ENCYCLOPEDIA OF

ENVIRONMENTAL ANALYSIS AND REMEDIATION

VOLUME 7

WILEY ENCYCLOPEDIA SERIES IN ENVIRONMENTAL SCIENCE

ROBERT A. MEYERS
ENCYCLOPEDIA OF ENVIRONMENTAL ANALYSIS AND REMEDIATION

ATTILIO BISIO AND SHARON BOOTS
ENCYCLOPEDIA OF ENERGY TECHNOLOGY AND THE ENVIRONMENT

EDITORIAL BOARD

ENCYCLOPEDIA OF ENVIRONMENTAL ANALYSIS AND REMEDIATION

VOLUME 7

Robert A. Meyers
Editor

A Wiley-Interscience Publication
John Wiley & Sons, Inc.
New York / Chichester / Weinheim / Brisbane / Singapore / Toronto

This book is printed on acid-free paper. ♾

Published simultaneously in Canada.

Library of Congress Cataloging-in-Publication Data:

Encyclopedia of environmental analysis and remediation / Robert A. Meyers, editor.
p. cm. — (Wiley encyclopedia series in environmental science)
Includes bibliographical references and index.
ISBN 0-471-11708-0
1. Environmental sciences—Encyclopedias. 2. Environmental pollution—Analysis—Encyclopedias. 3. Environmental protection—Encyclopedias. I. Meyers, Robert A. (Robert Allen), 1936– . II. Series.
GE10.E49 1998
628—dc21 97-41586
CIP

Printed in the United States of America

10 9 8 7 6 5 4 3 2 1

ENCYCLOPEDIA OF

ENVIRONMENTAL ANALYSIS AND REMEDIATION

VOLUME 7

RADON, THORON, AND DECAY PRODUCTS

EARL O. KNUTSON
Consultant ·
ANDREAS C. GEORGE
Consultant
Bayside, New York

In the past 10 years, there has been a great deal of concern and awareness about the health risks from exposure to radon/thoron and their progeny. The discovery of elevated indoor radon concentrations scattered throughout most of the U.S. prompted many federal, state, and local agencies and private individuals to test for radon. The high concentrations found in some regions helped to locate "hot" areas and promote testing by a wider section of the general population.

As of this writing the U.S. Environmental Protection Agency (EPA), Office of Radiation and Indoor Air, states that more than 12 million homes have been screened for radon of which about 10% need to be mitigated to reduce the radon levels to <150 Bq m^{-3}. However, about 85% of the residential buildings in the U.S. have not been tested for radon. In real estate transactions, only about 16% of the homes are tested for radon, and EPA is urging 100% testing.

This increased interest in measuring radon/thoron and their progeny has stimulated instrument research and development. The selection of the instruments and methods included here is primarily based on the experience of the authors and hundreds of users in the commercial sector, such as surveyors, inspectors, real estate professionals, mitigators, environmental firms, and researchers. This article emphasizes portable, commercial instruments, drawing on a recent paper of one of the authors (1). Other good reviews are available that emphasize classical, usually manual, methods (2–5).

MEASUREMENT CONSIDERATIONS AND APPLICATIONS

Purpose

This section provides background information for the professional person who plans and makes measurements of radon, thoron and their decay products. Included are definitions of important quantities and their units, typical average values and typical variations of these quantities, some information about health effects, etc. More information, including the relevant mathematical equations, is available in recent books (6–9) and review papers (10,11).

Sources of Indoor Radon

Radon, chemical symbol Rn, is a naturally occurring substance discovered in 1900 and isolated in 1908. Early names for this element were "emanation" and "niton." Radon's atomic number is 86 and its atomic weight is approximately 222. Chemically, it is a member of the noble gas family of elements. As such, it has very little chemical affinity for any other substance. Radon has several isotopes, of which three are worthy of mention: ^{222}Rn, ^{220}Rn, and ^{219}Rn. Although the correct name for all three isotopes is radon, the three also have common names, radon, thoron and actinon, respectively.

Radon is radioactive. Its atoms disintegrate according to the law of radioactive decay, each yielding an alpha particle and a residual atom (decay product) different from radon. In the case of ^{222}Rn, if there were no source of this substance, half the world's supply would disappear in 3.823 days, and half the remainder would disappear in another 3.823 days, etc, so the whole supply would be gone in a few weeks. The other radon isotopes have even shorter half-lives. However, there is a source, because radon is a decay product of certain metallic elements found in trace amounts in all rock and soil.

When the soil-bound radon precursor elements decay, some of the radon escapes into the air in the interstices of the soil, yielding "soil gas" rich in radon. On open ground, some soil gas and radon leaks into the atmosphere where it is rapidly diluted by mixing with air. Houses built on or into the ground often have an inside air pressure slightly lower that the atmospheric pressure outside. Thus soil gas migrates through leaks and cracks into these houses, carrying radon atoms with it. *This is the major source of indoor radon.* If the above-ground parts of the house are moderately air tight, the radon concentration inside reaches levels much higher than outdoors.

Among the other sources of indoor radon are radon-laden water drawn from deep wells and emanation of radon from building materials (masonry) that contain radon precursors. In the case of radon-laden water, most of the radon is released into the air when the water is agitated, as in washing clothes and showering.

Radioactive Decay, and Interactions with Aerosols and Ions

Table 1 shows the decay chains of ^{222}Rn, commonly called radon, and ^{220}Rn, commonly called thoron. The first decay product in each case is polonium, a metallic element. It is believed that Po oxidizes rapidly in air, so that most Po is found as the compound PoO_2. As shown in Table 1, polonium is just the first of a series of decay products. All of the nuclides between ^{222}Rn and ^{210}Pb have short half-lives. These nuclides are often collectively called the short-lived decay products of ^{222}Rn. Similarly, the nuclides between ^{220}Rn and ^{208}Pb are called the short-lived decay products of thoron. There is a branching point at ^{212}Bi in the ^{220}Rn chain, where 64% of these atoms decay by β emission and 36% decay by α emission. Note also that Po and Pb show up twice, as different isotopes, in each chain. Other terms used in place of decay products are "daughters" and "progeny." Colloquially, the term radon is often understood to include the short-lived decay products and the gas itself. For this reason, we often use the term "radon gas" when we want to focus on the gas and exclude the decay products.

Because the half-lives are quite short, the mix of decay products in a sample changes considerably even during sampling and certainly afterward. *Decay* of one nuclide and *ingrowth* of its product are an everyday fact of life in radon and radon decay-product measurements. Examples are shown in the figures following. Ingrowth and decay are well understood and are predicted accurately by a set of equations known as

Table 1. Decay Data for Radon and Thoron

Radon Series				
Nuclide	Half-Life	Alpha Energy, Mev	Beta endpoint energy, MeV	Gamma energy, MeV
^{222}Rn	3.8235 d	5.4897		
^{218}Po	3.04 min	6.00[a]		
^{214}Pb	27. min		0.67, 0.73	0.242, 0.295, 0.352
^{214}Bi	19.9 min		3.28	0.609, 1.120, 1.764
^{214}Po	163.7 μs	7.686		
^{210}Pb	22.3 yr			
Thoron Series (note that ^{212}Bi has two modes of decay: beta(64%) $\xrightarrow{212}$ Po $\xrightarrow{208}$ Pb; alpha(36%) $\xrightarrow{208}$ *Tl* $\xrightarrow{208}$ Pb)				
Nuclide	Half-Life	Alpha Energy, Mev	Beta endpoint energy, MeV	Gamma energy, MeV
^{220}Rn	55.6 s	6.288		
^{216}Po	145 ms	6.778		
^{212}Pb	10.64 h		0.28, 0.57	0.239
^{212}Bi	60.6 min	6.051(25%), 6.090(9.6%)	2.252(64%)	0.288, 0.727, 0.785
^{212}Po	0.298 μs	8.784		
^{208}Tl	3.053 min		1.28, 1.52, 1.796	0.277, 0.511, 0.583, 2.614
^{208}Pb	stable			

[a] This value from Ref. 5; all others from Ref. 91. Some values have been rounded.

the Bateman equations. A convenient recursive method for Bateman-type calculations has been given (12).

Unlike its parent, Po has a high affinity for other substances. Polonium atoms, or PoO_2 molecules, stick tightly to any solid or liquid surface they strike. This is true also of the other short-lived decay products. Indoors, Po atoms that impinge on walls or other surfaces stick there, and the remaining decays in the chain take place mostly on those surfaces. Typically, about half the short-lived decay products in a room are sequestered on the surfaces of the room.

Another important feature of the air is the small particles always present. The term *aerosol* denotes a more or less permanent mixture involving small solid or liquid particles and a gas that suspends them. The entire atmosphere, indoor and out, is such a mixture. The concentration of particles depends on activities, such as smoking, cooking, and automobile traffic, and ranges from about 10^3 to about 10^6 particles per cm^3 of air. In fact, aerosol particles far outnumber the atoms of short-lived decay products found in air, typically by 10^5 to 1. The particles have diameters ranging from a few nm to tens of μm, also far larger than the diameter of decay product atoms or molecules.

Although radon/thoron gas is not affected by airborne particles, the decay products are. *Attachment* is the process whereby Po atoms or PoO_2 molecules impinge on the surface of aerosol particles and stick there. These collisions result from the thermal motion of the airborne atom or molecule, characterized by a diffusion coefficient of about 0.05 cm^2s^{-1}. Once the atom or molecule attaches to an aerosol particle, it shares the motion of that particle, which also has thermal motion but with a diffusion coefficient several orders of magnitude smaller than the atom or molecule. The difference in motion is so dramatic that it is customary to speak of two different species, *unattached* decay products and *attached* decay products. Attachment is second only to decay as an important process affecting decay-product atoms.

The quantitative aspects of attachment resemble the process of radioactive decay. Both processes are described by a rate constant or by half-life. There is a difference, however, in that the rate constant for radioactive decay is a true constant, whereas the rate constant for attachment varies with the aerosol concentration and particle size. For example, the constant describing radioactive decay of ^{218}Po is ln(2)/3.04 = 0.228 min^{-1}, where 3.04 min is the half-life of ^{218}Po. The rate "constant" describing attachment varies in the range 0.1 to 10 min^{-1}, with a typical value of 1 min^{-1}. This means that a newly formed ^{218}Po atom is likely to become attached before it decays. Quite the opposite is true for ^{216}Po, which has a much higher rate of radioactive decay (half-life = 145 ms).

A third process (crucially important for some measurement techniques) is *electrical neutralization.* Alpha particles strongly ionize atoms and molecules by ripping out orbital electrons. When Rn decays, the emerging alpha particle carries off two units of positive charge, but it also rips out several electrons from the host atom, so that the decay product is left with a positive charge. However, this charge is normally lost in less than a second through one of three mechanisms (13). The rate constant for neutralization is largely determined by the concentration of certain air contaminants, including NO and NO_2.

Two other processes are important in principle, but not very important in practice, namely, *clustering* and *recoil.* Clustering is the accretion of various condensable molecules in the air onto Po or PoO_2, producing a cluster whose mass is significantly more (perhaps twice) than that of Po or PoO_2. Recoil is a process that affects the decay product of attached Po. When Po decays, the alpha particle departs in a random direction and the decay-product nuclide recoils in the opposite direction. The recoil energy is easily enough to dislodge the decay product from the surface of the aerosol particle. As a result, more than half of the ^{214}Pb starts life as unattached atoms, even if the parent ^{218}Po was attached.

Quantities and Units

The amount of radon is expressed in terms of the *number of atoms,* but there is another quantity widely used for radon and for any radioactive substance including radon decay products. That quantity is the *activity,* the number of disintegrations expected per unit time. The two quantities are related: the activity is the product of the number of atoms and the decay constant. For radon, the decay constant is 2.094×10^{-6} s^{-1}. Thus,

Table 2. Quantities and Units Important in Radon Measurements

Quantity	Symbol	SI unit	Older Unit	Relationship
Activity	A	becquerel, Bq (= 1/s)	curie, Ci	1 Ci = 3.7×10^{10} Bq
Activity concentration	A	Bq/m^3	pCi/L	1 pCi/L = 37 Bq/m^3
Potential alpha energy	PAE	joule, J		
Potential alpha energy concentration	PAEC	J/m^3	Working level, WL	1 WL = 20.8 μJ/m^3
Absorbed dose	D	gray, Gy (= 1 J/kg)	rad (= 100 ergs/g)	1 rad = 0.01 Gy
Equivalent dose	H	sievert, Sv	rem	1 rem = 0.01 Sv
Effective dose	E	sievert, Sv		

for example, if we have 10^6 atoms of radon, the activity is 2.094 disintegrations per second.

Table 2 shows the main quantities and units used in connection with radon, thoron, and their decay products. The international (SI) unit for activity is the *becquerel,* abbreviated Bq and defined as one disintegration per second, 1 s^{-1} (14). (This differs from the unit hertz, also s^{-1}, which pertains to periodic phenomena.) The older unit of activity is the curie (Ci), defined as 1 Ci = 3.7×10^{10} Bq. A submultiple, the picocurie (pCi; 1 pCi = 0.037 Bq) has long been used in radon work.

For airborne radon and decay products, the *activity concentration* is more important than the activity itself. The SI unit of activity concentration is Bq m^{-3}. The older unit is pCi L^{-1}, where 1 pCi L^{-1} = 37 Bq m^{-3}.

For mixtures of short-lived decay products, a special quantity called the *potential alpha energy* has been developed to describe the amount. This was developed in the early 1950s, when it was discovered that the principal health effects of radon were actually caused by the decay products, rather than radon itself. To understand potential alpha energy, imagine storing a sample of short-lived decay products until all these atoms have decayed. The potential alpha energy is simply the total energy released through alpha decays during this storage. The SI unit of potential alpha energy is the joule (J). As with activity, the concentration (potential alpha energy concentration, or PAEC) is more important than the potential alpha energy itself. The SI unit of PAEC is J/m^3. There is also an older unit, the *working level* (WL), that originated in the early 1950s; 1 WL = 2.08×10^{-5} J/m^3.

Equilibrium factor, or *equilibrium ratio,* is a useful dimensionless quantity for specifying the concentration of the short-lived decay products, at least those of ^{222}Rn. To understand this quantity, consider a hypothetical case in which the airborne ^{222}Rn, ^{218}Po, ^{214}Pb, and ^{214}Bi all have the same activity concentration. (This situation, called secular equilibrium, is never achieved in practice.) For this case, calculation shows that the PAEC in nJ m^{-3} is numerically equal to 5.56 times the radon concentration in Bq m^{-3}. The equilibrium factor is then defined as the ratio of the measured PAEC (in nJ/m^3) to $5.56A_0$, where A_0 is the radon concentration (in Bq m^{-3}). The equilibrium factor is also defined for thoron and its decay products. Here the numerical factor is 75.2.

Three quantities used to quantify the health effects of radiation are worth mentioning, although in the case of radon there is no instrument or technique available for directly measuring them. (Radon measurements are only a starting point, and dose determinations require lengthy calculations that bring in physiological and biological factors.) The quantities are *absorbed dose, dose equivalent,* and *effective dose equivalent.* Absorbed dose is a purely physical quantity, namely, the energy deposited in a substance divided by the mass of the substance. The SI unit of absorbed dose is J/kg, which has been given the special name gray (Gy). Empirically, it was found that the severity of the biological effect in body tissue depends on the absorbed dose and also on the type of radiation, α, β, γ. Dose equivalent and effective dose equivalent, which are multiples of the absorbed dose, were introduced to facilitate comparing the three types of radiation in terms of biological effect and risk to health. All three varieties of dose have the same physical units, J/kg. However, this unit is given special name gray (Gy) when referring to absorbed dose and the special name sievert (Sv) when referring to dose equivalent or effective dose equivalent. A clear discussion of these quantities and the relationships between them is available (15). Older units for absorbed dose and equivalent dose are *rad* and *rem,* respectively (see Table 2).

Health Effects

There is wide, if not unanimous, agreement that exposure to radon makes the person exposed more susceptible to cancer of the lung, especially if the person smokes (16). This is the health effect that has been, and continues to be, the driving force for virtually all measurements of radon gas and its decay products. The increased risk of lung cancer was first established in the 1950s from the experience of uranium miners (17). Uranium is a radon precursor, so that areas rich in uranium also have potential for high concentrations of airborne radon and radon decay products. Because mine ventilation was geared to clearing smoke and dust rather than radon, early miners often worked in atmospheres with elevated concentrations of radon.

In the early 1950s it was established that the injury to lung tissue is caused by alpha particles from radon's short-lived decay products, rather than from radon itself. When the person breathes, some of the inhaled decay products deposit on the lining of the airways in the lung. Many of the alpha particles from ensuing disintegrations deliver their full energy into the surface layer of lung tissue. Radon, a noble gas, does not accumulate on lung surfaces and is therefore less harmful than the decay products. Recognition of the key role of alpha particles from the short-lived decay products prompted the adoption of PAEC (see "Quantities and Units") as a measure of the potential for detrimental health effects. It was recognized also that a face mask blocking inhalation of the decay products provides a large measure of protection.

Cancer is an illness that develops slowly, over one or two decades, so it is impossible to determine the specific event that caused a given cancer. The concept of *risk* has been developed to describe the health effects of exposure to radon decay products. Risk is believed proportional to *exposure,* which is the time-integrated decay-product concentration. A recently published study found that the *lifetime excess risk* of lung cancer for uranium miners is 0.079 for each J h m^{-3} of exposure (18). This is crucial information, but it pertains to miners exposed in the 1950s and 1960s where conditions differed from those in modern mines or in homes.

To estimate lung cancer risk for current or future miners or for occupants of high-radon houses, the first requirement is to determine the exposure. This is done by the measurement methods described later in this article. However, many conditions in modern mines, or in houses, differ from those in 1950s–1960s mines. The differences include

- the mix of age and of gender;
- the level of exertion;
- the amount of fresh air;
- the number and size of aerosol particles.

Since about 1975, calculation methods have been developed to take these factors into account. One of the findings is that aerosol particles (last factor), acting through attachment, are an important modifier of the risk coefficient. Later in this chapter we describe methods for measuring the number and size of aerosol particles.

Exposure Limits

Various authorities and commissions in the United States and in many other countries have issued regulations or advice (guidelines) about exposure. In the early 1950s, the U.S. Public Health Service advised that radon decay-product concentration in mines should be reduced to 1 WL (= 20.8 $\mu J/m^3$), by ventilation if necessary. In 1969, after many public hearings and much expert testimony, the U.S. Department of Labor imposed a limit of 4 *working level months per year* for exposure to decay products (19). In this context, "month" was taken to be 170 hours, so in modern terms, the annual limit is 0.00354 J h/m^3. This limit is still in effect for exposure in workplaces.

Exposure in private houses became a matter of public concern after the 1984 discovery of a private home in eastern Pennsylvania with an exceedingly high concentration of radon (>80,000 Bq m^{-3}). Many states launched surveys that found high radon levels in a few percent of the housing stock. In response to the requests for advice about safe levels of radon, in 1986 the U.S. Environmental Protection Agency issued a guideline that remedial action should be taken if the long-term average radon concentration exceeds 148 Bq/m^3 (4 pCi/L) (20). One reason this guideline was expressed in terms of radon rather than decay products is that radon measurements are simpler and less expensive to perform.

The regulation and the guideline both continue in force as of this writing, but both are subject to review and revision.

Typical Concentrations

Table 3, adapted from a recent review article (11), gives typical indoor and outdoor concentrations of radon, thoron, and decay products. For outdoor radon, data on diurnal and seasonal patterns in the concentration of radon gas have been summarized (21). The average diurnal plots show the highest concentrations in the predawn hours and the lowest in the early afternoon, the highest about three times the lowest. There is also a seasonal pattern, such that the average of the highest month is about three times that of the lowest, but the exact pattern differs for different sites. Annual averages also change from year to year, in concert with year to year changes in weather.

Table 3. Typical Concentrations of Radon, Thoron, and Their Decay Products[a]

	Radon		Thoron	
	Outdoor	Indoor	Outdoor	Indoor
Gas, Bq/m^3	8	50	5	10
Decay products, nJ/m^3	28	100	15	38

[a] Adapted from Table 16 of Ref. 11.

Turning to indoor concentrations, Figure 1 is a summary based on several hundred of the earliest radon measurements made in U.S. residential housing. The frequency distribution is fitted by a log normal distribution with median 35 Bq m^{-3} and geometric standard deviation equal to 2.8. A more scientific survey done in 1989–1990 (22) generally supported the summary shown in Figure 1. Based on 4,658 measurements in single family homes, the statistics were average concentration equal to 54 Bq/m^3, median equal to 29 Bq/m^3, geometric standard deviation equal to 2.97. Other categories of housing were also measured.

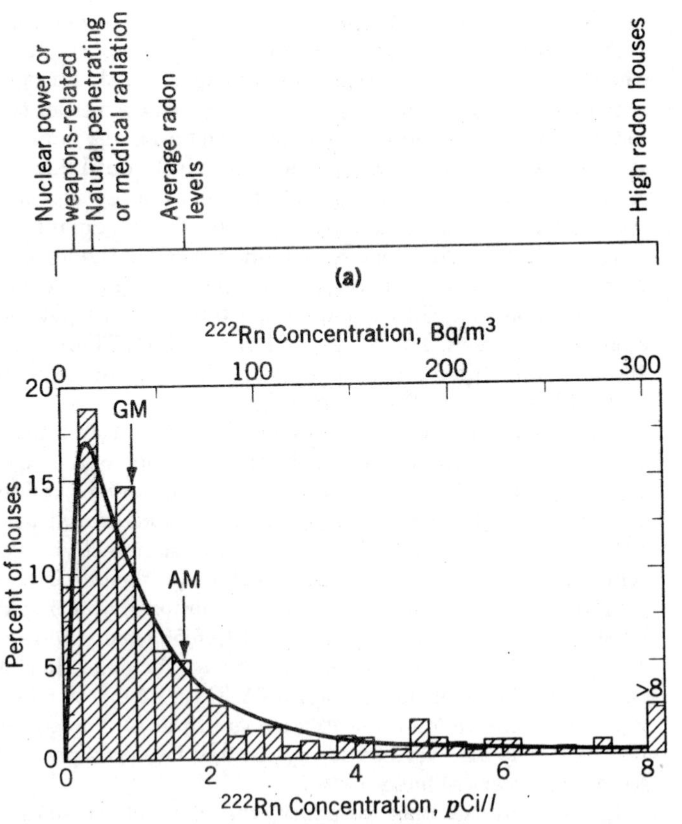

Figure 1. Probability distribution of ^{222}Rn in U.S. homes, based on early measurements in 552 homes. (**a**) Comparison to other radiation exposures; (**b**) the smooth curve is a log-normal distribution with GM = 35 Bq/m^3, GSD = 2.8, and AM = 61 Bq/m^3. Used with permission (6).

The concentration of indoor radon and its decay products show hourly, seasonal, and weather-related variations at least as large as for those outdoors, mentioned above. This variability can lead large errors when deducing long-term averages from short-term measurements.

Quality Assurance; Measurement Protocols

In one sense, data has quality if it is scientifically sound and its precision and accuracy are known (23). In a broader sense, quality is also judged by whether or not the data serves the intended purpose, which is true only if the data suffices to answer the questions that prompted the measurements in the first place. In this broader sense, quality depends as much on the type and amount of data as it does on its precision and accuracy. In environmental measurements, it is also important that the measurements are done right the first time. It is often unacceptable to make measurements first and determine data quality later.

Quality assurance (QA) is the process by which it is assured that a set of measurements will yield data with acceptable quality. QA must be an integral part of any measurement campaign, starting with the planning. The authors recommend the following:

- clearly state the objectives of the measurements;
- determine the type and amount of data required to meet the objectives;
- determine the precision and accuracy required of the individual measurements;
- select a measurement method and a sampling protocol to meet the type, amount, precision, and accuracy requirements;
- determine what steps, such as measuring blank and standard samples, are necessary to ensure that each datum meets the precision and accuracy requirements;
- determine how the data, both numeric and verbal, are initially recorded and how they are transcribed, stored and preserved (this applies to the quality-check data and the measurement data itself).

At this point it is usually necessary to estimate the cost of the measurement campaign. If not affordable, it is necessary to simplify the objectives and repeat the process.

Following are several reasons for measuring radon, thoron, or their decay products:

- screening or range-finding;
- diagnosing radon sources or entry points at a given site;
- testing a given site for compliance with regulations/guidelines;
- determining if a site needs immediate remedial action;
- determining an average concentration within a given geographical area (thousands of individual sites);
- detailed study of health hazard at one or a few sites;
- long-term measurements to determine an individual's average annual exposure.

Each of these study objectives has implications regarding the type of measurements and the amount of data needed. For example, high accuracy and precision are not needed in screening measurements. Rather, simplicity and speed are important. Conversely, testing a given site for compliance with a guideline requires a method which yields an accurate long-term average, good documentation, and quality assurance procedures so that, when the results are in, they are trustworthy.

Table 4 shows some terms used to differentiate among sampling methods. The first three entries are broad categories differing in the time sequence of sampling and signal registration. The bottom of the table shows two terms describing the power requirements.

Detailed QA recommendations for measuring indoor radon gas or decay products are in a recent US EPA document (23). For example, the document recommends a minimum two-day sampling period for any data used for mitigation decisions. Recommendations are also given as to where and when to place samplers in buildings (to avoid situations that bias the data). Beyond this, specific QA recommendations are given for each of 15 categories of measurement methods or instruments. For example, a five-part QA program is recommended for the alpha track method of measuring indoor radon. In abbreviated form, these five steps are as follows:

- *Calibration.* The calibration factor for every type of detector and analysis instrument should be determined initially and every 12 months thereafter.
- *Known exposure measurements.* Anyone providing measurement services with alpha track devices should arrange for exposing 3 out of every 100 devices (but no less than 3 per year and no more than 6 per month) in a known radon atmosphere.
- *Duplicate (collocated) detectors.* Anyone providing measurement services with alpha track devices should reserve 10% of the detectors (but no more than 50 per month) for placement side by side in typical measurement sites.

Table 4. Characterization of Sampling Methods

Type of Method	Characteristics
Instantaneous (or grab) sampling	An air sample is drawn through the sampler for a short period of time, not exceeding 10 min, followed by a (usually longer) period when the radiation signal is being recorded.
Continuous sampling	Sampling is continuous for at least several hours, with or without air flow, and the radiation signal is recorded at regular intervals, such as 0.25 h or 1 h.
Integrating sampler	Sampling takes place over a longer period, one day to one year, and the radiation signal accumulates at a rate proportional to the air concentration. Readout is at the end of the period and yields the average value during the measurement period.
Passive sampler	A sampler that operates without pumping air.
Active sampler	A sampler that involves drawing air through the sampler by means of an air pump. Usually this requires external electrical power.

- *Control detectors.* The laboratory should determine the background for each batch of alpha track devices by measuring unexposed detectors, and make this information available to those using the devices. Additionally, those providing measurement services should set aside 5% of the detectors (but no more than 25 detectors per month) and store them unopened in a low radon atmosphere.
- *Routine instrument checks.* Instruments used for analysis should be checked each day to ensure that they are in good working order.

The detectors identified in the second through fourth of these steps are to be analyzed in the same way as the detectors exposed in the field. This provides running data on the precision and accuracy of the detectors. One study (24) described experiences showing the need for this type of ongoing QA.

Statistics: Lower Limit of Detection (LLD)

Most techniques for measuring radon, thoron, or their decay products hinge on counting radioactive decay events. Watching counts ring up on a scaler gives an impression of exactitude, but this is deceptive. The central fact of radioactive decay is that it is governed by the laws of chance. The consequences for measurement are seen in Figure 2, which depicts the counts in consecutive 10 minute intervals from a midsized scintillation cell. The data was constructed by a Monte Carlo simulation that mimics the randomness of radioactive decay. It was assumed that the cell was initially filled with air containing 100 Bq/m^3 (Fig. 2**a**) or 10,000 Bq/m^3 (Fig. 2**b**) of radon, with 1 count min^{-1} of background noise in each case. Aside from the ingrowth period (which will be discussed later under "scintillation cell") the most striking feature is the scatter in the upper panel of the figure. The counts per 10 min vary from 45 to 90, so that individual 10 min counts are nearly worthless. In the lower panel where the simulated activity is 100 times higher, the scatter is relatively much less so that a single 10 min count seems adequate. The question is "How can we cope with this randomness?"

Fortunately, the statistical aspects of count data, such as those shown in Figure 2, are well understood (see, eg, 25,26). Provided that the counting period is short compared to the half-life of the radionuclide, the number of decay events is governed by the Poisson distribution. This is a single-parameter distribution for which the mean is equal to m and the standard deviation is equal to the $\sqrt{m}$, where m is the parameter. This property is a great time saver. Without it, the measurement uncertainty could be determined only by repeating the experiment ten or more times, then computing the standard deviation of the results. With it, a single count suffices. If n is the number of counts observed in a *single* time interval, the mean and standard deviation are estimated from n and $\sqrt{n}$, respectively. (The radon and thoron series nuclides include some short-lived species, see Table 1, so that the condition stated above for validity of the Poisson distribution is not always met in radon measurements. Areas of difficulty are pointed out as needed in the following.)

The $\sqrt{n}$ rule is the basis for most of the equations used to estimate the statistical component of uncertainty in radioactivity measurements. For example, the net count is determined from

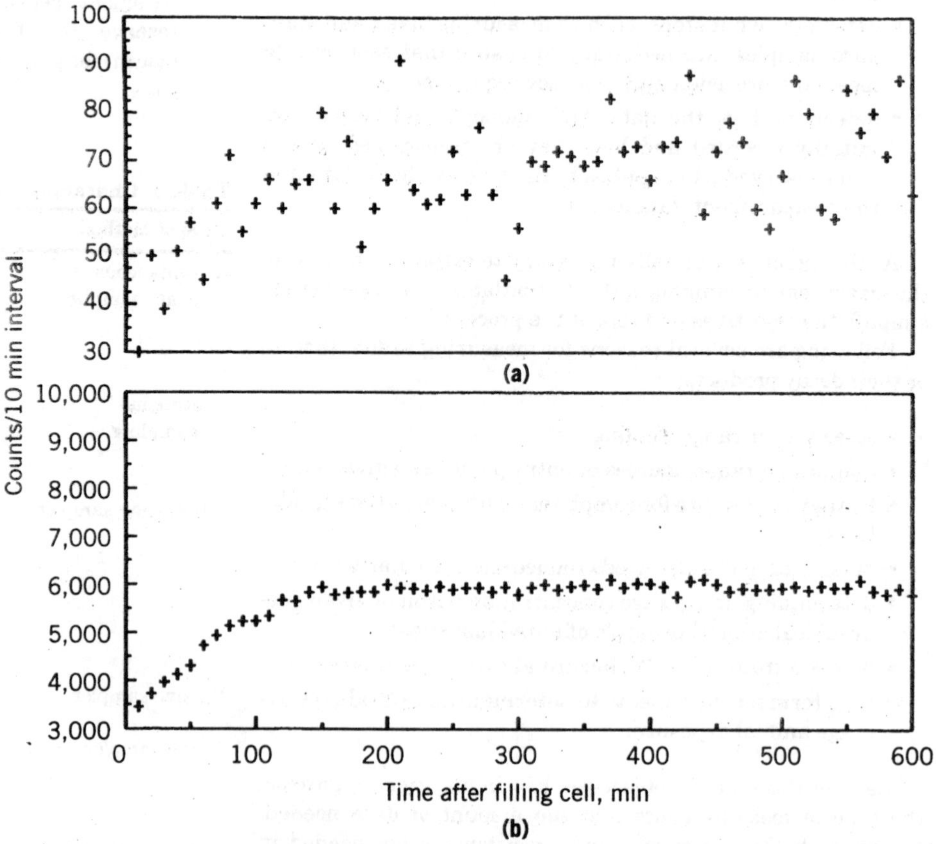

Figure 2. Illustration of randomness in alpha-count data; simulated ingrowth of count rate in a scintillation cell, where in (**b**) cell volume = 0.515 L, counting efficiency = 0.64, and cell background = 1 min^{-1}. Assumed radon concentration: (**a**) 100 Bq/m^3; (**b**) 10,000 Bq/m^3.

the following equation:

$$N = G - BT_g/T_b \quad (1)$$

where N is the net count; G is the gross count; B is the background count; T_g is the time used to determine the gross count; and T_b is the time used to determine the background count.

To determine the uncertainty of the net count, the $\sqrt{n}$ rule is applied separately to the gross count and the background count, and these are combined using the following propagation of errors formula:

$$\sigma_n = \sqrt{\sigma_g^2 + \sigma_b^2\left(\frac{T_g}{T_b}\right)^2} = \sqrt{G + B\left(\frac{T_g}{T_b}\right)^2} \quad (2)$$

where σ_n is the uncertainty (standard deviation) of the net count; σ_g is $\sqrt{G}$, the uncertainty of the gross count; and σ_b is $\sqrt{B}$, the uncertainty of the background count.

Similarly, the uncertainty (standard deviation σ_r) of the net count rate r (= $G/T_g - B/T_b$) is given by

$$\sigma_r = \sqrt{\frac{G}{T_g^2} + \frac{B}{T_b^2}} = \sqrt{\frac{g}{T_g} + \frac{b}{T_b}} \quad (3)$$

where $g = G/T_g$ is the gross count rate and $b = B/T_b$ is the background count rate.

Normally, net counts (or count rate) is a step along the way. Further equations are needed to arrive at the final result. The equation for estimating uncertainty in a given situation is based on the $\sqrt{n}$ rule, but also on the working equation(s) for the measurement method. Equations are given as needed in the following.

In some cases, as in measurements of radon progeny, T_g has to be kept small (a few minutes) because of the short half-lives of the radionuclide. However, there is no such limit on T_b. Often, matters are arranged so that the second term in equations 2 and 3 is small compared to the first, in which case $\sigma_n = \sqrt{G}$ and $\sigma_r = \sqrt{(g/T_g)}$ (Note, however, that the background should not be neglected in determining N from equation 1).

The *lower limit of detection* (LLD) is a useful term for describing the sensitivity of those measurement methods based on counting discrete events. The LLD is a statement of the lowest sample activity that is confidently distinguished from zero. The following equation (27) is widely used:

$$\text{LLD} = \gamma(k_\alpha + k_\beta)\sigma_0 \quad (4)$$

where γ is the calibration constant that converts net count rate into activity; α, β are the acceptable probabilities of type I and type II errors, respectively; k_x is defined by $N(k_x) = 1 - x$; N is the cumulative normal distribution function; and $\sigma_0 = \sqrt{(b/T_g + b/T_b)}$ is the limiting value of σ_r for very low net count rate.

Normally, α and β are both taken as 0.05, so that $k_\alpha = k_\beta = 1.645$. As an example, the data in Figure 2 pertains to a scintillation cell with $b = 1.0$ count per minute (cpm) and $\gamma = 16.9$ Bq m^{-3} per cpm. Assuming that 1 h was allowed for both the background and gross count, the LLD is 10 Bq m^{-3}.

Although LLD is a useful quantity in many cases, it is not useful in all cases. The ability to *detect* does not guarantee the ability to *measure with acceptable precision.* Thus in discussing radon progeny measurements, it is common to use the *minimum measurable concentration* (MMC), defined as the lowest concentration measured with a precision (relative standard deviation) of 20% (28).

Clearly, a 10-min count is not adequate for the case in Figure 2**a**, and this brings up an esoteric statistical point discussed in one study (29): for count intervals substantially larger than the half-life of the decay products, each radon decay event may be recorded more than once, with the consequence that the $\sqrt{n}$ rule is not strictly valid. The study shows that the relative uncertainty in net count rate is larger than that calculated from the ordinary $\sqrt{n}$ rule. The required correction factor is in the range 1 to $\sqrt{J}$, where J is the average number of counts expected per radon atom decay. For the cell considered in Figure 2, $J = 1.92$. In practice, this is not a large concern.

Those making measurements must maintain a healthy respect for the consequences of the stochastic nature of radioactive decay and calculate the uncertainty for each measurement, using the appropriate equations. Uncertainty is one of the important indices of data quality.

INSTRUMENTATION FOR MEASURING RADON AND THORON GAS

Virtually every measurement plan for any type of radon study should include and begin with measurement of ^{222}Rn. This assertion is true for two main reasons:

1. ^{222}Rn is centrally important. Even though the health impact is from its decay products rather than from ^{222}Rn itself, studies of the indoor environment have shown that, because of the interplay of several factors, ^{222}Rn measurements are just as good as decay-product measurements as a basis for estimating health risk (30).
2. Simple and inexpensive methods are now available for measuring ^{222}Rn. Simply put, the user gets more useful information per dollar spent by measuring ^{222}Rn than by measuring other nuclides.

For thoron gas ^{220}Rn, the situation is quite different from that for radon gas. Its half-life is so short that it decays before it has time to fully mix into the volume of the room. This complicates measurements and makes thoron gas a poor predictor of the concentration of its decay products. We return to this when discussing methods for measuring decay products.

Of course there are certain situations, indoor and more commonly outdoor, in which thoron gas measurements provide useful information. The techniques for measuring thoron gas are usually variations of those used for measuring radon gas, and these are discussed at appropriate points in the following.

Much development and testing of radon measurement techniques have been done since the middle 1980s, spurred by a growing market for this technology. This has resulted in simple, inexpensive, and reliable techniques and an industry that provides measurement services. When faced with a need to measure radon gas, one should look first to the service industry.

Table 5 shows key features of presently available commercial instruments. Physical principles and typical uses are discussed in the following, beginning with those methods which have been most thoroughly commercialized. Further information is available (1).

Adsorptive Collection of Radon

Although radon is a noble gas, it has a moderately high boiling point, −61.8°C (31–34). Thus it is trapped by adsorption. Activated charcoal, similar to that used in gas masks to trap poisonous gases, and in filters to purify water and air streams, is

Table 5. Commercially Available Instruments or Services for Measuring Radon or Thoron Gas

Manufacturer or Service Provider[a]	Manufacturer's Designation or Generic Type	Principle of Sample Collection	Principle of Detection	Lower Limit of Detection, Bq/m^3 at Exposure Period[b]	Size and Weight	Power Source	Cost in 1995, U.S. $
Integrating type: Readout done after exposure, service companies supply samplers and perform readouts							
See ads in *Health Phys.*	Charcoal	Diffusion/Adsorption	Gamma counting	7.5 at 4 d	25–200 g	None	$15–25
See ads in *Health Phys.*	Charcoal	Diffusion/Adsorption	Liquid scintillation (alpha + beta)	5.0 at 2 d	1–3 g 20 mL vial	None	$15–25
See ads in *Health Phys.*	Solid state	Diffusion	Track counting	5–10 at 90 d	10–100 mL	None	$15–25
Rad Elec	E-PERM	Diffusion	Discharging of electret	7 at 3 d short term 15 at 90 d long term	0.2 L	None	$75
RTCA	Ra-Dome	Diffusion	Discharging of electret	15 at 90 d long term	68 mL	None	$25–30
Continuous type: Data recorded at regular intervals on paper, or in internal memory for later retrieval							
femto-TECH	CRM-510	Diffusion	Ionization pulse counting	15 at 4 h	14 × 16 × 19 cm 2.3 kg	Battery or 115 V ac	$3,500
Genitron	ATMOS-12D	Air pump	Ionization pulse counting	15 at 1 h	22 × 50 × 37 cm 10 kg	115/230 V ac	$8,000
Genitron	AlphaGuard MC50	Diffusion	Ionization pulse counting	15 at 1 h	40 × 17 × 15 cm 4.5 kg	Battery (10 d) or 115 V ac	$10,000
Niton	RAD-7	Air pump/ Electrical	Semiconductor pulse counting	15 at 1 h	24 × 19 × 29 cm 5.5 kg	115 V ac or battery (4 d)	$3,000
Sun Nuclear	Model 1023	Diffusion/ Electrical	Semiconductor pulse counting	37 at 4 h	13 × 13 × 18 cm 0.9 kg	Battery or 115 V ac	$800
SARAD	RTM 2010	Air pump	Semiconductor pulse counting		115 V ac	ac/dc 7.5 V, 500 mA	
Grab or continuous type: Multiple operating modes available through front-panel programming							
Sun Nuclear	Model 1025	Air pump/ Electrical	Semiconductor pulse counting	7.5 at 1 h	28 × 14 × 13 cm 2.3 kg		$3,000
Pylon	Model AB-5	Air pump	Scintillation cell pulse counting	7.5 at 1 h	22 × 10 × 24 cm 3.7 kg	Battery, 48 h	$4,000
Eberline	RGM-3	Air pump	Scintillation cell pulse counting	7.5 at 1 h	41 × 25 × 30 cm 12 kg	Battery	$4,000

[a] See Table 14 for addresses.
[b] Where LLD values were not given by manufacturer, an estimate is given.

also effective in trapping radon, even at room temperature. This forms the basis for one of the most inexpensive and convenient methods available for measuring indoor radon. Analysis is done subsequent to sampling, either by gamma counting with a NaI detector or by alpha/beta counting using the liquid scintillation technique. Because radon decays even as it is being collected, the method is not useful for long-term sampling. However, it is well suited to short-term (2 to 7 days) measurements.

Solid-State Nuclear Track (SSNT) Detectors

Some substances sustain permanent damage upon exposure to alpha particles. For certain plastic substances, such as cellulose nitrate (LR-115) and allyl diglycol carbonate (CR-39), the damage is made visible as tracks by chemical or electrochemical etching, usually in NaOH solution (35). The track density is determined by a microscope, with either manual or automated counting, or by an electrostatic discharge technique. This is the basis for another widely used method for measuring radon concentration.

Because the alpha damage is permanent, the length of exposure is not limited by the half-life of radon. The method is therefore well suited for determining long-term averages. Exposure intervals of 3 months to 1 year are commonly used. The detector material is not sensitive to beta and gamma radiation, and correction for ambient background is not necessary.

However, correction is required for the background tracks that result from flaws in the plastic. Also, the calibration constant depends on altitude because alpha particle range increases with decreasing air density. For SSNT detectors utilizing the LR 115 plastic, an increase of 41% was found at an altitude of 1500 m, compared with sea level (36).

A passive integrating device for measuring both radon and thoron has been developed in Japan (37). This device consists of two hemispherical diffusion chambers in which two polycarbonate film detectors discriminate between the two radioactive gases. Exposures longer than 3 months are required for adequate sensitivity.

Ionization Chambers

From the first days of the atomic age, ionization of otherwise nonconducting materials has been a favorite method of detecting alpha particles. In 1908, Rutherford and Geiger used ionization to count the alpha particles in their determination of the activity of radium (38). In recent years, many devices based on ionization have been developed for both laboratory and field measurement of radon.

Ionization devices are used in many different ways. Similar to that used by Rutherford, alpha-particles are counted individually (pulse ionization chambers). Modern electronic amplifiers have greatly facilitated this method because pulses are counted without "gas amplification" as was necessary in early days. Another way of using ionization is to measure the average electric current from collecting ions caused by alpha particles.

A third way is to collect the ions by an electret, as ions accumulate, the effective surface charge of the electret decreases, and this is measured by a special voltmeter. In recent years, this last technique has been developed into a practical and low-cost method of measuring radon, so it will be presented first.

Electret Ionization Chamber. These devices consist of a chamber made of conducting plastic, containing an electret (39). After opening a seal, radon diffuses into the chamber through a filter. Different designs exist, some for short-term exposures of 2–7 days and others for long-term exposures ranging from 3 months to 1 year. The results are obtained in the field by reading the voltage with the special (electrostatic, noncontact) voltmeter. Care is needed to avoid touching the electret, because touching partly discharges it. If enough charge is left after one exposure, the device is reused. Because the devices also respond to gamma radiation, a correction must be applied for natural gamma background. Some designs use a second electret sealed into an air tight chamber to measure this gamma response, whereas, in other designs, corrections are derived on the basis of standard gamma radiation tables for different parts of the country.

Pulse Ionization Chambers. Various designs are available for these devices, which operate by electronically counting the burst of ionization caused by each alpha-particle (40,41). Some (not commercially available) are laboratory-based devices intended for measuring very low concentrations of radon, as found in the atmosphere over the oceans. These devices are also used as transfer standards for calibrating field instruments, thereby providing traceability to national standards laboratories.

At present, there are three models of commercially available, portable pulse ionization chambers (see Table 5). One uses a pump to bring the sample into the sensitive volume, whereas two depend on diffusion. All are continuous devices that display updated radon readings at intervals of 10 min to 1 h, and all provide a method of logging data either by a printer or by internal electronic memory. Some also measure and store the temperature, humidity, and barometric pressure. Two of the devices operate from rechargeable batteries. The ATMOS-12D has spectroscopic energy resolution capability, so, in principle, it distinguishes between ^{222}Rn and ^{220}Rn.

Current Ionization Chamber. The MC50 automatically switches to DC current if the radon concentration is so high that pulse counting is not possible.

Electrostatic Collection of ^{218}Po and ^{216}Po

According to a 1913 report, 88% of freshly formed (1-μs-old) ^{218}Po atoms have one unit of positive charge, and this number has stood the test of time (13). Presumably, it is true for ^{216}Po as well. These atoms are collected by a suitable electric field onto a sensitive surface where alpha particles from the subsequent decays are registered. Several instruments have been designed using this technique. In the early designs, a thermoluminescent dosimeter chip or a scintillator-photomultiplier was used as the alpha-particle detector. Modern designs use a semiconductor alpha-detector, which counts individual alpha particles and also differentiates (by energy) between alpha particles from the various nuclides of interest, ^{218}Po, ^{216}Po, ^{214}Po, and ^{212}Po. (Alpha-particles from ^{212}Bi in the ^{220}Rn chain have energy close to those from ^{218}Po, so that these two cannot be distinguished by energy. Fortuitously, the half-lives are so different that, at worst, the former acts as a slowly varying background in the determination of ^{218}Po.) One of the design issues with these instruments is that the electric field must be strong enough to collect ^{218}Po before it becomes electrically neutralized and to collect ^{216}Po before it decays.

As seen in Table 5, there are now three models of radon instruments using electrostatic collection. They are continuous monitors, although one is operable in a grab sample mode. Two use a pump to draw in the sample, and one uses diffusion. Two (RAD-7 and Sun Model 1025) measure ^{218}Po and ^{214}Po separately, so that the response time for changes in radon concentration is of the order of 3 min. For this reason, they are sometimes called radon sniffers.

Scintillation Cell

Along with ionization, *scintillation* is a classic method of detecting alpha particles. The famous 1913 experiment of Geiger and Marsden to test Rutherford's hypothesis about the atomic nucleus was done by counting light pulses from alpha particles striking a scintillator screen (42). Then as now, a thin layer of powdered, silver-activated zinc sulfide bonded to a flat surface makes a suitable scintillator. In the old days, the scintillations were counted by eye, with the help of a low power microscope. The modern version uses a photomultiplier tube to detect light flashes and electronic circuitry to do the counting.

Scintillation, in the form of a scintillation cell, is also one of the oldest and most widely used methods for measuring radon. In a 1957 article (43), a particular design was presented that was widely copied and became known as the Lucas cell. Most scintillation cells are hollow right-circular cylinders, made of metal, plastic or glass, with a window on one end and a flat or hemispherical cap on the other. The inside surface, but usually not the window, is coated with ZnS(Ag) scintillator powder. In the end opposite the window, there are one or two sampling tubes, each fitted with a valve. The vessel is rigid enough so that air can be pumped out of it.

Table 6 shows the performance characteristics of one scintillation cell built from a design used for about 15 years at EML (44,45). This information is offered as an aid in selecting from among the several commercial cells now available. In this cell, all internal surfaces are coated with the ZnS(Ag) powder. Note that the efficiency for counting alpha particles from ^{222}Rn is about 53%. The difference from 100% is caused mainly by geometry: both the diameter and height of this cell are larger than the range in air (7.2 cm) of alphas from ^{222}Rn, so that many of these alphas expend all their energy in air and never strike the scintillator surface. To produce a countable light pulse, alpha-particles must strike the surface with about 2 MeV of residual energy. The efficiency is larger for the more energetic alphas from the decay products, and the average efficiency of this cell for alphas in the ^{222}Rn chain is 64%. Clearly, the efficiency depends on the air pressure within the cell and on the threshold setting of the photomultiplier. Scintillation cells and photomultiplier tubes should be calibrated together. Specifications for building this cell are in the EML Procedures Manual (46), but there is little reason to build these cells because very good cells are now available commercially.

Measurement of Radon Gas. Scintillation cells are versatile and are used in several ways. The simplest way is to evacuate the cell, open it briefly at the place to be sampled, then store it to be counted with a photomultiplier tube at a later time. A filter at the inlet is used to ensure that only gas, not decay products, gets into the cell. (Another method, possible for cells with two valves, is to connect a small pump and draw air through the cell. To insure that the old sample is completely flushed out, the volume drawn should be about 10 times the cell volume.) Figure 2, already discussed in the section on statistics, shows simulated count data from the cell described in Table 6. In the first 3 h, the count rate increases because of the ingrowth of ^{218}Po and ^{214}Po. After three hours the average count per 10 minutes is about 70 in Figure 2a and about 6000 in Figure 2b. Thus, the common practice is to store cells for three hours before counting.

As seen in Table 5, scintillation cells are also used in some commercial radon monitors. These instruments consist of a scintillation cell, a photomultiplier tube, an air pump, and electronics for pulse counting, all incorporated into one portable package. The two instruments have multiple modes of operation, including continuous. Differences between the two include the capability for simultaneous measurement of radon decay products, temperature, humidity, and barometric pressure; the frequency of logging data; and memory available for storing data.

For continuously operating scintillation cells, there is a time lag in the response to changing radon concentration. Similar to Figure 2, three hours are needed to fully respond to a change in radon concentration. This has been discussed by several workers who have proposed a look-back calculation to improve the response (47–49). However, this calculation is seldom done.

Table 6. Main Characteristics of the Midsized EML Scintillation Cell

Cell volume, liter	0.515
Background, count/min	1
Efficiency for counting alphas	
^{222}Rn	0.534
^{220}Rn	0.599
^{218}Po	0.638
^{216}Po	0.658
^{214}Po	0.745
Average for ^{222}Rn, ^{218}Po, ^{214}Po	0.639

Measurement of Thoron Gas. The challenge in measuring thoron gas is to extract the thoron signal from an equally large, or often much larger, signal from radon. There are two ways that scintillation cells measure thoron gas. Both use the difference in half-life between the two series of nuclides.

One method uses the short half-life of thoron gas, and its first decay product, relative to radon. The grab sampling mode is used. In the first 3 minutes after filling a cell with a mixture of radon and thoron gas, there is a flurry of counts as the ^{220}Rn and ^{216}Po decay. If the activities are high as when measuring soil gas radon/thoron, one-minute counts yield good data. Thus, for example, if a one-minute count is taken in the second minute and another is taken in the sixth minute, mathematics are applied to calculate the concentrations of both radon and thoron. A complete discussion is available (50,51).

The second method uses the very short 145-ms half-life of ^{216}Po, so that an alpha particle from ^{220}Rn is followed in short order by another from ^{216}Po. The continuous flow mode of operation for the scintillation cell is used. The electronics are set up to record the time of arrival of each pulse, measured to the nearest ms or nearest 10 ms from an arbitrary starting point. If the average pulse rate is no more than a few per second (activity in the cell is no more than a few Bq), then the pulse-pairs from ^{220}Rn-^{216}Po can be resolved from the randomly spaced pulses from the decay of unrelated atoms. Details of this method are in the literature (45,52,53). The same is accomplished with hard-wired circuitry (54,55).

Two-Filter Tube

This method is now obsolete except in specialty applications. The method adapts quite easily to measuring thoron, so it is worth considering where both radon and thoron are to be measured.

In this method, the sample is taken by drawing air through a tube, usually cylindrical, with a filter on each end. The entrance filter allows radon and thoron to pass but blocks all decay products. If the airflow is sufficiently fast (residence time < 15 s), the concentration of radon and thoron inside the tube equals that outside. ^{218}Po atoms are formed inside the tube at the rate A_0V, where A_0 is the activity concentration of radon and V is the volume of the tube between the two filters. Depending on airflow and geometry, a large fraction of these ^{218}Po atoms are swept onto the exit filter. For thoron gas, the half-life of ^{216}Po is so short that the species collected is the next decay product ^{212}Pb. The fraction collected is calculated from diffusion theory (56), typically 90%. After a set time, sampling is stopped, and the exit filter is removed for alpha counting. The flat filter surface is ideal for alpha counting, so that the counting aspect is inherently more accurate than for gas measurements. Indeed, if done carefully, the two-filter method approaches a reference standard for measuring radon gas.

The two-filter method scales up quite well, so a high sensitivity is obtained by increasing the volume V. For example, a monitor with $V = 1$ m^3 and a sweep airflow of about 0.4 m^3/min is used to measure outdoor radon gas at Mauna Loa Observatory, where the levels are sometimes below 0.1 Bq/m^3 (57). This monitor has been automated. At fixed intervals, the filter is moved from the collection position to the counting position, and a fresh filter is installed for collection.

By using a detector that differentiates between alphas from the two decay chains, both thoron and radon are determined. Alternatively, the exit filter is recounted several hours later, after the short-lived radon decay products have decayed. The sensitivity for thoron, but not radon, is improved by extending the sampling time to 10 or 20 hours.

Summary: Practical Guide to Measuring Radon and Thoron Gas

For different measurement applications and recommended protocols, specific and practical instruments must be used to obtain the required information in timely fashion. EPA recommends (23) that the minimum measurement duration be 48 hours to ensure consistency among different groups and to include two daily radon cycles. Except for real estate transactions, most of the measurements for screening last from 2–7 days.

The choice of instrument depends on the measurement strategy and the purpose of the test. Detailed strategy recommendations for various indoor situations are in the above EPA document.

For *radon screening*, the present authors recommend either an activated carbon collector or a short-term electret ionization chamber. Both instruments are sensitive to environmental radon levels. The cost per exposure ranges from $15 to $50 and provides a single average value for the 2–7 day exposure. For *long-term* integrated measurements (3–12 months), we recommend long-term electret ionization chambers or nuclear alpha track detectors. The cost is from $25 to $50 per exposure.

For measurements before, during, and after *mitigation* and for *research measurements,* grab sampling methods with a scintillation cell or a flow-through activated carbon collector are recommended. The cost per measurement is $50 or more, because an operator must be at the site to make the measurement.

Many commercial sources have been identified in the forgoing text. For up to date information, the reader should consult advertisements that appear in the journal *Health Physics*.

INSTRUMENTATION FOR MEASURING DECAY PRODUCTS

As stated at the beginning of the last section, all measurement plans should include measuring radon gas. However, it is the decay products that are injurious to health, and in occupational settings the regulations are written in terms of the decay products. Therefore there are many cases in which measurement of decay products are either required or just make good sense.

The concentration of decay products is specified in terms of the activities (Bq/m^3) of the individual nuclides or as the potential alpha energy concentration (PAEC, nJ/m^3) which is a measure of the decay products in aggregate. PAEC is a good measure when health impact is the concern.

Table 3 shows that the PAEC from thoron is not typically negligible relative to that from radon. The following shows that it is relatively easy to measure the combined PAEC of radon and thoron. However, it is estimated that the health impact per unit of thoron PAEC is only about one-third that of radon PAEC (58). This is a good reason to make the extra effort of measuring the two PAECs separately.

As with radon, there are now several types of commercially available instruments for measuring the radon decay-product concentration, some of which also measure thoron decay products. In most measurement situations, there is a suitable commercial instrument. However, in the following, we discuss the manual methods first. (Although many protocols have been proposed and each has some merit, only the most widely used protocols are included here.) When used with care, the manual methods are more accurate and precise than instrument packages. Furthermore, the manual methods are the standard against which automated instruments are calibrated.

Manual methods require calculations that range from simple to complicated. The authors have developed a set of personal computer programs to do some of these calculations. This set, called EOKEMLUSDOE, is available from authors or from their former institution.

Table 7 illustrates the calculation of PAEC for an atmosphere containing a sample mixture of radon and thoron decay products. This is used as a basis for illustrating some of the measurement methods discussed in the following.

Sample Collection Using a Filter

Like most air sampling tasks, measurement of the airborne decay products of radon or thoron involves two parts: sample collection and sample assessment. For the collection step, the most commonly used method is *filtration,* which is the topic of this section. As seen in the last section, filtration is also one step in measuring radon and thoron gas (two-filter tube method). Other collection methods are available and are necessary for advanced measurements, such as determining the particle size of the airborne decay products. This is discussed in a later section.

As discussed earlier, decay products are continually formed in air by the radioactive decay of radon and thoron gas. Freshly formed decay products carry one unit of positive charge, which is usually lost in less than a second. In one minute or so, the decay product atom collides with and *attaches* to one of the aerosol particles always in the air. Even before this collision, the *unattached* decay products behave like very small ($\sim$1 nm) aerosol particles, that is, they stick to surfaces. Thus, the only practical way to collect a sample of decay products, attached or unattached, is to collect a sample of the aerosol particles. Ironically, only a tiny fraction of airborne particles, typically 1 in $\sim 10^5$, actually carry a decay-product atom.

Aerosol filtration is a broad subject that usually commands a chapter in books on measuring aerosols or air contaminants (59,60). Figure 3 (60,61) shows a sampling train that is commonly used for sampling radon/thoron decay products and aerosols in general. This train is for manual sampling, but similar components are used in automated instruments. Some prefer an alternative in which the flowmeter, usually a totalizing flowmeter, is placed after the pump. Either configuration is acceptable if used with care.

Table 7. Calculation of Potential Alpha Energy Concentration for a Hypothetical Mixture of Decay-Product Concentrations

Nuclide	Atoms per Bq	Potential Alpha Energy Per Atom, Mev	Potential Alpha Energy[a] Per Bq, nJ	Example Activity Concentration, Bq/m^3	Potential Alpha Energy Condentration, nJ/m^3
			Radon Series		
^{218}Po	263	13.69	0.579	100	57.9
^{214}Pb	2340	7.686	2.89	60	173.4
^{214}Bi	1723	7.686	2.13	40	85.2
^{214}Po	2.41×10^{-4}	7.686	2.98×10^{-7}	40	1.19×10^{-5}
Totals			*5.56*		*316.5*
			Thoron Series		
^{212}Pb	5.53×10^{4}	7.72	68.7	1	68.7
^{212}Bi	5250	7.72	6.52	0.1	0.652
^{212}Po	4.30×10^{-7}	8.784	6.08×10^{-10}	0.064	3.89×10^{-11}
^{208}Tl	264			0.03	0
Totals			*75.2*		*69.4*

[a] The conversion 1 MeV = 1.609×10^{-4} nJ was used.

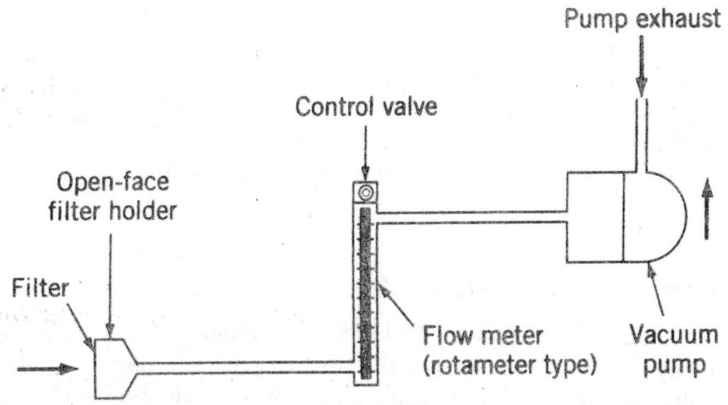

Figure 3. Radon–thoron decay product sampling train.

Filter. For decay product sampling, the filter most commonly used is a 47-mm diameter, cellulose ester porous membrane with nominal 0.8-μm pores (Type AA, Millipore Corporation, Bedford, Massachusetts). When operated at flow rates of 10 to 25 L min^{-1} (pressure drop about 4 to 10 kPa), this filter has an efficiency of at least 99.999% for collecting all sizes of particles. Although the membrane is about 100 μm thick, virtually all particles are caught in the first 10 μm. This thin deposit is ideal for subsequent alpha-counting. Other diameters of this type filter, 25 or 100 mm, are also used.

Another filter for decay-product sampling is a polyvinyl chloride-acrylonitrile porous membrane (Metricel DM-800, Gelman Scientific Company, Ann Arbor, Michigan). Its pressure drop and efficiency are similar to the cellulose ester filter, but it is much less prone to static electrification. This is an advantage because in certain circumstances (dry atmosphere and/or low particle concentration), a charged filter collects substantial amounts of charged ^{218}Po, even before the pump is started. This uncontrolled collection produces erroneous results.

A third filter type used in sampling decay-product aerosols is a thin (10-μm thick) polycarbonate sheet with 0.6- or 0.8-μm holes punched through (Nuclepore N060 or N080, Nuclepore Corporation, Pleasanton, California). This filter is so thin that alpha particles readily penetrate it, so that a large fraction of alphas from decay products caught on the front face can be counted from the back side. This is an advantage in automated instruments, largely offsetting the fact that the efficiency of this filter is only 80–90%, depending on flow rate and particle size.

Filter Holder. For decay-product sampling, it is important to use an *open-face* filter holder that places the filter as close as possible to the atmosphere being sampled. The reason is to minimize *plateout,* the loss of unattached decay products by deposition on the upstream parts of the filter holder. This is a real concern in atmospheres where the particle concentration is low, say less than 1000 particles cm^{-3}, so that the fraction unattached is large. Conversely, if the aerosol concentration is greater than 10^4 cm^{-3}, the design of the upstream part of the filter holder is not so important.

Additionally, the filter holder must (*1*) clamp the filter tightly so that there are no leaks around the edges; (*2*) support the filter so that it does not break because of the pressure; and (*3*) open easily so that the filter is removed quickly after sampling. Holders made of plastics are always suspect, because of the possibility of charging by friction. See *"Commercial Sources"* below for suggestions on filter holders and other components.

In radon progeny monitors designed to count during sampling, the filter, its holder and the detector must be designed as a unit. In most designs, the detector is mounted facing the front face of the filter, separated by a few mm, and the air sample must flow into this gap through a flow channel. All of these designs are prone to plateout. For the unattached decay products, this loss is very large (62). For a typical atmosphere where the unattached fraction of PAE is about 10%, plateout causes a loss of 10% of the signal.

Flow Meter. In Figure 3, a rotameter is used to monitor the airflow during sampling (be sure to use a rotameter that has the control valve *at the top,* as shown.) The rotameter is not a stand-alone flow measurement, however. Instead, a primary standard volume meter, such as a dry gas meter (similar to those used for residential heating gas service) or soap bubble meter, must be used periodically to calibrate the rotameter. An automated soap bubble meter (Gilibrator, Gilian Instrument Co., Wayne, New Jersey) is very convenient for this purpose. To calibrate, set up the flow train including filter exactly as it will be used, and then attach the primary standard flow meter to the *inlet* of the filter holder. (Often, this can be done with a rubber stopper that mates with the filter holder and a hose barb inserted in a hole in the stopper.) With the control valve, adjust the flow so that the rotameter ball is lined up with a selected scale mark. Then measure the flow rate with the primary standard. With a good quality rotameter, flow rate settings are repeatable within about 1%. The overall uncertainty in the measured flow rate, including calibration uncertainties, can be as small as about 3%. With a stopwatch to measure sampling time, the uncertainty in volume sampled can also be as low as 3%.

Automated samplers for decay products also use a flow train like that in Figure 3, but the rotameter is often replaced with an electronic flow meter. Although the better electronic flow meters are reliable, it is wise to check the calibration periodically with a primary standard as described above.

As an alternative to Figure 3, some workers prefer to eliminate the rotameter by permanently installing a dry gas meter at the exhaust of the pump. One concern is that, because of leaks, the volume of exhaust air is not the same as the volume of intake air. However, the method is certainly acceptable if care is taken to avoid leaks. The convenience gained by eliminating the periodic calibration of the rotameter is offset by the inconvenience of having to carry a dry gas meter everywhere.

The user should be on guard about the way airflow is expressed. Most commonly, the amount of air sampled is expressed in *ambient* liters or m^3, the actual volume sampled at the temperature and pressure at the sampling site. Occasionally, the amount sampled is expressed as standard liters or m^3. Most electronic flowmeters measure the *mass flow*, closely related to the volume measured in standard liters.

Sample Assessment

Scintillation Counters. The classical techniques for measuring radon decay products involve sampling on a filter, followed by one or more sequential measurements of gross alpha activity on the filter. The *drawer-type scintillation alpha counter* is one commonly used counter configuration (61) that is a benchmark against which other detectors are measured. The important features are as follows:

1. accepts filters (or other circular, flat substrates) up to 47-mm diameter;
2. 49% efficient for counting alpha particles, for all alpha particle energies of interest;
3. 0.1 min^{-1} typical background (0.01 min^{-1} can be achieved by special handling);
4. positive-going, TTL-compatible output pulses, suitable for standard pulse-counting circuits;
5. small, light (about 1 kg), and rugged enough for use in the field.

In this design, a 2.54-cm diameter, end-window photomultiplier tube is mounted window-down inside a lightproof housing with a drawer at the bottom. The housing also contains the power supply and pulse-shaping circuitry. The drawer has a 50-mm diameter circular recess which is visible when the drawer is opened. The exposed filter is placed into this recess and a scintillator disk consisting of ZnS(Ag) powder bonded to a Mylar sheet is placed on the filter. (It is crucial that the filter is placed "dirty side up" and the scintillator disk placed "shiny side up," so that there is close contact between the deposited decay products and the active side of the disk.) Closing the drawer seals the system against stray light and positions the filter-scintillator sandwich opposite the photomultiplier tube window. Counting may commence in about 15 s, after the residual effects of exposure to light have died away. In this configuration, the close contact between the filter and the scintillator gives an alpha-counting efficiency (item 2 above) close to the theoretical maximum 50%, independent of the alpha-particle energy. It has been found that light pulses from the scintillator easily reach the photomultiplier tube, in spite of a few mm of air gap.

Figure 4 shows simulated count data from a 10-L/min, 5-min sample taken in an atmosphere where the concentrations of ^{218}Po, ^{214}Pb, ^{214}Bi, ^{212}Pb, and ^{212}Bi are 100, 60, 40, 1, and 0.1 Bq m^{-3}, respectively. The counter efficiency was assumed to be 49% for all alphas, and the counter background was assumed to be 0.1 count min^{-1}. The first point plotted is for the second minute after the end of sampling. The most striking feature of the data is the scatter. A clear downward trend is seen in the first 15 min, followed by a level period for the next 25 min. At the low concentrations assumed, ^{212}Pb and ^{212}Bi have negligible impact on the data in Figure 4. The effect would be seen only after 200 min.

Other scintillation counter designs incorporate the scintillator into the counter itself, leaving an air gap between the filter being counted and the scintillator. In some designs, the scintillator is covered with an aluminized Mylar membrane to further shield the scintillator from stray light. Both the air gap and the light shield reduce the alpha-counting efficiency and make the efficiency energy-dependent. In this case, it is necessary to de-

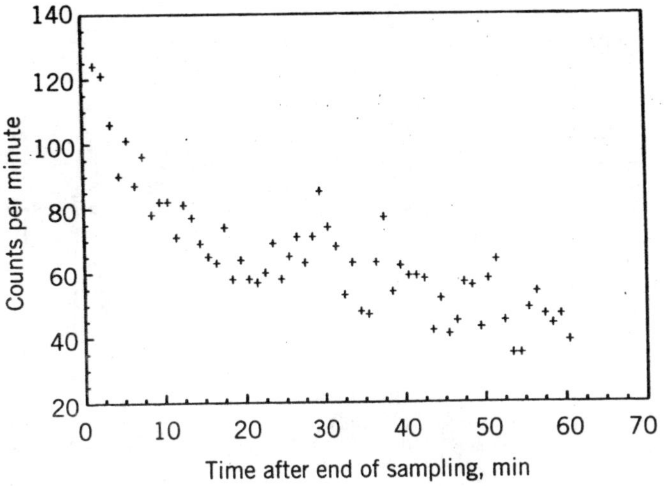

Figure 4. Alpha-count data from a simulated radon–thoron decay product sample where counting efficiency = 0.49 and counter background = 0.1 min^{-1}: 5-min sample at 10 L/min; counting efficiency 49% for all alphas; concentration of ^{218}Po, ^{214}Pb, ^{214}Bi, ^{212}Pb, 212Bi = 100, 60, 40, 1.0, and 0.1 Bq/m^3, respectively.

termine efficiency with standard alpha sources that have different energies, then interpolate or extrapolate to determine the efficiencies for decay-product alpha particles.

Semiconductor Alpha-Detectors. Starting in the late 1960s, semiconductor detectors have become available as alternatives to scintillation alpha-detectors. These detectors are smaller and more rugged than photomultiplier-based scintillation detectors. Early semiconductor detectors were small, expensive, and had low counting efficiency. Costs have dropped over the years, and detectors with sensitive areas 25 mm in diameter or larger are now quite common. These are the detectors chosen in commercially available automated instruments for radon decay-product measurement, although acceptance for use in manual sampling systems has not kept pace. Semiconductor detectors are also very good for spectrometry, because each alpha particle yields an output pulse with amplitude proportional to the particle's energy. Thus, alpha particles from ^{218}Po, ^{214}Po, and ^{212}Po are counted separately, facilitating determination of the mix of decay-product nuclides in the sample.

Figure 5 shows the pulse-height spectrum from a sample of radon decay products drawn onto a DM-800 filter at a relatively high flow rate (63). Separate peaks are seen for ^{218}Po and ^{214}Po alpha particles. The steep right edge reflects the fact that most of the aerosol particles are caught on the surface of the filter, and their alphas have a direct path to the detector. The tailing to the left side reflects the fact that some of the aerosol particles are drawn into the filter, so that their alpha particles have to penetrate through filter material to reach the detector. The alpha particles lose energy in penetrating through the filter. This spectrum was obtained after evacuating the spectrometer chamber. If the gap were filled with air, as in most field instruments, the peaks would be broader and overlap slightly.

Other Detection Methods. Solid-state nuclear track (SSNT) detectors, ionization chambers, and thermoluminescence chips

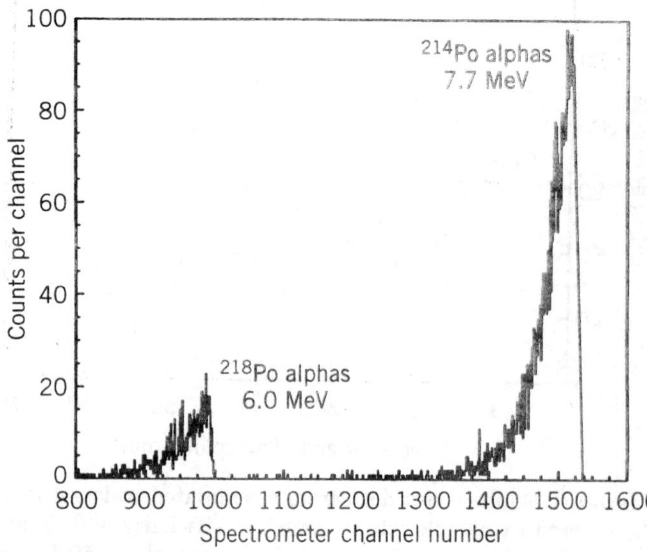

Figure 5. Example alpha-particle energy spectrum from a radon decay product sample using a DM800 filter, attached progeny, 25 L/min. The counting is done in a vacuum.

(TLD) have all been used to assess activity from filter samples of radon decay products. Measurement of beta and gamma radiation has also been used. Of these, only the SSNT detector is important now.

Specific Sampling/Counting Protocols

Kusnetz and Rolle One-Count Methods. In the summer of 1951, a simple field method was developed for measuring radon decay products in uranium mine atmospheres, expressed in terms of their potential alpha energy concentration (64,65). This basic method is still used today, especially in mining where simplicity and quick results are important. One study (66) found that the method is improved by selecting timing sequences that minimize the value of the maximum systematic error. Table 8 gives some numeric factors for use in the modified Kusnetz and Rolle methods.

For either the Kusnetz or Rolle method, the PAEC is calculated from the following equation:

$$\mathrm{PAEC} = \frac{r}{\epsilon T_s Q C_f} \tag{5}$$

where PAEC is the potential alpha-energy concentration (nJ/m^3); r is the net count rate (counts/min); ε is the fractional counter efficiency (counts/s) assumed to be the same for 6 and 7.7 MeV alphas; T_s is the sampling time (min); Q is the sampling flow rate (m^3/min); and C_f is the conversion factor (dpm per nJ collected), which depends on time as shown in Table 8.

The uncertainty caused by statistics is calculated from the following equation:

$$\sigma_p = \frac{\sigma_r}{\epsilon Q T_s C_f} \tag{6}$$

where σ_p is the uncertainty (nJ/m^3) in the PAEC and σ_r is the estimated standard deviation of the net count rate (from eq. 3). Beyond this, there is an inherent uncertainty in this method because the proportion of the different decay products is not known and may differ from that assumed in deriving equation 5. For the Rolle timing sequences in Table 8, it was shown that the maximum error from this source is 12%. The present authors think that 5% is typical for this uncertainty.

As an example, the simulated count data in Figure 4 has 671 counts in the period 35–45 min after the end of sampling. With this and the other values pertaining to Figure 4, equation 5 gives 379 nJ m^{-3}. The uncertainty from equation 6 is 15 nJ/m^3, and the inherent uncertainty is about 19 nJ/m^3. This result is shown in Table 10, as is an example calculation using the 5-, 6-, 5-min Rolle protocol.

To measure the PAEC of both radon and thoron, a two-count method can be used (67). As an example, for a 10-min sample,

$$\mathrm{PAEC_{Rn}} = \frac{r_1 - kr_2}{\epsilon T_s Q C_{f1}}, \text{ and } \mathrm{PAEC_{Tn}} = \frac{r_2}{\epsilon T_s Q C_{f2}} \tag{7}$$

where PAEC$_x$ is the PAEC (nJ/m^3) based on decay chain x, where x = Rn or Tn; r_1, r_2 = the net count rate (counts min^{-1}) based on (*1*) a 10-min count commencing 50 min after the end of sampling and (*2*) a 1-h count commencing 5 h after the end of sampling; C_{f1}, C_{f2} = 5.24, 0.64 counts/min/nJ, respectively; and k = 1.15. The other symbols were defined after equation 5. The statistical uncertainties are calculated from

$$\sigma_{Rn} = \frac{\sqrt{\sigma_1^2 + k^2\sigma_2^2}}{\epsilon T_s Q C_{f1}} \text{ and } \sigma_{Tn} = \frac{\sigma_2}{\epsilon T_s Q C_{f2}} \tag{8}$$

Table 8. Numeric Factors for the Modified Kusnetz and Rolle One-Count Methods for Measuring Radon Potential Alpha Energy Concentration

Modified Kusnetz method[b]		Rolle method[a]		
		Required Delay Time, min		
Delay Time, min	Conversion Factor, dpm/nJ	Counting Time, min	5-Min Sample	10-Min Sample
35	7.21	1	7.79	
45	6.25	2	7.33	
55	5.29	3	6.88	
65	4.32	4	6.46	
75	3.61	5	6.06	4.35
85	2.88	10	4.35	2.63

[a] Adapted from Table 1 of Ref. 66. (Conversion factor = 10.2 dpm/Bq in all cases)
[b] Adapted from Ref. 61; however, note the difference in definition of delay time. Delay time here means the interval between sampling and counting. (Sample time = 5 min, counting time = 10 min in all cases).

where σ_{Rn}, σ_{Tn} are the uncertainties (standard deviation, nJ/m3) in $PAEC_{Rn}$ and $PAEC_{Tn}$, respectively; and σ_1, σ_2 are the uncertainties in r_1 and r_2, respectively, from equation 3.

The study (67) states that equation 7 slightly overestimates $PAEC_{Tn}$ and supplies a graphed correction factor. For nominal $PAEC_{Rn}/PAEC_{Tn}$ = 3, 5 and 7, the correction factors are 0.96, 0.93 and 0.90, respectively.

Thomas-Tsivoglou and Nazaroff Three-Count Methods. In the experience of the authors, the most widely used manual method for measuring the concentration of radon decay products is Tsivoglou's method, as updated and refined (68). The Thomas-Tsivoglou method requires three sequential counts (in place of the one in the Kusnetz and Rolle methods) but yields more information, the concentrations for three nuclides: ^{218}Po, ^{214}Pb and ^{214}Bi. Later, a study (28) proposed alternate counting sequences more suited for measuring decay products at low concentrations.

Counting sequences and calculation coefficients for the Thomas-Tsivoglou and two Nazaroff protocols are given in Table 9. Each requires a 5-min sample. The first two allow 2 min for transferring the filter into the alpha counter, whereas the last allows 1 min. The following equations are needed to calculate the air concentrations (28):

$$A_i = \frac{1}{\epsilon Q}\sum_{j=1}^{3} K_{ij}N_j \text{ and } \sigma_i = \frac{1}{\epsilon Q}\sqrt{\sum_{j=1}^{3} K_{ij}^2 N_j} \quad (9)$$

where $i = 1, 2, 3$ denote ^{218}Po, ^{214}Pb, and ^{214}Bi, respectively; A_i is the activity concentration (Bq m^{-3}) of nuclide i; σ_i is the uncertainty (standard deviation; Bq m^{-3}) in A_i; K_{ij} are coefficients found in Table 9; j denotes the three count intervals; N_j are the net counts in interval j; ε is the fractional counting efficiency (counts s^{-1}Bq^{-1}), assumed to be the same for 6 and 7.7 MeV alphas; and Q is the sampling flow rate (m^3/min).

The PAEC and its uncertainty (σ_p) are also calculated directly from the net counts:

$$\text{PAEC} = \frac{1}{\epsilon Q}\sum_{j=1}^{3} L_j N_j \text{and} \sigma_p = \frac{1}{\epsilon Q}\sqrt{\sum_{j=1}^{3} L_j^2 N_j} \quad (10)$$

Terms involving the uncertainty of the background count rate are omitted because they are seldom significant.

To illustrate the use of these equations, the Thomas-Tsivoglou method is applied to the data of Figure 4. The gross counts in the three periods are 317, 1047, and 599, respectively. Thus, the calculation for ^{218}Po is as follows:

Table 9. Coefficients for Various Protocols for Determining ^{218}Po, ^{214}Pb, ^{214}Bi, and PAEC from Sampling on a Filter with Three Total Alpha-Counts Subsequent to Sampling

	Pattern of Coefficients (Nazaroff's Notation)[a]		
^{218}Po	K_{11}	K_{12}	K_{13}
^{214}Pb	K_{21}	K_{22}	K_{23}
^{214}Bi	K_{31}	K_{32}	K_{33}
PAEC	L_1	L_2	L_3
	Thomas-Tsivoglou Protocol (5, 7–10, 11–25, 26–35 min)[b]		
^{218}Po	0.006256	−0.00303	0.002856
^{214}Pb	0.000055	−0.000772	0.001828
^{214}Bi	−0.0008258	0.00122	−0.001383
PAEC	0.002016	−0.001383	0.003978
	Nazaroff Protocol I (5, 7–10, 13–30, 42–60 min)		
^{218}Po	0.004866	−0.00179	0.001208
^{214}Pb	−0.000452	−0.000103	0.000597
^{214}Bi	−0.000234	0.000646	−0.000548
PAEC	0.001011	0.000041	0.001255
	Nazaroff Protocol II (5, 6–9, 12–29, 40–60 min)		
^{218}Po	0.003876	−0.001429	0.000869
^{214}Pb	−0.000329	−0.000143	0.000551
^{214}Bi	−0.00024	0.000622	−0.000459
PAEC	0.00078	0.000082	0.001112

[a] Units: K_{ij}, Bq/min; L_j, nJ/min.
[b] That is, sample during time 0–5, and count during times 7–10, 11–25, 26–35, where all times are measured from the start of sampling.

$$N_1, N_2, N_3 = 316.7, 1045.6, \text{and} 598.1, \text{respectively};$$

$$A_1 = (0.006252 \times 316.7 - 0.00303 \times 1045.6 + 0.002856 \times 598.1)/(0.49 \times 0.01)$$

$$= 106.4 \text{ Bq m}^{-3};$$

$$\sigma_1 = (39.1 \times 10^{-6} \times 316.7 + 9.18 \times 10^{-6} \times 1045.6 + 8.16 \times 10^{-6} \times 598.1)^{0.5}/(0.49 \times 0.01)$$

$$= 33.4 \text{ Bq m}^{-3}$$

The calculations for ^{214}Pb and ^{214}Bi are done similarly, and the results are shown in Table 10. Also shown are the results of applying the Nazaroff I protocol to the data for Figure 4.

Raabe-Wrenn Weighted Least Squares Method. This method predates Thomas' modification of the Tsivoglou method (69).

Table 10. Computed Decay-Product Concentrations for the Data Shown in Figure 4

	Computed Concentration and Estimated Uncertainty (SD)[a]									
	Kusnetz		Rolle		Thomas-Tsivoglou		Nazaroff I		Raabe-Wrenn	
Nuclide	Value	SD	Value	SD	Value	SD	Value	SD	Value	SD
Po-218					106	33	121	23	112	15
Pb-214					62	10	61	4	62	5
Bi-214					38	11	33	6	43	3
Pb-212									1	0.3
Bi-212									−5	3
PAEC	379	15	334	37	320	23	316	9	370[b]	14

[a] Units: Bq/m^3 for individual nuclides, nJ/m^3 for PAEC.
[b] Includes an estimated 37 nJ/m^3 from thoron.

Table 11. Example Components for a Radon/Thoron Decay-Product Sampling System

Component	Source	Model Number
Primary flow standard	Gilian Instrument Co. Wayne, N.J.	Gilibrator System Kit
Filter	Gelman Sciences 1-800-521-1520	DM-800 (specify 47 mm)
Filter holder	Gelman Sciences 1-800-521-1520	1220 (47 mm)
Rotameter	Brooks Instrument Division, Emerson Electric Co. Hatfield, Pa.	1355HA2ABD5A (specify valve at top)
Pump	Gast Manufacturing Corp. Benton Harbor, Mich.	1531-107-G288
Interconnecting tubes	Flexible clear plastic tube with walls thick enough to resist collapse while pumping, selected to fit tightly on hose barbs.	
Drawer-type scintillation alpha counter	Ludlum Measurements, Inc. Sweetwater, Tex.	43-10 (specify no scintillator)
Scintillator disks	William B. Johnson & Associates, Inc. Ronceverte, W.Va. 24970	ASP-3 (specify 50 mm circles)
Power supply/Scaler	Ludlum Measurements, Inc. Sweetwater, Tex.	1000

However, it is more complicated in terms of amount of data needed and the calculations required and was not widely used until the advent of portable personal computers. The method uses total alpha-count data. Although the method is flexible as to the time schedule of counting, it is most commonly used with data recorded in one-minute increments like that shown in Figure 4. For the selected time schedule, the radioactive decay equations are used to construct a set of linear equations showing the possible time evolutions of the count data. A mathematical method called weighted least squares is then used to fit these candidate equations, in which the nuclide concentrations are the unknowns, to the data.

An example is shown in Table 10, where the Raabe-Wrenn calculation is applied to the simulated count data shown in Figure 4. Note that the Raabe-Wrenn method includes a calculation of the thoron decay products. To take advantage of this, is necessary to continue the counting for about 5 h. The results in Table 10 are therefore based on 294 one-minute counts, of which only the first 60 are shown in Figure 4. The calculated concentration of ^{218}Po is 112 ± 15 Bq m^{-3}. The uncertainly is smaller here than in the Thomas-Tsivoglou calculation, mainly because of starting the counting 1 min after sampling, rather than 2 min as in Thomas-Tsivoglou. The result for ^{212}Pb is usable, although the uncertainty is quite large. The result for ^{212}Bi is useless. It is virtually impossible to measure this nuclide without alpha spectrometry.

Duggan and Howell Method. An apparatus and procedure are described (70) for measuring individual radon decay products at the low concentration levels found in the open atmosphere. It features alpha spectrometric counting and the following protocol: 30 min sampling with simultaneous counting, 30 min wait, 30 min counting. Sensitivity was also helped by a high flow rate (80 L/min). Many variants have since been described (2–4).

The present authors have never found a compelling reason to adopt spectrometric counting for day-to-day manual sampling. This would change if a project came along that required frequent measurements of *thoron* decay products. Here the ability to separately count alphas from ^{212}Po in the thoron decay chain would be very helpful. Many of the commercial instruments described below use semiconductor detectors and alpha spectrometry. Further examples come up in the section on aerosol measurements.

Commercially Available Radon Progeny Apparatus and Instruments

Table 11 gives one set of components that the authors have used extensively for manually sampling radon and thoron decay products and are satisfactory. Comprehensive lists of sources for components are in the literature (71).

Table 12 contains key characteristics of the currently available commercial instruments. Further information is available in the literature (1) or by contacting the manufacturer.

Most commercial instruments for decay-product measurement operate continuously. Those using these instruments should be aware of the effects of ingrowth and decay. To illustrate these effects, Figure 6 is constructed to show the response of a hypothetical continuous monitor when sampling the sample atmosphere given in Table 7. The following

Table 12. Commercially Available Instruments or Services for Measurement of Radon or Thoron Decay Products

Manufacturer or Service Provider[a]	Manufacturer's Designation	Principle of Detection	Lower Limit of Detection at exposure period[b]	Size and Weight	Comments
	Integrating type: Readout done after exposure, service companies supply samplers and perform readouts				
CAIRS	Personal Alpha Dosimeter	Alpha track counting	4 nJ/m^3 at 1 week		Battery operated. Designed to be worn on person. Resolves thoron PAEC from radon PAEC; $50 per exposure.
CAIRS	Home Radon Monitor	Alpha track counting	4 nJ/m^3 at 1 week	0.43 kg	Resolves thoron PAEC from radon PAEC; $50 per exposure.
RAD Services	RAD/WL	Alpha track counting			
	Continuous type: Data recorded at regular intervals on paper, or in internal memory for later retrieval				
Eberline	WLM-1A	Semiconductor total alpha counting	10 nJ/m^3	15 × 12 × 20 cm; 2.6 kg	Battery operated. Resolves thoron PAEC from radon PAEC. WLM-1A: $3,000; WLR-1A programmer/reader unit: $4,000.
Thompson and Nielson	Radon WL Meter	Semiconductor total alpha counting	20 nJ/m^3	18 × 10 × 4.5 cm; 0.5 kg	$2,200. Printer/recorder options available.
Scintrex	WLM-30	Semiconductor spectrometric alpha counting	20 nJ/m^3	30 × 25 × 18 cm; 5.9 kg	Can resolve thoron PAEC from radon PAEC. $4,500.
Alpha Nuclear	Alpha Prism Model 560	Semiconductor total alpha counting	20 nJ/m^3	14 × 5 × 7 cm; 0.4 kg	Battery operation. Designed to be worn on person. Can resolve thoron PAEC from radon PAEC. $3,800 for sampler and microcomputer readout
SARAD	EQF 3020	Semiconductor spectrometric alpha counting	0.17 m^{-3} in 24 h		Measures unattached fraction
	Multiple operating modes available through front-panel or computer programming.				
Thompson and Nielson	Instant Radon Progeny Monitor	Semiconductor total alpha counting	20 nJ/m^3	21 × 20 × 18 cm 5.9 kg	Battery operated. Grab sampling. Rough PAEC in 5 min; better value in 22 min. $2,700.
Pylon	Model AB-5 monitor; Model AEP filter head	Scintillation pulse counting			Battery operated. Continuous or cyclic operation.
Pylon	WLx Working Level Monitor	Semiconductor total alpha counting	20 nJ/m^3	42 × 25 × 23 cm 6 kg	Battery operation (optional). Grab or continuous operation, multiple counting protocols. Resolves thoron PAEC from radon PAEC. $6,300.
Alpha-Nuclear	Alpha Smart 770A	Semiconductor spectrometric alpha counting	10 nJ/m^3	51 × 41 × 20 cm 20 kg	Cyclic operation, multiple counting protocols. Resolves thoron PAEC from radon PAEC; also individual nuclides. $35,000.

[a] See Table 14 for addresses.
[b] Where LLD values were not given by the manufacturer, an estimate is given.

values are assumed: sampling flow rate = 0.25 L/min; alpha-counts recorded in three energy bins, every 15 min; detector efficiency = 20% for all alpha-energies; inlet losses = 0; background = 0.033 cpm in each energy bin. Consistent with the recommended minimum, it is assumed that sampling lasts 48 h and counting continues 27 h after the end of sampling.

The data in Figure 6 were generated by a Monte Carlo procedure that mimics the statistics associated with radioactivity. However, no attempt was made to mimic the larger hour-by-hour, decay-product concentration changes, common both indoors and outdoors. The latter variability is more difficult to cope with than the statistical scatter.

The lesson from Figure 6 is that, with a continuous monitor, a 48-h sample is needed to assure fully measuring thoron decay products because about 40 h is needed for ingrowth of alpha activity from ^{212}Po. If the contribution of thoron decay products is small, 3 to 4 h is adequate to fully measure radon decay products. For both series of decay products, shorter sampling times are used if a mathematical correction is made for incomplete ingrowth.

To calculate the combined PAEC of radon and thoron decay products, continuous monitors use an equation of the following form:

$$\text{PAEC(Rn + Tn)} = E_{av}\frac{r_{sn}}{\epsilon Q} \text{ and } \sigma_p = E_{av}\frac{\sigma_{sn}}{\epsilon Q} \quad (11)$$

where PAEC (Rn + Tn) is the combined PAEC of radon and thoron (nJ/m^3); σ_p is the uncertainty (standard deviation, nJ/m^3) in PAEC (Rn + Tn); E_{av} is the average alpha-energy of the

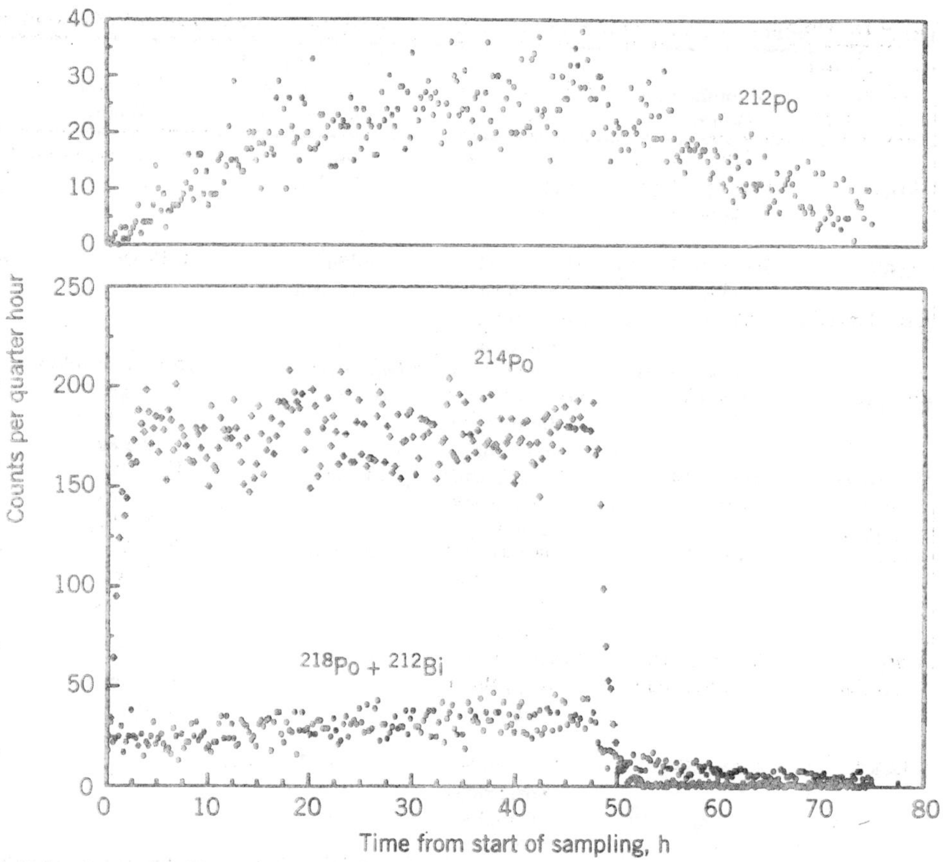

Figure 6. Simulated count data from a continuous radon–thoron decay product monitor where the flow rate = 0.25 L/min, counting efficiency = 20% for all alphas, counter background = 0.1 min^{-1}, and concentration of ^{218}Po, ^{214}Pb, ^{214}Bi, ^{212}Pb, and 212Bi = 100, 60, 40, 1.0, and 0.1 Bq/m^3, respectively.

mixture of alpha particles (nJ); r_{sn} is the steady-state, gross, alpha net count rate (counts/min); σ_{sn} is the uncertainty in r_{sn}, from equation 3; ε is the detector fractional efficiency; and Q is the sample flow rate (m^3/min).

As an example, the combined count rate for the data in Figure 6 is about (25 + 175 + 35) per 15 min, less 1.5 background counts per 15 min, so r_{sn} = 15.6 counts/min. E_{av} is computed from (8.8 × 25 + 7.7 × 175 + 6.0 × 35)/(25 + 175 + 35) = 7.5 MeV = 0.00122 nJ. Also, ε = 0.2 and Q = 0.00025 m^3/min. With these values, equation 11 yields PAEC (Rn + Tn) = 380 nJ/m^3, as compared to the true value 317 + 69 = 386 nJ/m^3 from Table 7.

For continuous monitors that measure only the total alpha count rate, E_{av} is not known in advance. However, it is known to be within a range of 6.84 MeV (for pure ^{218}Po, which yields 2 alpha particles) to 7.72 MeV (pure ^{212}Pb), that is, in the range 0.00110 to 0.00124 nJ. Both extremes are unlikely, so the true value is close to the middle of the range, that is, 0.00118 nJ.

In addition to the combined PAEC, many of the available monitors measure the separate contributions of the radon and thoron series. For monitors that count the alphas in separate energy bins, as in Figure 6, it is straightforward to separate the PAEC contributions. For monitors without this capability, the contribution of the thoron series is estimated by analyzing the decaying count rate after the end of sampling. As seen in Figure 6, the contribution of the radon series is largely gone after 4 h, so the count rate beyond this point is used to estimate the PAEC of the thoron series.

There are many other differences among the available decay-product monitors. Some separately measure the activity concentration of ^{218}Po, ^{214}Pb, ^{214}Bi, ^{212}Pb, and ^{212}Bi. Some separately measure the unattached fraction of decay products.

Summary: Practical Guide to Measuring Radon and Thoron Decay Products

Those charged with measuring radon decay-product concentration should look first at the several commercial instruments described in Table 12, and should check the advertisements in recent issues of the journal *Health Physics* for up-to-date information. These instruments combine all the necessary components into one or two packages, which are easier to handle and have a shorter learning curve than a manual system as described in Table 11. Furthermore, many of these instruments have a choice of operating modes, including grab sampling, and provide automated data handling, storage, and computation.

However, the manual methods will probably continue to be useful. A daily or twice-weekly sample with a manual method provides QA that a continuous instrument is working properly. Occasionally, there are cases where the 5- or 10-min time resolution offered by the manual method is advantageous compared with the few-hour time resolution inherent in continuous instruments. A word of caution: although the manual methods appear straightforward, understanding the fundamentals coupled with training and practice are needed to ensure consistently reliable results.

MEASURING RADON/THORON DECAY-PRODUCT PARTICLE SIZE

Importance of Particle Size

One of the permanent features of the atmosphere, both indoor and out, is particles. Suspended in each cm^3 of air there are from 10^3 to 10^6 diverse particles, some solid, some liquid. Most particles are spherical, but there are also fibers, rock fragments, flocs of soot, among others. The size of these particles, expressed as diameter (or *equivalent* diameter, if not spherical), cover more than a 1000-fold range from 10 nm to >10 μm. All of these particles are carried along by air currents, but they also move relative to the air. The largest particles settle by gravity onto horizontal surfaces. The smallest particles travel about by Brownian motion until they strike and stick to a surface, such as a wall, ceiling, or furniture. Thus if there were no sources, particles would gradually disappear from the atmosphere. But there are sources. Indoors, cooking and smoking of tobacco produce billions of particles of size 200–500 nm, and these gradually mix into the atmosphere. Gas cooking flames produce huge numbers of very small particles, about 10 nm in diameter. A leaky vacuum cleaner bag blows dirt particles from the floor back into the air. Consumer spray products produce a mist that contributes to the aerosol. Outdoors, another mix of sources comes into play. The number and size of particles in the air continually change because of the interplay between sources and losses. In addition to gravity settling and Brownian motion, other processes affecting (some) particles are shrinkage by evaporation or growth by absorption of water vapor. A whole branch of science dealing with aerosols has emerged in the past 35 years. Several books give a good general exposition of theory and practice (72,73). Specific information on methods of measurement are in the recent handbooks already cited.

As noted before, radon decay-product atoms become entangled with the aerosol. Before attachment, radon decay products behave like 1-nm aerosol particles, and once attached they take on the size characteristics of the host particle. Thus, in effect, the size range for radon decay products is 1 nm to >10 μm. Size is important for the reasons stated above (gravity settling, Brownian diffusion) and also because size affects the amount and location of deposition in the respiratory tract. Figure 7 shows the fractional deposition in the tracheobronchial region of the respiratory tract, the region where most radon-induced cancers are found. The figure pertains to adult males in light exercise, with oral breathing (74). The fraction of inhaled particles that deposits varies from about 55% for 3-nm particles to about 2% for 500-nm particles. In locations or situations where questions about the health hazard from radon decay products are pressing, it is necessary to include a measurement of the particle size distribution.

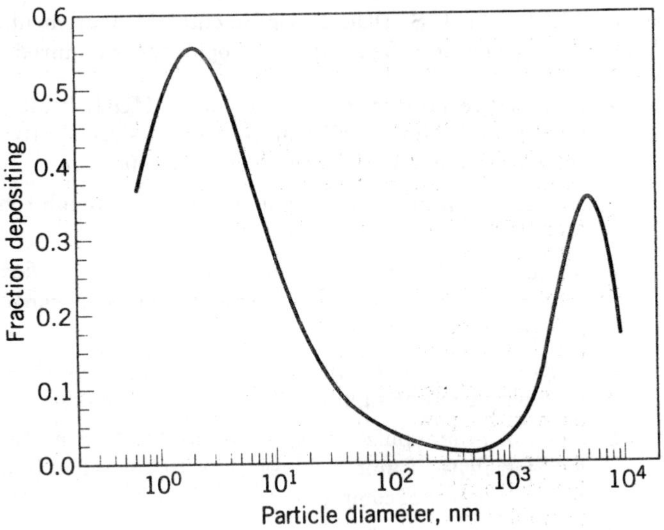

Figure 7. Deposition of aerosol particles in the tracheobronchial region of the lung of an adult male after light exercise, particle density = 1.5 g/cm³ (74).

Measurement of the Unattached Fraction

In the classical view, radon/thoron decay products comprise two distinct fractions, attached and unattached, that differ by roughly a factor of 100 in particle size. The two greatly differ in their tendency to deposit on surfaces, such as the bronchial epithelium. For this reason, simple techniques were developed to sample and analyze the fractions separately. Although cases have been found recently where the distinction between attached and unattached is blurry, the classical picture is still valid in many cases. Therefore, there is still value in measuring the unattached fraction.

The simplest and most commonly used method for determining the unattached fraction is the *wire screen method* (75). A circular piece of 60-mesh wire screen is mounted into an open-face filter holder and air is drawn through as in sampling on a filter. Tests show that with air flow of 11.5 cm s^{-1}, the screen collects 77% of unattached decay products but less than 1% of attached decay products. After sampling, the upstream face of the screen is alpha counted in the same manner as a filter. The counting efficiency is less than for a filter, because some of the activity on the screen is deposited on the downstream side. Tests show that the combined collection and counting efficiency of the above screen for unattached decay products is 49% of that for a filter. In summary, the method consists of sampling simultaneously with the screen and a conventional reference filter, then alpha counting the two simultaneously. The unattached fraction f is calculated from the following expression:

$$f = \frac{A_{screen}}{0.49 A_{total}} \tag{12}$$

where f is the unattached fraction; A_{screen} is the concentration as measured by the screen; and A_{total} is the concentration as measured by the reference filter.The quantity A refers to the activity concentration of any of the nuclides or to the PAEC.

This wire screen method is still in use, although different workers use different types of screen (76). Equations are now available for the collection (77,78) and counting (79) efficiency, so that the empirical calibration described above is no longer necessary. The graded screen array (GSA), mentioned below, is a descendant of the wire screen method.

Manual Methods for Measuring Decay-Product Particle Size

The steps involved in measuring decay-product particle size are explained by an example, namely, the manual method used since 1990 by the U.S. DOE Environmental Measurements Laboratory in New York. The following apparatus is required:

1. microorifice uniform deposit impactor (MOUDI; MSP Corporation, Minneapolis, Minnesota), a size-selective sampler that covers the size range of 50–5000 nm;
2. a suitable pump to draw air at 30 L min^{-1} through the MOUDI;
3. (optional) a diffusion-type sampler called the graded screen array (GSA), which extends the size range covered down to 0.5 nm;
4. a set of 10 drawer-type scintillation alpha counters, each with a power supply and pulse-shaping circuit;
5. a pair of 5-input pulse-counting boards (CTM-05; Keithley-Metrabyte, Taunton, Massachusetts) installed in a portable personal computer;
6. interconnecting cables;
7. software to manage the simultaneous collection of data from the 10 counters, to perform the Raabe-Wrenn calculation (described earlier) on each of the 10 sets of count data, and to calculate particle size distributions from the data and the known characteristics of the MOUDI.

Items 1, 2, 4, 5, and 6 are commercially available. If the GSA option is taken, it is necessary to use more counters and therefore a second computer with pulse-counting hardware. The software, developed with considerable effort over many years, is available from EML in a program set called EOKEMLUSDOE.

Table 13 shows the most important characteristics of the MOUDI. The air sample is drawn at 30 L/min through the eight stages, which are arranged in series (or *cascade*). At each stage, a sample is collected on a flat, circular, metal plate that is ideal for alpha-counting. As a good approximation, it can be said that Stage 3A collects all particles with diameter larger than 3200 nm (3.2 μm), that Stage 4A collects all particles with diameters in the range 1.8 to 3.2 μm, etc. The MOUDI also has a back-up filter that collects all particles smaller than 45 nm. Technical information about the MOUDI and the GSA are available in the literature (80,81).

After sampling for 5 to 10 minutes, the MOUDI is dismantled and the impaction plates and the back-up filter are transferred to the scintillation alpha counters. Two operators are needed for this step. With practice, the transfer is done within about 2.5 min, fast enough to collect at least some counts from the short-lived ^{218}Po. Counting proceeds exactly as in the radon/thoron decay-product measurement discussed in the previous section. Step-by-step procedures and some technical information are in the EML Procedures Manual (82).

Table 13. Characteristics of the MOUDI (Manufacturer's Data)

Stage	Cut Size, nm	No. of Nozzles
3A	3200	10
4A	1800	20
5A	1000	40
6A	560	80
7A	290	900
8A	173	900
B	97	2000
BB	45	2000

Automated Methods for Measuring Decay-Product Particle Size

Apparatus has been designed that does the sampling, counting, and much of the calculations automatically. The key design step was to eliminate the need for transferring the sample collection substrate, such as filters. Some advantages are obvious: around-the-clock, unattended, acquisition of particle size distribution data. A side benefit is increased sensitivity for measuring low concentrations, made possible by alpha counting during and after sampling. The disadvantage is that the time and effort required for the initial setup is greater than for a manual system.

A partly automated system based on a screen-type diffusion battery is described (83) that permits coverage of the size range 0.5 to 2000 nm (however, the median diameter cannot be over 400 nm). The flow rate through each unit of the battery was 2 m^3/h. The particles that penetrated each battery were collected on Nuclepore filters, and alpha particles from the filters were counted during and after sampling by surface-barrier semiconductor alpha-detectors. Typically, the sampling was continued for up to 1 h and counting was continued for 20 min longer. The authors indicate that measurements could be made at decay product concentrations down to 1 Bq/m^3. The basic system has four sampling/counting channels. For improved particle size data, consecutive samples are taken with different diffusion battery units attached to the channels.

A fully automated, semicontinuous system is described (84) also based on a screen-type diffusion battery. The system has six parallel flow channels and covers the size range 0.5 to 500 nm. The particles that penetrate the diffusion screens are collected on 25-mm diameter Millipore AA filters, and alpha particles are counted by a surface barrier detector mounted opposite each filter. In typical use, the flow rate in each channel is 15 L/min and the sampling period is 15 min. Two counting periods are used, the first a 15-min period coincident with sampling, and the second a 40-min period starting 20 min after the end of sampling. The detection limit is about 5 Bq/m^3. To allow time for the activity from the previous sample to decay, measurement cycles are started at 90 to 180 min intervals. A lot of data has been taken with this system.

A partly automated system is described (85) based on a nine-stage cascade impactor covering the size range 65 nm to 17.7 μm. The impaction surfaces consist of thin (4-μm thick) foil. Passivated, implanted planar silicon (PIPS) detectors are mounted below each impaction stage. The sampling flow rate is 5.1 m^3/h. For 30-min samples, the detection limit is about 4 Bq/m^3.

Sample Results from Measuring Particle Size

Figure 8 shows the results of one measurement of radon decay-product particle size, taken in February 1993 in the basement of a house in eastern Pennsylvania (86). The measurement was made with the MOUDI/GSA system described above. The plot shows how the PAEC is distributed with respect to particle size, in a standard format (87) constructed so that the area under the curve is equal to the total PAEC. The highest concentration is found attached to particles of 80–100 nm. This feature is typical. Fortunately for humans, this peak is at a size near the minimum for tracheobronchial deposition (see Fig. 7).

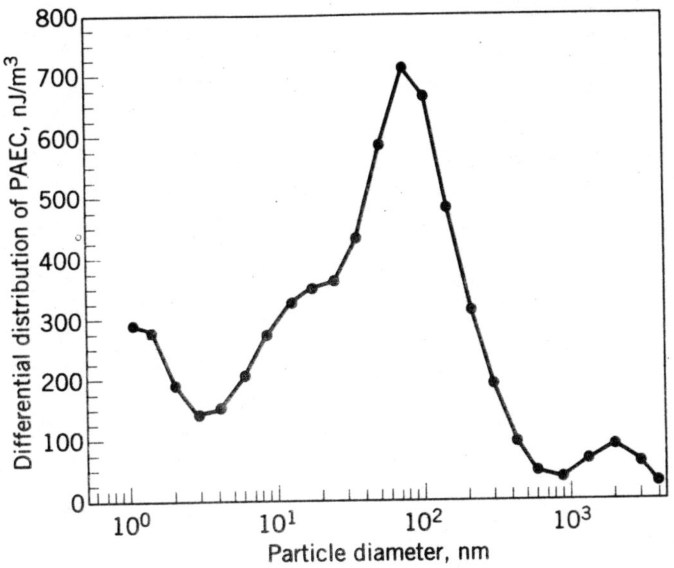

Figure 8. Radon PAEC particle size distribution measured indoors in an eastern Pennsylvania house, where radon = 1850 Bq/m^3, Rn DP = 1019 nJ/m^3, and aerosol = 2000 cm^{-3}.

If there had been smoking or cooking prior to this sample, the peak would have been found at 200–300 nm, even further into the minimum for deposition. There is also a substantial amount of PAEC at 1 nm (unattached decay products), which deposits readily in the tracheobronchial region, and a small but significant amount with a particle size larger than 1 μm.

Figure 8 is one case in which the classical, attached/unattached picture of decay products is inadequate. However, the house measured was untypical in several respects: the radon concentration was very high, about 12 times higher than the threshold level for remedial action; the PAEC was 1019 ± 15 nJ/m^3, 10 times the average as shown in Table 3; the aerosol concentration and the equilibrium ratio were low, 2000 cm^{-3} and 0.10, respectively. Hence, the spectrum in Figure 8 should not be accepted as typical.

RADON MITIGATION METHODS

It is assumed that any radon exposure causes some risk of lung cancer. EPA recommends reducing the radon concentration in any building if the results of a long-term radon measurement is greater than 150 Bq/m^3 (4 pCi/L). Obviously, buildings with very high concentrations of radon require prompt action.

Interim Steps

The risk from inhalation of high radon levels can be reduced by short-term or long-term mitigation methods. While waiting to decide on a long-term mitigation method, short-term mitigation measures can be taken to reduce the risk from radon. High levels of radon (>3700 Bq/m^3), require immediate attention. Limit the time spent in these spaces as much as possible, or ventilate the spaces with outside air. Mechanical ventilation may be needed, since merely opening windows may not be adequate. If there is a crawl space, check its radon concentration; if the concentration is high, ventilate the crawl space separately. The crawl space can be a major source of indoor radon.

Permanent Mitigation

For long-term or permanent mitigation, a qualified contractor with specific technical knowledge and special skills should be engaged. Usually, each state has a certified list of mitigators to provide the appropriate service. The cost of mitigating a building depends on its size and on the amount of radon reduction needed. The cost of mitigating a residence is typically in the range of $800–$1500.

The contractor normally performs a detailed investigation before recommending a permanent mitigation system. Radon reduction measures are usually aimed at preventing radon entry into a building. This is accomplished by diverting radon in the soil by depressurization, basement pressurization, sealing openings to the soil, and by isolating and ventilating areas in contact with the soil.

Following are the major radon reduction systems:

1. Active subslab suction or depressurization is the most utilized technique. More than 90% of the mitigations are of this type. It is also the most reliable radon reduction technique. It consists of installing one or more suction pipes through the floor slab into the soil or gravel bed underneath. The number of suction points depends on the air communication below the slab and on the strength of the radon source. The soil gas containing radon is usually moved outside with blowers of 2–7 m^3/min free-air capacity. Cracks and holes must be sealed first to achieve a uniform low-pressure field with the smallest blower and the least electric energy cost. The exhaust location should be located higher than the building or far away from the building to avoid radon reentry into the building.
2. A variant of (*1*) uses drain tiles around the periphery of the foundation as the gas collection system. In houses where the tile drains to a sump, the sump is capped and attached to an exhaust blower.
3. Passive subslab suction relies on natural air movement instead of a fan to draw soil gas and radon up from below the slab.
4. Block wall suction is used in basements with hollow block foundation walls. It is sometimes used in combination with subslab depressurization by suction in cases where the houses do not have perimeter drains.
5. In houses with a crawl space, radon concentration is reduced by ventilating the crawl space. Alternatively, the soil is covered with a heavy plastic sheet. A vent pipe and fan are then used to draw soil gas and radon from under the sheet and vent it outdoors.
6. Building pressurization. This method uses a fan to blow outdoor air into the basement. It creates positive pressure which is enough to prevent radon from entering the house. The effectiveness of this method is limited by house construction, climate, other appliances in the building, and occupant lifestyle.
7. A heat recovery ventilator (HRV) or air-to-air heat exchanger is used in some situations. The HRV system increases ventilation while using the heated or cooled air

Table 14. Addresses of Commercial Sources of Apparatus and Instruments.

Code Name	Full Name and Address	Telephone Number
AlphaNUCLEAR	AlphaNUCLEAR 1125 Derry Road East Mississauga, Ontario, Canada L5T-1P3	905-564-1383
CAIRS	Canadian Institute for Radiation Safety 102-110 Research Drive Saskatoon, SK, Canada S7N-3R3	306-975-0566
Eberline	Eberline/Thermo-Systems, Inc. P.O. Box 2108 Santa Fe, N.M. 87504-2108	505-471-3232
femto-TECH	femto-TECH, Inc. P.O. Box 8257, 325 Industry Drive Carlisle, OH 45005	513-746-4427
Genitron	Genitron Instruments Heerstrasse 149 D-60488 Frankfurt Germany	U.S. agent: 301-694-0011
Niton	Niton Corporation 74 Loomis Street Bedford, MA 01730	617-275-9275
Pylon	Pylon Electronics, Inc. 147 Colonade Road Ottawa, Ontario, Canada K2E-7L9	613-226-7920
Rad Services	R.A.D. Services and Instruments, Ltd 50 Silver Star Boulevard, Unit 208 Scarborough, Ontario, Canada M1V-3L3	FAX: 416-749-3398
Rad Elec	Rad Elec, Inc. 5714-C Industry Lane Frederick, MD 21704	301-694-0011
RTCA	Radon Testing Corporation of America 2 Hayes Street Elmsford, NY 10523	914-345-3380
SARAD	SARAD GmbH Dorfplatz 1 D-01705 Pesterwitz Germany	49-351-658-070 U.S. agent: 509-375-5157
Scintrex	Scintrex, Ltd. 222 Snidercroft Road Concord, Ontario, Canada L4K-1B5	905-669-2280
Sun Nuclear	Sun Nuclear Corporation 425-A Pineda Court Melborne, FL 32940-7508	407-259-6862
Thomson & Nielson	Thomson & Nielsen Electronics, Ltd. 25E Northside Road Nepean, Ontario, Canada K2H-8S1	613-596-4563

being exhausted to warm or cool the incoming air. HRVs also improve air quality indoors. However, there is a significant cost in cooling or heating with an HRV, and the reductions achievable are rarely more than 30%.

8. Radon in water reduction. There may be a significant radon supply into buildings using well water with radon concentrations >2 MBq/m^3 (50,000 pCi/L). There are two methods to remove radon from water:

   1. *Aeration.* This involves bubbling air in large volumes through the water to carry off the radon. This method is effective for very high concentrations of radon in water.

   2. *Granular activated carbon treatment.* The well water is filtered through a carbon bed where the radon is adsorbed on the activated carbon and the water leaves the bed with a reduced radon concentration. This system is about 90% efficient. However, the radon and decay products accumulated in the charcoal are a significant source of gamma radiation, which is objectionable in a domestic setting.

More information on the technical aspects of radon mitigation techniques is contained in documents from the Environmental Protection Agency (88–90).

BIBLIOGRAPHY

1. A.C. George, *Health Phys.* **70,** 451–463 (1996)
2. Nuclear Energy Agency, *Metrology and Monitoring of Radon, Thoron and Their Daughter Products,* Organization for Economic Cooperation and Development, Paris, 1985.

3. NCRP Scientific Committee 61, *Measurement of Radon and Radon Daughters in Air,* NCRP Report 97, National Council on Radiation Protection and Measurements, Bethesda, Md., 1988.
4. W.W. Nazaroff, "Measurement Techniques," in W.W. Nazaroff and A.V. Nero, Jr., eds., *Radon and Its Decay Products in Indoor Air,* John Wiley & Sons, New York, 1988, pp. 491–504.
5. B.S. Cohen, "Radon and Its Short-Lived Decay Product Aerosols," in K. Willeke and P.A. Baron, eds., *Aerosol Measurement-Principles, Techniques, and Applications,* Van Nostrand Reinhold, New York, 1993, pp. 799–815.
6. W.W. Nazaroff and A.V. Nero, Jr., eds., *Radon and Its Decay Products in Indoor Air,* John Wiley & Sons, Inc., New York, 1988.
7. D.J. Brenner, *Radon Risk and Remedy,* W.H. Freeman and Company, New York, 1989.
8. C.R. Cothern and J.E. Smith, Jr., eds., *Environmental Radon,* Plenum Press, New York, 1987.
9. D.G. Brookins, *The Indoor Radon Problem,* Columbia University Press, New York, 1990.
10. S.D. Schery, "Radon Isotopes and their Progeny in the Indoor Environment" in P.N. Cheremisinoff, ed., *Encyclopedia of Environmental Control Technology,* Gulf Publishing, Houston, Tex., 1989, Vol. 2.
11. J. Porstendorfer, *J. Aerosol Sci.* **25**(2) 219–264 (1994).
12. C. Samuelsson, *Nuclear Instrum. Methods* **A262**, 457–462 (1987).
13. P.K. Hopke, *Environ. Int.* **15,** 299–308 (1989)
14. D. Goldman and R. Bell, eds. *The International System of Units (SI)* NBS Special Publication 330, U.S. Department of Commerce, Washington, D.C., 1981.
15. A.C. James, "Lung Dosimetry", in W.W. Nazaroff and A.V. Nero, Jr., eds., *Radon and Its Decay Products in Indoor Air,* John Wiley & Sons, Inc., New York, 1988, pp. 292–299.
16. U.S. EPA, *A Citizen's Guide to Radon,* 2nd ed., U.S. Government Printing Office, Washington, D.C., 1994.
17. R.C. Ringholz, *Uranium Frenzy-Boom and Bust on the Colorado Plateau,* University of New Mexico Press, 1991, pp. 194–206.
18. P.K. Hopke and co-workers, *Environ. Sci. Technol.* **29,** 1359–1364 (1995).
19. R.C. Ringholz, *Uranium Frenzy—Boom and Bust on the Colorado Plateau,* University of New Mexico Press, 1991, pp. 246–253.
20. A.V. Nero, Jr., "Elements of a Strategy for Control of Indoor Radon," in W.W. Nazaroff and A.V. Nero, Jr., eds., *Radon and Its Decay Products in Indoor Air,* John Wiley & Sons, Inc., New York, 1988, pp. 467–468.
21. T.F. Gesell, *Health Physics* **45,** 289–302 (1983).
22. F. Marcinowski, R.M. Roberts, and W.M. Yeager, *Health Phys.* **66,** 699–706 (1994).
23. U.S. EPA, *Indoor Radon and Radon Decay Product Measurement Device Protocols,* EPA Report 402-R-92-004, National Environmental Information Service, Baton Rouge, La., 1992.
24. J.M. Matuszek, *J. Res. Natl. Inst. Stand. Technol.* **95,** 177–181 (1990).
25. National Council on Radiation Protection and Measurements, *A Handbook of Radioactivity Measurement Procedures,* 2nd ed., National Council on Radiation Protection and Measurements, Bethesda, Md., 1985, pp. 303–331, 508–516.
26. N. Tsoulfanidis, *Measurement and Detection of Radiation,* McGraw-Hill, New York, 1983, pp. 19–74.
27. B.S. Pasternack and N.H. Harley, *Nuclear Instr. and Methods* **91,** 533–540 (1971).
28. W.W. Nazaroff, *Health Physics,* **46**(2), 395–405 (1984).
29. H.F. Lucas and D.A. Woodward, *J. Appl. Phys.* **35,** 452–456 (1964)
30. H. Vanmarke and co-workers, "The Behavior of Radon Daughters in the Domestic Environment," in P.K. Hopke, ed., *Radon and Its Decay Products. Occurrence, Properties and Health Effects,* ACS Symposium Series 331, American Chemical Society, Washington, D.C., 1987, pp. 301–323.
31. A.C. George, *Health Phys.* **46,** 867–872 (1984).
32. A.C. George and T. Weber, *Health Phys.* **58,** 583–589 (1990).
33. B. Cohen and R. Nason, *Health Phys.* **50,** 457–463 (1986).
34. D.J. Gray and S.T. Windham, *NAREL standard operating procedures for radon-222 measurement using diffusion barrier charcoal canisters.* U.S. EPA Office of Radiation Programs, Report No. EPA 520/5-90-32 Washington, D.C., 1990.
35. H.W. Alter and R.L. Fleischer, *Health Phys.* **40,** 693–702 (1981).
36. L. Vasudevan and M.E. McLain, *Health Phys.* **66,** 318–326 (1994).
37. M. Doi, S. Kobayashi, and K. Fujimoto, *Rad. Prot. Dosim.* **45,** 425–430 (1992).
38. F.K. Richtmyer, E.H. Kennard, and T. Lauritsen, *Introduction to Modern Physics,* McGraw-Hill, New York, 1955, pp. 445–446.
39. P. Kotrappa, J.C. Dempsey, J.R. Hickey, and L.R. Stieff, *Health Phys.* **54,** 47–56 (1988).
40. I.M. Fisenne and H.W. Keller, *The EML pulse ionization chamber systems for ^{222}Rn measurements.* Environmental Measurements Laboratory, USDOE Report EML-437, New York, 1985.
41. P. Baltzer, K.G. Gorsten, and A. Backlin, *Nuclear Instruments and Methods in Physics Research* **A317,** 357–364 (1992).
42. F.K. Richtmyer, E.H. Kennard and T. Lauritsen, *Introduction to Modern Physics,* McGraw-Hill, New York, 1955, p. 146.
43. H.F. Lucas, *Review of Scientific Instruments* **28,** 680–683 (1957).
44. A.C. George, in P.L. Carson, W.R. Hendee, and D.C. Hunt, eds., *Proc. of Ninth Midyear Health Physics Symposium,* Denver, Colo. 1976, pp. 112–115.
45. E.O. Knutson, A.C. George, P. Shebell, and C.V. Gogolak, *Rad. Prot. Dosim.* **56,** 263–266 (1994).
46. *EML Procedures Manual,* 28th ed., HASL-300, U.S. Department of Energy, New York, Feb. 1997, Section 2.2.3.2.
47. J.W. Thomas and R.J. Countess, *Health Phys.* **36,** 734–738 (1979).
48. A. Busigin, A. Van Vooren, and C.R. Phillips, *Health Phys.* **37,** 659–667 (1979).
49. D.C. Ward and T.B. Borak, *Health Phys.* **61,** 799–807 (1991).
50. A.R. Hutter, *Health Phys.* **68,** 835–839 (1995)
51. *EML Procedures Manual,* 28th ed., HASL-300, U.S. Department of Energy, New York, Feb. 1997, Section 2.2.3.7.
52. R. Falk, H. More, and L. Nyblom, *Int. J Radiat. Appl. Instrum., Part A* **43,** 111–118 (1992).
53. J. Bigu and J. Elliot, *Nucl. Instrum. and Methods in Phys. Res.* **A344,** 415–425 (1994).
54. J. Bogen, M. Sappok, and G. Shumann, *Arch. Met. Geoph. Biokl. Ser. A,* **21,** 171–182 (1972).
55. Y. Li and S.D. Schery, *Radioactivity and Radiochemistry* **4,** 42–51 (1993)
56. J.W. Thomas and P.C. LeClare, *Health Phys.* **18,** 113–122 (1970)
57. A.R. Hutter, R.J. Larsen, H. Maring, and J.T. Merrill, *J. Radioanal. Nucl. Chem.* **193,** 309–318 (1995).
58. A.C. James, "Lung Dosimetry," in W.W. Nazaroff and A.V. Nero, Jr., eds., *Radon and Its Decay Products in Indoor Air,* John Wiley & Sons, Inc., New York, 1988, p. 301.
59. M. Lippmann, "Filters and Filter Holders," in B.S. Cohen and S.V. Hering, eds., *Air Sampling Instruments for Evaluation of Atmospheric Contaminants,* 8th ed., American Conference of Government Industrial Hygienists, Cincinnati, Ohio, 1995, pp. 247–278.

60. K.W. Lee and M. Ramamurthi, "Filter Collection," in K. Willeke and P.A. Baron, eds., *Aerosol Measurement: Principles, Techniques and Applications,* Van Nostrand Reinhold, New York, 1993, pp. 179–205.
61. *EML Procedures Manual,* 28th ed., HASL-300, U.S. Dept. of Energy, New York, Feb. 1997, Section 2.2.4.
62. T.S. Park and E.O. Knutson, *Rad. Prot. Dosim.* **67**, 121–128 (1996)
63. E.O. Knutson and K.W. Seo, *Evaluation of Selected Modern Filter Media for Use in Radon/Thoron Progeny Measurements,* Report EML-578, U.S. Dept. of Energy, New York, 1996.
64. R.C. Ringholz, *Uranium Frenzy: Boom and Bust on the Colorado Plateau,* University of New Mexico Press, 1991, p. 168.
65. H.L. Kusnetz, *Am. Ind. Hyg. Assoc. Q.* **17**, 85–88 (1956).
66. R. Rolle, *Health Phys.* **22**, 233–238 (1972)
67. E. Stranden, *Health Phys.* **38**, 73–76 (1980).
68. J.W. Thomas, *Health Phys.* **23**, 783–789 (1972).
69. O.G. Raabe and M.E. Wrenn, *Health Phys.* **17**, 593–605 (1969).
70. M.J. Duggan and D.M. Howell, *Int. J. Appl. Radiat. Isotopes* **19**, 865–870 (1968).
71. B.S. Cohen and S.V. Hering, eds., *Air Sampling Instruments for Evaluation of Atmospheric Contaminants,* 8th ed., American Conference of Government Industrial Hygienists, Cincinnati, Ohio, 1995, Chaps. 7, 12, 13, 22.
72. S.K. Friedlander, *Smoke, Dust and Haze—Fundamentals of Aerosol Behavior,* John Wiley & Sons, Inc., New York, 1977.
73. W.C. Hinds, *Aerosol Technology-Properties, Behavior, and Measurement of Airborne Particles,* John Wiley & Sons, Inc., New York, (1982).
74. G. Rudolf, R. Köbrich, and W. Stahlhofen, *J. Aerosol Sci.* **21**, Suppl 1, s403–s406 (1990).
75. A.C. George, *Health Phys.* **23**, 390–392 (1972).
76. C.R. Phillips, A. Khan, and H.M.Y. Leung, in W.W. Nazaroff and A.V. Nero, Jr., eds., *Radon and Its Decay Products in Indoor Air,* John Wiley & Sons, Inc., New York, 1988, pp. 203–258.
77. Y.S. Cheng, J.A. Keating, and G.M. Kanapilly, *J. Aerosol Sci.* **11**, 549–556 (1980).
78. M. Ramamurthi, R. Strydom, and P.K. Hopke, *J. Aerosol Sci.* **21**, 203–211 (1990).
79. S.B. Solomon and T. Ren, *Aerosol Sci. Technol.* **17**, 69–83 (1992).
80. V.A. Marple, K.L. Rubow, and S.M. Behm, *Aerosol Sci. Technol.* **14**, 434–446 (1991).
81. E.O. Knutson, A.C. George, and K. W. Tu, *Aerosol Sci. Technol.* **27**, 604–624 (1997).
82. *EML Procedures Manual,* 28th ed., HASL-300, U.S. Department of Energy, New York, Feb. 1997, Sections 2.2.2.5–2.2.2.8.
83. A. Reineking and J. Porstendörfer, *J. Aerosol Sci.* **17**, 873–879 (1986).
84. M. Ramamurthi and P.K. Hopke, *Aerosol Sci. Technol.* **14**, 82–92 (1991).
85. J. Kestin, G. Butterweck, J. Porstendörfer, and A. Reineking, *J. Aerosol Sci.* **22**, Suppl 1, s287–s290 (1991).
86. Dr. K.W. Tu, U.S. Dept. of Energy Environmental Measurements Laboratory, New York.
87. E.O. Knutson and P.J. Lioy, in B.S. Cohen and S.V. Hering, eds., *Air Sampling Instruments for Evaluation of Atmospheric Contaminants,* 8th ed., American Conference of Government Industrial Hygienists, Cincinnati, Ohio, 1995, pp. 121–138.
88. U.S. EPA, *Radon Technology for Mitigators,* U.S. Environmental Protection Agency, Office of Radiation Programs, Washington, D.C., Oct. 1993.
89. U.S. EPA, *Application of Radon Reduction Methods.* EPA/625/5–88/024, Washington, D.C., 1988.
90. U.S. EPA, *Consumer's Guide to Radon Reduction,* EPA/402-K92-003, Washington, D.C., 1992.
91. N.E. Holden, "Table of Isotopes" in D.R. Lide, ed., *CRC Handbook of Chemistry and Physics,* 75th ed., CRC Press, Boca Raton, Fla., 1994.

RAMAN SPECTROSCOPY. See FT/RAMAN SPECTROSCOPY IN FUEL ANALYSIS.

RAMAN SPECTROSCOPY IN WATER ANALYSIS

J. ERIC D. DAVIES
Lancaster University
Lancaster, England

Raman spectroscopy has undergone remarkable improvements in the 70 years since the effect was discovered by C.V. Raman in 1928 (1,2). This article will trace the development of Raman spectroscopy and attempt to show that it is finally attaining its long-promised potential in water analysis.

THE TECHNIQUE OF RAMAN SPECTROSCOPY

The Discovery

Having been elected a Fellow of the Royal Society in 1924, Chandrasekhara Venkata Raman was already well known for his studies of light-scattering phenomena when, in 1928, he discovered the effect which now bears his name.

Raman's original equipment was extremely simple: the source of radiation was sunlight, the detector was his own eye, and he used complementary filters in place of a spectrograph (Fig. 1). When both filters were placed in the incident beam no light reached the sample, but when they were placed in the positions shown in Figure 1, Raman found that the beam of sunlight could be seen traversing the liquid. This indicated that part of the beam was being scattered with modified wavelengths. It was necessary to determine whether the observed effect was a new effect or simply the well-known phenomenon of fluorescence (1).

> *Now the test with the complementary filters is precisely that ordinarily used for detecting fluorescence and indeed was first suggested by Stokes in his investigations on the subject. You may therefore rightly ask me the question how does this phenomenon differ from fluorescence? The answer to the question is, firstly, that it is of an entirely different order of intensity. A more satisfactory proof was however forthcoming when Mr Krishnan and myself examined the polarisation of this new type of radiation and found that it was nearly as strong as that of the ordinary light scattering in many cases, and is thus quite distinct from ordinary fluorescence which is usually unpolarised.*

Raman then used monochromatic radiation from a mercury lamp and when the scattered light was passed into a direct vision spectroscope he could see extra lines in the spectrum which were not present in the spectrum of the incident light.

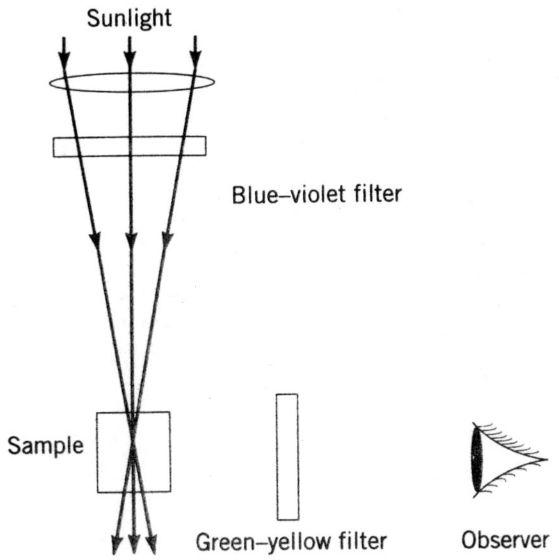

Figure 1. The arrangement used by Raman when discovering the Raman effect.

Photographic recording was used to obtain a permanent record. Raman correctly deduced that the observed effect was the optical analogue of the Compton effect and that the change in wavelength was a consequence of the excitation of molecular vibrations and rotations. Unknown to Raman, the effect had been predicted by Smekal in 1923 (3).

Shortly after the publication of Raman's paper, Landsberg and Mandelstam, working independently in Russia, reported a similar effect in crystalline quartz (4). Several workers soon confirmed Raman's discovery (5) and by the end of 1928 about 60 papers had been published on the Raman effect. In the following years Raman spectroscopy became the prime spectroscopic technique for studying molecular vibrations and rotations as recording infrared spectra at that time involved a laborious point by point plot.

International recognition of the value and importance of the discovery was swift: Raman was knighted in 1929 and awarded the Nobel Prize in Physics in 1930.

The Experimental Arrangement

Figure 2 shows the basic arrangement in Raman spectroscopy (6). Irradiation of the sample with monochromatic radiation of frequency ν_0 and energy $h\nu_0$ gives two types of scattered radiation. The Rayleigh radiation arises from elastic scattering and the Raman radiation from inelastic scattering of the photons by the molecule. The energy difference, $h\nu_m$, between the incident and Raman scattered radiation corresponds to the energy difference between molecular vibrational and/or rotational energy levels. Raman spectroscopy, like infrared (IR) spectroscopy, consequently gives vibrational, rotational, and vibration–rotation spectra.

Problematic features of this technique are:

1. The Raman effect is weak as the Rayleigh scattered radiation is much more intense than Raman scattered radiation. Using visible radiation on a liquid sample the intensities of the Rayleigh and Raman scattering are about 10^{-3} and 10^{-8} of the incident intensity. The technique thus requires sensitive detectors and good discrimination between the Rayleigh and Raman radiation. The latter has traditionally been achieved by using double and triple monochromators.
2. As visible radiation is normally used for excitation, photochemical and photophysical processes such as fluorescence and photodecomposition can be induced in addition to the Raman effect. Fluorescence is particularly problematic due to its intense nature and it can swamp the Raman radiation.
3. As the sample must not absorb any appreciable amount of the incident radiation, colored samples can be problematic due to the use of visible exciting radiation.

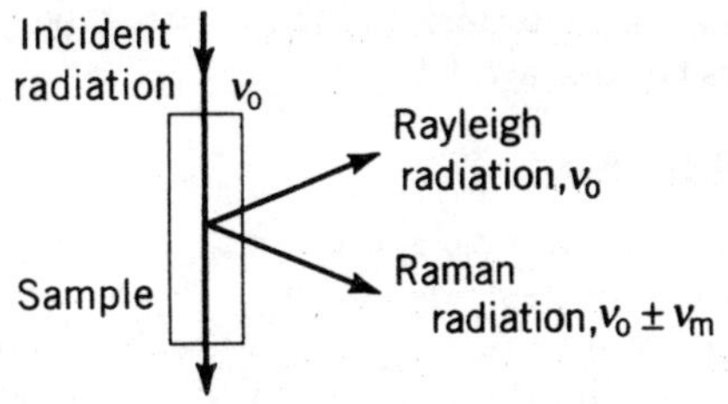

Figure 2. The basic experimental arrangement in Raman spectroscopy.

Advantageous features of the technique with regard to water analysis are:

1. The Raman spectrum of water is much weaker than its IR spectrum. This is, however, offset by the low intensity of the Raman scattering thus requiring much higher analyte concentrations in Raman than in IR spectroscopy.
2. The use of visible excitation radiation allows the use of glass sample cells and fiber optics.

The period 1930–1965

The early promise shown by Raman spectroscopy was vindicated during this period. It was the method of choice during the period 1930–1945 for studying the vibrational characteristics of molecules and was mainly applied to liquids and crystals. The weak nature of the scattering required the use of concentrated solutions but one early success with aqueous solutions was the detection of the Hg_2^{2+} species in solutions of mercury(I) nitrate (7). After 1945 the routine recording of infrared spectra became possible with the introduction of automatic infrared spectrometers. Raman spectroscopy thus lost its overall advantage as IR spectra could be easily recorded from gaseous, solution, and solid-state samples. Raman still, however, retained an advantage over IR in the study of aqueous solutions and low frequency vibrations.

Instrumental improvements during this period concentrated on increasing the intensity of the scattering by introducing high intensity sources (the mercury Toronto arc (8)) and introducing sources emitting in the yellow–orange–red region (radio-frequency powered rubidium and caesium lamps (9)) so that colored compounds could be examined. Photographic recording was replaced by direct recording using photomultipliers as detectors and the first commercial Raman spectrometers became available.

The Period Since 1965

The ideal source for Raman spectroscopy emits high intensity monochromatic radiation. In the early 1960s the first lasers were developed and were quickly tested as potential sources, first the pulsed ruby laser (10), then the continuous He–Ne laser (11). Both lasers emit red radiation, wavelengths at which the photomultiplier detector has decreasing sensitivity and the scattering has lower intensity due to its dependence on the fourth power of the exciting frequency. It was thus the development of tuneable Ar^+ (blue–green) and Kr^+ (blue–green–yellow–red) lasers which catalyzed further developments in Raman spectroscopy. Excimer lasers emitting UV radiation have been used in resonance Raman studies on colorless compounds and far-red and near-IR lasers are useful in overcoming problems of fluorescence. Detectors for visible radiation have been improved with the introduction of charge-coupled device (CCD) array detectors and the use of notch and edge filters has eliminated the need for double and triple monochromators. In addition to dispersive spectrometers, Fourier transform spectrometers utilizing near-IR excitation are now available commercially (12). All of these advances have significantly improved the applicability and sensitivity of Raman spectroscopy so that a wide range of compounds, irrespective of color or the presence of fluorescent material, is now amenable to the technique (13).

INTENSITY ENHANCEMENT TECHNIQUES

Conventional or *spontaneous* Raman spectroscopy, using the arrangement shown in Figure 2, has one major disadvantage with regard to the detection of low concentrations of analytes: it is an inherently weak scattering process. Two main methods can be used to enhance the intensity of the scattered radiation, both of which have been used in water analysis.

Resonance Raman Spectroscopy

The arrangement in *conventional* Raman spectroscopy is illustrated in Figure 3 where the frequency of the incident radiation (ν_o) is well removed from the frequency of an electronic (ie, UV/visible) absorption band (ν_e) in order to prevent absorption of radiation by the sample, ie, the frequency difference ($\nu_e - \nu_o$) is as large as possible. The intensity of a Raman band does, however, depend inversely on the magnitude of this quantity. The intensities of Raman bands, and consequently the sensitivity of the technique, can therefore be increased if ($\nu_e - \nu_o$) can be decreased. Decreasing ($\nu_e - \nu_o$) can, however, lead to the sample absorbing the radiation, resulting in its decomposition. This can usually be prevented by the simple procedure of flowing a liquid sample through the laser beam rather than using a static sample.

Having $\nu_o \sim \nu_e$ gives the technique of *resonance Raman spectroscopy* (RRS) where bands can have intensities 10^2–10^6 times greater than those obtained using conventional Raman spectroscopy. Sensitivities at the ppm and sub-ppm level are thus possible (14). Figure 4 illustrates the resonance technique in operation. The incident radiation (514.5 nm) is such that ($\nu_e - \nu_o$) is small for cyanocobalamin (vitamin B_{12}) but large for sodium nitrate. The former therefore displays a resonance effect but not the latter. Thus the cyanocobalamin band at 1504 cm^{-1} has about *half* the integrated intensity of the nitrate band at 1055 cm^{-1}, despite a concentration difference of *four orders of magnitude* (15). If the condition $\nu_o \sim \nu_e$ is not satisfied, but the excitation frequency falls within the absorption band profile, smaller intensity enhancements are achieved by the *preresonance Raman effect.*

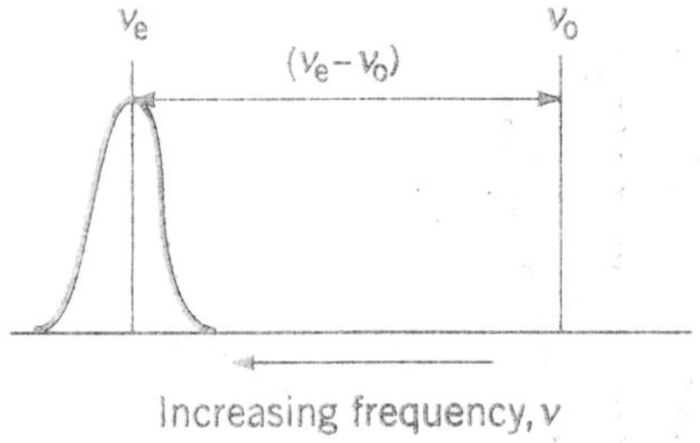

Figure 3. Relationship between the frequency of the incident radiation used in conventional Raman spectroscopy (ν_o) and the frequency of the sample absorption band (ν_e).

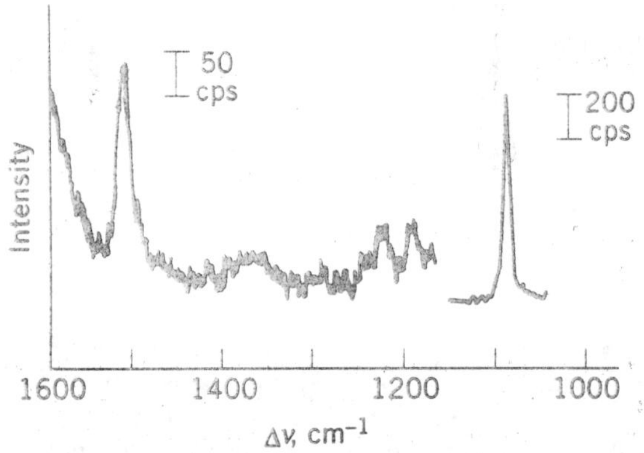

Figure 4. Raman spectrum of 5×10^{-6} *M* cyanocobalamin and 5×10^{-2} *M* sodium nitrate in 0.1 *M* HCl. Reprinted with permission (15).

Surface-Enhanced Techniques

It has been found that the Raman spectra of molecules adsorbed onto *roughened* metal surfaces, particularly silver, are intensity enhanced by a factor of 10^4–10^6 compared to the spectra of the liquid phase. This effect, *surface enhanced Raman spectroscopy* (SERS), is explained by a combination of physical (electromagnetic field) and chemical (charge or energy transfer) effects (16). SERS has been used extensively for the determination of a wide range of analytes at low concentrations (17). If the condition for resonance Raman spectroscopy (ie, $\nu_o \sim \nu_e$) is also satisfied, then both enhancement mechanisms are operative giving the technique of *surface-enhanced resonance Raman spectroscopy (SERRS).*

Other Raman-Related Techniques

The availability of laser sources, particularly high energy pulsed lasers, has led to the discovery of a number of new light-scattering phenomena related to the Raman effect, eg, the hyper-Raman effect, the stimulated Raman effect, the inverse Raman effect, and coherent anti-Stokes Raman spectroscopy (CARS). These effects are not discussed herein but, as some may be applicable to water analysis, further details are available (18).

APPLICATIONS

Raman spectroscopy has been used extensively to study analytes such as electrolytes and biological molecules in aqueous solutions. Studies of electrolyte solutions (19) give detailed information about the solvation of the species and the presence of different types of ion pairs, but they require relatively concentrated solutions. The studies on biological molecules (20) give information on conformations and can probe specific reaction sites. Time-resolved studies can follow the evolution of structure and elucidate the dynamics of biological processes. In this section studies on aqueous solutions, where the main emphasis has been detection of environmentally significant analytes at the ppm and sub-ppm levels, are reviewed.

Inorganic Analytes

As Raman spectroscopy is a *molecular* vibrational technique it cannot detect species such as metal cations and monatomic anions (Ce^-, Br^-, etc). The sensitivity of the technique thus depends on the intensity of the strongest Raman band of analytes such as NH_4^+, NO_3^-, and SO_4^{2-}.

Much efforts has been expended in determining the detection limits of inorganic analytes using conventional Raman spectroscopy. The studies can be divided into two classes depending on whether natural water samples or stock solutions, prepared using distilled/deionized water, is used. The former samples present a realistic assessment of the detection limits but natural waters usually contain photoluminescent material which gives a fluorescent background to the spectra. Although samples prepared in purified water generally give spectra free of a fluorescent background and thus have lower detection limits, they are not representative of actual natural water samples.

Using samples prepared in natural water, one study (21) used the 514.5 nm line of an Ar^+ laser to determine the detection limits of NO_3^-, PO_4^{3-}, SO_4^{2-}, and CO_3^{2-}. Using the strongest Raman line, that arising from the symmetric stretching mode, the detection limits were determined as 25, 50, 50, and 75 ppm, respectively.

A more detailed study of the experimental parameters which influence detection limits has been reported (22), but using solutions prepared with purified water. An equation linking the signal-to-noise ratio to analyte concentration, measurement time, spectral slit width, laser power fluctuations, and solvent background intensity was derived. The minimum detection limits, using the most intense Raman band, were found to be NO_3^-, 4.4 ppm; SO_4^{2-}, 7.3 ppm; CO_3^{2-}, 11.6 ppm; HCO_3^-, 21.1 ppm; HPO_4^{2-}, 15.7 ppm; $H_2PO_4^-$, 22.5 ppm; $CH_3CO_2^-$, 16.3 ppm; and CH_3CO_2H, 22.4 ppm. The detection limits for analytes in purified water are thus lower than those in natural water. Conventional Raman spectroscopy thus has detection limits in the range 25–75 ppm for typical anionic analytes in natural water (21).

The detection limit can be decreased substantially by using intensity enhancement techniques. This approach has been applied to the determination of nitrite in fresh-water and in seawater by using 507.1 nm argon ion excitation of the Raman spectrum (23). Resonance enhancement was achieved by converting the colorless nitrite into a colored azo product by using the chemical reactions shown in Figure 5. A further surface enhancement was achieved by adsorbing the azo compound onto a colloidal silver hydrosol giving a resultant increase in intensity due to the surface-enhanced resonance Raman (SERR) effect. The detection limit was determined to be 0.02 n*M* (0.9 ppb) in distilled water with a linearity range up to 30 n*M* (1.4 ppm).

SERS was also used in a study examining various aspects critical to the direct detection of trace cyanide in groundwater and in wastewater streams (24). The working surface consisted of a polished silver electrode which was electrochemically roughened with a single oxidation–reduction cycle in 0.01 *M* KCl. The magnitude of the signal enhancement was dependent on the applied potential (Fig. 6). Cyanide could be detected in 0.1 *M* KCl (in distilled water) with a linear response between 100 ppm and 10 ppb. The estimated limit of detection was about 8 ppb. The effects of pH, electrolyte level, and two common background ions, NO_3^- and SO_4^{2-}, were also studied. Cyanide response was found to be sensitive to pH, with optimal performance observed at neutral to basic pH. Electrolyte concentrations of 0.001 *M* reduced response to cyanide by a factor of five, whereas levels above 0.1 *M* had no significant effect. The addition of 10 ppm sulfate ion decreased response approximately 40%, whereas the presence of nitrate ion at concentrations up to 100 ppm had a negligible effect on SERS response. The results suggested that cyanide ion can be detected directly in high ionic strength aqueous solutions such as groundwater.

Although Raman spectroscopy can, in favorable circumstances, be used to detect inorganic analytes at the sub-ppm level, it is not the method of choice. Ion chromatography offers a much simpler sampling procedure and has the added advantage that monatomic anions can also be determined.

Organic Analytes

In contrast to the situation with inorganic analytes, much work has been carried out on applying Raman spectroscopy to the detection of organic analytes in water. The early studies used conventional Raman spectroscopy with later studies utilizing enhancement techniques to increase the sensitivity.

The first attempt at detecting an organic analyte at the ppm level used a He–Ne laser as the excitation source (25). Using a 5 mW laser, 50 ppm of benzene in distilled water could be detected using a 20 minute recording period. It was shown that the use of a more powerful krypton ion laser should enable the detection limit to be reduced to 1 to 10 ppm, utilizing the CH stretching mode band of pollutants such as benzene and motor oil (26). Attempts at the "remote sensing" of pollutants in water were also made (27). Using a 600 mW argon ion laser and the optical system shown in Figure **7a**, a detection limit of ~ 150 ppm for NO_3^- was obtained for a solution in distilled water located 6 m (21 ft) from the instrument (cf with a detection limit of 4.4 ppm in distilled water when using a conventional sampling arrangement (22)). Films of heating oil on water were also detectable (Fig. 7**b**).

As reported earlier for inorganic analytes, the detection limits of organic analytes can be lowered by using resonance and/or surface-enhanced resonance techniques. Colorless

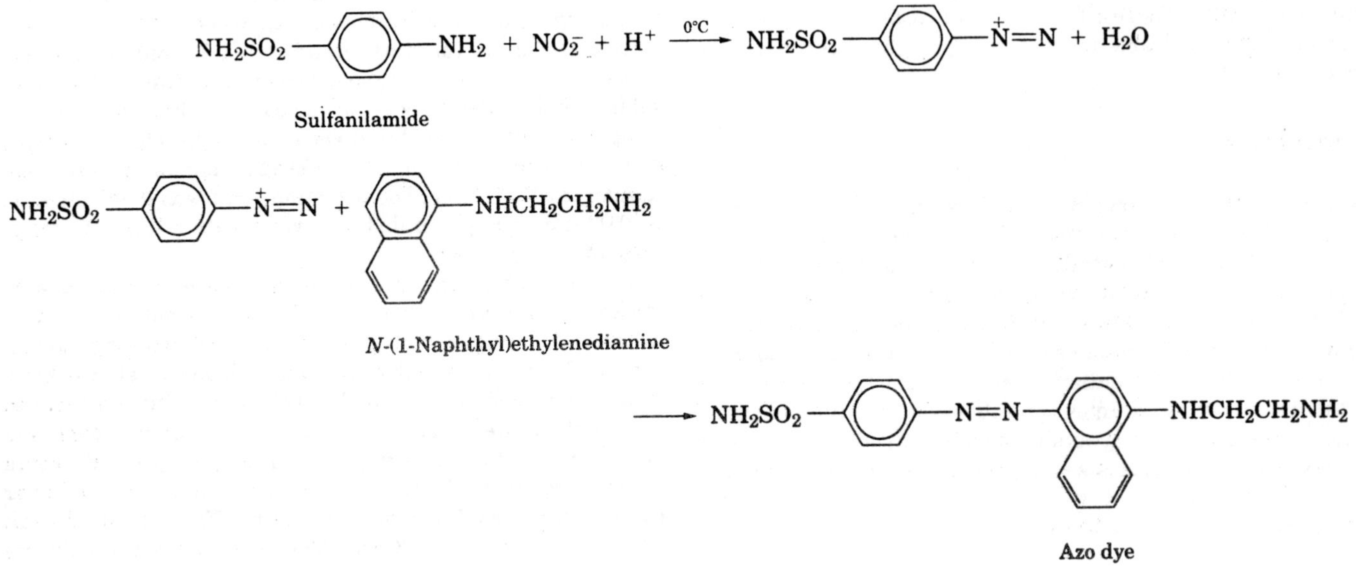

Figure 5. Reaction scheme for converting nitrite into the azo product.

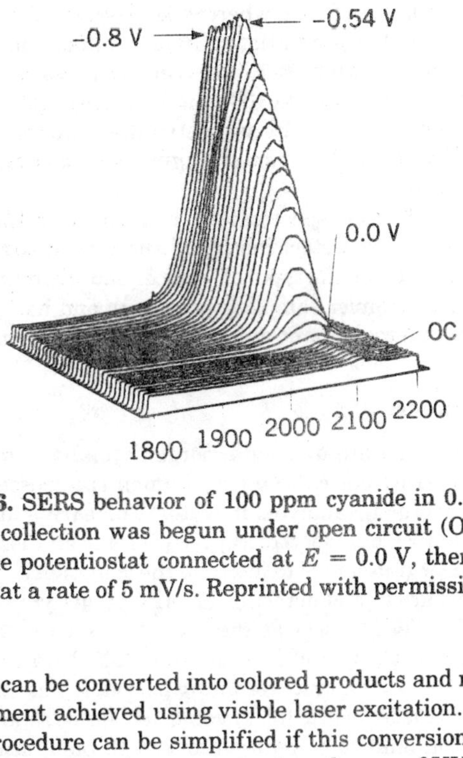

Figure 6. SERS behavior of 100 ppm cyanide in 0.1 *M* KCl. Spectral collection was begun under open circuit (OC) conditions. The potentiostat connected at $E = 0.0$ V, then ramped negative at a rate of 5 mV/s. Reprinted with permission (24).

analytes can be converted into colored products and resonance enhancement achieved using visible laser excitation. The analytical procedure can be simplified if this conversion step can be eliminated. This, however, requires the use of UV laser excitation for colorless compounds.

Fungicides and Pesticides. The spectra of fungicides and pesticides have been examined using the resonance effect on either colored compounds or by converting colorless compounds into colored products. Nitrophenol-based pesticides and fungicides have absorption bands in the violet or ultraviolet regions. Using argon ion 457.9 nm excitation, preresonance Raman intensity enhancement was observed which allowed detection at the ppm level (28). A further variable in determining the detection limit in these samples is that the enhancement can be changed by altering the pH of the solution. Figures 8, B and 8, C compare the Raman spectra of 31 ppm solutions of 2-nitrophenol at pH values of 4.9 and 10.9. The 2-nitrophenolate anion formed at the higher pH has a λ_{max} value closer to the excitation wavelength (488.0 nm) than 2-nitrophenol itself thus giving the increased intensity. The detection limits ranged from 0.5 ppm for 4,6-dinitro-2-*sec*-butylphenol to 7.0 ppm for *O,O*-dimethyl-*O*-4-nitrophenyl phosphorothioate in distilled water. 4,6-Dinitro-2-*sec*-butylphenol could be detected in river water spiked at a concentration of 3 ppm.

Colorless phenylamide-based pesticides were converted into azo derivatives in order to obtain resonance enhancement (29). The detection limits were $2-4 \times 10^{-8} M$ and the calibration curves were linear up to $1 \times 10^{-6} M$.

Phenols. The determination of phenols has been reported using a variety of techniques and samples, eg, conventional and resonance Raman spectroscopy on solutions in distilled and natural waters. The relative merits of different derivatization procedures have been studied with regard to the magnitude of the resonance enhancement, the thermal stability of the dyes in the intense laser beam (1 W power), and the occurrence and intensity of the fluorescence of the dyes formed (30). Using these criteria the best method was determined to be via the diazonium salt of 4-nitroaniline. The detection limits for phenol were found to be 20 ppb in distilled water and 50–300 ppb in natural waters depending on the source. The qualitative aspects of the method were investigated using phenol, *o*-cresol, and *m*-cresol. Comparison of the spectra shown in Figure 9 indicates that there are significant differences between the three spectra. The method is thus capable of giving both qualitative and quantitative results but it is limited as only a certain number of phenols react with diazotized 4-nitroaniline to give colored products.

The derivatization procedure using the diazonium salt of 4-aminobenzenesulfonic acid was used to form colored dyes using a spot test procedure utilizing 1 μL of the phenol solution (31). Test spectra using 1-naphthol-4-sulfonic acid showed that this phenol could be easily detected at the 4 ppm level, ie, using 4 ng of sample.

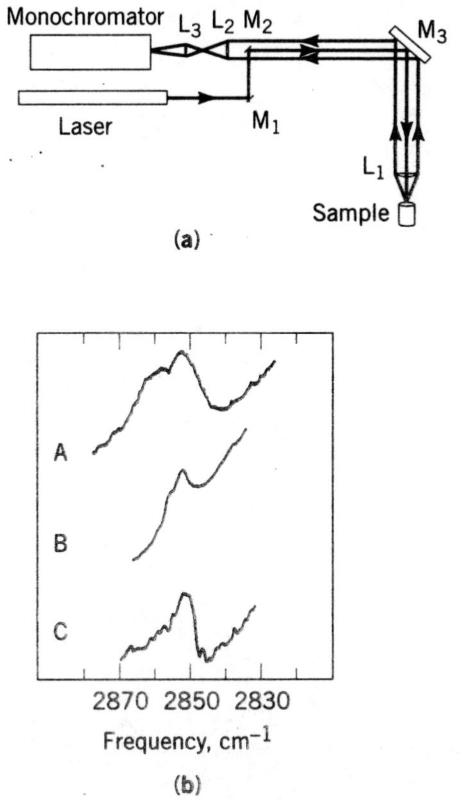

Figure 7. (**a**) Schematic of the optical system used for obtaining Raman spectra of samples located 6 m (21 ft) from the instrument; (**b**) Raman spectra of A, oil contained in a capillary tube located in the sample compartment; B, oil on the surface of water located in the sample compartment, and C, oil on the surface of water obtained with the "remote sensing" detection system. Reprinted with permission (27).

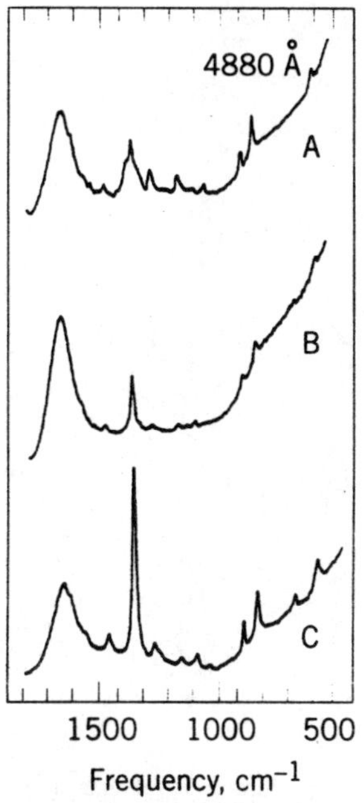

Figure 8. Raman spectra of 2-nitrophenol solutions in distilled water with 488.0 nm excitation: A 308 ppm, pH 4.4; B 31 ppm, pH 4.9; C 31 ppm, pH 10.9. Reprinted with permission (28).

A detailed study of the use of Raman spectroscopy in determining phenols in distilled water, using a range of chloro- and nitro-phenols and a variety of excitation wavelengths has been reported (32,33). As no derivatization was carried out, resonance enhancement was not observed. The detection limit for phenol using 514.5 nm excitation was determined to be 25 ppm, significantly less than the limit of 20 ppb obtained using the azo derivative (30). 2-Chloro-4-nitrophenol has a λ_{max} value of 400 nm and is expected to show some resonance enhancement. This is reflected in the wavelength dependence of the detection limit: 514.5 nm (0.92 ppm); 488.0 nm (0.42 ppm); and 457.9 nm (0.3 ppm) (32). The use of 457.9 nm excitation gave detection limits of 0.03, 0.4, 0.07, and 0.2 ppm for 2-nitro-, 3-nitro-, 4-nitro-, and 2-chloro-4-nitrophenol, respectively (33). Preliminary measurements using a UV laser (337.1 nm), which would give resonance enhancement for colorless compounds, indicated that such enhancements were observed. They did not, however, translate into lower detection limits as the optical system was not optimized for this wavelength.

Aromatic and Polycyclic Aromatic Hydrocarbons. Three techniques have been used for the determination of aromatics in water: SERS (34), RRS (35), and solid-phase microextraction (36). The SERS technique used a roughened silver surface coated with a self-assembled layer of octadecylthiol, which can partition the organics from the water. The adsorbed octadecylthiol gives a Raman spectrum and appropriate bands can be used as internal references for quantitative work. Sensitivity tests were conducted using aqueous benzene and naphthalene solutions. The detection limits were determined as 7.5 ppm for benzene and 2.3 ppm for naphthalene in purified water. On the assumption that the coating mimics octanol one factor determining the sensitivity is the octanol–water partition coefficient of the analyte. The octanol–water partition coefficients for polychlorinated biphenyls (PCBs) and polycyclic aromatic hydrocarbons (PAHs) are three orders of magnitude larger than for naphthalene. On this basis it was predicted that the detection limits for these species on the modified silver surface would be less than 10 ppb.

A previously reported resonance Raman study had used a tunable UV laser to detect PAHs such as naphthalene, anthracene, phenanthrene, and pyrene (35). By tuning the laser to the λ_{max} value of the PAH the detection limit was close to 20 ppb (10^{-7} *M* spectrum in Fig. 10). The vertical line in Figure 10 marks the position of the strongest Raman band at 1622 cm^{-1}, which has a S/N ratio >15 in the 10^{-7} *M* aqueous solution spectrum. The octadecylthiol-modified silver surface should therefore have greater sensitivity than the RRS method as well as being experimentally simpler.

The microextraction study used extraction of the aromatics onto solid poly(dimethylsiloxane) (36), which was subsequently examined using conventional Raman spectroscopy. The technique relies on the establishment of equilibrium between the analyte in the aqueous phase and the stationary phase. For the BTEX compounds (benzene, toluene, ethylbenzene, and xylenes) studied, 16–30 minutes were needed to reach equilibrium with a preconcentration enhancement of 2–3 orders of magnitude. The limits of detection using the most intense Raman bands were in the 1–4 ppm range with relative standard

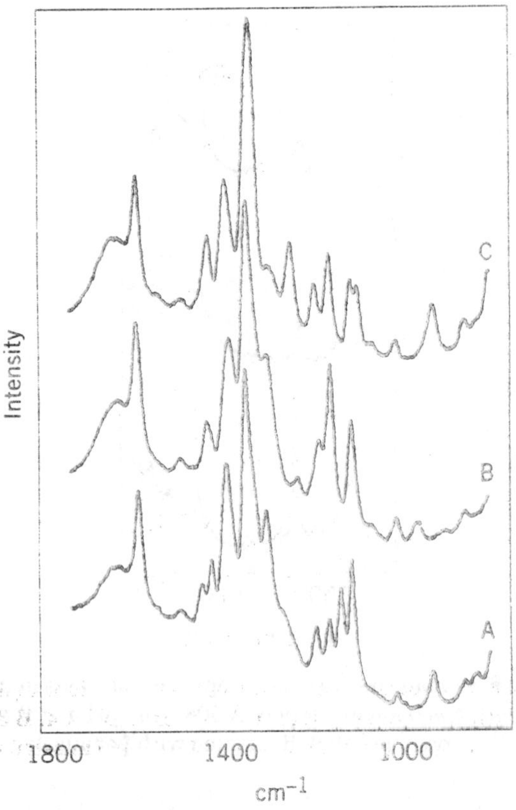

Figure 9. Resonance Raman spectra after derivatization with diazotized 4-nitroaniline of 1 ppm solutions in distilled water of: A phenol; B *o*-cresol; and C *m*-cresol. Reprinted with permission (30).

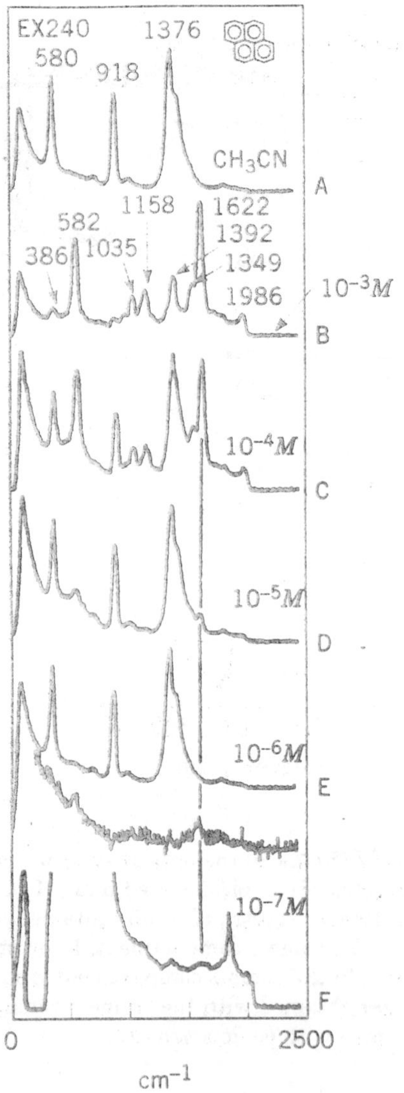

Figure 10. Concentration dependence of UV resonance Raman spectrum of pyrene using 240 nm excitation. Spectrum F is derived by subtracting the CH_3CN spectrum (A) from the 10^{-6} *M* pyrene spectrum (E). The bottom difference spectrum derives from subtracting a water Raman spectrum from the Raman spectrum of a 10^{-7} *M* aqueous pyrene solution. The vertical line indicates the position of the strongest pyrene Raman band at 1622 cm^{-1}. Reprinted with permission (35).

deviations of 3–9%. Preliminary studies using well water and river water spiked to 20 ppm concentration showed no significant interferences from the river and well water matrices.

Dyes. Dyes are ideally suited to analysis by resonance Raman spectroscopy using readily available visible laser emission lines. FD&C Red Dyes No. 2, 4, and 40 were studied using 300 mW of 488.0 nm argon ion radiation (37). The three dyes could be detected and identified at concentrations as low as 5 ppm in distilled water. Red Dye No. 40 could be detected in samples of cherry juice mix and in cherry soda (Fig. 11).

This technique has also been applied to the detection and identification of industrial fabric dyes (38). Two limits need to be defined: the identification limit and the detection limit. The former was defined as the concentration when at least three Raman bands with S/N ratio >3 were present. The detection limit required the presence of one band with a S/N ratio >3. These limits are given in Table 1, together with the detection limit when using spectrophotometry. The detection limits of the two techniques are comparable but resonance Raman spectroscopy has the added advantage of better identification capabilities. Resonance Raman spectroscopy has been used in a flow-through system for continuous monitoring of Superlitefast Rubine (39).

Aza-Arenes. Aza-arenes such as pyridine, quinoline, and acridine are of concern because of their carcinogenic and mutagenic properties. Sensitive and specific techniques are needed to determine *in situ* groundwater concentrations; prediction of contaminant transport requires an understanding of interfacial interactions between the analyte and the aquifer material.

The SERS technique, using adsorption on an electrochemically roughened silver electrode, was used to probe sensitivity and specificity (40). Using the electrode potential and the excitation wavelength as variables the detection limits in distilled water were found to be 700 ppb for quinoline and 4 ppm for pyridine. Natural water was simulated by adding humic material. The addition of 1 ppm of humic acid did not affect the S/N ratio of the spectrum of a 1580 ppm solution of pyridine. It should, however, be noted that the effect of humic acid on the spectra of lower concentrations of pyridine was not examined.

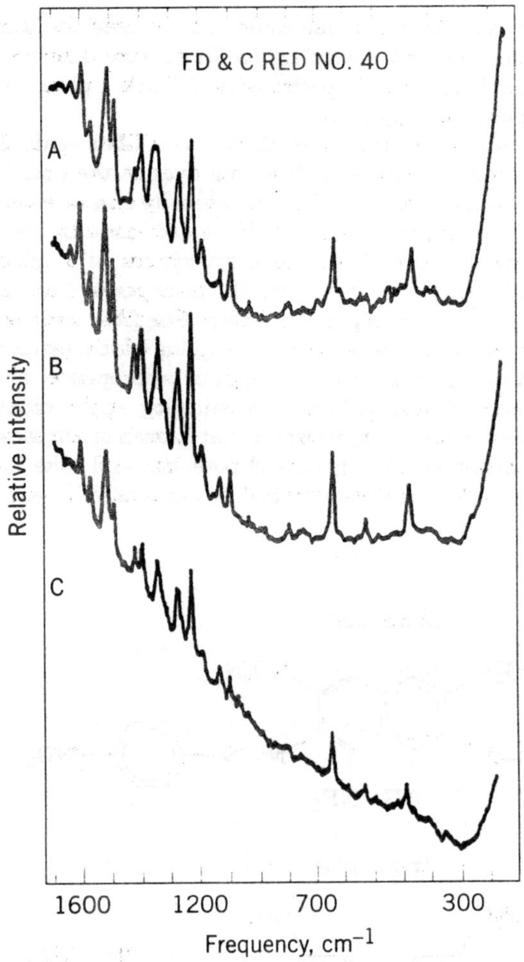

Figure 11. Resonance Raman spectra of A standard solution (5 ppm) of FD&C Red Dye No. 40; B Red Dye No. 40 in cherry juice mix; and C Red Dye No. 40 in cherry soda. Reprinted with permission (37).

Table 1. Minimum Detection (D) and Identification (I) levels (ppb) for Industrial Dyes[a]

	Raman		
	D[b]	I[c]	VS[d], D
Superfastlite Rubine	35	75	66
Procion Red	30	140	30
Lyrazol Fast Red	45	160	40
Direct Red 83	55	175	50

[a] Ref. 38.
[b] Detection limit at S/N >3 for the strongest band.
[c] Detection limit for three bands with S/N >3.
[d] Visible spectrophotometry using a 5-cm pathlength cell and a minimum detectable absorbance of 0.004.

Interfacial interaction was modeled by studying the effect of pH on the adsorption of aza-arenes from solutions containing electrolytes commonly found in groundwater ($NaClO_4$ and Na_2HPO_4) onto porous silica used as model aquifer material (41). The Raman studies indicated that the neutral aza-arene is hydrogen bonded to surface sites in aqueous media in which the pH is above the pK_a of the compound, whereas the interaction involves a protonated aza-arene/ClO_4^- ion pair when the pH is below the pK_a of the compound.

Chlorinated Hydrocarbons. The detection of chlorinated hydrocarbons in water using the SERS technique after adsorption onto electrochemically roughened (42) and octadecylthiol-modified (43) metal surfaces has been reported.

The study using electrochemically roughened copper electrodes reported sensitivities of <1 ppm for trichloroethylene, 10 ppm for CCl_4, and 15 ppm for $CHCl_3$ and perchloroethylene in high purity HPLC-grade water (42). Tests using two samples of trichloroethylene (TCE) contaminated groundwater showed that the TCE could be easily detected in the sample containing 280 ppm TCE, but not in the sample containing <1 ppm TCE.

The study using a 1-octadecylthiol-modified silver surface (43) used a technique similar to that used previously for the detection of aromatics (34). The detection limits are quoted as 12.6 ppm for perchloroethylene (PCE), 257 ppm for trichloroethylene (TCE), and 300 and 2086 ppm for *cis* and *trans*-dichloroethylene, respectively. The detection limit for PCE resembles that reported using the SERS study on copper (42) but that for TCE is two orders of magnitude greater than the value obtained using SERS (42).

DEVELOPMENTS AND PROSPECTS

Most of the above studies covering the period up to 1991 used a conventional scanning dispersive spectrometer. The main disadvantage is the time taken to record the spectrum at a typical scanning rate of about 5–50 cm^{-1}/min. The use of multichannel detectors (23,24,36,41,42) greatly reduces the spectral accumulation time. The methods are nevertheless still expensive, laboratory-based techniques, utilizing high power lasers with ppm and sub-ppm level sensitivity only attainable using intensity enhancement techniques (RRS, SERS, and SERRS). This approach requires colorless analytes to be converted into colored products or determined directly using UV lasers. Although the above studies have demonstrated that Raman spectroscopy is, in favorable circumstances, capable of determining analytes at low concentrations, it is far from being a routine analytical technique. Raman spectroscopy, unlike IR spectroscopy, can provide the entire vibrational spectrum for aqueous solutions. There are therefore substantial advantages to be gained from applying the technique as it has the ability to provide both qualitative and quantitative analysis of analytes.

A compact Raman spectrometer utilizing many of the recently introduced state-of-the-art devices has been used for quantitative detection of environmentally important dyes (44). The system (Fig. 12) is based on diode laser excitation (45) at 782 nm, fiber optic technology (46), an imaging spectrometer, and a state-of-the-art scientific CCD camera. The fiber optic probe is 3 meters in length and provides a much more versatile remote sensing capability than that described in (27) (see Fig. 7, A). As it requires only 1.5 V power at a few amps it potentially allows for portable operation. Some of the dyes examined (Fig. 13) are believed to be acutely toxic or genotoxic and may be found in wastewater discharges. As none of the dyes had appreciable absorption at the laser excitation wavelength of 782 nm there was no resonant enhancement of the spectral intensities. The four dyes give distinct Raman spectra

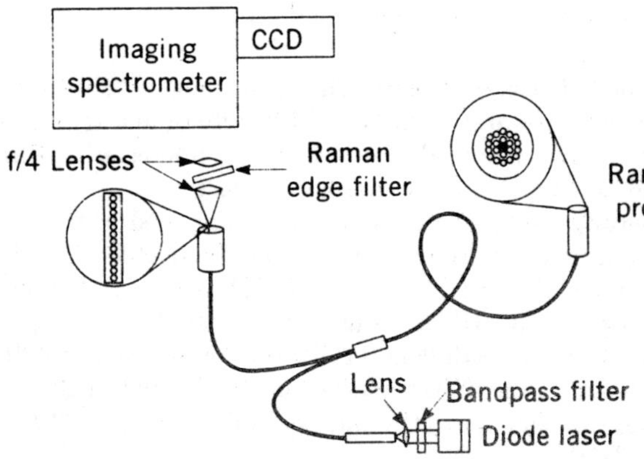

Figure 12. Schematic diagram of the diode laser/fiber optic Raman spectrometer. Reprinted with permission (44).

(Fig. 14) and the technique could thus be used for identification. Detection and identification limits were determined for each dye (Table 2) and spectra of Acid Black 1 down to 5 ppm are displayed in Figure 15.

This study showed that diode laser/fiber optic Raman spectroscopy is capable of detecting dyes at the ppm level in purified water without using any intensity enhancement techniques. The detection limit could be decreased by increasing the spectral accumulation time (maximum of 5 minutes in this study) and/or by increasing the laser power (maximum of 67 mW at the probe tip in this study). The fiber optic probe allows for remote as well as *in situ* sampling which, together with its portability, makes this technique appropriate for many environmental and industrial monitoring applications. This arrangement raises the prospect that Raman spectroscopy will finally attain its long-promised potential and take its place alongside other well-established water analysis techniques.

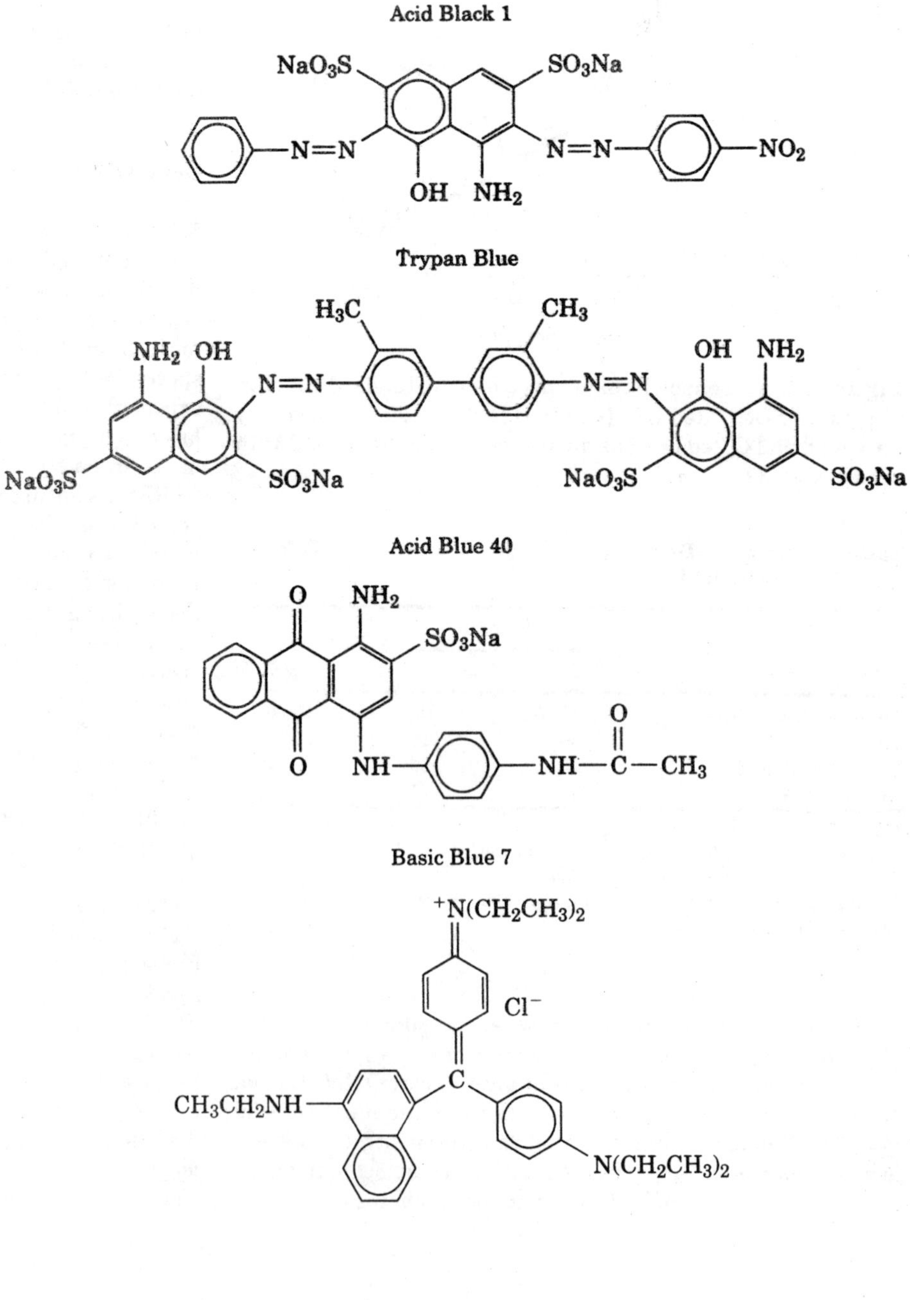

Figure 13. Structures of the dye molecules studied. Reprinted with permission (44).

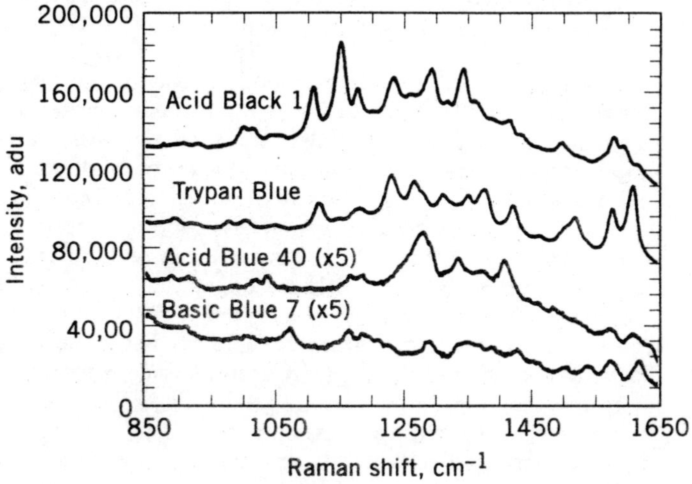

Figure 14. Raman spectra of the four dyes at 1000 ppm in distilled/deionized water. Reprinted with permission (44).

Table 2. Minimum Detection (D) and Identification (I) levels (ppm) Using Diode Laser/Optic Fiber Spectrometer[a]

	D[b]	I[c]
Tryptan Blue	0.4	5
Acid Black 1	0.2	2
Acid Blue 40	5	50
Basic Blue 7	25	100

[a] Ref. 44.
[b] Detection limit at S/N = 2 for the strongest band.
[c] Detection limit for six bands with S/N > 4.

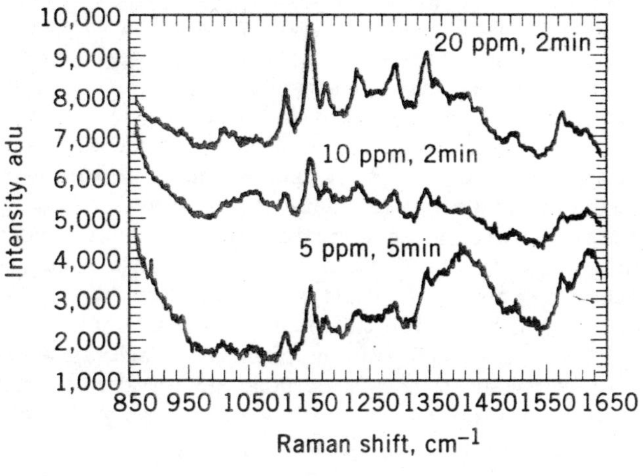

Figure 15. Raman spectra of Acid Black 1 at low concentrations: 20 ppm, 2 min integration time; 10 ppm, 2 min integration time; 5 ppm, 5 min integration time. Reprinted with permission (44).

Further developments and applications are awaited with interest.

BIBLIOGRAPHY

1. C. V. Raman, *Indian J. Phys.* **2,** 387–398 (1928).
2. J. R. Ferraro, *Spectroscopy* **11,** 18–25 (Mar./Apr. 1996).
3. A. Smekal, *Naturwiss.* **11,** 873 (1923).
4. G. Landsberg and L. Mandelstam, *Naturwiss.* **16,** 557 (1928).
5. R. W. Wood, *Nature* **122,** 349 (1928).
6. J. R. Ferraro and K. Nakamoto, *Introductory Raman Spectroscopy,* Academic Press, Inc., San Diego, Calif., 1994.
7. L. A. Woodward, *Philos. Mag.* **18,** 823–827 (1934).
8. H. L. Welsh, M. F. Crawford, T. R. Thomas and G. R. Love, *Can. J. Phys.* **30,** 577–596 (1952).
9. H. Stammreich, *Spectrochim. Acta* **8,** 41–45 (1956).
10. S. P. S. Porto and D. L. Wood, *J. Opt. Soc. Am.* **52,** 251–252 (1962).
11. H. Kogelnik and S. P. S. Porto, *J. Opt. Soc. Am.* **53,** 1446–1447 (1963).
12. P. Hendra, C. Jones, and G. Warnes, *Fourier Transform Raman Spectroscopy,* Ellis Horwood, Chichester, U.K., 1991; D. B. Chase and J. F. Rabolt, eds., *Fourier Transform Raman Spectroscopy,* Academic Press Inc., San Diego, Calif., 1994.
13. D. Pitt, *Spectrosc. Eur.* **8,** 34–35 (Feb. 1996); M. Pelletier and K. Davis, *Int. Lab.* **26,** 11D–11G (May 1996).
14. S. A. Asher, *Anal. Chem.* **65,** 59A–66A, 201A–210A (1993).
15. M. D. Morris and D. J. Wallan, *Anal. Chem.* **51,** 182A–192A (1979).
16. W. Krasser, *Chem. Brit.* **25,** 618–622 (1989); R. L. Garrell, *Anal. Chem.* **61,** 401A–411A (1989).
17. J. J. Laserna, *Anal. Chim. Acta* **283,** 607–622 (1993).
18. A. B. Harvey, ed, *Chemical Applications of Non-linear Raman Spectroscopy,* Academic Press, Inc., San Diego, Calif., 1981; D. A. Long, *Chem. Brit.* **25,** 592–596 (1989); J. P. Taran, *Chem. Brit.* **25,** 597–600 (1989).
19. D. E. Irish and T. Ozeki, in *Analytical Raman Spectroscopy,* eds., J. G. Grasselli and B. J. Bulkin, John Wiley & Sons, Inc., New York, 1991, pp. 59–106.
20. T. G. Spiro, ed, *Biological Applications of Raman Spectroscopy,* Vols. 1–3, John Wiley & Sons, Inc., New York, 1987–1988; T. G. Spiro, *Chem. Brit.* **25,** 602–605 (1989).
21. S. F. Baldwin and C. W. Brown, *Water Res.* **6,** 1601–1604 (1972).
22. K. M. Cunningham, M. C. Goldberg, and E. R. Weiner, *Anal. Chem.* **49,** 70–75 (1977).
23. K. Xi, S. K. Sharma, G. T. Taylor, and D. W. Muenow, *Appl. Spectrosc.* **46,** 819–826 (1992).
24. R. D. Shelton, J. W. Haas, and E. A. Wachter, *Appl. Spectrosc.* **48,** 1007–1010 (1994).
25. E. B. Bradley and C. A. Frenzel, *Water Res.* **4,** 125–128 (1970).
26. G. Bräunlich, G. Gamer, and M. S. Petty, *Water Res.* **7,** 1643–1647 (1973).
27. M. Ahmadjian and C. W. Brown, *Environ. Sci. Technol.* **7,** 452–453 (1973).
28. R. J. Thibeau, L. Van Haverbeke, and C. W. Brown, *Appl. Spectrosc.* **32,** 98–100 (1978).
29. S. Higuchi, O. Aiko, and S. Tanaka, *Anal. Chim. Acta* **116,** 1–6 (1980).
30. L. Van Haverbeke and M. A. Herman, *Anal. Chem.* **51,** 932–936 (1979).
31. H. Nakamura, K. Yoshimori, J. Itoh, M. Takagi, and K. Ueno, *Mikrochim. Acta* **2,** 131–142 (1981).
32. N. A. Marley, C. K. Mann, and T. J. Vickers, *Appl. Spectrosc.* **38,** 540–543 (1984).
33. N. A. Marley, C. K. Mann, and T. J. Vickers, *Appl. Spectrosc.* **39,** 628–633 (1985).
34. K. Carron, L. Peitersen, and M. Lewis, *Environ. Sci. Technol.* **26,** 1950–1954 (1992).
35. S. A. Asher, *Anal. Chem.* **56,** 720–724 (1984); C. R. Johnson and S. A. Asher, *Anal. Chem.* **56,** 2258–2260 (1984).

36. B.L. Wittkamp and D.C. Tilotta, *Anal. Chem.* **67**, 600–605 (1995).
37. C.W. Brown and P.L. Lynch, *J. Food Sci.* **41**, 1231–1232 (1976).
38. L. Van Haverbeke, P.F. Lynch, and C.W. Brown, *Anal. Chem.* **50**, 315–317 (1978).
39. L. Van Haverbeke and C.W. Brown, *Am. Lab.* **10**, 62–68 (1978).
40. M.M. Carrabba, R.B. Edmonds, and R.D. Rauh, *Anal. Chem.* **59**, 2559–2563 (1987).
41. R.A. Matzner, R.C. Bales, and J.E. Pemberton, *Appl. Spectrosc.* **48**, 1043–1053 (1994).
42. J.M.E. Storey, R.D. Shelton, T.E. Barber, and E.A. Wachter, *Appl. Spectrosc.* **48**, 1265–1271 (1994).
43. K. Mullen and K. Carron, *Anal. Chem.* **66**, 478–483 (1994).
44. D.A. Gilmore, D. Gurka, and M.B. Denton, *Appl. Spectrosc.* **49**, 508–512 (1995).
45. S.M. Angel, M. Carrabba, and T.F. Cooney, *Spectrochim. Acta* **51A**, 1779–1799 (1995).
46. P.R. Griffiths, I.R. Lewis, N.C. Chaffin, N.W. Daniel, and J.D. Jegla, *J. Mol. Struct.* **347**, 169–186 (1995); J. Lin, S.J. Hart, and J.E. Kenny, *Anal. Chem.* **68**, 3098–3103 (1996).

RECLAMATION OF DISTURBED LANDS

TERRENCE J. TOY
University of Denver
Denver, Colorado

W. LEE DANIELS
Virginia Polytechnic Institute and State University
Blacksburg, Virginia

Land disturbance and degradation resulting from human activities have persisted for millennia. It appears that early peoples set fires to drive herds of game animals and possibly to maintain grasslands. Deforestation and grazing caused severe erosion problems during the classic Greek and Roman periods (1). Various land conservation activities have been observed and recorded through the centuries. The Phoenician, Romans, Chinese, Incas, and Mayans constructed erosion-control terraces or contoured their fields (2). Specific efforts to repair or reclaim disturbed lands, however, are an enterprise only common within the past few decades. The recent interest in land reclamation is probably the consequence of population growth and the realization that planet earth is a finite resource; documentation of environmental damage resulting from disturbance; accumulated societal wealth and standards of living that permit long-term investment in the land; and the development of analytical and reclamation technologies capable of identifying and mitigating the environmental repercussions of land disturbance.

The National Academy of Sciences (3) defined three categories of remedial land treatment:

1. Rehabilitation: The land is returned to a form and productivity in conformity with a prior land-use plan including a stable ecological state that does not contribute substantially to environmental deterioration and is consistent with surrounding aesthetic values.
2. Reclamation: The site is hospitable to organisms that were originally present or others that approximate the original inhabitants.
3. Restoration: The condition of the site at the time of disturbance is replicated after the action.

According to these definitions, rehabilitation usually permits the greatest flexibility in future land use and incurs the least cost. Reclamation infers that the pre- and postdisturbance land uses are nearly the same. Restoration allows no land-use flexibility and incurs the greatest cost. There are various perspectives regarding the most appropriate goals for land treatment. Generally, industry favors rehabilitation; regulatory authorities favor reclamation, and many ecologists favor restoration. The central issue often is future land-use requirements and whether returning to the predisturbance use constitutes the best use of the land. It must be demonstrated, commonly, that alternative uses are attainable and warranted.

The terms rehabilitation, reclamation, and restoration have not been used consistently and connotations have varied through the years. The pertinent laws and regulations have been interpreted and enforced in different ways from time to time and from place to place. In the remainder of this discourse, we use the term "reclamation," but with the understanding that alternative land uses should remain possibilities. We define reclamation as the treatment of disturbed areas to create stable landforms and edaphic conditions to sustain predetermined land uses with minimal maintenance.

Reclamation Success

Reclamation goals have changed considerably over the years. It has long been recognized that establishing a vegetative cover is necessary to control erosion. This resulted in the "Green Grass Syndrome" which centered on the question "How high can we get grass to grow on the spoils piles?" (4). Although erosion control remains a fundamental objective, reclamation must achieve much more. The National Research Council (5) stated that an appropriate goal for reclamation is to ensure that society does not lose important land-use opportunities available before soil disturbance or generated in the reclamation process. Postreclamation land use increasingly became an important consideration, sometimes resulting in tensions between those advocating new or "better" uses and those supporting the return to past uses. As reclamation science progressed, the goals became more ambitious. As we learned to do better, we attempted to do more.

There are various tests of reclamation success. Erosion rates for reclaimed lands are compared with those of nearby, natural areas (6,7). Erosion rates are functions of several site characteristics including climate, topography, soil properties, and surface cover, and therefore serve as useful bases for site comparisons. Evidence of accelerated erosion is taken as evidence of hydrologic problems at a site (8). The vegetative cover, species diversity, and productivity of reclaimed areas are commonly compared with those of undisturbed reference areas. The "reference area" approach, however, is problematic, requiring replication of the pre-disturbance ecosystem, involving comparisons between mature and immature plant communities, and necessitating establishment of the original plant species on the disturbed soil (9). An alternative is the "utilitarian" approach that assesses the capability of the reclaimed land to properly function and thereby sustain its intended land use. For example, if grazing is the chosen land use, reclaimed lands should produce forage and support livestock gains appropriate to the region without detrimental effects to the ecosystem (9).

Purpose of Article

The reclamation process consists of several sequential steps: (*a*) site characterization, (*b*) reclamation planning, (*c*) material management, (*d*) topographic reconstruction, (*e*) replacement of topsoil or provision of a suitable substitute, (*f*)surface manipulation, (*g*) soil amendment, (*h*) revegetation, (*i*) irrigation, if needed, and (*j*) monitoring and maintenance. This article briefly discusses each of these components. Although most of the following discussion is based on our experiences with surface-mined lands, the principles are applicable to other types of land disturbances, such as quarrying, metal mining, or highway, industrial, and urban/residential construction.

TEMPORARY IMPACT MITIGATION

The reclamation of disturbed sites involves a variety of short- and long-term management practices, usually designed before disturbance, to minimize adverse impacts and maximize the site's future productive potential. It is important to realize, however, that some short-term effects, such as increased runoff, sediment yield, and wildlife displacement are inevitable from major land-disturbing activities. Thus, although long-term reclamation goals focus on landform reconstruction, overall hydrologic impacts, revegetation strategies, and post-reclamation land use, an active program of temporary impact mitigation is included in daily operations and contingency plans.

In general, water- and sediment-control measures, including routine facilities maintenance, are the most important practices for avoiding off-site environmental impacts from land disturbance. Intense thunderstorms falling on barren surfaces without appropriate erosion control result in extraordinarily high hillslope erosion rates, gullying, and even landslides. Barren sites must be protected from water run-on originating on upslope areas. Water flowing across the site must be routed through appropriate diversion channels and sediment detention structures as specified in applicable permits. Revegetation efforts must be concurrent with disturbance to minimize the total area exposed at any one time. Surface runoff ditches, sumps, and other temporary sediment ponds and structures require routine maintenance to ensure their effectiveness in controlling even minor storms.

In conjunction with the stormwater management practices, the material handling and placement plan is reviewed and adjusted as needed to prevent prolonged exposure of highly erosive or potentially toxic strata. As discussed later, the physical and chemical properties of all materials to be handled on site are rigorously analyzed before disturbance, and any materials posing short- or long-term water quality risks are identified. This level of material handling coordination requires daily adjustments to "cut/haul/fill" sequences and on-site water control.

Topsoil stockpiles are particularly susceptible to erosion or permanent saturation and require careful protection. These stockpiles are located away from traffic and material handling operations whenever possible. They are revegetated if the topsoil is stored for several months. The stockpile is identified with prominent signage and surrounded by a low berm of geologic material to divert lateral water flow and contamination with "nontopsoil" materials. Stockpiles are not used for waste disposal, especially contamination by petroleum products. Finally, the stockpiles are located for easy access during final topsoil load-out and respreading operations.

The short-term wildlife impacts of land disturbance are highly variable and difficult to predict, depending on the size of a disturbed site and its location. However, most disturbance results in the complete displacement of all habitat values and species for a finite period of time. Threatened or endangered species require special consideration and specific impact mitigation before the onset of the disturbance. Disturbance may be delayed until after the nesting season of raptors or other large predators. The movement of large animals (eg, deer or elk) through an active site is also controlled, particularly when large pits and highwalls are exposed. These hazards typically are fenced or surrounded by large berms to deter wildlife movement. Finally, some wildlife, especially certain rodents and birds, may interfere greatly with final revegetation efforts. Thus, specific animal control measures are often components of reclamation plans.

Effective reclamation is often measured by the attainment of long-term goals, including vegetative cover and diversity, hydrologic response, and the return of the site to productive use. These long-term goals are achieved through daily conformance to material handling and water control plans as necessarily adjusted and "fine-tuned" on-site while the disturbance progresses across the landscape. The geologic and hydrologic conditions encountered may differ considerably from those predicted in the original plans. Therefore, mine engineers and reclamation managers must be flexible and adopt new and more effective approaches to meet their long-term reclamation goals.

SITE CHARACTERIZATION

The physical and chemical properties of a disturbed area within a particular environmental setting profoundly influence reclamation planning, practice, and the probability of success. Site characterization usually entails analyses of climate, geology, topography, soils, vegetation, and hydrology. Although it is convenient to discuss these attributes individually, the environment functions as an open, process-response system with complex interrelations among components. Hence, changes in these site characteristics may have ramifications far-reaching in time and space.

Climatic Conditions

The climate of a site includes the mean condition of the atmosphere and the variability about mean values. Primary characteristics are temperature and precipitation. Important derived characteristics are growing-season length, potential evapotranspiration, and water surplus or deficit. Mean annual statistics usually are of little value in reclamation planning or practice. The temporal distribution of climate characteristics, however, is highly important. Short periods of unfavorable climatic conditions may severely impact reclamation success. For example, Kirby, Montana receives about 50 cm of precipitation annually but only about 15 cm during the growing season (10).

Temperature is a measure of heat energy within a system. Generally, there is a significant relationship between near-surface atmospheric and soil temperatures (11). Precipitation provides the mass input and, together with gravitational force, the kinetic-energy input into the hydrologic system operating at a site. Temperature and precipitation influence growing-season length, seed germination, potential evapotranspiration, water surplus or deficit, rock weathering, surface and subsurface flows, erosion rates, and the scheduling of reclamation practices. As temperature and precipitation affect these processes at a site, they influence landform stability, vegetative production, and reclamation success.

Climatic conditions vary considerably in time and space (10,12–14). Long-term, on-site records, therefore, are highly desirable for reclamation planning. It is necessary to assume that past conditions are indicative of future conditions. The annual distribution of temperature and precipitation largely determines the periods of greatest site vulnerability to runoff and erosion and the times most suitable for reclamation activities. The growing-season distribution of temperature and precipitation largely determines the opportunities and limitations for revegetation. Various climatic parameters are used in hydrologic and erosion models to estimate postreclamation process rates.

Long-term climate records often are available through federal or state agencies. These data are used for reclamation planning as long as the geographic setting at the meteorological station is similar to that of the reclamation site. Coal and metal mines commonly establish stations on-site to document prevailing climatic conditions during the several years of operation. Although hourly precipitation and temperature measurements often are desirable for process modeling, daily information may suffice.

Geologic Conditions

The engineering design and material handling plan for any disturbed site is usually dictated by the geologic conditions, including the weathered soil mantle. This is true regardless of whether the disturbance is caused by mining, highway development, or other construction activities. The major difference among these forms of disturbance is the depth of materials excavated and exposed. Mining operations typically require deep excavations, often extending into unweathered geologic strata. "Cut and fill" operations at construction sites, on the other hand, generally take place in the near-surface weathered zones of the soil-geologic column, except in the case of deep highway roadcuts. For the purpose of this discussion, we focus on mining because it results in the deepest and most severe disturbance of soil and geologic materials. Again, however, the basic principles discussed here apply to all major land-disturbing activities.

Mining companies are required to characterize the physical and engineering properties of the extractable resources and overburden because geologic conditions influence the economics of production and environmental considerations during mining and reclamation. Mining operations occur on the full range of lithologies and structures worldwide. The steps used in assessing geologic conditions are similar, however, and the procedures described here are applicable to most mine sites.

At the majority of surface mines the resource is covered by some thickness of overburden beneath a naturally developed mantle of soil. The overburden is indurated (competent) rock, partially weathered rock or "saprolite," unconsolidated sediments, or a combination of all three. Geologic materials between the soil mantle and the mined resource are called "spoils" following disturbance. Where mining operations remove multiple seams of a resource, the waste materials between the seams are called "interburden." The term "mining waste" is also used generically to describe all waste products of mining, including those generated by mining and associated processing.

The soil mantle is stripped or salvaged before overburden removal and stockpiled for use in final reclamation. At many sites, however, the native topsoil is scarce, of low quality, or was not saved previously. In these cases, reclamation uses topsoil substitutes derived from the overburden (15). Once these designated spoils are placed on a reclaimed surface and once they begin to weather and support plant growth, they are regarded as "mine soils." Although some choose to treat the soil mantle and the underlying overburden as very different materials, it is important to realize that they often reflect a weathering continuum from the surface downward. Soil reconstruction and topsoil substitution are discussed in subsequent sections.

In general, overburden materials comprise most of the materials handled and transported at any site, their volume and handling requirements largely determine the cost of mining. The physical and chemical properties of the overburden control hillslope stability in backfills and the geochemistry of surface and ground waters. Thus, it is essential that we understand (*1*) the premining geologic conditions, (*2*) the geologic conditions after mining, and (*3*) the dynamics of the physical and chemical processes affecting these geologic materials following disturbance. The geochemical characteristics of mining wastes are also strongly affected by the climatic conditions at the site, particularly as they affect leaching intensity.

In North America, mining takes place under a variety of geologic conditions. Surface mining of coal disturbs the largest land area, and most of our knowledge concerning mine reclamation was developed at coal mines. North American coal reserves are divided into three general geologic regions: (*1*) the Appalachian basin, (*2*) the Midwestern (Eastern) inland basin, and (*3*) the Western region. In the Appalachians, relatively thin coal seams (<3 m) are overlain and separated by nearly horizontal sequences of sandstones, siltstones, shales, and mudstones. The pyrite (FeS_2) content of the coals and the associated overburden is highly variable, but is the most important single geologic property from the perspectives of mineral processing and environment. The Midwestern coals generally are contained in less competent, sedimentary rocks than those of the Appalachian region, and the strata commonly are overlain by glacial tills and loess. In the Western region, coal seams are considerably thicker (up to 65 m) than those in the East and often covered by partially lithified or unconsolidated sediments and glacial materials. The Western coals and overburden are generally lower in pyrite content than their Eastern counterparts, because of differences in the paleoenvironment of formation (16). The lower pyrite content substantially reduces the processing costs of these coals and the potential water quality and soil chemistry impacts associated with mining these resources.

Likewise, a wide variety of sand and gravel, carbonate, and dimensional stone-mining operations take place throughout North America in a variety of geologic settings. In addition, there are numerous heavy and precious metal-mining operations in the West. In general, metal mining exploits various metamorphic and hydrothermal deposits that frequently contain sulfide compounds and other highly reactive minerals with the potential for severe environmental impacts.

Minimum Criteria for Assessing Geologic Conditions. As part of the permit application, a mining company completely maps, documents, and characterizes the lithologies and geologic structures at the proposed site, including their dip, thickness, and spatial extent. Other data, such as hardness and swell factor, are necessary to accurately estimate the volume of all materials to be handled. The most important step to accurately evaluate the potential environmental effects and reclamation challenges resulting from the mine operation is collectively referred to as "overburden analysis" (17) and is required for all geologic and soil materials found on-site before any disturbance is permitted. The results of premining overburden analyses are used to (*1*) identify potentially toxic strata and develop a plan for their isolation; (*2*) predict water quality impacts resulting from material handling and placement operations; (*3*) determine the geotechnical properties (eg, shear strength, swell, compaction characteristics) of all materials; and (*4*) ascertain which strata are suitable topsoil substitutes, where necessary.

Each stratum is sampled at several locations from drill cuttings and/or bulk samples if sediments are encountered. Care is taken to prevent contaminating the samples with drilling greases. The number and spacing of boreholes required to adequately characterize geologic conditions varies widely. The experience of the operator, consultants, and regulatory agencies must insure that representative samples are obtained for testing. Samples are taken from outcrops and shallow excavations for geotechnical analyses, and samples from drill cuttings are analyzed for pH, acid-base accounting, extractable nutrients, cations, and soluble salts. Other specialized analyses for heavy metals, such as arsenic (As), and selenium (Se), may be required because of local geologic conditions and permitting requirements. Once all of the materials at a site have been characterized, the potential environmental effects of the various materials are forecast and the limitations imposed by the physical or chemical properties of the spoils are addressed in mining, materials handling, and reclamation plans. The majority of adverse environmental impacts due to mining result from the lack of adequate initial overburden analyses and subsequent adjustment of mining plans and procedures.

Topographic Conditions

Topography refers to the surface configuration of an area, perhaps described as rugged, rolling, gentle, or smooth. Average hillslope gradient and relief (the difference between the highest and lowest elevations) are expressions of topography. Relationships between the shape of the land and geomorphic process rates have been documented extensively (18). As hillslope gradient increases, the propensity for mass movement, runoff, erosion, and sediment yield rates increase. The U.S. Environmental Protection Agency (19) stated that as hillslope steepness increases, there is a corresponding rise in the velocity of the surface runoff which in turn results in greater erosion. A doubling of the velocity of water, produced by increasing the degree (*gradient*) and length of the hillslope, enables water to move particles 64 times larger, allows it to carry 32 times more soil material, and makes the erosive power, in total, 4 times greater. The National Academy of Sciences (3) noted that overburden removal techniques need to account for the fact that "maximum vegetative stability" cannot be obtained on hillslopes steeper than 33% (3:1). "Optimum vegetative stability" requires slopes of less than 25% (4:1), and the use of agricultural machinery requires that slopes are no greater than 20% (5:1). Although the exact effect of hillslope gradient in a particular circumstance is debatable, the concepts involved remain valid. Gentle hillslope gradients usually facilitate other reclamation operations. For small drainage basins without significant floodplains, there is usually a direct relationship between hillslope gradient and also channel gradient.

Hillslope aspect or orientation (the direction that a surface faces), together with gradient, largely determines the quantity of solar radiation received at a given latitude. As solar radiation increases, soil temperature, the annual number of freeze-thaw and wet-dry cycles, and evapotranspiration rates increase, whereas soil moisture, organic matter, and vegetative production decrease.

The topography surrounding a disturbed site also influences reclamation plans and practices. The reconstructed surface must blend with the undisturbed landscape so that matter and energy fluxes smoothly traverse the reclaimed surface. Abrupt changes in hillslope or channel gradients frequently are the locus of accelerated process rates.

Topographic maps prepared and distributed through the U.S. Geological Survey generally are available showing predisturbance land surfaces at a scale of 1:24,000 with contour intervals of 3–6 m (10 or 20 feet). Likewise, aerial photographs or remotely sensed imagery often can be obtained through various government agencies showing the land at various times in the past. It is frequently necessary, however, to construct larger scale topographic maps from detailed surveys for operational and reclamation purposes.

Soil Conditions

The soil mantle at a proposed mine site usually develops from the uppermost geologic stratum. Subsequently this material is modified by pedogenic processes operating through time under specific conditions of climate, vegetation, and topography. The soil profile strongly influences hydrologic processes. In particular, rainfall is partitioned into infiltrating water and runoff based on the texture and structure of the A horizon, coarse-fragment content, surface roughness, and gradient. The soil's water-holding capacity, controlled by the combined factors of texture, aggregation, bulk density, and overall depth, directly influences plant productivity, leaching potential, and groundwater replenishment.

Soil chemical characteristics, including pH, nutrients, sodicity, salinity, and other metals, also directly influence plant productivity, species adaptation, and land-use capability. The topsoil layer (A + E horizons, as defined by the Office of Surface Mining) is enriched with nutrients and organic matter because of bioaccumulation and litter recycling. The thickness and humus content of soils are typically good indicators of overall soil quality. The organic matter or "humus" (*1*) buffers soil acidity, (*2*) provides nutrients through mineralization, (*3*) binds potentially toxic metals, and (*4*) enhances the aggregation, aeration, and water-holding capacity of the topsoil layer. Therefore, the accumulation and maintenance of organic matter commonly is considered the most important soil property affecting plant growth (20) and is a top priority for reclamation.

North American mining also occurs under a variety of soil conditions because of variations in geologic materials and climate. In the Appalachians, soil development commonly is limited by the steep hillslopes, typical of this intensively dissected region, and the resistance of siliceous rocks to weathering. The rugged terrain and thin soils severely limit the land-use potential, particularly for agriculture. The soils of the Midwestern inland basin develop on glacial tills and loess yielding deeper profiles than typical of the Appalachians. These soils and the gently rolling Midwestern landscape often support productive agriculture before mining. Because of the arid and semiarid climatic conditions, the soils of the Western coalfields are less leached and differentiated than those in the east and retain much higher soluble salt contents. Alkalinity, sodicity, salinity, and water deficiencies frequently limit their land-use potential.

As soils develop, water soluble salts and dispersible solids (clays, iron, and sometimes humus) are leached or "eluviated" from the topsoil layers and accumulate in the "illuvial" subsoil (B horizon), usually producing strongly differentiated profiles through time. Subsoils in the Eastern U.S. typically are much higher in clay, iron, and total acidity than their overlying topsoils, whereas Western subsoils typically are high in secondary carbonates or sulfates. The subsoil also is significantly more compact (higher in bulk density) than the topsoil because the pore spaces are filled with illuvial materials. Therefore, subsoil layers, particularly B horizons, typically are much less desirable plant growth media than topsoils, and lands where the topsoil has been degraded by erosion or scalping are much less productive. The geochemically altered zone below the depth of rooting and transpiration, called the "C horizon", constitutes a transition between soil and geologic materials. This zone may be quite thick, is often more oxidized and leached than underlying strata, and may be superior in plant growth potential to the overlying B horizon (21).

In addition to its ability to support plant growth, the physical and chemical properties of soils combine with climate and landform characteristics to influence hillslope stability and other engineering characteristics, such as shrink-swell potential, foundation suitability, and compactibility. Collectively, the land-use potential of a site is strongly affected by its predisturbance soil conditions, and the postdisturbance land-use potential is strongly affected by the properties of the reconstructed soils. Although it is not possible to completely recreate preexisting soil conditions after disturbance, a thorough understanding of the soil and its alteration by mining is essential to reclamation planning and success. It is also important to note that recreated mine soils are very dynamic systems, particularly when they are created from newly exposed geologic materials which rapidly change in the years following reconstruction (22).

Because of the importance of the soil to reclamation planning and success, it is essential to map and classify soils before mining or other disturbance. The scale of soil mapping should be sufficient to support correct interpretations and reclamation planning. Modern USDA-NRCS soil maps are compiled at a scale of 1:24,000 which is too generalized for detailed planning, particularly for topsoil handling plans. Because most final mine plan and permit maps are constructed at scales of 1:6,000 or larger, we suggest soil mapping or field correction at the same scale. Soil characterization includes the major soil types, associated horizons, depths, physical, and chemical properties. Particular attention should be given to topsoil (A + E horizon) thickness and the presence of any potentially toxic (acidic, sodic, or saline) subsoil horizons so that they can be isolated during material handling. Many subsoil horizons, especially in the Eastern United States, contain plastic clays that complicate mining and reclamation during wet weather because they are difficult to handle, compact, and stabilize.

As a part of the mapping and classification process, the landscape is also delineated into areas of similar landform, geology, drainage, and dominant soil type to identify individual landscape units. In fact, a well-executed soil map can be used to predict underlying geologic materials and near-surface hydrologic conditions. Therefore, the soil map reflects the overall surficial geomorphology of the site. When this map is then combined with data for the soil properties discussed above, it serves as an invaluable tool for reclamation planning and mine site planning and development.

Vegetative Conditions

The plant community of an area may include native and introduced, sensitive and tolerant, common and endangered species. The quantity, quality, and diversity of vegetation at a site reflect the entire environmental setting in addition to past and present human activities. Vegetation, however, is not merely a passive respondent to site characteristics but is important in determining the rates of rock weathering, soil development, mass movement, surface and subsurface flows, and erosion. The presence and absence of particular species indicates a site's potential to support plant life and should be carefully considered during reclamation planning.

It is very important to estimate accurately the premining productivity and diversity of the vegetation because most reclamation programs, especially for coal mining, require return of the site to equivalent predisturbance conditions. Without premining information, the postdisturbance productivity and diversity are compared to that of external reference areas, which may not be appropriate or reliable. The predisturbance vegetative assessment also renders important clues concerning adverse soil conditions (eg, dominance of halophytes) or drainage conditions (eg, dominance of hydrophytes) and should directly influence mine planning and development. As noted previously, threatened and endangered species must be carefully documented, and appropriate habitat mitigation and propagation protocols must be included in the permit application. In addition, when the reclamation plan requires the return to native vegetation, the predisturbance assessment should identify those species that could pose propagative problems. Specific predisturbance harvesting plans for both seed and live vegetation may be necessary. This is particularly true of semiarid plant communities with diverse native plant assemblages (23).

Hydrologic Conditions

The hydrologic conditions at a site include the quantity, quality, movement, and storage of water above and below the surface. Hydrology is determined by upslope and on-site characteristics of climate, geology, topography, soil, and vegetation. Hydrology, in turn, affects land form, soil, and vegetative processes. Climate provides the water input to the hydrologic system whereas the other characteristics determine the movement of water into and across the surface. Precipitation inputs immediately participate in the system whereas snow and ice inputs must await melting.

According to the Horton model (24), runoff commences when precipitation rates exceed the infiltration capacity of surface materials. Flow accumulates in the downslope direction, increases in depth and velocity, and follows the path of the steepest gradient to hillslope channels, such as rills and gullys, or valley channels, such as gullys and streams. The quantity or discharge of water and sediment determines channel characteristics (25). Mean, seasonal, and peak discharges are commonly measured or estimated because they are essential to water management. In addition, the surface of sediment particles provide vehicles for transporting adsorbed chemical compounds.

There is an alternative hydrologic model that couples surface and near-surface water flow. In humid areas with dense vegetative covers, buildup of organic matter in the soil, and high infiltration capacities, most precipitation passes into the soil and travels laterally (throughflow) to reemerge farther downslope (return flow). The combination of runoff and return flow from saturated areas, known as "saturated overland flow," is a major contributor to stream channels in humid regions.

Where unimpeded by impermeable soil horizons or geologic strata, water moves vertically through porous media to join the groundwater reservoir beneath the surface. Groundwater also flows downgradient or in the direction of least pressure. Should the land surface intersect the groundwater flow, water issues forth as seeps or springs. Groundwater discharge provides the lowest constant flow in perennial streams, known as "base flow." The measurement or estimation of groundwater discharge is also essential to water management. In addition, solutes carried in groundwater affect soil chemistry and water quality where they reach the land surface or stream.

The solute, sediment, and organic components of water determine water quality. As discharge passes over and through soils or geologic materials, it acquires a load of various dissolved and suspended compounds, derived from both natural and human sources. Concentrations of those substances may reach levels at which they become deleterious to the environment or unsuitable for particular uses, whereupon they are considered pollutants. Water quality measurements are important because pollutants severely affect natural or reclaimed ecosystems and human populations. Standard water quality tests usually are performed by selected laboratories using samples collected at designated locations throughout the disturbed site.

RECLAMATION PLANNING

The second step in the reclamation process is planning. Ideally, the reclamation strategy and tactics are finalized before land disturbance. This allows full consideration of the (*1*) probable and possible consequences of disturbance, (*2*) project management protocols to facilitate reclamation, and (*3*) evaluation of alternative reclamation practices. Most land-disturbing industries, such as coal mining, must obtain operating permits, and both temporary impact mitigation and permanent reclamation plans are necessary components of the permit applications. Reclamation planning, therefore, is driven by the environmental setting, consequences of disturbance, reclamation goals, and legal requirements.

Environmental Setting

Site characterization provides a detailed and quantitative definition of the environmental setting in which reclamation takes place. The physical attributes, hydrologic and geomorphic processes, on-site and off-site constraints, and any special opportunities or problems, such as unique habitats or endangered species, are identified and described. Baseline data requirements may be specified by regulations or suggested by guidelines and past experience. All desirable on-site data usually are not available. The acquisition of additional data necessitates an assessment of collection costs compared with the risks that could result from decisions made on the basis of incomplete information.

Consequences of Disturbance

Land-disturbing activities result in both common and unique impacts. Most activities at least disrupt vegetative cover and agitate upper soil horizons. Other activities invert and homogenize the soil profile in addition to the above impacts. Some activities also fragment geologic strata and structures, significantly altering site topography. Collectively, these modifications change the magnitude and frequency of surface and subsurface hydrologic and geomorphic processes.

Impacts also vary depending on methods of operation. The consequences of underground and surface mining differ considerably. Within the category of surface mining, the effects of open-pit, quarry, contour, mountaintop removal, and area strip mining also differ. Even within a category of surface mining, such as open-pit mining, consequences differ depending on the mineral produced. For example, low-grade ores commonly are processed on-site to minimize transportation cost. The waste products of processing contain various particle sizes, quantities of acids, heavy metals, and radioactive constituents (Fig. 1). Safe, long-term, disposal of such wastes warrants careful planning and special techniques (18). Clearly, reclamation planning requires complete understanding of the consequences of particular land-disturbing activities.

Reclamation Goals

The goals for reclamation can be viewed from theoretical and practical perspectives. Simply put, a reclamation project should aim to produce environmentally sound and stable conditions that ultimately integrate the disturbed area into the general ecosystem (26). Although directed toward surface-mine reclamation, the National Research Council (27) provided a useful general statement of reclamation goals, applicable to most types of land disturbance:

> *For most sites, reclamation is understood as rehabilitating the disturbed site by grading in a manner compatible with the surrounding landscape, or with the post-mining use, and establishing a self-sustaining vegetation cover. Reclamation is intended to control erosion, to restore the hydrologic balance, and to limit the off-site impacts, during and after mining. The ultimate purpose is to restore the mining site to productive use, for example, for agriculture, forestry, industry, housing, recreation, or fish and wildlife habitat. Desirably this use should be at least as beneficial as the capability of the site before mining although not necessarily for the same purpose.*

Figure 1. Coal mining on steep terrain in the Eastern United States.

From a practical perspective, the goals for reclamation include (*1*) compliance with laws and regulations, (*2*) attainment of land-use objectives, and (*3*) minimal long-term site maintenance. Laws, regulations, and reclamation activities are directed toward sustaining the postreclamation land use. Accordingly, a reclamation plan actually contains several component plans addressing the steps necessary to achieve these goals, including (*1*) material management, (*2*) topographic reconstruction, (*3*) topsoil replacement or substitution, (*4*) surface manipulation, (*5*) soil amendment (*6*) revegetation, (*7*) irrigation, and (*8*) site monitoring and maintenance. These steps are discussed in forthcoming sections.

Legal Requirements

Decades of uncontrolled land disturbance in the U.S. resulted in large areas of abandoned land no longer suitable for productive uses and often contributing to the degradation of adjacent lands. In recent years, society demonstrated its resolve to minimize and mitigate long-term environmental destruction through a series of laws and regulations at the federal, state, and local levels. Federal laws must be somewhat general to accommodate within the same statute the variety of situations encountered throughout the country. The regulations and guidelines promulgated, together with the manner of enforcement, add specificity concerning allowable activities and preferred reclamation practices. State laws and regulations are usually more exact, often mandating certain actions or practices. Local control, frequently administered through zoning ordinances, reflects conditions in the immediate vicinity of a site or the particular interests of a community. Usually, reclamation programs are obliged to meet the requirements of several regulatory programs simultaneously.

Many federal statutes influence reclamation, depending on the land-disturbing activity and site characteristics. It is only possible here to identify the most important laws:

1. River and Harbor Act (RHA; 28)
2. National Environmental Policy Act (NEPA; 29)
3. Clean Water Act (CWA; 30)
4. Water Quality Act (WQA; 31)
5. Surface Mining Control and Reclamation Act (SMCRA; 32)
6. Comprehensive Environmental Response, Compensation, and Liability Act (CERCLA; 33)
7. Resources Conservation and Recovery Act (RCRA; 34)
8. National Pollution Discharge Elimination System (NPDES; 35)
9. Contaminated Sediment Management Strategy (36)

The contents of these enactments are not static but periodically change through amendment or interpretation. Those responsible for both compliance and enforcement sometimes believe that they are "shooting at moving targets."

The Surface Mining Control and Reclamation Act (SMCRA) is rather unique because it focuses on a particular industry (coal) and one method of extraction (surface mining). Reclamation requirements are quite specific and variously perceived as providing efficient direction for reclamation programs or as being overly restrictive, stifling imagination and creativity. One controversial requirement, return of the land to the approximate original contour (AOC), is discussed in the section concerning topographic reconstruction. It seems that the regulation of other land-disturbing industries, such as metal mining, is evolving toward the requirements of SMCRA, although considerable variation from state-to-state remains.

MATERIAL MANAGEMENT

The costs of blasting, excavating, hauling, and placing overburden are the major costs of any land-disturbing operation that requires earth-moving. Engineers and reclamation planners strive to minimize material handling costs wherever possible. Modern mine operations, however, also are directed by environmental regulations requiring that certain overburden strata are selectively handled and placed to minimize adverse effects or to maximize their utility in achieving water quality and reclamation goals. It is imperative that material handling operations are carefully planned and designed before mining to ensure that the economic goals of the resource extraction plan mesh with regulatory compliance and societal demands for environmental protection. The incremental cost of regulatory compliance compared to the environmental benefits of very stringent standards is a major source of controversy in the mining communities of the U.S. (37), Australia, and most other countries where mining laws are implemented and enforced.

Material Volume and Disposal Requirements

As discussed earlier, one of the important components of site characterization is accurate estimation of the overburden or spoil volumes that must be moved and eventually placed as "fill." Virtually all rocks and sediments volumetrically expand or "swell" to some extent (typically 15 to 30%) when they are blasted, hauled, and replaced. The mine engineer, therefore, must anticipate "excess spoil" in designing material handling operations (38). In the Western U.S., mining often removes sufficient resource to compensate for the increase in volume, but this is seldom the case in Eastern coal mine operations. When mines generate excess waste materials, adjacent areas, such as hollows, are utilized for disposal or the reconstructed land forms are configured to accommodate the excess on the original "footprint" of the site. In the latter case, postmining land forms are often steeper and more erosive than the original land forms. Most low grade ores are processed on-site and generate additional solid wastes, tailings, or slimes which also must be handled in an environmentally responsible manner (18). These materials usually are placed in separate fills or impoundments away from the active mine site, but in some circumstances they are returned to the mine pit for codisposal with the overburden. The disposal of processing wastes influences the overall volumetric balance of spoil disposal and the need to disturb additional lands beyond the area of the actual resource deposit.

The volume of each material to be handled together with its physical and chemical properties are also carefully assessed because each affects the reclamation and final closure plan. For example, if a shale stratum slakes (disintegrates) readily, it is excluded from steep backfills to achieve mass stability. It is not within the scope of this article to address the engineering properties of mine spoils in detail, but the materials composing the backfills, hollow fills, and other postmining land forms must meet minimum standards of shear strength and compaction according to the placement and grading plan. In addition, if the mine is operating under a topsoil-substitution reclamation plan, the substitute strata must be available in sufficient volume to cover the entire reclamation surface. Finally, potentially toxic strata must be adequately buried or isolated using nontoxic materials. This may be a major constraint when a mine disturbs a preponderance of acid-forming materials. Certain areas in the Eastern U.S. have been proposed as "unsuitable for mining" because of their large volume of acid-producing strata (39).

Characterizing Mining Wastes for Revegetation and Water Quality

In addition to volumetric and material handling considerations, it also is important that all materials are evaluated carefully for their potential revegetative and water quality impacts. In general, the presence of potentially acid-forming pyrites is a major concern for mine and reclamation planners worldwide. Other chemical characteristics, such as sodicity, salinity, or the presence of selenium salts are locally important, especially in the Western U.S. In the following section we describe the sequential evaluation of mine spoils typical of the Eastern U.S. and note analyses appropriate for other mining regions. For this discussion, we assume that topsoiling is not possible. So we must evaluate other materials as potential topsoil substitutes. We also assume that all strata have been sampled adequately as discussed earlier.

Overburden Acidity, Alkalinity and Sodicity

Spoil chemical composition is a primary control of revegetative success and potential water quality interactions. The "equilibrium pH" is a measure of active soil acidity and is the most commonly used and misused indicator of spoil quality. In general, pH values between 6.0 and 7.5 favor plant growth, although native species may be adapted to soil pH values beyond this range. Virtually no plant materials are available, however, that tolerate soil pH values below 3.5 or higher than 9.0. Pyrite oxidizes to form sulfuric acid and drastically lowers pH to values less than 4.0, whereas carbonates (eg, $CaCO_3$) significantly increase pH as they weather. Spoils high in soluble carbonates typically have pH values near 8.0, whereas those containing appreciable sodium (Na) develop pH values in excess of 9.0. The pH values of unweathered mine spoils containing significant quantities of pyrites in excess of neutralizers (carbonates) often rapidly decrease upon exposure to water and oxygen at or near the Earth's surface. Preweathered and leached mine spoils and soils generally are not subject to the rapid pH changes of newly exposed spoils. Hence, pH is a more reliable indicator of long-term acidity levels in the pre-weathered and oxidized strata near the surface.

The oxidation of pyrite produces large quantities of soluble salts in addition to acidity. Pyritic spoils may take many years of leaching to reach chemical equilibrium. Afterwards, however, pyritic spoils are a suitable plant growth medium if the sulfate salts are removed and the pH rebounds to a value of approximately 4.0. Many abandoned coal-processing waste piles in the Appalachian region are invaded by native pioneer species after their surface spoil chemistry stabilizes (40). Care should be taken to protect or stockpile this weathered surface zone on older spoil piles during mining operations.

Acid-base accounting (ABA) (41) is the best common indicator of spoil quality. ABA measures the total or pyritic sulfur content of a sample and assumes complete oxidation to form sulfuric acid. Similarly, the total neutralization capacity of carbonates and any other base compounds in the sample is measured and assumed fully reactive. In the absence of carbonates and assuming complete reaction, each 1.0% of pyritic sulfur generates sufficient acidity to require 32 metric tons (t) of agricultural lime (finely ground $CaCO_3$) per 1000 t of spoil to achieve pH 7.0. This conversion also conveniently generates liming recommendations. The ABA method is subject to various sources of error (42), but it is a fairly reliable and readily available analysis that yields consistent and usually conservative liming prescriptions. Spoils with net alkalinity should weather to mine soils with pH values above 5.5 (above 7.0 in arid and semiarid areas) which are quite suitable for plant growth following fertilization. Spoils with net acidity levels in excess of 2.0 t of lime requirement per 1000 t should be avoided. Spoils with net acidities in excess of 5.0 t per 1000 tonnes by law must be handled as toxic materials because of their potential impact on water quality and revegetation.

In environments where annual precipitation is less than approximately 70 cm and evapotranspiration rates are high, including much of the Western U.S., spoils and soils are usually poorly leached and retain the soluble salts produced by weathering processes. The salt content of these materials, measured by the exchangeable sodium percentage (ESP) and electrical conductance (EC), are important chemical properties. Sodic soils with an ESP of 10 or more generally exhibit physical properties detrimental to revegetation. ESP and the sodium absorption ratio (SAR) are closely related (43). Many reclamationists feel, however, that ESP provides a better method than SAR for calculating the amendment rates of very sodic soils (ESP > 10) (44). Although soils with an EC of 4 mmhos/cm at 25°C are considered saline, some Western grasses are adapted to values as high as 7–10 mmhos/cm (45). Certain spoils in the Western U.S. also contain sufficient quantities of selenium and arsenic to cause concern for water quality and revegetation (46). In Eastern spoils, large quantities of soluble salts usually indicate active pyrite weathering and low pH conditions.

Extractable Nutrients

Soils must supply plants and microbes with adequate quantities of numerous macro- and micronutrients in aqueous form to maintain productivity rates. Reclamation plans usually require that soil media are similar to native soils in their capacity to supply these nutrients. Unfortunately, extractable quantities of phosphorus (P), calcium (Ca), magnesium (Mg), and potassium (K) have never been correlated on a regional basis with actual plant uptake and performance on mine soils. This fact, coupled with the variability of lithology and mineral solubility in mine spoils, makes it very difficult to accurately predict potential fertility. The quantities of nutrients reported from soil tests, however, may be used for general comparisons among mine spoils or soils. When spoil P, Ca, and Mg levels are low, it is assumed that the material is either very low in soluble primary minerals or has been extensively oxidized and leached, requiring lime and fertilizer applications for optimal plant production. Conversely, very high values of P, Ca, and Mg cannot be interpreted as providing adequate plant nutrients because many soil-testing extracts, particularly those employing strong acids, dissolve these elements from solid mineral phases that would not be available to plants until released by weathering.

Spoil Lithology and Physical Properties

Although most decisions regarding the use or disposal of a material are based on chemical properties, the physical properties of the spoil following placement strongly influence postdisturbance productivity. The lithology or sedimentary characteristics of spoil largely determine the physical properties of mine soils as they develop on reclaimed surfaces.

The plant-available water-holding capacity of a spoil is controlled by its particle size distribution inherited from the parent rock or sediment as modified by mining and weathering. Spoils derived entirely from sandstones, sandy sediments, and glacial tills are very coarse and droughty. Spoils derived entirely from fine siltstones, shales, loess, or finely textured sediments form surface crusts which impede water infiltration and percolation. Spoils composed of multiple lithologies are often superior to those composed of a single rock type because the likelihood of extremes in texture (eg, sands or clays) is reduced by mixing (47).

The bulk density of natural soils generally ranges from 1.1 to 1.5 g/cm^3 whereas the bulk density of compacted spoils commonly ranges from 1.8 to 2.2 g/cm^3, depending upon fragment size and sorting. Many reconstructed mine soils are highly compacted (bulk density > 1.5 g/cm^3) to a depth of several meters by the traffic of rubber-tired loaders and haulers, and to a lesser extent, bulldozers. Severely compacted mine soils (bulk density > 1.7g/cm^3), particularly those with less than 50 cm of effective rooting depth, cannot store sufficient plant-available water to sustain vigorous plant development through protracted droughts. Compacted zones also limit infiltration, perch water tables in unexpected locations resulting in saturation and deleterious anaerobic conditions within the rooting zone, and cause mass instability.

Weathered near-surface geologic materials with pH values less than 5 generally should not be used as topsoil substitutes. Quite often, these strata are preferred by the mine operators in the Eastern U.S. because they disaggregate into finely textured spoil when blasted, which makes them easy to handle and spread. An advantage of using weathered, finely textured spoils is their great particle surface area, capable of holding large quantities of plant-available water. These brownish red materials contain large quantities of iron oxides, however, which can diminish long-term P availability because they adsorb or "fix" applied P. Surprisingly, the majority of even unweathered sedimentary rocks in the Appalachians disintegrate into spoils that contain 30 to 50% soil-size fragments (<2 mm) upon blasting and supply adequate plant-available water as long as they are not compacted on the reclaimed surface.

Mining Fill Construction

Most spoils generated by mining are placed into fills for disposal as long as they are nontoxic and are not required as topsoil substitutes. These materials are (*1*) returned to the original mine pit, (*2*) placed into adjacent pits or topographic hollows, or (*3*) back-filled above grade to cover exposed highwalls and return the mine site to AOC. The economics of mining usually dictate that materials are handled only once with minimal haul distance, thus, careful planning is essential in developing an efficient and orderly material management plan. As required by geotechnical or volumetric concerns, the spoils are compacted in separate lifts (layers) to increase shear strength and decrease porosity. Despite compaction, almost all fills settle or subside to some extent from geostatic loading. Most of this settlement takes place during the first wetting cycle following placement (48), although long-term incremental settling also occurs. The depth of settlement depends on the thickness of the fill and the extent of compaction, but seldom exceeds 10% under controlled conditions. The bulk of the fill land form is constructed as various run-of-mine spoils are removed and hauled from the active mine pit, unless toxic materials are present. It is only as the surface approaches the final reclamation configuration that controlled overburden placement procedures are employed rigorously, followed by topsoil or the topsoil-substitute placement. Occasionally, all of the materials moved at a mine are suitable topsoil substitutes, hence, no discriminant material handling is necessary.

Isolation of Toxic Materials

Toxic materials in the Eastern U.S. are generally those with acid-forming potential (ABA > 5 parts per thousand acidity). Burial of potentially acidic materials at least 1 meter below the rooting zone generally is sufficient to eliminate revegetative problems. This strategy does little to prevent groundwater impacts, however, as water percolates below the rooting zone and becomes acidified. Similar problems result from toxic materials placed in mine pits or on benches of undisturbed clays

after coal removal. Water quality impacts are particularly severe when toxic materials are placed in zones that experience fluctuating water tables with episodic oxidation and leaching cycles. Modern toxic-fill designs in the East place toxic materials above the water table and provide a compact, impermeable cap above the toxic fill to "shed" percolating groundwater around the material. Although constructing these caps poses little problem for mines developed in fine, unconsolidated sediments, it may be difficult to accumulate sufficient quantities of finely textured materials for a suitable cap when mining is entirely in coarse-textured lithologies. These compacted caps also perch local water tables, saturating the fill, and possibly cause mass instability. More sophisticated "leaching barriers," coupled with subsurface leachate collectors, are placed above and below acid-forming, metal sulfide tailings in many parts of the U.S. and worldwide.

In arid and semiarid coal mining environments, such as the Western U.S., the acid-forming potential of sulfides is less problematic because of (*1*) low leaching rates, (*2*) low sulfide contents of geologic materials, and (*3*) the alkalinity of the overburden. The upward migration of salts and other soluble toxic elements results from high rates of evaporation that "wick" water with entrained ions toward the soil surface. When significant quantities of sodic, saline, or other phytotoxic materials are placed in fills near the surface, it is possible to construct compacted or coarse-textured salt-migration barriers at the base of the topsoil layer to protect plant productivity.

Topsoil-Substitute Construction

Many modern surface mines and construction operations utilize topsoil substitutes derived from spoils or unconsolidated overburden materials for at least some of their final reclamation surface. Careful coordination of operations is required so that the designated "topsoil-substitute" material actually becomes part of the reclaimed surface. Failure to control the placement of overburden materials results in lateral and vertical variability of spoil properties because various lithologies comprise the final surface. Although this may not be a problem when all materials are suitable for plant growth, patches of highly acidic, sodic, or saline spoils commonly decrease the percentage of vegetative cover sufficiently to prevent regulatory compliance and bond release. Severe compaction of otherwise chemically suitable spoils also decreases the vegetative density, particularly over the extended periods now required before final bond release.

One of the most important criteria for selecting a topsoil substitute is the cost and practicality of selectively handling the designated material within the mining plan. If an ideal stratum is thinner than the usual blasting lift thickness or placed inopportunely within the geologic column, its use may not be feasible. Quite often, two or more adjacent strata within the same blasting lift are identified as substitute materials permitting them to be handled and spread together. Accordingly, materials with dissimilar physical and chemical properties are mixed to yield physical and chemical properties more favorable than either of the individual materials.

About one meter of uncompacted material is usually sufficient to store adequate plant water through prolonged droughts. Shallow bedrock or coal seams, large boulders, or compacted layers commonly limit the rooting depth in mine soils. The position of rock outcrops or stoniness generally indicates the average depth and water-holding capacity of mine soil. The depth of material above bedrock is estimated by observing the contact between the mining bench and the outslope (15). The soil depth at this location may not reflect the entire area, however. The presence of swampy areas on a bench usually indicates shallow rock or compacted zones. When poor plant growth is encountered in soils with otherwise suitable physical and chemical properties, insufficient rooting depth is almost always the cause.

Summary

Careful material handling creates a suitable subsoil or base layer upon which topsoil is returned. Sometimes, overburden materials are used as topsoil substitutes. Regardless, well-planned and executed overburden selection and placement are necessary to ensure that the final reclamation surface is optimized for revegetation.

The following material handling steps are recommended for successful reclamation. First, when topsoil volumes are insufficient to cover the reclamation surface adequately, identify the best strata of substitute material, based on the following properties: thickness, rock type, acid-base balance, pH, and the Na, As, Se, or soluble salt content. Other elements, such as Ca, Mg, K, and P, should enter into the decision but are not critical because they are added easily as fertilizers. Second, operations should be coordinated so that the designated strata are separated and hauled to the proper areas for final grading. Excessively stony spoils (more than 70% >2 mm) should not be used as surface materials. Third, the selected spoil should be end-dumped on the final surface to provide a minimum thickness of 1.2 m of topsoil substitute for plant root development as generally required by regulations.

TOPOGRAPHIC RECONSTRUCTION

Most natural landscapes are composed of drainage basins which, in turn, consist of hillslopes and stream channels in an orderly spatial arrangement for efficiently conveying water and sediment. Drainage basins are open, process-response systems with integrated morphologic (shape) and cascading (matter and energy) elements. Water and sediment accumulate in the downgradient direction causing progressive adjustments in hillslopes and stream channels. Numerous statistical relationships among drainage basin form and geomorphic processes are summarized in a study (18). These relationships support the proposition that, under natural conditions, an approximate dynamic equilibrium exists between land forms and the processes that create them. Geomorphic work, such as erosion, takes place at low to moderate rates because there is an approximate balance between the forces operating on or beneath the surface and the resistances afforded by surface and near-surface materials. The specific character of the equilibrium depends on the prevailing climate, geology, topography, soil, vegetation, and hydrology.

Land disturbance violates the state of approximate dynamic equilibrium causing disparity between forces and resistances and, hence, accelerated process rates. Mass movements and erosion are mechanisms by which the equilibrium or balance is reestablished. Rates of adjustment are highest shortly after the disturbance and decrease through time (49). The consequences of disturbances and subsequent adjustments are transmitted through the system as "complex responses" (25) because of its integrated character.

From this perspective, the goal of reclamation is to recreate an approximate dynamic equilibrium between land forms and processes, between forces and resistances. Topographic reconstruction simulates the work of processes operating over millennia to create hillslopes and stream channels. A reconstructed topography that closely resembles the equilibrium landscape necessitates less "fine-tuning" by natural processes than one that departs significantly from the equilibrium configuration. The character of the postreclamation equilibrium differs from that of the predisturbance equilibrium because geologic and soil properties cannot be completely replicated. Topographic design, therefore, should be based upon expected properties following reclamation rather than predisturbance properties. The importance of topographic reconstruction cannot be overestimated because the resulting land forms are the foundation upon which other reclamation practices are executed and eventual land uses take place.

Regulation

The postreclamation configuration of landscapes and land forms is usually not mandated by federal laws or regulations but controlled through state or local guidelines and the reclamation plan approval process. For example, the Bureau of Land Management (50) recommends hillslope gradients no greater than 33% (3:1) for reclaimed waste rock disposal sites at metal mines. Departures from the guidelines receive careful scrutiny and may require substantial justification. Generally, it must be demonstrated that the proposed configuration will yield stable, minimum maintenance surfaces, capable of supporting the proposed land use over the long term. In this context, "stable" includes mass movement, erosion, and vegetative processes. State laws and local ordinances may be more specific.

The regulations promulgated under SMCRA advocate the return of the land to the approximate original contour (AOC). This has proven controversial. One study (51) states that a return to AOC does not necessarily ensure that adequate reclamation will occur and frequently results in long hillslopes and excessive soil compaction that accelerate erosion and create difficulties for revegetation. One comprehensive study found that many AOC backfills in the central Appalachian Mountains face long-term mass stability problems and concluded that AOC mining and reclamation techniques in steeply sloping terrain is not an appropriate form of reclamation (52). Original land form morphology may not be stable following fragmentation of the geologic strata and structures. Further, there are situations where configurations other than AOC permit more productive and desirable post-reclamation land uses.

Hillslope Reconstruction

A hillslope is a land surface extending from a drainage divide to a floodplain or stream channel. Under natural conditions, hillslopes are the product of geomorphic processes operating on soil and geologic materials through time. Reclaimed hillslopes are the product of grading and shaping practices. The hillslope profile, a transect extending from crest to base, is a simple expression of form. Changes in gradient along the profile length result in convex, straight, concave, or complex (convexo-concave) profiles. Within a landscape, assemblages of hillslopes also have three-dimensional geometry, as illustrated in Figure 2. Changes in shape along hillslope length and width result in valley head hillslopes (hollows), valleyside hillslopes (side slopes), and spur end hillslopes (noses).

Relationships between hillslopes and process rates are well documented. Usually (other factors being equal), as gradient and height increase, mass stability decreases. Convex profiles are more susceptible to mass movement than concave profiles.

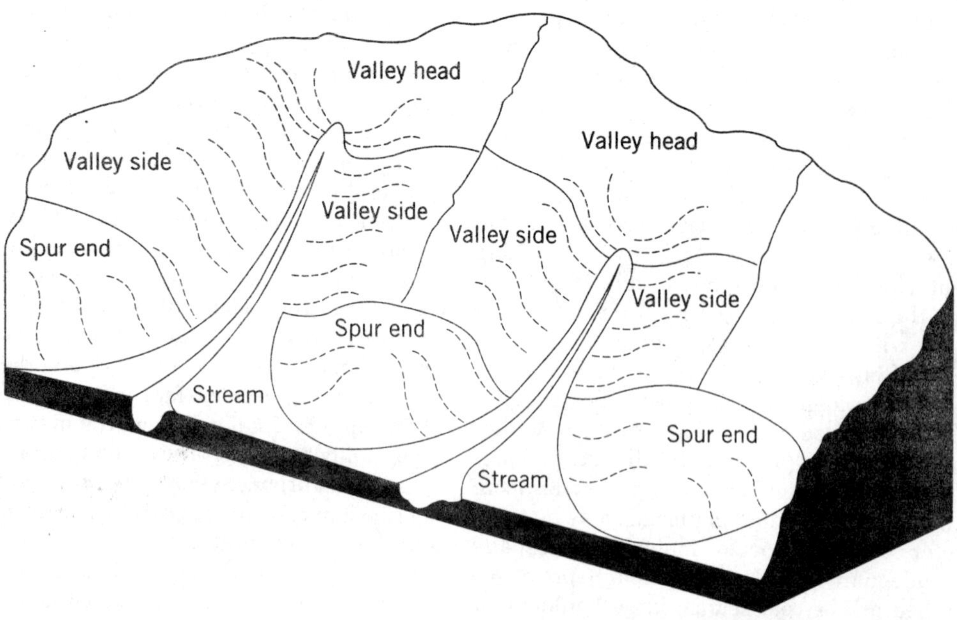

Figure 2. Location of the valley head, valley side, and spur end hillslopes in the landscape.

As gradient and length increase, erosion rates increase. A convex profile is more erodible than a straight profile, and a straight profile is more erodible than a concave profile (53). Erosion rates are highest on the straight midslope segment of complex profiles and least on the upper convex element, and either erosion or deposition occurs in the lower concave element (54).

Three-dimensional shapes also influence process rates. Valley head hillslopes are less susceptible to mass movement than valleyside or spur end hillslopes because they are supported laterally by adjacent valleyside hillslopes. The flowlines on valleyside hillslopes are generally parallel and the relationships between profile and erosion rates prevail as described above. The flowlines on valley head hillslopes, however, converge in the downslope direction and cause higher erosion rates than on valleyside or spur end hillslopes (55). The flowlines on spur end hillslopes diverge downslope and result in lower erosion rates than on valleyside or valley head hillslopes.

The foregoing suggests that the reclaimed hillslopes should be as short and gentle in gradient as feasible. Concave profiles are preferable, with complex, straight, and convex profiles progressively less desirable. In three dimensions, spur end hillslopes are preferable, with valleyside and valley head hillslopes progressively less desirable where erosion control is a primary consideration. Of course, the entire reclaimed landscape is unlikely to be composed of a single hillslope form. As designs depart from preferred shapes, erosion-control practices become increasingly important.

Stream-Channel Reconstruction

A stream channel is a trough in the landscape that conveys water and sediment. Runoff, throughflow, ground water, sediment, and solutes contribute to stream flow. Under natural conditions, streams are classified in various ways: (*1*) based on the material in which they are entrenched, hence, alluvial or bedrock channels, (*2*) based on the gain or loss of water from the groundwater reservoir, hence, influent or effluent, or (*3*) based on the frequency of flow, hence, perennial, intermittent, or ephemeral (18). On reclaimed lands, channels are also (*1*) temporary, for managing run-on, runoff, and stormwater, or (*2*) permanent, replacing streams destroyed or damaged by disturbance. Most of the principles of stream behavior are derived from research involving perennial, alluvial channels, or engineered conveyance channels.

Stream-channel form is a function of hydrologic processes operating upon geologic materials. The basic stream theory following was first articulated (56), verified by many investigations, and summarized thusly (57): (*1*) streams cut their own valleys; (*2*) the gradient of each stream is adjusted to the velocity, volume, and sediment load carried; and (*3*) a stream system is integrated by mutual adjustment of its constituent parts. Expressions of channel shape include the cross section, gradient, channel pattern, and drainage pattern. The cross section is a transect from bank to bank consisting of bankside slope, width, and depth. Channel gradient refers to the inclination of the channel along its course. Channel pattern is the plan-form sinuosity of the channel itself (eg, straight, braided, meandering). Drainage pattern is the plan-form geometry of the channel assemblages comprising the drainage network (eg, dendritic, trellis, radial).

Water and sediment discharges are the two primary independent variables determining channel form. One study (25) summarizes our understanding of channel-form adjustments resulting from changes in water and sediment discharges in the following relationships. The superscripts "+" and "−" indicate "increases" and "decreases," respectively:

$$Q^+ \approx b^+, d^+, \lambda^+, S^- \quad (1)$$

$$Q^- \approx b^-, d^-, \lambda^-, S^+ \quad (2)$$

$$Qs^+ \approx b^+, d^-, \lambda^+, S^+, P^- \quad (3)$$

$$Qs^- \approx b^-, d^+, \lambda^-, S^-, P^+ \quad (4)$$

where Q is the discharge of water; Q_s is the discharge of sediment; b is the channel width; d is the channel depth; λ is the meander wavelength; S is the channel gradient; and P is the channel sinuosity (ratio of channel to valley length). According to the first relationship, as the discharge of water increases, channel width, depth, and meander wavelength increase, whereas channel gradient decreases.

The study (25) acknowledged that water and sediment discharges usually change together and provided another set of relationships to express the probable adjustments resulting from such simultaneous changes. Drainage pattern is largely determined by geologic structure. A dendritic pattern is characteristic of drainage systems developed in horizontal strata composed of homogeneous lithology, reflecting the absence of geologic controls. A trellis pattern is characteristic of dipping strata of varying lithologies (58).

Channel forms on reclaimed lands are products of grading and shaping practices subsequently altered by hydrologic processes. There are two general philosophies concerning the channel design: geomorphic and engineering. The geomorphic approach advocates replication of the predisturbance channel characteristics based on the assumption that those forms were stable or equilibrium forms in the past and will be so again. The engineering approach employs "tractive-force" or "regime" theory to determine stable channel forms for flows of a selected magnitude and frequency (eg, 100-yr, 24-h storm).

The foregoing suggests that the first consideration in designing stream channels for reclaimed lands is channel type: (*1*) alluvial or bedrock, (*2*) influent or effluent, (*3*) perennial, intermittent, or ephemeral. The theory and practices for alluvial channels should be applicable when channels traverse well-fragmented geologic strata and structures. In the Eastern U.S., reclaimed channels are often effluent, perennial or intermittent, whereas in the Western U.S., reclaimed channels are often influent and ephemeral. Design channels should convey water and sediment from storms of moderate magnitude and frequency (100-yr, 24-h) under the expected geologic, topographic, soils, vegetative, and hydrologic conditions without erosion or deposition. Channel gradients should follow a smooth, gentle, concave-upward curve without abrupt changes along the course or at the confluences of reclaimed channels and undisturbed streams. The dendritic drainage pattern should be appropriate in the absence of geologic control.

Drainage Basin Reconstruction

A drainage basin (watershed, catchment) is the area contributing water and sediment to a drainage network, bounded upslope by a drainage divide and downslope by the confluence

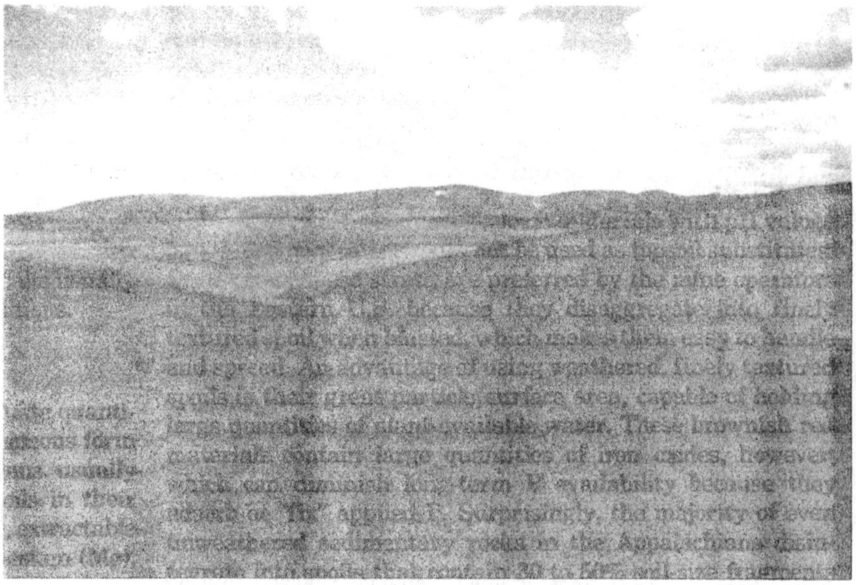

Figure 3. Reclaimed small drainage basin at a coal mine in the Western United States.

of the main channel with another drainage system. As previously noted, they are open, process-response systems achieving and maintaining an approximate state of dynamic equilibrium. Large reclamation projects are likely to require reconstruction of complete drainage basins (Fig. 3).

Drainage basin form is described in terms of (*1*) area, (*2*) relief, or (*3*) shape. Basin area refers to the surface between divides that receives precipitation, generates runoff and throughflow, and yields sediment as a result of erosion. As basin area increases, runoff increases, and water and sediment discharges increase, whereas sediment yield (discharge per unit area) decreases somewhat for small basins involved in reclamation.

Relief is the change in elevation between the upslope divide and the downslope channel confluence. This characteristic is expressed through the simple "relief ratio" (change in elevation/basin length along main channel), the "weighted mean slope," or the complex "hypsometric curve." As relief increases, runoff, erosion, and sediment yield increase.

Drainage basin shapes range from nearly linear to nearly circular. Shape influences the delivery and accumulation of runoff to channels. In long narrow basins, tributary flow reaches the main channel at different times. In wide fan-shaped or nearly circular basins, tributary flows arriving at about the same time result in high peak flow (59). Although the general relationship between basin shape and the storm hydrograph is accepted, quantification has proven difficult, usually precluding consideration in drainage basin design.

Drainage density (total channel length/basin area) is a measure of basin dissection by stream channels and is regarded as the most important characteristic of a channel system. Drainage density increases as mean annual runoff, mean annual flood, percent bare surface, and sediment yield increase. This basin characteristic, then, is the product of all of the environmental conditions that influence the flow of water and sediment through the system.

The foregoing suggests that reclaimed drainage basins should be as small in area and gentle in relief as feasible. A linear shape results in slower runoff concentration and lower peak flows. Adequate drainage density is essential. Otherwise, further dissection of the area with substantial sediment production is inevitable.

Summary

Topographic reconstruction creates the landscape and foundation upon which other reclamation practices and eventual land use take place. Following reclamation, the landscape and component landforms operate as an open, process-response, system with cascades of matter and energy across and through morphologic elements. Accordingly, the consequences of unsuccessful reclamation for any one component are transmitted through the system.

The water and sediment outputs from hillslope systems become the inputs to channel systems. The discharges of water and sediment largely determine channel characteristics. Drainage basin area and relief strongly influence the water and sediment flows within a climatic region and the complexity of the drainage network, represented by the drainage density. Consequently, estimates of water and sediment accumulations in the downgradient direction provide the basis for topographic reconstruction.

REPLACEMENT OF TOPSOIL AND SOIL RECONSTRUCTION

Revegetation of reclaimed surfaces requires a suitable growth medium. In most cases, topsoils have the necessary physical, chemical, and biological properties to sustain plant development, although the use of substitute geologic materials is often inevitable. In general, topsoil layers are higher in organic matter, microbial activity, and nutrients than the underlying subsoil or geologic material. Topsoil contains a significant seed bank that can be used to great advantage in revegetation. Stockpiling and reuse of topsoil almost always facilitates achievement of reclamation goals. When spoils or other coarse-textured materials are revegetated without topsoil covers, moisture stress induced by compaction, high coarse-fragment contents, salts, and high surface heat is usually the primary factor limiting growth. If pyrite or other sulfidic materials are placed near the surface, acidity or excess salt accumulation may be a major problem but is largely controlled by careful material placement and appropriate soil amendments.

Topsoiling

Most regulatory programs for land-disturbing activities require segregating and removing the topsoil (A + E horizons) and returning it to the land surface during reclamation, particularly when the lands supported agriculture. The topsoil layers are removed carefully from the disturbed area while leaving the deeper subsoils, as directed by accurate soil surveys. In many areas the topsoil cannot be effectively stripped until the woody vegetation, stumps, and roots are removed. Usually, the litter layers (0 horizons) are blended with the underlying topsoil horizons. Litter layers, however, can be separated and used as very effective mulching materials for revegetation, particularly when native plant communities are to be reestablished.

Because topsoil is a living resource, direct haul reclamation strategies, whereby topsoil is stripped from one area and immediately reapplied to another, are superior to stockpiling options. Direct haul topsoiling requires on-site coordination and generally is more feasible on gently rolling terrain than on steep hillslopes. Stockpiling topsoil for extended periods generally reduces microbial activity (60) and the viability of seed propagules. If topsoil is stockpiled too deeply or compacted causing anaerobic conditions to develop, its value as a plant growth medium is significantly reduced because the microbial assemblage declines and soil chemistry changes. If long-term topsoil storage is inevitable because of the mining plan, the stockpiles should not exceed a height of 2 m and should be placed in well-drained landscape positions to permit adequate gas exchange between the soil and the atmosphere. Long-term stockpiles should also be revegetated with deep-rooted plant species compatible with the postmining land use so that beneficial microbial activity is maintained and the topsoil is protected from wind and water erosion.

The topsoil should be reapplied at its original thickness whenever possible. Unfortunately, topsoil is often "lost" during mining operations, and remaining topsoil is frequently "stretched" to cover areas larger than those from which they were removed. A positive correlation exists between topsoil thickness and vegetative productivity for row-crop production in the Midwest (61) and for revegetation of various disturbed materials. A common result of topsoiling operations is thinner and more compact topsoils than were originally present. Many topsoiled areas require extensive ripping and tillage to alleviate compaction, particularly when rubber-tired trucks and scrapers are employed. Topsoil should not be regraded when it is too wet or too dry because this tends to degrade the soil structure and makes applying a uniform thickness quite difficult. Applying topsoil on wet or low-strength subgrades often leads to rutting and uneven soil thickness. Topsoil replacement, then, requires careful planning and on-site management to realize its maximum benefits.

Soil Reconstruction

In certain circumstances, such as disturbance of prime farmland in the Midwestern U.S. by surface mining coal, the entire soil profile is removed selectively in several layers and then sequentially returned to the site after mining. Natural subsoils (B horizons) are frequently more acidic and clayey than topsoils, which makes them less hospitable for plant growth. This is especially true when their natural structure is destroyed by the stripping process. Therefore, replacement of natural subsoil horizons is not always beneficial. Subsoils reconstructed from deeper, less-weathered layers (C horizons) or selected spoil materials result in a more productive plant-rooting medium when combined with topsoils. Field and greenhouse studies comparing complete soil reconstruction with alternative mine-soil layering strategies often show that a wide variety of soil and overburden materials is suitable for use in soil reconstruction, as long as their physical and chemical properties are understood and managed appropriately. Some subsoils produce even higher yields than topsoil (61). Recent work in Texas (62) found that mixing subsoil and topsoil increased the productivity of two clayey soils. These and other studies show that, although complete soil-profile reconstruction is a good practice in concept, it is not always the optimal alternative at a specific site.

The use of soil covers to reclaim materials with toxic chemical properties greatly improves the prospect of reclamation success. Even thin (<25 cm) layers derived from mixed subsoil horizons provide sufficient water-holding capacity and a suitable plant-rooting medium to establish grasses, legumes, and native shrubs. Thicker soil covers are necessary for long-term legume survival on acidic materials (40). The use of agricultural lime at the base of a soil cover is essential when thin layers are placed over acidic materials. The liming rate is based on the potential acidity of the underlying material and the soil-cover thickness.

SURFACE MANIPULATIONS

Following topographic reconstruction and topsoil replacement, the reclaimed surface is modified in various ways to facilitate water management and erosion control. Types of surface manipulations range from common agricultural practices, such as plowing, disking, and furrowing, to severe surface alterations, such as listering, gouging, and construction of small basins with the blade of a bulldozer. The physical effects of these activities are to break soil crusts, roughen the surface, and relieve soil compaction caused by equipment traffic. The hydrologic effects are increased infiltration capacity of the surface, increased soil moisture, and reduced volume and velocity of runoff. In climatic regions where soil moisture limits plant growth, surface manipulations commonly expedite revegetation (63). The geomorphic effect reduces the erosion rate and trap sediment in small surface depressions, holding the soil on-site. Descriptions of numerous manipulation practices are available elsewhere (19,64).

Where lands are returned to intensive uses requiring easy equipment passage, such as agriculture or managed turf grass, the final reclaimed surfaces must be very smooth. This necessitates extensive grading by bulldozers or motor scrapers with soil compaction as an inevitable side effect. These "traffic pans" may be quite deep (>50 cm), depending on soil-moisture conditions at the time of grading.

Compaction is often the limiting factor in successful reclamation and frequently occurs when multiple layers of soil are returned to a site during complete soil reconstruction (65). Compaction has also been identified as a major problem in reclaiming bentonite mine land in the Northern Great Plains (66). Generally, deep ripping is effective for remediating compaction on prime farmlands reclaimed in Illinois, Kansas, and Kentucky (65), but the results are occasionally inconsistent, especially when forage crops, such as alfalfa, are grown. Compaction of sandy tailings also results from wet settling combined with bulldozer traffic during regrading (67).

Minimal grading is actually superior to smooth grading for the establishment and long-term vigor of many vegetative communities, particularly forests (68) or unmanaged herbaceous and shrub communities. Rough-graded areas have higher infiltration rates, lower runoff and erosion rates, and provide a greater variety of microsites for diverse plantings to germinate and prosper than smooth-graded areas. Roughened surfaces effectively trap windblown plant propagules, fungal spores, and insects. The benefits of rough grading, particularly, are apparent on landscapes reclaimed without topsoil in the Appalachians and Midwestern U.S., but the potential value should not be overlooked for other regions.

The benefits of most surface manipulations are temporary. One study (63) found that the microtopographic relief created by furrowing, pitting, and ridging decreased by 40–60% during a 6-year study in Wyoming. Another study (69) found that the value of dozer basins and surface gouges generally lasted less than two years at sites in Montana and North Dakota. Climate, soil erodibility, topography, and rate of vegetative establishment control the effective longevity of these practices.

SOIL AMENDMENT

The final step in preparing the soil of a disturbed site for revegetation is the application of amendments, such as various combinations of agricultural lime, gypsum, inorganic fertilizers, and organic materials. The application of amendments depends on five factors: (*1*) the physical and chemical properties of the soil; (*2*) the local climatic regime; (*3*) the topography and accessibility of the site; (*4*) the vegetative type to be reestablished and postreclamation land use; and (*5*) the regulatory-compliance requirements.

The physical and chemical properties of the soil material are the most important factors in amendment selection. Where topsoils are returned to the site, amendments are prescribed by standard soil testing. Specific recommendations are usually available for most land uses. Topsoils are typically treated with lime and fertilizers as necessary to achieve prescribed pH and plant nutrient conditions. When regraded subsoils, spoils, or other geologic materials are to be revegetated without topsoil, determining the proper prescription is more complicated and is based upon the predicted soil conditions expected to develop through time as indicated by the overburden analyses discussed earlier. Because these materials lack the organic matter and plant-nutrient pools found in topsoil, they often require higher rates of fertilizer application and organic-matter amendments. Thus, the soil-amendment "package" for a particular site usually includes a combination of treatments specifically designed to produce optimal conditions for plant growth. In cases where productive native topsoils are seeded with native vegetation, minimal or no soil amendments are necessary.

The quantity and timing of amendment applications depends on the predicted vegetative demands which are largely determined by the climate, specific growth habits, and adaptations of the plant community. On mined lands, regulatory performance standards also control the use of supplemental amendments following establishment. For example, on coal-mined lands in the Eastern U.S. required by permit to control erosion and return the land to natural forest vegetative cover, supplemental amendments are specifically prohibited during the 5-year probation period for vegetative establishment and bond release. Agricultural uses, such as hayland and pastureland, on the other hand, allow periodic amendment during the probation period as long as the application rates are typical for that land use on unmined lands. In other cases, such as highway rights-of-way and construction sites, immediate erosion-controlling vegetative covers with minimal long-term maintenance typically are mandated for rural areas that require substantial initial applications of amendments but only infrequent subsequent applications. High-visibility urban areas are often managed very intensively for aesthetic reasons with frequent post-establishment applications of soil amendments.

Soil pH Control

The majority of environmental problems associated with coal and metal mining worldwide are the result of pyrite weathering and its effects on soil and water chemistry. Pyrite weathering is also a problem in any earth-disturbing activity where pyrite-bearing lithologies are exposed to atmospheric conditions. The pH value of such materials depends on their pyrite content, the length of exposure, and the acid-neutralizing capacity of the matrix around it (70). The rate of pyrite weathering and acid production is generally high in the oxygenated surface layer, which is also the zone utilized by plant roots.

Native and introduced plant communities are adapted to a range of soil pH values which, in turn, control nutrient availability, metal solubility, and microbial activity. There is no specific pH value that is "ideal" for reclamation other than that required to establish vegetative cover. In general, most agricultural row crops demand pH values between 6.0 and 7.5, whereas certain Eastern U.S. woody species and many grasses readily tolerate pH values below 5.0. Conversely, many Western species are adapted to higher (>7.5) pH values. Natural pH values in the Eastern U.S. typically are more acidic than in the Western states. Hence the use of agricultural lime to raise pH values is common in the East.

Soil pH values also are subject to change through time as geologic materials weather. Topsoils high in organic matter are buffered against major pH shifts, and their liming requirements are predicted reliably by standard soil-testing procedures. The liming requirement for sulfidic materials incorporates an estimate of the soil's buffering capacity and the desired pH modification as determined by the ABA approach or by other accepted methods (41). It is not uncommon to prescribe two or three times as much lime as predicted by ABA to ensure that the pH is controlled over long periods of time. This adds considerably to the cost of reclamation when toxic materials are involved. For example, coal-processing waste piles in the central Appalachians often require from 25 to more than 100 tonnes of lime per hectare for revegetation. In the absence of sulfidic materials, liming applications seldom exceed ten tonnes per hectare.

Liming materials must meet regulatory standards for agricultural use, including fineness, reactivity, and calcium carbonate equivalence. Liming is more effective when incorporated into the soil (to 15 cm) and several months elapse, allowing the lime to react completely. Hydroxide and oxide lime sources react much more quickly, but must be used carefully to avoid burning sensitive seedlings, or raising the soil pH to deleteriously high values. Coarse-textured lime particles become encapsulated with Fe (iron) or Mn (manganese) precipitates that severely limit their solubility. The high Mg content of dolomitic limes also complicates the revegetation of extremely acidic materials because $MgSO_4$ is much more soluble in soil solutions than gypsum ($CaSO_4$) and produces higher salinity.

A rapid decline in pH releases phytotoxic concentrations of acid-soluble metal ions into soil solutions and reduces the availability of many plant nutrients. Another problem caused by pyrite oxidation is the production of sulfate salts which accumulate to toxic concentrations in the root zone. When the lime added to sulfidic materials is completely reacted, the original pyrite is oxidized, the iron and other metals are complexed into various oxide forms, and the sulfur is precipitated as sulfates, particularly gypsum.

Agricultural lime is applied directly or as part of the hydroseeder slurry or with dry spreaders. Up to 5 tonnes per hectare of lime is applied to steep hillslopes through repeated hydroseedings, but higher rates must be applied by mechanical spreaders or blowers. Although the application of lime on flat or gently rolling terrain is relatively simple, application on steeply sloping topography is much more difficult. Further, it is virtually impossible to incorporate lime adequately into the soils of steep, rocky hillslopes with conventional tillage equipment. Generally, it is applied to the surface and "back-bladed" with a bulldozer. This usually mixes the lime to a depth of only 4–6 cm. Additional lime may be required after several years on many sites, especially if the soil is clayey and high in aluminum. Experience shows that repeated lime applications to extremely acidic soils are much more effective than one high-rate application because the time intervals between applications allow complete reaction of the liming compounds and precipitation of the secondary reaction products.

Many subsoils and lithologies in the arid and semiarid Western U.S. contain salts that generate high pH, sodic, and saline conditions, limiting revegetative success. The on-site treatment of these materials is often impractical because most known remedies require substantial irrigation for leaching, coupled with large applications of gypsum. Alternatively, these materials are buried beneath the rooting zone or blended with less alkaline, sodic, and saline materials.

Fertilizer Applications

Plants require six essential macronutrients (N (nitrogen), P, K, S (sulfur), Ca, Mg) and a variety of micronutrients including iron and boron. In general, sufficient quantities of Ca, Mg, and S are supplied by the weathering of geologic materials, limes, or atmospheric sources. Thus N, P, and K are the only elements routinely applied to reclaimed soils. Topsoils usually contain significant reservoirs of all essential elements, so topsoiling reduces fertilizer application rates. When native plant communities are reestablished, fertilizer applications may be detrimental to establishing and developing desired species. Inorganic fertilizers are applied by conventional spreaders or dissolved in a hydroseeder slurry.

When topsoil is returned to a reclamation site, fertilizer requirements for a particular vegetative cover are determined by standard soil-testing and prescription programs that account for organic and mineral-nutrient reserves in the soil and estimate the necessary supplements. When spoils or other geologic materials are revegetated, however, standard soil-testing procedures are difficult to apply and interpret. The soil's various nutrient pools must be rebuilt to assure availability through time. This commonly requires trial and error in a particular region, although a number of principles direct the planning process, as discussed below.

Mine-spoil materials and newly exposed geologic materials are virtually devoid of available N, so the N required to sustain plant growth must come from the initial fertilization and subsequent symbiotic N fixation by legumes infected with *Rhizobium* bacteria. In humid regions, usually less than 150 kg/ha of fertilizer N are added initially and much of this is subject to leaching. The majority of the N requirement, therefore, must come from N fixation and subsequent mineralization of organically combined N. Therefore, maintaining a vigorous legume component within the plant community is essential. Total annual biomass production in humid regions is controlled by the N availability of most reclaimed soils, particularly for herbaceous grasses and legumes. Nitrogen application rates are generally lower when revegetation specifies tree establishment to limit herbaceous-plant competition. Nitrogen rates are also lower in semiarid environments where annual soil and plant N fluxes are less than in humid regions.

Because N primarily is combined with the organic matter of soils, organic amendments greatly enhance total N availability. Sewage sludge is a particularly effective mine-soil amendment, according to numerous studies (71), but is not always available in sufficient quantities at remote sites. Local and state regulations frequently restrict the use of sewage wastes on disturbed lands. Sawdust and bark mulch also help increase the initial organic-matter content of soils but generally are low in N and immobilize free soil nitrates for an extended period after incorporation. Recently, yardwaste composts and other fully stabilized organic materials are increasingly available in many areas and may play an important role in mined-land reclamation.

The maintenance of plant-available, water soluble P in mine soils is hindered by two factors: (*1*) newly exposed geologic materials generally are low in plant-available P, and (*2*) as mine soils weather, they become enriched in Fe oxides at low pH (<6.0) and in secondary carbonates at high pH (>7.5) causing specific adsorption of P from solution. The propensity of mine soils to fix P actually may increase with time, especially for materials with high iron content. It is, therefore, essential to establish and build an organic P reservoir in the soil to supply long-term plant requirements through P mineralization. Sewage sludges and other compost products are particularly effective in restoring the organic P pool in disturbed soils. High application rates of P (150 to >350 kg/ha) are required to ensure that sufficient labile P is available during the first several years. Small quantities of P become available to the plant community as calcium phosphates in the parent material decompose, but this is insufficient to meet the needs of a vigorous plant community. Many mycorrhizal plant species are able to extract P from complex sources, such as partially crystalline oxides, and should be used whenever possible.

Most mine soils or geologic materials contain sufficient Ca, Mg, and K to sustain plant growth over extended periods. Older weathered and leached mine spoils, however, are deficient in these essential nutrients, particularly if the reclaimed land is managed as hayland or pastureland with periodic removal of biomass. Deficiencies of these nutrients are identified easily and remedied through standard soil-testing procedures and fertilization.

Assuming adequate initial soil-nutrient conditions, the long-term productivity of a plant-soil system depends on two major processes: (*1*) accumulation of organic matter and N, and (*2*) establishment of an organic P pool with minimal P fixation. Both of these depend, in turn, on introduction and operation of microbial communities through time. In most mine/soil systems where fertilization beyond the first growing season is permitted, it is likely that N first limits plant growth because of high demand and that P fixation becomes a limiting nutrient in later growing seasons as the mine soils weather. However, if plant production is limited by N, the transfer of fertilizer P into organic forms is limited, thus increasing P fixation losses. Similarly, low P levels also hinder N accumulation because symbiotic N-fixing bacteria demand substantial quantities of P. The development of an organic-matter pool is essential to long-term soil fertility. The essential plant nutrients must be managed together, not as independent entities.

With adequate initial liming and fertilization, the nutrient requirements of newly established vegetation are met easily on most disturbed areas. Regulatory standards, however, usually require that a vigorous plant community persist for a minimum period of at least five years at Eastern coal mines and 10 years at Western coal mines. Maintaining this goal is difficult when periodic nutrient augmentation is prohibited, especially on harsh sites. The effects of initial fertilization generally last for about two growing seasons. Thereafter, decreases in standing biomass and vegetation cover are common even on the best of mine soils. Thus, a program of periodic site inspection and maintenance is mandated by the prevailing regulatory program and incorporated into revegetation plans. It is important to reiterate that various vegetative communities require substantially different soil pH and fertility. Consequently, the soil-amendment package must be tailored carefully to the specific needs of the desired vegetation. For example, in the East, the herbaceous mixed grass/legume community required for hayland and pastureland demands high pH and fertility that inhibit the establishment of conifers and native hardwood trees. Material handling and soil amendments must be planned well in advance of disturbance to ensure recreation of appropriate soil conditions for revegetation and reclamation.

REVEGETATION

Revegetation is a principal goal of reclamation and results in many desirable secondary water quality and aesthetic benefits. Revegetative goals vary from simple erosion control to the full restoration of complex native communities. The approaches and protocols employed, therefore, are specific for region, site, and land-use. Although detailed discussion of revegetative practices is beyond the scope of this article, the basic principles are summarized.

Plant Material Selection

The plant species selected for revegetation are dictated by the overall reclamation plan, local site and soil conditions, the post-mining land use, and sometimes the pre-disturbance vegetative history. Plants used in reclamation must withstand short- and long-term changes in climate and soil. The importance of heat and drought tolerance cannot be overemphasized, and the soils at many sites also exert pH and nutrient stresses. In the East, acid-tolerant species are preferred whereas alkaline- and salt-tolerant species are preferred in much of the West. Adapted native species are usually preferred, but it is important to remember that disturbed soils are very different from natural soils. Consequently, exotic species are often more successful than native species, particularly when unusual soil conditions result from the disturbance. The native seed banks contained in direct hauled topsoil or litter layers are utilized to provide locally adapted plant materials. Forage and agronomic plant species (eg, tall fescue, *Festuca arundinacea*) frequently are selected rather than native species because large and consistent seed supplies are commonly available and relatively inexpensive. Certified seed from reputable sources should be used in any case. Many plant species cannot be established readily from seed, necessitating planting of cuttings or seedlings.

The seeds of many plant species require treatments to ensure germination including stratification (cold/moist incubation), scarification (physical abrasion of the seed coat), or exposure to fire. A number of species also resist germination for one or more seasons (known as "hard seed"). Land-use specific seeding recommendations are available for most localities from local regulatory agencies, universities, private consultants, and reference books.

Revegetative protocols usually include a rapidly growing annual cover to provide initial erosion control, shade, and mulch for slower growing perennials. The initial cover or "nurse" crop absorbs and retains essential plant nutrients reducing losses due to leaching and runoff and then recycles these nutrients to the soil as detritus. Once the perennial species are well established, usually by the second year, and plant/soil nutrient cycles are operational, the probability of long-term, successful reclamation is greatly improved.

Seedbed Preparation

Conventional tillage practices, such as plowing, disking, harrowing, and culti-packing are commonly employed to prepare the seedbed where agronomic crops or grasses are to be established on topsoils or other finely textured materials. Planting follows tillage immediately whenever possible to permit seed germination before surface crusts form and to limit fertilizer and seed losses due to erosion. Where coarse-textured or rocky spoils are to be revegetated, tillage generally is not possible, but site preparation begins well in advance of seeding. Where possible, it is advisable to allow spoils or mining wastes to lay exposed for at least six months before seeding to allow salts to leach and acids to react.

Site preparation is problematic on steeply sloping surfaces. Gradients in excess of 25% limit the use of conventional agricultural equipment. Bulldozers, traveling up and down the hillslopes, are used generally for site preparation, such as spreading and incorporating soil amendments. In humid areas, water flow through acidic spoil produces acid seeps or "hot spots" along the hillslope. These areas often appear chalky white because of salt accumulations and develop pH values of 1.5–3.0. These areas should be identified and treated with large quantities of lime to prevent future problems in plant development. Immediately before seeding, hillslopes should be "tracked" with a bulldozer or other suitable equipment. Tracking up and down the hillslope creates narrow microdepressions on the contours that retain water, seed, and fertilizer. These tracks usually are the first areas to exhibit seed germination and plant growth. Although grading and tracking are recommended for hay land, pastureland, or other management-intensive uses, it causes soil compaction, especially deleterious to forest establishment and long-term stand vigor. If forestry is the intended postmining land use, topographic reconstruction is limited to that required for mass stability, and the surface should remain "roughly graded." Extensive highway revegetative research (72) finds that rough-graded hillslopes also favor the establishment of herbaceous plant communities.

Seeding Techniques

The strategies and techniques to establish vegetation vary with site conditions and plant species mix. In the Eastern United States, hydroseeding is common on steep hillslopes and involves spraying a liquid slurry of lime, fertilizer, seed, and mulch onto the final graded surface. On the gently rolling lands reclaimed for agriculture in the Midwest, seeding utilizes conventional agricultural equipment, such as broadcast seeders and drills. In the Western U.S., broadcast seeders are used on steep hillslopes and seed drills with culti-packers are most often used on the gentle to moderate gradient hillslopes. As a general rule, broadcast seeding requires about twice the application rate of drill seeding to compensate for seed falling in locations unfavorable for germination. A number of innovative techniques have been developed in the Western U.S. and Australia (23,73) to harvest and reapply a "living mulch" of chopped vegetation and seed which are superior to conventional seeding techniques for reestablishing sensitive native plant communities. Woody forest and shrub species typically are hand planted from nursery-grown seedling stock, or occasionally directly seeded, as with oak acorns in the Eastern U.S. Certain woody species, such as black locust (*Robinia pseudoacacia*) are also established by hydroseeding. It is important to emphasize that seeding techniques vary with local conditions and reclamation objectives. Reestablishment of native plant communities often requires complex sequences of seed collection and treatment, soil replacement and amendment, site preparation, and postestablishment maintenance as permitted.

Success in establishing vegetation is influenced strongly by climate. Local guidelines for species mixtures and seeding practices typically vary seasonally. In some areas, the climatic "window" for seeding is narrow, requiring careful planning and scheduling to insure that topographic reconstruction, site preparation, and other logistical considerations are completed in advance. In other regions, such as the Southeastern U.S., seeding is successful throughout the year although the plant materials utilized vary considerably by season.

Mulching

Whereas plantings on agricultural lands do not require mulch applications, the majority of reclamation protocols include mulching treatments of some kind (74). Mulches protect newly graded, barren, hillslopes from erosion until the plant community is established and are often prescribed for temporary reclamation. Mulches also serve several longer term objectives: (*1*) insulating the soil surface from the diurnal heat fluctuations thereby keeping the soil/seed bank cooler during the day and warmer at night than if the surface was bare; (*2*) increasing rainfall infiltration and reducing evaporative water losses from the soil thereby increasing soil-water availability while reducing runoff and erosion, and (*3*) decomposing through time to supply energy, carbon, and a limited quantity of nutrients to the microbial biomass.

Straw, wood chips or sawdust, compost, and other materials are applied to directly seeded sites. Mulches are usually spread and anchored in place with chemical sprays or mechanical crimpers to prevent displacement by wind or water flow. Alternatively, mulches are included with the hydroseeder slurry. Where hydroseeding techniques are utilized, such as in the Eastern U.S., wood fiber or wastepaper fiber mulches are commonly applied at rates of 1000 to 3000 kg/ha. Care is taken to avoid thick mulch layers, however, that separate the germinating seed from the rooting medium or that inhibit seedling emergence. Likewise, care is taken in applying and incorporating high carbon mulches, such as straw, into soils that are inherently N and P deficient. Rapid immobilization and depletion of plant-available N and P ensues from decomposition by microbes.

IRRIGATION

The National Academy of Sciences (3) concluded that the best techniques for revegetating disturbed lands in the Western U.S. must include sparing irrigation during seedling establishment because of erratic climatic conditions. Much was written about the advantages and disadvantages of irrigation in the decade following this statement. For the semiarid northern Great Plains, one study (75) remarked that irrigation to supplement low quantities of natural precipitation (*1*) helps ensure rapid and successful perennial plant establishment; (*2*) permits planting later into the growing season, helping to control weedy species; and (*3*) provides some control of the final species composition in newly established stands. For the arid Southwest, another study (76) remarked that supplemental irrigation is necessary for stand establishment when the annual precipitation is less than 20 cm (8 in.).

Irrigation improves seed germination and plant establishment. Some suggest, however, that "die-off" following the termination of irrigation largely negates the initial benefits. In addition, supplemental water allows temporary survival of some less drought-resistant plant species and perhaps the development of shallow root systems. There are questions concerning available sources of irrigation water in semiarid and arid regions where virtually all water rights were allocated long ago. Irrigation is considered an expensive reclamation practice.

It appears that the trend is away from irrigation on reclaimed lands. In semiarid regions, most reclamation managers believe that irrigation is unnecessary and revegetation must endure climatic variability. In arid regions, opinions seem to vary somewhat by industry. Although irrigation is still used at a few places following the surface mining of coal, many reclamation managers have, or intend to, discontinue the practice. A few engaged in the reclamation of lands mined for metals advocate irrigation if the cost is not prohibitive and a water supply is available near the site. If irrigation has a place in the reclamation of semiarid and arid lands it is to replicate "normal" precipitation during the initial year. Generally, irrigation is not regarded as essential, but possibly as good insurance, given precipitation variability from year to year and growing season to growing season (10,12).

MONITORING AND MAINTENANCE

Land management after reclamation includes site monitoring and maintenance. The performance of the newly constructed environmental system and its integration into the surrounding, undisturbed area is evaluated by site monitoring. The assessment is usually based on observed indicators and collected data. Attention commonly focuses on (*1*) the quantity, quality, and control of surface and subsurface water; (*2*) the quantity and quality of vegetative cover; and (*3*) the rates of geomorphic processes, such as mass movement and erosion.

Most drastic land disturbance is governed by some regulatory program that mandates site monitoring and maintenance, but the extent and compliance period vary considerably by country, state, region, or locality. For example, surface mining of coal in the U.S. is regulated intensively from premining stages to final performance-standard enforcement and bond release. Highway and construction activities also must meet stringent short-term sediment and erosion-protection standards and then are typically released from oversight within months after the targeted vegetative standard is achieved. On the other hand, sand and gravel mine operations in the Eastern U.S. are typically required to meet only short-term sediment- and erosion-control standards and to return the site to minimum revegetative standards before permit release. If subsequent problems, such as acid seeps or barren areas develop on these noncoal mine sites, typically, it is left to the landowner to remediate rather than the contractor.

The reclamation programs required by SMCRA (32) for lands disturbed by surface coal mining in the U.S. are among the most rigorous for any land-disturbing activity in the world. These coal mines must meet stringent standards for surface and groundwater quality, land form reconstruction, and long-term revegetation in conformance with preselected land-use objectives. This regulatory program has been subject to considerable criticism and we are not necessarily endorsing its adoption for all environments. It is useful, however, as a model for examining the implementation and enforcement of monitoring and maintenance programs for many types of disturbed lands. An important component of SMCRA is the financial performance bond that is placed in escrow at the time the permit is issued and is returned to the permit holder incrementally as various components of the reclamation plan are completed. The final bond increment is retained until the permittee demonstrates that revegetation and water quality standards have been successfully met for five (Eastern U.S.) or ten (Western U.S.) consecutive years. In summary, the SMCRA regulatory package includes specific performance and monitoring standards for most potential environmental impacts caused by surface mining of coal. Noncompliance with the standards results in delays in the return of bond commitments or complete forfeiture if the company refuses to comply or terminates the business. These monies are used to hire private environmental consulting and contracting firms to provide the necessary site reclamation.

Surface Water

Surface-mining activities must not significantly degrade surface water quality or quantity in accordance with the regulations promulgated under SMCRA (32). Baseline monitoring of all surface waters potentially affected by the operation is required for at least six months before disturbance. Once mining commences, all surface water flowing across the mine site must be controlled and discharged through points permitted under the National Pollution Discharge Elimination System (NPDES) (35). Suspended and dissolved solids and a variety of chemical properties, such as pH, conductance, Fe and Mn, must meet established standards for contaminant concentrations and total mass discharge for the designated performance period. These regulations typically require multiple monitoring points throughout the mine site in addition to in-stream monitoring above and below discharge points. Monitoring frequency is determined by the regulatory authority but should be done at least monthly for most properties or during discharge if large sediment ponds with ephemeral discharge are part of the water management plan. Any areas yielding substandard water quality must be remediated by revegetation, chemical treatment, or reconstruction of water control or detention structures. The monitoring period may be extended by the regulatory agency to ensure long-term compliance.

Ground Water

Likewise, surface mining must not significantly degrade groundwater quality or quantity. Baseline data is gathered by well networks and hydrogeologic investigations before disturbance. Each significant water-bearing stratum is identified, and several monitoring wells are installed above and below the mine along the potentiometric gradient. Near-surface aquifers are differentiated from deeper, confined or partially confined aquifers in the monitoring plan. Potential acid-mine drainage impacts to ground water or aquifers used for drinking water are particular concerns. In addition to wells and piezometers, springs or seeps above or below the mine site and any mine-fill discharge points (underdrains) are also monitored monthly for groundwater quality. Several monitoring points are located directly in the mining path and are replaced immediately after backfilling. Any significant water quality degradation is remediated on-site, or if the impact is deemed irreversible, various mitigation measures ranging from drinking water replacement wells to monetary fines are imposed.

Vegetation

Vegetation directly influences surface water quality and long-term landscape stability through its capacity to substantially reduce erosion rates. Revegetative procedures must establish a permanent and diverse vegetative cover that supports the postmining land use for the designated performance period. Revegetative standards are enforced by site inspections. Cover and species diversity data are gathered and compared to minimum quantitative standards. For example, if the reclamation goal at a site is erosion control and return to native forest, the site must support an average living ground cover of 90%

and contain the specified number of woody species required by the permit. On the other hand, lands returned to hayland and pastureland must meet annual yield and forage-quality goals established as typical for the area.

When vegetative failure occurs, the probable cause is determined, such as (*1*) adverse soil conditions; (*2*) poor seed or plant material quality; or (*3*) adverse local site conditions, perhaps killing frosts or floods. Where adverse soil conditions are suspected, representative soil samples are taken from both the barren areas and nearby areas exhibiting acceptable plant growth. By comparing the soil-test results (eg, ABA, pH, various salts), the limiting soil factor(s) are identified and amended as necessary. If soil conditions are not the cause of plant failure, the seed or plant materials are examined carefully for germination and vigor. Finally, it may be necessary to reevaluate the choice of plant materials relative to site conditions. Although local climatic limitations cannot be managed, seed selection, seeding dates, mulching practices, and local water flow regimes are adjusted to improve the probability of successful establishment. If the size of the retreated area is significant, remedial actions reset the "bond clock" for that area or possibly for the entire permit. If all revegetative and water quality standards for the site are satisfied at the end of the performance period, the operator is released from further permit compliance and monitoring obligations, and all remaining performance bond commitments are returned.

Mass Movement and Erosion

Geomorphic processes are generally expected to operate at rates similar to those of nearby undisturbed lands where an approximate state of dynamic equilibrium prevails. The reclaimed land, therefore, should not have accelerated process rates. For mass movement processes, tension cracks, slumps, or hummocky topography indicate mass instability. For soil erosion, rilling or gullying indicate accelerated rates. For stream channels, bank erosion, bed entrenchment, or excessive in-channel deposition indicate instability. All of these phenomena suggest an imbalance between forces and resistances. Process rates for reclaimed and nearby undisturbed lands are measured and compared to furnish quantitative evaluations of successful reclamation (6,7).

Occasionally, site monitoring identifies places in the reclaimed landscape where reclamation practices have not performed as expected. This may be the result of faulty practice design or field application or the consequence of unexpected events, such as low-frequency, high magnitude rainstorms, overgrazing by wildlife, or vandalism. If the specific reason for the problem is recognized, then appropriate remedial measures are designed and implemented. Sometimes, repetition of earlier practices, such as reseeding, is all that is required. Other times, more drastic action, such as regrading, topsoiling, and revegetation, are necessary. The adage "a stitch in time saves nine" is apropos.

CONCLUSION

Throughout the U.S. during the past 30 years, the quality and quantity of land reclamation has dramatically improved. Society expressed its determination to cease land-use practices causing long-term environmental degradation through numerous federal, state, and local statutes. Industry accepted the responsibility to minimize and repair the damage to the land. Successful reclamation is reported in various journals and magazines each month. Unfortunately, misconceptions persist concerning environmental abuse by some industries based on past and now uncommon cases.

In addition to accumulating research results and experience, much of the improvement in reclamation is traceable to advances in (*1*) methods of impact assessment; (*2*) reclamation planning; (*3*) equipment design; and (*4*) available materials, including erosion-control products and seed sources. Impact assessment and reclamation planning benefitted from expansions of environmental databases and refinements of hydrologic, geomorphic, and engineering models. Computerization enhanced the accessibility and efficiency of these databases and models. A market for specialized equipment emerged as reclamation became part of standard business operations in several industries. Innovative erosion-control products were developed for special circumstances. The varieties and quantities of seed available for revegetation increased, especially for native species.

Nevertheless, uncertainties remain in reclamation science. Geologists have compiled evidence of climatic changes throughout Earth history and numerous investigations document temporal variabilities during the last few hundred years. Regardless of the causes, it is clear that climate has changed in the past and is expected to do so in the future. Usually, discussions of climatic change center on fluctuation in mean annual temperature and precipitation, although changes in the frequency distribution of rainfall intensities also deserve attention. The ramifications of climatic change on reclaimed lands depend on the present climatic conditions at a site together with the direction and magnitude of the changes. The effects of different climatic regimes may be anticipated with considerable accuracy for some processes based upon physical laws and existing environmental analogs. For example, increases in temperature and precipitation increase rock weathering rates. With other processes, such as runoff and erosion, however, the response reflects the combined influence of several factors. For example, accepting the general form of the Langbein and Schumm relationships between mean annual precipitation and sediment yield (77), increases in precipitation result either in increases or decreases in sediment yield, depending on the adjustments of vegetative cover to new temperature and precipitation conditions. Further, changes in runoff and sediment yield alter channel properties, as suggested earlier. Although we cannot predict the consequences of climatic change with certainty, we can consider various scenarios in reclamation planning.

There are also questions concerning the long-term stability of reclaimed landscapes. The relationships between forces and resistances vary through time as natural processes work on and within reconstructed landforms. Eventually, a storm and the runoff it generates exceed design criteria. Episodic land form failure, however, also is a part of natural systems as illustrated by landslide scars.

Successful reclamation depends on our understanding of critical site-specific conditions and implementation of the best techniques for reconstructing them for a particular land use. Whenever possible, natural topsoils are salvaged, stockpiled, and returned to the final surface. When this is not possible, a topsoil substitute is created from the best available subsoil or geologic materials to serve as a plant growth medium.

In a worst case scenario, and under extraordinary circumstances, such as abandoned mined lands, the reclamation planner must reclaim a site "as is" without any topsoiling or selected substitute media. Usually, the physical and chemical properties of soils and spoils, together with their fertility, are easily assessed and adjusted unless potentially toxic materials are encountered. Standard soil-testing procedures are useful for comparing various spoil or soil substitute materials but the results cannot be interpreted with the same accuracy as for natural soils. Potential acidity and alkalinity are the most important chemical properties to be estimated. Physical properties of spoils and soils detrimental to revegetation, such as high bulk density due to compaction and low water-holding capacity, are difficult to adjust after land disturbance. In fact, compacted soils or substitutes with minimal water-holding capacity are the most common factors limiting successful reclamation. Soil acidity, salinity, and other toxic chemical conditions also limit successful reclamation, but are more localized spatially than detrimental physical properties.

Whether a site is reclaimed with complete soil reconstruction or with a topsoil substitute, its long-term success depends on the reestablishment of essential soil processes, such as organic-matter accumulation and nutrient cycling. Fortunately, these processes are restored rapidly by a well-designed plan of soil material selection and placement together with the judicious use of soil amendments. The accumulation of organic matter and N through time, with minimal P fixation by soil iron oxides are important factors controlling the long-term vegetative productivity in humid regions, whereas water deficiencies and saline/sodic soil conditions are important in arid and semiarid environments. Many waste products, such as sewage sludge, organic composts, feedlot wastes, and coal fly ash are quite useful as mine-soil amendments with secondary recycling benefits. However, these materials must be evaluated and managed carefully to ensure that the geochemical environment into which they are introduced immobilizes toxic components.

The development of a stable, self-sustaining vegetative cover is a goal of most reclamation projects. Revegetative strategies vary widely with the type of ecosystem to be reclaimed. In general, rapidly growing annual species stabilize the site, retain nutrients, and control erosion and leaching, protecting the soil so that native species invade successfully and dominate with time. The compatibility of native and introduced species should be considered carefully together with long-term site management goals and biodiversity issues. In particular, the mix of species must be locally adapted, hardy, and resistant to long-term pH, nutrient, water, and disease stresses.

The plant community itself plays an important role in soil development and nutrient dynamics, particularly through organic matter and N accumulation and long-term concentration of essential nutrients into the plant biomass for microbial recycling and remineralization. The advantages of utilizing native seedbanks in topsoils and litter layers should be considered carefully in reclamation planning. The direct haul of these materials to the final reclamation surface is encouraged whenever possible. Although the selection and culture of appropriate plant materials is a critical component of reclamation plans, it cannot compensate for improper material handling and soil reconstruction practices. The revegetation strategy is developed in conjunction with the other components of the reclamation plan, not simply as an afterthought.

Careful planning, materials handling, and topographic reconstruction are keys to successful reclamation. It is seldom possible to design effective reclamation strategies without an intimate working knowledge of the land-disturbance practices responsible for the environmental impacts that require remediation. It is therefore necessary for reclamation managers to interact with and understand the diverse perspectives of en-

Figure 4. Reclaimed valley-fill hillslope at a coal mine in the Eastern United States.

gineers, geologists, soil scientists, hydrologists, biologists, and others trained in related disciplines, such as geomorphology. Similarly, reclamation strategies must be financially feasible and communicated clearly to appropriate decision makers.

Finally, the reclaimed site must merge into the larger landscape of which it will be a functional part (Fig. 4). The surrounding landscape provides reference areas for comparative research. Often, it is possible to utilize the reclamation process to produce more productive land forms than originally occupied the site. This is a distinct possibility when the predisturbance landscape was degraded previously by soil erosion, mass movement, or earlier mining.

RECOMMENDATIONS FOR FUTURE RESEARCH

Although there have been substantial advances in reclamation science during the past several decades, numerous challenges and opportunities remain for additional research. There are aspects of each component in the reclamation process that merit further investigation. The recommendations following constitute a partial list of possibilities based on our experiences and perceptions at this time.

Global positioning systems (GPS) and geographic information systems (GIS) are rapidly developing technologies by which the spatial distribution of site characteristics, such as topography, geology, soils, vegetation, and surface and subsurface hydrology, are recorded, mapped, and analyzed with considerable accuracy. Information is easily added, updated, and displayed as often as desired. The construction of "spatially oriented" databases facilitates every step to successful reclamation. In the future, process models will be coupled with geographic information systems to simulate and predict the dynamic nature of disturbed and reclaimed surfaces.

Substantial progress has been made in our understanding of hillslope hydrology and erosion processes since the 1930s. Research commonly focused on the environmental factors affecting process rates and erosion-control practices. The topographic influence on erosion rates considered hillslope length and gradient, and to a lesser extent, hillslope shape along the profile from crest to base. Much remains to be learned about the effect of three-dimensional hillslope geometry on hydrologic and erosion processes.

There are various models of hillslope evolution that may apply to reclaimed surfaces. Field investigations near Tucson, Arizona and Denver, Colorado suggest that the "gully gravure" process sequence could occur on disturbed lands where coarse-textured material overlie, or are mixed with, finely textured materials. Rill- or gully-erosion processes on hillslopes cause the coarse material to fall into and accumulate in channels. Fine materials are entrapped within the interstices of the coarse material, reducing the porosity and permeability of the in-channel deposit. Runoff is deflected toward the periphery of the deposit, shifting the loci of erosion laterally along the hillslope. Additional coarse material from the channel banks falls into the new channel. The processes continue yielding a surface at lower elevation composed of coarse, armoring material. This hillslope approximates an equilibrium with the environment of formation and possesses low erodibility due to the armoring. Documentation of the gully gravure process sequence on reclaimed surfaces appears warranted because the resulting land form characteristics strongly influence long-term process rates. Additional discussion of gully gravure is available (78).

Reclamation cannot replicate all of the geologic, soil, and vegetative properties that existed before disturbance. It is assumed that soil and vegetative characteristics develop through time and eventually return to a semblance of the previous equilibrium condition or perhaps attain a new equilibrium condition. The evolving soil and vegetative properties affect the hydrology and erodibility of hillslopes and the discharge and sediment load delivered to stream channels. Relationships 1–4 in the section "Stream Channel Reconstruction" express the expected changes in channel morphology from changes in discharge and sediment load. It should be possible to examine changes in soil, vegetation, and channel properties through a few decades at older reclaimed sites using the baseline data and aerial photographs obtained before disturbance together with current field measurements.

Standard soil-testing procedures and plant-nutrient uptake models created for natural or agricultural conditions should be examined, modified, and validated for disturbed soil and geologic materials under reclamation conditions. Specifically, development and calibration of accurate techniques to estimate liming requirements for sulfidic materials should be conducted under a wide range of site and climatic conditions. In addition, models of long-term N, P, and C accumulation in mine soils and their use in evaluating the evolution of equilibrium soil conditions also demand further study. This research should be coupled with detailed studies of soil microbial consortia.

BIBLIOGRAPHY

1. J. Hughes, *J. Forest Hist.* **26**(2), 60–75 (1982).
2. F. Troeh, J. Hobbs, and R. Donahue, *Soil and Water Conservation,* 2nd ed. Prentice–Hall, Engelwood Cliffs N.J., 1991.
3. National Academy of Sciences, *Rehabilitation Potential of Western Coal Lands,* Ballinger Publishing Co., Cambridge, Mass., 1974.
4. R.P. Hansen, in K. Vories, ed., *Reclamation of Western Surface Mine Lands, Workshop Proceedings,* Ecology Consultants, Inc., Ft. Collins, Colo., 1976, pp. 1–7.
5. National Research Council, *Surface Mining, Soil, Coal, and Society,* National Academy Press, Washington, D.C., 1981.
6. G.C. Lusby and T.J. Toy, *Earth Surf. Proc. and Landforms* **1**(4), 375–386 (1976).
7. T.J. Toy, *Earth Surf. Proc. and Landforms* **5**(14), 289–302 (1989).
8. W. Curtis, K. Dyer, and G. Williams. A Manual for Training Reclamation Inspectors in the Fundamentals of Hydrology. U.S. Dept. of Agriculture, Forest Serv., Northeastern Forest Exper. Sta., Berea, Ky. U.S. Government Printing Office, Washington, D.C.
9. G. Schuman, D. Booth, and J. Waggoner, *J. Soil and Water Conser.* **45**(6), 653–657 (1990).
10. T. Toy and B. Munson, Climate Appraisal Maps of Rehabilitation Potential of Strippable Coal Lands in the Powder River Basin, Wyoming and Montana, Misc. Field Studies, MF-932, U.S. Geological Survey, U.S. Government Printing Office, Washington, D.C., 1978.
11. T. Toy, A. Kuhaida, and B. Munson, *Soil Sci.* **126**(3), 181–189 (1978).
12. T. Toy, and B. Munson, Climate Appraisal Maps of Rehabilitation Potential of Strippable Coal Lands in the Four Corners Area, Arizona, Colorado, New Mexico, and Utah, Misc. Field Studies, MF-

1098, U.S. Geological Survey, U.S. Government Printing Office, Washington, D.C., 1979.

13. T. Toy, and D. Grim, Climate Appraisal Maps of Rehabilitation Potential of Strippable Coal Lands in the Green Rive Basin Area, Colorado, Utah, and Wyoming, Misc. Field Studies, MF-1212, U.S. Geological Survey, U.S. Government Printing Office, Washington, D.C., 1980.
14. T. Toy and D. Grim, Climate Appraisal Maps of Rehabilitation Potential of Strippable Coal Lands in the Williston Basin, Montana, North Dakota, and South Dakota, Misc. Field Studies, MF-1675, U.S. Geological Survey, U.S. Government Printing Office, Washington, D.C., 1984.
15. W.L. Daniels and D.F. Amos, *Environ. Geochem. and Health* **7,** 8–15 (1985).
16. F.T. Caruccio, J.C. Ferm, J. Horne, G. Geidel, and B. Baganz, Paleoenvironment of Coal and its Relation to Drainage Quality, USEPA Industrial Environmental Research Lab., Cincinnati, USEPA 600/7-77-067, 1977.
17. A.A. Sobek, W.A. Schuller, J.R. Freeman, and R.M. Smith. Field and Laboratory Methods Applicable to Overburdens and Minesoils, U.S. EPA Industrial Environmental Research Lab., Cincinnatti, U.S. EPA 600/2-78-054, 1978.
18. T. Toy and R. Hadley, *Geomorphology and Reclamation of Disturbed Lands.* Academic Press, Orlando, Fla., 1987.
19. U.S. EPA, Erosion and Sediment Control, Surface Mining in the Eastern U.S., Design. EPA Tech. Trans. Publ., EPA-625/3-76-006, U.S. Government Printing Office, Washington, D.C., 1976.
20. H. Jenny, *The Soil Resource,* Springer-Verlag, New York, 1980, pp. 113–144.
21. L.R. Hossner, S.G. Tipton, and D.G. Purdy, in R.E. Dunker et al., ed., Proc., 1992 National Symp. on Prime Farmland Reclamation, University of Illinois at Urbana, 1992, pp. 25–30.
22. K.C. Haering, W.L. Daniels, and J.A. Roberts, *J. Environ. Qual.* **22,** 194–200 (1993).
23. D.R. Brooks, in R.I. Barnhisel et al., ed., *Reclamation of Drastically Disturbed Lands,* American Society of Agronomy, Madison, Wisc., 1998 (in press).
24. R. Horton, *Bull. Geol. Soc. Am.* **56,** 275–370 (1945).
25. S. Schumm, *The Fluvial System,* John Wiley & Sons, Inc., New York, 1977.
26. G. Harwood, A Guide to Reclaiming Small Tailings Ponds and Dumps, U.S. Dept. of Agriculture, Forest Service, Gen. Tech. Rept. INT-57, U.S. Government Printing Office, Washington D.C., 1979.
27. National Research Council, *Surface Mining Non-Coal Minerals, Committee on Surface Mining and Reclamation,* National Academy of Sciences, Washington D.C., 1979.
28. U.S. Congress, The Rivers and Harbors Act of 1899 (RHA), PL 97–322, 1899.
29. U.S. Congress, National Environmental Policy Act (NEPA), PL 91–190, 1969.
30. U.S. Congress, Clean Water Act (CWA), PL 95–217, 1972.
31. U.S. Congress, Water Quality Act (WQA), PL 100-4, 1977.
32. U.S. Congress, Surface Mining Control and Reclamation Act of 1977, PL-95-87, 1977.
33. U.S. Congress, Comprehensive Environmental Response, Compensation, and Liability Act (CERCLA), PL 94-499, 1980.
34. U.S. Congress, Resource Conservation and Recovery Act (RCRA), PL 94–550, 96–482, and 98–616, 1982.
35. U.S. EPA, National Pollutant Discharge Elimination System (NPDES), 40 Code of Fed. Reg., Parts 121–125. U.S. Government Printing Office, Washington, D.C.
36. U.S. EPA, EPA's Contaminated Sediment Management Strategy, Office of Water, EPA 823-R-94-001. U.S. Government Printing Office, Washington, D.C., 1994.
37. N. Shover, D.A. Clelland, and J. Lynxwiler, *Enforcement or Negotiation, Constructing a Regulatory Bureaucracy,* SUNY Press, Albany, N.Y., 1986, pp. 107–123.
38. C.E. Zipper, W.L. Daniels, and J.C. Bell, *J. Soil and Water Conserv.* **44,** 279–283 (1989).
39. J.M. McElfish and A.E. Beier, *Environmental Regulation of Coal Mining: SMCRA's Second Decade,* Environmental Law Institute, Washington, D.C., 1990, pp. 237–252.
40. W. L. Daniels, K.C. Haering, and D.C. Dove, *Va. Coal Energy Qua.,* 45–59 (1989).
41. J.G. Skousen, J.C. Sencindiver, and R.M. Smith, *A Review of Procedures for Surface Mining and Reclamation in Areas with Acid-Producing Materials,* West Virginia Energy and Water Research Center, Morgantown, W.Va., EWRC871, 1987, p. 3.
42. E.F. Perry, in *Proc., Symp. on Surf. Mining, Hydrology, Sedimentology, and Reclamation,* University of Kentucky College of Engineering, Lexington, Ky., 1985, pp. 369–375.
43. U.S. Salinity Laboratory Staff, Diagnosis and Improvement of Saline and Alkali Soils, *Agricultural Handbook 60,* U.S. Department of Agriculture, Washington, D.C., 1954.
44. G.E. Schuman and J.L. Meining, *Soil Sci. Soc. Am. J.* **57,** 1083–1088 (1993).
45. J.A. Smith, E.J. DePuit, and G.E. Schuman, *J. Environ. Qual.* **15,** 427–435 (1986).
46. S.E. Fisher and F.F. Munshower, in *Reclaiming Mine Soils and Overburden in the Western United States,* Analytical Parameters and Procedures Soil Conservation Society of Amer., Ankney, Iowa, 1987 pp. 109–134.
47. J.A. Roberts, W.L. Daniels, J.C. Bell, and J.A. Burger, *J. Environ. Qual.* **17** 55–62 (1988).
48. R.D. Krebs and J.W. Dulin, *Residential Foundations for Reclaimed Land,* Virginia Tech Research. Div., Powell River Project, Blacksburg, Va., 1987, pp. 10–13.
49. W. Graf, *Am. J. Sci.* **277,** 178–191 (1977).
50. U.S. BLM, *Solid Minerals Reclamation Handbook,* BLM Manual Handbook H-3042-1, U.S. Government Printing Office, Washington D.C., 1992.
51. F. Brenner, *Mining Magazine* 211–217 (Sept. 1985).
52. J. Bell, W. Daniels, and C. Zipper, *Landscape and Urban Plann.* **18,** 127–138 (1989).
53. L. Meyer and M. Romkens, in *Proceedings of the Third Interagency Sedimentation Conference,* PB-245-100, Water Res. Council, U.S. Government Printing Office, Washington, D.C., 1976, p. 2-65–2-76.
54. R. Hadley and T. Toy, *J. Research U.S. Geol. Survey,* **5**(4) 487–490 (1977).
55. M. Carson and M. Kirkby, *Hillslope Form and Process,* Cambridge Univ. Press, New York, 1972.
56. J. Playfair, *Illustrations of the Huttonian Theory of the Earth,* William Creech, Edinburgh, Scotland, 1802.
57. R. Chorley, A. Dunn, and R. Beckinsale, *The History of the Study of Landforms.* V.1, Meuthen & Co. Ltd., London, England, 1964.
58. A.D. Howard, *Am. Assoc. Pet. Geol. Bull.* **52,** 2246–2259 (1967).
59. M. Morisawa, *Streams: Their Dynamics and Morphology,* McGraw Hill, New York, 1968.
60. D. Brown and R.G. Hallman, Reclaiming Disturbed Lands, USDA-USFS Equip. Dev. Center, Missoula, Mont., 1984, pp. 19–25.

61. I.J. Jansen and W.S. Dancer, in *Proc., Symp. on Surf. Mining, Hydrology, Sedimentology, and Reclamation,* University of Kentucky College of Engineering, Lexington, Ky., 1981, pp. 357–362.
62. J.A. DeMent, S. Cooney, and J. Denman, in R.E. Dunker and coworkers, eds., *Proc., 1992 National Symp. on Prime Farmland Reclamation,* University of Illinois at Urbana, 1992, pp 177–182.
63. G. Schuman, F. Rauzi, and G. Howard, *Reclamation and Revegetation Res.* **6,** 49–54 (1987).
64. R. Hodder, in J. Thames, ed., *Reclamation and Use of Disturbed Lands in the Southwest,* University of Arizona Press, Tucson, Ariz., 1977, pp 217–233.
65. R.I. Barnhisel and R.B. Gray, in J. Skousen, and co-workers, eds., *Proc., 1990 Mining and Reclamation Conf. and Exhib.,* West Virginia University, Morgantown, W.Va., 1990, pp. 143–149.
66. G.E. Schuman, in R. E. Dunker et. al., eds., *Proc., 1992 National Symp. on Prime Farmland Reclamation,* University of Illinois at Urbana, Urbana, Ill., 1992.
67. W.L. Daniels, M.H. Genthner, and R.L. Hodges, in *Proc., 1992 Annual Meeting of the Amer. Soc. for Surface Mining and Reclamation,* IRRB, Chisolm, Minn., 1992, pp. 37–47.
68. J.L. Torbert, Restoring Forest Production on Surface Mined lands, Ph.D. Dissertation, Virginia Polytechnical Institute, Blacksburg, Va., 1995.
69. D. Dollhopf, I. Jensen, and R. Hodder, Effects of Surface Configuration in Water Pollution Control on Semi-arid Mined Lands, Montana Exper. Sta. Res. Rept. 114, Bozeman, Mont., 1977.
70. N. Van Breemen, in Kittrick et al., eds., *Acid Sulfate Weathering,* Soil Science, Society of America Pub. 10, Madison, Wisc., 1982, pp. 95–108.
71. W. E. Sopper, *Municipal Sludge Use in Land Reclamation,* Lewis Publishers, Boca Raton, Fla., 1993.
72. J.N. Booze-Daniels, W.L. Daniels, J.M. Krouse, and D.L. Wright, in R.I. Barnhisel et al., eds., *Reclamation of Drastically Disturbed Lands,* American Society of Agronomy, Madison, Wisc., 1998 (in press).
73. E.F. Aldon and W.R. Oaks, eds., *Reclamation of Mined Lands in the Southwest, A Symposium,* Soil Conservation Society of America, New Mexico Chapter, Albuquerque, 1982.
74. B.M. Slick and W.R. Curtis, A Guide for the Use of Organic Materials as Mulches in the Reclamation of Coal Minesoils in the Eastern United States, USDA-USFS Northeastern Exp. Sta. Gen. Tech. Rep. NE-98, Berea, Ky., 1985.
75. R. Ries, *Farm Res.* **6,** 21–23. (1980).
76. E. Aldon and H. Springfield, in J. Thames, ed., *Reclamation and Use of Disturbed Lands in the Southwest,* University of Arizona Press, Tucson, Ariz., 1977, pp 229–237.
77. W.B. Langbein and S. Schumm, *Am. Geophys. Union Trans.* **39,** 1076–1084 (1958).
78. W. Osterkamp and T. Toy, *Bull. Geol. Soc. Am.* **106**(10), 1233–1241 (1994).

RECYCLING MUNICIPAL SOLID WASTE

GARRICK E. LOUIS
FRANCIS C. MCMICHAEL
Carnegie Mellon University
Pittsburgh, Pennsylvania

There are three sides to the system of recycling municipal solid waste (MSW). They are the policies, programs, and processes by which it is carried out. This article focuses on the policies and programs under which recycling is implemented in the United States. It presents the major issues associated with recycling and critically evaluates the nation's recycling policies. It also discusses the impact of recycling programs on the management of MSW across the nation.

The article does not cover the technical details of recycling processes or describe how to set up a recycling program. There are several excellent references in the literature that do so. Some of these are cited in the Appendices to this text.

This article begins with a background that provides a definition of recycling and of MSW. The background then considers the environmental and economic reasons for recycling. Different processes by which recycling may be categorized close out this section. The next section reviews the history and status of recycling in the United States. It presents national statistics on MSW recycling reported by the Environmental Protection Agency (EPA). It also presents statistics on state programs as reported in MSW trade journals. The next section summarizes federal and state recycling policies and programs. It reviews the evolving federal policy on MSW recycling and discusses the programs by which these policies are implemented. It then tabulates each state's recycling policy, its recycling goals, and the status of its recycling program. The following section looks at recycling MSW from a systems perspective. It discusses EPA's policy of integrated solid waste management (ISWM). This includes integrating recycling with waste-to-energy (WTE). It then considers the economic and environmental costs of recycling MSW. It closes with a relative cost evaluation of recycling including the Iron Law of Recycling. The next section compares mandated recycling to market-based approaches. It also discusses the issue of flow control and the possible impact of the Carbone decision on municipal recycling programs. The final section concludes the discussion. It revisits the rationale for recycling and the Iron law. It presents a 10-point list of things to consider when implementing MSW recycling policies. It also includes a flexible integrated solid waste management scheme which takes a systems approach to recycling.

BACKGROUND

Recycling Defined

No simple definition of recycling can do justice to the diversity of perceptions, expressions, and beliefs that are associated with this term. However, at least for consistency in this article, a concise, yet inclusive definition is presented. Recycling may be thought of as the process of reducing resource consumption by extending the useful life of a product or material through its reuse in the same or different applications. This process is illustrated in Figure 1. This definition includes the relatively short anthropocentric time scale, represented by streams R_1, R_2, and R_3, and the longer ecocentric time scale which allows for ecological decomposition and reformulation of materials, represented by stream R_0.

Thus recycling is a process rather than a single activity. It takes place at any stage in the life cycle of a material or lifespan of a product. This definition includes the common distinction between reuse and recycling (1,2). When defined separately, reuse refers to utilizing the product in its original form whereas recycling involves reprocessing and/or remanufacturing new products from the original component materials. The definition proposed here also includes energy recovery from MSW as

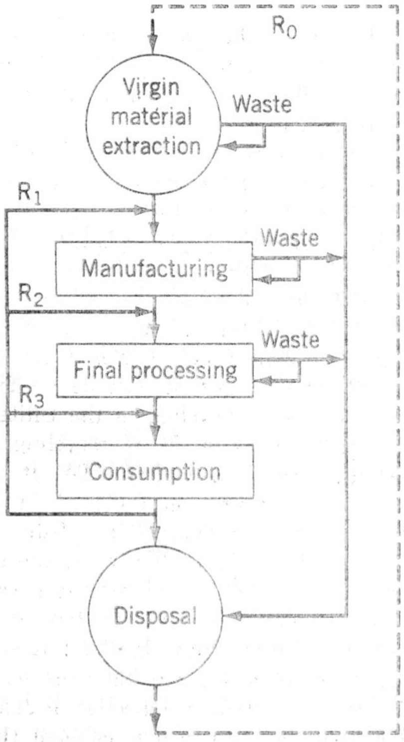

Figure 1. The recycling process: R_0 = biochemical degradation and synthesis into new materials, eg, petroleum from fossils; R_1 = reduction to raw materials and use in new products, eg, recycled paper; R_2 = cleaning, repairing, and reuse for same application, eg, a used car; and R_3 = reuse directly for same or different applications, eg, used furniture, waste-to-energy.

a form of recycling. This is by no means a consensus view. However, for the purposes of this article, generality is better served by including rather than excluding energy recovery.

MSW Defined

In this article MSW is defined in terms of postconsumer waste. It is the nonhazardous waste stream defined by Subtitle D of the Resource Conservation and Recovery Act (RCRA) minus all industrial residues, agricultural wastes, waste from mining, utility ash, construction and demolition debris (CDD), municipal sewage sludge (MSS), medical waste, and hazardous waste from small quantity generators. Table 1 summarizes this definition and shows the place of MSW in the Subtitle D waste stream.

This definition is similar to that used by EPA in its biennial Characterization of Municipal Solid Waste (3). The EPA reports a national average recycling rate of 22% in 1993 (3). Thus, when we talk about recycling, we are referring to the potential that lies in the MSW stream, of which 22% is currently exploited. Howevever, it is sobering, if not useful, to put this potential in the context of the larger stream of Subtitle D waste. In this comprehensive stream, MSW is a negligible component. When the full Subtitle D waste stream is considered, it can be seen that the environmental impacts of at least four components are orders of magnitude greater than those from MSW. Nevertheless, MSW is the subject of the present discussion.

Why Recycle?

In 1960 total MSW generation in the United States was 80 million metric tons or 1.2 kg (2.7 lbs) per person per day (PPD). By 1993, the EPA reported total MSW generation of 188 million metric tons or 1.9 kg (4.4 lbs) per person per day (3). Thus total MSW generation increased by 135% and generation per person by 63% in the span of one generation. Population growth accounts for only 30% of the increase in total MSW generation over the period. The remainder is from the increase in the per person or per capita MSW generation rate. In an age of environmental awareness, this burden of solid waste is a major source of public concern. Recycling is a waste management option which appears to offer a solution to the problem. Driven by an increasing quantity and rate of MSW generation, there are many reasons advanced for recycling legislation, policies, and programs. Resource conservation, cost effectiveness, avoided disposal, health and environmental benefits are some of the more common. In essence, recycling is attractive because of the potential for energy and resource saving it provides as part of a systems approach to waste and materials management. Such an approach includes the reuse of products, their reduction to raw materials for manufacturing, and the "green design" of products and processes to minimize resource consumption and facilitate recycling. Thus, the motivations for recycling are summarized in two major categories: environmental reasons and economic reasons.

Environmental reasons include resource conservation, avoiding the impacts of waste disposal, and conserving landfill space. These altruistic reasons (5) address the desire to conserve energy and natural resources, including virgin materials displaced by recycling. Such conservation translates into a sustainable use of resources that does not compromise them for future generations. Recycling also reduces the impacts of waste incineration and landfilling. These include the environmental impacts on air and water quality and the perceived health risks associated with these impacts. Given these considerations, recycling also seems and feels like the right thing to do.

Economic reasons include the dollar value of recycled materials and products, the avoided cost of disposal, and the value of extended landfill and incinerator capacity. The reasons are likely to appear more attractive if the price of materials used to make products includes their environmental costs. Environmental costs include the value of depleted natural resources, impacts from the extraction of materials, and the manufacturing, transportation, recycling and disposal of products. Because the environmental cost of materials, products and processes are set by society, we call it their social cost. The social cost here is really a net cost. It is the difference between society's valuation of the benefits and costs from the environmental and other impacts of materials, products, and processes. Economic cost refers to the dollar value of inputs, such as materials, labor, energy, and the time value of money (interest).

Ideally, society should recycle at a level at which the net social and economic cost of recycling the last unit was more attractive than disposal of that unit of MSW. "More attractive"

Table 1. MSW in the RCRA Subtitle D Waste Stream[a]

Rank	Component	Weight, million tons	(Million t)	Weight %	Weight %[b]
1	Oil gas[c]	822,000	(747,000)	99%	
2	Industrial nonhazardous waste (INW)	7,600	(6,910)	1%	
3	Agricultural[d]	2,000	(1,820)	0%	
4	Mining (excluding coal mining)	1,400	(1,270)	0%	
5	MSW	158	(144)	0%	55%
6	Utility ash residue	85	(77)	0%	29%
7	Construction and demolition debris (CDD)	31.5	(29)	0%	11%
8	Municipal sewage sludge (MSS)	10.4	(9.4)	0%	4%
9	Medical (MW)	2.1	(1.9)	0%	1%
10	Small quantity generator (SQG)	1.2	(1.1)	0%	<1%
	Total	*833,000*	*(757,000)*	*100%*	*100%*

[a] Adapted from Ref. 4.
[b] Excluding gas and oil exploration waste, agricultural and farm wastes not applied to the land, and mining waste.
[c] Oil mud and debris from drilling and pumping operations. Assumes a density of 8 lb/gal.
[d] EPA's estimate for wet manure. Does not include other farm wastes.

here means a lower net cost or higher net benefit. Similarly, when we recycle, we would choose the option for which this *marginal cost* was lowest. In strict economic terms we also assume that marginal cost is equal to society's marginal benefit from recycling at this point.

Deciding on this optimal level of recycling is not an easy task. First, it is necessary to know the economic and social costs (especially environmental costs) of each alternative. This includes the extraction of raw materials to manufacturing the product, recycling the materials, or disposing of them. The option with the lowest long term cost, ie, the sum of private and social costs, should be chosen. This choice is sensitive to concerns about environmental quality, concerns for future generations, and optimism about future technological change. If the prices of raw materials, products, recycling, and disposal all reflect social and private costs, the decision might be left to the market. However, because there are significant externalities associated with the extraction of raw materials and the disposal of solid waste, the current prices of materials and disposal lead people to make the wrong decisions from a social viewpoint.

The private costs facing a MSW authority are illustrated in Figure 2. The horizontal axis is the percentage of MSW handled, from 0 to 100%. The marginal cost of collection and disposal (MC_D) declines throughout most of the range because of economies of scale in collection. However, MC_D begins to rise because the last 10–20% of MSW is difficult to collect. This may be MSW from remote areas or items like white goods which require special collections. The marginal cost of collection and recycling (MC_R) is also shown. MC_R also drops initially because of economies of scale in collection. At some collection level in many cities, MC_R becomes negative indicating that there are some materials where there is a net gain to the city from recycling. However, it is assumed that the highest value materials are picked up first. The first 7% of the MSW is high value metals (6). After that, the next 1% of materials are high value plastics, such as polypropylene. After that comes other plastics, glass, and paper. Considering aluminum, steel, high quality office paper, polypropylene and clear glass, only the first roughly 10% of MSW is profitably recycled with present technology and markets, even if the

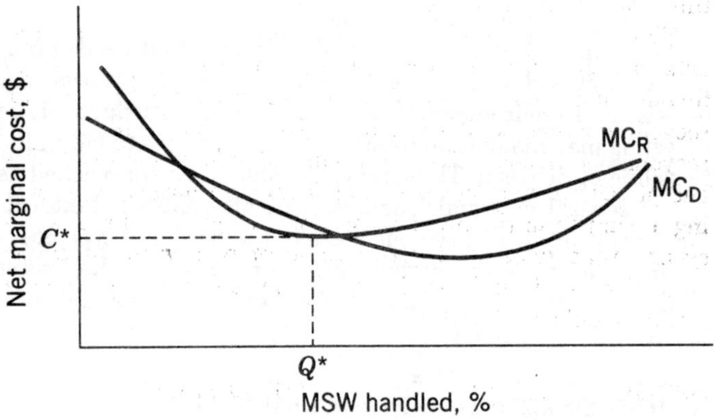

Figure 2. Comparing the net marginal costs of recycling and disposal: MCD = marginal cost of disposal (incineration of landfilling); MCR = marginal cost of recycling; Q^* = optimal level of recycling; and C^* = marginal cost of recycling at Q^* (7).

material were collected and separated. Thus, the value of the recyclable material declines steadily with the percentage of MSW that is recycled. The marginal cost of collection and separation increases as more products are added to the pile that must be sorted. The result is a large increase in MC_R above a critical recycling rate.

In this simplified example we compare materials recovery to either landfilling or incineration. In this case we would recycle at level Q^* which corresponds to the minimum C^* of the recycling cost curve.

This methodology for selecting the level of recycling is difficult. Given the current state of knowledge, a full-cost pricing basis is only an ideal for comparing recycling to disposal and for setting recycling policies. Still, let us use this ideal as a point of reference as we discuss recycling policies and programs through the remainder of the article.

High and Low Value Recycling

A broad definition of recycling includes the options of reuse, remanufacturing, and reprocessing. However, it is useful to

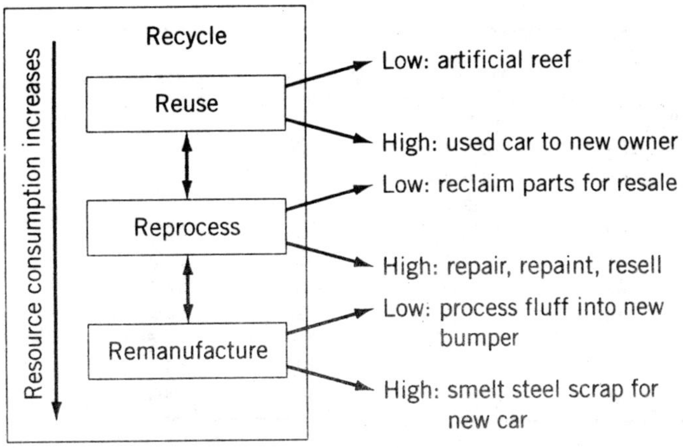

Figure 3. Low and high value recycling: a used car example.

differentiate between recycling which leads to high value uses and that which leads to low value uses (8). Figure 3 illustrates this distinction using the example of a used car.

The United States society, seeks recycling that yields products of high value. However, the labor, energy, process materials, and environmental costs associated with high value recycling may make it an unattractive alternative to low value recycling or disposal. Thus, recycling should be approached as one of several materials management alternatives. Reiterating a point from the previous section, the level and type of recycling that gives the highest value product with the lowest sum of economic and social costs from the use of available materials should be chosen.

HISTORY OF RECYCLING IN THE UNITED STATES

Overview

Organized recycling has a lengthy history in the United States (19). In the late nineteenth century, sanitary engineers extracted oils and fat from municipal refuse for soap and candle making. They also recovered rags for the manufacture of paper (10). The auto industry recycled scrap steel for the manufacture of new cars as early as the 1930s. Recycling was also a national priority during the years of World War II. However, recycling as a matter of public policy and legislation has its roots in the environmental movement of the 1960s.

The Solid Waste Disposal Act (SWDA) of 1965 (Public Law 89-272) initiated modern federal legislation focused on solid waste. The intent of SWDA was to facilitate the implementation of environmentally sound solid waste management and resource recovery systems. It provided technical and financial assistance to states and local governments and supported research to develop and implement these systems. *The 1968 National Survey of Community Solid Waste Practices* was commissioned by President Johnson, in part to further the objectives of the SWDA. It painted the first portrait of the nation's solid waste management system.

The SWDA was amended by the Resources Recovery Act (RRA) of 1970 (Public Law 95-512). This act shifted the national solid waste management focus away from disposal to recycling, including energy recovery from waste incineration. Provisions in the RRA for state solid waste management plans (SWMPS) stimulated state involvement in solid waste legislation and regulation. By 1975 Wyoming was the only state without an enacted solid waste management statute. In 1971, the State of Oregon enacted the nation's first bottle bill. This initiated government regulation of materials recycling from solid waste as it is known today.

The EPA was established in 1970 as the lead agency to control damage to the nation's air, water, and land resources from pollution. The Clean Air Act of 1970 (Public Law 91-604) established federal authority to regulate emissions from municipal waste incinerators, landfills (qv), and other waste management facilities. This led to the closure of many solid waste incinerators. Thus began a process whereby federal law, implemented by federal and state regulations, reduced the number of solid waste disposal sites. This created an indirect incentive to recycle. This pressure on solid waste disposal sites continued into the mid 1990s.

April 22, 1970 also marked the first Earth Day. This event stirred the public's awareness and concern for environmental issues. Solid waste was one of these issues. The public was favorably disposed to ideas like recycling, with its appeal for resource conservation and environmental protection. Federal and state governments soon responded to the public's interest with recycling legislation, policies, and programs. The fledgling EPA's report, Recovery and Utilization of Municipal Solid Waste, was the first to address recycling at the national level (11).

The Resource Conservation and Recovery Act (RCRA) of 1976 (public Law 94-580) established the Federal Government as a leading player in the management of solid waste. Intended to "promote the protection of health and the environment and to conserve valuable material and energy resources ...", RCRA's focus on implementation was a big step forward from the "evaluate and assess" model of RRA. RCRA established standards for sanitary landfills and guidelines for upgrading open dumps. It also set the stage for closure of thousands of open dumps and substandard landfills across the nation. This facilitated an increase in waste-to-energy (WTE) and materials recovery as solid waste management alternatives. The growth of WTE was also facilitated by the Public Utilities Regulatory Policy Act (PURPA) of 1978 (Public Law 95-617). PURPA guaranteed the sale of energy from WTE plants, operating as small energy producers, to electric utilities. However, high cost, competition from large, regional landfills and materials recovery facilities and public concerns about the health risks from WTEs curbed their increasing numbers by the early 1990s.

RCRA has driven solid waste management regulation and policy from 1976 to the present. In 1988, the EPA proposed the goal of reducing the nation's MSW stream by 25% through source reduction and recycling. In addition, in 1989 the EPA's hierarchy of integrated solid waste management (ISWM) gave priority to source reduction and recycling over combustion and landfilling (12). Many states adopted this hierarchy and mandated recycling in their solid waste laws. Thus, recycling was favored by public opinion, federal policy, and state laws. This led to a greater role for recycling in the management of MSW and a proliferation in the number of recycling programs. Now that two decades of program implementation have elapsed, the impacts of the nation's recycling policies can be assessed. This will enable lawmakers to refine these policies and make recycling a cost-effective component of solid waste and materials management systems.

Trends in the Recycling of MSW

Figure 4 summarizes the national trends in MSW management from 1960 to 1993 with projections to 1995. It shows that recycling (including composting) accounts for 7% (2.7 million metric tons) of the nation's MSW stream in 1960. This share rose to 22% (20 million metric tons) in 1993.

This is an average annual increase of 7% over the 33-year period. However, much of the increase is from the adoption of integrated solid waste management by the states since the mid 1980s. State ISWM policies had both mandatory and incentive-based recycling components. Thus the average annual increase in tons of material recycled was 15% for the period 1985 to 1993. This is more than twice the annual average for the 33-year period. Nationwide, the number of curbside recycling programs went from 1042 in 1988 to 7375 in 1995 (13). Thus recycling had a phenomenal growth in popularity across the nation and a variety of mandatory and voluntary programs now exist in communities in every state. To the extent that these programs are funded by direct user fees to participants, they represent the public's willingness to pay for recycling. However, most recycling programs are funded from general tax revenue. In this case, the public does not have a direct measure of the cost recycling. Thus the popularity of recycling programs does not directly reflect the public's willingness to pay for them.

An objective evaluation of recycling programs might ask three questions:

1. What and how much material is recycled?
2. What does it cost relative to other options for managing the materials?
3. How can the system be improved to conserve natural resources and protect the environment more cost effectively?

To answer these questions and see what can be learned about recycling MSW in the process, consider question *1,* What materials are recycled and how much? As noted earlier, the United States recycled roughly 22% of the 188 million metric tons of MSW generated in 1993. Figures **5a** and **5b** show the composition of the MSW stream and the quantity of materials recovered nationwide in 1993 (3).

Revenue from sales ranges from $48–$138 per ton for a typical bundle of materials in municipal recycling programs (9). Even after subtracting these revenues from the cost of collection and processing, recycling still has the highest dollar cost

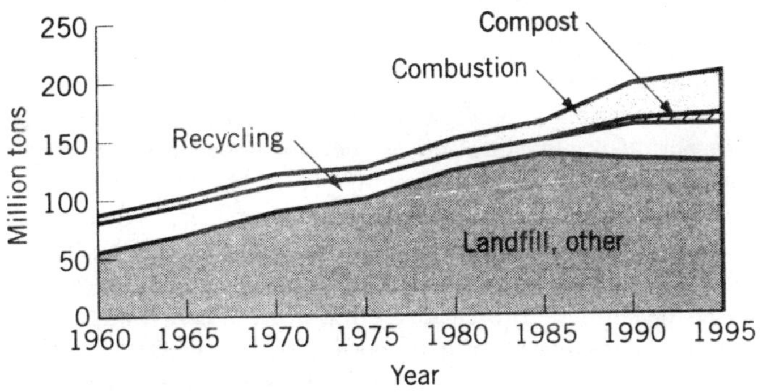

Figure 4. MSWM in the United States (1960–1993) (3).

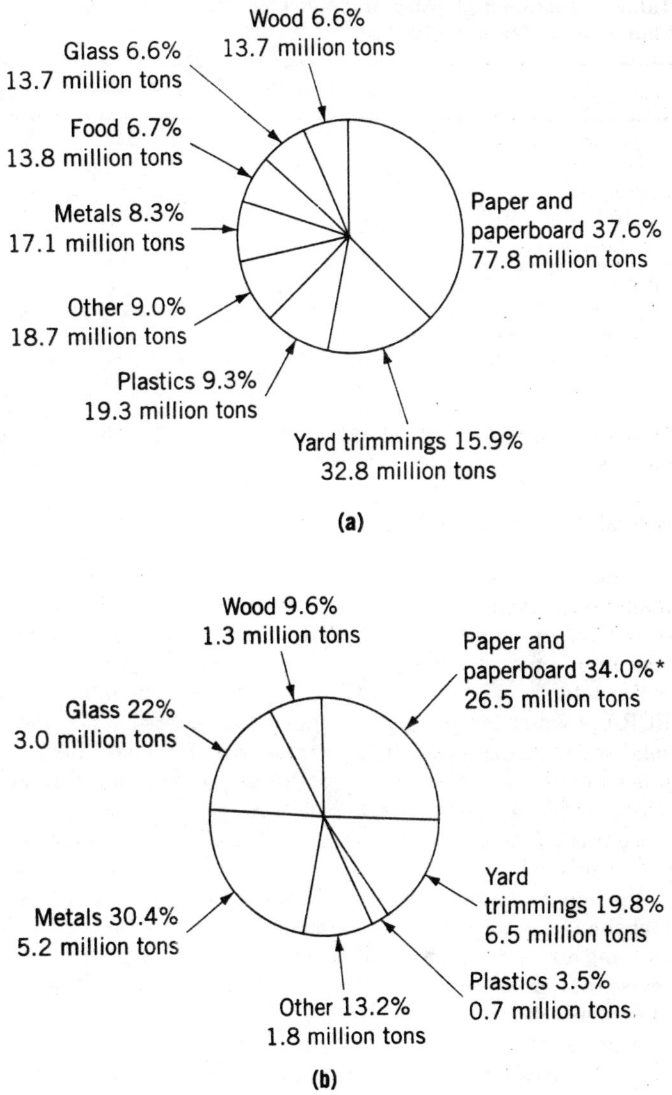

Figure 5. (**a**) Materials generated in and (**b**) recovered from MSW by weight, 1993 (3). Percentages are recovery as wt % of the material generated.

per unit of all the available waste management options. Table 2 compares the costs of the different technology options from the integrated solid waste management hierarchy. This partly answers the second questions, What is the cost of recycling relative to other options for managing the same materials? From Table 2 the answer is, it is the most expensive.

To delve a little deeper we note what when the prices of materials does not reflect their social costs, materials and energy recovery are the least attractive MSW management alternatives. Full-cost pricing of materials to reflect these unaccounted social costs, could increase their value when recovered. This could, in turn, raise the revenue from recycling and lower the net cost of recycling relative to other MSW management options. Until such full-cost prices are incorporated into all materials, consider other options for improving the attractiveness of recycling. Examine existing recycling policies and programs and in so doing seek to answer the third question, How can the system of recycling MSW be improved?

Table 2. Installed Capital and Net Operating Costs for MSW Management Options, 1994 $[a]

Technology	Net Capital + Operating Cost ($/t)[b]
Recycling (materials recovery)	$113–$410
Incineration	$105–$137
Landfilling	$90–$101
Composting	$82–$85, $137–$145[c]

[a] Ref. 9.

[b] Assumes a collection cost of $75/ton for nonrecyclables. Deducts the revenue from the sale of recovered materials.

[c] The first range is for yard trimmings; the second range is for sewage sludge.

FEDERAL AND STATE MSW RECYCLING POLICIES AND PROGRAMS

Federal Policies and Programs

Federal Policies. There is currently no federal law which mandates recycling in the United States. However, key pieces of federal environmental legislation advance "resource conservation" and "recycling" as parts of their objective. The most notable of these is RCRA (1976). Section 1003(a) of RCRA provides for technical and financial assistance for state solid waste management plans which includes "new and improved methods of collection, separation and recovery of solid waste..." Section 5001 directs the Secretary of Commerce to "encourage greater commercialization of proven resource recovery technology..."

President Clinton's Executive Order of October 20, 1993, established recycled content requirements for federal purchase of printing and writing paper. It also required federal agencies to revise their procurement practices to boost the recycled content of federally purchased products.

Federal policy on recycling includes policy guidelines from the EPA. Its policy of 25% reduction of the nation's solid waste stream, first proposed in 1988, and the integrated solid waste management hierarchy of 1989 (12) are the most notable for their lasting effects on the approach to recycling adopted by the states. In addition, the EPA also publishes numerous research results and reports which have policy implications for recycling programs. An example is its report on unit pricing of solid waste collections, which have implications for the funding of community waste management programs and their rates of materials recovery (14).

Federal Programs. Apart from grants for state solid waste management plans, the federal government also provides financial and technical support for recycling through a variety of programs. EPA's "Jobs Through Recycling" program funded the creation of Recycling Business Assistance Centers. These centers seek to promote economic development and create jobs by boosting capacity in the recycling industry.

State Policies and Programs

State Policies. Recycling policies at the state level are reflected in state recycling laws and implemented through regulations and programs administered by a variety of state agencies. In general these policies feature a mandate (43 states) or recommendation to reduce and recycle MSW by a target percentage and date. They are often accompanied by provisions to promote recycling through a variety of grants and financial incentives. Some states have procurement policies which favor products with recycled content.

The recycling mandates generally cover both residential and commercial sources and are implemented at the municipal level. Several state laws include bans on disposing of certain products and materials. These laws serve the dual purpose of promoting the recovery of the targeted materials and limiting their environmental and health impacts from disposal in landfills and incinerators. Additionally, diverting items like yard waste and white goods (eg, refrigerators, stoves, sofas) from landfills, extends the life of these facilities. Of course, alternatives must be found for handling the banned products. In considering these alternatives, assess their full cost (social economic) relative to disposal and choose the lowest cost option (all else being equal). Table 3 summarizes the types of state recycling laws and disposal bans.

The majority of state recycling laws seek to reduce, by a given percentage, the amount of MSW destined for disposal in incinerators and landfills. They pursue this goal through the recovery of materials from the waste stream, composting, and bans on the entry of products and materials into the MSW stream in the first place. This is essentially a *waste diversion* or *downstream* approach. It focuses on the waste stream after the MSW has been generated. In effect, this approach does not reduce the amount of waste generated, only the amount incinerated or landfilled.

A *source reduction* or *upstream* approach seeks to reduce the amount of MSW generated. Source reduction also seeks to reduce the toxicity of materials which enter the waste stream. Source reduction includes such elements as reducing the amount of packaging used in products and Green Design. Green Design seeks to reduce the resource consumption and toxicity of *the processes* used to manufacture products and of *the products* themselves. It also strives to design products that are easier to recycle and which yield a maximum amount of recyclable components when they are recycled. Table 4 summarizes the states' progress in achieving their recycling goals, their latest budget allocations, and their programs to promote recycling.

Column 7 of Table 4 gives a measure of the leverage or effectiveness of state expenditure on the recovery of recyclables. There is a wide variation across states in tons of material recovered per dollar of state funds budgeted for recycling. The range goes from a low of 0.03 metric tons (29 kg) per dollar in Wisconsin to a high of 24 metric tons (24,100 kg) per dollar in Utah. What lies behind this big difference across the states? Assuming that the budget reporting was consistent across states, could Utah really be that much more efficient at recycling than Wisconsin?

State spending on recycling programs has only an indirect effect on actual tons of material recovered. This is because the actual implementation of recycling operations is largely carried out by local governments. Thus, although state funds provide support for the purchase of equipment and program planning, regional and municipal budgets foot the bill for the tons of material recovered. Still, if "tons of material recovered" is a measure of the return on a state's investment in recycling, it is interesting to note the divergence in returns obtained across different states. Perhaps the results are highly dependent on local conditions in the state. However, it is also

Table 3. State Recycling Laws and Disposal Bans, 1995[a]

State	Recycling Law[b–d]	Disposal Bans[e]	State	Recycling Law[b–d]	Disposal Bans[e]
AL	Recycling plans	ad	MT		
AK		–	NE		abcdef
AZ	Opportunity to recycle	ab	NV	Recycling plans	d
AR	Opportunity to recycle	abc	NH	Recycling plans	acf
CA	Recycling plans	abde	NJ	Source separation	acf
CO		–	NM	Recycling plans	ad
CT	Source separation	ce	NY	Source separation	ab
DE	Opportunity to recycle	b	NC	Opportunity to recycle	abcdef
D.C.	Source separation	–	ND		ade
FL	Opportunity to recycle	abcdef	OH	Recycling plans	abc
GA	Opportunity to recycle	abc	OK	Recycling plans	b
HI	Recycling plans	abcef	OR	Opportunity to recycle	abdef
ID		ab	PA	Source separation	ac
IL	Recycling plans	abce	RI	Source separation	a
IN	Recycling plans	abc	SC	Opportunity to recycle	abcde
IA	Recycling plans	abcdf	SD	Recycling plans	abcdef
KS		ab	TN	Recycling plans	abd
KY	Recycling plans	sb	TX	Recycling plans	abd
LA		abe	UT		ab
ME	Source separation	a	VT	Recycling plans	abdef
MD	Recycling plans	bc	VA	Recycling plans	a
MA	Recycling plans	abcef	WA	Opportunity to recycle	ad
MI	Recycling plans	acd	WV	Source separation	abc
MN	Opportunity to recycle	abcdef	WI	Opportunity to recycle	abcdef
MS		a	WY		a
MO	Recycling plans	abcdef			

[a] Adapted from Ref. 15.
[b] Recycling plans: local government must prepare a recycling plan.
[c] Opportunity to recycle: local government must offer an opportunity to recycle via curbside collection, drop-off center, or by recovery from mixed MSW.
[d] Source separation: local government must mandate the separation of one or more components of the MSW stream.
[e] Disposal bans: a, vehicle batteries; b, tires; c, yard waste; d, motor oil; e, white goods; and f, other including portable batteries, scrap automobiles, glass and metal containers, antifreeze, newsprint, and cardboard.

possible that there are opportunities for learning from those states that effect relatively high materials recovery rates per dollar of state funds. This is certainly worth closer inspection.

State Programs. The are six major categories of state programs to promote and support recycling. They are tax incentives, grants, loans, recycling zones, economic development initiatives, and recycling market development councils or task forces. Let us briefly describe each of these in turn.

Tax incentives include the following options:

- Reduced state and local tax rates for recycling enterprises
- Property tax exemptions for recycling enterprises
- Tax credits for the purchase of recycling equipment
- Tax deductions for the purchase of postconsumer materials

State and local governments may offer reduced tax rates to businesses which are involved in materials or energy recovery. These may range from zero to several percentage points lower than the prevailing corporate tax rate. The reduced tax rates may be permanent, they may increase incrementally to the going rate, or they may be phased out after a given grace period. Beneficiaries may include independent materials recovery facilities (MRFs), waste-to-energy plants, and manufacturers who process recovered materials into recycled feedstock for new products. These programs also cover manufacturers using recycled materials.

Local governments may offer property tax exemptions and even donate property to attract recycling enterprises to particular locations. The desire to utilize abandoned former industrial sites (brown fields) aids these efforts. In general the tax exemptions are offered for a prescribed period after which the rate is gradually brought up to the prevailing rate.

Tax credits for equipment purchases are offered by both state and local governments. They cover the full range of equipment, from collection vehicles, to balers, sorters, and conveyors. Tax credits may be a flat amount per year or a percentage of the equipment's purchase price.

To promote the use of recycled content, states offer a tax deduction for the purchase of postconsumer materials. This is set at a percentage of the material's purchase price. Where disposal facilities are owned by the state, rebates on tipping fees may be offered to business which can demonstrate set rates of recycling their waste.

State recycling grants are generally made to regional and local governments or not-for-profit organizations. They are not commonly given to the private sector. In many states these grants are administered through the Department of Environmental Protection (or its equivalent). They may be awarded for

assisting in the preparation of regional solid waste management plans;

program implementation such as a rebate of installed capital cost for an MRF;

Table 4. State Recycling Goals, Achievement, Budget Allocations and Programs, 1995–1996[a]

State	Recycling Goal, %	Target Date	Rate, %[b]	1995 Recycled, Million Tons	1995–1996 Budget, $ million	Tons Recovered Per Budget $	Programs[c]
AL	25	None	15	0.81	0.27	3.00	bf
AK	10	1996	6	0.03	0.10	0.30	n/a[d]
AZ	None	None	10	0.45	1.20	0.38	abef
AR	40	2000	25	0.54	0.33	1.63	af
CA	50	2000	25	11.00	38.80	0.28	abcd
CO	50	2000	18	0.54	1.20	0.45	ace
CT	40	2000	23	0.70	2.40	0.29	f
DE	None	None	33	0.40	2.10	0.19	ae
DC	45	1994	15	0.04	0.30	0.13	n/a[d]
FL	30	1995	40	9.72	25.00	0.39	aef
GA	25	1996	n/a[d]	n/a[d]	n/a[d]	n/a[d]	f
HI	50	2000	20	0.40	2.40	0.17	b
ID	25	1995	n/a	n/a	n/a	n/a	n/a[d]
IL	25	1996	27	3.84	12.70	0.30	abcd
IN	50	2001	19	1.10	4.40	0.25	ace
IA	50	2000	28	0.89	0.19	4.66	ab
KS	None	None	8	0.28	1.50	0.19	ae
KY	25	1997	15	0.55	0.15	3.66	aef
LA	25	1992	6	0.29	n/a	n/a	ae
ME	50	1998	33	0.43	0.26	1.64	a
MD	20	1994	27	1.43	0.26	5.49	abcef
MA	46	2000	31	2.19	5.20	0.42	c
MI	25	2005	25	3.38	14.00	0.24	bce
MN	35[e],50[f]	1996	44	2.11	22.00	0.10	abcf
MS	25	1996	12	0.29	0.50	0.58	bef
MO	40	1998	18	1.35	3.80	0.36	bef
MT	25	1996	6	0.04	0.05	0.88	a
NE	50	2002	19	0.32	4.00	0.08	n/a[d]
NV	25	1994	12	0.31	0.23	1.34	a
NH	40	2000	12	0.13	0.225	0.56	bc
NJ	50	1995	42	3.57	1.80	1.98	abc
NM	50	2000	12	0.24	0.23	1.03	a
NY	50	1997	32	8.16	n/a	n/a	bce
NC	40	2001	30	3.00	1.10	2.73	aef
ND	40	2000	22	0.11	0.06	1.83	a
OH	25[e],50[f]	2000	34	8.26	3.50	2.36	bef
OK	None	None	12	0.30	0.28	1.07	a
OR	50	2000	32	1.10	2.50	0.44	abcf
PA	25	1997	17	1.53	36.00	0.04	bcef
RI	70	None	24	0.26	1.20	0.22	n/a[d]
SC	30	1997	16	1.11	5.40	0.21	f
SD	50	2001	30	0.25	n/a	n/a	n/a[d]
TN	25	1995	10	n/a	2.60	n/a	f
TX	40	1994	14	3.09	9.50	0.32	adef
UT	None	None	18	0.53	0.02	26.33	n/a[d]
VT	40	2000	35	0.21	n/a	n/a	bc
VA	25	1997	33	2.85	1.40	2.04	af
WA	50	1995	38	2.69	1.70	1.58	be
WV	30	2010	13	0.26	2.30	0.11	a
WI	None	None	28	1.52	47.50	0.03	abcf
WY	None	None	4	0.02	0.04	0.53	e

[a] Adapted from Refs. 17–19.
[b] For the year 1995.
[c] Ref. 20. a, tax incentive; b, grants; c, loans; d, zones; e, economic development (state agency involved in recycling market development); and f, market development councils.
[d] n/a = Not available.
[e] Rural.
[f] Urban.

- program staffing, eg, recycling coordinator, inspectors;
- performance rewards to municipalities for meeting or surpassing targets;
- household hazardous waste (HHW) collection programs; and
- public education, advertisement.

Loan programs generally target the private sector. They include

- low interest loans to small businesses;
- tax-exempt bonds to all recycling enterprises; and
- delayed payment of state fees and product surcharges (such as tire deposits).

Recycling zones are similar to industrial development zones. The state works with local governments to concentrate recycling businesses and markets in specified areas. These may target economically depressed areas and/or brown fields. A significant secondary benefit is often the provision of jobs. The zones facilitate the concentration of local resources on a single area for a specific purpose. Thus needed infrastructure is constructed with state funds and the programs benefit from the synergy of locating materials suppliers and users in the same area. California and Texas have strong programs for recycling market development zones.

Economic development refers to state programs that create and foster business and employment opportunities through recycling. They are generally broad and long term in design and include recycling market and industry development zones, small business loans, and grants. Rather than the environment or waste management divisions, these programs are housed in the State Department of Commerce, as in New York and Florida.

Market Development Councils and Task Forces are permanent bodies housed in a state agency. One example is the Office of Recycling Market Development (ORMD) in New York State. In addition, the state sponsors or supports such Councils through start-up and ongoing partial funding. The Councils and task forces are made up of waste management and marketing experts from the public, private and academic sectors. Their range of responsibilities covers acting as liaisons between the recycling industry and the state on market development, to fairly autonomous bodies actively creating new market opportunities for recycling within and outside the state.

Two key questions to ask in evaluating state recycling policies and programs are

- Why the particular targets (percent reduction and date) are objectives?
- How effective are the policies and programs at achieving their stated objectives?

Effectiveness is measured by the benefits produced (resource conservation, environmental protection) per unit of cost compared to other alternatives for managing the same materials. States generally do not include such quantitative evaluation of alternatives in their solid waste management plans. However, it is the result of just such analyses that could help states to frame effective recycling policies.

With respect to the question about targets as objectives, a rationale that systematically ties state recycling targets to resource conservation and environmental protection goals is not apparent. Furthermore, total data of federal, state, and local government annual expenditures on recycling are not available. The EPA reports that these expenditures yielded a 22% national average recycling rate in 1993. Unfortunately, there is no quantitative measure for the natural resources conserved or the environmental impacts avoided by this recycling, thus effectiveness of the various recycling policies and programs in achieving their objectives cannot be evaluated impartially. As a result the descriptions presented in the previous sections are limited.

SYSTEMS APPROACH TO RECYCLING

Integrated Solid Waste Management

As proposed in the Agenda for Action (12), the integrated solid waste management (ISWM) hierarchy is a ranked set of options for managing MSW. They were presented as a set of guidelines that could be adapted to local conditions to effect solid waste management so as to conserve resources conservation and to maximize landfill life. Source reduction was given the highest priority whereas combustion and landfilling were ranked equally and lower as disposal options. Figure 6 shows the original hierarchy.

Source reduction seeks to reduce the quantity and toxicity of material entering the waste stream. Quantity reductions target the mass and volume of the waste stream. It includes reformulating products, reducing the amount of packaging used, and the number of units consumed. Detergent serves as a good example. Since the 1970s, detergents have been reformulated to eliminate phosphorous which had significant negative environmental impacts. Detergents were densified and concentrated so that they could be sold in smaller, longer lasting containers. These containers were made from recycled material and also designed to be easier to recycle themselves. Reuse of products is another option for source reduction. The EPA and several community organizations have established extensive information and service networks for product reuse (1).

Recycling includes materials recovery and composting. As proposed in the Agenda for Action, it does not include energy recovery from incineration, as does the definition used in this article. The Agenda for Action proposed a 25% decrease in municipal discards through source reduction and recycling by the end of 1992. The EPA reported a national average recycling rate of 22% at the end of 1993 (3). Increases in the total and per capita amounts of MSW generation make it difficult to assess the impact of source reduction on the waste stream.

Combustion of MSW today largely involves resource recovery or waste-to-energy. It was given equal ranking with landfilling in the original hierarchy. However, on a hierarchy of volume reduction, it would rank just above landfilling because it can reduce the volume of material requiring final disposal by up to 95% with weight reduction of up to 90% (a range of 50–90%) of the original waste (21). There are strong opposing views about the incineration of MSW (22–24). Controversy notwithstanding, combustion, which we also call incineration or WTE, is appropriate for areas of high population density, where land is not available due to NIMBY (not-in-my-backyard) or other concerns.

Plastics (21–44 MJ/kg or 9,000–19,000 Btu/lb) and paper (12–28 MJ/kg or 5,000–12,000 Btu/lb) have high energy

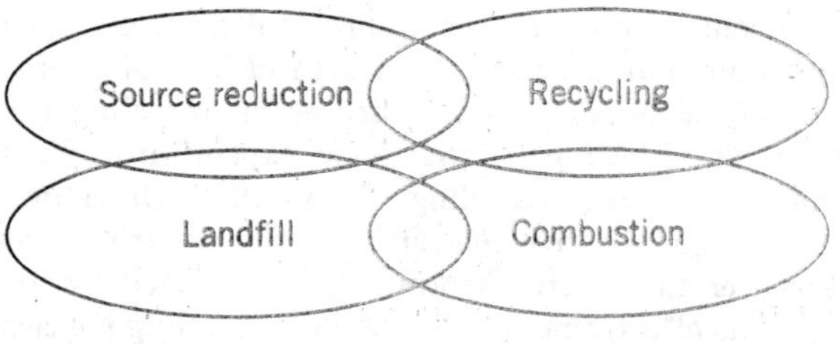

Figure 6. Integrated solid waste management.

contents relative to other components of the MSW stream (25). This makes them highly attractive to WTE facilities. High rates of recovery of these materials by aggressive recycling programs is a likely source of conflict with WTE in an integrated system employing both options. Operationally, this conflict has not materialized and WTE facilities operate successfully in communities with thriving materials recovery programs (26).

Landfilling is at the bottom of the ISWM hierarchy. On a tonnage basis it is the most commonly used MSW management option employed in the United States. In 1993, 62% of the MSW generated nationwide was disposed of in landfills. Large regional landfills in California, Ohio, Pennsylvania, and Virginia provide sufficient disposal capacity at fairly stable prices to allay the fears of a capacity shortage that reigned in the 1980s. Despite this, the nation's commitment to recycling remains high, indicating that the days of the "throwaway society" may indeed be over in the United States.

The ISWM hierarchy was, in part, a response to the "landfill crisis" of the 1980s. In fact if adopted as a sequence of operations, ISWM reduces the quantity of material destined for incineration and ultimate disposal in landfills as illustrated in Figure 7.

Integrated solid waste management incorporates the elements of the hierarchy into a cost effective system that is appropriate for local conditions. Thus, heavy expenditures on recycling are not appropriate for rural areas with widely dispersed, single family dwellings. The challenge of ISWM is to coordinate the different waste management options in a system that minimizes resource consumption and the health and environmental impacts of MSW. Technology options are available to reduce the risks associated with incineration and landfilling to whatever nonzero level proponents and critics are willing to afford. The real challenge lies in the systemwide changes required to make source reduction and recycling affordable and acceptable components of an integrated scheme for managing *materials* rather than simply *wastes*.

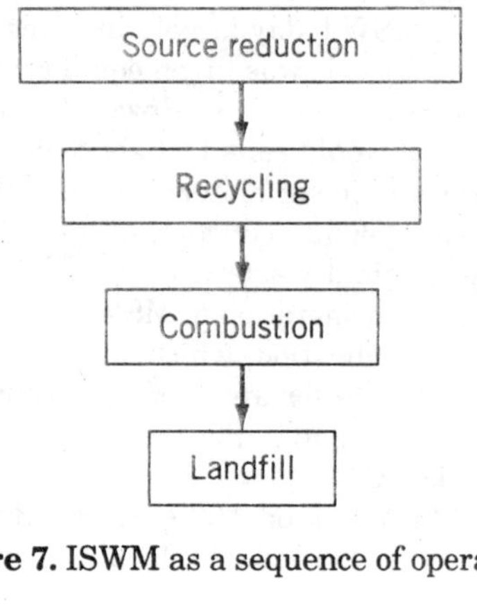

Figure 7. ISWM as a sequence of operations.

Life Cycle Assessment

Life cycle assessment (LCA) (27) is an analytical tool that is used to estimate the resource consumption and environmental impacts of products. LCA is useful for comparing the "greenness" of products or processes by tallying their respective resource requirements and environmental impacts (28). Working through the procedure can also help manufacturers choose materials and processes that are less resource intensive. These would be used to make products that are easier to recycle and have a higher content of recyclable components. If used correctly, LCA enables the policy-maker to assess the impacts of recycling materials at different rates from different products. Even with the most vigorous methodology, small differences in assumptions lead to different results in comparing products and materials. Furthermore, the differences calculated from these comparisons might not be large enough to warrant a confident recommendation of one over the other on the basis of "greenness." Thus LCA is a useful tool to inform policy decisions but may not be appropriate as a stand-alone basis for environmental decisions about products, materials or processes like recycling. If economic and social costs are attached to the resources consumed, the full costs and benefits of recycling may be evaluated. Figure 8 provides a simple illustration of the process. The procedure is a lot more complex than suggested by this example. Details are available (27).

Mass, materials, and energy balances around the system relate the net amounts of material and energy consumed as a product or a set of materials is taken from initial extraction of the raw materials to disposal and recycling. If it is assumed that the flow of mass and energy through the system is constant (steady state), then both Figures 8a and 8b reduce to the following equations:

Mass balance

$$\sum_{j=1}^{5} MI_j - \sum_{j=1}^{5} MO_j = 0$$

Materials balance

$$\sum_{j=1}^{5}\sum_{k=1}^{n} (MI_{jk} + MG_{jk}) - \sum_{j=1}^{5}\sum_{k=1}^{n} (MC_{jk} + MO_{jk}) = 0$$

Energy balance

$$\sum_{j=1}^{5} (EI_j + EG_j) - \sum_{j=1}^{5} (EC_j + EO_j) = 0$$

where j is the suffix for processes 1 through 5 (4 if recycling is excluded); k is the suffix for material species; MG_{jk} is the mass of material species k, generated in process j; MC_{jk} is the mass of material species k, consumed in process j; EG_j is the energy generated in process j; EG_j is the energy consumed in process j; and MI, MO are the same as defined in the notation of Figure 8.

The designer or decision maker can look at each resource stream and/or process to consider changes to their components. This alters the composition of the resulting products which, in turn, changes the economic and environmental impacts of their life cycle. In Figure 8**a**, processes 2 and 3 represent the life

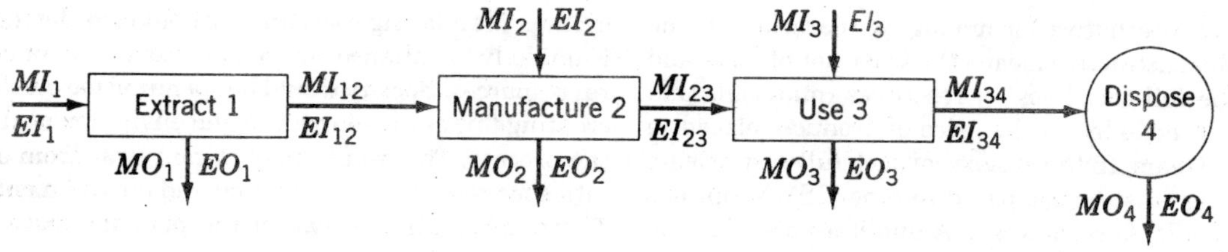

(a)

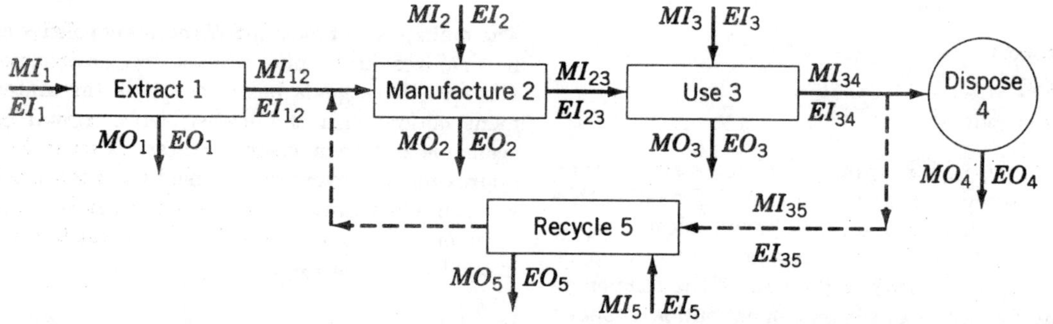

(b)

Figure 8. Use of life cycle assessment to evaluate recycling: life cycle (**a**) without and (**b**) with a recycle process. MI_1, mass input to process 1 (extraction of raw materials); EI_1, energy input to process 1; MO_1, mass output from process 1; EO_1, energy output from process 1; MI_{12}, mass output from process 1 to process 2; and EI_{12}, energy output from process 1 to process 2.

span of the product in a downstream, waste management approach to recycling. In Figure 8**b**, the materials management approach focuses on all processes in the material's life cycle. Only one recycle stream is shown in this simplified example. In reality, recycling streams may go between the output and input streams of any process or processes.

Recycling is feasible when the net material and energy consumption of the life cycle with recycling is less than that of the life cycle without recycling. A simple comparison of the quantities of material and energy saved by recycling could result in misleading conclusions. Instead, these quantities must be weighted by some measure of their quality to make a meaningful comparison. For example, a recycling process which saves 100,000 Btus of energy per ton of product in the form of steam at 100°C is not necessarily more desirable than one which saves 10,000 Btus of energy per ton of product in the form of superheated steam at 200°C. Attaching a price to the materials and energy saved is one way to weight them for quality. Thus, in comparing the life cycle with recycling to that without recycling, either the weighted value of materials and energy saved (such as the dollar value of the resource saved), or only materials and energy of the same quality across life cycles must be compared.

More aggressive criteria may require that all or some subset of the resource flow streams (MI_j, MO_j, EI_j, EO_j) in the life cycle with recycling be smaller than they are with no recycling. The disposal option may include incineration and/or landfilling. The recycle box may include materials and energy recovery as well as composting. Thus, the procedure illustrated in Figure 8 allows us to compare recycling to other ISWM options. When the preferences of society are used to attach prices to the resource flows in these figures, we approach an evaluation of recycling on the basis of full-cost pricing.

Full-Cost Pricing

The best way to inform consumers and producers and to motivate then to act in socially desirable ways is to establish a pricing mechanism for materials and products that reflects their full social cost, including resource depletion and environmental damage. Full-cost pricing (29) of raw materials would lead producers to make more desirable choices of materials and lead them to designs that are easier to reuse and recycle. A major problem with the current system is that product wastes in MSW arrive at the MRF having been manufactured with little or no thought for making them easy to recycle. Full-cost pricing would change the choice of materials and design so that processing at the MRF is an integrated part of the product's design. Figure 8 shows the points at which the designer could evaluate the process stream and make the necessary changes.

Unfortunately, more research is required to determine the full cost of materials. After that is done, it will be necessary to develop a means of implementing the concept. Neither task will be easy, but the alternative is to neglect environmental problems through incomplete policies or to attempt to regulate every decision involving materials and products that eventually has an environmental impact.

The Iron Law of Recycling

The conceptual basis for framing recycling policy is clear. Ideally, recycling should be carried out when it is the most socially attractive alternative for managing materials in the MSW stream. Most attractive means the least net of costs and benefits after the full social cost of resources (materials and energy) have been included in the price of products placed in the MSW waste stream. In the absence of such full-cost pricing methods, recycling must be compared to other ISWM options on the basis of available social costs. A qualitative evaluation of environmental costs may also accompany these comparisons and be used in making the final decision.

The major components of any MSW management option are

collection,
transfer/transport,
processing, and
disposition of output.

In recycling, disposition of output includes the sale of materials from MRFs to recycled materials markets, the sale of electricity from WTE plants to local utilities, or the sale of compost from composting facilities. Table 2 presents a cost comparison of the ISWM options based on the sum of the component costs listed above for each waste management option. It shows that recycling has the highest cost per unit of all the ISWM options. The Iron Law of Recycling (29) provides a rule for acting on these results. The Iron law states "Collection cost plus sorting cost minus the proceeds from selling the material should be less than the collection and disposal cost of MSW (29)." A simple example shows how the Iron law works.

It is fair to assume that collecting recyclables are $75/t more expensive than collecting MSW. The cost of sorting recyclables is about $150/t and landfill tipping fees are about $35/t (data for the City of Pittsburgh, 1993) (29). To satisfy the Iron Law of Recycling, sorted recyclables must sell for at least $190/t. More quotations on the cost of materials recovery are available (30). Note that we are not considering the environmental benefits of recycling. These could change the calculus. In this respect, the Iron law is not quite fair because recycling was initiated to save resources and improve environmental quality. Including the value of these benefits in the social cost of materials and products may overcome this shortcoming of the Iron law. However, recycling has economic and environmental costs of its own. Additional trucks, fuel and sorting facilities are needed. Every truck mile added for recycling collections and transportation adds carcinogenic diesel particles, carbon monoxide, organic compounds, oxides of nitrogen, and rubber particles to the environment. Collection in urban areas also increases traffic congestion and noise. These also carry a social cost. Thus the activities associated with recycling increase resource use and environmental discharges. They may, in fact, increase the cost of recycling.

In 1993, only aluminum, which sold at $750/t, satisfied the Iron law. Plastic sold at $100–$130/t, steel and bimetal cans at $80/t, clear glass at $50/t, and newsprint at $30/t. Satisfying the Iron law is difficult: if only metals were collected, sorting costs would fall and collection costs would rise (because less material would be collected at each house). If it is assumed that recycling metals satisfies the Iron law, the goal of 25% recycling of MSW would not be met.

When combined with the full-cost pricing of materials (to include the social valuation of resource depletion and the environmental impacts of their use), the Iron law gives a solid basis for making an informed economic decision about recycling. Mandating recycling without considering the Iron law is unlikely to enhance environmental quality or conserve natural resources. Society would be spending too much on recycling, creating further resource use and environmental discharges in the process. This would take money away from other opportunities for resource conservation and environmental protection. These, after all, are two of the primary motivations for recycling policy.

POLICY OPTIONS FOR MSW RECYCLING

The main goals of most MSW recycling policies are to conserve natural resources and protect the environment. Recycling achieves these goals best when it is the materials management option with the lowest net marginal cost when the economic and social costs of materials are included. As a first approximation, every dollar spent leads to a fixed level of environmental discharges. With this in mind, let us now review the common policy options employed to promote recycling of MSW in the United States.

Mandates

Recycling mandates are laws requiring the public's participation in some form of recycling. They may include designation of the types and quantity of materials to be recycled, the manner in which materials are to be prepared and collected for recycling, and the methods by which recycling program costs are to be paid. Mandated recycling is the law in at least 43 states. Local ordinances also add to the operational aspects of state recycling mandates. Table 5 presents some of the pros and cons of mandated recycling. The list is by no means complete and many would add and subtract points on either side of the table.

Germany's Green Dot program is an example of a nationally mandated recycling system. Though not a U.S program, an examination of the German program illustrates some strengths and weaknesses of a mandated system noted in Table 5. Germany's Green Dot Program refers to the nation's Packaging Regulation passed in April 1991. The regulation mandated recycling in response to a crunch in disposal capacity caused by earlier restrictions on new landfills and incinerators. It

Table 5. Some Advantages and Disadvantages of Mandated Recycling

Advantages	Disadvantages
Compliance enforceable	Noncompliance is costly; enforcement is costly
Recovered quantities relatively predictable	Markets unpredictable
Goals and schedules possible	Quantities not tied to markets
Politically popular	Public does not pay full share of cost directly
Permits banning troublesome waste from MSW stream	Often subsidized, eg, by surcharge on landfill fees
	Recovery rates may not be the level of lowest marginal cost
	Inattention to full costs can result in greater resource consumption and negative environmental impacts than not recycling

set target percentage rates and dates by which industry must collect and recycle different types of packaging. Packaging includes materials made from glass, metal, paper, plastics, or wood. Packaging products range from bottles, cans, cartons, cups, corrugated cardboard boxes, plastic bags and trays, and shrink wrap to wooden pallets. Companies selling packaged products in Germany receive a license from the Duales System Deutschland (DSD) to do so. These companies pay a royalty to DSD for a permit to place a green dot on their products. The dot indicates that the company's product has been licensed under the Packaging Regulation. This dot gave the program its name. Waste collectors then charge a fee to DSD for every package collected which carries the green dot. Companies which do not pay for the green dot permit must arrange to accept packages returned from their products.

The law bans WTE for recycling the returned packaging. Consumers are permitted to return the product packaging to the store from which it was purchased. The manufacturers then bears the cost of collecting and recycling the material. The cost for managing the roughly 4.1 million metric tons of packaging per year were estimated at $5 billion to start up and $1.2 billion for annual operating expenses (31). Industry picked up the initial costs for the program but likely passed these increased costs on to consumers in the form of higher prices (see MUNICIPAL WASTE-TO-ENERGY TECHNOLOGY).

The law resulted in very high recovery rates for recyclables. These overwhelmed industry's capacity to utilize the materials. A glut of recyclables, particularly plastic and paper was created. These materials soon began to appear on other European materials markets, as German industries dumped them to comply with the law. This depressed several recycling markets across Europe, thereby hampering recycling efforts in communities many borders removed from Berlin or Frankfurt. To date it is unclear what structural changes can be made to bring the supply of recyclables from consumer returns into alignment with true market demand from industry. Until that is done, it is likely that the glut of materials will remain, and economic dumping will continue to hurt recycling in places with less stringent regulations than Germany's which permit access to German recyclables.

In reviewing the Green Dot program based on the points in Table 5, the following may be observed:

Advantages:

- Compliance was enforceable on industry. The quantity of packaging used is thus likely to decrease. Was the comprehensive Green Dot law the best way to effect this change?
- The quantities of material likely to be recovered were predictable. Budgets and capacity requirements could therefore be estimated proactively.
- Goals and schedules were included in the law. This could further assist long range infrastructure planning.
- The public's response to the law was strong and positive.
- Incineration was banned as an option. This put further pressure on industry to recycle (or export).

Disadvantages:

- Information on administrative costs and the rate of noncompliance were not available. However, annual operating cost estimates were $1.2 billion for 4.1 million metric tons of material. This amounts to roughly $293/t of recyclables. It would be interesting to compare this to the total collection and disposal costs for the same materials via landfilling or incineration in Germany. If they are lower, then the Green Dot program violates the Iron Law of Recycling.
- Provisions to stimulate market demand for recyclables and stabilize prices were inadequate.
- Swamped manufacturers depressed prices as they dumped materials on the market.
- The public paid indirectly via price pass-alongs by manufacturers. There was no direct cost to the public as individuals for the benefits received from this legislation. They simply "recycled" as much packaging as they cared to return.
- Manufacturers paid royalties for the green dot license or for removal of their products. Also, a deposit system on certain packages was used to pay part of the collection and processing costs. These fees and royalties subsidized the revenues from recyclables which could not alone cover the cost of the program.
- The costs of the Green Dot program are borne primarily by industry. Thus it would be difficult to claim that Germany society as a whole is equating the marginal cost to the marginal benefit from recycling. In fact, the benefits accruing to the public are essentially realized without direct cost to them. Out of a sense of environmental goodwill, they are likely to recycle at levels above that which would be dictated if they faced rising marginal costs. Thus the program is not economically efficient and the optimal allocation of resources does not occur.

This review skirts finer issues as whether Green Dot and other manufacturer responsibility policies are truly "polluter pay" systems. Instead, the focus has been on the observable results of the legislation. This focus shows that the German Green Dot program has most of the key advantages and disadvantages outlined in Table 5 for a mandated recycling system. The disadvantages outweigh the advantages of the system as presently implemented. To the extent that these are truly generic to mandated systems, the lessons from the Green Dot program bear close learning. This is particularly true for those considering similar legislation in the United States.

Market-Based Approaches

Market-based approaches use economic incentives or fees to affect recycling behavior. In some cases, the economic policies target product manufacturers (supply side). In others, the policies target consumers (demand side). Full-cost pricing is an economic policy that affects all users of material and products, including manufacturers and consumers.

Unit Pricing. The principle of unit pricing is to charge each user for the amount of MSW generated. The charge is based on the weight or volume of material set out for collection.

The unit pricing program is implemented by requiring customers to purchase a specially marked bag in which to put their MSW at the curb. The user fee is reflected in the price of the bag. Alternatively, special containers are provided by the service provider which the customers must use to put out their MSW. Each housing unit then receives a monthly bill for its

MSW service. If more trash is put out than can fit into the container provided, the customer pays a higher rate on the additional amount.

Theoretically, the unit price, is set relative to the marginal cost of the system's waste management services. However, factors, such as household income or the type of waste, affect the actual price. For example, if the municipality's goal is to increase recycling, a higher unit price is charged for regular MSW than for recyclables.

Figure 9 shows how unit pricing affects a household's generation of MSW and the quantity it recycles. The figure assumes that there is no illegal dumping. Thus all MSW generated is set out for regular pickup or for recycling. Under the flat fee system, the marginal cost of MSW pickup to the household is zero. This is represented by the cost curve MC_1 which lies along the horizontal axis. The household's demand for MSW pickup service is represented by the demand curve D. The intersection of D and MC_1 gives the household's MSW generation level X_1 without unit pricing.

The household's cost curve under unit pricing is represented by MC_2. The intersection of D and MC_2 shows that unit pricing has reduced the household's MSW generation to level X_2, at a price P_2. This reduction in MSW generation (and pickup) is accomplished if the household avoids the purchase of goods with associated waste, such as excess packaging. Additionally, the household recycles materials that would otherwise go into the MSW stream at zero collection price.

Thus, where unit pricing is used in conjunction with a lower or no-cost recycling program, there are three primary expected benefits:

- Customers will reduce the amount of nonrecyclable MSW they set out.
- Customers will increase the amount of recyclables they put out for collection.
- The program will reduce system costs because of lower generation rates, or revenues from the program will cover the costs of MSWM.

In its 1990 study of unit pricing, the EPA reviewed programs in 17 communities across the nation. It made a detailed study of three of these communities. The relevant results are shown in Table 6.

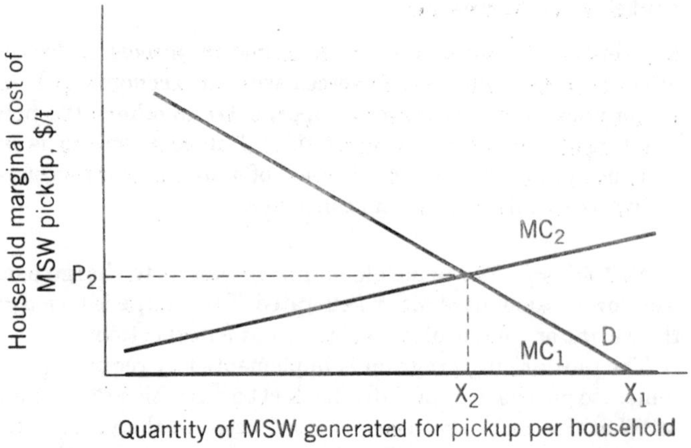

Figure 9. Impact of unit pricing on household MSW generation.

In 1993, R.W. Beck and Associates conducted a survey of 80 cities and counties across the nation to determine their use of private MSWM services and the way they financed their system. Table 7 presents the key relevant results reported in BioCycle (33).

In the survey of unit pricing programs in Portland, Oregon, Seattle, Washington, Tacoma, Washington, Wilkes–Barre, Pennsylvania, Bothell, Washington, and Minneapolis, Minnesota, similar results are reported (33). It was found that unit pricing was associated with a decrease in the amount of MSW set out for disposal and an increase in the rate of residential recycling. Like the EPA study, the effect on system costs or revenues could not be isolated. Also, as with the EPA results, a functional relationship between characteristics of the unit pricing program and the reported changes in consumer behavior was not established.

In a study of Charlottesville, Virginia, similar results are also reported. Under unit pricing, households reduced their weight of garbage by 14%, reduced the volume of garbage by 37%, and increased the weight of recyclable materials by 15.7%. It is estimated that illegal dumping accounts for 32% of the total reduction in weight of garbage presented for disposal at the curb (34).

There are many variables that determine how effective a unit pricing program is in reducing the amount of waste to be collected and increasing the rate of recycling. The key variables identified in the EPA study are summarized in Table 8.

With the loss of flow control, more MSWM administrators are likely to take a look at unit pricing schemes to reduce the tonnage of waste they must dispose of, lower their system costs, and increase their amount of recyclables. This is particularly true in those jurisdictions where state grants are available for recycling programs but not for MSW disposal. The results presented here suggest that reduced MSW generation and increased materials recovery may occur. However, the verdict is still out on any expected reduction in system costs.

Advance Disposal Fees. Advance disposal fees (ADFs) have been around for some time in the form of surcharges on items that are difficult to dispose of, such as automobile tires and lead-acid batteries. On October 1, 1993, they pioneered in Florida as a major market instrument to increase recycling from the supply side. The status of the program is now uncertain because it was not reauthorized by the legislature for 1996. Florida's ADF program departed from the common flat fee on materials with special disposal problems. It sought to promote recycling by targeting products manufacturers (and distributors). Such a program has significant implications for MSW recycling if widely adopted by other states. Thus the Florida system is described in greater detail here.

In the Florida scheme, the ADF targets containers made from a certain material type. The material types are steel, aluminum, glass, plastic and plastic coated paper. Containers made from a target material with less than 50% recycled content (25% for plastic and 35% for glass) statewide are assessed a one penny per container fee (36). Containers with greater than 50% recycled content of the target material Statewide are exempt from the one penny fee (two cents as of January 1, 1995). Recycled material content does not include the virgin scrap component generated during manufacture of the container.

The system provides an incentive for container manufacturers to increase the recycled content of their products.

Table 6. Effect of Unit Pricing on MSW Generation[a]

	Perkasie, Pa.[b]	Ilion, N.Y.[b]	Seattle, Wash.[c]
Year	1987–1988	1988–1989	1986–1987
Change in tons of regular MSW, %	−41	−51	−1.1
Change in tons of recyclables, %	193	132	6.7
Change in tons total MSW collected, %	−18	−44	0.4
System cost relative to flat rate	Net loss of revenue	NA	NA

[a] Ref. 32.
[b] New program; predominantly single family residences; curbside recycling.
[c] Existing program; mixed single and multifamily units; no curbside recycling at time of study.

Table 7. Prevalence of Unit Pricing in 80 Selected MSWM Systems[a]

City Use	Percentage
Public MSW collection, %	68
Private MSW collection, %	32
Property tax or other indirect charges, %	35
Direct disposal fees, %	65
Unit pricing for households, %	16

[a] Ref. 33.

Table 8. Variables Influencing the Effectiveness of Unit Pricing Programs[a]

Variable	Effect
Voluntary program	Low participation rate
Mandatory program	High participation rate
Curbside recycling also offered	Increased recycling and reduced disposal
Effective prohibitions on illegal dumping	Increased recycling and reduced disposal
Weight-based fee	Reduced weight of MSW disposed of
Volume-based fee	Reduced volume disposed of but weight may increase due to compaction in the container
Single family units	High compliance rate
Multifamily and commercial units	Low compliance rate

[a] Adapted from Refs. 33 and 35.

Containers which meet the 50% criterion can be sold for less than those that do not. Additionally, materials with a sustained recycling rate statewide that is greater than the recycled content goals are also exempt from the one penny fee. Thus, aluminum and steel cans have been exempt from the fee since the program's inception. Their respective industries have documented sustained recycling rates in excess of 50% for these materials in Florida. A take-back provision allows individual companies to claim an exemption if they cause target materials to be removed from the Florida waste stream and recycled in other products at a rate equal to the content goals. Thus glass and plastic from beverage containers can be recycled into "glassphalt" and plastic furniture. These latter applications have a longer useful product life than beverage containers and provide the benefit of delaying the material's entry into the waste stream.

Companies responded well to the program. Within six months of its inception there were 173 petitions for exemption from 105 companies. This resulted in the exemption of over 60% of candidate containers. With a sustained recycled content of well above 50%, beer bottles account for a significant proportion of the glass container exemptions. This is an important source since there were 2.3 billion glass containers sold in Florida in 1995 of which over 850 million were beer bottles (36).

The ADF program raised roughly $24 million and $30 million for Florida in the 1994/95 and 1995/96 fiscal years, respectively. These funds were designated to finance state environmental programs, including grants to counties for recycling programs. These revenues were expected to decline as companies responded by increasing the recycled content of their products. It is not clear whether the avoided cost of disposal of these recycled materials will compensate for the environmental revenue lost from declining ADF fees. That is the expectation if the fee is set at an economically efficient level. Companies would have initially absorbed the cost of recycling to provide the benefit of avoided disposal of materials. Eventually, these costs get passed on to consumers as higher prices, and society pays the cost for the environmental benefit of recycling. Implementing such an ideal is likely to be quite challenging. In a real system, inequities in the burden of costs and benefits are likely and approaching the ideal may require an iterative process of policy fixes.

Florida's ADF program uses a market incentive to achieve the state's recycling goals. It does so by stimulating manufacturers' demand for recycled material while allowing companies to choose how they will meet the standards of the program. The program has provided a substantial revenue stream for environmental initiatives and solid waste management in the state. This may make it attractive to other states wishing to adapt the program to their own goals and target materials. ADF programs are already in existence in a different form for tires, motor oil and batteries in at least 22 states (37). They are a potentially powerful instrument for increasing the recycling of materials (and thus reducing disposal) from the supply side. If widely adopted by the States, ADFs could affect the composition and quantity of the MSW stream. They could also provide a source of funding for MSW management.

Flow Control

The basic idea behind flow control is to direct the flow of MSW generated in a given region to a specific facility or facilities. For example, a municipality may wish to direct all the MSW generated within its jurisdiction to a particular materials recovery facility (MRF) for processing, all the residue from the MRF to a designated WTE plant for combustion and all the ash from the WTE plant to a specified landfill for disposal. The rationale behind flow control is to use the revenues obtained from the guaranteed waste stream to repay bonds and other expenses for managing the region's municipal solid waste. Municipalities may enter into agreements with private companies to construct and operate MSWM facilities financed by municipally guaranteed, tax exempt bonds. The assurance of a minimum tonnage of MSW to be processed by the facility at a given price is usually a requirement of the financing agreement. Failure on the part of the municipality to deliver these minimum gate receipts would result in compensation payments for the revenue lost from the deficient amount of MSW. This latter arrangement is referred to in the industry as "put or pay".

There are three types of flow control:

- legislative flow control
- contractual flow control
- economic flow control

Legislative Flow Control. Under enabling legislation from the state, the City Council or County Board of Governors passes an ordinance directing the flow of waste to specified waste management facilities. For example, in August 1989, the town of Clarkstown, New York, enacted the "Solid Waste Transportation and Disposal Ordinance" in response to a consent decree with the State Department of Environmental Protection over previously poor landfill practices in the town. The town agreed to close its old landfill and build a transfer station on the same site. The transfer station would accept bulk solid waste and separate its recyclable component. The town guaranteed minimum waste flows of 120,000 tons per year for five years to the private contractor who built the facility at a cost of $1.4 million. At the end of five years the town could purchase the facility for $1. The town would pay for any deficit in tonnage of gate receipts below the guaranteed minimum. The tipping fee was $81/t. The town's solid waste ordinance required all waste haulers operating in Clarkstown to utilize the transfer station for disposing of their MSW. Thus Clarkstown utilized legislative flow control to comply with its consent agreement with the State and to safely manage it MSW. This was to become a defining case for flow control and for MSW management as a whole.

Contractual Flow Control. This involves granting franchises, licenses, or contracts to private service providers by the municipality or county to manage its MSW. As a condition of the franchise, license, or contract, the municipality stipulates the way its waste is to be managed. Later we will show that this system avoids some of the legal problems of legislative flow control.

Economic Flow Control. Here a municipality or county could attract waste to its facility by lowering tipping fees or by increasing the fees for other types of services, such as the collection of recyclables. The lower tipping fees would have to be subsidized by funds from other sources, such as advanced disposal fees (on items like tires) or surcharges on MSW collection services. Economic flow control appears to be free of the constitutional challenges faced by legislative flow control.

Legal Status of Flow Control. During the 1980s and early 1990s, flow control was widely used by municipal and county governments to guarantee flows to newly built MRFs and WTE plants. Many of these were financed by bonds issued and rated on the basis of flow-control revenue streams. Unfortunately, the tipping fees at flow-control-protected facilities were often higher than fees at other facilities within the region. Additionally, owners of facilities not covered by flow control agreements felt that they were denied an opportunity to compete fairly for the waste that was preempted by flow control. One such private facility owner was C & A Carbone, Inc. in the town of Clarkstown, New York (38).

Carbone operated a waste sorting and shipping facility in the town of Clarkstown that was similar to the transfer station newly constructed by the town. Though Carbone was permitted to continue the operation, the town's ordinance required Carbone to send sorted nonrecyclables to the its transfer station. Carbone found that this was a redundant transaction for his already sorted waste. Furthermore, the $89 per metric ton tipping fee at the Town's transfer station was higher than Carbone's combined cost for shipping and disposal at a facility outside of Clarkstown. Carbone violated the town's ordinance by sending sorted waste to the out-of-town facility for disposal. The town was initially successful in forcing Carbone's compliance with the ordinance through the State Courts. However, on May 16, 1994, the United States Supreme Court, in a 6–3 decision, found that the town's flow control ordinance violates the United States Constitution's Commerce Clause. They reversed the State Court's decision. In coming to its decision, the Supreme Court drew on precedent in the case of Philadelphia vs New Jersey (39). In that case it deemed that solid waste is "an article of commerce". Thus solid waste is subject to the Constitutional provision that interstate commerce could be regulated only by the Federal Government. The Supreme Court found that the town of Clarkstown's ordinance:

1. regulated interstate commerce because its economic effects were interstate in reach;
2. discriminated against interstate commerce by allowing only the favored operator to process waste that is within the town's limits;
3. the town could address its MSWM concerns through nondiscriminatory alternatives.

This ruling, in favor of Carbone, was felt throughout the waste management industry. Though the exact number of systems and facilities operating under flow control has not been enumerated (40) (at least 26 States had expressly granted municipalities the power to implement flow control), the ruling essentially threatened all that functioned under legislated systems similar to that of Clarkstown. It put their municipal and county bonds in jeopardy of downgrading or forfeiture, should these public agencies now be unable to guarantee the minimum flows promised under flow control. Those municipalities and counties with put-or-pay contracts would now face severe financial consequences from any lost flows to their designated facilities. The case of New York State, in which the landmark case originated, illustrates the impacts of the Carbone decision.

"As of June 1, 1995, there were 37 New York municipalities (districts, towns, counties authorities) authorized by the State

Table 9. Impact of Flow Control on Selected New York State Solid Waste Planning Units[a]

Planning Unit	Municipal Revenue Bonds in Flow-Control Facilities[b]	Moody's Bond Rating, 5/95[c]
Town of Hempstead, Nassau County	$270MM	Unfavorable without flow control
Onondaga Co. Resource Recovery	$193MM	Unfavorable
Town of Huntington, Suffolk County	$170MM	Under review
Westchester County	$112MM	NA
Warren/Washington Counties	$87MM	NA
Town of Babylon, Suffolk County	$85MM	Under review
Town of N. Hempstead, Nassau Co.	$66MM	NA
Oneida–Herkimer County SWMA	$50MM	Unfavorable without flow control
Dutchess County	$38MM	NA
Town of Islip, Suffolk County	$33MM	NA
Ulster County	$31MM	Stable
Montgomery, Otsego, Schorarie SWMA	$29MM	NA
DANC (Jefferson and Lewis Counties)	$28MM	Unfavorable without flow control
Franklin County	$20MM	NA
Monroe County	$12MM	NA
St. Lawrence County	$8MM	Downgraded

[a] Ref. 43.
[b] MM = million.
[c] NA = not available.

legislature to enact flow control legislation. Almost 15 million of the States 18 million citizens reside in those municipalities, most of which have enacted such legislation (41)." The State, identified 16 waste planning units (regions) with a combined total of $1.2 billion (1995 dollars) in debt financing, primarily in bonds for WTE facilities. Table 9 summarizes the impact of the Carbone decision on these municipalities.

Thus, using New York State as an example, the loss of flow control can disrupt the plans of many municipal and regional solid waste management systems. Many could be forced to suspend all or part of their MSWM services to meet their put-or-pay obligations and to meet their municipal bond obligations.

In concurring with the majority decision in Carbone, Justice Sandra Day O'Connor wrote, "It is within Congress' power to authorize local imposition of flow control. Should Congress revisit the area, and enact legislation providing a clear indication that it intends the states and localities to implement flow control, we will, of course, defer to that legislative judgment (42)." This opened the door for a Federal fix of the potential mess created by the Carbone decision. Thus municipalities turned to the federal government for relief.

On May 16, 1995, Senate Bill S.534, "Interstate Transportation of Municipal Solid Waste Act of 1995," was approved by a 94–6 vote. This bill would grandfather flow control ordinances in place before the Carbone decision (42). However, the House version of the bill was defeated by a vote of 149–272 on January 31, 1996 (44). This effectively put the issue to rest for the remainder of the 104th Congress. Flow control failed to get Congressional relief not on its own weakness but because both the House and Senate bills coupled it to the more troublesome issue of state control over the interstate flow of MSW. Flow control failed to win approval because many Representatives voted against the interstate portion of the bill, rather than against flow control per se. Separate bills for flow control and interstate transportation of MSW will have to be developed for the 105th Congress if flow control is to pass the Congressional test on its own merits.

Flow control is not without its detractors. In a study sponsored by Browning Ferris Industries (BFI), released in May 1995 by National Economic Research Associates, flow control adds approximately 40% to the national average disposal charge (45). This added fuel to an earlier report by EPA (based on 1992 data) that there was "no evidence that flow controls were essential either for the development of new solid waste capacity or for the long term achievement of state and local goals for waste reduction, reuse and recycling (46). In addition, the Environmental Industry Association and its National Solid Wastes Management Association claim that flow control marks an attempt by towns and cities to create solid waste monopolies. Congress must weigh these findings and opinions against the reported $20 billion in municipal revenue bonds potentially at risk when it considers municipal flow control in 1997. Given the size of potential losses, it is likely that local flow control will be reprieved in some form under federal law.

For the present, municipalities and counties have three main options to legally address the flow control issue. First is the use of contractual flow control. Here the public body can enter into agreements for the collection and delivery of MSW to facilities with which it has a contract. Under this scenario, the public body is acting as a "market participant" rather than as a "market regulator". This passes the constitutional test according to the language of the Carbone decision. The second option is for the municipality to collect the MSW itself and deliver it to any facility it chooses for disposal. This may be achieved by municipal ownership of collection equipment or by the use of municipal personnel to pick up MSW on vehicles owned and operated by a private contractor. The third option is for municipalities to completely privatize their MSWM services. Under this system the entire contract for MSWM services would be let for competitive bid and paid for from municipal funds. None of these options effectively addresses municipal obligations under put-or-pay contracts or gate-receipt-guaranteed revenue bonds. However, in the absence of Congressional relief, these are the only legal options available.

Summary

Federal and state recycling policies and programs largely neglect a strict economic rationale in pursuing their objectives. Targets, mandates, and subsidies may actually misallocate resources to recycling and result in increased resource consumption and environmental damage if full costs are not taken into account.

Unit pricing, advance disposal fees, and even flow control offer only partial answers to the question of how to make recycling work. Full-cost pricing promises significant advances but the methodology and implementation strategies are still under development.

In essence, the challenge is to make recycling an economically efficient, cost-effective policy given the available means.

CONCLUSION

As a society, recycling continues because it is believed to help to conserve natural resources and protect the environment. The logic is simple; reusing the materials that would otherwise have been thrown away means less virgin materials and possibly less energy are used to manufacture new products. Additionally, generating less waste and diverting some of it from incinerators and landfills reduces the environmental impacts of these facilities.

Yet recycling itself requires the consumption of resources and creates environmental impacts. It costs time, money and energy to separate, collect, transport, sort, and process recovered materials. Additional collection trucks for recyclables generate pollution, consume more fuel, and create more traffic. Recycling processes consume water, electricity, and heat. The actual recycling processes employ chemicals like bleaching agents or acids, which may be discharged as wastes from processing plants. Thus we must approach recycling with a bit more than naive enthusiasm. The full effect of recycling and its role in the overall scheme of waste management should be recognized.

Two comparisons can be made in evaluating the benefits obtained from recycling MSW. First, compare the financial cost of recycling to the avoided cost of disposing of the same products and materials. Thus the collection, transportation, sorting, and other processing costs of recycling are added and the revenues from the sale of materials are subtracted. Compare this difference to the sum of the collection, transportation, and disposal costs of the same waste components if they were part of the regular MSW stream. This is the avoided cost of disposal. According to the Iron Law of Recycling, the sum of the economic costs for recycling must be less than that for disposal. If it is not, then recycling fails the economic cost test. It means that there must be other components of cost and benefits that we must consider in our evaluation. This leads to the second comparison.

In the second comparison, a systems approach to evaluating recycling is taken. Society's valuation of the environmental costs and benefits associated with the various products, materials, and processes is included. Thus a value is attached to the depletion of natural resources associated with the extraction of raw materials. Next manufacturers are charged for the environmental impacts associated with the processing of these raw materials into feedstock and finally into products for sale to consumers. Manufacturers will then include these charges in the price of their products. The process is continued by putting a tax on fuels for the pollution they cause for transporting products to markets. Distributors would probably add these taxes to their price for the product. Finally a charge is added to each product to cover the value of the cost of collecting, transporting, and disposing of it when it enters the MSW stream. Environmental costs from each of these operations may also be included in the charges.

In this comparison, at each step along their life span, users of materials and products pay their full economic and social price. Users would then seek the least costly alternative materials, products, and processes. As the prices of virgin materials increase under this scheme, manufacturers would be motivated to reduce the costs of recycling as they search for less costly sources of raw materials.

Compare the cost of recycling against that of disposal with the social costs included. The level of recycling at which these costs are lower than those for disposal represents the optimum level of recycling that society is willing to pay for. It will probably not be 100% of the MSW stream. Thus disposal will still be required. When full-cost pricing is considered, the optimal level of recycling may be less than the targets set by many state programs. This is no cause for alarm. It simply says that the targets set by the states have costs which exceed the benefits that its citizens are willing to pay for when full-cost pricing prevails. Figure 10 illustrates this point.

In Figure 10, MC_1 represents society's marginal cost curve without full cost pricing and MC_2, MC_3 with full-cost pricing. MB is its marginal benefit from recycling. If the inclusion of full-cost prices results in an overall increase in the net cost of recycling, all other things being equal, society would reduce its level of recycling from Q_1 to Q_2 percent of total MSW. On the other hand, if full-cost pricing increases the value of recovered materials and energy to such an extent that there is an overall reduction in the net cost of recycling, all other things being equal, society would increase its level of recycling from Q_1 to Q_3. This assumes that the total quantity of MSW generated remains unchanged. In fact, it is likely that higher materials costs under full-cost pricing would lead to a reduction

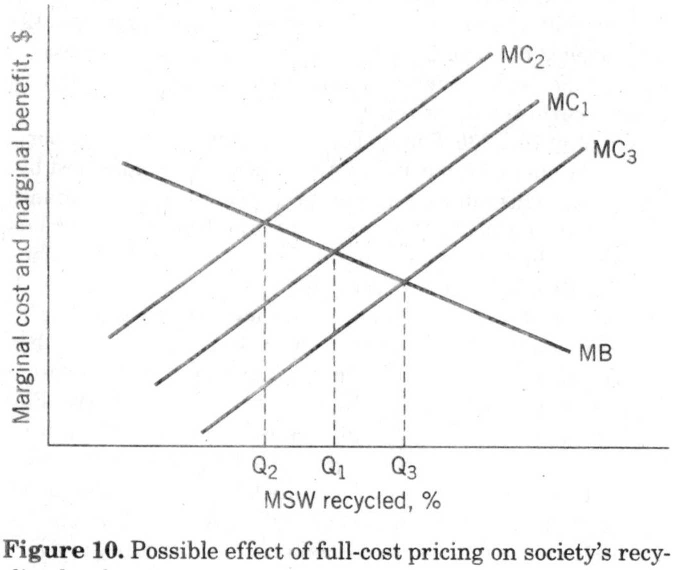

Figure 10. Possible effect of full-cost pricing on society's recycling level.

in the amount of packaging. This would reduce the amount of MSW generated. Thus the % recycled would be calculated on the basis of a smaller MSW stream. The result could be a higher calculated percentage of MSW recycled even though the actual tons of material recycled (and MSW generated) had decreased. It is important to recognize that full cost pricing would affect both the quantity of MSW generated and the rate of recycling. Although the quantity of MSW generated is likely to decrease, it is difficult to predict the way in which the rate of recycling would change. Actually, the effects of full-cost pricing are likely to be different for each material or process affected. The final overall rate of recycling will be the net of these differing results.

It appears that full-cost pricing can help us on the way to economically efficient recycling policies. Unfortunately, the methodology is not yet available. Until it is, it is preferred to stick as closely as possible to the full-cost pricing approach. A 10-point approach to improving the current MSW recycling system follows:

1. Eliminate subsidies on virgin materials where possible. This could make recyclables more competitive. Subsidies on virgin material can become a complex issue. Some may be tied to agriculture or trade policies. Their elimination to facilitate recycling may have significant public policy implications in these other areas. As a result this due consideration should be given in acting on this recommendation.
2. Keep source reduction at the top of the ISWM hierarchy. Green design is one option that offers promise. It seeks to reduce the resource consumption and toxicity of the materials going into products. It also focuses on designing materials to increase the ease of recycling them and the amount of materials recovered when they are recycled. Manufacturer take-back policies would encourage producers to make such product design changes. However, such policies should be carefully timed so as not to exceed the infrastructure and capacity of manufacturers. These options have been discussed further (8).
3. Make the fullest accounting of economic and social costs possible. Use these to compare the costs of recycling, composting, incineration, and landfilling. Apply the Iron Law of Recycling. Where environmental and other social costs (like aesthetic impacts from odorous facilities or increased traffic congestion) are not quantified, a qualitative rating scheme may be developed. Use this to reevaluate the results obtained after applying the Iron law. For each material, choose the level of recycling that minimizes marginal cost, equates it to marginal benefit, and is attainable given the market demand and recycling infrastructure for the material.
4. Approach a National Recycling Law with extreme caution. The German Green Dot program provides an example of some of the potential pitfalls. In addition, local preferences and conditions have a lot to do with the benefits people perceive from recycling. It is difficult to accommodate these local factors in blanket national legislation. However, there are some issues that could benefit from national attention. These include a quality standard for recyclables, a consistent method for calculating recycling rates, subsidies on virgin materials, and support for developing full-cost pricing methodology. However, federal legislation is not necessary to effect these changes. For example, the recycling industry and existing standards organizations can develop standards for recyclable materials. The EPA can prepare and make a convincing case to the states for a robust methodology for calculating recycling rates. Virgin material subsidy programs may be phased out where appropriate.
5. Use market mechanisms, such as user fees, where possible. These personalize the cost of waste generation and disposal and encourage recycling at a level where benefits are equated to costs.
6. Avoid the targets and goals often associated with mandates. These are not linked to markets and may lead to overspending on recycling at the expense of other opportunities for resource conservation and environmental protection.
7. Keep recycling as a part of a truly integrated solid waste management system. Let it compete equally with disposal on the basis of costs versus benefits. Recycle in those instances and at those amounts where it is the least cost (net of cost minus benefit) option.
8. In this integrated system, address the risks associated with different waste management technologies. Reduce them to operationally acceptable levels. Give society a chance to make informed decisions about the alternatives against the costs required for its acceptable levels of risk. Bear in mind that zero risk is not likely to be attainable with any technology.
9. Implement the system at the local level but make the necessary structural and policy changes at the regional, state and federal levels so that the system is effective. Be prepared to make iterative improvements to the system as we learn how to recycle from a systems perspective.
10. Keep in mind the place that MSW occupies relative to other forms of waste and issues that compete for spending on natural resource conservation and environmental protection. Spend what is appropriate on MSW in the light of other pressing problems. Use what is spent on MSW most effectively to conserve resources and protect the environment. Spend on recycling only what it takes to get its maximum contribution to resource conservation and environmental protection within an integrated solid waste management system.

Municipal Recycling in a Flexible ISWM System

One of the constraints under the current system of recycling is that materials are supposed to go to market whether or not their demand is strong. This is a major weakness because, with the exception of aluminum, the markets for all other commonly recycled components of MSW are highly unstable (47). Recycling is not truly integrated into a system of waste management alternatives when the amount of material to be recycled is rigidly set without consideration for prevailing market conditions or the economics of the other alternatives. Figure 11 presents a flexible integrated solid waste management system

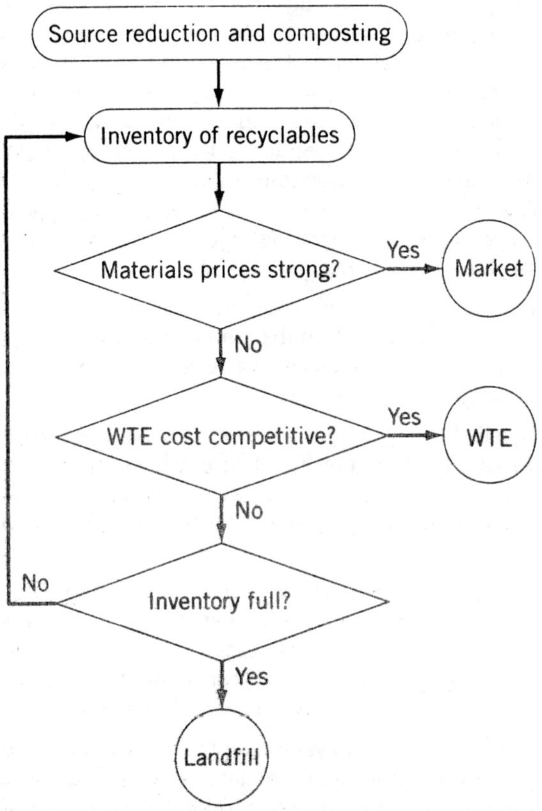

Figure 11. Recycling within a flexible ISWM hierarchy, where WTE = waste-to-energy.

that puts recycling on the same operational level as incineration and landfilling. It provides for continually assessing the net cost of all MSW management options. This information is used to decide how much of each material in the MSW stream to allocate the different options at any given time.

Assume that to the fullest extent possible the social and economic cost are included in the prices of materials and products and also in the cost of MSW management processes. Thus, it can be assumed that the level of recycling chosen reflects society's equating the marginal benefits of recycling to its marginal cost. This is done relative to the other MSW management options available. This was discussed in a previous section and reiterated in the 10 points to recycling above.

In the flexible ISWM scheme, source reduction first reduces the quantity of MSW generated. Composting then diverts an economical portion of organics from the waste stream. Voluntary reuse options, such as community-run product exchanges, remove additional portions of the MSW stream at this time. The MSW stream is then ready for collection, transportation, processing, storage, and disposition by any of the three remaining alternatives. The net cost of each alternative is compared. In addition, the anticipated future cost of each alternative is weighed in deciding how much of each material type to store in inventory for sale in the future as recyclables. Materials are kept in inventory, sold on recycled materials markets, sent to WTE plants or landfilled, depending on the lowest net present value of the total cost for each option. Work is currently underway to automate this decision model. Using it gives a municipality the opportunity to choose the combination of technologies within an integrated waste management system that minimizes the cost (social and economic) of managing its MSW.

Summary

This review of MSW recycling in the United States shows that current policies and programs largely ignore economic efficiency as they pursue target recycling and waste reduction rates. The efficiency criteria for recycling are that it be cost-competitive with other waste management options and that it occur at a level that matches society's benefit from recycling to its cost. The role that full-cost pricing plays in recycling (and materials management in general) must be placed on surer economic ground. The 10-point approach to recycling policy is sufficient until full-cost pricing schemes are tried and implemented.

The article departs from the standard how-to pieces on recycling and challenges the reader to question the assumption that recycling is always right. The references direct the reader to a selection of the wealth of available how-to manuals. It is the aim of this article to show how to approach such how-to manuals asking, if not knowing, *why* and when to recycle.

SOURCES OF INFORMATION

Federal Government Agencies

EPA headquarters:

United States Environmental Protection Agency (EPA)
Office of Solid Waste
401 M Street, SW
Washington, D.C. 20460-5301W
Tel: 703-308-8895, 1-800-424-9346.

EPA regional offices (15):

EPA Region 1 (CT, MA, ME, NH, RI, VT).
HER-CAN6JFK Federal Building
Boston, MA 02203-2211
EPA Region 2 (NY, NJ, Puerto Rico, the Virgin Islands).
II-AWM
26 Federal Plaza
New York, NY 10278
EPA Region 3 (DE, Washington, DC, MD, PA, VA, WV)
3HW53
841 Chestnut Street
Philadelphia, PA 19107
EPA Region 4 (AL, FL, GA, KY, MS, NC, SC, TN)
4WD-RCRAFF
345 Courtland Street, NE
Atlanta, GA 30365
EPA Region 5 (LL, IN, MI, MN, OH, WI)
HRP-8J
230 South Dearborn Street
Chicago, IL 60604
EPA Region 6 (AR, LA, NM, OK, TX)
68-HH

First Interstate Bank Tower
1445 Ross Avenue
Dallas, TX 75202-2733
EPA Region 7 (IA, KS, MO, NE)
RCRA-SPG
726 Minnesota Avenue
Kansas City, KS 66101
EPA Region 8 (CO, MT, ND, SD, UT, WY)
8HWM-RI
999 18th Street
Denver, CO 80202-2405
EPA Region 9 (CA, HI, Guam, American Samoa)
H-3-1
75 Hawthorne Street
San Francisco, CA 94105
EPA Region 10 (AK, ID, OR, WA)
HW-072
1200 Sixth Avenue
Seattle, WA 98101

State Solid Waste Agencies

Alabama
Alabama Department of Environmental Management (ADEM)
Solid Waste Branch
1751 W.L. Dickinson Drive
Montgomery, AL 36109-2608
Alaska
Alaska Department of Environmental Conservation and Environmental Health
The Solid Waste Program Manager
410 Willoughby St.
Juneau, AK 99801
Arizona
Arizona Department of Environmental Quality
Waste Programs Division
3033 N. Central Avenue
Phoenix, AZ 85012
Arkansas
Arkansas Department of Pollution Control & Ecology
The Regional Solid Waste Coordinator
P.O. Box 8913
Little Rock, AR 72119
California
California Integrated Waste Management Bureau
8800 Cal Center Drive
Sacramento, CA 95826
Colorado
Colorado Department of Public Health and Environment
HMWMD-SWIM-B2
4300 Cherry Creek Drive South
Denver, CO 80222-1530
Connecticut
Connecticut Department of Environmental Protection
Waste Management Bureau
79 Elm Street
Hartford, CT 06106-5127
Delaware
Delaware Department of Natural Resources and Environmental Control
P.O. Box 1401
Dover, DE 19903
Florida
Florida Department of Environmental Protection
2600 Blair Stone Road
Tallahassee, FL 32399-2400
Georgia
Georgia Department of Natural Resources
Environmental Protection Division
4244 International Parkway, Suite 104
Atlanta, GA 30354
Hawaii
Hawaii Department of Health
Solid & Hazardous Waste Branch
Environmental Management Division
919 Ala Moana Blvd #210
Honolulu, HI 96814
Idaho
Idaho Division of Environmental Quality
Air Resource and Prevention Bureau
1410 North Hilton
Boise, ID 83706
Illinois
Illinois Environmental Protection Agency
Solid Waste Management Section-Bureau of Land
2200 Churchill Road
P.O. Box 19276
Springfield, IL 62794-9276
Indiana
Indiana Department of Environmental Management
Office of Solid & Hazardous Waste Management-Regulatory Division
100 N. Senate Avenue
P.O. Box 6015
Indianapolis, IN 46206-6015
Iowa
Iowa Department of Natural Resources
Wallace Building
900 E. Grand Avenue
Des Moines, IA 50319
Kansas
Kansas Department of Health and Environment
Division of Environment
Forbes Field, Building 740

Topeka, KS 66620
Kentucky
Kentucky Department of Environmental Protection
Division of Waste Management
Solid Waste Branch
14 Reilly Road
Frankfort, KY
Louisiana
Louisiana Department of Environmental Quality
Solid Waste Division
P.O. Box 82178
Baton Rouge, LA 70884-2178
Maine
Maine Department of Environmental Protection
Division of Solid Waste
State House Station 17
Augusta, ME 0433-0017
Maryland
Maryland Department of the Environment
Waste Management Administration
2500 Broening Highway
Baltimore, MD 21224
Massachusetts
Massachusetts Department of Environmental Protection
One Winter Street
Boston, MA 02108
Michigan
Michigan Department of Environmental Quality
Solid Waste Management Unit
P.O. Box 30241
Lansing, MI 48917
Minnesota
Minnesota Pollution Control Agency
Groundwater and Solid Waste Division
520 Lafayette Road
St. Paul, MN 55155
Mississippi
Mississippi Department of Environmental Quality
Office of Pollution Control
P.O. Box 10385
Jackson, MS 39289-0385
Missouri
Missouri Department of Natural Resources
Division of Environmental Quality
P.O. Box 176
Jefferson, MO 65102-0176
Montana
Montana Department of Environmental Quality
Waste Management Division
Box 200901
Helena, MT 59620-0901
Nebraska
Nebraska Department of Environmental Quality
P.O. Box 98922
Lincoln, NE 68509-9822
Nevada
Nevada Department of Environmental Protection
333 West Nye Lane
Carson City, NV 89710
New Hampshire
New Hampshire Department of Environmental Services
Waste Management Division
6 Hazen Drive
Concord, NH 03301-6509
New Jersey
New Jersey Department of Environmental Protection
Division of Solid and Hazardous Waste-CN 414
120 S. Stockton Street
Trenton, NJ 08625-0414
New Mexico
New Mexico Department of Environment
P.O. Box 26110
1190 St. Francis Drive
Santa Fe, NM 87502
New York
New York State Department of Environmental Conservation
Division of Solid & Hazardous Waste
50 Wolf Road-Room 488
Albany, NY 12233-7250
North Carolina
North Carolina Department of Environmental Health and Natural Resources
Division of Solid Waste Management
PO Box 27687-7687
Raleigh, NC 27611
North Dakota
North Dakota Department of Health
Division of Waste Management
P.O. Box 5520
Bismarck, ND 58504-5520
Ohio
Ohio Environmental Protection Agency
Division of Solid & Infectious Waste Management
PO Box 1049
Columbus, OH 43216-1049
Oklahoma
Oklahoma Department of Environmental Quality
Waste Systems Planning Division
1000 N.E. 10th Street
Oklahoma City, OK 73117-1212
Oregon
Oregon Department of Environmental Quality
Waste Management Division

811 SW 6th Avenue
Portland, OR 97204

Pennsylvania

Pennsylvania Department of Environmental Protection
Bureau of Waste Management
14th Floor-RCSOB
PO Box 8472
Harrisburg, PA 17105-8472

Rhode Island

Rhode Island Department of Environmental Management
Waste Management Division
291 Promenade Street
Providence, RI 0908-57672

South Carolina

South Carolina Department of Health & Environmental Control
Solid Waste Management Division
2600 Bull Street
Columbia, SC 29201

South Dakota

South Dakota Department of Environmental and Natural Resources
Waste Management Program
523 E. Capitol Avenue
Pierre, SD 57501

Tennessee

Tennessee Department of Health and Environment
Solid Waste Management Unit
701 Broadway
Nashville, TN 37243

Texas

Texas Department of Health-TNRCC
Waste Planning and Assessment Division
P.O. Box 13087
Austin, TX 78711-3087

Utah

Utah Department of Environmental Quality
Division of Solid & Hazardous Waste
PO Box 144880
Salt Lake City, UT 84114-4880

Vermont

Vermont Department of Environmental Conservation
Solid Waste Management Program
103 South Main Street
Waterbury, VT 05671-0404

Virginia

Virginia Department of Environmental Quality
Waste Management Division
PO Box 10009
Richmond, VA 23240

Washington

Washington Department of Ecology
Solid & Hazardous Waste Management Program
PO Box 47600
Olympia, WA 98504-4700

West Virginia

West Virginia Department of Commerce, Labor & Environmental Resources
Office of Waste Management
1356 Hansford Street
Charleston, WV 25301

Wisconsin

Wisconsin Department of Natural Resources
Solid Waste Management Division
101 S. Webster Street
PO Box 7921
Madison, WI

Wyoming

Wyoming Department of Environmental Quality
Solid Waste Permitting & Corrective Action Program
250 Lincoln Street
Lander, WY 82520-2848

Other National Information Sources and Associations (15,25)

Aluminum Association, Inc. (AA), Washington, D.C.
American Paper Institute, Inc. (API), New York
Association of State and Territorial Solid Waste Management Officials, Washington, D.C.
Center for Plastics Recycling Research (CPRR), Piscataway, N.J.
Coalition of Northeastern Governors (CONEG), Washington, D.C.
Compost Council, Alexandria, Va.
Council for Solid Waste Solutions (CSWS), Washington, D.C.
Council on Plastics and Packaging in the Environment (COPPE), Washington, D.C.
Environmental Defense Fund (EDF), New York
Food Service and Packaging Institute, Arlington, Va.
Glass Packaging Institute (GPI), Washington, D.C.
Institute for Local Self Reliance (ILSR), Washington, D.C.
Institute of Scrap Recycling Industries, Inc. (ISRI), Washington, D.C.
Keep America Beautiful, Inc. (KAB), Stamford, Conn.
National Association for Plastic Container Recovery (NAPCOR), Charlotte, N.C.
National Association of Towns and Townships, Washington, D.C.
National Oil Recyclers Association, Washington, D.C.
National Recycling Coalition (NRC), Washington, D.C.
National Resource Recovery Association
The United States Conference of Mayors, Washington, D.C.
National Solid Waste Management Association, Washington, D.C.
National Soft Drink Association (NSDA), Washington, D.C.
Plastics Recycling Foundation, Washington, D.C.
Renew America, Washington, D.C.

Society of the Plastics Industry, Inc. (SPI), Washington, D.C.

Solid Waste Association of North America (SWANA), Silver Spring, Md.

Solid Waste Composting Council (SWCC), Washington, D.C.

Steel Can Recycling Institute (SCRI), Pittsburgh, Pa.

BIBLIOGRAPHY

1. N. Golbeck, and D. Goldbeck, *Choose to Reuse: An Encyclopedia of Services, Businesses, Tools and Charitable Programs that Facilitate Reuse,* Ceres Press, New York, 1995.
2. S.M. Selke, *Packaging and the Environment: Alternatives, Trends and Solutions,* Technomic, Lancaster, Pa., 1994, p. 119.
3. U.S. EPA, *Characterization of Municipal Solid Waste in the United States: 1994 Update,* EPA/530-R-94-042, Washington, D.C., 1994.
4. U.S. EPA Report to Congress, *Solid Waste Disposal in the United States,* EPA/530-SW-88-011B, Vol. 2. Washington D.C., Oct. 1988.
5. J. Vining, N. Linn, and R. Burdge, *Environ. Manage.* **16** (6), 785–797 (1992).
6. U.S. EPA, *Characterization of Municipal Solid Waste in the United States: 1995 Update-Executive Summary,* EPA/530-S-96-001, Washington, D.C., 1996.
7. T.E. Duston, *Recycling Solid Waste: The First Choice for Private and Public Sector Management,* Quorum Books, Westport, Conn., 1993, Chapt. 8.
8. L.B. Lave, C. Hendrickson, and F.C. McMichael, Environ. Sci. Technol., **28**(1), 19A–24A (1994).
9. G. Louis, "Regional Integrated Municipal Solid Waste Management in the Northeastern United States," Ph.D. Thesis, Carnegie Mellon University, Pittsburgh, Pa., 1996.
10. M. Melosi, *Garbage in the Cities: Refuse Reform and the Environment: 1880–1980,* Texas A&M Press, College Station 1981.
11. N.L. Drobny, H.E. Hull, and R.F. Testin, *Recovery and Utilization of Municipal Solid Waste,* United States EPA, SW-127, Washington, D.C., 1971.
12. U.S. EPA, *The Solid Waste Dilemma: An Agenda for Action,* EPA/530-SW-89-019, Washington, D.C., 1989.
13. *BioCycle,* "The State of Garbage in America", annual reports 1989–1996.
14. U.S. EPA, *Variable Rates in Solid Waste: Handbook for Solid Waste Officials,* EPA/530-SW-90-084, Washington, D.C., 1990.
15. D. Kimbal, *Recycling in America,* ABC-CLIO, Santa Barbara, Calif., 1992.
16. C. Miller, *Waste Age,* 92–98 (Mar. 1995).
17. R. Steuteville, *BioCycle,* 35–41 (May 1996).
18. R. Steuteville, *BioCycle,* 55–61 (Apr. 1996).
19. J. Heumann, E. Striano, and K. Egan, *Waste Age,* 35–47 (Aug. 1996).
20. R. Stueteville, *BioCycle,* 31 (May 1994).
21. M.J. Clarke, M. de Kadt, and D. Saphire, in S.R. Golden, ed., *Burning Garbage in the United States: Practice vs. State of the Art,* INFORM, New York, 1991.
22. R.A. Denison, and J. Ruston, *Recycling and Incineration: Evaluating the Choices,* Environmental Defense Fund, Washington D.C., 1990.
23. B. Commoner, *Making Peace with the Planet,* Pantheon Books, New York, 1990.
24. J.H. Alexander, *In Defense of Garbage,* Praeger, Westport, Conn. 1993.
25. G. Tchobanoglous, H. Thiesen, and S. Vigil, *Integrated Solid Waste Management, Engineering Principles and Management Issues,* McGraw–Hill, New York, 1993.
26. J.T. Aquino, ed., *Waste Age / Recycling Times' Recycling Handbook,* Lewis Publishers, New York, 1995.
27. M.A., Curran, *Environmental Life-Cycle Assessment,* McGraw–Hill, New York, 1996.
28. M.B. Hocking, *Science,* **251**(1), 504–505 (1991).
29. C. Hendrickson, L. Lave, and F. McMichael, *Chemtech,* Aug. 1995, pp. 56–60.
30. J.T. Aquino, ed., *Waste Age / Recycling Times' Recycling Handbook,* Lewis Publishers, New York, 1995, Chapt. 3.
31. J.T. Aquino, ed., *Waste Age / Recycling Times' Recycling Handbook,* Lewis Publishers, New York, 1995, Chapts. 3, 8.
32. EPA, *Charging Households for Waste Collection and Disposal...* EPA 530-SW-90-047, Washington, D.C., 1990.
33. R. Cuthbert, *BioCycle,* 63–65 (May 1994).
34. D. Fullerton, and T. Kinnaman, *Household Demand for Garbage and Recycling Collection with the Start of a Price Per Bag,* Departmental Study in the Heinz School of Public Policy and Management at Carnegie Mellon University, Pittsburgh, and the Department of Economics at University of Virginia, Charlottesville, 1993.
35. U.S. EPA, *Charging Households for Waste Collection and Disposal...,* EPA 530-SW-90-047, Washington, D.C., 1990.
36. R. Martin, *BioCycle,* 75–77 (Oct. 1994).
37. J. Glenn, *BioCycle,* 30–37 (May 1992).
38. D. Stickney, *University of Cincinnati Law Review,* **64,** 283–318 (1995).
39. 437 United States Supreme Court, 617, 624, as reported in the *United States Law Week,* 62 LW 4315.
40. B. Parker, *Waste Age,* 115–118 (Apr. 1993).
41. *The New York State Solid Waste Management Plan: 1995 / 1996 Update,* New York State Dept. of Environmental Conservation, Division of Solid & Hazardous Materials, Albany, N.Y., 1996.
42. R. Woods, *Waste Age,* 79–90 (May 1995).
43. Ref. 41, pp. 43–45.
44. J. Heumann, and R. Woods, *Waste Age,* 36 (Feb. 1996).
45. R. Woods, and J. Aquino, *Waste Age,* 10–14 (June 1995).
46. U.S. EPA, *Report to Congress on Flow Control and Municipal Solid Waste,* United States EPA, Washington, D.C., Mar., 1995.
47. P. McCreery, *Waste Age,* 67–76 (Mar. 1996).

Selected References on Implementing Recycling Programs

J. Carless, *Taking Out the Trash: A No-Nonsense Guide to Recycling,* Island Press, Washington, D.C., 1992.

T.E. Duston, *Recycling Solid Waste: The First Choice for Private and Public Sector Management,* Quorum Books, Westport, Conn., 1993.

EPA, Decision Maker's Guide to Solid Waste Management, 2nd ed., EPA530-R-95-023, Washington, D.C., 1995, particularly Chapt. 6.

H.F. Lund, ed., *The McGraw–Hill Recycling Handbook,* McGraw–Hill, New York, 1993.

M.J. Rogoff, and J.F. Williams, *Approaches to Implementing Solid Waste Recycling Facilities,* Noyes, Park Ridge, N.J., 1994.

Selected References on Recycling Operations and Processes

BioCycle staff, ed., *The BioCycle Guide to Maximum Recycling,* JG Press, Emmaus, Pa. 1993.

P.N. Cheremisinoff, ed., *Waste Minimization and Recycling,* Gulf, Tex., 1992.

L.F. Diaz, *Composing and Recycling Municipal Solid Waste,* Lewis Publishers, Boca Raton, Fla., 1993.

H.F. Lund, ed., *The McGraw–Hill Recycling Handbook,* McGraw–Hill, New York, 1993.

PEER Consultants and Cal Recovery, Inc., *Material Recovery Facility Design Manual,* CRC Press, Boca Raton, Fla., 1993.

G. Tchobanoglous, H. Theisen, and S. Vigil, *Integrated Solid Waste Management, Engineering Principles and Management Issues,* McGraw–Hill, New York, 1993.

T.J. Bratton, *World Wastes,* 51–55 (June 1994).

R.A. Denison, *Annu. Rev. Energy Environ.* **21**, 191–237 (1996).

U.S. EPA, *State and Local Solutions to Solid Waste Management Problems,* EPA/530/SW-89-014. Washington, D.C., 1992.

U.S. EPA, *The Consumers' Handbook for Reducing Solid Waste,* EPA530-K-92-003. Washington, D.C., 1992.

E.M. Kirschner, *C&EN,* 19–22 (Nov. 4, 1996).

REFINERY WASTEWATER, TOXICITY REDUCTION

W. WESLEY ECKENFELDER, Jr.
Eckenfelder Inc.
Nashville, Tennessee

Aquatic toxicity in refinery and petrochemical wastewaters may result from chemicals present in the feedstock, chemicals added or generated in the process, or chemicals generated during the wastewater treatment process, usually referred to as soluble microbial products (SMP). In most cases, the chemicals originally present or generated are biodegradable and can be removed in the biological treatment process. In some cases, a physical–chemical source treatment may be required. SMP generated through the wastewater treatment process are nonbiodegradable and are best handled by the application of powdered activated carbon (PAC) integrated into the activated sludge process.

The biodegradability of 37 organic substances found in petrochemical wastewater has been determined (1). As can be seen from Table 1, most of these are biodegradable under appropriate operating conditions. It should be noted that in some cases extended acclimation could have achieved greater reductions in COD. Other studies have shown that most refinery or petrochemical organics typically found in wastewaters are either stripped or biodegraded in a biological treatment process and therefore should not present toxicity problems in the effluent. It should be noted, however, that in some cases long sludge ages are required to achieve the desired result. The distribution of hydrocarbons in the activated sludge process treating refinery wastewaters is shown in Figure 1. As can be seen in most cases, greater than 90% removal is achieved through conventional biological treatment. In a comprehensive study at the Shell Martinez Refinery in California, toxicity due to naphthenic acids quantified as "soluble oil and grease", ammonia, process chemicals, and a water treatment flocculent were identified (2). The toxicity profile for this refinery is shown in Figure 2 for stickleback fish. As is true for most industrial wastewaters, variability in influent concentration and composition results in a highly variable effluent toxicity. It should be noted that toxicity is very species-specific, as shown in Figure 3. What may be a problem in one refinery may not be a problem in another refinery, depending on the specified species.

Naphthenic acids are present in heavy California crude and partition into the aqueous phase. They are removed in the activated sludge process when operated at high sludge ages (20–30 days). The 96-h LC_{50} for naphthenic acids is in the range of 5 mg/L. Toxicity is affected by pH. Decreasing the pH from 7.8 to 6.6 causes the stickleback survival rate to drop from 100 to 0%. Reducing effluent oil and grease to less than 10 mg/L should control the naphthenic acid concentration below the toxic effect level (2). Toxicity, due to specific organics that are biodegradable, is controlled by adjusting the operating sludge age to achieve a nontoxic concentration of that specific organic in the effluent. This is shown in Figure 4 for nonylphenol.

Ammonia is only toxic in the nonionized form (NH_3) which is a function of pH. As the pH is increased, the percentage of ammonia as NH_3 is increased, hence the toxicity is increased (Fig. 5). The ammonia toxicity to aquatic species in refinery effluent is shown in Figure 6. Ammonia is also toxic to nitrifiers in roughly the same order of magnitude as many aquatic species. This is a two-edged sword. Inhibition of nitrification results in higher effluent ammonia levels and hence aquatic toxicity. This probably also results in permit violations. Removal of residual ammonia from refinery effluent is usually achieved by biological nitrification. Many chemicals found in refinery and petrochemical effluents in low concentrations, as well as ammonia, are inhibitory to nitrification. This is shown in Figure 7 for a petrochemical effluent. At 22°C an SRT of 25 days is required to achieve nitrification whereas in municipal wastewater under these conditions an SRT of only 2–3 days is required. Inhibition has been reduced by changing plant operating conditions from conventional plug flow to step feed (Fig. 8) (1). It should be noted that if inhibition caused by ammonia exists, a complete mix basin configuration is preferred so that the ammonia concentration at all points in the basin is the same as the effluent. In any event, the nitrification rate in refinery effluents should be determined in order to define the SRT necessary to meet effluent requirements. Procedures for determining nitrification rates in industrial effluents have been outlined (3).

Flocculents are used for coagulation and flocculation prior to secondary clarification. Studies have shown that some of the polyelectrolytes in use as coagulants were acutely toxic to three-spine stickleback at relatively low levels (2). If a polyelectrolyte is to be used in the wastewater treatment plant an effluent toxicity check should be made. In some cases it may be necessary to change the polyelectrolyte to a nontoxic form.

Effluent toxicity can be caused by the cumulative effect of trace quantities of organic chemicals and by SMP generated in the process. The SMP can amount to 5–10% of the degradable COD and can contribute to effluent toxicity and inhibition of nitrification. These effects are schematically shown in Figure 9. As the COD is reduced, toxicity due to chemicals originally present in the wastewater, such as naphthenic acids, is reduced due to biooxidation of the organics. However, toxicity due to SMP can increase with increasing COD reduction. Because these organics cannot be specifically identified, toxicity is usually related to effluent COD. A toxicity COD correlation for a West coast refinery is shown in Figure 10 (4).

Table 1. Biodegradability of 37 Organic Substances[a]

No.	Test Substance	Molecular Weight	Method[b]	Concentration[c], mg/L	Acclimation, Days	Removal, % COD_{Mn}	TOC
1.	Epichlorohydrine	92.5	A	130	1	89	67
2.	Trichloropropane	147.5	A	220	1	55	76
3.	Formic acid	46	Res	100			95
4.	Methyl anhydrice	98	A	170	1	97	98
5.	Methyl alcohol	32	A	100	1	91	92
6.	*n*-Butyl alcohol	74	A	100	1	82	93
7.	*tert*-Butyl alcohol	74	Res	100			64
8.	Ethylene glycol	62	A	220	1	92	93
9.	Glycerin	92	A	220	1	97	94
10.	2-Chloroethyl alcohol	805	A	400	1	20	1
11.	Polyethylene glycol	414	A	200	1	4	10
12.	Poly(vinyl alcohol) (PVA)	>4200	a	140	1	5	20
13.	Toluene	92	A	27	1	27	60
14.	2-Methlaniline	107	A	100	1	92	83
15.	*N*-Methylaniline	107	Res	100			42
16.	2,4-Diaminotuloene	122	A	85	1	49	53
17.	2,4-Dinitrotoluene	182	A	60	1	0	1
			Res	100		0	0
18.	4-Chloro-2-methyl aniline	141.5	A	110	1	16	7
19.	2-Chloro-5-amino-*p*-toluene sulfonic acid	222.5	A	210	1	0	0
			Res	100		0	0
20.	*p*-Toluidine-*m*-sulfonic acid	187	A	230	1	0	0
21.	*p*-Toluene sulfonic acid	172	A	100	10		90
			Res	140		95	91
22.	Phthalic anhydride	148	A	65	1	33	88
23.	Phthalimido	147	A	120	1	63	88
24.	Benzoic acid	122	A	150	1	44	86
25.	Stylene glycol	38	A	200	3	75	76
26.	Bisphenol	228	A	58	1	72	57
27.	*p*-Meth-1-thio-phenol	140	A	100	14	0	0
28.	Acetoacetanilide	177	A	100	10	87	86
29.	Acetoacetic-*p*-toluidide	191	A	100	10	83	86
30.	*β*-Naphthol	144	A	43	1	20	8
			Res	100		96	97
31.	*β*-Oxynaphthoic acid	188	Res	100		81	82
32.	*N*-Methyl-2-pyrrolidone	99	A	150	1	94	92
33.	4,4-Dimethyl 1-1, 3-dioxane	116	A	150	1	14	26
34.	Dimethyl sulfoxide	78	A	200	14	0	0
35.	Methyl cellulose		A	186	1	8	6
36.	Soap (RCOONa)		A		1	84	83
37.	Silicon surfactant		A		1	13	10

[a] Ref. 1.
[b] A = activated sludge degradability test; Res = respiration meter method.
[c] Initial concentration of test substance.

A summary of their data is shown in Table 2. The increase in effluent COD and toxicity from the DAF effluent and the WWTP effluent can probably be attributed to SMP. In these cases, pretreatment or upstream controls are usually ineffective. The most effective treatment is activated carbon adsorption in which toxic organics and SMP are effectively adsorbed on the carbon. In most cases, the most cost-effective solution is the use of powdered activated carbon (PAC) which is added prior to, or directly into, the aeration basin. It has been shown that PAC reduces residual COD concentrations from 242 to 50 mg/L and TOC from 82 to 4 mg/L, with a resulting increase in the LC_{50} from 15 to <100% (also 100% survival) (4). This was achieved with a PAC dosage of 75 mg/L.

CONCLUSIONS

1. Toxicity in refinery effluents can be caused by chemicals present in the influent wastewater including ammonia, chemicals added in the treatment process, and by SMP generated in the treatment process.
2. Most chemicals in the influent wastewater exhibiting toxicity are biodegradable and will be removed in the biological process.
3. Effluent toxicity can be caused by the cumulative effect of trace organics and by SMP generated in the process. The most cost-effective solution for this problem is the addition of powdered activated carbon (PAC) to the biological process.

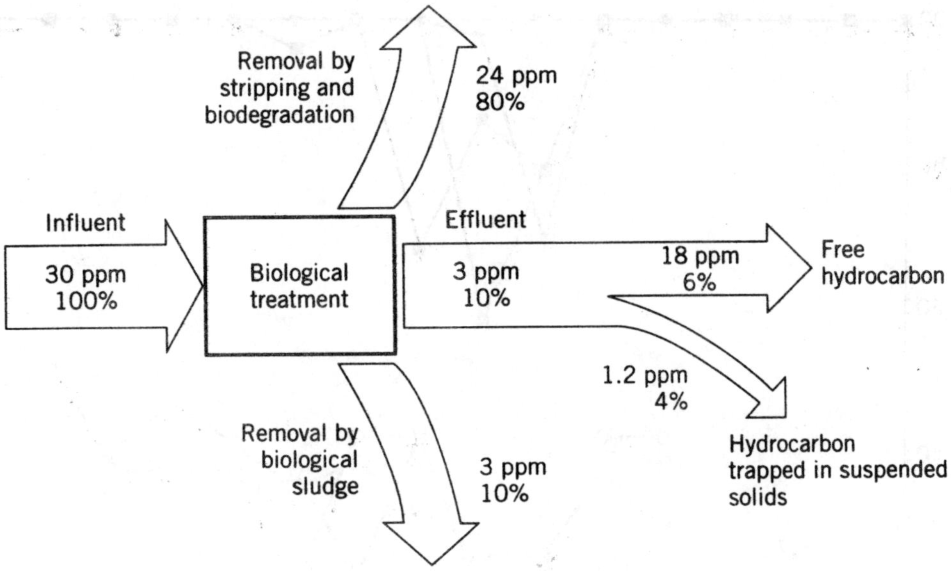

Figure 1. Removal of hydrocarbons in the activated sludge process.

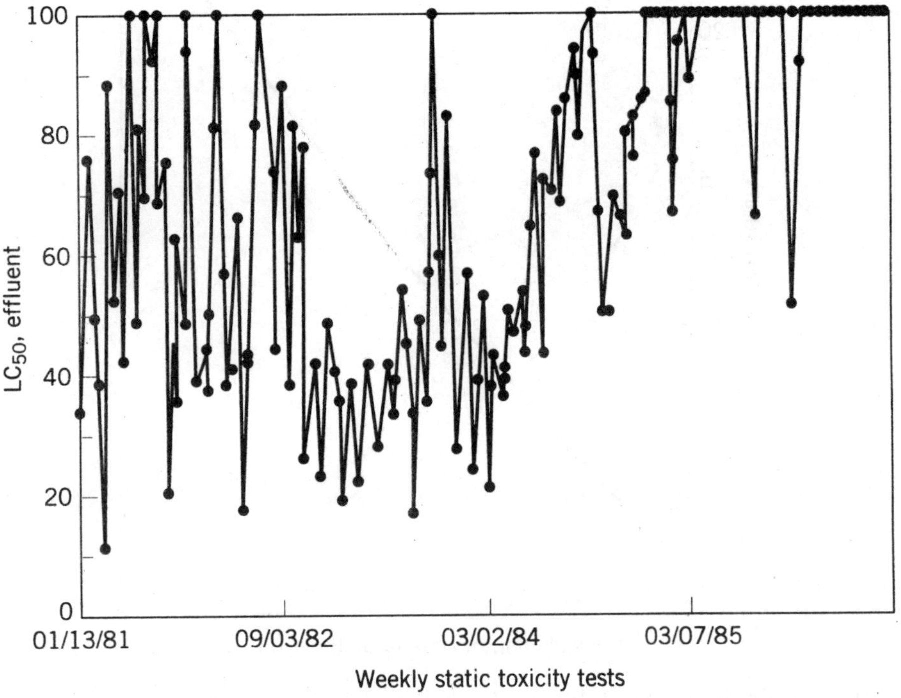

Figure 2. Acute toxicity of refinery effluent to stickleback.

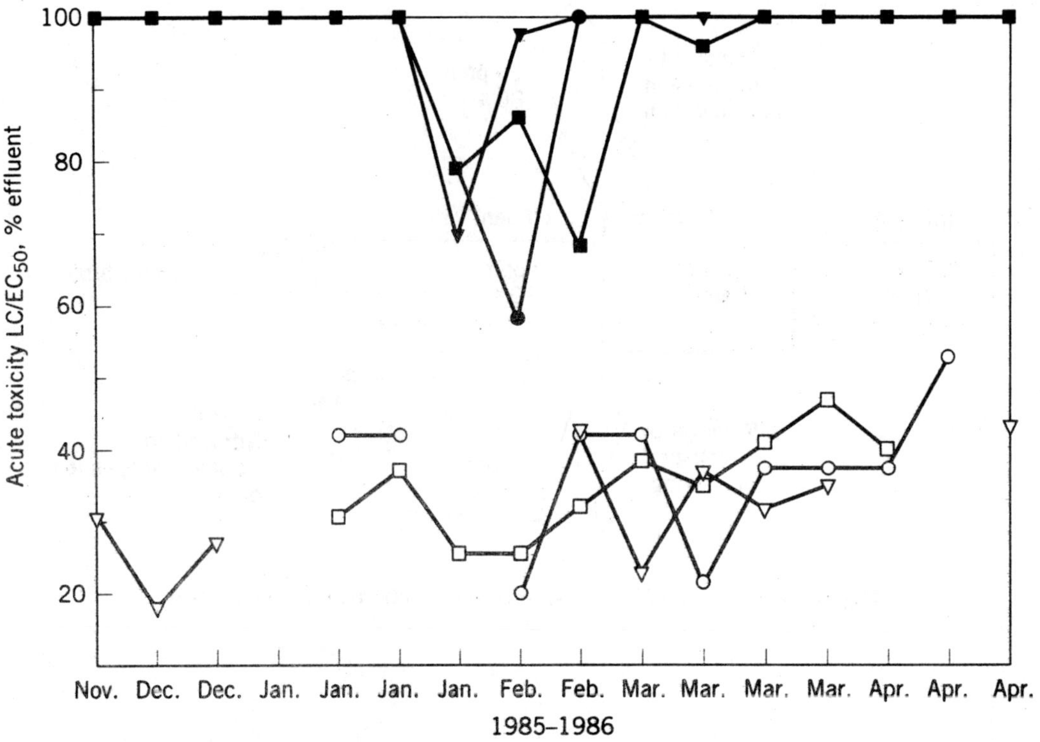

Figure 3. Acute toxicity of six species to refinery effluent: ●, sheepshead; ■, stickleback; ▼, Microtox; ○, algae; □, rainbow; and ▽, mysid.

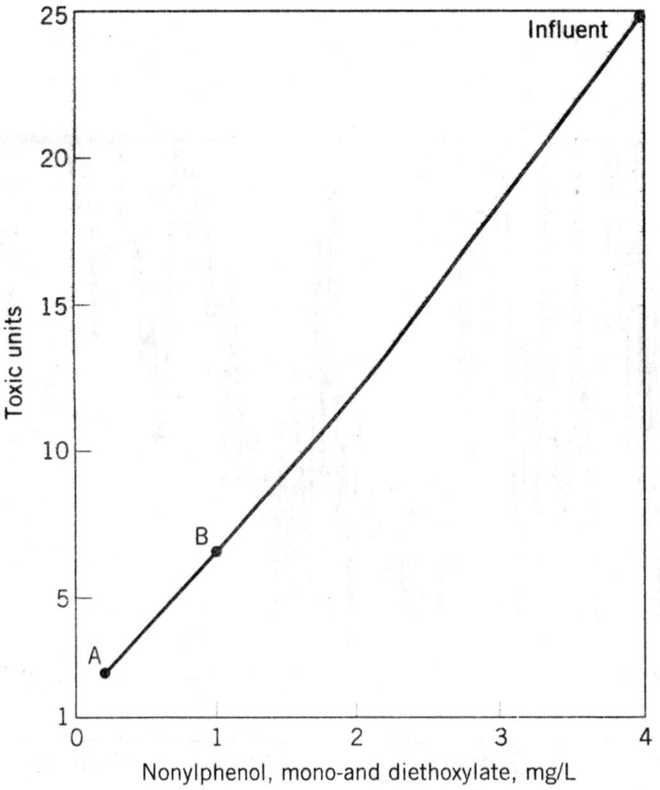

Figure 4. Effects of SRT on toxicity reduction for nonylphenolics, where A is SRT at 21 d; B, 4.5 d.

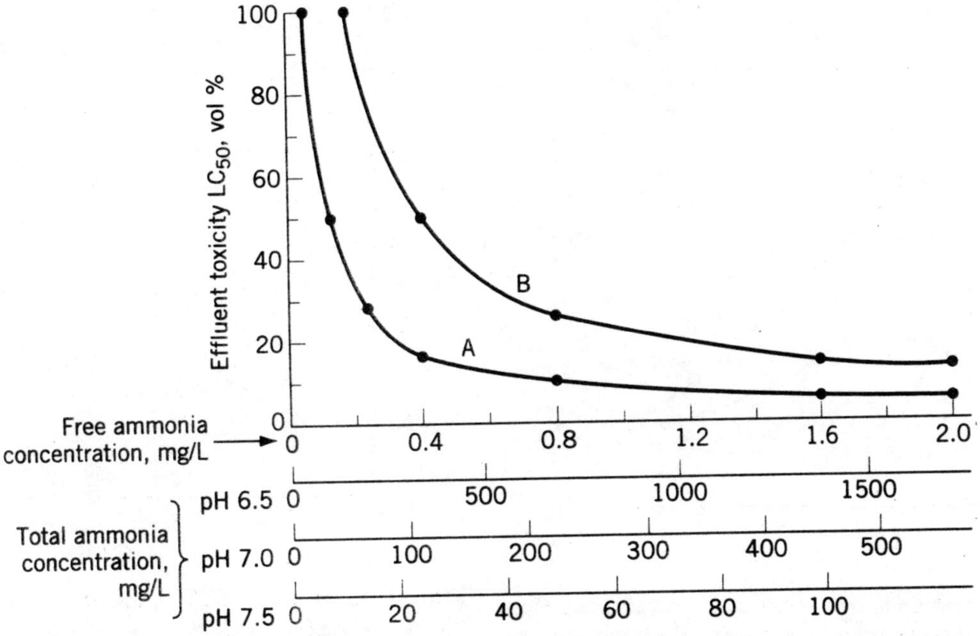

Figure 5. Toxicity of ammonia vs pH, where A shows LC_{50} = −0.05 mg/L and B, LC_{50} of 0.2 mg/L.

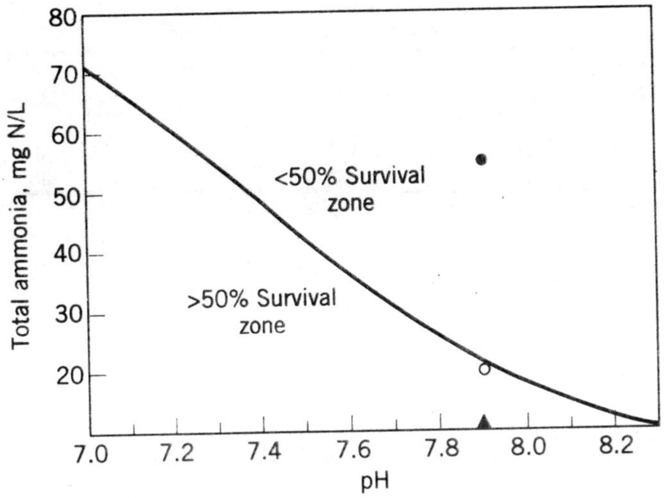

Figure 6. The effect of pH on ammonia toxicity for refinery effluent calculated from the Erickson equation: $LC_{50} = [1.0/1 + 10^{1.03(7.32-\mathrm{pH})}]K - a + [H^+]/K_a$, where • shows EAL 20°C 0% survival; EAL 20°C 0% survival and MMC 17°C 100% survival; and ○, MMC 17°C 100% survival.

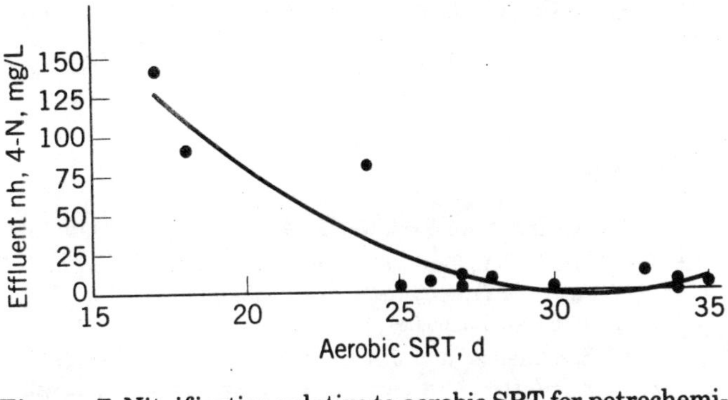

Figure 7. Nitrification relative to aerobic SRT for petrochemical wastewater at 22–24°C.

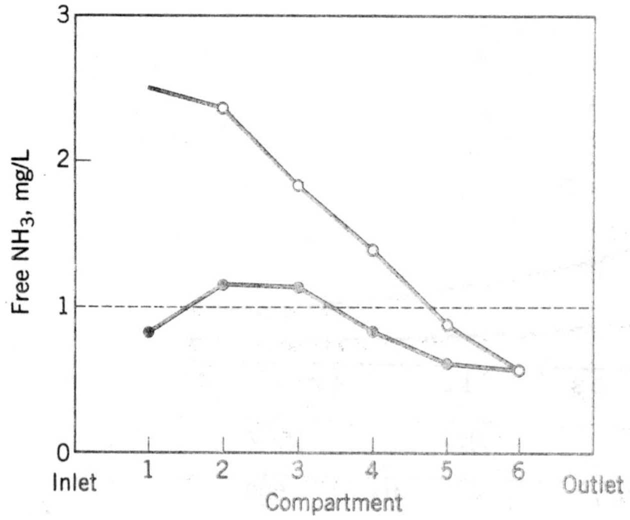

Figure 8. Concentration of free ammonia in the aeration tank where ● shows step aeration; ○ conventional aeration; and (---), Anthonisen's inhibitory level.

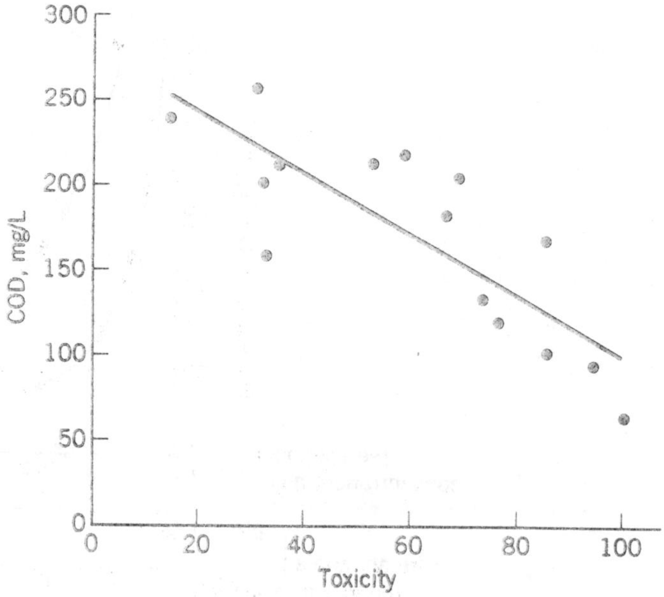

Figure 10. Toxicity/COD correlation for WWTP effluent, where $r = 0.821$.

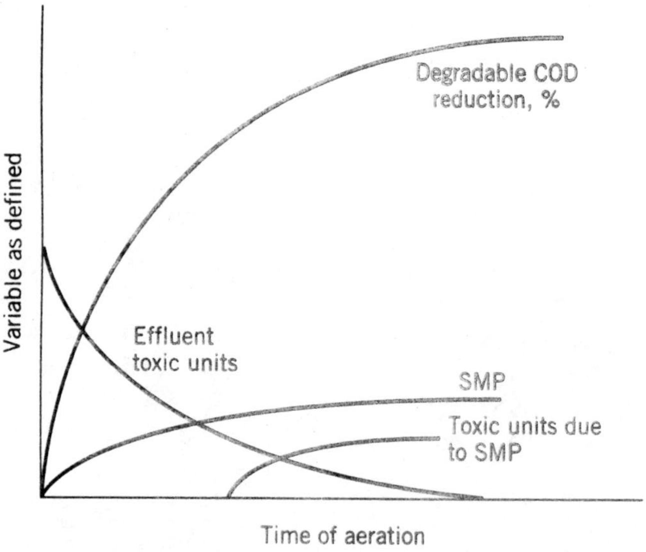

Figure 9. Toxicity relationships in refinery effluents.

Table 2. Organic Content and Toxicity of Process Wastewaters[a]

Process Water	Organic Content, mg/L COD	Organic Content, mg/L TOC	Toxicity asLC_{50}
Oxidizer product	5430	1700	6.1
Pretreated phenolic water	1180	380	22
WWTP effluent	198	65	26
DAF effluent	168	50	35

[a] Ref. 4.

BIBLIOGRAPHY

1. Y. Okawa and co-workers, *Wat. Sci. Tech.* **20**(10), 193 (1988).
2. P.B. Dorn, in D.L. Ford, ed., *Case Histories, The Petroleum Refining Industry in Toxicity Reduction Evaluation and Control,* Technomic Pub. Co., Lancaster, Pa., 1992.
3. W.W. Eckenfelder and J. Musterman, *Activated Sludge Treatment of Industrial Wastewaters,* Technomic Pub. Co., Lancaster, Pa., 1995.
4. J. Wong and co-workers, *Wat. Sci. Tech.* **25**(3), 221 (1992).

General References

N. Galil and co-workers, *Wat. Sci. Tech.* **20**(10), 21 (1988).

P.W. Lankford and W.W. Eckenfelder, *Toxicity Reduction in Industrial Effluents,* Van Nostrand Reinhold, New York, 1990.

REMOTE SENSING DATA SYSTEMS AND NETWORKS

PAUL M. MATHER
The University of Nottingham
Nottingham, U.K.

The establishment of an effective market in earth observation data and products is dependent to a considerable extent on the provision of computer and administrative systems which are capable of receiving, processing, cataloguing, and archiving the data collected by instruments carried by remote sensing satellites as they orbit around the earth. Also needed are systems that provide facilities for potential users to locate information and products derived from this remotely sensed data, as well as software appropriate to their specific applications or needs. The delivery of remotely sensed data to an user thus requires the design, implementation, and maintenance of complex systems capable of handling a data flow that is estimated will be of the order of 20×10^{12} bits per day in the early years of the twenty first century for the United States alone (1). The development of such systems to provide users with a browse and search facility to enable the location of suitable data sets, and of the infrastructure required to use those data sets effectively, is a major challenge which is being faced by national space agencies across the globe.

This article has three sections. The first section contains a summary of the nature of earth observation data obtained from remote sensing systems carried by earth orbiting satellites. The topic of ground systems for the receiving, processing, archiving, and distribution of these data is also considered. Finally, a summary of current and planned facilities for user interaction with data catalogues, particularly those that are available through the World Wide Web (WWW), is provided.

EARTH OBSERVATION DATA: NATURE AND APPLICATIONS

Earth observation (EO) data are acquired by instruments carried onboard orbiting spacecraft. These data can be of two kinds: image and profile. Image data can be considered as a two-dimensional digital "picture" of the earth's land, sea, and ice surfaces, such images being obtained for several spectral bands such as the visible red, green, and blue or the thermal infrared. Profile data consist of one-dimensional sequences of point measurements made along a line on the earth's surface lying directly below the path of the orbiting satellite. These profile measurements relate either to the point properties of the earth's land and sea surfaces, for example, ocean wave height, or of the atmosphere. For example, profiling instruments called altimeters are carried by the European ERS satellites and the U.S./French TOPEX/Poseidon mission, and are capable of measuring the height of ocean waves (Fig. 1). Although Figure 1 appears to be a two-dimensional image, it was produced by processing a sequence of profiles generated by the altimeter as the satellite traveled on successive (adjacent) orbit tracks across the earth's surface. Figure 1 shows waves as high as eight meters (26 ft) in the Southern Ocean.

Images are produced by a process of scanning. Currently, the most widely used satellite scanner systems are the Advanced Very High Resolution radiometer (AVHRR) carried by the U.S. National Atmospheric and Oceanic Administration (NOAA) satellites, the Thematic Mapper (TM) carried by the U.S. Landsat-4 and Landsat-5 satellites, and the High Resolution Visible (HRV) instrument carried by the French SPOT satellite. Other systems producing images in the visible, infrared, and microwave regions of the electromagnetic spectrum are carried by the Japanese ADEOS and JERS satellites, the Indian IRS-1 satellite, Canada's Radarsat, and the European ERS satellites. Figure 2 shows an image of the Camargue area of Southern France, collected by the SPOT HRV instrument on January 12, 1987. Since HRV operates in the near-infrared as well as the red and green regions of the visible spectrum, images do not appear in natural colors, for example, with vigorous vegetation shown in green. Instead, vegetated areas appear as shades of red, and the water may appear black. Water in lagoons may appear bluish on certain images, due to large amounts of suspended sediment. Clouds are white or whitish blue and plowed fields may have a greenish tinge.

The number of different spectral bands collected simultaneously by an imaging device is called the *spectral resolution* of the instrument. Of the systems mentioned above, the AVHRR collects images in either four or five spectral regions, depending on the type of sensor. There are two types. Both collect information in the visible, near-infrared, middle infrared, and thermal infrared. The second type includes an additional thermal infrared band. Landsat's TM has seven spectral bands spanning the visible to thermal infrared regions of the spectrum, whereas the SPOT HRV has three bands, two of which are located in the visible region of the spectrum with the third band measuring surface reflection in the near-infrared wavelengths. The choice of the number of bands to be scanned, and the location of the bands in particular regions of the electromagnetic spectrum, depends to a large extent on the applications in which the data are to be used. For example, systems designed to monitor vegetation cover and status operate in the visible and the near-infrared regions of the spectrum, whereas geological information is derived from the mid-infrared spectral bands.

Image and profile data obtained by remote sensing of the earth's surface and atmosphere are used in a variety of applications, including weather forecasting, wave height and sea ice monitoring, map production and revision, petroleum exploration, geological mapping, oceanography, agricultural crop monitoring, forest monitoring, and environmental science (3). A survey of the full range of earth observation data applications is beyond the scope of this article; however, details of U.S. and foreign programs can be found in NASA's WWW pages (3). A key characteristic of satellite-derived data is that temporal sequences of such data are obtained at regular intervals depending on the orbital characteristics of the particular satellite, thus enabling the study of change over periods ranging from days to decades. For example, the U.S. NOAA series of satellites has an orbit repeat cycle of 12 hours, so that images of a given area can be obtained at 12-hour intervals. Landsat 4 and 5, on the other hand, have a 16-day orbital repeat cycle. A satellite's orbital characteristics reflect the requirements of the applications of the data acquired by the instruments onboard that satellite. As the data from the AVHRR sensor carried by the NOAA satellites is used extensively in weather forecasting, a high temporal frequency is required. The main applications of image data from Landsat's Thematic Mapper instrument are crop and land cover monitoring, which require a lower temporal resolution than weather forecasting. The

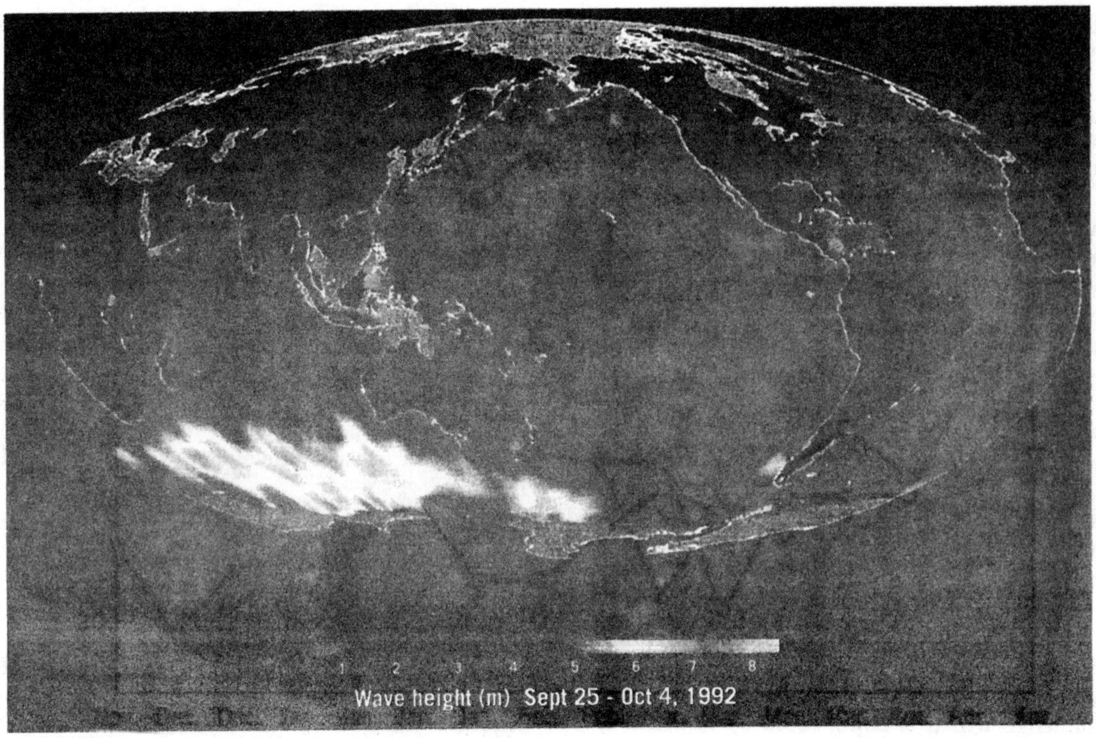

Figure 1. Image showing global wave heights (in meters) from NASA altimeter carried by the TOPEX/Poseidon mission (2).

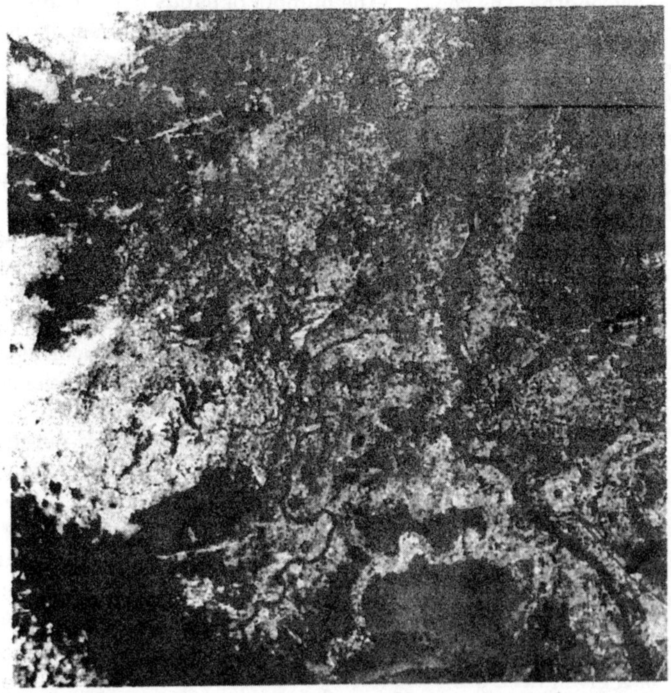

Figure 2. SPOT HRV image of the Camargue area of southern France. Courtesy of SPOT Image Corp.

repetitive nature of satellite imagery means that changes in the characteristics of the earth's land, sea, and ice surfaces as well as of the atmosphere can be measured and monitored at appropriate temporal scales. The number of repeat visits made by a sensing system over a given time period is called the *temporal resolution* of the system.

The *spatial resolution* of an imaging instrument, like the spectral and temporal resolutions, is also related to the requirements of a specific application. The highest available resolution image data will be provided by U.S. very high resolution (VHR) satellite systems developed from military technology that has been made commercially available following Presidential Directive 23. These systems (EarlyBird and Space Imaging are planning a 1997 launch; Quickbird is planned for 1998, and Orbview for 1999) provide accurate orbit tracking using GPS, with resolutions in panchromatic mode ranging from 1–3 meters (~3–10 ft). Such imagery will find applications in cartography at scales of 1:5000 and larger, as well as in land cover and disaster monitoring. An important attribute of VHR data is that it can also be acquired in stereoscopic mode, providing the opportunity for the development of digital elevation models (DEM). An example of a coarse-scale DEM of Yellowstone National Park derived by conventional means is shown in Figure 3. These elevation models are used in environmental modeling, landscape planning, highway construction, communications planning, and other similar applications.

Medium scale (10–30 meter) resolution imagery is provided by the Thematic Mapper carried by the Landsat satellites, the High Resolution Visible system onboard the French SPOT satellite. In addition, Japanese satellites such as the Advanced Earth Observing System (ADEOS) (5) and the Indian IRS-1 series provide multispectral imagery. ADEOS, for example, carries the Advanced Visible and Near-Infrared Radiometer (AVNIR), which provides imagery in four spectral bands with a spatial resolution of 16 meters plus a panchromatic band

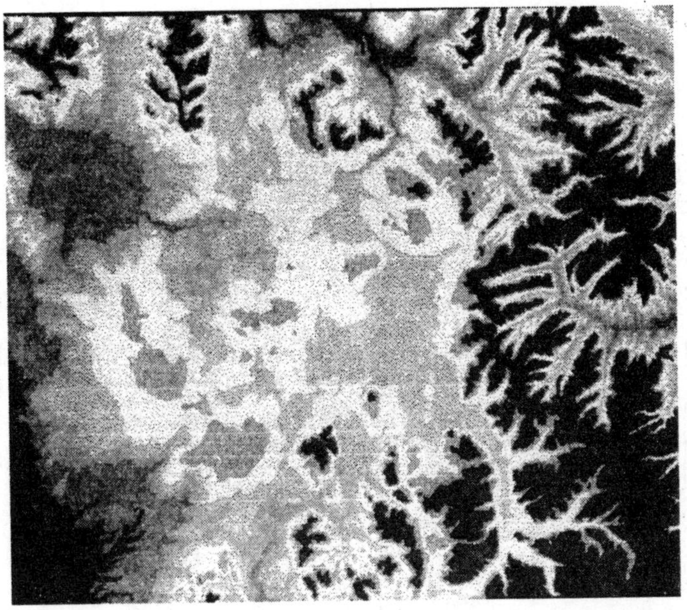

Figure 3. Digital elevation model (DEM) of the Yellowstone National Park. Scale ~1 : 250,000. Produced by the Defense Mapping Agency and distributed by the U.S. Geological Survey, Eros Data Center (Sioux Falls, S.D.) (4).

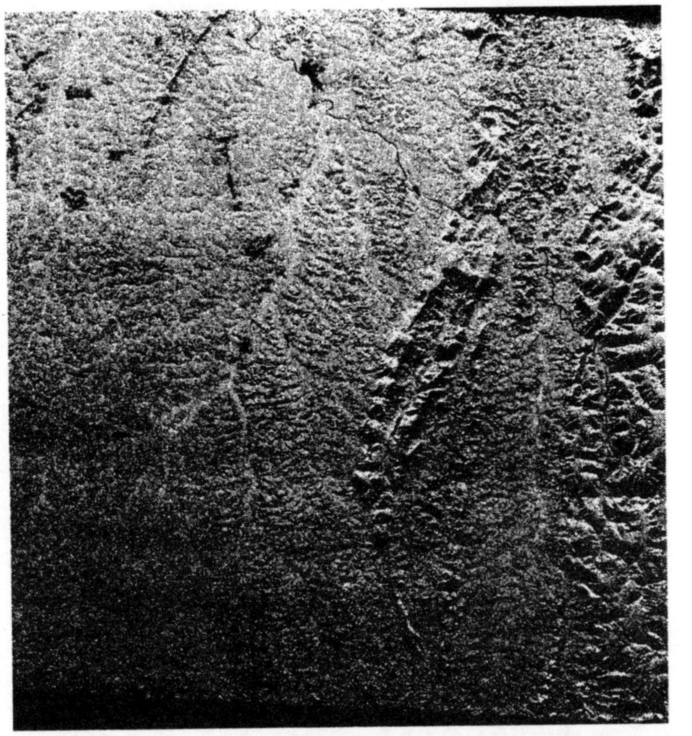

Figure 4. Synthetic aperture radar (SAR) image of part of the Brazilian rainforest, May 17, 1992. The image was acquired by the Japanese J-ERS-1 (Earth Resources Satellite 1) (6).

with a spatial resolution of 8 meters. Instruments for recording data from the oceans and the atmosphere are also carried by ADEOS. Data from the Indian IRS-1C satellite is marketed by the EOSAT Corporation (Lanham, Maryland), including 5-meter resolution panchromatic imagery and 25-meter resolution multispectral data.

EOSAT also markets data from the Japanese JERS-1 satellite which carries an imaging microwave radar sensor. The Canadian RADARSAT and the European ERS satellites also carry radar sensors, which have the advantage that they are day/night and weather independent, so that they can "see" through clouds or during the hours of darkness. The spatial resolution of the synthetic aperture radar systems carried by these satellites is of the order of 25 meters. An example of a microwave radar image is shown in Figure 4. This image was produced by the Japanese JERS-1 system in May 1992. It shows an area of Brazilian Amazonia. The drainage pattern as well as geological and topographic structures are very clearly seen, even though the area (which is covered by tropical rain forest) may have been hidden by cloud.

Coarse resolution data are obtained from systems such as the NOAA AVHRR, which provides imagery in four or five spectral bands ranging from the visible to the thermal infrared, and with a nominal resolution of 1.1 kilometers at nadir, that is, at the point on the ground immediately beneath the satellite. These data are captured on a daily basis and provide a means of observing global weather systems for forecasting purposes. Data from this sensor are also used to provide weekly summaries of the status of global vegetation based on a vegetation based on a vegetation index.

Future plans for satellite observing systems and the accompanying ground systems are focused on the Earth Observing System (EOS). Figure 5, derived from NASA's EOS Science Office, shows the satellite platforms scheduled for launch between 1997 and 2008, together with the instruments that will be carried by each platform. The non-U.S. platforms are shaded. The U.S. EOS contribution to a global observing system will begin with the Tropical Rainfall Monitoring Mission in 1997 and the AM-1 and Landsat-7 platforms in 1998. Together with systems such as those provided by the Japanese NASDA ADEOS II (1999), the French SPOT 4, and the European Space Agency (ESA) ENVISAT (also 1999) a considerable number of sensors operating in different regions of the electromagnetic spectrum and at different spatial and temporal resolutions will be providing increasingly voluminous amounts of data on the state of the various Earth systems, ie, atmosphere, biosphere, hydrosphere, and cryosphere.

EARTH OBSERVATION DATA: THE GROUND SEGMENT

Descriptions of existing and planned satellite data collection systems are described in the preceding section from which it is clear that, if present plans come to fruition, a very considerable volume of digital data will become available for a wide range of science and practical applications across a range of disciplines. In this section the mechanisms by which profile and image data are collected from orbiting platforms and prepared for the end user are considered.

The space-faring nations of the world have their own national space agencies, for example, the U.S. National Aeronautics and Space Agency (NASA), Canada's Centre for Remote Sensing (CCRS), France's Centre Nationale d'Etudes Spatiale (CNES), India's National Remote Sensing Agency which is part of the Indian Government's Department of Space, Japan's National Space Development Agency (NASDA), and Europe's Space Agency (ESA). In each of these countries and

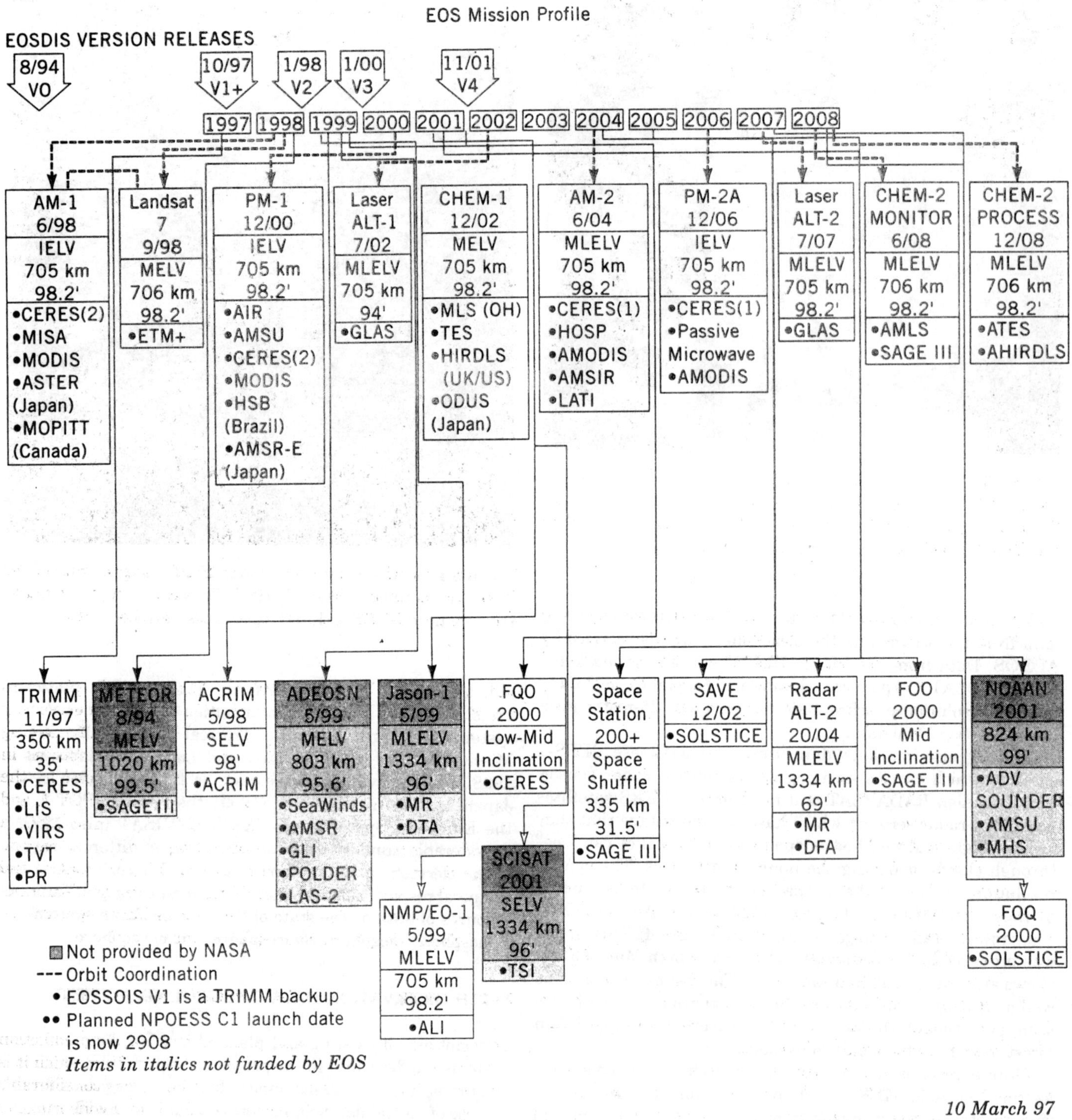

Figure 5. Earth observing system: missions planned by NASA, NOAA, NASDA (Japan), and the Russian Space Agency, 1997–2008 (7).

regions data reception facilities and data processing centers exist for the acquisition, preprocessing, and archiving of data obtained from satellite missions. Because the instruments carried by orbiting spacecraft view the entire globe (apart from areas around the North and South poles) there is international cooperation so that ground stations operate on a cooperative basis. Thus, data from the Indian IRS-1 satellite is collected by European and North American ground stations, and Landsat TM and MSS data are downlinked to ground receiving stations worldwide. In addition, Tracking and Data Relay Satellites (TDRS) are used to transmit data from U.S. remote sensing satellites such as Landsat to a reception station at White Sands, New Mexico, thus obviating the need to rely on foreign cooperation. An alternative is to use onboard tape recording systems to store data which are collected while the satellite is out of range of a ground station, and to replay the recording once in range of a suitable ground receiving station.

Data from earth-orbiting satellites are pre-processed in a number of ways before they are made available to the end user. Preprocessing and correction are almost invariably necessary before earth observation data can be used in practical applications. Raw data transmitted from the satellite to the ground station, either directly or via a communications satellite link, contain various systematic errors. For example, the Landsat Thematic Mapper instrument operates in both "forward" and "reverse" mode, so that data are acquired as the sensor's mirror swings from left to right and from right to left. Each sweep of the mirror collects 16 scan lines. A raw TM image would, therefore, appear to have the order of the pixels reversed within each group of 16 scan lines. These and other anomalies are generally removed from the data at the receiving station.

Radiometric and geometric correction can also be carried out at the receiving station. *Absolute* radiometric correction refers to the conversion of the data from counts to physical values. Uncorrected (raw) data are provided as counts, usually on a 0–255 scale representing 256 levels in a gray scale image, but the relationship between data values collected at different dates crucially depends on the calibration of the sensor. Because of variations in sensor calibration over time, as a result of sensor degradation, the physical meaning of a given count varies from time to time, causing problems if the data are to be compared. Physical values of radiance collected by the sensor can be retrieved from counts, usually by a function such as $radiance = offset + gain \times count$. The values of *gain* and *offset* change through time due to sensor degradation, so that the radiance associated with a given count will not be constant between two images separated by a time gap of months or years. Any attempt to infer change in the characteristics of the land or sea surface from a study of counts is therefore unreasonable.

Relative radiometric correction consists of "balancing" the outputs from each detector so that they have the same statistical properties. Landsat TM has 16 detectors per spectral band, and these detectors deteriorate over time. Hence, to avoid visually distracting variations in brightness down the image, the means and standard deviations of the data generated by each of the 16 detectors are equalized. Processing and archive centers generally supply relative radiometrically corrected data.

Geometric rectification (or geo-referencing) is needed to remove the effects of earth rotation, panoramic distortion, variation in the satellite platform's attitude (pitch, roll, and yaw) and, for electromechanical systems such as Landsat TM, the nonlinearity of the mirror motion. However, even if these effects are removed the resulting image does not fit any map projection, and cannot be overlaid on other geographic data sets, for example within a Geographic Information System (GIS). Geo-referencing refers to the precise adjustment of the image geometry so that the geometry conforms to that of a specified map projection. For existing satellite systems such corrections use the principle of point or feature matching which requires that matching control points are located on both image and map, so that a mathematical transformation can be derived based on the polynomial functions. With the advent of the Global Positioning System (GPS) and laser ranging it will be possible for the location of the satellite in its orbit relative to the earth's center of mass to be derived, and this information can be used in orbital models to calculate the coordinates of any pixel in the image. More elaborate forms of geo-referencing are required when the target area contains significant elevation variations. In such cases a digital elevation model (DEM) is used to provide (x, y, z) coordinates of the control points and the resulting transformation corrects for variations in elevation as well as in planimetric position.

Ground receiving stations may also have the function of archiving (storing) both raw and corrected data and products derived from such data. Alternatively, this function may be taken on by a data center or a commercial company. The EOSAT company operates both preprocessing and archiving facilities in the United States, as does SPOT Image. The Eurimage company takes on a similar role in Europe. Data centers may be associated with ground receiving stations or they may be in a physically separate location. The archiving function is closely linked to that of dissemination and data distribution, which is described in the final section of this article.

EARTH OBSERVATION DATA: THE USER SEGMENT

Data collected by earth-orbiting satellites are transmitted to ground receiving stations, where they may be preprocessed and transferred to a data center for distribution and archiving. During the first 20 or so years of satellite remote sensing activities, data centers in different countries operated more or less independently so that the products and services provided by each center were often incompatible. Data were generally provided on 26.6 cm (10.5 in.) magnetic tapes in a wide variety of formats. During this time a significant obstacle to the more widespread use of remote sensing was the apparent problem of reading data from magnetic tapes. The main U.S. data center at this time was the USGS EROS Data Center at Sioux Falls, South Dakota. Its equivalent in Europe was the European Space Agency's facility at Frascati, near Rome, Italy. These centers continue to operate today, but they form part of an international cooperative network, promoted by the Committee on Earth Observation Satellites (CEOS) and its working groups. CEOS was established in 1984 and has the following aims (8,9):

- to optimize the benefits of space-borne earth observations through cooperation of its members in mission planning and in the development of compatible data products, formats, services, applications, and policies;
- to aid both its members and the international user community by serving as the focal point for international coordination of space-related earth observation activities, including those related to global change;
- to exchange policy and technical information to encourage complementarily and compatibility among space-borne earth observation systems currently in service or development, and the data received from them;
- to address issues of common interest across the spectrum of earth observation satellite missions;
- to allow individual members of CEOS to use their best efforts to implement CEOS recommendations in their respective earth observation programs.

The CEOS International Directory Network (CEOS IDN) (10) is a cooperative international effort with the aim of providing access to information on available data sets in different parts of the world (the Americas, Europe, Japan, Russia, India, etc). Information about data sets is termed *metadata*. The IDN provides access via the World Wide Web (WWW) to scientific information in the areas of the earth sciences, plus space physics, solar physics, planetary science, and astronomy/astrophysics.

The U.S. node of the CEOS IDN is the NASA Global Change Master Directory (GCMD) which offers a comprehensive source of information about satellite and *in situ* earth science data, with broad coverage of the atmosphere, hydrosphere, oceans, solid earth, and biosphere (11). The GCMD provides user access to a search engine which submits the user query to a database containing summary information (metadata) describing data sets whose characteristics match the query specification. Some of these data sets can be downloaded using the file transfer protocol (FTP).

The NASA GCMD is one example of a data center that provides access to earth observation and related science data. The U.S. government's *Mission to Planet Earth* (MTPE) (12), the NASA Earth Observing System (EOS) (13), and the U.S. Global Change Research Program (14) are examples of other programs that support data archiving activities.

EOS is intended to provide a complete infrastructure for the ground and user segments. The Earth Science Data and Information System (ESDIS) at the NASA Goddard Spaceflight Center (Greenbelt, Maryland), is tasked with the following operations: spacecraft control, science data processing, and archiving, management, and distribution of earth science data from NASA missions. The last of these functions is carried out through the EOS Data and Information System (EOSDIS) (15). EOSDIS provides facilities for the user to access individual centers, called Distributed Active Archiving Centers (DAACS) as well as foreign centers.

Each DAAC is responsible for a specific area of science data, as shown in Table 1. In addition, EOSDIS provides access to the Japanese National Space Development Agency's (NASDA) Earth Observation Center (EOC) and the NASA Global Change Master Directory.

Table 1. DAACs Associated with U.S. EOSDIS

Distributed Active Archive Center	Typical Data Sets
Alaska SAR facility, University of Alaska, Fairbanks	Microwave radar data and products
Goddard Space Flight Center, Greenbelt, Md.	Ocean color, hydrology, and precipitation, atmospheric dynamics, atmospheric chemistry, land biosphere
Langley Research Center, Hampton, Va.	Earth Radiation Budget experiment, International Cloud Climatology project, NASA Water Vapor project
Global Hydrology Resource Center, Huntsville, Ala.	Tropical Rainfall Monitoring Mission Lightning data
NOAA Satellite Active Archive	AVHRR satellite data and products, Defense Meteorological Satellite data and products
National Snow and Ice Center, Boulder, Colo.	Snow and ice data sets
Oak Ridge National Laboratory, Oak Ridge, Tenn.	Biogeochemical dynamics data for terrestrial ecosystems, Carbon Dioxide Information Center
CIESIN, University Center, Mich.	Socioeconomic data and applications

The European Commission, with the agreement of the European Space Agency, has agreed to establish a distributed Centre for Earth Observation (CEO) which will be part of a future European Earth Observation System (EEOS) (16) responsible for:

1. Acquisition, cataloguing, and archiving of raw and pre-processed data from European and selected non-European satellites,
2. Distribution of preprocessed data using CEOS-recommended data formats and computer networks,
3. Development of applications of earth observation data,
4. Establishing a data policy to meet the needs of all categories of users, and
5. Providing an interface between European activities and developments taking place elsewhere in the world.

Point 5 leads to the difficult question of interoperability. Four major data and information systems are mentioned above: the NASA Global Change Master Directory, EOSDIS, the Japanese EOC, and the European CEO. Although it is possible for an individual user to interrogate all the available systems separately, in an ideal world it would be more acceptable to have the same query passed between different information systems and the result returned to the user's terminal. Interoperability is possible at a variety of levels (17). At the lowest level the user has access to a central node which allows a choice to be made from a list of available data centers such as the EOSDIS DAACs. A slightly more sophisticated version would pass some contextual information from the nodal system to the remote system. At level 3, the user would not be aware of the existence of separate systems, and queries would be passed automatically from one system to another, with results returned in the opposite direction. Finally, at the highest level of interoperability all of the associated systems would act like a distributed database so that relational operations could be carried out on data set descriptions held at separate locations. The CINTEX experiment, described in (17), represents an early attempt to link several data centers in an interoperable manner.

The provision of access to metadata describing data sets relating to earth observation is only one aspect of the user segment. Facilities need to be implemented to give access to descriptions of services and software, as well as tutorial systems and bibliographic databases, to allow a user to select appropriate data, methods, hardware, and consultancy services. Systems such as the NASA Global Change Master Directory, the CEOS IDN, and EOSDIS have been established for several years and are continually being expanded both in terms of capability (for example, the inclusion of free text searching as opposed to keyword-based search) and content. Other related systems, such as the CEOS WGISS Yellow Pages (8) (provided by the CEOS Working Group on Information Systems and Services, WGISS), give access to details of satellite sensor systems and to services offered by CEOS member agencies, such as the national space agencies, and international bodies, such as the United Nations Environment Program (UNEP) Global Resource Information Database (GRID), the United Nations Food and Agriculture Organization (FAO) World Agricultural Information Center (WAICENT), and the International Geosphere–Biosphere Program (IGBP) Data and Information System. The CEOS Yellow Pages are a good place to start if a user wishes to find suitable remotely sensed data for a particular application.

The European Commission Centre for Earth Observation has recently introduced its European-Wide Service Exchange (EWSE) on the World Wide Web (18). The EWSE aims to provide access not only to data and information services but to service providers (organizations such as consultants, universities, and educational institutions) and to sources of information on software. The CEO Algorithms, Models, and Software (AMS) database (19) contains information about software packages that provide functionality ranging from image processing and Geographic Information Systems (GIS) to scientific modeling. AMS also includes an extensive tutorial package to guide the novice user, and a bibliography of around 100 items. EWSE now contains links to a broad representation of applications of earth observation data to a range of environmental problems (2).

SUMMARY

The process of earth observation from space requires significant investment. In the early stages of earth observation, much of this investment went into the space segment (development of satellite platforms, instruments, and launchers) and into the ground segment, with relatively small amounts being allocated to the user segment. Since the mid-1980s it has been realized that the market for EO products and services requires a formal infrastructure (the user segment) which will give access to data, products, services, software, and tutorial assistance. Current international developments (NASA GCMD, EOSDIS, NASDA's EOC and the EC's CEO) are now in existence to provide this infrastructure, with international coordination provided by the Committee on Earth Observation Satellites (CEOS). The need for this expensive infrastructure will become clearly apparent once the next generation of earth observation systems is in place.

BIBLIOGRAPHY

1. R. Harris, in P.M. Mather, ed., *TERRA 2: Understanding the Terrestrial Environment: Remote Sensing Data Systems and Networks,* John Wiley and Sons, Ltd., Chichester, U.K., 1995, pp. 43–60.
2. *http://topex-www.jpl.nasa.gov/discover/images/first_image.gif*
3. NASA, *Mission to Planet Earth,* program is described at *http://www.hq.nasa.gov/cgi-bin/print_hit_bold.pl/office/pao/NASA/mtpe.html;* practical applications of earth observation data are given at *http://ewse.ceo.org/anonymous/case_studies/list_case_studies.pl;* the *Space Mission Acronym List and Hyperlink Guide* is located at *http://ranier.oact.hq.nasa.gov/Sensors_page/MissionLinks.html.*
4. *http://edcwww.cr.usgs.gov/glis/graphics/guide/1_dgr_dem/figure1.gif*
5. ADEOS details at *http://www.eorc.nasda.go.jp/ADEOS/*
6. *ftp://ftp.eorc.nasda.go.jp/pub/JERS-1/SAR/NASDA-007-1/SAR-Amazon_deforest/Tiff/Brazil.rain.forest-tiff*
7. *http://eospso.gsfc.nasa.gov/Images/mission.gif*
8. *http://www.smithsys.co.uk/yp/*
9. *http://gds.esrin.esa.it/CEOSinfo*
10. *http://gcmd.gsfc.nasa.gov/ceosidn/*
11. *http://gcmd.gsfc.nasa.gov/*
12. *http://www.hq.nasa.gov/cgi-bin/print_hit_bold.pl/office/pao/NASA/mtpe.html*
13. *http://eospso.gsfc.nasa.gov/eospso_homepage.html*
14. *http://www.usgcrp.gov/*
15. *http://spsosun.gsfc.nasa.gov/EOSDIS_main.html*
16. P. Churchill, G. Mancini, S. Ruddock, and J. Ellis, in Ref. 1, pp. 61–74.
17. B.D. Thomas, M.D. Elkington, and M.S. Hutchins, in Ref. 1, pp. 113–122.
18. *http://ewse.ceo.org/*
19. WWW URL as [19]-click *Algorithms and Software.*

REMOTE SENSING FTIR SPECTROSCOPY

DOOYONG PARK
Hansung University
Seoul, Korea

STEVEN P. LEVINE
University of Michigan
Ann Arbor, Michigan

GEORGE M. RUSSWURM
ManTech Environmental Technology Inc.
Research Triangle Park, North Carolina

In this article, the basics and field applications of FTIR and remote sensing (RS) are discussed. The remote sensing configurations of RS-FTIR are introduced. Following that, advantages and disadvantages of the system are presented. Various topics related to the field applications of RS-FTIR are discussed. This section includes target air pollutants, the process of path siting, the background spectrum, resolution effects, quality assurance and quality control, and actual field applications.

RS-FTIR instruments transmit an IR beam across a long open path through ambient air to monitor gaseous air pollutants. Air monitoring techniques for the past decades have been improved to increase simplicity, accuracy, and sensitivity. Additional improvements have been made to acquire real-time data, be less obtrusive, and have the ability to analyze multiple compounds simultaneously. The RS-FTIR technique has found increasing applications for environmental, indoor, and industrial air monitoring, because it addresses many of these criteria. The RS-FTIR technology was initially discussed more than 20 years ago (1–3). However, despite the success reported in these early manuscripts, the method was largely overlooked by the vast majority of the experts engaged in air monitoring until the late 1980s. Most commercial instrumental developments and their applications were made over the past five years. This was because of the recent growth in requirements to monitor a large number of air pollutants (4). An increased need for real-time monitoring in environmental arenas following the Bhopal incident was another driver for the development of these techniques.

FUNDAMENTALS OF RS-FTIR

A number of optical remote sensing (ORS) technologies have been developed and demonstrated over the past decades for applications in various environmental monitorings. Recently there has been a renewed interest because of the potentially large number of compounds that may be required to be monitored. Improvements in hardware and software techniques related to the ORS for the past decade have been made so that they can generally measure multiple compounds simultaneously and quickly. Various sampling strategies have been actively pursued to employ these techniques in environmental air monitoring. Thus, ORS technologies are expected to play a significant role in many industrial and environmental settings to monitor atmospheric pollutant gases.

A guidance document has been developed that includes a glossary of terms used in RS-FTIR (5). Remote sensing is generally defined as the measurement of gases, solids, and liquids *in situ* without the use of a closed sample cell in the instrument, or a sample storage medium. The optical remote sensor is generally set up to transmit a beam of radiation through a parcel of air that contains one or more species of interest. Remote sensors are generally diverse spectroscopic instruments that can use either dispersive or nondispersive optics. When the U.S. Environmental Protection Agency (EPA) developed a prototype of RS-FTIR system to measure environmental pollutants over long distances, it was initially called the remote optical sensing emission (ROSE) FTIR. Since the first ROSE FTIR air monitoring system was reported in the literature in 1979 (2), several terms have been used to indicate this type of FTIR: RS-FTIR, open-path FTIR, FTIR open path-monitoring, or FTIR open-path remote sensing. Although there is no consensus yet, all these terms are generally used to describe the same FTIR system.

Configurations of RS-FTIR System

RS-FTIR has logically evolved from the gas-cell FTIR. It has four main components: (*1*) the FTIR spectrometer, (*2*) the transmitting and receiving optics, (*3*) the electronics, and (*4*) the computer system.

There are two primary geometrical configurations for transmitting the IR beam along the path. One is a bistatic system and the other a monostatic system (Figs. 1 and 2). The bistatic system has the detector at one end of the beam path and the source at the other. The monostatic system has both the IR source and the detector at the same end of the path. Thus, in the bistatic system, the optical path length is equal to the physical path length between the source and the detector, whereas in the monostatic system, the optical path length is twice the physical path length. In remote systems, the gas cell of the classical FTIR is replaced by telescopes to collimate, transmit, and receive IR radiation through the ambient air. In the bistatic configuration, the IR source unit (and transmitting optics) is placed at one location, and the detector unit at another location. The optical path is one way from the IR source to the detector unit. In the monostatic configuration, the source and detector are situated together into a single unit. This system is placed at one end of the optical path and, to reflect the beam back to the detector, a mirror or multiple mirrors are used at the other end.

Some RS-FTIR systems are designed to use uncontrolled light sources which are already presented in the environment. This system measures naturally emitted IR energy by chemical species of interest. The configuration of this measurement technique is referred to as the passive mode in contrast to the active mode which uses its own IR source (5).

All IR sources that are available today for the use with an FTIR are heated elements. The ideal temperature of the source would be approximately 3000 K so that it would have the maximum power output at about 1.1 μm. However, sources which operate at this temperature are not available. Most currently used IR sources operate at a cooler temperature of from 1200 to 1500 K. The other limitation of the available sources is that their sizes are large compared to the focal length of the collimating optics. The beam can never be better collimated than the angle that the source subtends at the collimating device. Thus, all available FTIR systems have beam divergence. This implies that retroreflectors or receiving telescopes tend to be overfilled, which results in loss of the transmitted energy.

In most cases, the beam is collimated and collected by using a telescope before and after it is transmitted along the path. Two types of telescopes are mainly being used: the Cassegrain and the Newtonian telescopes. In the Cassegrain, the beam exits from the end of the telescope; in the Newtonian, it exits from the side.

Bistatic System

The bistatic configuration has only a single beam splitter in the interferometer, whereas the monostatic configuration requires a secondary beam splitter. Therefore, in the monostatic configuration, a large portion of the beam intensity is lost at the retroreflector, the receiving telescope, and the secondary beam splitter (see "Monostatic System" below). This is a fundamental advantage of a bistatic configuration over a monostatic configuration.

At the source unit, there must be some device to collimate the beam before it is transmitted. A beam collector is needed at the detector. Most commercial RS-FTIR instruments use the telescopes to transmit and receive the beam.

The interferometer system can be installed either in the source unit or the detector unit. Thus, there two configurations in the bistatic systems. One configuration places the interferometer in the IR source unit (see Fig. 1**a**). The other configuration places the interferometer in the detector unit, which is more commonly used for the bistatic systems (see Fig. 1**b**). In one system, the IR energy is transmitted through the interferometer before it traverses the path and is thus modulated and the incoming black body radiation can be suppressed. In the other, the black body radiation is modulated along with the source energy. The bistatic system is unable to distinguish the IR radiation between the active instrument and the background sources. However, the detector unit of this system can be operated in the passive mode without using an active IR source (see Fig. 1**c**).

When the bistatic system is used for ambient air monitoring in rough terrain or confined areas, practical problems may be encountered. The bistatic configuration requires power sources for both the source and detector units that are located at each end of the beam path. When the contaminated area is relatively small, a single pass of the beam between the source and detector may not be enough to measure the absorbance given the low concentration of typical ambient air contaminants. If air contaminants exist in a limited area, the absorbance by the target chemicals cannot be increased by simply further separating the source and detector. It is often impossible to have a sufficient

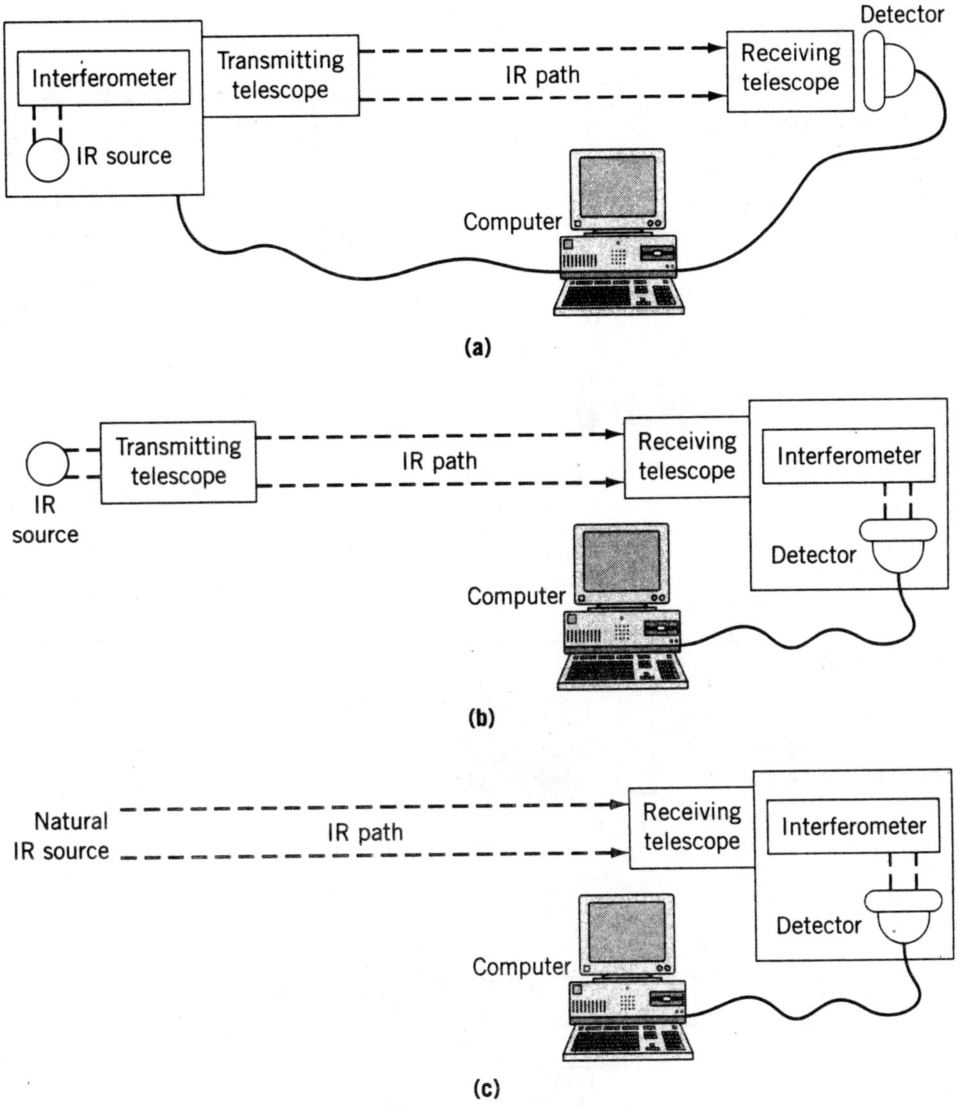

Figure 1. Schematic of the bistatic configuration of RS-FTIR: (**a**) interferometer placed in the IR source unit; (**b**) interferometer placed in the detector unit; (**c**) configuration of (**b**) can be used as a passive mode.

path length in indoor or workplaces, but in this case, intermediate mirrors may be used to increase the path length (Fig. 3).

Monostatic System

There are currently two configurations used in typical monostatic systems (see Fig. 2). Basically, all optical components are in a single module. In one configuration, offset corner-cube mirrors are used to translate the beam slightly for its return path to the offset detector (see Fig. 2**a**). The beam path originating to the mirrors is actually different from that returning to the detector. Thus, two telescopes are required: one to transmit and the other to receive the beam.

The second monostatic configuration uses a beam splitter that allows the source, interferometer, and detector to be housed in a single module equipped with a single telescope for sending and receiving (see Fig. 2**b**). If a mirror or retroreflector is used to reflect the beam at the end of the path, the incident and reflected beams are collinear. Thus, only a single telescope is required for transmitting and receiving the beam in this configuration. Usually, in field monitoring situations a retroreflector array is used to reflect the IR beam at the end of the path instead of flat mirror.

A corner-cube retroreflector is an arrangement of mirrors that reflects the beam so that the incident and reflected directions of propagation are collinear but opposite. It is usually made of three reflecting surfaces that are mutually orthogonal, the shape of which is similar to the corner of the room where the floor and two adjacent walls meet.

The bistatic and monostatic systems are different. Because of the second beam splitter, 75% of the energy is lost compared to the bistatic system. The loss of transmission affects the signal-to-noise (S/N) ratio. However, this configuration gives several advantages. The most significant advantage is that the IR beam is modulated before it is transmitted along the path. Therefore, the system is able to effectively reject any unmodulated energy entering the instrument. The other one is that the alignment with the use of a retroreflector is very easy. Because of the design of the orthogonal mirror configuration it is quite insensitive to small changes in alignment such as those caused by a wind or vibration.

This configuration has been used with a self-contained beam aiming device that allows translation in the horizontal and vertical axes. A mirror attached to a computer-controlled solenoid permits rapid beam movement to the directions defined by the parameters that are programmed into the data

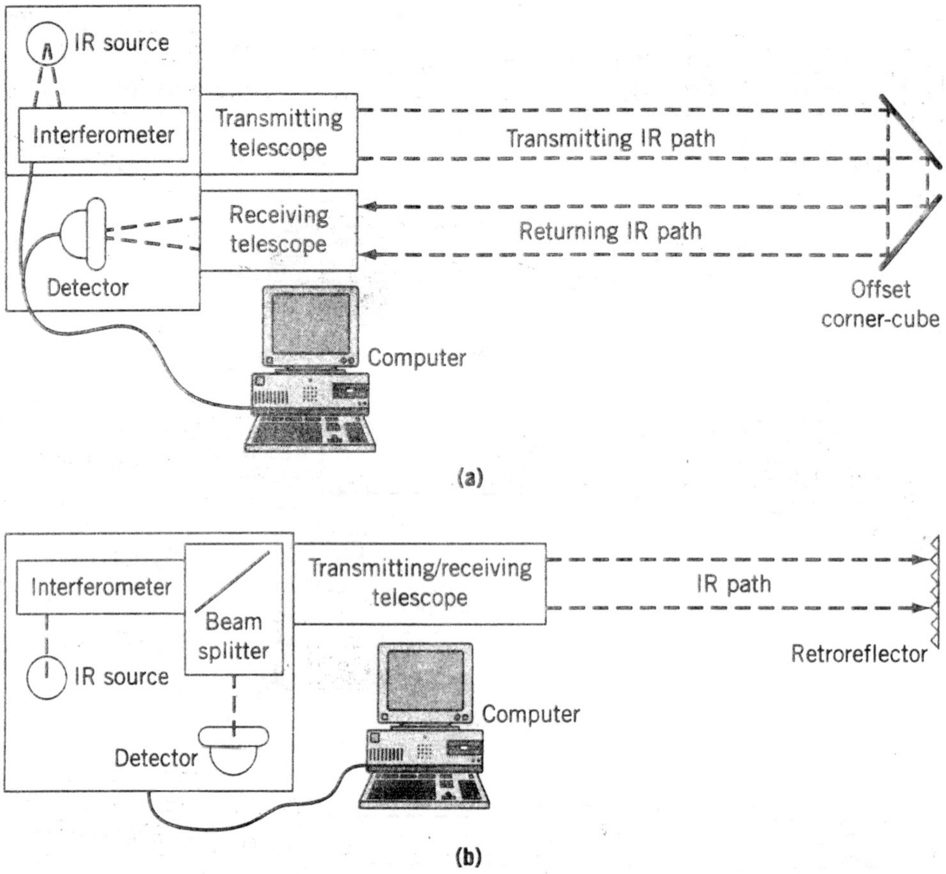

Figure 2. Schematic of the monostatic configuration of RS-FTIR: (**a**) beam path is translated using offset corner cube (two telescopes are required); (**b**) all components are placed in a single module (one telescope used to transmit and receive beam). An additional beam splitter is required to send the beam to the detector, thus maximum 12.5% of transmitted beam from the IR source would be detected.

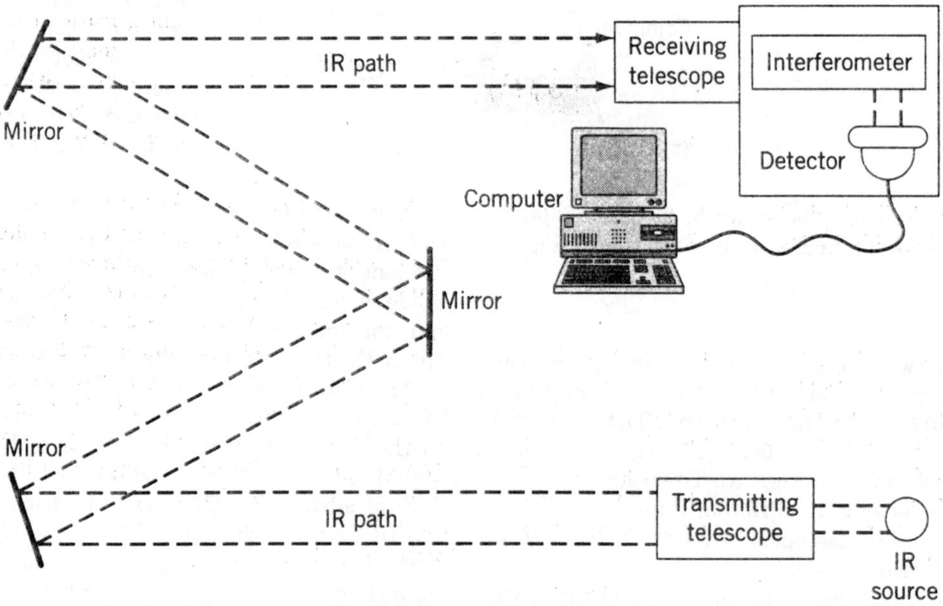

Figure 3. Schematic of a beam configuration using intermediate mirrors to extend the beam path length.

system. Thus, many path measurements can be obtained with the use of a large number of retroreflectors. This improvement allows the collection of the sample and background spectra frequently without moving the whole system, and the measurement of the emissions for many paths to cover an entire area. It is also suitable for plant fence line monitoring in many industrial settings. Recently, this system has been discussed for use in application of tomographic reconstructions of the air pollutant concentration field for a whole measurement area.

ADVANTAGES AND DISADVANTAGES OF REMOTE SENSING

Traditional air monitoring methods used to evaluate environmental air pollution are generally classified into three types: (*1*) direct reading, (*2*) those which remove the contaminant from a measured quantity of air using proper sampling media, and (*3*) those which collect a known volume of air.

Many of the direct reading instruments are capable of measuring only one or a limited number of air pollutants at a time, and can be used only to sample at individual points. It is also well known that certain classes of these instruments, because of their design and complexity, require frequent calibration.

The second and third methods usually require subsequent laboratory analysis. These traditional methods have several inherent limitations. For sampling and laboratory analysis methods, there is a time delay to get the measurement results from the laboratory. In general, these methods produce time-integrated measurement data that do not provide the temporal variation of concentrations for the sampling period. Temporal information is necessary for acute toxins and irritants, and in cases where the air contaminant concentrations must be causally linked to process state or meteorological factors. In addition, these methods are not suitable for polar or labile compounds because they can be easily lost during sampling, shipping, and storing. Also, point monitors sample large volumes; FTIRs sample large areas.

Ideally, an air monitoring instrument should be sensitive, specific, capable of identifying unknown components, fast, cheap, and easy to use. RS-FTIR does not satisfy all these requirements. However, some of these requirements can be achieved only with RS-FTIR techniques. Unlike traditional methods, with remote sensing no sampling container is used. Thus, sampling errors introduced during collection are eliminated because there is no contact during measurement. This allows the remote system to overcome one significant limitation of the traditional methods which may not be able to accurately measure polar and labile compounds because of the high and unpredictable losses in the sampling system. The cost and humanpower requirements should also be decreased because no sampling procedure is needed. It was also possible to monitor inaccessible or isolated areas with this measurement technique. An example of this is the monitoring of stack emissions from a location on the ground outside the fence line of a plant.

Measurements with the use of RS-FTIR are made in real time or near real time. Thus, it is appropriate for measuring transient contaminants, to detect leakage, and to warn workers or residents. It provides the measurement data with a good time resolution. In addition to the capability of simultaneous multicompound analysis, RS-FTIR system stores the entire IR spectrum.

RS-FTIR produces path averaged measurement data. This can be an advantage to evaluate emission fluxes over an environmental area. With FTIR not giving the maximum concentration and the point monitor perhaps missing the plume, both RS-FTIR and point monitors have weakness. Path averaged data are perfect for application of RS-FTIR in computed tomographic techniques which require essentially multiple path integrated data to reconstruct air contaminant concentrations over a measurement plane.

Several disadvantages have also been discussed for RS-FTIR applications. The primary problem facing users of open path optical methods is the acquisition of a clean background spectrum. The background is the spectrum obtained when no target compound is present. Thus, the background would be a spectrum taken in the absence of the particular absorbing species of interest with all other conditions being equal. With the use of a classical FTIR and a closed sample cell, a background spectrum can be easily obtained. However, with the use of a RS-FTIR instrument, it is extremely difficult to acquire a background spectrum in the field situation, because the air pollutants cannot be simply eliminated from the atmosphere.

A background spectrum is required because Beer's law is used in the calculation. Although several techniques have been developed to generate a background, there is no prescribed method to create an ideal background. Each method used to acquire a background spectrum is discussed in detail in the section "Acquisition of Background Spectrum" below.

Most RS-FTIR systems are in limited production and therefore are relatively expensive. Thus, the instrument cost may be a limitation. In many environmental monitoring situations, the path averaged data can be a significant limitation. RS-FTIR integrates the air contaminant concentration along the beam path. By dividing by the total path length, a path averaged concentration is obtained. Thus, there is no difference between 10 ppm of a well-mixed contaminant along a 10-m beam path, or 100 ppm contained within 1 m of a 10-m beam path and 0 ppm in the remaining 9 m. The same result of 100 ppm would be obtained from the measurement using RS-FTIR in these two scenarios. Therefore, it is necessary to develop proper sampling strategies to address this critical issue of poor spatial resolution.

FIELD APPLICATIONS OF RS-FTIR

Target Air Pollutants

The term "gaseous air pollutants" indicates gases and vapors which are present in the form of the gaseous phase in an ambient air matrix. Gases are generally understood to be in the gaseous state, and the liquid status of gas has a vapor pressure higher than 1 atm at room temperature. Vapors mean the gaseous phase of compounds which are in the liquid state, and have a vapor pressure less than 1 atm at room temperature. Gases and vapors, under ordinary environmental conditions, exist in the gaseous state even when present at high concentrations, and are considered to have the same behavior. One of the advantages of the IR techniques is that various gases and vapors can be measured simultaneously. Homonuclear diatomic molecules such as nitrogen, oxygen, and hydrogen and monatomic gases such as helium, argon, and radon do not absorb significant IR radiation over the relatively short paths of interest here. Most target chemicals in environmental monitoring are heteronuclear diatomic or polyatomic molecules which absorb IR radiation. Therefore, it is possible to measure almost all gaseous molecules with the use of IR spectroscopic methods. For these reasons, IR spectrometers have been widely used for more than 40 years.

Path Siting

Field monitoring strategies for the use of RS-FTIR depend on the monitoring situations such as objectives, target compounds, meteorological conditions, and instrument performance. The most important procedure in field monitoring is to set up a path so that the selected beam path is representative of the contaminated area. The path configuration must be based on the meteorological conditions. For short-term monitoring program, a flexible monitoring strategy using various path designs can be based on the meteorological conditions. However, the path in the long-term monitoring program must be set up to account for changes of the meteorological conditions, such as wind directions. In addition to the meteorological conditions, it is generally recommended that several factors be considered when designing the path configuration: instrumental parameters, such as the S/N ratio; the characteristics of the target compounds, such as concentration and absorption coefficients; concentrations of the interferences, such as H_2O and CO_2; characteristics of the sources and of the most sensitive receptors (target populations); and the appropriate sites to accommodate the instruments and peripherals (5).

Appropriate criteria for ambient air monitoring for use of RS-FTIR are included in Part 58 of Chapter 1 of Title 40 of the *Code of Federal Regulations* (40CFR58) which was being promulgated by the EPA (6). These amendments cover how the path is to be chosen in terms of obstructions, height above the ground, and changes in path height. They also describe the appropriate positioning of the path in relation to buildings, stacks, and roadways. The following considerations are included in these criteria:

- At least 80% of the path must be between 3 and 15 m above the ground.
- At least 90% of the path must be at least 1 m vertically or horizontally away from walls.
- If the path has to be near a building, then it must be on the windward side of the building.
- Buildings or other obstructions may possibly scavenge the gases of interest. At least 90% of the path must have unrestricted airflow and be located away from obstructions so that it is removed by at least twice the height that the obstacle protrudes above the path.
- At least 90% of the path must be at least 20 m from the drip lines of trees.
- When monitoring is done for ozone, 90% of the path must be at least 10 m from a road that carries fewer than 10,000 cars a day. This criterion changes to 250 m for heavily traveled roads (>110,000 cars a day).

Selection of Path Length. Determination of an appropriate path length is not a simple chore although a preliminary hint can be derived from Beer's law. Many factors that cause scattering and absorption of the IR light must be considered. *In situ,* the absorption mainly due to water vapor limits the maximum path length.

The shortest path can be calculated from Beer's law with knowledge of the absorption coefficient, α, and the minimum measurable absorbance. The absorption coefficient, α, can be calculated from the reference spectrum of the compound of interest. The minimum detectable absorbance is generally defined as 3 times the root mean square baseline noise as measured under normal operating conditions. For example, with a reference spectrum of benzene, the absorption coefficient, α, is 1.0×10^{-4} (ppm/m) at 1038 cm^{-1}. When the minimum detectable absorbance of the instrument is 3×10^{-4}, the minimum path integrated concentration (concentration $\times$ path length) can be obtained as follows:

$$A = \alpha CL$$

where A is the absorbance, α is the absorbance coefficient, C is the concentration, and L is the path length. Because the product of concentration and path length is the path integrated concentration, CL is the path integrated concentration. Thus,

$$\begin{aligned} CL &= A/\alpha \\ &= 3 \times 10^{-4}/1 \times 10^{-4} \text{ (ppm/m)} \\ &= 3 \text{ ppm/m} \end{aligned}$$

Therefore, if the minimum concentration desired to be measured is 0.01 ppm, the minimum path length would be 300 m.

When the path length is determined, it is also critical to consider the spatial distribution of the target analytes air contaminant concentration along the path. If the target analytes exist only in a limited region within the optical path, the beam path should not be increased. As the beam path increases, only the effects of spectral interferences will be increased while the IR beam travels the portion of the path which has zero concentration of the target compound. In this case a path longer than the minimum reduces the sensitivity of the measurement.

Consideration of Winds. In general, the beam path must be set up to traverse the area which is desired to be monitored. For example, the path may be chosen close to the physical boundaries of the plume when plumes of finite size are being measured. There is also no need to change the direction of the path when the concentrations of gases crossing a fence line are being monitored. However, most environmental monitoring using RS-FTIR depends on the wind to deliver the air pollutants being emitted from a source to the beam path. Information on the prevailing wind is often used to set up the path. However, wind direction and speed will change dramatically over small regions and time. Thus, in practice a knowledge of the historical prevailing winds does not provide reliable data to select a path. For example, the prevailing wind information cannot be used to calculate emission rates from the RS-FTIR data. It is desirable to measure the meteorological data such as wind speed and direction directly using a meteorological monitoring station when emission rates are required. A knowledge of the winds does provide reliable data for a long-term permanent installation where monitoring is being done daily over months or years of time.

Although path orientation is significant, there is no data available to address the variation of the concentration with altitude. No comparison was made between a slant and horizontal path to date.

Acquisition of Background Spectrum

When RS-FTIR is used for air monitoring, most users use absorption spectra for the quantitative data analysis although there are some people that use the interferogram directly. If absorption spectra is to be used, a background spectrum must be obtained as discussed earlier in the RS advantages and disadvantages section. However, there is no universally agreed on method for obtaining a satisfactory background spectrum to date. The method for choosing how to acquire a background is probably more dependent on the actual program requirements and not so much the site. Several techniques have been developed and used. However, all of them are potential sources of error. Therefore, the method to acquire a background spectrum must be chosen carefully on a site-by-site basis.

There are basically two approaches to obtain the background spectrum when monitoring in the field: (*1*) collection of a spectrum under similar conditions, and (*2*) manipulation of the spectrum. Three techniques are commonly used to obtain the background spectrum using similar conditions to those for actual data acquisition. The first is to acquire a background spectrum along a path that is upwind from the contaminant source. This is called the upwind background spectrum. The second is to collect a background spectrum when there are no source emissions which might result in a detectable concentration of the target analyte. The third is to collect a background spectrum over a short path length. This is called the short-path background spectrum.

The upwind background is generally collected along a path perpendicular to the sampling beam path or to the wind direction. As long as the wind does not transport contaminants across the path, a usable background can be obtained. When the area of the source is relatively small and its upwind side is accessible, this method is often used. It is necessary to be familiar with the information related to the wind directions of the monitoring area and the contaminant source emission characteristics to obtain satisfactory upwind backgrounds. These backgrounds are typically collected before and after monitoring to minimize the movement of the instrument. It is possible to collect background spectra as often as necessary with the use of the rapid beam aiming RS-FTIR system. It is also possible to acquire an upwind background spectrum by waiting until the wind shifts so that the sampling path is along an upwind side of the source. There must not be an emitting source in an upwind area to use this method.

There are some advantages to the collection of upwind background spectra. It is likely that the components in the air are the same for the sample and background spectra, except for the target analytes. The operating characteristics also are the same. If the beam path for the background can be set up along the side of the source area, a second retroreflector can be used, the background spectra can be taken frequently without moving the entire system. However, there are also some disadvantages and limitations. There still may be errors due to differences in alignment between the sampling and background paths. If the background is perpendicular to the sampling path, it is necessary to ensure that the pollutants to be monitored are not contained in wind crossing the background path. In general, this type of background is collected at the beginning and end of the monitoring period (each day). Therefore, it is possible that significant errors can be embedded due to a change in water vapor. An upwind background may simply not be available in some industrial situations. In addition, an upwind background will not be able to allow the operator to determine the variability in the gases coming across the data path from the upwind direction.

Another possible method to collect background spectrum in the fields is to bring the retroreflector or external IR source close to the detector. This method effectively eliminates the absorption caused by the contaminants. This is called the short-path background. The short-path and sample spectra can be easily configured to collect background frequently, if two retroreflectors are available.

The main drawback to the short-path background collection is the possibility of saturating the detector. This will result in nonlinear response of the detector. Thus, one difficulty arises in the specification of an appropriate distance of the short path. If the distance is long enough to avoid the saturation problem, the absorption by the contaminant may increase sufficiently to bias the results. It is possible to overcome the saturation problem with reduction of light intensity by rotating the retroreflector or using a wire mesh screen to attenuate the signal.

The second approach to obtain a background spectrum is data manipulation. Several techniques have been proposed to generate valid background spectra in this way (7,8). This is referred to as the synthetic background. One technique that is supplied in the software package with the commercially available instruments generates a synthetic background by fitting a polynomial function to data points selected from a single beam sample spectrum. The data points for the fit should never be selected at peaks or within an absorption region, otherwise distortions will be introduced into the intensity function. The synthetic background that is produced must have a smooth curve without artificial dips and peaks. A synthetic background cannot be constructed in the wave number regions where H_2O and CO_2 absorb strongly because of the number of overlapping peaks and strong absorption.

An iterative technique has been investigated to create a synthetic background (8). An initial I_0 spectrum is used with a single beam spectrum I to form an absorbance spectrum. This absorbance spectrum is analyzed, and concentrations found are subtracted from the absorbance spectrum. Then, the absorbance spectrum is converted to the I spectrum with the original I_0. This recreated I spectrum becomes the I_0 for the next analysis. The initial results using this technique were positive in limited tests although they are not conclusive due to the lack of reported data.

The average of several background spectra that were taken over a certain time period can be used for the entire data set for that period when the experimental conditions maintain constant. As long as the environmental conditions are constant, each individual background spectrum should have the same noise level. Thus, the final error in the average background should be smaller by a factor of $\sqrt{N}$ because the S/N ratio is proportional to the square root of the number of co-added single beam spectra. Here, N is the number of spectra that are averaged. However, the experimental conditions such as water vapor and CO_2 concentrations are continuously changing in most field situations. Thus, this method can be used only in limited conditions.

Currently, no unique ideal technique is available to obtain a perfect background under all conditions. Therefore, it is important for the operator to be aware of the advantages and disadvantages of each method prior to use. Generally, it is advised to use a synthetic background whenever possible (5). If the absorbing peaks are narrow, as they are for methane or hydrogen fluoride, a valid synthetic background can be successfully constructed. However, it is difficult to generate a valid synthetic background for broad absorbing features because with broad absorbing bands, even small changes in the curvature of the baseline can result in large errors. It is recommended that a short-path background be considered for use only when the compound to be monitored has a broad absorption feature. In all cases, two criteria for background spectra are crucial:

One is that the background should not contain any absorption features due to the target compounds. The other one is that the background spectrum must be valid for the time period over which it is used. For example, a new background must be acquired if the curvature chages because of water vapor and whenever any optical component (IR source, mirror, windows, telescope, etc) is changed in the instrument.

Resolution Effects

Insufficient resolution causes deviations from Beer's law. The linear relationship between IR absorbance and concentration is observed only when the spectrum is measured at a resolution that is better than the full width at half height (FWHH) of the band. If the FWHH of the band is narrower than the instrumental resolution, the measured spectrum is actually a convolution of the instrument line shape and the true band shape. Therefore, there will only be an approximately linear relationship between the measured absorbance and concentration. For accurate quantitation, better resolution is required. If higher resolution is used for all measurements, the ability to resolve spectral overlap is also improved. Thus, if the absorption features due to interfering species are found in the analytical bands of the target compound, it is better to obtain the data with resolution high enough to resolve the spectral features of the target compound from those of interferences. The most common and the strongest interferences that exist in the normal ambient atmosphere are water vapor and CO_2. Water vapor absorption lines are present in the spectrum over almost all regions and they affect the absorption features of almost every air pollutant chemical. Water vapor and CO_2 have IR absorption bands with band widths as narrow as 0.1 cm^{-1}. The absorption bands that make up the rotational fine structure of CO_2 have bandwidths of approximately 0.2 cm^{-1} according to the USF HITRAN-PC database (9). It is also likely that the atmospherically pressure broadened water vapor lines are 0.2 cm^{-1} wide. Therefore, a spectral resolution of 0.125 cm^{-1} is required to characterize the full features of spectra for atmospheric chemical species including water vapor and CO_2. However, as higher resolution is used, analysis time and data storage requirements are increased, and the S/N ratio is decreased as discussed earlier. The optimum resolution would provide adequate specificity with an appropriate balance between the S/N ratio and analysis time. Resolution ranges of 0.25–2 cm^{-1} have been suggested for use in RS-FTIR. However, there is no consensus as to what resolution is generally applicable (5,10). Therefore, the optimum resolution for monitoring air pollutants with the use of RS-FTIR must be chosen in the consideration of target chemicals and monitoring situations.

An FTIR system was evaluated for quantitative air monitoring of toxic chemicals at a semiconductor manufacturing facility. The system was designed and evaluated to determine the resolution that provided the optimum balance between analysis time, data storage space, S/N, accuracy of quantitation, and the ability of differentiation of compounds with overlapping spectra (11). This system was used to analyze part per billion (ppb) levels of organic vapors and metal hydrides such as arsine, phosphine, and diborane. Spectral measurements were taken at resolutions ranging from 0.5 to 8 cm^{-1} over a 20.25-m path in a multipass cell. In this study, acceptable quantitation was achieved at 8 cm^{-1} resolution for organic vapors. The higher resolutions were required for the metal hydrides because the absorption bands of them were overlapped with those of atmospheric interferences. For example, phosphine overlaps with CO_2 and arsine overlaps with water vapor. It was also found that the accuracy with quantitative results was affected with decreased resolution. When pure diborane was analyzed using CLS, only the 0.5 cm^{-1} resolution measurement showed a linear relationship between the absorbance and concentration. The effect of resolution on the quantitation by CLS analysis was also evaluated for overlapping compound mixtures of Freons 11, 13B1, and 22. These compounds have broad spectral features that overlap. Each of the individual compounds was accurately quantified at 0.5, 2, 4, and 8 cm^{-1} resolution, respectively, in a 1:1:1 mixture at concentrations of 10, 1, and 0.1 ppm. Resolutions of 1–2 cm^{-1} were found to be adequate for multicomponent quantitative analysis of VOCs, such as acetone, chloroform, toluene, methanol, 1,1,1-trichloroethane, methyl ethyl ketone, carbon tetrachloride, and the xylene isomers using CLS method (12). An adverse effect on the multicomponent analysis was observed at a resolution poorer than 4 cm^{-1}.

The optimum resolution for RS-FTIR applications has been addressed (13). Theoretical calculations describing the relationship between absorbance and concentration for a single Lorentzian line with a half-width of 0.1 cm^{-1} measured at resolutions from 0.01 to 0.1 cm^{-1} were presented. Under these conditions, linearity was observed only for the highest resolution case and at the lowest concentrations. The deviation from linearity most likely would be observed in small molecules such as HCl, CO, CO_2, and H_2O. On the other hand, larger compounds with broader absorption bands demonstrated a more linear relationship between the absorbance and concentration. Advantages and disadvantages of using low resolution measurement in atmospheric monitoring with the use of FTIR have been discussed (14). The smaller size and greater portability of the instrument, an improved S/N ratio, and a lower cost were discussed as advantages. A greater difficulty in visualizing the IR bands of the target compounds and a potential problem due to deviations from Beer's law were considered as disadvantages. A test case of measuring the xylene isomers at resolutions of 2, 4, 8, and 16 cm^{-1} showed good quantitative results using a partial least-squares (PLS) program. Recently, high resolution RS-FTIR measurements of VOCs have been studied at a variety of monitoring sites (15). With measurements of spectra at 0.125 cm^{-1} resolution, detection limits for benzene of 0.5 ppm m were achieved at path lengths between 60 and 100 m at a fuel storage area. The narrow benzene band at 674 cm^{-1} could be separated from the strong CO_2 absorption bands with the high resolution measurement.

Quality Assurance and Quality Control

Because errors are inherent in all measurement systems, valid data can be obtained only when the uncertainties are controlled within an acceptable range. Thus, the validity of data can be assured only when proper documents are available to ensure the precision and accuracy of the measurement system. A quality control (QC) program concerned with sampling and analysis is a systematic attempt to assure the precision and accuracy of future analysis by detecting determinate errors in analysis and preventing their recurrence. Confidence in the accuracy of analytical results and improvements in analysis precision are established by identification of the determinate sources of error. The precision will be governed by the indeterminate error inherent in the procedure, and can be estimated by statistical techniques. For a result to be accurate, the procedure must not only be precise but be without bias.

The QC program should cover instrumental control as well as total analytical control. In general, quality assurance (QA) includes those operations and procedures to ensure that measurement systems are in statistical control and performing within prescribed specifications. The measurement system includes sampling, analysis, quality control, data processing, and reporting. The definition of QA which has been formulated by the American National Standards Institute and the Institute for Electric and Electronic Engineers is "a planned and systematic pattern of all actions necessary to provide adequate confidence that the item or product conforms to established technical requirement" (16).

Although the need for QA procedures has been growing for RS-FTIR monitoring, there is currently no widely practiced QA program to validate RS-FTIR data. For environmental monitoring with RS-FTIR, true values or standards are only rarely attainable. It is also extremely difficult to compare the data taken with RS-FTIR to those taken with other methods because no comparable methods are applicable in practice. Thus, proper sampling strategy and instrumental maintenance are critical to obtain valid data for RS-FTIR monitoring. However, there is very little information in the literature that addresses QA issues.

The EPA policy requires that all EPA organizational units involved in environmentally related measurements participate in an agency-wide QA program. This requirement for QA applies equally to those monitoring and measurement activities supported or specified through contracts, grants, regulations, or other formal agreement (17).

The EPA has identified four different categories of QA project plans:

Category I projects are those that are designed so that their results can be used directly, without additional support, for the purpose of compliance or other litigation. As such, the data must be able to withstand legal challenge and therefore have the most rigorous and detailed requirements.

Category II projects are those whose data complement other projects. When combined with the output from other projects, these data can be used for rule making or policy making.

Category III projects are those producing data that allow the evaluation and selection of basic options for use in feasibility studies.

Category IV projects are those that are associated with research and development projects. The results are used to assess the basic or underlying assumptions or suppositions of other work.

The goal of these plans are to ensure that all data are scientifically sound, of known quality, and thoroughly documented. These plans are prepared to be fairly general and broad so that any project can fit into one of these categories. Category I applied to routine monitoring of air contaminants requires 16 items that must be addressed. The other categories require 13, 12, and 6, respectively. Each item of the QA requirements for Category I project plans is briefly discussed as follows.

Project Description. The project description includes a summary for the overall monitoring design and measurement parameters, such as sampling sites, sampling frequency and duration, target compounds, and interferences. This may be done in tabular form listing all parameters. For RS-FTIR monitoring, a brief description of the comparison methods such as the canister sampling and an analysis method is included in this section.

Project Organization and Responsibility. In this section, all individuals must be given their responsibilities and the relationships among them in order to obtain valid and usable data and proceed smoothly with the a monitoring program. The lines of authority must be clearly stated and any subcontractors must be described.

QA Objectives. This section is to define what levels of quality are required for the data. The data quality requirements are based on the intended use of the data. Once data quality requirements are established, the QA project plan must specify items that can be quantitatively defined. Limit of detection/quantitation, precision, accuracy, comparability, and completeness of data must be included in the QA objectives.

Site Selection and Sampling Procedure. For RS-FTIR monitoring, site selection is extremely important. Not only a suitable area but also an appropriate beam path must be selected to obtain valid data. A description of the site and the rationale for site selection must be addressed in this section. The plan must completely describe the sampling procedure including instrument operations and background collections. Flow diagrams or charts are often used.

Sample Custody. Proper procedures during sample storage and shipping from sample acquisition to sample analysis are referred to as chain of custody of sample. They are very important to use the data for documenting violations of regulations. For RS-FTIR monitoring, background and sample data are usually stored on disk. Thus, proper procedures for data management such as back-up must be completed.

Calibration Procedures and Frequency. In general, calibrations are one of the most critical procedure in the QA program for environmental monitoring program. However, there is no calibration procedure for RS-FTIR in the classical meaning because the spiked samples (known concentration air samples) cannot be prepared and applied in the atmospheric conditions. Although the instrument can be checked for the precision using a gas cell with a spiked air sample, this procedure is not routinely conducted in most field applications.

Analytical Procedures. For RS-FTIR monitoring, this section must specify an analytical method that is being used for quantitation. In addition, spectral band regions of each target analyte or procedures to choose proper spectral windows for quantitation must be included in this section. If any gas cell is used for obtaining reference spectra or checking the precision of the instrument, a detailed procedure for use should be addressed.

Data Reduction, Validation, and Reporting. In this section, any statistical procedures to handle data must be described including sample calculation. All procedures used from the data generated to the end user must be also described. This section also describes uniform and consistent data validation criteria that will be used to reject outliers.

Internal Quality Control Check. All internal QC checklists must be determined and described. Purposes and significance also must be included. Various QC checks can be made for RS-FTIR monitoring including daily checks for return signal magnitudes and the RMS baseline noise measurements.

Performance and Systems Audits. A performance audit is to determine whether the instrument is operating as it was described in other sections. A system audit consist of a review of the total data production process to check whether the proposed procedures and sampling protocols have been followed. Auditor, auditing protocol, and the acceptance criteria must be described in this section. The performance audit of the RS-FTIR instrument can be done by using a gas cell filled with for a known concentration of gas to measure the response. Generally, it is not conducted in field environmental monitoring.

Preventive Maintenance. This section is to assure that the instrumental functions are maintained adequately, usually using an equipment logbook which records the history of usage, maintenance, and repairs. The preventive maintenance and frequency must be described.

Calculation of Data Quality Indicators. This section must specify the data quality indicators such as precision, accuracy, control limits, and outliers. Detailed procedures to calculate them and all statistical routines must be included. There is no consensus for data quality indicators to be used for RS-FTIR monitoring.

Corrective Action. Each monitoring program must establish the total corrective action scheme which is designed to identify and eliminate defects caused during the entire project process. This section must describe how the corrective action is to be done. For example, identification of defect, tracing them to the source, planning and implementation of proper measures to correct them, and documentation for the results of the corrective process are included.

Quality Control Reports to Management. This section describes what contents will be included in the report and to whom and when they will go. The report must also address any results of ongoing performance and system audits and any problems related to QA plan with proposed corrective actions.

References. If any references are available related to the QA plan, they must be included in each section.

Other Items. Any voluntary items can be included to accomplish the QA goal.

At the time of this writing the items listed under these categories are being revised and expanded.

Field Applications

The unique advantages of the RS-FTIR technique have been exploited to monitor atmospheric pollutants over a variety of environmental conditions (4,18,19). A large portion of applications have been in measurements at landfills, waste sites, and in the vicinity of chemical industrial complexes. Many of these applications have been focused on measuring multiple air contaminants at very low concentration levels. Some of the applications have been aimed to assess emission rates from areas or stacks. Recently, RS-FTIR has been recognized as an emerging monitoring technique for workplaces and the indoor environment.

Technology in this area is progressing rapidly. Faster, lighter, and smaller RS-FTIR systems with higher throughput are expected to be developed. New sampling strategies will also be developed for the use of this technique, such as air pollutant concentration mapping using RS-FTIR coupled with computed tomographic techniques.

EPA's Requirement of Air Monitoring. The Clean Air Act Amendments (CAAA) and the Superfund Amendments and Reauthorization Act (SARA) have been major drivers to impel development and applications of RS-FTIR technology. Enactment of the new CAAA of 1990 resulted in the requirement that EPA must consider the measurement and reporting of atmospheric burdens, concentrations, and emission rate of a large number of chemicals and numerous types of emission sources. Title I of the CAAA will designate areas as attainment or nonattainment areas for ozone, and rank these areas from marginal to extreme. It requires that ozone reductions via reduction of VOCs and NO_x must be 3% or greater per year from sources classified as serious, severe, or extreme. It also requires enhanced monitoring including VOCs and NO_x for these areas. Under Title III of the CAAA, as many as 189 compounds may have to be monitored at the same location. It is also required that every major source (both new and existing) to implement maximum achievable control technology (MACT) standards to reduce hazardous air pollutants (HAPs) emissions. In addition, Title III provides for investigation of accidental releases. Title VIII requires that EPA has to review the emission factors every three years and develop emission factors for the 189 pollutants identified in the CAAA (20).

The traditional point-sampling methods may not be well suited for some types of sources and target air pollutants in these regulations. The CAAA indicated that the best available technology may be used in order to monitor for the presence of various atmospheric gaseous pollutants. Most of the toxic gases covered under the new CAAA have absorption spectra in the two IR regions of 700 to 1200 cm^{-1} and 2400 to 3000 cm^{-1}. Furthermore, the RS-FTIR have a number of unique advantages that may be able to fill gaps between the existing technology and monitoring requirements. Thus, the RS-FTIR system can be an ideal tool for monitoring the atmospheric pollutants required by the CAAA.

Ambient Air Monitoring. Since the first RS-FTIR instrument built by Herget and Brasher (21), a number of applications for monitoring atmospheric pollutants has been made with the use of RS-FTIR in various areas, such as urban areas, industrial sites, indoors, and hazardous waste sites (4,18,19). Various tests have also been conducted to assess the feasibility of the RS-FTIR systems for use ambient in air monitoring. The results support the idea that the RS-FTIR instrument is able to measure air contaminant concentrations for a wide range of gaseous atmospheric pollutants with an acceptable accuracy (22–25).

In-plant monitoring applications have been important to detect leakage or spills of toxic chemicals due to the advantage of real-time monitoring. The RS-FTIR system is suitable for monitoring the perimeter of distributed process or storage areas. Several demonstration tests have been applied to fence-line monitoring at petrochemical industries (13,26). Time series analysis and statistical process control methods have been explored to use this system for rapid detection of departures from defined normal operating conditions (27). Those studies showed that the RS-FTIR system has a high potential to be used to alert to residents in the vicinity of sources effectively against to the accidental leakage of the toxic chemicals.

Estimation of Emission Rates. There has been an increasing interest in developing the technique so that the RS-FTIR systems can be used to measure the emission rates of pollutant gases from hazardous waste sites or other sources (28–30). Because environmental control strategies are focused on reduction of emissions from sources under Title III of the CAAA, it is necessary to estimate the emission rates from sources. Determination of emission rates requires not only the air pollutant concentrations but also meteorological data such as wind direction and speed. In general, the path-integrated concentrations are measured across the plume downwind from sources and the meteorological data over the path are collected at the same time. Then, the emission rates are calculated using straightforward concentration ratio calculation or estimated with the use of tracers or plume dispersion models. By using the path-integrated concentrations of RS-FTIR rather than those of the point sampling data, the mathematical calculation can be simplified to estimate the emission rates from area or point sources (29). However, these techniques may not be satisfactory to estimate emission rates because both measurements of concentrations and wind characteristics are taken at the ground level.

Recently, a new approach referred to as the plane-integrated technique has been developed to overcome the uncertainties due to their heavy reliance on the ground-level measurement and assumed plume dispersion characteristics (30). A RS-FTIR system which is able to move the beam path rapidly was used to establish a measurement plane made up of several beam paths. The measurement plane was positioned in a plane downwind from a source such that a large fraction of the plume passes through the plane for direct measurement. Thus, the concentration profile with the plume was directly measured downwind from the source, allowing for a more direct and representative estimation of the emission rate. This methodology is still under development. However, the result of the preliminary test showed that the average emission rates were obtained reasonably close to the actual emissions released from the simulated volume source. Thus, RS-FTIR has a promising potential usage for estimation of the emission rates. However, a true picture of the concentration field coming from a plant complex may only be attainable with use of RS-FTIR computed tomography methods.

Stack Sampling. When Herget and Brasher developed the first RS-FTIR system, the detector unit including the interferometer could measure concentrations of hot stack gaseous pollutants without using an IR source (21). Extended studies have followed to determine temperature, composition, and concentrations of chemical species in plumes of gas stacks using IR emission measurements at a distance. However, the studies showed that errors in emission measurements can be much larger that those in absorption measurements because of uncertainties in temperature and concentration gradients from the source (31). Emissions from small building smoke stacks have been successfully detected by distinguishing natural gas from fuel oil based on the H_2O/CO_2 concentration ratio in exhaust gases (32).

Indoor and Workplace Air Monitoring. A great deal of effort has been made in development of the RS-FTIR technique for use in workplaces to measure workers exposure to air contaminants or to link process states to worker exposure. For applications of RS-FTIR in workplaces, a relatively small transportable system has been developed (7). Initial studies were focused on comparison tests between the averaged beam path concentration and multiple point data along the path to use the technique in workplaces (33,34). It has not been generally accepted for workplace monitoring. One of the main limitations is that the measurement data do not provide the spatial distribution of the air contaminant concentrations. Therefore, it is difficult to assess the exposure level to the hazardous chemicals for each individual worker from the path integral data. However, various sampling strategies are being investigated to address the spatial resolution problem with the use of multiple beam paths: bi-beam measurement, multisegmented beam path measurement, and computed tomographic technique.

Computed Tomographic Application. Optical remote sensing coupled with computed tomographic technique results in chemical concentration mapping over a measurement plane. The technique of the computed tomography (CT) also known as reconstructive tomography (RT) or computer-aided tomography (CAT) refers to a group of techniques to reproduce an original image from a number of integrated measurements along various paths in the plane (35). With the use of this technique, a number of the RS-FTIR measurement data across the measurement plane can be used to reproduce the chemical concentration distribution. This technique of air pollutant concentration mapping will be very useful for rapid detection of air pollutant releases, source identifications, tracing the air contaminants, and estimation of emission rates.

The first experimental study in a full room-sized chamber on the CT reconstruction of indoor air concentrations coupled with RS-FTIR was successfully demonstrated by Yost et al. (36). Experimental CT reconstruction studies were conducted in a full room-sized chamber with the use of a single RS-FTIR system which is able to aim the beam to the positions programmed into the computer (37,38). The possibility of coupling CT with RS-FTIR to map the outdoor concentrations of gaseous contaminants has been theoretically explored and discussed along with the technical and economical feasibility of this approach (39). This technique is in the research and development stages. However, this technology has a great potential to be able to provide both time and spatial resolution.

FUTURE

Technology related to both remote sensing and FTIR is rapidly progressing to the point that the system has better sensitivity, specificity, and high throughput. The future system will be small, rugged, and moderately priced with higher spectral resolution. Eventually, we can probably expect to use a hand-portable, tiny optical bench with no moving parts. Such an instrument will depend on an array of IR detectors operating at room temperature.

There are high expectations for great advances in the software to collect, manipulate, and analyze data. For example, quantitation software is expected to be developed based on a combination of pattern recognition or artificial intelligence and calculation-intensive fitting strategies. User-oriented computer programs will also be developed so that the system can be operated by nonexperts. Various sampling strategies will be further developed and modified to enable collection of the data that are in compliance with requirements. Software will display the data in the form of graphics, tables, or charts without manual manipulations.

BIBLIOGRAPHY

1. P.L. Hanst, *Appl. Spec.* **24**(2), 161–174 (1970).
2. W.F. Herget and J.D. Brasher, *Appl. Opt.* **18**(20), 3404–3420 (1979).
3. W.F. Herget, *Appl. Opt.* **21**(4), 635–641 (1982).
4. S.P. Levine and G.M. Russwurm, *Trends Anal. Chem.* **13**(7), 258–262 (1994).
5. G.M. Russwurm, and J.W. Childers, *FT-IR Open-Path Monitoring Guidance Document,* 2nd ed., ManTech Environmental Technology, Inc., Research Triangle Park, N.C., 1995.
6. U.S. EPA, *Fed. Reg.* **59**(159), 42541–42552 (1994).
7. H.K. Xiao, S.P. Levine, W.F. Herget, J.B. D'Arcy, R. Spear, and T.H. Pritchett, *Am. Ind. Hyg. Assoc. J.* **52**(11), 449–457 (1991).
8. H.K. Xiao and S.P. Levine, *Anal. Chem.* **65**(17), 2262–2269 (1993).
9. University of South Florida. USF HITRAN-PC, University of South Florida, Tampa, Fla., 1993.
10. L.S. Ying, and S.P. Levine, *Anal. Chem.* **61**(7), 677–683 (1989).
11. C.R. Strang, S.P. Levine, and W.F. Herget, *Am. Ind. Hyg. Assoc. J.* **50**(2), 70–77 (1989).
12. T.L. Marshall, C.T. Chaffin, M.R. Witkowski, V.S. Makepeace, S.E. Leggans, R.M. Hammaker, W.G. Fateley, P. Saarinen, K. Salonen, and J. Kauppinen, "The Effects of Resolution on the Multicomponent Analysis of Volatile Organic Compounds (VOCs) of Interest in Open Air Long Path Fourier Transform Infrared (FTIR) Spectrometry," *Presented at the 44th Pittsburgh Conference and Exposition on Analytical Chemistry and Applied Spectroscopy,* Atlanta, Ga., Paper No. 478, Mar. 1993.
13. R.L. Spellicy, W.L. Crow, J.A. Draves, W.F. Buchholtz, and W.F. Herget, *Spectroscopy* **8**(9), 24–34 (1991).
14. P.R. Griffiths, D. Qin, R.L. Richardson, and C. Zhu, "Atmospheric Monitoring with FT-IR Spectrometry: Strengths and Weaknesses of Measurement at Low Resolution," *Presented at the 44th Pittsburgh Conference and Exposition on Analytical Chemistry and Applied Spectroscopy,* Atlanta, Ga., Paper No. 1142, Mar. 1993.
15. H. Bittner, T. Eisenmann, H. Mosebach, M. Erhard, and M. Resch, "Measurements of Diffuse Emissions of Volatile Organic Compounds by High Resolution FTIR Remote Sensing," in *SP-89 Optical Remote Sensing for Environmental Monitoring,* Air & Waste Management Association, Pittsburgh, Pa., 1994, pp. 443–454.
16. Institute of Electrical and Electronics Engineers Standard, IEEE 730-1984, *IEEE Standard for Software Quality Assurance Plans,* 1984.
17. T.W. Stanley and S.S. Verner, "The U.S. Environmental Protection Angency's Quality Assurance Program," in J.K. Taylor and T.W. Stanley, eds., *Quality Assurance for Environmental Measurements, ASTM STP 867,* American Society for Testing and Materials, Philadelphia, Pa., 1985, pp. 12–19.
18. W.B. Grant, R.H. Kagann, and W.A. McClenny, *J. Air Waste Manage. Assoc.* **42**(1), 18–30 (1992).
19. M. Sigmonds, H. Xiao, and S.P. Levine, *Am. Ind. Hyg. Assoc. J.* **55**(10), 953–965 (1994).
20. B. Eklund, and J. LaCosse, *Summary Document for the Houston OPM Workshop,* Radian Corporation, Austin, Tex., 1992.
21. W.F. Herget, and J.D. Brasher, *Appl. Opt.* **18**(20), 3404–3420 (1979).
22. W.A. McClenney, G.M. Russwurm, M.W. Holdren, A.J. Pollack, J.D. Pleil, J.L. Varns, J.D. Mulik, K.D. Oliver, R.E. Berkeley, D.D. Williams, K.J. Krosk, and W.T. McLeod, *Superfund Innovative Technology Evaluation: The Delaware Site Study,* US EPA (AREAL-MRB), Research Triangle Park, N.C., 1991.
23. G.M. Russwurm, R.H. Kagann, O.A. Simpson, W.A. McClenny, and W.A. Herget, *J. Air Waste Manage. Assoc.* **41**(8), 1062–1066 (1991).
24. G.M. Russwurm and W.A. McClenny, "A Comparison of FTIR Open-Path Data with TO-14 Canister Data," in *Proceedings of EPA/A&WMA International Symposium on Measurement of Toxic and Related Pollutants,* (EPA-600/9-90-026; NTIS PB91-120279), Raleigh, N.C., 1990.
25. J.L. Hudson, M.J. Thomas, and J. Arello, "An Overview of the Intercomparability and Performance of Multiple Open-Path FTIR Systems as Applied to the Measurements of Air Toxics," in *Proceedings of the 1992 Air and Waste Management Association Specialty Conference on Optical Remote Sensing,* Houston, Tex., 1992.
26. American Petroleum Institute (API), *Remote Sensing Feasibility Study of Refinery Fenceline Emissions,* API, Washington D.C., 1994.
27. M. Simonds-Malachowski, S.P. Levine, G. Herrin, R.C. Spear, M. Yost, and Z. Yi, *J. Air Waste Manag. Assoc.* **44**(5), 673–682 (1994).
28. S.D. Piccot, S.S. Masemore, E.S. Ringler, and S. Srinivasan, *J. Air & Waste Manag. Assoc.* **44**(3), 271–279 (1994).
29. T.J. Lyons, and W.D. Scott, "Atmospheric Diffusion," *Principles of Air Pollution Meteorology,* CRC Press, Boca Raton, Fla., Chapt. 3, 1990.
30. S.D. Piccot, S.S. Masemore, W. Lewis-Bevan, E.S. Ringler, and D.B. Harris, *J. Air Waste Manag. Assoc.* **46**(2), 159–171 (1996).
31. W.F. Herget, *Appl. Opt.* **21**(4), 635–641 (1982).
32. R.C. Carlson, A.F. Hayden, and W.B. Telfair, *Appl. Opt.* **27**(23), 4952–4959 (1988).
33. H.K. Xiao, S.P. Levine, J.L. Nowak, and M. Puska, *Am. Ind. Hyg. Assoc. J.* **54**(9), 545–556 (1993).
34. M.G. Yost, H.K. Xiao, R.C. Spear, and S.P. Levine, *Am. Ind. Hyg. Assoc. J.* **53**(10), 611–616 (1992).
35. R.A. Brooks, and G.D. Chiro, *Radiology* **117** 561–572, (1975).
36. M. Yost, A. Drescher, Y. Zhou, P. Saisan, M. Simonds, W. Nazaroff, S. Levine, and A. Gadgil, *Am. Ind. Hyg. Assoc. J.* **55**(5), 395–402 (1994).
37. A.C. Drescher, D.Y. Park, M.G. Yost, A.J. Gadgil, S.P. Levine, and W.W. Nazaroff, *Atmos. Environ.* **31**(5), 727–740 (1997).
38. A.C. Drescher, A.J. Gadgil, and W.W. Nazaroff, *Atmos. Environ.* **30**(6), 929–940 (1996).

39. M.L. Fischer, A.C. Drescher, A.J. Gadgil, and M.G. Yost, "A System for Rapid Detection and Mapping of Gas Plumes on 100 m Scales: Examination of some Technical and Economic Issues," in *Proceedings of Optical Sensing for Environmental and Process Monitoring,* Orman A. Simpson, ed., A&WMA Vol. VIP-37(SPIE Vol. 2365), 1995, pp. 151–159.

REMOTE SENSING SATELLITE SYSTEMS.

See EARTH SURFACE MONITORING FROM SPACE.

REVERSE OSMOSIS

D. BHATTACHARYYA
W. C. MANGUM
University of Kentucky
Lexington, Kentucky

M. E. WILLIAMS
EET Corporation
Knoxville, Tennessee

Reverse osmosis (RO) is a fairly mature technology used in the area of seawater and brackish water desalination. Use of reverse osmosis as a separation tool is, however, a relatively young technology. In the late 1950s, it was shown that cellulose acetate RO membranes were capable of separating salt from water, although the water fluxes obtained were too small to be practical (1–4). Then in the early 1960s, a method for making asymmetric cellulose acetate membranes having relatively high water fluxes and separations was developed, making RO separations both possible and practical (5–7).

Since that time, the industrial development of thin-film composite membranes (8) and improvements in polymer materials have widened applications of RO from desalination to treatment of hazardous wastes, water purification uses, and material recovery in the electroplating industries. These membranes can tolerate wide pH ranges, higher temperatures than cellulose acetate, and harsh chemical environments. Moreover, water flux and solute separation characteristics have highly improved, resulting in many applications. For example, these RO membranes have found uses in wastewater treatment, production of ultrapure water, water softening, and food processing (9). Sales of RO membrane products were as high as $118 million annually in the late 1980s, and continued growth is expected (10).

The driving force for the development and use of RO membranes is the advantage of these systems over traditional separation processes such as distillation, extraction, ion exchange, and adsorption. Because reverse osmosis is a pressure-driven process, no energy-intensive phase changes or potentially expensive solvents or adsorbents are needed for RO separations. Reverse osmosis is a process that is inherently simple to design and operate compared to many traditional separation processes. Also, simultaneous separation and concentration of both inorganic and organic compounds are possible using the RO process. In addition, using nanofiltration, ie, loose RO, membrane-selective solute separations based on charge and molecular weight and size differences are possible. Finally, reverse osmosis technology can also be combined with ultrafiltration, pervaporation, distillation, and other separation techniques to produce hybrid processes that result in highly efficient and selective separations. Excellent reviews of RO technology and theories have been reported in the literature (7,9–15).

MEMBRANE MATERIALS AND MODULES

Reverse osmosis membrane separations are governed by the properties of the membrane used in the process. These properties depend on the chemical nature of the membrane material, which is almost always a polymer, as well as its physical structure. Properties for the ideal RO membrane include low cost, resistance to chemical and microbial attack, mechanical and structural stability over long operating periods and wide temperature ranges, and the desired separation characteristics for each particular system. However, few membranes satisfy all these criteria and so compromises must be made to select the best RO membrane available for each application. Excellent discussions of RO membrane materials, preparation methods, and structures are available (8,13,16–21).

Most commercially available RO membranes fall into one of two categories: asymmetric membranes containing one polymer, or thin-film composite membranes consisting of two or more polymer layers. Asymmetric RO membranes have a thin (~100 nm) permselective skin layer supported on a more porous sublayer of the same polymer. The dense skin layer determines the fluxes and selectivities of these membranes whereas the porous sublayer serves only as a mechanical support for the skin layer and has little effect on the membrane separation properties. Asymmetric membranes are most commonly formed by a phase inversion (polymer precipitation) process (16). In this process, a polymer solution is precipitated into a polymer-rich solid phase that forms the membrane and a polymer-poor liquid phase that forms the membrane pores or void spaces.

An excellent review of composite RO and nanofiltration (NF) membranes is available (8). These thin-film, composite membranes consist of a thin polymer barrier layer formed on one or more porous support layers, which is almost always a different polymer from the surface layer. The surface layer determines the flux and separation characteristics of the membrane. The porous backing serves only as a support for the barrier layer and so has almost no effect on membrane transport properties. The barrier layer is extremely thin, thus allowing high water fluxes. The most important thin-film composite membranes are made by interfacial polymerization, a process in which a highly porous membrane, usually polysulfone, is coated with an aqueous solution of a polymer or monomer and then reacts with a cross-linking agent in a water-immiscible solvent.

Although RO membranes have been formed and tested using a wide range of different materials and preparation techniques, cellulosic polymers such as cellulose acetate and cellulose triacetate, linear and cross-linked aromatic polyamide, and aryl–alkyl polyetherurea are among the most important RO membrane materials (8,10,20). Asymmetric cellulose acetate membranes continue to enjoy widespread use despite some disadvantages, for example, a narrow (4.5–7.5) pH operating range, because the polymer is subject to hydrolysis, is susceptible to biological attack, exhibits compaction (mechanical compression) at high pressures that results in reduced water flux, and has low (~35°C) upper temperature limits. Polyamide and polyurea composite membranes typically have higher water fluxes and salt and organic rejections, an ability to withstand higher temperature and larger (4–11) pH variations, and immunity to biological attack and compaction. However, these membranes tend to be less chlorine-resistant and more susceptible to oxidation as compared to cellulose acetate membranes. Figure 1 shows water flux and NaCl rejections for three different classes of RO membranes made from a variety of polymer materials. Selected solute rejections, both inorganic and organic, for a large number of RO membranes can be found in the literature (9).

Whereas the membrane material largely determines the water and solute fluxes in an RO process, the packaging of the RO membrane is also important for process feasibility. Requirements of a membrane module include the offering of mechanical support to the RO membrane even at high (up to 8 MPa (1160 psi)) operating pressure, a design that minimizes pressure drop across the module as well as fouling and concentration polarization, and a module that is relatively inexpensive and easy to replace in the membrane process. The most common commercially available membrane modules include plate-and-frame, tubular, spiral-wound, and hollow-fiber elements (11). For large-scale desalination applications, spiral-wound modules have been commonly used. Another newer type of membrane module, Disc Tube (Rochem), which has reduced fouling problems, has been applied to various environmental problems ranging in sizes from 10 to 2000 m^3/d (22,23). This system has also been reported to withstand operating pressures up to 27.6 MPa (4000 psi).

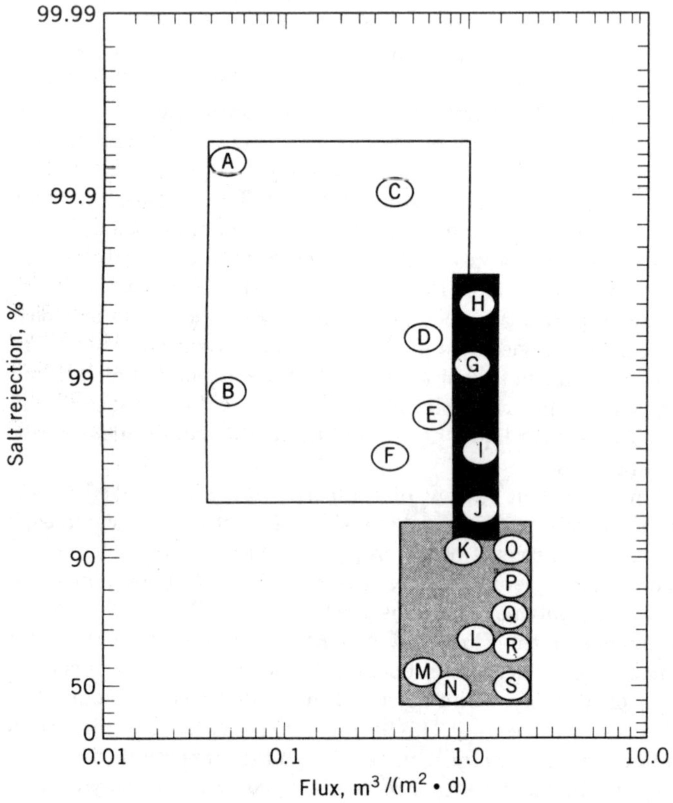

Figure 1. Water flux and NaCl rejection of several membrane types (10), where (□) represents seawater membranes, which operate at 5.5 MPa and 25°C; (■), brackish water membranes, which operate at 1500 mg/L NaCl feed, 1.5 MPa, and 25°C; and (■) nanofiltration membranes, which operate at 500 mg/L NaCl feed, 0.74 MPa, and 25°C. A represents cellulose acetate–cellulose triacetate; B, linear aromatic polyamide; C, cross-linked polyether; D, cross-linked fully aromatic polyamide; E, other thin-film composite membranes; F, asymmetric membranes; G, BW-30 (FilmTec); H, SU-700 (Toray); I, A-15 (Du Pont); J, NTR-739HF (Nitto-Denko); K, NTR-729HF (Nitto-Denko); L, NTR-7250 (Nitto-Denko); M, NF40 (FilmTec); N, NF40HF (FilmTec); O, UTC-40HF (Toray); P, NF70 (FilmTec); Q, UTC-60 (Toray); R, UTC-20HF (Toray); and S, NF50 (FilmTec). To convert MPa to psi, multiply by 145.

THEORETICAL ASPECTS

A reverse osmosis membrane acts as the semipermeable barrier to flow in the RO process, allowing selective passage of a particular species, usually water, while partially or completely retaining other species, ie, solutes such as salts. Chemical potential gradients across the membrane provide the driving forces for solute and solvent transport across the membrane. The solute chemical potential gradient, $-\Delta\mu_s$, is usually expressed in terms of concentration; the water (solvent) chemical potential gradient, $-\Delta\mu_w$, is usually expressed in terms of pressure difference across the membrane.

Measurable Process Parameters

The RO process is relatively simple in design. It consists of a feed water source, feed pretreatment, high pressure pump, RO membrane modules, and in some cases, post-treatment steps. A schematic of the RO process is shown in Figure **2a**.

The three streams and associated variables of the RO membrane process are shown in Figure **2b**: the feed; the product stream, called the permeate; and the concentrated reject stream, called the concentrate or retentate. The water flow through the membrane is reported in terms of water flux, J_w.

$$J_w = \frac{\text{volumetric or mass permeation rate}}{\text{membrane area}}$$

Solute passage is defined in terms of solute flux, J_s.

$$J_s = \frac{\text{mass permeation rate}}{\text{membrane area}}$$

Solute separation is measured in terms of observed rejection, R, defined as

$$R = 1 - \frac{C_P}{C_F} \tag{1}$$

where C_P and C_F are the permeate and feed solute concentration, respectively. The quantity of feed water that passes through the membrane, ie, the permeate, is measured in terms of water recovery, r, defined for a batch RO system as

$$r = \frac{\Sigma J_w A_w \Delta t}{V_F} = \frac{V_P}{V_F} \tag{2}$$

where A_m is the membrane surface area; Δt the time interval, and V_F the volume of the feed; and for a continuous system as

$$r = \frac{J_w A_m}{F_F} = \frac{F_P}{F_F} \tag{3}$$

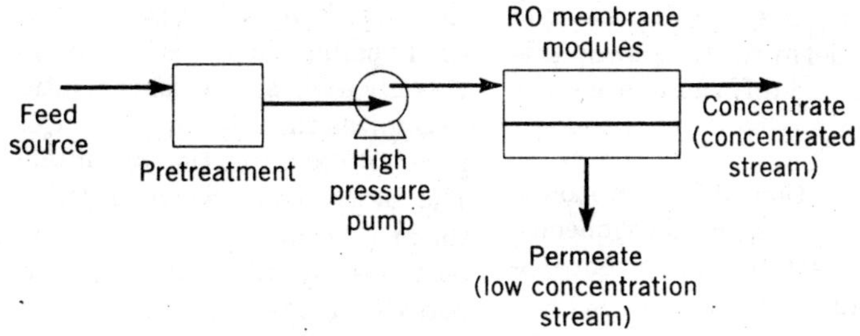

(a)

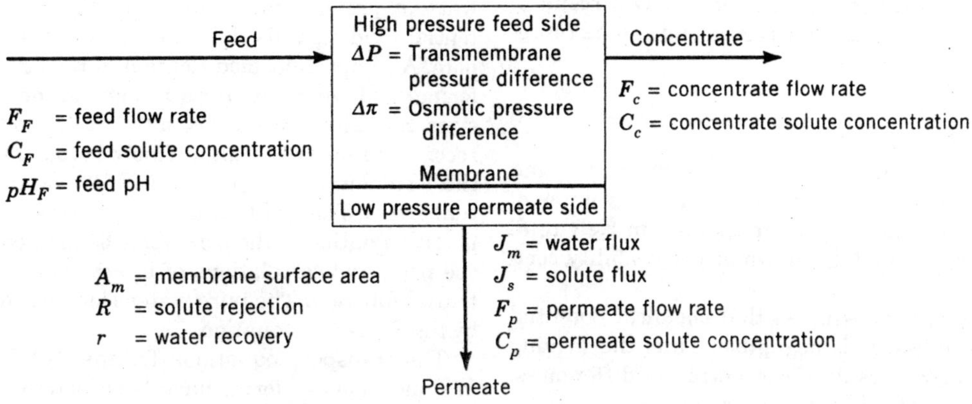

(b)

Figure 2. Schematic of (**a**) an RO membrane process, and (**b**) the RO process streams.

where F_F and F_P are the feed and permeate flow rates, respectively. In a batch membrane system, water is recovered from the system as the concentrate (retentate) is recycled to the feed tank. As a result, if the solute is rejected, the feed concentration, C_F, continuously increases over time. For a continuous membrane system, fresh feed is continuously supplied to the membrane.

Water flux is sometimes normalized according to the initial or pure water flux, J_{wo}, as J_w/J_{wo}, or as flux drop, defined by

$$\text{flux drop} = 1 - \frac{J_w}{J_{wo}} \tag{4}$$

The pressure difference between the high and low pressure sides of the membrane is denoted as ΔP; the osmotic pressure difference across the membrane is defined as $\Delta\pi$; the net driving force for water transport across the membrane is $\Delta P - \sigma\Delta\pi$, where σ is the Staverman reflection coefficient and $\sigma = 1$ means 100% solute rejection. The standardized terminology recommended for use to describe pressure-driven membrane processes, including that for reverse osmosis, has been reviewed (24).

Transport Models

Many mechanistic and mathematical models have been proposed to describe reverse osmosis membranes. Some of these descriptions rely on relatively simple concepts; others are far more complex and require sophisticated solution techniques. Models that adequately describe the performance of RO membranes are important to the design of RO processes. Models that predict separation characteristics also minimize the number of experiments that must be performed to describe a particular system. Excellent reviews of membrane transport models and mechanisms are available (9,14,25–29).

Reverse osmosis models can be divided into three types: irreversible thermodynamics models, such as Kedem-Katchalsky and Spiegler-Kedem models; nonporous or homogeneous membrane models, such as the solution–diffusion (SD), solution–diffusion–imperfection, and extended solution–diffusion models; and pore models, such as the finely porous, preferential sorption–capillary flow, and surface force–pore flow models. Charged RO membrane theories can be used to describe nanofiltration membranes, which are often negatively charged. Models such as Donnan exclusion and the extended Nernst-Planck include electrostatic effects. The transport models focus on the top thin skins of asymmetric membranes and of composite membranes because the skins determine fluxes and selectivities for most membranes. Also, most of the membrane models assume equilibrium, near-equilibrium, or steady-state conditions in the membrane.

A fundamental difference exists between the assumptions of the homogeneous and porous membrane models. For the homogeneous models, it is assumed that the membrane is nonporous, that is, transport takes place between the interstitial spaces of the polymer chains or polymer nodules, usually by diffusion. For the porous models, it is assumed that transport takes place through pores that run the length of the membrane barrier layer. As a result, transport can occur by both diffusion and convection through the pores. Whereas both conceptual models have had some success in predicting RO separations, the question of whether an RO membrane is truly homogeneous, ie, has no pores, or is porous, is still a point of debate. No available technique can definitively answer this question. Two models, one nonporous and diffusion-based, the other pore-based, are discussed herein.

Solution–Diffusion Model. In the solution–diffusion model, it is assumed that (*1*) the RO membrane has a homogeneous, nonporous surface layer; (*2*) both the solute and solvent dissolve in this layer and then each diffuses across it; (*3*) solute and solvent diffusion is uncoupled and each is the result of the particular material's chemical potential gradient across the membrane; and (*4*) the gradients are the result of concentration and pressure differences across the membrane (26,30). The driving force for water transport is primarily a result of the net transmembrane pressure difference and can be represented by equation 5:

$$J_w = \frac{D_{wm} C_{wm} \overline{V}_w}{R_g T \delta} (\Delta P - \Delta \pi) \equiv A(\Delta P - \Delta \pi) \qquad (5)$$

where D_{wm} and C_{wm} are the water membrane diffusivity and concentration, respectively; $\overline{V}_w$ is the partial molar volume of water; R_g, the gas constant; T, the temperature in Kelvin; δ, the membrane thickness; and A, the water permeability coefficient.

For the solute flux, it is assumed that chemical potential difference owing to pressure is negligible. Thus the driving force is almost entirely a result of concentration differences. The solute flux, J_s, is defined as in equation 6:

$$J_s = \frac{D_{sm} K_{sm}}{\delta} (C_{\text{wall}} - C_P) \equiv B(C_{\text{wall}} - C_P) \qquad (6)$$

where D_{sm} is the membrane solute diffusivity; K_{sm}, the solute partition coefficient for homogeneous membrane; C_{wall} and C_P, the feed solute concentration at the membrane wall and the permeate solute concentration, respectively; and B, the solute permeability coefficient. In the absence of concentration polarization, C_{wall} can be replaced by the bulk solute concentration, C_B, or by the feed concentration, C_F, for negligible water recovery. Using the relations for solvent and solute flux, solute rejection, R, for the solution–diffusion model can be expressed as in equation 7:

$$\frac{1}{R} = 1 + \left(\frac{B}{A}\right)\left(\frac{1}{\Delta P - \Delta \pi}\right) \qquad (7)$$

Equation 7 shows that as $\Delta P \rightarrow \infty, R \rightarrow 1$. The principal advantage of the solution–diffusion (SD) model is that only two parameters are needed to characterize the membrane system. As a result, this model has been widely applied to both inorganic salt and organic solute systems. However, it has been indicated (26) that the SD model is limited to membranes having low water content. Also, for many RO membranes and solutes, particularly organics, the SD model does not adequately describe water or solute flux (27). Possible causes for these deviations include imperfections in the membrane barrier layer, pore flow (convection effects), and solute–solvent–membrane interactions.

When it was recognized (31) that the SD model does not explain the negative solute rejections found for some organics, the extended solution-diffusion model was formulated. The SD model does not take into account possible pressure dependence of the solute chemical potential which, although negligible for inorganic salt solutions, can be important for organic solutes (28,29).

Surface Force–Pore Flow Model. The surface force–pore flow (SFPF) model (8,32) is a two-dimensional extension of the finely porous model. Whereas the finely porous model considers only axial solute concentration gradients, the SFPF model recognizes that the solute concentration in an RO membrane pore may be a function of radial as well as axial position (29). The SFPF model assumes that (*1*) water transport through the membrane occurs in pores by viscous flow; (*2*) solute transport takes place by diffusion and convection in the membrane pores; (*3*) transport of both water and solute through the membrane pores is determined by interaction forces, friction forces, and chemical potential gradients of the water and solute; (*4*) the pores of the membrane are cylindrical and run the length of the membrane barrier layer; (*5*) a molecular layer of pure water is preferentially sorbed on the pore wall; and (*6*) a potential field controls the solute distribution of the membrane pore. An important modification of the SFPF model has been formulated which assumes a distribution of membrane pore sizes. Some inconsistencies in the SFPF model similar to those recognized for the original finely porous model (33,34) led to the formulation of a modified SFPF correcting these conceptual errors. The SFPF models require the solution of a velocity profile equation in the pore, force balance equation of solute in the pore, and correlations of interaction parameters with potential function. The pure water flow rate has been described by the Poiseuille equation.

The transport equations for the SFPF model, expressed in dimensionless form, have been solved using a variety of numerical techniques. Liquid chromatography was employed to determine $\tilde{A}$, the measure of electrostatic repulsion force between the ionic solute and the membrane, or $\tilde{B}$, the measure of short-range van der Waals forces for a solute (7,33,35). Trial-and-error was then used to find the membrane pore radius, R_p. Alternatively, if R_p was specified, then $\tilde{A}$ or $\tilde{B}$ was varied to produce agreement in the predicted and measured permeate concentrations. Measured pore radius values and one experimental data point for permeate concentration was also employed to eliminate the need for trial-and-error solution of the transport equations (36). Both solution techniques indicate that the SFPF model gives excellent predictions of solute separation for a wide range of inorganics and organics under varying operating conditions. However, for some dilute organics that cause substantial decreases in water flux, the models do not adequately predict the water flux.

Concentration Polarization

Concentration polarization (CP) is the term used to describe the accumulation of rejected solute at the surface of a membrane, causing the solute concentration at the membrane wall to be much greater than that of the bulk feed solution. The enhanced solute concentration on the membrane surface causes a reduction of the water flux owing to increased osmotic pressure (see eq. 5), precipitation of sparingly soluble salts such as $CaSO_4$, reduced permeate quality, etc. A dramatic reduction in water flux for a $CaSO_4$ solution owing to CP is shown in Figure 3 (37). As soon as the mixing was stopped, CP increased so that the wall concentration of $CaSO_4$ exceeded saturation and the flux dropped steadily owing to precipitate formation.

Reviews of concentration polarization have been reported (14,38,39). Because solute wall concentration may not be experimentally measurable, models relating solute and solvent

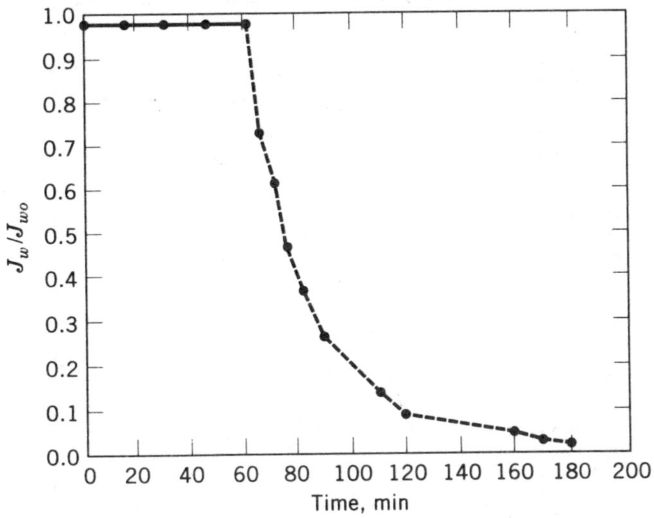

Figure 3. Comparison of flux for an aromatic polyamide (FT30-BW) membrane for (—) mixed and (---) unmixed solutions of $CaSO_4$, where the initial concentrations are 1.75 g/L and $\Delta P = 2.1 \times 10^6$ N/m^2 (2.1 MPa (304.5 psi)). Mixing was carried out for 60 min (see eq. 4).

fluxes to hydrodynamic parameters are needed for system design. The Navier-Stokes diffusion–convection equation has been numerically solved to calculate wall concentration, and thus the water flux and permeate quality (40).

A simplified model using a stagnant boundary layer assumption and the one-dimension diffusion–convection equation has been used to calculate wall concentration in an RO module. The integrated form of this equation, the widely applied film theory (41), is given in equation 8.

$$\frac{C_{\text{wall}} - C_P}{C_B - C_P} = \exp\frac{V_w}{k_s} \qquad (8)$$

Correlations for the mass-transfer coefficient, k_s, as the Sherwood number for various membrane geometries have been reviewed (39).

Using this simplified model, CP simulations can be performed easily as a function of solution and such operating variables as pressure, temperature, and flow rate, using software packages such as Mathcad. Solution of the CP equation (eq. 8) along with the solution–diffusion transport equations (eqs. 5 and 6) allow the prediction of CP, rejection, and permeate flux as a function of the Reynolds number, *Re*. To facilitate these calculations, the following data and correlations can be used: (*1*) for mass-transfer correlation, the Sherwood number, *Sh*, is defined as $Sh = 0.04 \cdot Re^{0.75}Sc^{0.33}$, where *Sc* is the Schmidt number; (*2*) osmotic pressure follows van't Hoff's equation, ie, $\pi = iCR_gT$, where i is the number of ions; (*3*) $A = 4.05 \times 10^{-7}$ cm/(s · kPa) (4.1×10^{-5} cm/(s · atm)) and $B = 1.3 \times 10^{-4}$ cm/s; (*4*) $\mu = 1$ mPa · s(= cP), NaCl diffusivity in water $= 1.6 \times 10^{-5}$ cm^2/s, and solution density $= 1$ g/cm^3. Figure 4 shows typical results of this type of simulation of salt water permeation through an RO membrane. Increasing the Reynolds number in Figure **4a** decreases the effect of concentration polarization. The effect of feed flow rate on NaCl rejection is shown in Figure **4b.** Because the intrinsic rejection, $R = 1 - C_P/C_{\text{wall}}$, is defined in terms of the wall concentration, theoretically R should be independent of the Reynolds number and in Figure **4b** the intrinsic rejection is essentially so. On the other hand, as C_{wall} increases, C_P should increase and thus the observed rejection is expected to be a strong function of feed flow rate. The maximum flow rate that can be used to minimize CP is controlled by the membrane module pressure drop. The effects of spacers, complex flow patterns, types of fluids, etc on CP behavior have been reported (40).

STUDY OF RO VARIABLES AND TYPICAL EXPERIMENTAL SETUP

Factors affecting RO membrane separations and water flux include feed variables such as solute concentration, temperature, pH, and pretreatment requirements; membrane variables such as polymer type, module geometry, and module arrangement; and process variables such as feed flow rate, operating time and pressure, and water recovery.

For simple, noninteracting inorganic salts, the trends in water flux and rejection behavior can be easily predicted (10). Most of the RO transport models (eq. 5) indicate that water flux, J_w, should increase linearly with the net pressure. Temperature affects variables such as viscosity and diffusivity. The result of a temperature increase is generally an increase in water (solvent) flux and a decrease in solute rejection. Water flux can also gradually decrease over operating time, measured in days or months of operation, as a result of fouling, membrane compaction, or other changes in membrane structure. Salt rejection (eq. 7) increases with the net pressure up to an asymptotic value. For ionizable organics, the rejection of ionized species is considerably higher than the nonionized species.

The intrinsic rejection and maximum obtainable water flux of different membranes can be easily evaluated in a stirred batch system. A typical batch unit (42) is shown in Figure 5. A continuous system is needed for full-scale system design and to determine the effects of hydrodynamic variables and fouling in different module configurations. A typical laboratory/pilot-scale continuous unit using computer control and on-line data acquisition is shown in Figure 6.

Fouling

One of the primary concerns in reverse osmosis processes is the irreversible fouling of the membrane surface. This can lead to product flux decline as well as possible changes in rejection and selectivity. Fouling is a broad term that refers to the deposition or association of feed stream constituents on the membrane surface or within the membrane pores, consequently leading to a decrease in the membrane performance. The cause of fouling can generally be traced to constituents in the membrane feed stream. Several general categories of foulants found in process streams can be identified in relation to reverse osmosis processes: biological fouling agents, colloids, scaling agents, organics, metal oxides, and suspended solids in general. Depending on the particular application, one or several of these categories may become a factor in the RO process. For example, in the desalination of seawater, problems with colloids, suspended solids, biological fouling, and scaling may occur. Wastewater reclamation, on the other hand, suffers primarily from fouling by dissolved organics and biological slime. Fouling must be taken into account when considering the use of RO on an industrial level.

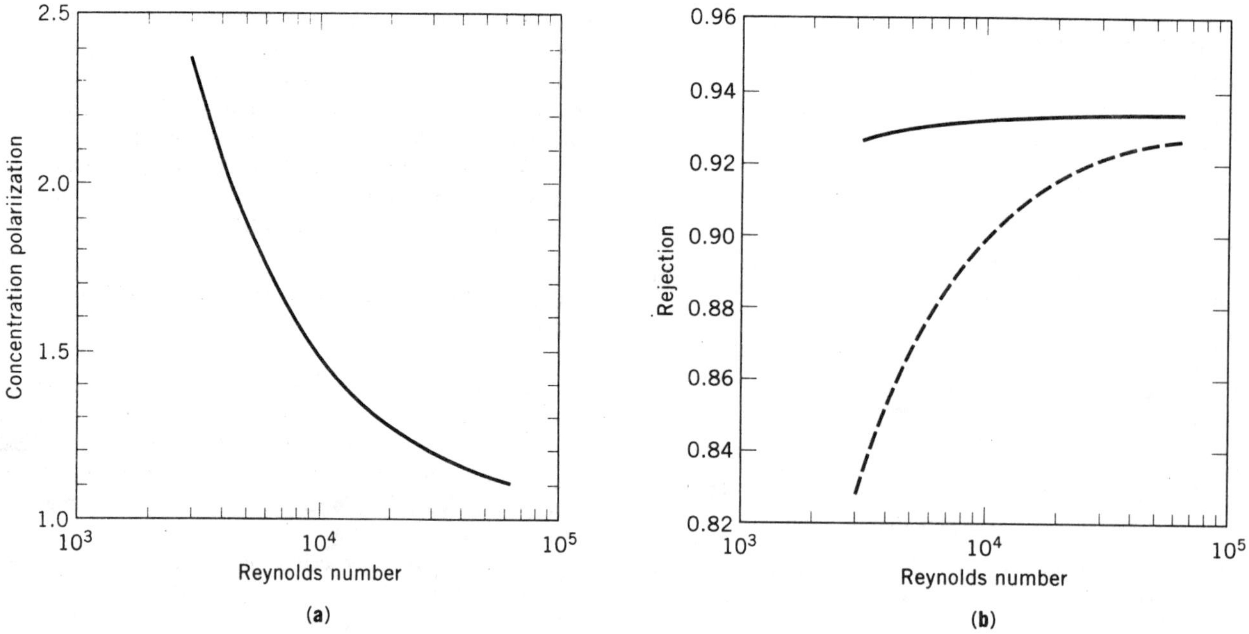

Figure 4. Mathcad simulations (C_F = 5000 mg/L) as a function of Reynolds number for a NaCl solution: (**a**) concentration polarization (CP), and (**b**) (---) observed and (—) intrinsic rejections using a simple film theory model.

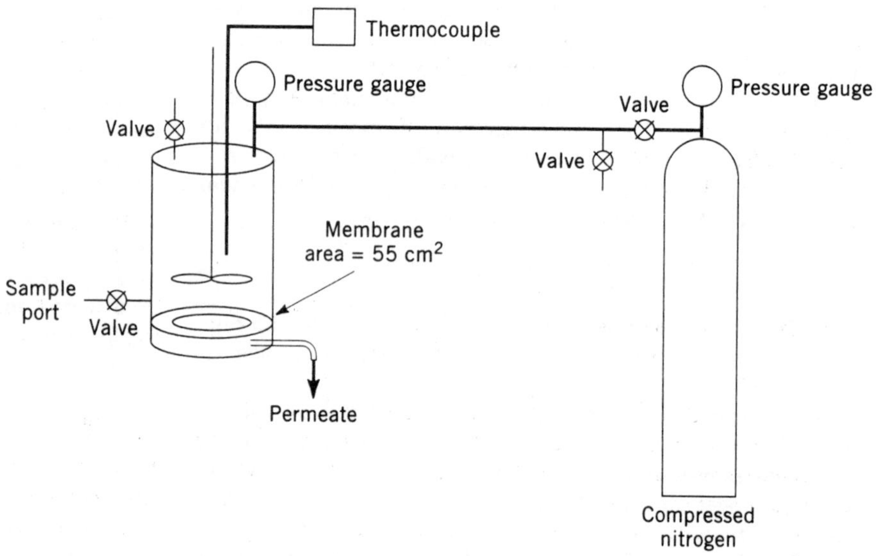

Figure 5. Schematic for a stirred batch RO unit pressurized by compressed nitrogen.

The manner in which the foulants decrease the membrane flux varies, depending on the particular foulant. Gel layers can be formed at the surface from a variety of constituents that do not enter the membrane but rather accumulate at the surface. In the presence of inorganic salts at high water recovery, solubility limits can be exceeded, causing the formation of a scale on the membrane surface. Various microorganisms can attach and grow on the membrane, thus creating a biological slime. Other components that can form gel layers are humic acids, proteinaceous material, and other large molecules. In the presence of fine colloids or suspended solids, the membrane pores may also become plugged.

The success of a reverse osmosis process hinges directly on the pretreatment of the feed stream. If typical process streams, without pretreatment to remove partially some of the constituents listed, were contacted with membranes, membrane life and performance would be unacceptable. There is no single pretreatment for all types of foulants. Pretreatment methods range from pH control, adsorption, to filtration, depending on the chemistry of the particular foulant. Some of the pretreatment methods for each type of foulant are as follow (43–45):

Fouling material	Pretreatment method
Biological fouling	Chlorination, ozonation, or UV light
Colloids	Coagulation/filtration
Scaling	Acid treatment, antiscale agents (SHMP), chelating agents (EDTA), or sand filtration

Organics	Coagulation/filtration, activated carbon, or oxidation
Metal oxides	Proper materials or acid treatment
Proteins	Microfiltration or ultrafiltration
Suspended solids	Cartridge filtration or screen filters

Depending on the number, nature, and concentration of the potential foulants in the feed stream, a logical pretreatment plan must be developed to ensure the success of the RO plant.

Biologicals. Biological components form slimes on a membrane surface that can cause a reduction in the membrane flux as well as membrane degradation (43,45–47). Once microbes attach to the membrane surface, they can multiply. Steps such as chlorination (0.5 mg/L) are generally taken to kill biological constituents before the feed enters the membrane module. Chlorine, however, causes damage to most RO membranes. Typical cellulose acetate membranes can withstand chlorine concentrations of 0.3–1.0 mg/L; polyamide membranes can withstand less than 0.05 mg/L. Therefore, any pretreatment that uses chlorine must have an additional step for chlorine removal, the most common of which is the use of sodium bisulfite. Too much residual chlorine after the removal step can lead to membrane damage, which in turn usually leads to an increase in flux accompanied by a dramatic decrease in rejection. If, however, all of the chlorine is removed while a few bacteria remain, the bacteria continue to grow. For this reason, it is desirable to have a chlorine-resistant membrane and prevent biological growth by allowing a low concentration of chlorine to be present in the RO module.

There has been considerable research on chlorine-resistant RO membranes (48–52). A poly(*trans*-2,5 dimethyl)piperazinthiofurazanamide used in the presence of low (3 mg/L) concentrations of chlorine resulted in a membrane life of three years (48). A copolyamide hollow-fiber membrane for use in desalination has been developed that is resistant to 0.5 mg/L chlorine (49). Millipore Corporation has also developed a sulfonated polysulfone member that has desirable chlorine-resistance properties.

Inorganics. Inorganic salts can precipitate on the membrane surface if the solubility limit of the salt is exceeded. This may occur upon increased water recovery. Depending on the types of salts present in the feed stream, different steps can be taken to prevent scaling. For waters containing bicarbonates and calcium, calcium carbonate, $CaCO_3$, presents a potential

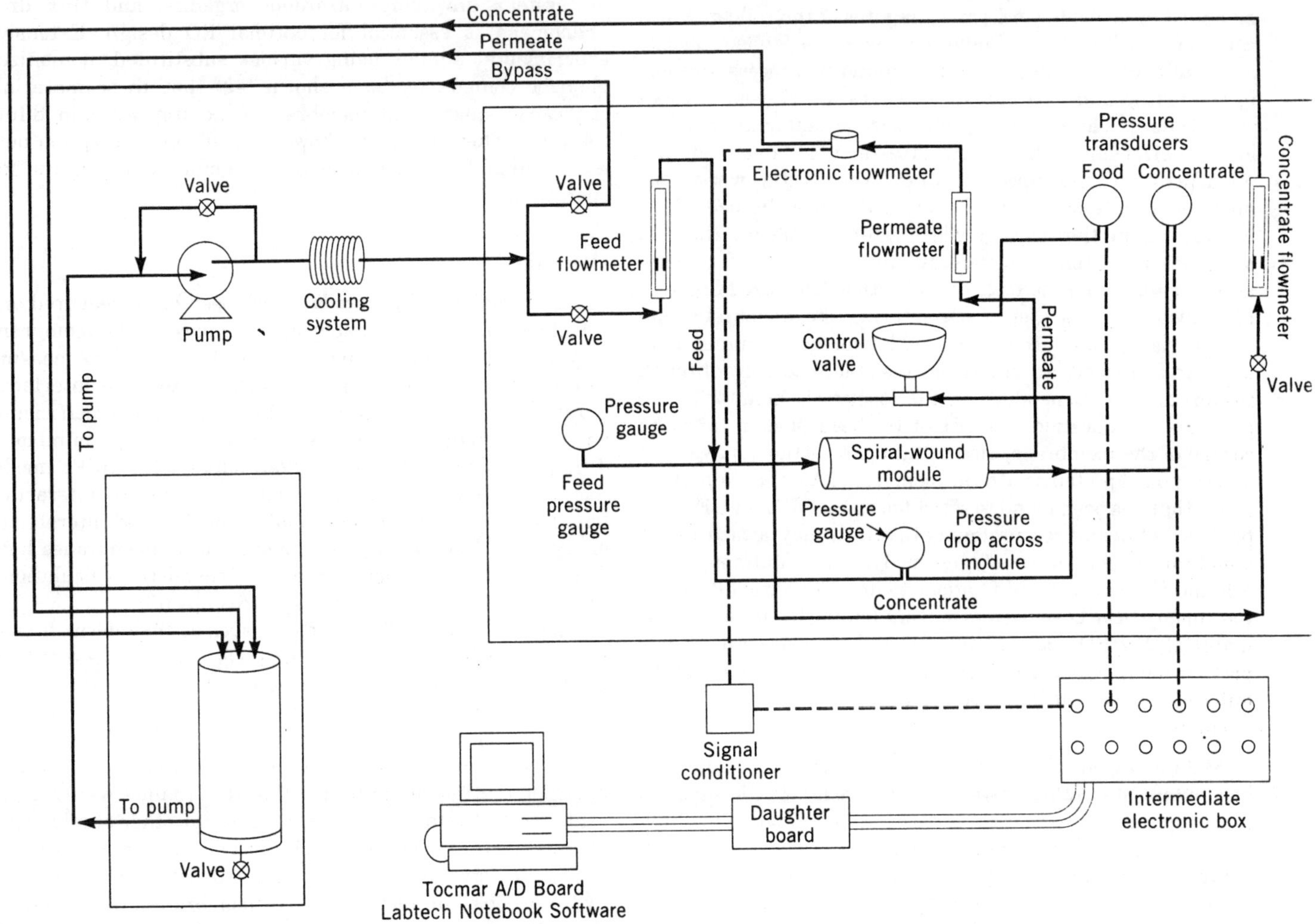

Figure 6. Schematic for a computer-controlled, continuous RO unit pressurized by a pump. A/D = analog to digital.

scaling problem. Treatment of the water with acid converts bicarbonate to carbon dioxide, thus eliminating the scaling problem (53). Calcium and magnesium can be removed by precipitation with lime and subsequent removal. Chelating agents such as ethylenediaminetetraacetic acid (EDTA) and polyphosphates are also frequently used to prevent precipitation.

Metal oxides found in RO feed streams typically originate from corroded pipes found in the RO process. These metal oxides can deposit on the membrane surface and decrease the membrane flux. This type of fouling can be prevented by using the proper materials of construction in the piping system to prevent corrosion.

Solids and Colloids. Suspended solids can accumulate at the membrane surface, creating an additional resistance to flow through the membrane as well as a possible feed channel, such as that for a spiral-wound module; plugging; and subsequently a decrease in flux. Prevention of this type of fouling lies in the removal of the suspended solids, which can be accomplished using filters and screens prior to arrival at the RO unit.

Colloidal materials present in surface waters can also plug RO membranes, causing a decrease in permeate flux. Colloidal plugging can be avoided by using one of several possible pretreatment steps. Ultrafiltration (UF) or microfiltration (MF), depending on the size of the colloid, can be used to filter out the colloidal material. Alternatively, a coagulant such as alum can be added to the water to form aggregates of the colloid, which can then be filtered in a similar manner as suspended solids.

Proteinaceous material is often found in process streams in the textile and food industries where RO is used to clean wastewaters. The effect of proteinaceous material on membrane performance has been investigated extensively for ultrafiltration and microfiltration. Very little work has, however, been done in the RO area. Because the use of RO is growing in the area of wastewater treatment, there is a significant interest in determining the cause of proteinaceous fouling and methods of prevention. The size of protein molecules prevents them from entering the membrane matrix. Instead, these accumulate at the surface, forming a gel layer and creating an additional resistance to flow through the membrane. Depending on the isoelectric point(s) of the protein(s) in question, the pH of the feed stream, and the charge of the membrane, electrostatic interactions may cause association of proteins with the membrane. Another important factor that has been found to affect fouling in MF and UF is the presence of denatured protein (44), which may attach to the membrane surface to satisfy thermodynamic requirements. In MF and UF, surface modifications to impart hydrophilicity on the surface have been used to prevent fouling by proteinaceous material. For RO applications, MF and UF can be used as a pretreatment step to eliminate the protein at the RO module, but there may still be a fouling problem in the MF or UF modules.

Multiple Agents. In most applications, there is more than one foulant present in a given process stream. The best option in the operation of a reverse osmosis plant would be to eliminate fouling through module design and feed pretreatment. However, economic considerations almost always prevent the complete elimination of fouling. Typically, an economically feasible pretreatment process is designed, and when membrane performance degrades to a predetermined level, the process goes through cleaning cycles.

Membrane Cleaning. Methods for cleaning membranes have been developed (54–56). Cleaning methods for RO membranes (54) can be divided into two classes. Physical cleaning methods include the use of sponge balls, backpressure flushing, and vibration. Chemical cleaning methods may employ EDTA or other chelating agents, surfactants, acids, or sodium dithionate for iron-containing scales. Physical cleaning can only be performed on flat-plate (scrubbing) or tubular membranes (sponge balls). For the hollow-fiber and spiral-wound modules, which offer the advantage of high surface area, physical cleaning is not an option. Chemical treatments must be used to remove the fouling layer in these types of modules. For example, calcium scales can be removed by the use of citric acid. Once again, knowledge of the feed stream and hence the foulants is of the utmost importance when cleaning membranes. Extensive flux declines may also occur in systems containing dilute low molecular weight organics owing to interactions with membrane polymers.

ORGANIC–MEMBRANE INTERACTIONS

Understanding the feed characteristics of membrane polymers and organics solute interactions (57), fundamental transport mechanisms and simplified predictive models particularly for systems involving hazardous organics, and flux drop phenomena is essential for optimal RO design. Extensive experimental studies using various substituted, nonionized phenolic compounds have shown (42) that these materials can cause substantial membrane flux drop even in dilute solutions. These organics showed significant adsorption onto a cross-linked aromatic polyamide membrane during the RO operation.

Interactions.

Solute–membrane interactions have been both measured and calculated. Liquid chromatography employing the membrane polymer as the column packing has been used to measure directly inorganic and organic solute–membrane interfacial parameters, such as equilibrium distribution coefficients, Gibbs free energy, and surface excess for a large number of different polymers (7,58,59). Additionally, the SFPF model has been used to calculate solute–membrane interaction forces, coulombic or van der Waals (7,33,36), and interactions of organic solutes with polybenzimmidazole membranes have been characterized using quantum chemistry calculations, liquid chromatography, and ir spectroscopy (60).

The effects of small halocarbons, such as chloroform, bromoform, and carbon tetrachloride, on several RO membranes have been considered (61). The organics were found to cause a reduction in the membranes' void spaces, representing a decrease in water content of the membrane, possibly because of substitution of the water in the membrane by the organic. Analysis of partition coefficients indicated that the organics were significantly sorbed into the membrane. Several other studies have also shown that even dilute solutions of some organic solutes, such as chloro- and nitrophenols, and polynuclear aromatic hydrocarbons interact with RO membranes, substantially changing water flux characteristics (9,62–64). Adsorption of the solute onto the membranes was thought to be related to the lower water flux observed.

Organic Rejections and Flux Drop Behavior.

Many studies have been performed on the separation of organics and organic pollutants by RO membranes, and these studies have identified some of the unique aspects associated with organic separation. Separation and flux data of cellulose acetate membranes have been compiled for a large number of organic compounds, including many organic pollutants (7,65). Organic separation can vary from 0 to 100%, depending on the characteristics of the organic material, eg, polarity, size, and charge, and the RO operating conditions, such as feed pH and operating pressure.

Several investigations have compared organic rejections of cellulose acetate to that of other membranes. Aromatic polyamide and composite membranes usually have organic rejections greater than those of cellulose acetate membranes. Several organic rejections of the Toray composite membrane PEC-1000 (polyfuran) have been listed (66). Most rejections were high, including 97% for acetone and 99% for phenol. Separation results for several polar organic solutes, eg, alcohols, phenols, carboxylic acids, amines, and ketones, and various phenolic derivatives have been reported (67,68) for a composite membrane. The main factors affecting rejection included molecular weight, molecular branching, polarity, and degree of dissociation for ionizable compounds. Both cellulose acetate and thin-film composite membrane separations of a wide variety of organic pollutants have been compared (69), the latter involving FT30 (FilmTec), a cross-linked aromatic polyamide. The composite membrane rejections, greater than 90% for most of the organics studied, and water fluxes were substantially better than those of the cellulose acetate membrane. However, adsorption of some of the organics on the membranes was noted. Polyamide membrane rejections of chlorinated hydrocarbons (>95%) have been observed to be much greater than those of cellulose acetate membranes (70). Separation results for four composite and two asymmetric membranes for a variety of single and multicomponent organic solutions, including many organic pollutants, have been reported (71). Rejections varied from only 25% up to >99%, depending on the solute, but generally the composite membrane rejections were higher.

The rejection and flux characteristics of aromatic polyamide, FT30-BW, membranes for various chlorinated organics have been investigated (42). For substituted phenols, the rejections of ionized ($pH > pK_a$) species were always considerably higher than the nonionized species. Rejections of several environmentally significant organics are shown in Table 1. In general, rejection increased with increasing substituent group addition for the chloro- and nitrophenols. Both specific and nonspecific interactions are possible. On the other hand, water flux drop was highly significant for certain substituted, nonionized ($pH < pK_a$) organics. Figure 7 shows the normalized flux results for several organics. With the exception of PCP, the water flux drop caused by the dilute (insignificant osmotic pressure) chlorophenol solutions increased in the order of increasing numbers of chlorines on the phenol. The hydrogen bonding of the phenol to groups on the membrane polymer, such as the carbonyl group of polyamide FT30 membrane, could be expected to cause the observed flux drop by reducing the membrane hydrophilicity. Figure 8 supports this, showing a significant decrease in water flux as solute acidity (decreasing pK_a) increases. For these compounds, flux declines were insignificant when the RO system operated at pH values greater than the corresponding pK_as, ie, the solutes were ionized.

Table 1. Rejections by Aromatic Polyamide Composite Membrane[a,b]

Compound	Rejection, %
Standard NaCl	98
Nonionized phenolics	
Nitrophenol	55
Phenol	60
Chlorophenol	73
Dinitrophenol	80
Trichlorophenol	90
Pentachlorophenol	99
Ionized phenolics[c]	94–99
Other organics	
Chloroaniline	>90
Benzene	93
Dichlorobenzene	96
Ethyl benzene	>97
Naphthalene	98
Trichlorobenzene	>99
Anthracene	>99
Phenanthrene	>99

[a] RO membrane FT30-BW. Compounds are in dilute aqueous solution.
[b] Pressure = 1.3–2.0 MPa (13–20 atm).
[c] For example, rejection of 2,4,6-trichlorophenol at pH = 10 is 99.4%

A nonspecific interacting compound, benzene (see Fig. 7), caused a smaller (<8%) water flux drop even though the benzene was more highly adsorbed (2.9×10^{-6} mol/cm^2) as compared to the specific interacting compound, 2,4,6-trichlorophenol (TCP), which was adsorbed to a lesser (9.2×10^{-7} mol/cm^2) extent but caused substantially more (almost 50%) flux decline. Flux and separations characteristics can be modeled upon modifying the SD equations by incorporating an organic sorption term (72) or by solving the general diffusion–convection equations using organic adsorption terms (42). A modified solution–diffusion (MSD) model can easily be obtained by assuming that the sum of organic and water concentrations is constant, ie, total concentration in membrane, $C_{tm} = C_m + C_{wm}$, and representing C_m/C_{tm} by a Langmuir adsorption term. The modified SD equations for water and solute flux are as in equations 9 and 10:

$$J_w = \left[\frac{1}{1 + b_0 C_F}\right] A * (\Delta P - \Delta\pi) \quad (9)$$

$$J_s = B * \left[\frac{b_0 C_F}{1 + b_0 C_F} - \frac{b_0 C_P}{1 + b_0 C_P}\right] \quad (10)$$

A simulation plot of normalized flux vs feed concentration, C_F, is shown in Figure 9 for a known A^*, a modified water permeability parameter, and various values of b_0, a measure of organic–membrane interaction parameter. Flux decreases significantly with increasing b_0. From experimental results using FT-30 membranes, the following values of b_0 were obtained for nonionized organics: 280 M^{-1} for 2-chlorophenol, 1380 M^{-1} for 2,4,6-trichlorophenol, and 4400 M^{-1} for 2,4-dinitrophenol. Thus RO membranes typically used for treatment of aqueous feeds are not as effective in the treatment of organic feeds. Development of novel RO membranes for the separation of organics would be of interest.

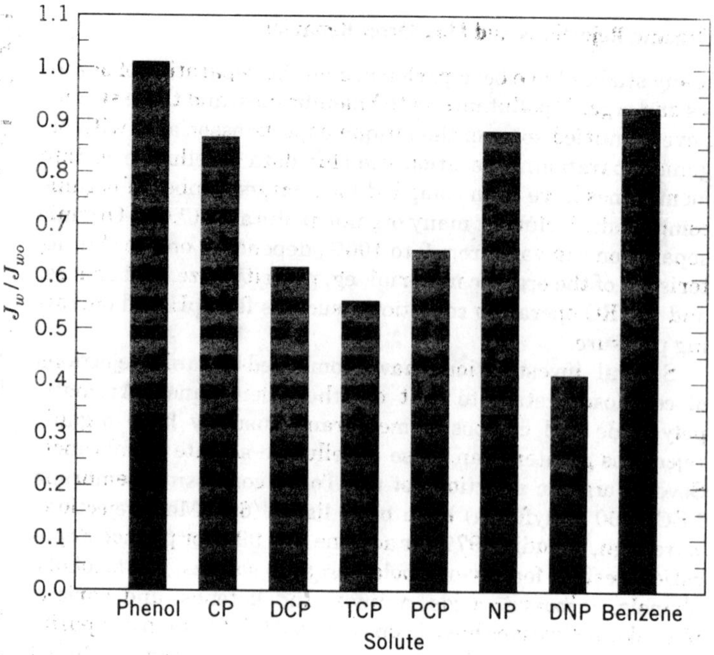

Figure 7. Normalized water flux for hazardous organic compounds using FT30-BW membranes at $T = 24°C$, $\Delta P = 1.38$ MPa (200 psi), pH of the feed = 3.0 − 4.0, and $r_f = 52 - 62\%$. C_F for phenol, 2,4,6-trichlorophenol (TCP), and 2,4-dinitrophenol (DNP) was 0.24 m*M;* for 2-chlorophenol (CP), 0.23 m*M;* for 2.4-dichlorophenol (DCP), 0.22 m*M;* for 2-nitrophenol (NP), 0.21 m*M;* for benzene, 0.19 m*M;* and for pentachlorophenol (PCP), 0.02 m*M.*

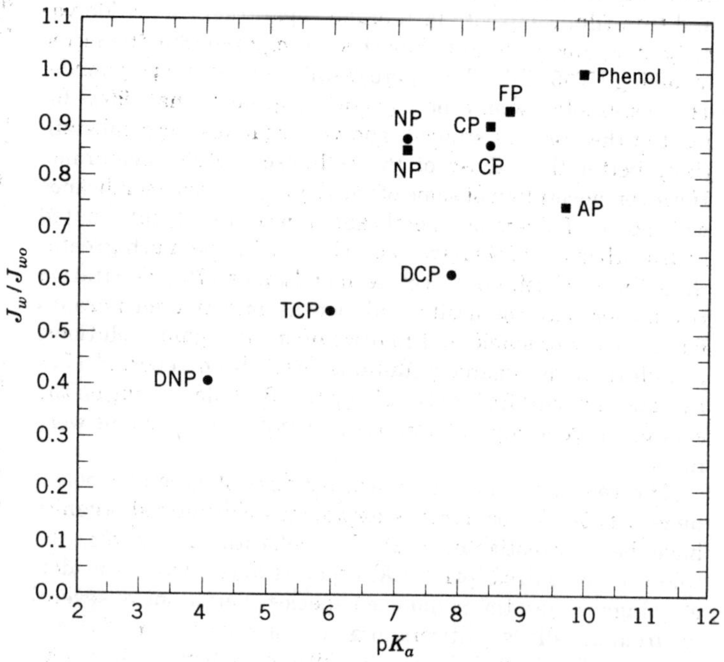

Figure 8. Water flux as a function of pK_a for various phenols using FT30-BW membranes at $T = 24°C$, $\Delta P = 1.38$ MPa (200 psi), pH of feed = 3.0 − 4.6, and $r_f = 34 - 62\%$, where (■) represents $C_F = 0.5$ m*M*; (●), ~0.2 m*M*. AP = 2 − aminophenol; FP = 2 − fluorophenol; for other abbreviations, see Fig. 7.

APPLICATIONS

Developments and advances in both membrane materials and reverse osmosis modules have increased the range of applications to which RO can be applied. Whereas the RO industry has developed around water desalination (9,53,73,74), RO has become a significant cornerstone in other industries. These include wastewater cleanup for electroplating (75–78), radioactive processing (79–82), landfill leachate (76,83), and municipal wastewater (84–87); ultrapure water production for electronics-grade (88,89), laboratory-grade (90), and pharmaceutical-grade (91) materials; and food processing (9).

Wastewater Applications

Significant advances have been made to reduce environmental pollution problems caused by the generation of hazardous wastes. Waste management approaches include abatement, minimization, recycle and reuse, end-of-pipe treatment, and environmentally safe disposal of effluents and residuals. Even using the implementation of waste reduction procedures, significant amount of wastes often require further treatment. These treatment technologies may include biological treatment, adsorption, stripping, chemical and photolytic oxidation, membrane separation (RO, NF, pervaporation, catalytic membranes), and supercritical water oxidation. For separation,

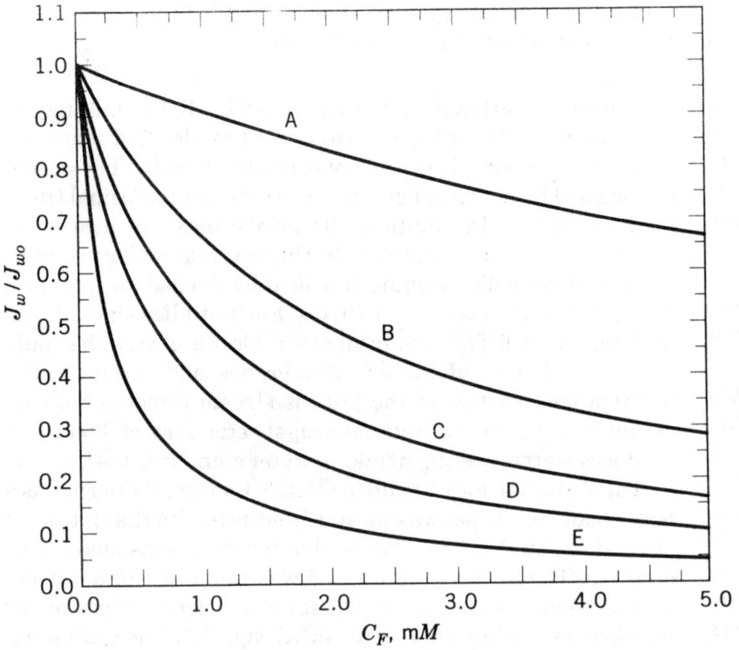

Figure 9. Parametric simulation using the MSD model of water flux vs feed concentration as a function of organic sorption coefficient, b_0, where for A, $b_0 = 100\ M^{-1}$; for B, $500\ M^{-1}$; for C, $1000\ M^{-1}$; for D, $1695\ M^{-1}$; and for E, $4000\ M^{-1}$.

material recovery, and destruction of hazardous organics, a combination of membrane processes with other technologies can provide the optimum treatment system. Many wastewater treatments use reverse osmosis.

Several advantages of the RO process make it particularly attractive for dilute aqueous wastewater treatment: (*1*) RO systems are simple to design and operate, have low maintenance requirements, and are modular in nature, making expansion of the systems easy; (*2*) both inorganic and organic pollutants can be removed simultaneously by RO membrane processes; (*3*) RO systems allow recovery/recycle of waste process streams but do not affect the material being recovered; (*4*) RO membrane systems often require less energy and offer lower capital and operating costs than many conventional treatment systems; and (*5*) RO processes can considerably reduce the volume of waste streams so that the resulting concentrated system can be treated more efficiently and cost-effectively by other processes, such as incineration or other destruction processes (75,92,93). In addition, RO systems can replace or be used in conjunction with other treatment processes, including oxidation, adsorption, stripping, or biological treatment, to produce a high quality product water that can be reused or discharged.

Applications reported for RO processes include the treatment of organic-containing wastewater, wastewater from electroplating and metal finishing, pulp and paper, mining and petrochemical, textile and food processing industries, radioactive wastewater, municipal wastewater, and contaminated groundwater (75,76,83,94–99). One of the newer applications involves the treatment of wastewater generated by the semiconductor industry, which generates wastewater containing large amounts of hydrofluoric acid (HF) as well as ionic impurities that render the HF solution useless. RO membranes that allow selective permeation of HF while rejecting the ionic impurities to produce a pure HF solution that can be reused have been reported (100).

Electroplating and Metal-Finishing Process Wastewaters. In most cases, process wastewaters from the electroplating and metal-finishing industries must be treated to remove heavy metals before being discharged. Reverse osmosis is ideal for many of these operations because it allows both recovery of the heavy metals and reuse of the product water in the process. The RO process has been used in the treatment and recovery of wastewater containing nickel, acid copper, zinc, copper cyanide, chromium, aluminum, and gold (75,97,98). Results for an industrial printed circuit board production plant have been summarized (77), demonstrating the effectiveness of RO for plating salts recovery and virtual elimination of wastewater discharge.

Cellulose acetate and polyamide (FT30) membranes were used at three Japanese plating shops employing nickel, chromium, and gold plating lines (78). Up to 80% water recoveries and high metal and total dissolved solids (TDS) (>95%) rejections were possible. The product water was recycled. The RO processes were found to be cost-effective in treating the wastewaters, and the compact nature of the RO system made it highly desirable to the customers because of space limitations. The RO treatment of effluent from an electrolytic polishing process for aluminum products has also been discussed (101). The streams contained phosphoric acid and aluminum from rinse water. DDS HR-98 membranes allowed 96–98% acid recovery, up to an acid concentration of 20%, and produced permeate water suitable for reuse. The membranes appeared to be stable to the feed even at the low (0.9–1.0) pH values found at high recoveries.

Radioactive Processing Wastewaters. Because of high rejection of inorganic compounds, RO membranes have been studied for treatment of radioactive effluents. These membranes have been used to treat uranium conversion process effluent containing toxic, corrosive, and radioactive compounds (79). The FT30 membranes had rejections of uranium of 99.5% for water recoveries up to 70%. Results indicated that the treated effluent was within regulatory discharge standards. In addition, a hybrid process consisting of nanofiltration (NF), reverse osmosis, and precipitation has been used to treat uranium effluents (80). The process removed both soluble and suspended uranium species. Up to 95% uranium recovery was possible, and the treated effluent met environmental standards. The FT30 membranes gave uranium rejections of >99%. Cellulose acetate membranes have been shown to remove 99% or uranium from effluents containing uranium nitrate when the uranium was complexed with EDTA (81). RO can be used to treat contaminated effluent from a uranium metal plant (82), whereby 95% by volume of the treated effluent is within discharge limits.

Leachates. The treatment of landfill leachates serves two purposes. It facilitates the production of a water stream that can be discharged while concentrating the hazardous material and reducing its volume. Several studies have been conducted on the treatment of landfill leachates using RO processes. Reverse osmosis has been shown to be effective, ie, high rejection at high recovery, in the removal of total organic carbon (TOC), TDS, chemical oxygen demand (COD), ammonia, heavy metals, and alkalinity from different types of leachates (102,103). In some cases, flux decline has been reported. However, proper pretreatment and membrane cleaning can alleviate this problem. In an early study, RO membranes were used to remove >91% of TOC from sanitary landfill leachate (102). In one case, RO treatment of landfill leachate resulted in removals of 94.5% alkalinity, 97% COD, 97% total solids, 92.1% volatile solids, and 96.6% ammonia (104). FT30 membranes were used to treat soil-wash leachates containing heavy metals and organic contaminants so that the treated water could be recycled back to the soil-washing step (64). TOC rejections as high as 80–85% and heavy metal (Pb, Zn, Ni, Cu) rejections of 94–98% were found. However, water flux decreases of up to 33% were noted. The effects of the addition of EDTA or surfactant and feed preozonation were also investigated. Feed preozonation was found to have substantially improved membrane water flux. Specific organic rejections included >98% for pentachlorophenol and 2,4-dinitrophenol, >97% for ethylbenzene, >81% for xylene, and >90% for chloroaniline. The treatment of landfill leachates containing organic acids has been reported (105). Good separations of volatile fatty acids were obtained and TDS removed sufficiently to allow discharge of the product water. A commercial leachate treatment site in Germany was operational for >10,000 h without membrane replacement (83). Tests on leachates from sites where the waste was segregated have been conducted (76). Waste containing a large amount of ash was found to be unsuitable for treatment by RO.

Drinking Water. The ability of RO membranes to remove both inorganic and organic compounds have made these systems attractive for the treatment of contaminated drinking water supplies (106). Reverse osmosis processes can simultaneously remove hardness, various inorganics such as bromate and bromide ions (107), color (108), many kinds of bacteria and viruses, and organic contaminants such as agricultural chemicals and trihalomethane precursors. RO treatment of drinking water sources has been reviewed (109), and successful removal of a wide variety of contaminants has been accomplished. Membrane processes are also finding applications for the removal of natural organic matter (NOM) and disinfection byproducts (DBPs). An excellent overview of selected processes, eg, adsorption, coagulation, RO, and NF, for removing NOM and DBPs has been reported (110). Membrane processes show the potential to achieve the highest removal of NOM and DBP precursors, but for surface water applications, high pretreatment are required to minimize fouling.

Municipal Wastewater. The application of RO membranes to the treatment of municipal wastewater has also had some success. Reverse osmosis can remove dissolved solids that cannot be removed by biological or other conventional municipal treatment processes. In addition, RO membranes can also lower organics, color, and nitrate levels. However, extensive pretreatment and periodic cleaning are usually needed to maintain acceptable membrane water fluxes. Early studies showed that high removals of TDS and moderate removals of organics could be achieved. Use of tubular membranes and a substantial pretreatment system of the process stream removes both inorganics and organics from municipal secondary effluent and produces water meeting drinking water standards (84).

The Water Factory 21 site in Orange County, California, is a large-scale municipal wastewater treatment site that has been studied in detail (85,86). The feed to the plant consisted of secondary effluent, and the process was composed of a variety of treatment systems, including several different types of RO membranes having a 1.9×10^4 m^3/d capacity. The process reduced TDS and organics to levels that allowed the effluent to be injected into groundwater aquifers used for water supplies. Use of several RO membranes to treat secondary effluent containing various salts and dissolved organic materials has been reported (87). TDS rejections of up to 99% and TOC rejections as high as 90% were found possible, and fecal coliform group rejections were >99.9%. Losses in water flux over time were noted but these could be partially restored by periodic cleaning.

Desalination

Desalination of seawater and brackish water has been and, as of the mid-1990s, is the primary use of RO. Driven by a need for potable water in areas of the world where there is a shortage, this industry has developed. Desalination involves the reduction of the total dissolved solids (TDS) concentration to less than 200 mg/L. RO offers several advantages over other possible desalination processes such as distillation, evaporation, and electrodialysis. The primary advantage of RO over the traditionally used method of distillation is the energy savings that is afforded by the lack of a phase change in RO.

Reverse osmosis processes for desalination were first applied to brackish water, which has a lower TDS concentration than seawater. Brackish water has less than 10,000 mg/L TDS; seawater contains greater than 30,000 mg/L TDS. This difference in TDS translates into a substantial difference in osmotic pressure and thus the RO operating pressure required to achieve separation. The need to process feed streams containing larger amounts of dissolved solids led to the development of RO membranes capable of operating at pressures approaching 10.3 MPa (1500 psi). Desalination plants around the world process both brackish water and seawater (15).

Most desalination plants use similar pretreatment and posttreatment methods in the desalting process (Fig. 10). Feed water is passed through multimedia, sand, and cartridge filters to remove particulates and suspended solids. In addition, the feed water is typically treated with chlorine to kill any microorganisms in the water, followed by coagulation and filtration to remove the dead cells. For most commercially available membranes, dechlorination using sodium bisulfite before entering

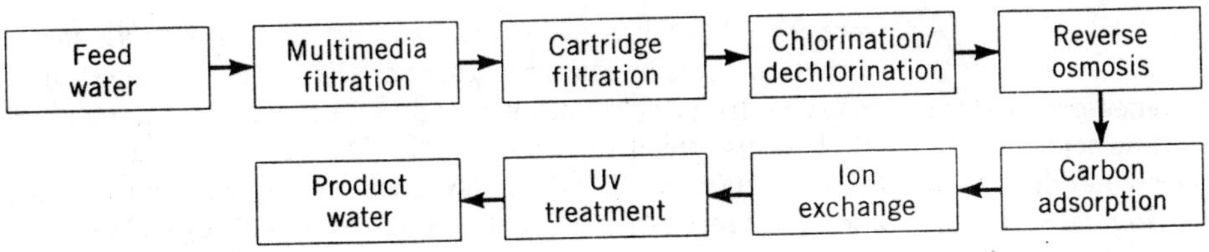

Figure 10. Flow diagram for the setup of a desalination plant employing reverse osmosis, including pre- and post-treatment steps.

the membrane module is necessary to prevent membrane damage. Chelating agents and acid are injected into the feed water to prevent precipitation and scaling on the membrane surface. Post-treatment of the permeate usually consists of the addition of lime to increase the pH along with chlorine to prevent biological growth. In seawater desalination, the concentrated brine is usually returned to the source. For brackish and well water desalination, disposal of the brine is necessary.

The largest desalination plant in the world is a brackish water plant in Yuma, Arizona having a capacity of 275,000 m^3/d (73). Spiral-wound cellulose acetate membranes produce permeate water having 200 mg/L TDS from intake water containing 3100 mg/L TDS. Seawater desalination requires a much higher pressure to achieve a given separation than brackish water. The largest seawater desalination plant in the world operates in Jeddah, Saudi Arabia. It has a capacity of 56,800 m^3/d, and the TDS content of the seawater is approximately 44,000 mg/L (74). Pretreatment modules for this plant include chlorine treatment, a dual media filter, and a cartridge filter.

Some desalination plants combine distillation with reverse osmosis to produce both power and water. Multistage flash (MSF) processes are used to produce both power and distilled water. The combination of RO and MSF and the advantages of such a combination have been reported (111). Distillation processes typically reduce the TDS concentration to levels well below the required specifications. Because the product water from the two processes is combined, the RO process can produce water at higher TDS concentrations and still meet the potable water specifications. In addition, the power produced from the MSF process can be used in the RO process, cutting energy costs.

Ultrapure Water

Reverse osmosis has been used to produce ultrapure water for a variety of applications. The electronics industry requires extremely pure water for the manufacture of semiconductors and other electronic components (88,89). For example, sodium should be present in no greater concentration than 0.001 mg/L and there can be no more than one microorganism per mL. Whereas RO removes the low molecular weight organics that pass through an ultrafiltration membrane, by itself, RO cannot produce water of this purity. Rather, RO is merely one step in a hybrid of processes designed to achieve this level of purity. The development of new membrane modules for ultrapure water production processes has been reported (112). These membranes showed high rejections of TOC and inorganic ions at low concentrations.

RO is also used to produce ultrapure water for many laboratory uses (90) as well as in the medical and pharmaceutical industries (91). As for the electronics industry, purity is achieved using a combination of processes. A typical hybrid process for the production of ultrapure water is shown in Figure 11. The order in which the various steps take place may vary from case to case.

Food Processing

One of the first applications of RO was in the food processing industry. The primary advantage of RO over the traditionally used processes in the food industry is that RO operates at low temperatures which can prevent the denaturation of some materials used in foodstuffs. Because high temperatures are not required, energy costs are reduced as well. Moreover, RO is relatively simple in terms of the equipment design. These factors lead ultimately to a reduction in capital and operating costs, accompanied by an increase in product quality.

RO, primarily used in the dairy industry, is expanding into other areas of food processing. RO can be used for a variety of operations, ranging from wastewater treatment and material recovery to clarification and concentration. Material recovery is advantageous for two reasons. By recovering valuable products, eg, proteins, from waste streams, profits can be increased while costs for waste disposal decreased. An excellent review of the different applications of RO in food processing is available (9).

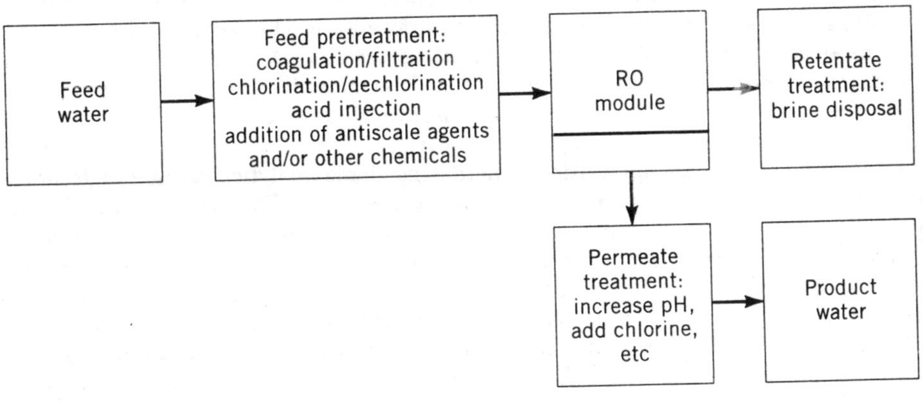

Figure 11. Schematic of membrane-based hybrid process for ultrapure water production.

NANOFILTRATION

Nanofiltration, also called loose RO, membranes are a more recent development in the field of RO membrane separations. These membranes typically have much higher water fluxes at low pressures when compared to traditional RO membranes. Nanofiltration (NF) membranes are usually charged, ie, have carboxylic groups, sulfonic groups, etc, and as a result, ion repulsion (Donnan exclusion) is the primary factor in determining salt rejection. More highly charged ions such as SO_4^{2-} are rejected more fully than monovalent ions such as Cl^- by a negatively charged nanofiltration membrane. Another characteristic of charged membranes is the tendency to foul in the presence of components having opposite charge to that of the membrane (113).

Nanofiltration membranes usually have good rejections of organic compounds having molecular weights above 200–500 (144,115). NF provides the possibility of selective separation of certain organics from concentrated monovalent salt solutions such as NaCl. The most important nanofiltration membranes are composite membranes made by interfacial polymerization. Polyamides made from piperazine and aromatic acyl chlorides are examples of widely used nanofiltration membrane. Nanofiltration has been used in several commercial applications, among which are demineralization, organic removal, heavy-metal removal, and color removal (116).

Nanofiltration has been used to remove both organics and inorganics in various wastewaters. The use of NF membranes to remove color-causing compounds from effluent containing lignins and high salt concentrations in a wood pulping process has been reported (117). Color removals were >98% at water recoveries up to 95%, whereas the inorganics were poorly rejected, allowing the use of low operating pressures. NF can give high separations of color-causing compounds such as lignin sulfonates in paper pulping wastewaters (118). NF removes >95% of chlorinated organic compounds from alkaline pulp and paper bleaching effluents having high water fluxes (119). The use of NF membranes to remove hardness and organics in textile mill effluents (120), the recovery of low molecular weight dyes from high salt concentration effluent (121), and the use of NF membranes in the treatment of food processing wastewaters (115,118) have been reported. Specific uses included the desalting of whey and the reduction of high biological oxygen demand (BOD) and nitrate levels in potato processing waters (122).

NF processes are also finding wide applications in drinking water production by removing natural organics from water that cause trihalomethane formation during coagulation (116). NF membranes have been used to remove organic matter from highly colored groundwater (123) as well as from river water (124). Nanofiltration has been used in a hybrid process employing biological treatment and chemical oxidation to treat leachate from a dumpsite (125). Increases of 9–17% in the elimination rate as compared to a process without nanofiltration have been observed.

NF40 (Dow/FilmTec) membranes were used to separate mixtures of cadmium and nickel (126). These membranes have also been examined with and without feed pretreatment by ozonation (127,128) for removal of various chlorophenols and chloroethanes. TOC rejections up to 90% were possible using ozonation pretreatment. The use of NF in a process for treating uranium wastewater where NF40 uranium rejections were 97–99.9% has been reported (80). The use of NF membranes to replace activated carbon filters for the removal of organics from offshore-produced water containing residual oils has also been discussed (129). The produced waters contained ~1000 mg/L soluble organics, mostly carboxylic acids, and high inorganic concentrations, ~15,000 mg/L Na^+ and ~25,000 mg/L Cl^-, as well as other dissolved ions. Organic rejections were suitable to meet discharge standards; inorganic rejections were low (<20%), allowing operation at low pressures.

DESIGN CONSIDERATIONS

To ensure the successful design of a reverse osmosis process, several factors should be considered. These considerations encompass the feed solution, the membrane module, and the use of other processes in the pre- and post-treatment processes. A thorough knowledge of the feed stream and its components is essential to the prevention of membrane damage and product impurities. Once the feed stream is characterized and the process objective is defined, design can be initiated.

Feed characterization, particularly for nondesalination applications, should be the first and foremost objective in the design of a reverse osmosis plant. This involves the determination of the type and concentration of the main solutes and foulants in the stream, temperature, pH, osmotic pressure, etc. Once the feed has been characterized, a realistic process objective can be defined. In most cases, some level of pretreatment is needed to reduce the number and concentration of foulants present in the feed stream. Pretreatment necessitates the design of processes other than the RO module, thus the overall process design should use the minimum pretreatment necessary to meet the process objective. Once the pretreatment steps have been determined and the final feed stream defined, the RO module can be selected.

Design of the membrane module system involves selection of the membrane material; the module geometry, eg, spiral-wound or hollow-fiber; product flow rate and concentration; solvent recovery; operating pressure; and the minimum tolerable flux (9,11). The effects of these variables can be obtained from laboratory or pilot experiments using different membranes and modules. The membrane module as well as the solvent recovery can be chosen to minimize fouling. Spiral-wound modules are widely used because these offer both high surface area as well as a lower fouling potential.

Prediction of reverse osmosis performance is useful to the design of RO processes. Simulation of RO processes can be separated into two categories. The first is the prediction of membrane module performance. The second is the simulation of a network of RO processes, ie, flow sheet simulations, which can be used to determine the optimum placement of RO modules to obtain the overall process objective.

Owing to the variety of situations encountered in RO applications, there is no single analytical technique to predict membrane module performance. The module and the feed stream, along with the operating parameters, determine system performance. To predict module performance, a model that governs transport through the membrane must be used. A wide range of models have been proposed, ranging from the solution–diffusion model to the irreversible thermodynamics model. Each model is based on several assumptions and is therefore limited in its range of applicability. Simulations that make use of these models have been developed for several systems. Data obtained from a pilot plant in the treatment of cooling tower water by tubular reverse osmosis have been used to simulate a full-scale plant (130). The finely porous model for transport through the membrane is employed. The solution–diffusion model has also been used to simulate the transport of a saline solution through the DuPont B10 permeator (131). A finite-element technique was used (40) to predict the concentration polarization in a parallel-plate membrane module, which involves the solution of convection–diffusion equation, given a mass-transfer model and boundary conditions. This model was used to determine the effect of the operating variables on the flux and rejection behavior of NaCl systems.

Given the first type of simulation, it is advantageous to be able to design a system of RO modules that can achieve the process objective at a minimal cost. A model has been integrated into a process simulation program to predict the stream matrix for a reverse osmosis process (132). In the area of waste minimization, the proper placement of RO modules is essential for achieving minimum waste at a minimum cost. Excellent details on how to create an optimal network of RO modules is available (96).

ECONOMIC ASPECTS

Economics plays a role in the design of any industrial process and RO is no exception (133–136). In 1990 sales of $118 million were attributed to the RO industry (10). Costs of RO and evaporative processes involving desalination of different feed water sources have been compared (133). RO was found to be the lowest in cost compared to multistage flash and multieffect evaporator having thermal vapor compression. The costs associated with RO processes have been reviewed (134). Capital costs include equipment, land, installation, and utilities. Equipment costs include the components for pretreatment of the feed, storage of the product, pipes, pumps, and the RO module. Equipment costs in terms of capacity corresponding to 1 m^3/d range from $400 for brackish water plants to $900 for seawater plants. Operating costs include the usual costs for an industrial process: energy, labor, and chemicals. However, for the RO process there are additional operating costs from membrane replacement, which range from $0.05/$m^3$ for brackish water plants to $0.10/$m^3$ for seawater plants. Energy consumption in seawater desalination reverse osmosis is usually the largest operating cost because of the high pressures required. Energy costs have been reported from $0.10/$m^3$ to $0.40/$m^3$.

Overall production costs range from $0.50–1.85/$m^3$, depending on the type, size, and location of the plant. Other factors contributing to costs are the extent of pretreatment necessary for acceptable performance, and the degree of solvent recovery. For applications other than desalination, economic calculations must also include the processing of retentate (the concentrate stream) and the benefits of material recovery and water reuse. If valuable materials can be recovered by an RO process, that added value decreases the total production costs. The integration of RO processes and other concentration methods, such as evaporation and pervaporation, can provide significant minimization of waste by simultaneous material recycling and water reuse.

Table 2. Capacities and Costs of Desalination Processes[a]

Desalination Feed	Capacity, m^3/d	Costs, $[b]
	Hollow-fiber module	
Seawater	18.9	322
Brackish water	60.6	63.7
	Spiral-wound module	
Seawater	15.1	101.3
Brackish water	28.4	42.3

[a] The membrane material is a thin-film composite in a 20-cm module.
[b] For each m^3/d.

The primary products of the RO industry are the membrane modules that determine both the geometry and specific membrane type. Therefore, the success of the RO industry is dependent on the production of inexpensive, long-lasting membrane modules. These modules must be mechanically stable at high pressures and provide good hydrodynamics to minimize fouling. Maximizing the membrane surface area per unit volume of the module is also a critical objective. The predominant modules used industrially are the spiral-wound module and the hollow-fiber module, both of which offer high membrane surface area. Costs and capacities of spiral-wound and hollow-fiber thin-film composite membrane modules are given in Table 2.

RO membrane modules are available from many manufacturers including, for hollow-fiber modules, DuPont and Dow/FilmTec Corporation, and for spiral-wound modules, UOP Inc., Millipore Corporation, Nitto-Denko America, Inc., Toray Industries Inc., Dow/FilmTec Corporation, and DuPont.

NOMENCLATURE

Symbol	Definition	Units
A	Water permeability coefficient	$L^3/(L^2 \cdot tP)$
A^*	Modified water permeability constant	$L^3/(L^2 \cdot tP)$
$\tilde{A}$	Measure of electrostatic force between solute and pore wall	L
A_m	Membrane surface area	L^2
B	Solute permeability coefficient	t^{-1}
$\tilde{B}$	Measure of solute–membrane pore wall interaction force	L^3
B^*	Modified solute permeability constant	mol/$(L^2 \cdot t)$
b_0	Sorption coefficient	L^3/mol
C_B	Bulk concentrate solute concentration	mol/L^3 or M/L^3
C^F	Feed solute concentration	mol/L_3 or M/L^3
C_m	Membrane solute concentration	mol/L^3
C_P	Permeate solute concentration	mol/L^3 or M/L^3
C_{tm}	Total concentration of water and solute in the membrane	mol/L^3
C_{wall}	Feed solute concentration at the membrane wall	mol/L^3 or M/L^3
C_{wm}	Membrane water concentration	mol/L^3
D_{sm}	Membrane solute diffusivity	L^2/t
D_{wm}	Membrane water diffusivity	L^2/t
E	Electrochemical potential	mV or V
F_F	Feed flow rate	L^3/t
F_P	Permeate flow rate	L^3/t

J_s	Solute flux	mol/($L^2 \cdot t$) or M/($L^2 \cdot t$)
J_w	Water flux	$L^3/(L^2 \cdot t)$
J_{wo}	Pure water flux	$L^3/(L^2 \cdot t)$
K_{sm}	Solute partition coefficient for homogeneous membrane	dimensionless
k_s	Mass-transfer coefficient	L/t
L	Length	cm or m
P	Pressure	kPa or Pa
ΔP	Transmembrane pressure difference	$M/(L \cdot t^2)$
r	Permeate water recovery	dimensionless
R	Solute rejection	dimensionless
Re	Reynolds number	dimensionless
R_g	Gas constant	$ML^2/(t^2 \cdot T)$ mol
R_p	Membrane pore radius	L
Sc	Schmidt number	dimensionless
Sh	Sherwood number	dimensionless
t	Time	s
T	Temperature	K
V_F	Feed volume	L^3
V_P	Permeate volume	L^3
V_w	Water permeation velocity	L/t
$\bar{V}_w$	Partial molar volume of water	L^3/mol
δ	Membrane thickness	L
$\Delta\pi$	Transmembrane osmotic pressure difference	$M/(L \cdot t^2)$
μ_s	Solute chemical potential	E/mol
μ_w	Water chemical potential	E/mol
σ	Staverman reflection coefficient	dimensionless

BIBLIOGRAPHY

1. C. Reid and E. Breton, *J. Appl. Polym. Sci.* **1,** 133 (1959).
2. P. Ferguson, *Desalination,* **32,** 5 (1980).
3. H. Lonsdale, *J. Membrane Sci.* **10,** 81 (1982).
4. L. Applegate, *Chem. Eng.* **91,** 64 (June 11, 1984).
5. S. Loeb and S. Sourirajan, *Adv. Chem. Series,* **38,** 117 (1962).
6. S. Loeb, in A. Turbak, ed., *Synthetic Membranes,* American Chemical Society, Washington, D.C., 1981.
7. S. Sourirajan and T. Matsuura, *Reverse Osmosis / Ultrafiltration Principles,* National Research Council of Canada, Ottawa, Canada, 1985.
8. R. J. Petersen, *J. Membrane Sci.* **83,** 81–150 (1993).
9. D. Bhattacharyya and co-workers, in W. Ho and K. Sirkar, eds., *Membrane Handbook,* Van Nostrand Reinhold Co., Inc., New York, 1992, pp. 263–390.
10. R. Riley, in *Membrane Separation Systems,* Vol. 2, U.S. DOE Report, DOE/ER/30133-H1, U.S. Dept. of Energy, Washington, D.C., 1990.
11. B. Parekh, ed., *Reverse Osmosis Technology,* Marcel Dekker, Inc., New York, 1988.
12. G. Belfort, ed., *Synthetic Membrane Processes: Fundamentals and Water Applications,* Academic Press, Inc., New York, 1984.
13. D. Lloyd and T. Meluch, *Selection and Evaluation of Membrane Materials for Liquid Separations,* American Chemical Society, Washington, D.C., 1985.
14. R. Rautenbach and R. Albrecht, *Membrane Processes,* John Wiley & Sons, Inc., New York, 1989.
15. C. Fell, in R. Noble and S. Stern, eds., *Membrane Separation Technology,* Elsevier Science, Inc., New York, 1995, pp. 113–140.
16. R. Kesting, *Synthetic Polymeric Membranes: A Structural Perspective,* Wiley-Interscience, New York, 1985.
17. I. Cabasso, in J. I. Kroschwitz, ed., *Encyclopedia of Polymer Science and Engineering,* 2nd ed., Vol. 9, John Wiley & Sons, Inc., New York, 1987, pp. 509–579.
18. W. Koros and co-workers, *Prog. Polym. Sci.* **13,** 339 (1988).
19. R. Baker, in Ref. 10.
20. H. Strathmann, in M. Porter, ed., *Handbook of Industrial Membrane Technology,* Noyes Publications, Park Ridge, N.J., 1990, pp. 1–60.
21. R. Petersen and J. Cadotte, in Ref. 20, pp. 307–348.
22. P. Stanford, *Proceedings of North American Membrane Society Conference,* Portland, Oreg., 1995.
23. T. Peters, *Desalination,* **83,** 159 (1991).
24. V. Gekas, *Desalination,* **68,** 77 (1988).
25. W. Koros, *Chem. Eng. Prog.* **91,** 68 (Oct. 1995).
26. M. Soltanieh and W. Gill, *Chem. Eng. Comm.* **12,** 279 (1981).
27. M. Mazid, *Sep. Sci. Technol.* **19,** 357 (1984).
28. W. Pusch, *Desalination,* **59,** 105 (1986).
29. J. Dickson, in Ref. 11, pp. 1–51.
30. H. Lonsdale, U. Merten, and R. Riley, *J. Appl. Polym. Sci.* **9,** 1341 (1965).
31. H. Burghoff, K. Lee, and W. Pusch, *J. Appl. Polym. Sci.* **25,** 323 (1980).
32. T. Matsuura and S. Sourirajan, *Ind. Eng. Chem. Proc. Des. Dev.* **20,** 273 (1981).
33. H. Mehdizadeh and J. Dickson, *J. Membrane Sci.* **42,** 119 (1989).
34. H. Mehdizadeh, "Modeling of Transport Phenomena in Reverse Osmosis Membranes," dissertation, McMaster University, Hamilton, Ont., Canada, 1990.
35. H. Mehdizadeh and J. Dickson, *Comp. Chem. Eng.* **14,** 157 (1990).
36. D. Bhattacharyya and co-workers, *Chem. Eng. Comm.* **42,** 111 (1986).
37. J. Siler, "Reverse Osmosis Membranes-Concentration Polarization and Surface Fouling: Predictive Models and Experimental Verifications," dissertation, University of Kentucky, Lexington, 1987.
38. E. Matthiasson and B. Sivik, *Desalination,* **35,** 59 (1980).
39. V. Gekas and B. Hallstrom, *J. Membrane Sci.* **30,** 153 (1987).
40. D. Bhattacharyya, S. Back, and R. Kermode, *J. Membrane Sci.* **48,** 231 (1990).
41. P. Brian, in U. Merten, ed., *Desalination by Reverse Osmosis,* MIT Press, Cambridge, Mass., 1966, pp. 161–202.
42. M. Williams, "Measurement and Mathematical Description of Separation Characteristics of Hazardous Organic Compounds with Reverse Osmosis Membranes," dissertation, University of Kentucky, Lexington, 1993.
43. H. Ridgway, in Ref. 11, pp. 429–481.
44. G. Belfort, R. Davis, and A. Zydney, *J. Membrane Sci.* **95,** 1–58 (1994).
45. H. Ridgway, M. Rigby, and D. Argo, *J. AWWA,* **77,** 97–106 (1985).
46. J. Lepore and R. Ahlert, in Ref. 11, pp. 141–183.
47. T. Osta and L. Bakheet, *Desalination,* **63,** 71–80 (1987).
48. S. N. Gaeta and co-workers, *Desalination,* **83,** 383 (1991).
49. K. Nita and co-workers, *Desalination,* **96,** 33–41 (1994).
50. R. Singh, *Desalination,* **95,** 27–37 (1994).
51. J. Glater, S.-K. Hong, and M. Elimelech, *Desalination,* **95,** 325–345 (1994).

52. G. Congjie, L. Xueren, and B. Zhiguo, *Desalination*, **83**, 271–278 (1991).
53. A. Ko and D. Guy, in Ref. 11, pp. 185–278.
54. S. Ebrahim, *Desalination*, **96**, 225–238 (1994).
55. S. Ebrahim and H. El-Dessouky, *Desalination*, **99**, 169–188 (1994).
56. A. Jaffer, *Desalination*, **96**, 71–79 (1994).
57. H. Mehdizadeh and J. Dickson, *AIChE J.* **39**, 434–445 (1993).
58. J. Jiang and co-workers, *Desalination*, **17**, 107 (1989).
59. S. Gao and Q. Bao, *J. Liq. Chromatog.* **12**, 2083 (1989).
60. J. Jiang and C. Jiayan, *Desalination*, **78**, 389 (1990).
61. R. Cheng and co-workers, *Desalination*, **85**, 33 (1991).
62. D. Bhattacharyya and co-workers, *EPA Report*, EPA/600/87/053, U.S. Environmental Protection Agency, Washington, D.C., 1987.
63. D. Bhattacharyya and M. Madadi, *AIChE Symposium Series* **84**(261), 139 (1988).
64. D. Bhattacharyya and A. Kothari, *EPA Report*, Cooperative Agreement No. CR814491, U.S. EPA, Washington, D.C., 1992.
65. S. Sourirajan, *Reverse Osmosis*, Academic Press, Inc., New York, 1970.
66. M. Kurihara and co-workers, *Desalination*, **38**, 449 (1981).
67. K. Koyama and co-workers, *J. Appl. Polym. Sci.* **27**, 2845 (1982).
68. K. Koyama and co-workers, *J. Appl. Polym. Sci.* **29**, 2929 (1984).
69. S. Lynch and co-workers, *EPA Report*, EPA/600/1-84/018, U.S. EPA, Washington, D.C., 1984.
70. J. Rickabaugh and co-workers, *Proceedings of the 12th Annual Hazardous Wastes Symposium*, Cincinnati, Ohio, 1986.
71. W. Pusch, Y. Yu, and L. Zheng, *Desalination*, **75**, 3 (1989).
72. R. Rautenbach and A. Gröschl, *Proceedings of the 203rd American Chemical Society National Meeting*, San Francisco, Calif., 1992.
73. E. Lohman, *Desalination*, **96**, 349–358 (1994).
74. Y. Ayyash and co-workers, *Desalination*, **96**, 215–224 (1994).
75. P. Cartwright, *Desalination*, **56**, 17 (1985).
76. K. Linde, A.-S. Jönsson, and R. Wimmerstedt, *Desalination*, **101**, 21–30 (1995).
77. C. Schmidt, I. White, and R. Ludwig, *EPA Report*, EPA/600/R-93/150, U.S. EPA, Washington, D.C., 1993.
78. K. Imasu, *Desalination*, **56**, 137 (1985).
79. G. Hsiue and co-workers, *Desalination*, **71**, 35 (1989).
80. M. Chu, C. Tung, and M. Shieh, *Sep. Sci. Technol.* **25**, 1339 (1990).
81. S. Prabhakar and co-workers, *Sep. Sci. Technol.* **27**, 349 (1992).
82. S. Prabhakar and co-workers, *Sep. Sci. Technol.* **29**(8), 1001–1010 (1994).
83. R. Rautenbach and R. Mellis, *Desalination*, **101**, 105–113 (1995).
84. H. Tsuge and K. Mori, *Desalination*, **23**, 123 (1977).
85. I. Nusbaum and D. Argo, in Ref. 12, p. 377.
86. M. Reinhard and co-workers, *J. AWWA* **78**(4), 163 (Apr. 1986).
87. Y. Suzuki and T. Minami, *Wat. Sci. Tech.* **23**, 1629 (1991).
88. C. Frith, in Ref. 11, pp. 279–310.
89. D. Ulieru, *Proceedings of the 19th Annual Technical Meeting of the Institute of Environmental Science*, Las Vegas, Nev., 1993.
90. Y. Egozy, J. Denoncourt, and G. Ganzi, in Ref. 11, pp. 311–346.
91. P. Parise, B. Parekh, and R. Smith, in Ref. 11, pp. 347–398.
92. P. Sinisgalli and J. McNutt, *J. AWWA* **78**(5), 47 (May 1986).
93. P. Cartwright, *Desalination*, **83**, 225 (1991).
94. C. Slater, R. Ahlert, and C. Uchrin, *Desalination*, **48**, 171 (1983).
95. A. Ghabris, M. Abdel-Jawad, and G. Aly, *Desalination*, **75**, 213 (1989).
96. M. El-Halwagi, *AIChE J.* **38**(8), 1185–1198 (1992).
97. J. Schoeman and co-workers, *Wat. Sci. Tech.* **25**, 79–93 (1992).
98. C. Slater, J. Zielinski, and R. Wendel, *J. Environ. Sci. Health*, **A27**(5), 1175–1193 (1992).
99. F. Awadalla, C. Striez, and K. Lamb, *Sep. Sci. Technol.* **29**, 483–495 (1994).
100. A. Kulkarni, D. Mukherjee, and W. Gill, *Chem. Eng. Comm.* **129**, 53 (1994).
101. T. Thorsen, *Desalination*, **53**, 217 (1985).
102. E. Chian and F. De Walle, *EPA Report*, EPA-600/2-77-186b (1977).
103. C. Slater, R. Ahlert, and C. Uchrin, *Env. Prog.* **2**, 251 (1983).
104. R. Kinman and D. Nutini, *Proceedings of the 1990 International Congress on Membranes and Membrane Processes*, Chicago, Ill., 1990.
105. J. Lepore and R. Ahlert, *Waste Manage.* **11**, 27 (1991).
106. AWWA Membrane Technology Research Committee, *J. AWWA* **84**(1), 59 (Jan. 1992).
107. T. Marhaba and S. Medler, *Proceedings of 1994 National Conference on Environmental Engineering*, 1994.
108. L. Tan and R. Sudak, *AWWA J.* **84**, 79–87 (1992).
109. T. Eisenberg and E. Middlebrooks, *Reverse Osmosis Treatment of Drinking Water*, Butterworth & Co., Boston, Mass., 1986.
110. J. Jacangelo and co-workers, *J. AWWA*, **87**, 64–77 (1995).
111. L. Awerbuch and co-workers, *Desalination*, **76**, 189–197 (1989).
112. T. Ikeda, H. Muragishi, and T. Uemura, *Desalination*, **98**, 391–400 (1994).
113. M. Nyström, L. Kaipia, and S. Luque, *J. Membrane Sci.* **98**, 249 (1995).
114. P. Eriksson, *Env. Prog.* **7**, 58 (1988).
115. J. Cadotte and co-workers, *Desalination*, **70**, 77 (1988).
116. L. Raman, M. Cheryan, and N. Rajagopalan, *Chem. Eng. Prog.* **90**(3), 68–74 (1994).
117. A. Bindoff and co-workers, *Desalination*, **67**, 453 (1987).
118. K. Ikeda and co-workers, *Desalination*, **68**, 109 (1988).
119. M. Afonso and co-workers, *Water Res.* **26**, 1639 (1992).
120. M. Simpson, C. Kerr, and C. Buckley, *Desalination*, **64**, 305 (1987).
121. M. Perry and C. Linder, *Desalination*, **71**, 233 (1989).
122. Anon., *Food Eng.* **60**, 124 (1988).
123. P. Fu and co-workers, *J. AWWA* **86**(12), 55–72 (1994).
124. K. Agbekdo and co-workers, *Rev. Sci. Eau*, **7**(2), 183–200 (1994).
125. R. Rautenbach and R. Mellis, *Desalination*, **95**, 171–188 (1994).
126. D. Bhattacharyya, R. Adams, and M. Williams, in D. Butterfield, ed., *Biological and Synthetic Membranes*, Alan R. Liss, New York, 1989.
127. M. Williams, R. Deshmukh, and D. Bhattacharyya, *Env. Prog.* **9**, 118 (1990).
128. D. Bhattacharyya and M. Williams, *EPA Report*, EPA/600/2-91/045 (1992).
129. C. Dyke and C. Bartels, *Env. Prog.* **9**, 183 (1990).
130. C. Brouckaert and C. Buckley, *Water SA*, **18**(3), 215–224 (1992).
131. A. Malek, M. Hawlader, and J. Ho, *Desalination*, **99**, 19–38 (1994).
132. H. Niemi and S. Palosaari, *J. Membrane Sci.* **84**, 123–137 (1993).
133. S. Talati, *Desalination*, **97**, 353–361 (1994).
134. R. Ray, in Ref. 9, pp. 355–390.

135. J. Birkett, in Ref. 11, pp. 484–504.

136. S. Ebrahim and M. Abdel-Jawad, *Desalination,* **99,** 39–55 (1994).

General Reference

D. Bhattacharyya, W.C. Mangum, and M.E. Williams, "Reverse Osmosis" in J.I. Kroschwitz and M. Howe-Grant, eds., *The Encyclopedia of Chemical Technology,* 4th ed., Vol. 21, John Wiley & Sons, Inc., New York, pp. 303–335.

RISK ASSESSMENT

R. A. FJELD
K. L. COMPTON
Clemson University
Anderson, South Carolina

Environment risk assessment is a quantitive estimate of the human health, ecological, or aesthetic effect of the potential or actual release of a contaminant to the environment. Ideally, it is an objective evaluation which, in combination with economic and societal factors, can form the basis for environmental regulatory or management decisions. Environmental risk assessment is a systematic, quantitative process that consists of four steps: release analysis, transport analysis, exposure analysis, and consequence analysis. The purposes of this article are to provide (*1*) the scientific and engineering basis for each of the four analysis steps, (*2*) mathematical models that can be used for screening level calculations, and (*3*) applications and limitations of the risk assessment process. Although the focus here is on the calculation of human health risk, many of the principles can be applied to ecological risk assessment as well.

Illustrated in Figure 1 are typical examples of environmental risk assessment problems. These problems generally have the following elements: (*1*) an actual or potential release of a contaminant to the environment, (*2*) environmental pathways through which the contaminant is subject to transport, (*3*) routes or mechanisms of human or other receptor exposure to the contaminant, and (*4*) the risk of adverse human health, ecological, or aesthetic effects. Most contaminants that are released to the environment either serve some purpose (such as protecting crops against insects and disease) or are the byproduct of some activity (such as the generation of electricity) that society considers to be beneficial. While contaminants are associated either directly or indirectly with some benefit to society, they also pose a risk that covers a spectrum from insignificant to the deaths of thousands of people. Some historical examples in which contaminant releases resulted in documented adverse effects are given in Table 1. Society is thus faced with the difficult task of protecting itself by placing restrictions on activities that pose unacceptable levels of risk and consequences such as those in Table 1, while at the same time allowing itself to benefit from activities that pose acceptable levels of risk. This can be accomplished through the risk management process, which is illustrated conceptually in Figure 2. This is a subjective process in which risks are balanced against benefits within social and economic constraints, subject to political feasibility. As illustrated in Figure 2, risk assessment is the technical tool that is used to provide scientifically based quantitative estimates of risk.

Risk Assessment Process

The risk assessment process involves the integration of information from disciplines as varied as hydrology, environmental chemistry, meteorology, toxicology, and demography. Due to the highly multidisciplinary nature of a risk assessment, it is typically divided into a number of steps, each corresponding to a particular area of expertise. For example, in an influential publication, the National Research Council (1) recommended that risk analysis be divided into four steps: hazard identification, dose–response assessment, exposure assessment, and risk characterization. Others have organized the process by four somewhat different steps: release assessment, exposure assessment, consequence assessment, and risk estimation (2). The American Association of Engineering Societies (3) recommends a similar scheme. The U.S. EPA guidance document for Superfund risk assessments suggests three steps: exposure assessment, toxicity assessment, and risk characterization.

All of these are substantially similar in that each evaluates the potential for the release of a contaminant, the transport of a contaminant through the environment, the rate at which humans are exposed to contaminated environmental media, the health effects of that exposure, and a final step integrating all of the above information to summarize the consequences of a release of contaminants to the environment. In this article, the analytical process of risk assessment is divided into four discrete assessment steps: release assessment, environmental transport assessment, human exposure assessment, and consequence assessment. The first step, release assessment, involves the identification of contaminants, an assessment of the likelihood of a release, and the quantification of the spatial and temporal release rate ($\dot{S}$, in mg/s or similar units) of a contaminant into the environment. In the second step, environmental transport analysis, pathways through which contaminants are transported in the environment are identified, and contaminant concentrations (C, in mg/m^3, mg/kg, or similar units) in air, food, and water are quantitatively estimated. Human exposure assessment, the third step, involves the identification of exposed populations (receptors) and exposure routes, and estimates the rate at which humans are exposed to an environmental contaminant. This step results in a quantitative estimate of contaminant dose (D, in mg/(kg $\cdot$ d) or Sieverts) to members of the exposed population. Consequence analysis, the final step, combines information on the toxicity of the chemical with estimates of the doses received by an exposed population to generate an estimate of the probability and severity of adverse human health effects. These steps are represented by the block diagram in Figure 3.

Each of the four blocks represents a mathematical model (or a collection of mathematical models) for the release, transport, exposure, and consequence components of the overall risk assessment model, and the output of one component is the input for the next. Each component requires parameters that must be specified for a given problem. Some of these parameters are subject to variability and are more appropriately characterized by distributions rather than discrete values. When parameter variability distributions are propagated through the steps of the risk assessment procedure, a risk curve such as that given in Figure 4 results.

The ordinate of the risk curve is exceedence probability which is the probability with which the number of effects

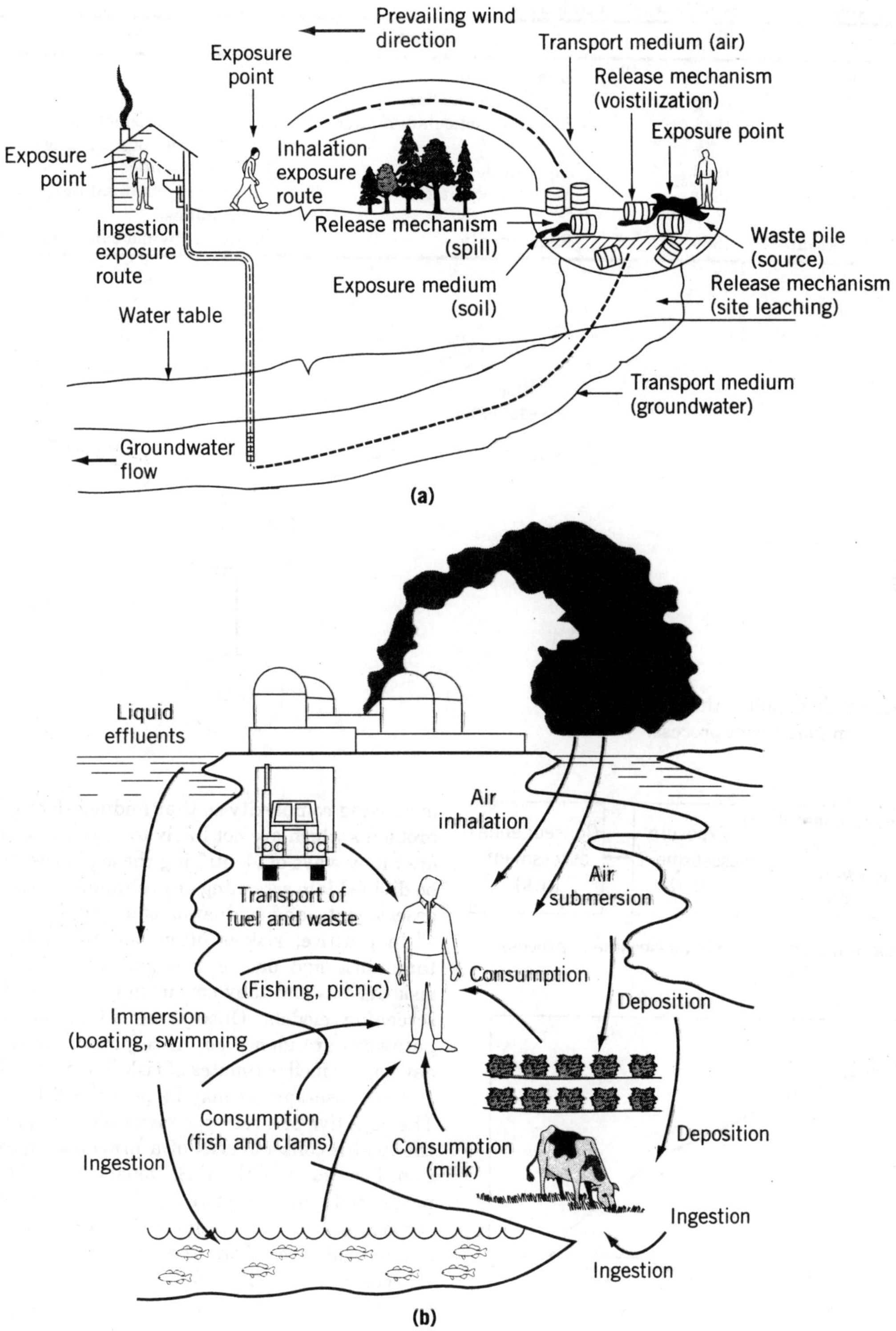

Figure 1. Typical risk assessment problems: (**a**) waste disposal (49); (**b**) operational facilities (adapted from Ref. 59).

exceeds the value on the abscissa. For example, the probability with which the number of effects exceeds 100 is approximately 10^{-6}. The risk curve is another name for the complementary cumulative distribution of risk. The relationships among the complementary cumulative distribution, the cumulative distribution, and the probability density are considered later in this article. Uncertainty in risk is due to lack of knowledge, leading to a family of risk curves. The risk curve in Figure 4 represents the median risk. Often a risk distribution such as that in Figure 4 is characterized either by a measure

Table 1. Adverse Effects of Historical Releases

Location	Date	Contaminant	Effect
London, U.K.	1852	Human waste	Cholera
Ducktown, Tenn.	Early 1900s	SO_2 from a smelter	Death of vegetation
Donora, Pa.	1948	SO_2 and particulate matter	21 fatalities in population of 14,000; 6,000 cases of respiratory distress
New York, N.Y.	1953	Air pollution	200–400 fatalities
Bhopal, India	1984	Methyl isocyanate	~1,754–10,000 fatalities; extensive casualties
Chernobyl, Ukrain	1986	Radioactivity	Increased thyroid cancer in children

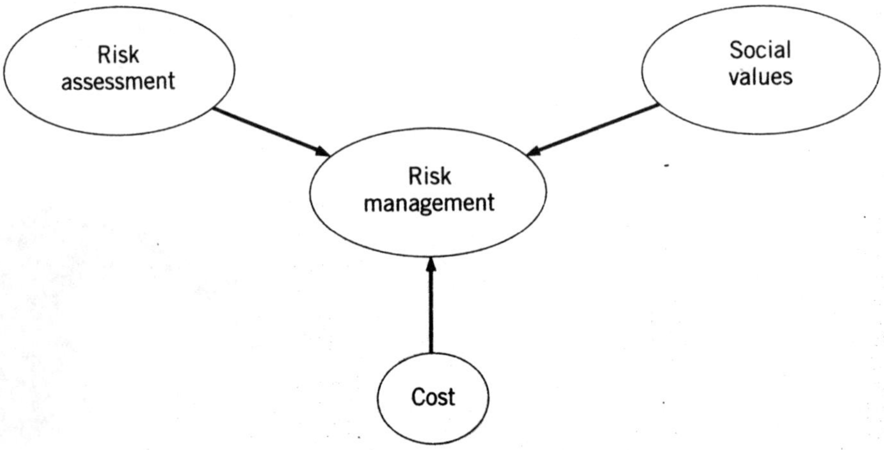

Figure 2. Diagram of the risk management process.

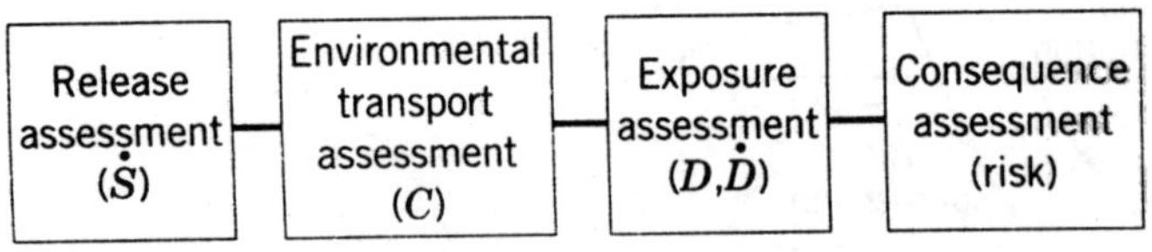

Figure 3. Block diagram of the risk assessment process.

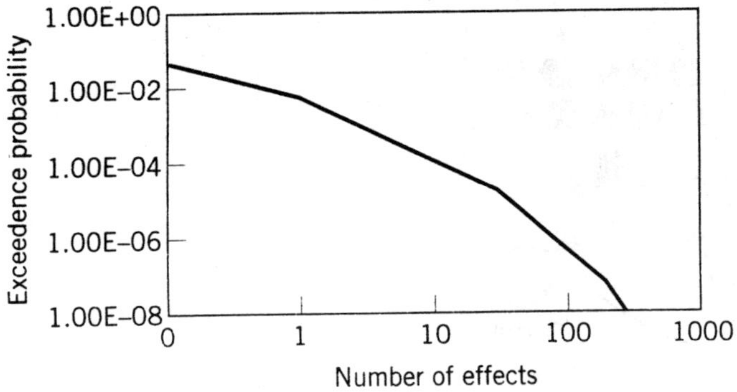

Figure 4. Exceedence probability versus number of adverse effects.

of central tendency, such as the mean or mode, or by a fractile ($R_{50\%}$, $R_{95\%}$, etc).

A complete environmental risk assessment of a given activity would involve all potential contaminants and all potential pathways. However, resource limitations usually dictate that risk assessments be conducted in progressive phases of increasing complexity so that undue effort is not expended on problems which are not likely to be of concern. Although there are many ways of identifying these phases, they can generally be divided into screening or preliminary assessments, detailed assessments, and evaluation of alternatives.

In practice, risk is often dominated by one or two contaminants and one or two pathways. These are known as risk drivers and they can usually be identified using simple screening models. Once these dominant contaminants and pathways are identified, more precise models can be used to develop refined estimates of risk in a detailed assessment.

Risk assessments may be performed for various purposes. The objective of a retrospective risk assessment is to estimate the health consequences of a previous exposure of a population. For example, the U.S. Department of Energy (DOE) has performed retrospective risk assessments, called dose reconstructions, to evaluate the effect of their nuclear production facilities from the 1940s through the 1980s. This is accomplished through historical records of release and environmental measurements, mathematical models of environmental transport, and epidemiological studies of exposed populations. The objectives of risk assessments performed in support of environmental remediation activities are to provide an estimate of the risk posed by a contaminated site and to provide input into selecting a remedial alternative. The objective of a prospective risk assessment is to investigate the potential for health effects from planned facilities and to ensure that the facilities will not pose an undue burden on nearby populations. Risk assessments are often used by regulatory agencies to establish health-based environmental quality criteria, such as the development of drinking water standards by the EPA under the Safe Drinking Water Act. Risk assessments are also sometimes used in litigation where damages to exposure to a contaminant is alleged (toxic torts) and can be used by either the defendant or the plaintiff to support their side of a case.

In this article, environmental risk assessment is presented as a collection of mathematical models. In practice, some portions of the process may be accomplished through direct measurement or by expert judgment. For example, if measurements of contaminant concentration are available, if the measurement error is small, and if there is adequate temporal and spatial resolution of the measurements, release and transport calculations might not be necessary. More realistically, a limited number of measurements might be available. These can be used to validate or calibrate the models. Similarly, if measurements of dose or dose surrogates are available, they can be used to supplement exposure calculations.

As indicated above, risk assessments are a critical component of the environmental decision-making processes. Applications requiring an absolute estimate of risk, such as regulatory decisions or toxic torts, generally are very demanding and push the risk assessment process to its technical limits. Applications such as the establishment of cleanup priorities for a number of waste sites which require only a relative measure of risk are generally not as technically demanding.

RELEASE AND TRANSPORT ASSESSMENT

Release Assessment

The first step of the risk assessment procedure is to characterize the release of the contaminant to the environment. Release assessment (sometimes referred to as source term assessment) has three components: identification of the contaminants, characterization of their probability of release (which is sometimes trivial, being 100% for existing or planned releases), and quantification of their spatial and temporal emission rates. Ideally, characterization would be accomplished by direct measurement. Direct measurement could be used, for example, for the analysis of routine releases of easily measured contaminants from known emission points. However, for preoperational assessments of proposed facilities, assessments of hypothetical accidents, or retrospective risk assessments of historical releases, direct measurement of the release is not possible. A second approach for characterizing releases is through process knowledge. This involves identifying and quantifying the various processes that occur within a facility which can result in a contaminant release. This includes knowledge of contaminant generation and loss within the facility, release pathways, and the effectiveness of natural and man-made controls. For example, the release of gaseous effluent from a coal-fired power plant requires knowledge of the rate at which coal is burned, the composition of the coal, the production of gaseous and particulate contaminants in the combustion process, subsequent gas-phase reactions and physical removal processes that alter the contaminant composition, and the efficiency of air pollution control devices. Process knowledge of this type forms the basis for a material mass balance from which the gaseous effluents are identified and their releases are quantified. A third approach is based on environmental measurements. The measurements might be of the contaminant itself, the effects of the contaminant, or reaction/degradation products of the contaminant. If a suspected contaminant is found in the vicinity of a facility and there are no other plausible sources, then the environmental measurement can be used for identification. However, if there are multiple potential sources or if the contaminant is not measured directly, identification through environmental measurements becomes problematic. Quantification of the release is achieved by back-calculation through an environmental transport model. In practice, some combination of effluent measurements, environmental measurements, process knowledge, and modeling is used to characterize releases.

Contaminant Identification. For purposes of human health risk assessment, it is convenient to place a contaminant into one of five general categories.

Organic compounds: these can be further subdivided into halogenated and nonhalogenated. Halogenated organic compounds include solvents such as carbon tetrachloride, methylene chloride, trichloroethane (TCA), trichloroethylene (TCE), and tetrachloroethylene (PCE); pesticides such as aldrin or dieldrin, chlordane, DDT, diazinon, ethylene dibromide (EDB), heptachlor, Kepone, and lindane; herbicides such as 2,4-D and atrazine; and organic acids such as pentachlorophenol. Nonhalogenated organic compounds include solvents (eg, acetone, benzene, toluene, and xylene); carbamate fungicides (eg, mancozeb, maneb, and zineb); pesticides (eg, malathion); and herbicides (eg, paraquat). The highly toxic organic compound dioxin, in contrast with the types of chemicals discussed above, is not intentionally produced but may be present as a contaminant in other chemical products and in emissions from incinerators.

Metals: although metals occur naturally in the environment, human activities often result in high concentrations of metals in areas where humans or animals may be exposed. Many metals are toxic at sufficiently high concentrations. Some metals of environmental concern include arsenic, barium, cadmium, chromium, lead, mercury, selenium, silver, nickel, beryllium, and cobalt.

Inorganic gases: included here are gases that present both localized risks and those that have the potential for contributing to regional or global effects. Inorganic gases capable of posing direct health threats include Cl_2, H_2S, HCl, HF, and CO. Gases implicated in regional air quality problems and possibly in global climate change include CO_2, NO_x, and chlorofluorocarbons. Sulfur dioxide, SO_2, is a common by-product of coal-fired electricity generation and fits into both groups.

Radioactive materials: this category includes both anthropogenic and naturally occurring radioactive elements (radionuclides). Human-made radionuclides are often separated into fission products such as ^{3}H, ^{85}Kr, ^{90}Sr, ^{131}I, and ^{137}Cs; activation products such as ^{14}C, ^{55}Fe, $^{58,60}Co$, ^{63}Ni, and ^{134}Cs; and transuranic elements such as ^{237}Np, $^{238,239}Pu$, ^{241}Am, and ^{244}Cm. The principal naturally occurring radionuclides are ^{40}K and the members of the uranium, actinium, and thorium decay series, the most prominent being ^{226}Ra and ^{222}Rn and its short-lived decay products.

Radionuclides are typically characterized by their activity and their half-life. Activity is an indication of the rate at which radiation (either in the form of subatomic particles such as electrons, neutrons, or helium nuclei, or electromagnetic radiation such as gamma- or X-radiation) is emitted by the material. Over time, the activity decreases as the radionuclide decays into a stable, nonradioactive isotope. The rate at which this occurs is characterized by the half-life, which is the amount of time required for the activity to decrease to half of its original value.

Others: it is impossible to capture all contaminants of potential concern in a small number of discrete categories, and this category contains those that do not fit into one of the other four. It includes, for example, viral and bacterial pathogens, particulate matter, and fibers such as asbestos, and genetically engineered organisms.

Specification of the physical/chemical form can be important in the characterization of a release, since the form of a contaminant can affect its transport and its effects. For example, a given mass of plutonium in the form of respirable-sized particulate matter poses a greater hazard than the same mass in the form of a single piece of plutonium metal; chromium(VI) is more toxic than chromium(III); the mobility of most metals in the subsurface depends to a large degree on the metal's oxidation state; PCBs have different chemical forms known as congeners and toxicity is congener-dependent. In addition, many chemicals are transformed in the environment, and the fate of a contaminant is highly dependent on the initial chemical form.

Release Probability. Most accidental contaminant releases occur as the result of an unlikely event (such as an earthquake, tornado, airplane crash, etc) or a sequence of unlikely events (such as a series of component failures or human error in a chemical plant or a nuclear power plant). The probabilistic nature of accidental releases is another source of variability and is commonly characterized by an exceedence probability plot such as that illustrated in Figure 5.

Plots such as these are generated by a process formally known as probabilistic risk assessment (PRA) (4,5). In PRA, event trees and fault trees are used to model probabilistic releases. A fault tree is a logic diagram that depicts the failure of components which can lead to system failure. The fault tree is constructed by identifying all possible failures which may lead to a particular accident. For example, a loss of coolant accident in a nuclear power plant can occur in a variety of ways, such as a rupture in a pipe or the sticking of a pressure relief valve. Each of the possible events leading to a loss of coolant is listed. Subsequently, each of the events leading to these events is listed. Eventually, a tree is generated which gives all possible events leading to the undesired outcome. The probability of the final event can then be estimated using the probabilities of each event along with Boolean logic. An event tree, on the other hand, is a graphical means of identifying the possible outcomes of a single initiating event.

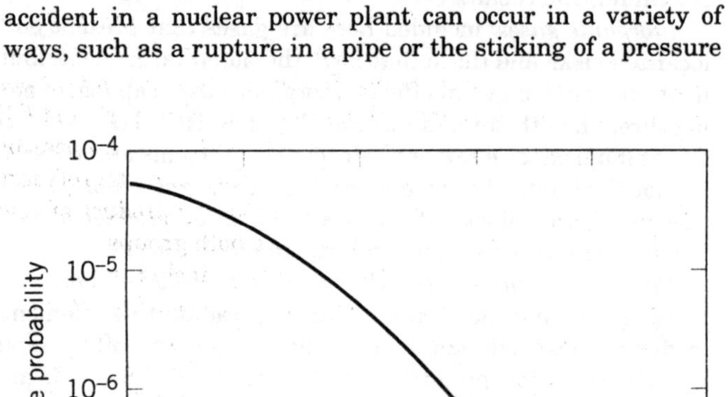

Figure 5. Exceedence probability for an accidental release.

Contaminant Emission Rate. Given that a release has occurred or will occur, the next step is to determine the spatial and temporal dependence of the emission rate, $\dot{S}(\bar{r}, t)$. Although the temporal and spatial dependence of emission rate is often complex, many of the closed-form analytical contaminant transport models are, by necessity, based on very simple spatial and temporal dependencies. As a consequence, actual emission rates are often replaced by simple approximations. For example, an emission from a stack or a process sewer outfall might be approximated as a point source; emission of atmospheric pollutants from a highway or leaching of wastes buried in a long, shallow trench might be approximated as a line source; and release from contaminated surface soils of landfills might be approximated as a geometrically simple area source such as a disk or a rectangle.

Temporally, instantaneous and step approximations are common. Instantaneous emissions are those that occur over a period of time which is very short compared to the time required to reach a receptor. For example, upset conditions such as a fire or the failure of a pollution control device in an industrial facility may result in an abnormally high emission rate for a short period of time. Step emissions are those which are relatively constant after they begin; for example, emission of a pollutant from an industrial stack can be approximated as a step source, as can the leaching of wastes buried in a landfill.

Emission rate approximations are formed such that the total amount of contaminant released under the approximation is equal to the actual amount released. Mathematically, this is expressed as

$$\int\int \dot{S}_{\text{app}}(\bar{r},t)d\bar{r}dt = \int\int \dot{S}_{\text{act}}(\bar{r},t)d\bar{r}dt \qquad (1)$$

where $\dot{S}_{\text{app}}$ is the approximate emission rate and $\dot{S}_{\text{act}}$ is the actual emission rate. The mathematical forms of the common emission rate approximations are given in Table 2 and illustrated in Figure 6.

Environmental Transport Assessment

Contaminant transport pathways for two typical risk assessment problems were shown in Figure 1: (Fig. **1a**) subsurface releases from buried waste and (Fig. **1b**) atmospheric releases from an industrial facility. The second step of the risk assessment process, environmental transport analysis, involves the determination of contaminant concentrations in air, food, and water as a result of the relevant transport pathways. For some problems, such as estimating current contaminant concentrations due to a pre-existing waste disposal site, it might be possible to determine the concentrations through a network of field measurements. More commonly, measurements are either not practical or not possible, and the concentrations must be based on transport models. The contaminant transport problem is complex because of the inherent complexity of environmental systems. In addition to the hydrological processes

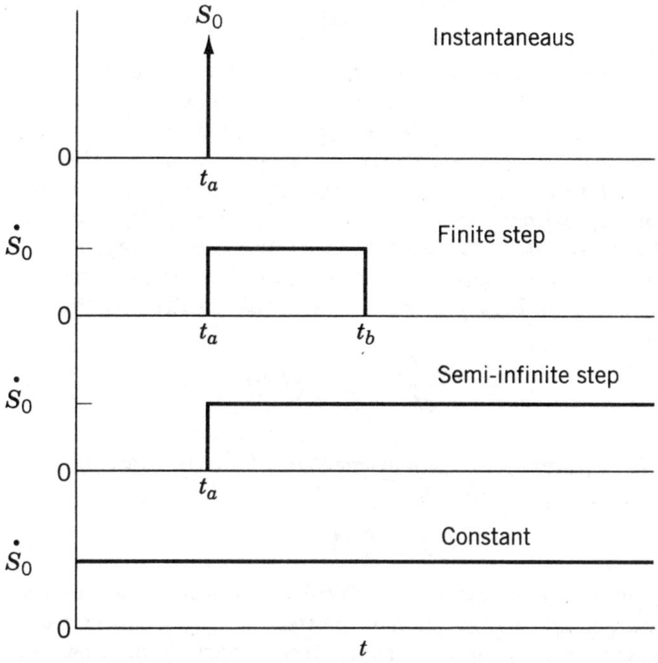

Figure 6. Temporal approximations for emission rates.

that govern fluid transport, additional physical, chemical, and biological processes can also be important. These processes are not always well understood and they can depend on many factors. These factors, in turn, may be poorly understood or highly variable. Nonetheless, by combining and interfacing empirical data for processes that are poorly understood with mathematical theory for processes that are well understood, it is possible to develop models for predicting contaminant concentrations in air, water, and food. Even though the empirical and theoretical base of knowledge is dispersed across many disciplines, a unified approach to transport modeling can be taken. In this unified approach, all of the empirical and theoretical transport models are based on a single governing equation called the contaminant transport equation. The contaminant transport equation is the continuity equation from fluid mechanics modified to apply to a contaminant that is intimately mixed with the fluid through dissolution, entrainment, or suspension.

Introduction to Environmental Transport. The contaminant transport equation is the means through which the various transport processes can be taken into account in a quantitative way. The sophistication of the mathematical description of a process depends on the level of theoretical understanding of the process or the degree to which the process has been characterized empirically. Complex but poorly understood or characterized processes are typically represented by very simple models. Conversely, well-understood or characterized processes generally have more complex or more rigorous mathematical descriptions. There are several recent introductory texts in the area of environmental modeling (6,7).

Transport Processes. The contaminant transport equation is a differential equation for the time rate of change of contaminant concentration in a fluid medium such as air, surface water, or groundwater. It takes into account the processes that can cause the amount of contaminant contained in a volume element of the medium to change. Generically, the processes are (*1*) transport across the surface (either from the outside to the inside or vice versa), (*2*) generation inside the volume, or (*3*) destruction inside the volume. Processes that cause a contaminant to be transported across the surface do not change the total amount of contaminant contained in all volume elements and are therefore referred to as conservative processes (Fig. 7). There are two conservative processes: advection and dispersion. Advection is the transport of a contaminant with the mean flow of the fluid. Dispersion is a general term that is used to describe a variety of processes that cause contaminant movement to deviate from the mean flow of the fluid. These processes include molecular diffusion, turbulent diffusion, and advective inhomogeneities. Molecular diffusion is the spreading of contaminants due to Brownian motion. It might be important, for example, in the migration of contaminants in slow moving groundwater or, in a slightly different context, their migration within a soil particle following contaminant sorption on the surface. Turbulent diffusion is the spreading of a contaminant due to turbulence of the medium. It is responsible for the spreading of a contaminant plume in the atmosphere and contributes to spreading in streams and rivers. Advective inhomogeneities refer to contaminant spreading that occurs as a result of nonuniform velocity distributions of the medium. For example, in a laminar flow channel (ie, a slow-moving river or stream) the fluid velocity is zero at the interface between the fluid and the sides and bottom of the channel. Thus, contaminant molecules near the sides and bottom travel slower than those near the center, resulting in spreading of the contaminant in the direction of flow. Advective inhomogeneities also contribute to dispersion in the subsurface for a variety of reasons to be discussed later.

Contaminant Transport Equation. A derivation of the contaminant transport equation is beyond the scope of this article, but it is analogous to the derivation of the mass continuity

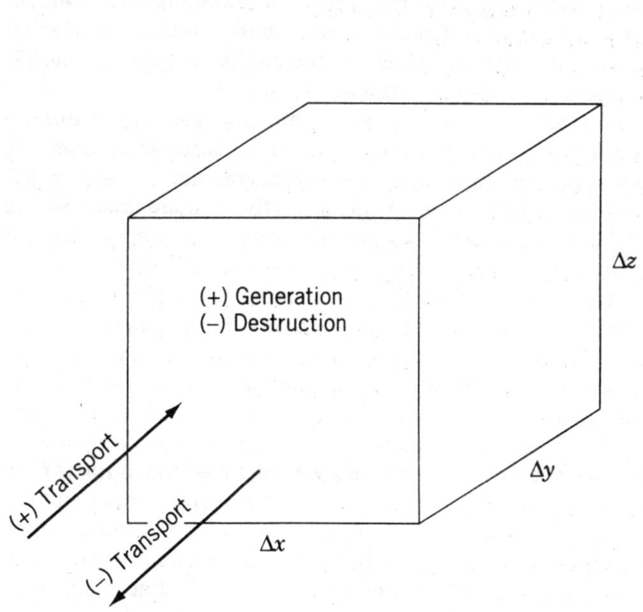

Figure 7. General processes responsible for the contaminant concentration to change in a differential volume.

equation in fluid mechanics (8). In word form,

$$\begin{bmatrix}\text{Time rate of}\\ \text{change of}\\ \text{contaminant}\\ \text{concentration}\end{bmatrix} = \begin{bmatrix}\text{Net rate of}\\ \text{increase due}\\ \text{to dispersion}\end{bmatrix} + \begin{bmatrix}\text{Net rate of}\\ \text{increase due}\\ \text{to advection}\end{bmatrix} + \begin{bmatrix}\text{Generation}\\ \text{rate}\end{bmatrix} - \begin{bmatrix}\text{Destruction}\\ \text{rate}\end{bmatrix} \quad (2)$$

The first two terms on the right-hand side of the equation represent the conservative processes and the last two represent the nonconservative processes. In mathematical form, equation 2 becomes

$$\frac{\partial C(\bar{r},t)}{\partial t} = \nabla \cdot D\nabla C(\bar{r},t) - \nabla \cdot C(\bar{r},t)\bar{v} + g(\bar{r},t) - d(\bar{r},t) \quad (3)$$

where $C(\bar{r},t)$ = contaminant concentration, mg(c)/m^3; $\bar{r}$ = position vector; t = time, s; $\bar{v}$ = fluid velocity vector, m/s; D = dispersion coefficient, m^2/s; $g(\bar{r},t)$ = contaminant generation rate, mg(c)/cm^{3}s; and $d(\bar{r},t)$ = contaminant destruction rate, mg(c)/cm^{3}s.

Equation 3 is general in the sense that models used to determine concentrations in air, food, and water can be derived from it. For example, it is the basis for simple nondimensional compartmental models of the type used for contaminant uptake by vegetation from contaminated soil. At the same time, it is the basis for complex three-dimensional models used for groundwater, surface water, and atmospheric transport. Of course, it must be remembered that equation 3 and its solutions are macroscopic approximations for natural processes that are discrete on a microscopic scale. In addition, when applied to a given transport problem, the accuracy of predictions based on the contaminant transport equation depends on the accuracy of transport parameters (such as dispersion coefficients, reaction rate constants, partition coefficients, etc) used to quantify specific processes.

Solutions of the Contaminant Transport Equation. The contaminant transport equation is the fundamental basis for most mathematical models commonly used to make estimates of the spatial and/or temporal dependence of contaminant concentration in air, water, or food. Presented in this section are selected general solutions on which mathematical models of specific transport processes or pathways are based.

Homogeneous Compartments The homogeneous compartment is one in which the contaminant concentration does not vary spatially. Environmental compartments are either advective or nonadvective. A nonadvective compartment is one in which contaminant movement across the surface can be approximated as occurring in the absence of fluid flow. Typical examples would be the partitioning of a sorbable, soluble contaminant between the aqueous phase and suspended particulate matter in a surface water; the uptake of a contaminant into the human body by ingestion of contaminated vegetation; or the absorption of a gaseous contaminant, such as SO_2, into rain drops. In some nonadvective compartments, the contaminant might actually be carried by a fluid medium, but the flow rate is negligible. An example would be contaminant uptake by aquatic animals. An advective compartment is one in which the contaminant is carried into the compartment by the fluid. An example would be a well-mixed reservoir that is contaminated by the river that feeds it.

For a nonadvective compartment, the dispersion and advection terms in are zero and equation 3 reduces to

$$\frac{dC(t)}{dt} = g(t) - d(t) \quad (4)$$

where the generation and destruction terms include nonadvective transfer into and out of the compartment. In the application of equation 4 (as well as equation 3) to environmental processes, generation is often approximated as zero-order (ie, $g(t)$ = constant) and destruction processes are often approximated as first-order (ie, $d(t) = k \cdot C(t)$), where k is the first-order rate constant (s^{-1}). When both occur simultaneously, the "constant uptake/first-order removal" model, illustrated in Figure 8, results. For this model, equation 4 becomes

$$\frac{dC(t)}{dt} = g - kC(t) \quad (5)$$

The solution for a compartment in which the concentration initially is zero ($C(0) = 0$) is

$$C(t) = \frac{g}{k}(1 - e^{-kt}) \quad (6)$$

Equation 6 is displayed graphically in Figure 9, where the contaminant concentration is plotted as a function of time. Initially, the concentration increases linearly with time, but the rate of increase continually declines, and the concentration asymptotically approaches steady state. The concentration is within 1% of its steady-state value after seven half-lives, t_h, where $t_h = \ln 2/k$. If the initial concentration is nonzero, the same steady-state concentration obtains, but the time required to reach steady state depends on the half-life and the initial concentration. Later in this article, the constant uptake/first-order removal model is applied to vegetation (constant deposition on foliage and first-order removal by

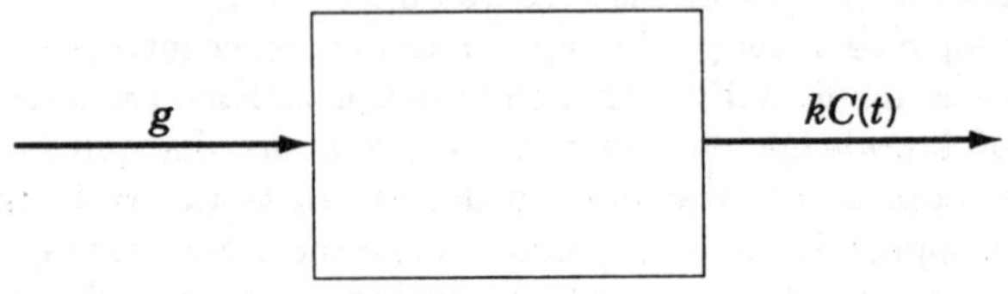

Figure 8. Constant uptake/first-order removal model.

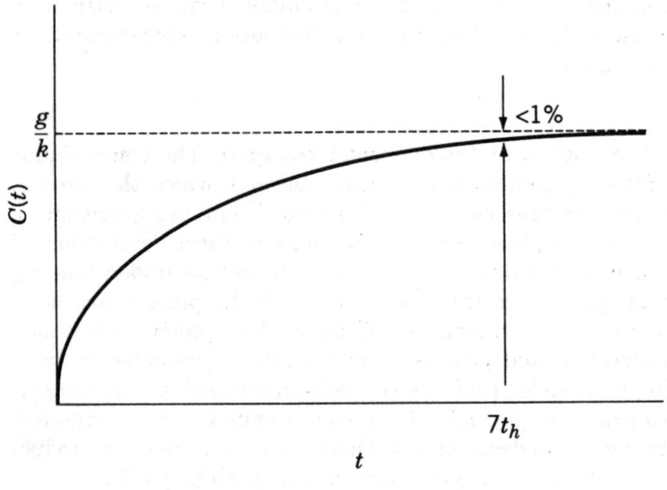

Figure 9. Concentration vs time for a homogeneous nonadvective compartment.

various processes) and to animals (constant uptake by foraging contaminated food and first-order biological removal of the contaminant from the animal).

In some problems, the time scale for uptake is small compared to the time scale of interest, and the contaminant can be considered to undergo instantaneous uptake. This is illustrated in the instantaneous partitioning model shown in Figure 10. In this model, the concentration in compartment 2 is linearly related to the concentration in compartment 1 through a partition factor,

$$C_2 = \mathrm{PF} \cdot C_1 \quad (7)$$

where C_1 is concentration in compartment 1, C_2 is concentration in compartment 2, and PF is the partition factor. The dashed arrow in Figure 10 is used to indicate that the transfer from compartment 1 to compartment 2 is often considered reversible. The instantaneous partitioning model is applied to transfer from water to suspended sediment, soil to vegetation, and water to aquatic plants and animals. The units of the partition factor depend on the concentration units for compartments 1 and 2. For example, if compartment 1 is water and compartment 2 is suspended sediment, then the partition factor has units of L/kg; if compartment 1 is soil and compartment 2 is vegetation, then the units are kg(s)/kg (v).

An advective compartment with first-order degradation is illustrated in Figure 11. The advective flow rate into and out of the compartment is Q, the contaminant concentration in the influent fluid is C_0, and the initial contaminant concentration in the compartment is zero. The compartment volume, V, is constant. For this problem, equation 5 takes on the following form,

$$\frac{dC(t)}{dt} = \frac{C_0 Q}{V} - \frac{C(t)Q}{V} - kC(t) \quad (8)$$

Mathematically, equation 8 is the same equation 5 with g being replaced by C_0Q/V and k being replaced by $Q/V + k$. The solution is

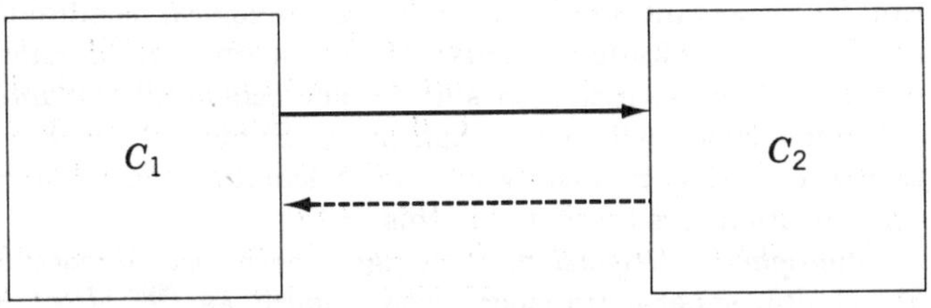

Figure 10. Instantaneous partitioning model.

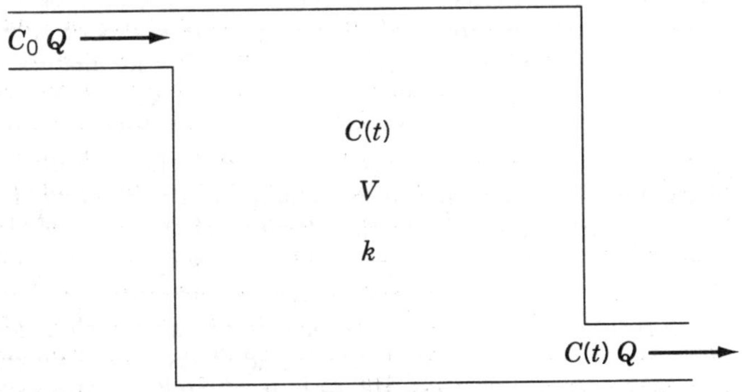

Figure 11. Advective compartment, first-order degradation.

$$C(t) = \frac{C_0 Q}{V\left[\frac{Q}{V} + k\right]}\left[1 - e^{-\left[k + \frac{Q}{V}\right]t}\right] \quad (9)$$

and the transient behavior of contaminant concentration is analogous to that depicted in Figure 9 where $t_h = \frac{\ln 2}{k + Q/V}$. This type of model can be used for well-mixed lakes and for discrete segments of a river.

One-Dimensional Transport. A common contaminant transport problem involves the release from an industrial facility to a river or stream as illustrated in Figure 12**a.** The one-dimensional approximation for this problem is illustrated in Figure 12**b.**

In this approximation, the contaminant is uniformly distributed across the channel and becomes a function of x and t only, and equation 3 reduces to

$$\frac{\partial C(x,t)}{\partial t} = D\frac{\partial^2 C(x,t)}{\partial x^2} - u\frac{\partial C(x,t)}{\partial x} + g(x,t) - d(x,t) \quad (10)$$

Presented in Table 2 are advective and advective/dispersive solutions to equation 10 with first-order degradation during transport ($d(x,t) = kC(x,t)$) for a point (spatial) release and the four temporal approximations given in Table 2.

Graphical depictions of the solutions are either in the form of concentration vs distance or concentration vs time. A concentration vs distance plot is a snapshot of the spatial extent of concentration at a fixed time. A concentration vs time plot shows the temporal behavior of contaminant concentration at a fixed location.

Instantaneous Release. Advection/dispersion solutions for an instantaneous release are shown graphically in Figure 13. The concentration follows a Gaussian distribution in space. The peak of the distribution is transported at the mean fluid velocity, and as the contaminant is carried away from the source ($x = 0$), it spreads due to dispersion. The degree of spreading is quantified through the standard deviation of contaminant concentration, which increases with time (and, equivalently, with distance) according to

$$\sigma = \sqrt{2Dt} = \sqrt{2D\frac{x}{u}} \quad (19)$$

If the distribution is Gaussian, approximately 68% of the contaminant is contained within one standard deviation (ie, locations between $(ut - \sigma)$ and $(ut + \sigma)$), and approximately 95% is contained within two standard deviations. Concentration versus time at given locations is illustrated in Figure 14.

Concentration versus time is asymmetric, characterized by a rapid rise and a more gradual decline. This occurs because dispersion and advection are acting in concert before the peak arrives and in opposition after it leaves. The gradual decline of concentration is referred to as tailing and the portion of the curve to the right of the peak concentration is called the dispersion tail. The relative importance of advection compared to dispersion at a given location is quantified by the Peclet number Pe, which is given by $Pe = ux/D$. In Figure 14 the Peclet number is 50.

Semi-Infinite Step Release. Concentration versus distance profiles for a semi-infinite step emission are shown in Figure 15. The effects of dispersion are evident in the shape of the leading edge, ie, it is sharper at t_1 than it is at t_2. Dispersion effects are more clearly illustrated in Figure 16, where curves of concentration versus time at a given distance are shown for Peclet numbers of 1, 100, and 1000. At $Pe = 1000$, dispersion is negligible and the solution reduces to the purely advective form.

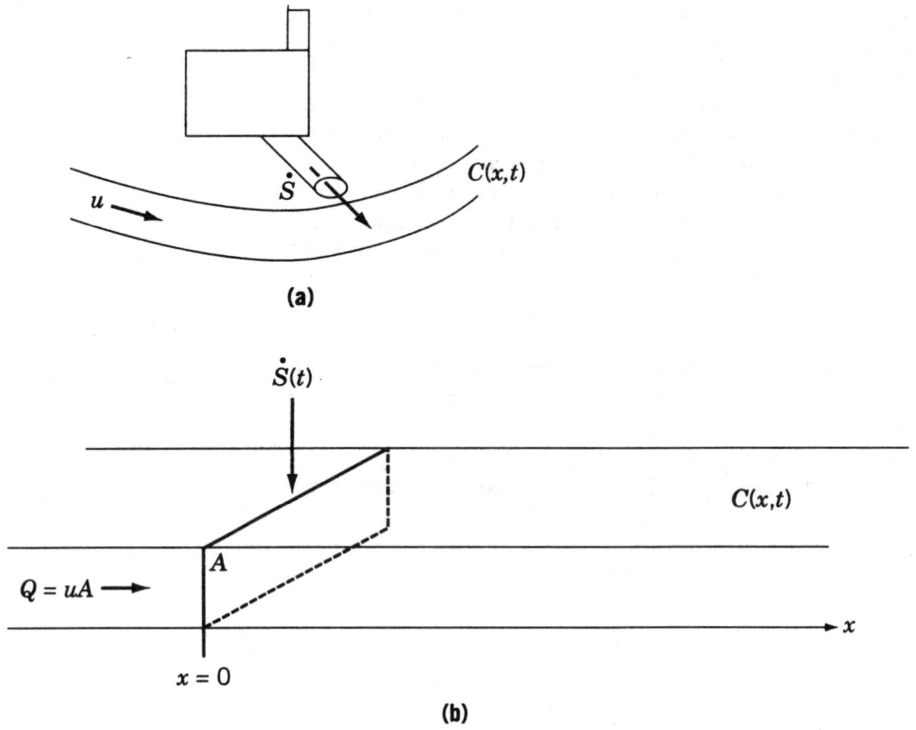

Figure 12. One-dimensional transport.

Three-Dimensional Advection/Dispersion. There are many important environmental transport problems for which one-dimensional approximations are not appropriate. For example, a two-dimensional approximation might be necessary for a wide, shallow river or a confined aquifer, and three-dimensional approximations are usually needed in the atmosphere, large lakes and oceans, and unconfined aquifers. Mathematically, multidimensional transport is analyzed by adopting the convention of aligning the x-axis with the direction of advective flow. Applying this to equation 3 yields

$$\frac{\partial C}{\partial t} = D_x \frac{\partial^2 C}{\partial x^2} + D_y \frac{\partial^2 C}{\partial y^2} + D_z \frac{\partial^2 C}{\partial z^2} - u \frac{\partial C}{\partial x} + g - d \qquad (20)$$

where D_x is longitudinal dispersion coefficient, D_y is transverse dispersion coefficient in the y (usually horizontal) direction, D_z is transverse dispersion coefficient in the z (usually vertical) direction, and $C = C(x, y, z, t)$. The mathematical form of the solution to equation 20 depends on the geometry of the medium and the spatial and temporal dependence of the release. For a point, instantaneous release into a medium where boundary effects are negligible, the solution is

$$C(x, y, z, t) = S_0 \frac{e^{\frac{-(x-ut)^2}{4D_x t}}}{\sqrt{4\pi D_x t}} \frac{e^{\frac{-y^2}{4D_y t}}}{\sqrt{4\pi D_y t}} \frac{e^{\frac{-z^2}{4D_z t}}}{\sqrt{4\pi D_z t}} \qquad (21)$$

At any given time, the contaminant concentration profile is Guassian in the x-, y-, and z-directions as illustrated in Figure 17.

Up to here, the contaminant transport equation has been applied to problems that were generic in the sense that the transport medium was not specified. In subsequent sections, the equation is applied specifically to atmospheric, surface water, groundwater, and food chain pathways. A common feature of these models is that they each contain one or more empirical transport parameters. A critical part of transport modeling is specifying appropriate values for these empirical parameters.

Atmospheric Transport. A typical risk assessment problem involving atmospheric transport is illustrated in Figure 18. A contaminant is emitted from a facility, either routinely or accidentally, at an emission rate, $\dot{S}(t)$. For a release of a given magnitude and duration (ie, for a given $\dot{S}(t)$, atmospheric contaminant concentration at ground-level depends primarily on wind direction, wind speed, release height, and atmospheric dispersion. However, removal or transformation mechanisms can be important under certain meteorological conditions (such as precipitation events) or for select contaminants (such as fluorine, hydrogen sulfide, photochemical oxidants, or radionuclides with short half-lives). Simple atmospheric transport models commonly used for calculating contaminant concentrations are presented in this section.

Atmospheric Dispersion. Although there are important global atmospheric transport issues (such as the effect of CO_2 on temperature or the effect of chlorofluorocarbons on ultraviolet radiation) and regional atmospheric transport issues (such as the effect of SO_2 released in one state or one country on acid precipitation in neighboring states or countries), the focus here is on effects within a few kilometers of the release. Localized transport occurs in a region of the atmosphere adjacent to the surface of the earth known as the planetary boundary layer, which varies in thickness from 200 to 2000 m. The dispersion of contaminants in the planetary boundary layer depends on the stability of the atmosphere. Atmospheric stability, in turn, depends largely on the vertical temperature gradient. The vertical temperature gradient which results solely from the decrease in pressure with height above the surface is known as the dry adiabatic lapse rate and is approximately −1°C per 100 m (−5.4°F/1000 ft). However, the actual temperature profile generally differs from the dry,

Table 2. Common Transport Equation Approximations and Solutions

Rate	Functional Form	Equation
	Emission Rate Approximations	
Instantaneous at $t = t_a$	$\dot{S}_{app} = S_0\delta(t - t_a)$	
Semi-infinite step beginning $t = t_a$	$\dot{S}_{app} = \dot{S}_0 h(t - t_a)$	
Finite step from $t = t_a$ to $t = t_b$	$\dot{S}_{app} = \dot{S}_0 h[h(t - t_a) - h(t - t_b)]$	
Constant	$\dot{S}_{app} = \dot{S}_0$	
	Advective Point Source Solution	
Instantaneous, $S_0\delta(t)$	$C(x,t) = \frac{S_0}{Q}\delta\left(t - \frac{x}{u}\right)e^{-k\frac{x}{u}}$	(11)
Semi-infinite, $\dot{S}_0 h(t)$	$C(x,t) = \frac{\dot{S}_0}{Q}h\left(t - \frac{x}{u}\right)e^{-k\frac{x}{u}}$	(12)
Finite step, $\dot{S}_0[h(t) - h(t - T)]$	$C(x,t) = \frac{\dot{S}_0}{Q}\left[h\left(t - \frac{x}{u}\right) - h\left(t - T - \frac{x}{u}\right)\right]e^{-k\frac{x}{u}}$	(13)
Constant, $\dot{S}_0$	$C(x,t) = \frac{\dot{S}_0}{Q}e^{-k\frac{x}{u}}$	(14)
	Advective/Dispersive Point Source Solution	
Instantaneous, $S_0\delta(t)$	$\frac{S_0}{A\sqrt{4\pi Dt}}e^{-\frac{(x-ut)^2}{4Dt}}e^{-k\frac{x}{u}}$	(15)
Semi-infinite, $\dot{S}_0 h(t)$	$\frac{\dot{S}_0}{2Q}\left[erfc\left(\frac{x - ut}{\sqrt{4Dt}}\right) + e^{\frac{ux}{D}}erfc\left(\frac{x + ut}{\sqrt{4Dt}}\right)\right]e^{-k\frac{x}{u}}$ [a]	(16)
Finite step, $\dot{S}_0[h(t) - h(t - T)]$	$C(x,t) = \frac{\dot{S}_0}{2Q}[g(t) - h(t - T)g(t - T)]e^{-k\frac{x}{u}}$	(17)
Constant, $\dot{S}_0$	$C(x,t) = \frac{\dot{S}_0}{Q}e^{-k\frac{x}{u}}$	(18)

[a] Where $g(t') = \left[erfc\left(\frac{x-ut'}{\sqrt{4Dt'}}\right) + e^{\frac{ux}{D}}erfc\left(\frac{x+ut'}{\sqrt{4Dt'}}\right)\right]$.

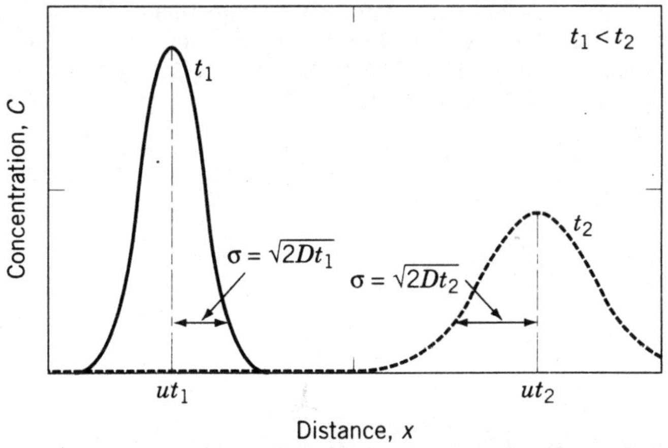

Figure 13. Concentration vs distance for an instantaneous release.

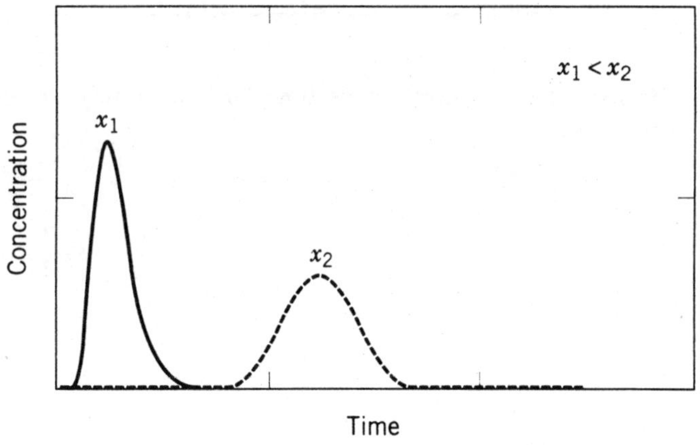

Figure 14. Concentration vs time for an instantaneous release.

adiabatic rate for any number of reasons including the localized effects of winds and topography, heat transfer from the air to the earth's surface or from the earth's surface to the air, or the movement of large air masses. Under dry, adiabatic conditions, the atmosphere is considered to be neutral with respect to contaminant dispersion, ie, vertical motions are neither enhanced nor suppressed. However, if the temperature profile is subadiabatic (either it does not

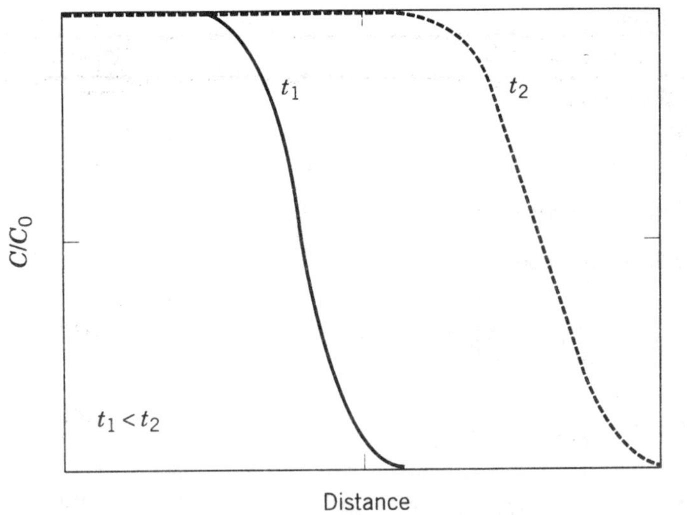

Figure 15. Concentration vs distance for a semi-infinite step release.

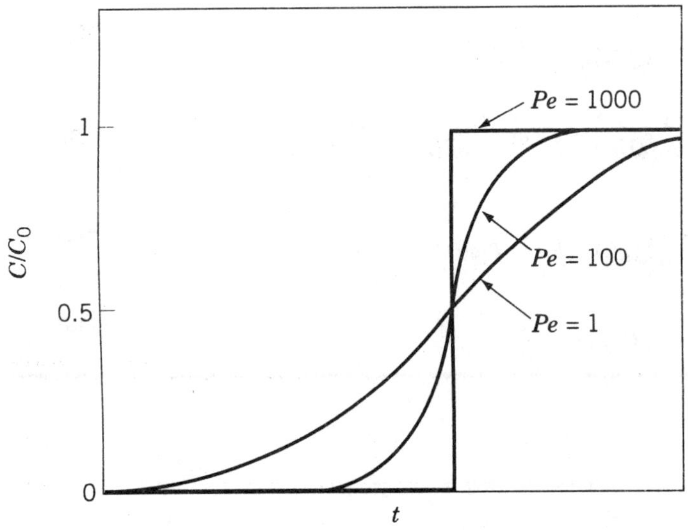

Figure 16. Concentration vs time for a semi-infinite step emission.

decrease with height as rapidly as it does under dry, adiabatic conditions or it increases with height), vertical motions are suppressed. This occurs because the temperature of a parcel of air undergoing a vertical displacement follows the dry, adiabatic lapse rate. Consequently, a parcel that is displaced upward has a lower temperature than the surrounding air (as illustrated in Fig. 19) and a corresponding higher density. It thus experiences a force that pushes it down toward its original position. Similarly, a parcel that is displaced downward has a higher temperature than the surrounding air and a corresponding lower density. It experiences a force that pushes it upward toward its original position. Thus, under these conditions, the atmosphere is inherently stable and dispersion is low. Conversely, if the atmospheric temperature profile is superadiabatic (eg, decreases more rapidly than under neutral conditions), vertical motions are enhanced, the atmosphere is inherently unstable, and dispersion is high.

For purposes of dispersion modeling, it is customary in the United States to characterize atmospheric stability in terms of the Pasquill-Gifford stability classification system. The various stability classes and corresponding temperature gradients are given in Table 3.

Gaussian Plume Model. For modeling purposes, many atmospheric transport problems can be approximated by the simplified problem depicted in Figure 17. Consistent with the conventions discussed in the previous section, the x-axis is aligned with the wind direction and the z-axis is aligned with the vertical direction. The release occurs at $z = h$. In the atmosphere, dispersion is quantified by the eddy diffusivity (9), K. Replacing D in equation 20 by K yields

$$\frac{\partial C}{\partial t} = K_x \frac{\partial^2 C}{\partial x^2} + K_y \frac{\partial^2 C}{\partial y^2} + K_z \frac{\partial^2 C}{\partial z^2} - u \frac{\partial C}{\partial x} + g - d \quad (22)$$

When applied to a constant, elevated release without generation or destruction, equation 22 yields

$$C(x, y, z) = \frac{\dot{S}_0}{2\pi u} \frac{e^{\frac{-y^2}{4K_y \frac{x}{u}}}}{\sqrt{2K_y \frac{x}{u}}} \frac{e^{\frac{-(z-h)^2}{4K_z \frac{x}{u}}}}{\sqrt{2K_z \frac{x}{u}}} \quad (23)$$

In its present form, equation 23 cannot be used in practice because there are neither theoretical nor empirical equations for eddy diffusivity that are satisfactory. The useful operational form of equation 23 is known as the Guassian

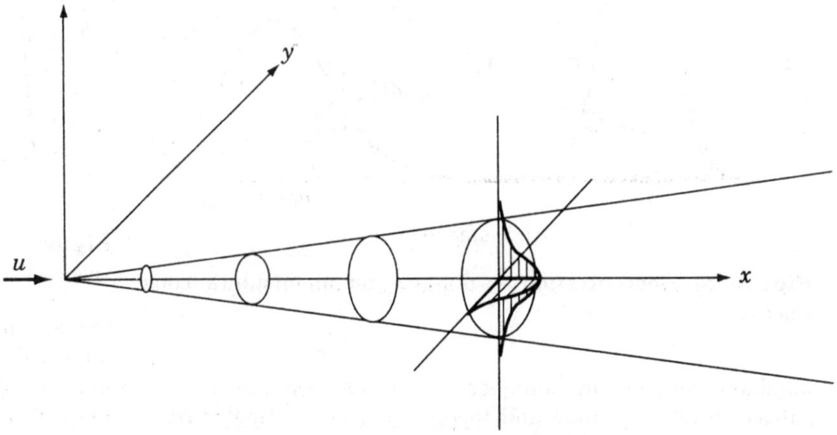

Figure 17. Three-dimensional advection dispersion.

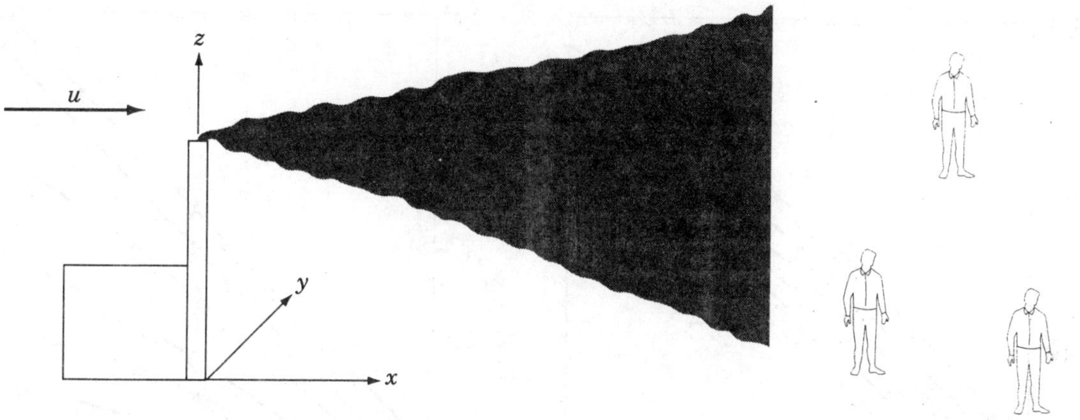

Figure 18. A typical atmospheric transport scenario.

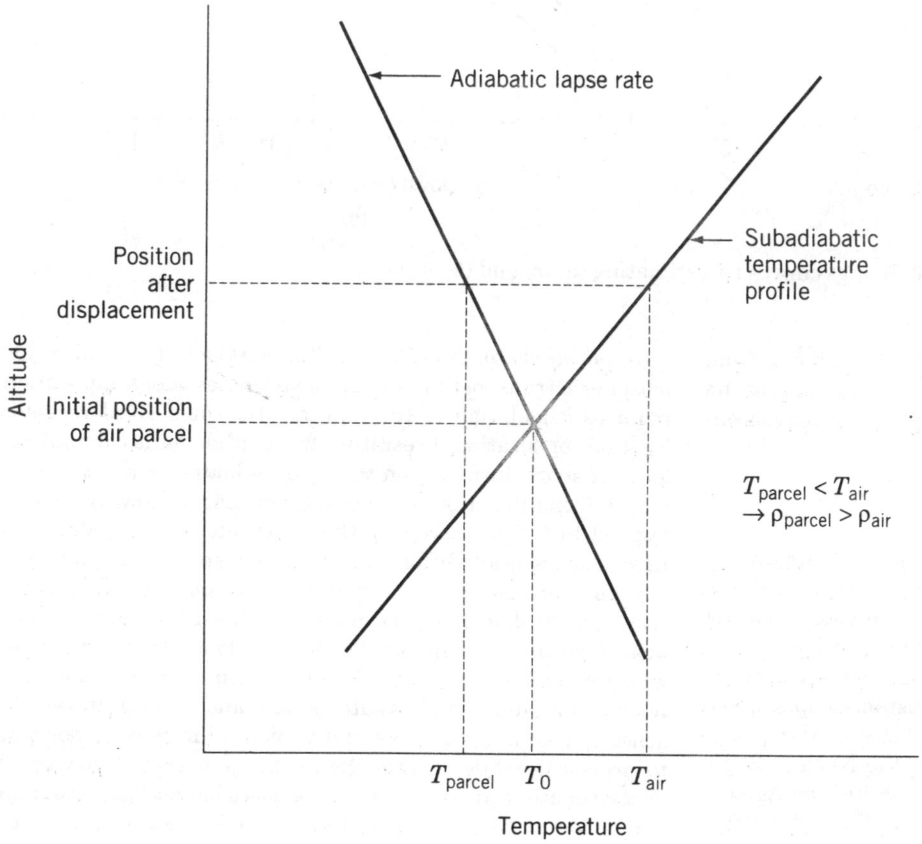

Figure 19. Suppression of vertical motion by a subadiabatic temperature profile.

Table 3. Pasquill-Gifford Stability Classes

Stability Class	Description	Temperature Gradient °C per 100 m
A	Extremely unstable	$\frac{\Delta T}{\Delta z} \leq -1.9$
B	Moderately unstable	$-1.9 < \frac{\Delta T}{\Delta z} \leq -1.7$
C	Slightly unstable	$-1.7 < \frac{\Delta T}{\Delta z} \leq -1.5$
D	Neutral	$-1.5 < \frac{\Delta T}{\Delta z} \leq -0.5$
E	Slightly stable	$-1.5 < \frac{\Delta T}{\Delta z} \leq 1.5$
F	Moderately stable	$1.5 < \frac{\Delta T}{\Delta z} \leq 4.0$

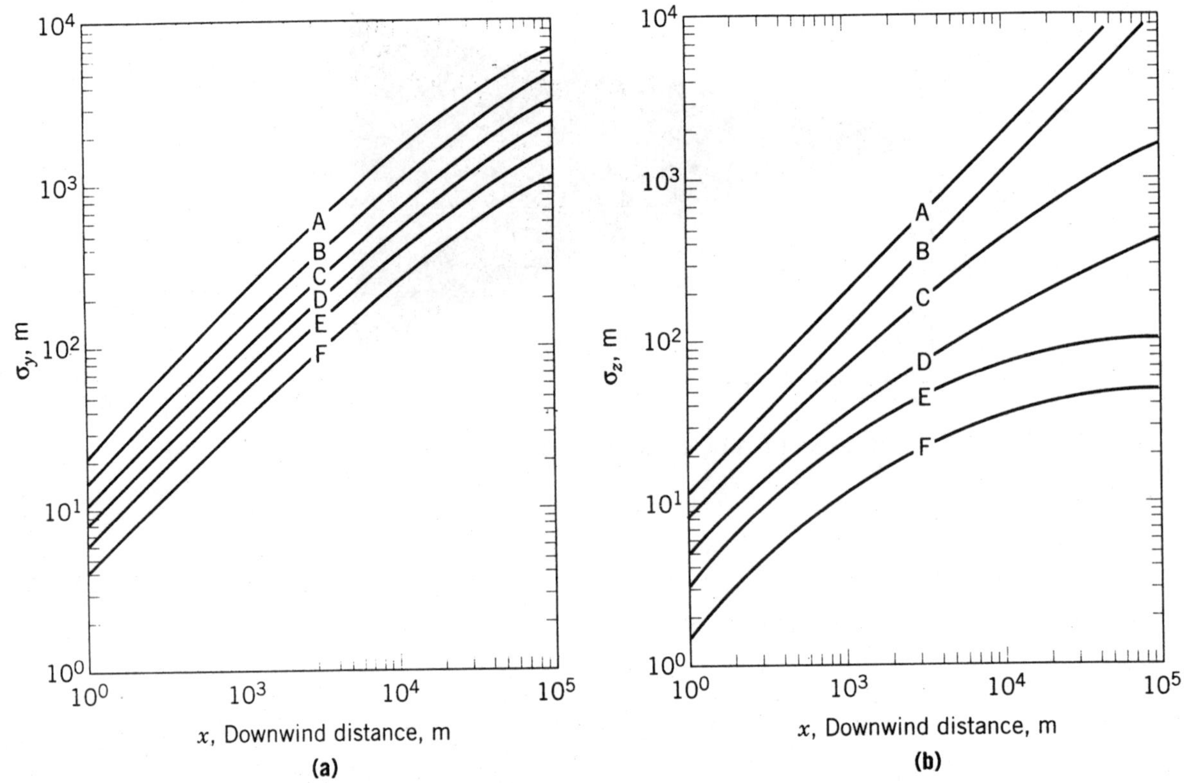

Figure 20. Briggs curves for estimating (**a**) σ_y and (**b**) σ_z (10).

plume model. It is obtained by substituting $\sigma = \sqrt{2Kx/u}$ (from eq. 19) and assuming complete reflection of contaminant at the ground. The resulting expression for the ground-level concentration is

$$C(x, y, 0) = \frac{\dot{S}_0}{\pi u \sigma_y \sigma_z} e^{\frac{-y^2}{2\sigma_y^2}} e^{-\frac{h^2}{2\sigma_z^2}} \quad (24)$$

This equation requires values of σ_y and σ_z, which are standard deviations of contaminant concentration in the y- and z-directions, respectively. A number of schemes are available for specifying σ_y and σ_z. In Figure 20 are Briggs curves for estimating these parameters (10). These parameters are applicable for flat terrain and downwind distances up to about 10 km. Fits to these curves are given in Table 4. Many state and federal regulatory agencies, including the Nuclear Regulatory Commission and the Environmental Protection Agency, use a different set of parameters known as the Pasquill-Gifford curves (8,9).

Table 4. Equations Recommended by Briggs[a]

Stability Class	σ_y, m	σ_z, m
A	$\frac{0.22x}{\sqrt{1+0.0001x}}$	$0.20x$
B	$\frac{0.16x}{\sqrt{1+0.0001x}}$	$0.12x$
C	$\frac{0.11x}{\sqrt{1+0.0001x}}$	$\frac{0.08x}{\sqrt{1+0.0002x}}$
D	$\frac{0.08x}{\sqrt{1+0.0001x}}$	$\frac{0.06x}{\sqrt{1+0.0015x}}$
E	$\frac{0.06x}{\sqrt{1+0.0001x}}$	$\frac{0.03x}{1+0.0003x}$
F	$\frac{0.04x}{\sqrt{1+0.0001x}}$	$\frac{0.016x}{1+0.0003x}$

[a] See Fig. 20.

Applications of the Gaussian Plume Model. The typical atmospheric transport problem in a risk assessment application requires calculation of the concentration at a fixed receptor location (or locations) resulting from either an accidental release of short duration (on the order of hours) or a normal release of long duration (on the order of years). However, when applied to a fixed receptor, the Gaussian plume model with Briggs or Pasquill-Gifford dispersion parameters is valid only for time periods on the order of tens of minutes. This is because (*1*) the dispersion parameters are based on observations made over this time span, (*2*) wind directions tend to fluctuate, which causes the receptor's position with respect to the centerline of a plume to fluctuate as well, and (*3*) even in the absence of fluctuations in wind direction, plumes meander due to large-scale eddies. Consequently, the Gaussian plume model overestimates concentrations for releases longer than about an hour. One of two averaging techniques is normally used for longer time periods, both of which require hourly measurements of wind direction, wind speed, and temperature gradient (or, equivalently, stability class).

The first technique is referred to here as the summation of Gaussian plumes. The Gaussian plume model is used to calculate the concentration for each hourly period, and the average concentration over the entire exposure period is given by

$$\overline{C} = \frac{\sum_{i=1}^{N} C_{g,i}}{N} \quad (25)$$

where $C_{g,i}$ is the Gaussian plume concentration for the ith set of hourly meteorological measurements and N is the total number of measurements. In prospective analyses performed to support permitting or licensing applications, meteorological data are usually collected for a year and $N = 8760$ (corresponding

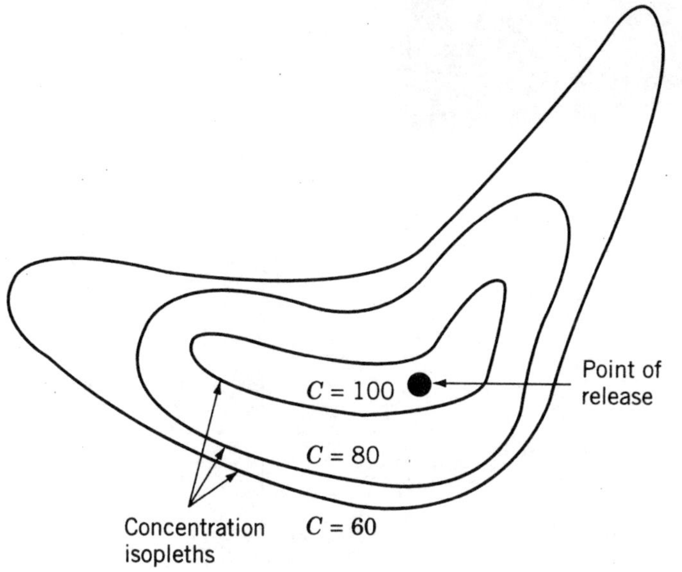

Figure 21. Concentration isopleths in the vicinity of a release point.

to 365 days of hourly measurements). If equation 25 is applied to each node on a grid surrounding the release point, a map of concentration isopleths, such as that illustrated in Figure 21, can be produced.

The second technique, which is less computationally intensive than the summation of Gaussian plumes and is used by the Nuclear Regulatory Commission (11) in the licensing of nuclear power plants, is the sector average approximation. The region surrounding the release is divided into 16 sectors, each centered on one of the 16 compass points (N, NNE, NE, ENE, etc) used for specifying wind direction. For each hourly period that winds are blowing into a given sector, the concentration is taken to be constant across the sector. This average concentration is defined such that the total contaminant mass equals that predicted by the Gaussian plume model, as illustrated in Figure 22.

The resulting sector averaged concentration, $\overline{C}_{\mathrm{SA}}$, is

$$\overline{C}_{\mathrm{SA}} = \frac{2fS_0}{u\sigma_z x}\exp\left(-\frac{h^2}{2\sigma_z^2}\right) \tag{26}$$

where f is the fraction of time that the wind is blowing into the sector. When used in conjunction with hourly meteorological data, the sector averaged concentration is calculated by

$$\overline{C}_{\mathrm{SA}} = \frac{2\dot{S}_0}{x}\sum_{i=1}^{I}\frac{1}{u_i}\sum_{j=A}^{F}\frac{f_{i,j}}{\sigma_{z,j}}\exp\left(-\frac{h^2}{2\sigma_{z,j}^2}\right) \tag{27}$$

where I = number of wind speed groups and $f_{i,j}$ = fraction of time winds blow into the given sector under stability j and with speed u_i.

Other Factors. The Gaussian plume model incorporates the principal factors, ie, wind direction, wind speed, atmospheric stability, and release height, that typically dominate contaminant concentrations in air. However, other factors can be important in certain situations. Although many of these other factors are complex and difficult to model accurately, simple expressions are available for making estimates of their impact on atmospheric concentrations. These simple expressions are given below.

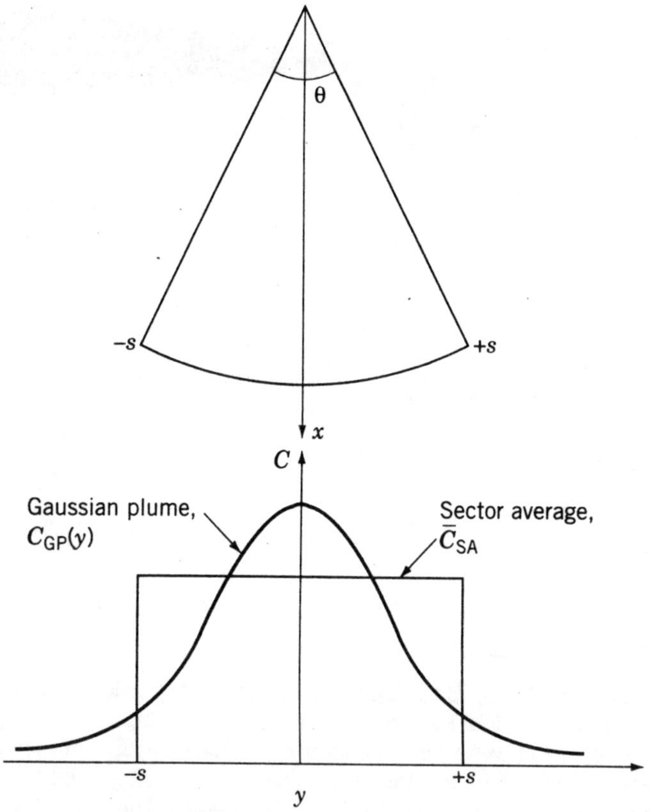

Figure 22. Sector average approximation.

Plume Rise. Plume rise refers to the effect of the momentum and buoyancy of a plume of gases emitted from a stack (Fig. 23). This can cause the effective height of a release to be as much as a factor of 10 greater than the physical release height. The expression for plume rise h_{pr}(m) given in equation 28 (12) is presented here because it requires specification of only two parameters, heat emission rate Q_H (MW) and wind speed u (m/s).

$$h_{pr} = 530 \cdot \frac{Q_H^{0.25}}{u} \tag{28}$$

More accurate estimates require additional parameters such as effluent velocity, effluent and ambient temperature, stack diameter, and atmospheric temperature gradient.

Building Wake. Building wake refers to the turbulence that develops as air passes around and over a building. This is important because it can increase near-field dispersion. The preferred methods for predicting the effects on contaminant concentrations are field measurements or wind tunnel simulations. However, these approaches are rarely practical and in their absence empirical adjustments to the dispersion parameters are made. The following has been given for σ_y and σ_z (13):

$$\sigma_y^* = \left(\sigma_y^2 + \frac{cA}{\pi}\right)^{0.5} \tag{29}$$

$$\sigma_z^* = \left(\sigma_z^2 + \frac{cA}{\pi}\right)^{0.5} \tag{30}$$

where A is the cross-sectional area of the building normal to the wind direction and c is a constant between 0.5 and 0.67. Since the dispersion parameters increase with downwind distance, the effect of the adjustment decreases with distance

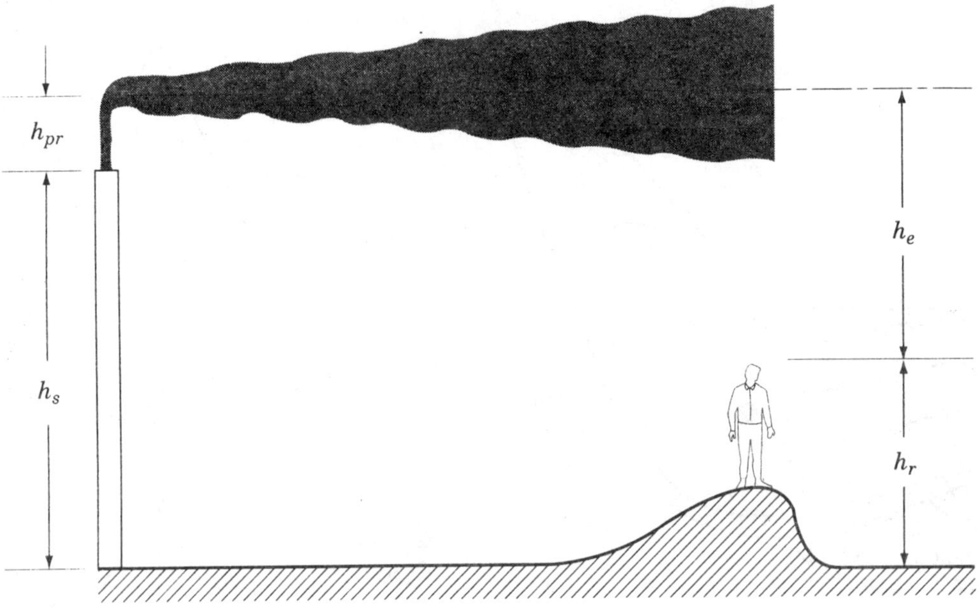

Figure 23. The effects of plume rise and receptor location on effective release height.

from the release. Building wake can also affect release height. A release that is less than the height of the building is captured in the wake and is effectively released at ground level. As the release height increases above the building height, the effect of building wake is difficult to predict. The following is recommended: for a release greater than 2.5 times the building height, the release is unaffected by building wake; for a release below the height of the building, a ground-level release is assumed; for a release between the building height and 2.5 times the building height, a ground-level release is assumed for part of the time and an elevated release is assumed for the remainder of the time (14).

Effective Release Height. Ground-level concentrations are particularly sensitive to release height, as seen in equations 24 and 26. The effective height of a release can differ from the physical release height because of the effects of plume rise, building wake, and receptor elevation. The effects of plume rise and receptor elevation on effective release height are illustrated in Figure 23 and are given by

$$h_e = h_s + h_{pr} - h_r \tag{31}$$

where h_s is the physical release height and h_r is the elevation of the receptor relative to the ground elevation at the release.

Removal Mechanisms. In the derivation of the Gaussian plume model, nonconservative processes such as contaminant generation, transformation, decay, degradation, deposition, washout, or rainout were all neglected. Contaminant generation, such as the production of photochemical oxidants, is far too complex for consideration in this article as are rigorous treatments of removal mechanisms. However, by approximating removal mechanisms as first-order processes, their effect can be estimated by

$$C = C_{GP} e^{-k\frac{x}{u}} \tag{32}$$

where C_{GP} is the concentration calculated from the Gaussian plume model and k is the first-order rate constant for the removal mechanism of interest. For degradation, the rate constant is the relevant chemical reaction rate constant and for radioactive decay it is the radioactive decay constant. For deposition and washout, the rate constants are v_d/z_{eff} and $R\omega_v/z_{eff}$, respectively, where v_d is contaminant deposition velocity (m/s), R is precipitation rate (m/s), ω_v is volumetric washout factor, and z_{eff} is effective height. Values of deposition velocities and volumetric washout factors are available (15). Deposition velocities are discussed further in the section on food-chain transport. For a ground-level release, $z_{eff} = \sqrt{\pi/2}\sigma_z$ and for an elevated release, $z_{eff} = \sqrt{2\pi} e^{h^2/2\sigma_z^2} \sigma_z$.

The Gaussian plume model should not be used for rough terrain or for distances farther than 10 km from the point of contaminant release. An alternative is the variable trajectory model which takes into account spatial and temporal variations of air flow and should provide more accurate estimates of atmospheric concentrations. In such a model, individual "puffs" are tracked over time intervals on the order of 30 minutes. However, it requires detailed data on wind direction in the vicinity of a site that are normally not available.

Groundwater Transport. The groundwater pathway usually represents the principal route of exposure for risk assessments involving spills, buried wastes, or underground storage tanks. Contaminants that are released from the surface or near-surface zones can migrate downward under the force of gravity until they reach an aquifer. The groundwater in an aquifer typically moves laterally and can thus carry contaminants to off-site wells or outcrops. Discussed in this section are basic models for predicting the fate and transport of contaminants in groundwater.

Subsurface Formations. In the context of contaminant transport in groundwater, subsurface media are categorized as one of three types: porous, fractured porous, and fractured nonporous. In porous media, the flow of water and solutes occurs through interstitial spaces between the soil particles. Typical porous media are sand, silt, and clay. Fractured, porous media are characterized by rapid flow through fissures and slower flow through the porous matrix. Subsurface media subject to fractures include sandstone, shale, and clay. In fractured, nonporous media such as rock, flow does not occur through the matrix and is confined to the fractures. Although fractures in porous and nonporous media are important because they represent short circuits for contaminant transport, their identification and characterization is extremely difficult and simple models for fractured flow transport do not exist. Porous media are of two types: saturated and unsaturated. In saturated media, all of the interstitial spaces are filled with water and the driving force for fluid transport is the hydrostatic pressure. In unsaturated media, the interstitial spaces contain both water and air. Water flow occurs as a result of both the pressure gradient and capillary action. Air flow occurs as a result of diffusion and atmospheric pressure.

A generic subsurface formation is depicted in Figure 24. The vadose (or unsaturated) zone extends from the land surface to the water table, which forms the interface between the saturated and vadose zones. Subsurface media are further characterized according to their ability to transmit water under normal pressure gradients. An aquifer is a saturated region that can transmit significant quantities of water and an aquitard is a poorly permeable region that impedes groundwater movement. An aquifer that lies between two aquitards is referred to as confined and one whose upper boundary is the water table is referred to as unconfined.

For the purposes of flow and transport modeling, the physical characteristics of subsurface media are quantified by two parameters: bulk soil density, ρ_B, and moisture content, θ. They are defined for a representative volume, V_T, of the medium. Bulk soil density is dry soil mass per unit volume and moisture content is the ratio of liquid volume to total volume. In the saturated zone, moisture content is equal to porosity, n, which is the ratio of the interstitial volume to the total volume. In the vadose zone, moisture content is less than porosity because of the presence of air. In some saturated media it is important to distinguish between porosity and effective porosity, n_e, which is the ratio of mobile water volume to total volume. The mobile water volume can be less than the interstitial volume because of water that is trapped and immobilized in dead end pores. In the remainder of this article, the effective porosity and porosity are taken to be equal.

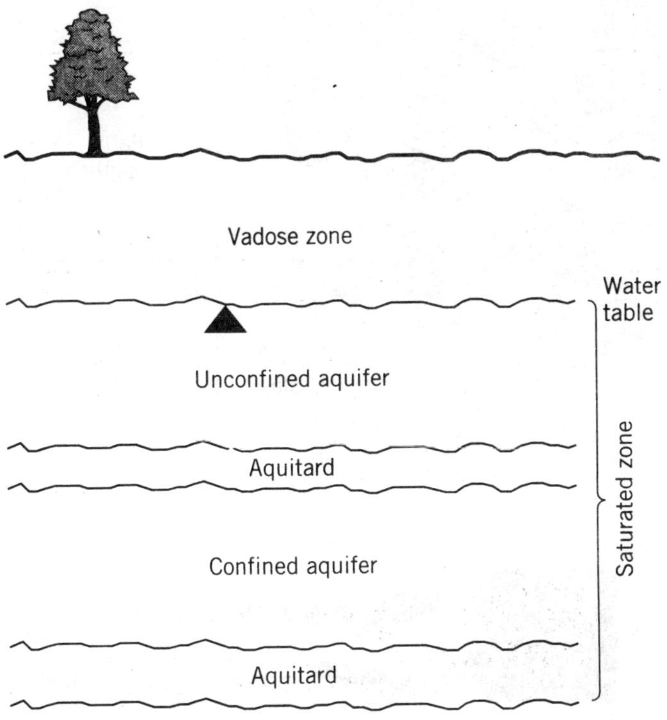

Figure 24. Characterization of subsurface formations.

On a microscopic scale, the flow of water around and between the soil particles in a porous medium is very complex. Some of these complexities are discussed later in this section as they relate to dispersion mechanisms. However, on a macroscopic scale, the average speed of the water between two points (the mean linear velocity) can be estimated using Darcy's law. Darcy's law was developed originally for calculating the volumetric flow rate of water in an aquifer. However, it is also useful in contaminant transport because it provides a means of calculating mean linear velocity. In one dimension, Darcy's law takes the following form,

$$\frac{Q_w}{A} = -k_H \frac{dH}{dx} \tag{33}$$

where Q_w is volumetric flow rate (m^3/s), k_H is hydraulic conductivity (m/s), dH/dx is the pressure gradient with pressure expressed in units of hydraulic head (m), and A is cross-sectional area of the aquifer (m^2). Mean linear velocity, u, obtains by dividing both sides of equation 33 by θ, yielding

$$u = -\frac{k_H}{\theta}\frac{dH}{dx} \tag{34}$$

Thus, the prediction of mean linear velocity requires knowledge of the hydraulic head.

The values of parameters such as the hydraulic conductivity and the effective porosity are strongly dependent on the type of soils composing the subsurface. In the absence of field measurements, an approach to estimating these parameters is to use the mean particle diameter to estimate the porosity, bulk soil density, and hydraulic conductivity. A relationship reported in the EPA's groundwater screening document (16) between porosity and mean particle diameter is

$$\theta = 0.261 - 0.385 \cdot \ln(d) \tag{35}$$

where d is the mean particle diameter in centimeters. The relationship between the density of the soil particles, the porosity, and the bulk soil density can be easily derived and is

$$\rho_B = \rho_P \cdot (1 - \theta) \tag{36}$$

where ρ_p is the particle density (often taken as 2.65 g/cm^3). Finally, the Karmen-Cozeny equation (17) can be used to estimate the hydraulic conductivity for soils with low clay content:

$$k_H = 478 \cdot \left[-\frac{\theta^3}{(1-\theta)^2} \cdot d^2\right] \tag{37}$$

However, the use of these estimating equations or of representative data are not a substitute for measured values. Modeling of conditions at a specific site should be done with site-specific measured parameters, as these values can vary widely.

Saturated Zone Transport. *Dispersion.* Dispersion in the subsurface is due to the combined effects of diffusion and advective inhomogeneities. Advective inhomogeneities, which are usually referred to as mechanical dispersion in the groundwater literature (18), generally dominate, and they occur at both the pore level and at the aquifer level. At the pore level, there are three basic mechanisms (19): (*1*) nonuniform velocity distributions across the pore channels, (*2*) variability in the flow paths that water molecules follow, and (3) variability in the size of the pore channels. At the aquifer level, variability in hydraulic conductivity and porosity due to heterogeneities of the porous medium causes additional variability of groundwater velocity and a corresponding increase in dispersion. Many subsurface media are highly heterogeneous, containing regions where water flow is high and regions where it is low. For example, a sand medium may contain clay lenses. Water flow through the clay lens is slower than it is in the surrounding sand, resulting in the spreading of contaminants suspended in the water. The pore level and aquifer level causes of dispersion are illustrated in Figure 25.

Mechanical dispersion coefficients are taken to be proportional to the average linear velocity, and the proportionality constant is the dispersivity, α. Diffusion and mechanical dispersion are combined linearly to yield the following:

$$D_L = D' + \alpha_L u \qquad (38)$$

$$D_T = D' + \alpha_T u \qquad (39)$$

where D' = molecular diffusion coefficient (m^2/s), D_L = dispersion coefficient parallel to the direction of flow (m^2/s), D_T = dispersion coefficient perpendicular to the direction of flow (m^2/s), α_L = longitudinal dispersivity (m), and α_T = transverse dispersivity (m).

Dispersivities have been measured both in laboratory column experiments and in field studies. The following provides a fit to the data (applicable for $x < 3500$ m) (21):

$$\alpha_L = 0.0175x^{1.46} \qquad (40)$$

where the distance along the flow direction, x, and dispersivity are both in meters.

Sorption. There are a number of processes through which contaminants become associated with particulate matter. Collectively known as sorption, they include ion exchange, complexation/hydrolysis, oxidation/reduction, and colloid/polymer formation. When an aqueous phase contaminant and a solid phase come into contact (either through the addition of the solid phase to a solid-free solution or addition of a contaminant to a solution containing solids), some of the contaminant becomes associated with the solid phase. Consequently, the solid-phase concentration increases and the aqueous-phase concentration decreases until equilibrium is achieved between the two phases. The concentrations at equilibrium are commonly expressed through an empirical quantity known as the equilibrium distribution coefficient, K_D, which is the ratio of solid to aqueous-phase contaminant concentration:

$$K_D = \frac{C_s}{C_l} \qquad (41)$$

where C_s is solid-phase contaminant concentration, mg(c)/kg, and C_l is aqueous (ie, liquid) phase concentration, mg(c)/L. The units of the distribution coefficient are thus L/kg. Distribution coefficients depend on the chemical form of the contaminant, the physical and chemical characteristics of the solid phase, and the chemical characteristics of the aqueous

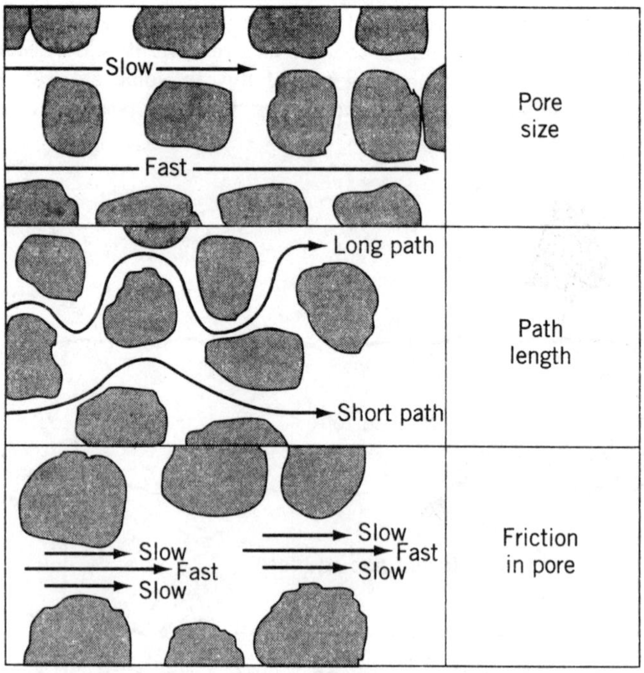

(a)

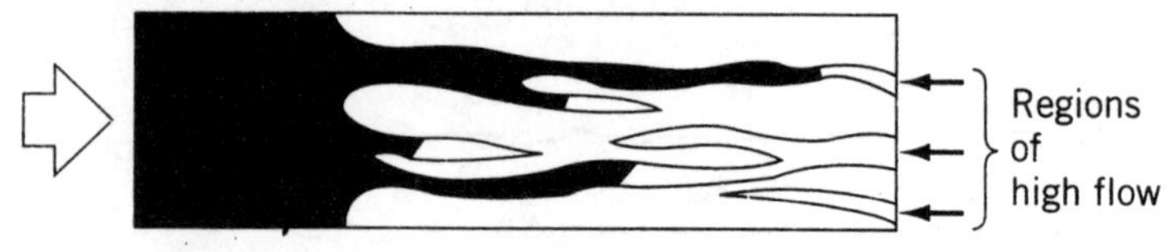

(b)

Figure 25. Illustration of mechanical dispersion processes: (**a**) pore level; (**b**) aquifer level (20).

Table 5. Distribution Coefficients of Elements for Different Soils[a]

K_D	Sand	Loam	Clay	Organic
<1	H,Tc	Tc		
1–10	I,Mo,Np	I	I,Tc	Tc
10–100	Ag,Cd,Co,Cr,Mn,Ru,Sr,U	Cd,Cr,Np,Sr,U	Mo,NP	I,Mo
100–1,000	Ce,Cs,Fe,Ni,Pb,Po,Pu,Ra,Zn	Ag,Fe,Mn,Po	Ag,Cd,Co,Fe,Mn,Ni,Sr	Cd,Co,Cr,Cs,Sr,U
1,000–10,000	Am,Cm,Th	Am,Ce,Co,Cs,Pu,Zn	Am,Cs,Pu,Ra,Th,U,Zn	Ni,NP,Pu,Zn
>10, 000		Cm,Pb,Ra	Ce	Ag,Am,Pb,Ru,Th

[a] Ref. 22.

Table 6. Sorbed Fraction as a Function of Liquid Volume Ratio and Distribution Coefficient[a]

Liquid Volume Ratio, θ	Solid to Liquid Mass Ratio	Fraction Sorbed	
		K_D = 10 L/kg	K_D = 0.1 L/kg
1	10^{-6}	10^{-5}	10^{-7}
1	10^{-4}	10^{-3}	10^{-5}
0.996	10^{-2}	0.09	10^{-3}
0.71	1	0.41	0.07
0.2	10	0.68	0.12
0.024	100	0.71	0.16

[a] Particle density = 2.5 g/cm^3.

phase (principally pH and ionic strength). Presented in Table 5 are typical distribution coefficients for various elements in different types of soil.

Table 5 serves to illustrate the dependence of distribution coefficient on the type of soil and the variability from element to element. Most organic contaminants do not sorb readily to minerals and thus have low distribution coefficients, although polychlorinated biphenyls (PCBs) are a notable exception. However, organic contaminants can sorb to other organic matter in the soil. Most soils contain organic matter. In some cases, it is possible to estimate a distribution coefficient for organic compounds sorbing onto soil organic carbon through the use of the K_{oc} value of the contaminant and the fraction of organic carbon in the soil as follows:

$$K_D = f_{oc} \cdot K_{oc} \tag{42}$$

This expression is only valid if the amount of organic carbon in the soil is great enough to sorb the contaminant. At low levels of f_{oc}, (ranging from 0.002 for dichloroethane with a K_{ow} of 62 to 0.000002 for pyrene with a K_{ow} value of 209,000), there may be insufficient organic matter to sorb the contaminant. In general, the critical value of f_{oc} is inversely related to the K_{ow} value of the contaminant.

The fraction of the contaminant sorbed to an immobile phase can be estimated as

$$f_s = \frac{\frac{\rho_B}{\theta} K_D}{1 + \frac{\rho_B}{\theta} K_D} \tag{43}$$

In modeling the groundwater transport of contaminants, sorption must be considered unless the distribution coefficient is very low. This is because of the large surface area available for sorption in the subsurface. In Table 6, sorbed fractions are given for various saturated zone moisture contents and distribution coefficients of 10 and 0.1 L/kg.

At a saturated zone moisture content of 0.2, the sorbed fraction is almost 70% for K_D = 10 L/kg, and it is greater than 10% for K_D = 0.1 L/kg. In contrast, even at a high surface water suspended solids concentration (10^{-2} kg/L (10,000 mg/L)), the sorbed fraction is less than 10% for K_D = 10 L/kg and is 0.1% for K_D = 0.1 L/kg.

As a consequence of the importance of sorption in ground water transport, the contaminant transport equation (eq. 3) is typically modified to explicitly incorporate its effect on aqueous-phase concentrations. In the linear, equilibrium model presented here, sorption is considered to occur instantaneously (which, in practical terms, means quickly compared to advection, dispersion, or reaction time scales), and the solid-phase concentration is linearly related to the aqueous-phase concentration as in equation 41. This yields the following one-dimensional form for subsurface contaminant transport:

$$\frac{\partial C}{\partial t} = \frac{D}{R}\frac{\partial^2 C}{\partial x^2} - \frac{u}{R}\frac{\partial C}{\partial x} + \frac{g_T}{\theta R} - kC \tag{44}$$

where C is aqueous-phase contaminant concentration, g_T is total (aqueous and solid phase) contaminant released to the subsurface, and R is retardation factor. The retardation factor, which is given by

$$R = 1 + \frac{\rho_B K_D}{\theta} \tag{45}$$

is a measure of the importance of sorption. If the contaminant does not sorb (ie, if $K_D = 0$), $R = 1$. As K_D increases, R increases. Physically, $1/R$ is the fraction of contaminant in the aqueous phase. Comparing the dispersive and advective transport terms in equation 44 to those in equation 10 it is seen that the forms are similar, with D being replaced by D/R and u being replaced by u/R. Since $R > 1$ for a sorbable contaminant, the net effect of sorption is to make dispersive and advective transport of the contaminant to be slower than that of the water by the factor R. Thus, the mean linear contaminant velocity, u_c, is

$$u_c = \frac{u}{R} \tag{46}$$

Table 7. Effect of Sorption on Contaminant Velocity in Saturated Zone[a]

K_D, L/kg(s)	R	u_c/u
0	1	1
0.1	1.5	0.65
1	6.3	0.16
10	54	0.02
100	5330	0.002

[a] $\theta = 0.2$; bulk soil density = 1500 kg(c)/m^3.

The effect of retardation on contaminant velocity is illustrated in Table 7, where the ratio of contaminant velocity to water velocity is given as a function of distribution coefficient.

Sorption can have a dramatic effect on the advection of a contaminant. For example, at a distribution coefficient of 10, which is at the low end of the range for metals in a subsurface medium containing clay, the contaminant moves only 1/50th as fast as the water. The reason is that only 1/50th of the contaminant is associated with the aqueous phase and is mobile; the remainder is associated with the solid phase and is immobile. The subscript T is included with the generation term in equation 44 to emphasize that it is the total amount of contaminant that is released per unit of subsurface volume. The fraction that partitions into the aqueous phase is $1/R$, and the fraction of the total volume occupied by the aqueous phase is θ.

Colloidal Transport. A complicating feature is added to transport of contaminants in the subsurface due to colloids. Colloids are very small particles (diameter <1 micrometer), which, due to their small size, may be able to migrate through porous media. Colloidal particles can include organic macromolecules, suspensions of insoluble liquids, bacteria, or mineral particles. If contaminants sorb onto these colloidal particles, they can form a second mobile phase which can be transported much faster than would be expected from Table 7; in fact, they can sometimes move even faster than the average groundwater linear velocity. Since a high sorption coefficient will increase both the immobile sorbed phase concentration and the mobile colloidal fraction, the effects of a high K_D value can be difficult to interpret properly if colloidal transport is suspected. Modeling the effect of colloidal transport can be very difficult, as it is dependent on the type and concentration of colloids in subsurface waters.

Transformation. Depending on the type of contaminant present, transformation processes can significantly affect the concentration of a contaminant at a point down-gradient of the source. These processes are nonconservative and can result in either an increase or, more commonly, a decrease in the level of a contaminant present. They can be divided into two categories: biotic or abiotic. Biotic processes are those which are catalyzed by biological processes of microscopic fauna in the subsurface (typically bacteria). Abiotic transformation processes are those which are due to physical processes such as radioactive decay or chemical reactions between the contaminant and surrounding media. Although the basic processes underlying the transformations are often fairly well understood, rigorous application of the relevant models is often difficult. Thus, transformation processes are often modeled as simple first-order processes. Care should be taken in modeling transformation: modeling the degradation of one compound without accounting for the resulting generation of the transformation product can lead to an underestimate of risk, since the transformation product may be as toxic as the parent compound.

Biological transformations are typically the result of subsurface microbes which utilize organic contaminants as an energy source. These can occur under aerobic conditions, or under anaerobic conditions if an electron acceptor such as nitrite is present. Although the modeling of biodegradation can be highly complex, certain simplifying assumptions can be made to approximate the process as first-order. These include the assumption that the concentration of the compound is very low and that the amount of microbial activity and oxygen are relatively constant.

Some common physical transformations include radioactive decay, hydrolysis, dehydrohalogenation, oxidation, and reduction. Hydrolysis is a process in which water reacts with a compound and substitutes a hydroxyl ion for an end group. In general, compounds with low numbers of end groups are rapidly hydrolyzed, and those with large numbers are hydrolyzed at slower rates. However, the converse is true for dehydrohalogenation, in which a halogen and a hydrogen are removed from carbons, leaving a double bond between the carbon atoms. This type of process proceeds rapidly for highly halogenated compounds, and slows as the degree of halogenation decreases. Oxidation and reduction can cause changes in the speciation of the contaminant, which can strongly affect its transport properties.

Solutions. Solving equation 44 or its three-dimensional correlate can be complex. As discussed in the section on the contaminant transport equation, there are two methods for solving the equation: numerical and analytical. Numerical techniques typically involve the use of finite element or finite difference models. These techniques can be highly sophisticated and are not covered in this article. Analytical solutions to equation 44 for a wide variety of initial conditions and boundary conditions are available (23). Two problems are considered here: an instantaneous release from a point source and a finite step release from a point source. Both of these were considered in the general discussion of advective/dispersive transport. As applied to groundwater modeling, the instantaneous release solution is

$$C = \frac{S_0/\theta RA}{2\sqrt{\pi D_L t/R}} \exp\left(-\frac{(Rx - ut)^2}{4D_L Rt}\right) \quad (47)$$

Equation 47 is plotted in Figure 26 as a function of distance at a given time after the release. With increasing retardation, the distance traveled by the pulse decreases and the total mass of contaminant also decreases. Both of these occur because the fraction of contaminant in the aqueous phase decreases with increasing R.

The initial and boundary conditions for the semi-infinite release are

$$C(x,0) = 0 \quad x \geq 0 \quad (48)$$

$$C(0,t) = C_0 \quad t \geq 0$$

$$C(\infty,t) = 0 \quad t \geq 0$$

The solution is identical to equation 16 with $R = 1$ and $C_0 = \frac{\dot{S}_0}{\theta uA}$. In most groundwater transport problems, the Peclet

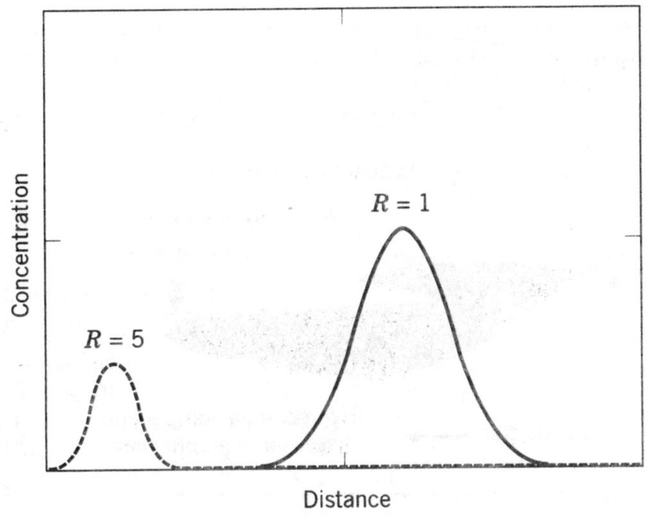

Figure 26. Concentration vs distance for an instantaneous release and $R = 1, 2,$ and 10.

number is large, and the second term in equation 16 can be neglected, yielding

$$C = \frac{C_0}{2} erfc\left(\frac{Rx - ut}{2\sqrt{D_L Rt}}\right) \tag{49}$$

Plots of C/C_0 as a function of distance are given in Figure 27 for various values of R. As retardation increases, the distance traveled by the contaminant front decreases. Here again, this occurs because the fraction of contaminant in the aqueous phase decreases with increasing R, and the aqueous phase must travel slower to maintain the equilibrium solid-phase concentration.

One-dimensional solutions are useful in estimating the behavior of a transient plume, but they provide conservative over-estimates of steady-state concentrations because transverse dispersion is neglected. This is particularly important at distances that are equal to or greater than the cross-sectional dimensions of the contaminated region. The vertical and horizontal spread (VHS) model is used by the EPA to include the effect of transverse dispersion. In this model, the contaminant is assumed to have reached its steady-state concentration and, therefore, longitudinal dispersion is not important. The boundary conditions are

$$C(0, y, z) = \begin{cases} C_0 & \text{for} \begin{cases} 0 < z < Z \\ -\frac{W}{2} < y < \frac{W}{2} \end{cases} \\ 0 & \text{otherwise} \end{cases} \tag{50}$$

where W and Z are the width and depth, respectively, of the contaminated region. The solution is

$$\frac{C}{C_0} = \frac{1}{4}\left\{erf\left[\frac{y + \frac{W}{2}}{2\sqrt{\alpha_H x}}\right] - erf\left[\frac{y - \frac{W}{2}}{2\sqrt{\alpha_H x}}\right]\right\} \cdot \left\{erf\left[\frac{z + Z}{2\sqrt{\alpha_V x}}\right] - erf\left[\frac{z - Z}{2\sqrt{\alpha_V x}}\right]\right\} \tag{51}$$

where α_H and α_V are transverse dispersivities in the horizontal and vertical directions, respectively. The database for transverse dispersivities is more sparse than for longitudinal dispersivities and the uncertainty associated with specifying them is large. In the absence of site-specific data, the following are recommended (24):

$$\alpha_H \approx 0.3\alpha_L \tag{52}$$

$$\alpha_V \approx 0.033\alpha_L \tag{53}$$

An approximate three-dimensional model can be formed by multiplying equation 49 by equation 51. The effect of first-order degradation/decay on concentration can be approximated by estimating the travel time of the contaminant to the point x as xR/u and then using an exponential multiplicative factor:

$$C' = Ce^{-k\frac{xR}{u}} \tag{54}$$

where k is the first-order rate constant, C is the concentration from equation 47, equation 49, or some other conservative model, and C' is the concentration corrected for degradation, decay, etc.

Vadose Zone Transport. The modeling of water flow and contaminant transport in the vadose zone is more complex and difficult than in the saturated zone because of the presence of both air and water in the pore spaces. Although a rigorous treatment is beyond the scope of this article, estimates of contaminant travel time can be made using a simple compartmental model. The average percolation rate, P, is obtained from a water balance on the top soil layer or the root zone (the average depth of vegetation roots) over some period of time, typically a year. P has units of $m^3/(m^2 \cdot yr)$, or simply m/yr. A water balance is essentially an accounting technique, which involves estimating water inputs and losses to the topsoil layer. Water can enter the soil through precipitation and can be lost from overland runoff to surface water bodies, evapotranspiration to the atmosphere, or percolation into deeper soil. Changes in the moisture content within the soil layer can also affect the amount of water available for percolation down to the groundwater. The process is shown in Figure 28.

The average linear velocity through the vadose zone, ν, is

$$\nu = \frac{P}{\theta} \tag{55}$$

The contaminant velocity thus becomes

$$\nu_c = \frac{P}{R_v \theta} \tag{56}$$

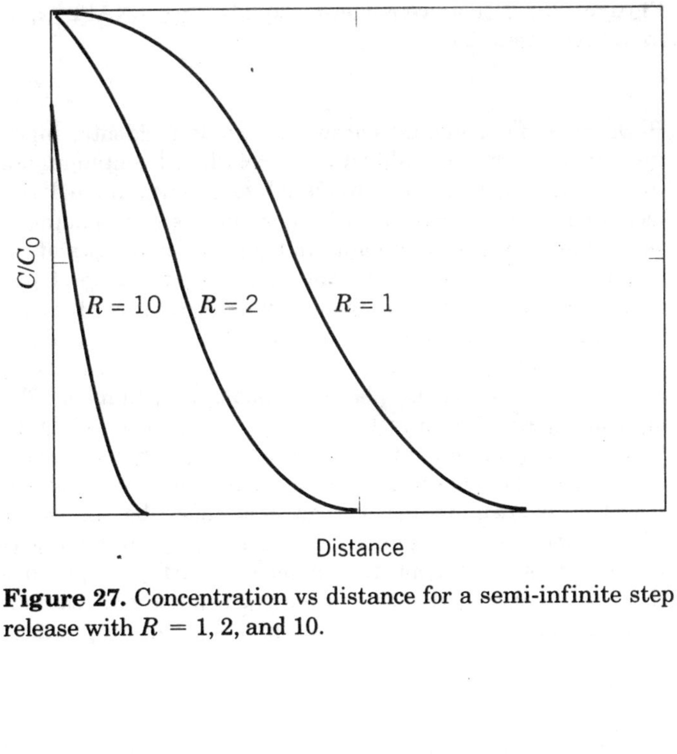

Figure 27. Concentration vs distance for a semi-infinite step release with $R = 1, 2,$ and 10.

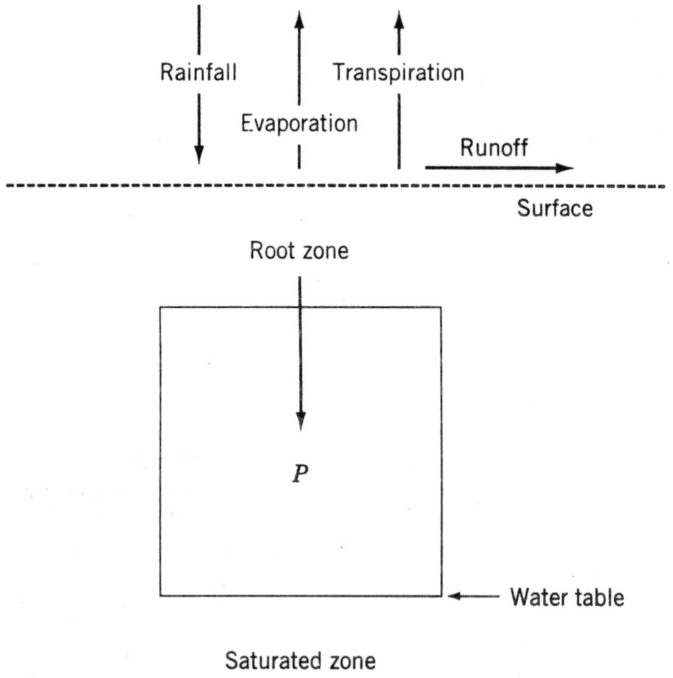

Figure 28. Water balance for the top soil layer.

where ν_c is contaminant velocity in the vadose zone and R_v is retardation factor for the vadose zone soil.

Nonaqueous-Phase Liquids. Nonaqueous-phase liquids pose a particular problem for the groundwater modeler. The preceding discussion has focused on contaminants in an aqueous phase. However, many contaminants are relatively immiscible in water. Such contaminants are known as nonaqueous phase liquids (NAPLs). The density of these compounds can either be greater or less than that of water. NAPLs which are lighter than water are known as light nonaqueous phase liquids (LNAPLs) and those which are denser than water are dense nonaqueous phase liquids (DNAPLs). The effects of NAPL spills are illustrated in Figure 29.

Since some of the liquid dissolves, the modeling of the dissolved phase may be handled by the techniques discussed above. LNAPL spills result in a pool of liquid which floats on top of the saturated zone, and is the source of a constant input of contaminant into the groundwater. DNAPL spills sink through the saturated zone, pooling at the lower confining layer. The path of the DNAPL through the saturated zone gives rise to a vertical line source extending over the depth of the saturated zone, with a pool at the lower layer dissolving into the groundwater. However, since the nonaqueous phase constitutes a continuing and variable source of dissolved phase contaminants, the application of simple approaches is problematic. Quantitative modeling of the movement of the nonaqueous phase liquid itself is quite complex and beyond the scope of this article. However, the theory is quite well developed, as the problem has been of interest to petroleum geologists for many years. Introductory groundwater modeling texts can provide greater detail on the modeling of NAPL transport (19).

Surface Water Transport. Usually, the principal surface water pathway of interest is ingestion of drinking water.

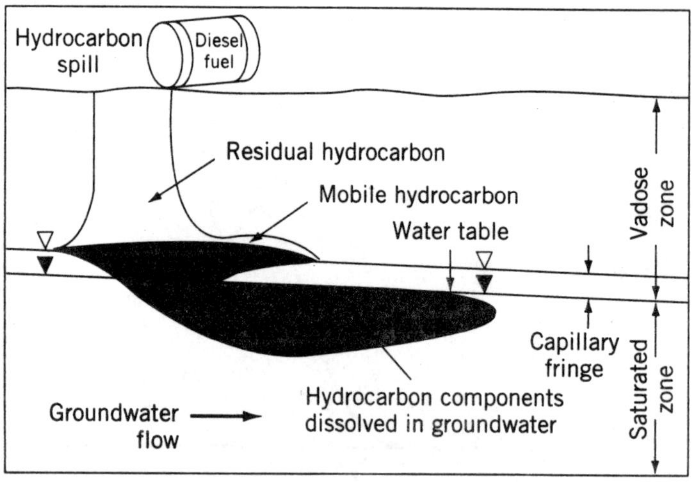

(a)

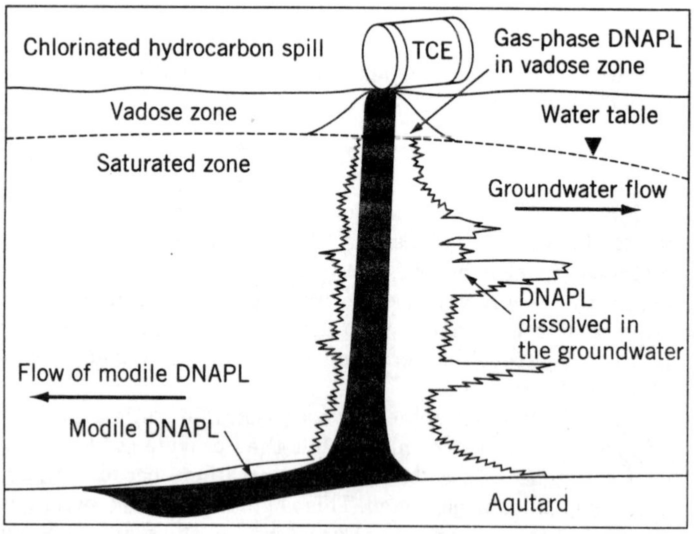

(b)

Figure 29. Nonaqueous-phase liquids: (**a**) LNAPL spill; (**b**) DNAPL spill (19).

However, other potential pathways include incidental ingestion while swimming, inhalation of volatilized contaminants, consumption of food crops irrigated with contaminated surface water, and consumption of fish or shellfish. Depending on the pathway, the contaminant, and the characteristics of the water body, suspended sediment may impact transport processes. Consequently, it is important to distinguish between contaminants in the aqueous and solid phases. Aqueous-phase contaminants are in dissolved form and are readily available for assimilation by plants, animals, and humans. The aqueous-phase concentration is bounded by the solubility. Solid-phase contaminants are sorbed to or otherwise associated with the solid phase and are generally less available; however, particulate matter can sometimes be an important vector for contaminant transport. To account for both phases, it is convenient to subdivide a surface water into three physical compartments: water, suspended sediment, and bottom sediment. It is sometimes necessary to couple these

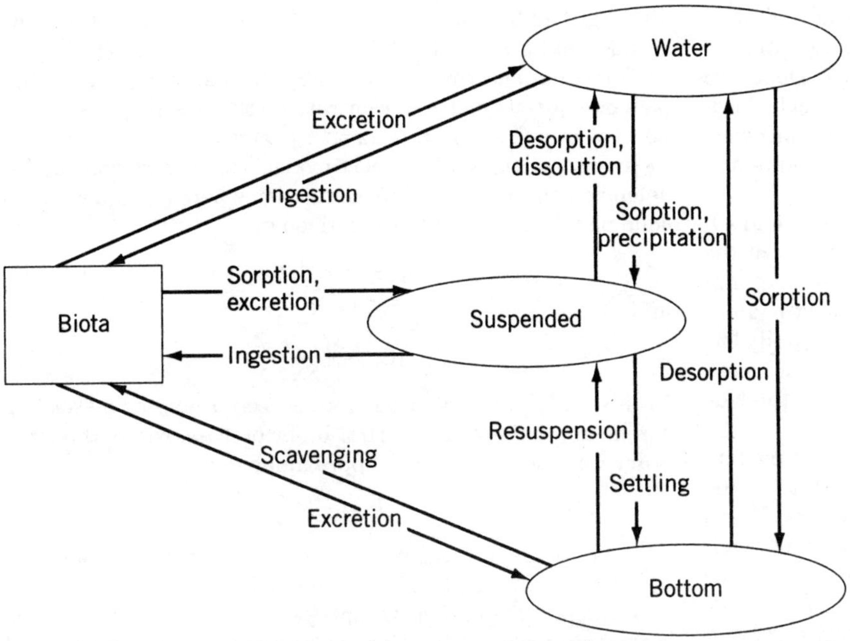

Figure 30. Illustration of surface water compartments, potential coupling with biotic food-chain compartments, and possible mechanisms for contaminant transport among the compartments.

compartments with biotic compartments in food-chain pathways. The compartments and processes that may be important in contaminant transport among them are illustrated in Figure 30.

Figure 30 is a simplified conceptualization of the processes that may play a role in surface water transport because it neglects intracompartmental processes. For example, aqueous-phase reactions such as hydrolysis, oxidation–reduction, or complexation could yield multiple chemical species of an aqueous-phase contaminant in the water compartment; and the suspended sediment compartment encompasses a spectrum of particle origins, sizes, and morphologies ranging from colloidal precipitates to sand particles. Fortunately, it is rarely necessary for intracompartmental processes to be included in surface water transport models, and they are not considered here.

Depending on the contaminant and the characteristics of the water body, it may be possible to neglect one or two of the compartments and many, if not all, of the transfer processes. For a nonreacting, highly soluble contaminant in a fast-flowing river or stream, ingestion of drinking water is likely to be the most important pathway, and attention can be restricted to the aqueous phase of the contaminant and the water compartment. For a contaminant that is strongly sorbed to particulate matter in a lake, ingestion of bottom-feeding fish and benthos may be the most important exposure route, and all three compartments and multiple transfer processes might need to be considered. Surface water transport is generally a more difficult problem than atmospheric transport because the nonconservative processes often assume greater importance and advection/dispersion processes are dependent on the geometry of the receiving body. Thus, it is convenient to divide surface water bodies into the following categories: rivers and streams, lakes, reservoirs on rivers, estuaries, and oceans. The general characteristics and relative importance of dissolved phase transport and solid (sediment) phase transport can differ greatly for these different systems.

Rivers and Streams. Advection and dispersion can usually be approximated by open channel flow, and the residence time of the aqueous phase is on the order of days or weeks. For highly soluble contaminants, the importance of the solid phase depends on the suspended solids concentration and the affinity of the contaminant for suspended particles. The residence time of the solid phase varies because the majority of solid-phase contaminants are associated with small particles. During periods of low flow, these particles settle to the bottom. Depending on the topography of the stream bed, sediment may be immobilized in traps or subject to resuspension during periods of high flow. If the sediment is resuspended, some fraction of the contamination may dissolve or desorb back into the water compartment.

Lakes. Small lakes can often be treated as homogeneous compartments. Large lakes (such as the Great Lakes) have rapid initial dispersion in large-scale turbulence and wind-driven currents. Ultimately, dispersion is limited by the dimensions of the lake. Since flow rates through lakes are small compared to lake volumes, residence times are long, typically years to tens of years. Virtually all sediment is trapped and, except for contaminants that enter the food chain through bottom-feeding fish or become exposed when lake levels drop, bottom sediment becomes an effective sink for sorbed contaminants.

Reservoirs on Rivers. A river/reservoir system, formed by one or more dams, is a hybrid between a river and a lake. Thus, the description above for rivers applies to riverine sections and that for lakes applies to reservoir sections. Reservoirs can sometimes be treated as a series of homogeneous compartments. The dividing line between compartments can be arbitrary or can correspond to a natural geographic feature such as a junction with a major tributary or a human-made feature such as a causeway. The residence time for aqueous-phase contaminants is longer than in rivers, but removal by settling of the solid phase may be significant. Solid-phase contaminants are more likely to be resuspended than in a lake.

Estuaries. Advection and dispersion are dominated and enhanced by tidal flows. Aqueous-phase concentration at a given location can exhibit sinusoidal behavior corresponding to tides. Sediment has net downstream movement in upper portions of an estuary, but lower portions serve as sediment traps with virtually all of the sediment deposited. Sorption in brackish water is usually less than in fresh water. Dissolved phase contaminants are unavailable except through food-chain pathways because there is no drinking water pathway. Generally, availability of contaminants is high for plants and animals that are part of a food-chain pathway

Oceans. Coastal turbulence causes rapid dispersion, and advection can be affected by winds and ocean currents. On the Continental Shelf, residence times for aqueous-phase contaminants are on the order of hundreds of days. Sediment is a minor concern because sediment loadings offshore are usually low and sorption is generally low in salt water. Contaminants are only available through food-chain pathways because of the lack of a drinking water pathway.

Sorption. Sorption, previously considered in the context of groundwater transport, can play a significant role in the transport and fate of contaminants in surface waters. It can reduce aqueous-phase concentrations, thereby reducing availability through the ingestion pathway but at the same time increasing availability through some food-chain pathways. The effect of sorption on aqueous-phase concentration is often small, yet it is still significant from the perspective of food-chain concentrations.

When the time required to reach equilibrium (often it is on the order of hours or less) is short compared to the travel time of the contaminant from release to the point of interest, the equilibrium distribution coefficient discussed previously is useful in quantifying sorption in surface water transport models. The fraction of contaminant associated with the solid phase, or the fraction sorbed, f_s, is calculated by

$$f_s = \frac{K_D SS}{1 + K_D SS} \tag{57}$$

where SS is suspended solids (sediment) concentration (kg/L). Figure 31 is a plot of fraction sorbed as a function of the $K_D SS$ product. This plot can be used to determine if sorption affects the aqueous-phase concentration.

For $K_D SS > 0.1$, the fraction sorbed is less than 10% and represents a reduction of the aqueous-phase concentration of less than 10%. In clear water, suspended solids concentrations are typically less than 10^{-5} kg/L (10 mg/L) and sorption is therefore significant only for $K_D > 10^4$ L/kg. Although suspended sediment concentrations can reach 10^{-2} kg/L (10,000 mg/L) in streams subject to heavy erosion, it rarely exceeds 10^{-3} kg/L (1,000 mg/L), in which case sorption would be significant for $K_D > 10^2$ L/kg.

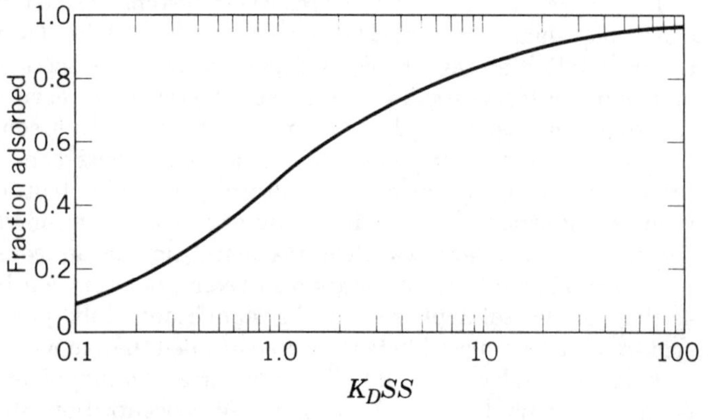

Figure 31. Fraction sorbed as a function of the product of distribution coefficient and suspended sediment concentration.

For situations where either sorption has a significant impact on aqueous-phase concentrations or the solid phase is important in a food-chain transport pathway, it is necessary to include sorption in a transport model. This can be done by developing a model for total concentration where the aqueous and solid phase concentrations are given by

$$C_l = (1 - f_s)C_T \tag{58}$$

and

$$C_s = \frac{f_s C_T}{SS} \tag{59}$$

In this model, removal of solid-phase contaminants by settling can be approximated as a first-order process, where the first-order rate constant may be approximated as

$$k = \frac{v_s f_s}{Z} \tag{60}$$

where v_s is the average settling velocity of sediment particles and Z is average depth of the water body.

Advection/Dispersion. Quantifying advection and dispersion for all of the five surface water systems is beyond the scope of this article. The discussion here is confined to two simple approximations: a one-dimensional, open-channel approximation that can be applied to rivers and streams, and a homogeneous compartment approximation that can be applied to well-mixed lakes or reservoirs.

One-Dimensional, Open-Channel Approximation. The one-dimensional, open-channel approximation is appropriate for narrow rivers or at downstream locations in wide rivers, since the plume will typically spread quickly across the width and depth of the channel. Dispersion in an open channel is due primarily to turbulence and nonuniform velocity distributions. This is illustrated in Figure 32. Here, spreading of an instantaneous release across a cross section of the channel at $t = 0$ is illustrated for uniform and nonuniform velocity distributions. If the velocity profile across the channel were uniform, as in Figure 32**a**, spreading of the contaminant at $t = t_1$ and $t = t_2$ would be due solely to turbulence. However, velocity profiles in channels are not uniform. In a rectangular channel with laminar flow, the velocity has a parabolic distribution horizontally across the channel and a logarithmic distribution vertically. The effect of a nonuniform profile in the vertical direction is illustrated in Figure 32**b**, where the maximum velocity is set equal to the velocity in Figure 32**a.** The surface spreading of the contaminant is the same as in Figure 32**a;** however, below the surface the water velocity decreases with depth and is zero at the bottom. Thus, contaminant molecules near the bottom of the channel travel only a short distance. Superposed onto this is the effect of a nonuniform velocity profile across the channel, illustrated in Figure 32**c**. Turbulent mixing mitigates these effects by causing slow moving parcels of water near the boundaries to mix with the faster moving ones in the middle of the channel. The net effect has been examined (25), and the following empirical expression for the dispersion coefficient in the direction of flow has been derived:

$$E_x(\text{m}^2/\text{s}) = 0.011 \frac{u^2 W^2}{u^* H} \tag{61}$$

where H is channel height (m), W is channel width (m), s is channel slope (m/m) and u^* is shear velocity (m/s),

$$u^* = \sqrt{gHs} \tag{62}$$

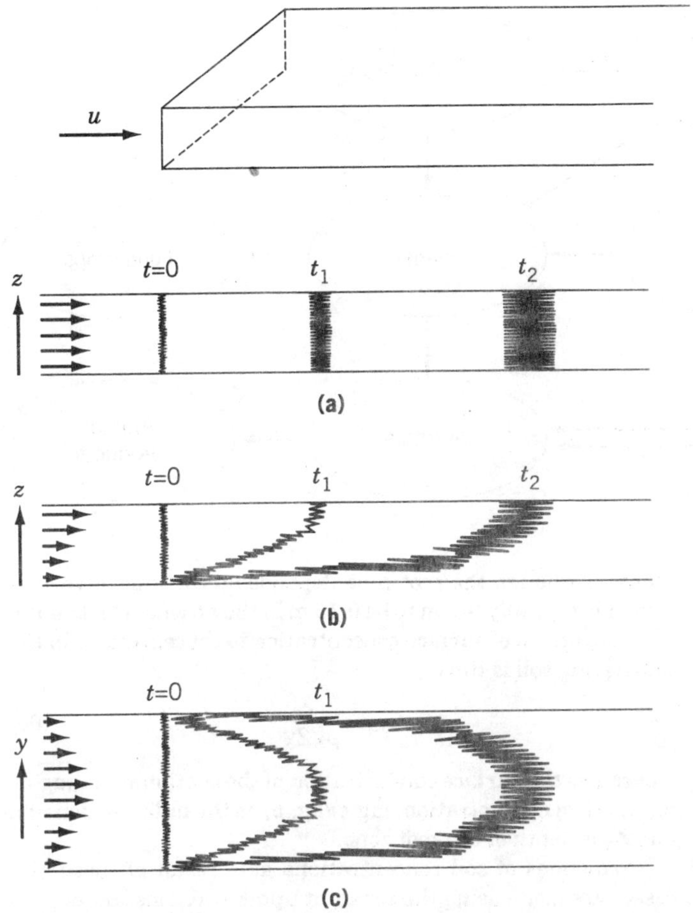

Figure 32. Dispersion due to turbulence and nonuniform velocity distributions in an open channel: (**a**) uniform, (**b**) logarithmic (vertical); and (**c**) parabolic (horizontal).

For an instantaneous release uniformly distributed across a 1-D river the solution is

$$C(x,t) = \frac{S_0}{WH\sqrt{4\pi E_x t}} \exp\left[-\frac{(x-ut)^2}{4E_x t}\right] \quad (63)$$

Equation 63 is simply equation 15 with $k = 0$ and D replaced by E_x. Thus, the concentration vs time and concentration vs distance plots are the same as those presented previously in Figure 13 and Figure 14. At a given location, the maximum concentration occurs at

$$t^* = \frac{-E_x + \sqrt{E_x^2 + u^2x^2}}{u^2} \quad (64)$$

Homogeneous Compartment Approximation. The homogeneous compartment approximation is widely used for estimating contaminant concentrations in lakes. A typical problem would be a semi-infinite step release of a contaminant to a well-mixed lake at $t = 0$. The release could be associated with flow into the lake, as illustrated in Figure 11, or not. The solution (which can be obtained from a differential equation similar to equation 3) for contaminant concentration as a function of time following the initiation of a nonflow associated release, $\dot{S}_0$, is

$$C_T = \frac{\dot{S}_0}{(Q + V\sum_i k_i)}\left\{1 - \exp\left[-\left(\frac{Q}{V} + \sum_i k_i\right)t\right]\right\} \quad (65)$$

where the k_i are first-order rate constants for removal by degradation, decay, settling of solid phase contaminant, etc. The aqueous- and solid-phase concentrations are obtained from equations 58 and 59. Stratified lakes, river/lake systems, and rivers can be modeled as a set of discrete, homogeneous compartments in series.

Food-Chain Transport

In estimating human exposures to contaminants, consumption of contaminated food products such as produce, meat, milk, fish, and shellfish may be significant pathways. These pathways may be relatively simple, such as ingestion of seafood contaminated as a result of a contaminated surface water body, or may involve multistage exposure pathways, such as air/water → soil → plants → animals → milk. Presented in Figure 33 are the principal pathways through which contaminants can be incorporated into food.

Food-chain transport involves the determination of the concentration of contaminants at various points in the food chain. In human health risk assessments, the final point is typically determined by the type of food raised and consumed in the contaminated region. Although physical processes usually control the initial release and dispersion of a contaminant, biological processes become important in the food chain. In addition, ecological risk assessments may have as an endpoint the level of contamination of a species of interest, such as protected fish or wildlife.

A significant difference between a food-chain compartment and an atmospheric, surface water, or groundwater compartment is the potential for increasing the contaminant concentration. This can occur as a result of both biological and physical processes. If this occurs in living organisms, it is termed bioaccumulation. Although all contaminants do not bioaccumulate, those that do pose a concern due to the steady increase in contaminant levels due to long-term, low level releases. In addition, due to the characteristic of ecosystems that higher organisms prey on lower organisms, a process known as biomagnification may also occur, in which the concentration of contaminant in animals increases as the contaminant moves up the food web. Biomagnification and dilution are illustrated conceptually in Figure 34.

A classic example of biomagnification is DDT transport through aquatic ecosystems, where concentrations at the top of the food chain (in ducks, ospreys, hawks, etc) were found to be as much as four orders of magnitude higher than at the bottom (in lake sediment). Contaminants which are lipid soluble, for example, may build up in fatty tissues of organisms, reaching concentrations higher than those in the food which the organism ingests. Contaminants with high lipid solubilities, as measured by the octanol–water partition coefficient (K_{ow}), tend to bioaccumulate. The octanol–water partitioning coefficient is a physical constant of a compound which indicates its lipid solubility. It is similar to the K_D value but is the equilibrium partitioning ratio between a contaminant dissolved in water and in octanol. K_{ow} values greater than 4.3 to 5 (26) indicate a significant bioaccumulation potential. This occurs because the contaminants successively accumulate from sediment to benthic organisms, from benthic organisms to fish, and from fish to birds.

Contaminant transport through food-chain pathways can involve a variety of complex physical, chemical, and biological processes. Even though the basic mechanisms for some of these processes are known, the systems are so complex and

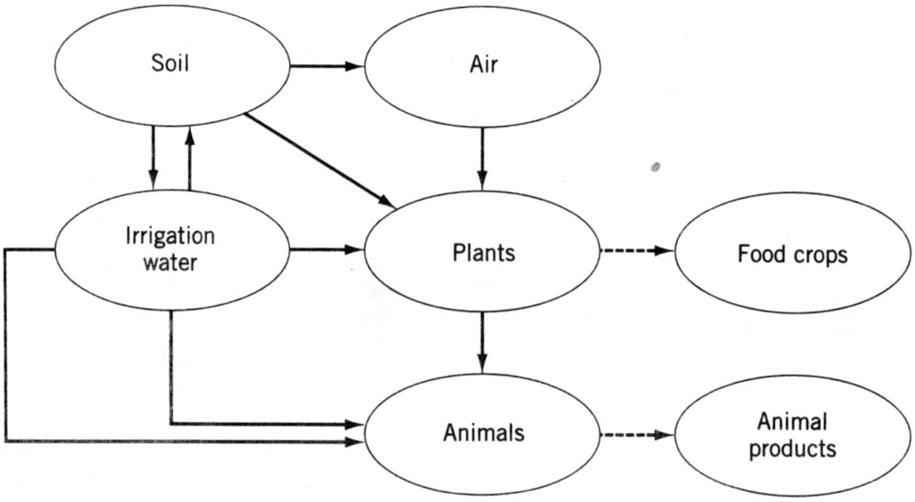

Figure 33. Diagram of the principal food-chain transport pathways.

Figure 34. Contaminant dilution ($C_{out}/C_{in} < 1$) or biomagnification ($C_{out}/C_{in} > 1$) in a food-chain compartment.

depend on so many variables that integrated theoretical models either do not exist or require data that are not readily available. Consequently, food-chain transport models used in risk assessments are largely empirical, and they rely on the base of contaminant specific laboratory and field data that exist. The base of existing data is substantial for radiological contaminants, but it is limited for chemical contaminants. The discussion in this section is therefore based on the approach used in radiological assessment, but it is equally applicable to chemical contamination, provided that data are available.

For modeling purposes, these data are almost universally expressed in terms of nonadvective homogeneous compartmental models. Mathematically, these are either instantaneous partitioning models or constant uptake models. Presented in this section are the models that are commonly used to predict contaminant concentrations in the soil, vegetation, and animal compartments in Figure 33.

Concentration in Soil. Because a major pathway for food contamination involves uptake from contaminated soil by plants, determination of soil concentrations is often a first step in a food-chain transport assessment. There are three principal deposition pathways through which surface soils can become contaminated: dry or wet deposition from the atmosphere, direct deposition from a leak or spill, and irrigation with contaminated water. Removal can occur by degradation, decay, resuspension to the atmosphere, runoff from the surface, and leaching to deeper soil.

For modeling purposes, contaminants are considered first to be deposited on the soil surface and then to be uniformly mixed to the depth of the root zone. The root zone is the depth to which the roots of the vegetation extend and hence is the volume of contaminated soil available for transfer of contamination to the plants. Although the root zone depends on the type of vegetation, it is typically taken to be 0.15 m in the absence of site data. The conversion of surface concentration to concentration in the underlying soil is thus

$$C_s = \frac{\chi}{\rho_B Z_R} \tag{66}$$

where χ is the surface concentration of the contaminant, mg(c)/m; C_s is soil concentration, mg(c)/kg, ρ_B is the bulk soil density, and Z_R is depth of the root zone.

Predictions of soil concentrations as a result of these processes are made using the constant uptake (ie, constant deposition), first-order removal model. When applied to soil, equation 5 becomes

$$\frac{dC_s}{dt} = \frac{\dot{\chi}}{\rho_B Z_R} - k_s C_s \tag{67}$$

where $\dot{\chi}$ is deposition rate to the surface, mg(c)/(m^{2}s), and k_s is a first-order rate constant for removal from soil. Depending on the contaminant and the soil, the rate constant may actually be the sum of rate constants for degradation, decay, resuspension, etc.

The solution to equation 67 is

$$C_s(t_{s,e}) = \frac{\dot{\chi}}{k_s \rho_B Z_R}(1 - e^{-k_s t_{s,e}}) \tag{68}$$

where $t_{s,e}$ is soil exposure time. Equation 68 is typically used for long-term exposures such as those due to routine atmospheric releases from a facility. The limiting (ie, steady-state) concentration is

$$C_s = \frac{\dot{\chi}}{k_s \rho_B Z_R} \tag{69}$$

This is the concentration resulting from a continuous long-term release. For exposures resulting from short-term, accidental releases, removal processes can often be neglected and the soil concentration can be estimated by

$$C_s(t_{s,e}) = \frac{\dot{\chi}}{\rho_B Z_R} \cdot t_{s,e} \tag{70}$$

Applying equation 69 or equation 70 to a specific scenario involves determining the appropriate form for the deposition and removal rates. These are discussed below.

Table 8. Resuspension Factors[a]

Condition	Resuspension Factor Range, m^{-1}	Comments
Wind stress	2×10^{-11} to 8×10^{-9}	Bare soils, ^{91}Y aqueous chloride
	9×10^{-8} to 1×10^{-7}	^{210}Po as oxide
	9×10^{-8} to 5×10^{-7}	U_3O_8
	1×10^{-4} to 1×10^{-9}	Time-dependent model for plutonium in soil
	9×10^{-11} to 3×10^{-4}	Literature review
	1.8×10^{-13} to 6.1×10^{-10}	Nevada Test Site, plutonium aerosols
	$<2 \times 10^{-9}$	Nuclear weapons test debris, 13 yr after deposition
	$<5 \times 10^{-10}$	Nuclear weapons test debris, 22 yr after deposition
	3.6×10^{-9} to 4.9×10^{-8}	Chernobyl ^{137}Cs deposition: initial factor
Vegetation	2.9×10^{-8} to 6×10^{-7}	Fluorescent powder
	3×10^{-8} to 5×10^{-5}	U_3O_8
Mechanical disturbances	1.5×10^{-6} to 3×10^{-4}	Plutonium
	4.8×10^{-5} to 1.1×10^{-2}	ZnS, per disturbance
	1×10^{-5} to 2×10^{-4}	ZnS, per disturbance
	1×10^{-10} to 4×10^{-2}	Literature review

[a] Ref. 29.

Atmospheric Deposition. Dry and wet surface deposition rates are calculated from

$$\dot{\chi} = (1 - f_v)v_d C_a \tag{71}$$

where C_a is contaminant concentration in air, f_v is the fraction of the contaminant intercepted by vegetation (taken as 0.25 in the absence of site-specific data), and v_d is deposition velocity. The deposition velocity is an empirical quantity which depends on the contaminant and the type of surface. Values of deposition velocity, which range from 10^{-5} m/s for strontium on grass to 10^{-1} m/s for particulate matter, have been tabulated (27).

Irrigation Deposition. Deposition by irrigation can be calculated from

$$\dot{\chi} = \dot{I} C_w \tag{72}$$

where $\dot{I}$ is the irrigation rate (m/yr) and C_w is contaminant concentration in the irrigation water. Irrigation rates vary depending on the water needs of the crop being raised and can range from 0.23 m/yr to 1.4 m/yr. The default agricultural irrigation rate used by the NRC is 0.76 m/yr. Deposition due to a spill can be considered a special case of deposition by irrigation, in which the irrigation proceeds for a very short time (ie, the time over which the spill occurs).

Atmospheric Resuspension. Surface soils can be resuspended into the atmosphere through the action of wind and mechanical disturbance. Resuspension is quantified through the resuspension factor, κ (m^{-1}), which is defined by

$$\kappa = \frac{k_{s,r}}{v_d} \tag{73}$$

where $k_{s,r}$ is the resuspension rate constant. Physically, the resuspension factor is the fraction of contaminant resuspended per meter in the direction of the wind. It ranges from 10^{-5} to 10^{-10} m^{-1}, with typical values being on the order of 10^{-8} m^{-1}. Resuspension is affected by many factors, including the type of soil, the atmospheric conditions, and the level of mechanical disturbance (eg, plowing, vehicular traffic, gardening, etc). The range of possible values is indicated in Table 8.

Concentration in Vegetation. The next step in the process is the determination of the contaminant concentration in vegetation as a result of (*1*) uptake from contaminated soils through the roots and resuspension of contaminated soil onto plant surfaces and (*2*) direct deposition or absorption from the air through foliage or flowers. Removal processes include transpiration, wind removal, and wash-off by rain, as well as radioactive decay if modeling radionuclides. These processes are illustrated in Figure 35.

The concentration of a contaminant in vegetation is thus a balance between these uptake and removal processes. The total concentration is calculated as the linear combination of the contributions from deposition and uptake, or

$$C_v = C_v^d + C_v^u \tag{74}$$

Removal Processes. Similar to soil, the removal rate constant may be the sum of rate constants for more than one process. Contaminants may be removed by wind removal, washoff by precipitation, and a variety of other processes. Rate constants for these processes are not well known. In the absence of data, a value of 0.05/d is recommended (29) for k_v, which corresponds to a half-time of 14 days.

Direct Deposition. Direct deposition is modeled through the constant uptake, first-order removal model, which takes on the following form:

$$\frac{dC_v^d}{dt} = v_d C_a f_v \frac{T_v}{Y_D \frac{FW}{DW}} - k_v C_v^d \tag{75}$$

where $C^d{}_v$ is concentration in vegetation due to deposition, mg(c)/kg(v, wet); T_v is the translocation factor, Y_D is dry yield, kg_v(dry)/m^2; FW/DW is the fresh weight to dry weight ratio of the vegetation, kg(v,wet)/kg(v,dry); and k_v is a rate constant for removal from vegetation by either biophysical processes or radioactive decay. Physically, $v_d C_a$ is an areal deposition rate. This is then multiplied by the fraction of the deposited contaminant which is intercepted by the vegetation and by the translocation factor, yielding a total areal deposition rate into the vegetation. Conversion to a mass concentration in the vegetation is then achieved using the yield or mass per area of the crop. Translocation factor is the fraction of externally deposited contaminant that is transferred to the edible part of the plant. The value is taken to be 1 for produce such as leafy vegetables and 0.1 for most other crops. Dry yield is the dry mass of a crop produced per unit area of land. This dry yield is

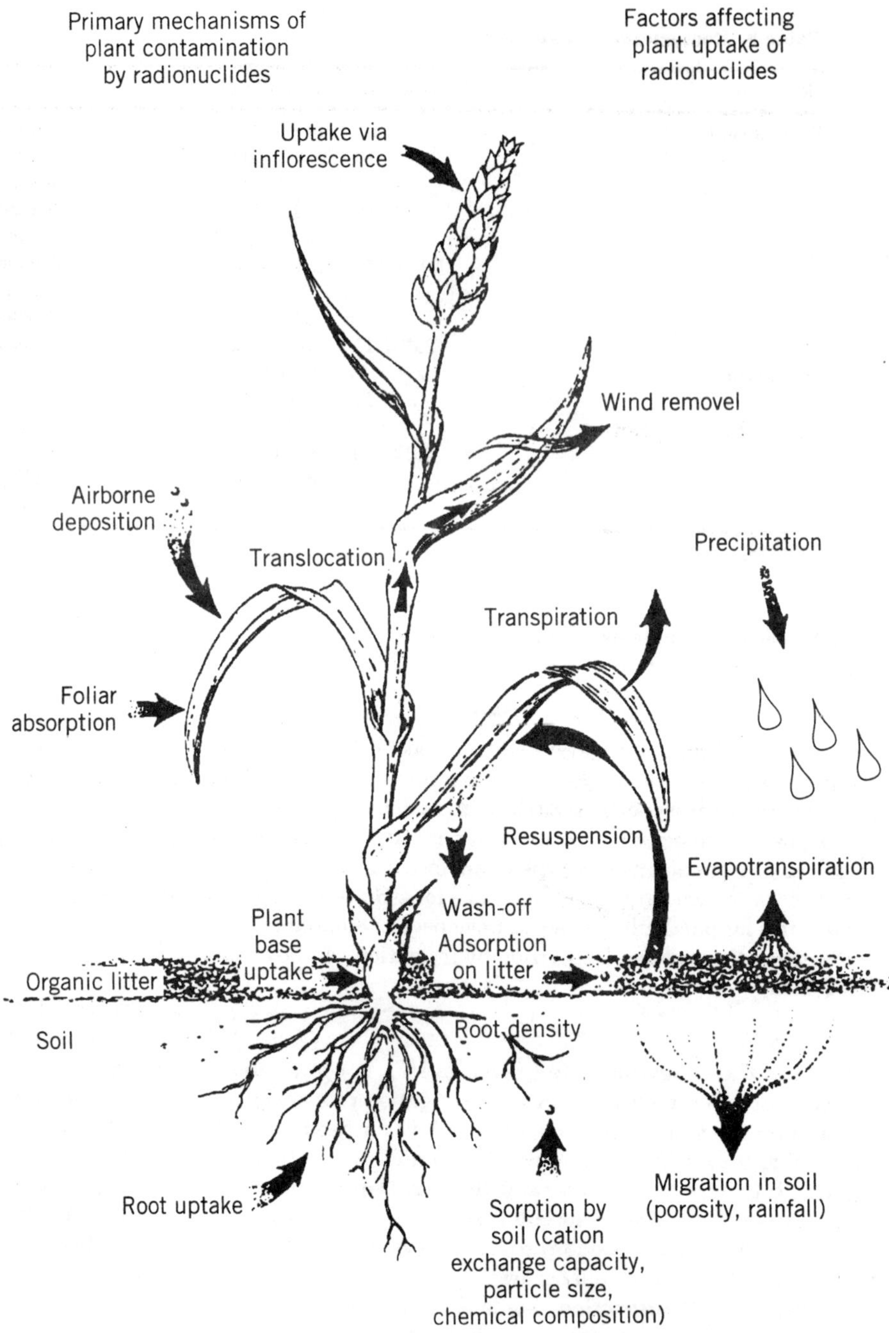

Figure 35. Principal pathways affecting contaminant concentration in food crops (15).

then corrected to a wet yield by a ratio of fresh (or wet) weight to dry weight, which can range from near 1 for grains to 20 for leafy produce such as lettuce. Alternatively, the wet yield may also be used use directly. Nominal values for food crop wet yields range from 4 kg(w)/m^2 for vegetables to 2 kg(w)/m^2 for fruits and leafy vegetables to 1 kg(w)/m^2 for grains (28).

The solution to equation 75 is

$$C_v^d(t_{v,e}) = \frac{f_v v_d C_a T_v}{Y_D \frac{FW}{DW} \sum_i k_{v,i}} (1 - e^{-k_v t_{v,e}}) \qquad (76)$$

where $t_{v,e}$ is the exposure time for vegetation. For annual crops, the maximum exposure time is the length of the growing season, or about 90 days.

For a short-term exposure, with removal mechanisms neglected, equation 76 reduces to

$$C_v^d(t_{v,e}) = \frac{f_v v_d C_a T_v}{Y_D \frac{FW}{DW}} t_{v,e} \qquad (77)$$

Uptake from Soil. Contaminant uptake from soil is a highly complex biophysical process and is affected by a number of factors. These include the physicochemical form of the contaminant, the plant species and the level of translocation within the plant, characteristics of the soil, levels of nutrients and trace elements already present in the soil, and the spatial distribution of the contaminant in the soil (15). Unfortunately, a general model for predicting uptake as a function of all of these variables is not currently available. Instead, the results of experimental studies have been carried out based on the instantaneous partitioning model. This model assumes that the level of the contaminant in the soil and in the plant instantaneously achieve equilibrium. The form of the model is thus:

$$C^u_v = B_v C_s \tag{78}$$

where $C^u{}_v$ is concentration in vegetation due to uptake from the soil, mg(c)/kg(v), and B_v is the bioaccumulation factor. For some contaminants and some plants, separate bioaccumulation factors are available for roots, leaves, and fruit. They may be expressed either on a wet or dry weight basis. Although a rigorous model would account for all of the factors given above, bioaccumulation factors are typically given only as a function of the contaminant and the plant.

Concentration in Animals. Animal products of concern for human exposure include fish, other aquatic organisms, and terrestrial grazing animals. Aquatic organisms such as fish may be exposed to contaminants in the water column while bottom-feeding fish and benthic organisms (animals that live in the sediments at the bottom of a surface water body) may be exposed to contaminants from the sediments and the water.

The instantaneous partitioning model is used to calculate contaminant concentrations in fish and other aquatic animals through exposure to aqueous-phase contaminants. Just as in equation 78, the form is:

$$C_f = B_f C_w \tag{79}$$

where C_f is the concentration in fish, mg(c)/kg(f); C_w is the concentration in the aqueous phase, mg/L; and B_f is bioacccumalation factor for fish, L/kg(f).

Grazing animals take up contaminants through contaminated feed or water. The constant uptake, first-order removal model is used for grazing animals such as cows or goats, and it is applied to either milk or meat. For milk, the differential equation takes on the following form:

$$\frac{dC_{\text{milk}}}{dt} = C_v \dot{U} \frac{f_{\text{milk}}}{V_{\text{milk}}} - kC_{\text{milk}} \tag{80}$$

where C_{milk} is concentration in milk, mg(c)/L; $\dot{U}$ is uptake rate of vegetation, kg(v)/d; f_{milk} is the fraction of contaminant transferred to milk; V_{milk} is milk volume, L; and k is a first-order removal rate constant. The solution is

$$C_{\text{milk}}(t) = C_v \dot{U} \left[\frac{f_{\text{milk}}}{kV_m} (1 - e^{-kt}) \right] \tag{81}$$

The term in the brackets in equation 81 is defined as the milk transfer factor, F_{milk} (d/L), which physically is the fraction of daily intake of contaminant that appears in the milk per unit milk volume. Thus, the calculation of contaminant concentration in milk simplifies to

$$C_{\text{milk}} = \dot{U} C_v F_{\text{milk}} \tag{82}$$

An analogous approach is used for the calculation of contaminant concentration in the meat of grazing animals, yielding

$$C_{\text{meat}} = \dot{U} C_v F_{\text{meat}} \tag{83}$$

where C_{meat} is contaminant concentration in meat, mg(c)/kg(meat), and F_{meat} is the meat transfer factor, d/kg(meat).

Determination of the uptake rate is important in determining the total contaminant uptake of the animal, since animals may not take up all of their food from a contaminated pasture. Uptake rates for different animals and animal products are shown in Table 9.

Since these are total intake rates, they must be reduced if animals are grazed or fed on uncontaminated crops for part of the time. A final consideration is the time required for food products to reach the consumers. This time, known as a holdup time, can range from 20 days for beef to 14 days for fruits and vegetables, to one day for leafy vegetables and dairy products. For crops exposed to releases of relatively short-lived radionuclides, holdup times may significantly reduce the amount of contaminant taken up by humans. In such cases, the appropriate decay correction should be applied.

Exposure Assessment

For a contaminant to pose a risk to human health, there must be some pathway whereby the individual takes the contaminant into the body. Exposure is the contact of humans with a contaminant, either through physical contact with a contaminated medium, inhalation of contaminated air, or ingestion of contaminated water or food. Exposure assessment is the process of quantitatively characterizing the route and duration of exposure.

Because of the differences in chemical toxicity due to the exposure duration (discussed in greater detail in subsequent sections), exposure duration is typically qualitatively characterized as either acute, subchronic, or chronic. There is no hard and fast rule as to the dividing lines between these types of exposures; however, acute exposures are either instantaneous or occur over a short time (on the order of a few days). Subchronic exposures are protracted, either continuous or intermittent, but the duration is not a significant fraction of an entire lifetime. The duration of a subchronic human exposure can range from several months to several years. Chronic exposures occur on a continuing, long-term basis. A convention used by the EPA to distinguish between subchronic and chronic exposures is to define a subchronic exposure as less than 10% of total lifetime and chronic as greater than 10%. Thus, an exposure duration of seven years (corresponding to an average life span of 70 years) or less can be considered subchronic, and more than seven years can be considered chronic.

Contaminant Dose. Dose is the quantitative measure of exposure, and dose rate, $\dot{D}$, is the dose received per unit time. Dose is a function of the concentration of the contaminant in environment media, the contact rate with the contaminated media, the body weight of the exposed individual, the duration of exposure, and the absorption of the contaminant into the body. In radiological assessments, contact rate is known as uptake rate. Dose and dose rate are related through

$$D = \int_{t=0}^{t_e} \dot{D}(t)dt \tag{84}$$

where t_e is the length of actual exposure. In quantifying the health effects of chemical exposures, the average dose rate is commonly used. It is related to the dose and instantaneous dose rate by

$$\bar{\dot{D}} = \frac{\int_{t=0}^{t_e} \dot{D}(t)dt}{t_e} = \frac{D}{t_e} \tag{85}$$

where t_e is the period over which the exposure is averaged. There are two different traditions in which doses calculated in human health risk assessments have developed. Toxicologists typically quantify chemical dose as mass of contaminant per unit body mass (ie, mg of contaminant per kilogram of body weight). For estimating the biological effects of radiation, the health physics community has defined a concept known as the

Table 9. Animal Feed and Water Intake Rates kg/(w)/d[a]

Intake Media	Beef Cattle	Dairy Cattle	Poultry (Meat)	Poultry (Eggs)
Fresh forage	27	36	0.13	0.13
Stored hay	14	29	n/a	n/a
Stored grain	3	2	0.09	0.09
Water	50	60	0.3	0.3

[a] Ref. 28.

radiological equivalent dose, H, which is a measure of the biological damage to the organism caused by exposure to ionizing radiation. The SI unit of radiological equivalent dose is the Sievert (Sv).

Chemical Dose. Doses due to exposure to chemicals can be expressed either on the basis of administered chemical dose (D_{ADM}), which is the amount of chemical to which the organism is exposed; absorbed chemical dose (D_{ABS}), which is the amount of chemical which actually passes through an exchange boundary such as the gut lining or the skin; or the effective chemical dose (D_{EFF}), which is the concentration of the biologically active form of the chemical in the target organ. However, due to wide variation in the distribution and metabolism of chemicals in the body, effective chemical dose is very difficult to quantify. Thus, chemical dose is usually expressed either as administered dose or absorbed dose.

Chemical administered dose is a function of the concentration in the contaminated medium, the rate of exposure to the contaminated media (henceforth termed the contact rate), the effective time of exposure, and the body weight. It is given by

$$D_{ADM} = \frac{C \times CR \times t_e}{BW} \tag{86}$$

where C is the concentration in the contaminated media, CR is the contact rate, t_e is the effective time of exposure, and BW is the body weight.

Chemical absorbed dose is determined from administered dose by

$$D_{ABS} = D_{ADM} \times f_e \tag{87}$$

where f_e is the absorption efficiency. Absorption efficiency is the fraction of the chemical which is absorbed into the body. It can be highly variable, being a function of the chemical and physical form of the contaminant, the pathway through which the chemical is administered (eg, ingestion, inhalation, dermal, etc), age, and diet.

Determination of the chemical effective dose as a function of the chemical absorbed dose is complex. It involves modeling the distribution and transformation of the contaminant in the body. This is performed with the use of pharmacokinetic models. However, for a majority of contaminants, data are not available to properly construct a pharmacokinetic model. Thus, most environmental risk assessments use the administered or absorbed dose rather than the effective dose.

In studies of the chronic or subchronic toxicity of a chemical, dose is typically expressed as the average daily dose over the period of the test. In order to express human doses due to environmental exposures in a comparable manner, the dose (either administered or absorbed) is averaged over a period of time corresponding to the conditions of the test, known as the averaging time, t_{avg}. For example, most carcinogenicity studies are performed over the life-span of the animal. Thus, the equivalent human dose is averaged over the average human life-span, often taken as 70 years. This is expressed in fashion similar to equation 75:

$$\bar{\dot{D}} = \frac{\int_{t=0}^{t_e} \dot{D}(t)dt}{t_{avg}} = \frac{D}{t_{avg}} \tag{88}$$

Use of $t_{avg} = 70$ years results in a lifetime average daily dose. For noncarcinogens, dose is averaged over the total period of exposure, so that $t_{avg} = t_e$

Radiological Dose. Biological damage due to radiation exposure depends primarily on the amount of energy deposited per unit mass of tissue and the type of radiation. Since some types of radiation can travel large distances in air and can pass through a significant thickness of tissue, radiation dose can arise from either internal or external sources of radiological contaminants (radionuclides). The basic dosimetric quantities are radiation absorbed dose, D_R; equivalent dose, H; and effective dose, E (30). Radiation absorbed dose is the amount of energy absorbed in tissue from the radiation per unit mass. It has units of J/kg, which is given the special name Gray (Gy). The combined effect of absorbed energy and type of radiation on biological damage is taken into account by the equivalent dose through

$$H = \omega_R D_R \tag{89}$$

where ω_R is the radiation weighting factor. The radiation weighting factor is 1 for gamma-rays and electrons, 5 for protons, 20 for alpha-particles, and ranges from 5 to 20 for neutrons depending on energy. Equivalent dose, which has units of Sieverts (Sv), is used for the quantification of the class of effects defined in a subsequent section as deterministic. Effective dose, which also has units of Sieverts, is used for the quantification of the class of effects termed stochastic, which includes cancer and inherited effects. This is because the relative sensitivity of body tissue to the induction of stochastic effects varies from organ to organ. Effective dose is essentially a risk weighted dose equivalent, and it is given by

$$E = \omega_T H \tag{90}$$

where ω_T is the tissue weighting factor.

The equivalent dose and the effective dose are measures that are independent of the specific radionuclide. This permits the consequence of exposure to multiple radionuclides to be combined by summation of the individual doses.

The determination of radiation dose due to the uptake of radioactivity is quite complex because it involves the calculation of both radionuclide transport in the body and radiation transport/energy absorption. The health physics community has simplified this task through the development and publication of radiation effective dose factors for virtually all radionuclides of technological interest. The dose factor is the effective dose (Sv) resulting from the instantaneous uptake of one Becquerel of radioactivity. Separate values are given for uptake by inhalation and ingestion. Effective dose due to uptake of radioactivity can thus be calculated by

$$E = C \times U \times t_e \times DF \tag{91}$$

where DF is the radiation effective dose factor (Sv/Bq) for the individual radionuclide.

Quantification of Exposure Routes. There are three major ways in which a contaminant can enter the body and exert its toxic effect: inhalation, ingestion, and absorption through the skin. For radioactive contaminants an additional exposure pathway, that of direct external exposure to ionizing radiation, may also need to be considered. The following discussion focuses primarily on quantifying the contact rate due to different exposure routes.

Inhalation. Airborne contaminants subject to inhalation are present either in particulate or gaseous form. Contact rate for inhalation of contaminated air is equal to the inhalation rate, which is also known as the pulmonary ventilation rate. Although inhalation rate is a function of age, weight, sex, activity level, and physical condition, it is rare that all of these factors are taken into consideration. Values range from 0.4 m^3/h for children at rest to 4.8 m^3/h for adult males engaged in heavy activity. Default values given in EPA's Superfund guidance documents are 1.25 m^3/h for a suggested upper bound and 0.8 m^3/h as an average for adults.

Ingestion. Exposure via the ingestion pathway is conceptually similar to that of the inhalation pathway. Contact rate for ingestion is the ingestion rate and is the amount of contaminated media ingested over a given time. Contaminated media which may be ingested are water, food, and soil.

For contaminated water, uptake occurs due to ingestion of drinking water and incidental ingestion of water while swimming. For adults, values for ingestion of drinking water range from 0.3 to 2.8 L/d with averages ranging from 1.2 to 1.6 L/d (31). Children consume less water than adults and for them the range is 0.5 to 0.8 L/d (32). The actual amounts vary depending on the individual level of physical activity and the temperature of the individual's environment. Sources include water direct from the tap and beverages made with tap water such as concentrated juices and coffee or tea.

For food, ingestion rate is a function of the type of food and the age and sex of the individual. Ingestion of different foodstuffs are typically determined from dietary surveys. Major food groups used in environmental risk assessments which may be subject to contamination are meat, dairy products, eggs, fish and shellfish, and vegetables. In determining dose, the ingestion rate of concern is that of contaminated food. Since much of the total diet of an individual may not be locally produced, the fraction of the total diet which is affected by contamination must be specified. This is usually food produced (or caught) by the receptor for personal consumption, although situations may exist (such as Chernobyl) in which contaminated food is distributed commercially. Ingestion rate for foods is quantified by

$$CR_i = IR_i \times f_i \tag{92}$$

where IR_i is the daily consumption of the type of food and f_i is the fraction of the particular food which is contaminated. The subscript i indicates the type of food under investigation, since different types of foods are ingested at different rates and are locally produced in varying ratios. Typically, the percentage of locally produced food is highest for individuals living in rural areas and lowest for those residing in cities, with suburban residents lying between the two. In addition, individuals engaging in hunting and fishing are likely to consume higher fractions of locally produced game, fish, and shellfish and may have higher than average consumption rates of game and seafood.

The final ingestion pathway of concern involves contaminated soils. Under normal circumstances, this is due to either ingestion of airborne particulate matter that has been trapped in the mouth, soil from homegrown food products, or dust from handling food or cigarettes with unwashed hands. Normally, young (less than seven years old) children can be expected to ingest, on average, between 100 and to 200 mg of soil per day, although ingestion rates of up to 1000 mg/day are possible for normal ingestion by children (33). However, a phenomenon known as pica may cause soil ingestion rates to be abnormally high for some subpopulations. Pica can be of significant concern if it leads to an abnormal level of direct ingestion of contaminated soil. Unfortunately, it is very difficult to accurately quantify the amount of soil that pica children ingest, as it can vary widely.

Dermal Contact. Dermal exposure involves the absorption of a contaminant through the skin. Although the skin is an effective barrier to the entry of contaminants, some are capable of penetrating this barrier. Contact rates used for estimating dermal dose can be more difficult to quantify than those for the ingestion or inhalation pathway. Contact rate can be estimated using either of two approaches. The first, represented in equation 93, is an empirical approach conceptually similar to that used for ingestion and inhalation, with an absorption fraction giving the percent of the contaminant absorbed through the skin:

$$CR = \frac{MCR}{\rho} \cdot f \tag{93}$$

where MCR is the mass contact rate of an individual with the contaminant, expressed as g/day, ρ is the density of the contaminated media, and f is the fraction of the contaminant absorbed.

The second is represented by equation 94 and is based on simple Fickian diffusion through the skin. It thus requires an estimate of the permeability (PC) of the skin for different types of contaminants and an estimate of the total area of the body exposed to the contaminated media:

$$CR = PC \times SA \tag{94}$$

For dermal contact, the amount of body surface area exposed to the contaminated medium has a significant effect on the total rate at which the contaminant is absorbed. Body surface area is a function of age, sex, and exposure route. For example, individuals swimming or bathing in contaminated water may expose anywhere from 80–100% of their body to contaminated water, whereas individuals handling contaminated media may only expose their hands, or about 5–6% of their total surface area.

CONSEQUENCE ASSESSMENT

Basic Toxicology

Here, the impact of contaminant exposures on human health is considered. The ultimate goal is to learn how to make quantitative estimates of health effects. In this section the mechanisms of cellular toxicity and the clinical effects that can result are addressed.

Biological Basis of the Deterministic/Stochastic Distinction. At a fundamental level, most health effects of chemicals or radiation are due to alteration of the structure and/or function of molecules in the body. This can lead to the disruption of normal cellular and physiological function. If cellular damage is not repaired, it may (*1*) cause cells to malfunction, die, or not reproduce or (*2*) modify the genetic material (DNA) in cells without affecting their ability to reproduce. These two types of cellular damage ultimately yield the fundamentally different types of effects introduced earlier. Deterministic effects (those for which the *severity* is a function of dose) are the result of the first type of damage. As the number of affected cells increases (due to increases in the contaminant dose), the severity of the effect increases. An example is illustrated in Table 10, which shows that the clinical symptoms of breathing SO_2 become more severe as the SO_2 concentration increases.

The effect is quite different if the genetic material in the cell is modified. If the cell survives, the genetic change is passed on to future cell generations, and the number of affected cells doubles with each new generation. If the initial damage is in a somatic cell, the ultimate result can be cancer. If it is in a germ cell in a reproductive organ, the ultimate result can be a hereditary effect in descendants. These effects are referred to as *stochastic* (those for which the probability is a function of dose). Since the effect is binary, ie, either it occurs or it does not, the severity is not related to the dose. Examples of stochastic effects include cancer and some teratogenic effects. Contaminants that are known carcinogens in humans are asbestos (lung and pleura), benzene (leukemia), cigarette smoke (lung, bladder, larynx, and esophagus), and radiation (leukemia, bone, thyroid, lung, and others).

Biological Basis of Thresholds. Many, if not most, tissues and organs can continue to function normally even if they lose a large number of cells. For these, there exists a toxic threshold below which adverse effects do not occur. Such a threshold would also exist for stochastic effects that require multiple alterations of DNA in space or time by more than one contaminant interaction. Such effects are referred to as threshold effects. Effects that can result from the action of a single molecule of a contaminant or radioactive emission are nonthreshold effects since any exposure, however small, carries some increased risk. Nonthreshold stochastic effects include cancers that can be initiated by a single interaction with DNA, such as might occur with alpha-radiation in epithelial layer cells in the lung; or teratogenic effects that can result from the alteration or destruction of a single bud cell in a developing fetus, such as occurred with the drug thalidomide. Although nonthreshold deterministic effects may exist, most deterministic effects occur in systems with sufficient reserve capacity to prevent the damage from impacting the health of the exposed individual. Thus, there is a threshold for the health impact resulting from exposure, and deterministic effects are considered to be threshold effects.

Understanding the deterministic/stochastic and threshold/nonthreshold distinctions is crucial for purposes of quantifying human health effects. For threshold effects, either deterministic or stochastic, there are two methods available for quantifying risk: a margin of safety approach, based on the ratio of the dose to the threshold dose, and a fractional response approach, in which the fraction of a population responding at a given dose is estimated. Nonthreshold effects can only be characterized using the fractional response method.

Table 10. Example of Deterministic Effect: SO_2 Toxicity

Concentration, ppm	Duration	Effect
0–0.06	1 Week	None
0.15–0.25	1–4 d	Cardiorespiratory effects
1–5	1 h	Tightness in chest
5–10	1 h	Choking
10	1 h	Severe respiratory distress

Toxicology Fundamentals. Toxicology is the science that deals with "poisons" and their effects. The concept of a poison, or toxin, is thus central to the study of toxicology. This has been well described (34): "One could define a *poison* as any agent capable of producing a deleterious response in a biologic system. This is not, however, a useful working definition for the very simple reason that virtually every known chemical has the potential to produce injury or death if present in sufficient amount. Paracelsus (1493–1541) phased this well when he noted, 'All substances are poisons; there is none which is not a poison. The right dose differentiates a poison and a remedy.'" This is illustrated in Table 11, where the dose which is fatal to 50% of the rats exposed is given for substances ranging from table sugar to aflatoxin.

An elementary knowledge of toxicology is necessary for the evaluation of the consequences of exposure to chemical or radiological contaminants. Presented in the following sections are some basic mechanisms whereby contaminants interfere with normal bodily functions. This background can assist in evaluating the consequences of contaminant exposures. Because most contaminants exert their toxic effect at the cellular

Table 11. Approximate Oral LD_{50} in Rats[a,b]

Chemical	LD_{50}, mg/kg body weight
Sucrose (table sugar)	29,700
Sodium chloride (common salt)	3,000
Vanillin	1,580
Aspirin	1,000
Copper sulfate	960
Chloroform	908
Caffeine	192
DDT	113
Nicotine	53
Strychnine	16
Sodium cyanide	6
Aflatoxin B_1	5

[a] Most values were taken from NIOSH Registry of Toxic Effects of Chemical Substances, 1979. Results reported elsewhere may differ, as will results for species other than rats.

[b] Chemicals are listed in order of increasing toxicity in rats, ie, sucrose is the least and aflatoxin B_1 is the most toxic.

level, a discussion of normal cell physiology and function is presented first. This is followed by a discussion of some of the mechanisms of contaminant disruption of normal cellular function.

Basic Cellular Physiology. The human body is composed of a variety of specialized cells performing specific tasks necessary for maintaining normal physiological functioning. Although these cells vary widely in structure and function, they almost all have certain elements in common. The primary components of the cell are the cell membrane, which serves to contain the cellular contents and to control the entry and excretion of material from the cell; the nucleus, which is the control center for cell function and division; and various organelles, which are subunits of the cell that provide special functions within the cell. The cellular membrane is composed of two layers of phospholipid molecules which are oriented such that the external surfaces of the membrane are hydrophilic and the internal surfaces are hydrophobic. Such a structure protects the cell against foreign substances by greatly impeding the diffusion of compounds outside the cell through the cellular membrane. The membrane also contains a variety of membrane proteins which assist in transferring materials into and out of the cell and in receiving "messages" from extracellular messenger compounds such as hormones. The nucleus contains strands of DNA which serve as a blueprint for the specialized proteins which the cell manufactures. For normal functioning of succeeding cell generations, the DNA must remain intact and be replicated when the cell divides. Important organelles include the mitochondria, in which oxidation of small molecules provides energy for cellular processes and the endoplasmic reticulum, which manufactures proteins and lipids. In addition, an internal network of fibrous proteins known as the cytoskeleton provides structure to the cell.

General Mechanisms of Toxicity. Most chemicals exert their toxic effect through reactions at the cellular level. There are several basic mechanisms through which a chemical compound interferes with normal cellular function (34), including,

- Interference with receptor–ligand interactions
- Interference with excitable membrane functions
- Interference with cellular energy production
- Binding to biomolecules
- Interference with calcium homeostasis
- Toxicity from selective cell death

These mechanisms are discussed in greater detail below.

Interference with Receptor–Ligand Interactions. Many biological processes in the body, particularly in the nervous system, are mediated by receptor–ligand interactions. Receptors are components of cells with which a chemical (ligand) interacts to produce its characteristic biological effects. Ligands produced naturally in the body, such as hormones or neurotransmitters, interact with the receptor to produce a normal physiological response. Chemical contaminants, however, are exogenous ligands which can act either to activate or to block the response. Since many receptor–ligand interactions are highly dependent on chemical structure, small differences in chemical structure between contaminants can have a dramatic impact on the different toxic responses elicited.

Interference with Excitable Membrane Functions. Excitable membranes are cellular membranes which regulate the electrical potential across the cellular membrane as a part of normal cell function. In the nervous system, such membranes are responsible for the transmission of electrical impulses along the cell. Chemicals which interfere with excitable membranes can significantly reduce the functioning of these tissues. The effects of DDT on the nervous system (by blocking open a sodium gate on the axonal membrane, thus degrading the ability of the cell to achieve a normal resting state) is a prime example.

Interference with Cellular Energy Production. Energy is produced in the cell via the oxidation of carbohydrates to produce adenosine triphosphate (ATP). This stored energy is used in the cell in a variety of biochemical reactions which require the input of energy, such as active transport of biologically required chemicals through the cell membrane and the synthesis of proteins. Interference with ATP production and utilization can result in cellular degradation and death. Chemicals which interfere with oxidative transport, such as hydrogen sulfide, can also lead to a degradation of cellular energy production.

Binding to Biomolecules. Within the cell, certain biomolecules (proteins and enzymes) are critically important for normal cellular function. Binding of contaminants can interfere with these functions. For example, carbon monoxide exerts its toxic effect by binding to iron in hemoglobin, thus impairing the ability of the blood to deliver oxygen to tissues. Lipids, the primary component of cellular membranes, can be damaged by peroxidative attack, leading to ruptures in the cell membrane and consequent cell death. Such attacks on lipids are often combined with depletion of molecules (intracellular thiols) which assist the cell in withstanding oxidative stress. Finally, nucleic acids such as DNA and RNA are frequent targets of contaminants. They may bind to sites on DNA or RNA, creating adducts. These adducts can inhibit the expression of critical gene products and interfere with critical enzyme production.

Interference with Calcium Homeostasis. Calcium homeostasis refers to the ability of the cell to maintain the proper concentration of calcium ions. Calcium is required as part of normal cell function both in the maintenance of the cytoskeleton and as a second messenger in the regulation of intracellular functions. Interference with calcium homeostasis can lead to cell death through disruption of the cytoskeleton and degradation of cellular structure. Contaminants such as nitrophenols, dioxins, halogenated alkanes and alkenes, and some metal ions are known to interfere with calcium homeostasis.

Toxicity from Selective Cell Death. Some compounds are selectively cytotoxic, causing cell death only in certain types of tissues. The neurotoxicity of manganese stems from its toxic effect on certain brain cells, and radioactive isotopes of iodine are especially toxic to the thyroid because iodine concentrates there. Selective cytotoxins are of particular concern during embryonic development because loss of a relatively limited number of cells in the developing fetus can have dramatic consequences. The destruction of limb bud cells by thalidomide taken during a critical stage of pregnancy is an example.

Clinical Effects. It is convenient to separate the clinical effects of contaminants into four categories: systemic, carcinogenic, hereditary, and teratogenic effects. Systemic effects refer to deterministic effects in body organs (liver, kidneys, brain, etc) or systems (immune, respiratory, digestive, etc); carcinogenic effects refer to the induction of malignant tumors; hereditary effects refer to adverse effects observed in offspring or in future generations as a result of parental

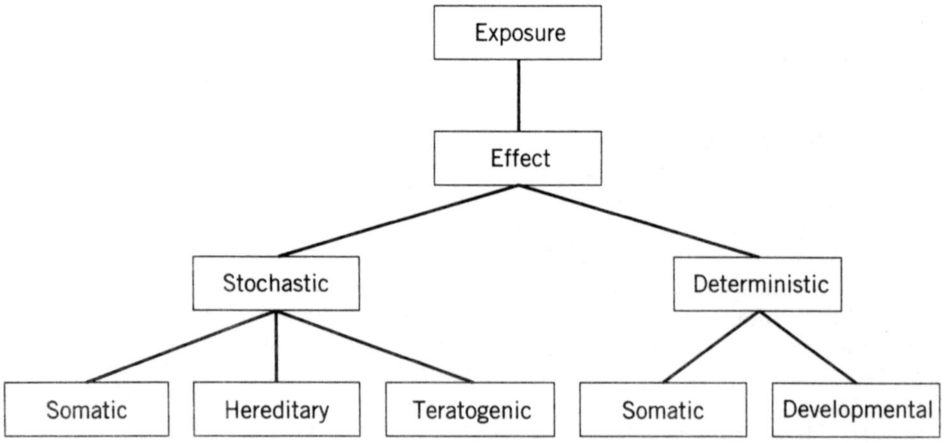

Figure 36. Classification of health effects.

exposure prior to conception; and teratogenic effects refer to developmental abnormalities induced between conception and birth. The relationship between deterministic/stochastic and systemic/carcinogenic/teratogenic/hereditary effects is illustrated in Figure 36.

Systemic Effects. This section is intended to provide an overview of normal systemic function and ways in which specific chemicals can impair the functioning of vital organs. The systems discussed are the blood, kidney, liver, immune system, and nervous system. Although there are many other vital organ systems in the body, this group is the site of action of many environmental contaminants. The deterministic effects addressed in this section are independent of stochastic carcinogenic effects.

Blood. Blood serves several vital functions in the body, primarily by acting as a chemical transport and communication system. Among the most important functions are oxygen transfer by hemoglobin in red blood cells and resistance to disease associated with the white blood cells. Most blood cells are produced in the bone marrow, and the blood system is dependent on the bone marrow to replace cells destroyed as a result of natural processes. The precursors of mature blood cells are constantly being produced in the bone marrow. Thus, damage to the bone marrow typically results in clinical effects being observed in the affected individual's blood chemistry.

Toxic insults to the blood can occur in a variety of ways. Some chemicals affect the ability of the hemoglobin in red blood cells to transfer oxygen. Carbon monoxide, which readily binds to hemoglobin, is well known for reducing oxygen transfer by the blood. Other chemicals act directly on the bone marrow and reduce the production of red and white blood cells. Ionizing radiation, benzene, lindane, chlordane, and arsenic have all been shown to depress the production of blood cells (resulting in a condition known as *pancytopenia*) at sufficient levels of exposure (35).

Kidney. The kidney serves several functions. One important function is selective filtration of the blood and plasma to remove waste products for elimination from the body while retaining water, salts, sugar, amino acids, and other essential components. Another important function is the secretion of enzymes which regulate the volume of extracellular fluid in the body and the electrolyte composition of bodily fluids. Toxic insults to the kidney can result in either decreases in the blood flow, reduction in the removal of wastes, excessive elimination of required chemicals, or an alteration in enzyme production. A variety of contaminants including metals such as lead, mercury, and uranium, halogenated hydrocarbons such as chloroform, TCE, and PCE, and organic solvents such as ethylene glycol and toluene, are known to induce renal (ie, kidney) dysfunction (36).

Liver. The liver is a versatile organ that serves a variety of functions, the most important of which revolve around its role in metabolism through the manufacture, storage, and/or processing of carbohydrates, lipids, and proteins. It also secretes bile, destroys red blood cells, and stores iron and vitamins. Because of the liver's many functions, contaminants can act in a variety of ways to cause liver injury. The principal ones include necrosis (tissue destruction), cirrhosis (hardening of tissue), and the accumulation of abnormal amounts of fat. A particularly important function of the liver from the standpoint of contaminants is the detoxification of compounds by transforming them into a more water-soluble form. Whereas this is an important protection mechanism for some compounds, others which are relatively benign in their administered form can be transformed into more toxic species. This process, known as metabolic activation, is attributed to metabolizing enzymes. For example, allyl alcohol can be metabolized to the neurotoxin acrolein by the alcohol dehydrogenase enzyme. Although the enzymes responsible for metabolic activation are present throughout the body, they are particularly abundant in the liver. Electrophilic intermediate metabolites that are produced can react with DNA, leading to the initiating step in carcinogenesis, as discussed later.

Nervous System. The nervous system consists of millions of nerve cells known as neurons which are responsible for receiving and processing sensory information and then controlling muscles to respond to external stimuli. It is divided into two primary components; the central nervous system (CNS) and the peripheral nervous system.

Disorders of the neuron can be classified by the area which is affected. Along with their other effects, pesticides are typically acute neurotoxins. DDT, for example, interferes with the excitable membrane of the axon, preventing it from achieving a resting state. Organophosphate and carbamate pesticides can bind to neurotransmitter receptor sites, causing repetitive firing of nerve cells.

Immune System. The immune system is primarily responsible for providing the body with the ability to recognize and eliminate harmful agents. The key components of the immune system are the white blood cells, which encapsulate foreign substances for their subsequent destruction or excretion. A wide array of environmental contaminants have been implicated in interfering with immune system function. Benzene, halogenated hydrocarbons, dioxins, pesticides, and metals are all known to suppress the immune system. Organic compounds such as vinyl chloride, TCE, and PCE, and metals such as gold and mercury can cause an autoimmune response (ie, an immune response to an individual's own tissue). Many contaminants, including organic solvents and metals such as beryllium and mercury, can induce allergic responses.

Carcinogenic Effects. In the United States, approximately 30% of the population will suffer from cancer and half of those will die from it. Along with heart attacks and strokes, it is one for the three leading causes of death. Although life style choices such as the use of tobacco products, alcohol consumption, and diet are thought to be responsible for a majority of cancers, it is also known that cancer can result from exposure to radiation and to some chemicals. In fact, cancer is usually the primary stochastic effect of concern in environmental health risk assessments. Cancer is of particular concern because it can be induced at doses far below (and, possibly, at any nonzero dose) the level required to induce an observable systemic effect. Thus, controlling exposures to prevent systemic effects may be ineffective in providing an acceptable level of protection against cancer.

Carcinogenesis. Cancer is a class of diseases, all characterized by an uncontrolled growth of cells. Carcinogenesis is the process whereby normal cells begin to divide abnormally, leading to an abnormal cellular mass in the body. If the growth of cancerous cells is not checked by some combination of radiation therapy, drug therapy, and surgery, the affected individual is likely to die. The cancerous cells consume most of the body's energy and starve healthy cells, they secrete digestive enzymes that destroy healthy cells, and they suppress the immune system, making the individual susceptible to secondary infection.

Although the mechanisms for the induction of cancer are not known precisely, there is evidence to suggest a multistage process consisting of at least two conceptual steps: initiation and promotion. Initiation is the mutation of DNA in a single cell by chemicals or radiation (which can be either of natural origin or products of human activities). This mutation proliferates through succeeding cell generations. Except for the presence of the mutated gene, these cells are normal in that they remain differentiated and their growth is regulated. In the promotion step, which is not well understood, the cells begin to grow and divide abnormally, creating a tumor (also termed a neoplasm). In some tumors, the cells retain most of the characteristics of their differentiated ancestor cells and are contained by the surrounding tissues. These tumors are termed benign (although the abnormal growth of a benign tumor can sometimes lead to significant health effects). However, in some tumors, the cells lose their differentiating characteristics, reverting to immature cells. Whereas mature, differentiated cells are destroyed by the immune system if they are found outside of their normal tissue, immature cells are not. They can migrate away from the site of the primary tumor through the blood and lymphatic circulatory systems and start new tumors in distant tissues and organs. This process is known as metastasis, and tumors which have the capability to metastasize are known as malignant.

Carcinogens are often divided into two classes: mutagenic (or genotoxic) carcinogens, which can initiate a cancer by transforming the DNA of the target cell, and epigenetic carcinogens, which exert their carcinogenic effects without interacting with genetic material. About 90% of known carcinogens are mutagenic. Although all of the factors that cause a chemical to be mutagenic are not clear, it appears that most mutagenic carcinogens are either highly electrophilic compounds or those which have electrophilic metabolites. If these compounds enter the cell, their chemical characteristics allow them to form a covalent bond with DNA. This structure is known as an adduct. Although DNA is regularly monitored and repaired by special enzymes, these processes are not all error free. Unrepaired genetic damage or inaccurate repair can lead to genetic mutations which may be passed on to cells derived from the mutated cell. Mutations of somatic cells may lead to cancer and mutations of germ cells may lead to hereditary disorders. Other forms of DNA damage include strand breaks and cross-links between DNA strands, caused by free-radical attack on genetic material or by the absorption of energy from ionizing radiation. If unrepaired, this damage can lead to mutations in the cell as it divides and replicates its DNA. In addition, some DNA repair processes themselves can introduce mutations when attempting to repair damaged DNA. There is considerable evidence for the key role of mutagenicity in cancer, including the monoclonal nature of most cancers and the fact that most carcinogens are also mutagenic.

Epigenetic carcinogens are those which do not directly affect DNA but instead act to promote the growth of a tumor or suppress natural defenses against cancer. This includes promoters, hormone-modifying compounds, cytotoxic compounds, and immunosuppressive agents. Often, these compounds induce DNA synthesis and cellular proliferation. Unlike genotoxic carcinogens, most epigenetic carcinogens require relatively high and prolonged exposure to manifest their carcinogenic effect.

Identifying Carcinogens. There are two screening methods used in the initial evaluation of the carcinogenic potential of a chemical. The first is a chemical structure analysis, in which the chemical structure of a compound is compared to that of known carcinogens. It is thought that similar chemical structures may have similar carcinogenic properties. The other method is a direct evaluation of the mutagenic potential in an *in vitro* test. The standard mutagenicity assay is known as the Ames test, named after the biochemist who developed it. The test involves exposure of a certain strain of Salmonella bacteria to the suspected carcinogen. The bacteria are placed in a medium which inhibits their growth; however, if a mutation occurs at a certain genetic location, the bacteria are enabled to grow and divide. The number and size of the bacterial colonies give an indication of the mutagenic potential of the compound.

Known Carcinogens. Of the many chemicals present in the environment, only a few have been conclusively determined to cause cancer. However, of these chemicals, only a small percentage have been extensively tested, and it is likely that as the number of chemicals tested increases, the list of known carcinogens will expand. A partial list of environmental contaminants and industrial processes for which there is convincing (Group 1) evidence for human carcinogenicity is presented in Table 12.

Teratogenic Effects. Teratogenic effects are those induced between conception and birth. Although a teratogenic effect is rigorously defined as a birth defect (either a malformation or retardation), embryonic or fetal death can occur from the same mechanisms that cause birth defects. Approximately 3% of live-born infants have major teratogenic (also referred to

Table 12. Known Human Carcinogens[a]

Aflatoxins	Underground hematite mining with exposure to radon
Alcoholic beverages	Iron and steel founding
Aluminium production	Isopropyl alcohol manufacture (strong acid process)
4-Aminobiphenyl	Manufacture of magenta
Arsenic and arsenic compounds	Melphalan
Asbestos	8-Methoxypsoralen (Methoxsalen) plus ultraviolet radiation
Auramine manufacture	Mineral oils, untreated and mildly treated oils
Azathioprine	MOPP and other combined chemotherapy for cancer
Benzene	Mustard gas (sulfur mustard)
Benzidine	2-Naphthylamine
Beryllium and compounds	Nickel and nickel compounds (essentially sulfate and sulfide)
Betel quid with tobacco	Nonsteroidal estrogens (not necessarily all in group); includes diethylstilbestrol
Bis(chloromethyl)ether and chloromethyl methyl ether (technical grade)	Steroidal estrogens (not all in group)
Boot and shoe manufacture and repair (occupational exposure)	Combined oral contraceptives and sequential oral contraceptives
Butanediol dimethanesulfonate	(Myleran)
Cadmium and compounds	Painter (occupational exposure)
Chlorambucil	Phenacetin (analgesic mixtures containing)
Chlornaphazine	Rubber industry
(2-Chloroethyl)-3-(4-methylcyclohexyl)-1 nitrosourea	Salted fish, Chinese-style
Chloromethyl methyl ether (technical)	Solar radiation
Chromium compounds (hexavalent)	Shale oils
Coal gasification	Soots
Coal-tar pitches	Sulfuric acid (occupational exposures to strong inorganic acid mists of sulfuric acid)
Coal tars	Talc containing asbestiform fibers
Coke production	Thiotepa
Cyclophosphamide	Tobacco products (smokeless)
Cyclosporin	Tobacco smoke
Erionite	Treosulfan
Ethylene oxide	Vinyl chloride
Furniture and cabinet-making	

[a] Ref. 37.

as congenital) malformations that are recognized in the first year of life. The total frequency of major and minor malformations that ultimately occur is estimated to be 16%. As is the case with other health effects, it is difficult to distinguish between the incidence of effects caused by a given contaminant exposure and the background incidence. Nonetheless, there is a substantial body of epidemiological data to establish a substantial list of known teratogens.

Teratogenesis. Teratogenic action predominates during the embryonic stage (first eight weeks) of development when rapid and complex changes are occurring. Because of the dramatic changes that are taking place, there are critical windows of development for the various organs and body systems and the timing of an exposure plays a critical role in the potential effect. For example, organ development occurs between 21 and 56 days after conception, the neural plate develops between 18 and 21 days, and lower limbs between 31 and 32 days. There are three general mechanisms through which teratogenic effects occur: alteration or destruction of embryonic cells, alteration of differentiation processes, or alteration of the mother's system. The destruction of a small number of cells by low doses early in a critical period may result in a normal, but growth-retarded organ; whereas high doses late in the period may yield a malformed or nonfunctional organ. The explanation is that if the damage is early, then the surviving cells will partly compensate by hyperplasia (enhanced cell production). If it is late, there is no time for cell replacement. However, the effects may be nonspecific in that there is a general inhibition of cellular processes in all tissues, resulting in overall growth retardation or death, with the latter resulting in death of the embryo. The alteration of specific cells during a very specific time in their differentiation can result in a well-defined anomaly or malformation. For example, thalidomide affects the development of limb bud cells during the sixth or seventh week of pregnancy, yielding rare malformations of the arms. Whereas the first two mechanisms impact the embryo directly, the third acts through the mother. The most common problem is nutritional deficiency which is ultimately manifested in effects such as growth retardation, thyroid deficiencies, and delays in maturation of the central nervous system that are not reversible.

Examples of Teratogens. Presented in Table 13 is a partial list of known human teratogens. It must be recognized that human data are very limited, and many toxicologists believe that any contaminant can alter embryonic development if exposure occurs at the appropriate dose and time. Over 900 agents are known to cause teratogenic effects in animals.

Hereditary Effects. Hereditary effects are those that are expressed in the immediate or remote descendants of the exposed individual. Whereas carcinogenic effects occur as a result of damage to genetic material in somatic cells, hereditary effects occur as a result of damage to genetic material in germ cells in the reproductive system. There are two general types of damage: alterations of DNA (gene mutations) and alterations in the structure or number of chromosomes (chromosomal aberrations). Dominant mutations can appear in either parent, whereas recessive mutations must be present in both parents for the effect to be expressed. Recessive mutations yield few ac-

Table 13. Known Human Teratogens (Drugs and Chemicals)[a]

Androgenic hormones	Trimethadione
Aminopterin	Courmarin anticoagulants
Cyclophosphamide	Valproic acid
Busulfan	Tetracyclines
Thalidomide	13-*cis*-Retinois acid
Mercury (organic)	Lithium
Chlorobiphenyls	Methimazole
Diethylstilbestrol	Radiation
Diphenylhydantoin	

[a] Ref. 38.

tual effects in the first few generations, but they contribute to the general pool of genetic damage in subsequent generations.

Approximately 30% of all spontaneous embryonic and fetal deaths and 6% of stillbirth and infant deaths are attributed to genetic damage. Among live births, the severity of inherited effects varies over a wide range with approximately one-third to one-half characterized as serious. It is estimated that 2–4% of newborns have a serious genetic abnormality, and 10% of people ultimately experience a serious genetic disorder. Dominant mutations can result in hereditary disorders such as polydactyly (extra fingers and toes), achondroplasia (short-limbed dwarfism), Huntington's chorea (progressive involuntary movements and mental deterioration), muscular dystrophy, anemia, and retinoblastoma (an eye cancer). Recessive mutations are known to cause hemophilia (failure of blood clotting), color blindness, phenylketonuria (PKU, a type of mental deficiency), Tay-Sachs disease (blindness and death in the first few years of life), sickle cell anemia, and cystic fibrosis.

Although there is no human epidemiological evidence of hereditary effects due to environmental contaminants, there is ample evidence in animal studies for contaminant induced hereditary effects.

Dose Response

The previous section focused solely on identifying the potential responses to contaminant exposures. However, risk assessments usually require a quantitative estimate of the incidence of the effect in an exposed population (or, similarly, of the risk that a particular individual will be adversely affected). This requirement is the motivation for quantitative dose response modeling.

Concepts in Dose Response. ***Assumptions in Dose Response.*** Critical to the understanding of dose response modeling is an understanding of the three critical assumptions (34) upon which dose response studies are based. The first is that the response is due to the contaminant to which the individual is exposed. In effect, this assumption requires a causal mechanism for the contaminant to create a response. Mere observance of a correlation between a toxic response and a particular chemical exposure does not, in itself, justify a quantitative estimate of dose response. For example, an observed response attributed to a particular contaminant in a laboratory study may actually be due to a different agent, eg, an impurity in the test solution. In animal studies, as discussed in a following section, there is a great deal of control over the administered contaminant, and hence this is not typically a significant source of uncertainty. However, in epidemiological studies, it is considerably more difficult to associate a given effect with a given cause, although epidemiological studies can be buttressed by animal experimentation to identify the mechanism of action.

The second assumption in dose response modeling is that the response is related to the dose in a quantitative fashion. In order for this to hold, there must typically be three elements. The first is that there is a molecular or receptor site with which the contaminant interacts to produce the response. The second is that the production and degree of the response are related to the concentration of the contaminant at the reactive site. The final element is that the concentration of the contaminant at the site is proportional to the exposure of the individual. This element is essentially the relationship between administered dose and effective dose discussed in the section on exposure assessment.

Finally, the third key requirement in quantitative dose response analysis is that there must be a precise and quantitative means of expressing the toxic endpoint. Without a replicable and directly measurable toxicity endpoint, quantitative analysis cannot be performed. The toxic endpoint can be expressed in a variety of ways and can be either discrete or continuous. Discrete endpoints are those which are either present or absent in an individual; for example, the presence or absence of cancer in a particular animal. Discrete endpoints are thus binary; ie, they only have two possible values. Continuous endpoints are those which vary in an individual in a fashion proportional to the dose. Mathematically, they may take on any value. This distinction corresponds roughly to the stochastic/deterministic distinction introduced in previous sections, where stochastic responses are discrete whereas deterministic endpoints are continuous. Quantification of continuous responses in a *population,* however, is very difficult. The usual approach is to select a particular point of the continuous response and then treat the data as quantal.

The most common endpoint is a fractional population response at a given dose level. Such an endpoint is obtained by estimating the fraction of an exposed population exhibiting some quantal response. The dose response models discussed in this section are all based on some measure of fractional response.

Factors Affecting Response. There are a variety of factors affecting the degree of response due to exposure to a contaminant. Although the total dose received is generally the most important, other factors such as dose rate and the route and site of exposure can also be important.

As discussed previously, the administered dose of a chemical contaminant is the total amount of the contaminant, over a specified period of time, to which the individual is exposed. Dose is usually measured in mg of contaminant per kg of body weight for chemical contaminants and in Sieverts for radiological exposure. Often, doses are converted to average daily dose rate by averaging the total dose received over an averaging time to yield units of mg/kg/d.

For many contaminants, the dose rate can have a significant effect on the toxicity of the exposure. The administered dose rate for chemical contaminants is simply the rate at which the contaminant is taken into the body as a result of exposure. In general, a dose received over a short period of time is more toxic than the same dose received over a protracted period. This is often due to the time dependence between the administered dose and the effective dose at the target organ, usually associated with the ability of the body to eliminate the contaminant or repair toxic insults. If the body can quickly degrade or eliminate the contaminant, or rapidly repair injuries, partitioning of the exposure may lead to reduction or elimination of the toxic response. However, if the natural protective mechanisms of the body are overwhelmed, either by a large dose or by inefficient repair mechanisms, a toxic response may be observed. Elimination of the contaminant from the body causes the administered dose to be different from the effective dose, whereas repair of biological damage is more purely an issue of the dose-rate dependence of toxicity. This is illustrated in Figure 37, where contaminant concentration at some target organ is plotted on the left y-axis and exposure time is plotted on the x-axis.

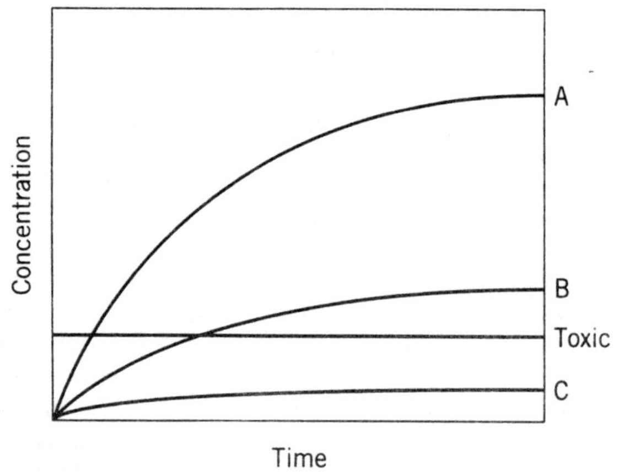

Figure 37. Dose rate dependence of toxicity.

In this example, the body is assumed to detoxify the contaminant at a rate proportional to the amount of contaminant present in the body. Contaminant A is administered much faster than it can be detoxified. It steadily accumulates at the active site, and it ultimately exceeds a toxic concentration. Contaminant B, administered at a lower rate, accumulates at a slower rate but eventually reaches a toxic level. Contaminant C, on the other hand, is administered at such a low rate that the detoxification processes prevent it from reaching a toxic level. An analogous example could be developed to illustrate the effects of repair mechanisms, in which the y-axis represents the level of biological damage and the rate constant represents the rate of repair rather than the rate of detoxification.

Most contaminants have one or more target organs at which they exert their most deleterious effect. The route through which a contaminant is administered may significantly affect the concentration at the target organ and thus the degree of toxic response. Again, this is a distinction made necessary due to the use of administered doses rather than effective doses in dose response analysis. Substances which undergo detoxification or metabolic activation may have significantly different toxicities when exposure is through inhalation rather than through ingestion. This is because most chemical contaminants pass through the liver when taken orally. Since the liver is a site of a great deal of metabolic activity, there is a high chance of either activation or detoxification prior to entering systemic circulation. Inhalation of contaminants, on the other hand, results in the entry of the substance to systemic circulation almost immediately. In addition, contaminants are absorbed into the body with different efficiencies when administered via different routes. The effectiveness of the skin as a barrier to foreign substances tends to cause dermal toxicity to be fairly low for systemic toxins. Inhalation toxicities are typically the highest, followed by ingestion and dermal toxicities.

Hazard Identification and Dose Response Testing. Quantitative dose response information comes from two sources, epidemiological studies on exposed human populations and controlled laboratory experiments with animals. However, given the large number of potentially toxic chemicals and the multiplicity of factors affecting toxicity, the base of data for quantitative dose response analysis is far from complete. The data that do exist are generally limited to situations involving high doses and dose rates, and they are for limited and often nonrepresentative populations. Unfortunately, the majority of risk assessments are concerned with effects of low doses due to environmental exposures to general populations. This requires an extrapolation from high dose to low dose and, in the case of animal studies, from animals to humans.

Most dose response analysis (particularly for deterministic effects) lacks a firm foundation of biologically based models. The models which do exist are often highly simplified abstractions of the actual physiological processes involved in contaminant toxicity. In practice, such considerations dictate that effects be extrapolated from tests with similar durations and routes of exposure. If the dose rates or routes of contaminant exposure are substantially different than those used to develop dose response data, then the data must either be adjusted to reflect the changed conditions of exposure or the risk estimates must be qualified to reflect the uncertainties associated with the extrapolation.

As discussed previously, dose response testing usually provides data in the form of the cumulative population response at different dose levels. Such data are the goal both for epidemiological studies of human exposures and for controlled laboratory experiments on animals. A study might yield hypothetical data such as those in found in Table 14.

These data typically yield an S-shaped (or sigmoidal) curve when the cumulative response is plotted against the log of the dose. A plot of these data, with a dose-fitted response function, is given in Figure 38. Data and curves such as those given above can be generated for both the deterministic and stochastic effects of contaminants. If the contaminant has a population threshold (dose at which the most sensitive member of a population first exhibits an effect), there is a region where no response is observed. As the dosage is increased above the threshold, there is a region of the curve at relatively low dose levels where a few susceptible individuals show a response. Following this is a region where a large fraction of the population exhibits symptoms of exposure. As the doses increase still further, the most resistant individuals in the population finally show the effects of exposure. The type of data obtained is similar for both the stochastic and deterministic effects of contaminants; however, in the current practice of risk assessment the data are analyzed differently for the two types of effects. For deterministic effects, the population threshold is found. For stochastic effects, which are often assumed to lack a threshold, a dose response function is obtained.

Dose–Response Interpretation. *Threshold Identification.* Evaluation of the deterministic effects of contaminant exposures typically focuses on the identification of a population threshold. However, since the doses are increased discretely

Table 14. Hypothetical Dose–Response Data

Dose, mg/kg/d	Fractional Response
1	2.2
3	7.3
10	19.8
30	38.2
100	61.9
300	80.3
1000	92.7

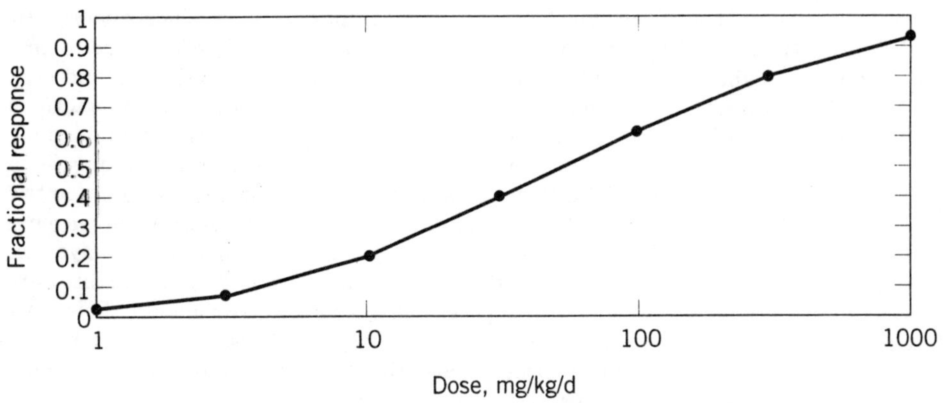

Figure 38. Hypothetical dose–response curve.

rather than continuously, and since the entire population is not tested, the actual true population threshold is not observable. Instead, there is a dose at which no statistically significant effects are observed in the exposed population, and the next highest dose yields detectable effects. The highest dose at which no statistically significant effects are observed in the exposed population is known as the no observed adverse effects level (NOAEL), and the lowest dose which results in detectable effects is known as the lowest observed adverse effects level (LOAEL). These two points bracket the true population threshold in the exposed population, as shown in Figure 39.

Dose–Response Functions. For nonthreshold stochastic effects, the response at low dose levels is difficult to distinguish from the normal background incidence of the effect. Thus, an appropriate dose response model is chosen and fitted to the data at high doses. This model is then used to estimate low dose reponse. Model types will be discussed in greater detail in a following section.

Epidemiological Data. Human evidence for a causal link between a contaminant and a health effect or human data for developing a quantitative dose–response relationship are developed from epidemiological studies. These studies are of two general types, cohort and case control. In a cohort study, a contaminant exposure is identified, and the objective of the study is to identify health effects in the exposed population. This is accomplished by comparing the incidence of disease in the exposed population to that in a similar, unexposed control population. A cohort study of the survivors of Hiroshima and Nagasaki has been, and is still being, used to develop dose response relationships for the induction of cancer by radiation. Cohort studies are prospective in the sense that the exposure is identified first, and the exposed population is monitored for some period of time in the future to infer if effects occur.

In a case control study, a group of subjects suffering from a specific disease is identified, and the objective of the study is to identify the cause, which could be a contaminant exposure, lifestyle, heredity, etc. For a suspected contaminant, this is accomplished by comparing the exposure history of the affected population with that of an unaffected control population. A case control study established the link between diethylstilbestrol (DES) and vaginal cancer. Case control studies are retrospective in that the effect is identified first and historical data are used to infer the cause.

The principal advantage of epidemiologial studies is that they provide data and information on humans. This eliminates uncertainties associated with the extrapolation of animal data to humans. However, epidemiological data are subject to uncertainty, primarily due to uncertainties in the estimated doses, the difficulty in obtaining appropriate control and experimental populations, and the presence of confounding factors such as other chemical exposures or lifestyle risk factors. In addition, data on humans are limited to a relatively small number of contaminants. For these reasons, dose response relationships for most chemical contaminants must be derived from animal studies.

Animal Studies. Several types of animals tests are used to quantify the toxicity of chemical contaminants. These are typically classified by their duration and the effect for which they test.

Acute lethality tests establish the lethality of the contaminant under conditions of acute exposure. The contaminant is typically given in a single exposure, and the test animals are observed for 14 days. The typical measure obtained from such

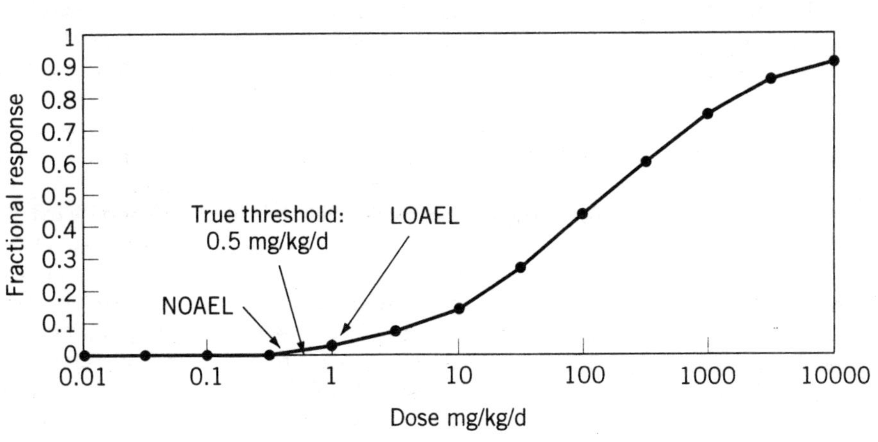

Figure 39. Identification of toxic thresholds.

a test is the median lethal dose, or LD_{50}. This is defined as the single dose of a contaminant which results in the death of 50% of the test species under a specified set of conditions. A similar test can be performed to established the effects of protracting the same total dose over a short period of time, typically 14 days.

If a contaminant is selected for more extensive testing, subchronic studies may be performed to identify cumulative effects. These tests are typically conducted over a 90-day period and serve two functions: they generate estimates of a threshold dose if the contaminant is thought to have a threshold level of effect, and they allow for design of subsequent chronic tests by identifying the maximum tolerated dose (MTD). The MTD is defined as the highest dose of a contaminant given over a specified period of time which produces minimal nonlethal signs of toxicity. In practice, a commonly used indicator of toxicity is a slight suppression of body weight.

Chronic tests, which range anywhere from six months up to the lifetime of the animal, are used to assess the deterministic effects of cumulative toxicity and to evaluate the carcinogenic potential of a contaminant. Information from chronic toxicity tests for deterministic effects include a NOAEL and a LOAEL, as well as dose response data used for fitting dose response models. For carcinogenicity or other stochastic effects assumed to have no dose threshold, chronic tests provide dose–response data which are used to calibrate a mathematical model for estimating responses at low levels of exposure.

Selection of dose levels for animal testing is bounded by two competing goals. On one hand, doses are kept as low as possible to ensure that the toxicity of the chemical does not cause premature deaths. The MTD derived from the subchronic tests is typically the highest dose used. Other doses are typically fractions of the MTD. However, if dose levels are too low, it may be difficult to identify the level of response because of poor statistical resolution. Lowering the dose requires a corresponding increase in the numbers of animals required to provide a statistically adequate estimate of the response due to the chemical. Typically, animal studies use 50 animals at each dose level. A major factor driving the size of a test is the cost. Obviously, the more animals which are needed, the greater the costs involved in procuring the animals, caring for them, and performing the examinations. For example, subchronic and chronic studies using rats can range from hundreds of thousands to millions of dollars. With 50 animals per test group, a dose capable of causing an 8% ($p < 0.025$) increase in contaminant-induced tumors could result in the absence of tumors in the test group. To date, the largest study conducted is known as the ED01 study and was performed in the late 1970s by the National Center for Toxicological Research for the U.S. Food and Drug Administration. Designed to clarify the shape of the dose–response function at low levels, it used over 24,000 test animals and it was designed to detect a 1% increase in tumor incidence. From this, it is clear that animal testing at the regulatory target range of 10^{-4} to 10^{-6} is not feasible.

Health Effects Modeling. The goal of health effects modeling is to make a quantitative prediction of the response of a human population exposed to environmental contaminants. Environmental exposures to chemical and radiological contaminants are often chronic at low dose rates. Direct data on human health impacts are extremely limited for such exposures, since the effects may be statistically indistinguishable from the natural variability of the effect. Consequently, the quantitative estimation of human health effects due to environmental exposures usually requires two critical (and often controversial) extrapolations. If reliable data on health effects in human populations are not available from adequate epidemiological studies, health impacts in humans must be inferred from animal studies, with the implicit assumption that animals are adequate models of human toxicological response. For stochastic responses, particularly those which have very low or nonexistent thresholds, a second extrapolation is required which is potentially even more problematic. This is the extrapolation of effects at very high doses or dose rates to the very low doses or dose rates characteristic of environmental exposures.

Animal to Human Extrapolation. Due to the large differences in size between humans and test animals, a given mass of contaminant does not have the same effect in humans as it does in animals. This necessitates the adjustment of the doses administered to laboratory animals to an equivalent human dose. The equivalent human dose is the dose which yields the same level of effects in human populations as that in animal populations. There is considerable debate as to the appropriate method for inferring equivalent human doses from animal doses. Early studies indicated that the LD_{50} was approximately equal in humans and animals when the dose was divided by the body weight. This observation led to the traditional unit of chemical dose as milligrams of contaminant per kilogram of body weight (39). However, this relationship does not always hold true. The LD_{50} values are more nearly equal between species when a relationship based on body surface area is used. However, by tradition, dose data are reported in units of mass per body weight. A mass scaled dose can be converted to a surface area scaled dose by noting that surface area is proportional to $BW^{2/3}$. The equivalent human daily dose, D_{Human}, corresponding to an animal daily dose, D_{Animal}, is given by

$$D_{\mathrm{Human}} = D_{\mathrm{Animal}}\left(\frac{BW_{\mathrm{Animal}}}{BW_{\mathrm{Human}}}\right)^{\frac{1}{3}} \tag{95}$$

where BW is body weight in kilograms and D is dose in mg/kg/d.

If the data are available, a pharmacokinetic model may provide a more accurate estimate of the equivalent human dose. Pharmacokinetic models are typically compartmental models of contaminant transport in humans and animals, as illustrated in Figure 40. The compartments can be empirically defined for mathematical convenience, or they can be chosen to simulate actual organ systems characterized by physiological parameters such as organ weight, organ specific metabolic rates, etc. The transport rate from one compartment to another is approximated as either a constant or linear function of the concentration in the compartments. The result is a set of coupled differential equations which relate changes in contaminant concentration in various compartments due to chemical transfer and metabolic transformation.

Dose Extrapolation Models. Models used for low dose extrapolation can be divided into two types: tolerance distribution models, used to estimate the distribution of individual toxic thresholds in a population, and mechanistic models, used in estimating the probability of developing a given stochastic effect as a result of exposure to a chemical or radiological contaminant. Tolerance distribution models are essentially properly parameterized statistical distributions, whereas

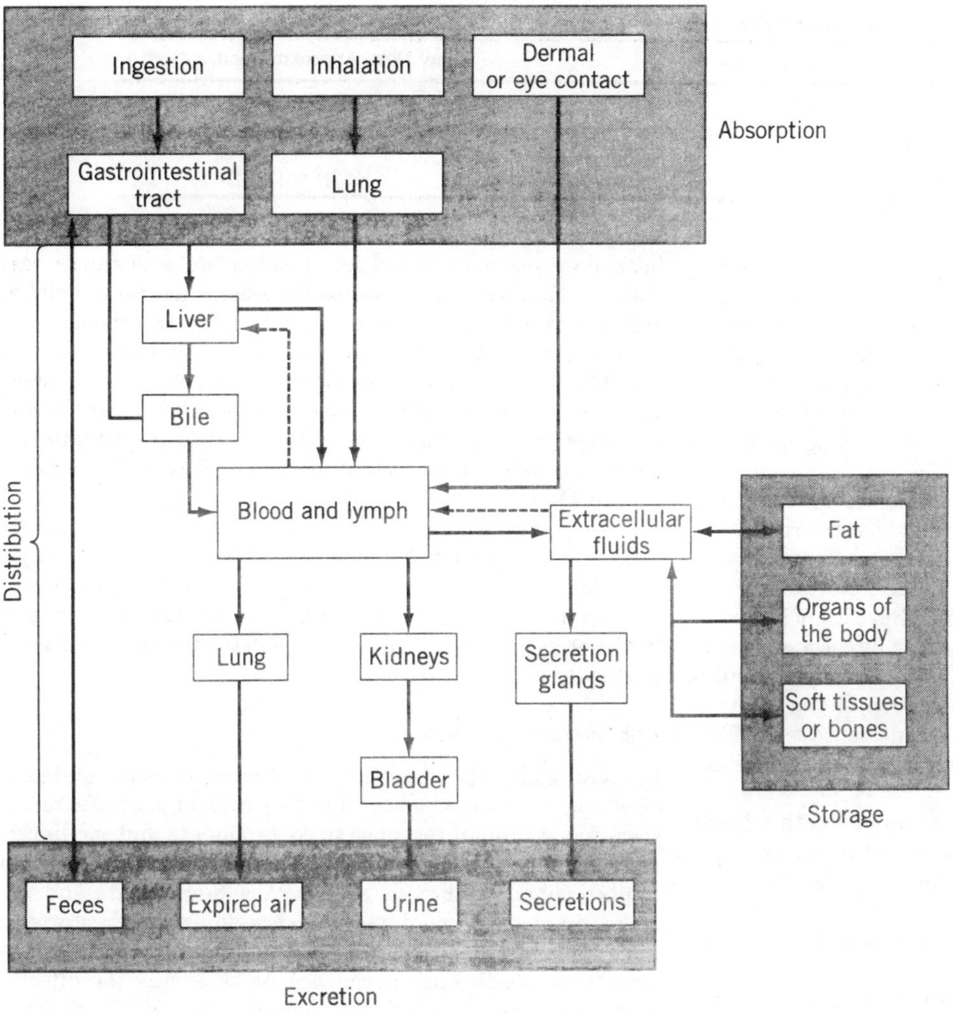

Figure 40. Pharmacokinetic models (40).

mechanistic models draw on a theoretical understanding of the biology of the stochastic effect being modeled (usually cancer) to predict the incidence. The primary theoretical difference is the underlying assumption of tolerance distribution models that sensitivities are distributed in a population in a statistical fashion, whereas mechanistic models of cancer induction typically assume that all individuals are equally sensitive to cancer induction, and the probability of cancer is related to the dose received.

Tolerance Distribution Models. For deterministic, threshold effects, statistical models such as the probit (log–normal), the Weibull, and the log–logistic (logit) distributions, and modifications of these can be used to describe the distribution of thresholds in a population. When applied to a population, these models yield the expected number of individuals who exceed the population threshold as a result of a particular dose. The mathematical form of three common tolerance distribution models is given in Table 15.

Table 15. Statistical Tolerance Distribution Models

Model	Functional Form
Log–normal (log–probit)	$R(D) = \frac{1}{\sqrt{2\pi}} \int_{-\infty}^{\rho_0+\rho_1 \cdot \ln D} e^{-\left(\frac{u^2}{2}\right)} du$
Log–logistic (logit)	$R(D) = (1 + e^{-(\rho_0+\rho_1 \cdot \ln D)})^{-1}$
Weibull	$R(D) = 1 - e^{-(\rho_0+\rho_1 D^m)}$

The log–normal distribution is probably the most common statistical distribution in toxicology because it has been shown to fit experimental results for a wide array of contaminants. It is sometimes modified by forcing the curve to be linear at low doses, producing the Mantel-Bryan model. The Weibull and the logit models are similar to the log–normal in their general shape; however, the logit model allows fitting data with a higher degree of variability. Since the Weibull and logit distributions predict threshold doses lower than those of the probit, they are typically considered to be more conservative.

Cancer Models. Mechanistic models are often used in predicting the incidence of cancer. These are based upon a biologically simple theoretical explanation of the mechanisms of carcinogenesis. Although such models can become mathematically complex, they typically incorporate a number of very simple biological assumptions. The most common models are the one-hit, multihit, and multistage models. Presented in Table 16 are the general mathematical forms for these models.

The one-hit model, which was first proposed in 1950 (41), is based on the hypothesis that exposure to even a single molecule of a carcinogenic substance (or a single particle of ionizing radiation) has the potential to cause cancer. At low doses, the probability of cancer increases approximately linearly as a function

Table 16. Mechanistic Models of Carcinogen Dose–Response

Model	Functional Form	Low Dose Approximation, $\rho D \ll 1$
Single hit	$R(D) = 1 - e^{-(\rho D)}$	$R(D) = \rho D$
Multihit	$R(R) = 1 - \sum_{j=0}^{k} [e^{-\rho D}(\frac{(\rho D)^j}{j!})]$	$R(D) = \rho D^k$
Multistage	$R(D) = 1 - e^{-(\rho_0 + \rho_1 D + \cdots + \rho_k D^k)}$	$R(D) = (\Sigma \rho_i)D$

of dose. Such a model is consistent with the theory that one molecule of a carcinogenic substance is capable of interacting with DNA and causing a mutation that leads to cancer. However, it is well known that repair mechanisms are present in the nucleus which protect the DNA against damage. Multihit models (42), therefore, are based on the assumption that two or more interactions must occur within a given time period to initiate a transformation.

Multistage models (43) of carcinogenesis are based on the hypothesis that cancer induction is typically not the result of a single transformation event, but that several different events must occur in a fixed sequence prior to the induction of cancer. Multistage models are useful in that they fit a wide array of experimental data sets, as well as having a reasonably solid biological basis. One variation of this is the linearized multistage model (44) in which the linear portion of the curve is constrained to lie at its 95th percentile confidence level. This is intended to provide a conservative estimate of the cancer potency at low doses.

A final mechanistic model, which accounts for both cellular transformation and cellular proliferation, is the Moolgavkar-Venzon-Knudson (MVK) model (45,46). In this model the transformation of a single cell is sufficient to initiate a cancer and cellular progression is irreversible and independent of all other initiated cells.

A generic dose response model proposed for exposure to radionuclides is

$$R(D) = \gamma_0[1 + f(D)g(\beta)] \quad (96)$$

where D is the dose in Sieverts, γ_0 is the age-specific background risk of death due to a given type of cancer, $f(D)$ is the basic dose response model for radiation induced cancer, and $g(\beta)$ is a function that accounts for other important variables such as sex, age, age at exposure, and time since exposure (47).

Selection of an appropriate models is problematic. Current scientific knowledge is generally not available to conclusively identify the correct model. Often, several models can fit high dose experimental data equally well, yet yield dramatically different estimates of the risk at low doses. This is illustrated in Figure 41.

Selection of an appropriate model thus often requires the exercise of professional judgment on the part of the analyst. Such factors as biological plausibility, statistical goodness of fit, and degree of conservatism need to be taken into account in the selection of a model used to estimate the effects of a particular exposure.

Risk Characterization

Unfortunately, the quantitative characterization of health effects is constrained and dictated by current scientific understanding of human response to contaminants and the limited base of existing epidemiological and animal data. The principal limitations are that there are very few human data and that there are virtually no data at the low doses representative of environmental exposures. A related limitation is the fact that since most studies are performed by observing the effect of contaminants on populations rather than individuals, the characterization of risk usually involves the estimation of impacts on populations. Risks to specific individuals are extremely difficult to estimate and can typically only be characterized by the relation of the individual to the entire population. Another

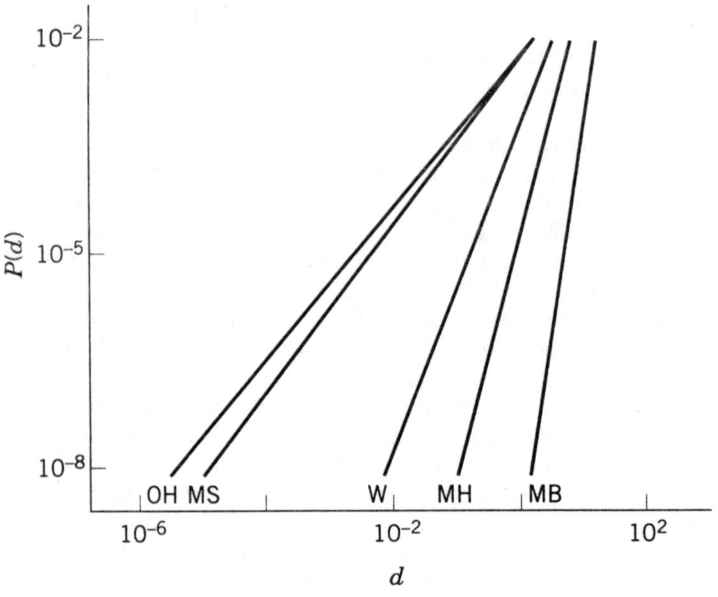

Figure 41. The effect of different models on extrapolation from high dose to low dose (48).

limitation is that dose response data are usually given in a quantal form (eg, as the fraction of subjects who experience a discrete symptom such as a 25% increase in liver weight) rather than a graded form (eg, as a graph of increase in liver weight as a function of dose).

For purposes of risk characterization it is useful to distinguish between those risks which exhibit a toxic threshold and those which do not. The former are referred to as threshold effects and the latter are referred to as nonthreshold effects. The toxic threshold, or threshold dose, is the dose below which no member of an exposed population is expected to display a clinical response. Doses less than the threshold dose do not cause an effect. For nonthreshold effects, there is the theoretical possibility of an effect at any nonzero dose. From the perspectives of regulatory policy and contemporary practice in risk assessment and risk management, all deterministic effects are categorized as threshold effects and all stochastic effects are categorized as nonthreshold effects. However, from the perspective of science, the characterization of health effects associated with low doses is still an open question and is a topic of intense interest and study because of its implications with respect to regulatory policy.

General Metrics. There are two general metrics used in characterizing the risk of exposure to contaminants. One applies to nonthreshold effects and the other applies to threshold effects. For nonthreshold effects, the metric is fractional response, which is the fraction of the population expected to exhibit the effect of interest. Although conceptually the metric could be applied to both stochastic and deterministic effects, in current practice it is used solely for stochastic effects as noted in the previous paragraph. For threshold contaminants, the metric is the margin of safety, which is the ratio of the threshold dose to the actual dose.

Fractional Response. Fractional response is obtained by estimating the dose and then determining, from the appropriate dose response model, the fraction of the population which is affected. This was illustrated in Figure 38. For example, a dose of 1 mg/kg/d results in a response being observed in 50% of the population. In practice, this is not done graphically; rather, the dose response models presented previously are used to estimate the response at a given dose. A significant challenge in risk characterization lies in the selection of the appropriate model used to estimate the response. Ideally, the appropriate mechanistic model that relates response to dose would be selected. Unfortunately, the mechanisms that lead to a physiological response are not always precisely known, particularly at the very low exposures characteristic of environmental contamination. Thus, the selection of a model is a mixture of scientific judgment and policy judgment. Typically (particularly for chemical carcinogens) the models and model parameters are conservative; ie, biased to produce an over-estimate of the risk.

Cancer. Although cancer is not the only possible stochastic effect of contaminant exposure, it has become the effect of greatest concern, and most work on characterization of stochastic health effects has been associated with cancer. For this reason, the focus is on cancer in this section. The concepts involved in carcinogenic risk characterization can be applied to noncarcinogenic stochastic effects as well.

A number of theoretical models for cancer were given previously. These are based on postulated mechanisms for carcinogenicity. A significant characteristic of several of the models is that at low doses, they may be approximated as linear functions of dose, ie:

$$R = \rho D \tag{97}$$

where R is the fraction of the population responding, ρ is the cancer slope factor ($(\text{mg/kg/d})^{-1}$) or radiation risk factor (Sv^{-1}), and D is the dose (mg/kg/d or Sv). The approximation is valid for low values of the product ρD. However, sufficiently low doses will usually cause this product to be small, regardless of the value of ρ.

Estimating the total incidence of the stochastic effects in an exposed population is straightforward. Given the fractional response due to a given dose and the total population exposed at that dose, the total incidence is simply

$$R_n = RP = \rho DP \tag{98}$$

where R_n is the total number of effects in the exposed population, R is the fractional response at dose D, and P is the total population receiving the dose.

Estimating the risk to a specific individual in the exposed population is difficult, for the reasons discussed previously. Common practice in risk assessment is to assume that the risk faced by a given individual is the same as the fractional response given by equation 97. This approximation holds given that no additional information is known about the individual which might indicate their susceptibility to cancer relative to the entire exposed group. Hence, the risk is simply the chance that an individual selected at random from that population has cancer as a result of the exposure; ie, if every year, n individuals out of a population of N contract cancer, then the individual risk is n/N.

Deterministic Effects. Although deterministic effects are not usually characterized using the fractional response method, it is possible to perform such characterization in a method exactly analogous to that used for modeling stochastic effects. Threshold deterministic effects exhibit a distribution of thresholds across a population. If the average contaminant dose and distribution of thresholds were known, the fraction of the population experiencing adverse effects could be estimated.

Margin of Safety. If there is a threshold dose associated with exposure to a given contaminant, then it is possible to reduce exposures to a level protective of the entire population. The population margin of safety has been defined previously as the ratio of the threshold dose to the actual dose, or

$$\text{MOS} = \frac{\text{Threshold dose}}{\text{Actual dose}} \tag{99}$$

If the margin of safety is greater than one, the threshold dose is greater than the actual dose, and no toxic effects are expected in the exposed population. If the margin of safety is less than one, there is a possibility that effects may be observed. However, without knowledge of the shape of the dose response curve, it is not possible to calculate a percentage of the population expected to display the toxic effect. Quantitative evaluation of risk when the margin of safety is less than one would require the use of an appropriate tolerance distribution model.

One of the most significant limitations in the application of the margin of safety method of risk characterization is estimation of the threshold dose. It is extremely rare that data are available which give a reliable estimate of the threshold dose, as discussed in a previous section. One way of dealing with this is to use an experimental NOAEL or LOAEL adjusted with the use of safety factors to estimate the threshold dose. In brief, the idea is that a NOAEL or LOAEL is observed as a result of epidemiological studies of human populations or is estimated from animal data. This estimate is used as a proxy for the threshold and is then reduced by the use of safety factors. This is expressed as:

$$\text{MOS} = \frac{\text{Threshold dose}}{\text{Actual dose} \cdot \prod \text{Safety factors}} \tag{100}$$

The effect of including safety factors is to provide a more conservative estimate of the safety margin than would be achieved with a NOAEL or LOAEL alone.

Aggregating Multiple Chemical Exposures. In the previous sections, the ways in which risk associated with contaminant exposure could be characterized quantitatively were examined. These were done for exposures to single contaminants. However, such situations are quite rare. A more typical example would involve exposure of a population (or populations) to a mixture of contaminants. This presents a challenge because information is needed on contaminant interactions in the body, but data on toxic interactions are extremely limited. Conceptually, three types of interactions may take place:

Interaction	Effect
Additive	$R(\Sigma D_i) = \Sigma R(D_i)$
Synergistic	$R(\Sigma D_i) > \Sigma R(D_i)$
Antagonistic	$R(\Sigma D_i) < \Sigma R(D_i)$

An additive interaction is one in which the effect of simultaneous exposure to multiple contaminants is simply the some of the individual effects. Unless these is information to suggest that synergistic or antagonistic reactions between contaminants may occur, responses are typically assumed to be additive. A synergistic interaction is one in which the effect of multiple contaminants is greater than the sum of the effects taken individually. Synergistic responses are particular concern because of the potential to underestimate the risk posed by a mixture of contaminants. An antagonistic interaction is one in which the effect of multiple contaminants is less than the sum of the individual effects. Antagonistic interactions may result in an overestimate of risk, leading to potentially unnecessary remediation. The professional judgment of a trained toxicologist is required when modeling contaminant mixtures that may have nonadditive interactions.

In most risk assessments, similar effects (those with similar mechanisms of toxicity) are assumed to be additive. Thus, for multiple contaminants with similar stochastic effects, the risk is added. However, the aggregation of contaminants under the margin of safety approach is a bit more complex. One method is through the hazard quotient, which is the reciprocal of the margin of safety. Thus, a margin of safety greater than one, indicating that the threshold dose is not exceeded, corresponds to a hazard quotient of less than one. The hazard quotient (HQ) is often used, rather than the margin of safety, since if summing hazard quotients from exposure to different chemicals yields a composite HQ of less than one, then there cannot be any individual exposures with a HQ greater than one. This makes the HQ useful as a screening tool to ensure that no individual contaminant exposure will result in toxic effects in the exposed population. However, if a composite HQ is greater than one, more detailed analysis is required to determine the potential incidence of an adverse affect.

Applications. In this section, one of the more common applications of risk assessment is addressed and the application of the concepts introduced above in the Environmental Protection Agency's Superfund risk assessment program discussed. Although this is not definitive guidance on performing Superfund risk assessments, the basic ideas are presented and explained. Guidance for performing Superfund risk assessments is provided in the EPA publication *Risk Assessment Guidance for Superfund* (49), often known as the RAGS manual.

In general, EPA assessments divide contaminant effects into two classes: noncarcinogenic and carcinogenic. These categories correspond roughly to the deterministic/stochastic distinction, although it is important to recognize that there are significant noncarcinogenic stochastic risks. Carcinogenic risks are characterized using the fractional response technique, and noncarcinogenic risks are characterized using a form of the margin of safety.

Hazard Indices and Hazard Quotients for Noncarcinogens. As discussed earlier, the determination of a population threshold for the response to a contaminant is fraught with uncertainty. To deal with this uncertainty, the EPA has defined a quantity known as the Reference Dose (RfD), which is defined as the dose which is expected to be without appreciable effects over a lifetime of exposure to the most sensitive member of a population. It is typically derived from a NOAEL or LOAEL, modified by the application of uncertainty factors, as follows:

$$\text{RfD} = \frac{\text{NOAEL}}{\prod_I UF_I} \tag{101}$$

where RfD is the reference dose and *UF* are uncertainty factors to account for differences between test conditions and conditions of human exposure. For consistency. EPA recommends the use of standardized uncertainty factors. These are given in Table 17. A similar approach is used for risks due to inhaled substances. It involves replacing the no-effects dose with the no-effects atmospheric concentration. This yields a quantity known as the reference concentration (RfC).

Although reference doses per se are not defined for radionuclides, an analogous approach can be used based on threshold doses for different deterministic effects. Acute thresholds for deterministic effects are 0.5 Sv to the bone marrow for depression of homeostasis, 0.5–2 Sv to the lens of the eye for

Table 17. Default EPA Uncertainty and Modifying Factors

Type of Factor	Value
Animal to human extrapolation	10
Use of subchronic data to calculate chronic RfD	10
Use of LOAEL when NOAEL is not available	10
Interindividual human variability	10
Professional judgment (modifying factor)	>0–10

Table 18. Reference Doses for Selected Contaminants[a]

Contaminant	Target Organ	RfD, mg/kg/day
Heptachlor epoxide	Liver	0.000013
Aldrin	Liver	0.00003
Chlordane	Liver	0.00005
DDT	Liver	0.005
Carbon tetrachloride	Liver	0.007
Aldicarb	Nervous system	0.0013
Epichlorohydrin	Kidney	0.002
Paraquat	Chronic pneumonitis	0.0045
Beryllium	Growth and body weight	0.005
Chloroform	Liver	0.01
Tetrachloroethylene	Liver, kidney	0.02
Bis(2-ethyl-hexyl)phthalate	Liver, female mortality, kidney	0.02
Pentachlorophenol	Liver, kidney	0.03
Methylene chloride	Liver	0.06
Fluoride	Teeth (objectionable dental fluorosis)	0.06

[a] Ref. 50.

detectable opacity and 5 Sv for cataracts, 0.15 Sv to the testes for temporary sterility and 3.5–6 Sv for permanent sterility. However, since these levels of radiation imply a high probability of stochastic effects, a reference dose is typically not calculated for the deterministic effects of radioactivity. However, RfDs are developed for the chemical toxicity of many radioactive elements because some radionuclides, such as U-238, may be present in the environment at sufficient concentrations to cause deterministic effects due to their chemical toxicity, even when the radiological effects are relatively minimal. Presented in Table 18 are reference doses for selected chemical contaminants.

The quantitative measure of noncarcinogenic effects for Superfund assessments is the hazard quotient (HQ) as defined previously,

$$HQ = \frac{D}{RfD} \qquad (102)$$

where D is the average daily dose (for either subchronic or chronic exposures) in mg/kgd and RfD is the corresponding subchronic or chronic reference dose.

Aggregating risks from multiple noncarcinogenic contaminants results in a hazard index (HI), which is the sum of the hazard quotients for multiple contaminants and/or multiple exposure pathways. EPA practice is to segregate the HQs according to exposure duration (ie, acute, chronic, and subchronic exposures) and then report a separate hazard index for each exposure duration (49). As discussed earlier, a composite hazard index of less than one indicates that no effects are likely to be observed. If the composite HI is greater than one, the situation is a bit more complicated. It may be possible to segregate contaminants by critical effect (first toxic response observed as dosage is increased) and then sum HQs only for contaminants with similar critical effects. However, such an estimation may be time-consuming and complex, and if performed improperly, may misrepresent the actual toxic response. In a case where a composite HI of greater than one exists, the judgment of a trained toxicologist is necessary to properly interpret the effect of multiple chemical exposures.

Estimating Cancer Incidence. Cancer incidence is currently estimated using a linear, nonthreshold approximation. The key parameter in the equation is ρ, the cancer slope factor or cancer potency factor. For chemical carcinogens, virtually all of the slope factors are obtained from animal data. An appropriate dose response curve is fit to the data, and the slope factor is the slope of the curve at low doses. Since EPA recommends and publishes slope factors that correspond to the upper 95% confidence level, their values are likely to produce conservative overestimates of the actual cancer risk.

If cancer risks are high (ie, over 0.01), the dose response curves typically depart from linearity. In such cases, the one hit model may be used to estimate the risk. Fractional response according to the one-hit model is given by

$$R = 1 - e^{-\rho \cdot D} \qquad (103)$$

where R is fraction of the population responding, ρ is potency factor or slope factor, and D is dose. For radiation, the slope factor is called a probability coefficient (30). The probability coefficient for fatal cancer incidence is $5 \times 10^{-2} Sv^{-1}$.

Presented in Table 19 are slope factors from EPA's Integrated Risk Information System (IRIS) for selected chemical contaminants. IRIS is a database that contains summaries of human and animal evidence on the toxicity of contaminants and values of slope factors and reference doses.

Aggregating carcinogenic risks is straightforward in Superfund risk assessments. For each receptor population, the risks due to each contaminant and exposure pathway are summed. If these risks are high, the risks may be summed as follows:

$$R = 1 - e^{-\sum_i \rho_i \cdot D_i} \qquad (104)$$

where i is an index variable which ranges over all combinations of exposure pathways and all contaminants.

Summary. In this section, the methods available for characterizing the risks associated with contaminant exposure have been reviewed. The distinction between threshold and nonthreshold effects (and therefore deterministic and stochastic effects) will inform the choice of the appropriate metric. Some variant of the fractional response measure and the margin of safety measure are present in almost all risk assessments. Estimating maximum individual risks using reasonable maximum values for parameters and population incidence using average values for the parameters is typically performed through the use of these simple equations and metrics.

Table 19. Cancer Slope factors for Selected Contaminants[a]

Contaminant	Slope Factor, $(mg/kg/day)^{-1}$	Contaminant	Slope Factor, $(mg/kg/day)^{-1}$
Acrylonitrile	0.54	Hexachlorobutadiene	0.078
Aldrin	17	α-Hexachlorocyclohexane	6.3
Bis(chloroethyl) ether	1.1	β-Hexachlorocyclohexane	1.8
Carbon tetrachloride	0.13	Hexachlorodibenzo-*p*-dioxin	6200
1,2-Dichloroethane	0.091	Hexachloroethane	0.014
Epichlorohydrin	0.0099	1,1,2,2-Tetrachloroethane	0.2
Heptachlor epoxide	9.1	2,4,6-Trichlorophenol	0.011

[a] Ref. 50.

As the careful reader may have noticed, many of the decisions to be made in characterizing risk are severely hampered by lack of data. In addition, policy considerations often dictate the preference for certain measures over others. An accurate, scientifically grounded characterization of risk is extremely difficult to obtain even under the best of circumstances. Thus, the task of the risk analyst is to ensure that the risk manager clearly understands the strengths and weaknesses of the methods chosen.

UNCERTAINTY ANALYSIS

Background

From the preceding discussions, it is clear that there is a great potential for variability and uncertainty in assessments of human health risk. They are present in each step of the process. For example, uncertainty and variability in levels of contaminant inventory, emission rates, lack of accurate knowledge of site-specific environmental transport mechanisms, questions regarding the toxicity of the compounds, and lack of knowledge about the receptor populations may all contribute to variability and uncertainty in the final result. The process of characterizing the variability and uncertainty associated with the risk from a contaminant or contaminants is known as uncertainty analysis. An uncertainty analysis can be used for various purposes. The principal purpose is to provide an estimate of the distribution of the risks associated with a given scenario and the level of confidence which can be placed in the risk results. A second purpose is to allow a user to evaluate a risk assessment in the context of other factors in a risk management decision. A clear expression of the uncertainty in the risk calculation is necessary in the process of balancing risk against the other factors that must be taken into account such as cost, political acceptability, and prevailing social norms and priorities. Finally, by providing a means of identifying factors which dominate the variability and uncertainty and, therefore, which may affect decisions in the event of new information, uncertainty analyses can provide insight into areas where additional research would be most useful.

Uncertainty analysis is currently a highly active field of research, and several introductory texts and publications are available for further reading (51,52,53).

Characterizing Uncertainty. There are two types of uncertainty (52). Type A uncertainty refers to the variability of a parameter spatially, temporally, or across a population. Type B uncertainty refers to lack of knowledge of the true value of a fixed parameter, of the distribution of a variable parameter, or of the validity of a mathematical model. For example, the body weight of adults in the U.S. population is subject to Type A uncertainty, or variability, as there is a wide range of adult body weights throughout the population. Based on statistical sampling, it is possible to estimate the frequency distribution of different body weights and thus characterize the Type A uncertainty. However, the true mean of the distribution of body weights is not known and thus subject to Type B, or lack of knowledge, uncertainty. In the remainder of this article, the term variability is used to describe Type A uncertainty and lack of knowledge is used to describe Type B uncertainty.

Variability is represented by a frequency distribution, or a function which represents the relative frequency with which different values of a parameter are observed. If necessary, variability can be reduced by reducing the scale of the question, or by desegregating the inputs. For example, the variability in fish consumption can vary widely across the entire population, due in part to the presence in the overall population of subsistence fishermen who consume a great deal of fish. However, a site-specific survey might reveal that there is a generally low level of fish consumption in an inland rural community, and therefore the variability in fish consumption rates in that subpopulation would be smaller than in the entire U.S. population.

Lack of knowledge is typically represented using a probability distribution; that is, a mathematical function which assigns a certain probability of occurrence to each particular value of a parameter. Such a distribution may be based on an underlying frequency distribution; for example, in the fish consumption example, if an individual is picked at random from a population, then she or he has a single value of average fish consumption which is unknown. However, it is possible to assign a probability distribution to this uncertainty by assuming that since the individual is picked at random from the population, the uncertainty can be characterized by the underlying frequency distribution. Uncertainty can only be reduced by additional information through research. In addition, it is typical that the distributions describing natural variability are subject to uncertainty.

Distributions. Uncertainty and variability are characterized by distributions. Distributions can represent either the frequency with which a certain result occurs or the probability that a certain result occurs. Frequency distributions are used to characterize parameter variability and probabilities are used to characterize lack of knowledge of a parameter. Both probability and frequency distributions follow the same

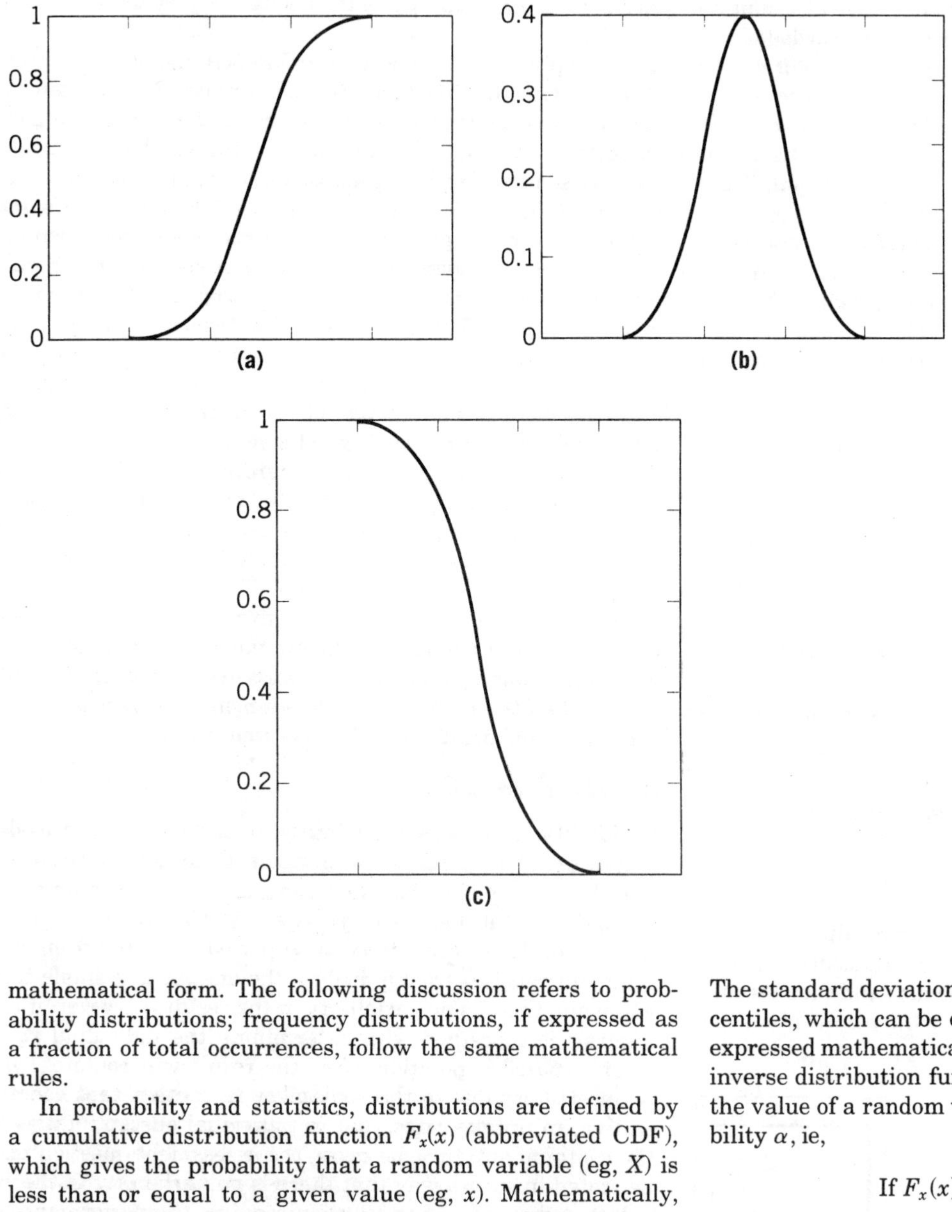

Figure 42. (**a**) Cumulative distribution function, (**b**) probability density function, and (**c**) complementary cumulative distribution function.

mathematical form. The following discussion refers to probability distributions; frequency distributions, if expressed as a fraction of total occurrences, follow the same mathematical rules.

In probability and statistics, distributions are defined by a cumulative distribution function $F_x(x)$ (abbreviated CDF), which gives the probability that a random variable (eg, X) is less than or equal to a given value (eg, x). Mathematically, this is represented by the expression $F_x(x) = P\{X \leq x\}$. An example is illustrated in Figure 42. The complementary cumulative distribution function $F_x^c(x)$ (abbreviated CCDF) is $P(X > x)$ and is equal to $1-F_x(x)$. A distribution can also be described by the probability density function, $f_x(x)$, (PDF), which is the integral of the CDF. The relationships between these different, ways of describing a distribution are

$$F_x^c(x) = 1 - F_x(x) \tag{105}$$

and

$$f_x(x) = \frac{d}{dx} F_x(x) \tag{106}$$

It is sometimes convenient to characterize a distribution using select descriptive statistics that are obtained from either the PDF or the CDF. The arithmetic mean, or first moment of the distribution, is given by

$$\mu_x = \int x \cdot f_x(x) dx \tag{107}$$

and the variance, which is the second moment of the distribution, is

$$\sigma_x^2 = \int (x - \mu_x)^2 \cdot f_x(x) dx \tag{108}$$

The standard deviation is the square root of the variance. Percentiles, which can be obtained graphically from the CDF, are expressed mathematically through the inverse of the CDF. The inverse distribution function $F_x^{-1}(\alpha)$ is a function which gives the value of a random variable associated with a given probability α, ie,

$$\text{If } F_x(x) = \alpha, \text{ then } F_x^{-1}(\alpha) = x \tag{109}$$

Thus the αth percentile is given by $F_x^{-1}(\alpha)$. Important percentiles include the median or 50th percentile ($\alpha = 0.5$) and upper bound percentiles such as the 95th ($\alpha = 0.95$) and 99th ($\alpha = 0.99$).

Analyzing Uncertainty and Variability. A quantitative uncertainty analysis is an attempt to characterize the magnitude and significance of variability and lack of knowledge uncertainty in an assessment endpoint which arises due to variability and lack of knowledge uncertainty, respectively, in the inputs. As discussed earlier, the analysis of variability is a problem distinct from that of the analysis of lack of knowledge uncertainty, although similar techniques are used for both. The following sections present some methods for the quantitative analysis of variability and lack of knowledge uncertainty.

Expert Elicitation

The quantitative specification of variability and lack of knowledge uncertainty is oftentimes subjective, requiring the scientific judgment of specialists to estimate the distributions of either model inputs or model outputs. This process is referred to as expert elicitation. The result is a subjective distribution characterizing either the variability or lack of knowledge uncertainty in a model parameter. Subjective assessment can be used at any stage in the uncertainty analysis process. It can be used to develop distributions for individual parameters in a general risk model, which can then be propagated using the mathematical techniques discussed in the next section; it can be used to develop distributions for intermediate values, such as the contaminant concentration at a particular location; or it can be used to determine the distribution of risk itself, provided that the risk metric can be defined in a clear and unambiguous way.

NCRP Commentary 14 (53) summarizes the 10 steps of a formal elicitation:

- Selection of issues and associated parameters
- Selection of experts
- Preparation of background information
- Training of the selected experts in the elicitation method
- Presentation of issues, questions, and background information to experts
- Analysis of issues, questions, and background information by experts
- Discussion of analyses of issues and background information by experts
- Elicitation of judgments from the experts
- Consolidation (and possibly aggregation) of the elicitation results
- Review and communication of the elicitations

It is important to realize that different results may be obtained on the same issues from different assessment groups. Obtaining results from independent groups and then reconciling the differences can lend strength to the results of an elicitation. Further discussion of the elicitation process is available (54,55).

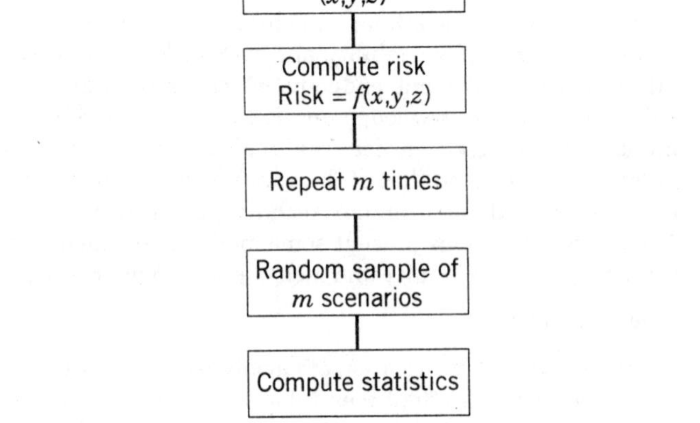

Figure 43. Diagram of the Monte Carlo process (53).

Three protocols have been used for eliciting subjective probability judgments (51). Of these, the Stanford/SRI Assessment Protocol was the first to be developed, and it provides a good example of the sequence of events used in conducting subjective assessments. This protocol was developed by decision analysts at Stanford University during the 1960s and 1970s and consists of five stages; motivation, structuring, conditioning, encoding, and verifying. The motivational stage is an introductory step in which the problem is presented and the possibility of a motivational bias is assessed. The structuring stage seeks to establish a clear, unambiguous definition of the value to be assessed. The conditioning phase aims to elicit the experts' cognitive processes and make them transparent to the subject matter expert and the analyst. This assists in eliminating potential cognitive biases such as overconfidence, in which an expert does not assign a sufficiently large variability to their belief, and anchoring, in which the estimate of risk is biased towards the initial estimate. The fourth phase, encoding, is the actual process of generating a probability distribution from the expert's knowledge. This typically involves the establishment of extreme values and then filling in the probabilities of values within the bounds. Finally, the results are verified in the fifth stage, when the results of the assessment are discussed with the expert and are checked for consistency.

Analysis of Variability

Variability, or Type A uncertainty, in the results of a model output are strictly due to the framing of the assessment question. For example, if the assessment endpoint is contaminant concentration at a given location at a given time or the total cancer fatalities over a fixed time period resulting from contaminant releases from a facility, the answer is a single true value and there is no variability in the result. However, if the analysis is concerned with determining the individual risks to an exposed population, then the result will contain Type A uncertainty due to the variability in contaminant concentration, in uptake rates, and in individual susceptibilities to the contaminant. In some cases, the assessment question may be stated in such a way that there is no variability in the result, but the underlying question requires the performance of a variability assessment. For example, the average individual risk across a population is a single value, but determination of the average requires a variability analysis.

There are a variety of techniques available for propagating the variability in the inputs through the risk model. Mathematical methods include numerical simulation techniques such as Monte Carlo analysis and analytical methods used either to approximate the parameters of the output distribution or to determine a closed form expression for the distribution of risk.

Monte Carlo Simulation. The Monte Carlo method is probably the most widely used technique for propagating variability through calculations of environmental risk. Since the advent of fast, inexpensive computing power, highly sophisticated numerical methods are available for users of desktop computers. In this section, the basic principles of Monte Carlo methods are considered. The process, which is illustrated in Figure 43, is conceptually simple.

The combinations of the input values are obtained by random sampling from the distributions assigned to the input variables. Each combination of input values defines a scenario. These values are used in the risk model to estimate the risk associated with that scenario. By repeating this process, a random sample is obtained from the distribution of risk. Standard statistical techniques may then be used to estimate parameters of this risk distribution, such as the mean, variance, and percentiles. In addition, the accuracy of these estimates may be evaluated using standard statistical techniques.

There are essentially five steps in a Monte Carlo analysis as applied in uncertainty analysis:

1. Computer generation of a pseudo-random number simulating a uniformly distributed random variable.
2. Transformation of the uniform sample to the desired random variable.
3. Sampling from the distributions for each of the input parameters to generate the desired number of input scenarios and calculation of the risk for each input scenario.
4. Calculation of descriptive statistics of the output distribution such as the mean, variance, or fractiles
5. Estimation of the accuracy of the statistics.

Several variations of the basic technique are available, designed to reduce the number of scenarios which must be generated and thus reduce the required computing effort. The basic technique is known as simple or crude Monte Carlo, in which a value for each of the n input variables is obtained by random sampling from the n input distributions. These inputs are then used to determine a value of the output by performing a deterministic calculation. This process is repeated m times, resulting in a sample of size m. Standard statistical techniques are then used to estimate descriptive statistics of the output distribution, along with the associated uncertainty about those statistics. Crude Monte Carlo is the technique illustrated in Figure 43.

Importance sampling is a method of concentrating attention on specific areas of the distribution which may be of interest. For example, in an analysis of low probability, high consequence accidents, interest may be focused on the extreme tails of the distributions rather than the behavior of the mean. In importance sampling, values of particular importance (eg, tails of distributions) are sampled more often. This is done by artificially inflating the probability of the values of interest through weighting. After the sampling is performed, the probability is generated by reducing the value of the output samples proportional to the amount by which they had been weighted.

In stratified sampling, the input distributions are divided into intervals, and each of these intervals is sampled separately. A common form of stratified sampling is known as Latin hypercube sampling (LHS). In LHS, each input distribution is divided into m equiprobable intervals. Each of these intervals is then sampled, yielding m values of the input variables. A scenario is then generated by sampling one value at random from each of the m values of the n inputs without replacement. This yields m scenarios, with each value from each input being used only once. Because the sample scenarios tend to be more spread out over the input domain, the values for the moments of the output are more exact than with unstratified sampling. However, the statistics for estimating the accuracy of the moments are more difficult to compute and may yield an underestimate of the true precision.

Analytical Approximations. In some cases, analytical approximations are available to allow derivation of the moments of an output distribution. Although these techniques are generally not as accurate as Monte Carlo simulations, they can be useful in screening analyses for the rapid computation of descriptive statistics for the output distribution and identification of potential major contributors to uncertainty. In cases where the model is strictly linear with respect to the inputs, the results are exact. The following discussion focuses on a technique known as error propagation.

The error propagation technique is based on representing the function under consideration as a polynomial derived from a Taylor series expansion. This expansion is taken about the means of the input variables. Thus, this approximation substitutes a polynomial surface for the more complex response surface. The first-order approximation yields a hyperplane tangent to the response surface at the point defined by the vector of input means. It then is clear that this approximation is fairly accurate about the point of expansion, but that the error induced by the expansion may be significant away from the point of expansion.

To derive estimates of the mean and variance, the first order Taylor series approximation is substituted into equations 107 and 108 resulting in the following expressions for the approximate mean and variance:

$$\mu_z = f(\mu_i) \tag{110}$$

$$\sigma^2 = \sum_i \sigma_i^2 \left(\frac{\partial f}{\partial x_i} \Big|_{\mu_i} \right)^2 \tag{111}$$

It can be seen from equation 110 that the error propagation variance in the model output is simply a weighted average of the variances of the model inputs.

The error propagation method can be quite useful under certain conditions. However, it does have some fairly significant limitations. The accuracy of an error propagation estimate is a function of the linearity of the model and the variance in each of the inputs. It is often thought that taking higher order terms in the Taylor series expansion yields a more accurate approximation. However, due to the local nature of the expansion, and the global nature of the integration, this does not hold true in general. If the model is approximately linear in the region where the inputs are likely to lie, then the method can produce good results. However, for a highly nonlinear model, the estimates can be significantly in error, possibly by several orders of magnitude.

Analytical Methods. Analytical methods can potentially yield the most powerful and flexible tools available for performing propagation of uncertainty and variability, but they can involve highly intricate calculations. Development of formal expressions for transformation of random variables requires some mathematical background in probability theory, as it can require methods such as integral transform techniques. However, if analytic techniques can be used, the generality of the results make them the most powerful of all of the methods. The result of this method is a derived distribution of the variability of the model output. The following section presents some basic theorems used in formally computing output distributions. Among the conceptually simpler techniques, several operate directly on the cumulative density function and the probability density function. Two of these are the transformation of CDFs and the transformation of PDFs.

The transformation of cumulative distribution functions is the most general method, requiring no restrictive assumptions. Provided that the risk model is a continuous function, the basis for this technique is given by

$$F_Y(y) = P\{g(X_1, X_2) \le y\} = \int\int_{\{(x_1,x_2):g(x_1,x_2)\le y\}} f_{x_1,x_2}(x_1, x_2)dx_1dx_2 \quad (112)$$

Essentially, the joint density of the input variables is integrated over the response surface generated by the model. Although this method does not require any restrictive assumptions, it can often be difficult to set up the limits of integration properly. Except in problems with one or two variables and a simple functional relationship between the inputs and outputs, this method can rapidly become intractable. Under certain conditions, transformation of the probability density functions directly offers a potentially simpler approach, which does not require complex integration. For a function of a single variable, this method can be fairly straightforward, reducing to the relationship

$$f_Y(y) = f_x(g^{-1}(y))\left|\frac{d}{dy}g^{-1}(y)\right| \quad (113)$$

Essentially, the inverse of the risk model (ie, the value of a parameter which yields a given risk) is substituted into the probability density function of the input variable, and this is multiplied by the absolute value of the derivative of the inverse.

Unfortunately, few risk equations are known to be dominated by the variability in a single variable, and the technique becomes more complex with multivariate expressions.

Alternative Techniques. In addition to the techniques discussed above, there are a variety of other methods for propagating variability. These include, for example, response surface methodologies (56) and discrete probability distributions (57,58). These have basic similarities with the techniques discussed above. For example, response surfaces are polynomial approximations of a model output generated by fitting a polynomial to the results of a model output, rather than using a Taylor series approximation to approximate the output. Discrete probability distributions are conceptually similar to Monte Carlo methods, except that the combinations of potential scenarios are explicitly identified and conditional probabilities used to estimate the result.

Analysis of Lack of Knowledge Uncertainty

Lack of knowledge (Type B) uncertainty arises from a variety of sources. These can be grouped under the following areas:

1. Lack of knowledge of correct model form
2. Lack of knowledge of true parameter values
3. Variability in model parameters
4. Analyst errors

Quantitative risk assessment, like all ventures in mathematical modeling of the physical world, inevitably introduces errors in reducing complex biological and physical systems to mathematical models. Regardless of the mathematical complexity of a model, it is an abstract simplification of the real world. In some cases, the physical and biological processes may be relatively well characterized, with broad scientific consensus on the correct approach to model a given phenomenon. In these cases lack of knowledge uncertainty is primarily a question of proper parameterization of the model and in the agreement between the assumptions of the model and the real physical circumstances. In other cases, however, there is not such agreement regarding the proper modeling approach. For example, there is a great deal of controversy about modeling the effects of exposure to contaminants at the low doses characteristic of environmental exposure. In such cases, professional judgment is required, and a thorough documentation of the rationale for selecting a particular model

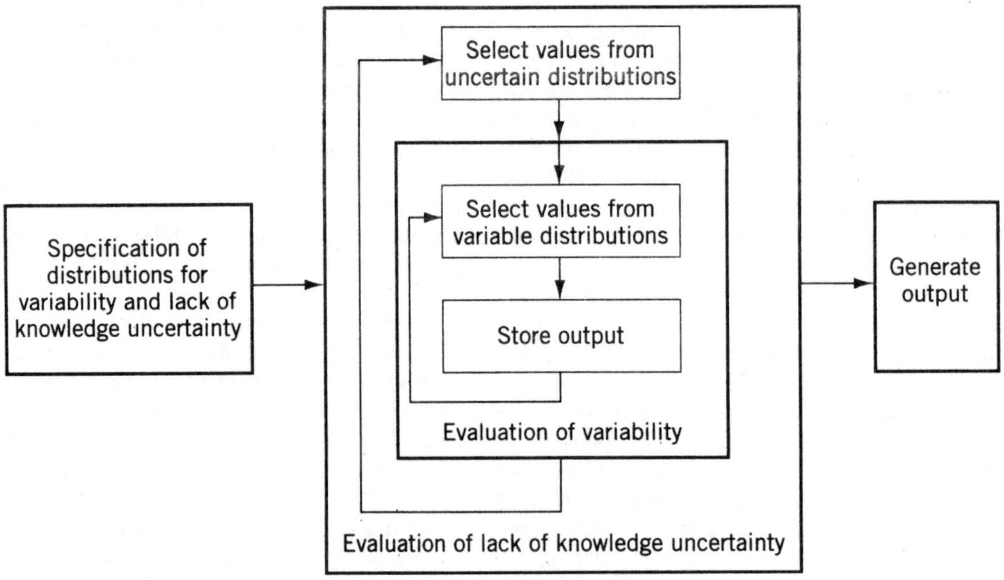

Figure 44. Two-stage Monte Carlo assessment of uncertainty and variability.

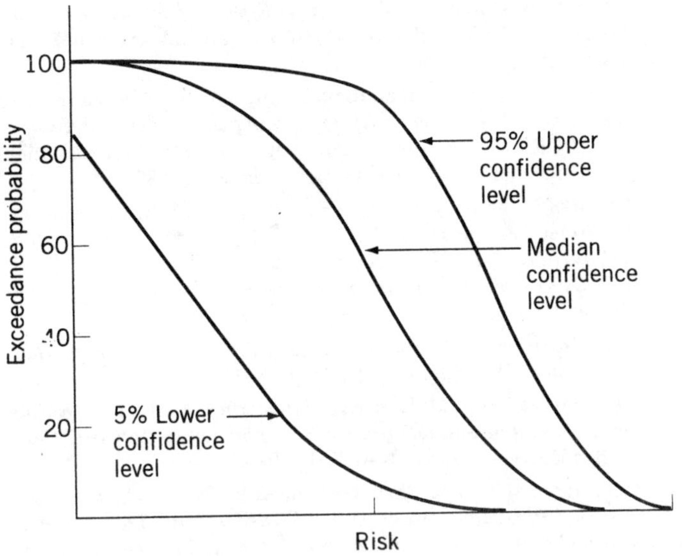

Figure 45. Presentation of results of a combined Type A/Type B uncertainty assessment.

is necessary to aid the risk manager in interpreting the results of the assessment.

Estimating the overall uncertainty, due to both variability and lack of knowledge uncertainty, in the results of an analysis is difficult. Current practice involves the use of the methods discussed previously for estimating variability. In Monte Carlo assessments, for example, the uncertainty in model results is estimated using a two-step Monte Carlo procedure, in which the inner loop is the estimation of variability and the outer loop is a Monte Carlo sampling of the uncertain parameters of the input frequency distribution. This is illustrated in Figure 44.

The results of such an analysis are shown in Figure 45, reflecting both the variability and the lack of knowledge uncertainty in the final result. An alternative technique is to use expert elicitation to determine the overall uncertainty in an assessment endpoint, using the techniques discussed previously.

BIBLIOGRAPHY

1. National Research Council, *Risk Assessment in the Federal Government: Managing the Process,* National Academy Press, Washington, D.C., 1983.
2. V. Covello and M. Merkhofer, *Risk Assessment Methods—Approaches for Assessing Health and Environmental Risks,* Plenum Press, New York, 1993.
3. American Association of Engineering Societies, *Risk Analysis: The Process And Its Application,* Institute for Regulatory Science, Columbia, Md., 1996.
4. N. Rasmussen, "The Application of Probabilistic Risk Assessment Techniques to Energy Technologies," in T. Glickman and M. Gough, eds., *Readings in Risk,* Resources for the Future, Washington, D.C., 1990.
5. H. Kumamoto and E. Henley, *Probabilistic Risk Assessment and Management for Engineers and Scientists,* IEEE Press, New York, 1996.
6. M. Clark, *Transport Modeling for Environmental Engineers and Scientists,* John Wiley & Sons, Inc., New York, 1996.
7. J. Schnoor, *Environmental Modeling: Fate and Transport of Pollutants in Water, Air, and Soil,* John Wiley & Sons, Inc., New York, 1996.
8. P. Gerhart and R. Gross, *Fundamentals of Fluid Mechanics,* Addison-Wesley Publishing Co., Reading, Mass., 1990.
9. S. Kao, "Theories of Atmospheric Transport and Diffusion," in D. Randerson, ed., *Atmospheric Science and Power Production,* DOE/TIC-27601, National Technical Information Service, Springfield, Va., 1984.
10. S. Barr and W. Clements, "Diffusion Modeling: Principles and Applications" in D. Randerson, ed., *Atmospheric Science and Power Production,* DOE/TIC-27601, National Technical Information Service, Springfield, Va., 1984.
11. U.S. Nuclear Regulatory Commission, *Regulatory Guide 1.111—Methods for Estimating Atmospheric Transport and Dispersion of Gaseous Effluents in Routine Releases from Light Water Reactors,* USNRC, Washington, D.C., 1977.
12. G. Briggs, I. Van der Hoven, R. Engelmann, and J. Halitsky, "Processes Other Than Natural Turbulence Affecting Effluent Concentrations," in D. Slade, ed., *Meteorology and Atomic Energy 1968,* TID-24190, National Technical Information Service, Springfield, Va., 1968.
13. F. Gifford, "An Outline of Theories of Diffusion in the Lower Layers of the Atmosphere" in Ref. 12.
14. H. Brenk, J. Fairobent, and E. Markee, "Transport of Radionuclides in the Atmosphere" in J. Till and H. Meyer, eds., *Radiological Assessment,* NUREG/CR-3332, National Technical Information Service, Springfield, Va., 1983.
15. H. Peterson, "Terrestrial and Aquatic Food Chain Pathways" in Ref. 14.
16. Office of Solid Waste, U.S. EPA, *Background Document: Groundwater Screening Procedure,* U.S. Environmental Protection Agency, National Technical Information Service, Springfield, Va., 1986.
17. R.G. Shepherd, *Ground Water,* **27**(5), 633 (1989).
18. J. Bear, *Dynamics of Fluids in Porous Media,* American Elsevier Publishing Co., New York, 1972.
19. C. Fetter, *Contaminant Hydrogeology,* Prentice Hall, Upper Saddle River, N.J., 1992.
20. C. Fetter, *Applied Hydrology, 2d ed.,* Macmillan Publishing Co., New York, 1988.
21. S. Neuman, *Water Resourc. Res.* **26**(8), 1749–1758 (1990).
22. M. Sheppard and D. Thibault, *Health Phys.* **59**(4), 473 (1990).
23. M. van Genuchten and W. Alves, *Analytical Solutions of the One-Dimensional Convective-Dispersive Solute Transport Equation,* Technical Bulletin No. 1661, U.S. Dept. of Agriculture, 1982.
24. P. Domenico and V. Palciauskas, *Groundwater* **20,** 303–311 (1982).
25. H. Fischer, E. List, R. Koh, J. Imberger, and N. Brooks, *Mixing in Inland and Coastal Waters,* Academic Press, Inc., San Diego, Calif., 1979.
26. G. Suter, ed., *Ecological Risk Assessment,* Lewis Publishers, Chelsea, Mich., 1993.
27. G. Sehmel, "Deposition and Resuspension" in Ref. 9.
28. Nuclear Regulatory Commission, *Residual Radioactive Contamination from Decommissioning* (NUREG/CR-5512, Vol. I), National Technical Information Service, Springfield, Va., 1992.
29. Nuclear Regulatory Commission, "Regulatory Guide 1.109—Calculation of Annual Doses to Man from Routine Releases of Reactor Effluents for the Purpose of Evaluating Compliance with 10 CFR 50, Appendix I" Nuclear Regulatory Commission, Washington, D.C., 1977.

30. International Commission on Radiation Protection, *1990 Recommendations of the International Commission on Radiation Protection (ICRP Publication 60),* Pergamon Press, New York, 1991.
31. M.E. Gillies, and H.V. Paulin, *Int. J. Epidemiol.* **12**(1), 45–50 (1983).
32. International Commission on Radiation Protection, *Report of the Task Group on Reference Man,* Pergamon Press, New York, 1981.
33. E.J. Calabrese, P.T. Kostecki, and C.E. Gilbert, "How Much Soil do Children Eat? An Emerging Consideration for Environmental Health Risk Assessment", *Exposure Factors Handbook,* PB90-106774, U.S. Environmental Protection Agency, Washington, D.C., 1989.
34. C.D. Klaasen, and D.L. Eaton, "Principles of Toxicology," in M.O. Amdur, J. Doull, and C.D. Klaasen, eds., *Casarett and Doull's Toxicology: The Basic Science of Poisons,* 4th ed., Pergamon Press, New York, 1991.
35. J.W. Harris, and J.W. Kellermeyer, *The Red Cell- Production, Metabolism, and Destruction: Normal and Abnormal,* rev. ed., Harvard University Press, Cambridge, Mass., 1970.
36. W.R. Hewitt, R.S. Goldstein, and J.B. Hook, "Toxic Responses of the Kidney" in Ref. 34.
37. International Agency for Research on Cancer, monographs Vols. 1–60, WHO, 1972–1994; U.S. Dept. of Health and Human Services, *Sixth Annual Report on Carcinogens,* National Toxicology Program, Washington, D.C., 1991.
38. J. Manson and L.D. Wise, "Teratogens" in Ref. 34.
39. R.L. Dedrick, *J. Pharmacokin. Biopharmaceut.* **1,** 435 (1973).
40. Environ Corp., *Elements of Toxicology and Chemical Risk Assessment,* rev. ed., Environ Corp., Arlington, Va., 1988.
41. S. Iverson and N. Arley, *Acta Pathol. Microbiol. Scand.* **27,** 773 (1950).
42. J. Cornfield and N. Mantel, *Biometrics* **33,** 1 (1977).
43. P. Armitage and R. Doll, *Brit. J. Cancer* **8,** 1 (1954).
44. K.S. Crump, "Mechanisms Leading to Dose Response Models", in P. Ricci, ed., *Principles of Health Risk Assessment,* Prentice Hall, Englewood Cliffs, N.J., 1985.
45. S.H. Moolgavkar and D.J. Venzon, *Mat. Biosci.* **47,** 55–77 (1979).
46. S.H. Moolgavkar and A.G. Knudson, *J. Nat. Cancer Inst.* **61,** 49–52 (1981).
47. National Research Council, *Health Effects of Exposure to Low Levels of Ionizing Radiation (BEIR V),* National Academy Press, Washington, D.C., 1990.
48. National Council on Radiation Protection and Measurements, *Comparative Carcinogenicity of Ionizing Radiation and Chemicals (NCRP Report Number 96),* National Council on Radiation Protection and Measurements, Bethesda, Md., 1989.
49. Office of Emergency and Remedial Response, U.S. Environmental Protection Agency, *Risk Assessment Guidance for Superfund, Vol. I: Human Health Evaluation Manual (Part A) Interim Final,* EPA/540/1-89/002, U.S. Environmental Protection Agency, Washington, D.C., 1989.
50. U.S. Environmental Protection Agency, *Integrated Risk Information System,* Washington, D.C., 1996.
51. M.G. Morgan, and M. Henrion, *Uncertainty: A Guide to Dealing with Uncertainty in Qualitative Risk and Policy Analysis,* Cambridge University Press, New York, 1990.
52. International Atomic Emergency Agency, *Evaluating the Reliability of Predictions Made Using Environmental Transfer Models (Safety Series 100),* International Atomic Emergency Agency, Vienna, 1989.
53. National Council on Radiation Protection and Measurements, *A Guide for Uncertainty Analysis in Dose and Risk Assessments Related to Environmental Contamination (NCRP Commentary No. 14),* National Council on Radiation Protection and Measurements, Bethesda, Md., 1996.
54. E.J. Bonano and co-workers, *Elicitation and Use of Expert Judgement in Performance Assessment for High Level Radioactive Waste Repositories (NUREG/CR-5411),* Sandia National Laboratory, Albuquerque, N.M., 1990.
55. M. Meyer and J. Booker, *Eliciting and Analyzing Expert Judgement: A Practical Guide,* Academic Press, Inc., New York, 1991.
56. D.J. Downing, R.H. Gardner, and F.O. Hoffman. *Technometrics,* **27**(2), 151–163 (1985).
57. S. Kaplan *Risk Anal.* **1,** 198–198 (1981).
58. S. Ahmed and co-workers, *Nucl. Technol.* **59,** 238 (1982).
59. U.S. Atomic Energy Commission, *The Potential Radiological Implications of Nuclear Facilities in a Large Region in the USA in the year 2000 (WASH-1209),* Washington, D.C., 1973.

S

SALT WASTE IN OIL AND GAS PRODUCTION, REMEDIATION

LLOYD E. DEUEL, JR.
Soil Analytical Services, Inc.
College Station, Texas

This article is intended for all audiences interested in the effects and impact of salt on the natural environment, as it relates to the petroleum extraction industry, and measures necessary for remediation.

Hazardous wastes, including heavy metals, hydrocarbons, and other substances, are perceived by the public as a greater environmental threat associated with the oil and gas exploration and production industry than salt, and thus receive greater public scrutiny and corporate attention. However, in terms of actual damage to the environment and capital outlay necessary for remediation, all other constituents pale in comparison to salt. This article

- examines salt sources and concerns for an adverse impact
- defines salinity effects on soils
- defines sodium effects on soils
- describes means available for characterizing contaminated soils
- provides practical methods for reclaiming contaminated soils
- defines passive remediation and conditions that may warrant the wasting of land resources

QUANTIFYING SALT IMPACTED SOIL

The exploration and production (E&P) function of the petroleum industry generates liquid and solid wastes from a variety of activities. Produced water constitutes the vast majority of liquid wastes. The American Petroleum Institute (API) estimates about 98% of the total E&P wastes are produced water (1). Solid wastes are primarily drilling and completion fluids and cuttings. E&P generates about 300 million barrels of drilling waste annually (2). Many of these waste streams contain large amounts of salt.

SOURCES AND CONCERNS FOR AN ADVERSE IMPACT

The primary source of salt in the oil and gas industry is produced water. This is often referred to as *oilfield brine*. It is comprised of a complex mixture of dissolved constituents called ions. The ions are equally balanced between positively charged elements called cations (eg, calcium, Ca^{+2}; magnesium, Mg^{+2}; sodium, Na^{+1}; potassium, K^{+1}) and negatively charged anions (eg, sulfate, SO_4^{-2}; carbonate, CO_3^{-2}; bicarbonate, HCO_3^{-1}; chloride, Cl^{-1}). The primary constituent common to produced water is sodium chloride (NaCl), averaging about 38,000 mg/liter in combined mass. Some are saturated with concentrations in excess of 280,000 mg/liter, as NaCl. Principal constituent composition for produced waters sampled in Oklahoma, Louisiana, Michigan, Texas, and Alaska are given in Table 1. Average NaCl concentration for the produced waters represented in Table 1 is 72,901 mg/liter.

Salts associated with E&P solids emanate from additives used in drilling fluid formulations and geologic strata circulated out of the hole as cuttings. These solids are usually commingled and disposed of in the reserve pit constructed for this purpose. Salinity levels measured for 1,198 reserve pit samples analyzed by the author averaged 38.3 dS/m. This is equal to 18,548 mg/kg total salt calculated for E&P waste solids at the average saturated paste moisture equivalent of 79% and conversion constant of 613 mg/L/dS/m (3). The total salt equivalent for 300 million barrels of E&P waste solids at this concentration is approximately 1,000,000 tons.

Impact thresholds and effects that salt has on both soil and water resources is well defined in the scientific literature. Salt affects soil by rendering soil moisture unavailable to plants, disrupting the normal nutrient balance, and causing physical deterioration. Salt contaminating water resources can make it unfit for irrigation, and human or animal consumption. Of these, irrigation water is the most sensitive to salt.

MEASURING THE EFFECT OF SALT ON SOILS

The effect of salt on soils is measured in terms of salinity (soluble salts in soil solution) and sodicity (sodium adsorbed or bound to soil particles). Salinity defines the concentration of ions dissolved in soil water, and is measured directly by electrical conductivity (EC). Soil water is held under tension or suction and as such is not available for analysis. To compensate for this, the standard procedure is to measure soil EC on a water extract. Soil scientists use what is called a saturated paste soil water extract. A saturated paste is prepared by adding pure water to a soil sample until it is saturated (the void space is full). Soil at the saturated moisture equivalent has a paste-like consistency, glistens at the surface, and flows slowly when tipped on its side (4). The paste is stirred gently and allowed to stand overnight, then extracted by vacuum and analyzed for specific conductance.

The saturated paste method provides an estimate of salt in the soil at normal field moisture equivalents. Saturation percentage is about four times the 15-atmosphere moisture equivalent or permanent wilting point (irreversible plant desiccation), and approximately two times the 0.33 atmosphere moisture equivalent or field capacity (moisture held in soil against drainage by gravity). Soil with a saturated paste EC of 4 desiSeimens per meter (dS/m) is considered saline by agricultural standards and can inhibit growth of sensitive plants. The general crop response to soil EC is given in Table 2. Specific crop responses are presented by agronomic class in Figures 1, 2, and 3.

MEASURING THE EFFECT OF SODIUM ON SOILS

Two parameters are used to measure and evaluate the effects of excess sodium in soils (4). These are the sodium adsorption ratio (SAR) and the exchangeable sodium percentage (ESP). SAR is a parameter used to describe the composition of irrigation water or existing soil solution and is used to predict the

Table 1. Comparative Analysis of Produced Water

	Values by Source[a]				
Parameter	Oklahoma	Louisiana	Michigan	Texas	Alaska
pH, su	7.1	7.2	6.4	6.8	6.9
EC, dS/m	61.9	115	384	215	38.8
Sodium, mg/L	13,050	19,070	49,200	42,869	7,775
Calcium, mg/L	730	1,129	25,500	13,848	213
Magnesium, mg/L	318	253	4,400	2,437	206
Chloride, mg/L	20,412	28,980	80,000	88,147	15,000
Sulfate, mg/L	2,021	3,504	na	3,504	na
Bicarbonate, mg/L	207	579	na	579	na

[a] na = Not analyzed.

Table 2. General Crop Response as a Function of EC[a]

EC, dS/m	Effect on Crop Yield
0–2	None
2–4	Slight
4–8	Many crops affected
8–16	Only tolerant crops yield well
>16	Only very tolerant crops yield well

[a] Ref. 4.

ESP. SAR is defined by the following equation:

$$\mathrm{SAR} = \mathrm{Na}^{+}/[(\mathrm{Ca}^{+2} + \mathrm{Mg}^{+2})/2]^{1/2}$$

where cation concentrations are expressed in milliequivalents/liter. ESP is either measured directly as the exchangeable sodium divided by the cation-exchange capacity (CEC) × 100, or estimated from SAR by the following equation:

$$\mathrm{ESP}/(100 - \mathrm{ESP}) = 0.015 \times \mathrm{SAR}$$

This equation assumes an equilibrium status and may not reflect the true condition if the soil is not in equilibrium with the irrigation water or soil water extract reflecting a recent spill event. A soil is considered sodic if the ESP exceeds 15% of the CEC. CEC refers to the negative charge of soil colloids (clay size particles) and the attraction of this negative charged surface for the positively charged cations (6). Capacity refers to the total number of cations that must be adsorbed to achieve electrical neutrality. This is a surface phenomenon and the cations adsorbed are continually being desorbed and replaced by other cations in soil solution, thus the term cation-exchange capacity. The dominance of a given cation in soil solution favors its adsorption in an exchange reaction. Too much sodium in soil solution results in a sodic soil.

Excess sodium is toxic to plants and causes deterioration of soil structure. Deterioration of soil structure is closely associated with the processes of swelling and dispersion. Calcium-saturated clay particles are comprised of four or more parallel clay platelets called tactoids (Fig. 4). Tactoids clump together to form particles and aggregates. As sodium associated with brine leaks or spills replaces calcium on the external exchange sites, particles within aggregates separate form each other and become dispersed (Fig. 5). Dispersed particles can move from one layer to the next becoming lodged in finer pores, and eventually clog up the soil. As sodium increases and replaces calcium in the interlayer the tendency to swell increases. Multivalent cations provide electrostatic bridges between platelets. Sodium is not only monovalent but also has a large

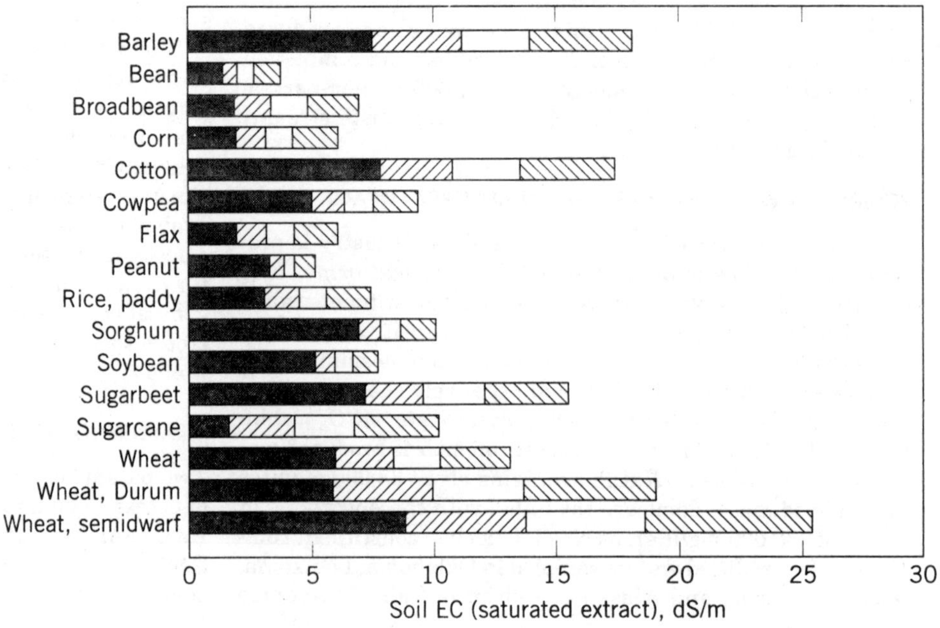

Figure 1. Yield decrease of fiber, grain, and special crops due to soil salinity, where ▧ is 30–50%; □, 15–30%; ▨, 0–15%; and ■ 0% (5).

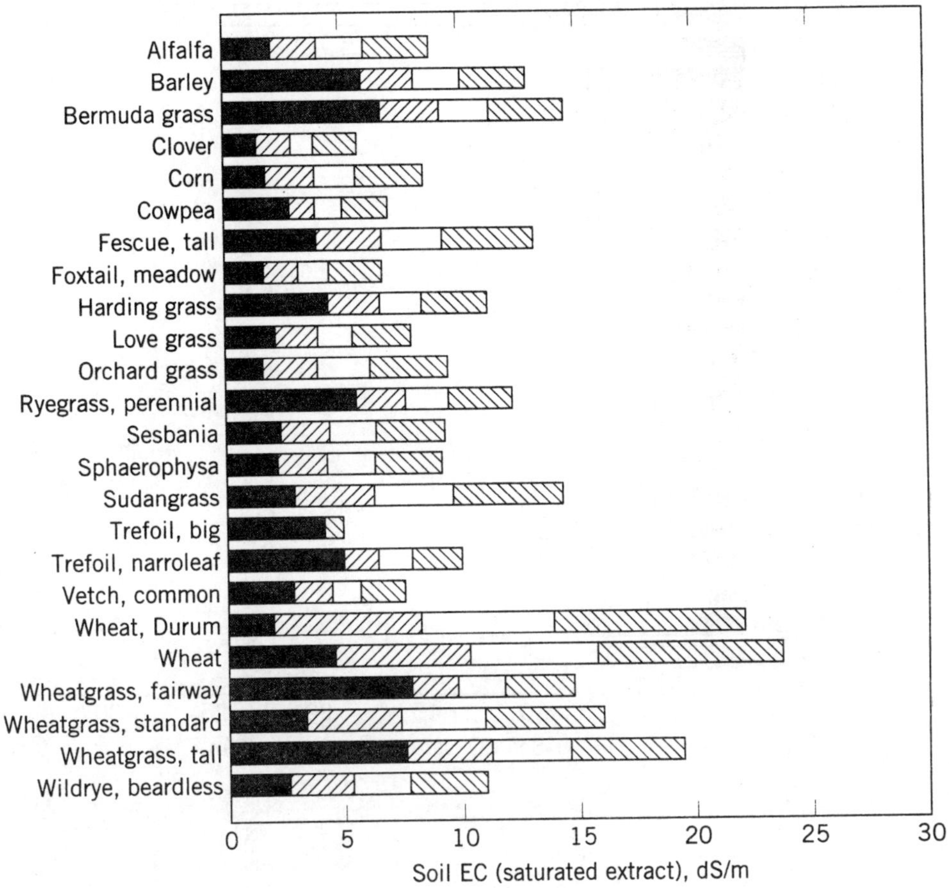

Figure 2. Yield decrease of grasses and forage (barley, corn, cowpea, wheat, and wheat (Durum)), crops due to soil salinity, where ▧ is 30–50%; □, 15–30%; ▨ 0–15%; and ■ 0% (5).

hydrated radius that further increases the distance between platelets. The deterioration of tactoid structure is generally complete at an ESP of about 50%. This condition provides the greatest challenge in reclamation of salt-affected soils.

REMEDIATION OF SALT IMPACTED SOILS

Remediation of Saline Soils

As stated previously, a saline soil is one with a saturated paste EC in excess of 4 dS/m. Reclamation is indicated when the salt level limits or alters the intended land use. The only way to reclaim a saline soil is to leach it with fresh water. Soluble salts are carried with the water to a maximum depth of penetration that will prevent the return of salt by capillary action. Irrigation is required if the evapotranspiration exceeds precipitation (7).

Soil structure is generally not a limiting factor for saline soils. However, a textural or structural impediment to leaching requires some form of drainage if the restriction is above the required depth of penetration. The quantity of water needed for reclamation is dependent on the required depth of penetration and texture which controls the effective pore volume. The requirement may be as great as 1.5 to 2 pore volumes. For a heavy clay soil the depth of water required for leaching is roughly equivalent to the required depth of penetration. In arid regions a heavy clay soil could require up to 1.5 m (5 ft) of fresh water per acre of affected land.

Sometimes a moderately saline soil (EC $<$ 25 dS/m), comprised of coarse or medium texture, can be reclaimed to support salt-tolerant crops, by adding up to 10% low salt manure. The addition of organic matter increases the cation-exchange capacity (Table 3) of the affected soil. A 10% treatment can adsorb up to 3175 kg (7000 lb) of salt per acre.

Salt not adsorbed is diluted in soil solution because organic matter dramatically increases the available water content of the soil. Available water refers to the moisture held in soils between field capacity and its wilting point (Fig. 6). The addition of 10% organic matter is expected to increase the available moisture about 4.8 cm (1.9 in.) per acre. Some fertility adjustment may be necessary by adding fertilizer to replace plant nutrients lost by the leach process required to remove excess salt.

REMEDIATION OF SODIC SOILS

Reclaiming a sodic soil is far more difficult than reclaiming a saline soil. The condition of sodicity (ESP $>$ 15%) requires the addition of calcium amendments, time for them to react with the soil, and percolation of water through the soil to remove excess sodium (8). Whether or not this can even be done depends on the degree of dispersion and swelling.

The usual practice is to apply sufficient calcium to replace adsorbed sodium from the top 30 cm (12 in.) of the impacted soil. Calcium is normally obtained from either high ionic strength, soluble calcium sources (eg, $CaCl_2$ or $Ca(NO_3)_2$), or gypsum ($CaSO_4 \cdot 2H_2O$). Acids are also effective in combination with dolomitic lime ($CaMgCO_3$). Lime alone is of no benefit unless the soil is acidic. This would be an anomaly since most sodic soils are strongly alkaline with a pH in excess of 8.5. Dry powder or granular amendments must be mechanically mixed

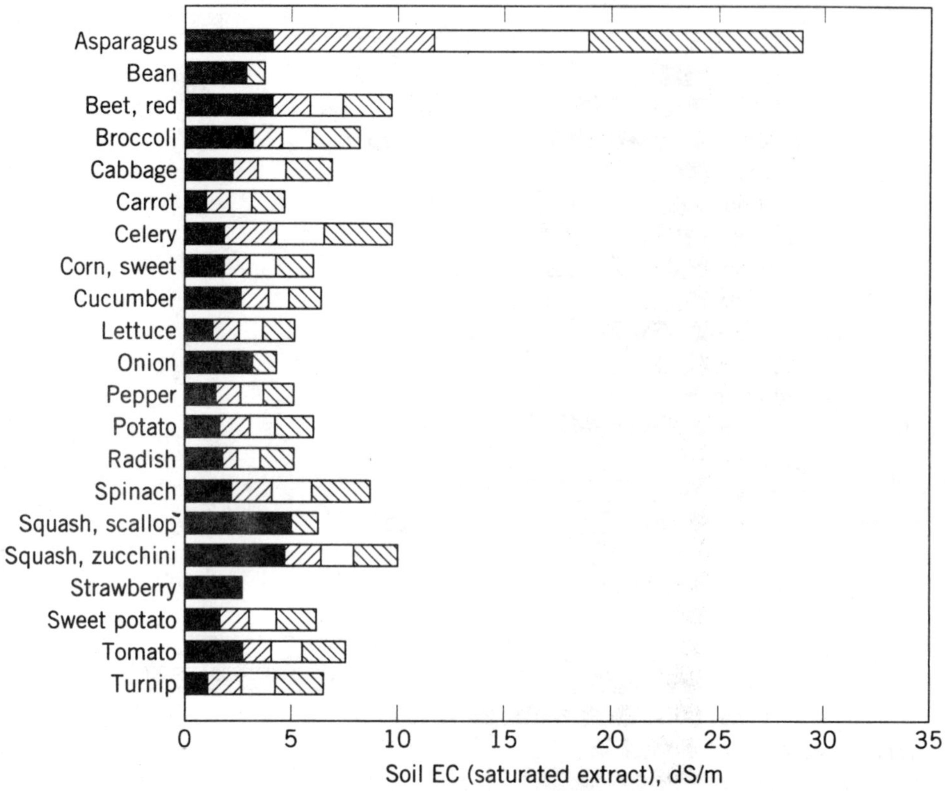

Figure 3. Yield decrease of vegetable and fruit crops due to soil salinity, where ▧ is 30–50%; □, 15–30%; ▨, 0–15%; and ■ 0% (5).

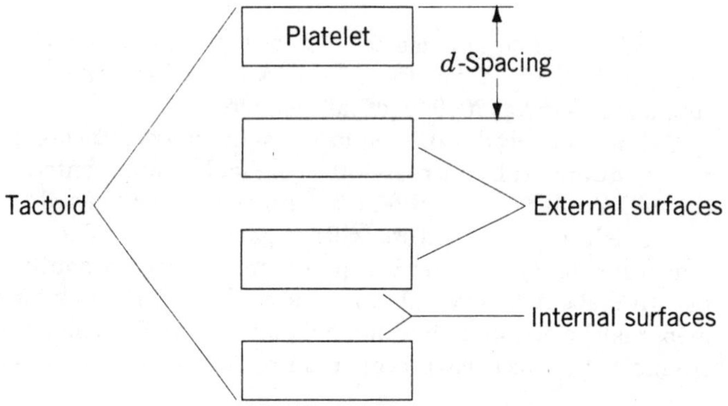

Figure 4. Surface features of clay platelets forming tactoids.

in the soil and require a long time (3–5 yr) to react. Liquid products are generally faster reacting (3–6 months) and do not require mechanical operations unless a strong surface crust, hardpan, or other near-surface cemented layers restrict infiltration and penetration of the treatment into the soil.

As a general rule, when time is a factor or mechanical means are not available or practical for product distribution, then soluble calcium sources are indicated. Soluble calcium sources are particularly effective on recent brine spills that have not equilibrated to a sodic soil condition. Agricultural-grade gypsum is an effective calcium amendment when time is not a factor and the material can be mechanically blended with the impacted soil.

Poor physical condition and resulting low hydraulic properties of the soil often impede the exchange of sodium by calcium

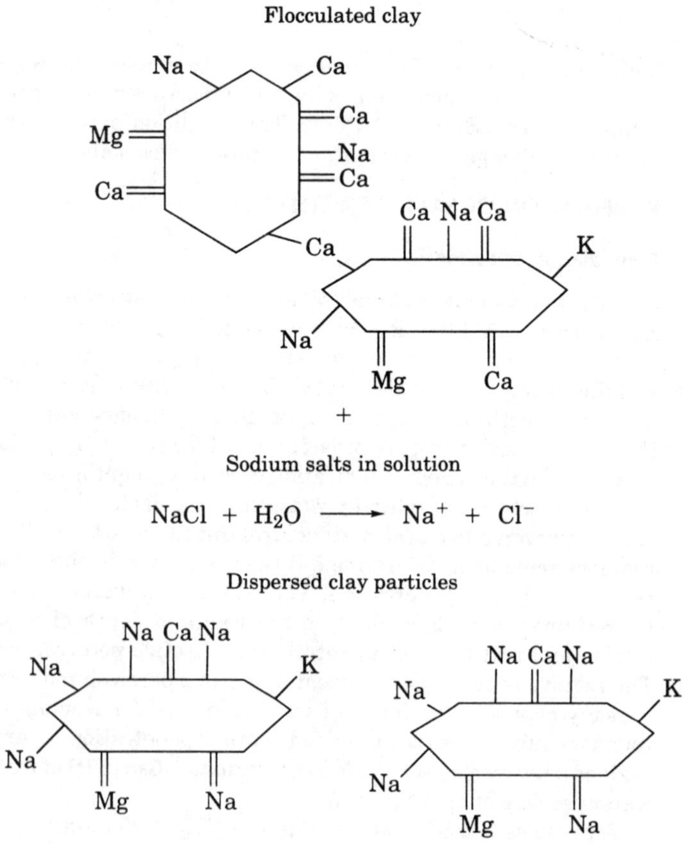

Figure 5. Dispersion of flocculated clay with sodium.

Table 3. CEC of Common Exchange Materials in Soils

Material	CEC, meq/100 g
Organic matter	100–300
Vermiculite	100–150
Montmorillonite	60–100
Chlorite	20–40
Illite	20–40
Kaolinite	2–16
Sesquioxides	0

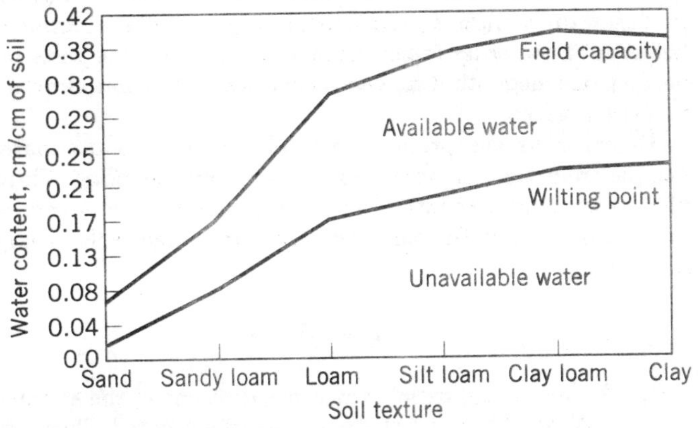

Figure 6. Available water as influenced by soil texture.

without the aid of structuring amendments, wetting agents, and/or mechanical processes. Structuring amendments aid by increasing soil aggregation and are particularly effectual in increasing the infiltration rate. Some of the more common structuring agents include poly(vinyl alcohol), polyacrylamide, and natural plant polymers.

Mechanical processes physically loosen the soil and increase the surface area for reaction with chemical amendments and water. Disking is an operation that provides good mixing of amendments and soil. Disking is normally limited to an effective depth of 10–15 cm (4–6 in.). Chiseling or deep plowing is necessary where sodic materials occur at greater depths. The goal in treating sodic soils is to reclaim the surface 3 meters (1 ft) to a favorable structural status. Amendments eventually work deeper into the soil profile and increase the effective root zone.

If amendments and mechanical processes are ineffectual, the only option, save wasting the land, is to bury the undesirable material under a cover with suitable soil. Sodic materials have no impact in a buried state.

REMEDIATION OF SALINE–SODIC SOILS

Saline–sodic soils are defined as having a saturated paste EC > 4 dS/m, and ESP >15%. These soils are often easier to reclaim than sodic soils alone because they are in better physical shape. However, if the excess salt is removed without addressing the sodicity problem, the soil properties could change noticeably and become as difficult to reclaim as a sodic soil. To prevent this from happening calcium is added to the soil prior to the leach process or added as an amendment to the leach water (8).

As with saline soils, excess salts should be leached below the maximum depth of penetration to prevent resalinization of the root zone. Sodic materials are treated to a depth of 3 m (1 ft).

PASSIVE REMEDIATION AND WASTING LAND RESOURCES

Passive treatment is differentiated from wasting of land resources in that amendments are applied to the soil and even incorporated by mechanical means. Mulches may also be used to conserve water. Time and nature are allowed to take their course. Passive implies there will be some lessening of an adverse impact over time. Passive treatments are ideally suited for historic salt-damaged soils in humid or subhumid climates with annual precipitation (>25 in.).

Wasting of land resources must be considered an option when treatment cannot reclaim the property for its intended land use without the expenditure of water and other soil resources exceeding the value of the property being reclaimed. Treating a part of the land for aesthetic reasons and planting halophytes where practical may be considered.

BIBLIOGRAPHY

1. API, *Oil and Gas Exploration and Production Wastes,* Document No. 417-01-09, American Petroleum Institute, Washington, D.C., 1987.
2. API, *Basic Petroleum Handbook,* 13:3, American Petroleum Institute, Washington, D.C., Sept. 1993.
3. L.E. Deuel, Jr., and G.H. Holliday, *Soil Remediation for the Petroleum Extraction Industry,* Pennwell, Tulsa, Okla. 1994, pp. 35, 130.
4. U.S. Soil Salinity Lab Staff, *Diagnosis and Improvement of Saline and Alkali Soils,* USDA Handbook 60, U.S. Printing Office, Washington, D.C., 1954.
5. E.V. Maas, *Appl. Agric. Res.* **1,** 12–26 (1986).
6. N.C. Brady, *The Nature and Properties of Soils,* 10th ed., MacMillan Publishing Co., New York, 1990, pp. 200–204.
7. F.M. Sandoval and W.L. Gould, in F.W. Schaller and P. Sutton, eds., *Reclamation of Drastically Disturbed Lands,* American Society of Agronomics, Madison, Wisc. 1978, pp. 485–501.
8. R.J. Prather and co-workers, *Soil Sci. Amer. J.* **42,** 82–786 (1978).

SAMPLE SIZE FOR SITE INVESTIGATION

NAGARAJ K. NEERCHAL
University of Maryland
Baltimore, Maryland

TIM LEGORE
Boeing Computer Services
Richland, Washington

HERBERT LACAYO
N. PHILLIP ROSS
U.S. Environmental Protection Agency
Washington, D.C.

A current problem in environmental remediation work is the lack of a satisfactory process or paradigm for the overall site investigation and remediation process that relates the major part of the process in a systematic and quantitative way. A generic process has been created under the RCRA and Superfund laws but is frequently vague on the relationship between the various parts of the process and especially the relationship between the conflicting interests of the various parties (private and public) who have legitimate stake in the site cleaning. In practice, this results in gaps on how to proceed. A number of individual tools have been developed to deal with parts of the site investigation and remediation process, but very little has been done to connect these parts into a contiguous whole in which site investigation parameters (eg, sample plan design) can be clearly and traceably related to the identified interests of the stakeholders. One reason for this state of affairs is the enormous complexity of the problem which causes each remediation to appear hopelessly unique.

This article is an attempt to remedy a portion of that problem. One way will be by considering the interests of all parties in achieving the "desired post-remediation condition of a site." These interests and concerns can be illuminated by focusing on how much data must be gathered, usually by sampling, in order to test the hypothesis that remediation has been successful. Simply stated, enough samples are taken after cleanup to convince us that a good job was done.

The remediation process is divided into the following five parts:

1. Describing contamination of the site. This is usually done by gathering sufficient samples from within and around the site to get a good description of the spatial distribution of the contamination.
2. Decision rule. A rule or procedure that will be used to decide whether or not the site is dirty. This can be used both before and after remediation to decide whether or not one has successfully cleaned up the site. The rule is usually based on information obtained from site-sampling data.
3. Post-closure risk. After having decided on a sampling plan and taking a sample, what risks are being taken by the interested parties under the alternatives of the decision rule.
4. Error rates. Those responsible for making decisions associated with site remediation want to have a good estimate of the probability that they made a mistake in their decision.
5. Determine a sampling plan and sample size. How many samples are needed must be determined in order that error rates and the associated post-closure risk are "reasonably low." The type of sampling plan to use must also be determined.

This article does not concern ways to estimate risk costs. It will be assumed that the methodology is defined and costs usually increase with concentration. For those who are unfamiliar with site cleanup, a brief description of a hypothetical site is included, as well as the usually differing viewpoints of the government and the parties responsible for the site.

A PARADIGM FOR DECISION-MAKING

The fundamental decision to be made at a waste site is whether or not to remediate the site. Or, if a site has been remediated, is further cleanup necessary? Secondary considerations include choosing a specific remediation technology, where to remediate, and how much to remediate. Once the basic mechanism for arriving at the fundamental decision is established, the secondary considerations can be addressed as problems of optimization or of compromises in the choice of parameters. In what follows, the most complicating factors that arise in the course of every remediation are simplified in order that the overall analysis not be lost in details.

Describing Contamination of the Site

The U.S. EPA describes some statistical tools that may be used for designing sampling plans and identifying contamination at facilities (1). In order to better understand why these tools are needed, one model that may be used to describe contamination at a site is given:

Define ϵ as the probability of obtaining a contaminated sample from the site under simple random sampling. Think of the site as divided into a very large number, N, of sampling units (Fig. 1). Let the number of contaminated cells in Figure 1 be N_c.

$$\epsilon = \frac{N_c}{N}$$

Let $X_1, X_2, \ldots, X_N$ denote the concentrations of the samples 1, 2, ... N, whether or not they are contaminated. Then the average concentration of the site is approximately

$$\delta = \frac{\sum_{i=1}^{N} X_i}{N}$$

In addition to the above description of contamination, the uncontaminated parts of the facility are assumed to contain as

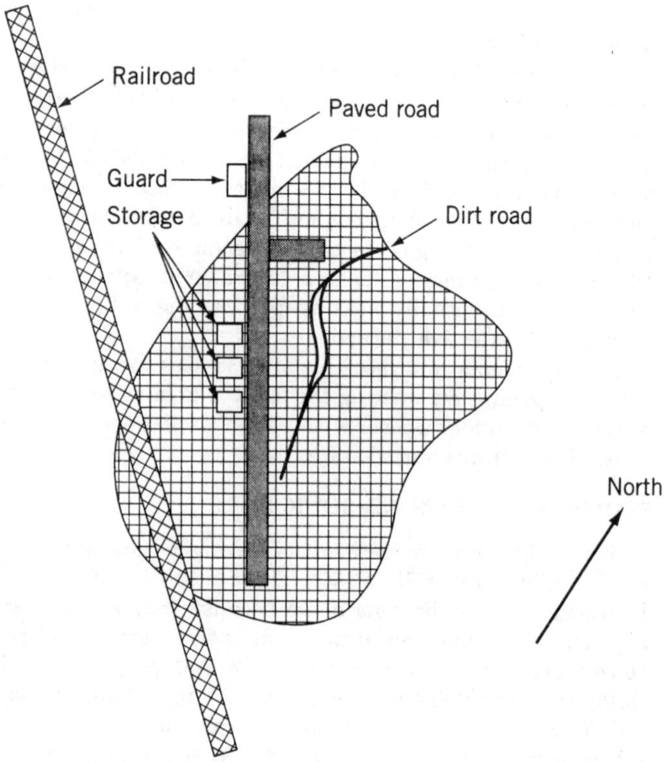

Figure 1. Hypothetical site.

background the analytes of concern that vary naturally across the facility. Note that the values $X_1, X_2, \ldots, X_N$ form a histogram with the properties that the mean of the histogram is δ. If ϵ is the fraction of the contaminated sites, however, then $100\epsilon\%$ of the upper right-hand area of this histogram represents these sites. Therefore, the area under the histogram above the value δ_0 is $100\epsilon\%$ (Figs. **2a** and **2b**). Note also that as the sampling grid is made finer (ie, $N \to \infty$), the histogram can be approximated by a continuous density curve with the following properties:

$$\int_{\delta_0}^{\infty} f(x)dx = \epsilon \qquad \int_0^{\infty} xf(x) = \delta$$

Since risk is a function of ϵ, δ, and δ_0, a direct linkage can be made between the desired post-remediation risk and these critical values. If ϵ and δ are high, then very likely the site is badly contaminated and very likely the decision to remediate will be made (ie, the answer is clear: there is no danger of any decision error and no sophisticated machinery is needed for the decision process). Thus, the crux of the matter lies in identifying the critical values of ϵ and δ. Note that δ_0 is usually given as a health-based standard or as a background value for a contaminant. In fact it had to be known in order to calculate ϵ. Specifically, N_c could not be calculated without knowing δ_0. This introduction was to display its inter-relationship of δ, ϵ, and the contamination distribution in Figures **2a** and **2b.**

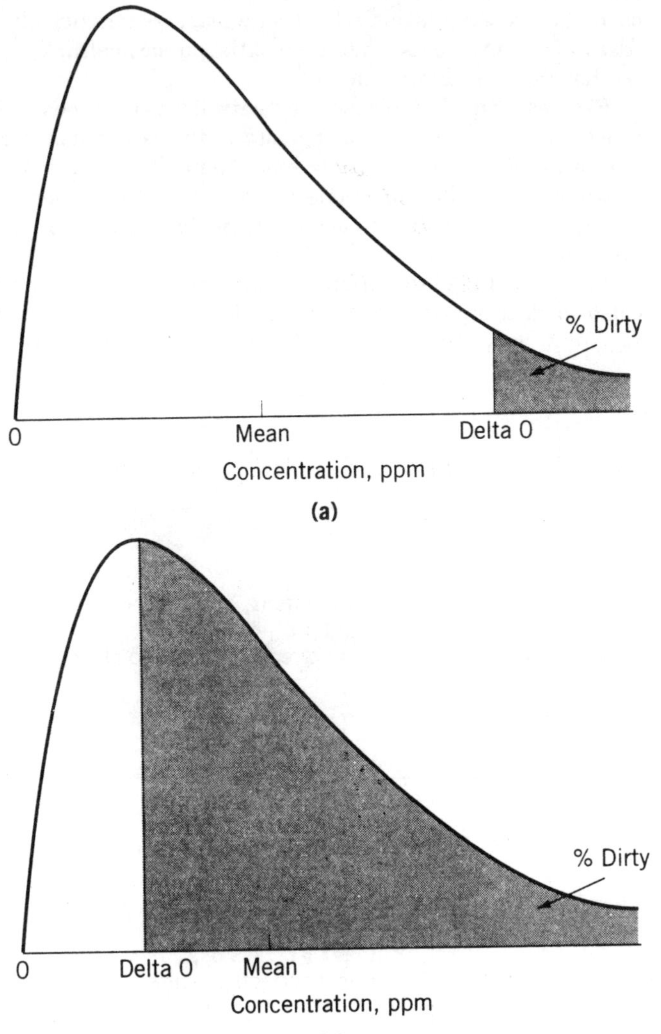

Figure 2. (**a**) Model for measurements from a clean site; (**b**) model for measurements from a dirty site.

Decision Errors and Loss Functions

The decision process for deciding whether or not to remediate is usually based on a numerical decision rule that requires sampling data (ie, a statistical test) that may or may not yield a "correct" decision for the waste site. In a site investigation, the accuracy of a decision is measured in terms of the probabilities associated with the two possible decision errors: (*1*) the error against the public: deciding that the site is clean when in fact it is contaminated, and (2) the error against the site owner/responsible party (RP): deciding that the site is dirty when in fact it is clean and hence the responsible party must pay for an unnecessary cleanup.

Managers, decision-makers, and politicians are all concerned with the probable consequences (ie, risks) of the decisions they make. The process of evaluating the risks requires listing various cleanup scenarios. For simplicity, these scenarios have been summarized by the four end points described below (Table 1).

$L_{d,d}$: The loss incurred when the decision is made that the site is dirty (denoted by d) and, indeed, the site is dirty (also denoted by d). This is a good end point in the sense that the decision was correct and hence remedial action is correctly taken. However, there is a loss in the sense that cleanup costs will be incurred by the government or the party responsible for the site (ie, responsible party). The cleanup costs usually increase with increasing contamination.

$L_{d,c}$: The loss incurred when the decision is made that the site is dirty (d), when in fact the site is clean (denoted by c). This is of course an error and it results in the cost of remediation when it is unnecessary. It is useful to think of this as an error against the responsible party. *This is frequently referred to as Type I Error, but this terminology will not be used herein because it is uninformative and varies according to whether the de facto assumption is "not clean" or "not dirty." In a site investigation, de facto assumption of "clean or dirty" varies according to the economic interests of the interested parties or where the they may be in the remediation process or in what the courts mandate.*

Table 1. End Points of Decisions

True Site Condition	Decision Taken: Site is Clean, $t \le t^*$	Decision Taken: Site is Dirty, $t > t^*$
Site dirty, $\delta > \delta_0$	Error against government resulting in loss, $L_{c,d}$	Correct decision resulting in loss, $L_{d,d}$
Site clean, $\delta \le \delta_0$	Correct decision resulting in loss, $L_{c,c}$	Error against RP resulting in loss, $L_{d,c}$

$L_{c,d}$: The loss incurred when the decision is that the site is clean, when in fact the site is dirty. This is an error of not cleaning/remediating when it should be done. That error can result in large costs in terms of increased health problems or in the transport of contamination to other areas. The responsible party may face legal problems due to being accused of negligence of public safety. It is useful to think of this as an error against the public health. This is sometimes called a Type II Error, but for reasons mentioned above, this terminology will not be used.

$L_{c,c}$: The loss incurred when the decision is made that the site is clean and in fact it is clean. Here the costs are usually the planning, sampling, and testing needed to obtain enough information for the decision-maker to make an informed decision. Usually, this planning cost is incurred before the decision of whether or not to remediate is taken. Thus, this cost is in fact included in each case, described above, as a component.

These various endpoints ($L_{d,d}$, $L_{d,c}$, $L_{c,d}$, $L_{c,c}$) are illustrated in Figure 3. In Figure 3 the four cases are related to a decision rule, like that given below, where it is decided that the site is dirty or clean according to whether the sampled value t is greater or less than a given reference value t^*. Thus, the decision rule states:

If $t > t^*$ Decide site is dirty and clean up

If $t \leq t^*$ Decide site is clean and do not clean up

In the above decision rule, t^* is computed based on a number of things including the background level δ_0, the variability (σ) in the measurements, the level of decision errors tolerated by the stakeholders, and the modeling distribution used.

The purpose of this article is to reach a wide audience including nonspecialists, therefore the relationship between the decision rules, the true state of the site, and the consequent losses are discussed more fully.

Consider the first quadrant (Q_I) of Figure 3. This quadrant illustrates what causes the remediation process to end in $L_{d,d}$. In Q_I, $\delta > \delta_0$, so the site is in fact dirty. That is, the true average concentration of the contaminant is greater than δ_0, the maximum allowable contamination level. δ_0 is called the health standard if it comes from health considerations; it is called the background standard if it is decided that the site contamination should not exceed the usual background for that geographical area.

Also in Q_I, note that $t > t^*$. This means that based on the evidence gathered as result of carrying out a sampling plan, value of t has been calculated from the sample values, and the value of t is greater than the reference value t^*. Then, as suggested by the decision rule, it is decided that the site is dirty, and so arrive at end point $L_{d,d}$, ie, the loss that will be incurred if the site is dirty and it has been correctly decided that the site is dirty.

The same sort of reasoning holds in quadrants II, III, and IV. Now consider Q_{II}. The points in Q_{II} can be denoted by (t, δ), where $t \leq t^*$ and $\delta > \delta_0$. Also in Q_{II} note that $t \leq t^*$ (where t is the value calculated from the sample and t^* is the reference value), the decision rule indicates that the site is clean. Since $\delta > \delta_0$, the site in reality is dirty. Thus, the points in Q_{II} correspond to situations where it has been decided that the site is clean, but in fact it is dirty. In other words, Q_{II} contains all the cases where an error is made against the public health. Q_{III} and Q_{IV} have similar interpretations.

Figures 4 and 5 represent hypothetical loss functions which illustrate how the losses may depend on the extent of contamination in the site. Every effort must be made to obtain or approximate these loss functions in order to better understand the consequences of the decision to remediate or not to remediate a site.

In Figure 4 the loss suffered as a consequence of having decided to clean up (ie, $t > t^*$) is considered. The graph depicts $L_{d,d}$ for $\delta > \delta_0$ and $L_{d,c}$ for $\delta \leq \delta_0$. Note that $L_{d,c}$ remains con-

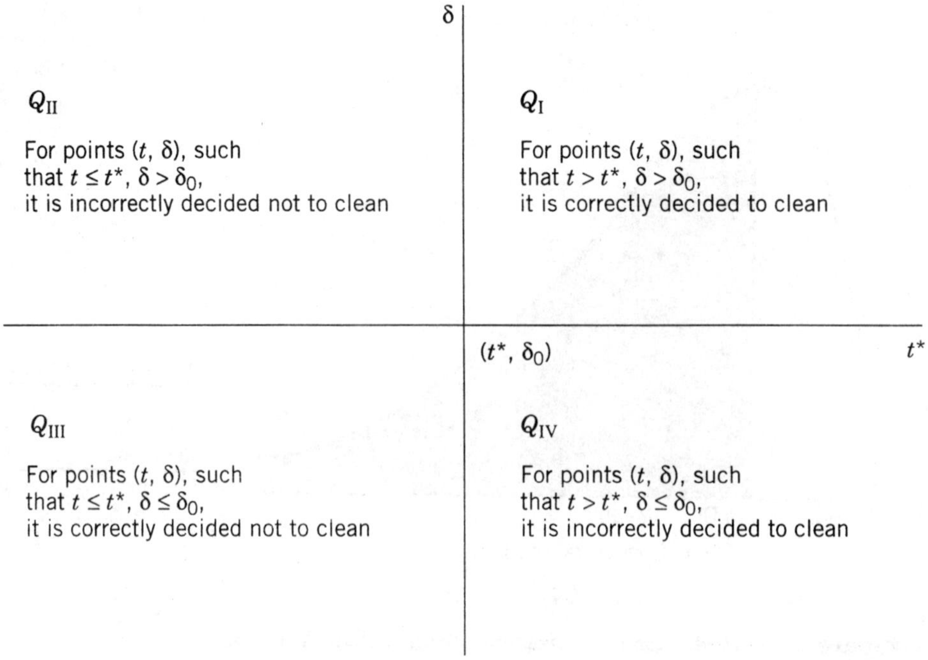

Figure 3. The various decision end points.

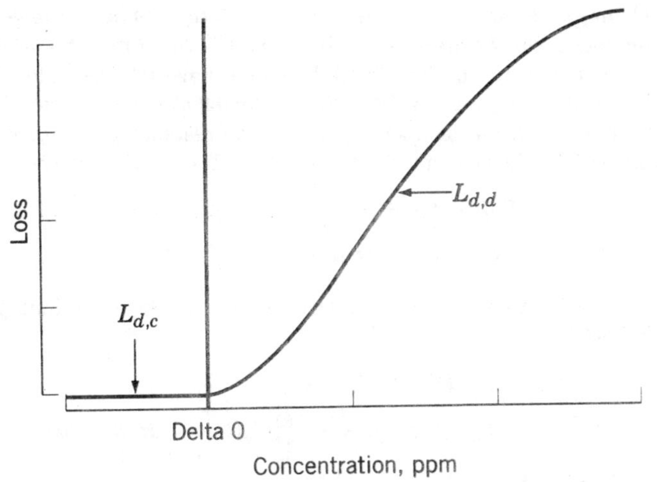

Figure 4. Loss as a function of contamination when the decision taken was that the site is dirty ($t > t^*$) and needs cleanup.

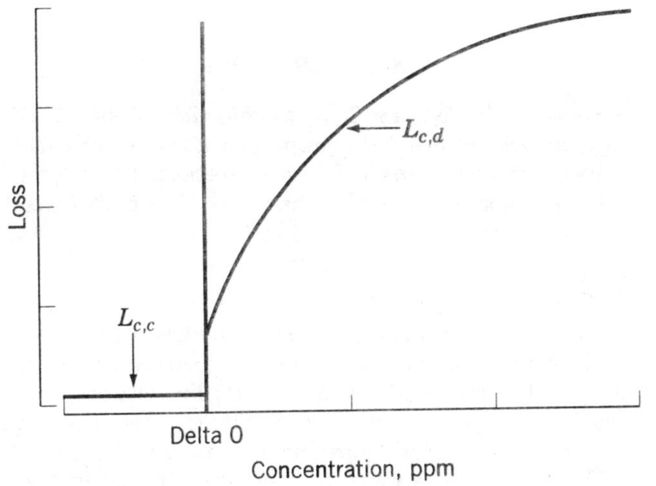

Figure 5. Loss as a function of contamination when the decision taken was that the site is clean ($t \leq t^*$) and does not need any further cleanup.

stant since the only cost incurred is the cleanup cost which should be the same for all $\delta \leq \delta_0$. Realities of the situation indicate that the needless cleanup will go smoothly and at a minimal cost. However, for $\delta > \delta_0, L_{d,d}$ increases with δ and convex down indicating a higher cost of cleanup for higher contamination of the site and a decrease in marginal increases in cost. It may be realistic to assume that in the case of extremely high levels of concentration, a maximal cost of cleaning up the site will be incurred by having to excavate the existing soil and refilling with clean soil.

Figure 6 depicts the loss suffered as a consequence of having decided to leave the site as is ($t \leq t^*$). Here again, $L_{c,c}$ is a constant for all values of $\delta \leq \delta_0$, since there is no cleanup and costs are only for analysis and sampling. However, if $\delta > \delta_0$, the cost to the public health may be high and increase rapidly with increasing δ.

Note that the realities of site cleanup dictate the amount of investigation and/or estimation of the likelihood of all end points, but risk estimation should be made an explicit part of the site analysis along with the data quality objectives (DQO)

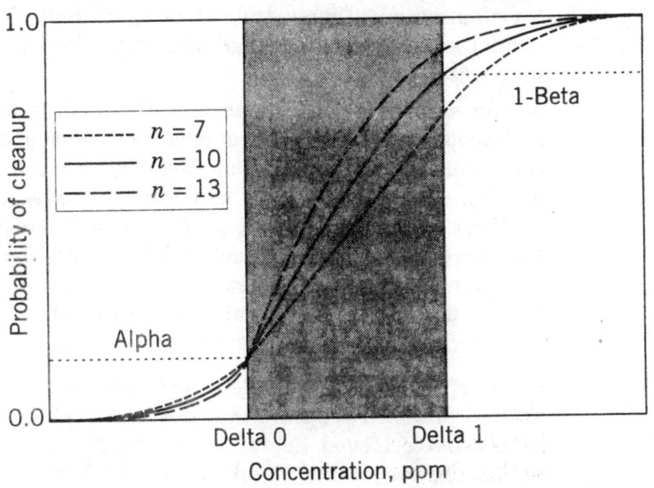

Figure 6. Illustration of how the sample size determined in the DQO paradigm. The levels decision errors, α and β, are specified by the RP and the government, respectively. The minimal sample which will achieve these levels is then sought.

process. Also note that Figures 4 and 5 give essential features of loss functions, but they are by no means the only types of loss functions that can be used in practice.

Risk Functions

Now that the various components of risk have been identified, additional pieces of information are needed to define the expected loss or risk. (*Risk* is used in a sense different than its meaning in "health risk" or "environmental risk assessment". Its meaning in statistical literature is similar to its use in "risk in investment"). These pieces are δ_0, δ, α, and β.

δ_0: This is either a health-based standard of what is an acceptable concentration of a pollutant or it is a background concentration. δ_0 can be treated as a constant or variable depending on the nature of the problem. For example, it is sometimes known, assumed, or agreed upon that the health-based standard will be a given constant, δ_0, in parts per million. This happens when toxicologists and the health professionals have enough experience with the pollutant to propose such a constant. On the other hand, it may have been agreed upon by the various groups involved (responsible party (RP), EPA, the local Department of Justice attorney, etc) that site cleanup should proceed until the land is as clean as the surrounding uncontaminated land. Then, δ_0 is a variable that must be estimated by sampling the uncontaminated surrounding. Here, δ_0 is treated as a constant.

δ: This is the true but unknown concentration of the contaminant at the site ($0 \leq \delta \leq 1$).

P(decision error against public or government): This is the probability that it is decided that the site is clean when in fact the site is dirty. This usually comes about as follows:

1. The site under consideration has come under scrutiny because expert opinion or a preliminary examination indicates that the site may be dirty. At this point (2) the working assumption or null hypothesis is that the site is dirty (ie, $\delta > \delta_0$).

2. Information is gathered and sampling is done. The results are concentration readings from n locations in the site.
3. A way of deciding if the site is clean or dirty (ie, a decision rule) has been agreed upon based on some formula or calculation involving sampling $X_1, X_2, ..., X_n$ values. For example, t = average of $X_i's$, t = max of the $X_i's$, or t = some complex function of the $X_i's$. The standard form of the decision rule is the same as before:

If $t > t^* \Rightarrow$ decide site is dirty and clean up the site
If $t \leq t^* \Rightarrow$ decide site is clean and do not clean up

where t^* is a reference value usually obtained from statistical tables to ensure that the data quality objectives are achieved. In practical terms, t^* is chosen so that the probability of making a wrong decision is suitably low.

4. Suppose that $t \leq t^*$. Then, according to the decision rule it is decided that the site is clean.
5. On the other hand, suppose that the site is dirty. The sampling variability has led to a mistake. The probability of ending up with this mistake is P(decide that the site is clean | site is really dirty), or simply

$$P(t \leq t_0 \mid \delta > \delta_0)$$

P(Decision error against the responsible party): probability of deciding that the site is dirty when in fact the site is clean. This usually comes about as follows:

Steps 1 through 3 are the same as above in the description of α.

4. Suppose $t > t^*$ and it is decided that the site is dirty.
5. If the site is clean, and the sampling variability has led to a mistake, the probability of this mistake is P (decide that the site is dirty | site is really clean) or simply

$$P(t > t^* \mid \delta \leq \delta_0)$$

Now the expected loss (also denoted by E(loss) or risk) can be expressed, based on the above terminology and the information in Figures 3 and 4. Since the concentration δ is unknown, it is prudent to examine various values of this parameter.

Case (1) $\delta \leq \delta_0$. If the decision was to clean up, then the loss suffered by the responsible party (RP) is $L_{d,c}$ and there is no loss suffered by the public or the government (see Figs. 4 and 5). If the decision was not to clean up, then the RP suffers a minimal loss due to sampling costs. Thus, the risk to the responsible party (if the contamination level is δ) is

$$R_{RP} = L_{d,c}P_\delta(t > t^*) + L_{c,c}P_\delta(t \leq t^*)$$

where the subscript δ is used to emphasize the fact that these probabilities depend on the extent of true concentration.

In this case, the public does not suffer any loss, regardless of the outcome of the decision rule and therefore $R_G = 0$ (ie, risk to government = R_G).

Case (2) $\delta > \delta_0$. Once again two possibilities are considered (ie, $t > t^*$ or $t \leq t^*$) (see Figs. 4 and 5). If the decision is to clean up (the correct decision) the loss, $L_{d,d}$, basically includes the cleanup cost and there will be no loss to the public or the government. On the other hand, if an error is committed by deciding not to clean up, then the public/government suffers a loss of $L_{c,d}$ and the loss to the RP is $L_{c,d}$ in terms of potential damage claims via litigations. Thus, risk to the responsible party and the risk to the government are given by the expressions below:

$$R_{RP} = L_{d,d}P_\delta(t > t^*) + L_{c,d}P_\delta(t \leq t^*)$$

$$R_G = L_{c,d}P_\delta(t \leq t^*)$$

The expressions for the two cases may be combined as follows:

$$\begin{aligned} R_{RP} &= L_{d,c}P_\delta(t > t^*) + L_{c,c}P_\delta(t \leq t^*) \ \text{if}\ \delta \leq \delta_0 \\ &= L_{d,d}P_\delta(t > t^*) + L_{c,d}^{RP}P_\delta(t \leq t^*) \ \text{if}\ \delta > \delta_0 \end{aligned}$$

Similarly,

$$\begin{aligned} R_G &= 0 \ \text{if}\ \delta \leq \delta_0 \\ &= L_{c,d}P_\delta(t \leq t^*) \ \text{if}\ \delta > \delta_0 \end{aligned}$$

Finally,

$$R_{Total} = R_G + R_{RP}$$

Note that R_G, R_{RP}, and R_{Total} are functions of the concentration δ and the probabilistic properties of the statistic (t) used in the decision rule. In the following sections, the risk for two typical decision problems that may come up in a site investigation is computed.

An Example

The description of this risk problem is fairly general. All site problems cannot fit the above classical form because risk scenarios and situations are so rich in their diversity and complexities that from time to time new situations appear that require new viewpoints, and perhaps this framework will not fit.

The following shows how a site optimization problem (3) fits into the framework described herein. The purpose is to show how risk depends on the choice of the decision rule function d (estimators) and the sample size. A part of that work is summarized below and the equivalent concepts of the current framework shown in brackets.

For this example, n items are sampled (n soil samples, or n pints of water) from a large population. The outcomes of testing the samples $X_1, ..., X_n$ are:

$$X_i = \begin{cases} 0 & \text{if sample is clean} \\ 1 & \text{if sample is dirty} \end{cases}$$

Three estimators (to be evaluated) of the proportion of the site that is dirty are given by

$$d_1 = \frac{\sum_{i=1}^{n} X_i}{n}$$

$$d_2 = \frac{\sum_{i=1}^{n} X_i + 1}{n + 2}$$

$$d_3 = \frac{\sum_{i=1}^{n} X_i + \sqrt{(n)}/2}{n + \sqrt{(n)}}$$

As before, a decision rule follows:

$$\text{Clean up } if\ d_1 > CS$$

$$\text{Do not clean up } if\ d_1 \le CS$$

where CS denotes a cleanup standard recommended by experts. In this example δ_0 is called CS and a simple choice was made for t^*, namely CS. Similar decision rules may be formed based on d_2 and d_3 as well. The statistical model for the observations $X_1,\ldots,X_n$ assumes that X_is are independent and

$$P(X_i = 0) = 1 - P(X_i = 1) = p$$

Consider a concrete example to clarify the terminology. Suppose a site is sampled (see Fig. 1) to estimate what fraction of the site is contaminated. Specifically, p is estimated, the contamination (in this example, $p = \delta$), by one of the possible estimation formulas, d_1, d_2, or d_3 (that can be used in the decision rule) and compared with cleanup standard CS (which is 2/12). First consider d_1.

Take a sample of size 12. Suppose the result of testing the samples is (000100010100). Then $d_1 = (X_1 + X_2 + +X_{12})/n = 3/12 = 1/4$. Since $1/4 > 2/12 = CS$, it is decided that the site is dirty by applying the decision rule. Consider the four possible situations that may occur and the losses under each:

$$\text{If } p \le CS \text{ and } d_1 \le CS \text{ then loss} = 0$$

$$\text{If } p \le CS \text{ and } d_1 > CS \text{ then loss} = C_1$$

$$\text{If } p > CS \text{ and } d_1 > CS \text{ then loss} = C_1$$

$$\text{If } p > CS \text{ and } d_1 \le CS \text{ then loss} = C_2$$

The above loss structure results in the risk function

$$\begin{aligned} R_d(p) &= C_1P(d > CS) && \text{if } p \le CS \\ R_d(p) &= C_1P(d > CS) + C_2P(d \le CS) && \text{if } p > CS \end{aligned}$$

Note that the loss structure given above is obtained as a special case of the framework, by taking $p = \delta, t = d_1, CS = t^*, CS = \delta_0, L_{c,c} = 0, L_{c,d} = C_1, L_{d,d} = C_1, L_{c,d} = C_2$. These risks for the different estimators (d_1, d_2, d_3) are plotted in Figure 6 for comparison with the other loss functions considered in this article.

Note that the purpose was to show that the choice of decision rules depends both on sample size and the particular objectives of the decision-maker (3). This is unfortunate because all the parties together represent the people. Having them sit at the table with everyone's concerns and risks shown would be helpful to a fair decision process. The approach given in this paper provides a framework for such a discussion.

Note that each estimate d_i is optimal in some sense: d_1 is unbiased and the maximum likelihood estimator; d_2 is a commonly used Bayes estimator; d_3 is the minimax estimator; but there is no overall winner among these three (3). The choice, if it were to arise, would be a matter of which choice supported current policy.

Relationship of Risk to the DQO Process

The classical statistical hypotheses testing paradigm may be considered in setting up data quality objectives. These DQOs are then used to determine the sampling plan and sample size. For the discussion here, assume that the site is clean and evidence against this hypothesis is sought. This is a reasonable scenario provided the RP has finished cleaning the site and is asking for permission to release the land for public use.

The DQO team, as is required by the DQO process, has gone through an extensive analysis of various sources of variability in the measurements. Suppose that σ measures the sum total of all the variability. For simplicity it is assumed that $\sigma = 1$. Since the assumption is that the site is clean, the following null (clean: H_0) and alternative (dirty: H_1) hypotheses are formed.

$$H_0 : \delta \le \delta_0 \text{ and } H_1 : \delta > \delta_0$$

The decision rule used in this context is based on either the z-test or Wilcoxon rank sum test. To keep the discussion simple, consider the z-test.

Decision rule:

$$\begin{aligned} &\text{Clean up site again} && \text{if } t > z_{1-\alpha} \\ &\text{Accept site for human use} && \text{if } t \le z_{1-\alpha} \end{aligned}$$

where $z_{1-\alpha}$ is the 100 $(1-\alpha)$ percentile of the standard normal distribution, and α is the maximum allowable probability of a decision error against RP, as agreed upon during the DQO process. That is:

$$P(\text{decide to clean up when site is in fact dirty})$$

$$= P(t > z_{1-\alpha} \mid \delta \le \delta_0) \le \alpha$$

The current process of determining the sampling plan and sample size rests on the DQO team, through extensive negotiations among the stakeholders, the RP, EPA, local or state governments, and comes up with the following key factors: (*1*) a maximum acceptable level for the probability of decision error against the responsible party (α), and (*2*) a critical level of contamination (δ_1) above the safe background standard (ie, $\delta_0 < \delta_1$) and desired level assurance (β) such that:

$$P(\text{decision error against public when level of contaminant is } \delta_1)$$

$$= P(t \le z_{1-\alpha} \mid \text{true concentration is } \delta_1) \le \beta$$

Note that requirements (*1*) and (*2*) may rephrased as follows:

$$P(\text{deciding to clean up}) = P(t > z_{1-\alpha} \mid \delta \le \delta_0) \le \alpha$$

$$P(\text{deciding to clean up}) = P(t > z_{1-\alpha} \mid \delta = \delta_1) \le 1 - \beta$$

These requirements are illustrated in Figure 6. Thus, between δ_0 and δ_1 the probability of a decision to clean up is not restricted and gives room to minimize sampling cost. For the decision rule given above (2), the minimum sample size needed to achieve these DQOs is given by:

$$n = \frac{\sigma^2(z_{1-\alpha} + z_{1-\beta})^2}{(\delta_0 - \delta_1)^2}$$

Under the framework developed in this paper, the risk of that decision rule is:

$$\begin{aligned} Risk(\delta) &= L_{d,d}(\delta)P_\delta(t > z_{1-\alpha}) + L_{c,d}(\delta)P_\delta(t \le z_{1-\alpha}) \text{ if } \delta > \delta_0 \\ &= L_{d,c}(\delta)P_\delta(t > z_{1-\alpha}) + L_{c,c}(\delta)P_\delta(t \le z_{1-\alpha}) \text{ if } \delta \le \delta_0 \end{aligned}$$

The use of subscript δ in P_δ emphasizes that the variability in the decision rule depends on the true concentration level at the site. The function P_δ, depends on the the probabilistic properties of the statistic used in the decision rule, which in turn is a function of the sample size, and the extent of decision errors tolerated (α, β). For example, in the case of z-test,

$$P_\delta(t \le z_{1-\alpha}) = \Phi\left(z_{1-\alpha} + \frac{\delta_0 - \delta_1}{\sigma/\sqrt{n}}\right)$$

where Φ is the cumulative distribution function of the standard normal distribution. Thus, risk can be written, conceptually, as a function of δ, α, β, n, δ_0, δ_1, and σ. That is,

$$Risk = R(\delta, \alpha, \beta, n, \delta_0, \delta_1, \sigma)$$

In practice, the decision rules may be based on statistical tests whose distributions are less tractable than that of z-tests.

DISCUSSION OF DECISION PROCESS

Risk is frequently a matter of viewpoint, a matter of what the individuals involved believe that they have to lose or gain. The classical modeling of risk is formulated in terms of a number of parameters. These may include the cleanup standard (δ_0), the contamination level (δ), the decision rule (a function of t and t^*), the sample size (n), and lastly the error to the public and error to the responsible party.

This modeling allows the risk to be displayed (total risk, risk to government, risk to responsible party) in a variety of ways. The first example of this is Figure **7d** (3). Here the loss functions are as given in Figures **7a** and **7b,** and the probability of $t > t^*$ is given in Figure **7c.** Figure **7d** displays risk to the government (R_G) and the risk to the responsible party (R_{RP}) as a function of the contamination level δ. In the risk plots of Figure **7d,** it is worth noting that the risk to the responsible party is rather low until the contamination level gets fairly high, but the government undergoes a very large risk for medium range values of δ. It may not be in the public interest to accept such a large risk. If the government suspects that δ is in that medium range, they would be well advised to increase the sample size or change δ_0 or t^* if possible. If the government strongly believes $\delta < \delta_0$ or δ is very high, they might be willing to sit pat (ie, accept the sample size $n = 7$). Naturally, the truth or, more exactly, the usefulness of the above modeling depends on the accuracy of the modeling in depicting reality.

Figures **7e, 7f, 7g,** and **7h** are the results of the modeling when using cost data (3). This shows that even the most simple formulations of this problem lead to instabilities in the risk outlook in the δ_0 region (see Fig. **7h**).

Figures **8a–8c** display the risk as a function of sample size n and site contamination level δ. This is shown on the surfaces illustrated in Figures 8**a,** 8**b,** and 8**c.**

The cleanup standard δ_0 and the decision point t^* are fixed such that:

$$P(t > t^* \mid \delta = \delta_0) = 0.001$$

Figures **9a–9c** and **10a–10c** are similarly defined except that

$$P(t > t^* \mid \delta = \delta_0) = 0.2 \text{ and } 0.05$$

respectively. These are discussed together.

First note that Figure 8**a** gives the risks to the government as a function of δ and n. Similarly, Figures 8**b** and 8**c** give the risk to the RP and the total risks (government + RP).

There are certain features about these risk surfaces that are worth noting. Suppose that both the government and RP do not know the true contamination levels, and in fact do not have any good estimates of δ. Suppose further that the government is not in total ignorance of δ, and they have reason to believe that the true contamination δ_T is in some range, $0 < \delta_T < \delta_U$, where δ_U is some upper bound of likely contamination. In such a situation (see Fig. 8**a**) the government is at high risk for low values of the sample size n when δ is high. They naturally will want a high n. On the other hand, if sampling costs are low, the RP may want a large sample size as an insurance against unnecessary cleanup if it is suspected that δ is close to but greater than δ_0.

As seen by these remarks on Figure 8**a** and Figure 8**b,** the government and RP are frequently in adversarial roles. In theory (but not in practice) if the government and the RP are the same, then the analysis of the situation only involves the total risk, R, as shown in Figure 8**c.** There are many such examples where the government and RP may the same (eg, the RP is the U.S. Navy whereas the interests of the government are represented by EPA or DOJ). If they have reason to believe that δ_T is low, ie, near or below δ_0, then the government may be more willing to accept a sample size n.

Now consider the risk from the viewpoint of the RP. If the RP knows or believes that the true concentration δ_T, is high, ie, $\delta_T > 2$, then (as seen in Fig. 8**b**) they may want a low sample size in order to lower their own risk. This assumes that either they do not know the government risk, R_G, or do not care.

If the RP believes that δ_T is low and sampling is cheap, then the RP may not care if the sample size is large or small since the risk to them is fairly low. Lastly, note that for a relatively high value of $P(t > t^* \delta = \delta_0) = 0.2$ (see Figs. **10a–10c**) the government is poorly protected against high risk for low values of n. The Government is reasonably protected, however, for middle values of n for $\delta_T > \delta_0$. However, the RP will almost always have to clean up, for all except small values of n. The situation is complex enough so that the RP and government interests may or may not coincide on the sample size.

The nature of this cleanup problem permits a look at the risk structure in the following special way. The sample size needed to achieve the desired probabilities of decision errors is given by

$$n = n(\alpha, \beta, \delta_0, \delta_1, \sigma) = \sigma^2 \left[\frac{z_{1-\beta} + z_{1-\alpha}}{\delta_0 - \delta_1}\right]^2$$

where

$$\alpha = P(\text{government decides site is clean when in fact } \delta = \delta_1)$$

and

$$\beta = P(\text{RP cleans site when in fact } \delta = \delta_1)$$

Figures **11a–11d** show the relationship between the sample size, α and β, for various values of δ_1. For example, given any combination of α and β, n can calculated as seen in Figure **11a** where the surface n vs α and β is given when δ_1 is 0.1 SD (standard deviation) above δ_0.

Figures 12**a** and 12**d** are more complex than the foregoing. First recall that in the model, risk is a function of the various

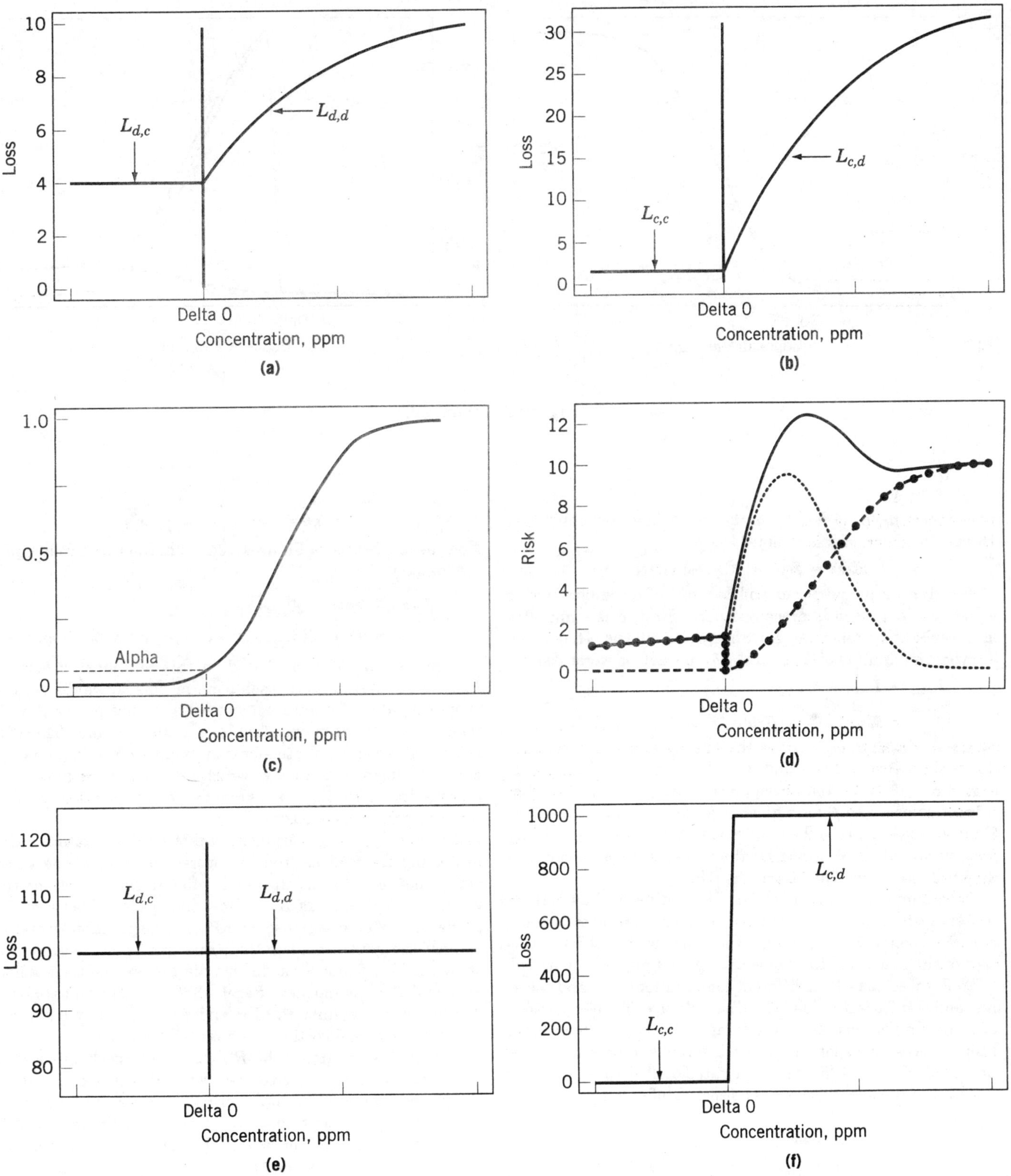

Figure 7. (**a**) Loss function for case $t > t^*$; (**b**) loss function for case $t > t^*$; (**c**) $P(t > t^*)$ as function of concentration; (**d**) for $\delta < \delta_0$, the total risk consists only of risk to RP and for large δ the total risk consists mostly of risk to RP (**d–d–d** is R_G, -------- is R_{RP}, — is R_{Total}); (**e**) plot loss function for case $t > t^*(3)$; (**f**) plot loss function for the case $t \leq t^*(3)$; (**g**) $P(t > t^*)$ as function of contaminant concentration; and (**h**) for $\delta < \delta_0$, the total risk consists only of risk to RP and for large δ the total risk consists mostly of risk to RP.

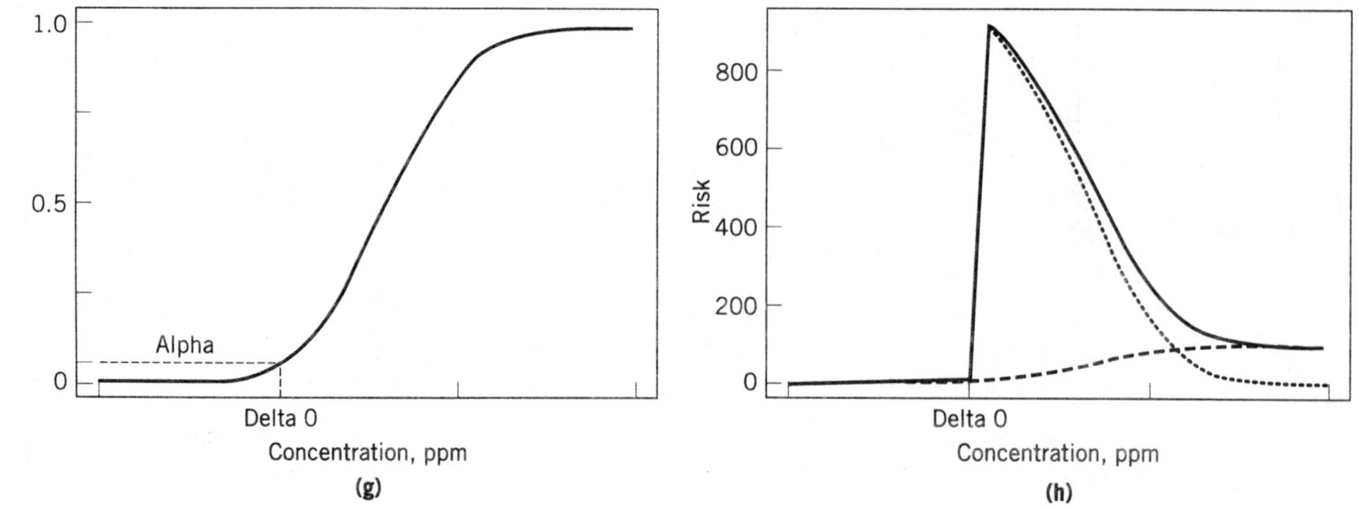

Figure 7. *(continued)*

parameters, α, β, δ, δ_0, δ_1, n, and σ various cost constants. Hence, as earlier, risk is denoted by:

$$Risk = R(\delta, \alpha, \beta, n, \delta_0, \delta_1, \sigma)$$

For illustrative purposes and because of its importance, a worst-case approach is presented here. First, δ is eliminated as a variable by replacing it with R_{max} as follows. That is, for a given combination of $(\alpha, \beta, n, \delta_0, \delta_1, \sigma)$, define worst risk by:

$$R_{max} = R_{max}(\alpha, \beta,)$$

$$= Max_\delta R(\delta, \alpha, \beta, n(\alpha, \beta, \delta_0, \delta_1, \sigma), \delta_0, \delta_1, \sigma)$$

where $n = n(\alpha, \beta, \delta_0, \delta_1, \sigma)$ is the sample size which realizes the desired error rates α and β for δ_0, δ_1, and σ. Having generated $R_{max}(\alpha,\beta)$ by the above process, R_{max} can be plotted as a function of α and β for a given combination of δ_0, δ_1, and σ. Figures 12**a**–12**d** give the relationship between α, β, and R_{max} for several values of δ_1 and for the two sets of loss structures depicted in Figures 7**a**–7**b** and 7**g**–7**h.**

Proceeding in this manner (ie, eliminating δ, thus making the analysis easier) is somewhat arbitrary and sometimes a matter of policy. For example, if the policy with respect to this type of site is to avoid the worst case, then it might be justified.

In Figures 12**a**–12**d,** different combinations of α and β correspond to different values of the sample size. To see the effect of increasing or decreasing the sample size, segments of these risk surfaces corresponding to the different sample sizes can be produced. Figures 13**a**–**h** are generated to illustrate this point. Recall that the sample size n is a function of δ, α β, δ_0, δ_1, and σ. That is,

$$n = n(\delta, \alpha, \beta, \delta_0, \delta_1, \sigma^2)$$

or

$$n = n(\alpha, \beta)$$

Therefore, for a given sample size, the corresponding β for a fixed combination of the remaining parameters can be found, that is

$$\beta = \beta(n, \alpha, \delta_0, \delta_1, \sigma)$$

In view of this risk may be written as a function of n, eliminating β as follows:

$$Risk = R(\delta, \alpha, n, \delta_0, \delta_1, \sigma^2)$$

Now, as in the case of Figures 12**a**–12**d,** the worst risk is plotted, namely,

$$R_{max} = R_{max}(\alpha, n, \ldots)$$

$$= Max_\delta R(\delta, \alpha, \beta, (n, \alpha, \delta_0, \delta_1, \sigma), n, \delta_0, \delta_1, \sigma)$$

Figures 13**a**–13**h** are based on this relationship between R_{max} and α. Note that at each point on this curve there corresponds a value of β since all variables or parameters that determine β have been specified ($n = 25$, $\delta_0 = 0$, etc). Therefore this β value can be plotted for every point on the R_{max} (α) vs α curve, giving a β vs α curve which may be read as follows: for every point on the R_{max} (α) vs α curve, go up (or down) until intersecting the β vs α curve.

Frequently, the government and the RP are interested in protecting themselves from the worst disaster, or the worst-risk situation. That is, they would like to choose options that minimize the value of R_{max}. How the trade-off between the probability of error against the RP, and the probability of error against the government or the public can be achieved will be shown in this framework. In this discussion, the terminology of α and β is as defined. Recall that α is the probability of decision error against the RP and β is the probability of a decision error against the government (or public).

Figure 13**a** is simply the R_{max} vs α, the probability of an error against the RP, for fixed n. From this it is easier to see the trade-off between α and β for fixed n when it is eliminated as a variable by focusing on R_{max}. How can Figure 13**a** help either the government or RP? In this particular situation, the R_{max} goes down as the error against the RP goes up, and, of course, as the error to the government goes down. Therefore if the Government believes Figure 13**a** then they should push for a low β and force a high probability of error against the RP. This is due to the high value of R_{max}, indicating potential harm to the public if the government makes a mistake.

Also note that in Figure 13**a** the maximum total risk is examined, broken down, and plotted separately as the maximum risk to the RP and the maximum risk to the government.

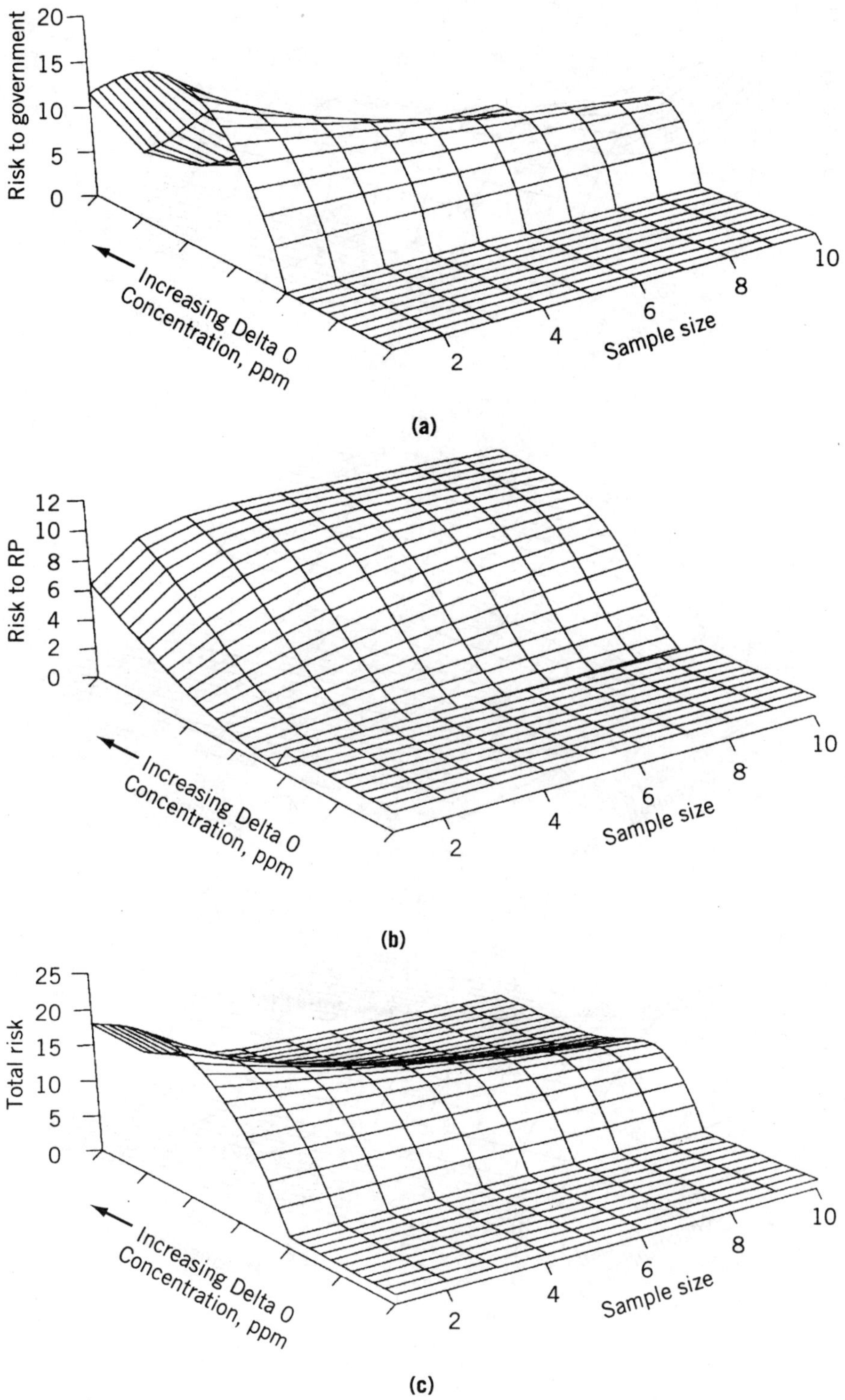

Figure 8. (**a**) Risk to government, R_G; (**b**) risk to RP; and (**c**) total risk, as a function of contamination level δ and sample size n when error against RP is 5%.

The filled blank squares represent the relation between α and β where α is read on the x-axis and β is read on the right-hand side y-axis. The next two curves, the dotted line and the empty squares, represent the risk to the government and the responsible party, respectively. The last curve (shown by empty squares with a line through them) is the total risk (ie, the sum of the risk to the government and the risk to the responsible party). The potential risk to the government is so high that they might pick up some or all of the cost of cleanup if the RP cannot. Figure 13**b** is the same as Figure 13**a** except that δ is 1 instead of 0.5. Figure 13**c** and 13**d** are similar to Figures 13**a** and 13**b** except that the sample size is 25. Figures 13**a** through 13**d** are based on the loss functions depicted in Figures 7**a** and 7**b**. Similarly, Figures 13**e** through 13**h** are constructed using the loss function given in Figures 7**e** and 7**f**.

CONCLUSIONS

Negotiation between the government and responsible party are played out in a much broader context of human interactions,

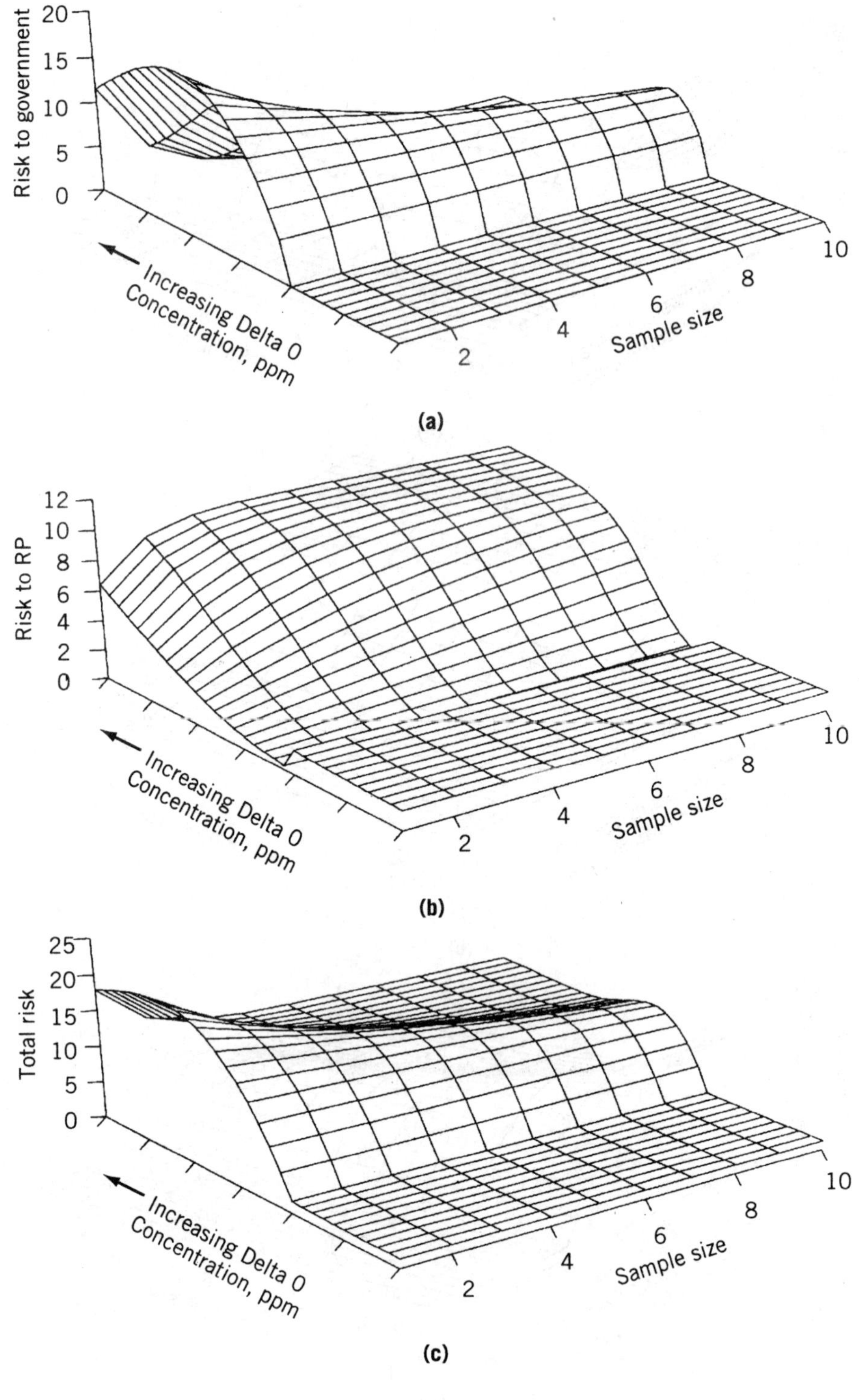

Figure 9. (**a**) Risk to government; (**b**) risk to RP; and (**c**) total risk, when error against RP is 0.01%.

traditions, and local and national politics. It should be stressed that the sound modeling approach proposed in this article increases understanding of the interplay of risks between concerned parties. Such knowledge facilitates negotiations even in the face of real-world complications. The mathematical framework discussed herein which incorporates the concerns and risks of the government and responsible party (RP) into the decision is a step in that direction.

Implementing the traditional data quality objectives (DQO) framework also requires negotiations between the government and RP to determine the levels of decision errors that can be tolerated. Each party comes with a sampling scheme hoping that it will be acceptable to the other. If there is a gap in the expectations, a DQO facilitator can intervene to work out a compromise. The framework discussed herein provides a forum for a bipartisan discussion. Both sides are encouraged to put

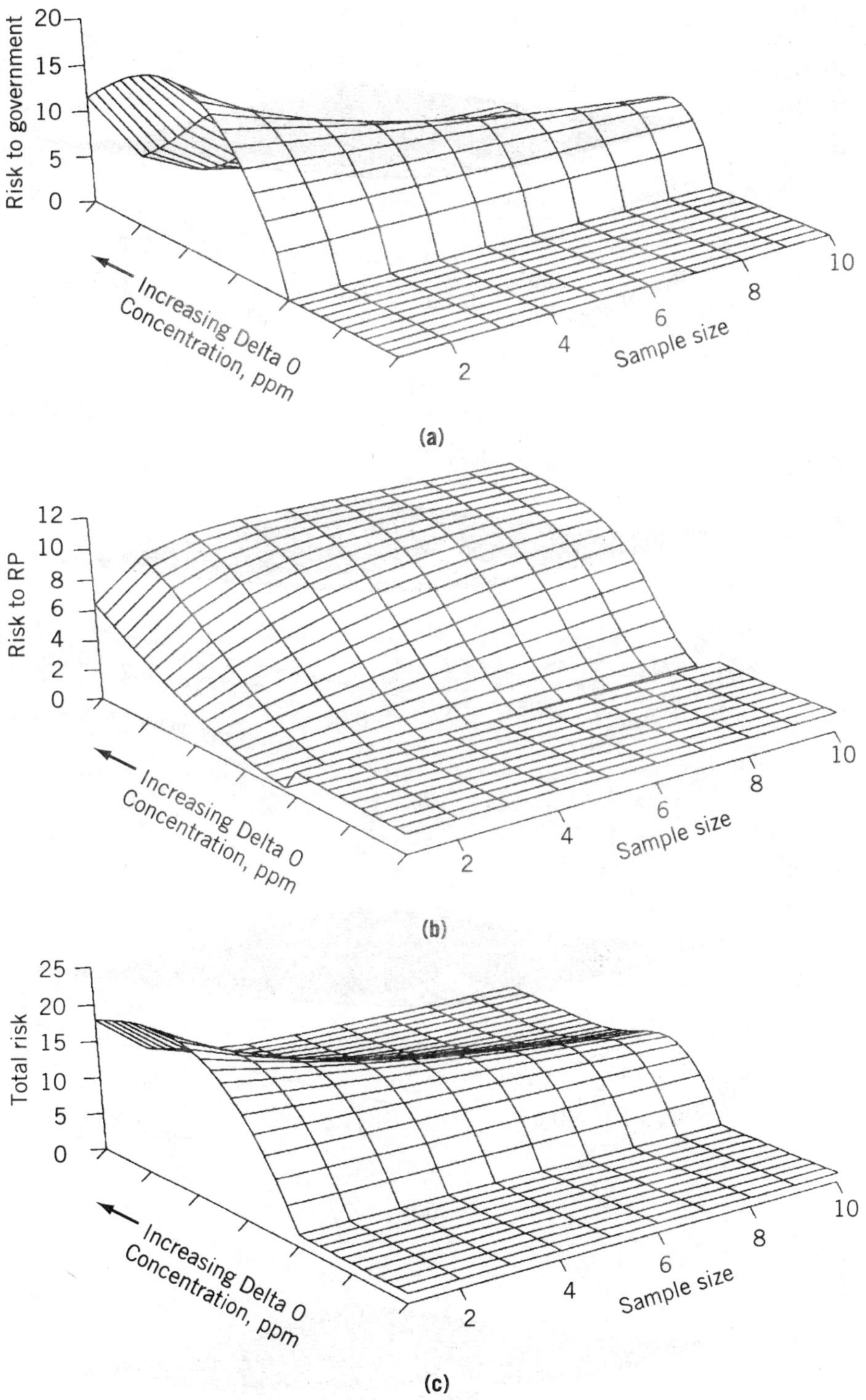

Figure 10. (**a**) Risk to government, (**b**) risk to RP, and (**c**) total risk, when error against RP is 20%.

their cost structure into the loss functions so that some delicate issues are brought to the fore. Thus, risk estimates and feasible levels of decision errors are related to the cost structures of both parties. The loss functions can be made known to all, giving a sense of full disclosure and creating an atmosphere conducive to compromises.

The outcomes or decisions that result from an analysis of this type are heavily dependent on the cost/loss functions. Sometimes the loss functions are difficult to quantify, or even worse, there is strong disagreement both quantitatively and qualitatively on their forms. It would be misleading to give the impression that the calculation of the decision errors and the associated risks incurred by the concerned parties (government and RP) are the only or even the major consideration in a site remediation process. In fact, in a real-world scenario, things like statutory requirements, public perception of the danger of the site to their health, economic impact to the community, and even the political ideology of the government officials are all involved. For example, public or political reaction to the problem may be so strong that remediation is not

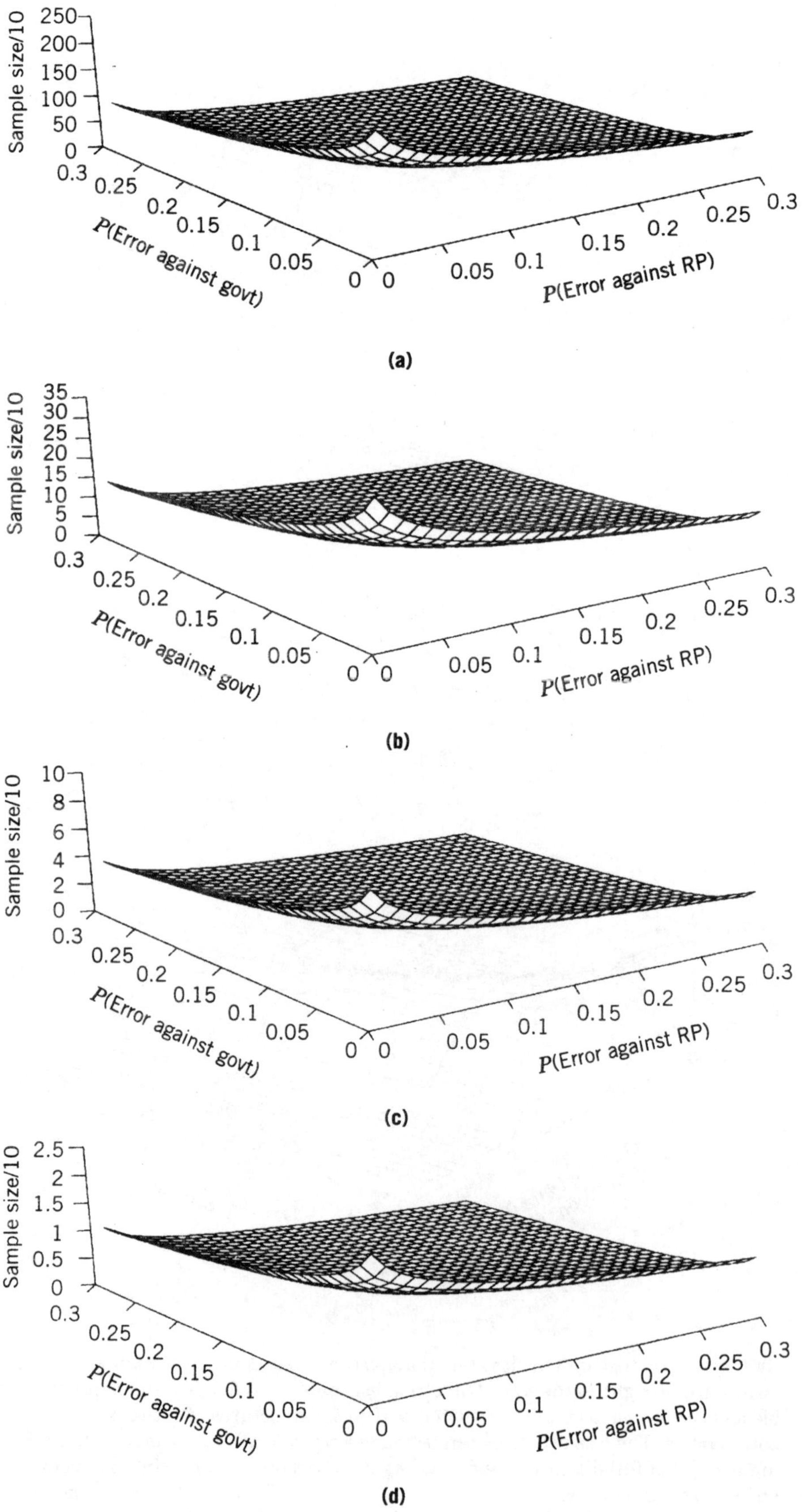

Figure 11. (**a**) Sample size for various values of decision errors: (**a**) delta 1 is 0.1 SD > delta0; (**b**) delta 1 is 0.25 SD > delta0; (**c**) delta 1 is 0.5 SD > delta0; and (**d**) delta 1 is 1 SD > delta0.

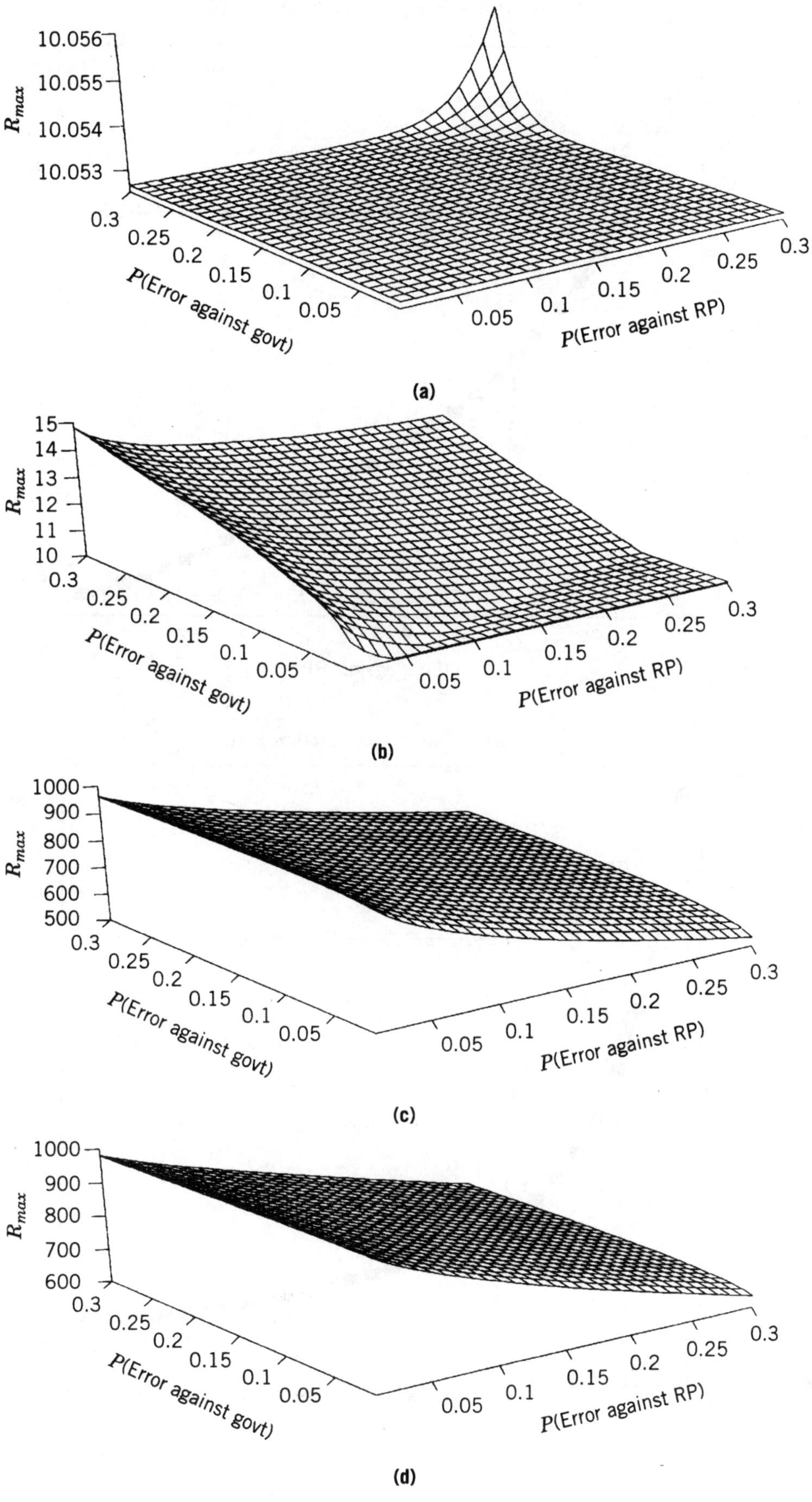

Figure 12. (**a**) Maximum total risk to government and RP (loss structure as in Figs. 7**a** and 7**b**): (**a**) delta 1 is 0.5 SD > delta0; (**b**) delta 1 is 1 SD > delta0; loss structure as in Figs. 7**g** and 7**h** (**c**) delta 1 is 0.5 SD > delta0; (**d**) delta 1 is 1 SD > delta0.

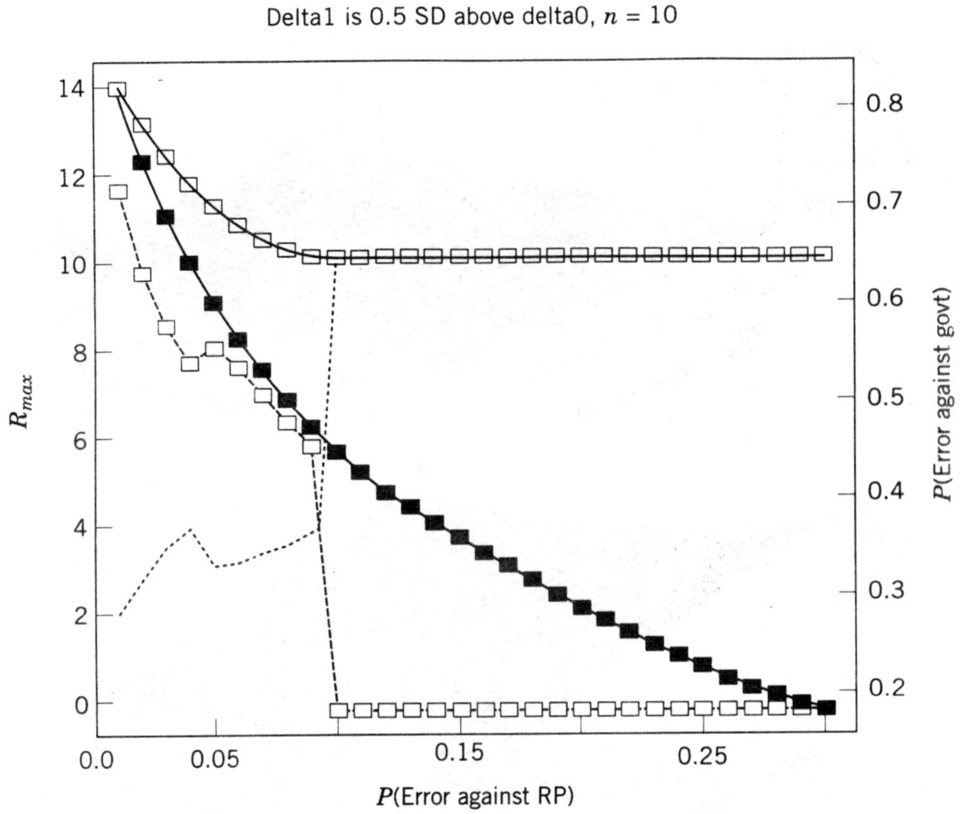

(a)

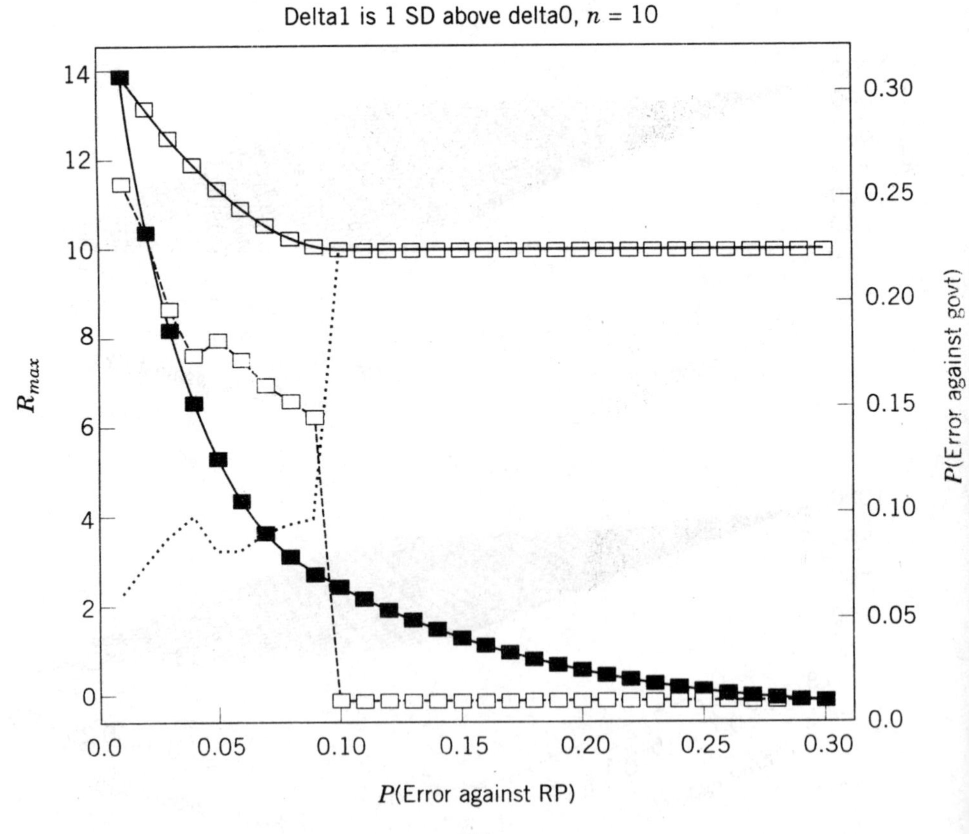

(b)

Figure 13. Maximum risks and decision errors for the loss structure in: (**a**) Figs. 7**a** and 7**b**, sample size 10, δ_1 is 0.5 σ above δ_0, (**b**) Figs. 7**a** and 7**b**, sample size 10, and δ_1 is 1 σ above δ_0; (**c**) Fig. 7**a** and 7**b**, sample size 25, δ_1 is 0.5 ς above δ_0; (**d**) Figs. 7**a** and 7**b**, sample size 25, and δ_1 is 1 σ above δ_0; (**e**) Figs. 7**g** and 7*h*, sample size 10, δ_1 is 0.5 ς above δ_0; (**f**) Figs. 7**g** and 7**h**, sample size 10, δ_1 is 1 σ above δ_0; (**g**) Figs. 7**g** and 7**h**, samples size 25, δ_1 is 0.5 σ above δ_0; and (**h**) Figs. 7**g** and 7**h**, sample size 25, and δ_1 is 1 σ above δ_0.

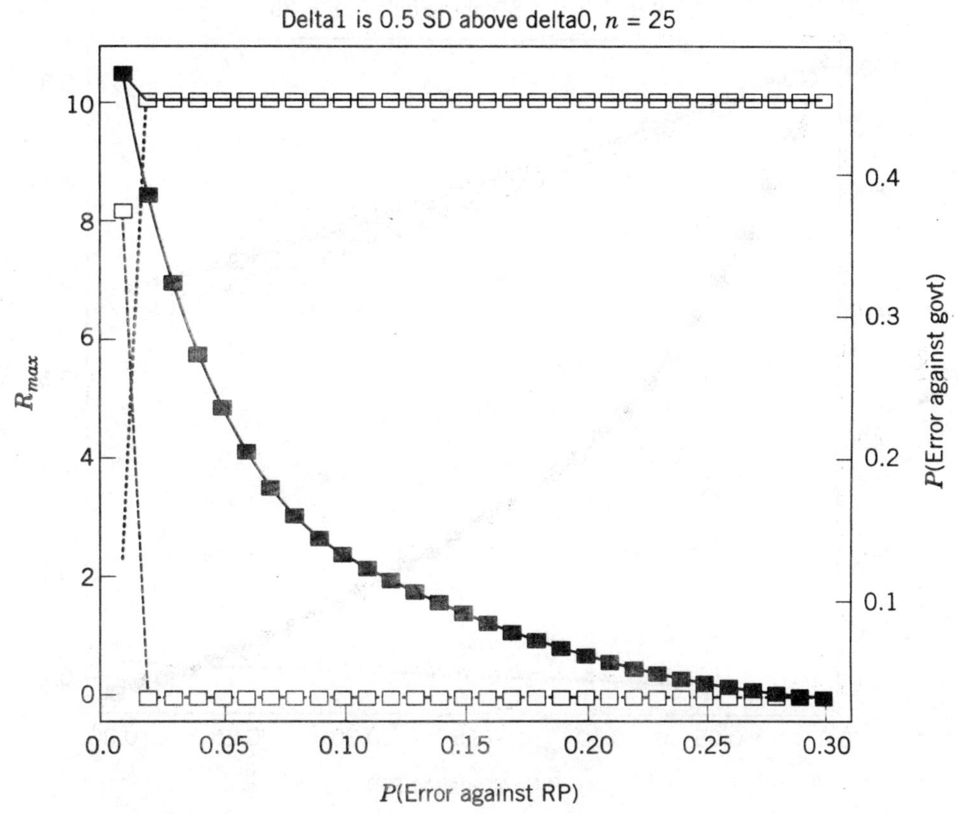

(c)

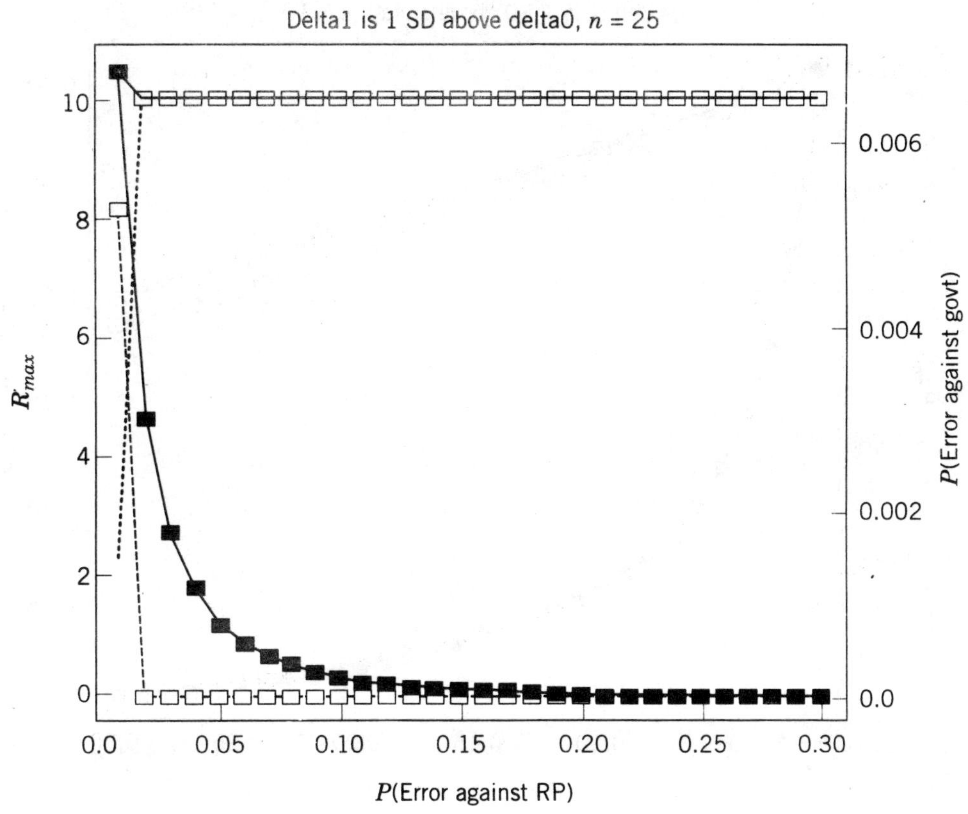

(d)

Figure 13. *(continued)*

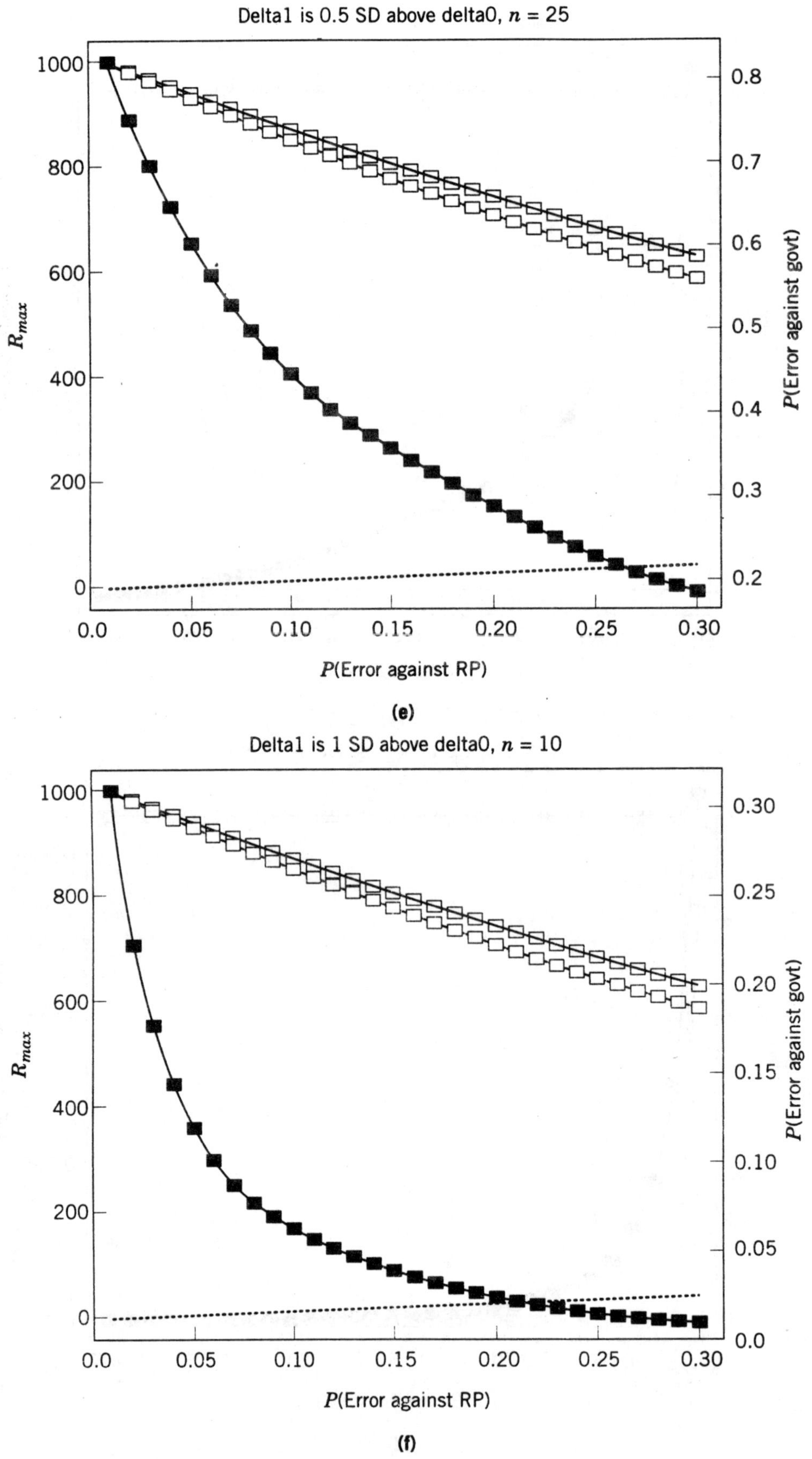

Figure 13. (*continued*)

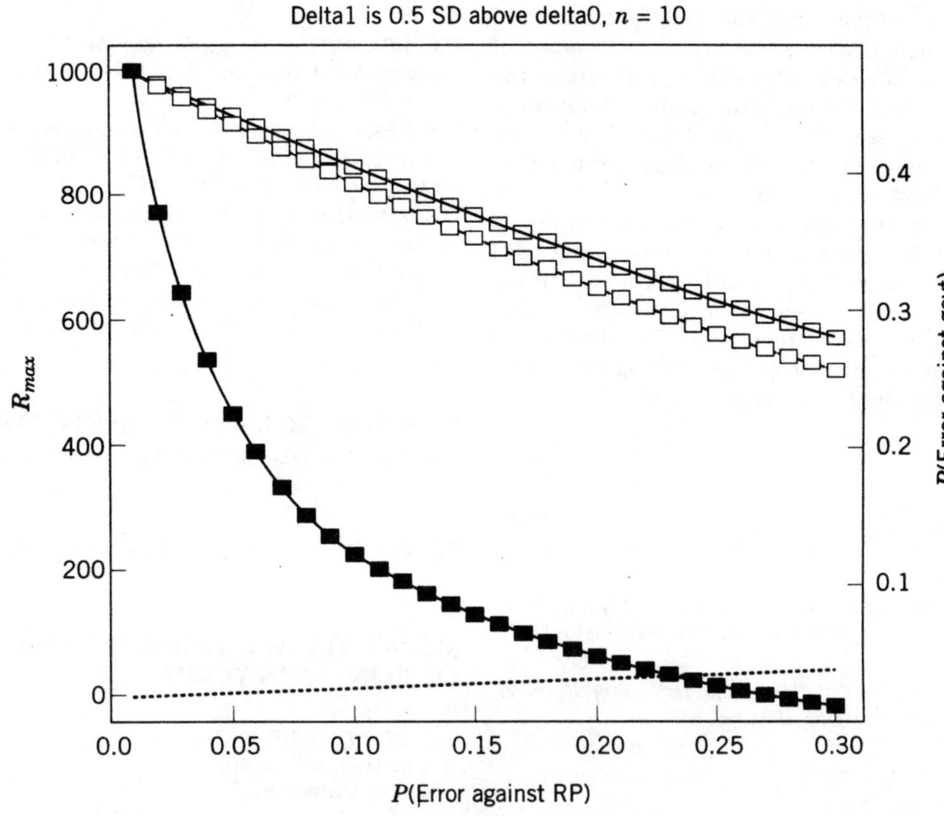

(g)

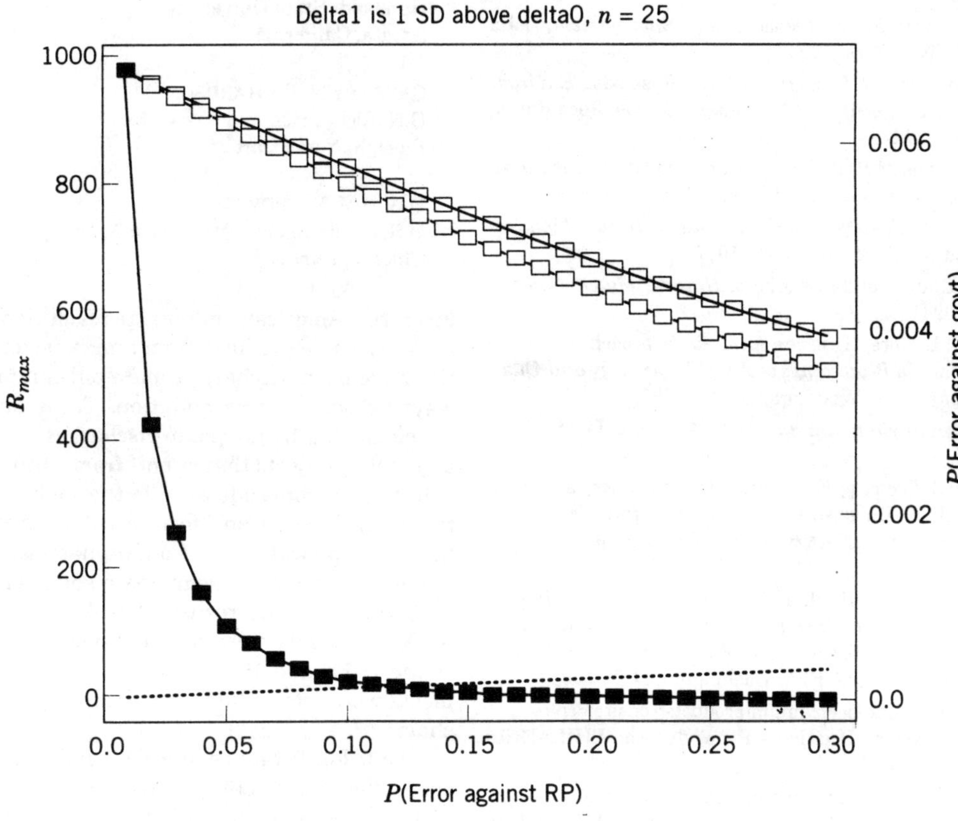

(h)

Figure 13. *(continued)*

seriously considered. The classical example for this was the government's reaction to Love Canal where all the residents of that area were relocated as opposed to serious reclamation of the land. Remediation, a less costly solution, was lost in the concern for public reaction. Another example is a situation so overwhelming, such as the toxic site near Richland, Washington, that the suggested remediation solutions appear to be beyond current scientific or economic abilities.

Each site investigation is unique in its own way. However, in this article a standard statistical approach has been used to quantify risks. A strong effort toward such a quantification of risks of all concerned parties will bring out the various issues for discussion in a well-defined way. This should lead to enhanced cooperation among the concerned parties and a smoother closure of the site remediation.

BIBLIOGRAPHY

1. U.S. EPA, *Guidance for Data Quality Assessment: Practical Methods for Data Analysis,* EPA QA/G-9, EPA/600/R-96/084, Washington, D.C., July 1996.
2. U.S. EPA, *Methods for Evaluating the Attainment of Cleanup Standards,* Vol. 2, *Ground Water,* EPA 230-R-92-014, Washington, D.C., 1992.
3. H. Lacayo, In G.P. Patil and C.R. Rao, eds., *Handbook of Statistics,* Vol. 12, North Holland, New York, 1994, pp. 881–895.

General References

J.O. Berger *Statistical Decision Theory and Bayesian Analysis,* Springer, New York, 1985.

V. Barnet and K.F. Turkman, *Statistics for the Environment,* John Wiley & Sons, Inc., New York, 1993.

H.T. David and S. Yoo, In C.R. Cothern and N.P. Ross eds., *Environmental Statistics, Assessment, and Forecasting,* Lewis, Boca Raton, Fla., 1994, pp. 361–372.

W. Feller, *An Introduction to Probability Theory and It's Applications,* 3rd ed., John Wiley & Sons, Inc., New York, 1968.

R. Gilbert, *Statistical Methods for Environmental Pollution Monitoring,* Van Nostrand Reinhold, New York, 1987.

M.J. Messner, C.A. Claton, and co-workers, *Hazard. Mater. Control,* **3,**(3), 123–151 (1994).

N.K. Neerchal and T. LeGore, "Decision Analysis Approach to Sample Size Determination," in *Proceedings of Waste Technology and Quality Assessment Conference,* Washington, D.C., 1995.

W.R. Ott, *Environmental Statistics and Data Analysis,* Lewis, Boca Raton, Fla., 1995.

G.P. Patil, "Statistical Ecology, Environmental Statistics, and Risk Assessment," in H. A. David, *Advances in Biometry: 50 Years of the International Biometric Society,* John Wiley & Sons, Inc., New York, 1997.

H. Lacayo, R.J. Nadeau, G.P. Patil, and L. Zaragoza, *Proceedings of the Workshop on Superfund Hazardous Waste: Statistical Issues in Characterizing a Site: Protocols, Tools, and Research Needs,,* Statistical Policy Branch, U.S. EPA, Washington, D.C., 1993.

htpp//www.epe.gov/superfund/oerrr/impm/products/rodsites/rpaphise.htm, for records of Decision—Pennsylvania: PHILADELPHIA to SEVEN VALLEYS.

htpp//www.epe.gov/superfund/oerrr/impm/products/rodsites/rcasan01.htm, for records of Decision—California: SAN JOSE to SAN JOSE.

N.P. Ross, and S. Harris, "From Model Building to Risk Management: Evolving Standards of Professional Responsibility," in W.A. Wallace, eds., *Ethics of Modeling,* Elsevier Science Inc., New York, 1994.

U.S. EPA, *Methods for Evaluating the Attainment of Cleanup Standards,* Vol. 1, *Soils and Solid Media,* EPA 230/02-89-042, Washington, D.C., 1988.

U.S. EPA, *Guidance for the Statistical Analysis of Groundwater Monitoring Data at RCRA Facilities,* EPA/530/R-93/003, (NTIS: PB89-151026), Washington, D.C., 1992.

U.S. EPA, *Methods for Evaluating the Attainment of Cleanup Standards,* Vol. 3, *Referenced-Based Standards for Soils and Solid Media,* EPA 230-R-94-004, Washington, D.C., 1994.

U.S. EPA, *Guidance for Data Quality Objectives Process,* (EPA QA/G4), EPA/600/R-96/055, Washington, D.C., 1994.

SAMPLING SOIL WATER AND GROUNDWATER. See SOIL WATER AND GROUNDWATER SAMPLING.

SCWO. See SUPERCRITICAL WATER OXIDATION.

SEDIMENTS, MANAGEMENT AND REMEDIATION OF CONTAMINATION

KENNETH R. DEMARS
University of Conneticut
Storrs, Conneticut

RONALD C. CHANEY
Humboldt State University
Arcata, California

GREGORY N. RICHARDSON
G.N. Richardson and Associates
Raleigh, North Carolina

DAVID M. PETROVSKI
U.S. Environmental Protection Agency
Chicago, Illinois

From the American industrial revolution (circa 1830) to the present, waste products from mines, factories, and municipalities have been discharged and spilled into the nations waterways as point-source pollution. These discharges have been supplemented by nonpoint discharges of fertilizer, pesticides, and other products that runoff from rural areas. Much of these pollutants become adsorbed to the surface of fine-grained particles which settle and form a contaminated sediment at the bottom of the waterway. Contaminated sediments may be subsequently buried by clean sediments, remobilized by erosion or disturbance, and resuspended by dredging. These contaminated sediments were long considered benign until research in the 1960s and 1970s revealed that some chemicals in sediments were bioavailable and could bioaccumulate up the food chain.

The federal/state regulations and criteria which affect the excavation, treatment, and disposal of contaminated sediments and dredged materials are continuously changing to encorporate the results of new research data and experiences. Prior to the 1960s, dredged materials were not considered to be problematic, thus disposal alternatives were typically evaluated on the basis of economics or lowest cost rather than environmental impact. This generally meant open water disposal of harbor and channel sediments, and land (or wetland) disposal of river sediments. During the mid- to late 1960s, questions were raised about the effects of dredging on water quality. In 1971 the U.S. Environmental Protection Agency (EPA) issued the report "Criteria for Determining Acceptability of Dredge Spoil Disposal to the National Waters" as the first effort to identify and classify contaminated sediments in the nations waterways for purposes of controlling dredging and disposal operations. The major concern at that time was to limit nutrient loadings such as nitrogen and phosphorous. Thus, sediment quality criteria (SQC) were established from limited scientific data to define the type and quantity of pollutants associated with dredged material. Sediments that failed these criteria were considered unsuitable for open water disposal and required placement in a confined disposal facility (CDF). Since that time the EPA criteria for classifying contaminated sediments has expanded to include numerous heavy metals and PCBs with the sediment quality criteria undergoing frequent updating. Although the development of these criteria was expected to have a positive effect on human health and well being, they have significantly slowed the process and increased the cost of dredging clean sediments from the nations waterways.

It is obvious from the nature of the problem that the process of remediating a site with contaminated sediments is very difficult, complex, and multidisciplinary in nature and that existing regulations, in many respects, often complicate the problem. Although regulations specify the criteria for defining a hazardous material, it is not clear how to sample and test a submerged site to define the spatial extent of pollutants, how to select and operate the appropriate dredge, how to treat or contain and isolate the contaminated sediments, and finally how to monitor and verify that target levels have been attained. There have been a small number of contaminated sediment remediations and demonstrations to date, however, this field is still in its infancy and is likely to increase in importance in the future.

In 1992, U.S. Congress passed the Water Resources Development Act which requires EPA, in consultation with the National Oceanic and Atmospheric Administration and the U.S. Army Corps of Engineers, to conduct a national survey of data regarding sediment quality in the United States. The goals of the National Soil Inventory (NSI) are to assess the national extent and severity of sediment contamination, to identify areas in need of further assessment and to identify high risk areas for appropriate remedial action. The NSI is to be updated every two years with the first report to congress in summer 1995. The effort to date (1) is based on over 3.4 million records from water body segments around the country including sediment chemistry and biological sample measurements of tissue residue, toxicity, benthic abundance, and histopathology. These preliminary analyses have identified over 1700 water body segments as potential areas of concern with many additional areas likely to be added in the future. The first step in remediating these areas must be to identify and control the source followed by a concerted effort to clean the contaminated sediments.

A listing of industries associated with particular sediment contaminants is presented on Table 1. Such contamination is typically historical and much of it predates regulatory control. Additionally, ongoing releases from combined sewer overflows (CSOs) can be locally significant. Sediments are also contaminated by nonpoint sources such as agricultural runoff. A portion of the sediments are so highly contaminated that they require remedial action. For example, about one-half of the total amount of sediments dredged in the Great Lakes (approximately $1.5 \times 10^6\ m^3$) are sufficiently contaminated to preclude their unconfined release to the environment. In this article, such materials are referred to as problematic sediments. The management and possible remediation of these dredged materials is the focus of this article. These contaminated sediments require special consideration during dredging and disposal operations because of potentially adverse impacts on water and air quality and aquatic organisms. Sound planning and design of dredging operations and either disposal facilities or remediation processes are necessary to protect the environment, while keeping these activities economically viable. The remaining half of the sediments are generally classified as clean and suitable for unconfined/untreated disposal.

The regulatory requirements for the disposal of dredged material are determined by both the type and level of the contaminants associated with the dredged material, as well as the extent to which the contaminants could potentially be released from the sediments to proximal air, groundwater, or surface water. To date, regulatory concern with most contaminated dredged material disposal projects has been focused primarily on containment of release routes to water, but as decisions need to be made on the proper handling of some of the more contaminated materials, potential contaminant loss pathways like air emission and seepage of free oil-phase organic contamination are being examined more closely. This is reflected in Table 2, which depicts several pertinent relationships between the level of sediment contamination, the degree of contaminant partitioning to the water associated with the sediments, in conjunction with three conceptual categories under which sediment disposal can occur and the significant disposal regulations. As depicted on Table 2, these three conceptual approaches to dredged material disposal are labeled "beneficial use or open water disposal," "solids retention," and "hydraulic isolation."

Regulation of Dredging Activity

CWA–COE 404 Program. Section 404 of the Clean Water Act (CWA) regulates the disposal of dredged or fill material into navigable water by granting the U.S. Army Corps of Engineers (U.S. ACE) the authority to designate disposal areas and the authority to control their use. Navigable waters now include all waters which may be susceptible to use in interstate commerce, or the use, degradation, or destruction of which could affect interstate commerce. The CWA regulates the impact of effluent discharge and dredged material disposal activities on navigable waters. As such, the CWA typically addresses the disposal or impact of materials having a high liquid and low solid content. As an example, all states within EPA Region 5 (Great Lakes) are authorized under the CWA and have implemented water quality standards (WQS) that are as strict or stricter than those in the CWA, and these states must certify or waive WQS Certification of all proposed dredged material disposal into the US waters in order to comply with 404 permit requirements.

Table 1. Historical Sources of Sediment Contaminants[a]

CONTAMINANTS / INDUSTRIES	Aluminum	Ammunitions	Antifouling paints	Automobile	Batteries	Chemical manufacturing	Commercial farming	Corrosion metallurgy	Dairy	Detergents/surfactants	Dye	Electrical	Explosive	Flat glass	Fruits and vegetables	Leather/tanning	Meat products	Metal finishing refining	Metallurgical processes	Nitric acid manufacturing	Oxide manufacturing	Perfume	Pesticides	Petroleum refining	Phosphate mining	Phosphorus	Photographic	Pigments/inks	Plastics	Printing plates	Pulp and paper mills	Rubber	Steam power	Steel/iron	Sulfuric acid	Textiles	Utilities	Valuable mineral mining	Wastewater treatment plants	Potential nonpoint sources
Acenaphthene							+				+												+						+											
Aldrin							+																+																	
Ammonia						+	+		+				+		+		+			+			+	+				+						+	+	+			+	
Anline											+			+				+				+						+						+	+					
Arsenic				+														+	+							+	+						+	+						+
Benzo(*a*) anthracene																																		+						
Benzo(*a*)pyrene				+														+															+	+		+				+
Cadmium	+			+		+						+						+															+							+
Chlordane																																							+	
Chlorpyrifos																																								
Chromium	+			+		+		+								+		+			+						+						+						+	+
Copper		+	+	+				+				+	+					+			+												+						+	+
Cyanide								+										+																+						
DDE																																								+
DDT							+																																	+
Dieldrin							+																+																	
Endrin																							+																	
Ethyl Parathlon																						+																		
Fluoranthene																																								+
Heptachlor																							+																	
HCB																							+																	
HCBD																																								
HCCPD					+													+																						
Lead			+		+			+				+						+						+				+	+				+	+				+	+	+
Mercury			+			+						+																+			+						+		+	+
2-Methylnaphthalene						+																																		
Nickel	+							+				+						+	+												+		+						+	+
Oil and grease				+		+												+	+												+			+						+
PCBs				+								+																						+					+	
Phenanthrene																																		+			+			
Phosphorus		+					+		+	+					+								+		+	+	+			+				+			+			
Pyrene																																		+					+	+
Selenium						+						+		+										+			+	+				+					+			
TCDD																								+							+									
TCDF																								+							+									
Toxaphene						+	+																+																	
Zinc		+		+				+		+		+						+	+		+									+		+		+			+	+	+	+

[a] Ref. 2.

Table 2. CDF Design Criteria Based on Contaminant Level and Partitioning[a]

	Problematic Sediments			
Minor Sediment Contamination Minimal Partitioning	Minor Partitioning	Moderate Partitioning	Extensive Partitioning	RCRA/TSCA Sediments
Beneficial use or open water disposal	Solids Retention approach	Hybrid approach	Hydraulic isolation approach	Hydraulic isolation CDF MTG design
	Increasing Degree of sediment isolation	Hydraulic isolation during sediment disposal only	Increasing degree of hydraulic isolation	CWA 402 permit for a point discharge
	Filter intake CDFs			
CWA 401 certification for dredged material discharge	CWA 401 certification for diffuse discharge	CWA 401 cerfitication for point or diffuse discharge	CWA 401 certification for point discharge	EPA minimum
	← 87 COE CDF Design manual →			
	Shoreline and in-lake CDF designs			Technology guidance
	← Shoreline and in-lake CDF designs →			
		Upland CDF designs →		
	← Region 5 CDF guidance designs →			
← CWA Section 404 and NEPA, EPA/COE framework →				
← Increasing degree of sediment contamination →				

[a] Ref 3.

In approving a discharge site for a 404 permit, the ACE applies guidelines developed jointly by ACE and EPA in a Memorandum or Agreement (4) as codified in 40 CFR Part 230. These guidelines require that dredged or fill material should not be discharged into an aquatic ecosystem unless it can be demonstrated that the discharge will not have an unacceptable impact either individually or in combination with other activities that could affect the ecosystem. A primary emphasis of this review is the ability of the disposal operation to be carried out to satisfy the four requirements found in Section 230.10. These requirements are as follows. 230-10 (a): The discharge site must represent the least damaging, practical alternative; 230.10 (b): Requires compliance with established legal standards (eg, State Water Quality Certification); 230.10 (c): No significant degradation of the aquatic ecosystem; and 230.10 (d): All practical means must be made to minimize or mitigate adverse environmental impact. The concept of a "mixing zone" in the vicinity of the point of discharge is introduced in 404(b)(1) guidelines presented in 40 CFR 230 and may also appear in state water quality regulations. This concept allows contaminant concentrations to exceed state WQS only within a predefined mixing zone that is immediately adjacent to the point of discharge or dredging activity. The mixing zone analysis is being eliminated by many States in favor of WQS limits based on biological concentrating contaminants (BCC).

CWA–EPA 404 Program Review. Under CWA 404(c), the EPA Administrator is given authority to veto a Section 404 permit issued by U.S. ACE if the administrator determines the discharge of dredged or fill materials into the specified disposal area will have an unacceptable effect on municipal water supplies, shellfish beds, fishery areas, wildlife, or recreational areas. EPA has promulgated regulations that define the environmental restrictions associated with these disposal areas (40 CFR Part 231).

CWA–State 401 Program Review. As stated above, Section 401 of the CWA provides that an application for a 404 permit must include a certification from the state in which the discharge originates that the discharge will comply with the following sections of the CWA:

301:Effluent Limitations

302:Water Quality Related Effluent Limitations

303:Water Quality Standards and Implementation Plan

306:National Standards of Performance

307:Toxic and Pretreatment Effluent Standards

Primary emphasis in this review is maintenance of state WQS beyond the mixing zone at the point of discharge. The State must establish procedures for public notice for all applications for 401 Certification. A 404 permit cannot be granted, and a USACE maintenance project cannot proceed if certification is denied by the State.

CWA–State 402 Program Review. The continuing discharge of rainwater and/or seepage from a confined disposal facility (CDF) is also subject to control under the CWA. Such discharges may be required to adhere to discharge limitations, based on State 401 WQS Certification and authority. Some confusion has occurred in the past because regulations under Section 402, the National Pollutant Discharge Elimination System (NPDES) program also seem to be appropriate. EPA Great Lakes Region 5 has expressed interest in using Section 402 to regulate certain types of discharges from CDFs, such as rainwater, that are not specifically exempted from Section 404. However, Section 402 has never been used for this purpose, and Section 401 has equivalent power. A state may, when appropriate, issue a 401 Certification letter that could be virtually indistinguishable in content from a typical NPDES permit. However, because most CDF releases, which are usually diffuse and intermittent, do not usually violate state WQS outside of an approved mixing zone, state 401 Certifications may range from simple waiver letters to permit-like letters requiring specific pollution monitoring and limits, groundwater monitoring requirements, fish migration timing restrictions, and even plan modifications that may or may not be acceptable to the ACE or the permittee. Any project consisting of problematic sediments or dredged materials is likely to require considerable effort coordinating for an acceptable 401 Certification from the State.

Marine Protection Research and Sanctuaries Act (MPRSA). Section 102 of MPRSA requires EPA, in consultation with ACE, to develop environmental criteria for all ocean disposal. Section 103 of MPRSA provides that U.S. ACE must specifically authorize all ocean disposal of dredged materials.

National Environmental Policy Act (NEPA). Dredged material disposal activities must comply with applicable NEPA requirements regarding identification and evaluation of alternatives associated with the dredging project. Sections 102(2) and 102(2)(E) of NEPA require the examination of reasonable alternatives to the actions proposed by the lead federal agency. All alternatives, even those discarded, must be analyzed in an environmental assessment (EA) or environmental impact statement (EIS) (qv). For federal dredging projects, the ACE is responsible under NEPA for developing alternatives for the discharge of dredged material. The U.S. ACE is also responsible for preparing NEPA documents for Section 404 permit decisions. NEPA does not define performance objectives, but does require the process evaluation procedure.

Contaminated Sediments

Regulation of Contaminated Dredged Materials. For the purposes of this document, dredged materials are considered contaminated when the ambient or leachable concentration of metals or organic compounds exceed federal RCRA or TSCA regulatory limits or when contamination exists in high enough concentrations and are sufficiently available to affect human and/or ecosystem health. Contaminated dredged material may threaten human health by direct exposure or when contaminated water or organisms are consumed. Consequently, an acceptable disposal option for such sediments will entail some form of isolation from the ambient environment. Contaminants in sediments can be from point sources (municipal or industrial effluent), nonpoint sources (eg agricultural runoff or airborne sources), and from other sources such as spills or leaks.

RCRA and TSCA regulations provide for the land disposal of contaminated materials in cells that hydraulically isolate the waste material from proximal ground and surface waters. Wastes placed within such landfill cells must not contain free liquids as defined by the paint filter test (EPA Method 9095).

RCRA and TSCA land disposal regulations are therefore directed at wastes that are very high in solids and have little or no free liquids. Conversely, CWA regulations are directed at the disposal of waters containing very small concentrations of suspended solids and partitioned contaminants. With large portions of both water and solids, environmentally acceptable disposal options for dredged sediments combine aspects common to both of these regulatory perspectives.

In general, the potential for sediments to hold or bind contaminants increases with increasing percentages of fine-grained sediments (silts and clays) and organic carbon content. Fine-grained sediments have a high surface area per unit mass and significant ionic forces that attract and retain many classes of contaminants. The presence of organic matter increases the affinity of sediments for metals and organic contaminants.

Sediment pH and oxidation/reduction status also affect the mobility and biological availability of contaminants. Strongly acidic (low pH) sediments slow microbial activity and increase the soluble level of toxic metals. Weakly acidic to slightly alkaline sediments result in less mobile metals. The oxidation/reduction potential of a sediment, as measured by the redox potential of the sediment, has a major effect on the capture or release of metals by a sediment. Frequently, the sediment contaminants are chemically stable under the geochemical conditions existing within the *in situ* sediments. However, changes in the pH or redox potential of a sediment during dredging, remediation or disposal will substantially affect contaminant mobilization. This influences the partitioning of the contaminant between the sediment solids and the pore water. EPA is currently updating equilibrium sediment quality criteria (SQC) that normalize nonionic organic contaminants to organic carbon and metal contaminants to acid volatile sulfide (AVS). The role of long-term equilibrium contaminant partitioning is reflected in the regulatory scheme shown on Table 2. Contaminant mobilization mechanisms in dredged materials, SQC, and remediation of contaminated materials are discussed in greater detail in a later section.

Problematic Sediments. In this article, problematic dredged materials are defined as materials which are sufficiently contaminated to preclude their unconfined release to the environment, eg open water disposal or beneficial reuse. Problematic dredged material are commonly the result of nonpoint source releases and spills or point-source releases which occurred prior to the effective date of the CWA as well as most other environmental regulations. Because the unrestricted release of problematic dredged materials to the environment is unacceptable, the Adequate Environmental Performance and Protection (AEPP) disposal criteria for these materials will require containment to some extent. The extent of containment required is dependent on the nature, concentration, and partitioning potential of the specific contaminants. Provided an acceptable level of suspended solids is achieved in the effluent, the AEPP criteria for the disposal of sediments associated with low levels of tightly sorbed contamination may require only physical isolation of the dredged material. AEPP criteria for the disposal of sediments with higher levels of contaminants that partition to the aqueous phase at concentrations of concern will generally require hydraulic isolation of the dredged material as well as any derived effluent. These two approaches to the disposal of dredged materials have been labeled the "sediment retention approach" and the "hydraulic isolation approach," respectively. As discussed below, the disposal of sufficiently contaminated problematic dredged material can be regulated by the Resource Conservation and Recovery Act, Subtitle C (RCRA-C) and/or Toxic Substances Control Act (TSCA). AEPP criteria for the disposal of such highly contaminated materials which meet the regulatory requirements for land disposal will require a hydraulic isolation design based on RCRA-C Minimum Technical Guidance (MTG).

TSCA Sediments. Enacted in 1976 by U.S. Congress in response to the discovery that DDT and other pesticides were impacting waterfowl and that polychlorinated biphenyl (PCBs) were present in large numbers of birds and fish, TSCA required EPA to test certain existing chemicals to determine if they posed an unacceptable human health or environmental risk. While the 29 sections of TSCA cover a wide range of considerations, TSCA has become fundamentally associated with the control of PCBs at concentrations equal to or greater than 50 ppm. The use, control, and disposal of PCBs under TSCA are covered in 40 CFR Part 761, as issued in 1982. Subpart D specifies that, in general, PCBs at concentrations equal to or greater than 50 ppm must be disposed of in an incinerator. However, 40 CFR Part 761.60 (a)(2) also allows PCBs at concentrations less than 500 ppm to be disposed of in a chemical waste landfill as defined in 761.75. TSCA specifically addresses dredged sediments contaminated with PCB concentrations equal or exceeding 50 ppm in 761.60(a)(5). This section provides three methods of PCB contaminated sediment or sludge disposal: (*1*) incineration, (*2*) placement in a TSCA chemical waste landfill, and (*3*) using a disposal method approved by the Regional Administrator. Application for the (*3*) must demonstrate that disposal in an incinerator or chemical landfill is not reasonable and appropriate, and that the alternative disposal method will provide adequate protection to health and the environment. TSCA also allows for the temporary storage of wastes containing PCBs at concentrations equal to or greater than 50 ppm for a time period not to exceed one year in an approved storage facility.

A chemical waste landfill under 40 CFR Part 761.75 (TSCA) must satisfy well-defined technical liner requirements. These technical liner requirements are based on hydraulic isolation of the PCB contaminated material and include the following:

1. As a minimum, the landfill must have 4 feet of *in situ* natural soils or 3 feet of compacted soils having a hydraulic conductivity equal to or less than 1×10^{-7} cm/s. A synthetic membrane liner must be used when, in the judgment of the Regional Administrator, the hydrogeologic or geologic conditions at the site require it to achieve an equivalence to these criteria.
2. The landfill design must include a leachate collection/monitoring system.
3. The bottom of the liner system must be at least 50 feet from the historic high water table. Floodplains, shorelands, and groundwater recharge areas shall be avoided.
4. The landfill must be protected from the 100 year flood.
5. The landfill site must have low to moderate relief to minimize long-term erosion.

6. The landfill must be equipped with a monitoring system that allows both ground and surface water to be monitored.

These regulatory requirements, although not consistent in minor design details with those required by RCRA-C for a hazardous waste landfill, are consistent with the full containment concepts of RCRA. The TSCA chemical landfill requires a greater thickness of vertical buffer between the liner and seasonal high groundwater table. From a dredged material disposal perspective, the 50 feet above the historic high water table requirement would virtually eliminate all CDFs for TSCA waste in the United States if not for alternative (*3*) eg, disposal method approved by the Regional Administrator. The conceptual basis for the TSCA regulations is the hydraulic isolation of the waste from surface and groundwaters. This is identical to the conceptual basis for RCRA-C and RCRA-D regulations.

RCRA Subtitle C Sediments. As promulgated in 1976 and amended by Hazardous and Solid Waste Amendments (HSWA) in 1984, RCRA-C is a national regulatory mechanism that controls the treatment, storage, and disposal of hazardous waste as defined under 40 CFR 261 and thereby minimizes the present and future threat to human health and the environment. 40 CFR 261 specifies that a waste is hazardous if it is not excluded from regulation as a hazardous waste and it meets any of the following criteria: (*1*) is named as a hazardous waste and is included in one of three lists under 40 CFR 261, (*2*) exhibits a hazardous waste characteristic, (*3*) results from the mixture of a waste with listed hazardous waste, (*4*) is derived from a listed hazardous waste or exhibits a hazardous waste characteristic, and (*5*) is an environmental media which contains a listed hazardous waste. The first list under 40 CFR 261 contains hazardous wastes from nonspecific sources (eg, spent halogenated solvents, such as perchloroethylene). The second list identifies hazardous wastes which commonly result from manufacturing or industrial processes (eg, bottom sediment sludge from the treatment of wastewater by wood preservers). The third list contains commercial chemical products or chemical manufacturing intermediates which, when discarded, must be treated as hazardous waste. If a waste is not listed as hazardous, it is still covered by RCRA if it exhibits one of four hazardous waste characteristics: ignitability, corrosivity, reactivity, or toxicity. For example, under the definition provided for toxicity, a solid waste is hazardous if the extract obtained from the sample using the Toxic Characteristic Leaching Procedure (TCLP) test contains a hazardous constituent at or above the regulatory threshold provided for that constituent. To date, most RCRA-C sediment characterization efforts in the Great Lakes area (EPA Region 5) have focused on toxicity as defined by the TCLP test.

Of major potential significance to the disposal of contaminated dredged materials is the land disposal ban enacted by HSWA. These regulatory requirements mandate that hazardous wastes or Subtitle C media must be properly and adequately treated prior to land disposal. Such treatment would include destruction of organic constituents and/or solidification of the metal constituents in the waste such that it will pass a TCLP test. In 1986, EPA banned the land disposal of untreated hazardous waste containing dioxin and solvents (see DIOXIN EMISSIONS INVENTORY). Next, in 1987, EPA banned the disposal of certain untreated hazardous wastes that had previously been banned by California. Third, EPA published a ranking of all other hazardous wastes based on their hazard or volume. From 1988 to 1990 EPA enacted a ban on the land disposal of these wastes without treatment. By banning the direct land disposal of such hazardous wastes, Congress sought to reduce the rate of hazardous waste production and the long-term risks associated with hazardous waste disposal.

The treatment requirements under RCRA-C for contaminated environmental media such as sediment, depend on the type of RCRA-C contaminant associated with the sediment. Conceptually, sediments can be contaminated with one or more listed hazardous waste or can exhibit one or more hazardous waste characteristics. In practice however, the identification of sediments which require compliance with the RCRA-C regulations for listed waste has been rare. This is due to the RCRA-C exclusions under 40 CFR 261.4(a) for hazardous waste mixed with domestic sewage and waste released from industrial point-source discharges subject to Section 402 of the CWA. An overview of the treatment and disposal regulations for sediments requiring compliance with RCRA-C is provided below.

The treatment and disposal of sediments contaminated with a listed hazardous waste can be handled in one of two ways. Where feasible, such sediments can be treated to the extent that the sediments no longer contain the listed waste. The threshold levels for this determination are provided by the EPA Region in which the activities occur. Treatment procedures which reduce the listed RCRA-C wastes from the sediment to concentrations below these thresholds render the sediments nonhazardous. Subsequent disposal of the sediments is then not regulated under RCRA-C. In contrast, if the treated sediments still contain listed hazardous wastes, but meet the treatment standards provided under 40 CFR 268, the disposal of the sediments would continue to be regulated under RCRA-C. Consequently, such treated sediments would need to be dewatered until the sediments pass the paint filter test and subsequently placed in a RCRA-C disposal facility.

Similarly, the treatment requirements for dredged sediments which exhibit a RCRA-C hazardous waste characteristic can also be addressed through the following scenario. RCRA-C requires treatment of sediment to the standards provided in 40 CFR 268. Once adequately treated, disposal of the sediments would no longer be regulated under RCRA-C.

RCRA-C differs from TSCA in that many specific requirements for treatment, storage, or disposal (TSD) facilities are not codified. Technical requirements for TSDs have been codified in 40 CFR 264 as Minimum Technology Requirements (MTR) and are given in Minimum Technology Guidance (MTG) documents prepared by EPA. Site location requirements are codified with respect to seismic considerations and floodplains (40 CFR 264.18). RCRA does not contain site restrictions that would eliminate in-lake or near-shore TSDs or CDFs. Under Subpart N, the hazardous waste regulations simply specify a liner system designed to prevent any migration of wastes out of the landfill and a leachate collection system immediately above the liner that will prevent the leachate depth from exceeding 30 cm over the liner. These specifications form the basis for the worst-case hydraulic isolation CDF design criteria.

The groundwater protection standards of Subtitle C involve the use of EPA-approved health-based standards for evaluating compliance with approved environmental performance standards. Such standards include the Maximum Contaminant Levels (MCL) established by the CWA. When no MCLs exist, the reference doses (RFDS) for threshold constituents, and carcinogenic potency factors (CPFs) for nonthreshold constituents are used. CPFs measure the ability of a chemical to cause cancer assuming a risk level of 10^{-6} for Class A and B carcinogens and 10^{-5} for Class C carcinogens (5).

RCRA Subtitle D Sediments. As promulgated on October 9, 1991, Subtitle D (40 CFR 258) defined minimum landfill standards for those facilities receiving MSW and for landfills codisposing sewage sludge with MSW. The codisposal regulations were copromulgated under the authority of RCRA-D and CWA. Separate regulations are being developed for sewage sludge monofills under CWA Authority. The RCRA-D regulations are the first to clearly establish performance objectives for the landfills. A default composite liner is provided in Subtitle D, however the general performance criteria for a MSW landfill requires that groundwater quality be maintained such that all constituents are at concentrations below their MCLs at a regulatory specified point of compliance adjacent to the landfill.

Remediation of Contaminated Sediments. Remediation of contaminated dredged material to remove or lower the level of contaminants may involve combining several treatment processes to achieve cleanup objectives. Treated sediments, once cleaned, can be used as beneficial fill material outside of a CDF or as designed components within the CDF. For example, cleaned fine-grained sediments could be used to line a CDF. The Assessment and Remediation of Contaminated Sediments (ARCS) program in EPA Region 5 has performed pilot- and bench-scale demonstrations of selected technologies. A full discussion of these technologies is presented in a later section. It should be noted that, at present, these technologies have not been widely implemented at full scale.

The containment of contaminated sediments in CDFs is generally viewed as the most cost-effective remediation tool. This is consistent with a parallel trend in Superfund sites where containment of the contaminated soils or wastes has been the selected alternative in a clear majority of existing Record of Decisions (RODs). This containment policy is expressed in recent revised Remedial Investigation/Feasibility Study (RI/FS) procedures for contaminated municipal solid waste landfills (6). These revised RI/FS procedures make containment of the waste the presumptive remedy. As with dredging projects, the quantities of contaminated solids and relatively low contamination levels associated with MSW landfills frequently makes current waste remediation technologies prohibitively expensive.

Confined Disposal of Problematic Sediments

Nationwide CDFs were built to dispose of clean dredged materials; within the Great Lakes Region (EPA Region 5), however, all existing CDFs were designed for the disposal of problematic dredged materials. Such facilities are designed to retain dredged solids while allowing the carrier water to be released from the containment area. The general conceptual drawing for this process is shown on Figure 1. The principal functional components of the CDF are the containment dikes that hold the dredged materials and a weir, or filter wall, or treatment system that allows the carrier water to decant from the CDF. The design of containment dikes is site-specific and must consider the following factors: dike stability with respect to shear strength, settlement, seepage, erosion, and available building materials or construction equipment. The influent dredged material may be a slurry if hydraulic dredging processes are used or a water/soil–clod matrix if bucket excavators are used. This dredged material influent and the decanted effluent are normally characterized by the suspended solids concentration, suspended particle size gradation, type of carrier water (fresh or saline), and rate of flow.

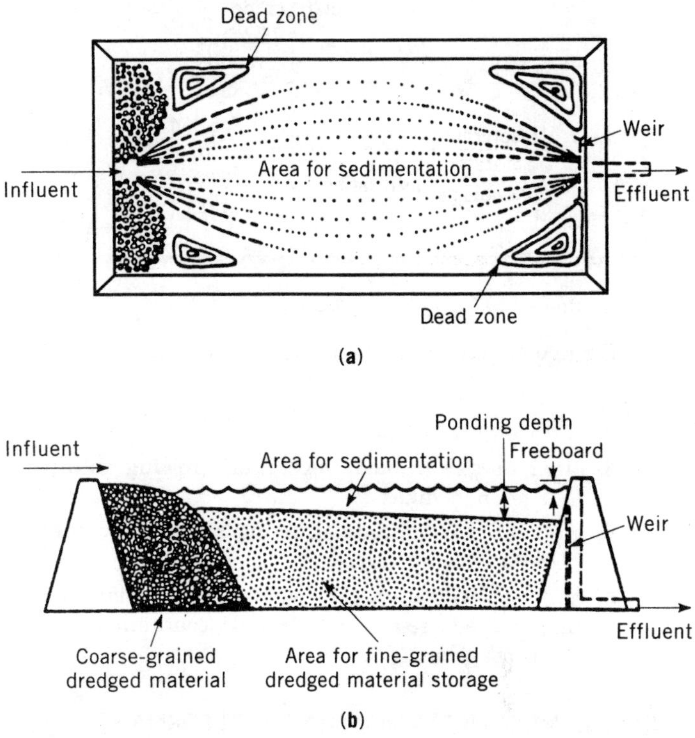

Figure 1. Conceptual diagram of a CDF: (**a**) plan (**b**) cross section.

The containment basin must be sized to provide an adequate volume to contain the dredged slurry and to provide sufficient retention time for the associated water to allow solids to fall out of suspension. Flocculent may be used to increase the rate at which solids fall out of suspension if the site area is restricted. Weirs are designed such that the water depth above the solids can be maintained constant as the CDF fills with sediment.

The Great Lakes CDFs were designed so that the effluent leaving the CDF has a concentration of suspended solids less than 1 to 2 grams per liter for freshwater conditions (EM 1110-2-5027). Typical effluent requirements in EPA Region 5 are usually in terms of tens of milligrams per liter and are enforced at those CDFs having a measurable point discharge.

CDF Contaminant Pathways

The design of the CDF must be concerned with potential pathways by which the contaminant can leave the containment system. For an upland or typical island CDF such pathways are shown on Figure 2 and include the following:

1. effluent drained through the weir or dikes
2. leachate drained from the sediments that may move into the groundwater

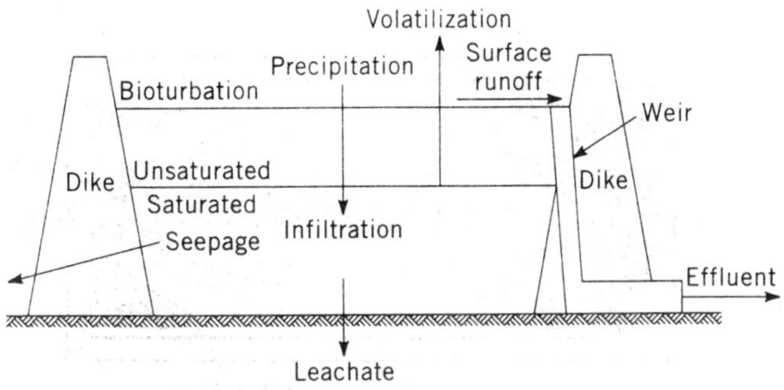

Figure 2. Contaminant pathways for upland CDF (7).

3. surface water runoff
4. plant uptake through vegetation growing within the CDF or on its dikes
5. animal uptake in animals living or nesting within the CDF
6. from volatilization or wind borne migration of fine-grained soils and bound contaminants into the atmosphere.

DEVELOPMENT OF ENVIRONMENTAL PERFORMANCE CRITERIA

In 1971, Region 5 of the EPA issued "Criteria for Determining Acceptability of Dredge Spoil Disposal to the National Waters" as a first effort to identify and classify polluted sediments in the nation's waterways for purposes of controlling dredging and disposal operations. These ambient concentration thresholds were published before there was adequate scientific data to define the type, quantity, and partitioning potential of contaminants which are associated with dredged material which impact water and sediment quality at the dredging or disposal site and, thus, become bioavailable. Sediments that failed this evaluation were considered unsuitable for aquatic (open water) disposal and required confined disposal or remediation. In 1977, EPA Region 5 began to use sediment pollution criteria defined by the Great Lakes guidelines shown on Table 3.

During this same time period the U.S. ACE was developing an approach for evaluating the pollution potential and bioavailability of contaminated dredged materials during dredging and disposal operations. The U.S. ACE position about establishing regulatory criteria for the disposal of clean and contaminated dredged materials in the oceans and other aquatic bodies such as lakes and rivers has been discussed (8). It was noted that the criteria for defining environmental protection should:

1. be meaningful and based on the best possible knowledge
2. not precede the current technical state of the art
3. be based on laboratory procedures that can be performed satisfactorily at routine testing laboratories
4. attainment and implementation should not be prohibitively expensive

The first guidelines for the discharge of dredged materials into waters of the United States were issued in 1975 and were quickly followed by interim guidance manuals for discharge into navigable waters and into ocean waters (9). The latter manual is referred to as the Green Book. In 1991, EPA and U.S. ACE jointly published an update of the 1977 Green Book for the discharge of dredged material into ocean waters (7). This update was intended to address the differences between the regulations governing disposal under CWA and MPRSA and is specifically intended for CWA 404(b)(1) compliance determinations. EPA and U.S. ACE are currently drafting a guidance manual for evaluation of material proposed for discharge in inland and near coastal waters (Inland Testing Manual). Because the Green Book applies to ocean and coastal waters it has not been used by EPA Region 5 for the Great Lakes. EPA Region 5 has instead used a case-by-case approach using the 1977 Great Lakes Guidelines (see Table 3), International Joint Commission guidelines, and other criteria to make the decision of whether the dredged materials are contaminated and thus inappropriate for open water disposal. While the proposed Inland Testing Manual is national in scope, the U.S. ACE North Central Division and Region 2, 3, and 5 are also developing a Great Lakes Testing Manual.

Joint U.S. ACE/EPA Sediment Management Strategy

Based on the USACE decision-making framework (10) as applied to Commencement Bay, Washington (11), a joint USACE/EPA guidance document for evaluation of environmentally acceptable dredged material management alternatives was developed. The document is intended to serve as the technical framework for U.S. ACE and EPA personnel in evaluating the environmental acceptability of dredged material management alternatives. Three dredged material management alternatives are considered in the joint framework: open water disposal, confined (dike) disposal, and beneficial use application. The joint EPA/USACE dredged material management program is summarized in the flow chart on Figure 3. This proposed management program is divided into five major elements as follows.

Evaluation of Dredging Project Requirements. The need for the dredging project must be established under NEPA, CWA, and/or MPRSA. Under NEPA, the initial impact assessment for the dredging project relates to the purpose and need for the proposed action in the case of new work and the continued viability in the case of an existing project. The needs determination under CWA or MPRSA is specifically concerned with a justification of the necessity for dredged material disposal in waters of the U.S. or ocean, respectively.

Identification of Alternatives. Under NEPA, the environmental impact of disposal options such as CDFs, open water, and beneficial uses must be considered. The NEPA document must discuss all reasonable alternatives even if they are beyond the capability of the applicant or lead agency making the application. Note that NEPA does not specify performance standards but simply requires the evaluation of alternatives.

Initial Screening of Alternatives. Reasonable dredged material management alternatives include those that are practical or feasible given environmental, technical, and economic constraints. All potential alternatives (ie, CDF, open water, beneficial use, etc) are evaluated with respect to the availability of the required and suitable disposal site, design limitations, climatic conditions, dredging equipment availability, physical

Table 3. EPA Guideline Values for Harbor Sediment Classification[a]

Compound/Element, Ranking	Nonpolluted	Moderately Polluted	Heavily Polluted
Volatile solids %	<5	5–8	>8
COD	<40,000	40,000–80,000	>80,000
TKN	<1,000	1,000–2,000	>2,000
Oil and grease	<1,000	1,000–2,000	>2,000
Lead	<40	40–60	>60
Zinc	<90	90–2000	>200
Ammonia	<75	75–200	>200
Cyanide	<0.10	0.10–0.25	>0.25
Phosphorus	<420	420–650	>650
Iron	<17,000	17,000–25,000	>25,000
Nickel	<20	20–50	>50

[a] Value in mg/kg dry weight unless otherwise specified.

and chemical aspects of the sediments, local interests or concerns, and known environmental regulatory (eg, endangered species) and economic concerns. The initial screening is based on available sources of sediment contamination information.

- Available results prior physical, chemical, and biological tests of the materials proposed to be dumped.
- Available results of prior field characterization studies of the materials proposed to be dumped (eg, physical characteristics, organic–carbon content, and grain size).
- Available information describing the source(s) of the material to be dumped which would be relevant to the identification of potential contaminants of concern.
- Existing data contained in files of either the EPA or COE or otherwise available from public or private sources. Examples of sources from which relevant information might be obtained include:
  - Selected Chemical Spill Listing (EPA)
  - Pesticide Spill Reporting System (EPA)
  - Pollution Incident Reporting System (USCG)
  - Identification of In-Place Pollutants and Priorities for Removal (EPA)
  - Hazardous waste sites and management facilities reports (EPA)
  - COE studies of sediment pollution and sediments
  - Federal STORET, BIOS, CETIS, and ODES databases (EPA)
  - Water and sediment data on major tributaries (USGS)
  - NPDES permit records
  - Fish and Wildlife Service
  - CWA 404(b)(1) evaluations
  - Pertinent and applicable research reports
  - MPRSA 103 Evaluations
  - Port Authorities
  - Colleges/Universities
  - Records of state agencies (eg, environmental, water survey)
  - Published scientific literature
  - CERCLA/RCRA Facility Files

Detailed Assessment of Alternatives. Environmental acceptability of a dredged material management alternative must include the following:

1. evaluation of the adequacy and timeliness of existing data
2. evaluation of the physical characteristics of the sediments
3. evaluation of sediment contamination
4. perform appropriate CWA testing and assessments
5. evaluation of sediment management/control options

Once a detailed assessment of the alternatives is completed and the CDF alternative has been selected, the flow chart on Figure 4 is used for the detailed assessment of the CDF alternative to include testing and related evaluations. This requires a site assessment of all potential CDF locations including an evaluation of physical impacts, site capacity, and possible management options. The primary concern of this flow chart is the short- and long-term impact of contaminated dredged material placement into the CDF and the need for additional control measures.

Alternative Selection. The selection of a preferred/proposed alternative during the NEPA EIS phase is based on environmental acceptability, technical feasibility, costs, and other factors.

Evaluation of Contaminant Pathways

The evaluation of the impact of the disposal of contaminated dredged materials on the local environment is based on a detailed laboratory testing protocol that considers each of the possible contaminant pathways. Based on the test results and analysis, the laboratory testing program will determine if additional control measures are required for the proposed CDF. Additional control measures consist of liners, covers, and filters.

Effluent Discharge Quality. The large volumes of fluids generated during dredging may contain undesirable levels of dissolved contaminants and suspended fine solids with adsorbed contaminants. The modified elutriate test (12) is the current method which the U.S. ACE uses to predict effluent

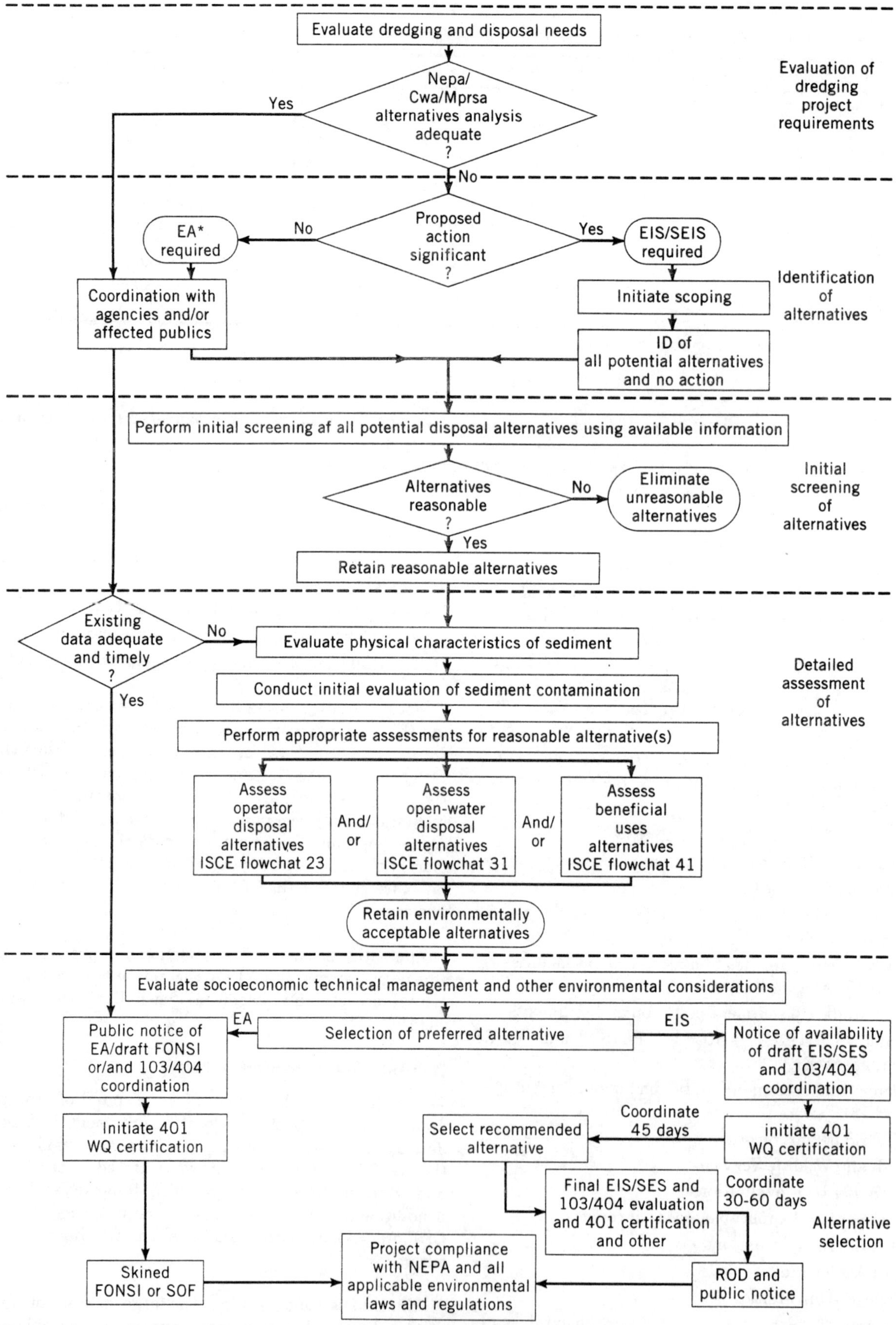

*If at any time in the EA process, the federal action is released as being significant, EIS scoping is inflated.

EA • Environmental assessment
EIS/SES • Environmental impact statement/supplement
FONSI • Finding of no significant impact

ROD = Record of decision
SOF = Statement of findings

Figure 3. Flow chart illustrating framework for determining environmental acceptability of dredged material disposal alternatives (7).

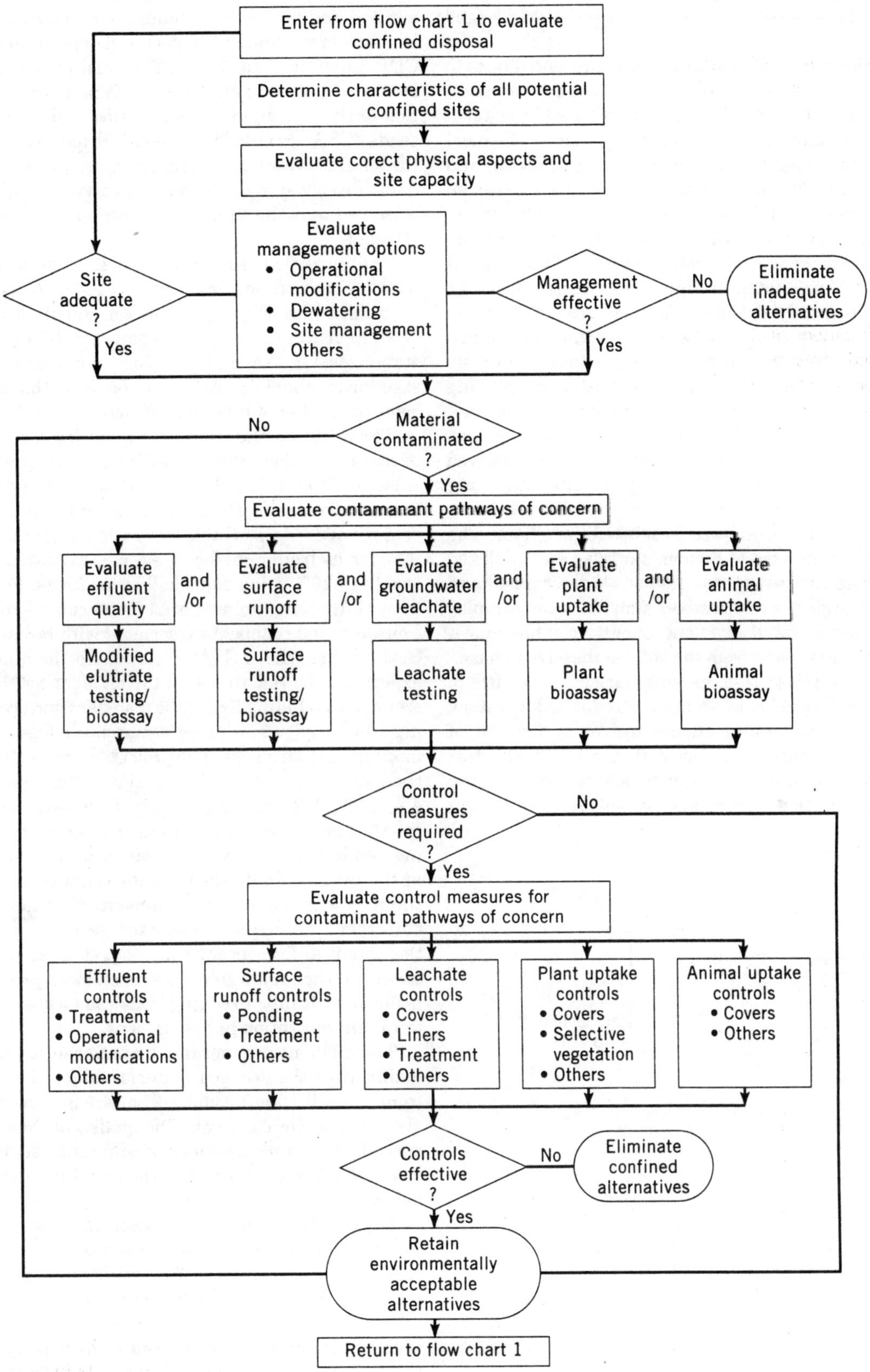

Figure 4. Flow chart illustrating framework for testing and evaluation for confined (diked) disposal (7).

quality from a CDF. This is a modification of the standard elutriate test used for assessment of aquatic disposal of dredged material.

A review of the standard elutriate test is presented here to aid in understanding the transition represented by the "modified" elutriate test. The U.S. ACE has always based its criteria on the change in ambient water quality (dissolved chemical concentrations) that might be expected during dredging and disposal in a CDF. The compliance of a dredged material operation with state water quality standards (WQS) is assessed by the elutriate test which is designed to simulate the dredging/disposal process. A schematic diagram of the original elutriate test (8) is presented in Figure 5 as designed for the disposal of hydraulically dredged material. As noted in the diagram, four volumes of disposal site water and one volume of contaminated dredged material are shaken vigorously in a beaker for 30 minutes to simulate hydraulic dredging and to allow for the exchange of contaminants from the solid phase (adsorbed organics and metals) to the dissolved phase. Also, prior to mixing the water with sediments, the dissolved concentrations of pollutants in the disposal site water are measured and listed as X_1. After shaking the sediment/water suspension, the mixture is allowed to settle for one hour. The supernatant is then filtered to remove particles and analyzed for the same chemical constituents previously determined and listed as X. Although the test method simplifies the complex interaction between polluted sediment and water, it has shown that there is little migration from the solid to dissolved phase.

The modified elutriate test, as summarized in Figure 6, was developed by U.S. ACE to be a more realistic laboratory simulation of effluent quality during hydraulic disposal of dredged material in open water or within a CDF. The test involves mixing contaminated sediment and water from the dredge site in proportion to the expected influent concentration as influenced by the dredging method. The mixture is then aerated and settled in a column. Modifications to the test can be made to simulate transport and deposition for an estimated CDF retention time (13,14). The unfiltered supernatant (effluent) is subjected to chemical analysis to define the concentration of the contaminants associated with the total suspended solids (TSS). Similarly, a chemical analysis is performed on the filtered water to determine the concentrations of contaminants dissolved in the water. This allows prediction of both the dissolved and the total concentration of contaminants in the effluent.

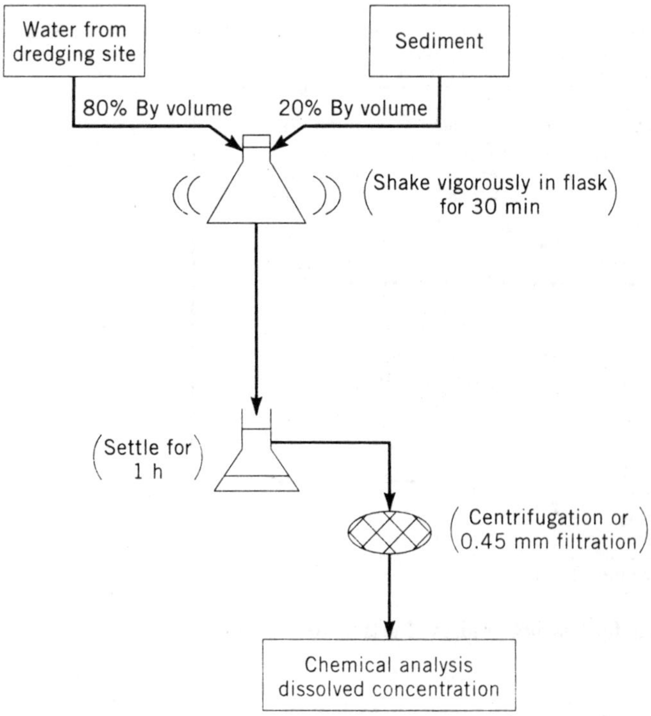

Figure 5. Elutriate test (8).

The modified elutriate test is generally used with a parallel column settling test to predict actual effluent TSS as outlined in Figure 7. The estimated total concentration of effluent contaminants is compared to state water quality standards. If the effluent does not meet the state WQS standards, the U.S. ACE can propose the use of a mixing zone to further dilute the effluent.

The modified elutriate test provides conservative estimates of contaminants within the effluent; bulk quantities (mass released) of contaminants released are highly dependent on the volume of dredging fluids generated during dredging. The percentage of dredging fluids by weight can range from more than 90% for hydraulic methods, eg, using a cutterhead dredge, to less than 20% using clamshell excavations. For a hypothetical annual dredging volume of 200,000 cubic yards of the highly contaminated sediments associated with Indiana Harbor in the Great Lakes, the U.S. ACE estimated the quantities of PCBs expected to be released with the effluent for three CDF filling scenarios including hydraulic transfer from scow, direct pumping from dredge site using a matchbox-type dredgehead, and mechanical placement. The projected annual PCBs released for these three methods relative to a common size CDF are 6.3, 4.2, and 0.0027 kg, respectively. Ironically, methods of dredging that minimize contaminant transport at the site of dredging can lead to a maximum potential for contaminant release at the point of discharge from the containment facility and ultimately into the local environment. For example, the use of a cutterhead dredge minimizes the generation of turbidity and the potential for contaminant migration at the dredging site. However, the high water to solids effluent produced by the cutterhead hydraulic dredging does produce a potential for contaminant migration in the effluent.

The effluent is generated by both the supernatant that remains after the dredged material falls out of suspension and from rainfall runoff coming from dredged material that rises above the water elevation. The quality of the effluent is influenced by type and quantities of pollutants, sediment mineralology, and organic content, and the dissolution kinetics discussed below.

Organic Contaminants. An alternative approach to evaluate the relationship between sediments and bioavailability of toxic organic contaminants is the equilibrium partitioning (EqP) method. The EqP approach relies on established water quality criteria to assess sediment toxicity. The first basic assumption of the EqP approach is that sediment toxicity is correlated to the concentration of the contaminants in the interstitial water and not to the total sediment concentration. The second basic assumption is that contaminants partitioned between the interstitial water and the sediment sorbents (ie, organic carbon) are in equilibrium. Such equilibrium concentrations are significantly larger than developed in the elutriate tests.

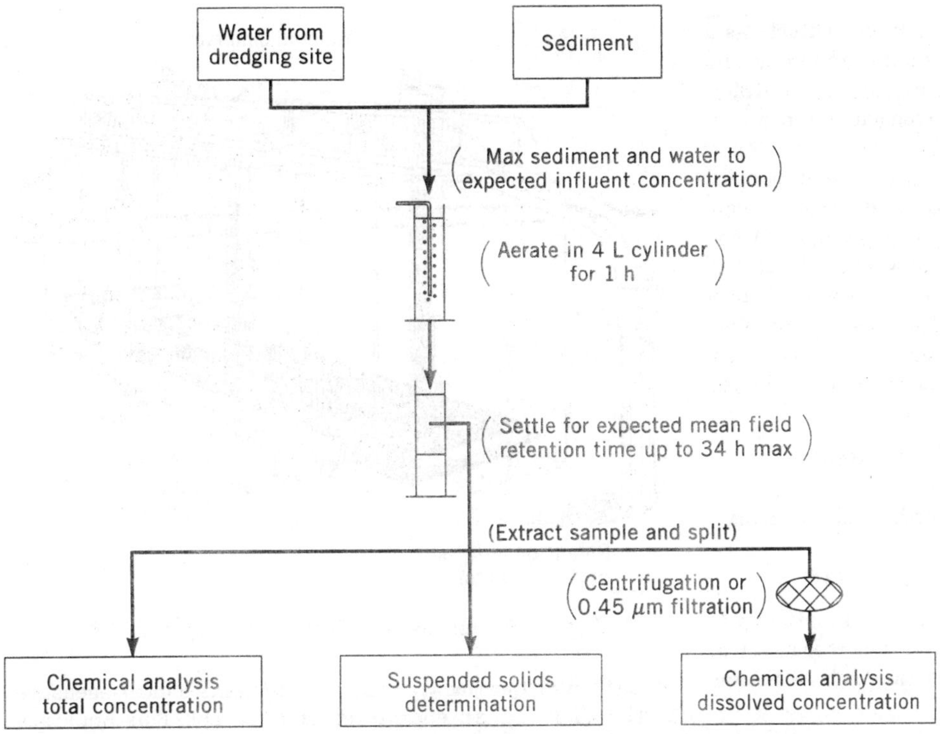

Figure 6. Modified elutriate test (12).

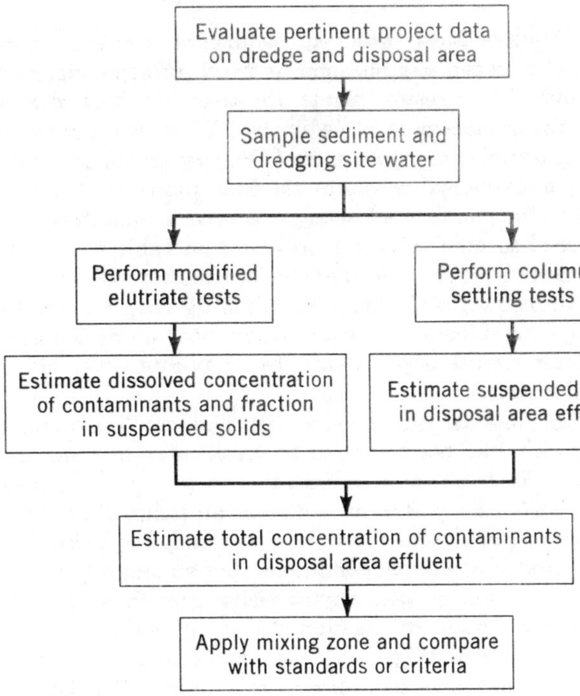

Figure 7. Effluent quality predictive technique.

Therefore, for a given contaminant, if the total sediment concentration, the concentration of sorbent(s), and the partitioning coefficient are known, then the interstitial pore water contaminant concentration can be calculated. Given the third assumption that organic carbon is the significant sorbent, the contaminant concentration can be predicted based on the total organic compound (eg, TOC normalization).

Metal Contaminants. Of particular recent interest is research into the relationship between sediment sulfide content and bioavailability of toxic metals. Studies have shown that sulfide content of freshwater and marine sediments is a predominant determinant of metal toxicity (15–17). This is based on the fact that many toxic metals form insoluble sulfides in anoxic or reduced environments (eg, cadmium, copper, lead, nickel, and zinc) and most freshwater and marine sediments have significant sulfide contents. Note that this includes uncontaminated sediments. These researchers examined acid volatile sulfide (AVS) content of sediments, the solid-phase sediment sulfides (FeS and MnS, H_2S, HS, and S_2^{-2}) that are soluble in cold acid, and the concentration of simultaneously extracted metals (SEM) using the ratio (SEM)/(AVS). AVS binding of metals reduces their bioavailability by decreasing metal solubility and is a precursor to formation of more insoluble pyritic sulfides (eg, FeS_2). Formation of pyritic metal sulfides is one mechanism of potential permanent removal of toxic metals. AVS of marine sediments in relation to cadmium toxicity to three species of amphipods has been measured, and it was found that AVS is the sediment phase that determines the LC_{50} for cadmium and also that correlations occurred between mortality and interstitial water metal activity (15). Cadmium and nickel were studied in estuarine sediments in 10-day exposure tests with an amphipod and oligochaete species (17). In this study, molar [SEM (cadmium and nickel)]/[AVS] ratios greater than one were consistently toxic to the amphipod (*Hyallela azteca*) and ratios of less than one were not. Metal availability was apparent through bioaccumulation in the oligochaete worm (*Lumbriculus variegatus*) when [SEM]/[AVS] ratios exceeded one. Freshwater sediments were spiked with cadmium and tested by a 10-day exposure with the same species of oligochaete worm, and the snail *Helisoma* sp (16). Here again, toxicity was not observed when the cadmium/AVS ratio was less than or equal to one.

Metal bioavailability is decreased by AVS, which acts as a sink for metals as long as the sediment remains anaerobic and the redox potential is too low to allow for oxidation of sulfides. Decreasing solubility of metal sulfides coincides with an increased AVS affinity for metals, that is Mn < Fe < Ni < Zn < Cd < Pb < Cu < Hg (15). These studies suggest an important sediment and interstitial water metal toxicity assessment technique which may be useful in developing sediment quality criteria. Note that potential effect associated with dredging and dredge material disposal alternatives can significantly impact the beneficial role played by AVS. The importance of microbial processes to sediment [AVS] should be noted, considering this a critical need in future research (17). General rules of thumb for sediments are:

$$[\text{metal}]/[\text{AVS}] < 1 \text{ sediment has little toxicity}$$

$$1 \leq [\text{metal}]/[\text{AVS}] \leq 10 \text{ sediment has little to high toxicity}$$

$$[\text{metal}]/[\text{AVS}] > 10 \text{ sediment is highly toxic}$$

Copper may be an exception to this rule due to its affinity to also bind to organic compounds, and most of this research was performed on marine and lake sediment (ie, high in Fe and aluminum oxides).

Surface Water Runoff Quality. Concerns regarding surface water runoff from the dredged material in the CDF begin when the elevation of the dredged material exceeds that of the water in the CDF. This concern can be mitigated by an adequate residence time for the supernatant (eg, water draining from the dredged materials) and runoff waters in the CDF such that the fines settle out and are not carried out of the CDF by the effluent. Final contours of sediment within the CDF can exceed the elevation of perimeter dikes. Thus, there is the long-term potential for sediment and contaminant mobility with the runoff water if the dredged material is not properly managed.

The quality of surface runoff water within a CDF can be expected to vary with time as a result of the changing physicochemical condition of the dredged material. When initially placed in a CDF, the dredged material is saturated, reduced, anaerobic, and of neutral to slightly basic pH. However, dredged material placed above the zone of saturation within the CDF are subject to drying. This promotes aerobic microbial degradation or volatilization of organic components, oxidation of some metals and sulfides, and changes in pH (usually lower and more acidic depending on the mineral composition of the dredged material). This reduction in pH can increase the solubility of metals and organic contaminants, leading to release of the contaminants.

A laboratory testing protocol has been developed by the U.S. Army Waterways Experiment Station (WES) to simulate the complex process of rainfall runoff (18,19). The WES surface runoff simulator is shown in Figure 8. The test protocol requires a large sediment sample from the dredge site to be placed in the soil lysimeter in the saturated reduced state. At various stages of the drying process, rainfall events are applied to the dredge material and surface runoff water samples are collected for analysis of water quality parameters. Rainfall simulations are performed several times over the full range of air drying. These results have been shown to accurately predict surface runoff water quality which can be compared to water quality standards and thus determine the need for runoff water treatment or the need to apply control measures to the CDF (eg, an engineered final cover). This apparatus can also be used to evaluate control measures such as lime treatment, soil covers, and plant stabilization.

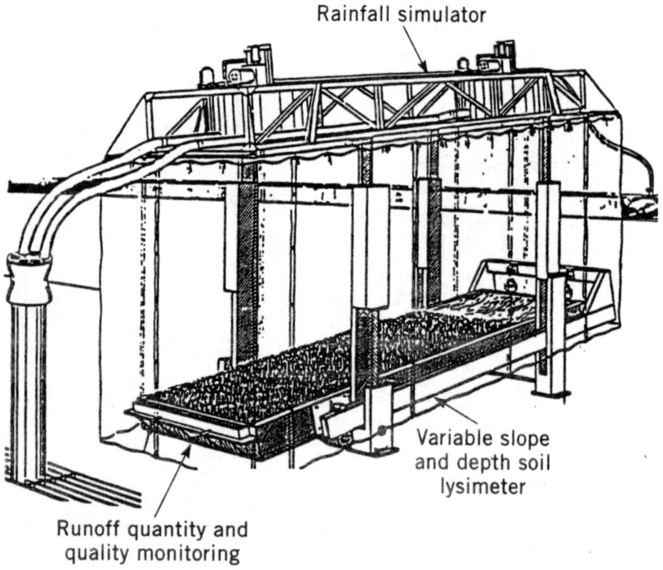

Figure 8. Surface runoff evaluation system (19).

Plant Uptake. Most dredged sediments contain large amounts of nitrogen and phosphorus which promote vigorous plant growth. As a result, plants are easily established in a CDF following placement of sediments. When contaminated dredged materials are placed in the CDF, there is the potential that contaminants will move into the food chain from bioaccumulation in the plants. A summary of sources and pathways of toxic trace elements to plants is presented on Table 4. Organic contaminants typically have large molecules that will not pass through the cell walls of plants, thus limiting the plant uptake of organics. Also, because most contaminated dredged materials contain a suite of pollutants they may interact to affect biological systems.

The U.S. ACE has developed a plant bioassay test protocol (19,20) which has been applied to freshwater and marine sediments. For freshwater sediments, the test protocol uses *Cyperus escalantus* (yellow nutsedge) as an index plant. This plant is grown in buckets filled with contaminated sediments from the dredge site under conditions that simulate the CDF environment. Plant growth, phytotoxicity (growth reduction), and bioaccumulation are monitored during the test in the apparatus shown in Figure 9. A chemical analysis is performed on plants harvested from the test buckets. The level of chemicals present in the plants can be compared to allowable threshold levels such as U.S. FDA crop limits. When limits are exceeded, the sediment management strategy invokes control measures such as restricted placement of polluted dredge materials in the CDF to minimize exposure, such as placement of clean surface layers, limiting plant growth to desirable plant species, etc. Dredged materials producing excessive plant uptake require designed control measures (eg, an engineered final cover).

Animal Uptake. Both upland and inlake CDFs are invaded and colonized by many animal species after each phase of

Table 4. Sources and Pathways of Toxic Trace Elements of Plants[a,b]

Uptake	Sb	As	Be	B	Cd	Cr	Co	Cu	Pb	Hg	Ni	Se	Sn	V
By roots														
Soil or groundwater	x	x	x	x	x	x	x	x	x	x	x	x	x	x
Fallout to soil from air pollution	x	x	x	x	x	x	x	x	x	x	x	x		x
Sewage sludge soil amendments		x			x	x		x	x	x	x	x		
Biocides applied to soil and/or seed		x		x	x			x	x	x			x	
Surface water contamination		x				x			x					x
Fertilizers		x			x	x		x		x	x		x	
Industrial pollution		x			x	x		x		x	x		x	
By leaves and stems														
Pollutant fallout from industrial sources	x	x	x		x	x	x	x	x	x	x	x		
Pollutant fallout from auto emissions									x					
Biocide applications to plants		x		x	x			x	x	x			x	
Pollution fallout from incineration of fossil fuels and refuse	x	x	x	x	x	x	x	x	x	x	x	x	x	x

[a] Sb, antimony; As, arsenic; Be, beryllium; B, boron; Cd, cadmium; Cr, chromium; Co, cobalt; Cu, copper; Pb, lead; Hg, mercury; Ni, nickel; Se, selenium; Sn, tin; V, vanadium.
[b] Ref. 19.

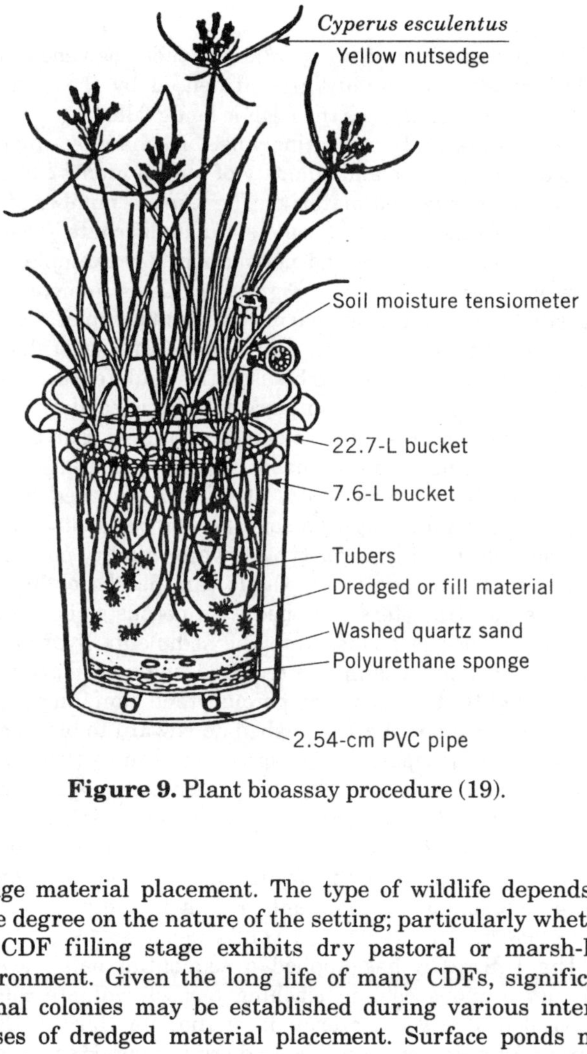

Figure 9. Plant bioassay procedure (19).

dredge material placement. The type of wildlife depends to some degree on the nature of the setting; particularly whether the CDF filling stage exhibits dry pastoral or marsh-like environment. Given the long life of many CDFs, significant animal colonies may be established during various interim phases of dredged material placement. Surface ponds may exist during the operational life of a CDF. Such ponds can support the presence of fish and benthic organisms. Bioavailable metals and organic compounds can then be transferred from the dredged material to fish or benthic organisms. The contaminants then can move quickly up the food chain. Such interim uptake problems can be reduced by the application of clean dredged material to the surface of problematic dredged material. Hierarchal food chain development is dependent on the lowest form of plant and animal species. Contaminants that are bioavailable at small concentrations to the lowest species may be magnified during migration up the food chain. The more recent view of biomagnification is that the levels of organic chemicals accumulating in animals has more to do with lipid content than the species position in the food chain. A summary of the sources and pathways of toxic trace elements to animals is presented on Table 5. The test protocol used by the U.S. ACE to evaluate animal uptake of contaminants from dredged sediments was borrowed from the European Economic Community (EEC). This test was used by the EEC to determine the hazardous nature of manufactured chemicals prior to sale in Europe.

As adapted by the U.S. ACE, the EEC test involves placement of ordinary earthworms in a container filled with polluted dredge materials as shown of Figure 10. The material is kept in moist, semimoist, and air-dried conditions to simulate the expected range of CDF environments. For a period of 28 days, the earthworms are monitored for toxicity and bioaccumulation of contaminants. It has been noted that this test protocol can indicate potential environmental effects for dredged material placed in upland environments (19). It has also been shown useful for identifying bioavailable metals and organic contaminants. Contaminant problems are identified by comparing body burdens with allowable FDA limits. Contaminated sediments producing excessive uptake in the earthworms will require CDF control measures for the dredge materials (eg, an engineered final cover).

Groundwater Leachate Quality. When a dredged material is placed in an upland CDF or above the water elevation in an in-lake CDF, a downward hydraulic gradient develops as the slurry consolidates. This gradient forces leachate into the underlying strata. The degree of leachate penetration into the underlying strata is influenced by the site hydraulics. For instance, CDFs located at near-shore settings may be within a

Table 5. Sources and Pathways of Toxic Trace Elements to Animals[a,b]

	Sb	As	Be	B	Cd	Cr	Co	Cu	Pb	Hg	Ni	Se	Sn	V
Terrestrial														
Breathing contaminated air	x	x	x	x	x	x	x	x	x	x	x	x	x	x
Eating contaminated plant or animal tissue	x	x	x	x	x	x	x	x	x	x	x	x	x	x
Drinking contaminated water	x	x	x	x	x	x	x	x	x	x	x	x	x	x
Licking or preening fur or leathers									x					
Receiving therapeutic drugs (domestic animals)		x												
Eating biocides or poison baits		x											x	
Aquatic														
Metal in water	x	x	x	x		x	x	x	x	x	x	x	x	x
Runoff and fallout		x			x	x		x	x	x	x	x	x	x
Sewage and industrial waste outfalls	x	x		x	x	x	x	x	x	x	x		x	
Mine tailings or smelter waste leachate		x			x	x		x	x	x			x	
Contaminated plants, animals, or sediment		x	x	x	x	x	x	x	x	x	x	x	x	x
Biocides or runoff		x		x				x					x	
Lead shot									x					

[a] Sb, antimony; As, arsenic; Be, beryllium; B, boron; Cd, cadmium; Cr, chromium; Co, cobalt; Cu, copper; Pb, lead; Hg, mercury; Ni, nickel; Se, selenium; Sn, tin; and V, vanadium.
[b] Ref. 19.

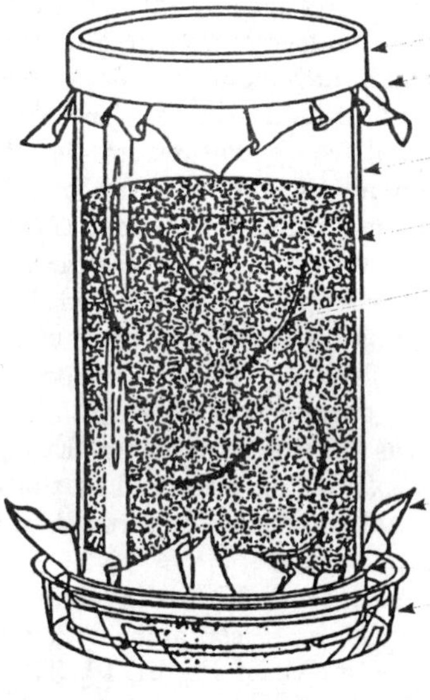

Figure 10. Earthworm bioassay procedure (19).

zone of groundwater discharge that would limit or prevent leachate penetrations. Following consolidation, water seepage through the sediment may continue to generate leachate due to surface infiltration of rainwater. Movement of contaminants due to a hydraulic gradient is termed advective transport. Simultaneously with the advective transport, the contaminants undergo molecular diffusion. This process is more significant through fine-grained sediments due to their low hydraulic conductivity. Diffusion of contaminants occurs due to contaminant concentration gradients and typically at rates so slow that it can be neglected. The quantity of leachate can be accurately predicted from advective flow theory if the site and CDF hydrogeological parameters are known. However the quality of leachate is affected by the type and quantities of pollutants, sediment mineralogy and organic content, equilibrium concentration of contaminants and pH of the pore fluid, and dissolution kinetics to name a few. Many of these parameters that influence leachate quality are influenced by the geochemical factors that change as the CDF is being filled.

There is presently no routinely used leachate testing protocol to define the type and quantity of contaminants carried by leachate from dredged material. There are a number of standardized test methods that could simulate leachate generation from both consolidation and percolation. For example, a soil/sediment press as shown in Figure 11 (21) is commonly used to extract pore fluids from marine sediment samples. Contaminant concentrations in such pore waters will generally be at equilibrium. This test could be used to simulate leachate generation from soluble contaminants during consolidation. Additionally, the soil press sample can be extracted from the press for use in a flexible wall permeameter cell to simulate leachate generation due to percolation. The fluids removed from the dredged material by either a soil press or permeameter could be subjected to analysis for typical water quality parameters. Field samples of pore water fluids can also be obtained using cone penetrometers equipped with porous stone sampling ports. The cone is pushed into the sediments hydraulically. Once at the depth a sample is required, the cone shaft is withdrawn slightly to expose the porous stone sampling part. At this time, an inner shaft is pushed downward to break a rubber septum that separates the porous stone and a glass vacuum sample tube. A sample of the pore water is then drawn by the vacuum into the glass tube. The pore water sample is then recovered by the withdrawal of the cone penetrometer. Pore water samples obtained in this fashion are filtered and represent the actual soluble contaminant present in the subgrade (see CONE PENETROMETER-DEPLOYED CHEMICAL SENSORS).

The U.S. ACE has applied a sequential batch leaching procedure and a column leaching test to evaluate confined disposal of dredged material from Indiana Harbor; Everett Harbor, Washington; and New Bedford Harbor, Massachusetts (12,22). The U.S. ACE has not yet endorsed a leachate test for routine testing and continues to perform research and evaluation testing on leachate generation.

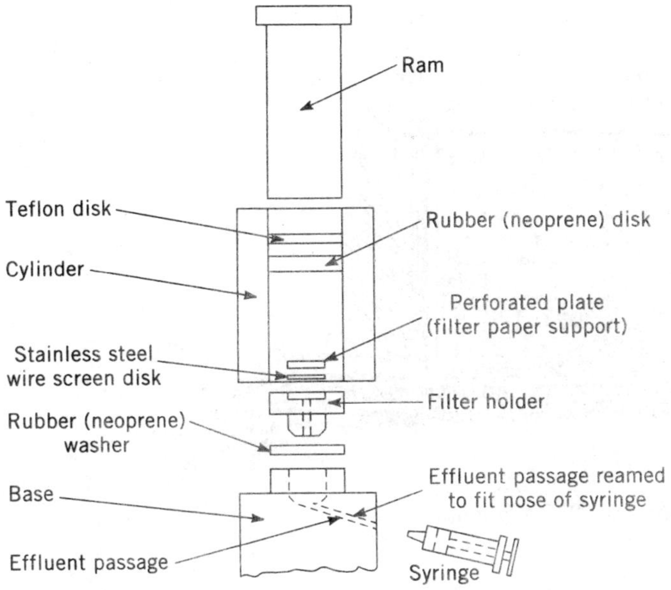

Figure 11. Soil press for sediment leachate sampling (21).

Water Column Discharge Quality. During much of the operational life of an in-water CDF, the surface of the dredged material is below the water level within the CDF. The surface water can act as a host to aquatic animals and provides a contaminant uptake pathway. An interim cover of clean dredged material may be required to limit this short-term contaminant pathway. If required, an interim cover placed over the contaminated sediments must be able to limit the advective movement of contaminants vertically through the interim cover and biological uptake of the contaminants. It is assumed that the sediments and cover are saturated and in direct communication with the water column. The impact of contaminants moving through the interim cover have been assessed using small and large-scale laboratory water column discharge reactor units (23). These units were designed to assess the effectiveness of subaqueous capping at open water sites, but the same principles apply for overlying soil covers for CDFs. The effectiveness of the cover is assessed by following the movements of chemical contaminants and microbial spores found in the contaminated sediments into the overlying water column and by monitoring the biological uptake of chemical contaminants by clams and polychaetes. Such tests on Dutch Kills sediments from New Jersey indicated that as little as 10 cm of sand cover was effective in isolating the water column but 50 cm was required to eliminate biological uptake. No test standards currently exist for the water column uptake reactor test.

ADEQUATE ENVIRONMENTAL PERFORMANCE AND PROTECTION (AEPP) EVALUATION CRITERIA FOR CDFS

Prior to reviewing AEPP criteria for potential contaminant pathways from a CDF, it is important to understand the difficulties associated with even the identification of problematic dredged materials. The current confusion regarding sediment testing guidelines and criteria were expressed by the U.S. General Accounting Office as follows (24):

According to EPA, some scientists believe that the 1975 testing guidelines EPA developed in conjunction with the Corps under section 404(b)(1) of the Clean Water Act—guidelines the Corps is currently using for disposal decisions on dredged material—are technically inadequate. However, Corps headquarters officials maintain that the guidelines have evolved substantially, have been tested thoroughly, and are adequate. Both the Corps and EPA have conducted detailed research on contaminated sediment, but the agencies have not always agreed on how the research should be used to determine which sediment needs confinement. The lack of agreement occurs generally because of the difficulty in establishing clear cause-and-effect relationships between the contaminant concentration in sediment and a biological impact on humans and wildlife.

In 1991, EPA and U.S. ACE jointly published testing guidance for the discharge of dredged material into ocean waters. Currently EPA and U.S. ACE are developing a similar testing guidance for evaluation of material proposed for discharge to inland and near coastal waters (referred to as Inland Testing Manual).

The environmental performance of a CDF should be verified by a monitoring program that assesses the success of control measures for each of the contaminant pathways that are identified during the testing protocol. Evaluation criteria are required to serve as the standard by which the test results from laboratory simulations and field monitoring are evaluated. Although finalized evaluation criteria are not currently available for all contaminant pathways, there are sufficient data in the literature from EPA, FDA, the European Economic Community, and research data to help with the decision-making process. Table 6 summarizes the type of existing information that might be useful for evaluating each contaminant pathway. The contaminants or nutrients of concern must be selected on a case-by-case basis. Contaminants of interest are those that have FDA limits, are the subject of state fisheries advisories, or have EPA water quality criteria.

Effluent Criteria

The factors influencing effluents from CDFs are shown on Figure 12. An unacceptable effluent discharge is one in which the concentration of any dredged material constituent, after allowance for mixing, exceeds applicable state water quality standards (WQS). Effluents from a CDF may be discharged as a point-source through a weir system or filter cells, or as a nonpoint-source flow through the dikes. Assuming that the sediments have been established as potentially contaminated, the modified elutriate and column settling tests are used to

Table 6. Evaluation Criteria for Contaminant Pathways[a]

Pathway	Criteria
Effluent	EPA water quality standards (Table B1) State water quality standards CWA mixing zone criteria State mixing zone criteria
Runoff	EPA water quality standards (Table B1)
Plant uptake	Research results (Table B2)
Animal uptake	FDA and European action levels for foodstuffs (Table B3)
Leachate	EPA drinking water standards (Table B4)
Air emissions	Pending EPA air quality standards

[a] Ref. 3.

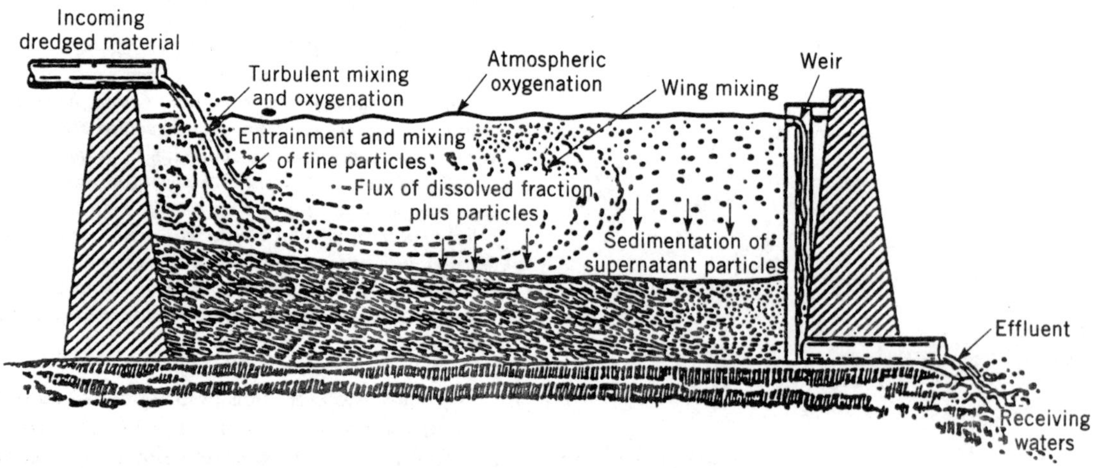

Figure 12. Factors influencing CDF effluents.

predict effluent quality as shown on Figure 7. The standards used for final comparison include turbidity criteria from state WQS as required under Section 401 of the CWA.

Surface Water Runoff Criteria

The final closure of the CDF must ensure that water runoff from the facility meets the same criteria discussed above for effluent discharge. Additionally, many states require a sediment control plan and potential discharge control structures, eg sedimentation ponds, to limit erosion and siltation of adjoining bodies of water. As discussed previously, current CDFs do not have a standard approach to closure but are consistently very flat in profile. Such minimal grades provide for control of water runoff related erosion at the expense of increased infiltration potential. Site specific testing can be performed using the test device shown on Figure 8 (18).

Plant Uptake Criteria

The potential for movement of unacceptable levels of contaminants from the dredged material into the environment through plants and eventually into the food chain must be restricted. Appropriate management strategies therefore are formulated to place problematic dredged material to minimize plant uptake. The management strategies control and manage plant species on the site so that desirable plant species that do not uptake and accumulate contaminants are allowed to colonize the site, while undesirable plant species are removed or eliminated. Plant uptake can be limited during interim periods of the CDF life when a significant surface area of problematic dredged material may be exposed for years. Therefore control of plant uptake is considered throughout the life of the CDF and not just after final closure.

Animal Uptake Criteria

A potential for animal uptake of unacceptable levels of contaminants from dredged material may exist. Appropriate management strategies therefore need to be formulated that consider the ultimate environment in which the dredged material is placed, the anticipated ecosystem developed, and the physicochemical processes governing the biological availability of contaminants for animal uptake. Interim periods during the operational life of the CDF may be more critical than after the final closure. Surface ponds created during the filling of a CDF may support aquatic life that will provide the basis for contaminant uptake through the food chain via visiting avian and terrestrial species.

Groundwater Leachate Criteria

Pollutants that are dissolved in the excess water of the dredged sediments can contaminate the underlying groundwater. Such migration of contaminants may be caused by pressure gradients that develop in the pore water during consolidation of the dredged sediments or at sites that are natural groundwater recharge zones and result in advective transport of the contaminant to the groundwater or by diffusion of the contaminant caused by concentration gradients.

Contamination of underlying groundwaters is also a design consideration in common landfills. For instance, the recent promulgation of RCRA part 257.3-4 provides groundwater contamination criteria for solid waste landfills. These revised regulations state that the solid waste facility shall not contaminate underground drinking water sources beyond its boundary or point of compliance. For purposes of this requirement, contamination is defined as concentrations of substances exceeding maximum contaminant levels (MCLs) developed by EPA under Section 1412 of the Safe Drinking Water Act (SDWA) or existing ambient groundwater quality. The list of current MCLs is given on Table 7. EPA is in the process of amending this list so adjustments to the MCLs can be anticipated. The point of compliance is usually taken as the water table at the nearest monitoring well. In addition, many states have nondegradation and cleanup criteria that may be orders of magnitude more stringent than MCL levels. Such small levels of acceptable contaminant presence are commonly due to biologically concentrating contaminants (or biomagnification) that result in toxicity concerns.

SELECTION OF REMEDIAL OPTIONS

The Clean Water Act (CWA) requires the survelliance and montioring of effluents and water quality. Since many contaminants have very low solubilities in water, the monitoring

Table 7. Maximum Contaminant Levels[a]

Chemical	CAS Reg. No.	MCL, mg/L
Chromium (hexavalent)	[*7440-47-3*]	0.05
2,4-Dichlorophenoxy acetic acid	[*94-75-7*]	0.1
1,4-Dichlorobenzene	[*106-46-7*]	0.075
1,2-Dichloroethene	[*107-06-2*]	0.005
1,1-Dichloroethylene	[*75-35-4*]	0.007
Endrin	[*75-20-8*]	0.0002
Fluoride		4.0
Lindane	[*58-89-8*]	0.004
Lead	[*7438-82-1*]	0.05
Mercury	[*7438-87-6*]	0.002
Methoxychlor	[*72-43-5*]	0.1
Nitrate		10.0
Selenium	[*7782-49-2*]	0.01
Silver	[*7440-22-4*]	0.05
Toxaphene	[*8001-35-2*]	0.005
1,1,1-Trichloroethene	[*71-55-6*]	0.2
Trichloroethylene	[*79-01-6*]	0.005
2,4,5-Trichlorophenoxy acetic acid	[*93-76-5*]	0.01
Vinyl chloride	[*75-01-4*]	0.002

[a] Section 1412 of the Safe Drinking Water Act.

of effluents and water quality may not indicate the presence of contaminants in sediments. Sediment particles reach the water body by a variety of routes. The particles vary in chemical composition and in their physical properties. The constituents of sediments such as clay, organic matter, hydrated iron, manganese oxides, and associated characteristics such as particle size distribution, pH, oxidation–reduction condition, and salinity of the water body all affect the interaction between the individual sediment particles and the contaminants present in the surrounding water. As an example the behavior of particles having a high surface area to mass ratio (ie, specific surface) will be controlled predominantly by surface forces, whereas a particle having a low specific surface will be controlled by body forces (ie, gravity). As a consequence materials with high specific surfaces will contain more contaminants. Thus sediment particle size influences the presence of contaminants with the sediment and the potential for contaminant migration. The finer the particle size the greater the potential for higher concentrations of contaminants. Smaller particles remain suspended for longer periods of time, are easily resuspended in high tides and floods, and travel further from the source of contamination. Although contaminants are generally found associated with fine particles in any water body, their distribution is not uniform. They typically are found in pockets and in the deeper water. In addition, the vertical profile of contaminants are strongly affected by benthic organisms that are present in the water body. In summary, the particle size, organic content, and mineralogy can affect the selection of a remediation technology. The selection of the appropriate remediation technology is described in detail in a later section.

Contaminants typically found in sediment can be classified as follows:

Polynuclear aromatic hydrocarbons (PAHs)
Pesticides
Chlorinated hydrocarbons
Mononuclear aromatic hydrocarbons (such as benzene)
Phthalate esters
Metals
Cyanides and organo-metals + others

These contaminants enter the water body from various sources and contact the sediment particles by a either direct sinking or solubilization of the organic matter and subsequent adsorption on the sediment particles.

In most aquatic systems the suspended sediments and the top part of the soil column contain a higher concentration of contaminant than the overlying water column. The sediment thus becomes a sink of contaminants which can become remobilized back into the water column. In addition, the contaminants that are bound to sediments also have the potential to undergo chemical changes and thus evolve into potentially different chemical species. For example oxidation of organic matter in sediment can lead to conditions favorable to the release of bound metals into the water column (25). Other bound trace metals, especially mercury, can be methylated or converted to other organo-metallic forms by the action of micro-organisms. These organo-metals have been shown to bioaccumulate in fish (26).

CONTAMINATED SEDIMENT TREATMENT

Treatment of contaminated sediment usually involves combining several appropriate technologies into an overall scheme for achieving a cleanup objective. The treatment of dredged contaminated sediment and subsequent placement in a confined disposal facility (CDF) typically include the following steps: removal, transport, interim storage, pretreatment, treatment, disposal, and effluent (including surface runoff) or leachate treatment. In this section the pretreatment and treatment steps will be discussed. All processes in this article were recommended for study in the ARCS program (27,28) or have been used in the mining industry (29).

Soil Chemistry

Properties of Sediments Affecting Contaminants. Sediments requiring remediation can vary widely in terms of physical and chemical properties. The primary physical characteristic is texture (ie, the distribution of sand, silt, and clay-sized particles). In general, sandy sediments have little attraction for either toxic metals or synthetic organics (pesticides and industrial organics). In contrast, fine-textured sediments such as silt and clay have a much greater affinity for all classes of contaminants. Fine-textured material at the sediment–water interface and suspended silt and clay particles effectively remove contaminants from the water column. These fine-textured materials have large surface areas when compared to either their volume or mass (eg, specific surface). This large surface area serves as a site for colloidal chemistry to occur. These particles tend to accumulate in more quiescent reaches of waterways. Separation of the less contaminated gravel and sandy fractions from contaminated sediments may be possible, yielding material clean enough for disposal without restriction, while also reducing the volume of the contaminated sediment requiring treatment.

A second important physical property is the organic matter content. Fine-textured sediments generally contain from one to several percent naturally occurring humic material derived from microbal transformation of plant and animal detritus. Sandy sediments typically have less than 1% humic material. Humic material may be present as discrete particulates or as coatings on either clay or fine sandy sediments particles and is important in two respects: (*1*) the humic material greatly increases the affinity of sediments for metals (eg, bind, chelate, and complex) and nonpolar organic contaminants and (*2*) it serves as an energy source for sediment microbial populations.

Measurement of *in situ* water content is a third physical property of sediments usually important to remediation decisions because it is a measure of the amount of void space present in the sediment and as a result the density. The chemical properties of sediments also greatly affect the mobility and biological availability of contaminants. Sediment acidity and oxidation–reduction status are two important chemical parameters. Strongly acidic (low pH) sediments can in general slow microbial activity and increase the soluble levels of toxic metals. Weakly acidic, neutral, and slightly alkaline conditions (higher pH) favor metal immobilization processes. The oxidation–reduction status of a sediment, measured as redox potential, has a major effect on the retention or release of a number of metals, either directly or as a result of the different reactions of metals with oxidized and reduced sediment constituents. Changes in pH and redox potential of contaminated sediments from their initial *in situ* condition at a dredging site to different conditions at a disposal site can substantially affect contaminant mobilization/immobilization processes. Other important chemical properties of sediments include salinity conditions, sulfide content, the amount and type of cations and anions, and the amount of potentially reactive iron and manganese.

Clay minerals have a high specific surface and associated electrical charge. Specific surface is defined as the surface area of a particle divided by either its mass or volume. The electrical charge on this surface is predominately negative with some positive sites on the ends of the clay mineral. The charge distribution is created by the atomic lattice structure that makes up the clay mineral. The behavior of water-cation systems adjacent to this surface can be described by the Gouy-Chapman theory developed from colloidal chemistry. The distribution of these cations in terms of distance from the colloid surface is termed the double layer. These cations distribute themselves along the surface of the clay platlet but are not permanently attached to minerals. The effect of various parameters on the double layer thickness and resulting soil structures is presented in Table 8. The Gouy-Chapman theory developed an equation for the double layer thickness that can be expressed as follows: $t = \frac{\epsilon KT}{8\pi n e^2 v^2}$ where t = double-layered thickness, ϵ = dielectric constant, k = Boltsmann constant, T = temperature, n = electrolyte concentration, e = elementary charge, and v = cationic valance.

The dielectric constant as utilized in the Gouy-Chapman equation is a measure of the pore fluids ability to act like an insulator and influence the strength of the clay minerals electric field. A reduction of the clay minerals electric field reduces the thickness of the double layer and hence, the colloidal clay behavior. As the dielectric constant increases, the thickness of the double layer increases causing a tendency for particle dispersion and the pore fluid acts more like a polar molecule. Therefore, if the liquid within the pore spaces is not polar (such as a petroleum product), then the diffuse double layer will not develop. A table of typical values of dielectric constant for a variety of different fluids is presented in Table 9.

Forms and Reactions of Contaminants. Typical trace and toxic metal contaminants in dredged sediments commonly include copper, zinc, cadmium, lead, chromium, nickel, arsenic, mercury. These elements are usually present in soils and sediments at low concentrations from natural sources. It is when one or more of these contaminants is present in elevated concentrations that they pose a potential problem. Toxicity may be experienced if these excess metals are released to the water column or are present in forms readily available to plants and animals that come in contact with the sediment material.

Metals dissolved in the water column or pore water are available to most aquatic organisms. Metals bound to clay minerals and humic material by cation exchange processes are available due to equilibrium partitioning between these bound metals and dissolved metals. On the opposite extreme, metals bound within the crystal lattice structure of clay minerals are generally much less bioavailable.

Metals may be mobilized or immobilized if the chemical environment (eg, pH and Eh) of the sediment or dredged material changes. Therefore, understanding the influence of the *in situ* sediment chemical environment, and the potential of alteration in these parameters at the disposal site are important to the selection of disposal alternatives for contaminated sediments. The fate of potentially available metals as sediment conditions change is presented in Table 10. In addition there may be a complex interaction between some of these processes as the pH or oxidation status of a sediment is altered. As metals are released from one form, they may be immobilized again by another process.

Organic contaminants can vary widely in water solubility depending on their molecular composition and functional groups. Like metals in a sediment–water system, most organic contaminants tend to become strongly associated with the sediment solid phase, particularly the humic fraction. Thus, at most sites, the distribution of organic contaminants between dissolved and solid phases is a function of their water solubility and the percent of naturally occurring humic materials in the sediment. However, at heavily contaminated sites, organic contaminants may also be highly associated with petroleum-based or sewage-based organics.

Unlike metals, however, organic contaminants do degrade. Though all organic contaminants degrade at some rate, some have half-lives on the order of several decades. Chemical half lives are measures of persistence, or how long a chemical will ramain in various environmental media. The half lives are the result of all removal processes (eg, phase transfer, chemical transformation, and biological transformation) acting together. A summary of typical half lives of organic contaminants found in contaminated sediments is presented in Table 11. Some organics are subject to enhanced degradation rates under certain sediment chemical conditions.

Sediment Toxicity Assessment. Assessing the toxicity of sediments and any potential threat they pose to human health and the environment is an important step in identifying materials that need to undergo a remediation process. Presently,

Table 8. Effect of Various Parameters on Double-Layer Thickness and Resulting Soil Structures[a]

	Double-Layer Thickness		Soil Structure	
Parameter	Increase	Decrease	Dispersed	Flocculated
Electrolyte concentration	Decrease	Increase	Decrease	Increase
Ion valence	Decrease	Increase	Decrease	Increase
Dielectric constant	Increase	Decrease	Increase	Decrease
Temperature	High	Low	Increase	Decrease
Size of hydrated ion	Large	Small	Increase	Decrease
pH	High	Low	Increase	Decrease
Anion adsorption	Increase	Decrease	Increase	Decrease

[a] As shown, a change in any particular parameter (increase or decrease) has a related affect on the double-layer thickness and the soil structure.

Table 9. Typical Values of Dielectric Constants for Various Fluids

Chemical Name	Dielectric Constant
Cyclohexane	2
Xylene	2
Benzene	2
Aniline	6
Propanol	20
Acetone	22
Ethanol	26
Methanol	32
Water	80

several different kinds of tools are available to use in making decisions concerning sediment assessment and desired levels of remediation. The primary tools include environmental regulations based on bioaccumulating contaminants and sediment assessment methods. Environmental regulations are the most important and have previously been covered in a previous section (Regulation of Dredging Activity). In the following, sediment assessment methods will be discussed.

Although sediments are an extension of the water column, assessment of sediment toxicity is much more complex than assessment of water quality. Due to the nature of sediment chemistry, presence of contaminants does not necessary mean that the sediment is toxic. For example, contaminants may be present but complexed by humic material in the sediment and thus relatively unavailable.

The EPA has developed a "Sediment Classification Methods Compendium" (31) to serve as a reference for methods that could be used to assess the quality of contaminated sediments. This compendium describes the various methods, as well as their advantages, limitations, and existing applications. The sediment quality assessment methods are presented in Table 12. Each method either directly or indirectly attempts to delimit levels of contamination within sediments such that above those levels either (*1*) acute and/or chronic toxicological effects become manifest or (*2*) some amount of bioaccumulation occurs. The sediment quality assessment methods described can be classified into two basic types: quantitative or qualitative (see Table 12). Quantitative methods are chemical-specific and can be used to generate numerical sediment quality values. Descriptive methods are qualitative and cannot be used alone to generate numerical sediment quality values for particular chemicals.

Currently, the EPA is working toward the development of nationally applicable sediment quality criteria (SQC). The SQC represents the best recommendation of sediment contaminant concentrations that will not unacceptably affect benthic organisms or their uses. SQC would eventually be developed for each high priority contaminant of concern.

Equilibrium partitioning (EqP) theory is the EPA's selected method to establish national SQC. The EqP approach relies on established water quality criteria to assess sediment toxicity. The first basic assumption of the EqP approach is that sediment toxicity is correlated with the concentration of the contaminants in the interstitial water and not to the total sediment concentration. The second basic assumption is that contaminants partitioned between the interstitial water and the sediment sorbents (eg, organic carbon) are in equilibrium. Therefore, for a given contaminant, if the total sediment concentration, the concentration of sorbent(s), and the partitioning coefficient are known, then the interstitial contaminant concentration can be calculated. The interstitial contaminant concentration can then be compared to established water quality criteria to assess sediment toxicity. A flow chart describing the process is presented in Figure 13.

Due to variation in the specific sediment sorbent(s) that different classes of contaminants sorb to, methodologies for deriving SQC vary with different classes of compounds. For nonionic organic chemicals the methodology requires normalization to organic carbon. For metal contaminants a methodology has been developed that is based on normalization to acid volatile sulfide. A flow chart describing the process is presented in Figure 14.

Pretreatment

Pretreatment technologies prepare dredged material for treatment or disposal. Objectives of pretreatment processes are largely dependent on the treatment or disposal options following in the treatment process train. The objectives of this step are usually one or more of the following:

- To provide a suitable material for further treatment and/or disposal operations
- To enhance or accelerate settling of the dredged materials solids
- To reduce the water content of the dredged materials solids

Table 10. Typical Fate of Potentially Available Metals In a Changing Chemical Environment[a]

Metal Type	Initial Condition	Environmental Changes	Result
Carbonates, oxides, and hydroxides	Salts in the sediment	Reductions of pH	Release of the metals as the salts dissolve
Adsorbed on iron oxides	Adsorbed in sediment	Sediment becomes reducing or acidic	Iron oxides become unstable and release metals
Chelated to humic	Chelated in sediment	Strongly immobilizes metal in both reducing and oxidizing sediments (however, there is some indication that the process is less effective if a reduced sediment becomes oxidized)	
Sulfides	Very insoluble precipitate	Sediment becomes oxidized	Sulfides become unstable, oxidize to sulfates, and release the metals

[a] Ref. 30.

- To separate coarser, potentially cleaner, solids from the fine-grained, more contaminated solids (particle classification)
- To reduce the overall cost for the remedial action

Pretreatment technology processes include (*1*) dewatering, (*2*) particle classification (separation), and (*3*) slurry injection.

Dewatering. Dewatering technologies are processes used to reduce the moisture content of slurries or sludges to expedite the handling and to prepare the material for further treatment or disposal. The water generated during dewatering generally contains contaminants as well as suspended solids. Types of dewatering processes include mechanical, evaporative, and settling ponds. Dredged material dewatering has traditionally been accomplished in ponds or CDFs, which rely on a combination of seepage, drainage, consolidation, and evaporation. Evaporative processes by themselves alone are not considered a viable alternative for use in the Great Lakes region due to the climate. Most mechanical dewatering processes reduce the moisture of the feed material to a level comparable to *in situ* sediment (about 50 wt %) and work best on homogeneous waste streams.

Implementability for dewatering processes other than settling ponds is poor primarily because of the high processing rates required for most dredging operations. The primary purpose for dewatering to minimize the loss of leachate through the dikes and bottom of the CDF. A summary of recommended dewatering techniques for use in the ARCS program (27) is presented in Table 13.

Particle Classification. Separation by grain size is important in the management of soils and sediments contaminated with toxic materials since the contaminants tend to sorb primarily onto fine-grain clay and organic matter. The most appropriate solids separation technology for a given site depends on the volume of contaminated sediments; composition of the sediments (particle size distribution of the feed); specific gravity and chemical analysis of the soil; mineralogical composition; characteristics of the soil; percentage of sand, humus, clay, or silt; composition of the organics in each soil fraction; moisture content; pH; characterization of the contaminants; types of dredging or excavation equipment employed; site location and site surroundings (Fig. 15). Particle classification options include screening processes that depend on size, processes that depend on a combination of size and density or density alone, and processes that depend on conductive or magnetic properties of the particles. A summary of recommended particle classification technologies evaluated by EPA are shown in Figure 15, by the ARCS program (27) in Table 14, and mining industry techniques in Table 15 (29).

Slurry Injection. Slurry injection technologies are used as a pretreatment step either to add chemicals that condition the sediment for further treatment and/or accelerate the settling of the suspended solids or to provide nutrients or microbes that enhance biodegradation of organics. The process involves the injection of chemicals or microorganisms into the dredged material slurry to take advantage of the mixing process available in the pipeline from a hydraulic dredge. A summary of recommended slurry injection techniques for use in the ARCS program (27) is presented in Table 16.

Treatment

This section discusses the various treatment technologies available for the decontamination/detoxification of contaminated sediment. Many of the process options are not stand-alone processes, but are components of a system that may involve multiple treatment steps to address multiple contaminant problems. Specific treatment objectives include the following:

- Destruction of toxic organic contaminants by conversion to nontoxic end products
- Removal of heavy-metal or organic contaminants from contaminated dredged material, thereby concentrating toxic material into a media of smaller volume (extraction), and/or to reduce the volume of solids for further treatment or disposal
- Reduction of the mobility of contaminants in dredged material to a level compatible with acceptable risks
- Compatibility with removal and final disposal options
- Compliance with acceptable capital and operating costs
- Minimization contamination of other environmental media
- Avoidance of the addition of potentially toxic materials or the production of toxic materials during the treatment process

In addition, most of the processes outlined below will also require one or more pretreatment processes.

Table 11. Half-Life Range of Typical Contaminants[a]

	Half-Life Range in Surface Water, days		
Chemical Name	Low	High	Other Half-Life Ranges
PCBs	2	12.9	
Acenaphthene			
Acenaphthylene	0.125		
Anthracene			
Benzo(*a*)anthracene	0.10	5.00	
Benzo(*a*)pyrene	0.40		
Benzo(*b*)fluoranthene	1.00	2.00	
Benzo(*ghi*)perylene			
Benzo(*k*)fluoranthene			
Chrysene	0.20		
Dibenzo(*a,h*)anthracene	0.02		
Fluoranthene	1.00	2.00	
Fluorene			
Indeno(1,2,3-*cd*)pyrene	0.02	2.1	
Naphthalene			
Phenanthrene	0.38	2.00	
Pyrene			
Dieldrin	723 d		
Chlordane	Very persistent		
Vinyl chloride	25 min		
Chloroform	15 months		Blue gill, 1 d
1,1,1-Trichloroethane	5–9 months		
Carbon tetrachloride			Blue gill, 1 d
Dibromochloromethane	274 yr		
Bromoform	686 yr		
Chlorobenzene	5.8 h		
Bis(2-chlorethyl)ether			Blue gill, 4–7 d
2-Chlorophenol			Blue gill, 1 d
2,4-Dimethylphenol	1 d		
2,4-Dichlorophenol	8–9 d		
Hexachlorocylcopentadiene	11 d		
2,4,6-Trichlorophenol			Rat blood, 20 h
2-Chloronaphthalene	Biomagnified in water		
2,4-Dinitrophenol	No significant microbial degradation in 64 d		
2,6-Dinitrotoluene	>64 d		
Benzidine			Dogs and rats, 68–88 h
3,3′-Dichlorobenzidine			Long life in soil
Bromoform	686 yr		
Chlorobenzene	5.8 h		
Bis(2-chlorethyl)ether			Blue gill, 4–7 d
2-Chlorophenol			Blue gill, 1 d
2,4-Dimethylphenol	1 d		
2,4-Dichlorophenol	8–9 d		
Hexachlorocylcopentadiene	11 d		
2,4,6-Trichlorophenol			Rat blood, 20 h
2-Chloronaphthalene	Biomagnified in water		
2,4-Dinitrophenol	No significant microbial degradation in 64 d		
2,6-Dinitrotoluene	>64 d		
Benzidine			Dogs and rats, 68–88 h
3,3′-Dichlorobenzidine			Long life in soil

[a] Refs. 5 and 31.

Biological Treatment. Biological degradation technologies use bacteria, fungi, or enzymes to break down PCBs, pesticides, and other organic constituents into less toxic compounds. The microorganisms may be indigenous microbes, conventional mutants, or recombinant DNA products. Biological degradation proceeds with slower reaction rates than thermal or chemical techniques. A summary of recommended biological techniques evaluated by the ARCS program (27) is presented in Table 17.

Chemical Technologies. The objective of chemical processes, such as chelation, oxidation, or reduction, is to change the form of a toxic material to render it less toxic or to change its solubility, stability, separability, or other properties affecting

Table 12. Sediment Quality Assessment Methods[a]

Method	Type: Numeric	Descriptive	Combination	Concept
Bulk sediment toxicity	x			Test organisms exposed to sediments containing unknown quantities of potentially toxic chemicals; at end of specific period, response of test organisms is examined in relation to specified biological endpoint
Spiked-sediment toxicity	x			Dose response relationships established by exposing test organisms to sediments spiked with known amounts of chemicals or mixtures of chemicals
Interstitial water toxicity	x			Toxicity of interstitial water is quantified and identification evaluation procedures are applied to identify and quantity chemical components responsible for sediment toxicity; procedures are implemented in three phases: (*1*) characterization of interstitial water toxicity, (*2*) identification of the suspected toxicants, and (*3*) confirmation of toxicant identification
Equilibrium partitioning	x			Sediment quality value for given contaminant is determined by calculating sediment concentration of contaminant that corresponds to interstitial water concentration equivalent to EPA water quality criterion for contaminant
Tissue residue	x			Safe sediment concentrations of specific chemicals established by determining sediment chemical concentration that results in acceptable tissue residues; methods to derive unacceptable tissue residue based on chronic water quality criteria and bioconcentration factors, chronic dose response experiments or field correlations, and human health risk levels from the consumption of freshwater fish or seafood
Freshwater benthic community structure		x		Environmental degradation measured by evaluating alterations in freshwater benthic community structure
Marine benthic community structure		x		Environmental degradation measured by evaluating alterations in marine benthic community structure
Sediment quality triad	x	x	x	Sediment chemical contamination, sediment toxicity, and benthic infauna community structure measured on the same sediment; correspondence between sediment chemistry, toxicity, and biological effects used to determine sediment concentrations that discriminate conditions of minimal, uncertain, and major biological effect
Apparent effects threshold	x		x	AET is sediment concentration of contaminant above which statistically significant biological effects (eg, amphipod mortality in bioassays, depressions in abundance of benthic infauna) is always expected; AET values empirically derived from paired field data for sediment chemistry and range of biological effects indicators

[a] Ref. 28.

handling or disposal. Chemical treatment technologies use chelating agents, acid or base addition, chlorine displacement, oxidation, or reduction in the destruction, detoxification, or removal of contaminants found in contaminated sediments. A summary of recommended chemical technologies reviewed/evaluated by the ARCS program (27) is presented in Table 18.

Extraction Technologies. Extraction is the removal of a contaminant (organic/inorganic) from soils or sediment by dissolving it in a fluid that is later recovered and subsequently treated to remove the contaminant or recycled. Extraction is often used to remove volatile organics from permeable soils and can also be used to treat soils and sludges contaminated with the following: metals, inorganics, organics (PCBs, TPH, halogenated solvents, aromatics), amines, ethers, and anilines. Extracting agents that are employed vary according to the contaminants to be treated. They include water, acids, bases, complexing and chelating agents, surfactants, and certain reducing agents.

There are two basic methodologies employed utilizing extraction processes: (*1*) containerized extraction, and (*2*) *in situ* extraction. In the containerized extraction process soil/sediment is excavated and placed in a container to allow contact with the extracting liquor. The extracting liquor is then injected into the soil, and contaminants are extracted from the soil and concentrated in the washing solution. The washing solution leaches the contaminants out of the soil and is collected and recycled by a specific treatment technology appropriate for the type of contaminants involved. For *in situ* extraction processes, the contaminated site is typically flooded with an appropriate extraction solution and the elutriate leaching from the site is collected either by a series of shallow well points or subsurface drains. The leachate is then treated and recycled back into the site.

A summary of recommended extraction technologies reviewed/evaluated by the ARCS program (27) is presented in Table 19. The technologies presented in Table 20 include (*1*)

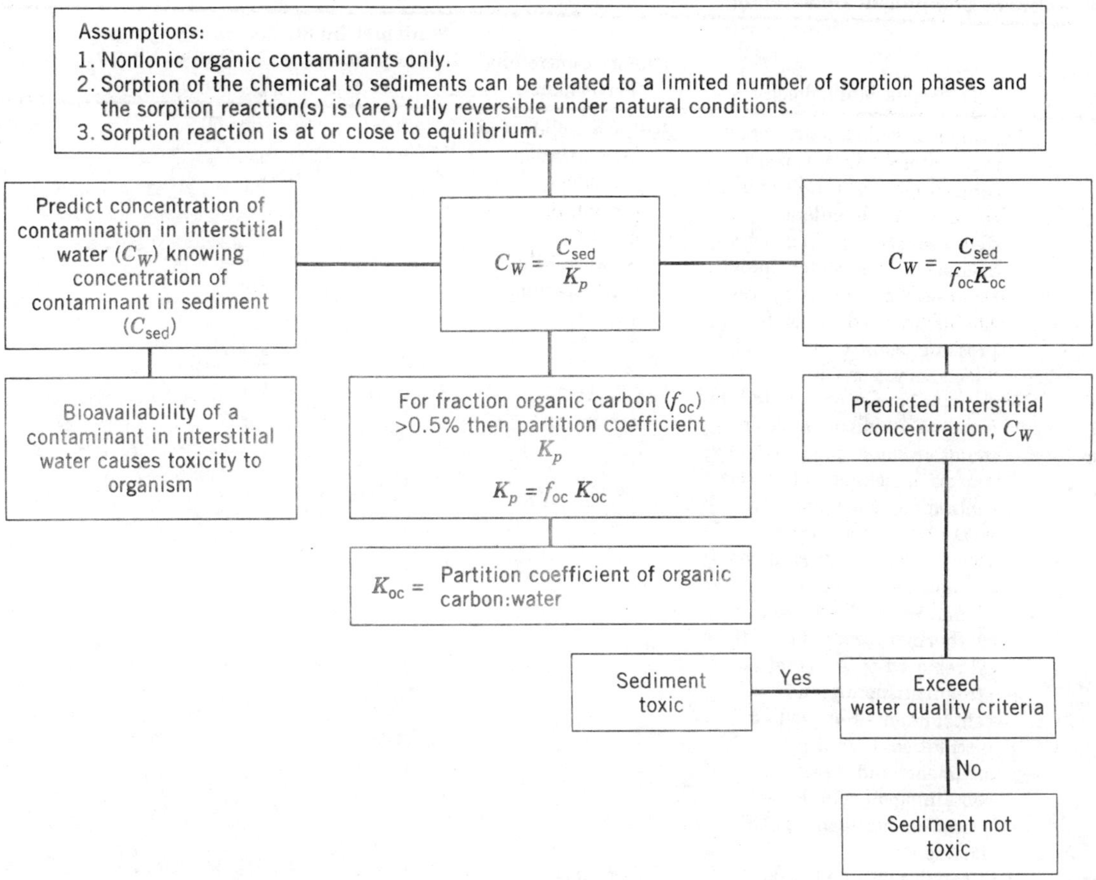

Figure 13. Summary of EPA's EqP model for estimating toxicity of nonionic organic chemicals in sediments.

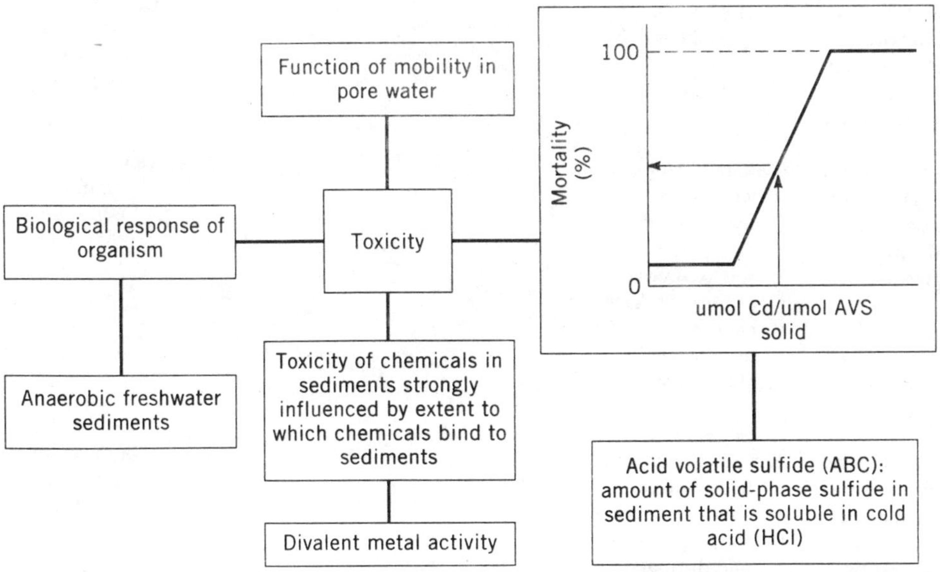

Figure 14. Summary of acid volatile sulfide method for determination of toxicity of cadmium in sediments.

acid leaching, (*2*) BEST, (*3*) C. F. Systems-Propane, (*4*) Low Energy (acetone/kerosene), and (*5*) surfactants.

Immobilization Technologies. The purpose of immobilization technologies is to limit the mobility of contaminants that are present in sediments. Most immobilization techniques take the form of solidification or stabilization (S/S) processes performed prior to placement of the dredged material in the CDF. The objectives of S/S are generally to improve the handling and physical characteristics of the material, decrease

Table 13. Pretreatment Component Dewatering

Process Option	Treatment Methodology	Factors Controlling Performance	Minimum Input Slurry Range, wt %	Output Slurry Range, wt %	Cost[a]
Belt filter press	Use single or double moving belts to dewater sludges. Consists of three stages: (*1*) conditioning by adding of flocculent or thickening drum; (*2*) gravity drainage of free water, and (*3*) compression zone. Compression zone is provided by top belt pressing against bottom belt which serves as filter.	Sediment solids feed concentration, flocculent, amount of pressure between rollers, filter opening size	1–40	50–60	Capital cost for 500,000 yd^3 facility \$6 million; operation and maintenance \$20–\$100 per yd^3
Carver-Greenfield evaporation	Food-grade carrier oil injected into waste as fluidizing agent to maintain liquid phase as solids content increases. Oil-soluble contaminants extracted from waste by carrier oil, and volatile compounds stripped out and condensed with carrier oil and water. Water removed by thermal treatment. Carrier oil removed by evaporation and stream stripping, and contaminants removed for distillation. Useful for sediments and wastes containing oil-soluble contaminants such as PCBs, dioxins.	Temperature, carrier oil	8		
Centrifugation	Uses rapid rotation of fluid mixture inside rigid vessel to separate components based on mass. Types are solid bowl, basket, and disk. Concentrates soil/sediments from fine gravel to silt.	Particle size; efficiency drops drastically for particles <10 μm	Disk = 1	Solid basket = 15–40; basket = 9–25; disk = 6	Capital cost for 50 lb/h dry \$500,000; operation and maintenance \$20–\$100 per yd^3
Chamber filtration	Utilizes rigid individual filtration chambers operated in parallel under high pressure to dewater slurry. Used in situations requiring large area of filtration in minimal area.	Pressure plate warpage and deterioration of plate gasket.		>50	Operation and maintenance cost \$20–\$100/yd^3
Gravity thickening	Operates on differences in specific gravity between solids and water to accomplish separation. Process accomplished in circular vessel similar to conventional clarifier.	Maximum flow, waste type, volume of solids per day, percent solids, specific gravity, surface chemistry, maximum particle size, percent solids required in underflow.			Operation and maintenance \$20–100/yd^3
Primary settling (CDF)	Functions like large settling pond. Effective settling area reduced by wind, turbulence, and short-circuiting.	Efficiency controlled by hydraulic characteristics of settling pond and drainage mechanism.		~50	Operation and maintenance <\$20/yd^3

Table 13. (*continued*)

Process Option	Treatment Methodology	Factors Controlling Performance	Minimum Input Slurry Range, wt %	Output Slurry Range, wt %	Cost[a]
Subsurface drainage (CDF)	Consists of network of perforated pipes or filter material placed under or around perimeter of CDF. Pipes or filter material drain to series of sumps where water is withdrawn.	Permeability of dredge and filter material, time.			Operation and maintenance <$20/yd^3
Vacuum filtration	Consists of rotating cylindrical drum mounted horizontally and partially submerged in trough containing slurry. Vacuum supply applies negative pressure to inside of drum while the moist solids adhere to filter.	Vacuum, degree of drum submergence, drum speed, filter media, porosity, feed type, solids, concentration, chemical conditioning, management.		20–40	Operation and maintenance $20–$100/yd^3
Wick drains (CDF)	High permeability polymeric strips to promote both radial and vertical drainage. Normally placed on 1.5 m (5 ft.) centers to depths of 12 m (40 ft.) Surcharge used to promote dewatering. Have been used on CDFs.	Wick spacing, depth of CDF, material type, amount of surchage.			Operation and maintenance cost $20–$100/yd^3

[a] 1 m^3 ≈ 0.77 yd^3.

the surface area of the sediment mass across which transfer or loss of contaminants can occur, and/or limit the solubility of contaminants by pH adjustment or sorption phenomena. Immobilization techniques are typically used for heavy-metal contaminants because they cannot be destroyed and extraction from soils is a difficult process. Solidification or stabilization is also effective for limiting the volume of leachate leaving the site. It should be noted that placement of contaminated dredged material in a low permeability landfill covered with a low permeability cap is in itself a physical immobilization technique. Thus immobilization techniques can include chemical techniques that modify the dredged materials and containment techniques that simply prevent the movement of water to or from the waste. A summary of immobilization techniques for use in the ARCS program (27) is presented in Table 20.

Thermal Technologies. Thermal technology processes include low temperature thermal desorption, incineration, pyrolysis, vitrification, supercritical and wet air oxidation, and other processes that require heating the sediment up to thousands of degrees above the ambient temperature. Thermal processes are generally the most effective options for destroying organic contaminants. As a first step, low temperature thermal desorption may be needed to change the contaminant to the vapor phase so that the gas can be collected. Incineration typically achieves destruction/removal efficiencies greater than 99% for organic contaminants. Heavy metals in the incineration processes generally pass through the process, except that some of the most volatile metals (such as lead and mercury) can volatilize from the higher temperature processes. Pyrolysis in contrast involves heating the material in the absence of oxygen. Volatile materials are driven off and collected or destroyed by secondary processes, and metals, salts, and other nonvolatile materials melt into molten glass. Vitrification processes use high voltage graphite electrons to provide the primary heat source. A summary of thermal techniques reviewed/evaluated by the ARCS program (27) is presented in Table 21.

SUMMARY AND CONCLUSIONS

Sediments serve as "sinks" for contaminants discharged into receiving waters. If these contaminants are toxic, they constitute a significant problem since these contaminants have the potential for bioaccumulation and for infecting the overlying waters if they are released from the sediments. Determination of the persistence and fate of these toxic pollutants in contaminated sediments requires a knowledge of the contaminant types within the sediment, and the manner in which these are "held".

The strategies for management of contaminated sediments generally pertains to the stewardship of natural and financial resources or the development of cost effective solutions. The terms *contaminated* and *clean,* as they pertain to sediments, are somewhat arbitrary. At the national level, agencies such as EPA, U.S. ACE, and Environment Canada continually research and develop guidelines for determining the levels of various pollutants in sediments that cause adverse ecological and human health effects. Sediment quality criteria have been evolving for the last two decades by using a wide range of biological

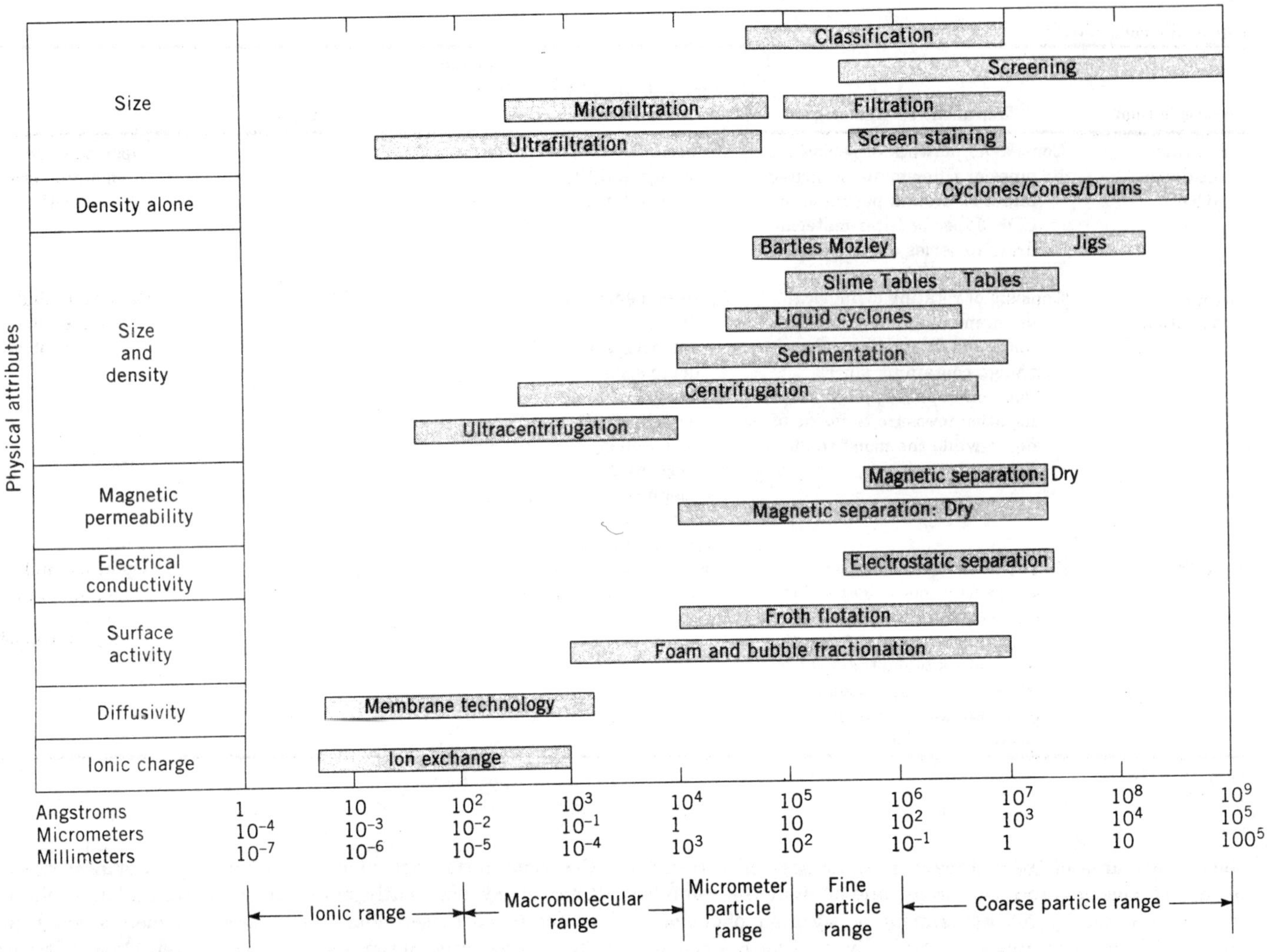

Figure 15. Physical separation technology and particle size.

and chemical assessment techniques for evaluation of both water column and benthic impacts. Although it is arguable that these criteria are adequately defined, the primary remediation problem is the high cost of field sampling and laboratory testing to define that portion of a site that fails the criteria and requires remediation from the portions which are clean and those which are at a problematic contaminant level, ie, higher than the remediation criteria. Several hundred core samples may be required for a single site.

For navigation dredging projects the U.S. ACE and U.S. EPA have developed a technical framework for environmental evaluation of dredged material disposal alternatives. The framework uses a tiered approach which on a site specific basis evaluates the alternatives including (*1*) no action, (*2*) contain in place, (*3*) treat in-place, (*4*) remove and contain, and (*5*) remove and treat.

For contaminated sediment remediation projects, management strategies are not well established but the U.S.ACE and U.S.EPA are presently cooperating to develop and implement a strategy. The legal authorities for remediation and enforcement are provided in legislation of the Clean Water Act, CERCLA/Superfund, RCRA, TSCA, the Oil Pollution Act of 1990, and many state regulations.

Technical and scientific evaluations would be expected to dominate any remediation strategy, however other factors are also important, such as public support for the project; consensus of all affected groups such as business, industry, commerce, and the general public is necessary. The common goal is to develop an environmentally sound plan which reflects the views of the community and satisfies the overlapping jurisdictions of the many national, regional, and local agencies involved.

Containment and isolation are important components in the regulation of contaminated sediment disposal in the United States. The basic design considerations are to control all potential contaminant migration pathways. The applicable regulations covering the disposal of contaminated sediments is shown to be a function of both the concentration of contaminant and the partitioning coefficient for the contaminant in water. Those sediments that have very low levels of contamination or are contaminated with constituents having a very low solubility in water, eg, having a low partitioning coefficient can

Table 14. Pretreatment Particle Classification Technologies

Process Option	Size of Particle Handled	Treatment Methodology	Equipment Requirements	Advantages	Limitations	Size Requirement	Cost[a]
Flotation	Fines	Physiochemical process to concentrate contaminants. Material is ground finer than 48–65 mesh so that contaminants are able to float by air bubbles. After grinding, suspended particles treated with chemicals to cause particles to be air-avid and water repellent. Mixture agitated with air and contaminants attached to air bubbles, float to surface, forming froth. Concentrated froth skimmed off leaving clean product.	Mechanical flotation device uses upright impeller for agitation and aeration. Pneumatic device uses air blowers.	Corrosion resistance	Dependent on surface chemistry, particles size, shape distribution of feed, contaminant distribution with particle size, characteristic of soil, specific gravity and chemical analysis of soil, mineralogical analysis, solids to liquids ratio, pH.		Capital cost \$24–\$160K. Operation and maintenance \$20–\$100 yd^3.
Grizzlies	Debris to gravel, >1.0 in.	Vibrating or fixed separation units used to remove oversized material and reduce slurry velocity prior to subsequent processing.	Device consists of parallel frame mounted bars spaced 1–5 in. apart. Units can be used in series or parallel.	Improves the reliability and efficiency of separation technologies. Maintenance costs minimal.	None.	Typically 6–9 ft wide and 12–18 ft long.	Operation and maintenance cost <\$20/yd^3

Table 14. Pretreatment Particle Classification Technologies

Process Option	Size of Particle Handled	Treatment Methodology	Equipment Requirements	Advantages	Limitations	Size Requirement	Cost[a]
Hydraulic classifiers	Gravel to fine sand	Physical separation and classification by grain size of sand and gravel from fines by settling velocity through water.	Typically V-shaped rectangular trough or cone in which slurry of dredge material is directed. Sand and gravel particles fall to bottom and recovered through discharge valves; fine-grained material carried off by water flow.	Units typically designed for mobile system application with quick startup and shutdown time. Fairly simple maintenance requirements.	Removal and classification of solids from 3/8 in. to ~74 μm. Some fines removed with sand and gravel		Capital cost varies with size and capacity. Typical range \$30–\$76K. Operation and maintenance \$20–\$100/yd^3
Hydrocyclones	Separates clay/silt from coarser particles	Separates solids (gravels and sands) from slurry by vortex-generated centrifugal forces.	Cone-shaped vessel with tangential feed inlet at top. Discharge pipe located short distance into vessel to form vortex.	Applicable to sites with limited space. Wide variety of sizes and can be operated in series or parallel.	Not effective with highly viscous, high clay content slurries, slurries with solids concentration >30%, or particles with specific gravity ~2.5–3.2.	Small size	Capital cost varies according to size and number. Typical range \$1–\$5K. Operation and maintenance <\$20/yd^3.
Impoundment basins	Gravel to fine silt	Physical separation and classification by grain size of gravel to fine silt by settling velocity through water	Diked area or CDF	Handles large dredging operations.	Requires large area. No solids removal system. Needs backhoe or clamshell dredge. Minimal control when used employed as particle separator	Large area	Operation and maintenance <\$20/yd^3
Magnetic/electrostatic	≤100 μm	Separates slurry based on difference in magnetic and conductive properties between contaminant and soil.	Slurry forced through magnetized steel wool containing 95% void space. Intensity of magnetic field can vary from 1–100 kG.	Soils may be dry or in slurry form.	Particles must be <100 μm. If contaminants finely disseminated throughout soil, unlikely to work.		Operation and maintenance \$20–\$100/yd^3

Moving screens	1/8–6 in.	Plain screening surface in rectangular frame. Typically nested from coarse to fine at slope of 20° to allow for a cascading effect. Screens may be vibrated, revolving, or gyrated. Vibrated most common.	Vibrated rectangular framed screen arranged in nest. Usually placed at 20° slope to allow particles to cascade from one screen to next.	Can typically handle 300–950/t of material.	Soil must be dry or washed through screen with water. Screen replacement frequent.	3–10 ft wide and 6–30 ft long.	Capital cost for screen 10′ long, 5′ wide, 5′ deep with a capacity of 200 t/h ~$25K. Operation and maintenance $20–$100/yd^3.
Shaking table	Wide range from slimes to rocks	Feed enters long upper edge of table. Wash water and shaking action spread feed over table. Particles sorted according to size by arrangement of riffles that decrease in height along length toward the discharge end.	Rectangular table with a slope 0–6°. Surface smooth with riffles. Table capable of being shaken.	Versatile gravity separation.			Operation and maintenance $20–$100/yd^3.
Spiral classifiers	Sand and gravels ≤3/8 in.	Settled solids fed from hydraulic classifier into spiral classifier by rotating screw placed on incline. Fines carried to bottom of incline while sand and gravel carried up incline by screw.	Rotating screws mounted in inclined rectangular vessel with capability of washing, dewatering, and sediment. Require pretreatment by hydraulic classifier.	Removes clay and silt adhering to sand and gravel. Designed for mobile systems. Capable of processing up to 950 t/h.		Typically 22–34 ft long, 8–19 ft wide and 8–12 ft deep.	$14–$77K for single screw; $37–$150K for double screw. Operation and maintenance from $20–$100/yd^3.

Table 14. Pretreatment Particle Classification Technologies *(continued)*

Process Option	Size of Particle Handled	Treatment Methodology	Equipment Requirements	Advantages	Limitations	Size Requirement	Cost[a]
Stationary screens	Separates fines from sands and gravels.	Screen surface continuously curved. Slurry passed over screen with velocity high enough to produce centrifugal force to keep slurry in contact with screen.	Curved screen surface. The screen consists of frame mounted parallel bars spaced close together for finer particle separation. Pressurized water spray can be added to increase efficiency.	No moving parts; wedge bar screen more resistant to abrasion.	Less efficient than vibrating screen.	Small size.	Operation and maintenance \$20–\$100/yd^3.

[a] 1 m^3 ≈ 0.77 yd^3.

Table 15. Screen Types[a]

Distinguishing Characteristics	Classifications	Description	Motion	Speed, Amplitude	Applications	Advantages/ Disadvantages
			Stationary grizzly			
Heavy-duty surface of fixed bars	Conventional	Heavy bars running in flow direction, sloped to allow gravity transport. Bars may spread along length to minimize blinding	Stationary surface (vibrating grizzlies also available: bar vibrating screens)		Scalping before crushers	Simple, robust, probability form-blinding, resistant
	Probability	Bars divergent in vertical plane				
			Roll grizzly			
Surface of rotating rolls		Stationary screen surface, but nonuniform shape of rolls conveys material.			Coarse separations before crushing, primarily conveyor.	Conveying action allows near horizontal operation in low head room situations.
			Sieve bends			
Slurry feed, fixed bar surface	Straight or curved surface	Stationary, parallel bars at right angles to slurry flow. Surface may be straight (with steep incline) or curved to 300°.	Stationary		Separations in range 2 mm to 45 μm, or those too coarse for hydrocyclone, or where density effects make classifier unsuitable, dewatering.	Relatively high efficiency and capacity. Sharpness of cut less than true screen. Separation slightly affected by mineral density. Excessive dewatering can be a problem.
			Revolving screens			
	Trommel	Slightly inclined cylindrical screen, may have concentric surfaces	Below critical speed (ball mill).	15–20 rpm	Wet or dry separations 60 to 10 mm if dry, smaller if wet.	Simple, useful for scrubbing or rough size separations. High wear, low surface utilisation.
Screen surface rotating around cylinder axis	Centrifugal	Vertically mounted cylindrical screen; centrifuges particles through screen.	Operates above critical speed. Also has vertical action of 800–1000 cycles/min.	60–80 rpm	Wet or dry separations 12 mm to 400 μm; dewatering.	High wear
	Probability	Particles drop through "surface" formed by bars radiating out like wheel spokes.	Radiating bars rotate about vertical axis. Speed of rotation determines cut size.		Developed for separating coal <6 mm	Relatively high capacity with fine separations, cut size easily changed and controlled by varying speed
			Vibrating screens			

Table 15. Screen Types[a] *(continued)*

Distinguishing Characteristics	Classifications	Description	Motion	Speed, Amplitude	Applications	Advantages/ Disadvantages
			Stationary grizzly			
	Inclined (sub-classified by vibrator mecha-nisms)	Inclined rectangular screening surface which allows material to flow with aid of vibrations.	Mechanical vibrations give circular motion at center: elsewhere it depends on vibrator Elec-tromagnetic vibrators may give linear vibration at center.	600–7000 rpm Low <25 mm	Wide applications, generally down to 200 μm in mineral industry, but down to 38 μm in chemical industry, using the high speeds.	Relatively high efficiency and capacity, but capacity generally inadequate below 200 μm.
High speed motion, designed primarily to lift particles off surface.	Horizontal	Horizontal rectangular screening surface. Linear vibration must have horizontal component to convey material along screen.	Linear motion, with vertical component to provide lift, and horizontal component for conveying.	600–3000 rpm Low <25 mm	Similar to inclined screens.	Similar to other vibrating screens, but can also be used where head room is restricted.
	Probability	Series of relatively small inclined screen surfaces: separates by statistics rather than physical constraint.			Similar to inclined screens	Generally superior to conventional vibrating screen. High capacity and efficiency for given space: low noise, low power. Low efficiency at low loading.
			Shaking screens			
Slow linear mo-tion essentially in plane of screen.		Usually slightly inclined. May have several surfaces in series with dif-ferent apertures.	Linear motion, essentially in plane of screen; parti-cles tend to remain in contact with screen surface.	30–800 rpm 25–1000 mm	Down to 12 mm for coal prepara-tion and nonmetallic minerals. Higher speeds may size down to 250 μm.	Low headroom and power require-ments. May be used for convey-ing and sizing. Accurate for large sizes. High maintenance cost, low capacity.
			Rotary sifters			
	Reciprocating	Rectangular screen surface with slight (~5°) incline.	Circular motion is applied at feed end, and produces re-ciprocating motion at dis-charge end.			
Circular motion applied to screen surface	Gyrating	Circular screen surface.	Circular motion over most of the screen surface.	500–600 rpm Low <25 mm	Generally used for finer separa-tions (12 mm to 45 μm, wet or wet or dry) in non-metallurgical industries.	Suitable for finer separations, but with low capacity.

Table 15. *(continued)*

Distinguishing Characteristics	Classifications	Description	Motion	Speed, Amplitude	Applications	Advantages/ Disadvantages
	Gyrating	Circular screen surface.	Screen moves with circular motion, but also has oscillating vertical component.			

[a] Ref. 28.

Table 16. Pretreatment: Slurry Injection

Process Option	Size of Particle Handled	Treatment Methodology	Equipment Requirement	Advantages	Limitations	Costs[a]
Chemical clarification	Silt and clays	Ust of conditioners to promote coagulation or flocculation of smaller size colloidal size particles. Process increases rate of sedimentation. Coagulants include inorganic chemicals (salts of iron and aluminum), organic polyelectrolytes. Polyelectrolytes recommended for dredged material.	Pipeline transport of dredged material slurry required to mix conditioner to mix conditioner. Method to inject conditioner required.	Polyelectrolytes are less dependent on pH and require lower dosages compared to inorganic chemicals.	Settling process required after injection of conditioners. Prolonged periods of mixing reduce effectiveness.	Operation and maintenance <$20/yd^3.
Microbe addition	Silt and clays	Use of microorganisms acclimated to degrade toxic materials. Microorganisms injected into pipeline slurry, mixing through sediment mass.	Pipeline transport of dredged material slurry required to mix microorganisms. Method to inject microorganisms required.		Different microorganisms required for different types of toxics.	Operation and maintenance \$20–\$100/yd^3

[a] 1 m^3 ≈ 0.77 yd^3.

be controlled by designing the CDF for solids retention. For higher concentrations of contaminants, or for contaminants having high partitioning, the CDF must be designed based on hydraulic isolation of the waters released by the sediments.

Dredged material contaminants placed within a CDF can leave the system using the following pathways: (*1*) waterborne pathways including effluents drained from the waste, effluent seeping through the dikes, leachate seeping into the groundwater within the foundation of the CDF, and surface water runoff; (*2*) via plant or animal uptake; and (*3*) from airborne emissions. Each contaminant pathway may be controlled by engineered components, ie, with a liner or by selective placement of the dredged materials.

Waterborne contaminants moving through the dikes or leaching into the subgrade can be controlled either with engineered barriers such as clay or synthetic liners or by placement of clean fine-grained dredged materials initially on the bottom of the CDF to provide hydraulic isolation. Similarly, water runoff, animal and plant uptake, and air emissions can be controlled either through the use of an engineered cover system or the placement of clean dredged material over the contaminated dredged material. A number of testing protocols are available to evaluate the need for pathway control.

Remedial technologies can be broadly divided into either active remediation or in-place remediation. Active remediation involves dredging of the contaminated sediments and then treating them *ex situ* under highly controlled conditions. In contrast, in-place remediation involves processes such as isolation by natural burial, capping and armoring or bioremediation. A treatment/remediation method is selected after a

Table 17. Contaminant Sediment Treatment: Biological

Process	Applicable Contaminants	Mechanism	Factors Affecting	Demonstrated for Remediation of Sediments	Typical Cost[a]	Comments
Aerobic bioreclamation	Organic materials	Microorganism are encouraged to grow in aerobic environment utilizing the organic contaminant as food source.	Nutrients for growth of micro-organisms, oxygen, temperature, soil moisture, pH.	Bench	Operation and maintenance \$20–\$100/yd^3	Methodology can be used to treat soils and groundwater contaminated with organic materials
Anaerobic bioreclamation	Organic contaminants such as halogenated compounds, various aromatics, some pesticides.	Process uses anaerobic metabolism that includes anaerobic respiration using nitrate or sulfate as terminal electron acceptor and fermenting methanogenic processes using methanogenic consortium (strictly anaerobic bacteria).	Temperature, redox potential of ≤ -250 mV, no oxygen, nitrates, or sulfates present.	Bench	Operation and maintenance \$20–\$100/yd^3.	Process is slower and fewer compounds can be degraded in comparison to aerobic bioreclamation.
Bioreactors	Organic materials	Aerobic biodegradation technology that functions to increase availability of contaminants and nutrients to microbes within tank. Slurry dewatered and resultant water treated, if necessary.	Nutrients, oxygen, pH.	No	Operation and maintenance \$20–\$100/yd^3.	Limitations of technology include inhibition of microbial metabolism by chloride substitution and heavy metals, and temperatures outside the optimum range.

[a] $1\ m^3 \approx 0.77\ yd^3$.

preliminary estimate of the magnitude and extent of sediment contamination and the associated risks to human health and the environment.

The nature of contaminated sediments effects the selection of treatment technologies. The volume of sediments present at contaminated underwater sites tend to be quite high, while the concentrations of contaminants in the sediments tend to be relatively low in comparison to the concentration of contaminants in soils at hazardous waste cleanup sites. In addition, the majority of contamination is typically associated with the fine-grained material. Because most of the contaminants are associated with the finer grained material, substantial cost savings can be achieved for active remediation by first applying extractive technologies, thus reducing the volume of material requiring further treatment by more expensive destructive methods. Once the volume of material has been reduced, a treatment strategy for the fine-grained material can then be devised. In addition dredged materials must first undergo a dewatering stage prior to treatment because of the high initial water content.

Treatment technologies can be broken down into two broad categories: (*1*) those that work on the sediments *in situ* and (*2*) those that process sediment after dredging. In general, *in situ* treatment technologies for contaminated sediments are less developed than the technologies that can be applied to dredged materials. Any decision to leave sediments in place is highly dependent on an evaluation of the relative risks posed by the sediments left untreated on the bottom, the risks of performing a treatment operation on *in situ* sediments, and the risks associated with the removal and subsequent disposal or treatment of the contaminated dredged materials.

Table 18. Contaminant Sediment Treatment: Chemical

Process	Applicable Contaminants	Mechanism	Factors Affecting	Demonstrated for Remediation of Sediments	Cost[a]	Comments
Chelation	Metals	Chelating molecule (ligand) forms multiple bonds with metal ions resulting in a heterocyclic (ringed) structure. Chelating agents either increase metal mobility for removal or immobilize by sorbing onto clay materials.	pH, chelate dosage, chelating agent, metal type	No	Operation and maintenance of \$100–\$200/ yd^3.	Chelating agents must be used according to their affinity for particular metals.
Nucleophilic substitution	PCBs, acids, oils, thiols, dioxins, chlorobenzenes.	Chemically removes chlorine from organic (aromatic) compounds using electron donator principle.	Requires alkaline conditions (high pH level), temperature, reaction time, reagent dose, moisture content.	Yes	Operation and maintenance \$150–\$200/ yd^3.	Two primary processes: APEG, alkali metal hydroxides in polyethylene glycol: KPEG, potassium–polyethylene glycol.
Oxidation of organics	Oxidation organics such as aldehydes, phenols and benzidine. Limited applications for slurries, tar, and sludge.	Oxidation state of atom increased by removal of electrons. Oxidation transforms, degrades, or immobilizes contaminants in soil system.	Oxidant concentration, pH, oxidation potential.	Yes	Operation and maintenance \$150–\$200/ yd^3.	Toxic products may form if reaction not complete. Process not applicable for high strength complex waste systems, nor highly halogenic organics. Methodology can be employed *in situ*.

[a] $1\ m^3 \approx 0.77\ yd^3$.

Treatment technologies for dredged material have been evaluated and demonstrated by a number of groups (27). A review of the literature indicates that all treatment technologies typically leave some degree of contamination in the sediment (3). No single technology has been shown to be effective for all contaminants. Typically, technologies are designed to deal with either the organic contaminants (such as PCBs) or heavy-metal contaminants. Some of the available technologies (eg, particle separation and solidification) may effectively treat both types of contaminants, but their application is limited to sediments with specific characteristics that are not present at all sites. Complications have been shown to arise due to the presence of certain contaminants in application of some technologies, and volatile contaminants may be lost unintentionally during the application of thermal treatment processes. In addition, treatment technologies are effective on only some types of sediment. These limitations suggest that multiple step treatment processes will be necessary to treat sediment.

The following conclusions apply to contaminated dredged materials:

1. Contaminated sediment remediation projects are multidisciplinary in nature and are driven by federal and state regulations. Standards test methods and data are needed to provide a basis for improved communication and understanding between the involved parties.
2. Site investigations to characterize the nature and extent of contamination are difficult and costly. A tiered-testing approach has been proposed by U.S.ACE/ U.S.EPA to identify suspect sites.
3. There remains an urgent need to develop *in situ* chemical sensors such as the polyacrylamide gel probe for discrete data or possibly fiber optic sensors with telemetry

for continuous data monitoring before, during, and after disposal.

4. There is a need to develop a number of analytical and stochastic models to support scientific and engineering evaluations of a remediation option such as the mass-balance analysis of contaminants during dredging, placement in a CDF and isolation.
5. Natural sediments, by nature, are soft and difficult to sample and handle. Dredged sediments have entrained water and exhibit fluid-like behavior.
6. Contaminated sediments usually contain oil and grease or other organic components with the potential to generate gas and odor following dredging. These may require special attention during dredging and remediation.

Table 19. Contaminant Sediment Treatment: Extraction

Process	Applicable Contaminants	Mechanism[b]	Factors Affecting	Remediation of Sediments, Dredged Materials	Typical Cost[c]
Acid leaching	Metals, amines, anilines, ethers, phenols.	Steam and/or weak acids (aqueous solutions of sulfuric, hydrochloric, nitric, phosphoric, and carbonic acid) used as dissolving agents for basic metal salts, amines, anilines and ethers processes. *In situ* treatment may be performed using weak acid solutions (acetic acid and sodium hydrogen phosphate). Basic solutions (NaOH) used as flushing agents for soils contaminated with metals (zinc, tin, lead, and phenols).		Pilot, no	Operation and maintenance \$63–\$200/ yd^3.
Basic extraction sludge treatment (BEST)	PCBs, oils, creosote	Mixture of 1 part free water sediment and 1–7 parts amine-based solved, usually TEA. Single liquid phase forms and solid matter removed by centrifugation and dried to remove residual TEA. TEA is then processed to remove contaminants.	Treatment must be sealed from atmosphere and operated under nitrogen blanket. Temperature must be below critical solution temperature. Alkaline conditions (pH 10) required.	Demonstrated, bench	Operation and maintenance \$60–\$300/ yd^3.
C.F. Systems-Propane	PCBs and organics	Contaminated sediment fed to top of extractor and makes nonreactive contact with condensed propane solvent flowing up through the extractor at a pressure of 250 psi. Typically used in a series of three extractors. Contaminated propane is fed to a separator where it is vaporized and recycled by recompression.	Propane, pressure, temperature.	Pilot/pilot	Operation and maintenance \$100–\$200/ yd^3.
Low energy acetone–kerosene[a]	PCBs and organics	This process separates sediment into liquid and solid, then leaches solid portion with hydrophilic solvent (acetone). Acetone is steam-stripped from sediment. PCBs treated with hydrophobic solvent (kerosene).	Water content >50%	Conceptual, bench	Operation and maintenance \$41–\$200/ yd^3.

Table 19. *(continued)*

Process	Applicable Contaminants	Mechanism[b]	Factors Affecting	Remediation of Sediments, Dredged Materials	Typical Cost[c]
Surfactants	Hydrophobic organics, nonsoluble organics (PCBs, crude oil, tertiary oils).	Surfactants injected into soil to flush contaminants. Typically use mixture of anionic and nonanionic surfactants.		Demonstrated, no	Operation and maintenance \$100–\$200/ yd^3.

[a] Field and laboratory have exhibited mixed results.
[b] TRA = triethylamine.
[c] $1\ m^3 \approx 0.77\ yd^3$.

Table 20. Contaminant Sediment Treatment: Immobilization

Process	Applicable Contaminants	Mechanism	Factors Affecting	Remediation of Sediments, Dredged Materials	Typical Cost[a]	Comments
Chloranan encapsulation	Organic compounds, heavy metals, oil, grease.	Injection of cement, water, and chloranan into contaminated solids or sludges. Chloranan encapsulates organic contaminants to prevent interference with solidification process. Contaminants immobilized from soils in concrete-like leach-resistant matrix.	Water content, type of solid/ slurry density of soil.	Demonstrated, no	Operation and maintenance \$100–\$200/yd^3.	
In situ encapsulation	Organic and inorganic materials (metals, esticides, phenols, and PCBs)	Injection and/or mixing of solidification/stabilization reagents with contaminated material. Mixing can be accomplished using agricultural spreaders and tillers, special augers, backhoes, or clamshells.	Type of material, contaminant, and reagent. Water content.	Demonstrated, no	Operation and maintenance \$100–\$200/yd^3.	
Lime-based pozzolan	Organics and low levels of metals.	Contaminated materials mixed with carefully selected reactive, pozzolanic additive (typically bituminous coal fly ash) to pasty consistency and subsequently blended with hydrated lime to form solid mass.		Demonstrated, bench-scale testing for dredged material	Operation and maintenance \$100–\$200/yd^3.	Only tolerates low levels of mercury and moderate levels of lead. Greater leachability than cement-based processes.
Portland cement-based	Arsenic, lead, zinc, copper, cadmium, nickel, PCBs.	Portland cement added as binding agent along with pozzolanic product (fly ash) to react with free calcium hydroxide and thus improve strength, handling characteristics, and chemical resistance of stabilized sediment.	Type of contaminant, water content.	Demonstrated, bench-scale testing for dredged material	Operation and maintenance \$100–\$200/yd.	Cement reduces mobility of heavy metals due to conversion to insoluble hydroxides or carbonates because of elevated pH of cement. Other sorbents added to improve characteristics.

Table 20. Contaminant Sediment Treatment: Immobilization *(continued)*

Process	Applicable Contaminants	Mechanism	Factors Affecting	Remediation of Sediments, Dredged Materials	Typical Cost[a]	Comments
Proprietary processes	Typically metals and higher molecular weight organics.	Proprietary reagent typically mixed with contaminant, then combined with cement product to form stabilized mass. (large number of processes on market).	Type of contaminant, soil type, reagent type.	Demonstrated, bench-scale testing for dredged material	Operation and maintenance \$100–\$200/yd.	
Sorption	Organic and inorganics.	Involves adding solids material to soak up free liquid in soil or waste to produce product that is easier to handle. Sorbents include natural materials such as fly ash, klin dust, vermiculite, and bentonite, as well as synthetic materials such as activated carbon, resins, Hazorb, Locksorb, etc	Quantity of sorbent for removal of liquid varies according to the nature of liquid, solids content of waste, and chemical reactions that may occur.	Demonstrated, no	Operation and maintenance \$100–\$200/yd.	Sorbents inexpensive and plentiful, but often required in large amounts, producing disposal problem.

[a] $1\ m^3 \approx 0.77\ yd^3$.

Table 21. Contaminant Sediment Treatment: Thermal

Process	Applicable Contaminants	Mechanism	Factors Affecting	Remediation of Sediments, Dredged Materials	Typical Cost[a]	Comments
Advanced electric reactor	PCBs, dioxins, heavily halogenated organics, nerve gas	Vessel with porous carbon core surrounded by carbon electrides. Radiation heat shield surrounds core and electrodes. Nitrogen acts as gaseous blanket that isolates reactants from core and heat-transfer medium between carbon electrodes and porous core. Two post-reactor zones connected in series.	Temperature, residence time	Pilot/no	Operation and maintenance >\$200/yd^3.	Transfers energy to the waste through radiation.
Circulating bed combustor	Halogenated, nonhalogenated hydrocarbons.	High air velocity to circulate solids and create larger and more agitated combustion zone. Dry limestone added to feed material to react with acid gas so that wet scrubber not necessary.	Temperature, residence time	Demonstrated, yes	Operation and maintenance \$100–\$200/yd^3.	Uses relatively low temperatures (871°C) due to high degree of turbulence.

Table 21. *(continued)*

Eco-logic	Chlorinated wastes, organic wastes containing PCBs, halogenated benzen, phenols, cycloalkane, alkanes, dioxin, dibenzofuran.	Thermochemical destruction process relies on ability of hydrogen to dechlorinate organic compounds at high temperatures.	Temperature, residence time	Pilot/pilot	Operation and maintenance \$100–\$200/yd^3.	Mobile unit
Fluidized-bed incineration	Halogenated and non-halogenated organics, phenols, methyl methacrylate, oil refinery wastes.	Process consists of a cylindrical, vertical, refractory-lined vessel that contains inert granular material, usually sand, on perforated metal plate. Combustion air introduced at bottom of incinerator and bubbles through inert bedding material, causing bedding material to become fluidized an agitated. Waste pumped into vessel and combusted within bubbling material.	Temperature, residence time	Demonstrated, no	Operation and maintenance >\$200/yd^3.	
High temperature slag incineration	Waste including low level radioactive wastes and most stable chlorinated aromatics.	Transforms waste into mechanically strong and chemically stable basalt-like material in granular or bulk form. Homogenous waste stream fed into combustion chamber. Burner heats top of waste into layer of molten slag. Slag droplets flow to granulator where they are quenched, causing them to burst into granules.	Temperature, residence time	Demonstrated, no	Operation and maintenance >\$200/yd^3.	
Infrared incineration	Halogenated and non-halogenated organics, dioxins, spent activated carbon.	Waste prepared to consistency and size required. Feed fed on wire mesh conveyor belt into infrared primary chamber. Material stirred gently by rotary rakes, or cake breakers.	Temperature, residence time, waste layer thickness, combustion air flow rate.	Pilot, no	Operation and maintenance >\$200/yd^3.	

Table 21. Contaminant Sediment Treatment: Thermal *(continued)*

Process	Applicable Contaminants	Mechanism	Factors Affecting	Remediation of Sediments, Dredged Materials	Typical Cost[a]	Comments
In situ vetrification	Organics, inorganics, heavy metals, PCBs	Process converts waste into chemically inert and stable glass and crystalline product. Mechanism involves placing 4 large electrodes vertically inserted in square arrangement in soil. Graphite placed on surface to complete circuit between electrodes. Sand, glass frit, or soda ash can be added to soil to improve process. Combustion gases need to be processed.	Current, electrode spacing, water content.	Pilot, bench	Operation and maintenance >$200/yd^3.	
Low temperature thermal stripping	Volatile organics and PCBs	Only enough heat applied to volatize contaminants.				

[a] 1 m^3 ≈ 0.77 yd^3.

ACKNOWLEDGMENTS

The authors appreciate the support of U.S. EPA Region V (contract 68-CO-0068-43) for the opportunity to work on this project. This document does not represent the position or policies of the U.S. EPA.

BIBLIOGRAPHY

1. C. Fox, EPA Office of Science and Technology, Washington, D.C., personal communication, 1995.
2. *The Merck Index,* 11th ed., Rahway, N.J., 19–; *An Overview of Sediment Quality in the U.S.,* June 1987; *Principles of Water Quality Management,* 1980.
3. G.N. Richardson, R.C. Chaney, and K.R. Demars. *Design, Performance and Monitoring of Dredged Material Confined Disposal Facilities in Region 5,* EPA 905-B-96-003, U.S. Environmental Agency, Chicago, Ill., Aug. 1996.
4. *Fed. Reg.* **55,** 9211 (Mar. 12, 1990).
5. *Superfund Public Health Evaluation Manual,* EPA/540/1-86/060, Office of Emergency and Remedial Response, Washington, D.C., 1986.
6. *Conducting Remedial Investigations / Feasibility Studies for CERCLA Municipal Landfill Sites,* EPA/540/P-91/001, Office of Emergency Remedial Response, Washington, D.C., 1991.
7. *Evaluation of Dredge Material Proposed for Ocean Disposal,* EPA-503/8-91/001, Washington, D.C., 1991.
8. J.W. Keeley and R.M. Engler, *Discussion of Regulatory Criteria for Ocean Disposal of Dredged Material: Elutriate Test Rationale and Implementation Guidelines,* U.S. Army Engineers, MP D-74-14, Waterways Experiment Station, Vicksburg, Miss., 1974.
9. *Implementation Manual for Section 103 of Public Law 92-532 (Marine Protection, Research, and Sanctuaries Act of 1972),* Environmental Effects Laboratory, Waterways Experiment Station, Vicksburg, Miss., 1977.
10. N.R. Francinques and co-workers, *Management Strategy for Disposal of Dredged Material: Contaminant Testing and Controls,* Misc. Paper D-85-1, Waterways Experiment Station, Vicksburg, Miss., 1985.
11. R.K. Peddicord, C.R. Lee, M.R. Palermo, and N.R. Francingues, *General Decisionmaking Framework for Management of Dredged Materials, Example Application to Commencement Bay, Washington,* Misc. Paper MP-D-86, U.S Army Waterways Experiment Station, Vicksburg, Miss., 1985.
12. M.R. Palermo and co-workers, *Evaluation of Dredged Material Disposal Alternatives for U.S. Navy Homeport at Everett, Washington,* Technical Report EL-89-1, U.S. Army Engineer Waterways Experiment Station, Vicksburg, Miss., 1989.
13. D.E. Averett, M.R. Palermo, and R. Wade, *Verification of Procedures for Dredged Material Containment Areas for Solids Retention,* Technical Report D-88-2, Waterways Experiment, Vicksburg, Miss., 1988.
14. M.R. Palermo and E.L. Thackston, *Refinement of Column Settling Test Procedures for Estimating the Quality of Effluent From Confined Dredged Material Disposal Areas,* Technical Report D-88-9, Waterways Station, Vicksburg, Miss., 1988.
15. D.M. DiToro, J.D. Mahoney, D.J. Hansen, K.J. Scott, M.B. Hicks, S.M. Mayr, and M.S. Redmond, *Environ. Toxicol. Chem.* **9,** 1489–1504 (1990).
16. A.R. Carlson, G.L. Phipps, V.R. Mattson, P.A. Kosian, and A.M. Cotter, *Environ. Toxicol. Chem.* **10,** 1309–1319 (1991).
17. G.T. Ankley and co-workers, *Environ. Toxicaol. Chem.* **10,** 1299–1307 (1991).
18. C.R. Lee, and Skogerboe, "Quantification of Soil Erosion Control by Vegetation on Problem Soils," *Proceedings, International Conference on Soil Erosion and Conservation,* Soil Conservation Society of America, St. Louis, Mo., 1983.
19. C.R. Lee and co-workers, *General Decisionmaking Framework for Management of Dredged Material: Example Application to Com-*

merce Bay, Wisconsin, Misc. Paper D-91-1, U.S. Army Engineer Waterways Experiment Station, Vicksburg, Miss., 1991.

20. B.L. Folsom, C.R. Lee, and D.J. Bates, *Influence of Disposal Environment on Auxiliary and Plant Uptake of Heavy Metals in Dredged Material,* Technical Report EL-81-12, Waterways Experiment Station, Vicksburg, Miss., 1981.
21. *Annual Book of ASTM Standards,* American Society for Testing and Materials, Philadelphia, Pa., 1993.
22. T.E. Myers and J.M., Brannon, *New Bedford Harbor Superfund Project, A Cushnet River Estuary Engineering Feasibility Study of Dredging an Dredged Material Disposal Alternatives: Report 5, Evaluation of Leachate Quality,* Technical Report E-88-15, U.S. Army Engineer Waterways Experiment Station, Vicksburg, Miss., 1988.
23. J. Brannon and co-workers, *Effectiveness of Capping in Isolating Dutch Kills Sediment From Biota and the Overlying Water,* Misc. Paper D-86-2, U.S. Army Engineer Waterways Experiment Station, Vicksburg, Miss., 1986; J.M. O'Connor and S.G. O'Connor, *Evaluation of the 1980 Cappino Operations at the Experimental Dump Site, New York Bight Apex,* Technical Rep. D-33-3, U.S. Army Engineer District, New York, Oct. 1983.
24. *Water Resources: Future Needs for Confining Contaminated Sediment in the Great Lakes Region Report to Chairman,* Subcommittee on Water Resources, Committee on Public Works and Transportation, House of Representatives, GAO/RCED-92-89, United States General Accounting Office, July 1992.
25. J.C. Luand and K.Y. Chen, *Environ. Sci. Technol.,* **11,** 174–181 (1997).
26. G.R. Phillips and R.C. Russo, *Metal Bioaccumulation in Fishes and Aquatic Invertebrates: A Literature Review,* EPA-600/3-78-103, Environmental Research Laboratory, U.S. Environmental Protection Agency, Duluth Minn., Dec. 1978; R. Eisler, *Mercury Hazards to Fish, Wildlife and Invertebrates: A Synoptic Review,* U.S. Fish and Wildlife Service Biological Report, 85 (1.1), Contaminant Hazard Reviews, Apr. 1987.
27. D.E. Averett, B.D. Perry, E.J. Torrey, and J. Miller, *Review of Removal, Containment, and Treatment Technologies for Remediation of Contaminated Sediment in the Great Lakes,* Misc. Paper EL-90-25, U.S. Environmental Protection Agency, and USAE Division, Chicago, Ill., 1990.
28. *Handbook: Remediation of Contaminated Sediments,* EPA/625/6-91/028, Cincinnati, Ohio, Apr. 1991.
29. *E.G. Kelly, and D.J. Spottiswood, Introduction to Mineral Processing, John Wiley & Sons, Inc., New York, 1982.*
30. Technical Resource Document: Design, Construction, and Operation of Hazardous and Non-Hazardous Waste Surface Impoundments, EPA/530/SW-91/054, Washington, D.C., 1991.
31. D.N. McNelis and co-workers, *Exposure Assessment Methodologies for Hazardous Waste Sites,* University of Nevada, Las Vegas, 1984.

General References

Standard Methods for the Analysis of Water and Waste Water, 17th ed., American Public Health Association, American Water Works Association, Water Pollution Control Federation, Washington, D.C., 1989.

F.A.J. Armstrong, and A.L. Hamilton, in P.C. Singer, ed., *Trace Metals and Metal-Organic Interactions in Natural Waters,* Ann Arbor Science Publishers, Ann Arbor, Mich., 1973, pp. 131–156.

Assessment and Remediation of Contaminated Sediments (ARCS), *1991 Work Plan,* U.S. Environmental Protection Agency, Great Lakes National Program Office, Chicago, Ill., 1991, 65p.

R.J. Auerlich, R.K. Ringer, and S. Iwamoto, *Arch. Environ. Contam. and Toxicol.* **2** (1), 43–51 (1974).

J.M. Brannon, *Comprehensive Analysis of Migration Pathways (CAMP): Contaminant Migration Pathways at Confined Dredged Material Disposal Facilities,* Misc. Paper 0-90-5, Dept. of the Army Waterways Experiment Station, Vicksburg, Miss., 1990.

T. Bursztynsky, *Wetlands and Pollutant Treatment Mechanisms,* Technical Memorandum No. 69, San Francisco Bay Area Water Quality Planning Program, Association of Bay Area Governments, Berkeley, Calif., 1981, 56 p.

L.M. Walsh, ed., *Application of Sewage Sludge to Cropland: Appraisal of Potential Hazards of the Heavy Metals to Plants and Animals,* CAST Report No. 46. Council for Agricultural Science and Technology, Ames, Iowa, 1976, 63 pp; reprinted as EPA 430/9-76-013 (MCD-33).

R.L. Chaney "Plant Accumulation of Heavy Metals and Phytotoxicity Resulting from Utilization of Sewage Sludge and Sludge Composts on Cropland," *Proceedings of the National Conference on Composting Municipal Residues and Sludges,* Information Transfer, Inc., Rockville, Md., 1978.

R.L. Chaney, *Potential Effects of Waste Constituents on the Food Chain in Land Treatment of Hazardous Wastes,* Noyes Data Corp., Park Ridge, N.J., 1983.

R.L. Chaney, A.M. Decker, and C.C. Shaeffer, personal communication, 1980.

D. Chisholm, *Can. J. Plant Sci.* **52,** (4) 583–588 (1972).

Ecological Evaluation of Proposed Dredged or Fill Materials Into Navigable Waters: Interim Guidance for Implementation of Section 404(b)(1) of Public Las 92-500 (Federal Water Pollution Control Act Amendments of 1992), Misc. Paper D-76-17, Waterways Experiment Station, Vicksburg, Miss., 1976.

Productive Land Use of Dredged Material Containment Areas: Planning and Implementation Considerations, Technical Report No. DS-78-20, Waterways Experiment Station, Vicksburg, Miss., 1978.

Confined Disposal of Dredged Material, Engineer Manual EM 1110-2-5027, Dept. of the Army, Washington, D.C., Sept. 1987.

Disposal Alternatives for PCB-Contaminated Sediments from Indiana Harbor, Indiana, Misc. Paper EL-87-9, Vols. 1 and 2, Waterways Experiment Station, Vicksburg, Miss., 1987.

Sebewaing Harbor Confined Disposal Water Quality Monitoring, U.S. Army Engineering District, Detroit, Mich. Nov. 1988.

D'Angremond, *Assessment of Certain European Dredging Practices and Dredged Material Properties,* Adriaan Volker Dredging Co., Rotterdam, the Netherlands, 1978.

R.D. Davis and P.H.T. Beckett, *New Phytology,* **80** (1978).

R.D. Davis, *Plant Soil,* **49** (1983).

K.R. Demars, G.N. Richardson, R.N. Yong, and R.C. Chaney, eds., *Dredging, Remediation and Containment of Contaminated Sediments,* ASTM STP 1293, American Society for Testing and Materials, Philadelphia, Pa., 1995, 333 pp.

D.M. DiToro, *Toxicity of Cadmium in Sediments: The Role of Acid Volatile Sulfide,* 1989, 38 pp.

K.H. Domsch, K. Grabbe, and J. Fleckenstein, "Heavy Metal Content in the Culture Substrate and in the Mushroom, *Agaricus bisporus,* Grain in Composts Mixed with Municipal Waste and Sewage, (in German), 2 Pflanzenern, Boden K., 1976, pp. 467–501.

R. Eisler, *Fourth Annotated Bibliography on Biological Effects of Metals in Aquatic-Environments,* U.S. Environmental Protection Agency, Washington, D.C., 1979.

R. Eisler, *Cadmium Hazards to Fish, Wildlife and Invertebrates: A Synoptic Review,* U.S. Fish and Wildlife Service Biological Report, 85 (1.2), Contaminant Hazard Reviews, July 1985.

R. Eisler, *Chromium Hazards to Fish, Wildlife and Invertebrates: A Synoptic Review,* U.S. Fish and Wildlife Service Biological Report, 85 (1.6), Contaminant Hazard Reviews, Jan. 1986.

R. Eisler, *Arsenic Hazards to Fish, Wildlife and Invertebrates: A Synoptic Review,* U.S. Fish and Wildlife Service Biological Report, 85 (1.12), Contaminant Hazard Reviews, July 1988.

M. Elster, *A Summary of Contaminated Sediment Remedial Actions in Region V of the U.S. Environmental Protection Agency,* EPA 171 R-92-007, Apr. 1992, 42 pgs.

R.M. Engler, T. Wright, C.R. Lee, and T.M. Dillon, *Corps of Engineers' Procedures and Policies on Dredging and Dredged Materials Disposal,* EEDP-04-8, Waterways Experiment Station, Vicksburg, Miss., 1988.

Design and Construction of Covers for Solid Waste Landfills, EPA/600/2-79/165, Municipal Environmental Research Laboratory, Cincinnati, Ohio, 1979.

An Overview of Sediment Quality in the United States, U.S. Environmental Protection Agency, Office of Water Regulation and Standards, Washington, D.C., June 1987.

Handbook: Groundwater, Technology Transfer Document, EPA/625/6-87/016, Cincinnati, Ohio, 1987.

Technical Guidance Document: Geosynthetic Design Guidance for Hazardous Waste Landfill Cells and Surface Impoundments, EPA/600/S2-87/097, Washington, D.C., 1987.

Technical Guidance Document: Final Covers on Hazardous Waste Landfills and Surface Impoundments, EPA/530-SW-89-047, Washington, D.C., 1989.

Report on Great Lakes Confined Disposal Facilities, U.S. Environmental Protection Agency, EPA/905/9-90/003, Region 5, Aug. 1990.

Design and Construction of RCRA/CERCLA Final Covers, EPA/625/4-91/025, Office of Research and Development, Washington, D.C., 1991.

Guidance on Selecting Techniques for Remediating Contaminated Sediment, Draft, Risk Reduction Engineering Laboratory, Cincinnati, Ohio, 1991.

Technical Guidance Document: Construction Quality Management for Remedial Action and Remedial Design Waste Containment Systems, EPA/540/R-92/073, Washington, D.C., 1992.

Draft Technical Guidance Document: Quality Assurance and Quality Control for Waste Containment Facilities, Hazardous Waste Engineering Research Laboratory, Cincinnati, Ohio, 1992.

Evaluating Environmental Effects of Dredged Material Management Alternatives—A Technical Framework, EPA842-B-92-008, Washington, D.C., 1992.

J.C. Evans, R.C. Chaney, and H.Y. Fang, *Influence of Pore Fluid on Clay Behavior,* Fritz Engineering Laboratory Report No. 384.14, LeHigh University, Bethlehem, Pa., 1981.

W.B. Ezell, Jr., ed., *An Investigation of Physical, Chemical, and/or Biological Control of Mosquitos in Dredged Material Disposal Areas,* Technical Report D-78-48, U.S. Army Engineer Waterways Experiment Station, Vicksburg, Miss., 1978.

O.G. Fitzhugh, A.H. Nelson, E.P. Laugh, *Arch. Ind. Hyg. Occup. Med.* **2,** 433–442 (1980).

A.K. Furr, G.S. Stoewsand, C.A. Bache and D.J. Liske, *Archive Environ. Health,* **26,** 67–91 (1976).

L.P. Gough, *Element Concentrations Toxic to Plants, Animals, and Man,* U.S. GPO, Geological Survey Bulletin, Washington, D.C., 1979, p. 1466 (as cited in ATSDR 1989).

A. Haliburton, *Guidelines for Dewater/Densifying Confined Dredged Material,* Technical Report No. DS-78-11, Waterways Experiment Station, Vicksburg, Miss., 1978.

Handbook-Remediation of Contaminated Sediments, EPA/625/6-91/028, Center for Environmental Research Information, Office of Research and Development, U.S. Environmental Protection Agency, Cincinnati, Ohio, Apr. 1991, 44p.

J.E. Harrison and L.C. Chisholm, *Identification of Objectionable Environmental Conditions and Issues Associated with Confined Disposal Areas,* Contract Report D-74-4, U.S. Army Engineer Waterways Experiment Station, Vicksburg, Miss., 1974.

W. Harrison, and co-workers, *Abatement of Malodors at Confined Dredged Material Disposal Sites,* Contract Report D-76-9, U.S. Army Engineer Waterways Experiment Station, Vicksburg, Miss., 1976.

D.E. Johnson, E.W. Kienholz, J.C. Baster, E. Sprangler, and G.M. Ward, *J. Anim. Sci.* **52** (1), 108–114 (1981).

A. Kabata-Pendias and H. Pendias, *Trace Elements in Soils and Plants,* CRC Press, Boca Raton, Fla., 1984, 315 pp.

M. Kelly, *Mining and the Freshwater Environment,* Elsevier Applied Science, New York, 1988.

R.J. Krizek and Salem A.M., *Behavior of Dredged Material in Diked Containment Areas,* Technical Report No. 5, Dept. Civil Engineering, Northwestern University, Evanston, Ill., 1974.

R.J. Krizek and co-workers, *Investigation of Effluent Filtering Systems for Dredged Material,* Northwestern University, Evanston, Ill., 1976.

R.J. Krizek and F. Somogyi, "Perspectives on Modeling Consolidation of Dredged Materials," *Proceedings of Symposium Sediment Consolidation Models—Predictions and Validations,* ASCE, New York, 1984.

D.S. Landeen, *Animal Intrusion Status Report for Fiscal Year 1990,* WHC-EP-0398, Westinghouse Hanford Co., Richland, Wash., 1991.

H.V. Leland and J.S. Kuvabara, "Trace Metals" in G.M. Rand and S.R. Petrocelli, eds., *Fundamentals of Aquatic Toxicology: Methods and Applications,* Hemisphere Publishing Corp. New York, 1985.

A.J. MacLean and A.J. Dekker, *Can. J. Soil Sci.* **58,** 381–369 (1976).

Consolidated Fact Sheet, Background and Technical Reference, National Nutrients, Las Vegas, Nev., 1988.

M.R. Palermo, *Interim Guidance for Predicting the Quality of Effluent Discharged from CDFs,* Tech Note EEDP-04-1, U.S. Army Waterways Experiment Station, Vicksburg, Miss., 1984.

M.R. Palermo and E.L. Thackston, *J. Environ. Eng.* **114**(6), (Dec. 1987).

F.P. Perera and A.K. Ahmed, *Respirable Particles, Impact of Airborne Fine Particles on Health and Environment,* Ballinger Publishing Co., Cambridge, Mass., 1979.

S.A. Pranger and P.R. Schroeder, *Dye Tracer Studies at the Kenosha, Manitowoc, Milwaukee, and Kewaunee Harbors Confined Disposal Facilities,* Misc. Paper D-86-4, Dept. of Army Waterways Experiment Station, Vicksburg, Miss., 1986.

P.R. Schroeder, *Chemical Clarification Methods for Confined Dredged Material Disposal,* Technical Report D-83-2, Dept. of Army, Waterways Experiment Station, Vicksburg, Miss., 1983.

P.R. Schroeder and co-workers, *Verification of the Hydrologic Evaluation of Landfill Performance (HELP) Model Using Field Data,* EPA/600/2-87-050, EPA Hazardous Waste Engineering Laboratory, Cincinnati, Ohio, 1987.

P.R. Schroeder, and co-workers, *Hydrologic Evaluation of Landfill Performance (HELP) Model: Documentation for Version 2,* DOA Waterways Experiment Station, Vicksburg, Miss., 1988.

R.P. Sharma, J.C. Street, M.P. Verma, and J.L. Shupe, *Environ. Health Perspec.* **28,** 59–66 (1979).

C. Stohr and co-workers, *Photogram. Eng. Remote Sens.* **53**(11), (1987); American Society for Photography and Remote Sensinf, Falls Church, Va.

U.S. Dept. of Agriculture (USDA), Peer Review, *Standards for the Disposal of Sewage Sludge,* U.S. EPA Proposed Rule 40, Parts 257 and 503, Washington, D.C., 1969.

W. Van Driel and co-workers, *Comparison of the Heavy-Metal Uptake of Cyperus Esculentus of Agronomic Plants Grown on Contaminated Sediments,* Misc. Paper D-83-1, U.S. Army Engineer Waterways Experiment Station, Vicksburg, Miss., 1983.

P.H. Williams, J.D. Shenk, and D.E. Baker, *J. Environ. Qual.* 7(3), 450–454 (1978).

SELENIUM IMMOBILIZATION IN WASTEWATER REMEDIATION

BRUCE A. MANNING
U.S. Department of Agriculture
Riverside, California

Remediation of wastewaters that contain high levels of selenium (Se) and other potentially toxic elements is important for protecting water quality. Selenium can be a contaminant in coal fly ash waste, oil refinery effluent, and saline agricultural drainage water and exists as soluble selenate (Se(VI) or SeO_4^{2-}) and/or selenite (Se(IV) or $HSeO_3^-$) oxyanions. After leaching from soils or solid wastes, the mobility of soluble Se is controlled by redox environment and adsorption reactions at the mineral–water interface. The Se(IV) species is strongly retained in soil by adsorption, whereas Se(VI) tends to be problematic because it is less strongly adsorbed and highly mobile. Common soil mineral components which adsorb anions from solution with high affinity include both crystalline and poorly crystalline Fe and Al oxides.

The work presented here is an investigation of the effectiveness of *in situ* precipitation of ferric oxyhydroxide (FeOOH) in removing Se(VI) from solution. Batch experiments containing Se(VI) were conducted to test the immobilization characteristics of *in situ* precipitation of FeOOH from both Fe(II) and Fe(III) starting compounds. The solid-phase FeOOH precipitates were then selectively extracted and dissolved in order to evaluate the stability of the resulting FeOOH–Se complexes. Precipitation of FeOOH from ferric chloride ($FeCl_3$) removed nearly 100% of Se(VI) in 6.3×10^{-6} *M* Se(VI)/L solutions at Fe(III) starting concentrations greater than 10^{-3} *M*. Selenate which was removed from solution by *in situ* FeOOH precipitation was resistant to extraction by phosphate (PO_4) indicating that Se(VI) had been bound internally within FeOOH particles (occluded) during the precipitation process. Ferrous iron starting materials ($FeSO_4$ and $FeCl_2$) partially reduced Se(VI) to selenite (Se(IV)), although this process did not account for all the Se(VI) removed. Both Se(VI) and Se(IV) were recovered from the FeOOH solid phase by PO_4 extraction and $HClO_4$/HCl acid dissolution after *in situ* precipitation of FeOOH from Fe(II). The Se(VI) removal process by Fe(II) oxidation was a combination of Se(VI) occlusion, Se(VI) reduction to Se(IV), and adsorption.

FACTORS CONTROLLING SELENIUM IMMOBILIZATION IN THE ENVIRONMENT

Selenium is a potentially toxic trace element which occurs naturally in soils and geologic source materials. Levels of Se tend to be highest in marine shale parent material (1), and can be mobilized by oxidation of metal selenosulfides (eg, FeSeS) to form soluble Se(VI) (2,3) which is then leached from the soil or sediment profile. Agricultural drainage waters, coal fly ash leachate, and certain oil refinery wastewaters that contain adverse concentrations of dissolved Se(IV) and Se(VI) can threaten surface and groundwater quality because of the potential toxicity of these ions to both wildlife and humans.

The overall mobility of Se in soil and sediment depends on the prevailing redox potential of the system (4,5). In oxidizing environments, such as well-aerated soil and surface waters, the Se(VI) species predominates. Mildly reducing conditions favor the stability of Se(IV), which is water soluble but strongly bound to soil and sediment mineral surfaces. Redox transformations between the soluble Se(VI) and Se(IV) species such as

$$SeO_4^{2-} + 4\,H^+ + 2\,e \rightleftharpoons H_2SeO_3 + H_2O \qquad (1)$$

are common in soils and surface waters, but are dependent on such factors as pH, kinetic limitations, and microbial activity in the system. In strongly reducing environments, which can occur in flooded soils, anoxic sediments, and groundwaters, insoluble species of Se such as elemental or metallic Se(0) and selenide (Se(−II)) predominate. Selenide associates strongly with dissolved ferrous iron to form solid-phase selenopyrite-like compounds by the following reaction:

$$HSe^- + Fe^{2+} \rightleftharpoons FeSe\downarrow + H^+ \qquad (2)$$

Selenium is also incorporated into plant biomass by the uptake of soluble Se followed by the formation of selenium-containing amino acids such as selenocysteine and selenomethionine (6). Reducing conditions and plant uptake of Se are conditions under which Se is temporarily immobilized.

REMEDIATION OF SOLUBLE SELENIUM IN WASTEWATER

Several remediation techniques have been tested at various scales to remove soluble Se from wastewaters. These techniques include phytoremediation, microbial volatilization, microbial reduction, and *in situ* precipitation. Phytoremediation and microbial volatilization take advantage of the ability to biotransform dissolved Se oxyanions into volatile species such as dimethylselenide (DMSe) and dimethyldiselenide (DMDSe) which are subsequently outgassed from the system. Microbial reduction involves biotransformation of Se oxyanions to insoluble inorganic Se solid phases such as Se(0) or FeSe. The use of *in situ* or chemical treatments attempts to minimize the migration of soluble Se by immobilization.

Plants are capable of volatilizing soil Se at rates as high as 16.2 g Se/(acre · day) (7) although investigations of the use of plants for remediation of Se in wastewater are still preliminary. The formation of gaseous Se in plants which accumulate Se such as *Astragalus racemosus* was first shown in the 1960s (8). Later work (9) identified the gas formed by Se accumulators as DMDSe, although nonaccumulator plants such as cabbage produce primarily DMSe (10). The proposed biochemical mechanism of DMSe formation in the plant leaf is the enzymatic cleavage of Se–methyl selenomethionine selenonium salt ($DMSe^+$–R). More recent work (11) has shown that plant roots are the primary site of production of volatile Se which may involve a coupled plant–microbial process in the root rhizosphere.

Microbial volatilization of Se has been extensively tested on a field scale and has been applied to remediation Se-laden soil and evaporation pond water (12,13). The microorganisms primarily responsible for volatilization of Se are soil fungi (14,15). The *in situ* biomethylation technique for maintaining the soluble Se concentration below the hazardous waste criterion of 1.0 ppm was tested at Kesterson Reservoir, a 520-hectare evaporation pond in California's Central Valley which is now drained (16). It was concluded (12,13) that ammendment with protein such as casein increased the rate of Se volatilization. However, more work is necessary to determine the optimum conditions for microbial volatilization of Se such as ammendment rates with co-enzymes and methyl donors, aeration, and microbial innoculants.

Anaerobic bacterial reactors were tested on a pilot scale and were successful in lowering Se concentrations in drainage water from 300 to 500 ppb down to 14 to 40 ppb (16). Anaerobic bacterial and microalgal–bacterial reactors are based on the principle that bacterial respiration in anoxic sludge blankets causes the reduction of Se(VI) to less soluble Se(IV) and Se(0). Often, bacterial and microalgal treatments which are designed to reduce Se(VI) to Se(IV) and Se(0) are combined with a final polishing step by ion exchange or *in situ* coprecipitation with ferric hydroxide formed from $FeCl_3$. Abiotic methods have also shown promise in removing soluble Se from drainage water including reverse osmosis (qv) and ion exchange.

The use of Fe oxide to remove Se from contaminated waters is of particular interest because of its simplicity, and it has been successfully used in other remediation applications such as attenuating heavy metals in soil and mine wastes (17,18). Coprecipitation of Se(IV) and Se(VI) with FeOOH has potential for remediating adverse concentrations of Se in wastewaters and has been applied to Se immobilization in evaporation pond sediments on a bench scale (19). Removal of soluble Se(VI) by reaction with ferrous hydroxide ($Fe(OH)_2$) has been extensively investigated in a series of reports submitted to the San Joaquin Valley Drainage Program by U.S. Bureau of Reclamation Scientists (2220–24). These workers concluded that the reaction between $Fe(OH)_2$ and Se(VI) produces elemental Se, Fe oxide, sodium hydroxide, and water. Elemental Se is then captured within the Fe oxide precipitate. Field-scale application of this technique concluded that the efficiency of the removal/immobilization reaction was sensitively dependent on wastewater components which inhibit the reaction, such as dissolved oxygen, nitrate, silicate, bicarbonate, hardness, trace metals, and phosphates. Precipitation of Fe oxide is a process which occurs naturally in soil and sediment under conditions of Fe(II) oxidation and Fe(III) hydrolysis as well as in zero-valent iron reactive barriers.

REACTIONS OF SELENIUM WITH FERRIC OXYHYDROXIDE

Of the three polymorphs of FeOOH, ie, goethite (α-FeOOH), akaganite (β-FeOOH), and lepidocrosite (γ-FeOOH), goethite is the most common in soils and forms from both the oxidation of Fe(II) and hydrolysis of Fe(III) (25). Hydrolysis of dissolved $Fe(H_2O)_6^{3+}$ can form any of several Fe(III) oxide, hydroxide, or oxyhydroxide phases of varying crystallinity, depending on such factors as the starting Fe concentration, temperature, pH, anion environment, and precipitation time (25,26). Goethite forms predominantly when the starting Fe(III) concentration is $<0.03\ M$ Fe(III) (26) and during rapid oxidation of Fe(II) (27). During the *in situ* FeOOH precipitation process, Se(IV) adsorbs on fresh surface and may become bound on internal FeOOH surfaces (occlusion) as subsequent layers of FeOOH are precipitated (28). Intraparticle diffusion followed by adsorption is a dominant mechanism of $HSeO_3^-$ sorption in porous alumina (29) suggesting that $HSeO_3^-$ binding to internal sites in oxide particles is also an important process.

Adsorption of the Se(IV) on minerals, especially the oxides of Fe, has been extensively investigated (30–34). These studies have concluded that Se(IV) is adsorbed on Fe and Al oxide minerals by a pH-dependent inner-sphere ligand-exchange reaction. Inner-sphere complexes contain covalent chemical bonds which form between ions in solution and the solid surface when the ion penetrates the coordination shell of the surface metal ion (in this case Fe):

$$\equiv Fe{-}OH_2^+(s) + HSeO_3^-(aq) \rightleftharpoons \equiv Fe{-}O{-}SeO_2H(s) + H_2O \tag{3}$$

In this reaction $\equiv Fe{-}OH_2^+(s)$ represents a protonated hydroxyl group on the FeOOH surface. Electrostatic or outer-sphere surface complexes retain a water monolayer between the ion and the oppositely charged surface:

$$\equiv Fe{-}OH_2^+(s) + HSeO_3^-(aq) \rightleftharpoons \equiv Fe{-}OH_2^+(s)\cdots HSeO_3^- \tag{4}$$

Both IR (33) and extended X-ray absorption fine structure (EXAFS) spectra (35) indicate that Se(IV) adsorbed on Fe oxide surfaces forms binuclear bridging complexes, such as

$$2{\equiv}Fe{-}OH_2^+(s) + HSeO_3^-(aq) \rightleftharpoons Fe{-}OSe(O)OFe(s) + 2\,H_2O + H^+(aq) \tag{5}$$

Iron oxides exhibit variable surface charge by undergoing surface protonation and deprotonation reactions similar to weak acids:

$$\equiv Fe{-}OH(s) \rightleftharpoons \equiv FeO^-(s) + H^+(aq) \tag{6}$$

$$\equiv Fe{-}OH(s) + H^+(aq) \rightleftharpoons \equiv Fe{-}OH_2^+(s) \tag{7}$$

This partially accounts for the pH-dependence of ion adsorption on oxide surfaces. Anions such as $HSeO_3^-$ and SeO_4^{2-} undergo stronger electrostatic attraction to the positively charged surface $\equiv Fe{-}OH_2^+$ sites at low pH. The H_2O molecule is also a better leaving group than OH^- following the specific adsorption reaction (as in eq. 3).

Adsorption reactions of Se(VI) with soils and reference minerals have also been investigated under controlled laboratory conditions. The Se(VI) species is weakly adsorbed on both soils (36,37) and Fe minerals (30,31), although the chemical mechanisms which explain the apparent weak association of Se(VI) with these materials are still not clear. Using EXAFS it has been shown that the SeO_4^{2-} ion retains its primary waters of hydration when it forms an electrostatic, outer-sphere surface complex with goethite (α-FeOOH) (35):

$$\equiv Fe{-}OH_2^+(s) + SeO_4^{2-}(aq) \rightleftharpoons \equiv Fe{-}OH_2^+(s)SeO_4^{2-} \tag{8}$$

In contrast, in another study (38), which also used EXAFS data, it was concluded that Se(VI) forms inner-sphere, binuclear bridging complexes with α-FeOOH and hydrous ferric oxide surfaces similar to $HSeO_3^-$, HPO_4^{2-}, and arsenate ($HAsO_4^{2-}$). Most studies, however, have concluded that the SeO_4^{2-} ion is more weakly bound to mineral surfaces and more readily leached from soils than the SeO_3^{2-} ion and thus is the primary species of concern for remediation technology.

The main objectives of the experimental work presented in this article were to determine the partitioning and identity of Se reaction products after allowing *in situ* precipitation of FeOOH to proceed in solutions containing Se(VI). Both Fe(II) and Fe(III) were used as Fe reactants in an attempt to separate the effects of chemical reduction of Se(VI) by Fe(II) and the effects of adsorption and occlusion which were expected to occur in both the Fe(II) and Fe(III) systems. Possible Se reaction products in the Se(VI)/Fe(II)/$NaClO_4$/H_2O system were soluble Se(IV), adsorbed Se(IV) and/or Se(VI), occluded Se(IV) and/or Se(VI), and precipitated Se(0). The speciation of the Se reaction products was investigated by selective extraction of the solid-phase FeOOH formed. Occluded Se is resistant to exchange with competing adsorbates such as HPO_4^{2-} and is desirable from the standpoint of environmental remediation because it is immobilized in the solid phase.

EXPERIMENTAL SECTION

General Information

All chemicals in this study were reagent grade. Distilled deionized (DI) water was produced using Megapure System DF deionizing cartridges. Experiments were carried out as batch reactions in 40 mL polycarbonate Oak Ridge-type centrifuge tubes. Separate Fe(II) and Fe(III) starting solutions were made from $FeCl_2 \cdot 4H_2O$, $FeCl_3 \cdot 6\,H_2O$, $FeSO_4 \cdot 7\,H_2O$, and $Fe_2(SO_4)_3 \cdot nH_2O$ salts dissolved in 0.1 *M* $NaClO_4$ to make 0.1 *M* Fe. Iron(II) solutions were unstable and visible evidence of Fe(III) precipitation occurred in 2–3 h. In all experiments, a 1,10-phenanthroline colorimetric method (39) was used to measure residual Fe(II) after *in situ* FeOOH precipitation. Crystalline goethite was prepared according to a modified method (40) by titration of a 0.45 *M* $Fe(NO_3)_3$ solution with KOH to pH > 12.5 and reacting for 7 d. The solid material was washed with DI water and stored as a suspension. An aliquot of the suspension was dried and the identity of the solid was found to be goethite by X-ray diffraction.

Selenium Analysis

In all experiments, Se was determined by hydride generation atomic absorption spectrophotometry (HGAAS) by established methods (3,41,42). Analyses were performed on a Perkin-Elmer Model 2100 spectrophotometer at a wavelength of 196.0 nm with an electrodeless discharge lamp operating at 4.5 watts. Hydrogen selenide (H_2Se) was stripped from solution with Ar (flow rate 45 mL/min) after reaction of Se(IV) with sodium borohydride ($NaBH_4$) using a Varian vapor generation accessory (VGA 76) and atomized by a quartz tube heated with an acetylene flame. Reagent and sample flow rates were 1 mL/min (0.16 *M* $NaBH_4$), 1 mL/min (10 *M* HCl), and 7 mL/min (sample). The sensitivity of the technique was 0.5 mg/L Se(IV) in sample solutions.

Speciation of Se was determined by subdividing sample solutions into selective treatments for Se(IV) and total Se (Se(IV + VI)). Selenite was determined by acidifying aliquots of sample to 4 *M* HCl and analyzing directly by HGAAS within 1–2 h. Solutions of Na_2SeO_3 and Na_2SeO_4 were used as standards. Preliminary work verified that Se(VI) did not react with $NaBH_4$ to form H_2Se and, thus, the technique was selective for Se(IV). To determine Se(IV + VI), 5 mL sample aliquots solutions and 5 mL concentrated HCl were transferred to 50 mL graduated glass digestion tubes along with 0.2 mL of 2% ammonium persulfate ($NH_4S_2O_8$). The tubes were heated in a digestion block at 100°C for 30 min to first oxidize all Se to Se(VI) followed by quantitative reduction of Se(VI) to Se(IV). The digest was brought to volume with 4 *M* HCl for HGAAS analysis. Total Se was then determined by HGAAS and Se(VI) was calculated as the difference between the Se(IV + VI) and the Se(IV) analyses.

Experiments with Se(VI) Solutions

Typical Se(VI) concentrations in agricultural drainage water between 400 and 1400 mg/L have been reported (3) and thus a starting concentration of 500 mg/L (6.3×10^{-6} *M* Se(VI)) was chosen. In centrifuge tubes, solutions of 30 mL total volume containing 6.3×10^{-6} *M* Se(VI) in 0.1 *M* $NaClO_4$ with varying Fe(II) or Fe(III) concentration (10^{-4} to $10^{-1.1}$ *M* Fe) in 30 mL total volume were shaken for either 14 d (Fe(III)) or 28 d (Fe(II)). In order to induce FeOOH precipitation, microliter quantities of 1 *M* $NaHCO_3$ were pipetted into the tubes daily. The quantity of $NaHCO_3$ added was 3 mol H^+/mol Fe, equivalent to the expected proton production during Fe(III) hydrolysis:

$$Fe(H_2O)_6^{3+} \rightleftharpoons Fe(OH)_3 \downarrow + 3\,H_2O + 3\,H^+ \qquad (9)$$

After the reaction was complete the tubes were briefly centrifuged at low rpm (1300 g) to minimize compaction of FeOOH particles. Supernatant solutions were filtered (0.2 μm) and designated extract #1, which was analyzed for Se(IV), Se(IV + VI), Fe(II), and pH. The FeOOH particles were rinsed once with approximately 2 mL of DI water and the rinsate added to extract #1. The FeOOH particles were then resuspended in 30 mL of 0.01 *M* K_2HPO_4 (pH 8). After shaking for 20 h, the tubes were briefly centrifuged, and the solution filtered (0.2 μm) and analyzed for Se(IV) and Se(IV + VI). This extract was designated extract #2 which contained PO_4-desorbable Se. The FeOOH particles were then dissolved in 4 *M* $HClO_4$/2 *M* HCl. After 2 h the mixtures were centrifuged (12,500 g, 10 min) to create extract #3 in which Se(IV) and Se(IV + VI) were determined as above.

RESULTS AND DISCUSSION

The amount of Se(VI) removed from solution by *in situ* precipitation of FeOOH depended on the Fe starting compound (Fig. 1). Selenate was removed from the reaction solution (extract #1) most effectively by $FeCl_3$. Over the $10^{-3.5}$ to $10^{-2.5}$ *M* Fe treatment range, FeOOH precipitation removed Se(VI) from solution at a rate of 3 mmol Se(VI) mol/Fe. Below starting concentrations of 10^{-2} *M*, the Fe(II) compounds did not remove appreciable Se(VI). The Se(VI) removal pattern for $FeCl_2$ was similar to that for $FeCl_3$, though the maximum concentration of Se(VI) associated with solid-phase FeOOH obtained from $FeCl_2$ was 0.3 mmol Se(VI) mol/Fe. Solid FeOOH formed from $FeSO_4$ resulted in 0.2 mmol Se(VI) mol/Fe. Unreacted Fe(II) measured in extract #1 of both the $FeCl_2$ and $FeSO_4$ systems was below the 1 mmol Fe(II)/L detection limit of the colorimetric procedure. The solubility of Fe(III) in equilibrium with goethite at pH 6 is less than 10^{-18} *M* (43) and thus the pH values of the final reaction mixtures for both the Fe(II) and Fe(III) starting materials (Fig. 2) were too high to maintain soluble Fe(III).

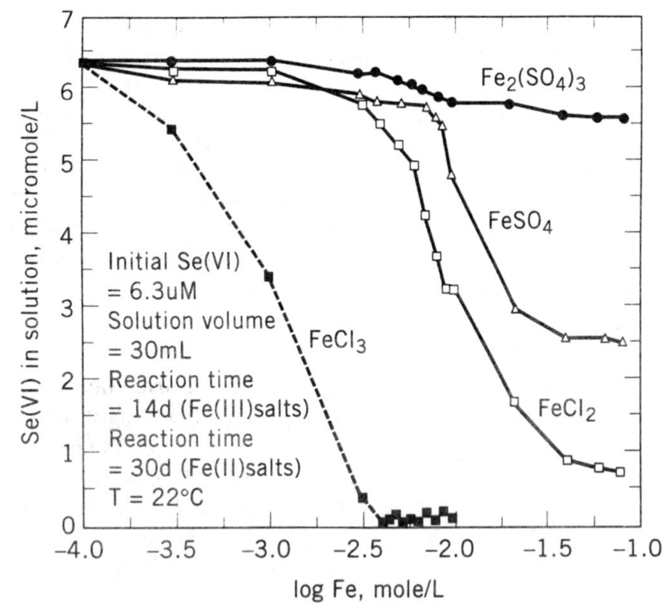

Figure 1. Effect of Fe starting concentration on the removal of soluble Se(VI) by *in situ* formation of FeOOH from Fe(III) and Fe(II) starting materials.

The system was considered to be at equilibrium when all Fe(II) had oxidized, hydrolyzed, and precipitated, and no further removal of Se(VI) was expected. Ferric sulfate did not remove appreciable Se(VI) (see Fig. 1) and for this reason was not further analyzed for the identity and location of Se reaction products. Nearly all of the original Se(VI) was accounted for and remained unreacted in the extract #1 solution over the range of $Fe_2(SO_4)_3$ treatments. The chemical similarity of the SeO_4^{2-} and SO_4^{2-} ions suggests that SO_4^{2-} competition inhibits SeO_4^{2-} adsorption and coprecipitation during FeOOH precipitation. The formation of $FeSO_4^+$ complexes at high SO_4^{2-} concentrations has been shown to interfere with polymerization of ferric oxyhydroxides (26).

The FeOOH solids formed in the $FeCl_3$, $FeCl_2$, and $FeSO_4$ treatments were further extracted using 0.01 *M* K_2HPO_4 to determine the exchangable Se reaction products. Total Se in the 0.01 *M* K_2HPO_4 extract (extract #2) of solid-phase FeOOH formed from $FeCl_3$, $FeCl_2$, and $FeSO_4$ is given in Figure 3. About 40% of the original Se(VI) was exchangable with the HPO_4^{2-} ion in the $FeCl_3$ system at the $10^{-2.5}$ *M* Fe(III). No Se(IV) was detected after determination of Se(IV)/Se(VI) speciation in extract #2 from the $FeCl_3$ system. Figures 4 and 5 indicate that chemical reduction of Se(VI) to Se(IV) had occurred in the Fe(II) systems and that both SeO_3^{2-} and

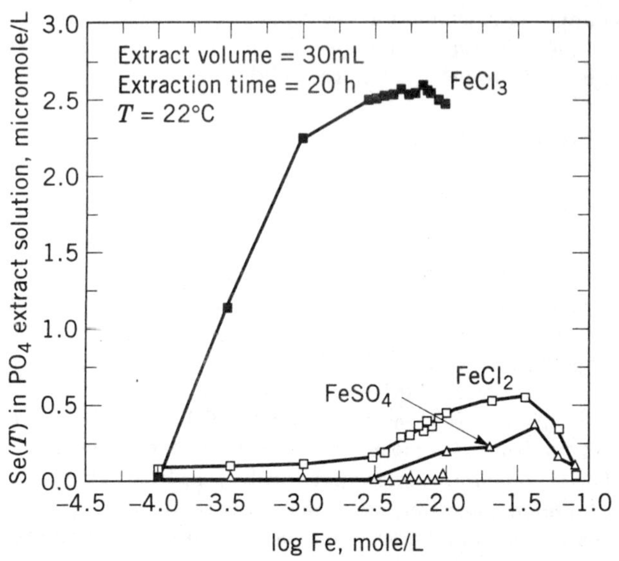

Figure 3. Total Se in phosphate extracts (extract #2) of FeOOH solids formed after *in situ* precipitation of FeOOH from Fe(III) and Fe(II) starting materials.

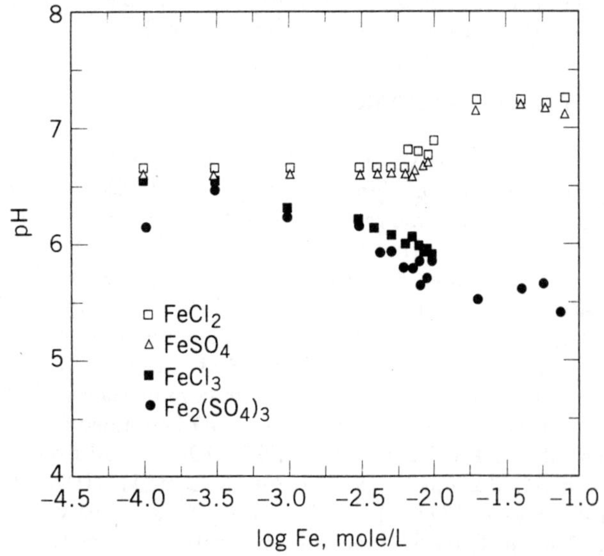

Figure 2. pH measured in Se(VI)/Fe reaction solutions (extract #1) after *in situ* formation of FeOOH from Fe(III) and Fe(II) starting materials.

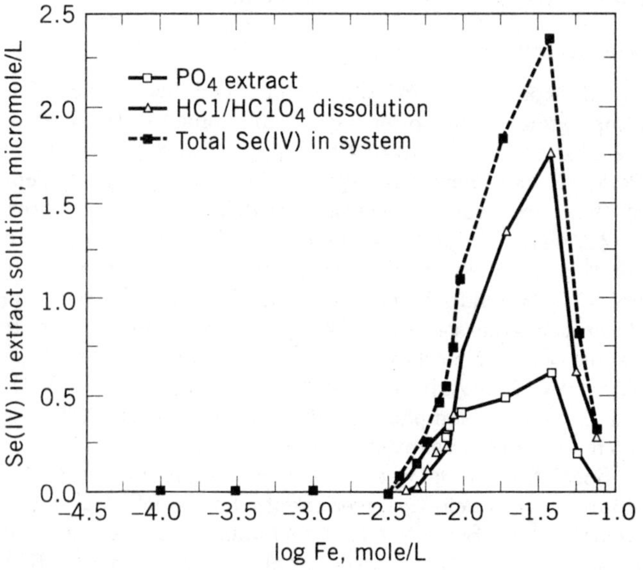

Figure 4. Selenite (Se(IV)) in selective extracts of FeOOH solids formed after *in situ* precipitation of FeOOH from $FeCl_2$.

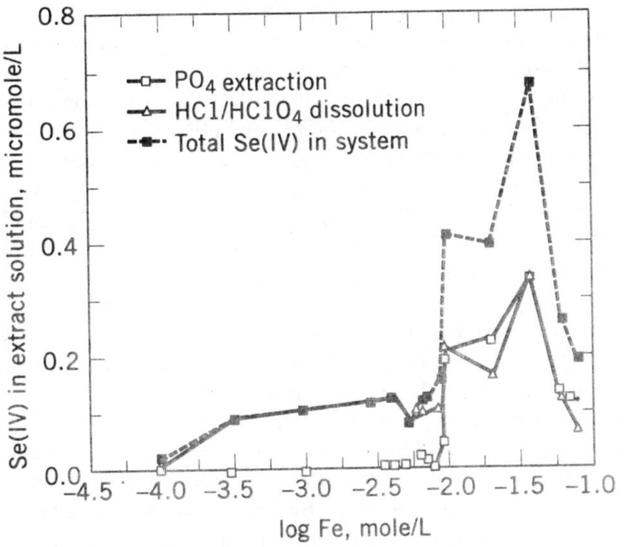

Figure 5. Selenite (Se(IV)) in selective extracts of FeOOH solids formed after *in situ* precipitation of FeOOH from $FeSO_4$.

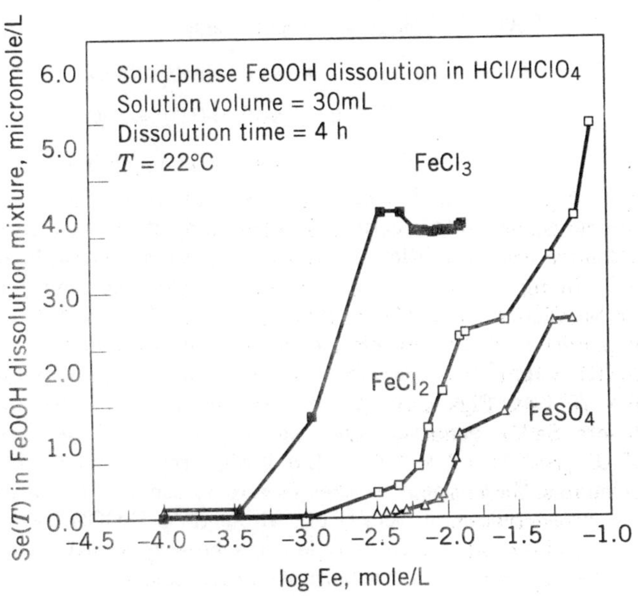

Figure 6. Total Se in 4 *M* $HClO_4$/2 *M* HCl dissolution mixtures (extract #3) of FeOOH solids formed after *in situ* precipitation of FeOOH from Fe(III) and Fe(II) starting materials.

SeO_4^{2-} ions were desorbed by the HPO_4^{2-} ion. The extraction efficiency reached a maximum at $10^{-1.5}$ *M* Fe and decreased with higher Fe concentrations suggesting that some Se was not available to exchange with HPO_4^{2-}. In both Fe(II) and Fe(III) treatments, a decline in PO_4-extractable Se in the high Fe treatments ($>10^{-2.5}$ *M*Fe) also indicated that occlusion within FeOOH solids had occurred. Ferrous sulfate reduced a maximum of 10% of the original SeO_4^{2-} in the $10^{-1.5}$ *M*$FeSO_4$ treatment (see Fig. 5). Sharp declines in Se(VI) recovered as Se(IV) occurred above the $10^{-1.5}$ *M* Fe concentration in both Fe(II) salts.

Total Se in acid dissolution mixtures (extract #3) is shown in Figure 6. The FeOOH dissolution process was rapid in the acid mixture and no solids were visible after 1.5 hours. Total Se in extract #3 corresponded with the Se(VI) removal profiles (see Fig. 1) for both Fe(II) and Fe(III) treatments. The $FeCl_2$ system (see Fig. 4) contained a large proportion of the original Se(VI) associated with the solid phase as a mixture of Se(VI) and Se(IV). About 30% of Se in extract #3 from the 10^{-2} *M* $FeCl_2$ system was Se(IV) suggesting that Fe(II) partially reduced Se(VI) to Se(IV) in this system. A small amount of Se(IV) (0.3 μM Se(IV)) was measured in extract #3 of the 10^{-2} *M* Fe(III) treatment (about 5% of the original Se(VI)). The Se(IV)/Se(VI) speciation in the extract indicated that greater than 95% of Se in the dissolution mixture was Se(VI) suggesting that Se(VI) was occluded within FeOOH precipitated from $FeCl_3$.

Large fluctuations in analytical recoveries of Se were associated with Fe treatments greater than $10^{-1.5}$ *M* Fe, reflecting differences in the FeOOH precipitation environment at varying Fe starting concentrations. As indicated in Figures 4 and 6, Se(IV) measured in extracts decreased rapidly above starting concentrations of $10^{-1.5}$ *M* Fe. Above Fe starting concentrations of about 10^{-2} *M* Fe, Se removal proceeded, but it became increasingly difficult to recover all of the original Se added. High concentrations of dissolved Fe(III) in extract #3 may also have complicated the HGAAS analysis of Se by interfering with the formation of H_2Se in the vapor generation reaction coil. Precipitation of FeOOH from $FeCl_3$ removed Se(VI) from solution and formed a surface-bound component of Se(VI) which was recovered from the solids by 0.01 *M* K_2HPO_4 extraction. Selenate that remained within the solid phase was operationally defined as occluded Se(VI). At Fe treatments concentrations greater than $10^{-2.3}$ *M* Fe, occlusion processes were more pronounced and effectively shielded surface-bound Se(VI) from exchange with the HPO_4^{2-} ion. No evidence was found for the reduction of Se(VI) to Se(0). Therefore, it was postulated that 6.3 μM Se(VI) solutions were partially reduced to Se(IV) by millimolar concentrations of Fe(II). Reduction of Se(VI) to Se(IV) by chemical coupling with Fe(II) oxidation was not, however, the dominant process in Fe(II)/Se(VI) systems. Occlusion of Se(VI) during the precipitation process accounted for the majority of Se(VI) removal from solution regardless of the oxidation state of the starting Fe compound.

Reduction of Se(VI) to Se(IV) is desirable because of the greater affinity of Se(IV) for the FeOOH surface. The standard Gibbs free energy change ($\Delta G°$) for the Se(VI) → Se(IV) reduction reaction in solution is strongly dependent on pH (44). When the Fe(II) oxidation half-reaction is coupled, Se(VI) reduction under alkaline conditions, the net reaction is thermodynamically unfavorable (44,45):

$$2\ Fe(II) \leftrightharpoons 2\ Fe(III) + 2\ e^- \ \Delta G° = +(35.51\ kcal/mol)rxn$$

$$SeO_4^{2-} + H_2O + 2\ e^- \leftrightharpoons SeO_3^{2-} + 2\ OH^- \ \Delta G° = -(2.31\ kcal/mol)rxn$$

$$2\ Fe(II) + SeO_4^{2-} + H_2O \leftrightharpoons 2\ Fe(III) + SeO_3^{2-} + 2\ OH^- \ \Delta G° = (+33.21\ kcal/mol)rxn$$

When coupled with Fe(II) oxidation half-reaction under acidic conditions, the calculated $\Delta G°_{rxn}$ for the Se(VI) → Se(IV) reduction is thermodynamically favorable (44,45)

$$2\ Fe(II) \leftrightharpoons 2\ Fe(III) + 2\ e^- \ \Delta G° = (+35.51\ kcal/mol)rxn$$

$$SeO_4^{2-} + 4\,H^+ + 2\,e^- \rightleftharpoons H_2SeO_3 + H_2O \quad \Delta G^\circ = (-53.04\ \text{kcal/mol})\text{rxn}$$

$$2\,Fe(II) + SeO_4^{2-} + 4\,H^+ \rightleftharpoons 2\,Fe(III) + SeO_3^{2-} + 2\,OH^- \quad \Delta G^\circ = (-17.52\ \text{kcal/mol})\text{rxn}$$

The acidity produced from the hydrolysis of Fe(III) cations during experiments was titrated with alkalinity (as $NaHCO_3$), although acidic conditions were probably established at intervals during the reaction period causing favorable conditions for Se(VI) reduction. The formation of Se(IV) from Se(VI) also depended on Fe(II) concentration and was greatest at $10^{-1.5}$ *M* Fe(II) where the Fe(II)/Se(VI) ratio was approximately 6×10^3 (see Figs. 4 and 5). The lowest Fe(II) concentration where Se(VI) reduction was observed was $10^{-2.5}$ *M* Fe(II) (Fe(II)/Se(VI) = 5×10^2) in the $FeCl_2$ system. Most metal selenates, for example $Fe_2(SeO_4)_3$, are too soluble to control the concentration of Se(VI) (4), although a FeOOH–Se(VI) outer-sphere adsorption complex has been proposed as the initial step in the Se(VI) adsorption mechanism (46).

Evidence for reduction of Se(VI) to Se(IV) by Fe(II) was shown by the Se(IV)/Se(VI) speciation in sequential extracts from FeOOH formed from Fe(II) starting compounds. Reduction of Se(VI) resulted in specifically adsorbed Se(IV) which was recovered in PO_4 extracts of the solids formed from Fe(II) (see Figs. 4 and 5). Selenite formed on reduction of Se(VI) was most likely adsorbed on FeOOH by an inner-sphere mechanism forming mono- and bidentate surface species. Increased Se(IV) adsorption on FeOOH at pH values below the point of zero charge of the oxide solid is due to anion attraction to positively charged surface $Fe{-}OH_2^+$ functional groups.

The crystal morphology of FeOOH tends to vary depending on precipitation environment (18). Iron oxides such as lepidocrosite (γ-FeOOH), one of the ferrihydrite minerals (eg, $Fe_5(HO_8)4H_2O$), or hematite (α-Fe_2O_3) may form under different remediation conditions. In the experiments presented here, there was evidence for the formation of poorly crystalline α-FeOOH, although the actual identity of the FeOOH phases which formed were not determined in all treatments because both Fe concentration and anion were varied. Crystal growth kinetics were also probably different between the high and low Fe starting concentrations as well as pH, although the final pH of the reaction was fairly uniform (see Fig. 2).

SUMMARY

It should be emphasized that selenium immobilization by *in situ* precipitation of FeOOH has not yet been tested on a large scale. Though bench-scale studies have investigated the chemical reactions of Se(IV) and Se(VI) with various Fe oxide phases, field-scale applications of this technology are still needed. The work presented here took a pure system, bench-scale approach to determine the mechanism of Se(VI) removal by *in situ* precipitation of FeOOH. Four different Fe salts ($FeCl_2 \cdot 4H_2O$, $FeCl_3 \cdot 6H_2O$, $FeSO_4 \cdot 7H_2O$, and $Fe_2(SO_4)_3 \cdot nH_2O$) were examined for their ability to remove Se(VI) from solution by forming solid-phase FeOOH from both dissolved Fe(II) and Fe(III). Attempts to neutralize and equalize pH in the reaction mixtures were successful and no detectable Fe remained in solution at the time of Se analysis.

Selenate removal from solution by *in situ* precipitation of FeOOH depended on the Fe/Se ratio, the Fe oxidation state, and the accompanying anion in the Fe starting material. Precipitation of FeOOH from $Fe_2(SO_4)_3 \cdot nH_2O$ did not remove Se(VI) from solution due to the large excess of SO_4^{2-} ion in the system which competed with SeO_4^{2-} for retention sites on FeOOH. Partial removal of Se(VI) by $FeSO_4$ was accomplished at Fe(II) starting concentrations greater than 10 *mM* Fe. The Cl^- ion in the $FeCl_3$ and $FeCl_2$ systems inhibited Se(VI) removal less than the SO_4^{2-} ion. It was concluded from selective extracts of the Fe(III) solids from $FeCl_3$ that Se(VI) had been occluded during the precipitation process.

Coupling of Fe(II) oxidation with Se(VI) reduction to Se(IV) accounted for approximately 10 and 35% of Se(VI) removal at $10^{-1.5}$ *M* Fe(II) starting concentration in the $FeSO_4$ and $FeCl_2$ systems, respectively. The remainder of the original Se(VI) that was removed from solution was bound in solid-phase FeOOH as Se(VI). There was no evidence in this study for the reduction of Se(VI) to Se(0) by chemical coupling with the Fe(II) oxidation reaction. Both Se(VI) and Se(IV) were removed from the Fe(III) solids by exchange with HPO_4^{2-} and acid dissolution of the solids with $HClO_4$/HCl. This suggested that Fe(II) is a weak reducing agent with respect to Se(VI) reduction and that the Se(VI) removal process was a combination of Se(VI) reduction to Se(IV) and Se(VI) occlusion.

BIBLIOGRAPHY

1. J. M. McNeal and L. S. Balistrieri, in L. W. Jacobs, ed., *Selenium in Agriculture and the Environment,* SSSA Special Publication No. 23, Soil Science Society of America, Madison, Wis., 1989, pp. 1–13.
2. G. R. Bradford and D. Bakhtar, *Environ. Sci. Technol.* **25,** 1704–1707 (1991).
3. T. S. Presser and I. Barnes, *Water Resources Investigations,* Rep. 84-4112, U.S. Geological Survey, Menlo Park, Calif., 1984.
4. M. A. Elrashidi and co-workers, *Soil Sci.* **144,** 141–152 (1987).
5. P. H. Masscheleyn, R. D. Delaune, and W. H. Patrick, Jr., *Environ. Sci. Technol.* **24,** 91–96 (1990).
6. A. Shrift, *Ann. Rev. Plant Physiol.* **20,** 475–494 (1969).
7. N. Terry and co-workers, *J. Environ. Qual.* **21,** 341–344 (1992).
8. B. G. Lewis, C. M. Johnson, and C. C. Delwiche, *J. Agric. Food Chem.* **14,** 638 (1966).
9. C. S. Evans, C. J. Asher, and C. M. Johnson, *Aust. J. Biol. Sci.* **21,** 13 (1968).
10. B. G. Lewis, Ph.D. dissertation, University of California, Berkeley, 1971.
11. A. M. Zayed and N. Terry, *J. Plant Physiol.* **143,** 8–14 (1994).
12. W. T. Frankenberger and U. Karlson, *In Situ Volitilization of Selenium from Evaporation Ponds,* a report prepared for the San Joaquin Valley Drainage Program, Calif., 1990.
13. U. Karlson and W. T. Frankenburger, Jr., *Soil Sci. Soc. Am. J.* **53,** 749–753 (1989).
14. R. W. Fleming and M. Alexander, *Appl. Microbiol.* **24,** 424–429 (1972).
15. L. Barkes and R. W. Fleming, *Bull. Environ. Contam. Toxicol.* **12,** 308–311 (1974).
16. U.S. Dept. of the Interior, *A Management Plan for Agricultural Subsurface Drainage and Related Problems on Westside San Joaquin Valley,* Final Report of the San Joaquin Valley Drainage Program, California Dept. of Water Resources, Sacramento, Calif., 1990.
17. N. C. Scrivner and co-workers, *Ground Water Monit. Rev.* **6,** 53–58 (1986).

18. D.W. Blowes and co-workers, *Geochim. Cosmochim. Acta* **55,** 965–978 (1991).
19. B.A. Manning and R.G. Burau, *Environ. Sci. Technol.* **29,** 1526–1533 (1995).
20. C.D. Moody and co-workers, *Bureau of Reclamation Experimental Results for Removing Selenate from Agricultural Drainage Waters,* a report submitted to the San Joaquin Valley Drainage Program, USBR Engineering Research Center, Calif., 1988.
21. C.D. Moody, G.L. Howard, and A.P. Murphy, *Selenium Removal from Agricultural Drainage Waters by Chemical Reduction,* a report submitted to the San Joaquin Valley Drainage Program, USBR Engineering Research Center, Calif., 1988.
22. A.P. Murphy, *Ind. Eng. Chem. Res.* **27,** 187–191 (1988).
23. M.K. Price and R.J. Eisenhauer, *Selenium Process Testing Report,* a report submitted to the San Joaquin Valley Drainage Program, USBR Engineering Research Center, Calif., 1988.
24. L.H. Rowley, *Selenium Pretreatment Study Report,* a report submitted to the San Joaquin Valley Drainage Program, USBR Engineering Research Center, Calif., 1989.
25. U. Schwertmann and R. Taylor, in J.B. Dixon and S.B. Weeds, eds., *Minerals in Soil Environments,* 2nd ed., Soil Science Society of America, Madison, Wis., 1989, pp. 379–438.
26. S. Music, Z. Orehovec, and S. Popovic, *J. Mat. Sci.* **29,** 1991–1998 (1994).
27. R. Taylor and U. Schwertmann, *Clay Min.* **10,** 299–310 (1974).
28. B.A. Manning, Ph.D. dissertation, University of California, Davis, June 1993.
29. L. Papelis, *Environ. Sci. Technol.* **29,** 1526–1533 (1995).
30. L.S. Balistrieri and T.T. Chao, *Geochim. Cosmochim. Acta* **54,** 739–751 (1990).
31. L.S. Balistrieri and T.T. Chao, *Soil. Sci. Soc. Am. J.* **51,** 1145–1151 (1987).
32. F.J. Hingston, A.M. Posner, and J.P. Quirk, *Adv. Chem. Ser.* **70,** 82–90 (1968).
33. R.L. Parfitt, R.L., and J.D. Russell, *J. Soil Sci.* **28,** 297–305 (1977).
34. M.A. Anderson and D.T. Malotky, *J. Colloid Interface Sci.* **72,** 413–427 (1979).
35. K.F. Hayes and co-workers, *Science* **238,** 783–786 (1987).
36. R.H. Neal and G. Sposito, *Soil Sci. Soc. Am. J.* **53,** 70–74 (1989).
37. M.H. Alemi, D.H. Goldhamer, and D.R. Nielsen, *J. Environ. Qual.* **17,** 608–613 (1988).
38. A. Manceau and L. Charlet, *J. Colloid. Interface Sci.* **168,** 87–93 (1994).
39. *Standard Methods for the Examination of Water and Wastewater,* 16th ed., American Public Health Association, Washington, D.C., 1985, pp. 215–221.
40. R.J. Atkinson, A.M. Posner, and J.P. Quirk, *J. Inorg. Nuc. Chem.* **30,** 2371–2381 (1968).
41. O. Weres, A.R. Jaouni, and L. Tsao, *Appl. Geochem.* **4,** 543–563 (1989).
42. R. Fujii, S.J. Deverel, and D.B. Hatfield, *Soil Sci. Soc. Am. J.* **52,** 1274–1283 (1988).
43. W.L. Lindsay, *Chemical Equilibria in Soils,* John Wiley & Sons, Inc., New York, 1979, p. 132.
44. C.E. Cowan, *Review of Selenium Thermodynamic Data,* Electric Power Research Institute, EPRI EA-5655, Menlo Park, Calif., 1988.
45. W. Stumm and J.J. Morgan, *Aquatic Chemistry,* 2nd ed., John Wiley & Sons, Ltd., Chichester, U.K., 1980, pp. 418–503.
46. P. Zhang and D.L. Sparks, *Environ. Sci. Technol.* **24,** 1848–1856 (1990).

SELENIUM SPECIATION IN NATURAL WATERS

KATTA J. REDDY
University of Wyoming
Laramie, Wyoming

Selenium (Se) in natural waters (eg, soil water, groundwater) exist in the form of free ionic species (eg, selenate (SeO_4^{2-}) and selenite (SeO_3^{2-})), inorganic complexes (eg, $CaSeO^0{}_4$, $MgSeO^0{}_3$), and dissolved organic carbon complexes (DOC–Se). Speciation is defined as the partition of dissolved Se in natural waters into free ionic species and complexes. Knowledge of the speciation of dissolved Se in natural waters is fundamental for understanding Se biogeochemical processes such as biological uptake (eg, uptake by microorganisms and plants), adsorption, precipitation, mobility, and remediation. For example, studies have shown that SeO_4^{2-} is mobile than SeO_3^{2-} in natural waters because it is not adsorbed by the mineral surfaces. The purpose of this article is to review the present knowledge of the speciation of dissolved Se in natural waters, particularly soil waters and groundwaters. This article emphasizes Se sources, importance of dissolved Se speciation, factors affecting dissolved Se speciation, and methods for analyzing the dissolved Se speciation in natural waters.

BACKGROUND INFORMATION

Globally selenium (Se) is an important elements because it has wide application. Selenium was discovered by Berzelius in 1817. The Se atomic number is 34 and it is in the Group VI(A) elements of the Periodic Table. The Se atomic weight is 78.96. Selenium is a required micronutrient for animals and humans; excessive Se, however, can be toxic. Its requirement for plants is not clearly understood. The range between Se toxicity and deficiency for animals and humans is rather narrow. Selenium toxicity problems have been documented as early as the twelfth century. For example, Marco Polo traveling in China in 1295 was probably describing toxicity symptoms of Se in livestock. In 1560, Father Simon Pedro described what was probably Se toxicity symptoms in humans in Colombia. In 1857, T.W. Madison, U.S. Army Surgeon, described the suspected Se toxicity symptoms in horses near Fort Randall, South Dakota. In China, two types of Se endemic human diseases, cardiomyopathy (Se deficiency) and selenosis (Se toxicity) were reported (1).

Selenium occurs naturally in soils and groundwaters. The main geological source of Se in soils is derived from cretaceous-age shales. However, soils derived from igneous rocks are expected to contain low Se. Selenium is expected to be found in higher concentrations in semiarid and arid environments of western United States, China, Mexico, and Colombia. In some humid environments of eastern United States, China, United Kingdom, Japan, India, and New Zealand, Se concentration is expected to be lower. The common range of total Se concentrations in most soils is between 0.01 and 2 mg/kg. However, in some soils of the world, total Se concentrations can be as high as 1200 mg/kg (2). Such soils are commonly known as seleniferous soils.

Groundwater is an important natural drinking water resource for many worldwide. Many parts of the world also use groundwater for agriculture and industry. In recent years, the use of groundwater has increased more rapidly than use of surface water. In many rural areas of the world groundwater is the only source of drinking water. Therefore, any contamination of groundwater could lead to severe consequences. Because a narrow range between deficiency and toxicity exists, the Se concentration in drinking water is regulated not to exceed 10 μg/L (3).

Very limited information regarding the dissolved Se concentrations in natural waters is available. Dissolved Se concentrations in fresh water may range between 0.1 and 400 μg/L (4). Dissolved Se concentrations between 16 and 231 ng/L for river waters in Japan have been reported (5). Dissolved Se concentrations ranging between 60–70 ng/L in Finnish groundwaters, rivers, and lakes have also been reported (6).

Human activities may introduce Se into soil–water systems. Burning fossil fuels (coal), disposal of coal residues, mining activities, and application of fertilizers put Se into soil–water systems. Anthropogenic activities such as coal-burning power plants and smelting and refining industries are the main sources of Se (7). Worldwide total Se input to fresh waters from coal-burning power plants is $6–30 \times 10^4$ metric tons of Se per year. Smelting and refining industries contribute $3–20 \times 10^4$ metric t of Se per year to fresh waters.

Anthropogenic activities that impact soil–water systems with Se are well documented. As an example, disposal of agricultural drainage water into the wetlands of Kesterson National Wildlife Refuge in California, caused bioaccumulation of Se by plants, fish, and waterfowl at levels that were harmful (8). In another case, dissolved Se concentrations as high as 333 μg/L were reported in shallow post-coal mining groundwater in Wyoming (9).

IMPORTANCE OF SELENIUM SPECIATION IN NATURAL WATERS

The chemistry of Se in natural waters is complex because it exists in different oxidation states (+6, +4, 0, and −2). These oxidation states consists of a number of species. Some of these species are listed below:

$$Se(+6) : SeO_4^{2-}, HSeO_4^-, H_2SeO_4^0, CaSeO_4^0, \text{and } MgSeO_4^0 \quad (1)$$

$$Se(+4) : SeO_3^{2-}, HSeO_3^-, H_2SeO_3^0, CaSeO_3^0, \text{and } MgSeO_3^0 \quad (2)$$

$$Se(-2) : Se^{2-}, HSe^-, H_2Se^0, CaSe^0, \text{and } MgSe^0 \quad (3)$$

Among these, the SR(+6) and Se(+4) oxidation states are commonly found in most soil–water enviroments. However, under reducing conditions Se(0) and Se(−2) oxidation states may be expected.

As shown in Figure 1, dissolved Se in natural water undergoes different biogeochemical processes. As water infiltrates through the soil, Se solid phases dissolve and release SeO_4^{2-} and SeO_3^{2-} ionic species into soil water. The SeO_4^{2-} and SeO_3^{2-} ionic species are also released from the mineral surfaces into soil water through the desorption process. The SeO_4^{2-} and SeO_3^{2-} ions in soil water interact with metals to form complexes (eg, $CaSeO^0{}_4$, $MgSeO^0{}_4$, $CaSeO^0{}_3$, $MgSeO^0{}_3$). Soil water also contains dissolved organic carbon (DOC) due to the decomposition of vegetation and biological activity; therefore, DOC–Se complexes are expected.

Studies have shown that Se uptake and toxicity to microorganisms depend on its chemical form (10). Certain microorganisms can volatilize inorganic Se species in soil water into dimethyl selenide ($(CH_3)_2Se$) (gas) (11). Similarly, Se uptake and toxicity to plants also depend on its chemical form (12).

The adsorption process of Se in soil water depends on the free concentrations of SeO_4^{2-} and SeO_3^{2-} ionic species. For Example, it has been shown that SeO_4^{2-} is not adsorbed by the mineral surfaces when compared with SeO_3^{2-} (13–15). Under reducing (anoxic) conditions a solid phase such as FeSe is shown to precipitate in sediments (16). These Se biogeochemical processes (speciation, biological uptake, adsorption, and precipitation) in turn control the mobility of Se species in soil waters. Similar Se biogeochemical processes are expected in groundwater environments.

Based on the thermodynamic properties, it has been predicted that SeO_4^{2-} and SeO_3^{2-} should be found dominate dissolved Se in alkaline soil waters (17). Under acidic conditions $HSeO_3^-$ should be found. Experimental studies have confirmed the presence of both SeO_4^{2-} and SeO_3^{2-} species in fresh waters (18,19). DOC–Se complexes have also been reported in natural waters (20–23). In addition to SeO_4^{2-}, SeO_3^{2-}, and DOC–Se, metal–Se complexes (eg, $MgSeO_4^0$) in groundwaters of the Powder River Basin, Wyoming, have been suggested (24).

The above review suggests that Se biogeochemical processes in natural waters depend on the chemical form of the dissolved Se, and the considerable research suggests natural waters contain not only SeO_4^{2-} and SeO_3^{2-} but also metal–Se complexes and DOC–Se complexes. If dissolved Se contains different species, then its partition into free ionic species, metal complexes, and DOC–Se complexes is important both for understanding Se deficiency and toxicity processes as well as developing effective remediation methods.

SELENIUM SPECIATION IN NATURAL WATERS

The importance of understanding dissolved Se speciation in biogeochemical processes and in remediation of Se in natural environments has been discussed. In this section, the factors affecting Se speciation, advantages and limitations of current Se speciation methods, and a new method for analyzing Se speciation in natural waters are reviewed. Information regarding the remediation of Se in contaminated soils and waters is described elsewhere in this *Encyclopedia* (see SELENIUM IMMOBILIZATION IN WASTEWATER REMEDIATION; GROUNDWATER AND SOIL REMEDIATION).

Factors Affecting Se Speciation in Natural Waters

The most important factors that affect dissolved Se speciation in natural waters include the pH and the redox potential (Eh). It has been suggested that Se(−2) species should be found in reducing environments, Se(+4) species in moderately oxidizing environments, and Se(+6) species in oxidizing environments (17). Transformations of Se as affected by sediment pH and Eh have been examined (16). The Eh varied from −200 to 450 mV and the pH ranged between 6.5 and 9. Results suggest that Eh and pH have a major effect on the speciation and transformations of dissolved Se in sediments.

Although general trends for Eh conditions controlling Se speciation have been observed, experimental evidence is lacking. Theoretical solubility and Eh criteria for Se in soils have been developed (17,25). Predictions of soluble Se speciation can

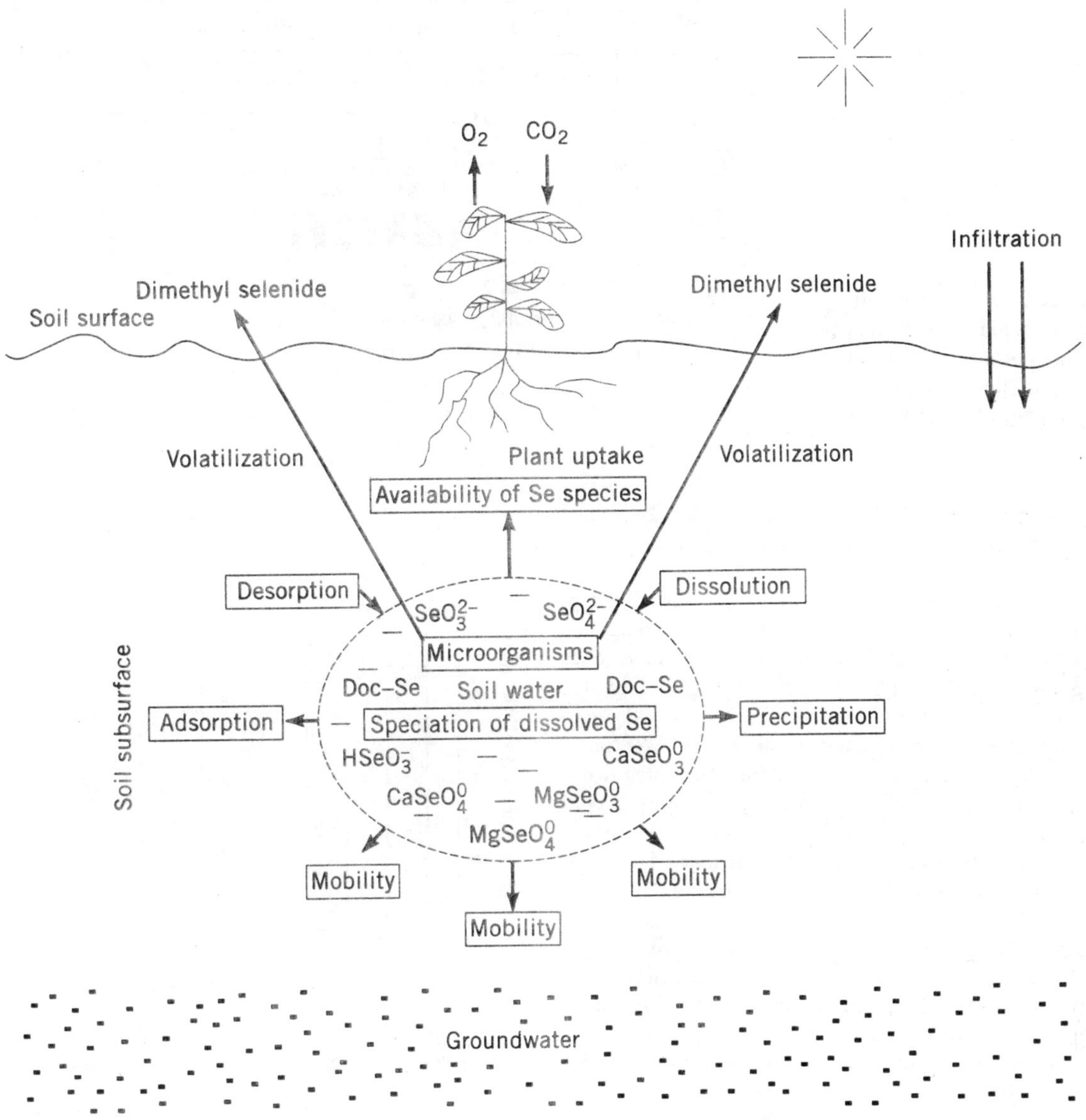

Figure 1. Schematic representation of a hypothetical soil subsurface system and various chemical and biological processes that occur in soil waters (not to scale).

be made based on Eh versus the Se mole fraction (17). However, because of uncertainty in the measurement of Eh and the application of Eh measurements, experimental evidence verifying redox–Se speciation predictions is lacking.

A simple system to control Eh and pH in soil waters has been reported (26). An Eh and pH controlling system to examine iron and manganese solubility and availability in soil waters has been employed (27–29). An improved Eh-controlling system has been developed to determine iron solubility in soil waters (30). A Eh-controlling apparatus similar in concept to that used in Reference 30, with a few modifications, has been developed (31). Unlike previously described systems, this system does not require the addition of organic matter to provide energy for microorganisms to reduce soil water or groundwater. Decomposition of organic matter may release organic acids, which may complex with metals.

The design of the redox controlling apparatus (RCA) (31) consists of a sample chamber containing pH, platinum (Pt), and reference electrodes connected to the pH/Eh meter (Fig. 2). The pH/Eh meter and oxygen and hydrogen gas tanks were connected to a control box to maintain the desired Eh in the sample chamber. The oxygen gas tank contained 95% oxygen and 5% carbon dioxide and the hydrogen gas tank contained 3% hydrogen, 5% carbon dioxide, and 92% argon. The purpose of carbon dioxide in the gas mixture is to maintain a constant pH (5.0 ± 0.1) in the samples. A peristaltic pump was used to circulate samples in the chamber. The gas outlet was connected to a fume hood.

The RCA control box was designed to read mV signals from the platinum and reference electrodes and the signals were compared to the minimum and maximum set values on the oxygen and hydrogen dials. If the mV signal was less than the set range, the controller switched on a solenoid that released oxygen gas into the sample chamber to increase the mV reading to the set value. Similarly, hydrogen was switched on if the mV reading was greater than the upper limit. The relay (control box) was calibrated to operate from +1.2 to −1.2 V.

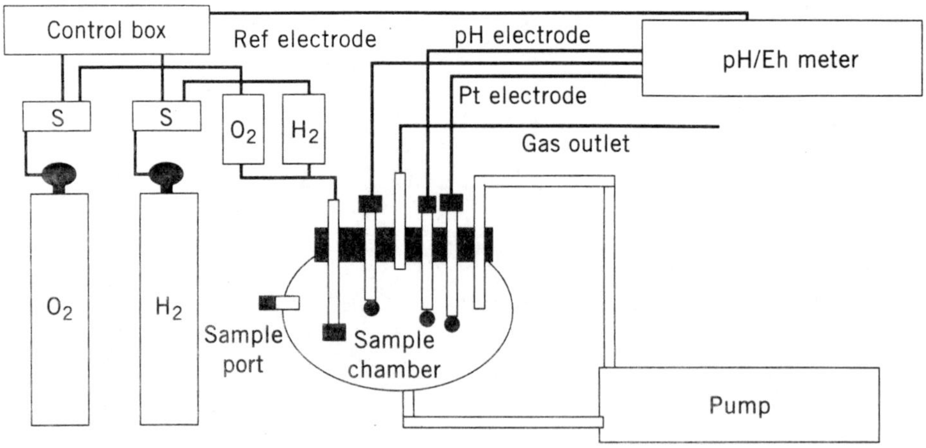

Figure 2. Redox potential controlling apparatus (RCA) used to examine the effects of redox potential (Eh) on the speciation of dissolved Se in soil water and groundwater samples.

Two samples were tested (31): 1 L of distilled deionized water mixed with 250 grams of soil and 1 L of groundwater alone, in the sample chamber to determine the effect of Eh on dissolved Se speciation. Samples were amended with 0.5 g of carbon black containing 10% platinum to improve the response of the electrode. The soil water sample was amended with 5.0 mg/L of SeO_4^{2-} and 5.0 mg/L of SeO_3^{2-}. The groundwater sample was amended with 8.0 mg/L of SeO_4^{2-} and 8.0 mg/L of SeO_3^{2-}. Samples were reacted from 72 to 170 hours to equilibrate with a particular Eh setting. Samples were withdrawn from the sample chamber through a sample port with a syringe and were analyzed immediately with an IC, using manual injection through a 0.45-μMillipore filter for SeO_4^{2-}, SeO_3^{2-}, nitrate (NO_3^-), nitrite (NO_2^-), chloride (Cl^-), and sulfate (SO_4^{2-}).

The dissolved Se in aqueous environments could also be analyzed with hydride generation atomic absorption spectrophotometry (HG/AAS) and fluorometry. Other methods such as neutron activation, mass spectrometry, and X-ray fluorescence are also available for the analysis of dissolved Se. More details regarding these methods are discussed later.

The RCA was successful in controlling the Eh of the soil water and groundwater samples at a given measured Eh value. The addition of 5% CO_2 to the gas mixture maintained a constant pH 5.0 ± 0.1 in the groundwater sample. The pH of the soil water sample also remained constant and was more representative of the native pH of the sample. However, the soil water sample resisted Eh changes caused by adding hydrogen and oxygen. This was evidenced by the longer time periods required for the Eh to stabilize at a particular setting.

Oxidation of SeO_3^{2-} to SeO_4^{2-} was not observed in the groundwater sample at an Eh of 444 mV (Fig. 3). Reduction of SeO_4^{2-} to SeO_3^{2-} in the groundwater sample also was not observed in an Eh range of 444 to −280 mV. However, the reduction of SeO_3^{2-} in the groundwater sample occurred much more readily. Selenite in the groundwater began to reduce after 72 hours at an Eh of 213 mV and the concentration diminished by 99% at an Eh of −280 mV. When the Eh in the groundwater sample was increased to +603 mV, very little or no SeO_3^{2-} was recovered, indicating that SeO_3^{2-} probably was volatilized or converted to hydrogen selenide (H_2Se) gas.

The SeO_4^{2-} in the soil water sample decreased initially by 30% after 72 hours at 455 mV and decreased by 80% after a total elapsed time of 288 hours at −160 mV. This decrease

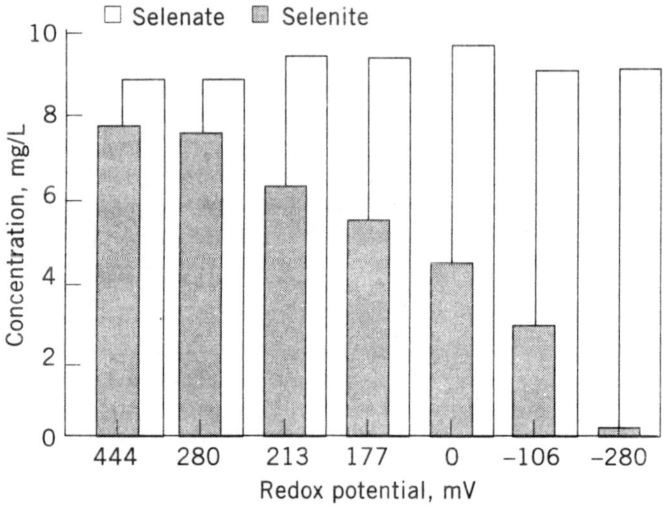

Figure 3. Effect of redox potential on the dissolved concentration of SeO_4^{2-} and SeO_3^{2-} in groundwater sample.

in SeO_4^{2-} concentration may be attributed to the reduction to SeO_3^{2-} due to both chemical and biological processes. However, SeO_3^{2-} in the soil water sample was apparently adsorbed by iron and manganese oxides and clay minerals in the soil; therefore, speciation changes caused by Eh could not be determined.

To help substantiate measured Eh values in relation to particular Eh conditions, this study (31) also monitored the concentrations of other anions participating in oxidation and reduction transformations. As an example, NO_3^- concentration in the groundwater sample decreased at 236 mV and was not detected after 120 hours. A concomitant increase in NO_2^- concentration was observed with the decrease in NO_3^-. However, after another 72 hours, NO_2^- concentration also diminished. Sulfate concentrations remained constant and appeared unaffected by the Eh.

The reduction of NO_3^-, NO_2^-, and SeO_3^{2-} indicates that appropriate electron activity for the rapid reduction of SO_4^{2-} and SeO_4^{2-} should be present in groundwater sample. NO_3^- was reportedly highly correlated with Se in groundwater and maintains the Eh above what is required for SeO_4^{2-} reduction (32). When oxygen and NO_3^- are removed from the system, the Eh decreases and SeO_4^{2-} is reduced. This effect, however, was not observed for groundwater sample in this study (31). Selenite in solution is susceptible to reducing conditions, but did not appear to be oxidizable in the groundwater sample under the experimental conditions employed. These results could be caused by the lack of labile electrons to participate in the particular redox couples in the groundwater sample or could be caused by kinetic constraints. For example, oxidation of SeO_3^{2-} to SeO_4^{2-} in soil waters has been reported to require from 48 to 240 hours (16,33).

The response of the SeO_4^{2-} and SeO_3^{2-} conversions to the imposed redox conditions of the redox controlling apparatus in the groundwater sample did not completely follow the thermodynamic predictions. As stated previously, the rates of Se redox transformations have been reported to be slow and the study in Reference 31 focused on relatively short reaction periods; thus, the nonoccurrence of some of the predicted redox transformations could be caused by a lack of equilibrium conditions in the redox chamber.

The reduction of SeO_4^{2-} in the soil water followed thermodynamic predictions somewhat more closely than in the groundwater sample. Reduction of SeO_4^{2-} in the soil water occurred at Eh values similar to those reported elsewhere (16,34). The absence of SeO_4^{2-} reduction to SeO_3^{2-} in groundwater has important implications. Because of the difficulty in obtaining reliable measurements of Se speciation in soil water or groundwater samples, total Se is often the only measurement determined. The speciation of dissolved Se into SeO_4^{2-} and SeO_3^{2-} generally is calculated on the basis of mathematical models. Results obtained in one study (31) suggest that in groundwater, SeO_4^{2-} and SO_4^{2-} remains stable under reduced conditions when thermodynamic predictions indicate otherwise. A comparison between groundwater and the soil water with respect to the reduction of SeO_4^{2-} indicates the importance of chemical and biological properties in the oxidation and reduction processes.

Methods for Analyzing Se Speciation in Natural Waters

Different approaches are available for the speciation of dissolved metals in natural waters. These speciation approaches are broadly divided into two groups: (*1*) analytical approaches and (*2*) mathematical modeling. The analytical approaches include anodic stripping voltametry, ion-selective electrodes, hydride generation atomic absorption (HG/AAS), ion chromatography (IC), ion-exchange techniques, filtration, and dialysis. More details including advantages and limitations of these methods, are discussed elsewhere (35). For in-depth information readers should review *Complexation Reactions in Aquatic Systems: An Analytical Approach* (36), *Speciation of Metals in Water, Sediment and Soil Systems* (37), and *Metal Speciation: Theory, Analysis and Application* (38).

The HG/AAS and IC are the most commonly used methods to analyze the speciation of dissolved Se in natural waters. The HG/AAS method measures total dissolved Se in natural water as Se(+4), from which Se(+6) and DOC–Se can be determined by altering the pretreatment steps as follows (39,40). The concentration of Se(+4) in a sample is measured by generating H_2Se with a $NaBH_4$ solution and 7 *M* HCl (undigested). Another aliquot of sample is heated for 20 minutes at 85°C with 7 *M* HCl to reduce Se(+6) to Se(+4). The concentration of Se in this solution is considered as the sum of Se(+6) and Se(+4) (digested). The difference between the concentration of Se in digested and undigested samples is considered as the concentration of Se(+6). The total Se concentration in samples is determined by oxidizing organic matter with H_2O_2 for 20 minutes at 85°C and then digesting with 7 *M* HCl for another 20 minutes at 85°C. The difference between total Se and digested Se concentrations is assumed to be the concentration of dissolved organic Se species.

The concentration of Se(+6) and Se(+4) determined with HG/AAS in soil water and groundwater samples consists of SeO_4^{2-} and SeO_3^{2-} species and their solution complexes. If SeO_4^{2-} and SeO_3^{2-} could be isolated and removed directly from natural waters, then their concentrations could be measured using HG/AAS. The advantage with HG/AAS is that it can detect as low as 1 μg/L of Se in aqueous solutions.

The IC method can measure the concentration of SeO_4^{2-} and SeO_3^{2-} simultaneously in aqueous solutions. However, SO_4^{2-}, a common anion of natural waters (eg, soil waters or groundwaters), interfere in SeO_4^{2-} and SeO_3^{2-} separation with IC (41,42). Furthermore, limit of the IC quantification for Se analysis has been reported to be 181 μg/L (33).

Often mathematical models (eg, MINTEQA2, GEOCHEM, WATEQFC) are used to calculate the speciation of dissolved metals in natural waters. Chemical data such as dissolved concentrations of metals and ligands, pH, and Eh of the soil water and the groundwater samples are entered into these models in order to calculate the chemical speciation (ie, activity of free ions and complexes) by solving a series of mass–balance equations through an iterative procedure.

There are three important issues when mathematical models are applied to determine the chemical speciation of soil water and groundwater samples. (*1*) Models are based on the assumption of equilibrium. Therefore, soil waters and groundwater to be tested should be close to a steady-state condition (ie., the condition in which little change has occurred in the major cations and anions involved). (*2*) The mass–balance equations for each dissolved species should contain all possible solution species to ensure accurate calculation of the speciation. Omission of any significant solution species from the mass–balance equation will cause overestimation of the activity of dissolved free ions and complexes. (*3*) Variations occur in the equilibrium constants for solution species. All these factors could lead to the misinterpretation of the speciation of dissolved chemical in natural waters.

Few studies actually have examined the speciation of dissolved Se in soil waters or groundwaters, despite its importance in understanding Se availability, toxicity, and mobility in natural waters. This is partly due to the difficulty involved in the separation of Se(+6) and Se(+4) oxidation states without destroying their natural distribution.

A method has been developed to separate dissolved Se in natural waters into Se(+6), Se(+4), Se(0), and Se(−2) with electron capture detection (ECD) gas chromatography using 1,2-diamino-3,5-dibromobenzene without preconcentration (5). It was reported that dissolved Se in river and seawater samples from Japan were dominated by Se(+6) species. Passing groundwater samples through a column of XAD-8 resin at pH 1.6–1.8 can remove organic compounds without altering the chemistry of Se(+6) and Se(+4) species (43). A procedure has been described to separate Se(+6) and Se(+4) from organic-rich groundwater samples. In this procedure, dissolved organic compounds were separated by passing groundwater samples through a XAD-8 anion-exchange column at pH 1.6–1.8 and resulting solutions were determined for Se(+6) and Se(+4) using HG/ASS.

Different methods for the speciation of dissolved Se in soil water samples have been evaluated (21). It was found that XAD-8 resin separation is a useful method for determining the concentrations of Se(+6), Se(+4), and hydrophobic organic Se in soil water samples. An isotope dilution mass spectrometry (IDMS) method for the determination of dissolved Se species in natural waters has been developed (22). Dissolved Se in environmental samples was separated into Se(+6) and Se(+4).

Inherent problems associated with the IDMS method as well as other methods have been discussed (23), and a sample treatment scheme to determine the speciation of dissolved Se in natural waters presented. In this method, dissolved Se in natural waters was separated into Se(+6), Se(+4), and dissolved organic Se. The dissolved organic Se was further speciated into hydrophobic base, acidic, and neutral Se fractions. However, in this study (23) the concentrations of the sum of the different dissolved Se species were 11 to 21% below the total dissolved Se concentrations and they attributed these differences to the Se associated with hydrophilic base, acidic, and neutral fractions. Nevertheless, these differences may also be due to metal Se complexes (eg, $CaSeO_4^0$, $MgSeO_4^0$, $CaSeO_3^0$, $MgSeO_3^0$).

In summary, current dissolved Se speciation methods are useful for laboratory characterization of natural waters, however, they are:

- time consuming,
- require extensive sample pretreatment steps,
- not practical for field applications, and
- cannot isolate and extract dissolved SeO_4^{2-} and SeO_3^{2-} ionic species directly from aqueous solutions.

New Method for Analyzing Se Speciation in Natural Waters

In earlier sections the importance of Se speciation was discussed, as well as factors affecting Se speciation, and current methods for analyzing Se speciation in natural waters and their advantages and limitations. In this section, a new method to isolate and extract dissolved SeO_4^{2-} and SeO_3^{2-} species from natural waters is described.

Discovery of the CuO Method for Dissolved Se Speciation The conditions which led to the discovery of CuO methods for dissolved Se speciation is described here. Studies have shown that the common anion SO_4^{2-} of natural waters interference with the separation of dissolved Se during the IC analysis (41,42). To remove SO_4^{2-} from groundwater samples, various procedures have been used (24). For example, $BaCl_2$ and $Ba(OH)_2$ solutions were used to precipitate SO_4^{2-} as $BaSO_4$; however, SO_4^{2-}, SeO_4^{2-} and SeO_3^{2-} also coprecipitated.

Subsequently, Cu^{2+} (cupric) was selected to precipitate SO_4^{2-} from aqueous solutions because Cu^{2+}–Se solution species are not significant and Cu^{2+}–Se solid phases are highly soluble (17). The $CuCl_2$ solution was added to aqueous solutions containing SO_4^{2-}, SeO_4^{2-}, and SeO_3^{2-}. Then, SO_4^{2-} was precipitated as cupricoxysulfate by adding NaOH to the solutions to increase the pH to 12.0 (44). The color of this precipitate was sky blue. This procedure appeared to work for some samples; on further investigation, addition of $CuCl_2$ and NaOH to solutions was not completely effective in removing SO_4^{2-}. After one to two hours, the color of the cupricoxysulfate precipitate changed from sky blue to black and at this point SeO_4^{2-} and SeO_3^{2-} could not be detected in solutions. The pH of the solutions had dropped to around 6.0. When the pH was increased to 12.0, then SeO_4^{2-} and SeO_3^{2-} were detected in solutions. X-ray diffraction analysis of the black precipitate suggested that it was tenorite (CuO).

Overall, these experiments indicated that increasing the pH of $CuCl_2$ solutions by adding NaOH precipitated CuO. Excessive Cu^{2+} in solutions slowly lowered the pH to around 6.0 through hydrolysis reactions. At this point, SeO_4^{2-} and SeO_3^{2-} were not detected in solutions because CuO adsorbed SeO_4^{2-} and SeO_3^{2-} in solutions; when the solution pH was increased to 12, SeO_4^{2-} and SeO_3^{2-} were detected because of the desorption of these species from the CuO surface. These results suggested that the CuO particles selectively adsorbed SeO_4^{2-} and SeO_3^{2-} at pH 6.0 and desorbed them at pH 12.0 in the presence of other anions.

Further investigation suggested that the CuO particles in aqueous solutions show distinct adsorption and desorption phenomena for SeO_4^{2-} and SeO_3^{2-} ionic species at different pHs. As shown in Figure 4 the CuO particles adsorb 100% SeO_4^{2-} at pH 6 and desorb 100% at pH 11. For SeO_3^{2-}, CuO particles adsorb 100% at pH 9.5 and desorb 100% at pH 13. A possible explanation is that the zero point of charge (ZPC) for CuO occurs at pH 9.5 (45); below this pH the surface of the CuO is positive, and above pH 9.5 the surface is negative.

Analysis of Se Speciation in Soil Waters and Groundwaters with CuO

The dissolved Se speciation procedure (24) consists of the following six steps:

1. Filtration of groundwater samples through 0.45-μMillipore filters.
2. Separation of dissolved Se into Se(+6), (Se+4), and organic Se with HG/AAS.
3. Isolation and extraction of SeO_4^{2-} and SeO_3^{2-} with CuO particles.
4. Analysis of SeO_4^{2-} and SeO_3^{2-} with HG/AAS.
5. Determination of concentrations of metal SeO_4^{2-} and SeO_3^{2-} complexes by subtracting the concentrations of SeO_4^{2-} and SeO_3^{2-} from Se(+6) and Se(+4) concentrations, respectively.
6. Speciation of dissolved Se in groundwater samples.

The CuO procedure was tested to determine the dissolved Se speciation in three groundwater samples (24). Subsequently, the CuO procedure was applied to determine the dissolved Se speciation in three soil water samples. The soil water and groundwater samples were reacted with CuO particles for four hours for adsorption and 24 hours for desorption.

The selected chemical data of soil water and groundwater samples are shown in Table 1. The SO_4^{2-} concentrations in soil water and groundwater samples prevented direct determination of SeO_4^{2-} and SeO_3^{2-} with IC, except for soil water 1. However, dissolved Se concentration in this sample was below the detection limit of IC. Dissolved Se speciation in soil water and groundwater samples, as determined by HG/AAS, is summarized in Table 2. Also included in Table 2 are the SeO_4^{2-} and SeO_3^{2-} concentrations extracted with CuO particles.

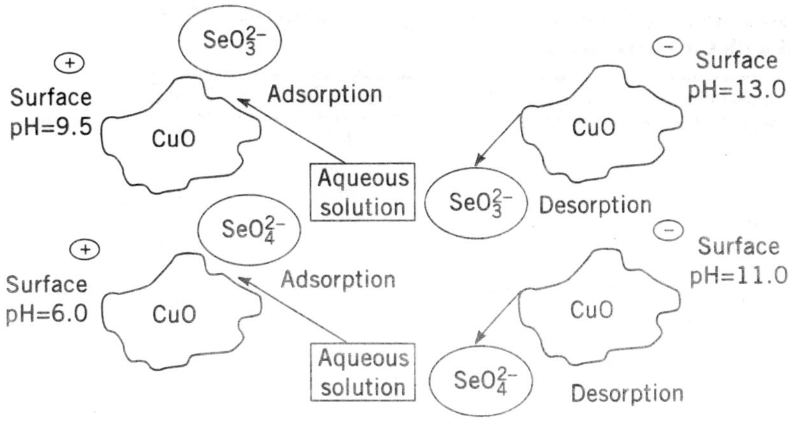

Figure 4. Schematic representation of SeO_4^{2-} and SeO_3^{2-} adsorption and desorption by the CuO particles in aqueous solution.

Table 1. Selected Chemical Data of Soil Water (SW) and Groundwater (GW) Samples, mg/L[a,b]

	SW 1	SW 2	SW 3	GW 1	GW 2	GW 3
Calcium	214	243	275	470	427	545
Magnesium	48	63	76	693	2,080	391
Sodium	31	18	31	339	1,180	347
Potassium	14	14	9.4	18	38	33
Sulfate	166	960	1,130	4,980	11,300	2,960
DOC	60	19	7.6	14	88	87
Se, μg/L	21	69	120	22	151	107
pH	7.9	6.1	5.8	5.5	7.1	6.3

[a] Unless otherwise indicated

[b] Courtesy of the American Chemical Society (24).

The total dissolved Se concentrations in soil water samples ranged between 21 and 120 μg/L (see Table 2). For soil water 1, 62% of total dissolved Se is in the form of Se(+6) and 19% is in the form of Se(+4). The dissolved organic Se is 19% of the total dissolved Se. The total dissolved Se concentrations in soil water 2 and 3 were dominated by Se(+6) and organic Se. Dissolved Se concentrations in groundwater samples ranged between 22 and 151 μg/L. Results in Table 2 show that 6 to 36% of the total dissolved Se in groundwater samples is in the form of Se(+4). The Se(+6) concentrations ranged between 41 and 79% of the total dissolved Se. Organic Se concentrations ranged between 14 and 22% of the total dissolved Se.

Results from the CuO extraction method show that the removal rate for SeO_4^{2-} ranged between 69 and 82% when compared with Se(+6) concentrations. The Se analysis of the CuO supernatant solutions suggested that 18 to 31% of SeO_4^{2-} was left in the solutions. However, for SeO_3^{2-} the removal rate is 90 to 100% except, for soil water 1, when compared with Se(+4) concentrations. These results also suggest that other anions such as SO_4^{2-} and DOC did not interfere with the SeO_3^{2-} removal process by CuO particles. The observed 69 to 82% removal rate of SeO_4^{2-} by the CuO particles from the soil water and groundwater samples could be due to the presence of other metal–Se complexes (eg, $MgSeO_4^\circ$, $CaSeO_4^\circ$), which may not be adsorbed by the CuO particles.

To determine the effect of divalent cations on the concentration of SeO_4^{2-} in aqueous solutions, different concentrations of Mg^{2+} were reacted with 1.0 mg/L of SeO_4^{2-} for 30 minutes (24) (Fig. 5). Then solutions were filtered through 0.45-μMillipore filter and solutions were analyzed for SeO_4^{2-} with IC. Data obtained in this experiment suggested that as Mg^{2+} concentration increased, the SeO_4^{2-} concentration decreased from 1.0 to 0.76 mg/L, which suggest that Mg^{2+}–SeO_4^{2-} solutions probably consists of $MgSeO^\circ_4$ complex. The observed removal rate of 69 to 82% for SeO_4^{2-} by the CuO particles in soil water and groundwater samples may be attributed to the presence of 18 to 31% of metal–SeO_4^{2-} complexes (eg, $MgSeO^\circ_4$, $CaSeO^\circ_4$) because oxides may not adsorb strongly bonded complexes (46). As an example, it was demonstrated (47) that formation of $CdCl^\circ_2$ complex lowers the concentration of Cd^{2+}, which in turn decreases the adsorption and increases the mobility of Cd in soils amended with wastes containing high concentration of chloride.

The speciation of dissolved Se concentrations in soil water and groundwater samples is presented in Figures 6 and 7. These results show that total dissolved Se concentrations in soil water samples were dominated by SeO_4^{2-} (43–72%). The dissolved organic Se concentrations ranged between 6 and 28% and the concentrations of other species (metal–Se complexes) were between 14 to 24%. For groundwater 1, dissolved Se was dominated by SeO_3^{2-} (36%) and SeO_4^{2-} (32%), followed by organic Se (23%) and metal–Se complexes (9%). For the other two groundwater samples, dissolved Se was dominated by SeO_4^{2-} (58–65%), followed by organic Se (14–20%), metal–Se complexes (14–16%), and SeO_3^{2-} (6–7%). The dissolved organic Se concentrations can be further separated into hydrophobic base, acidic, and neutral fractions by following procedure (22). However, further research to separate the dissolved Se associated with hydrophilic base, acidic, and neutral fractions is needed.

The most important factors that affect the CuO method include reaction time, pH, concentration of CuO particles, and size of the CuO particles. Although some studies focused on longer reaction times (4 hours for adsorption and 24 hours for desorption), recent studies (48) suggested that short reaction times are involved in adsorption and desorption of SeO_4^{2-} and SeO_3^{2-} by the CuO particles. For example, only less than 15 minutes required for 100% adsorption and desorption of SeO_4^{2-} by the CuO particles. For SeO_3^{2-} less than 5 minutes for adsorption and less than 30 minutes for desorption were sufficient.

Table 2. Speciation of Dissolved Se in Soil Water (SW) and Groundwater (GW) Samples with HG/AAS and Concentrations of SeO_4^{2-} and SeO_3^{2-} extracted with CuO particles, $\mu g/L$[a]

Sample	pH	DOC, mg/L	Se(+6)SeO_4^{2-}[b]	Se(+4)SeO_3^{2-}	Organic Se	Total Se
SW 1	7.9	60	13	4	4	21
			9[b]	3[b]		
SW 2	6.1	19	50	<1	19	69
			40[b]	<1		
SW 3	5.8	7.6	112	1	7	120
			86[b]	1[b]		
GW 1	5.5	14	9	8	5	22
			7[b]	8[b]		
GW 2	7.1	88	110	10	31	151
			87[b]	9[b]		
GW 3	6.3	87	85	7	15	107
			70[b]	7[b]		

[a] Courtesy of the American Chemical Society (24).
[b] Extracted with CuO.

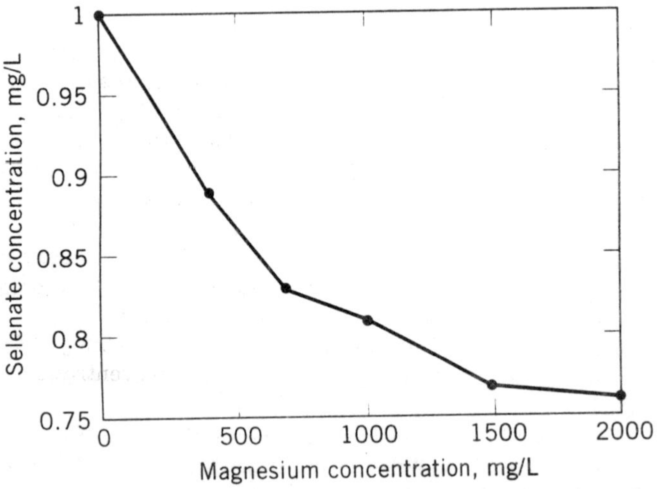

Figure 5. Effect of Mg^{2+} concentration on Se(+6) concentration in aqueous solution. Adapted with permission from the American Chemical Society (24).

The zero point of charge for CuO occurs at pH 9.5; below pH 9.5 the surface of CuO particles is positive, and above pH 9.5 the surface of CuO particles is negative. Therefore, accurate pH adjustments in isolation and extraction of SeO_4^{2-} and SeO_3^{2-} ions from soil water and groundwater samples is important. To extract SeO_3^{2-}, acidification of natural waters is not necessary because the pH of most natural waters are below 9.5.

The concentration of CuO particles had a greater effect on the adsorption of SeO_4^{2-}ions than SeO_3^{2-} ions because with 10 g/L of CuO particle concentration and 2.0 mg/L of SeO_4^{2-}, only 40% adsorption at pH 6.0 was observed. For SeO_3^{2-} ions, under similar conditions, 100% adsorption was observed at pH of 8.5. These results suggest that the concentration of CuO will be important in extracting SeO_4^{2-} and SeO_3^{2-} ions from natural waters.

The particle size of the CuO may also effect adsorption and desorption of SeO_4^{2-} and SeO_3^{2-} ions in natural waters. The pH of commercial CuO and laboratory synthesized CuO (estimated surface area is 52 m^2/g) in SeO_3^{2-} aqueous solutions has been compared. The commercial CuO had larger particles size than laboratory-synthesized CuO. The SeO_3^{2-} concentration ranged between 0 and 11 mg/L. The laboratory prepared CuO showed greater changes in pH than commercial CuO. This suggests that laboratory prepared CuO particles have a larger surface area and consequently have higher adsorption capacity than commercial CuO.

SUMMARY

Studies have shown that dissolved Se in soil waters or groundwaters exist in inorganic and organic forms. These include free ionic species (SeO_4^{2-} and SeO_3^{2-}), metal–Se complexes ($MgSeO_4^0$, $CaSeO_4^0$, $MgSeO_3^0$, $CaSeO_3^0$), and organic Se (hydrophobic base, acidic, and neutral Se). If dissolved Se exists in different forms, then isolation, extraction, and measurement of individual Se species in natural waters are fundamental for understanding Se biogeochemical processes such as availability, toxicity, adsorption, precipitation, and mobility. Additionally, such information is also required for developing effective remediation methods.

The speciation of dissolved Se in natural waters is not well understood. This is partially due to the difficulty associated with separation of dissolved Se species without changing their natural distribution. Although current Se speciation methods such as HG/AAS, IC, IDMS, and XAD-8 are useful for laboratory characterization of natural waters, they are time consuming and not practical for field application. The CuO method described (24) isolated and extracted SeO_4^{2-} and SeO_3^{2-} ionic species directly from soil water and groundwater samples in the presence of other anions. The CuO method can extract as low as 1 $\mu g/L$ of SeO_4^{2-} and SeO_3^{2-} from natural waters, and its detection limits are determined by the method of Se analysis (ie, HG/AAS, fluorometry, or IC).

The CuO method is rapid, simple, and effective. This method can potentially be used to extract SeO_4^{2-} and SeO_3^{2-} species from different environmental samples. The CuO particles could be taken to the field to extract SeO_4^{2-} and SeO_3^{2-} ions from natural waters, preserve these oxidation states, and later these species could be desorbed and analyzed with HG/AAS, fluorometry, or IC, depending on the concentrations of Se.

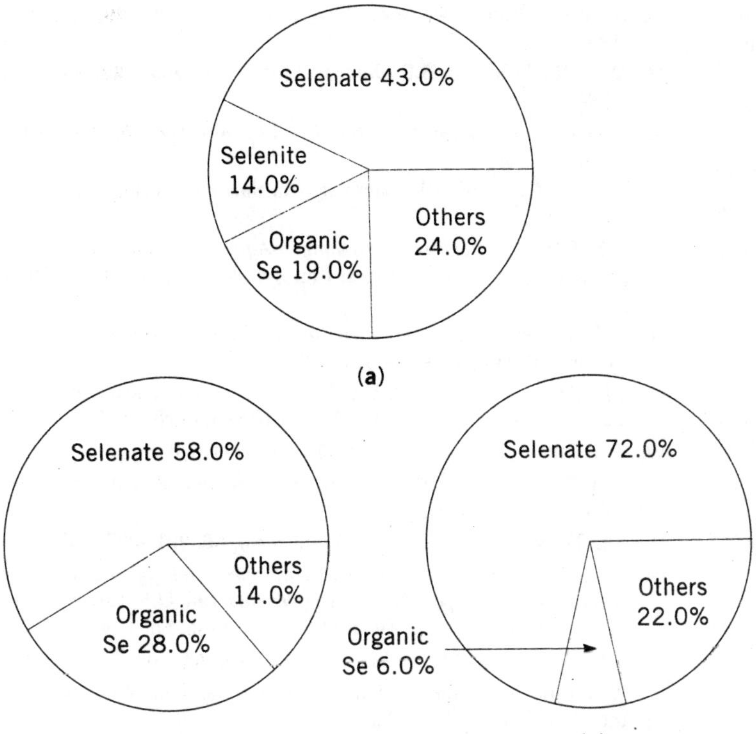

Figure 6. Speciation of dissolved Se in soil waters: (**a**) soil water 1; (**b**) soil water 2; (**c**) soil water 3. Others mean metal–Se complexes such as $MgSeO_4^0$, $CaSeO_4^0$, $MgSeO_3^0$, $CaSeO_3^0$.

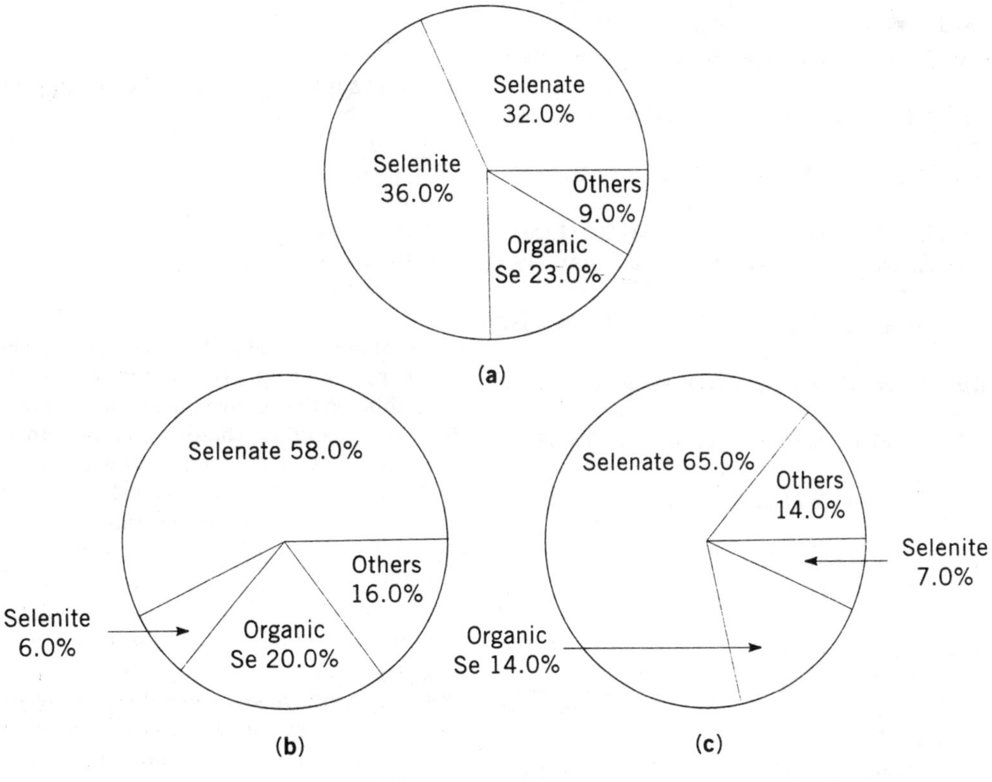

Figure 7. Speciation of dissolved Se in groundwaters: (**a**) groundwater 1; (**b**) groundwater 2; (**c**) groundwater 3. Others mean metal–Se complexes such as $MgSeO_4^0$, $CaSeO_4^0$, $MgSeO_3^0$, $CaSeO_3^0$.

Finally, further research to isolate DOC–Se in natural waters into hydrophilic base, acidic, and neutral fractions is needed.

BIBLIOGRAPHY

1. G. Yang and co-workers, *Amer. J. Clin. Nutr.* **37,** 872–881 (1983).
2. D.C. Adriano, *Trace Elements in the Terrestrial Environment,* Springer-Verlag, New York, 1986, pp. 390–420.
3. World Health Organization (WHO), *Guidelines for Drinking-Water Quality,* Geneva, Switzerland, 1984, p. 130.
4. D.J. Swaine, *Trace Subst. Environ. Health* **12,** 129–134 (1978).
5. H. Uchida and co-workers, *Environ. Sci. Technol.* **14,** 541–544 (1980).
6. D. Wang and co-workers, in A. Aitio, A. Aro, J. Jarvisalo, and H. Vainio, eds., Royal Society of Chemistry, London, 1991, pp. 50–56.
7. O.J. Nriagu and J.M. Pacyna, *Nature* **333,** 134–139 (1988).
8. H.M. Ohlendorf, in L.W. Jacobs, ed., *Soil Science Society of America,* Madison, Wis. 1989.
9. D.L. Naftz and J.A. Rice, *Appl. Geochem.* **4,** 565–575 (1989).
10. R.J. Shamberger, *Biochemistry of Selenium,* Plenum, New York, 1983.
11. J.W. Doran, *Adv. Microb. Ecol.* **6,** 1–32 (1982).
12. I. Rosenfeld and O.A. Beath, *Selenium,* Academic Press, Inc., New York, 1964.
13. J.S. Ahlrich and L.R. Hossner, *J. Environ. Qual.* **16,** 95–98 (1987).
14. L. Balistrieri and T. Chao, *Soil Sci. Soc. Amer. J.* **51,** 1145–1151 (1987).
15. R.H. Neal and G. Sposito, *Soil Sci. Soc. Amer. J.* **53,** 70–74 (1989).
16. P.H. Masscheleyn and co-workers, *Environ. Sci. Technol.* **24,** 91–96 (1990).
17. M.A. Elrashidi and co-workers, *Soil Sci.* **144,** 141–152 (1987).
18. G.A. Cutter, in M. Ihnat, ed., CRC Press, Boca Raton, Fla., 1989, pp. 243–262.
19. K.W.M. Siu and S.S. Berman, in M. Ihnat, ed., CRC Press, Boca Raton, Fla., 1989, pp. 263–293.
20. T.D. Cooke and K.W. Bruland, *Environ. Sci. Technol.* **21,** 1214–1219 (1987).
21. J.L. Fio and R. Fujii, *Soil Sci. Soc. Amer. J.* **54,** 363–369 (1990).
22. D. Tanzer and K.G. Heumann, *Anal. Chem.* **63,** 1984–1989 (1991).
23. D. Wang and co-workers, *Environ. Sci. Technol.* **28,** 383–387 (1994).
24. K.J. Reddy and co-workers, *Environ. Sci. Technol.* **29,** 1754–1759 (1995).
25. H.R. Geering and co-workers, *Soil Sci. Soc. Amer. J.* **32,** 35–40 (1968).
26. W.H. Patrick and co-workers, *Soil Sci. Soc. Amer. J.* **37,** 331–332 (1973).
27. C.N. Reddy and co-workers, *Agron. J.* **68,** 987–989 (1976).
28. A.P. Schwab and W.L. Lindsay, *Soil Sci. Soc. J.* **47,** 201–205 (1983).
29. F.W. Boyle and W.L. Lindsay *Soil Sci. Soc. Amer. J.* **50,** 588–593 (1986).
30. E.W. Brennen, M.S. thesis, Colorado State University, Fort Collins, Colo., 1988, p. 64.
31. K.J. Reddy and co-workers, "Effects of Redox Potential on the Speciation of Selenium in Groundwater and Coal Mine Backfill Materials," in G.E. Schuman and G.F. Vance, eds., Princeton, W.V , 1995, pp. 230–236.
32. H. Weres and co-workers, *Water, Air, Soil Poll.* **49,** 251–272 (1990).
33. M.J. Blaylock and B.R. James, *J. Environ. Qual.* **22,** 851–857 (1993).
34. G. Sposito and co-workers, *Soil Sci. Soc. Amer. J.* **55,** 1597–1602 (1991).
35. G.M.P. Morrison, in L. Landner, ed., Springer-Verlag, New York, 1986, pp. 55–73.
36. J. Buffle, *Complexation Reactions in Aquatic Systems: An Analytical Approach,* Ellis Horwood Limited, West Sussex, U.K., 1988, p. 692.
37. L. Landner, *Speciation of Metals in Water, Sediment and Soil Systems,* Springer-Verlag, NY, 1986.
38. J.R. Kramer and H.E. Allen, *Metal Speciation: Theory, Analysis and Application,* Lewis Publishers, Chelsea, Mich. 1988, p. 357.
39. G.A. Cutter, *Anal. Chem. Acta* **98,** 59–66 (1978).
40. S.M. Workman and P.N. Soltanpour, *Soil Sci. Soc. Amer. J.* **44,** 1331–1333 (1980).
41. T.B. Hoover and G.D. Yager, *Anal. Chem.* **56,** 221–227 (1984).
42. J.A. Oppenheimer and co-workers, *Speciation of Selenium in Ground Water,* Report No. EPA-600/S2-84-190, U.S. Environmental Protection Agency, Washington, D.C., 1985, pp. 1–4.
43. R.D. Roden and D.E. Tallman, *Anal. Chem.* **54,** 307–309 (1982).
44. W.L. Lindsay, *Chemical Equilibria in Soils,* John Wiley & Sons, Inc., New York 1979, p. 226.
45. G.A. Parks, *Chem. Rev.* **65** (1965).
46. K.J. Reddy and S.P. Gloss *Appl. Geochem. Suppl. Issue* **2,** 159–163 (1993).
47. P.M. Giordano and co-workers, *Environ. Sci. Technol.* **17,** 193–198 (1983).
48. J. P. Lin and K. J. Reddy, *Agron. Abstr.* **88,** 208 (1996).

SITE CHARACTERIZATION, EXPEDITED

G. A. STENBACK
B. H. KJARTANSON
A. BEVOLO
J. D. WONDER
Iowa State University
Ames, Iowa

Environmental restoration is the process of determining whether a potentially contaminated site represents a threat to the public or the environment and, if necessary, remediating it to an acceptable condition or monitoring it to determine whether the threat warrants remedial action in the future. As illustrated in Figure 1, site characterization is the process within environmental restoration that develops the information and analysis necessary to assess the risk posed by the site. Site characterizations are conducted where hazardous wastes have been, may have been, or might be released into the environment which might then impact the health of humans or ecosystems. The product of a site characterization is a conceptual site model describing the geology, hydrogeology, contaminant source and spatial distribution, contaminated media, transport pathways and mechanisms, exposure routes, and potential or actual receptors with sufficient accuracy and precision to be able to perform a reliable risk assessment. The objective of the site characterization is to reduce the

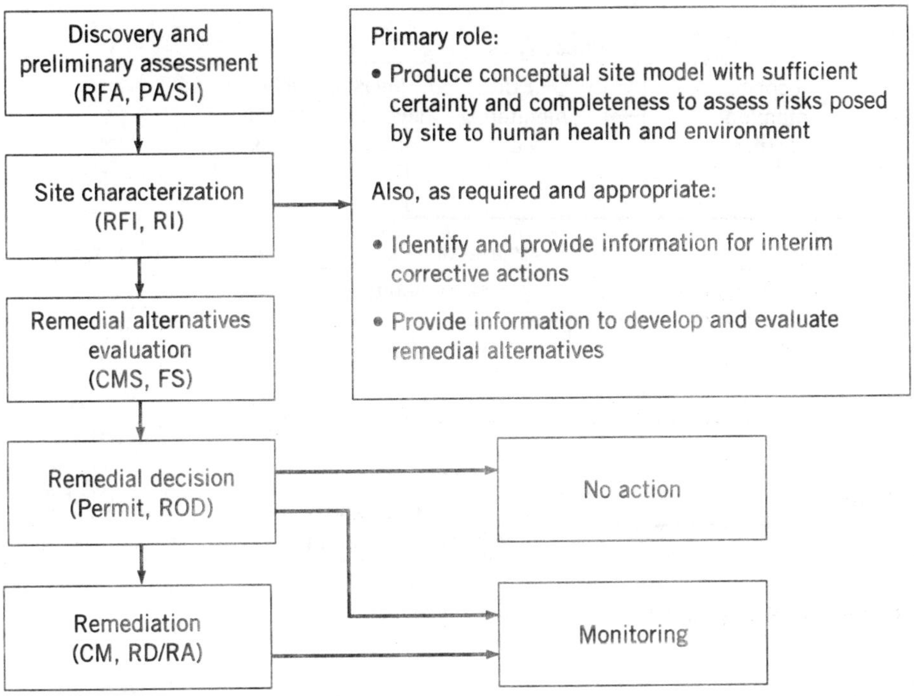

Figure 1. Basic environmental restoration process.

uncertainty of the conceptual site model sufficiently so that a correct decision regarding the need for remedial action and appropriate future land use can be made.

Conventional Approach to Site Characterization

A conventional environmental site characterization typically involves use of drilling equipment for the installation of many soil borings and monitoring wells for the collection of soil and groundwater samples. This is followed by an off-site contract laboratory program (CLP) chemical analysis of soil and/or water samples with subsequent data analysis attempting to define the type, spatial extent, and temporal extent of contamination. After a field mobilization, data collection, and data analysis event, a typical outcome may be that sufficient uncertainty exists to warrant establishing additional sample points for collection and analysis of more environmental samples. As illustrated in Figure 2, this process may be repeated several times with successive field mobilizations until sufficient data are obtained to perform a risk assessment and feasibility study. Each subsequent phase will typically be about 6 to 12 months long. The conventional method may include the use of some geophysical techniques such as ground penetrating radar, electromagnetic surveys, magnetometric surveys, gravity surveys, and radiological surveys.

The conventional approach has several cost, time, and methodological limitations. Installation of many monitoring wells used to define a contaminant plume is not only expensive, but results in the placement of unnecessary monitoring wells in areas which are free of contamination. Sequential tasks conducted by independent contractors using junior personnel in the field, with multiple field mobilizations is expensive, time consuming, and may lead to lack of adequate data integration. Use of full suite chemical analyses by off-site laboratories is expensive and leads to slow turnaround time so that the field investigator is effectively working without up-to-date information to guide the field work. The use of rigid work plans based on inadequate data often results in the placement of monitoring wells in the wrong locations and/or using incorrect screen intervals. Expedited site characterization (ESC) was developed in an effort to overcome many of these limitations.

Expedited Site Characterization

Expedited site characterization is a site characterization methodology which focuses on the synergistic use of nonintrusive and minimally intrusive investigation techniques to efficiently and effectively characterize hazardous waste sites. Key aspects of ESC include maximizing the use of existing site historical and physical data and the use of rapid data collection, interpretation, and visualization technologies to update the conceptual site model as the investigation proceeds. The central principle of ESC is effective selection, organization, and use of information to make fast and accurate decisions. The ESC process was developed and tested (2) at the DOE's Argonne National Laboratory and has been further developed by the DOE's Ames Laboratory. ESC has been successfully applied at over 10 sites since 1989 (1). Sites include U.S. Department of the Interior, Bureau of Land Management landfills (1,3), U.S. Department of Agriculture, Commodity Credit Corporation former grain storage facilities (4–6), the DOE Pantex weapons facility (7), a former manufactured gas plant site (described in this article), a DOE uranium processing waste site, an oil refinery site (8) and oil seepage basins at the DOE Savannah River Site (described in this article). ESC was developed in an effort to reduce site characterization time and costs while simultaneously providing better site characterizations than the conventional approach. ESC is made

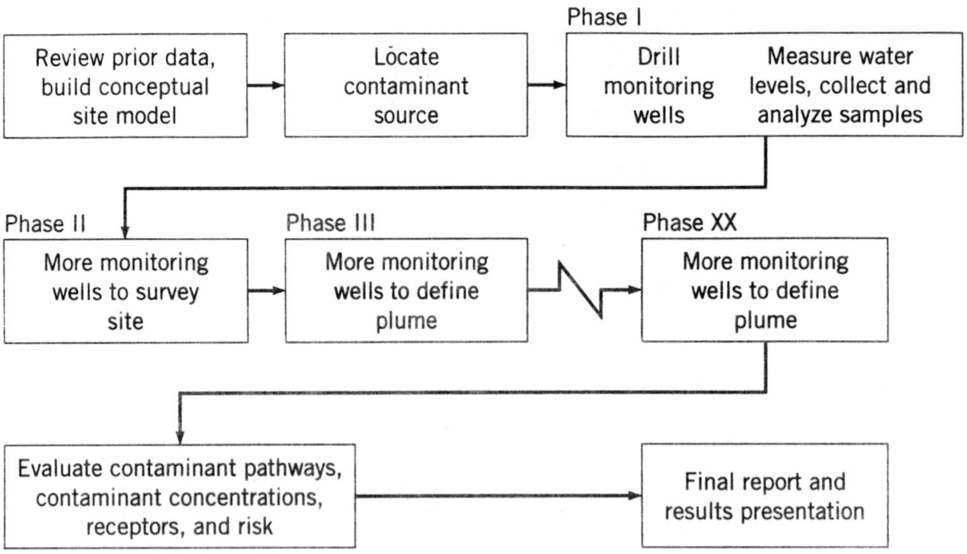

Figure 2. Conventional approach to site characterization (1).

possible because of recent advances in site characterization technologies including near-surface geophysical and direct push technologies, chemical field analytical technologies, and data management, analysis, and visualization technologies.

As illustrated in Figure 3, an ESC field investigation is conducted in two phases, called Phase I and Phase II, which are each typically two to six weeks in duration. An initial planning phase precedes field Phase I whereas a final analysis and reporting phase follows field Phase II. Throughout the investigation, ESC personnel maintain effective communication with the appropriate regulatory authority. A compilation and evaluation of prior data, including information regarding the historic land use, geologic and hydrologic reports, local well data, prior site investigations, soil surveys, and aerial photographs, is used to develop a preliminary conceptual site model prior to Phase I. The Phase I work then focuses on characterization of potential geologic and hydrogeologic contaminant migration pathways using several complimentary methodologies, location of contaminant sources, and development of a list of site-specific contaminants of potential concern. Many rapid analytical measurements with predefined quality control specifications are then made in the field during Phase II with a real understanding of the problem based on the information gained during Phase I so that the spatial distribution of the contaminants and contaminant fate can be determined. The most important and critical feature of ESC is the use of an integrated, multidisciplinary team of experienced professionals that conduct all field activities using a dynamic work plan which incorporates the flexibility for on-site decision-making during the investigation process. Data integration, interpretation, and visualization are performed on-site as new data become available to support the on-site decision-making process.

This article describes the fundamental principles, processes, and advantages of the ESC methodology as it has been applied by Ames Laboratory. Two examples of the application of the ESC process are given: one at a former manufactured gas plant (FMGP) site in the U.S. midwest and another at oil seepage basins at the Savannah River Site in South Carolina.

THE ESC PROCESS

An ESC investigation is a methodology for conducting a RCRA facility investigation (RFI) or CERCLA remedial investigation (RI). The ESC process is carried out by a core technical team of professionals with expertise in geologic systems, hydrologic systems, and chemical systems and is supported by personnel with expertise in data management, statistics, health and safety, quality assurance and quality control, logistics, remediation engineering, risk assessment, community relations, and legal regulations. The ultimate team composition is specific to the site and may include disciplines other than those mentioned above. Under the supervision of a core technical team leader, the team members work together in data integration, analysis, and interpretation to ensure that ambiguities arising from conflicting data and data gaps are properly resolved. All work is carried out by a multidisciplinary team of experienced experts. The team that plans the site work also manages the site work. Emphasis is given to the use of a dynamic work plan, coordination of nonintrusive geophysical and minimally intrusive geologic and hydrogeologic sampling, on-site sample collection and analysis of screening and/or definitive type (see Ref. 9 for a definition of data types) chemical contaminant data, and on-site data management, integration, interpretation, and visualization to aid further development of the conceptual site model as the investigation proceeds.

Initial Planning and Program Development Phase

A site characterization begins with some identified reason to suspect a potential threat to human health and/or the environment due to the release or potential release of environmental contaminants. The landowner or responsible parties may enter into a contractual agreement with the ESC technology provider to characterize a site. This marks the first stages of the initial phase of the ESC process (see Fig. 3) whereby a core ESC team is established.

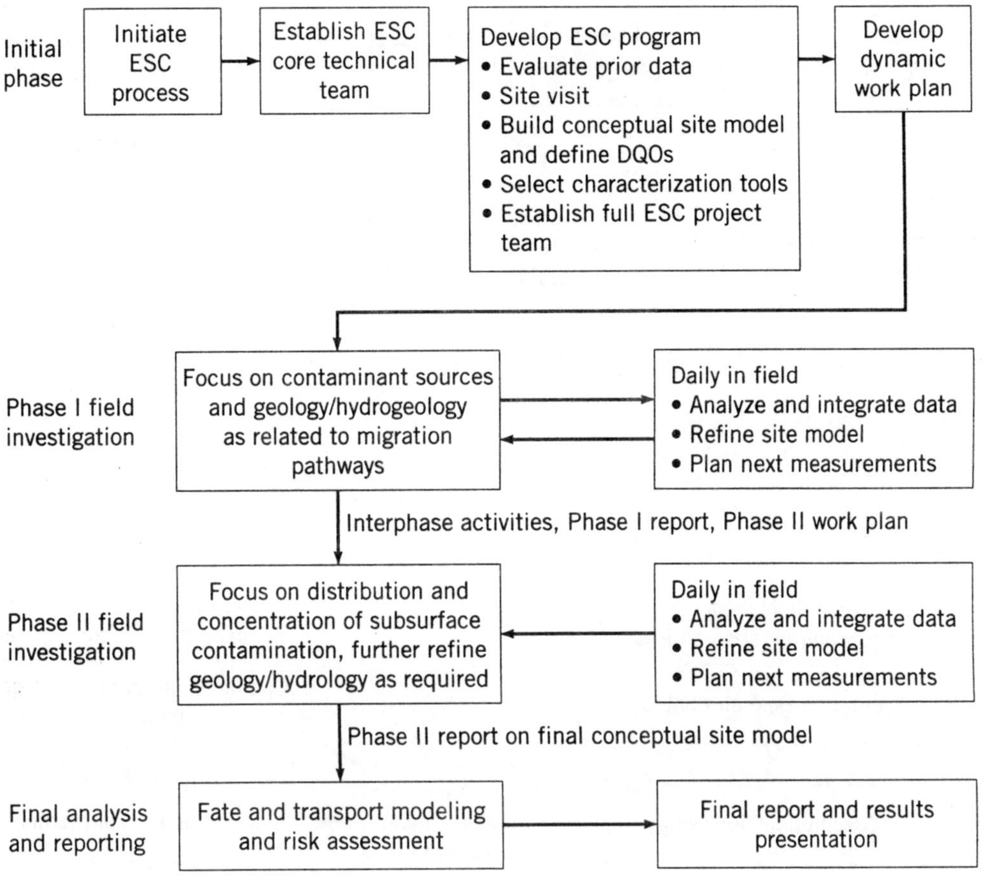

Figure 3. Flow chart of the ESC process (2).

In developing the ESC program (Fig. 4), the core team begins to collect and evaluate any relevant site information, including historic land use, regional and local geological and geohydrological reports, soil surveys, aerial photographs, local municipal well data, site soil boring logs, and site soil and/or water sample analyses. Interviews with current or former site workers, property owners, or nearby residents may be conducted. Data gathered during this phase is critically evaluated to assess its accuracy and usefulness, to assess data conflicts or data gaps, and to begin to develop a preliminary conceptual site model. The boundaries of the site to be investigated are clearly defined, often according to property lines, but also through negotiation with the site owners, federal, state, or local regulatory agencies, and possibly adjacent property owners. One or more visits to the site are made to observe the site surface conditions including terrain, vegetation, signs of vegetative stress, areas of discolored soil, and work obstructions such as buildings, fences, material storage areas, power lines, trees, wetlands, etc.

Uncertainties in the conceptual site model are defined and data needs and decision rules (such as the EPA data quality objectives) (9) are developed. An assessment of potential characterization tools is made and the full ESC project team is organized. A dynamic site investigation work plan which describes the site conceptual geologic and contamination models and the Phase I field work to be conducted is then developed. The work plan must allow for some flexibility regarding decisions made by the site ESC manager during the on-site investigation. Ames Laboratory has sought to involve appropriate regulatory agencies as soon as possible in this process to insure that their concerns are understood and incorporated into the work plan right from the start.

Tools and methods (eg, analytical, geophysical, sampling strategy, etc) used in the ESC process are chosen as appropriate for each specific site; as new tools are developed they may be utilized and compared with existing tools, validated, and eventually incorporated into the ESC process. In this regard, the ESC process itself is evolving in an effort to find the most efficient and cost-effective methods to perform adequate site characterization, and seeks to achieve regulatory acceptance of tools which can be shown to provide reliable data. The primary classes of tools for ESC are aerial remote sensing, surface and borehole geophysics, subsurface direct push and drilling technologies, chemical analytical methods, and data management and visualization. Figure 5 illustrates the general application of tools for an ESC investigation. The ESC strategy for applying characterization tools is to begin with the easy measurements (samples from existing wells) and wide area measurements (aerial photography and geophysical surveys) and later proceed to the more difficult and expensive ones, such as methods that deeply penetrate the subsurface. There is an emphasis on selecting geological and hydrogeological characterization tools that corroborate each other, especially if one of the tools is developmental or relies on innovative technology. Generally, only one analytical method is used to define the contaminant spatial distribution at the

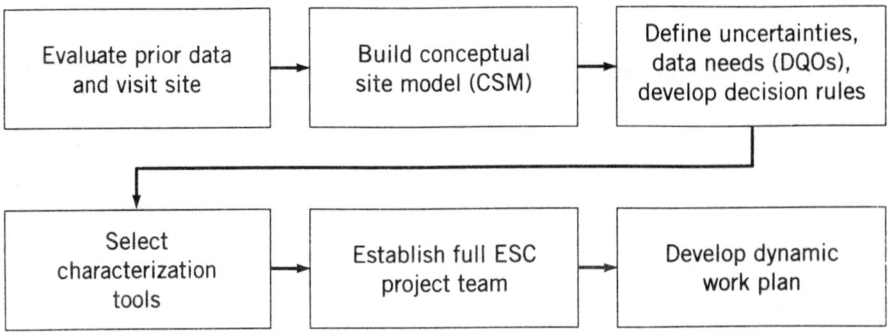

Figure 4. Activities undertaken by the ESC core team to develop the ESC program.

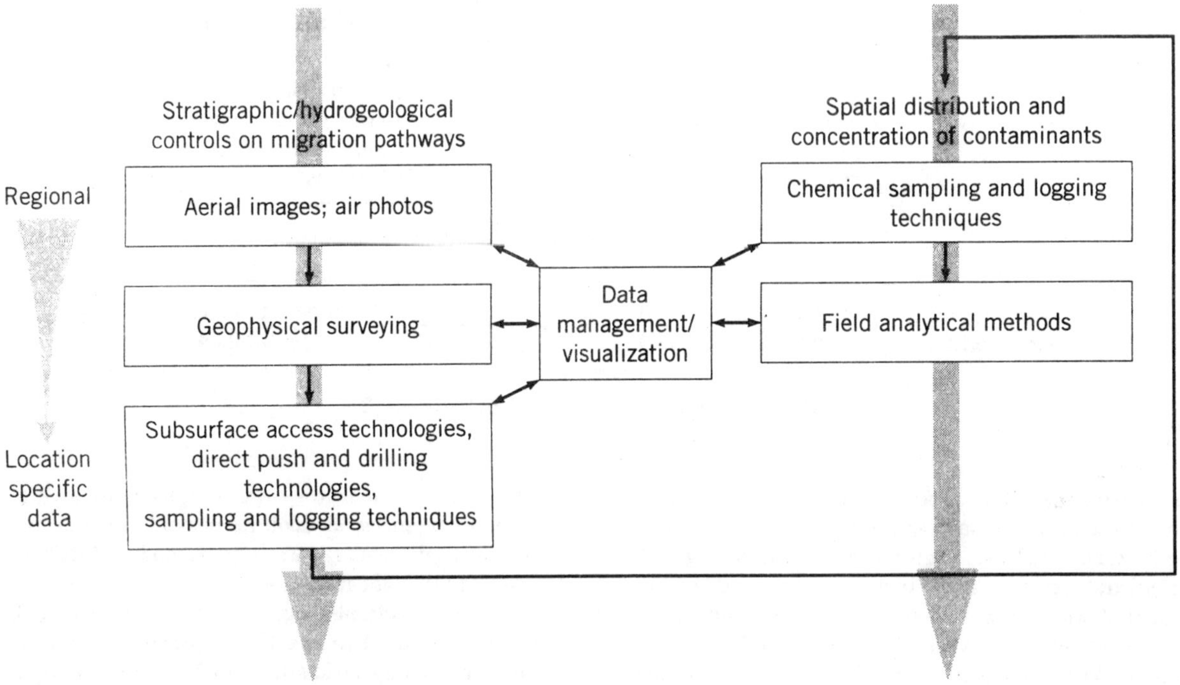

Figure 5. Application of tools for an ESC investigation.

prespecified level of analytical quality as appropriate to the investigation needs.

Phase I Field Investigation

The major focus of the Phase I investigation is to develop the geologic and hydrogeologic models sufficiently well to understand potential contaminant migration pathways to be used as a guide to the Phase II investigation, and to develop a list of contaminants of potential concern which will be tracked during Phase II. For characterizing the geological and hydrogeological properties of the site, ESC first seeks to understand the regional geologic setting of the site followed by focusing on smaller scale details. For example, is the site in an alluvial geologic environment, coastal plain, glacial, etc? Regional geologic studies and reports can be most useful. In addition, the study of air photos by experienced personnel may reveal clues to the regional setting and possibly indicate the presence of larger scale local features (eg, buried channels, karst, faults) that would be less apparent from direct observation at the site. Geophysical survey methods would be applied next in combination with sampling of groundwater from existing monitoring wells, surface soils, waste materials, or other media as appropriate. Geophysical methods inexpensively provide up to 100% aerial site coverage and are extremely useful in delineating anomalies and mapping the continuity of strata, such as aquifers and/or aquitards, on a site-wide scale. A limitation with geophysical methods, however, is that measurement results may have multiple interpretations. Thus, geophysical measurements must be calibrated and corroborated to determine the correct subsurface structure giving rise to the measured signal.

Direct push technologies (DPT) are generally more productive and less intrusive than drilling technologies. DPT methods can provide depth discrete soil, groundwater, and soil gas samples as well as continuous soil samples for both geological characterization and chemical analysis. In addition, on-board sensors may be deployed by the DPT systems for both stratigraphic logging and chemical detection. Once the geology and hydrogeology of the contaminant pathways have been sufficiently defined, the updated site geological and hydrogeological model can be used to forecast the locations of the contaminant plumes. Field analytical methods, such as the use of an on-site mobile laboratory capable of producing required data quality from screening to definitive data, coupled with on-site computerized data management and visualization tools, permits frequent updating of the spatial distribution and concentration of the contaminants in the conceptual site model. Current contaminant data are effectively used in conjunction with knowledge of geologic and hydrogeologic conditions controlling contaminant migration to optimize placement of subsequent samples to efficiently delineate the extent of contamination in one field mobilization.

Geology A suite of geophysical tools, such as ground penetrating radar, seismic reflection and refraction, borehole geophysics, electromagnetic sensors, and radiological sensors may be considered for potential use at a site, according to need and their applicability to existing site conditions. Geophysical tools have been used to locate buried utilities, buried wastes (including metal or plastic objects), to develop soil stratigraphy, depth to bedrock, or search for subsurface regions with elevated organic or inorganic concentrations (see for example 10 for a review of geophysical methods). If there is not already one or more continuously logged boreholes drilled through the soil units of interest at the site, then several "calibrating" boreholes should be drilled and both visually and geophysically logged to develop a general site stratigraphic column. This will be used to correlate and calibrate the direct push sensor output and to aid the interpretation of the geophysical data collected during Phase I. While the electromagnetic and borehole geophysical data may be interpreted in a preliminary sense while on-site, many of the geophysical tools require about one to four weeks of off-site processing for analysis and interpretation. Major benefits of the geophysical methods are their ability to provide widespread coverage of the site rapidly and cost effectively and their ability to indicate anomalous conditions.

Hydrogeology. If groundwater information is lacking, the team may decide to install temporary piezometers to better understand the groundwater system (these activities may actually be initiated prior to Phase I and monitoring may continue throughout the ESC process). Together with historic data from any existing piezometers or monitoring wells at the site, this data may provide seasonal and longer term natural variations in water table depth and the prevailing direction and variations of groundwater flow. Piezometer nests may be installed to evaluate vertical hydraulic gradients. Slug and/or bail tests may be performed in piezometers and/or monitoring wells to obtain data to estimate the hydraulic conductivity of the screened soil units (see HYDROGEOLOGY).

Contaminants. If volatile or semivolatile contaminants are suspected, a grid of passive and/or active soil gas stations may be installed to estimate the location and extent of contamination. Soil and/or water (groundwater and/or surface water) samples from both background (locations on-site or nearby which are not affected by site contaminants) and from expected source and other potentially contaminated areas of the site may be collected and sent to an off-site CLP laboratory for analysis of a full suite of potential contaminants including metals, pesticides, inorganic compounds, organic compounds, and radioactive wastes, as appropriate. Analytical screening methods, such as chemiluminescence, active and passive soil gas, and immunoassay, have been used to locate the main body of plumes of volatile or semivolatile compounds. While some of these technologies may provide contaminant concentrations in the field, in some cases samples and sample collectors are sent to an off-site laboratory for analysis requiring about two to four weeks unless special analytical services yielding a faster turn-around time are requested.

Refine Conceptual Site Model and Identify Nature and Scope of Contamination. The outcome of the Phase I activities includes a site-specific list of contaminants known to be present at the site, refined site geologic and contaminant conceptual models, and a work plan for the Phase II site investigation activities. The site specific list of contaminants is developed on the basis of a pre-established decision rule; eg, comparison of average or maximum observed contaminant concentrations with background concentrations and/or applicable or relevant and appropriate requirements (ARARs). The concentration levels and spatial distribution of the site-specific contaminants will be further investigated during Phase II; note that generally only the site-specific contaminants will be investigated during Phase II thereby eliminating costs associated with unnecessary full suite analyses. Any uncertainties in the geologic environment causing uncertainties in the conceptual model should be identified during Phase I or the post-Phase I interim analysis so that a work plan can be developed to address these issues during the Phase II investigation. The nature and uncertainties of the problem are brought into clear focus so that only data that are needed by the remedial action decision-makers are collected during the Phase II investigation.

Phase II Field Investigation

The Phase II activities comprise an intensive field investigation according to the work plan developed following Phase I. The Phase II field work is managed by the same ESC team that performed the Phase I work and developed the Phase II work plan. The intended output of the Phase II activities include an updated site geologic and contaminant model with sufficient information to allow remedial decisions to be made with relatively high confidence. Although the focus of Phase II is mapping contaminants, as illustrated in Figure 3, further refinement of the geology may form an integral part of the Phase II field work. The ESC team and a variety of sampling and subsurface exploration tools are brought to the site. A mobile analytical laboratory equipped to test for each contaminant of concern in each media of concern yielding screening or definitive type data, as needed, is brought to the site. Data are generally made available from the analytical and geologic assessment tools within hours of the sampling event. As new data become available, they are incorporated into the conceptual site model which is used to aid the selection of subsequent sample locations. Initial sample placement may be near the expected source of contamination with subsequent sample locations for contaminant analyses chosen according to the current understanding of contaminant migration pathways on the basis of available geologic and hydrogeologic data and prior contaminant concentration measurements. Sample location decisions are based on input from the ESC team members with primary authority and ultimate decisions made by the core team leader.

Geology. The primary tools for Phase II characterization of the site geology are the minimally intrusive direct push technologies (DPTs) such as the percussion probe and cone penetrometer. The percussion probe has been used to push a soil electrical conductivity probe into subsurface soils to develop a depth versus soil electrical conductivity profile for stratigraphic logging; Geoprobe Systems has developed this technology (11). The cone penetrometer has been used to develop a soil behavior classification profile versus depth using methods previously described (12–17). Both percussion probe and cone penetration test (CPT) initial pushes are performed adjacent to several (preferably two or more) continuously logged boreholes to correlate and calibrate the direct push output with the actual soil type. Both of these direct push technologies have the ability to take soil samples allowing a visual classification of the soil; however, samples are generally taken for contaminant analysis and not solely for a soil classification. Nevertheless, this sampling capability often provides additional information regarding soil classification and soil mass characteristics, such as fractures in clays, root holes, etc, that may not be apparent from the probe sensor output. Both direct push technologies can collect groundwater samples as well. Additional probe sensors exist, such as the CPT laser induced fluorescence (LIF) sensors to detect petroleum and polycyclic aromatic hydrocarbons (18–21), the piezocone to sense the dynamic pore fluid pressure during a CPT push, and the percussion probe soil electrical conductivity sensor. Lastly, these DPT technologies can generally seal the probe hole by injecting grout into the hole as the probe rods are removed. This grouting feature greatly reduces the chance of cross-contamination of geologic units and groundwaters that are hydraulically separated by aquitards which may be penetrated by the DPT probes.

Hydrogeology. Cone penetrometer piezocone data may provide a rough estimate of depth to groundwater. In addition, a series of pore pressure dissipation tests may be conducted to enable an estimation of the horizontal coefficient of consolidation and the hydraulic conductivity of the fine grained soils using methods previously described (22,23). These pore pressure dissipation tests may be carried out long enough to allow the soil pore pressure to stabilize so that a static pore pressure versus depth assessment can be made. With a properly calibrated piezocone, static pore pressures at depth may be used to characterize vertical hydraulic gradients in subsurface soil units.

Contamination. Contamination in soil and/or groundwater is assessed on-site in mobile analytical laboratories using EPA SW-846, or other methods approved by the regulatory authorities. Soil and groundwater sampling has been performed using the percussion probe as the method of choice. A variety of small diameter percussion probe groundwater sampling devices are available (24). Samplers can be used to collect groundwater grab samples whereby a sampler is installed at a specified depth, the sample collected, followed by removal of the sampler. Additionally, permanent or semipermanent monitoring wells may be installed for longer term groundwater monitoring. Several samplers for the collection of discrete soil samples at a specified depth for visual inspection and/or chemical analyses are described by Kejr Engineering (25,26). The percussion probe is mounted on the back of a pick-up truck, and is relatively fast, produces little contaminated waste material to contain, and is relatively maneuverable over uneven terrain and in tight spaces.

Data Analysis. The data analysis and integration computer programs provide relational database capability and are used to archive the data and provide the ability to integrate data from a variety of sources. The computer programs allow a site plan to be overlaid onto any plot, can produce surface or 2-D and 3-D iso-contour plots with rotation of 3-D images, fence (cross-section) diagram generation, and possibly geostatistical analysis of data. Spreadsheets or custom programs may be used for data handling and analysis as needed.

Phase II Output. The output of Phase II is site geology and contamination models developed to the extent that risk assessment may be undertaken and remedial options, if remediation is required, may be screened in a preliminary sense. A report describing the final analysis and site characterization results is written following the Phase II investigation. Presentation of results is carried out as appropriate.

Key Guiding Principles Behind ESC

ESC is distinct from conventional site characterization approaches through the dynamic on-site decision-making nature of the investigation which is carried out by a multidisciplinary team of experienced professionals. The dynamic investigation is made possible by the technological advancements in and availability of near-surface geophysical, DPT probing and sampling tools, mobile field analytical laboratories, and data integration and analysis tools. Using the ESC approach, the field team can integrate and interpret data from multiple sources while in the field in near real-time allowing near optimal placement of subsequent sample locations. The result of this is that fewer samples are collected in uncontaminated regions, contaminant migration pathways are better defined, and the contaminant spatial distribution can be characterized with one mobilization of the field team. These inherent advantages of ESC over the conventional site characterization approach have resulted in better, faster, and safer site characterizations.

EXAMPLES OF ESC APPLICATION

Former Manufactured Gas Plant Site

The DOE's Ames Laboratory conducted a demonstration of the ESC process at a former manufactured gas plant (FMGP) site in the midwestern United States. The main aspect of this investigation was to demonstrate the ESC process and compare innovative site characterization technologies with state-of-the-practice characterization technologies. Immunoassay, chemiluminescence, active and passive soil gas, geophysical borehole logging, ground penetrating radar, electromagnetic offset logging, seismic refraction and reflection, minimally intrusive direct push technologies, and on-site mobile analytical laboratories were all used at the FMGP site. While one objective of this demonstration was to field, compare, and contrast several differing technologies, it should be noted that cultural conditions at the site, including a rail switching yard immediately south of the site, metal fences and storage sheds on-site, on-site vehicle traffic, overhead power lines, and subsurface utilities, severely impacted the applicability and interpretation of results of some of the geophysical methods used. This illustrates that a particular technology should be chosen on the basis of its applicability to the problem at hand with consideration of the existing site conditions under which the technology will be utilized.

FMGP Site Background. The FMGP site occupies about 2.5 acres and is shown in plan view in Figure 6. A manufactured gas plant operated at the site from the middle 1870s until 1946. Major by-products of this process were coke, tar, ash, and spent purifiers. The final disposition of these by-products on the site is unknown.

At the FMGP site Pleistocene glacial till, glaciolacustrine deposits, fluvial deposits, and loess lie unconformably over discontinuous layers of limestone and dolomite. Prior to the ESC demonstration, 16 borings and 7 monitoring wells were drilled and visually logged at the FMGP site. Soil and groundwater samples were collected from these borings and monitoring wells and sent off-site for analytical analyses for contaminants. Data from these boring logs indicate that the surface soils are stratified and may be broadly grouped into three basic soil units: an upper cohesive unit of clayey and silty material overlying a granular unit composed of sands and gravely sands, in turn overlying a lower, discontinuous cohesive unit of clayey till. The site soils are overlain by about 0.3 to 3 m (1–10 ft) of fill materials of variable consistency including soil materials, glass, metal, bricks, coal, and wood fragments. The depth to bedrock varies from about 6 to 12 m (20–40 ft) or more. A typical stratigraphic profile is illustrated in Figure 7.

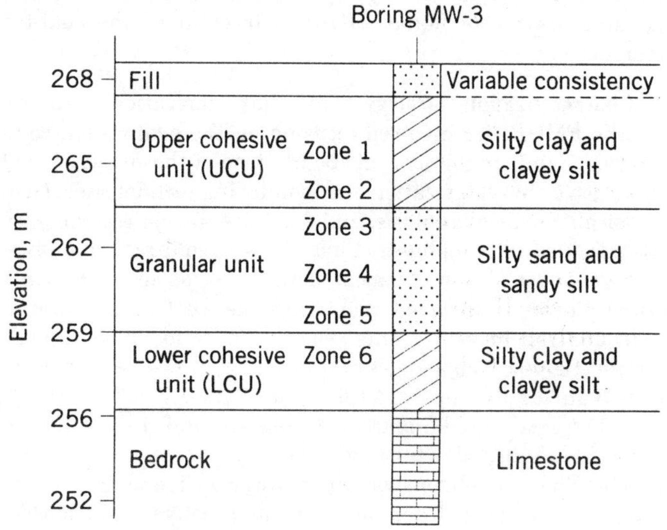

Figure 7. FMGP monitoring well 3 stratigraphy showing stratigraphic zones.

For the purpose of developing and assessing the site contaminant model, the site stratigraphy was subdivided into six zones as noted in Figure 7. This zonation simplified the collection, integration, and interpretation of contamination data as related to the stratigraphic controls on potential migration pathways.

Nature of Contamination. Visible coal tar or fuel contamination was present in the soil sampled from several prior borings at the site. Soil samples submitted for analysis showed contaminants including benzene, toluene, ethylbenzene, xylene, phenols, and polycyclic aromatic hydrocarbon (PAH) compounds. The primary target contaminants in this ESC in-

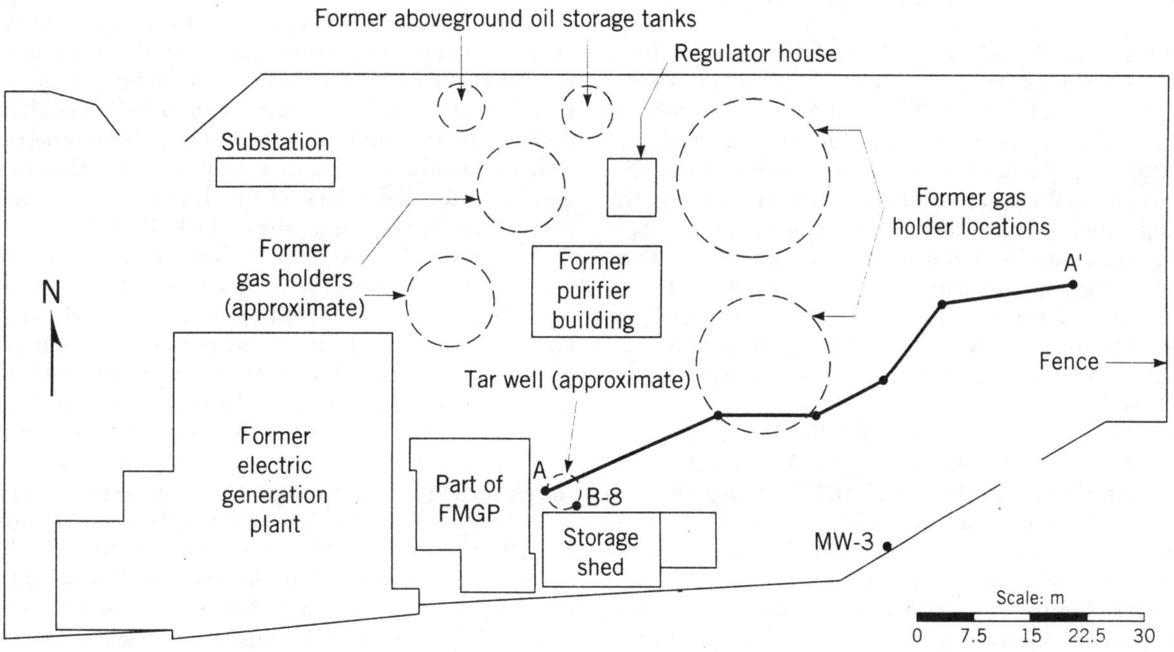

Figure 6. Plan view of the former manufactured gas plant site.

vestigation were the PAH compounds contained in the dense nonaqueous phase liquid (DNAPL) fraction of the coal-tar residue.

Characterization Strategy. The characterization strategy for the FMGP site involved focusing on the evaluation, comparison, and promotion of both state-of-the-practice and innovative characterization and monitoring technologies. Geophysical, chemical analysis, and data integration technologies were fielded in two phases. Phase I involved largely noninvasive geophysical and chemical screening-type measurements while Phase II involved more invasive DPT and contaminant analysis measurements. The Phase I and Phase II work for the fielded technologies were conducted over a five-week period during the spring of 1994. The technologies fielded, approaches used in fielding the technologies, and the results are described in the following sections.

The Phase I demonstration strategy included the use of chemiluminescence, immunoassay, and active and passive soil gas screening technologies to begin to define the location and extent of the PAH contamination in the subsurface soils. This was followed by a suite of geophysical methods, including seismic reflection and refraction, borehole geophysics, ground penetrating radar, and electromagnetic offset logging, utilized to define both the geology and subsurface contamination.

Given the site stratigraphic characteristics and the nature of the subsurface contaminants, the basic objectives for the ESC demonstration were to:

- Define the site soil stratigraphy.
- Define the topography of the lower cohesive unit. This is especially important due to the DNAPL transport characteristics (27) and potential remediation schemes, such as free product pumping.
- Define the location and magnitude of soil PAH contamination in the subsurface.
- Demonstrate efficient and effective integration of multiple technologies, including DPTs, geophysical methods, analytical methods, and data management and analysis.

Phase I Significant Results. In terms of defining the bedrock surface, the seismic refraction survey results compare reasonably well (usually within about 0.6 to 0.9 m (2–3 ft)) with the 16 borehole and 3 on-site monitoring well log data which had been collected at the site prior to the ESC investigation. The borehole geophysical logging results were also verified by the pre-existing borehole logs. The borehole geophysical methods utilized included natural gamma, gamma–gamma, neutron, and induction logging. The borehole logging results added information to the interpreted interface between weathered and unweathered bedrock and were useful for calibrating and verifying some of the other intrusive techniques discussed in following sections.

The 300 MHz ground penetrating radar (GPR) data appeared to locate shallow foundations and services; however, for the deeper penetrating 100 MHz GPR survey there is significant potential error in trying to determine the depth to bedrock. This is due to the presence of relatively high electrical conductivity shallow clayey soils (the upper cohesive unit) which severely limited the penetration of the radar signal. Moreover, the heterogeneous nature of the surficial fill materials would tend to disperse and attenuate the radar signal. The electromagnetic offset logging did not provide useful data at this site primarily due to the significant cultural features present at the site, including chain-link fences, stacks of steel piping, steel storage sheds, and vehicle traffic. Qualitative comparisons of stratigraphic profiles produced by the seismic reflection interpretations with those produced by intrusive data showed poor matches.

Phase I contaminant screening data was obtained from three immunoassay (IMA) kits, a chemiluminescence technique, and active and passive soil gas measurements. Ordinarily, this many screening techniques would not be utilized at one site during an ESC project; however, as previously mentioned, the intent of this ESC demonstration was to compare and contrast a variety of methodologies. Each method was applied at a depth of about 1.5 to 2.1 m (5–7 ft) and again at a depth of about 3 to 4.6 m (10–15 ft) at the sample locations shown in Figure 8. The results of the Phase I contaminant screening technologies generally compare well with each other, as well as with data obtained from the 16 on-site soil borings and three on-site monitoring wells.

IMA is a technology that was developed by the medical testing community and is based on very high selectivity of enzyme activity that permits sensitive analysis of organics in the presence of many potentially interfering compounds. The IMA kits require little training or expertise and give results on-site in a matter of minutes using color coded liquid extractants. They detect a weighted average of the PAHs of interest and can only be used in a screening mode. The IMA kits used in this investigation were the Ensys and Millipore kits, which gave a gravimetric measurement in ppm by weight, and a Quantix kit, which gave a volumetric measurement in ppm by volume (see IMMUNOASSAY, ENVIRONMENTAL).

The chemiluminescence technology was being marketed in Europe at the time of this FMGP investigation and was selected for use at the FMGP site because of its simplicity and potential for quantitative analysis. After a proprietary liquid-phase extraction from a soil sample, a chemical was added to the extract and the luminescence was monitored by a simple solar cell-based detector that converts the emitted light signal into an electric current. Calibration factors, using the results of HPLC analyses on representative site samples, are required to convert the measured intensity of the light to concentration. Results were reported in terms of total PAH (TPAH) less naphthalene. In the analysis below, the chemiluminescence data were converted to TPAH by noting that naphthalene typically accounted for about 16% of the TPAH value on the basis of Phase II soil sample analyses at the FMGP site.

As shown in Figure 9, the shallow (corresponding roughly to zone 1 in Fig. 7) Phase I screening-type soil contaminant plume constructed from the chemiluminescence TPAH data is generally centered around the former purifier building (see Fig. 6) and extends toward the storage shed at the southern edge of the site. As the detection limit for the chemiluminescence data used to generate the TPAH contours in Figure 9 was 10 ppm (corresponding to one on the log scale), Figure 9 indicates that the eastern and western edges of the site at the 1.5 to 2.1 m (5–7 ft) depth range have low levels of PAH contamination. Figure 10 shows the TPAH contours for the shallow 1.5 to 2.1 m depth range constructed on the basis the Ensys IMA kit. Note that the contours shown in Figures 9 and 10 are in generally good agreement with respect to the locations of the main contaminated soil region and those areas which are relatively

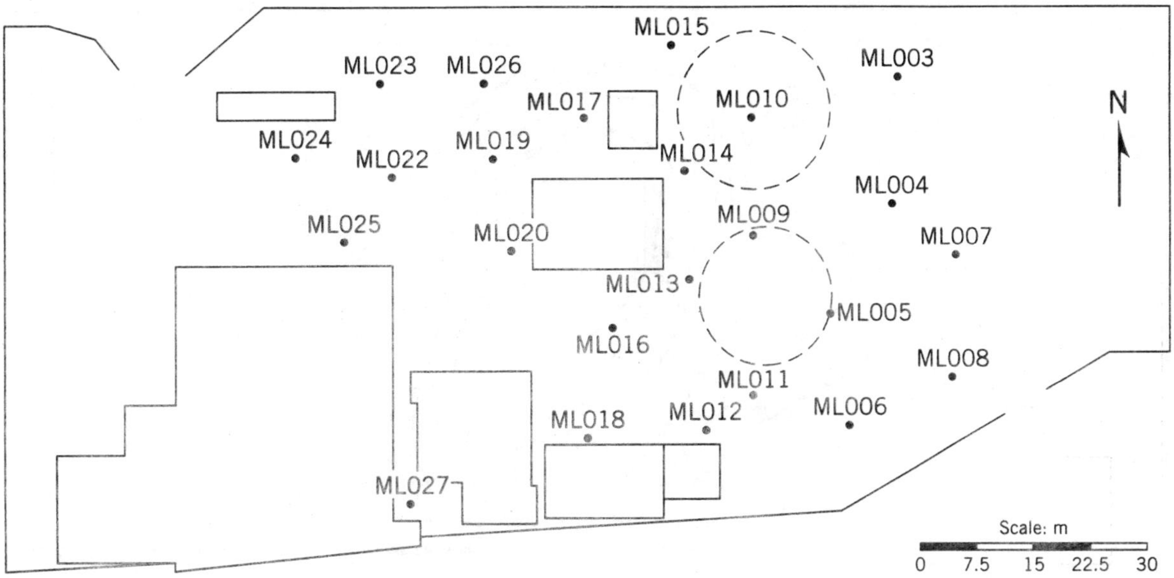

Figure 8. FMGP ESC Phase I screening data sample locations.

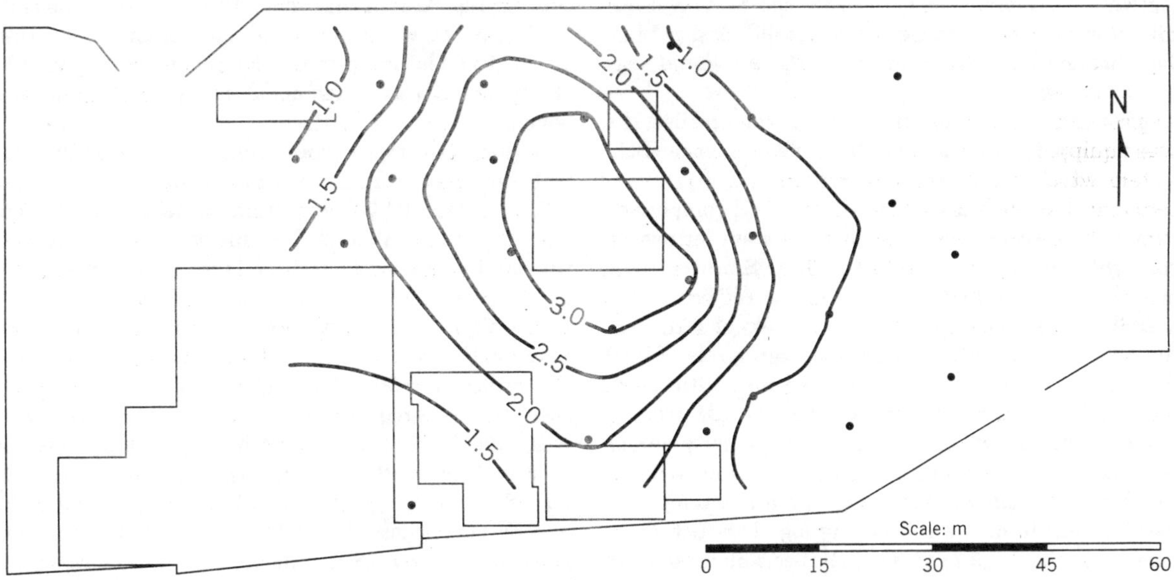

Figure 9. Logarithm of total PAH based on chemiluminescence data (mg/kg).

clean. This information is useful in updating the contaminant conceptual site model and in planning the Phase II ESC sampling activities.

Phase II Significant Results. The Phase II work at the FMGP site consisted of two minimally intrusive, direct push characterization technologies together with percussion probe-based soil sampling and on-site mobile analytical laboratory chemical analysis of soil samples for PAH compounds. An innovative cone penetrometer test (CPT) system, the site characterization and analysis penetrometer system (SCAPS) (18), and an innovative Geoprobe direct sensing soil conductivity sensor (11) were both fielded during this investigation. Further information is given on the SCAPS and its use at this FMGP site (28). The objectives of this phase of the investigation were to confirm and further refine the site geologic and contamination conceptual models as defined through the Phase I activities (following the ESC approach) and to compare and evaluate the CPT and percussion probe systems fielded. A key objective for the direct push technologies was to define the surface topography of the lower cohesive unit. These direct push technologies, however, are limited in providing direct data to define the surface of the bedrock. The first task of the ESC Phase II DPT program was to carry out CPT and percussion probe

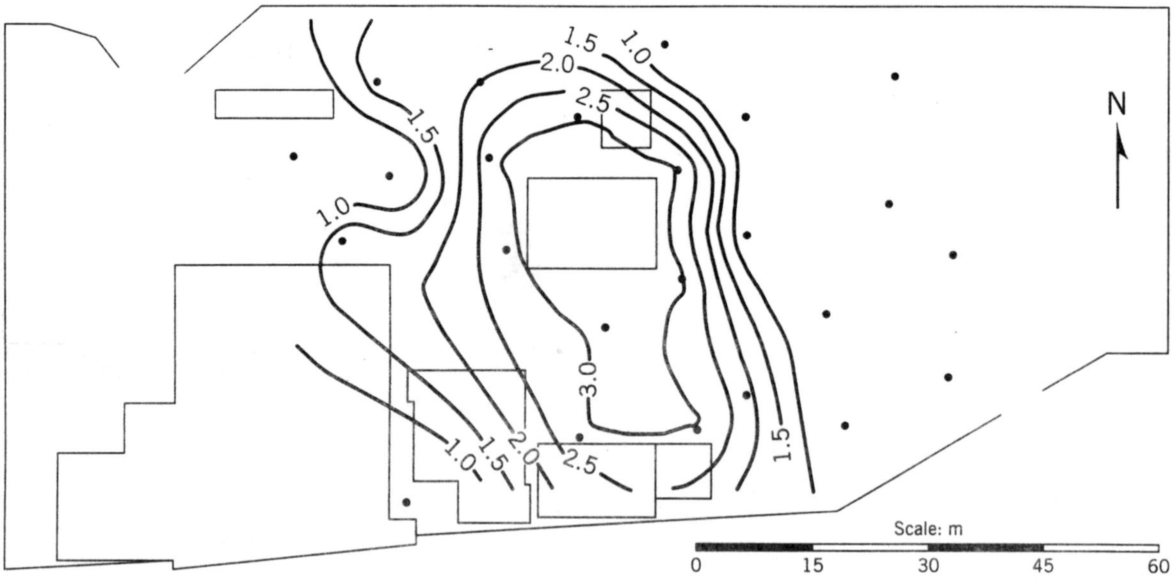

Figure 10. Logarithm of total PAH based on Ensys IMA data (mg/kg).

electrical conductivity pushes adjacent to several existing monitoring wells to compare and calibrate the SCAPS interpreted soil behavior types (stratigraphic profile) and percussion probe conductivity measurements with the visual logs from the soil borings.

Besides providing continuous stratigraphic soil profiles, the SCAPS was equipped with a laser induced fluorescence (LIF) sensor system which could detect petroleum, oil, and lubricant contaminants in the subsurface. As the PAH compounds show a strong fluorescence response to the ultraviolet excitation source light used in the SCAPS, the SCAPS proved to be an extremely valuable technology to delineate regions of the subsurface which had nonexistent to low levels of PAH contamination from regions with moderate to high levels of PAH contamination. A continuous log of LIF response with depth was provided at each push location. An example LIF trace is illustrated in Figure 11 where fluorescence intensity versus depth data collected about 4.6 m (15 ft) laterally from soil boring B-8 (see B-8, Fig. 6) are plotted along with a visual log (material classification) from the soil boring log. This LIF trace shows evidence of about 0.9 m (3 ft) of PAH contaminated soil directly overlaying the lower cohesive clayey soil at about 10 m (35 ft) below ground surface along with evidence of PAH contaminated soil from about 0.9 to 7 m (3–23 ft) below ground surface.

The percussion probe conductivity sensor provided a continuous log of soil conductivity with depth at a push location that generally clearly delineated the clayey from the sandy soils units at the site. As the resistivity of a soil will increase (conductivity decrease) with increasing hydrocarbon contamination in its pores, the percussion probe conductivity logging could also be used as a contaminant screening tool for the low electrical conductivity PAH contaminants. Several soil conductivity logs showed a marked downward dip in conductivity at the base of the granular unit. This was underlain directly by the lower cohesive unit, as interpreted from the logs. Because soils contaminated with PAH compounds will have a reduced conductivity, it was surmised that the dip in the conductivity log may be due to the presence of free-phase coal-tar residue containing PAHs in that zone. An example of this is illustrated in Figure 11 which shows a conductivity trace taken about 3.7 m (12 ft) laterally from a logged soil boring together with a fluorescence trace taken about 4.6 m (15 ft) from the same soil boring.

Figure 11 shows a typical erratic conductivity in the top five feet indicating surficial fill materials. At 5.7, 6.8, and 7.4 m (19, 22.5, 24.5 ft) below ground surface (bgs) the conductivity trace dips below 40 milli-Siemens per meter (mS/m) adjacent to the sand seams and sandy soils with free oil noted at 18 and 20 ft bgs in the nearby boring log. From 5–16.5 ft bgs relatively high (>90 mS/m) conductivities exist indicative of the upper cohesive zone soils. This is underlain by soils with conductivities near 50 mS/m indicative of the middle granular unit soils; the elevated peak at 6.4 m (21 ft) bgs may be associated with the trace of clay noted in the nearby boring log. From 10.2 to 10.6 m (33.5–35 ft) bgs the conductivity dips below 40 mS/m adjacent to the free oil contamination indicated in the nearby boring log. Below 10.9 m (36 ft) bgs the conductivity rises to about 80 mS/m indicating that the lower cohesive unit had been penetrated.

Another valuable source of data for comparison with the CPT and Geoprobe conductivity data and for further development of the site geologic and contamination models is the visual description logs associated with each of the 123 soil cores (which resulted in 127 soil samples) collected from 29 map locations for quantitative chemical analysis. The soil samples for this effort were collected by a two person crew operating a Geoprobe® Model 8-MU sampling system. Core samples 0.6 m long (2 ft) by 2.54 cm in diameter were collected from six subsurface zones at specified depths. A detailed descriptive log of each of the 127 samples collected was kept and data on core recovery, date, time, sample number, depth, soil description, water content, and visual/odor evidence of contamination were recorded.

The CPT and conductivity soil profiles show significantly more detail than do the boring logs and show that the CPT and conductivity soil profiles were in close agreement with

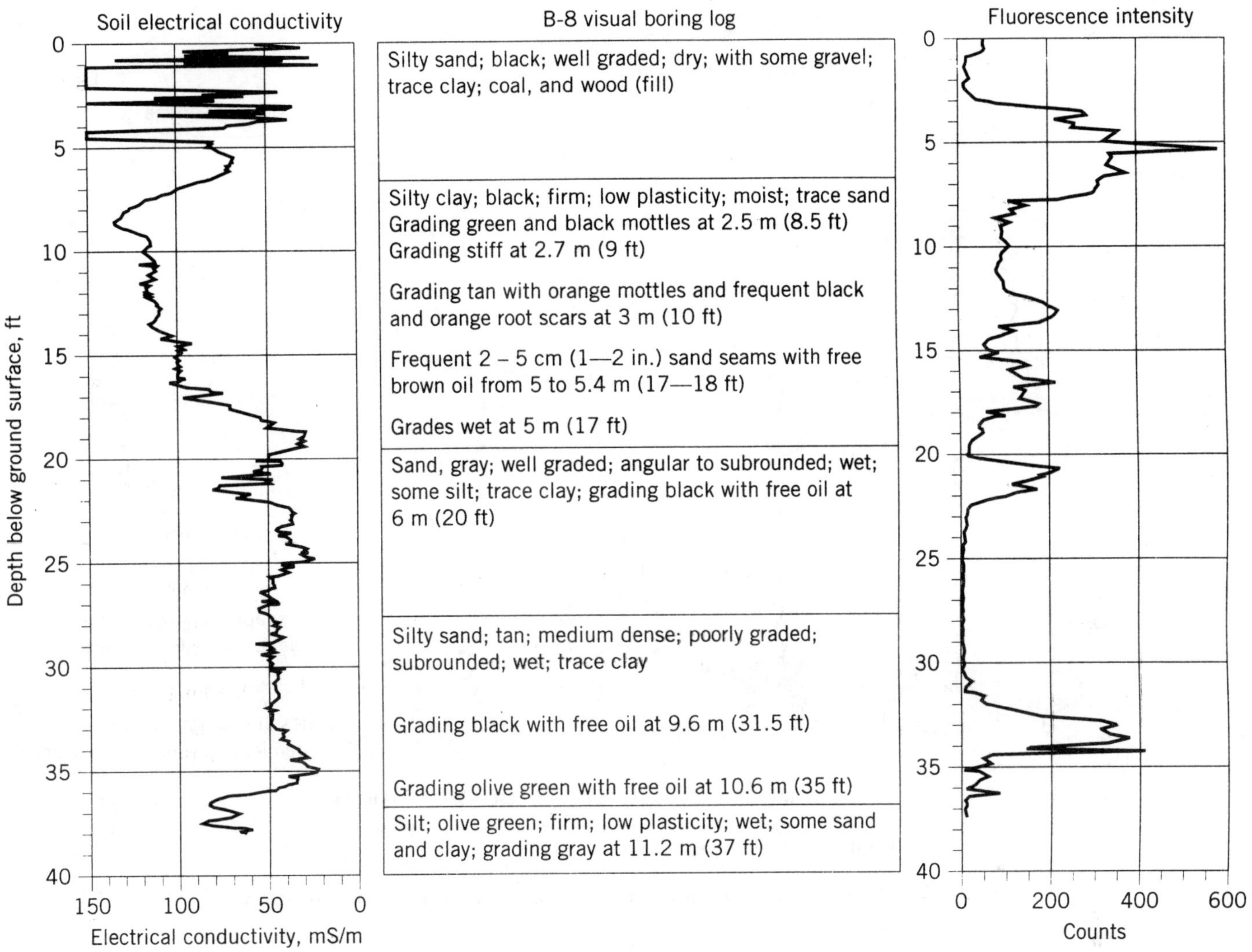

Figure 11. Electrical conductivity profile, soil boring log, and LIF trace north of the storage shed shown in Figure 6 (29).

the boring logs. The strategy for subsequent CPT and conductivity deployment was to fill in data gaps to verify and/or enhance the geologic and contamination conceptual models. Figure 12 shows a stratigraphic cross section prepared using the six conductivity traces shown (with scale increasing from left to right) as the primary source of stratigraphic information. Stratigraphic contacts and contamination data taken from several nearby boring logs (B-6 and B-8) and CPT data (ML-77 and ML-109) were added along the cross section. Figure 12 illustrates the development, and generally good agreement, of both stratigraphic and contaminated regions of the subsurface on the basis of the soil boring, CPT, and conductivity data.

Figure 13 shows contour plots of the surface elevation of the lower cohesive unit and available data locations for four selected days: day 2, 3, 5, and 10 of the on-site DPT pushing. These plots are based on all of the DPT data available at the end of each day during Phase II of the ESC investigation and illustrate the use of data to update the conceptual site model as the investigation proceeds. Comparison of the day 2 and day 5 maps shows some major changes in the surface model, particularly on the western portion of the plots where little data were collected during the first few days on-site. Comparison of the four maps in Figure 13 shows that the general features of the surface were reasonably well defined after the third day of sampling and that additional data points beyond day 3 primarily identified only minor fluctuations in the surface topography model. Because no data were collected under the large buildings near the southern edge of the site, there is some uncertainty regarding the topographic surface elevation in this region. The fence along the southern edge of the site identifies the property boundary beyond which samples could not be taken.

Figure 14 shows contours (log scale) of the CPT LIF fluorescence intensity in stratigraphic zone 5 (see Fig. 7) which directly overlays the lower cohesive soil unit, the surface topography of which is shown in Figure 13. At the FMGP, fluorescence intensities below 100 counts were generally associated with low contaminant levels (<100 mg/kg TPAH) while fluorescence intensities greater than about 200 counts were typically associated with moderate (>100 mg/kg TPAH) to high (>500 mg/kg TPAH) levels of contamination (28). Accordingly,

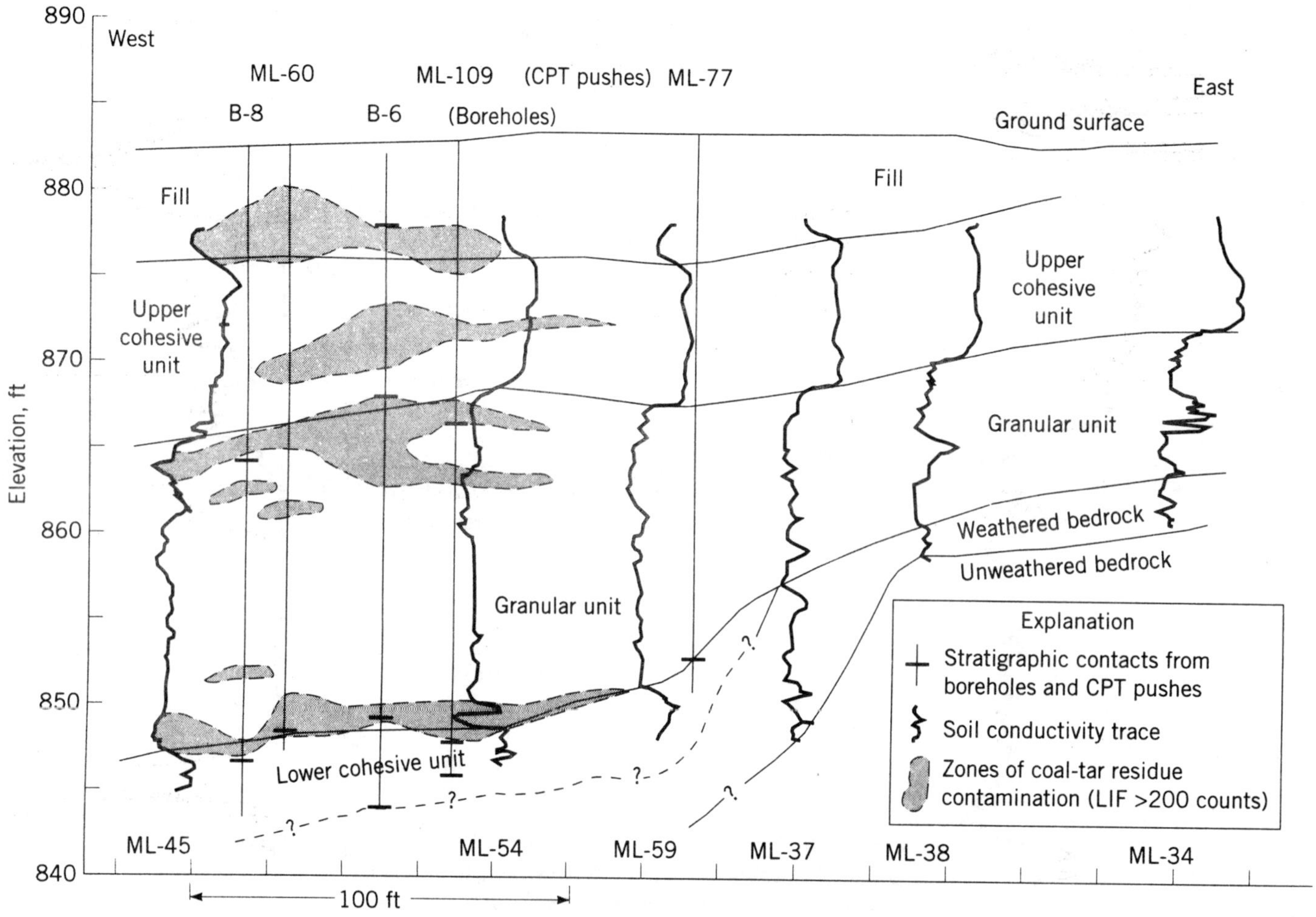

Figure 12. Stratigraphic cross section A–A′ from Figure 6 (29) (1 ft = 0.3048 m).

the LIF probe was an excellent screening tool to indicate the presence of PAH contaminants in the subsurface. The LIF contours in Figure 14 indicate the presence of PAH contamination under the central portion of the site and trending off-site toward the south. This is consistent with DNAPL migration downgradient along the upper surface of the lower cohesive soil unit (see Fig. 13) which acts to retard the downward migration of the DNAPL as described previously (27). Such LIF information often formed the basis for placement of soil samples to confirm the presence or absence of PAH contamination.

Analytical Data. In general, soil samples were only taken at locations and depths where information content was expected to be maximum based on all prior evidence, including pre-ESC data, percussion probe conductivity logging, CPT-LIF, and CPT soil stratigraphic data. Occasionally a 0.6 m (2 ft) section was requested from a depth with the expectation that two contiguous zones would contribute equally to the 0.6 m core. In these cases the prior stratigraphic evidence was of such high quality that nearly all of these types of cores met our stringent expectations and two distinct samples could be generated from one core. A total of 127 soil samples were selected from the core samples and analyzed on-site for the 16 target PAH compounds. Twenty-one additional samples were so contaminated (by visual inspection) that they were simply archived and qualitative GC/MS fingerprinting of the free product phase was done.

A major effort at the FMGP was the fielding of five separate on-site extraction schemes for the semivolatile PAHs in soils. The extraction methods used were sonication, supercritical fluid extraction (SFE), thermal desorption, microscale, and microwave extraction. These five extraction schemes were coupled to three gas chromatography/mass spectrometry (GC/MS) instruments. Three subcontractors were involved in this rather unique activity; successful implementation of this activity required their complete cooperation in the field. The sonic solvent system consisted of a Tekmar Sonic Disrupter coupled to an HP Series II 5890 GC with an HP 5970B MS detector. The SFE system included an HP Model 7680T SFE Module coupled to an HP Series II GC with an HP 5972 MS detector. The thermal extraction unit consisted of a Thermex extractor from Ruska Instruments with the same GC/MS unit as was used for the SFE. A CEM Corporation Model 1000 Microwave Extractor was used. Eighteen samples were also sent off-site and analyzed using Method 8270B for SVOCs. Only 16 targeted PAHs were measured by the off-site CLP laboratory. Autosamplers were used to permit overnight analysis. The sonication and microscale extraction analyses were done on one GC/MS instrument, the supercritical fluid and microwave extraction analyses were done on a second GC/MS instrument, and the

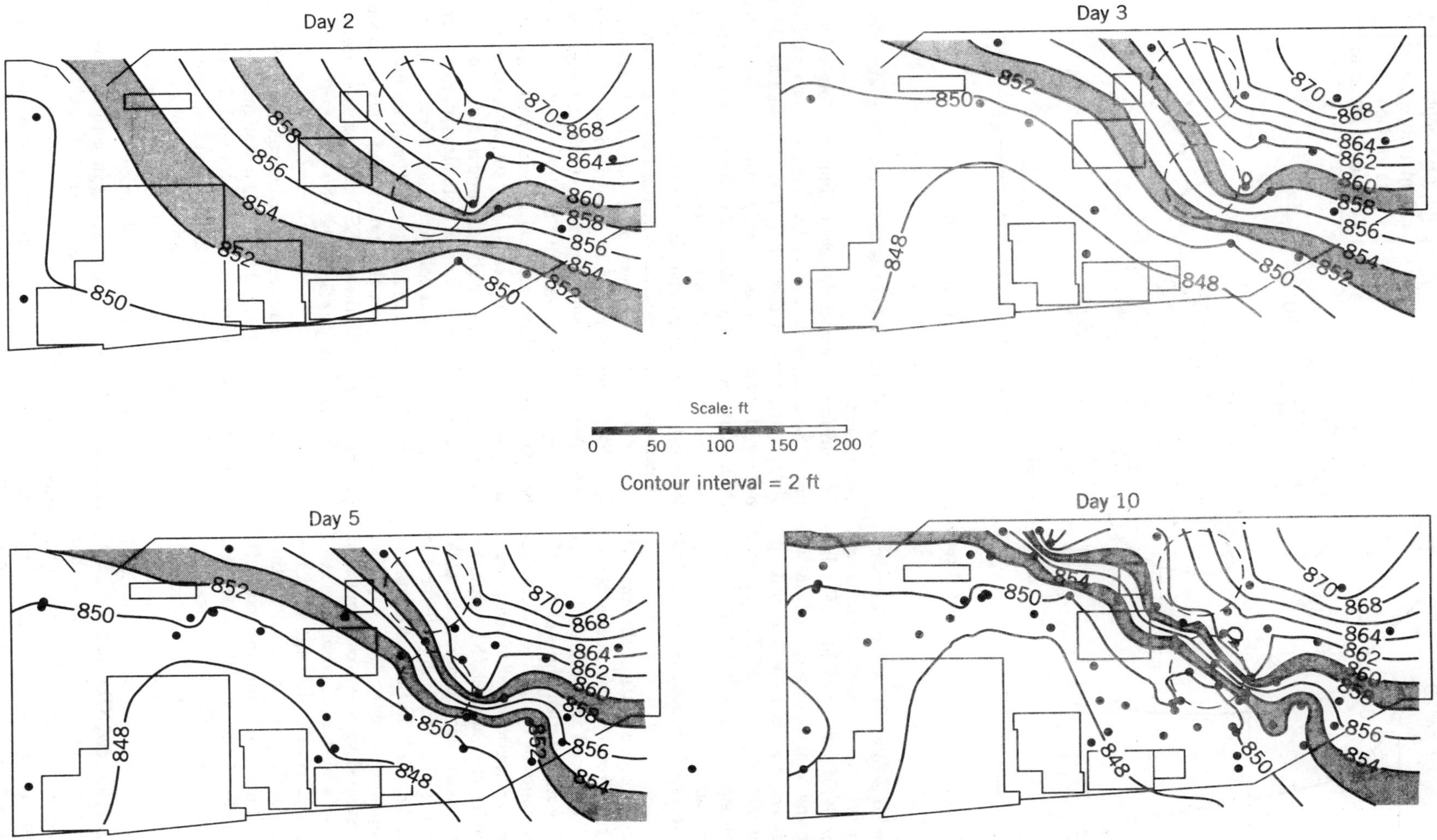

Figure 13. Daily lower cohesive unit surface elevation (feet above mean sea level, msl) with sample locations shown (one foot = 0.3048 m) (29).

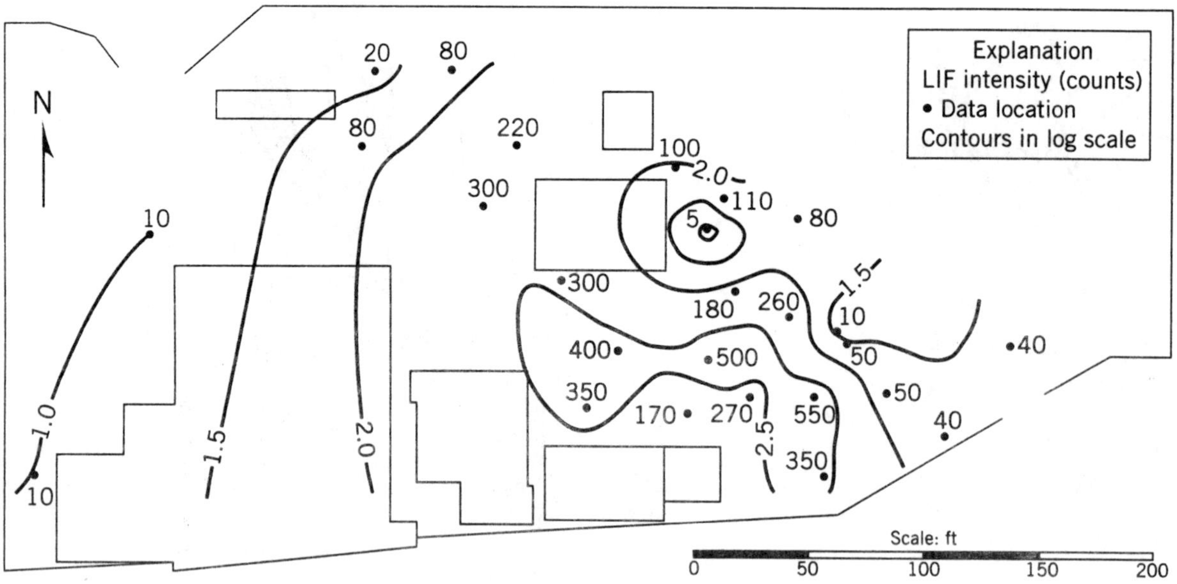

Figure 14. CPT fluorescence intensity in stratigraphic zone 5 (see Fig. 7).

thermal extraction was done on a third GC/MS instrument. Samples prepared by microwave and microscale extractions were analyzed by whichever GC/MS was available. For homogenized soil samples split and analyzed using each of the five extraction methods, differences as large as one order of magnitude occurred in about 20% of the cases. The thermal extraction data tended to be significantly lower while the microscale and microwave extraction data tended to be significantly greater than the average of the results for the PAH measurements.

In terms of the site characterization goals of this project, samples were selected starting from outside the expected plume area followed by moving toward the suspected plume with the intent of using the low quantitation limits of the GC/MSs to map the action level limits of the plume. This plan conflicted with the technology evaluation goal of the project because it led to a large number of nondetect (ND) samples that could not be used for the technology comparison task. Nevertheless, this site sampling plan was followed for the first several days of soil sampling. In addition, estimates of the sample analysis throughput were lower than expected and if this approach were continued very few contaminated samples would be available for a systematic comparison of the five extraction methods. As a result, after the first few days the sampling and analysis plan was modified so that only samples judged to be contaminated were automatically sent to all five extraction modules. Other samples were first analyzed using the sonication method. These samples were then sent to the other extraction systems only if the sonication results confirmed that PAH contamination was present. Even with this revised system implemented, however, a significant number of the samples analyzed returned ND results.

A fortunate circumstance was realized after a few days that made the criteria for selecting the contaminated samples for further analysis relatively easy. It was noticed that soil samples that were characterized as having a slight but definite odor turned out to have PAH concentrations in the 1 to 10 mg/kg range. This helped make good estimates as to which samples to send for analysis without incurring any idle GC/MS instruments.

A uniform method detection limit of 1 mg/kg was determined for each extraction method and for each analyte. Some of the samples were diluted 5-, 10-, or even 100-fold prior to testing; if such a diluted sample returned a nondetect result (ND, below the method detection limit), then it was recorded as <5, <10, or <100 mg/kg, respectively (depending on the level of dilution). Dilution of samples at this level causes severe limitations in the usefulness of the resulting data whenever a nondetect value is observed because the detection limit for total PAH may be raised to a level above the target action level. The spatial prediction method (30) is useful for spatial interpolation and uncertainty analysis with nondetect data, but the method does not deal directly with the problem of multiple detection limits caused by the use of several levels of sample dilution in the analytical preparation process.

As this ESC project was a technology demonstration and comparison, and not a full site characterization effort, adequate analytical data to map the PAH contaminated soils were not obtained. However, those data that were obtained are in good agreement with the pre-ESC data, as well as with the soil screening technologies fielded at this site.

Discussion. In making the conclusions noted above and detailing the lessons learned from this investigation which follow, it must be kept in mind that a primary goal of the ESC project at the FMGP site was to compare and evaluate state-of-the-practice technologies and innovative technologies; the goal was not to carry out a full ESC style characterization of the site. In this context, in terms of geological characterization technologies, a lesson learned from this investigation is that while geophysical survey techniques are an integral part of the ESC process, all techniques may not be appropriate for all sites. In this case, the difficult stratigraphic conditions and significant cultural interferences, along with, in some cases poor data collection and analysis/reporting procedures, led to significant error and uncertainty in the geophysical survey results and severely limited their useful application. The potential limitations of geophysical methods need to be carefully considered on a site specific basis.

The application, versatility, and the high quality of data from direct push technologies was demonstrated and proven at this site. The SCAPS CPT demonstrated the reliability and efficiency of both the stratigraphic logging capabilities and LIF sensing capabilities in developing the site stratigraphic model and delineating areas of PAH contamination. The capabilities of the percussion probe soil conductivity system were also proven. It had more maneuverability than the CPT and was also more operationally efficient. With proper calibration, it can be reliably used to fill in stratigraphic data between two locations with known stratigraphic profiles. In addition, again with proper calibration and verification, it can be used to enhance the site contamination model. Both direct push technologies have the capability to provide much more detailed stratigraphic information than conventional auger boring, which is important when considering contaminant fate and transport. On the down side, both direct push technologies had difficulties detecting weathered bedrock and unexpected bedrock or large gravel/cobbles resulted in several broken probes. More rugged equipment for getting through upper heterogeneous fill layers, which could be present on many developed sites, could be considered.

There was a great deal of soil sampling in the east and northeast areas of the site. Once it was established that contamination in this area was low or nonexistent, this sampling activity should have been shifted west to where higher levels of contamination were detected. The eastern boundary of the contaminant plume was defined with relatively high accuracy; however, there is a large uncertainty regarding the location of the western boundary of the plume. This is due in large part to several large buildings in the southwest corner of the site; however, it may have been possible to obtain several more samples at selected locations adjacent to these buildings. As one of the primary objectives of the Ames Laboratory ESC project was to collect enough data to allow direct comparison of state-of-the-practice and innovative technologies, this does not impact on the success of project.

Lastly, this study indicates the potential for significant variation of chemical analysis results for PAHs in soils, even when following regulatory approved SW-846 procedures. The uncertainty and potential variability associated with soil matrix effects, sample selection, preparation, and extraction procedures far outweigh inaccuracies in the chemical analysis methodologies themselves. These results are consistent with other studies on the relative contribution of sampling and analysis to the total variance (31,32). Additionally, it was found that the measurement error standard deviation was roughly proportional to the magnitude of the measured contaminant concentration. In contrast, the CPT-LIF was found to be an effective contaminant screening tool that gives a continuous profile of contamination at a specific location with results that are largely operator independent.

D-Area Oil Seepage Basins: Savannah River Site

The Savannah River Site (SRS) ESC program was a unit assessment RI activity carried out within the RCRA/CERCLA Facility Investigation/Remedial Investigation (RFI/RI) at the D-Area Oil Seepage Basin (DOSB), SRS, South Carolina. Prior to the ESC investigation, risk assessment concerns were potential transport of unit-related contaminants to surface and subsurface soils, groundwater, and off-site transport to surface water bodies. If in the future, institutional control of the area were lost and no further remedial action were taken, onsite construction workers and residents could become potential receptors. Under the requirements of CERCLA, the DOSB was slated for additional assessment and perhaps, environmental remediation (33). An assessment of the environmental media was accomplished using the ESC methodology. The Phase I and Phase II ESC activities described below were conducted during the late spring and summer of 1995.

Site History and Existing Information. The DOSB was constructed in 1952 as a series of at least three unlined 2.4 m (8 ft) deep trenches separated by berms to dispose of waste oils and other fluids not suitable for burning (33). A plan view of the DOSB is shown in Figure 15. Waste oils including machine cutting oil and transformer and other shop fluids were transported to the basin in drums, opened, and dumped into the trenches. These waste oils were periodically burned along with general office and cafeteria waste. No historical evidence of overflow of the basin exists. This practice continued until 1973 when open burning at the site ceased. The basin received waste oils until 1975 when it was removed from service and back-filled with soil. Covering the DOSB with soil has reduced, but not eliminated, the potential for exposure of humans and wildlife to environmental contaminants. An unknown number of 208 L (55 gal) drums, possibly containing waste oil, remained in the basin. The basin remains inactive and was covered by bushes and grasses and surrounded by trees during the ESC activities.

A preliminary evaluation of the unit was carried out between 1983 and 1988. This phase involved the installation of four monitoring wells (DOB-1 to DOB-4 in Fig. 15) in 1983 and 1984 and a literature review and unit reconnaissance. The preliminary evaluation indicated that the unit has received hazardous substances. Next, a unit screening/limited field investigation was carried out between 1988 and 1994. This phase involved a series of geophysical and geochemical (analyses of soil samples) surveys to identify areas potentially impacted by hazardous substance release. ESC was integrated into the remedial program unit assessment phase for the DOSB in late 1994 and early 1995.

General Site Geology. The site is known to be underlain by a glauconitic marine clay referred to as the "green clay". Prior to the ESC investigation the green clay was thought to exist at a depth of about 20 m (65 ft), however, little was known regarding its surface topography, thickness, or continuity. The green clay is overlain by alluvial sands, silts, and clays (33). Radioactive waste has been studied at the SRS H-area about eight miles northeast of the D-area, where the green clay acts as a strong confining layer, or aquitard, and while the green clay is likely to be largely continuous, a major source of discontinuity exists in the form of erosion by channeling, which may have occurred during any of several periods of unconformity that have occurred since the green clay was deposited (34). The green clay is dissected along a number of present-day stream channels at the SRS.

Four U.S. Army Corps of Engineers SCAPS pushes were performed along the perimeter of the DOSB site during 1991

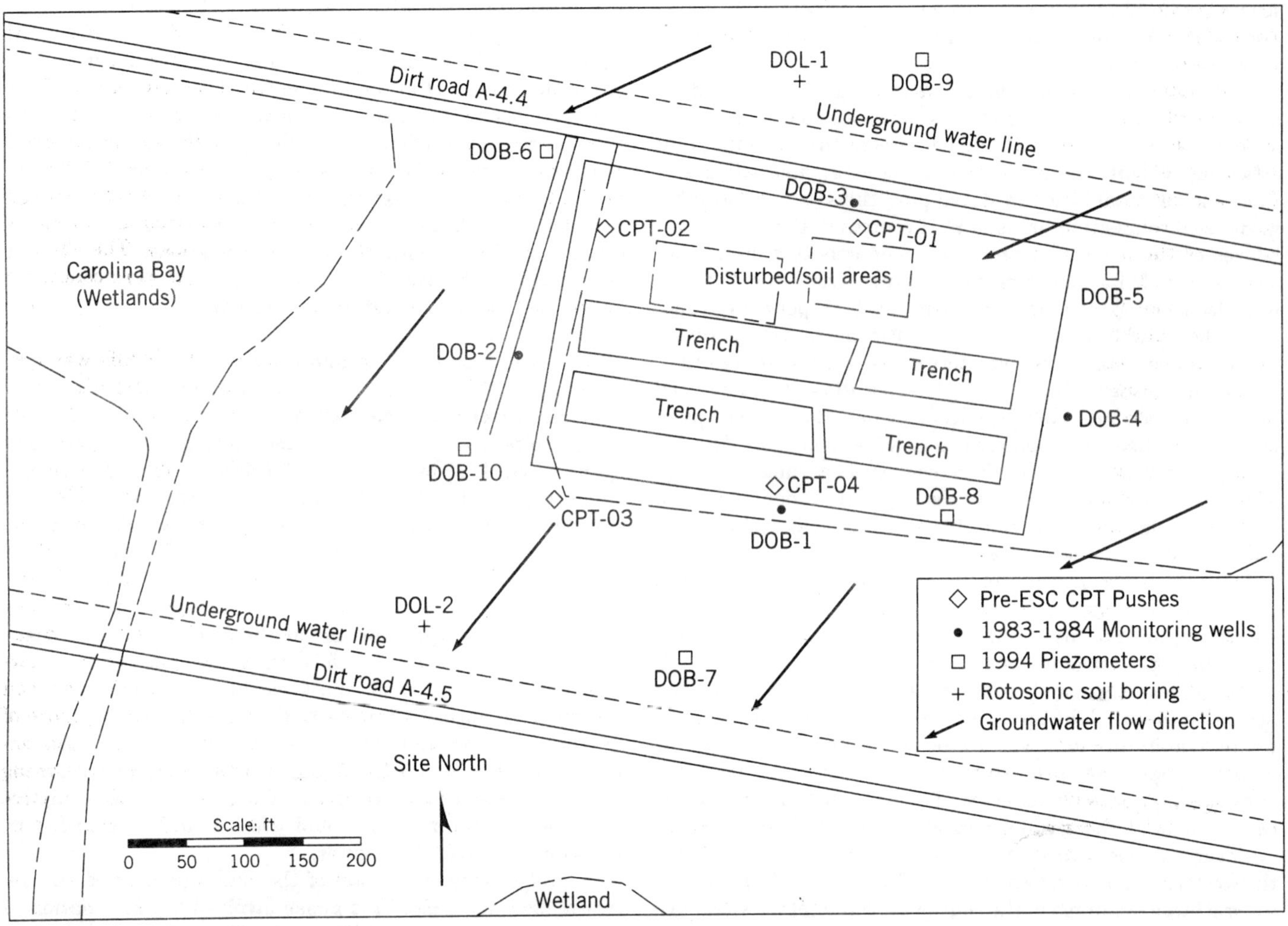

Figure 15. SRS DOSB (trench and disturbed soil area) and surrounding area (one foot = 0.3048 m).

(18) (see Fig. 15 for locations). Three of the CPT pushes were made to a depth of 15.2 m (50 ft) and one to a depth of 9.1 m (30 ft). The CPT was equipped with a LIF sensor (described for the work at the FMGP in the preceding sections) (18). The soil classifications from these CPT pushes generally indicate sandy soils overlying silty and clayey soils at a depth of about 1.5 to 4.6 m (5–15 ft). This sequence overlays about 3 m (10 ft) of sandy soils overlaying more silty and clayey soils at a depth of about 7.6 to 9.1 m (25–30 ft). These soils overlay interbedded sands, silts, and clays. Because of a large amount of apparent interbedding and significant variation between push locations, it was not possible to confidently ascribe continuity of any soil unit between the push locations, nor was it clear whether the green clay had been penetrated by any CPT push. This significant uncertainty highlighted a need for geophysical and additional DPT data to develop the site geologic model and assess the nature and continuity of the soil units underlying the site. The LIF logs do show elevated fluorescence intensities over the 3 to 6 m range (10–20 ft), with some lesser "hits" near 9.1 and 13.7 m (30 and 45 ft) bgs indicating the possible presence of a fluorescent class of organic contaminants in the subsurface.

An initial conceptual site model geologic cross section summarizing the initial state of knowledge of the site stratigraphic conditions prior to ESC Phase I is shown in Figure 16. This east–west general cross section shows the site underlain by sediments consisting primarily of sands, silt sands, and clayey sands. Clayey/silty layers, lenses, or stringers with unknown continuity could be present at depths of about 3 m (10 ft) and 7.6 to 9.1 m (25–30 ft). Regional data and local experience indicates that the sands could be locally cemented and that a shell matrix may intermittently overlay the green clay. A best estimate of the depth to the green clay was about 20 m (65 ft); however, it was not known for certain if the green clay was even present at the site. The boundaries of two parallel trenches and two areas of disturbed soil had been approximately defined by prior geophysical measurements. The trenches were estimated to be about 2.4 m (8 ft) deep. Water table depths as high as 1.2 m (4 ft) below ground surface had been recorded indicating that groundwater has come into contact with waste materials remaining in the trenches.

Prior Investigation Data. Analysis of groundwater samples from on-site monitoring wells over a period of about 10 years has indicated the presence of tetrachloroethylene (PCE), trichloroethylene (TCE), and vinyl chloride (VC) (33). In 1988

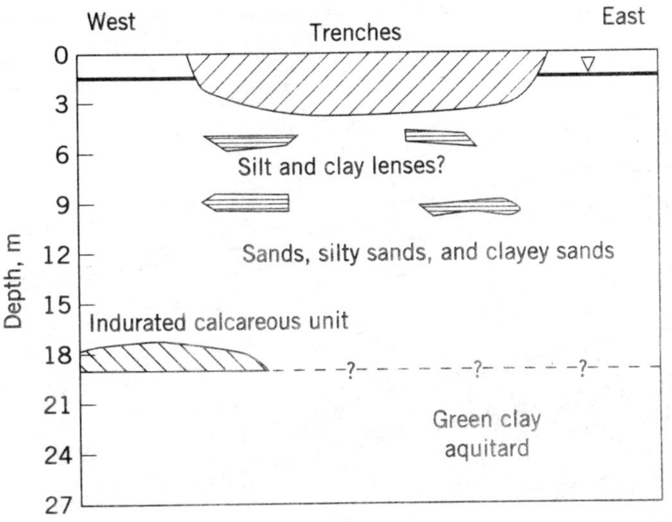

Figure 16. SRS DOSB initial conceptual site model geologic cross section (35).

a drum was punctured during a soil sampling drilling activity and was found to contain a liquid with high concentrations of several organic compounds including TCE and PCE. Contaminants detected in soil samples collected from beneath the unit include PAHs, alkyl benzenes, phthalate, lead, chromium, and antimony.

Selected soil gas measurements were made over the trench area in 1991. The soil gas data for PCE, TCE, chloroform, ethane, Hg, propane, *I*-butane, *N*-butane, ethylene, and propylene all indicated elevated regions and localized "hot" spots for these contaminants. Soil gas measurements for VC, *trans*-1,2-dichloroethylene, 1,1,1-trichloethane, and carbon tetrachloride were largely near or below the detection levels. PCE had the largest area of elevated soil gas measurements; the soil gas data for PCE are shown in Figure 17. Figure 17 shows that the elevated PCE soil gas measurements follow the trench area very closely, with the exception of four slightly elevated measurements in the extreme northeast of the sampled area. Extreme elevated TCE measurements are confined to the southwest trench area whereas elevated chloroform and mercury measurements were found between the DOSB area and the Carolina Bay (see Fig. 15). While soil gas has been used as a screening method to evaluate the extent of a plume of groundwater contaminated with volatile organic compounds (36), the PCE and TCE soil gas measurements here all show NDs to the south and west of the DOSB area in the direction of the near surface groundwater flow. As the Phase II data (described in the following sections) show groundwater PCE and TCE contamination trending southwest of the trenches in the direction of the groundwater flow, the soil gas PCE and TCE are probably associated primarily with source material in the trenches and not with the groundwater plume. The TCE soil gas data does not indicate the presence of any source areas outside of the trench area.

A second soil sampling event in 1993 collected samples at 0.6–1.2, 1.8–2.4, and 3.6–4.2 m (2–4, 6–8, 12–14 ft) depth intervals at 14 sample locations within the DOSB area. This investigation discovered oil-stained soils underlying much of the trench area and analytical results indicating the presence of dioxins, furans, naphthalene, α-benzene hexachloride (α-BHC), β-BHC, α-chlordane, γ-chlordane, PAHs, DDE, DDT, PCBs, and metals (Al, Sb, As, Be, Cd, Ca, Cr, Co, Cu, Fe, Pb, Mg, Mn, Hg, Ni, Ag, V, and Zn).

Three monitoring wells were installed in May 1983, and a fourth in June 1984, to characterize the geologic conditions and monitor the groundwater level and groundwater quality of the basin. Locations of these wells, DOB-1 to DOB-4, are shown in Figure 15. These four poly(vinyl chloride) (PVC) monitoring wells are screened from about 3 to 12 m (10–40 ft) bgs. Quarterly data (1st quarter 1985 to 2nd quarter 1994) collected from these wells show water table depths varying between 1.2 to 4.3 m bgs (4–14 ft); however, potentiometric data is inconclusive with respect to groundwater flow direction (33). In September 1994, six PVC water table piezometers were installed surrounding the DOSB (see DOB-5 to DOB-10 in Fig. 15) with 6 m screened zones extending from 0.9 to 7 m (3–23 ft) bgs. Data from the six piezometers indicates that the near surface groundwater flows generally to the southwest.

Characterization Strategy. The objectives for the DOSB ESC Phase I investigation were to:

- Define the site geology and hydrogeology as related to the system of contaminant migration pathways. In particular, the presence, continuity, and thickness of the apparent 3 m (10 ft) and 7.6 m (25 ft) deep silty/clayey layers and the green clay aquitard.
- Define the site-specific contaminants in groundwater, shallow soils, and surface water sediments that will be tracked in the ESC Phase II investigation.

The data needs defined by these objectives, coupled with relatively undisturbed nature of the site soils (apart from the trenches themselves) and few cultural interferences, led to the following selection of characterization tools. Shallow (medium frequency) GPR was selected to characterize source materials in and around the trenches. Deep (low frequency) GPR was selected to characterize stratigraphic units beneath the unit and the surrounding wooded area. The shallow GPR data was collected along tightly spaced parallel lines, while the deep GPR data was collected along much longer, widely spaced lines that passed over the unit and extended into wooded areas and along roads. An electromagnetic survey (EM) was conducted to measure soil conductivity over several depth ranges across the site.

Rotasonic drilling was used to drill boreholes for two new piezometers. Rotasonic drilling (37–39) was chosen because it produces high quality soil cores with almost no waste and can easily drill through flowing sands. One purpose of these boreholes was to provide detailed stratigraphic description logs down to and into the green clay (which were not available from any other boring on the site) for correlation and calibration of subsequent measurements. A second purpose for these boreholes was for hydraulically testing the lower portion of the aquifer overlying the green clay, which had not been isolated by any of the previous wells or piezometers.

Soil, wetland sediment, and groundwater samples were collected using hand augering and DPT tools and sent to off-site fixed-base laboratories for analyses for 173 Phase I target analytes. A total of 92 Phase I samples were collected. After Phase I the list of contaminants of potential concern was reduced to eight analytes in groundwater. This made

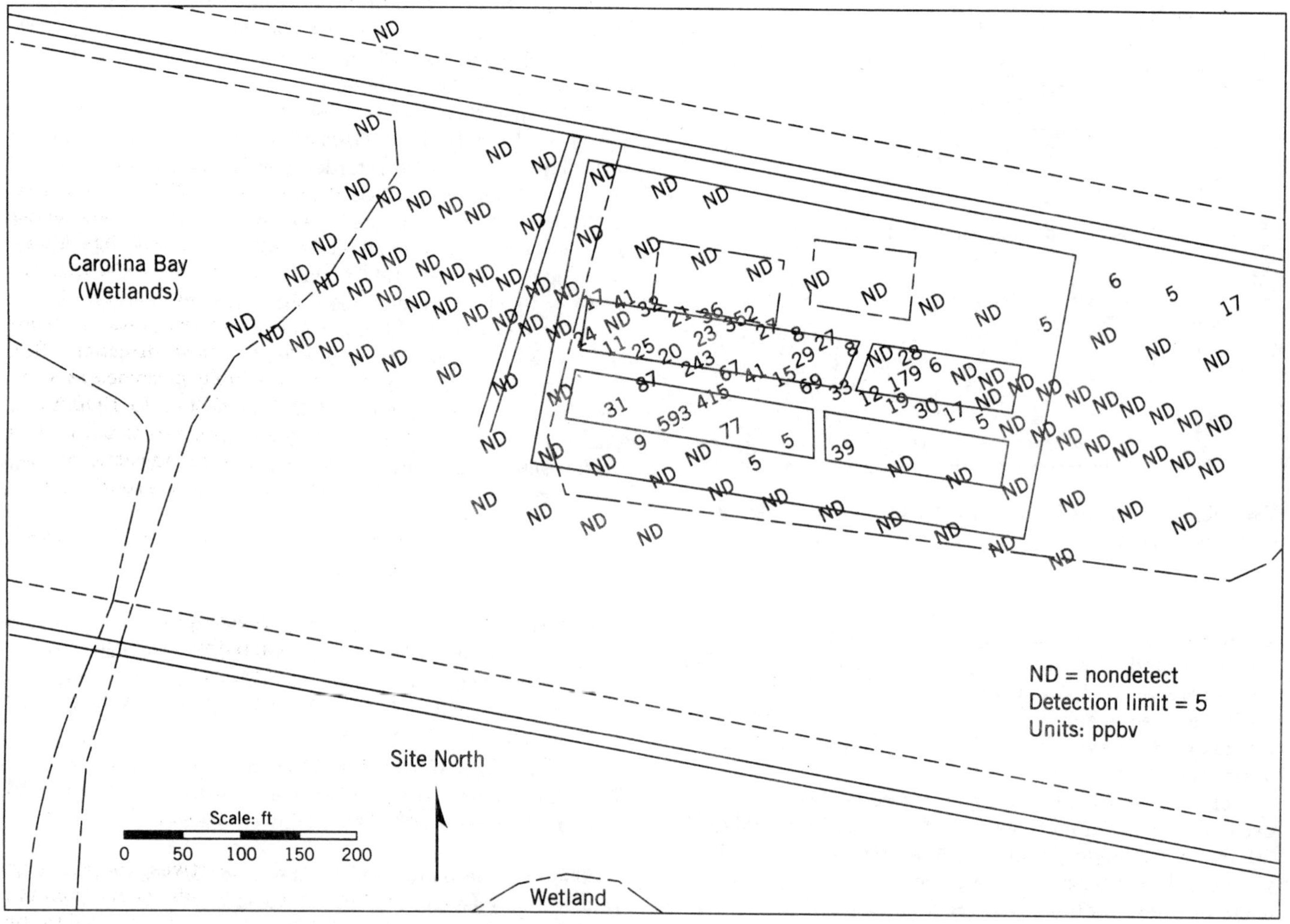

Figure 17. DOSB 1991 soil gas tetrachloroethylene (PCE) measurements (one foot = 0.3048 m).

deployment of a mobile lab at the field site economical for the ESC Phase II investigation.

ESC Phase and Phase II Timeline. Figure 18 shows a timeline of the approximately three-month period during which the ESC Phase I and Phase II investigations took place. The most significant aspect is the large amount of information gained during the short time required for the ESC investigation compared with the what was learned in the approximately 12 years of prior site investigation activities.

Phase I Significant Results. ***Geology.*** Geophysical activities that were carried out in ESC Phase I (see Fig. 18) include GPR, EM38, EM31, and EM34 electrical conductivity measurements, and geophysical logging of two boreholes. The objectives of these activities were to map the aquifer stratigraphy (GPR, 40 MHz to map to a depth of about 18 to 21 m (60–70 ft), and 100 MHz to map to a depth of about 6 m (20 ft)), including the green clay confining layer, and map soil conductivity for several depth ranges over the region (EM38 to a depth of about 1.5 m (5 ft), EM31 to about a 6 m (20 ft) depth, and EM34 to about a 15 m (50 ft) depth). Two boreholes were drilled with the innovative rotasonic drilling method to obtain continuous soil cores from the ground surface to a depth of about 3 m (10 ft) into the green clay. The rotasonic drill locations are referred to as DOL-1, drilled north (upgradient) of the DOSB area, and DOL-2, drilled to the southwest (downgradient) of the DOSB area. The locations of DOL-1 and DOL-2 are shown in Figure 15. DOL-1 was drilled to a depth of 19.8 m (65 ft) and DOL-2 to a depth of 16.8 m (55 ft). PVC piezometers were installed in both DOL-1 and DOL-2 with a 3-m screen located within the sandy soils overlying the green clay. Both borings were drilled and the piezometers set in one day. The rotasonic boreholes were logged with natural gamma, gamma–gamma (density), neutron–neutron (porosity), and induction (conductivity) probes.

Continuous electromagnetic (EM) measurements giving excellent lateral resolution were collected along lines spaced 1.8 m (6 ft) apart over most of the DOSB area. The EM measurements show elevated soil conductivity around the trenches with elevated conductivity values emanating from the trench area and extending to the west and southwest of the trenches in general agreement with the direction of groundwater flow

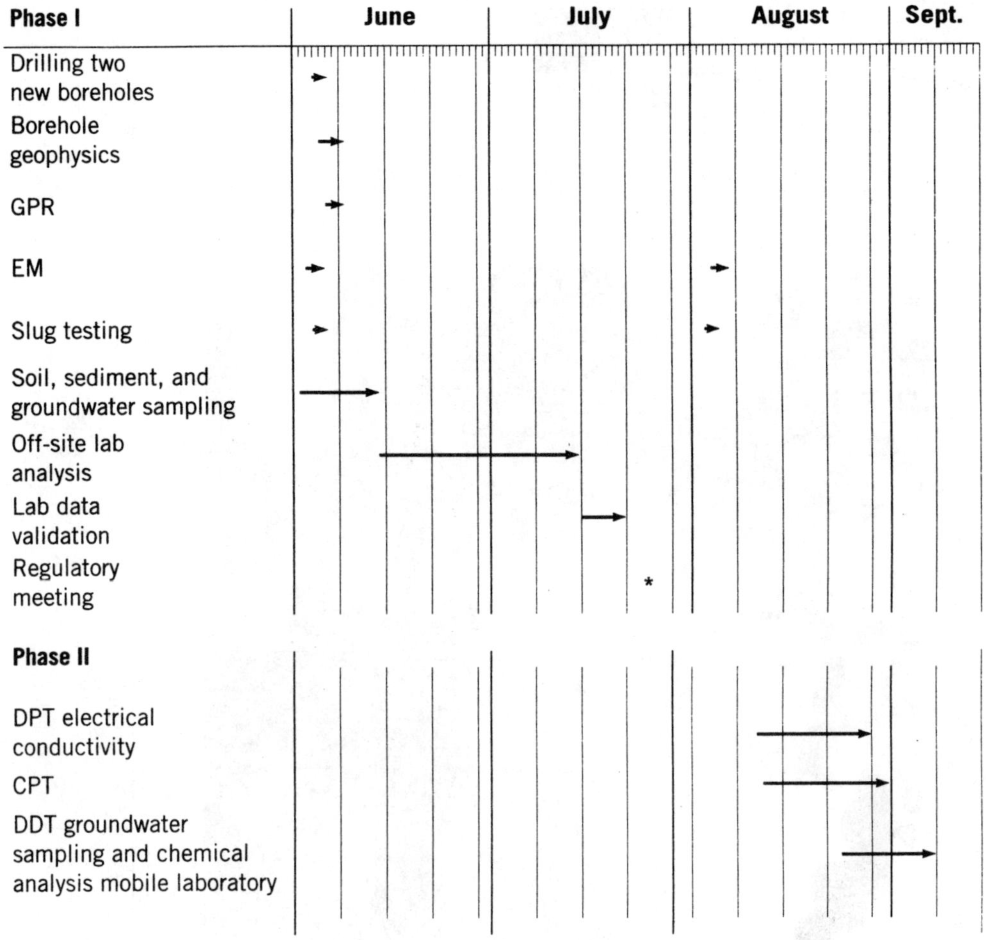

Figure 18. SRS DOSB 1995 ESC Phase I and Phase II investigation timelines.

indicated by the piezometers (Fig. 19). Elevated conductivity values were associated with buried metallic and nonmetallic materials and an inorganic groundwater plume. Downgradient wells DOB-2 and DOL-2 show elevated conductivities at a depth of about 7.6 m (25 ft) that were interpreted as possible zones of inorganic contamination. The GPR data indicate uniform shallow soil conditions, disturbance in the trench areas, areas with local buried metallic and nonmetallic targets, and large magnitude radar reflectors below the trenches and water table possibly associated with DNAPLs. The green clay was found at a depth of 14.3 m (47 ft) in DOL-1 and 11.3 m (37 ft) in DOL-2 (35).

The GPR results indicate the existence of a clay layer between 6 and 9 m (20 and 30 ft) bgs that is continuous over much of the site and a thin aquitard at a depth of about 3 m (10 ft) over much of the site. In addition to facies changes, absence of these intermediate aquitard reflectors in some of the GPR data may be due to either masking by above-ground reflectors or to high ground conductivity above the aquitards caused by contaminants in the aquitards absorbing the radar signal. GPR results indicate that the green clay occurs at depths of about 11 to 14 m (35–45 ft), but may not be continuous over the site. The lack of resolution of the green clay aquitard using radar data is partly due to its gradational contact with the aquifer above it. Because of the interbedded nature of the soils, the exact depth at which the green clay begins was also difficult to identify in the DOL-1 and DOL-2 monitoring well drill cores. The GPR data did confirm the location of the boundaries of the trenches and areas of disturbed soil shown in Figure 15.

Analytical Data. Phase I analytical data collected in the late spring and early summer of 1995 (see Fig. 18) were measured by an off-site laboratory using EPA SW-846 methods with definitive type data quality. An off-site lab was used during Phase I because it was uneconomical to use an on-site mobile lab for full suite analyses on a small set of samples and the waiting period for the analytical results coincided with the waiting period for the detailed geophysical results. This activity therefore could be integrated into the period between Phase I and Phase II without negatively impacting the investigation.

The surface electromagnetic geophysical measurements taken during the Phase I investigation detected an area of elevated electrical conductivity to the southwest of the site. As a result, several additional Phase I groundwater sampling locations were added in the area to the southwest and several previously planned sample locations were moved to positions where contamination was more likely. This is an example of the type of flexibility in the ESC dynamic work plan whereby on-site decision making based on current field data was used to optimize subsequent data collection.

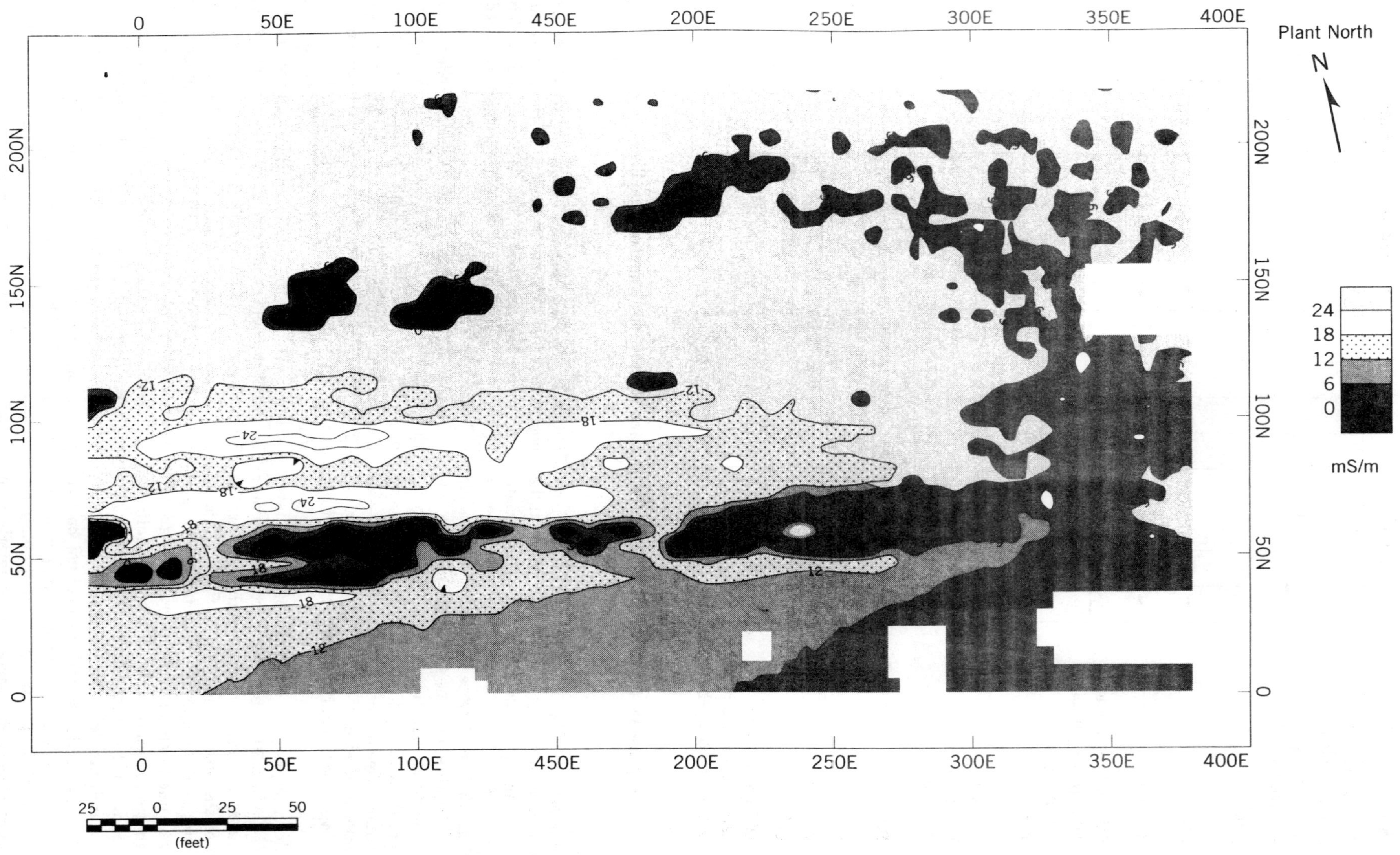

Figure 19. DOSB trench and disturbed soil area (see Fig. 15) EM31 conductivity contours (35) (one foot = 0.3048 m).

During July 1995 (see Fig. 18) there was a regulatory meeting with the ESC core team, SRS personnel, EPA Region IV, and the South Carolina Department of Health and Environmental Control to review the Phase I geophysics and define the list of SSCs to track during Phase II. The outcome of this meeting included a need to further define the elevated conductivity plume to the southwest with additional GPR and EM surveying and soil conductivity probing. Within about one week the geophysics crew was back out on the site to conduct further GPR and EM surveys during a Phase IB investigation during the first week of August 1995 (see Fig. 18). This again illustrates the importance of the dynamic, flexible aspect of the work plan. It was decided at the regulators meeting that only groundwater contamination would be tracked during Phase II; vadose soils were contaminated but already had sufficient data to perform a risk assessment. It was agreed that the contaminants listed in Table 1 would be tracked as the SSCs of potential concern during the ESC Phase II investigation.

Figure 20 shows a contour plot illustrating the outcome of the additional conductivity measurements. Note the elevated conductivity readings extending to the south and west of the trench area in the general direction determined for the movement of the shallow groundwater. It was hypothesized that this electrical conductivity anomaly was due to inorganic contaminants in the groundwater. This hypothesis was tested by measuring the specific conductance of groundwater samples from upgradient monitoring well DOL-1 and downgradient monitoring well DOL-2 (see Fig. 15). The average specific conductance of samples from the downgradient well DOL-2 was four times more than that from the upgradient well DOL-1.

Table 1. SRS DOSB ESC Phase II Site-Specific Contaminant List

Contaminant Type	Species
Metals	Sb, As, Be, Mn
Volatiles	PCE, TCE, vinyl chloride
Pesticides	α-Benzene hexachloride

On the basis of the Phase I results, the objectives for the subsequent Phase II investigation were determined to be:

- Confirm and refine the site geology and hydrogeology models as related to the system of contaminant migration pathways with particular focus on the continuity, depth and thickness of the 3 m, 7.6 m, and green clay aquitards.
- Assess the lateral and vertical extent of the site specific contaminants (listed in Table 1) in the groundwater.

The Phase II investigation was guided by the GPR and EM measurements taken during the Phase IB investigation, which had been integrated into the site conceptual model. Figure 18 shows that about one week had elapsed between the Phase IB geophysical work and the start of the Phase II investigation.

Phase II Significant Results. The Phase II investigation was carried out during an approximate four-week period as shown in Figure 18. DPT equipment and mobile analytical laboratories were brought to the site to conduct the investigation. For calibration purposes, the first DPT pushes were carried out

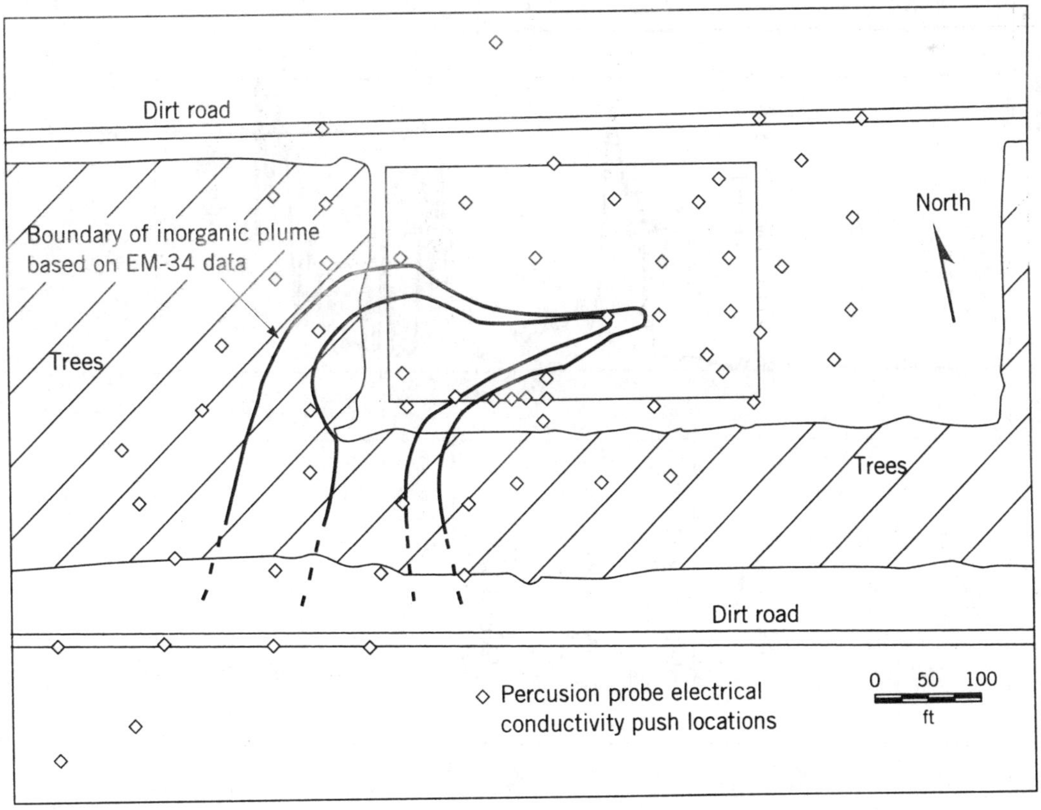

Figure 20. Percussion probe push locations and inorganic plume boundary based on EM34 data (35) (one foot = 0.3048 m).

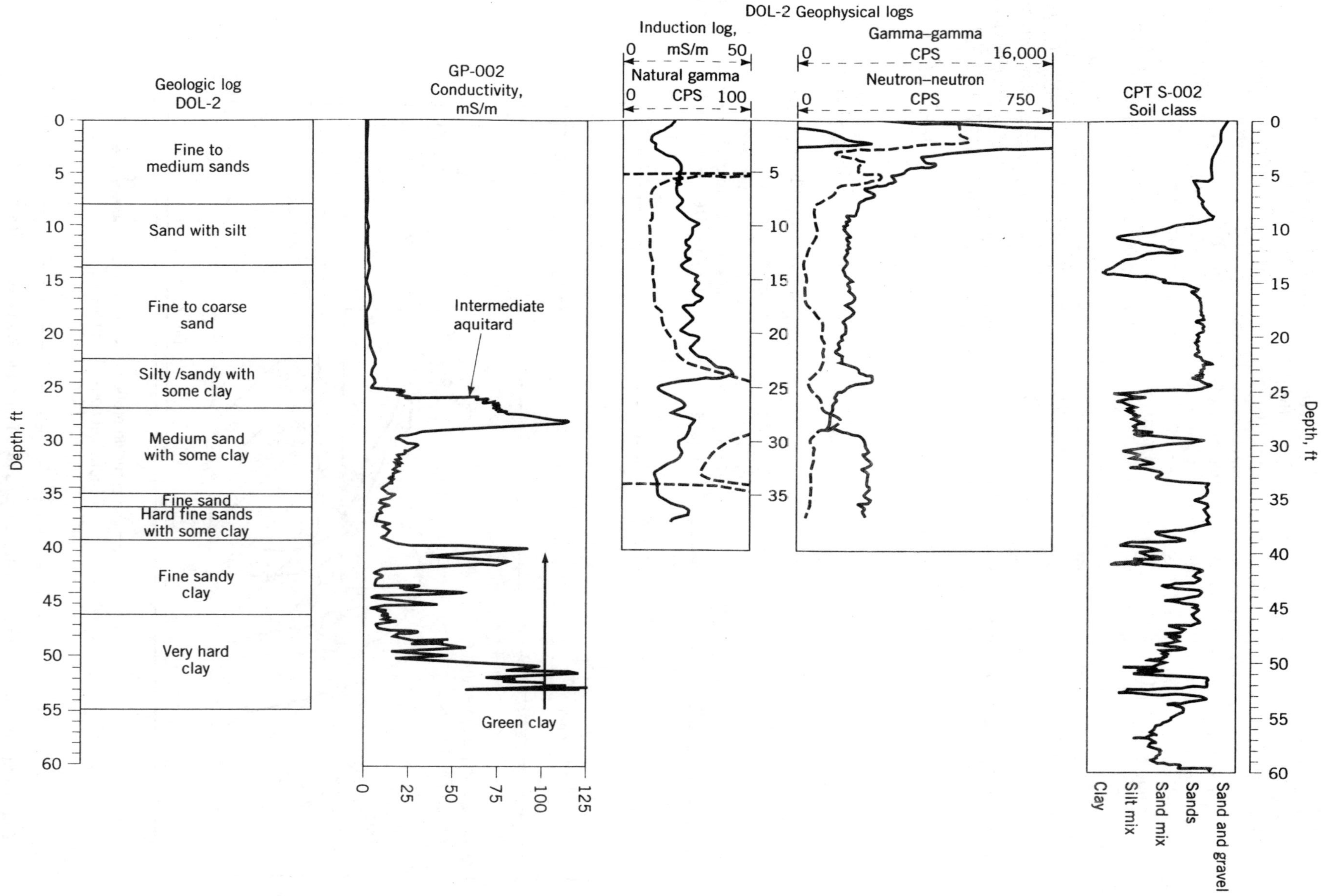

Figure 21. Comparison of DOL-2 borehole geologic log, percussion probe electrical conductivity, borehole geophysical logs, and CPT stratigraphic log. CPT S-002 soil class information from Ref. 16.

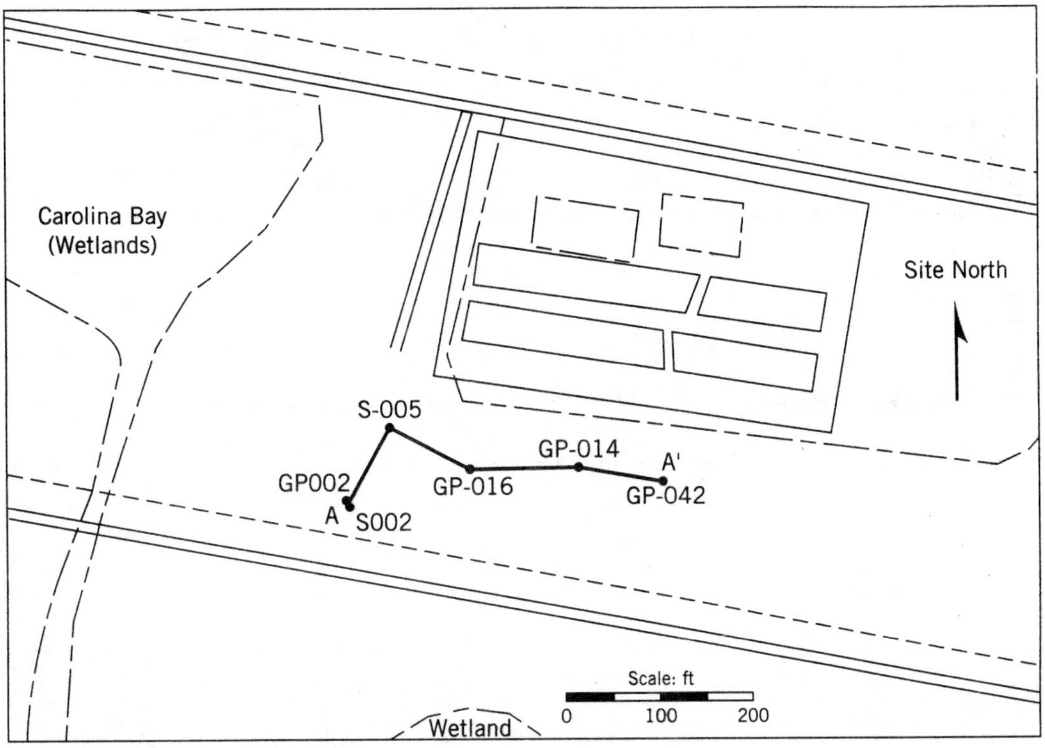

Figure 22. SRS DOSB map showing the location of the fence diagram in Figure 23 (one foot = 0.3048 m).

adjacent to monitoring wells DOL-1 and DOL-2. Two CPT rigs with cone tip stress and sleeve stress sensors were used to further define the site stratigraphy. Soils were classified according to soil behavior types using modified empirical results (16). Additionally, the CPT rigs both had pore pressure sensors which were used to measure both the dynamic pore pressure during each push and to perform pore pressure dissipation testing within the fine-grained soils that developed significant positive dynamic pore pressures during a push. The pore pressure dissipation test data was used to estimate the horizontal coefficient of consolidation and hydraulic conductivity of the fine grained soil units at the site using previously known methodology (22,23). Percussion probing equipment was used to measure soil electrical conductivity to further define areas of high and low electrical conductivity. High electrical conductivity soils generally correlate to elevated clay content and/or soils contaminated with inorganic materials. CPT stratigraphic logging was able to determine whether soils had high clay content or not, and thus were able to reduce uncertainty associated with the electrical conductivity probe data. The smaller, pick-up truck mounted percussion probe was much more maneuverable within the wooded area than the 20 ton CPT rigs which were generally confined to working in cleared spaces and along roadways.

Figure 21 illustrates the use of multiple technologies to assist the interpretation of responses and reduce uncertainty in the conceptual site model. Figure 21 shows the visual soil description log taken from the downgradient monitoring well DOL-2, along side of the percussion probe conductivity log, the borehole geophysical logs, and a CPT soil behavior-type classification based on previous work (16). Note that the so-called green clay which occurs below about 12 m (40 ft) in these logs is a clayey/silty/sandy soil. This green clay was often overlain by interbedded soils and a sharp contact delineating green clay was not apparent in many of the electrical conductivity and CPT logs; however, there was never a definitive indication that the green clay was absent in any of the logs. Figure 23 shows a series of CPT soil classifications and electrical conductivity traces taken along the line shown in Figure 22. Note the apparent continuity of the 7.6 m (25 ft) weak aquitard across this portion of the site clearly seen in both the conductivity and CPT logs. The 3-m unit is seen clearly in the CPT logs but does not show a strong characteristic response in the conductivity logs. Soils classified as green clay are also shown in Figure 23, where the characteristic green clay response appears to be interbedded sands, silts, and clays with either abrupt or steadily increasing conductivity readings below about 10.7 to 12 m (35–40 ft) below ground surface. On the basis of interpretations such as illustrated in Figure 23 and the DPT locations shown in Figure 24, Figure 24 shows the surface topography of the green clay. Recall that, as mentioned previously, the green clay contact appears to be gradational in nature with interbedded sands, silts, and clays in overlying soils.

Percussion probing equipment was used to collect groundwater samples for chemical analysis conducted in on-site mobile laboratories. Data from the analytical labs was generally available within several hours of sample collection and

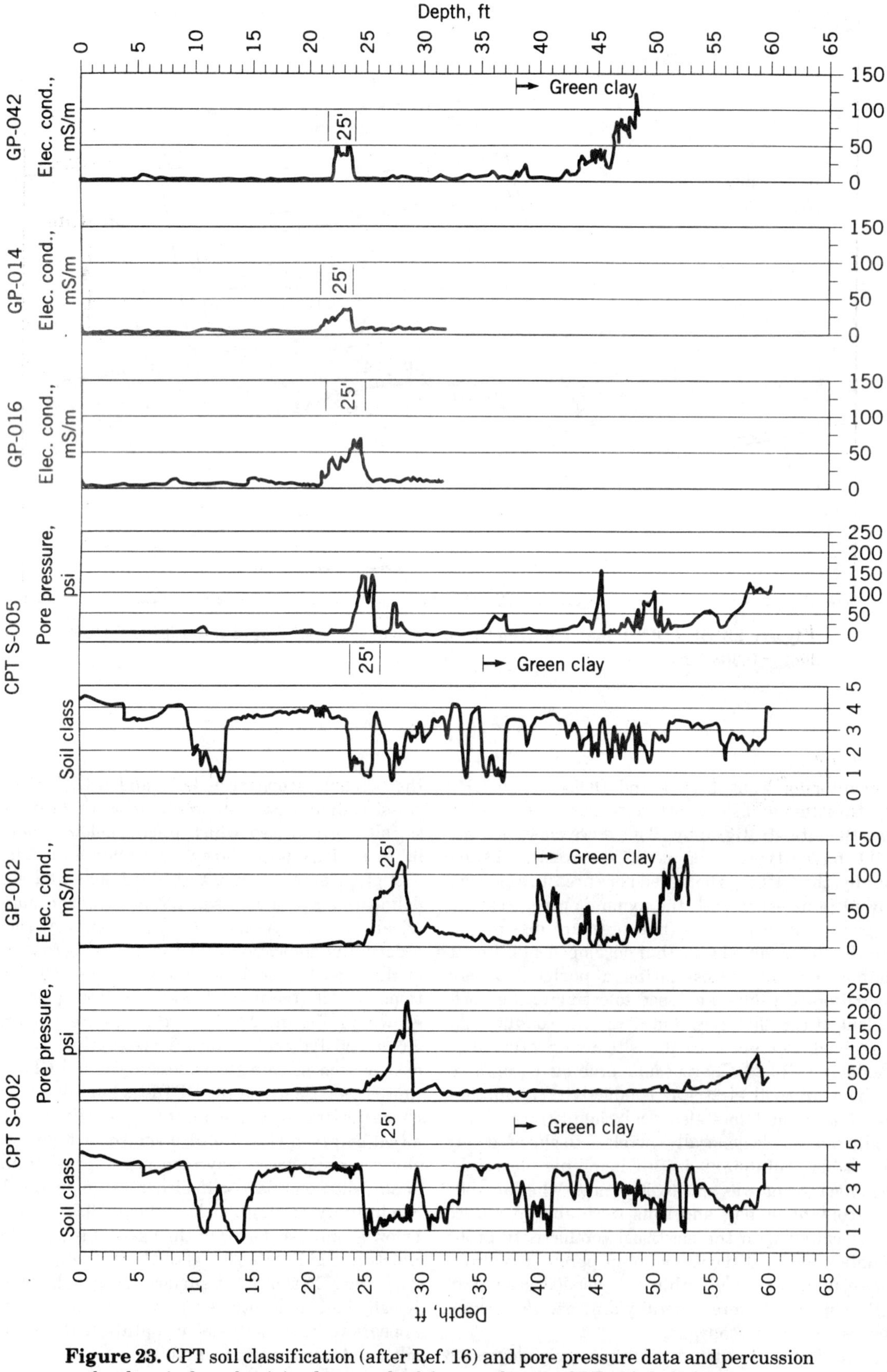

Figure 23. CPT soil classification (after Ref. 16) and pore pressure data and percussion probe electrical conductivity data south of the trench area. CPT soil classification: <1 = clays, 1–2 = silt mix, 2–3 = sand mix, 3–4 = sands, and 4–5 = sand and gravel (one foot = 0.3048 m); one psi = 6.89 kPa).

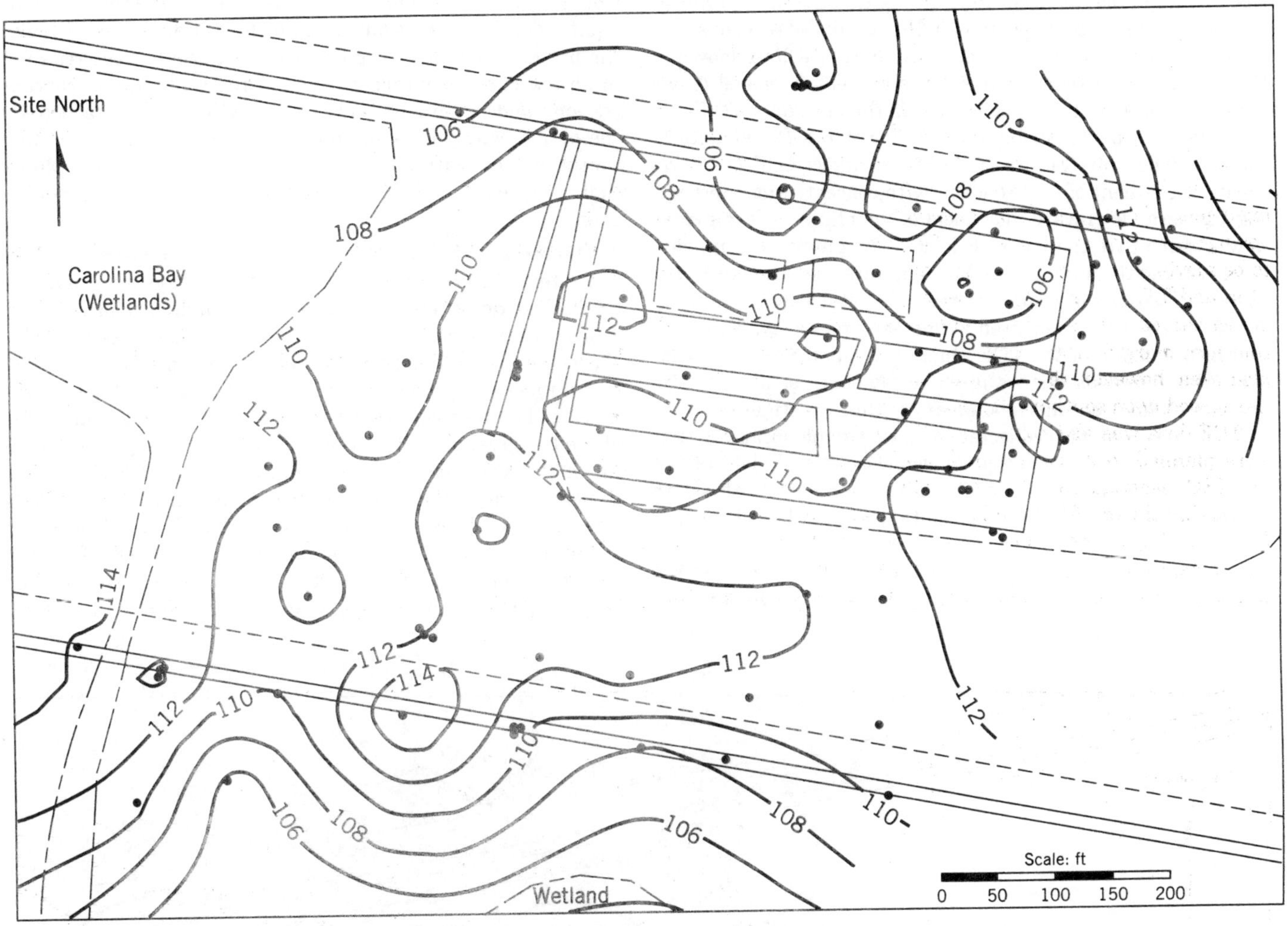

Figure 24. Green clay surface elevation (feet above msl) with DPT push locations shown (one foot = 0.3048 m).

was integrated with other site data on computers housed in a separate mobile unit on-site. Data management and visualization software was used to analyze the data, refine the geologic and contaminant conceptual site models, and determine locations for subsequent samples while in the field. Sampling depths were guided by the geological and hydrogeological models which were updated periodically as the investigation progressed. The goal was to track each SSC until nondetects bound each plume in all three dimensions. Sampling began at the source area and then radiated out. Sample locations were initially taken close to the trenches to detect drum-sized sources and using these results subsequent locations were moved away from the trenches with each action level for each SSC defining the stopping rule. At each sampling location, samples were taken from potentially seven depths: above and within the shallow (3 m) aquitard, top, middle, and bottom of the aquifer between the shallow and intermediate aquitards, within the intermediate (7.6 m) aquitard, and within the aquifer directly overlying the green clay. As sampling progressed downgradient, confidence grew that specific plumes were well defined and the analytical program was modified by dropping some analytes from the SSC list. For example, about halfway throughout the sampling program all pesticide analyses were reported as below the method detection limit and the action level and so pesticide analyses were stopped. This illustrates the dynamic aspect of the on-site decision making process that is integral to ESC.

Figure 25 shows all of the Phase II groundwater sample locations along with the TCE plume determined by contouring the maximum concentration observed over the individual sample depths at each sample location. Note that the contaminant plume trends away from the site in roughly the same location and direction as the elevated EM34 conductivity station measurements shown in Figure 20. The TCE and EM34 plumes (Figs. 25 and 20, respectively) do not overlap exactly, however, the EM34 measurements are responding to inorganic (metals) contaminants which are moving with the groundwater but may have a different source within the site. The EM data was particularly valuable in locating the contaminant plume at this site. The plume was unusually narrow, making it difficult to detect; it would not have been detected using the conventional methodology in place before the ESC investigation. The narrow size of the plume suggests that the contamination originates primarily from a small location within one of the trenches.

A cross section along the line A–A′ in Figure 25 is shown in Figure 26. The cross section shows a conceptual model of the soil units based on DPTs together with the measured TCE concentrations in parts per billion with ND shown for nondetects (MDL = 0.61 ppb). Figure 26 shows nondetect TCE measurements upgradient of the trenches, with concentrations indicating a general downward movement of the plume as it migrates downgradient (to the left in Fig. 26). The plume also appears to be moving through the shallow and intermediate aquitards (3 m and 7.6 m soil units); however, in most cases, only nondetect values were measured within the green clay. Note that southern margin of the TCE plume extends beyond the sampled area, however, the wetlands to the south of the site was the agreed upon southern boundary of this investigation.

TCE data was also collected at those sample locations that were planned in a conventional monitoring well field before any ESC investigation was contemplated. Analysis of this data shows that the pre-ESC planned data would not have been sufficient to define the plume's northern or southern edge, and that additional wells would have been needed to define the plume's extent and configuration. Also, the preplanned data would not have been able to detect the plume's presence at mid-aquifer depths to the south toward the wetlands. Accordingly, the pre-ESC planned sampling locations would have resulted in an inaccurate characterization of the contaminant concentrations and extent. In contrast, the ESC TCE data bounds the plume (with the exception of the intentionally overlooked southern edge) with nondetect data and defines a TCE plume that is consistent with the hydrogeologic site model and source area.

Discussion. The SRS ESC investigation demonstrated effective integration of Phase I and Phase II technologies and met the goals of defining the site geology and hydrogeology as they impact the transport and fate of the site-specific contaminants. The contaminant plumes were well defined and demonstrated to originate within the trench area and migrate to the southwest in the direction determined for the flow of groundwater above the green clay. The EM data was especially instrumental in the development of the conceptual site models for the geology, geohydrology, and contaminant spatial distribution. The Phase I EM data fed directly into the Phase II planning and investigation. The interpretation of geologic data from the minimally intrusive investigation technologies demonstrated that, while this site is stratigraphically complex, these tools

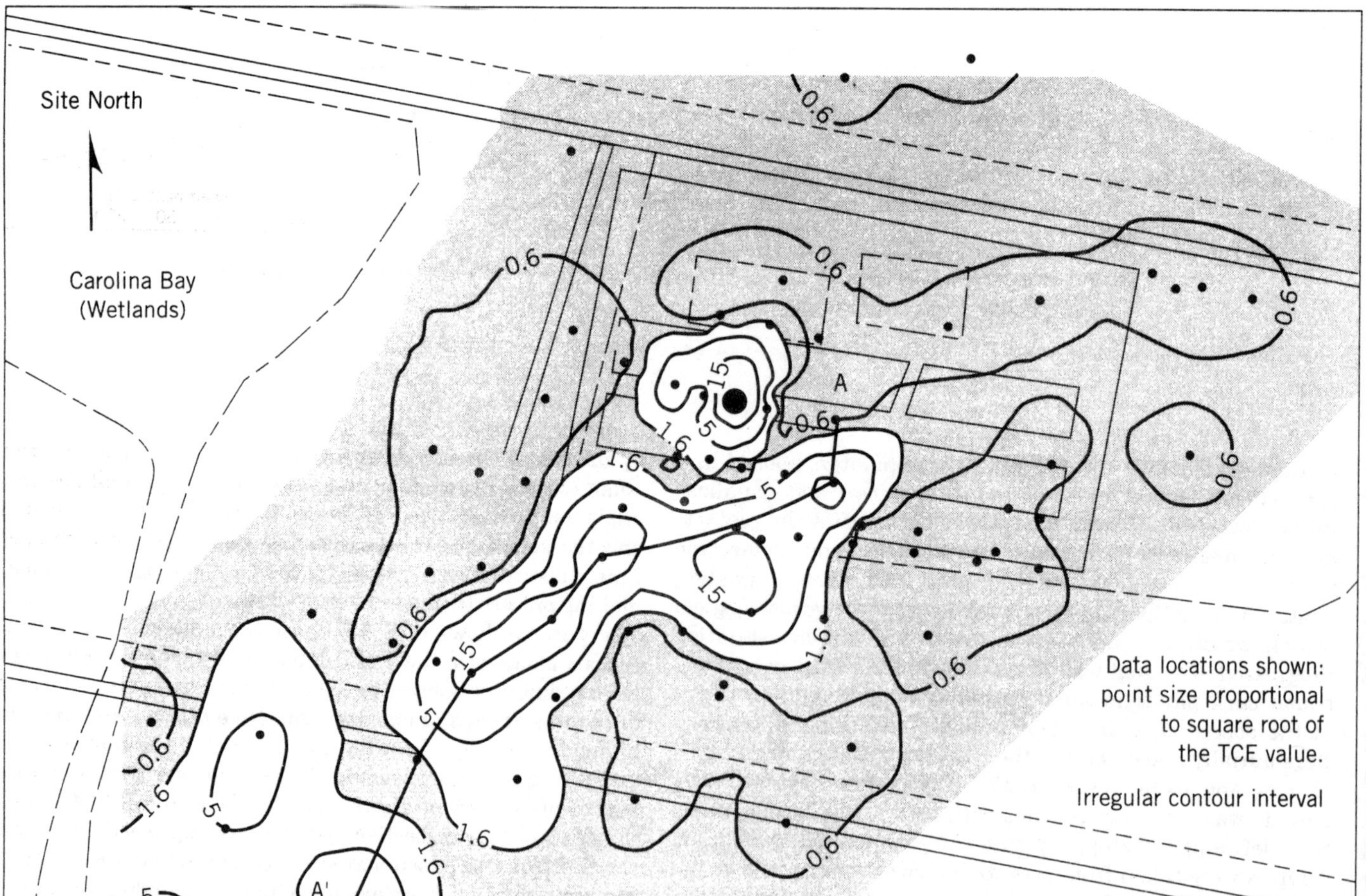

Figure 25. Groundwater TCE concentration contours (ppb). Light shading shows area predicted to be less than the RBC of 1.6 ppb (one foot = 0.3048 m).

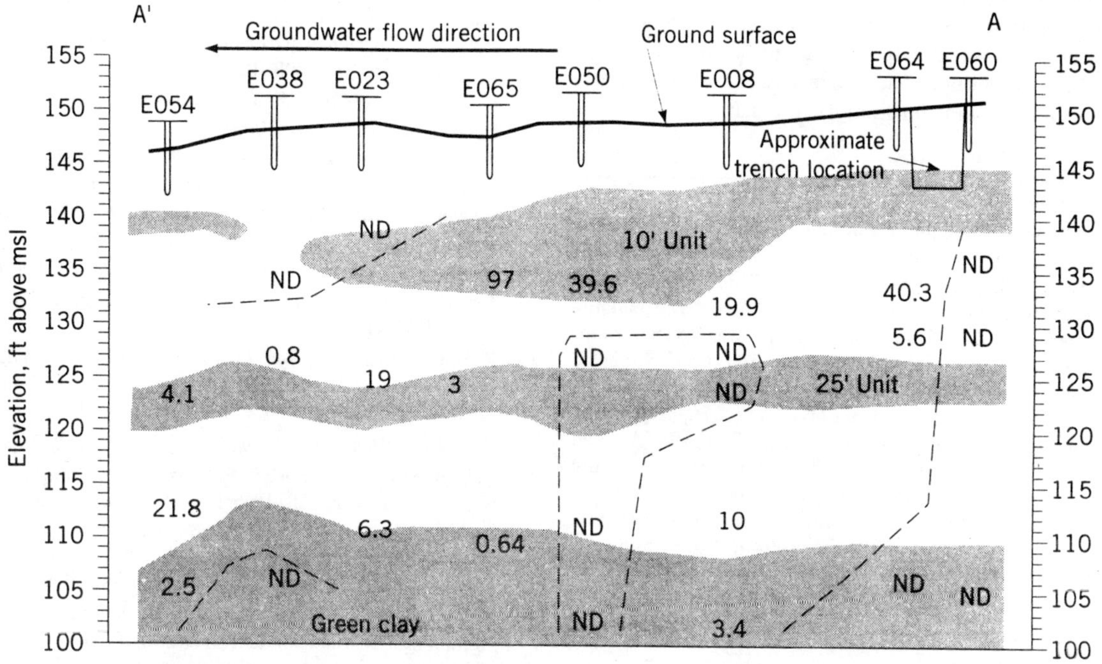

Figure 26. Cross section showing major soil units and measured groundwater TCE concentrations (ppb). Dashed lines separate nondetect (ND) results (one foot = 0.3048 m).

played a vital role in the development of the stratigraphic models, and in particular, the determination of the green clay continuity and general depth across the site. Finally, percussion probing made it possible to collect a large number of groundwater samples in just a few weeks, while on-site analytical laboratories with short analytical turn around times using definitive data methods made possible the use of current data to form a basis for the placement of subsequent sample locations to optimally define the contaminant spatial distributions.

SUMMARY AND CONCLUSIONS

ESC offers a better, faster, and safer way to characterize hazardous waste sites than the conventional approach. In an ESC project all prior and existing data is evaluated and integrated into the conceptual site model to plan the next phase to be conducted and/or measurements to be collected. Measurements are collected with specific goals in mind. In general, the first phase of an ESC investigation focuses on locating the source and nature of contamination and defining the geology and hydrogeology so that a proper assessment of contaminant fate and transport can be made. The second phase generally focuses on defining the contaminant plume(s) spatial distribution in each media of concern (soil, groundwater, surface water, etc). The contaminant spatial distribution is determined quickly and efficiently by use of direct push technology sampling equipment and on-site mobile analytical laboratories which produce analytical results generally within hours of sample collection. Prior analytical results, taken together with contaminant pathways, are used to optimize the placement of subsequent sample locations while in the field during the investigation.

The field work of an ESC program must be flexible to allow for changes in the work plan made on the basis of actual data and the developing conceptual site model. The field work is conducted by a multidisciplinary team of experienced professionals. Tool and technology selections are made on the basis of the specific site questions that need to be answered and on the applicability of the tools to the specific site conditions. Particular emphasis is given to nonintrusive and minimally intrusive technologies which are proven to provide reliable data, are operationally safe, and minimize potential exposure of site workers to site contaminants.

At the SRS where past cultural impact was minimal and the site geology was relatively complex, the EM and GPR geophysical surveys played a significant role developing the initial conceptual site models and were instrumental in planning the Phase II investigation by providing nearly complete coverage of the site and identifying areas with anomalous readings. The subsequent use of the percussion probe electrical conductivity sensor and the CPT rigs during Phase II enhanced the development of the geologic model, and in particular, the location and continuity of the green clay and the two intermediate aquitards.

At the FMGP site where there were many cultural interferences, the geophysical data did not provide much useful information, apart from indicating that the bedrock surface was weathered and highly variable in depth. However, at the FMGP site, the geology was relatively simple and the percussion probe electrical conductivity sensor and CPT rig were clearly sufficient to characterize the significant aspects of the site geology. Additionally, at the FMGP site, the CPT LIF, and to some extent, the electrical conductivity probe, were extremely useful as screening tools to identify soils contaminated with PAH compounds; these areas were subsequently targeted as sample locations for analytical testing.

ACKNOWLEDGMENTS

We gratefully acknowledge the participation of Global Environmental, Inc., McLaren/Hart Environmental Engineering Corp., Geoprobe Systems, Inc., Kansas City District, U.S. Army Corps of Engineers, PACE Environmental Laboratories, Inc., CEM Corp., D'Appolonia Environmental Services, Inc., The GEHM Corp. and Entech Engineering, Inc., Technos, Inc., Coleman Energy and Environmental Systems-Blackhawk Geosciences Division, Resolution Resources, Inc., and Dynamic Graphics, Inc. in the FMGP site ESC project. We gratefully acknowledge the participation of the DOE SCAPS, Applied Research Associates, Technos, Inc., Zebra Environmental Corp., and McLaren/Hart Environmental Engineering Corp. in the ESC investigation at the SRS D-Area Oil Seepage Basins.Ames Laboratory is operated for the U.S. Department of Energy by Iowa State University under contract No. W-7405-Eng-82. This work was supported by the Office of Environmental Technology and Development.

BIBLIOGRAPHY

1. C. B. Purdy, R. Benson, A. Bevolo, T. P. Starke, and S. B. Wyrick, "DOE's Expedited Site Characterization (ESC) Course," presented at *National Ground Water Association's Outdoor Meeting,* Las Vegas, Nev., May 1, 1995.
2. J. C. Burton, J. L. Walker, T. V. Jennings, P. Aggarwal, B. Hastings, W. T. Meyer, C. M. Rose, and C. L. Rosignolo, *Superfund XIV,* Vol. II, Hazardous Materials Research and Control Institute, Greenbelt, Md., Nov. 30–Dec. 1, 1993, pp. 809–826.
3. J. C. Burton, *Proceedings of the 1992 Federal Environmental Conference and Exhibition,* Hazardous Material Control and Research Institute, Greenbelt, Md., 1992, pp. 242–251.
4. P. K. Aggarwal, J. C. Burton, and C. R. Rose, *Proceedings FER III & WM II Conference and Exhibition,* Rockville, Md., 1994, pp. 1311–1317.
5. J. C. Burton, *Proceedings FER III & WM II Conference and Exhibition,* 1994, pp. 1407–1415.
6. B. Hastings, G. Hildebrandt, T. Meyers, W. Saunders, and J. C. Burton, "Integration of Geophysics With the Argonne Expedited Site Characterization Program at a Site in the Southern High Plains," in *1995 Symposium on the Application of Geophysics to Engineering and Environmental Problems* (SAGEEP '95), Orlando, Fla., 1995.
7. J. C. Burton, J. L. Walker, P. K. Aggarwal, and W. T. Meyer, "Argonne's Expedited Site Characterization: An Integrated Approach to Cost- and Time-Effective Remedial Investigation," paper presented at *88th Annual Meeting,* Air and Waste Management Association, San Antonio, Tex., 1995, 27 pp.
8. Ames Laboratory, *Report on the Results of the Expedited Site Characterization at the Chzechowice Refinery,* Ames Laboratory, Ames, Iowa, 1996.
9. U.S. EPA, *Data Quality Objectives Process for Superfund: Interim Final Guidance,* EPA/540/G-93/071, Publication Number 9355.9-01, Washington, D.C., Sept. 1993.
10. R. C. Benson, in D. E. Daniel, ed., *Geotechnical Practice for Waste Disposal,* Chapman & Hall, London, 1993, pp. 311–357.
11. C. D. Christy, T. M. Christy, and V. Wittig, *A Percussion Probing Tool for the Direct Sensing of Soil Conductivity,* Technical Paper No. 94-100, Geoprobe® Systems, Salina, Kans., Mar. 1994, 14 pp.
12. B. J. Douglas and R. S. Olsen, "Soil Classification Using the Electric Cone Penetrometer," *Symposium on Cone Penetration Testing and Experience, Geotechnical Engineering Division,* ASCE, St. Louis, 1981, pp. 209–227.
13. R. G. Campanella, P. K. Robertson, and D. Gillespie, *Can. Geotech. J.* **20,** 23–35 (1983).
14. A. C. Meigh, *Cone Penetration Testing: Methods and Interpretation,* CIRA Ground Engineering Report, *In-situ* Testing, Butterworths, Adlard & Son Ltd., Letchworth, Hertfordshire, U.K., 1987, 141 pp.
15. R. S. Olsen and J. V. Farr, *Proceedings of the ASCE Conference on Use of In-Situ Testing in Geotechnical Engineering,* ASCE, New York, 1986, pp. 854–868.
16. R. S. Olsen, in J. L. Von Thun, ed., *Earthquake Engineering and Soil Dynamics II—Recent Advances in Ground-Motion Evaluation,* ASCE, June 27–30, (1988), pp. 374–388.
17. P. K. Robertson, *Can. Geotech. J.* **27,** 151–158 (1989).
18. J. P. Koester, L. T. Lee, R. S. Olsen, D. H. Douglas, G. D. Comes, S. S. Cooper, and J. F. Powell, *Field Trials of the Site Characterization and Analysis Penetrometer System at the Savannah River Site (SRS),* Technical Report GL-93-16, U.S. Army Corp of Engineers, Waterways Experiment Station, July 1993.
19. S. H. Lieberman, S. E. Apitz, P. G. Malone, and S. S. Cooper, "Real-Time In Situ Measurements of Fuels in Soil: Comparison of Fluorescence and Soil Gas Measurements," *Proceedings of the Third International Symposium, Field Screening Methods for Hazardous Wastes and Toxic Chemicals,* Las Vegas, Nev., Feb. 1993.
20. R. W. St. Germain and G. D. Gillispie, *Proceedings of the Fourth International Symposium on Field Screening Methods for Hazardous Wastes and Toxic Chemicals,* Las Vegas, Nev., Feb. 1995, pp. 467–477.
21. G. D. Gillispie and R. W. St. Germain, *Proceedings of the Fourth International Symposium on Field Screening Methods for Hazardous Wastes and Toxic Chemicals,* Las Vegas, Nev., Feb. 1995, pp. 478–484.
22. P. K. Robertson and R. G. Campanella, *Guidelines for Geotechnical Design Using the Cone Penetrometer Test and CPT with Pore Pressure Measurement,* 4th ed., Hogentogler & Co., Inc., Baltimore, Md., Nov. 1989, pp. 105–123.
23. P. K. Robertson, J. P. Sully, D. J. Woeller, T. Linne, J. J. M. Powell, and D. G. Gillispie, *Can. Geotech. J.* **29,** 539–550 (1992).
24. *A Guide to Geoprobe® Tools for Groundwater Samplers and Permanent Monitoring Devices,* Technical Bulletin 95-1000, Kejr Engineering, Inc., Salina, Kans., Oct. 1995, 32 pp.
25. *Geoprobe Large Bore Soil Sampler,* Technical Bulletin 93-660, Kejr Engineering, Inc., Salina, Kans., Sept. 1996, 16 pp.
26. *Geoprobe Macro-Core® Soil Sampler,* Technical Bulletin 95-8500, Kejr Engineering, Inc., Salina, Kans., Sept. 1996, 36 pp.
27. R. G. Luthy, D. A. Dzombak, C. A. Peters, S. B. Roy, A. Ramaswami, D. V. Nakles, and B. R. Nott, *Environ. Sci. Technol.,* **28**(6), 266–276 (1994).
28. G. A. Stenback, B. H. Kjartanson, A. Bevolo, D. Wonder, and K. Older, *Proceedings of the Fourth International Symposium on Field Screening Methods for Hazardous Wastes and Toxic Chemicals,* Las Vegas, Nev., Feb. 1995, pp. 450–457.
29. G. A. Stenback, B. H. Kjartanson, A. Bevolo, and D. Wonder, "The Use of Direct Push Technologies in Expedited Site Characterization," *Proceedings of the Uncertainty in the Geologic Environment: From Theory to Practice,* ASCE, Uncertainty '96, July 31–Aug. 3, 1996, 15 pp.
30. M. L. Stein, *J. Comput. Graph. Stat.* **1**(1), 91–110 (1992).
31. O. R. West, R. L. Siegrist, T. J. Mitchell, and R. A. Jenkins, *Environ. Sci. Technol.,* **29**(3), 647–656 (1995).

32. U.S. EPA, *Soil Sampling Quality Assurance User's Guide,* EPA/600/8-89/046, Washington, D.C., Mar. 1989.
33. WSRC, *Phase II RCRA Facility Investigation / Remedial Investigation Work Plan for the D-Area Oil Seepage Basin,* Westinghouse Savannah River Co., prepared for U.S. DOE, May 1995.
34. B. R. James and R. A. Freeze, *Water Resources Res.* **29**(7), 2049–2065 (1993).
35. Technos, *Final Report: Geologic, Hydrologic and Geophysical Work at the D-Area Seepage Basin,* Vol. 1, Technos Project No. 95-108, Savannah River Site Plant, Savannah, Ga., prepared for Ames Laboratory TIP by Technos, Inc. Dec. 7, 1995.
36. C. W. Fetter, *Contaminant Hydrogeology,* Macmillan Publishing Co., New York, 1993, pp. 383–384.
37. T. D. Ault, S. Logan, and A. J. Madaj, *Proceeding of the National Groundwater Association Seventh Annual Outdoor Action Conference and Exposition,* Las Vegas, Nev., 1994, pp. 621–636.
38. J. C. Barrow, *Ground Water Mon. Rev.,* 153–160 (Spring 1994).
39. R. Stephan, *Proceedings of the National Groundwater Association, Eighth Annual Outdoor Action Conference and Exposition,* Las Vegas, Nev., 1995, pp. 325–337.

SITE CLEANUP, COST AND RISK CONTROL

DANIEL GOODMAN
Montana State University
Bozeman, Montana

STANLEY BLACKER
MAC Technical Services, Co.
Alexandria, Virginia

The project life of a site cleanup, from first detection of a problem to declaration of a closure that is satisfactory to the standards of the regulators and the conscience of the responsible party, can span a large amount of time and encompass many activities. Some of the steps may be mandated by the regulatory process, as for example the Superfund project milestones of Preliminary Assessment, Site Investigation, Remedial Investigation, Feasibility Study, Remedial Design, and Remedial Action. Regardless of regulatory strictures, practical considerations require that the cleanup involve, in addition to the engineering steps of the remediation itself, data collection to define the scope of the contamination problem, data collection to inform major strategy decisions for the cleanup, and data collection to guide the details of implementation of the remediation. These various activities all take time and money, so overall efficiency in the cleanup project can only be achieved if all the steps are coordinated in such a manner that each step adds value in moving toward the ultimate project goal, and each step provides the information and site preparation necessary for effective operation of the next steps, without investing more resources in any one step than is needed for that step to serve its role in the larger scheme of the cleanup.

This article is concerned with the logic and the technical tools for coordinating all the steps of a site cleanup in order to achieve overall efficiency. The relevant concepts and technical tools employ ideas and vocabulary that are encountered in many different, but partially overlapping, disciplines and schools of thought, such as *total quality management, value engineering, experimental design, data quality objectives, cost–benefit analysis, observational approach, adaptive management, decision theory, operations research,* etc. On the one hand it is a strength to have such a wealth of disciplines to draw on, but for purposes of communication, the multiplicity of vocabularies is a liability. This article will mostly rely on the terminology of statistical decision theory and control theory, which are, comparatively, the fundamental disciplines.

CONCEPTUAL FRAMEWORK: MINIMIZATION OF COST SUBJECT TO RISK CONSTRAINTS

Problem Definition: The Big Picture

An environmental site becomes a candidate for cleanup when it is *perceived* as a problem. In the early days of this perception, it may not be clear exactly what is the crux of the problem at this site; and for the same reason it may not be exactly clear what specific actions would make the "problem" go away. Needless to say, it would be the gravest mistake to start in with *technical* planning at a stage in the project when the problem definition is so vague, for this is an invitation to wasteful false starts and a confusion between tactics and strategy that might obstruct a real search for efficiency.

It is often helpful to take the broadest possible view of an environmental contamination problem when attempting to formulate a site-specific problem definition that will encourage, rather than hinder, technical optimization. Generally, a contamination problem is, at root, a problem of potential exposure: there is contamination at the site that could come in too much contact with people or with some sensitive ecological feature. Considered at the same level of generality, there are three approaches to solving the problem: (*1*) the contamination could be removed, destroyed, or neutralized, (*2*) the contamination could be left in place but sealed off in such a way as to prevent exposures, or (*3*) the contamination could be partially contained and any leakage could be intercepted by continued active measures to prevent exposure. Initially, all relevant solution approaches should be on the table.

At some sites it maybe most effective to begin, for example, with an approach that relies primarily on *interception* of a moving contamination front, then shift to *containment,* and finally shift to a *removal* strategy. At other sites, other combinations or sequences of remediation actions may be superior. The possible efficiencies of these transitions in remedial approach will not be available if options are foreclosed without due consideration.

At a complicated site there may be several kinds of contamination in several different media (buildings, tanks, soils, water, etc). There may well be cascading effects such that cleanup activities directed at one contaminant in one medium will have consequences on other contaminants and other media. A premature focus on the minutiae of one small aspect of the cleanup runs the risk of initiating activities that interfere with some other aspect of the cleanup, or of missing an opportunity to let low priority portions of the cleanup get a free ride on the coattails of high priority remediation activities that will be undertaken in any case.

These cautionary comments argue for the importance of significant, early, in-depth communication among the stakeholders in the cleanup, in order to get a clear perspective on the real problem, and possible solutions, well before any technical planning begins. Blunders left uncorrected at this phase of problem definition will hamper the entire course of the project. Early success in problem definition will set the stage for later technical successes.

Because problem definition depends so much on communication, a "recipe" cannot be offered for this phase of the endeavor the way that it can be for the technical optimization steps. Success in problem definition is a matter of good-faith brainstorming (as distinct from grandstanding), careful listening, organized attention to all issues that arise, and willingness to re-examine preconceptions about the site, the problem, and the possible solutions.

Total Project Cost and the Time Factor

The many steps of a site cleanup project (remediation activities as well as data collection activities, decision-making activities, project administration, and response to regulatory action, legal action, penalties, etc) all bear costs. The goal of project integration is to minimize the *sum* of these costs. Because these costs are incurred at different times during the course of the project, a *discount rate* is applied to compute a time-weighted sum that can allow valid comparison of different project plans with different timelines. The discount rate reflects the effect that the time of payment has on the burden to the paying party.

Cost and Uncertainty: Total Expected Project Cost

The actual total cost of a project can only be computed after completion of the cleanup. For planning purposes, a means to calculate total cost *in advance* is necessary, but this calculation must take account of the reality that there is uncertainty about the course the project may take. For example, there may turn out to be greater or lesser amounts of contamination than initial surveys indicated, and so a larger or smaller investment in remediation may be required to attain compliance. A remediation technology that initially seemed promising may prove inadequate for the particular site, requiring additional investments that were not included in the original budget which assumed that everything would go as planned. The best planning guide, given this kind of uncertainty, is the *total expected cost:* this is the weighted sum of the total costs of each possible path to project completion, where the total cost of each path is weighted by the probability of that path.

By comparison, planning based on best-case, worst-case, or "best" estimates can be very misleading, since these particular scenarios neglect the probability of other outcomes that bear other costs. The labeling of a particular estimate as "best" can be especially deceptive, since that label may suggest to some customers that this estimate should be acted on as if it were certainly true, whereas it actually may be far more prudent to adopt a plan that hedges its bets in case the "best" estimate turns out to be mistaken and the "worst" is so unlikely that it is not the appropriate fall-back. The single most probable outcome can still be less probable than the collection of all alternatives.

It is unfortunate, but true, that in the notation of *decision theory,* the uncertainty about cost is itself termed a risk, that is, the risk of incurring various costs. This risk in dollar cost is not what people commonly mean when discussing the risk in an environmental cleanup. Consistent with the common sense usage, for purposes of this article, the term *risk* is reserved for uncertain environmental and human health outcomes, and the summary of uncertainty about the dollar cost of the project is referred to in the language of dollar costs of outcomes, probabilities of outcomes, and the total expected project cost, as defined above.

Environmental Risk, Health Risk, and Uncertainty

From a regulatory and public welfare perspective, the goal of an environmental cleanup is reduction of *risk* to an acceptable level, where risk here refers to adverse human health and ecological outcomes that are uncertain, but probable enough to be a concern. With respect to human health, the level of risk thought acceptable corresponds to a very small probability of the feared outcome. In some cases the standard is as low as a one in one million probability of cancer death during a lifetime of exposure.

Nevertheless, the risk goal must be understood as a *constraint* to be satisfied, as distinct from *risk minimization.* Risk minimization is too open-ended a goal to provide practical guidance for a cleanup, since it is always possible to invest more to reduce risk incrementally, even though zero risk cannot be achieved. For this reason, the risk standard is fixed as a matter of policy.

In general, the risk standards set by policy do reflect a rationale. The risk of very bad outcomes (such as death) is usually set low enough to be comparable to other risks that people commonly ignore or voluntarily accept. In setting these standards, responsible agencies give thought to the reasonableness of the costs of attainment, although this aspect of policy-making is not conducted in as systematic a way as some constituencies prefer.

For cleanups with a long duration, the risk standards may operate at two levels, one defining the range of allowable situations during the course of the cleanup and the second defining a condition that will constitute *closure.*

Project Optimization as a Minimization of Expected Cost Subject to Risk Constraints

Even though the risk goal may be set as a matter of policy, the way to achieve that goal is largely a matter of practical choices among various alternatives. The choices may be alternative remediation technologies, alternative data collection designs, and alternative strategies for using data to guide the remediation. These alternatives leave plenty of scope for different cleanup plans for a given site: all of these plans might be adequate to satisfy the risk standard and achieve site closure, but they would do so for different costs. From the standpoint of the regulator, these different plans might be equivalent, but from the standpoint of the party paying the bills there is an obvious preference for the plan with the lower price tag. *In this article the problem is to find a plan that minimizes the total expected cost while satisfying the regulatory risk standards.*

The advantage to posing the problem in this way is that this constitutes a technically well-defined optimization problem for which there is an abundance of powerful mathematical and computer tools that help in finding a solution. It is especially important, in posing the problem in this way, that there is only one quantity, ie, total expected cost, that will be minimized. Formulations that attempt to minimize (or maximize) more than one quantity are mathematically ambiguous, and the technical tools that can be brought to bear in solving them are much less powerful.

How do Risks of Catastrophe Fit the Paradigm?

Although many Superfund cleanups, and similar regulation-driven cleanups, are guided by explicit programmatic policy that sets the acceptable risk standard, some one-of-a-kind projects may appear to present risks of a sort that are difficult to fit into such a formula. For example, complicated engineering projects might raise the specter of a disaster scenario, such as a massive explosion or leak that seems incommensurate with standards expressed in terms of a very low acceptable probability of an individual's death from a lifetime exposure.

Because of various sources of initial uncertainty about the site or uncertainty about the engineering technology for the remediation, the probability of the disaster, as assessed from the perspective of adopting a fixed project plan while the uncertainty is still large, may well be large enough to cause concern. If the data to resolve the important site and engineering uncertainties, in advance, can be collected relatively cheaply, then this is the obvious solution to control the risk. If the cost of such information would itself be overwhelming, however, then there could appear to be an impasse.

In fact, such an impasse is generally the result of a mistaken limitation of perspective. There really is no need to assess the risk (including the probability of an eventual catastrophe) from the perspective of a fixed project plan that must be decided on while the uncertainty is still high. In the face of such uncertainty, the sensible engineering approach is to adopt a *flexible* project plan that has built into it enough monitoring of danger signs, and enough backup systems for dealing with whatever incipient failures are signaled by the monitoring, so that the probability of the catastrophe is reduced to so negligible a level that the risk is commensurate with the more mundane scenarios of a risk standard dealing with very low probabilities of death or injury to an individual. When the catastrophe scenario is engineered out of the picture, researchers are left on more familiar ground, where the assessment questions are simply those of confirming that the proposed monitoring systems and backup plans are adequate to provide the desired level of security. Then the project optimization would consist in comparing the costs of different, but adequate, packages of monitoring system and backup plans, to find the package with the lowest expected total cost.

In other words, a perspective with a disaster scenario that threatens to paralyze the analysis can often be replaced by a tame cost–benefit analysis that simply weighs the costs and probabilities of remediation activities following correct guesses versus the costs and probabilities of damage control necessitated by incorrect guesses, under various investments in site characterization data, monitoring data, and engineering feasibility studies. Much of the *art* in facilitating site remediation lies in restating initial concerns so that they become manageable technical problems.

REGULATORY FRAMEWORK: THE CONSTRAINTS

Understanding Risk-Based Standards

The risk goal for a cleanup is stated in terms of *probabilities* of outcomes. For example, the risk goal may be a probability of one cancer per million, but this is not to say that of one million people exposed to those conditions, exactly one of them will develop cancer; and it assuredly is not to say that of one million people exposed in sequence it would be the one millionth person who developed cancer. What the probability statement does mean is that, in any given sample of a million people with this exposure, there is a probability of zero cancer deaths, a probability of one cancer death, a probability of two cancer deaths, etc, and the *average* of this distribution is one cancer death. In other words, there remains an acknowledged uncertainty, within the language of the set standard, as to what will actually happen in a given trial of the tolerated exposure.

The reason the goal is stated in terms of a probability is that the outcomes of interest genuinely are uncertain. Some of the sources are uncertainty about the dose response curve, uncertainty about individual responses to a given dose, uncertainty about exposures that may occur, and uncertainty about the actual (as opposed to target) residual contamination level at closure.

Neither the uncertainties nor the contamination can be reduced to zero, but expenditures can be made to reduce both incrementally. Uncertainties can be reduced by investment in data, and contamination levels can be reduced by remediation activities. The best *balance* between investment in data and investment in remediation is one of the concerns in this article.

Obtaining Consensus on Regulatory Standards for the Site

Some aspects of the regulatory standard are almost entirely the prerogative of the regulator, within limits set by program policy. There is little point in attempting to renegotiate this sort of policy or regulatory call on a site by site basis. These *constraints* on the cleanup are dealt with by obtaining, from the regulators, a definitive statement of what they are, and then managing the cleanup so as to comply.

The goal of a specified residual risk level, expressed in terms of a probability of cancer per lifetime exposure, is an example of such a regulatory standard that is simply one of the rules of the game. Another example would be specific regulatory limits on emissions and discharges.

Other aspects of the regulatory standard applicable to a site can depend strongly on decisions that may originate from stakeholders other than the regulators. For example, future site-use decisions can have large effects on what sorts of exposure scenarios are likely, with corresponding influence on what residual concentrations of contamination are consistent with the risk goal.

The status of exposure models and dose response curves lies somewhere in between that of the allowable risk, which is almost pure policy, and that of the future site use, which is whatever is agreed to among the property owners and the zoning authority. Exposure models and dose response curves are matters of fact, so in principle these could be renegotiated with the latest data, and fine-tuned to the peculiarities of the particular site; but in practice it generally is not worth attempting this negotiation. The inherent scientific uncertainties in most of the relevant dose response curves and exposure models are usually quite great, but these uncertainties (having to do with low dose extrapolation and vagaries of contaminant transport, human use patterns, and contaminant uptake mechanisms) still reflect the state of knowledge after synthesizing what is usually a considerable body of scientific effort. For this reason it is not likely that one more examination of this information, or one more experiment, will substantially alter the balance of evidence. Therefore, except in cases where the current working consensus is egregiously wrong and not defended with any conviction, it usually is best to accept the current agreement on these uncertain matters of fact, rather than provoke a confrontation over an issue where the prospects for satisfactory resolution are low.

Translating Regulatory Standards to Operational Constraints on Residual Contamination and Uncertainty

The level of residual risk that is deemed acceptable for closure is set by regulators. The future site use is defined by landowners and local zoning authorities. The exposure scenario is determined by the future land use and the nature of the contaminant and the contaminated medium. With these items given, an exposure model and dose response curve is applied to determine what amount and pattern of residual contaminant will be acceptable at closure. The acceptable residual contamination constitutes a contamination scenario that results in a risk no higher than the risk goal, when run through the exposure model and dose response curve appropriate for the situation. This usually is expressed as a maximum allowable concentration of the contaminant computed as an average over a spatial unit (and perhaps over some temporal unit as well).

For example, the operational statement might be, "no more than 5 ppb as an average concentration over a 2500 cu ft volume of surface soil representing a rectangular prism 1 ft deep and 50 ft on a side." The dimensions defining the volume of averaging must make sense in the context of the exposure scenario. In order to justify a volume of particular dimensions, the exposure mechanisms themselves must average the individual's exposure to a similarly proportioned volume. In the above example, it might reasonably be asked whether the exposure mechanism is consistent with averaging contamination to a depth of one foot, and whether use patterns do average exposure to a 50 × 50 ft area.

When there is substantial variation in the concentrations of contamination over the site, the averaging greatly affects practicalities of the cleanup, by defining how *small* a hot spot must be searched for to preclude unacceptable risk, and by defining how *large* a hot spot can be tolerated if it is offset by neighboring areas of lower contamination. The larger the averaging volume, the less expensive the cleanup, all other things being equal; so the rationale for the averaging volume merits careful scrutiny. By the same token, the length of temporal unit over which effective exposure is averaged determines allowable temporary exceedances which could offer important scope for minor releases during the course of the cleanup. A case study presented elsewhere describes a multimillion dollar Superfund cleanup where resolution of ambiguities about the role of the spatial averaging volume proved crucial to an optimization that halved the total project cost (1).

Translating Policy Into Operational Constraints on Intermediate Phases of Cleanup.

Some aspects of the constraints on the course of a cleanup are matters of almost pure policy. For example, the interim level of risk that will be allowed during the course of a cleanup is set by the regulators. Similarly, there may be regulations *directly* governing allowable discharges that will apply during the course of a cleanup. As with the regulatory criteria for closure, these policy-driven standards governing the course of a cleanup are constraints to be worked with, rather than negotiable elements of the remediation plan.

In addition to these matters of definitive policy, there are common "soft" policies embodied in the regulators' usual preferences for quicker rather than slower remediations and preferences for not allowing contamination to spread. While these preferences are understandable, they are not essential to controlling risk.

For example, a slower remediation in some circumstances may be more efficient (especially, if it utilizes a less costly remediation technology); and with appropriate monitoring and backup systems (and bonding to remove uncertainty about ability to pay for the maintenance of these systems), the slower remediation need not add to the risk. Similarly, some degree of temporary spatial extension of a contamination plume will not add any risk if there is no exposure route and if there is good enough monitoring and backup technology to ensure that the plume can be contained short of boundaries beyond which its extension could cause exposures. Tolerating such a temporary extension of a contamination plume could allow for a less expensive remediation that still reaches the desired conditions for closure.

For these reasons, there may be great rewards from careful examination of the rationale behind some of the soft policies, and perhaps reconsidering some of those soft policies on a case by case basis, in the interests of greater technical flexibility in a cleanup, while still adhering strictly to the essential risk standards.

TECHNICAL FRAMEWORK: THE COST MINIMIZATION

The Central Strategy

Early, careful communication with the regulators will have resulted in translation of the regulatory standards to a form usually expressed as (*1*) a condition for closure defined as a probability for not exceeding a maximum allowable residual concentration of contaminant computed as an average over a specified unit of space and time, and (*2*) possible interim side conditions defined as probabilities for not exceeding a maximum allowable release computed as an average over a specified unit of space and time.

The standards continue to include a statement of probability in order to link the desired hypothetical condition to the actual observations made (monitoring, surveys, sampling) in an attempt to estimate whether the desired condition really has been attained. For example, it may never be known for sure whether the *true* average concentration of contaminant is below a stipulated threshold, but it will be known whether the average of the *measurements* made on the *samples* taken is below that threshold, and by how wide a margin, and then on that basis the probability that the true average is below the threshold can be calculated.

With the standards stated in this way, everything else about the remediation (in particular *how* to get the average contamination below the desired level, and *how* to make measurements to confirm the goal) is a set of technical options that can be selected from on technical grounds. The *efficiency* of the project depends on the ability to make these selections in such a way as to minimize cost.

The preceding section has suggested that there is merit to carefully reducing the regulatory constraints to the absolute minimum necessary for stipulating a risk standard in order to maximize the scope for consideration of technical alternatives in a cleanup. This is not to recommend that flexibility, for its own sake, automatically leads to the desired efficiency in a cleanup. Flexibility increases the opportunities for expensive mistakes, unless the exploitation of that flexibility is very disciplined.

Basic Rules for Disciplined Use of a Flexible Approach

1. At any decision point, it is necessary to know the cost of every outcome that might reasonably eventuate, and its probability. The probability of any outcome is the frequency with which that outcome would take place in repeated circumstances with identical information. This probability is not an estimate, and it is not a test of a statistical hypothesis. This probability is obtained by an objective calculation utilizing the available information.
2. The best that can be done at each decision point in a cleanup is to choose the option with the lowest expected cost. The expected cost of each option can only be calculated from the correct knowledge of the cost of each outcome and its probability. If the probability used in the assessment is incorrect, the decision may be incorrect. Correct decision here means a decision that would be best in the long run. In any given instance the consequences of the decision may turn out one way or the other. Work with the law of averages, which operates on the long run.
3. Never choose an option associated with a non-negligible probability of an outcome for which the damage is catastrophic and the cost of damage control would break the bank. Always make sure that the uncertainty associated with the option chosen is manageable through monitoring for incipient engineering failure and triggering affordable backup actions to prevent that failure.
4. A cleanup is not the result of only one decision. A cleanup involves a sequence of decisions. The decision-maker for each decision must look ahead, in the way that a good chess player evaluates the sequence of possible moves and countermoves that could follow the move currently in play. Weighing options at any decision point during the course of the cleanup must consider every possible path from that decision point until closure is achieved. An exclusive focus on the immediate consequences of the current step can create a vulnerability to choices that may lead to costly consequences several steps further on, just as a superficially attractive move in chess could be a trap.

Analytical Tools for Effective Decision Support

For uncomplicated cleanups where the cost structure is one sided (eg, the cost of enough data to resolve most of the uncertainty is small compared to the cost of a remediation mistake resulting from making a decision with insufficient data, or vice versa), common sense and a well organized nonmathematical assessment of the relevant factors often suffices for making decisions that achieve reasonable efficiency. As the complexity of the decision sequence increases, however, and as the various trade-offs become closer calls (eg, what if data and remediation mistakes are *both* very expensive?), intuitive methods may prove inadequate. The number of combinations to consider can be extensive. Then, recourse to rigorous, computerized, quantitative analysis is the only hope for successful application of the optimization approach described in the preceding sections.

A rigorous, formal technical framework for optimization can be obtained by combining elements of (*1*) *statistical decision theory* to set up the problem of minimizing total expected cost, (*2*) a particular version of *Bayesian statistical inference* to compute the probabilities of outcomes of actions and the probabilities of outcomes of future sampling, and (*3*) the operations research methods of *dynamic programming* to structure an efficient search for the least costly remediation plan when the number of combinations to be evaluated in a decision tree is too great for a simple exhaustive search.

Each of these three technical tools (decision theory, statistical inference, and dynamic programming) for cleanup optimization is a well-defined technical subject, each with a considerable literature and a substantial history of applications to many kinds of practical problems. In this article, some of the technical tools are described, and some portions of the analysis for a particular case study are illustrated.

There are no user-friendly computer packages currently available that put these three tools together for simple solutions of general cleanup optimization. At present, optimization of a complicated cleanup requires a customized analysis.

Statistical Decision Theory. The goal of minimizing the total expected cost of a remediation is a concept drawn from *statistical decision theory,* where, in the usual notation of that discipline, decisions are made so as to maximize the *expected utility* (2). In this formulation, the *utility* of an outcome for a remediation is the negative of the *cost,* where the *constraints* control the human health and environmental risk. Statistical decision theory provides a framework for evaluating the alternatives in terms of the sum of the utility of each possible consequence multiplied by its respective probability. This framework is largely one of common sense. The setup for computing total expected cost as a sum of costs weighted by probabilities is illustrated in some detail in the case study that follows.

Much of the technical content of statistical decision theory has to do with understanding worthwhile strategies for stating alternative decisions and *decision rules,* and with the calculation of the *probabilities* that are used in the formula for expected utility (3).

A decision rule is an exact statement of how data will be processed mathematically to arrive at a decision. Generally, in the engineering design of a remediation a decision process can be readily specified that would make correct decisions *if* perfect information about properties of the site (such as amount, location and movement of contamination, characteristics of soils, aquifers, and containment structures, and future meteorological events) were available. Actual information about the relevant properties is not perfect, is usually incomplete (as a result of sampling), and somewhat incorrect (as a result of measurement error).

A practical decision rule must therefore be stated in terms of operations on the data actually obtained, rather than in terms of the hypothetical perfect information. The literature on statistical decision theory provides guidance on using the *statistics* of actual data to formulate decision rules that make best use of the information content of data, while properly hedging the decision according to a quantification of the uncertainty about the true conditions.

Bayesian Statistical Inference. The probabilities used in the formula for expected utility are probabilities of various possible true conditions, as calculated from imperfect data. These probabilities of true conditions based on fixed data are rather different from the calculations of *confidence* delivered by conventional statistical hypothesis testing. The calculations of *confidence* are actually calculations of the probability of obtaining *different data* under hypothetical repetition of the data collection process, assuming *fixed hypothetical true conditions* (4). There are many attractions of conventional statistical hypothesis testing, but the quantities calculated by this approach are not the quantities needed in order to proceed with a decision theoretic analysis.

The approach to statistical inference that does deliver probabilities of unknown true conditions, based on given fixed data, is *Bayesian statistical inference.* The Bayesian approach is not the most commonly used sort of statistical inference, in part because, before the days of personal computers, the complexity of the calculations (which do not lend themselves to tabulation of key quantities, in contrast to the hypothesis testing calculations) was a practical obstacle, and in part because the Bayesian calculation for each application requires the *prior distribution,* ie, the probabilities of unknown possible true conditions *without reference to the data for the case in question.* The need for a prior distribution is a nuisance and poses a conceptual hurdle that bears on the mathematical and physical meaning of a probability.

For most of this century, discussion of Bayesian statistics was mired in controversy over whether probability was objective or subjective. Fortunately, in the 1990s (simultaneously with widespread access to computers that alleviated much of the numerical difficulty), many practitioners have set aside the philosophical debate, and there is increasing pragmatic use of Bayesian inference in the mainstream statistical literature (5).

In these applications, the controversial aspects of the Bayesian approach are avoided by remaining in situations where there are sufficient comparative data from related situations (ie, data on other locations within the same site, data on other similar sites, etc) that simple empirical compilations of the comparative data will suffice for a convincing, objective, approximate distribution that can successfully be used as an approximation for the prior distribution. The theoretical foundations of this particular approach to obtaining a prior distribution were established during the era of active philosophical debate (6). In recent years these methods for using data to estimate a prior have come to be known as *empirical Bayes* or *heirarchical Bayes* and there is a substantial literature on practical applications (7–9).

The Bayesian calculations have a transparent logic that lends itself to modern Monte Carlo computer methods (10). For complicated applications, however, such as the application illustrated in the case study later in this article, the straightforward computational approaches may be much slower than desired. In these circumstance the calculations can be greatly accelerated by use of less transparent closed-form mathematical solutions to some of the integrals (11) as part of the computational strategy.

Dynamic Programming. The decision theory framework, linked to Bayesian calculations of the probabilities of uncertain outcomes, provides an unambiguous mathematical formula for the expected cost of each option, but the numerical calculations to find the plan with the lowest expected cost can be overwhelming if there is much depth to the search in a multistep remediation. The difficulty is that decisions made in one step can affect the outcomes of decisions that will be made in all later steps, so the mathetical problem is of adjusting all the decision rules that will be used in all the interdependent steps in the sequence (taking account of the multiple possibilities in each step, owing to uncertainty), to arrive at a sequence of decision rules that minimizes the total cost for the entire sequence. The simultaneous adjustment of many decision rules creates a number of possible combinations that defies simple exhaustive evaluation.

The remarkable breakthrough accomplished by dynamic programming (12) is to reformulate the analysis of the decision in a given step into *components* separating the contribution of the decision in that step to (*1*) the costs and rewards accrued in that step and (*2*) the contribution of the decision in that same step to the costs and benefits that will be accrued in all subsequent steps. This leads to a *recursive* formulation that expresses the cost–benefit analysis of the decision in step n as an additive function of the status entering step n (which depends only on starting conditions and previous decisions) and the resulting status for step $n + 1$. Thus what was originally an optimization problem involving n interdependent decisions gets transformed to n separable optimization problems of one decision each. The recursion formula also allows for a backward stepwise analysis that often simplifies the problem, since in the last stages of a cleanup (when there is the least remaining uncertainty and when the decision is about the last remnants of contaminant) the decision is generally the simplest.

Even when the restructuring accomplished by a dynamic programming formulation does not yield a direct solution, the simplification brought about by the recursion often yields a problem that offers good prospects for efficiently structuring the final numerical search for the least costly plan. This may be accomplished by a variety of computer-intensive methods including exhaustive search, hill-climbing methods (13), and simulated annealing (14).

Conceptually, the key to reformulating a remediation problem according to the dynamic programming recursion is to recognize the kinds of contributions that activities in step n of the cleanup make to later steps in the cleanup. The activities in step n may include direct remediation actions (such as excavation of a contaminated volume of soil, or pump-and-treat of a contaminated volume of groundwater), containment actions (such as installation of a barrier), or data-gathering actions. Interestingly, these all have interlinked consequences for later steps in the cleanup, in that contamination remediated now will not need to be removed later, a layer excavated now exposes the next layer for easier data collection or easier remediation if needed, an extraction well put in place now facilitates future monitoring at that location, data collected now may resolve some questions that otherwise would have to be faced later, etc. The dynamic programming approach makes these linkages explicit and states them in a way that allows rigorous analysis of the merits of various strategies for remediation at the site.

Informally, the linkages between present remediation activity and future uncertainty motivate a flexible remediation approach, in which the details of the remediation plan are made up as the remediation proceeds, in the hopes that remediation decisions based on guesses made early in the cleanup will pay off either in terms of remediation accomplished (because these guesses proved right) or in terms of uncertainty reduction (even if early guesses proved wrong). A personal intuitive approach to such complicated iterative project planning has been promoted as the *observational approach,* has been promoted (15).

Because flexibility without rigorous analysis multiplies the opportunities for spectacularly expensive errors, in a high stakes project, a formal analysis of a flexible strategy, according to the dynamic programming formulation, offers an insurance policy that deserves careful consideration. The formalities of trading off investing in further information versus investing in action based on the information already in hand is treated in the literature on adoptive control (16). An especially relevant application is found in the literature on decisions about investment in mining exploration versus investment in extraction of known reserves (17). The formalities of using experimental manipulations (as distinct from pure data collection) to deliberately reveal information about an incompletely known system are treated in a literature on *probing* (18). There is a considerable literature applying the dynamic programming approach to water resources systems (19).

REMEDIATION DECISIONS FOR 179 HAZARDOUS WASTE STORAGE TANKS AT THE HANFORD NUCLEAR RESERVATION

Background to the Problem

This example presents a simplified version of a real situation, at the U.S. DOE facility at Hanford, Washington (20). On the nuclear reservation there are 179 buried tanks, ranging in capacity from half a million to a bit over a million gallons each, containing mixtures of wastes that constitute a potential chemical as well as radioactive hazard. Of the 179 tanks, 149 are at least 40 years old, and are of single-shell construction, without provisions for containing leaks from the primary vessel, and with only crude systems for detecting leaks. Some of these old tanks are already suspected of leaking. The remaining 30 tanks are newer, double-shelled tanks with better monitoring and containment features. There is a similar situation, but with a smaller number of tanks, at the DOE facility at Savannah River, South Carolina.

There is a regulatory agreement that all of the old single-shelled tanks at Hanford should be emptied. The detailed plan for removal, treatment, and disposal of the contents has not yet been decided. There are several possible methods for retrieving material from the tanks. Some of the waste volume can be reduced by separating out some of the water in an evaporator facility. Other processes can separate other components of the waste. The less radioactive material can be disposed of in a form of chemically stabilized cement, in relatively uncomplicated burial vaults. One option for disposal of the highly radioactive fraction is vitrification (heat fusion) to form a solid glass material, for eventual underground storage in highly shielded, elaborately monitored facilities.

The various engineering designs for retrieval have different respective tolerances for the physical properties of the material. Each form of disposal has its own requirements for maximum and minimum concentrations of various constituents. Feedstock that is out of specification can cause damage to treatment plants, and result in product that does not have the correct properties (eg, leaching rates and radiation levels) for the intended type of disposal. The vitrification form of disposal is the most expensive, therefore it is desirable to minimize the volume processed by this path.

There are a number of different waste types in the various tanks. Deliberate blending from different tanks has the potential for creating efficient mixes of feedstock for the respective treatment, separation, and disposal processes. Mistaken mixes have a potential for generating severe problems owing to possibilities of chemical explosion or propagating exothermic reactions from several possible incompatible combinations of waste, the potential to amass excessive amounts of radioactive material resulting in problems of too much heat, radiolysis, and possible criticality, and damage to the physical plant. Mistaken mixes that do not cause any of these severe problems still can impose a serious cost by generating suboptimal feedstock. A further constraint on actually carrying out a blending program is that there is limited free tank volume for moving material and for temporary storage of interim mixes.

Because of the chemical and radiation hazards, all operations with this waste material, including sampling, are extremely expensive. The tanks have very restricted access ports, precluding detailed, representative sampling of any horizontal spatial inhomogeneity in the wastes. There is known vertical inhomogeneity in the wastes. Horizontal inhomogeneity is suspected.

There are extensive, but imperfect, records of approximate volumes of transfers into and out of each tank, and the approximate compositions of many of the original waste types that went into each tank are thought to be known. A few liquid samples and sample cores have been collected and analyzed from many of the tanks, and more sampling is under consideration. Statistical analysis of results from intensive replicated sampling of two of the tanks suggests considerable within-tank heterogeneity, and for some of the tank constituents there has also been enough analytical variation to contribute significantly to the uncertainty.

The existing results of sparse sampling of the tanks, in combination with historical records concerning material transfers into and out of the tanks, provide rough estimates of the composition and physical properties that are important to decisions that will be made in the remediation. Because of the heterogeneity of the material in the tanks, each additional sample will reduce the uncertainty only incrementally. The high cost of sampling, in conjunction with the limited access ports for sampling, make it likely that no affordable amount of additional sampling will be sufficient to resolve *all* of the practically significant uncertainty while the waste material is still in its current form in the tanks.

In summary, the contents of the Hanford waste tanks will be treated in a long sequence of operations during the course of achieving a satisfactory disposal. The sequence of operations presents many decision points where choices can be made about the remediation. The choices made will greatly affect costs in a program that will probably spend several tens of billions of dollars to reach closure. The correct choices at each decision point depend on information that at present is known only imprecisely. Increasing the precision of the desired information for making decisions will itself be expensive.

The Big-Picture Statement of the Problem

The overall problem is to design an optimal set of decision rules governing how much sampling is done at each step, and defining what decision should be made based on data obtained at each step. The range of decisions at any step can include more sampling, transferring waste material between tanks, mixing waste material from different tanks, feeding waste material to various separation processes, treating waste material by chemical or physical modification, and feeding waste material to various disposal processes. The test of optimality is whether the total expected cost for reaching closure (acceptable disposal of all the material in all the single-shelled tanks) is as low as possible, while conforming to all risk constraints during the course of the project. The total expected cost includes the expected costs of data, costs of implementing correct decisions, and costs of dealing with the aftermath of incorrect decisions.

Because of the length and degree of branching in the decision tree describing the road map from the present status to eventual closure, this optimization is a prime candidate for application of the dynamic programming framework to structure the analysis. In that framework, the overall optimization is made manageable by recasting it as a sequence of much smaller optimization problems. The crucial link between the overall optimization and the component optimizations is that the analysis for each component decision assesses the costs and benefits of that decision in terms of *all* the possible consequences of that decision, from that point in the process until closure. Working backward in a backward recursion dynamic programming approach may greatly simplify the search process, because the depth of tree to be searched at each step is then only one layer, as the later layers (later in real time as the actual remediation proceeds) have already been resolved in an analysis that started at the end.

Perspective from Analysis of One Component Step

At a given step in dealing with a given tank there is a range of choices of what to do next with the material that is currently in the tank, where the range of choices includes collecting more data. At this step there is existing information (process and sampling) about the tank contents, which may provide sufficient accuracy for making decisions about some tanks, but not about others whose rough estimates are too close to some crucial threshold concentration for a comfortable conclusion as to which side of the threshold they actually lie on. One hypothetical constituent, whose mean concentration in the tank is the crucial quantity for making the decision about the next operation on the material in that tank at that step, is considered here. The threshold value is designated as T and the true (but unknown) average concentration in the tank as A.

Because the number of tanks is large enough to average out much of the uncertainty about the respective individual tanks, the existing information, considered over the entire collection of tanks, is adequate to give a reasonably reliable picture of the distribution of concentrations over the set of tanks and the distribution of within-tank variability (quantified as the variance between samples, within-tank) over the set of tanks. The existing information does not give as reliable an estimate of the mean (or the variance) for any *particular* tank, because there are far fewer data from any one tank than for the entire set of tanks. Put another way, the combined existing information from the collection of tanks is probably adequate to give a good estimate of *how many* of the tanks are above threshold (because that is an aggregate property, for which there are many data points), but it is not adequate to pin down exactly *which* of the tanks are above threshold (because that is a property of the individual tank, for which there are few data).

Because of the nature of the hazards posed by imaginable contingencies that could arise in the remediation operations and interim maintenance of this set of tanks, some of the costs of mistakes are very high. Because of all the precautions that are required, the costs of correct decisions and of sampling are also very high. Almost all the unit costs are measured in millions of dollars.

Given the large number of tanks about which decisions need to be made, there are likely to be tanks for which, at a given step, the estimates from existing data are sufficient to conclude that the tank is well above threshold; there are likely to be tanks for which the estimates from existing data are sufficient to conclude that the tank is well below threshold; and there are likely to be some tanks where the estimates from existing data leave enough uncertainty that the collection of more samples from that tank at that time is warranted.

To simplify the illustration, assume there are only two choices, in addition to collection of more data, as to the next action: choice O is appropriate if the tank concentration is *over* the threshold T, and choice U is appropriate if the tank concentration is *under* the threshold T. To further simplify, assume that the logistics of sample collection and analysis for a particular step make it impractical to conduct the sampling *sequentially* (taking one sample, analyzing that sample, considering the decision between choice O and U, and again deciding whether one more sample is warranted before the decision, etc). Finally, assume that the consequences of each choice, O or U, are of only two kinds, corresponding to whether the decision turns out to be right or wrong for that step. In the present case, genuine limitations on access ports leave no worthwhile scope for sophisticated evaluation of *where* to sample in each tank: the obvious answer is to sample wherever access permits, preferentially selecting ports that have not yet been sampled.

In reality, there may be more options to choose from at a given step, so there will be more categories than simply over- or under-threshold for dealing with a tank. The costs of decision consequences (especially the consequences of a mistake, where the mistake may be based on a large or small error in the estimate of concentration A) may also be more graded. There may also be times when sequential sampling is a practical option, and at many cleanup sites there is the question of *where* to sample, in addition to the question of *how many* samples. All these complications can be accommodated by extending the same logic as presented in this simplified illustration.

Preview of the Notation

The analysis that follows is concerned with a *decision* about whether a tank is over- or under-threshold. Since the decision is made under conditions of uncertainty, the tank could *actually* be over- or under-threshold, regardless of the decision. Once the decision is made, subsequent events and ongoing monitoring will reveal the relationship between the decision made and the actual tank status, but at the time of the decision the actual tank status is unknown, so meaningful reference can be made only to the *probabilities* of the consequences of any decision, based on the probabilities of the actual tank status as computed from the data in hand at that time.

When a decision is made about collecting additional data, those additional data themselves are not yet in hand, so it is possible then to quantify only the *probabilities* of what those future data will show, the *probabilities* of how those data, when obtained, will cause revised calculations about the *probabilities* that the tank is over- or under-threshold, the *probabilities* of the various *decisions* to be made once the additional data are in hand, and the *probabilities* that the future decisions will later turn out to have been right or wrong. Obviously, careful mathematical notation will be necessary to keep track of actual but unknown tank status, decisions (that may or may not prove correct) about the tank, decision consequences that result from each possible combination of decision made and actual tank status, data in hand, data not yet in hand, and probabilities of all of the above.

Optimizing the Decision about Over- or Under-Threshold in a Given Component Step

Strategy. The decision strategy is to minimize the total cost of all the possible decision outcomes, each weighted by its respective probability. In the present example there are two decision categories (over- or under-threshold) and each has two categories of outcome (right or wrong), so there are a total of four outcomes to consider.

Cost Structure. The costs of each outcome are designated as follows:

C_{to}: cost of implementing a decision that the tank is over-threshold, where this decision turns out to be correct (*to* for this situation derives from the label "true over" paralleling the common terminology of "true positive")

C_{tu}: cost of implementing a decision that the tank is under-threshold, where this decision turns out to be correct (*tu* is "true under," parallel to the common terminology of "true negative")

C_{fo}: cost of implementing a decision that the tank is over-threshold, then implementing corrective action when it turns out that this decision was mistaken (*fo* is "false over," parallel to the common terminology of "false positive")

C_{fu}: cost of implementing a decision that the tank is under-threshold, then implementing corrective action when it turns out that this decision was mistaken (*fu* is "false under," parallel to the common terminology of "false negative")

Each cost includes all the costs, from the time when the decision is made until closure is attained. This cost to reach closure depends on decisions made at later steps in the remediation as well as the immediate decision that is being weighed, but because a dynamic programming framework is used to structure the analysis as a backward recursion, the decisions in the subsequent steps are already determined, contingent on whatever data are obtained up until the time of each step. Thus the four respective costs (true over, true under, false over, false under) weighed in the present step can be partitioned into costs attributable to immediate actions/consequences and costs attributable to later stages. To the extent that the later stages are the same following the actions (including damage control, if any) in this step (ie, if the paths converge regardless of the present decision), the costs of the immediate actions/consequences will drive the present decision. If the present decision includes a branch point for *distinct* paths to closure, the subsequent steps will also differ in the options and costs they present, so the cost contributions of the subsequent steps will affect the present decision.

The need to consider costs of subsequent steps, in addition to the immediate consequences of the present decision, introduces an unwelcome complication to the cost assessment, but this is the link that ensures that the step-by-step optimization will optimize the entire remediation. An intuitive appreciation of this linkage can be gained by understanding this as creating a dual perspective, whereby the decisions within a given step in the remediation are evaluated both in terms of the amount of remediation accomplished in this step *and* in terms of the contribution of this step to "positioning" the remediation to enhance the effectiveness of actions in later steps.

Probabilities of Outcomes. There are only two relevant categories of actual status of the tank contents: actually over-threshold (*ao*) or actually under-threshold (*au*). Depending on the decision category chosen, nominally over-threshold (*O*) or nominally under-threshold (*U*), these determine the outcome (true over, true under, false over, or false under). The probabilities of the two relevant categories of actual status are designated as follows:

P_{ao}: probability, given the data, that the tank contents are over-threshold

P_{au} probability, given the data, that the tank contents are under-threshold

Note that $P_{au} = (1 - P_{ao})$, since these exhaust the possibilities.

Probabilities, given the data, are technically different from the significance or confidence levels computed by conventional statistics intended for tests of hypotheses. Probabilities, given the data, are obtained by Bayesian inference. In order for the probabilities to be correct long-run frequencies of occurrence, as is required, empirical Bayes methods often must be used to avoid the difficulties posed by subjective theories of probability. In the empirical Bayes approach, the *prior* probability is based on objective data, and provided there are enough such data, the calculated probabilities that result will have the desired properties. In the present example, the existing data (historical process information as well as sampling data) for the entire collection of tanks forms the empirical basis for estimating prior probabilities.

In the present case, assume that the within-tank heterogeneity gives rise to a normal distribution in the samples from a given tank, where both the true mean, *A*, and the true variance, σ^2, of that distribution are unknown, but where the mean and variance of the observed samples from that tank give some information on the true mean and variance. The decision about the tank could be made without error if the true mean concentration *A* were known for that tank. As it is, statistical operations with the data will yield a joint probability distribution of the unknown tank mean, *A*, and the unknown within-tank variance, σ^2. The variance σ^2 influences the interpretation of the relation of the true mean to the average of the samples already in hand, and it influences the appraisal of what is likely to be learned from a few more samples.

In the Bayes formulation, the distribution for the uncertain quantities, A and σ^2, given the data specific to the instance in question, x, is

$$p(A, \sigma^2|x) = kp(A, \sigma^2)p(x|A, \sigma^2) \quad (1)$$

where the result $p(A, \sigma^2|x)$ is called the *posterior distribution*, and $p(A, \sigma^2)$ is called the *prior distribution* based on all available information for related instances. The term $p(x|A, \sigma^2)$ is a standard statistical function, called the *likelihood function*, relating the probability of the observed data to hypothetical values of the parameters of the distributions from which the data were drawn. In the present case, the likelihood function is simply the textbook distribution of samples from a normal distribution with mean A and variance σ^2. (If some other distribution, for example a log–normal, better describes the distribution of sample values within a tank, that would be used for the likelihood function.) The value k is a normalizing constant obtained from

$$k = \frac{1}{\int_0^\infty \int_0^\infty p(A, \sigma^2)p(x|A, \sigma^2)dAd\sigma^2} \quad (2)$$

Assume, for the calculations in this illustration, that the joint prior distribution $p(A, \sigma^2)$, describing the distributions of within-tank means, and within-tank variances, among all 149 single-shelled tanks, is a kind of two-dimensional distribution, where one dimension describing the distribution of means is a family of normal distributions, and the second dimension describing the distribution of variances is based on a gamma-distribution (specifically, the reciprocal of the variance has a gamma-distribution). The joint prior distribution is specified in terms of four parameters: the grand mean (average of the by-tank means over all the tanks), the variation in means (standard deviation of the collection of by-tank means), the mode of the distribution, over all the tanks, of the within-tank standard deviations (as the measure of within-tank inhomogeneity), and the variation (standard deviation) of the distribution of within-tank standard deviations.

Figure 1 shows the relationship between this two-dimensional distribution, the four parameters used to define it, and respective within-tank means and within-tank standard deviations for every individual tank for which preliminary information is available.

In reality, the joint prior distribution almost certainly has a more complicated shape that does not correspond nicely to a familiar mathematical function such as Figure 1. There are methods for dealing with this complication; however, for purposes of this discussion, a simple distribution has the advantage that it can be summarized in terms of four basic parameters that have some intuitive meaning, so it can be shown how changes in each of these understandable parameters leads to changes in the results of the analysis.

The quantity of interest for the decision, the mean A for the tank in question, is itself one of the elements of the joint posterior distribution of tank parameters computed from applying Bayes' formula to the data. The distribution of the decision quantity A is obtained as a marginal of the joint distribution of the two uncertain parameters.

$$p(A|x) = \int_0^\infty p(A, \sigma^2|x)d\sigma^2 \quad (3)$$

The probability, given the data, that the true mean in the tank in question exceeds the threshold T is then computed simply by determining what fraction of the distribution of A, for this tank, exceeds T:

$$P_{ao} = \int_T^\infty p(A|x)dA \quad (4)$$

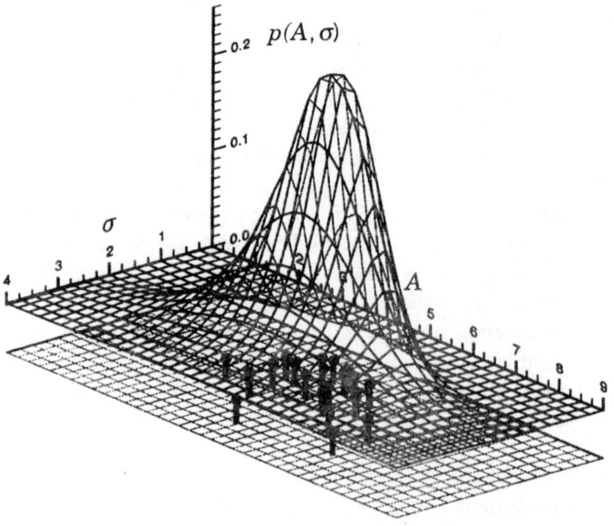

Figure 1. The joint distribution of within-tank averages (A) and within-tank, between-sample standard deviations (σ) in concentrations of the constituent of interest considered over the set of waste tanks (20 tanks), in relation to sample data from the respective tanks. The lower plane, which shares the same A-axis and σ-axis scaling with the three-dimensional coordinate system, summarizes the available information for each tank, where each tank is represented by the (A, σ) pair corresponding to the position of one of the ball-on-stick symbols. That is, each ball-on-stick symbol indicates that there is a tank with data showing a mean concentration that is the value of A corresponding to that location on the lower plane and showing a standard deviation (heterogeneity of samples from that tank) that is the value of σ corresponding to that location on the lower plane. The three-dimensional coordinate system adds a vertical axis for probability, corresponding to the *density* of occurrences of tanks with A and σ on the lower plane. The gridded surface in the 3-D coordinate system describes a smooth probability distribution fitted to the ragged histogram of actual occurrences below. The mathematical 2-D distribution relative to the A-axis is a family of normal probability functions. Each vertical section through the surface parallel to the A-axis is a Gaussian, where all the Gaussians have a common position of their mean on the A-axis, but with different respective variances. The mathematical form relative to the σ-axis is such that a projection onto the (σ, p) plane yields a transformed gamma-distribution (the reciprocal of σ^2 is gamma-distributed). The smooth surface is fitted to the data on density of occurrences by matching three parameters (the grand mean on the A-axis, the standard deviation of the projection on the A-axis, and the standard deviation of the projection on the σ-axis), then fitting the location of the mode (highest point of distribution) the projection on the σ-axis. In the example drawn, the mean on the A-axis is 4.5, the standard deviation on the A-axis is 1.5, the mode on the σ-axis is 1.5, and the standard deviation on the σ -axis is 1.0. These parameter values are used as inputs in scenario (b) solved in Table 1.

as shown graphically in Figure 2.

The calculation of P_{au} follows immediately.

Probability-Weighted Costs of Outcomes and The "Best" Decision. Compare the expected cost of the consequences of deciding that the tank is nominally above threshold versus the cost of deciding that the tank is nominally below threshold. In either case, the costs will be the sum of the probability-weighted costs of the consequences when it turns out that the decision was right or wrong.

D_o designates the expected cost of a decision to treat the case as nominally over-threshold

D_u designates the expected cost of a decision to treat the case as nominally under-threshold

Forming the probability-weighted sums of the costs of the possible outcomes gives

$$D_o = P_{ao}C_{to} + P_{au}C_{fo} = P_{ao}C_{to} + (1 - P_{ao})C_{fo} \quad (5)$$

$$D_u = P_{au}C_{tu} + P_{ao}C_{tu} + P_{ao}C_{fu} = (1 - P_{ao})C_{tu} + P_{ao}C_{fu} \quad (6)$$

The break-even point for deciding one way or the other occurs when the costs of either decision are the same, which, equating the right-hand sides of equations 5 and 6, and rearranging algebraically, gives:

$$P^*_{ao} = \frac{C_{tu} - C_{fo}}{C_{to} - C_{fo} + C_{tu} - C_{fu}} \quad (7)$$

where P^*_{ao} is the break-even point, the critical level for the probability of exceedance.

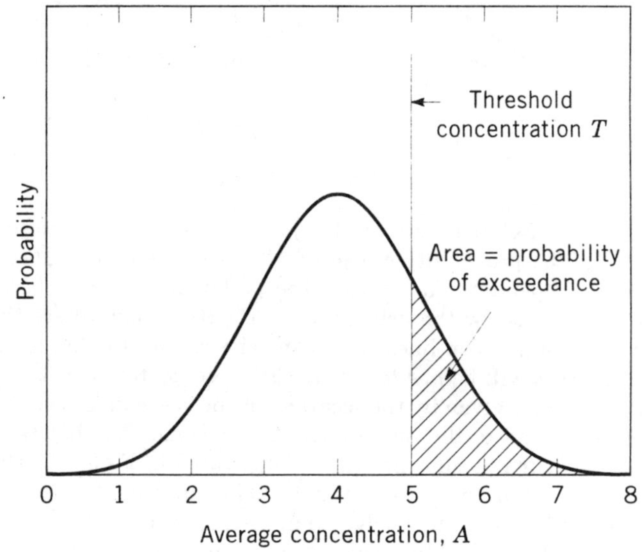

Figure 2. Relation of the distribution of the unknown true average concentration, A, to the probability that the tank is in exceedance of the threshold concentration T, for a case where the threshold concentration is 5. The bell-shaped curve (t-distribution) is the marginal posterior distribution for the unknown true average concentration, A, in the tank, inferred from sample data from this tank and comparative information from other tanks, as per equations 1 through 3. The probability of exceedance, as per equation 4, is the area under the curve of the tail of the distribution to the right of the threshold T.

In other words, the decision about whether the tank is over (O) or under (U) threshold (T) is optimized in a given component step, based on existing data, by applying the following rule: select the course of action associated with deciding that the tank is over-threshold whenever the calculated probability that the tank is over-threshold exceeds the critical probability solved in equation 7, and otherwise select the course of action associated with deciding that the tank is under-threshold.

Formally, then, the condition for justifiably deciding that the tank is over-threshold is

$$P_{ao} > \frac{C_{tu} - C_{fo}}{C_{to} - C_{fo} + C_{tu} - C_{fu}} \quad (8)$$

Note that the degree of certainty required for deciding that the tank is over-threshold varies as a function of the respective costs of all four possible outcomes. Figure 3 shows how graphs of the cost of deciding (O) or (U), as functions of the current calculation of the probability of exceedance, P_{ao}, define a crossover point, P^*_{ao}, such that when the probability of exceedance exceeds this value (as per equation 8), the cost of deciding (O) will be less than the cost of deciding (U).

Optimizing the Decision about Collecting More Data in a Given Component Step

Strategy. The decision strategy is to minimize the total cost of all the possible decision outcomes, each weighted by its respective probability. In the present example there are two general decision categories: (*1*) to proceed without additional data, using the rule developed above for optimizing the decision about over- or under-threshold in a given component step, or (*2*) collect more data, then decide about over- or under-threshold, using the same rule developed above but with the larger data set. Under category (*1*) the expected cost has already been calculated in the preceding section: it is the lower of D_o or D_u (as given by eq. 5 and 6). Under category (*2*) the expected cost is more complicated, because it includes the cost of new data, plus the sum of the probability-weighted costs of the outcomes that will follow upon making the future decision using the new data.

Limiting the Number of Sample Sizes to be Evaluated: The Cost of Uncertainty. In determining the number of additional samples to collect, the cost of new information is balanced against the value of new information. The cost of new information is simply the cost of obtaining and analyzing the specified number of new samples: this is the unit cost of an additional sample times the number of new samples. The value of information is the cost reduction expected from making the decision with more (and therefore, on the average, better) information.

The decision to collect more samples really consists of a large family of decisions: one more sample, two more samples, three more samples, etc. To simplify the search, first restrict it by asking whether there is an easy way to identify an upper limit to the number of samples that could be justified. If a cost could be assigned to the uncertainty associated with the existing information, this would set an upper bound to the costs for additional data that could possibly be offset by the cost reductions that the information in those data confers.

Based on existing data, the respective probabilities that the mean concentration in the tank is above or below threshold are P_{ao} and P_{au} (computed from eq. 4). This means that with

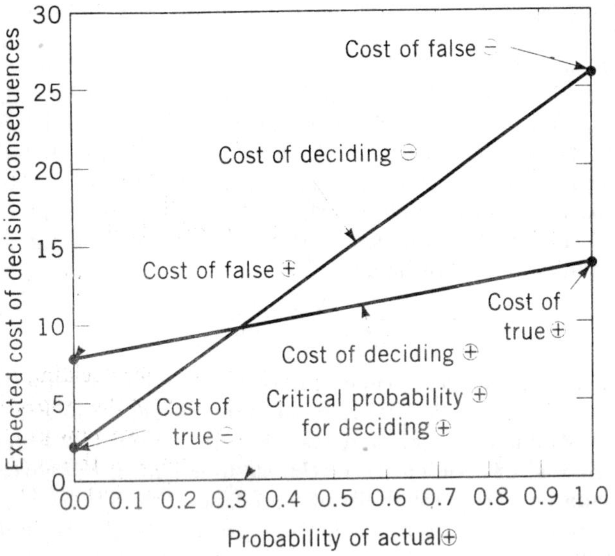

Figure 3. Relation between the expected cost of deciding *positive* or *negative* and the *probability of actual positive* (that is, P_{ao}, the probability of exceedance). Equation 5 shows why the expected cost of deciding *positive* describes a straight line connecting the unit costs of *false positive* and *true positive* outcomes. Equation 6 justifies the straight-line relationship for the expected cost of deciding *negative.* The point where the two decision cost lines cross defines the *critical probability of exceedance* (that is P^*_{ao}). If the probability of exceedance for a given tank is larger than the critical level, the expected cost of deciding *positive* is lower than the expected cost of deciding *negative,* and therefore, in the absence of further information, the correct choice would be to decide *positive.*

unlimited additional data there is a probability P_{ao} that the data would show with no uncertainty that the tank is above threshold, and there is a probability P_{au} that the data would show with no uncertainty that the tank is below threshold, which exhausts the possibilities. So the expected cost of the decision, D_∞, not including the cost of the data themselves, if there were no uncertainty, is the probability-weighted sum of the costs of the two possible correct decisions that would result:

$$D_\infty = P_{ao}C_{to} + P_{au}C_{tu} \tag{9}$$

The difference between this cost (with no uncertainty) and the cost with the present uncertainty (the lower of D_o or D_u) is therefore the cost of the present uncertainty. So the cost of the present uncertainty, I, is the lower of

$$I_o = D_o - (P_{ao}C_{to} + P_{au}C_{tu}) \tag{10}$$

and

$$I_u = D_u - (P_{ao}C_{to} + P_{au}C_{tu}) \tag{11}$$

If the unit cost of an additional sample is C_s, then the maximum number of additional samples worth considering is the largest whole number in the ratio formed from the cost of the present uncertainty divided by C_s. Thus the expected total cost only needs to be searched using additional sample sizes from 1 up to this easily calculated maximum, n_{max}, the smaller of

$$n_{max_o} = integer(I_o/C_s) \tag{12}$$

and

$$n_{max_u} = integer(I_u/C_s) \tag{13}$$

Figure 4 shows graphically how the cost of uncertainty relates to the other unit costs (C_{to}, C_{fo}, C_{tu}, C_{fu}) as a function of the current calculation of the probability of exceedance, P_{ao}.

Expected Cost of Decision After More Data Are Collected. The strategy for this calculation is readily understood, but the calculation itself is very involved. The strategy is to use the distribution already obtained for the uncertain parameters, A and σ^2, based on the existing information, to compute the probabilities of all possible future data that could be obtained if a particular number of additional samples were collected. The inferences that would be drawn from those future data, and the probabilities and costs of making right or wrong decisions based on those inferences, are then systematically calculated and averaged for the expected cost of all possible outcomes of collecting the additional data including the cost of the new sam-

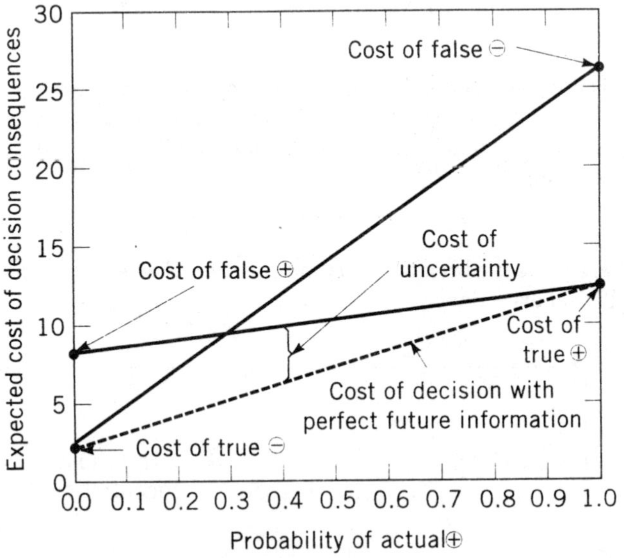

Figure 4. Relation between the *probability of actual positive* (ie, P_{ao}, the probability of exceedance) and the cost of uncertainty. Depending on whether the probability of exceedance is above or below the critical level, P^*_{ao} (the point where the two cost of decision lines cross), the choice with the lower expected cost will be *positive* or *negative,* respectively, with an expected cost given by the location on the respective cost line corresponding to the probability of exceedance for that tank with the current information, as illustrated in Figure 3. The expected cost of the decision that would result from having sufficient information to resolve all uncertainty about the tank is given by the dashed line connecting the unit costs of *true positive* and *true negative* as per equation 9. The *cost of uncertainty,* at a given value of probability of exceedance, is the difference between the cost of a decision with perfect information (dashed line) and the lower of the lines describing the decision cost for a *positive* or a *negative* decision at this location on the probability axis. The cost of uncertainty is zero where there is no present uncertainty (if the probability of exceedance is zero or one) and is greatest at the critical probability of exceedance defining the crossover point for deciding *positive* or *negative* with that cost structure.

ples themselves. Whichever number of new samples is associated on this basis with the lowest expected cost, is the best number of new samples. If that expected cost is lower than the expected cost with existing information, that number of new samples is the correct decision; if that expected cost is higher than the expected cost with existing information, then the correct decision is to proceed with selection of an action without collecting additional data at this step.

More formally, the quantity of interest is:

Dn the expected total cost, including the cost of the new samples, of using the decision criterion from inequality 8 after collecting n new samples.

This requires calculation of the probabilities of the conclusions that will be drawn (correctly, and incorrectly) from the new data. Let $p_n(P_{ao})$ designate the probability that when n new samples are collected the new probability assigned to the conclusion that the case is above threshold is P_{ao}. Note that for any particular value of P_{ao}, the anticipatory $p_n(P_{ao})$ also gives the probability that the new data will lead to the inference that the probability of no exceedance is P_{au}, since $P_{au} = 1 - P_{ao}$.

For a given pair of values of the uncertain parameters, A and σ^2 (whose joint probability is already defined, given the existing data, from eq. 1), the likelihood function (as used in eq. 1) can be used to give the probability of future data. The probability of each possible configuration of future data in turn allows recursive application of equation 1 and equation 3 (where the posterior distribution from the inference with the existing data serves as the prior distribution applied in eq. 1 for inference from the future data). In the terminology of statistical decision theory, this calculation of probabilities of inferences from data not yet in hand is called a *preposterior analysis*.

The distribution of probabilities of the decision quantity resulting from the preposterior analysis is then submitted to equation 4, to obtain a distribution, valid before the collection of the future data, for the probability $p_n(P_{ao})$ that the future data will lead to a conclusion that the probability of the mean, A, being above threshold is the value P_{ao}.

The probability function $p_n(P_{ao})$ completes the information needed for evaluating the expected worth of future data. Integrating over all possible values of P_{ao}, weighting each value by its probability, using inequality 8 to define the decision that will be made from inference on the new data, using the value of P_{ao} itself to calculate the probabilities that this future decision will later turn out to be correct or to require damage control, and using the unit cost information, then the expected total cost for taking n new samples and making the remediation decision for this step on the basis of the new data is

$$Dn = nC_s + \int_0^{P^*_{ao}} p_n(P_{ao})\,[(1 - P_{ao})C_{tu} + P_{ao}C_{fu}]dP_o + \int_{P^*_{ao}}^1 p_n(P_{ao})\,[P_{ao}C_{to} + (1 - P_{ao})C_{fo}]dP_{ao} \qquad (14)$$

where the critical probability, P^*_{ao}, is the break-even point defined in equation 7.

The value of D_n computed from equation 14 gives the expected total cost of collecting n new samples and then using the information they yield to make the decision to select the action, in this step, consistent with the new calculation of the probability that the tank average concentration, A, is over the threshold T. This cost is then compared with the expected cost for other numbers of new samples, $D_1, D_2, .., D_{nmax}$, and with the expected cost for selecting an action without additional data. The design with the lowest cost is the correct choice for this component step of the remediation.

The expected cost D_n is a net cost. Equation 14 is readily partitioned to reveal the components of this net cost: the value of future information and the cost of future data, for a given number of additional samples. The term nC_s is the gross cost of the additional data, and the sum of the two integrals gives the expected cost of the consequences of the future decision using the additional data. Subtracting the expected cost of future decision consequences from the smaller of D_o or D_u (the expected cost with no additional samples) gives the expected gross cost reduction attributable to the n additional samples. This is the *value of information*, shown graphically in Figure 5. The value of information exhibits a pattern of diminishing returns (which makes sense, since the value of future information cannot exceed the current cost of uncertainty, no matter how many additional samples are proposed). The cost of the additional data is proportional to the number of new samples proposed, so the maximum expected net benefit (value minus cost) is achieved at an intermediate number of additional samples.

Information and Calculation Requirements

The input information to use this method is not at all esoteric: it is the same information on costs and feasibility and sampling data that would be used for any common sense engineering

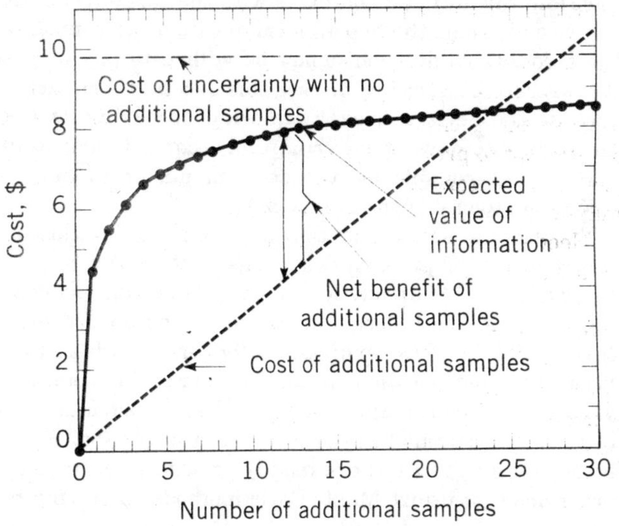

Figure 5. Value of future information and cost of future information as functions of the number of additional samples. The solid curve shows the expected value of information (expected gross reduction of future cost of decision) from additional samples, calculated from scenario (a) of Table 1, using the partitioning of equation 14. The dotted line shows the gross cost of the additional samples when the cost per new sample is 0.35. The net benefit of the additional samples is the difference between the value of information and the cost of the samples. In the present case, the net benefit is maximum at four additional samples (optimal number of new samples for this scenario). The dashed line is the cost of uncertainty with no additional samples, so the expected value of information will asymptotically approach this ceiling with increasing numbers of additional samples.

analysis, and the calculations are natural candidates for standard computer methods.

The input quantities necessary for the analysis are nothing more than the various unit costs of defined activities or consequences (C_s, C_{to}, C_{tu}, C_{fo}, C_{fu}), the sampling data (x) bearing on the tank in question, and the statistics (in this case the four parameters that define the prior distribution of comparative information) that describe the distributions of by-tank mean concentrations and within-tank heterogeneity over the entire collection of tanks.

Quantifying the cost of sampling (C_s) and the costs of responding to a correct selection of action (C_{to} or C_{tu}) should present no special difficulty. Quantifying the costs of what turned out to be "false over" decisions (C_{fo}) or "false under" decisions (C_{fu}) is a little more complicated.

The most favorable setting for weighing the costs of false positive ("false over") and false negative ("false under") decisions occurs where the choices are essentially among engineering alternatives. Then the costs of "errors" are quantified from an understanding of how an incorrect decision at a particular stage will be detected at some later stage, and understanding what recovery and replacement actions will be undertaken in consequence of the detection of that error, as described in a previous section.

The most challenging setting for weighing false positive and false negative decisions arises where it is believed that the error will go undetected indefinitely, but where there is nevertheless a reluctance to let the error be uncontrolled. One way of handling this difficult context is with the seemingly artificial exercise of posing the hypothetical question: How much would the accountable decision maker be willing to pay to prevent the error from occurring? In fact, this question is not artificial, because the decision maker *will* wind up paying for data in an attempt to prevent the error from occurring, so it would be perfectly reasonable for the decision maker to have some maximum tolerable price tag in mind.

Needless to say, the mathematics of obtaining closed form solutions for the probabilities in this analysis can be difficult. Equations 1–14 already look rather daunting, owing to the integrations. These forms of the equations do not show the further details of the expressions for terms such as the likelihood functions, or the joint distribution of the mean and the variance in the prior and posterior of Bayes' formula, or the recursive application of Bayes' formula to solve for $p_n(P_{ao})$. Nevertheless, all the steps are readily implemented in computer programs that utilize Monte Carlo methods for solving Bayes' formula.

The Monte Carlo numerical methods are computer-intensive. For example, a pure Monte Carlo strategy for solving the example presented here required, for each scenario, two weeks of execution time on a work station (DEC 5000-260). But, whatever cost or delay this represents is minor compared to the many millions of dollars that can be saved as the result of the two-week calculation.

For cases where there is a need to explore multiple scenarios in a shorter amount of time, it is possible to use a variety of explicit (but complicated and problem-specific) mathematical solutions to replace some of the Monte Carlo calculations for a much faster solution. By investing considerable mathematical expertise in the computer program used for this example, a personal computer version was developed that solves each scenario in a matter of seconds.

Example Solutions of Scenarios

Table 1 shows solutions for three example scenarios that could be plausible cost structures and data for some of the Hanford tank decisions.

Scenario (a) has a comparatively high cost of false positives and false negatives. It demands a higher probability of exceedance before deciding *positive* with existing data. The data on this tank are sparse, but indicate that this tank is barely over the critical probability for exceedance, triggering a positive decision, under circumstances where the expected costs of a positive or negative decision are almost identical. Because the probability of exceedance is so close to the break-even point, and given the high costs of false positive and false negative outcomes, the cost of the uncertainty with the present data is high. Since the cost of additional samples is not too high relative to the cost of uncertainty, additional samples are worthwhile. The optimal number of additional samples reduces the expected cost by about one-third, which is a substantial amount when costs are reported in millions.

Scenario (b) has a less threatening cost structure, in particular the cost of false positives is comparatively low, so it demands less certainty of exceedance to trigger a positive decision. The data are still sparse, but they make a more convincing case that the tank is below threshold. So much so that the probability of exceedance is below the critical probability, and the decision with existing data is that the tank is negative. Because of the less threatening relative costs of mistakes, and because the current probability of exceedance is not so near to the break-even point, the cost of uncertainty is lower. Because the cost of additional samples is high, and the cost of uncertainty is modest, no additional samples are justified.

Scenario (c) has the same cost structure as scenario (b) for the costs of decision outcomes (and therefore the same critical probability of exceedance for triggering a positive decision), but it has a vastly cheaper unit cost for additional samples. Scenario (c) is also distinguished by a distribution of comparative information for the entire set of tanks that creates a default presumption that the tank is over-threshold (rather than the reverse presumption in scenarios (a) and (b), above). Despite the slightly stronger data for this tank (three samples rather than the two samples from scenario (b), but with the same average), the presumption of exceedance, owing to the comparative information, leads to a decision with the existing data that the tank is a positive. If the decision were made, instead, that the tank is a negative, the expected costs would be about 25% higher. Because of the very low cost of additional samples, the optimal number of additional samples is substantial, and these additional samples would reduce the expected cost by about 10%.

Intuitive Appraisal of Response to Inputs

The method described previously uses mathematical formulas for identifying the best decision within any given step of the remediation. The appearance of mathematical formulas can create an impression of opacity to intuition, but there is an understandable, logical flow to the calculation, and, despite the math, it is shown here that in each stage of the calculation there is an intuitively reasonable relationship between changes in the various inputs and the resulting changes in the calculated best decision.

Table 1. Decision Theory Solutions for Three Example Scenarios[a]

Solution	Scenario (a)	Scenario (b)	Scenario (c)
Input Specifications			
Costs of decision consequences			
Cost of true positive C_{to}	20.0	10.0	10.0
Cost of false positive C_{fo}	15.0	4.0	4.0
Cost of true negative C_{tu}	2.0	2.0	2.0
Cost of false negative C_{fu}	60.0	25.0	25.0
Other Unit Costs			
Cost of additional sample C_s	0.35	1.0	0.05
Comparative information			
Grand mean of concentrations	4.50	4.50	5.50
Standard deviation of means	1.50	1.50	1.50
Modal within-tank std dev	1.00	1.50	1.50
Std dev of distribution of std devs	0.25	1.00	1.00
Engineering specifications			
Concentration action level T	5.0	5.0	5.0
Sampling data for this tank			
Number of existing observations	1	2	3
Observation No. 1 x_1	4.40	3.00	3.00
Observation No. 2 x_2		4.00	4.00
Observation No. 3 x_3			3.50
Decision Theory Results			
Results of cost structure			
Critical Probability of exceedance P^*_{ao}	0.245	0.118	0.118
Decision consequences with existing data			
Current Probability of exceedance P_{ao}	0.246	0.109	0.191
Expected cost of deciding positive D_o	16.2	4.7	5.1
Expected cost of deciding negative D_u	16.3	4.5	6.4
Best action decision (O or U)	O	U	O
Cost of current uncertainty I	9.8	1.6	1.6
Evaluation of worth of additional data			
Optimal number of additional samples n	4	0	7
Expected cost with additional samples D_n	11.1	4.5	4.7

[a] Std dev = standard deviation.

The costs of all possible outcomes of the decision are compared in order to identify the critical probability level for the decision. In the case of the binary decision (over or under) there were four possible outcomes whose costs are taken into account, and one critical probability level, the critical value for the probability of exceedance, P^*_{ao}, such that when the probability of exceedance, as calculated from data on the tank, is that large or larger, the action appropriate for the tank being "over" should be selected. This critical value is determined by a calculation shown in Figure 1, where the individual costs of all four possible outcomes served to anchor the two straight lines whose crossover point defined the value of the critical probability P^*_{ao}. Thus, changing any one of those unit costs will change the critical probability P^*_{ao} and accordingly influence the decision.

For example, if the cost of a false negative (a "false under") is made larger, this will raise the right-hand end of the line describing the cost of a negative decision in Figure 1, so the place where the two lines cross will move to the left, resulting in a lower value for P^*_{ao}. A lower value for P^*_{ao} means that *less* certainty of an exceedance is required to trigger to the decision that the tank is "over." Which makes sense, because the cost change imposes a higher penalty for mistakenly deciding that the tank is "under." More generally, when the cost of a particular outcome increases, the critical probability level for triggering a decision shifts *away* from the decision that could lead to that more costly outcome.

The existing information is used to compute a probability for the tank composition. The existing information consists both of sampling data specific to the tank in question (encapsulated in the data x) and comparative data from other tanks (encapsulated in the prior distribution $p(A, \sigma^2)$). Both these sources of information enter equation 1 as distribution-like functions: the prior distribution is explicitly a distribution of parameter values, and the likelihood function is an un-normalized probability of the data considered over the full range of parameter values. The normalized *product* of these two distribution functions gives the probability distribution for the parameters describing tank status, so whichever of the two input distributions, the prior or the likelihood, is narrower will dominate the product that constitutes the result. A narrower distribution arises from higher information content (more certainty). Therefore, in this view of equation 1, as the amount and quality of data from the tank in question increases, the inference (posterior distribution) becomes more and more driven by those data, and less influenced by the comparative information on the other tanks (prior distribution). This is as it should be, because the information on the other tanks serves only as a default in the absence of adequate information on the tank of interest.

The uncertainty in the conclusion about the concentration in the tank of interest is reflected in the breadth of the posterior distribution inferred for the tank's average concentration; the broader that distribution, the greater the uncertainty. Factors that contribute to the breadth of the respective distribution-like functions representing the sampling data for this tank and the comparative description of other tanks are also the factors which will contribute to the breadth of the distribution for the inferred tank average. These factors are sparsity of sampling data for this tank, heterogeneity in sampling data for this tank, high frequency of large within-tank variation in the comparative distribution, and large variation between tanks in the comparative distribution. A final factor leading to uncertainty about the concentration is disagreement between the comparative distribution and the actual data for the tank in question.

The inferred probability distribution for the average concentration in this tank is used to compute a probability for tank status, which is then compared to the critical probability. The probability of tank status, that is, the probability that the true tank average is over the threshold, is computed as the area of the posterior distribution that exceeds this threshold. When the observed tank average is under the threshold, the portion of the distribution that exceeds the threshold is a kind of "spillover" that increases with the breadth of the distribution, as shown in Figure 2. In other words, when the observed average is under-threshold, the calculated probability of exceedance increases with the uncertainty about the true tank average. Therefore, as the uncertainty about the tank average increases, under these circumstances, the probability of exceedance can increase until it passes the critical level, changing the decision from "under" to "over."

The cost of uncertainty with the present data puts a ceiling on the cost of additional data that could possibly be worthwhile. The cost difference between making the better or worse, "over" or "under", decision, given the uncertainty, diminishes the closer the probability of exceedance is to the critical level (where the cost difference is zero) (see Fig. 3). For the same reason, as Figure 4 shows, the cost of the uncertainty increases, the closer the probability of exceedance is to the critical level (because near the critical level, there is no better decision to choose). As a result, the closer the probability of exceedance (with the existing data) is to the critical level, the larger the number of additional samples that can be justified (for a fixed cost per additional sample).

Of course, as the cost of additional samples declines, the number of additional samples that can be justified increases. If the cost of additional samples is negligible, it will be justifiable to collect however many samples are required to remove all the effective uncertainty. Then the cost of the decision will be the lower dashed line shown in Figure 4, which is simply the probability-weighted cost of making the correct decision (true positive or true negative). Given a cost-ceiling on justifiable investment in additional samples, the relationship between cost per sample and optimal number of additional samples is obvious. The relationship between other factors and the optimal number of additional samples is less self-evident.

The optimal number of additional samples is found by comparing the cost of additional data to the cost reduction expected from having those data. Another way to understand the relationship between the optimal number of new samples and the probability of exceedance with the existing data is that in order for the new samples to make a difference, the new samples must be capable of changing the decision (from "over" to "under", or vice versa). The greater the potential for the new data to change the decision, the larger the optimal number of new samples, all other things being equal.

The general factors which contribute most consistently to the potential for the new data to change the decision are being close to the critical probability of exceedance (close to the crossover point in Fig. 3), and sparse existing data. Two, more complicated factors, that can work at cross purposes (because they affect P_{ao} as well) are low information content in the comparative distribution, and high information content in future data. The detailed factors that can contribute to low information content in the comparative distribution are a mean concentration near the threshold value T, and a large standard deviation in the distribution of the mean. High information content in the new samples is encouraged by low within-tank heterogeneity, as signaled by a comparative distribution for the within-tank variability that has a narrow spread and a mode at a low value, and it is encouraged by existing data that exhibit little variation. The effect of high information content in future data will be clearest when there are fewest existing sampling data from the tank in question.

Comparison with Conventional Statistical Hypothesis Testing

The example presented illustrates use of decision theory coupled with Bayesian probability calculations. Most technically trained researchers are more familiar with a different statistical tradition, based on *hypothesis testing* and using a different definition (variously called *classical, frequentist,* or *sampling theoretic*) for the underlying probabilities to represent uncertainty. The two approaches do not generally give the same solution; when they differ, the Bayesian decision theory calculation, in the kind of application considered here, will give solutions that lead to a lower total expected cost. The reasons for this difference are rather deep, and will not be discussed in any detail here. What is presented in this section is an illustration of the differences by carrying out the calculations both ways in a few scenarios.

A conventional hypothesis testing approach considers the two types of errors, false positives and false negatives, and controls them in different ways. One is treated as the Type 1 statistical error (mistakenly accepting a false null hypothesis), and is controlled by use of a confidence limit (or significance test) with a stipulated alpha-level; the other is treated as the Type 2 statistical error (mistakenly rejecting a true null hypothesis), and is controlled by adjusting the sample size until a *power* calculation indicates that a stipulated beta-level has been satisfied.

In a regulatory setting, where a "positive" signifies that regulatory action is needed, and a "negative" signifies that regulatory action is not needed, the hypothesis testing approach conventionally controls false negatives with the confidence limit mechanism, and it controls false positives with the power mechanism. The logic of the hypothesis testing approach is just as legitimate if the mechanisms are switched, and the confidence limit is used to control false positives and the power calculation adjusts the sample size to control false negatives. But the operating characteristics for the procedure with the switched mechanisms will not be identical.

In a conventional application of hypothesis testing to the Hanford tank example, a confidence interval would be computed on the *estimate* of the mean from a given tank, and the upper confidence limit would be compared to the threshold value T to make the decision. Use of the upper confidence limit would control the probability of false negatives to a frequency α, by creating a margin of safety favoring a positive decision, where the size of the margin tracks the uncertainty in the estimate of the mean. This aspect of the error control system will work regardless of the sample size, for as the sample size changes, the size of the margin of safety provided by the confidence limit changes accordingly.

In the conventional application of the hypothesis testing approach, the identification of the desired sample size arises through the stipulation of a very particular false positive error scenario. This can be expressed, for example, as the situation where the true concentration in the tank is $(T - g)$, and stipulating an allowable probability for the mistaken conclusion under this circumstance that the concentration is above the threshold T. The positive quantity g then represents the width of a "gray area" within which false positives are tolerated with frequency β, even though the true concentration is below the threshold. The power calculation finds a sample size that satisfies this criterion, when the decision rule is determined by the upper confidence limit mechanism selected previously.

Two immediate difficulties in the application of this approach, compared to the decision-theoretic framework, are that (*1*) no specific guidance is provided on how to select the probabilities α and β that define the error tolerances, and (*2*) the role of the particular error scenario chosen for defining the false positive situation (ie, the value of g in the scenario where the true negative value, that is mistaken for a positive, is $T - g$), and the relation of the probability of that false positive scenario to other false positive scenarios, is not clear. It is common, both in textbook examples and often in practice, to set the alpha error rate at 5%, the beta error rate is often set at 10 or 15%, and the magnitude of the false positive error margin g is often rationalized in terms of a region of indifference, near the threshold T, where the customer is not greatly concerned to discriminate whether the true value is slightly below threshold by that amount versus above threshold by any amount. These choices of specifications for the hypothesis testing approach are not firm, and there is no theoretical reason they should be, even though these specifications strongly affect the results.

Also, there may be ambiguity in the choice of a value for the within-tank heterogeneity (specified as a variance, σ^2) that is used in the power calculation. When there are enough existing data on the tank in question to compute a trustworthy variance among the samples, this will provide an obvious value for σ^2. Of course, where there are enough existing data to calculate a secure estimate for the variance, uncertainty about the quantity A that is driving the decision will probably be so slight that the decision is academic. If there are no existing samples from this tank, or if the number of existing samples is too small for the estimate of the variance to be secure, then it becomes a matter of judgment to select a value for σ^2 in the hypothesis testing approach.

Obviously, if there is this much room for judgment in the setting of the driving specifications for the hypothesis testing approach, and there is no formal linkage between the setting of these specifications and the costs and probabilities of various outcomes, the practical performance in the long run is unpredictable, for these input specifications can determine the result. A final and more esoteric shortcoming of the classical hypothesis testing approach, is that the probabilities computed in the confidence and power mechanisms are averages over all possible data configurations, and not specific to the data that are actually in hand, so these probabilities are not interchangeable with the probabilities, such as the computed probability of exceedance P_{ao}, used in the decision-theoretic approach.

Here the comparative performance is illustrated, where traditional values of 5 and 15% are used for the alpha and beta error rates, and an arbitrary value of 5% of T is selected for g, the width of the indifference region.

Table 2 shows three example scenarios with their hypothesis testing solutions. These three scenarios duplicate the conditions of the scenarios illustrating decision theory solutions in Table 1, but much of the input information used by the decision theory calculation is not used by the hypothesis testing approach (so this information does not appear in Table 2), and the hypothesis testing approach requires some additional input specifications that are listed in Table 2. As for the information *used* by the hypothesis testing approach, the three scenarios illustrated in Table 2 are identical except for the existing sampling data from the tank. Accordingly, there is not much response of the hypothesis testing results to the different scenarios, except that scenario (*c*) has enough data to trigger a decision that the tank is a *negative* and no additional samples are required. (Under the hypothesis testing approach, where only false *positives* are controlled by the power calculation determining the required sample size, a *negative* decision automatically leads to a conclusion that no additional data are required.)

A comment is in order about the specification of the standard deviation (assumed heterogeneity) in Table 2. In scenario (*a*) there are insufficient data to compute a standard deviation at all. In scenario (*b*) the data lead to a calculated standard deviation estimate of 0.71, but the 95% confidence limits on this estimate are 0.32 to 26.7, which does not create much of a sense

Table 2. Hypothesis Testing Solutions for Three Example Scenarios

Solution	Scenario (*a*)	(*b*)	(*c*)
Input Specifications			
Error tolerances			
False negative rate α	0.05	0.05	0.05
False positive rate β	0.15	0.15	0.15
False positive scenario			
Lower border of gray area $(T - g)$	4.75	4.75	4.75
Assumed heterogeneity			
Standard deviation	0.50	0.50	0.50
Engineering specifications			
Concentration action level T	5.0	5.0	5.0
Sampling data for this tank			
Number of existing observations	1	2	3
Observation No. 1 x_1	4.40	3.00	3.00
Observation No. 2 x_2		4.00	4.00
Observation No. 3 x_3			3.50
Hypothesis Testing Results			
Decision with existing data			
Action decision (O or U)	O	O	U
Decision on need for additional data			
Required number of additional samples	30	29	0

of security in the number. In scenario (*c*) the calculated estimate of the standard deviation in the data is 0.5, and the 95% confidence limit on this estimate is 0.26 to 3.1. The specified value 0.5 is plausible for all three scenarios.

Using the cost information and comparative information in Table 1, the machinery of statistical decision theory can be employed to calculate the expected costs of implementing decisions according to the recommendations generated by the hypothesis testing approach in Table 2. The results, comparing the expected costs (and the recommendations) of the two methods are shown in Table 3. There the two methods generate quite different conclusions: they can differ (in both directions) on the decision with existing data as to whether the tank is a *positive,* and they can differ (again in both directions) as to the number of additional samples that should be collected. The decision theory approach, outlined in previous sections, in no case performs worse than the hypothesis testing approach, and generally performs better, often by a very wide margin, as for example in scenario (*b*).

In general, the decision theory solution responds in an intuitively satisfying way to changes in the respective input variables. The hypothesis testing approach never performs better and often performs markedly worse than the decision theory solution, but in a pattern that is rather difficult to predict (because of the arbitrary nature of some of the input specifications in the hypothesis testing approach, and because the hypothesis testing approach does not use cost information or comparative data as explicit formal inputs).

CONCLUSION

An integrated approach has been presented for minimizing the cost of a cleanup while complying with regulatory risk constraints. This approach is *integrated* in that it optimizes the project as a whole, provides a framework for translating verbal issues (including regulatory goals) into quantitative operational language that can guide the cleanup, considers all the costs in pursuing the project to completion, takes account of the respective probabilities of uncertain outcomes, and evaluates trade-offs between different remediation activities and between remediation actions and data collection.

Table 3. Comparison of Decision Theory Results and Hypothesis Testing Results for Three Scenarios

Scenario[a]	Decision Theory	Hypothesis Testing
(*a*)		
Decision consequences with existing data		
Action decision	*O*	*O*
Expected cost of decision	16.2	16.2
Evaluation of worth of additional data		
Recommended number of additional samples	4	30
Expected cost with additional samples	11.1	18.7
(*b*)		
Decision consequences with existing data		
Action decision	*U*	*O*
Expected cost of decision	4.5	4.7
Evaluation of worth of additional data		
Recommended number of additional samples	0	29
Expected cost with additional samples	4.5	32.2
(*c*)		
Decision consequences with existing data		
Action decision	*O*	*U*
Expected cost of decision	5.1	6.4
Evaluation of worth of additional data		
Recommended number of additional samples	7	0
Expected cost with additional samples	4.7	6.4

[a] From Tables 1 and 2.

Guidance on achieving this integration is a mixture of seemingly nontechnical common sense advice having to do with problem definition, semitechnical advice on translating regulatory goals into unambiguous constraints, and very technical advice on optimization of all the component decisions that then govern the cleanup. It would be a mistake to assume that the significance of the various ingredients in this mixture of advice increases in proportion to the degree of technicality. In practice quite the opposite is true.

Because of the human factor, the precise capture of the key issues in the problem definition and operationalization of regulatory concerns is often the most difficult part of optimizing a real cleanup. It can also be the most important part, because it sets the stage for all the details that follow. No amount of math and statistics will salvage a project plan for a misstated problem; with a misstated problem all that the math and statistics will provide is an elegant answer to the *wrong* question. By contrast, successful problem definition states the problem in such a way that all the remaining decisions for optimization are *merely* technical, which is a relief because there is a large portfolio of proven, relatively mechanical formulations (some of which have been illustrated here) that can be applied for technical optimization.

The rewards are high for adopting this integrated approach. Skillful, systematic application of the approach described here can greatly cut the cost of a cleanup compared to the project plan considered before an integrated optimization analysis was done.

The sky is the limit, in some sense, as to how much money and time can be saved, since, in a complicated project, there is hardly any ceiling on possible *in*efficiencies that are accessible via the alternative of informal, improvisational project organization. Even at the micromanagerial level of optimizing within a single step of a project, the scope for savings can be large. The cost–benefit analysis approach presented for setting decision error tolerances shows that, depending on the relative costs of the respective decision outcomes, the correct critical probability level for the decision can range from 0 to 100%, so reliance on a traditional upper 95% confidence limit can result in incorrect decisions with an arbitrarily high frequency, and can lead to arbitrarily large over-investment or under-investment in additional data. Depending again on the relative costs of decision outcomes and the unit costs of additional samples, the cost penalties, from the excess probability of wrong decisions and from the misinvestment in additional data, can be arbitrarily high. There is always a more costly way to do a cleanup, and the integrated approach outlined here helps avoid it.

BIBLIOGRAPHY

1. S. Blacker, and D. Goodman. *Environ. Sci. Technol.* **28,** 471A–477A (1994).
2. J.O. Berger, *Statistical Decision Theory and Bayesian Analysis,* Springer-Verlag, New York, 1985.

3. M.H. DeGroot, *Optimal Statistical Decisions,* McGraw-Hill, New York, 1970.
4. E.L. Lehmann, *Testing Statistical Hypotheses,* 2nd ed., John Wiley & Sons, Inc., New York, 1986.
5. A. O'Hagen, *Kendall's Advanced Theory of Statistics,* Vol. 2B, *Bayesian Inference,* Edward Arnold, London, 1994.
6. R. von Mises, *Ann. Math. Statist.* **13,** 156–165 (1943).
7. J.S. Maritz, and T. Lwin, *Empirical Bayes Methods,* 2nd ed., Chapman and Hall, London, 1989.
8. A. Gelman, J.B. Carlin, H.S. Stern, and D.B. Rubin, *Bayesian Data Analysis,* Chapman and Hall, London, 1995.
9. B.P. Carlin, and T.A. Louis, *Empirical Bayes Methods for Data Analysis,* Chapman and Hall, London, 1996.
10. A.F.M. Smith, and A.E. Gelfand, *Am. Statist.* **46,** 84–88 (1992).
11. H. Raifa, and R. Schlaifer, *Applied Statistical Decision Theory,* Division of Research, Harvard Business School, Boston, Mass., 1961.
12. R. Bellman, *Adaptive Control Processes: A Guided Tour,* Princeton University Press, Princeton, N.J., 1961.
13. S.L.S. Jacoby, J.S. Kowalik, and J.T. Pizzo, *Iterative Methods for Non-linear Optimization Problems,* Prentice-Hall, Englewood Cliffs, N.J., 1972.
14. S. Kirkpatrick, C.D. Gelatt, and M.P. Vecchi, *Science* **220,** 671–680 (1983).
15. L. Bjerrum, A. Casagrande, R.B. Peck, and A.W. Skempton, eds., *From Theory to Practice in Soil Mechanics: Selections from the Writings of Karl Terzaghi with Bibliography and Contributions on His Life and Achievements,* John Wiley & Sons, Inc., New York, 1960.
16. S.J. Yakowitz, *Mathematics of Adaptive Control Processes,* American Elsevier Publishing Co., New York, 1969.
17. K.J. Arrow and S.S.L. Chang, in P.T. Liu, ed., *Dynamic Optimization and Mathematical Economics,* 1980, Plenum, New York, pp. 105–116.
18. Y. Bar-Shalom, *IEEE Trans. Autom. Control* **26,** 1184–1195 (1981).
19. Y.Y. Haimes, *Hierarchical Analyses of Water Resources Systems,* McGraw-Hill, New York, 1977.
20. *Statistical and Cost–Benefit Enhancements to the DQO Process for Characterization Decisions,* DOE/EM-0316, NTIS, 1996, provides an earlier version of this analysis.

SLUDGE REMEDIATION. See SUPERCRITICAL WATER OXIDATION FOR WASTEWATER AND SLUDGE REMEDIATION.

SMOLDERING BIOMASS FUELS, MEASURING AND MODELING

LISA MCKENZIE
URS-Grenier, Inc.
Denver, Colorado

Biomass fires release a complex mixture of compounds, including hydrocarbons, oxygenated hydrocarbons, halogenated organic compounds, polycyclic aromatic hydrocarbons (PAHs), sulfur compounds, and nitrogen compounds, into the atmosphere (see POLYCYCLIC AROMATIC HYDROCARBONS IN THE ENVIRONMENT). Some of these compounds, such as formaldehyde and carbon monoxide, are locally significant due to their toxicity (see FORMALDEHYDE IN THE ENVIRONMENT). These compounds could be hazardous to people who are exposed to smoke from biomass fires. For example, wildland firefighters made approximately 12,000 medical visits for respiratory problems during the 1988 wildfires in Yellowstone National Park Wyoming, (1). Some of the emissions are also globally significant as greenhouse gases (eg, methane and carbon dioxide), as photochemically reactive trace gases in the troposphere (eg, nonmethane hydrocarbons and nitrogen oxides), and as precursors to ozone destroying radicals (eg, methyl chloride and methyl bromide).

The burning of biomass is widespread, with more than 1800 Tg dm/yr burned in extra tropical zones (U.S., Canada, Europe, former USSR) (2). This large mass includes biomass burned in wildfires, prescribed fires, domestic fires, and fires of agricultural residues. The biomass burned in extra tropical zones accounts for 23% of the biomass burned globally. Currently, most biomass burning (77%) occurs in the tropics, where the use of fires has significantly increased during the past two decades because of rapid changes in land use practices and increases in population. Fires in the tropics are widely used for deforestation, shifting cultivation, domestic energy consumption, pest control, and clearing of agricultural wastes. The emissions may create serious air quality and health problems. Recently, the importance of wildfires in the Northern Hemisphere boreal forests has become of concern. The boreal forests contain between one-quarter and one-third of all carbon stored in terrestrial biomass, much of it in the duff layer (organic subsurface forest floor) (3). These intense fires are presently a significant factor in boreal forest ecology, with 5–10 million hectares burned annually (4). Fire frequency may increase by up to 40% as conditions become hotter and drier in the boreal forests with global climate change (5).

The emissions from smoldering combustion are especially important considering that wildland firefighters are heavily exposed to smoldering emissions during mopping operations and people in general are exposed to smoldering emissions as a result of domestic heating and cooking fires. Smoldering emissions are prevalent in duff fires which are particularly relevant in boreal forest fires as discussed previously. The haze attributed to wildfires in Alaska in 1988 had a chemical signature consistent with emissions from smoldering combustion and notably different from emissions from flaming combustion (6). This article summarizes terminology and methodology used to investigate emissions from smoldering biomass both in laboratories and in the field and reviews the findings from several studies.

SMOLDERING COMBUSTION

Definition of Smoldering Combustion

A fire that is 100% efficient in the combustion of organic material emits all carbon as carbon dioxide. Most fires are not this efficient and emit carbon in other forms, such as carbon monoxide, as a result of incomplete combustion. Therefore, the completeness of biomass combustion or combustion efficiency can be expressed by comparing carbon dioxide emissions to emissions of incomplete combustion products. There are several expressions that have been used to describe combustion efficiency, some of which are summarized in Table 1. The dominant presence of visible radiation from flames and black smoke characterize flaming fires, whereas

Table 1. Expressions for Describing Combustion Efficiency

Expression	Smoldering Combustion, %
$CO_2/1834$[a]	<90
$CO_2/(CO + CO_2)$[b]	<92
CO/CO_2[b]	>8

[a] Maximum possible CO_2 emissions (g/kg), assuming a vegetation carbon content of 50% (7).
[b] Molar ratios (8,9).

sparse or no flames and the presence of dense white smoke characterize smoldering fires or the smoldering phase of fires.

Physical and Chemical Processes that Produce Emissions During Smoldering Combustion

Smoldering combustion is not a process in and of itself, but rather a composite of several unique physical and chemical processes. These processes, summarized in Figure 1, include volatilization, pyrolysis, oxidation, and reduction. During a biomass fire, all smoldering combustion processes may be occurring simultaneously, along with flaming combustion processes. Several emissions result from more than one process

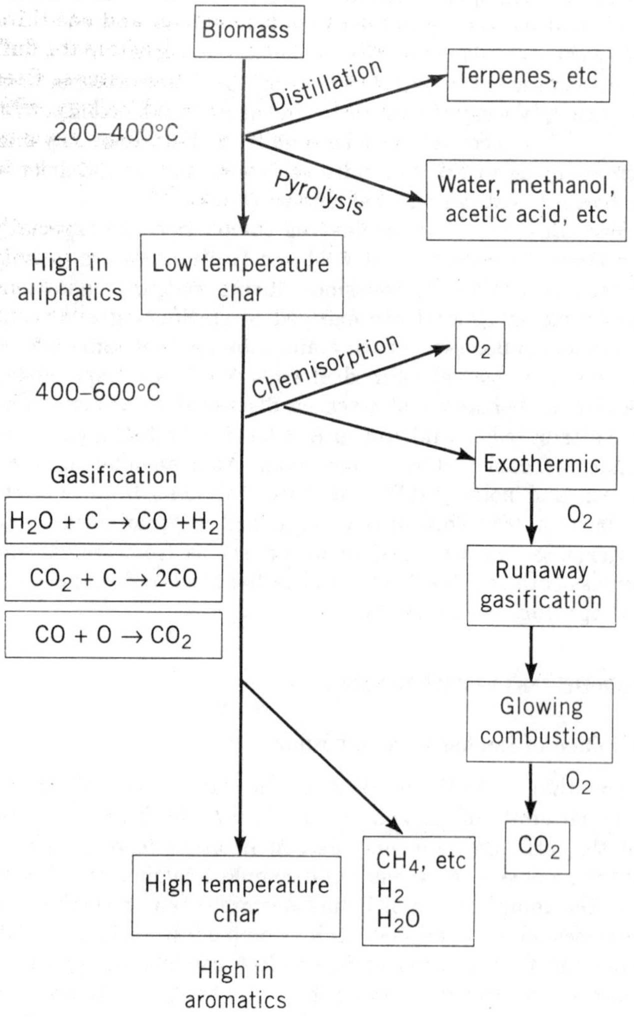

Figure 1. Some processes in smoldering combustion of biomass (10).

(eg, methane, methyl halides, carbon dioxide). Periods of oxygen depletion or thermal quenching of oxidation reactions, characteristic to large forest fires, may result in an increase in smoldering combustion emissions (11).

Studies of effects of heat flow geometry in layers of smoldering newspaper (a lignocellulosic material) insulation (12) and of the influence of inorganic compounds on the combustion of wood (13,14) are very relevant in elucidating some of the processes associated with smoldering combustion in biomass. A description of these processes follows (15). The smoldering of a fuel such as duff (the forest floor subsurface layer consisting of partially decayed organic matter) occurs by way of relatively slow advance of a glowing combustion front. As this front advances, biomass adjacent to the front gains heat, plant products such as terpenes are distilled off, and the biomass is converted to char. This process produces mostly water and occurs in a relatively oxygen-deficient environment because of the oxidative processes in the adjacent glowing combustion front. Thus, the low temperature (200–400°C) char formation phase produces volatile pyrolysis products. These pyrolysis products are known to arise from the pyrolysis of the polysaccharides and lignins which compose biomass. For example, acetol is a major pyrolysis product from a range of polysaccharides, especially in the presence of inorganic components, and guaiacol is a major product of the pyrolysis of lignin. Both acetol and guaiacol have been observed in the emissions of smoldering combustion. Secondary reactions of pyrolysis products catalyzed by the low temperature char may also occur, such as appears to be the case for methyl halide formation. Subsequently, the char layer is heated further as the front advances, reaching 450–500°C where oxygen chemisorption on the char is maximized, especially in the presence of inorganic components. This chemisorption is the major exothermic process in smoldering combustion of the char and with suitable fuel geometry results in the generation of heat more rapidly than the removal of heat with consequent "runaway" combustion, which is a major source of carbon monoxide and carbon dioxide and probably of hydrocarbons and PAHs.

MEASUREMENT OF EMISSIONS

Emissions from smoldering combustion have been measured in several different scenarios. Investigators have measured emissions from fires set in laboratories, where several experimental parameters, such as fuels and fire conditions, can be controlled. Laboratory fires can range from bench-scale fires of a few grams of biomass to combustion chamber fires of a few kilograms of biomass. Fires in the field have included experimental fires set at prescribed burn sites and for clearing fields and agricultural residues and wildfires. Investigators have also measured emissions from wood stoves, barbeques, and fireplaces. Smoke has been sampled from near fires, which results in concentrated samples with complex matrices. The complexity of the matrix, which is in large part water, can result in complex chromatograms containing many peaks that can be difficult to separate. Smoke has also been sampled from plumes several kilometers from fires, which can result in dilute samples. Then, the challenge is detecting the compounds of interest. Table 2 summarizes the methods used to collect emissions from smoldering combustion.

Table 2. Methods Used to Collect Emissions from Smoldering Combustion

Method	Target Compounds
Evacuated steel cansiters	Hydrocarbons, arenes, halogenated hydrocarbons, fixed gases
Tedlar gas sampling bags	Hydrocarbons, fixed gases
Cold traps[a]	Oxygenated organics, nitrogenated organics
Mist chambers[b]	Organic acids, amines, ammonia
Sorbant tubes[c]	Arenes, oxygenated organics
Adsorbant traps containing Tenax, molecular sieve, and/or carbon[d]	Hydrocarbons, oxygenated organics, halogenated hydrocarbons
Glass fiber filters[e]	High molecular weight oxygenated organics, PAHs

[a] Refs. 15 and 16.
[b] Refs. 17–21.
[c] Ref. 22.
[d] Refs. 23–26.
[e] Refs. 27–34.

Biomass Fires in the Laboratory

Several investigators have examined the emissions from biomass burning through scaled-down laboratory fires. Laboratory fires offer several advantages in the investigation of biomass emissions. The fuel can be well characterized both physically and chemically in a laboratory fire and experimental variables, such as fuel type, size, packing ratios and moisture content, and wind, external temperature, and humidity, can be controlled. It is relatively easy to sample emissions directly from the fire. Exact measurements of mass fuel consumed and volumes of smoke can be obtained, and smoldering and flaming emissions can be sampled separately. Laboratory fires have disadvantages in that it is difficult to reproduce to exact physical state of fuel found in the field (ie, living tissue vs dead tissue, mass of fuel burned), and it is difficult to reproduce the fire behavior found in large field fires (ie, large biomass fires produce their own weather).

Beech wood pieces (200 mg) were combusted in a VCI oven at 400, 600, and 800°C in air (23). The resulting emissions were pulled through two Tenax TA cartridges and one molecular sieve cartridge connected in series. Although the experiments did not attempt to mimic biomass burning conditions, they can give insight into the processes, especially pyrolysis processes, associated with smoldering combustion of biomass materials. Small bench-scale fires, as shown in Figure **2a** and 2**b,** have been used to characterize the emissions from smoldering temperate forest fuels, such as ponderosa pine and Douglas fir tissues (15,16,24,35). Approximately 20 g of fuel ground to pass a 1 mm sieve was placed in a ceramic dish. Smoldering was started either by heating the fuel with a 10 A electrical radiant heating coil (in place throughout the fire) or by lighting the fuel with a match. An inverted glass funnel (32 cm diameter) was placed over the fuel with a gap (5 × 0.5 cm) to allow for air intake and the neck of the funnel was connected to a sampling device(s) with Teflon lines. The sampling device could be glass traps immersed in an ethylene glycol bath at −45°C (cold trap), an adsorbent Tenax-GR/Graphpac trap, or a gas sampling bag or canister. Smoke was pulled through the bench-scale apparatus either with an aspirator or air sampling pumps at approximately 2 L/min. The final products of the combustion were a mixture of ash, char, and some unburned fuel, a creosote or tar on the inside of the funnel, either condensed volatile compounds in the cold trap or material collected on the adsorbent trap, and noncondensible or untrapped gases in the Tedlar bag. These products were then analyzed by various methods which are described later. Small bench-scale fires such as these are an efficient and inexpensive way to screen the emissions from various biomass fuels. The bench-scale fire apparatus has been successfully adapted to sample portions of emissions from larger laboratory fires in combustion chambers. The glass funnel in Figure 2 was replaced with a stainless steel funnel attached to a stainless steel pipe, which was connected to the Teflon tube leading to the condensing tubes shown in Figure **2a.** The bench-scale apparatus was placed on a cart with no other changes. Intact forest-floor squares were placed in a rectangular (122 × 41 cm, 6 kg fuel) basket of steel wire mesh (0.64 × 0.64 × 122 cm) and lit at one end with a propane torch. As the flame front moved across the litter layer, the thick white smolder smoke behind the flame front was sampled through the stainless steel funnel. The flames ceased after approximately 10 minutes, leaving behind smoldering litter and duff, and sampling continued. The adapted bench-scale apparatus could also be used to sample portions of emissions from fires in the field.

Large experimental fires have been performed on burning tables (17). In these fires, samples of unaltered fuels were placed on a 60 × 60 cm burning table that rested on top of a high resolution analytical balance. The table could be angulated to 45° to simulate various wind directions, as in heading and backing fires. A stack (120-cm diameter inverted stainless steel funnel) was located 50 cm above the burning table. All major mechanical parts were stainless steel and the entire apparatus was mounted in a mobile trailer. An electrical fan in the top of the funnel prevented drafts and drew the emissions through the funnel. The flow rate of emissions through the stack was measured with a differential pressure sensor. A fan wheel-anemometer originally used for the same purpose was too sensitive to dirt. Stack and fuel temperatures were determined with nickel–chromium temperature sensors. Emissions were sampled through various stainless steel tubes placed in the neck of the funnel. A portion of the emissions were pumped through a stainless steel tube connected to optical analyzers via several flow controllers, a cold trap (−40°C), a glass fiber filter, and a stainless steel sinter–filter (to remove water, particles, and tar) for *in situ* analysis of carbon monoxide and carbon dioxide. Exhaust gas was pumped through a Teflon tube, a cold trap (−20°C), and a Teflon filter (0.5 μm) (to remove water and big particles) into an analyzer for *in situ* determination of nitrogen oxides. Grab samples were collected via a probe from the stack by sucking the emissions through a glass fiber filter into an evacuated stainless steel canister (1 L) equipped with two bellows valves. A portion of the emissions were bubbled through a solution of 30 mL 0.1 *N* sulfuric acid in polyethylene bottles at 8.5 L/min for subsequent ammonia analysis. Another portion of the emissions were collected in a mist chamber for sampling ammonia and amines.

Even larger laboratory fires have been carried out in combustion chambers and wind tunnels. Fuel is burned in continuously weighed fuel beds and smoke is drawn into inverted funnels and stacks, as described previously. Collection devices and *in situ* instrumentation can be located on the

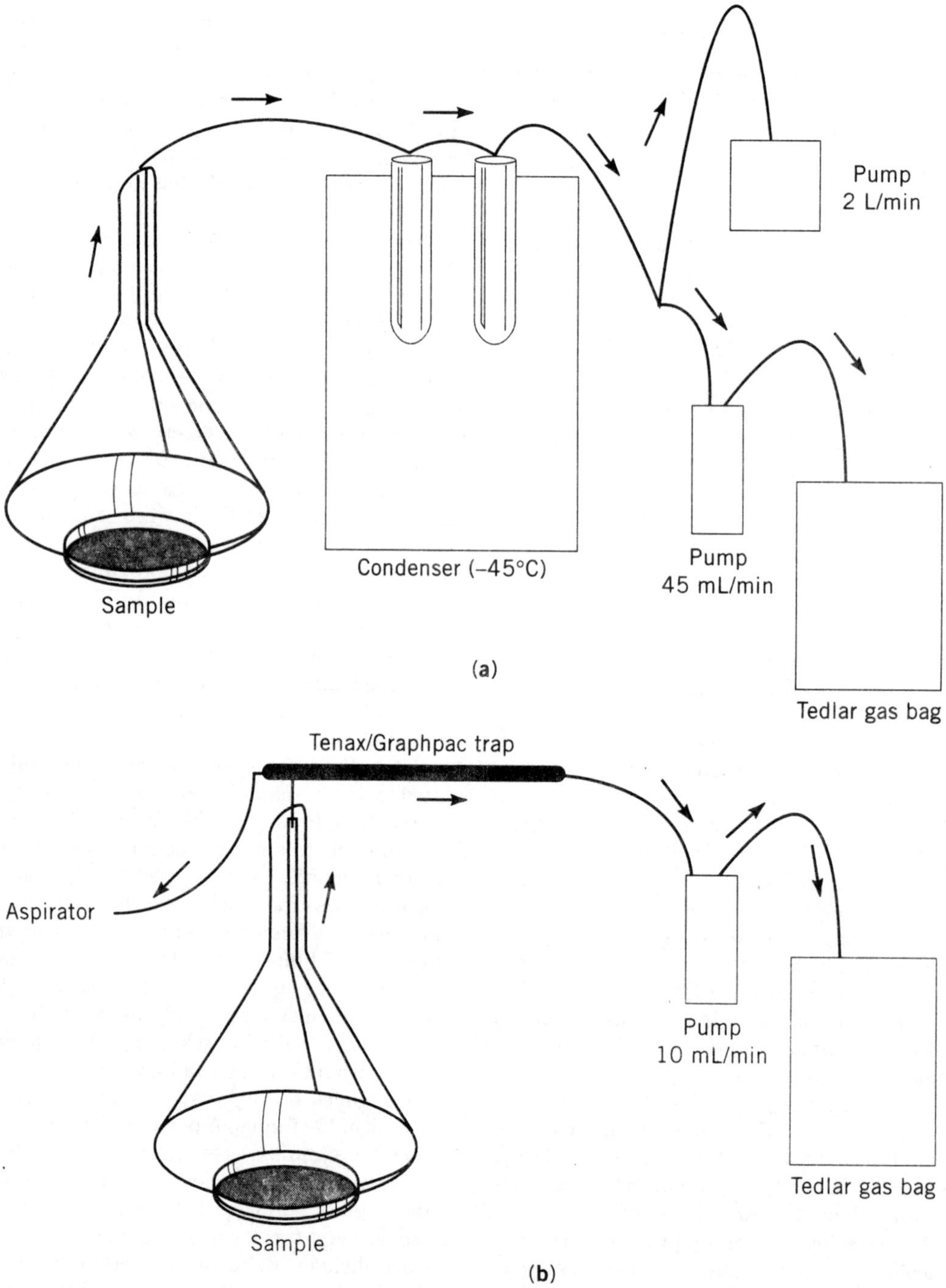

Figure 2. (**a**) Schematic for bench-scale fire apparatus with cold trap; (**b**) schematic for bench-scale fire apparatus with Tenax trap (24).

ground or on sampling platforms in the stacks above the fire. Fires in combustion chambers and wind tunnels are advantageous in that field conditions can be mimicked more closely. The disadvantage is the expense and time required to prepare each fire. Homogenous Douglas fir sticks (1.2 × 1.2 × 20 cm) packed into fuel beds have been burned, as have undistributed litter and duff layers, and logging slash fuel beds on a 0.5 m^2 burning platform. Emissions were drawn through a 40-cm diameter stainless steel duct by a fan (36). A type-S pitot tube was used to measure velocities in the duct and an inertial impactor was used to exclude particles larger than 2.0 μm. Fine particulates (PM 2.0 μm) were collected on 47-mm Teflon filters and the emissions were cooled and collected in five-layer gas sampling bags. Burns have also been carried out in an open-circuit, forced draft wind tunnel consisting of a fuel feed table, inlet section, combustion test section, and stack sampling section (37). The fuel is loaded onto a conveyer belt and transferred to a stainless steel rod conveyor, located above a refractory brick floor, at the front of the combustion test section. A track-mounted traversing sampling platform is located 9.8 m above the fuel bed in the combustion test section.

Investigators have also used wood burning stoves, fireplaces, open cooking fires, and barbeques both in the laboratory and in the field to investigate emissions from smoldering combustion. *In situ* measurements and grab samples are usually obtained from probes within or at the top of the chimney. These types of experimental fires represent and successfully mimic a major source of smoldering emissions, domestic wood burning. Their application to open burning conditions is limited because fires in stoves are often oxygen-starved.

Straw barley and birch and pine wood chips have been burned in ceramic pots (1 L) (25,26). A larger ceramic pot was placed over this pot and samples obtained by pumping the emissions, from a hole in the bottom of the larger pot, through a triple-layer sampling cartridge. Emissions of the open burning of twigs placed on a gridiron were sampled by the same method directly from the smoke plume and smoke samples from a wood stove were collected as they exited the chimney. Pine and oak wood has been burned in a traditional brick fireplace. Emissions were collected continuously over the entire course of the fires (3 h) with a dilution stack sampler. Spruce and birch wood has been burned in small air-tight horizontal residential wood stoves (64 × 54 × 32.5 cm), with a typical heat output of 4–5 kW (27). This was burned with both an air supply to produce flames and without an air supply, inducing smolder conditions. Emissions were sampled from the flue with a glass quartz probe 1 m above the stove. Particles were collected on a Gelman glass fiber filter held at 125°C and the remaining emissions were passed through several impingers cooled gradually from 0 to −60°C to remove water. The emissions were finally adsorbed onto a 5 × 5 cm column of Amberlite XAD-2. Coniferous wood and duff was burned in a residential wood stove, installed with a high volume air sampler with glass filter paper 4.9 m above the stove (28). Emissions passed through the flue and into a vertical tower (containing the high volume air sampler) by natural draft.

Emissions have been measured from improved mud, biogas, and charcoal cookstoves placed on floors in East Indian kitchens (38). Fuelwood and dung cake were burned in the mud cookstoves. Emissions were sampled with battery-operated carbon monoxide personal monitors at intervals of 10 min for 1 h at squatting (50 cm) and standing (150 cm) positions above the stoves. Suspended particulates were collected on glass filters at 1.5 L/min and analyzed for PAHs. Emissions from wood, charcoal, and kerosene stoves have been collected in households in Manila (50). Because none of the stoves had flues or chimneys, samples were collected into evacuated stainless steel canisters (850 mL) directly above lit stoves at a point where the temperature of the emissions had cooled to 65 ± 5°C.

Biomass Fires in the Field

Most field fire emissions have been sampled from specially equipped aircraft flying through smoke plumes from full-scale biomass fires. Fires in remote locations can be investigated from aircraft with relatively little lead time and without disturbing burning conditions. The disadvantage of aircraft sampling is that sampling occurs relatively far from the fire, allowing more opportunity for secondary reactions to occur in the smoke plume. Some investigators have developed ground-based sampling protocols.

A fire atmosphere sampling system (FASS) has been developed to measure carbon emissions and fire parameters in the field, in real time (9,39). FASS units are ground-based with an 8 to 14 m tower for gas and particulate sampling, as shown in Figure 3. Each unit contains a nondispersive infrared gas monitor for measuring carbon dioxide, electrolytic cells to measure carbon monoxide and nitrogen oxide, and sensors to measure temperature and wind velocity. Gas sampling canisters and particulate filters are also contained in each FASS unit. Individual on-board computer systems control the data acquisition. In the field, FASS units were placed 70 to 200 m from the line of ignition and the sampling was triggered when the temperature 50 m from the unit reached 67°C. This field sampling approach has the advantage of obtaining grab and integrated smoke samples from very near a fire. This protocol can be used effectively to sample emissions from prescribed and agricultural fires, but would be more difficult to use in measuring emissions from wildfires because they occur unexpectedly and often in remote locations.

Wildland firefighters have used equipment to sample specific emissions from fires as they were being fought (22). Individual firefighters were fitted with a 4 kg (9 lb) sampling apparatus containing three personal sampling pumps which drew air into a gas sampling bag; sorbant tubes for benzene, formaldehyde, and acrolein; and a cycle and filter assembly. Smoke was sampled directly from the firefighters breathing zone. This protocol allows for sampling directly at the point of firefighter exposure, but is limited in the number of emissions that can be sampled at one time.

In Situ Measurements of Emissions from Biomass Fires

Open-path Fourier transform infrared spectroscopy (FTIR) has been used to measure a variety of emissions, including carbon dioxide, carbon monoxide, methane, C_2 hydrocarbons, formaldehyde, formic acid, acetic acid, methanol, ammonia, nitrogen oxide, nitrogen dioxide, hydrogen cyanide, sulfur dioxide, and carbonyl sulfide from biomass fires in combustion chambers and in the field. Complete descriptions of the methodology is available, of which a brief summary follows (10,40). A Fourier transform spectrometer with optics for collecting infrared radiation from a collimated IR source up to 200 m distant in the field or 3 m distant in the combustion chamber stack was mounted on a trailer for field experiments or on the sampling platform in combustion chamber experiments. Smoke from the fires passed through the IR beam between the source and the spectrometer, and IR spectra were recorded continuously before, during, and after a fire. The resulting IR absorption spectra were then analyzed using the MALT/Classic least squares method to determine simultaneous concentrations of many gas-phase species. MALT is a program which calculates calibration spectra from HITRAN absorption line parameters, such that the calculated spectra closely match real experimental spectra over a range of gas concentrations. More traditional methods are required for species for which spectral line parameters are not available. Smoke plume measurement via *in situ* FTIR offers several advantages over measurements made from grab and

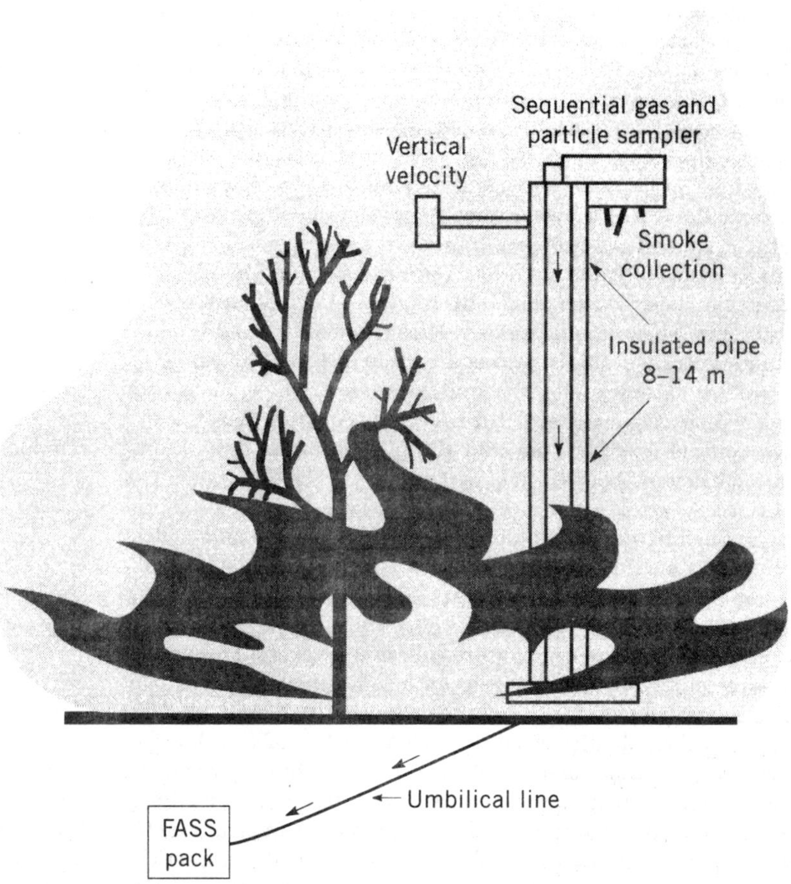

Figure 3. Schematic for FASS pack (9,39).

integrated samples of smoke plumes. Simultaneous real-time measurements of a wide array of gas-phase species can be made and since the measurements are integrated over a long path through the smoke plume, the measurements are not subject to small-scale local variations. There also are no uncertainties associated with sample storage, and reactions on the walls of sampling tubes, pumps, or vessels. It is also possible to monitor the real-time emissions of trace gases throughout the fire. Disadvantages are that the equipment is expensive, not all emissions absorb in the IR, and not all emissions can be detected from the overlapping IR spectra.

Carbon monoxide and carbon dioxide can be measured *in situ* by drawing smoke through separate nondispersive infrared detectors. One IR detector is calibrated for carbon monoxide detection; the other is calibrated for carbon dioxide detection (19,17). Electrolytic cells have also been used as *in situ* nitrogen oxide and carbon monoxide sensors (9). Ammonia and nitrogen oxides ($NO + NO_2$) can be measured *in situ* by various chemiluminescence techniques. Nitrogen oxide can be chemically oxidized to nitrogen dioxide; nitrogen dioxide is then detected after a chemiluminescence reaction with a solution of Luminol (41). Alternatively, nitrogen dioxide can be reduced on a molybdenum thermal converter (350°C) and nitrogen oxide is then detected via chemiluminescence. Any peroxide, nitrogen dioxide, nitrous acid, or peroxynitrites present could also be reduced to nitrogen oxide; therefore this method actually measures NO_y. A second parallel stainless steel thermal converter (850°C) oxidizes ammonia to nitrogen oxide, which is also detected via chemiluminescence. The concentration of ammonia is then determined by calculating the difference between the two converters. Because nitriles and amines can also be oxidized to nitrogen oxide by the same method, ammonia could be overestimated.

COLLECTION OF EMISSIONS FROM BIOMASS FIRES

Traps and Filters

Smoke from laboratory and field biomass fires has also been collected by various means for subsequent laboratory analysis. Emissions from laboratory fires have been pulled through cold traps as described previously. The resulting condensate is then extracted into methylene chloride. The extract or the nonextracted condensate can then be injected into a gas chromatographer/mass spectrometer (GC/MS) and analyzed for oxygenated organic compounds. Water-soluble acidic gases (eg, formic and acetic acid, ammonia, and amines) have been collected using a mist chamber technique which has been described in detail (18,42). Briefly, a fine mist of water droplets strips water-soluble trace gases from large volumes of sampled air or smoke and concentrates them to a small aqueous volume (10–20 mL). The mist is created by pumping the sample through one nebulizing nozzle and drawing deionized water through another nebulizing nozzle into a glass chamber. The droplets are then trapped by a Teflon filter and drip down into a deionized water reservoir. Efficient removal of water-soluble trace gases from the sample stream is ensured by the large surface-area-to-volume ratio of mist droplets. The deionized water solution can then be analyzed by colormetric or ion chromatographic (IC) techniques to quantify ions of interest (eg, acetate and formate ions). Several investigators have used the mist chamber technique to collect field-fire smoke from aircraft and smoke from laboratory fires.

Another method used to trap emissions from biomass fires is to pull smoke through a trap containing some type of adsorbent material. The adsorbent material is then either solvent-extracted for GC analysis or directly desorbed onto a GC. Smoke has been pulled with an air pump at a rate of 10–50 mL/min through triple-layer glass cartridges with separate layers of Carbosieve S-III, Carbotrap, and Tenax TA (25). The cartridges were then placed in a desorption oven in line with a liquid nitrogen cold trap and a gas chromatographer (GC) in series. The traps were desorbed at 235°C with a helium flow of 20–40 mL/min onto either a gas chromatographer/flame ionization detector (GC/FID) or a GC/MS for analysis of hydrocarbons and furans. Emissions were collected on a Tenax-GR/Graphpac (30.5 × 0.254 cm ID) trap at 8.8–10.2 mL/min with a portable air sampling pump (24). The traps were desorbed at 100°C while being flushed with nitrogen at 40 mL/min. The eluents were cyrofocused with liquid nitrogen and then introduced onto a gas chromatographer/electron capture detector (GC/ECD) for analysis of methyl halides. Smoke emissions have been collected on two Tenax-TA cartridges and one molecular sieve cartridge connected in series (23). The Tenax-TA cartridges were extracted with ethanol and analyzed the eluates for phenols, arenes, and PAHs by GC/MS and gas chromatography/infrared detection (GC/IRD). The molecular sieve cartridges were transferred into headspace vials with 2 mL of frozen 1 m hydrochloric acid and 4 m sodium chloride desorption solution. The vials were placed in a headspace sampler coupled to a GC/MS and analyzed for volatile aldehydes, ketones, alcohols, and hydrocarbons.

Various types of glass fiber filters have been used to collect smoke from smoldering combustion. Glass fiber filters can either be heated (400–750°C) or solvent-rinsed to remove any possible organic contaminants. After sample collection, the fibers are either Soxhlet or ultrasonically extracted with various combinations of organic solvents. The fine aerosols in smoke have been collected by first passing the smoke through several parallel cyclone separators, and then through several parallel quartz fiber filters (43). The filters were then combined and ultrasonically extracted in hexane and benzene/2-propanol. The extracts were concentrated, filtered, and analyzed by GC/MS for oxygenated organic compounds (C_8 compounds and above). High volume polyurethane foam (PUF) plugs located behind quartz fiber filters have been used to trap smoke emissions (29–31). The PUF plugs were then extracted into acetone by sonification for 6 h. The extracts were concentrated under nitrogen and analyzed by GC/MS for phenols, methoxyphenols, alkanes, and PAHs. Several investigators have used high volume air samplers to pull smoke through glass fiber filters for PAH analysis. The glass fiber filters can then be ultrasonically extracted with cyclohexane–methylene chloride (32,33), Soxhlet extracted with acetone–hexane (32,33), methylene chloride (28), or acetone (27). The extracts are then concentrated and analyzed by GC/FID, GC/MS, or high performance liquid chromatography/fluorescence (HPLC/fluorescence). XAD-2 resin has been used with glass fiber filters to collect gaseous PAHs. The XAD-2 resin is then extracted with the glass fiber as described above (32,34).

All of these trapping methods (cold traps, adsorbents, and filters) result in a concentrated smoke sample. These methods can therefore be advantageous for detecting compounds present at ppb and ppt levels (eg, PAHs). Using these methods, small integrated portions of smoke can be continuously sampled throughout an entire burn. For small fires, almost all the smoke from the fire can be circulated through a trap (eg, the cold trap). The emissions factors determined from this type of method represent a sum of the emissions of a particular compound from the various processes occurring in smoldering combustion (ie, distillation and pyrolysis). A disadvantage to collecting smoke emissions on traps and filters is that the traps can become quickly saturated or plugged, therefore traps should also be tested for breakthrough volumes of analytes.

Canisters and Bags

Many investigators have collected emissions from biomass fires in evacuated stainless steel canisters. The interior of the canisters are typically beam welded and electropolished to ensure a smooth surface with no scratches or holes, which can act as diffusion traps or reaction sites. The successful containment of trace gases depends heavily on the nature of the inner surfaces of the canisters. Great care is taken at all steps in canister preparation to avoid contaminating the inner walls of the canisters. A detailed description of canister preparation, including construction, electropolishing, and cleaning is available (44). Various manifold/pumping systems are used to feed previously filtered (glass fiber or Teflon filters) smoke into canisters. Canisters have been used to collect integrated and grab samples of smoke from aircraft, the ground, and laboratory fires. Integrated smoke samples have also been pumped into nonreactive Tedlar bags. The sample is then either kept in the Tedlar bag or transferred to stainless steel canisters. Aliquots of samples are then removed from the canister or Tedlar bag and analyzed for trace gases by a variety of techniques.

Analysis of Emissions

Carbon dioxide and carbon monoxide have been determined using the *in situ* methods already discussed. They have also been determined from samples collected in gas canisters or bags either by infrared absorption, or reduced to methane with hydrogen and a nickel catalyst to 375°C and determined by GC/FID techniques. Carbon dioxide has also been measured by gas chromatography/thermal ionization detection (GC/TID). Carbon monoxide, along with molecular hydrogen, has also been measured by the hot mercury oxide reduction technique.

Table 3 summarizes various instrumental methods used to analyze smoke from smoldering combustion of biomass for organic compounds. GC/FID has been used extensively to analyze smoke for hydrocarbons (including methane) and arenes, and it has been used to analyze smoke for halogenated organic compounds. As hydrogen atoms are replaced with other functional groups such as carbonyls and alcohols, FID becomes less sensitive. It is a good general method for quantitating

Table 3. Organic Compounds Identified in Smoldering Biomass Emissions

Compound	Method(s)	Fuel(s)	Refs.
		Hydrocarbons	
Methane	GC/FID	Boreal forest, African savanna, tropical forest, Brazilian savanna, temperate forest, wetlands, wood in stoves, laboratory fires (savanna grass, wood, hay, straw), laboratory fires (ponderosa pine wood needles, bark, litter, duff, humus)	44,39,45,9,46,47,11,48,49,50,17,15,16
	FTIR	Australian savanna, laboratory fires (sagebrush, temperate forest slash), laboratory fires (pine needles, temperate forest duff)	51,10,40
	IR	Wood in stoves	27
Ethane	GC/FID	Boreal forest, African savanna, temperate forest, laboratory fires (savanna grass, wood, hay, straw), laboratory fires (ponderosa pine wood needles, bark, litter, duff, humus), laboratory fires (straw, spruce twigs, birch wood)	44,39,45,11,48,17,15,16,25
	FTIR	Australian savanna, laboratory fires (sagebrush, temperate forest slash), laboratory fires (pine needles, temperate forest duff)	51,40,10
Ethene	GC/FID	Boreal forest, African savanna, temperate forest, laboratory fires (savanna grass, wood, hay, straw), laboratory fires (ponderosa pine wood needles, bark, litter, duff, humus), laboratory fires (straw, spruce twigs, birch wood)	44,39,45,48,17,15,16,25
	FTIR	Laboratory fires (sagebrush, pine needles, temperate forest duff)	10
Ethyne	GC/FID	Boreal forest, African savanna, temperate forest, laboratory fires (savanna grass, wood, hay, straw) laboratory fires (straw, spruce twigs, birch wood)	44,39,45,11,48,17,25
	FTIR	Australian savanna, laboratory fires (sagebrush, temperate forest slash, pine needles, temperate forest duff)	51,10
Propane	GC/FID	Boreal forest, African savanna, temperate forest, laboratory fires (savanna grass, wood, hay, straw), laboratory fires (straw, spruce twigs, birch wood)	44,39,45,11,48,17,25
Propene	GC/FID	Boreal forest, African savanna, temperate forest, laboratory fires (savanna grass, wood, hay, straw), laboratory fires (straw, spruce twigs, birch wood)	44,39,45,11,48,17,25
Propyne	GC/FID	Boreal forests, laboratory fires (straw, spruce twigs, birch wood)	44,25
Propadiene	GC/FID	Laboratory fires (straw, spruce twigs, birch wood)	25
n-Butane	GC/FID	Boreal forest, African savanna, laboratory fires (savanna grass, wood, hay, straw), laboratory fires (straw, spruce twigs, birch wood)	44,39,45,17,25

Table 3. *(continued)*

i-Butane	GC/FID	Boreal forest, African savanna, laboratory fires (straw, spruce twigs, birch wood)	44,39,25
trans-2-Butene	GC/FID	Boreal forest, laboratory fires (savanna grass, wood, hay, straw), laboratory fires (straw, spruce twigs, birch wood)	44,17,25
1-Butene	GC/FID	Boreal forest, African savanna, laboratory fires (savanna grass, wood, hay, straw), laboratory fires (straw, spruce twigs, birch wood)	44,39,45,17,25
i-Butene	GC/FID	Boreal forest, African savanna, laboratory fires (savanna grass, wood, hay, straw), laboratory fires (straw, spruce twigs, birch wood)	44,39,45,17,25
cis-2-Butene	GC/FID	Boreal forest, African savanna, laboratory fires (savanna grass, wood, hay, straw) laboratory fires (straw, spruce twigs, birch wood)	44,39,17,25
1,3-Butadiene	GC/FID	Boreal forest, African savanna, laboratory fires (savanna grass, wood, hay, straw), laboratory fires (straw, spruce twigs, birch wood)	44,39,17,25
2-Butyne	GC/FID	Laboratory fires (straw, spruce twigs, birch wood)	25
Butenyne	GC/FID	Laboratory fires (straw, spruce twigs, birch wood)	25
1-Butyne	GC/FID	Laboratory fires (straw, spruce twigs, birch wood)	
Methylbutane	GC/FID	Boreal forests, laboratory fires (straw, spruce twigs, birch wood)	44,25
n-Pentane	GC/FID	Boreal forest, African savanna, laboratory fires (savanna grass, wood, hay, straw), laboratory fires (straw, spruce twigs, birch wood)	44,39,17,25
Cyclopentene	GC/FID	African savanna	39
	GC/MS	Laboratory fires (straw, spruce twigs, birch wood)	25
3-Methyl-1-butene	GC/MS	Laboratory fires (straw, spruce twigs, birch wood)	25
trans-2-Pentene	GC/MS	Laboratory fires (straw, spruce twigs, birch wood)	25
2-Methyl-2-butene	GC/MS	Laboratory fires (straw, spruce twigs, birch wood)	25
1-Pentene	GC/FID	Boreal forest, African savanna	44,39
	GC/MS	Laboratory fires (straw, spruce twigs, birch wood)	25
2-Methyl-1-butene	GC/FID	African savanna	39
	GC/MS	Laboratory fires (straw, spruce twigs, birch wood)	25
cis-2-Pentene	GC/MS	Laboratory fires (straw, spruce twigs, birch wood)	25
Isoprene	GC/FID	Laboratory fires (savanna grass, wood, hay, straw)	17
	GC/MS	Laboratory fires (straw, spruce twigs, birch wood)	25
Cyclopentadiene	GC/MS	Laboratory fires (straw, spruce twigs, birch wood)	25
trans-1,3-Pentadiene	GC/MS	Laboratory fires (straw, spruce twigs, birch wood)	25
Hexane	GC/FID	Boreal forest, African savanna, laboratory fires (straw, spruce twigs, birch wood)	44,39,25

Table 3. Organic Compounds Identified in Smoldering Biomass Emissions (*continued*)

Compound	Method(s)	Fuel(s)	Refs.
1-Hexene	GC/FID	Boreal forest, African savanna, laboratory fires (straw, spruce twigs, birch wood)	44,39,25
1-Heptene	GC/FID	Laboratory fires (straw, spruce twigs, birch wood)	25
1-Octene	GC/FID	Laboratory fires (straw, spruce twigs, birch wood)	25
		Arenes	
Benzene	GC/FID	Boreal forest, African savanna, temperate forest, laboratory fires (savanna grass, wood, hay, straw), laboratory fires (straw, spruce twigs, birch wood), wood in stoves,	44,39,22,17,25,27,51
	FTIR	Australian savanna	
Toluene	GC/FID	Boreal forest, African savanna, laboratory fires (savanna grass, wood, hay, straw), laboratory fires (straw, spruce twigs, birch wood, wood in stoves	44,39,17,25,27
Ethylbenzene	GC/FID	Laboratory fires (straw, spruce twigs, birch wood)	25
Styrene	GC/FID	Laboratory fires (straw, spruce twigs, birch wood)	25
Dimethylbenzenes	GC/FID	African savanna, laboratory fires (savanna grass, wood, hay, straw), laboratory fires (straw, spruce twigs, birch wood)	39,17,25
		Polycyclic Aromatic Hydrocarbons	
Naphthalene	GC/MS	Oak wood in stove, ponderosa and lodgepole pine duff in stove, laboratory fires (ponderosa pine wood and cottonwood)	34,28,35,32
	HPLC/fluorescence	African savanna	
Triphenyl	GC/MS	Ponderosa and lodgepole pine duff in stove	28
Acenaphthene	GC/MS	Oak wood in stove, ponderosa and lodgepole pine duff in stove	34,28
Dibenzofuran	GC/MS	Ponderosa and lodgepole pine duff in stove	28
Benzofuran	GC/MS	Laboratory fires (cottonwood)	35
9*H*-Fluorene	GC/MS	Ponderosa and lodgepole pine duff in stove	28
t-Butylnaphthalene	GC/MS	Ponderosa and lodgepole pine duff in stove	28
9-Methylenefluorene	GC/MS	Ponderosa and lodgepole pine duff in stove	28
Phenylanisole	GC/MS	Ponderosa and lodgepole pine duff in stove	28
Dihydrophenanthrene	GC/MS	Ponderosa and lodgepole pine duff in stove	28
Phenanthrene	GC/MS	Oak wood in stove, lodgepole pine wood, ponderosa, and lodgepole pine duff in stove, wood in stove	34,28,29
	HPLC/fluorescence	African savanna	32
Anthracene	GC/MS	Oak wood in stove, lodgepole pine wood, ponderosa, and lodgepole pine duff in stove, wood in stove	34,28,29
	HPLC/fluorescence	African savanna	32
Phenylnaphthalene	GC/MS	Ponderosa and lodgepole pine duff in stove	28
Fluoranthene	GC/MS	Oak wood in stove, lodgepole pine wood, ponderosa, and lodgepole pine duff in stove, wood in stove	34,28,29
	HPLC/fluorescence	African savanna	32

Table 3. *(continued)*

Benzacenaphthylene	GC/MS	Ponderosa and lodgepole pine duff in stove	28
Pyrene	GC/MS	Oak wood in stove, lodgepole pine wood, ponderosa, and lodgepole pine duff in stove, wood in stove	34,28,29
	HPLC/fluorescence	African savanna, Australian native vegetation	32,33
Dodecahydrochyrsene	GC/MS	Ponderosa and lodgepole pine duff in stove	28
Benzofluorene	GC/MS	Ponderosa and lodgepole pine duff in stove	28
Benzo(*ghi*)fluoranthene	GC/MS	Ponderosa and lodgepole pine duff in stove, wood in stove, oak wood in stove, lodgepole pine wood, ponderosa, and lodgepole pine duff in stove	28,29,34
Benz(*a*)anthrancene	GC/MS	Oak wood in stove, lodgepole pine wood, ponderosa, and lodgepole pine duff in stove, wood in stove	34,28,29
	HPLC/fluorescence	Australian native vegetation	33
Benzo(*b*)fluoranthene	GC/MS	Oak wood in stove, ponderosa and lodgepole pine duff in stove, wood in stove	34,28,29
	HPLC/fluorescence	African savanna, Australian native vegetation	32,33
Benzo(*e*)pyrene	GC/MS	Oak wood in stove, lodgepole pine wood, ponderosa, and lodgepole pine duff in stove	34,28
	HPLC/fluorescence	African savanna	32
Benzo(*a*)pyrene	GC/MS	Oak wood in stove, ponderosa and lodgepole pine duff in stove, wood in stove	34,28,29
	HPLC/fluorescence	African savanna, Australian native vegetation	32,33
Perylene	GC/MS	Ponderosa and lodgepole pine duff in stove	28
Dihydrodibenzanthrancene	GC/MS	Ponderosa and lodgepole pine duff in stove	28
Indeno(*1,2,3-cd*)pyrene	GC/MS	Ponderosa and lodgepole pine duff in stove	28
Benzo(*ghi*)perylene	GC/MS	Ponderosa and lodgepole pine duff in stove	28
	HPLC/fluorescence	Australian native vegetation	33
Anthanthrene	GC/MS	Ponderosa and lodgepole pine duff in stove	28
Retene	GC/MS	Oak wood in stove, laboratory fires (pine and oak wood)	34,43
Chrysene	GC/MS	Oak wood in stove, lodgepole pine wood, ponderosa, and lodgepole pine duff in stove, wood in stove	34,28,29
	HPLC/fluorescence	African savanna, Australian native vegetation	32,33
Benzo(*k*)fluoranthene	GC/MS	Oak wood in stove, ponderosa and lodgepole pine duff in stove, wood in stove	34,28,29
	HPLC/fluorescence	African savanna, Australian native vegetation	32,33
Coronene	HPLC/fluorescence	African savanna, Australian native vegetation	32,33
		Oxygenated Organic Compounds	
Methanol	GC/MS	Laboratory fires (ponderosa pine wood, needles, bark, litter, duff, humus)	15,16
	GC/MS/IR	Laboratory fires (beech wood)	23
	FTIR	Laboratory fires (pine needles, pine duff)	10
Formaldehyde	DNPH derivitation/ HPLC/UV	Laboratory fires (jack pine, red oak, cedar, and ash wood), temperate forest	52,22

Table 3. Organic Compounds Identified in Smoldering Biomass Emissions *(continued)*

Compound	Method(s)	Fuel(s)	Refs.
	GC/FID	Spruce and birch wood	27
	FTIR	Australian savanna, laboratory fires (sagebrush, pine needles, temperate forest slash, and duff)	51,10,40
Formic acid	GC/MS	Laboratory fires (ponderosa pine wood, needles, bark, litter, and duff)	15,16
	Mist chamber/IC	Boreal forest, African savanna, laboratory fires (hardwood, brush, and leaves)	19,20,18
Ethanal (acetaldehyde)	DNPH derivitation/ HPLC/UV	Laboratory fires (jack pine, red oak, cedar, and ash wood)	52
	GC/MS/IR	Laboratory fires (beech wood)	23
	FTIR	Australian savanna	51
Acetic acid	GC/MS	Laboratory fires (ponderosa pine wood, needles, bark, litter, duff, and humus)	15,16
	FTIR	Laboratory fires (pine needles and duff)	10
	Mist chamber/IC	Boreal forest, African savanna, laboratory fires (hardwood, brush, and leaves)	19,20,18
Propanone (acetone)	DNPH derivitation/ HPLC/UV	Laboratory fires (jack pine, red oak, cedar, and ash wood)	52
	GC/MS/IR	Laboratory fires (beech wood)	23
	GC/FID	Spruce and birch wood	27
2-Propenal	DNPH derivitation/ HPLC/UV	Laboratory fires (jack pine, red oak, cedar, and ash wood), temperate forest	52,22
	GC/MS/IR	Laboratory fires (beech wood)	23
3-Hydroxypropanal	GC/MS	Laboratory fires (ponderosa pine wood)	15,16
1-Hydroxy-2-propanone (acetol)	GC/MS	Laboratory fires (ponderosa pine wood, needles, bark, litter, duff, and humus)	15,16
Propionic acid	GC/MS	Laboratory fires (ponderosa pine wood, needles, bark, litter, and duff)	15,16
Propanal	DNPH derivitation/ HPLC/UV	Laboratory fires (jack pine, red oak, cedar, and ash wood)	52
2-Oxopropanal (pyruvic aldehyde)	GC/MS	Laboratory fires (ponderosa pine wood, needles, and litter)	15,16
2-Oxopropanoic acid (pyruvic acid)	Mist chamber/IC	African savanna	20
trans-2-Butenal (crotonaldehyde)	DNPH derivitation/ HPLC/UV	Laboratory fires (red oak, cedar, and ash wood)	52
Butanal	DNPH derivitation/ HPLC/UV	Laboratory fires (jack pine, red oak, wood, and ash)	52
trans-2-Butenoic acid (crotonic acid)	GC/MS	Laboratory fires (ponderosa pine wood, needles, bark, and litter)	15,16
Vinyl acetate	GC/MS	Laboratory fires (ponderosa pine wood, needles, bark, litter, and duff)	15,16
Butyric acid	GC/MS	Laboratory fires (ponderosa pine bark, litter, and duff)	15,16
2-Oxopropionic acid, methyl ester	GC/MS	Laboratory fires (ponderosa pine wood)	15,16
2-Hydroxy-3-oxobutanal	GC/MS	Laboratory fires (ponderosa pine wood)	15,16
3-Oxobutanoic acid, methyl ester	GC/MS	Laboratory fires (ponderosa pine wood, needles, bark, litter, and duff)	15,16
Hexanal	DNPH derivitation/ HPLC/UV	Laboratory fires (red oak, cedar, and ash wood)	52
3-Furfural	GC/MS	Laboratory fires (ponderosa pine wood)	15,16

Table 3. (*continued*)

2-Cyclopenten-1-one	GC/MS	Laboratory fires (ponderosa pine wood, needles, bark, litter, duff, and humus)	15,16
2-Furfural	GC/MS	Laboratory fires (ponderosa pine wood, needles, bark, litter, duff, and humus), laboratory fires (ponderosa pine and cottonwood)	15,16,35
	DNPH derivitation/ HPLC/UV	Laboratory fires (red oak, cedar, and ash wood)	52
2-Furanmethanol	GC/MS	Laboratory fires (ponderosa pine wood, bark, litter, and duff), laboratory fires (ponderosa pine and cottonwood)	15,16,35
2-Cyclopenten-1,4-dione	GC/MS	Laboratory fires (ponderosa pine wood)	15,16
Dihydro-2(3*H*)-furanone (γ-butyrolactone)	GC/MS	Laboratory fires (ponderosa pine wood, needles, bark, and litter), laboratory fire (cottonwood)	15,16,35
2-(5*H*)-Furanone	GC/MS	Laboratory fires (ponderosa pine wood, needles, bark, litter, and duff)	15,16
2-Methyl-2-cyclopenten-1-one	GC/MS	Laboratory fires (ponderosa pine wood)	15,16
2-Acetylfuran	GC/MS	Laboratory fires (ponderosa pine wood, needles, bark, litter, and duff), laboratory fires (ponderosa pine and cottonwood)	15,16,35
Methylfuranone (angelicalactone)	GC/MS	Laboratory fires (ponderosa pine wood)	15,16
5-Methyl-2-furfural	GC/MS	Laboratory fires (ponderosa pine wood, needles, bark, litter, duff, and humus), laboratory fires (ponderosa pine and cottonwood)	15,16,35
Phenol	GC/MS	Laboratory fires (ponderosa pine wood, needles, bark, litter, duff, and humus), laboratory fires (ponderosa pine and cottonwood), hard and softwood in stoves and fireplaces	15,16,35,30
2-Methylphenol (*o*-cresol)	GC/MS	Laboratory fires (ponderosa pine wood, needles, bark, litter, and duff), laboratory fires (ponderosa pine and cottonwood), hard and softwood in stoves and fireplaces	15,16,30,35
3/4-Methylphenol (*m*/*p*-cresol)	GC/MS	Laboratory fires (ponderosa pine wood, needles, bark, litter, and duff), laboratory fires (ponderosa pine and cottonwood), hard and softwood in stoves and fireplaces	15,16,30,35
Benzaldehyde	GC/MS	Laboratory fires (ponderosa pine and cottonwood)	35
	DNPH derivitation/ HPLC/UV	Laboratory fires (jack pine, red oak, cedar, and ash wood)	52
o-Hydroxybenzaldehyde	GC/MS	Laboratory fires (ponderosa pine wood, needles, litter, and duff), laboratory fires (ponderosa pine and cottonwood)	15,16,35
3,4-Dimethoxybenzaldehyde (valeraldehyde, veratraldehyde)	DNPH derivitation/ HPLC/UV	Laboratory fires (red oak, cedar, and ash wood)	52
		Laboratory fires (pine and oak wood)	43
2-Methylbenzaldehyde (*p*-tolualdehyde)	DNPH derivitation/ HPLC/UV	Laboratory fires (jack pine, red oak, cedar, and ash wood)	52
4-Methoxybenzaldehyde (anisaldehyde)	GC/MS	Laboratory fires (pine wood)	43

Table 3. Organic Compounds Identified in Smoldering Biomass Emissions *(continued)*

Compound	Method(s)	Fuel(s)	Refs.
4-Hydroxy-3-methoxy-benzaldehyde (vanillin, 4-formylguaiacol)	GC/MS	Laboratory fires (ponderosa pine wood, needles, bark, litter, and duff), laboratory fires (ponderosa pine and cottonwood), hard and softwood in stoves and fireplaces, laboratory fires (pine wood)	15,16,35,30,43
2-Methoxyphenol (guaiacol)	GC/MS	Laboratory fires (ponderosa pine wood, needles, bark, litter, and duff), laboratory fires (ponderosa pine and cottonwood), hard and softwood in stoves and fireplaces	15,16,35,30
4-Methylguaiacol	GC/MS	Laboratory fires (ponderosa pine wood, needles, bark, litter, and duff), laboratory fires (ponderosa pine and cottonwood), hard and softwood in stoves and fireplaces, laboratory fires (pine and oak wood)	15,16,30,35,43
4-Ethylguaiacol	GC/MS	Laboratory fires (ponderosa pine wood), laboratory fires (ponderosa pine and cottonwood), hard and softwood in stoves and fireplaces, laboratory fires (grass)	15,16,30,35,42
4-Vinylguaiacol	GC/MS	Laboratory fires (ponderosa pine wood), laboratory fires (cottonwood)	15,16,35,
4-Propylguaiacol	GC/MS	Hard and softwood in stoves and fireplaces, laboratory fires (pine and oak)	30,43
4-Acetylguaiacol	GC/MS	Hard and softwood in stoves and fireplaces	30
4-Propenylguaiacol	GC/MS	Laboratory fires (ponderosa pine wood), Hard and softwood in stoves and fireplaces	15,16,30
4-Prop-1-en-3-oneguaiacol	GC/MS	Laboratory fires (ponderosa pine wood)	15,16
4-Allylguaiacol (eugenol)	GC/MS	Laboratory fires (ponderosa pine wood and bark), laboratory fires (ponderosa pine and cottonwood), hard and softwood in stoves and fireplaces	15,16,35,30
Acetonylguaiacol	GC/MS	Hard and softwood in stoves and fireplaces, laboratory fires (pine and oak wood)	30,43
2,6-Methoxyphenol (syringol)	GC/MS	Laboratory fires (cottonwood), hardwood in stoves and fireplaces	35,30
4-Methylsyringol	GC/MS	Laboratory fires (cottonwood), hardwood in stoves and fireplaces	35,30
4-Ethylsyringol	GC/MS	Laboratory fires (cottonwood), hardwood in stoves and fireplaces	35,30
4-Allylsyringol	GC/MS	Hardwood in stoves and fireplaces	30
4-Propylsyringol	GC/MS	Laboratory fires (cottonwood), hardwood in stoves and fireplaces	35,30
4-Propenylsyringol	GC/MS	Hardwood in stoves and fireplaces	30
4-Acetylsyringol	GC/MS	Hardwood in stoves and fireplaces, laboratory fires (oak wood)	30,43
4-Acetonylsyringol	GC/MS	Hardwood in stoves and fireplaces, laboratory fires (oak wood)	30,43
1-(4-Methoxyphenyl)propane	GC/MS	Laboratory fires of grass	43
1,2-Hydroxybenzene (catechol)	GC/MS	Hard and softwood in stoves and fireplaces	30

Table 3. *(continued)*

4-Methylcatechol	GC/MS	Hard and softwood in stoves and fireplaces	30
3-Methoxycatechol	GC/MS	Hard and softwood in stoves and fireplaces	30
C_2-Catechol	GC/MS	Hard and softwood in stoves and fireplaces	30
Benzoic acid	GC/MS	Laboratory fires (ponderosa pine wood, needles, and litter), laboratory fires (ponderosa pine and cottonwood)	15,16,35
3,5-Methoxy-4-hydroxybenzoic acid	GC/MS	Laboratory fires (oak wood)	43
4-Methoxybenzoic acid (anisic acid)	GC/MS	Laboratory fires (pine and oak wood and grass)	43
(4-Methoxyphenyl)acetic acid (homoanisic acid)	GC/MS	Laboratory fires (grass)	43
4-Hydroxy-3-methoxybenzoic acid (vanillic acid)	GC/MS	Laboratory fires (pine and oak wood)	43
1-(3,4-Dihydroxy-5-methoxyphenyl)ethanone	GC/MS	Laboratory fires (oak wood)	43
4-Hydroxy-3,5-dimethoxybenzaldehyde (syringaldehyde)	GC/MS	Hardwood in stoves and fireplaces, laboratory fires (oak wood)	30,43
3,4-Dimethoxyphenylpropanone	GC/MS	Laboratory fires (pine and oak wood)	43
3-Methoxy-4-hydroxyphenylacetic acid (homovanillic acid)	GC/MS	Laboratory fires (pine and oak wood)	43
3,4-Dimethoxybenzoic acid (veratric acid)	GC/MS	Laboratory fires (pine and oak wood)	43
3,4-Methoxyphenylacetic acid (homoveratric acid)	GC/MS	Laboratory fire (pine wood)	43
1-(3,4,5-Trimethoxyphenyl) ethanone	GC/MS	Laboratory fires (oak wood)	43
1-(3,4,5-Trimethoxyphenyl) propanone	GC/MS	Laboratory fires (oak wood)	43
3,4,5-Trimethoxybenzoic acid	GC/MS	Laboratory fires (oak wood)	43
1,2-(4-Methoxyphenyl)ethane (dianisyl)	GC/MS	Laboratory fires (grass)	43
1,2-(3-Methoxy-4-hydroxyphenyl)ethane (divanillyl)	GC/MS	Laboratory fires (pine and oak wood)	43
Bis(3,4-dimethoxyphenyl) methane	GC/MS	Laboratory fires (pine and oak wood)	43
1,3-(3-Methoxy-4-hydroxyphenyl)propane (divanillylmethane)	GC/MS	Laboratory fires (pine and oak wood)	43
1,2-(3,4-Methoxyphenyl)ethane (diveratryl)	GC/MS	Laboratory fires (pine and oak wood)	43
1,4-(3-Methoxy-4-hydroxyphenyl)butane (1,2-divanillylethane)	GC/MS	Laboratory fires (pine wood)	43
3-Methoxy-4-hydroxyphenyl-3,5-methoxy-4-hydroxyphenylmethane (bis(guaiacylsyringyl))	GC/MS	Laboratory fires (oak wood)	43
Dehydroabietic acid	GC/MS	Laboratory fires (pine and oak wood)	43
(IR)-7β-Ethenyl-1,2,3,4,4a,4bα,5,6,7,9,10,10aα-dodecahydro-1, 4aβ,7-trimethyl-1α-phenanthrenecarboxylic acid (pimaric acid)	GC/MC	Laboratory fires (pine wood)	43
Sandaracopimaric acid	GC/MS	Laboratory fires (pine wood)	43

Table 3. Organic Compounds Identified in Smoldering Biomass Emissions *(continued)*

Compound	Method(s)	Fuel(s)	Refs.
(IR)-7-α-Ethenyl-1,2,3,4,4a,4bαm-5,6,7,8,10,10a-dodecahydro-1,4aβ,7-trimethyl-1α-phenanthrenecarboxylic acid (isopimaric acid)	GC/MS	Laboratory fires (pine wood)	(43)
1,2,3,4*a*,4*b*,4,6,10,10*a*-Decahydro-1,4*a*-dimethyl-7-(1-methylethyl)-1-phenanthrenecarboxylic acid (abietic acid)	GC/MS	Laboratory fires (pine wood)	43
7-Oxodehydroabietic acid	GC/MS	Laboratory fires (pine wood)	43
1,2-(3,5-Methoxy-5-hydroxyphenyl)ethane (disyrinyl)	GC/MS	Laboratory fires (oak wood)	43
Tetrahydro-3,4-(3-methoxy-4-hydroxy-α-tolyl)furan (tetrahydro-3,4-divanillylfuran)	GC/MS	Laboratory fires (pine and oak wood)	43
Tetrahydro-3-(3-methoxy-4-hydroxy-α-tolyl)-4-(3,4-methoxy-α-tolyl)furan (tetrahydro-3-vanillyl-4-veratrylfuran)	GC/MS	Laboratory fires (pine and oak wood)	43
Dihydro-3,4 (3-methoxy-4-hydroxy-α-tolyl)-2(3*H*)-furanone (matairesinol)	GC/MS	Laboratory fires (pine wood)	43
Bis(3,4,5-trimethoxy-phenyl)ethane	GC/MS	Laboratory fires (oak wood)	43
Tetrahydro-3,4,-(3,4-methoxytolyl)furan (tetrahydro-3,4-diveratrylfuran)	GC/MS	Laboratory fires (pine and oak wood)	43
Dihydro-3,4-(3,4-methoxytolyl)-2-(3*H*)furanone (dihydro-3,4-diveratryl-2(3*H*)-furanone)	GC/MS	Laboratory fires (pine wood)	43
Dihydro-3-(3,5-methoxy-4-hydroxytolyl)-4-(3-methoxy-4-hydroxytolyl)-2(3*H*)-furanone (dihydrovanillyl-syringyl-2(3*H*)-furanone)	GC/MS	Laboratory fires (pine wood)	43
Stigmast-6-en-3β-ol (β-sitosterol)	GC/MS	Laboratory fires (pine and oak wood and grass)	43
Dihydro-3(2,3,4-methoxytolyl)-4-(3,4-methoxytolyl)-2(3*H*)-furanone (dihydro-3(2′,3′,4′-trimethoxyphenyl)-4-veratryl-2(3*H*)-furanone)	GC/MS	Laboratory fires (pine wood)	43
		Nitrogenated Organic Compounds	
Acetonitrile	FTIR	Australian savanna	51
	GC/TID	Laboratory fires (savanna grass, wood, hay, and straw)	17
Acrylonitrile	GC/TID	Laboratory fires (savanna grass, wood, hay, and straw)	17
Propionitrile	GC/TID	Laboratory fires (savanna grass, wood, hay, and straw)	17
Methylamine	GC/TID	Laboratory fires (savanna grass, wood, hay, and straw)	17
Dimethylamine	GC/TID	Laboratory fires (savanna grass, wood, hay, and straw)	17
Ethylamine	GC/TID	Laboratory fires (savanna grass, wood, hay, and straw)	17

Table 3. *(continued)*

Trimethylamine	GC/TID	Laboratory fires (savanna grass, wood, hay, and straw)	17
2-Methyl-1-butylamine	GC/TID	Laboratory fires (savanna grass, wood, hay, and straw)	17
n-Pentylamine	GC/TID	Laboratory fires (savanna grass, wood, hay, and straw)	17
Hexamethylenetetraamine	GC/MS	Laboratory fires (ponderosa pine duff and humus)	15,16
Pyridine	GC/MS	Laboratory fires (ponderosa pine litter, duff, and humus)	15,16
Benzonitrile	GC/MS	Laboratory fires (ponderosa pine litter, duff, and humus)	15,16
		Halogenated Organic Compounds	
Dichloromethane (methylene chloride)	GC/FID, GC/ECD	African savanna, Boreal forest	53,44
Chloromethane (methyl chloride)	GC/FID, GC/ECD	African savanna	54,53
	GC/ECD	Laboratory fires (Douglas fir wood, slash and duff), laboratory fires (oak leaves, pine needles, straw, cornstalks), temperate forest, laboratory fires (Douglas fir wood, needles, bark, and litter)	22,55,48,24
	GC/FID	Laboratory fires (savanna grass, wood, hay and straw), Australian bush	17,56
Trichloromethane (chloroform)	GC/ECD	African savanna	53
Carbon tetrachloride	GC/ECD	African savanna, boreal forest	53,44
Trichloroethane	GC/ECD	African savanna, boreal forest	53,44
Trichloroethene	GC/ECD	African savanna	53
Tetrachloroethene	GC/ECD	African savanna, boreal forest	53,44
Trichlorotrifluoroethane	GC/ECD	African savanna	53
Dichlorodifluoromethane (CFC 12)	GC/ECD	Boreal forest, temperate forest	44,48
Trichlorofluoromethane (CFC 11)	GC/ECD	Boreal forest	44
1,1,2-Trichloro-1,2,2-trifluoroethane (CFC 113)	GC/ECD	Boreal forest	44
Iodomethane (methyl iodide)	GC/ECD	Laboratory fires (Douglas fir needles, wood, bark, and litter)	
	GC/ECD, GC/MS	African savanna	24,54
Bromomethane (methyl bromide)	GC/MS	Chaparral, African savanna, boreal forest, and laboratory fires (savanna grass)	57
	GC/ECD, GC/MS	African savanna	54
	GC/ECD	Laboratory fires (Douglas fir needles and litter)	24

organic samples, especially those with high water content as in smoke, but is limited in its ability to conclusively identify compounds. Tenative identifications are made by matching sample retention times with retention times of standards. Smoke has been most often analyzed for halogenated organic compounds by GC/ECD, which is a specific method for compounds containing electronegative atoms such as halogens. It is a good method for the quantitation of halogenated organic compounds, but it is less effective for samples with high water content. It has the same compound identification limitations as GC/FID. Several investigators have used GC/MS to analyze smolder smoke for hydrocarbons, PAHs, oxygenated organic compounds, organic nitrogen compounds, and halogenated organic compounds. GC/MS is a superior universal method which can offer conclusive compound identification, based on a mass spectrum and retention time, but it is not as quantitative as GC/FID or GC/ECD. Gas chromatography/mass spectrometry/infrared (GC/MS/IR) has been used to identify oxygenated organic compounds. This method offers even more conclusive identifications than just GC/MS by also providing an IR spectrum for each compound. Organic nitrogen compounds have also been analyzed by GC/TID. The TID is selective for organic compounds containing nitrogen and phosphorus, although compound identifications are tentative because they are made based on retention times. Details of the individual GC columns, conditions, and chromatograms are given in the references in Table 3. Some compounds do not pass through GC columns and must be determined by other methods. Several hydrocarbons and some oxygenated organic compounds and organic nitrogen compounds have been identified with *in situ* FTIR, as described previously. PAHs in smoke have been determined using HPLC/fluorescence. Aldehydes and ketones have been determined by *in situ* derivatization of smoke with 2,4-dinitrophenylhydrazine (DNPH) in an acetonitrile solution. The aldehyde and ketone derivatives are then separated via HPLC equipped with an UV detector. The aqueous solutions collected in the previously described mist chambers have been analyzed for carboxylic acids by IC. Individual experimental details are given in the references Table 3.

CHEMICAL CHARACTERIZATION OF FUELS

Studies of smoldering emissions from burning biomass have revealed that emission of individual compounds are dependent on fuel chemistry, which can vary considerably (15,16,24). Therefore, a few methods for chemical fuel characterization are presented. Carbon, hydrogen, and nitrogen content of biomass can be determined by combusting ground samples in oxygen at 925°C, converting carbon, hydrogen, and nitrogen to carbon dioxide, water, and nitrogen oxides, respectively. Nitrogen oxides are then reduced to nitrogen, and the gases are separated by steady wave front chromatography and detected with a thermal conductivity detector. Total inorganic content can be determined by ashing samples at approximately 550°C for several hours. Individual metal concentrations, such as calcium and potassium, are most often determined by acid-digesting an ashed sample and analyzing the digest for cations by ion-coupled plasma emission spectrometry. Chloride concentrations in biomass materials can be determined by extracting ground samples into aqueous sodium nitrate and using a chloride ion electrode to measure chloride in the extract. Chlorine and bromine can be determined by ashing an aqueous paste of ground sample and calcium oxide, resulting in all chlorine and bromine being in the forms of calcium chloride and calcium bromide. Chloride and bromide are extracted from the ash with hot water, and the extracts are analyzed by IC (58). Chlorine, bromine, and iodine can also be determined in biomass tissues by instrumental neutron activation analysis. Several methods for determining cellulose and polysaccharide content of woody materials have been described (59), as have several methods for determining lignin content in woody materials (60).

CALCULATION OF EMISSION FACTORS AND EMISSION RATIOS

Concentrations of compounds within a smoke plume depend on several factors, such as how near the fire the smoke was sampled, dilution of smoke within the plume, and the phase of combustion (ie, flaming or smoldering). This makes it difficult to make direct comparisons of emissions between studies and to determine the contribution of biomass burning to the global flux of compounds. It is therefore necessary to express emissions in terms of emission factors (EFs) and/or emission ratios (ERs).

Direct Method for Calculating Emission Factors

Emission factors (the mass of compound emitted per dry mass of fuel consumed) have been determined by several methods, depending on the parameters measured during a fire. When it is possible to measure the mass of fuel consumed by a fire and the total flow of air over a fire (as in small laboratory fires), emission factors can be calculated directly. The mass of a particular analyte emitted is divided by the mass of fuel consumed to give an emission factor in mg/kg or g/kg. The mass of an analyte measured in units of concentration (ie, mg/L) is determined from the total flow rate (L/min) of air over the fire and total sampling time (min). When it is not possible to measure the total fuel consumed by a fire or the total flow of air over a fire (as in large laboratory fires and field fires), indirect methods have been used to estimate emission factors.

Carbon Balance Method for Calculating Emission Factors

The carbon balance method assumes that all fuel carbon is emitted during a fire, that most of the carbon emitted from a fire is in the forms of carbon dioxide and carbon monoxide, and that the ER of an analyte to carbon dioxide and carbon monoxide remains constant throughout a fire. The fuel consumption can be determined by summing the mass of carbon in carbon dioxide and carbon monoxide emissions, along with the mass of carbon in any other measured emissions, and dividing the sum by the carbon content of the fuel, often assumed to be 0.5 or 0.45×10^3 g/kg:

$$W = (\Sigma C_{\text{emit}})/R$$

where W = fuel consumed (g/m^3), ΣC_{emit} = carbon fraction of the emission (g/m^3), and R = the carbon fraction of the fuel. The emission factor is then calculated by dividing the concentration of a compound in the smoke (g/m^3) by the fuel consumed (g/m^3) (7,48,61). For example, the emission factor for carbon dioxide (EF_{CO_2}) would be:

$$EF_{CO_2} = \Delta CO_2/W \tag{1}$$

where ΔCO_2 = the concentration of carbon dioxide measured (g/m^3) minus background carbon dioxide.

The weakness of the carbon balance method lies in the assumptions. Not all of the carbon in the fuel is released during a fire; much of it remains behind in the char and fuels vary in their carbon content. The method is strengthened if several samples are collected throughout each phase of a fire, thus allowing for a more integrated approach. With these assumptions removed, W simply equals C_{fuel}, the mass of fuel carbon burned. The emission factor for carbon dioxide would then be:

$$EF_{CO_2} = \Delta CO_2/C_{\text{fuel}} = (\Sigma C_{\text{emit}}/C_{\text{fuel}})/(1 + \Delta CO/\Delta CO_2 + \Delta CH_4/\Delta CO_2 + \ldots) \tag{2}$$

where ΔCO = the concentration of carbon monoxide measured (g/m^3) minus background carbon monoxide and ΔCH_4 = the concentration of methane measured (g/m^3) minus background methane. The sum in the denominator should also include concentrations of any other measured organic emissions, such as particulates and nonmethane hydrocarbons. The expression $\Sigma C_{\text{emit}}/C_{\text{fuel}}$ can be estimated from measuring the carbon content of the fuel before and after a fire. If all the carbon-based emissions in a fire were measured, the sum of the emission factors should equal $\Sigma C_{\text{emit}}/C_{\text{fuel}}$.

Calculating Emission Ratios

Several investigators have found positive correlations between incomplete combustion products and carbon monoxide, whereas these products are weakly correlated with carbon dioxide (17,19,22). This implies that constant ratios may exist between products of incomplete combustion. The EF of compounds emitted from smoldering combustion can be derived from the ER to carbon monoxide as follows:

$$ER_x = \Delta X/\Delta CO \quad (3)$$

$$EF_x = (\Delta X/\Delta CO) \times (nEF_{CO}) \quad (4)$$

where n = the number of carbon atoms in species X (51,62,63). Equations 3 and 4 can be used to estimate emission ratios for nitrogen and sulfur-based species by dividing them by the mean molar ratio of fuel nitrogen or sulfur to fuel carbon. Calculating emissions factors from emission ratios rely on the same assumptions employed in the carbon balance method, especially the assumption the ER to carbon monoxide remains constant throughout the fire. However, this may not be true for all compounds emitted during smoldering combustion, because many processes occur during smoldering combustion that produce various compounds to varying degrees. Some investigators have observed that ER of compounds, such as methyl bromide and methanol, to carbon monoxide and carbon dioxide vary throughout a fire (10,57).

The concept of dividing fires into two phases, flaming and smoldering, separated in time is not an accurate description of what actually occurs during a fire (10). Emissions have been measured from combustion chamber fires in real time throughout burns using the FTIR techniques already described. Emissions that result from pyrolysis and distillation (eg, methanol, acetic acid, formic acid, and terpenes) occur during times dominated by flaming combustion and in times dominated by smoldering combustion. Poor temporal correlations between emissions that result from pyrolysis and distillation processes and the emissions of both carbon monoxide and carbon dioxide have also been observed (10). There are at least three processes that should be considered in a fire: flaming, smoldering, and pyrolysis/distillation; a representative sample cannot always be taken at any arbitrary time during a fire. It should also be noted that few fires are neatly separated into processes. All these processes and several others often occur simultaneously. Therefore, the assumption that smoldering combustion emissions, especially those that result from pyrolysis and distillation processes, are correlated with carbon monoxide emissions could result in over or under-estimation of the EF for some of these compounds.

EMISSIONS FROM SMOLDERING COMBUSTION OF BIOMASS

Figure 4 illustrates the partitioning of carbon from the smoldering combustion of ponderosa pine wood. Char is defined as the material left after the burn; tar is defined as a brown viscous liquid which collects on the walls of the sampling apparatus. Levoglucosan predominates the composition of the tars. As can be seen, most of the fuel carbon is partitioned in the char during smoldering combustion, followed by partitioning into carbon dioxide and carbon monoxide. The rest of the carbon is either partitioned into the tar or emitted as various trace components including oxygenated organic compounds and hydrocarbons.

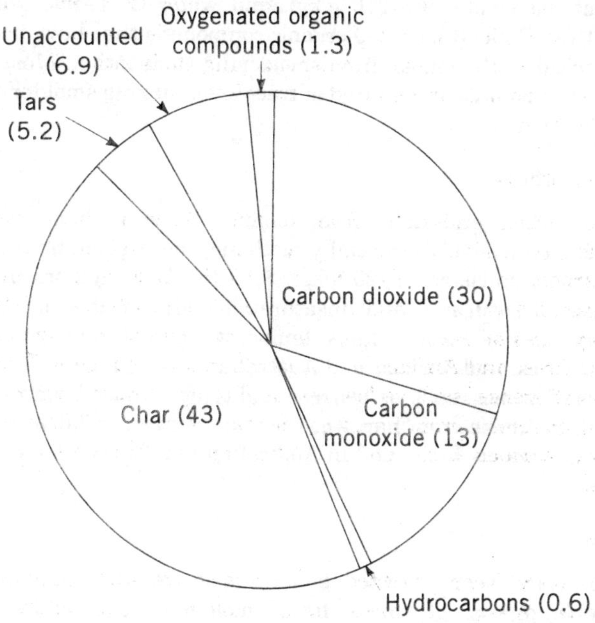

Figure 4. Carbon partitioning from smoldering ponderosa pine wood fires (16). Numbers in parentheses are percentages of total carbon.

Figure 5 illustrates the proportions of trace gas emissions measured from the smoldering combustion of a duff bed collected from a temperate forest. Over half of the trace emissions are hydrocarbons, of which methane is most significant. Oxygenated organic compounds comprise one-third of the trace emissions, followed by inorganic nitrogen compounds, and sulfur dioxide. Halogenated organic compounds and PAHs comprise less than 1% of trace emissions from biomass burning.

The predominate emissions released from the smoldering combustion of biomass are carbon dioxide (430–1600 g/kg),

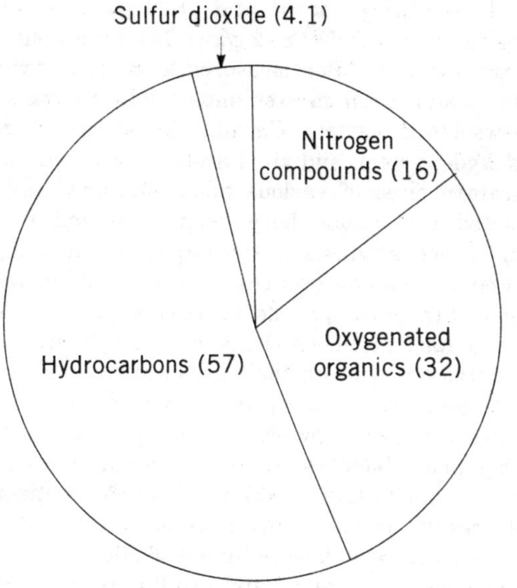

Figure 5. Proportions of trace gas emissions from smoldering duff fires (10). Numbers in parentheses are percentages of total trace gas emissions.

carbon monoxide (90–270 g/kg), and water (234–990 g/kg) (9,10,15). Table 3 lists 200 organic compounds that have been identified in the smoke from smoldering combustion. Most of these compounds are emitted at trace levels during smoldering combustion.

Hydrocarbons

Hydrocarbon emissions from biomass burning have been studied extensively, especially methane, the predominate hydrocarbon emission (3–20 g/kg) (8,10,15). Investigators have measured saturated and unsaturated hydrocarbons in laboratory fires of various fuels, temperate, boreal, and tropical forest fires, and African and Australian savanna fires. Trace levels of arenes, such as benzene and toluene, have been measured in smoke from hardwood burning stoves, in laboratory fires of various fuels, and in Australian and African savanna fires.

PAHs

PAHs have been detected both in gaseous and condensed particle phases of smoke from smoldering combustion of biomass fuels. Naphthalene, acenaphthylene, phenanthrene, benzo(*a*)pyrene, pyrene, and fluoranthene are emitted in the greatest amounts (27,28,32–34). Total PAH emissions were 50 times lower than total volatile organic compound emissions in smoke sampled from wood burning in residential fireplaces and no PAHs with more than six rings were detected (34). PAH emissions increased under oxygen-starved conditions. Hardwood burned in a residential stove emitted 20–32 mg of total PAHs per kg of dry fuel under oxygen-starved conditions (27). PAHs are most likely formed on the char at high smolder temperatures as indicated in Figure 1.

Oxygenated Organic Compounds

Considerable work has been done to identify and quantify oxygenated organic compounds from burning biomass. The oxygenated organic compounds include some of the major trace species emitted in the smoldering combustion of biomass: methanol (1–5 g/kg), acetic acid (1–8 g/kg), vinyl acetate (0.2–3 g/kg), and acetol (0.2–2 g/kg) (15). Emissions of formic and acetic acid have been measured from laboratory fires of many fuels and in air masses impacted by biomass burning over central and eastern Canada. Emissions of methanol, formaldehyde, acetol, and vinyl acetate have been measured in laboratory fires of various fuels. Methoxylated phenols and related compounds have been measured in particulates and gaseous phases in the plumes emitted from wood stoves, and in urban air over cities where wood burning occurs (31). Some of these compounds, eg, phenol, guaiacol, and their derivatives, have also been identified in smoke from smoldering combustion of various fuels in bench-scale fires. Most of the oxygenated organic compounds emitted from smoldering combustion are known pyrolysis products of macromolecules which comprise biomass, cellulose, hemicelluloses, lignin, phytosterols, and resins (16,43). For example, acetic acid most probably results from the pyrolysis of acetate ester groups known to exist in xylan hemicelluloses (64,65) and the phenolic compounds, such as guaiacol and vanillin, result from the pyrolysis of lignin (66,67). Some of the higher molecular weight oxygenated compounds simply vaporize and subsequently condense unaltered into aerosols (43).

Nitrogen and Sulfur Compounds

Several investigators have measured nitrogen compounds in smoke from laboratory fires of various fuels and in smoke from temperate forest and African and Australian savanna fires. The volatilization of fixed nitrogen from plants and soil during fires has been termed *pyrodenitrification* (51,68). Three nitrogen compounds, pyridine, benzonitrile, and hexamethylenetetraamine, have been detected in the smoke from ponderosa pine litter (top surface of forest floor), duff, and humus (forest floor subsurface layer of decayed organic material), although they these compounds were not detected in smoke from ponderosa pine sapwood, needles, and bark (15). Markedly higher nitrogen to carbon ratios were found in the forest floor materials than in the sapwood or bark, but the highest ratio was in the needles. It was also determined that nitrogen is concentrated in the chars of the floor layers and diluted in the chars of the needles during burning, suggesting nitrogen losses from the needles during burning in forms not detected, such as ammonia or nitrogen oxides. Hexamethylenetetraamine must therefore be formed in the char smolder zone and/or in the smoke plume when formaldehyde and ammonia, precursors to hexamethylenetetraamine, emissions are high.

Carbonyl sulfide and sulfur dioxide have been observed in smoke from laboratory fires and from African savanna fires (10,69,70). Biomass burning accounts for approximately 10% of the global flux of carbonyl sulfide (69).

Halogenated Organic Compounds

It has been estimated that biomass burning contributes up to 50% of total atmospheric methyl chloride and up to 30% of atmospheric methyl bromide (54,71). Several investigators have measured elevated levels of methyl chloride in smoke plumes from biomass fires and in smoke from controlled laboratory fires. More recently, elevated levels of methyl bromide have been measured in smoke plumes from biomass fires. Methyl iodide has also been recently measured in plumes from biomass fires. Several other chlorinated organic compounds and a few chlorofluorohydrocarbons have been measured in smoke plumes from burning biomass in equatorial Africa and boreal forests at concentrations orders of magnitude lower than methyl chloride (44,53). Biomass fires may emit methyl chloride and methyl bromide as a result of at least two processes. Most of the methyl chloride and methyl bromide emitted results from a secondary low temperature smolder process occurring on the char. A small percentage (<10%) of the emissions may result from prefire methyl chloride and methyl bromide volatilizing from the fuel as heating begins (24).

FACTORS INFLUENCING EMISSIONS AND MODELING

Combustion Efficiency

Emissions of smoldering combustion products tend to increase as the combustion efficiency decreases. The combustion efficiency is dependent on fire response to fuel arrangement and weather influences. In some cases, the combustion efficiency can be used to predict the emission of a particular compound. For example, the emission of 2-furaldehyde from fires of various ponderosa pine fuels can be predicted fairly well from the combustion efficiency ($CO_2/(CO_2 + CO)$) as

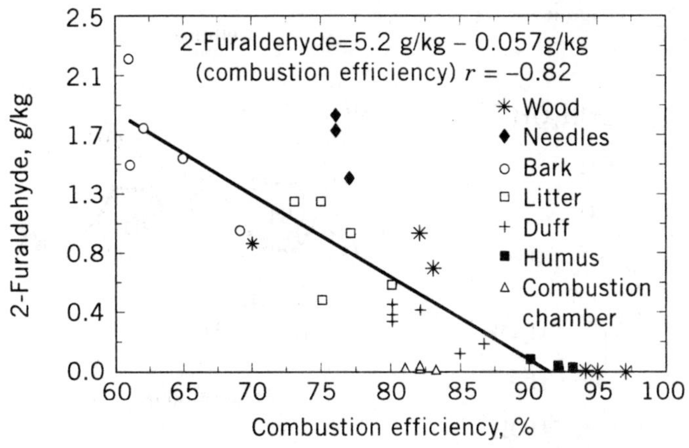

Figure 6. 2-Furaldehyde vs combustion efficiency (24).

shown in Figure 6 (15,24). Combustion efficiency has been used to predict particulate ($r^2 = 0.59$), carbon monoxide ($r^2 = 0.99$), methane ($r^2 = 0.87$), and molecular hydrogen ($r^2 = 0.91$) emissions from fires in Amazonian savannas and forests (9). Good negative linear correlations were found between combustion efficiency and emissions of methanol ($r^2 = 0.95$, $n = 6$), formaldehyde ($r^2 = 0.91$, $n = 9$), and acetic acid ($r^2 = 0.93$, $n = 6$), although the models are only predictive for combustion efficiencies greater than 0.89 (10). Models have been developed for predicting methane emissions based on combustion efficiency (8). This data set included measurements from fires in Amazon forests and savannas, African savannas, and North American chaparral, wetlands, and boreal forests. It has been found that emissions of methane not only depend on combustion efficiency but also on the type of fuel burned. Therefore, three separate models are presented for predicting methane emissions from combustion efficiency: (*1*) forests with little duff ($r^2 = 0.71$, $n = 24$), (*2*) forests with duff ($r^2 = 0.77$, $n = 46$), and (*3*) grass-like material ($r^2 = 0.80$, $n = 31$). Fires of forests with little duff produced the most methane, followed by fires of forests with duff, for the same combustion efficiency. The models could be used to reasonably predict methane emissions from combustion efficiency for fires of these types of fuels. The combustion efficiency has also been found to be more predictive of emissions from smoldering combustion when the type of fuel being burned was considered. The linear relationship between combustion efficiency and emissions of smoldering combustion products was observed to diminish at low combustion efficiencies (<80%). Combustion efficiency is a convenient parameter to measure in field experiments and, as has been discussed, can be used to predict emissions of some compounds from smoldering combustion, with a degree of caution. The emission of many compounds also depend on the chemistry of the fuel being burned, and unless the chemical composition of the fuel is considered in a model, the combustion efficiency is not very predictive of these emissions and could be misleading.

Fuel Chemistry

The chemical composition of fuels burned in biomass fires is quite diverse. Although three polymers, cellulose, hemicellulose, and lignin, shown in Figure 7**a**, primarily comprise biomass, variations occur in the monomers that are part of the hemicelluloses and lignins. For example, hardwood (angiosperm) lignins contain syringol not found in softwood (gymnosperm) lignins and grass lignins contain *p*-coumaryl alcohol not found in woody materials. Various monomeric units are than pyrolyzed as these fuels burn. Examples of differing monomeric lignin precursors are shown in Figure 7**b.** Smoke from softwoods, hardwoods, and grasses are enriched in products from coniferyl alcohol, syrinapyl alcohol, and *p*-coumaryl alcohol, respectively (43). The percentage of cellulose, hemicellulose, and lignin varies from plant type to plant type, between tissue types within a species, and even with season and location within a tissue. The wood of Douglas fir has three times the polysaccharide (cellulose and hemicellulose) content of the needles and six times the polysaccharide content of the bark (72). A principal components analysis on 26 oxygenated organic compounds (most known to be pyrolysis products of cellulose, hemicellulose, and lignins) emitted from smoldering fires of ponderosa pine wood, needles, bark, litter, duff, and humus revealed distinct components which could be described as polysaccharide, hemicellulose, or lignin components (15,24). Softwoods also contain diterpenoid acids which directly volatilize during smoldering combustion and which can be used as unique markers for softwood burning (43).

Biomass materials vary in nitrogen content depending on season, species, and tissue type. Foliage and floor layers tend to have the highest nitrogen concentrations (15,17,41,51). Emissions of nitrogen oxides from African savanna fires are linearly dependent on fuel nitrogen content ($r = 0.93, n = 14$) (62). Nitrogen oxide emissions from open cooking fires in Africa were higher when fuels expected to have high nitrogen concentrations, eg, dung, were burned than when low nitrogen content fuels, ie, wood, were burned (73).

Biomass materials also display remarkable variability in trace chemistry depending on species, tissue type, and location. Chlorine and bromine concentrations, along with molar ratios of chlorine to bromine, can vary by at least two orders of magnitude. Grasses and foliage have the highest chlorine concentrations and woods have the lowest. Litter and duff have the highest bromine concentrations, while wood and grasses have the lowest. Chlorine and bromine concentrations in Douglas fir tissues tend to increase with proximity to the Pacific Ocean, a large source of chloride and bromide (74). Mineral concentrations also vary considerably by species and tissue. Minerals concentrations are highest in metabolically active foliage (24,75). Some minerals, such as potassium and calcium, catalyze oxygen chemisorption on char, resulting in elevated levels of carbon dioxide emissions during smoldering combustion (76,77).

Predictive models for methyl chloride and methyl bromide emissions based on fuel chlorine and bromine content along with other parameters have been developed. Methyl chloride emissions have been modeled from laboratory fires of temperate forest fuels to fuel chloride content, reaction intensity (net heat release rate per unit area of fuel bed), and combustion efficiency (CO_2 emitted/CO_2 maximum) ($r^2 = 0.78$, $n = 30$) (36). Methyl chloride emissions increased with decreasing combustion efficiency and reaction intensity and increasing fuel chloride concentrations. Methyl chloride and methyl bromide emissions have been modeled from laboratory fires of Douglas fir tissues and temperate fuels and field fires of African

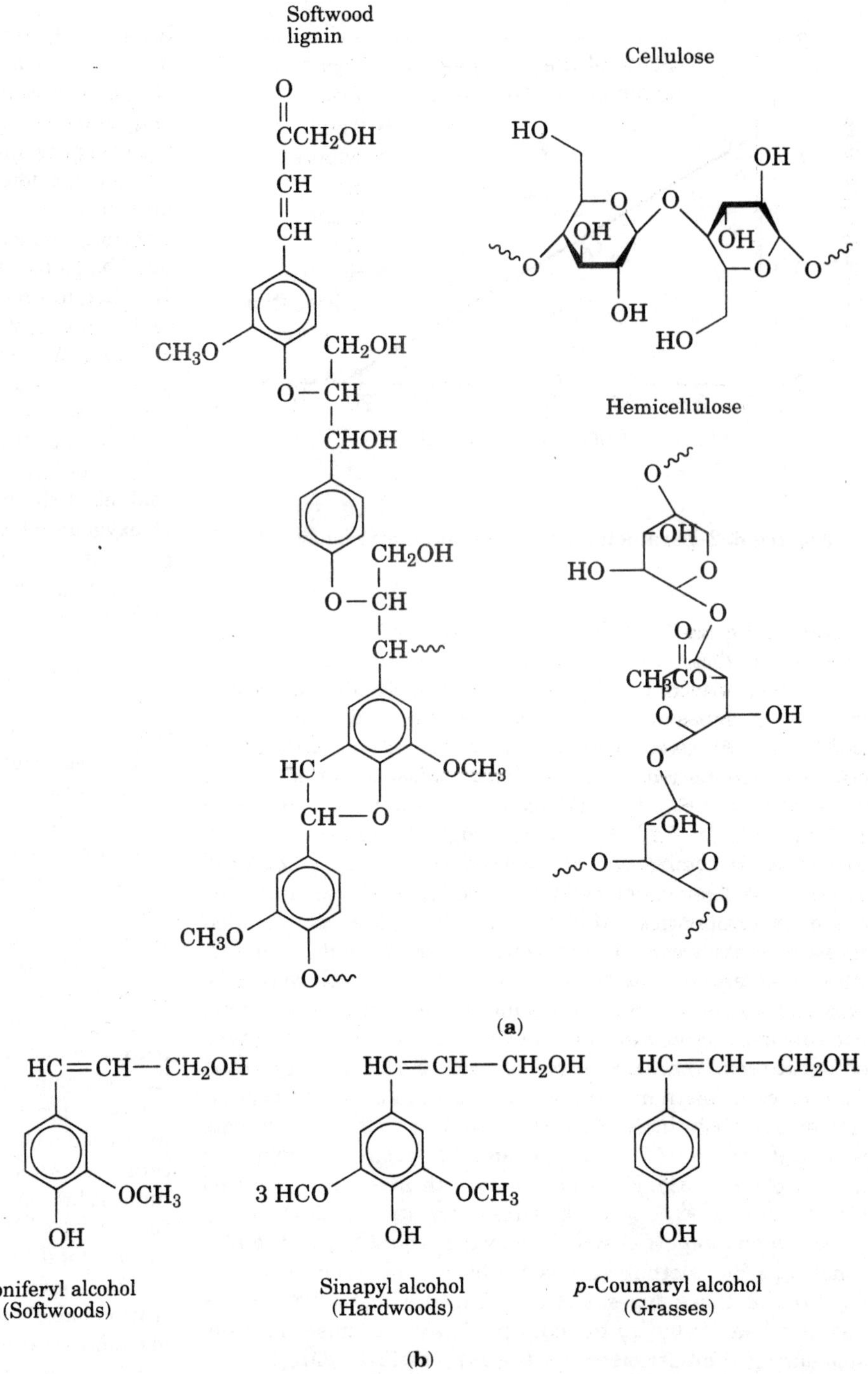

Figure 7. (**a**) Major polymers in biomass tissues; (**b**) possible monomeric units in lignins.

savannas to fuel chlorine and bromine content, fuel carbon content, carbon monoxide emissions, and carbon dioxide emissions (24). Both methyl chloride and methyl bromide emissions were primarily dependent on fuel chlorine and bromine content and secondarily on combustion efficiency. Methyl chloride emissions increased with increasing fuel chlorine and emissions of carbon monoxide and decreasing emissions of carbon dioxide ($r^2 = 0.91$, $n = 44$). Methyl bromide emissions increased with increasing fuel bromine concentrations and emissions of carbon dioxide and decreasing emissions of carbon monoxide ($r^2 = 0.81$, $n = 24$).

Smoldering Combustion Emissions as Tracers for Wood Burning

Air pollution (qv) is a serious problem in many urban areas, with a multitude of potential sources, such as industrial processes, automobiles, and wood burning stoves and fireplaces. One of the challenges of atmospheric science is to identify these sources and quantify their contribution to overall air pollution. Investigators have sought to identify unique chemical tracers from wood smoke in order to quantify its contribution to air pollution in various cities. Methyl chloride has been proposed as one such chemical tracer (78,79). The gaseous tracer model was developed to relate concentrations of methyl chloride emitted to the burning of wood to the mass of fine particulates, a measure of air pollution, from wood burning. The model was used to determine the contribution of wood burning to air pollution in Portland, Oregon in the winter of 1991. A disadvantage to this model is the varying background levels of methyl chloride. The ocean and wood rot fungi are sources of methyl chloride which vary with location, even within a city. Methoxylated phenols, lignin pyrolysis products, have also been proposed as chemical tracers for wood burning (31). Guaiacyl derivatives have been used as markers to distinguish wood smoke in urban pollution and syringol as a unique marker for the burning of hardwoods. Emissions of guaiacol (2-methoxyphenol) and its derivatives correlate well with emissions of PM-10 carbon. Methoxylated phenol concentrations were highest in residential areas with high wood burning activity in both Minneapolis, Minnesota and Salt Lake City, Utah. Background concentrations of methoxylated phenols can be expected to be very low.

Physical Parameters

The mass of fuel present for consumption in a smoldering fire can influence emissions during the smolder. This is an important consideration when extrapolating from small bench-scale fires to larger field fires. Small amounts of fuel (20 g) burned in bench-scale fires emitted less methane, ethane, propionic acid, and butyric acid and more 2-furaldehyde than fires of similar fuels burned in larger combustion chamber fires (6 kg) (24). It is probable that higher temperature smolder processes occurred in the combustion chamber fires, because of an increase in opportunity for the formation of insulated heat-trapping areas in the larger masses of fuel. High temperature (>400°C) smolder processes, illustrated in Figure 1, favor the formation of methane and ethane through secondary reactions of pyrolysis products eg, 2-furaldehyde, on the char. It may be that propionic acid and butyric acid formation is also favored at higher smolder temperatures.

Ratios to Carbon monoxide

The concentration of a compound measured in a smoke plume depends on how near the fire the sampling occurs and the extent of dilution within the smoke plume. In order to compare values determined in different studies, it is necessary to either calculate EFs or ERs to a reference gas, such as carbon monoxide or carbon dioxide. Reference gas ratios are also useful in estimating trace gas emissions from fires of similar fuels. Trace gas emissions are not commonly measured in field fires, whereas carbon monoxide and carbon dioxide are routinely measured. Several investigators have found good positive correlations between smoldering combustion products and carbon monoxide, whereas these products are weakly correlated with carbon dioxide (17,19,22). Therefore, carbon monoxide is the best reference gas to use when comparing or estimating emissions from smoldering combustion. There are limitations to using carbon monoxide ratios which have been previously discussed, but given the complexity of the emissions from smoldering combustion, the variability of fuels, and the feasibility of measuring all emissions in any given fire this method is useful in obtaining general estimations.

Reference gas ratios have been used to estimate wildland firefighter exposure levels to smoldering combustion emissions. Concentrations of 2-propenal, formaldehyde, benzene, and carbon monoxide to which a group of wildland firefighters were exposed have been measured (22). The emissions of 2-propenal, formaldehyde, and benzene were found to correlate well with carbon monoxide emissions, and ERs to carbon monoxide were calculated. These ERs were then used to estimate wildland firefighter exposure levels to these compounds, based on carbon monoxide emissions. The same method has been used to determine wildland firefighter exposure levels to several oxygenated organic compounds (15,24). This method has also been used to estimate benzene emissions from biomass fires in the United States and globally (80,81).

ENVIRONMENTAL IMPLICATIONS

Human Health

People come in contact with smoke from smoldering combustion in several ways. They may be exposed to these emissions while cooking or heating with wood stoves, clearing agricultural residues or land, or fighting fires. Residents in the vicinity of large biomass fires may also be exposed to the emissions of smoldering combustion. Wildland firefighters may experience significant eye and respiratory tract irritation, which may cause a slight decrement in lung function and could lead to increased incidents of respiratory illness. They may also be exposed to elevated levels of carbon monoxide, leading to adverse health effects from insufficient tissue oxygenation (22). High ambient concentrations of carbon monoxide were observed in the states of Para and Rondonia, Brazil, mostly caused by smoke from fires burning Amazon forests (9). Carbon monoxide emissions have been measured from cookfires of acacia wood, dung cake, charcoal, and biogas at the breathing level height in squatting and standing postures in East Indian kitchens. Dung cake emitted the most carbon monoxide, followed by fuel wood. These emissions were four to five times the World Health Organization safe limit of 10 ppm (38). Wildfires occasionally result in poor local air quality, such as in northern California and southern Oregon in 1987 (80), near Yellowstone National Park (Wyoming) in 1988 (80), and in northwestern Canada during the very active burning season of 1994 (82). Emissions from woodstoves account for as much as 80% of the fine-particulate loading in several U.S. communities (30). In California, burning of agricultural wastes such as wheat straw and tree prunings has resulted in increased levels of carbon monoxide and particulates (37).

Table 4 lists exposure levels to some of the compounds produced during the smoldering combustion of biomass based on a carbon monoxide exposure level of 54 ppm. Exposure levels of formaldehyde exceed the TWA limit by a factor of 30 and exposure levels of 2-propenal and benzene begin to approach the TWA limit. 2-Propenal, better known as acrolein, is a strong irritant. Formaldehyde is also an irritant and benzene exposure has many health effects, including aplastic anemia. The other compound exposure levels listed are well below the allowable TWAs based on a peak exposure to carbon monoxide of 54 ppm when considered individually. It should be noted, however, that the synergistic effects and TWAs of some or all of these compounds have not been determined. Many of the compounds listed are known irritants, which may help to explain some of the 12,000 visits firefighters made to medical facilities for respiratory problems during the 1988 wildfires in Yellowstone National Park (1). Other compounds, such as

Table 4. Exposure Levels for Compounds Found in Smoldering Biomass Emissions

Compound	Exposure Levels[a], ppb	TWA[b], ppb, and Effect[c]
2-Furaldehyde[d]	2.6 ± 1.6[d]	2000; Suspected carcinogen, irritant
5-Methylfuraldehyde[d]	10 ± 4.6[d]	Irritant
2-Acetylfuran[d]	1.5 ± 0.37[d]	Irritant
Phenol	17 ± 11[d]	5000; Irritant
o-Cresol	4.9 ± 0.65[d]	2000; Irritant
m/*p*-Cresol	9.2 ± 1.7[d]	2000, Irritant
Guaiacol	16 ± 2.5[d]	Irritant
4-Methylguaiacol	9.2 ± 1.4[d]	Irritant
Vanillin	1.1 ± 0.086[d]	Irritant
Acetol	65 ± 92[d]	No toxicological data given
Vinyl acetate	150 ± 50[d]	Suspected carcinogen, irritant
2-Cyclopenten-1-one	3.9 ± 1.3[d]	Irritant
Acetic acid	700 ± 170[d]	10,000; Irritant
Formic acid	110 ± 46[d]	5000; Irritant
Propionic acid	86 ± 12[d]	10,000; Irritant
Methyl 3-oxobutanoate	16 ± 7[d]	Irritant
Methanol	110 ± 53[d]	200,000; Irritant
Formaldehyde	500[e]	16; Carcinogen, irritant
2-Propenal	70[e]	100; Irritant
Benzene	60[e]	100; Carcinogen, anemia, immune system depressant
Total irritants	*1770 ± 360*	

[a] Based on a 54 ppm CO exposure ±1 standard deviation.
[b] Time weighted average from NIOSH based on a 40-h work week.
[c] As listed on material safety data sheets for inhalation.
[d] Ref. 24.
[e] Ref. 36.

2-furaldehyde, formaldehyde, benzene, and vinyl acetate, are either known or suspected human carcinogens.

Several of the PAHs identified in smoke from smoldering combustion are also known or suspected human carcinogens, such as benzo(*a*)pyrene (83). Ames Salmonella tests have been performed with PAHs extracted from wood smoke. The tests were positive for mutagenicity, which increased in tests with extracts from oxygen-starved fires (oxygen-starved fires produced higher concentrations of PAHs) (27).

Photochemical Smog

The emissions from smoldering combustion contain all the necessary compounds needed for the formation of photochemical smog, ie, nitrogen oxides and unsaturated hydrocarbons. Chemical and meteorological evidence both suggest biomass fires as the source of photochemical smog layers over the Atlantic Ocean and African tropical forests. The global ozone production from biomass fires is estimated to be comparable in strength to the stratospheric input (84) (see Ozone control and voc reactivity). Locally elevated ozone is significant due to its toxicity to humans and plants. Ozone can cause inflammation of the eyes and upper respiratory tract (85) and tissue damage in plants. Sharp increases in ozone with altitude have been observed in the tropics, therefore the likelihood of vegetation damage by ozone from biomass burning would be especially high in tropical mountainous regions (2).

Acid Deposition

Emissions from biomass burning can also result in acid rain (qv), which has been observed in Venezuela, Brazil, Africa, and Australia (2). The acidity of the rain has been mainly attributed to organic acids (especially formic and acetic acid) and nitric acid. Formic and acetic acid emitted from biomass burning and nitric acid photochemically formed from biomass burning emissions of nitrogen oxides dissolve in aqueous aerosols, resulting in acidic rain or dry acid deposition (pH = 3.8–6.2). Approximately 23 to 42% of rain acidity in South Africa has been attributed to biomass burning (86). Acid deposition can threaten ecosystems by direct damage to plants, amphibians, and insects, and through soil acidification.

Perturbation of Nutrient Cycles

Detailed explanations of how biomass burning perturbs nutrient cycles are available (2,71). A summary of the explanations follow.

Biomass burning volatilizes nitrogen from ecosystems. Over 50% of the volatilized nitrogen is in the form of molecular nitrogen and is not redeposited into the ecosystem, resulting in a significant loss of nutrient nitrogen through pyrodenitrification. Long-range transfer of nitrogen emitted in other forms, such as nitrogen oxides and ammonia, also depletes nitrogen from ecosystems that are repeatedly burned. Long-term perturbations in nitrogen nutrient cycle also occur as a result of biomass burning, with elevated soil nitrogen oxide and nitrous oxide emissions continuing for up to six months after a burn. Other nutrients, such as potassium, phosphorus, magnesium, and sulfur may also be depleted from ecosystems through repeated biomass burning.

Biomass burning results in the sequestration of carbon as charcoal, constituting a significant sink for atmospheric carbon dioxide and therefore a source of atmospheric oxygen. The charcoal is not subject to microbial oxidation. On geological time scales atmospheric oxygen levels may increase, which increases the risk of fires and thus establishes a positive feedback loop for the buildup of atmospheric oxygen.

Ecosystems that are not burned may benefit from the nutrient volatilizations during the burning of other ecosystems. In the central Amazon Basin, as much as 90% of the nitrogen and sulfur deposited are from external sources. One possible external source may be the long-range transport of these nutrient plumes of smoke from biomass burning in other areas. Of course, this long-range transport could also result in acid deposition and increased levels of ozone in unburned ecosystems.

Atmospheric Chemistry

Emissions from biomass burning could disrupt the oxidizing potential of the atmosphere (71). The atmosphere has a "self-cleaning" mechanism which photochemically oxidizes pollutants, such as hydrocarbons, to carbon dioxide, preventing their buildup in the atmosphere. The key oxidizer in the troposphere is the hydroxyl radical. The hydroxyl radical also provides a major sink for methane and carbon monoxide. When nitrogen oxide concentrations are low, hydroxy radicals are consumed as they oxidize these pollutants through a series of chain reactions. An increase in global hydrocarbon, methane, and carbon monoxide concentrations, through biomass burning, could result in a greater decrease of hydroxyl radicals. Global consumption of hydroxyl radicals could result in further global increases in carbon monoxide and methane through a positive feedback mechanism. Observed global increases in carbon monoxide and methane indicate this fundamental change in tropospheric chemistry (see TROPOSPHERIC AIR POLLUTION MONITORING). This effect reverses at localized biomass burning sites. As discussed previously, biomass fires emit nitrogen oxides. Hydrocarbon oxidation in the presence of large amounts of nitrogen oxides creates additional hydroxyl radicals and ozone. The regional limitation is due to the relatively short residence time of nitrogen oxides compared to methane and carbon monoxide. Although hydroxyl radical concentrations may increase regionally in areas of biomass burning, the global increases in methane and carbon monoxide, largely from biomass burning and especially from smoldering combustion, could lead to a global decrease in hydroxyl concentrations. Therefore, many pollutants that are normally oxidized by hydroxyl radicals could build up, leading to significant changes in global atmospheric chemistry.

Climate Change

Several of the emissions from biomass burning, such as methane and carbon dioxide, absorb infrared radiation and are therefore potential greenhouse gases. Biomass burning contributes to global increases in carbon dioxide by direct emission of carbon dioxide and through deforestation (elimination of trees that are a sink for carbon dioxide). Biomass burning vies with industrial sources as a source of carbon dioxide, contributing to 18–40% of the global greenhouse effect (2). Biomass burning and smoldering combustion in particular are significant sources of methane, contributing approximately 6% of the global flux. The contribution of biomass burning to the global flux of methane is expected to increase in the next decade as use of fuel wood increases and time between savanna fires and fallow cycles in shifting agriculture become shorter (8).

Biomass burning also produces aerosols which could have an effect on global climate. Aerosol lifetimes are on the order of days and their climatic effects are initially on a regional scale. However, regional effects in equatorial areas could lead to global effects, since equatorial regions are extremely important in receiving solar energy and then vertically redistributing the heat through the atmosphere. Any change in the redistribution could have significant global implications. Aerosols reflect radiation back into space, reducing the amount of the sun's energy that reaches the earth's surface, whereas black soot aerosol absorbs radiation, intensifying the sun's energy at the earth's surface. Significant daytime temperature drops have been observed below smoke plumes from forest fires (2). Aerosols also provide nuclei for the formation of cloud droplets, which then form clouds. The large number of aerosols in biomass-burning plumes results in the formation of many small cloud droplets, which are effective in reflecting solar radiation.

Stratospheric Effects

Biomass burning emits methyl chloride and methyl bromide, especially during smoldering combustion, and may contribute up to 50% of total atmospheric methyl chloride and 30% of atmospheric methyl bromide. Methyl chloride is the second largest source of stratospheric chlorine, and methyl bromide is the largest source of stratospheric bromine (54). The lifetimes of both these methyl halides are sufficiently long (1–2 yr) for them to reach the stratosphere. Once in the stratosphere, they are photolyzed, releasing chlorine and bromine radicals that enter into catalytic ozone-depleting reactions. Bromine radicals are 20 to 60 times more efficient in depleting stratospheric ozone than chlorine radicals (87). Methane emitted from biomass burning also interferes with the stratospheric ozone balance, acting as a sink for stratospheric chlorine radicals.

FUTURE RESEARCH DIRECTIONS

Many investigators have collected and analyzed the emissions from smoldering combustion. Although they have identified well over 200 compounds smoldering biomass emissions, there are many unknown compounds that need to be identified. Some of these unknown compounds could be toxic or have significant effects in the atmosphere.

In addition to identifying and quantifying compounds in smoldering biomass emissions, research to investigate mechanisms and kinetics of formation of these compounds is needed. For example, to what extent do formic acid emissions result from pyrolysis of polysaccharides in biomass and to what extent do formic acid emissions result from oxidation of formaldehyde in smoke plumes? What compounds result from secondary reactions on chars and what compounds result from secondary reactions in smoke plumes? It is becoming increasing clear that smoldering biomass emissions depend on the chemistry of the fuel being burned (which can vary considerably), burning conditions, and the extent to which various processes occur during a particular smoldering fire. Ecosystems prone to biomass burning and the individual plants in them need to be chemically characterized, as do other biomass fuels. Procedures are needed that can better investigate burning conditions and compound formation processes for biomass fires, either in the laboratory or in the field. These factors must all be considered in models that seek to predict global contributions of biomass burning to fluxes of individual compounds and human exposure to individual compounds.

Most large-scale field studies have occurred in tropical Africa and South America (Brazil). It cannot be assumed that fires in other regions will have the same emission profiles. Therefore, many other regions also need to be studied, for example Asia, western South America, and the boreal forests. As climate changes occur, biomass regional burning patterns may change. Conditions in the boreal and temperate forests of the Northern Hemisphere may become hotter and drier, causing an increase in fires in this region. Population growth could result in an increase in the extent of fuel consumed in domestic cooking and heating fires and therefore an increase in human exposure to smoldering biomass emissions. If wood becomes scarce, fuels such as dung may become more prevalent. Emissions from charcoal making, an oxygen-starved process that involves high temperatures, have yet to be characterized.

These are just a few areas in which more research is needed to better understand the impact of the smoldering combustion emissions on regional and global atmospheric chemistry and on human health. Biomass fires are incredibly complex and diverse systems and there is still much to understand about them.

BIBLIOGRAPHY

1. D. E. Ward, "Air Toxics and Fireline Exposure," *10th Conference on Fire and Forest Meteorology,* Ottawa, Canada, 1989.
2. M. O. Andreae, in R. S. Oremland, ed., *Biogeochemistry of Global Change: Radiatively Active Trace Gases* Chapman & Hall, New York, 1993, pp. 113–150.
3. E. S. Kasischke, N. H. F. French, L. L. Bourgeau-Chavez, and N. L. J. Christensen, *J. Geophys. Res.* **100,** 2941–2951 (1995).
4. B. J. Stocks, D. R. Cahoon, W. R. I. Cofer, and J. S. Levine, *Monitoring Large-Scale Forest Fire Behavior in Northeastern Siberia Using NOAA-AVHRR Satellite Imagery,* Biomass Burning and Global Change, Williamsburg, VA., 1995.
5. M. D. Flannigan and C. E. Van Wagner, *Can. J. For. Res.* **21,** 61–72 (1991).
6. S. C. Wofsy and co-workers, *J. Geophys. Res.* **97,** 16731–16746 1992.
7. D. E. Ward, R. A. Susott, A. P. Waggoner, P. V. Hobbs, and J. D. Nance, "Emission Factor Measurements for Two Fires in British Columbia Compared with Results for Oregon and Washington," *29th Annual Meeting of the Pacific Northwest International Section of the Air and Waste Management Association,* Bellevue, Wash. 1992.
8. W. M. Hao and D. E. Ward, *J. Geophys. Res.* **98,** 20657–20661 (1993).
9. D. E. Ward and co-workers, *J. Geophys. Res.* **97,** 14601–14691 (1992).
10. R. J. Yokelson, D. W. T. Griffith, and D. E. Ward, *J. Geophys. Res.* (submitted, 1996).
11. L. F. Radke and co-workers, in J. S. Levine, ed. *Global Biomass Burning: Atmospheric, Climatic, and Biospheric,* MIT Press, Cambridge, Mass. 1991, pp. 209–224.
12. T. J. Ohlemiller and F. E. Rogers, *Combust. Sci. Technol.* **24,** 139–152 (1981).
13. W. P. Pan and G. Richards, *J. Anal. Appl. Pyrol* **16,** 117–126 (1989).
14. W.P. Pan and G.N. Richards, *J. Anal. Appl. Pyrol* **17,** 261–273 (1990).
15. L. M. McKenzie, W. M. Hao, G. N. Richards, and D. E. Ward, *Environ. Sci. Technol.* **29,** 2047–2054 (1995).
16. L. M. McKenzie, W. M. Hao, G. N. Richards, and D. E. Ward, *Atmos. Environ.* **28,** 3285–3292 (1994).
17. J. M. Lobert and co-workers, in Ref. 11, pp. 290–304.
18. R. W. Talbot, M. O. Andreae, H. Berresheim, D. J. Jacob, and K. M. Beecher, *J. Geophys. Res.* **95,** 16799–16811 (1990).
19. B. L. and co-workers, *J. Geophys. Res.* **99** (1994).
20. G. Helas, H. Bingemer, and M. O. Andreae, *J. Geophys. Res.* **97,** 6187–6193 (1992).
21. W. R. Hartmann, Ph.D. dissertation, University of Mainz, Germany, 1990.
22. T. E. Reinhardt, "Smoke Exposure at Prescribed Burns," *Draft Final Report to the USDA Forest Service,* Pacific Northwest Research Station, University of Washington, Dept. of Environmental Health, Seattle, 1994.
23. A. Hauk, M. Sklorz, G. Bergmann, and O. Hutzinger, *J. Anal. Appl. Pyrol* **28,** 1–12 (1994).
24. L. M. McKenzie, Ph.D. dissertation, University of Montana, Missoula, 1996.
25. G. Barrefors and G. Petersson, *J. Chromatogr.* **710,** 71–77 (1995).
26. G. Barrefors and G. Petersson, *Chemosphere* **30,** 1551–1556 (1995).
27. T. Ramdahl, I. Alfheim, S. Rustad, and T. Olsen, *Chemosphere* **11,** 601–611 (1982).
28. Y. L. Tan, J. F. Quanci, R. D. Borys, and M. J. Quanci, *Atmos. Environ.* **26A,** 1177–1181 (1992).
29. S. B. Hawthorne, D. J. Miller, R. M. Barkley, and M. S. Krieger, *Environ. Sci. Technol.* **22,** 1191–1196 (1988).
30. S. B. Hawthorne, M. S. Krieger, D. J. Miller, and M. B. Mathiason, *Environ. Sci. Technol.* **23,** 470–475 (1989).
31. S. B. Hawthorne, D. J. Miller, J. J. Langenfeld, M. S. Krieger, *Environ. Sci. Technol.* **26,** 2251–2262 (1992).
32. P. Masclet, H. Cachier, C. Liousse, and H. Wortham, *J. Atmos. Chem.* **22,** 41–54 (1995).
33. D. J. Freeman and F. C. R. Cattell, *Environ. Sci. Technol.* **24,** 1581–1585 (1990).
34. N. R. Khalili, P. A. Scheff, and T. M. Holsen, *Atmos. Environ.* **29,** 533–542 (1995).
35. L. A. Edye and G. N. Richards, *Environ. Sci. Technol.* **25,** 1133–1137 (1991).
36. T. E. Reinhardt and D. E. Ward, *Environ. Sci. Technol.* **29,** 825–832 (1995).
37. B. M. Jenkins and co-workers, *Environ. Sci. Technol.* **27,** 1763–1776 (1993).
38. J. B. Kandpal, R. C. Maheshwari, and T. C. kandpal, *Energy Conserv. Manag.* **36,** 1073–1079 (1995).
39. W. M. Hao, D. E. Ward, G. Oblu, and S. P. Baker, *J. Geophys. Res.* **101,** 23577–23584 (1996).
40. D. W. T. Griffith, W. G. Mankin, M. T. Coffey, D. E. Ward, and A. Riebau, in Ref. 11, pp. 230–230–239.
41. R. Delmas and co-workers, *J. Atmos. Chem.* **22,** 175–193 (1995).
42. R. W. Talbot, K. M. Beecher, R. C. Harriss, and W. R. Cofer III, *J. Geophys. Res.* **93,** 1638–1652 (1988).
43. B. R. T. Simoneit and co-workers, *Environ. Sci. Technol.* **27,** 2533–2541 (1993).
44. D. R. Blake, T. R. J. Smith, T.-Y. Chen, W. J. Whipple, and F. S. Rowland, *J. Geophys. Res.* **99,** 1699–1719 (1994).
45. B. Bonsang, C. Boissard, M. F. Le Cloarec, J. Rudolph, and J. P. Lacaux, **22,** 149–162 (1995).
46. W. R. I. Cofer and co-workers, *Atmos. Environ.* **27A,** 1903–1907 (1993).

47. Y. J. Kaufman and co-workers, *J. Geophys. Res.* **97**, 14581–14599 (1992).
48. K. K. Laursen, P. V. Hobbs, L. F. Radke, and R. A. Rasmussen, *J. Geophys. Res.* **97**, 20687–20701 (1992).
49. W. R. I. Cofer and co-workers, *J. Geophys. Res.* **95**, 1865–1870 (1990).
50. K. R. Smith and co-workers, *Chemosphere* **26**, 479–505 (1993).
51. D. F. Hurst, D. W. T. Griffith, and G. D. Cook, *J. Geophys. Res.* **99**, 16441–16456 (1994).
52. F. Lipari, J. M. Dasch, and W. F. Scruggs, *Environ. Sci. Technol.* **18**, 326–330 (1984).
53. J. Rudolph, A. Khedim, R. Koppmann, and B. Bonsang, *J. Atmos. Chem.* **22**, 67–80 (1995).
54. M. O. Andreae and co-workers, *J. Geophys. Res.* **101**, 23603–23613 (1996).
55. R. A. Rasmussen, L. E. Rasmussen, M. A. Khalil, and R. W. Dalluge, *J. Geophys. Res.* **85**, 7350–7356 (1980).
56. S. Tassios, D. R. Packham, *J. Air Pollution Control Assoc.* **35**, 41–42 (1985).
57. S. Manö, *Science* **263**, 1255–1256 (1994).
58. T. P. Gaines, M. B. Parker, and G. J. Gascho, *Agron. J.* **76**, 371–374 (1984).
59. J. W. Green, in R. L. Whistler, ed., *Methods in Carbohydrate Chemistry,* Vol. III, Academic Press, Inc., New York, 1963, pp. 9–20.
60. Y. Z. Lai and K. V. Sarkanen, in K. V. Sarkanen and C. H. Ludwig, eds., *Lignins Occurrence, Formation, Structure, and Reactions,* Wiley-Interscience, New York, 1971, pp. 165–230.
61. L. F. Radke and co-workers, in P. V. Hobbs, and M. P. McCormick, eds., *Aerosols and Climate,* Hampton, Va., 1988, pp. 411–422.
62. R. Delmas, J. P. Lacaux, and D. Brocard, *Environ. Mon. Assess.* **38**, 181–204 (1995).
63. D. F. Hurst, D. W. T. Griffith, J. N. Carras, D. J. Williams, and P. J. Fraser, *J. Atmos. Chem.* **18** (1994).
64. R. J. Evans and T. A. Milne, *Energy and Fuels* **1**, 123–137 (1987).
65. G. D. McGinnis, Ph.D. dissertation, University of Montana, Missoula, 1970.
66. O. Faix, D. Meier, and I. Fortmann, *Holz als Roh- und Werkstoff* **48**, 281–285 (1990).
67. A. D. Pouwels, A. Tom, G. B. Eijkel, and J. J. Boon, *J. Anal. Appl. Pyrol* **11**, 417–436 (1987).
68. J. M. Lobert, D. H. Scharffe, W. M. Hao and P. J. Crutzen, *Nature* **346**, 552–554 (1990).
69. B. C. Nguyen, N. Michaloplulos, J. P. Putaud, and B. Bonsang, *J. Atmos. Chem.* **22**, 55–56 (1995).
70. P. J. Crutzen, L. E. Heidt, J. P. Krasnec, W. H. Pollock, and W. Seiler, *Nature* **282**, 253–256 (1979).
71. P. J. Crutzen and M. O. Andreae, *Science* **250**, 1669–1678 (1990).
72. F. Shafizadeh and W. F. DeGroot, *Fuels and Energy From Renewable Resources,* Acamdemic Press, Inc., New York, 1977, pp. 93–114.
73. J. Ludwig, M. O. Andreae, G. Helas, L. Marufu, and F.X. Meixner, "Emission of CO_2, NO and CO form Domestic Biomass Burning in Africa," *European Geophysical Society XXI General Assembly,* The Hague, The Netherlands, 1996.
74. L. M. McKenzie, D. E. Ward, and W. M. Hao, in J. S. Levine, ed., *Biomass Burning and Global Change,* MIT Press, Cambridge, Mass., 1997, pp. 241–248.
75. C. W. Philpot, *Forest Sci.* **16**, 461–471 (1970).
76. F. Y. Hshieh and G. N. Richards, *Combust. Flame* **76**, 49–56 (1989).
77. F. Y. Hshieh and G. N. Richards, *Combust. Flame* **76**, 37–47 (1989).
78. M. A. K. Khalil, S. A. Edgerton, and R. A. Rasmussen, *Environ. Sci. Technol.* **17**, 555–559 (1983).
79. S. A. Edgerton, M. A. K. Khalil, and R. A. Rasmussen, *Environ. Sci. Technol.* **20**, 803–807 (1986).
80. D. E. Ward, J. Peterson and W. M. Hao, "An Inventory of Particulate Matter and Air Toxic Emissions from Prescribed Fires in the USA for 1989," *1993 Annual Meeting and Exhibition of Air and Waste Management Association,* Denver, Colo., 1993.
81. D. E. Ward and W. M. Hao, "Air Toxic Emissions from Burning of Biomass Globally—Preliminary Estimates," *85th Annual Meeting and Exhibition of the Air and Waste Management Association,* Kansas City, Mo., 1992.
82. J. E. Dibb and co-workers *Atmos. Environ.* **30**, 553–561 (1996).
83. R. C. James, in P. L. Williams and J. L. Burson, eds., *Industrial Toxicology,* Van Nostrand Reinhold, New York, 1985, pp. 230–259.
84. G. Helas and co-workers, *J. Atmos. Chem.* **22**, 163–174 (1995).
85. M. G. Duffell, in Ref. 83, pp. 162–193.
86. C. R. Turner and J. J. Pienaar, "Biomass Combustion and Rain Chemistry in the South African Highveld Region," *European Geophysical Society XXI General Assembly,* The Hague, The Netherlands, 1996.
87. R. J. Cicerone, *Science* **263**, 1243–1244 (1994).

SOIL CHARACTERIZATION. See PARTITIONING TRACERS FOR SOIL CHARACTERIZATION.

SOIL AND GROUNDWATER REMEDIATION. See GROUNDWATER AND SOIL REMEDIATION.

SOIL AND WATER QUALITY IN AGRICULTURE, ASSESSING AND IMPROVING

DARRELL J. BOSCH
MARY LEIGH WOLFE
Virginia Polytechnic Institute and State University
Blacksburg, Virginia

Agriculture is one of the world's major economic activities. Agriculture's importance is increasing because of the need to produce more food for a growing population, expand incomes for rural populations, and generate foreign exchange through trade in agricultural products. Agricultural output has increased dramatically through better crop varieties and livestock breeds, improved plant fertilizers, pesticides to protect crops from weeds, insects, and diseases, and increased use of irrigation. Beef, poultry, pork, and dairy enterprises are increasingly produced in confinement systems where production conditions are carefully controlled to enhance livestock growth and output. Although increased output has raised producers' incomes and kept prices low for consumers, it has resulted in increased pollution. Pollution damages to soil and water quality are a concern because of the large scope of agriculture and because society requires continued increases in agricultural output to meet growing world food needs. This chapter summarizes the major pollutants associated with agriculture, their sources, and damages to soil and water quality attributed to these pollutants. Agricultural methods to reduce soil and water quality damage from pollutants are described. Reasons why farmers may be reluctant to adopt methods to control pollution are discussed. Policies to induce farmers to adopt pollution control practices are also compared.

PRIMARY AGRICULTURAL POLLUTANTS CAUSING SOIL AND WATER QUALITY DAMAGE

The major pollutants associated with agriculture include sediment, nutrients (especially nitrogen, N, and phosphorus, P), pesticides, salts, and pathogens.

Sediment

Sediment refers to the soil particles that are detached and transported by soil erosion, then deposited elsewhere on the land or in surface water. Erosion is a natural process caused by water (raindrop impact and flowing water) and wind. Significant quantities of sediment are moved without any human influence. The type of erosion of concern to agriculture is accelerated erosion caused by land-disturbing activities. Table 1 indicates relative erosion rates per unit area from various land uses. On nonfederal land in the U.S., including cropland, pastureland, and rangeland, about 4.1 billion metric tons (4.5 billion tons (Mg)) of soil are moved annually by soil erosion (1), about 40% by water, and 60% by wind. Approximately 66% of the water erosion and about 34% of the wind erosion occur on cropland.

Soil erosion by water consists of three processes: detachment, transport, and deposition. Detachment of soil particles occurs when the shear forces exerted by flowing water or raindrop impact exceed the soil's ability to withstand those forces. Detached particles move when the transport capacity of the water exceeds the sediment load. Deposition of sediment occurs when the transport capacity is less than the sediment load. Transport capacity decreases because of factors, such as reduced velocity when water flows into vegetation and reduced volume of flow due to infiltration.

Erosion and the resulting sediment are detrimental to both the land resource that is eroded and the water body to which the sediment is delivered. Erosion depletes the land resource by removing top soil, leaving a shallower, less fertile soil. Detrimental effects of sediment in receiving waters include deterioration or destruction of aquatic habitat, deterioration of aesthetic value, loss of storage capacity in reservoirs, and accumulation of bottom deposits that inhibit normal biological life. Sediment destroys spawning areas and food sources and directly harm fish and other aquatic wildlife. In addition, pollutants, such as nutrients and pesticides, are often transported with sediment. The detrimental effects of those pollutants are described in the following sections. Annual in-stream damage by sediments has been estimated at between $3.2 and $13 billion (3).

Nutrients

Nutrients, such as N and P, and micronutrients, are needed by all living matter. Typically, in aquatic ecosystems, the quantities of micronutrients are abundant. However, N, P, or sometimes carbon are in short supply. The nutrient in the shortest supply relative to the needs of plants and organisms is called the limiting nutrient because its quantity affects the rate of production (growth) of organic matter. Most surface water bodies are either P (most of the inland lakes) or N (some estuaries and rivers) limited (4). When excessive amounts of nutrients are added to a water body, however, negative impacts result and those nutrients become pollutants. The nutrients of most concern with respect to water quality are N and P.

The major impact of excessive nutrients on water quality in surface waters is accelerated eutrophication. Eutrophication is a natural process by which a body of water progresses from its origin, eg, a glacial lake, to its extinction, dry land. Eutrophication is caused by a net increase of organic matter (net productivity), ie, organic matter production (primary productivity), nourished by nutrients and minerals, exceeds organic matter loss by respiration, decay, grazing by higher organisms, and outflow. The rate of eutrophication is variable, depending on a number of factors. During investigations of eutrophication problems in Madison, Wisconsin lakes, it was noted that algal blooms occur when concentrations of inorganic P and inorganic N exceed 0.01 mg P/L and 0.3 mg N/L, respectively (5).

Some of the related physical and chemical changes caused by advanced eutrophication include pH variations, oxygen fluctuations or lack of oxygen, waste excretions by some algae, and increased turbidity (4). These changes interfere with recreational and aesthetic uses of the body of water and also cause a shift in fish and shellfish populations from better quality fish to rough fish. In addition, taste and odor problems caused by algae make water less suitable for potable use and contact recreation. When the concentration of phytoplankton algae exceeds a certain threshold nuisance value during the late summer period, the situation is termed algal bloom.

Nitrogen. Various forms of N exist in soils, plants, animals, and the atmosphere. The two major pools of soil N are organic and inorganic. The largest quantity of N in soil occurs as organic N. Most organic N is not directly available to plants, whereas inorganic forms, particularly ammonium (NH_4^+) and nitrate (NO_3^-), are available. N is transformed between organic and inorganic forms through processes that constitute the N cycle (Fig. 1).

The N cycle in soil is divided into immobilization and mineralization. Immobilization of N in soils results from physical-chemical attractions, chemical precipitation from insoluble N salts, biochemical reactions, and N uptake by soil organisms and plants (4). Immobilized N is temporarily held and is not free to follow additional N cycle pathways. Mineralization is the conversion of organically bound N to inorganic mineral forms (NH_4^+ and NO_3^-). Microbial decomposition of organic matter results in producing NH_4^+ (ammonification), which, with good aeration and suitable temperatures, is oxidized to nitrite (NO_2^-) and then to NO_3^- (nitrification). Unless the nitrate-forming bacteria is inhibited, practically no nitrite accumulation occurs. Products of mineralization are susceptible to further transformations including volatilization, denitrification, and leaching, resulting in losses or outputs of inorganic N from the soil. Nitrate is reduced to gaseous N compounds and lost to the atmosphere when poor aeration and suitable carbon sources are available for denitrification. Ammonium ions are held to the cation exchange sites in soils (ammonium exchange) and are available for plant use. Nitrate ions are soluble in soil water.

Table 1. Representative Rates of Erosion from Various Land Uses[a]

Land Use	Metric Tons per km^2 per Year	Tons per mi^2 per Year	Relative to Forest = 1
Forest	6.3	18	1
Grassland	63	180	10
Abandoned surface mines	630	1800	100
Cropland	1260	3600	200
Harvested forest	3150	9000	500
Active surface mines	12600	36000	2000
Construction	12600	36000	2000

[a] Adapted from Refs. 1 and 2.

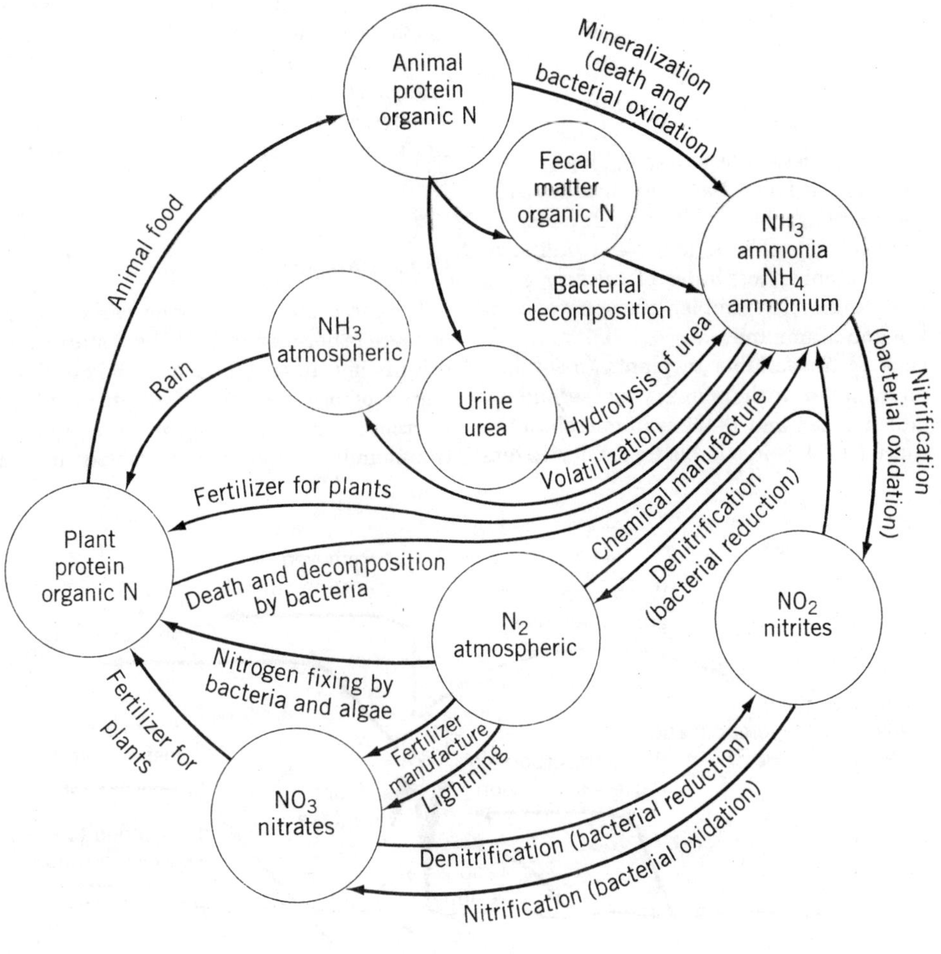

Figure 1. The nitrogen cycle (8).

Mineralization occurs most commonly when plant and animal residues are added to soil. Although only 2–3% of immobilized N is mineralized annually, the release of N to inorganic forms has long supplied a significant portion of crop needs and is about 60 kg N/ha per year for a representative surface soil in humid temperate regions (6). In arid soils, mineralized quantities are less, except in irrigated areas that receive large inputs of plant residues.

NO_3^- is the form of N of greatest concern for water quality in the soil because it leaches and moves with the soil water. Losses of organic N are related to surface runoff. NO_3^- in surface runoff and water leaching through the soil pollute drinking water sources, both surface and ground water, and stimulate unwanted plant and algal growth in lakes and reservoirs. Health hazards associated with NO_3^- result mainly from the bacterial conversion of ingested NH_4^+ to NO_2^- and then to NO_3^-. NO_3^- induces methemoglobinemia in infants (blue baby syndrome), a reduction in the oxygen-carrying capacity of the blood. Nitrogen inputs to the soil are in the form of aerial deposition, biological N_2 fixation, animal and green manures, synthetic nitrogen fertilizers, and municipal sludges. The N in biological residues is primarily in the organic and ammonium form. Fertilizer N is usually in the ammonia, ammonium, nitrate, or urea forms.

Phosphorus. Phosphorus (P) is also essential for plant growth. Some of the significant plant functions affected by P are photosynthesis, biological N_2 fixation, crop maturation, and root development (6). The P cycle (Fig. 2) depicts the interactions of various forms of P with microbes, plants, and animals.

Phosphorus in soil and surface water is classified as sediment-bound or soluble and as organic or inorganic. Soluble P includes orthophosphates, inorganic polyphosphates, and some organic forms. The organic P fraction in soils varies from 20 to 80% depending on microbial activity, clay mineralogy, pH, and the addition of soil amendments (6). The inorganic forms are mainly iron and aluminum phosphates in acid soils and calcium phosphates in alkaline soils. All inorganic forms of P in soils are extremely insoluble, and concentrations of P in soil water solutions are low, generally less than 0.2 mg/L.

Sediment-bound P is classified into the following groups (6): (*1*) adsorbed: labile and exchangeable P; (*2*) organic: phospholipids, nucleic acids, inositol phosphates, and other unknown forms; (*3*) precipitates: insoluble phosphates of calcium, iron, and aluminum; and (*4*) minerals: simpler phosphate compounds of calcium, iron, and aluminum.

The bulk of native P is unavailable to plants (insoluble) because 98–99% is associated with primary or secondary minerals and soil organic matter, and 1–2% is in microbial tissue. Because only about 0.01% P is in soluble forms, fertilizers and manures containing soluble P are applied to supplement available soil P for plant uptake (6). Much of the available P thus supplied to soil is chemically converted to forms unavailable to plants. As a result, crops take up only 10 to 25% of the fertilizer-applied P in the same year. Though crops utilize a portion of the accumulated P in subsequent years, the buildup of P is a source of environmental concern due to the transport of P in sediment-bound and dissolved forms by surface runoff to receiving waters. The sediment-bound P delivered to surface waters is a long-term concern due to the interchange of P from bottom deposits with the water column. Forty-five percent of all P loadings to the Chesapeake Bay are attributed to fertilizers. Total P concentrations up to 300 mg/L have been measured in feedlot runoff in the Great Plains whereas total P losses from well-drained coastal plains in Virginia are in the range of 9–40 mg/L (4). It was reported that Lake Okeechobee in south central Florida was in danger of becoming hypereutrophic because of P enrichment from agricultural activity (7).

The other less-documented aspect of P buildup involves the transport of P into ground water. There is evidence of dissolved P movement into ground water through sandy and peat soils in Florida and the Netherlands. Total P losses from subsurface tile drainage range from 3 to 10 kg/ha per year (4).

Organic Matter

All organic matter contains carbon in combination with one or more other elements. All substances of animal or vegetable origin contain carbon compounds and, therefore, are organic. Decay of organic matter is effected by naturally occurring microorganisms that convert the organic matter into simpler compounds. When organic matter is added to water, micro-

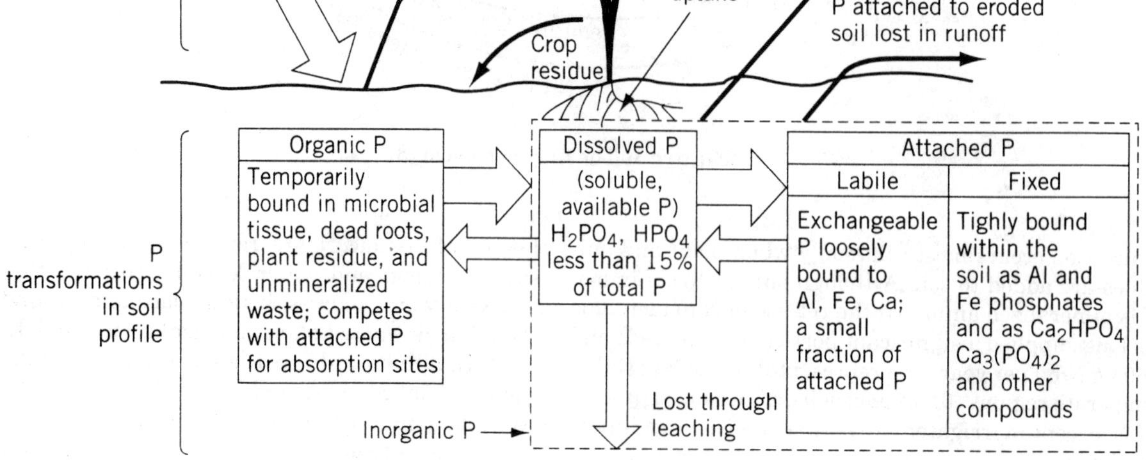

Figure 2. Abbreviated phosphorus cycle (8).

organisms, principally bacteria, begin to consume and convert the organic material. As the organisms consume the organic matter, they also consume free oxygen. The level of oxygen depletion depends primarily on the amount of organic matter added; the size, velocity, and turbulence of the stream; the initial dissolved oxygen levels in the organic matter and in the stream; and the temperature of the water (8).

Adding organic matter to a stream lowers oxygen levels to such an extent that fish and other aquatic life are forced to migrate from the polluted area or die because of lack of oxygen. Biochemical oxygen demand (BOD) is the amount of oxygen per unit volume of water utilized by microorganisms to biochemically oxidize the organic matter during a period of time at a specified temperature (9). BOD is a qualitative test rather than an absolute measure of oxygen. Five-day BOD, BOD_5, is commonly reported.

Pesticides

Pesticides are inorganic or organic chemicals used to control weeds, insects, and diseases. The use of modern synthetic organic chemicals is credited with substantially increasing yields of agricultural crops and assisting in controlling pests and diseases, such as malaria. Pesticides include herbicides used for weed control, insecticides for control of insects, and fungicides to control fungus, particularly in orchards and vegetable farms.

Residual pesticides in soil or on crops is transported from fields to surface and ground waters. Pesticides negatively affect water quality (10). By definition, pesticides are poisons. If they are released to the aquatic environment before degradation, they kill nontarget and other aquatic species. Pesticides hinder photosynthesis in aquatic plants affect reproduction, respiration, growth, and development of aquatic species, reduce their food supply, and destroy their habitat. Pesticides lower aquatic organisms' resistance and increase susceptibility to other environmental stresses. Some pesticides bioaccumulate in tissues of fish and other species, thus posing a health risk to humans and other species that consume the fish. There are also human health risks in consuming contaminated water. Some pesticides are carcinogenic and mutagenic and/or teratogenic. Pesticides also reduce commercial/sport fishing and other recreational values of water bodies.

The impact of a pesticide on water quality is related to its toxicity and persistence, whereas its transportability is a function of its solubility and ability to adsorb to soil particles. Toxicity is defined by the LD_{50}, the dose that kills 50% of the test animals to which it is administered under experimental conditions, expressed as milligrams (mg) of toxicant per kilogram (kg) of body weight (11). The LD_{50} is measured in terms of oral (fed to, or placed directly in the stomach), dermal (applied to the skin), and respiratory toxicity (inhaled). The amount of pesticide required to kill a human being is correlated with the LD_{50} of the material to animals in the laboratory.

The persistence of a pesticide is expressed by its half-life. Degradation of pesticides begins at formulation and continues for hours, days, weeks, months, years, or decades, depending on the chemical and the conditions. Degradation processes are classified as physical, chemical, and biological. Physical degradation processes are caused primarily by light and heat. Chemical degradation is promoted through hydrolysis, oxidation, and reduction (12). Biological degradation is facilitated by microorganisms including bacteria, fungi, and actinomycetes. The relative importance of each degradation process depends to a great extent on the specific pesticide, including its use pattern, physical properties, and chemical structure (13) and on environmental conditions, such as soil moisture, oxygen content, temperature, organic matter content, pH, and light intensity.

For most pesticides, degradation ultimately results in universally present materials such as carbon dioxide, ammonia, water, mineral salts, and humic substances (14). However, different chemicals are formed before pesticides are completely degraded. Some of the degradation products are more persistent, more soluble, and contaminate water and food resources. In some situations, pesticides (eg, aldoxycarb, promecarb) form as degradation products of other pesticides (14). Degradation products of a number of commonly used pesticides, including atrazine, cyanazine, and carbofuran, are detected in groundwater. The fate and significance of the degradation products of many currently used pesticides are not clearly understood.

Pesticides are transported from agricultural fields to surface and ground waters by atmospheric transport, surface runoff, erosion, and leaching. Pesticides also reach surface and ground waters through accidental spills or dumping of excess chemicals, particularly at mixing sites.

Atmospheric transport includes drift during application and volatilization both during application and from soil and vegetative surfaces after application. The amount of drift depends on the method of application and the formulation of the pesticide. Small droplet or dust applications increase losses, whereas granular applications decrease losses. Climatic conditions at the time of application are also important, especially wind conditions.

Volatilization depends on the vapor pressure, water solubility, adsorption coefficient of the pesticide, and on the moisture content and temperature of the soil (15). Soil incorporation, cover crop, high clay and organic matter content, and deep plant penetration reduce volatilization. Volatilization increases pesticide loading to the atmosphere and the possibility of eventual redeposition to streams and lakes via precipitation.

Transport of pesticides by runoff and leaching are determined by the adsorption and solubility properties of the pesticides, the characteristics of the site, and the timing, mode, placement, and rate of chemical application. The most common mode of transport for soluble and weakly adsorbed pesticides is leaching through the soil profile. However, intense rainfall, steep topography, or saturated soil conditions cause such pesticides to be lost via runoff. Moderately adsorbed pesticides are primarily transported in surface flow but are also lost with eroded sediment. Strongly adsorbed pesticides are relatively immobile in the soil profile, generally remaining close to the surface where they were applied. Because the soil surface is the area of greatest biological activity and is where pesticide degradation rates are greater, adsorption of a pesticide to soil particles allows for degradation and potential reduction of the pesticide's toxicity. Strongly adsorbed pesticides are transported primarily on eroded sediment.

Each pesticide has unique characteristics. Although information about the toxicity, persistence in soil, and the predominant transport mode is available for many herbicides, insecticides, and fungicides (eg, 16–18), many details are still unknown (see PESTICIDES, ENVIRONMENTAL FATE).

Salts

Salts have negative impacts on both soil and water quality. Sodic and saline soils contain salts sufficient to impair crop growth. Sodic soils contain sodium (exchangeable-sodium percentage ≥15), whereas saline soils contain soluble salts. The major solutes of saline soils are sodium, calcium, magnesium, sulphates, and chlorides. Minor amounts of bicarbonates and potassium are also present. Sodic soils are poorly aggregated, crusted, and hard when dry, and sticky and plastic when wet (19). They are often dark brown or black due to surface coatings of dispersed organic matter. Sodic soils reduce infiltration of water into the soil, thus negatively affecting crop growth. Saline soils impair uptake of water by plants. Saline soils occur naturally, but many of the soil salinity problems in the western U.S. have occurred because of poor quality (high salt content) irrigation water. Many western U.S. irrigation waters contain from 0.7 to 37 metric tons of salt per ha-m (0.1 to 5 tons of salt per ac-ft) of water (20). In the southwestern desert areas, as much as 50 tons of salt is applied to one acre during a 12-month growing season (112 metric tons of salt per ha). In many instances, insufficient drainage and lack of adequate leaching has left excess salts in the soil profile.

When salts are drained or leached from saline soils, it is likely that those salts are delivered to a water body. Salts affect the types and population of fish and aquatic wildlife that are present, favoring salt-tolerant species (10). Fluctuations in salinity cause greater problems than the absolute level of salinity. Salts destroy habitat and food source plants for fish species. Salinity causes skin and eye irritation, thus reducing the recreational value of water bodies. The water body also becomes less desirable as a drinking water supply.

Saline seeps affect crop production under dry land conditions. Saline seeps result from a combination of geologic, climatic, hydrologic, and cultural (land-use) conditions. The primary cause is a change from grassland or forest to a cropping system, such as crop-summer fallow rotation, which utilizes less moisture from the soil profile than the previous land use. Excess water percolates down to the water table and then moves laterally to lower soil surfaces, dissolving salts in the material through which it flows. Subsurface conditions eventually force the flow to the surface, where it evaporates and results in a saline seep. A seep reduces or eliminates crop growth in the discharge area because of increased concentrations of soluble salt in the root zone. Saline seeps have been identified in the Great Plains in the U.S. and in Australia, India, Iran, Turkey, and Latin America (21).

Pathogens

The primary source of pathogens in agriculture is livestock production. The excreta from warm-blooded animals contain countless microorganisms, including bacteria, viruses, parasites, and fungi. Some of the organisms are pathogenic (disease causing), and many of the diseases carried by animals are transmittable to humans. Pathogens are introduced to surface water through runoff originating on feedlots or flowing over grazed pastures or fields on which manure is applied. Pathogens are also discharged from livestock and poultry processing plants. Pathogens are leached to groundwater from feedlots, grazed pastures, and fields on which manure is applied. Water contaminated by pathogens is a human health hazard by both consumption and contact. Therefore, the recreational value of the water body and the drinking water quality are impaired. The presence of pathogens in water increases the treatment costs for drinking water.

SOURCES OF AGRICULTURAL POLLUTION

Pollutant loads from agricultural production operations depend on many factors, including land use, type of crops or animals, crop rotation, soils on which crops are grown, topography, climatic conditions, farming technology, irrigation and drainage, timing and location of chemical or fertilizer application, and cultivation practices. Potential environmental quality problems due to agricultural operations more often depend on the production practices and waste management techniques utilized by farmers than the size of the operation, the number of animals, or the amount of waste.

There are several ways to classify sources of agricultural pollution. One approach is to identify the primary pollutants associated with different agricultural land uses. Table 2 contains a list of the most prevalent pollutants transported by runoff and subsurface water from various agricultural land uses. Pesticides result primarily from cropland and orchards or nurseries. Salts are delivered from irrigated areas, including both fields and orchards or nurseries. The primary sources of pathogens are those operations that involve animals.

In addition to pollutants transported by runoff and subsurface water, agricultural land-use activities in or adjacent to streams have negative impacts on water quality. Areas located immediately adjacent to streams, including the stream banks, are called riparian areas. Livestock permitted uncontrolled access to streams and riparian areas shear and slough stream bank soils and eliminate stream bank vegetation. Crop production systems that plow fields to the edge of the stream bank destabilize the banks causing them to collapse or causing gully headcuts to form. The material from degraded stream banks and active gullies smothers remaining riparian vegetation and deposits in the adjacent stream channel. Composition and richness of aquatic organism species change as a result of loss of cover, food source, and changes in flow, temperature, and dissolved oxygen (DO).

Draining low-lying areas (generally wetlands) for crop production has profoundly affected the physical state of the landscape (4). Loss of riparian vegetation through clearing, loss of a natural high water table from drainage, and loss of the flood buffering capacity of low lying areas result in a stream system that is more flood-prone when wet and more drought-prone when dry. Loss of consistent stream base flow is one of the most detrimental changes affecting the reproductive habits of most aquatic organisms.

A second approach to classifying sources of agricultural pollution is on the basis of the type of pollution generated by the source. The two types of pollution, point source (PS) and nonpoint source (NPS) differ in method and frequency of discharge to water. PSs discharge directly to surface waters from a definable pipe or ditch and to ground water at a specific recharge or well site. In contrast, NPSs include land surfaces, such as cropped fields and parking lots. Discharges from PSs are generally the result of relatively continuous human activities, such as industrial processes, wastewater treatment, and food processing. In contrast, the frequency of NPS pollutant discharge depends on precipitation and climatic conditions that lead to

Table 2. Pollutants from Various Agricultural Land Uses[a]

	Pollutant[b]						
		Nutrients		Pesticides			
Land Use Category	Sediment	Adsorbed	Dissolved	Adsorbed	Dissolved	Salts	Pathogens
Dryland cropland	x	x	x	x	x		
Irrigated cropland	x	x	x	x	x	x	s
Pasture land	x	x	x	s	s		x
Range land	x	x	x	o	o		x
Confined animal feeding operations	x	x	x			x	x
Orchards or nurseries			x		x	x	x

[a] Adapted from Ref. 4.
[b] x = common; s = sometimes; o = occasionally.

runoff and water percolating through the soil profile. These characteristics of PS and NPS pollution influence the methods used to control each type of pollution. PS controls generally include biological, physical, and chemical treatment processes for polluted waters. NPS controls include management, vegetative, and structural measures on the land. PSs are often regulated by discharge permits, whereas NPS control is primarily voluntary.

Agricultural activities include both point and nonpoint sources of pollution, although the majority of agricultural sources are nonpoint. This article focuses primarily on sources of NPS pollution, with brief descriptions of point sources and their control.

Point Sources

Point sources of pollution from agriculture occur primarily in the food and fiber processing sectors and from pesticide and fertilizer handling. In addition, some animal production operations are classified as point sources. The wastes produced by agriculture vary in quantity and quality. Wastes from food processing are low strength, high volume liquids and solids, whereas those from livestock operations are high strength, low volume wastes.

Food Processing. Food processing wastes result from washing, trimming, blanching, pasteurizing, and juicing raw materials; the cleaning of processing equipment; and the cooling of the finished product. In most plants, cooling waters have low contamination and are reused to wash and transport the raw product. Waste characteristics have been compiled from various processing operations, including vegetables, fruits, and wine; meat-packing, slaughterhouses, packinghouses, and rendering; poultry processing plants; milk plants; and fish processing (eg, 22,23). Generally food processing wastes are low in N, have high BOD and suspended solids, and undergo rapid decomposition. Some wastes, such as those from beet processing, are highly colored. Fresh wastes have a pH close to neutral; during storage the pH decreases. In addition to the organic content, fruit and vegetable processing wastes contain other pollutants, such as soil, lye, heat, and insecticides.

Livestock Production. Concentrated animal feeding operations (beef, dairy, swine, sheep, and poultry) are classified as point sources of pollution by federal legislation and are subject to National Pollutant Discharge Elimination System (NPDES) permits in the U.S. (8). The definition of a concentrated animal feed operation is a function of the number of animal units and the frequency of pollutant discharge. Many operations do not meet the criteria and, therefore, do not require permits. State regulations that are more stringent supersede the federal regulations.

The quantity of runoff from feedlot areas depends on the permeability and infiltration characteristics of the lot, antecedent moisture conditions, the number of animals on the lot, the surface storage characteristics, and rainfall intensity. The surface of the feedlot has much lower infiltration rates and permeability than the undisturbed soil because animals compact the surface and add organic matter from animal manure. The elevated organic content of the soil surface crust in a feedlot protects against erosion and consequently sediment yields from feedlots are lower than from other disturbed soil surfaces.

Pollutants are also leached from feedlots to ground water. Nitrate and bacteria are the primary constituents of animal waste that affect ground water quality.

Nonpoint Sources

Cropland. On cropland, pollutants are transported to surface and ground water by surface runoff, interflow, and percolating water. Often the reduced pollution in one component causes an increase in another component. For example, decreasing surface runoff increases percolation, thus moving the pollutants to groundwater rather than surface water. Erosion and sediment transport by surface runoff and nitrate leaching into groundwater are considered predominant sources of pollution from cropland. Tillage is a primary cause of erosion on cropland due to soil surface disturbance. Conversely, tillage increases hydrologic surface storage and permeability that sometimes balance the increased erosion potential by decreasing runoff. The amount of runoff affects the amount of both dissolved and sediment-bound pollutants.

The majority of N and P lost in runoff from cropland is adsorbed to sediment. For example, in one study, 92% of the total N and 62–80% of the total P measured in runoff was adsorbed to sediment (24). In a second study, 82% of the N and 93% of the P transported by runoff was sediment-bound (25). In those studies, average annual dissolved N and P losses in runoff never exceeded 1% of the annual fertilizer application, but the ammonium concentration frequently exceeded the

0.5 mg/L lower limit considered necessary to stimulate algal growth in impounded water. One study focused on cotton with N fertilizer applied, but no P fertilizer was applied. Dissolved P losses in 77% of the runoff exceeded 0.1 mg/L (25).

Pesticides are detected in many surface and groundwaters. Atrazine, the pesticide most often detected, is an herbicide used to control broadleaf weeds and some species of annual grasses and is widely used in the corn belt. The EPA estimated that atrazine was the most heavily used pesticide in the U.S. in 1991, at the rate of 80 million pounds, primarily on corn (11). Atrazine has potential for significant loss in runoff because it is relatively highly soluble in water (33 mg/L) and is surface applied (26). Atrazine losses in runoff of less than 10% are most common (26), but some studies have shown higher losses. For example, one study reported losses as great as 60% in 11 watersheds during storms (27). In west Tennessee, during a 12-month period following pesticide application, approximately 1.5% of the applied atrazine was lost in runoff (26). Concentrations as high as 0.25 mg/L were detected. By the fourth storm after application, atrazine concentration was below detection limits. Atrazine was detected in the 10 to 20 cm soil depth after the first rainfall, but was not detected below 20 cm at any time during the study.

Atrazine concentrations measured in wells ranged from less than 0.01 to 88 mg/L (28). Peak concentrations were observed in shallow wells downgradient from irrigated fields at the end of the irrigation season. This study monitored two other pesticides, alachlor and dieldrin, and found extremely low concentrations, less than 0.1 mg/L, in all but one water sample (28). Atrazine was also investigated in irrigation wells in Nebraska and in irrigation tailwater recovery (reuse) pits (29). Seepage from the pits in late spring and early summer increased NO_3 and atrazine concentrations in ground water. Measured atrazine concentrations ranged from 1.02 to 23.1 mg/L in the pits and from less than 0.005 to 6.96 mg/L in the groundwater. During the rest of the year, NO_3 and atrazine levels were much lower in the pits.

Pasture and Rangeland. Overgrazing (grazing an area until vegetation damage occurs) and permitting livestock to approach and enter channels and streams are the major pollution activities on pastures and rangelands. It is reported that sediment yield from pastures in the state of Washington was minimized when vegetation cover remained greater than 50%, regardless of animal trampling disturbances (30). Another study investigated nutrient and sediment losses from unimproved pastures, that is, pastures not treated with fertilizer, weed control, or selected forage species, and found insignificant differences in nutrient losses from grazed compared with ungrazed areas, although sediment losses were greater in grazed than in ungrazed areas (31).

Livestock Production. Animal manure and waste from beef, dairy, swine, and poultry production facilities are sources of nonpoint pollutants. The terms manure and animal waste are often used synonymously. However, it is useful to consider manure as combinations of feces and urine only, whereas waste includes manure plus other material, such as bedding, soil, wasted feed, and water used for sanitary and flushing purposes. Litter is a specific form of poultry waste that results from "floor" production of birds after an initial layer of a bedding material, such as wood shavings, is placed on the floor at the beginning of and during the production cycle.

Livestock manure and waste produced in confined and semiconfined facilities are of primary concern because the waste is applied to fields for disposal and utilization. Depending on the rate and timing of application, waste constituents are transported from the field by runoff and leaching. The potential for delivery of pollutants to surface waters is greatest when rainfall occurs soon after manure application. The potential for leaching to ground water is a function of both the amount and timing of application relative to crop needs. The principal constituents of animal waste that affect surface water are organic matter, nutrients, and fecal bacteria, whereas nitrates and bacteria are the primary constituents that affect groundwater quality.

Manure from unconfined animals and poultry, such as those on range or pasture, are of less concern than that from confined operations because nutrient deposition in animal manure is less than nutrient removal by the grazed pasture, thereby minimizing the potential for nutrient loss to ground or surface water.

POLLUTION CONTROL

Approaches for controlling pollution are divided into prevention, mitigation, and treatment. Prevention is controlling pollution at the source, ie, eliminating (or reducing) generation of the pollutant. Mitigation focuses on reducing the amount of pollutant delivered to surface or groundwaters, often by inserting measures between the pollutant source and the receiving water. Treatment typically involves collecting and treating the polluted water to achieve some level of water quality that depends on the intended use of the water. For example, water for drinking must meet higher quality standards than water for some industrial processes.

The most appropriate control measures are a function of the particular pollution situation, including the pollutant(s), the source, and the intended end use of the water. For example, the approach used to control many PSs is to reduce the generation of the pollutant(s) (elimination is often impossible), then treat the polluted water before it is discharged into receiving waters. In contrast, NPS control tends to focus more often on prevention and mitigation.

Point Source Pollution Control

In general, the waste treatment and disposal methods used in other industries can be used for point source control in agriculture. To determine which methods are appropriate, the characteristics of the particular waste must be known. For example, biological treatment is appropriate for a liquid waste containing dissolved organic solids, whereas composting and incineration are alternatives for solid wastes with high organic content. Wastes from food processing plants are highly variable, both between plants and temporally within plants. Therefore, plant-specific information is required in determining appropriate pollution control measures. An additional consideration related to agricultural point sources is the frequency of discharge. Fruit and vegetable processing are seasonal, yielding intermittent discharges. This factor must be considered in designing treatment and disposal methods. In addition to treatment and disposal, point source control is facilitated by decreasing the amount of waste produced, reducing the strength of the waste produced, and reutilizing some of the waste.

Food Processing. The wastes from food processing plants are largely organic and can be treated in municipal waste treatment plants. The flow and organic load fluctuations must be adequately evaluated when these wastes are allowed to enter municipal treatment facilities. Fruit and vegetable processing can cause overloading in these facilities because of the variable characteristics and short processing season. Meat, poultry, and milk processing are less seasonal. The quantity and strength of wastewaters requiring treatment and disposal is reduced by methods such as (*1*) reducing freshwater requirements by recycle systems; (*2*) segregating strong wastes for separate treatment; (*3*) modifying processes to minimize waste generation; and (*4*) educating plant personnel about pollution control and water conservation.

Livestock Waste Management. Animal waste management systems are combinations of practices for storing and treating manure and other wastes, controlling runoff from feedlots, managing removed waste and runoff, and disposing of animal carcasses, or other waste. Manure storage facilities are usually classified by type and form of stored manure and/or construction of storage, for example, above or below ground liquid manure tank, earthen storage basin, and solid manure storage. The forms of manure range from solid to liquid, depending on the solids and moisture content and are defined as follows (32). Solid manure has a liquid content low enough that the mixture piles or stacks, with a solids content usually greater than about 15%. This condition occurs when some bedding or soil is added or sufficient air drying has occurred to raise the solids content. Liquid manure has a liquid content high enough that the mixture will flow and pump relatively easily. The solids content is usually less than about 13%. Semisolid and slurry forms are between solid and liquid. The transition between categories is not sharply defined.

Solid manure and other dry waste or dead animals (especially poultry and swine carcasses) is treated by composting, the biological degradation of organic solids under aerobic conditions to a relatively stable humus-like material called compost (32). Composting solid wastes reduces the bulk of organic material to be handled, improves its handling properties, and destroys pathogens. The compost usually has a lower N content than the original waste.

Liquid manure is treated in lagoons. Anaerobic lagoons are widely accepted in the U.S. for treating animal waste (8). The primary purpose of lagoons is to treat and stabilize livestock waste through biological activity. If properly managed, nutrient and BOD content are reduced. A lagoon operates more effectively if loading is by small, frequent (daily) inflow, rather than large, infrequent slug loads. A lagoon is constructed to avoid leakage and ground water pollution. The effluent (treated wastewater) from lagoons is land applied to minimize water pollution problems, as described in the section on nutrient management.

Runoff-control systems for feedlots and manure storage are governed by two basic principles. First, all clean water originating outside the feedlot or storage area is diverted so that the water does not come in contact with the contaminated soil. Second, the water originating inside the feedlot is disposed of to minimize its pollution potential. There are five basic components for feedlot runoff management: (*1*) diversion: to prevent unpolluted water from running on the feedlot; (*2*) drainage: the feedlot surface must be properly drained; (*3*) debris basins: prevent solids from reaching detention (holding) ponds; (*4*) detention ponds: hold runoff; and (*5*) disposal: to land or evaporation.

Clear water diversion systems include terraces and diversions, which direct water from upland watersheds away from the feedlot or barnyard, and gutters with downspouts on buildings, which prevent roof drainage from entering the area. When a natural waterway (creek) crosses a feedlot, it is necessary to pass it through a culvert or relocate the feedlot.

Collecting runoff with a system of curbs, gutters, or terraces prevents it from directly entering defined watercourses and concentrates polluted runoff for treatment before utilization or disposal. Runoff is contained for short periods in settling channels or basins, vegetative filters, or infiltration areas. Holding ponds provide short-term storage.

It is estimated that livestock waste management systems (all practices combined) reduce pollutant loadings as follows (averages) (4): bacteria 74%, sediment 64%, total N 62%, total P 21%. Individual practices in the system vary widely. One study investigated the effectiveness of a two-stage anaerobic lagoon treatment system with 81- to 118-day hydraulic retention times (33). Significantly reduced concentration in dairy wastewater from milking parlors of 80–82% for volatile solids, 90–93% for chemical oxygen demand (COD), and 55–73% total Kjeldahl nitrogen (TKN) were found. Another study stated that a waste treatment lagoon (and proper land application) reduces bacteria by 65–100%, dissolved P by 10–69%, and N by 32–91% (4). Most recommendations suggest applying manure to satisfy crop N needs. Recently, concern has been raised that this practice results in applying excess P.

Nonpoint Source Pollution Control

NPS pollution is controlled through the use of best management practices (BMPs). A BMP is defined as a practice or combination of practices that is determined by a state (or designated area-wide planning agency) after problem assessment, examination of alternative practices, and appropriate public participation to be the most effective practicable (including technological, economic, and institutional considerations) means of preventing or reducing the amount of pollution generated by nonpoint sources to a level compatible with water quality goals.

The selection and application of BMPs are related to the pollutants that need to be controlled, the type of agricultural activity contributing such pollutants, the load reduction required, and to site-specific climatic, topographical, and managerial considerations. There is no single technology or BMP to control all NPSs. For example, erosion control practices also control sediment-bound pollutants, but may not affect dissolved pollutants. Effectiveness of BMPs is site-specific, depending on characteristics, such as topography, soil type, and climate. Variability in BMP effectiveness can be extreme.

One study summarized the steps involved in the identifying and implementing BMPs as follows: (*1*) identify the water quality problem—What pollutants are present? What load reductions are required? (*2*) determine sources that may be contributing to the problem; (*3*) determine the type of control needed on the basis of the problem and sources; (*4*) identify and select the most appropriate BMPs; (*5*) implement BMPs; and

(6) monitor the effectiveness of the BMPs in achieving water quality goals (22). The most appropriate BMPs are those that are agronomically effective, environmentally effective, economically achievable, socially acceptable, and implementable, where those terms are defined as follows (22):

agronomically effective: practices that promote productivity and commodity quality and that conserve, preserve, and, if possible, enhance soil productivity;

environmentally effective: practices that prevent, reduce, or control pollutant loads (dissolved or bound on particulates) entering surface or groundwater to a level compatible with water quality goals; interaction between surface and ground water and air should be considered, ie, reducing a pollutant in one medium may increase it in another;

economically feasible: practices whose economic implications are compatible with water quality and competing goals;

socially acceptable: practices that are viewed as reasonable and are supported by the public;

implementable: practices that are legal; institutions are in existence or can be formulated to provide flexibility and simplicity in administration to achieve implementation.

BMPs are generally classified as vegetative, management, or structural. In general, there are BMPs in each class to control different pollutants. Individual BMPs are often somewhat more effective in controlling certain types of pollutants than others. For example, BMPs that reduce erosion reduce delivery of nutrients and pesticides adsorbed to sediment, but may have little or no impact on dissolved pollutants. Specific BMPs and their effectiveness are described in the following sections.

Vegetative BMPs. The primary objective of vegetative (and many management) BMPs is to reduce erosion and thus sediment yield. The single best method of preventing soil erosion is to protect the soil surface with vegetative cover. A number of practices stress maintenance of vegetative cover during critical times. Reducing sediment yield generally also reduces the delivery of sorbed pollutants, so many BMPs focus on erosion control.

There are many different types of vegetative BMPs and many variations of individual types. The following paragraphs describe some of the most commonly used BMPs.

Cover Crops. Cover crops are close-growing grasses, legumes, or small grain crops that cover the soil during the critical erosion period. Cover crops reduce erosion by decreasing soil detachment. The growing vegetation also retards movement of any detached particles and their adsorbed pollutants. When legumes are used as a cover crop or as part of a crop rotation, they provide a N source for the subsequent crop, thus minimizing the need for commercial fertilizers. Cover crops also use nutrients that are otherwise lost through leaching during fallow periods. Cover crops are 40–60% effective in reducing sediment and 30–50% effective in removing total P (4).

Crop Rotation. Crop rotation is a system of periodically changing the crops on a particular field. Crop rotations reduce erosion and associated adsorbed pollutant loads by improving soil structure, which decreases detachment. Rotations also reduce the need for pesticide and nutrient applications because pest reproductive cycles are disrupted by changing host crops and nutrient needs are more correctly balanced for a moderate level of production. The most effective crop rotations for water quality protection involve at least two years of grass or legumes in a four-year rotation (4).

The effectiveness of a crop rotation system is difficult to measure. Often the best estimate is an average annual load reduction throughout the rotation, so a higher reduction in one year coupled with a low reduction in another gives a moderate level for the entire rotation. It is estimated that some rotations reduce annual losses of N by 50% and of P by 30% (4).

Strip-Cropping. Strip-cropping is the practice of growing crops that require different types of tillage, such as corn and alfalfa, in alternate strips. Typically, strips of grass or other closely growing crops are alternated with strips of row crops. The primary purpose is to reduce soil losses. Strip-cropping systems provide up to a 75% load reduction in sediment and up to a 50% load reduction in total P (4).

Vegetated Filter (Buffer) Strips. Filter strips utilize strips of closely growing vegetation, such as sod or bunch grasses or small grain crops, with the primary purpose of water quality protection. Filter strips are generally placed between the agricultural land (for example, cropped fields and animal waste application areas) and the body of water to be protected. Strips are designed to remove sediment and other pollutants from runoff by slowing water velocity and allowing sediment and sorbed pollutants to deposit. If filter strips are wide enough (detention time is adequate) and planted with appropriate vegetation (high-nutrient-demanding species), they remove some nutrients from runoff and shallow subsurface flow.

Filter strips remove 35–90% of sediment and sediment-bound N, but are much less effective in removing P, fine sediment, and soluble nutrients, such as nitrate (14% removal) or orthophosphate (5–50% removal) (4). Nutrient removal decreases as the number of runoff events passing through the filter increases (34,35). During the active growing season, some filter strips are effective in removing P. The most effective systems rely on perennial grasses. Some research indicates that deep-rooted native bunch grasses in strip-cropping systems are also effective in "recovering" leached N that escapes below the roots of conventional row crops (4).

A review details studies of trapping mechanisms in grass filter strips and presented the following observations (36). As runoff enters grass filter strips, its velocity is reduced leading to reduced sediment transport capacity, and thus depositing excess sediment and adsorbed chemicals in the upslope edge of the filter. Slowing the runoff also provides an opportunity for chemicals to adsorb on vegetation, litter, or the surface layer of the soil. Infiltration in grass filters is enhanced by the presence of macropores created through grass root activity. This infiltration traps fine sediment and dissolved chemicals in the soil matrix. Chemicals trapped in the soil surface are converted in time by biological and chemical transformations and plant uptake. Coarser sediment fractions are trapped near the upslope edge of the filter, whereas finer particles are trapped within the filter.

The long-term effectiveness of grass filters is investigated in two studies (35,37). It was observed that, as sediment is trapped at the filter edge, berms form that force inflows from subsequent storms to flow parallel to the filters until channels are encountered that allow flow to cross the vegetated filter strip as concentrated flow (35). This flow channelization reduces the effectiveness of the grass filter.

Field Borders. Like filter strips, field borders utilize strips of closely growing vegetation, such as sod or bunch grasses or small grain crops, with the primary purpose of water quality protection. Field borders normally consist of perennial vegetation planted at the edge of fields to control erosion, regardless of their proximity to water. Field borders are effective in preventing detachment of soil particles in the areas covered by the border, but have little effect on controlling erosion, pollutant detachment, or transport from the fields they surround. Field borders are most effective in intercepting and trapping sediment which otherwise leaves the field. Pollutant reductions for borders alone have not been fully evaluated.

Grassed Waterways. Grassed waterways are designed and managed to transport water at nonerosive velocities from fields to ditches or other outlets. Grassed waterways often serve as outlets from terrace systems and other practices.

Riparian Buffer Zones. Riparian buffer zones are vegetated areas adjacent to drainage channels, streams, or rivers. The vegetation includes grasses, shrubs, and/or trees. Zones are complete ecosystems or function as transition areas between aquatic and terrestrial ecosystems, while maintaining a distinct set of vegetative and soil characteristics. The grasses and low vegetation in the buffer zone filter both surface and subsurface flow, while the roots of taller vegetation take up and transform pollutants and nutrients from shallow ground water. In summarizing seven riparian buffer studies, reductions were noted of 79–98% N and 50–85% P in surface flow and 68–100% N and −114–33% P in subsurface flows (36). The negative reduction in P indicates the variable effectiveness of these zones, depending on site, vegetative, and soil conditions. Various width zones have been investigated without conclusive results as to optimum widths for particular pollutant removal efficiencies. Pesticide removal has been documented but not adequately quantified (4).

Management BMPs. ***Conservation Tillage.*** Conventional tillage is defined as the method of preparing a seedbed by completely inverting the soil and incorporating all residues with a moldboard plow. The soil surface is tilled further with a disc harrow to provide a smooth seedbed free of clods. After planting row crops, a cultivator is sometimes used to control weeds. The result is a bare soil surface during the early spring, a time of high rainfall and runoff erosivity in many locations.

Tillage practices have changed drastically in the past 25 years from the conventional tillage described above because of developments, such as herbicides to control most major weeds at reasonable costs; decreased availability of labor on the farm; increased fuel costs; and public concern about soil erosion as a source of water pollution. These developments have stimulated farmers to adopt reduced-tillage methods, which allow less erosion than conventional tillage. Reduced-tillage practices are known collectively as conservation tillage.

Conservation tillage is any tillage system that leaves at least 30% of the soil surface covered with crop residue after planting. Conservation tillage controls erosion and sediment by decreasing soil detachment, which also helps control the loss of adsorbed pesticides and nutrients. Residue management, a significant component of reduced-tillage systems, has developed as an approach to control erosion (38) and increase soil fertility (39) by leaving crop residue on the soil surface (40). Two studies (41,42) conclude that residue cover reduces erosion in four ways: (*1*) dissipation of the energy from raindrop impact; (*2*) slowing runoff, and increasing flow depth, which in turn reduces the impact of raindrops; (*3*) absorption of some of the forces from runoff that are usually applied to the soil surface; and (*4*) creation of small reservoirs of ponded runoff causing deposition. As slope steepness increases, crop residues are less effective in controlling soil erosion (41). Residues partially incorporated into the soil are more effective in reducing erosion than residues loose on the surface (42).

Another study (43) differentiates the effects on soil erosion of flat vs standing residue. Residue lying prone on the soil surface (flat residue) helps reduce water erosion by lessening the impact of raindrops on the surface and retarding the rate at which water can leave the field through increased infiltration of surface water. Residue left upright (standing residue) is less effective at reducing water erosion than flat residue. However, standing residue is generally more effective in controlling wind erosion than the equivalent mass of flat residue alone, in some cases by an order of magnitude. Standing residue reduces the velocity shear stress near the soil surface and intercepts soil and other material transported by the wind within the residue. Flat residue reduces the soil surface area exposed to the wind and, therefore, reduces the detachment rate of soil particles from the surface. Also, unanchored flat residue by itself is susceptible to removal by wind, potentially nullifying any mitigating effects on wind erosion.

A study (44) evaluates the surface-residue cover reductions from 30 combinations of primary and secondary tillage in soybean residue and 18 combinations in corn residue. It was found that only one of the combinations in soybean residue left more than 30% cover, which supports the need for no-till planting or cover crops following soybeans. More combinations in the corn residue left more than 30% cover. Through comparison of the results with published values, it was concluded that site-specific conditions produce results varying considerably from the published values of residue cover remaining because of particular tillage operations (44). Published values are useful in planning but actual reductions in cover should be determined with a test strip in actual field conditions prior to the field operation. Specific cover amounts vary with soil type, moisture content, crop yield, travel speed, and depth of tillage for a particular implement.

Conservation tillage is highly effective in reducing sediment yield but not in controlling soluble nutrients and pesticides (4). Reduced sediment losses of 30–90%, in total P loss of 35–85%, and in total N loss of 50–80% are found. However, pollution potential increases (and reduction efficiencies decline) if fertilizers are not incorporated in the soil in a conservation tillage system. Some of the conservation tillage systems and their effectiveness are described in the following paragraphs.

No-tillage (no-till, zero tillage) is a method of planting crops that involves no seedbed preparation other than opening a small slit or punching a hole into the soil to place the seed at an intended depth. There is usually no cultivation during crop production. Chemical weed control is normally used. No-till farming generally consists of leaving the previous year's crop residue and planting into the undisturbed surface. No-till is considered the most effective erosion control practice applicable to agricultural lands. However, a greater use of herbicides and sometimes lower yields are experienced in some soils.

Advantages of long-term no-till farming compared with conventional methods include reduced soil loss (38,45,46), increased surface residue, improved soil drainage (47), and reduced labor costs (48). Some studies have focused on no-till management (42,49,50). In a field study, it was found that sediment losses are significantly reduced when long-term no-till is the management practice (40). It was concluded that long-term no-till practices are effective and practical in reducing rill erodibility and sediment loss on the two sites investigated in north central Illinois.

A ten-year study reports that no-till management controls erosion within acceptable limits on highly erodible graze-out wheat land (51). Conservation of nutrients in soil water is another benefit of no-tillage systems (52). Another study reported that no-till did not significantly affect subsurface biological activity and denitrification (53).

Other types of conservation tillage systems include ridge tillage, strip tillage, mulch tillage, and reservoir, or basin, tillage. In ridge tillage, the soil is undisturbed prior to planting, which is done on ridges 10–15 cm higher than row middles. Residues are moved aside or incorporated on about one third of the soil surface. Herbicides and cultivation are used to control weeds. In strip tillage, the soil is undisturbed prior to planting. Narrow and shallow tillage is done in rows using rotary tiller, in-row chisel, etc. Up to one third of the soil surface is tilled at planting time. Herbicides and cultivation are used to control weeds. In mulch tillage, the soil surface is disturbed by tillage prior to planting, but at least 30% of residues are left on or near the soil surface. Tools such as chisels, field cultivators, disks, and sweeps are used (eg, stubble mulch). Herbicides and cultivation are used to control weeds. In reservoir, or basin tillage, a large number of depressions or small reservoirs or basins are formed to retain rain or sprinkler-applied irrigation water. The reservoirs prevent runoff, decrease erosion, and increase water conservation.

Contour Farming. Contour farming is field operation, such as plowing, planting, cultivating, and harvesting, conducted across a slope, in contrast to up and down hill.

Livestock Exclusion. Permanent or temporary fences are used to exclude livestock from areas where grazing or trampling cause erosion of stream banks and lowering of water quality by livestock activity in or adjacent to the water. Keeping animals out of streams reduces deposition of fecal material, and thus availability of nutrients and bacteria. Turbidity from in-stream trampling is eliminated because particles are not detached. Stream banks are not denuded, so that instead of being a pollutant (sediment) source, the banks provide a filter for runoff thus reducing delivery of sediment and adsorbed materials to the stream. In addition, if the riparian area is large enough, the vegetation removes some of the nutrient load, decreasing the availability of dissolved pollutants. Reductions of 50–90% of suspended solids and total P are reported (4).

In addition to fencing, livestock exclusion is accomplished with herding and by providing off-stream water sources for the livestock. Field observations of cattle behavior indicate that 92% of the time cattle prefer to drink from an off-stream water source rather than from the adjacent stream (54). In the same study, installation of an off-stream water source reduces the total time that cattle spend in the stream by 58% and the amount of stream bank erosion by 76%. Pollutant concentrations are also reduced, total suspended solids by 90%, total N by 54%, sediment-bound N by 68%, total P by 81%, and sediment-bound P by 75%. Bacterial concentrations are reduced by approximately 50–75%.

Range and Pasture Management. Rangeland is land generally planted with native grass and managed as range. Pastures are normally seeded with an improved grass variety and managed by agronomic practices (liming, fertilization, irrigation). Range and pasture management includes seeding and reseeding; brush management (mechanical, chemical, physical, or biological); proper stocking rates and proper grazing use; and deferred rotational systems that protect the vegetative cover on improved pasture and native rangelands. Keeping land permanently covered in high quality (closely spaced) vegetation decreases soil loss to negligible amounts (detachment and transport are interrupted), and sediment and adsorbed pollutants are not lost from the land surface. If correctly managed, range and pastures function like filters to reduce the load of N, P, and sediment up to 80% (4).

Nutrient Management. Nutrient management focuses on managing N, P, and potassium (K) on an annual basis for both expected crop production and to protect water quality. The goals of farm nutrient management planning are to minimize adverse environmental effects, primarily on water quality, and to maintain or improve long-term farm net returns by avoiding unnecessary nutrient applications. The dual goals of nutrient management are easier to achieve for field crops for which fertilizer expenses are significant compared to high value crops for which fertilizer expenses are minor.

For successful nutrient management, all sources and sinks of nutrients on the farm must be considered, with the goal of matching the sources and sinks as closely as possible. Soil testing is used to determine residual nutrient levels in the soil. The results of the soil test are used in conjunction with the productivity of the soil for the particular crop, crop nutrient requirements, and the leaching potential of the soil to determine the amount of each nutrient needed to meet the goals of nutrient management. The farmer should request the appropriate quantities of each nutrient to be applied, rather than using a specific fertilizer grade. A fertilizer dealer can supply those nutrients at least cost using fertilizer materials and/or grades in stock.

The timing and placement of nutrients directly affects crop utilization and losses of nutrients to the environment. Nutrient management emphasizes applying mobile nutrients, such as N, in split applications, thus providing the nutrients when they are most needed by the crop and not providing excess nutrients that are leached. Placement of fertilizers is more important with regard to soil-immobile nutrients such as P. Because P is more susceptible to transport with runoff, surface applications increase the potential for P losses to surface water. Incorporation of fertilizer reduces the potential for surface losses of nutrients. P is most efficiently applied in a band, reducing and delaying P bonding to soil particles and increasing the time P is available to growing plants.

Manure is an environmentally sound and economic nutrient source. Variation in characteristics makes manure more difficult to manage than commercial fertilizers, but advantages in economics and improvements in the physical characteristics of soil are gained. Additional considerations related to the utilization of manure include storage and hauling costs, storage losses of N, and relatively low N, P, and K in manure compared to commercial fertilizer. These issues must be evaluated on a site-specific basis, as must all other aspects of nutrient management, to determine the best approach for a particular site.

Integrated Pest Management. Integrated pest management (IPM) is a combination of biological, cultural, and chemical practices to control crop pests (insects, weeds, diseases) in an economically efficient and ecologically sound manner (55). IPM uses traditional practices, such as crop rotation and selection of resistant crop varieties, along with sophisticated pesticide application management, such as pest scouting, minimal application rates, most judicious timing, and selection of the least toxic and least persistent chemical. Pest management technologies include the following (56):

biological control: using beneficial organisms against pest organisms;

cultural control: using rotations and cultivations to reduce pest problems (57,58);

chemical control: judiciously using pesticides and other chemicals in a responsible manner (59);

legal control: abiding by state and federal regulations that prevent the spread of pest organisms.

IPM practices reduce pollution mainly by decreasing the applied amounts of pesticide or crop-protection chemical and by selecting the least toxic, least mobile, and/or least persistent material (4). IPM may not always reduce the amounts of pesticides applied. Complete studies on the effectiveness of IPM show varying results depending on the exact type of pesticide, soil, moisture, and crop conditions. One study reviewed economic evaluations of IPM programs for a number of commodities, including cotton, soybeans, vegetables and flowers, fruits and nuts, peanuts, tobacco, corn, and alfalfa (56). The programs evaluated involved insect, disease, weed, and nematode IPM. Based on the results of 61 studies (summarized in Table 3), it was noted that farm-level evaluation of IPM benefits and costs indicates generally lower pesticide use, crop production costs, and economic risk, and higher net returns to producers.

Structural BMPs. Structural BMPs typically require some type of construction, often including earthmoving. Generally, structural BMPs are more expensive to implement and maintain than vegetative and management BMPs. Some of the most common structural practices are described in the following sections.

Terraces. A terrace is an embankment or combination of an embankment and channel constructed across a slope to control erosion by reducing the slope and by diverting or storing surface runoff instead of permitting it to flow uninterruptedly down the slope. Terraces are level or graded. Level terraces store or pond water for infiltration or outlet through an underground conduit. Graded terraces divert water to a suitable outlet.

Terraces change the effective slope of the land, which decreases runoff velocity and sediment detachment. Soil particles and adsorbed pollutants are thus not transported from the field. Terraces that retain all water applied to the field are also effective in reducing the loss of dissolved pollutants. Graded terraces or those with outlets allow runoff to leave the field and, therefore, do not control dissolved or suspended pollutants well. Level terraces remove up to 95% of sediment, up to 90% of its associated adsorbed nutrients, and between 30 and 70% of dissolved nutrients (4).

Diversions. Diversions are a form of graded terrace but are defined as channels constructed across a slope with a supporting ridge on the lower side. Diversions prevent water from flowing across areas where pollutants are available, such as highly erodible soils, feedlots, or areas where fertilizers or pesticides are mixed, stored, or applied at high rates. Diversions must be provided with a stable outlet, such as a grassed waterway or buffer zone. Diversions reduce sediment movement by 30–60% and adsorbed nutrients by 20–45% (4).

Grade-Stabilization Structures. Grade-stabilization structures control the grade and gully head-cutting in natural and artificial channels and provide an outlet for other conservation practices. Structures are constructed of metal, wood, rocks, concrete, or earth. The area surrounding the structure must be suitably stabilized.

Grade-stabilization structures reduce water velocity, thus preventing additional sediment detachment, and decrease the transport capacity of the water. They also often act as traps, reducing additional sediment transport. Sediment-adsorbed nutrients are thus controlled much more effectively than dissolved pollutants. Grade-stabilization structures result in 5–75% reduction in suspended solids leaving the unstable area, but data on the area below the structures are unavailable (4).

Critical Area Planting. Critical area planting involves planting suitable vegetation (such as grasses, trees, and/or shrubs) on highly unstable sites, such as gullies, denuded streambanks, dams, and embankments where usual planting methods are not effective. Critical area planting is classified as a structural rather than a vegetative practice because intensive methods are often needed to establish vegetation in critical areas. Examples of intensive methods include erosion control blankets (natural or sÿnthetic matting that protects the soil surface); gabions (wire cages filled with rocks or rocky soil); hydromulching (applying mulch hydraulically); increased seeding rate; and hand planting.

Sedimentation Basins. Sedimentation basins are designed and constructed to intercept sediment laden runoff from earth disturbance sites and to detain the water long enough for sediment to settle and remain on the bottom of the basin, while discharging clean water to the uncontrolled environment (60). In addition, basins must be designed to provide sufficient storage capacity for the sediment removed. Sedimentation basins preserve the storage capacity of downstream reservoirs, canals, etc, and often serve an entire small watershed. Sedimentation basins remove 40–87% of the incoming sediment, up to 30% of the adsorbed N, and 40% of the total P (4). Soluble pollutants are not controlled effectively by sedimentation basins.

Detention Ponds. Detention ponds collect runoff from agricultural fields for storage and pollution control. Detention ponds are not drained and often support aquatic vegetation. Available P and N are taken up by these plants and pesticides are degraded. Detention ponds, especially those with aquatic vegetation, are generally more effective than sedimentation basins in removing sediment and nutrients.

Stream Bank Stabilization. Stream bank stabilization is achieved by structural or vegetative methods to protect the stream bank from erosion. Riprap, concrete, wood, or rock gabions are used, but vegetative stabilization is most effective for pollution control (4). Eroding stream banks contribute sediment to the waterway and streambed and also cause increased water temperatures because of the loss of riparian vegetation that shades the stream. Increased water temperatures negatively affect aquatic organisms. Loss of floodwater storage or hydrologic assimilative capacity to regulate flow is

Table 3. Summary of Results of Farm-Level Economic Evaluations of IPM Programs[a]

Commodity	States	Number of Studies	Average % Change in Pesticide Use[b]	% Change in Production Cost with IPM	% Yield Change with IPM	% Change in Net Returns Per Acre	Level of Risk with IPM
Cotton	TX, GA, MS, NC, SC, LA, MO, TN, AZ, NM, CA, AR	18	−15	−7	+29	+79	Decreased
Soybeans	NC, VA, MD, GA, IN	7	−35	−5	+6	+45	Decreased
Corn	IN, IL, and 10 other states	3	+20	+3	+7	+54	–
Vegetables and flowers	CT, CA, MA, TX, FL, OH, NY, HI	15	−43	Quality increased in 4 studies remained the same in the others			
Fruits	NY, MA, WA, NJ, CA, CT	8	−20	0	+12	+19	–
Peanuts	GA, TX, OK, NC	5	−5	−5	+13	+100	–
Tobacco	NC	2	−19	–	0	+1	–
Alfalfa	OK, WI, Northwest	3	−2	–	+13	+37	Decreased
Unweighted average[c]			−14.9	−2.8	+11.4	+47.8	Decreased

[a] Ref. 56.
[b] For those producers who adopted the specified IPM practices compared to those who did not.
[c] Weighting is not possible without an accurate accounting of the acreage affected for each commodity in each state.

also an effect. Stabilization methods eliminate detachment of soil particles and restore one part of the original hydrology of the system.

Effective stabilization reduces sediment loading by 90%, but is highly dependent on the type of vegetation and the stability and width of the reclaimed area. When combined with riparian area restoration, removal rates of 50–70% for P and 80–90% for N are found. When planted in conjunction with a grass filter strip, load reductions of 80% of total suspended solids (TSS), 60–90% of total N, and 30–90% of total P are possible (4).

Irrigation Management. Irrigation water management (IWM) is a combination of practices that control irrigation water and chemigation (application of pesticides and/or fertilizers through the irrigation system) to prevent pollution and reduce water loss (4). IWM includes proper scheduling, efficient application, efficient transport systems, utilization and reuse of tailwater and runoff, and management of drainage water.

Seepage-control practices (lining ditches with concrete or converting to pipe conveyances) greatly reduce leaching losses and decrease the availability of salts in canals for application (and runoff) from fields. Tailwater pits (which are usually lined to catch rainwater runoff or drainage from furrows at the end of a field) prevent any pollutants from entering any bodies of water. IWM increases the efficiency of water application and prevents excess runoff that carries with it dissolved nutrients and pesticides. IWM is particularly effective in reducing N and pesticide loadings to ground water and salt loading to surface waters (4). Tailwater pits are about 50% effective in sediment removal and are highly effective in nutrient and pesticide pollution control as long as the collected water is reused and not discharged.

BARRIERS TO SOIL AND WATER QUALITY PROTECTION

The negative effects of certain agricultural practices on soil and water quality raise the question of why farmers engage in such practices. What are the barriers that prevent farmers from adopting alternative practices to reduce environmental damage? A number of barriers have been identified.

Property Rights

Contamination of water with manure, pesticides, and sediment from crop and livestock production is an unintended consequence called an externality (61). Pollution of water renders it less desirable or unfit for other uses such as drinking, recreation, and wildlife as described earlier. Property owners often assume the implicit right to pollute water and are usually not penalized for doing so. Property owners, who incur costs to reduce such pollution and improve water quality, gain to the extent that the improved quality affects water uses on their land, such as drinking and irrigation. However, much of the improved-quality water leaves their property and is enjoyed by others. Landowners do not have property rights to water once it leaves their land and do not receive a return from improvements in water quality enjoyed by others. Thus, the structure of property rights encourages property owners not to account for pollution resulting from their activities.

Variations in property rights in different societies have implications for soil and water quality damage. In many, if not most, countries, rights to agricultural land are privately held. Landowners have an incentive to maintain land productivity and value. Landowners who wish to maximize the value of their land attempt to conserve soil as long as the marginal costs of conserving an additional soil unit are less than the present value of the net benefits of increased productivity from that added soil unit over time (62). Present value refers to the discounted sum of net income earned from the unit of soil in each time period where the discount rate used is the relevant interest rate for the farmer. Landowners' efforts to protect soil productivity reduce erosion and benefit water quality as well.

In some societies, particularly less developed countries (LDCs), property rights to certain lands are ill defined or defined in ways detrimental to maintaining land productivity. For example, in several African and Latin American countries, rights to forested lands are obtained by clearing and farming the land. If pesticides and fertilizers needed to increase productivity on existing cultivated lands are expensive, farmers are better off leaving their plots and cutting and burning new forest lands to exploit the natural fertility of newly cleared land. These actions maximize returns on their labor but lead to deforestation, increased erosion, and degradation of soil and water quality (63).

Profitability

Farmers are concerned about profitability of investments to protect soil and water quality. Many soil and water quality protection practices reduce profits because environmental protection is not rewarded in the marketplace. Farmers' losses from practices to reduce water quality damage depend on several factors. First, the size of region over which the reduced emission requirement is enforced is important. If the requirement is enforced on a small production region, it has relatively little impact on output price with the result that more losses are absorbed by producers in the region. For example, if the requirement results in reduced yields, the losses to producers are approximately equal to the price times the yield loss plus or minus any changes in production costs. If the requirement is applied to a large region, the reduced production would cause prices to rise reducing farmers' losses at the expense of consumers. If demand for the product is sufficiently inelastic or unresponsive to price, the price increases might actually raise farmers' incomes. For example, if all major wheat-producing countries require following stringent soil conservation measures in wheat production, wheat production costs would rise and supplies would decline. Given the importance of wheat in consumers' diets, the price of wheat might rise by a percentage larger than the decline in supplies, causing farmers' incomes to rise.

Second, the length of time allowed for market forces to adjust to the policy affects its impact on profitability. As more time elapses after reduced emissions are required, market forces may be set in motion to reduce farmers' costs of complying. Farmers who incur losses are willing to pay more for information or products that reduce emissions at lower cost. Firms and/or public agencies observe these profit opportunities and invest more in research and product development and promotion. These investments increase the likelihood of new products and technologies that reduce emissions at lower costs. Thus, in many instances the long run costs to farmers may be below the short run costs. For example, livestock producers might be required to restrict manure application rates to crops to levels so that all P in applied manure is removed by crops (P standard). This requirement would lessen potential P runoff into surface water, but would tend to reduce manure application rates, increase the distance manure has to be transported for disposal, and increase producers' costs of disposal. New technologies such as microbial phytase could be developed to make livestock more efficient in using P in their diets, which would reduce P in manure (64). Farmers would be willing to pay for such products, which lower farmers' costs of manure disposal under the P standard.

Third, the type of policy used to reduce emissions has a large impact on farmers' costs. Some policies, such as regulations and taxes, place all of the burden on farmers, whereas voluntary policies, such as subsidies, may largely spare farmers and pass most or all costs on to taxpayers. In the next sections, we evaluate the profitability of conservation practices aimed at three types of pollutants: (*1*) sediment, (*2*) pesticides, and (*3*) nutrients.

Profitability Impacts of Controlling Sediment Loss. When sediment erodes, the most productive topsoil layers are lost first. Conserving soil protects its productivity and crop yield potential. However, soil-conserving practices may have other negative impacts on yields, or may require substituting less profitable crops or other costly structural practices, such as terraces. For example, no-till crop production may result in a poorer stand or less favorable root growth with certain types of soil and weather conditions compared with conventional practices. The overall impact of reduced till or no-till on profitability is difficult to generalize. Almost an equal number of studies find that reduced or no-till systems are more profitable than conventional till or vice versa (65). Important factors affecting profitability of conservation tillage are original depth of topsoil, weather patterns, cropping patterns, management, and prices of inputs (65). For example, the relative profitability of no-till increases where topsoil is shallow and where climate conditions are dry, conditions which place moisture and soil conservation at a premium (66,67). Also, higher fuel costs make conservation tillage systems more attractive.

The length of time the farmer plans to farm the land and the effects of erosion control on land sale values are also important. Soil conserving investments yield a return many years in the future. If the farmer does not plan to farm many more years, future returns from more distant years are not considered, reducing the profitability of soil-conserving investments. However, if the sale value of land is increased by erosion control efforts, even farmers who plan to sell their land soon capture the benefits of erosion control (68).

Although profitability is enhanced and soil loss reduced by conservation tillage, further reductions in soil loss reduce farmers' net returns. It is estimated that reducing soil losses on highly erodible land to a T-level reduces farmers' net returns by \$37 per hectare (\$15 per acre) per year (69). A T-level is the amount of annual soil erosion from a site which just equals the amount of new soil created through natural processes each year. Another study estimates that farmers' net returns are reduced from $2.62 to $7.76 per metric ton ($2.38 to $7.04 per short ton) of soil saved per year depending on soil type (70). These estimates consider soil-conserving BMPs, such as reduced tillage and substitution of less erosive crops or pasture. Reductions in farmers' profits per ton of soil saved are likely to increase with the desired reduction in soil loss (65).

Profitability Impacts of Controlling Pesticide Losses. Pesticide losses to the environment can be reduced while increasing returns to producers. For example, a survey of 61 scientific studies of IPM on cotton, soybean, corn, vegetables, fruits, peanut, tobacco, and alfalfa revealed an average reduction in pesticide applications of 15% (56). Thirteen of the studies (21%) found that pesticide use actually increased with IPM. Overall, the studies found reduced pesticide use resulted in decreased production costs of nearly 3%. Yields increased on average by about 11% resulting in an average increase of 47% in net returns. Few of the studies found that net returns declined with IPM. IPM strategies introduce more flexibility into production and allow using pesticides more cost effectively and environmentally safely.

For the foreseeable future it is likely that many, if not most, crop enterprises will require some pesticide use for profitable crop production (71). Banning pesticides is likely to reduce farmers' net income although the extent of the reduction depends on several factors. First, the number of pesticides banned (65) affects the potential losses. Bans of a single pesticide generally have less effect because the product would be replaced by a substitute (72). Bans on all pesticides would reduce income by a greater amount. Second, the amount of time allowed for markets to adjust to the ban affects its costs. When a ban is instituted, firms have incentives to develop alternative technologies to replace the banned pesticide(s). These new technologies include alternative pesticides, new plant varieties that are less susceptible to the pest, and alternative forms of biological control. The new technologies may allow crop production as profitable as before the pesticide ban. Third the area to which the ban applies affects losses. Policymakers can reduce aggregate costs by targeting a chemical ban at areas which are most susceptible to leaching and/or runoff. Targeting a ban minimizes the impacts on commodity supplies and reduces increases in food prices to consumers. However, farmers affected by the ban may suffer large losses unless government compensation is provided.

Profitability Impacts of Controlling Nutrient Losses. Frequently some nutrient (N and P) loss is avoided while actually increasing farm net income. For example, strategies such as splitting N applications and testing soil N levels to guide N applications help farmers reduce N applications and losses and save fertilizer costs without losing yields. Farmers whose soils test high in P can forego additional P applications without harm to yield. Reducing P applications reduces future soil P levels and the amount of P in eroded sediment. These "win–win" changes benefit the environment without loss of farm income and should be sustainable once farmers see the benefits of the changes.

It is questionable whether the desired reductions in nutrient pollution to groundwater and/or surface water can be achieved using these "win–win" types of strategies alone. Achieving further reductions in nutrient losses to meet water quality goals will result in loss of net income to farmers because of yield losses or planting less profitable crops (65,73,74). In the long term, some of these losses are passed on to consumers in the form of higher prices if nutrient reductions are required over large production regions (75).

Lack of Information

The fact that "win–win" opportunities are found raises the question why farmers did not discover these opportunities on their own. Acquiring information about these opportunities is costly for farmers in time and/or money. If totally borne by farmers, these costs may exceed the increased profits from the practices, and the practices are not adopted. For example, farmers may be unaware of opportunities to utilize alternative nutrient sources, such as legume carryover and manure applications, as a substitute for commercial fertilizer (76). If the government provides the information to farmers as a free service, farmers may adopt the recommendations and realize a profit. For example, in the U.S., the Cooperative Extension Service and/or other state agencies provide free manure nutrient testing to farmers.

Farmers frequently underestimate the potential for water quality damage from agricultural practices on their land or the susceptibility of sites to water quality damage (77). Farmers' production practices may not change greatly as the leaching or runoff potential of the cropped site changes, perhaps because farmers fail to recognize which sites are most vulnerable to leaching or runoff (78). NPS pollution is a slow, diffuse process difficult to observe without expensive monitoring equipment. Effects of NPS pollution occur at great distances from the source of pollution. Sediment, the most visible agricultural pollutant, has much more damaging effects on waterways in which the sediment collects than on the productivity of the eroded land. Continued education is needed to help farmers recognize the impacts of their practices on water quality.

Risk

Many farmers are willing to take steps to reduce net returns risk even if these steps reduce their profit potential. Uncertainty about crop water availability, nutrient requirements, or potential damage by pests may induce farmers to apply larger amounts of irrigation water, fertilizer, or pesticides than are required to maximize profits.

IPM practices allow producers to reduce both risk and pesticide use (56,65). IPM practices, including monitoring of field pest conditions, timely applications of pesticides, and alternative control methods, allow the producer to maintain or improve crop protection against pests, often while reducing pesticide use. Some pesticides are usually required with these strategies, however, and restricting or eliminating pesticides altogether would increase the risks of yield and net returns to levels unacceptable to many producers. Producers with greatest concerns about reducing income risk would be most severely affected by pesticide restrictions (79).

Crediting organic N in manure, soil organic matter, and legume carryover may allow farmers to reduce inorganic fertilizer applications. However, greater dependence on organic N sources may increase farmers' risks because the mineralization rate of organic N is a function of weather. If mineralization is slower than anticipated, crop yields may be reduced. Soil N and plant tissue tests combined with split applications of N fertilizer should reduce these risks by providing farmers with more information on plant nutrient status. However, splitting applications to the crop presupposes that field conditions allow the farmer to make a second application, which may not be the case. Some farmers may prefer to rely on one application of inorganic N at planting to minimize the risk of yield loss caused by inadequate nutrients from less reliable organic sources or inability to sidedress N (80,81).

Soil loss control programs, which require farmers to reduce sediment losses or to adopt selected BMPs, may also increase farmers' risks relative to returns (82). As a result, producers who prefer low risk strategies may be more reluctant than profit maximizers to adopt such practices.

Other Financial and Institutional Barriers

Water quality improvements may require large capital investments, such as manure storage structures, alternative types of tillage equipment, and soil conservation structures. Bankers may be reluctant to lend funds for these investments. Farmers with high debt to asset ratios are less able to make investments to protect water quality.

Historically, the U.S. and numerous other countries have developed commodity programs, quotas, or marketing orders to protect prices of selected crops and livestock. To preserve their bases or quotas, farmers may be constrained to produce fixed areas of crops that intensively use polluting inputs or practices. For example, U.S. producers of corn, wheat, cotton and other commodity program crops had to produce their base (the land area of commodity which program participants can produce and for which subsidies are received) or lose the base. The amount of subsidy was often tied to the farm's "program yield", a proven per hectare (acre) yield, which encouraged intensive use of inputs to increase yields and often increased pollution potential. Producers who attempted to rotate less chemical intensive crops, such as legumes, with program crops risked loss of base.

The Federal Agriculture Improvement and Reform Act of 1996 (FAIR) increased production flexibility. U.S. producers can enter into seven-year production flexibility contracts to become eligible for government payments for program commodities (83). Producers may plant their total contract acreage to almost any crop except fruits and vegetables and still receive government payments. There is no requirement to maintain a proven yield for program commodities. The new program encourages flexibility and potentially reduces incentives to plant erosive program crops or apply yield-increasing chemicals and nutrients. However, more work is needed to assess actual environmental benefits.

Agricultural Trade

World agricultural trade is an important determinant of prices farmers receive in many countries. In 1994, the total value of world exports was $4.2 trillion. When the effects of inflation are removed, this amount represents a 55% increase compared to 1980 (84). Many countries depend on agricultural trade to generate income and foreign exchange. Most countries impose constraints on trade preventing products from being traded on the basis of purely comparative advantage. Comparative advantage refers to a country's relative advantage in producing a product when all costs of producing the product including externalities are considered. Protection mechanisms include quotas and tariffs on agricultural imports, support prices and/or target prices for domestically grown commodities, and subsidies for inputs, such as irrigation water, pesticides, and fertilizer. The amount of protection varies by country. The lowest levels occur in less developed countries (LDCs), Australia, and New Zealand, which are major agricultural exporters. The U.S. and Canada have more protection, while the most protection is given in Europe, Japan, and South Korea (85). In 1989, the average producer subsidy equivalent was 19% in the U.S. and 30% in Europe but only 6% in Australia (86). A producer subsidy equivalent is a measure of the value of transfers to producers from taxpayers and consumers resulting from import quotas, tariffs, input subsidies, price supports, and other policies. Many LDCs use national marketing systems to keep agricultural prices low for the benefit of urban consumers. LDCs also use input subsidies for fertilizer, irrigation water, and agricultural credit to partially compensate for low prices and encourage the use of modern technology.

Experts debate whether current trade protection or more liberalized trade is best for water quality and the environment. Those who argue that the current pattern is preferable do so on three grounds. First, liberalized trade would increase production, consumption, and the pressure on scarce resources. Second, liberalized trade would produce a greater flow of agricultural production to LDCs where environmental standards are lower and result in increased environmental damage on a global scale. Finally, liberalized trade will produce pressure for harmonizing environmental standards so that the lowest common denominator of standards prevails (87).

In spite of the previous arguments, free trade is likely to produce water quality benefits in many but not all areas. Because the world population is growing rapidly and not likely to stabilize for decades, there is little choice but to expand food production. Free trade allows expanding output efficiently with less pressure on resources. In developed countries (DCs) with high levels of protection, government commodity price supports stimulate intensive use of chemicals and fertilizer which increases pollution potential (87,88). Subsidies may also encourage the cultivation of marginal land that is more prone to pollution (89,90). Removal of subsidies would lower fertilizer and chemical use and intensive livestock feeding and reduce the potential for pollution. However, in DCs with less protection, liberalization might worsen water quality. Higher prices would stimulate more land cultivation and use of chemical inputs.

It is not clear whether free trade will increase agricultural pollution in LDCs. Higher world prices and elimination of policies designed to keep agricultural prices low in LDCs will tend to shift crop and livestock production toward LDCs (85). Higher output might be obtained through expanded use of physical capital and labor on existing land, rather than clearing of new lands (91,92). Chemical and fertilizer use intensity would increase; however intensity of chemical use is far below that in DCs. Also, if trade liberalization included removal of subsidies for chemicals, fertilizer, and irrigation water in LDCs, the trend to increased input use would be lessened. Livestock production in LDCs would be less intensive than in DCs, creating fewer problems for environmentally safe manure disposal compared to that in DCs. Increased demand for labor to expand agricultural output would raise the cost of labor. Because harvesting wood for fuel is highly labor intensive, the cost of wood fuel would increase with higher labor costs. Alternatives to wood for fuel, such as solar and hydroelectric power and petroleum would become more competitive. Reduced reliance on firewood would reduce deforestation (85). Higher commodity prices would increase demand for land and give farmers more incentives to invest in soil conservation structures (89).

Nonetheless, increased pressure on environmental resources is likely in some areas through more intensive chemical use and increased cropping of environmentally vulnerable lands. Higher prices may stimulate research and development activities that are skewed toward export promotion at the expense of water quality (93). If the research is successful in increasing the country's relative advantage in chemical-intensive crops, water quality may be further damaged.

Increased pressure on natural resources and potential for water quality damage could be counteracted with environmental protection policies. However, it is not clear that LDCs would implement these policies because of their political and economic costs. It is hypothesized that LDCs place a relatively lower value on clean water than richer countries because other needs, such as adequate food and shelter, take precedence (94). Deforested areas and parks and protected areas as indices of environmental quality provide empirical support for this hypothesis (95). It was found that the income elasticity of demand for environmental quality was significantly lower in low income countries (95). The income elasticity of demand refers to the percentage increase in environmental quality demanded as income increases by one percentage point. This evidence suggests that increased income from liberalized agricultural trade in LDCs would not translate into large increases in demand for environmental protection within those LDCs.

POLICIES TO OVERCOME BARRIERS

Agricultural NPS pollution has a major impact on soil and water quality. Numerous barriers discourage farmers from voluntarily adopting less polluting practices. Therefore it is not surprising that a number of alternative policies are proposed to reduce agricultural NPS pollution. Some of the more important types of policies are listed in Table 4 and discussed below.

Regulation

Regulatory policies require that producers conform to rules regarding the types and/or amounts of inputs, outputs, or pollutants. In this section, we discuss the advantages and disadvantages of various types of regulatory policies.

Performance Standards. Performance standards establish a ceiling on allowable pollution that farmers must not exceed (96). For example, a standard may be set for allowable groundwater NO_3^- concentration in a watershed. Alternatively, limits could be placed on NO_3^- runoff or leaching from individual farms. Performance standards give the farmer freedom to choose how to meet the standard best. Thus, farmers' control costs should be less than if government dictates how the farmer should meet the standard. Farmers' control costs are any type of direct or indirect cost which results in a lower net return to the farmer. For example, if reducing pollution requires that a farmer plant a less profitable crop, then the control cost is measured by the reductions in profits from including a less profitable crop in the rotation. Allowing pollution control measures to differ among farms may be a more effective way to meet the objective than requiring that all producers implement a given set of practices. Performance standards also encourage research and innovation to develop lower cost methods of reducing pollution.

Requiring that all farmers adhere to the same standard may not be efficient if farmers' costs of control differ. Allowing the standard to vary so that farmers with lower costs achieve greater reductions will result in lower overall costs of control. Another drawback of performance standards is their high administrative cost caused by the difficulty of measuring or estimating pollution. Performance standards might be more applicable to PSs than NPSs because by definition NPS pollution is widely dispersed and subject to delays between the polluting practice and the effect of the pollutant. Performance standards could be applied to NPS pollution by using mathematical models to estimate pollutant flows resulting from agricultural practices (97). Models require input data and skilled personnel and may not represent reality. However, great strides are being made in methods for acquiring and manipulating input data and in developing credible NPS models.

Design Standards. Design standards require that farmers follow a set of prescribed production practices for the amounts of inputs, the types and/or amounts of crops or livestock, or the techniques used to produce enterprises (96). In the United States, the 1990 Amendments to the Coastal Zone Management Act (PL 101-508) describe management practices that farmers in coastal areas must follow to control NPS pollution. States must implement plans to ensure compliance. States may overlook agriculture or certain categories of agriculture if they are not significant contributors to nonpoint pollution. States may also substitute voluntary programs for regulatory programs if they achieve NPS pollution control objectives of the regulations (98). Measures include cropland erosion protection, nutrient and pesticide management to minimize potential for runoff and/or leaching, and irrigation management (99). States have their own regulations. For example, producers on sandy soils in some parts of Nebraska must refrain from applying N during fall or winter, must conduct soil N tests, and must keep records of N applications (100). Some countries, including the Netherlands, Denmark, and Norway, limit the number of livestock per hectare so as to limit manure production to the potential of crops grown on the farm to utilize the nutrients in the manure (101).

Land use controls prohibit certain types of agricultural activity. These controls are usually applied at local levels, rather than statewide or nationally. For example, zoning ordinances may forbid concentrated animal feeding operations in areas near surface water bodies or with high potential to contaminate groundwater aquifers. Farmers in parts of the Chesapeake Bay drainage are required to leave uncultivated buffers adjoining surface water bodies.

Compliance with design standards is easier to monitor than compliance with performance standards. Design standards save farmers the cost of determining the most efficient

Table 4. Advantages and Disadvantages of Policies to Reduce Agricultural Pollution

Policy Type	Advantages	Disadvantages
Performance standard	1. Gives farmers flexibility to choose the cheapest way to reduce pollution	1. High administrative cost
	2. Encourages innovation	2. Does not allocate pollution reduction to farms with lowest control costs
Design standard	1. Easier to monitor compliance than performance standards	1. Does not allow farmers to choose the cheapest way to reduce pollution
	2. Environmental impacts more predictable than other policies	2. Discourages innovation
Licensing	1. Lower administrative cost than other regulatory options	1. Banning products is costly to farmers if no close substitutes exist
Pollution tax	1. Allows farmers to choose the cheapest way to reduce pollution	1. Difficult to measure farm pollution
	2. Encourages innovation	2. Difficult to estimate tax level that will achieve desired pollution reduction
	3. Generates government revenue	3. Politically unpopular with farmers
	4. Allocates pollution reductions to farmers with lower control costs	
Input tax	1. Easy to administer	1. May have small impact on input use and pollution
	2. Generates government revenue	2. Does not allocate pollution reduction to farms with lowest pollution control costs
		3. Politically unpopular with farmers
Pollution reduction subsidy	1. Allows farmers to choose cheapest way to reduce pollution	1. Difficult to estimate pollution reduction
	2. Allocates pollution reductions to farmers with lower control costs	2. Costly to taxpayers
	3. Encourages innovation	
Practice subsidy	1. Easy to administer	1. Does not allocate subsidies to farms with highest potential to reduce pollution
	2. Politically popular	2. Does not allow farmers to choose cheapest way to reduce pollution
		3. Discourages innovation
		4. Costly to taxpayers
Liability rules	1. Allows farmers to choose the cheapest way to reduce pollution	1. Politically unpopular
	2. Allocates more pollution reductions to farms with more pollution potential	2. Responsibility for pollution damage may be difficult to assess
	3. Encourages innovation	
Research and development	1. Easy to administer	1. Difficult to predict if research will be successful
	2. High payoff if research is successful	2. Time and cost required to develop new technologies
	3. Politically popular	
Moral suasion	1. Easy to administer	1. Limited long-term effectiveness
	2. Low cost to taxpayers	2. Does not allocate pollution reduction to farmers with the lowest pollution control costs
	3. Quick impact in emergency situations	
Education	1. Identifies opportunities to increase farm profits and reduce pollution	1. May not reach farms with the lowest costs of reducing pollution
	2. Stimulates innovation and adoption of pollution reduction techniques	2. Not effective if farmers' profits are reduced
Pollution reduction trading	1. Lowers costs of reducing pollution	1. Difficult to monitor compliance
	2. Low cost to taxpayers	2. Uncertain effectiveness of practices adopted to reduce nonpoint source pollution
	3. Encourages innovation	
Targeting	1. Increases cost effectiveness of government funds to control pollution	1. May be politically unpopular
	2. Allocates pollution reductions to farms with lower costs of pollution control	2. Administrative cost of identifying farms to be targeted
Free trade	1. Shifts production to land with less potential to pollute	1. Possible increased pollution in some LDCs
	2. Increases production in countries with low labor costs which encourages substitution of labor for chemicals	2. Politically unpopular in countries which lose agricultural production
Trade agreements	1. Require improved environmental standards as a condition of increased trade flows	1. Limited environmental scope
	2. Protect environments in LDCs	2. Difficult to enforce

strategy for reducing pollution. Also, impacts of design standards on pollution may be more predictable, assuming that farmers comply with the regulation. However, design standards place greater limits on farmer options and increase farmer costs of compliance. Uniformly applied standards do not consider variations in farm control costs. Overall costs of farmer compliance could be reduced by focusing more on farms whose profits are reduced relatively less by the design standard and focusing less on farms where the design standard results in large profit reductions. Design standards stifle research and innovation in new technologies, because standards give little leeway to differing methods for complying with regulations. Thus, in the long term, costs may be significantly higher for this type of regulation compared to other policies.

Licensing and Registration. Requiring licenses or registration is a method of regulating the manufacture and/or use of selected products. Licensing may reduce the use of a product and disseminate information about the product to users (101). Permission to produce or use a product may be withheld when its perceived negative impact on the environment cannot be justified in the view of the regulator. Licensing has a lower administrative cost compared to other regulatory policies, because the regulator deals with a relatively small number of manufacturers rather than many farms. The cost of such regulation depends on the restrictiveness of the registration and licensing procedure including how long registration is valid, how much information must be provided by the manufacturer and/or user in the registration procedure, and how strict the criteria are for awarding a license. The costs of withholding registration (product banning) depend on the availability of close substitutes for the product in question. Where close substitutes are available at low additional cost, banning may be a low-cost and effective environmental protection option.

Although the U.S. and most other developed countries require registering pesticides, there is variation in the requirements for obtaining approval and the period of approval. For example, in the Netherlands, registration expires after a period of two to ten years, requiring renewal by the chemical manufacturer. By contrast, pesticide registration in the U.S. is permanent. Registration of pesticides is supposed to be reviewed every five years, but the burden of review is placed on the regulatory agency, and consequently few are actually reviewed (101).

Landowner Rights and Responsibilities. In many if not most countries, landowners implicitly have a right to pollute whereas the burden is placed on governments to compensate landowners for reducing environmental damage. Governments could take actions to place more burden on landowners to manage their lands in ways that reduce soil and water quality damage (71). For example, Iowa has instituted legislation that requires farmers to follow soil conservation practices. Clarification of landowner responsibilities could be accomplished through state and/or federal laws, as requirements for receiving government program benefits (such as commodity cross compliance), through land easements (such as the Conservation Reserve Program or CRP), by educational programs, and by landowner certification programs (71). Clarification of landowner responsibilities would greatly enhance the effectiveness of other voluntary and mandatory soil and water quality protection programs by shifting the burden of compliance from governments to private individuals.

Regulatory Compliance. A disadvantage of regulations is the high cost of monitoring compliance with regulations by a large number of farms. However, in many countries, the number of farms is declining and agricultural production is becoming more vertically integrated. For example, in the U.S. a few integrated poultry firms control production, processing, and marketing of most poultry. Integrators contract production to farms but exercise control of production conditions and reimbursement. These trends may reduce costs of monitoring and enforcing regulatory compliance. Laws could be passed to make integrators responsible for pollution from contracted production. For example, if processors who own poultry are responsible for safe disposal of poultry waste, then regulations regarding safe disposal of waste will be reflected in contracts with producers (71). In this way the impacts of regulations directed at only a few firms would be reflected in the behavior of many farmers. The costs of regulatory monitoring and enforcement are passed from taxpayers to private firms and/or consumers of the firms' products.

Economic Incentives

Incentive strategies attempt to improve water quality by making it in the self-interest of polluters to change their behavior rather than requiring them to do so. Incentive strategies include subsidies, taxes, and liability for damages resulting from pollution.

Taxes. Taxes are negative incentives that increase production costs to reflect the additional costs of pollution to society. Implicitly, the levy of taxes presumes that society has the right to clean water and that the polluter must pay for the right to pollute. Tax revenues can be used by government to invest in pollution control. However, a tax scheme that achieves water quality objectives without unduly penalizing agriculture may be complex to design and may require periodic revision to adapt to changing economic conditions. Taxes also have high political costs because they are viewed as reducing the value of farmers' property to provide a public good (clean water). In this view, the tax is a partial taking of private property by the government for which compensation is required (102). Numerous forms of taxes can be levied; two of the most important are taxes on pollution and on chemical inputs.

Pollution Tax. A pollution tax could be levied per unit of emission from farms. Direct measurement of emissions could be done for PSs but would be difficult if not impossible for NPSs. Simulation models would probably be used to estimate emissions resulting from specified practices. Taxes on emissions would impact farmers directly in proportion to their contribution to the problem. Taxes induce farmers with lower control costs to reduce pollution more compared to those with higher costs. These taxes give farmers the flexibility to determine the least-cost method for reducing pollution (103). Emission taxes provide incentives for research and innovation in ways to reduce pollution. Administrative costs of such a tax are likely to be high because of the difficulty of determining farmers' emission levels. Also, determining the level of a tax to achieve the desired reduction in pollution is complicated by the difficulty of predicting farmers' responses.

Input Tax. A tax on chemical inputs would induce farmers to reduce chemical use. Consequently, pollution is likely to be reduced. For example, Iowa levies a tax on pesticides and N fertilizer and invests receipts in water quality protection programs (104). An input tax is easier to administer than an emissions tax because inputs are easier to observe. However, this tax does not account for variations in potential pollution among sites and in costs of controlling pollution. As a result, an input tax does not reduce use enough in areas with high potential for pollution and reduces use too much in areas with low potential (102). Farmers' use of chemical inputs may not respond greatly to price increases, in which case the impact of taxes on pollution is slight (105,106). Compared with emission taxes, input taxes somewhat reduce the ability of farmers to select the most cost-effective method for reducing pollution. Input taxes stimulate research on ways to use the higher cost inputs more effectively. Tax proceeds are often also used to fund research on methods to improve water quality.

Subsidies. In contrast to taxes, subsidies give positive signals to producers to change behavior to account for pollution. The subsidy concept implies that farmers have the right to pollute and that society must pay farmers to protect water quality. Subsidies have many of the same economic impacts as taxes in changing farmer behavior, and they are likely to be effective in reducing water quality damage. Subsidies should be politically acceptable to farmers and landowners. Subsidy programs are costly to governments, making the programs unpopular with nonfarming taxpayers.

Pollution Reduction Subsidies. Subsidies could be paid to farmers for reducing their pollution emissions. Emissions are difficult to measure directly. Thus, models would probably be used to estimate emissions resulting from alternative sets of practices for given soil and weather situations. For two reasons this type of subsidy is likely to minimize farm costs of reducing pollution (103). First, farms with the lowest costs of achieving such reductions are likely to take greater advantage of the subsidy compared to higher cost farms. Second, farmers can use specialized knowledge of their farming operations to select least cost practices for achieving reductions. This subsidy would also give firms incentives to develop new technologies and practices that reduce pollution emissions more cheaply. Pollution subsidies induce research and innovation in ways to achieve pollution reductions at lower cost.

Reduced emission subsidies would be expensive to administer. Application to NPSs requires detailed modeling of farm practices using varying soil and weather conditions to estimate pollution emissions. Another problem is determining baseline emissions for use in estimating reductions. Farmers would have an incentive to overstate their baseline emissions to qualify for more reduced emission credits.

Practice Subsidies. Farmers could be paid for adopting BMPs likely to reduce emissions. Practices could include reducing input use or taking land out of production altogether. Practice subsidies are perhaps the most widely used type of incentive for agriculture and water quality. These subsidies first assumed importance in soil erosion control efforts, but have since been extended to controlling pesticide and nutrient losses. Subsidized BMPs may be structural, for example, installing terraces or a sedimentation pond to control erosion runoff or installing manure storage to reduce manure application in periods when application is likely to result in runoff and leaching. Nonstructural practices may also be subsidized, such as scouting for IPM to reduce pesticide applications, changing crop rotations, or nutrient management planning.

In the U.S., the President's Water Quality Initiative (107) established special project areas where subsidies are paid for identified practices. Under provisions of the 1996 Federal Agriculture Improvement and Reform Act, the Federal government has authorized $1.3 billion for seven years to assist crop and livestock farmers in making environmental and conservation improvements. The program provides cost share or incentive payments for land management and structural practices (83). A number of states have subsidy programs for practices including manure storage construction, stream bank protection, tree planting, installation of contour stripcropping, and IPM scouting.

Practice subsidies are likely to be cheaper to monitor and administer compared with emission subsidies, because practices are easier to observe than emissions, and because it is easier to predict farmers' responses to a practice subsidy. However, practice subsidies reduce farmers' flexibility in choosing the most cost effective way to reduce pollution. The amount of subsidy paid for a practice is the same regardless of the amount of pollution reduction achieved. For that reason, overall costs of achieving pollution reductions are not minimized, because too many practices are subsidized on land with low potential to pollute and too few on land with high pollution potential. Practice subsidies are also less likely to induce innovation and research compared with emission subsidies and taxes, because the subsidy is tied to a defined practice which limits flexibility.

An indirect form of practice subsidy is the cross-compliance requirement for government program benefits. U.S. farmers who wish to qualify for commodity subsidies included in the farm program must comply with specific requirements on highly erodible land to minimize erosion. These requirements may vary by farm, but could include practices such as contour stripcropping, changing crop rotations, and installation of terraces. Further, farmers must refrain from converting native prairie land to cropland that is sensitive to erosion (sodbuster requirement) and from converting wetlands (swampbuster). Conservation compliance is efficient to administer because it is tied to an existing program, the farm clientele has already been identified, and much is already known about the farms involved. Conservation compliance is politically unpopular with farmers, adding requirements to an existing subsidy program. In general taxpayers benefit from obtaining additional water quality protection benefits for the given government expenditures.

The long-term effectiveness of conservation compliance in protecting water quality is uncertain because farmers' decisions about participation are based on their economic benefits from the program, rather than on the water quality gains. In periods of high prices, farmers may not participate in the program and may not comply with protection measures (103). Farmers who participate in government programs may not be those with the greatest pollution problems; thus conservation compliance may not reduce pollution on farms with the most potential for low cost pollution control. Also, farmers who do comply are not given flexibility in choosing the least cost ways to reduce pollution which increases costs. The lack of flexibility in this strategy also reduces incentives to invest in research in new technologies to protect water quality at lower cost.

The Conservation Reserve Program (CRP) idles lands with high potential to damage water quality while supporting crop prices by reducing crop production. Farmers submit bids to the government indicating the payment required for removing land from production for periods of ten years. If accepted, they receive annual payments in exchange for idling land using a conserving plant cover. By June 1992, 14.8 million hectares (36.5 million acres) had been enrolled at an average annual cost of $1.8 billion or nearly $126 per hectare ($51 per acre). Initially the objective of the program was to reduce soil erosion on targeted lands with water quality enhancement a secondary goal; however with the 1990 farm bill (PL 101-624), water quality enhancement became a major goal. Administrative costs of the program are relatively low considering the length of time land is idled. A major potential benefit of the program is the ability to target lands based on their contribution to water quality objectives. A weakness of the program is that water quality benefits are lost when CRP contracts expire, again releasing land to potential production. This weakness could be partly overcome by purchasing long-term easements on environmentally sensitive lands to limit farming practices to those with low potential for pollution (71).

Liability Rules. Subsidies and taxes should be implemented before actual damage from the polluting activity occurs. Liability is an ex post facto incentive in that the farmer is held liable for damages that result from polluting activities. Alternatively, the input supplier/manufacturer may be held liable in place of the farmer. For example, Connecticut farmers are not liable for damage to groundwater from chemical application as long as they follow label directions. Chemical manufacturers are held liable for groundwater damage resulting from use of their products. Liability may be established explicitly by administrative actions, civil lawsuits, or criminal prosecutions by state or federal agencies. Private individuals may also establish liability through civil lawsuits in state or federal courts. A finding of liability may result in civil or criminal penalties and/or an injunction requiring the defendant to desist from the polluting activity (108). The effect of liability rules is to make producers aware ex ante of the potential damages of their activities so that they prevent pollution to avoid paying for these damages. Farms that pollute more are more likely to be held liable for damages and, consequently, have more incentive to reduce pollution. Also, firms are given flexibility to reduce pollution in the cheapest way. Liability also gives firms incentives to invest in developing new technologies and practices that farmers can use to reduce emissions and potential damages. Liability rules may be politically unpopular because they force those suffering damages to take the polluter to court and prove damages before receiving payment (96). The responsibility for emissions may be difficult to assess, because NPS pollution is diffuse. Also, there may be long delays between pollution emissions and actual damages to groundwater.

In the U.S., a number of states have enacted groundwater protective legislation setting standards for allowable levels of various contaminants. Frequently, state standards are based on the EPA's standards for contaminants and characteristics of drinking water. Farmers or agricultural processing firms who violate these standards may be held liable for damages. However, a number of states have also acted to limit farmers' liability for pesticide contamination of groundwater, if farmers follow good farming practices and are not in violation of water pollution laws (108).

Research and Development

Research and development is used to identify alternative technologies that produce agricultural products with less environmental damage (103). Interest in funding such research has increased with awareness of the damages from agricultural practices. For example, Low Input Sustainable Agriculture (LISA) funding was set at $40 million in the 1990 farm bill (PL 101-624). One of the primary objectives of this research is environmental protection from agricultural practices. More research is needed on economic and social factors that affect producers' decisions to adopt new technologies. A better understanding of these factors can enhance the development of new technologies with more potential for adoption and the design of more effective policies to control agricultural pollution (71).

Agricultural research is politically attractive to farmers particularly if the costs of funding research are shared by taxpayers in general. Benefits of technological breakthroughs that are economically viable and voluntarily adopted by farmers are reaped for many years and have low administrative costs. Assuming that the new technologies are economically viable and that researchers are given flexibility in how they conduct research, new technologies will be adopted on farms having the greatest potential to reduce pollution at least cost.

However, prospects for research success in developing profitable technologies are uncertain, and new technologies take years to develop, which makes research unlikely to be an effective short-term water quality protective strategy (103). Funding research imposes burdens on taxpayers. It may be possible to use receipts from taxes on inputs, such as pesticides and fertilizers, to fund research programs as is being done in Iowa and some other states and in some European countries. In this way, the cost of new research is borne by those most likely to be responsible for pollution.

Moral Suasion

Moral suasion involves attempting to convince polluters to cease pollution voluntarily because it is "the right thing to do." Moral suasion may be effective if it is used to convince firms that voluntary compliance with environmental goals is needed to avert future regulations that may have greater costs. Because farmers comply voluntarily, administrative costs are low, there is little political resistance from farmers, and the burden on taxpayers is small. Moral suasion is most effective as a temporary measure in an emergency situation where quick action is needed to avert environmental damage. Moral suasion is less effective in the long run particularly among farmers subject to strong domestic and international competition. Farmers who voluntarily raise their costs without compensation to achieve water quality goals may find they are less competitive. Even if effective, there is no guarantee that farmers who comply voluntarily will be those with the lowest costs of pollution control. Hence overall control costs may be higher than with other strategies. This strategy does not produce incentives to invest in new technologies for controlling pollution at lower cost.

Education

Policymakers may provide farmers with information about the impacts of their practices on pollution and the opportunities to reduce pollution while increasing net returns. Often the information is provided free of charge to farmers at taxpayers' expense. Educational strategies may involve use of mass media, such as television, radio, published handouts, and leaflets. Demonstration plots allow farmers to learn by example. In the U.S., 16 federally funded Water Quality Demonstration Projects were initiated in 1990 and 1991 to disseminate information about water quality protection to wide areas using field demonstrations, farm tours, and mass media (109).

An alternative strategy is to educate farmers on a one-on-one basis. For example, many states in the U.S. have nutrient management programs whereby trained specialists develop nutrient management plans for farmers. Frequently such plans result in fertilizer cost savings for farmers and less nutrient losses. Farmers learn more about nutrient management which provides benefits beyond the immediate impacts of the plan. At the federal level in the U.S., this one-to-one approach is best typified in the 74 Hydrologic Unit Area Projects and 110 Water Quality Special Projects begun in the early 1990s. Technical experts funded by the federal government provide assistance to farmers in selecting and implementing BMPs with the greatest water quality benefits for the farm.

Education is a benefit to farmers because this information subsidy often increases profits. Frequently much or all of educational costs are borne by taxpayers. It may be possible to fund programs from receipts of taxes on potentially polluting inputs, such as pesticides and fertilizers. Overall, education is likely to be effective in protecting water quality to the extent that "win-win" opportunities exist, ie, opportunities for farmers to increase profits at the same time as water quality damages are reduced. Educational programs will not be effective if they promote practices that reduce farmers' net returns without compensation. Educational programs that are broad and encompass most of the enterprises on the farm will help identify least cost ways to reduce pollution losses. Voluntary educational programs will not necessarily identify farms with the lowest cost opportunities to reduce pollution. Farmers' willingness to be trained in pollution reducing methods depends more on personal attributes and progressiveness than on the farm's opportunities for low cost pollution reduction. Education should stimulate research and development of new technologies for water quality protection. Educated farmers are more aware of the need to protect water quality and more receptive to adoption of new technologies.

Existing educational approaches in soil and water pollution control have been criticized for lack of sophistication in audience identification and information delivery (71). Program effectiveness could be increased by tailoring information content according to type of farm enterprise or service. For example, educational programs aimed at agricultural input dealers, crop soil consultants, or bankers may be needed in addition to those aimed at farmers. An example is the Certified Crop Advisor (CCA) program, which trains and certifies crop advisors. The goal of the program is to assure that crop advisors have the basic knowledge to help growers produce crops efficiently while protecting environmental quality. The voluntary program is conducted under the direction of the American Society of Agronomy in cooperation with agribusiness and various government agencies.

Pollution Reduction Trading

Frequently, agricultural NPSs of pollution are not regulated whereas point sources are regulated. Tradeable discharge permits allow PSs to reduce their costs of complying with the discharge standard by trading with the unregulated NPSs (110,111). The trades presume that the NPS has a right or permit to pollute and the PS does not. A trade allows a PS to exceed its allowable standard by a specified amount in exchange for the NPS agreeing to reduce its discharges. The agreement might require the NPS to install a set of pollution control practices whose average pollution reduction is some agreed amount.

Trading lowers total costs of achieving environmental goals because farmers who accept trades normally have lower costs of reducing pollution than PSs who initiate trades. For example, in the Tar-Pamlico watershed of North Carolina, it was estimated that upgrades for plants to control N and P emissions would cost $50 to $100 million. These goals could be met with NPS improvements primarily for agriculture costing $11.2 million (112). Trades allow farmers the flexibility to choose least cost pollution control practices. Nutrient trading also provides incentives to PSs to further reduce control costs through innovation (113). If PSs can develop new technology that helps them reduce emissions more cheaply than via existing trades, they can do so (113). Trading is voluntary, which should increase its acceptability among farmers. Finally, trading does not require large government expenditures.

In spite of its advantages, nutrient trading has been little used for several reasons. First, effectiveness of NPS pollution reduction measures is uncertain (primarily due to weather and site specific conditions), whereas control effectiveness of PS measures is relatively certain. Second, monitoring compliance with NPS measures is difficult and expensive because of large numbers and wide dispersion of NPS polluters. Third, trading requires an implicit acknowledgment by PSs that NPSs have a right to pollute while they do not. PSs may feel that NPSs have equal responsibility to control pollution, and NPSs may not wish to draw attention to themselves as polluters. Fourth, the location of pollution may be important particularly with nonconservative pollutants that degrade. In that case, pollution reduction by NPS polluters may not achieve the same water quality improvement as an equivalent reduction at a PS.

Targeting

Given the high costs of reducing NPS pollution, targeting control efforts to watersheds and/or farms within watersheds with the greatest potential to reduce pollution cheaply is an attractive option. Targeting is not an independent policy option, such as those discussed above, but rather a strategy for implementing one or more of the previously described control options. For example, targeting can be a mechanism for prioritizing watersheds to receive government subsidies for agricultural pollution control efforts. Three factors determine a watershed's relative ranking in efforts to control agricultural pollution (114). First is the extent to which agricultural pollution reductions make a difference in improving the quality of the water. For example, if the goal is to make surface waters in a watershed fishable and swimmable and reductions in agricultural pollution will achieve that goal, then agriculture in the watershed may receive priority for pollution control efforts. Second, the potential demand for water quality improvement in the watershed is also important. Heavily populated watersheds may receive higher priority than sparsely populated watersheds, because more people are potentially affected by the water quality improvement. The value that individuals place on obtaining the improvements is also important. One study shows that allocation of soil erosion control funds among regions in the U.S. would be altered considerably after taking into account demand for reducing off-site damages compared to the allocation obtained when only on-farm costs of erosion are assessed (115). Third, the overall costs of reducing agricultural pollution in the watershed must be considered. Watersheds with generally lower costs would receive higher priority for water quality protective efforts.

The rules that should be followed in allocating limited resources to agricultural water quality control are (114,115) (*1*) eliminate watersheds where feasible agricultural pollution controls cannot achieve water quality objectives for watershed; (*2*) allocate funds to the watershed where the value of water quality improvement per dollar of public expenditure on agricultural control programs is highest and continue to do so until the value of control per dollar of expenditure is higher in another watershed; (*3*) at that point shift control dollars to another watershed whose value per dollar is now highest; and (*4*) continue to follow this principle until all funds for agricultural control are exhausted.

Targeting can also be done within watersheds at farms with relatively low costs of pollution control because of their economic and physical characteristics (116). For example, farms with large livestock enterprises and no existing manure storage may be able to reduce nutrient losses by utilizing storage and spreading manure in seasons when runoff and leaching are less likely. A study found that per-ton costs of erosion reduction in highly erodible west Tennessee watersheds were 34% lower than the national average (117). Operator characteristics, such as education and knowledge about low-input farming techniques, affect which farmers are likely to successfully adopt water quality protective techniques. Farmers with better education, who are more willing to take on risks or who are more concerned about protecting water quality, may be more likely to accept new technologies that reduce water quality damage. Farm location may be an important factor. For example, farms near surface water may be able to achieve larger reductions in pollutants delivered to surface water at a given cost than more distant farms.

Targeting increases the cost effectiveness of pollution control expenditures by the government, and it can allocate pollution reduction responsibilities to farms having the lowest control costs. However, costs of determining which farms are likely to have low control costs must be considered as part of total pollution control costs. These costs include conducting studies with sample farms to determine what factors are associated with low control costs (116). Once these factors are identified, their distribution among the population of farms must be determined to implement targeting. These costs may in some cases exceed the potential cost savings realized from selecting low cost farms to achieve pollution control objectives. In addition, political considerations may make it difficult to selectively target farms for pollution control efforts if targeting is perceived to be unfair. Political resistance is perhaps less of a problem with targeting subsidies, which compensate farmers for their costs of controlling pollution.

Programs that retire lands for environmental purposes can be made more effective by targeting watersheds and sites within watersheds with the greatest water quality problems and the most potential to pollute ground and surface waters. For example, in the U.S., targeting lands for enrollment in the Conservation Reserve Program based on the potential to protect ground and surface water rather than minimizing erosion can increase the relative cost effectiveness of government funds by 50% (118). In particular, benefits could be enhanced by focusing on lands bordering surface water where pollution is high (119). Since 1990, environmental benefits per dollar of government cost have been considered in evaluating lands for enrollment in the program.

Agricultural Trade Reforms

Promote Free Trade. Promotion of free trade in agricultural products benefits water quality by encouraging the movement of agricultural production to areas having the greatest comparative advantage in agriculture. Livestock and crop production would tend to move away from the DCs and toward LDCs particularly Latin America (85). Crop production would tend to move to areas with more productive and less erosive soils. Soil and water quality would improve, because LDCs use fewer machinery and chemical inputs and more labor because of their lower labor costs relative to DCs. Livestock production would rely more on grazing and less on spatially concentrated confinement with fewer problems for environmentally safe manure disposal. Although free trade may increase agricultural pollution in some LDCs, the net global effect should be a reduction in agricultural pollution. Reductions in agricultural pollution in countries that currently subsidize agriculture should offset pollution increases elsewhere. Other trade reforms to be discussed in the following sections can encourage environmental protection in LDCs.

Trade Agreements. Increased agricultural production may result in greater pollution in some LDCs. However, LDCs are likely to place lower priority on environmental quality than DCs because of their urgent need to generate foreign exchange and produce the basic necessities for their populations. LDCs do not have the infrastructure and government budgets that DCs have to implement policies that reduce agricultural pollution. Trade agreements can include provisions to protect environmental quality in both DCs and LDCs from damage resulting from liberalized trade. A side agreement to protect the environment was included in the 1993 North American Free Trade Agreement (NAFTA) between the U.S., Canada, and Mexico. NAFTA provisions to promote environmental quality include (88) (*1*) that international environmental agreements (such as the Basel Convention on the Control of Transboundary Movements of Hazardous Waste) take precedence over NAFTA provisions; (*2*) that members are allowed to enforce environment-related performance requirements on foreign investors; and (*3*) that members are exhorted to enforce their own environmental laws. In addition a Commission for Environmental Cooperation (CEC) is funded to (*1*) monitor implementation of NAFTA environmental provisions; (*2*) work to harmonize and raise environmental standards of the three member nations; and (*3*) review cases of nonenforcement of environmental standards by the members.

Although NAFTA is the most important example to date of a "green" trade agreement, some critics question the efficacy of the NAFTA environmental agreements because of (*1*) the limited scope of environmental nonenforcement actions that may be challenged; (*2*) the limited enforcement power of the CEC;

and (*3*) failure to give transboundary environmental problems highest priority (88).

The Uruguay Round Agreements (URA) of the General Agreement on Trade and Tariffs (GATT), which were concluded in 1993, contain several provisions to protect environmental quality including (88) (*1*) explicit mention in the preamble of the need to pursue sustainable development; (*2*) development of codes for allowable sanitary, phytosanitary, and technical barriers to trade designed to pursue legitimate health, public safety, and environmental objectives; (*3*) allowance of environmental subsidies; and (*4*) a more public dispute settlement process. Panels convened to settle trade disputes may seek scientific advice on environmental issues. Regarding point (*2*), sanitary and phytosanitary provisions allow imposing trade measures to protect human, animal, or plant life from pests and diseases and from contaminants and additives in food. Technical barriers to trade allow developing product standards to protect the environment and human, animal, and plant health and safety.

Environmental Trade Measures. As of 1994, 20 international environmental agreements including the 1985 International Code of Conduct on the Distribution and Use of Pesticides and the 1989 Basel Convention on the Control of Transboundary Movements of Hazardous Wastes and Their Disposal utilized trade measures, such as import/export restrictions and trade sanctions to enforce environmental objectives (120). The most effective trade measures are those that are applied multilaterally; however multilateral negotiation is a slow and costly process. Unilateral measures invoked in support of the agreed upon environmental standards are less likely to be effective because (*1*) they may be overturned by trade dispute resolution panels; and (*2*) they are easy to circumvent via transhipment of goods from the nation whose goods are being embargoed to other countries which do not subscribe to the embargo.

Global Environmental Management Institutions. International institutions could be devised to (*1*) insure that environmental objectives are not compromised by trade measures; and (*2*) advance environmental objectives more generally. It is proposed that environmental staffing in the World Trade Organization of the GATT be increased to pursue the first objective (120). To pursue the second objective, it is suggested that an Intergovernmental Panel on Trade and Environment outside the GATT be set up to further environmental reform internationally. This panel could develop an information base on environmental implications of development that could help develop consensus for international environmental reform. For example, such a panel could develop a blueprint for management assistance to LDCs to help them anticipate and respond to environmental damages resulting from increased economic development due to liberalized trade.

Harmonization of Standards. Countries may harmonize their product standards, production and processing methods, and environmental performance standards. Harmonization prevents some nations from acquiring a trade advantage by ignoring environmental costs. Harmonization may reduce the costs of companies that export agricultural products to many countries because exporters meet only one set of standards rather than many. Harmonization of environmental standards also has advantages when the actions of one country affect the level of the environmental amenity in another country. Multilateral organizations, such as the United Nations Food and Agriculture Organization, the International Monetary Fund, and the World Bank and international trade and environmental agreements, could be used to obtain and enforce such standards (121).

Harmonization ignores countries' differences in preferences for environmental amenities. Lower income countries may prefer a lower level of environmental quality to produce adequate food for their populations. Also, countries may differ in the capacity of their environments to absorb agricultural wastes and convert them into harmless or even ecologically beneficial products. Forcing all countries to adopt a consistent set of standards is unlikely to be economically efficient (122). Significant political and economic costs may be required to negotiate harmonized standards. Common standards may not be politically feasible because some nations will resist perceived encroachments on their sovereignty and refuse to comply with standards. For these reasons, countries should proceed cautiously and harmonize standards or products where the benefits clearly outweigh costs. In cases where harmonization makes sense, wealthier nations should be prepared to subsidize improved water quality standards in lower income countries. Subsidies may be justified because actions to protect water quality in low income countries may have environmental benefits for wealthier countries, and wealthier nations may be more willing to pay for environmental amenities (123).

SUMMARY AND CONCLUSIONS

Primary Agricultural Pollutants

The primary pollutants associated with agriculture include sediment, nutrients (especially N and P), pesticides, salts, and pathogens. Sediment is the result of soil erosion, a naturally occurring process that is accelerated by land-disturbing activities. The primary activity in agriculture that generates sediment is crop production. Tillage of the soil, to prepare a seedbed, to harvest a crop, or to control weeds, increases the susceptibility of the soil to erosion. Soil erosion and the resulting sediment are detrimental to the eroded land and to the water in which the sediment is deposited. The loss of top soil decreases soil fertility and adversely affect crop yields. Soil formation is a slow process; in many areas the rate of erosion exceeds the rate of soil formation resulting in a net loss of soil. Sediment impacts physical, biological, and chemical characteristics of receiving waters. Sediment decreases the storage capacity of reservoirs. Sediment deteriorates or destroys aquatic habitat by decreasing the amount of light that penetrates the water and by depositing on the bottom of water bodies, smothering or inhibiting aquatic organisms and vegetation. Sediment also negatively affects water quality by transporting other pollutants, such as nutrients and pesticides, that sorb to the sediment.

Although nutrients are needed by all living matter, they become pollutants when added to water in excessive amounts. The main water quality concern related to nutrients is accelerated eutrophication, caused primarily by excessive amounts of nitrogen (N) and/or phosphorus (P). Eutrophication occurs when there is a net increase of organic matter. Accelerated eutrophication occurs when excessive nutrients increase the rate of organic matter production, particularly algae and other aquatic plants. The increased production utilizes oxygen causing a shift in fish and shellfish populations from better quality fish to rough fish. Taste and odor problems caused by algae can also make water less suitable for potable use and contact recreation.

The use of pesticides to control weeds, insects, and diseases has substantially increased crop yields and assisted in controlling diseases such as malaria. Pesticides are transported to surface and groundwaters through several modes. Drift occurs during application. Pesticides are leached to groundwater and are delivered to surface water dissolved in runoff or adsorbed to sediment. Pesticides negatively affect water quality. By definition, toxic pesticides kill non-target and other aquatic species. Pesticides negatively affect essential processes in aquatic plants, including photosynthesis, reproduction, respiration, growth, and development. Some pesticides bioaccumulate in fish tissues and other species, thus posing a health risk to humans and other species that consume the fish. There are also human health risks associated with consuming contaminated water. The water quality impact of a specific pesticide is related to its toxicity, persistence, solubility in water, and ability to adsorb to soil particles. Each pesticide has unique characteristics.

Salts become pollutants when concentrations are sufficient to negatively affect soil and water quality. The presence of salts in soils decreases crop growth. Some saline soils occur naturally, but many have resulted from irrigation using water with high salt content. Salts in water bodies affect the types and populations of fish and aquatic wildlife that are present. Fluctuations in salinity cause greater problems than the absolute level of salinity. Salts destroy habitat and food source plants for fish species.

Pathogens differ from the other pollutants described because pathogens in any concentration present a human health hazard. The primary source of pathogens in agriculture is livestock production and processing. Pathogens are delivered to surface waters by runoff from feedlots or from fields fertilized with manure. Pathogens are also discharged from livestock or poultry processing plants.

Sources of Agricultural Pollution

Sources of agricultural pollution are classified as PSs and NPSs. PS pollution results from pipes or other definable discharge points. NPS pollution is caused by rainfall and runoff and occurs throughout a diffuse area, such as a field. There are some PSs in agriculture, primarily processing plants and feedlots, but most of the sources are nonpoint, particularly those associated with crop production.

Pollution occurs when two conditions are satisfied. First, the pollutant must be available. Some of the pollutants, including sediment and salts, are available naturally. Others, such as pesticides, are supplied through farming operations. Nutrients occur naturally and are applied by manure and inorganic fertilizer applications. Second, a method of transporting the pollutant from the source to receiving waters must be present. In NPS pollution, the method of transport is runoff or percolating water, initiated by rainfall or irrigation.

In general, primary pollutants are associated with different agricultural land uses. Sediment, nutrients, and pesticides are associated with dryland agriculture. In addition to sediment, nutrients, and pesticides, salts are associated with irrigated agriculture. The main pollutants associated with pasture and range land are sediment and nutrients. Confined animal production facilities are potential sources of sediment, nutrients, salts, and bacteria. Dissolved nutrients, pesticides, salts, and pathogens are associated with orchards or nurseries.

The pollutant loads that result from agriculture depend on many factors, including land use, type of crops or animals, crop rotation, soils on which crops are grown, topography, climatic conditions, farming technology, irrigation and drainage, timing and location of chemical or fertilizer application, and cultivation practices. Because so many factors are involved, wide ranges of pollutant loadings occur from the same land use. Potential environmental problems due to agricultural operations often depend more on the production practices and waste management techniques utilized by farmers than the size of the operation, the number of animals, or the amount of waste.

Pollution Control

The most successful approach to pollution control is to prevent pollution from occurring in the first place. However, prevention is often not a viable economic alternative. The second best approach is to reduce the amount of pollution that occurs. Pollution reduction occurs at the source or between the source and the receiving waters. If pollutants reach receiving waters, then treatment is the only alternative to correcting the problem. In many cases, treatment is not feasible because of the size of the polluted water body or groundwater.

Pollution control measures differ for PS and NPS pollution. PS control is effected through effluent limitations established by a permitting agency. A PS discharger can utilize a variety of options, from prevention through treatment, to achieve effluent limitations. Typically, biological, physical, and chemical treatment processes are applied to all wastes.

NPS pollution control is achieved by best management practices (BMPs). The term BMP is used to indicate a single practice or a combination of practices that is an effective practicable means of preventing or reducing the amount of pollution generated by NPSs. BMPs are selected and implemented on the basis of the water quality problem, the water quality goals to be achieved, site-specific conditions, and the type of operation contributing the pollutants. The effectiveness of BMPs is site-specific, depending on factors, such as climate, topography, and soils.

BMPs are classified as vegetative, management, and structural. Vegetative BMPs provide vegetative cover on the soil surface, the best method to prevent soil erosion. Reducing soil erosion decreases the amount of sediment and sediment-adsorbed pollutants delivered to surface waters. Some vegetative BMPs include cover crops, crop rotations, strip-cropping, vegetated filter strips, and grassed waterways.

Practices related to soil manipulation, use and application of water, manure storage and application, fertilizer application, pesticide application, and livestock management are often considered to be management BMPs. One of the primary management BMPs is conservation tillage. Crop production depends on the soil resource. Tillage is the practice of manipulating the soil to enhance crop production. Tillage also increases soil erosion and delivery of sediment and sediment-adsorbed pollutants to surface waters. Conservation tillage refers to tillage practices that leave at least 30% of the soil surface covered with residue. Residue management is a concept that goes hand-in-hand with conservation tillage. The presence of residue on the soil surface decreases soil erosion.

Management BMPs for manure, fertilizer, and pesticide application focus on the amount and timing of application and on the application technology. The main concept is to use application rates that minimize the potential for water quality problems. For fertilizers and pesticides, emphasis is on applying just the amount needed; this is beneficial for both economic and environmental reasons. For manure, from a disposal viewpoint, it is desirable to use as much as possible, but at rates unlikely to cause water quality problems. Practices that address manure and fertilizer application are considered nutrient management. Integrated pest management (IPM) is a concept that draws many pest management practices together, including biological, cultural, and chemical practices.

Structural BMPs include practices that require some type of construction, often including earth-moving. Examples of structural BMPs include terraces, diversions, grade-stabilization structures, critical area planting, detention ponds, and stream bank stabilization. Structural BMPs are usually the last choice, following vegetative and management practices, because structural practices are more expensive to implement and maintain. Often, vegetative and management practices are used in conjunction with structural practices.

Effectiveness of different practices is site-specific. Thus, each NPS problem must be addressed individually to determine the appropriate BMPs. In addition, caution must be exercised in applying practices to ensure that the problem is solved, not transferred. For example, practices that reduce runoff and, therefore, yield of sediment and adsorbed pollutants, may increase the leaching of dissolved pollutants to ground water because, in general, if water does not run off, it infiltrates.

Many BMPs have been developed to control NPS pollution. Many studies have been conducted to determine the effectiveness of different practices. Research continues to further develop understanding of the mechanisms involved in different practices and to develop new practices, but many practices have already proven effective. Although a number of pollution control practices are available, adoption of those practices is not widespread enough to have solved the pollution problem.

Barriers to Adoption

Several factors encourage agricultural practices that degrade soil and water and inhibit the adoption of more favorable practices. Frequently property rights are defined so that farmers are not penalized for farming practices that damage water quality through pesticide, sediment, and nutrient pollution. Farmers do not hold property rights to water after it leaves their farms; thus they are unable to earn a return from actions to reduce pollution and improve the usefulness of water for other uses, such as drinking, recreation, and wildlife.

Some practices reduce pollution and increase profitability whereas others reduce profits. The profitability effects of practices depend on the scale at which the practices are carried out, the length of time allowed for markets to adjust to pollution reduction requirements, and the type of policy. Pollution reduction practices vary in their effects on farmers' profits. Lack of information may be a barrier to adoption of pollution control practices. Farmers may be unaware of opportunities to increase profits and reduce pollution at the same time. Acquiring such information entails costs in terms of time and/or money. Farmers are also concerned about the effects of practices to control pollution on income risk. For example, greater reliance on organic nutrients in place of inorganic fertilizer may increase yield and income risks. Other financial and institutional factors may also inhibit adoption of pollution control. For example, structural practices such as terraces or manure storage that require large amounts of capital may be too expensive for farmers with limited capital and/or access to credit. Current agricultural trade policies encourage agricultural pollution. Developed countries (DCs) subsidize their agricultural sectors, increasing agricultural production and encouraging use of agricultural chemicals and fertilizer and intensive livestock production with resulting manure disposal problems. Less developed countries (LDCs) exploit their agricultural sectors by keeping prices below world levels, which encourages the movement of labor out of agriculture. At the same time, LDC governments may attempt to compensate for low prices by subsidizing chemicals, fertilizer, and irrigation water, which increases their use and potential for pollution.

Policies to Overcome Barriers

Policies to reduce agricultural pollution include regulation, economic incentives, research and development, moral suasion, pollution reduction trading and trade reforms. Targeting policies at watersheds and farms within watershed may be used to improve their effectiveness. Because barriers to adoption of pollution reduction practices vary by economic and social characteristics of the farm and farm manager, a mix of policy strategies is needed to reduce soil and water quality damage (71). For example, educational and technical assistance may be adequate for producers who are willing to adopt less polluting practices but are unaware of them. Market incentives, regulations, or other policies may be needed for those who are aware of such practices, but unwilling to adopt them. Policy instruments may also complement each other; for example, research strategies may accelerate the development of more effective technologies for reducing soil and water quality damage that can subsequently be integrated into voluntary or regulatory programs.

Regulatory policies require farmers to conform to specified practices or to achieve specified reductions in pollution. Design standards that dictate how pollution reductions are to be achieved are less costly to enforce than other policies, but increase costs by stifling research and innovation and reduce farmers' flexibility in choosing the most cost-effective way of pollution control. Performance standards give farmers flexibility in how to reduce pollution and can reduce farmers' costs of compliance. However, performance standards have high costs of monitoring compliance. Product registration can reduce use of products having adverse water quality impacts, and convey information to producers and consumers about the products' environmental impacts.

Incentives in the form of taxes, subsidies, and liability rules may be imposed to make farmers account for costs of pollution. An input tax which discourages use of a polluting input is easy to administer and generates government revenue. However, it may have little impact on use of the polluting input. A pollution tax or a pollution subsidy gives farmers more flexibility to reduce pollution in the lowest cost way, but this policy is hindered by difficulty in measuring farmers' pollution. Practice subsidies are preferred because they are politically popular and easy to administer. However, they discourage innovation and do not allocate subsidies to farms with the highest potential to reduce pollution at low cost. Liability rules encourage farmers to reduce pollution in the least costly way, allocate pollution reduction to farms with lower control costs, and encourage innovation to reduce pollution control costs. However, they are politically unpopular because they require those harmed by pollution to take costly court action to obtain judgments against polluters.

Research and development yields new technologies that increase farmers' returns and reduce soil damage and/or loss of pollutants to the environment. Social research enhances technology diffusion by showing why farmers adopt or fail to adopt new technologies. New technologies have low administrative costs if adopted voluntarily by farmers and provide benefits for many years. However, research requires years to discover and refine new technologies, making it unlikely to provide short-term water quality benefits. Also paying for research places a burden on taxpayers in general, unless ways are developed to tax users and producers of polluting inputs to raise funds.

Moral suasion seeks to convince polluters to cease pollution out of a sense of moral obligation to protect the environment. This strategy may produce quick results in emergency situations, but is unlikely to be effective in the long term. Those who voluntarily reduce pollution may lose profits and have difficulty competing with those who do not comply.

Pollution reduction trading allows PS polluters with high costs of pollution control to trade with agricultural NPS polluters with lower control costs. PSs pay NPSs to reduce pollution at lower costs than if the PSs achieved the reductions themselves. Trading remains largely unexploited because of the uncertain effectiveness of NPS pollution controls and the difficulty of monitoring NPS compliance.

The effectiveness of pollution reduction programs is enhanced by targeting watersheds and/or farms where pollution control is achieved most cheaply. Overall costs of achieving watershed water quality objectives are reduced because only the farms with the lowest costs of reducing pollution are involved. However, care must be taken that costs of acquiring information required to identify these farms do not exceed farm savings. Targeting may also create a perception of unfairness in singling out farmers for pollution control efforts, although perceived unfairness is less of a problem with subsidy programs.

Trade reforms to liberalize world trade should benefit water quality in many DCs because agricultural production will shift to LDCs. Pollution from intensive use of chemicals in crop production and manure disposal from intensive livestock production operations will lessen by such a shift. Agricultural pollution could increase in some LDCs although removal of subsidies for fertilizers, chemicals, and irrigation water could lessen the undesirable impacts of expanded production there. Trade agreements, environmental trade measures, global environmental management institutions, and harmonization of standards can also help insure protection of soil and water with expanded trade. DCs should be prepared to subsidize the environmental protection efforts of LDCs recognizing that environmental protection in LDCs has benefits for DCs and that LDCs may legitimately have lower environmental protection objectives than do DCs.

Some policy tools produce "win–win" outcomes in which farmers' profits are increased and pollution is reduced. For example, education may encourage farmers to adopt soil N testing, which reduces N applications and N pollution and increases farmers' profits. However, most policy tools reduce pollution at a cost to farmers, taxpayers, and/or consumers. Further reductions in agricultural pollution will depend on how much society values a cleaner environment and is willing to pay for pollution reduction.

ACKNOWLEDGMENT

The authors express appreciation to D. Orden and J. Pease for helpful comments on an earlier draft. Any errors of fact or interpretation are the sole responsibility of the authors.

BIBLIOGRAPHY

1. USDA, "1992 NRI Summary Statistics," Natural Resources Conservation Service, URL http://www.nhq.nrcs.usda.gov/nritables.html, Sept. 1995.
2. U.S. EPA, *Erosion and Sediment Control Surface Mining in the Eastern U.S.,* EPA-625/3-76-006, U.S. EPA, Washington, D.C., 1976.
3. E.H. Clark, Jr., J.A. Haverkanp, and W. Chapman, *Eroding Soils—The Off-Farm Impacts,* The Conservation Foundation, Washington, D.C., 1985.
4. V. Novotny and H. Olem, *Water Quality: Prevention, Identification, and Management of Diffuse Pollution.* Van Nostrand Reinhold, New York, 1994.
5. C.H. Sawyer, *New Engl. Water Works Assoc.,* **61** 109–127 1947.
6. N. Brady, *The Nature and Properties of Soils,* Macmillan Publishing Co., New York, 1990, pp. 315–350.
7. J. Zhang, C.T. Haan, T.K. Tremwel, and G.A. Kiker, *Trans. ASAE* **38**(3), 767–773 (1995).
8. USDA-Soil Conservation Service, *Agricultural Waste Management Field Handbook,* U.S. Government Printing Office, Washington, D.C., 1992.
9. E.P. Taiganides, "Animal Waste Management and Wastewater Treatment," in D. Strauch, ed., *Animal Production and Environmental Health,* Elsevier Science Publishers B.V., Amsterdam, 1987, pp. 91–153.
10. U.S. EPA, *Report to Congress: Nonpoint Source Pollution in the U.S.,* United States EPA, Office of Water Program Operations, Water Planning Division, Washington, D.C., 1984.
11. G.W. Ware, *The Pesticide Book,* Thompson Publications, Fresno, Calif., 1994.
12. D. Bennett, "Evaluation of the Fate of Pesticides in Water and Sediment," in D.H. Hutson and T.R. Roberts, eds., *Environmental Fate of Pesticides,* John Wiley & Sons, Ltd., Chichester, U.K., 1990, pp. 149–173.
13. J.R. Coats, "Pesticide Degradation Mechanisms and Environmental Activation," in L. Somasundaram and J.R. Coats, eds., *Pesticide Transformation Products,* American Chemical Society, Washington, D.C., 1991, pp. 10–30.
14. L. Somasundaram and J.R. Coats, "Pesticide Transformation Products in the Environment," in L. Somasundaram and J.R. Coats, eds., *Pesticide Transformation Products,* American Chemical Society, Washington, D.C., 1991, pp. 2–9.
15. D.J. Arnold and G.G. Briggs, "Fate of Pesticides in Soil: Predictive and Practical Aspects" in D.H. Hutson and T.R. Roberts, eds., *Environmental Fate of Pesticides,* John Wiley & Sons, Chichester, England, 1990, pp. 101–122.
16. Weed Science Society of America, *Herbicide Handbook of the Weed Science Society of America,* The Weed Science Society of America, Urbana, Ill., 6th ed., 1989.

17. C.R. Worthing and R.J. Hance, eds., *The Pesticide Manual,* 9th ed. The British Crop Protection Council, Surrey, U.K., 1991.
18. CPCR, *Crop Protection Chemicals Reference,* Chemical and Pharmaceutical Press, John Wiley & Sons, Inc., New York, 1993.
19. J.D. Rhoades, "Overview: Diagnosis of Salinity Problems and Selection of Control Practices," in K.K. Tanji, ed., *Agricultural Salinity Assessment and Management,* ASCE, New York, 1990, pp. 18–41.
20. C.H. Pair, W.H. Hinz, K.R. Frost, R.E. Sneed, and T.J. Schiltz, *Irrigation,* 5th ed., The Irrigation Association, Arlington, Va., 1983.
21. A.D. Halvorson, "Management of Dryland Saline Seeps," in K.K. Tanji, ed., *Agricultural Salinity Assessment and Management,* ASCE, New York, 1990, pp. 372–392.
22. R.C. Loehr, *Pollution Control for Agriculture,* 2nd ed. Academic Press, Inc., Orlando, Fla., 1984.
23. CAST, "Processing Wastes," in *Waste Management & Utilization in Food Production and Processing,* Council for Agricultural Science and Technology, Ames, Iowa, 1995, pp. 67–96.
24. E.E. Alberts, G.E. Schuman, and R.E. Burwell, *J. Environ. Qual.* **7**(2), 203–208 (1978).
25. L.L. McDowell, G.H. Willis, and C.E. Murphree, *Agric. Ecosys. Environ.* **25**(2–3), 119–137 (1989).
26. S.J. Klaine, M.L. Hinman, D.A. Winkelmann, K.R. Sauser, J.R. Martin, and L.W. Moore, *Environ. Toxicol. Chem.* **7**(8), 609–614 (1988).
27. R. Frank and G.J. Sirons, *Sci. Tot. Environ.* **12,** 223–239 (1979).
28. G.A. Junk, R.F. Spalding, and J.J. Richard, *J. Environ. Qual* **9**(3), 479–483 (1980).
29. R.F. Spalding, M.E. Exner, J.J. Sullivan, and P.A. Lyon, *J. Environ. Qual.* **8**(3), 374–383 (1979).
30. M.D. Smolen, G.D. Jennings, and R.L. Huffman, "Impact of Nonpoint Sources of Pollution on Aquatic Systems: Agricultural Land Uses," in *Nonpoint Source Impact Assessment,* Rep. No. 90-5, CH2M–Hill, Water Environment Research Foundation, Alexandria, Va., 1990.
31. L.B. Owens and co-workers, *J. Environ. Qual.* **18,** 232 (1989).
32. ASAE, "Uniform Terminology for Rural Waste Management," ASAE S292.5 OCT94 in *ASAE Standards,* 42nd ed., ASAE, St. Joseph, Mich., 1995, pp. 523–526.
33. J.M. Sweeten and M.L. Wolfe, *Trans. ASAE* **37**(4), 1145–1154 (1994).
34. W.L. Magette, R.B. Brinsfield, R.E. Palmer, and J.D. Wood, *Trans. ASAE* **32,** 663–667 (1989).
35. T.A. Dillaha, R.B. Reneau, S. Mostaghimi, and V.O. Shanholtz, *Trans. ASAE* **32**(2), 491–496 (1989).
36. S.P. Inamdar, *Investigation of Hydrologic and Sediment Transport Processes on Riparian Hillslopes,* Ph.D. Dissertation, Virginia Polytechnic Institute and State University, Blacksburg, Va., 1996.
37. J.C. Hayes and J.E. Hairston, *Modeling the Long Term Effectiveness of Vegetative Filters as Onsite Sediment Controls,* ASAE Paper No. 83-2081, St. Joseph, Mich., 1983.
38. G.F. McIsaac, J.K. Mitchell, and M.C. Hirschi, *Trans. ASAE* **33**(5), 1541–1550 (1990).
39. S.C. Geiger, A. Manu, and A. Bationo, *Soil Sci. Soc. Am. J.,* **56**(1), 172–177 (1992).
40. K.W. King, D.C. Flanagan, L.D. Norton, and J.M. Laflen, *Trans. ASAE* **38**(1), 159–164 (1995).
41. L.G. Brenneman and J.M. Laflen, *Trans. ASAE* **25**(5), 1245–1250 (1982).
42. N.P. Cogo, W.C. Moldenhauer, and G.R. Foster, *Soil Sci. Soc. Am. J.* **48**(2), 368–373 (1984).
43. L.E. Wagner and R.G. Nelson, *Trans. ASAE* **38**(2), 419–427 (1995).
44. L.C. Brown, R.K. Wood, and J.M. Smith, *Appl. Eng. Agric.* **8**(3), 333–339 (1992).
45. A.R. Pesant, J.L. Dionne, and J. Genest, *Canadian J. Soil Sci.* **67,** 835–843 (1987).
46. C.A. Lembi, M.D. Bri Hon, and M.A. Ross, *Evaluation of Nitrogen Application Technique and Tillage System on Nitrogen Runoff from an Erodible Soil,* Technical Report 174, Purdue Univ., West Lafayette, Ind., 1985.
47. W.M. Edwards, M.J. Shipitalo, L.B. Owens, and L.D. Norton, *Geoderma* **46,** 73–84 (1990).
48. A. Weersink, M. Walker, C. Swanton, and J.E. Shaw, *J. Soil Water Conserv.* **47**(4), 328–334 (1992).
49. J.C. Zhu, C.J. Gantzer, S.H. Anderson, E.E. Alberts, and P.R. Beuselinck, *Soil Sci. Soc. Am. J.* **53,** 1210–1214 (1989).
50. J.M. Laflen and T.S. Colvin, *Trans. ASAE* **24**(3), 605–609 (1981).
51. W. A. Berg and co-workers, *J. Soil and Water Conserv.* **43,** 407 (1988).
52. B.R. Stinner and co-workers, *Agric. Ecosystems. Environ.* **20,** 81 (1988).
53. T.B. Parkin and S.E. Meisinger, *J. Environ. Qual.* **18,** 12 (1989).
54. R.E. Sheffield, *Off-Stream Water Sources for Grazing Cattle as a Stream Bank Stabilization and Water Quality BMP,* M.S. Thesis, Virginia Polytechnic Institute and State University, Blacksburg, Va., 1996.
55. R. Heitefuss, *Crop and Plant Protection: The Practical Foundations,* Ellis Horwood Limited, West Sussex, U.K., 1989.
56. G.W. Norton and J. Mullen, *Economic Evaluation of Integrated Pest Management Programs: A Literature Review,* Publication 448-120, Virginia Cooperative Extension, Virginia Polytechnic Institute and State University, Blacksburg, Va., 1994.
57. T.H. Coaker, "Cultural Methods: The Crop," in A.J. Burn, T.H. Coaker, and P.C. Jepson, eds., *Integrated Pest Management,* Academic Press Inc., San Diego, Calif., 1987, pp. 69–88.
58. H.F. Van Emden, "Cultural Methods: The Plant," in A.J. Burn, T.H. Coaker, and P.C. Jepson, eds., *Integrated Pest Management,* Academic Press Inc., San Diego, Calif., 1987, pp. 27–68.
59. I.J. Graham–Bryce, "Chemical Methods," in A.J. Burn, T.H. Coaker, and P.C. Jepson, eds., *Integrated Pest Management,* Academic Press Inc., San Diego, Calif., 1987, pp. 113–159.
60. B.W. Engle and A.R. Jarrett, *Trans. ASAE* **38**(2), 435–439 (1995).
61. A. Randall, *Resource Economics: An Economic Approach to Natural Resource and Environmental Policy,* 2nd ed. John Wiley & Sons, Inc., New York, 1987.
62. American Agricultural Economics Association Task Force, *Soil Erosion and Soil Conservation Policy in the United States,* Occasional Paper No. 2, American Agricultural Economics Association, Ames, Iowa, 1986.
63. B. Knerr, "Agricultural Policies and Deforestation in Sub-Saharan Africa," in M. Loseby ed., *The Environment and the Management of Agricultural Resources: Proceedings of the 24th Seminar of the European Association of Agricultural Economists,* European Agricultural Economics Association, Viterbo, Italy, 1992, pp. 70–90.
64. E.T. Kornegay, "Nutritional, Environmental, and Economical Consideration for Using Phytase in Pig and Poultry Diets," in E.T. Kornegay, ed., *Nutritional Management of Food Animals to Enhance and Protect the Environment,* CRC Press, Boca Raton, Fla., 1996, pp. 227–302.
65. G. Fox, A. Weersink, G. Sarwar, S. Duff, and B. Deen, *Northeastern J. Agric. Res. Econ.* **20**(1), 124–142 (1991).

66. D.J. Walker, *Am. J. Agric. Econ.* **64**(4), 690–699 (1982).
67. D.B. Taylor and D.L. Young, *Western J. Agric. Econ.* **10**(1), 63–76 (1985).
68. G. Fox and S.J. Taff, "Topsoil, Incentives, and Sustainability," in *Sustainable Agriculture: Its Policy Effects on the Future of Canada's and Ontario's Agrifood System,* Proceedings of a Conference Sponsored by the George Morris Centre, University of Guelph, Ontario, Canada, May 1990, pp 85–99.
69. A.J. Barbarika and M.R. Dicks, *Agric. Econ. Res.* **40**(3), 12–20 (1988).
70. P.P. Setia and C.T. Osborn, *North Central J. Agric. Econ.* **11**(1), 95–103 (1989).
71. National Research Council, *Soil and Water Quality: An Agenda for Agriculture,* Committee on Long-Range Soil and Water Conservation, Board on Agriculture, National Academy Press, Washington, D.C., 1993.
72. R.O. Burton and M.A. Martin, *North Central J. Agric. Econ.* **9**(2), 181–194 (1987).
73. G.L. Casler and J.J. Jacobs, *J. Northeastern Agric. Econ. Council* **4**(2), 174–184 (1975).
74. G.H. Pfeiffer and N.K. Whittlesey, *Western J. Agric. Econ.* **1**(2), 264–267 (1977).
75. E.O. Heady and G. Vocke, *J. Environ. Qual.* **8**(2), 143–148 (1979).
76. P.E. Norris and L.A. Shabman, *Amer. J. Altern. Agric.* **7**(4), 148–156 (1992).
77. J.A. Pease and D.J. Bosch, *J. Soil Water Conserv.* **49**(5), 477–483 (1994).
78. D.J. Bosch, J.W. Pease, S.S. Batie, and V.O. Shanholtz, *Crop Selection, Tillage Practices, and Chemical and Nutrient Applications in Two Regions of the Chesapeake Bay Watershed,* VPI-VWRRC-BULL 176, Virginia Water Resources Research Center, Virginia Polytechnic Institute and State University, Blacksburg, Virginia Nov. 1992.
79. E. Feinerman, D. Holtkamp, and S.R. Johnson, "Farm-Level Evaluation of Agricultural and Environmental Policies with an Integrated Modeling System," Working Paper 89-WP48, Center for Agricultural and Rural Development, Iowa State University, Ames, Iowa, 1989.
80. W.T. McSweeney and J.S. Shortle, *Northeastern J. Agric. Res. Econ.* **18**(1), 1–11 (1989).
81. E. Feinerman, E.K. Choi, and S.R. Johnson, *Am. J. Agric. Econ.* **72**(4), 975–984 (1990).
82. W.T. McSweeney and R.A. Kramer, *Land Econ.* **62**(2), 159–173 (1986).
83. USDA, *Agricultural Outlook Supplement,* Economic Research Service, Washington, D.C., 1996.
84. FAO (Food and Agriculture Organization of the United Nations), *FAO Trade Yearbook 1994,* FAO Statistics Series No. 127, Rome, 1995.
85. K. Anderson, *World Econ.* **15**(1), 153–171 (1992).
86. USDA, *Estimates of Producer and Consumer Subsidy Equivalents: Government Intervention in Agriculture, 1982–1992,* Statistical Bulletin No. 913, Economic Research Service, Commercial Agriculture Division, Washington, D.C., (Dec. 1994). USDA, *Estimates of Producer and Consumer Subsidy Equivalents: Government Intervention in Agriculture, 1982–1986,* Economic Research Service, Washington, D.C., 1988.
87. C. Harold and C.F. Runge, *Am. J. Agric. Econ.* **75**(3), 789–793 (1993).
88. U.S. Congress, Office of Technology Assessment, *Agriculture, Trade, and Environment: Achieving Complementary Policies,* OTA-ENV-617, U.S. Government Printing Office, Washington, D.C., 1995.
89. M.D. Young, "International Trading Arrangements, the Intensity of Resource Use, and Environmental Quality," in J.B. Braden and S.B. Lovejoy, eds., *Agriculture and Water Quality: International Perspectives,* Lynne Rienner Publishers, Boulder, Colo., 1990, pp. 197–215.
90. C.F. Runge, J.P. Houck, and D.W. Halbach, "Implications of Environmental Regulations for Competitiveness in Agricultural Trade," in J.D. Sutton, ed., *Agricultural Trade and Natural Resources: Discovering the Critical Linkages,* Lynne Rienner Publishers, Boulder, Colo., 1988, pp. 195–216.
91. D. Cavallo, "Agriculture and Economic Growth: The Experience of Argentina, 1913–84," in A. Maunder and A. Valdes, eds., *Agriculture and Governments in an Interdependent World,* Dartmouth for the IAAE, London, 1989.
92. M. Lopes, *The Mobilization of Resources from Agriculture: A Policy Analysis for Brazil,* Ph.D. Dissertation, Purdue University, West Lafayette, Ind., 1977.
93. A. Schmitz, G.C. Van Kooten, and W.H. Furtan, "Issues in Commodity Trade: Implications for Natural Resources," in J.D. Sutton, ed., *Agricultural Trade and Natural Resources: Discovering the Critical Linkages,* Lynne Rienner Publishers, Boulder, Colo., 1988, pp. 195–216.
94. V.W. Ruttan, *Am. J. Agric. Econ.* **53**(5), 707–717 (1971).
95. J.M. Antle and G. Heidebrink, *Econ. Devel. Cultur. Change* **43**(3), 603–625 (1995).
96. P. Bohm and C.S. Russell, "Comparative Analysis of Alternative Policy Instruments," in A.V. Kneese and J.L. Sweeney, eds., *Handbook of Natural Resource and Energy Economics,* North-Holland, New York, 1985, pp. 395–460.
97. D.G. DeCoursey, *J. Soil and Water Conserv.* **40**(5), 508–513 (1985).
98. R.E. Heimlich and C.H. Barnard, *Economics of Agricultural Management Measures in the Coastal Zone,* Agricultural Economic Report No. 698, Natural Resources and Environment Division, Economic Research Service, USDA, Washington, D.C., 1995.
99. U.S. EPA, Office of Water, *Guidance Specifying Management Measures for Sources of Nonpoint Pollution in Coastal Waters,* 840-B-92-002, U.S. EPA, Washington, D.C., 1993.
100. D. Williamson, "Implementation of the Nebraska Nitrate Control Legislation," in *Nonpoint Pollution: 1988—Policy, Economy, Management, and Appropriate Technology,* American Water Resources Association, Minneapolis, Minn., 1988, pp. 133–139.
101. G.D. Anderson, A.E. De Bossu, and P.J. Kuch, "Control of Agricultural Pollution by Regulation," in J. B. Braden and S. B. Lovejoy, eds., *Agriculture and Water Quality: International Perspectives,* Boulder, Colo. 1990, pp. 63–104.
102. K. Segerson, "Incentive Policies for Control of Agricultural Water Pollution," in J.B. Braden and S.B. Lovejoy, eds., *Agriculture and Water Quality: International Perspectives,* Lynne Rienner Publishers, Boulder, Colo., 1990, pp. 39–62.
103. D.G. Abler and J.S. Shortle, *Northeastern J. Agric. Res. Econ.* **20**(1), 53–60 (1991).
104. L. Mosher, *J. Soil Water Conserv.* **42**(5), 333–335 (1987).
105. K. Kumm, "Incentive Policies in Sweden to Reduce Agricultural Water Pollution," in J.B. Braden and S.B. Lovejoy, eds., *Agriculture and Water Quality: International Perspectives,* Lynne Rienner Publishers, Boulder, Colo., 1990, pp. 105–116.
106. A. Dubgaard, "Programs to Abate Nitrate and Pesticide Pollution in Danish Agriculture," in J.B. Braden and S.B. Lovejoy, eds., *Agriculture and Water Quality: International Perspectives,* Lynne Rienner Publishers, Boulder, Colo., 1990, pp. 105–116.

107. USDA, "United States Department of Agriculture's Water Quality Initiative 1992 Work Plan," Working Group on Water Quality, Washington, D.C., 1991.
108. T.A. Feitshans, *J. Soil and Water Conserv.* **45**(2), 211–215 (1990).
109. USDA, *Agricultural Resources: Cropland, Water, and Conservation Situation and Outlook,* AR-30, Economic Research Service, Resources and Technology Division, Washington, D.C., 1993.
110. A. S. Malik, B. A. Larson, and M. Ribaudo, *Water Res. Bull.* **30**(3), 471–479 (1994).
111. U.S. General Accounting Office, *Water Pollution: Pollutant Trading Could Reduce Compliance Costs if Uncertainties Are Resolved,* GAO/RCED-92-153, Report to the Chairman, Committee on Public Works and Transportation, House of Representatives, U.S. Government Printing Office, Washington, D.C., 1992.
112. E.M. Jacobson, D.L. Hoag, and L.E. Danielson, "The Theory and Practice of Pollution Credit Trading in Water Quality Management," in P.E. Norris and L.E. Danielson, eds., *Economic Issues Associated with Nutrient Management Policy: Proceedings of a Regional Workshop,* SRDC Publication No. 182, Southern Rural Development Center, Mississippi State University, Mississippi State, Mississippi, 1994, pp. 69–92.
113. D. Letson, *Natural Res. J.* **32**(2), 219–232 (1992).
114. W. Van Vuuren, C. Giraldez, and D.P. Stonehouse, "Importance of Benefit Identification in Evaluating Water Pollution Control Programs," Working Paper WP94/08, Department of Agricultural Economics and Business, University of Guelph, Guelph, Ontario, 1994.
115. M.O. Ribaudo, *Land Econ.* **65**(4), 320–332 (1989).
116. C.L. Carpentier, *Value of Information for Targeting Agro-Pollution Control: A Case Study of the Lower Susquehanna Watershed,* Ph.D. Dissertation, Virginia Polytechnic Institute and State University, Blacksburg, Va., 1996.
117. W.M. Park and D.G. Sawyer, "Targeting Soil Erosion Control Efforts in a Critical Watershed," ERS Staff Report No. AGES850801, Economic Research Service, USDA, Washington D.C., 1985.
118. T. Osborn, *J. Soil Water Conserv.* **48**(4), 271–279 (1993).
119. C.L. Lant, S.E. Kraft, and K.R. Gillman, *J. Soil Water Conserv.* **50**(2), 193–200 (1995).
120. D.C. Esty, *Greening the GATT: Trade, Environment, and the Future,* Institute for International Economics, Washington, D.C., 1994.
121. C. F. Runge, "Environmental Regulation and Agricultural Trade Competitiveness," in J.B. Braden and S.B. Lovejoy, eds., *Agriculture and Water Quality: International Perspectives,* Lynne Rienner Publishers, Boulder Colo., 1990, pp. 183–196.
122. M.A. Marchant and N. Ballenger, *J. Agric. Appl. Econ.* **26**(1), 108–128 (1994).
123. J.L. Seale, Jr. and G.F. Fairchild, *J. Agric. Appl. Econ.* **26**(1), 97–107 (1994).

SOIL AS AN ECOSYSTEM

MOHAN K. WALI
FRED P. MILLER
The Ohio State University
Columbus, Ohio

WHAT IS SOIL?

Soil is the sustenance of life, and water its elixir. An old saying, attributed to the Chinese, sums it up: "Man—despite his artistic pretensions, his sophistication, and his many accomplishments—owes his existence to a six-inch layer of topsoil and the fact that it rains." This truism is as old as human knowledge of the use of soils. For more than 150 years and from many continents, studies have documented the interplay of the geologic substrates, climate and organisms and their interactions that produce a rich diversity of soils that, in turn, support the myriad mosaic of life. This article presents an integrated and holistic view of the ecological, biological, and engineering dimensions of the soils in light of the rich past and recent work.

During its development, soil science has proceeded along two distinct lines. Pedology, the study of soil development as a phenomenon in itself, incorporates primarily soil–landscape relationships and morphologic, physical, chemical, and mineralogic features. Edaphology, on the other hand, emphasizes soil–plant relationships in agricultural systems, forests and all land masses that were or are capable of supporting life (1).

There are many definitions of soils, but three capture both the essence and the progression in thinking. First is the edaphic concept of soil: "The surface stratum of earthy material, as far as the roots of plants reach" (2). Second, in a totally revised soil classification system proposed and adopted by soil scientists of the United States Department of Agriculture (USDA) a century later, the definition retained the plant growth function: "Soil ... is the collection of natural bodies on the earth's surface, containing living matter, and supporting or capable of supporting plants" (3).

Of late, three distinct and diverse trends have become apparent: (*1*) numerous technological advances, in conceptual development in other disciplines and in instrumentation, have made it possible to study soils in remarkably fascinating ways; (*2*) many potentially arable soils brought under cultivation have been degraded by management practices; and (*3*) soils have been used as a repository for society's residuals and waste products, which often results in short- and long-term toxicities and pollution of groundwaters and adjacent ecosystems. Thus the recent, and third, definition of soil incorporates the classical and the modern, and the use and abuse conundrum (4):

> *The thin skin at the earth's surface known as soil is the primary reason life is possible on this planet—this is the life-sustaining pedosphere. This biologically active, porous, and structural medium is an effective integrator and dissipator of mass flux and energy, the medium through which biomass productivity is sustained, the foundation upon which structures are built, the fabric to which organisms are anchored and housed (habitat), the repository of solid and liquid wastes and the living filter for bioremediation of waste products and water supplies. Soils are the long-term capital on which a nation builds and grows. It is a basic component of ecosystems and is one of the most vulnerable to degradation through mismanagement.*

THE ECOSYSTEM CONCEPT—DEFINITION AND USEFULNESS

An *ecosystem* has been visualized as an interacting complex of organisms with "the whole complex of physical factors" (5). It has been emphasized that although our interest may be in living organisms, "we cannot separate them from their special environment, with which they form one system. It is the systems so formed ... that are the basic units of nature on the earth" (5). Ecosystems may be of any size, of any kind, and in a relatively stable equilibrium.

Many modifications of this definition are now part of the ecological literature, but two may be worthy of note. An ecosystem was noted to be "a system composed of physical–chemical–biological processes active within a *space–time* unit of any magnitude" (emphasis added) (6). Furthermore, it is emphasized that an ecosystem is "any area of nature that includes living organisms and non-living substances interacting to produce an *exchange* of materials between the living and non-living parts" (emphasis added) (7). Both the space *and* time relationships and the exchange of materials are crucial to theoretical and practical considerations.

Structurally and functionally, an ecosystem, or its Russian analog, the *biogeocoenose* (8), comprises the abiotic components of climate and soils; the producer components of green plants and chemosynthetic bacteria; the consumer components of herbivores, carnivores, and omnivores; and the decomposer component of detritivores. Bacteria and fungi that decompose dead organic matter are fundamental to the cycling of nutrients (Fig. 1). From an ecologic standpoint, the only producers on earth are the green plants and chemosynthetic bacteria (thus autotrophs or primary producers); and all animals (including humans) are heterotrophic consumers. The ecosystem has a number of structural components (Fig. 2) (9), and a rich base of information exists on these structural components.

- Ecosystem
  - Biocoenose
    - Plant population A
    - Plant population B...
    - Plant community A
    - Plant community B...
    - Animal population A
    - Animal population B
    - Animal community A
    - Animal community B...
    - Microbial population A
    - Microbial population B...
    - Microbial community A
    - Microbial community B...
  - Climatope
    - Incoming solar radiation, light
    - Air temperature
    - Precipitation
    - Wind, snow
    - Air pollution...
  - Eda(pho)tope
    - Soil particle size
    - Soil water
    - Soil temperature
    - Soil air
    - Soil biota
    - Soil toxins

Figure 2. Structural components of an ecosystem. Modified from Ref. 9.

System Interconnectedness

The elegance of the ecosystem concept is that it emphasizes the interconnectedness of ecosystem variables (ie, of climate, soils, plants, animals, and microbes), of the system as a whole both in space and in time. Six attributes of the concept of ecosystem attest to its strength (10,11): structure, function, complexity, interaction and interdependence, temporal change (ecological succession), and flexible spatial dimensions (allowing experimental flexibility).

Though the ecosystem concept originated in Europe, much of its development occurred in the United States (12): "The ecosystem story is largely the American tale ... the ecosystem concept appeared to be modern and up to date. It concerned systems, involved information theory, and used computers and modelling. In short, it was a machine theory applied to nature." The visualizing of the ecosystem as composed of functional compartments makes it amenable to mathematical modeling (Fig. 3). The ecosystem concept is now ever-present both in scientific literature and in nonscientific writing. In a survey of its members on the importance and relative ranking of ecological concepts, the British Ecological Society found that the ecosystem concept was at the top of the 50 concepts rated (14).

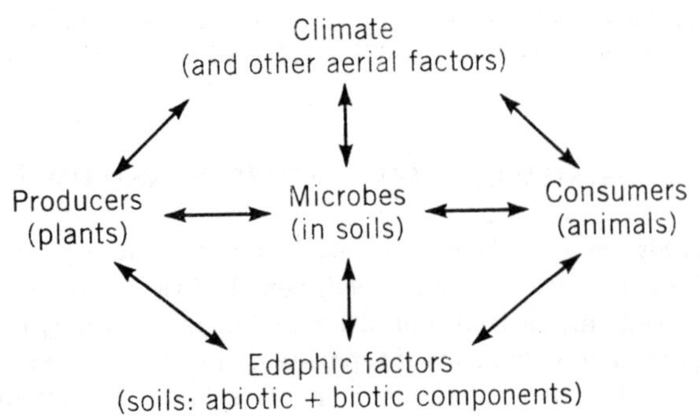

Figure 1. Functional and interacting components of an ecosystem.

Soils form the base of the ecological pyramid in terrestrial ecosystems. The primary productivity of plants, and in turn, the sustenance and welfare of all animal consumers, including humans, depends on the fertility, unstressed root environment, and stability of the soils. Thus, by reinforcing the interconnectedness of system variables and parameters, the ecosystem concept is at the core of regional, national, and global sustainability issues.

THE SOIL ECOSYSTEM

The Principle of Limiting Factors

A soil ecosystem may be defined as a system of organisms—plant and animal—that is in intimate interaction with the soil's physical characteristics (particle size and water relations), chemical factors (prevalence, abundance, and exchange of many chemical elements), thermal relations or heat budgets (acquisition and loss), air relations (dynamics of oxygen, carbon dioxide, and other gases), and biota (from consumers to decomposers). It is the collective expression of these system variables that determines the extent of energy flows and hydrologic and biogeochemical cycles, which, in turn, dictate the dynamics of the presence, abundance, and distribution of biotic communities.

Like all constituent ecosystems processes, the dynamics of organisms in soil ecosystems follow Liebig's law of the minimum: when a process is governed by a number of factors, the rate of the process is determined by the factor in

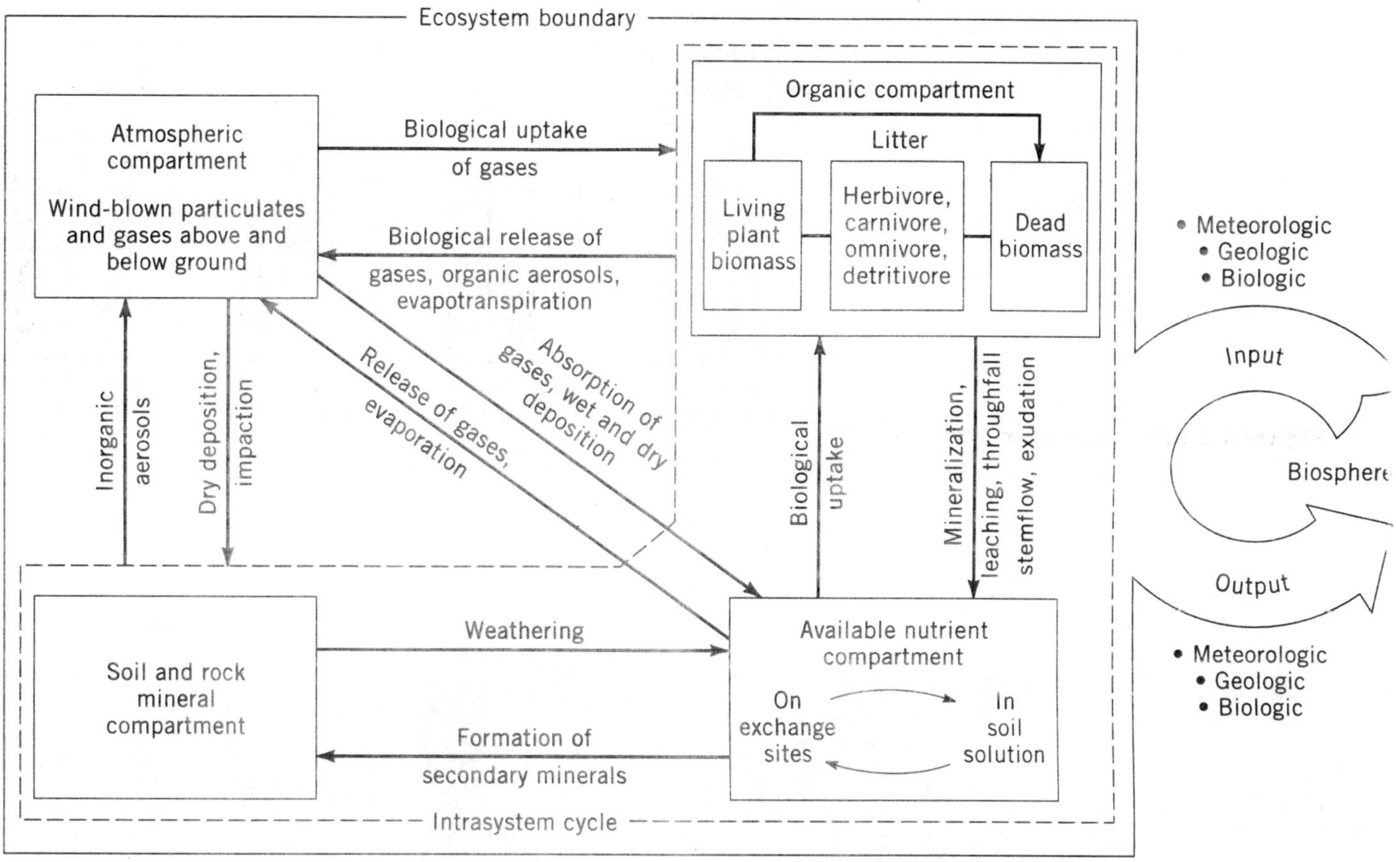

Figure 3. Major components of an ecosystem. Note that at least three of the major compartments within the ecosystem boundary are directly related to soils (13).

shortest supply (Fig. 4). An obvious corollary of the law of the minimum is the law of toleration (16), ie, when a factor is present in abundance (eg, salts), organisms may not be able to tolerate it. Both of these laws, have been codified: for "physiological phenomena, assimilation, respiration, growth, and the like, which have a varying magnitude under varying external conditions of temperature, light, supply of materials, ... it is customary to speak of three cardinal points, the *minimal* condition below which the phenomenon ceases altogether, the *optimal* condition at which it is exhibited to its highest observed degree, and the *maximal* condition above which it ceases again" (17). Under special but brief conditions, a factor may compensate for the short supply of another (18). But in ecosystem processes, compensation seldom implies substitution, and few essential factors can be replaced.

Soil Physical Components

Particle Size and Structure. The primary particle constituents of the soil are sand (very coarse to very fine 2.00–0.05 mm), silt (0.05–0.002 mm), and clay (<0.002 mm). The relative percentage proportions of these particles is referred to as particle size distribution, or soil texture. The universally used textural triangle aids scientists in grouping soils into textural classes (Fig. 5). Soil structure (which may be platty, blocky, angular, or prismatic) is the aggregation of these primary particles under field conditions into primary and secondary structural units (peds). Micropores (small spaces or voids) are formed as the primary particles clump together into primary aggregates. Furthermore, as primary aggregates form secondary aggregates, the spaces or voids between structural units are larger, and hence macropores are formed. Together, the two kinds of spaces or voids (micropores and macropores) make up the total soil porosity.

Soil porosity is intimately related to the mass–volume relationship of soils in their field state (referred to as bulk density, or apparent specific gravity). Bulk density provides a good estimate of the physical resistivity that roots will encounter as they penetrate deeper into the soil. Thus the lower the bulk density, the greater the void space to accommodate roots. The average bulk density of agricultural soils is in the range of 1.1–1.5 g/cm^3. For organic soils, bulks densities are generally <1.0; for soils that are compacted, either naturally or by human activities (eg, through cultivation, machinery traffic, snow-mobiling), bulk densities are >1.5 g/cm^3. Conversely, natural phenomena (eg, freezing–thawing) and activities of soil animals (eg, ants, termites, earthworms) may considerably increase the soil porosity and decrease bulk density. Actual pore space may be estimated by considering both bulk density and particle size density (true specific gravity):

$$1 - (\text{Bulk density}/\text{Particle density}) \times 100$$

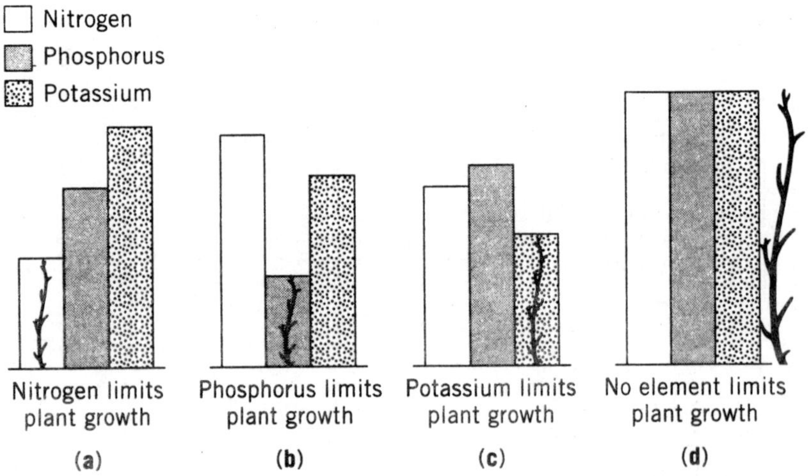

Figure 4. A simple illustration of the law of the minimum (15).

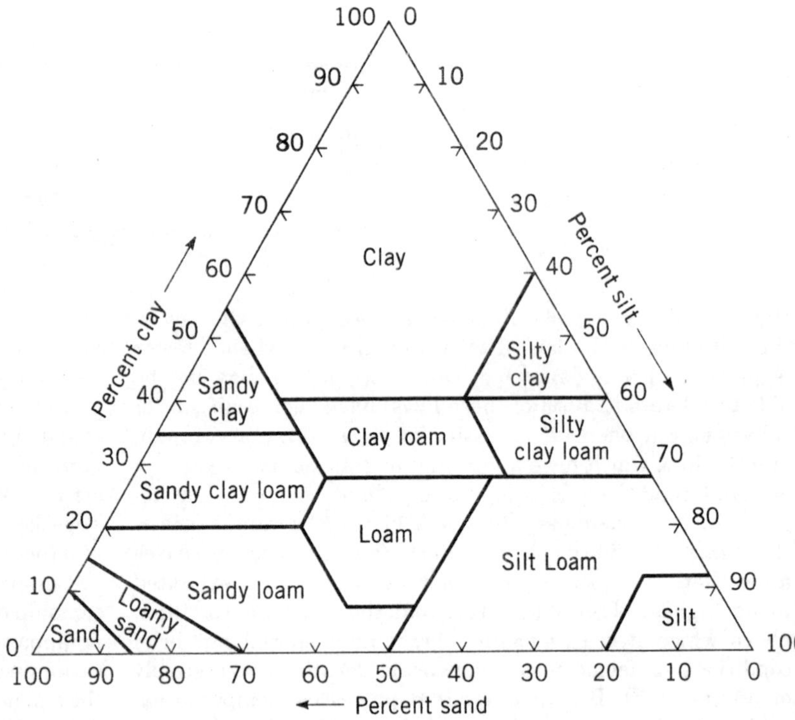

Figure 5. A diagram to determine soil textural class based on soil particle size distribution (3,19).

Together, soil texture, particle mineralogy, and structure determine gas exchange, water retention properties, nutrient availability and mobility, temperature relations, and the distribution and abundance of soil biota.

Many ecological studies (since at least 1929) have shown a relationship between texture (silt and clay content) and site indices of forests, reflecting the water-holding capacity of soils and availability to trees; early examples include the growth of *Pinus resinosa* in Connecticut and the aspen site index in Minnesota and Wisconsin (20). Perhaps the most rigorous use of particle size was in its relation to the distribution of species in the Danish oakwoods (21); the texture gradient comprised 13 classes, and each class had a 2% differential of silt and clay content. A quantified soil particle size gradient based on the silt and clay content, bulk density, depth of soil genetic horizons, and stoniness produced effective species optimalities in the boreal zone of British Columbia (20). Besides forest ecosystems, texture is used extensively to evaluate soil fertility in agriculture and rangelands.

Soil Water. The amount of water remaining in the soil after a precipitation event (rain, snow, hail) determines the growth, abundance, and distribution of plant and animal life. Where water is in short supply, as in the arid regions of world, vegetation is sparse. Where human practices accelerate the loss of vegetation in such arid areas, desertification can result. In high rainfall areas, exposed soil and its component particles, especially the fine particles and nutrients, are prone to loss. In many tropical rain forest regions where vegetation is burned and cleared to make land available for farming, the fields lose their fertility through combined crop removal and erosion-leaching losses in 5–10 years. Known as slash-and-burn agriculture, this practice has also been referred to, somewhat surprisingly, as desertification (22).

Advances in soil physics and plant physiology have made it possible to assess the energy status of water in soil. Alone and in combination with other factors, the energy status of water has been the subject of numerous studies in plant ecology. The roles of water in nutrient availability, ion mobility, and soil amelioration and as a metabolic requirement of growth for plants are well known. Most earlier ecological studies of water, which dealt with the distribution and abundance of plants, were mainly qualitative. Assessment of the energy status of water relations in the soil–plant–atmosphere continuum (SPAC) is much more appropriate than the consideration of any single phase system. For the soil alone, the total soil water stress has the major components of matric potential and osmotic potential.

$$\Psi_{soil} = \Psi_m + \Psi_s + \Psi_g + \Psi_p$$

where Ψ_m is the matric potential produced by capillary and surface binding forces, Ψ_s is the osmotic potential produced by solutes in the soil water, Ψ_g is the gravitational forces operating on soil water, and Ψ_p is the external pressure (which is negligible because the pressure is near atmospheric pressure in the root zone) (23).

Several factors regulate the availability of soil water to plants, but the two most significant are the energy status of water and its transmissibility. The mechanical resistance to roots by soil horizons of high bulk density and the prevailing water table are also important. Chemical toxicity (eg, high aluminum concentrations, pH extremes, salinity–alkalinity) also limits the rooting environment and thus limits the available water for plants.

As soil water tension, or the tenacity with which water is held by the soil increases, less and less water is available to plants. The availability of water along increasing tensions depends on the particle size distribution characteristics of the soil as well as the varying ability of the plants to use it (which varies from species to species). There is a general consensus that a soil's available water capacity—the difference between the soil's water content at −30 kPa (field capacity) and −1500 kPa (permanent wilting point) soil water tension—can be used as an approximate value for the ecological grouping of plants (24). Some considerations in such an assumption should be emphasized: neither field capacity nor permanent wilting point are equilibrium values but represent only gross generalizations conveying approximate equivalence of the respective water states. Water is not equally available to plants within this range, nor do all plants wilt at −1500 kPa (24).

The aspect of transmissibility involves water transfer by viscous flow in the liquid phase, molecular diffusion, and possibly migration in the adsorbed phase (25). Saturated flow depends on the macroporosity of the soil. The differences in both the energy status and transmissibility of water are not significant <−100 kPa, being directly a function of soil structure between 0 and <−100 kPa. Unsaturated flow assumes great significance above a tension value of <−100 kPa and is a function of the particle size (the total surface of the natural soil peds or structural blocks) and bears a direct relationship with soil water tension. Water movement is further impeded if clay–sodium complexes abound.

Vegetation distribution as a function of water gradient has been illustrated for a number of ecosystems (20,26–30). However, soil water characterization when used alone in ecological studies has limitations. The direct interception of precipitation by tree crowns and of dew may be cited as two such examples. Thus in several studies the soil water measurements as above have been used together with internal water stress measurements of plants that have provided good correlations with plant distribution (31–33). A newly proposed geographic information system-derived, integrated moisture-predicting species composition in forests (34,35) is a major step and holds promise in elucidating ecological relationships of natural communities. The importance of soil water in human welfare is beyond question, and the recent work on a broad regional scale of the Middle East region emphasizes the importance of soil water in the rise and fall of civilizations (36). The dynamics of soil water, therefore, play a critical role in soil–plant behavior and the response of soil microbes in the soil ecosystem.

Soil Air. The rich diversity of organisms that abound in soils require both aerobic and anaerobic conditions. The aerobic organisms require oxygen supply. Small fauna, like ants and termites, can increase the pore space substantially by churning the soils (37). Soil air dynamics, or aeration, bear an intimate relationship with water. As water infiltrates soils and percolates into and through soil voids, it drives out the gases. Once saturation occurs and persists, chemical processes must take place under anaerobic conditions. Such conditions, depend on the reduction–oxidation potential, and alter the forms in which ions are available (Table 1). However, oxygen depletion produces changes that are complex; eg, flooding may produce a sequence of different oxidation states of ions over time (Fig. 6). Thus specialized plant populations and communities adapted to these oxygen-depleted conditions abound in bogs, fens, and other wetland areas. Anaerobic processes occur in soil aggregates when capillary pores are filled with water.

Thermal Relations. Like all chemical and biochemical processes, soil processes are intimately related to energy transfers and temperature. In the European ecological literature, temperature relations are referred to as "heat economy" or as "heat budgets." It is generally agreed that up to optimality, the biological processes follow Vant Hoff's law, ie, the rate of chemical reactions doubles for every 10°C increase in temperature:

$$Q_{10^\circ C} = 2$$

In general, soil temperatures show fluctuations similar to air temperatures. However, a major distinction is that there is a diurnal lag in temperature across the atmosphere–soil interface. This lag results in a general dampening of diurnal temperature extremes and does not subject plant roots and micromacrobial life to wide fluctuations. This temperature dampening effect increases with soil depth to the point at

Table 1. Chemical States of Essential Nutrients in Aerobic and Anaerobic Conditions[a]

Element	Oxidized Form	Reduced Form	Redox Potential for Transformation, mv
Nitrogen	NO_3	N_2O, N_2, NH_4	220
Manganese	Mn^{4+}	Mn^{2+}	200
Iron	Fe^{3+}	Fe^{2+}	175
Sulfur	SO_4	S^{2-}	−75 to −150
Carbon	CO_2	CH_4	−250 to −350

[a] Ref. 38.

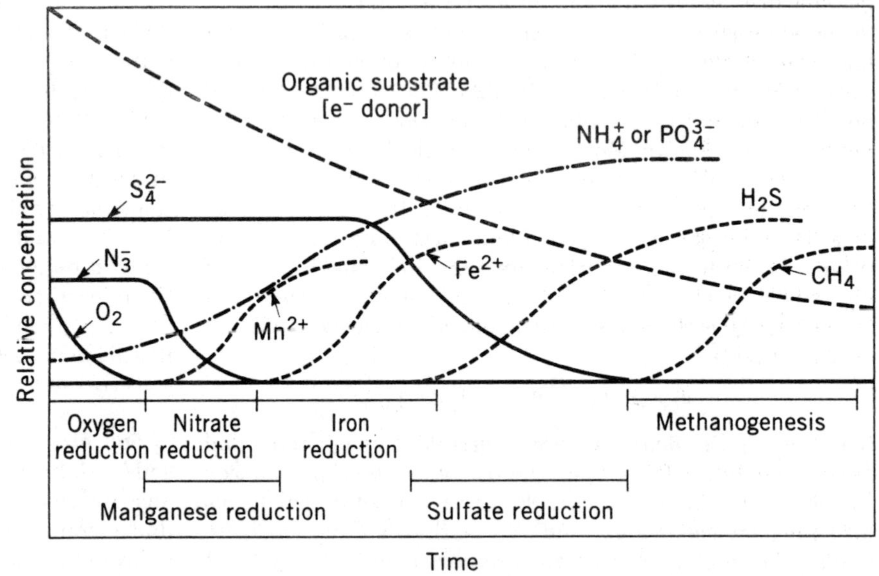

Figure 6. The effect of flooding and onset of anaerobic conditions on soil processes (after Reddy, see Ref. 38).

which constant diurnal, and eventually constant annual, temperatures are reached, as modified by the influence of groundwater and bedrock.

Soil temperature is related to soil particle size, soil color, and water content. Soils with dark-colored horizons and low water content absorb heat readily and are warmed quickly by the sun. Although, sandy soils are highly reflective, especially when low in organic matter, the temperature of the surface layer of such soils is high during the day; at night these soils lose heat quickly. Hence, these soils show a wide diurnal temperature range. Soils with a higher water content (eg, wetlands) require greater energy to warm up, since the heat constant for water is about five times greater than soil. Thus, even though sandy soils low in organic matter are highly reflective, their tendency to rid themselves of water by rapid drainage means they warm rapidly, which allows for earlier planting than do heavier textured (more silt and clay) and/or poorly drained soils, which require much more solar energy to warm the soil–water matrix.

Soil Chemical Relations

Inorganic Components and Plant Nutrition. Cations (positively charged) and anions (negatively charged) in soils form the ingredients from which the plants build their biomass. How plants acquire these ions from soils is one of the most fascinating marvels of nature. Plant roots, with their sensitive surficial membranes and colloid-like properties, exchange ions with soil and organic colloids—clays and finely decomposed organic material (humus). There are several types of clays (also called clay colloids, clay minerals, and silicate clay minerals) that can be structurally represented in three major categories. Kaolinite has one tetrahedral sheet (Si) to one octahedral sheet (Al) and is hence referred to as a 1:1 type crystal lattice (Fig. 7).

Structural formula	$(OH)_8Al_4Si_4O_{10}$
Oxide formula	$4\ H_2O.2\ Al_2O_3.4\ SiO_2$

Its size range is 0.5–5.0 μm and its net negative charge ranges from 2 to 5 $cmol_c/kg$. Montmorillonite, on the other hand, has an octahedral (Al) sheet between two tetrahedral (Si) sheets and hence is called a 2:1 type crystal lattice (Fig. 7).

Structural formula	$(OH)_4Al_4Si_8O_{20}.nH_2O$
Oxide formula	$3\ H_2O.2\ Al_2O_3.8\ SiO_2$

These clay mineral types have a size range of 0.01–2.0 μm and a net charge range of 60–120 $cmol_c/kg$. The third, illite, also has a 2:1 type structure, but unlike montmorillonite it has a nonexpanding intralayer space similar to the nonexpanding kaolinite (Fig. 7).

$$KAl_2(Si_3Al)O_{10}(OH)$$

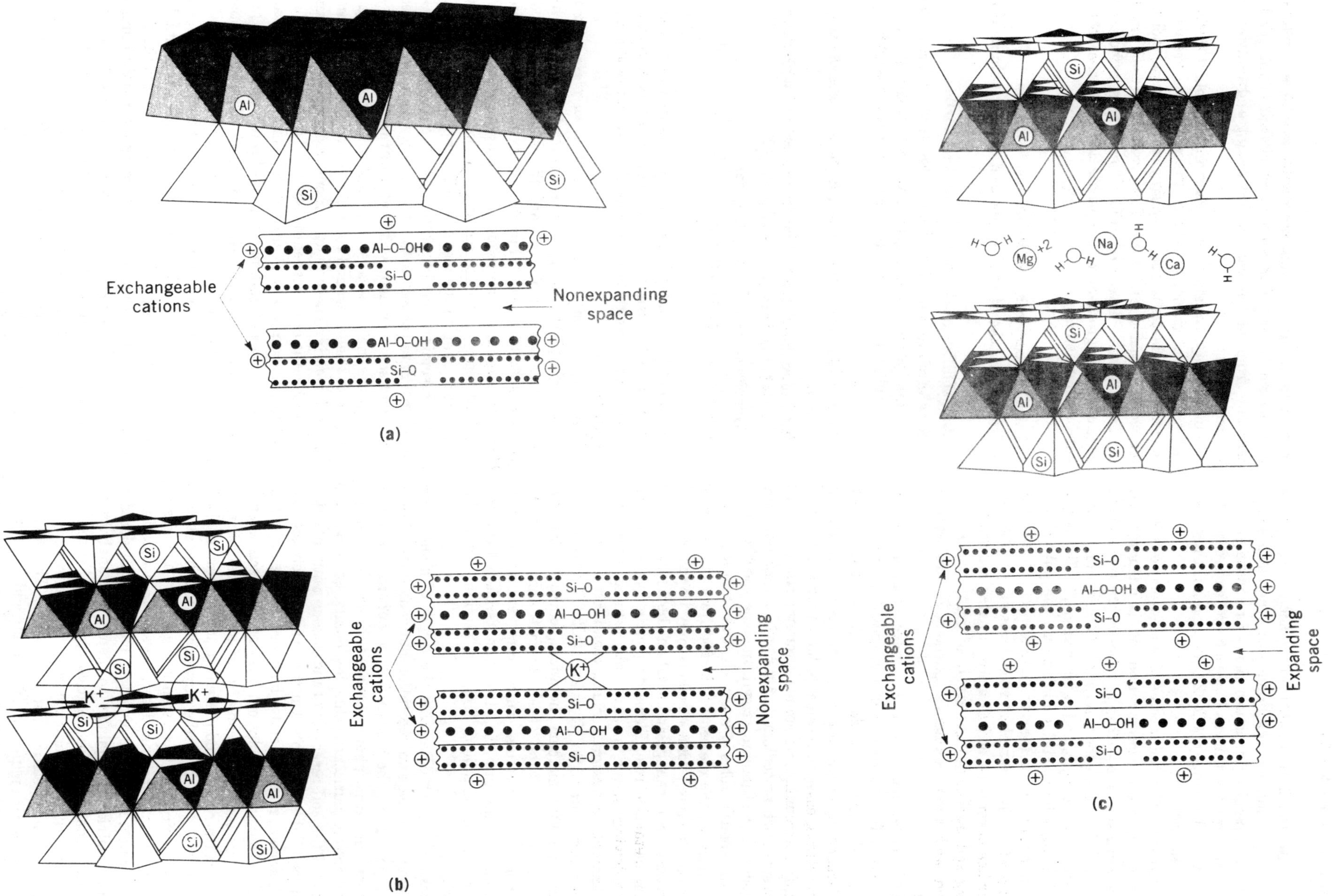

Figure 7. The structure of clay minerals: (**a**) kaolinite, (**b**) illite, and (**c**) montmorillonite (15).

In both kaolinite and illite (in that order), the nonexpanding structure limits the number of exchange sites whereas in montmorillonite, the expanding structure allows for a much greater number of exchange sites, thus its large ion exchange capacity. The illites have a size range of 0.2–2.0 μm and a net negative charge of 15–40 $cmol_c/kg$.

Plant acquisition of nutrient ion exchange from soils is not simply a function of ion concentration but is influenced by valence and ionic size. Ion exchange between the two colloids (soil and root) is based on the type and concentration of ions, charge density and strength, and surface area (39). The larger the number of smaller particles making up the soil volume, the larger the number of surfaces per unit area or volume; thus the exchange of ions between plant roots and soils is a phenomenon of surface chemistry as influenced by clay mineralogy. Quantifying soil exchange capacity is of great importance; both cation and anion exchange phenomena are equally important.

Physiologic experiments with a number of plant species have ascertained the indispensability of specific nutrient ions for plant growth, and hence these ions have been termed essential. Depending on the relative quantities of each element required by plants, essential elements have been divided into two groups: macronutrients and micronutrients (Table 2). Some elements such as sodium, lithium, and aluminum have also been shown to stimulate growth in some specific cases. Although the specific roles of many elements and their ionic form are unknown, the elements recorded from plants show a phenomenal diversity; some elements may be taken up by plants through ion exchange and solution chemistry without playing any metabolic role. Elements recorded in plants include the following: silver, arsenic, gold, barium, beryllium, bismuth, bromine, cadmium, cerium, cobalt, chromium, dysprosium, erbium, europium, fluorine, gallium, gadolinium, germanium, hafnium, mercury, holmium, iodine, lanthanum, lithium, lutetium, niobium, lead, promethium, polonium, radium, rubidium, rhenium, ruthenium, antimony, scandium, selenium, samarium, tin, strontium, terbium, titanium, thulium, uranium, vanadium, tungsten, yttrium, ytterbium, and zirconium.

Five basic processes determine the origin of plant nutrients and drive the soil and vegetation nutrient systems (44): nutrient inputs from the weathering of minerals; nutrient inputs from the atmosphere; nutrient losses by leaching and volatilization; nutrient cycling by plants, animals, and microbes; and the soil reservoir of indigenous nutrients that are readily available to plants. On the basis of their physicochemical behavior, nutrient states have been classed into four groups: dissolved in soil solution, adsorbed in exchangeable form, bound in nonexchangeable form in mineral particles and organic matter (OM), and present in soluble and insoluble forms in the OM (45). It is the physicochemical behavior of nutrients, ie, their form of existence, solubility, and change from one form to another, that determines their availability to plants (45). But a multitude of factors must be considered in determining nutrient availability to plants. Soil analyses for all essential elements reflect relative amounts of potentially available plant nutrients and must be correlated directly with plant response and specific plant populations. The ability of plant roots to absorb a multitude of elements and ions that are soluble only in the strongest solvents illustrates the fallacy of the claim sometimes made that particular elements extracted by a certain type of solvent are readily available to plants. It is important to note that without correlating plant uptake of nutrients from the soil medium, soil testing and analysis alone is meaningless.

Ion mobility, uptake, and/or exchange by plant roots and the subsequent transport within the plant are arduous yet fascinating subjects in soil chemistry and plant physiology. A classical diffusion and mass flow concept of soil nutrient availability was developed (46). The factors affecting the concentration gradient of a specific ion with the root and surrounding soil are

- Initial concentration of the ion in soil solution.
- Rate of uptake of the ion per unit area or volume of root surface.
- Rate of diffusion of the ions to the root surface.
- Rate of movement of the ions to the root surface by mass flow in water moving toward the root along the water gradient (which results from water being absorbed by the roots).
- Rate of diffusion of the ion along the surfaces of soil particles.
- Rate of replenishment of the ions associated with other ions in solution from ions held by the soil.
- Capacity of the soil to replenish ions.
- Movement or growth of root.

Although some misgivings have been expressed on some aspects of mass flow and diffusion (47), substantial experimental evidence has recently been provided in support of the original hypothesis (48).

The role of roots is equally important in plant nutrition. A single rye plant (*Secale cereale*) grown for 4 months had a root surface area of 639 m^2, and the combined length of all roots was 623 km (40). Thus the ability of roots to scavenge for nutrients can be enormous. Depending on genetic and physiological traits, roots are generally surface feeders, but they may reach deep to the water table. Some plants that inhabit arid and semiarid environments, like the grasses, have even more intimate contact with soils, a fact well known to ecologists and soil scientists who find it difficult to separate soils from roots. That is because roots and soil bacteria form a mucilaginous sheath around roots that extends into the immediate soil matrix. These mucilaginous sheaths are also known as soil sheaths or rhizosheaths (49,50). The formation of a rhizosheath requires water; since water is not readily, if at all, available in dry environments, it has been argued that the water must have "come from the plant itself... The plant must exude some moisture back into the soil through the roots" (50). This energy expenditure keeps the plant roots from desiccating and continuing to scavenge for water and nutrients. While there is both intraspecific and interspecific competition for resources in nature, there is also a remarkable partitioning of resources for which roots provide an excellent example (Fig. 8). They may also exude compounds that prevent roots of other species to grow, a phenomenon called allelopathy.

A number of workers have used single-factor measurements and indirect synthetic gradients as parameters for measuring or indexing soil fertility and soil nutrient status and have demonstrated the distribution of vegetation (populations

Table 2. Elements of Significance in Plant Nutrition and Their Roles[a]

Element	Confirmed or Suspected Role
Macronutrients	
Nitrogen	Proteins, nucleic acid, coenzymes
Sulfur	Amino acids, cystine, cysteine, methionine, thiamine, biotin, sulfolipids
Phosphorus	Energy metabolism, adenosine triphosphate, nucleic acids, coenzymes, pyrophosphorylases
Potassium	Activitor of numerous enzymes, starch synthase
Calcium	Lamella of cell walls as pectates, amylase
Magnesium	Chlorophyll, phosphorylation (+), ribulose diphosphate–carboxylase (+)
Micronutrients	
Iron	Ferredoxin, cytochromes
Manganese	Activator of several enzymes, Hill reaction, amino paptidases,
Zinc	Metalloenzymes, including several dehydrogenases, carbonic anhydrase, alkaline phosphotase, hexokinase, ribonuclease
Copper	Ascorbic acid oxidase, cytochrome oxidase, plastocyanine
Boron	Sugar transport, cell wall synthesis, lignification
Chlorine	Enzyme action in photosystem II in photosynthesis
Molybdenum	Several metalloenzymes, key role in nitrate reductase
Nickel	Metal component in urease, many hydrogenases
Beneficial Nutrients	
Sodium	C_4 photosynthetic pathway
Silicon	Cell walls, protection against pathogens
Cobalt	Vitamin B_{12} coenzyme, nitrogen-fixing systems
Selenium	Lowering susceptibility to insects
Aluminum	Stimulation of root cap size

[a] Refs. 40–43.

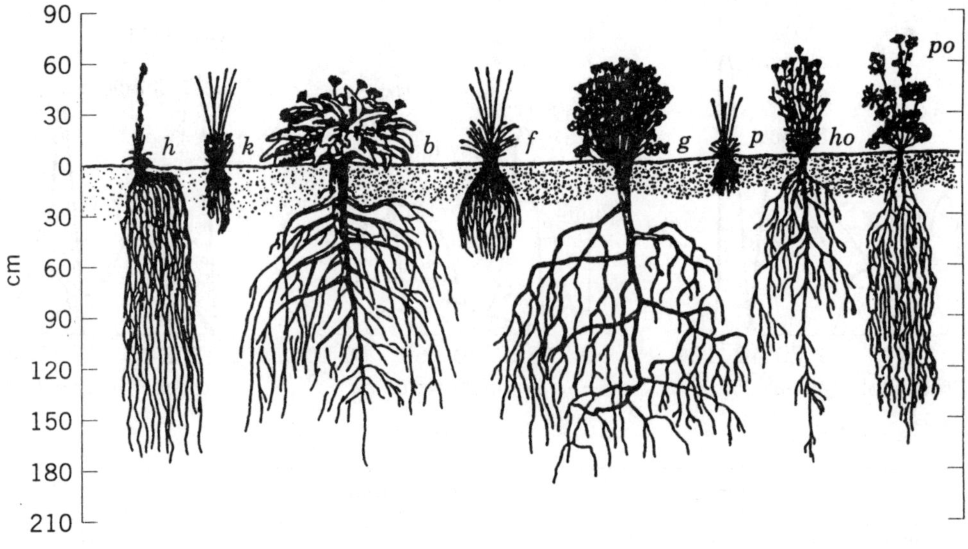

Figure 8. Differences in depth and spread of root systems, showing resource partitioning (51).

and communities) along these nutrient gradients. Calcium, potassium, nitrogen, pH, and electrical conductivity have been widely used for single-factor approaches for assessing a soil's fertility status. While most early European ecologists used pH and base saturation measurements, Scandinavian ecologists emphasized electrical conductivity measurements as a rough approximation of soil fertility. In many ecological studies, nutrient gradients have been based on soil calcium and nitrogen.

Based on field data, two-dimensional nutrient–water regime grids have been used to illustrate the ecological responses of forest trees (52). Triangular models, much like the textural triangle, have been used effectively to demonstrate tree growth in relation to forest soil variables, such as a combination of soil water, particle size, and macronutrients (53). Comprehensive and quantitatively based ecological gradients based on quantitative measurements of a number of soil attributes (such as depth of genetic horizons, bulk density, and effective rooting depth) have provided valuable information on the ecological modalities of species (Fig. 9).

The focus of studies examining the role of micronutrients and other trace elements on plant growth has predominantly been on agricultural crops and horticultural cultivars (see Table 2). Consideration of these elements in the distribution of natural communities, and in adverse environmental impacts (Table 3) is relatively recent. The forms in which these elements exist, their availability and mobility in plant growth, and their downward and lateral movement in soil formation have been explored.

Evidence suggests that most of these metals are present in a chelated form or as metal–organic matter complexes or both (55). Fertilizer trials using trace elements in a chelated form have shown strong indications that the organically bound trace elements are taken up by plants. Roots have been shown to have as strong an ability to chelate metals from the substrate, as do some of the strongest synthetic chelating agents (56). This is not surprising, because soil microbiological activity results in the production of tartarates, malates, citrates, and other such substances that have great chelating capacity (57). It has been found that most of the principal chelating donor groups present in the soil are involved in the complexing of metals by OM (55). Besides polymers that contain many ligand groups, there are the amino, imino, keto, hydroxy, thioether, carboxylate, and phosphonate groups. The presence of these chelating agents in soils may help explain why trees are able to absorb enormous quantities of nutrients from relatively insoluble compounds (57).

Soil Biota

The diversity of life that exists in the soil and the staggering numbers of both identified and unidentified species are phenomenal. The respective roles of soil biota in the functioning of ecosystem processes is indispensable.

The soil biota belong to several groups. The microbes are predominantly bacteria and fungi, along with some algae, cyanobacteria, and some faunal elements. The bacteria are primarily responsible for the decomposition of organic material, nitrogen fixation, and denitrification. The major bacterial contribution to ecosystem processes in the soil medium is in nutrient amelioration, particularly the oxidative processes of the nitrogen cycle, which converts ammonium to nitrite and nitrate by nitrifying bacteria; nitrate is the preferred ion of many plants. These bacteria, members of the genus *Rhizobium,* are responsible for nitrogen fixation. They reside in nodules—gall-like structures on the roots of legumes (eg, peas, beans, clovers, alfalfa)—a highly beneficial and synergistic relationship between the plants and bacteria. Wherever they grow, legumes ameliorate soil nitrogen levels considerably, a property that is the basis of production-sustaining crop rotations in agriculture, although synthetic nitrogen fertilizers have supplanted this practice in many areas.

Besides legumes, there are a number of species that are also nodulated and have the capability of fixing atmospheric nitrogen. These include *Alnus* in the cool regions of the Northern Hemisphere; *Ceanothus* and *Shepherdia* in North America; *Elaeagnus* in Asia, Europe, and North America; *Myrica* in the temperate regions of the world; *Coriara* in Japan, New

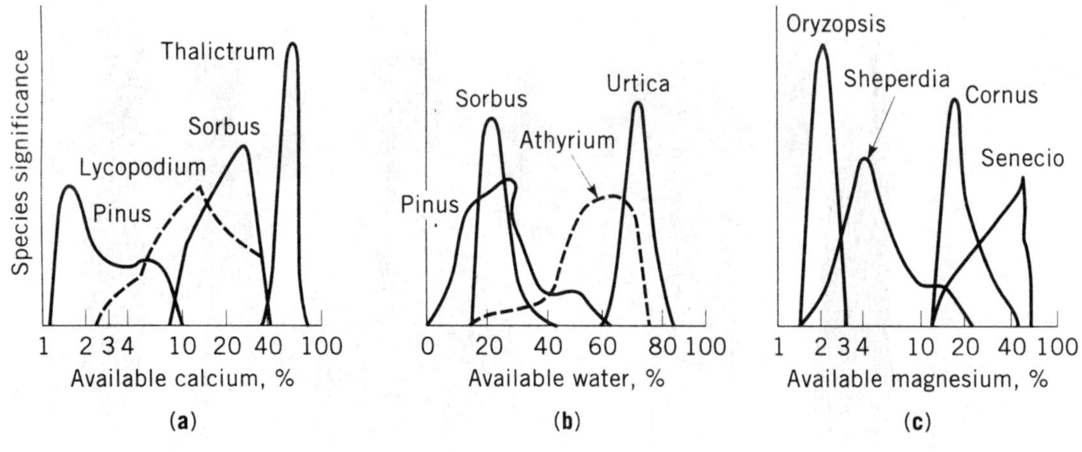

Ecological Group	Calcium Group Limit	Calcium Group No. Species in Group	Water Group Limit	Water Group No. Species in Group	Magnesium Group Limit	Magnesium Group No. Species in Group
I	<2.3	14	≤15	8	<1.5	14
II	2.3–4.5	31	16–30	49	1.5–3.0	30
III	4.6–9.0	34	31–45	56	3.1–6.0	19
IV	9.1–18	26	46–60	29	6.1–12	32
V	18–36	10	61–75	15	12–24	31
VI	36–73	20	≥76	9	24–48	10
VII	>72	6			>48	9

Figure 9. Selected plant population distributions along quantified gradients of (**a**) soil calcium, (**b**) soil water, and (**c**) soil magnesium (10,11).

Table 3. Commonly Used Terms for Elements in Soil Ecosystems[a]

Description	Comments
Metal	Metals may be defined by the physical properties of the elemental state, but are better identified by consideration of their chemical properties; the term is used indiscriminately to refer to both the element and compounds (eg, "the uptake of copper by ..." does not distinguish the form in which the metal is absorbed)
Heavy metal	A hopelessly imprecise and thoroughly objectionable term that originates from a categorization by density, hardly biologically significant; the term is now often used to mean any metal with atomic number >20, but there is no general concurrence
Essential metal	Broadly, one that is required for the complete life cycle of the organism and whose absence produces specific deficiency symptoms relieved only by that metal and whose effect is shown by a dose–response curve; the term is often used misleadingly, since it should be accompanied by a statement of which organisms show a requirement for the element
Beneficial metal	An older term, now largely not used; implies that a nonessential metal could improve vigor
Toxic metal	*All* metals are toxic, but the degree of toxicity varies greatly from metal to metal and from organism to organism; toxicity, like essentiality, may be defined by a dose–response curve
Abundant metal	The term usually refers to the proportion of the element in the earth's crust, although it may be defined in terms of other regions, eg, oceans or fresh water.
Available metal	One that is found in a form that is easily assimilated by living organisms
Trace metal	A metal found in low concentration (parts per million or less) in some specified source, eg, soil, plant, tissue, or groundwater; sometimes this term has confusing overtones of low nutritional requirement (by a specified organism)
Micronutrient	More recent term used to describe the second of the meanings of trace metal

[a] Ref. 54.

Zealand, Central and South America, and the Mediterranean region; and *Casuarina* in the tropics and subtropics from East Africa to the India Archipelago, Pacific Islands, and Australia. So successful was the practice of legume-based crop rotations in agriculture that it prompted several scientists to suggest similar legume-based rotational systems for those forest ecosystems where soils were nitrogen-deficient for tree growth. In the tropics, particularly in areas where rice is cultivated, substantial nitrogen fixation comes from cyanobacteria, particularly of the taxa *Anabena* and *Nostoc*.

Given the significance of the statement that "countless millions of microbes can be placed on the head of a pin" (58), investigators have appropriately concentrated their attention on the contributing roles of microbes in ecosystem function. In the grassland ecosystems of the western United States, it has been estimated that the annual aboveground biomass productivity (what we see) was 240 g/m^2, whereas the belowground root–microbial biomass productivity (what we do not see) was 360 g/m^2 (58). Standing crop (previous years' standing dead plus current year's growth) estimates have shown even more striking differences: aboveground 275 g/m^2; belowground 1625 g/m^2. Thus the magnitude and importance of the role of microbial populations cannot be understated.

Recent estimates indicate that >95% of all land plants depend to some degree on a symbiotic relationship between their roots and certain fungi, termed mycorrhizae (*myco* = fungus; *rhiza* = root) (59,60). There are four major types of mycorrhizal relationships, which vary both in the taxonomic groups of the plants and fungi involved and in the nutritional benefits that the partners derive from the relationships (61).

Endomycorrhizae. The most widespread and common form of mycorrhizal symbiosis is termed the endomycorrhizae, or vesicular–arbuscular, mycorrhizae (VAM). Almost all angiosperm families and some ferns, fern allies, and mosses form VAM relationships with a small group of primitive fungi related to bread molds (62). As a consequence of the differences in the diversity between the plant hosts (250,000–300,000 described species) and fungal symbionts (400–500 species in five genera), there is little specificity between fungus and plant in this relationship (62). The few families that do not form VAM symbioses include the cosmopolitan mustard family (Cruciferae or Brassicaceae) and the Australian protea family (Proteaceae). In addition, there are several other families, notably the Chenopodiaceae and the Amaranthaceae, in which annual and short-lived perennial species are not mycorrhizal but long-lived perennials are (63). Fossils of the earliest land plants show strong VAM colonization of their roots, indicating that this relationship is probably primitive to the origin of land plants (61).

The two common names for this relationship are both based on the anatomical and morphological attributes of the infected roots. In all forms of mycorrhizae, there is a fungal hyphal network that grows throughout the intercellular spaces of the root cortex and also ramifies broadly throughout a volume of soil outside the root, ranging up to 7–10 cm from the root epidermis. In VAM mycorrhizae, there are also highly specialized branched structures called arbuscules, which develop within the cell wall (but outside the cell membrane) of certain cortical cells. Transfer of materials between plant and fungus occurs where the arbuscule abuts the cortical cell membrane (64). The prefix *endo* has been applied to these mycorrhizal fungi, because of the position of the transfer organ inside the cell wall. In addition, the fungi produce intercellular storage structures, termed vesicles. Hence vesicular–arbuscular or endomycorrhizae.

VAM-forming fungi are capable of taking up and transporting phosphorus and many trace metals from the soil to the plant (64). In addition to taking up soluble phosphate from the soil, the enzyme systems of the VAM fungi are able to take up phosphorus mineral forms that are not available to roots (65). In semiarid and arid regions, the growth enhancement achieved by VAM-infection is as much due to increased water flow to the roots as it is from improved nutrition (66). Because of the highly branched nature of the hyphal network of VAM fungi, the volume of soil from which the plant can forage for water and phosphorus is 1,000–10,000 times greater than that available to an uninfected root (64).

Fungi, in turn, receive fixed carbon from the plants, typically as 3-carbon sugars. In rapidly growing crop plants, as much as 7% of the daily photosynthetic gain may be transferred to the mycorrhizal fungus (61). However, the growth enhancement from having the VAM relationship is so large that growth and yield of mycorrhizal plants are typically much greater than those of uninfected plants, despite this carbon drain.

Almost all major agronomic crops depend on VAM mycorrhizal relationships, including the major forage crops: wheat, corn, soybeans, oats, and rice (Fig. 10). Among the commercially important fruits and vegetables, only those in the families listed above (eg, canola, rape, spinach, sugar beets) depend to some degree on VAM mycorrhizae (62) Similarly, most shrubs and trees of natural ecosystems (others than those discussed below) depend on VAM, with the degree of dependence proportional to plant life span. In particular, trees typical of ecosystems in which the soils have moderate to high inorganic nitrogen availability (especially as nitrate) but that may be phosphorus limited strongly depend on VAM (67).

Ectomycorrhizae. The second most common mycorrhizal relationship is termed ectomycorrhizae (ECM). ECM relationships are formed by certain species of higher fungi (Basidiomycetes and Ascomycetes) and certain groups of woody plants, including several families of conifers (pines, spruces, and firs) and angiosperms (in oaks, hickories, eucalypts, and acacias) (61). The number of species of plant and fungus involved in the ECM relationship is much smaller, and plant–fungus specificity is much greater than VAM. Evidence from the fossil record and from current studies of fungal evolution indicate that this relationship arose independently a number of times in the late Mesozoic and Cretaceous periods (61).

Overall, data show that mycorrhizal associations considerably enhance plant growth by playing an important role in carbon dynamics, nitrogen, and phosphorus cycling (61). Furthermore, "in addition to N and P, mycorrhizal fungi take up and translocate any element required for plant growth, and mycorrhizae also enhance the transport of water from soil to plant" (61).

Soil Fauna. Likewise, soil fauna are numerous and diverse. Soil fauna are generally classified on the basis of size into three groups (Fig. 11): microfauna, generally >200 μm, include mostly protozoa and a few small members of mites, nematodes, and rotifers; mesofauna, generally 200 μm to 1 cm long, include millipedes, most mites, and most nematodes; and macrofauna, several centimeters in length, include earthworms, large arthropods, and vertebrates that live in the soils (68). These organisms exhibit much taxonomic diversity (68) and affect soil ecosystems in diverse ways (61,68,69). They include detritivores, herbivores, and carnivores. Their diversity is as important as are their numbers and biomass (Table 4).

The roles of two or three groups of organisms and their affects on soil will illustrate the importance of soil fauna. The earthworms, whose role in soils made an impression even on Charles Darwin (70), provide a multitude of functions: soil turnover; incorporation of organic matter, fertilizer, and lime; improvement of soil aeration; acceleration of humification; conversion of organic nitrogen and phosphorus into plant-assimilable forms; stimulation of respiratory and enzyme activities of soil; destruction of phytopathogenic fungi; increase in yield of roots, grass, and other crops; and preservation of soil structure through production of polysaccharides (70). The useful role of earthworms in organic waste management, in soil amelioration, and in land reclamation and as bioindicators of environmental contamination has also been emphasized (71).

Ants are also abundant in some soils, and ant mounds or hills are dispersed over large areas in many parts of the world. By one estimate 20 anthills/ha pile up over 3800 kg earth on the surface. In the northern Great Plains, 1000 mounds were estimated in an approximately 4-ha area; there may be as many as 50 colonies/ha, with a total of >13,800 individuals. Ant activity, by tunneling and moving the earth, was shown to increase the porosity of soil by >30%, thereby considerably altering water transport and nutrient mobility. In two contrasting situations showing the effects of ant activity, the low pH values of pine soils in Poland increased, whereas the high

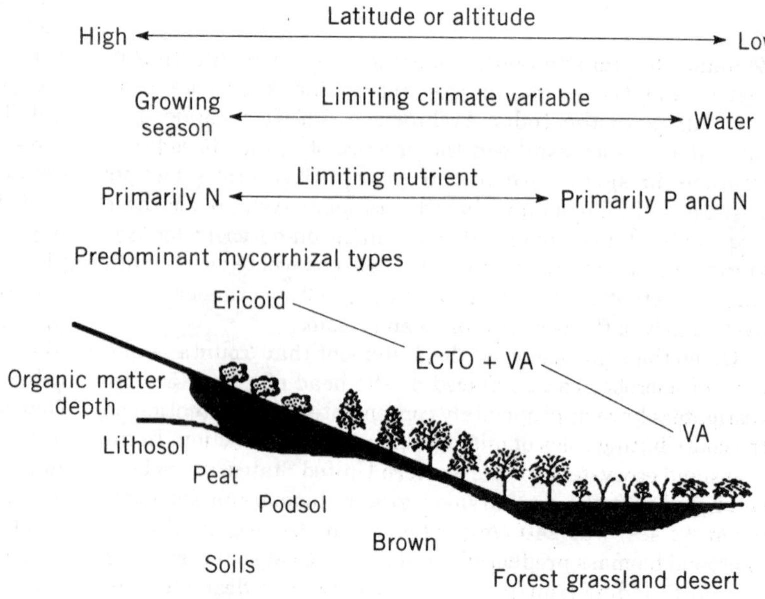

Figure 10. The distribution of mycorrhizal fungal associations along the gradients of climate, soil types, and vegetation (61). ECTO, ectomycorrhizae; VA, vesicular-arbuscular mycorrhizae.

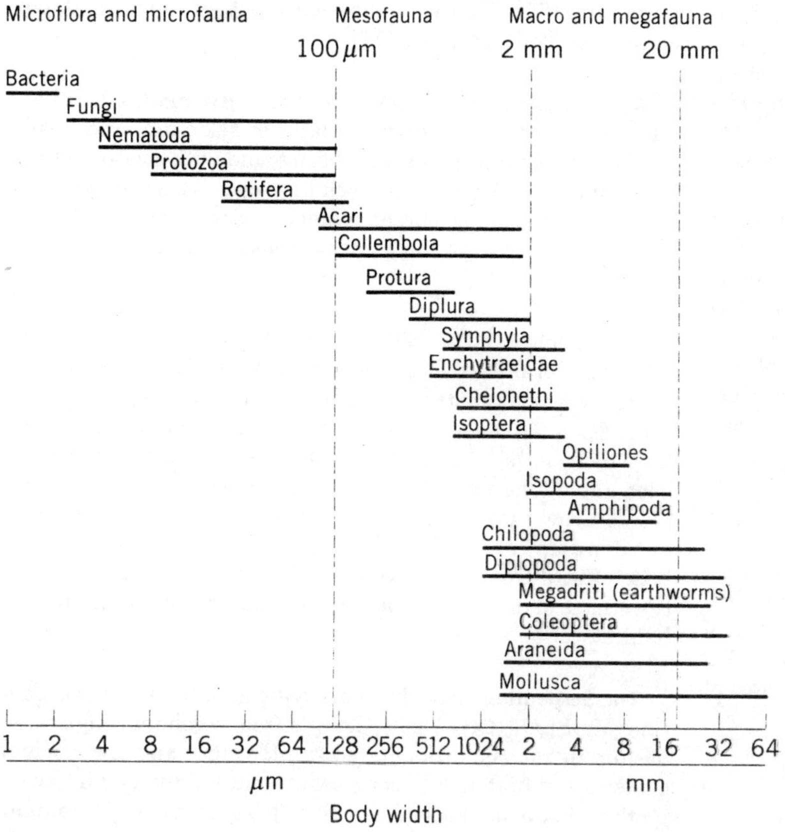

Figure 11. The size classification of organisms in decomposer food webs by body width (68).

Table 4. Numbers and Estimated Biomass of Soil Fauna in Conventional Tillage (CT) and No Tillage (NT) Agroecosystems at Horseshoe Bend[a]

Fauna	CT, ×m^{-2}	NT, ×m^{-2}	CT, mg dry wt/m^2	NT, mg dry wt/m^2
Nematodes[b]				
Bacterivores	1,836	909[c]	237	117
Fungivores	227	500[c]	14	31
Herbivores	945	1,064	93	104
Subtotal	*3,008*	*2,473*	*344*	*252*
Microarthropods[d]				
Mites	41,081	78,256[c]	118	303
Collembola	6,244	14,684[c]	17	40
Insects	2,105	2,548		
Subtotal	*49,430*	*95,488*	*135*	*343*
Macroarthropods[e]				
Ground-beetles	7	33[c]	6	30
Spiders	1	17[c]	1	14
Others	6	28[c]		
Subtotal	*14*	*78*	*7*	*44*
Annelids[f]				
Earthworms	149	967[c]	3,129	20,307
Enchytraeids	1,867	520	592	17
Subtotal	*1,986*	*1,487*	*3,188*	*20,963*
Total	*54,438*	*99,526*	*3,674*	*20,963*

[a] Ref. 68
[b] Means of samples from June to October 1983; numbers are $\times 10^{-3}$.
[c] For numbers of organisms, tillage treatments differ significantly at $P = 0.05$.
[d] Means of samples from May to December 1983.
[e] Means of samples from April to June 1983.
[f] Means of samples from April 1983.

soil pH values in a northern Plains grassland decreased; in both cases, there were significant changes in resulting plant community composition (37). The role of termites, which is on a larger spatial scale than that of ants, has been shown to be important in tropical soils, with some termite mounds exceeding 2–3 m in height and consisting of hundreds of kilograms of soil.

In addition, gopher mounds, prairie dog mounds, and other small vertebrates affect soil properties. For example, it was found that prairie dogs and other herbivores remove large quantities of nutrients, likely as much as 75% of those translocated aboveground annually, and hence may substantially influence nutrient cycling (72). A higher nitrogen content was found in areas colonized by prairie dogs than in those not colonized. Studies of the impact of large vertebrates on soil properties and vegetation composition are relatively recent. Both field studies and simulation modeling have documented the effect of large mammals in African ecosystems, particularly in nitrogen cycling (73).

Overall, it is believed that compared to microbial activity, every other faunal component has a much smaller role (58). This statement was substantiated by estimating the microbial biomass at 100 g/m^2 in contrast to invertebrates at 0.1–1.0 g/m^2, small vertebrates at 0.2–2 g/m^2, and large vertebrates at 1–5 g/m^2.

DYNAMIC PROCESSES OF THE SOIL ECOSYSTEM

Plant Population and Community Dynamics

The strong influence of soil on the occurrence, abundance, and distribution of plants and elemental uptake by plants has been directly implicated in population and community differentiation at several levels. This has been referred to as edaphic specialization. Concentrations of particular elements in soils and the specificity of plants to be able to take up these elements are equally important in edaphic specialization. Many studies show high accumulations within certain species of plants of a number of elements (eg, nitrates, aluminum, copper, molybdenum, nickel, and selenium). These accumulators are referred to as indicator plants. The latter have been used in geochemical prospecting, in geological and soil mapping, and in groundwater surveys (74). The role of indicator species is extremely important in discussions on the preservation of rare, endangered, and threatened species. Plant responses ranging from localized and small populations to large areas inhabited by multiple species provide evidence of edaphic specialization. Some examples are provided below.

Evolution in Closely Adjacent Populations. Evolution in closely adjacent populations has shown how selection pressures produce ecotypes tolerant of high elemental concentrations (ecotypes are genetically based physiological races). The ecotypic differentiation in *Anthoxanthum odoratum* and *Festuca ovina,* tolerant of high zinc concentrations (75), and the selection of copper- and cadmium-tolerant *Athyrium yokocenscens* ecotypes on copper tailings (76) are illustrative.

Single Species Populations Occupying a Large Area. *Arctostaphylos myrtifolia* has been shown to grow in dense stands in the foothills of the Sierra Nevadas in California on Eocene laterite and sericitic schists, which are low in cations and fertility, have pH values of 2–3.95, and contain high soluble aluminum (77). This specialization has been termed edaphically controlled endemism (77).

Accumulations in Genera. The genus *Astragalus* has been found to be an obligate accumulator of selenium, and nearly all members of this genus have been found with concentrations exceeding 10,000 ppm (78). Such high concentrations not only indicate seleniferous parent materials and soils but also the necessity for screening such plants before animals use them as forage.

The Calcicole (Calciphilous) and Calcifuge (Calciphobous) Plants. Several British ecologists have elucidated the mechanisms by which plants are able to inhabit limestone areas. The calcicoles have the ability to take up large amounts of calcium, for which, it has been speculated, these plants have both a need and a tolerance. Studies further suggest that while these plants grow in a pH range of 7–8.5, they must be efficient in taking up iron. Calcifuge species, in contrast, show strong toxic effects from high levels of calcium and bicarbonate and the effects of deficiencies of nutrients that are inherent in such habitats.

The Serpentine Soils. In many regions of the world, serpentine ($Mg_6Si_4O_{10}(OH_8)$) soils, derived from secondary high magnesium ultrabasic rocks, have low pH values and low calcium contents but high nickel, magnesium, chromium, and iron contents and support specialized plant taxa, resulting in edaphic endemics (79,80). The flora of these unusual areas is so distinct that it has been suggested that "serpentine vegetation is not a seral stage in the vegetation of other soils but a distinct climax of its own distinctive successions" (81).

Local Floras. Diploid and polyploid plants were found to be distributed closely in relation to soil environmental gradients in the Ogotoruk Valley in northwestern Alaska (82). Gradients included those of soil texture, soil moisture, soil temperature, permafrost, and disturbance. For a large number of species (85%), the patterns of correlations were consistent. Plants represented in the local and sparse flora of shale barrens have been shown to require unusually high amounts of sulfur and to belong to the genera that are common around sulfide deposits. Perhaps the most investigated are the serpentine floras.

Soil Control of a Continental Flora. The morphological adaptations such as xeromorphy and sclerophylly in plants and indeed the origin of a large segment of the continental flora in Australia can be explained by a single major essential element: phosphorus (83). Migration and evolution of diversified taxa found in the xeric components of Australian floras from rain forest families are promoted by low phosphate and not aridity. The evolutionary response under the influence of low phosphate is an increase in the frequency of two vegetative characters: xeromorphy and sclerophylly. These traits can be experimentally reversed by the addition of phosphate and nitrate (83).

Energy Flow

The Relationship of Thermodynamics to Soil Ecology. Thermodynamics governs the processes of energy capture within and transmission through biological systems, and these thermodynamically governed processes result in both matter being produced by and energy exiting from the biological system (84). Ecosystems, therefore, are the conductors and transmitters of energy and material that drive both soil formation and biological sustenance. The quantity and duration of energy transport manifests itself through biological energy capture and climate, which, in turn, determine the character of the biosphere that imparts a stamp of unique morphology on the resulting soil (85).

The application of thermodynamic principles and models to soil science dates back at least to 1935 when the energy relationship of water in soil was examined (86); soil genesis models based on thermodynamics were developed later (87). Water flow was used as the energy currency for transforming soil parent material into a more ordered soil profile, ie, reducing entropy (87). This model, however, was based on thermodynamic principles more appropriate to *closed systems* (88).

More recently, it was argued that a form of thermodynamics that deals with flows of matter and energy in *open systems* is necessary (89). This approach leads to nonequilibrium thermodynamics that describe open systems, such as soil-plant ecosystems, that tend toward steady states characterized by minimum production of entropy. Ecosystem sustainability can be defined on the basis of this thermodynamic approach, ie, the balance between ordering and dissipative processes.

Table 5. Ordering and Dissipative Processes in Soil–Plant Systems Categorized as Biological or Physical[a]

Ordering Processes, Entropy Decreases	Dissipative Processes, Entropy Increases
Biological	
Photosynthesis	Respiration
Growth	Senescence
Formation of humus	Decomposition of humus
Physical	
Water flow (profile development)	Water flow (erosion, leaching)
Flocculation	Dispersion
Aggregation	Disaggregation
Structure development	Breakdown of structure
General characteristics	
Larger units	Smaller units
Fewer of them	More of them
More ordered	Less ordered

[a] Ref. 85. The pairs are not necessarily exact opposite.

Sustainability of Soil Ecosystems and Agriculture—Ordering and Dissipative Processes. Solar energy flow provides the energy currency that sustains the biosphere, including soil ecosystems, and drives the self-organization (ie, ordering or the reduction of entropy) of the biosphere (90). This solar energy flow can result in effective work or products, such as biological production, soil ecosystems, and their sustenance, only if infinite isothermal reservoirs at high (solar energy source) and low (thermal sink, ie, outer space) temperatures exist (89).

Biological processes provide excellent models of these ordering and dissipative processes (89). Ordering processes such as photosynthesis and its complementary processes result in high molecular weight compounds (eg, carbohydrates, cellulose, proteins) that have complex, ordered structures derived from smaller molecules of CO_2, H_2O, NH_3 and various elements and inorganic substances. Thus agriculture can be defined thermodynamically as the building of large molecules (eg, carbohydrates and proteins in grains, meat, milk, etc) from small molecules and elements. Similarly, soil development and soil ecosystems are driven and sustained by ordering processes, giving rise to unique soil profiles and ecosystems derived from various parent materials and biological processes and biological components.

Dissipative processes are also at work within the biosphere, even under conditions of a steady state of minimum entropy production. These dissipative processes are exemplified by respiration and senescence that degrade photosynthesized molecules and structures back into smaller molecules such as CO_2, and less complex inorganic compounds and structures. Ordering and dissipative processes in the soil–plant ecosystem can be categorized (Table 5). Soil degradation occurs when the dissipative processes become excessive, eg, soil profile destruction by erosion.

Sustainable agriculture and sustainable soil ecosystem management can be defined using this thermodynamic-ordering-dissipative model. The goal of sustaining agriculture and soil ecosystems is to strive for lessening of entropy, ie, minimization of dissipative processes caused by soil perturbations such as excessive erosion, nutrient losses, soil compaction, soil structure deterioration, and salinization. This model can be used also to understand the concept of ecosystem resilience when human-induced ecological perturbations push an ecosystem away from its steady state condition.

The cutting of the eastern U.S. deciduous forests and the cultivation of these forest soils and subsequently the prairie soils were major perturbations of these ecosystems. But the biological capacity of these soil ecosystems was not destroyed. Under proper management, these soil ecosystems have resulted in a virtual agricultural cornucopia while maintaining a suitable balance of ordering and dissipative processes. One advocated criterion of sustainability is that there should be the maintenance of a proper balance between ordering and dissipative processes, ie, maintaining entropy at the lowest level by managing agriculture (and soil) systems that permit the establishment of steady states (88).

> *Continuous agriculture, or even monoculture, may therefore be one of the more sustainable options available, but with four provisos: (1) The system must achieve a steady state. (2) The flows within the system, particularly that of nitrogen, must be of appropriate magnitude. (3) The capacity for self-organization must be retained in the system; this means making sure that the biological potential is maintained. (4) The maintenance of the steady state must not involve excessive expenditure of energy, and entropy production, outside the system, in fertilizer manufacture, for example (89).*

Such steady states can be monitored by how much dissipative by-products flow out of the system. Therefore, water quality (as measured by turbidity–sediment, nitrate, phosphorus) can be the flow indices of ecosystem health as measured by dissipative product exhausts to show how well a biological or soil ecosystem is functioning (84).

Biogeochemical Cycles

Materials in nature, including water, are cycled and reused. This is not true for energy, which flows in one direction only. But energy and material cycling have a dynamic flux from a source (from which energy and materials flow) to a sink (to which they flow). Although material cycles are also referred to as nutrient cycles, in nature all elements cycle, including toxic elements. Three groups of biogeochemical cycles are recognized: the hydrological, or water cycle; gaseous; or true cycles, which have gas, liquid, and solid phases (carbon, nitrogen, oxygen, sulfur), and sedimentary cycles (eg, calcium, magnesium, phosphorus), through which elements are taken up from the soil by plants, passed on to the herbivores and carnivores, and returned to the soil. As the name implies, the fluxes of biogeochemical cycles are determined by several factors: climatic variables (particularly temperature and precipitation), the nature of the geological material, the type of soil, and the types of plants and animals.

The flow of solar energy is the engine that powers the hydrologic cycle, drives the earth's climate patterns, and energizes the biogeochemical cycles. These cycles are critical to the sustenance of life and are crucial to the formation of soils and the maintenance of soil ecological processes that are requisite for sustaining soil quality. The amount, residence time in a particular pool, and rate (flux) of transfer of material from one pool or reservoir to another varies dramatically with the chemical character (solubility, volatility, reactivity, pool composition) of the material. For example, water, which occurs in all three states (gaseous-vapor, liquid, solid) within a relatively narrow and common temperature range, can be cycled in large volumes rather quickly. Calcium and silicon, on the other hand, are chemical elements that are commonly tied up in sediments and rocks. Thus they may take many years or millennia to be solubilized or physically moved (by erosive water, wind, glacial action, or volcanic eruption) from one pool or location to another. Carbon, nitrogen, and oxygen are major elemental building blocks of the biosphere and constitute major components of the atmosphere. They can cycle quickly through the biosphere, atmosphere, and hydrosphere as well as become lodged in sediments as part of the lithosphere.

Nearly three-quarters or 70.9% (361,743,640 km^2) of the earth's surface is water. The volume of the oceans is 1,285,467,390 km^3 in contrast to 35,012,730 km^3, or 2.72% of fresh water. The water cycle is also referred to as a closed cycle, since water lost to the atmosphere through both evaporation and transpiration must return to earth through precipitation (Fig. 12). In contrast, the elemental cycles are open.

Among the gaseous cycles, carbon and nitrogen have received most attention. Carbon is stored in a variety of material from animal tissues to tree trunks to soil humus, and it is continuously trapped by green plants through the process of photosynthesis and continuously released by all life through respiration (Fig. 13). Thus carbon is exchanged between the atmosphere, the oceans, the terrestrial biosphere, and more slowly, with sediments and sedimentary rocks (92), making atmospheric carbon dioxide a linchpin in the cycling phenomenon. Much concern has been expressed in recent years on the increasing concentration of carbon dioxide in the atmosphere as a result of human activities.

Because of its limiting nature, particularly in agronomic crops, research on the nitrogen cycle has been going on for over a century. In well-aerated cropland, nitrogen mobilization takes place in a number of steps (Fig. 14), from the decomposition of organic matter to the formation of ammonium ions through the processes of aminization and ammonification. Since a majority of the plants investigated prefer to take up nitrogen in the form of nitrates, the oxidation steps mediated by nitrifying bacteria are, therefore, crucial. These bacteria are aerobic and prefer a circumneutral pH. Plants in wetland areas, where conditions are anaerobic, have the ability to use nitrogen in the form of ammonium. Rice plants transport oxygen to roots, exude it to the rhizosphere, and oxidize NH_4^+ to NO_3^-.

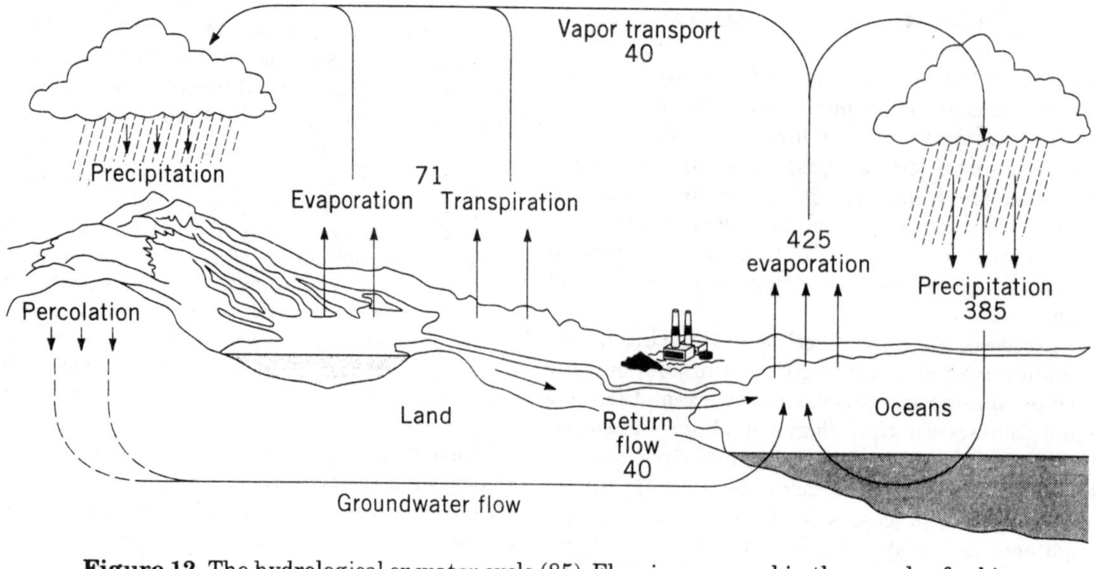

Figure 12. The hydrological or water cycle (85). Flow is measured in thousands of cubic kilometers per year.

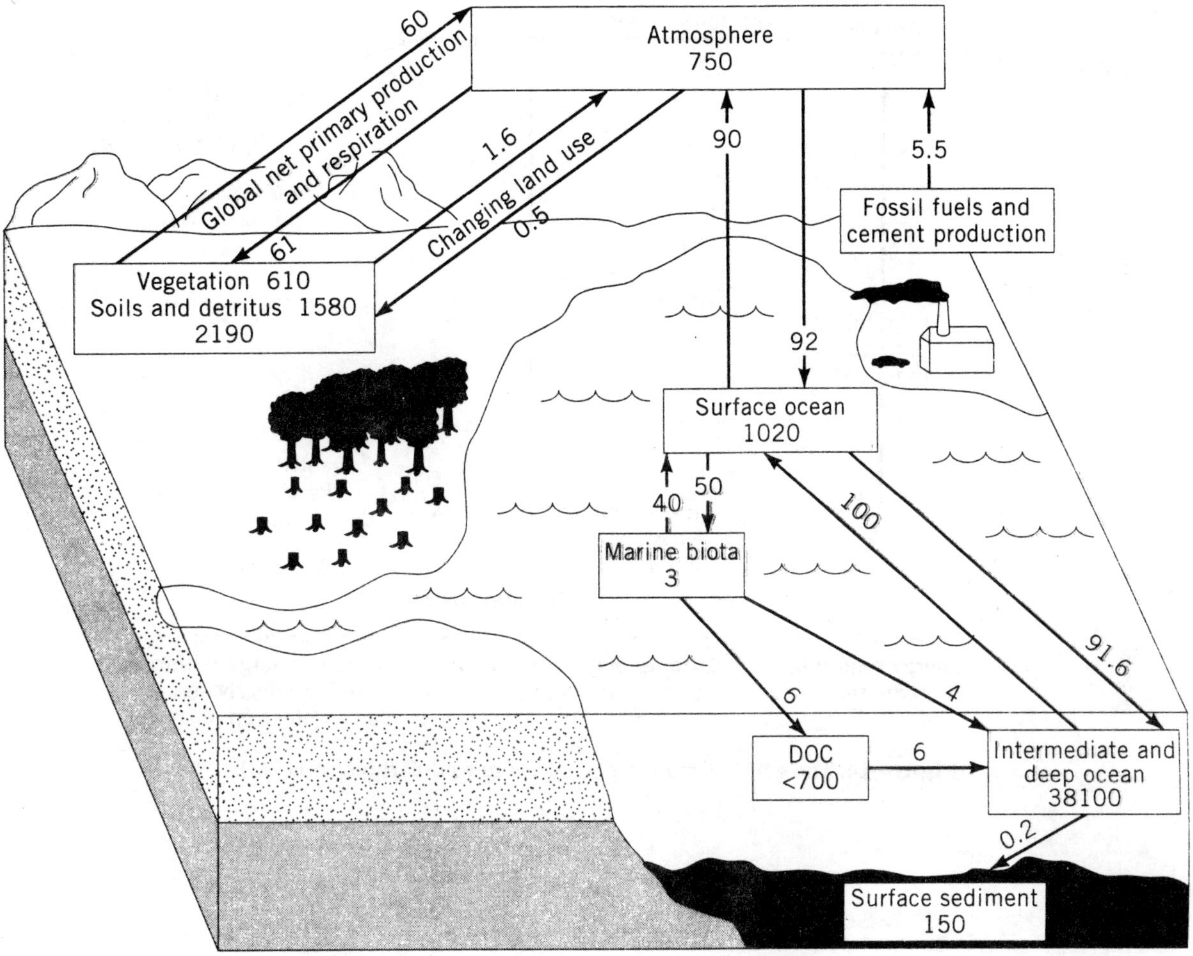

Figure 13. The global carbon cycle (92). (Units: GtC for reserviors; GtC/yr for fluxes; all averages for period 1980–1989).

Intensive work on the role of mycorrhizal associations, the possibility of nodule association in nonleguminous plants (using biotechnology tools), and the mechanisms by which natural plant populations are able to obtain nitrogen all hold promise. Recent reports suggest that the production of polyphenols by pine litter in strongly acid and infertile soils may help plants obtain nitrogen directly from dissolved organic forms (94) (Fig. 15). Quantifying the role of consumers for all nutrients in general and altering nitrogen relationships in particular, is also of great importance. Simulation models have been provided on the role of large mammals in nutrient cycling in African ecosystems (73); the models applies generally to temperate grasslands as well (Figs. 16 and 17). Carbon and nitrogen are both intimately related to biological activity (96,97). Thus both are indicators not only of soil fertility but also of ecosystem productivity. Nutrient uptake and sequestration are equally important in forest ecosystems, and partitioning is differential in different parts of trees (Fig. 18) (98).

Sedimentary cycles include all other nutrients such as calcium, phosphorus, magnesium, and a number of micronutrients like iron, manganese, and copper (see Table 2). Generally, these cycles have fewer steps than the nitrogen cycle; however, the same factors apply: climate, type of soil, biota, and particularly the microorganisms. But no matter how important these cycles are to the perpetuation of life, there is a downside. Elements that may be considered toxic (or compounds that are not biodegradable) get increasingly concentrated as they pass from one trophic level to the next. Two such examples are the concentration of ^{90}Sr and DDT. In the first case, ^{90}Sr was captured directly from the atmosphere by arctic lichens (reindeer moss) from where it was passed on to reindeer and humans. During passage from one trophic level to the next, the concentrations increased by several orders of magnitude. Being in the same elemental group as calcium and reacting similarly, ^{90}Sr in lactating (calcium-rich, milk-producing) mammals was particularly problematic. Similarly, food chain concentration of DDT occurred through accumulation from aquatic or terrestrial systems to animals and birds. This phenomenon has been referred to as bioconcentration, bioaccumulation, or biomagnification.

Ecosystem Resilience

Ecosystem stability, resilience, and diversity have been topics of much discussion recently. Does ecosystem stability lead to diversity or does diversity enhance stability of the ecosystems? "Stability (*sensu stricto*) is the propensity of a system to attain

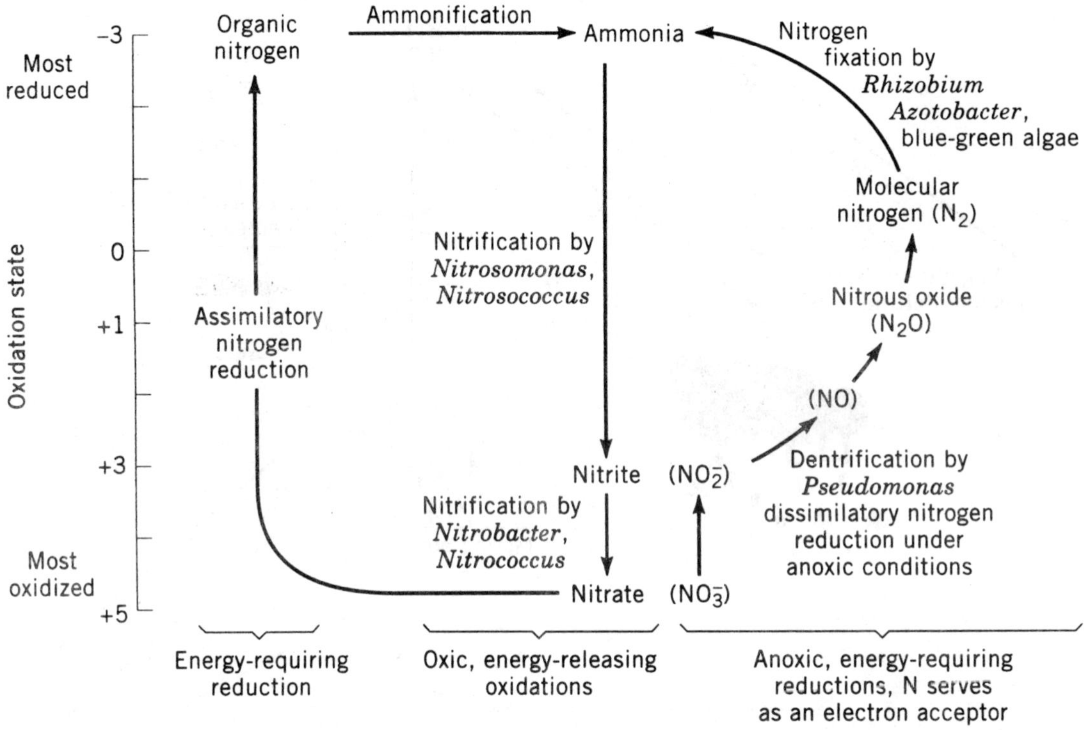

Figure 14. The generalized nitrogen cycle in well-aerated soils (93).

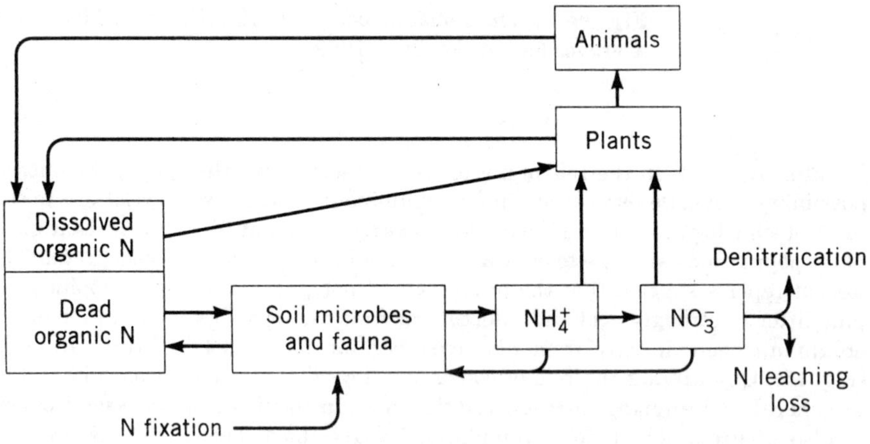

Figure 15. The nitrogen cycle depicting uptake of nitrogen by plants from the organic pool (94).

or retain an equilibrium condition if, [in] steady state or steady oscillation... Resilience, on the other hand, is the ability of a system to maintain its structure and patterns of behavior in the face of disturbance" (99). For soil ecosystems, stability and resilience are central to how much disturbance can be tolerated before soils are irretrievably degraded; some of these considerations, important as they are, are only now receiving attention (100).

"Is variety only the spice of life or is it a necessity for the long life of the total ecosystem comprising man and nature" (7)? Some long-term experiments, the results of which have been published recently, suggest that biological diversity stabilizes soil ecosystems (101).

Soil Formation

Relative to human time scales, soil formation is a slow process. Indeed, soils have been characterized as the least renewable of our natural resources. Soils are derived from rocks (igneous, sedimentary, and metamorphic) and their breakdown products where geological and climatic forces have acted on them. If the rocks and their breakdown products have stayed in place,

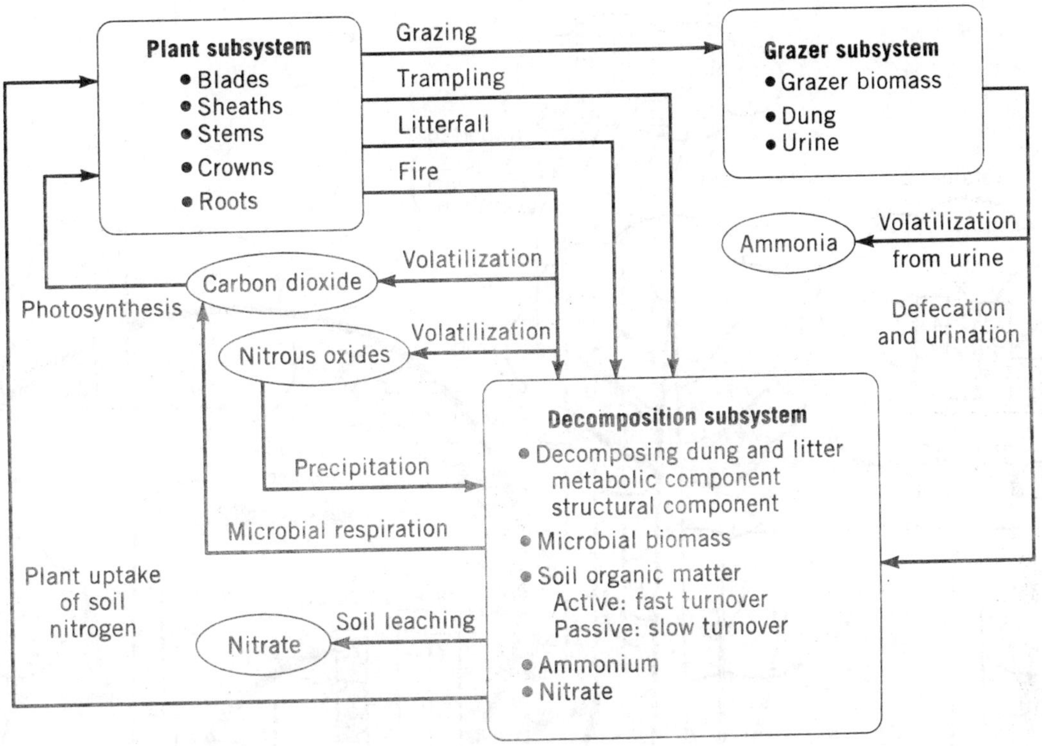

Figure 16. Nutrient cycling pathways in grasslands that support abundant large animals (73).

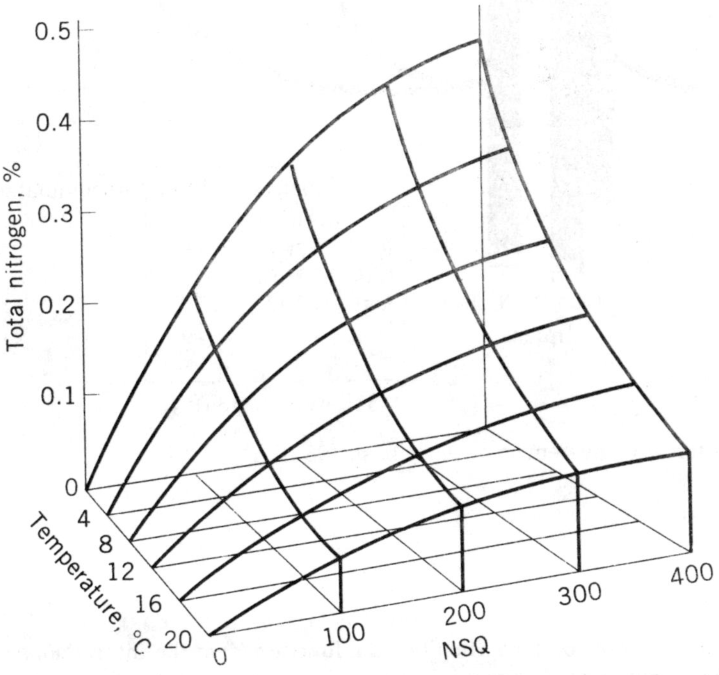

Figure 17. Generalized trends in soil total nitrogen content in relation to mean annual temperature and mean annual Niederschlag Saatigungs quotient (NSQ). NSQ is calculated by dividing precipitation (in millimeters) by absolute saturation deficit of air (in millimeters of mercury). The model depicts the Great Plains area from Canada to Mexico (back to front curves), and from the desert to the humid regions (left to right curves) (96).

we refer to them as residual parent materials. However, in many cases the parent materials have been reconfigured and transported by several natural forces; thus parent materials are glacial if they were transported by glacial ice, alluvial if moved by water, and eolian if moved by wind. Lake sediments now exposed are the lacustrine parent material. Similarly, exposed oceanic sediments become marine parent materials.

Rocks and their fragments are degraded by the forces of weathering. Thermal expansion and contraction; the erosional forces of ice, water, and wind; and the abrasive influences of plants and animals all constitute physical weathering. Mineral hydration by water, oxidative, and acidic processes constitute the forces of chemical weathering. In nature, both physical and chemical processes proceed simultaneously,

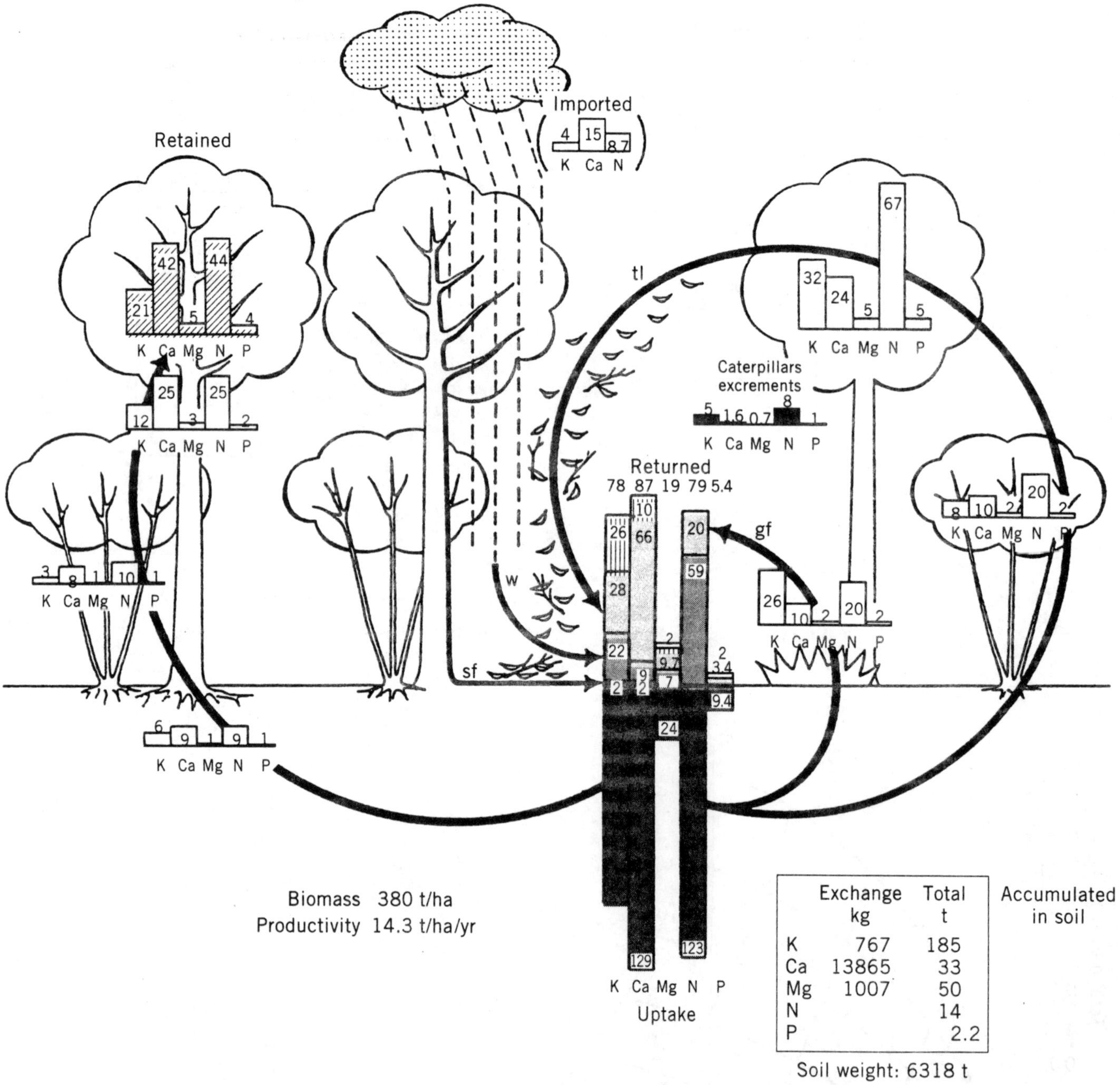

Figure 18. Nutrient cycling in a forest ecosystem on a yearly basis (98).

though the influences of one may be more predominant over the other in given cases. For example, the tundra parent materials are overtly acted on by the forces of physical weathering, whereas those of the tropical rain forest are much more influenced by chemical weathering. Temperature, quantity of precipitation, and the type of vegetation in a given area will determine how quickly soil formation can proceed.

Building on the original Russian soil science concepts, an "environmental formula of soil forming factors" (also referred to as the state factor approach) was proposed (96):

$$s = f(cl, o, r, p, t)$$

where soil formation (s), is a function (f) of the interactions of climate (cl), organisms (o), relief or topographic features (r), parent materials (p), and time (t). It was emphasized that assessing soil-forming factors requires considering factors in reference to time zero, in so far as possible. This does not suggest, as some have erroneously interpreted, that other factors are less important. Thus, with emphasis on different state factor variables, several sequences can result: climosequences, biosequences, toposequences, lithosequences, and chronosequences, by which the dominant influences are climate, biota, relief, parent materials, and time, respectively. Each may be evaluated in sequences in which the other four factors are constant (although they are rarely so in nature).

It has been appropriately argued that the state factors for soil genesis are indeed the same as those that control vegetation development (102):

$$v = f(cl, o, r, p, t)$$

where v is vegetation development. Because of its importance as a significant ecological factor, the variable fire (py) was added to the state factor equation (103). Subsequent work has elaborated considerably on the state factor approach (96). Though some of the theoretical and practical aspects in genetic modeling of soils (104,105) and their implications in ecology (106) have been critically examined, the state factor approach has provided a framework for some elegant work nonetheless. For ecological studies, the state factor approach "is a dominant conceptual approach ... and a crucial background for experimental analyses" (107).

Soil Ecosystems of the World

Soil varies in both the horizontal and vertical dimensions. The dynamic nature of the soil-forming processes, a product of climatic, geological, and biotic interactions results in general patterns that may be characterized thus: as water percolates down through the soils, it carries with it soluble nutrients and clay removed from the upper horizons; this removal process is called eluviation. These materials accumulate in the deeper horizons by the process of illuviation. Two other major processes that contribute to soil formation are biocycling and churning of soil by plant or animal activity (pedoturbation). Given the regionalclimate and the indigenous flora and fauna, these four processes result in soils characteristic of a region (Table 6).

Over time, scientists have organized many of these attributes not only to understand the genesis of soil systems but also to decipher general patterns that could be used to better manage soils for agriculture and forestry. Thus soil classification schemes have evolved that have incorporated the prevailing understanding and knowledge of soil systems. Such classification systems exist in a number of countries depending on the tradition in which the science of soils developed (eg, United States; Soviet Union, France, Canada, and a system in development at the UN Food and Agriculture Organization).

The U.S. system currently in use was introduced in 1960 (3); until then the soil taxonomic system was biased in favor both of Russian terminology and concepts of soil genesis. The system introduced in 1960 was the culmination of many decades of work by leading soil scientists; the soil classification was labeled "comprehensive." Indeed it was, for it incorporated not only meaningful rationale behind the concepts but also both qualitative and quantitative information. It has gone through several revisions since its inception. At the broadest level of integration, it is divided into 11 orders (formative elements of the names are provided in parentheses): alfisol (*alf,* "pedalfer"), andisol (*ando,* "dark soil"), aridisol (*id, aridus,* "dry"), entisol (*ent,* "recent"), histosol (*ist, histos,* "tissue"), inceptisol (*ept, inceptum,* "beginning"), mollisol (*oll, mollis,* "soft"), oxisol (*ox,* "oxide"), spodosal (*od, spodos,* "wood ash"), ultisol (*ult, ultimus,* "last"), and vertisols (*ert, Verto,* "turn"). Worldwide, these 11 orders have 55 suborders, 238 great groups, 1,243 subgroups, 7,504 families, and 18,807 series. A broad grouping of climatic, vegetation, and soil types, with particular attention to temperature–evapotranspiration (TE) and precipitation–evapotranspiration (PE) indices has been provided (109) (Fig. 19). Figure 20 shows the distinction among the orders, and Figure 21 shows the distribution of major soils in the world (110).

USE, RESPONSE, AND MANAGEMENT

Soil Ecology and Landscape Ecology

The relationships between the geosphere and landscape components are intimate (Fig. 22). In its broadest context, land includes soil, geologic substrata, topography, relief, aspect (spatial orientation), ground water, surface water (up to the arbitrary point where water bodies assume their own definition or identity), minerals and fossil fuels, air, and climatic attributes.

As people occupy and develop land, natural landscapes are eventually converted to cultural landscapes, ie, the landscape ecology is more and more impacted and characterized by human-induced activities and uses. While various attributes of land serve as the impetus for people to occupy and use land (eg, aesthetic appeal, mineral extraction, access to river transportation, and adjacent harbor), the most extensive worldwide use of land is devoted to agriculture, livestock grazing, and forestry. Soil, therefore, is among the primary attributes of land that permits and sustains this agrarian land use (the other attributes are solar energy and water, which define the resource trinity for agriculture).

Landscape ecology is an interdisciplinary science dealing with the relationships between human society and its living space, including both open and built-up landscapes. It is a relatively recent spin-off of applied ecology and geography, having been conceived by geographers and ecologists in central Europe following World War II (112). Because most soils on human-occupied landscapes have been disturbed and are under continuous management, the disciplines of soil ecology and landscape ecology have much in common and rest on the foundation of understanding soil and landscape processes and behavior under various use and management scenarios and social interventions. Human impacts on ecosystems have been both direct and indirect, establishing humans as the *ecological dominants* (113). Therefore, soil ecology is more and more focused on human-induced impacts on soil and the response of soil to many uses and to a spectrum of management actions.

Major Uses of Soil and Land

Changes in land use in the last few centuries have been phenomenal. In the past 300 years

> *the world's forests and woodlands diminished by 1.2 billion ha, or 19% of the [year] 1700. Grasslands and pastures have declined by 580 million ha, or 8% of the [year] 1700 estimate. Croplands brought into cultivation show a net increase of 1.2 billion ha, for a 466% increase in less than three centuries. Agricultural expansion and depletion of forests and grasslands were greater in absolute terms over the 30 years between 1950 and 1985 than in the 150 years between 1700 and 1850 (114).*

An approximation of the world's land use compiled from several sources along with the average population density (115)

Table 6. Some Processes of Soil Formation That Are Complexes of Subporcesses and Reactions[a]

Term	Fourfold Categorization	Brief Definition
Eluviation	3	Movement of material out of a portion of a soil profile as in an albic horizon
Illuviation	3	Movement of material into a portion of soil profile as in an argillic or spodic horizon
Leaching (depletion)	2	General term for washing out or eluviating soluble materials from the solum
Enrichment	1	General term for addition of material to a soil body
Erosion, surficial	2	Removal of material from the surface layer of a soil
Cumulization	1	Aeolian, hydrologic and man-made additions of mineral particles to the surface of a soil solum
Decalcification	3	Reactions that remove calcium carbonate from one or more soil horizons
Calcification	3	Processes including accumulation of calcium carbonate in Cca and possibly other horizons of a soil
Salinization	3	The accumulation of soluble salts such as sulfates and chlorides of calcium, magnesium, sodium, and potassium in salty (salic) horizons
Desalinization	3	The removal of soluble salts from salic soil horizons
Alkalization (solonization)	3	The accumulation of sodium ions on the exchange sites in a soil
Dealkalization (solodization)	3	The leaching of sodium ions and salts from natric horizons
Lessivage	3	The mechanical migration of small mineral particles from the A to the B horizons of a soil, producing in B horizons relative enrichment in clay (argillic horizons)
Pedoturbation	3	Biologic, physical (freeze–thaw and wet–dry cycles) churning and cycling of soil materials, thereby homogenizing the solum in varying degrees
Podzolization	3,4	The chemical migration of aluminum and iron and/or organic matter, resulting in the concentration of silica (ie, silication) in the layer eluviated
Desilication (ferrallitization, ferritization, allitization)	3,4	The chemical migration of silica out of the soil solum and thus the concentration of sesquioxides in the solum (goethite, gibbsite, etc), with or without formation of ironstone (laterite; hardened plinthite) and concretions
Decomposition	4	The breakdown of mineral and organic materials
Synthesis	4	The formation of new particles of mineral and organic species
Melanization	1,3	The darkening of light-colored mineral initial unconsolidated materials by admixture of organic matter (as in a dark Al or mollic or umbric horizon)
Leucinization	3	The paling of soil horizons by disappearance of dark organic materials either through transformation to light-colored ones or through removal from the horizons
Littering	1	The accumulation on the mineral soil surface of organic litter and associated humus to a depth of <30 cm
Humification	4	The transformation of raw organic material into humus

[a] Ref. 108.

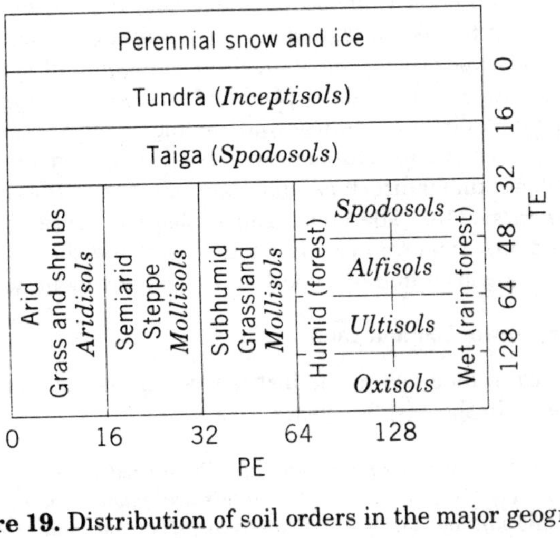

Figure 19. Distribution of soil orders in the major geographic regions of the world (109).

provides some useful insights. About one-third of the earth's land area consists of mountains, deserts, tundra, and lands not conducive to habitation while cropland, pasture-grasslands, and forest-woodlands account for 11, 26, and 30%, respectively. Population densities vary by an order of magnitude, with Asia having the greatest population density and Oceania (including Australia) having the lowest density. Density of population has dramatic effects on the kind and intensity of land and soil use (Table 7).

The rate of land use change between 1979–1981 and 1989–1991 (Table 8) (116), indicates that while the percentage figures are relatively small, they represent millions of hectares, reflecting changing human effects on the earth's landscapes. For example, South America's large increase in cropland and pastureland over this decade of comparison came out of its forestland resources; whereas Europe and North America's cropland and pastureland use actually decreased over this time due to more stable populations, increasing agricultural production from the soil and land resources; and political action (eg, crop production control programs).

For the United States, data on the major uses of land in both the conterminous 48 states and all 50 states (Table 9), show that the extent of cropland varies annually in response to federal cropland diversion programs, weather conditions, market forces, and farm consolidations. Urban land use in the United States increased by 340,000 ha annually between 1980 and 1990, accounting for 22.6 million ha in 1990. Most of the changes in cropland between 1982 and 1992 (Fig. 23) can be accounted for by the Conservation Reserve Program.

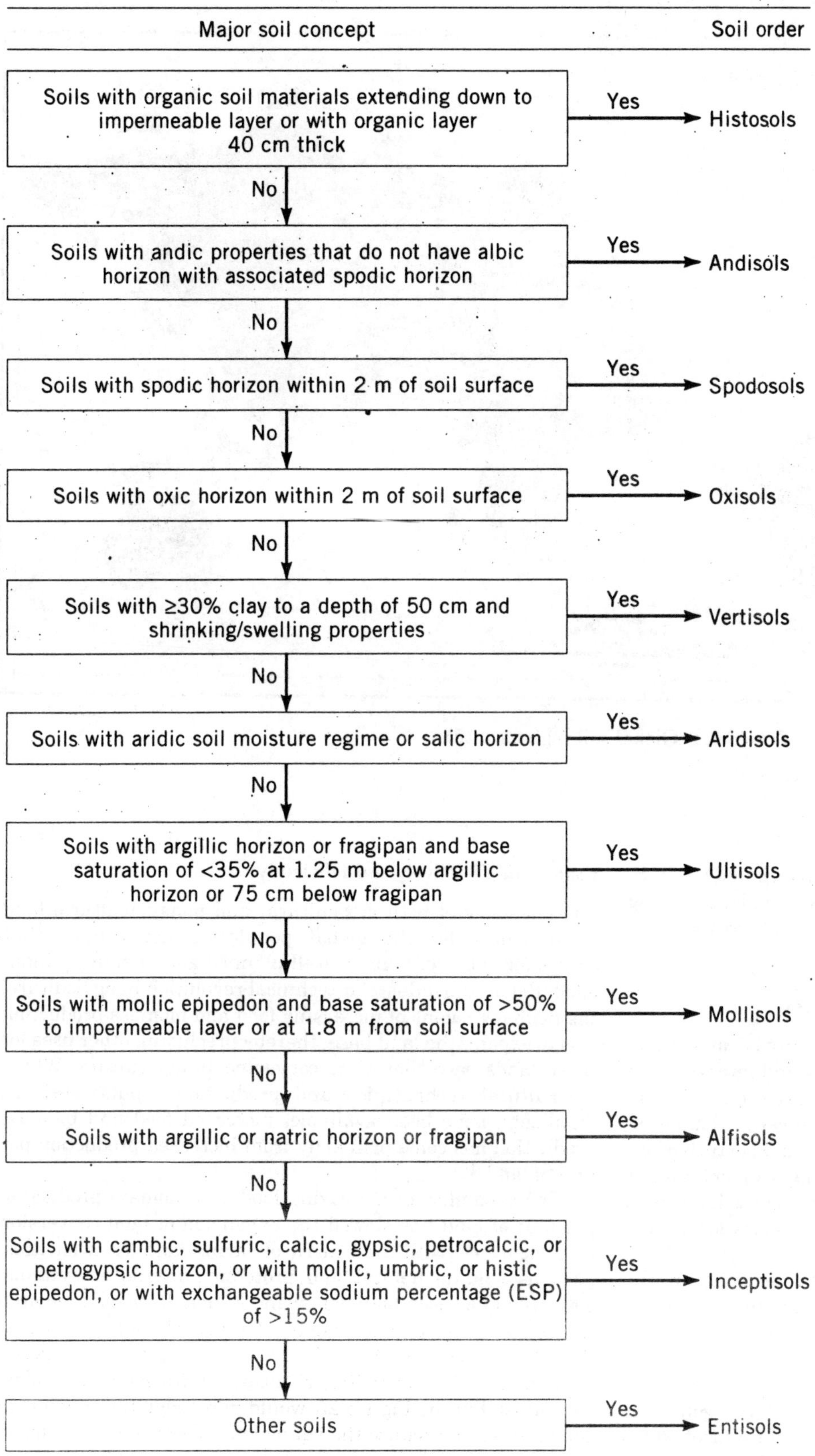

Figure 20. Simplified key showing the central concept of the 11 orders in soil taxonomy (19).

While human needs for agricultural land, grasslands, forestland, and land for urban uses are extensive and obvious, humans use and demand land for a variety of other uses. As economic affluence and availability of leisure time increase, people demand that land be devoted to a variety of uses, from national and state parks to animal preserves and wetlands. Growth in many outdoor recreation activities and visitation numbers at national and state parks continue

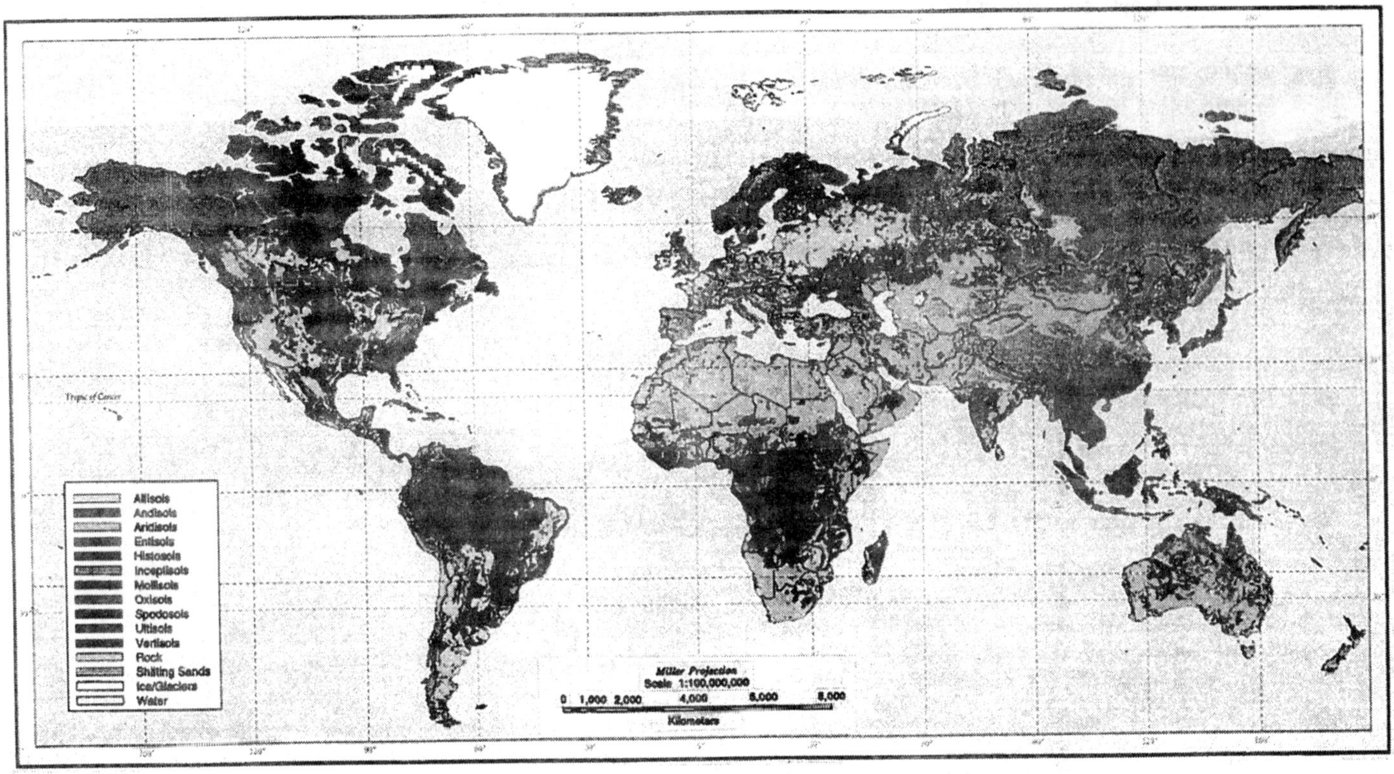

Figure 21. Global soil regions (110).

to climb (Fig. 24). In 1994, 270 million people visited the U.S. National Park System, equivalent to each U.S. citizen visiting a national park at least once a year; with a 15% increase expected by the year 2000.

Consumption of Terrestrial Biomass by Humans

Land use data for both the world and the United States attest to the tremendous impact human demands and interventions have had on terrestrial ecosystems. There have been major impacts on soil ecology, including accelerated erosion, changes in hydrologic response, altered soil chemistry, and perturbations of biodiversity and soil biology. These changes in soil ecology affect soils' behavior and response under various land uses, including the capacity of soils to produce biomass for human consumption.

Because the human population totally depends on the photosynthetic process through primary production for its energy and nutrition, it follows that the most extensive use of the earth's land and soil resources are devoted to agriculture, forestry, and biomass production for human needs. It has been estimated that annual global primary productivity generates about 224×10^9 t dry weight of biomass, about 60% of which is produced in terrestrial ecosystems (119). Through direct human consumption (eg, food and fiber), indirect use (eg, feed for animals), and forfeited primary production (eg, urban and infrastructure land uses), 35–40% of the earth's annual primary production ($46-53 \times 10^9$ t dry weight biomass) from terrestrial ecosystems is consumed or foregone by human intervention.

Using Land Effectively and Efficiently

The adequacy of both agricultural and nonagricultural land requirements for the global population necessitates that ecosystem interventions be well planned and effectively managed. Before the industrial-technical revolution in agriculture, the primary means of increasing food and biomass production was to expand the land base, thereby precluding other uses for these lands once they were converted to agriculture. Where agricultural technologies and production inputs, such as fertilizers, have been available, increased food and biomass production has come primarily from increased production per unit of land.

This capability of increasing food and biomass production per unit of land has slowed the expansion of land conversion to agriculture, with the benefit of saving land for other uses, such as wildlife habitat and protecting fragile ecosystems. This trend has been both global and regional. Global land area harvested for grain has remained relatively stable at about 600 million ha since 1960 (Fig. 25). If the global population growth had relied on 1960 agricultural technology until today, the upper line in Figure 25 would represent the amount of land needed to produce the equivalent of agricultural production in 1990. Globally, the land spared since 1960 by raising grain yields, which accounts for over half of all calories, equals an area equivalent to the Amazon basin. In India, increased wheat yields allowed India agriculture to spare 42 million ha over the 30 year period of 1961–1991, an area equivalent to the size of Sweden or California. Similarly, the same trend occurred in the United States for the period 1950–1990 (Fig. 26).

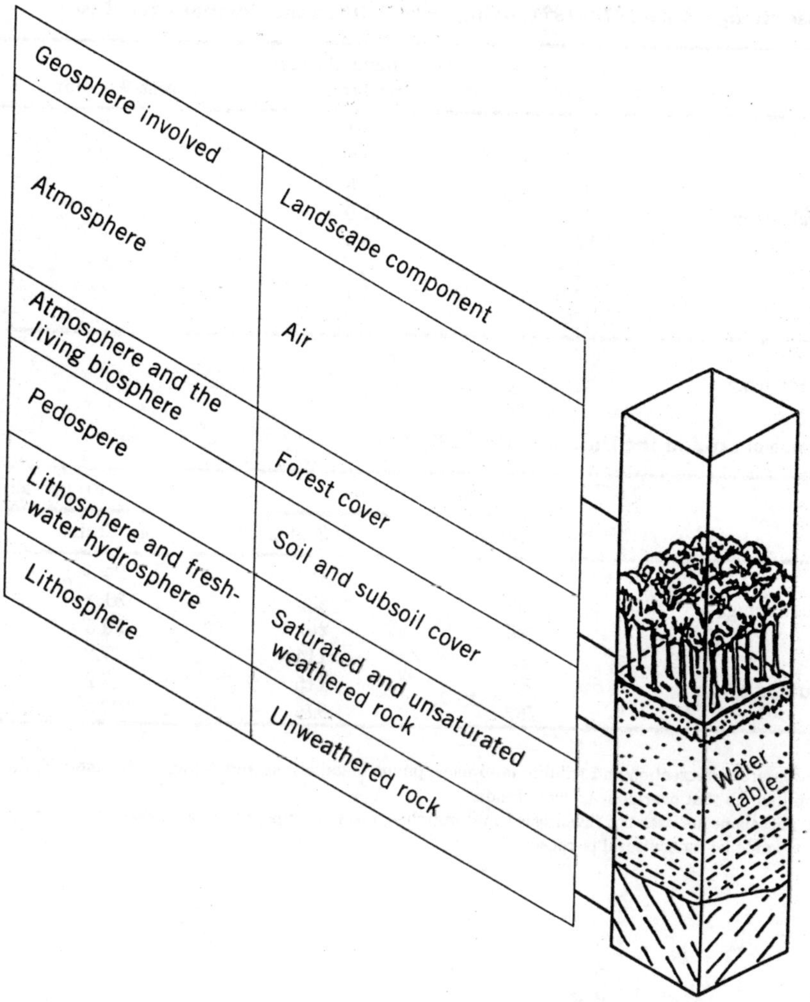

Figure 22. A landscape prism showing the relationships between the geosphere and landscape components in a forest ecosystem (111).

Table 7. World's Land Area, Uses of Land, and Population Density[a]

World and Continents[b]	Land Area, $\times 10^6$ ha	Land use, $\times 10^6$ ha: Cropland	Permanent Pasture Grassland	Forest-Woodland	Other	Population Density, per ha
World	13,042	1,441	3,357	3,898	4,345	0.43
Africa	2,964	181	900	685	1,198	0.24
Asia	2,679	457	759	532	931	1.23
North and Central America	2,138	272	362	710	794	0.19
South America	1,753	114	494	829	316	0.18
Europe	473	138	83	157	94	1.09
Former USSR	2,190	229	327	828	806	0.13
Oceania	845	50	431	157	207	0.03

[a] Ref. 116.
[b] Does not include Antarctica.

If 1950 technology were used today, the 1990 U.S. cropland requirement would double. Technological advances, therefore, have had dramatic effects on both land use and soil ecology.

Soil Disturbance and Degradation

Under natural conditions, soil–plant ecosystems tend toward steady states, characterized by minimum production of entropy (89). Through natural ecological processes (eg, energy flow, biogeochemical cycling, population dynamics) inherent within soil–plant ecosystems, the balance of ordering and dissipative processes still results in some level of "process exhausts" being released from ecosystems. An ecosystem delineated by a watershed, for example, will discharge over time a given quantity of gases, soluble ions, and compounds, and particulate

Table 8. Land use change since 1979–1981, using 1989–1991 as the Comparative Base[a]

	Cropland, %	Permanent Pasture Grassland, %	Forest-Woodland, %	Other, %
World[b]	1.8	2.4	−7.8	5.5
Africa	5.0	0.9	−3.8	0.9
Asia	1.3	9.5	−4.9	−4.6
North and Central America	−0.7	1.0	0.3	−0.5
South America	12.7	4.4	−5.1	3.5
Europe	−1.8	−3.7	0.9	4.6
Former USSR	−1.0	1.7	−22.2	40.7
Oceania	9.9	−4.8	−0.1	9.1

[a] Ref. 116.
[b] Does not include Antarctica.

Table 9. Major use of land in the United States, 1992[a]

	Area, $\times 10^6$ ha		Proportion of Total, %	
Land use	48 States	50 States	48 States	50 States
Cropland	186	186	24.3	20.3
Grassland, pasture, range	239	239	31.1	26.1
Forest-use land	226	262	29.5	28.6
Special uses[b]	55	114	7.2	12.4
Other land, including urban[c]	61	115	7.9	12.5
Total	*767*	*916*	*100.0*	*100.0*

[a] Ref. 116.
[b] Rural transportation areas, recreation and wildlife land uses, public installations and facilities, farmsteads, farm roads, including approximately 36 million ha that overlap with forestlands.
[c] Includes areas in miscellaneous uses not inventoried and marshes, open swamps, bare rock areas, desert, tundra, and other land generally having low value for agricultural purposes.

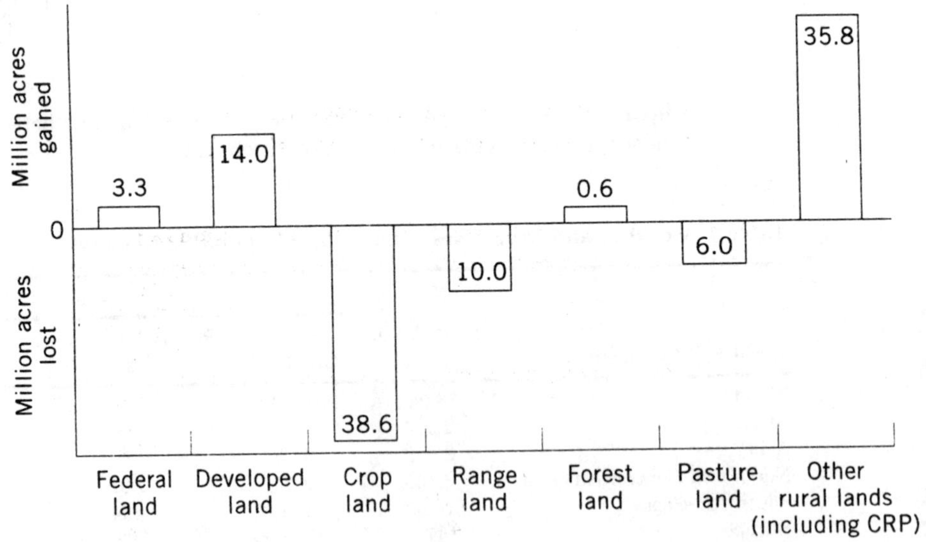

Figure 23. Land use changes, 1982–1992 (116).

materials. Natural landscapes are slowly dissolved and eroded under the relentless effects of the hydrologic cycle and other ecological processes. The character and quantity of the dissolved and particulate material load of the earth's streams, lakes, estuaries, and oceans testifies to the magnitude of these ecosystem "exhausts."

As these natural landscapes and ecosystems become inhabited by people, ecosystems, and their component soils become disturbed. The balance between ordering and dissipative processes is shifted such that more "exhausts" are emitted, either from disturbance of the ecosystem (eg, increased sediment discharges from accelerated erosion, solubilized nitrogen from accelerated mineralization of the disturbed soil organic matter) or through additions to the ecosystem (eg, with fertilizers, pesticides, manures, sludges, and other materials). Significant areas are also under massive networks of roads and highways, parking lots, and airports.

Each landscape or ecosystem segment, be it an agricultural field, farm, or watershed, will exhibit a characteristic exhaust response pattern related to its ecosystem properties and degree of unmanaged disturbance. This land use response concept is illustrated in Figure 27 for a Mid-Atlantic landscape

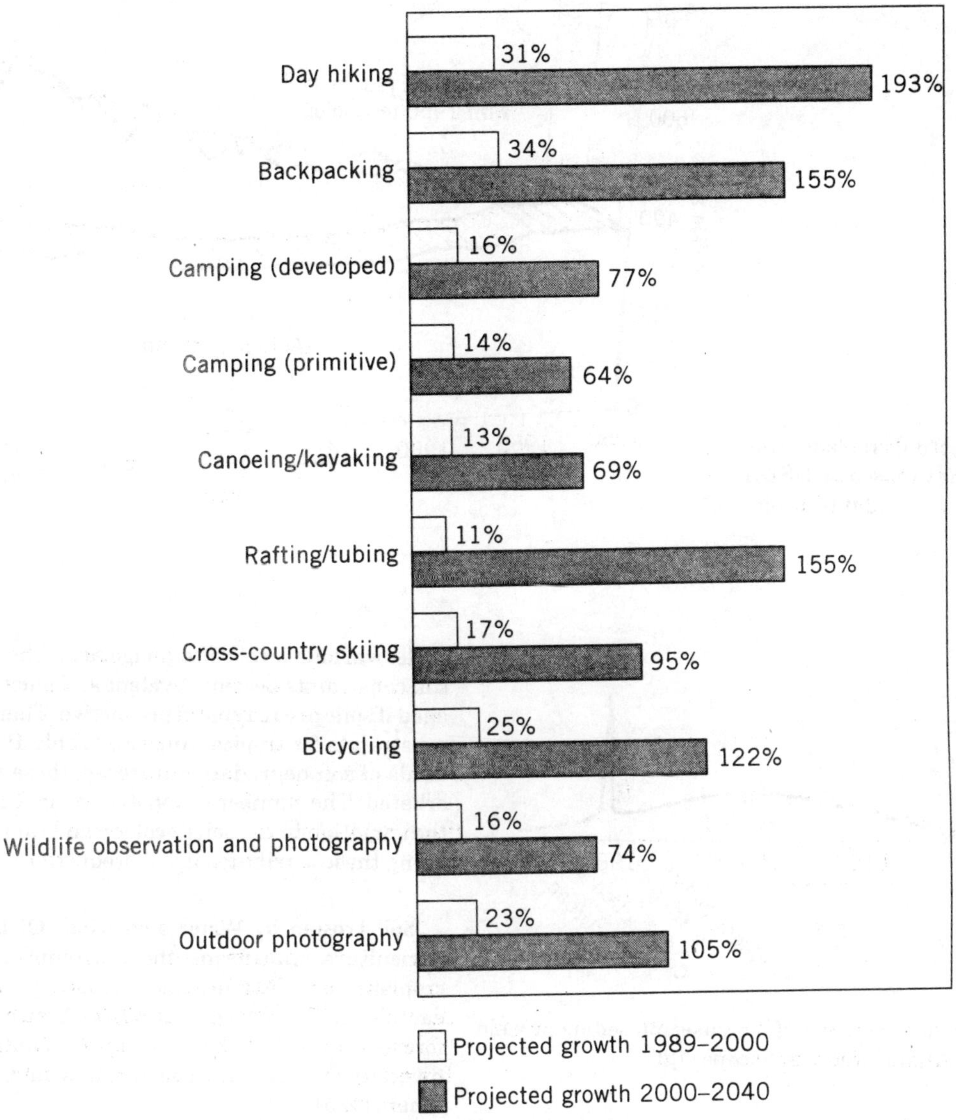

Figure 24. Projected growth in outdoor recreation activities through the year 2040 (117,118).

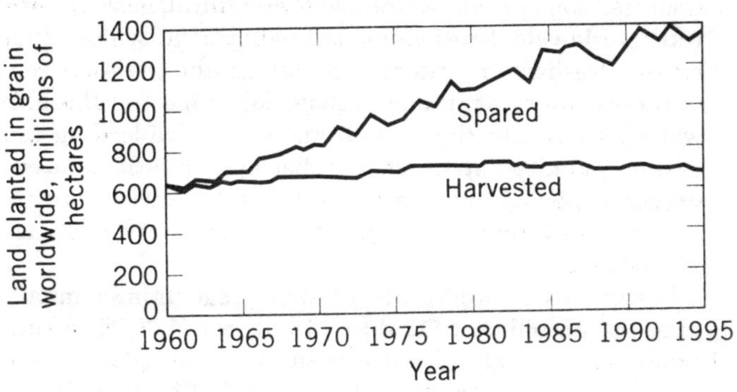

Figure 25. Improved yields have allowed the global area actually harvested for grain to remain relatively stable (117,118).

using sediment discharge as the measured exhaust. During its indigenous mixed deciduous forest cover, the natural geologic erosion around 1800 was estimated at a sediment discharge rate to streams of 10–20 t/km^2 annually (121). As the forests were cut and the landscape converted to cropland, human-induced erosion increased, as did the sediment discharge to the streams, peaking in the early 1900s. As some cropland reverted to pasture and/or back to woodlands and as better soil conservation management systems were implemented on this landscape, the erosion rate and sediment discharge rate decreased through 1960. At that time, much of the landscape underwent clearing for urban construction (in proximity to Baltimore, Maryland), leaving the landscape's soils exposed to the full brunt of the erosive forces of the hydrologic cycle. The result was a brief period, or spike, of extremely high erosion and sediment discharge to the landscape's streams; and ultimately, a portion was discharged to the Chesapeake Bay. Once the urban development was stabilized with pavement and landscaping, the erosion and sediment discharge rate subsided to near the predisturbance rates, even though more runoff occurred because of increased impervious area.

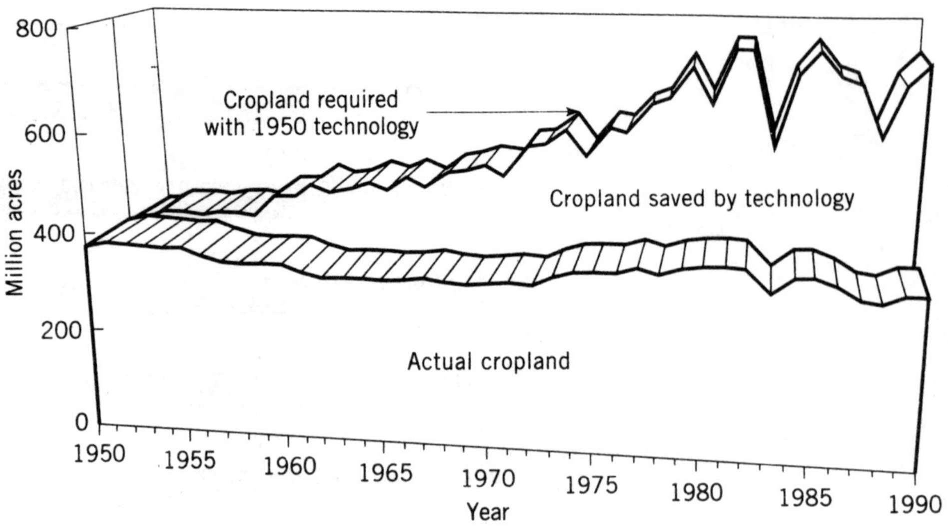

Figure 26. Actual cropland area saved by technology (based on USDA data) (118b).

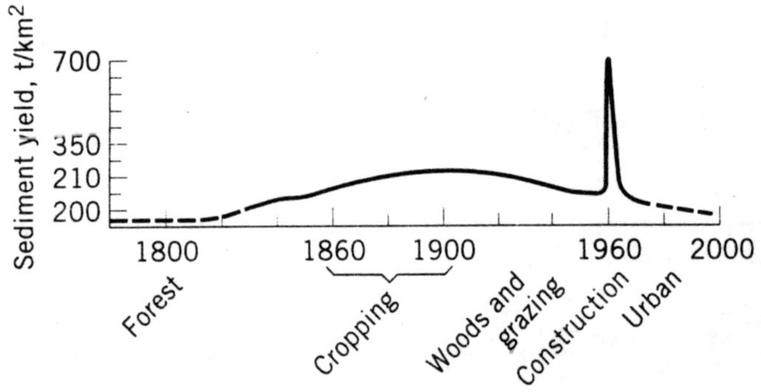

Figure 27. Schematic sequence of land use and sediment yield from a fixed Mid-Atlantic area landscape (120).

The concept illustrated in Figure 27 can be extended to other landscape exhausts, such as dissolved minerals and ions, biological oxygen demand (BOD) loads, chemical oxygen demand (COD) loads, suspended organic matter, microbial species, xenobiotics, and other materials. The point is that each ecosystem exhibits a characteristic response via a discharge or exhaust print under natural, undisturbed conditions. And when disturbed, the discharge or exhaust print changes in relation to the character of the ecosystem and the kind and degree of disturbance. Soil ecosystems, because of their extensive use through agriculture and managed plant-animal systems, are major drivers of landscape ecosystem responses to ecosystem disturbance and land use change. Thus it is necessary that soils and landscapes be managed such that soil landscape disturbance does not destroy the productive capacity of soil ecosystems and/or damage the adjacent ecosystems that receives the exhausts (eg, sediment, nutrients, and BOD–COD loads) when the ordering–dissipative balance has been severely tipped.

The extreme variation of soils with respect to their physical, chemical, and biological properties results in a variety of constraints for plant growth. As soils are disturbed for agriculture under various levels of management, the significance of these soil constraints becomes evident and must be overcome or managed if soils are to remain productive. There are a variety of soil constraints for tropical America (Table 10). As erosion or other kinds of soil degradation increase, these constraints are exacerbated. The number of constraints in Table 11 reminds us of the variation in the soils' ecology and ramifications of not managing these attributes if soil productivity is to be sustained.

Soil Erosion by Water and Wind. Of all human activities, agriculture transforms the environment most widely, with cropland and pastureland accounting for one-fifth of the earth's land surface (125,126). Agriculture has consumed forests, drained wetlands, changed habitats, and reduced biodiversity by favoring some plants and animal species over others (125).

Soil erosion by both water and wind is among the most ubiquitous human-induced dissipative processes affecting soil and land ecology worldwide. The extensive use of soil for agriculture, grazing, and forestry coupled with other human-induced land disturbances (eg, urban construction, highways) has accelerated soil erosion severalfold over natural, geologic erosion rates worldwide. Erosion-induced sediments discharged to the earth's freshwater systems account for the greatest volume of water-quality degradation material in most of these fresh water systems. In the United States, the five leading causes of water quality impairment are bacteria, siltation, nutrients, oxygen-depleting substances, and metals. Agriculture ranks as the major source of water quality impairment for U.S. rivers and lakes (127).

Estimates of worldwide erosion range from annual soil losses of 33 billion t (126) to 75 billion t (127). The ratio of human-induced global soil erosion to natural geologic erosion has been estimated at 3–4 to 1 (126,128). Thus 28–30% (126) of an estimated annual 33 billion t soil erosion for the earth's terrestrial ecosystem are attributed to natural geologic erosion. In a critique of these human-induced erosion figures it was noted that both data sets rest on shaky underpinnings (129).

Table 10. Geographical Extent of Major Soil Constraints in tropical America (23°N–23°S) in Regions Dominated by Acidic, Infertile Soils[a]

	Tropical America, 1493 10^6 ha		Acidic, Infertile Soil Region, 1043 10^6 ha	
Soil Constraint	10^6	Percentage of Total Area	10^6 ha	Percentage of Total Area
N deficiency	1332	89	969	93
P deficiency	1217	82	1002	96
K deficiency	799	54	799	77
High P fixation	788	53	672	64
Al toxicity	756	51	756	72
S deficiency	756	51	745	71
Zn deficiency	741	50	645	62
Ca deficiency	732	49	732	70
Mg deficiency	731	49	739	70
H_2O stress >3 months	634	42	299	29
Low H_2O holding capacity	626	42	583	56
Low ECEC[b]	620	41	577	55
High erosion hazard	543	36	304	29
Cu deficiency	310	21	310	30
Waterlogging	306	20	123	12
Compaction hazard	169	11	169	16
Laterite hazard	126	8	81	8
Fe deficiency	96	6	?	?
Acidic sulfate soils	2	0	2	0
Mn toxicity	?	?	?	?
B deficiency	?	?	?	?
Mo deficiency	?	?	?	?

[a] Ref. 122.
[b] ECEC = Exchangeable Al + Exchangeable Ca + Exchangeable Mg + Exchangeable K.

Table 11. Global Extent of the Degree of Soil Degradation for the Major Types of Soil Degradation[a]

Factor	Light, mha	Moderate, mha	Strong and Extreme, mha	Total, mha
Water	343	527	224	1094
Wind	269	254	26	549
Chemical degradation	93	103	43	239
Loss of nutrients	52	63	20	135
Salinization	35	20	21	76
Pollution	4	17	1	22
Acidification	2	3	1	6
Physical degradation	44	27	12	83
Total	749	911	305	1965

[a] Refs. 100, 123, and 124.

In assessing the area of degraded lands worldwide, researchers arrived at the figure of 1.97 billion ha degraded land, 84% of which was attributable to water and wind erosion (100,123,124). Table 11 lists the global extent and degree of soil degradation and the major causative factors; Table 12 outlines the relative contributions of the causative factors by type of soil degradation (100,123,124). Figure 28 shows the degree of the world's vegetated land disturbance.

Soil erosion data for the *total* continguous U.S. land area (Table 13) show that about 60% of the annual 6.27 billion t soil erosion loss is caused by rainfall and movement of water over the soil surface. The annual average U.S. *cropland* erosion rate (water and wind caused) in 1992 was 12.5 t/ha; 55% percent of which was caused by ranfall and runoff induced sheet and rill erosion. The nation's 155 million ha cropland coupled with 161 million ha range land make up 40% of the land surface of the contiguous United States; this area accounted for 62% of the total estimated erosion in 1992 (131).

U.S. cropland erosion data by erodibility class for the years 1982 and 1992 (Table 14) show that most erosion of cropland ocuurs on highly erodible land (HEL), which totals 42.5 million ha, or less than 27% of the total U.S. cropland in 1992. This 27% of highly erodible cropland, however, accounted for nearly 55% of total sheet, rill, and wind erosion of U.S. cropland in 1992, averaging nearly 25 t/ha compared to 7.6 t/ha of the non-HEL.

Soil Productivity Impairment by Erosion-Degradation. Does past and projected soil erosion pose a threat to the world's soil resources and their capacity to sustain production for an

Table 12. Types of Soil Degradation and Causative Factors Worldwide[a]

	Causative Factor, mha				
Type of Soil Degradation	Deforestation of Natural Vegetation	Overexploitation of Natural Vegetation	Overgrazing	Agricultural Activities	(Bio)industrial Activities
Water erosion	471	36	320	266	–
Wind erosion	44	85	332	87	–
Chemical degradation	62	10	14	133	23
Physical degradation	1	+	14	66	–
World	579	133	679	552	23

[a] Refs. 100, 123, 124.

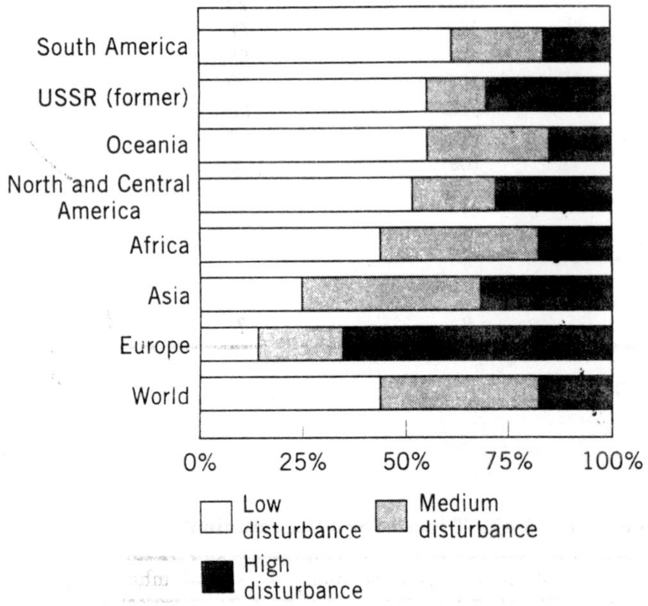

Figure 28. Degree of human disturbance of the vegetated land area worldwide (130).

increasing population? Have the dissipative processes and resiliency for these erosion-prone soils to sustain their ecological functions for crop production been exceeded? For many soil areas, the answer is yes. Excessive erosion has simply truncated the soil profile to the extent that the rooting zone necessary for sustaining plant growth no longer exists, or if some portion remains, its volume and/or physical-chemical attributes preclude an environment conducive to optimum plant growth. Estimates of the global magnitude of strong to extreme degradation (Table 11) caused by water and wind erosion is estimated at 250 million ha, or 15% of the total area of the earth's erosion-impacted soils. The global annual losses in crop and livestock production from land degradation of drylands has been estimated to be $42 billion (132).

While the magnitude of the earth's degraded lands is impressive, if not alarming, there remains the question of how impaired the world's soil resources are as a result of this degradation. A weighted average soil productivity loss was calculated at 17% over the 45 year span between World War II and 1990, implying an average annual soil productivity loss of 0.4% for the world's degraded-eroded lands (129,133–136). Although this estimated productivity loss is, at best, a rough approximation, it is still well below the technology-driven annual yield, production increases of 1–3% for much of the world's agricultural areas, a situation that tends to lessen the concern about soil erosion and land degradation.

Projections of the world's agricultural productivity are many and varied (134). While population projections into the twenty first century are expected to peak and level within the range of 9–12 billion, the world's arable land production capability—although damaged and likely to be further damaged—is still capable of producing more.

Despite the declining total grain area and per capita grain area harvested, total world grain production continues its upward trend (Fig. 26). Furthermore, grain carryover stocks have not changed drastically in the last 25 years (137), although sharp changes can occur from time to time. World meat and milk production have continued to rise during the last 40 years, as have pulses, fruits, vegetables, soybean, and oil crops (137,138). Many of the developed countries have to limit their production capacity to keep from glutting national markets. Thus world food production is still rising, having increased by 24% between 1979–1981 and 1989–1991; production in developing countries as a group increased by 39%, and in Africa, by 33% in the same period (122).

When total food production is measured vis à vis the demands of the growing population, the figures indicate that the world is maintaining supply for the demand. However, during the 1980s, per capita food production increased about 5% for the decade, or one-fifth the rate of total world food production increase (24%). In developing countries, the per capita increase was 13% in the same period, although three-quarters of Africa's countries produced less on a per capita basis at the end of the 1980s than at the beginning (139). Many other countries in the Near East and Latin America have been barely able to match production with population growth in the past 20 years (138). Thus the interpretation of per capita food production data (Fig. 29) requires caution, since there is great disparity among countries and regions. Population growth, coupled with poverty, subverts efforts to devise sustainable-production agriculture, thereby exacerbating environmental degradation (115). Nevertheless, there is potential for increasing the world's per capita food production. In 1993, global per capita grain production was projected to be 315 kg (137), an increase of 22 and 6% over the 1950 and 1970 per capita production figures, respectively.

Higher production projections for the future are based on the premise that no global increase in temperature will occur. Currently, more than 10 million people are classified as environmental refugees, with their nomadic behavior directly attributable to unproductive land (140). That number may be as

Table 13. Estimated Erosion by Type and Source in the Contiguous United States, 1992[a]

Type and Source	Land Area, Million acres	Annual Erosion[b], Billion tons	Erosion/Acre/Year, Tons
Sheet and rill			
Cropland	382	1.20	3.1
Pasture	126	0.13	1.0
Range	399	0.40	1.2
Forest[c]	395	0.32	0.8
Other rural nonfederal[c]	89	0.47	5.3
Developed land[d,e]	92	0.25	2.7
Water areas and federal[f]	457	0.44	1.0
Subtotal sheet and rill	1940	3.21	1.7
Ephemeral	NA[g]	NA	NA
Streambank and gully[e]	NA	0.85	NA
Total water caused	1940	4.06	2.1
Wind			
Cropland	382	0.93	2.5
Pasture	126	0.01	0.1
Range	399	1.75	4.4
Forest	395	NA	NA
Other rural nonfederal[c]	89	0.16	1.8
Developed land, water, and federal	549	NA	NA
Total wind caused	1940	2.85	1.5
Total all types	1940	6.91	3.6

[a] Ref. 131.
[b] Preliminary.
[c] Erosion data not available for 1992. Assumed to be at the same per acre level as estimated in the 1987 National Resources Inventory. Includes lands in the Conservation Reserve Program.
[d] Erosion data not available for 1992. Assumed to be the same total tonnage as estimated in the 1982 National Resources Inventory.
[e] Includes roadsides and construction sites.
[f] Not available for 1992. Assumed to be the same total tonnage as estimated in the 1977 National Resources Inventory (stream-bank was 0.55 million t and gully was 0.30 million t).
[g] Not available.

Table 14. U.S. Cropland and Cropland Erosion by Erodibility Class, 1982 and 1992[a]

Item and Class	1982 Million acres	1982 Percent	1992 Million acres	1992 Percent
Cropland				
HEL	125	29.7	105	27.5
Non-HEL	296	70.3	277	72.5
Total	421	100.0	382	100.0
	Million tons		Million tons	
Cropland erosion				
HEL	1,801	57.0	1,160	54.5
Non-HEL	1,361	43.0	969	45.5
Total	3,162	100.0	2,129	100.0
		Tons/acre		
Cropland erosion/acre				
HEL	14.4		11.0	
Non-HEL	4.6		3.4	

[a] Ref. 131.

high as 25 million (141), and it has been emphasized that it is environmental degradation that makes people migrate: "with poverty and life on the environmental limits as the main motivating force, it matters little whether the migrants are labelled environmental or economic refugees" (141). It was added that in a changed global climate, an increase in temperature will lessen the supply of water in soils and the number of refugees may rise sixfold, to a high of 150 million. While it is recognized that many of the models of global climate change reflect more the nature of changing scenarios than future predictions, the following point must be underscored: most environmental changes are slow and subtle. Once a major change, such as an increase in global temperature occurs, it might take decades to reverse it, if ever. This unpredictability of the effect of a climate change on crop production and its policy implications has been examined (142,143). An ecological-systems approach has been

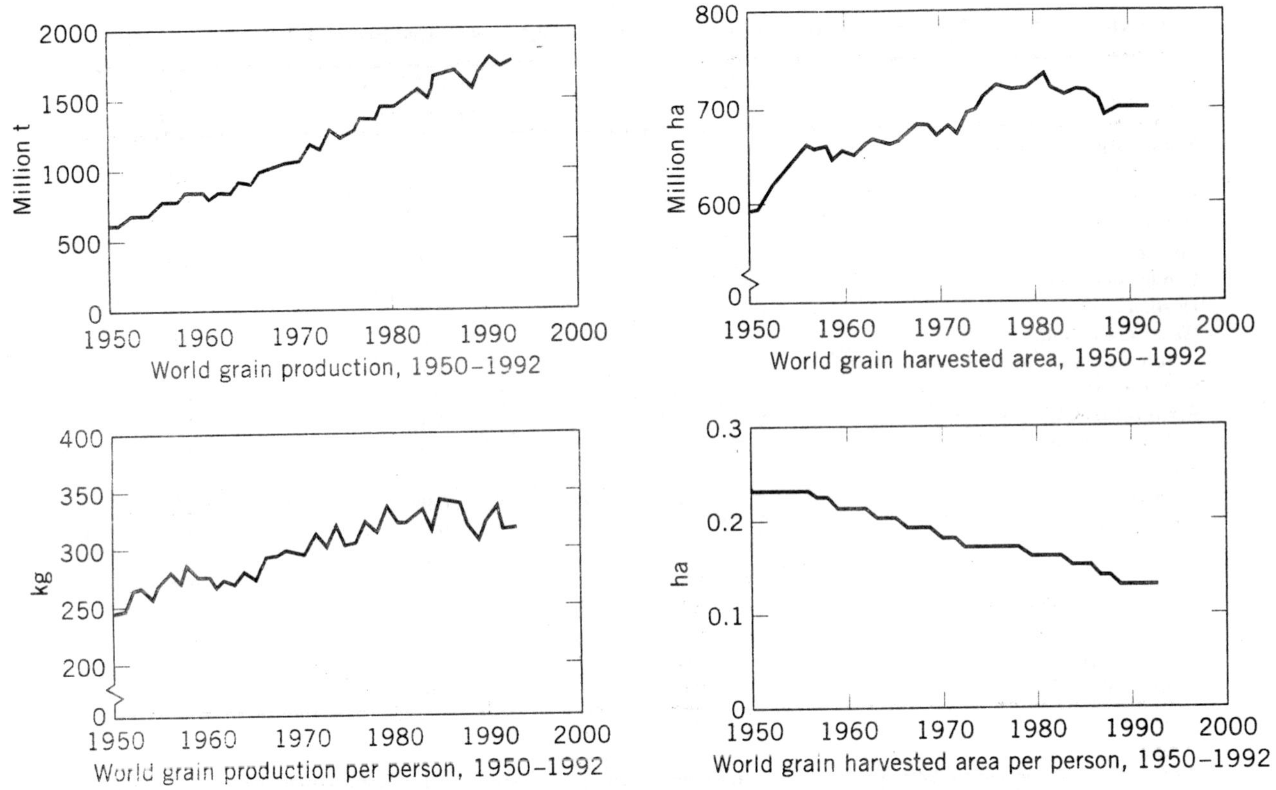

Figure 29. World grain production trends for the period 1950–1992 (137).

suggested (144) to assess agroecosystem behavior in relation to climate change. Because of the complexity of the issues, it is not surprising that many professionals emphasize the need to slow the rate of human population increase and economic development—complex global issues directly related to soil ecology.

The impact of soil erosion in the United States on the nation's soil productivity has been evaluated through a number of assessments (117,118,127,131). In the United States, estimates of lost productivity due to soil erosion indicate that while soils and cropping systems vary in their response to erosion the annual productivity loss from all U.S. cropland is in the range of 0.1% (145). Although this figure is small, this lost productivity translates to the equivalent of millions of hectares of U.S. cropland lost in the future. It has been projected that based on 1977 erosion rates in the United States the equivalent of 3.8 and 7.7 million ha of cropland will be lost over the next 25 and 50 years, respectively, due to forfeited productivity resulting from 1977 levels of erosion (145). Using the erosion productivity impact calculator (EPIC) model developed by USDA, soil scientists (146) calculated that 1982 levels of soil erosion on U.S. cropland over the next 100 years with no change in technology, will reduce the nation's corn, cotton, and soybean yields 3–4% below what they would otherwise be if there were no erosion.

Cutting-edge science invested in inherently poor or degraded land, will result in mediocre yields and a compromised return on the science investment. The impacts of soil erosion and land degradation are largely masked by modern agricultural technology, and the full potential of such technology is sacrificed when it is applied to degraded or poorly managed soils. For example, in the highly erodible Palouse area of Washington in the northwestern United States, wheat yield losses from erosion have been estimated at 670 kg/ha, which is a 15–20% reduction in yield, compared to noneroded land (147). Still, improved wheat varieties and other technologies have allowed Palouse-area farmers to continually increase their wheat yields in spite of the erosion-induced drag on yields.

Despite poor land stewardship, technological advances have resulted in a continually increasing grain output since World War II, while the actual harvested area has declined in the last 10 years (Fig. 29). However, world grain production has leveled off on a per capita basis and has recently tapered downward for many countries. The grain area harvested per capita has steadily declined during the entire period. It is difficult to determine how much of this downward trend is the result of forfeited production capacity from land degradation as opposed to a shift in land use for other crops. However, it is clear that the trend is real (147) (Tables 11 and 12). While biotechnology, coupled with improved breeding, will certainly lead to better yields, society should not burden scientists by siphoning off technology's yield-increasing potential as a substitute for better soil stewardship or use damaged soils without rehabilitating them.

The economic impact of erosion in the United States has been estimated through a variety of assessments. In one study using the EPIC model, the annual average gross on-farm costs of 100 years of U.S. cropland erosion at 1982 erosion rates was \$252 million (117,118). In a later study, the annual gross cost of erosion-induced on-farm losses of productivity in the United States was estimated at \$500–600 million (148).

Once degraded, can soil ecosystems be repaired? The key point for ecosystem resiliency is that the biological system can retain its capacity for self-organization if the perturbations are not too rapid or push the system too far from its steady state (89). For example, soils may be severely degraded, but if their biological potential and ecological processes, though damaged, can be maintained, then the soil system can rebound. The time required for such soil system resilience, however, will differ greatly from soil to soil, depending on the severity of the perturbation and ecological conditions governing the rate of self-reorganization such as soil development.

Socioeconomic Impacts of Soil–Land Degradation. The ultimate impact of human-induced land degradation is the concurrent or subsequent degradation of the quality of life. Thus degraded land ecology and soil ecology translate to socioecological impacts (Fig. 30). "As society's dependence on resources wrought from ecosystems deepens and therefore, the stress under which those ecosystems are put increases, the need to assure ecological sustainability has become widely recognized" (150). This recognition has led to efforts to forestall degradation, rehabilitate badly degraded ecosystems, and protect areas that have important ecological value (150). The ecological, economic, social, and aesthetic considerations to be considered in ecosystem rehabilitation can be conceptualized in the interactive schematic matrix shown in Figure 31.

Agriculture

The essence of agriculture is the capture of solar energy through selected plants, the manipulation of nutrient stocks, and the conversion of complex organic compounds to degradable forms. Management of soil–plant–animal systems involves removing stresses, eg, inadequate or excessive water, weeds, diseases, and insect pests. Expanding this concept to a more comprehensive understanding of the commonalities involved in agroecosystems, one can schematically illustrate the key processes and functional relationships in a traditional production agricultural ecosystem (Fig. 32). Human interventions in natural ecosystems result in the selection and manipulation of each component of the processes and functional relationships shown in Figure 32. While climate cannot be controlled, species selection can be made for the terrestrial ecosystem earmarked for agriculture. Likewise, soil resources are selected and managed as are the farming practices, nutrient cycling methods, and biodiversity components.

Production agriculture systems usually occur within adequate water–temperature (climate) gradients. Based on aboveground net primary production, a gradient indicator can be drawn for the grassland regions of the United States (Fig. 33). Such ecological gradients are so pronounced that they can be translated to economic value (Fig. 34), once again illustrating the interplay between natural landscapes and cultural landscapes.

The key to sustaining a productive agroecosystem, therefore, is to manage the system so that those ecological processes and functional relationships that sustain the soil system are protected (Fig. 32). To illustrate the point, compare an automobile factory to a farm-both enterprises use production resource inputs and generate product outputs (automobiles or crops). While the similarity of the resource inputs and product outputs of the enterprise is obvious, there is a crucial difference in the way these inputs are processed. The automobile plant relies on a sequence of specific functions using machines and mostly nonrenewable resources. The physics and engineering principles employed in the design, operation, and maintenance

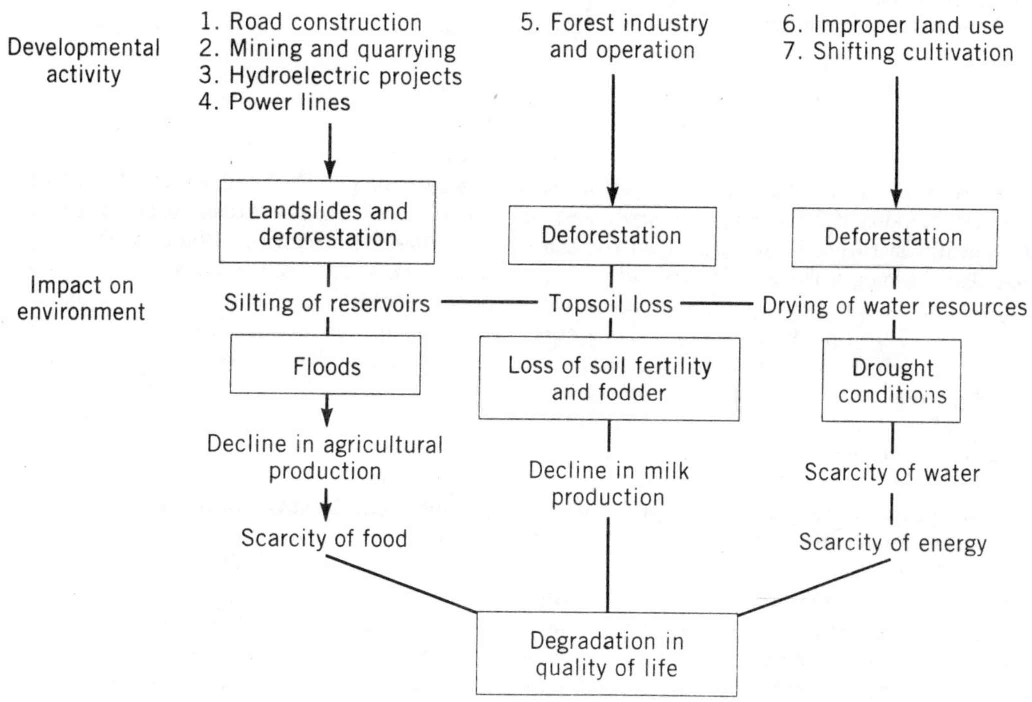

Figure 30. Socioenvironmental effects of deforestation on the quality of life (149).

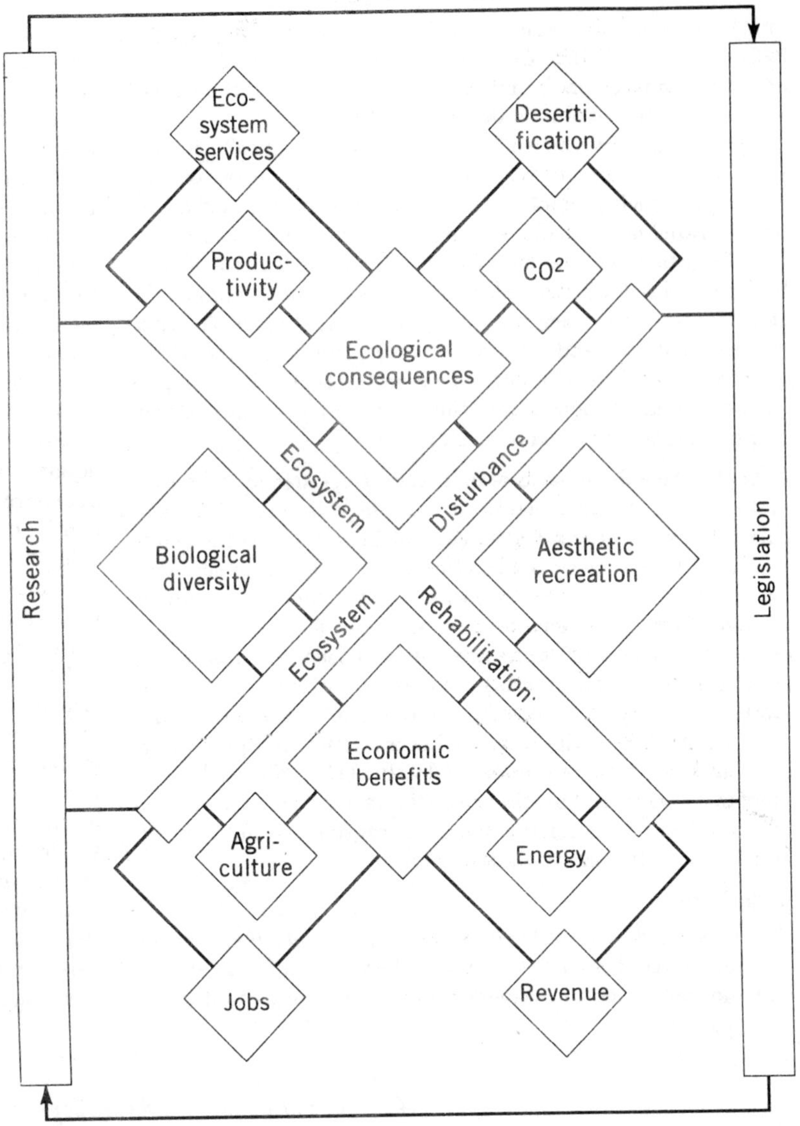

Figure 31. Ecosystem disturbance and need for rehabilitation in a matrix of ecological, economic, social, and aesthetic considerations (149).

of these machines are well understood. But, the agricultural (or forestry or fisheries) analog of the factory analogy "does not consist of known machines; rather it consists of all the complex and diverse rules of ecology. These rules are not yet well known to science, yet they are, in principle, the 'machines' inside the 'factory' of agricultural production" (154). Thus agriculture, grasslands management, forestry, and other managed agroecosystems are driven, regulated, and sustained by the machinery of ecological processes. The soil resource and thus soil ecology are crucial components of the ecological machinery of agriculture, its management, and sustenance.

Regardless of the type of agriculture, its location or the management style and philosophy of the farmer, there are certain requisites, tenets, and/or premises on which all agriculture is based or to which all agriculture must adhere (Fig. 32). These agricultural requisites, tenets, and premises must be addressed, regardless of whether the farming system is a primitive, subsistence agroecosystem in a developing country or a high-tech farm operation using the latest science and technology. Each agricultural production system along this spectrum must address each of these factors.

Natural Resource Base. Plant-based agriculture relies on the resource trinity of soil, solar energy, and water. The combination of tillable soil with adequate available water and sunlight exists on only about one-fourth of the earth's land surface with an even smaller fraction qualifying as prime agricultural land, ie, deep, relatively level, medium textured soils. The world's most productive soils have developed from relatively young geologic parent materials such as alluvium, volcanic ash, glacial deposits, and wind-blown silts (loess). It is, therefore, not surprising to find today's most productive and highly developed agroecosystems operating on these young materials where the other two components of the resource trinity (solar energy and water) exist. Agriculture can function on poor soils and steep lands (terraces in Asia), but only with intensive management and investments that are difficult to sustain, particularly in many developing countries where production resources are scarce and where poverty drives production at any

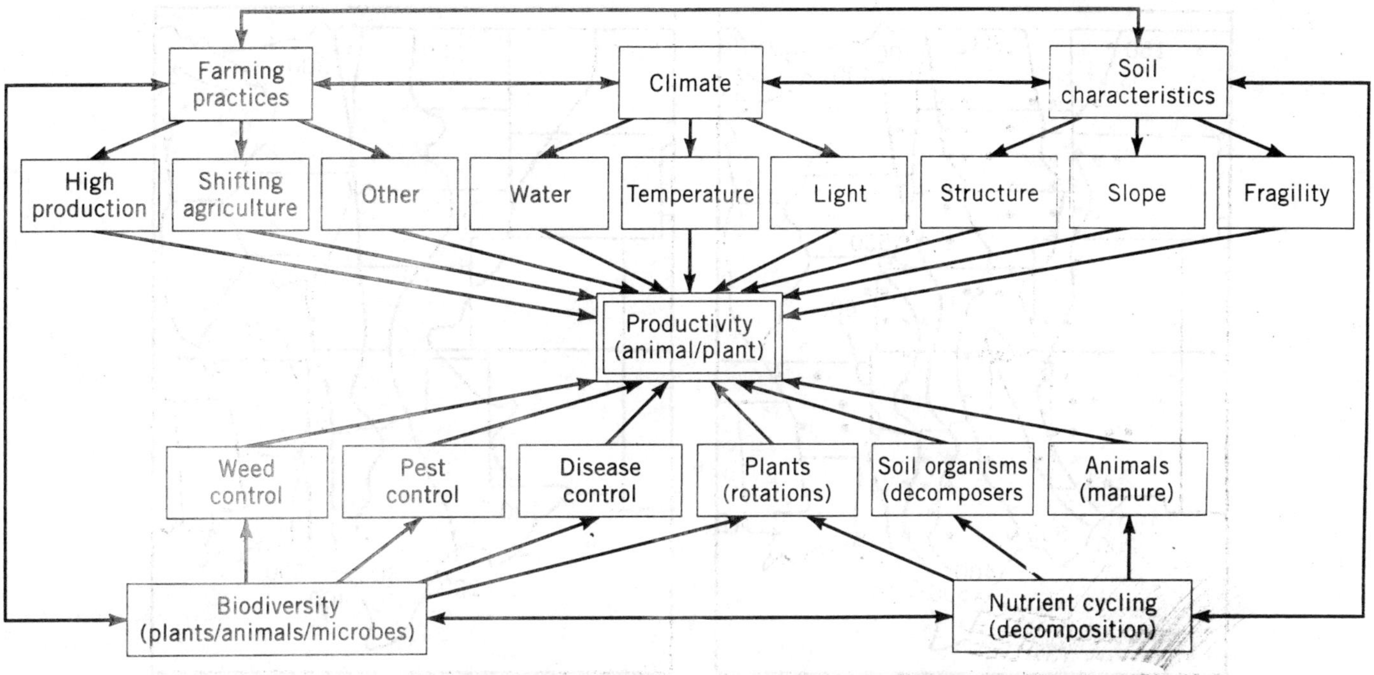

Figure 32. Key processes and functional relationships in a traditional agroecosystem (151).

environmental cost. The United States has been well endowed with the natural resources necessary to sustain an economically viable, highly productive agriculture.

Crop-Product Selection. Economic viability can only be sustained if markets exist for crops and/or animal products. While there are approximately 80,000 plant species that produce some part that is edible, just 50 are actively cultivated to any extent, with 30 of the most commonly produced crops accounting for most of the food on which humans rely (Fig. 35). Likewise, <10 animal species account for most of our animal-derived foods. While many plant and animal species can be grown and efficiently produced in many parts of the world, factors beyond plant–animal adaptability govern why certain agroecosystems produce such a limited portion of the potential crop–animal spectrum. Thus the development of markets (including subsidy programs) and infrastructure (processing, storage, transportation, system) for certain crops and animal products are major factors limiting the kinds of commodities produced. Dietary preferences and cultural traditions are also significant drivers of the limited biodiversity of agriculture. Thus it would seem that farmers have little choice in what they produce, unless they are willing to develop their own niche markets for specialty crops and/or animal products. However, such exotic or new crops might serve as a vector for diseases or insects that attack the more common crops, illustrating the trade-offs and risks associated with most production options.

Providing Seed–Soil Contact. Crop seed must have soil contact to germinate and sustain seedling growth through root establishment. There are many tillage systems that can produce satisfactory seedbeds, including moldboard plowing, disking and harrowing an entire field, chisel plowing, and no-tillage systems (the ancient jab-stick and modern slit-planters). While each of these tillage systems satisfies the requisite soil–seed contact criterion, there are economic, environmental, and social trade-offs. Excessive soil disturbance can result in the deterioration of soil tilth and the soil's biological integrity. Soil erosion can be severe for certain soils under various combinations of landscape character (slope, length of slope), climate, and tillage system. Furthermore, certain tillage systems require specific equipment and/or chemical treatments that are costly or may subject adjacent ecosystems or groundwater to contamination. These systems require trade-offs in energy and human resources, ranging from total reliance on human power in hand cultivation to horse-drawn implements to diesel power. Tillage systems are chosen, therefore, based on a combination of factors such as availability of resources (eg, equipment), their cost compared to the cost of human labor and other options, soil management objectives (eg, erosion potential, ease of soil to till), time available for seedbed preparation, existing power, and size of operation. A cash grain farmer employed at a local industry will probably choose a different tillage system than will an Amish dairyman or an organic farmer specializing in certain crops. While one system may be more environmentally sound, it may not satisfy the needs of other farmers with different lifestyles and management objectives.

Maintain Soil Quality. Soil husbandry must protect soils from erosion and degradation through compaction and loss of tilth (structure), organic matter loss, salinization, water logging, and nutrient depletion. Soil management systems should strive to maintain the soil's biological and physical integrity, thereby facilitating natural ecological processes (eg, nutrient cycling, biological maintenance, and renewal) rather than unnecessarily interrupting these processes. Those attributes

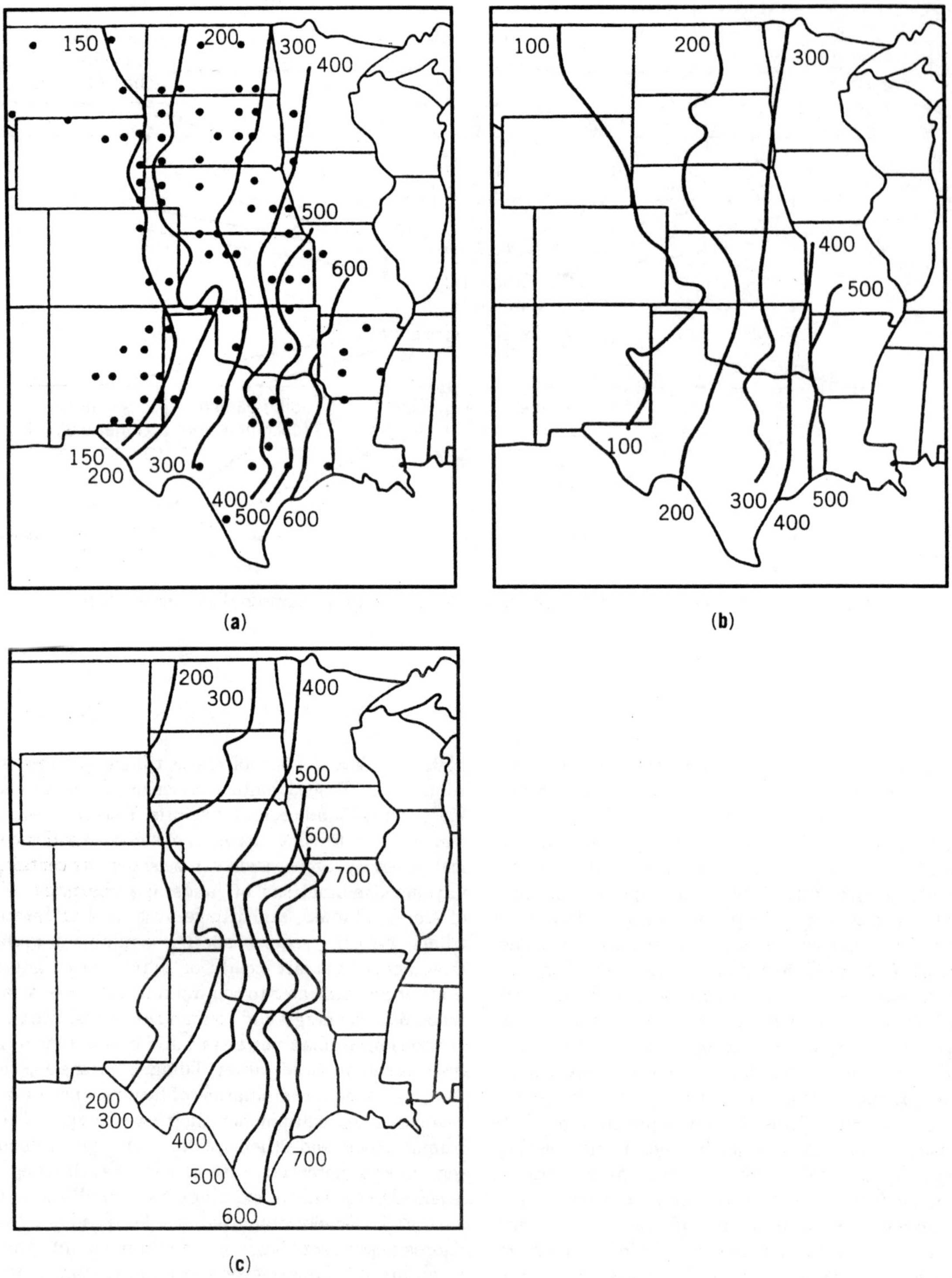

Figure 33. Aboveground net primary production (in grams per square meter) for the central grassland region (**a**) druing years of average precipitation, (**b**) during unfavorable years, (**c**) during favorable years (152).

that distinguish soil from inert geologic material are its porosity, organic-biological components, ability to hold, release, and exchange nutrients, and its water-holding capacity. Good soil management protects the soil from erosion and maintains the soil's capacity to sustain its biological properties through such practices as residue management and growing cover crops. The impact of two management scenarios on the physical, chemical, and biological properties of a soil in the Palouse region of Washington illustrates this (Table 15). Each management option has trade-offs with respect to management

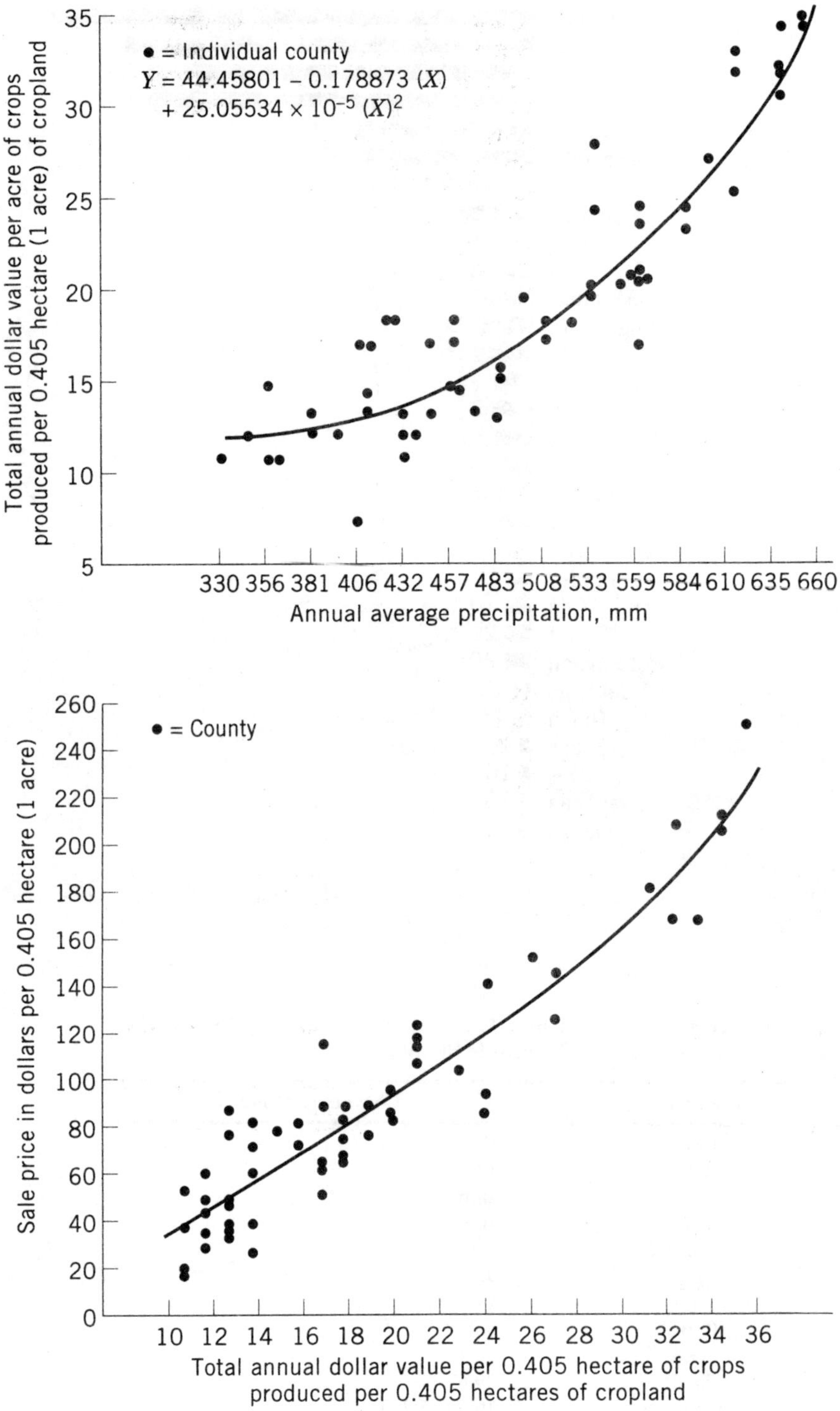

Figure 34. Relationship between total annual dollar value of crops, annual precipitation, and sale price of land (153).

intensity (time), costs, equipment requirements, environmental impacts, soil quality maintenance and lifestyle choices. Land management has a dramatic effect on the shape and quantitative character of the time line plot shown in Figure 27. Similar time-line plots can be made of nutrient runoff and other production resources that have the potential to become nonpoint source pollutants.

Provide Nutrient Management. Plants require 17 elements for their metabolic functions, 14 of which are derived from the soil (most of the oxygen, carbon, and hydrogen are derived from the atmosphere and water via the photosynthetic process). As plant products (eg, grain, stover, forage) and animal products are removed from the farm and cannot be recycled back onto the land from which they originated, the soil's inherent nutrient stock is ultimately diminished. Unless replenished through combinations of commercial fertilizers, manures, composts, sludges, or legume crops, nutrient deficiencies will ultimately limit production. Depending on the type of farm (eg, cash grain vs an animal-based operation), the nutrient management system options will vary. It makes no difference to the plant or plant-consuming herbivore if these nutrients are derived from commercial fertilizers or organic

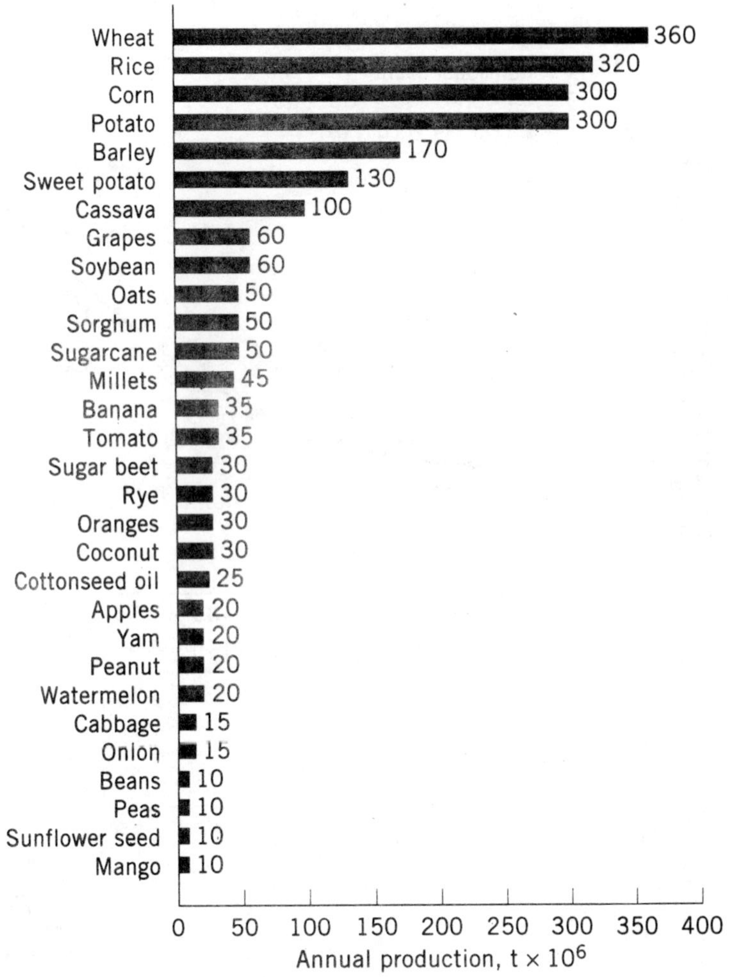

Figure 35. The relative array of the 30 most commonly used crop plants. Note that there are about 80,000 plant species with edible parts (155).

Table 15. Soil Physical, Chemical, and Biological Properties of a Naff Silt Loam Soil in the Palouse Region of Washington State Managed Under Conventional and Organic Strategies[a]

Soil Property	Organic Farm	Conventional Farm
Organic matter (g/kg) (0–10cm)	27[b]	17
Polysaccharides (mg/kg) (0–10 cm)	1.13[b]	1.00
CEC ($cmol_c$/kg) (0–10 cm)	17.3[b]	15.6
Total N (mg/kg) (0–15 cm)	1179[b]	1066
Ext. P. (mg/kg)	5.5	5.3
Ext. K ($cmol_c$/kg)	1.2[b]	0.7
Bulk density (Mg/m)	1.0	1.0
Modulus of rupture (MPa)	0.016[b]	0.2
Surface horizon thickness (cm)	39.8[b]	36.7
Depth to Bt horizon (cm)	55.6[b]	39.8
Color (dry)	10 YR 4/2	10 YR 5/2, 5/3
Consistency (dry)[c]	Slightly hard	Hard to very hard
Consistency (wet)[c]	Friable	Firm
Structure	Granular	Subangular blocky parting to granular

[a] Ref. 156.
[b] Significant at the 0.05 and 0.01 levels, respectively.
[c] Average of summer and fall 1981 and spring 1982 samples. All other samples taken July 1985.

sources. Once certain nutrients are built up in the soil, a cropping system may not need specific nutrients for many years. But the principle of limited nutrient stocks still holds—no amount of management or any elixirs can produce or extract nutrients that are not indigenous to the soil or added through other means. The soil's chemical and biological components of its ecology are strongly influenced by such factors as temperature, moisture content, and soil acidity and alkalinity (measured as pH). The impact of pH on microorganisms and nutrient availability is significant, thus farmers must manage the pH range of their soil through liming and acidification (Fig. 36). The optimum pH is specific to plant species that the grower plants.

Provide Pest Management. Native plant ecosystems have evolved through natural selection to the point that communities of plants have adapted to their environmental niches, including indigenous insects and diseases that would destroy or eliminate plants that were not tolerant or resistant to these pests. When non-native plants (crops) are introduced into agricultural production systems, they are usually susceptible to a variety of native as well as introduced diseases and insects. Pest management, therefore, is a requisite for any agricultural system. There are a variety of pest management strategies, including developing and selecting pest-resistant or -tolerant germplasm; maximizing biodiversity through the rotation of different crops; and using biological controls, cultural practices—burning or plowing under residues, managing planting dates (eg, delayed wheat planting to protect against Hessian fly), fallowing—and pesticides. Each pest management strategy or combination has trade-offs with respect to management intensity (convenience, timing), available equipment, cost, efficacy and duration, environmental impacts, and social acceptance.

Provide Weed Control. Most farmers rate weed control as their top priority management problem. Problem weeds usually can out-compete crops for space, light, soil moisture, and nutrients. Weed control strategies include mechanical cultivation, crop rotations, cultural practices such as narrowing row widths to gain quicker canopy closure (to reduce sunlight getting to germinating weeds), limited biological controls and herbicides. Regardless of the cropping system used, weed management strategies are necessary, and their trade-offs must be weighed with respect to cost, convenience, availability of equipment, erosion risk, runoff/leaching (pollution) risk, management objectives and social acceptance.

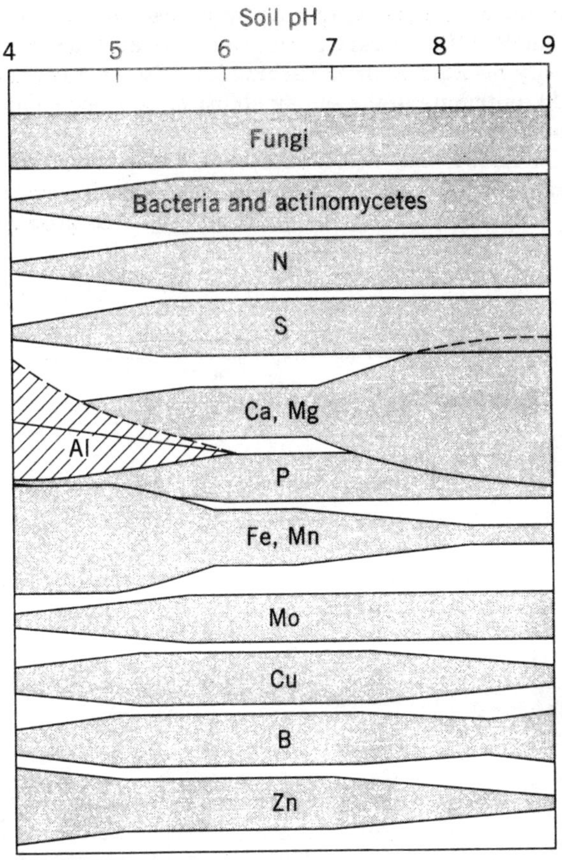

Figure 36. Relationships between pH, microorganisms, and plants nutrients (19).

Provide for Water Management. Agricultural production systems are commonly exposed to water stress during some portion of the growing season—either from too much or from too little water. Water management strategies such as drainage, irrigation, time of planting, and selection of soil type and germplasm are commonly employed. Each strategy has economic, practical, environmental and social trade-offs, including policy implications for irrigation and/or drainage schemes.

Lifestyle Choice. The human element in any agricultural production system controls the menu of options and trade-offs for each of the aforementioned requisites. One of the objectives of managing an agricultural system is risk aversion. Farmers vary in their willingness to accept risk; therefore, certain management options are eliminated. Likewise, environmental stewardship levels vary among farmers. Lifestyle choices have increased under today's management options. In many parts of the developed world, a farmer can choose to work off-the-farm to generate more income to accommodate his or her lifestyle and still operate a sizable farm. This would not have been possible many years ago, since the necessary equipment, chemicals, and other production resources were not available. Thus agricultural sustainability is tied not only to economic and environmental issues but also to social (public acceptance, lifestyle choices, etc) considerations. Farmers with different farming systems or opportunities may be quite happy with their lifestyles, but their options for risk aversion, exercising their degree of environmental stewardship, and economic flexibility might be quite limited and different. Government policies and economic conditions play major roles in the sustainability of these lifestyles as well as their environmental implications.

These agricultural production requisites and tenets constitute the basis for sustaining future agricultural production for the world's current and future populations. The world's soil resources are crucial to sustaining this production. As noted, much of the earth's soil resources has been degraded, but there is still a tremendous capacity to produce more. Soil management that maintains the soil's ecological integrity must be both a goal and a requirement for future human well-being.

Grasslands

The world's grasslands and pastures account for about one-quarter of the earth's land use (Table 7). Many of the world's great grassland zones, notably the North American Prairies, the South American Pampas, and the eastern European Steppes, have had much of their better-watered areas converted to arable land with grazing often relegated to marginal land (157). Still, great expanses of grazing areas exist (eg, Tibetan Plateau, African savannas).

In the United States, pasture and range land accounted for about 325 million ha in 1992, nearly 36% of the nation's land area (116). Three major types of grazing land are included in this total: cropland pasture (3%), grassland pasture and range

(92%), and grazed forest land (5%). About two-thirds of this land is privately owned or under state or local government control (115).

These vast grassland areas have undergone significant alterations by human activity. Domesticated cattle have replaced native herbivores. Nearly 100 million cattle have replaced the estimated 4 million bison on the U.S. western prairie. The world's cattle population is about 1.2 billion. The rate at which the large native herbivores were replaced by domesticated herbivores on these grassland ecosystems has been unequaled since the Pleistocene extinctions (158).

Nearly all the world's rangelands except those in the Arctic have been degraded to some degree by livestock overgrazing, the introduction of exotic species, fuelwood harvesting, suppression of the natural fire cycle, wildlife depredation, and conversion to cropland or housing (115). Estimates of the world's range degradation since 1945 are shown in Figure 37.

A picture of human pressure on native grasslands is particularly poignant in the Sahel of Africa. For centuries, the grasslands of this region have provided adequate biomass for the traditionally nomadic local pastoral societies' grazing needs. As human and livestock population pressure has increased, the carrying capacity of the ecosystem has been exceeded. Livestock rearing has become a high risk operation because of the uncertainty of climate prediction in this already arid region (100,123,124).

Reasons for the degradation of the world's pastoral lands and the mitigation necessary to sustain these grasslands have been proposed (157) and include encroachment of cultivation, improved animal health, population growth, breakdown of traditional systems, subsidies on grain and feed, opportunistic mentality, and lack of alternatives to invest profits. Future emphasis in sustainably managing the world's grasslands must include an understanding the socioeconomic issues, grazing patterns, long-term ecological studies and dynamics of systems, effects of ecological disturbances, field testing of plant genetic resources already available, and enhanced resource capture (157).

The concept of herbivore carrying capacity cannot be based solely on botanical considerations but must take into account the management objectives of range land users (159). It has been asserted that there is no single biologically optimal stocking rate that can be defined independently of the different management options associated with various forms of animal exploitation (159). This assertion affirms the concept of landscape ecology by which social values and peoples' relationship to the land greatly modify the terrestrial capacity to support various management scenarios.

The soil resources undergirding the world's grasslands and pastures are vital components of these vast plant ecosystems. These soil resources have developed under and assumed many of their properties and much of their morphological character from the native plant communities constituting the grassland–range land biomes. Harvesting these biomes through grazing, fuel scavenging, and other uses and management interventions changes the composition of the native plant communities and, in turn, affects the soil ecosystem. Soils become susceptible to degradation through erosion by water and wind. These human-induced changes in plant community composition also have implications on the nutrient regimen, nutrient leaching, and nutrient use patterns of soil–plant communities.

Ecologists studying grassland–range land ecosystems have debated this diversity–sustainability hypothesis, ie, that the sustainability of soil nutrient cycles and thus sustainability of soil fertility depends on biodiversity (101). Under controlled conditions, it has been shown that plant productivity and resource use were significantly greater at higher plant diversity in both developing grassland ecosystems and native grasslands (101). Greater nutrient use at higher species diversity resulted in lower leaching loss of soil nitrogen, which should contribute to the sustainability of nutrient cycling and soil fertility.

It is known that soil fertility and productivity influence diversity. It has been demonstrated that the converse is also true: plant diversity has a significant effect on productivity,

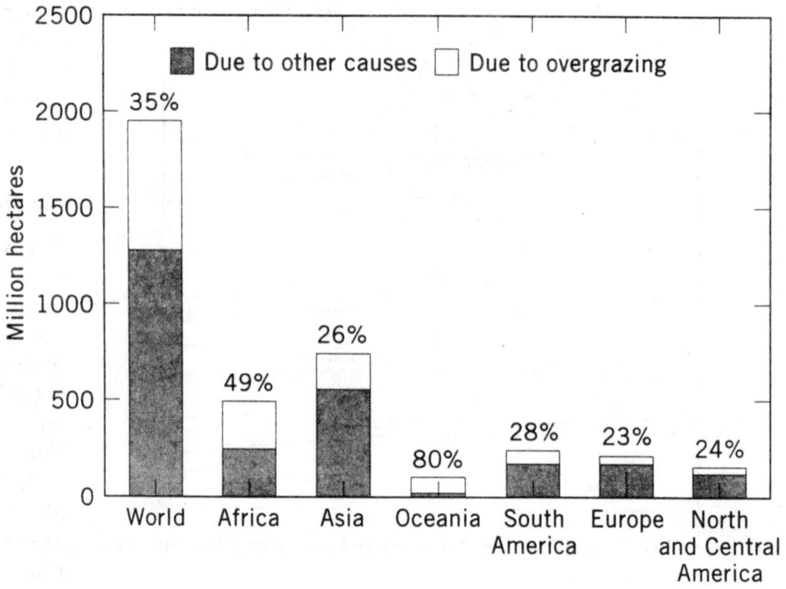

Figure 37. Degradation of rangelands worldwide since 1945 (138).

nutrient use, and nutrient retention (101). Thus biodiversity influences ecosystem productivity, sustainability, and stability. It follows, therefore, that soil ecology, including the physical, chemical, and biological character of soils, is greatly influenced by the kind and intensity of use, management, and biodiversity of the world's grasslands and range lands. If these soil resources and the biomes they support are to be sustainably managed, attention must be given to the effects of ecological disturbance and long-term ecological dynamics of these systems, as has been proposed (157).

Forest Lands, Forestry

The area devoted to forests and woodlands is about 30% of the earth's terrestrial area (see Table 7). Forests once covered >40% of the global land surface (160). Considering only tropical forests (1.76 billion ha) and forests in industrialized countries (1.43 billion ha), the world's estimated forest resources exceed 3.4 billion ha (115). The former USSR, with 755 million ha of forest, and Brazil, with 561 million ha of forest, accounted for nearly 40% of the global natural forest cover in 1990 (115).

Recent FAO reports indicate that between 1980 and 1990, tropical forest areas shrank an average of 15.4 million ha (0.8%) per year (115). Thus during the 1980s, 154 million ha of tropical forests—an area three times the size of France—were cut and converted to other uses. These tropical forests have lost half their original expanse in the past 50 years, the most rapid vegetation change in human history (160). Forest areas in industrialized countries increased slightly through the 1980s, a trend caused by increased agricultural technology that resulted in higher yields per unit of land and, therefore, less land needed for agriculture. Some of this land relinquished from agriculture was converted to forests.

Why are forests, with their exceptional potential to support humanity, allowed to decline? More than half of all tropical deforestation is the result of slash-and-burn agriculture by displaced landless peasants. These impoverished people, making up several hundred million of the world's 1.3 billion people living in absolute poverty, are driven to migrate into the forests by economic and population pressures, land hunger, and other reasons (160). Thus the source of most tropical deforestation (and associated degradation of soil ecosystems) lies in an amalgam of factors that are usually far removed from the forests and lie outside the purview of traditional forestry measures.

Seldom in human history has a continent been transformed as rapidly and extensively as North America during the great American westward migration and settlement during the 1800s. The clearing of the eastern forests and breaking of the vast North American prairies by the plow occurred with unprecedented speed and extent; the changes in cropland and forest land areas between 1850 and 1980 are shown in Figure 38. Between 1850 and 1910, American settlers and farmers cleared about 77 million ha of forest, a rate averaging 3525 ha/day for 60 years! By 1920, clearing forests for agricultural land had largely halted (161).

About 33% of the U.S. land area (298 million ha) is forest, amounting to about two-thirds of the area that was forested around 1600 (421 million ha) (Fig. 38). About 66% of U.S. forest land is classified as timberland—forest land capable of commercial production—70% of which is in the east (161). In the early 1900s timber harvest in the United States exceeded timber growth by a 2:1 margin. By the 1940s, timber growth

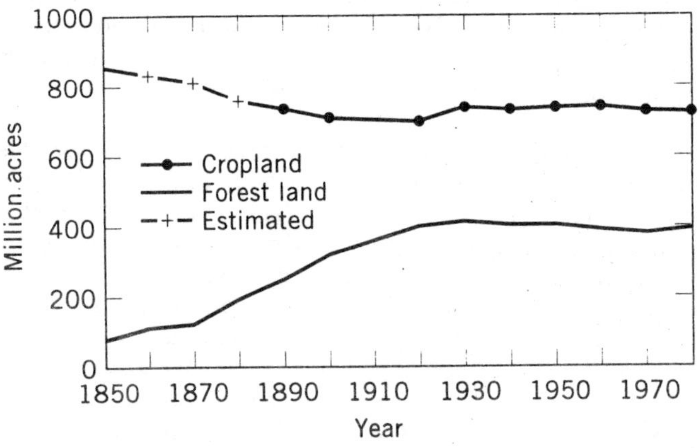

Figure 38. U.S. cropland and forest land areas from the 1850s to the 1970s (161).

and harvest rates came into balance as forest growth rates improved and harvest rates declined modestly. Net timber growth exceeded harvest by 54% in 1976, declining to 33% in 1991 because growth rates did not increase as rapidly as in the past and harvest levels also continued to increase. Today, annual forest growth in the United States is about 3.5 times what it was in 1920 (161).

Waste Loading, Cycling, Assimilative Capacity

Waste Streams. Humans as biological beings generate and discharge bodily wastes that must be treated, recycled, and/or assimilated by ecological processes in the environment. Communities of humans, from farmers and herdsmen to city dwellers and manufacturers, generate a variety of waste streams, potentially toxic chemicals, and hazardous materials that are either discharged or dispensed into the environment as anthropogenic exhausts, buried in landfills, or recycled to some degree.

In the United States, nearly 200 million t of solid waste are generated annually, about 37% of which consists of paper and paperboard. Yard wastes, metals, plastics, glass, food wastes and other materials compose the balance. Most of this waste (>80%) is not recycled or reused.

Industry, municipalities, and agriculture generate even greater volumes of waste loads and by-product materials, ranging from fly ash and residues from coal-fired electric utilities and manufacturing processes to raw and processed sewage discharges and agricultural and food processing wastes. As environmental consciousness and standards increase, greater volumes of certain wastes are precipitated to prevent or minimize waste discharges to the atmosphere (eg, scrubbing stack-flue gases) and the hydrosphere (eg, enhanced sewage treatment, resulting in greater volumes of sludges). About one-third of U.S. housing units still use on-site sewage disposal systems, accounting for annual discharges of 375 billion L of effluent that must be absorbed by the soil–geologic substrate ecosystem.

Repositories for Wastes—The Soil Medium. There are only three repositories for waste streams: the atmosphere, the hydrosphere, and the biogeopedosphere or terrestrial ecosystem. As more stringent limits are being placed on waste discharges to the atmosphere and hydrosphere, the soil–land ecosystem is being used more intensively as a waste repository and/or treatment medium. Soil ecology, therefore, is much influenced by anthropogenic waste loads and is a determinant of the assimilation capacity of these waste streams (162).

Soil, as a porous medium with a large surface or sorptive area, provides a medium for filtration. It also provides a medium for diffusing oxygen. The surface area of the clay minerals and organic matter is chemically active, providing chemical exchange sites. Dynamic and varied biological populations in the surface soil horizons provide biologically active properties to soils that can deactivate and decompose various wastes. Because of these properties and functions, the soil medium possesses a natural capacity to accept and renovate certain amounts and kinds of both solid and liquid wastes. The soil system is most effective in renovating wastes that contain filterable solids, nontoxic organic and inorganic material, microorganisms, and simple compounds and salts (including nitrogen and phosphorus).

Liquid Wastes. Because soils differ, the liquid waste loads that they can accommodate vary as well as the degree of waste remediation. If environmental impacts are to be minimized, the soil's hydraulic and assimilative–remediation capacity must not be exceeded.

A large volume of liquid wastes that lend themselves to soil disposal and remediation are household or domestic wastes. Rural electrification led to pressurized domestic water supplies, which permitted even the most remote homes and villages to have indoor plumbing facilities. Thus the old pit privy was replaced. This much-maligned privy was really a rather efficient waste disposal system. It depended on a relatively simple biological decomposition process, which was efficient, cheap, and required little space and hardly any management. The waste volume was confined solely to what was excreted daily by the human body. But with the advent of pressurized water systems, the few liters of human waste handled by the privy each day became diluted in large volumes of water along with detergent salts and other domestic wastes. This large volume of liquid waste requires disposal and renovation.

Domestic sewage is a complex brew, consisting of all manner of things that go down drains or are flushed down toilets (163). Since water is merely the vehicle for flushing and transporting these wastes to their disposal or treatment site, the resulting sewage commonly contains 99.9% water by weight. Average daily sewage discharges from pressurized water systems range from about 190 to >250 L/person. The components of the sewage include both suspended and dissolved organic and inorganic substances in addition to bacteria, viruses, and other microorganisms. Health officials are primarily concerned about the proper disposal of these microorganisms, because of potential health hazards if water and food supplies are contaminated. The composition of raw sewage and effluent is shown in Table 16.

Sewage consists of organic matter, some of which is microbially digested (carbon released as CO_2, CH_4, and soluble organic matter) and contains a host of microbial species, both pathogenic (disease causing) and nonpathogenic to humans. Some microbes provide a valuable asset in the sewage decomposition and treatment process while others are passive.

The amount of organic matter in sewage is indexed by its BOD, which is a measure of sewage concentration, or strength, and its potential for microbial and chemical oxygen consumption as carbon is used by these organisms as an energy source or as oxygen is consumed through direct chemical oxidation reactions. The BOD loads discharged to the soil must be assimilated by the soil medium. Therefore, the soil, in combination with sewage digestion via the septic tank or aerobic tank, provides secondary sewage treatment. Efficient primary (removal of solids) and secondary treatment (microbial digestion) can yield clear effluents that still contain significant amounts of organic and inorganic compounds. These salts and dissolved materials can be removed by a variety of processes, collectively referred to as tertiary treatment or advanced wastewater treatment (AWT).

Remediating septic tank effluents by passing them through the soil medium is a form of tertiary treatment. By the time an effluent has passed through an adequate volume of soil, the processes of filtration, sorption, oxidation, microbial digestion, and dilution have resulted in a purified effluent equivalent to those produced by tertiary treatment in a sewage treatment plant.

Site Requirements for On-Site Waste-Disposal and Renovation. The primary requirements for a successful on-site wastewater remediation system are adequate hydraulic conductivity, adequate soil volume, and lack of other site constraints.

Adequate Hydraulic Conductivity. Even the most impervious soils, as determined by the most hydraulically limiting horizons within soil profiles, have some finite capacity to absorb and transmit liquids. The problem for soil scientists and sanitarians is to determine what is adequate for an on-site disposal system, using the soil as the medium for waste disposal and remediation.

The lower the soil's hydraulic conductivity, the greater the area required to dispose of the effluent. Because it is nearly impossible to construct large sorption areas and load them uniformly and because the cost and land requirements are impractical for such large disposal areas, a compromise between hydraulic conductivity and soil area must be made.

For a soil to function effectively as a waste treatment medium, it must accommodate not only the percolate from natural precipitation but also the waste load imposed on top of the rainfall percolate fraction. Even many well-drained soils may be at capacity with respect to their natural loading or hydraulic conductivity. The point to be emphasized is that many soils can accommodate their natural loadings but may not have much capacity to accept more. Therefore, an on-site waste disposal system installed in a well drained soil may contribute more liquid than the soil's hydraulic capacity can tolerate.

To further complicate the situation, loading a soil continuously with either septic tank effluent or water will result in a striking decrease in hydraulic conductivity over time. This reduction of a soil's hydraulic conductivity at the effluent loading surface or soil–effluent interface is due to a combination of physical, chemical, and biological clogging or crusting processes. The impact of these processes on the hydraulic conductivity of a soil when submerged is shown in Figure 39.

It is this crusting process that renders the percolation test ineffective as a design criterion. The percolation test estimates saturated hydraulic conductivity. However, the system eventually operates in the soil medium at unsaturated hydraulic

Table 16. Raw Sewage and Affluent Constituent Composition of Typical Domestic Liquid Wastes[a]

Constituent	Raw Sewage Entering Septic Tank, mg/L	Effluent Leaving Septic Tank, mg/L
BOD	150–300	80–160
COD	300–900	174–400
Suspended solids	150–400	20–50
Nitrogen (total)	20–40	20–40
Organic nitrogen	15–30	5–10
NH_4–N	15–30	20–35
NO_3–N	<1	<1
Phosphorus (total)	5–20	3–10
Ortho	1–7	2–9
Dissolved/solids	400–800	400–800
Chloride	20–100	20–100
Fecal coliform	10^8–10^{10}[b]	10^5–10^7[b]
Fecal streptococci	10^5–10^7[b]	10^3–10^5[b]

[a] Ref. 163.
[b] Most probably number per 100 mL.

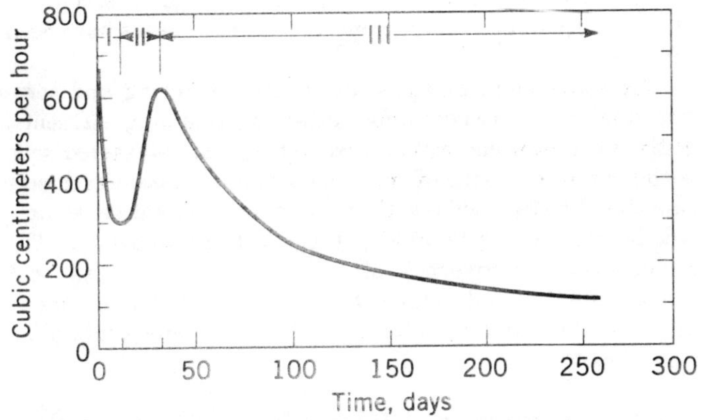

Figure 39. Typical percolation rate curve for prolonged water spreading on a soil core occurring in three phases (163).

conductivities as governed by the infiltration rate at the clogged surface. And it is this phenomenon that has taxed the ability of the soil scientist, engineer, and sanitarian to come up with a procedure whereby the soil percolation test can be correlated with the eventual crusted hydraulic conductivity once the system is operational.

Adequate Soil Volume. A soil's final equilibrium hydraulic conductivity alone does not ensure proper effluent treatment, for the remediation of the effluent is as important as its disposal. Too often, the concept of out of sight, out of mind is relied on. Proper effluent treatment requires not only adequate hydraulic conductivity for effluent disposal but also adequate soil volume for remediation of the effluent before it discharges to the groundwater. A practical limit that yields a minimal health or environmental hazard suggests that at least 1 m of unsaturated, medium-textured soil should be available below the point of effluent discharge and above the water table, creviced bedrock, or impervious layer. An uneven bedrock contact with the soil or creviced bedrock with local open channels can result in untreated or partially treated effluent discharges, or short-circuiting, to the groundwater. Greater soil volume is necessary for sandy or coarse-textured soils.

Lack of Site Constraints. Having located soils with adequate hydraulic properties and adequate remediation capabilities, the site must be examined for other constraints that might jeopardize the functioning of an on-site waste treatment system. The configuration of the landscape should not allow flowing and excess surface or subsurface runoff to inundate the system, once installed. Excessive slope can conspire with gravity to intercept the effluent, causing it to break out at the surface before it has traveled far enough through the soil to effect complete remediation.

Mechanisms Operating in Soil That Influence Effluent Remediation. The biological and chemical constituents of waste effluent must be renovated through the combined remediation mechanisms and processes available in soils. These soil ecology processes act on waste effluent components to effect at least some degree of remediation. The remediation capacity of soils is determined by the ability of their infiltrative surface and underlying horizons to (*1*) mechanically filter the largest effluent constituents (eg, particulate matter and colloidal suspension), (*2*) maintain at least a minimum oxygen diffusion rate within the porous medium, (*3*) expose large amounts of surface area to the infiltrating effluent so that constituents may be sorbed either physically or electrostatically, and (*4*) the presence of a well-adapted biologically active flora and fauna that form a biological environment that is competitive with and antagonistic to the biological contaminants in effluents.

Filtration. The mineral skeleton of the soil and the subsequent crust formation serve as an effective physical filter. This filtration capacity of the soil serves not only to remove suspended solids but also to retain microorganisms and facilitate biological treatment of both dissolved and suspended organic matter. As noted, the infiltrative surface takes the brunt of the filtration impact. The slowing of the hydraulic conductivity with time aids in the remediation process. The increased contact time of the effluent with the soil enhances filtration and the probability of sorption, BOD reduction, and bacteriological die-off.

Bacteria behave like other particulate matter in soils. They are large enough (0.5–3 μm) to be removed by straining, sedimentation, entrapment, and adsorption. Thus the filtration process is a major mechanism in the renovation of particulate matter and bacteria in effluents passing through the soil medium. Most particulate matter is removed within the crust or clogging zone and seldom penetrates beyond 10–15 cm in the soil system.

Sorption. Sorption is the binding of one substance by another through the mechanisms of *ab*sorption (the taking in or reception by molecular or chemical action) and *ad*sorption (a binding of substances on a surface). The large surface area within the soil's pore space provides a great capacity for sorption of suspended solids and soluble materials in effluents passing through the medium. Furthermore, this surface area is neither chemically nor electrostatically inert.

The iron and aluminum oxides and hydrous oxides coating the subsoil clay minerals and the magnesium-hydroxy clusters or coatings on the weathering surfaces of ferromagnesian minerals provide especially effective sorption sites. Elements and compounds such as boron, phosphates, fluorides, and many of the metals are sorbed from the effluent by these coatings. The electrostatic properties of the clay minerals and organic matter provide an ionic exchange capacity that is capable of sorbing ionic and biological constituents from the effluent.

Oxidation. Effluents contain a variety of natural and synthetic organic compounds, the concentration or "strength" of which are collectively expressed in terms BOD, COD, and total organic carbon (TOC).

The well-drained soil with an adequately designed and installed effluent disposal field is capable of providing an adequate oxygen supply for BOD and COD reduction. This oxygen supply is furnished via the unsaturated soil pores by diffusion and/or dissolved oxygen percolating with rain water into the zone of the oxygen demanding substrate. The aerobic bacteria using the carbon in the organic material as an energy source require oxygen for the decomposition process.

Thus a well-aerated, medium-textured soil provides a medium whereby the BOD and COD load in the effluent percolates slowly through the soil. This process may provide enough time for the heterotrophic bacteria to effect remediation by consuming the carbon energy supply and reducing the BOD and COD load.

Microbial Antagonism. Pathogens generated from human waste encounter environmental conditions outside the body that are not conducive to their reproduction and longevity. Coupled with this situation and the competition of the already existing and adapted soil microbial community, the conditions for pathogenic remediation by soils are good. The survival times for many pathogens often exceed several days to several months once they are discharged to the soil environment. The key to effective pathogenic remediation is to have adequate soil volume and sorptive capacity so that the combined mechanisms of filtration and sorption will retain the organisms long enough for die-off and slow their travel time through the soil.

Fate and Disposition of Effluent Biological Contaminants in Soil. As the microorganism-laden effluent moves into the soil, the physical, chemical, and microbiological properties of the soil act to render effluents nontoxic. As a general rule, pathogenic organisms are removed rapidly as they pass through all but the coarsest soils. The soil system contains vast numbers of microorganisms that are capable of tremendously diverse and complicated reactions. It is a complex and competitive system that greets the incoming pathogens and allows only the best adapted and most competitive microorganisms to survive.

Pathogen survival in soil systems is determined by such factors as soil texture, organic matter levels, pH, moisture content, temperature, and competition from native soil microflora; soil pathogen survival data in soils have been compiled (Table 17). The movement of pathogens in soil systems by percolation appears to be related not only to survival times of the pathogen but also to the soil type and hydraulic considerations.

Larger microbes such as pathogenic worm ova and protozoa cysts generally exceed the mean pore diameter of soils and are considered to be removed near the soil–effluent interface by the physical process of filtering. Bacteria and viruses are small enough to move through pores in soil, and their movement is related to the hydraulic conductivity and adsorption processes in soil.

In both field and laboratory studies, several factors appear to be important in determining viral adsorption. Soil factors such as pH values, type of cations on the exchange complex, type and amount of clay and amount of organic matter are important in determining adsorption. Viral movement is primarily restricted to the top few centimeters of the soil column, although greater distances of movement within soils have been reported.

The soil system provides an effective filtering and sorptive medium for remediating pathogen-containing effluents. Pathogen movement will be greatest in coarse-textured soils with large pores, monovalent cations, and low cation-exchange capacity. Shallow soils with inadequate attenuation volume will be most likely to yield pathogens to groundwater. The remediation effectiveness of soil with depth shows that the greater the soil volume that the pathogen must travel through, the less the probability of groundwater contamination (Fig. 40).

Fate and Disposition of Effluent Chemical Contaminants in Soil. One of the most important functions of soil microorganisms is the decomposition of organic compounds by common metabolic pathways. Because anaerobic metabolism occurs at slow rates and is not complete, aerobic conditions in the soil are preferred.

When ammonia-containing effluent encounters the aerobic filter field, the process of nitrification occurs, resulting in oxidation of the ammonia to nitrate. A family of four discharges ~32–34 kg (70–75 lb) nitrogen annually to a septic system. In a well-designed system, most of this nitrogen is nitrified in the aerobic zone. Coupled with NO_3–N contributions from other sources (eg, mineralization of soil organic matter), high density septic system installations contribute significant amounts of nitrate to the groundwater each year. For a septic system serving a family of four, the minimum area necessary

Table 17. Survival Times of Various Pathogenic Organisms in Soils[a]

Organism	Group	Survival Times
Ascaris ova	Parasitic worms	2.5–7 years
Entamoeba cysts	Protozoa	40–170 days
Salmonella	Bacteria	1–85 days
Coliform group	Bacteria	100–150 days
Mycobacterium	Bacteria	5–15 months
Enteroviruses	Viruses	?

[a] Ref. 163.

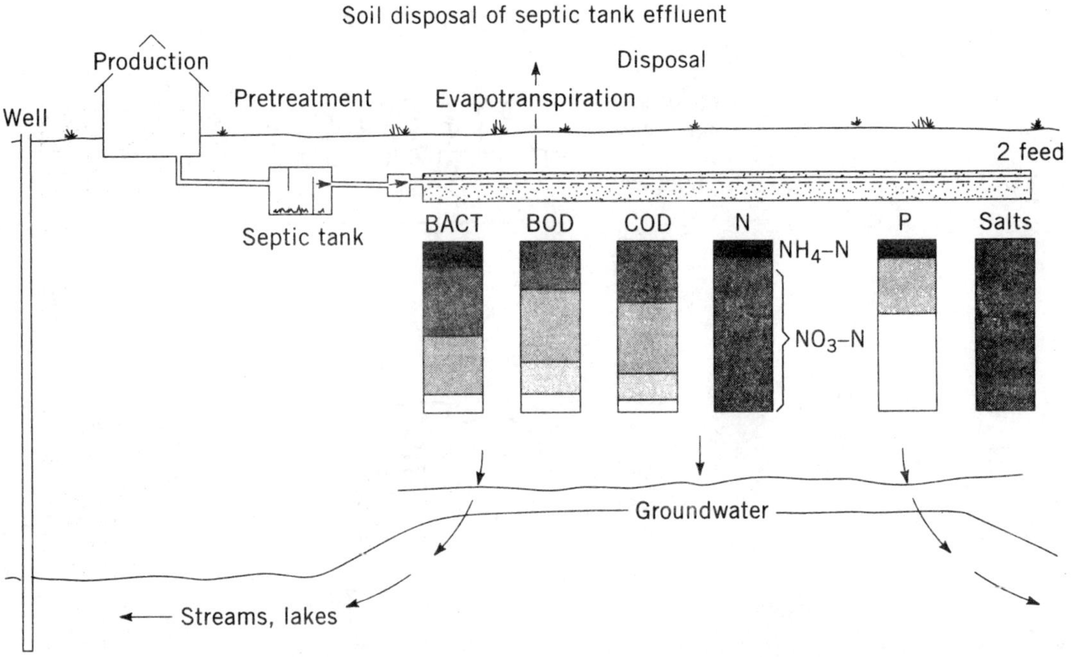

Figure 40. Schematic representation of relative attenuation of effluent constituents as they percolate through the soil profile (163).

to accommodate the necessary dilution to concentrations of <10 ppm NO_3-N in the groundwater ranges from 0.4 to 0.1 ha (1 to 0.25 acre), depending on contributions from other sources and rainfall.

The average family of four contributes 4–4.5 kg (9–10 lb) of total phosphorus to the septic system annually. Less than half of the total phosphorus is soluble inorganic phosphorus compounds. The soil medium is well suited for phosphorus removal from effluents and provides an effective means of effluent renovation with respect to this nutrient. The major mechanisms involved in phosphorus retention by soils are precipitation reactions and adsorption. Precipitation is controlled by a reaction between an anion (orthophosphate) and cation, such as iron, aluminum, and calcium. The sorption mechanism is controlled by the amount of surface area and the chemical nature (oxides, hydroxy clusters, carbonates, exchange capacity) of this area. Soils, therefore, have a large but finite capacity to sorb phosphorus. However, continued loading of the soil medium with phosphorus can lead to saturation of the system, while renovation of the phosphorus-containing effluent does not occur.

Most phosphorus is sorbed within a few centimeters of the loading surface (Fig. 40), although there are reports of some soils providing inefficient phosphorus renovation. Thus soils with iron and aluminum contents in humid regions and calcium contents in more arid regimes provide effective media for phosphorus remediation.

The dissolved solid load in effluents is 200–400 mg/L. These effluents contain a variety of salts, many of which form ions in solution; including Ca^{2+}, Mg^{2+}, Na^+, K^+, Cl^-, SO_4^{2-}, $H_2PO_4^-$, HPO_4^{2-}, NO_3^-, HCO_3^-, CO_3^{2-}, as well as other and more complex ionic molecules. Soils possess a finite exchange capacity and, therefore, have a fixed capacity per unit weight to remove salt ions. The soil eventually becomes saturated and eventually contributes little toward effluent renovation with respect to these salt constituents.

The soluble anions (SO_4^{2-}, Cl^-, NO_3^-) are not removed from the effluent (unless the NO_3-N is denitrified) and contribute to the mineralization of the groundwater. Furthermore, the renovation of the salts, including the cationic species, in the effluent decreases with time. As the soil ion exchange sites become saturated, the ionic composition of the renovated water will eventually be the same as that of the effluent. Hydraulic dilution will eventually render this salt flux harmless unless the disposal system density is too great.

Land Treatment of Wastewaters. Land treatment involves the application of liquid wastewater on the land surface instead of discharging the wastes within the soil medium beneath the soil surface. Where communities cannot afford high standard secondary or tertiary treatment plants, land treatment of wastewater may offer an economical and effective sewage treatment alternative.

The objective of sewage treatment is to return the renovated sewage carrier (water) to either surface waters or groundwaters with minimal impact on these ecosystems. Since raw sewage cannot be applied to the land surface because of health risks, some degree of sewage pretreatment is necessary before land application can be considered in the United States. There are three major types of land treatment systems. The primary differences in these systems involve the mechanisms by which the effluent is delivered to the land and their application rates (Fig. 41).

Land Reclamation by Solid Waste Amendments. Land reclamation is the process by which soil ecology is artificially

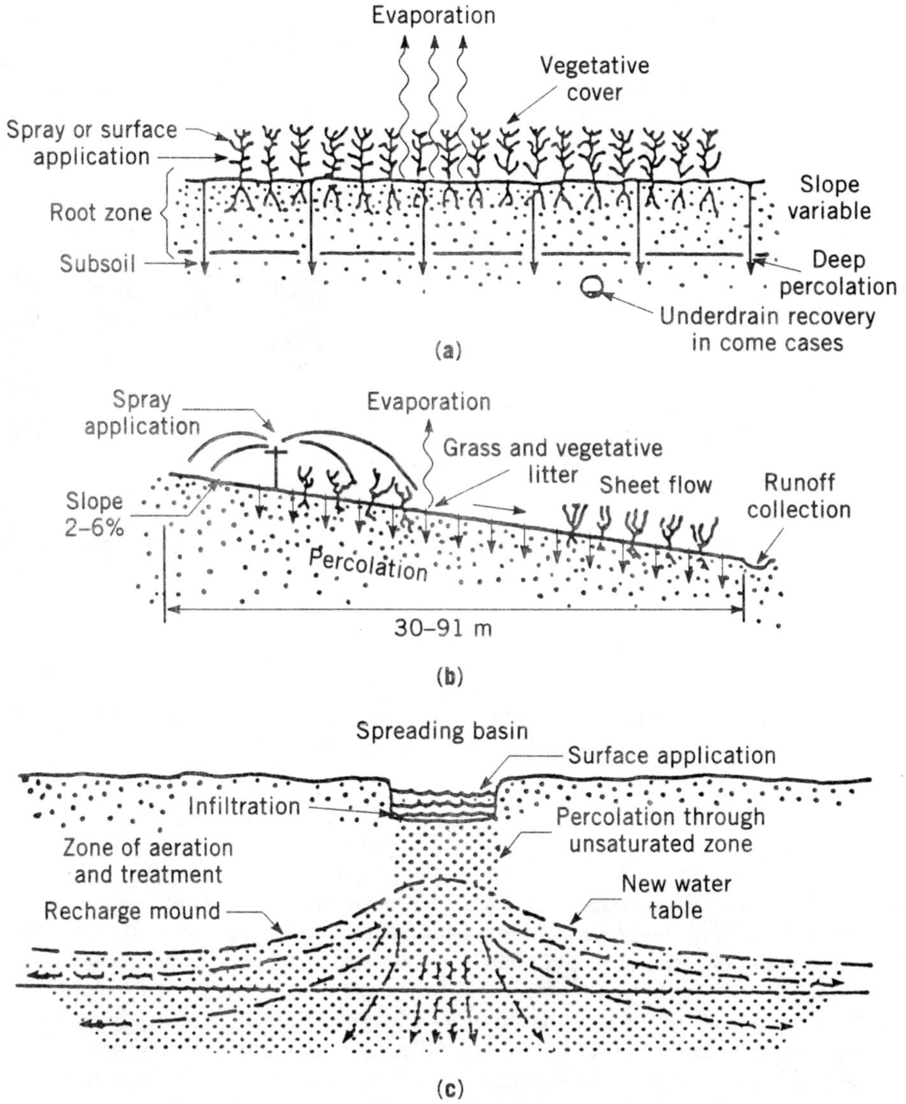

Figure 41. Land treatment distribution systems: (**a**) spray irrigation, (**b**) overland flow, (**c**) rapid infiltration (163).

enhanced and productive land areas are created by the addition of foreign matter, such as stabilized sludges, solid wastes, or dredge spoils. Reclamation of strip mines spoils is one category of land that can accommodate solid wastes, such as sludges and other mineral and organic solid wastes. Other categories of land reclamation by solid wastes include urban expansion by filling in of marginal lands. Some sanitary landfills can be considered reclamation projects, since their ultimate usage is often a park, recreation, or conservation site.

Reclaiming or altering any of these ecosystems with solid wastes poses the problem of potential contamination of the groundwater or adjacent ecosystems. Overloading these sites with nitrates, pathogens, metals, or other potentially toxic materials must be avoided. The carrying capacity of these lands with respect to solid waste loads and their leachates is finite. Therefore, site evaluations must be carried out and the proposed waste characteristics must be known so that an environmentally safe recommendation can be made.

Land Spreading of Wastes. Spreading certain waste residuals on agricultural land has been practiced for decades. Wastes that lend themselves to such recycling management include manures, sludges, and certain other solids such as garbage and septage. With poor management and/or overload, however, surface and groundwaters can become contaminated with nitrates, phosphates, and toxic metals from sludges. Pollution of receiving water bodies is also a concern.

For proper management, all organic wastes spread on the land should be stabilized, conditioned, or chemically treated before application to minimize odors and pathogenic hazards. Manure can be liquified and digested in either aerobic or anaerobic lagoons. Sewage sludge is stabilized by anaerobic digestion. Loading rates of such materials are usually determined by their nitrogen and/or phosphorus content. Nitrogen applications should not exceed the sum of losses through volatilization, denitrification, and plant uptake. A gross generalization suggests that nitrogen loadings should not exceed twice the amount of plant uptake. Phosphorus can ultimately be the nutrient that limits waste loadings, since its removal from landscapes occurs primarily by crop removal. For certain sludges, the toxic metal content and/or pathogen content may be the constraining loading criterion.

Sanitary Landfilling. Land filling accounts for about half of all urban and industrial generated solid waste and about 90% of all collected municipal refuse. Landfilling consists of confining solid wastes to the smallest practical volume and covering each day's accumulations with a layer of soil to prevent nuisance, health, and safety hazards. Soil and earth materials are, therefore, used to contain, cover, and revegetate landfill sites. Soil material must also filter and renovate waste leachates before they enter the underlying water table. The design of a landfill is site-specific and must include consideration of soil and geology, hydrology, topography, and climate in addition to the quantity and quality of waste to be accommodated.

Sewage Sludge Trenching. Where agricultural land and/or reclamation projects are not adequate or suitable for accepting large quantities of municipal wastewater sludges, a modified landfill technique has been developed to handle these solid wastes. Trenches 0.6–1.2 m (2–4 ft) deep and 0.6 m (2 ft) wide are excavated and filled with either digested or raw-limed dewatered sludge. Loading rates are in the range 3100–3400 wet t/ha (620–580 dry t/ha) of dewatered filter cake sludge. The soil is replaced over the sludge where continued anaerobic digestion occurs. Stabilization and continued dewatering may take up to 5 years. Nitrate formation is slow and is restricted to the aerobic perimeter of the trench walls.

Engineering Uses of Soils

In terms of volume, soil ranks as the leader among all building materials. Soil material is used for road embankments, earthen dams, dikes, and the supporting medium for building foundations. Soil landscapes can be cut, filled, and molded by earth-moving equipment to provide the desired landscape configuration for buildings, houses, highways, recreation areas, and numerous other uses. The character of soil materials determines their load-supporting capability and stability for such uses. Some soils can accommodate higher loads than others. Inadequate knowledge of soils has resulted in spectacular failures. Expansive soils (those with high expansive clay contents) cause an estimated $6 billion worth of structural damage annually in the United States, accounting for the nation's most costly natural hazard (164).

The engineering uses of soils and the behavior of soils in response to loads properly belong to the disciplines of soil engineering and soil mechanics and are part of the broad domain of soil ecology. Soil mechanics consists of studying soil problems and analyzing the soil's response and stability to various loadings, using the principles of statistics, mathematics, and applied mechanics. Soil engineering encompasses a broader scope than does soil mechanics. It is the practical science of applying both fundamental engineering principles and theoretical methods in solving problems related to soil behavior under various applied loads or forces.

Soil: The Three-Phase System. It is convenient for the engineer to think of soil as a three-phase system consisting of solids, water, and air. A cubic meter of soil may be partitioned into the three phases.

Soil Consistency Limits. Adding water to dry soil can change the consistency of the soil from a solid to a semisolid state to a plastic state, eventually arriving at a soil consistency possessing the character of a liquid as more water is added. This generalized relationship between water content and resulting volume can be expressed as a proportion of the original oven-dry soil volume. The points marking the consistency state change are known as the shrinkage limit (SL), plastic limit (PL), and liquid limit (LL). These consistency parameters are termed the Atterberg limits, after their originator, who proposed and published the concepts in 1911 and 1912.

The SL is defined as the water content at which the volume change is no longer linearly proportional to a change in water content. This is the only one of the three parameters that can be rigorously defined in terms of a water content volume change. The liquid and plastic limits are determined by standardized laboratory procedures. The numerical difference between the liquid and plastic limits is the plasticity index (PI). The plasticity index is a measure of the relative amount of fines or silts and clays in a sample and an indicator of clay mineralogy.

Particle Size Classes. Particle size has a tremendous influence on soil properties. Chemical, physical, and biological properties of soils are greatly influenced by the particle size distribution within soil horizons. Likewise, hydraulic conductivity, soil temperature, aeration, and bulk density are determined in large part by the particle size distribution of soils. Particle size distribution in soils influences the strength and compressibility of soils. These properties are critical in the consideration of bearing strength in soils and their stability for engineering purposes.

Each soil material or soil horizon has a characteristic particle size distribution, or gradation, curve (Fig. 42). A well-graded or well-sorted soil is one containing significant amounts of various particle size fractions, yielding a gradual curve with a low gradient (Fig. 42a, curve B). A poorly graded soil contains a predominance of one particle size class, yielding a steep distribution curve (Fig. 42, curve A). Engineers refer to the distribution and size of grains in a soil as its gradation. Over the years, the various branches of earth sciences and engineering have developed their own particle size classes. Several particle size scales are shown in Figure 43.

Engineering Soil Classification. To the engineer, the loading and stability of soil are of primary concern. The empirical correlation of soil physical properties with their observed behavior under a variety of conditions has allowed engineers to develop their own engineering soil classification systems. It is important to be aware that several engineering soil classification systems exist, eg, the Unified System and the American Association of State Highway and Transportation Officials (AASHTO) system.

Engineering Properties of Soils. The engineer is interested in only the most important soil characteristics that influence the capacity of a soil to support a load, remain stable, or resist movement or settlement. These soil characteristics include the major components of the soil's structural strength (internal friction and cohesion), elasticity, density, compressibility, potential for volume change, permeability, capillarity, and corrosivity. The engineering indices used to characterize the behavior of soils represent the integration of soil properties that soil scientists use as a diagnostic criterion in the pedologic classification of soils. One must distinguish between determinant soil properties (eg, particle size distribution, pore space, mineralogy) and behavioral indices. The latter result from measuring the response of soil to a prescribed kind of manipulation. These determinant and behavioral soil attributes combine to yield the mechanical and hydraulic properties that determine the suitability of soils for engineering uses.

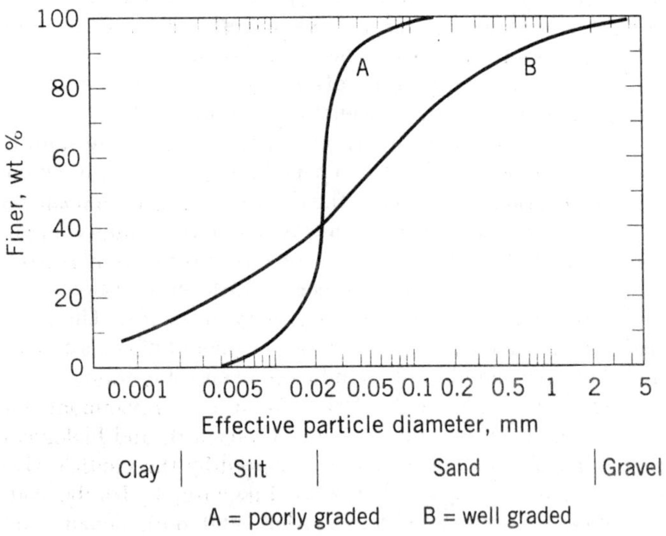

Figure 42. Particle-size distribution curve (118).

Soil Compaction. Compacting a soil with a specific compactive force at a given moisture content will yield a certain density. By changing the moisture content under the same compactive force, a different density will be obtained. Likewise, increasing or decreasing the compactive effort will result in higher or lower soil densities for each moisture content. These relationships of compactive force and moisture are best illustrated by way of a soil moisture–density curve (Fig. 44).

As water is added, the individual mineral particles slide together more readily, increasing the soil density as the water begins to have a lubricating effect. At the maximum density point, the soil fines are close to the plastic limit, and the air voids are completely surrounded by water (166). The moisture content corresponding to the maximum density is termed optimum moisture. The optimum moisture is a compromise: there is enough water to permit the soil grains to distort and reorient themselves into a smaller volume but not so much water that the voids are filled (167).

Maximum densities of soil range from 1360 to 2320 kg/m^3 (85 to 145 lb/ft^3) on an oven-dry basis. The higher the maximum density, the lower the optimum moisture content in soils (Fig. 44). The maximum density of a soil provides at least approximate information on the soil's gradation, whereas the optimum moisture yields approximate information on the clay and silt content. The shape of the moisture–density curve can also provide valuable information about the influence of moisture on the load-supporting value of the soil. A relatively flat or broad curve indicates a soil that possesses the same load-supporting capability over a wide range of moisture contents (167).

Adding more water beyond the optimum moisture point results in moving the soil particle grains farther apart, thus increasing the proportion of the soil–water volume occupied by lower density water. Continued water additions ultimately change the soil material to a liquid, with no soil strength and low density. With more prolonged or greater compactive force, the maximum density is reached at a lower moisture content, and the compaction curves move up and to the left as shown in Figure 44.

Structural Strength of Soils. In considering the load-bearing capacity and stability of soils for engineering purposes, two conditions must be investigated. One is the shear strength and the other is the soil's deformation characteristic under a load—a process referred to as consolidation. Therefore, a soil will support a load only when its shearing resistance is not exceeded or if excessive consolidation does not occur (168).

Shear Strength. Many construction materials such as steel and concrete have considerable compressive and tensile strength. The engineer and architect can specify the strength of these building materials when designing a structure. But the soil on which the building load is supported must be used as is. Soil exhibits only a fraction of the strength of these building materials. Therefore, in designing the foundation

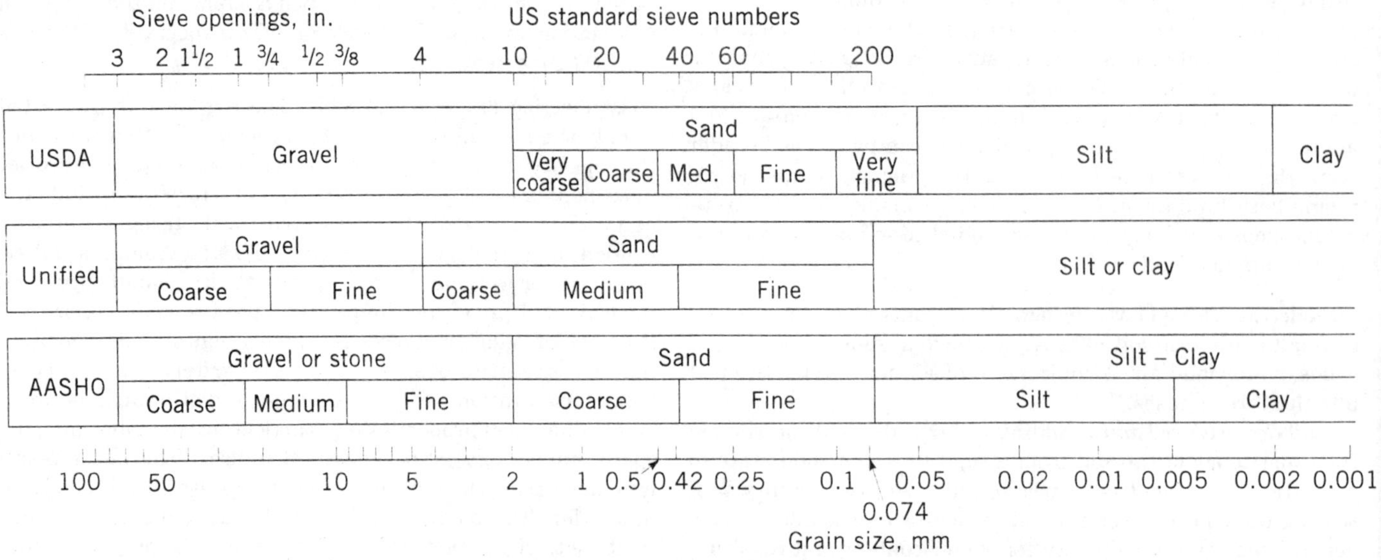

Figure 43. Comparison of particle size classes (165).

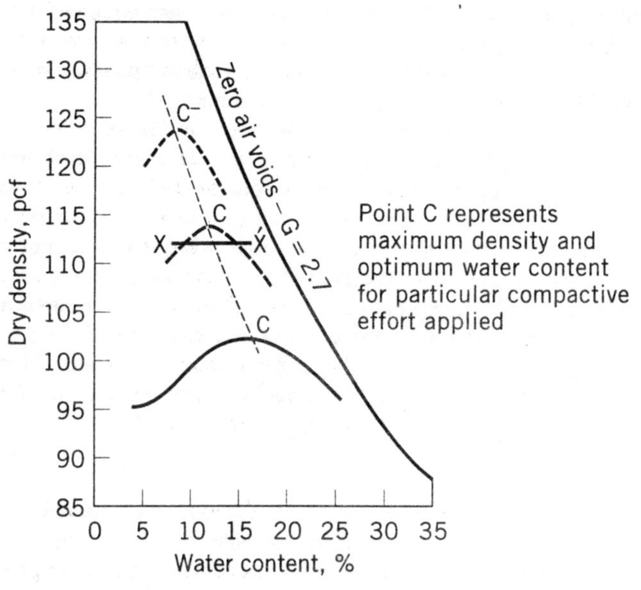

(a)

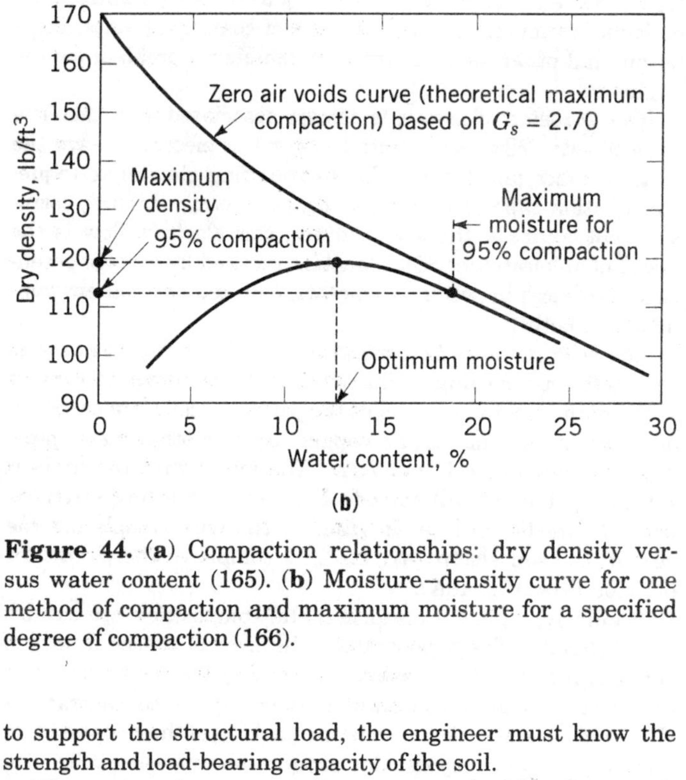

(b)

Figure 44. (**a**) Compaction relationships: dry density versus water content (165). (**b**) Moisture–density curve for one method of compaction and maximum moisture for a specified degree of compaction (166).

to support the structural load, the engineer must know the strength and load-bearing capacity of the soil.

The engineering computations concerned with soil strength deal primarily with its shear strength, ie, the resistance to sliding of one soil mass against another (169). It is the shear strength of soil that allows it to remain in equilibrium even though its surface is not level. Accordingly, it is soil shearing resistance that allows it to support a load. Water does not have any shear strength. Thus the continued addition of water to a soil will cause the soil mass to attain the properties of a liquid (liquid limit), resulting in reduction and eventual loss of the soil's shear strength, which is the cause of many landslides. A soil with a slope gradient is exposed to a lateral gravitational force. Loss of shear strength coupled with the extra load due to the excess water often results in failure or a landslide.

The shear strength of a soil is derived primarily from two sources. The first is the internal friction of the soil mass caused by resistance of the individual soil grains to sliding due to forces or pressure of the soil grains against each other. Internal friction is the predominant source of shear strength for coarse textured or granular soil materials, also referred to by the engineer as cohesionless soils. Internal friction increases with increasing loads or external pressure on the soil and is relatively unaffected by moisture content.

The second source of shear strength is cohesion: the molecular forces exhibited by the mutual attraction of small particles coupled with the presence of thin water films. Unlike granular soils that exhibit internal friction, the cohesive forces in clayey soils (and thus the soil strength) depend on the moisture content. From the standpoint of soil strength, therefore, the engineer speaks of two idealized concepts of soil material. One is granular soil, or cohesionless soil, for which internal friction is the source of strength, and the other is cohesive soil, for which the forces of cohesion impart the primary component of structural strength.

Consolidation and Compression. A soil may fail under a structural load because either the soil's shear strength was exceeded, because of soil consolidation, or because of a combination of these factors. The engineer must consider both factors in the design of foundations. Soils also exhibit characteristics that are not encountered with other building materials; namely, volume reduction and settlement under a load. When a load, such as a pavement, building, or earthen embankment, is placed on a soil, the supporting soil may deform or exhibit a volume reduction. The load on the soil rearranges the soil grains and compresses the soil, causing void volume reduction and settlement.

Where structures cannot be anchored to inelastic bedrock, the problem of volume change usually faces the engineer whenever soil materials are loaded. Cracking of buildings and structural distortions are caused through excessive and/or uneven settlement. The Leaning Tower of Pisa stands as testimony to the severity of the volume change and deformation problem in soils.

In most cases, the volume reduction of soils under loads is not due to lateral soil displacement. If this were the case, the displaced soil would result in bulges of soil around the structure. The confining soil around the loaded soil usually prevents displacement. While such displacement can and does occur, most volume reduction is due to void reduction within the soil. If the soil is dry, air within the voids is compressed when the soil is loaded. In most cases, this compressed void air can be expelled relatively fast. The resulting volume change or settlement is nearly instantaneous or occurs within a short time. This quick soil volume change under a static load has often been referred to as compression.

If a loaded soil is at or near saturation, the water in the voids must be expelled before the soil can experience volume reduction. If the soil is quite permeable, the water can be expelled quickly. Under this condition, the loaded soil will settle rapidly, thus resembling the compression phenomenon. But if the saturated soil is slowly permeable, the water in the voids must flow out of the soil mass before the soil grains can be rearranged. The slow settlement resulting from water draining out of the voids within loaded soils is termed consolidation. Again some engineering references refer to consolidation as simply the rate of volume change or settlement per unit time, while others view consolidation as a special form of compression, caused by water draining from soil voids (170).

The time–settlement curves for a sand under compression and a saturated clay experiencing consolidation, using the definitions based on settlement–time lag are illustrated in Figure 45. The time lag to reach equilibrium may extend to several years before equilibrium is attained and settlement ceases. Structures built on fine-textured or cohesive soils may continue to settle at a decreasing rate for the duration of their life (166). The Leaning Tower of Pisa, built in 1174, is still settling.

Elasticity and Volume Changes. Most soil materials remain compressed after compaction or consolidation. Removal of the load or compactive force from such soil materials reveals only slight rebound. Most soil materials are compressible and present no special design problems to the engineer (167).

Other soil materials, however, are elastic and rebound considerably after a compactive force has been applied or after a load has been removed. Soils with flexible components, such as micaceous soils or soils high in organic matter, commonly exhibit elastic properties. Such soils make poor subgrades for flexible (eg, asphalt) pavements, because they will compact or compress and then expand or rebound with the passage of every vehicle. Soon the pavement is torn apart (167).

Another problem facing the engineer, is the expansion of certain types of clay soils in the presence of water. Most soils have a tendency to shrink and swell as moisture contents change. Such volume changes in soil can cause serious problems to structures if the volume change is large and, as noted, contribute to the most costly natural hazard in the United States (171). The principal causes of soil expansion or swelling in the presence of water are the amount and type of clay, the exchangeable cations, and the organic matter content. Maximum expansion would be expected to occur in a saturated, high clay content soil with a high sodium content.

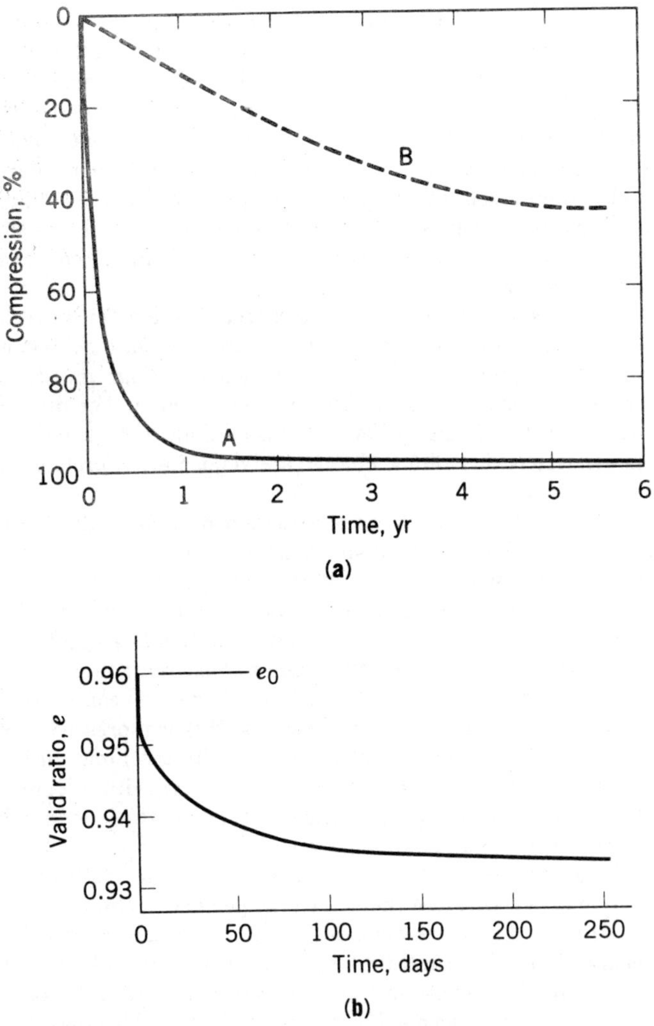

Figure 45. (**a**) Typical time–compression curve for sand (curve A) and time–consolidation curve for cohesive soil (curve B) (165). (**b**) Time–settlement curve for an increment in stress (166).

Soil Corrosion. A variety of metallic materials used by society are buried or driven into the soil. Pipelines transport much of our energy needs, water, chemicals, and other products. Electric utility lines and communication cables are often buried. Metal pilings are driven into the soil to support various loads. While attempts are made to protect these multibillion-dollar investments through the use of coatings, paints, insulation, and other means, corrosion remains a problem for the engineer.

The corrosion of metals is always associated with the flow of electricity. Wherever electric current or electrons leave the metal surface and flow to the surrounding electrolyte (aqueous solution capable of conducting an electric current), corrosion will occur. In the soil medium, this electron flow is the result of an electrochemical reaction induced by voltage potential differences between two metals, between two soil environments, or both.

Soil with its attendant moisture and dissolved ions acts as an electrolytic medium. Whenever the pipe comes in contact with another metal, such as at the junction (coupling or splice) of a galvanized pipe with a copper, iron, or other metal pipe, a simple galvanic cell will form. The metal with the highest solution potential will corrode because it is losing electrons from its anodic surface. In place of the wire completing the circuit, the soil electrolytic medium completes the circuit for electron flow (Fig. 46**A**).

Similarly, a pipe of uniform metal composition can exhibit two different voltage potentials wherever it passes from one soil to another. This situation is caused by the two soils having different electrolyte concentrations or oxygen concentrations (Fig. 46**B**). Local differences in the packing of the soil and in its moisture content may develop oxygen concentration cells. The area with the least oxygen is anodic where corrosion is most likely to occur, compared areas where oxygen is more readily accessible (172). Even in the same soil, a concentration cell can form around a buried pipe if the moisture content, degree of aeration, or ionic composition of the soil solution is different on top of the pipe than it is on the pipe bottom.

When a buried metallic pipe crosses from one soil to another, both galvanic and concentration cell corrosion can occur. The pipe shown in Figure 47 has been spliced with a different metal thereby establishing a galvanic cell. The electrolyte concentration, chemical composition, and degree of aeration from concentration cells change as the pipe passes from one soil to

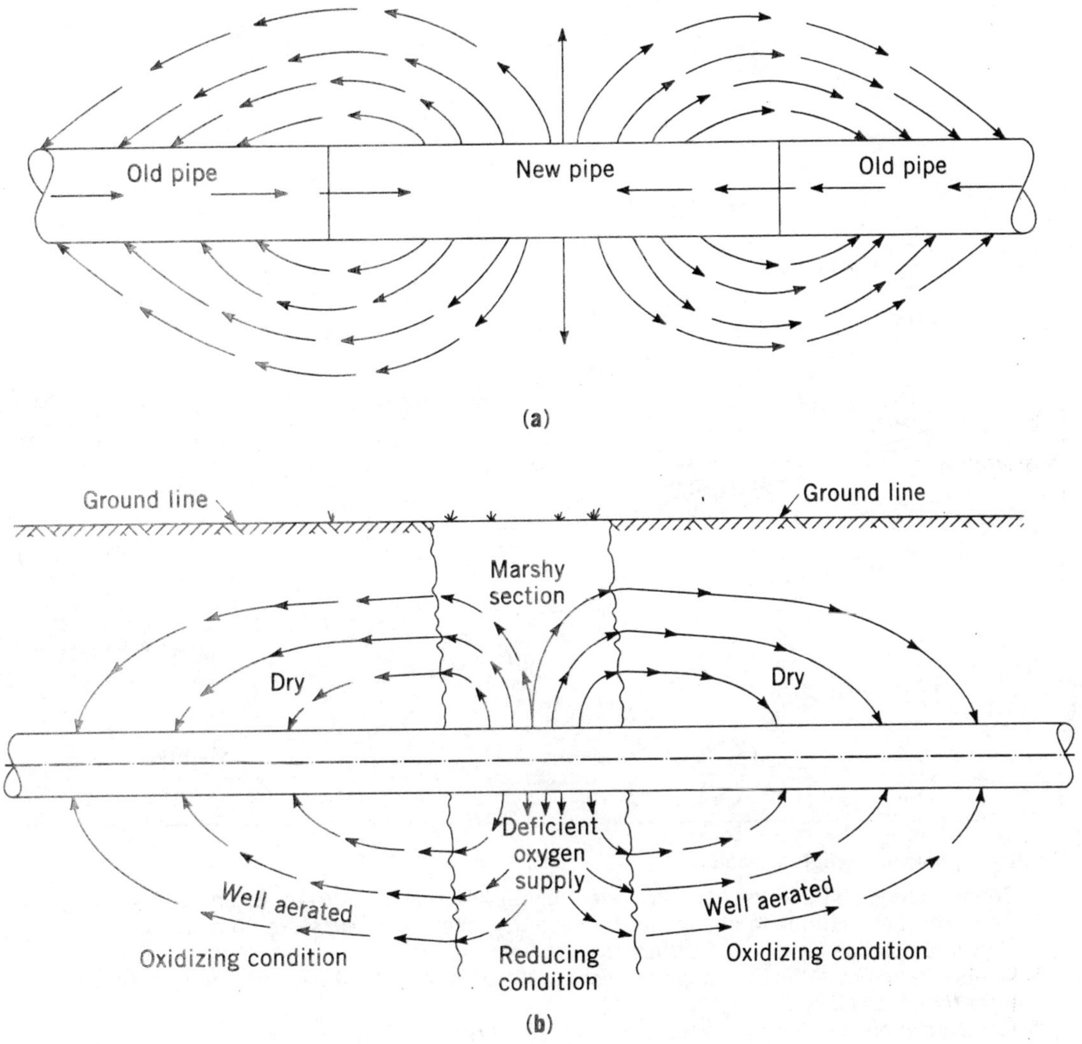

Figure 46. (**a**) Corrosion due to dissimilar pipe. (**b**) Corrosion caused by pipeline passing through marshy section. H. V. Beezley and G. R. Olson, *Petrol. Eng.* **29,** May D31–D34, June D28–D34 (1957).

another and from one horizon to another; electrolyte concentrations not the same above and below the pipe.

Corrosion engineers use a variety of indices and measurements to assess the corrosion potential of soils. Resistivity–conductivity measurements of soils provide an index of the soil's electrolytic strength. Resistivity integrates both the moisture and salt content of the soil. Several classes of corrosivity for uncoated steel are recognized based on resistivity and total acidity (Table 18).

Redox potential provides an index of oxygen concentration and diffusivity in the soil as measured by the oxidation–reduction potential. The latter determines the aerobic–anaerobic environment in soils, which is critical to corrosion reactions. However, the redox measurements are difficult to make. Their use as an index of corrosivity in the natural soil medium has been successful.

Other soil properties used to assess the corrosion potential of soils are permeability, soil texture, and total acidity. Soil reaction (pH) does not correlate well with corrosion potential, except under acidic conditions ($pH < 4.0$).

For most soil corrosion problems, the primary cause of corrosion is nonuniformity in the distribution of oxygen and moisture. In general, the most corrosive soils are those with high salt contents with attendant low resistivities and high conductivities and soils with high hydrogen ion concentrations. Soils with variable oxygen concentrations also form oxygen concentration cells, which promote electrochemical voltage differences and current flow. Since pipelines are contained in a disturbed soil volume, such cells can form due to air pockets, organic debris, and stones when back-filled. The presence of sulfides or easily weatherable sulfide minerals (eg, pyrite) produces not only strong electrolytes but soluble anode area corrosion salts. Such soluble salts keep fresh metal exposed to the driving electrochemical force and result in a highly corrosive environment. Both sulfur-reducing and oxidizing bacteria also can play a role under these conditions.

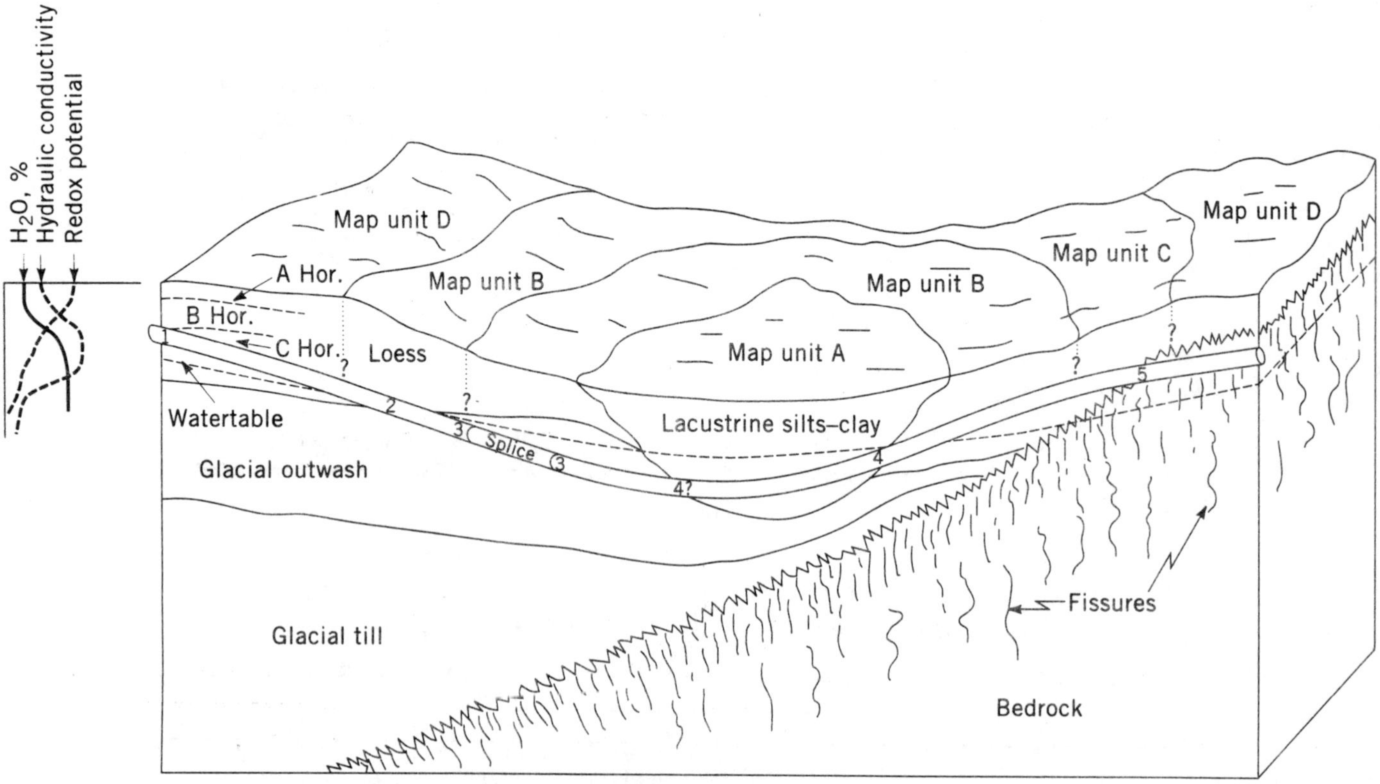

Points of potentially high corrosion:

1. Concentration cell due to chemistry (electrolyte) differences between B/C horizons
2. Concentration cell due to change in permeability, chemistry, moisture, oxygen
3. Galvanic cell due to splice of different metal into pipe
4. Concentration cell due to change of permeability, oxygen, moisture, chemistry; perhaps no corrosion if pipe remains below water table
5. Concentration cell due to change of electrolyte, oxygen, moisture

Figure 47. Cross-section of soil landscape showing metallic pipeline and points of potentially high corrosion risk (172).

Concrete deterioration or corrosion is also a serious problem in some soils. High rates of concrete deterioration generally occur under low pH (<5) and high concentrations of water soluble sulfate (>7000 ppm). Hence problems can occur with concrete structures aboveground in industrial areas with acid rain.

GLOBAL CHANGE AND SOIL ECOSYSTEMS

Human activities have considerably altered the nature of resource use. As a result, the processing of raw materials, the accelerated use of energy both in material production and transportation, and rapidly rising needs of a growing human and livestock population have brought with it much environmental degradation. In addition, waste products in the forms of gases and aerosols have made their assault on the atmosphere. Environmental degradation of land resources is local, that of rivers and lakes is local and regional, but atmospheric changes are global.

In all likelihood, however, negative impacts on the environment will be exacerbated as a result of the synergism of a number of environmental factors. Forest decline, or Waldsterben, was initially attributed either to elevated ozone levels or to high ozone levels and atmospheric deposition combined rather than any single factor. But some see forest decline as a spiral, the cumulative conspiracy of predisposing, contributing, and inciting factors that results in forest decline (173) (Fig. 48). It is believed that most environmental insults work that way. And now, it appears, that the increase in atmospheric carbon dioxide concentration may also trigger responses in other variables as well.

It has been noted that soils are only part of the total ecosystem, and that some changes in soils as a result of climate change (changes in the mineral composition) might not be as important in the short term as other changes (eg, changes in the composition and quality of vegetation and water supply) (174). However, other soil parameters, such as soil organic matter content and soil moisture capacity, can potentially affect vegetation, erosion, salt content, and so on.

The most immediately visible effects of climate change will almost certainly be on plant communities, especially their composition and distribution. These changes will be mediated partly by changes in soils, but the latter will themselves undergo other changes because of plant community change. The system is dynamic and requires dynamic, interactive modelling. This topic is seen as one of the most important targets for future research (174).

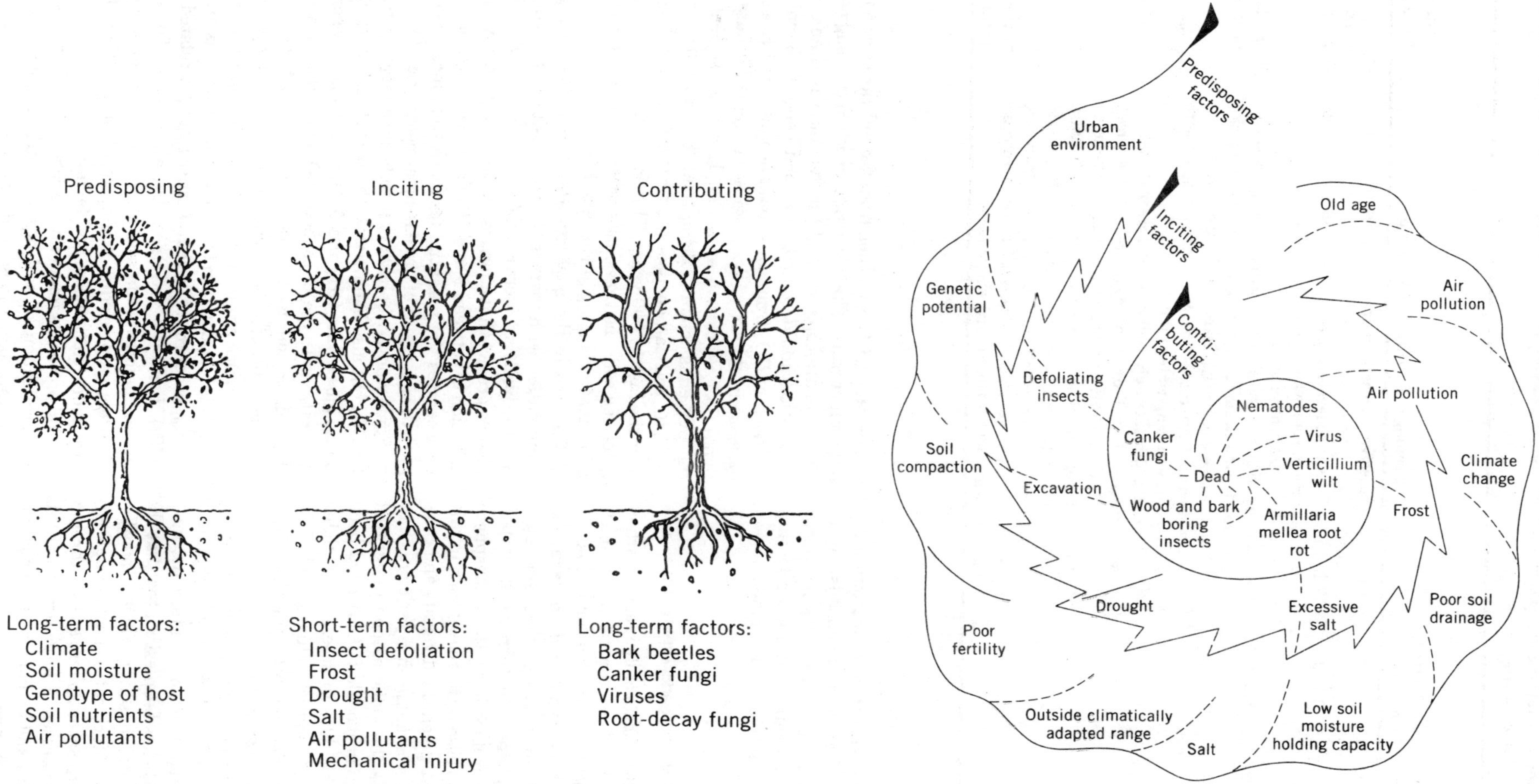

Figure 48. The many factors involved in the decline of trees; the soil factors of water, nutrients, and toxic materials have a central role (173).

Table 18. Soil Corrosivity Classes for Uncoated Steel Based on Total Acidity, Resistivity–Conductivity of the Saturation Extract, and Drainage–Texture–Aeration Relationships[a]

Soil Corositivity Class	Total Acidity, mEq/100 g	Resistivity, ohm-cm	Conductivity, mmho/cm	Drainage–Texture Relationship	Water–Air Permeability
Very low	<4	>10,000	<0.1	Somewhat excessive; excessively drained coarse-textured soils	Rapid to very rapid
Low	4–8	5000–10,000	0.1–0.2	Well drained with moderately coarse and medium-textured control section; somewhat poorly drained with coarse-textured control section	Moderate to rapid
Moderate	8–12	2000–5000	0.2–0.4	Well drained with moderately fine-textured control section; moderately well drained with medium-textured control section; somewhat poorly drained with moderately coarse-textured control section; very poorly drained with high, nonfluctuating water table	Moderately slow to slow
High	12–16	1000–2000	0.4–1.0	Well drained and moderately well drained fine-textured soils; moderately well drained, moderately fine-textured soils; somewhat poorly drained, medium and moderately fine-textured control sections; poorly drained with coarse to moderately fine-textured control sections; very poorly drained soils; water table fluctuates with foot of surface	Slow and very slow; saturated
Very high	>16	<1000	>1.0	Somewhat poorly to very poorly drained fine-textured soils; mucks, peats with a fluctuating water table	Very slow; saturated

[a] Ref. 172.

It is firmly believed that while scientific investigations must be done at an accelerated pace on whole ecosystems, it is important to equally emphasize the point that if a significant global change occurs, the effects of global change will have a lasting effect on soils and thus will be disruptive to all human support systems, including food production.

Increasing Carbon Dioxide Concentration and Temperature

Most recent work on assessing the effects of increasing carbon dioxide concentration have focussed on the carbon budget, which refers to the balance between sources and sinks of carbon dioxide in the atmosphere, expressed in terms of anthropogenic emissions and fluxes between the main reservoirs—the oceans, the atmosphere, the terrestrial carbon pool—and the buildup of carbon dioxide in the atmosphere (93). It is important, therefore, to assess the role of soils in this overall budget. A word of caution: given the diversity and complexity of the earth's individual ecosystems, it is far more challenging to quantitatively assess the globe as one integrated ecosystem. Thus global circulation models (GCM) have been generated with the best available data to assess their global warming potential (GWP). Carbon sources, fluxes, and sinks have been the topic of several studies recently, and the following discussion is based on these assessments.

There is disagreement about the quantity of carbon in the soil, but Table 19 shows carbon in the top 1 m of soil, organized by major soil orders. As expected, histosols contain the most carbon and aridisols the least. It has been estimated that the carbon content in the organic matter in soils worldwide now lies between 1400 and 1500 Pg (175). Much of this pool is relatively labile and could be returned to the atmosphere as CO_2 through decomposition and other soil processes. A large portion of this pool is held in the soils of tundra and boreal forests, which are extremely vulnerable to climate change.

The rate of CO_2 emission from the soil varies as a function of soil temperature. There is much evidence that suggests warming of the atmosphere might induce increased amounts of soil respiration and decomposition, and hence CO_2 emission. The emission of CO_2 from soils could be quite significant, as a 1% increase in the CO_2 flux from soil could contribute and additional 68.7 Pg carbon annually. This is equivalent to 14% of the CO_2 emissions resulting from the combustion of fossil fuels. The soil carbon pool (and organic matter) has increased over the last 18,000 years. This increase is greater in formerly glaciated soils, as these soils were essentially bereft of soil organic matter during glaciation (175).

Changes in the quantity and distribution of vegetation in combination with increased decomposition of soil organic matter are likely to increase CO_2 emissions to the atmosphere. Models predict that changes in the above factors will lead to the emission of between 9 and 181 Pg carbon. None of these models incorporates the potential feedback from increased vegetative growth as a result of increasing atmospheric CO_2.

Some workers believe that soils are a major source and sink for CO_2 but also believe that the potential flux of other greenhouse gases, like CH_4 and N_2O, is much less important (175). The rate and type of carbon (CO_2) emission or sequestration depends on soil use and management practices. Soils hold approximately twice as much carbon as the atmosphere and three times as much as global biota. About 80% of the nitrogen stored in soils is found within the top 10 cm of soil. This nitrogen is the principle source of N_2O and other nitrogen-related compounds (NO_x). Land use activities that alter carbon fluxes from soil include deforestation and afforestation, biomass burning, cultivation, crop residue management, application of fertilizers, rice paddy cultivation, and farming systems.

Soil processes important to carbon flux from the soil include mineralization of soil organic carbon, soil erosion, nitrogenous transformations, and anaerobiosis. The potential for a

Table 19. Organic Carbon Content in the World's Soils by Soil Orders of the U.S. Classification System[a]

Soil Order	Global Area, 10^3 km^2	Organic Carbon[b] in Upper 100 cm			Organic Carbon[b] in Upper 15 cm	
		Mg/ha	Global Pg[c]	Percent of Global	Range, %[d]	Typical, %[d]
Entisols	14,921	99	148	9	0.06–6.0	—[e]
Inceptisols	21,580	163	352	22	0.06–6.0	—
Histosols	1,745	2,045	357	23	12–57	47
Andisols	2,552	306	78	5	1.2–10	6
Vertisols	3,287	58	19	1	0.5–1.8	0.9
Aridisols	31,743	35	110	7	0.1–1.0	0.6
Mollisols	5,480	131	73	5	0.9–4.0	2.4
Spodosols	4,878	146	71	5	1.5–5.0	2.0
Alfisols	18,283	69	127	8	0.5–3.8	1.4
Ultisols	11,330	93	105	7	0.9–3.3	1.4
Oxisols	11,772	101	119	8	0.9–3.0	2.0
Miscellaneous land	7,644	24	18	1	—	—
Total	*135,215*		*1,576*	*100*		

[a] Ref. 108.
[b] Organic matter may be roughly estimated as 1.72 times this value. Organic nitrogen may also be estimated from organic carbon values by dividing by 12 for most soils. For aridisols and arid region soils this factor is somewhat lower, about 10. For histosols and other humid, wetland soils this factor is somewhat higher, about 20 (21).
[c] Petagram = 10^{15} g.
[d] Percent on mass basis (ie, g/100 g).
[e] These soils are too variable to suggest a typical value.

significant feedback mechanism exists for soils in that a 3°C increase in global mean temperature could release 50–100 Pg carbon over the following 50 years. Improved land use practices on cultivated land in addition to the restoration of degraded lands could sequester 0.1–1 Pg carbon per year (176).

Decomposition rates of soil organic matter by microbial activity are influenced "by the physicochemical conditions and the quality of organic matter." Organic matter quality is influenced by "hydrothermal conditions, chemical composition of vegetation, and the mineralogical composition and texture of the soil" (177). Soil organic matter is not a homogeneous mixture of material. The potential CO_2 flux depends on the substance and fractional composition of the organic matter. In addition to the physical makeup, different microbial populations alter the emission of CO_2 (177).

According to one study (178), soil carbon is contained in one of two pools: soil organic carbon (SOC), the relatively shallow pool (<1 m deep; 1550 Pg carbon), and the deeper soil inorganic pool (SIC) (>1 m deep; 1700 Pg carbon). The soil nitrogen pool is estimated to contain as much as 95 Tg nitrogen. During the last 140 years, increased agricultural and other land use activities have led to the emission of 90–120 Pg carbon from soils and vegetation. The annual rate of carbon loss is one order of magnitude greater for tropical soils than it is for temperate soils (2 Pg carbon/year versus 0.2 Pg carbon/year).

While there has been much interest in the CO_2 fertilization effect (ie, enhanced photosynthesis thus biological productivity), there has not been any conclusive evidence that suggests it is occurring (the lack of growth is most likely to be the result of nutrient deficiencies). Changes in climate may also lead to increased methanogensis from anaerobic conditions in soils (wetland soils) (179). There now appears to be conclusive evidence that in the past 100 years, the concentration of carbon dioxide in the atmosphere has increased mostly, if not entirely, as the result of anthropogenic activities. Models simulating net primary productivity under doubled-CO_2 conditions predict a 16.3% increase in net primary productivity globally. When looking at individual ecosystem types, there is a wide disparity in the net primary productivity response to climate change (from an increase of 50% for deserts to no increase for *selected* northern ecosystems). The lack of increase for the northern ecosystems is a result of the nitrogen-limiting conditions (180).

In the Northern Hemisphere, organic soils, such as peatlands, are expected to be significantly affected by the predicted changes in climate. Like many natural communities, the most obvious response of peatlands to increasing temperature will be to deteriorate at their southern margin with peatland-conducive environments migrating farther north. This transition may become complicated by a lack of summer precipitation, leading to increased occurrences of drought and fire at the southern boundary of the zone. The net effect will be that the northward migration of peatland-producing conditions will not keep pace with peatland degradation toward the southern edge.

The net flux of CO_2 from peatlands may be more difficult to estimate, as the increased potential for tree growth might at least partially offset CO_2 emissions as a result of enhanced decomposition rates from increasing temperatures and lowered soil moisture levels (lowered water table). However, lower water tables will decrease the emission of methane as anaerobic conditions cease to exist.

The effects that the predicted climate change could have on northern permafrost regions could be quite serious. With even a modest increase in warming, areas not characterized by permafrost will rapidly change, thus significantly altering the ecology and biogeochemistry of any affected area. Many successional communities may not be able to migrate or adapt to a rapid change in climate. Despite these potential changes, the long-term effects of a melting permafrost are rather unclear. Changes in the erodability of soils, water table level and plant successional patterns have the ability to both contribute and sequester CO_2 and CH_4 (179).

In tropical and dry temperate ecosystems, the increase in net primary productivity as shown by the model simulation under a climate change scenario is mostly a result of a CO_2 fertilization effect. In contrast, the increase in productivity in northern and moist temperate zones is thought to be the result of increasing the nitrogen availability in the soil.

Water Relations

There is convincing evidence from simulation models that an increase in temperature will result in increased evaporation rates and transpiration by plants. This is particularly true in arid and semiarid regions, where water is already in short supply. For example, when simulation models of the effect of increased carbon dioxide alone and of increased carbon dioxide in conjunction with aerosols on monsoon rainfall in China and India were simulated, the differences in the two scenarios was staggering: a 7–14% drop in the average rainfall of 88 cm over the 120-day period was noted (181). That was considered to be a serious threat to agriculture in China and India.

Many climate change phenomena occur at relatively large spatial scales over long periods of time. Much research has focused on these types of changes (eg, global circulation models), yet there has been little effort to examine change that occurs on a short time scale over a relatively small area (ie, erosion). When looking at the potential impacts that climate change may have on erosion, the following factors must be addressed:

> *Precipitation intensity; precipitation pattern, or frequency distribution of precipitation events over the year; changes in the level and frequency of extreme precipitation events; wind velocity during a precipitation event; frequency of frost-thaw cycles in the soil surface, which affect shear soil strength; and soil water content at the start of an event, determining the amount of surface runoff (182).*

Overall Effects on Productivity

When one typically thinks of global change, there is a tendency to consider only the large-scale processes and cycles (183). However, soil scientists must scale down the information from global circulation models to create datasets that portray daily and seasonal weather patterns. As noted, global change effects on soils will be on a localized level (183). That is to say, the effect on soils resides at a local level, but it is the frequency of the local effects that give changes in soil global significance. Vegetation and climate are important factors in soil formation that should not be examined separately, as changes in these factors will influence soils both directly and indirectly. Historically, one view holds, soils have displayed an enormous amount of variability during periods of climatic change. As these periods of change were slower than the current projected change, it is thought that soils will remain out of equilibrium throughout the period of change. Most important will be changes in inorganic mineral content, erosion, element cycles, soil organic matter, and emissions of greenhouse gases.

Ecosystems that are important in the overall storage of carbon and may be significantly affected by climate change include tundra, forests, prariries, wetlands, rice paddies, and oceans. Significant processes and practices that contribute to the overall carbon flux from soils include soil tillage, biomass burning, and the depletion of soil fertility.

An intensive look at 19 GCM and other models showing specific effects on agricultural production revealed many effects on individual crop species and large shifts in what are now productive regions (184). One model suggested that for every 1°C, the U.S. corn belt would shift 144 km to the north and 100 km to the east. This amounts to about a 175 km shift toward the northeast. Another model reported a decrease in winter wheat, spring wheat, and soybeans up to about 10% in the United States (175). From all evidence available, it appears that the only beneficiaries of such a warming scenario will be the northern regions—Canada, Russia, and those regions adjacent to their boundaries. This last conclusion seems also to be held by the recent report of the Intergovernmental Panel on Climate Change (185).

Agricultural systems that contribute to the flux of carbon to the atmosphere include "mechanized methods of deforestation, plow-till farming, continuous cropping on marginal lands and ecologically-sensitive ecoregions, low-input and resource-based shifting agriculture and related bush fallow systems, substance farming leading to fertility depletion and soil degradation, and overstocking and overgrazing" (178). Uncertainties and barriers to research include unstandardized methods, data quality and reliability, and missing or incomplete data.

THE CHALLENGES AHEAD

On the research front, exciting times lie ahead. The furtherance of knowledge both in soil science and other basic sciences, particularly enhanced by the revolutionary computational abilities unavailable just a few years ago, promises to advance the frontiers of soil ecosystems. An example is the wide recognition of such concepts as chaos theory—more appropriately, the science of complexity, which emphasizes that complex systems evolve in random patterns that are difficult to predict (186,187). The use and application of geographic information systems is another area of promise, though somewhat indirectly, for soil properties must be inferred from landscape attributes, in particular those of vegetation. The use of appropriate genetic stock that emphasizes more fitting of the plant to the soil rather than the current approach of fitting soils to plants is an area of needed research.

But from an applied and human perspective, there are some immediate serious concerns. While our knowledge base on global change and related areas is limited, there is sufficient information at smaller scales to predict the likely outcomes of contemporary damage to and degradation of soil ecosystems. In many cases, both theoretical and practical prescriptions have been suggested (188). The larger question is, How can those suggestions be conveyed to convince the decision makers and policymakers of the acceptance, legitimacy, and urgency of environmental issues in real terms so that prudent actions can be put in place?

But the excitement of discovery remains. The soil ecosystem can be compared to "a merry-go-round with all its joys, expectations, worries, and surprises of the riders and spectators, one finds that it is not an ordinary carousel. It is rather a three-dimensional type, its path being spiral, as it turns it climbs, thereby getting a broader view of the landscape. When a turn is completed, the sight, though familiar, is vastly different from the starting point" (189).

Acknowledgments
Our warm thanks to Drs. Ralph E. J. Boerner and Frank L. Himes; to S. E. Watts and T. O. West for a review of the manuscript, and to D. Pant, M. K. Jacobs, and B. J. Hunt for excellent logistical help.

BIBLIOGRAPHY

1. N. P. Remezov and P. S. Pogrebnyak, *Forest Soil Science,* Israel Program for Scientific Translations, Jerusalem, 1969.
2. E. W. Hillgard, *Soils,* Macmillan, New York, 1911.
3. U.S. Dept. of Agriculture Soil Conservation Service, Soil Survey Staff, *Soil Classification—A Comprehensive System: 7th Approximation,* U.S. Government Printing Office, Washington, D.C., 1960.
4. L. P. Wilding, *Agron. News,* 4 (Apr. 22, 1994).
5. A. G. Tansley, *Ecology,* **16,** 284–307 (1935).
6. R. L. Lindeman, *Ecology,* **23,** 399–418 (1942).
7. E. P. Odum, *Fundamentals of Ecology,* W. B. Saunders, Philadelphia, 1959, 1971.
8. V. N. Sukachev and N. Dylis, *Fundamentals of Forest Biogeocoenology,* Oliver & Boyd, London, 1964.
9. R. F. Daubenmire, *Plant Communities,* Harper & Row, New York, 1967.
10. J. P. Kimmins, *Forest Ecology,* Macmillan, New York, 1988.
11. J. P. Kimmins, *Forest Ecology: A Foundation for Sustainable Management,* 2nd ed., Prentice-Hall, Englewood Cliffs, N.J., 1996.
12. F. B. Golley, *A History of the Ecosystem Concept in Ecology,* Yale University Press, New Haven, Conn., 1993.
13. G. E. Likens, F. H. Bormann, R. S. Pierce, J. S. Eaton, and N. M. Johnson, *Biogeochemistry of a Forested Ecosystem,* Springer-Verlag, New York, 1977.
14. J. M. Cherrett, ed., *Ecological Concepts: The Contribution of Ecology to an Understanding of the Natural World,* Blackwell Scientific, Oxford, UK, 1989.
15. R. L. Donahue, J. C. Shickluna, and L. S. Robertson, *Soils. An Introduction to Soils and Plant Growth,* Prentice-Hall, Englewood Cliffs, N.J., 1971.
16. V. E. Shelford, *J. Morphol.,* **22,** 551–618 (1911).
17. F. F. Blackman, *Ann. Bot.,* **19,** 281–295 (1905).
18. E. Rübel, *Ecology,* **16,** 336–341 (1935).
19. N. C. Brady and R. F. Weil, The Nature and Properties of Soils, 11th ed., Prentice-Hall, Englewood Cliffs, N.J., 1996.
20. M. K. Wali and V. J. Krajina, *Vegetatio,* **26,** 237–381 (1973).
21. M. Køie, *Dansk Bot. Arkiv.,* **14,** 1–164 (1951).
22. P. S. Ramakrishnan, *Shifting Agriculture and Sustainable Development,* UNESCO, Paris, 1992.
23. P. J. Kramer and J. S. Boyer, *Water Relations of Plants and Soils,* Academic Press, San Diego, Calif., 1995.
24. R. O. Slatyer, *Plant–Water Relationships,* Academic Press, New York, 1967.
25. J. R. Philip, *Ann. Rev. Plant Physiol.,* **17,** 245–268 (1966).
26. O. Loucks, *Ecol. Monogr.,* **32,** 137–166 (1962).
27. R. H. Waring and J. Major, *Ecol. Monogr.,* **34,** 167–215 (1964).
28. T. B. Mowbray and H. J. Oosting, *Ecol. Monogr.,* **38**(4), 309–344 (1968).
29. D. A. Wikum and M. K. Wali, *Ecol. Monogr.,* **44,** 441–464 (1974).
30. K. T. Killingbeck and M. K. Wali, *Oikos,* **29,** 31–60 (1978).
31. F. P. Scholander, H. T. Hammerl, E. D. Bradstreet, and E. A. Hemminsen, *Science,* **148,** 339–346 (1965).
32. R. H. Waring and B. D. Cleary, *Science,* **155,** 1248–1254 (1968).
33. R. W. Ruess and M. K. Wali, *Oecologia,* **47,** 200–203 (1980).
34. L. R. Iverson, C. T. Scott, M. Dale, and A. M. G. Prasad, paper presented at the IUFRO 20th World Congress, Tampere, Finland, Aug. 1995.
35. L. R. Iverson, M. E. Dale, C. T. Scott, and A. Prasad, *Landscape Ecol.,* **12,** 331–348 (1997).
36. D. Hillel, *Rivers of Eden,* Oxford University Press, New York, 1995.
37. M. K. Wali and P. B. Kannowski in M. K. Wali, ed., *Prairie: A Multiple View,* University of North Dakota Press, Grand Forks, 1975.
38. W. J. Mitsch and J. G. Gosselink, *Wetlands,* 2nd ed., Van Nostrand, New York, 1993.
39. H. Jenny in E. Truog, ed., *Mineral Nutrition in Plants,* University of Wisconsin Press, Madison, 1951.
40. E. Epstein, *Mineral Nutrition of Plants, Principles and Perspectives,* John Wiley & Sons Inc., New York, 1972.
41. H. G. Gauch, *Inorganic Plant Nutrition,* Dowden, Hutchinson, & Ross, East Stroudsburg, Pa., 1972.
42. J. F. Stevenson, *Cycles of Soil, Carbon, Nitrogen, Phosphorus, Sulfur, Micronutrients,* John Wiley & Sons Inc., New York, 1986.
43. H. Marschner, *Minerals Nutrition of Higher Plants,* 2nd ed., Academic Press, San Diego, Calif., 1995.
44. S. T. Trudinger, *Soil and Vegetation Systems,* 2nd ed., Clarendon Press, Oxford, UK, 1988.
45. L. Wiklander in W. Rühland, *Handbuch der Pflanzenphysiologie,* vol. 4, Springer-Verlag, Vienna, Austria, 1958.
46. S. A. Barber, *Soil Sci.,* **93,** 39–49 (1962).
47. P. H. Nye and P. B. Tinker, *Solute Movement in Soil-Root System,* University of California Press, Berkeley, 1977.
48. S. A. Barber, *Soil Nutrient Bioavailability, A Mechanistic Approach,* John Wiley & Sons Inc., New York, 1995.
49. M. E. McCulley and M. J. Canny, *Plant Soil,* **111,** 159–170 (1988).
50. M. Watt, C. G. Topp, M. E. McCulley, and M. J. Canny, *Physiol. Plant* (in press) (1997).
51. P. J. Kramer, *Water Relations of Plants,* Academic Press, New York, 1983.
52. V. J. Krajina, *Ecol. West. North Am.,* **2,** 1–147 (1969).
53. E. V. Bakuzis, in G. M. Van Dyne, ed., *The Ecosystem Concept in Natural Resource Management,* Academic Press, New York, 1969.
54. D. A. Phipps, in N. W. Lepp, ed., *Effect of Heavy Metal Pollution on Plants, Vol. 1: Effect of Trace Metals on Plant Function,* Applied Science Publishers, Englewood Cliffs, N.J., 1981.
55. J. L. Mortensen and F. L. Himes in F. E. Bear, ed., *Chemistry of the Soil,* 2nd ed., Reinhold Publishing, New York, 1964.
56. J. C. Brown, L. O. Tiffin, and R. S. Homes, *Plant Physiol.,* **35,** 878–886 (1960).
57. S. A. Wilde, *Forest Soils,* Ronald Press, New York, 1958.
58. F. E. Clark in Ref. 37.
59. D. S. Hayman, *Can. J. Bot.,* **61,** 944–963 (1983).
60. E. I. Newman and P. Riddell, *New Phytol.,* **106,** 745–751 (1987).

61. M. F. Allen, *The Ecology of Mycorrhizae,* Cambridge University Press, New York, 1991.
62. E. I. Newman and P. Riddell in Ref. 37.
63. B. G. DeMars and R. E. J. Boerner, *Flora,* **191,** 179–189 (1996).
64. D. S. Hayman in Ref. 37.
65. R. E. J. Boerner, *Mycorrhiza,* **1,** 169–174 (1992).
66. G. R. Safir, J. S. Boyer, and J. W. Gerdemann. *Science,* **172,** 581–582 (1971).
67. K. A. Vogt, D. A. Publicover, and D. J. Vogt, *Agri. Ecosys. Environ.,* **35,** 171–190 (1991).
68. D. C. Coleman and D. A. Crossley Jr., *Fundamentals of Soil Ecology,* Academic Press, San Diego, Calif., 1995.
69. D. L. Dindal, ed., *Soil Biology Guide,* John Wiley & Sons, New York, 1990.
70. R. Hartenstein, *Adv. Ecol. Res.,* **15,** 379–409 (1986).
71. C. A. Edwards, *Biology and Ecology of Earthworms,* 3rd ed., Chapman & Hall, New York, 1995.
72. A. D. Whicker and J. K. Detling, *BioScience,* **38,** 778–785 (1988).
73. S. J. McNaughton, R. W. Ruess, and S. W. Seagle, *BioScience,* **38,** 794–800 (1988).
74. H. L. Cannon, *Taxon,* **20,** 227–256 (1971).
75. J. Antonovics, A. D. Bradshaw, and R. G. Turner, *Adv. Ecol. Res.,* **7,** 1–85 (1971).
76. H. Usui, in M. K. Wali, ed., *Ecology and Coal Resource Development,* Vol. 2, Pergamon Press, New York, 1979.
77. R. Gankin and J. Major, *Ecology,* **45,** 792–808 (1964).
78. J. M. Kingsley, *Poisonous Plants of the United States and Canada,* Prentice-Hall, Englewood Cliffs, N.J., 1964.
79. A. R. Kruckeberg, *California Serpentines,* University of California Press, Berkeley, 1984.
80. G. Stebbins and J. Major, *Ecol. Monogr.,* **35,** 1–35 (1965).
81. R. H. Whittaker, *Ecology,* **35,** 275–288 (1954).
82. A. W. Johnson and J. G. Packer, *Science,* **148,** 237–239 (1965).
83. N. C. W. Beadle, *Ecology,* **47,** 992–1007 (1966).
84. R. E. Ulanowicz, *Growth and Development—Ecosystem Phenomenology,* Springer-Verlag, New York, 1986.
85. T. M. Addiscott in R. B. Bryant and R. W. Arnold, eds., *Quantitative Modelling of Soil Forming Processes, Special Publication 39,* Soil Science Society of America, Madison, Wis., 1994.
86. R. K. Schofield, *Trans. Third Int. Cong. Soil Sci.,* **2,** 37–48 (1935).
87. E. C. A. Runge, *Soil Sci.,* **115,** 183–193 (1973).
88. M. R. Hoosebeek and R. B. Bryant, *Geoderma,* **55,** 183–210 (1992).
89. T. M. Addiscott, *Eur. J. Soil Sci.,* **46,** 161–168 (1995).
90. H. J. Morowitz, *Entropy for Biologists,* Academic Press, New York, 1970.
91. J. W. Maurits La Riviere, "Threats to the World's Water" in *Readings from Scientific American Magazine,* W. H. Freeman Co., New York, 1990.
92. D. Schimel, I. G. Enting, M. Heimanm, T. M. L. Wigley, D. Raynaud, D. Alves, U. Siegenthaler and co-workers, in J. T. Houghton, L. G. Meiro Filho, J. Bruce, H. Lee, B. A. Callander, E. Haites, N. Harris and K. Maskell, eds., *Climate Change 1994: Radiative Forcing of Climate Change and an Evaluation of the IPCC IS92 Emission Scenarios,* Cambridge Univ. Press, New York, 1995, pp. 35–71.
93. R. E. Ricklefs, *Ecology,* 3rd ed., Freeman, New York, 1990.
94. R. R. Northrup, Z. Yu, R. A. Dahlgren, and K. A. Vogt, *Nature,* **377,** 277–229 (1995).
95. F. S. Chapin III, *Nature,* **377,** 199–200 (1995).
96. H. Jenny, *Factors of Soil Formation,* McGraw-Hill, New York, 1941.
97. H. Jenny, *The Soil Resource: Origin and Behavior,* Springer-Verlag, New York, 1980.
98. P. Duvigneaud and S. Denaeyer-De Smet in D. E. Reichle, ed., *Analysis of Temperate Forest Ecosystems,* Springer-Verlag, New York, 1970.
99. C. S. Holling in W. C. Clark and R. E. Munn, eds., *Sustainable Development of the Biosphere,* Cambridge University Press, New York, 1986.
100. L. R. Oldeman in D. J. Greenland and I. Szabolcs, eds., *Soil Resilience and Sustainable Land Use,* CAB International, Oxford, UK, 1994.
101. D. Tilman, D. Wedin, and J. Knops, *Nature,* **397,** 718–720 (1996).
102. J. Major, *Ecology,* **32,** 392–412 (1951).
103. J. Major in R. Knapp, ed., *Vegetation Dynamics, Handbook of Vegetation Science 8.* Junk Publishers, The Hague, the Netherlands, 1974.
104. D. H. Yaalon, in D. H. Yaalon, ed., *Paleaopedology-Origin, Nature and Dating of Paleosols,* Israel University Press, Jerusalem, 1971.
105. N. E. Smeck, E. C. A. Runge, and E. E. MacKintosh in L. P. Wilding, N. E. Smeck, and G. F. Hall, eds., *Pedogenesis and Soil Taxonomy, Vol. 1: Concepts and Interactions,* Elsvier Science, Amsterdam, The Netherlands, 1983.
106. D. W. Billings, *Q. Rev. Biol.,* **27,** 251–265 (1952).
107. P. J. Vitousek, *Factors of Soil Formation: A Fiftieth Anniversary Perspective, Special Publication 33,* Soil Science Society of America, Madison, Wis., 1994, pp. 87–97.
108. S. W. Buol, F. D. Hole, and R. J. McCracken, *Soil Genesis and Classification,* 2nd ed., Iowa State University Press, Ames, 1980.
109. R. V. Ruhe, *Catena* **2** 309–320 (1975).
110. U.S. Dept. of Agriculture, *Keys to Soil Classification,* 6th ed., Soil Conservation Service, Govt. Printing Office, Washington, D.C., 1994.
111. J. A. C. Fortescue, *Environmental Geochemistry: A Holistic Approach,* Springer-Verlag, New York, 1980.
112. Z. Naveh and A. S. Liegerman, *Landscape Ecology, Theory and Application,* Springer-Verlag, New York, 1994.
113. J. Tivy and G. O'Hare, *Human Impact on the Ecosystem,* Oliver and Boyd, Edinburgh, 1981.
114. J. F. Richards, "Land Transformation" in B. L. Turner II, W. C. Clark, R. W. Kates, J. F. Richards, J. T. Mathews and W. B. Meyer, eds., *The Earth as Transformed by Human Action,* Cambridge Univ. Press, New York, 1990, pp. 163–178.
115. World Resource Institute, *People and the Environment,* Oxford Univ. Press, New York, 1994.
116. A. B. Daugherty, *Major Uses of Land in the United States, 1992,* USDA, Agricultural Economic Report, 1995.
117. U.S. Dept. of Agriculture, Forest Service, *An Analysis of the Outdoor Recreation and Wilderness Situation in the U.S.: 1989–2004,* 1989.
118. U.S. Dept. of Agriculture, Soil Conservation Service, *Summary Report, 1992 National Resources Inventory,* U.S. Government Printing Office, Washington, D.C., 1992; L. Tweten and C. zulauf, *Ohio Chal.,* **6**(1), 3–5 (1993).
119. P. M. Vitousek, P. R. Ehrlich, A. H. Ehrlich, and P. A. Matson, *BioScience,* **36,** 368–373 (1986).
120. M. G. Wolman, *Georg. Ann.,* **49a,** 385–395 (1967).
121. U.S. Environmental Protection Agency, Office of Water, *National Water Quality Inventory: 1994 Report to Congress. Executive Summary,* U.S. EPA, Washington, D.C., 1995.

122. P. A. Sanchez and J. G. Salinas, *Adv. Agron.*, **34**, 279–406 (1981).
123. L. R. Oldeman, R. T. A. Hakkeling, and W. G. Sombroeck, *World Map of the Status of Human-Induced Soil Degradation: An Explanatory Note,* International Soil Reference and Information Centre, Wageningen, the Netherlands, 1990.
124. L. R. Oldeman, V. W. P. van Engelen, and J. H. M. Pulles in Ref. 121.
125. J. H. Ausubel, *Amer. Sci.* **84,** 166–178 (1996).
126. R. Lal in R. Lal and B. A. Stewart, eds., *Soil Degradation,* Springer-Verlag, New York, 1990.
127. D. Pimentel and co-workers, *Science,* **267,** 1117–1122 (1995).
128. S. Judson, *Am. Sci.,* **56,** 356–374 (1968).
129. P. Crosson, *Science,* **269,** 461–464 (1995).
130. World Resources Institute, *World Resources, 1994–95,* Oxford University Press, New York, 1994.
131. R. Magleby, C. Sandretto, W. Crosswhite, and C. T. Osborn, "Soil Erosion and Conservation in the United States: An Overview," *Agriculture Info. Bull. 718,* USDA, Washington, D.C., 1995.
132. H. Dregne and co-workers, "A New Assessment of the World Status of Desertification," *Desertification Control Bulletin 20,* UN Environment Programme, Food and Agricultural Organizaiton, Rome, 1992.
133. *Science,* **261,** 423 (1993).
134. L. R. Brown, H. Kane, and E. Ayers, *Vital Signs: The Trends That Are Shaping Our Future,* Norton, New York, 1993.
135. W. J. Hudson in D. Pimentel and C. W. Halls, eds., *Food and Natural Resources,* Academic Press, San Diego, Calif., 1989.
136. N. Meyers and J. Simon, *Scarcity or Abundance? A Debate on the Environment,* Norton, New York, 1994.
137. U.S. Dept. of Agriculture, Foreign Agricultural Service, "World Grain Situation and Outlook," *Circular Series F69-93,* USDA, Washington, D.C., 1993.
138. World Resource Institute, *World Resources, 1992–1993,* Oxford University Press, New York, 1992.
139. P. Pinstrup-Andersen, "World Food Trends and How They May Be Modified," paper prepared for the *CGAIR International Centers Week, Oct. 25–29, 1993,* Washington, D.C., 1993, 44 pp.
140. J. L. Jacobson, "Environmental Refugees: A Yardstick of Habitability," *Worldwatch Paper 86,* Worldwatch Institute, Washington, D.C., 1988.
141. N. Myers, *BioScience,* **43,** 752–761 (1993).
142. P. A. Oram, *Climate and Food Security,* International Rice Research Institute, Manila, Philippines, 1989.
143. P. A. Oram in Ref. 142.
144. G. Uehara, *Trans. 15th World Congress Soil Sci.* **6a** 4–19 (1994).
145. W. E. Larson, F. J. Pierce, and R. H. Dowdy, *Science* **219** 458–465 (1983).
146. R. Strohbehn and K. Alt, "An Economic Perspective of Soil Conservation Policy," *Agricultural Resources: Cropland, Water, and Conservation Situation and Outlook Report AR-8,* USDA, Washington, D.C., 1987.
147. M. D. Jennings, B. C. Miller, D. F. Bezdicek, and D. Granatstein, *J. Soil Water Conserv.,* **45,** 75–80 (1990).
148. P. R. Crosson in W. C. Clark and R. E. Munn, eds., *Sustainable Development of the Biosphere,* Cambridge University Press, New York, 1986.
149. M. K. Wali in M. K. Wali, ed., *Ecosystem Rehabilitation: Preamble to Sustainable Development,* Vol. 1, SPB Academic, The Hague, the Netherlands, 1992.
150. S. Woodley, J. Kay, G. Francis, eds., *Ecological Integrity and the Management of Ecosystems,* St. Lucie, Boca Raton, Fla., 1993.
151. C. E. Edwards, T. L. Grove, R. R. Harwood, and C. J. Pierce Colfer in C. A. Edwards, M. K. Wali, D. J. Horn, and F. Miller, eds., *Agriculture and the Environment,* Elsevier Science, Amsterdam, the Netherlands, 1993.
152. E. J. Kormondy, *Concepts of Ecology,* 4th ed., Prentice Hall, Englewood Cliffs, N. J., 1995.
153. F. C. Westin, M. Stout, Jr., D. L. Bannister, C. J. Frazee, *Soil Sci. Soc. Amer. Proc.* **37,** 606–611 (1973).
154. J. H. Vandermeer paper presented at the *Symposium on Enhancing the Future of the Land Grant System,* Apr. 3–4, 1992.
155. N. D. Vietmeyer, *Science,* **232,** 1379–1384 (1986).
156. J. P. Reganold, *Am. J. Altern. Agri.,* **3,** 144–154 (1989).
157. F. Riveros, paper presented at the *17th International Grassland Congress,* Palmerson North, New Zealand, Feb. 8–21, 1993.
158. H. A. Mooney, paper presented at the *17th International Grassland Congress Palmerson North,* New Zealand, Feb. 8–21, 1993.
159. M. Hadley, paper presented at the *17th International Grassland Congress, Palmerston North,* New Zealand, Feb. 8–21, 1993.
160. N. Myers, *Science,* **268,** 823–824 (1995).
161. D. S. Powell, J. L. Faulkner, D. R. Darr, Z. Zhu, and D. W. MacCleery, *Forest Resources of the United States, 1992,* General Technical Report RM-234, USDA, Forest Service, (1993).
162. J. Taradellas, G. Bitton, and D. Rossel, *Soil Ecotocicology,* Lewis, Boca Raton, Fla., 1997.
163. S. N. Goldstein and W. S. Moberg, Jr., *Wastewater Treatment Systems for Rural Communities,* Commission on Rural Water, Washington, D.C., 1973.
164. R. N. Yong and B. P. Warkentin, *Introduction to Soil Behavior,* Macmillan, New York, 1966.
165. R. H. Karol, *Soils and Soil Engineering,* Prentice-Hall, Englewood Cliffs, N.J., 1960.
166. G. B. Sowers and G. F. Sowers, *Introductory Soil Mechanics and Foundations,* Macmillan, New York, 1961.
167. Portland Cement Association, *PCA Soil Primer,* Portland Cement Association, Chicago, Ill., 1973.
168. E. J. Yoder in J. R. Schultz and A. B. Cleaves, eds., *Geology in Engineering,* John Wiley & Sons Inc., New York, 1955.
169. U.S. Dept. of the Interior, *Earth Manual,* 2nd ed., U.S. Government Printing Office, Washington, D.C., 1974.
170. D. E. McCormack, *Properties of Canfield and Geeburg Soils, and Their Engineering Behavior,* Ph.D. dissertation, Ohio State University, Columbus, 1973.
171. U.S. National Academy of Sciences, U.S. National Academy of Engineering, *Confronting Natural Disasters,* National Academy Press, Washington, D.C., 1987.
172. M. Romanoff, "Underground Corrosion," *National Bureau of Standards Circular 579,* U.S. Government Printing Office, Washington, D.C., 1957.
173. P. D. Manion, *Tree Disease Concepts,* Prentice-Hall, Englewood Cliffs, N.J., 1991.
174. J. P. Legros, P. J. Loveland, and M. D. A. Rounsevell in M. D. A. Rounsevell and P. J. Loveland, eds., *Soil Responses to Climate Change,* Vol. 23, Springer-Verlag, New York, 1994.
175. W. H. Schlesinger in G. M. Woodwell, ed., *Biotic Feedbacks in the Global Climatic System: Will the Warming Feed the Warming?* Oxford University Press, New York, 1995.
176. R. Lal, J. Kimble, E. Levine, C. Whitman in R. Lal, J. Kimble, E. Levine, and B. A. Stewart, eds., *Soils and Climate Change,* CRC Lewis, Boca Raton, Fla., 1995.
177. D. M. Howard and P. J. A. Howard, *Soil Biol. Biochem.,* **25,** 1537–1546 (1993).
178. R. Lal, J. Kimble, E. Levine, C. Whitman in Ref. 177.

179. J. M. Melillo, A. D. McGuire, D. W. Kicklighter, B. Moore, C. J. Vorosmarty, and A. L. Schloss, *Nature,* **363,** 234–240 (1993).
180. E. Gorham in Ref. 175.
181. G. Mudur, *Science,* **270,** 1922 (1995)
182. P. Botterweg, in Ref. 174.
183. P. B. Tinker and J. S. I. Ingram in Ref. 174.
184. B. Smit, L. Ludlow, and M. Brkaeich, *J. Environ. Qual.,* **17,** 519–527 (1988).
185. R. Houghton in *The Encyclopedia of Energy, Technology, and the Environment,* Vol. 1, John Wiley & Sons, New York, 1995.
186. J. Gleick, *Chaos: Making a New Science,* Penguin Books, New York, 1987.
187. M. M. Waldrop, *Complexity 7. The Emerging Science of the Edge of Order and Chaos,* Simon & Schuster, New York, 1992.
188. F. P. Miller and M. K. Wali, *Can. J. Soil Sci.,* **75,** 413–422 (1995).
189. H. Jenny, *Soil Sci. Soc. Am. Proc.,* **25,** 428–432 (1961).

SOIL DECONTAMINATION BY SUPERCRITICAL WATER OXIDATION

BRIAN S. KOCHER
The University of Akron
Akron, Ohio

SOIL DECONTAMINATION

The contamination of soil is a potential environmental disaster that must be addressed in order to ensure safety. Supercritical water oxidation is well suited to destroy hydrocarbons and organic materials in a totally enclosed system without harmful side products produced by other thermal processes. However, issues such as final permissible contamination levels and economics will determine if the process should be considered in the remediation of soil. Other variables affecting consideration are the extent of contamination, type (inorganic, organic, chlorinated, etc), soil type (clay, etc), and proximity to inhabitation, etc. These issues will determine the best method of addressing remediation of the site.

Soil Preparation

Supercritical water oxidation involves the *ex situ* processing of soils or sludges to either on-site or off-site plants through transportation by truck, rail car, and/or pipelines. Once the area to be remediated is determined, the contaminated soil is removed from the ground and slurried.

The soil must be made into a pumpable form of either a slurry or sludge. Factors affecting the ability of the soil to form a slurry in water are hydrophobicity, particle size distribution, and morphology. The properties of water change throughout the process as temperature and pressure approach and surpass the critical point. Although these properties change dramatically over the course of operation, a soil can be slurried in such a manner as to create a stable slurry over these ranges.

An important factor in slurrying any solid is the size of the particles. The size determines to a large extent the slurrability and the degree of stabilization of the slurry. Many different slurry types exist and they are determined by their particle size as follows: fine (<1 mm), coarse (50–150 mm), stabilized (<0.2 mm fine and <50 mm coarse), and colloidal (30–150 μm). Coarse particle slurries are only stable when dynamically stressed (stirred or pumped). Otherwise, the particles tend to settle in tanks or pipelines. Stabilized slurries use additive packages such as surfactants, electrolytes, and polymers to enhance stability. In the supercritical water oxidation process, these additives would be either destroyed (surfactants and polymers) or precipitated out as solids (electrolytes). In general, the finer (smaller) the particle the more stable the slurry. However, this can dramatically increase sizing cost.

Another factor determining the ability of the soil to form a slurry is the hydrophobicity of the soil. In contaminated soil, this is determined by the type and extent of the contamination. The more hydrophobic the soil is the less water it holds and the more likely it is to form a stable slurry. However, the hydrophobic nature of the soil is determined to a high degree by the contamination level. In contaminated soil sites, the contamination is highly variable which may make it difficult to produce a slurry with consistent rheological properties.

The morphology of the soil (size and shape) is determined by the nature of the soil (sandy loam to clay) and contamination type (polar, nonpolar, and aromatic). The morphology affects the inter-particle interactions which determine slurry concentrations that can be used through processing parameters, such as viscosity.

Remediation of Soil

Soil contamination is classified into two broad categories, organic and inorganic. Inorganic contamination consists of heavy metals (lead, mercury, etc), usually in the form of salts. Organic contamination is typically in the form of liquid or solid hydrocarbons, aromatics, and organics containing heteroatoms (chlorinated organics, nerve gas, etc).

Hydrocarbon constituents in the soil can easily be removed and destroyed by supercritical water oxidation. Supercritical water oxidation shows the ability to destroy a wide range of hydrocarbon chemical species producing only water and carbon dioxide, greatly reducing the formation of hazardous products of incomplete combustion. This indicates that supercritical water oxidation can be used to treat almost any organically contaminated soil or sludge.

Supercritical water solubilizes the hazardous organic species contained in the soil into the highly oxidative atmosphere where supercritical water oxidation is able to destroy the compounds that contaminate the soil. However, in order to be processible the soil slurry must contain levels of between 1 and 20% hydrocarbon. Fuel in the form of methane or fuel oil must be added to the system to ensure self-sustaining operation (autothermal) if the organic content is below 10% (1).

The inorganic materials are in the form of heavy metals which are a common soil contaminate. They are produced by many industrial processes and are usually disposed of improperly. By definition, a heavy metal is any cation having an atomic weight greater than 23 (or greater than sodium). By this it is gathered that a tremendous amount of heavy metals are present in the environment of which some are toxic and/or carcinogenic. The most important heavy metals are cadmium, chromium, copper, strontium, lead, and mercury. The presence of these metals in the environment can lead to long-term health problems. They are usually found in the form of salts as sulfates, phosphates, and hydroxides (2).

Typical removal methods involve the use of pH adjustment to solubilize and precipitate out the heavy metals. This is usually done in two- or three-stage processes. Ion-exchange resins are also used as a method to remove them from water. Ion-exchange resins are selective in their removal of these. Another treatment is the use of oxidation–reduction reactions to reduce toxicity and solubility (2).

In supercritical water, heavy metals are not soluble to any great degree and precipitate out of solution. Depending on conditions they may participate in oxidation reactions (cyanide) or reduction reactions (chromium, mercury, lead, etc) producing less toxic and less soluble species. Heavy metals remaining in the process effluent are removed in the cleansing step through the use of ion-exchange resins.

The Fate of Remediated Soil

The remediated soil is recovered from the system at either the salt separation section and/or after depressurization depending on the operating conditions, soil type, and particle size. Supercritical water oxidation remediates organically contaminated soil to a more environmentally acceptable level, in many cases exceeding 99.99 + %. Supercritical water removes the contaminants from the soil matrix enabling oxygen to attack the hydrocarbons converting them into carbon dioxide, water, and salts. However, supercritical water does not selectively remove only hazardous organic materials from the soil, but all soluble organic species. All organic species are removed and destroyed in this system. The process produces a sterile soil that is decontaminated to an acceptable level.

Example: Remediation of PAH-Contaminated Soil. As with other hazardous pollutants, the contamination of air, soil, water, and groundwater from PAHs is not a new problem. Pollution from polycyclic aromatic hydrocarbons (PAHs) was first documented in Germany during the early 1800s. These compounds are produced by a wide range of industrial processes involving high temperature operations. However, the main source of PAH contamination in Europe and North America is from abandoned town gas sites, or manufactured gas plants. Manufactured gas plants, which started after 1850 and peaked around 1930, utilized coal gasification for the production of a low Btu gas. This gas was used as the primary source of heating and lighting in both the industrial and private sectors. The Btu value of the coal gas was enriched by the addition of light petroleum stocks. A by-product of this process was the production of coal tar, an unwanted liquid rich in PAHs. Like most unusable by-products of the time, the liquid was disposed of on-site by dumping it into unlined pits. As the availability and use of electricity increased, the need for town gas diminished and the sites were abandoned. Over the years the exact number and location of these sites have been lost. Currently, there are an estimated 2500 contaminated town gas sites across North America.

The need to remediate town gas sites has led to the development of technologies that detoxify or destroy the contaminants instead of disposal in underground storage tanks or landfill (qv). Typically, the contamination from town gas sites has migrated over an extensive area. In some instances, both the subsurface soil and groundwater have been contaminated. PAHs have been shown to have low removal efficiencies with traditional remediation methods. This is due in part to the hydrophobic and nonvolatile nature of the compounds. Remediation of town gas sites is further complicated by the dependency on the type of soil matrix, extent of contamination, and the distribution of compounds present, making it an expensive process. Therefore, the development of treatments that can be applied on-site and/or *in situ* is crucial.

The removal efficiency of hydrocarbon contaminants from the soil is presented in Table 1 and shown in Figure 1 (3). These results are encouraging; a decontamination efficiency of greater than 99 + % was achieved at oxygen space velocities of 0.09, 0.5, and 1 cc/cc/min. The maximum removal percentage was achieved by introducing 2.5 liters (STP) of oxygen at a space velocity of 0.5 cc/cc/min which resulted in a 99.5% removal of the contaminants. The remaining contaminants were not soluble in the supercritical water–oxygen mixture and thus were not removed/destroyed by the oxidation process.

The extraction/destruction percentage of PAHs increases as the residence time is increased. This behavior is expected because the destruction of the PAHs into carbon dioxide and water allows more PAHs to be extracted until equilibrium is again reached in the extractive atmosphere.

The high solubility of hydrocarbons in supercritical water allows for almost total removal of hydrocarbon contamination from the soil matrix by removing them into the supercritical water/oxygen phase. The complete miscibility of oxygen in supercritical water allows the hydrocarbon contaminants to be brought into intimate contact. This contact allows for rapid and complete oxidation of the hydrocarbon contaminants. The reactor effluent included products such as carbon dioxide and water. The hydrocarbon contaminants are removed/destroyed in a single step, leaving clean soil.

Table 1. Removal Percentage of Hydrocarbon Contaminants from Town Gas Soil[a]

Oxygen, L	Space Velocity, 0.09 cc/cc/min	Space Velocity, 0.5 cc/cc/min	Space Velocity, 1.0 cc/cc/min
0.09	85.3		
0.18	87.3		
0.27	93.8		
0.36	93.9		
0.45	90.2		
0.50		75.4	
1.00		85.0	93.6
1.98	99.2		
2.00		89.1	90.8
2.50		99.5	99.3

[a] Ref. 3.

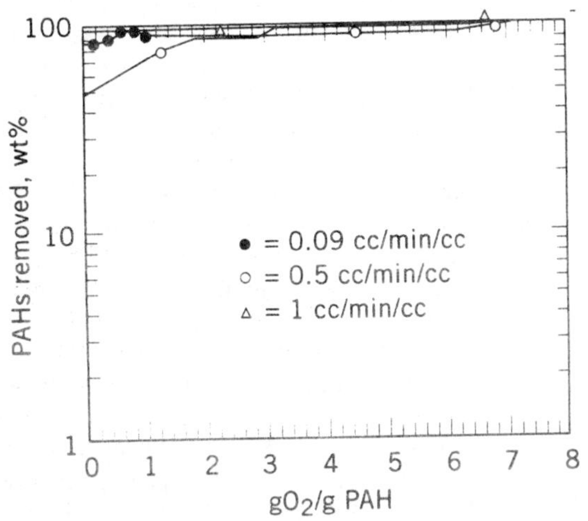

Figure 1. PAH removal/destruction from contaminated soil (3).

BIBLIOGRAPHY

1. D. Kodra and V. Balakotaiah, *Ind. Eng. Chem. Res.* **33**, 575–580 (1994).
2. R. A. Corbitt, *Standard Handbook of Environmental Engineering*, McGraw-Hill Publishing Co., New York, 1990.
3. F. O. Azzam, "Application of Supercritical Water Oxidation the Complete and Partial Oxidation of Chemical Wastes," Dissertation at the University of Akron, Ohio, 1993.

SOIL QUALITY ANALYTICAL METHODS

M. R. CARTER
Agriculture & Agri-Food Canada, Research Centre
Charlottetown, Prince Edward Island, Canada

Soil is a very complex and varied material composed of interacting mineral, organic, water, and air components. It has strong biological features. Knowledge about the chemical, physical, and biological properties of soil can only be fully obtained by recourse to analytical methods. The need for soil analytical methods is underlined, by both the scientific and general public, in the ongoing concern with soil and environmental quality protection, and the sustainability of the biosphere. Soil is used for various purposes such as a medium for plant growth, the recycling of numerous waste products, and as a construction material. Thus, methods are required that can characterize soil material for a specific use or purpose. In addition, there are a wide range of disciplines that have an interest in soil analysis, such as agriculture, forestry, earth sciences, civil engineering, and environmental science. For all these disciplines, the need to describe and characterize soil material requires the continued development of analytical and testing methods. This article describes the basic concept and approach to laboratory analysis of mineral soil material, and provides a select description of some common chemical, physical, and biological methods, with an emphasis on methods of analysis for soil quality.

SOIL QUALITY

Soil is the "excited skin" of the subaerial part of the earth's surface. As naturally formed bodies, soils are components of terrestrial ecosystems, but they are also themselves an ecosystem. The soil body provides important functions, such as: supporting plant and animal life; providing a store of biogenic energy and nutrients; partitioning the flows of energy and water; mitigating toxic elements; providing a store for greenhouse gases and mitigating their accumulation in the atmosphere; and acting as an environmental buffer. Soil forms at a relatively slow rate; therefore, soil material utilized for agriculture and other uses should be considered a nonrenewable natural resource.

Soils occur naturally in the landscape. Thus, they are a component of land which can be viewed as the natural integrated entity of soil, water, and vegetation. Soil is composed of several recognizable components, such as the mineral fraction, organic matter, soil water, and soil air. Analysis of soil material in the laboratory can provide information on soil properties (1), but it should be realized that some of these properties may not be always equivalent to the same property measured in situ, because the removal of the material will also remove the influence of adjacent soil and the ambient environment.

Defining Soil Quality

Interest in assigning quality to soil evolves around the various functions that soils perform in ecosystems (2). Soil quality is considered a key to sustainable land management and a key indicator in environmental science. Placing a value on soil in regard to a specific function, purpose, or use leads to the concept of soil quality. However, in contrast to water and air where the function can be directly related to human and animal consumption, the function placed on soil is often diverse and usually not directly linked or involved with human health. Table 1 provides a list of common soil functions.

Soil quality can only be assessed by measuring properties. However, analysis of soil properties alone cannot provide a measure of soil quality unless the properties evaluated are calibrated or related against a designated role or function of the soil. Thus, implicit in any definition of soil quality is an understanding of the stated function of the soil, or what the soil does. Soil quality can be simply defined as the fitness of a soil for a specific use. However, to accommodate the concepts of sustainability and ecosystem, and to stress the need to protect the environment, definitions have been expanded and developed, as indicated by the following:

> *Soil quality* is the capacity of a soil to function, within ecosystem and land-use boundaries, to sustain biological productivity; maintain environmental quality; and promote plant, animal, and human health (2).

Inherent versus Dynamic Soil Quality

Soil quality has two parts: an intrinsic soil part which covers a soil's *inherent* or natural capacity for a specific function (eg, crop growth), and a *dynamic* part influenced by the soil user or how the soil is managed. This has important implications for the approach to soil analysis.

> Inherent soil quality properties, such as mineralogy and particle size distribution, are mainly viewed as almost static and usually show little change over time.

Assessment of inherent soil quality can be achieved by utilizing national land resource inventories. Such databases can be analyzed in a computerized geographic information system (GIS) to develop broad regional assessments of inherent soil quality. Many present databases cover climatic information (eg, growing degree days), soil texture, and depth. Land resource inventories also often include data on soil porosity, nutrient retention, and both physical and chemical rooting conditions (eg, physical and chemical barriers to root growth).

> Dynamic soil quality properties that are responsive to soil management and manipulation are subject to change over relatively short time periods.

For example, total organic matter may change over a period of years to decades, whereas pH and labile organic matter fractions may change over a period of months to years. In comparison, microbial biomass and populations, soil respiration, nutrient mineralization rates, and macroporosity can change over a period of hours to days. Thus, maintenance or improvement of dynamic soil quality deals primarily with those attributes or indicators which are most subject to change, loss, and depletion.

Evaluating Soil Quality

> A useful framework to evaluate soil quality is based on the following sequence: functions, processes, attributes or properties, attribute indicators, and methodology (2).

Soil quality is evaluated on the basis of the function in question. Functions deal with what the soil does, or what the soil is asked to do. Each function can be characterized by specific soil processes that support the function which is being imposed on the soil. Soil quality attributes can be defined as measurable soil properties that influence the capacity of the soil to perform a specific function. Generally, attributes describe a critical soil property involved with the process. In many cases the specific attribute or property may be difficult to measure directly, so an indicator can be used to serve as an indirect, practical measure of the attribute. Indicators can represent a single attribute or may also represent a set of attributes. It is generally acknowledged that indicators have the following characteristics:

- sensitive to variations in soil management (but not overly sensitive)
- relatively low sampling error and
- easily measured with readily available methodology.

Once a property is identified for a specific soil type or situation, information is needed in regard to soil quality standards for a given set of conditions. Development of soil quality standards, especially defining the limits or critical range of the property, and outlining standard methods of analysis is as yet not well developed. In most cases several properties are needed to adequately characterize a soil process or function. This allows the establishment of minimum data sets. Such data sets are composed of a minimum number of soil properties/indicators that will provide a practical assessment of one or several soil processes of importance for a specific soil function.

SOIL SAMPLE

Laboratory analysis of soil is dependent on obtaining a suitable soil sample. However, soil properties can exhibit both temporal and spatial variability, and also vary with depth. Thus, information about soil properties usually involves a sequence of steps and procedures as follows:

- description of soil sampling site
- selection of optimum sampling time, procedure, and number of soil samples required
- storage and preparation of samples
- selection of suitable analytical method
- calibration of the method and quality control
- statistical analysis and
- interpretation of analytical results.

In many cases some information about soil sampling site characteristics is needed because soil type can vary in relation to its position both within a soil horizon, soil profile, and landscape (1). Even in the same soil type, the various sources of soil variation necessitate that an adequate sampling scheme be adopted so that valid inferences can be made about the soil population from the sample analyzed. Selection of a specific sampling site can be judgemental if information is already available about the soil or location in question, or if visible differences are evident. In most cases little is known about the soil location or site, so random or probability sampling schemes need to be utilized. To accommodate different situations, various soil sampling schemes and methods have been developed for both disturbed and undisturbed soils (1,3). The following are some common sampling procedures:

- simple random sampling: sample points are not influenced by other points
- systematic or grid sampling: sample points are at regular distances from each other (Fig. 1)
- composite sampling: sample points are grouped and the samples within each group composited to give well mixed, but fewer samples and
- spatial structure sampling: sample points at regular distances to estimate relationships between distance and soil variation (2)

These sampling procedures allow some measure of soil variation to be calculated for the measured variables (x) in the number (n) of soil samples collected. The most basic statistic is the mean ($\bar{x}$), the variance (s^2), and the standard deviation (s). They are calculated as follows:

$$\bar{x} = \Sigma x/n \quad (1)$$

$$s^2 = \Sigma(x-\bar{x})^2/n-1, \text{ and } s = \sqrt{s^2} \quad (2)$$

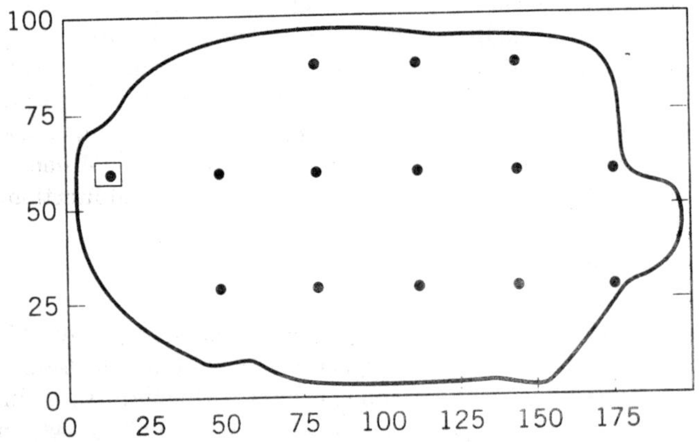

Figure 1. A systematic or grid sampling scheme showing soil sampling points separated by a fixed distance. Reprinted with permission (1). Copyright 1993, CRC Press, Inc.

Dividing s by the $\sqrt{n}$ gives the standard error of the mean. To assess the relative degree of variability among the measured variables, the coefficient of variation (CV) can be determined. The CV can vary among soil properties from a relative low value (<10%) for soil bulk density and water filled pore space, to a medium value (10% > CV < 50%) for soil particle size distribution and water content at low to high tension, and a high value (>50%) for soil water and air permeability. The calculation for CV is as follows:

$$CV = (s/\bar{x})(100) \tag{3}$$

Soil variation with depth caused by changes in the thickness and density of soil horizons often necessitates that samples are obtained on a mass per unit area basis, rather than on the basis of depth alone, for valid comparisons among sampling sites (Fig. 2).

In addition to adequate soil sampling procedures, attention must also be given to soil handling and preparation prior to analysis. These quality control considerations are a prerequisite to minimize error in all phases of soil analysis, especially those presented by both soil population and measurement variability. Soil handling concerns involve such things as the protocol for drying field moist soil samples and the degree of grinding and screening that soil samples require, prior to analysis (1,3,4). In addition to this, storage of soil samples before analysis requires attention to both state of the soil (eg, dry or moist), temperature, and the type of container used (1,5). Special precautions and conditions apply for processing soil samples for biological analyses (5). Quality control is also an important aspect of the actual laboratory analysis procedure. This includes the use of quality control procedures such as calibration of measuring instruments, use of both certified (obtained from international organizations) and internal reference samples where possible (ie, for total concentrations of elements), and statistical quality control (ie, routine evaluation of the $\bar{x}$ and the s of the analytical values over time).

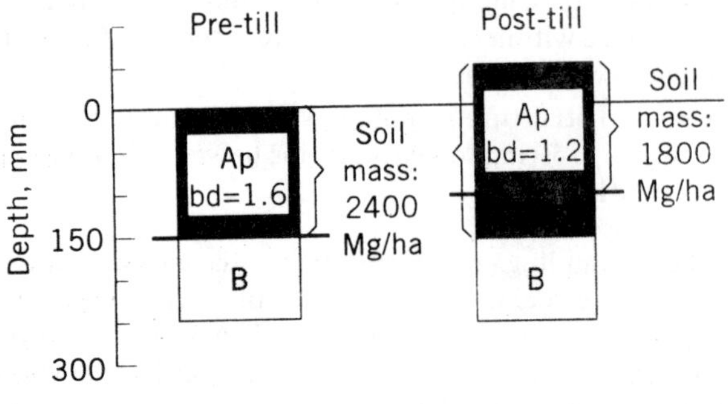

Figure 2. Changes in soil sampling depth needed to obtain a similar mass of soil by accommodating differences in soil bulk density Db. Reprinted with permission (15). Copyright 1995, Agricultural Institute of Canada.

In most cases, expressing soil analysis results on a volume basis is preferred over a concentration basis. Soil cores obtained using metal cylinders are well suited for soil volume sampling, and are utilized for many of the methods used in this article. Various core samplers are available to obtain vertical or horizontal cores at different depths. The sampling procedure used for cores is as follows (1). Cylinders used to obtain cores should exceed 7 cm diameter to obtain representative volume of soil and to minimize edge disturbance during sampling. Cores should be obtained at near-soil field capacity moisture levels to prevent shattering or compaction. Excavated cores should be trimmed flush with the end of the cylinder, capped, and wrapped in plastic to prevent evaporation and to provide protection during transport. Cores can be stored at 0 to 4°C to reduce faunal activity.

SOIL CHEMICAL METHODS

Soil is a three-phase system of solid, liquid, and gaseous components in an ever shifting equilibrium. Chemical analysis can involve characterization of all three phases. The solid phase consists of various soil minerals. Generally, most analysis is concerned with specific constituents of the solid and liquid phase that have importance for a certain soil function (eg, soil as a medium for plant growth). The interface between the solid (generally clay sized, <2 μm, secondary phyllosilicate minerals) and liquid phase is usually a net negatively charged particle surface surrounded by a diffuse layer of hydrated counter-ions merging into the soil solution or liquid phase. In many cases, the chemical composition of both the diffuse layer and the soil solution is of interest. Analysis of the soil air or atmosphere is often conducted to determine soil aeration, and physiological active gases such as ethylene, toxic gases, and radiatively active gases. For plant growth studies, the availability of soil nutrients and chemical elements that impact on plant productivity are a major concern for soil test determinations.

Mineralogical Properties

Soil mineral analysis is generally undertaken to characterize the nature and proportions of minerals in soil material (1,3). The gravel (>2 mm), sand (0.02–2.0 mm), and silt (0.002–0.02 mm)-sized fractions in soil are mainly composed of primary minerals (eg, quartz, feldspars, micas) inherited from the parent material (eg, igneous and metamorphic rocks), whereas the clay (<0.002 mm)-sized fraction is mainly composed of secondary minerals (eg, crystalline layer silicates, carbonates). The latter are formed by physical and chemical reactions during soil formation or inherited from sedimentary rocks. The aggregated nature of soil components presents a situation where crystalline minerals of varying size are associated with noncrystalline material, as well as other materials such as organic matter and salts. Thus, before mineral analysis can occur the soil material needs to be dispersed to isolate or separate the mineral fraction. Pretreatments are usually required to destroy or remove aggregating agents such as iron oxides and organic matter. Soluble salts also may need to be removed because they can prevent the needed soil dispersion. Selective chemical dissolution methods can also be used to selectively remove noncrystalline material. In some cases use of sonification can be used to disperse the soil.

Because there is usually a range of mineral particle sizes (eg, from <1 μm to 2 mm) in the dispersed and pretreated soil sample, it is necessary to first separate the soil material into major size fractions by sieving or centrifugation. Clay (<2 μm) and silt (2–50 μm)-sized mineral fractions are usually identified by X-ray diffraction (1,3), although other techniques are often used under specific situations such as IR spectrometry, magnetic methods, and thermal analysis (3). Carbonates can also be determined by X-ray diffraction (1). Sand (>50 μm)-sized mineral fractions are separated by specific gravity or magnetic susceptibility and identified on the basis of their crystallographic, optical, physical, and chemical properties.

Chemical Properties

Methods for analysis of total elemental concentrations in soil were developed from the procedures used for analysis of rocks and minerals (6). The two basic procedures used for decomposing soil material into soluble forms are as follows:

- fusion in Na_2CO_3 followed by dissolving the fused sample in HCl and
- acid digestion using hydrofluric acid (HF), or HF in combination with sulfuric (H_2SO_4) and perchloric ($HClO_4$) acid.

The derived solutions can be analyzed for a range of elements (eg, Si, Al, Ca, Mg, Fe, K, Na, Ti, and other elements) using AAS or flame emission spectrometry for K and Na (6–8). Emission spectrometry is now considered a superior method for analysis of dissolved elements (6).

Analysis of soil organic matter is common in environmental studies. Organic matter (50–58% C) is usually estimated from the concentration of soil organic C and ranges in concentration from 0.4 to 10% in temperate-region mineral soils, and is considered to be a store of organic elements that are important plant nutrients (eg, N, P, and S). These latter elements are usually found in relatively small concentrations in soil organic matter (<5% for N; <2% for P and S). In soil, the term *organic* implies that the element is mainly associated with organic matter (ie, C bonded element). For example, organic C is mainly found in the soil organic matter, whereas inorganic C is mainly found in carbonate minerals. Methods used to determine total concentrations of the previously mentioned organic elements usually involve digestion or combustion techniques to liberate the element or convert it to an inorganic form (1,4,6). Generally, the following basic approaches can be used to determine organic C.

- wet oxidation using acid dichromate followed by back titration of the remaining dichromate with ferrous ammonium sulfate
- wet oxidation as stated but collection and determination of the evolved CO_2 and
- dry oxidation in a furnace and collection and determination of the evolved CO_2 (Fig. 3).

Modification of these methods can include a separate treatment to measure or compensate for carbonates so that the total soil C content can be determined. Figure 3 shows the apparatus used in a dry oxidation technique. Oxygen is used to ensure uniform combustion at a temperature of 900°C, and KI to remove oxides of N and S. The evolved CO_2 is trapped in NaOH and determined by titration. Total soil N is also measured by wet oxidation (eg, Kjeldahl method) and dry combustion (eg, Dumus method) methods. In the former, soil samples are digested with H_2SO_4 to convert organic N to NH_4^+ followed by distillation with strong alkali (NaOH) to liberate NH_3. The Dumas method utilizes high temperatures and hot Cu to convert organic N to N_2 and is readily automated. Soil P can reside in both organic and inorganic forms, thus methodology for P usually employs the following sequence:

- acid digestion or fusion in Na_2CO_3 to determine total P
- acid/base (eg, HCl/NaOH or H_2SO_4/NaOH) extraction to measure inorganic P and
- extraction (eg, HCl/NaOH/$HClO_4$) or ignition to estimate organic P.

In most cases, organic P is measured by difference between total and inorganic P. Soil sulfur can exist in both inorganic (including several valence states) and organic forms, with the latter predominating in soil surface horizons. Methodology for determination of total S generally involves converting all S forms to a single form using a variety of wet digestion and dry combustion techniques. Alkaline digestion (NaOBr) to convert all S to sulfate, then application of a hydriodic acid reducing mixture to convert sulfate to sulfide followed by colorimetric determination of the sulfide is one recognized procedure (1).

Biochemical Properties

Soil is a dynamic medium for many organisms that transform and cycle organic materials (ie, mainly plant residues) and chemical elements. These chemical processes result in the formation of organic compounds that are important for the functioning of the soil. Many of these biochemical compounds and their related properties can be measured and characterized. Organic matter C is composed of 10–20% carbohydrates, 20% amino sugars and amino acids, 10–20% fatty acids or lipids, with the remainder as aromatic humic compounds. The latter are a complex mixture of organic compounds (9). Some organic matter fractions (eg, particulate or macroorganic matter, and light fraction) that represent fragments of recently added plant material and nonhumified material (Fig. 4) can be separated from the soil using physical methods (ie, sieving and density separation) prior to chemical analysis (1), as follows. For macroorganic matter, an air-dried 2-mm soil sample is dispersed using Na hexametaphosphate and the soil suspension passed through a 50-μm sieve. The sand plus organic matter on the sieve is dried and analyzed for organic C. For light fraction, an air-dried 2-mm soil sample is place in a centrifuge tube along with NaI (1:2, soil:NaI). After shaking,

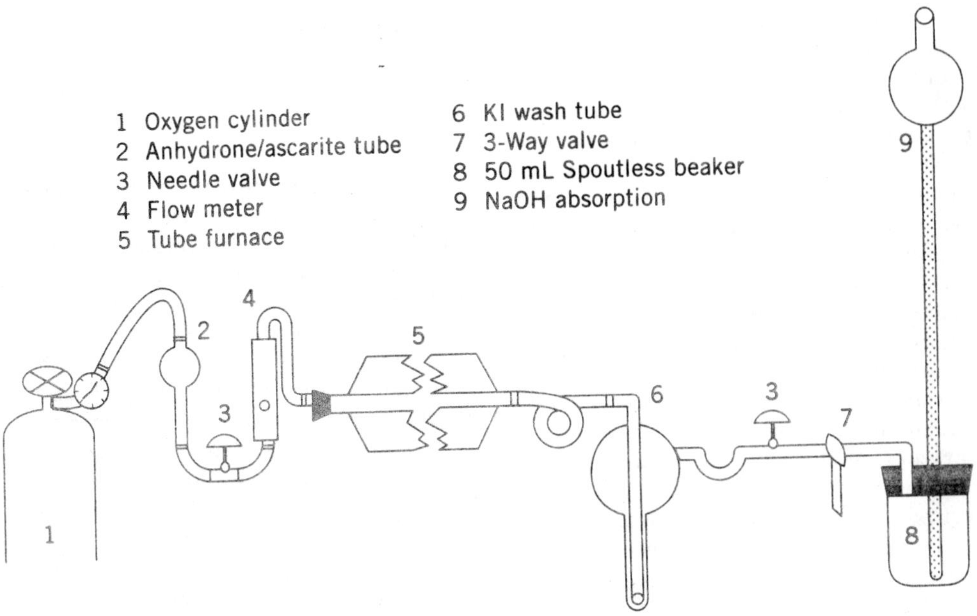

Figure 3. A tube furnace and gas absorption train for determining total or organic carbon in soil samples. Reprinted with permission (1). Copyright 1993, CRC Press, Inc.

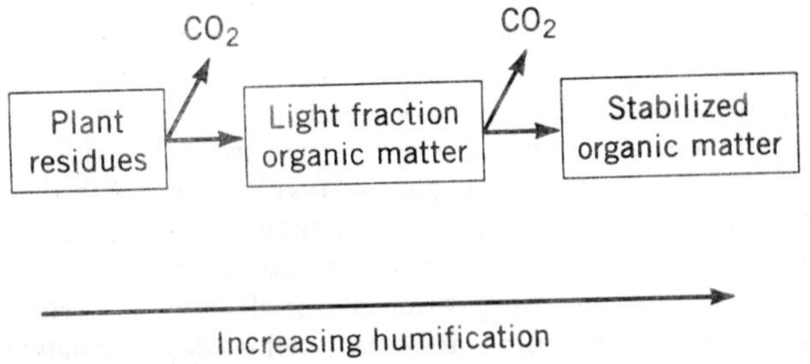

Figure 4. Simple model for decomposition of plant residues indicating the transitory pool of light fraction between residues and stable soil organic matter. Reprinted with permission (11). Copyright 1996, CRC Press, Inc.

the tubes are centrifuged and the organic matter floating on the dense NaI is removed, dried, and analyzed for organic C.

Soil carbohydrates are generally extracted by acid hydrolysis (eg, conc. H_2SO_4) to release saccharide monomers and the hydrolysate analyzed for individual monosaccharides by chromatography, or total saccharides by colorimetric methods (1,4). Because of the great variability in soil carbohydrate complexity and resistance to hydrolysis (eg, cellulose), and the need to extract carbohydrates that are implicated in soil aggregation processes, dilute acid (eg, 0.5 M H_2SO_4) is often used as the extractant. Nitrogen compounds (eg, amino sugars or acids) are also extracted by acid hydrolysis (eg, using conc. HCl) and the neutralized hydrolysate analyzed for the organic form of N using the Kjeldahl steam distillation procedure or by colorimetric methods (1,6). Soil lipids are extracted using organic solvents (eg, chloroform) which separates the lipids from their association with carbohydrates and proteins, and the extract analyzed for lipid composition by chromatographic fractionation (6). In many cases, an acid pretreatment is needed to separate lipids bound by clay minerals, or iron and aluminum salts.

The traditional approach to analyze soil humic compounds is by extraction with alkali (NaOH) followed by acidification (HCl) of the extract to obtain the humic, fulvic, and humin fractions (Fig. 5). This procedure is useful in comparing organic matter dynamics in soils from different environments or zones (1). The three fractions can be analyzed by various chemical methods to obtain their overall chemical composition. However, because humic compounds are not discrete substances, but rather heterogenous and complex in nature and cannot be defined in specific chemical terms, information on their total chemical compositions alone is some what limited in use. Most interest is directed to characterizing the major functional groups (ie, carboxyl, alcohol, phenolic hydroxyl, and carbonyl) that provide information on the structural composition of humic compounds and how they function in soil and water ecosystems. Several instrumental methods are now available to study the structure of humic substances such as ^{13}C-NMR spectroscopy, electron spin resonance spectroscopy, pyrolysis-MS, and supercritical gas extraction (9).

Diffuse Layer and Soil Solution

Fine mineral particles and organic matter can possess an electrostatic charge. Isomorphous atomic substitution in the lattices of clay minerals results in a permanent negative charge, whereas hydrolysis reactions at broken bonds at the edges of clay lattices, and dissociation of acidic functional groups in organic compounds can result in a pH-dependent negative charge. In some acidic soils, the permanent negative charge can be neutralized by aluminum polymers or iron

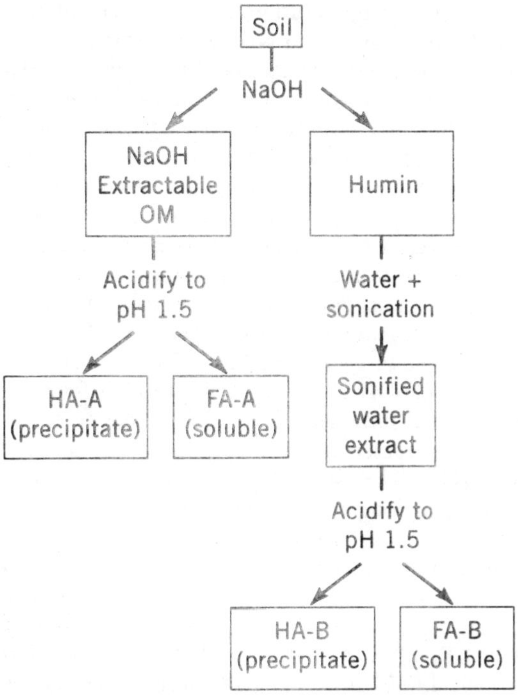

Figure 5. Analytical flow diagram for the classical alkali extraction and fractionation of soil organic matter. Reprinted with permission (1). Copyright 1993, CRC Press, Inc.

oxides. Further soil acidification can cause deprotonation of oxo- or hydroxyl-groups associated with Al and Fe, or acidic functional groups in organic compounds, resulting in a positive charge as the pH declines. This in turn results in the potential for an anion exchange capacity (AEC). The presence of negative charges results in the attraction of a balancing or exchangeable cations, which in their totality represents a soil's cation exchange capacity (CEC). For most soils, the CEC is much greater than the AEC, thus the former is more commonly measured. Methods to measure CEC, which for the reasons given is not an independent or single-valued property, mainly involve displacement of the exchangeable cations using a saturated salt solution followed by summation of the displaced cations (ie, summation method), or replacement of the adsorbed cations by a saturating salt solution containing an index cation that can be subsequently displaced and measured (ie, displacement method) (1,4,6). Generally, the saturating salt solution is buffered to the soil pH so that the CEC obtained reflects ambient conditions. The choice of replacing cation (eg, Na, NH_4, or Ba) mainly depends on the presence of specific competing ions and soluble salts. The following sequence is used for determination of exchangeable cations and CEC by the ammonium acetate method at pH 7.

A 2-mm soil sample (10.00 g) is added to a centrifuge tube along with 40 mL of 1*M* NH_4OAc (adjusted to pH 7). The tube is shaken and the suspension filtered into a 250-mL flask. The flask is made up to volume with 1*M* NH_4OAc, and a portion analyzed for Ca, Mg, K, and Na (eg, C, cmol(+) L^{-1}). The residual NH_4OAc is removed from the soil by washing with isopropanol, then the soil is leached with four 50-mL portions of 1M KCl into a 250-mL flask, and made up to volume with distilled water. The leachate is analyzed for NH_4 using an autoanalyzer. Identification and quantification of the soil exchangeable cations are routinely analyzed by AAS or emission spectrometry.

The exchangeable cations (M, cmol(+)kg^{-1}) and CEC (cmol(+)kg^{-1}) are calculated as follows:

$$M = C(0.25\ \text{L/g soil})1000 \qquad (4)$$

$$CEC = [(\text{mg N L}^{-1})(1\ \text{cmol}(+)/140\ \text{mg})] \times (0.25\ \text{L/g soil}) \times 1000 \qquad (5)$$

The amount of exchangeable acidity (ie, exchangeable H^+, Al^{3+}, Al–OH, and weak acids), which is mainly a function of soil pH and the composition (ie, percent base or Ca, Mg, K, Na saturation) of the exchangeable ions, is often measured in acid agricultural soils to assess lime requirement (1,4).

Analysis of the soil solution is of interest for many disciplines such as plant nutrition and environmental studies. Many soil chemical and biological processes are controlled by the composition of the soil solution. Direct extraction of the soil solution is the preferred method but is time consuming. Therefore, various extractant:soil ratios are employed to provide indirect estimates of the true soil solution. Examples of these methods are saturated pastes, where the soil is moistened to the saturation level and the solution extracted by suction, and fixed (eg, 1:5) soil:water extracts (1,6). Such extracts allow measurement of total soluble salts (usually by electrical conductivity), individual ions, pH (usually a 0.01 *M* $CaCl_2$ solution is used in place of water), sodicity indices (eg, sodium adsorption ratio and critical Ca ratios), and organic compounds such as soluble C and phenolic materials. Direct methods used to isolate the soil solution include column displacement (eg, using liquids such as ethanol or saturated calcium sulfate) and centrifugal filtration (relative centrifugal force of 10^3 will drain pores of 1 μm or greater). The analytical sequence for column displacement is as follows: In portions of 30 g, 6-mm sieved moist soil samples are placed into a plastic column and gently packed until 400 to 500 g of soil have been added. One centimeter of sand is added to top of soil column. About 100 mL of displacement solution (eg, saturated $CaSO_4$ containing KSCN) are added to the top of the column, and over time 5-mL aliquots of the leachate are retained until SCN^- is detected (by using $FeCl_3$). The first 2–3 mL of leachate are discarded, and the remaining SCN^- free soil solution analyzed for cations, anions, and other constituents. Displaced soil solutions can be analyzed using AAS, emission spectrometry, or ion chromatography. The activities and chemical species in the solutions can also be calculated (1).

Composition of Soil Air

The composition of the soil air reflects the various internal soil biological and chemical processes, as well as root respiration, and it is regulated by levels of soil moisture and the rate of exchange between soil and the atmosphere. In addition to the major gases in the atmosphere (N_2, O_2, and CO_2), soil air can contain a wide array of other gases such as methane (CH_4), ethylene (C_2H_4), nitrous oxide (N_2O), ammonia (NH_3), and various sulfur gases. Generally, the major impediment to obtaining satisfactory analysis of the soil air is the sampling process. In the field, various air wells can be used that allow the soil atmosphere to be withdrawn by vacuum after a time of equilibrium (1), or chambers can be inserted into, or placed over the soil (6). For loose soil samples, incubation flasks can be utilized in the laboratory. GC utilizing an ultrasonic detector is the most versatile method for the separation, identification, and determination of individual gases in soil air (6). However, other detectors provide greater sensitivity for specific gases using GC (1,8). Determination of gases dissolved in the soil solution, which is not possible using GC, can be accomplished using electrochemical (eg, platinum microelectrode to measure diffusion of O_2) and amperometric (eg, O_2 electrode) sensors (Fig. 6).

Soil Test Analysis

The utilization of the soil as a medium for plant growth generates a need for measurement of nutrient elements in soil. In contrast to total chemical analysis, the concern of soil test analysis is to characterize both nutrient element availability to the plant and assess the presence of toxic elements. Soil test methodology involves selecting chemical extraction techniques where the extracted nutrient or nutrients can be statistically correlated with plant tissue concentration, and plant response or productivity (1,10). In most cases this involves the use of mild extractants, that somewhat reflect the absorbing capability of plant roots, by removing nutrients in the soil solution and displacing easily exchangeable nutrients. Measurements for plant available nutrients are often soil and climate specific, and directed mainly to agronomic situations where a relatively high level of plant nutrients are required. In reality for many elements there is no permanent form or pool of plant available elements in soil, rather a dynamic equilibrium among solution-exchangeable-precipitated forms. Under natural ecosystems where plant available elements can be low, the internal cycling of nutrients is probably more important than the maintenance of measurable levels of nutrients.

There are a wide range of extractants available, especially for water-soluble elements. The need to estimate exchangeable nutrients (eg, K, NH_4) requires the utilization of an exchange cation in the extracting solution. For N, where plants utilize nitrogen only in the NO_3 and NH_4 form, both KCl and K_2SO_4 are common extractants (1). Ammonium acetate (NH_4OAc) is employed to extract water-soluble and readily exchangeable Ca, Mg, and K. For more strongly adsorbed K, which is only slowly exchangeable, boiling HNO_3 and HCl are often used. In some soils, sodium bicarbonate ($NaHCO_3$) is used to extract available P (P as PO_4) as well as K and N (NO_3, and NH_4). Both P and S require some form of sequential extraction to provide a measure of plant available forms, because they usually reside in both organic and various inorganic (eg, Al, Fe, and Ca) forms. A sequence of extractants for P is as follows: resin strip for exchangeable P, $NaHCO_3$ for mainly soil solution P, NaOH for Fe– and Al–P, dilute HCl for Ca–P, and concentrated HCl for residual P (Fig. 7). For S, soil solution SO_4-S can be extracted using a weak $CaCl_2$ solution, whereas SO_4–S adsorbed on soil surfaces (if soil pH < 6.5) or associated with insoluble compounds (eg, gypsum and carbonates) can be extracted with a phosphate buffer solutions.

Multielement soil extractants, such as the following Mehlich III extractant, extract many available plant nutrients. The extractant is composed of 0.2 *M* CH_3COOH, 0.25 *M* NH_4NO_3, 0.015 *M* NH_4F, 0.013 *M* HNO_3, and 0.001 *M* ethylene diamine tetraacetic acid (EDTA). P is extracted by the CH_3COOH and NH_4F. Exchangeable K, Ca, Mg, and Na are extracted by NH_4NO_3 and HNO_3. Micronutrients Cu, Zn, Mn, and Fe are extracted by NH_4 and the chelating agent EDTA.

Many nutrient elements that are only required in small amounts for plant growth can present a potential for toxicity at higher concentrations. Also, other elements such as heavy metals some of which are essential elements for plant growth can at certain concentrations adversely affect plants and human health. For these elements selective extractants are required (1,6). For example, micronutrient cations (eg, Fe, Mn, Cu, and Zn) and certain heavy metals (eg, Cd, Cr, Ni, and Pb) that can be derived from natural sources or pollutants can be extracted from the soil using chelating agents such as diethylene triamine pentaacetic acid (DTPA) or EDTA. Other extractants include dilute acids (eg, HCl, HNO_3). Micronutrient anions (eg, B, Mo, and Se) can be extracted with various extractants such as hot water and ammonium bicarbonate–DTPA.

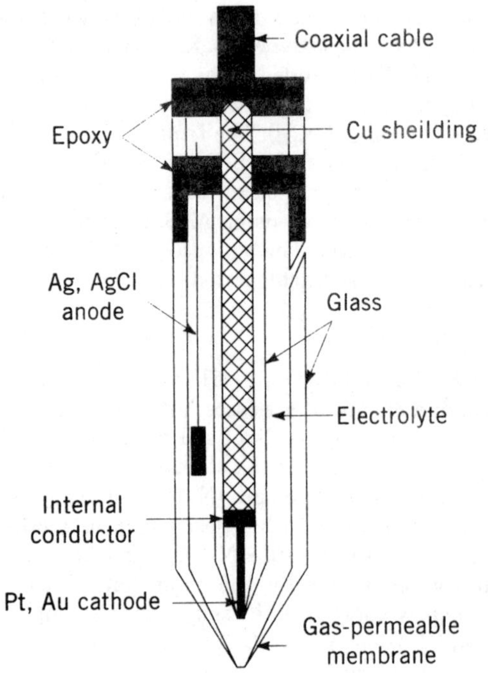

Figure 6. Basic components of a O_2 microelectrode. Reprinted with permission (1). Copyright 1993, CRC Press, Inc.

Soil Chemical Quality

Soil chemical quality is important for many soil functions such as medium for plant growth, regulation of energy, and buffering or filtering (see Table 1). Specific processes involved are storage of nutrients; acceptance, retention, and release of nutrients; and reduction in adverse chemical conditions. The main soil properties utilized to characterize these processes are given in Table 2.

Nutrient supply can be viewed from the concept of quantity/intensity, where quantity includes the total amount of nutrient available in the soil organic matter, exchange complex, and soil solution. Whereas intensity reflects the concentration of nutrient in the soil solution at any one time. Clay type and amount reflects the ability of the soil to adsorb and retain plant nutrient cations, and also buffer or regulate release of nutrients. Organic matter contains a store of plant nutrients and energy-rich compounds that on microbial decomposition provide a ready source of plant available nutrients. In addition,

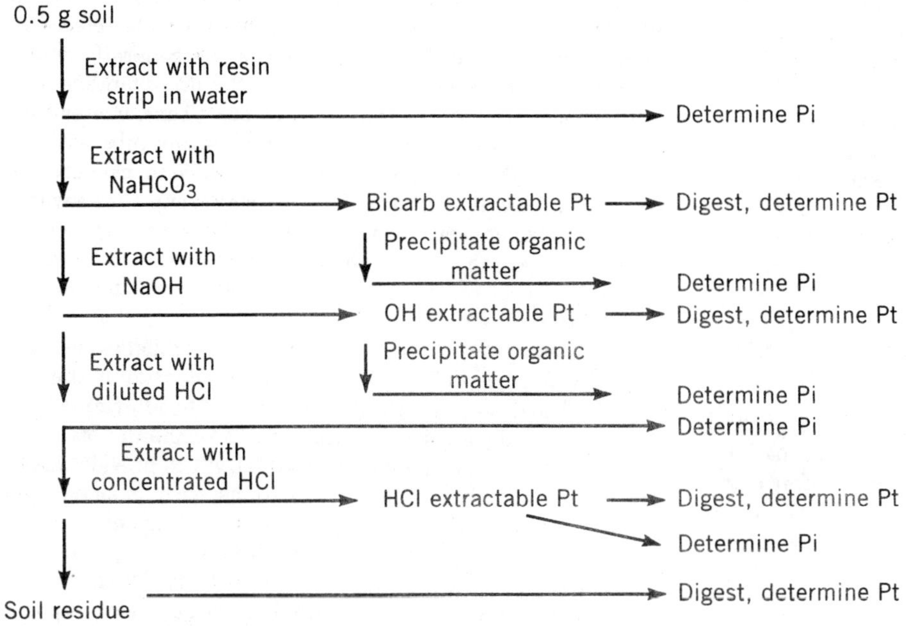

Figure 7. A flow chart indicating the sequential extraction of soil phosphorus. Reprinted with permission (1). Copyright 1993, CRC Press, Inc.

Table 1. Function Components of a Soil in Regard to Crop Production[a]

Function Component	Function Characteristics/Processes
Medium of plant growth	Suitable medium for seed germination and root growth
	Absence of adverse chemical conditions (acidity, salinity, sodicity)
	Supply balance of nutrients
	Suitable medium for microbes (nutrient cycling, decomposition)
	Promote root growth and development
Regulate water	Receive, store, and release moisture for plant use
	Adequate water retention to buffer and reduce effects of drought
	Adequate infiltration and storage capacity to reduce runoff
Regulate gases	Accept, hold, and release gases
	Permeable for adequate air movement and exchange with atmosphere
Regulate energy	Store and release (recycle) energy-rich organic matter
Buffer or filter	Accept, hold, and release nutrients
	Sequester energy compounds and/or biotoxic elements
	Detoxify substances harmful to plants

[a] Adapted from Ref. 2.

Table 2. Soil Properties for Characterizing Soil Chemical Quality

Soil Property	Contribution to Soil Chemical Quality
Clay type and content	Reflect ability of soil to adsorb and retain plant nutrients, and also buffers and regulate release of nutrients
Organic matter	Provides a store of plant nutrients; functional groups retain cations and pesticides
Cation exchange capacity	Indicates number of cations adsorbed on the diffuse layer, and ability of soil to retain and buffer changes
pH and pE	Control element solubility and adsorption at acidic functional groups
Salinity and sodicity	Excess salts influence plant growth and ion balance; sodicity influences soil structure and permeability
Trace metal elements	Indicate deficiency or toxic levels for plants

the functional groups of the humic component of soil organic matter can retain cations and remove pesticides from the soil solution. The cation exchange capacity reflects the contribution of both clay minerals and organic matter to the total number of cations that are weakly adsorbed on the diffuse layer. Both pH and pE of the soil solution can have a controlling influence on element solubility, mineral dissolution,and element adsorption at acidic functional groups. Changes in soil pH can have major impacts on the soil as a medium for plant growth. Acid soils often have nutrient deficiencies in Ca, Mg, K, and P, and toxic concentrations in Mn, Al, and trace element cations, whereas alkaline soils can present deficiencies in Zn, Fe, and P. Excess salts in the soil solution, as found in saline soils, can limit the growth of plants by decreasing water absorption and in many cases causing an imbalance of nutrients or toxicity. In some cases the presence of certain trace elements can present a phytotoxic situation. Sodicity is related to relatively high levels of exchangeable Na, which can cause swelling of soil microaggregates and clay dispersion, resulting in reduced movement of water.

SOIL PHYSICAL METHODS

Soil physics deals mainly with the properties of the solid soil phase, especially the framework it supplies for the storage of water and the movement of water and air. Physical properties of soil have many aspects that range from semipermanent (eg, clay mineral content, retention of soil water) to dynamic properties (eg, porosity, structure). The latter tend to change under the influence of climate and management, and tend to be soil and site specific. Generally, the methodology for soil water content and porosity is based on clear or accepted physical principles. In contrast, soil structure measurements are very method-dependent and subjective. However, even though this complex situation often confounds the development of quantitative methodology, it has produced a relatively small number of reference methods that utilize a common approach to soil structure assessment (11,12).

Properties of Soil Water

The water content of soil is one of the most commonly measured parameters in soil analysis. A simple approach to estimate total water content is to determine the loss of mass after oven-drying a soil sample at 105°C to a constant mass. The analysis can be expressed as the ratio of the mass (or the volume) of water and dry soil as shown in the following gravimetric method:

$$\%H_2O(\text{mass}) = [(\text{mass of wet soil, g}) - (\text{mass of dry soil, g})/(\text{mass of dry soil, g})]100 \quad (6)$$

$$\%H_2O(\text{volume} = [\%H_2O(\text{mass})] \times \text{soil bulk density}) \quad (7)$$

In many cases the ratio of the volume of water and dry soil is used, especially if the soil is a swelling soil. Soil water content can range from near zero to a variable level associated with the volume of the soil pore space. In addition to oven-drying procedures to calculate the mass of soil water, various instrumental techniques are available (1,3). Time-domain reflectometry (TDR) is based on the electrical properties of the water molecule specifically the dielectric constant of water, which is much higher than that of soil solids or air. Thus, the dielectric constant of soil is related to the water content. The TDR method is useful for both laboratory and field use. For mainly field studies, or in use with large bulk soil samples, the neutron thermalization method and the gamma ray attenuation method can be used. The former uses high-energy fast moving neutrons, which are scattered and slowed to the level of thermal energy by collisions with nuclei of similar mass and size, such as the H nuclei. Because most H in soil is in water, the intensity of thermal neutrons provides a measure of soil water content. Gamma ray attenuation uses gamma rays which are passed through the soil and absorbed by soil particles and water. Thus, under conditions where soil solids are maintained at constant density the degree of gamma ray attenuation or reduction in intensity is related to soil water content.

Water in soil is held in a certain physiochemical state that can be characterized in terms of free energy per unit mass or potential (1,3). The potential (Φ) is related to the sum of the various forces in soil that hold onto water (eg, matrix, osmotic, gravitational). Matrix potential arises from the attraction of the soil solids for water, whereas the osmotic potential relates to the influence of dissolved solutes on water activity. For methodology purposes, water potential relates to the amount of work expressed in terms of pressure (ie, megapascals) required to remove water from the soil system. Matrix potential can be measured using resistance blocks, tensiometers, and water desorption curves. Resistance blocks contain electrodes embedded in a matrix of porous material such as gypsum, fiber glass, or nylon. After equilibration in soil, the electrical conductivity of the block can be related to soil matrix potential. Resistance blocks are generally better suited to dry soil conditions and have the disadvantage of a slow response time to changes in soil water. Tensiometers are composed of a porous ceramic cup attached to a tube with an airtight cap and a side arm containing a device to measure the pressure of the water in the tube. The porous cup allows water to move freely between the soil and tube; thus, the hydrostatic pressure within the tube can be related to the soil matrix potential. Use of tensiometers is generally restricted to wet soil conditions.

Water desorption curves, or soil water retention curves, relate soil water content to the soil matrix potential as the soil dries (1). The desorption curves can be used to determine flow of water and soil pore size distribution. The combination of soil cores (ie, soil sampled in metal cylinders), which provide undisturbed and controlled boundary conditions, with tension tables and pressure vessels is the approach most commonly used to characterize desorption curves in the laboratory (Fig. 8).

A matrix potential down to −0.3 MPa is dependent on the soil macrostructure, or the specific arrangement of pores and solids; thus, the necessity for undisturbed soil cores. However, at lower potentials desorption is mainly related to the composition of the soil matrix, rather than pore-solid arrangement; therefore, disturbed soil samples (eg, subsample removed from the core) can be used. The procedure is as follows. Soil cores with the bottom end covered with a nylon cloth (mesh 53 μm) are first saturated in a water tray, then, with a slight twist, placed on the tension table to obtain a firm contact with the tension medium. Choice of the tension media, which can consist of fine sand, silica flour, glass beads, and aluminum oxide powder, is dependent on the degree of tension used. The tension table can be equipped with a hanging water or regulated vacuum to provide the required tension. After equilibrium has been obtained at a desired tension the core is removed, weighed, and returned to the tension table and hydraulic contact reestablished. This procedure can be repeated to give a stepwise decrease in matrix potential to −0.05 MPa. For soil water potentials <−0.05 MPa, the cores can be placed on a ceramic plate in a pressure vessel where the bottom side of the plate is open to the atmosphere. Hydraulic contact between the core and plate is established by first saturating the plate with water, removing the excess, then spreading a layer of hydraulic contact medium (ie, tension medium) on the cloth-covered bottom of the core. A sequence of pressures can then be applied within the vessel, which decreases the matrix potential of the water in the ceramic plate, resulting in movement of the water from the soil through the plate. At the termination of the tension table and pressure plate procedures, the soil core is oven-dried at 105°C.

Calculation of the soil volumetric water content (θ_v, % v/v) at each matrix potential is determined by subtracting the mass of moist soil core (W_p, g) measured at each matrix potential from the mass of oven-dried soil core (W_d, g), and dividing the result by the volume (V_t, mm^3) of the core as follows:

$$\theta_v = (W_\rho - W_d)/(V_t)0.01 \quad (8)$$

In agronomic and plant growth studies, the maximum level of water held by the soil after natural drainage (termed field capacity) and the level associated with permanent wilting of

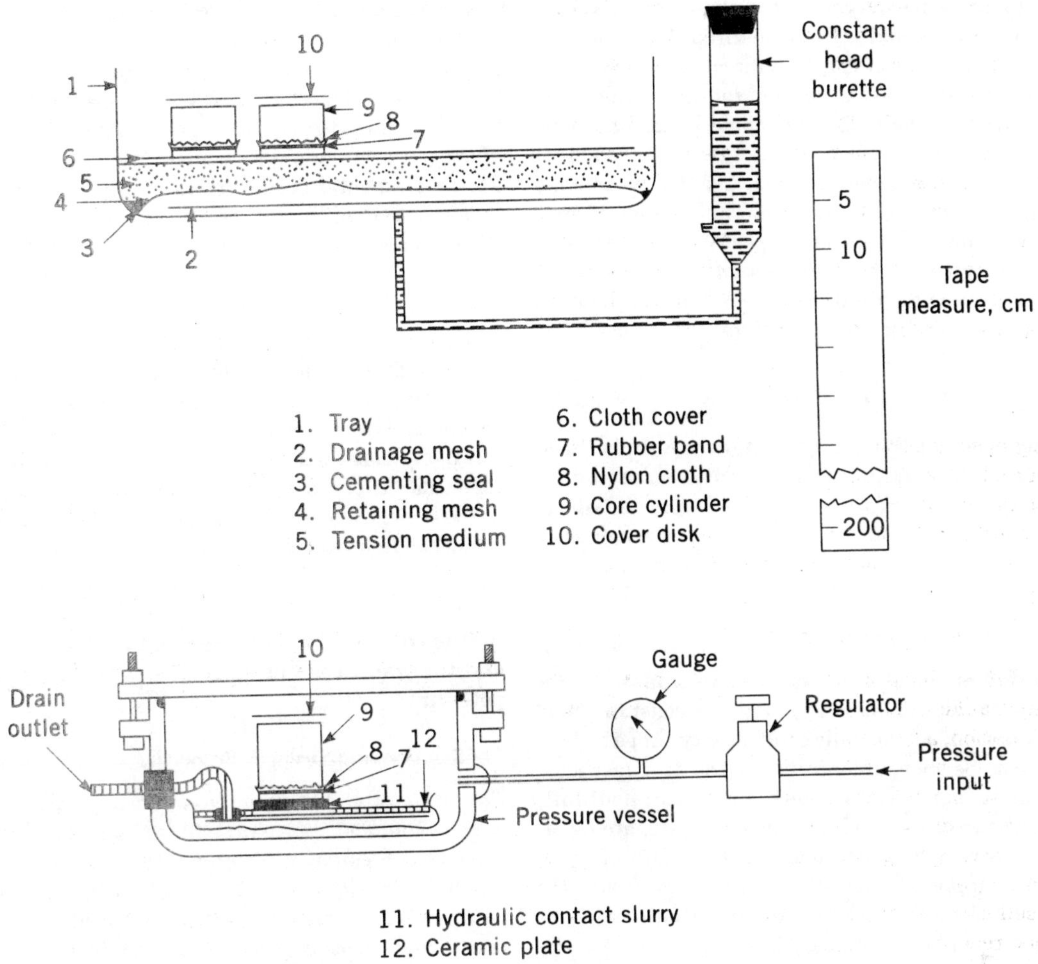

Figure 8. Schematic diagrams of the components of a tension table and a pressure vessel for determination of desorption curves on soil cores. Reprinted with permission (1). Copyright 1993, CRC Press, Inc.

plants (termed the permanent wilting point) are of interest, even though both terms are not constants (1,3). The difference between field capacity and wilting point, which are both related to the type and amount of clay and level of organic matter, provides a rough measure of the soil water available to plant roots (ie, soil water availability). They are commonly measured in the laboratory (on sieved soil or intact soil cores). The following tensions and tension media, and equilibrium times are applicable for 75-mm high cores in the determination of field capacity and wilting point (1).

For field capacity, a tension table at equilibrium pressures of −0.005 MPa, a tension media of glass beads (dia 30 μm), and an equilibrium time of 3 days are used. If the traditional equilibrium pressure of −0.033 MPa is used, a tension media of aluminum oxide (diam. 9 μm) and an equilibrium time of 10 days can be used.

For wilting point, pressure plates at equilibrium pressures of −0.2−1.5 MPa, and an equilibrium time of 12 to 25 days are used.

In addition to these terms, the water content associated with the transition of the soil from the solid state (ie, shrinkage limit) to the semisolid state (ie, plastic limit), and to the liquid state (ie, liquid limit) is of interest especially for tillage and engineering manipulation of soil. The shrinkage limit is determined by placing a wet, remoulded soil sample in a mold and determining the degree of shrinkage after drying. Both the plastic and liquid limit, which are related to soil clay and organic matter contents, are determined or based on shear tests (1). The plastic limit is the moisture content at which a rolled thread of moistened soil will begin to crumble, whereas the liquid limit is the moisture content at which a groove formed in a moistened soil sample will begin to flow and close under a specific number of standardized impacts.

Water Movement in Soil

Water flow in soil is controlled by the temperature of the water, the size distribution of the soil particles, the form and continuity of the soil pores, and the ability of soil structural aggregates to resist the unstabilizing effect of water or the dispersive effect of certain ions in the diffuse layer and soil solution. In the laboratory, water flow or the hydraulic conductivity can be characterized under both saturated and unsaturated conditions (1,3). Saturated hydraulic conductivity measures the ability of a water saturated soil to transmit water. Because of the potential interference of bubbles of entrapped air within the soil on water flow, it is important that the soil saturation procedure is conducted slowly and that the water passed through the soil is first de-aired. Two common approaches to estimate saturated hydraulic conductivity in the laboratory are the constant head method and the falling head method. Both utilize soil cores. In the constant head method, a constant hydraulic head gradient is maintained throughout the duration of the measurement period, usually by keeping a constant water level on top of the core. The volume (V, mL) of water collected from the core of a specific cross-sectional area (A, cm^2) and length (L, cm), over a specific time period (T, s) and at a set hydraulic head difference (H, cm) can be used to calculate saturated hydraulic conductivity (K_s, cm s^{-1}) as follows:

$$K_s = VL/(ATH) \tag{9}$$

In the falling head method, the constant hydraulic head gradient is replaced by a falling head of water on top of the core. The hydraulic conductivity is calculated from the initial (H_1, cm) and final (H_2, cm) hydraulic head difference using the cross-sectional area (a, cm^2) of the standpipe containing the falling head of water as follows:

$$K_s = (aL/AT)ln(H_1/H_2) \tag{10}$$

Overall, the falling head method is more applicable for soils with low hydraulic conductivity than the constant head method. In combination, a hydraulic conductivity range of 10^{-1} to 10^{-7} cm s^{-1} can be measured utilizing the two methods. Generally, in most agricultural soils saturated hydraulic conductivity ranges from 10^{-5} (clay or poorly aggregated soil) to 10^{-2} (sandy or well aggregated soil) cm s^{-1}. Both of these methods are also applicable for determining the hydraulic conductivity of soil clods or blocks of soil, in situations where soil cores are not readily obtainable (1). A schematic for the apparatus for the falling head soil core method, conducted under water, is given in Figure 9.

For water movement above the water table, which mainly occurs as unsaturated flow, information is required on soil unsaturated hydraulic conductivity. This information can be estimated from the S-shaped curve for the desorption curve and certain pore size distribution concepts, using the empirical van Genuchten equations (1). Direct laboratory measurements can be obtained using the tension infiltrometer (tension permeameter), which can provide measurements of near-saturated (ie, hydraulic conductivity at matrix potentials in the range of 0 to −0.002 MPa) hydraulic conductivity (1). Unsaturated hydraulic conductivity is very sensitive to soil texture and especially differences in the structure (ie, size, shape, and stability of soil aggregates and pores) of the soil. Thus, for laboratory measurements soil cores are required to ensure that the original soil structure is maintained. The tension infiltrometer is composed of a water reservoir tower which can be closed to the atmosphere and a bubble tower which sets the tension on a porous membrane (Fig. 10).

The infiltrometer measures both the initial (ie, sorptivity) and steady-state (ie, unsaturated hydraulic conductivity) infiltration into the soil through the porous membrane which is maintained at a constant negative water pressure (ie, tension). The procedure is as follows. The infiltrometer, set to the required potential by adjusting the water level in the bubble tower, is placed on top of the soil core. Sand is used as a contact medium between the core and the infiltrometer. At the early stage of water infiltration or sorptivity phase, the movement of water into a dry soil is controlled by soil capillary forces of the soil matrix.

Sorptivity (S, cm s$^{-1/2}$) is measured by recording the time (T, s) for each incremental drop in the water reservoir (cm) until water starts to drip from the bottom of the core. By using the above data, the cumulative infiltration (I, cm) is determined for each incremental drop in the water reservoir utilizing the internal radius of the infiltrometer reservoir tower (R, cm) and the soil core (r, cm) as follows:

$$I = R^2/r^2 \tag{11}$$

Each value for I is plotted against $T^{1/2}$ and the slope of the straight line portion of the graph is equal to S. Continuing this procedure will achieve steady-state infiltration, where set incremental drops in the water reservoir (H, cm) occur at relatively constant time (T, s) intervals. This allows the determination of unsaturated conductivity (K, cm s^{-1}) for the specific negative soil water potential used in the procedure, using the following calculation:

$$K = H(R^2)/T(r^2) \tag{12}$$

The soil core can be oven-dried to obtain soil bulk density, and the water content associated with the specific soil water potential.

Compaction and Strength Properties

The state of soil compactness can be evaluated by soil bulk density and some measure of maximum compaction. The core method is a common technique to estimate both soil dry and wet bulk density (1). Soil cores of known volume (V, cm^3) and weight ($W1$, g) are pressed into a moist soil (to prevent soil shattering), carefully removed, and the soil trimmed to the edge of the core, and the core sealed in a plastic bag. The core is weighed ($W2$, g), oven dried at 105°C, and then weighed again ($W3$, g). The bulk density (Mg m^{-3}) expressed as either dry bulk density (D_b) or wet bulk density (D_{bw}) is calculated as follows:

$$D_b = (W3-W1)/V \tag{13}$$

$$D_{bw} = (W2-W1)/V \tag{14}$$

For partially filled cores, the proportion of the core not filled with soil can be estimated by filling the soil core with very fine glass beads, and comparing the weight or volume of beads utilized to that used for a full core (1). As an alternative to cores, the bulk density of clods or large aggregates can be measured by coating the clod with a water-repellent substance and weighing the clod in air and then in a liquid of known density (3). This procedure can only be used for soils that produce stable and relatively large clods. Bulk density varies according to soil particle size distribution and organic matter level; therefore, it is not possible to compare values across soil types. To address this situation, the bulk density can be expressed relative to some measure of maximum soil bulk density (2). The latter can be established for a specific soil by applying a standard amount of compaction to a soil core. The ratio of the resulting bulk density (ie, maximum bulk density, D_{bm}) and the actual bulk density gives a relative measure of soil compaction as follows:

$$\text{Relative compaction}(\%) = (D_{bm}/D_b)100 \tag{15}$$

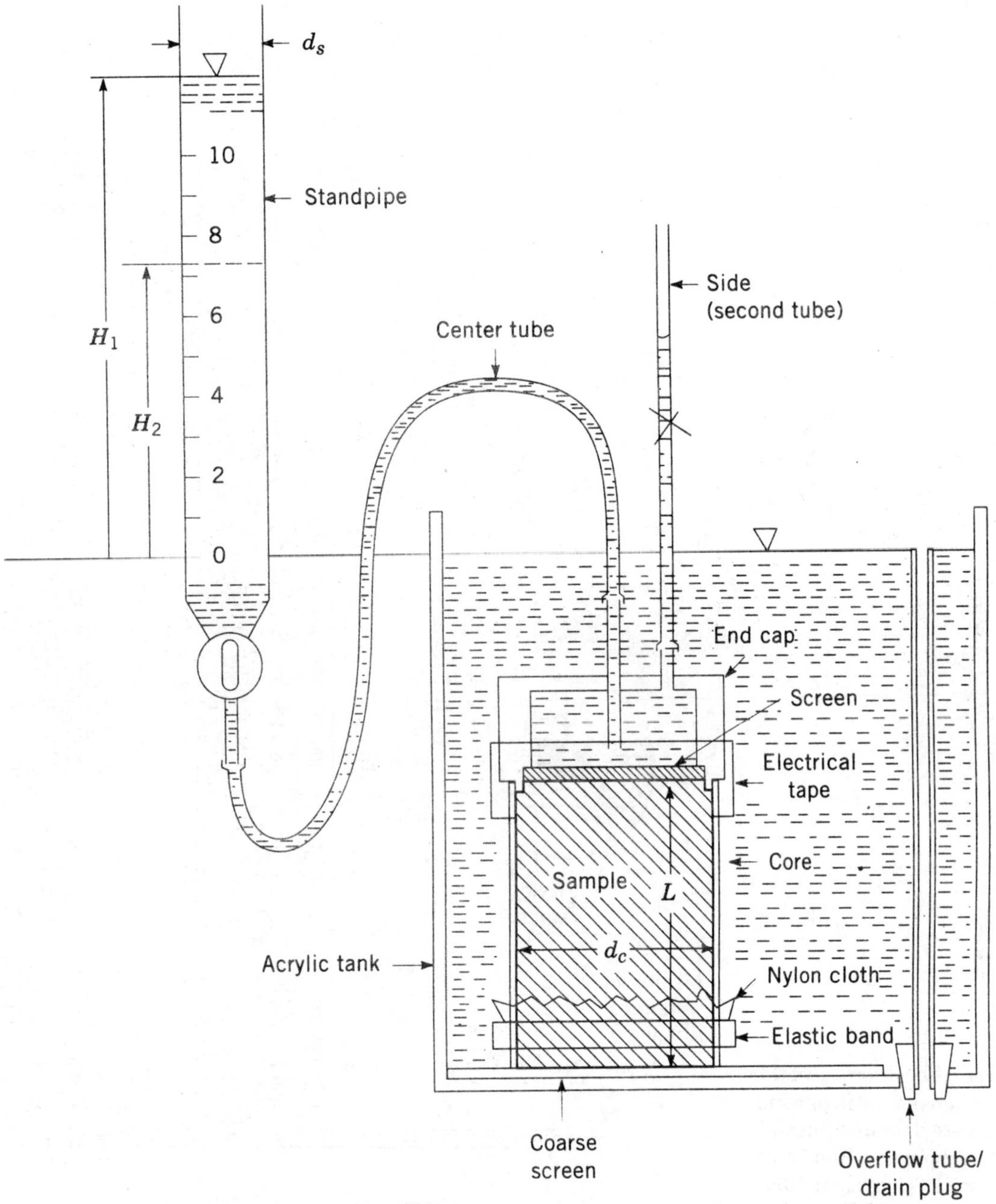

Figure 9. Schematic diagram of the components of a falling head soil core method for hydraulic conductivity. Reprinted with permission (1). Copyright 1993, CRC Press, Inc.

Soil strength can be characterized using penetration resistance and shear strength (2). Penetrability is concerned with the ease or resistance that occurs as a metal push-rod (ie, penetrometer) is be pushed into a soil (3,12,13). The main type of penetrometer used in field studies is the cone penetrometer, which measures the resistance to penetration developed by a cone attached to a push-rod as it is pushed into the soil. Penetrometer factors that influence the resistance to penetration are cone angle, diameter, roughness, and the rate of penetration. In comparison, soil factors that affect resistance to penetration are water content and potential, strength, and bulk density. For laboratory measurement on soil cores (ie, at least 20 × cone diameter), small diameter (eg, <3 mm) cones have been developed that can be utilized under relatively constant conditions, such as specific rate of penetration (eg, 0.05 cm/min) and at a known water content and bulk density level (1,3,12). In many cases, penetration resistance expressed in MPa can be related to limiting soil strength for plant root growth.

Shear strength reflects the cohesive and frictional properties of soil that are related to the forces that hold soil particles together and the roughness and shape of the particles (2). Measurement of shear strength can be accomplished using a shear vane, which consists of several metal plates or vanes on a

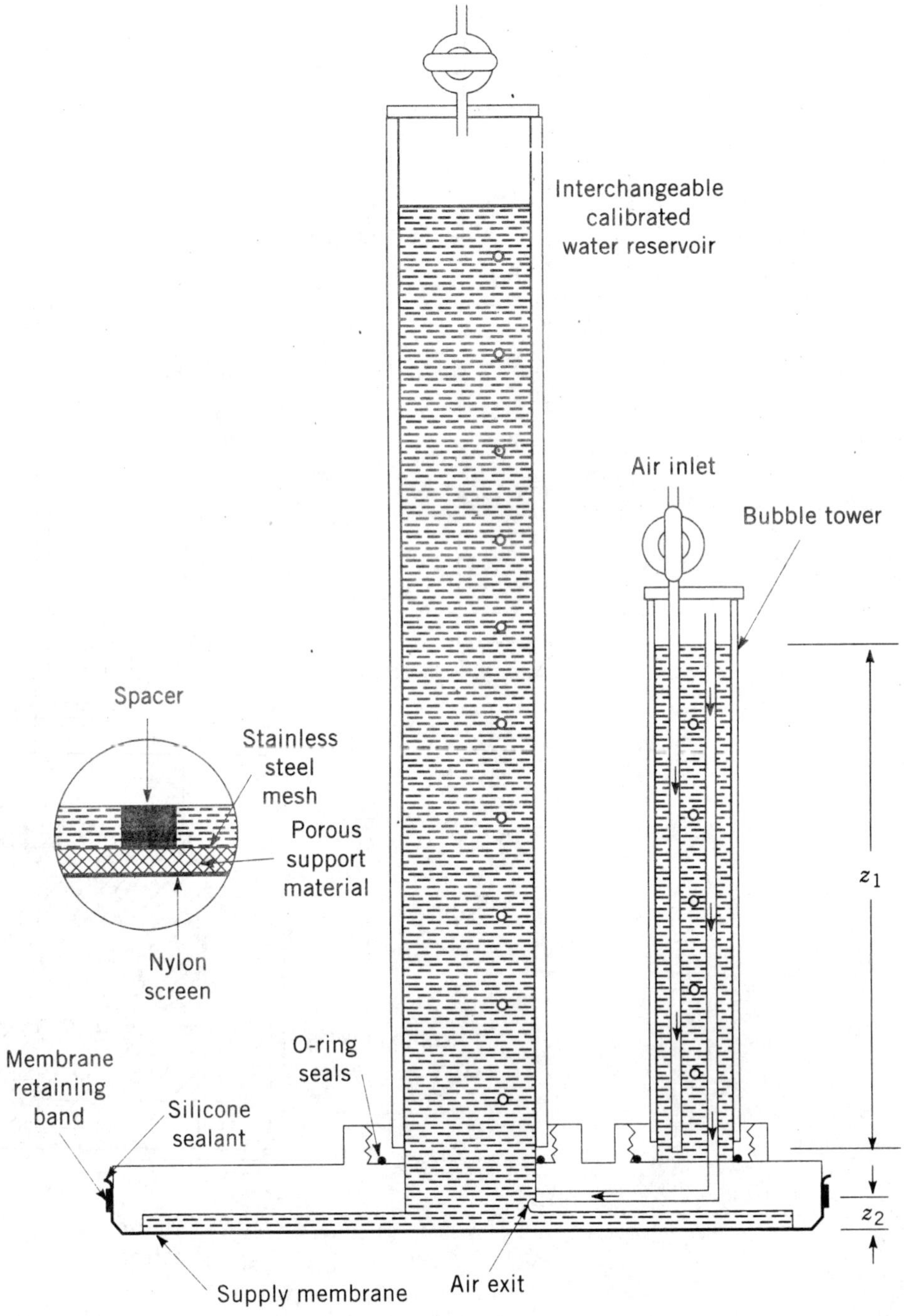

Figure 10. Schematic diagram of the components of a tension infiltrometer (disk permeameter) for unsaturated hydraulic conductivity. Reprinted with permission (1). Copyright 1993, CRC Press, Inc.

ring. The vanes on the shear vane are pushed into the soil and the shear vane rotated, and the rotational force required to rotate the vanes measured (ie, MPa). As with all strength apparatus, the specific dimensions of the vanes are recorded and the soil properties. Small shear vanes are available for use on soil cores.

Porosity and Aeration Properties

Soil porosity parameters reflect the character of soil as a three-phase system of solids, liquids, and gases. Porosity measurements are useful indicators of the soil physical condition, especially the distribution of the total pore space into different categories of relatively large diameter pores that influence the movement of water and air. For laboratory measurements of soil porosity, the use of soil cores provide relatively undisturbed soil of a known volume and weight.

Total soil porosity (S_t) is a measure of the volume percentage pore space and can be estimated from soil dry bulk density (D_b) and soil particle density (D_p) as follows:

$$S_t = 1-(D_b/D_p) \tag{16}$$

The D_b parameter can be considered as the ratio of the mass of oven-dried soil solids to the bulk volume of the solids plus pore space, whereas D_p is often assumed to approximate the

D_p of quartz (ie, 2.65 Mg m^{-3}). Greater accuracy for D_p can be obtained by the use of a pycnometer (3). In moist soils there is usually a need to further characterize the total pore space into air-filled porosity and water-filled porosity, and to obtain a measure of pore-size distribution (1). Air-filled porosity (*Fa*) is the fraction of the soil bulk volume occupied by air at a certain moisture content, or preferably soil matrix potential. It can be measured as follows.

A saturated soil core (weight = W_s) of a known volume (V_t) is placed on a tension table at a specific water potential. The weight of the equilibrated core (W_p) is recorded, and the following calculation is used:

$$F_a = (W_s - W_p)/V_t \quad (17)$$

The water-filled porosity parameter (F_w), often called relative saturation, expresses the volume of water in soil relative to S_t. It can be simply calculated from S_t and F_a as follows: $F_W = S_t - F_a$. Generally, a F_w over 65–70% indicates anaerobic conditions.

Pore-size distribution can be determined from desorption curves that relate soil water content to matrix potential (1,3). By using the soil core and tension table method, as described earlier under the section on "Properties of Soil Water", saturated soil cores are placed on a tension table and a stepwise decrease in matrix potential is imposed. Equation 8 can be used to calculate the volumetric water content for any one matrix potential. To determine the volume of water removed between two potentials, the weight of the soil core at the higher potential (W_{p1}, g) is subtracted from the weight of the soil core at the lower potential (W_{p2}, g) as follows:

$$\theta_{v1} - \theta_{v2} = (W_{p1} - W_{p2})/(V_t)0.01 \quad (18)$$

On the basis of the assumptions of the capillary model, the amount of water removed as the soil is progressively drained by a decreasing matrix potential corresponds to a change in the energy status of the soil water and the pore size. Water molecules in soil pores are held by the attractive force of other water molecules and the attractive force of the soil matrix, which is related to the radius or diameter (D, m) of the soil pore, and the angle of contact between the water and soil matrix (assumed to be zero in soil). The energy per unit area, characteristic of the above forces, is expressed by the surface tension of water (γ, J m^{-2}). Opposing these attractive forces are the density of water (ρ, Mg m^{-3}), the acceleration due to gravity (g, m s^{-2}), and the mass and length (m) of the water column in the soil pore (ie, equivalent to the negative water potential, h). These various terms can be arranged to give the upper limiting diameter of pores that remain full of water at a certain tension or negative water potential, as follows:

$$D = 4\gamma(\rho g h)^{-1} \quad (19)$$

This equation can be simplified to give D (μm), at a temperature of 20°C, of the smallest pore drained at a specific potential (h, kPa) by the following: $D = 300/h$. Therefore, the volume of water removed between two water potentials can be expressed also as the volume of pores between those potentials, and the potentials used to calculate the specific pore diameters involved using equation 19. Thus, the volume of water removed between two specific pore size diameters would equal the volume of the pore space for that pore size range. It should be noted that these pore diameters are approximate values because soil pores are not constant in shape or size. Also, the method is dependent on using a decreasing matrix potential because the desorption curve does not retrace itself in the reverse direction (due to a phenomenon termed hysteresis). Further, the method presumes that the soil volume does not change and that saturation can be obtained. This can be a problem for fine-textured soils because shrinkage can influence the soil volume over time, whereas in very sandy soils low water adhesion can cause gravitational loss of water. For clay soils, the desorption curve method to estimate pore size distribution can be used at relatively high water potentials (eg, 0 to −0.001 MPa). At lower tensions the mercury intrusion method is applicable (3). Generally, the water desorption method is applicable for use at high water potentials (ie, −0.001 to −0.02 MPa). This can be extended to lower potentials by use of pressure plate extraction and pressure membrane extraction (1). In most cases, in place of the complete pore size distribution, the amount of pore space larger than a certain size is of interest. Several studies have characterized the large pore space (eg, >50 μm), often termed the macropore space, although no specific size range is universally accepted (1).

Generally, these parameters of soil porosity tend to be static because they provide little information about the movement of air or water in the soil. In many cases, both the size and continuity of pores is of interest. The latter has a main influence on air and water movement in soil. Micromorphological methods can be used to directly determine the shape and continuity of pores >100 μm (1,13). Soil cores can be sectioned and visible soil pores segregated or classified (eg, macrostructure of pores >200 μm) on the basis of size and shape to provide a semi-quantitative analyses. However, the best method of assessing pore continuity is use of a gas at low pressures. In contrast, use of water to assess pore continuity may significantly change the soil pore structure over time. Air permeability can be measured in the laboratory, using soil cores, by the constant pressure gradient method (1,3,13). In this method an air flux, obtained from a pressurized air bottle and regulated by a pressure gauge, is passed through a soil core of a certain length (L, cm) and cross-sectional area (A, cm^2) at a constant low pressure (eg, 2.5 cm water) as measured with a water manometer, and the volume of air (V, cm^3 s^{-1}) passing through the core measured with a flow meter. A low air pressure is used to prevent turbulent flow in the soil core which may change the pore size distribution as the air flows through the soil. Various apparatus can be designed to hold the core as shown in Figure 11. The air permeability coefficient (K, cm^2), termed the intrinsic permeability to air, is calculated as follows:

$$K = (V)(L/A)(7.4 \times 10^{-8}) \quad (20)$$

The constant (7.4×10^{-8}) represents air viscosity at 22°C and a constant air pressure of 2.5 cm water (0.25 kPa). Air permeability can be calculated at various soil moisture contents. For example, cores placed on a tension table to drain macropores (eg, at −0.006 MPa) can then be subject to air permeability measurements to characterize macropore continuity in the core.

A useful method to assess soil aeration is the platinum microelectrode for measurement of the oxygen diffusion rate (ODR) through the liquid phase of the soil (1,3). The microelectrode (cathode) is connected by an electrical circuit to a reference Ag–AgCl electrode (anode) which has a KCl

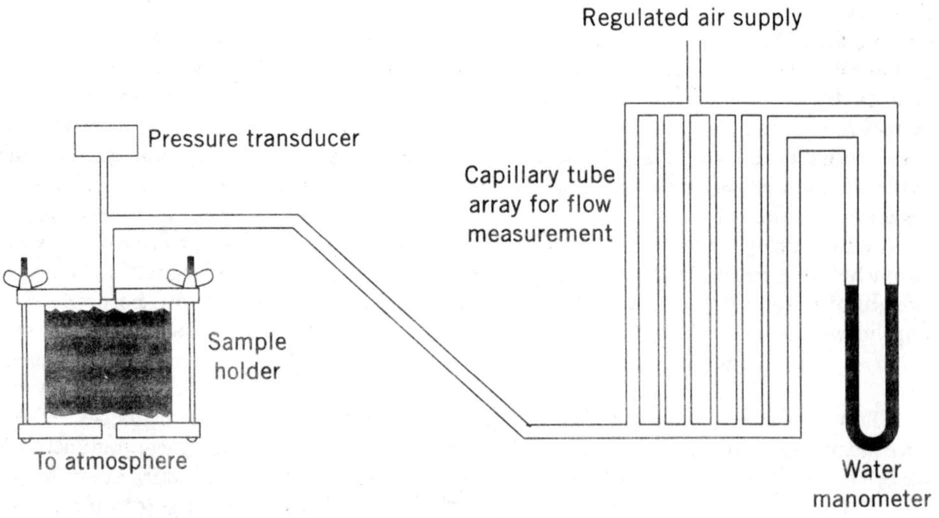

Figure 11. Schematic diagram of the components of the apparatus for measurement of air permeability. Reprinted with permission (16). Copyright 1990, Blackwell Science Ltd.

saturated agar gel salt bridge. The procedure is as follows. The platinum microelectrode (A surface area, m^2) is inserted into the soil or soil core along with the reference electrode which is connected to the soil via the salt bridge. A voltage is then applied (ie, usually −0.65 V or less) between the electrodes and the steady-state current (i, μA) measured. The latter is proportional to the flux of O_2 at the surface of the platinum microelectrode. The ODR ($\mu g\ m^{-2}\ s^{-1}$) is then calculated as follows:

$$\text{ODR} = (i/A)(8.3 \times 10^{-5}) \qquad (21)$$

The constant (8.3×10^{-5}) reflects the molecular weight of O_2 and the Faraday constant. The microelectrodes should be periodically cleaned to remove soil colloidal material or salts that can influence the potential and subsequently the output current. The method is conducive to sequential measurement of multiple microelectrodes, which can be retained in the soil at different depths. The ODR can be related to indices of soil compaction such as dry bulk density (Fig. 12).

The platinum electrode, as a combined unit with the Ag–AgCl anode and KCl electrolyte, can also be used as an electrode to measure directly O_2 concentration in both the gaseous and solution phases of the soil (see Fig. 6).

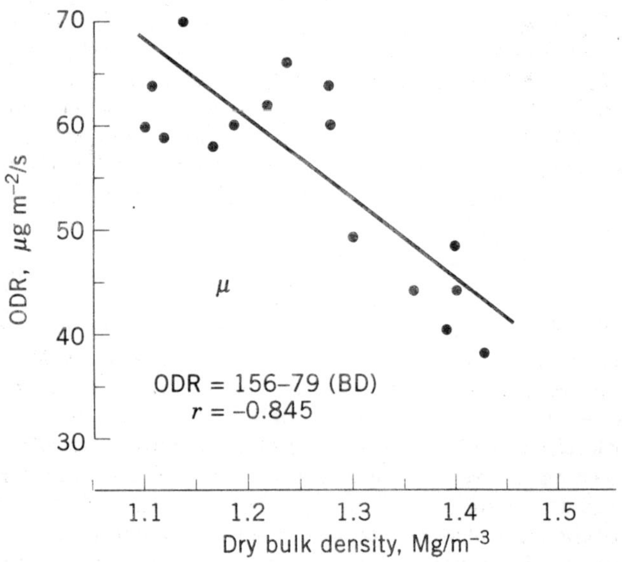

Figure 12. Relationship between soil dry bulk density and oxygen diffusion rate in a fine sandy clay loam.

Soil Structure

Soil structure has been variously defined but in the broadest sense can be described as the spatial arrangement of both soil particles and voids. In this regard, soil structure has significant effects on air–water relationships in soil (2,11,13). The structure of soil secondary particles can be placed into three broad categories as follows:

- single grained (sometimes referred to as structureless), completely unattached particles as in sands
- massive, tightly packed particles (eg, compacted clay soil) and
- aggregated, where primary particles associate to form secondary particles or aggregates.

Soil aggregates formed under natural conditions (called peds), without the aid of soil tillage or other forms of soil disturbance, are characterized by their shape, arrangement, and size. However, the shape and size of aggregates in cultivated soils can be greatly modified by agricultural practices. Thus, the emphasis is placed not only on the size and arrangement of aggregates, but also on the stability of the aggregate arrangement.

The formation of soil aggregates is dependent on both abiotic and biotic factors, the former being mainly related to soil clay content and the capacity for natural (eg, alternating shrink-swell, freeze-thaw, wet-dry) structure forming processes. The aggregation process is not totally random but occurs with some degree of order, in that for many soil types where biotic factors are the main aggregating agents small soil particles are apparently stabilized in large particles (11,14). Although soil specific, this ordering or arrangement of particles and aggregates can be viewed conceptually as an hierarchy consisting of three main orders (11):

- clay microstructures (<2 μm diameter)
- microaggregates (2–250 μm)
- macroaggregates (>250 μm)

In soils (eg, Oxisols) where inorganic compounds (ie, Fe and Al oxides) are the main aggregating agents the above hierarchal order may not be evident. In clay microstructures, clay–organic matter complexes are stabilized by humic acids and inorganic ions (eg, Ca). Microaggregates are stabilized directly by microbial materials such as polysaccharides, hyphal fragments, and bacterial cells or colonies (11). In comparison, the formation of macroaggregates and their temporary stabilization is the result of a combination of mechanisms related to plant roots and the activity of soil fungi and fauna. Generally, the study of soil structure is usually not related to soil structure itself (ie, spatial heterogeneity of the different components or properties of soil) but rather to the functionality of soil structure: that is, how an aspect of structure affects specific soil processes. A useful approach to describe soil structure in agricultural soils, which accommodates both aggregates and pores, is to characterize structure on the following basis (14):

- structural form: arrangement and size of the pore space
- structural stability: ability to retain the distribution and size of its aggregates after exposure to various stresses (eg, external forces of impact, shear, abrasion, and slaking) and
- structural resiliency: ability of a soil to recover its pore space arrangement after the removal of a specific stress (eg, compaction).

Soil structural form can be characterized by soil bulk density at the macroscopic scale. Pore-size distribution, as described earlier, provides a more detailed measure of structural form than bulk density. Pore space can be categorized on the basis of the dominant water process (ie, water transmission or storage pores). For example, the volume of pores >60 μm equivalent diameter (ie, macropores or transmission pores) can be compared to the volume of 60–0.2 μm pores (ie, available water or water storage pores). In many cases the volume of transmission pores, often termed macropores, can serve as a useful guide to soil structural form.

Soil structural stability can be addressed from both the pore or aggregate viewpoint: however, the latter usually dominates through ease of measurement. Generally, the distribution of aggregates, either in their natural field state or after being subject to disruptive forces, and the stability of a certain aggregate size or class are the main approaches to assess soil structural stability. Sieving is the main direct method to obtain aggregate distribution, whereas water is the dominant disruptive agent used to assess aggregate stability. In some cases sonification, dispersion, or shaking may precede the sieving procedure. Also, sieving can be replaced by turbidimetric, densimetric, and dispersive techniques, whereas other reagents such as alcohol or benzene are used in addition to water (11).

Wet sieving is the most commonly utilized technique to assess aggregate distribution and stability to slaking forces, and provide a measure of aggregate stability to water (1,3,11,13). There are multiple permutations of the technique due to differences in sample preparation, wetting technique, and sieving procedure. In addition, the moisture status of the aggregates at sampling and moisture status during storage prior to analysis can influence aggregate stability. A common reference method is the use of wet sieving to assess the wet aggregate stability of air-dry 1–2-mm aggregates (1,3). The following sequence is used. A specific weight ($W1$, g, oven-dried weight basis) of aggregates (either air dry or rehumidified) is placed on a single 250-μm sieve, and the sieve is placed in a wet sieving apparatus. The sieve containing the aggregates is raised and lowered, under water, over a specific distance for a set time period and rate. After wet sieving, the water-stable aggregates remaining on the sieves are washed into a flask, oven-dried, and weighed ($W2$, g), then treated with Na-hexametaphosphate to disperse the aggregates and separate any primary particles. The dispersed aggregates are returned to the sieve and washed. The primary particles remaining on the sieve are oven-dried and weighed ($W3$, g).

This correction for the presence of sand particles is usually employed so that the stability data represent true aggregates. The percent wet aggregate stability (%WAS) is calculated as follows:

$$\%\text{WAS} = 100(W2-W3)/(W1-W3) \qquad (22)$$

This method has been widely used and under standard procedures has relatively low coefficients of variation (3). In many cases, use of a single aggregate size may mask the dynamics of aggregate stability in relation to soil and crop management differences. In response to such situations use of whole soil samples, in contrast to specific aggregate sizes, can be used. For example, wet sieving can be used to separate water-stable microaggregates (<250 μm) from water-stable macroaggregates (>250 μm). The latter are generally sensitive to soil and crop management inputs (11). In other situations, the size distribution of water-stable aggregates is of interest (1) as demonstrated by the following technique. A weighed ($W1$, g, oven-dried weight basis) moist or dry soil sample is spread evenly over the top of a nest of sieves of decreasing opening diameter (eg, 5, 2, 1, 0.5, and 0.25 mm). The nest of sieves is then placed in the wet sieving apparatus until the water is level with the top sieve. The continuously immersed sieves are lowered and raised over a set height, for a set number of times, and for a specific time period. After wet sieving the stable aggregates are removed from the sieve, dried, and weighed ($W2$, g). The stable aggregates are then dispersed, as described earlier, and returned to their specific sieve size, and the weight ($W3$, g) of primary particles is determined. The proportion of water stable aggregates (WSA) in each size fraction (WSA_x) is calculated as follows:

$$\text{WSA}_x = (W2_x - W3_x)/(W1 - \text{total sand}) \qquad (23)$$

where x = number of size fractions (1,2,3, ..., n).

The mean weight diameter (MWD, mean intersieve diameter × proportion of aggregates in each size fraction including those which passes the finest sieve) is a common indice to characterize the size distribution of water stable aggregates.

At a lower scale microaggregation, the level of dispersed clay can also provide a useful indice of soil structural stability, especially for processes involved with reduced permeability and soil crusting (1,11). Turbidimetric techniques that measure changes in particle size distribution, specifically dispersed clay-or silt-sized particles, have proven useful to characterize the influence of cropping systems on soil structural stability, sodicity, and crusting on soil permeability. Generally, the methods used have the following sequence: some form of soil dispersion (eg, soil shaking), followed by measurement of suspended clay or clay plus fine silt particles using light transmission. Water stability of microaggregates can be directly related to field infiltration rates and the development of surface seal formation and subsequent runoff (11).

Sieving air-dry soil to obtain dry aggregate stability to external forces of impact, abrasion, and shear is a common technique, especially to assess a soil's susceptibility to wind erosion. Laboratory methods have utilized a rotary sieve to obtain a complete distribution of dry aggregate sizes (1,3). The technique has the following sequence: Soil samples are air-dried and then placed in the rotary sieve, which is operated until all the soil passes through the sieve. The data obtained can be used to supply simple indices of dry soil aggregate distribution, such as mean weight diameter (MWD). The percentage of soil aggregates >0.84 mm diameter provide a commonly used indice of the wind erodibility of soil.

Soil structural resiliency is characterized by assessing a soil's ability to self-generate structure through natural processes. Biological processes, such as the activity of soil fauna, and physical processes such as freezing and thawing, and wetting and drying are the main agents involved in structure recreation at the soil surface (14). These processes however do not have a similar effect on all soils. Planar pores, for example, generated by physical processes are usually unstable. In contrast, some soils form a friable mass of small (<5 mm) stable aggregates under the influence of physical processes. These self-mulching soils are mainly clay-to-clay loams that undergo sufficient swelling to allow structure formation. They can be characterized by three main properties: high clay content; tendency to slake after puddling, and stability to dispersion of the slaked material.

Soil Physical Quality

Soil physical quality is important for many soil functions such as medium for plant growth, and regulation of water and air (see Table 1). Specific processes involved are suitable medium for root growth and development, acceptance, retention, and release of water; and permeability for air and water movement.

The main attributes or soil properties utilized to characterize these processes are given in Table 3. Many of these properties reflect the ability of soil to store and transmit water, solutes and gases. Inherent soil properties, such as clay content, influence the magnitude of the soil water holding capacity. This is also augmented by the organic matter content. Dynamic soil properties, such as soil density and porosity, control the soil storage capacity for water and air. Pore-size distribution is considered to be a good indicator of the soil structural condition and has proven useful for predicting water infiltration rates, water availability to plants, soil water storage capacity, and soil aeration status. Macroporosity, air permeability, and oxygen diffusion rate provide a measure of soil macropore continuity and organization. Soil strength and aggregate stability reflect a soil's ability to resist compaction and other stresses that can lead to loss of structure.

Table 3. Soil Properties for Characterizing Soil Physical Quality

Soil Property	Contribution to Soil Physical Quality
Clay and organic matter	Controls a soils ability to retain and store water, and determines soil water availability
Hydraulic conductivity	Reflects soil permeability to water
Pore size distribution	Indicates the size and structure of the soil pore space
Macropore space and organization	Indicates level of large pores and soil aeration capacity
Oxygen diffusion rate	Reflects soil aeration status
Bulk density	Indicates compactness of soil
Soil strength	Reflects ability of soil to resist compaction
Aggregate stability	Indicates a soil's ability to resist stress due to water and abrasion, and maintain structural organization

SOIL BIOLOGICAL METHODS

Although soil is a three-phase system of solid, liquid, and gaseous components, it also contains living organisms or a biotic component. Soil biological properties are related to these organisms and the biochemical substances they produce, and the soil processes they control. The importance of microbes to soil ecology is reflected in the numerous analytical techniques available in the area of soil microbiology (1,5,6). Common methods to characterize soil biological properties emphasize techniques that can isolate and enumerate soil organisms, and monitor their activity. In recent years, molecular techniques have been developed to monitor microbial populations.

Microbial Populations

The soil microbial community is very diverse and includes various physiological types, such as heterotrophs that use organic compounds as a C and energy source, and autototrophs that obtain energy from oxidizing inorganic compounds (1,5,6). The main microbes of interest are bacteria, actinomycetes, and fungi. Their abundance in soil can range from 10^5 to 10^{10} g^{-1} dry soil. Methodology is also available for soil microorganisms such as algae and cyanobacteria, and viruses (5). Much effort has been applied to selectively enumerate and isolate specific soil microbes as evidenced by the wide range of enumeration procedures and culture mediums currently available. Generally, many of these use the plate count technique or the most-probable number (MPN) technique. The underlying principles involved can be described by the following sequence: dispersion of a soil sample in a suitable diluent, applying an aliquot of the dispersed sample to a selective growth medium, incubating the inoculated growth medium under specific conditions, and enumerating the developed microbial colonies (1). The selective growth medium is composed of components that favour or allow the growth of a desired microorganism. This procedure can be illustrated by the following technique used to isolate specific microorganisms involved in soil C transformations, which ultilizes the spread plate counting method (1). The selective growth culture media would consist of the specific C source (eg, cellulose, chitin, or starch), common nutrients for microbial growth, and agar a polysaccharide derived from marine algae that serves as a jelling agent for the culture medium.

Soil Sample. Composite soil samples are obtained according to an accepted sampling design to provide some measure of variability. Soil cores provide a soil volume measurement. The samples are maintained moist and at ambient temperature or cooler. The samples are passed through a 2-mm sieve and mixed throughly.

Dilution Preparation. Moist soil subsamples are weighed (eg, 10 g) and placed in a dilution jar (200 mL) containing 95 mL of diluent (eg, distilled water or specific salt/buffer solution), to give an approximate 10^{-1} dilution. The jar is shaken vigorously for 20 min, then a 10-mL sample is transferred to a new dilution jar containing 90 mL to give a 10^{-2} dilution. This step is repeated to provide a sequence of dilutions to 10^{-7}.

Spread Plate. Sterilized media is prepared and poured into sterile plastic petri plates and the agar allowed to solidify. An aliquot (eg, 0.1 mL) of these dilutions (starting with the highest dilution) is transferred onto the agar surface of a plate and the suspension spread using a sterile glass spreader. The plates are placed in an incubator (eg, in dark at 24–28°C) and over time individual colonies of organisms are counted on plates containing 30–300 colonies. The number of colony forming units (CFU) are determined by adjusting the weight of the soil sample to an oven-dry basis and using one of the colony counts as shown by the following calculations:

$$\%\text{Water} = [(\text{soil wet wt} - \text{soil oven-dry wt})/(\text{soil oven-dry wt})]100 \quad (24)$$

$$\text{Dry wt of soil} = \text{wet wt}[1 - (\%\text{water}/100)] \quad (25)$$

$$\text{CFU/g soil} = (\text{mean plate count})(\text{reciprocal of the dilution})/\text{soil dry wt} \quad (26)$$

In many cases, microbial population can be estimated by the MPN, which is based on a process-related attribute and the concept of dilution to extinction (1,5). In this method, the aliquots of dilutions are inoculated into a test tube containing a liquid medium that serves as a substrate for a unique characteristic associated with growth of the organism (eg, ability to turn media turbid or production of detectable product). Presence of the specific characteristic after incubation of the dilutions is evidence of microbial growth. Number of microbial cells is estimated from the reciprocal of the dilution (eg, if the highest dilution exhibiting growth is 10^{-5}, then 10^4 to 10^5 cells are present). Generally, MPN does not depend on the evidence from one single aliquot or dilution but uses the pattern of positive and negative results and subjects them to a mathematical analysis. This method is commonly used in soil studies to enumerate nodule forming rhizobia, vesicular–arbuscular mycorrhizal propagules, autotrophic nitrifying bacteria, and sulfur oxidizing bacteria.

Microbial Biomass

The magnitude of the soil microbial biomass is of interest, especially for ecological-based studies concerned with the biomass as a sink and source of nutrients. Light microscopic methods have been developed to estimate bacteria and fungal biomass by using cell length and width dimensions, and a dry weight/cellular volume conversion factor (5). However, such methodology is time consuming. Several techniques have utilized the change in the soil respiration rate, as a result of exposing the soil to a fumigant or by adding a C substrate (eg, glucose), to estimate microbial biomass (5,6). In the former, a soil sample is fumigated under specific conditions with a fumigant (eg, $CHCl_3$ vapor) to kill the living microbes. The resulting increased decomposition, in comparison to an unfumigated soil, as measured by the release of CO_2 over a specific period of incubation (ie, 10 days) is directly proportional to the size or amount of C in the biomass. The release or flush of CO_2 after fumigation is the result of germinating microbial spores utilizing the C substrate provided by microbial cells killed by the fumigant. Fumigation techniques are suitable for estimating the C, N, P, and S content of the soil microbial biomass (1,5,6). Replacing the incubation step by directly extracting the soil immediately after fumigation provides a relatively rapid chemical method to estimate the size of the biomass. The following gives the analytical sequence for determining microbial biomass C using $CHCl_3$ fumigation–extraction method.

Soil Sampling and Preparation. Soil cores are obtained according to some acceptable sampling scheme to provide a measure of soil variability. The field-moist samples can be stored at 4°C (for a week) or stored overnight at 15°C. Soil is passed through a 4–6-mm sieve prior to analysis. Three soil subsamples (eg, 15–50 g) are placed in glass beakers one of which is immediately extracted with K_2SO_4.

Fumigation Treatment. Under a fumehood, the two remaining soil subsamples are placed in a desiccator, along with a beaker containing purified $CHCl_3$ (50 mL) and boiling chips. A vacuum is applied until the $CHCl_3$ boils vigorously, then the sealed desiccator is maintained under vacuum in the dark at 25°C for 24 h.

Extraction of Microbial Biomass C. The vacuum in the desiccator is carefully released, the $CHCl_3$ removed (and disposed as a hazardous waste), the beakers containing the soil samples removed, and 0.5*M* K_2SO_4 (to give 1:2 to 1:5 ratio of soil to extract) added to each beaker. After shaking for 1 h, the soil suspension is filtered and the filtrate analyzed for organic C.

After adjusting soil weight to an oven-dry basis (using equation 24) and correcting for the total volume of the solution in the extracted soil (ext mL = soil wet wt − soil dry wt + mL of K_2SO_4 added), the total weight of extractable C in both fumigated (CF) and unfumigated (CUF) soil, and the microbial biomass C are calculated using the following equations:

$$\text{CF or CUF}(\mu\text{g/g soil}) = \text{extractacle C}(\mu\text{g/mL}) \times (\text{ext mL})(\text{g dry soil}) \quad (27)$$

$$\text{Microbial biomass C}(\mu\text{g/g soil}) = (\text{CF} - \text{CUF})/0.25 \quad (28)$$

where 0.25 is the efficiency of extraction (k_{ec}) of microbial biomass C based on *in situ* calibration techniques (1).

Microbial Processes

Microbial activity can influence the mineralization and transformation of many soil elements, such as C and N, and influence the composition of the soil air. The majority of methods devised to characterize microbial processes in soil are concerned with estimating C and N transformations. However, methods are also available to assess other microbial mediated transformations such as the oxidation and reduction of S, Fe, and Mn (5). Carbon mineralization involves the biological oxidation of organic C, where CO_2 is released to the soil atmosphere from metabolizing soil microbes and fauna. Soil respiration can be defined as the sum of the soil's metabolic activities that produce CO_2 (5). Laboratory methods to measure soil respiration generally involve use of incubation chambers, where CO_2 evolved from a soil sample is trapped in alkali and the resulting acid–base neutralization produces CO_3, which can then be measured by titrimetric, gravimetric, or conductimetric analysis. For analysis of CO_2 in gaseous samples, chromatography or IR gas analysis can be used. The static method provides a simple means of measuring soil respiration in the laboratory as follows. A soil sample (W, g) in a beaker is placed in an incubation jar, along with a beaker containing 1M NaOH (use 10 mL per 25 g of soil), and the jar is sealed. A water blank and an NaOH blank are also incubated separately. After a period of time the NaOH beaker is removed from the jar, and an excess of 1.5 M $BaCl_2$ with a few drops of phenolphthalein indicator are added to the beaker. The solution is then titrated with 1 M HCl (S). Both the water and NaOH blank (B) are also titrated. The following equation is used to calculate mg CO_2-C.

$$\text{mg CO}_2\text{–C} = (\text{mL HCl used in } B - \text{mL HCl used in } S) \times 6 \quad (29)$$

Nitrogen mineralization involves the conversion of the various organic N components in soil to inorganic N. Methods to characterize this phenomenon usually employ incubation techniques similar to that used for C mineralization. The importance of N for plant growth has created a need to estimate the N mineralization potential in soil. This potential is determined by incubating a soil sample under standard conditions and periodically leaching the sample (eg, using 0.01 M $CaCl_2$) and measuring the NO_3–N and NH_4–N in the leachate over time, until the cumulative N mineralized approaches a plateau (1,5). Both the amount of cumulated N and the duration of incubation are used to calculate the N mineralization potential.

Biological denitrification is the stepwise process where NO_3 is reduced to N_2. It can be estimated in the laboratory on soil cores obtained from the field, using the acetylene block technique. The acetylene blocks the last step in the denitrification process (ie, $N_2O \rightarrow N_2$). The accumulated nitrous oxide (N_2O), measured by GC, provides a measure of the soil denitrification rate (1,5). The specific sequence for the method is as follows. Soil cores are weighed (W, g) and placed in a glass jar sealed with serum stoppers. A 10-mL sample of gas is removed from the sealed jar and replaced with 10 mL of high purity acetylene. After time to allow acetylene diffussion, a gas sample is removed from the sealed jar and the N_2O concentration measured using GC. In most cases, the amount of dissolved N_2O in the soil water is also calculated to give a measure of the total N_2O produced.

Microbial processes in soil are mediated by enzymes that serve as biological catalysts. They control organic matter decomposition and their activities reflect several aspects (ie, soil moisture and other environmental parameters). Generally, laboratory methods used to measure enzyme activities in soil involve the incubation of a soil sample with a specific substrate, under standard conditions, followed by determining the rate at which the substrate is converted to a measurable product (5,6).

Soil Fauna

Various soil animals use the soil as a permanent or part-time habitat. Generally, they can be classified on the basis of size as follows: microfauna (<200 μm long), eg, soil protozoa and nematodes; and macrofauna (>200 μm long), eg, microarthropods and worms. Protozoa are unicellular, mainly aerobic organisms that can be classified into three groups: flagellates, amoeba, and ciliates (5). They are active in the soil pore space, in moisture films, and concentrate near high populations of bacteria, around roots, and labile organic matter. Populations of protozoa can range from 10^3 to 10^6 g^{-1} dry soil. Many protozoa feed on bacteria and fungi, and are involved in a wide range of soil processes. Similar to bacteria and fungi, they respond quickly to changes in the soil environment. Methods of enumeration include direct observation (eg, dried soil smears, staining, and fixation) and indirect observation (eg, dilution and MPN, density centrifugation). Generally, choice of method depends on the objective of the analysis, the group or species of protozoa under investigation, and the type of soil and soil environment (5).

Nematodes, the most abundant multicellular soil animal, are a diverse group of unsegmented roundworms that are categorized on the basis of food source, physiology, and method of feeding (1,5). Their abundance in soil can range from 10^4 to 10^7 m^{-2}. Although the total soil biomass of nematodes is much less than that found for microbes, their metabolic activity is higher. They all feed on living material (eg, plant roots, microbes) and, in soil systems, tend to be viewed as regulators of microbial populations. Various methods are available to enumerate nematode populations in soil. Direct methods involve staining and fixation (5). Extraction of nematodes from soil can be accomplished by several methods such as sieving and decanting, density centrifugation, filtration, and elutriation (1,5). The following sequence is used in the centrifugal–flotation method. A soil sample is gently sieved (6-mm sieve) and mixed, and a subsample (100 g) is placed in water (1:6, soil:water). After stirring, the soil is allowed to settle for up to 60 s. The suspension is decanted through two sieves (500- and 38-μm diam. openings). The contents of the 38-μm sieve are transferred to centrifuge tubes and the tubes centrifuged at 420 g. The water is discarded, and the centrifuge filled with a sucrose solution (1.18 specific gravity) and recentrifuged to remove soil. The sucrose solution, which includes the suspended nematodes, is passed through a 26-μm sieve to remove the nematodes.

Soil arthropods mainly consist of centipedes, millipedes, and certain insects. The latter consist of certain microarthropods such as springtails and mites. Soil populations can reach 10^4 m^{-2}. Microarthropods play a significant role in organic matter decomposition through their interactions with microbes, especially by increasing the rate of microbial biomass turnover (1,5). Extraction methods are mainly twofold: behaviour modification or trapping techniques that require the participation of the arthropod; and mechanical separation based on body size, density, or hydrophobicity (1,5). A typical example of the first is to apply a temperature or moisture

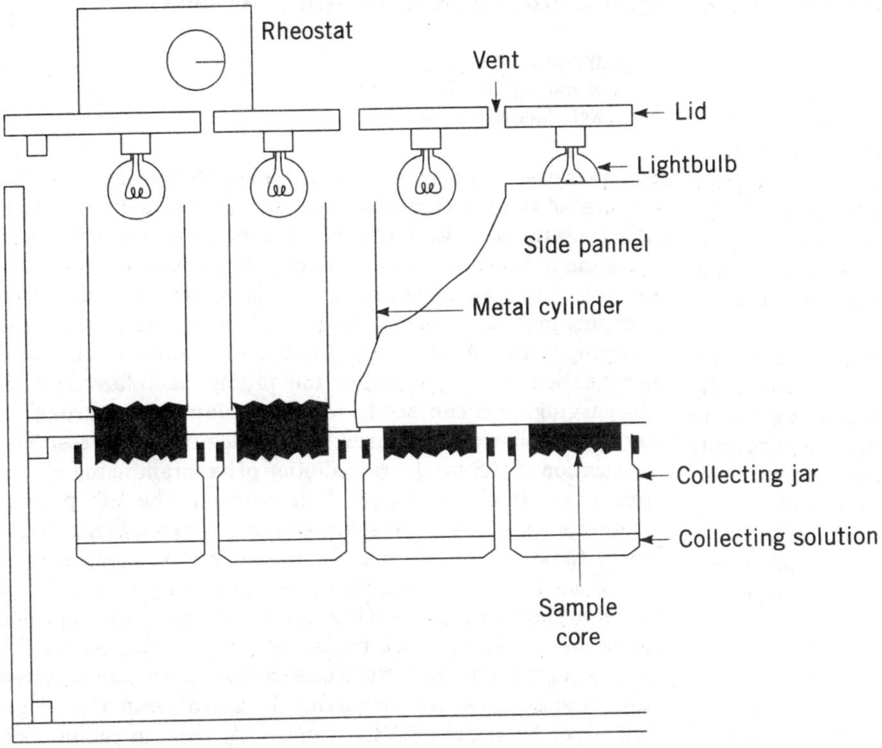

Figure 13. Front view of an extractor to assess microarthropod populations in soil. Reprinted with permission (1). Copyright 1993, CRC Press, Inc.

gradient to the soil, which causes the arthropod to move away from the adverse condition (ie, desiccation), and consequently allows them to be trapped in some container (eg, collecting jar containing a preservative). Mechanical methods consist of flotation in salt solutions, elutriation, and centrifugation. The following is the description of the stages used in a behaviour modification method, which utilizes heat (Fig. 13).

A soil sample or core is placed inside the extractor, which consists of a plywood box containing a heat source (light bulbs), and a shelf to hold the soil samples and collection jars and which separates the box into an upper hot zone (>20°C) and lower cool zone (<15°C). Metal cylinders are placed over the soil core, to concentrate the heat from the light bulbs, and a collecting solution (eg, polyethylene glycol) is placed into the jars, which are positioned under the soil samples or cores. The lower level of the extractor is maintained cool by use of refrigerated air. Gradual desiccation of the top of the soil sample causes a downward migration of the fauna. Extracted arthropods are removed from the jars and stored in ethanol.

Earthworms can have a strong influence on soil physical properties, movement of organic matter, and soil mixing through the production of casts. Generally, soil worms can be separated into two main groups: true earthworms (*Lumbricidae*) and the smaller potworms (*Enchytraeidae*). The population of earthworms in soil can be patchy and variable, but can range from 0 to 10^3 m^{-2}. Methods used to sample earthworms, similar to those for arthropods, can be grouped as behavioural, physical or mechanical, and indirect (1). The latter is based on physical evidence of earthworm presence. A simple and common mechanical method is to sample a soil area (eg, 0.1 m^2 and 0.1-m deep) and sort the soil by hand. Larger soil samples can be subject to a washing and sieving procedure as follows. Soil samples are collected using large cores or a spade, and soaked two days in a solution of Na-hexametaphosphate (2%, w/v, to disperse the soil clay particles) and formaldehyde (4%, v/v, to kill and fix earthworms). Soil samples are then placed in a large water container,

Table 4. Soil Properties for Characterizing Soil Biological Quality

Soil Property	Contribution to Soil Biological Quality
Organic matter	Energy source for soil organisms and nutrient source for plants. Influences many soil processes and properties.
Macroorganic matter and light fraction	Labile fractions of soil organic matter that respond quickly to soil disturbance and C inputs.
Microbial biomass	Both a sink and source of nutrients. Major agent in soil energy flow, and organic matter transformations, and responds quickly to environmental and management changes in soil.
Soil respiration	Provides a sum of a soil's metabolic activities, especially decomposition of organic compounds.
Nitrogen mineralization	Reflects microbial activity, and the ability of a soil to supply mobile forms (ie, NO_3) of N.
Enzymes	Indicate changes in microbial and biochemical activity.
Fauna	Involved in development and maintenance of soil biopore system, and mixing of organic and mineral material. Sensitive to soil pollution and degradation.

broken apart, then flushed with water and sieved to separate earthworms.

Soil Biological Quality

Soil biological quality is important for several soil functions, such as a medium for plant growth, regulating soil water, and regulating energy flows (see Table 1). Specific processes involved are supply balance of plant nutrients, suitable medium for microbes and fauna, influence soil water storage and movement, and store and release (recycle) energy-rich organic matter.

The main attributes or soil properties utilized to characterize these processes are given in Table 4. Organic matter is the prime energy source for soil organisms and is also involved in many soil processes. Organic matter dynamics are especially reflected by changes in the labile light fraction and macroorganic matter. The microbial biomass is both a sink and a source for nutrients and is the main agent for organic matter recycling. The fauna also are involved in a host of soil processes. Soil respiration, N mineralization, and enzymes are measures of microbial activity.

BIBLIOGRAPHY

1. M.R. Carter, ed., *Soil Sampling and Methods of Analysis,* Lewis/CRC Press, Baco Raton, Fla., 1993.
2. E.G. Gregorich and M.R. Carter, eds., *Soil Quality for Crop Production and Ecosystem Health,* Elsevier Science Publishers, Amsterdam, the Netherlands, 1997.
3. A. Klute, ed., *Methods of Soil Analysis, Part 1, Physical and Mineralogical Methods,* 2nd ed., American Soc. Agron., Madison, Wis., 1986.
4. P.R. Hesse, *A Textbook of Soil Chemical Analysis,* John Murray, London, 1971.
5. R.W. Weaver et al., eds., *Methods of Soil Analysis, Part 2, Microbiological and Biochemical Properties,* American Soc. Agron., Madison, Wis., 1994.
6. A.L. Page, ed., *Methods of Soil Analysis, Part 2, Chemical and Microbiological Properties,* 2nd ed., American Soc. Agron., Madison, Wis., 1982.
7. L.M. Walsh, ed., *Instrumental Methods for Analysis of Soils and Plant Tissue,* Soil Sci. Soc. America, Madison, Wis., 1971.
8. K.A. Smith, ed., *Soil Analysis: Modern Instrumental Techniques,* Marcel Dekker, New York, 1991.
9. P. MacCarthy, C.E. Clapp, R.L. Malcolm, and P.R. Bloom, eds., *Humic Substances in Soil and Crop Sciences: Selected Readings,* American Soc. Agron., Madison, Wis., 1990.
10. L.M. Walsh and J.D. Beaton, eds., *Soil Testing and Plant Analysis,* Soil Sci. Soc. America, Madison, Wis., 1973.
11. M.R. Carter and B.A. Stewart, eds., *Structure and Organic Matter Storage in Agricultural Soils,* Lewis/CRC Press, Baco Raton, Fla., 1996.
12. K.A. Smith and C.E. Mullins, eds., *Soil Analysis: Physical Methods,* Marcel Dekker, New York, 1991.
13. W. Burke, D. Gabriels, and J. Bouma, eds., *Soil Structure Assessment,* A.A. Balkema, Rotterdam, The Netherlands, 1986.
14. B.D. Kay, *Advances Soil Sci.,* **12,** 1–52 (1990).
15. B.H. Ellert, and J.R. Bettany, *Canadian J. Soil Sci.* **75,** 529–538 (1995).
16. P. Blackwell, *J. Soil Sci.* **41,** 215–228 (1990).

SOIL REMEDIATION BY AIR SPARGING

CLIFFORD J. BRUELL
University of Massachusetts
Lowell, Massachusetts

Air sparging is an *in situ* remediation technology used for removal of volatile organic compounds (VOCs) and biodegradable contaminants from groundwater aquifers. Generally, air sparging is used for the remediation of contamination caused by petroleum hydrocarbons or chlorinated solvents. The sparging process is also known as *in situ* aeration or *in situ* air sparging (IAS). When IAS is used to facilitate the microbial degradation of contaminants the process is referred to as biosparging. IAS can act to remove contaminants through a variety of physical, chemical, and biological processes. The application of IAS for the remediation of a contaminated source area is illustrated in Figure 1. In principle the IAS process is quite simple. Clean pressurized air is injected beneath the water table within the area of contamination using slotted or screened wells and the rising air induces mass transfer of VOCs to the vapor phase while adding oxygen to the aqueous phase to promote aerobic biodegradation. Contaminated vapors rising from beneath the water table enter the unsaturated zone (vadose zone) where further biodegradation can occur. Soil vapor extraction (SVE) is generally used in conjunction with IAS to control the migration of contaminated vapors within the vadose zone (1).

IAS is often used as an alternative to pump and treat for the remediation of contaminants located below the water table. The major drawback to pump and treat is that most hydrocarbon contaminants have relatively low water solubilities; therefore, excessively large quantities of clean water are required to solubilize and mobilize these contaminants (2). Subsequently this water (now contaminated) must be treated before reinjection into an aquifer or release to sewer. IAS avoids the production of large volumes of contaminated water by treating the water in place. Furthermore, the addition of oxygen can stimulate *in situ* biodegradation of both dissolved and immobile sorbed contaminants.

Air sparging has also been applied to control the movement of contaminated groundwater plumes in the form of sparged gate wells, sparging trenches, or sparging curtains (3,4). A sparged gate well is illustrated in Figure 2. When using gate wells, barriers such as driven interlocking sheet-piling are used as cutoff walls to direct the contaminated groundwater across a relatively narrow sparge zone located within a gate well. Sparging trenches are either open trenches, or trenches that have been back-filled with coarse gravel or pebbles. These trenches are positioned to intercept contaminant plumes. Sparging curtains use a series of sparge wells to intercept a plume. In all of the aforementioned configurations either stripping of VOCs or biodegradation occurs as contaminated water passes through the sparged zone.

APPLICABILITY OF IAS AND BASIC OPERATING PRINCIPLES

Typical Contaminants Removed Via Air Sparging

IAS is conventionally used for the remediation of groundwater contaminants that are volatile or aerobically biodegradable (5). Usually these contaminants originate from fuels, fuel

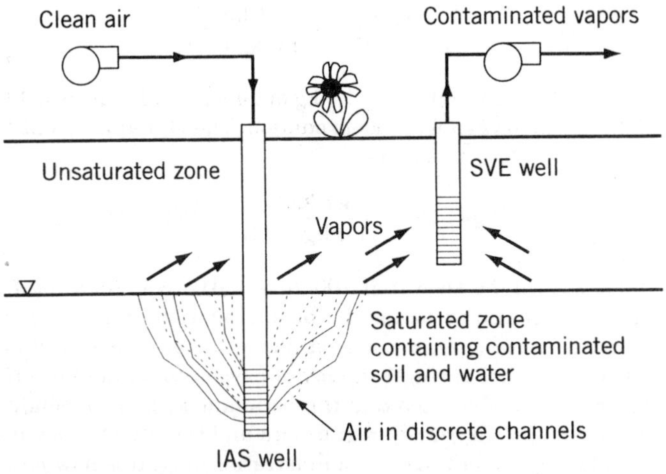

Figure 1. IAS used in conjunction with SVE for the remediation of a contaminated groundwater aquifer.

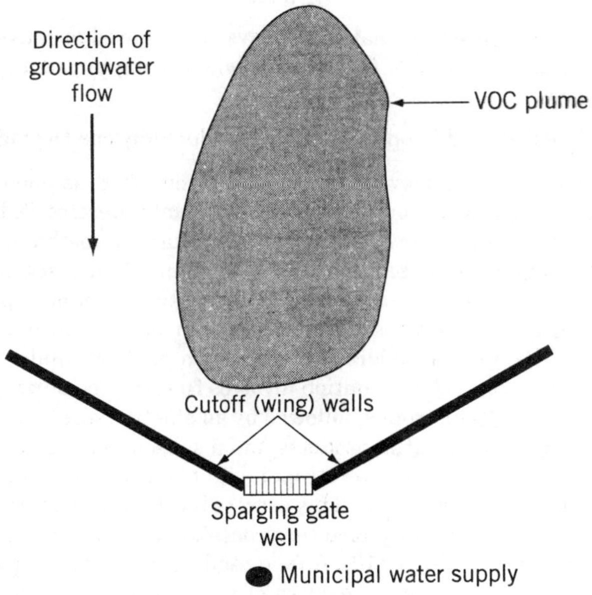

Figure 2. Use of IAS in gate well for plume containment.

oils, and chlorinated solvents that have been improperly disposed, leaked, or spilled into an aquifer. Typical VOCs from gasoline hydrocarbons include benzene, toluene, ethyl benzene, and xylenes (BTEX). The gasoline additive methyl-*t*-butyl ether (MTBE) is also often found in conjunction with gasoline contamination. Common chlorinated solvents include trichloroethylene (TCE) and perchloroethylene (PCE). The use of IAS for the removal of mineral oil and grease (6) and polycyclic aromatic hydrocarbons (PAH) such naphthalene (7) has been documented.

IAS can be employed for the removal of nonaqueous phase liquid (NAPL) contaminants situated beneath the water table that are immobilized on the solid phase (eg, smeared). To avoid the possibility of spreading NAPLs, alternate technologies such as skimming techniques or soil vapor extraction should be used first, to remove potentially mobile NAPLs, before implementing IAS.

Contaminant Volatilization Potential

IAS provides intimate contact between injected gas and the contaminated liquid phases. This enables the mass transfer of volatile contaminants, across the liquid–gas interface, from the aqueous phase to the vapor phase. This process is sometimes referred to as *in situ* volatilization or *in situ* air stripping. The potential of mass transfer of a contaminant between an aqueous phase and a vapor phase is dependent on the partitioning of the contaminant under equilibrium conditions as described by the Henry's law constant (*H*). Henry's law constant describes the ratio of a chemical's vapor pressure to aqueous solubility as follows:

$$H = \text{vapor pressure(atm)/aqueous solubility(mol/m}^3\text{)}$$
$$= \text{atm} \cdot \text{m}^3\text{/mol}$$

Henry's constants for common groundwater contaminants are provided in Table 1 (8–10). Contaminants with larger values of the Henry's constants have a greater potential for stripping from water. To be efficiently removed from an aquifer via IAS a contaminant must have a Henry's constant greater than 10^{-5} atm · m^3/mol (11). It has been estimated that in the initial weeks or months immediately following the initiation of air sparging that volatilization accounts for a greater portion of hydrocarbon removal than does biodegradation (12).

Contaminant Biodegradability

The biodegradability of a contaminant is dependent on its chemical structure and environmental conditions. Environmental contaminants that have been demonstrated to be aerobically biodegradable can be aerobically biosparged. Most gasoline constituents are moderately to highly biodegradable. Chlorinated solvents are somewhat refractory; however, selected solvents (eg, TCE) have been demonstrated to degrade slowly under the proper environmental conditions.

Table 1. Henry's Law Constants for Selected Chemicals

Chemical	Henry's Law[a] Constant, *H* (atm · m^3/mol)	Ref.
	Gasoline Constituents	
Benzene	5.43×10^{-3}	8
Toluene	6.61×10^{-3}	8
Ethyl benzene	7.9×10^{-3}	8
m-Xylene	6.91×10^{-3}	8
o-Xylene	4.94×10^{-3}	8
p-Xylene	7.01×10^{-3}	8
Styrene	2.28×10^{-3}[b]	8
	Gasoline Additives	
tert-Butyl alcohol (TBA)	1.20×10^{-5}	9
Diisopropyl ether (DIPE)	9.97×10^{-3}	9
Methyl-*t*-butyl ether (MTBE)	5.72×10^{-4}[c]	10
	Chlorinated Solvents	
Carbon tetrachloride	2×10^{-2}[b]	8
Chloroform	3.75×10^{-3}	8
Trichloroethylene (TCE)	8.92×10^{-3}	8
Tetrachloroethylene (perchloroethylene)	2.27×10^{-2}	8

[a] Values are at 25°C unless otherwise specified.
[b] Values is at 20°C.
[c] Value is calculated based on published data (10).

Aerobic biodegradation is generally must faster than anaerobic decomposition. The flux of oxygen into groundwater aquifers under natural conditions is very slow and constitutes the rate-limiting step for aerobic microbial biodegradation of aqueous phase contaminants. Therefore, aquifers that have been contaminated with petroleum hydrocarbons are usually found to be anoxic. IAS promotes the mass transfer of oxygen from the vapor phase to the aqueous phase. IAS increases groundwater dissolved oxygen (DO) levels and promotes the aerobic biodegradation of selected contaminants in both soil and groundwater. This process is termed as biostimulation. At one IAS site, a three-order-of-magnitude increase in bacterial population was observed, within the capillary fringe and groundwater, in conjunction with gasoline hydrocarbon degradation (13).

IAS Oxygen (O_2) Delivery

IAS is capable of delivering large volumes of gas to an aquifer. However, when IAS is used to oxygenate an aquifer the rate of oxygen mass transfer into an aquifer is often the rate-limiting step controlling aerobic biodegradation. Estimates of oxygen transfer efficiency range from 0.05 to 0.5% (12). It should be noted that estimates of oxygen transfer capability are highly dependent on site conditions and system design. Assuming all oxygen goes to biodegradation, a 0.5% transport efficiency translates to a hydrocarbon degradation rate of 4 kg/yr per m^3/h (15 lb/yr per standard ft^3/min (scfm)) of air flow. A 1.2-m (4-ft) thick smear zone, with an 2.4-m (8-ft) radius contaminated with 1,000 ppm hydrocarbons, sparged at 11 m^3/h (6.5 scfm) with air, would require one year to biodegrade (14).

Fine-grained soils with low air permeabilities can severely restrict the flow of sparging gas. If IAS flow is limited to less than 3.4 m^3/h (2 scfm), sparging with air may be inapplicable because of the limited amount of O_2 transferred to the aquifer under these low flow conditions. However, sparging with pure oxygen may make sparging feasible. Air contains approximately 20% oxygen (O_2) by volume. The highest dissolved oxygen (DO) levels experienced within water that is maintained in equilibrium with air, at typical groundwater temperatures, is approximately 8 mg/L. This gas/liquid partitioning is governed by Henry's law; therefore, when using pure oxygen (100% O_2) as a sparging gas, it is possible to achieve DO levels five times as high or approximately 40 mg/L. Use of pure O_2 as a sparge gas may result in higher O_2 mass transfer and biodegradation rates, possibly enabling the use of sparging in tight soils.

Sample Calculation: Oxygen Requirements for Biodegradation

The minimum amount of DO required for aerobic biodegradation of contaminants can be estimated via a stoichiometric mass balance. Gasoline is a complex mixture containing hundreds of hydrocarbons. Toluene a gasoline constituent can be employed as a gasoline surrogate to simplify the calculation. The dissolved oxygen requirement for the aerobic mineralization of 200 kg of toluene (C_7H_8) to carbon dioxide and water is as follows:

$$\underset{92\text{ kg}}{C_7H_8} + \underset{288\text{ kg}}{9\ O_2} \Rightarrow \underset{308\text{ kg}}{7\ CO_2} + \underset{72\text{ kg}}{4\ H_2O}$$

Based on this stoichiometric relationship the O_2 requirement can be expressed on the basis of kg O_2 per kg C_7H_8:

$$\frac{288\text{ kg }O_2}{92\text{ kg }C_7H_8} = \frac{3.1\text{ kg }O_2}{1.0\text{ kg }C_7H_8}$$

It can be seen that every 1.0 kg of toluene will require 3.1 kg of dissolved oxygen for biodegradation. The O_2 requirement for 200 kg of C_7H_8 is as follows:

$$200\text{ kg }C_7H_8 \times \frac{3.1\text{ kg }O_2}{1.0\text{ kg }C_7H_8} = 626\text{ kg }O_2$$

Now it can be seen that 200 kg of C_7H_8 requires a total of 626 kg O_2 minimum for aerobic biodegradation. It has been estimated that a sparging system using air as a sparging gas is capable of delivering 3 kg/day of oxygen to an aquifer (15). Again it should be noted that this estimate is highly dependent on site conditions and system design and the use of this value (ie, 3 kg/day) is for the sake of illustration only. It is now possible to estimate the time required to deliver the oxygen required for aerobic biodegradation as follows:

$$626\text{ kg }O_2 \times \frac{\text{day}}{3\text{ kg}} = 209\text{ days}$$

Therefore, an estimated 209 days are required to deliver the minimum amount of O_2 required when using air as a sparging gas.

IAS Methane Addition to Promote Trichloroethylene Degradation

The chlorinated solvent trichloroethylene (TCE) is one of the most ubiquitous groundwater contaminants. Aerobic TCE oxidation can be achieved using mixed cultures of methanotropic and heterotrophic bacteria. When methane is utilized as the sole carbon source for energy and growth, methanotropic organisms produce the enzyme methane monooxygenase (MMO) (16). MMO can epoxidize alkenes, such as TCE, under aerobic conditions. TCE oxidation involves first the epoxidization of TCE by methanotrophs, followed by an abiotic hydrolysis of the epoxide to nonvolatile products and a subsequent degradation of these products by heterotrophic organisms to carbon dioxide, chloride, and water (17). This process has been observed in the laboratory in an air phase that contains 0.6% natural gas (ie, methane) by volume (18). A field study of IAS natural gas addition, to stimulate the growth of methanotrophs to facilitate TCE degradation, has been conducted at the U.S. Department of Energy (DOE) Savannah River site (19).

Behavior of Injected Gas in Porous Media: Laboratory Studies

In IAS systems, the behavior of gases after injection into a water-saturated porous media is complex and it is not completely understood. However, laboratory and field studies have provided valuable insights into this process. It should be noted that the behavior of injected air within porous media is highly site specific. Therefore, a comprehensive investigation of site stratigraphy and a field-scale IAS evaluation are required to enable effective IAS system design.

To study air sparging, visualization studies of injected gas behavior in a two-dimensional model aquifer using 0.2–4-mm diameter spherical glass beads as a water-saturated porous media have been conducted (20). In 4-mm beads, which corresponds to a medium or coarse gravel, when air was injected into the bottom of the tank, a plume of discrete bubbles 1–3 bead diameters in size would move upward, apparently under the influence of buoyant forces, in a stumbling motion. In 0.75-mm or smaller beads, which corresponds to fine sands, silts, or clays, air flow occurred in continuous pore-scale air channels formed inside the medium. These channels were discrete and continuous. In response to increased air flow, new channels of similar size were formed. These channels appeared at the edge of the air plume and as increasing numbers of new channels formed within the plume. When air flow was increased or decreased, new channels would grow or collapse accordingly. In the plume, it was estimated that 50% of the pore space was occupied by air and 50% water. A transition between flow regimes was observed when using 2-mm beads, where both bubbles and channels were observed. In mixed bead experiments, used to simulate natural soils, air channels were shown to be sensitive to even slight medium heterogeneities. It was theorized that these distortions are as a result of minor variations in permeability and air entry pressures which result from pore-scale heterogeneity. Layers of beads of different sizes and permeabilities were used to simulate stratifications (macroscale heterogeneous layering). When injected air encountered the bottom of a lense of smaller beads with a lower permeability, a stagnation of high air saturation occurred and lateral migration of air was observed. The area above the low permeability lense was not reached by the injected air.

Model-scale IAS experiments using water-saturated soils demonstrated that in coarse-grained soils air could rise solely as a result of buoyant forces. However, in fine-grained soils additional external pressure was required to overcome capillary forces. The dividing line between these two behaviors was determined to be a coarse sand (d_{50} = 0.8 mm) (21).

Based on these experiments it is expected that in a field setting, air flow resulting from IAS will move in discrete pore-scale channels and that distorted air plumes resulting from pore-scale heterogeneity will predominate. Additionally, when macroscale heterogenous layering is present, that lateral migration of injected air will occur.

A two-dimensional laboratory-scale physical model was used to investigate the effects of air injection rate and groundwater flow rate on IAS oxygen mass transfer rates (22). In one set of experiments, groundwater flow was held constant and air flow was varied. At relatively low air injection rates, increases in air injection rates lead to the formation a higher density of air channels which resulted in increased interfacial area and subsequently increased mass transfer. This trend continued until an optimal air flow rate was reached beyond which decreases in mass transfer were observed. Air flow rates beyond optimal did not create new air pathways, but instead, resulting higher air pressures induced reductions in relative permeability causing water to circumnavigate the air flow zone. This was observed by dye tracer studies. In a separate set of experiments where air flow was held constant and groundwater flow was varied, increases in groundwater flow rates led to significant increases in oxygen transport, over the range of flow rates investigated. It was therefore deduced that in the field, advective flow of groundwater may be a significant factor controlling mass transfer of oxygen. However, increases in air injection rates will not always result in increased oxygen mass transfer rates.

Behavior of Injected Gas in Porous Media: Field Studies

To characterize the distribution of air resulting from IAS within an aquifer, technologies such as electrical resistance tomography (ERT) and neutron probes have been used. A typical distribution of air as characterized using neutron probe technology is shown in Figure 3 (23). Neutron probe measurements have revealed that in the vicinity of an air sparging well, air saturations can range from 30 to >50% and these values decrease radially away from the sparging wells. It is postulated that in the core area near the sparging well, where the percent air saturation is similar to the vadose zone, that volatilization and biodegradation will approximate that obtained within the vadose zone. The potential of volatilization within the core area of high air saturation applies to both dissolved VOCs and NAPLs (24). At lower levels of air saturation, biodegradation, and volatilization is probably going to be diffusion limited because the air channels are likely to be more widely spaced (23).

IAS Zone of Influence

IAS wells must be properly positioned for successful remediation of a contaminated aquifer. The area influenced by an IAS well is often referred to as a radius of influence (ROI). The ROI has been defined as "the distance from an air sparging well that defines a volume where air flow can be detected or the affects of air contact, groundwater mixing or groundwater oxygenation are detected and consistent" (25). An ideal uniform ROI is depicted in Figure 4. However, when IAS is attempted using heterogeneous soils with layering of silts or clays, a uniform ROI is rarely experienced. Under most real-world conditions the ROI is rather irregular as shown in Figure 5. As a result of this behavior the term zone of influence (ZOI) is often used to describe the area of IAS activity.

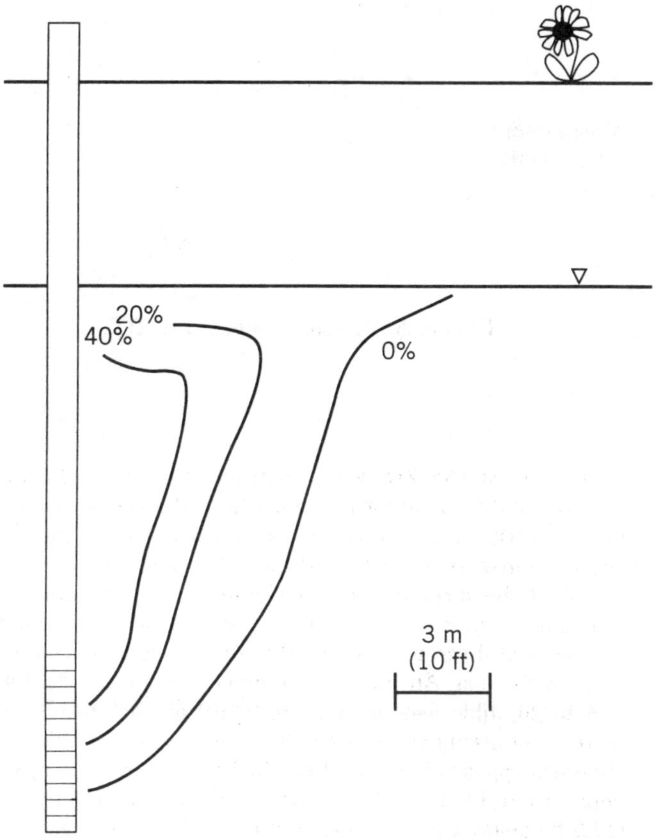

Figure 3. Typical distribution of air within an aquifer under steady-state conditions as measured using neutron probes (23).

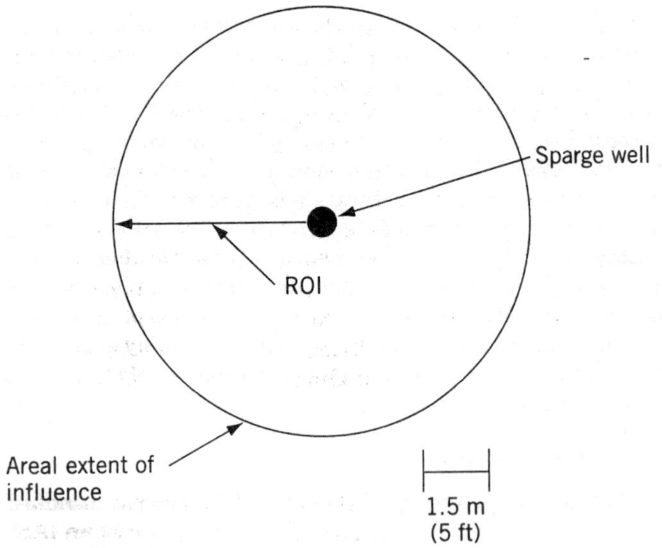

Figure 4. Ideal uniform IAS ROI.

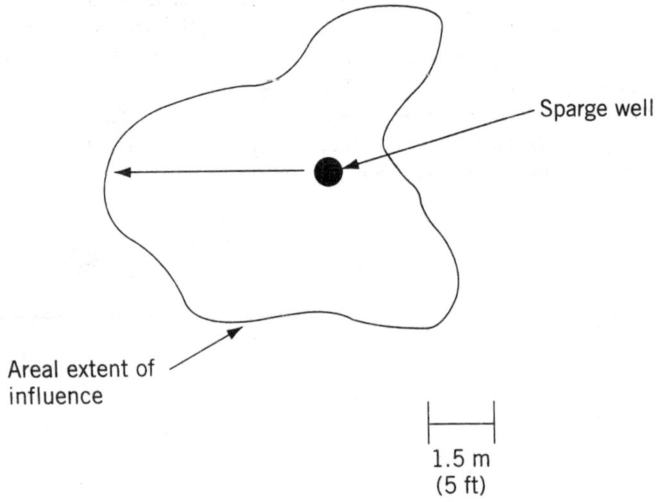

Figure 5. Typical irregular IAS ZOI.

A study of IAS ZOI was conducted at a field site, in an area of relatively uniform sand, where the soil surface was covered with shallow standing water which allowed the visual observation of areas where air channels would form. The areal extent of the formation of air channels was examined as a function of flow rate and depth of injection points. Air channels appeared at discrete locations that did not seem to change or move with time. An increase in air flow from 4.2–34.0 m^3/h (2.5 to 20 cubic feet per minute (cfm)) did not result in an increase of areal extent of air channels but instead additional channels appeared in the established zone. Increased injection depth from 1.8 m (6 ft) below the water surface to 5.3 m (17.5 ft) below water surface did result in an increase of the areal extent of channel formation from a radial distance of 1.8–4.9 m (6–16 ft) (26).

Preferential Flow In the Field

A homogenous soil is classified as anisotropic if its properties vary with direction. In the field horizontal permeability can often be 10 to 1000 times greater than vertical permeability as a result of geologic sedimentation (27). Soil anisotropy can potentially influence the movement of air in both the saturated and unsaturated zones, and influence the uniformity of groundwater treatment (24).

If the IAS air injection rate exceeds the rate of vertical or horizontal migration of air from an aquifer, large pockets of air can form within the aquifer (28,29). These pockets are detected by the disappearance of water in previously submerged saturated zone monitoring wells. Additionally, pockets of air can serve as pressure sources, driving water out of a monitoring well with great force for prolonged periods of time, following the discontinuation of air injection.

It is postulated that injected air in the field will behave in the same fashion as was seen in laboratory experiments. In water-saturated soils, injected air will migrate vertically until it encounters a confining layer of lower permeability causing it to migrate laterally. The horizontal migration of air as a result of confining layers is depicted in Figure 6.

In some cases, IAS activity may be detected in selected outlying monitoring points, whereas monitoring points in relatively close proximity to a sparging well will show no evidence of sparging. This behavior is termed as preferential flow and is believed to be as a result of heterogenous and anisotropic soil conditions such as extensive confining layers of low permeability strata. The phenomenon of preferential flow has been documented at closely monitored field sites using various monitoring parameters including tracer gases (30,31). Figure 7 depicts IAS preferential flow to an outlying region. When preferential flow conditions exist, the effectiveness of the IAS system is compromised. Additionally, if preferential flow or any lateral movement of air and water occurs due to confining layers of low permeability or a large horizontal to vertical permeability ratio, then the potential for lateral spreading of contaminants exists in the field (5,28,32).

Groundwater Mounding

Groundwater mounding is the upward movement of the water table which occurs as a result of injecting large vol-

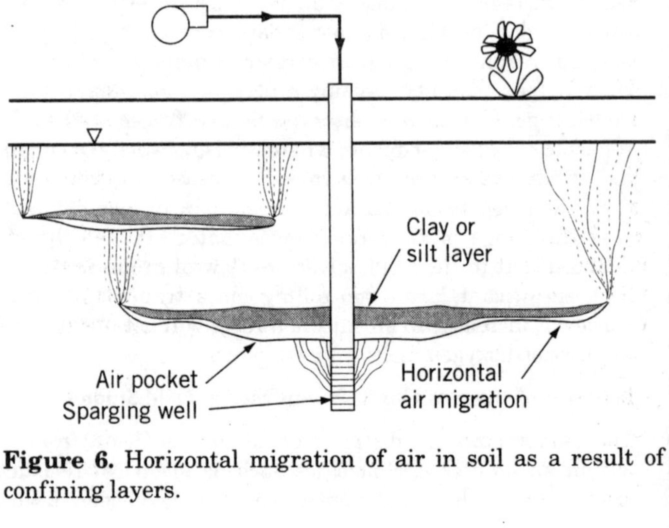

Figure 6. Horizontal migration of air in soil as a result of confining layers.

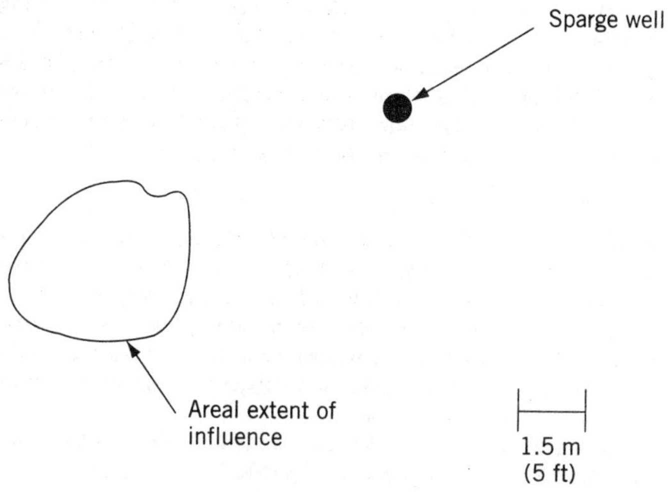

Figure 7. IAS preferential flow detected in an outlying region as a result of a confining Layer.

umes of air into the saturated zone. The water is displaced vertically by the injected air and an increase in water table elevation is observed. Mounding is depicted in Figure 8. The magnitude of mounding is highly dependent on site conditions but it may range from a fraction of an inch (cm) to several feet (m). Excessive mounding caused by IAS can result in the lateral movement of water and contaminants, and therefore should be monitored carefully. Mounding is generally transient in nature. At the onset of sparging the rate of air flow injected into the saturated zone is greater than air flow leaving the saturated zone, and the large volume of air beneath the water table results in a localized rise in water table elevation. Water migrates radially away from the IAS well. When the air establishes pathways to the unsaturated vadose zone the mounding begins to collapse and mounding will often dissipate as steady-state conditions are achieved. This process may take several hours to reach steady state. When an IAS system is turned off, the water table will often become depressed in the vicinity of the IAS well because air now escapes from beneath the water table faster than the displaced water can flow back. The mounding process has been simulated using

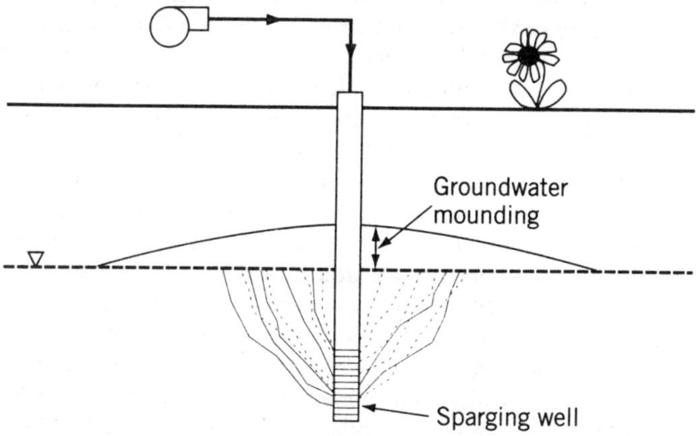

Figure 8. Groundwater mounding in the vicinity of an IAS well.

numerical techniques and confirmed in closely monitored field studies (33).

IAS air flow can reduce the permeability of an aquifer to the flow of water disrupting the movement of natural groundwaters. This behavior can result in a persistent rise in groundwater elevations up-gradient of sparge wells and contribute to lateral movement of groundwater flow (33). This must be considered when designing sparging curtain systems to ensure that contaminated groundwater flows through the curtain and not around it.

Pulsed Operation of IAS Systems and Groundwater Mixing

Pulsed IAS operation is cycling the air injection system on and off for periods ranging from several hours to several days. This practice is used to enhance IAS system performance resulting from induced groundwater mixing. An additional benefit of pulsed IAS operation is a reduction of wear on mechanical equipment and conservation of energy.

When IAS is applied to most natural soils, a pattern of discrete air channels is established within the saturated zone. Dissolved-phase VOCs and NAPLs partition only into air channels with which they are in intimate contact. Therefore, when IAS is first initiated, volatilization rates are often relatively high. However, as time progresses, volatilization rates decrease in what often resembles a first-order type of decay until volatilization rates become asymptotical. This behavior is probably as a result of diffusion limitations: ie, contaminants must diffuse to air channels from progressively greater distances. The diffusion distances are a function of the density of air channels or the maximum distance between air channels. Additionally, oxygen transport occurs as the result of the dissolution of oxygen from air-filled channels into water within the aquifer. Therefore, some of the same diffusion limitations that apply to IAS volatilization also apply to IAS induced in-situ biodegradation.

IAS air flow will occur in channels that have the lowest air entry pressures. These channels are formed from a series of connected pores with the largest pore diameters. As a result of this behavior air channels probably reestablish themselves in the same locations when sparging systems are turned on and off (24). However, cycling an IAS system on and off can induce groundwater mixing, bringing additional dissolved-phase contaminants into contact with air-filled channels. During nonsteady-state conditions this mixing likely results from the physical displacement of water by invading air and capillary interactions of air and water. During IAS, injected air physically displaces groundwater from selected soil pores and increases pore pressures. When sparging is ceased, subsequent water imbibition can result in bidirectional mixing of the aquifer. This process functions over distances ranging from inches (centimeters) to site-scale distances depending on stratigraphy (34).

Generally, pulsing has less of an effect on residual NAPL movement and distribution. Because the position of air channels and residual NAPLs remain essentially the same, it is postulated that volatilization of NAPLs will not be substantially enhanced as a result of pulsing (24).

The use of IAS/SVE pulsing to enhance volatilization rates has been demonstrated at field sites. Volatilization rates resulting from IAS are quantified by monitoring hydrocarbon content and flow rates within SVE exhaust gases. Figure 9

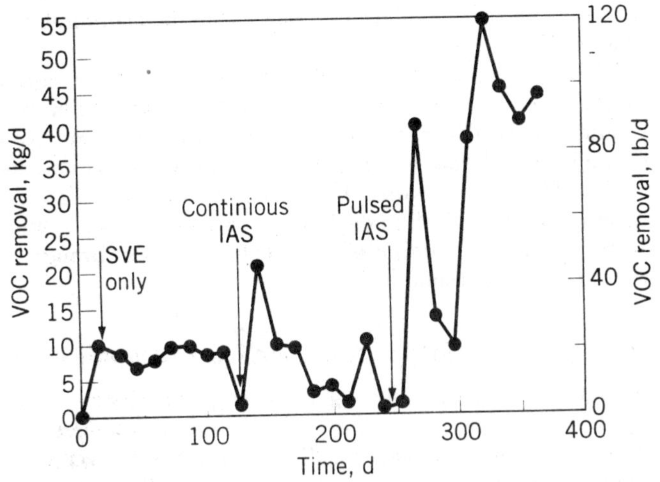

Figure 9. Impact of IAS pulsing on performance (34).

illustrates typical hydrocarbon removal rates at an IAS/SVE site which used multiple sparging wells (34). The sequence of remedial activities is clearly illustrated. An initial period of 126 days of only SVE was conducted to remove any potentially mobile contaminants from the vadose zone alone. Subsequently, a combination of continuous IAS/SVE was conducted from the period of 126 days to 252 days until overall hydrocarbon removal rates had decayed to low levels. Pulsed operation was then implemented on day 252. When individual wells showed diminished removal rates they were shut off for periods of from 12 to 24 hours before reactivation. The net result of pulsing was a three- to fivefold increase in VOCs removal when pulsing versus continuous IAS (34).

The benefits of pulsing of IAS to enhance oxygen delivery and *in situ* biodegradation are not well defined. This is because the volatilization process differs from the biodegradation process. When air injection is halted, significant quantities of gaseous and dissolved oxygen remain trapped within the soil matrix, enabling oxygen transport and utilization to continue. Conversely, volatilization effectively ceases when air flow stops. Therefore, it has been theorized that air injection pulsing frequencies to optimize oxygen transport, will be lower than those required to optimize volatilization (22).

Recommended Soil Conditions for IAS

It has been reported that a soil hydraulic conductivity of 0.001 cm/s or greater is required for IAS (35,36). The most desirable soil conditions for the application of IAS are uniform, medium- to coarse-grained sands or gravels, where air flow will be primarily due to buoyancy. Under these conditions air flow should be predictable and display a more uniform distribution in the vicinity of the sparging well impacting a higher percentage of the soil pores (25,28). However, in highly permeable soils, injected air will predominately travel in the vertical direction limiting the zone of influence.

If confining layers of fine-grained soils such as silts and clays restrict the vertical movement of air, lateral transport of potentially contaminated vapors may occur (5,28,32.) When silts and clays predominate, it may impossible to achieve adequate IAS flow rates without use of excessive pressures. However, if air injection pressures exceed overburden pressure, soil heaving and rupture can occur (28,37). A pressure-induced crack will result in a short-circuited air flow and diminished IAS system performance (28,38). Therefore, use of IAS may be inappropriate when fine-grained soils predominate or are present as confining layers.

Modeling of IAS

Efforts to model the IAS process take many forms. Analytical and numerical modeling techniques which are run on computers have been used to better define the interrelationships of pertinent site conditions and operational parameters. Modelling of IAS-induced air flow, groundwater flow, contaminant mass transfer or biodegradation all provide a better understanding of the IAS process.

An early modeling attempt utilized a three-dimensional (3-D) numerical model known as SPARG1 (Envirogen, Inc.) for describing pressure distributions which were subsequently used to simulate the flow field resulting from IAS, provided valuable insights into the influence of air flow and hydrogeological conditions on IAS behavior (39). A model used to describe the behavior of IAS trenches that uses local equilibrium assumptions has been used to investigate the effects of air flow, groundwater flow, aquifer thickness, number of theoretical transfer units, Henry's constant, and initial VOC concentration on system performance (40). Simple steady-state and nonsteady-state models describing the removal of dissolved VOCs by IAS has been modeled (41). A model of IAS DNAPL droplet removal which assumes diffusional transport from a spherical DNAPL droplet through a thick stagnant layer to advecting mobile water has been used to investigate the dependency of cleanup times on the model parameters (42). The removal of dissolved and NAPL VOCs has been investigated; the model assumes that the air moves through the porous media in persistent channels and that NAPL dissolves and moves to these channels by diffusion and dispersion processes (43); further modification to the model permits the simulation of removal via biodegradation (44). A multiphase, multicomponent simulator know as TETRAD (DYAD 88 Software, Inc.) which employs standard two-phase flow theory, including considerations of the compressibility of water and air, was used to investigate how air distribution within an aquifer changes with space and time (45); additionally this model has also been used to investigate IAS mounding (33). A finite-difference based, three-dimensional, three-phase and compositional numerical simulator known as INVAD (University of Houston) which includes the effects of capillarity, gravity, and relative permeability was used to investigate the influence of small-scale heterogeneity and fingering on IAS patterns of air distribution (46). A finite-element multiphase model known as MAGNAS3 (HydroGeoLogic, Inc.) capable of modeling active vapor, water, and NAPL in 3-D was used to simulate a variety of spatial and temporal parameters including aquifer pressures, contaminant (ie, benzene and xylene) cleanup, and oxygen addition (47).

Models provide valuable insights into the IAS process and serve as a first approximation for the purposes of design. For simplicity most of these models assume highly idealized and well-defined aquifer conditions. Therefore, most models have the limitations and do not adequately address the complexity of most field sites without significant calibration. Calibration generally is in the form of a pilot-scale field evaluation.

DETERMINATION OF IAS ZOI IN THE FIELD

No reliable theoretical method exists to predict ZOI resulting from IAS. Therefore, it is necessary to conduct a field evaluation or pilot test to determine the ZOI at a site. No standard method exists to determine ZOI in the field. Determination of IAS ZOI is an ongoing area of IAS research. As new technologies to determine IAS zone of influence evolve, the suitability of existing methods come into question. Generally, numerous parameters are monitored during a pilot test and analysis of the combined data set is used to determine the ZOI.

Bubbling in an open monitoring well is an indication of the presence of an air channel. A hand-held mirror can be used to reflect sunlight into the well to aid in the visual observation of air bubbling. Low volume flow meters can be used to quantify air flow from individual wells. However, an open monitoring well may constitute a path of least resistance to air flow. Air flow that occurs along preferential pathways may result in bubbling at locations that far exceed the effective ZOI as measured by other parameters. Therefore, bubbling that occurs in individual wells far beyond the effective ZOI should be investigated because this behavior indicates the that transport of potentially contaminated vapors is occurring (25).

Groundwater mounding has been used as an indication of ZOI (1,28). Groundwater mounding is not a direct indication of ZOI because of the transient nature of the process and because the extent of groundwater mounding generally exceeds the area of air injection (5,33).

Monitoring of saturated zone pressures can be accomplished by using wells or piezometers screened exclusively in the saturated zone. However, due to the nature of propagation of pressure from a source, measurements of saturated zone pressures often result in over-estimates of ZOI. Additionally, the saturated zone pressure response is encountered in conjunction with water table mounding in the vicinity of the sparge well. Therefore, both mounding and saturated zone pressures are transient in nature.

Vadose zone pressures have been used to determine ZOI. Often IAS is used in conjunction with soil vapor extraction (SVE). SVE results in a negative pressure distribution in the vadose zone. When an IAS system is turned on, the air flowing from beneath the water table will result in a decrease in the magnitude of negative pressure measured within the vadose zone. Comparisons of plots of vadose zone pressures versus radial distance from the IAS well, made while continuously running the SVE system and toggling the IAS system on and off, have been used to estimate IAS. However, it should be noted that reliance on unsaturated zone gas pressures to determine ZOI is plagued by some of the same problems as was encountered when using saturated zone pressures.

Cross-borehole electrical resistance tomography (ERT) is an experimental technique that has been used for the determination of air distribution patterns in the saturated zone resulting from IAS activities at a limited number of sites (48). The technique utilizes a series of electrodes located at various elevations that are positioned in adjacent monitoring wells. An electrical potential is imposed between electrodes in the two wells and corresponding measurements of current flow and voltage drop are made and analyzed. ERT data provide a direct and accurate representation of the movement of air in the saturated zone in the form of a series of two-dimensional plots depicting percent water saturation. These plots provide a relatively clear picture of the shape and size of the air-filled regions in the vicinity of a sparging well. It should be noted that ERT does not provide an indication of the movement of dissolved air. Comparisons of ROIs measured by ERT versus conventional monitoring techniques indicate that use of conventional techniques may lead to sizable overestimates of ZOI. The effective use of ERT requires considerable data analysis and interpretation; therefore, ERT cannot be used in a real-time mode. Because of the cost and time requirements, ERT should be considered primarily as a research tool.

Neutron probes have been used to provide near-real-time measurements of the degree of air saturation resulting from air sparging (23). A neutron probe contains a fast neutron radioactive source (eg, americium 241-beryllium) and a slow thermal neutron detector housed in a cylindrical probe which is lowered into access wells on a cable to make measurements. Fast neutrons released from the source collide with nuclei of similar mass (eg, hydrogen) and are slowed or thermalized. The thermal neutron counter is used to monitor the process and the result is related to water content via calibration curves. The resolution of the technique is suitable to provide two-dimensional plots of air saturation surrounding a sparging well and provide estimates of IAS ZOI in relatively short time periods.

Groundwater DO content is often used to delineate ZOI. In areas where aerobic biodegradation of contaminants is being conducted by indigenous microorganisms, DO levels as low as 0.2 mg/L are often encountered. IAS activity can increase DO content as a result of direct aeration of groundwater, mixing of aerated and unaerated groundwaters, and to a limited extent diffusional transport of dissolved oxygen to oxygen-deficient zones. Increased mixing is often observed when air sparging systems are cycled on and off every several hours (34). Following brief periods of air sparging DO content is often observed to rise in areas directly impacted by air sparging (48,49). Regardless of the transport mechanism, elevated DO levels resulting from air sparging are an indication of IAS ROI and are especially significant when biodegradation is the prime removal mechanism. Measurement of DO levels is one of the most popular methods of determining IAS ROI (25). It should be noted that conventional monitoring wells that are open to the atmosphere and intersect an air channel may constitute a path of least resistance to air flow. In this case bubbling in a monitoring well would tend to strip volatile contaminants and increase the DO levels of that well's contents, making DO or groundwater quality measurements questionable (37). Therefore, DO measurements should be made via pumping samples to the surface or using DO probes, in wells that are sealed at the surface.

Tracer gas tests are conducted by injecting a suitable tracer gas into the pressurized line that carries air to the IAS well. Gas is usually injected as a slug for several minutes. To avoid advective transport of tracer gases in the vadose zone, tests are usually conducted with the SVE system off. A series of gas sampling probes that are positioned a short distance above the saturated zone and laid out in a radial grid pattern are used to sample the gas phase and determine ZOI (31). The detection of tracer gases that are not normally present at appreciable background-levels serves as a positive indication of the influence of ZOI.

Gases such as helium (He), sulfur hexafluoride (SF_6), argon (Ar), and methane (CH_4) have been used as tracer gases in numerous field studies. Permits for their use are generally not required or are easily obtained. Helium has a low background concentration in the atmosphere, it is light and inert, and has a relatively low aqueous solubility. Helium has been used to characterize gas distribution in air channels within the saturated zone and vapor transport in the vadose zone (13). Portable helium detectors (Mark Model 9820) have been used to detect helium in the field at relatively low levels (29). SF_6 is inert and is not normally found in background measurements. It can be detected at extremely low levels using instruments such as electron capture detectors (ECDS) or portable halogen detectors (TIF Model 5550) (30,31).

Tracer gas tests are also used to confirm the ability of an SVE system to contain and recover IAS gas rising from the water table into the unsaturated zone. To determine capture, the SVE system is left on while a slug of tracer gas is introduced into the IAS well; the SVE exhaust is then sampled over time and a mass balance is conducted to determine the percent tracer gas recovered by the SVE system.

It should be noted that under conditions of preferential flow it is difficult to determine the ZOI using measurements at only a limited number of radially located sample locations. Therefore, the use of a comprehensive sampling network positioned in a series of concentric rings around the sparging well is most desirable.

DESIGN AND CONSTRUCTION OF IAS SYSTEMS

Layout of IAS Well Fields

The layout of an IAS well field is dependent on the extent, distribution, and mobility of the target contaminant. Often IAS is used to remediate residual contamination trapped beneath the water table when the contaminant is behaving as an immobile persistent source. In this situation, multiple IAS wells are installed within the source area to remediate the remaining contaminant. If the ZOI experienced in an aquifer is a uniform radius as shown in Figure 4, then the design of a remediation system is simplified. On the surface, the lay out of IAS wells would appear as slightly overlapping circles providing coverage of the contaminated aquifer as illustrated in Figure 10. However, field testing should be conducted to confirm that adequate coverage is achieved.

Intersecting a mobile contaminant plume with an IAS system is a technique that is often used to protect sensitive downgradient receptors. If a relatively mobile contaminant plume exists in a relatively narrow saturated zone (ie, <3-m (10-ft) deep) which is located close to the soil surface, then it is possible to use a sparging trench to intersect the plume and strip or biodegrade the contaminant within the trench area. The sparging trench is positioned normal to the direction of groundwater flow and usually consists of a horizontal perforated or slotted pipe for an air supply which is located at the bottom of the trench and is back-filled with a coarse gravel. The layout of a sparging trench is illustrated in Figure 11.

A sparging curtain is a series of sparging wells used to intersect a mobile plume. A sparge curtain might be used when the contaminant plume is relatively deep and a sparging trench is impractical. Layout of typical IAS sparging curtain is illustrated in Figure 12.

It should be noted that the introduction of air into a previously water-saturated aquifer will possibly alter the local hydraulic gradients. Therefore, in all sparging layouts (especially sparging curtains) monitoring should be conducted to ensure that contaminated plumes migrate through the ZOI of the IAS wells and not around them.

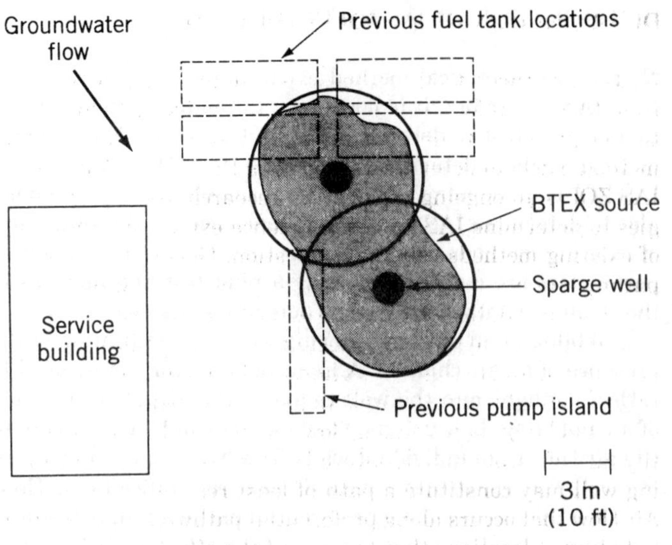

Figure 10. Layout of IAS wells to remediate a contaminated source area.

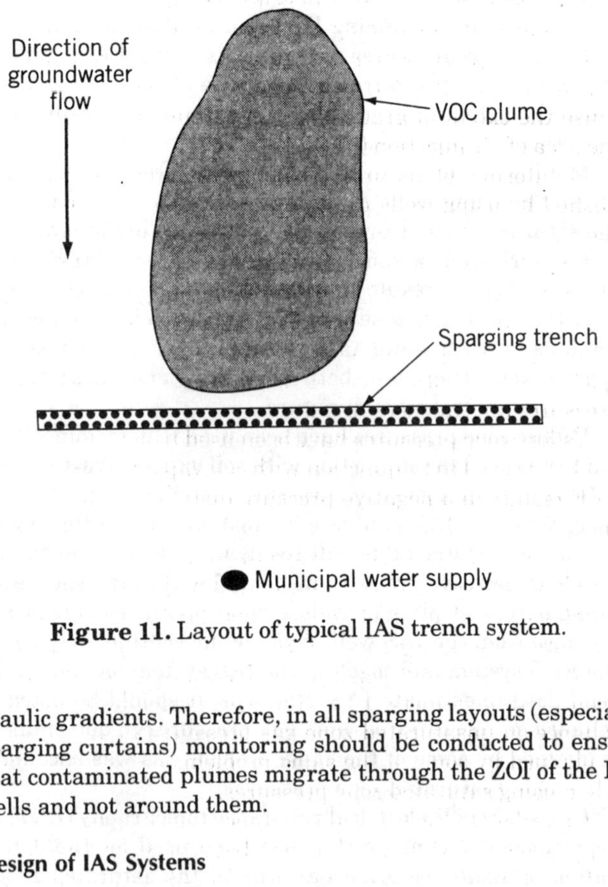

Figure 11. Layout of typical IAS trench system.

Design of IAS Systems

A typical IAS system consists of a compressor, manifold, and wells. Table 2 provides typical design and operation parameters that were determined from the analysis of 59 sites within a database assembled by the American Petroleum Institute (API) (25). Additionally, an examination of ROIs (ZOIs) at 37 IAS sites was conducted by the API. The results are presented in Figure 13 (25). The IAS-ROI was 1.5–6.1 m (5–20 ft) at the majority of the sites. The report of ROIs greater than 12.2 m (40 ft) were considered to be questionable. This

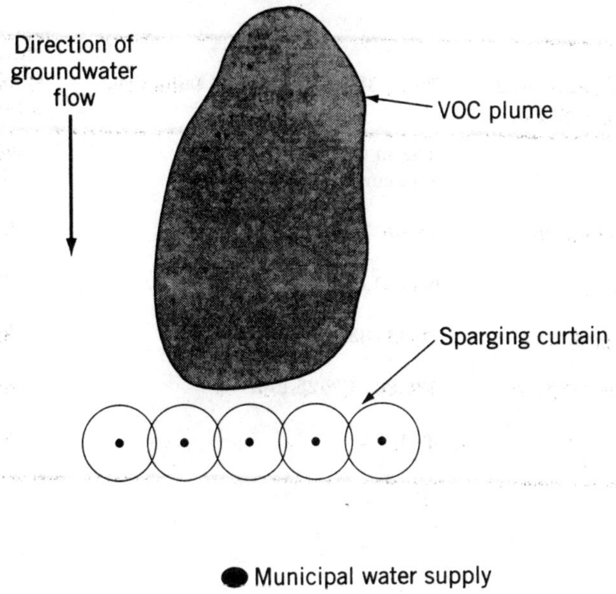

Figure 12. Layout of typical IAS sparging curtain.

data can serve as a starting point for IAS design efforts. Actual site conditions may necessitate alternate designs. To ensure the safety of personnel and equipment designers always consult manufacturers concerning the proper application of their products.

Materials of IAS system construction should be selected with care. PVC piping is often rated for nonshock liquid pressures in excess of 690 kPa (100 psig) and temperatures of 60°C (140°F). However, when carrying air which is a compressible fluid, a sharp blow to the piping could cause the pipe to burst and result in injury. Therefore, PVC pipe is often used when operating pressures and temperatures do not exceed 276 kPa (40 psig) and 38°C (100°F), respectively. To protect manifold piping from excessive temperatures 6.1–12.2 m (20–40 ft) of galvanized pipe should be used to separate the compressor from the PVC manifold. When higher pressures and temperatures are expected, the use of cleaned galvanized steel pipe or other corrosion-resistant metal piping should be considered. Use of solvents, glues, and various pipe sealants, on manifold or well components situated below the groundwater table, should be avoided to prevent potential groundwater contamination. Use of threaded fittings sealed with Teflon tape is recommended.

IAS wells are generally drilled using hollow stem auger (HSA) methods. A 5.08-cm (2-in.) diameter well is commonly employed to allow the use of readily available drilling techniques and hardware components. A 0.61-m (2-ft) long screen is generally used which provides more than sufficient contact area with the soil. In practice the majority of sparging air will exit in a relatively small area at the top of the well screen where the exit pressure is at a minimum. Because IAS systems are designed to convey relatively small volumes of compressed air, smaller well diameters and screen lengths may be successfully employed. Driven wells may be adequate under certain conditions.

A typical IAS well is shown in Figure 14. In drilled wells the selected filter pack should approximate native soils. Well screens with a 0.25–0.5-mm (0.010–0.020-in.) slot size are generally used. The screen can be slotted pipe such as flush threaded schedule 40 or 80 PVC pipe or wire-wound galvanized screen. Microporous bubblers have also been used in lieu of well screens (50). Observations of IAS well behavior using ERT revealed the importance of grouting of wells in the area above the well screen to avoid vertical short-circuiting of injected air within permeable well packing material (48). Suitable grouting is required to avoid the potential of ejection of a pressurized well from the ground, jeopardizing the safety of personnel in the vicinity of the well and rendering the well unusable.

The top of the well screen should be located beneath the lowest layer of contamination so that rising air will pass through the contaminated zone. For gasoline hydrocarbons and other contaminants that are lighter than water, the residual smear zone is usually encountered within the vicinity of the seasonal low, water table elevation. Contamination resulting from chlorinated solvents, that are heavier than water, will often be found at the bottom of the saturated zone overlying relatively impermeable soils or bedrock. In this case the well screen is positioned as close as practical to the contaminated soils without severely compromising the flow of sparge gas.

A piping manifold is necessary to convey and regulate air flow within an IAS system. A typical IAS system for a single well is shown in Figure 15. Pressure regulators, gauges, and a gate valve are used to regulate air flow. An automatic pressure relief valve should be used to protect system components by venting excess pressure. When using IAS, a large bubble of pressurized air will often form within the aquifer. If the air compressor is turned off and piping is disconnected, air and water may blow-back under great pressure for extended periods of time. A check valve prevents accidental blow-back and a manual pressure relief valve allows controlled system pressure bleeding. Ports to add tracer gases and for the installation of a flow meter are options that may be useful for research purposes. Hard piping or quick-connect flexible high-pressure hoses and couplings may be used to make connections. A metallic wire should be used to trace plastic piping to enable the location of buried components when using metal detectors. Compression of air often results in the production of condensate which may accumulate in piping and manifold components. Therefore, condensate bleed ports should be provided. Heat tracing should be utilized if freezing temperatures are anticipated.

The behavior of an IAS system is highly site specific; therefore, a pilot-scale evaluation using a portable compressor is necessary to evaluate pressure and flow specifications for blower sizing (14). Compressor specifications should require only oil-less type compressors to avoid introducing hydrocarbons into the aquifer. Reciprocating compressors are generally used if sparging pressures in excess of 103.5 kPa (15 psig) are expected. Oil-less reciprocating compressors often use Teflon components to avoid the use of lubricants and are capable of achieving pressures in excess of 621 kPa (90 psig). A receiver (ie, air tank) should be properly sized to allow continuous flow to IAS wells at specified flow rates. Receivers should be equipped with an automatic or manual drain valve to allow condensate removal. Heat tracing should be provided to avoid winter freezing. Compressors generate heat and are quite noisy. Therefore, when using compressors near residential areas, compressors should be situated in properly ventilated, sound-insulated enclosures. Compressor inlets should not be

Table 2. Typical Design and Operation Parameters for IAS Wells[a]

Parameter and Range	Most Often Used Value (No. of Sites)	Second-Most Often Used Value (No. of Sites)	Third-Most Often Used Value (No. of Sites)	Total Number of Sites
Screen length, 0.15–3.05 m	0.61 m (16)	0.91 m (8)	1.52 m (7)	40
Well diameter, 2.54–10.16 cm	5.08 cm (17)	10.16 cm (7)	2.54 cm (5)	37
Overpressure, 2.41–125.67 kPa[b]	2.41–34.45 kPa[b] (14)	34.45–68.90 kPa[b] (9)	68.90–103.35 kPa[b] (5)	31
Well screen depth below water table, 0.61–8.08 m	1.52–3.05 m (10)	3.05–4.57 m (8)	0.61–1.52 m (6)	31
In-situ sparging flow rate, 2.21–67.96 m^3/h	2.21–8.50 m^3/h (16)	8.50–16.99 m^3/h (9)	25.48–33.98 m^3/h (5)	39
In situ sparging pressure, 24.11–172.25 kPa[b]	34.45–68.90 kPa[b] (17)	68.90–103.35 kPa[b] (8)	137.80–172.25 kPa[b] (6)	40
(SVE ROI)/(IAS ROI) ratio, 0.16–7.42	1–2 (12)	0.16–1 (6)	3–4 (3)	26

[a] Ref. 25.
[b] To convert kPa to psi, multiply by 0.145.

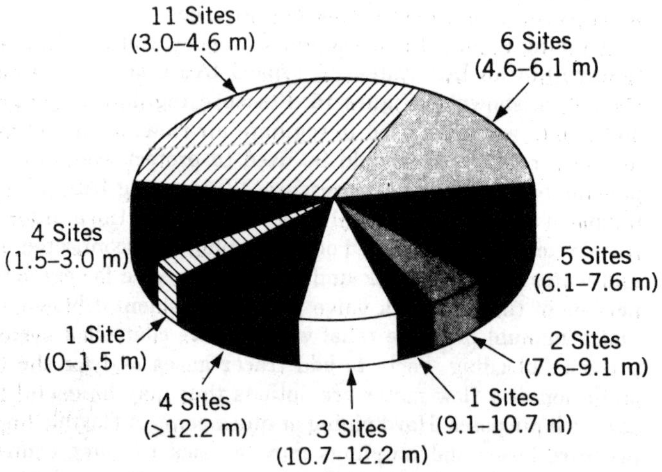

Figure 13. Distribution of experimentally determined ROIs found at field sites (25).

located in service stations or near SVE stacks to avoid the introduction of airborne contaminants.

Sample Calculation: Maximum Operating Pressure To Avoid Soil Rupture

Excessive sparging pressures may result in soil rupturing and cracking, allowing sparging gas to short-circuit to the soil surface. This scenario is likely to render the sparging system ineffectual. It is possible to estimate the maximum air pressure that can be safely applied without the production of soil fractures. The calculation is based on an estimate of total overlying pressure from the weight of the soil and water column situated above the top of the IAS well screen. Subsequently, a safety factor is utilized which limits applied pressure to 60–80% of total overlying pressure. This calculation should be considered only as an estimate because field conditions may vary. A sample calculation follows (25,51,52).

Assume that soil is unconsolidated medium-size sand with a 35% porosity (0.35) and a particle density of 2.65. The water table is located 6.40 m (21 ft) below the soil surface and the top of the sparging well screen is located 8.23 m (27 ft) below the soil surface or 1.83 m (6 ft) below the water table as shown in Figure 16.

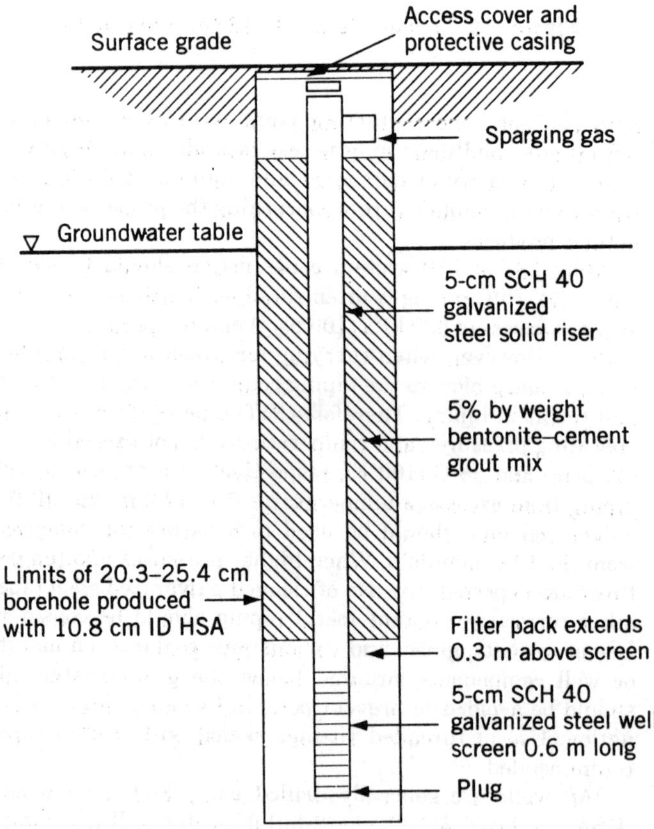

Figure 14. Typical IAS well design.

Calculations are as gauge pressure.

$$\text{Water pressure} = 1.83\ \text{m} \times 0.35 \times 9.81\ \text{kN/m}^3$$
$$= 6.28\ \text{kN/m}^2 \text{ or } 6.28\ \text{kPa}$$

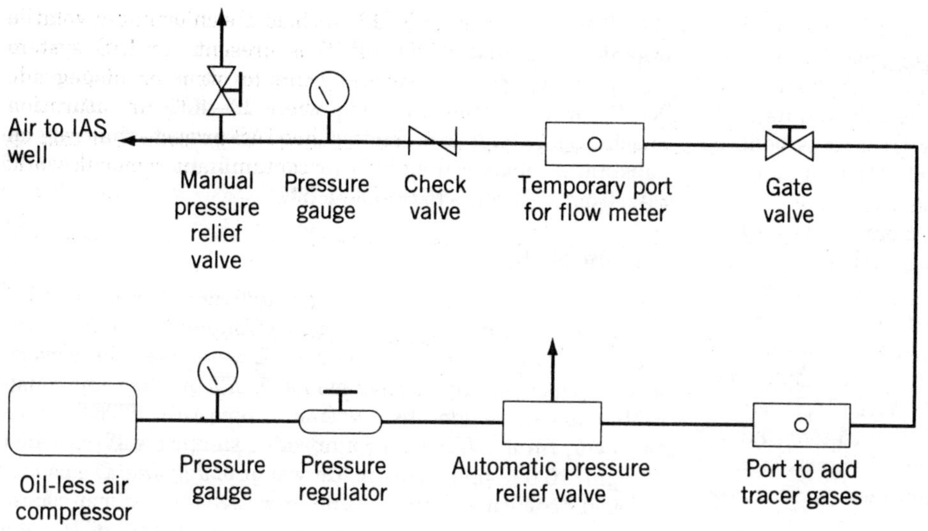

Figure 15. Typical IAS piping manifold.

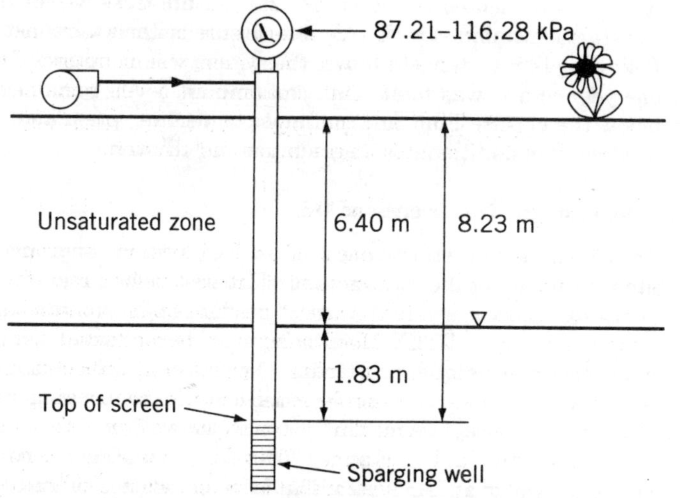

Figure 16. Layout used to illustrate the estimation (via calculation) of the maximum IAS well pressure to avoid soil rupture. To convert kPa to psig, multiply by 0.145.

$$\text{Soil pressure} = (6.40 + 1.83)\text{m} \times 2.65 \times (1{-}0.35) \times 9.81\ \text{kN/m}^3 = 139.07\ \text{kPa}$$

$$\text{Total pressure} = (6.28 + 139.07) = 145.35\ \text{kPa}$$

Maximum pressure to avoid soil fractures

$$= (0.6 \times 145.35) = \boldsymbol{87.21\ kPa} \text{ to } (0.8 \times 145.35) = \boldsymbol{116.28\ kPa}$$

Initial IAS pressures required to establish flow generally exceed pressures required to maintain steady-state air flow. Initial IAS activity will displace overlying water, resulting in lower pressures under steady-state flow conditions. The pressure required to maintain steady-state flow is the sum of the pressure required to overcome hydrostatic pressure at the top of the well screen and overpressure. Overpressure can be defined as the pressure in excess of that required to overcome hydrostatic head at the top of the well screen that is experienced under steady-state flow conditions (25). Overpressure is related to soil type. Tighter, fine-grained soils usually exhibit larger overpressures. Assuming an operating overpressure of 34.5 kPa (5.0 psi) for 8.49 m^3/h (5 cfm) of air flow in a medium sand, the pressure required to maintain flow in the system can be estimated as follows:

$$\text{Hydrostatic head} = 1.83\ \text{m} \times 9.81\ \text{kN/m}^3 = 18.0\ \text{kN/m}^2 \text{ or } 18.0\ \text{kPa}$$

$$\text{Pressure required to maintain flow} = 18.0\ \text{kPa} + 34.5\ \text{kPa} = 52.5\ \text{kPa}$$

It can be seen that this pressure is much less than half of the maximum pressure to avoid soil fractures. Pressure required initially to establish flow will be slightly higher, eg, 10–20 kPa (1.5–2.9 psi).

EFFECTIVENESS OF IAS

IAS Rebound

Since the inauguration of IAS as a remediation technique there has been considerable controversy concerning the short- versus long-term effectiveness of the technology. The efficacy of air sparging is largely due to the extent and duration of contact between injected air and contaminants within an aquifer. IAS can achieve significant, lasting reductions in groundwater contaminant levels if applied appropriately for sufficient periods of time.

Often, following a period of sparging (eg, several months), groundwater contaminant levels as measured in conventional monitoring wells will be substantially reduced. However, when IAS is terminated, contaminant levels often rebound to levels approaching or exceeding presparging values. Rebound may occur within weeks to months. In some cases rebound may correspond to the spring high groundwater level. It is believed that rebound is due to diffusion or mobilization of contaminants from areas not directly influenced by IAS.

A classical example of contaminant rebound was encountered when IAS was used to remediate a source area containing NAPL petroleum solvents consisting primarily of hexane and mineral spirits and approximately 1–2% of PCE, TCE, and other chlorinated solvents (24). Figure 17 shows a typical response of dissolved-phase VOC levels, monitored in an operating water extraction well located amongst three IAS wells, to IAS. Air sparging was applied for 5 months until contaminant levels were below 250 μg/L. However, when IAS was stopped the NAPL source remained and contaminant levels rebounded nearly back to presparging levels.

A study focusing on the effectiveness of air sparging was conducted which examined 21 IAS sites (6 chlorinated and 15 petroleum hydrocarbon) covering a wide range of sparging system parameters and site conditions (14). All sites selected for the study had groundwater quality data from time periods before IAS was initiated immediately before IAS was terminated, and in the months following IAS shutdown to observe contaminant rebound. Rebound occasionally occurred many months (eg, 2–16 months) after IAS system shutdown. Petroleum-contaminated sites showed more rebound than sites contaminated with chlorinated compounds. High air flows (ie, >17 m^3/h(>10 scfm) per well) and a network of closely spaced IAS wells (ie, <6.1 m(<20 ft)) positioned throughout the source area minimized rebound. More rebound was observed when treating source areas containing smeared or residual NAPL than when treating dissolved phase plumes. This demonstrates the importance of NAPL removal before implementing IAS and the importance of intimate contact between contaminants and sparge gas.

Based on a study of IAS system performance it is recommended that IAS systems treating source areas should be operated for at least 10 months to avoid rebound, although dissolved plumes may be addressed in periods of 4–6 months (14). However, when an NAPL such as the chlorinated volatile organic compound (CVOC) PCE is present, an IAS system will probably require several years to strip or biodegrade NAPL located beyond the zone where 20–40% air saturation is achieved (24). It is estimated that IAS projects that rely on substantial biological activity for contaminant removal would take from 1–3 years to complete (53).

IAS Case Study

Sparging was used to accelerate groundwater cleanup at a former gasoline station in southern New England. Mobile NAPL recovery, pump and treat, and SVE were used for several years to successfully remove mobile NAPL product and remediate the vadose zone to less than 5 ppm total BTEX levels. However, total BTEX in groundwater samples still exceeded 10 ppm (10,000 ppb). This result was probably due to gasoline trapped beneath the water table in relatively uniform coarse to very fine sands. An IAS system, utilizing a total of 14 wells operated at injection flow rates of 3.4–10.2 m^3/h (2–6 scfm) per well, was operated intermittently over a period of 60 days. Total BTEX levels in groundwater versus time observed at the site is presented in Figure 18 (25). Within three weeks of the start of combined SVE/IAS, site-closure criteria were met. Following IAS system shutdown the system was monitored for one year and it was found that contaminant levels remained below the closure limit and continued to decline, presumably as a result of additional biodegradation and dilution.

Monitoring the Effectiveness of IAS

To determine the effectiveness of an IAS system, appropriate monitoring of the contaminant of interest before and after sparging is necessary. In the case of gasoline spills, monitoring usually focuses on BTEX. Monitoring must be conducted using appropriate techniques and timing. Conventional drilled monitoring wells may serve as preferential conduits to sparging air if a large air pocket forms that contacts the well or if they intersect an individual air channel. Therefore, sampling of a conventional well of an IAS system that is in operation is of limited

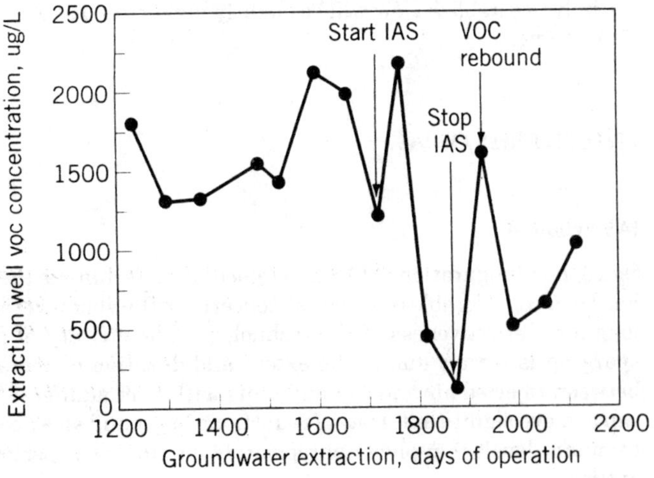

Figure 17. Evidence of rebound of contaminant levels at an IAS site in the vicinity of a NAPL source (24).

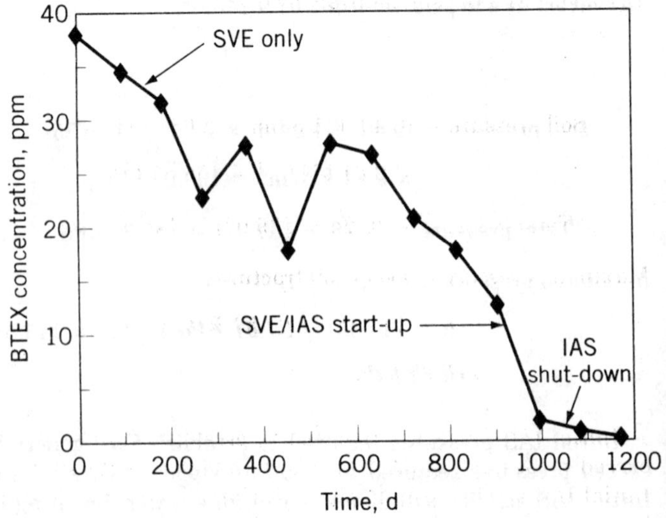

Figure 18. Groundwater total BTEX concentrations versus time at an IAS site (25).

value. In areas where monitoring wells are directly impacted by the flow of injected air, a waiting period of greater than 1 month may be required before representative samples can be obtained. During IAS, direct-push (Geoprobe) discrete implant soil samplers appear to be more representative of aquifer conditions than drilled conventional monitoring wells (29). Reboring and sampling of both soils and groundwaters in areas outside the immediate influence of the sparge wells are suggested (54).

SUMMARY

IAS is a remediation technology used to remove dissolved and immobilized NAPL contaminants such as gasoline hydrocarbons or chlorinated solvents from groundwater aquifers. The IAS process utilizes wells to inject clean air beneath the groundwater to strip volatile contaminants, and to add oxygen to facilitate aerobic biodegradation of contaminants. Injected air moves through the porous soil matrix in discrete pore-scale channels within the ZOI of an IAS well. When stratified soils containing silts and clays inhibit the vertical migration of sparging gas, lateral preferential flow of potentially contaminated vapors can occur, and the effectiveness of the IAS system is compromised. IAS first removes contaminants that are in intimate contact with air channels. Subsequent contaminant removal becomes diffusion limited. Pulsed operation of IAS wells results in aquifer mixing which can offset some diffusion limitations. Neutron probes, electrical resistance tomography, use of tracer gases, and measurements of dissolved oxygen levels are some of the methods used to determine the ZOI of a sparging well.

Behavior of injected air is highly site specific. Therefore, a field pilot-scale investigation is required to enable effective design. IAS systems can be constructed using conventional hardware but care must be taken to avoid introducing contaminants from construction materials or hardware. IAS systems should be carefully designed and operated to avoid soil rupture making the system ineffective. If IAS systems are not properly designed, operated and sampled rebound of contaminant levels is likely to occur.

BIBLIOGRAPHY

1. R.A. Brown, F. Jasiulewicz, *Pollut. Eng.*, 52–55 (July 1992).
2. R.E. Hinchee, D.C. Downey, and E.J. Coleman, *Proceedings of the Conference on Petroleum Hydrocarbons and Organic Chemicals in Ground Water: Prevention, Detection and Restoration,* API/NWWA, Houston, Tex., 1987, pp. 147–164.
3. J.F. Pankow, R.L. Johnson, J.A. Cherry, *Ground Water* **31**(4), 654–663 (1993).
4. M.C. Marley, E.X. Droste, F. Li, and R.J. Cody, "Design of an In Situ Sparging Trench", in *Proceedings of the Conference on Petroleum Hydrocarbons and Organic Chemicals in Ground Water: Prevention, Detection and Restoration,* API/NGWA, Houston, Tex., 1994.
5. E.K. Nyer, and S.S. Suthersan, *Ground Water Monit. Remed.,* **13**(4), 87–91 (1993).
6. D. Lord, J. Lei, M-C. Chapdelaine, J-L. Sansregret, and B. Cyr, in R.E. Hinchee, R.M. Miller, and P.C. Johnson, eds., *In Situ Aeration: Air Sparging, Bioventing, and Related Remediation Processes,* Batelle Press, Columbus, Ohio, 1995, pp. 121–134.
7. B.D. Symons, R. Linkenheil, D. Prichard, C.A. Shanke, and D. Seep, in R.E. Hinchee, R.M. Miller, and P.C. Johnson, eds., *In Situ Aeration: Air Sparging, Bioventing, and Related Remediation Processes,* Batelle Press, Columbus, Ohio, 1995, pp. 135–143.
8. EPA, *Subsurface Contamination Reference Guide,* Office of Emergency and Remedial Response, Washington, D.C. 20460, EPA/540/2-90/011, 1990.
9. J.H. Montgomery, *Groundwater Chemicals Desk Reference,* Vol. 2, Lewis Publishers, Chelsea, Mich., 1991.
10. Merck & Co., Inc., *The Merck Index,* 10th ed., Merck & Co., Inc., Rahway, N.J., 1983.
11. R.A. Brown, C. Herman, and E. Henry, "The Use of Aeration in Environmental Clean-Ups", Presented at *Haztech International Pittsburgh Waste Conference,* Pittsburgh, Pa., 1991.
12. P.M. Boersma, F.S. Petersen, P. Newman, and R. Huddleston, "Use of Groundwater Sparging to Effect Hydrocarbon Biodegradation," In *Proceedings of the Conference on Petroleum Hydrocarbons and Organic Chemicals in Ground Water: Prevention, Detection and Restoration,* API/NGWA, Houston Tex., 1993.
13. M. Javanmardian, J.S. Huber, C.B. Olson, W.P. Schwartz, C.A. Masin, and V.J. Kremesec, in R.E. Hinchee, R.M. Miller, and P.C. Johnson, eds., *In Situ Aeration: Air Sparging, Bioventing, and Related Remediation Processes,* Batelle Press, Columbus, Ohio, 1995, pp. 191–202.
14. D.H. Bass and R.A. Brown, in *Proceedings of the Conference on Petroleum Hydrocarbons and Organic Chemicals in Ground Water: Prevention, Detection and Restoration,* API/NGWA, Houston, Tex., 1995, pp. 621–636.
15. R.A. Brown, "Oxygen Sources for Biotechnological Application", Presented at *Biotechnology Work Group,* Monterey, Calif., Feb. 21–23, 1989.
16. L. Semprini, P.V. Roberts, G.D. Hopkins, and P.L. McCarty, *Ground Water* **28**(5), 715–727 (1990).
17. L. Semprini, G.D. Hopkins, P.V. Roberts, D. Grbic-Galic, and P.L. McCarty, *Ground Water* **29**(2), 239–249 (1991).
18. J.T. Wilson and B.H. Wilson, *Appl. Environ. Micro.* **49**(1), 242–243 (1985).
19. DOE, "Cleanup of VOCs in Non-Arid Soils-The Savannah River Integrated Demonstration," U.S. Dept. of Energy, WSRC-MS-91-290, Rev 1., 1991.
20. W. Ji, A. Dahmani, D. Ahlfeld, J. Lin, and E. Hill, Jr., *Ground Water Monit. Remed.,* **13**(4), 115–125 (1993).
21. K. Wehrle, in F. Arenak, M. Hinsenveld, and W.J. Van den Brink, eds., *Contaminated Soil '90,* Kluwer Academic Publishers, Netherlands, 1990, pp. 1061–1062.
22. K. Rutherford and P.C. Johnson, in *Proceedings of the Conference on Petroleum Hydrocarbons and Organic Chemicals in Ground Water: Prevention, Detection and Restoration,* API/NGWA, Houston, Tex., pp. 129–143, 1995.
23. L.J. Acomb, L.J., D. McKay, P. Currier, S.T. Berglund, T.V. Sherhart, and C.V. Benediktsson, in R.E. Hinchee, R.M. Miller, and P.C. Johnson, eds., *In Situ Aeration: Air Sparging, Bioventing, and Related Remediation Processes,* Batelle Press, Columbus, Ohio, 1995, pp. 47–61.
24. P.M. Boersma, K.R. Piontek, and P.A.B. Newman, in R.E. Hinchee, R.M. Miller, and P.C. Johnson, eds., *In Situ Aeration: Air Sparging, Bioventing, and Related Remediation Processes,* Batelle Press, Columbus, Ohio, 1995, pp. 39–46.
25. M.C. Marley, and C.J. Bruell, *In Situ Air Sparging: Evaluation of Petroleum Industry Sites and Considerations for Applicability, Design and Operation. API Publication 4609,* American Petroleum Institute, Washington, D.C., 1995.
26. A. Leeson, R.E. Hinchee, G.L. Headington, and C.M. Vogel, in R.E. Hinchee, R.M. Miller, and P.C. Johnson, eds., *In Situ Aeration: Air*

Sparging, Bioventing, and Related Remediation Processes, Batelle Press, Columbus, Ohio, 1995, pp. 215–222.

27. R.A. Freeze and J.A. Cherry, *Groundwater,* Prentice Hall Inc., Englewood Cliffs, N.J., 1979, pp. 30–33.
28. M.C. Marley, D.J. Hazebrouck, and M.T. Walsh, *Ground Water Monit. Rev.* Spring 1992, pp. 137–145.
29. P.C. Johnson, R.L. Johnson, C. Neaville, E.E. Hansen, S.M. Stearns, and I.J. Dortch, in R.E. Hinchee, R.M. Miller, and P.C. Johnson, eds., *In Situ Aeration: Air Sparging, Bioventing, and Related Remediation Processes,* Batelle Press, Columbus, Ohio, 1995, pp. 1–20.
30. R.S. Baker, M.E. Hayes, and S.H. Frisbie, in R.E. Hinchee, R.M. Miller, and P.C. Johnson, eds., *In Situ Aeration: Air Sparging, Bioventing, and Related Remediation Processes,* Batelle Press, Columbus, Ohio, 1995, pp. 63–73.
31. J.M. Fischer, T.P. Wilson, and A.C. Kane, in *Proceedings of the Conference on Petroleum Hydrocarbons and Organic Chemicals in Ground Water: Prevention, Detection and Restoration, API/NGWA,* Houston, Tex., 1996, pp. 601–619.
32. L.M. Martin, R.J. Sarnelli, and M.T. Walsh, in *Proceedings of the National R&D Conference on the Control of Hazardous Materials,* San Francisco, Calif., 1992, pp. 318–328.
33. P.D. Lundegard, in R.E. Hinchee, R.M. Miller, and P.C. Johnson, eds., *In Situ Aeration: Air Sparging, Bioventing, and Related Remediation Processes,* Batelle Press, Columbus, Ohio, 1995, pp. 21–37.
34. W.S. Clayton, R.A. Brown, and D.H. Bass, in R.E. Hinchee, R.M. Miller, and P.C. Johnson, eds., *In Situ Aeration: Air Sparging, Bioventing, and Related Remediation Processes,* Batelle Press, Columbus, Ohio, 1995, pp. 75–85.
35. M.E. Loden and C-Y Fan, in *Proceedings of the National Conference on the Control of Hazardous Materials,* Hazardous Materials Control Research Institute, San Francisco, Calif., 1992, pp. 328–334.
36. A.C. Middleton, and D. Hiller, "In Situ Aeration of Ground Water: A Technology Overview," in *Proceedings of 1990 Environment Canada Montreal Conference,* Montreal, Quebec, 1990.
37. R.L. Johnson, P.C. Johnson, D.B. McWhorter, R.E. Hinchee, and I. Goodman, *Ground Water Monit. Remed.* **13**(4), 127–135 (1993).
38. H. Gudemann and D. Hiller, "In Situ Remediation of VOC Contaminated Soil and Groundwater by Vapor Extraction and Water Aeration," in *Proceedings of the Third Annual Haztech International Conference,* Cleveland, Ohio, 1988.
39. M.C. Marley, F. Li, and S. Magee, "The Application of Models in the Design of Air Sparging Systems," In *Proceedings of the Conference on Petroleum Hydrocarbons and Organic Chemicals in Ground Water: Prevention, Detection, and Restoration,* API/NGWA, Houston Tex., 1992.
40. D.J. Wilson, S. Kayano, R.T. Mutch, Jr., and A.N. Clarke, *Sep. Sci. Technol.,* **27**(8 & 9), 1023–1041 (1992).
41. D.J. Wilson, *Sep. Sci. Technol.* **27**(13), 1023–1041 (1992).
42. L.A. Roberts and D.J. Wilson, *Sep. Sci. Technol.* **28**(5), 1127–1143 (1993).
43. D.J. Wilson, R.D. Norris, and A.N. Clarke, *Sep. Sci. Technol.* **31**(5), 915–939 (1996).
44. D.J. Wilson, R.D. Norris, and A.N. Clarke, *Sep. Sci. Technol.* **31**(10), 1357–1376 (1996).
45. P.D. Lundegard, and G. Andersen, in *Proceedings of the Conference on Petroleum Hydrocarbons and Organic Chemicals in Ground Water: Prevention, Detection, and Restoration,* API/NGWA, Houston, Tex., pp. 461–476.
46. R.E. Hinkley, M-R. Chen, and J.E. Killough, in *Proceedings of the Conference on Petroleum Hydrocarbons and Organic Chemicals in Ground Water: Prevention, Detection and Restoration,* API/NGWA, Houston, Tex., 1995, pp. 223–237.
47. G.D. Beckett, D. Huntley, and S. Panday, in *Proceedings of the Conference on Petroleum Hydrocarbons and Organic Chemicals in Ground Water: Prevention, Detection and Restoration,* API/NGWA, Houston, Tex., 1995, pp. 207–221.
48. P.D. Lundergard and D. LaBrecque, *J. Contam. Hyd.* **19,** 1–27 (1995).
49. M.C. Marley, E.X. Droste, and R.J. Cody, "Mechanisms that Govern the Successful Application of Sparging Technologies," Presented at the *87th Annual Meeting of the Air and Waste Management Association,* Cincinnati, Ohio, June 19–23, 1993.
50. W.B. Kerfoot, "Slantwell Sparging," in *Proceedings of the Tenth Annual Conference on Contaminated Soils,* University of Massachusetts, Amherst, Mass., 1995.
51. Wisconsin Dept. of Natural Resources (WDNR), *Guidance for Design, Installation and Operation of In Situ Air Sparging Systems,* Wisconsin Dept. of Natural Resources, Emergency and Remedial Response Program, PUBL-SW186-93, 1993.
52. R.R. DuPont, C.J. Bruell, D.C. Downey, S.G. Hulling, M.C. Marley, R.D. Norris, and B. Privitz, *Innovative Site Remediation Technology, Bioremediation: Design and Application,* Phase II Monograph, William C. Anderson, ed., WASTECH® and The American Academy of Environmental Engineers, 1998.
53. H. McCleary, *Bioremed. Report* **3**(6), 1–3 (1994).
54. J.F. Billings, J.E. Griswold, and B.G. Billings, in R.E. Hinchee, R.M. Miller, and P.C. Johnson, eds., *In Situ Aeration: Air Sparging, Bioventing, and Related Remediation Processes,* Batelle Press, Columbus, Ohio, 1995, pp. 111–120.

SOIL SAMPLING AT HAZARDOUS WASTE SITES

BRIAN SCHUMACHER
J. JEFFREY VAN EE
EVAN J. ENGLUND
U.S. Environmental Protective Agency
Las Vegas, Nevada

The sampling of soil, or any heterogeneous media, requires an understanding of the spatial and temporal scales of interest to the decision makers. Depending on how the media is sampled and analyzed, and how the data are processed, almost any "valid" contaminant concentration can be obtained. In the absence of other data, data from the sampling and analysis of soil (or hazardous waste) are frequently assumed to represent the actual contaminant concentration in the vicinity of the sampling point when emphasis is placed on characterizing the distribution of a contaminant in the soil. When additional data are obtained in the vicinity of that sampling point, differences in contaminant concentrations almost always occur. The dilemma, then, is to try to determine the true concentration of the contaminant within a given area or volume of soil using the most efficient sampling design and minimizing the errors that occur during sample collection.

This article will provide a basic understanding of the processes and factors involved in the sampling of soils at a hazardous waste site. These processes and factors can be divided into several general categories, which include sample planning, documentation, quality assurance/quality control (QA/QC), sampling network design, site characterization, sample collection, and sample handling and preparation concerns. Basic guidance on dealing with factors influencing accuracy, precision, representativeness, and comparability of data and how to develop concise statements of the sampling effort objectives will be emphasized. Interpretation of the collected data will be discussed as a means for not only answering the basic objectives of the sampling effort, but also as a means for determining whether the quality of the data is sufficient to meet the needs of the user. Information on how to answer some of the most frequently asked questions, such as, "How many samples are needed and where should they be taken?" will be provided. Regrettably, no simple answers exist. Once the planning phase is completed and the soil sampler goes to the field, discussion of site features that influence the distribution and ultimate management of the contaminants at the site and how to correctly collect the soil sample will be presented. Unfortunately, no checklists are capable of addressing all of the important issues and factors that influence the sampling and analysis of soils and heterogeneous models. Site characterization and soil sampling are multimedia, multidimensional, multidisciplinary efforts that require vigorous communications among all of the principal parties involved. A genuine grasp of the basic concepts and critical definitions is extremely important to perform efficient and effective soil sampling for characterizing hazardous waste sites.

Some of the concepts presented in this article may be new and appear to conflict with existing guidance and procedures. In this article, the basic foundation for the efficient, best scientific investigation of environmental media is the U.S. Environmental Protection Agency (EPA) new data quality objectives (DQO) process. Nothing presented in this article conflicts with that process—a process that focuses on refining the critical questions that drive the decision-making process and on the data that are necessary to provide answers to those questions. The material presented in this article may cause people to rethink how they have been characterizing soils at hazardous waste sites.

SAMPLING PLANNING

Typically, at the beginning of a site investigation, historical information is sought to allow hypotheses and conceptual models of the site to be developed. All potential pathways of contaminant movement and exposure routes to living organisms should be identified and addressed. For example, if the contamination is in the soil, one pathway of concern may be the migration of the contaminant to the groundwater. The project planners should use the limited historical information about the site to determine what additional data are needed and where to collect them to address the critical questions of how much risk the site poses to humans and the environment.

Data are collected from a variety of sources and from a variety of locations to develop a conceptual model of the site. There may be no one complete conceptual model that exists for a given site and other models may need to be developed as the site investigation progresses and new information becomes available. A variety of mathematical models, and sometimes physical models, may be employed to estimate data among collected data points and to predict data at other times, at critical receptors, and in other media.

Far too often, efficient collection of samples and data is difficult because critical definitions and well-defined problem statements are often lacking at the beginning of an investigation even though the conceptual models may have been well defined. Usually, the controlling factors regarding the amount, quality, and type of data collected are time and money and not whether the results will adequately represent the site, fill the data gaps, or allow for a defensible decision to be made. The traditional thinking is that the more data that can be collected, the better the final conceptual model and decision-making process become.

Data that are collected in a site investigation are often viewed in a spatial context. The increasing popularity of computer-based geographical information systems (GIS) permits the rapid display of both two- and three-dimensional data. Although data from a site investigation are frequently analyzed using classical statistics, geostatistics is increasingly being used. None of these tools, however, can properly be used to answer critical questions about a site unless the critical questions have been properly highlighted and defined. Frequently, the generally accepted critical question that drives an investigation is whether an action level for one, or a variety of contaminants, is being exceeded. This question, by itself, is meaningless unless the *support* (1) or spatial volume that accompanies the action level is prescribed. Sampling heterogeneous materials, such as soil, can lead to a wide range of valid contaminant concentrations from which comparisons with an action level may be made depending on the volumes of material that are being represented. For example, the concentration of a contaminant in a 5-g sample may exceed the action level, whereas the concentration of contaminant in a soil core sample from which the 5-g subsample was obtained may not exceed the action level.

Terminology

A number of spatial units need to be considered and defined early in the investigation of a hazardous waste site in order to permit the efficient collection of data and the answering of critical questions (Fig. 1). All of these units are used in the development of the conceptual model for a site.

The first unit to consider is the distribution unit. The distribution unit is basically a volume of soil that has been contaminated. A large site may consist of a number of distribution units (ie, contaminated areas). Different processes may have been involved in the distribution unit for distributing

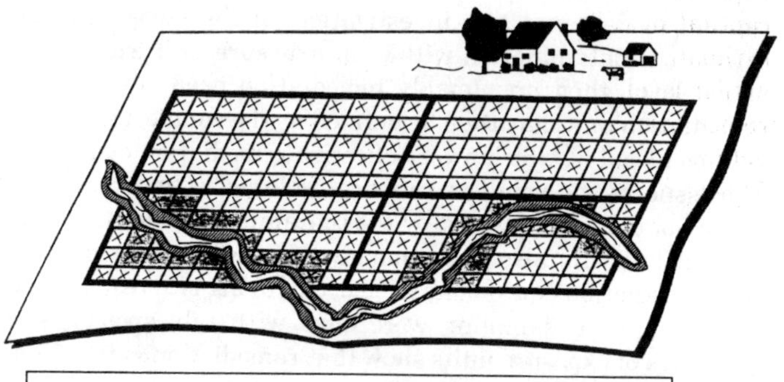

Figure 1. A site may be broken up into a variety of units to allow for critical questions to be addressed in the DQO process.

the contaminant. The contaminant may have been uniformly distributed as one might expect in a settling pond. Conversely, a contaminant may be distributed quite irregularly if leaking drums were stored in the contaminated area. It should be noted that distribution units may be described differently in different regulatory programs, eg, operable units. No matter what distribution units may be called, knowing in advance how the contaminant may have been distributed at a site should influence the decision-making process for determining how and where samples must be collected to answer critical questions about the site. The sampling procedures and number of samples needed to characterize an area where leaking drums were stored can be expected to be different than for the bottom of a settling pond. In statistical terms, the process of subdividing a site into different distribution units that may be sampled differently is called stratification.

The second unit to consider before an investigation progresses significantly is the exposure unit. The exposure unit is the volume of material, in this case soil, for which an action level or threshold concentration applies. Unfortunately, action levels are typically listed without the accompanying exposure unit. Action levels may be listed for humans for residential or industrial exposures, but usually the spatial area for which a person may be exposed to the action level (for some period of time and experience little risk) is seldomly defined precisely. Exposure units need to be defined as precisely as possible before extensive data collection begins if the exposure unit is going to serve as the foundation for the decision-making process. It may be necessary to spend some time to examine the basis and data for the action level to determine the type of exposure, underlying assumptions, and area for which actual exposure and risk of contact with the contaminant is of concern. For example, does the residential exposure unit assume 1/4 acre lots for which an individual may be exposed over a lifetime, on average, to unacceptable risks from the contaminant? Alternatively, are hot spots within the exposure unit of greater concern because the exposure and dose to children are expected to be higher? These are potential questions that must be considered when defining the exposure unit. The establishment of the exposure unit, together with the action level for a contaminant, is important in the development of the conceptual model and the associated decision-making process in a site investigation.

The third unit, the remediation unit is related to the exposure unit through the decision-making process and conceptual models in a site investigation. If the measured and estimated contamination within an exposure unit exceeds an action level, then presumably remediation needs to occur. A remediation unit may then be defined as the smallest practical volume of soil at which a cost-effective remediation can occur. The issue of the smallest practical volume is important because not only is it defined by the contaminant concentrations present but it is also operationally defined by the sampling and remediation equipment to be used during site restoration. For example, if sampling were done with tablespoons, and estimates of exposure units show that remediation is required, backhoes, rather than tablespoons, would more likely be used to remediate the area. Although samples were collected with a small support and decisions were based on those data, decisions would likely be made and based on a larger support. If only a small fraction of the exposure unit were contaminated (ie, a hot spot), would it be necessary to remediate the entire volume of the exposure unit, or can smaller areas be remediated? It may be possible to make decisions that are less costly on the basis of remediation units, eg, backhoe volumes, rather than entire exposure units. Addressing the volume of the remediation unit at the beginning stages of a site investigation allows for development of decision-rules that can guide the decision-making process and the subsequent selection of sampling procedures, analytical methods, and concentration estimation methods. (A decision-rule is described in the following section on EPA's new, revised DQO process.)

The problem of hot spot identification and the required, optimal sampling network becomes easier to address if the remediation units have been defined at an early stage. If hot spots were identified through the use of tablespoon samples at a particular grid spacing, it would not be practical to precisely define the boundaries of the hot spot at a relatively small spatial scale if a backhoe were used to remove contaminated areas. High-resolution sampling for hot spots may not be cost effective if the area to be effectively and economically remediated is on a larger scale. Definition of the remediation unit prior to the collection of a significant number of samples from the field permits the costs in obtaining data at various spatial densities (to address a decision-rule) to be balanced against the costs in remediating the area with no further information or data being required.

Data Quality Objectives

All three previously defined units, the distribution, exposure, and remediation units, may be used in the DQO process to achieve an optimization of the site characterization process. The DQO process fosters communications and the development of critical decision-rules for the study of soils and hazardous waste sites (Fig. 2). A stepwise, iterative process, involving major participants in the decision-making process for a site, is used to focus attention and resources on the critical questions or hypotheses that need to be tested. A variety of questions may be answered in an investigation but there are

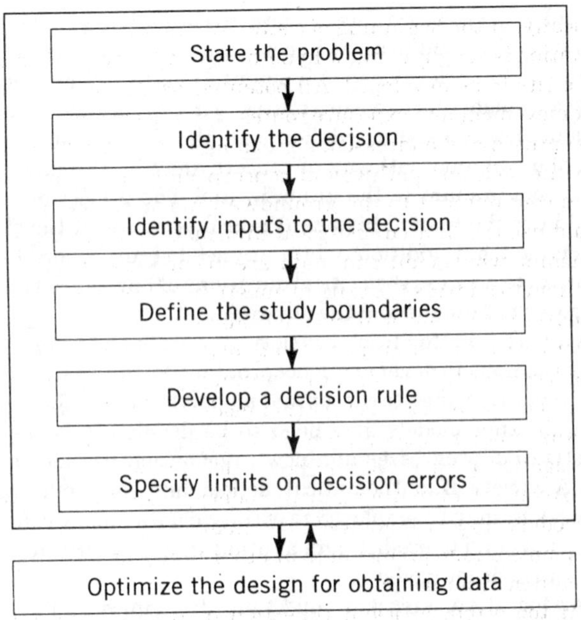

Figure 2. The data quality objective process.

usually just a few contaminants and questions that drive the decision-making process for a site investigation. An example of a decision-rule would be, if the mean concentration of arsenic exceeds the action level within the exposure unit by 10%, then the unit should be remediated by progressively removing those remediation units that contribute to the exceedance.

The DQO process has been used in the past to establish objectives for precision, accuracy, representativeness, completeness, and comparability for the reported data. The major focus of quality assurance and quality control efforts has been on the analytical phases of the sample collection and measurement process. Current DQO guidance from the EPA attempts to focus more efforts toward examining the sensitivity of the decision-making process to all key inputs, rather than just those related to sample analysis. In other words, what are the effects on the basic decision-making process (ie, the decision-rule) from imprecise, inaccurate, nonrepresentative, incomplete, and incomparable data from a variety of sources? This change in emphasis in the DQO process recognizes that there are numerous, and often significant, sources of variability outside the analytical process. There is variability in the process that is used to estimate contaminant concentrations and other physical parameters between measured points. There are errors introduced in the sample collection, subsampling, sample preparation, and sample transportation steps. All of these steps can introduce errors that are significantly greater than those encountered in the analytical process. A well-executed DQO process considers, in parallel, the tolerance for error in the basic decision-making process as well as the tolerance for error in the sample collection and measurement processes.

Further information on the DQO process may be found in the *Guidance for the Data Quality Objective Process* (2) and *The Data Quality Objectives Process for Superfund: Interim Final Guidance* (3) documents. A related document, *Guidance for Data Quality Assessment* (4) describes statistical methods for evaluating data to determine whether the DQOs have been met.

QUALITY ASSURANCE/QUALITY CONTROL

The overall goal of the QA/QC program in any investigation is to assess and ensure that the variability in the data is attributable, as much as possible, to the media and environment that is being measured and not to the measurement (including sampling) process. A properly designed and implemented QA/QC program can achieve that goal. In the field, QA/QC samples may be placed in the sample collection and preparation processing stream at several different locations. Depending on their placement, a variety of different errors can be assessed. However, the determination of the number, type, and placement of QA/QC samples within the sample stream is as problematic for a site investigation as the determination of the number, type, and location of routine samples that must be collected. All too often the number and type of samples to be used in both situations are determined by the projected budget for the site investigation and by estimated analytical costs and not by a judicious consideration of the acceptable error rates in making basic decisions (ie, false positives and false negatives).

A rationale has been developed for how to determine the number, type, and placement of QA/QC samples in soil (5). The decision-making process for determining how extensive a QA/QC program needs to be for a particular investigation is not simple. A stepwise process needs to be developed in which data from the QA/QC effort are progressively examined and compared to the sample data and to DQOs. Basic concepts need to be understood and applied to ensure that the data collected meets the needs of the decision-maker.

Data that are collected as part of an environmental investigation are going to exhibit variability. The expectation is that variability associated with the measurement process is sufficiently low that variability in the data can be attributed to the measured parameters or population. QA/QC programs are implemented to assess and control measurement variability, but these efforts are frequently insufficient or unable to assess or control all of the sources of bias and variability throughout the measurement process. Often, the QA/QC programs are targeted toward the analytical phase of the measurement process, but considerable bias and variability may be present in other areas.

As reported in the Rationale document (5):

$$\sigma_t^2 = \sigma_m^2 + \sigma_p^2$$

where σ_t^2 = total variability, σ_m^2 = measurement variability, and σ_p^2 = population variability. Also:

$$\sigma_m^2 = \sigma_s^2 + \sigma_h^2 + \sigma_{ss}^2 + \sigma_a^2 + \sigma_b^2$$

where σ_s^2 = sampling variability (standard deviation), σ_h^2 = handling, transportation, and preparation variability, σ_{ss}^2 = preparation variability (subsampling variability), σ_a^2 = laboratory analytical variability, and σ_b^2 = between batch variability. (It is assumed that the data are normally distributed or that a normalizing data transformation has been performed.)

Bias can change over time and become intertwined with imprecision—both of these contribute to variability in the measurement process. A variety of QA/QC samples may be used in a rigorous manner to try to assess the magnitude and source of errors in the sampling of soils at hazardous waste sites. Some of the errors may never be assessed because the true distribution and concentration of contaminants in the environment can never be truly measured; they can only be estimated. The best approach is to try to minimize bias and imprecision at the outset by carefully considering and selecting the sample collection, sample preparation, subsampling, and analytical techniques that are apt to produce representative data with low measurement error.

Some of the most common QA/QC samples used during the collection of soils include field evaluation samples, field blanks, field duplicates, and preparation splits. Field evaluation samples (eg, performance evaluation materials or site-specific soil quality assurance materials) are soil samples of known concentration that are subjected to the same manipulations as routine samples. The field evaluation samples should be introduced in the field at the earliest stage possible. These samples can then be used to estimate total measurement error and, if used in duplicate, precision estimates can be made.

Field blanks provide a measure of various cross-contamination sources, decontamination efficiency, and other potential sources of error introduced from sources other than the sample. Three common types of blanks used in the field include the field rinsate blanks, preparation rinsate blanks, and trip blanks. Rinsate blanks (ie, equipment blanks) are used to assess the efficiency of the decontamination process. These blanks are obtained by pouring deionized/distilled water over the sampling equipment (for the field rinsate blank) or preparation equipment (for the preparation rinsate blank) after it has been decontaminated and then collecting the rinsate for analysis. Trip blanks are generally used during the collection of volatile organic compounds. A trip blank consists of the sample containers filled with deionized/distilled water, solvent (eg, methanol), or evacuated vials used during headspace analyses. For some sampling efforts (depending on the DQOs), a portion of the trip blanks will be opened in the field (excluding evacuated vials) to assess if there is any contamination that may have occurred during sampling, preparation, or shipment to the analytical laboratory. The remainder of the trip blanks will not be opened in the field and can be used to assess contamination due to incomplete sealing of volatile organic analysis (VOA) vials and sampling jars or improperly cleaned glassware.

The field duplicate sample is an additional sample taken near a routine field sample and may be used to determine total within-batch measurement variability although differences in the two samples may also be attributable to short-range spatial variability in the soil. If the data from field duplicates are significantly different from the corresponding routine sample, then there is good cause to question the representativeness of the routine sample data and to employ data from other QA/QC samples to determine whether the problem is with short-range spatial variability. If the problem seems to be associated with short-range variability, then a larger support, or compositing of the samples on a small, spatial scale, may be all that is necessary to obtain more representative samples.

Preparation split samples are soil samples that are collected after the soil has been properly homogenized and, thus, all portions of the sample have theoretically equal concentrations. The preparation split samples are used to estimate the sum of subsampling, analytical, and data-handling error variations (ie, total within-batch error minus the error that occurs during sampling, handling, transportation, and preparation/homogenization).

SAMPLING NETWORK DESIGN

A successful sampling design accomplishes the objectives of the investigation at minimal cost. The DQO process is one approach that provides a sound framework for sampling network design. The DQO process accomplishes three critical tasks. First, the DQO process develops a clear, quantitative statement of the decision that must be made. This is generally in the form of an if-then-else decision rule: If the sample measurements exhibit characteristic x, then take action a; else, take action b. Second, by specifying tolerable error limits on the decision, the DQO process allows for the fact that the true concentration in volume v is never precisely known. Finally, the design-performance diagram can be used to evaluate the expected performance of alternate sampling design options and to select from among the acceptable designs the one with the lowest associated cost.

In the DQO approach, error limits are established across the entire range of possible contaminant concentrations, and can be plotted against true concentration to form a design-performance diagram (Fig. 3). The acceptable range of performance is the area including the gray region and the tolerable false positive and false negative regions. Each sampling scheme has an associated performance curve that shows the probability of making a correct or incorrect decision at any given true concentration. The performance curve (also known as a power curve) shown as the solid line in Figure 3 represents a sampling scheme that has acceptable performance because it falls within the specified limits. This particular curve also represents a minimum cost design because it meets

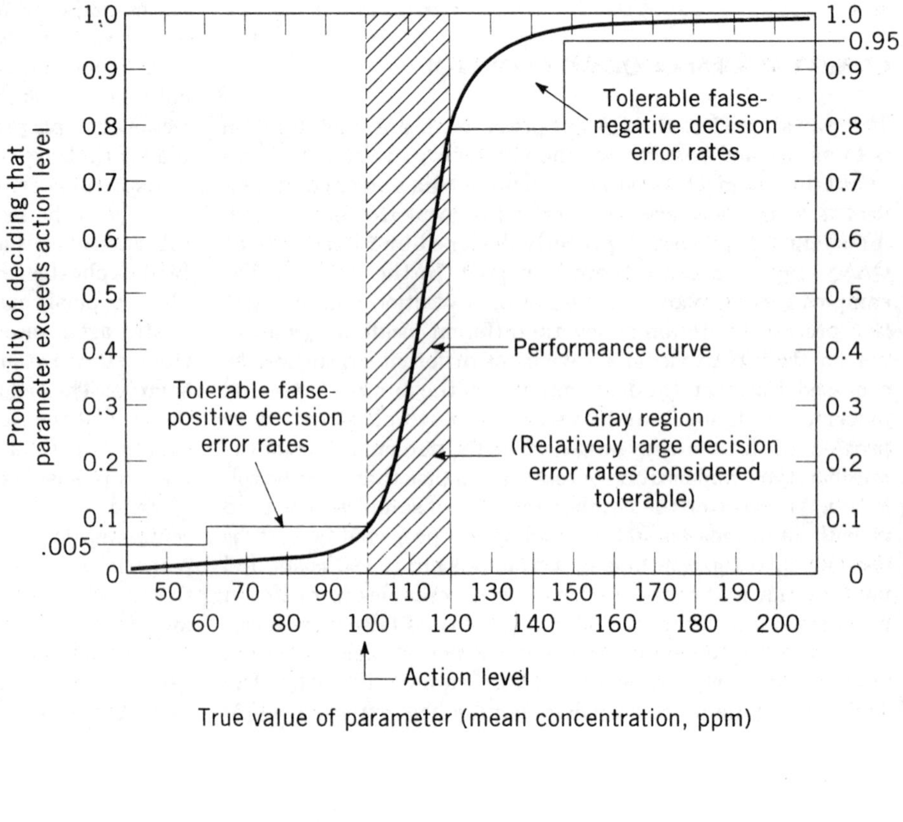

Figure 3. A typical design performance curve used to evaluate the expected performance of alternative sampling design options and to select, from among the acceptable designs, the one with the lowest associated cost.

but does not exceed the specifications. A sampling scheme with fewer samples or less accurate measurement would have higher error rates and be unacceptable, whereas a scheme with more samples or higher quality measurements would be acceptable, but unnecessarily expensive.

It is not easy to develop design performance criteria. Determining acceptable decision error rates is a subjective process that is difficult and uncomfortable for many people to perform. There is a tendency to initially set the error rates so low that they lead to unacceptably high sampling costs, in which case the process must be repeated to obtain a more reasonable sampling design with lower overall associated costs.

Basic Spatial Sampling Strategies

To illustrate some of the basic spatial sampling strategies, two common decision scenarios in soil remediation will be examined. Additionally, alternative sampling strategies under the same decision scenarios will be discussed.

The first decision scenario involves the estimation of the mean concentration of contaminant in a specified volume of soil. Assume that a single distribution unit, or stratum, has been identified, such as the sediment in an abandoned waste lagoon, and that the decision maker will have to remediate the sediments if the mean concentration exceeds a specified threshold. From a design-performance diagram it has been determined that a sampling design will be acceptable if it the standard error (*se*) of the mean is less than or equal to 4, or equivalently, a variance (V_e) less than or equal to 16 ($se^2 = V_e$). Further, assume that the measurement error variance (V_m) for collecting and analyzing a soil core is 10, and the population variance (V_p) is 90 (ie, 10% of the total variance is due to measurement error including sample collection, preparation, and analysis, whereas 90% is due to natural variability at the site). The total variance (V_t), including measurement error, is:

$$V_t = V_p + V_m = 100$$

For this scenario, a soil core costs \$25 to collect in the field and \$300 to analyze in the laboratory. The question that needs to be answered is, "What is the best sampling design?"

The classic approach to estimating a population mean is by the arithmetic mean of a random sample. Random sampling is always a valid method, even though it may not always be the most efficient for spatial sampling. In the random sampling approach, it is simple to calculate the standard error of the mean and evaluate the costs of alternate sampling schemes.

For a random sample of N measurements, the variance of the error of the mean (V_e) is simply equal to V_t/N. The design process now involves nothing more than finding the smallest value of N that will meet our example design requirement, namely, $V_e \leq 16$. In this case, $N = 7$, making $V_e = 100/7 = 14.3$ which is <16. The final cost of the random sampling design is $7 \times \$325 = \$2275$.

The bulk of the cost in the case of this random sampling design is due to laboratory analysis, so let us next consider a less expensive alternative. Assume that an alternate laboratory method is available at a cost per analysis of \$100, but that this method is less precise, with $V_m = 40$ instead of 10. If the initial study objectives were to get high quality measurements of individual soil cores, this approach would not be useful. By taking 4 separate measurements of a single core sample, V_m could be reduced to $V_m/4 = 40/4 = 10$ and equal the precision/quality of a single measurement by the original method. But the cost of the replicated analyses would be $4 \times \$100 + \$25 = \$425$ per sample—more expensive than the original method selected. However, the original decision rule has nothing to do with measuring individual cores, only with estimating the mean of the sediments in the waste lagoon. For this purpose, $V_t = V_p + V_m = 90 + 40 = 130$. To meet the required objective, nine measurements would be needed so that $V_e = 130/9 = 14.4 < 16$. The cost of this design alternative becomes $9 \times \$125 = \$1125$, a substantial financial improvement from the \$2275 cost of the initial random sampling design with the more precise (but more expensive) analytical method.

Suppose that with the initial measurement method, where $V_m = 10$, the population variance is much higher than the measurement variance. In such cases, composite sampling can be a cost-effective alternate approach. In composite sampling, individual core samples are physically combined to make a single unique sample from which one or more measurements of the sample can be made. The overall estimation error is: $V_e = V_p/N_c + V_m/N_a$, where N_c is the number of cores or increments that make up a composite sample, and N_a is the number of analyses performed on a single composite sample. There will also be an additional error component and an added cost introduced by the process of homogenizing and subsampling the composite, but for this example, it is assumed that this added error and cost are negligible. The problems associated with subsampling a composite sample are far from trivial and will be discussed in detail in the subsequent section on particulate sampling theory. In the terminology of that theory, a composite sample is a multiple-increment sample taken to reduce the effects of short- and long-range heterogeneity.

The composite sampling design process becomes one of evaluating combinations of N_c and N_a and selecting the lowest cost combination that meets the initial design requirements. In this case, the best result is obtained when $N_c = 15$ and $N_a = 1$ (ie, a single analysis of a 15-core composite sample). The error associated with the alternate approach would be: $V_e = 90/15 + 10/1 = 6 + 10 = 16$, and the associated cost is $(N_a \times \$300) + (N_c \times \$25) = \$675$. This achieves the original objectives at a lower cost than with individual samples.

From these examples, it has been shown that an improvement on the original design can be achieved by using a lower cost analytical method or by compositing. Can the project planners do even better by combining both methods and analyzing composite samples by the cheaper method? Interestingly, in this particular example, the answer turns out to be no. With a single composite sample, measurement error is again a significant factor, and the cost of reducing V_m through multiple analyses becomes cost-prohibitive. The best option to meet the original goal of $V_e \leq 16$ is with $N_c = 15$ and $N_a = 4$ (where $V_e = V_p/N_c + V_m/N_a = 90/15 + 40/4 = 6 + 10 = 16$), for a total cost of \$775 [equals $(15 \times \$25) + (4 \times \$100)$]. Although this approach costs slightly more than the single, more precise analysis (\$400 vs \$300), the four replicate analyses on the composite sample provide at least some measure of quality assurance in terms of precision (laboratory analytical repeatability).

This example was intended to illustrate the type of economic analysis of sampling alternatives that is part of the DQO process, not to provide a blanket endorsement of composite sampling or cheap but imprecise analytical methods. With different costs, variances, and objectives the conclusions could be quite different.

Estimating Variances

Although this analysis is relatively simple, the results are only strictly true if *all* of the assumptions are correct. The major assumptions were:

- the variance of the population distribution and the measurement error distribution for each measurement alternative are known,
- the true costs of sample collection and measurement are known, and
- all measurements are independent of each other.

In practice, these assumptions are at best only approximately true, so it is worth looking at how they affect the analysis of sampling alternatives. The relative magnitude of the variance of the population versus the variances of the alternative measurement methods and the relative costs of the alternatives are all important in determining the overall sampling strategy. Generally, the relative costs of measurement methods can be estimated reasonably accurately. The more difficult problem is estimating the error variances. The true variance of a contaminant in soil at a site or within a stratum will always remain unknown. The best way to estimate it is through the analysis of sample data, but this poses a dilemma. It is more difficult to estimate variances precisely than it is to estimate means. It is likely that by the time one collects enough samples to get a satisfactory estimate of variance, there will be far more data than needed to efficiently estimate the mean. The situation becomes worse if composite sampling or another alternate measurement method were evaluated because an estimate of the measurement variance(s), in addition to the total variance, is needed. Analytical variances are often well-known, but measurement variance includes many sampling and subsampling error components that are site-specific. Estimating measurement variance from sample data requires the analyses of multiple field replicates. Estimating these variances can easily cost many times more than simply estimating the mean.

The alternative to estimating variances from data is to estimate them through educated guesswork (ie, best professional judgement). Usually, at least a few measurements will be available from previous investigations which can in turn be used to estimate the measurement variance. Analogies can be made with data sets from other similar sites. Measurement variance can likewise be estimated from quality assurance data collected at similar sites, or it can be estimated by combining the analytical variance from laboratory studies with estimates of preparation and subsampling variances from particulate sampling theory (as described later in this article).

The question of which approach to take to estimate the variances should be addressed as part of the data quality objectives process. In addition to determining the decision to be made and how accurately it must be made, it is also necessary to know how accurately the decision maker needs to know how accurate the decision is. Unfortunately, there is no standard formula for determining these requirements. In the previous example, if the single-composite, single-analysis option is chosen, there will be no way to evaluate whether or not the desired precision is achieved. The driving factor will be the magnitude of the consequences of decision errors. Suppose in the previous example that the variances were preliminary estimates based on prior investigations of similar sites. Suppose, also, that the decision involved is relatively minor; for example, whether to leave slightly contaminated soil in place or remove it to a municipal land fill at a cost of $5,000 to $10,000. In this situation, the described analysis might well be sufficient, and one of the compositing schemes discussed would be chosen. However, if the decision were between disposal in the municipal landfill versus in a hazardous waste facility at a typical cost of $100,000, it is reasonable to spend more money on sampling in order to be more confident about the quality of the decision. Again, there is no unique simple strategy for such cases; nonetheless, some form of sequential sampling may be appropriate.

In sequential sampling, an initial set of data is collected and evaluated. The decision to be made is now threefold: yes, remediation is required; no, the soil is clean and can be left *in situ*; or, collect more data to help better define the situation. The initial date is used to estimate the mean, variance, and standard error, and if the standard error is low enough, the decision is made. If the standard error is too high, the variance is used to estimate how many additional samples are needed in the second sampling phase. The process is repeated until a confident yes or no decision can be made. The number of samples collected in the initial set is determined by economics and logistics. If measurements are being made in the field, it may be practical to re-evaluate the data after each measurement. This approach minimizes the total number of samples required. However, if the samples must be sent to a laboratory for measurement, there are time delays and mobilization costs each time a sampling crew returns to the field. This may make it advantageous to make a conservative initial estimate of variance and attempt to over-sample the initial phase to avoid further sampling.

Sequential sampling can be used with either individual or composite samples. Unfortunately, economic analysis of the alternatives is very difficult, especially when the population is not normally distributed. Real-world distributions of contaminated soil measurements are often highly skewed and contain nondetects (ie, data below the instrument or method detection limit). In such cases, the only practical method for comparing the cost-effectiveness of alternative sequential designs is through rather elaborate computer simulations of the sampling process. Composite samples may well have an advantage in that the composite samples will be more normally distributed than the original population. The occurrence of a more normally distributed population makes it more likely that normality-based tests, such as Students *t*-test, can be used with a small number of measurements.

Spatial Dependence

In the previous examples, it was assumed the data were independent. However, the fate and transport of contaminant in soil is determined by physical and chemical processes that do not operate at random. This is also true of natural soil parameters such as clay content, porosity, etc, that can influence the spatial distribution of contaminant. In general, it is expected that soil measurements taken a few centimeters apart would be more similar in concentration than when the measurements are taken kilometers apart. Quantifying such spatial dependence and how it affects sampling and estimation is the subject matter of geostatistics. The reader is referred to

Reference 6 for a practical introduction to geostatistics. Additionally, Reference 7 provides an excellent discussion of the specifics involved in spatial dependence during soil sampling applications. Only a brief discussion of the basic concepts and a few rules-of-thumb for sampling and estimation will be provided in the following text.

Spatial dependence can be quantified in the form of a semivariogram (Fig. 4), which shows the variance of measurements as a function of the spatial vector (distance and direction) separating the points. More precisely, the semivariogram shows the expected value of $(z(x)-z(x+h))^2/2$, where z is the analyte concentration at points x and $x + h$, and sampling location $x + h$ is separated from sampling location x by the vector h. The semivariogram incorporates all of the components of the total variance. The y-intercept at $h = 0$ includes all of the independent measurement variance components, such as field and laboratory preparation, subsampling, and analytical error. The semivariogram at very small h values (ie, very small distances from the original sampling location) would reflect the variance of collocated samples. The shape of the semivariogram curve defines the spatial dependence structure. The mean value of the semivariogram over all possible sample pairs at all possible separation vectors is simply the total population variance. The semivariogram provides a very useful model of both independent and dependent variance components, but like any of the components, creating a semivariogram entirely from data can be very expensive.

When data are spatially dependent, the arithmetic mean is no longer the best (minimum error variance) estimate of the population mean. Instead, a weighted mean should be used. In geostatistics, a weighting method called kriging is frequently used, in which the semivariogram is used to compute sample weights that minimize the estimation variance. When two or more samples are clustered very closely together (ie, sampling distances between two points is small), whether by chance or by design, they are effectively providing duplicate information. This duplication is indicated by low semivariogram values among neighboring samples and smaller weights are assigned to such samples. Conversely, samples with no near neighbors receive higher weights. By down-weighting clustered data that are duplicative, kriging provides better estimates than simple averaging.

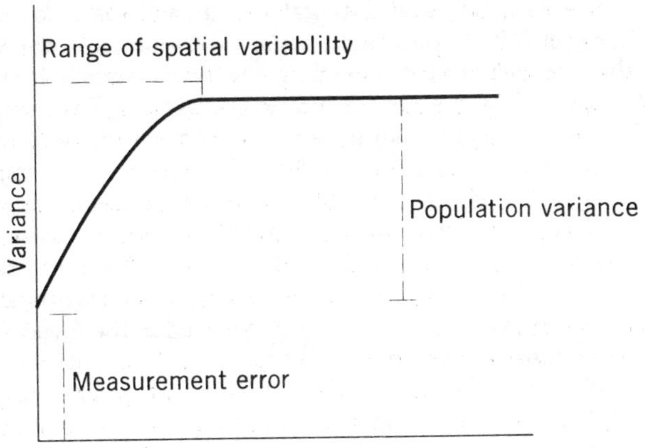

Figure 4. A typical semivariogram showing the variance of measurements as a function of the spatial vector (distance and direction) separating the data points.

The most efficient sampling design, when spatial dependence exists, is one that minimizes the duplication of information; or equivalently, maximizes the distance from each sample to its nearest neighbor. The solution to this problem is simple: use a systematic sampling scheme and sample at regular sampling grids. It has been shown (7) that grid sampling can be significantly better than random sampling in some cases. They also reported that the arithmetic mean of a regular grid is essentially equivalent to the kriged mean for the same data set. This suggests that regular grids are also optimal for composite sampling where the physical mixing of the soil cores is equivalent to arithmetic averaging. Triangular grids are the most efficient (in two-dimensions) but square grids are very nearly as good and are generally more convenient in the field. Because it is usually safe to assume the existence of some form of spatial dependence in soil contamination, regular grid sampling is recommended, preferably with a random origin and orientation.

Sampling for Local Estimation

The purpose of local estimation is usually to classify a site into higher concentration areas that need some form of remedial action and lower concentration areas that need none. When it is assumed that the site contains high and low concentration areas, it is the same as saying that there is spatial dependence (ie, similar values tend to be grouped together). This spatial dependence becomes a significant factor in sampling network design. Similar to global estimation, sampling on regular grids is more efficient for local estimation.

Local estimation and decision-making can become very complex. It is critical to establish a very clear decision rule at the beginning that must include the scale of the decision. The simplest approach to local estimation is to subdivide the site into an array of discrete units. These discrete units may be as large as exposure units or as small as remediation units. Generally, high remediation costs lead to the consideration of smaller decision units in an attempt to be more selective in which units must undergo the expense of being remediated. Each unit will be classified based on an estimate of its mean concentration. Subdividing a site into smaller remediation units means that many decisions will be made instead of just one and that the consequences of error for any given decision become much smaller. The DQO process should reflect this by establishing broader error tolerances for smaller units.

Composite sampling remains an attractive option for local estimation because compositing increases the area represented by each sample and reduces the sample variance. Compositing is particularly effective when individual core concentrations vary greatly over distances that are small compared to the scale of decision, and when this short range variability is high relative to measurement error. Generally the larger the area over which a composite is taken, the greater the reduction in sample variability; however, the composite area should not be larger than the decision unit.

Local estimation differs significantly from global estimation because the local unit being estimated is a part of the larger population. As a result of spatial dependence, each unit is correlated to some extent with its neighboring units. It becomes possible to estimate the concentration of a unit by interpolation from nearby data, even when there are no samples in the unit itself. In this case the design problem is no longer one of adequately sampling each unit, but rather one of sampling the population area in order to adequately estimate each unit.

There are two basic approaches to optimizing sampling designs for local estimation. Both require an estimate of the semivariogram. The first approach is based on the fact that when estimates of remediation units are made by kriging, estimates of the corresponding standard errors are also produced. The basic assumptions involved in kriging result in the standard error estimates being determined by only the semivariogram and the sample locations, not by the data values, so it becomes possible to calculate kriging standard errors for hypothetical sampling network designs. The performance of alternate designs can be compared to the established decision error limits. The approach has been detailed (8,9). The major problem with this approach is that kriging estimates of kriging standard errors are quite sensitive to errors in estimating the semivariogram, and to errors in the basic assumptions.

A promising alternative approach to the problem of developing complex designs for local estimation has been described (10,11). This approach involves first creating a detailed computer model of the soil contamination distribution that is consistent with both the semivariogram and with any available data using a geostatistical technique called conditional simulation. The resulting model contains a very dense grid of data points (typically, 10,000 or more points). A few of these grid points will contain actual measured data values whereas the remainder of the grid points will contain simulated measurements that are both reasonable and realistic in detail. This approach is somewhat less sensitive to errors in estimating the semivariogram because the model is forced to fit the actual data values. After the detailed model has been constructed, it is possible to simulate the entire process of sampling, estimation, and decision-making for any particular sampling design and to compare the cost-effectiveness of alternative designs. The basic procedure for evaluating a sampling design by conditional simulation is as follows:

1. Overlay a grid of decision units over the modeled area. Each unit will contain many simulated sample values.
2. Calculate the mean concentration of all the simulated values in each unit. These are the true concentrations of the units. Comparing these means to a specified action level determines what the correct remediation decision for each unit should be.
3. Select a sampling design scheme, such as 50 samples at randomly selected locations over the model area.
4. Determine a set of sample locations and draw the simulated sample concentration values from the model.
5. Estimate the mean concentration of each decision unit, apply the decision rule, and determine whether the decision is correct by comparison with the true concentration from step 2.
6. Repeat steps 4 and 5 at least 100 times with different sample sets generated by the same sampling scheme and keep track of all decisions obtained.
7. On a design performance diagram, plot the proportion of correct decisions versus the concentration for each of the decision units. If all of the points fall within the design performance limits, the sampling scheme is deemed acceptable.

The conditional simulation procedure is complex and clearly best suited to large, expensive remediation projects. Its primary advantages include the flexibility to evaluate alternate spatial designs, such as regular grids and composites; to tabulate detailed performance statistics, such as costs of sampling and remediation; and to quantify the amount of contaminant that would remain unremediated following the selected sampling design.

This simulation approach can also be used to evaluate the cost-effectiveness of the overall sampling/remediation strategy. In a design-performance diagram like that in Figure 3, the right-hand boundary of the acceptable performance zone limits the probability that highly contaminated soil will remain unremediated, whereas the other boundary limits the probability that relative uncontaminated soil will be unnecessarily remediated. The former is of primary interest to the regulator because it protects against risk to humans and the environment. The latter boundary, however, is primarily an economic choice on the part of the party responsible for remediation. The trade-off is between sampling and remediation costs. In practice, the actual decision level will occur at estimated concentrations near the center of the gray region. Moving the left-hand boundary farther to the left lowers the effective decision level and increases the amount of remediation to compensate for the greater uncertainty due to less sampling. Moving it to the right raises the effective decision level and increases the amount of sampling required. Simulation can be used to evaluate these trade-offs and find the most cost-effective solution that meets the regulatory requirements.

DOCUMENTATION

Accurate documentation is essential for the success of a field sampling program. Documentation should occur in all phases of a soil sampling program including: planning, sample collection, and laboratory analysis. Three documents usually required are the field sampling plan (FSP), quality assurance project plan (QAPP), and health and safety plan (HASP). The FSP provides guidance for all fieldwork by defining in detail the sampling and data-gathering methods to be used on a project (12). Topics that should be discussed in the FSP are the site background, sampling objectives, sample location and frequency by matrix (including QA/QC samples), sample designation, sampling equipment and procedures with standard operating procedures (SOPS), if available, and sample handling and analysis. The QAPP describes the policy, organization, functional activities, and QA/QC procedures necessary to achieve the DQOs defined for the project. The detail to be included in QAPPs depends on the category of investigation being undertaken. Most investigations under the Superfund program have compliance or litigational implications and thus require discussion of the following 14 points: project description; project organization and responsibilities; quality assurance objectives; site selection and sampling procedures; sample custody; calibration procedures and frequency; analytical procedures; data reduction, validation, and reporting; internal quality control checks; performance and system audits; preventative maintenance; calculation of data quality indicators; corrective action; and quality control reports to management (13). If both a FSP and QAPP are required for the same project, to avoid duplicative efforts for sections that contain the same information, it is recommended to simply reference the overlapping section in the one document from the other document. The HASP is prepared as a support document to protect the health and safety of the soil sampler and to meet OSHA guidelines and regulations (12). Topics to be discussed in a HASP include: identification of key health and safety personnel (eg, site health and safety officer); hazard risk analysis by task and operation; employee training requirements; personal protective equipment; medical surveillance requirements; frequency and types of air monitoring; site control measures; decontamination procedures; standard operating procedures for the site; emergency response procedures; and confined space entry should sampling be performed in a pit or hole.

During sampling and analysis, it is only through proper documentation that a soil sample can be linked with a sampling location (including depth) and the date and time of collection. Use of detailed field and laboratory logbooks to document routine information (eg, sampling date, weather conditions, sample numbers, QA/QC sample use, air monitoring data, etc), problems encountered during sampling (eg, inability to penetrate to desired depth), sampling location changes, site characterization information (to be discussed), soil characteristics, production of the site map (to be discussed), or analytical procedural changes. The use of standardized forms greatly enhances field documentation by ensuring the collection of complete information required at the site. Other required documentation associated with sample identification and shipment include: sample labels; chain-of-custody forms, custody seals, and shipping airbills (14).

SITE CHARACTERIZATION

Whenever a soil sampler, sampling team, or management team approaches a new site that needs to be evaluated, one of the initial steps performed should be site characterization (ie, site description). A good site characterization program consists of two major steps, namely, site observation and subsequent site mapping. Site characterization may be limited to a reconnaissance survey or involve a full-scale (or definitive) survey. A reconnaissance survey is performed to confirm historical data through the collection of a few selected soil samples for analysis or to identify site features that may indicate contaminant sources, pathways, affected populations, and potential monitoring/sampling locations. A full-scale survey is undertaken to obtain soil samples for analysis with their results being used in the final decision-making process (15).

For both survey types, a basic suite of characteristics should be noted and recorded during the site visit. This suite includes information and observations on climate and weather, slope, surface erosion and erodibility, surface runoff potential, vegetation, and the presence of macro- and meso-fauna. Eventhough these parameters are readily observed, they should be recorded because they may provide explanations for observed trends in the data and possible considerations for long-term planning and remediation efforts at the site.

In addition to the site characteristics, numerous soil characteristics should be examined and recorded in the field because they provide information on potential subsurface transport pathways, variability of soil properties, location of contaminants, and identify features that may influence remedial operations. More complete descriptions and information on site characteristics are available (16,17) (see SITE CHARACTERIZATION, EXPEDITED).

Site Observations

Climate and weather are site characteristics that often influence the collection of soil samples. Climate, the average condition of the weather over a period of years, influences the time of year that sampling and site remediation efforts can occur (16). Clearly, soil sampling during the winter months in the northeastern United States may be restricted due to cold weather and frozen ground conditions, yet in contrast, winter sampling in the southwestern United States may be preferable to avoid the extreme high temperatures commonly encountered during the summer months. Weather conditions, the state of the atmosphere during the field investigation, influences the specific time of sampling. Specific conditions to note include temperature, precipitation, wind speed and direction, and humidity. While often noted, but not recorded, these conditions generally influence the health and safety of the sampler and have little direct influence on soil sample collection with the exception of precipitation. For example, when the wind is blowing, samplers should always position themselves upwind to avoid inhalation of dust and other particles that may contain the contaminants of concern. During time of high winds, sampling should be avoided to prevent the spread of the contaminants sorbed on fugitive dust particles leaving the waste site.

Noting the slope of the soil surface is important because it influences the rate and amount of runoff and erosion of water, soil, and associated contaminants (16,18). Slope features to be noted include gradient, length, shape, and topographic position. Simple terms such as flat, moderate, or steep can be used to describe the gradient. If greater accuracy is required, slope gradients may be recorded as a slope percentage (the change in elevation per horizontal distance between two points) or a slope angle (measured in degrees from horizontal). Slope length, usually measured in meters or feet, influences the behavior of water and the potential for erosion. In general, longer slopes have greater runoff and erosion potential than shorter slopes (18). Slope shape (eg, convex, concave, or flat) and topographic position (eg, summit, shoulder, sideslope, or floodplain) influence the movement of water and soil on the surface and in the subsurface. For example, steep convex shoulder positions are more likely to experience erosion than flat or concave floodplains in which soil deposition is likely to occur (19).

Site characterization of surface erosion is important to assess soil loss or deposition in the past and to assess future erosion potential. The determination of surface erosion is easily done visually by comparison of observed changes in soil texture and color between surface and subsurface horizons (16). If subsurface colors or textures are noted at or near the surface, then that area has undergone and is prone to erosion. The presence of rills, gullies, or other erosional features also indicate erosion is occurring at the site and mark potential pathways for movement of contaminated soils within and from the site. Valuable information on surface erosion and erodibility of soils at a site can be found in county soil surveys published by the U.S. Department of Agriculture (USDA)–National Resource Conservation Service (formerly the USDA–Soil Conservation Service).

Surface runoff potential is used to evaluate the potential for transport of contaminants at the soil surface to surface streams or other water bodies (16). The runoff potential is controlled by slope and saturated hydraulic conductivity. Saturated hydraulic conductivity may be defined as the rate of movement of water through a soil or as a measure of the ability of a soil to transmit water under saturated conditions (20). Runoff is directly proportional to slope (ie, as slope increases, surface runoff potential also increases) and inversely proportional to the saturated hydraulic conductivity (ie, as the hydraulic conductivity increases, the runoff potential decreases). Similar to surface runoff, valuable information on soil hydraulic conductivities and surface runoff may be found in soil surveys published by the USDA-National Resource Conservation Service.

Vegetation serves as an indicator of site history, erosion potential, and contaminant location. During site characterization, the sampler should observe the nature (eg, hardwood forest), kind (eg, maple trees with mixed understory), extent (eg, dense, scattered, sparse, or bare), and distribution of the site vegetation. The presence of stunted vegetation, luxuriant plant growth in comparison to surrounding areas, discolored leaves, and burn spots can all indicate the toxic effects of contaminants in the soil. In areas of heavy metal contamination, sampling of vegetation along with soil sampling may be desirable to assess potential metals exposure through bioaccumulation (16).

Macro- and mesofauna, like vegetation, can provide an indication of the presence of contaminants. Additionally, during their routine daily activities, fauna create potential pathways to enhance vertical movement and leaching of contaminants to depth within a soil, mix the soil, and are capable of moving large quantities of subsurface materials to the surface. Macrofauna are those animals that can be measured in centimeters (or inches) such as burrowing rodents, earthworms, and insects (16). In contrast, mesofauna are the smaller animals including other insects, arthropods, nematodes, and smaller worms. Simple terms, such as many, common, few, and none, can be quickly recorded to indicate the presence of indigenous fauna. A lack of any visible fauna may indicate an area of high contaminant concentrations.

Site Mapping

The second step in the site-characterization process is the production of a site map. Site mapping generally occurs in two phases: initial and final mapping. The site map is important because it allows for planning sampling locations and allows for the subsequent plotting of the contaminant spatial distribution. The initial site map is a crude sketch of the sampling site. Features that should be drawn on the site map include both permanent and temporary objects. Permanent objects to be mapped include playgrounds, buildings, roads, railroads, boulders, and bodies of water (eg, streams or rivers). Temporary objects include trees, fences, power poles, and telephone poles. Drainage ways (eg, gullies or depressions) are sometimes plotted because they represent potential pathways of contaminant migration. Other features that may be included on the initial site map are bare spots, percent vegetative cover, and unusual surface features such as soil discolorations and surface staining.

Once the first draft sketch of the site has been completed, the approximate locations of the sampling points can be plotted based on the sampling design. More exact locations of the sampling points can be made using surveying equipment or global positioning systems (GPS) units depending on the needs of the program. It will be from these marked points that the samplers will collect the soil samples. If changes in sampling location need to be made due to unmapped surface features or unforeseen subsurface obstacles (eg, large tree roots, buried rocks, etc), the sampler should indicate the type of obstacle encountered, move approximately 30 cm (12 inches) in any direction away from the obstacle, and try to recollect the sample. This process may have to be repeated several times in different directions from the initial point or at greater distances until a satisfactory location is found. The new sampling point should be marked on the site map to indicate the sampling location change.

After sampling has been completed, the final location of each sampling point needs to be accurately determined. Exact locations of the final sampling points can be made through a variety of different techniques ranging from measuring the distance and direction from a fixed reference point (or points) with tape measures, through the use of field survey equipment and GPS units. Each sampling point should be reconfirmed as to its approximate spatial location in the field and on the site map. A more formal site map may then be produced using this collected information in conjunction with, or separate from, the analytical results.

SAMPLE COLLECTION

After the site has been adequately mapped and all important site observations have been noted, the next step is to collect the soil samples. It is often quoted that it is during sample collection that a majority (80% or more) of the total measurement error occurs and yet it is during this phase of a program that the least attention to detail is paid by the decision-makers, project managers, chemists, etc. The following text provides guidance on how to properly collect a soil sample taking into account all the potential pathways for error to enter the system, provides means to minimize these errors, provides information on common concerns during the selection of sampling equipment, and discusses the advantages and disadvantages of commonly available sampling tools.

Particulate Sampling Theory

A particulate sampling theory was developed by Pierre M. Gy for the mining industry to provide a more accurate and precise means for identifying heterogeneous ore grades. The theory is based on the relationship that exists between the variability of the material, the particle sizes within the material, the distribution of the component of interest (eg, pollutant), and the size of the sample collected. Particulate sampling theory is based on sampling correctness, which in turn is based on a property intrinsic to the material itself and the equipment used to extract the sample (21). Gy defines a correct sample as "A sample in which all particles in a randomly chosen sampling unit have the same probability of being selected for inclusion in the sample". The importance of this theory to environmental soil sampling is that the theory identifies various sources of error that can influence the final determined contaminant concentration and presents the means for controlling these errors. The following text is a very brief synopsis of Gy's particulate sampling theory as it pertains to soil sampling. A more complete and in-depth discussion of all aspects of the sampling theory is available (21).

Two models are described in Gy's particulate sampling theory: a continuous model which deals with variables through time and space, and a discrete model which addresses sampling populations of fragments. Because soil sampling is more closely related to sampling populations of discrete fragments, only those errors associated with the discrete model will be discussed. The errors associated with the discrete model include the fundamental error; grouping and segregation; short-range heterogeneity; increment delimitation; increment extraction; preparation; and analytical errors.

Sources of Variation and Sampling Error. One of the key concepts of Gy's particulate sampling theory and its determination of the fundamental error is that the relationship between the maximum particle size in a collected sample and the weight of the sample to be collected must be clearly defined. By following the formulae (to be presented), the sampler can select a particle size range that will reduce the relative variance in the final results to an acceptable level (as defined in the DQO process) and ensure that enough soil is collected to present an unbiased sample to the laboratory for analysis.

The fundamental error (FE) is associated with the natural variability inherent (ie, the heterogeneity that is inherent to the composition of each fragment or particles making up the lot) in the composition of every particle making up the lot to be sampled (21). A lot is defined by Gy as the batch, volume, or sampling unit from which increments and samples are collected. Lots may range in size from a bottle of soil at the laboratory, to a dump truck load of soil, through the entire hazardous waste site under investigation. The fundamental error is the only error that cannot be eliminated; however, it can be reduced by comminution of the maximum particle size or by taking larger sample sizes in the lot. Mathematically, the fundamental error is defined as:

$$s_{\mathrm{FE}}^2 = (1/M_{\mathrm{S}} - 1/M_{\mathrm{L}}) IH_{\mathrm{L}}$$

where s_{FE}^2 = relative variance of the fundamental error; M_{S} = mass of the sample (g); M_{L} = mass of the lot (g), and IH_{L} = constant factor of constitution heterogeneity. When the mass of the lot, M_{L}, is large in relation to the sample weight to be collected, M_{S}, the formula can be simplified to:

$$s_{\mathrm{FE}}^2 = IH_{\mathrm{L}}/M_{\mathrm{S}}$$

IH_{L} can be estimated using the following relationship:

$$IH_{\mathrm{L}} = fgc\ell d^3$$

where f = shape factor (dimensionless); g = particle-size (or granulometric) factor (dimensionless); c = mineralogical factor (g/cm^3), ℓ = liberation factor (dimensionless), and d = diameter of the largest particle (cm).

The shape factor (f) addresses error due to all fragments not being perfect cubes and thus not fitting perfectly through the square holes in sieves used to screen the soil (21). The shape factor is influenced by the number of particles passing through two consecutive upper and lower sieves, average particle diameter, average particle density, and the mass of the fraction to be analyzed. Mathematically,

$$f = M/pd^3\lambda$$

where M = mass of fraction collected on the lower sieve (g); p = number of individual particles; d = average diameter of particles passing through the two sieves (cm), and λ = average density of collected fraction (g/cm^3). For all practical purposes, a quick examination of the sample under a microscope is sufficient to determine the shape factor based on visual observation and comparison with the following approximate shape factor values. The shape factor used in calculating the fundamental error can be approximated by using values of 0.5 or slightly less for most minerals and soils, 0.1 for flat minerals such as micas, and ≥ 1 for acicular minerals, such as asbestos and tourmalines. Shape factors as large as 10 may be used in the presence of very long, thin, needle-shaped minerals.

The particle size distribution factor (g) addresses error due to all fragments not being exactly the same size and thus not being the coarsest fragments in the sample (21). Mathematically,

$$g = \Sigma M_{\mathrm{L}\alpha}/M_{\mathrm{L}}$$

where $M_{\mathrm{L}\alpha}$ = mass of the size fraction (α) of concern (g), and M_{L} = mass of the lot (g). Two general assumptions for g that can made in calculating the fundamental error are that for most soils, g is approximately equal to 0.25 and that if the soil has been screened to a particular size fraction, then g can be approximated by a value of 0.55.

The mineralogical factor (c), also known as the chemical or mineralogical composition factor, accounts for the maximum heterogeneity within the lot when the mineral (or constituent of interest, eg, a pollutant) is completely liberated from the other material (gangue). It results from differing particle densities and associated differing concentrations (21). Although the full mathematical formula accounts for the average concentration of the lot and the varying densities of the fraction of concern and the gangue or rest of the material, for most soils the following formulas may be used:

$$c = \lambda_{\mathrm{M}}/a_{\mathrm{L}} \text{ if } a_{\mathrm{L}} < 0.1 \qquad c = (1 - a_{\mathrm{L}})\lambda_{\mathrm{g}} \text{ if } a_{\mathrm{L}} > 0.9$$

where λ_{M} = density of pure mineral or constituent of concern (g/cm^3), a_{L} = critical content (proportion of the constituent of interest in the lot) of the lot (dimensionless), and λ_{g} = density of gangue or background materials (g/cm^3). A particle density of 2.6–2.65 g/cm^3 is often used to represent the density of the soil in these formulae. The critical content (a_{L}) of the lot is usually an approximation based on historical data about the proportion (or concentration) of the contaminant of concern at the site.

The liberation factor (ℓ) is a correction factor taking into account that c is a measure of the maximum possible heterogeneity where the contamination occurs as completely separated (liberated) discrete particles within the soil matrix (21). Thus ℓ, whose values lie between 0 and 1, adjusts for heterogeneity of the sample or lot. Mathematically,

$$\ell = (a_{\max} - a_{\mathrm{L}})/(1 - a_{\mathrm{L}})$$

where $a_{\max}$ = maximum proportion of the pollutant associated with the largest contaminated particles in the lot, and a_{L} = average proportion of the pollutant in the lot. The liberation factor can be estimated depending on the degree of visual heterogeneity within the sample using values of 0.8, 0.4, 0.2, 0.1, and 0.05 for matrices that appear to be very heterogeneous, heterogeneous, average, homogeneous, and very homogeneous, respectively. At most hazardous waste sites where soil contamination is the result of a spill, ℓ values range between 0.05 and 0.2. In contrast, where discrete contaminant particles have been spread out at a site (eg, ground Pb plates from used automotive batteries), then a liberation factor of between 0.8 and 1 may be appropriate.

The final and perhaps most influential factor used in the determination of the fundamental error is the diameter (d) of the opening of a square mesh retaining no more than 5% of the sample (21). The largest particles strongly influence the fundamental error because they have the greatest inherent heterogeneity among particles.

By knowing the factors that control the FE, the sampler can minimize, but never eliminate, the fundamental error. However, the fundamental error is not the only source of error or bias that must be considered during the collection of a soil sample. The sampler must be careful to collect a correct sample and thus avoid introducing the other biases into the final results. These biases are the result of the grouping and segregation, increment delimitation, increment extraction, preparation, and analytical errors. Fortunately, these errors can be minimized or eliminated through the careful collection of the soil sample.

The grouping and segregation error (GE) results from the distributional heterogeneity, the heterogeneity that is inherent to the manner in which separate and distinct particles (or aggregates) are scattered or spread out within the lot to be sampled (21). The grouping factor reflects the number of increments (ie, the group of particles extracted from the lot in a single operation of a sampling device) making up the sample compared to the number of fragments (ie, particles) making that sample. The grouping factor approaches zero as the number of increments approaches the number of fragments. The segregation factor accounts for the segregation of different types of particles and the natural range between the minimum and maximum distributional heterogeneity within a sample. The segregation factor ranges between zero and one. The major factor affecting GE is gravity. GE can occur due to gravitational separation that results from differences in:

- particle density
- size
- shape (eg, round particles tend to move or roll easier in the sample or pile than flat particles)
- magnetic properties (eg, magnetite particles will preferentially tend to stick to themselves or steel sampling utensils)
- electrostatic charge (eg, charges on the sides of plastic bottles or bags can cause the adhesion of charged clay particles)
- moisture which tends to clump soil particles together
- air turbulence during sample splitting or homogenization can cause loss of the fines and
- vibration (eg, during transport from the site, on a laboratory cart, or from a vibrospatula used to obtain subsamples) in which heavy particles will slowly settle to the bottom of the sample container.

To minimize the grouping error, the collection of as many small increments as practically possible is recommended. This process of collecting numerous small increments increases the probability of collecting all types of groups for inclusion in the sample and thus reduces the grouping error. To minimize the segregation error, homogenization of the lot before sampling is recommended taking care that the homogenization process itself is free from the factors that increase GE.

The short-range heterogeneity error (CE_1) is a random, discontinuous error that is influenced by FE and GE (CE_1 = FE + GE) and thus, can only be minimized (21). CE_1 is what most soil samplers are referring to when they discuss the natural variability of a soil through time or space and what geostatisticians call the nugget effect. Because CE_1 is a random and discontinuous error, it can only be minimized by performing correct sampling and, thus, minimizing the factors that influence FE and GE.

Sampling is a selection process. The selection process can either be probabilistic (ie, sampling with a random selection component) or nonprobabilistic. Nonprobabilistic sampling (eg, grab sampling) can never be strictly correct. Probabilistic sampling can be correct or incorrect. Probabilistic sampling is correct when all of the constituents of the lot to be sampled have an equal probability of being sampled and when all constituents that are not of the lot have no probability of being selected. The error associated with the selection/sampling process is known as the materialization error (ME) and is the combination of the two relative variances, the delimitation error (DE) and the extraction error (EE); that is, ME = DE + EE. The materialization error must be minimized to approach the notion of correct sampling. (Note: although the preparation error (to be discussed) is technically part of the materialization error, it is usually treated separately.)

The increment DE can be a major source of sampling bias and results from incorrectly defining the boundaries or limits of the volume of material to be extracted and physically collected from the sample (21). In order to define the proper shape, the dimensionality of the lot, which ranges from 0 to 3, must be defined. A zero-dimensional lot exists when the whole lot is used for analysis and, thus, no DE exists, or when the order (through time or space) of the sampling has been lost or is irrelevant (eg, a processed laboratory sample). Similar to a zero-dimensional lot, a one-dimensional lot may be defined as a thin, continuous, elongated pile or stream, except that the order of the sample is important. For zero- or one-dimensional lots, a correct sample is obtained by collecting all of the soil between two parallel planes cut across the entire sample. An example of a one-dimensional lot may be a core sample from which a slice is cut for the analytical subsample.

A two-dimensional lot is correctly sampled by a cylinder with a constant cross-section and consists of a mass with an upper and lower boundary. The third dimension, although existent, for a two-dimensional lot should be insignificant when compared to the other two dimensions. Typical examples of two-dimensional lots in a soil is a horizon (or a designated sampling layer) in which the thickness of the horizon is negligible compared to its horizontal distribution or a core sample collected through the entire soil horizon.

A three-dimensional lot is a sample in which all three dimensions are important and significant. Examples of three-dimensional lots include: large waste piles, truck loads of soil, or an entire waste site. Three-dimensional lots are nearly impossible to sample correctly because the proper shape of the sample is a sphere. Three-dimensional samples are typically reduced to two-dimensional samples for correct subsampling purposes. The one- and two-dimensional lots are most frequently encountered during soil sampling. The two-dimensional lot is collected in the field, whereas the one-dimensional lot is sampled during sample preparation and analysis.

The other important aspect of DE is that the sampling device must be able to physically collect all particles (or portions of particles) that lie within the edges of the sampling device (Fig. 5). If a particle or fraction of particle lies outside the sampling device, then it must be excluded from the collected sample. However, collection of only those portions of particles that lie within the edges of the sampling device is neither practical nor possible because the sampling device cannot cut soil particles. To account for the cutting error associated with the sampling device as it delimits the sample, the increment extraction error (EE) has been defined. To properly extract a sample, the rule of the center-of-gravity must be respected (21). The rule of the center-of-gravity simply states that if the center-of-gravity for a particle lies within the sampling device boundaries, then the fragment must be included with high probability in the sample (Fig. 6). Conversely, if the particle's center-of-gravity lies outside the sampling device boundaries, yet part of the particle lies within the sampling device's edges, then the particle must be excluded with high probability from the collected sample (Fig. 7). Factors that can affect the center-of-gravity rule during soil sampling are the:

- Straightness of the cutting edge: The tips of the cutting edge should be superposable to each other by simple lateral translation (ie, both edges of equal length).
- Shape of the cutting edge: The cutting edge should form an angle between the outer wall of the sampler, and the line parallel to the cutting edge should be less than 45° or equal to 90° (a flat cutting edge). If the cutting edge has two beveled edges from the tip, then a right angle should be formed at the cutter tip between the two sides.

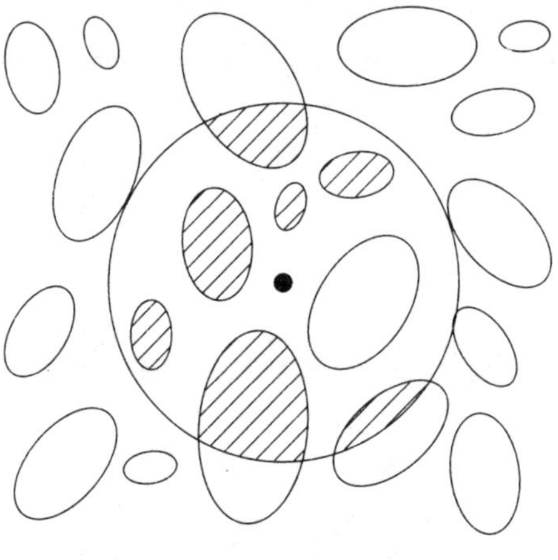

Figure 5. Illustration of proper increment delimitation. Reprinted with permission from Ref. 21. Copyright 1993. CRC Press.

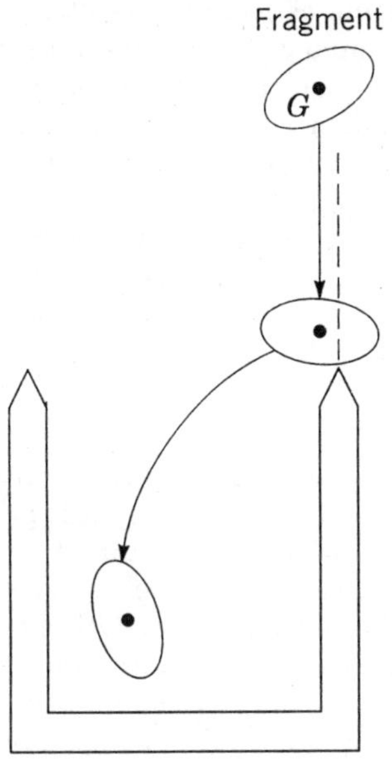

Figure 6. Illustration of the rule of the center of gravity: G = the center of gravity of the fragment or soil particle. Reprinted with permission from Ref. 21. Copyright 1993. CRC Press.

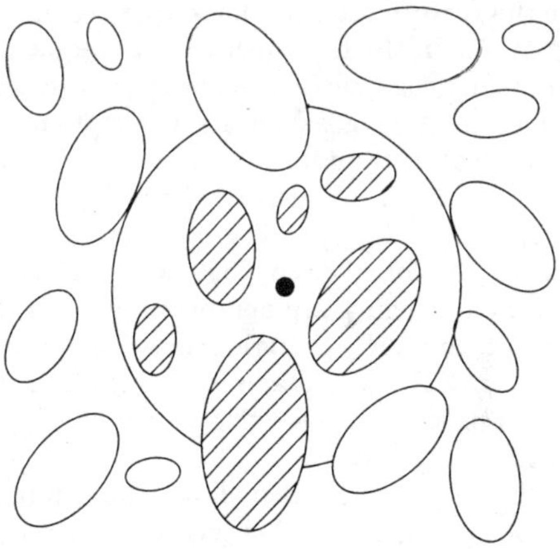

Figure 7. Illustration of proper increment extraction. Reprinted with permission from Ref. 21. Copyright 1993. CRC Press.

- Width of the cutting mouth: For $d \geq 3$ mm, the width of the cutting mouth should be at least $3 \times d$. If $d < 3$ mm, then the width should be at least 10 mm in diameter.

One possible means to minimize the extraction error is to select a larger diameter coring device.

The last two errors, the preparation (PE) and analytical errors (AE), are the result of physical sample manipulation or from noise in the system (21). The major sources of preparation error include contamination, alteration of chemical form (eg, volatilization or redox reactions), human error, and loss of the sample via misplacement or spillage. Preparation errors may be unintentional and occur during sample grinding, sieving, and storage, or through other mistakes such as mislabelling or improper sample handling. Occasionally, intentional error (eg, fraud or sabotage) may be encountered such that pollutants of concern are intentionally lost during preparation or sampling. Analytical error is the error of which most analysts are cognizant. Unfortunately, it is in reducing, controlling, and quantifying analytical error that most quality assurance/quality control time, effort, and budget is expended. As presented throughout the discussion of particulate sampling theory, without giving proper attention to the actual soil sample collection, the sample received at the analytical laboratory may have already been so severely biased that controlling the analytical error may hardly be worth the time and expense.

Application of Particulate Sampling Theory to Soil Sampling. One of the keys to Gy's theory is that it provides a means to determine the quantity of soil to collect and submit a representative sample with an acceptable and set relative variance of the fundamental error to the analytical laboratory based on the maximum particle diameter. Additionally, if a large sample is submitted, it provides a comminution (ie, particle-size reduction) protocol to reduce the sample size yet maintain the fundamental error below the acceptable relative variance as defined in the DQO process (22). Of course, this assumes that the sample is collected correctly using all possible means to control and minimize the GE, DE, EE, PE, and AE so that these errors are negligible compared to FE. Two approaches used to determine the required sample mass are to calculate the mass by solving for M_S in the fundamental error equation or to create a sampling nomograph (Fig. 8). Following the calculational pathway, for example, based on the assumptions of an allowable percent relative standard deviation for FE of 15% (or a relative variance of FE of 0.0225); $M_L \gg M_S$; d = 2 mm; f = 0.5; g = 0.25; ℓ = 0.2; λ = 2.65 g/cm^3; and contaminant concentrations (a_L) = 50, 100, 250, 500, and 1000 ppm, then the required sample weights are approximately 471, 236, 94, 47, and 24 g, respectively. By changing the allowable relative standard deviation of FE to 5% and following the same assumptions, the required sample weights would be 4240, 2120, 848, 424, and 212 g, respectively.

A sampling nomograph also allows the sampler to control sampling and sample processing such that no step exceeds the DQO established for allowable fundamental error. The nomograph provides a visual interpretation of steps that may be necessary to reduce particle and sample size as well as providing the required sample mass for a given maximum particle size. The construction of a nomograph simply involves solving and plotting the equation for the fundamental error. The fundamental error was defined as:

$$s_{FE}^2 = (1/M_S - 1/M_L) f g c \ell d^3$$

This equation can be rearranged to identify a sampling constant, C, by assuming constant values for f, g, c, and ℓ and assuming M_L is much larger (at least 10-fold) than M_S (21). The formula then becomes:

$$s_{FE}^2 = Cd^3/M_S$$

where $C = fgc\ell$. In order to put all the pertinent information on a single nomograph, a logarithmic coordinate system is used and thus the formula becomes:

$$\log s_{FE}^2 = \log C + 3 \log d - \log M_S$$

For a given particle size d and a sampling constant C, the value of log s^2_{FE} is directly proportional to $-\log M_S$ (21). The derivative of log s^2_{FE} with respect to M_S equals -1. Then, the line representing the log s^2_{FE} as a function of $-\log M_S$ for a given particle size d and a sampling constant C, has a slope of -1 on the nomograph. Therefore, to plot the nomograph for a given d with a constant C, the log s^2_{FE} is plotted on the y-axis, log M_S is plotted on the x-axis, and log d is plotted as a family of parallel lines with a slope of -1 (see Fig. 8).

To identify a minimum required sample size given the largest particle diameter and acceptable relative variance limit, the sampler can follow the diagonal particle size line to the point where it intersects the horizontal variance line. At the point of intersection, the value on the x-axis equals the minimum sample size required to collect a representative sample.

Alternately, if a large sample is collected whose mass must be reduced to a smaller size (eg, to the size required for an extraction procedure), the nomograph provides information on how to reduce the sample mass without exceeding the acceptable DQO relative variance level. To identify the steps

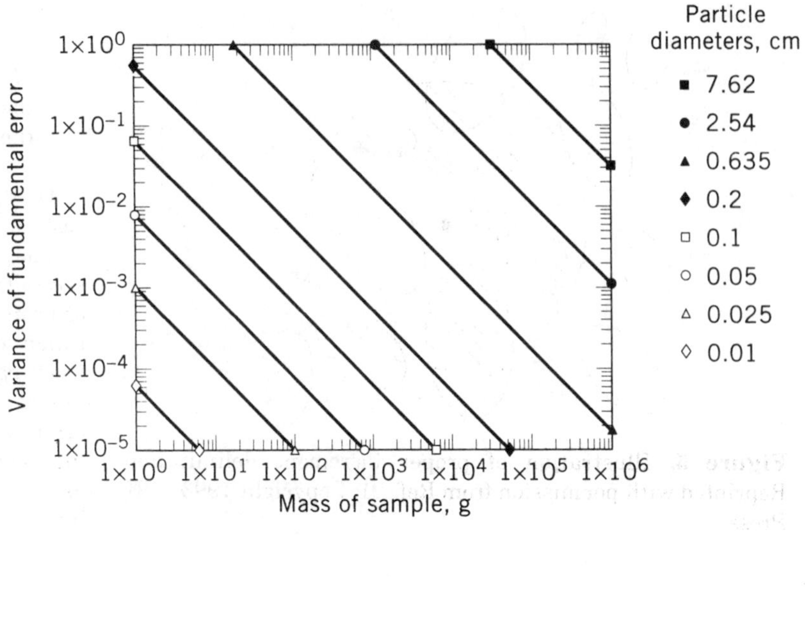

Figure 8. Typical sampling nomograph for soils where $g = 0.25, f = 0.5, \ell = 0.2$, $\lambda_M = 2.65$ g/cm^3, and $a_L = 1000$ μg/g (ppm). See text.

necessary (if any) to properly reduce the sample mass, the sampler needs to know the acceptable relative variance limit as well as the starting and final sample weights. Following the same process for determining the required sample size, if the intersection of the particle size and variance lines results in a sample size smaller than the final required weight, then no particle size reduction is necessary, and the sample can be collected bearing in mind to minimize the other sampling errors. If not, then some form of particle size reduction (ie, comminution) must be taken. The reduction in particle size can be performed by grinding, correct splitting, or by screening the sample to remove larger particle sizes (22). The removal of the coarsest particle sizes should be based on the assumption that the largest particles do not significantly contribute to the contaminant concentration. To select an appropriate particle reduction scheme, simply follow the maximum particle diameter line from its intersection with the starting weight line until it intersects the acceptable variance line. At this point, a vertical line should be drawn until it intersects a smaller particle diameter line that falls below the variance line for the final required sample size. Once an appropriate smaller particle diameter line has been identified, then the whole sample needs to be reduced to have the coarsest particles equal to or smaller than that selected diameter. Several steps may be required to properly reduce the maximum particle sizes but the key is to not go above the acceptable relative variance limit.

Selection of Sampling Equipment

The selection of sampling equipment should be performed with care since the sampling equipment must be able to collect a representative and correct sample and because the sampling equipment comes in direct contact with the soil. Criteria that should be considered during the selection of the appropriate sampling tool are: chemical and physical compatibility; matrix effects; volume capacity; physical requirements for equipment use; ease of operation; time requirement; decontamination and reuse potential; cost; and soil type to be sampled (15,23,24). Further, the appropriate sampling devices selected must be correct with respect to delimitation and extraction errors previously discussed. Each of these factors will be discussed in greater detail in the following text.

Because the soil comes into direct contact with the sampling tool, the sampling tool must be compatible both chemically and physically with the soil and the contaminants present in the soil (23). Chemical compatibility is a concern primarily when organic contaminants are present. The sampling tool should not be a source of additional sample contamination and should be free from (or at least resistant to) chemical degradation that may occur due to interactions with the samples. For example, plastic or acetate sleeves commonly used as liners in coring equipment should be avoided when sampling for organics because they can bleed phthalates and other organic compounds into the sample as well as degrade due to the solvent effect of some organics that may be present in the contaminated soil. Conversely, certain organic contaminants may partition into the plastic resulting in lower contaminant concentrations and cross contamination, if the plastic sampler is used to obtain another sample. Physical compatibility is a consideration in terms of the physical strength of the sampler to resist deformation when collecting the soil and for the possibility of sample contamination due to physical abrasion of the sampler with the separated particles becoming part of the analytical sample. For example, the use of stainless steel sampling equipment for sampling chromium-contaminated soils should be avoided because chromium is a major elemental component in stainless steel.

The effect of the sampling device on the soil matrix is an important concern because the design of the device influences the representativeness of the collected sample (23). As previously discussed, if the sampling device excludes a given particle-size fraction (Gy's increment extraction error) or improperly collects the sample by influencing which part of the whole sample is collected (Gy's increment delineation error), then major sources of error enter into the sampling and analytical stream. Further, sample disturbance is a primary consideration when sampling soil for VOCs. An intact core is highly preferred (versus a disaggregated sample obtained from devices such as augers) during collection of VOC-contaminated soils because the minimization of atmospheric exposure of the sample and consequent VOC loss from the sample is imperative.

Most sampling devices provide adequate sample volume (23). However, sampling device volumes should be compared to the volume necessary for all required analyses, quality assurance purposes, and provide sufficient excess sample for archiving and re-analyses purposes. If the device does not provide adequate volume, the following options should be considered: collecting multiple increments/samples in very close proximity to the original location, using a similar device with increased capacity, using an alternate device with increased capacity, or modifying the existing device.

The physical requirements, ease of operation, and time required for use of the sampling equipment all relate to the transport and operation of the selected sampling device (23). Physical requirements to be considered are the device's size and weight because the sampler may have to carry the device to the sampling location and manually collect the sample. If the device is power driven, then the power source and ancillary equipment (eg, drill rigs, trucks, etc) are a concern in selecting these devices. Ease of operation concerns involve the training of personnel to effectively and properly use the sampling device. Fortunately, most manually operated sampling equipment is relatively simple to use, and representative samples can be collected as long as proper care is taken during sample collection. A factor influencing both the ease of operation and time requirement for sampling is the labor requirement. Most power-driven sampling tools require multiple-person teams to effectively and safely use the equipment but have the advantage of shorter sample collection times, especially when depths exceed one meter (39 inches). In contrast, manual sampling tools, although allowing a single sampler to quickly and efficiently collect surface samples, require much longer times to collect samples at depth than power-driven tools.

Decontamination, reuse potential, and cost are concerns when sampling equipment may be used for multiple sampling programs (23). Three aspects of decontamination, and thus the device's reuse potential, to consider are the procedure's ease, success (ie, ability to eliminate or minimize the potential for sample cross-contamination), and time involved. Ideally, the decontamination procedure should be simple, quick, and successful. Most soil sampling equipment can be effectively decontaminated with a simple soap and water wash followed by water rinse. However, where certain organic compounds are of concern, a solvent rinse may also be necessary. (Note: caution should be taken during the selection and use of solvents to rinse sampling equipment to avoid cross-contamination of the samples, health and safety issues, and spillage such that the solvents become part of the problem at the site.) If successful decontamination procedures are not possible due to the soil matrix or contaminant (eg, soil contaminated with heavy oils or tars), the use of disposable sampling tools should be considered to avoid sample cross-contamination. Cost considerations should be contemplated and compared to the life expectancy of the equipment and use expectancy (ie, will the equipment be used for multiple projects within the organization or should it be leased for a short one-time project).

Several soil types make the routine collection of soil samples difficult and thus require specialized (or modified) sampling tools. These soil types include stony soils, noncohesive soils, and saturated (or nearly saturated) soils (15,23). Sampling difficulties are encountered in stony soils due to the blocking or plugging of the device's opening thus preventing the collection of the sample. Large stones may also prevent the penetration of the sampling device to any significant depth. Noncohesive soils, such as loose sands, require special sampling equipment that retains the soil within the sampler and thus prevents the soil from flowing out the open end of the sampling device. In contrast, saturated soils (eg, mucks and muds) are generally difficult to collect and remove from the sampling tool especially if the soils have high clay contents. Under these circumstances, specialized sampling tools are required that will allow for the simple removal of the soil from the sampling device.

Soil Sampling Tools

Soil sampling tools can be divided into two main categories based on the depth to which the samples can be collected. Surface samplers are designed to generally collect the sample within the upper 30 cm (12 in.) with a single pass (ie, single insertion into the soil) of the sampling device. Subsurface devices are generally designed to sample to a depth of 1.5 m (60 in.) in a single pass but can usually collect deeper samples with multiple passes. Table 1 presents commonly available sampling tools and information on the selection criteria previously discussed. Project-specific parameters, such as compatibility, matrix effects, decontamination requirements, and cost will need to be determined on a per-project basis.

Although spoons, scoops, trowels, and shovels (or backhoes) are perhaps the most readily available and widely used sampling tools for the collection of surface samples, they also tend to introduce the greatest sampling error. The most common cause of the sampling error is the improper increment delimitation in which part of the sample is unintentionally excluded from the sample to be analyzed (21). For example, when using a shovel, rarely is the hole dug such that the sidewalls are parallel to each other and perpendicular to the surface. The natural tendency is to dig a hole with sidewalls at an angle to the surface (Fig. 9, C). If the sample collected from the soil is removed from the hole, the lower portion of the sample is underrepresented because less sample is collected at that depth than at the wider surface mouth. Under these circumstances, if the contaminant has been leached to a depth within a given sampling increment, a lower contaminant concentration may be obtained due to sample dilution with the cleaner overlying material. Conversely, if the contaminant is present at or near the surface, then the reported concentrations may overestimate the actual concentration due to inclusion of a greater proportion of surface versus subsurface soil. To overcome this problem, it is recommended to dig a hole to the required depth and collect the sample from the walls of the hole being careful to collect a uniform sample thickness from the entire length of the sampling stratum. Alternately, care must be taken to dig the hole properly to include equal proportions of all fractions within the sampling increment (Figs. 9, A and B).

The same general principle is the basis for the error introduced when using spoons, scoops, and trowels. Most of these samplers have a rounded bottom which when used to collect soil preferentially collects less of the lower portion of the sample than the surface and thus biases the results (Fig. 10, A). Additionally, because none of these sampling utensils has sidewalls capable of holding the entire sample within the sampler, as the sample is removed and transported to the sample container, portions of the sample may be lost by falling off the sampler (Fig. 11, A). In some cases, both the upper and lower portion of the sample may be lost with the greatest sampled portion being collected from the middle of the sampling stratum. To properly collect a sample with these types of tools would require equipment modification to have parallel side walls and a flat bottom capable of holding the entire sampling increment (Fig. 10, B and Fig. 11, B).

Tube-type samplers, including probes, punches, corers, tube samplers, barrel samplers, continuous samplers, and zero contamination samplers, tend to have a distinct advantage over other forms of samplers because an intact soil core is collected. The cylindrical shape of the intact core is the correct shape for a two-dimensional lot and, thus, reduces the increment delimitation error as long as care is taken to collect the entire layer or depth defined as the sampling increment (Figs. 9, A and B) (21).

When more sample is collected than required for the sampling increment (eg, a 25-cm (10-in.) core is collected and subsampled to collect the top 10 cm (4 in.)), care must be taken to delineate the proper increment by cutting the new boundary parallel to the surface of the sample at the desired depth. For the thinner diameter tube-type samplers (<2.5 cm ID; 1 in.), the increment extraction error may be a concern due to exclusion larger particles at the mouth of the sampling tube.

Augers, whether used to collect surface or subsurface samples, with the exception of the hollow stem auger, collect samples that are disaggregated by the cutting bit or within the auger threads. This cutting action destroys natural soil structure and makes it difficult to accurately delineate soil horizon boundaries or sampling depths. This disaggregating property of augers also makes them unsuitable for the collection of soils contaminated with VOCs because the sample disturbance allows for the rapid loss of VOCs to the atmosphere. Increment delimitation error may also occur with augers because the lowest portion of the sample (ie, that portion where threads do not overlap and have an upper and lower thread to hold the sample) cannot be collected.

Thief and triers are two types of sampling tools that were not designed for the collection of soils but can be used when necessary. Thiefs are usually pushed into the sample and rotated to open windows and allow the sample to flow into the central chamber/barrel to be collected (23). This type of filling

Table 1. Guide to Sampling Equipment Selection[a]

Type of Sampler	Typical Lengths/ ID[b]	Relative Sample Size	Liners	Core Samples	Suitable Soil Types: Cohesive	Suitable Soil Types: Moisture	Suitable Soil Types: Stony	Manual or Power	Labor
Surface Samplers									
Spoons, trowels, and shovels	Var[c]	Var	No	No	All	All	Yes	Manual	Single
Soil punches and sampling tubes	6–10/1–3	Small	Yes	Yes	Coh	All	No	Manual	Single
Probes and tube samplers	6–24/1	Small	No	Yes	Coh	All	No	Manual	Single
Soil corers	2–12/1–3	Small	Yes	Yes	Coh	All	No	Manual	Single
Split core/barrel samplers	6–12/1–2	Small	Yes	Yes	Coh	Int	No	Manual	Single
Screw augers	6–12/1–2	Small	No	No	Coh	Int	No	Manual	Single
Soil recovery augers	8–12/2–3	Large	Yes	Yes	Coh	Int	No	Manual	Single
Subsurface Samplers									
Augers									
Power augers	36–60/2–6	Large	No	No	All	Int	No	Power	Multi
Solid stem flight augers	36–60/2–6	Large	No	No	All	All	Yes	Power	Multi
Hollow stem augers	36–60/2–4	Large	No	Yes	All	All	Yes	Power	Multi
Regular bucket augers	6[d]/2–4	Large	No	No	Coh	Int	No	Manual	Single
Sand bucket augers	6[d]/2–4	Large	No	No	Non	Int	No	Manual	Single
Mud bucket augers	6[d]/2–4	Large	No	No	Coh	Wet	No	Manual	Single
Dutch augers	6[d]/3	Large	No	No	Coh	Wet	No	Manual	Single
Screw augers	12[d]/1–2	Small	No	No	Coh	Int	No	Manual	Single
Post-hole auger	6[d]/2–8	Large	No	No	Coh	Wet	No	Manual	Single
Tube-type samplers									
Shelby/thin wall tube sampler	30/2–4	Large	No	Yes	Coh	Int	No	Both[e]	Multi
Subsoil probes	36–48/1	Small	No	Yes	Coh	All	No	Manual	Single
Split spoon/tube samplers	12–60/2–4	Large	Yes	Yes	Coh	Int	No	Both[e]	Multi
Continuous sampler	48–60/2–5	Large	Yes	Yes	Coh	All	No	Power	Multi
Veihmeyer sampling tubes	48–72/1	Large	No	Yes	Coh	Int	Yes	Manual	Single
Zero contamination sampler	12–24/1	Small	Yes	Yes	Coh	Wet-Int	No	Both[e]	Multi
Ring-lined barrel sampler	24–60/2–4	Large	Yes	Yes	All	Wet-Int	Yes	Both[e]	Multi
Miscellaneous Samplers									
Triers	6–60/1–3	Var	No	No	Coh	Int	No	Manual	Single
Thiefs	6–72/1	Var	No	No	Non	Dry	No	Manual	Single
Backhoes	Var	Large	No	No	All	All	Yes	Power	Multi

[a] This table does not purport to be exhaustive. Only the general names of the sampler types are presented. Varying names for the same sampler or same type of sampler with slight modifications are commonly identified in the literature and sales catalogs.
[b] Approximate lengths and internal diameters (ID) are presented in inches. To convert inches to meters, multiply by 2.54×10^{-2}.
[c] Var = variable depending on length of sampler or size of hole dug; Coh = cohesive; Non = noncohesive; Int = intermediate wetness (ie, not dry nor wet); Multi = two or more persons required to operate effectively.
[d] Manual auger lengths given are just for portion of the sampler where the actual sample is held. Most bucket augers have handles and extensions that allow for greater sampling depths with multiple passes.
[e] Manually operated versions of these samplers can be operated by one person.

action leads to the concerns about incorrect sample delimitation because: (*1*) the sample below the windows, although truly a part of the sample, cannot be collected; (*2*) only those particles that flow easily are collected; (*3*) particles larger than the window diameter are excluded, thus biasing the collected sample to the finer particle sizes; and (*4*) the proper cylindrical shape for a two-dimensional lot is not guaranteed. Triers, on the other hand, although nearly capable of collecting the proper sample cylinder, do not have sidewalls to hold the sample once collected. This lack of sidewalls can lead to sample loss during sample removal and transfer operations resulting in potentially biased analytical results.

SAMPLE HANDLING AND PREPARATION

Once the sample has been collected, the next steps that occur prior to preparing subsamples for analysis, are sample handling and preparation. Various concerns and steps are involved which are briefly presented here to remind the sampler of their final responsibilities. These steps include sample compositing, homogenization, preservation, and storage.

Sample compositing is the process of combining several distinct subsamples to create a single sample for analysis (25). The combined subsamples are then homogenized to make up the final sample submitted for analysis. The ad-

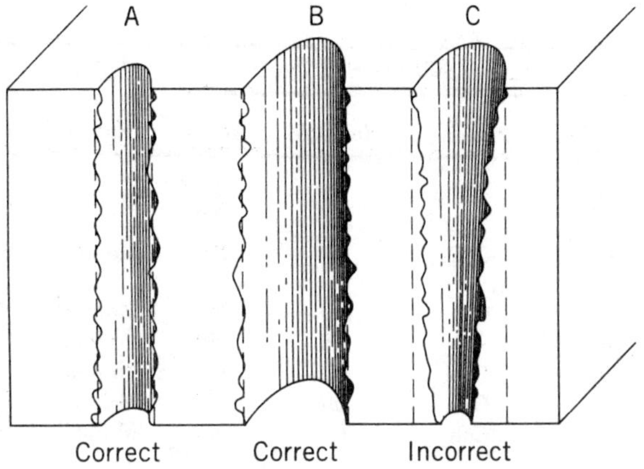

Figure 9. Illustration of correct and incorrect sample delimitation using a coring device (A) and shovels (B and C). Reprinted with permission from Ref. 21. Copyright 1993. CRC Press.

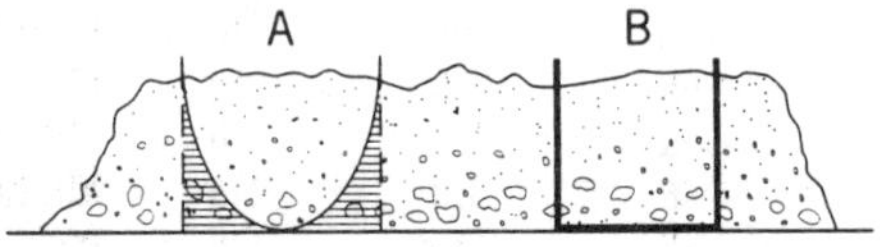

Figure 10. Incorrect increment delimitation introduced by a scoop with round bottom. Reprinted with permission from Ref. 21. Copyright 1993. CRC Press.

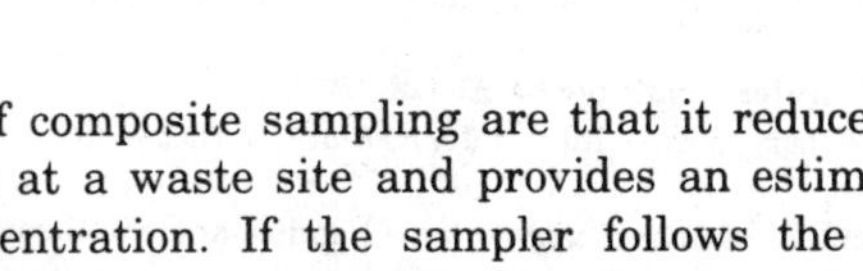

Figure 11. Incorrect increment delimitation introduced by scoop with no side walls to prevent material from falling. Reprinted with permission from Ref. 21. Copyright 1993. CRC Press.

vantages of composite sampling are that it reduces the cost of analysis at a waste site and provides an estimate of the mean concentration. If the sampler follows the principles of Gy's particulate sampling theory, then sample compositing after the collection of numerous small increments is an excellent way to reduce GE. The disadvantages of sample compositing are a loss of information about variation within the sampling area (26), loss of sensitivity because of sample dilution (27), and the time required to homogenize the samples prior to laboratory subsampling. In situations where rapid, field-portable methods are used (eg, field portable X-ray fluorescence spectrometry), compositing may not be necessary because individual samples can be analyzed in real time to provide a more definitive answer concerning the distribution of a contaminant at the waste site.

Numerous techniques have been used to homogenize soils prior to the removal of the analytical subsample (28,29). Some of these techniques include bottle shaking, stirring, sheet mixing or rolling, tumbling, mechanical mixing (eg, cement mixers or V-blenders), riffling, cone-and-quartering, and sectorial splitting. Homogenization is performed to diminish GE within the sample or lot. When performing sample homogenization, perhaps the most frequently asked question is "When has homogeneity been reached?" Although there is no patented answer to this question, most investigators will homogenize their samples for a fixed time, a fixed number of passes, or until visual homogeneity is reached without ever testing the degree of sample homogeneity. Investigators should always test their homogenization technique prior to full-scale usage especially because some homogenizers will actually lead to resegregation of the sample.

In one experiment to determine the effectiveness (degree of homogeneity) and the efficiency (time consumption) of riffling versus cone-and-quartering as homogenization techniques, it was found (29) that using a closed-bin riffle splitter (which also reduced visual fine particle loss) was more consistently effective and efficient than cone-and-quartering, open-bin riffle splitting, or random sampling, when the sample was passed through the riffle splitter five times. In this experiment, homogenization was performed on bulk soils (approximately 5 kg each with textures ranging from sand through clay) by either pouring the soil collected in both bins back through the riffle splitter, or by forming a new cone as the individual quarters are removed during cone-and-quartering. Although homogenization can help overcome grouping and segregation problems that occur during shipment and handling, properly performed incremental sampling can also produce the same benefits without the added time and expense (22).

Sample preservation for most soil samples is simply to cool the sample to 4°C or colder (30). For soils contaminated with VOCs, preservatives such as methanol or other biocides are commonly used, in conjunction with cooling, to help reduce biological degradation and volatilization of the VOCs. Sample storage in either plastic or glass containers is generally acceptable for soils contaminated with inorganics. In contrast, organically contaminated soils should be stored in glass containers (with Teflon-lined caps for VOC containing soils). It should be noted that some compounds are light-sensitive, and those samples should be stored in the dark to avoid any photochemical reactions.

ACKNOWLEDGMENT

The U.S. Environmental Protection Agency, through its Office of Research and Development (ORD), funded and performed the research described here. It has been subjected to the Agency's peer review and has been approved as an EPA publication. The U.S. government has the right to retain a nonexclusive, royalty-free license in and to any copyright covering this article. Mention of trade names or commercial products does not constitute endorsement or recommendation for use.

BIBLIOGRAPHY

1. T.H. Starks *Mathematical Geol.* **18**(6), 529–537 (1986).
2. U.S. EPA, *Guidance for the Data Quality Objective Process, EPA QA/G4,* Quality Assurance Management Staff, Washington, D.C., 1994.
3. U.S. EPA, *The Data Quality Objectives Process for Superfund: Interim Final Guidance,* EPA/540/R-93/071, Quality Assurance Management Staff, Washington, D.C., 1993.
4. U.S. EPA, *Guidance for Data Quality Assessment—Practical Methods for Data Analysis, EPA QA/G-9,* EPA/600/R-96/084, Quality Assurance Management Staff, Washington, D.C., 1996.

5. U.S. EPA, *A Rationale for the Assessment of Errors in the Sampling of Soils,* EPA/600/X-89/203, Environmental Monitoring Systems Laboratory, Las Vegas, Nev., 1989.
6. E.H. Isaaks and R.M. Srivastava, *An Introduction to Applied Geostatistics,* Oxford University Press, New York, 1989.
7. R. Webster and M.A. Oliver, *Statistical Methods in Soil and Land Resource Survey,* Oxford University Press, New York, 1990.
8. A.B. McBratney, R. Webster, and T.M. Burgess, *Comp. Geosci.* **7**(4), 331–334 (1981).
9. A.B. McBratney and R. Webster, *Comp. Geosci.* **7**(4), 335–365 (1981).
10. E.J. Englund and N. Heravi, *Environ. Ecol. Statistics.* **1,** 247–263 (1994).
11. E.J. Englund and N. Heravi, in A. Soares, ed., *Geostatistics Troia '92,* Kluwer, Dordrecht, 1993, pp. 613–624.
12. U.S. EPA, *Guidance for Conducting Remedial Investigations and Feasibility Studies under CERCLA,* EPA/540/G-89/004, U.S. EPA, Washington, D.C., 1989.
13. G.F. Simes, *Preparing Perfect Project Plans,* EPA/600/8-89/087, US EPA, Cincinnati, Ohio, 1989.
14. M.E. Byrnes, *Field Sampling Methods for Remedial Investigations,* Lewis Publishers, Boca Raton, Fla., 1994.
15. D.S. Barth, B.J. Mason, T.H. Starks, and K.W. Brown, *Soil Sampling Quality Assurance User's Guide,* 2nd ed. EPA 600/8-89/046, US EPA, Las Vegas, Nev., 1989.
16. R.E. Cameron, *Guide to Site and Soil Description for Hazardous Waste Site Characterization, Vol. 1 -Metals,* EPA/600/4-91/029, US EPA, Las Vegas, Nev., 1991.
17. J.R. Boulding, *Description and Sampling of Contaminated Soils—A Field Pocket Guide,* EPA/625/12-91/002, U.S. EPA, Cincinnati, Ohio, 1991.
18. Soil Survey Staff, *Soil Survey Manual,* U.S. Dept. Agr. Hbk. **18,** (1951).
19. L.M. Thompson and F.R. Troeh, *Soils and Soil Fertility,* McGraw-Hill Book Co., New York, 1978, pp. 35–39, 127–128.
20. A. Klute and C. Dirksen, "Hydraulic Conductivity and Diffusivity: Laboratory Methods," in A. Klute, ed., *Methods of Soil Analysis, Part I—Physical and Mineralogical Methods,* 2nd ed., American Society Agron. Madison, Wis., 1986, pp. 687–734.
21. F. F. Pitard, *Pierre Gy's Sampling Theory and Sampling Practice,* 2nd ed., CRC Press, Boca Raton, Fla., 1993.
22. B.J. Mason, *Preparation of Soil Sampling Protocols: Sampling Techniques and Strategies,* EPA/600/R-92/128, U.S. EPA, Las Vegas, Nev., 1992.
23. ASTM, *Draft Standard Guide for Selection of Sampling Equipment for Waste and Contaminated Media Data Collection Activities,* American Society for Testing and Materials, W. Conshohocken, Pa., 1995.
24. T.E. Lewis, A.B. Crockett, R.L. Siegrist, and K. Zarrabi, *Soil Sampling and Analysis for Volatile Organic Compounds,* EPA/540/4-91/001, U.S. EPA, Las Vegas, Nev., 1991.
25. ASTM, *Draft Standard Guide for Composite Sampling and Field Subsampling for Environmental Waste Management Activities,* American Society for Testing and Materials, W. Conshohocken, Pa., 1996.
26. J. Crépin and R.L. Johnson, in M.R. Carter, ed., *Soil Sampling and Methods of Analysis,* Lewis Publishers, Boca Raton, Fla., 1993, pp. 5–18.
27. L.H. Keith, *Environmental Sampling and Analysis—A Practical Guide,* Lewis Publishers, Boca Raton, Fla., 1991.
28. R.E. Cameron, G.B. Blank, and D.R. Gensel, *Sampling and Handling of Desert Soils,* NASA Tech. Rep. 32-908, Jet Propulsion Lab, Calif. Inst. of Technology, Pasadena, Calif., 1966.
29. B.A. Schumacher, K.C. Shines, J.V. Burton, and M.L. Papp, *Soil Sci. Soc. Am. J.* **54,** 1187–1190, (1990).
30. U.S. EPA, *Test Methods for Evaluating Solid Waste, SW-846,* 3rd ed, Office of Solid Waste and Emergency Response, Washington, D.C., 1986.

SOIL SORPTION OF ORGANIC POLLUTANTS AND PESTICIDES

CARY T. CHIOU
U.S. Geological Survey
Denver, Colorado

Soil may be defined as a collection of natural bodies synthesized in profile form from a variable mixture of broken and weathered minerals and decaying organic matter that cover the earth surface (1). Discounting the water content in soil, which varies between soils from different geographic locations and between different horizons of a vertical soil profile, the principal soil components are mineral matter and organic matter. The minerals are composed of aluminosilicates and oxides in various crystalline and amorphous forms and vary considerably in size, ranging from very small colloidal clay particles ($<2\ \mu m$) to relatively large sand particles ($>50-60\ \mu m$). The soil organic matter, which originates primarily from biologically degraded plant tissue and becomes part of the underlying horizons by infiltration or physical incorporation, consists of a heterogeneous makeup of organic constituents (such as lignins, carbohydrates, protein, fats, and waxes); it contains a large fraction of operationally defined *humic substance,* which is yellow-to-brown in color and moderately refractory to biological degradation (2). Although the molecular structures of humic substances have not been well characterized, they are known to be high molecular weight, amorphous materials (frequently referred to as *humic polymers*) with significant polar-group contents. In ordinary soils, more than 90% of the dry organic matter is made up of carbon, hydrogen, and oxygen, with minor amounts of nitrogen, sulfur, and phosphorus. Except for relatively few organic soils, ordinary soils are rich in minerals and are referred to as mineral soils. The organic matter content for a large number of mineral soils falls between 0.5 and 3.0% by weight.

Studies of the mechanism of organic compound sorption on soil began in the 1940s following the advent of pesticides in agricultural practice, which called for assessment of their effectiveness and persistence in soil. In the 1970s, research in this field became more active after more varieties and quantities of pesticides and industrial organic wastes were found widely spread in the environment, raising public concern over the long-term environmental contamination. Through the effort over the years, the research in this field has now come to the point where the mechanistic behavior of the soil (or sediment) in terms of its organic matter and mineral components in sorption of organic pollutants and pesticides can be placed in a relatively good perspective (3). This enables one to make reasonable predictions of the sorption behavior of these compounds under different system conditions and to evaluate the old and new data for consistency.

In the present description and discussion of the mechanistic roles of soil organic matter and minerals in uptake of organic compounds and pesticides by soil, some terminology is used frequently in referring to the mechanism involved. The term *sorption* is commonly used to denote the uptake of a contaminant (solute or vapor) by soil or sediment without reference to a specific mechanism. The term *adsorption* refers to the condensation of vapor or solute (both being commonly referred to as *adsorbate*) on the surfaces or interior pores of a solid (*adsorbent*). Although no explicit distinction is made between different types of adsorption in this article, it is understood that physical adsorption results from London–van der Waals forces (the same forces that are responsible for adhesion of a molecule to a bulk solid) between the adsorbate and adsorbent. When the forces leading to adsorption are related to chemical bonding, the adsorption is referred to as *chemisorption*. The distinction between these two types of adsorption is not always a sharp one; adsorption of polar molecules onto polar solids may fall under either classification depending on the adsorption energy.

The term *partition* or *partitioning* is used to describe a process in which the sorbed material dissolves in an organic phase by forces common to solution (eg, by van der Waals forces). This is analogous to the extraction of an organic solute from water into an organic solvent. When the organic phase is a solid (eg, soil organic matter), partition is distinguished from adsorption by the homogeneous distribution of the sorbed material throughout the entire volume of the solid phase, analogous to, for example, the partitioning (solubilization) of benzene into polystyrene. In adsorption, the adsorbate occupies only the surface of the solid and does not permeate its entire volume. A powerful adsorbent invariably has a high surface area. By contrast, the equilibrium partition coefficient of a compound between two separate phases (eg, an organic phase and water) is a function of its solubilities in the two phases, and the magnitudes of the individual solubilities are strongly influenced by interactions of the functional groups between the solute and individual solvents. As a general rule, a solute exhibits a high solubility in a solvent when the solute and solvent possess similar compositions and polarities (ie, like dissolves in like). The surface area of the solvent has no bearing on solute solubility.

The term *isotherm* is used to describe the relationship of the concentrations of a solute between two separate phases under equilibrium at a constant system temperature. Hence, the *adsorption isotherm* of a solute or a vapor expresses the relation of the amount of solute or vapor adsorbed on a solid (adsorbent) to the concentration of the solute (or vapor) in solution or in the vapor phase at equilibrium. By contrast, the *partition isotherm* of a solute or a vapor describes the relation of the solute concentrations between two solution phases or between a solution and a vapor phase at equilibrium. Similar to the term *sorption*, the term *sorption isotherm* is used to describe the isotherm without reference to a specific uptake mechanism.

The term *organic matter* is used to refer to the bulk of the organic content in soil (or sediment). Although differences in composition between humic and nonhumic organic matter in soil can influence the overall behavior of the soil organic matter, separate treatments of their contributions to soil sorption are not possible because practically all past sorption studies have been carried out with intact soils. In general, except for top surficial soils where a significant amount of undecomposed (or poorly decomposed) plant litter may exist, soil organic matter normally contains mainly well-humified organic material. The sorption data presented are the equilibrium (or near-equilibrium) values. Whereas the concentrations of given contaminants between phases in natural systems deviate frequently from those at equilibrium, the equilibrium data serve as a useful guide to the direction of contaminant movement at a particular point in time and to the likely earlier contamination event. A comparison of the field data with equilibrium values also enables one to recognize whether a compartment (such as soil or sediment) functions as a sink (for receiving a given contaminant) or as a source (for releasing a given contaminant) under a specified condition. Such information is often sought in the characterization of a contamination site.

This article presents an overview of the sorption interactions of nonionic contaminants (organic pollutants and pesticides) with soils (or sediments), the correlations of sorption coefficients with properties of contaminants and soils, and the impacts of sorption interactions on contaminant activities. The sorption mechanism of ionic organic pollutants and pesticides, which constitute a small fraction of all contaminants, is generally better understood. Cationic pesticides (such as paraquat and diquat) are known to sorb primarily by exchange of cations on soils and clays (4–7) and organic anions usually show insignificant sorption to soil (4,8). Because of the unique roles of soil organic matter and mineral matter in the sorption of nonionic compounds, detailed consideration will be given to the mechanistic functions of these two basic soil components and to the effect of water on their respective functions.

HISTORICAL DEVELOPMENT IN SOIL SORPTION STUDY

Influences of Mineral Matter, Organic Matter, and Water

The complex and heterogeneous nature of soil (or sediment) compositions has been a major retarding factor in the development of soil sorption theory, especially with regard to the individual and combined functions of soil organic matter and mineral matter. In earlier studies of pesticide–soil interactions, the soil was considered either as a single adsorbent or as a mixed adsorbent of some kind, analogous to some other well-defined adsorbents. Although this view accounted largely for the sorption of pesticides to relatively dry soils and minerals, it ran into difficulties in explaining the sorptive behavior of water-saturated soils. The latter system is of special importance in assessing the transport and impact of organic pollutants and pesticides in soil–water systems.

The adsorptive character of dry and water-unsaturated soils was well demonstrated, for instance, in earlier studies of vapor uptake of such pesticides as methyl bromide (9), chloropicrin (10) and ethylene dibromide (11,12) by relatively dry mineral soils and clays in which the uptake of the organic vapors was suppressed by moisture content. Similarly, the uptake of parathion and lindane from hexane solution by soils was found to be suppressed by soil moisture (13,14). These observations indicate that organic compounds and water compete for adsorption on soil minerals, which provide most of the surface area in ordinary soils. In addition, the uptake of organic compounds from the vapor phase and from hexane solution on dry soils and clays gives nonlinear isotherms characteristic of the adsorptive nature of the soil and clay.

By contrast, the uptake of nonionic compounds and pesticides by soils in aqueous systems displays a set of unique features. Most notably, the extent of soil uptake for given organic compounds shows a strong dependence on the organic matter content of the soil (15–31). The uptake of organic vapors by wet soils shows a similar effect (12,32). The predominant effect of soil organic content in this case is demonstrated by the relative invariance of the sorption coefficients of given organic compounds among soils, or size fractions of soil, when the coefficients are normalized to the organic matter content of the soil (17–19,27,29,33–35). The sorption isotherms of nonionic compounds on water-saturated soils are all relatively linear (12–14,20,25,26,28,32,34–38), and are not strongly temperature dependent, showing only small exothermic heats (13,14,19,25,39–41). Moreover, soil uptake of binary nonpolar solutes from water shows no apparent competition between the solutes (14,26,42), in contrast to the strong competitive effects that are found in sorption by dry soil from the vapor phase and from organic solvents (9–14,40,43–49).

In spite of the fact that the observed relationship between soil uptake and organic matter content had greatly simplified predictions of the uptake of nonionic organic compounds from water by soils, there was no single widely accepted view regarding the sorptive mechanism with soil organic matter in earlier published works. Over the years, many researchers assumed that soil organic matter functions as a high surface area adsorbent (16,50,51) capable of adsorbing nonionic organic compounds by hydrophobic interactions (23,52–57). The hydrophobic adsorption concept, however, is not in keeping with general adsorption criteria and with soil sorption data in aqueous systems. Moreover, the high surface area for soil organic matter (550–800 m^2/g) as reported earlier by Bower and Gschwend (50) was based on the amount of ethylene glycol (EG) retained by soil organic matter, which was later shown to be an artifact of the enhanced solubility of polar EG in relatively polar soil organic matter (58–60). Using high organic-content soils (peat and muck) as a model for soil organic matter, the measured surface area by the standard Brunauer-Emmett-Teller (BET) method with nitrogen as the adsorbate was only about 1 m^2/g (58,60,61), which is nearly three orders of magnitude lower than the earlier assumed value.

To uphold the idea that soil organic matter is a high-surface-area adsorbent, some investigators (51) suggested that the large difference between the reported surface areas of soil organic matter by polar-solvent retention method and standard BET (N_2) method represents the internal surface of the soil organic matter. It should be noted, however, that such *internal surface* has a trivial meaning, because it is impervious to low molecular weight gases such as N_2 (58,59). Whereas many porous solids have large internal surfaces, in the sense that they are internal to the outer granule boundary of the solid (as in activated carbon), they are nevertheless external to the bulk solid (59,62). These internal surfaces make sense only if they are accessible to gases of low molecular weight.

In consideration of the high uptake of monuron (a herbicide) on muck soil in aqueous systems (15), Hartley (63) speculated that a solvent action of the *oily constituent* in soil organic matter might be responsible for the enhanced pesticide uptake (while making no effort to distinguish this effect from surface adsorption). This view was further emphasized by Goring (17) in assessing the distribution of organic toxicants between soil water and soil solids. Although it is reasonable to expect that the oily constituent in organic matter would exhibit such a solvent effect, the role of soil organic matter as a whole (or that of the nonoily components) remained to be defined. Lambert (64) later considered the role of soil organic matter to be analogous to that of a solvent medium in partition chromatography and suggested a correspondence between sorption coefficient and solvent–water distribution coefficient. It has been noted (36) that the soil–water equilibrium isotherm of parathion was linear over a significant range of equilibrium concentrations in water and suggested that part of the parathion might be adsorbed as a liquid dissolved in the organic fraction of the soil similar to the partition process in liquid–liquid extraction. Further confirmation of the partitioning of the pesticide between organic matter and water was not available at that time, however, and the validity and significance in soil uptake by partitioning into soil organic matter in aqueous and nonaqueous systems was not duly recognized.

As a means of estimating contaminant sorption coefficients between soil (or sediment) and water, numerous studies illustrated the linear relations between the sorption coefficients, normalized either to soil organic matter content (K_{om}) or to soil organic carbon content (K_{oc}), and the corresponding octanol-water partition coefficients (K_{ow}) for different classes of compounds (19,25,26,28,29,33,34,54,65–68).

Soils as Dual Sorbents for Organic Compounds

In a series of studies, Chiou et al. (14,25,26,44,69,70) advanced the idea that the sorption of low polarity nonionic organic compounds from water to soils with significant organic contents consists primarily of solute partition (solubilization) into the soil organic matter. In addition to the recognized dependence of soil sorption on organic matter content, they found the equilibrium isotherms of relatively nonpolar solutes in water to be essentially linear from low to high relative concentrations (ratios of equilibrium concentrations to solute solubilities). Moreover, the equilibrium heats of sorption for solutes are less exothermic than heats of solute condensation from water (ie, the reverse heats of solution), and the system shows a lack of solute competition. The inability of the soil mineral fraction to significantly adsorb nonionic organic compounds from water is attributed to the strong dipole interaction of minerals with water, which suppresses the adsorption of these compounds on minerals. In keeping with solute partitioning into soil organic matter, it was found that solutes (as liquids or supercooled liquids) with higher normalized sorption coefficients (K_{om}) usually exhibit lower limiting sorption capacities on soil organic matter. By application of the Flory–Huggins model to account for solute solubility in (amorphous) soil organic phase, Chiou and co-workers (26) established a partition equation to account for the magnitudes of the observed sorption coefficients. This analysis led to the recognition that the primary factor affecting the sorption coefficients of slightly water-soluble organic compounds is the solubility of the compounds (as liquids or supercooled liquids) in water. The frequently observed empirical correlation between sorption coefficient (K_{om} or K_{oc}) and octanol-water partition coefficient (K_{ow}) was recognized to be the consequence that the solute solubility in water is the major determinant of both K_{om} and K_{ow} values (26,71). That the soil organic matter functions essentially as a partition medium for organic solute uptake is further in keeping with the later finding of its low surface area (about 1 m^2/g) (58), which is too small to account for the solute uptake by soil organic matter in terms of an adsorption process.

The differences in sorption characteristics of nonionic organic solutes on soil in aqueous and nonaqueous systems are reconciled with the postulate that the soil (or sediment) behaves as a dual sorbent, in which the mineral matter functions as a conventional solid adsorbent and the organic matter as a partition medium (14,26,44,69). The linear isotherms and other characteristics as observed in aqueous systems are attributed to solute partitioning into the organic matter, with a concomitant suppression of adsorption by water on mineral matter. The nonlinear isotherms and higher sorption capacities on dry soils are ascribed to adsorption on minerals, which predominates over the simultaneous partition in soil organic matter. The schematic plots in Figure 1 are used here to depict the relative contributions of mineral matter and organic matter in sorption of a mineral soil having a moderate organic matter content (say, 1–2%). The scales for these two contributions are not exact but are drawn to stress the dominant effect of either mineral adsorption (for dry soil) or partition in organic matter (for water-saturated soil). For dry soil, as shown in Figure 1**a**, the much greater nonlinear adsorption on mineral matter than the linear partition into organic matter results in a high and nonlinear total soil sorption (the sum of the two contributions). For the water-saturated soil, as shown in Figure 1**b**, the mineral adsorption is greatly suppressed by water and the partition in organic matter predominates to give a relatively low and linear total sorption.

The specific roles of soil organic matter and mineral matter as described provide the point of departure for understanding the diverse and seemingly contradictory sorption data of various organic contaminants on soil from water solution, from organic solvent solution, and from the vapor phase. These topics and related systems will be treated in a logical sequence. Because there is a continuum of the organic matter content in soil, the present analysis of sorption data in aqueous systems is confined only to those soils having more than 0.1–0.2% organic matter so that the effect of soil organic matter on soil uptake would be significant and could be reliably quantified. Prior to the analysis of specific data, we first consider some important features associated with adsorption and partition processes to better appreciate their mechanistic differences. The symbols used in this article for description of contaminant interactions and processes are listed in the Nomenclature.

PARTITION AND ADSORPTION CHARACTERISTICS

There are a number of useful textbooks and articles dealing with adsorption theory and with mathematical models for interpreting interactions of organic compounds with various types of adsorbents in vapor and solution systems (62,72–76). A succint description of the fundamentals of adsorption of vapors and solutes on solids (adsorbents) is available elsewhere in this *Encyclopedia* (see ACTIVATED CARBON ADSORPTION). The influences of solid surface area and surface heterogeneity on adsorption and the aspects of adsorption heat, isotherm shape, and adsorptive competition are placed in good perspective.

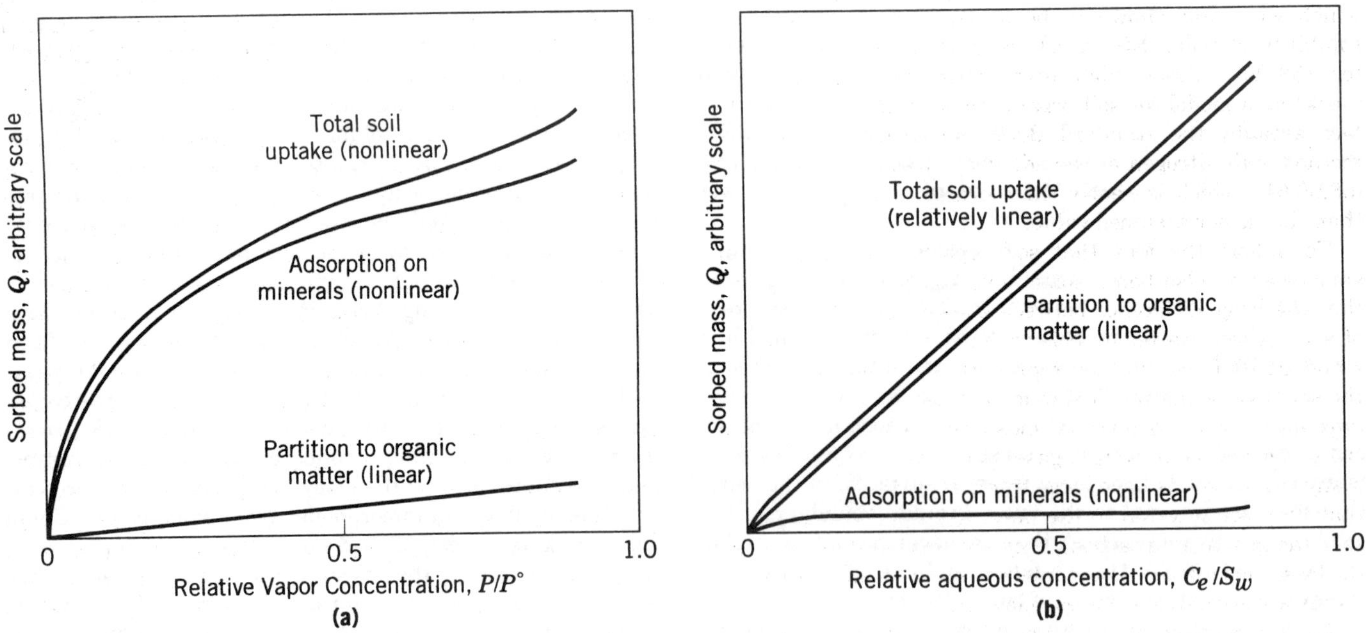

Figure 1. Schematic plots of contributions by adsorption on soil mineral matter and by partition in soil organic matter to uptake of organic compounds by a mineral soil at a system temperature. (**a**) Uptake from vapor phase on dry soil; P and P^o are the equilibrium pressure and saturation vapor pressure of the compound, respectively. (**b**) Uptake from water solution on wet soil; C_e and S_w are the equilibrium concentration and water solubility of the compound, respectively.

It is beyond the scope of this article to address the various adsorption models, therefore, the reader is advised to consult the cited literature. Here we consider only distinct features of adsorption and partition equilibria. An account of the solute partition effect is essential to defining the soil sorption in aqueous systems.

Because our common notion of the partition process comes largely from experience with the solvent extraction of solutes, it is desirable to elucidate herein the basic parameters that influence solute partition coefficients in solvent–solvent systems. To keep our treatment in order, we first consider the partition equilibrium of an organic solute in a two-phase solvent–water mixture, in which the solute partition coefficient can be reasonably accounted for by the conventional solution theory (Raoult's law). We consider later a similar two-phase system where the organic phase is macromolecular in nature and the solute partition coefficient is accounted for by the Flory–Huggins solution theory. The latter system is immediately applicable to the partition of organic compounds from water to soil organic matter.

The partition coefficient of a slightly water-soluble organic compound in a solvent–water mixture, where the solvent exhibits a small solubility in water, can be derived by application of conventional solution theory to a two-phase equilibrium mixture (71), giving

$$K = C_o/C_w = (\gamma_w^*/\gamma_o^*)\cdot(\overline{V}_w^*/\overline{V}_o^*) \tag{1}$$

which may be transformed with the assumption of $\overline{V}_w^* = \overline{V}_w$ to a logarithmic form as

$$\log K = -\log S_w - \log \overline{V}_o^* - \log \gamma_o^* - \log(\gamma_w/\gamma_w^*) \tag{2}$$

where K is the solute partition coefficient; C_o is the (equilibrium) solute concentration in the solvent phase (mol/L); C_w is the solute concentration in water; γ_w is the solute activity coefficient in water (Raoult's convention); γ_w^* is the solute activity coefficient in solvent-saturated water, γ_o^* is the solute activity coefficient in water-saturated solvent; $\overline{V}_w^*$ is the molar volume of solvent-saturated water (L/mol); $\overline{V}_o^*$ is the molar volume of water-saturated solvent phase; $S_w = 1/(\gamma_w \overline{V}_w)$ is the water solubility of solute as a liquid or supercooled liquid (mol/L); and $\overline{V}_w$ is the molar volume of water. For a solute that is sparingly soluble in water, all the activity-coefficient terms are practically constant. Likewise, the molar volume term ($\overline{V}_o^*$) is constant, which is determined by the molar volume of the pure solvent corrected for water miscibility. The log (γ_w/γ_w^*) term, which corrects for the enhancement of solute solubility in water due to saturation of the solvent, is relatively insignificant for solutes with high S_w in systems where the solvent has a low solubility in water. The use of supercooled-liquid solubility rather than solid solubility for solid solutes in equation 2 is to correct for the melting-point effect that affects only the solute solubility, but not the partition coefficient because this effect cancels in the two solvent phases (71,77). The supercooled liquid solubility of a solid solute in a solvent at a temperature below its melting point can be calculated from the solid solubility adjusted for the difference in fugacity of the solid (f_s) and its supercooled liquid (f_l). The adjustment for solute solubility in water gives

$$S_w^{(l)} = S_w^{(s)}\left(\frac{f_l}{f_s}\right) \tag{3}$$

with

$$\ln(f_l/f_s) = \frac{\Delta \overline{H}_f}{R}\left(\frac{T_m - T}{T\cdot T_m}\right) \tag{4}$$

where $S_w^{(s)}$ and $S_w^{(l)}$ are solubilities of the solid and its supercooled-liquid, respectively; T is the system temperature (degrees Kelvin); T_m is the solid's melting temperature; $\Delta\overline{H}_f$ is the solid's molar heat of fusion; and R is the gas constant. The term f_s/f_l (<1) is called the solid activity, which gives the ideal solubility of the solid (in mole fraction) in any solvent at T. The term $\Delta\overline{H}_f/T_m$ is called the molar entropy of fusion of the solid ($\Delta\overline{S}_f$).

Many important points associated with the partition equilibria of relatively water-insoluble organic solutes in solvent–water mixtures are worth noting. First, the magnitudes of K for organic solutes are determined to a large extent by the reciprocals of S_w values, rather than by the solubilities in the organic solvent (ie, the magnitudes of γ_o^*). The effect of γ_o^* on K is generally small relative to that of S_w for organic solutes because of their high degrees of miscibility with the organic solvent, as is typically found for nonionic organic compounds in octanol–water systems. As a result, one finds a relatively linear relationship between log K_{ow} (octanol–water) and log S_w for a wide range of chemical classes (71,78–80). A list of the log K_{ow} and log S_w values for some substituted aromatic solutes (liquids and supercooled liquids) is given in Table 1. A plot of log K_{ow} against log S_w for these solutes is presented in Figure 2 along with the *ideal line,* which is defined as $\log K_{ow}^{o} = -\log S_w - \log \overline{V}_o^*$, with $\log \overline{V}_o^* = -0.92$ for water-saturated octanol (71). The regression of log K_{ow} against log S_w gives

$$\log K_{ow} = -0.862 \log S_w + 0.710 \tag{5}$$

with $n = 36$ and a correlation coefficient $r^2 = 0.988$. The difference between log K_{ow}^{o} and log K_{ow} provides a measure of $\log \gamma_o^* + \log(\gamma_w/\gamma_w^*)$, or essentially of $\log \gamma_o^*$ for most solutes, since $\log(\gamma_w/\gamma_w^*)$ is significant only for those solutes that are extremely insoluble in water (such as DDT, some PCBs, and hexachlorobenzene) (71).

Second, the partition coefficient (K) of a slightly water-soluble solute in a solvent–water mixture is largely independent of the solute concentration below the limit of the solute solubility. That this is true is attributed to the fact that the activity coefficient of the solute in aqueous solution (which is inversely related to S_w) remains practically constant if the solubility is small and that the γ_o^* value is generally small and not sensitive to the solute concentration in the organic phase (because of high compatibility between solute and solvent (71)). It is thus expected that there is a relatively linear relation between solute concentrations in organic and water phases to give a virtually constant partition coefficient over a wide range of solute concentration relative to the solute's solubility. For example, the experimental K_{ow} of fenuron (a herbicide) remains constant over three orders of magnitude in aqueous concentration (1–3000 mg/L) (81). By contrast, the adsorption isotherm of a solute on an adsorbent (eg, activated carbon) in single-solute systems is linear only in the very low relative concentration region (the Henry's law region). Thus the isotherm linearity provides a useful basis for distinguishing partition from adsorption.

The third distinct feature in partition equilibria is the relative independence of the individual solute partition constants in binary or multiple solute systems, as is generally experienced in the solvent–solvent extraction of multiple solute

Table 1. Octanol–Water Partition Coefficients (K_{ow}) and Liquid or Supercooled Liquid Solubilities (S_w) of Substituted Aromatic Compounds[a]

Compound	Log S_w, mol/L[b]	Log K_{ow}
Aniline	−0.405	0.90
o-Toluidine	−0.817	1.29
m-Toluidine	−0.853	1.40
N-Methylaniline	−1.28	1.66
N,N-dimethylaniline	−2.04	2.31
o-Chloroaniline	−1.53	1.90
m-Chloroaniline	−1.37	1.88
Benzene	−1.64	2.13
Toluene	−2.25	2.69
Ethylbenzene	−2.84	3.15
n-Propylbenzene	−3.30	3.68
Isopropylbenzene	−3.38	3.66
1,3,5-Trimethylbenzene	−3.09	3.42
t-Butylbenzene	−3.60	4.11
Fluorobenzene	−1.80	2.27
Chlorobenzene	−2.36	2.84
Bromobenzene	−2.55	2.99
Iodobenzene	−2.78	3.25
o-Xylene	−2.72	2.77
m-Xylene	−2.73	3.20
p-Xylene	−2.73	3.15
Diphenylmethane	−4.07	4.14
1,2-Dichlorobenzene	−2.98	3.38
1,3-Dichlorobenzene	−3.04	3.38
1,4-Dichlorobenzene	(−3.03)	3.39
1,2,4-Trichlorobenzene	−3.57	4.02
Biphenyl	(−3.88)	4.09
2-Chlorobiphenyl	(−4.57)	4.54
3-Chlorobiphenyl	−5.16	4.95
Naphthalene	(−3.08)	3.36
2-Methylnaphthalene	(−3.69)	4.11
Phenanthrene	(−4.48)	4.57
Anthracene	(−4.63)	4.54
Pyrene	(−5.24)	5.18
Hexachlorobenzene	(−5.57)	5.50
p,p′-DDT	(−6.74)	6.36

[a] Ref. 71.
[b] Values at 20 to 25°C. Numbers enclosed with parentheses are the supercooled liquid solubilities estimated according to methods described (71).

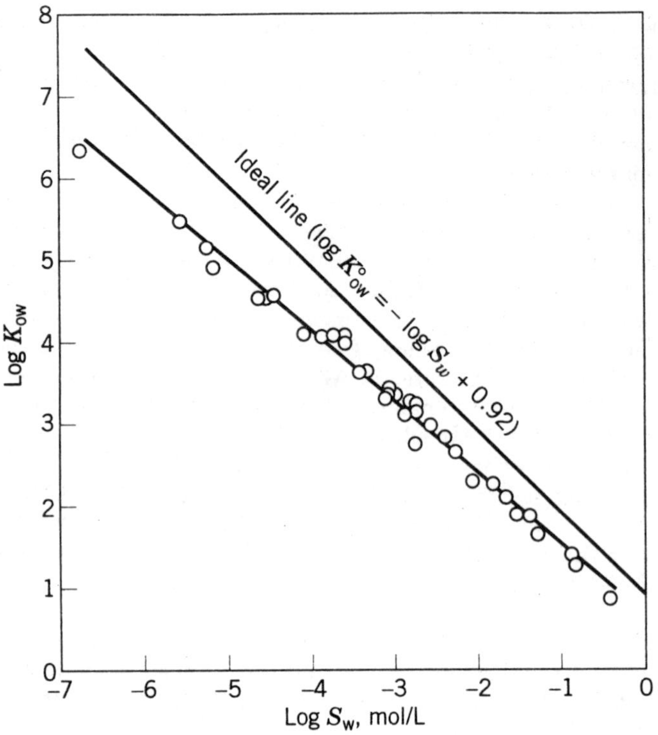

Figure 2. Correlation of octanol–water partition coefficients (K_{ow}) and water solubilities (S_w) of substituted aromatic compounds from Table 1. Reproduced with permission (71).

components. The partition constant of a solute would be significantly affected by the presence of other solutes only when the solvency of the solvent is significantly affected by very high concentrations of the other solutes. Such cases, however, are relatively rare. In the normal concentration range it has been shown, for example, that the individual solute partition coefficients in octanol–water systems determined with a combination of solutes are essentially the same as when determined singly (42). In adsorption, the competition between solutes for adsorbent surfaces or specific sites of the adsorbent leads to reduction of the adsorption capacities of the individual solutes in a multisolute system relative to their capacities as single solutes (74,76). This reduction is generally very marked for the solutes in a multisolute system at the concentration range in which the individual single-component adsorption capacities are relatively comparable.

Another unique characteristic that is associated with the partitioning of organic compounds into an organic phase is the observation that the limiting sorption capacities for such systems tend to be smaller with compounds exhibiting higher partition coefficients. For a series of nonionic solutes, the high partition constants (and low water solubilities) tend to correlate strongly with molecular weight; therefore, solutes with high molecular weights tend to have lower limiting sorption capacities (although the effect is considerably less than directly proportional). The inverse relationship between limiting capacity (at water saturation) and molecular weight is opposite to what one would expect for adsorption on a solid, where one would expect either a constant limiting adsorption capacity (for pore filling, as on activated carbon) or an increase in adsorption capacity (for open surfaces). By contrast, the reduction of solubility in the organic phase with increasing molecular weight of the solute (which magnifies any incompatibility between solute and organic solvent) is just what one would expect for a partition process. This point is illustrated by the comparison of the solubilities and octanol–water partition coefficients of DDT and benzene. The K_{ow} value of DDT is approximately four orders of magnitude greater than that of benzene, whereas the S_w of DDT as a supercooled liquid is approximately five orders of magnitude lower than that of benzene. Whereas DDT has a solubility in water-saturated octanol of approximately 32 g/L (or 0.090 mol/L) as a solid at 24°C (and approximately 10 times as large for supercooled DDT), benzene is completely miscible with the solvent (71).

Lastly, it is useful to consider the equilibrium heat associated with the partitioning of a solute. The molar heat of transfer ($\Delta\overline{H}$) of a solute from water to an organic phase by partitioning is simply equal to the difference of the heats of solution of the solute in the two solvent phases (25,27). That is,

$$\Delta\overline{H} = \Delta\overline{H}_o - \Delta\overline{H}_w \qquad (6)$$

where $\Delta\overline{H}_o$ is the molar heat of solution in the organic phase and $\Delta\overline{H}_w$ is the molar heat of solution in water, both of which are usually positive. Low solubilities of organic compounds in water (ie, poor compatibilities with water) are commonly associated with large $\Delta\overline{H}_w$ values. The corresponding $\Delta\overline{H}_o$ terms should be much smaller for organic compounds because of their enhanced compatibilities with organic solvents. Equation 6 applies for both solid and liquid compounds. For solids, $\Delta\overline{H}_o$ and $\Delta\overline{H}_w$ include molar heats of fusion ($\Delta\overline{H}_f$) as a part of heats of solution, which are constant for given solid compounds. Therefore, $\Delta\overline{H}$ reflects only the difference of the heats of incompatibility of the dissolved liquid or supercooled liquid solute in the solvent media. According to equation 6, the heat of partition of an organic solute would be relatively small and less exothermic than its heat of condensation from water ($-\Delta\overline{H}_w$). This expectation is in keeping with the fact that the partition coefficients of organic compounds in octanol–water (and other solvent–water) systems show relatively small inverse dependence on temperature (27). By contrast, adsorption of a solute from dilute solution on an adsorbent normally exhibits a higher exothermic heat that results both from exothermic condensation of the solute on the surface and from exothermic interaction of the solute with the surface. Therefore, $\Delta\overline{H}$ for adsorption is usually more exothermic than the corresponding heat of solute condensation from the solution (ie, the reverse heat of solution); otherwise, the adsorption would be extremely weak. The difference in equilibrium heat between partition and adsorption is expected to be especially pronounced for solid compounds that have large $\Delta\overline{H}_f$ values, because $\Delta\overline{H}_f$ adds to the heat of solution, which becomes part of the heat of adsorption but has no effect on partitioning (as $\Delta\overline{H}_f$ cancels in partition equilibria). In a later section, we shall analyze the heat effect associated with the sorption of DDT (which has a large $\Delta\overline{H}_f$) from water by soil or sediment to substantiate the mechanism involved.

Up to this point we have considered many characteristics pertinent to the partition equilibria of nonionic organic compounds in a solvent-water system. Although these features are common, we should note that equation 2 expresses rigorously the magnitude of the partition constant (K) only for systems where the size disparity between solute and solvent is relatively small. Equation 2 loses its rigor when the size disparity between solute and solvent is large, because the mole–fraction-based Raoult's law no longer serves as a logical basis for the activities of solute and solvent (82–85). The problem with equation 2 may be noted from its improper implication that the partition constant of a solute should greatly decrease with a large increase of the solvent molar volume (or molecular weight) without a corresponding change in composition. This is, however, not consistent with the relatively high athermal solubilities of small organic compounds in (amorphous) polymeric substances (82–84) that are outside the predictions by Raoult's law. As a consequence, the application of Raoult's law for solute partitioning in a polymer would lead to $\gamma_o^* \ll 1$ for systems that show no evidence of specific solute–polymer interaction.

In a macromolecular solvent phase, a more logical reference basis for solute activity is the volume fraction rather than mole fraction. According to the Flory–Huggin theory (82–84), equation 2 must be corrected for the effect of size disparity on the solubility of a solute in an amorphous macromolecular phase (such as soil organic matter). A modified partition equation has been developed by applying the Flory–Huggins theory for solute solubility in an amorphous polymeric phase (26). On the assumption that the molar volume of solute is negligibly small compared to that of the polymer, the result is

$$\log K = -\log S_w\overline{V} - \log\rho - \frac{1+\chi}{2.303} - \log\left(\frac{\gamma_w}{\gamma_w^*}\right) \qquad (7)$$

in which $\overline{V}$ is the molar volume of the solute (L/mol); ρ is the density of the organic phase, introduced to express the solute concentration in the organic phase on a more convenient mass-to-mass basis; and χ is the Flory–Huggins interaction parameter (dimensionless), the sum of excess enthalpic (χ_H) and excess entropic (χ_S) contributions to the incompatibility of the solute in a polymer. The value of χ_S depends on the polymer network that affects the accessibility and flexibility of the polymer segment and is usually determined empirically. It has been suggested (86) that $\chi_S \cong 0.25$ for high polymers, and this value was adopted by Chiou et al. (26) for soil organic matter. The $\chi_H/2.303$ term in equation 7, equivalent to the $\log\gamma_o^*$ term in equation 2, expresses the extent of solute incompatibility with the macromolecular phase. The term $\log(\gamma_w/\gamma_w^*)$, as defined earlier, expresses the magnitude of solubility enhancement of the solute in water resulting from solubilization in water of a fraction of the macromolecular organic substance. As before, an *ideal line* is established by letting $\log K^\circ = -\log S_w\overline{V} - \log\rho - (1+\chi_S)/2.303$ and the difference between $\log K^\circ$ and $\log K$ gives a measure of the effect of $[\chi_H/2.303 + \log(\gamma_w/\gamma_w^*)]$, relative to $-\log S_w\overline{V}$, on $\log K$. If the molar-volume ratio between solute and solvent ($\overline{V}/\overline{V}_o^*$) is not negligible, a more general partition equation should be used instead (87).

Equation 7 has been applied to analyze the sorption (partition) coefficients of nonionic organic compounds between soil organic matter and water (26). A more detailed discussion of the results will be given later. It is proper to point out, however, that analysis of the partition data by application of equation 7 leads to essentially the same conclusion, as recognized in ordinary solvent–water systems, that S_w is the major factor affecting the partition coefficients (K) of relatively nonpolar solutes between the soil organic phase and water.

SORPTION BY SOILS AND SEDIMENTS IN AQUEOUS SYSTEMS

General Soil– and Sediment–Water Equilibrium Characteristics

The sorption data for relatively nonpolar organic compounds (solutes) are presented first and then the data for relatively polar organic compounds, because of some characteristic differences in their behaviors as influenced by water. Whereas the classification of polar versus nonpolar compounds is not simple, the polar compounds are considered the ones that possess significant polar groups in their molecular structures and the nonpolar compounds are the ones that contain little or no polar groups. Some common strong polar groups are –OH, $-NH_2$, –COOH, –CO–, and $-NO_2$. Polar groups in compounds enhance molecular interactions by polar forces and hydrogen bonding with each other and with other polar compounds. In general, the effect of a polar group in a molecule is more pronounced for small molecules than for large molecules. Polar organic solutes generally exhibit low partition from water to a water-immiscible (or partially miscible) organic phase relative to nonpolar solutes because of the increased affinity for water of the former. For instance, the log K_{ow} (octanol–water) values of highly polar organic solutes are generally <2.

In water solution the equilibrium sorption isotherms for relatively nonpolar organic compounds on soils or sediments are usually virtually linear, as has been demonstrated in a number of studies (13,18,25,26,28,30,36,39,41,44,56,66). Similar isotherm linearity has been found in the soil uptake of relatively nonpolar pesticides from the vapor phase onto moist (water-saturated) soils (32,40). In some studies, where slight isotherm curvatures were shown (eg, 20,56), the extent of the reported isotherm nonlinearity (either concave upward or downward) appears to be within the normal range of data scatter that cannot be distinguished from a linear plot. This is especially true in sorption studies of low organic content soils in which the solute uptake is calculated by the difference in solute concentrations in water before and after equilibration; here the experimental error in detecting small concentration changes is expected to be relatively large. Because the curvatures in allegedly nonlinear isotherms for some relatively nonpolar pesticides (eg, ethylene dibromide and lindane) (56) are quite small and show no consistent patterns, the results show no evidence of strong solute adsorption over the studied concentration range.

Although in previous studies the linear sorption isotherms presented were often confined to a low concentration range representative of practical situations, such linear isotherms also extend to high relative concentrations (C_e/S_w) for sparingly water-soluble nonionic solutes; C_e and S_w are the equilibrium solute concentration and solute solubility in water, respectively. Figure 3 shows typical linear isotherms at 20°C for the sorption of benzene, 1,3-dichlorobenzene, and 1,2,4-trichlorobenzene from water on a Woodburn soil, which

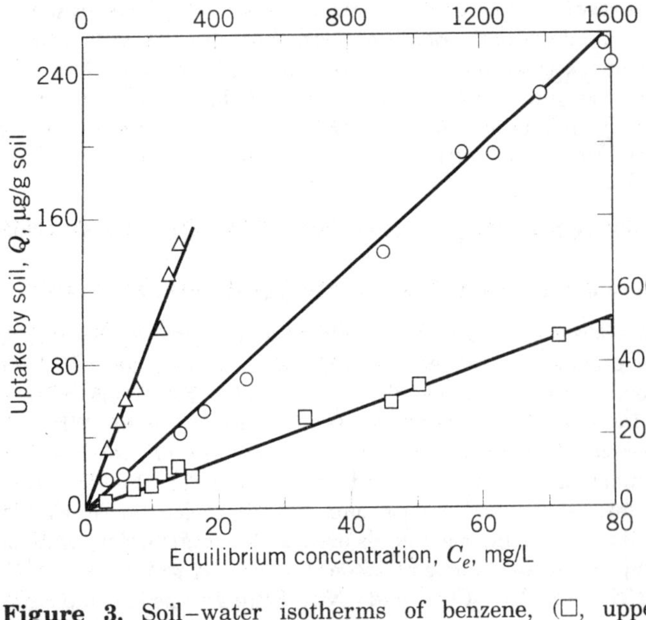

Figure 3. Soil–water isotherms of benzene, (□, upper and right scales), 1,3-dichlorobenzene (○), and 1,2,4-trichlorobenzene (△) on Woodburn soil (f_{om} = 0.019) at 20°C. Reproduced with permission (26).

contains 1.9% organic matter (f_{om} = 0.019) (26). The benzene isotherm is linear with C_e/S_w up to about 0.90. Similar linear isotherms for many halogenated organic liquids on a Willamette silt loam (fine-silty, mixed, mesic pachic Ultic Argixeroll) (f_{om} = 0.016) are shown in Figure 4, in which, for example, 1,2-dichlorobenzene shows linear sorption with C_e/S_w up to 0.95. The normalized K_{om} values of these halogenated solutes (ie, $K_{om} = K_d/f_{om}$, where K_d is the soil–water distribution coefficient) and their water solubilities at 20°C are given in Table 2. The wide isotherm linearity and the recognized dependence of solute sorption on soil organic matter in aqueous systems give evidence that soil sorption of relatively nonpolar organic compounds from water occurs mainly by partition in (rather than by adsorption on) soil organic matter. The low sorption capacities of these solutes on mineral soils result from both the low organic matter content in soil and the low partition efficiency of low-polarity solutes with relatively polar soil organic matter.

Although solute partitioning to soil organic matter was suggested earlier (36) as a possible mechanism for parathion sorption by soil from water, it did not gain wide acceptance in the absence of other supporting data. As a matter of fact, there has been some misconception about the implication of linear sorption isotherms. The isotherm linearity, as frequently observed for sparingly water-soluble organic compounds and pesticides on soil, was often thought to be a manifestation of the solutes' low concentrations in water that restrict soil adsorption to a linear range (56,88). Although adsorption isotherms of single solutes (and vapors) would be linear at very low relative concentrations (C_e/S_w), commonly referred to as the *Henry's law region*, the observed isotherm linearity that extends over a wide range of relative concentration (as with soil uptake from water) should not be confused with the linear portion of an overall nonlinear adsorption isotherm. To make this point evident one may, for example, compare the adsorption isotherms of 1,2-dichlorobenzene, 1,1,1-trichloroethane, 1,2-dichloropropane, and 1,2-dibromoethane on an activated carbon (Fig. 5) with their sorption isotherms on a Willamette silt loam (see Fig. 4). It is evident that the isotherms of the compounds on activated carbon are linear only at very low equilibrium concentrations relative to their water solubilities as shown in Table 2, whereas the sorption isotherms on soil show no obvious indication of curvature even at concentrations approaching saturation.

An important corollary of linear isotherms on soils in aqueous systems is the implication that the molar heat of sorption of the compound is constant and independent of its loading on soil. This result may be readily illustrated by the calculated equilibrium heat of sorption of a compound using its linear isotherms at two temperatures (13,25,39,40). Figure 6 gives a schematic plot of the linear isotherms of a compound at temperatures T_1 and T_2 (in degrees Kelvin), with $T_2 > T_1$. Let Q be the mass of the compound sorbed by a unit mass of soil (or, more closely, by a unit mass of soil organic matter) and C_e be the aqueous concentration of the compound in equilibrium with a given Q on soil. Each linear isotherm is assumed to extend over a wide range of equilibrium concentrations relative to solubility at the system temperature. The isotherms are drawn such that the soil uptake at T_1 is greater than that at T_2, as is usually observed for most organic compounds and pesticides; however, a reverse temperature dependence can

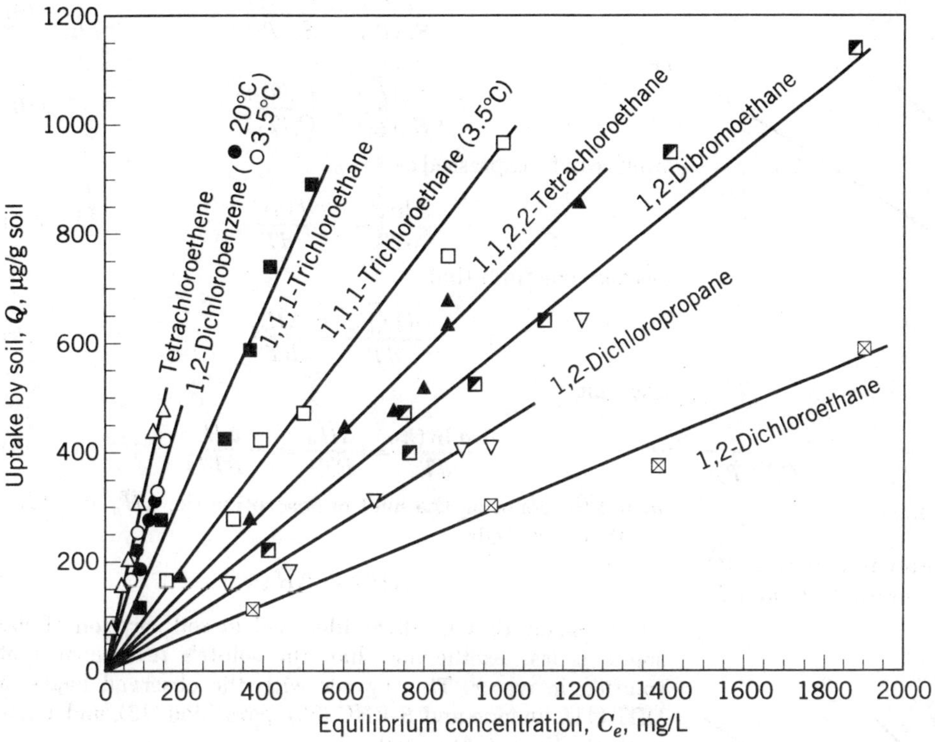

Figure 4. Soil–water isotherms of halogenated organic liquids on Willamette silt loam (f_{om} = 0.016) at 20°C. Reproduced with permission (25).

Table 2. Normalized Sorption Coefficients of Halogenated Organic Liquids from Water on Willamette Silt Loam (K_{om}) and Corresponding Liquid Solubilities in Water (S_w) at 20°C[a]

Compound	S_w, mg/L	K_{om}
1,2-Dichloroethane	8450	19
1,2-Dichloropropane	3570	27
1,2-Dibromoethane	3520	36
1,1,2,2-Tetrachloroethane	3230	46
1,1,1-Trichloroethane	1360	104
1,2-Dibromo-3-chloropropane	1230	75
1,2-Dichlorobenzene	148	180
Tetrachloroethene	200	210

[a] Ref. 25.

take place if the compound shows abnormal (ie, exothermic) heat of solution in water, as shown in Figure 4 for 1,1,1-trichloroethane (25).

The molar isosteric heat of soil sorption at a given uptake capacity can be obtained by use of the Clausius-Clapeyron equation. At capacity Q_A, for example, the equation gives

$$\Delta\overline{H}(Q_A) = -R\frac{\ln[C_e(A,T_2)/C_e(A,T_1)]}{1/T_1 - 1/T_2} \quad (8)$$

where R is the gas constant, and $C_e(A,T_2)$ and $C_e(A,T_1)$ are the equilibrium concentrations corresponding to Q_A at temperatures T_2 and T_1, respectively. The molar heat of sorption at capacity Q_B, that is, $\Delta\overline{H}$ (Q_B), can be similarly determined by substituting $C_e(B,T_2)$ for $C_e(A,T_2)$ and $C_e(B,T_1)$ for $C_e(A,T_1)$ in equation 8. Because the isotherms are linear, $C_e(A,T_2)/C_e(A,T_1) = C_e(B,T_2)/C_e(B,T_1)$, and hence $\Delta\overline{H}(Q_A) = \Delta\overline{H}(Q_B)$. By repeating the same calculations at any

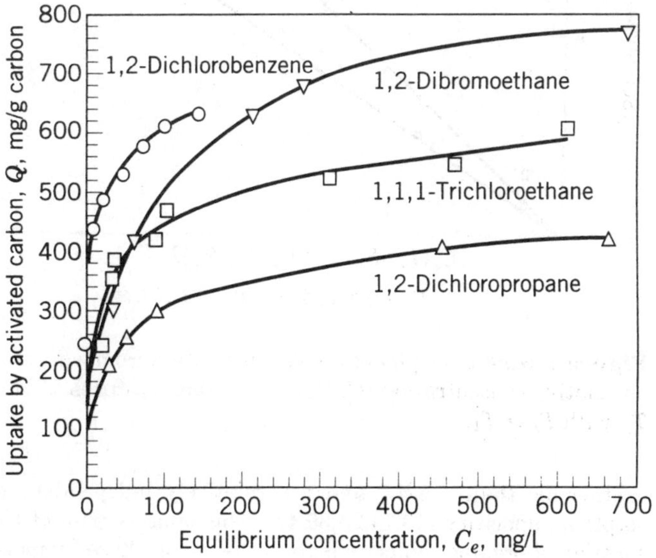

Figure 5. Adsorption isotherms of selected halogenated organic liquids from water on Pittsburgh CAL (12 × 40) activated carbon at 20°C. Reproduced with permission (27).

loadings, one thus concludes that the molar heat of sorption is constant and independent of the loading capacity. Because of the linearity of the isotherms, the concentration ratio in equation 8 is also equal to the ratio of the sorption coefficient (the slope of the isotherm) at T_1 to that at T_2.

The calculated molar heats of sorption for most organic compounds and pesticides on soil in water are generally less

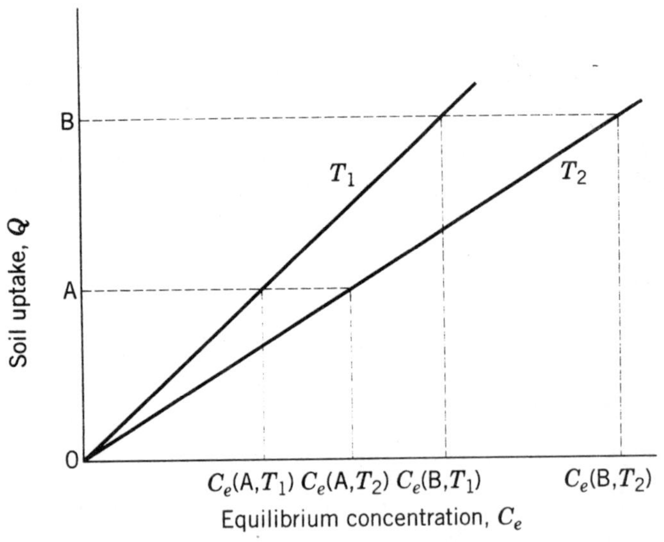

Figure 6. Schematic plot of linear soil sorption from water (Q) vs equilibrium concentration (C_e) at temperatures T_1 and T_2, with $T_2 > T_1$.

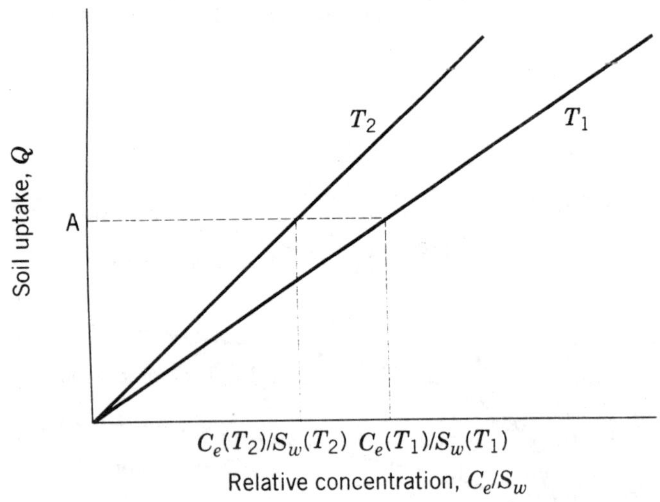

Figure 7. Schematic plot of linear soil sorption from water (Q) vs relative concentration (C_e/S_w) at two temperatures T_1 and T_2, with $T_2 > T_1$.

exothermic than $-\Delta\overline{H}_w$ and are relatively independent of sorption capacities (13,14,25,39,41). The same is true of the sorption of organic compounds from the vapor phase by moist (water-saturated) soils, such as found for ethylene dibromide (12) and lindane (40), in which heats of sorption are less exothermic than heats of vapor condensation ($-\Delta\overline{H}_v$). These results are inherently consistent with the suggested partition uptake of nonionic organic compounds by the soil organic matter of water-saturated soils.

The relation between $\Delta\overline{H}$ and $\Delta\overline{H}_w$ for soil sorption in aqueous systems can be readily explained by the temperature dependence of the normalized isotherms, as shown in Figure 7, in which the relative solute concentration (C_e/S_w) is used for the abscissa. By such a normalized plot, one usually finds a reverse temperature dependence for organic solutes in soil sorption (13,39); that is, at a given loading on soil

$$\frac{C_e(T_1)}{S_w(T_1)} > \frac{C_e(T_2)}{S_w(T_2)} \tag{9}$$

or

$$\frac{S_w(T_2)}{S_w(T_1)} > \frac{C_e(T_2)}{C_e(T_1)} \tag{10}$$

which can be expressed as

$$\frac{d\ln S_w}{dT} > \frac{d\ln C_e}{dT} \tag{11}$$

Because one finds that

$$\frac{d\ln S_w}{dT} = \frac{\Delta\overline{H}_w}{RT^2} \tag{12}$$

and that

$$\frac{d\ln C_e}{dT} = \frac{\Delta\overline{H}_d}{RT^2} = -\frac{\Delta\overline{H}}{RT^2} \tag{13}$$

with $\Delta\overline{H}_d$ denoting the heat of desorption (ie, $\Delta\overline{H}_d = -\Delta\overline{H}$), one therefore finds

$$\Delta\overline{H} > -\Delta\overline{H}_w \tag{14}$$

which means that the heat liberated in soil sorption of the solute is less exothermic than the solute's reverse heat of solution in water. This agrees with the observed heats of DDT (41), lindane and β-BHC (39), parathion (13), and 1,1,1-trichloroethane (25) on soil.

Equation 14 is consistent with the expectation for the partition process shown in equation 6. Because the heats of solution of organic compounds in an organic phase ($\Delta\overline{H}_o$) are generally small due to their improved compatibitilities, $\Delta\overline{H}$ will be small for compounds with low $\Delta\overline{H}_w$ values and may even become positive (endothermic) for compounds with abnormal (negative) $\Delta\overline{H}_w$ values. It has been shown (25), for example, that the $\Delta\overline{H}$ for 1,2-dichlorobenzene sorption by soil from water is nearly zero because of its low $\Delta\overline{H}_w$ and the $\Delta\overline{H}$ for 1,1,1,-trichloroethane is positive because of its negative $\Delta\overline{H}_w$ in the temperature range 3.5–20°C. One may thus conclude from these data that in systems where the $\Delta\overline{H}$ values are negative, such exothermic heats derive primarily from condensation of the solutes from water ($-\Delta\overline{H}_w$) and that, with $\Delta\overline{H}_o$ being normally positive, the interactions between soil organic matter and solute ($\Delta\overline{H}_o$) are normally endothermic (as usually is for the heat of solution). One finds small and nearly constant exothermic heats ($\Delta\overline{H}$) (and hence small temperature coefficients) in solute partition equilibria as a result of the partial cancellation in heat between $\Delta\overline{H}_o$ and $\Delta\overline{H}_w$.

The sorption data of p,p'-DDT are especially worth noting because DDT has a relatively high heat of fusion ($\Delta\overline{H}_f$), about 25 kJ/mol (89), and is highly incompatible with water, which should make $\Delta\overline{H}_w$ much greater than 25 kJ/mol. The sorption coefficient of DDT with soil or sediment expressed on the basis of the organic matter content of soil or sediment (K_{om}) is approximately 1.5×10^5 (41,90), whereas the heat of sorption ($\Delta\overline{H}$) is about -8.4 to -16.8 kJ/mol (41). Based on these values, one can calculate the standard entropy change for the transfer of DDT from water to the soil organic matter as

$$\Delta\overline{G}^o = -RT\ln K_{om} \tag{15}$$

and

$$\Delta\overline{S}^o = \left(\Delta\overline{H}^o - \Delta\overline{G}^o\right)/T \tag{16}$$

where $\Delta\overline{G}^{\circ}$ is the (molar) standard free energy change for the transfer of one mole (or a unit mass) of the solute from water at unit concentration to the organic matter phase at unit concentration, and $\Delta\overline{H}^{\circ}$ and $\Delta\overline{S}^{\circ}$ are the corresponding standard enthalpic and entropic changes. Now, if one takes $K_{om} \cong 1.5 \times 10^5$ and $\Delta\overline{H}^{\circ} = \Delta\overline{H} \cong -12.6$ kJ/mol for DDT at $T = 298$ K, one gets $\Delta\overline{S}^{\circ} \cong 58$ J/K · mol. Although the calculated $\Delta\overline{S}^{\circ}$ for DDT is subject to some uncertainty because of the inaccuracy of the $\Delta\overline{H}$ used, the magnitudes of the data would nonetheless show a gain in entropy for the transfer of DDT at the stated standard state. Such a gain in entropy deviates from the expectation for a decrease in entropy for adsorption of a trace component from solution. Later the heat effect associated with soil sorption of organic compounds in nonaqueous systems will be considered.

Accountability of solute partitioning in soil organic matter is further substantiated by the estimated magnitude of the solute solubility in soil organic matter. Since the isotherm is practically linear, the solubility of the solute in the organic matter may be determined by

$$S_{om} = S_w \cdot K_{om} \tag{17}$$

where S_{om} is the solubility in organic matter and S_w is the solubility in water. For solid DDT with $S_w = 5.5$ μg/L (91) and $K_{om} \cong 1.5 \times 10^5$ at 25°C, one therefore gets $S_{om} \cong 830$ mg/kg, or 0.83 g/kg. By comparison, the solubility of DDT in pure octanol is approximately 42 g/L (71), which is some 50 times its solubility in soil organic matter. The relatively low estimated solubility of DDT in soil organic matter is expected for a relatively nonpolar solid compound in a relatively polar macromolecular amorphous material. It is evident from these data that the very high sorption coefficient of DDT is primarily due to its extremely low solubility in water, which results in the high partition coefficient.

On the premise of solute partitioning, one expects organic compounds with higher water solubility to give lower K_{om} but higher S_{om} values because these compounds are usually also more compatible with organic solvents. For example, benzene having $S_w = 1780$ mg/L gives $K_{om} \cong 18$ (26) and $S_{om} \cong 32$ g/kg according to equation 17, which is approximately 40 times greater than the S_{om} value of solid DDT (due partly to the fact that benzene is a liquid) and is reconcilable with the observation that benzene is completely miscible with octanol and most organic solvents. Thus, although the S_{om} values for given solutes may vary somewhat among soils or sediments due to a compositional difference in organic matter, the magnitudes of the S_{om} values illustrated are largely in the range expected for low polarity organic compounds in relatively polar organic polymers. To rationalize differences in soil uptake of organic compounds from water, it has been suggested (56) that the less polar the compound the more it will tend to adsorb on a hydrophobic surface (soil organic matter) from a polar solvent (water), while removing solvent molecules from that surface. This *hydrophobic adsorption* model is in obvious contradiction to the results illustrated, namely that the limiting capacity of DDT in soil organic matter is far less than that of benzene. Furthermore, the finding that polar organic liquids exhibit much higher uptake than nonpolar organic liquids on a high organic content peat soil (92) is consistent with solute partition rather than with hydrophobic adsorption.

The sorption characteristics of individual solutes from a binary or multisolute system provides another distinction between partition and adsorption. In their study of the simultaneous sorption of pyrene and phenanthrene from aqueous solution by river sediments, Karickhoff and co-workers (28) contended that there was no discernible sorptive interference between the two compounds (although no explicit comparison was made between single-solute and binary-solute isotherms). Chiou and co-workers (14,26) found no apparent sorptive competition between *m*-dichlorobenzene and 1,2,4-trichlorobenzene and between parathion and lindane as binary solutes from water on soils. The isotherm data for parathion and lindane are illustrated in Figure 8. Thus, for these relatively nonpolar solutes on soil, the system does not manifest apparent solute competition, whereas a strong competitive effect would be expected if adsorption were the dominant mechanism (74,76). The absence of significant sorptive competition between nonpolar solutes simplifies the determination of individual solute sorption coefficients in multisolute systems.

The apparent noncompetitive effect for relatively nonpolar solutes reflects the dominance of solute partition in soil organic matter, as a result of the strong adsorptive competition of water for soil minerals. The low solubilities (ie, poor compatibilities) of the organic compounds in water promote their partitioning (solubilization) in the soil organic matter phase, which in effect accounts for the large difference between the sorption coefficients of DDT and benzene as discussed earlier. The ineffectiveness of the sorption of relatively nonpolar compounds on minerals from water has been documented, for instance, for lindane and dieldren on sand (93), lindane on calcium bentonite (70), and DDT on montmorillonite (41). Even for the sorption of more polar solutes, such as parathion on clay (20), phenol on geothite (94) and 2,4-D on montmorillonite (95), the sorption also tends to be quite weak. Because a trace amount of

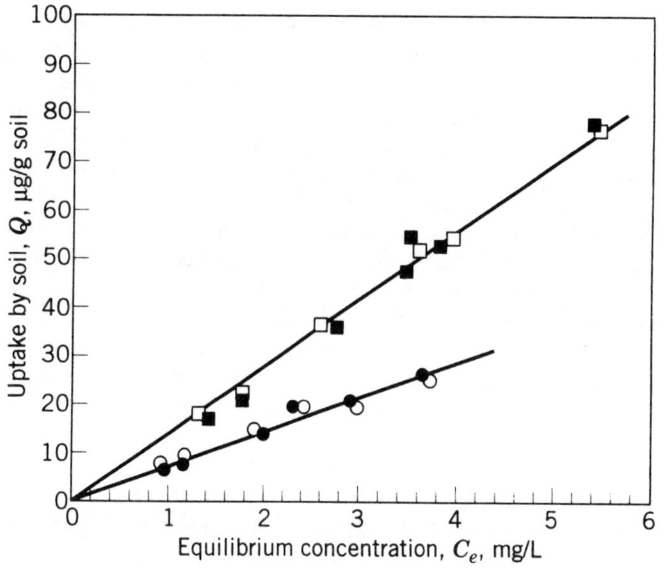

Figure 8. Sorption isotherms of parathion (□, ■) and lindane (○, ●) as single and binary solutes from water on Woodburn soil ($f_{om} = 0.019$) at 20°C. Open symbols are for single-solute data and solid symbols for binary-solute data. Reproduced with permission (14).

organic matter in unpurified clay minerals would have a significant impact on uptake by minerals from aqueous solution, the neglect of a small organic content in clays in sorption studies could lead to improper interpretations of the results. The enhanced uptake of 2,4-D by unpurified illite clay (95) might partly be an artifact of the organic impurities in the clay.

Despite the fact that the sorption data of low polarity compounds in water are illustrative of the dominance of solute partition in soil and sediment organic matter, the situation with relatively polar organic contaminants may be complicated under some conditions. For example, a close examination of the isotherms of 2-chlorophenol and 2,4-dichlorophenol from water on soils (96) reveals that the isotherms display a significant concave-downward curvature at low relative concentrations but a good linearity at high relative concentrations. In other words, the sorption coefficient is enhanced at low relative concentration but remains relatively constant at high relative concentrations. A similar effect was also documented for several polar substituted ureas (herbicides) on soils (81,97), where the lowering of solute relative concentrations by approximately three orders of magnitude to levels of $10^{-4}-10^{-5}$ resulted in an increase of the sorption coefficients by a factor of ~3.5. It was found (98) that the sorption coefficients of some phenolic compounds at low relative concentrations in binary- and ternary-solute systems were reduced by some 10–30% relative to those measured in single-component systems.

On the findings of substituted ureas on soils, Spurlock and Biggar postulated (81,97) a general partition (solubility) model to account for the concentration dependence of the sorption coefficient. The enhanced sorption coefficients at low relative concentrations (primarily at $C_e/S_w < 0.01$) were attributed to an enhanced solute partition resulting from a specific (exothermic) interaction of polar ureas with polar soil humic groups. This specific interaction is assumed to be most powerful at low concentrations due to *site* availability. At high relative concentrations, with the saturation of these specific sites, the polar solutes then show the usual partition in soil organic matter. This unusual bimodal partition model with soil organic matter is yet to be substantiated. Whereas it is possible for some polar solutes to exhibit specific interactions with certain active sites of the soil organic matter, it is not at all clear whether such specific interactions may be appropriately termed a solute partition (solubility) process when they occur.

By considering the dual mechanistic functions of the soil, it is possible that the unsuppressed adsorption of polar solutes on minerals could also lead to the enhanced sorption coefficient at low relative concentrations (99). This is because water may not suppress as effectively the adsorption of relatively polar organic solutes as it does the adsorption of low-polarity solutes on polar soil mineral matter. The consequence of this unsuppressed mineral adsorption or of the specific interaction with organic matter as mentioned on soil sorption coefficient may be better appreciated by reference to the plot in Figure **1b**. Since the partition to organic matter is essentially linear and the unsuppressed adsorption on minerals or specific interaction with organic matter is nonlinear (of a concave-downward shape), the partition is favored at high relative concentrations and disfavored at low relative concentrations. To the extent that the mineral adsorption or specific interaction is important, the total uptake (ie, the sum of the two contributions) on a soil would be dominated by partition at moderate to high relative concentrations and by unsuppressed mineral adsorption or specific interaction with organic matter at very low relative concentrations. In this case, the isotherm would be nonlinear at low relative concentrations but virtually linear at moderate to high relative concentrations. The solute concentration at which the linear partition and the nonlinear contribution cross is presumably a function of the organic matter content, the magnitude of K_{om} or K_{oc}, and the mineral type and content, which determine the relative effects of partition and adsorption.

The observed isotherm nonlinearity for certain polar solutes on soil, due either to a specific solute interaction with soil organic matter or to an unsuppressed solute adsorption on minerals, raises the adsorption-partition transition point, making the effect more readily noticed. At this time, the relative merit of all mechanisms contributing to the concentration dependence of the polar-solute-sorption coefficient remains to be settled by further experimental work. The knowledge of polar solute sorption to soil and sediment at low concentrations is important, because the concentrations of many polar solutes in natural water (usually at low μg/L to low mg/L) fall into a low relative concentration range because of their usually high water solubilities. The laboratory results obtained for polar solutes at high relative concentrations may not properly reflect the solute behavior in natural systems. The problem with low polarity solutes would be less serious in consideration of their wide isotherm linearity and their generally low water solubilities (which raise their relative concentrations). In the present discussion of the contaminant sorption behavior, consideration will be limited mainly to low polarity solutes or to polar solutes at appreciable relative concentrations.

Effect of Organic Matter Polarity and Composition on Solute Partition

An important subject to be addressed in the study of solute sorption coefficients (K_{om} or K_{oc}) is how much the soil or sediment organic medium property varies between soils and between sediments to affect the sorption coefficient (K_{om} or K_{oc}) of an organic pollutant. This information will determine if soils or sediments from dispersed geographic locations need to be studied separately (if K_{om} or K_{oc} values vary widely) or can be treated rather indiscriminately (if K_{om} or K_{oc} values are relatively invariant). Although some data are available for a few selected solutes (16,17), the K_{oc} data for these solutes were taken from a relatively small combined set of soil samples analyzed by different experimental procedures. Based on the combined K_{oc} data for a few selected solutes from different reports, it was found (33) that the K_{oc} variation between soils is generally less than a factor of three or four. Others (56) indicated that the K_{oc} could vary by as much as a factor of 10 or greater, based on selected K_{oc} values of some pesticides. Utilizing a correlation of K_{oc} with (O + N)/C weight ratio of natural organic matter, it has been estimated (38) that the K_{oc} variation of nonpolar solutes between soils to be less than a factor of three, based on the range of (O + N)/C values assumed for common soil organic matter.

Because different experimental procedures contribute to K_{om} or K_{oc} variation, especially in studies involving very low soil organic contents, a more definitive account of the K_{om} or K_{oc} variation between soils or sediments could only be established by data from a large set of soils and sediments of significant organic contents from diverse geographic sources using consistent and rigorous analytical methods. With this consideration, researchers (35) studied the K_{oc} values of two relatively nonpolar solutes, carbon tetrachloride (CT) and 1,2-dichlorobenzene (DCB), on 32 "normal soils" and 36 "normal bed" sediments collected from diverse geographic regions in both the United States and the People's Republic of China. The partition data of low polarity solutes with soil and sediment organic matter should best detect the differences in organic matter polarity and composition, if any, because the solubility of nonpolar solutes is sensitive to organic medium polarity (38) and because the adsorption of such solutes on soil/sediment minerals should be most effectively suppressed by water (99).

The sources of soils and bed sediments, the sample surface areas (SA) as determined by BET-(N_2) method, the organic carbon contents (% OC), and the measured K_{oc} values for CT and DCB (35) are shown in Table 3. Isotherm plots for CT and DCB on representative soils and bed sediments are shown in Figures 9 and 10. The linearity of the isotherms is typical of solute partition in soil and sediment organic matter, as expected. With the reported uncertainty of K_{oc} values being about $\pm 8\%$ and that of the organic-carbon mass fraction in soil or sediment (f_{oc}) being $\pm 5\%$, the observed differences between soil K_{oc}s (or between sediment K_{oc}s) for both CT and DCB are relatively small. The mean K_{oc} for CT on 32 normal soils is 60 (SD = ± 7) and the mean K_{oc} for DCB is 290 (SD = ± 42). The K_{oc}s for both CT and DCB on 36 normal bed-sediments are generally higher and show about the same variation as the K_{oc}s on soils. The mean K_{oc} on bed sediments for CT is 102 (SD = ± 11) and for DCB is 502 (SD = ± 66); they are greater by a factor of 1.7 than the mean K_{oc}s for CT and DCB on soils. This difference is more than the deviation of the means and is illustrated in Figures 11 and 12 by the sorption data of CT and DCB. The finding that the K_{oc}s for DCB are a factor of four to six greater than respective K_{oc}s for CT on all soils and sediments is approximately what one would expect according to equation 7 from the different solubilities of CT (800 mg/L) and DCB (154 mg/L) in water and their similar solubilities in soil organic matter (38).

The high degree of invariance of the K_{oc} values of CT and DCB in most soils or in most bed sediments is striking. The normalized sorption coefficients (K_{oc}s) for both solutes show little dependence on soil or sediment OC contents (eg, 0.16–6.09% for soils) and on (dry) soil or sediment surface areas (eg, 1.07–54.0 m^2/g for soils), as shown in Table 3. This finding reveals the similarity in organic-matter polarity/composition between soils and between bed sediments and the predominance of solute partitioning in soil/sediment organic matter over mineral adsorption; if mineral adsorption were important, the sorption coefficient would be affected by the sorbent's surface area. The range of variation for the soils is much less than that reported in earlier studies, which is based on a smaller set of combined data determined by different analytical procedures (33,56). The extreme K_{oc} values for soils (or sediments) differ by less than a factor of two. The relative invariance in K_{oc} suggests that the properties of the (humified) soil and sediment organic matter that control nonpolar solute solubility are quite similar for a wide variety of uncontaminated soils and for relatively pristine bed sediments. There does not appear to be a large variability in organic polarity and composition among well-weathered soils from diverse geographic locations. In view of the relative invariance in K_{oc} between soils or between sediments as illustrated, the use of average soil (or sediment) K_{oc} values for assessing the sorption of low polarity contaminants to different soils (or sediments) would seem sufficient in most environmental applications. However, an account of the difference in sorption to soil and sediment of low-polarity contaminants would seem warranted.

The fact that most soil K_{oc}s are distinct from bed-sediment K_{oc}s suggests that the process that turns eroded soils into bed sediments brings about a noticeable change in the property of the organic constituent. A possible cause for this change is that the sedimentation process fractionates soil organic constituents, such that the more polar and more water-soluble organic components in soil organic matter (eg, fulvic and humic acid fractions), or soil particles with more polar organic components, are separated out to form dissolved organic matter and colloids in water, with the less-polar soil organic constituents preserved in the bed sediment. The time scale to bring about a complete soil-to-sediment conversion should be a function of hydrodynamics. Another possible cause for this change would be the biological influence. However, this effect would seem small, because the difference between bed-sediment and soil K_{oc} values, while statistically significant, is not large.

Part of the variation in K_{oc} within bed sediments may reflect the extent of conversion of the eroded soils to bed sediments. Consider, for example, the relatively low K_{oc} values of CT and DCB with sediments 1, 6, 7 (see Table 3). Sediment 1 is a U.S. EPA sample taken from the mouth of Isaacs Creek at its junction with the Ohio River near Ripley, Ohio. That the K_{oc} values on sediment 1 are significantly lower than the rest but are very similar to soil K_{oc}s suggests that this sample could be a recently eroded soil which retains most of its soil organic composition. The somewhat lower K_{oc} values with sediments 6 and 7 relative to the average sediment K_{oc} may again be a result of incomplete conversion of eroded soils to bed sediments. The K_{oc} data suggest that bed sediments from most large rivers and lakes are relatively comparable in their organic polarities and compositions, probably because they are more aged and contain less recently eroded soils.

The difference between soil and bed-sediment K_{oc} values as detected by relatively nonpolar solutes lends a basis for identifying the source of suspended solids in rivers. In one study (35), the suspended solids from the Mississippi, Missouri, and Yellow Rivers were collected during high river flows, and the sample from the Illinois River was collected during a low-to-normal river flow. Here the K_{oc}s of CT and DCB are typical of those of soils for the former but are more representative of bed sediments for the latter (see Table 3). One may infer from these data that the suspended solids during high water flows in these three rivers consist mainly of newly eroded soils and the suspended solids from the Illinois River under low-to-normal water flow consists largely of resuspended bed sediment. The Yellow River suspended solid, which shows its origin as an eroded soil, is in keeping with the river's high carrying load of eroded soils during the high flow season. In contrast, a bed sediment collected from the Yellow River (sediment 22) gives K_{oc} values typical of those for other bed sediments. Thus the sorption data serves as a simple indicator of the source and time history of the suspended solids.

The relatively low K_{oc} values of CT and DCB, which are about one order of magnitude lower than their log K_{ow} values, suggest that soil and sediment organic matter must be relatively polar in nature to limit the partition (solubility) of these relatively nonpolar organic solutes. To investigate the effect of soil organic matter composition and polarity on solute

Table 3. Sources of Soils, Bed Sediments, and Suspended Solids and Their Surface Areas (SA) (m^2/g), Organic Carbon Contents (OC), and Measured Partition Coefficients (K_{oc}) of Carbon Tetrachloride (CT) and 1,2-Dichlorobenzene (DCB)[a]

No.	Source	SA	% OC	K_{oc}, CT	K_{oc}, DCB
	Soils				
1.	Burleigh Co., North Dakota (U.S. EPA reference soil 2)	7.85	2.40	52	263
2.	Oliver Co., North Dakota (U.S. EPA reference soil 3)		1.43	53	277
3.	Pierre, South Dakota (U.S. EPA reference soil 7)	22.4	2.21	63	319
4.	West-central Iowa (U.S. EPA reference soil 10)	8.84	2.04	58	248
5.	Manchester, Ohio (U.S. EPA reference soil 12)	9.38	2.25	57	230
6.	Columbus, Kentucky (U.S. EPA reference soil 19)	3.75	1.73	67	308
7.	Anoka, Minn.	1.07	1.08	61	261
8.	Piketon, Ohio	7.77	1.49	53	263
9.	Marlette soil, East Lansing, Mich.	3.99	1.80	45	223
10.	Spinks soil, East Lansing, Mich.	1.51	1.03	65	318
11.	Elliot, Illinois (International Humic Substances Society reference soil)	9.79	2.90	49	252
12.	Woodburn soil, Corvallis, Ore.	11.2	1.26	65	296
13.	Renslow soil, Kittitas Co., Wash.	11.6	2.40	59	340
14.	Sanhedrin soil, Mendocino Co., Calif.	7.88	6.09	68	344
15.	Cathedral soil, Fremont Co., Colo.	5.58	3.12	74	407
16.	Wellsboro soil, Otsego Co., N.Y.	5.73	3.47	68	383
17.	Fangshan District, Beijing, China	4.96	5.61	54	262
18.	Anda, Heilongjiang, China		2.83	67	288
19.	Jinxian Co., Jiangxi, China		0.34	53	327
20.	Nanjing, Jiangsu, China		1.08	61	236
21.	Changshu, Jiangsu, China		1.77	55	253
22.	Xuyi Co., Jiangsu, China	54.0	0.67	53	257
23.	Jinhu Co., Jiangsu, China		4.02	64	306
24.	Hongze Co., Jiangsu, China	22.8	0.81	70	313
25.	Dushan Co., Guizhou, China	8.20	2.54	61	315
26.	Gangcha Co., Qinghai, China	4.21	1.12	62	295
27.	Xinghai Co., Qinghai, China	2.86	0.16	59	264
28.	Luochuan Co., Shanxi, China		0.46	66	315
29.	Yishan Co., Guangxi, China	40.2	0.66	66	275
30.	Yangchun Co., Guangdong, China		0.83	64	293
31.	Xuwen Co., Guangdong, China		0.64	55	257
32.	Qiongzhong Co., Hainan, China	4.85	0.34	62	304
	Bed Sediments				
1.	Isaacs Creek at Ohio River, near Ripley, Ohio (U.S. EPA reference sediment 11)	20.2	1.50	66	301
2.	Mississippi River near Columbus, Ky. (U.S. EPA reference sediment 18)	22.1	0.79	103	476
3.	Illinois River, near Lacon, Ill. (U.S. EPA reference sediment 22)	3.39	2.20	116	572
4.	Kaskaskia River, Ill. (U.S. EPA reference sediment 25)	7.60	0.99	90	444
5.	Mississippi River (pool 2), St. Paul, Minn.	5.90	1.50	94	441
6.	Mississippi River (pool 11), Guttenburg, Iowa	4.86	1.13	87	370
7.	Mississippi River (pool 26), Alton, Ill.	15.5	1.40	91	387
8.	Mississippi River, Helena, Ark.	12.8	1.60	109	534
9.	Yazoo River, Vicksburg, Miss.	19.7	0.58	119	499
10.	Mississippi River, St. Francisville, La.	8.09	0.40	119	549
11.	Lake Charles, adjacent to the Calcasieu River Lake Charles, La.	13.3	1.97	112	536
12.	Marine sediment from Suisin Bay, northern San Francisco Bay, site 408.1	15.7	1.17	105	516
13.	Marine sediment from Suisin Bay, northern San Francisco Bay, site 416	21.6	1.48	107	532
14.	Marine sediment from Suisin Bay, northern San Francisco Bay, site 433	21.3	1.78	106	532
15.	Tangwang River, Yichun, Heilongjiang, China	12.8	4.73	106	551
16.	Songhuajiang River, Majiadukou, Jiling, China		1.12	117	553
17.	Tumen River, Helong Co., Jiling, China	4.93	1.99	93	420
18.	Xuanwu Lake, Nanjing, Jiangsu, China		4.12	103	557
19.	Guchen Lake, Gaochun Co., Jiangsu, China		1.24	94	554
20.	Lake Hongze, Sihong Co., Jiangsu, China	29.9	1.04	101	455
21.	Zhujiang River, Guangzhou, Guangdong, China		3.37	95	545
22.	Yellow River, Zhengzhou, Henan, China		0.11	112	589

Table 3. *(continued)*

No.	Source	SA	% OC	K_{oc}, CT	K_{oc}, DCB
	Soils				
23.	Yinghe River, Lushan Co., Henan, China	1.85	0.11	101	477
24.	Ziya River, Ci Co., Hebei, China	5.83	2.19	106	599
25.	Ganjiang River, Ruijin Co., Jiangxi, China	5.32	0.70	112	542
26.	Zishui River, Chengbu Co., Hunan, China	8.97	2.82	92	437
27.	Liuyanghe River, Liuyang Co., Hunan, China		0.29	116	555
28.	Youshui River, Xuanen Co., Hubei, China	11.9	1.22	92	451
29.	Niqu River, Louhuo Co., Sichuan, China	4.84	0.39	96	474
30.	Huaihe River, Bengbu, Anhui, China	17.6	0.50	112	535
31.	Huaihe River, Huainan, Anhui, China	8.21	0.45	91	466
32.	Jinghe River, Jingyuan Co., Ningxia, China	12.1	0.73	108	584
33.	Sangonghe River, Fukang Co., Xinjiang, China	4.00	0.38	103	499
34.	Yaluzangbu River, Lazi Co., Tibet, China	4.94	0.45	107	526
35.	Lake Pumo, Langkazi Co., Tibet, China	3.87	1.94	101	539
36.	Niyanghe River, Gongbujiangda Co., Tibet, China	3.12	0.54	93	487
	Suspended Soils				
1.	Mississippi River, Thebes, Ill.		1.82	60	296
2.	Mississippi River, St. Louis, Mo.		1.78	58	283
3.	Illinois River, Hardin, Ill.		2.60	89	423
4.	Missouri River, Herman, Mo.		2.87	49	231
5.	Yellow River, Zhengzhou, Henan, China		0.38	63	300

[a] Ref. 35.

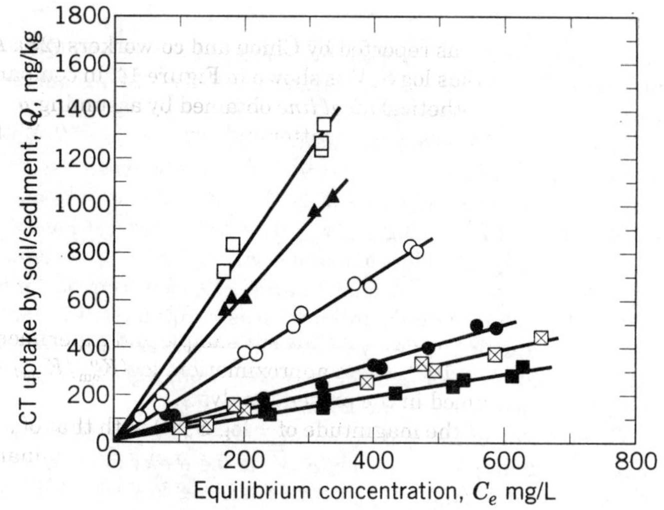

Figure 9. Sorption isotherms of carbon tetrachloride (CT) on representative soils and bed sediments from Table 3: (□) soil 14, (▲) sed.21, (○) sed.8, (●) soil 12, (⊠) soil 20, and (■) sed.10. Reproduced with permission (35).

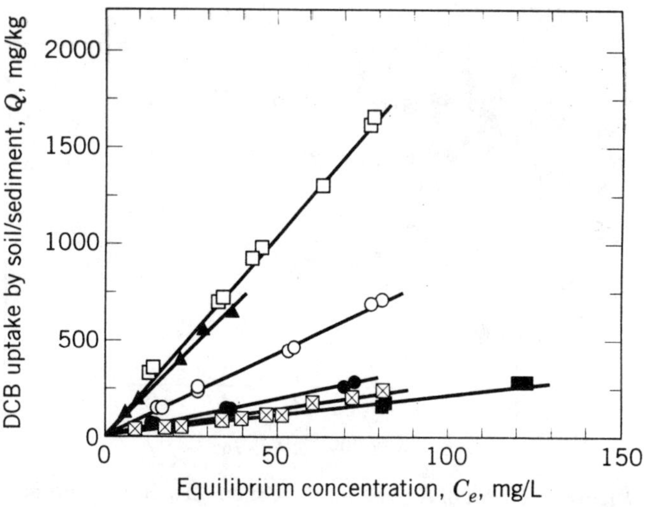

Figure 10. Sorption isotherms of 1,2-dichlorobenzene (DCB) on representative soils and bed sediments from Table 3: (□) soil 14, (▲) sed. 21, (○) sed. 8, (●) soil 12, (⊠) soil 20, and (■) sed. 10. Reproduced with permission (35).

partition, researchers (38) measured the partition coefficients of two relatively nonpolar solutes, benzene and CT, in relation to the elemental compositions of relatively ash-free natural organic matter: cellulose, muck, peat, and treated peat (peat washed by 0.1 *N* NaOH to lower the oxygen content). The weight ratio of oxygen (O) plus nitrogen (N) to carbon (C) of the natural organic matter [i.e., (O + N)/C] was used as an approximate polarity index of the sample, giving the relative polarity order: cellulose > muck > peat > treated peat. The K_{oc} (or K_{om}) for both CT and DCB shows an inverse relation with the (O + N)/C value of the organic matter sample. The results for CT, as shown in Figure 13, illustrate several points of interest. First, the partition of a nonpolar solute to a natural organic matter is sensitive to the organic matter polarity (or composition). The small K_{oc} of CT with cellulose relative to that with humified organic materials (such as muck or peat) results from a poor match in polarity between a nonpolar solute and highly polar cellulose, which makes cellulosic materials a poor partition phase for nonpolar contaminants. Second, the (O + N)/C values for normal soils should fall into a relatively narrow range, in view of relatively constant K_{oc}s of CT on soils from diverse sources (mean K_{oc} = 60; SD = ±7), which would place them somewhere between the (O + N)/C values of Houghton muck (0.777) and Florida peat (0.657). Third, the low (O + N)/

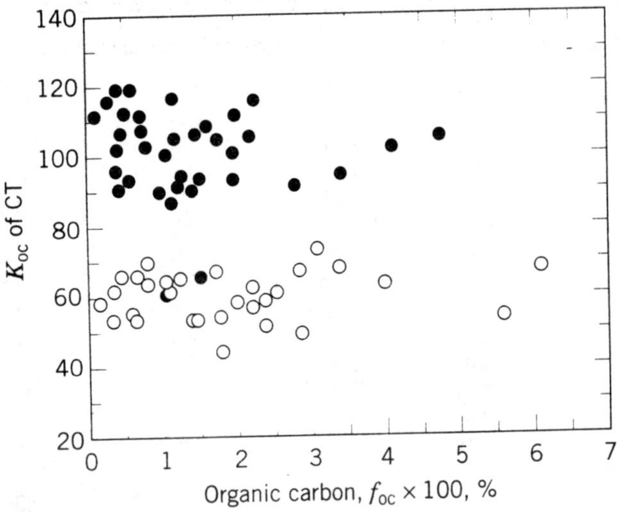

Figure 11. Plot of K_{oc} values of CT on 32 soils (○) and 36 bed sediments (●) vs their organic carbon contents from Table 3. Reproduced with permission (35).

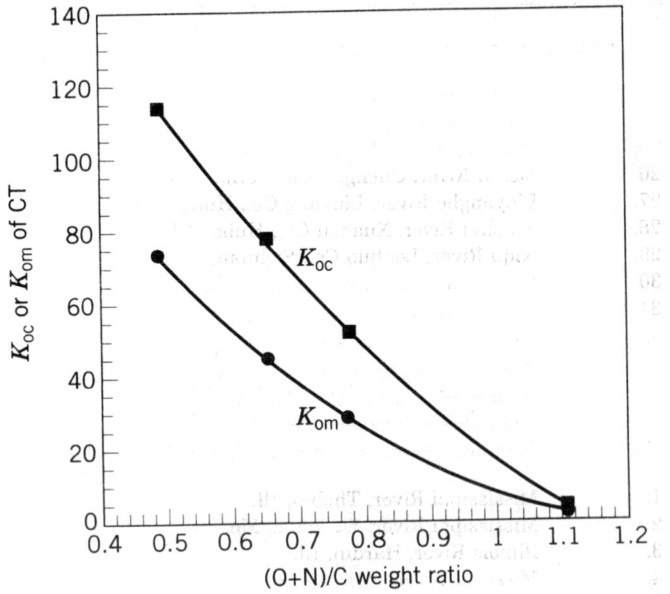

Figure 13. Plot of K_{om} and K_{oc} values of CT against (O + N)/C values of natural organic matters (38).

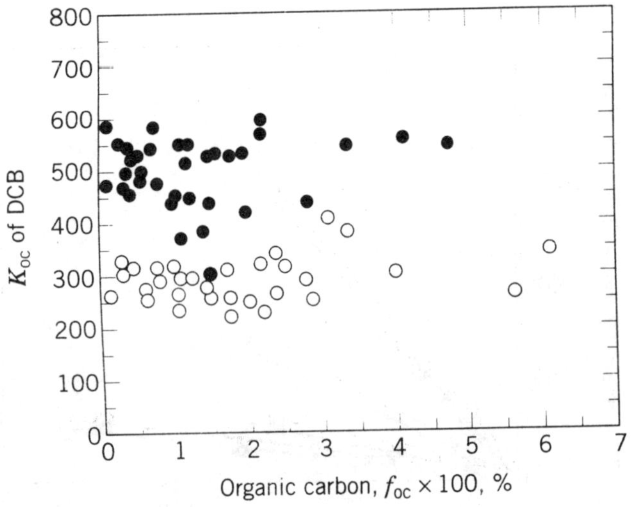

Figure 12. Plot of K_{oc} values of DCB on 32 soils (○) and 36 bed sediments (●) vs their organic carbon contents from Table 3. Reproduced with permission (35).

C value of the treated peat (0.488) and the observed K_{oc} of 115 for CT with the treated peat, which exceeds the K_{oc}s with normal soils but resembles the K_{oc}s with bed sediments, substantiate the contention that the sediment organic matter has a generally lower polar-group content than soil organic matter.

Effect of Solute Solubility on Partition Coefficient

We now consider in more detail the effect of solute water solubility (S_w) on the solute partition (sorption) coefficient with organic matter (K_{om}) in relation to model equation 7 and the relation of K_{om} with solvent (octanol)–water partition coefficient. Table 4 shows a list of the log K_{om} values of 12 substituted aromatic compounds with Woodburn soil (f_{om} = 0.019), the respective log S_w values, molar volumes ($\overline{V}$), and log K_{ow} values of the compounds, as reported by Chiou and co-workers (26). A plot of log K_{om} versus log $S_w\overline{V}$ is shown in Figure 14, in comparison with the hypothetical *ideal line* obtained by assuming ρ = 1.2 and χ_S = 0.25 for organic matter and $\log(\gamma_w/\gamma_w^*) = 0$. With given log $S_w\overline{V}$ of a compound, the difference between log K_{om}° from the ideal line and experimental log K_{om} defines the magnitude of $\chi_H/2.203 + \log(\gamma_w/\gamma_w^*)$. The magnitude of $\log(\gamma_w/\gamma_w^*)$ is relatively small for compounds with log $K_{om} \le 3$, assuming that the amount of organic matter released from soil into water after soil–water equilibration is <100 mg/L in water (the actual amount varies with the soil sample and experimental conditions) (70,100). As an approximation, $\log(K_{om}^{\circ}/K_{om}) \cong \chi_H/2.203$ is assumed in the present analysis.

Comparison of the magnitude of $-\log(S_w\overline{V})$ with that of $\log(K_{om}^{\circ}/K_{om})$ indicates that $-\log(S_w\overline{V})$ is the major determinant of log K_{om} for the organic compounds, leading to a highly linear correlation of log K_{om} with $\log(S_w\overline{V})$ as shown in Figure 14. For the 12 compounds (mainly chlorinated benzenes and PCBs) on Woodburn soil, the regression equations gives

$$\log K_{om} = -0.813 \log(S_w\overline{V}) - 0.993 \qquad (18)$$

with r^2 = 0.995, where for the solid solutes the S_w values (mol/L) used are the estimated supercooled liquid solubilities for reasons stated in the earlier section. Here a good correspondence exists between the correlation of log K_{om} versus $\log(S_w\overline{V})$ in equation 18 and the correlation of log K_{ow} vs log S_w in equation 5 for similar substituted aromatic solutes. The increased deviation between log K_{om}° and log K_{om} relative to that between log K_{ow}° and log K_{ow} is ascribed to the increased incompatibility of these low polarity solutes with relatively polar soil organic matter over that with the less-polar octanol. For more polar solutes, the deviations between log K_{om}° and log K_{om} should be less because of their enhanced partition in soil organic matter, and the relation of log K_{om} to log $S_w\overline{V}$ (or to a related property) may change accordingly.

Table 4. Water Solubilities (S_w), Molar Volumes ($\overline{V}$), Octanol–Water Partition Coefficients (K_{ow}), and Soil Organic Matter–Water Partition Coefficients (K_{om}) of Substituted Benzenes and Polychlorinated Biphenyls (PCBs)[a]

Compound	Log S_w[b], mol/L	$\overline{V}$, L/mol	Log $S_w \overline{V}$	Log K_{om}	Log K_{ow}
Benzene	−1.64	0.0894	−2.69	1.26	2.13
Anisole	−1.85	0.109	−2.82	1.30	2.11
Chlorobenzene	−2.36	0.102	−3.35	1.68	2.84
Ethylbenzene	−2.84	0.123	−3.75	1.98	3.15
1,2-Dichlorobenzene	−2.98	0.113	−3.98	2.27	3.38
1,3-Dichlorobenzene	−3.04	0.114	−3.98	2.23	3.38
1,4-Dichlorobenzene	(−3.03)	0.118	−3.96	2.20	3.39
1,2,4-Trichlorobenzene	−3.57	0.125	−4.47	2.70	4.02
2-PCB	(−4.57)	0.174	−5.33	3.23	4.51
2,2′-PCB	(−5.08)	0.189	−5.57	3.68	4.80
2,4′-PCB	(−5.28)	0.189	−5.97	3.89	5.10
2,4,4′-PCB	(−5.98)	0.204	−6.67	4.38	5.62

[a] Ref. 26.

[b] Values at 20 to 25°C; numbers in parentheses are supercooled liquid solubilities estimated according to methods described (71).

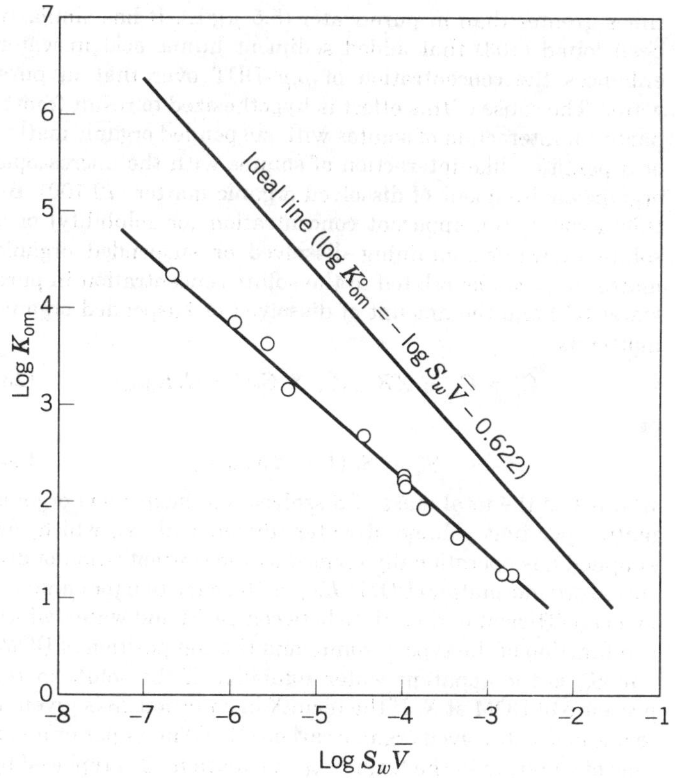

Figure 14. Plot of log K_{om} vs log ($S_w\overline{V}$) for substituted aromatic compounds from Table 4. Reproduced with permission (26).

Since the variation of $\overline{V}$ among solutes is quite small compared to that of S_w, a linear relation should also exist between log K_{om} and log S_w, as has been widely recognized (25,26,28,29,33,54,65,66). Using the data in Table 4, the correlation equation is

$$\log K_{om} = -0.729 \log S_w + 0.001 \quad (19)$$

with $n = 12$ and $r^2 = 0.996$. That water insolubility is the major determinant of K_{om} is expected in the partition equilibria of organic compounds between an organic phase and water, despite the fact that soil organic matter is not nearly as good a partition medium for nonpolar solutes as normal organic solvents such as octanol. For this reason, the log K_{om} value would be linearly related to log K_{ow}, as illustrated for various organic compounds and pesticides with different soils and sediments, as mentioned earlier. The data in Table 4 give

$$\log K_{om} = 0.904 \log K_{ow} - 0.779 \quad (20)$$

with $n = 12$ and $r^2 = 0.989$. For compounds with log K_{ow} = 2 − 5, the K_{om} values are here about one order of magnitude lower than the corresponding K_{ow} values. When the sorption coefficient is expressed in terms of soil organic carbon (K_{oc}), equivalent correlation equations are obtained by use of $K_{oc} \cong 1.72\ K_{om}$; the conversion factor of 1.72 is obtained by assuming that soil organic matter contains about 58% carbon (19). (Based on the mean K_{oc} of CT on soils and its standard deviation and the relation of K_{oc} to the organic matter composition in Figure 13, an assumed carbon content of about 54% in soil organic matter seems more appropriate. This would change the conversion between K_{om} and K_{oc} to $K_{oc} \cong 1.85\ K_{om}$.)

Since the magnitude of K_{oc} (or K_{om}) reflects the difference in solute solubilities in soil (or sediment) organic matter and water, which varies with the solute, the slopes and intercepts in equations 19 and 20 would depend on the compounds studied. A single correlation equation would suffice for different classes of solutes only when solutes with equal S_w or K_{ow} exhibit nearly the same (supercooled-liquid) solubility in soil or sediment organic matter (S_{om} in eq. 17), which however is not the case. The slope reflects the change in log S_{om} with the change in log S_w or log K_{ow} of the solute, whereas the intercept is the log K_{oc} (or log K_{om}) value of a hypothetical reference solute with log S_w or log K_{ow} equal to 0. As the S_{om} value changes with solute polarity in a different manner for different series of solutes, the slope and intercept of the correlation change accordingly. For example, Briggs (65) measured the log K_{om} values of a wide range of polar compounds and pesticides (anilines, anilides, phenols, nitrobenzenes, substituted ureas, carbamates, organophosphates, and others) on soil and found the correlation between $\log K_{om}$ and $\log K_{ow}$ as

$$\log K_{om} = 0.52 \log K_{ow} + 0.64 \quad (21)$$

with $n = 105$ and $r^2 = 0.90$. The majority of compounds and pesticides studied have $\log K_{ow} < 3$, however. As shown, equation 21 gives a smaller slope but a much larger intercept than equation 20. At $\log K_{ow} < 3$, as for most polar solutes, the $\log K_{om}$ values estimated by equation 21 are comparable in magnitude with respective $\log K_{ow}$ values and are thus considerably higher than the $\log K_{om}$ values estimated by equation 20.

The smaller slope in equation 21 than in equation 20 indicates that, whereas highly polar solutes (ones with very low $\log K_{ow}$) exhibit relatively high S_{om} values, the S_{om} decreases more rapidly with increasing K_{ow} as the solute polarity decreases within the selected series in comparison with the variation of S_{om} within a series of relatively nonpolar solutes. It has been shown that polar organic liquids exhibit a considerably higher partition (solubility) than nonpolar liquids in dry soil organic matter, due to more powerful molecular interactions by polar forces and hydrogen bonding over that by London (dispersion) forces (92). Therefore, unlike the $\log K_{ow}$ vs $\log S_w$ correlation, which is relatively insensitive to solute polarity (71), the $\log K_{om}$ vs $\log K_{ow}$ (or $\log K_{om}$ vs $\log S_w$) correlation is much more sensitive to solute polarity. In this sense, octanol (or K_{ow}) is not a good general model for soil organic matter (or K_{om}).

A notable difference in $\log K_{oc}$ vs $\log K_{ow}$ correlation also occurs between different classes of low polarity solutes. As an example, the correlation of $\log K_{oc}$ with $\log K_{ow}$ for a group of low polarity solutes consisting largely of polycyclic aromatic hydrocarbons (PAHs) on river sediments is given by Karickhoff and co-workers as follows (28):

$$\log K_{oc} = 1.00 \log K_{ow} - 0.21 \quad (22)$$

with $n = 10$ and $r^2 = 1.00$. Here, the intercept value in equation 22 is quite different from that in equation 20, but the range of $\log K_{ow}$ covered by both equations is about the same. At equal K_{ow}, equation 22 gives much higher K_{oc} for PAHs than does equation 20 for substituted benzenes and PCBs. In the range of $\log K_{ow} = 4{-}6$, the K_{oc} values predicted by equation 22 are approximately four to six times as large as predicted by equation 20. The slope of 1 in equation 22 suggests that the (supercooled-liquid) solubilities of different PAHs in sediment organic matter remain relatively unchanged with increasing $\log K_{ow}$. Because the correlation of $\log K_{ow}$ with $\log S_w$ shows no significant disparity between PAHs and chlorinated compounds (see Fig. 2 and Table 1), the difference between equation 20 and equation 22 stems supposedly from different solubilities of these solutes in soil or sediment organic matter (ie, from different S_{om} values according to equation 17). Indeed, PAHs, like other nonpolar solutes, are found to exhibit higher solubilities in sediment organic matter than in soil organic matter (101). In addition, the supercooled-liquid solubilities of a series of PAHs in soil organic matter are more enhanced and stay relatively constant with increasing K_{ow}, whereas those of substituted benzenes and PCBs decrease considerably with increasing K_{ow} (101).

From the above analysis, the accurate prediction of $\log K_{oc}$ (or $\log K_{om}$) values for a class of solutes on soil (or sediment) can be achieved only by use of the correlation for the same or similar class of solutes on soil (or sediment). The extrapolation of $\log K_{oc}$ (or $\log K_{om}$) vs $\log K_{ow}$ (or $\log S_w$) from a correlation of solutes to regions far beyond the original set of data could also lead to uncertain results.

INFLUENCE OF DISSOLVED AND SUSPENDED ORGANIC MATTER ON SORPTION COEFFICIENT

Effect of Dissolved and Suspended Natural Organic Matter

For many highly water-insoluble compounds, the term $\log (\gamma_w/\gamma_w^*)$ often has a more significant impact on $\log K_{om}$ in soil–water systems (eq. 7) than, say, on $\log K_{ow}$ in octanol–water systems (eq. 2). In soil–water mixtures, the term γ_w/γ_w^* expresses the enhancement of solute concentration (or solubility) by the amount of dissolved and/or suspended (colloidal) organic matter in water, that is, $\gamma_w/\gamma_w^* = C_e^*/C_e$. High molecular weight humic material in dissolved or suspended form in water at trace quantities is known to significantly enhance the water solubility of otherwise extremely insoluble organic compounds. It has been shown (102) that the apparent solubility of DDT in 0.5% soil sodium humate solution is more than 200 times greater than in pure water (5.5 μg/L). It has similarly been found (103) that added sediment humic acid in water enhances the concentration of *p,p'*-DDT over that in pure water. The cause of this effect is hypothesized to result from a partition interaction of solutes with suspended organic matter or a partition-like interaction of solutes with the microscopic organic environment of dissolved organic matter (70,100). By this account, the apparent concentration (or solubility) of a solute in water containing dissolved or suspended organic matter (C_e^*) can be related to the solute concentration in pure water (C_e) and the amount of dissolved or suspended organic matter as

$$C_e^* = C_e + XK_{dom}C_e = C_e(1 + XK_{dom}) \quad (23)$$

or

$$S_w^* = S_w(1 + XK_{dom}) \quad (24)$$

where X is the total mass of dissolved and suspended organic matter per unit volume of water (dimensionless), which, for simplicity, is operationally termed as the concentration of dissolved organic matter (DOM), K_{dom} is the partition (or enhancement) coefficient of the solute between DOM and water, which is a function of the type of solute and the composition of DOM, and S_w^* is the apparent water solubility of the solute in the presence of DOM at X. If the term X in equation 23 is given in terms of the dissolved organic carbon (DOC) mass per unit volume of water, then the term K_{dom} in equation 23 is replaced by K_{doc}. Relating C_e^* with solute concentration in soil (Q) at equilibrium results in an apparent solute distribution coefficient (K_d^*) as

$$K_d^* = Q/C_e(1 + XK_{dom}) = K_d/(1 + XK_{dom}) \quad (25)$$

where $K_d = f_{om}K_{om}$, as defined before. Normalization of K_d^* to f_{om} leads to

$$K_{om}^* = K_{om}/(1 + XK_{dom}) \quad (26)$$

in which K_{om}^* and K_{om} may be substituted by K_{oc}^* and K_{oc}, respectively, if K_d^* and K_d are normalized to soil or sediment organic carbon fraction (f_{oc}).

The magnitude of K_{dom} (or K_{doc}), from the standpoint of DOM, should depend on the polarity and size of DOM. For truly dissolved organic matter (not suspended organic matter), the DOM molecules must in principle be sufficiently large and must possess a sizable intramolecular nonpolar moiety to promote partition-like interaction with the solute. Dissolved low-molecular-weight organic matter (solvent) is not expected to be equally effective for promoting this type of interaction because of the size limitation. For the solutes, the important properties for large solubility enhancement are very low water solubility and significant compatibility with the organic phase.

An analysis of the functional relationship of K_{dom} (or K_{doc}) with the water solubility of organic compounds (solutes) and with the source and composition of DOM was presented by Chiou et al. (104,105), using a series of solutes with vastly different S_w values and of DOMs with varied compositions and structures. The solutes used and their S_w values at room temperature were: *p,p'*-DDT, $S_w = 5.5$ μg/L; 2,4,5,2',5'-PCB, $S_w = 10$ μg/L; 2,4,4'-PCB, $S_w = 115$ μg/L; lindane, $S_w = 7.8$ mg/L; and 1,2,3-trichlorobenzene (TCB), $S_w = 16.3$ mg/L. The various DOMs used were Sanhedrin soil humic acid (SSHA), Sanhedrin soil fulvic acid (SSFA), Suwannee River humic acid (SRHA), and Suwannee River fulvic acid (SRFA), extracted from soil and stream, and manmade phenylethanoic acid and polyacrylic acid. The high purities of extracted natural DOM samples provide sufficiently accurate elemental data (shown in Table 5) as indices of their polarities. Plots of the S_w^* values against DOM concentrations by use of equation 24 for selected solutes with SSHA as DOM are given in Figure 15. The corresponding plots with SRHA, phenylethanoic acid, and polyacrylic acid as DOMs are presented in Figures 16, 17, and 18, respectively.

From the results in Figures 15–18, the SSHA shows the greatest enhancing effect on solute solubility, whereas SSFA, SRHA, SRFA, phenylethanoic acid, and polyacrylic acid exhibit less or no effects. For the solutes, the effect decreases progressively from *p,p'*-DDT to 2,4,5,2',5'-PCB and to 2,4,4'-PCB with increasing S_w, and becomes negligible for relatively water-soluble lindane and TCB over the studied DOM concentrations (0–100 mg/L for SSHA, SSFA, SRHA, SRFA, and polyacrylic acid; 0–700 mg/L for phenylethanoic acid). The linear relationship between S_w^* and DOM concentration, the decrease of solubility enhancement with increasing S_w, and the lack of interference between binary solutes (for 2,4,5,2',5'-PCB and 2,4,4'-PCB with SSHA) are in keeping with the postulated partition-like interaction between the solute and DOM for the observed solubility enhancement. The different effects for a set of solutes with a given DOM are attributed to their different K_{dom} (or K_{doc}) values. The calculated log K_{dom} and log K_{doc} values of the solutes with SSHA, SSFA, SRHA, and SRFA are tabulated in Table 6. For a given DOM, because the magnitude of K_{dom} (or K_{doc}) should be inversely related to solute solubility in water (S_w), the log K_{dom} (or log K_{doc}) values are essentially linear with respective log K_{ow} values.

That small-sized DOMs (eg, phenylethanoic acid) are unable to produce as strong an effect as humic and fulvic acids agrees with the required favorable DOM size and polarity in such a partitionlike interaction. Note that if the size of DOM were not important, phenylethanoic acid would likely exhibit a greater enhancement effect than other DOMs because of its less-polar composition (70.6% C, 5.9% H, and 23.5% O). In situations where the solubility enhancement is caused by a specific interaction, one would expect such an effect to be more related to the acidity or polarity of the DOM, and therefore phenylethanoic acid should be a strong enhancer.

The relatively large solubility enhancement for DDT and PCBs with SSHA suggests that the enhancing SSHA must have a substantially larger molecular weight than those of the solutes, which is supported by the proposed molecular weights of 2,000–20,000 for soil humic acids (106). The kind of partition interaction that is envisioned here for a relatively insoluble solute with high molecular weight DOMs is mechanistically similar to solute solubilization in surfactant micelles, in which a microscopic organic phase is formed through aggregation of surfactant monomers. The much smaller enhancement effects with soil fulvic acid (SSFA), aquatic humic acid (SRHA), and aquatic fulvic acid (SRFA) may be ascribed in part to their smaller molecular sizes and to a greater extent to their greater polarities as measured by (O + N)/C. The suggested molecular weights are 1000–5000 for soil fulvic acids (107,108), 1,000–10,000 for aquatic humic acids (109), and 500–1000 for aquatic fulvic acids (110).

The comparable effects as found with SRHA, SRFA, and SSFA, despite some differences in their sizes, suggest that the DOM polarity has a greater impact on solubility enhancement in this situation. Furthermore, polyacrylic acid (with MW = 2,000 and 90,000) as a linear polyelectrolyte shows no enhancement effect, notwithstanding that its C (50%), H (5.6%), and O (44.4%) contents are comparable with those of SSFA. Here the inability of polyacrylic acid to enhance solute solubility may be ascribed to the frequent and orderly attachment of hydrophilic carboxyl groups to the carbon chain and to an extended chain structure, which do not permit the formation of a sizable intramolecular nonpolar environment. Thus, although the molecular size of DOM is essential, this property alone is not the deciding factor in solubility enhancement. To be an effective solubility enhancer, DOM must possess both a favorable size of nonpolar moiety and a relatively large molecular weight.

By the results in Table 6, the K_{dom} values with dissolved soil humic acid (SSHA) are about half as large as the K_{om} values with bulk soil organic matter in concentrating organic compounds on a unit weight basis. On the other hand, the suspended (colloidal) organic matter resulting from soil or sediment should give $K_{dom} \cong K_{om}$. In soil–water mixtures, the total amount of dissolved and suspended organic matter in water (X) usually increases with increasing soil-to-water ratio; at high X, the resulting K_d^* may then be less than K_d for certain solutes. Hence, if the aqueous phase contains, say, $X = 30$ mg/L of soil humic acid, the apparent K_{om}^* value of DDT with log $K_{dom} = 4.8$ would be approximately three times lower than the intrinsic K_{om}. Gschwend and Wu (100) investigated the apparent K_d^* values of highly insoluble 2,4,5,2',5'-PCB and 2,3,4,5,6,2',5'-PCB with sediments in relation to total organic carbons of suspended sediment microparticles. By use of equation 25, they showed that when precautions are taken either to eliminate or to account for the suspended microparticles in water the calculated intrinsic log K_{oc} (or log K_{om}) values stay constant for both compounds over a wide range of sediment-to-water ratios, whereas the apparent log K_{oc}^* value decreases with the amount of suspended microparticles. It would thus appear that the

Table 5. Percent Elemental Contents of Soil and Stream Organic Matter Extracts on Moisture-Free, Ash-Free Basis[a]

Organic Matter Extract	C	H	O	N	S	P	Ash
Sanhedrin soil humic acid (SSHA)	58.0	3.64	33.6	3.26	0.47	0.10	1.19
Sanhedrin soil fulvic acid (SSFA)	48.7	4.36	43.4	2.77	0.81	0.59	2.25
Suwannee River humic acid (SRHA)	54.2	4.14	39.0	1.21	0.82	0.01	3.18
Suwannee River fulvic acid (SRFA)	53.8	4.24	40.3	0.65	0.60	0.01	0.68

[a] Ref. 104.

relatively large variation of the log K_{om} values as commonly observed for highly water-insoluble solutes with different amounts of soils and sediments (30,111) could be ascribed in part to their different extents of water solubility enhancement by dissolved and/or suspended organic matter. It would also appear that the significant enhancement of the apparent K^*_{om} value of DDT with soils following ether and alcohol extraction of the soils (90) might be more reasonably explained by the reduction of the soluble or suspended soil organic matter, rather than by the postulate that the applied solvents remove lipids from organic matter and hence enhance the accessibility of DDT to soil organic matter.

By equations 23–26, the effect of DOM on solute solubility and hence on solute sorption coefficient would be less pronounced for solutes with higher water solubility because of the expected decrease in K_{dom} or K_{doc}. Consequently, relatively water-soluble compounds, such as lindane and TCB, show no detectable solubility increases even at quite high DOM concentrations (104,112). This consequence is predicated by equation 24. For example, for a solute–DOM pair to achieve $S^*_w = 2S_w$, it is required that $X = 1/K_{dom}$. Because K_{dom} increases with decreasing (supercooled liquid) S_w, a higher X is needed for a more water-soluble solute than for a less soluble solute to produce the same result. Thus, with soil humic acid as DOM, DDT with $K_{dom} = 6.6 \times 10^4$ would give $S^*_w = 2S_w$ at $X = 1.5 \times 10^{-5}$ (or 15 mg/L), whereas the more soluble TCB with $K_{dom} = 6.3 \times 10^2$ can achieve this result only at $X = 1.6 \times 10^{-3}$ (or 1600 mg/L); other relatively soluble solutes (most polar solutes) would similarly show no observable effect at $X < 100$ mg/L.

If one expresses $X = S_w/(S_w K_{dom})$ at $S^*_w = 2S_w$ and expresses $S_w K_{dom}$ in units of grams of solute per gram of DOM (ie, dimensionless), a more explicit relation is obtained between S_w and required X for solubility enhancement. For nonpolar solutes, the term $S_w K_{dom}$ is small because the corresponding $S_w K_{om}$ term, which expresses the solute S_{om} (eq. 17), is small (113), and because $K_{om} \geq K_{dom}$ for soil-derived DOM; the S_{om} values vary somewhat among nonpolar liquids and supercooled liquids (approximately 0.03 for benzene and <0.01 for supercooled DDT). Hence, for a solute to achieve $S^*_w = 2S_w$ with soil humic acid as DOM ($K_{dom} \cong 0.5\ K_{om}$), the required condition is that $X \geq 60S_w$, where S_w is for liquids or supercooled liquids (eqs. 3 and 4). This factor would increase with increasing polarity of DOM (as for SRHA and SRFA) because of their reduced K_{dom}. A general rule is that the concentration of natural DOM in water must be more than approximately two orders of magnitude greater than the solute's supercooled liquid S_w in order to produce an enhancement factor of 2. Since natural DOM seldom exceeds 100 mg/L, only those organic contaminants having $S_w < 1$ mg/L as supercooled liquids would possibly exhibit a significant increase in their concentrations in water by DOM. However, as discussed later, some man-made DOM, such as surfactant micelles, may produce a much greater effect than natural DOM at low concentrations.

In natural systems, the K_{dom} (or K_{doc}) values for given solutes would vary with the sources. Based on limited data available, the K_{dom} values for natural (not manmade) DOMs in rivers and streams appear to fall between values defined by soil humic acid (SSHA) and highly acidic aquatic fulvic acid (SRFA) (105), which vary by a factor of approximately five. The difficulties in extracting relatively pure aquatic humic materials for solute-DOM interaction studies have led some investigators to the use of *commercial humic acids* as surrogates for aquatic humic materials. Because the elemental compositions of some commercial humic acids used are vastly different from those of aquatic and soil-derived humic substances (105,114), the resulting data would not serve as a practical reference for the impact of aquatic humic materials. Here, for example, the exceptionally high K_{dom} (or K_{doc}) values as observed with Aldrich humic acid, which are approximately three times higher than that of soil humic acid and some four to twenty times higher than those of aquatic humic-acid and fulvic acid extracts, are consequences of its unusually high carbon (65.3%) and low oxygen (25.1%) contents (on an ash-free basis) in the purified sample (114). The use of K_{dom} (or K_{doc}) values derived from commercial humic acids would thus grossly overestimate the actual impact of aquatic humic materials on solute behavior in natural waters.

As noted, both suspended (colloidal) and dissolved organic matter could affect the apparent solute solubility or solute sorption coefficient. Whereas the effect of suspended organic matter may be properly termed the *third-phase effect* (100), the similar effect caused by van der Waals forces between solutes and an intramolecular (microscopic) organic environment of dissolved organic matter of both natural and manmade origins cannot be ascribed to that of a third-phase in the normal sense. In natural systems, both dissolved and suspended organic matters may be expected to influence the behavior of solutes to varying degrees, depending on their sources, concentrations, and compositions.

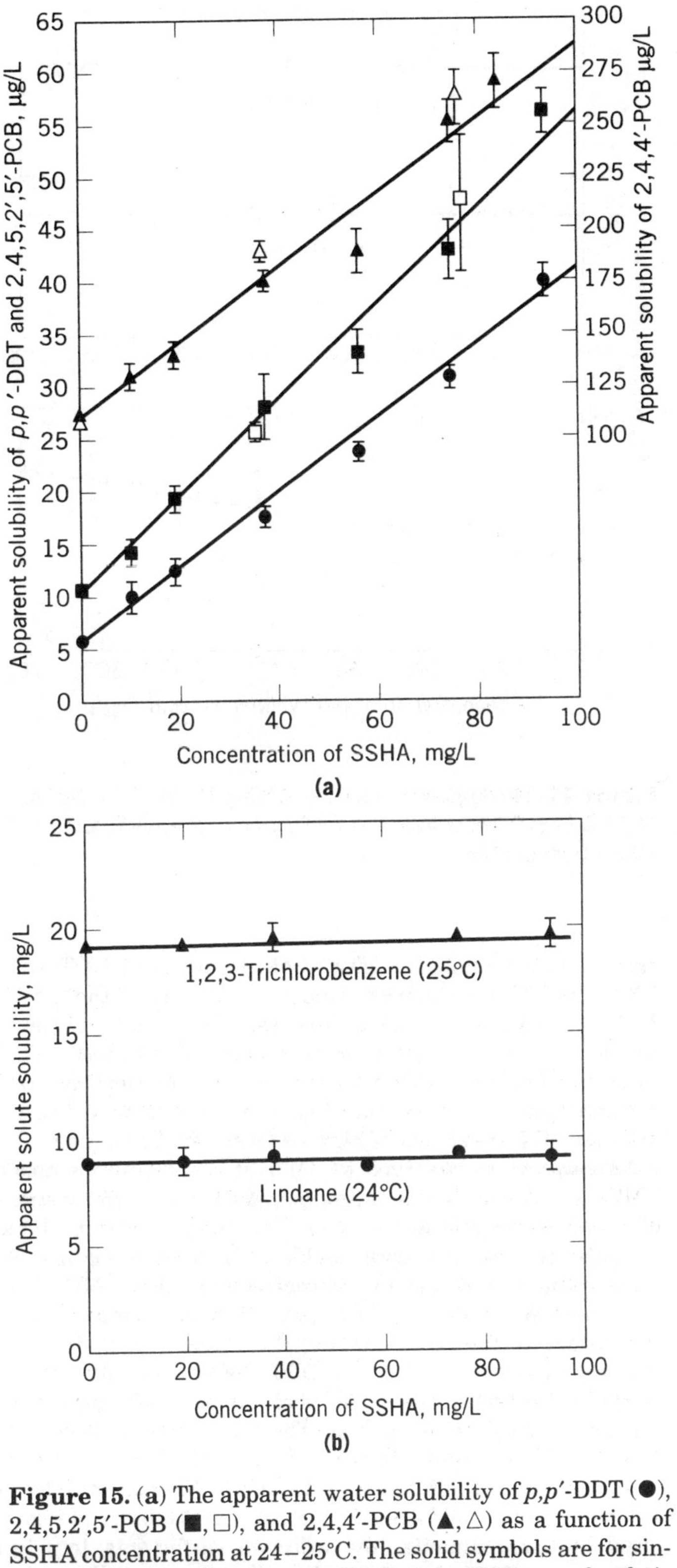

Figure 15. (**a**) The apparent water solubility of *p,p′*-DDT (●), 2,4,5,2′,5′-PCB (■, □), and 2,4,4′-PCB (▲, △) as a function of SSHA concentration at 24–25°C. The solid symbols are for single solutes, the open symbols for the two PCBs are for their binary mixtures. (**b**) The apparent water solubility of lindane and 1,2,3-trichlorobenzene as a function of SSHA concentration (104).

An alternative mechanism has been suggested by others to explain the observed inverse relation between the solute sorption coefficient (K_d^*) and the ratio of soil or sediment to water for relatively insoluble solutes. It has been hypothesized (115) that collisions between solute-loaded suspended soil or sediment particles induce desorption of the solute from the particles, the extent being related to the suspended particle concentration, particle organic carbon fraction (f_{oc}), and solute K_{oc}. However, the idea that the solute sorption coefficient is affected by particle collision is without precedent, as it violates the principle of equilibrium that the net amount of solute sorbed ought to be independent of particle collisions. From the kinetic theory, the root-mean-square velocity of a substance (suspended particle or solute) is inversely related to the square root of its molecular weight; therefore, the important collision would be between suspended particles and solute molecules rather than between suspended particles themselves. Solute–particle collision affects solute concentrations (in both solution and particle phases) only before the system reaches equilibrium, but poses no net effect on concentration once equilibrium is reached. Although the particle collision model gives a mathematical fit to equation 25, it lacks a physical basis to explain the influence of dissolved and suspended organic matter on solute solubility and the subsequent effect on the solute sorption coefficient.

Effect of the Man-Made Surfactants and Microemulsions

The discharge of household and industrial organic wastes into the environment may produce complex mixtures of dissolved and suspended organic matter in natural water. One kind of manmade dissolved and colloidal organic matter of special interest is surface-active agents (called surfactants) because of their large discharged quantities and unique molecular structures, enabling them to form stable aggregates (micelles) at certain concentrations. Whereas some surfactants or their degradation products could become potential pollutants in surface and ground water, their presence may also influence the solubility and partition behavior of other pollutants. The ability of a surfactant to form micelles is characterized by its *critical micelle concentration* (CMC). At concentrations below CMC, the dissolved surfactant is all in monomeric form; above CMC, the amount in excess of CMC forms micelles. Unlike ordinary dissolved organic substances that exist in monomeric form, the surfactant micelle provides a relatively large microscopic nonpolar organic environment into which solutes may partition. This property enables the micelle to greatly promote the (apparent) aqueous solubility of relatively insoluble solutes as compared with ordinary dissolved organic substances. As such, it also provides a potential means of removing pollutants from soil in remediation treatment, provided that the surfactant sorbed to the soil does not strongly enhance contaminant sorption.

A general equation similar in form to equation 24 may be written for the solubility enhancement of a solute by a surfactant in water (116):

$$S_w^*/S_w = 1 + X_{mn}K_{mn} + X_{mc}K_{mc} \qquad (27)$$

where S_w^* and S_w are as defined before; X_{mn} is the concentration of the surfactant as monomers in water (dimensionless); X_{mc} is the concentration of the surfactant in micellar form in water (dimensionless); K_{mn} is the partition (enhancement) coefficient of the solute between surfactant monomers and water; and K_{mc} is the partition coefficient of the solute between micelles and water. The X_{mn} and X_{mc} values are assigned as follows. If the

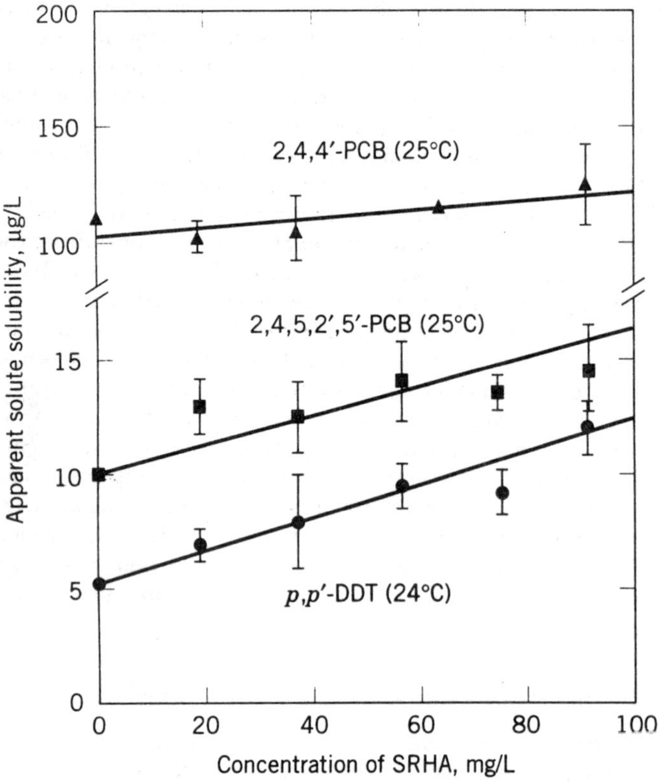

Figure 16. The apparent water solubility of *p,p′*-DDT, 2,4,5,2′,5′-PCB, and 2,4,4′-PCB as a function of SRHA concentration (104).

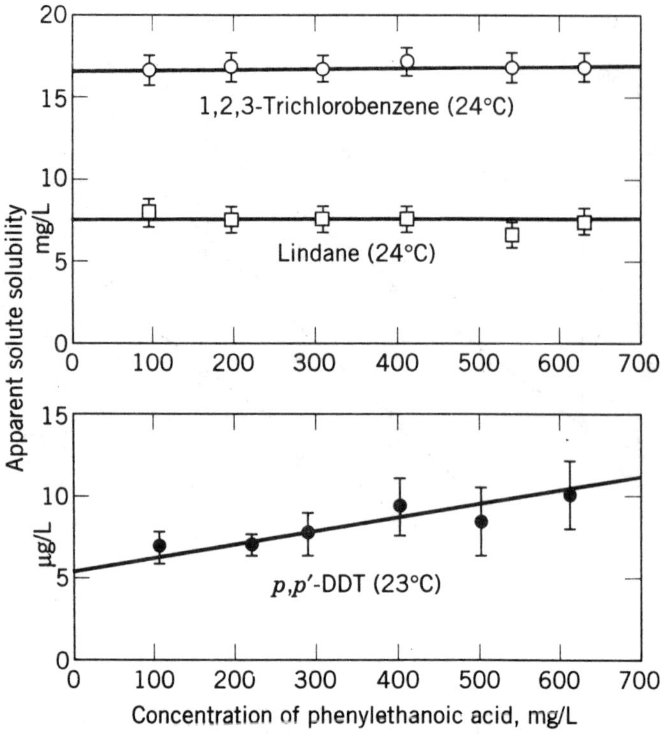

Figure 17. The apparent water solubility of *p,p′*-DDT, lindane, and 1,2,3-trichlorobenzene as a function of phenylethanoic acid concentration (104).

surfactant concentration $X \le CMC$, then $X_{mn} = X$; if $X >$ CMC, X_{mn} = CMC and $X_{mc} = X -$ CMC.

By equation 27, a plot of the apparent solute solubility (S_w^*) versus the surfactant concentration, extended over its CMC, will be bilinear, yielding a straight line with a slope of K_{mn} from $X = 0$ to $X =$ CMC, followed by another straight line of a much higher slope, K_{mc}, at $X \ge$ CMC, provided that the surfactant is molecularly homogeneous and has a single monomer-micelle transition point (ie, a single CMC). If the surfactant is molecularly heterogeneous, a succession of micellization may take place for different molecular fractions of the surfactant and the resulting plot of S_w^* versus X would not show a single sharp transition. (In this case, the measurement of surface tension versus X, as commonly used to detect CMC, gives a breadth of monomer-micelle transition rather than a single sharp break.)

Researchers studied (116) the effectiveness of surfactants to enhance solute water solubility by use of DDT and TCB as model solutes and several nonionic surfactants (Triton series and Brij 35), an anionic surfactant (sodium dodecyl sulfate, SDS), and a cationic surfactant (cetyltrimethylammonium bromide, CTAB) as reference surfactants. The structures and properties of these surfactants are given in Table 7. Triton and Brij 35 surfactants are molecularly heterogeneous in that the lengths of polar head groups, ie, the number of ethylene oxide (EO) units, are average rather than single values. By contrast, SDS and CTAB are virtually molecularly homogeneous.

For DDT, large solubility enhancements and sharply rising slopes are recognized as the concentrations of SDS and CTAB exceed their (nominal) CMCs. Enhancements of DDT with TX100 and TX114 show less sharp transitions and those with BJ35 and TX405 show a gradual transition, due presumably to their different degrees of molecular heterogeneity. TCB exhibits a similar solubility transition but a considerably lower enhancement, due to its much higher water solubility. The results for DDT are shown in Figures 19 and 20. In all cases, the enhancements as observed for DDT at concentrations above CMCs are tremendously large, manifesting the effectiveness of pseudophase micelles to solubilize highly water-insoluble nonpolar solutes. The more soluble TCB shows no significant enhancement at surfactant concentrations below CMCs. The calculated K_{mc} values of DDT and TCB are comparable in magnitude to the corresponding K_{ow} values (Table 8) (116). The calculated K_{mn} values for DDT, using solubility data at surfactant concentrations near zero, are some 40 times lower than corresponding K_{mc} values. The K_{mn} values are relatively comparable in magnitude with the K_{oc} values of the solutes with dissolved aquatic humic substances. Whereas micelles of all surfactants would greatly enhance the solubility of highly insoluble contaminants, the effective surfactants from an environmental standpoint are those that form micelles at low concentrations (ie, the ones with low CMCs).

When surfactants are combined with oils in certain ratios to form stable microscopic mixtures (called microemulsions), they become especially effective in enhancing the water solubility of nonpolar organic contaminants due to solute solubilization in the suspended microscopic oil phase. The stabilization of microemulsions is achieved through the interaction with

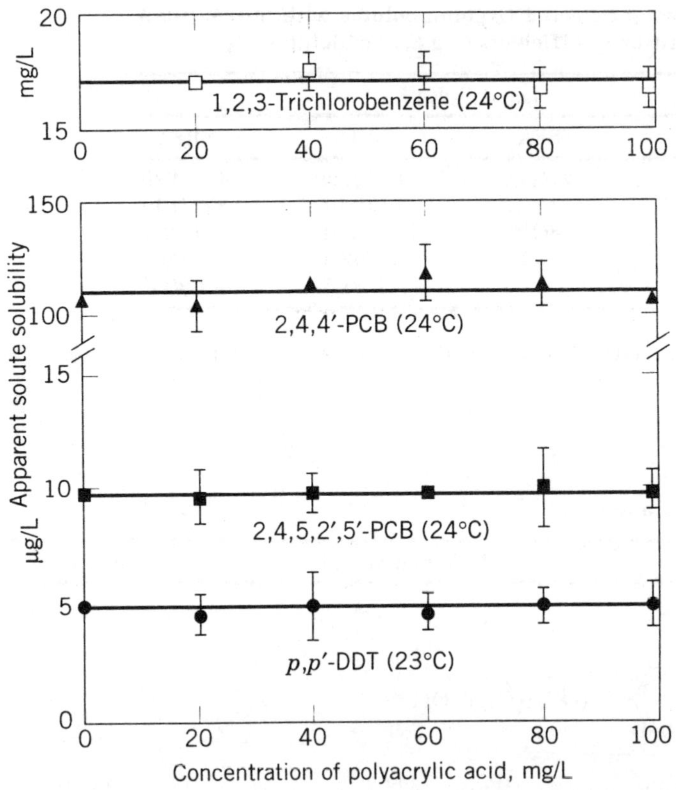

Figure 18. The apparent water solubility of *p,p′*-DDT, 2,4,5,2′,5′-PCB, 2,4,4′-PCB, and 1,2,3-trichlorobenzene as a function of polyacrylic acid (MW = 2000) concentration (104).

water of the polar head groups of surfactants attached to the oil particles. Whereas such microemulsions are sometimes marked as surfactants because of their abilities to reduce the surface tension of water, they exhibit a separate-phase property even at very low concentrations and hence they display no monomer-micelle transitions (ie, no CMCs). An example of emulsion-forming commercial products is the class of petroleum sulfonate-oil (PSO) surfactants, formulated by mixing free mineral oil with surface-active petroleum sulfonate in certain ratios, as shown in Table 9 (117). The enhancement of solute solubility in water by microemulsions can be described as

$$S_w^*/S_w = 1 + X_{em}K_{em} \tag{28}$$

where X_{em} is the concentration of microemulsion in water (dimensionless) and K_{em} is the solute partition coefficient between the emulsified phase and water. The K_{em} values determined for DDT and TCB with PSO microemulsions are about the same as the K_{mc} values, illustrating the separate-phase property of microemulsions (117). In comparison with normal surfactants, the emulsified oil phase can enhance the solute solubility much more effectively at low concentrations that are below the CMC of normal surfactants. In natural water near waste-discharge sites, the possible formation of an emulsified oil phase from components of untreated organic wastes could have a strong impact on the fate of certain organic pollutants.

The strong enhancement of the solute solubility by surfactants offers a potential means of remediating contaminated soils or sediments by use of surfactant solutions. If a surfactant (or microemulsion) solution is applied to remove contaminant(s) existing as a separate phase in the soil (or sediment), the result may be expected to be more straightforward based on equations 27 and 28, provided that the solution is readily accessible to the surfaces and void spaces of soil particles. Here the sorption of surfactants to the soil should not have a strong impact on the contaminant removal. On the other hand, if the contaminant exists only at subsaturated levels in the soil (ie, no excess contaminant phase is formed), the applied surfactant solution may produce different results, depending on the system, as described below.

For an organic contaminant at subsaturation in soil–water systems with a surfactant, the sorption coefficient may be expressed as

$$K_d^* = \frac{K_d + f_{sf}K_{sf}}{1 + X_{mn}K_{mn} + X_{mc}K_{mc}} \tag{29}$$

where K_d^* and K_d are as defined earlier; f_{sf} is the fraction of sorbed surfactant in soil; and K_{sf} is the solute (contaminant) sorption coefficient between sorbed surfactant and water. The surfactant concentration in water determines the relative contributions of $X_{mn}K_{mn}$ and $X_{mc}K_{mc}$. Here K_{sf} is a function of f_{sf} because the aggregation state of a sorbed surfactant on soil may vary with its loading (118). Similarly, the sorption coefficient for a contaminant in soil–water systems with a microemulsion would be

$$K_d^* = \frac{K_d + f_{em}K_{em}^{(s)}}{1 + X_{em}K_{em}^{(w)}} \tag{30}$$

where f_{em} is the fraction of sorbed emulsified material in soil; $K_{em}^{(s)}$ is the solute sorption coefficient between sorbed emulsified material and water; and $K_{em}^{(w)}$ is the corresponding coefficient between aqueous microemulsion and water. By equations 29 and 30, the prerequisite condition for soil remediation ($K_d^* < K_d$) requires that $f_{sf}K_{sf}/K_d$ be smaller than $X_{mn}K_{mn} + X_{mc}K_{mc}$ if a surfactant is used or that $f_{em}K_{em}^{(s)}/K_d$ be smaller than $X_{em}K_{em}^{(w)}$ if a microemulsion is involved. The relative values of K_d and K_d^* are usually a function of surfactant type and its concentration and the soil-sediment property.

To characterize K_{sf} as a function of f_{sf} (or $K_{em}^{(s)}$ as a function of f_{em}), the information on sorption of various surfactants on a variety of soils must be available to account for the amount and the state of sorbed surfactants. At this point, there is very limited knowledge in this area of research. Because of the unique molecular properties of surfactants (strong polar head groups with long nonpolar chains) and their large molecular weights, the sorptive behavior of surfactants on soil could differ significantly from that of ordinary nonionic organic solutes.

The sorption of cationic surfactants on clays and soils having cation-exchange sites is known to take place by a cation exchange (119–124). Incorporation of long-chain alkylammonium cations to charge-exchange sites either on clay or on soil organic matter creates an additional nonpolar organic environment for solutes to partition (ie, to give large $f_{sf}K_{sf}$), which makes $K_d^* \gg K_d$ for nonpolar organic solutes (125,126). Therefore, the addition of cationic surfactants to soil–water systems would normally increase (rather than decrease) the uptake of organic pollutants on soil and hence reduce pollutant migration through a soil profile. This would have a negative effect for soil remediation. Anionic surfactants with high charges

Table 6. Comparison of Log K_{dom} and Log K_{doc} Values of Selected Organic Solutes with SSHA, SSFA, SRHA, and SRFA with Respective Octanol–Water Partition Coefficients (log K_{ow}) of Solutes[a]

		Log K_{dom}/log K_{doc}			
Solute	log K_{ow}	SSHA	SSFA	SRHA	SRFA
p,p'-DDT	6.36	4.82/5.06	4.27/4.58	4.12/4.39	4.13/4.40
2,4,5,2′,5′-PCB	6.11	4.63/4.87	3.81/4.12	3.80/4.07	3.83/4.10
2,4,4′-PCB	5.62	4.16/4.40	3.58/3.89	3.27/3.54	3.30/3.57
1,2,3-Trichlorobenzene[b]	4.14	~2.8/3.0	~2.0/2.3	~1.7/2.0	~1.7/2.0
Lindane[b]	3.70	~2.5/2.7	~1.5/1.8	~1.2/1.5	~1.2/1.5

[a] Ref. 104.
[b] Approximate log K_{dom} and log K_{doc} values are obtained from linear extrapolations of log K_{dom} and log K_{doc} against log K_{ow} of *p,p'*-DDT, 2,4,5,2′,5′-PCB and 2,4,4′-PCB.

Table 7. Structures and Properties of Selected Commercial Surfactants[a]

Surfactant	Structure	Molecular Weight	CMC, mg/L
Triton series:	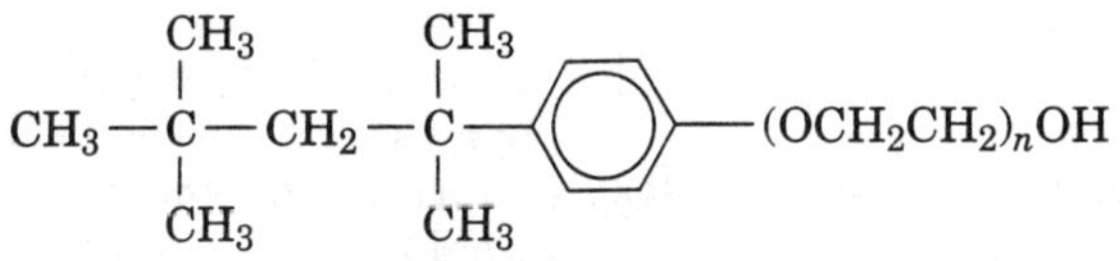		
Triton X-100 (TX100)	n = 9.5 (average)	628	130
Triton X-114 (TX114)	n = 7.5 (average)	536	110
Triton X-405 (TX 405)	n = 40 (average)	1966	620
Brij 35 (BJ35)	$C_{12}H_{25}(OCH_2CH_2)_{23}OH$	1200	74
Sodium dodecyl sulfate (SDS)	$C_{12}H_{25}{-}OSO_3^- \cdot Na^+$	288	2100
Cetyltrimethylammonium bromide (CTAB)	$C_{16}H_{33}{-}N(CH_3)_3^+ \cdot Br^-$	364	361

[a] Ref. 116.

usually show weak sorption on mineral soils (127). This would make anionic surfactants desirable for soil remediation. At this point, however, no adequate data are available to compare the effects of anionic surfactants and other surfactants on the sorption coefficients of organic contaminants in soil–water systems.

The influence of emulsified material on contaminant sorption coefficients has been investigated using PSO surfactants as reference materials (128). By use of naphthalene, phenanthrene, and 2,4,5,2′,4′,5′-PCB as model solutes and a low organic content soil (f_{oc} = 0.0010) as a model soil, Sun and Boyd (128) found that PSO reduces the K_d^* of 2,4,5,2′,5′-PCB more effectively than it does the K_d^* of phenanthrene, but imposes little effect on the K_d^* of more water-soluble nathphalene (Fig. 21). It appears that $f_{em}K_{em}^{(s)}/K_d$ is smaller than $X_{em}K_{em}^{(w)}$ for relatively insoluble solutes (high K_d), but the difference decreases for solutes of higher water solubility (low K_d). For an emulsified material in soil–water systems, one would expect $K_{em}^{(s)} \cong K_{em}^{(w)}$, since the sorbed material on solid surfaces would likely be in a separate phase similar to that in water. Therefore, if the material exhibits only a low-to-moderate sorption on a solid (ie, if f_{em}/X_{em} is not very large), $f_{em}K_{em}^{(s)}/K_d$ may be less than $X_{em}K_{em}^{(w)}$ (ie, $f_{em}/K_dX_{em} < 1$) for solutes with large K_d. For solutes of low K_ds, the f_{em}/K_dX_{em} increases and appears to approach 1 for naphthalene with a mineral soil. Overall, the addition of emulsified materials to water for remediation of soils with subsaturated contaminants would be most effective for highly insoluble contaminants.

The effect of nonionic surfactants on K_d^* of organic solutes has been studied with TX100 as a model surfactant (118,129). The K_d^* shows an apparent dependence on solute (contaminant) and soil properties and on surfactant concentration. Sun and co-workers (118) found that the K_d^* values of all the solutes (DDT, 2,4,5,2′,4′,5′-PCB and 1,2,4-trichlorobenzene) on a mineral soil (f_{oc} = 0.0010) show a marked initial rise with aqueous surfactant concentration, reach a maximum near or exceeding the CMC, and decline thereafter at high concentrations (Fig. 22). The initial rise of K_d^* may be ascribed to small $X_{mn}K_{mn}$ at low surfactant concentrations (where $X_{mc}K_{mc} \cong 0$ below the CMC) relative to $f_{sf}K_{sf}/K_d$. The decline of K_d^* at surfactant concentrations beyond CMC is ascribed to large $X_{mc}K_{mc}$ relative to $f_{sf}K_{sf}/K_d$ as the sorption of surfactants to soil (f_{sf}) approaches saturation. On the other hand, Deitsch and Smith (129) found that in the presence of TX100 the K_d^* of trichloroethylene (TCE) on an organic-rich soil (f_{oc} = 0.24) exhibits no increases over K_d up to moderately high surfactant concentrations (300 mg/L) and that the K_d^* shows a significant reduction relative to K_d only at surfactant concentrations much greater than CMC. By these differences, it appears that a surface-aggregation state is more readily formed by the sorbed surfactants on a mineral

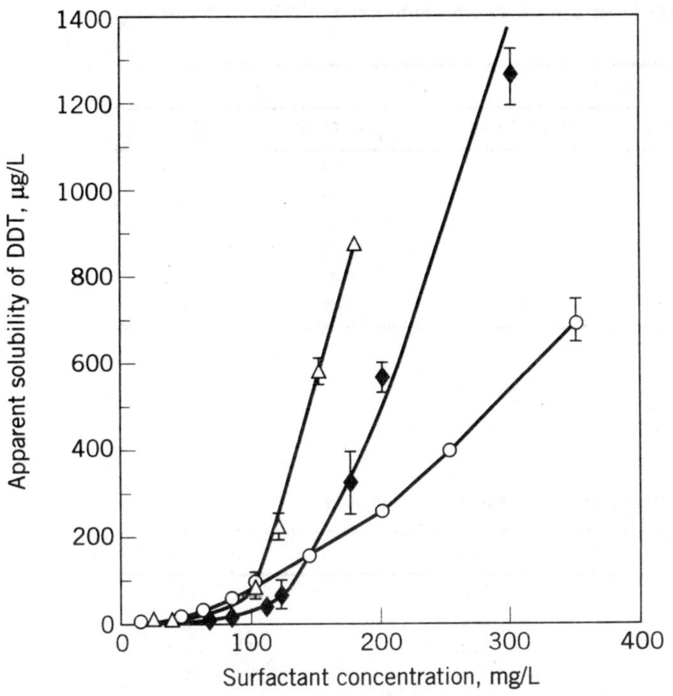

Figure 19. The apparent water solubility of *p,p′*-DDT as a function of the aqueous concentration of TX100 (♦), TX114 (△), and BJ35 (○) (116).

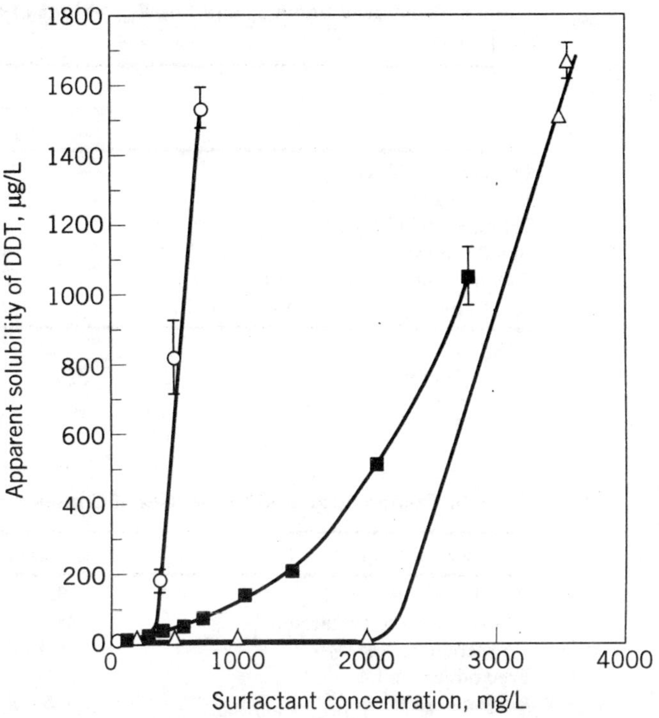

Figure 20. The apparent water solubility of *p,p′*-DDT as a function of the aqueous concentration of TX405 (■), SDS(△), and CTAB (○) (116).

soil than on an organic soil, because a greater amount of sorbed surfactants on the organic-rich soil may partition to SOM and because the generally higher surface area (or porosity) of a mineral soil may more effectively promote surface micellization (aggregation) of sorbed surfactants.

At the present time, the lack of detailed information on the sorption of various surfactants on soil in relation to soil properties, and the lack of an account of the magnitude of K_{sf} with respect to f_{sf} prohibit *a priori* prediction of the K_d^* of contaminants with the application of surfactants. Prior to the use of surfactants for remediation of subsaturated contaminants in soil, it is prudent to establish the relation between K_d^* and surfactant concentration as a guide to the selection of a proper surfactant and its concentration for maximum performance.

SORPTION BY SOILS IN ORGANIC SOLVENT SYSTEMS

Effect of Solvent Polarity on Solute Sorption

There are relatively few reported studies on the sorption of organic compounds and pesticides by soil from organic solvent solution. Results from such studies are useful for illustrating soil sorptive behavior in general and for establishing the basis for selection of suitable solvents to recover contaminants from soil, either for laboratory analysis or for remediation. Obviously, the best solvents to recover contaminants are those that effectively suppress the contaminant sorption to soil (or sediment). On the basis of the behavior of soil minerals and organic matter as discussed, it is anticipated that the partition of nonionic organic compounds and pesticides from organic solvents into soil organic matter should be weak because of the good solvency of the solution phase (14,69). Thus the extent of soil uptake would be controlled mainly by the efficiency of the compounds to compete with the solvent for adsorption on soil minerals. Because of the inherent polarity of inorganic minerals, the adsorptivity of a compound (solute or solvent) would be strongly influenced by the extent of polar interactions with the mineral.

Hance (130) reported experimental data that highlight the difference in the sorption of diuron from aqueous and petroleum solutions by an oxidized soil ($f_{om} \leq 0.015$) and an organic soil ($f_{om} = 0.76$). The sorption of diuron to the oxidized soil was remarkably greater from petroleum solution than from water solution, whereas the sorption to the organic soil was remarkably greater from water solution than from petroleum solution. By these findings, it was concluded that there is a competition between diuron and water for adsorption sites in soil under aqueous slurry conditions and that diuron competes more effectively for soil organic matter than for soil mineral surfaces. The high uptake of diuron from petroleum solution by the oxidized soil is due apparently to the strong (competitive) adsorption of the polar solute (diuron) on soil minerals at the expense of the relatively weak competition of a nonpolar solvent (petroleum spirit) for adsorption on minerals. On the other hand, the high uptake of diuron by the organic soil in water solution is in keeping with the solute partitioning into the organic matter, with concomitant suppression by water of adsorption on soil minerals.

Similarly, it was demonstrated (39) that there were major differences in the sorption of lindane by Venado clay (50% montmorillonite; $f_{om} = 0.06$) and Staten peat muck ($f_{om} = 0.22$) from aqueous and hexane solutions. The sorption capacities in aqueous systems are largely proportional to

Table 8. Measured Log K_{mn} and Log K_{mc} Values of *p,p′*-DDT and 1,2,3-Trichlorobenzene (TCB) on Selected Surfactants[a]

	Log K_{mn}		Log K_{mc}	
Surfactant	*p,p′*-DDT	1,2,3-TCB[b]	*p,p′*-DDT	1,2,3-TCB
Triton X-100 (TX100)	4.26	ND	6.15	3.82
Triton X-114 (TX114)	4.59	ND	6.18	3.95
Triton X-405 (TX405)	3.92	ND	5.56	—
Brij 35 (BJ35)	4.18	ND	5.75	3.31
Sodium dodecyl sulfate (SDS)	2.68	ND	5.38	3.54
Cetyltrimethylammonium bromide (CTAB)	3.54	ND	5.88	3.80

[a] Ref. 116.
[b] ND = Not detectable.

Table 9. Composition and Properties of Commercial Petroleum Sulfonate Surfactants[a]

Composition/Property	Petronate L	Petronate HL	Pyronate 40
Petroleum sulfonate %	62	62	41
Fraction as hydrocarbon	46.9	48.0	28.6
Fraction as $NaSO_3$	15.1	14.0	12.4
Free mineral oil, %	33.0	32.5	12.0
Inorganic salts, %	0.5	0.5	8.5
Water, %	4.5	4.5	38.5
Total hydrocarbon (water-free basis), %[b]	83.7	84.3	66.0
Petroleum sulfonate, MW	415–430	440–470	330–350

[a] Ref. 117.
[b] Sum of hydrocarbons in petroleum sulfonate and free mineral oil.

the organic matter content in the soil samples in accord with the partition model. In hexane solution, the sorption to the oven-dried Venado clay is markedly enhanced over that to dry peat muck. Moreover, whereas the equilibrium molar heats of lindane sorption in aqueous systems with both soil samples are less exothermic than the reverse heat of solution of lindane in water ($-\Delta\overline{H}_w$), the observed heats of sorption from hexane solution are much more exothermic than the reverse heat of solution in hexane. Here, again, the presumably weak adsorption of relatively nonpolar hexane on minerals allows lindane to compete favorably for adsorption on mineral matter from hexane (while partitioning with organic matter is minimized), and consequently the Venado clay shows a much greater sorption capacity.

Researchers studied (13) the soil uptake of parathion from a wide range of organic solvents and the effect of soil water content on parathion uptake from these systems. Parathion shows a high uptake on dry soils from hexane, a lower uptake from benzene, and virtually no uptake from such polar solvents as methanol, ethanol, acetone, chloroform, ethyl acetate, and dioxane. Although the uptake from hexane by dry and nearly dry soils is remarkably higher than from water, such uptake is strongly suppressed by humidity and approaches zero when the soils become water-saturated (whereas parathion exhibits a definitive uptake on the same soils from water). The saturation water content ranges from about 2 wt % for the sandy Mivtahim soil (6% clay; f_{om} = 0.003) to 17 wt % for the clay-rich Har Barqan soil (56% clay; f_{om} = 0.019). These observations led the authors to suggest that different mechanisms govern the parathion "adsorption" in aqueous soil suspensions and in hydrated soil–organic solvent systems. They assumed that in soil–water systems, the solvent (water) is preferentially adsorbed but that small amounts of parathion diffuse through water films and are adsorbed as they approach the colloid surfaces.

It is evident from the parathion sorption data of Yaron and Saltzman (13) that the (dominant) mechanism of the soil sorption in organic solvent systems is different from that in aqueous systems. Results found in aqueous systems are reconcilable with the assumed solute partition into soil organic matter. The high parathion uptake from hexane on dry soils is attributable to adsorption on soil minerals (mainly clays), on which the specific interactions of parathion's polar groups reduce the adsorptive competition of nonpolar hexane (whereas the partition of parathion into the organic matter is minimized by the good solvency of hexane). The parathion uptake from hexane would be depressed by humidity because of the strong adsorptive competition of water for minerals (in this case, water is considered as a competing solute), which leads eventually to a nearly complete suppression of parathion uptake when the soils become fully water saturated. By comparison, one notes that parathion exhibits a definitive uptake on soil from aqueous solution because the poor solvency of water (in addition to its suppression of mineral adsorption) makes partitioning of parathion into the organic matter a favorable process. The failure of the soils to sorb parathion from polar organic solvents is due appar-

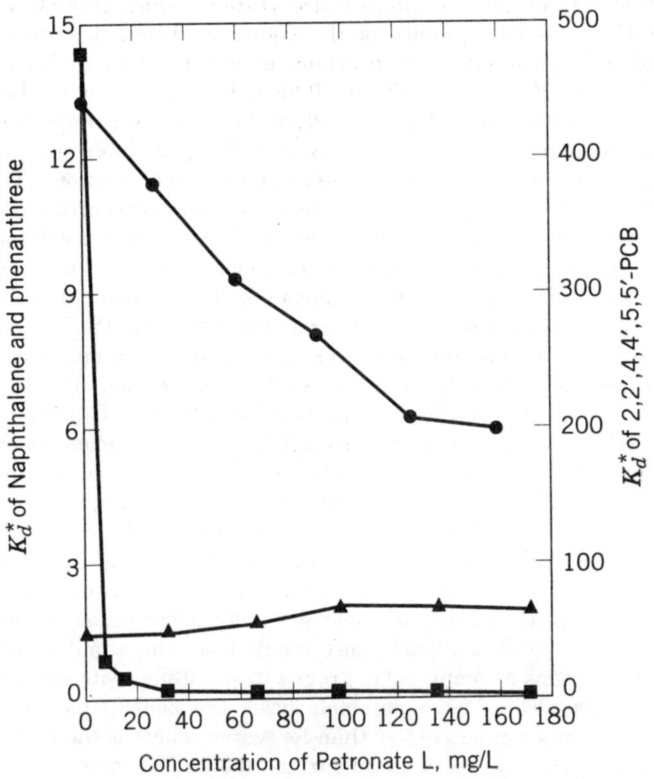

Figure 21. Plot of apparent K_d^* values of 2,2′,4,4′,5,5′-PCB (■), phenanthrene (●), and naphthalene (▲) with a soil (f_{oc} = 0.0010) versus petroleum sulfonate (Petronate L) concentrations in water. Reproduced with permission (128).

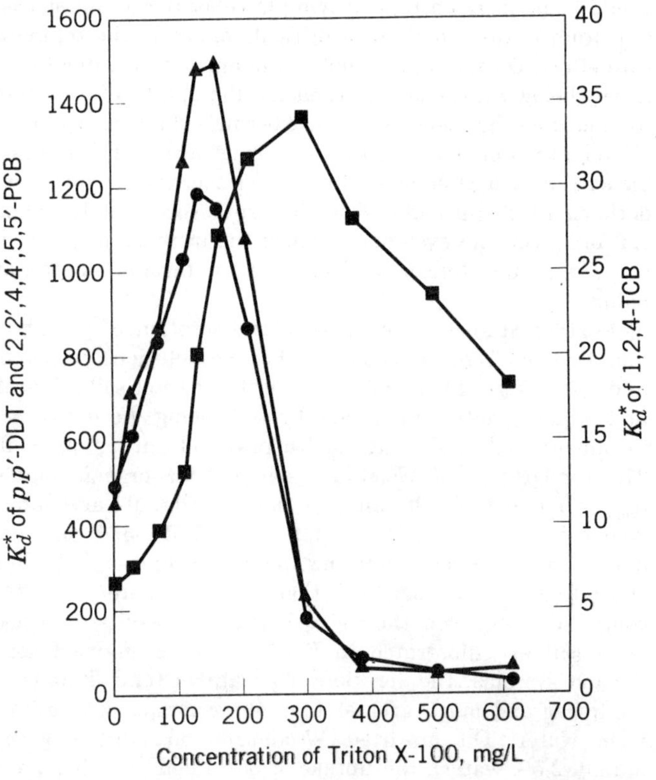

Figure 22. Plot of apparent K_d^* values of p,p'-DDT (▲), 2,2′,4,4′,5,5′-PCB (●), and 1,2,4-trichlorobenzene (■) with a soil (f_{oc} = 0.0010) versus Triton X-100 concentrations in water. Reproduced with permission (118).

ently to the fact that such solvents minimize solute (parathion) adsorption on minerals because of their polarity and reduce the solute partition into organic matter by their good solvency. Therefore, polar solvents are much more effective than nonpolar solvents in recovering nonionic organic contaminants from soil.

Effects of Temperature, Moisture, and Solute Polarity

Another interesting aspect that was recognized in an earlier study (13) is the increasing parathion uptake from hexane with increasing temperature on partially hydrated soils, as opposed to the decreasing sorption with increasing temperature in aqueous systems. Such differences led researchers (56) to suggest that the heat of adsorption of a solute from solution can be either exothermic or endothermic and consequently that the associated entropy change for solute adsorption can be either negative or positive.

The observed temperature dependence of parathion sorption from hexane with partially hydrated soils warrants careful consideration. As mentioned, a significant adsorption of single vapors or single solutes from a solution (to the extent that the concentration of the solute on the adsorbent surface is significantly enhanced over that in the solution phase) is usually accompanied by high exothermic heat. Although it is possible for a solute in single-solute systems to exhibit an endothermic heat of adsorption in competition with the solvent for adsorbent, adsorption of the solute in this case must be very weak because of the strong adsorption of the solvent (which makes the solute adsorption endothermic). This is supposedly the reason for the very weak adsorption of relatively nonpolar organic solutes on pure minerals from water. However, in certain binary-solute systems, the adsorptive competition of one solute against the other for adsorbent surfaces may give rise to an anomalous temperature effect if their heats of adsorption per unit area as single solutes are relatively comparable and if the displacement of one solute by the other produces a large gain in entropy of the system.

In parathion uptake from hexane on partially hydrated soils (13), the amounts of water in soils and in hexane solution are below saturation. Since parathion contains many polar groups, it should have a relatively high heat of adsorption per unit area with dry soil minerals. In addition, parathion is considerably bigger in size than water. Adsorption of one mole of parathion by displacing water from mineral surfaces of a partially hydrated soil (where there is enough water to cover all mineral sites) would release many moles of water into hexane solution. Although this displacement may be moderately endothermic for parathion, the gain in system entropy by releasing much more water from surface sites to unsaturated hexane solution may be high enough to offset the enthalpic deficiency to make the process favorable. This entropic driving force diminishes as the hexane solution becomes more saturated with water. When the water content in soil exceeds saturation, there would be no more entropic driving force for parathion adsorption from hexane, making it difficult to detect the temperature effect. On the other hand, if the amount of water in soil is way below the monolayer capacity, the uptake of parathion from hexane should be exothermic because of its adsorption on abundant water-free mineral sites. Thus, although it is possible for the adsorption of a solute to be either exothermic or endothermic (56), the endothermic adsorption occurs only for special binary-solute systems in which the increase in total entropy is enough to compensate for the unfavorable enthalpic balance.

Further studies were made of the sorption of parathion and lindane from aqueous and hexane solutions by Chiou and co-workers (14) to substantiate the roles of soil minerals and organic matter in uptake by soil, using Woodburn soil (a mineral soil) and Lake Labish peat soil (an organic soil). The composition of Woodburn soil is 1.9% organic matter (f_{om} = 0.019), 68% silt, and 21% clay and that of Lake Labish peat soil is 51% organic matter (f_{om} = 0.51), 36% silt, and 3.5% clay. In aqueous systems, both parathion and lindane show linear isotherms, and there is no apparent sorptive competition between the two solutes. The results on Woodburn soil are illustrated in Figure 8. In oven-dried soil–hexane systems, the sorption of parathion (and lindane) is nonlinear and much greater than the corresponding uptake from water. On air-dried Woodburn soil (with approximately 2.5% water), the uptake of parathion from hexane is significantly reduced; on water-saturated Woodburn soil (approximately 5% water), no detectable uptake is observed. The results are shown in Figure 23**a**. Parathion shows a similar endothermic uptake from hexane on air-dried Woodburn soil; the temperature effect is only qualitative because of a large scattering of the data.

As mentioned, comparison of soil uptake from solution is more appropriately made on the basis of relative concentration (rather than absolute concentration) which corrects for differences in solubility of the solute in different solvents. Since the solubility of parathion in hexane is much higher (5.74×10^4 mg/L at 20°C and 8.56×10^4 mg/L at 30°C) than in water (approximately 12 mg/L at 20°C) (14), it is important in comparing isotherms to extend the parathion sorption from hexane to sufficiently high absolute concentrations. The data in Figure 23**a** show that over a relative concentration of parathion between 0 and 0.01 at 20°C, the isotherm exhibits marked curvature, with capacities more than two orders of magnitude greater than in aqueous systems. Such curvature is not evident in the study of Yaron and Saltzman (13), because their study was limited to very low relative concentrations, which fell within the limiting Henry's-law region. The isosteric heat of parathion uptake calculated from 20°C and 30°C isotherms with dry Woodburn soil is highly exothermic (more than the reverse heat of solution of parathion in hexane, $-\Delta\overline{H}_h$) and varies with parathion loading capacity, as shown in Figure 23**b**. These characteristics support the contention that parathion adsorption on soil minerals is the dominant mechanism in sorption from hexane, as previously discussed. The high net exothermic heat per mole of parathion adsorption (up to 20 kcal/mol) may result from the simultaneous interactions of many polar groups in parathion with mineral surfaces. Thus, on a per mole basis the heat of parathion adsorption can be greater than for water, whereas the heat of adsorption per unit mineral surface should be greater for water to account for the suppression of parathion adsorption by water.

The sorption data of lindane from hexane on dry and partially hydrated (Woodburn) soils exhibit essentially the same patterns as with parathion, except that the uptake of lindane is considerably lower at equal relative concentration and is more sharply reduced by the water content (Fig. 24); the solubility of lindane in hexane at 20°C is 1.26×10^4 mg/L (relative

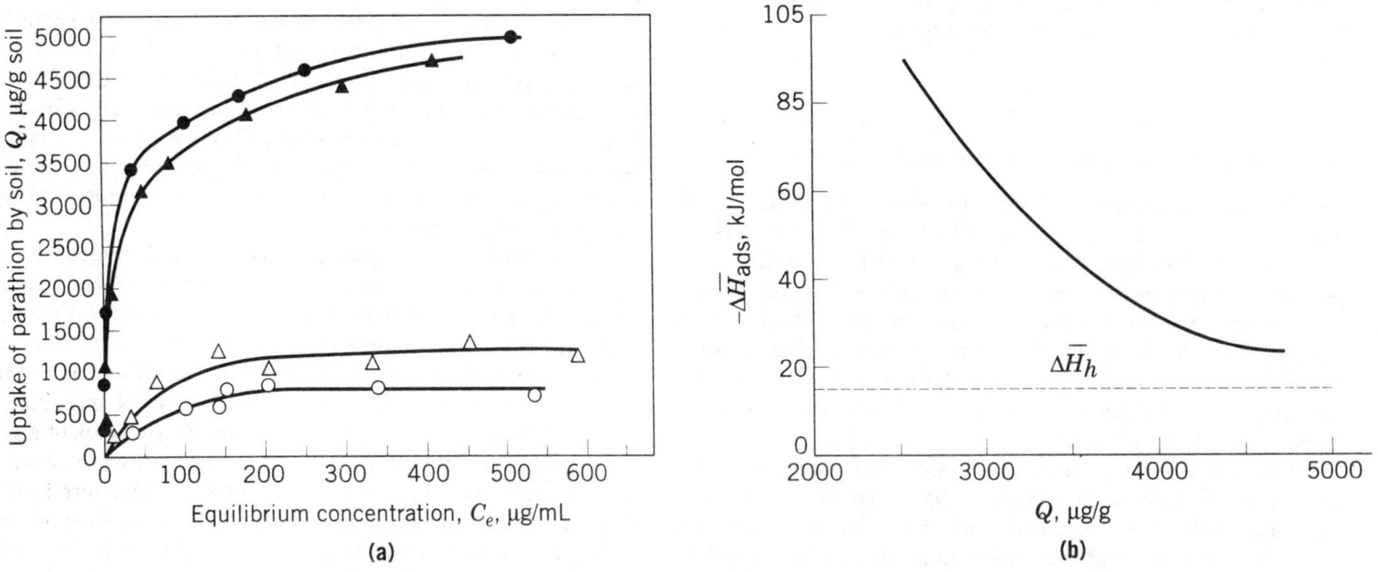

Figure 23. (**a**) Sorption isotherms of parathion from hexane on oven-dried (●, ▲) and air-dried (○, △) Woodburn soil at 20 and 30°C, respectively. The air-dried soil contains 2.5% water. (**b**) Isosteric heats of parathion sorption on oven-dried Woodburn soil calculated from the isotherms. Reproduced with permission (14).

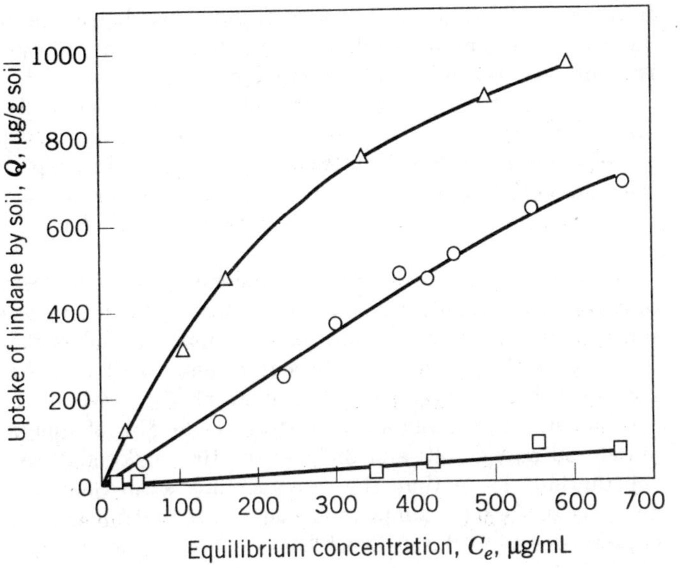

Figure 24. Sorption isotherms of lindane from hexane on oven-dried and partially hydrated Woodburn soil at 20°C: (△) lindane on oven-dried soil; (○) lindane on 5 mg water/g soil; (□) lindane on 25 mg water/g soil. Reproduced with permission (14).

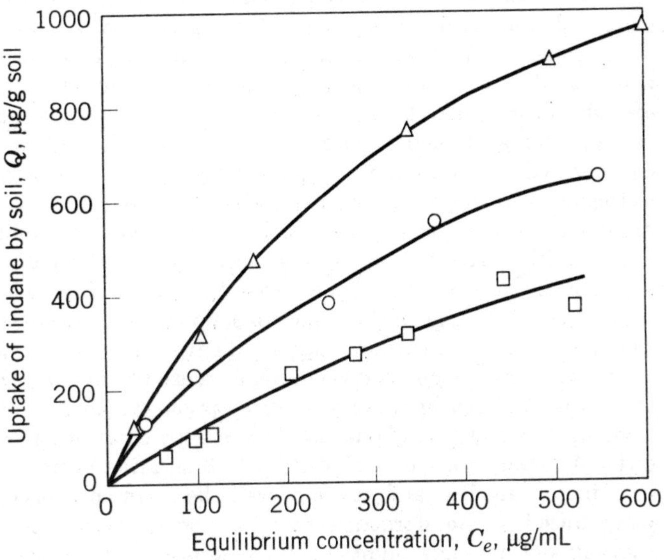

Figure 25. Sorption isotherms of lindane from hexane in the presence and absence of parathion as a competing solute on Woodburn soil at 20°C: (△) lindane only on soil; (○) lindane on 0.5 mg parathion/g soil; (□) lindane on 3 mg parathion/g soil. Reproduced with permission (14).

to 7.8 mg/L in water at 25°C). With approximately 2.5% water in soil, the lindane uptake is reduced nearly 25 times relative to the capacity with dry soil, which is much more than with parathion, and further addition of water to the saturation point suppresses lindane uptake to a nondetectable level. These differences are consistent with the low polarity of lindane relative to parathion, making lindane a less potent adsorbate, and thus a much weaker competitor than parathion against water for mineral adsorption. Being a relatively nonpolar adsorbate, lindane must have a considerably smaller heat of adsorption per unit area than water on dry soil minerals and consequently cannot effectively displace water from a partially hydrated soil. Thus, when all mineral surfaces are covered by water, as for Woodburn soil with approximately 2.5% water, the adsorption of lindane from hexane is likely to take place on less energetic and more uniform adsorbed-water surfaces, resulting in a weak and essentially linear uptake. This feature is in contrast to that found for parathion under the same system conditions.

Whereas the water content in soil suppresses the uptake of parathion and lindane from hexane, a similar competitive effect also occurs between parathion and lindane in their simultaneous sorption from hexane on dry Woodburn soil, as illustrated in Figure 25, where the uptake of lindane decreases with an increase of parathion uptake. Similarly, parathion and lindane show a considerably higher uptake from hexane on mineral-rich Woodburn soil than on organic-rich Lake Labish peat soil, which is the opposite of what is found in water solution, because of the dominance of solute adsorption on mineral matter from a nonpolar solvent and of the dominance of solute partition in organic matter from water (14).

SORPTION ON SOILS FROM VAPOR PHASE

Earlier interest in the sorption of organic compounds by water-unsaturated soils resulted from the need to evaluate the stability and persistence of fumigants and pesticides under field conditions. Shipinov (131) found that the isotherms for vapor uptake of hydrogen cyanide on dry soils were nonlinear (BET Type II isotherm). Similar nonlinear isotherms were reported for the vapor uptake of chloropicrin on dry soils and a close correlation was found between amounts of uptake and soil clay contents (10); the chloropicrin sorption decreased with increasing soil-moisture content. It has been found (11) that at a moisture content substantially below the wilting point, ethylene dibromide (EDB) was strongly sorbed by the soil, but that sorption decreased sharply to a minimum near the wilting point. Wade (12) studied the EDB sorption by three soils having significantly different clay vs organic carbon contents at moisture contents corresponding to the usual field range. Over the field moisture range, the EDB sorption by the soils was unaffected by soil–water content and the EDB isotherms were found to be linear, with the soil sorption capacities being closely proportional to the respective soil organic carbon contents; no correlation was found with clay content. Much greater EDB sorption was also noted on dry soils (12); moisture sharply suppressed the EDB sorption up to the point corresponding to the saturation water content of the soil and a further increase in moisture content beyond that point produced little additional effect.

The vapor uptake of EDB and 1,2-dibromo-3-chloropropane on dehydrated clays (which still retained small amounts of water) was studied in relation to clay types and exchanged cations (132–134). The extent of sorption was related to the external clay surface areas of specific clays, and the sorption data fit the BET adsorption model. The BET monolayer capacity of EDB increased from 17.5 g/kg for nonexpanding calcium kaolinite (1.0% water) to 72.1 g/kg for expanding calcium montmorillonite (2.3% water). The dependence on relative humidity (RH) of EDB vapor uptake on several soils and a calcium montmorillonite has also been studied (135). On dry soils and soils equilibrated with low ambient humidity (RH < 10%), the EDB isotherms were Type II in shape, in which clay soils displayed greater sorption capacity than soils low in clay content. An increase of RH from 0 to 50% progressively suppressed the EDB sorption (no data were available beyond an RH of 50%) with a concomitant change of the isotherm shape toward linearity. The isotherms with calcium montmorillonite were similar to those with soils, except that when RH increased from 0 to 10% there was a sharp increase in EDB sorption, which was attributed to expansion of the clay layers with creation of additional surface. Beyond RH of 10%, EDB sorption decreased with increasing RH, and at RH of 90% the sorption capacity was reduced to a small fraction of the sorption on dry clay. That the soils did not show additional EDB sorption on increasing RH from 0 to 10% suggests that the clay minerals in the studied soils were mainly of nonexpanding types. By these observations, Call (135) suggested that sorption of EDB on soils and clays containing low humidities resulted from adsorption on soil (mineral) surfaces, whereas sorption on wet soils and soils containing high humidities was attributed to dissolution in soil water or adsorption onto a water surface.

The vapor uptake of *cis*- and *trans*-1,3-dichloropropene was determined on three types of soils—humus sand ($f_{om} = 0.55$), peat sand ($f_{om} = 0.18$), and peat ($f_{om} = 0.95$), with moisture contents of 17%, 41%, and 120% of dry soil, respectively (32). The isotherms for all three soils were highly linear, with the soil-to-vapor distribution coefficients again being proportional to the respective organic matter contents; the calculated sorption coefficients based on soil organic content were largely independent of the soils for each vapor. The normalized sorption coefficients (K_{om}) of both compounds also exhibited a relatively small temperature dependence, with $\Delta\overline{H}$ being nearly constant and less than 1 kcal/mol exothermic (19). The data suggest that the soils used by Leistra were already water-saturated (note that the amount of water required to saturate the soil in sorption is obviously much smaller than the field water saturation capacity and appears to be close to the water content at the wilting point of the soil. The saturation water content in the present discussion should not be confused with the field water saturation capacity).

In light of the described vapor sorption data, it is evident that dry and partially hydrated soil minerals (especially, clays) are powerful adsorbents for organic compounds and that the contribution by clay adsorption greatly exceeds the concurrent partitioning into the organic matter on dry mineral soils. Apparently, at saturation-water contents, the adsorptive power of soil minerals for organic compounds is largely lost because of strong competitive adsorption of water (17,27), leaving partitioning with soil organic matter as the dominant mechanism. The fact that the organic-vapor uptake by dry soils is closely related to clay content rather than organic matter content suggests that dry clay is more powerful per unit weight in uptake of organic compounds by adsorption than organic matter in uptake by partitioning, whereas the reverse is true with the hydrated soils. The observed suppression of vapor uptake by moisture is essentially the same as found in the depression by water of parathion and lindane uptake from hexane solution. The only difference is that the organic solvent (hexane) also minimizes partitioning in the organic matter, which makes the solute uptake approach zero at full water saturation. Such partitioning is largely unaffected by moisture in sorption from the vapor phase (although the presence of moisture in organic matter may affect the partition uptake of organic compounds).

Results relating the equilibrium lindane and dieldrin vapor concentrations in soil with soil–water contents (40,43) further substantiate the relative roles of soil minerals and organic matter in soil sorption. At soil water contents of less than 2.2% on Gila silt loam (Typic Torrifluvent, $f_{om} = 0.006$), the equilibrium vapor densities of lindane (at a loading of approximately 50 mg/kg soil) and dieldrin (at 100 mg/kg soil) were substantially lower than the corresponding saturation vapor densities of the pure compounds, suggesting that the amounts of pesticides applied were much below the saturation limits on the soil. An increase in soil water content to more than 3.9%, however, resulted in a sharp increase of equilibrium vapor concentrations, which became equal to the saturation vapor concentrations of the pure compounds and stayed unchanged with further increasing water content up to the field capacity (17%). Results with dieldrin at 30 and 40°C are shown in Figure 26.

The fact that applied dieldrin at 100 mg/kg Gila soil and lindane at 50 mg/kg Gila soil show subsaturation vapor concentrations when soil water content is low and display saturation vapor concentrations when the soil is wet illustrates distinct mechanisms of sorption of soil minerals and organic matter. The low vapor concentrations with dry soil are ascribed to strong mineral adsorption, which overrides the effect of partitioning with organic matter. Upon wetting, water displaces most pesticide from soil minerals by adsorptive competition and, as a result, the amount of pesticide in soil becomes more than is required to saturate the organic matter, giving rise to saturation vapor concentrations. At water saturation, one notes that 100 mg dieldrin/kg Gila soil (with $f_{om} = 0.006$) corresponds to 17 g dieldrin/kg organic matter, and this loading is far greater than the solubility of solid dieldrin in soil organic matter at 25°C, that is, $S_{om} = 1.5$ g dieldrin/kg organic matter, calculated by equation 17 with $S_w = 0.20$ mg/L (91) and $K_{om} = 7400$ (65). Similarly, at water saturation, a loading of 50 mg lindane/kg Gila soil corresponds to 8.3 g lindane/kg soil organic matter, which is greater than the calculated $S_{om} = 3.0$ g lindane/kg organic matter for solid lindane at 25°C by equation 17 with $S_w = 7.8$ mg/L (91) and $K_{om} = 380$ (14). Thus, for the amounts of dieldrin and lindane applied to water-saturated Gila soil, the pesticide vapor concentrations will remain unchanged at saturation with increasing soil water content until the water content is high enough to dilute excess pesticides to below-saturation levels. The above analysis further accounts for the findings that the sorption capacity of lindane on the hydrated Gila soil is essentially the same as that on the same soil with 3.9 or 10% water (40). At 3.9% water on Gila soil, which corresponds to about the point of water saturation, the heat of lindane vapor sorption calculated from the temperature dependence of the isotherms was less exothermic than the reverse heat of lindane vaporization ($-\Delta\overline{H}_v$), due presumably to the dominance of vapor partition in soil organic matter of a water-saturated soil.

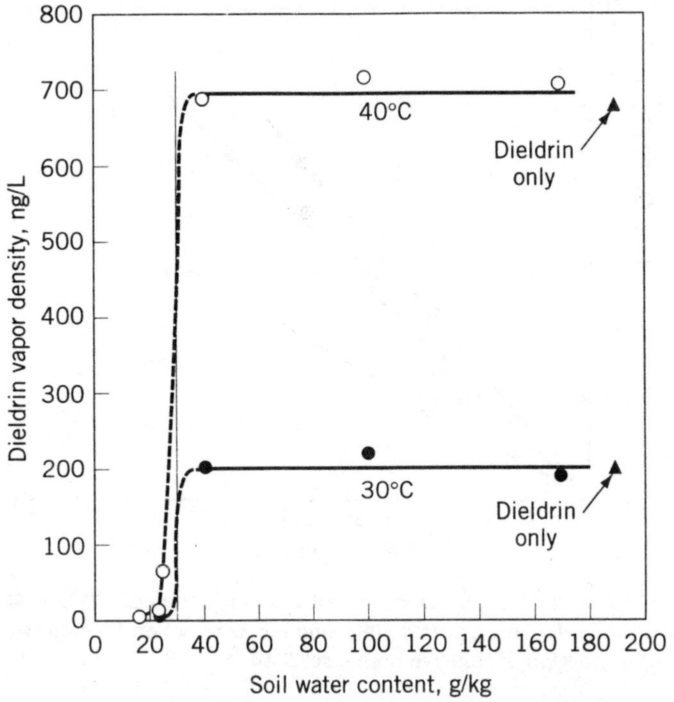

Figure 26. Effect of soil–water content on vapor density of dieldrin applied to Gila silt loam at 100 mg dieldrin/kg soil. Reproduced with permission (43).

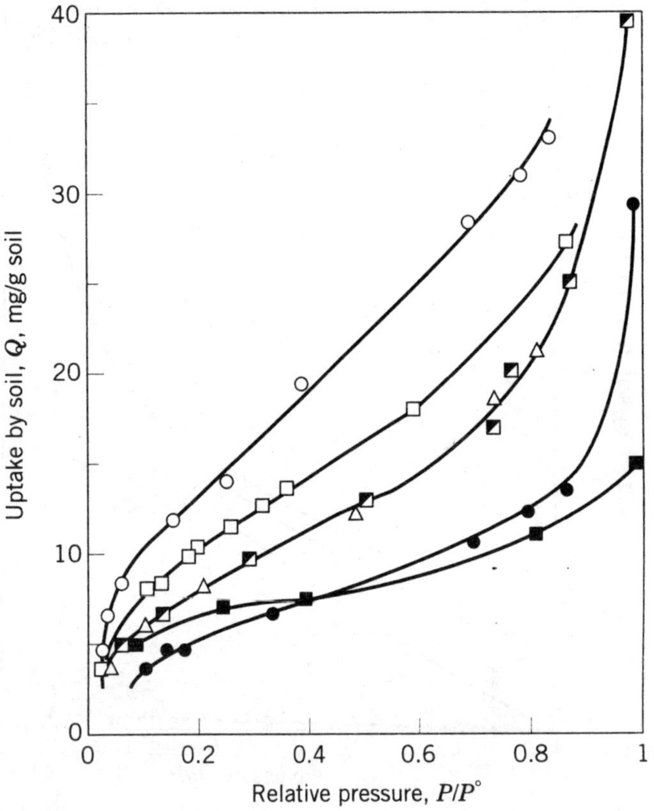

Figure 27. Uptake of vapors of organic compounds (□, 1,2,4-trichlorobenzene; △, chlorobenzene; ◩, *m*-dichlorobenzene; ●, benzene; ■, *p*-dichlorobenze) and water (○) on dry Woodburn soil vs relative vapor pressure at 20°C. Reproduced with permission (44).

To substantiate the effect of humidity on the mechanism and capacity of soil sorption of organic compounds, Chiou and Shoup (44) determined the vapor sorption of benzene, chlorobenzene, *m*-dichlorobenzene, *p*-dicholorobenzene, 1,2,4-trichlorobenzene, and water on dry Woodburn soil, and of benzene, *m*-dichlorobenzene, and 1,2,4-trichlorobenzene as functions of relative humidity. Isotherms for all compounds on dry soil in a normalized plot of Q vs P/P^o (where P and P^o are equilibrium and saturation vapor pressures, respectively) were distinctly nonlinear, with water showing the greatest capacity (Fig. 27). The observed data closely fit the BET adsorption model. In addition to the isotherm nonlinearity, the sorption capacities of organic vapors on dry soil were about two orders of magnitude greater than those of the same compounds from water on the same soil shown in Figure 3 (when the data therein are normalized in the form of Q vs C_e/S_w). The markedly greater sorption on dry soil is attributed to strong adsorption on soil minerals, which predominates over the simultaneous uptake by partitioning into the organic matter. The difference in sorption capacity between the dry and hydrated Woodburn soil samples illustrates the powerful suppression by water of adsorption of organic compounds on soil minerals.

The sorption of benzene, *m*-dichlorobenzene, and 1,2,4-trichlorobenzene by initially dry Woodburn soil was progressively depressed by increasing relative humidity (44). The results for *m*-dichlorobenzene are shown in Figure 28. In addition to the reduced capacities in the presence of water vapor, the isotherms for these relatively nonpolar vapors at RH $\geq$ 50% assume a practically linear shape at $P/P^o \leq 0.5$, which is attributed to adsorptive displacement of organic compounds by water at more energetic mineral surfaces. At high RH, the vapor uptake should consist largely of adsorption on adsorbed-water surfaces (since these vapors are not able to effectively displace adsorbed water) and of partition into the soil organic matter, producing relatively linear isotherms up to moderate P/P^o; at high P/P^o, the isotherms may exhibit nonlinearity due to multilayer vapor adsorption on water surfaces, if a significant amount of surface area remains (which decreases with the amount of water adsorbed). As adsorption of the compounds on water surfaces is further decreased by increasing amounts of water, the contribution to soil uptake by partition into the organic matter becomes increasingly more important and the isotherm linearity is extended to higher P/P^o. At approximately 90% RH, the sorption capacities of the compounds fall into a range close to those found in aqueous systems. A similar suppression by humidity of the vapor sorption is reported for trichloroethylene (TCE) on soil (46), TCE on clays and oxides (47), *p*-xylene on soil and silica gel (48), and chlorobenzene and toluene on soil (49). These results are consistent with the assumed functions of soil minerals and organic matter in the sorption of organic contaminants.

In contrast to the highly nonlinear isotherm and strong suppression by water of organic vapor sorption on a mineral

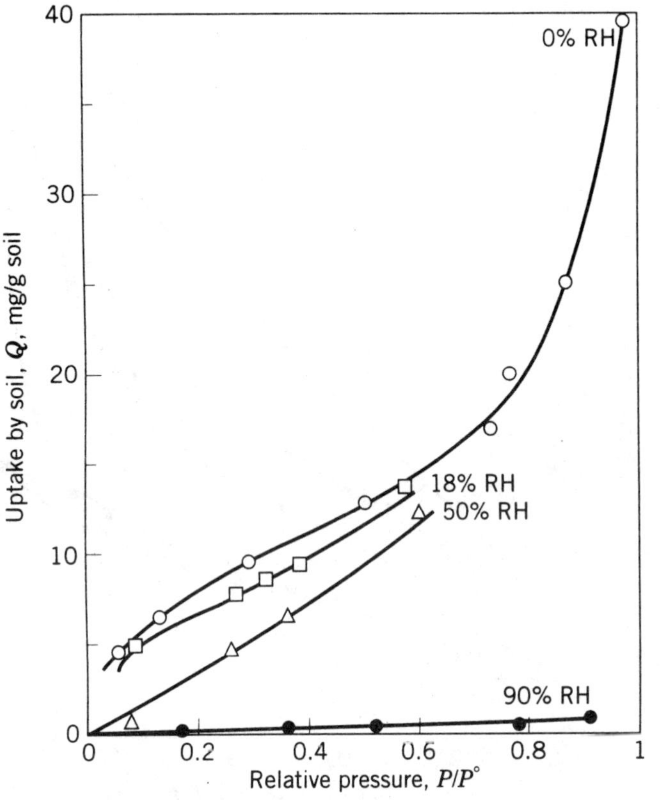

Figure 28. Uptake of *m*-dichlorobenzene vapor on Woodburn soil as a function of relative humidity at 20°C. Reproduced with permission (44).

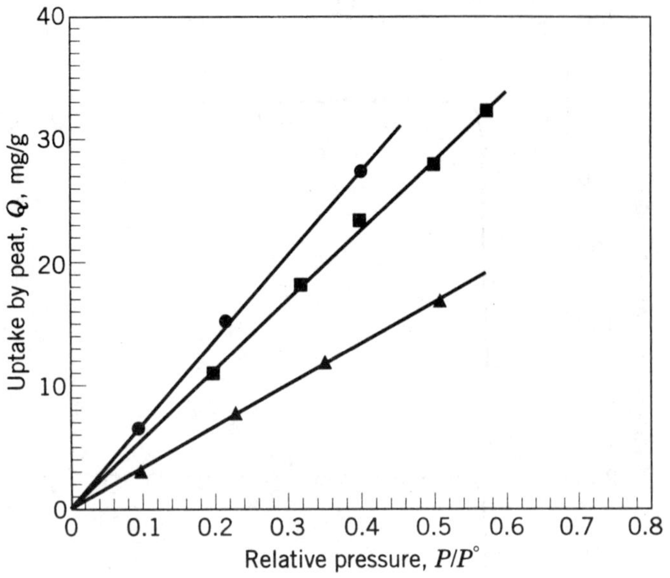

Figure 29. Uptake of vapors of trichloroethylene (TCE) (●), carbon tetrachloride (CT) (■), and benzene (▲) on Florida peat as a function of relative pressure (136).

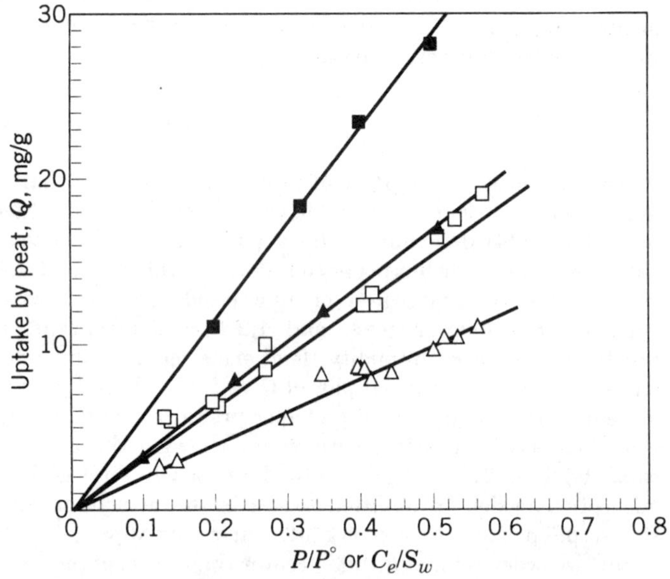

Figure 30. Comparison of uptake capacities of CT (■, □) and benzene (▲, △) from the vapor phase on dry Florida peat (Q vs P/P^o) and from water solution (Q vs C_e/S_w) on wet Florida peat. Solid symbols are for vapor-phase data and open symbols for water-phase data (136).

soil, the vapor sorption on a high-organic-content soil should be more linear and less sensitive to humidity because of the increased contribution of vapor partition in soil organic matter. It has been found (136) that vapor isotherms of benzene, carbon tetrachloride (CT), and TCE on a high organic content peat (f_{om} = 0.864) and muck (f_{om} = 0.815) are essentially linear and exhibit only moderate suppression by water. The results on dry peat are shown in Figure 29. The high isotherm linearity conforms to the predominance of vapor partition to organic matter; the observed capacity on dry peat is orders of magnitude too large to be accounted for by surface adsorption on the small surface area of the sample (1.5 m^2/g). The small mineral content in soil may be deeply covered by soil organic matter and hence be inaccessible to vapors. To determine the influence of water on vapor partition, the sorption isotherms from vapor phase (Q vs P/P^o) are compared with those from aqueous solution (Q vs C_e/S_w), because $P/P^o = C_e/S_w$ for sparingly soluble compounds. The data for benzene and CT on peat are shown in Figure 30. The saturation of organic matter by water is found to reduce the vapor partition by some 40–50% on peat and 30–40% on muck (136). This reduction is attributed to an increased polarity of soil organic matter upon water saturation, which reduces the partition efficiency of low polarity compounds (due to a greater mismatch in polarity between solute and organic matter). This effect, however, is small in comparison with the powerful suppression by water of the adsorption of nonpolar vapors and solutes on soil mineral matter.

The strong dependence of vapor sorption by a mineral soil on ambient humidity (or soil water content) provides a basis to account for the variation of the contaminant vapor concentration above the soil that may potentially occur in the field with changes in soil humidity (especially with surface soils). As illustrated in Figure 28, the relative pressure (ie, the chemical activity) of a contaminant in soil depends not only on the amount of contaminant in soil but also on the RH of the system because it affects the limiting (saturation) sorption capacity of the soil for the contaminant. For *m*-dichlorobenzene on Woodburn soil at, say, 0.4 mg/g soil, the relative pressure (P/P^o) of the compound is <0.01 at RH ≤ 50% but increases to approximately 0.8 at RH = 90%, which corresponds to nearly a hundredfold increase in vapor concentration (or chemical activity) due to a change in the soil's sorption capacity by water. The effect of RH on vapor activity is in conformity with results from previous studies (40,43) on the large variation of lindane and dieldrin vapor concentrations with soil–water content ranging from subsaturation to saturation levels. In natural environments, the transport of organic contaminants across soil–air interfaces may hence be influenced strongly by a drastic change of the ambient humidity.

IMPLICATIONS FOR CONTAMINANT ACTIVITY AND FATE

In the preceding sections, we have analyzed the dependence of sorption by soil (or sediment) on the properties of organic contaminants (pollutants and pesticides), the composition of soil (or sediment), and ambient factors such as temperature and humidity. The mechanistic roles of soil organic matter and soil minerals and the influence of soil water on their respective functions as elucidated provide the point of departure for understanding the diverse sorption behavior under various system conditions. In aqueous systems, the important factors for the sorption of relatively nonpolar compounds are the organic matter content of the soil (sediment), the water solubilities of the compounds, and, for highly water-insoluble contaminants, the amount and identity of dissolved and suspended organic matters. In nonaqueous systems, the important factors for sorption are the mineral type and content, the ambient humidity (or soil moisture content), and the solute (contaminant) and solvent polarities. Given the concentration of a relatively nonpolar contaminant in water and the soil organic matter content, one can make a reasonable estimate of the amount of sorption (partition) to soil at equilibrium from its K_{om} (or K_{oc}) value. The magnitude of K_{om} of a contaminant can in turn be estimated from the relation of K_{om} with water solubility (S_w) or with the octanol–water partition coefficient (K_{ow}) for a similar class of compounds.

Because the uptake of low-polarity contaminants by water-saturated soils is controlled mainly by partition into the organic matter (in which the solute has relatively small solubility), the contaminant mass sorbed to field soil or sediment from water would only be a small fraction of the soil/sediment organic matter content regardless of the source contaminant level, except for possible situations where the organic matter content is extremely small. One study (137), for instance, showed that concentrations of chlorinated hydrocarbons (DDT, DDE, and PCBs) in sediments from a marine site correlate positively with the organic carbon contents of the sediments; total chlorinated hydrocarbon concentrations ranged from 0.3 to 3.5 mg/kg sediment, whereas total organic carbon contents ranged from 4.5 to 17 g/kg sediment on a dry weight basis. Similarly, it was found (138) that sandy aquifer sediments (having trace amounts of organic matter) in a creosote-contaminated groundwater site exhibited little retention of substituted phenols and polyaromatic hydrocarbons, despite the fact that significant contaminant concentrations were found in groundwater.

The activity of a contaminant may be expected to be most sensitively influenced by a changing soil–water content from above to below the saturation capacity in the field because this drying-wetting cycle would cause a sharp change in chemical activity of the pesticide, as illustrated. The very top layer of surface soils supposedly undergoes a drying-wetting cycle to an extent depending on the change of ambient humidity and field practice. Under relatively dry conditions, strong adsorption by soil minerals (particularly, high-surface-area clays) in addition to partitioning into organic matter lowers the activity (as measured by P/P^o) of the contaminant in soil. Upon wetting, the chemical activity would rise sharply as a result of the displacement by water of those species previously adsorbed on

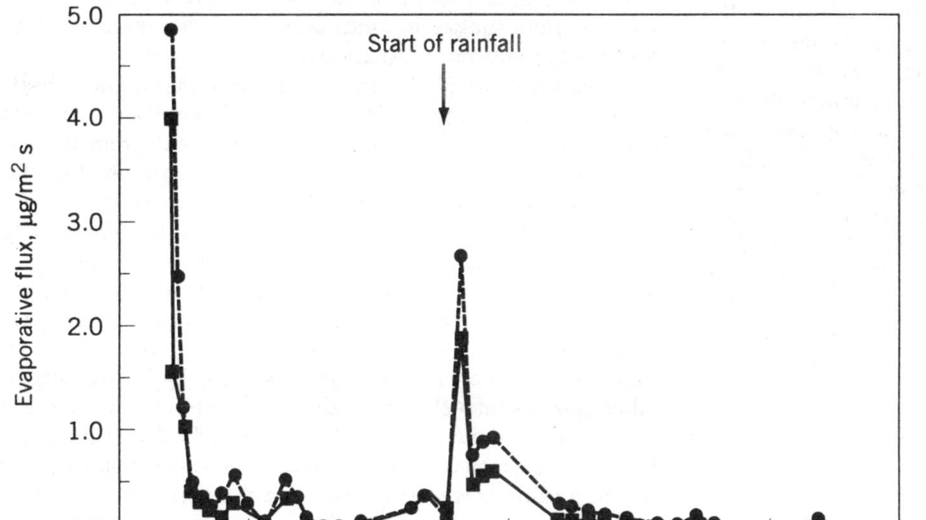

Figure 31. Evaporative fluxes of triallate (●) and trifluralin (■) herbicides with time showing the effect of a rainfall after the herbicide spray to a field site in Ottawa, Canada. Reproduced with permission (151).

soil minerals (17,44). Volatile pesticides originally sorbed by dry soils and clays are thus readily released by wetting the soil and clay (9–12,17,44,139–147). In fields to which pesticides have been applied, the rainfall event and the dew deposition on soil surfaces trigger large and sudden increases of pesticide concentrations in the air (148–151); when the soil surface is dry, the evaporative fluxes of pesticides are quite small. An example illustrating this phenomenon is given in Figure 31. Whereas many relatively nonvolatile pesticides and contaminants may not show significant vapor losses from soil, their chemical activities are equally affected by moisture in soil as illustrated in Figures 26 and 28.

Modifications of the chemical activity of pesticides in soil by soil–moisture content invariably lead to a change in the apparent pesticidal toxicity. Upchurch (152) found that diuron was more toxic to cotton (*Gossypium hirsutum* L.) under moist than under dry soil conditions. Barlow and Hodaway observed (153) that chlorinated hydrocarbon insecticides (lindane, DDT, and dieldrin) were inactivated by dry clay but were reactivated under high humidities for control of mosquitoes. Harris (154) reported empirical correlations of the insecticide toxicities of heptachlor, DDT, diazinon, V-C 13 (dichlofenthion), and parathion on a sand (f_{om} = 0.0052) and muck (f_{om} = 0.65) with soil–moisture content. The results are shown in Table 10. Heptachlor was 7.8, DDT 9.9, parathion 24.4, diazinon 132, and V-C 13 189 times more toxic to crickets when the sand was moist (5.5% water on a dry-weight basis) than when it was dry. By contrast, moisture in muck soil (162% water) had only a marginal effect on the insecticide toxicity. Comparison of insecticide toxicity between the two soil types yielded further striking results. At field moisture capacity DDT was 38.5, heptachlor 61.6, diazinon 65.7, V-C13 73.3, and parathion 92.0 times less toxic in the muck as compared with the sand. In moist soil all the insecticides were found to be strongly inactivated by the organic content of the soil, with the extent of inactivation being dependent on the specific insecticide used; in dry soil there was no obvious correlation between organic content and toxicity of the two soils. It was thus concluded that inactivation of the insecticides in moist soils was proportional to the organic content of the soil, whereas with dry soils inactivation was related to the adsorptive capacity of the mineral fraction (154).

The observed magnitude of the effect of soil moisture on the apparent toxicity of soil-incorporated organic pesticides is very closely correlated with the recognized effect of soil moisture on the individual functions of soil minerals and organic matter in soil uptake. Moreover, the remarkable effect of moisture on adsorption of the organic compound by soil minerals also agrees with our common experience that the air *in the field* gives a pleasant fragrance following a rain shower that succeeds a long period of drought. This phenomenon does not take place if the field has been wet prior to the rain shower. By these recognitions, we now have a better perspective of the role of humidity in influencing the activity of soil-incorporated contaminants and of the distinct mechanisms through which the soil organic and mineral matters function in sorption.

Table 10. Influence of Soil Type and Soil Moisture on Toxicity of Insecticides in Soils to First Instar Nymphs of Common Field Cricket (*Gryllus pennsylvanicus* (Burmeister))[a]

		LD_{50} mg/kg	
Insecticide	Soil Type[b]	Moist	Dry
Heptachlor	Plainfield sand	0.068	0.53
	Muck	4.19	5.39
DDT	Plainfield sand	1.75	17.3
	Muck	67.2	99.8
Diazinon	Plainfield sand	0.26	34.1
	Muck	17.0	11.5
V-C 13	Plainfield sand	3.80	717
	Muck	279	165
Parathion	Plainfield sand	0.25	6.00
	Muck	22.6	9.10

[a] Ref. 154.
[b] Plainfield sand (mixed, mesic Typic Udipsamment).

SUMMARY

This article considers mechanistic functions of soil organic matter and minerals in sorption of nonionic organic compounds by soil to account for the soil behavior under varying system conditions. The soil is perceived as a dual sorbent, in which the mineral matter functions as a conventional solid adsorbent and the organic matter as a partition medium. In aqueous suspension and for a normal soil, adsorption of organic compounds by mineral matter is suppressed by water due to strong polar interactions of water with minerals, and the soil sorption of low polarity contaminants consists primarily of solute partitioning into soil organic matter. The presumed solute partition with organic matter is supported by the small surface area of soil organic matter, the dependence of solute sorption on soil organic content, the linearity of soil–water isotherms, the absence of sorptive competition, the generally low heat of sorption, and the compliance of the sorption coefficient with the partition model derived for a macromolecular organic phase (soil organic matter). The theoretical basis underlying the empirical relationships of K_{om} (or K_{oc}) to water solubility (S_w) and to octanol-water partition coefficient (K_{ow}) is explained. The importance of soil and sediment organic matter and of dissolved and suspended organic matter in controlling the contaminant sorption coefficient in aqueous suspensions is emphasized.

The significance of mineral adsorption with dry and subsaturated soils is illustrated by data on soil sorption of organic compounds from nonpolar organic solvents and from the vapor phase. In these systems, the extent of adsorption of a compound by soil minerals in a soil decreases with an increase in moisture content and becomes minimal when the soil is fully saturated with water. The markedly greater sorption of a compound by the dry or subsaturated soil relative to that by the saturated soil is ascribed to mineral adsorption, which predominates over the simultaneous partition uptake by soil organic matter. As a result, a drying–wetting cycle in a normal soil causes a sharp change in the soil's sorption capacity and the chemical activity of a soil-incorporated contaminant. A change in soil moisture content below the saturation capacity therefore exhibits a strong impact on the apparent toxicity of soil-incorporated pesticides and on the release of organic contaminants across the soil–air interface. Specific examples are given to demonstrate these effects.

NOMENCLATURE

Symbol	Definition
C_e	Equilibrium solute concentration in water or solvent
C_e^*	Apparent C_e in solution containing dissolved organic matter
C_o	Solute concentration in organic–solvent phase
C_w	Solute concentration in water
f_{em}	Fraction of sorbed microemulsion in soil or sediment
f_l	Fugacity of a liquid solute or a supercooled liquid solute at system temperature
f_{oc}	Fraction of organic carbon in soil or sediment
f_{om}	Fraction of organic matter in soil or sediment
f_s	Fugacity of solid solute at system temperature
f_{sf}	Fraction of sorbed surfactant in soil or sediment
$\Delta\overline{G}^o$	Molar free energy change in the sorption of a solute at the standard state; see definition after eq. 16
$\Delta\overline{H}$	Molar heat of sorption of solute from solvent onto sorbent (soil or sediment)
$\Delta\overline{H}_d$	Molar heat of desorption of solute from sorbent into solvent ($\Delta\overline{H}_d = -\Delta\overline{H}$)
$\Delta\overline{H}_f$	Molar heat of fusion of solid compound
$\Delta\overline{H}^o$	Molar enthalpic change in sorption of solute at standard state
$\Delta\overline{H}_o$	Molar heat of solution of solute in organic phase
$\Delta\overline{H}_v$	Molar heat of vaporization of compound
$\Delta\overline{H}_w$	Molar heat of solution of solute in water
K	Solute partition coefficient between organic phase and water
K^o	Solute partition coefficient as defined by the hypothetical "ideal line"; see definition after eqs. 4 and 7
K_d	Solute distribution coefficient between soil (or sediment) and water
K_d^*	Apparent K_d, where dissolved organic matter is present in water
K_{doc}	Solute enhancement (or partition) coefficient between dissolved (or suspended) organic carbon and water
K_{dom}	Solute enhancement (or partition) coefficient between dissolved (or suspended) organic matter and water
$K_{em}^{(s)}$	Solute partition coefficient between sorbed microemulsion and water
$K_{em}^{(w)}$	Solute partition coefficient between aqueous microemulsion and water
K_{mc}	Solute partition coefficient between aqueous surfactant micelle and water
K_{mn}	Solute enhancement coefficient between aqueous surfactant monomer and water
K_{oc}	Solute partition coefficient between soil (or sediment) and water, normalized to f_{oc}
K_{oc}^*	Apparent K_{oc}, where dissolved organic matter is present in water
K_{om}	Solute partition coefficient between soil (or sediment) and water, normalized to f_{om}
K_{om}^*	Apparent K_{om}, where dissolved organic matter is present in water
K_{ow}	Solute partition coefficient between *n*-octanol and water
K_{sf}	Solute partition coefficient between sorbed surfactant and water
P	Equilibrium partial pressure of compound at system temperature
P^o	Saturation vapor pressure of compound at system temperature
Q	Uptake of compound (contaminant) by unit mass of soil or sediment
R	Gas constant
S_w	Solubility of solute (liquid or solid) in water
S_w^*	Solubility of solute in water in presence of dissolved organic matter
$S_w^{(l)}$	Solubility of liquid solute or (hypothetical) supercooled liquid solute in water at system temperature
$S_w^{(s)}$	Solubility of solid solute in water at system temperature
S_{om}	Solubility of solute in soil (or sediment) organic matter
$\Delta\overline{S}^o$	Molar entropic change in sorption of solute at standard state
T	System temperature (Kelvin)
T_m	Melting point of solid solute (Kelvin)
$\overline{V}$	Molar volume of solute
$\overline{V}_o^*$	Molar volume of water-saturated organic phase
$\overline{V}_w^*$	Molar volume of solvent-saturated water phase
X	Concentration of dissolved natural organic matter or surfactant in water
X_{em}	Concentration of microemulsion in water
X_{mc}	Concentration of surfactant micelle in water
X_{mn}	Concentration of surfactant monomer in water
γ_o^*	Solute activity coefficient in water-saturated solvent phase (Raoult's law convention)
γ_w	Solute activity coefficient in water
γ_w^*	Solute activity coefficient in solvent-saturated water phase
χ	Flory-Huggins interaction parameter for solute in polymer phase
χ_H	Enthalpic contribution to χ
χ_S	Entropic contribution to χ
ρ	Density of soil/sediment organic matter

ACKNOWLEDGMENT

The use of trade and product names in this article is for identification purposes only and does not constitute endorsement by the U.S. Geological Survey. The author thanks Ms. C. M. Rubin for assistance in manuscript preparation and Mr. D. W. Rutherford for assistance in figure illustration.

BIBLIOGRAPHY

1. N. C. Brady, *The Nature and Properties of Soils,* 8th ed., Macmillian, New York, 1974, pp. 11–14.
2. F. J. Stevenson, in G. R. Aiken, D. M. McKnight, R. L. Wershaw, and P. MacCarthy, eds., *Humic Substances in Soil, Sediment, and Water,* Wiley-Interscience, New York, 1985, pp. 13–29.
3. C. T. Chiou, in P. MacCarthy, C. E. Clapp, and R. L. Malcolm, eds., *Humic Substances in Soil and Crop Sciences: Selected Readings,* Soil Science Society of America, Madison, WI, 1990, pp. 111–160.
4. J. B. Weber, P. W. Perry, and R. P. Upchurch, *Soil Sci. Soc. Am. Proc.* **29,** 678–688 (1965).
5. J. B. Weber, *Am. Mineral.* **51,** 1657–1670 (1966).
6. S. A. Tahoun and M. M. Mortland, *Soil Sci.* **102,** 248–254 (1966).
7. S. B. Weed, *Am. Mineral.* **53,** 478–490 (1968).
8. C. I. Harris and G. F. Warren, *Weeds* **12,** 120–126 (1964).
9. R. C. Chisholm and L. Koblitsky, *J. Econ. Entomol.* **36,** 549–551 (1943).
10. F. L. Stark, Jr., *New York (Cornell) Agr. Exp. Sta. Mem.* **178,** 1–61 (1948).
11. W. J. Hanson and R. W. Nex, *Soil Sci.* **76,** 209–214 (1953).
12. P. I. Wade, *J. Sci. Food Agric.* **5,** 184–192 (1954).
13. B. Yaron and S. Saltzman, *Soil Sci. Soc. Am. Proc.* **36,** 583–586 (1972).
14. C. T. Chiou, T. D. Shoup, and P. E. Porter, *Org. Geochem.* **8,** 9–14 (1985).
15. H. R. Sherburne and V. H. Freed, *J. Agric. Food Chem.* **2,** 937–939 (1954).
16. G. W. Bailey and J. L. White, *J. Agric. Food Chem.* **12,** 324–332 (1964).
17. C. A. I. Goring, *Annu. Rev. Phytopathol.* **5,** 285–318 (1967).
18. S. M. Lambert, *J. Agric. Food Chem.* **16,** 340–343 (1968).
19. J. W. Hamaker and J. M. Thompson, in C. A. I. Goring and J. W. Hamaker, eds. *Organic Chemicals in the Soil Environment,* Vol. I, Marcel Dekker, New York, 1972, pp. 49–143.
20. S. Saltzman, L. Kliger, and B. Yaron, *J. Agric. Food Chem.* **20,** 1224–1226 (1972).
21. K. E. Savage and R. D. Wauchope, *Weed Sci.* **22,** 106–110 (1974).
22. A. B. Filnow, L. W. Jacobs, and M. M. Mortland, *J. Agric. Food Chem.* **24,** 1201–1204 (1976).
23. M. G. Browman and G. Chesters, in I. H. Suffet, ed., *Fate of Pollutants in the Air and Water Environments,* Part I, John Wiley and Sons, New York, 1977, pp. 49–105.
24. A. Felsot and P. A. Dahm, *J. Agric. Food Chem.* **27,** 557–563 (1979).
25. C. T. Chiou, L. J. Peters, and V. H. Freed, *Science* **206,** 831–832 (1979).
26. C. T. Chiou, P. E. Porter, and D. W. Schmedding, *Environ. Sci. Technol.* **17,** 227–231 (1983).
27. C. T. Chiou, in J. Saxena and F. Fisher, eds., *Hazard Assessment of Chemicals: Current Developments,* Academic Press, New York, 1981, pp. 117–153.
28. S. W. Karickhoff, D. S. Brown, and T. A. Scott, *Water Res.* **13,** 241–248 (1979).
29. S. W. Karickhoff, *Chemosphere* **10,** 833–846 (1981).
30. J. C. Means, S. G. Wood, J. J. Hassett, and W. L. Banwart, *Environ. Sci. Technol.* **16,** 93–98 (1982).
31. P. Nkedi-Kizza, P. S. C. Rao, and J. W. Johnson, *J. Environ. Qual.* **12,** 195–197 (1983).
32. M. Leistra, *J. Agric. Food Chem.* **18,** 1124–1126 (1970).
33. E. E. Kenaga and C. A. I. Goring, in J. C. Eaton, P. R. Parrish, and A. C. Hendricks, eds., *Aquatic Toxicology,* Am. Society Testing and Materials, Philadelphia, PA, 1980, pp. 78–115.
34. R. P. Schwarzenbach and J. Westall, *Environ. Sci. Technol.* **15,** 1360–1367 (1981).
35. D. E. Kile, C. T. Chiou, H. Zhou, H. Li, and O. Xu, *Environ. Sci. Technol.* **29,** 1401–1406 (1995).
36. A. R. Swoboda and G. W. Thomas, *J. Agric. Food Chem.* **16,** 923–927 (1968).
37. S. Sun and S. A. Boyd, *J. Environ. Qual.* **20,** 557–561 (1991).
38. D. W. Rutherford, C. T. Chiou, and D. E. Kile, *Environ. Sci. Technol.* **26,** 336–340 (1992).
39. A. C. Mills and J. W. Biggar, *Soil Sci. Soc. Am. Proc.* **33,** 210–216 (1969).
40. W. F. Spencer and M. M. Cliath, *Soil Sci. Soc. Am. Proc.* **34,** 574–578 (1970).
41. R. H. Pierce, Jr., C. E. Olney, and G. T. Felbeck, Jr. 1974, *Geochim. Cosmochim. Acta* **38,** 1061–1073 (1974).
42. K. Schellenberg, C. Leuenberger, and R. P. Schwarzenbach, *Environ. Sci. Technol.* **18,** 652–657 (1984).
43. W. F. Spencer, M. M. Cliath, and W. J. Farmer, *Soil Sci. Soc. Am. Proc.* **33,** 509–511 (1969).
44. C. T. Chiou and T. D. Shoup, *Environ. Sci. Technol.* **19,** 1196–1200 (1985).
45. H. P. Poe, K. T. Valsaraj, L. J. Thibodeaux, and C. Springer, *J. Hazard. Mater.* **19,** 17–32 (1988).
46. J. A. Smith, C. T. Chiou, J. A. Kammer, and D. E. Kile, *Environ. Sci. Technol.* **24,** 676–683 (1990).
47. S. K. Ong and L. W. Lion, *J. Environ. Qual.* **20,** 180–188 (1991).
48. K. D. Pennell, R. D. Rhue, P. S. C. Rao, and C. T. Johnston, *Environ. Sci. Technol.* **26,** 756–763 (1992).
49. C. Thibaud, C. Erkey, and A. Akgerman, *Environ. Sci. Technol.* **27,** 2373–2380 (1993).
50. C. A. Bower and F. B. Gschwend, *Soil Sci. Soc. Am. Proc.* **16,** 342–345 (1952).
51. K. D. Pennell and P. S. C. Rao, *Environ. Sci. Technol.* **26,** 402–404 (1992).
52. J. B. Weber and S. B. Weed, in W. D. Guenzi, ed., *Pesticides in Soil and Water.* American Society of Agronomy, Madison, WI, 1974, pp. 223–256.
53. R. D. Carringer, J. B. Weber, and T. J. Monaco, *J. Agric. Food Chem.* **23,** 568–572 (1975).
54. J. J. Hassett, W. L. Banwart, S. G. Wood, and J. C. Means, *Soil Sci. Soc. Am. J.* **45,** 38–42 (1981).
55. F. J. Stevenson, *Humus Chemistry,* Wiley-Interscience, New York, 1982, pp. 403–419.
56. U. Mingelgrin and Z. Gerstl, *J. Environ. Qual.* **12,** 1–11 (1983).
57. E. M. Murphy, J. M. Zachara, S. C. Smith, J. L. Phillips, and T. W. Wietsma, *Environ. Sci. Technol.* **28,** 1291–1299 (1994).
58. C. T. Chiou, J.-F. Lee, and S. A. Boyd, *Environ. Sci. Technol.* **24,** 1164–1166 (1990).
59. C. T. Chiou, J.-F. Lee, and S. A. Boyd, *Environ. Sci. Technol.* **26,** 404–406 (1992).

60. C. T. Chiou, D. W. Rutherford, and M. Manes, *Environ. Sci. Technol.* **27,** 1587–1593 (1993).
61. K. D. Pennell, S. A. Boyd, and L. M. Abriola, *Soil Sci. Soc. Am. J.* **59,** 1012–1018 (1995).
62. S. Brunauer, *The Adsorption of Gases and Vapors,* Princeton University Press, Princeton, NJ, 1945.
63. G. S. Hartley, in E. K. Woodford and G. R. Sagar, eds., *Herbicides and the Soil,* Blackwell, Oxford, 1960, pp. 63–78.
64. S. M. Lambert, *J. Agric. Food Chem.* **15,** 572–576 (1967).
65. G. G. Briggs, *J. Agric. Food Chem.* **29,** 1050–1059 (1981).
66. J. C. Means S. G. Wood, J. J. Hassett, and W. L. Banwart. *Environ. Sci. Technol.* **14,** 1524–1528 (1980).
67. P. J. McCall, D. A. Laskowski, R. L. Swann, and H. J. Dishburger, *Residue Rev.* **85,** 231–244 (1983).
68. D. A. Dzombak and R. G. Luthy, *Soil Sci.* **137,** 292–308 (1984).
69. C. T. Chiou, L. J. Peters, and V. H. Freed, *Science* **213,** 684 (1981).
70. C. T. Chiou, P. E. Porter, and T. D. Shoup, *Environ. Sci. Technol.* **18,** 295–297 (1984).
71. C. T. Chiou, D. W. Schmedding, and M. Manes, *Environ. Sci. Technol.* **16,** 4–10 (1982).
72. J. J. Kipling, *Adsorption from Solutions of Nonelectrolytes,* Academic Press, London, 1965.
73. A. W. Adamson, *Physical Chemistry of Surfaces,* 2nd ed., Wiley-Interscience, New York, 1967.
74. M. Manes, in I. H. Suffet and M. J. McGuire, eds., *Activated Carbon Adsorption of Organics from the Aqueous Phase,* Vol. I, Ann Arbor Science, Ann Arbor, MI, 1980, pp. 43–63.
75. S. J. Gregg and K. S. W. Sing, *Adsorption, Surface Area and Porosity,* 2nd ed., Academic Press, London, 1982.
76. D. H. Everett, in R. H. Ottweill et al., eds., *Adsorption from Solution,* Academic Press, London, 1983, pp. 1–20.
77. D. Mackay, *Environ. Sci. Technol.* **11,** 1219 (1977).
78. C. T. Chiou, V. H. Freed, D. W. Schmedding, and R. L. Kohnert, *Environ. Sci. Technol.* **11,** 475–478 (1977).
79. D. Mackay, A. Bobra, W. Y. Shiu, and S. H. Yalkowsky, *Chemosphere* **9,** 701–711 (1980).
80. M. M. Miller, S. P. Wasik, G.-L. Huang, W.-Y. Shiu, and D. Mackay, *Environ. Sci. Technol.* **19,** 522–529 (1985).
81. F. C. Spurlock and J. W. Biggar, *Environ. Sci. Technol.* **28,** 996–1002 (1994).
82. P. J. Flory, *J. Chem. Phys.* **10,** 51–61 (1942).
83. P. J. Flory, *Principles of Polymer Chemistry,* Cornell University Press, Ithaca, NY, 1953, pp. 495–520.
84. M. L. Huggins, *Ann. N.Y. Acad. Sci.* **43,** 1–32 (1942).
85. C. T. Chiou and M. Manes, *J. Chem. Soc. Faraday Trans. 1,* **82,** 243–246 (1986).
86. R. L. Scott, *J. Chem. Phys.* **17,** 268–279 (1949).
87. C. T. Chiou, *Environ. Sci. Technol.* **19,** 57–62 (1985).
88. W. G. MacIntyre and C. L. Smith, *Environ. Sci. Technol.* **18,** 295 (1984).
89. C. Plato and R. Glasgow, *Anal. Chem.* **41,** 330–336 (1969).
90. Y.-O. Shin, J. J. Chodan, and A. R. Wolcott, *J. Agric. Food Chem.* **18,** 1129–1133 (1970).
91. L. Weil, G. Dure, and K.-E. Quentin, *Z. Wasser Abwasser Forch.* **7,** 169–175 (1974).
92. C. T. Chiou and D. E. Kile, *Environ. Sci. Technol.* **28,** 1139–1144 (1994).
93. R. F. Boucher and G. F. Lee, *Environ. Sci. Technol.* **6,** 538–543 (1972).
94. E. C. Yost and M. A. Anderson, *Environ. Sci. Technol.* **18,** 101–109 (1984).
95. R. Haque and R. Sexton, *J. Colloid Interface Sci.* **27,** 818–827 (1968).
96. S. A. Boyd, M. D. Mikesell, and J.-F. Lee, in B. L. Sawhney and K. Brown, eds., *Reactions and Movement of Organic Chemicals in Soils,* Special Publ. 22, Soil Science Society of America, Madison, WI, 1989, pp. 209–228.
97. F. C. Spurlock and J. W. Biggar, *Environ. Sci. Technol.* **28,** 1003–1009 (1994).
98. S. A. Boyd, *Soil Sci.* **134,** 337–343 (1982).
99. C. T. Chiou, *Environ. Sci. Technol.* **29,** 1421–1422 (1995).
100. P. M. Gschwend and S.-C. Wu, *Environ. Sci. Technol.* **19,** 90–95 (1985).
101. C. T. Chiou, S. E. McGroddy, and D. E. Kile, *Environ. Sci. Technol.* **32,** 264–269 (1998).
102. R. L. Wershaw, P. J. Burcar, and M. C. Goldberg, *Environ. Sci. Technol.* **3,** 271–273 (1969).
103. C. W. Carter and I. H. Suffet, *Environ. Sci. Technol.* **16,** 735–740 (1982).
104. C. T. Chiou, R. L. Malcolm, T. I. Brinton, and D. E. Kile, *Environ. Sci. Technol.* **20,** 502–508 (1986).
105. C. T. Chiou, D. E. Kile, T. I. Brinton, R. L. Malcolm, J. A. Leenheer, and P. MacCarthy, *Environ. Sci. Technol.* **21,** 1231–1234 (1987).
106. M. Schnitzer and S. U. Kahn, *Humic Substances in the Environment.* Marcel Dekker, New York (1972).
107. J. Buffle, P. Deladoey, and W. Haerdi, *Anal. Chim. Acta* **101,** 339–357 (1978).
108. A. W. Underdown, C. H. Langford, and D. S. Gamble, *Anal. Chem.* **53,** 2139–2140 (1981).
109. E. M. Thurman, R. L. Wershaw, R. L. Malcolm, and D. J. Pinckney, *Org. Geochem.* **4,** 27–35 (1982).
110. G. R. Aiken and R. L. Malcolm, *Geochim. Cosmochim. Acta* **51,** 2177–2184 (1987).
111. S. W. Karickhoff, *J. Hydraul. Eng.* **110,** 707–735 (1984).
112. G. Caron, I. H. Suffet, and T. Belton, *Chemosphere* **14,** 993–1000 (1985).
113. C. T. Chiou, D. E. Kile, and R. L. Malcolm, *Environ. Sci. Technol.* **22,** 298–303 (1988).
114. R. L. Malcolm and P. MacCarthy, *Environ. Sci. Technol.* **20,** 904–911 (1986).
115. D. M. Di Toro, *Chemosphere* **14,** 1503–1538 (1985).
116. D. E. Kile and C. T. Chiou, *Environ. Sci. Technol.* **23,** 832–838 (1989).
117. D. E. Kile, C. T. Chiou, and R. S. Helburn, *Environ. Sci. Technol.* **24,** 205–208 (1990).
118. S. Sun, W. P. Inskeep, and S. A. Boyd, *Environ. Sci. Technol.* **29,** 903–913 (1995).
119. C. T. Cowan and D. White, *Trans. Faraday Soc.* **54,** 691–697 (1958).
120. B. K. G. Theng, D. J. Greenland, and J. P. Quirk, *Clay Miner.* **7,** 1–17 (1967).
121. R. M. Barrer, *Zeolites and Clay Minerals as Sorbents and Molecular Sieves,* Academic Press, London, 1978, pp. 407–486.
122. J.-F. Lee, J. R. Crum, and S. A. Boyd, *Environ. Sci. Technol.* **23,** 1365–1372 (1989).
123. S. Xu and S. A. Boyd, *Soil Sci. Soc. Am. J.* **58,** 1382–1391 (1994).
124. S. Xu and S. A. Boyd, *Environ. Sci. Technol.* **29,** 312–320 (1995).
125. S.-A. Boyd, M. M. Mortland, and C. T. Chiou, *Soil Sci. Soc. Am. J.* **52,** 652–657 (1988).
126. J. A. Smith, P. R. Jaffé, and C. T. Chiou, *Environ. Sci. Technol.* **24,** 1167–1172 (1990).

127. J. D. Rouse, D. A. Sabatini, and J. H. Harwell, *Environ. Sci. Technol.* **27**, 2072–2078 (1993).
128. S. Sun and S. A. Boyd, *Environ. Sci. Technol.* **27**, 1340–1346 (1993).
129. J. J. Deitsch and J. A. Smith, *Environ. Sci. Technol.* **29**, 1069–1080 (1995).
130. R. J. Hance, *Weed Res.* **5**, 108–114 (1965).
131. N. A. Shipinov, *Bull. Plant Protection (U.S.S.R.)* No. **1–2**, 192–199 (1940).
132. J. J. Jurinak, *Soil Sci. Soc. Am. Proc.* **21**, 599–602 (1957).
133. J. J. Jurinak, *J. Agric. Food Chem.* **5**, 598–600 (1957).
134. J. J. Jurinak and D. H. Volman, *Soil Sci.* **83**, 487–496 (1957).
135. F. Call, *J. Sci. Food Agric.* **8**, 630–639 (1957).
136. D. W. Rutherford and C. T. Chiou, *Environ. Sci. Technol.* **26**, 965–970 (1992).
137. W.-W. Choi and K. Y. Chen, *Environ. Sci. Technol.* **10**, 782–786 (1976).
138. D. F. Goerlitz, D. E. Troutman, E. M. Godsy, and B. J. Franks, *Environ. Sci. Technol.* **19**, 955–961 (1985).
139. C. T. Schmidt, *J. Econ. Entomol.* **40**, 829–837 (1947).
140. W. F. Meggitt, H. A. Borthwock, R. J. Aldrich, and W. C. Shaw, *Proc. Northeast. Weed Control Conf.* **8**, 21 (1954).
141. R. L. Gantz and F. C. Slife, *Weeds* **8**, 599–606 (1960).
142. C. R. Harris and E. P. Lichtenstein, *J. Econ. Entomol.* **54**, 1038–1045 (1961).
143. J. M. Deming, *Weeds* **11**, 93–95 (1963).
144. M. C. Bowman, M. S. Schechter, and R. L. Carter, *J. Agric. Food Chem.* **13**, 360–365 (1965).
145. R. A. Gray and A. J. Weierich, *Weeds* **13**, 141–147 (1965).
146. J. V. Parochetti and G. F. Warren, *Weeds* **14**, 218–284 (1966).
147. R. B. Taylorson, *Weeds* **14**, 294–296 (1966).
148. R. Grover, L. A. Kerr, and S. U. Khan, *J. Agric. Food Chem.* **29**, 1082–1084 (1981).
149. R. Grover, A. E. Smith, S. R. Shewchuk, A. J. Cessna, and J. H. Hunter, *J. Environ. Qual.* **17**, 543–550 (1988).
150. D. E. Glotfelty, A. W. Taylor, B. C. Turner, and W. E. Zoller, *J. Agric. Food Chem.* **32**, 638–643 (1984).
151. M. Majewski, R. Desjardins, P. Rochette, E. Pattey, J. Seiber, and D. Glotfelty, *Environ. Sci. Technol.* **27**, 121–128 (1993).
152. R. P. Upchurch, *Weeds* **5**, 112–120 (1957).
153. F. Barlow and A. B. Hadaway, *Bull. Entomol. Res.* **46**, 547–559 (1955).
154. C. R. Harris, *Nature (Lond.)* **202**, 724 (1964).

SOIL THERMAL DECONTAMINATION: FUNDAMENTALS

HIROSHI H. SAITO
JACK B. HOWARD
WILLIAM A. PETERS
Massachusetts Institute of Technology
Cambridge, Massachusetts

VERÓNICA BUCALÁ
Planta Piloto de Ingenieria Química, 12 de Octubre 1842
Bahía Blanca, Argentina

Technologies based on thermal treatment are already established for remediating contaminated soils. Per ton costs of soil thermal decontamination vary from as low as \$35 to several hundred dollars. Public concerns over off-specification performance of, and adverse emissions from, environmental cleanup technologies have motivated research on thermal cleaning of contaminated soils. Diverse equilibrium and kinetic processes affect soil thermal decontamination because of the chemical and physical complexity of soil itself and of soil-contaminant mixtures, especially as they undergo thermal treatment. Prior research has investigated some of these processes and elucidated mechanisms for contaminant release and by-product generation. Despite these valuable contributions, there is need for a better understanding of engineering fundamentals for the design, operation, monitoring, and diagnostics of *in situ* and aboveground soil thermal treatment processes. This article selectively reviews recent experimental and theoretical studies that contribute to this understanding.

INTRODUCTION TO SOILS AND THERMAL DECONTAMINATION OF SOIL

Soil (1–9) is a finely divided rock-derived material mixed with organic matter. Soil generally occurs in nature as thick or thin horizons (or layers), more or less parallel to the surface. Each horizon differs from its neighbors in one or more properties, such as color, texture, structure, consistency, porosity, and chemical reactivity. Soil is formed by the accumulation of unconsolidated rock fragments (or parent rock) and subsequent formation of horizons. Natural processes and human intervention affect the formation and composition of soil and soil horizons. Natural phenomena arise from the action of glaciers and other geological processes, climate, plants and animals, groundwater, wind, and gravity. They change soil physically (eg, decrepitation, suspension in, or inertial deposition from, aqueous media), and chemically (eg, by oxidation due to weathering, mineral dissolution in, or precipitation from, aqueous solutions). Soil composition and structure depend on the intensity and duration of natural processes and on human interactions with the soil. Human interactions entail beneficial utilization (eg, agriculture, construction), unintended abuse (eg, accidental spills of hazardous material), or intentional misuse (eg, premeditated dumping of wastes).

Soil composition and morphology varies significantly from site to site and within a site. Figure 1 summarizes the major classes of organic and inorganic compounds frequently encountered in soil. The mineral and organic fractions of soils are complex mixtures of a large number of chemical compounds reflecting their genesis in natural materials and processes. Soil organic matter is chemically complicated, which is in part a consequence of its origin in metabolic products and tissues of plants, animals, and microorganisms. Soil inorganic matter typically contains several compounds belonging to a range of rock and mineral classes, although various silicate minerals often dominate. Morphologically, soil particles vary widely in size. Individual particles exhibit a broad range of pore sizes, because they are formed from unconsolidated rock fragments. Organic and inorganic ingredients of soil decompose or otherwise react at sufficiently high temperatures (see below) or in reactive atmospheres.

Thermal treatment is already an established technology for removing organic contaminants from soils and shows significant promise for future applications to aboveground and *in*

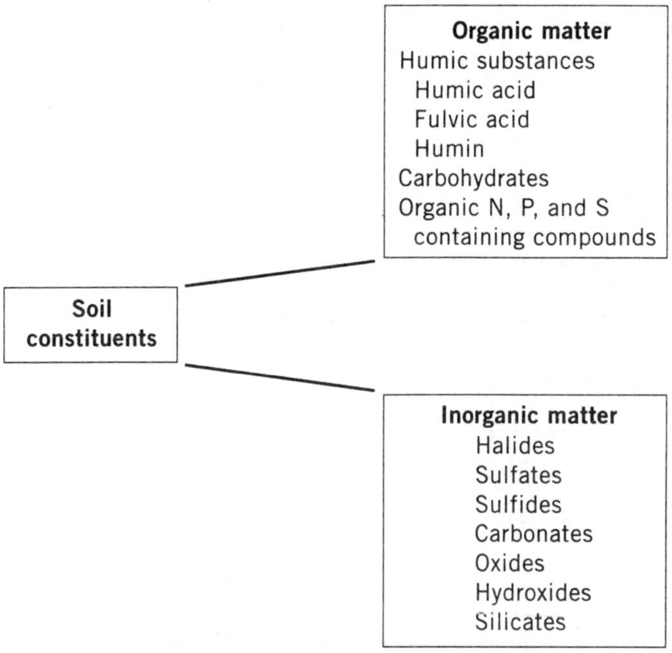

Figure 1. Typical constituents of soils. Reprinted with permission (1–9). Courtesy of the American Chemical Society.

situ soil cleanup. Thus the underlying details of how clean and contaminated soils behave when they are heated have practical and scientific relevance. Of particular interest are the fundamental chemical and physical processes responsible for contaminant separation or destruction. In the United States, polluted soils are found at various locations including federal and state Superfund sites and RCRA sites. In a 1991 document, the U.S. EPA projected an increase in the number of Superfund sites by 50–100 annually (10). By 1994 federal and state Superfund sites totalled over 1320 (11).

Versatility in handling a wide range of organic contaminants and extensive past experience with similar systems are two of the factors which have led to selecting thermal treatment for remediation at numerous contaminated sites. Well-designed and operated thermal destruction systems like incineration provide high overall levels of destruction for many hazardous waste streams (12). Destruction or removal efficiencies (DRE'S) as high as 99.9999% (13,14) are achievable for specific hazardous organics. Per ton costs of soil thermal treatment vary from $35 to $300–$700 (15). Public concerns about the health and environmental impacts of process emissions (16), about contaminant retention in the soil residues, and about the generation of other toxicants during normal operations and from process upsets (17,18) have contributed to poor public acceptance of incineration. Similar worries may create or sustain public resistance to other thermal destruction methodologies. Soil cleanup technology is evolving. As more sites have been evaluated and cleaned up using a variety of techniques, more has been learned about the cleanup process. However, imperfect technical understanding and surprises encountered at cleanup sites make evaluating and adopting remediation technologies challenging (13). The quests for additional technical understanding, lower costs, and broader public acceptance, motivate research to better understand the chemical and physical foundations of soil thermal treatment. Such understanding will enrich the scientific and engineering bases for system design, scaleup, operation, monitoring, and diagnostics. Potential performance benefits include more sustainable and efficient operation, better emissions management, and lower treatment costs.

FUNDAMENTAL CHEMICAL AND PHYSICAL PROCESSES IN THERMAL DECONTAMINATION

Qualitative Introduction to Particle-Based Phenomena

Regardless of flow sheet and equipment, the chemical or physical removal of foreign pollutant species from various locations within individual soil particles is common to all thermal remediation processes. The ability to achieve high levels of contaminant removal from each soil particle without generating hazardous by-products is an important metric of a technically successful treatment process. Thus, a length scale related to soil particle size proves useful in studying the fundamental processes of soil thermal decontamination. From this vantage point, the analysis can be expanded to account for more complex effects from clumps, beds, or other ensembles of individual soil particles and to process-related issues, such as mixing and heat transmissions over length scales comparable to process vessel dimensions. Contaminant transport, desorption, readsorption, and chemical reaction in packed soil beds (above-ground or *in situ*) are examples of multiparticle phenomena.

Each soil particle can be viewed simply as a porous solid with macropores and micropores. The pore walls and particle material consist of a complex mixture of organics and inorganics, some of which degrade at elevated temperatures, changing particle structure and chemical composition during the treatment process. Adding or removing contaminants (either single component or complex mixtures) modifies soil and perhaps complicates soil composition and morphology. Acid leaching of metals from soil, by acidic precipitation is well known (19,20). In some studies it has been assumed that low inventories of up to a few 100 ppm of low molecular weight contaminants, such as BTXs or trichlorethylene, cause minimal chemical or physical disturbance of the original soil. However, it is known that contaminant distribution within individual soil particles depends on the contamination route (eg, spilling vs adsorption from groundwater) and on the length of exposure. If the time between a spill and cleanup is short, the contaminant(s) have only enough time to diffuse into the macropore network of the soil particles. In longer exposures, the contaminant may distribute throughout the particles in both micro- and macropores. For high levels of contamination by high molecular weight organic materials (eg, coal tars), it is possible that the pore structures in soil particles contain and also are occluded by the contaminant. Such effects are reasonable in the light of reported concentrations of organic contaminants in soil, ranging from hundreds of ppm (21) to several weight percent (22). Hyper-contamination of soil has been found at former gas manufacturing plants as free-standing pools of coal-derived tars (23).

A simplified but instructive picture of soil decontamination is obtained by considering the behavior of a pollutant molecule initially adsorbed on a pore wall deep within a spatially isothermal soil particle. Increasing temperature and ensuing gradients in pollutant concentration (rigorously chemical potential) desorb the molecule from the pore wall and then release it from the particle surface via transport through one or more interconnected pores. Various complications arise of course. For example, a contaminant may undergo multiple re-adsorptions and desorptions on pore walls before escaping the particle. As the soil particle temperature increases, chemical transformation of the contaminant is more likely. A favorable result is contaminant breakdown to less toxic and or more volatile substances. More volatile substances are less prone to readsorption on pore walls and more easily transported within and away from the particle. Contaminant-derived volatiles facilitate transport of unreacted contaminant from the particle by creating a higher intraparticle driving pressure for bulk flow or by acting as a sweep gas. Contaminants reactions may also have adverse consequences. For example pyrolysis or other reactions (in the gas phase or during contact with the pore wall) can transform the contaminant into less volatile, higher molecular weight (MW) species, which are harder to expel from the soil interstices. Molecular weight growth also produces tars or even solid carbon which, if generated in sufficient amounts, blocks pores and hinders further decontamination. Carbon is formed by thermal reactions of aromatics, even at rather mild temperatures (300–400°C) if alkaline-earth metal oxides are present to act as promoting agents (24–26). Evidence for thermal decomposition of organics facilitated by calcium oxide and by clays and other minerals at temperatures and intraparticle residence times similar to soil thermal treatment is also found in the literature on the effects of native and exogenous mineral matter on coal pyrolysis (27–30). Another, beneficial complication is chemical or physical processes transforming the particle morphology to a more open structure, presenting less internal resistance to contaminant mass transfer.

A summary of many of the above phenomena is presented in Figure 2. As noted, potentially confounding factors are catalytic effects from soil minerals, modifications in soil particle morphology, and spatial temperature gradients. These last may arise when treating larger soil particles or using rapid surface heating rates (31). Decontamination modeling must therefore account for intraparticle variations in temperature-sensitive physicochemical properties, such as transport coefficients, phase equilibria (eg, vapor–solid partition coefficients), and chemical rate constants. It may also be necessary to solve coupled equations of change for heat, mass, and momentum transfer.

In practical thermal treatment systems, operating conditions depend on the equipment and soils being treated. Temperatures range from approximately 150°C in thermal desorption units to over 1200°C in rotary kiln incinerators. Operating pressures range from subambient in vacuum extraction systems to high pressure (eg, ≥25 MPa) in supercritical water oxidation systems. The contaminated material being treated is soil or sediment within a wide range of natural organic content and physical porosity. Organic contaminants are complex mixtures spanning wide ranges of molecular weight and volatility. As each factor potentially influences contaminant removal, the relative contributions of the above processes depend on treatment conditions, soil type, and contaminant properties. For example, when treating beds of soil, readsorption of volatilized contaminants from one layer to the next, effects of soil bed agitation and of soil particle breakage should be considered.

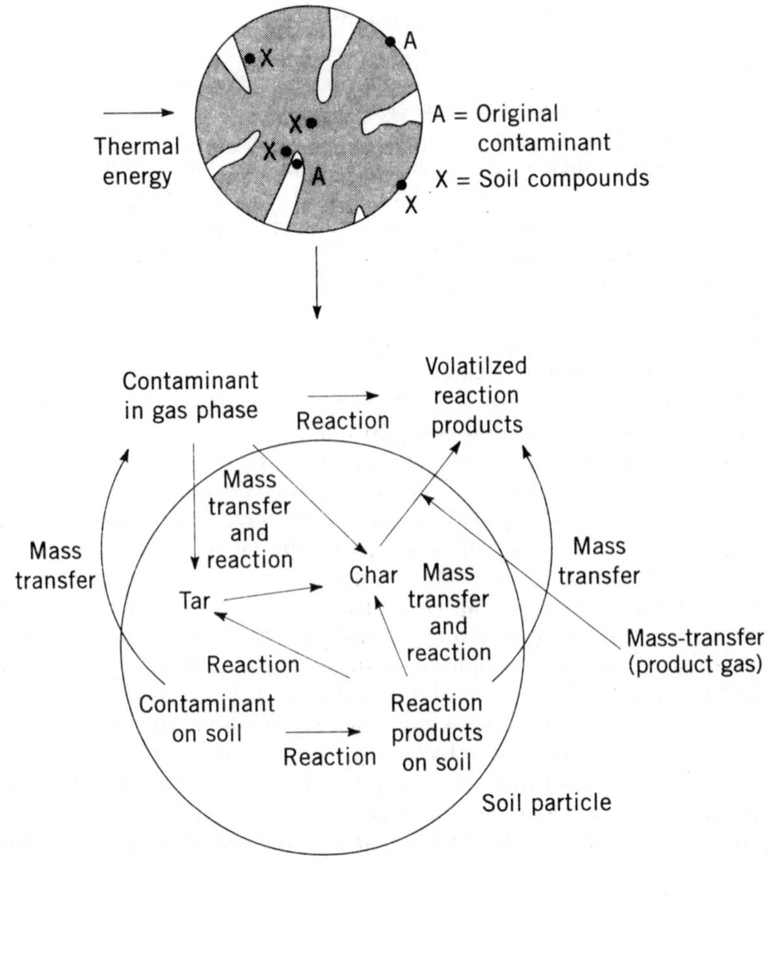

Figure 2. Possible pathways for transport and reaction of soil contaminants during thermal treatment.

Brief Overview of Inorganic and Organic Soil Constituents

Soils containing less than 12–20% organic carbon are termed mineral, and the rest are denoted organic soil. Although not a complete speciation, the inorganic fraction of soil is largely composed of four major mineral classes (32). One group consists of halide, sulfate, and carbonate minerals characterized by simple chemical structures, physical softness, and, for some components, solubility in water. Halite (NaCl), gypsum ($CaSO_4 \cdot 2H_2O$), calcite ($CaCO_3$), and dolomite ($CaMg(CO_3)_2$) are important entries in this class. The second group is sulfides of which pyrite (FeS_2) is the most common. Pyrite is usually not present in large quantities in most soils. Pyrite is unstable in oxidizing environments. Its weathering products include the minerals jarosite ($KFe_3(SO_4)_2(OH)_6$) and gypsum and sulfuric acid, which accounts for the acidic runoff from pyritic residues of mining and drainage of sulfide-containing areas. The third mineral group is comprised of oxides, hydroxides, and oxyhydroxides, eg, gibbsite ($Al(OH)_3$), geothite (FeOOH), hematite (Fe_2O_3), manganese oxides and hydroxides, and titanium dioxide, formed from cations and anions released by breakdown of other minerals during weathering. The fourth mineral group consists of silicates. This is a very large and important class, which accounts for nearly 40% of the common minerals, well over 90% of the Earth's crust, and the bulk of most soils. In sand and silt (75-μm–4.75-mm and 2–20-μm sized fractions of soils, respectively) minerals, such as olivines, pyroxenes, amphiboles, micas, feldspars, and quartz preponderate, whereas phyllosilicate minerals dominate the clay fraction ($\leq 2\ \mu$m).

Organic concentration varies from 0.1% in desert soils to nearly 100% in some organic soils (33) and is affected by site-specific factors such as drainage and clay content (34). As a benchmark, the average organic content of most good agricultural soils is about 1–5% (35). Soil organic matter, or humus, is one of the more chemically complex materials found in nature (1). Humus contains plant and animal residues at various stages of decomposition, cells and tissues of soil organisms, and substances synthesized by soil microbes (35). Humus is formed through numerous complex biochemical processes characteristic of decomposing organic matter (2,35), beginning simply with fracturing of the cell walls (2). Humus is generally a complex mixture of organic compounds consisting of two major groups of polymers: (*1*) "humic substances" ie, the more resistant fats, waxes, lignins, and oils derived from residues of higher plants, and (*2*) microorganism-synthesized compounds, such as polysaccharides and polyuronides. Humic substances constitute 70–80% of soil organic matter, and polysaccharides account for most of the remainder (36). Carbohydrates, proteins, peptides, amino acids, resins, pigments, more biochemically active fats and waxes, and other low molecular weight organic substances are also present but in very small quantities because they are easily attacked by microorganisms in soil and have relatively low survival rates (37).

Humic substances have been studied by soil scientists for 200 years (37–39), but much remains to be discovered regarding their origin, synthesis, chemical structure, reactions, and functions in the environment (33,34). Because soil organics are often tightly bound in soils, various techniques are employed to extract these substances and to analyze the extracts (33,38,39). Structures now hypothesized for humic substances are based on a classical extraction technique using dilute base and acid solutions. The extractions classify humic substances into three categories: humic acid, fulvic acid, and humin. The soil is initially extracted with an alkali to separate out the humin (unextractable fraction, tightly bound to soil), and the extract is treated with an acid to separate the humic acid and fulvic acid fractions. Scanning electron micrographs show that, in aqueous solutions, humic and fulvic acids are elongated, multi branched filaments 20–100 nm in width (33,36). Chemical analyses and studies of these fractions by a wide variety of analytical techniques, eg, elemental analysis, infrared spectroscopy, X-ray analysis, size-exclusion gel filtration, chemical oxidation, pyrolysis-gas chromatography, biodegradation (32,33,38,39) have led to proposals for the chemical structures of humic (38,40) and fulvic (33,38) acids. However, there is not broadly based agreement on the comprehensive chemical structures of humic and fulvic acids and of humin. Elemental analysis and functional group chemical analyses have yielded some information on humin (37). Optical examination (2) shows a substantial amount of the humin enclosed in microaggregates in the soil, the presence of polysaccharides and other biosynthesized products such as resins, waxes, and lipids that account for more than 10% of the humin, and the presence of products irreversibly bound to soil constituents, such as clays.

Thermal Chemistry of Soil

Pyrolysis occurs in all thermal treatment processes. Thus the literature on soil pyrolysis is a reasonable starting point for understanding soil behavior in practical thermal treatment technologies. Oxygen effects, when important are considered in terms of (*1*) the influence of oxygen on pyrolysis itself and/or (*2*) the impact of oxygen on the products of pyrolysis. One study (3) measured total weight loss and yields of tars and selected light gases (CO, CO_2, C_1–C_2 hydrocarbons) from rapid heating (nominal rates 1000°C/s) of an uncontaminated EPA Synthetic Soil Matrix in the temperature range 350–1050°C under 122 kPa (3 psig) helium. Carbon dioxide, CO, and tars were the major volatiles. It was hypothesized that (*1*) decomposition of carbonate minerals was an important source of CO_2; (*2*) back reaction of carbonate-derived CO_2 with organic carbon present in the soil contributes to CO production; and (*3*) the modest quantities of hydrocarbons are produced from the small organic fraction in the soil. Material and carbon balances in this study were good. Oxygen balances were fair and hydrogen balances were poor, possibly because yields of chemical water were not determined. This work further reveals that a multiple independent parallel reaction (MIPR) kinetic model satisfactorily describes total weight loss as affected by temperature and that release rates of several individual volatiles are quite well described by a single reaction, or for CO_2, two, independent, parallel, first-order reaction, Arrhenius kinetic models.

Understanding of soil pyrolysis is also obtained from pyrolysis-mass spectrometry studies of the structure of soil organic matter. Whole samples of an agricultural soil and a forest soil and of their organic fractions (humic acid, fulvic acid, and humin), were pyrolyzed in another study (40,41). Using a field ionization mass spectrometer, the study identified compound classes (eg, phenols, fatty acids, methyl naphthalenes) in the soil and in individual soil organic fractions. The research (42) identified over 200 compounds in the humic acid fraction of these soils, including *n*-alkanes, *n*-alkenes, aromatics and their alkylated derivatives, alcohols and phenols, and ketones. Earlier work (43) with humic soils also suggests the formation of organic acids, esters, and lignins during pyrolysis. Using analytical pyrolysis–gas chromatography/mass spectrometery (44), over 300 compounds were identified from pyrolisis of soil organic fractions from a Spanish brown soil, including light gases (carbon monoxide, carbon dioxide, methane, ethylene, ethane, propane, and higher molecular weight aliphatic hydrocarbons), alcohols (*n*-alkyl, phenols, aromatic), aromatics (unsubstituted, alkylated), fatty acids, and fatty acid esters among many others. Analytical pyrolysis studies of humic substances have been performed (45,46). Soil pyrolysis data are relatively limited, and appropriate diligence should be exercised in adapting these data to different equipment and treatment conditions and to other soils or contaminants (47).

Many common minerals, such as carbonates and clays, undergo chemical transformations at elevated temperatures. For example, calcite and dolomite evolve carbon dioxide to form their respective oxides (CaO and CaO · $(MgO)_x$) at elevated temperatures. At thermodynamic equilibrium, the partial pressure of CO_2 from the dissociation of calcite and dolomite is 1 atm at temperatures of 898 and 402°C, respectively (48). However these minerals dissociate at lower temperatures under CO_2 pressures below 101.3 kPa (1 atm). Clays also undergo chemical changes over a wide temperature range (49), eg, loss of chemical water below 400°C, dehydroxylation between 400 and 750°C, phase transitions above 750°C, and metal ion oxidation reactions (even without external oxygen) above 400°C, depending on specific mineral composition.

Thermal Chemistry of Selected Organic Contaminants of Soils

Chemical reactions of exogenous contaminants are important when the characteristic reaction times are comparable to or less than the residence time of the pollutant within the particle. Thus chemical reaction effects are increasingly significant with increasing temperature, eg, for lower volatility (ie, >two-ring) aromatic hydrocarbons as temperatures rise above about 600°C. Nonnegligible chemical effects can arise at significantly lower temperatures (eg, down to 300°C) if soil minerals or other soil constituents (including inorganic contaminants) catalyze or otherwise accelerate reactions (see discussion above). Space limitations preclude a comprehensive review of all relevant chemistry. To illustrate possible effects, reactions of two relatively simple prototypical soil contaminants, ie, toluene (a low MW, relatively high volatility single-ring aromatic hydrocarbon) and pyrene (a high MW, low volatility, polycyclic aromatic hydrocarbon (PAH) are discussed. The gas-phase (toluene and pyrene) and liquid-phase (pyrene) pyrolysis information used in these illustrations is instructive for soil because similar reactions could occur within soil pores and on soil surfaces.

Gas phase pyrolysis of toluene occurs at temperatures encountered in soil thermal treatment, eg, at temperatures as low as 700°C (50). Toluene thermally decomposes to a phenyl radical and a methyl radical or to a benzyl radical which also decomposes to produce C_2–C_5 compounds and radicals. PAHs and soot arise from addition of acyclic fuel fragments, such as acetylene (51), and by addition of aryls (52). Early experimenters, such as Berthelot, identified benzene, naphthalene, bibenzyl and anthracene as products of toluene pyrolysis (53). These products as well as styrene and phenanthrene were observed later on by Ferko (53). The formation of methane and hydrogen during toluene pyrolysis was confirmed (54). In an investigation (53), the pyrolysis of toluene at 700°C produced a single tar phase in which 23 products were identified in addition to toluene. Reaction pathways, such as dimerization, cyclodehydrogenation, and rearrangements explain the presence of many of the high MW products. Effects of pressure (1.333×10^{-6}–0.1333 kPa (10^{-5}–1 torr)) studied at low pressures to temperatures up to 1800°C showed products lower in MW than toluene (CH_3, C_2H_2, C_3H_3, C_4H_4, C_5H_5, C_6H_6, and C_7H_7) (55). At somewhat higher pressures, higher MW compounds up to at least $C_{20}H_{12}$ were observed. Toluene was pyrolyzed in a drop tube furnace in the 800–1500 K temperature range, and soot and tars exiting the reactor were collected in a filter (56). In the tar extracted from the filter by methylene chloride, PAHs ranged from C_{10} to C_{26} (aromatics smaller than C_{10} could not be collected with this apparatus). The residual solid remaining in the filter after extraction was soot. Based on these data, aromatic growth pathways were hypothesized (Fig. 3). At 1100 K, pyrolytic growth was quenched at an early stage, where 90% of the aromatic products had 16 carbons or less. Bibenzyl was the most abundant product identified. Small amounts of dimerization were observed, leading to the formation of benzyl toluenes and dimethyl biphenyls, which could then cyclodehydrogenate to products, such as anthracene and phenanthrene. As the temperature increased, aromatic compounds with 14, 16, 18, and 20 carbons were abundant. Addition of C_2 and C_4 carbon fragments were proposed as the pathways leading to these products (56). Isomerization of PAHs also appeared to be important. Aryl–aryl reactions were proposed to account for the significant numbers of PAHs with odd numbers of carbons that were also observed. Kinetic parameters were experimentally obtained for reaction pathways relevant to toluene pyrolysis: toluene and benzyl dissociation and the reaction of H atoms with toluene at temperatures from 1200 to 1800 K (57). These results were combined with lower temperature data to derive kinetic parameters for the 600–1800 K temperature range (57).

Pyrolysis of higher MW PAHs, such as pyrene, is now considered. One study (58) reported on the pyrolysis of liquid-phase pyrene in sealed glass capillary tubes at 475°C for one hour. Pyrene exhibited little breakdown to lower MW compounds like benzene, and no heavy residues were found in the tubes following heating. A small amount of gaseous products formed (moles gas collected to moles pyrene initially charged was 0.0011), but the amount was too small to be analyzed. In a later study by (59) using a comparable technique under identical heating conditions, pyrene pyrolysis products were identified by mass spectrometry. Soot and heavy residues, such as sludge-like heavy tars or products associable with aromatic ring-breaking reactions, eg, benzene, naphthalene, were not observed. Gaseous products, which made up 0.01 wt % of the initial charge, were predominantly hydrogen, and the remainder consisted of methane, ethane, ethylene, propane, and propylene. The dominance of hydrogen in the product gases supports the proposition that intermolecular condensation to eliminate hydrogen and form bipyrenes is a major pyrolytic reaction. Dihydropyrene and dipyrenyl were tentatively identified in the residues, which constituted 2.2% (wt) of the initial charge. Another study (60) observed a large number of products grouped into bipyrenes, bipyrene condensation products, fragmentation products, and soot from pyrolysis of pyrene in a

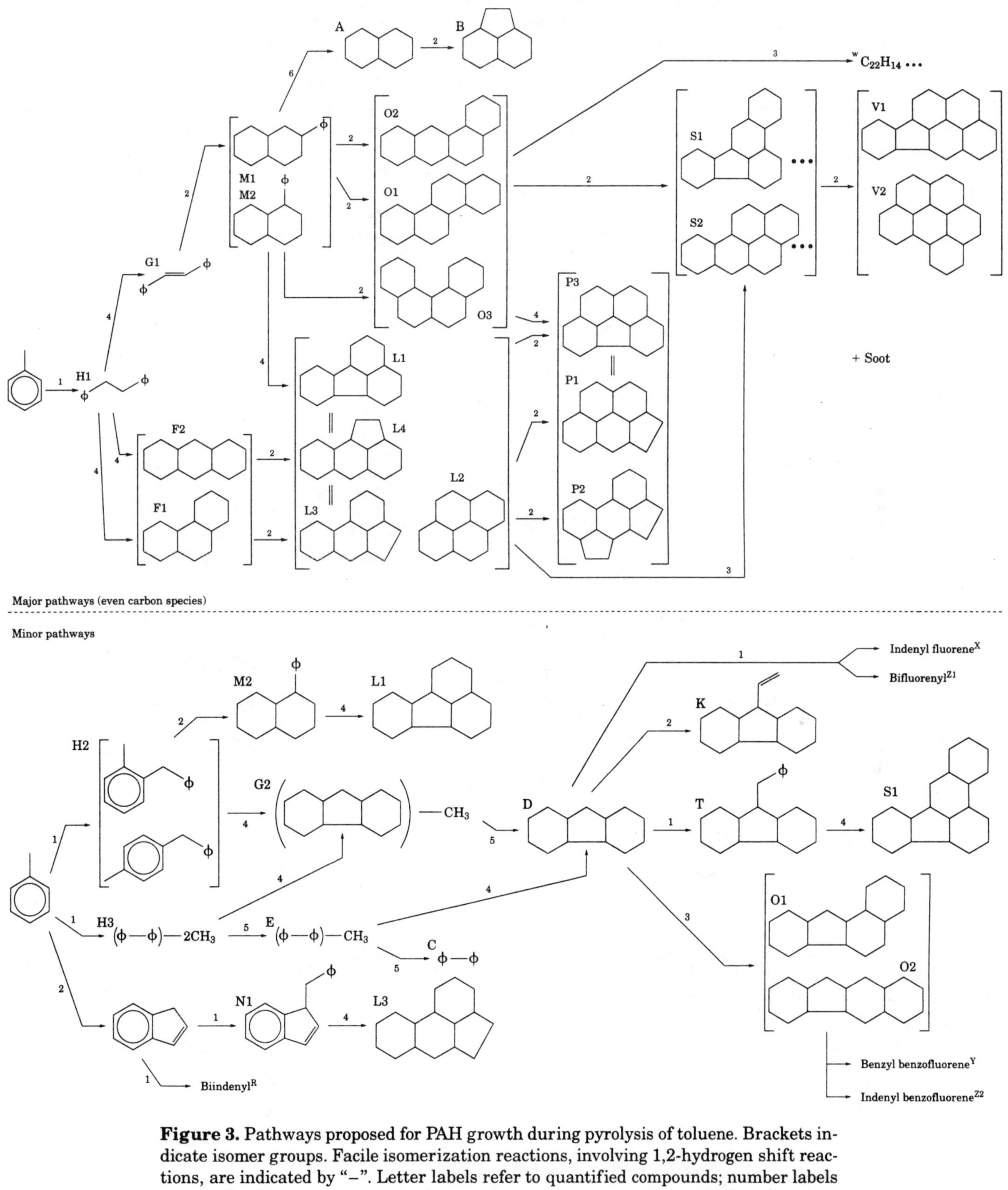

Figure 3. Pathways proposed for PAH growth during pyrolysis of toluene. Brackets indicate isomer groups. Facile isomerization reactions, involving 1,2-hydrogen shift reactions, are indicated by "–". Letter labels refer to quantified compounds; number labels refer to the following reaction pathways: 1 = aryl, 2 = two-carbon addition, 3 = four-carbon addition, 4 = hydrogen loss, 5 = methyl loss, and 6 = phenyl loss. Reproduced with permission (56).

drop tube furnace at 1213, 1303, 1393, and 1483 K. Bipyrenes (Fig. 4**a**) and bipyrene condensation products (Fig. 4**b**) are briefly discussed here. At 1213 and 1303 K, bipyrenes were the major product with all 6 isomers detected (Fig. 4**a**). The distribution of these isomers is explained by a statistical/steric argument based on the degeneracy of sites around the pyrene ring. Formation pathways for condensation products were deduced from the observation that their yields increase during bipyrene depletion.

It is also important to scrutinize possible effects of native and exogenous minerals and other inorganics on the thermal reaction behavior of the soil and its contaminants. Studies along these lines of coal pyrolysis and pyrolysis of various aromatic and aliphatic organic compounds are informative in the present context. Thus one group (24) studied the effects of calcium oxide on the pyrolysis of *n*-heptane, benzene, toluene, and 1-methylnaphthalene from 550 to 950°C, by passing vapor of the pure compound through packed beds of calcium oxide/quartz mixtures or of quartz in control experiments. Calcium oxide significantly increases the rates and extents of pyrolysis of the three aromatics. The temperature required for a given percentage conversion of a compound decreases by 140–170°C in the presence of CaO compared with quartz alone. Global activation energies derived from assumed first-order kinetics show that the activation energy over CaO is almost a factor of two lower than over quartz. Interestingly, CaO produces much

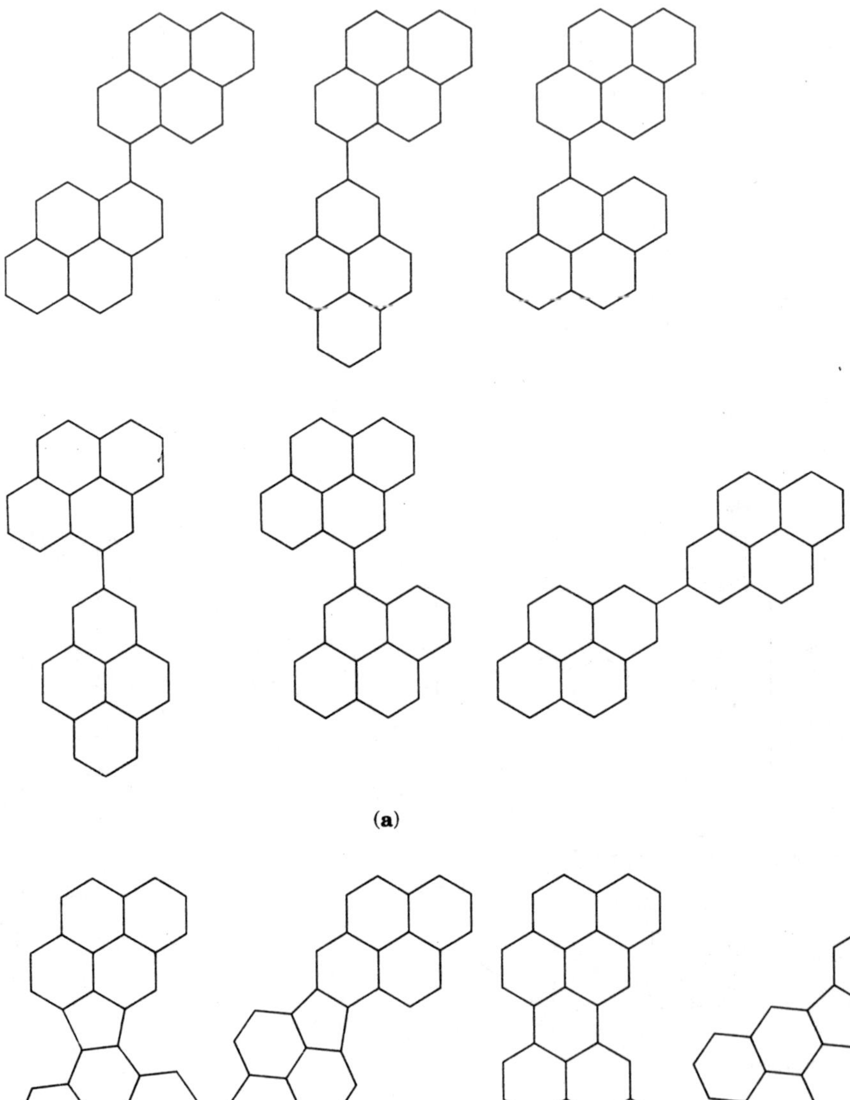

(**a**)

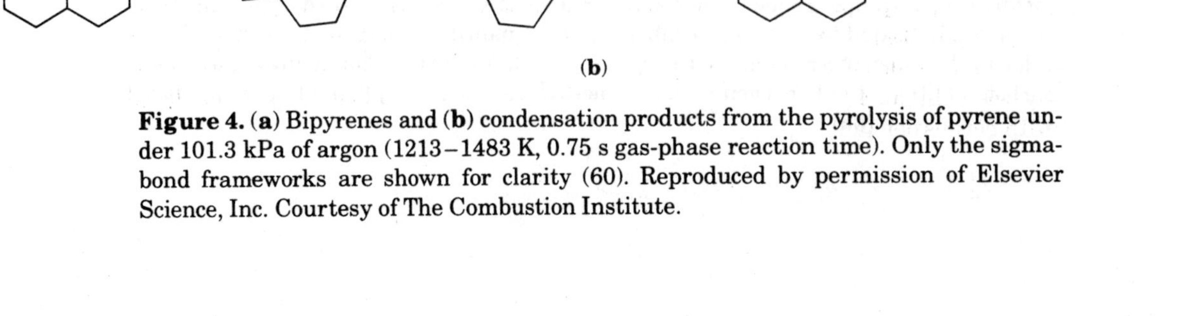

(**b**)

Figure 4. (**a**) Bipyrenes and (**b**) condensation products from the pyrolysis of pyrene under 101.3 kPa of argon (1213–1483 K, 0.75 s gas-phase reaction time). Only the sigma-bond frameworks are shown for clarity (60). Reproduced by permission of Elsevier Science, Inc. Courtesy of The Combustion Institute.

smaller increases in conversion of the aliphatic compound *n*-heptane. A study (25) of the effects of CaO on the pyrolysis of benzene, 1-methylnaphthalene, 1-methylanthracene, and 9-methylanthracene shows that CaO significantly increases overall rates and extents of aromatics pyrolysis and reduces the temperature for a specific level of conversion by 140°C or more. The magnitude of the effect increases with the molecular size of the compound. Coke is a major product of these reactions suggesting that condensation reactions are important over CaO.

The effects of minerals on the pyrolysis of coal have also been studied, for example, the effect of the addition of high calcium quicklime and of calcined dolomitic stones on coal pyrolysis in a dense phase fluidized bed (61). Coal pyrolysis with and without each mineral reveals that calcium oxide or calcined dolomite lowers tar yield and increases total gas, hydrogen, methane, and C_2 + hydrocarbon yields. Other studies (28,29) extracted native minerals from coal with HF and HCl, and back-added known amounts of minerals found in the coal, eg, calcite and clays. Demineralized coal treated with calcite or lime exhibits lower weight losses and higher char yields upon rapid pyrolysis compared to the demineralized sample (28). Tar and C_4–C_6 hydrocarbon yields are also reduced whereas yields of carbon oxides greatly increase. The same group also systematically studied the effects of adding clays, quartz, and other minerals (29). Pyrite, montmorillonite, and kaolinite reduce volatile yields by a statistically significant amount. Montmorillonite reduces tar yield by a statistically significant but slight extent. Kaolinite suppresses yields of light liquid hydrocarbons that were mainly benzene–toluene–xylene hydrocarbons. Calcium minerals increase light hydrocarbon yields more than other minerals, apparently by more pronounced secondary cracking of newly formed coal tar pyrolysis. Studies of C_3 to C_8 hydrocarbons (30) indicate that C_3 hydrocarbons are unaffected, except in the case of calcite and lime where yields are reduced by 20%. Calcite and lime lower yields of C_4–C_6 hydrocarbons. Montmorillonite, kaolinite, and calcite reduce C_6–C_8 volatile yields, whereas pyrite has less of an effect. Using similar techniques the effects of ion-exchanged alkali and alkaline-earth cations on rapid pyrolysis of a United States subbituminous coal were studied (27). Tar yields with ion-exchanged Ca, Na, and K are lower than those for raw coal at temperatures above 700–900°C, depending on the ion.

These examples show that, even rather volatile organic contaminants chemically react at elevated temperatures or at milder conditions with the catalytic or other chemical intervention of soil minerals or other inorganic substances. As discussed above, decontamination efficiencies are favorably or adversely impacted by some of these reactions. Some of the products of contaminant or other reactions, if improperly managed, can be more problematical than the original pollutant. Thus it was observed (26) that coal-tar mutagenicity to bacterial cells increases after vapor-phase pyrolysis of tar at temperatures above 700°C. From a practical perspective it is important to understand the thermal chemistry of contaminants and of soil itself, for the process conditions of interest. If serious potential for forming other toxicants is indicated by studies in research apparatus or by modeling calculations, processes can be designed with appropriate countermeasures, eg, improved mixing of volatiles with oxygen, increased temperature and time for primary oxidation, and use of afterburners. Thus the effects of oxygen on the thermal reactions of soil, soil decomposition products, and contaminants also need to be understood.

Chemical Reactions Between Soil and Organic Contaminants

Given the intimate contact between soil and the contaminant, it is reasonable to expect that soil, contaminants, and/or products of soil thermal decomposition, under some conditions of practical relevance, react to form additional species. Interactions between individual contaminants and/or their reaction products are a related possibility. Such effects further complicate the governing chemistry of soil thermal treatment. Evidence for soil-contaminant interactions is provided by one study (62) which identified benzene, toluene, ethylbenzene, styrene, benzaldehyde, benzofuran, chlorostyrenes, and chloroethylbenzenes in the volatiles evolved upon heating in air a bed of soil contaminated with 0.5 wt % ethylbenzene. The analytical system employed an on-line vapor sampler connected to a portable GC/MS facility (63). The study (62) interpreted the diversity of compounds identified as evidence that different reactions occur when heating contaminated soil in air, eg, degradation, oxidation, and production of chlorinated products by reaction of organic compounds with the hot upper layer of the bed of soil, which contained 0.12% chloride.

Other literature also suggests soil-related reaction phenomena. Another study (64) mentions that several organic chemical reactions occur in soil at ambient conditions: hydrolysis, substitution, elimination, oxidation, reduction, and surface-catalyzed hydrolysis and oxidation. It is thought that, as soil temperature increases, the reaction rates accelerate under the appropriate conditions. At least at low pressures, conventional aqueous-phase reactions like hydrolysis, substitution, and elimination (65) contribute little in the absence of free-standing water. Surface-catalyzed hydrolysis occurs at high temperature (66), and the reaction is enhanced as moisture is removed from the surfaces of clays like kaolinite. Some soils, clays, and minerals catalyze oxidation of many organic chemicals at ambient conditions and at higher temperatures and pressures (67). Reduction is the least studied and least understood of the reactions listed above (65), and its role in soil reactions is thus rather unclear. Further indications that chemical reactions occur between soil minerals and organic compounds are found in a new approach for removing polychlorinated biphenyls (PCBs) from soil, devised at Batelle's Pacific Northwest Laboratories (68). In this process, sodium bicarbonate is mixed with the contaminated soil, and the mixture is heated to 350°C. The hydrogen from the sodium bicarbonate replaces chlorine and other halogens in the PCBs yielding NaCl and a biphenyl lacking the toxicity of PCBs. Similar reactions might arise naturally during thermal treatment of some soils, because sodium bicarbonate is present in the minerals nahcolite and trona.

Clearly, if soil or contaminant constitutents or their degradation products react separately or with each other during thermal treatment, additional species and effluent production chemistries must be accounted for in modeling process performance.

Heat-Transfer and Mass-Transfer Issues

Nonuniform heating of solids creates internal temperature gradients. In soil thermal treatment, temperature nonuniformities may persist long enough to affect the rates and the extent of decontamination. In such cases the soil particle or bed is considered thermally thick and the kinetics are influenced or even limited by heat transfer. Quantitative analyses of heat-transfer (and mass-transfer) effects on solids thermal processing are complex and outside our scope. Useful information for soils can be obtained from the fuels literature. Directly pertinent references are numerous, as are helpful starting points for more in-depth study (69,70). The latter cites a number of investigations of the pyrolysis of thermally thick solid fuels and describes a mathematical model for deriving intrinsic rates of volatiles production in heat transfer controlled pyrolysis of solids and a method for computing simultaneous drying and pyrolysis within a solid fuel particle. One study (31) discusses an approach to estimating the effects of intrasample temperature gradients on apparent average rates and the extent of pyrolysis of externally heated fuel pieces.

Simplified heat-transfer analyses are useful for delineating the effects of practically relevant conditions and for estimating thermal response (lag) times, eg, the time necessary to equilibrate a temperature gradient throughout the entire soil particle. For example, with all other variables fixed, an increase in soil particle size or a decrease in its thermal diffusivity increases the thermal response time. Thus for soils composed largely of low thermal diffusivity minerals, significant spatial nonisothermalities persist for appreciable times, even for quite small particles. The magnitude of the intrasample spatial temperature gradient increases with an increasing surface heating rate. More quantitatively, Figure 5 shows how estimated thermal lag times are affected by particle size for a spherical soil particle heated instantaneously from 25 to 700°C. The calculations approximate soil as crystalline silicon dioxide and define lag time as the time for the difference between the temperature of the particle surface and the particle center to fall to 5% of the initial difference in these two temperatures. Other assumptions are that soil is a chemically inert, solid sphere with spatially isotropic, temperature-invariant thermal properties. For an order of magnitude estimate, a soil particle is assumed to be spatially isothermal if its thermal lag time is less than 10% of the time to heat the particle to the final temperature. Figure 5 then indicates that ~1 cm diameter particles are isothermal at heating rates below 10°C/min, that ~1 mm particles are isothermal for heating at ≤10°C/s, and that ~100 μm particles remain isothermal for rates ≤1000°C/s. Actual soil particles may contain significant internal void spaces, moisture, and reactive components which, respectively, impede intraparticle heat penetration owing to the lower thermal diffusivity of gas in the voids, the enthalpy of drying, and the enthalpy of chemical reactions (for endothermic reactions). Thus, more accurate estimates of soil thermal response times should account for effects of temperature variations in thermal physical properties, void volume, moisture, and heat absorption or release by chemical reactions and for the combined effects of intra- and extraparticle resistances on particle heating (71). Despite its simplifying assumptions, Figure 5 is useful for orientation estimates of single particle thermal response times and shows the importance of considering soil particle size and heating environment in evaluating thermal treatment schedules.

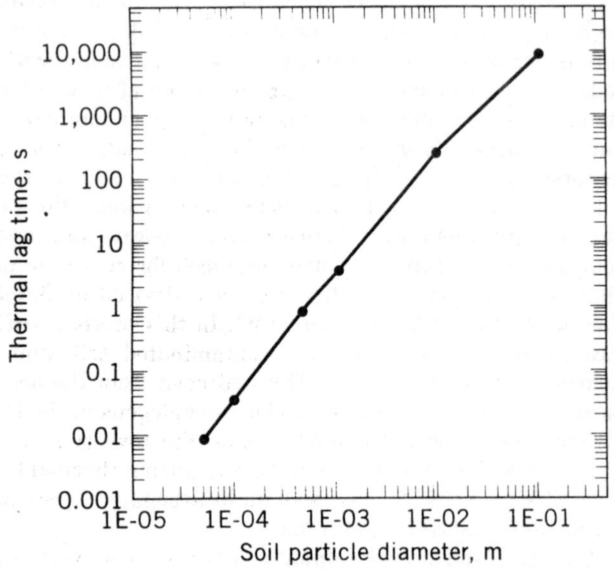

Figure 5. Estimated thermal lag times for the center of spherical soil particles as affected by soil particle size. Thermal lag time is defined as the time for the difference between the temperature of the particle surface and particle center to fall to 5% of the initial difference in these two temperatures when the surface temperature rises infinitely rapidly to its new value. A crystalline SiO_2 sphere heated from 25 to 700°C is assumed for this calculation.

Many mechanisms contribute to contaminant mass transfer within and away from individual soil particles, eg, diffusion through the solid soil matrix, Knudsen diffusion, bulk diffusion in pores, and forced convection within pores (see also the section on modeling). Pores are the primary transport channel because of the extremely low mass diffusivities for vapors within solids. Forced convection arises from rapid expansion of gas within the pores at high heating rates or from production of significant amounts of gas from decomposing soil or contaminants or from rapid volatilization of moisture. Chemically or physically induced modifications of the soil physical structure can slow or augment mass transport by restricting or opening the internal porosity. Characteristic diffusion times for contaminants within a soil particle are estimated from particle size and pore size data and knowledge of the mass diffusivities of the contaminants. However, the calculated values underestimate actual intraparticle diffusion times if readsorption/desorption of the contaminant on the pore walls is the rate controlling factor. Another consideration is the temperature dependence of mass transport parameters such as diffusivity. Significant intraparticle temperature gradients complicate the analysis of experimental data on mass-transfer effects (see the introduction above to heat-transfer effects).

For beds of soils, heat and mass transfer analyses must consider phenomena specific to individual particles, to particle ensembles, and to the bed itself. Quantitative approaches are discussed in the section on modeling. Among the factors relevant to decontamination are the effects of externally imposed gas flows through the bed on gas-to-particle heat and mass transfer and on accelerating contaminant mass transfer from the particle or bed surfaces. Other issues include temperature variations throughout the bed including possible readsorption of contaminants on cooler particles. Such effects arise even in rather shallow packed beds and potentially in dense phase fluidized beds if upsets degrade mixing and heat transfer. Kiln heaters are intentionally rotated to improve soil heating and gas-particle–volatiles mixing. The resulting soil bed and particle motions and particle decrepitation complicate quantitative analyses of heat and mass transfer, and their effects on decontamination (72).

TYPICAL EXPERIMENTAL SYSTEMS FOR RESEARCH SCALE STUDIES

Packed Beds

Two types of packed beds have been used for thermal adsorption/desorption studies (73–76) (Fig. 6). The first apparatus (Fig. **6a**), is designed to investigate the effects of the entire bed on decontamination. It consists of a soil bed placed in a heated pan. The pan is located on a balance so that bed weight is continuously measured. Preheated carrier gas passes across the upper face of the bed. This equipment has been used at temperatures up to 315°C, with air at ambient pressure as the carrier gas. The carrier gas is used to deposit or remove contaminants from the soil bed. One study (62,63) used an on-line GC/MS sampler to identify organic products from heating soil beds in this reactor. The simplicity of the reactor design and the ability to complement weight change data with gas analysis make this equipment versatile and informative. A further advantage is that this equipment allows quantitative study of thermally thick soil beds (typical depths are 6–8 cm). Effects of spatial gradients in temperature and species concentrations normal to the bed face heated are taken into account in interpreting and modeling bed performance. Bed temperatures are measured by *in situ* thermocouples (Fig. **6a**) (see section on Modeling).

The second apparatus (see Fig. **6b**) is designed to probe the effects of individual soil particles in decontamination. Shallow beds of soil (≤2.5 cm deep) are heated while passing preheated gas vertically upward through the bed. Gas flows have been used for adsorption/desorption experiments at temperatures up to 270°C. Contaminant equilibrium between the solid soil and the gas phase is described by a Freundlich isotherm (see section on Modeling).

Electrodynamic Balances

The electrodynamic balance (EDB) (Fig. 7), which employs an electric field to suspend a particle in a gas stream, has been used for contaminant adsorption/desorption experiments on soil particles (12,77–79). Particle weight is continuously determined from measured changes in the electric field strength required to suspend the particle at a fixed position. A suspended particle is heated to temperatures as high as 600–700°C without loss of charge (see below) using a focused laser beam. Particle temperature is continuously monitored by two-color optical pyrometry. Experiments at elevated pressures (eg, up to 50 atm or more) are possible, and a wide range of ambient gases is studied. Smaller diameter particles (of order 90–180 μm) are generally preferred to avoid intraparticle temperature gradients. Heating rates estimated to be as high as 10^5°C/s are attainable for smaller particles. To derive high-quality weight loss data requires spherical particles, an apparatus constant for the particle of interest, and no significant loss of electrical charge from the particle surface (79). Volatiles can be characterized by off-line analysis (eg, by GC) or by on-line techniques, although the small sample mass may require multiple runs to overcome the detection limits of analytical instruments.

Thermogravimetric Analyzers

Thermogravimetric analyzers (TGAs) continuously measure the weight of a sample undergoing heating, usually according to a preselected temperature-time history provided by a controlled heating element that is part of the apparatus. A variety of instruments are available commercially. A typical apparatus consists of a sensitive electronic balance, eg, a microbalance, in which one side suspends a pan containing the sample in a vertical tube furnace. Temperatures up to 1000°C and heating rates of up to 100°C/min are attainable, and various gaseous environments are used. Sample sizes vary, and some instruments have capacities up to 150 g. Much previous work on soils has been performed with samples weighing on the order of a few mg. Smaller samples allow study of more shallow beds which permit the use of higher heating rates while avoiding excessive intrasample temperature gradients. However, spatial concentration gradients may not be negligible, even in quite shallow beds of particles (80). Good knowledge of soil temperature is obtained for different heating rates by locating a thermocouple close to the specimen. This derives kinetic information from the measured weight changes. Soil-related TGA investigations have included studies of contaminant adsorption and/or desorption phenomena (80–82). Volatiles are characterized by analyzing the TGA effluent gas stream or by on-line techniques, eg, mass spectrometry (79). When specimens of small mass are examined, it may be necessary to run multiple experiments to generate sufficient quantities of products for analysis. For some specimens or operating conditions, vapor residence times at elevated temperature may be long enough that more thermally labile volatiles, eg, some tars, become susceptible to chemical reactions. These secondary reactions could modify the composition of volatiles exiting the TGA chamber and should be accounted for in estimating the identities and release rates of volatiles exiting the soil surface.

Chromatographic Methods

The effective diffusivities and adsorption equilibrium constants for chlorinated solvents were measured with a cylindrical packed soil pellet placed in a gas chromatograph oven (Fig. 8) (83). Pellets with a 1.43-cm^2 cross-sectional area and 0.29 cm high were heated to treatment temperatures in the range of 35 to 90°C in a helium environment. Contaminant was introduced as a concentration pulse on one side of the pellet and allowed to diffuse axially through the bed. Equal pressure was maintained on both sides of the pellet to eliminate forced convection of gas or contaminant through the bed. Moments of the profile of contaminant concentration vs time at the bed exit were analyzed to obtain pellet diffusivity, tortuosity, contaminant equilibrium constants, and adsorption rate constants for single contaminants (see modeling section). With quite straightforward experimental modifications, this informative technique should be applicable at higher temperatures to study contaminants of higher MW and lower volatility. Above some temperature, the effects of chemical reactions of contaminated soil must be accounted for. With suitable modifications in the analysis, this technique should also be readily applicable to mixtures of contaminants.

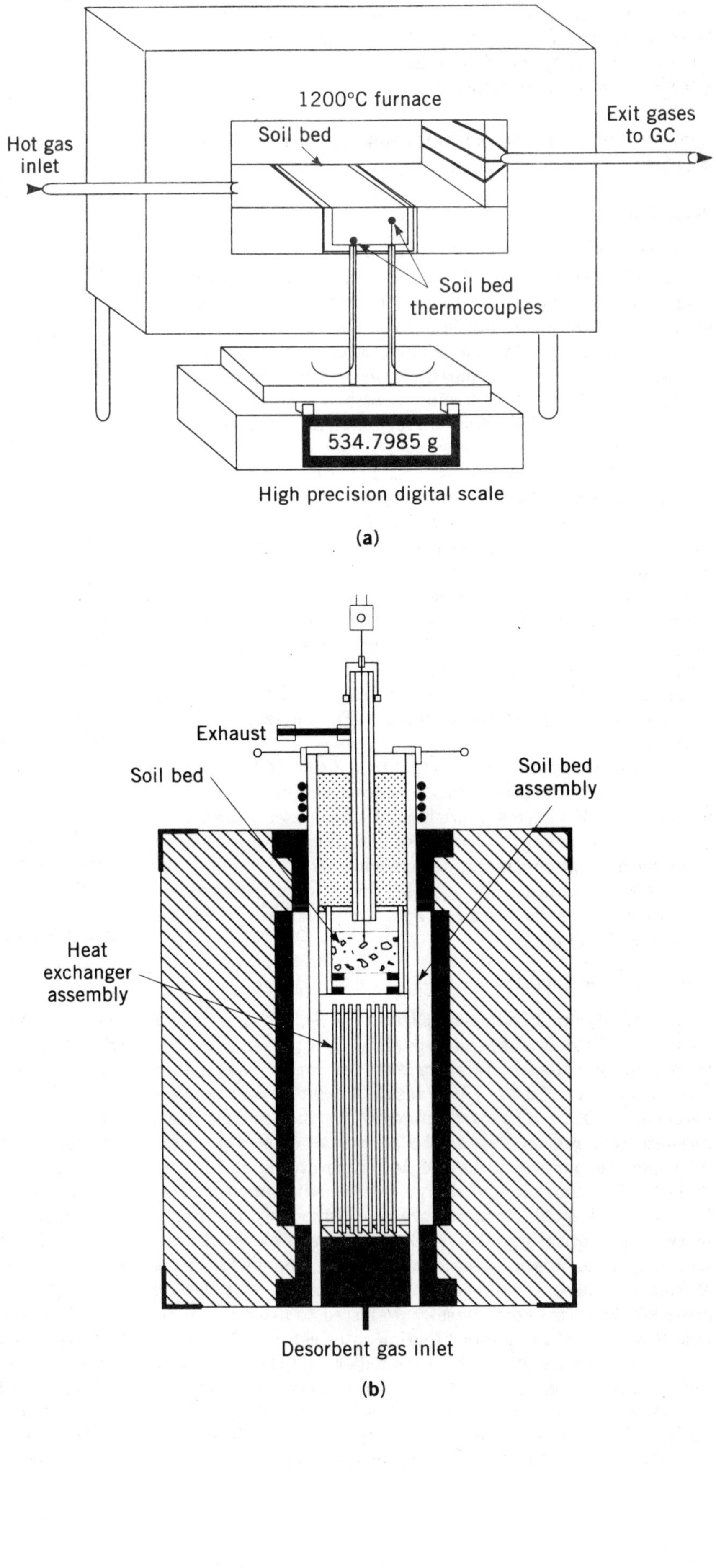

Figure 6. Apparatuses used for studies of the adsorption and desorption of organic contaminants from soils: (**a**) bed characterization reactor (BCR); (**b**) particle characterization reactor (PCR). Reprinted with permission of Elsevier Science Ltd. (74).

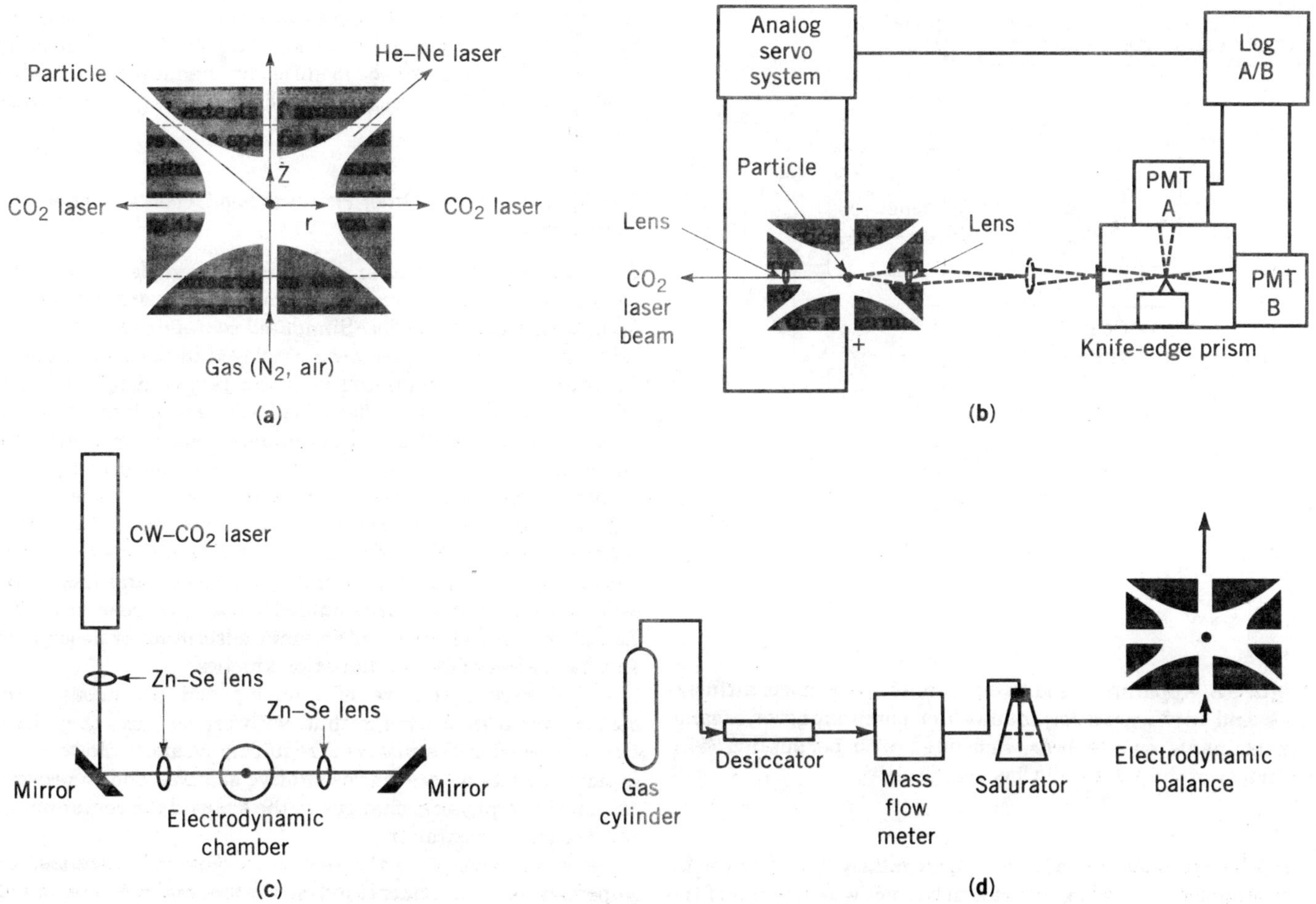

Figure 7. Schematic views of the electrodynamic balance used for studies of contaminants adsorption and desorption on soils (12,77,78): (**a**) cross-sectional view of the electrodes; (**b**) schematic of the position control system (PMT = photomultiplier tube); (**c**) schematic of the laser heating system; and (**d**) schematic view of the gas flow system. Reprinted with permission (12). Courtesy of the American Chemical Society.

Contaminant deposition was studied on a packed bed of soil in a modified gas chromatograph with a chromatographic response model to derive equilibrium constants for contaminants partitioning between the vapor phase and the adsorbed phase on the soil, heats of adsorption, axial dispersion coefficients, and intraparticle diffusion coefficients (84,85). Only slight modification of a standard gas chromatograph achieved desired control over temperature and heating rate with the chromatograph oven. With this technique, cumulative effects of sorption and desorption are inferred by measuring the initial and final mass of the soil column, and contaminant removal is detemined by chemical analysis of the column effluent. Experiments at elevated temperatures should be straightforward, but must account for the effects of chemical reactions of soil and contaminants.

Captive Sample Reactors With Electrically Heated Hot Stages

In these reactors, captive soil samples of known initial weight (typically ~50 mg), contaminated or neat (ie, uncontaminated) in control runs, are heated by passing electrical current through a hot stage (3,86–89). The apparatus (Fig. 9) (89) is similar to the electrical screen heater reactor used to study the rapid pyrolysis of coal (90,91) and biomass (31), except for two important modifications. To prevent the loss of friable soil particles, the heating screen was replaced with a 0.00254-cm (0.001-in.) thick stainless steel foil with a 2.01-cm-ID, 0.28-cm deep cylindrical well to confine the soil. The foil was mounted between two large aluminum electrodes within a cylindrical Pyrex chamber. The second modification inhibits evaporation of volatile contaminants during flushing of the reactor ambient gas in preparation for a soil heating experiment. A movable cylindrical loading rod (see Fig. 9) (89) is used to add soil to the foil after purging the reactor chamber. This rod consists of three 1.9-cm diameter aluminum sections connected by on-axis screws and threads. Soil is placed in the middle (1.9-cm long) section of the rod in a 0.32-cm blind hole drilled along a diameter of the rod to a depth of 1.6 cm. Soil is added to the foil by pushing the rod into the reactor chamber through two seals mounted on a side flange (see Fig. 9), until the rod is directly above the depression in the foil, and then rotating the rod about its axis to dump the soil onto the foil well. The soil

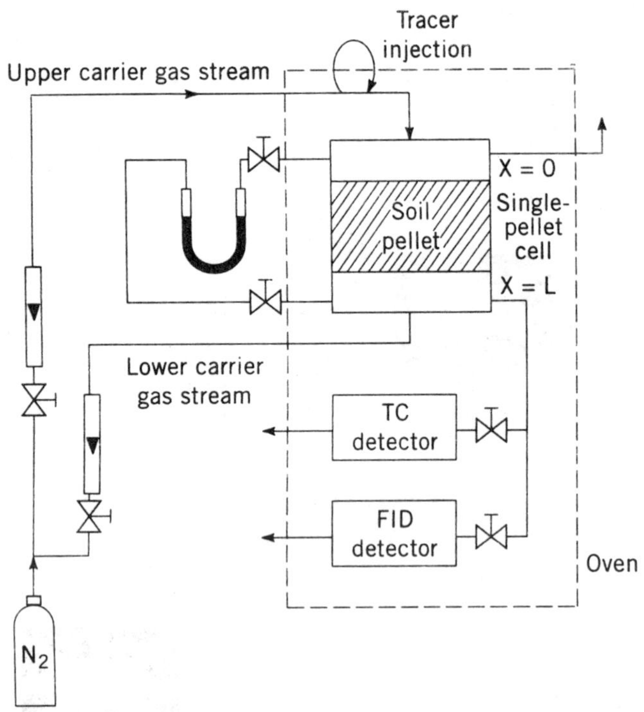

Figure 8. Apparatus used to measure effective mass diffusivities and coefficients for equilibrium partitioning of organic contaminants on soil beds. Reprinted with permission (83). Courtesy of the American Chemical Society.

particles are then spread into approximately a monolayer by agitating the foil with a vibrator in contact with the top of the reactor chamber. The middle section of the rod has relatively low mass and is separable from the other sections for weighing. The amount of soil used in a run is determined as the weight reduction in this piece after dumping the soil.

Heating rate (100–6000°C/s), final temperature (up to 1050°C), and time at final temperature (0–20 s) are independently selected for each experiment by adjusting variable transformers and timers that determine power levels to the heater foil. Samples cool down naturally, mainly by radiation and/or convection, at initial rates between 400 and 600°C/s (at ~101.3 kPa) depending on peak temperature. Soil temperature-time histories are measured throughout each experiment with a rapid response (0.00127-cm thick junction) type K thermocouple placed beneath the foil. Heat transfer calculations imply that the thermocouple closely tracks the bulk mean temperature of the soil particles (eg, conservatively to within about 10°C at a nominal heating rate of 1000°C/s). Experiments using two thermocouples show that, for heating rates up to 6000°C/s (88) particle temperatures are within about 30°C of the temperature measured with a thermocouple in contact with the bottom of the foil.

Products are collected for detailed analysis. Weight loss is determined from the weight change in the foil plus soil before and after heating. Material and carbon balance closures within 5–10% are reported for uncontaminated soil (3). Kinetic parameters for weight loss, products yields, and contaminants release (3,86–88) are obtained by regressing observed and predicted yields using the measured temperature–time histories for the soil. Newly formed volatiles are rapidly diluted and cooled by ambient gas upon exiting the soil sample, which lessens the opportunity for secondary reactions. Nevertheless it is important to be aware that more thermally labile volatiles could still be modified by chemical reactions before they are collected, especially when using higher operating temperatures (31,92).

Sample Selection and Other Pitfalls in Soil Decontamination Experiments

Care must be exercised in selecting soil samples and in applying various measurement techniques to quantify thermal decontamination behavior. Simulated contaminated soils and soils from field sites have been employed in research studies. Simulants for contaminated soils are prepared by adding a known contaminant to either a real soil sample from the field or to a model substrate. Contaminants used previously in simulated contaminated soils include volatile solvents (eg, toluene, carbon tetrachloride, chlorobenzenes, ethylbenzene), polycyclic aromatic hydrocarbons (eg, pyrene), and complex mixtures, such as No. 2 fuel oil. Substrates include clays, gas chromatographic packings, sand, glass beads, and industrial sorbents. Simulated contaminated soils have been used for fundamental studies of contaminant adsorption or desorption and for studies of decontamination kinetics.

An attractive feature of studying soil specimens from contaminated field sites is that, with proper care, they more closely resemble the material requiring cleanup. However, it is important to obtain representative samples and to prevent chemical and physical changes in the soil and the contaminant prior to experimentation.

As was discussed in the section on chemical reactions, an important issue in understanding the thermal behavior of soil samples is the extent to which contaminants interact with the soil or its reaction products during heating. The magnitude of such interactions may differ between actual and simulant samples. Moreover, these interactions may confound attempts to infer the extent of decontamination by simple subtraction of the weight loss behavior of "neat", ie, uncontaminated soil, from that of a contaminated specimen of the same soil (88), because contaminants directly or indirectly affect the weight loss of the soil itself. For example, weight loss is increased if the contaminant promotes additional volatilization, eg, by converting metal oxides in the soil minerals to CO and lower valence state metals. Weight loss is diminished if the contaminant decreases the volatility of soil pyrolytic products. This may arise when chemical reactions between the contaminant and soil volatiles produce higher molecular weight products. Contaminants may promote, or partipate in, coke deposition, constricting or blocking pores, hindered release of volatiles, and increased secondary conversion of volatiles to less mobile species. Providing contaminant–soil interactions are properly accounted for, studies of uncontaminated soil under similar conditions are useful control experiments for thermally treating contaminated specimens of the same soil. When interactive effects are known or suspected to be important (eg, studies at elevated temperatures, measured weight loss for contaminated soil exceeding the weight loss of uncontaminated soil plus the initial weight of contaminant), the decontamination behavior is more definitively characterized by determining yields and compositions of volatiles and residual products. This allows a direct measure of the identities and amounts of contaminants

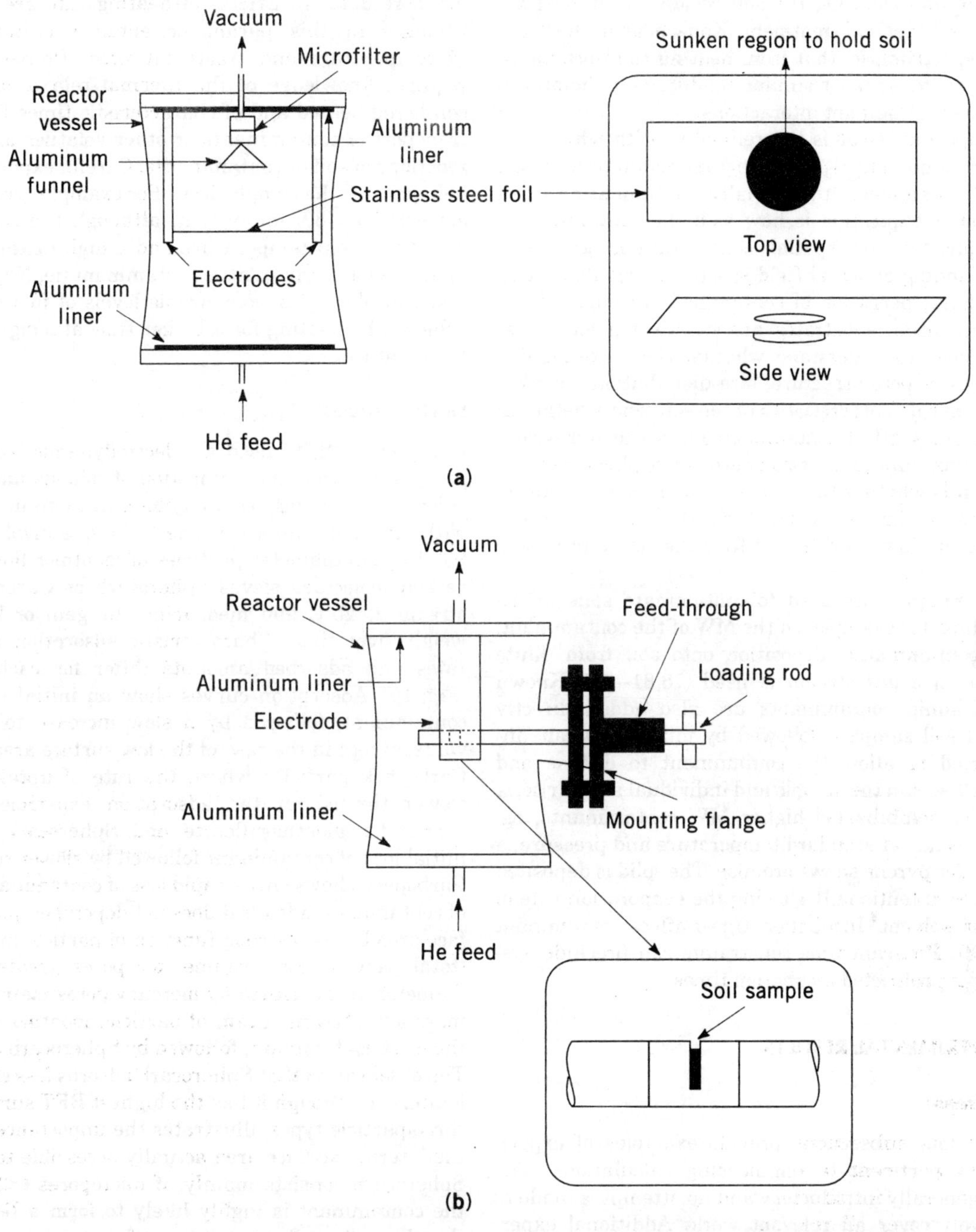

Figure 9. Electrically heated foil captive sample reactor used to study soil thermal decontamination at high heating rates (3,87–89): **(a)** front view showing foil hot stage and various stations for collection of tars and other condensables; **(b)** side view showing the loading rod positioned to drop soil particles onto the foil. Reprinted with permission (89). Courtesy of the American Chemical Society.

and other volatiles released from the soil. Creation or destruction of contaminants can be corrected for by isotopically labeled tracers. Solvent extraction of the soil after thermal treatment assays unevolved contaminant. However, because soil pollutants or their by-products may be poorly extracted by common solvents, this method may overestimate the true extent of contaminant removal. Further, the extraction solvent or the process itself may modify the soil or the contaminant residue.

To cross-check decontamination data, it is wise, when sample size and resources permit, to obtain measured material and elemental balances by analyzing untreated soil, and gaseous, liquid, and solid (ie, the soil residue after heating) products from soil thermal treatment. Knowledge of the "base line" product spectrum, ie, that from heating the uncontaminated soil under the same or similar conditions, is helpful in unraveling soil–contaminant interactions.

Another important issue is the reliability with which contaminated soil simulants represent actual contaminated soil in the field. For example, in artificially contaminating clean soils, an important question is how well the contamination technique replicates the spatial concentrations and soil-contaminant binding of actual field specimens, eg, deposition in soil micropores, perfusion of contaminant throughout the particle. When model substrates are substituted for actual soil, there is need to understand whether there are significant differences in pore structure, size distribution, or other physical or chemical characteristics of the soil, and whether or not such differences affect contaminant mass transfer within the soil or contaminant adsorption/desorption phenomena. A related concern is whether the process of artificially contaminating a natural or simulant soil disturbs the soil's physical or chemical structure in ways different from the actual processes of contamination in the field.

Various techniques are used to contaminate soils artificially, depending, for example, on the MW of the contaminant. For volatile contaminants, deposition onto soil from dilute concentrations in a gas stream is used (78,81–83). Known quantities of liquid contaminants are also added directly to preweighed soil samples, followed by mixing and an "incubation" period to allow the contaminant to diffuse and distribute itself within the sample and individual soil particles (73–76,89). For solubilizable higher MW contaminants, eg, those that are solids at standard temperature and pressure, a technique (88) for pyrene shows promise. The solid is deposited from solution by intentionally slowing the evaporation rate of a volatile host solvent. Incubation times affect contaminant desorption (74). Program time constraints can preclude systematic study of protracted incubation times.

SELECTED EXPERIMENTAL RESULTS

Summary of Trends

The following four subsections provide examples of experimental results pertinent to engineering calculations. The treatment is generally introductory and no attempt is made to comprehensively cover all relevant work. Additional experimental results are presented in the section on modeling to compare model predictions to observations. The objectives here are to illustrate that methodical experimental investigations of soil thermal decontamination elucidate important chemical and physical phenomena contributing to contaminant release; and to show that such experiments provide quantitative databases and parameterized equilibrium and kinetic models for orientation calculations and as inputs for more elaborate engineering design and environmental analysis computations.

With these points in mind, we summarize here some of the trends that emerge from the works cited. Thus, contaminant removal efficiency increases with increasing temperature or treatment time, provided that temperatures are not so high as to trigger chemical reactions that interfere with decontamination, eg, by transforming contaminant to coke and gas (see the section Fundamental Chemical and Physical Processes for more details). Effects of heating rate are more complex. Changes in this parameter entail simultaneous changes in temperature and treatment time. Proper interpretation requires knowledge of the thermal relaxation times of the soil being heated and of characteristic times for intraparticle transport of contaminants or other volatiles and for chemical reactions of soil or contaminants. Careful experimentation can help unravel the complexities. For example, controlled heating experiments (89) imply that, although the combination of a high treatment temperature and a high heating rate desorbs most of a modestly volatile contaminant (ie, No. 2 fuel oil) from a soil in about 1 s, comparable levels of fuel oil removal are achieved by heating for a longer time at a significantly lower temperature.

Single Particles

One group (12,78) used an electrodynamic balance (EDB) to study adsorption and desorption of toluene and carbon tetrachloride from a gas stream onto or away from single particles with different pore sizes. The technique involved suspending 90–190 μm diameter particles of montmorillonite clay or the carbon molecular sieves Spherocarb or Carbopack in a gas stream at 25°C and measuring the gain or loss in particle weight over time. Characteristic adsorption and desorption rates and adsorbed amounts differ for each particle type (Fig. 10). Adsorption curves show an initial rapid uptake of contaminant, followed by a slow increase to a steady-state value, except in the case of the low surface area, high porosity Carbopack particles where the rate of uptake is relatively slower throughout the adsorption experiment. Desorption curves for montmorillonite and Spherocarb show a steep initial loss of contaminant followed by slower release, whereas Carbopack shows only a rapid loss of contaminant. The amount of contaminant adsorbed does not depend on particle BET surface area but is a strong function of particle intrusion volume (total particle pore volume, for pores greater than 30 nm diameter) as measured by mercury porosimetry. Based on the mass adsorbed per gram of particle, montmorillonite adsorbs the most contaminant, followed by Spherocarb and Carbopack. The observation that Spherocarb adsorbs less than montmorillonite, even though it has the highest BET surface area of the three particle types, illustrates the importance of considering the internal surface area actually accessible to contaminants. Spherocarb consists mainly of micropores (<3 nm) in which the contaminant is highly likely to form a liquid phase and thereby restrict further access of contaminant to the particle internal surface area (12,78). This reasoning is supported by the results on Carbopack, which has large pores, low surface area, the smallest intrusion volume, and the lowest uptake of contaminant. Desorption studies at 0 and 25–30°C show that a significant fraction of contaminant is irreversibly adsorbed onto the particle. An order of magnitude analysis revealed that the amount of contaminant irreversibly adsorbed corresponds to about a monolayer coverage of the soil internal area.

TGA and Heated Foil Studies of Shallow Beds of Soil

One study (86) investigated the efficiency of pyrene removal from a clay soil with an electrical screen heater reactor. Soil specimens, artificially contaminated, or neat in control runs, were separately heated to temperatures between 200–800°C at a nominal heatup rate of 1000°C/s. Temperature strongly

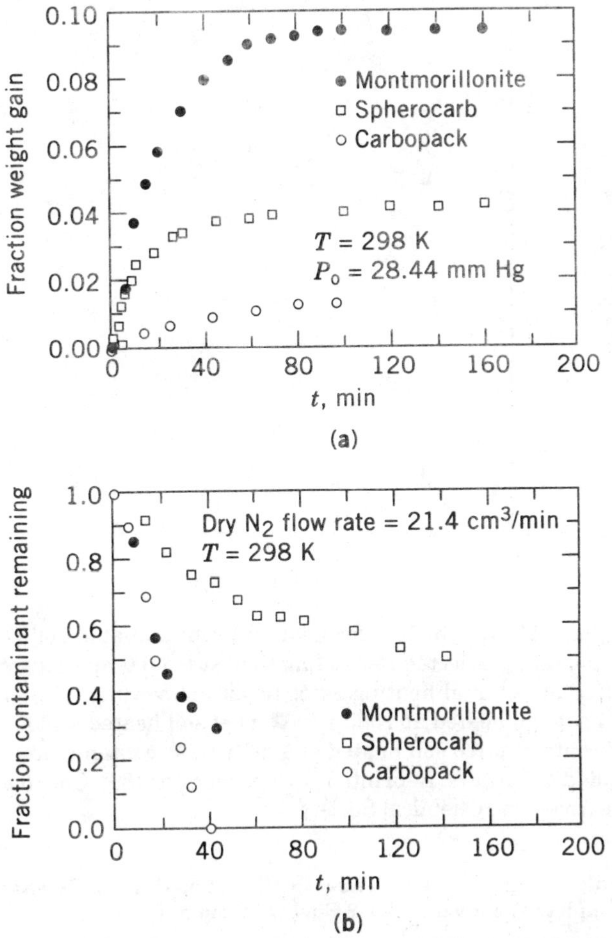

Figure 10. Data for toluene (**a**) adsorption on and (**b**) desorption from particles of various materials. Measurements were performed with an electrodynamic balance (see Fig. 7). Reprinted with permission (12). Courtesy of the American Chemical Society.

influences soil weight loss and contaminant removal. Contaminant removal is determined from the measured differences in soil weight loss between runs with contaminated and neat soil. The experiments indicate that pyrene is completely removed at temperatures of about 600°C or greater. A multiple, independent, parallel reaction model was used to fit the weight loss data to obtain global rate parameters for pyrene removal. The resulting fitted mean activation energy for pyrene removal from the soil is 155.3 kJ/mol (37.1 kcal/mol). This fitted quantity is best viewed as an overall or global kinetic parameter. It is not easily related to elementary chemical reaction steps and interestingly is about three times the heat of vaporization of pyrene at its boiling point (58.2 kJ/mol). This study (86) was exploratory. The experiments may have been affected by some loss of soil particles during heating and by nonuniform mixing of the pyrene with the soil. These two effects can artifactually increase rates of weight loss and of pyrene release deduced from the laboratory data. This work (86) created a solid foundation for follow-on studies by demonstrating that essentially total removal of pyrene from a clay soil is achieved in times of the order of 1s by heating shallow beds of soil to temperatures well below 1000°C; and that this experimental technique, modified to promote more spatially uniform contamination of the soil and to reduce or eliminate particle loss, provides useful kinetic data on soil decontamination by rapid heating.

Using the heated foil reactor and experimental techniques described previously, the effects of heating rate (200–6100°C/s), final temperature (300–700°C), holding time (0–25 s), and contamination level (4 and 7 wt %) on removing No. 2 fuel oil from a synthetic soil matrix were systematically studied (89). Fuel oil removal was determined from the difference in measured weight loss of samples of neat and contaminated soil (89). It was inferred that thermal treatment substantially eliminates No. 2 fuel oil from contaminated soil, with temperature, heating rate, heating time, and holding (soak) at final temperature each affecting the rates and extent of pollutant removal (Figs. 11–14), respectively. Fuel oil removal efficiency increases strongly with increasing peak temperature from 300–500°C, and then rises more slowly with further temperature increases, culminating in essentially complete decontamination at 700°C (see Fig. 11). The extent of oil removal at a fixed final temperature of 300°C increases with decreasing heating rate (see Fig. 12) probably because more time is available for contaminant release with slower heating. Different combinations of heating conditions remove the fuel oil virtually completely (Table 1).

Thus for ranges of conditions of interest in practical thermal treatment technologies, it is possible to obtain equivalent and indeed very high levels of decontamination by trade-offs in heating rate, temperature, and holding time. Changes in initial contamination levels from 7 to 4 wt % have little effect on the fractional extent of decontamination with 1000°C/s heating to 300°C for various holding times (89). The study also

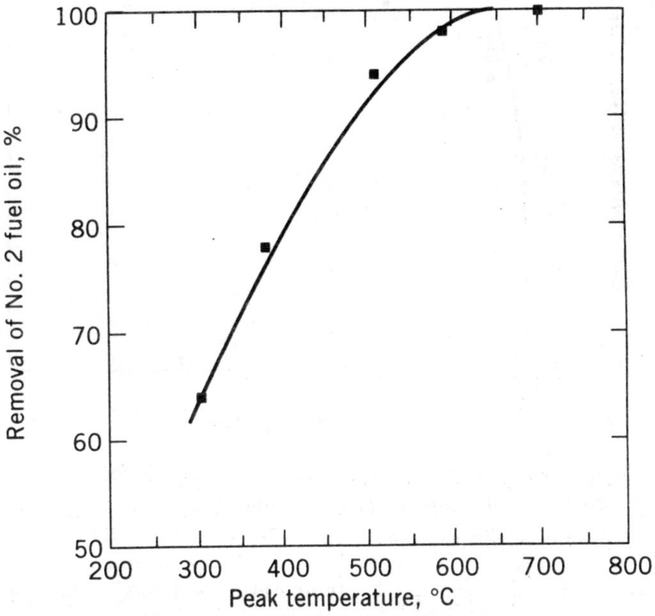

Figure 11. Percentage removal of No. 2 fuel oil as affected by peak temperature calculated from data on weight loss of contaminated and neat soil: (nominal heating rate = 1000°C/s; zero holding time at peak temperature; initial contamination 7 wt %). Curve: trend line. Reprinted with permission (89). Courtesy of the American Chemical Society.

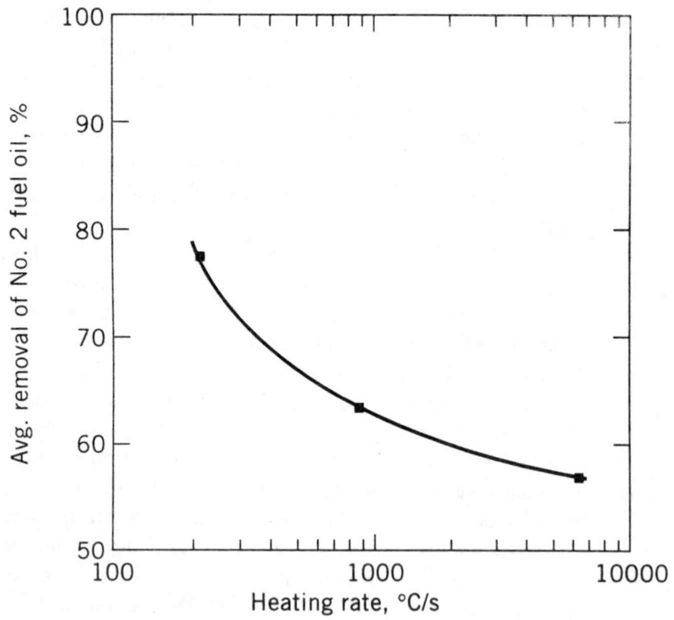

Figure 12. Average percent removal of No. 2 fuel oil as affected by nominal heating rate (peak temperature = 300°C; initial contamination = 7 wt % fuel oil; zero holding time at peak temperature). Curve: trend line. Reprinted with permission (89). Courtesy of the American Chemical Society.

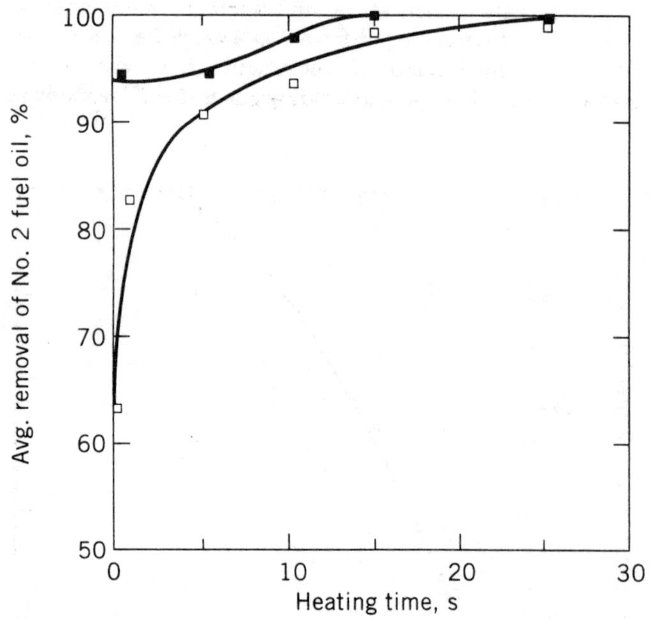

Figure 13. Average percent removal of No. 2 fuel oil as affected by total heating time, ie, time require to attain final (peak) temperature plus time held at final temperature: (nominal heating rate = 1000°C/s; initial contamination = 7 wt % fuel oil). □, peak temperature = 300°C; ■, peak temperature = 500°C. Curve: trend line. Reprinted with permission (89). Courtesy of the American Chemical Society.

concludes that the fuel oil undergoes significant chemical transformations at temperatures above 500°C with 1000°C/s heating rates (89).

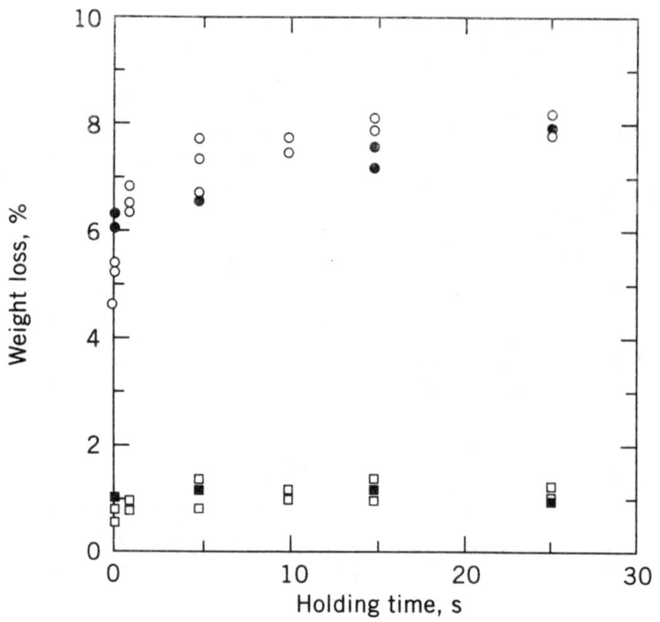

Figure 14. Weight loss for neat and contaminated soil (7 wt % fuel oil) as affected by holding time at final temperature for different nominal heating rates (peak temperature = 300°C). □, neat soil heated at 1000°C/s; ■, neat soil heated at 200°C/s; ○, contaminated soil heated at 1000°C/s; ●, contaminated soil heated at 200°C/s. Reprinted with permission (89). Courtesy of the American Chemical Society.

Table 1. Temperature–Time Profiles Leading to Essentially Complete Removal of No. 2 Fuel Oil from Soil[a]

Peak Temp, °C	Nominal Heating Rate, °C/s	Holding Time at Peak Temp, s
700	1000	0
300	1000	25
500	1000	15
300	200	25

[a] Ref. 89.

Another study (88,93) used the same apparatus to investigate removal efficiencies of a high MW, low volatility PAH contaminant (pyrene) from a synthetic soil matrix and chemical transformations of the contaminant, as affected by treatment temperature and heating rate. In contaminating the soil specimens with pyrene, a slower solvent evaporation rate was used than in (86) providing smaller pyrene crystals and improved soil–pyrene contact. Pyrene removal was calculated from the difference in weight loss of contaminated vs neat soil. Contaminant reactions were inferred from appreciable increases in CO yields from heating contaminated vs neat soil (Fig. 15). These studies (88,93) found that temperature strongly affects the rates and extent of contaminant removal and lead to the conclusion that prescribed thermal treatment substantially reduces and potentially can virtually eliminate pyrene contamination in soils. This indicates that controlled thermal treatment can also provide high levels of removal from soils of other PAHs similar in volatility and chemical reactivity to pyrene. The findings (88) on heating rate effects

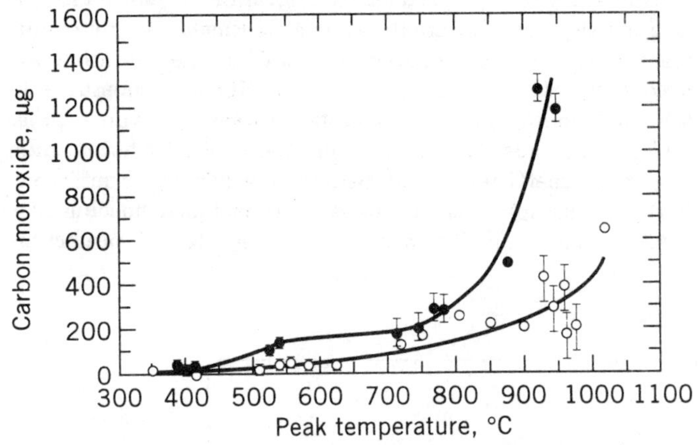

Figure 15. Carbon monoxide release for neat (○) and 4.75 wt % pyrene-contaminated soil (●) as affected by peak temperature (88,93) (nominal heating rate = 1000°C/s; zero holding time at peak temperature). Curves are trend lines (88).

on contaminant removal are more difficult to interpret. Yields of several light gases, eg, CO (Fig. 16), acetylene (Fig. 16), methane, and possibly ethylene from pyrene-contaminated soil, pass through maxima with increasing heating rate (in the 100–5500°C/s range). Heating rates between approximately 100 and 1000°C have little or no effect on total weight loss from neat and contaminated soils, but weight loss declines when the heating rate increases to 5500°C (88) (Fig. 17). These experiments allowed cooldown to begin immediately after attaining final temperature. Thus the decline in weight loss at the higher heating rate (Fig. 17) may reflect the shorter overall treatment times of these runs. Further research is needed to better understand heating rate effects under the conditions studied (88,93).

Adsorption and desorption of toluene, monochlorobenzene, and dichlorobenzene onto montmorillonite and Spherocarb particles were studied with a thermogravimetric analyzer (TGA) (81). The desorption process depends on soil type and especially on whether the soil is significantly microporous. This study (81) concludes that micropores are responsible for the slow desorption of the final monolayer of adsorbed contaminant after an initial period of rapid desorption. It was also determined that complete desorption of contaminant is achieved by prescribed heating, but that the minimum temperature required varies widely for different pollutant-particle combinations. For soil particles like Spherocarb with extensive microporosity, modeling results lead to the conclusions that surface diffusion is the main transport mechanism for contaminant release from small pores and that modeling contaminant transport only in terms of pore diffusion does not adequately describe desorption.

Using a TGA, a study (82) investigated the adsorption and desorption of benzene, trichlorobenzene, and water vapor onto and from sand and two dried soils (an EPA synthetic soil matrix and a Superfund site soil). A porous sphere taking into account intrapore diffusion coupled with a nonlinear Freundlich isotherm models the observed trends in both sorption and desorption quite well, using an effective intraparticle diffusivity as the only fitted parameter. Surface diffusion is not incorporated into the model. A parallel pore model, taking into account diffusion in both small and larger pores, improves the fit to the experimental data, but the inferred tortuosity of the pores required for the fit is too small, ie, well below typical values expected for porous particles. A plausible explanation is that relatively shallow contaminant penetration of the particle during adsorption makes contaminant removal easier during desorption, thereby the model calls for a smaller tortuosity.

Another group (80) used a TGA to study the effects of temperature, time, heating rate, and other variables on the removal of pyrene (initial loading 8 ± 0.4 wt%) from shallow piles (10 mg) of clay soil. Extensive decontamination (determined gravimetrically as 85–90%) occurs upon heating from ambient to temperatures between 200°C (at a heating rate of 5°C/min) and 300°C (at 50°C/min). The study concludes that evaporation of pyrene from the soil particles is important in removing 70% or more of the contaminant, but also that, even for the shallow beds and slow heating of the TGA, overall pyrene release rates from soil are substantially below the rates of evaporation of pure pyrene in the same apparatus.

Soil Bed Characterization

Soil thermal decontamination phenomena were studied (73–76,94) using three different research apparatuses designed to reflect important features of the behavior of soil particles and soil beds in large-scale thermal treatment equipment: a Particle Characterization Reactor (PCR), a Bed Characterization Reactor (BCR), and a small-scale rotary kiln. The BCR and PCR (Figs. **6a** and **6b**, respectively) are described in the section on Packed Beds. PCR desorption studies of four model soils show that nonporous materials, such as sand and glass beads, desorb *p*-xylene much more quickly than porous materials, such as peat and clay (74). The desorption of *p*-xylene from porous materials takes place in two stages: an initial region of rapid release which probably reflects external surface evaporation of the pollutant, followed by a much slower and rate-controlling process suggestive of intraparticle transport limitations or possibly of limited desorption rates. Experiments show that increasing temperature leads to greater evolution rates in evaporation and diffusion-desorption and to lower residual amounts of contaminant in the soil after treatment. When the PCR is used to partition *p*-xylene from the gas phase, the contaminant adsorption follows a Freundlich isotherm (74,75). This has important modeling implications for larger packed beds. Contaminant readsorption can occur if a thermal desorption wave thinner than the bed depth penetrates the bed at a rate allowing already decontaminated layers of soil to cool and then encounter contaminant vapors from deeper within the bed. These results show the importance of understanding evaporation and adsorption-desorption effects on the overall decontamination rates, even when the thermal treatment protocol allows for long duration heating of the soil (74).

PCR experiments on a gasoline-contaminated soil (74) also show the importance of contaminant volatility, with a more volatile component, *p*-xylene, desorbing ahead of a less volatile pollutant (naphthalene). Soil moisture increases contaminant desorption rates, probably because of steam distillation of the contaminant from the soil, or water occupying sites that otherwise bind contaminant (94). BCR data confirm that local bed temperature affects the desorption rate and the total amount of contaminant evolved (75). Comparison of PCR and BCR data with data from the small-scale rotary kiln shows that the contaminant evolution rate in the kiln is lower than that from the PCR but higher than that from the BCR. This finding, in light of other PCR and BCR results, suggests that in practical process equipment, such as rotary kilns, heating rate, gas–solid contacting, and possibly solid–solid contacting from the rotationally induced tumbling of the soil all affect the localized and cumulative evolution of contaminants (75).

One study (84,85) used chromatographic response analysis (see section on Chromatographic Methods) to determine various thermal parameters relevant to the pollution and decontamination of soil beds, eg, equilibrium constants for the partitioning of contaminants between the vapor phase and a phase adsorbed on the soil, heats of adsorption, axial dispersion coefficients, and intraparticle diffusion coefficients (84). The samples for the study (84) included specimens of surface soils from Newark, New Jersey. Model predictions

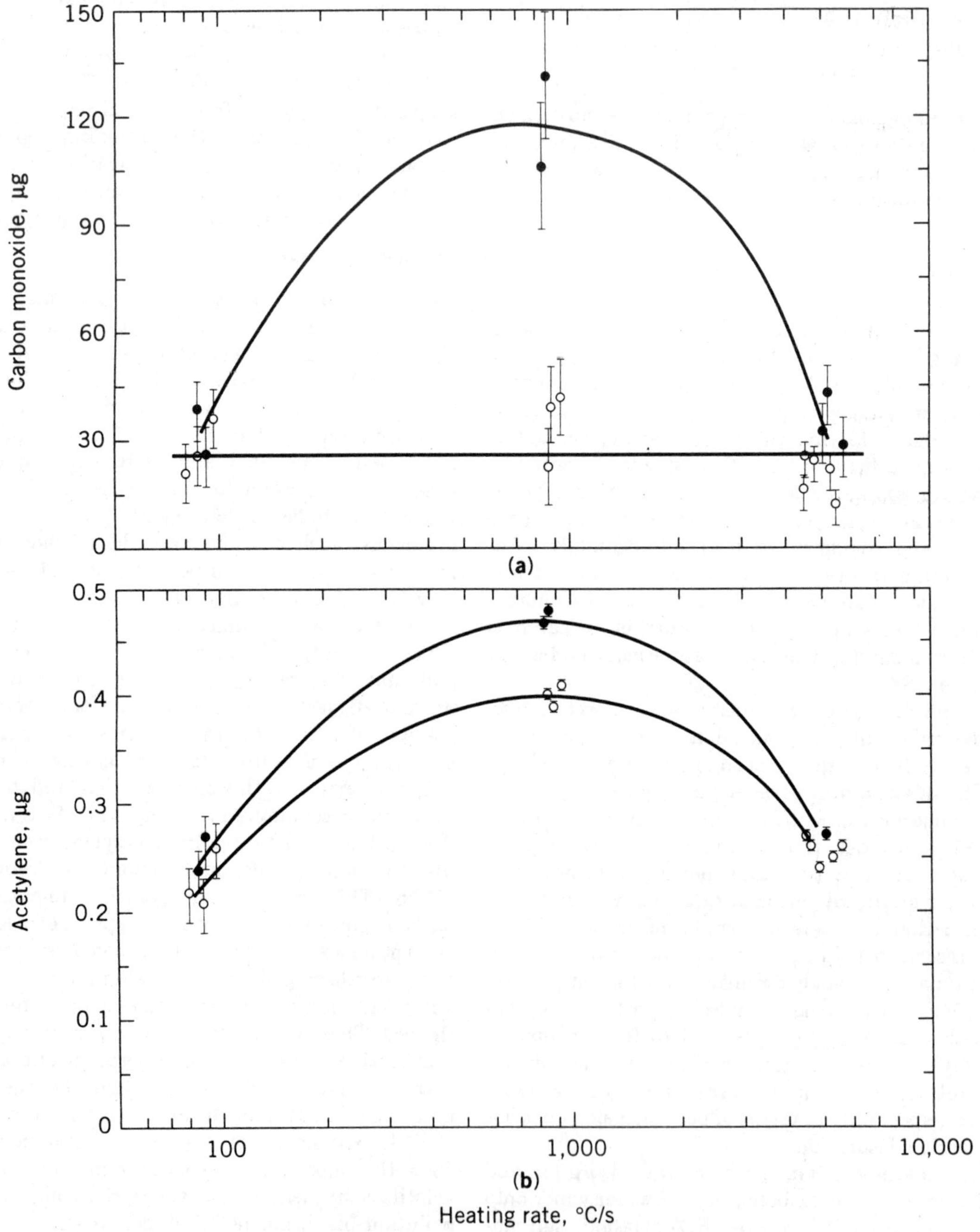

Figure 16. (**a**) Carbon monoxide and (**b**) acetylene release from neat (○) and 4.75 wt % pyrene-contaminated soil (●) as affected by nominal heating rate (88,93) (nominal peak temperature = 550°C; zero holding time at peak temperature). Curves are trend lines (88).

from experimentally determined parameters match experimental adsorption profiles reasonably well. Later work shows that soil particle size down to 0.225 mm does not affect the soil–contaminant equilibrium constants or heats of adsorption (85). For very small particles, augmentation of contaminant vapor pressure because of the curvature of the soil surface (the Kelvin effect (95)) must be accounted for. The effective diffusivities and adsorption equilibrium constants were measured (83) for monochloroethane, 1,2-dichloroethane, 1,1,1-trichloroethane, and 1,1,2-trichloroethane diffusing axially through a cylindrical packed soil pellet (see section on Packed Beds and Fig. 8). In these experiments, equal pressure was maintained on both sides of the pellet to prevent forced convection of carrier and contaminant. Moments of contaminant exit concentration vs time profiles were analyzed to derive the physical properties stated (see Packed Bed below).

Volatilizable Products From Heating Contaminated Soils

An on-line vapor sampler connected to a GC/MS system was used (62,63) to identify specific compounds evolved during thermal treatment of packed beds of soil. For example, benzene, toluene, ethylbenzene, styrene, benzaldehyde, benzofuran, chlorostyrenes, and chloroethylbenzenes were detected in the volatiles evolved upon heating in air a bed of soil contaminated with 0.5 wt % ethylbenzene. The chlorinated products are attributed to reaction of organic compounds with hot upper layers of the bed of soil, which contained 0.12% chloride (62).

Evidence was sought (88,93) for chemical transformations of an externally added contaminant (pyrene) during rapid heating of thin layers of a synthetic soil matrix in a captive sample, heated foil reactor (see "Captive Sample Reactors" below and Fig. 9). Contaminant reactions were inferred from differences in volatile yields from neat and contaminated soil. At temperatures above 530°C (at a nominal heating rate of 1000°C/s), levels of decontamination, calculated by comparing the weight loss of contaminated soil (corrected for weight loss from the soil itself) to the initial weight of contaminant, exceed 100% (Fig. 18). This behavior was attributed (88,93) to chemical reactions between the soil and the contaminant or contaminant reaction products causing additional volatilization of the soil. The observations that yields of CO (see Figs. 15 and 16) and C_2H_2 (Fig. 16) were greater from contaminated vs neat soil further support the proposition that pyrene and presumably other polycyclic aromatic compounds of similar volatility and chemical reactivity undergo chemical modifications during thermal desorption from soil at elevated temperatures, eg, at or above 550°C at a nominal heating rate of 1000°C/s. The different behavior of CO yields at 550°C, with changes in heating rate for contaminated vs neat soil (see Fig. 16), suggests that one or more of the proposed soil–contaminant reactions depends on heating rate. An important scientific and practical consequence is that soil–contaminant reactions may cause gravimetric measurements to overestimate the true level of pollutant removal. Thus it is prudent to test and, where necessary correct, for soil–contaminant effects, eg, by supplementing measurements of product yields with determinations of elemental and total material balances. The threshold temperature for such effects clearly depends on soil and contaminant reactivity and probably varies with heating rate.

THEORETICAL ANALYSES AND MATHEMATICAL MODELING

Accurate mathematical models to describe basic physicochemical phenomena occurring during the heating of contaminated

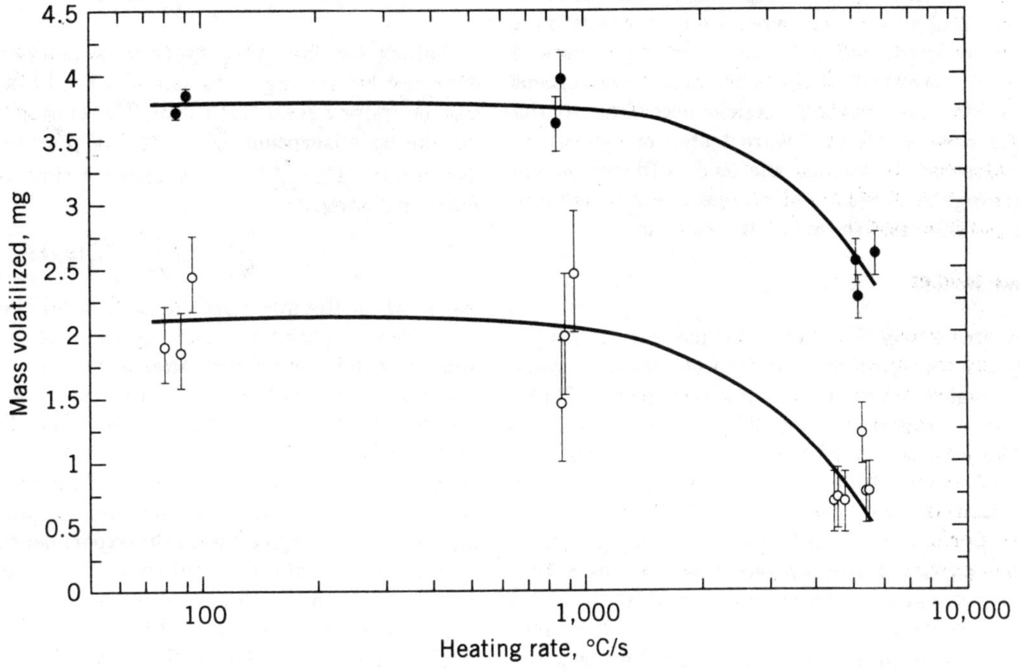

Figure 17. Weight loss for neat (○) and 4.75 wt % pyrene-contaminated soil (●) as affected by nominal heating rate (88,93) (nominal peak temperature = 550°C; zero holding time at peak temperature). Curves are trend lines (88).

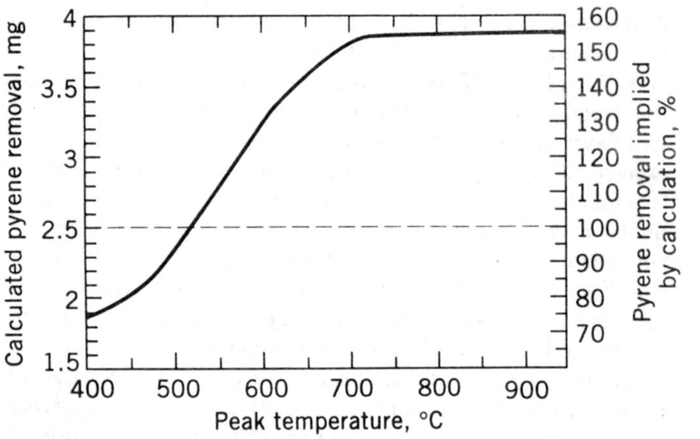

Figure 18. Effect of peak temperature on implied pyrene removal determined as the weight loss difference between neat and contaminated soil, assuming no interactions between pyrene and soil (88,93) (nominal heating rate 1000°C/s; zero holding time at peak temperature) (88).

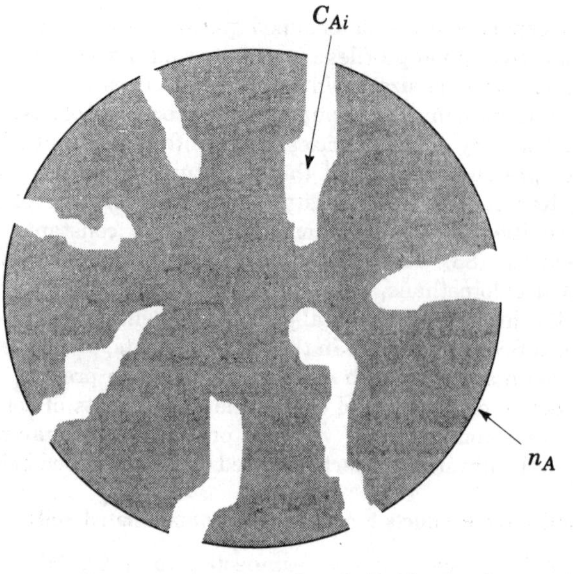

Figure 19. Conceptualization of an individual soil particle as a porous sphere (82).

soils are valuable tools for design, scale-up, simulation, monitoring, control, and optimization of thermal cleanup processes. Several authors have focused on the development of mathematical models to predict the required operating conditions for efficiently removing contaminants. On a phenomenological level, physical separation of the contaminants from the soil is a common goal of all thermal treatment technologies and sometimes is the dominant global rate process for soil cleanup. Depending on operating conditions, a variety of physicochemical phenomena contribute to the underlying details of decontamination. For example, many studies have focused on analyzing contaminant mass transfer through heated soil beds or within and away from individual soil particles. However, above a certain temperature, which in general depends on the contaminant, heating rate, and possibly soil type, chemical reactions of the contaminant are important and should also be accounted for in mathematical modeling. It therefore proves convenient to first classify the mathematical models according to the temperature range for which they were tested or developed. Then for each temperature domain, the models (based on the experimental arrangement of the soil sample) are grouped into those for single particles and those for packed beds.

Low Temperature Studies

Soil Particles. One group (82) used a thermogravimetric analyzer to study the transport and sorption of volatile organic compounds and water vapor within dry soil grains. Three different soils were employed: sand (30–60 mesh) (nominal opening 250–590 μm), a U.S. EPA synthetic soil matrix (20–30 mesh) (nominal opening 590–840 μm), and a field specimen of soil (20–30 mesh) (nominal opening 590–840 μm). The experiments were performed at 20°C. For modeling purposes, an individual soil particle was conceived of as a porous sphere (Fig. 19). The model couples intergrain diffusion with a nonlinear Freundlich isotherm ($n_A = AC_{Ai}^B$) to describe the partitioning of sorbates between the soil surface and the gas phase in the pores. Assuming that transport occurs only by gas-phase diffusion (surface diffusion is neglected), a mass balance on a small volume element within the grain is given by the following:

$$\epsilon_p \frac{\partial C_{Ai}}{\partial t} + \rho_p \frac{\partial n_A}{\partial t} = \frac{1}{r^2} \frac{\partial}{\partial r}\left(r^2 D_{ep} \frac{\partial C_{Ai}}{\partial r}\right) \quad (1)$$

Because the sorbed state is strongly favored (82), the first term in equation 1 is neglected. Using the Freundlich equation, equation 1 is rewritten as

$$\frac{\partial n_A}{\partial t} = D^* \frac{1}{r^2} \frac{\partial}{\partial r}\left[r^2 \left(\frac{n_A}{n_{A0}}\right)^{\frac{1-B}{B}} \frac{\partial n_A}{\partial r}\right] \quad (2)$$

Equation 2 is expressed in dimensionless form as follows:

$$\frac{\partial Q}{\partial \theta} = \frac{1}{x^2} \frac{\partial}{\partial x}\left[x^2 (Q)^{\frac{1-B}{B}} \frac{\partial Q}{\partial x}\right] \quad (3)$$

Values for the quantity Q as a function of x and θ are obtained by solving equation 3 with the appropriate initial and boundary conditions (82). The cumulative relative mass uptake by adsorption $\frac{M_\theta}{M_\infty}$ or the cumulative mass release by desorption $(1 - \frac{M_\theta}{M_\infty})$ at any specific time is found from the following integral:

$$\frac{M_\theta}{M_\infty} = 3 \int_0^1 x^2 Q(\theta) dx \quad (4)$$

where M_θ is the mass uptake at dimensionless time θ and M_∞ is the mass uptake at steady state (ie, as $\theta \to \infty$). The porous sphere model was used to interpret the experimental data using plots of relative mass gain (or loss) M_θ/M_∞ $(1 - M_\theta/M_\infty)$ vs the square root of dimensionless time (Fig. 20). The diffusion coefficient D^* was varied (which is equivalent to varying D_{ep}) to determine the best fit of the experimental data to the model. As shown in Figure 20, the porous sphere model agrees reasonably well with the experimental data. To further analyze the reliability of this model, a tortuosity factor τ was calculated from the combined diffusivity (calculated as a combination of the molecular and Knudsen diffusivity) and the effective diffusivity (calculated from the D^* value) (82). Similar τ values were found for different compounds in the same soil matrix (82). This result was expected because the contaminants have identical diffusion pathways. The porous sphere model

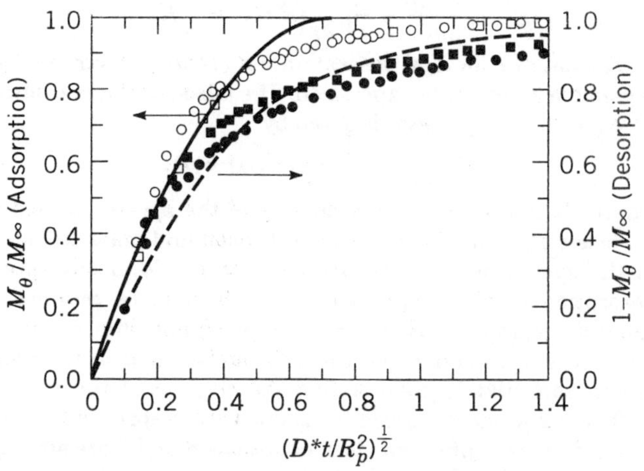

Figure 20. Kinetics for water vapor adsorption onto and desorption from a field soil. Circles represent data at 68% relative humidity and squares are for 85% relative humidity. Open and filled symbols denote adsorption and desorption data, respectively. Predictions of a porous sphere diffusion model, using the best-fitted values of the diffusion coefficient D^*, are shown as a solid line for adsorption and as a dashed line for desorption. Reprinted with permission (82). Courtesy of the American Chemical Society.

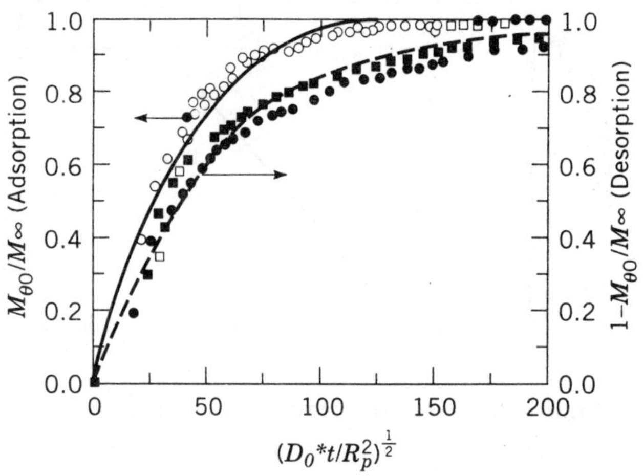

Figure 21. Kinetics for water vapor adsorption onto and desorption from a field soil. Circles represent data at 68% relative humidity and squares are for 85% relative humidity. Open and filled symbols denote adsorption and desorption data, respectively. Predictions of a fitted parallel-path pore model are shown as a solid line for adsorption and as a dashed line for desorption. Reprinted with permission (82). Courtesy of the American Chemical Society.

captures well the major features of the adsorption and desorption kinetics using a reasonable and self-consistent physical description of soil grains. However, there are some differences between the model predictions and the experimental data, eg, at longer times. These may reflect the modeling assumption that a single effective diffusivity applies to all pores. This issue was addressed by also interpreting the kinetic data with a modified version of a parallel pore model, which explicitly incorporates experimental data on pore size distribution. Details of this model can be found in (82). This model yields a better fit to the data than the porous sphere model, including for longer times (Fig. 21). However, the tortuosity inferred using effective diffusivities from this model was significantly less than unity. This finding suggests that it may also be necessary to incorporate further details of soil structure into the modeling, eg, intragranular sorption of contaminant.

Another group (81) studied the adsorption and desorption of toluene, monochlorobenzene, and dichlorobenzene onto and from particles of montmorillonite (45–53 and 90–125 μm) and Spherocarb (125–150 μm). A TGA was used as a single-layer, multiparticle reactor to measure the soil particle weight change during adsorption of organic vapor. By subjecting samples to different temperature-time histories, the effects of heating rate and of final temperature (up to 300°C) on the desorption process were determined. A desorption model was used to interpret the experimental data in terms of a probability density function *f(E)* that mathematically describes the range of sorbate–sorbent binding energies that a solid surface can exhibit, eg, because of surface heterogeneities (96). For the desorption of a monolayer of organic molecules from the solid surface, the desorption rate is assumed proportional to the amount of adsorbed molecules remaining on the surface:

$$\frac{dO_E}{dt} = k_0 \exp(-E/RT) O_{E,t} \Delta E \qquad (5)$$

where $k_E = k_0 \exp(-E/RT)$ is the rate constant, $O_{E,t}\,\Delta E$ is the number of adsorbed organic molecules remaining on the solid surface with activation energy between E and $E + \Delta E$, and O_E is the amount of organic molecules desorbed from the $O_{E,t}$ molecules with energy between E and $E + \Delta E$ during time period $0-t$. The following equation was derived to represent the total rate of desorption (96):

$$\frac{dO}{dt} = O_{tot}\,\frac{dE^*}{dt}\,f(E^*) \qquad (6)$$

where dO/dt is the desorption rate, O_{tot} is the total adsorbed amount of organic molecules, dE^*/dt is the rate of change with time of a quantity E^*, defined as the critical activation energy and satisfying (96) $\exp\{-\int_0^t k_0 \exp[\frac{-E^*}{RT(t)}]dt\} = e^{-1}$. To use this model for predictive purposes, it is necessary to calculate the energy distribution function from equation 6:

$$f(E^*) = \frac{1}{O_{tot}}\,\frac{dO/dt}{dE^*/dt} \qquad (7)$$

The quantities dO/dt and O_{tot} are measurable during temperature-programmed desorption, and E^* and dE^*/dt are calculated from the experimental data using previously described methods (81,96). If the *f(E)* computed from equation 7 is the "true" energy distribution function, it should be an intrinsic property of the soil and insensitive to the temperature-time history used in the temperature-programmed desorption. Thus, once this distribution is calculated, the model, in principle, can deal with any temperature program. Applying this model to the Spherocarb particles and using the typical temperature-time history shown in Figure 22, it was found that *f(E)* is virtually insensitive to the temperature program (Fig. 23). Further, the shape of *f(E)* for Spherocarb approximates that of a Gaussian probability density function (PDF). It was therefore concluded that the heterogeneity of solid Spherocarb adsorption energies is quite reliably modeled with a Gaussian PDF (81).

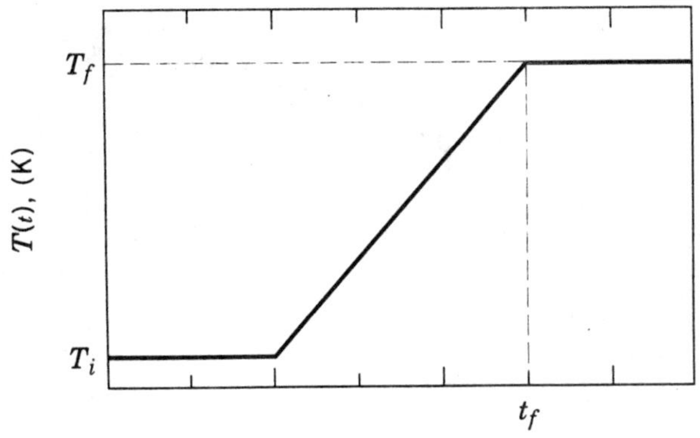

Figure 22. Typical temperature–time program (81).

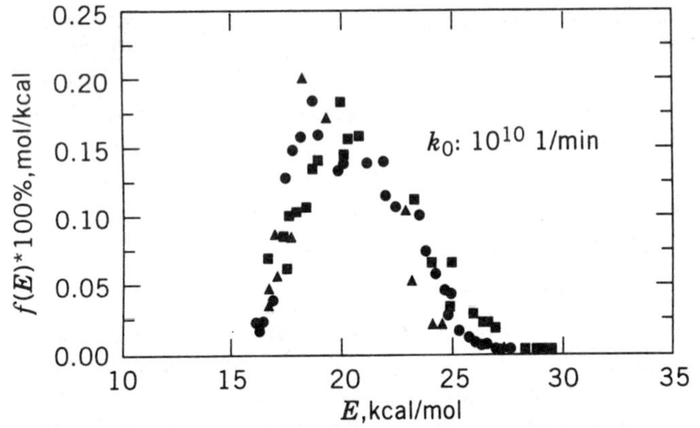

Figure 23. Energy distribution determined for desorption of dichlorobenzene from 125–150 μm particles of Spherocarb (■, slow heating ≈ 4.6°C/min; ●, fast heating ≈ 13.8°C/min; ▲, three-step heating which first raises the temperature from room temperature to 373 K, maintains it at this value for 35 min, then increases the temperature again to 473 K, maintains it for 40 min, and finally increases the temperature to 573 K (81) (1 kcal/mol = 4.186 kJ/mol).

To generate data for comparison with a proposed mathematical model, experiments were performed on desorption of contaminants from single pellets of soil (97). The single-particle reactor consisted of a clay pellet suspended inside a Pyrex tube, positioned inside the oven of a gas chromatograph. The pellets were constructed from fine clay particles (≤38 μm) using special molds. Two different montmorillonite clays were used. Adsorption isotherms for toluene were obtained for different temperatures. During the desorption experiments, the clay pellet was subjected to heating rates from 0.5 to 1.8°C/s and final temperatures at the pellet center up to ~180°C. To fundamentally assess the desorption process at an individual particle level, a detailed model was formulated incorporating several contaminant release processes, ie, heat transfer into the pellet, local desorption, bulk flow, pore diffusion, surface diffusion, and external mass transport. Figure 24 depicts the mass transport phenomena of this model in schematic form. Provided that the temperature profile within the spherical pellet is known, the heat balance is not needed. The vapor-phase species continuity equation for a contaminant species A within the spherical particle is given by

$$\epsilon_p \frac{\partial C_{Ai}}{\partial t} + \frac{1}{r^2}\frac{\partial}{\partial r}(r^2 N_A) = -R_A \qquad (8)$$

The net rate for local desorption $-R_A$ is often described by a reversible kinetic rate expression. The adsorbed layer continuity equation for species A is given by

$$\rho_p \frac{\partial n_A}{\partial t} + \frac{1}{r^2}\frac{\partial}{\partial r}(r^2 S_A) = R_A \qquad (9)$$

where S_A is the molar flux density of the species caused by surface diffusion. If the rates of diffusion and mass transport in the system are fast with respect to the local adsorption/desorption kinetics, equations 8 and 9 must be retained as separate equations. However, if local equilibrium exists between the contaminant on the clay surface and in the vapor phase immediately surrounding the clay, ie, if the rates of diffusion and mass transfer are slow with respect to the local adsorption/desorption kinetics, equations 8 and 9 are summed to eliminate the adsorption rate:

$$\epsilon_p \frac{\partial C_{Ai}}{\partial t} + \rho_p \frac{\partial n_A}{\partial t} + \frac{1}{r^2}\frac{\partial}{\partial r}[r^2(N_A + S_A)] = 0 \qquad (10)$$

The variables C_{Ai} and n_A are related by using an adsorption isotherm. The study (97) used the Freundlich isotherm $n_A = AC_{Ai}^B$. To integrate equation 10, the following constitutive equations for N_A and S_A were used (97):

$$\frac{1}{D_{AB}(\frac{\epsilon_p}{\tau})}[N_A - y_A(N_A + N_B)] + \frac{1}{D_{KA}}\left[N_A + \frac{0.36 C_{Ai} R_p^2 \epsilon_p}{\mu\tau}\nabla P\right] = -\nabla C_{Ai} \qquad (11)$$

$$S_A = -\beta \exp(-E_s/RT)\rho_p \frac{\partial n_A}{\partial r} \qquad (12)$$

Assuming that the ideal gas law applies, equation 10 is rewritten as follows:

$$\left(\epsilon_p C_i + \rho_p \frac{\partial n_A}{\partial y_A}\right)\frac{\partial y_A}{\partial t} + \left(\rho_p \frac{\partial n_A}{\partial T} - y_A \frac{\epsilon_p P}{RT^2}\right)\frac{\partial T}{\partial t} + \left(\rho_p \frac{\partial n_A}{\partial P} + y_A \frac{\epsilon_p}{RT}\right)\frac{\partial P}{\partial t} + \frac{1}{r^2}\frac{\partial}{\partial r}[r^2(N_A + S_A)] = 0 \qquad (13)$$

Summing equation 13 over species A and (nonadsorbing) B for a binary system, the total continuity equation is expressed as follows:

$$\left(\rho_p \frac{\partial n_A}{\partial y_A}\right)\frac{\partial y_A}{\partial t} + \left(\rho_p \frac{\partial n_A}{\partial T} - \frac{\epsilon_p P}{RT^2}\right)\frac{\partial T}{\partial t} + \left(\rho_p \frac{\partial n_A}{\partial P} + \frac{\epsilon_p}{RT}\right)\frac{\partial P}{\partial t} + \frac{1}{r^2}\frac{\partial}{\partial r}[r^2(N_A + S_A)] = 0 \qquad (14)$$

Provided that the temperature profile within the spherical pellet is known, equations 13 and 14 are the governing partial differential equations (PDEs) that describe the thermal desorption process, assuming local equilibrium. The unknown dependent variables are y_A and P. The boundary and initial conditions used in evaluating the governing equations are detailed (97). External mass transfer enters the model as an algebraic boundary condition. In solving the governing equations, not all of the quantities involved in the model were known a priori. The unknown quantities include τ, α, and E_s. Initially, the values of these three unknown parameters were determined simultaneously by adjusting them to maximize the agreement of predictions from the mathematical model with the experimental data. The problem encountered with this approach was that one transport mechanism dominates the overall model for any given set of parameters. Consequently, consistent values for the model

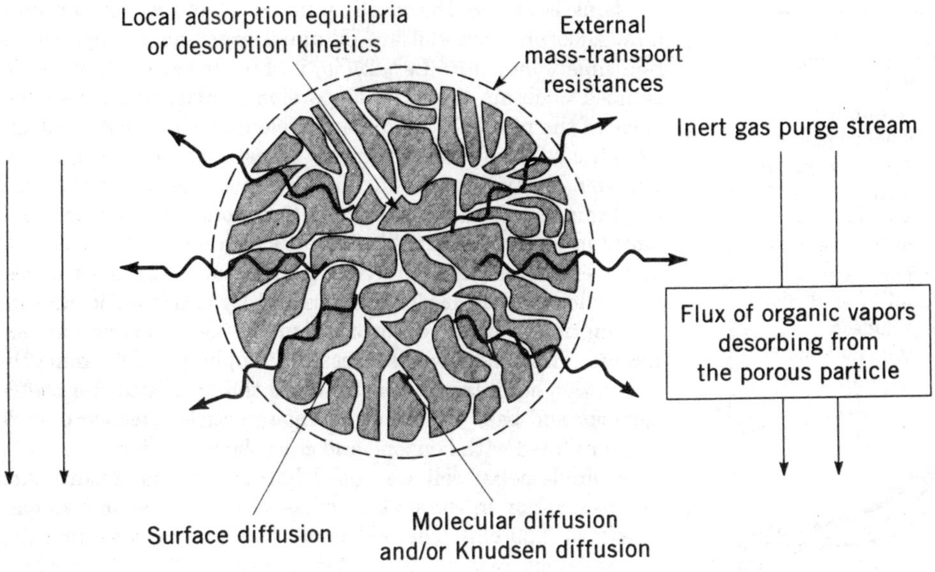

Figure 24. Schematic of mass-transport phenomena included in a soil thermal decontamination model. Reprinted with permission (97). Courtesy of the American Chemical Society.

parameters could be determined only for dominant transport mechanisms. Because this initial approach was not perfectly satisfactory, the study (97) used a simpler modeling strategy to determine which mechanism acting by itself best correlates the experimental data. Each of the five fits in Figure 25 was obtained with a single mechanism controlling the overall desorption rate. These results indicate that any one of the five mechanisms correlates the experimental data reasonably well. However, not all five mechanisms were judged equally probable as the dominant mechanism. Based on experimental evidence local pellet surface-to-vapor desorption and extrapellet mass transfer were discarded as plausible controlling mechanisms (97). Because the three remaining mechanisms (ie, molecular diffusion, Knudsen diffusion, and surface diffusion) are related to intrapellet mass transfer, the rate-controlling mechanism was associated with intrapellet diffusion.

Packed Beds. Chromatographic response factors (measured with a modified gas chromatograph) were used (84) to elucidate mass transfer of volatile organic substances (chloroform, methylene chloride, carbon tetrachloride, 1,1,1-trichloroethane, benzene, toluene, chlorobenzene, and 1,2,4-trichlorobenzene) through heated soil matrices. Organic soil taken from the ground surface in Newark, New Jersey, was sieved and No. 35/40 mesh particles were packed in a stainless steel tube 14.2 cm long. Depending on the contaminant, experiments were performed at temperatures ranging from 100 to 260°C. A mathematical model was derived (84) to relate the chromatographic behavior of organic contaminants in the packed soil column to the following basic mass-transfer processes responsible for that behavior: axial dispersion of components along the column; diffusion in a film surrounding the particle; and diffusion within the particle. The model assumes that adsorption and desorption of components between the gas phase of the intraparticle pore space and the particle surface are determined by thermodynamic equilibrium. Figure 26 is a schematic drawing of the packed soil bed system studied (84).

A mass balance for the gas phase in the column is given by the following

$$D_{ez}\frac{\partial^2 C_A}{\partial z^2} - v_s\frac{\partial C_A}{\partial z} - \frac{3(1-\epsilon_B)}{R_p}N_{R_p} = \epsilon_B\frac{\partial C_A}{\partial t} \tag{15}$$

where

$$N_{R_p} = D_{ep}\left(\frac{\partial C_{Ai}}{\partial r}\right)_{r=R_p} = k_f\left(C_A - C_{A-R_p}\right) \tag{16}$$

and the mass balance inside the particles is expressed by

$$D_{ep}\left(\frac{\partial^2 C_{Ai}}{\partial r^2} + \frac{2}{r}\frac{\partial C_{Ai}}{\partial r}\right) = \epsilon_p(1+K_a)\frac{\partial C_{Ai}}{\partial t} \tag{17}$$

Equation 17 reflects the linear equilibrium assumed between the solid-phase and gas-phase concentrations of contaminant, ie, $n_A\rho_p/C_{Ai}\varepsilon_P = K_a$. Using the appropriate boundary and initial conditions and a Laplace transform method, the study (84) developed expressions for contaminant concentrations at the column outlet. Applying moment analysis techniques, solutions were determined for the first absolute moment μ_1 (physically this is the average retention time) and the second central moment μ'_2 (physically the variance of the chromatographic response). In addition, the following equations were derived for corrected versions of μ_1 and μ'_2, which are more useful for practical computations:

$$\overline{\mu} = \frac{L\epsilon_B}{v_s}\left[1 + \frac{(1-\epsilon_B)}{\epsilon_B}\epsilon_p(1+K_a)\right] \tag{18}$$

$$\overline{\sigma}^2 = \frac{2LD_{ez}\epsilon_B^2}{v_s^3}\left[1 + \left(\frac{1-\epsilon_B}{\epsilon_B}\right)\epsilon_p(1+K_a)\right]^2 + \frac{2LR_p^2(1-\epsilon_B)\epsilon_p^2}{15v_sD_{ep}}(1+K_a)^2 + \frac{2LR_p(1-\epsilon_B)\epsilon_p^2}{3v_sk_f}(1+K_a)^2 \tag{19}$$

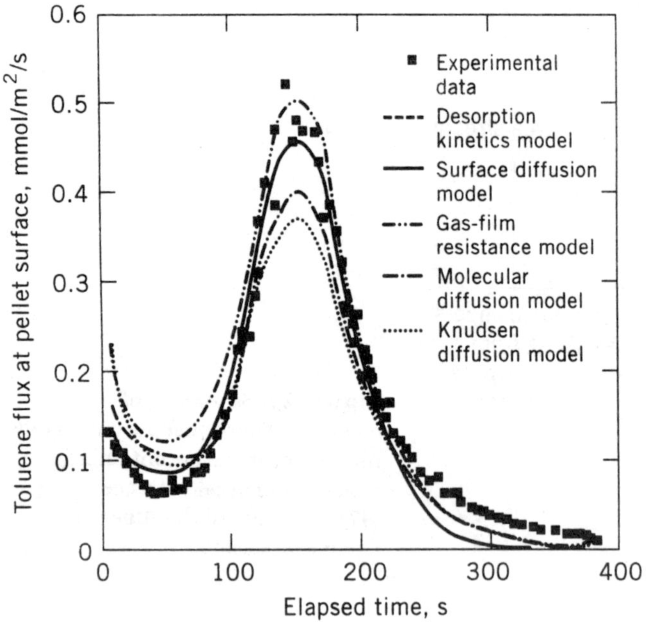

Figure 25. Comparison of rates of toluene release from a single clay pellet predicted from five single-mechanism desorption models. Reprinted with permission (97). Courtesy of the American Chemical Society.

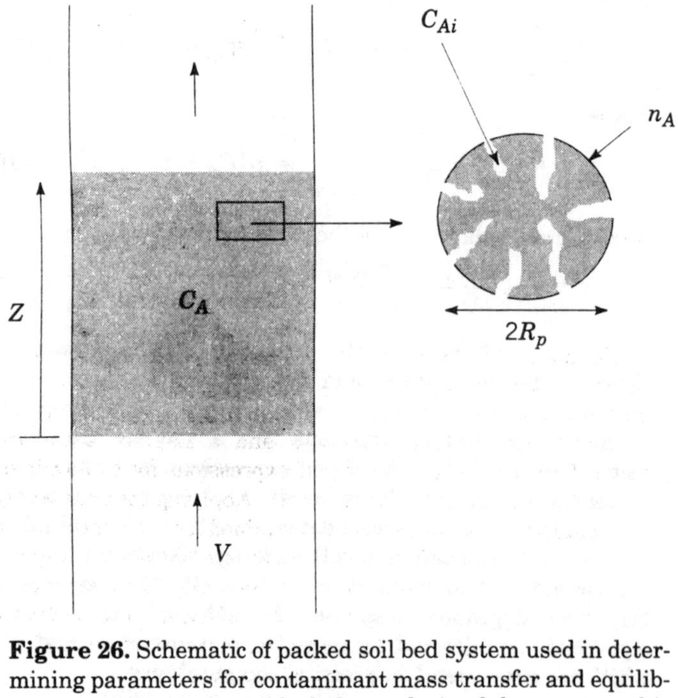

Figure 26. Schematic of packed soil bed system used in determining parameters for contaminant mass transfer and equilibrium partitioning in soil beds by analysis of chromatographic response factors (84).

Combining equations 18 and 19,

$$\frac{\overline{\sigma}^2 L\epsilon_B^2}{2\overline{\mu}^2 v_s} = \frac{D_{ez}}{v_s^2} + \frac{1-\epsilon_B}{\epsilon_B}\left(\frac{R_p^2}{15D_{ep}} + \frac{R_p}{3k_f}\right) \times \left[\frac{\epsilon_B}{(1-\epsilon_B)\epsilon_p(1+K_a)+1}\right]^{-2} \tag{20}$$

Using equation 18, equilibrium constants K_a for contaminant adsorption on soil are obtained from the average retention time. Plots of $\overline{\sigma}^2 L\epsilon_B^2/2\overline{\mu}^2 v_s$ vs $1/v_s^2$ from equation 20 are utilized to obtain the axial dispersion coefficient D_{ez} from the slope of the plot and the intraparticle diffusion coefficient D_{ep} from the value of intercept, given an estimate of the film mass transfer coefficient. As an illustration of this analysis, data on eliminating toluene contamination from a soil bed are compared in Figure 27 with the numerical solution of this model, using estimated mass-transfer parameters. The good agreement shown in Figure 27 lends credibility to this model and the assumption of linear adsorption under the experimental conditions. This model was successfully applied to determine the adsorption equilibrium constant and the mass-transfer coefficient for soil beds (different average particle sizes were used) contaminated with benzene and chlorobenzene (85).

A single-pellet cell was used (83) to investigate diffusion and adsorption in soil pellets (cross-sectional area of 1.43 cm^2, length of 0.29 cm). The soil pellet was prepared by pressing the soil (obtained from the Afyon region of Turkey) into a ring-shaped stainless holder and then placing it into the single pellet cell, located in the oven of a gas chromatograph. Pulse-response experiments with inert gas or chlorinated hydrocarbons passing over both end faces of the soil pellet were performed. To eliminate forced convective transport through the pellet, the pressures of the carrier gas streams in the upper and the lower chambers of the cell were adjusted to the same value (83). The soil pellets were heated up to final temperatures in the range 35 to 90°C. The single porous pellet concept used in this group's (83) model for decontaminating a soil bed is shown schematically in Figure 28. The vapor-phase species continuity equation for contaminant A within the porous pellet is written as follows:

$$\epsilon_p \frac{\partial C_{Ai}}{\partial t} + \frac{\partial N_A}{\partial z} = -R_A \tag{21}$$

The adsorbed layer continuity equation for species A, neglecting surface diffusion, is given by

$$\rho_p \frac{\partial n_A}{\partial t} = R_A \tag{22}$$

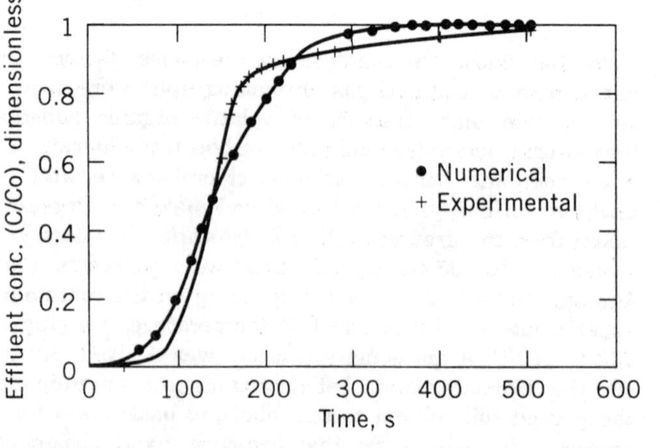

Figure 27. Comparison of experimental data and model predictions (see text) for cumulative elution of toluene contamination from a soil bed at 160°C as affected by time. Reprinted with permission (84). Courtesy of Gordon and Breach Publishers.

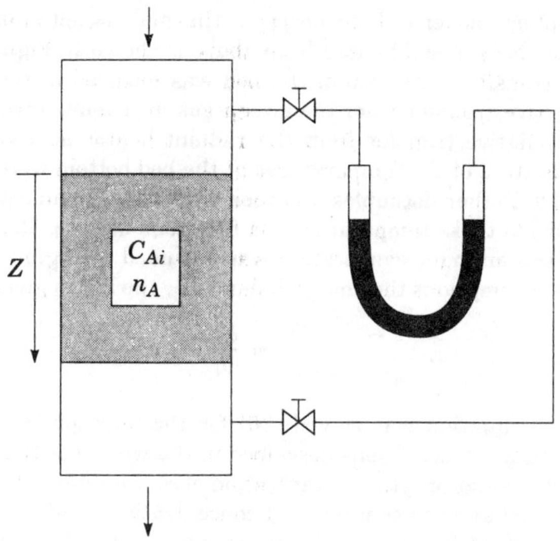

Figure 28. Schematic of a porous pellet system used to derive parameters for contaminant mass transfer and equilibrium partitioning in soil beds by analysis of chromatographic response factors. Reprinted with permission (83). Courtesy of the American Chemical Society.

For an assumed first-order, reversible vapor–solid partitioning process, the rate of contaminant adsorption on the soil surface is given by

$$R_A = k_a\left(C_{Ai}\epsilon_p - \frac{n_A\rho_p}{K_a}\right) \tag{23}$$

Equations 21 and 22 are solved by Laplace transforms using the initial and boundary conditions previously described (83). Applying moment analysis techniques (83), solutions were determined for the first absolute μ_1 and the second central μ'_2 moments. The resulting mathematical expressions are as follows:

$$\mu_1 = \frac{L^2\epsilon_p}{6D_{ep}}\left(1 + \frac{K_a}{\epsilon_p}\right)\left(\frac{3\frac{S}{L}D_{ep} + F}{\frac{S}{L}D_{ep} + F}\right) \tag{24}$$

and

$$\mu'_2 = \left(\frac{L^2\epsilon_p}{D_{ep}}\right)^2\left(1 + \frac{K_a}{\epsilon_p}\right)^2\left[\frac{(\frac{S}{L}D_{ep})^2 + \frac{2}{5}(\frac{S}{L}D_{ep})F + \frac{1}{15}F^2}{6(\frac{S}{L}D_{ep} + F)^2}\right]$$

$$+ \left(\frac{L^2K_a^2\epsilon_p}{3D_{ep}k_a}\right)\left[\frac{3(\frac{S}{L}D_{ep})^2 + 4(\frac{S}{L}D_{ep})F + F^2}{(\frac{S}{L}D_{ep} + F)^2}\right] \tag{25}$$

By setting $K_a = 0$ in equation 24, the first absolute moment for an inert tracer is obtained. Applying this model, it is possible to determine the values of D_{ep}, k_a and K_a. In fact, for high stream flow rates, equation 24 becomes

$$\mu_{1,\infty} = \frac{L^2\varepsilon_p}{6D_{ep}}\left(1 + \frac{K_a}{\varepsilon_p}\right) \tag{26}$$

To obtain K_a from equation 26 the quantity D_{ep} must be known. D_{ep} (from eq. 26) was determined from experiments performed with a nonadsorbing, ie, $K_a = 0$, tracer (helium) (83). The tortuosity factor was evaluated from D_{ep} and the calculated combined diffusivity (ie, $D_{ep} = \epsilon_p D_p/\tau$). Experiments were performed at different temperatures, but no significant variation of tortuosity with temperature was observed (83). Knowing D_{ep}, K_a values are determined from equation 24. With the effective diffusion coefficient and adsorption equilibrium constant obtained from the first moment analysis (and considering that the second moment is evaluated from experimental pulse-response curves), the rate constant k_a is the only unknown in equation 25. By using the described single-pellet technique, the adsorption equilibrium and rate constants are easily determined.

Experiments were designed to characterize intraparticle transport of contaminants, where the contaminant concentration and the temperature are each well known at the particle surface and to characterize interparticle transport, where the concentration and temperature are each known at the top of a bed of particles (76). A PCR (see Fig. **6b**) was used to evaluate intraparticle resistances, and a BCR (Fig. **6a**) was used in one-dimensional mass-transfer and heat-transfer experiments to consider resistances within a soil bed swept by a desorbent gas passing across the top of the particle bed (see also Packed Beds above). The soil used in the experiments was a 10% moisture clay with quartz, illite, and montmorillonite mineralogy. The average particle diameter was 838 μm, and *p*-xylene was the contaminant.

The PCR consisted of a bed of soil (up to 2.5 cm thick) through which a preheated desorbent gas was passed. The exhaust gas was analyzed with a gas chromatograph. After initiating desorbent flow, the soil bed reached 90% of its final operating temperature in approximately 10 min. The maximum temperature for the sweep gas was 270°C. By performing a mass balance on the gas-phase concentration of *p*-xylene within the bed and assuming negligible radial gradients and axial dispersion, the following expression was obtained for the transient concentration of contaminant (76):

$$-v_s\frac{\partial C_{Ai}}{\partial z} - \rho_B\frac{\partial n_A}{\partial t} = \epsilon_B\frac{\partial C_{Ai}}{\partial t} \tag{27}$$

It is assumed that contaminants on the soil particle surfaces and in the vapor phase above the soil particles are in local equilibrium and that the quantities n_A and C_{Ai} are related by the Freundlich isotherm $n_A = AC_{Ai}^B$. Because the isotherm parameters A and B depend on temperature, it is necessary to solve equation 27 together with the energy balance for the bed. It is assumed that conduction through the soil is negligible (the soil is primarily heated via convection from the gas) and that the effect of heat loss from contaminant evaporation is small (76). Results obtained from the energy balance are shown in Figure 29. By solving the energy and mass balances simultaneously, the gas-phase contaminant concentration C_{Ai} is evaluated. The contaminant concentration n_A in the solid phase is then determined by the Freundlich isotherm equation. In Figure 30 the model predictions of the average concentration of *p*-xylene remaining on the soil (calculated by integrating the predicted *p*-xylene concentrations remaining at each bed depth over the entire bed) after desorption for one hour are compared with the experimental results. Figures 29 and 30, respectively, show that the model agrees well with the experimental data on bed temperature and bed decontamination. The results indicate that, for the soil, contaminant, and treatment conditions studied, intraparticle transport of contaminant is not controlling.

The BCR (see Fig. **6a**) was used to determine the heat-transfer and mass-transfer resistances through a bed of soil.

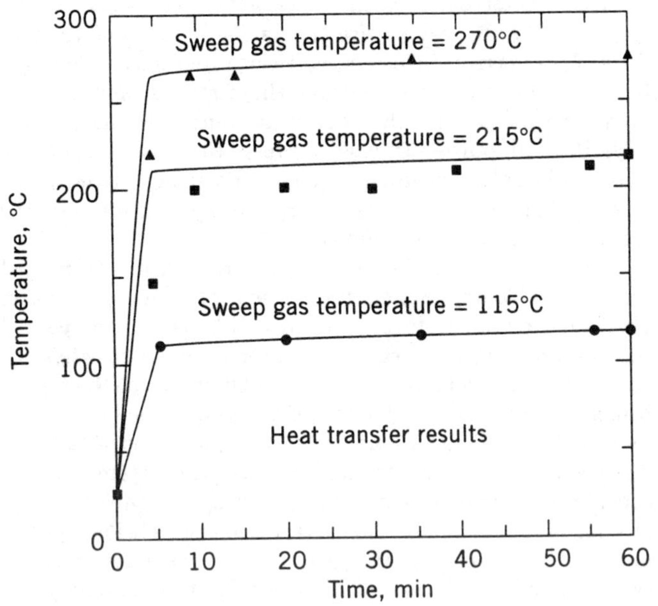

Figure 29. Comparison of model predictions (smooth curves, see text) and experimental measurements of the bed center temperature (symbols) for a bed of clay soil heated by passing preheated sweep gas upward through the bed. Initial inlet temperatures of the sweep gas are indicated. Reprinted with permission (76). Courtesy of the American Chemical Society.

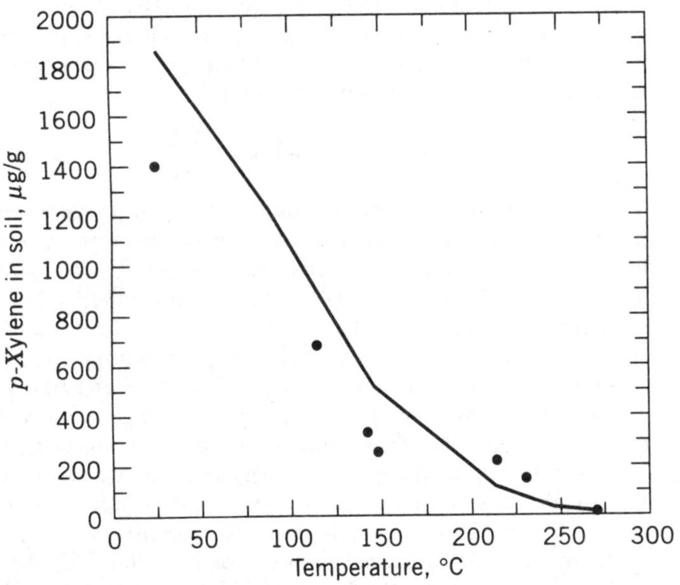

Figure 30. Comparison of model predictions (curve, see text) with experimental data (symbols) for the concentration of p-xylene remaining on a bed of soil after thermal desorption by a preheated sweep gas for one hour. Reprinted with permission (76). Courtesy of the American Chemical Society.

The bed was heated with a radiant heater located above the reactor, while desorbent gas was passed over the top of the bed. Temperatures within a 7.6-cm deep bed of soil were measured at different depths, and the weight loss of the bed was also recorded. A schematic of the processes analyzed in the bed desorption model (76) to describe thermal decontamination of thick beds of soil heated from above is shown in Figure 31. Heat transfer to the top of the bed was modeled in terms of convective transfer from the sweep gas at a temperature T_g, and radiative transfer from the radiant heater at a surface temperature of T_s. Temperatures at the bed bottom were measured with thermocouples and then were fitted to an equation to provide these temperatures as functions of time. Heat loss from contaminant evaporation was assumed negligible. With these assumptions the energy balance for the bed is given by

$$\alpha \frac{\partial^2 T}{\partial z^2} = \frac{\partial T}{\partial t} \tag{28}$$

This equation was solved (76) for the appropriate initial and boundary conditions described in the work. For the mass-transfer equations, the concentration of contaminant at the top of the bed and the contaminant concentration gradient at the bottom of the bed were both assumed to be zero. The contaminant concentration within the particle was represented by an average value n_A. Mass balances for the solid-phase and gas-phase concentrations (76) are given by

$$\frac{\partial n_A}{\partial t} = \frac{12 D_{AB}(1 - \epsilon_B)}{d_p^2 \rho_B}(C_A - C_{A,sat}) \tag{29}$$

and

$$\frac{\partial C_A}{\partial t} = \frac{1}{\epsilon_B}\frac{\partial}{\partial z}\left(D_{ez}\frac{\partial C_A}{\partial z}\right) - \frac{\rho_B}{\epsilon_B}\frac{\partial n_A}{\partial t} \tag{30}$$

The gas phase contaminant concentration $C_{A,sat}$ at the surface of the particle was found by using the Clausius–Clapeyron equation and the heat of adsorption from the Freundlich isotherm.

The assumption of negligible heat loss due to evaporation decoupled heat and mass transfer, and allows solving the heat transfer equation 28 independently of the mass transfer equations. The effective thermal conductivity (involved in the thermal diffusivity parameter α) was adjusted to fit the experimental data. The results of the heat transfer model were compared with measured data for three gas temperatures. As shown in Figure 32, for a 7.6-cm deep bed of clay particles, the model predicts the time and spatial variations of bed temperature quite reliably for all three gas temperatures. The mass transfer equations 29 and 30 were solved with the temperature profiles obtained from the heat transfer model and the appropriate initial and boundary conditions (76). A parameter involved in calculating $C_{A,sat}$ (76) was adjusted so that the resulting model predictions fit the 240°C data (76). Figure 33 compares resulting model predictions of p-xylene evolution vs time with experimental data for a 7.6-cm deep bed of clay particles and sweep gas temperatures of 175, 240, and 315°C. The mass transfer model accurately predicts that the contaminant evolution rate remains essentially constant at a given temperature for a lengthy period of heating and then tapers off at longer times. Figure 34 shows the model predictions for p-xylene desorption from a somewhat more shallow bed (5.1 cm deep) at a sweep gas temperature of 240°C. Again, the model accurately predicts the evolution of contaminant vs time.

Recently, a mathematical model was described (80) for removing pyrene contamination from a shallow pile of soil undergoing nonisothermal heating in a laboratory scale thermobalance (TGA). The analysis describes pyrene release in

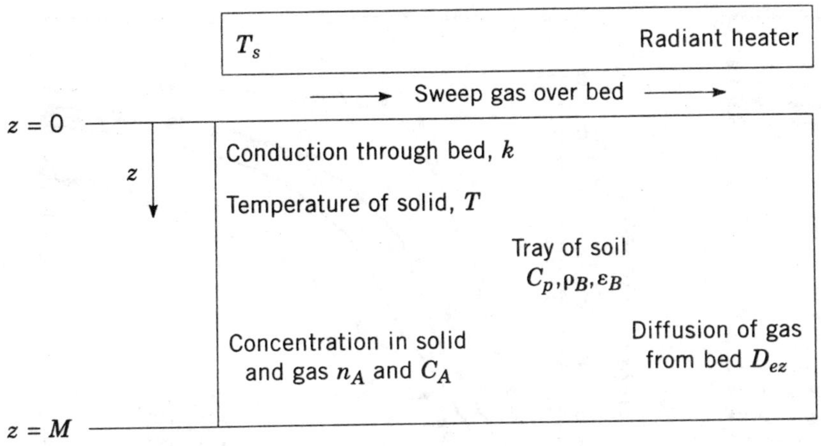

Figure 31. Schematic of processes analyzed in a bed desorption model developed to describe thermal decontamination of deep beds of soil heated from above, as in the bed characterization reactor (BCR). Reprinted with permission (76). Courtesy of the American Chemical Society.

terms of pyrene evaporation from a liquid sheath surrounding each soil particle, followed by nonreactive diffusive transport of pyrene through the soil pile and then through a nonreactive boundary layer occupying the volume between the top surface of the soil pile and the open top of the TGA crucible. For heating from near ambient temperature (ca 30°C) up to as much as 250°C, the model reasonably well predicted observed pyrene removal efficiencies as affected by heating rate, ie, 10–80%, 5–85%, and 2–70% ranges of decontamination at heating rates of 5, 25, and 50°C/min, respectively. The study (80) notes that two phenomena not treated in this model, pollutant chemical reactions and pollutant desorption from the soil, respectively, must be accounted for at higher temperatures and when contamination levels are comparable to roughly monolayer coverages of the soil. Follow-on modeling is in progress (102).

High Temperature Treatment

Thin Beds Of Soil. One study (3) reported yields, compositions, and global release rates of major products from heating a synthetic soil matrix. The soil specimens were mixtures prepared for the U.S. EPA to reflect certain attributes of soils encountered at Superfund sites in the United States. Because this study focused on determining how the soil decomposition itself affects thermal cleanup, exogenous contaminants were not added to the soil. The experiments involved heating 49–54-mg batches of 63–125-μm neat soil particles, thinly spread on a 0.001-in thick stainless steel foil, at nominal rates of 1000°C/s, to temperatures in the 350–1050°C range using the apparatus depicted in Figure 9. Above 500–600°C, total weight loss and the CO_2 yield increased strongly with increasing temperature. Significant weight loss, and by implication significant chemical decomposition of the soil, was observed at high temperatures. The data on total weight loss were fit to a multiple independent parallel reaction kinetics (MIPR) model (90,99). This global model assumes that the release of the devolatilization products occurs by a large number of independent parallel first-order reactions. The rate of product evolution by any individual reaction i is assumed to be first order in the amount of the product yet to be generated by that reaction:

$$\frac{dV_i}{dt} = k_i(V_i^* - V_i) \tag{31}$$

where V_i is the cumulative amount of product generated up to time t from reaction i, V^*_i is the ultimate amount of product generated by reaction i ($V_i \rightarrow V_i^*$ as $t \rightarrow \infty$), and k_i is the rate constant for reaction i, assumed to have the usual Arrhenius temperature dependence ($k_i = k_{0i}\exp(-E_i/RT)$). The total amount of product yet to be released by reaction i after reaction from zero time to a time t_f is then computed by integrating equation 31 over the period of devolatilization (ie, *0* to t_f)

$$(V_i^* - V_{if}) = V_i^* \exp\left(-\int_0^{t_f} k_i dt\right) \tag{32}$$

The model may be simplified by two assumptions (90): (*1*) that all reactions have the same preexponential factors ($k_{0i} = k_0$, for all i); and (*2*) that the number of reactions is large enough to permit expressing E as a continuous distribution function $F(E)$, representing the fraction of the ultimate weight loss from all reactions with an activation energy $\leq E$. Then $f(E) = dF(E)/dE$ is the probability density function for E, such that V^*_i is a differential part of V^*, the ultimate amount of product from all reactions. Then the quantity V^*_i can be written

$$V_i^* = dV^* = V^*dF = V^*f(E)dE \tag{33}$$

where $f(E)$ is normalized in the domain $0 \leq E \leq \infty$. The total amount of volatiles yet to be released $(V^* - V_f)$ is computed by summing the contribution from each reaction, ie, by integrating equation 32 over all values of E using equation 33:

$$(V^* - V_f)/V^* = \int_0^{\infty} \exp\left\{-k_0 \int_0^{t_f} \exp[-E/RT(t)]dt\right\} f(E)dE \tag{34}$$

It is recognized that the probability density function $f(E)$ is not a function of time because it describes ultimate yields, which, by definition, are independent of time. In general, $F(E)$ is any continuous distribution function, and $f(E)$ is the corresponding probability density function. In the present application, $f(E)$ is taken as a Gaussian (90):

$$f(E) = [\sigma(2\pi)^{1/2}]^{-1}\exp[-(E - E_0)^2/2\sigma^2] \tag{35}$$

With equation 35 the model utilizes three adjustable parameters: k_0, the mean E_0, and σ the standard deviation of the probability density function. The ultimate weight loss V^*

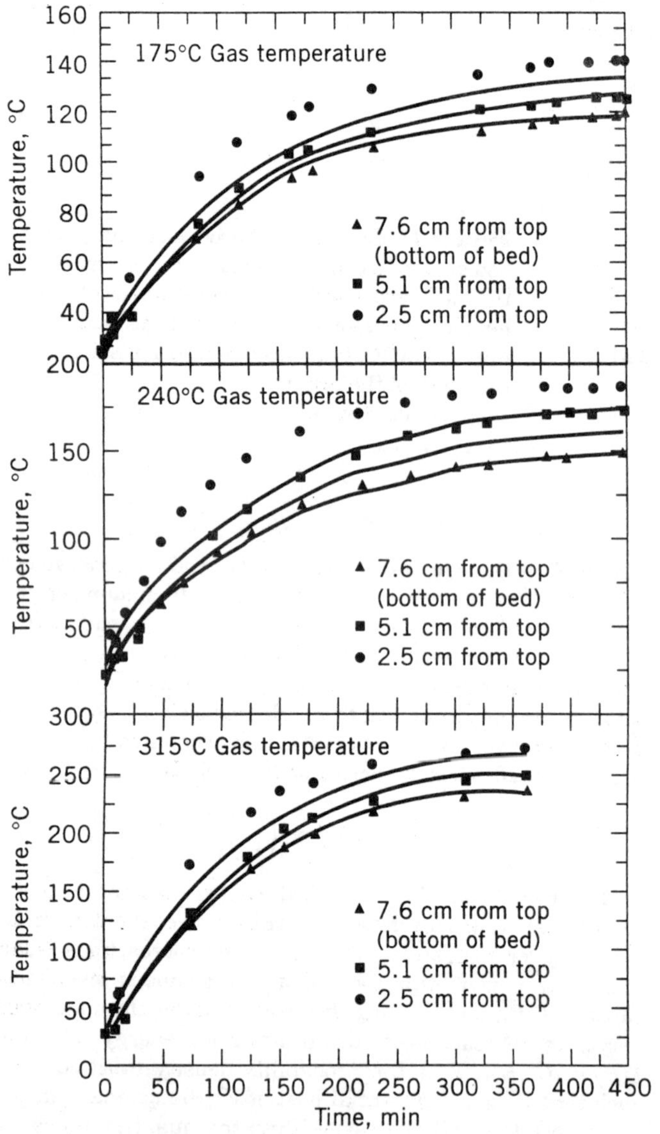

Figure 32. Comparison of temperatures at various depths in a 7.62-cm deep bed of clay particles predicted from a heat-transfer model (curves, see text) and experimentally measured (symbols), as affected by time for various sweep gas temperatures (sweep gas passed over top of bed). Reprinted with permission (76). Courtesy of the American Chemical Society.

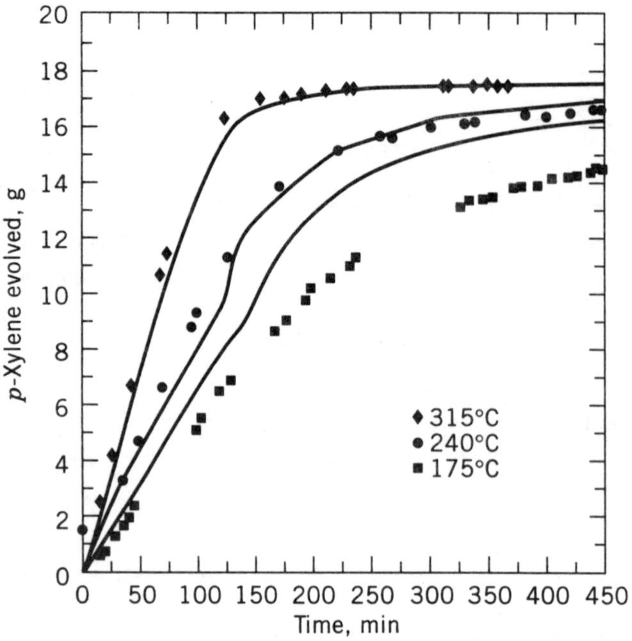

Figure 33. Comparison of quantity of p-xylene contaminant evolved from a 7.62-cm deep bed of clay particles predicted from a mass-transfer model (curves, see text) and experimentally measured (symbols), as affected by treatment time for various sweep gas temperatures (sweep gas passed over top of bed). Reprinted with permission (76). Courtesy of the American Chemical Society.

was obtained from experiments at high temperature and long residence times (3). The satisfying agreement between the predicted and measured weight losses implied in Figure 35 illustrate the good performance of this kinetics correlation. The data on yields of several individual products generated during the heating were successfully correlated with a single-reaction, first-order decomposition model (eq. 32) (3). Because CO_2 evolved by step-like behavior with increasing temperature, its release rates were modeled in terms of two, independent, parallel, first-order reactions.

Using the same apparatus and similar experimental methods, another study (88) compared, for the same neat soil, predictions of a MIPR model to trends observed in soil weight loss and acetylene production as affected by temperature (350–1050°C) at a nominal heating rate of 1000°C/s and as affected by nominal heating rate (100–5500°C/s) for a peak temperature of 530°C. When only heating rate data for neat soil weight loss were best fitted to the model, the model predicted decreasing weight loss with increasing heating rate, whereas the data show that weight loss is constant for heating rates from 100–1000°C/s, and then declines when the heating rate increases to about 5500°C/s (see Fig. 17). When the temperature data are added to the fit, the model predicts an even stronger decrease in weight loss with increasing heating rate.

To improve the quality of the data fits, two other global kinetic models were tested (88). One was a competitive first-order reactions model which assumed that, depending on the magnitudes of the relevant rate constants, a certain fraction of the soil is converted entirely to an observable volatile product P, entirely to a nonvolatile side product S, or to some mixture of P and S. The yields of one or more of the specific substances composing P may be quantified by experiment. For a constant volume soil particle, the differential equations to be solved for this model are

$$-\frac{dM_i(t)}{dt} = (k_s + k_p)M_i(t) \tag{36}$$

$$\frac{dM_s(t)}{dt} = k_sM_i(t) \tag{37}$$

$$\frac{dM_p(t)}{dt} = k_pM_i(t) \tag{38}$$

where M_i, M_S, and M_P are respectively the masses of volatilizable material, side product, and observed product, and k_S and k_P are the Arrhenius rate constants for the production of side product and observed product. Initial conditions are

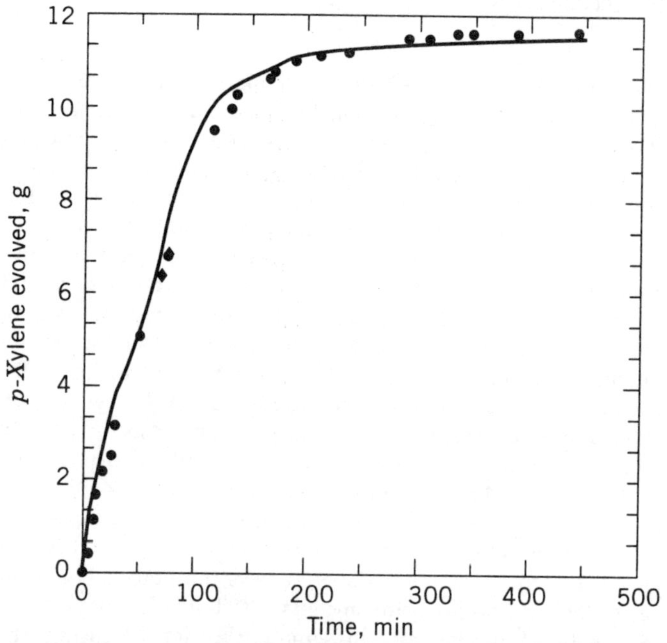

Figure 34. Comparison of quantity of *p*-xylene contaminant evolved from a 5.1-cm deep bed of clay particles predicted from a mass-transfer model (curves, see text) and experimentally measured (symbols), as affected by treatment time for a sweep gas temperature of 240°C (sweep gas passed over top of bed). Reprinted with permission (76). Courtesy of the American Chemical Society.

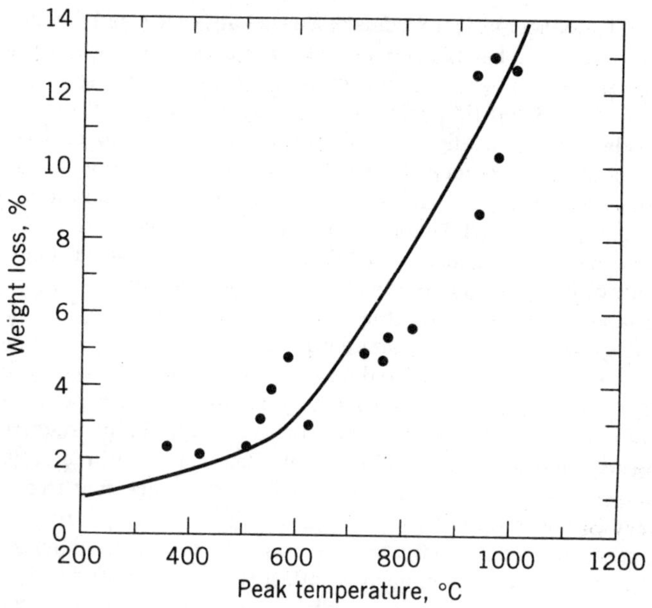

Figure 35. Experimental data points and model-generated curve for total weight loss from a synthetic soil matrix (see text) heated from room temperature to the peak temperature specified on the abscissa, at a nominal rate of 1000°C/s. Cooling of the soil at a nominal initial rate between 400 and 600°C/s began immediately after reaching peak temperature. Reprinted with permission (3). Courtesy of the American Chemical Society.

$M_S(0) = M_P(0) = 0$, and $M_i(0) = M_i^*$. This model utilizes up to five adjustable kinetic parameters: the Arrhenius preexponential factors k_{0S} and k_{0P}, the Arrhenius apparent activation energies E_S and E_P, and M^*_i the maximum amount of soil that can be converted to S, P, or $S + P$.

The study (88) also tested a sequential first-order reaction model in which a certain fraction of the soil was assumed to be converted to volatile product P, which then survives the experiment and is quantified, or further reacts to form nonvolatile side product S. For constant soil particle volume, the rate equations for this model are

$$-\frac{dM_i(t)}{dt} = k_P M_i(t) \quad (39)$$

$$\frac{dM_S(t)}{dt} = k_S M_P(t) \quad (40)$$

$$\frac{dM_p(t)}{dt} = k_P M_i(t) - k_S M_P(t) \quad (41)$$

where the symbols and initial conditions are the same as for equations 36–38.

When the parallel reaction model is best fit to heating rate data for acetylene, the resulting prediction is an increase in acetylene production with increasing heating rate, compared to the maximum in acetylene yields at intermediate heating rates exhibited by the data (see Fig. 16). Inclusion of temperature data in the fit results in poorer correlations and incorrect predictions of the data trends with temperature and heating rate.

When the sequential reaction model is best fitted only to the acetylene heating rate data, the model predicts a maximum in acetylene production at intermediate heating rates with the absolute values obtained, agreeing reasonably with the experimental data in Figure 16. However, the model is very sensitive to temperature. This leads to wide dispersion in the model predictions because of scatter in the experimental temperature–time histories, reasonable, but not perfect approximations of which were used in the modeling computations.

These comparisons between data and predictions show that none of these three models adequately captures all the kinetic phenomena of interest. Order of magnitude calculations indicate that mass transfer limitations are probably not responsible for this disagreement, suggesting a need to test other chemical kinetic models.

Closure

The behavior of individual soil particles, of shallow soil beds, and of deeper beds of soil at low and at high temperature provides a useful framework for categorizing mathematical models for removing organic contaminants from soil by thermal treatment. For all three soil configurations at lower temperatures, significant headway has been made in developing models and in testing those models for soil or soil simulants (3,12,73–76,78,80–85,87–89,93,94,96,97), particularly with regard to predicting rates and extent of release or uptake of primary contaminants, as affected by temperature and treatment time. Modeling capabilities are less well advanced for soil particles and soil beds (shallow and deep) at more elevated temperatures, especially where chemical reactions of contaminants, the soil, or both, are important.

The mathematical models discussed above are soil-focused, but they can be utilized in reactor models and flow sheet simulators for soil thermal cleanup processes, eg, see inter alia (72,98,100,101,103). In this application, soil-focused models are typically one of several subsystems formulated to deal with various complex phenomena including mixing and heat transfer, oxidation of volatilized contaminant and other soil-derived volatiles, and perhaps movement of the soil itself, eg, as in rotary kilns. Further, a range of length and timescales and steady-state and upset conditions require analysis. Recognizing that soil-focused models need to be integrated into larger simulation platforms, there is value in considering how to balance realism and universality in these models against user-friendly attributes like conceptual and mathematical simplicity, tractable data input requirements, and computational economy. Realism and universality entail the ability to reliably predict the major features of soil behavior important to engineering design, ie, rates and extent of contaminant release and by-products identities and generation rates, each as affected by temperature, time, heating rate, contaminant loading, pressure, soil particle size, moisture content, soil chemical and physical constitution, reactive atmospheres (oxidative, reducing), and contaminant physicochemical properties, eg, composition, volatility, chemical reactivity, etc. It may never be possible to achieve such versatility in one soil-focused model, although these model performance goals elucidate opportunities for further research (see below). There is merit in capitalizing fully on the capabilities of existing models and in seeking ways to extend them to different soils, contaminants, and operating conditions.

Fruitful areas for soil-focused modeling research include the high temperature behavior of soil particles and beds. Coupled effects of chemical reaction and physical transport are expected to be important, and thus extensions of models developed for contaminants transport under nonreactive conditions to elevated temperatures are promising embarkation points for this problem. An important related area is how to model soil decontamination chemistry in sufficient detail to predict the identities and release rates of important by-products. Several subsystem chemistries may be involved, for example, soil decomposition, primary and secondary reactions of the contaminant or soil products. Reaction effects are expected to become more significant with increasing temperature. Modifications in the physical structure and morphology of soil beds and soil particles impact decontamination behavior. For example, bed or particle porosity could change during the course of thermal treatment, altering effective diffusion paths or other transport resistances. Particle fragmentation as a mechanism for contaminant release from solids in a rotary kiln has been modeled (72). Additional work on modeling the physical structure of soil and soil beds during thermal treatment would be useful.

The need to develop soil-focused models that balance realism and computational tractability is noted above. Also needed is further research on how to model effects of soil type and contaminant properties, including studies to test, further evaluate, and extend already promising models to other soils, contaminants, and operating conditions. Effects of reactive atmospheres, both oxidative and reducing, should also be studied, as should interactions between organic and inorganic (eg, toxic metals) contaminants.

GAPS IN CURRENT UNDERSTANDING; RESEARCH OPPORTUNITIES

The last 15 years have witnessed significant advances in understanding the engineering fundamentals of soil thermal decontamination. Progress has occurred in experimental methods, mathematical modeling, generation of useful data, and elucidation of underlying chemical and physical phenomena. However, there are still substantial opportunities to improve the knowledge base of pertinent engineering fundamentals. For example, there is need to better understand decontamination behavior at higher temperatures, where chemical reactions of the contaminant and soil influence soil cleanup (see also section on Closure). Several technologies would expose soil and contaminants exiting the soil to elevated temperatures, eg, rotary kilns, fluidized beds, plasma arcs, vitrification, treatment in molten liquids, hydrothermal oxidation, etc. More generally, data and models are available for a comparatively small number of organic pollutants and soil types directly relevant to contaminated sites. A broader base of information on soil thermal decontamination behavior, eg, databases, engineering models, mechanistic understanding, would improve current capabilities for assessing the generality of already reported trends and elucidate heretofore undiscovered problems and opportunities for improving thermal methods of soil cleanup. Thus, valuable information would derive from further systematic studies of the effects of individual treatment conditions of practical interest (eg, temperature, time, heating rate, contaminant loading, oxidative and reducing atmospheres), soil properties (eg, particle size, bed geometry and mass, porosity, surface area, moisture content, chemical reactivity, mineral content), and contaminant attributes (state, composition, volatility, chemical reactivity) on decontamination dynamics including rates and extent of removal of original contaminants and yields and generation rates of by-products, especially toxic substances. Soils in the form of individual particles, shallow beds, deep beds, and moving beds that simulate practical treatment equipment are all of interest for further systematic study. Effects of moisture, native inorganic matter, and metallic or other inorganic contaminants in the soil on the removal of organic contaminants should also be studied. Relevant equilibrium and transport properties, eg, soil-contaminant partition coefficients and heats of adsorption and contaminant diffusivities in soil particles and soil beds, should be determined. Specific research opportunities in mathematical modeling of individual soil particles and soil beds are presented in the Closure section. This section elaborates on one of the major challenges in modeling soil thermal decontamination, ie, developing realistic, computationally efficient, soil-focused models that can be integrated into larger architectures for simulating reactors and entire processes.

In planning research on engineering fundamentals, investigators should consider how new data and models may be utilized to advance current and emerging technologies for aboveground and in situ thermal treatment of soil. Potential applications include simulation, monitoring, control, and optimization of reactors and processes and detection and diagnosis of system pathologies that could upset or even terminate process operability. Feedback from field tests and other larger-scale operations, can provide valuable insights on the strengths, limitations, and ranges of applicability of new engineering information and identify topics for further research.

CONCLUSIONS

Thermal treatment is already an established technology for removing organic contaminants from soil, and methodologies based on thermal treatment show significant promise for future applications, aboveground and *in situ*. Major challenges are to establish and maintain broadly based public acceptance of particular technologies in the light of concerns over adverse process emissions and/or of system failures that lower destruction efficiencies for targeted toxicants or create hazardous by-product effluents. Process economics are also important. Cleanup costs per ton of soil range from $35 up to several hundred dollars (15). Better understanding of the engineering fundamentals of soil thermal treatment can improve the technical foundations for current and new generations of soil thermal treatment technologies. In particular, such understanding can assist in the design, scale-up, operation, monitoring, and diagnostics of existing and emerging processes and could lead to new technology concepts for more efficacious soil decontamination. These advances should, in turn, improve public confidence that specific technologies can be relied upon to fulfill stated performance goals at tractable costs. This review shows that there has been significant headway in providing better understanding of the engineering fundamentals of thermally removing organic contaminants from individual soil particles and from shallow and deep beds of soil, including pertinent databases and mathematical models. This progress has been more substantial for systems at lower temperatures, where mathematical models showing satisfying agreement with observations, have been developed to predict rates and the extent of release or uptake of primary contaminants, as affected by temperature and treatment time. Further testing of these models for additional soils and contaminants and for conditions, where soil or pollutant chemical reactions are operative, are research opportunities. Research is also needed (see also sections on Closure and Research Opportunities) on soil behavior at elevated temperatures, effects of soil and contaminant chemical reactions, and effects of treatment conditions and soil and contaminant properties, on pollutant removal efficiencies, and on the generation of by-products, especially toxic or otherwise hazardous substances. Further integration of soil science (eg, 104, 105) and soil decontamination process fundamentals is also recommended.

NOMENCLATURE

Symbol	Definition	Unit
A	Freundlich isotherm parameter	$mol^{1-B}\ (m^3_{gas})^B/kg_{solid}$
B	Freundlich isotherm parameter	dimensionless
C_A	Gas-phase concentration in the interparticle space in the column	mol/m^3_{gas}
C_{Ai}	Gas-phase concentration inside the particle (or the pellet)	mol/m^3_{gas}
$C_{A,sat}$	Gas-phase concentration at the surface of the particle	mol/m^3_{gas}
C_i	Total gas-phase concentration inside the particle	mol/m^3_{gas}
D_{AB}	Molecular diffusivity	m^2/s
D_{ep}	Effective diffusivity for transport of contaminant through the particle (or pellet)	$m^3_{gas}/(m_{particle}s)$
D_{ez}	Effective diffusivity in the axial direction in a packed bed	$m^3_{gas}/(m_{bed}s)$
D_{KA}	Knudsen diffusivity	m^2/s
D_p	Combined diffusivity	m^2/s
D^*	$\frac{D_{ep}A^{-1/B}n_{AO}^{\frac{1-B}{B}}}{\rho_p B}$	m^2/s
D^*_0	Diffusion coefficient defined (82) for the parallel pore model	m^2/s
E	Desorption (or activation) energy	J/mol
E_i	Activation energy for reaction i	J/mol
E_P	Activation energy	J/mol
E_0	Mean of the probability density function	J/mol
E_s	Activation energy for surface diffusion	J/mol
E_S	Activation energy	J/mol
E^*	Critical activation energy	J/mol
$f(E)$	Probability density function for activation energy	$(J/mol)^{-1}$
$F(E)$	Distribution function for activation energy	
F	Stream volumetric flow rate	m^3_{gas}/s

k_a	Adsorption rate constant	s^{-1}
k_E	Desorption rate constant	s^{-1}
k_f	Mass-transfer coefficient in the external film of a particle	$m^3_{gas}/(m^2_{particle}s)$
k_i	Rate constant for reaction i	s^{-1}
k_P, k_S	Rate constants	s^{-1}
k_0, k_{0S}, k_{0P}	Preexponential factors	s^{-1}
k_{0i}	Preexponential factor for reaction i	s^{-1}
K_a	Adsorption equilibrium constant	dimensionless
L	Length of packed column	m_{bed}
M_i	Mass of initial volatilizable material	kg
M_P	Mass of observed product	kg
M_S	Mass of side product	kg
M_θ/M_∞	Cumulative mass gain relative to that at equilibrium	dimensionless
n_A	Solid-phase concentration	mol/kg_{solid}
n_{A0}	Solid-phase concentration in equilibrium with the imposed gas-phase concentration	mol/kg_{solid}
N_j	Gas-phase diffusion flux for species j	$mol/(m^2_{particle}\ s)$
O	Desorbed amount of organic molecules	kg
O_E	Desorbed amount of organic molecules	kg
O_{tot}	Total adsorbed amount of organic molecules	kg
P	Total pressure	Pa
Q	n_A/n_{A0}	dimensionless
r	Radial coordinate in the particle	$m_{particle}$
r_{p0}	Pore radius	m
R	Gas constant	J/(mol · K)
R_A	Net local rate of adsorption of species A	$mol/(m^3_{particle}\ s)$
R_p	Radius of the soil particle	$m_{particle}$
S	Area of end face of pellet	$m^2_{particle}$
S_A	Surface diffusion flux for species A	$mol/(m^2_{particle}\ s)$
t	Time	s
t_f	Final time	s
T	Temperature	K
T_g	Sweep gas temperature	K
T_s	Surface temperature	K
v_s	Superficial velocity	$m^3_{gas}/(m^2_{bed}\ s)$
V^*	Ultimate amount of product generated from all reactions	wt %
V_f	Cumulative amount of product generated up to time t_f from all reactions	wt %
V_i	Cumulative amount of product generated up to time t from reaction i	wt %
V_{if}	Cumulative amount of product generated up to time t_f from reaction i	wt %
V^*_i	Ultimate amount of product generated from reaction i	wt %
y_A	Gas-phase mole fraction of species A	dimensionless
x	r/R_p	
z	Axial coordinate	m
α	Thermal diffusivity	m^2/s
β	Preexponential factor in the Arrhenius-type expression for surface diffusivity	$m^2_{particle}/s$
ε_p	Particle (or pellet) porosity	$m^3_{gas}/m^3_{particle}$
ε_B	Bed porosity	m^3_{gas}/m^3_{bed}
θ	$\frac{tD^*}{R_p^2}$	dimensionless
λ	Preexponential factor for surface diffusion	m^2/s
μ	Fluid viscosity	kg/m s
$\overline{\mu}$	Corrected average retention time	s
μ_1	First absolute moment	s
μ'_2	Second central moment	s^2
ρ_B	Bed density	kg_{solid}/m^3_{bed}
ρ_p	Particle (or pellet) density	$kg_{solid}/m^3_{particle}$

ρ_s	Solid density	kg_{solid}/m^3_{solid}
σ	Standard deviation	J/mol
$\overline{\sigma}$	Standard deviation	
τ	Tortuosity factor	dimensionless

Acknowledgments

Financial support of this research by NIEHS Grant ES04675-06 (MIT-Superfund Hazardous Substances Basic Research Program) and by the North East U.S. EPA Hazardous Substance Research Center is greatly appreciated. V. Bucalá gratefully acknowledges the fellowship support of the Consejo Nacional de Investigaciones Científicas y Técnicas (CONICET) of Argentina, during her stay at MIT. We appreciate helpful comments from Mr. A. Modestino on research apparatus for soil thermal treatment. In connection with the authors' research on soil thermal treatment, we thank Dr. S. Rosenthal from Foster Wheeler Enviroresponse Inc., Dr. R. M. Frederick from the U.S. EPA, and Professors J. Bozzelli of the New Jersey Institute of Technology and J. Lighty and G. Silcox of the University of Utah for providing soil samples and/or information about soil specimens. We also thank Dr. G. Darivakis, Professor M. Flytzani–Stephanopoulos, Mr. E. Freeman, Professor P. Gilot, Ms. S. Han, Dr. M. Koch, Dr. A. Lafleur, Mr. J. Lee, Mr. P. Marrone, Professor A. Sarofim, Professor J. Tester, and Professor G. S. Yang for their collaboration in soil thermal decontamination research.

BIBLIOGRAPHY

1. J.L. Mortensen and F.L. Himes, "Soil Organic Matter," in F. E. Bear, ed., *Chemistry of the Soil,* 2nd ed., Reinhold Publishing, New York, 1964.
2. M. Bonneau and B. Souchier, *Constituents and Properties of Soils,* Academic Press, New York, 1982, Chapts. 6 and 9.
3. V. Bucalá, H. Saito, J.B. Howard, and W.A. Peters, *Ind. Eng. Chem. Res.* **35,** 2725 (1996).
4. M.L. Jackson, in F.E. Bear, ed., *Chemistry of Soils,* American Chemical Society Monograph Series, Reinhold Publishing New York, 1964, Chapt. 2.
5. C.G. Kowalenko, in M. Schnitzer and S.U. Khan, eds., *Soil Organic Matter,* Elsevier Science Publishing New York, 1978, Chapt. 3.
6. L.E. Lowe, in M. Schnitzer and S.U. Khan, eds., *Soil Organic Matter,* Elsevier Science Publishing, New York, 1978, Chapt. 2.
7. M. Schnitzer, in M. Schnitzer and S.U. Khan, eds., *Soil Organic Matter,* Elsevier Science Publishing, New York, 1978, Chapt. 1.
8. D.G. Schulze, in J.B. Dixon and S.B. Weed, eds., *Minerals in Soil Environments,* Soil Science Society of America, Madison, Wisc., 1989, Chapt. 1.
9. R.N. Yong, A.M.O. Mohamed, and B.P. Warkentin, *Principles of Contaminant Transport in Soils,* Elselvier Science Publishers, Amsterdam, the Netherlands, 1992.
10. "Superfund: Focusing on the Nation at Large - 1991", Publication #9200.5-701A, Document PB919-212-02, U.S. EPA, Office of Emergency and Remedial Response, Office of Program Management, Washington, D.C., 1991.
11. "GAO's Analysis of Superfund Cleanups Shows Uneven Progress", *Chem. Eng. News* **72**(44), 12 (1994).
12. L. Tognotti, M.F. Flytzani–Stephanopoulos, A.F. Sarofim, H. Kopsinis, and M. Stoukides, *Environ. Sci. Tech.* **25**(1), 104 (1991).
13. J.S. Hirschhorn and K.U. Oldenburg, *Chem. Eng. Prog.* **84**(12), 55 (1988).
14. L.J. Amend and P.B. Lederman, *Environ. Prog.* **11**(3), 173 (1992).
15. J.J. Cudahy, and W.L. Troxler, *J. Air Waste Manage. Assoc.* **42**(6), 844 (1992).
16. G.M. Long, *Chem. Eng. Prog.* **89**(5), 58 (1993).
17. W.P. Linak, J.D. Kilgroe, J.A. McSorley, J.O.L. Wendt, and J.E. Dunn, *J. Air Poll. Control Assoc.* **37**(1), 54 (1987).
18. W.R. Seeker, and C.P. Koshland, *Comb. Sci. and Tech.* **74,** i (1990).
19. G.E. Likens, C.T. Driscoll, and D.C. Buso, *Science* **272,** 244–246 (1996).
20. J. Kaiser, *Science* **272,** 198 (1996).
21. VISITT 3.0 (Vendor Information System for Innovative Treatment Technologies, version 3.0), IBM PC-based database of vendor information, U.S. EPA, Washington, D.C., 1994.
22. W.L. Troxler, J.J. Cudahy, R.P. Zink, J.J. Yezzi Jr., and S.I. Rosenthal, *J. Air and Waste Manag. Assoc.* **43**(11), 1512 (1993).
23. R.G. Luthy, D.A. Dzombak, C.A. Peters, S.B. Roy, A. Ramaswami, D.V. Nakles and B.R. Nott, *Environ. Sci. Technol.* **28**(6) 266A (1994).
24. D.L. Ellig, C.K. Lai, D.W. Mead, J.P. Longwell, and W.A. Peters, *Ind. Eng. Chem. Process Des. Dev.* **24**(40), 1080 (1985).
25. C.-K.S. Lai, Sc.D. Thesis, Massachusetts Institute of Technology, Cambridge, Mass., 1986.
26. C.C.-S. Chang, Ph.D. Thesis, Massachusetts Institute of Technology, Cambridge, Mass., 1986.
27. H.D. Franklin, R.G. Cosway, W.A. Peters, and J.B. Howard, *Ind. Eng. Chem. Process Design Dev.* **22,** 39 (1983).
28. H.D. Franklin, W.A. Peters, F. Cariello, and J.B. Howard, *Ind. Eng. Chem. Process Des. Dev.* **20**(4), 670 (1981).
29. H.D. Franklin, W.A. Peters, and J.B. Howard, *Fuel* **61**(2), 155 (1982).
30. H.D. Franklin, R.D. Cosway, W.A. Peters, and J.B. Howard, *Fuel* **61**(12), 1213 (1982).
31. M.R. Hajaligol, J.B. Howard, J.P. Longwell, and W.A. Peters, *Ind. Eng. Chem. Process Design Dev.* **21,** 457 (1982).
32. J.B. Dixon and S.B. Weed, eds., *"Minerals in Soil Environments"*, Soil Science Society of America, Madison, Wisc., 1989, Chaps. 1 and 3.
33. M. Schnitzer and S.U. Khan, *Soil Organic Matter,* Elsevier Scientific Publishing, New York, 1978.
34. *Soil Chemistry, Encyclopedia of Science and Technology,* 6th ed., McGraw–Hill, New York, 1987, Vol. 16.
35. *Soil, Encyclopedia of Science and Technology,* 6th ed., McGraw–Hill, New York, 1987, Vol. 16.
36. *Humus, Encyclopedia of Science and Technology,* 6th ed., McGraw–Hill, New York, 1987, Vol. 8.
37. M. Schnitzer and S.U. Khan, *Humic Substances in the Environment,* Marcel Dekker, New York, 1972.
38. F.J. Stevenson, *Humus Chemistry: Genesis, Composition, Reactions,* John Wiley & Sons, New York, 1982.
39. G.R. Aiken, D.M. McKnight, R.L. Wershaw, and P. MacCarthy, *Humic Substances in Soil, Sediment, and Water: Geochemistry, Isolation, and Characterization,* John Wiley & Sons, Inc., New York, 1985.
40. H.R. Schulten, *J. Anal. Appl. Pyrolysis* **25,** 97 (1993).
41. M. Schnitzer and H.R. Schulten, *Soil Sci. Amer. J.* **56,** 1811 (1992).
42. H.R. Schulten and M. Schnitzer, *Soil Science* **153**(3), 205 (1992).

43. R. Hempfling and H.R. Schulten, *J. Anal. Appl. Pyrolysis* **13,** 319 (1988).
44. C. Saiz–Jimenez and J.W. De Leeuw, *J. Anal. Appl. Pyrolysis* **9,** 99 (1986).
45. F. Martin, F.J. González–Vila, J.C. del Rio, and T. Verdejo, *J. Anal. Appl. Pyrolysis* **28,** 71 (1994).
46. G. Chiavari, G. Torsi, D. Fabbri, and G.C. Galletti, *Analyst* **119**(6), 1141 (1994).
47. C. Saiz–Jimenez, *Environ. Sci. Technol.* **28**(11), 1773 (1994).
48. R.S. Boynton, *Chemistry and Technology of Lime and Limestone,* 2nd. ed., John Wiley & Sons, Inc., New York, 1980.
49. G.W. Brindley and J. Lemaitre, "Thermal Oxidation and Reduction Reactions of Clay Minerals," in A.C.D. Newman, ed., *Chemistry of Clays and Clay Minerals,* John Wiley & Sons, Inc., New York, 1987.
50. G.M. Badger and T.M. Spotswood, *J. Chem. Soc.* **IV**(858) 4420 (1960).
51. M. Frenklach, D.W. Clary, W.C. Gardiner, Jr., and S.E. Stein, *Twentieth Symposium (International) on Combustion,* The Combustion Institute, Pittsburgh, Pa., 1984, p. 887.
52. L.D. Pfefferle, "Higher Hydrocarbon Formation During the Pyrolysis of Allene", presented at the *Mechanisms and Models of Soot Formation International Workshop,* Sept. 19–Oct. 2, Heidelberg, 1991.
53. G.M. Badger, "Pyrolysis of Hydrocarbons" in S.G. Cohen, A. Streitwieser, Jr., and N. W. Taft, eds., *Progress in Physical Organic Chemistry,* Interscience, New York, 1965, Vol. 3, pp. 1–40.
54. H. Blades, A.T. Blades, and E.W.R. Steacie, *Can. J. Chem.* **32,** 298 (1954).
55. R.D. Smith, *J. Phys. Chem.* **83,** 1553 (1979).
56. J.A. Mulholland, Ph.D. Thesis, Massachusetts Institute of Technology, Cambridge, Mass., 1992.
57. H. Hippler, C. Reihs, and J. Troe, *Z. Phys. Chem. (Munich)* **167,** 1 (1990).
58. J.J. Madison and R.M. Roberts, *Ind. Eng. Chem.* **50**(2), 237 (1958).
59. A.G. Sharkey Jr., J.L. Shultz, and R.A. Friedel, *Carbon* **4,** 365 (1966).
60. J. Mukherjee, A.F. Sarofim, and J.P. Longwell, *Combustion and Flame* **96,** 191 (1994).
61. Y.D. Yeboah, J.P. Longwell, J.B. Howard, and W.A. Peters, *Ind. Eng. Chem. Process Des. Dev.* **19**(4), 646 (1980).
62. W.H. McClennen, N.S. Arnold, K.A. Roberts, H.L.C. Meuzelaar, J.S. Lighty, and E.R. Lingren, *Comb. Sci. and Tech.* **74**(1–6), 297 (1990).
63. W.H. McClennen, N.S. Arnold, J.S. Lighty, E.G. Eddings, E.R. Lingren, K.A. Roberts, and H.L.C. Meuzelaar, "Monitoring the Evolution of Organic Compounds from the Thermal Treatment of Contaminated Soil Samples Using Short Column GC/MS", *American Chemical Society Symposium,* Division of Fuel Chemistry, Sept. 10–15, Miami, Fla., Preprint Extended Abstract, **34**(3), 1028 (1989).
64. J. Dragun, *J. Hazar. Mat.* **26,** 343 (1991).
65. J. Dragun, *The Soil Chemistry of Hazardous Materials,* Hazardous Materials Control Research Institute, Silver Spring, Md., 1988, Chapt. 8.
66. S. Saltzman, U. Mingelgrin, and B. Yaron, *J. Agric. Food Chem.* **24**(4), 739 (1976).
67. J. Dragun and C.S. Helling, *Soil Science* **139**(2), 100 (1985).
68. M. DiChristina, *Popular Science* **243**(2), 25 (1993).
69. J. Saastamoinen and J.–R. Richard, *Comb. Flame* **106,** 288–300 (1996).
70. D.A. Frank–Kamenetskii, *Diffusion and Heat Transfer in Chemical Kinetics,* translated by J.P. Appleton, Plenum Press, New York, 1969.
71. O. Levenspiel, *Engineering Flow and Heat Exchange,* Plenum Press, New York, 1984, Chapt. 11.
72. J.O.L. Wendt and W.P. Linak, *Comb. Sci. Tech.* **61,** 169 1988.
73. J.S. Lighty, E.G. Eddings, E.R. Lingren, D. Xiao, R.M. Winter, and W.H. McClennen, *Comb. Sci. and Tech.* **74,** 31 (1990).
74. J.S. Lighty, D.W. Pershing, V.A. Cundy, and D.G. Linz, *Nucl. Chem. Waste Manage.* **8,** 225 (1988).
75. J.S. Lighty, G.D. Silcox, D.W. Pershing, V.A. Cundy, and D.G. Linz, *Environ. Prog.* **8**(1), 57 (1989).
76. J.S. Lighty, G.D. Silcox, D.W. Pershing, V.A. Cundy, and D.G. Linz, *Environ. Sci. Technol.* **24**(5), 750 (1990).
77. E. Bar–Ziv, and A.F. Sarofim, *Progress in Energy & Combustion Science* **17,** 1 (1991).
78. M. Flytzani–Stephanopoulos, A.F. Sarofim, H. Kopsinis, and M. Stoukides, "Incineration of Contaminated Soils in a Electrodynamic Balance", *Emerging Technologies in Hazardous Waste Management II,* American Chemical Society, Washington, D.C., 1990.
79. A. Modestino, Department of Chemical Engineering, Massachusetts Institute of Technology, Cambridge, Mass., personal communications, May 17, 1996.
80. P. Gilot, J.B. Howard, and W.A. Peters, *Environ. Sci. Technol.* **31,** 461 (1997).
81. S. Han, M.S. Thesis, Massachusetts Institute of Technology, Cambridge, Mass., 1994.
82. T–F. Lin, J.C. Little, and W.W. Nazaroff, *Environ. Sci. Technol.* **28**(2), 322 (1994).
83. C. Cabbar, G. Dogu, T. Dogu, B.J. McCoy, and J.M. Smith, *Environ. Sci. Technol.* **28**(7), 1312 (1994).
84. Y.G. Wu, J. Dong, and J.W. Bozzelli, *Combustion Science and Technology* **85,** 151 (1991).
85. Y.G. Wu and J.W. Bozzelli, *Hazar. Waste Hazar. Mat.* **11**(1), 227 (1994).
86. G.S. Darivakis, *Work on Soils-Report,* Report to the MIT Energy Laboratory, 1990.
87. V. Bucalá, H. Saito, J.B. Howard, and W.A. Peters, in *Geo-Environmental Issues Facing The Americas,* Geotechnical Special Publication N° 47, 99–104, Geotechnical Division, American Society of Civil Engineers, Washington D.C., 1995.
88. H.H. Saito, Ph.D. Thesis, Massachusetts Institute of Technology, Cambridge, Mass. 1995.
89. V. Bucalá, H. Saito, J.B. Howard, and W.A. Peters, *Environ. Sci. Technol.* **28**(11), 1801 (1994).
90. D.B. Anthony, J.B. Howard, H.C. Hottel, and H.P. Meissner, "Rapid Devolatilization of Pulverized Coal", Proceedings Fifteenth Symposium (International) on Combustion, The Combustion Institute, Pittsburgh, 1975, pp. 1303–1317.
91. E.M. Suuberg, W.A. Peters, and J.B. Howard, *Ind. Eng. Chem. Process Des. Dev.* **17,** 37 (1978).
92. M.R. Hajaligol, J.B. Howard, and W.A. Peters, *Combustion and Flame* **95,** 47 (1993).
93. H.H. Saito, V. Bucalá, J.B. Howard, and W.A. Peters, to be submitted to I&EC Research, 1997.
94. J.S. Lighty, G.D. Silcox, D.W. Pershing, and V.A. Cundy, "On the Fundamentals of Thermal Treatment for the Cleanup of Contaminated Soils", *Air Pollution Control Association 81st Annual Meeting,* Dallas, Tex., 1988, Paper #88-17.5.
95. S.K. Friedlander, *Smoke, Dust and Haze,* Wiley–Interscience, New York, 1977, pp. 221–231.

96. Z. Du, A.F. Sarofim, and J.P. Longwell, *Energy & Fuels* 4(3), 296 (1990).
97. B.R. Keyes and G.D. Silcox, *Environ. Sci. Technol.* **28**, 840 (1994).
98. T.W. Lester, V.A. Cundy, A.N. Montestruc, C.B. Leger, S. Acharya, and A.M. Sterling, *Comb. Sci. Tech.* **74**(1–6), 67 1990.
99. J.B. Howard, "Fundamentals of Coal Pyrolysis and Hydropyrolysis," in M.A. Elliott, ed., *Chemistry of Coal Utilization,* 2nd Suppl. Vol. 1981, John Wiley & Sons, Inc., New York, 1981, Chapt. 12, pp. 665–784.
100. G.D. Silcox, F.S. Larsen, W.D. Owens, and M. Choroszy-Marshall, *Waste Management* **15**(5,6), 339 1995.
101. J.O.L. Wendt, W.P. Linak, and P.M. Lemieux, *Haz. Waste and Haz. Mats.* **7**(1), 41 1990.
102. P. Gilot, Universite de Haute Alsace, Ecole Nationale Superieure de Chimie de Mulhouse, Mulhouse, France, work in progress, 1997.
103. W.D. Owens, G.D. Silcox, J.S. Lighty, X.X. Deng, D.W. Pershing, V.A. Cundy, C.B. Leger, and A.L. Jakway, *Comb. Flame,* **86**(1–2), 101 1991.
104. G. Sposito, *The Chemistry of Soils,* Oxford University Press, New York, 1989.
105. G. Sposito, *Chemical Equilibria and Kinetics in Soils,* Oxford University Press, New York, 1994.

SOIL VAPOR EXTRACTION

RALPH S. BAKER
ENSR Consulting & Engineering
Acton, Massachusetts

DAVID J. BECKER
Hazardous, Toxic, and Radioactive Waste Center of Expertise
U.S. Army Corps of Engineers
Omaha, Nebraska

Soil vapor extraction (SVE) is the process of inducing airflow through unsaturated soil for the purpose of removing volatile organic compounds (VOCs). One or more blowers or vacuum pumps is employed to exert a vacuum and withdraw air from subsurface wells or trenches situated above the water table. The resulting pressure gradient causes air to flow through pores in the soil, from locations of higher to those of lower pressure. Along the way, VOCs partition into the air and are carried with it to the surface. Extracted air is often treated above ground prior to discharge. In systems consisting of many wells it is common to inject air into some of them to enhance and redirect subsurface airflow.

Soil vapor extraction can be among the most effective and cost efficient methods of removing VOCs from the vadose zone. Although the technique fully emerged only within the last decade, it has since become widely used at sites contaminated with petroleum hydrocarbons such as gasoline and with hazardous wastes including chlorinated solvents. Often SVE is used as a sole technology, but in many instances it is employed in conjunction with other *in situ* technologies such as bioventing (BV), air sparging (AS) and multiphase extraction of water and/or nonaqueous-phase liquid (NAPL). It is also employed in *ex situ* applications such as above-ground soil piles. As the use and successful application of SVE multiply, so do an appreciation of the practical limits of its applicability and efforts to overcome such limitations. Considering the rapid pace of innovation in the development and demonstration of technology enhancements, SVE can be expected to see continued widespread use.

APPLICABILITY

Contamination of groundwater by petroleum products and organic solvents has become a serious concern in industrialized countries. A substantial portion of the problem originated with underground storage tanks (USTs), which are ubiquitous at automotive service stations as well as commercial and industrial facilities. The U.S. Environmental Protection Agency (EPA) estimates that of the 2 million USTs in the United States, 295,000, or more than 10%, are leaking (1). Additional sources of hydrocarbon contamination in groundwater include surface spills, pipeline leaks, and releases from pits, ponds, and lagoons.

When a release of organic liquid occurs at or near the land surface, soil is nearly always the primary medium through which it first flows. The immediate effects of spills are commonly mitigated by pumping of liquids and/or excavation of the most contaminated soil. Thereafter, residual VOCs left in the soil frequently constitute an ongoing source of contamination as they volatilize and diffuse into the soil gas and the lower atmosphere and as they infiltrate and dissolve into the soil pore water and the groundwater. Emphasis has recently been placed on using relatively inexpensive *in situ* technologies such as SVE to remove this long-term contamination source.

Often SVE is a relatively short-term remedial activity, on the order of one to two years in duration for treatment of many gasoline- or solvent-contaminated sites. The actual duration and progress of treatment can vary considerably, however, because they depend on the setting, the ease with which air can be brought into contact with the contaminants, and the nature of the contaminants, including especially the starting concentrations and the remedial goals. Unfortunately there have been some instances in which SVE has been imposed upon the wrong site or in an inappropriate way. Proper application requires knowledge of site characteristics, adoption of attainable objectives, thorough consideration of remedial alternatives, and proper testing, design, and implementation of the technology.

Relationship between SVE and Bioventing

It is important to recognize at the outset the close relationship between SVE and bioventing. Bioventing resembles SVE in that it too makes use of subsurface airflow to treat contaminated soil. Bioventing differs from SVE, however, in that its focus is on biologically treating the contaminants in the soil rather than physically transferring them above ground. Under aerobic conditions, soil microorganisms are able to biodegrade large classes of organic contaminants, including many of the alkanes and monoaromatic compounds that comprise fuels. The products of aerobic biotransformations are innocuous: carbon dioxide, water, and cell biomass. To some extent this natural process occurs even without induced airflow; however, it is common for oxygen levels in hydrocarbon-contaminated soils to become depleted and for the process to become sluggish. In such cases, microbial activity and hence degradation rates can be significantly enhanced by active aeration, either by blowing air into or withdrawing it from the soil. The aim of bioventing is to induce sufficient airflow, delivered to the required places, to meet the oxygen demand of the hydrocarbon-degrading microorganisms. If this can be done, treatment can occur entirely below ground, without transfer of the contaminants to the surface. Bioventing thereby can circumvent costs associated with above-ground treatment of off-gas. In addition, unlike SVE, bioventing does not rely upon volatilization. As a result, it can be effective on both semivolatile and volatile organic compounds that are aerobically biodegradable.

When it comes to the treatment of soils contaminated with petroleum hydrocarbons, SVE and bioventing tend to be inseparable, because enhanced biodegradation nearly always accompanies aeration. By contrast, SVE is often preferred over bioventing for treatment of soils contaminated with chlorinated hydrocarbons, for which aerobic biodegradation occurs only under a relatively specialized set of conditions (2). The two related technologies can be contrasted further in that SVE airflow rates are typically much higher than those for bioventing. Consequently, there are instances where initially a system treating relatively high subsurface vapor concentrations may be operated in an SVE mode to accelerate their removal; but where later, as concentrations diminish and off-gas treatment is no longer needed, a transition to a slower-paced and less costly bioventing mode may be desirable (3).

Economic Importance

Site remediation in the United States is approximately a $10 billion-a-year industry (4). It is not known what fraction of this total is expended to test, design, construct, and operate SVE systems. Whatever the actual amount, the overall economic importance of SVE is still larger, however, because devalued properties become more valuable assets again upon cleanup, degraded natural resources are enhanced, and costs associated with more expensive and disruptive technologies are avoided. Another indication of the significance of SVE is the frequency of use. Although accurate numbers are difficult to obtain, applications at gasoline-contaminated sites in the United States number at least in the thousands. Also SVE enjoys widespread usage in Europe.

BACKGROUND

Soil Vapor Extraction

Soil vapor extraction has its antecedents in earlier technologies such as those developed for collection of gaseous hydrocarbons from petroleum reservoirs, mines, and landfills. The goals of such technologies were exploitation of economically valuable fossil and biogenic fuels and prevention of safety hazards, not remediation of subsurface contamination. As early as 1972, Knopik began applying SVE for cleaning up leaked gasoline from a UST at his service station in Forest Lake, Minnesota. By 1982, Knopik had employed his then-patented system at approximately 100 installations throughout the United States (5). Others may have developed the technology independently (6,7). Many of the early applications of SVE were not closely monitored; however, its use was examined (8) at a service station for gasoline-contaminated soil, and 350 gallons (1330 liters) of gasoline were removed in 100 days, achieving clean-up levels of 3 ppm or less in soil vapor and nondetectable concentrations in soil. Among the early developers of SVE systems in the late 1970s and early 1980s were Oil Recovery Systems, Exxon Company USA, Shell Oil Company, and the American Petroleum Institute.

In the mid-1980s, the Upjohn Company and others began to extend the use of SVE to hazardous waste sites. By the end of the 1980s, there were dozens of such sites (9), and by the early 1990s, hundreds. Focusing upon CERCLA sites alone, SVE made up 135, or 19% of the 697 source control treatment technologies selected by the U.S. EPA for remedial actions through fiscal year 1994 (Fig. 1) (10). None of the other technologies that the EPA categorizes as *innovative* were selected more often than SVE. Considering all treatment technologies, both established and innovative, SVE was second only to solidification/stabilization (selected in 202 instances, or 29% of the time). Reliance on SVE at hazardous waste sites increased in the early 1990s (Fig. 2), as did published information on its use (11,12).

As of August 1995, of the 135 SVE projects at Superfund sites, 52 were in predesign/design; 70 had completed designs, were being installed, or were operational; and 13 were completed (10). In 1993, the U.S. EPA designated SVE as a presumptive remedy for CERCLA sites contaminated with VOCs (13). The U.S. EPA lists nine Superfund Remedial Actions at which SVE has been completed, including a >20,000 yd^3 (>15,300 m^3) portion of the Rocky Mountain Arsenal in Colorado contaminated with trichloroethylene (TCE), and 42,000 yd^3 (32,100 m^3) at Fairchild Semiconductor in California contaminated with chlorinated hydrocarbons (14).

Bioventing

Bioventing emerged from SVE in the early 1980s, with reports by the Texas Research Institute (15,16). In pilot-scale studies conducted in a soil-filled tank, they found that SVE stimulated biodegradation, which may have accounted for as much as 38% of the removal of gasoline from the vented soils. The U.S. Air Force conducted field-scale research on BV at Hill, Tyndall, and other Air Force bases in the late 1980s and early 1990s. In May 1992, they began a nationwide initiative, employing a standardized protocol (17) to test BV at 55 contaminated sites, extended in December 1992 to over 135 sites. The U.S. Air Force lists three sites at which BV had been completed as of February 1994, including a 25,000-gal (95,000-L) spill of JP-4 (a jet fuel) to a depth of approximately 60 ft (18 m) at Hill AFB in Utah (18). The initiative recently yielded a manual of BV principles and practices (19). As with SVE, applications of BV began with fuel hydrocarbons and were later extended to other organic contaminants.

TECHNOLOGY DESCRIPTION

Application Environment

Soil vapor extraction is intended to remove contaminants from soils above the water table. Such contaminants are almost exclusively organic and present as a residual separate phase, dissolved in soil moisture, and in a vapor phase. Removal of the air, including the contaminant vapors, from the pore spaces around a well or trench induces flow of additional air toward

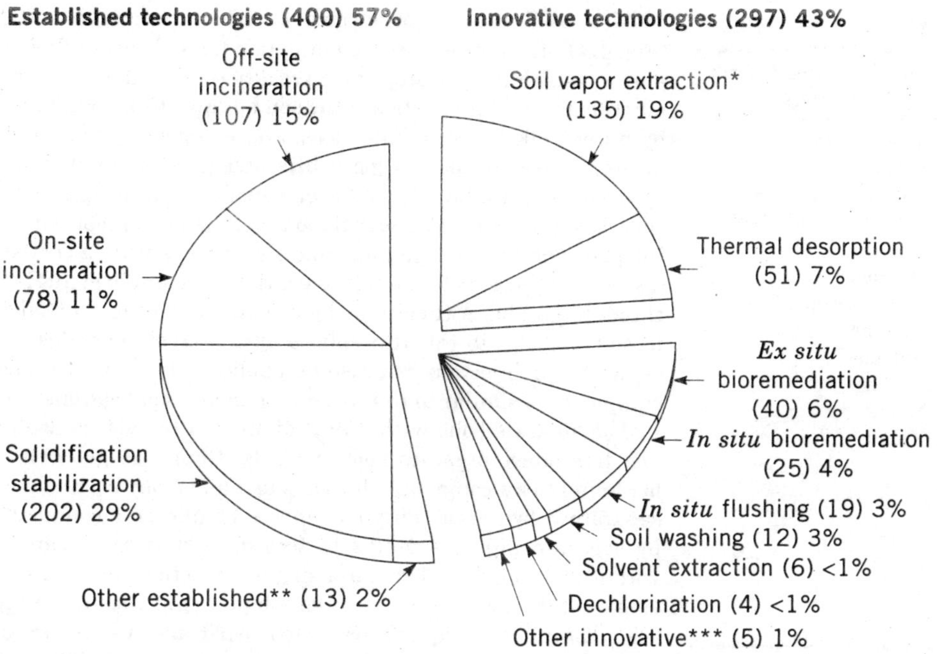

Figure 1. Superfund remedial actions—summary of source control treatment technologies selected through fiscal year 1994. Data derived from Records of Decision for fiscal years 1982–1994 and anticipated design and construction activities as of August 1995. A site may use more than one technology. Number in parens indicates number of times this technology was selected or used; *, includes two dual-phase extraction projects also listed as *in situ* groundwater technologies; **, other established technologies are soil aeration, open detonation, and chemical neutralization; ***, some other innovative technologies are phyto-treatment, contained recovery of oily wastes, cyanide oxidation, *in situ* vitrification, and plasma high temperature metals recovery (10).

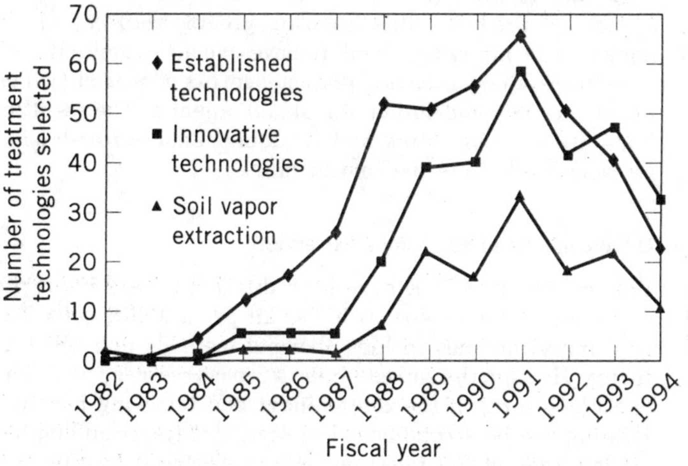

Figure 2. Superfund remedial actions—trends for number of times SVE was selected in comparison to total number of established and innovative treatment technologies for source control. Data derived from Records of Decision for fiscal years 1982–1994 and anticipated design and construction activities as of August 1995. A site may use more than one technology (10).

the well and disrupts the equilibrium between the liquid or dissolved contaminants and the air in the pores. More contaminant then volatilizes into the vapor phase and the process continues until the contaminant mass is removed or processes retard effective volatilization.

Contaminants must be able to volatilize into the air-filled pore space from both the separate phase and dissolved phase to achieve efficient removal. Contaminants most often addressed by SVE are fuels including gasoline, jet fuel, and some diesel constituents, and common halogenated solvents such as tetrachloroethylene, TCE, 1,1,1 trichloroethane, and freons (20). Table 1 lists the compounds typically considered amenable to SVE.

Soils exhibit a wide range of ability to transmit air. The soils must allow adequate airflow for volatilization of separate-phase and dissolved contaminants in an acceptable time frame without incurring large operating costs trying to get the air to move. Unsaturated sands and gravels are well suited to SVE because air moves freely through such soils. Clays often present significant barriers to airflow, just as they retard water flow, and may not be suited for SVE application. Mixtures of sands, silts, and clays vary greatly in the transmission of air, and the economic limit of applicability of SVE lies somewhere in these fine-grained mixtures (21).

The subsurface rarely consists of a single soil type. Evaluating soil heterogeneity is vital to designing an effective SVE system. Lenses of sand or gravel in otherwise fine-grained material have a dramatic effect on the movement of air, as do clay layers in sands.

Soil Vapor Extraction System Configuration

Soil vapor extraction has four standard components; subsurface collection point(s) (ie, wells or trenches), piping to transmit the vapors, condensate controls to remove condensed moisture to prevent damage to the blower, and an air-mover (a blower or vacuum pump). Additional components may include monitoring points for vacuum and soil gas concentration measurements, a treatment process for the extracted vapors, surface covers, and air injection points. Refer to Figure 3 for a schematic of a typical SVE system.

The air mover removes air from the subsurface and establishes a vacuum gradient in the soil which induces flow of air toward the extraction point. Monitoring points allow direct evaluation of the vacuum distribution, and thus the flow

Table 1. VOCs Considered Amenable to SVE[a]

Halogenated Volatile Organics	Nonhalogenated Volatile Organics
Carbon tetrachloride	Ketones/furans
Chlorobenzene	Acetone
Chloroethane	Methyl ethyl ketone
Chloroform	Methyl isobutyl ketone
1,1-Dichloroethane	Aromatics
1,1-Dichloroethylene	Benzene
1,2-Dichlorobenzene	Ethyl benzene
1,2-Dichloroethane	Styrene
1,2-Dichloroethylene	Toluene
1,1-Dichloropropane	*m*-Xylene
1,4-Dichlorobenzene	*o*-Xylene
1,1,1-Trichloroethane	*p*-Xylene
1,1,2-Trichloroethane	
1,1,2,2-Tetrachloroethane	
Ethylene dibromide	
Methylene chloride	
Tetrachloroethylene	
Trichloroethylene	
Vinyl chloride	

[a] Other compounds that have physical or chemical characteristics similar to the compounds listed may also be addressed by the presumptive remedy process (14).

paths of the air. Monitoring points are often set at different depths to determine vertical as well as horizontal vacuum gradients. Samples of soil gas taken from such points may also allow an indirect evaluation of the changes in the contaminant concentrations in soils.

Treatment units either capture or destroy the contaminant vapors and greatly reduce the amount of contaminant discharged to the atmosphere. Treatment units designed to capture the contaminants can be placed before or after the air mover, but units meant to destroy contaminants, usually by high temperature, are almost always placed downstream of the air mover.

Air removed from the subsurface is replaced by atmospheric air infiltrating into the subsurface. Surface covers are intended to prevent the infiltration of atmospheric air in the vicinity of the extraction points, thus making the air travel farther through the contaminated soil. These longer flow paths often allow more uniform treatment of soil with fewer extraction points. Covers also reduce the infiltration of precipitation or surface water which may reduce air permeability. Surface covers are often applied in locations with shallow water tables or high vertical air permeability. Covers may consist of nearly air-tight paving, although such paving is difficult to construct, or geomembranes covered by protective cover soil.

Airflow should be focused through the contaminated soil for best SVE effectiveness. In many cases, the coupling of air injection with SVE can create such focused paths. Air injection can also reduce the required vacuum needed at extraction points to achieve required air throughput. Air can be passively injected (see Fig. 1). In this case, the air is drawn into the subsurface through a vent open to the atmosphere solely by the vacuum exerted by a distant extraction point. Flow rates are usually low in such configurations. The air can also be injected at higher rates into an injection point by a blower or compressor.

Also SVE can be performed *ex situ* on excavated soil (Fig. 4). This may be a treatment option for soils excavated as part of tank removal projects. Excavation often improves the air transmission properties of the soil but may be accompanied by uncontrolled contaminant loss. Soil is normally placed on an impermeable liner in loose lifts and is often surrounded by a berm to contain leachate or runoff. Horizontal, slotted collection piping is set under the base of the pile and/or within the pile. The collection piping extends only to within a certain distance of the edge of the pile to provide more uniform airflow through the pile. A cover may be used to prevent uncontrolled loss of volatiles to the atmosphere during treatment. Passive or active air injection can also be applied. The above-ground equipment is similar to that used with *in situ* applications.

The simultaneous withdrawal of air and liquids (ie, water and free-phase organics such as fuels) from a single well is becoming very common. The various fluids can be removed separately by liquid pump(s) and a blower and conducted by separate piping systems to treatment devices or can be withdrawn together through a single extraction tube using a single total fluids pump, such as a liquid ring vacuum pump (22). The air and liquids are either separated just before or after the vacuum pump and treated as necessary. Figure 5 illustrates the concept. The simultaneous removal of fluids, also called multiphase extraction, provides several benefits. The application of vacuum reduces the need for physical drawdown of the liquid levels and thus reduces the amount of isolated product while allowing greater recovery of the liquids. The air withdrawal removes volatile contaminants from the capillary zone and promotes influx of oxygen-rich air to enhance bioremediation of residual organics. The resulting integration of bioventing and vacuum-enhanced free-product recovery has been termed bioslurping.

Enhancements to Soil Vapor Extraction

Enhancements to SVE have been directed toward improving contaminant mass recovery in low-air-permeability soils. Two primary techniques include thermal methods and soil fracturing. Heating the soil can improve mass recovery by raising vapor pressures of the contaminant and removing moisture. Heating can be accomplished in several ways, including hot-air injection, steam injection (23), or electrical heating (24). Hot-air injection may be used at sites where vapors are treated by thermal destruction and waste heat is available. Because of the high heat capacity of soils and low heat capacity of air, this can be a difficult way to achieve site-wide heating. Steam injection transmits more heat energy but increases soil moisture. Volatiles may be driven ahead of the steam condensation front toward extraction points. Electrical heating preferentially heats the soils with the highest resistivity, which are usually the soils such as clays that have the highest moisture content and the lowest air permeability.

Soil fracturing procedures use high-pressure air or liquids to induce tensile fracturing of low air-permeability soils and are most often applied to massive clays. Such fracturing increases the airflow through the soil matrix and shortens diffusion paths for the contaminants in the intact blocks of clay. Initial recovery rate increases of several orders of magnitude have been reported following fracturing (25).

Also SVE can be applied while using large augers or other soil-mixing devices to disrupt, but not excavate, the soil.

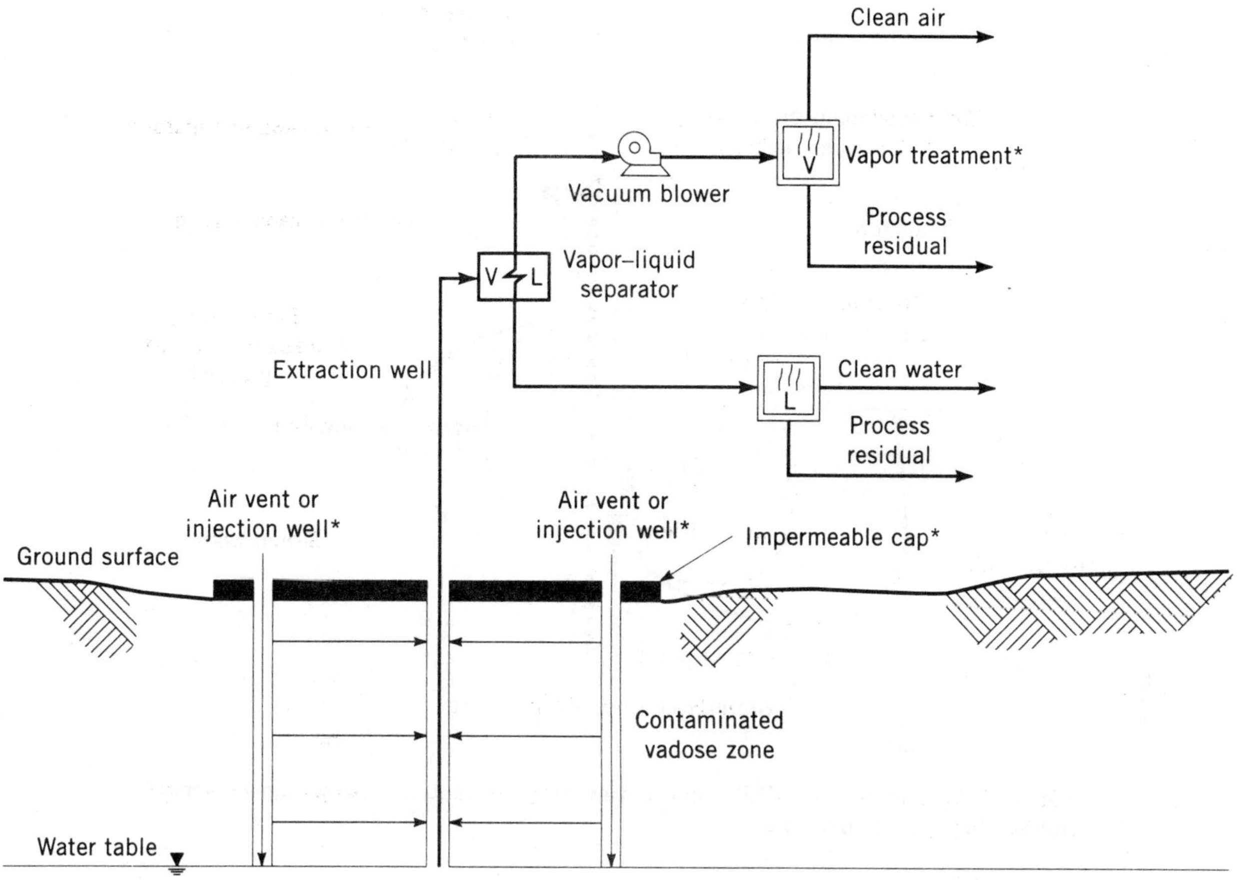

Figure 3. Generic soil vapor extraction system (20), where * denotes optional components.

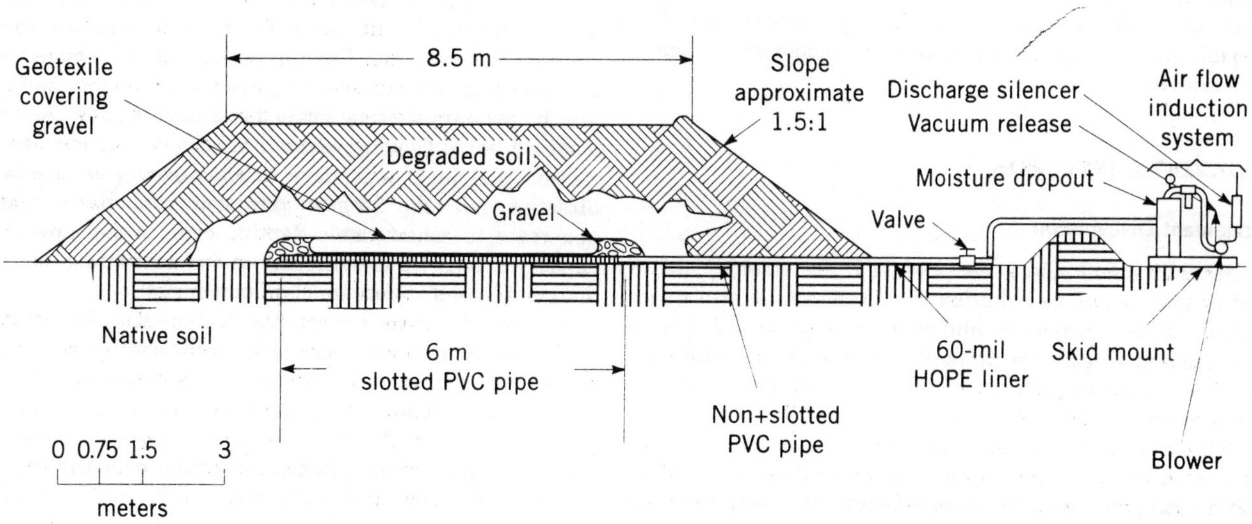

Figure 4. Above-ground soil pile (cross section) (3).

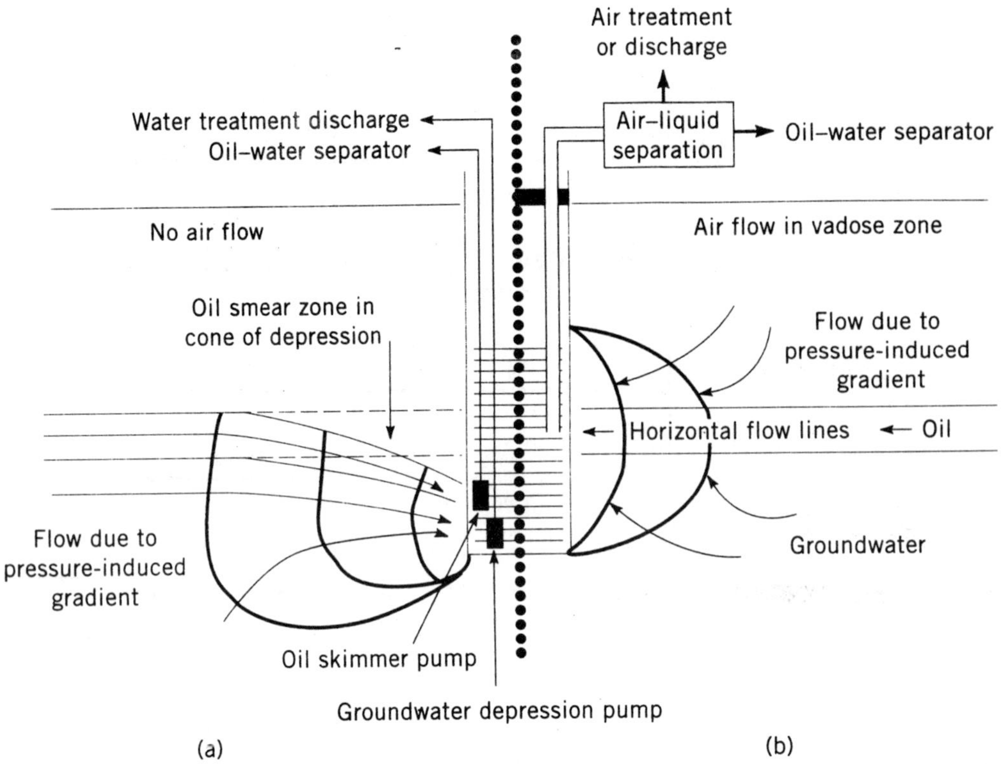

Figure 5. Comparison of LNAPL remediation using conventional two-pump system (**a**) and bioslurper system (**b**) (22).

Mixing *in situ* also improves airflow through the soil. Such mixing can be coupled with thermal enhancements.

Lastly, SVE can be conducted without the use of a mechanical air mover. Using check (one-way) valves, falling atmospheric pressure can, to some extent, be harnessed to extract air from the subsurface. In high air-permeability material, such techniques could possibly remove air from a significant area.

FUNDAMENTAL PRINCIPLES

Contaminant Distribution

The suitability and effectiveness of SVE are strongly influenced by the spatial distribution of contamination at a given site, both at the macroscale, and at the pore scale. It is therefore important to gain an understanding of the general patterns of soil contamination by VOCs, relative to the feasibility of moving air through the subsurface using SVE.

VOCs exist in the vadose zone primarily in the following forms: as free organic vapor in the soil pore gas; as dissolved organics in soil pore water; as adsorbed organics associated with solid surfaces; and as residual NAPL retained by capillary forces between solid particles. Residual NAPL will tend to constitute the largest mass of soil contamination (other than any pools of free-phase liquid that might be present), and therefore requires special attention. The relative importance of each of these forms of soil contamination depends on the nature of the release events, the properties of the contaminants of interest, and the nature and properties of the soil environment.

Nature of the Release

Patterns of soil contamination within the unsaturated zone can be very different for bulk spills of organic liquids as opposed to slow leaks. During infiltration of copious volumes of liquid from a point source, pressure gradients and gravity are the primary driving forces for saturated flow. As a result, a spilled liquid flowing within a relatively uniform and transmissive soil layer will move predominantly in a downward direction, via the largest interconnected pores, and may penetrate to considerable depths, even to the groundwater. Light NAPL (LNAPL) will float on the water table surface and distribute itself within the capillary fringe.

If along its path toward the water table the infiltrating NAPL intercepts even more permeable soil layers, structural cracks, macropores, or channels such as underground utilities, it will readily follow them downward in a process known as *preferential flow*. As more restrictive layers or barriers to flow are encountered, lateral movement over the top of such layers can become significant. Some of the NAPL may become perched above such lower permeability or higher moisture content zones. Toward the end of the spill event, as infiltration ends, unsaturated flow becomes the dominant flow process, causing some of the liquid to slowly redistribute itself for a limited distance into the adjacent soil, whereas the rest will remain along the initial flow pathways as residual NAPL. For such scenarios SVE may be a viable technology, because air flows preferentially within the larger soil pores, where some NAPL may still reside.

With release events involving slow leakage, by contrast, unsaturated flow is the dominant flow process from the outset, and a broader and shallower zone of contamination results. Thus within a relatively uniform and moderately transmissive soil layer, lateral movement can be equal in magnitude to downward flow. This *wicking* of NAPL tends to occur through medium-sized pores in the soil rather than through the largest pores. Under unsaturated flow the liquid will not, in fact, even be able to fill macropores that it encounters, so preferential flow is limited. Most of the residual NAPL will remain within the initially affected portions of the soil. Under such a scenario SVE will be more of a challenge, because air tends to flow preferentially through the larger pores and therefore may not come into direct contact with as much residual NAPL as in the former case. It is clear from this discussion that soil stratigraphy and morphology decide to a large extent how NAPL (and air) move through the subsurface. Other important determinants of the patterns of NAPL distribution that result from releases of different kinds include density and viscosity of the free-phase liquid, surface topography, the distribution of soil moisture, fluctuations in the water table elevation, and the degree of soil heterogeneity.

Contaminant Properties

The ability of the SVE process to impinge upon and extract contaminants is strongly influenced by their physical and chemical properties. These properties affect how contaminants are distributed within the soil, whether dissolved in pore water, adsorbed on the surface of particles, present as vapor, and/or occurring as NAPL (Fig. 6) (3). Knowing the following properties can make it easier to predict the behavior of a contaminant of interest, when subjected to SVE: (*1*) *solubility* indicates the degree to which a solid or NAPL, at equilibrium, will dissolve in water; (*2*) *soil adsorption coefficient* is the tendency of the dissolved compound, in turn, to adsorb to soil particles; and (*3*) *vapor pressure* and the *Henry's law constant* indicate how readily a dissolved compound, at equilibrium, will volatilize into the soil air.

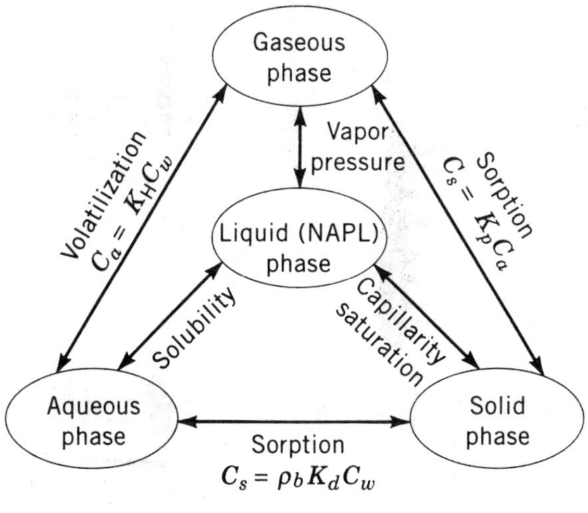

Figure 6. Partitioning of VOCs where C_a, C_w, and C_s = concentration of VOC component in air, water, and solid; K_H is Henry's constant; K_p, partition coefficient; K_d, distribution coefficient; and ρ_b, soil bulk density (3).

Many contaminants exist in the soil not by themselves, but as mixtures (eg, petroleum products) of individual compounds whose chemical and physical properties differ widely. When released within the environment, or subjected to SVE, the individual compounds within the mixture weather or respond to differing degrees, and, as they do, their distribution among the various phases changes. In the early stages of SVE, the more volatile and soluble fractions are more readily removed. The less soluble and volatile fractions have a greater tendency to remain in the soil. To some extent their extractability can be improved under elevated soil temperatures, which produce increases in both solubility and volatility. Biodegradation rates also increase with increases in soil temperature, up to approximately 40°C.

Henry's law determines the extent of volatilization of a contaminant or mixture dissolved in water and therefore applies to regions of the unsaturated zone where there is pore water but no immiscible phase is present (21). In such regions, the aqueous solubility of a contaminant restricts the attainable vapor concentration, ultimately limiting SVE's effectiveness, especially if remedial targets are quite stringent. Henry's law no longer applies, by contrast, at regions where air can come into direct contact with NAPL. In accordance with *Raoult's law*, vapor concentrations can be elevated in such a soil gas, even saturated with respect to the compound. Thus during the early stages of remediation in a source area containing NAPL, if some airflow can be induced, SVE recovery rates initially tend to be high.

Density governs whether a free-phase organic liquid is a dense NAPL (DNAPL) that, being denser than water, can sink below the water table, or a LNAPL that cannot. It must be remembered that where both types are present together at a site, they may be found cosolved within each other. The term *spreading coefficient* is related to density and interfacial tension in that it describes the degree to which a droplet of NAPL will spread itself over an air–water surface. In the unsaturated zone, even in dry soils, films of soil moisture coat the surfaces of soil particles, and in moist soils some of the soil pore spaces are water-filled. A droplet of LNAPL will tend to spread as a sheen over the water–air interface, positioning it to be exposed to airflow. A droplet of DNAPL, by contrast, is a nonspreading liquid and can *ball-up* within the aqueous phase, where it is not in position to be directly exposed to the soil air. It must first partition into the soil water and thence into the soil air. Thus in moist soils, residual DNAPL may be less susceptible to SVE than LNAPL (26).

Contaminant Transport and Removal during Soil Vapor Extraction

The removal of VOCs by SVE is carried out by several processes. In the more air-permeable regions of the soil, contaminants that volatilize into the soil gas are swept along with the bulk movement of the air in a process known as advection. Advection through lower permeability regions is relatively slow. In regions where little or no advection occurs because of low permeability, high soil moisture, or distance from flowing air, contaminant transport can still occur by molecular diffusion, which is much slower than advection. Diffusion is driven by concentration gradients, for example, between

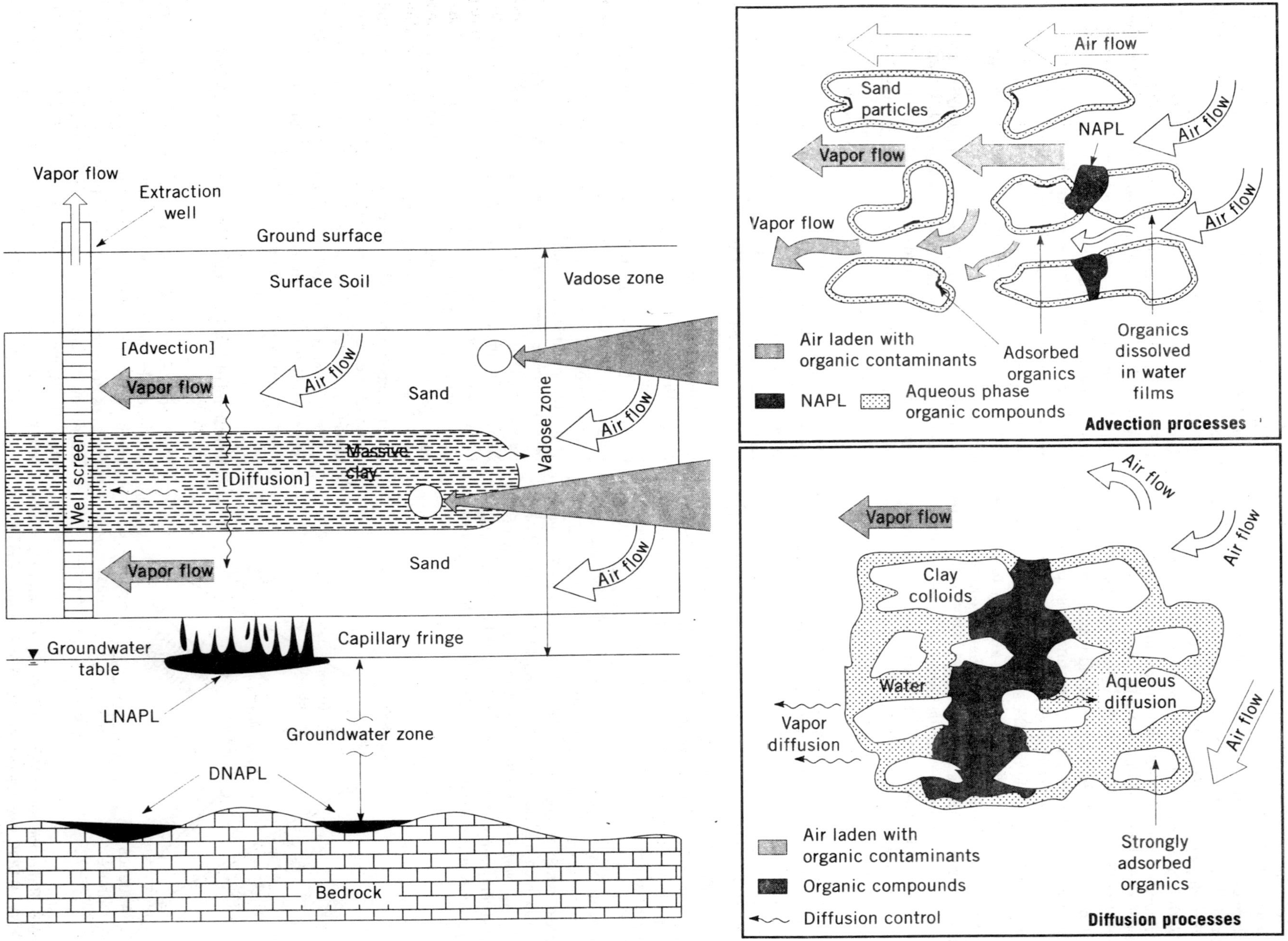

Figure 7. SVE transport processes (3).

contaminated locations that are not directly in advective pathways, and adjacent locations where vapor concentrations are diluted by advection. The greater the distance between advective pathways, the smaller the corresponding diffusion gradients and the slower and less effective the diffusion process. The complementary SVE transport processes of advection and diffusion are depicted in Figure 7.

Diffusion is not the only process that limits the rate of removal of contaminants during SVE. As described earlier, only a small fraction of the VOC mass at a site will be present in the soil gas at any time. Most of the contaminant mass must first partition into the air phase from sorption sites, soil water, and especially NAPL before it can be carried by airflow through the soil. The rates of desorption, dissolution, and volatilization each determine how fast contaminant mass is released into the advective airflow.

Contrary to the assumptions of many partitioning models, soil–fluid–vapor partitioning processes are often neither in equilibrium nor reversible. For example, sorbed compounds may not desorb as readily as soil adsorption coefficients would predict. This may occur if, over time, the sorbed compounds become strongly associated within inaccessible sorption sites or resistant soil fractions. Release of sorbed compounds from dead-end micropores is viewed as diffusion-limited and therefore is a nonequilibrium process (27). If, in the unsaturated zone, bodies of fine-textured soil are present in an otherwise coarse-textured matrix, their tendency to retain water rather tenaciously within numerous micropores ensures that mass transfer of contaminants from the aqueous to the vapor phases will tend to be very slow and diffusion-limited. Thus, compounds may not be as susceptible to volatilization or leaching, nor as bioavailable, as would be expected if their fate was not desorption or diffusion limited. As a consequence, compounds can prove to be more persistent during SVE treatment than would be expected otherwise. The greater the mass of contaminant that is positioned in space away from airflow pathways, the more the effectiveness of SVE will be limited by the slow rates of desorption or diffusion. Thus it is extremely important that SVE be carried out so as to distribute airflow as effectively as possible to contaminated locations in the soil.

Soil Properties

The description of soil particle size distribution, the *soil texture,* is a common basis for evaluating the applicability of SVE to a site. It is generally believed (and observed) that the coarser-textured the soil, the more likely it is for air to pass easily through it.

In actuality, the movement of air is not dependent on the particle sizes, but rather on the intervening pore sizes and the amount of pore space available to conduct air. *Porosity* is the dimensionless fraction, typically expressed as a percent, of a total volume of soil that is not taken up by solids. This porosity is normally filled above the water table by both air and water, possibly also by residual organic contaminant. Although there is a correlation between the pore size distribution and the particle size distribution (the larger the particles, generally, the larger the pore sizes), it is difficult to determine, particularly for mixtures of sands, silts, and clays. Such a correlation would be useful, because it is much easier to measure the distribution of grain sizes (eg, by ASTM method D422 (28)), than the pore size distribution.

The fraction of the total pore space that is not air filled is called the *saturation* and is usually expressed as a percent. The lower the saturation, the more air-filled porosity is available to conduct air. Thus, soil with a lower saturation will more readily conduct air than a comparable soil at a higher saturation. Although the saturation usually refers to soil moisture, residual organic contaminant (product) will also block pore space. Total saturation is the sum of the fractions of the pore space occupied by water and NAPL.

Water movement in unsaturated soils is affected by capillary forces that strongly restrict water movement through very small pore spaces but have much less overall effect on water movement through large pores. By contrast, the movement of air in unsaturated soils is not directly impeded by capillary forces on any scale. The pore size distribution however, does affect the amount of water that will be held tightly by capillary forces and block the flow of air. Thus, air will find it difficult to move through soils with small grain and pore sizes that retain more water.

Capillary pressure is defined as the difference between the soil fluid pressures of the nonwetting and wetting phases: $P_c = P_n - P_w$. In most soils, the wetting phase is the soil moisture and the nonwetting phase is the air filling the rest of the pore space. At the water table, the water pressure, by definition, equals the atmospheric (air) pressure. Below the water table, the water pressure increases as a function of depth. Above the water table, the pressure of the water phase is less than atmospheric; as a result, the capillary pressure is positive. This capillary pressure can be thought of as a suction force capable of holding water above the water table much like a sponge holds water. Capillary pressure can be expressed as a head by dividing it by the density of the fluid and the acceleration of gravity. Capillary pressure head is, by convention, considered positive if the pressure head in the wetting phase is less than atmospheric.

Heterogeneous soils can be thought of as having many sizes of pore spaces (or capillary-like tubes). Capillary pressures are largest (most positive) in the smallest pores (i.e., the water is held most tightly) and as a result water will drain from the larger pores before draining from the smaller pores. If it has not rained recently, and internal drainage has diminished, soils above the water table tend to exhibit a gradual decrease in saturation with increasing height above the water table. The large pores higher in the soil column will have drained, leaving moisture retained only in the smaller pores. Near the water table, more of both larger and smaller pores are filled. Figure 8 illustrates one relationship between capillary pressure head and saturation (29). When flow has ceased and soil moisture is at equilibrium, capillary pressure becomes equivalent to height above the water table, and the figure exhibits the profile of saturation with depth. The greater the capillary pressure, the smaller are the pores that remain water-filled.

For some distance (equivalent to a head h_b) above the water table, all pore spaces are saturated with water held at less-than-atmospheric pressure. This zone is commonly referred to as the capillary fringe and can be measured by the *bubbling pressure* or air pressure at which water can first be expelled from, or air bubbled through, the largest pores in initially saturated soils.

When applying SVE at a depth near the water table, the water table will rise locally in response to the applied vacuum (30). This rise, if significant, may isolate contaminated

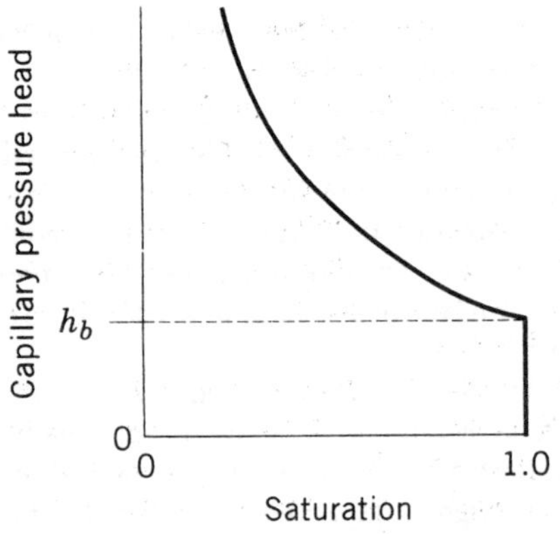

Figure 8. Capillary pressure head–saturation curve. Many soils exhibit a relationship of this general form. The bubbling pressure head, h_b, is the point at which water will first drain from an initially saturated soil's largest pores, making it possible for them to conduct air (3).

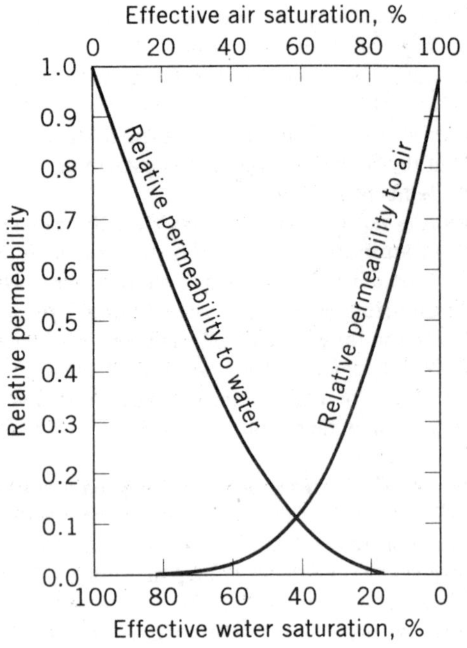

Figure 9. Relationship between water saturation (S_w) and relative permeability to air (3); $S_e = (S_w - S_r)/(1 - S_r)$, where S_e is effective and S_r residual water saturation.

soil from the airflow. This is particularly problematic at sites with a floating product smear zone. In addition, the relationship between saturation and capillary pressure predicts that the application of vacuum will reduce the capillary pressure head (the nonwetting phase pressure is reduced so the capillary pressure head also decreases) and thus, as long as there is a supply of water, the saturation of the soil will increase. This reduces the ability of the soils above the water table to transmit air. At sites with high moisture content or a shallow water table, this may significantly impede air flow.

The ability for a soil or other porous medium to transmit a fluid, independent of the characteristics of the fluid, is called the *intrinsic permeability* (k_i), and is measured in units of length squared. Saturated hydraulic conductivity (K_s) is the measure of the ability of the porous medium to transmit water when it is saturated, defined as:

$$K_s = \frac{k_i \rho_w g}{\mu_w} \tag{1}$$

where ρ_w is the water density, g is the acceleration of gravity, and μ_w is the dynamic viscosity of water. Similarly, an *air conductivity* can be defined with the appropriate values of air density and viscosity, which are functions of pressure and temperature and may be affected by high levels of contaminant vapors and humidity.

Most importantly, air conductivity is a function of saturation. Expressed in units of length squared, *air permeability* is independent of the air properties, but varies markedly with saturation as well as the nature of the soil. Relative (air) permeability is defined as the air permeability at a given saturation, divided by the intrinsic permeability (the air permeability at zero saturation). Figure 9 shows a typical relationship between saturation and relative (air) permeability.

Predicting Subsurface Airflow Using Soil Vapor Extraction Modeling

The laminar flow of fluids through porous media is described by *Darcy's law:*

$$q = \frac{k_i \rho}{\mu} \nabla \Phi \tag{2}$$

where q is the flow per unit area, k_i is the intrinsic permeability, ρ is the fluid density, μ is the fluid dynamic viscosity, and Φ is the total fluid potential. This equation is universally applied to predict water flow in soils; however, its applicability to predict air flow in the subsurface is less commonly recognized. For air flow, the intrinsic permeability is replaced by the air permeability, k_a, a function of saturation.

The fluid potential gradient is related to the pressure gradient by the Bernoulli equation:

$$\phi = gz + \frac{v^2}{2} + \int_{P_0}^{P} \frac{1}{\rho} dP \tag{3}$$

where z is the elevation, v is the fluid velocity, and P is absolute pressure. The first term on the right-hand side is negligible for elevation changes that are encountered in almost all SVE applications and the second term is negligible for laminar flow. This leaves only the third term, which can be expressed in a differential form:

$$\nabla \phi = \nabla \frac{P}{\rho} \tag{4}$$

It must be recognized that air density is not independent of pressure since gases are, unlike water, compressible. It has been suggested (31) that for absolute pressures of between 0.08 and 0.12 MPa (0.8 and 1.2 atm), the compressible nature of air does not contribute significantly to prediction errors for pressure and laminar flow.

Airflow in porous media also differs from water flow in that air velocities are nonzero at the pore space boundaries. This is termed slip flow and was first noted by Klinkenburg (32). This causes the airflow to be somewhat greater than predicted by Darcy's law. This is a significant effect for soils with pore spaces smaller than 10^{-3} mm (31).

A mathematical description of radial flow of air has been developed that accounts for pressure-dependent density and gas slippage (33):

$$\nabla \cdot (k_a \nabla P^2) = 2n_a \mu \frac{\partial P}{\partial t} \quad (5)$$

where n_a is the air-filled porosity and the other parameters have been described previously. This partial differential equation is nonlinear and can be solved analytically for a few boundary conditions or more generally using numerical (computer modeling) techniques.

Under steady-state conditions (no change with time), the right side of the equation reduces to zero. If k_a is assumed to be independent of pressure, which is not unreasonable for many settings where the saturation is not dramatically affected by the vacuum, the equation reduces to a form of the Laplace equation in P^2 for which useful analytical solutions have been developed. The solutions are based on specific assumptions about the geometry of the system or the boundary conditions. To apply the solutions to a specific problem, the assumed boundary conditions must reasonably approximate the site conditions. Steady-state conditions are usually quickly established at SVE sites under consistent operation, so these solutions are particularly relevant to predictions of long-term operation. An analytical solution has been developed for radial flow to a well that fully penetrates the vadose zone beneath a tight surface cover (33). This solution requires an estimate of the radius of pressure influence. This radius of pressure influence is typically estimated by plotting the observed vacuums versus the logarithm of the distance from each measurement point to the extraction well and extrapolating to a zero vacuum (Fig. 10).

Where air is drawn from a partially penetrating vertical well at an open or imperfectly covered site, there is leakage of air from the surface, resulting in two-dimensional radial flow. The solution under steady-state conditions is complex and requires that the effect of a point sink be integrated to simulate the finite vertical line sink (well). It also requires that the *method of images* be used to simulate both the no-flow boundary at the water table and the constant pressure boundary at the ground surface. The method of images relies on the principle of superposition or the fact that the pressure at any point can be determined by the sum of all external pressure effects. Researchers (34) provide a solution for radial flow at uncovered sites and provide useful nomographs for use of this relationship. A similar solution was provided with slightly different derivation (35). Figure 11 presents a cross section of the potential function and stream lines for a typical configuration. Given the nature of the solution in terms of pressure squared, the predicted pressure deviations from atmospheric pressure for extraction do *not* differ only by sign from the deviations predicted for injection. Although such equivalence exists for water (incompressible) injection and extraction, it does not exactly exist for air extraction and injection.

The effects from individual wells in a multiple well system can be summed to predict the pressure distribution throughout the volume to be remediated. If a large number of wells are present, this analysis is best done using numerical techniques.

In many instances, the travel time for air through a certain volume may be of interest, particularly if throughput of air is a criterion for design of an SVE system. Generally, travel time can be computed by integrating the change in distance by the seepage velocity at that point for the entire flow path of interest:

$$t = \int \left(\frac{1}{q_s}\right) ds \quad (6)$$

where t is the travel time, q_s is the seepage velocity, computed from Darcy's law but divided by the air-filled porosity, and ds is the incremental distance. For compressible flow, the travel times usually must be computed through numerical techniques; however, researchers (34) provide nomographs for travel times from the surface to an extraction well with several configurations of well screen relative to the surface and the water table. Vertical travel times to the surface for air injected at a well are also provided (35).

Numerical models are available for predicting SVE performance. Common models include those that can simulate pressure and airflow, predict contaminant or oxygen concentrations in soil gas, or do both. Such models may be very useful when (*1*) predicting airflow paths and travel times at heterogeneous sites or when multiple wells are included in the system, (*2*) optimizing well placement, (*3*) predicting offgas concentrations, or (*4*) estimating the time to remediation. Various references describe typical models (3,36).

Models require site data and should not substitute for actual pilot testing. Model boundary conditions must be a

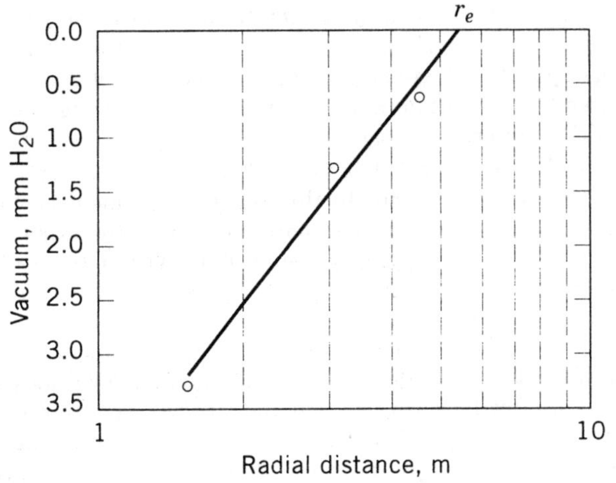

Figure 10. Use of distance-drawdown graphs of measured vacuum with distance from an SVE well to determine the radius of pressure influence, r_e (3) (1.0 mm H_2O = 9.8 Pa).

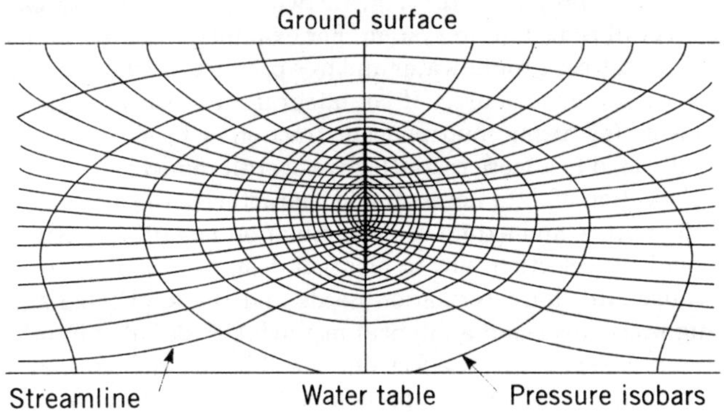

Figure 11. Cross section illustrating streamlines and pressure isobars during steady-state two-dimensional flow to a vertical SVE well beneath a leaky ground surface (3).

reasonable approximation of the site conditions. For pressure and flow models, input parameters normally include the air permeability distribution, porosity, surface and water table elevation, and soil gas properties (density, viscosity, and temperature). Models for predicting soil gas composition or contaminant transport normally require contaminant properties, including partitioning coefficients, information about the contaminant mass and composition (eg, constituents of fuels), dispersivity values, and degradation rates. Refer to the U.S. EPA (36) for more information on the considerations for modeling subsurface airflow.

Contaminant transport and composition models usually assume local equilibrium conditions so that equilibrium partitioning relationships can be used. This assumption may not be realistic at sites where kinetic limitations exist due to diffusion or where high airflow rates are used. Other parameters may be difficult to determine, such as dispersivity or total mass at the site.

If vacuum levels are relatively low, less than 0.2 atm gauge, one study (31) has shown that the relationships used to predict incompressible flow can be used without incurring unacceptable error. In this case, ground water flow models can be used with appropriate modifications for air conductivity and head values. In any case, the model assumptions must be reviewed for appropriateness given the site conditions and project objectives.

TECHNOLOGY SCREENING, BENCH-SCALE, AND PILOT-SCALE TESTING

Technology Screening Strategy

It is clear from the earlier section, Technology Description, that when faced with a specific site, one may choose from a number of options within the family of SVE technologies. The first steps in deciding whether SVE is a viable candidate technology involve weighing the primary decision criteria: the air permeability of the soil and the volatility and biodegradability of the contaminants of interest (Fig. 12). Among the related decision criteria are depth of the contamination, depth to water table, thickness of the capillary fringe, presence of barriers to airflow or preferential pathways, and moisture content. Along with technologies involving SVE, a host of other technologies also should initially be screened. One should avoid considering each candidate technology in isolation, however. It is often strongly preferable to integrate two or more technologies in the overall remediation system. For example, the system may need to address groundwater and free product recovery, as well as a vadose zone source. If so, integration of these remedial elements (eg, using multiphase extraction or bioslurping) will usually lead to a more cost-effective overall remedy.

Collection of data necessary to characterize the site as well as the nature and extent of contamination is a prerequisite to technology selection, not to mention design of SVE systems. Because site characterization needs and the applicable sampling procedures and analytical methods are detailed in other references (eg, 3, 21), the focus here will be to emphasize guiding principles and thought processes that come into play.

As discussed throughout the preceding section, Fundamental Principles, there are a number of physical, chemical, and biological site conditions that have a significant impact on the effectiveness of SVE. All too often, site characterization

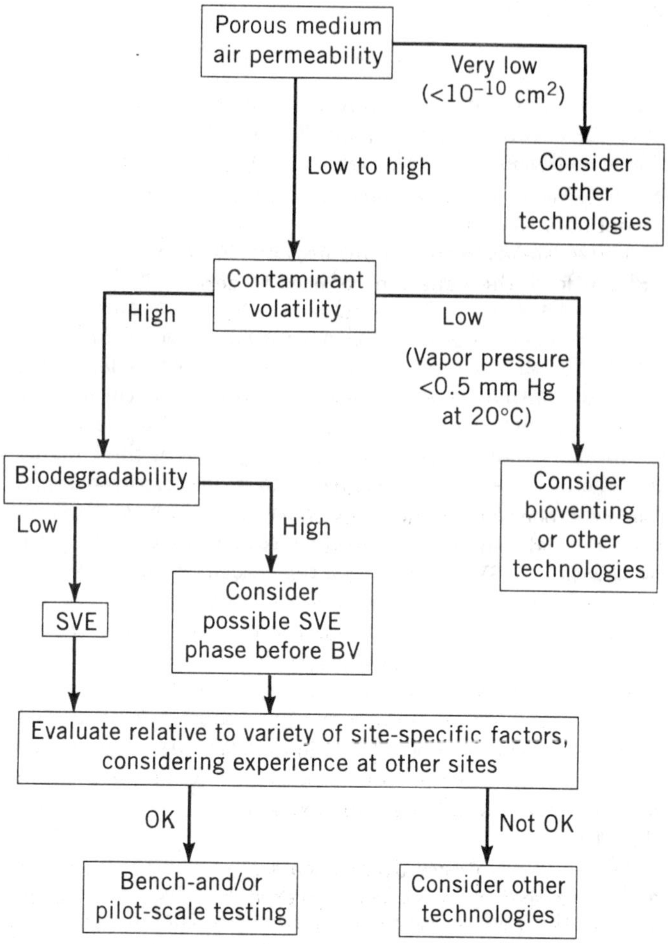

Figure 12. Technology screening decision tree (3).

data that could shed light on these conditions are not collected during initial site investigations. If, for example, personnel responsible for logging soil borings or observation pits were instructed to be aware of and systematically record features that might serve as barriers to airflow or preferential pathways, much duplication of effort could be avoided. It is not uncommon for unexpected contaminants to be extracted during startup of SVE systems. More care should be taken to characterize all volatile species, not just those of regulatory interest, during site characterization. In this regard, it is insufficient to characterize the vadose zone contamination at the expense of evaluating the underlying groundwater. Within the vadose zone itself, it is important to assess the three-dimensional distribution of contamination. If only surface soils are affected, SVE may not be a strong candidate. Moreover, if a small volume of relatively shallow soil is contaminated, SVE may not be as cost-effective as excavation and off-site disposal. Finally, if the presence of DNAPL is suspected in the unsaturated zone but not in the groundwater, there may be concerns that the installation of vent wells and/or the operation of SVE might increase the risk of downward mobilization of DNAPL, especially if DNAPL resides in fractured bedrock above the water table.

Bench-Scale Testing

It may be advantageous to conduct empirical tests at the bench-scale to ascertain whether SVE is an appropriate technology for a site. Although not performed as frequently as pilot tests, bench-scale tests can provide the only practical way to seek answers to certain kinds of questions under controlled conditions. For example, if airflow is induced through a soil, how will contaminant concentrations change with time as a function of flow rate, temperature, or moisture content? How realistic are the partition coefficients being adopted for predictive purposes? Can certain cleanup goals be attained? If they cannot be attained in representative soil cores under the best-case relatively confined conditions of the laboratory, it is doubtful that they will be attainable at the field scale.

One must be mindful that individual measurements made at the bench-scale do not necessarily represent behavior at larger scales. Measurements of air permeability, for example, reflect SVE system behavior more reliably if made at the field scale than at the column scale. Unfortunately, it can be expensive to perform a large number of *in situ* field-scale air permeability tests at a site. By collecting a representative number of soil cores, however, and subjecting each to relatively inexpensive laboratory air permeability testing, it can be possible to correlate the results of a single field-scale air permeability test with order-of-magnitude results from a number of soil cores collected at the same location. Additional soil cores collected site-wide can then be used to infer behavior of locations that cannot be tested at the field scale. Measurements of soil cores can be used as well to obtain certain key soil physical properties, such as capillary pressure–saturation curves (37), that cannot readily be obtained at larger scales. Both SVE column studies and measurements of soil properties should be performed on soil cores collected with as little disturbance as possible.

Pilot Testing Procedures and Analysis

Pilot tests for SVE are done to determine (*1*) air permeabilities, (*2*) relationship between vacuum and flow rate at the well head (a "system" curve), (*3*) vacuum or pressure distribution in the subsurface and the air flow paths, (*4*) off-gas concentrations, both initially and/or in the long term, and (*5*) the response of the water table or saturation to application of vacuum.

Equipment needed for a pilot test include (*1*) a power supply, (*2*) blower or vacuum pump (including associated equipment; these are often portable units), (*3*) extraction well (or trench) with sample port, (*4*) piping and valving (including an atmospheric bleed valve), (*5*) observation points (soil gas or vacuum probes), (*6*) vacuum, flow, and temperature monitoring equipment, (*7*) vapor (contaminant, oxygen, CO_2) concentration measurement equipment or device, and (*8*) offgas treatment device (if required). Many of these components are discussed in the Design section. Figure 13 schematically shows the layout of the typical pilot system. Regulatory requirements for offgas treatment during pilot testing vary greatly and should be verified prior to testing.

Vacuum monitoring should be conducted in probes installed at varying distances, typically from 1.5 to 15 m (5–50 ft), from the extraction well. Monitoring points are usually simple small-diameter probes with short screens set or pushed to the desired depths. Multiple depths should be monitored to determine the three-dimensional nature of airflow and to determine the degree of surface leakage.

The testing usually begins with a step test that determines the relationship of vacuum to flow from the well. This is a significant part of the system curve used extensively in design and is usually not a linear relationship, perhaps due to redistribution of liquids (saturation) near the well. Multiple flow rates (minimum of three) are tested by extracting at a single rate until vacuum at the well head stabilizes. The flow rate is then increased. Flow control is usually provided by regulating an atmospheric bleed valve. Flow measurement must be made upstream of the bleed valve. Figure 14 shows a typical system curve developed from a step test.

A constant rate test is usually performed on the well after the pressures have reequilibrated from any previous testing. Air is extracted at a constant rate approximating the rate at which air would be extracted under full-scale remediation. Data collected during the test includes vacuum response at the well head and in all monitoring points, and concentrations of contaminant, oxygen, and carbon dioxide in the extracted gas. Vacuum measurements are conducted on a logarithmically increasing schedule; for example, starting at intervals as short as 10 sec up to intervals of an hour or more later in the test. Vapor concentrations should be tested on a more regular frequency, generally every 2 to 4 h or at least three times during the test (one at the start, one in the middle, and one at the end of the test). The response of the water table may be monitored using either a sounder in an air-tight well or a pressure transducer or bubbler referenced to the air pressure in a sealed well (not the atmosphere) (3).

Duration of the test depends on the objectives, but may range from less than 1 h if air permeability is the only data need (eg, at a highly air permeable site), to several weeks if the goal is projecting longer-term off-gas concentration behavior. Air permeability tests should continue until vacuums in distant monitoring wells show negligible change. A more detailed description of pilot test procedures is provided elsewhere (3).

Data analysis is similar to analysis of groundwater pump tests. If the site is covered with a reasonably tight surface, a transient solution (33) for one-dimensional compressible radial flow can be used to estimate the apparent gas permeability. A graph is prepared of $(P/P_{atm})^2$, where P is the absolute pressure observed at the monitoring point, and P_{atm} is the absolute pressure of the atmosphere, versus $\ln(r^2/t)$ where r is the distance from the extraction well and t is the time of the measurement. The plot should be linear over the later times and the linear slope is used to calculate the air permeability using

$$k_a = -\frac{RT\mu Q_m}{2\pi b M P_{atm}^2 \text{ slope}} \tag{7}$$

where R is the gas constant, T is absolute temperature, Q_m is the mass flow rate, b is the unsaturated zone thickness, and M is the molecular weight of the gas. Researchers (38) also provide a similar and commonly used transient solution assuming one-dimensional radial flow.

If the site is uncovered, one study (34) provides a means to evaluate the air permeability from transient data. If applied (gauge) vacuums are less than 0.2 atm, incompressible solutions typically applied to groundwater pump test data can be used, including the Cooper-Jacob straight-line method or leaky aquifer solution (39). Later time data can be used to perform a pseudo-steady-state analysis for air permeability. Although

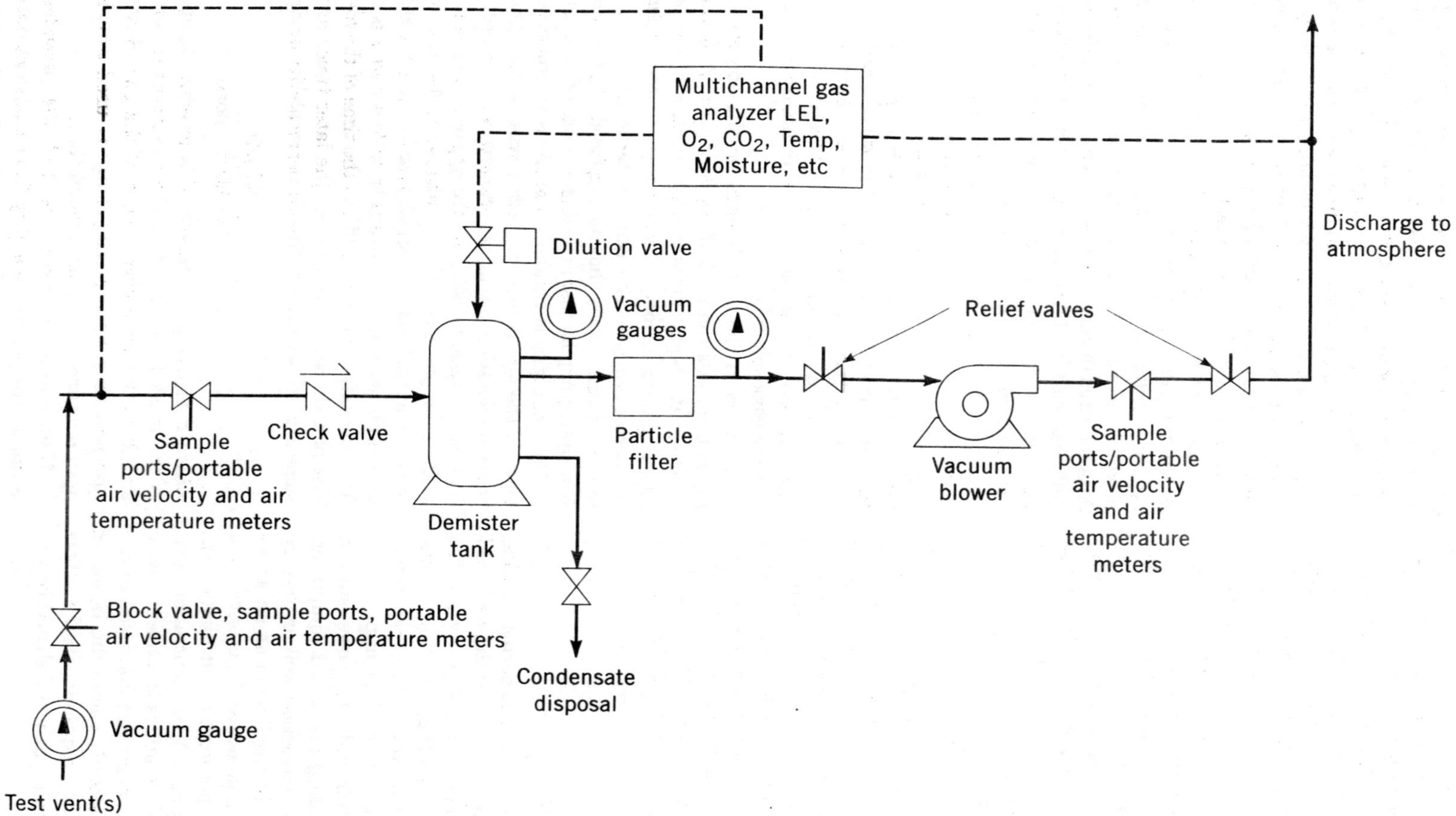

Figure 13. Simplified process flow diagram for conducting a typical SVE or bioventing system performance test (3).

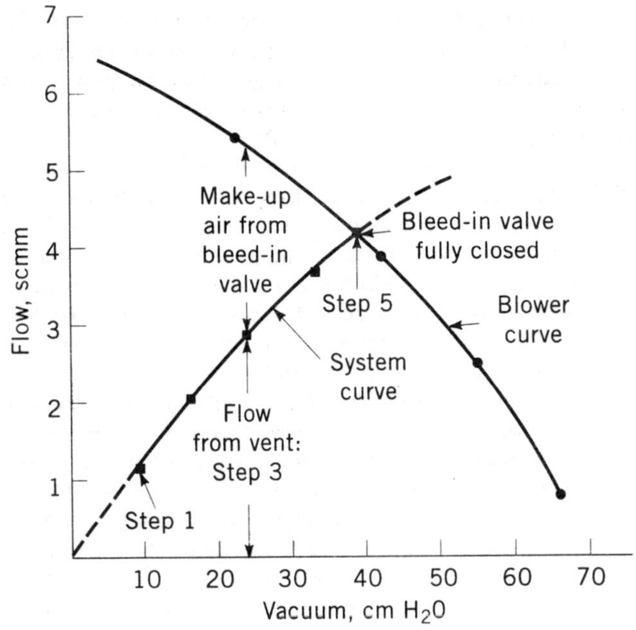

Figure 14. Example of system curve construction from stepped rate test (3).

pressure may still be changing, the time rate of change in pressure at later times becomes independent of distance from the well and air permeability can be calculated according to

$$k_a = \frac{Q_v P^* \mu}{\pi b} \frac{\ln(r_2/r_1)}{P_1^2 - P_2^2} \quad (8)$$

where Q_v is the volumetric flow rate, P^* is the pressure at the point of flow measurement, and r_1 and r_2 are the radial distances to the observation points where absolute pressures P_1 and P_2, respectively, are measured.

Researchers (40) provide a way to diagnose short-circuiting, a common problem with pilot tests. The observed vacuums in monitoring points are normalized by the extraction vacuum, the distance to the monitoring points is normalized by the depth to water table, and the data are plotted in a manner similar to Figure 15. A line is also depicted representing the theoretical vacuum versus distance relationship for an isotropic permeability site, where $K_h/K_v = 1$. Short-circuiting or another anomalous condition is indicated if a majority of the observations fall below the theoretical line (Fig. 15**c** and **e**).

Other field techniques possibly conducted as part of pilot testing include the use of tracer gas tests and down-hole flow-measurement devices. Tracer gas tests can be conducted with conservative tracers such as helium or sulfur hexafluoride to verify travel times and air permeability. Such tracers are released in individual observation points and monitored in the off-gas (41). Instruments have also been developed to thermally measure very low air-flow velocities in wells. These measurements can verify velocities calculated using gradients and air permeability estimates.

DESIGN AND APPLICATION

Design Strategy

If site characterization, technology selection, and bench- and/or pilot-scale testing have produced favorable results, the next step is to design the full-scale system. The overall objective of the design process is to attain certain remedial goals within an appropriate (preferably short) length of time and at an acceptable (hopefully modest) level of expenditure. The detailed design objectives include specifying the number and spacing of vent wells, flow rate and vacuum requirements, vent well and monitoring point construction, piping, method of vapor treatment, instrumentation, and process controls. Although these overall and specific objectives are not new, recently released U.S. Army Corps of Engineers guidance on SVE design (3) offers numerous opportunities to improve upon earlier design strategies. For example, there is a widespread tendency to overdesign SVE systems, using higher venting flow rates than necessary (42,43). Such systems are commonly designed to handle the initially high vapor concentrations associated with the most readily extracted fraction of the contaminant mass concentrations that are not sustainable very long. At many sites, because of diffusion or desorption-limited mass transfer, maintaining these higher flow rates does not improve removal but does increase the costs of moving air and treating the offgas. To design SVE systems as economically as possible, venting flow rates should instead attempt to maximize contaminant concentrations in the offgas over time, thereby minimizing the cost per mass of contaminant removed. Ideally the rate of airflow should match the rate of transfer of volatile contaminants from the soil and soil water into the soil air. Beyond a certain point, increasing the airflow rate will only dilute the contaminants being released from the soil, while increasing the quantity of air that has to be moved and treated.

Mathematical models are highly useful and often are indispensible tools during SVE design. They reduce the uncertainties inherent in the design process and permit the designer to ask a variety of "what if ..." questions. Even with a thoroughly analyzed design, however, there is no substitute for a healthy dose of empiricism. The nature of the subsurface is nearly always too complex to replicate in a model, nor would it be cost-effective to attempt to do so fully. Therefore it is important to build enough flexibility into the design so that it can be adjusted to changing conditions and data observed during operation.

Disciplines Involved

A project team with the appropriate background increases the likelihood of project success. Disciplines usually involved include mechanical/chemical engineer, hydrogeologist/geologist, geotechnical engineer, soil scientist, environmental engineer, electrical engineer, chemist, industrial hygienist or safety specialist, cost engineer, and regulatory specialist. Although the majority of the work may be done by one or a few individuals, consultation with other specialists improves design quality and avoids problems. Actual roles may vary, but a team with these skills should participate in the project. Close interaction, especially between those responsible for subsurface and aboveground design, is critical.

Elements of Design

There are many elements that may, depending on the circumstances, make up a complete SVE design package. Possible subsurface elements include well locations, well and trench construction, and construction of monitoring point locations;

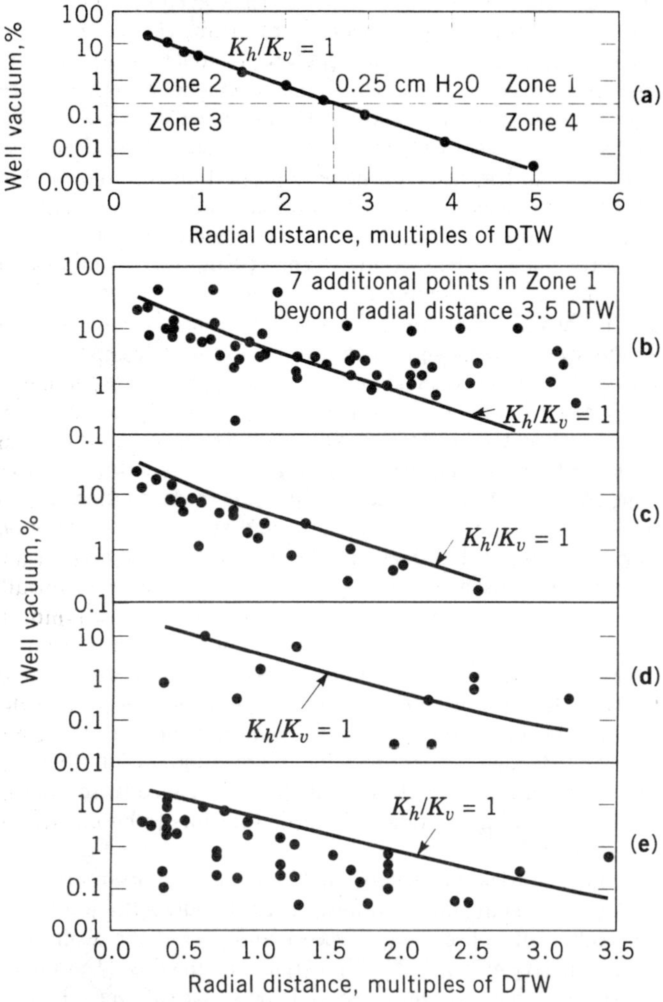

Figure 15. Field criteria for estimating venting feasibility, and evaluation of data from 22 pilot tests: (**a**) vacuum distribution zones for pass/fail criteria (pilot test observations falling within Zone 1 are considered passing); (**b**) high permeability sites passing field criteria; (**c**) high permeability sites failing field criteria; (**d**) low permeability sites passing field criteria; and (**e**) low permeability sites failing field criteria. K_h and K_v are the horizontal and vertical permeabilities, respectively, and DTW is depth to water table (40).

provisions for water and NAPL extraction (or simply control of upwelling); integration of related technologies such as bioventing, air sparging, or bioslurping; and incorporation of enhancements such as *in situ* heating, hydraulic fracturing, or perhaps oscillatory flow. Possible above-ground design elements include piping and valves, condensate control and disposal, particulate filters, blowers or vacuum pumps, silencer, off-gas treatment, instrumentation and process controls, electrical design, equipment support and housing, surface covers, and a review of process safety. With so many potential design elements, there are obviously a number of viable approaches to implementation and management of a design program. Several axioms however, can be elucidated. A design should not be commissioned before remedial goals of some kind have been established (eg, if the system is started up, when can it be shut off?). Operation of a pilot, or of the full-scale system in a pilot mode initially, produces data that can usually aid in system optimization. Nevertheless, it is not possible to accurately predict long-term SVE behavior on the basis of short-term performance (43).

A first stage in the design process is to review the site (and off-site) characterization data, plus data from bench- and/or pilot-scale studies, and identify any critical data gaps. Having a clear delineation of the contaminated zone is important. It may prove less costly to conduct a focused predesign investigation than to push forward with flawed data or adopt overly conservative design assumptions. Commonly *in situ* air permeability tests, at a minimum, are conducted. Sometimes a full review of site data can even lead to the conclusion that the site is not well suited to SVE, or that integration with another technology is necessary. A related step in the process is to consider the contaminant release mechanism that is apt to control the rate of remediation and to adopt an air exchange rate that is sufficient to match that rate of release. Depending on how limiting it is, one may decide that some form of enhancement of SVE is needed, or simply recognize that cleanup goals cannot be attained quickly.

A second phase in the design process is to design the ventilation rate in accordance with the estimated release rate limitations, and select pressure and vacuum requirements to attain the required airflow. It is important to consider at this point how one will be controlling the SVE equipment. If the site is remote, for example, frequent on-site monitoring and adjustment will not be practical. Here one will also want to consider whether horizontal trenches or vertical wells (or a combination) will be more effective, how many and where, screened at what depths, and whether a surface seal is needed. The result of this work is a tentative well layout. Airflow models can aid in this process, and will suggest ways in which the well layout can be improved. With due consideration to the relative importance of capital as well as operation and maintenance costs at the particular site, the blower/s (or vacuum pump/s), and the associated vapor treatment system, if any, should be designed. These can be the most expensive components in the overall SVE system. For example, if offgas treatment is not needed because the contaminants can be treated *in situ* by bioventing, overall costs can be halved (44). In some cases the off-gas can be redirected back into the ground. At this point it is not uncommon for feedback from the above-ground designers to send the subsurface designers back to their drawing boards in search of a less expensive, more effective overall approach.

Finally, the system components must be linked by suitable piping and valving. A preliminary piping and instrumentation diagram is created to show the relationships among many of the pertinent components (Fig. 16). Pneumatic calculations are performed based on system and blower curves and on manifold, valving, and instrumentation configurations to properly size equipment and, for example, to anticipate the effects of shutting off one or more wells of a multiwell system. The electrical connections for instruments, equipment, and process controls are established. Liquids handling must be addressed, as well as matters such as acoustics, aesthetics, equipment housing, and site security. Last, but not least, a process safety review needs to be conducted to ensure that any potential hazards or deficiencies are identified and corrected prior to system startup.

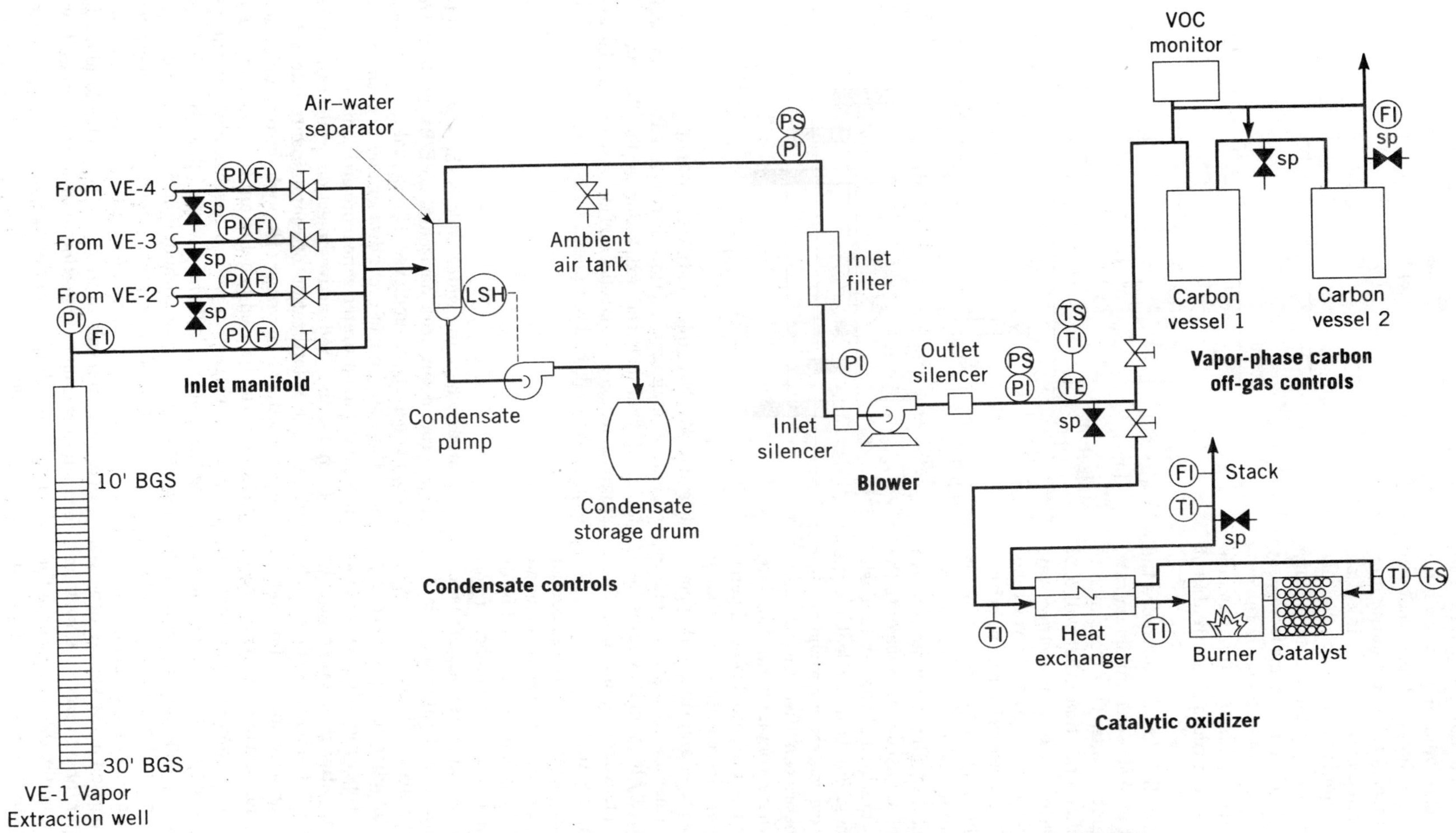

Figure 16. Typical SVE preliminary piping and instrumentation diagram (3): TE, temperature element; TI, temperature indicator; TS, temperature switch; FI, flow indicator; LSH, level switch high; PS, pressure switch; PI, pressure indicator; sp, sampling port; ⧖, manual/vacuum relief valve.

Well Layout

In the past, designers have often used a *radius of influence* approach to choosing well spacing. The radius of pressure influence has typically been defined as the distance from a vent well to locations where some small but measurable vacuum (or pressure) level can be detected at a given extraction (or injection) rate. It is assumed that since vacuum is detectable at that distance from the vent well, then air is moving and the soil is being treated. Well spacing would then be chosen on some factor, say 1.5 times the estimated radius of pressure influence at the projected flow rate. As was explained (45), this unfortunately fails to consider the actual flow rates at the edges of the treatment zone.

A more relevant approach to well layout is to require the initiation of an air velocity that exceeds some minimum rate, everywhere within the contaminated zone. This translates to a rate of air exchange (pore volume per time) that will lead to adequate cleanup in an allotted time. Put another way, the velocities need to be sufficient to result in travel times through the contaminated zone that are less than some critical travel time. Determination of this velocity (or travel time) is, however, problematic. Column studies of contaminant removal versus pore volumes of air passed through the column can give some indication of the amount of air needed to remediate the soil. There are several considerations in applying this analysis, however. The required air throughput is dependent on the initial soil concentrations (lower concentrations would require less air). Also, mass transfer kinetics can affect the efficiency of the removal, particularly where contaminants are only present in dissolved form in soil moisture or in dead-end pores. As previously discussed, optimum air velocities may change over time because of these phenomena and thus well spacing requirements may change over time also. Note that column test results, because of the relatively high airflow rates, may overestimate the air required to remove contaminants (refer to Ref. 3 for more information on this approach).

Analyses of air velocities and travel times to the extraction wells at various flow rates are required to verify adequate spacing (34,46). Required travel times can be estimated by dividing the time frame for remediation by the number of pore volumes required to remove a significant percentage of the contaminants. Computer models can be used for this analysis.

At sites requiring multiple wells, areas of little or no airflow are established where the effects of adjacent wells cancel out. This can be overcome either through operation of nearby wells at varying flow rates to move the stagnation point over time or by the use of air injection wells (Fig. 17). Computer models can be used to project the effect of passive or active air injection wells.

It is important to choose extraction and injection well screen intervals to focus airflow through the contaminated zone. If floating product is encountered, the greatest concentrations of vapors will often be found in the smear zone immediately above the water table. In this case, the screened sections of the wells should be placed in proximity to the water table for optimal removal efficiency (but with some portion of the vent screen extending far enough above the water table to prevent upwelling from occluding the screen). Additionally, the placement of the well screen deeper in the soil column has been shown, both analytically and empirically, to maximize the radius of influence of a given extraction well (34). Flow models are useful for selecting screen depths at complex sites.

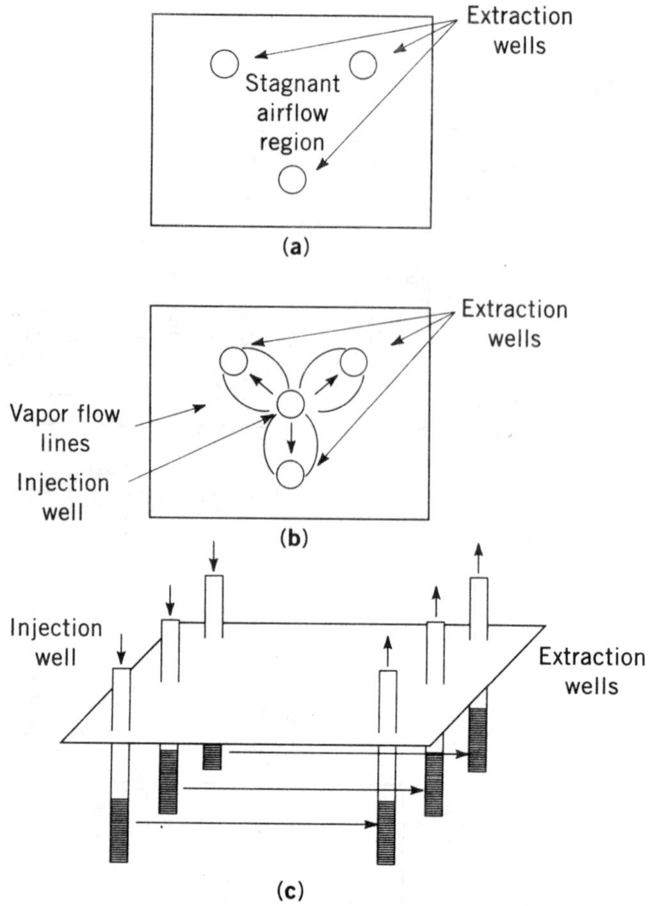

Figure 17. Example venting well configurations (11): (**a**) region of stagnant airflow between extraction wells; (**b**) injection well between extraction wells eliminates former stagnation zone; (**c**) parallel airflow streamlines between row of injection wells and row of extraction wells.

In areas where the water table is shallow (ie, <10 ft (3 m) below ground surface), horizontal SVE/BV wells or trenches should generally be employed. Horizontal SVE/BV wells minimize the upwelling of groundwater and in such cases extract air from the unsaturated zone more efficiently.

Both passive and active injection wells can be used to enhance the volatilization of contaminants into the soil air and control the air movement through the contamination zone. These wells can be used to help limit the radius of influence of a particular extraction well. A passive inlet system installed along the property boundary would allow SVE/BV to proceed at one of the properties without inducing migration of contaminants from the other property. The inlet wells may need to be quite closely spaced to create an effective boundary condition.

Screened intervals in injection wells should allow for uniform airflow from the injection to the extraction wells with little breakthrough of air to the atmosphere. Injection wells are usually installed vertically outside the edge of the contaminated area. A well-designed soil venting system allows vents to act interchangeably as extraction, injection, and/or passive inlet wells.

The disadvantages of installing passive inlet or active injection wells at a site include the added cost associated with the construction of additional wells and the additional energy cost of operating the compressor or blower for active injection wells. The addition of clean air to the subsurface may also dilute vapor phase contaminant concentrations, thereby increasing offgas treatment costs. This is true if poor well layout and choice of screened intervals allow preferential paths for clean air to move from the injection to the extraction well without adequately contacting contaminants or if the combined air flow rates are too high to account for mass transfer (diffusion) limitations.

Numerical models can be used to simulate more complex multiwell geometries and predict vacuum distributions. Such models use site data as input and are calibrated to observed behavior under pilot test or other known vacuum conditions. Once calibrated, the numerical model can be used to evaluate well placement options to determine the optimum geometry that meets air throughput criteria at a minimum total flow. The model can also be used later to evaluate changes in operations and should be recalibrated with observed full-system behavior. More information on computer modeling for SVE design is available elsewhere (36).

Monitoring Point Locations and Features

Monitoring probes are installed in the treatment area. The monitoring probes can be used to determine the vacuum, soil gas concentrations, or temperature at any one point. Such observations are necessary for verifying predicted airflow due to system operation and can be an indicator of progress toward remediation. The number of the probes must be adequate to determine the air flow throughout the contaminated zone. Special attention is given to representative areas near possible stagnation points equidistant from extraction wells.

The monitoring probes are usually small (eg, 2-in or 5-cm diam) wells with short screened intervals installed by driving or drilling so as to eliminate atmospheric air leakage (refer to ref. 3, Sec. 5–4 for construction guidance). Vacuum measurements are made using vacuum gauges or pressure transducers attached to the top of the probe. Soil gas concentrations are normally obtained using a portable vacuum pump to extract air from the probe. The pump must be able to overcome the induced vacuum in the subsurface. Subsurface temperatures can be estimated by measuring the temperature of extracted soil gas or by the placement of thermocouples in the boreholes with the monitoring probes. Soil moisture content can be monitored during SVE using a variety of methods including neutron depth moisture meters and tensiometers (37,47).

Extraction and Injection Well Construction

Vertical extraction (and injection) wells are generally constructed of plastic pipe and screen. Metal wells may be needed at sites with high levels of solvents or where thermal enhancements may be considered. The diameter of the well and the screen design are normally chosen to maximize the amount of open area to allow air to enter and travel up the well without significant pressure drop. The well diameter is also a function of the drilling method used to install the well. Drilling methods must not use liquids such as bentonite muds which may plug the soil pores. Two to six inch wells are normally employed and can be installed by auger methods (refer to ref. 3 for more detailed information on well installation).

In those cases where extraction and injection trenches are installed, the construction is very similar to the installation of any subsurface piping. The screen or slotted pipe is laid on a coarse bedding material in a trench. More of the same bedding material is placed over the piping and then low-permeability material is placed and compacted to original grade. A good seal in the trench is vital to prevent short circuiting from the atmosphere. In many cases where trenches are used, an air-tight surface cover is installed over the site.

Blower Selection

A specific blower or vacuum pump is selected based on mechanical, electrical, and pneumatic considerations. The first step in selecting a blower or vacuum pump is to conduct a pilot test to develop a pneumatic relationship between applied vacuum and air yield of a well. This was discussed in a previous section.

The second step is to calculate the friction loss for the system components and piping. The pneumatic calculations involved are described elsewhere (3). When the friction losses from subsurface and surface components are summed at each level of applied vacuum, the resulting positive relationship between vacuum and flow is termed a *system curve*. Finally a manufacturer's blower curve, which typically exhibits an inverse relationship between vacuum and flow, is superimposed over the system curve (see Fig. 14). A fixed-speed blower is capable of operating only along its blower curve, with an increase in applied vacuum being accompanied by a reduction in airflow, and vice versa. The point at which the system curve and a blower curve intersect represents the highest airflow that the blower is capable of extracting from the subsurface through the given piping and vent. Lesser amounts of airflow can be extracted using the same blower by admitting outside air into the well header pipe, usually by means of an intake valve (also known as a *bypass* or *bleed*). Greater amounts of airflow, however, cannot be extracted from the network except by changing to a larger blower.

The other major considerations in selecting a blower are its mechanical and electrical characteristics. Mechanical categories of common SVE blowers include regenerative blowers, rotary lobe blowers, and liquid ring vacuum pumps. These blower types are most applicable for low, medium, and high vacuum applications, respectively. Their modes of operation are described elsewhere (3). Vendors will typically offer several models of a given type of blower, each with a different flow capacity and blower curve. Blowers differ as to how efficiently they convert electrical energy to mechanical energy. The less efficient the conversion, the more power is lost to heating the airstream. Some rise in temperature can be useful, incidentally, as for example in reducing the relative humidity of offgas entering a carbon adsorption unit, which increases the efficiency of adsorption of VOCs. Too much temperature rise can, however, reduce the capacity of the carbon to adsorb the contaminant, and may present a fire risk. The designer must select the most efficient blower within the desired operating range of flows and pressures, while minimizing cost and maximizing flexibility of operation.

Off-Gas Treatment Options

To prevent exposure to extracted VOCs, treatment of SVE offgas may be required, although in some instances it may not be necessary. The method(s) selected must be capable of treating the contaminants of interest over the full range of influent concentrations that may occur during the course of system operation. This may entail several orders of magnitude of vapor concentration between the initial high levels produced upon startup and the low levels common during diffusion-limited tailing. Other considerations in offgas treatment selection include the required removal efficiency, flow rate range, secondary wastes produced, availability of utilities, and the required degree of control, monitoring, and automation. The cost of off-gas treatment can have a very significant effect on the overall site remediation costs. Given that off-gas treatment associated with SVE has many elements in common with emission control in general, only a relatively brief treatment of the subject is offered below.

Vapor-phase carbon can remove many classes of organic compounds including aromatics, aliphatics, and halogenated hydrocarbons and is employed in many SVE systems. Often two carbon canisters are placed in series to prevent breakthrough. Carbon usage is expensive when high concentrations of contaminants are encountered. Offsite regeneration or disposal is common, but onsite regeneration may be more cost effective for sites experiencing relatively higher long-term carbon usage.

Catalytic oxidation makes use of a catalyst, often platinum, to lower the activation energy of an oxidative reaction, allowing it to proceed at a relatively low temperature (290–370°C). Although most nonmethane hydrocarbons can be treated, catalytic oxidation of halogenated hydrocarbons produces acidic vapors, requiring installation of scrubbers. Many inorganic compounds including metals can substantially shorten catalyst longevity. In addition, the catalyst can overheat if the fuel content of the vapor stream is too high.

Thermal oxidizers heat the air stream high enough (900–1600 K) to combust the contaminants. Since there is no catalyst required, there are no concerns about poisoning of the catalyst. Very high energy efficiencies are possible, as is energy recovery, especially when vapor concentrations remain elevated. As concentrations diminish, however, thermal oxidizers become less efficient because they have to be supplemented with fuel.

Internal combustion engines can potentially serve both to perform the vacuum pump function and to provide offgas treatment. At low concentrations, supplemental fuel is required. Permitting the engine exhaust may increase monitoring requirements and thus operating costs. These units can accommodate only a relatively narrow, low range of airflows. Such engines can also tend to be sensitive to abrupt changes in soil moisture conditions.

Biofilters are increasingly attractive options for offgas treatment due to their relatively low initial and operating costs. A variety of support media are available, including several specially produced coated types that control biomass buildup and can handle shock loadings better than more conventional media such as soil, peat, and compost.

Other methods of off-gas treatment include resin adsorption, condensation of vapors, and, for very high vapor concentrations, flares.

Related Equipment

An SVE design is not complete unless various ancillary items are specified. Among these are the piping, valves, and manifold system; condensate collector, treatment and disposal system; particulate filter; blower silencer; instrumentation and process controls; electrical system; equipment housing; and surface cover. Although design considerations for each of these items are presented elsewhere (3), several are of particular interest.

Because soil vapor is typically close to or at 100% relative humidity, a sudden decrease in temperature or increase in pressure in the extracted vapor will cause condensation. If the water table is close to the vent screen, water droplets can become entrained within the vapor stream, or free water may even be drawn into the above-ground equipment. To prevent unwanted liquids from accumulating in piping, blowers, or air emission control devices, condensate controls are needed. Condensation can be minimized by burying or heat-tracing piping, and situating the above-ground SVE system components within a heated building (eg, 65°F or 18°C) so that the relative humidity of the vapor does not exceed saturation. Another approach, commonly employed during short-term applications, is to provide a physical or inertial type of air–water separator sized to hold a sufficient volume of condensate. A somewhat related item is a particulate filter to protect the blower from being damaged by particles that may enter the airstream.

Instrumentation and process controls ensure that the various SVE system components are coordinated and operate efficiently. Process control can be local, centralized, or remote. Local control is easy and inexpensive to install and operate. Remote control, by contrast, can be complex and expensive to install, but allows various levels of automation to be built in, enabling potential savings in labor, especially at inaccessible sites. Another important consideration is safety of operation, since potentially hazardous conditions can exist during SVE operations. Electrical codes must be followed, and adequate ventilation must be provided, especially where flammable mixtures could occur either during normal operation or in the event of a failure.

EFFECTIVENESS, CONSTRAINTS, AND CHALLENGES

Monitoring Performance

Ultimately an SVE system is judged successful when it has attained remedial objectives. Therefore, performance is measured primarily against the cleanup criteria that have been established, although practical considerations such as timeliness and cost and intangibles such as community relations also must be factored in. Typically, analysis of soil concentrations is performed to verify cleanup; however, a common method is to measure off-gas or soil gas concentrations, which are far easier to monitor than changes in soil concentrations. Unfortunately the early removal rates exhibited by an SVE system are seldom an indication of effectiveness, since extracted vapor concentrations tend to be highest near the outset of operation, but can seldom be sustained. It is typical for such concentrations to drop and approach an asymptotic level, while soil concentrations remain elevated.

A useful supplemental approach is to employ monitoring points to observe induced vacuum or pressure at several locations and depths of interest within the vicinity of each vent. Tracer gas tests, moreover, can indicate soil gas travel times from various locations to recovery wells (41). If early during operation the SVE system is not able to induce desired physical conditions, such as well-distributed airflow and adequate rates of air exchange within the contaminated zone, the negative impacts of any mass transfer limitations that might exist will only be exacerbated. In addition, performance will be severely compromised if upwelling or soil moisture occlude airflow. Monitoring of soil moisture levels and integration of groundwater controls during SVE can be essential ingredients for successful operation. Other valuable physical indicators of performance include even the relatively mundane monitoring of blower amperage, run-time, and on–off cycles. Short-term effectiveness is often measured in terms of adherence to air and condensate discharge requirements.

Rebound of Contaminant Levels and Probable Causes

When subsurface contaminant transport is limited by diffusion, desorption, or partitioning, SVE vent gas concentrations can appear deceptively low even when significant contamination remains in the nonvapor phases. In some cases, low off-gas concentrations may simply mean that venting flow rates are higher than necessary. Soil gas contaminant concentrations are typically monitored after SVE system shutdown to enable rebound to be detected. Rebounding concentrations indicate that a contaminant source is still present, and that its removal is mass-transport limited. If after a period of one to two weeks of shutdown, the vent gas concentrations are rechecked and remain above agreed-upon action levels, operation must resume until rebound levels no longer exceed action levels.

Site Limitations

There is no question that SVE is effective in removing contaminant mass from many sites. At marginally suited sites, however, SVE may not be capable of attaining remedial goals within a realistic time frame. Airflow at marginal sites is often nonuniform, bypassing contaminated zones, and leaving long diffusion pathlengths between advective channels. A very common site constraint is the presence of LNAPL within a thick capillary fringe, as occurs in medium- and fine-textured soils. Airflow through such a zone is minimal. Other common constraints include heterogeneous soils, perched water table zones or other regions of elevated soil moisture and low air permeability, fractured bedrock, and buried utilities. In the absence of adequate site characterization data, it is not wise to adopt the convenient assumption that the subsurface is homogeneous, because this is seldom the case. Whenever there is reason to suspect the presence of site constraints to SVE, practitioners are advised to exercise care throughout the application process (ie, in site characterization, technology screening, bench- and/or pilot-scale testing, design, and operation).

Cost

The definition of an innovative technology is one for which reliable cost and performance data are not yet available (10). Although SVE is more mature than many innovative technologies, it remains in this category. Costs for SVE are typically quoted within the range of $15 to $80/cubic yard ($12 to $60/m^3), depending on type and degree of contamination, scale of the project, site characteristics, and so on. Recently, however, the Federal Remediation Technologies Roundtable issued guidelines on cost and performance, which have been endorsed and are being implemented by the U.S. EPA, every branch of the Department of Defense, and the Department of Energy (48). As standardized cost and performance data begin to be recorded and compiled for a large number of sites, it will become possible to speak with far more certainty concerning the cost, applicability, and effectiveness of SVE relative to alternative technologies.

Soil Vapor Extraction Patents

A number of patents have been issued that relate to SVE (3). Although the central claim of a key SVE patent was invalidated (5), that decision is, as of this writing, under appeal. Given the changing status of the patent arena with respect to SVE, the reader is advised to consider the ramifications of the pertinent patents on their site activities.

Future Directions

Even while the acceptance and reliance on SVE for treatment of VOC-contaminated soils grows in the United States and abroad, some of the unbridled enthusiasm that led to earlier claims that SVE could be made to work in nearly all soils has fortunately been checked. A more sober and balanced view has emerged, that recognizes both the tremendous value of this simple and elegant remediation technology and its obvious limitations. Unknown even to some practitioners, methods to test, design, model, and implement SVE have advanced dramatically since its inception, and continue to do so today. Standardization of practice will probably become more prevalent in the future. The various available enhancements to SVE hold promise to enlarge its potential domain, as will its increasingly common integration with compatible technologies such as bioventing, air sparging, and multiphase extraction.

CONCLUSIONS

Soil vapor extraction is the most commonly used of all *in situ* remediation technologies, and enjoys wide application for clean up of soils contaminated with fuels and solvents. The standard of practice has advanced considerably over recent years, while understanding of the technology's limitations has also grown. The complex nature of the mechanisms involved, and the diverse skills called upon to implement SVE, demand the interaction of multidisciplinary teams throughout each stage of application. It is no wonder that soil vapor extraction remains a fascinating and fertile subject for students, researchers, and practitioners alike.

BIBLIOGRAPHY

1. U.S. EPA, *Cleaning Up the Nation's Waste Sites: Markets and Technology Trends.* EPA/542/F-93/003. Fact Sheet and Order Form. Office of Solid Waste and Emergency Response (05-110W), Washington, D.C., 1993.
2. J. T. Wilson and D. H. Kampbell, in *Handbook of Bioremediation*, Lewis Publishers, Boca Raton, Fla., 1994, pp. 117–129.

3. U.S. Army Corps of Engineers, *Soil Vapor Extraction and Bioventing Engineer Manual,* EM 1110-1-4001, ENSR Corp. for U.S. Army Corps of Engineers, Washington, D.C., 1995.
4. *Env. Bus. J.* 8(6), 12–16 (1995).
5. U.S. District Court, Court Order May 12, 1994 *James J. Malot v. Roy F. Weston, Inc. and Weston Services, Inc.,* Court Order May 12, 1994, Civil Action No. 1:92-CV-628-RLV. U.S. District Court for the Northern District of Georgia, Atlanta Division, 1994.
6. American Petroleum Institute, *Examination of Venting for Removal of Gasoline Vapors from Contaminated Soil,* API Publication No. 4429, Washington, D.C., 1980.
7. American Petroleum Institute, *Forced Venting to Remove Gasoline from a Large-Scale Model Aquifer,* API Publication No. 4431, Washington, D.C., 1984.
8. G. E. Hoag and B. Cliff, in E. J. Calabrese and P. T. Kostecki, eds., *Soils Contaminated by Petroleum: Environmental and Public Health Effects,* John Wiley & Sons, Inc., New York, 1985.
9. N. J. Hutzler, B. E. Murphy, and J. S. Gierke *State of Technology Review—Soil Vapor Extraction Systems,* EPA/600/2-89/024, Michigan Tech University, Houghton, Mich., 1989.
10. U.S. EPA, *Innovative Treatment Technologies: Annual Status Report,* 7th ed., EPA 542-R-95-008, Office of Solid Waste and Emergency Response, Washington, D.C., 1995.
11. T. A. Pederson and J. T. Curtis *Soil Vapor Extraction Technology Reference Handbook,* EPA/540/2-91/003, Camp Dresser & McKee, Inc. Office of Research and Development, Cincinnati, Ohio, 1991.
12. U.S. EPA, *A Technology Assessment of Soil Vapor Extraction and Air Sparging,* EPA/600/R-92/173, Risk Reduction Engineering Laboratory, Office of Research and Development, Cincinnati, Ohio, 1992.
13. U.S. EPA, *Presumptive Remedies Site Characterization and Technology Selection for CERCLA Sites with VOCs in Soils,* EPA/540/F-93/048, Washington, D.C., 1993.
14. U.S. EPA, *Innovative Treatment Technologies: Annual Status Report,* 6th ed., EPA 542-R-94-005, Office of Solid Waste and Emergency Response, Washington, D.C., 1994.
15. Texas Research Institute, *Laboratory Scale Gasoline Spill and Venting Experiment,* API Interim Report No. 7743-5:JST, Washington, D.C., 1980.
16. Texas Research Institute, *Forced Venting to Remove Gasoline Vapor from a Large-Scale Model Aquifer,* API Final Report No. 82101-F:TAV. Washington, D.C., 1984.
17. R. E. Hinchee, S. K. Ong, R. N. Miller, D. C. Downey, and R. Frandt, *Test Plan and Technical Protocol for a Field Treatability Test for Bioventing.* U.S. Air Force Center for Environmental Excellence, Brooks Air Force Base, Tex., 1992.
18. Air Force Center for Environmental Excellence (AFCEE), *Bioventing Performance and Cost Summary,* Brooks Air Force Base, Tex., 1994.
19. A. Leeson and R. E. Hinchee, *Principles and Practices of Bioventing,* Tyndall AFB, FL, USEPA, Cincinnati, OH, and US Air Force Center for Environmental Excellence, Brooks AFB, TX by Battelle Memorial Institute, Columbus, Ohio, 1995.
20. U.S. EPA, *Guide for Conducting Treatability Studies Under CERCLA: Soil Vapor Extraction,* EPA/540/2-91/019A, Office of Emergency and Remedial Response, Washington, D.C., 1991.
21. P. C. Johnson, A. Baehr, R. A. Brown, R. Hinchee, and G. Hoag, *Vacuum Vapor Extraction,* American Academy of Environmental Engineers, Annapolis, Md., 1994.
22. J. A. Kittel, R. E. Hinchee, R. Hoeppel, and R. Miller, in *Proceedings: Petroleum Hydrocarbons and Organic Chemicals in Ground Water.* NGWA/API, Dublin, Ohio, 1994.
23. L. A. Smith and R. E. Hinchee, *In Situ Thermal Technologies for Site Remediation,* Lewis Publishers, Boca Raton, Fla., 1993.
24. P. A. Gauglitz, T. M. Bergsman, S. M. Caley, W. O. Heath, M. C. Miller, R. W. Moss, J. S. Roberts, R. Shalla, M. H. Schlender, T. R. Jarosch, C. A. Eddy-Dilek, and B. B. Looney, *Six-Phase Soil Heating for Enhanced Removal of Contaminants: Volatile Organic Compounds in Non-Arid Soils Integrated Demonstration, Savannah River Site.* Office of Technology Development, U.S. Department of Energy, PNL-10184, US-406, Battelle Pacific Northwest Laboratory, Richland, Wash., 1994.
25. U.S. EPA, *Hydraulic Fracturing Technology, Technology Demonstration Summary,* EPA/540/SR-93/505, 1993.
26. U.S. EPA, *Laboratory Investigation of Residual Liquid Organics from Spills, Leaks, and the Disposal of Hazardous Wastes in Groundwater,* EPA/600/6-90/004, New Mexico Institute of Mining and Technology, Socorro, N.M., 1989.
27. J. J. Pignatello, in B. L. Sawhney and K. Brown, eds., *Reactions and Movements of Organic Chemicals in Soils,* Soil Science Society of America and American Society of Agronomy, Madison, Wis., 1989, pp. 45–80.
28. American Society for Testing and Materials *Methods for Particle-Size Analysis of Soils,* ASTM D422-63, 1990.
29. R. H. Brooks and A. T. Corey, *Irrigation Drainage* **92,** 61–88 (1966).
30. P. C. Johnson, C. C. Stanley, M. W. Kemblowski, D. L. Byers, and J. D. Colthart, *Ground Water Monitoring Rev.* **10**(2), 159–78 (1990).
31. J. W. Massmann, *J. Env. Eng.* **115**(1), 129–49 (1989).
32. L. J. Klinkenberg, *American Petroleum Institute Drilling and Production–Practice,* 1941, pp. 200–213.
33. D. B. McWhorter, *J. Contaminant Hydrol.* **5**, 297–314 (1990).
34. C. Shan, R. Falta, and I. Javandel, *Water Resources Res.* **28**(4), 1105–1120 (1992).
35. L. G. King, *Mathematical Models for Underground Injection of Gaseous Wastes into the Vadose Zone,* Battelle Technical Report No. BNRL-945. Battelle Pacific Northwest Laboratory, Richland, Wash., 1968.
36. U.S. EPA, *Review of Mathematical Modeling for Evaluating Soil Vapor Extraction Systems,* EPA/540/R-95/513, Washington, D.C., 1995.
37. R. S. Baker and J. Bierschenk, *J. Soil Contam.* **4**(1), 57–76 (1995).
38. P. C. Johnson, M. W. Kemblowski, and J. D. Colthart, *Groundwater* **28**(3), 413–429 (1990).
39. D. B. McWhorter and D. K. Sunada, *Groundwater Hydrology and Hydraulics.* Water Resources Publications, Littleton, Colo., 1977.
40. T. R. Peargin, and D. H. Mohr, in *Proceedings: Petroleum Hydrocarbons and Organic Chemicals in Ground Water.* NGWA/API, Dublin, Ohio, 1994, pp. 337–350.
41. M. C. Marley, in E.J. Calabrese and P. T. Kostecki, eds., *Hydrocarbon Contaminated Soils.* Vol. 3, Lewis Publishers, Chelsea, Mich., 1993, pp. 379–401.
42. F. Payne, in E.J. Calabrese and P. T. Kostecki, eds., *Abstracts of Eighth Annual Conference on Contaminated Soils,* School of Public Health, University of Massachusetts. Amherst, MA, 1993.
43. A. N. Clarke and D. J. Wilson, *Natl. Env. J.* 24–28 (Mar.–Apr. 1996).
44. H. J. Reisinger, E. F. Johnstone, and P. Hubbard, Jr., in R.E. Hinchee, B.C. Alleman, R.E. Hoeppel, and R. N. Miller, eds., *Hydrocarbon Bioremediation* Lewis Publishers, Boca Raton, Fla., 1994, pp. 40–57.
45. P. C. Johnson and R. A. Ettinger, *Ground Water Monitoring Rev.* 123–128 (Summer 1994).
46. R. W. Falta, K. Pruess, and D. A. Chesnut, *Groundwater* **31**(6), 1011–1120 (1993).

47. R. S. Baker and J. T. Wiseman, in *Proceedings: International Symposium on In Situ Treatment of Contaminated Soil and Water,* Air and Waste Management Association, Pittsburgh, Pa., 1992, pp. 26–41.
48. Federal Remediation Technologies Roundtable, *Guide for Documenting Cost and Performance for Remediation Projects,* EPA-542-B-95-002, Washington, D.C., 1995.

SOIL WATER AND GROUNDWATER SAMPLING

NEAL E. WILSON
Minnesota Pollution Control Agency
St. Paul, Minnesota

Contaminant regimes must be adequately characterized to protect human health and the environment. Moreover, site characterizations should be cost-effective. Over the last two decades the equipment and techniques used to obtain soil water or groundwater samples have become more refined. Environmental programs need to incorporate scientifically validated advances in groundwater sampling, so that potentially costly decisions about site management are based on sound data.

Some of the more commonly used methods for collecting groundwater samples have for the most part been found to be wanting. The following sections are intended to provide the information necessary to enable individuals to prepare and execute thorough and scientifically sound sampling and analysis plans. The data quality objective (DQO) methodology is presented to assist in preparing sampling and analysis plans, commensurate with the goals of the project.

SOIL WATER SAMPLERS

It may be desirable to collect water samples in the unsaturated (vadose) zone to establish an "early warning network", or because of the difficulties in installing monitoring wells in bedrock or a clayey regolith. The vadose zone begins at the earth's surface and extends downward to the first principal water-bearing aquifer. Water moving downward in the vadose zone does so within the soil matrix, or within macropores. Macropores in a soil matrix include worm channels, cracks, or root channels. Flow in the vadose zone changes from various degrees of partial saturation to preferential saturated flow (1).

Drive point or push-type samples are still for the most part limited to the saturated zone. As such, permanently installed soil water samplers are necessary to collect water samples from the vadose zone. The two principal types of soil water samplers (also called lysimeters) used to intercept water in the vadose zone are the pressure/vacuum type and the percolate or free-drainage type. A vacuum soil water sampler utilizes a vacuum to draw water into the sampler and thus collects interstitial water, and also water from immediately adjacent macropores. Percolate soil water samplers (pan or wick types) obtain samples predominantly from macropore flow derived from precipitation or irrigation wetting fronts. Each type of sampler has its own applications and limitations that should be evaluated prior to their use.

Pressure/Vacuum Soil Water Samplers

Pressure/vacuum samplers consist of a porous ceramic, stainless steel or Teflon tip attached to a poly(vinyl chloride) (PVC) or stainless steel tube. Inlet and outlet tubes exit the backplate of the device, and two chemically inert (usually Teflon) tubing lines run from the device to the ground surface. Prior to the installation of the pressure/vacuum samplers, they must be carefully cleaned and inspected for leaks. In accordance with the appropriately approved site safety and health plan (including a utilities check) a hole is drilled into the ground. The device is then placed into the hole, sometimes with silica flour placed around the tip to enhance hydraulic conductivity. Native fill is then carefully placed around the device, a bentonite plug is placed immediately above the device, and native fill is tamped back into the hole. A sufficient amount of time (optimally a year) should elapse subsequent to installation, to allow the soil to redevelop some structure.

To obtain a sample, a vacuum is applied from the ground surface into the device. The vacuum draws interstitial water that is held in the soil under tension into the sampling device. The vacuum is introduced with either a hand pump, a vacuum pump, or a vacuum tank. Optimally the vacuum should not decline over time. A constant vacuum produces more sample volume and reduces the potential for sample alterations (2,3).

Pressure/vacuum soil water samplers are limited to a depth of approximately 15 m, but special high pressure samplers equipped with check valves and a separate sampling reservoir may be used to depths of over 90 m. Numerous limitations are associated with pressure/vacuum soil water samplers (4):

1. The tip of the sampler may become plugged or may add or subtract chemical constituents.
2. Alterations of the sample may occur due to changes in redox potential, pH, or from volatilization.
3. The soil water may be in disequilibrium with the surrounding media.
4. Reservoirs (dead space) of between 30–80 mL of water may remain in the sampler.
5. Leaks may develop in the sampler or the hydraulic connection between the sampler and the surrounding soils may be difficult to obtain or may be lost.
6. Natural flow paths in the vadose zone may be altered by installing the device.
7. A nonconstant vacuum may skew the chemistry of the sample.
8. A sufficient sample may be difficult to obtain.
9. Holding times may be exceeded when obtaining the sample.
10. The flow directions associated with the sample may be uncertain.
11. Regulatory agencies may view soil water differently than groundwater.
12. Upward of 100 samples may be required to obtain a 70% confidence level.

Several reasons may account for the failure of soil water samplers to deliver a sufficient volume of soil water. These reasons include having a poor hydraulic connection between the soil and the sampler, leakage in the lines or sampler, a fractured tip, and freezing or clogging of the sampler and lines. In addition, the soil itself may not have enough moisture to pull a sample.

A tensiometer or gypsum block should be installed and used with the sampler. This is to ensure that an adequate amount of soil water is available to obtain an adequate sample volume. If a sufficient amount of water is available in the soil then observing the decay in vacuum over time (usually hours) or drawing a known volume of deionized water through the sampler should assist in trouble-shooting the problem.

Percolate Soil Water Sampler

In well-structured soil the downward migration of soil water is dominated by macropore flow. The optimal type of soil water sampler to intercept macropore flow is the pan or wick percolate (free-drainage) soil water sampler. A pan percolate soil water sampler is a nonreactive pan filled with fine-grained packing material. Sheet metal or glass blocks (5), aluminum (6), or plastic (7) have been used to fabricate pan samplers.

The bottom of the pan is sloped to a corner or the center of the device where the outlet is located, and inert tubing conveys the soil water to a collection vessel. Wick samplers are essentially pan samplers with an inert wick (usually fiber glass) incorporated into the design. Wick samplers pull water under tension into the sampler, in addition to collecting macropore flow.

Prior to installation, free-drainage soil water samplers need to be adequately cleaned. In accordance with an appropriately approved site safety plan (which includes a utilities check) a trench is excavated and a bench is cut into the trench at the desired sampling depth. The sampler is then pressed upward into the ceiling of the bench and is held in place with bracing. All edges of the sampler must be in firm contact with the sampler to ensure that the soil water does not bypass the sampler. Sometimes shelters are built over the trench to gain access to the sampler subsequent to installation. For the pan-type or percolate soil water sampler to function properly, enough water must move downward in a wetting front to overcome the capillary and adsorptive forces that retard downward migration of soil water. After each wetting front has passed the sample must be removed to prevent overflowing or cross-contamination between sampling events.

Summary

Soil water samplers may provide useful information about water chemistry and contaminant transport. The correct sampler needs to be selected and be properly installed and operated if it is to provide useful information. Moreover, samples obtained from soil water samplers must be evaluated in terms of their inherent limitations.

DESIGN AND INSTALLATION OF MONITORING WELLS

Locations of Monitoring Wells

The spatial configuration of a groundwater monitoring system is based on several factors. These factors include the type and quantity of the contaminants, the hydrogeologic characteristics of the site, and regulatory requirements. Whenever possible monitoring wells should be installed only after a sufficient amount of information about the soils, geology, hydrogeology, and geochemistry of the site has been compiled and assessed.

Monitoring Well Installation Work Plans

Existing information augmented with borings, piezometers geophysics, soil gas, or Geoprobe work assists in preparing a work plan for monitoring well installation. A work plan for monitoring well installation needs to specify the proposed well location, depth, materials, and construction techniques. An appropriately approved health and safety plan (including a utilities check) needs to be part of the work plan for monitoring well installation.

Soil Sampling

The locations and depths of borings undertaken before monitoring well installation need to reflect the spatial orientation of the site (8). When installing a monitoring well the borehole is usually logged simultaneously with well installation. Well logs provide geologic and hydrogeologic information necessary to further characterize the site. At minimum, samples from borings should be collected continuously for the first 4.5 m (15 ft) (where the most rapid lithologic changes usually occur), and then every 1.5 m or with each change in lithology. Borings are usually classified in the field using ASTM Method D2488, and a Munsell color chart. Further characterization of samples at contaminated sites, such as using a organic vapor monitor or Geiger counter, may also be necessary.

Split spoon samplers or Shelby tubes are most commonly used for obtaining samples in unconsolidated materials. Sometimes it may be appropriate to obtain both split spoon and Shelby tube samples in the same borehole. A split spoon sampler is a hollow metal tube (usually with 5 cm OD and 3.5 cm ID) that after sampling, opens along its major axis for sample recovery. Split spoon samplers are attached to sampling rods and are lowered into the borehole. A 63 kg (140 lb) hammer is dropped on the rod assembly either down-hole or at the surface (9). The number of blows required for each 15-cm increment of the sample rod is then recorded. Samples obtained from a split spoon sampler are used for field classification and for obtaining grain size distributions in a laboratory. Split spoon samplers may be lined with several close-fitting brass inserts. Brass inserts facilitates field classification while providing undisturbed samples for testing vertical permeabilities in a laboratory.

Shelby tube samplers are used in clays, silts, and fine-grained sands. Shelby tubes are stainless steel tubes, commonly 76 cm long with a 7-cm diameter. The tube is attached to a rod and is placed in the bottom of the borehole. The tube is pushed into the sample zone by a continuous rapid motion imparted by the drill rig's hydraulic ram, and the length of advance is determined by the resistance and condition of the formation (10). When the tube is removed from the borehole the bottom 2.5 cm of sample is removed for field classification, and the remainder is shipped to the laboratory in an upright position for vertical permeability testing.

Logging boreholes based on the return (drill cuttings) is generally not recommended because this method has limited lithologic accuracy. Logging based on the return may, however, be the only viable method available, as when drilling in a well-lithified formations.

Land Disposal Restrictions

Land disposal restrictions may apply to investigative derived wastes (IDW). IDW includes purge water or drill cuttings (11). If it is suspected that IDW is a hazardous waste it needs to be properly contained to await laboratory analysis. If the laboratory analysis confirms that the IDW are a hazardous waste, then the requirements of RCRA Subtitle C must be met. These requirements may include adequate characterization, placarding, using manifests, and disposal at a permitted hazardous waste disposal facility.

Field Documentation

Field notes describing the drilling are recorded by the professional in charge of the field work. These notes should include the project name, hole number, date started and completed, geologist's and driller's names, hole location (a narrative, and located on a map), rig type and drilling method, sampling equipment used, classification scheme, and description of salient features in proximity to the borehole. The borehole log should include the following information:

- texture (percent sand, silt, and clay)
- grain size, sorting, and degree of grain rounding (angular to spheroid)
- vertical extent of soil zones
- primary and secondary voids
- structural features such as bedding planes, cross-beds, graded features, lineations, solution channels, discontinuities, bioturbations, and respective orientations
- moisture content (wet, moist, dry)
- color (refer to a Munsell chart) and presence of discoloration or mottling
- observations on blow counts, advance rate, and sample recovery
- depth of borehole and of each water-bearing unit
- depth, location, and identification of any contaminants encountered
- OVM or Geiger counter readings of samples and explosivity meter readings from borehole, depending on site
- surveyed horizontal location and elevation of the top of borehole
- other comments or observations, such as deviations from drilling plan or method

Field classification of soil samples are undertaken to identify formation changes, to facilitate changes to the work plan, and to ensure that appropriate samples are sent to the laboratory for further analyses. All (deeper) borings not used for monitoring point construction should be properly sealed with bentonite or neat cement grout to minimize the potential for future pollutant migration along the borehole.

Drilling Methods

Hollow Stem. Hollow stem continuous flight augers are the most commonly used drilling tool to simultaneously log a borehole and install a monitoring well in unconsolidated materials. Each hollow stem flight is 1.5 m long, with large external threads. Hollow stem augers provide access for split spoon or Shelby tube sampling as drilling progresses. They also act as a temporary casing, facilitating the installation of the screen and casing string, filter pack, and annular seal.

Pilot bits are recommended when drilling with hollow stem augers. Pilot bits are used to reduce the potential for obtaining unrepresentative boring cores, and reduce the problems associated with heaving sands. When drilling in unconsolidated formations below the water table, tricone bits are sometimes used inside hollow stem augers. Tricone bits allow mud (a combination of variably viscous bentonite clay and water) to be pumped into formation. Drilling mud reduces the likelihood that formation materials would flow up into the hollow stem, locking up the pilot bit and auger. Drilling mud also reduces the potential that the next soil sample obtained is material drawn up into the auger. Drilling muds should be judiciously used, as they may chemically alter in the surrounding media and may also greatly reduce the hydraulic conductivity in proximity to the well. Drilling muds are difficult to remove from the borehole and are generally discouraged for installing monitoring wells whenever possible.

If the depth of the well or the nature of the formation materials precludes the use of a hollow stem auger, an alternative method of drilling must be selected such as using a cable tool, air percussion, air or mud rotary, or rotosonic drilling.

Cable Tool. A cable tool drill rig repeatedly raises and drops a heavy weight (the drill bit) to pound a hole into the ground. Casing is driven simultaneously with drill bit advancement if the formation is unconsolidated. Cuttings must be periodically removed from the borehole with a special bailer, and water must be added if it is not naturally present. Cable tool drilling is relatively slow, but may reduce smearing when compared to other drilling methods.

Air Percussion. Air percussion drilling may be selected when drilling in refractory glacial tills containing boulders, or when drilling in hard bedrock. Air percussion drilling utilizes compressed air to rapidly raise and lower a downhole hammer attached to the drill bit. Carbide buttons on the working face of the bit add durability. Air hammer drill bits incorporate holes and channels into the bit design so compressed air can be blown into the bottom of the borehole. Compressed air forced down through the drill rods and out of the bit cools the bit and forces cuttings to the surface. Obtaining representative formation samples with a split spoon or Shelby tube using air hammer (or air or mud rotary) drilling entails breaking down the drill string each time a sample is required, which can be time consuming.

Air or Mud Rotary. Air or mud rotary drilling uses a rotary action and downward force to act on the attached drill bit. Casing may be driven simultaneously with air or mud rotary drilling. In unconsolidated formations the borehole is kept open and cuttings are brought to the surface by using high pressure air (air rotary), or pressurizing clean water or a bentonite slurry (mud rotary) down through the drill rods, out through the bit, and back to the surface through the borehole. The air, water, or mud also lubricates and cools the bit, prevents inflow of aquifer materials, and reduces the potential for cross-contamination between aquifers. Bentonite slurry (mud) may be difficult to remove subsequent to drilling and may alter the chemistry of the groundwater (12). Adequate air filters must be used on the compressor when using an air rotary or air percussion drilling to prevent lubricating oils from invading the formation. Drilling foams that contain organic chemicals should be avoided to prevent altering the chemistry of the aquifer.

Rotosonic. Rotosonic drilling utilizes a sonic head that imparts high frequency sinusoidal waves to the drill bit. Rotosonic drilling is fast, and provides limited cuttings and excellent lithologic resolution (13). Rotosonic drilling may be more expensive than other drilling methods, and heaving sands may present a problem when using them.

Cleaning Drilling Equipment

Drilling and soil sampling equipment needs to be cleaned between boreholes. A cleaning plan should be part of the installation plan (14). The degree of decontamination needed between holes depends on the site. Steam cleaners, low and high pressure hot water washers, and combination units are all used to decontaminate drilling equipment. Steam cleaners use less water than high pressure washers, but they may form a mist. Mists at some sites may be a health and safety concern. A decontamination station where rinse water is collected, tested, and properly disposed of may be needed at some sites. A decontamination report should be part of the field notes. The drill rig itself should be visibly free of hydraulic fluid, grease, oils, or fuel, and work areas kept relatively clean and free of obstructions.

Materials Used for Monitoring Well Construction

A typical monitoring well consists of a screen, filter pack, riser, annular sealant, and surface completion details. The materials most commonly used for monitoring well screens are rigid PVC or stainless steel (SS) 304 or 316.

PVC and low carbon steel are the most commonly used riser materials. Sections of metal casing are either welded or joined together with threaded couplings. A stainless steel screen may be attached to length(s) of stainless steel riser. Black pipe attached to the stainless steel riser then runs up to the ground surface to reduce materials costs.

Stainless Steel. For most monitoring situations SS 304 is usually selected because it costs less than SS 316. Poor quality SS 304 screens (with bends or broken welds) may leach significant quantities of trivalent chromium and nickel due to crevice corrosion (15), so good quality screens need to be selected. Good quality SS 316 should be considered if the groundwater has low pH, redox, or dissolved oxygen concentrations, contains elevated chloride or sulfur-containing species, or high total dissolved solids (16).

Poly(Vinyl Chloride). PVC screens and casing may be joined together using solvents and glues, flush threading, or pipe threading. If a monitoring well is to be used for sampling organic compounds then solvents and glues must *not* be used to join PVC materials. Solvents and glues may leach a significant quantity of numerous organic compounds for many years subsequent to well installation. Flush threading (two pieces form a flush joint) may be preferable to pipe threading (an external coupling). This is because if an external coupling is required, it limits the annular space between the casing and borehole wall, making grouting more difficult.

Flush threads reduce the tensile strength of the casing string by about 30% as compared to a solvent-cemented joint (17). A 5-cm nominal diameter schedule 80 pipe has an inside diameter of 4.925 cm and a wall thickness of 0.553 cm. Schedule 40 pipe has an inside diameter of 5.250 cm, and a wall thickness of 0.391 cm. Schedule 80 PVC is much stronger than schedule 40 PVC pipe. If the sampling equipment requires a minimum inside diameter of 5 cm, then using 5-cm nominal diameter schedule 80 pipe may not be possible. Using 10-cm diameter schedule 80 casing instead may make wells more versatile, as they could be used in the future for remediation wells.

The maximum hanging weight of a casing string for 5-cm schedule 40 PVC pipe is 238 kg (525 lbs) (18). Therefore the maximum depth that a 5-cm nominal diameter schedule 40 PVC well could be installed without an offsetting buoyant force (from groundwater in the hole) is 199 m (656 ft). If the hydrostatic pressure of the annular seal (grout) is greater than the collapse strength of the casing material the casing will fail prior to grout hardening. The hydrostatic pressure acting on a casing string can be calculated using the following formula:

$$\text{Hydrostatic pressure (in psi)} = (\text{Height of fluid column (in ft)} \times \text{Fluid weight (in lb/gal)} \times (0.052))$$

The collapse strength for a 5-cm nominal diameter schedule 40 PVC riser pipe is 147 psi, and neat cement grout weighs approximately 15 lb/gal. The maximum depth using the above formula before collapse would occur would be 57 m (188 ft). Adding the grout in stages and letting the grout harden before the next addition of the grout has been suggested if an alternative design cannot be accommodated. (19). The maximum pressure that schedule 40 PVC pipe joints can withstand is 1.72 bar (25 psi).

National Sanitation Foundation-approved PVC pipe may represent a reasonable compromise for monitoring lower levels or the early detection of organic contaminants and/or monitoring for trace metals. PVC pipe is considerably less expensive than SS 304, stainless steel may bias trace metal concentrations in groundwater samples, and PVC has a limited adsorption potential with respect to lower concentrations of organic compounds.

PVC casing may deteriorate (solvate) in the presence of elevated concentrations of low molecular weight ketones, amines, aldehydes, and chlorinated alkenes and alkanes in the groundwater or surrounding soils (17). PVC may also become brittle with time or may crack due to shifting or freezing ground. PVC is also less forgiving when hit by heavy equipment. It may therefore be appropriate to limit the use PVC well construction materials to short-term (≤5 years) and relatively shallow (<15 m) water table monitoring situations.

PVC joints must be watertight and should never be overtightened. Teflon tape or chemically inert O-rings are required for joining sections of PVC pipe. The weight of the casing string should not be allowed to rest on the bottom of the hole until the grout sets up, or the compressive force may cause the screen or joints to structurally fail.

Keeping Monitoring Well Construction Materials Clean

Monitoring well screens must be either certified as factory cleaned and delivered to the site in a sealed plastic bag, or adequately steam-cleaned or high pressure washed in the field immediately prior to installation; the same is required for each section of riser. Bags of filter pack sand and bentonite and screen and riser materials must be kept away from sources of contamination while being delivered to the site. Sources of contamination include gasoline cans, gasoline-powered compressors, oils, hydraulic fluids, etc. During storage and transport clean plastic sheets should be placed under and over bags of filter pack sand and bentonite until ready for use.

Selecting a Filter Pack and Screen Size

For short-term presence/absence monitoring a 10 slot 20/40 filter pack combination, commonly called for irrespective of grain size distribution of sediments, may deliver data sufficient to meet qualitative DQO types. However, for longer-term monitoring of trace contaminant concentrations it becomes necessary to select the correct filter pack and screen. Selection is determined by first performing a grain-size analysis (in the field or in a laboratory) of the finest portion of the sediments within the proposed screened interval. The grain-size distribution of the sediments is plotted on 2–3 cycle semilog paper, and the D-70 of the sediments is obtained from the graph. (D-70 is equivalent to the sieve size that is 70% (or retains 30%) of the formation material.) Multiplying the D-70 by a multiple of 4–6 (20) may be used to determine the optimal filter pack. The screen size is then selected to retain between 90–95% of the filter pack. If sieving is undertaken in the field a range of filter pack and screen sizes may be required on the supply truck. Furthermore, laboratory confirmation of field sieving may be required. Filter pack materials need to be chemically inert and consist of well-rounded grains of uniform grain size because the well screen slots have uniform openings. The ideal uniformity coefficient should be 1 and the uniformity coefficient should not exceed 2.5 (21).

Monitoring Well Installation

Monitoring wells need to be installed in accordance with an approved work plan that includes an appropriately reviewed and approved safety and health plan. An explosivity meter and an organic vapor monitor (OVM) may need to be employed if explosive or dangerous levels of gases may be encountered when installing the well. Extreme care needs to be taken to avoid causing an explosion when drilling or welding, or by the improper use of a cutting torch. Well installation also needs to be in accordance with local well codes. Monitoring wells constructed through more than one aquifer may be required to have multiple casings, each casing sealed off from the other. Multiple casings are needed to reduce the potential for a introducing a hydraulic connection between aquifers.

Clean disposable vinyl gloves should be worn by the driller if holding the screen during installation. Water from an acceptable (potable) source of known chemistry must be used for the drilling processes. After drilling to the prescribed depth, the screen and casing sections are joined together and lowered into the hole section by section. Casing centralizers at the bottom of the well screen are sometimes used to evenly distribute the filter pack and the annular seal. Centralizers sometimes make it more difficult to use a tremmie, and to emplace a uniform filter pack and annular seal.

Monitoring wells need to be installed as straight as possible. The maximum deviation should be 1 degree or less for every 15 m (50 ft) of depth. When the hydraulic gradient is small a deviation survey should be considered because water levels taken in out of plumb wells may produce inaccurate determinations of groundwater flow direction and rate.

The volume displaced by casing, filter pack, and grout should be calculated prior to placing the filter pack and annular seal. The volume of filter pack sand and grout should be measured while the well is being constructed, and the amounts used recorded. If a significant disparity is noted between the calculated and actual volume of materials used the well may have to be immediately decommissioned.

Between 2 and 10% of the filter pack should be placed in the bottom of the borehole, prior to lowering the screen and casing string into the hole. A decontaminated 3.175 cm diameter tremmie pipe rotated in the hole should be used to install the filter pack. Clean water added to the sand may be necessary to keep the sand moving downward in the tremmie. A weighted measuring line is needed to gauge the progress of emplacement of the filter pack and the annular seal and to ensure complete coverage.

The filter pack is usually extended 0.6 m above the top of the screen, unless another aquifer is directly above the screen. The drill flights or temporary casing is pulled upward in the borehole while the filter pack, annular seal, and grout are emplaced and its progress is gauged using a weighted tape.

The Annular Seal

Bentonite (sodium montmorillonite, a clay) is usually used as the annular sealant. Bentonite may be added dry (pellets or chips) or as a slurry. A bentonite slurry may selected if the area above the screen is unsaturated. A bentonite slurry (following manufacturer instructions) is prepared by mixing 6.8 kg (15 lbs) of dry bentonite powder to 26.5 (7 gal) of water to produce 0.09 m^2 (1 ft^2) of bentonite slurry. A 0.6 m thick layer of fine sand should be placed on top of the filter pack if bentonite slurry is used. Fine sand on top of the filter pack reduces the potential that the bentonite slurry will invade the filter pack.

If bentonite pellets or chips are used then water must be added from the surface, or a sufficient quantity of water must be present in the formation or to allow the bentonite to properly hydrate. Between one to four hours should elapse (as specified by the manufacturer) between bentonite emplacement and subsequent well construction activities to allow for hydration and expansion. Adequately hydrated bentonite swells 10 to 15 times the volume of dry bentonite. Bentonite will not properly hydrate if the mix water has greater than 500 mg/L of total dissolved solids, or contains elevated chloride concentrations (17). Furthermore, if alcohols, ketones, and other polar organic solvents are present, bentonite may also not properly hydrate (21).

Grouting the Well

Grouting between the annular space and the casing is undertaken to provide a hydraulic seal, reduce downward migration of contaminants, and to increase the strength of the well. Grouting involves mixing and injecting either neat cement, bentonite, or neat cement with bentonite into the annular space between the casing and the drilled hole. Improperly emplaced grout (intrusion into the filter pack) may greatly bias groundwater chemistry (take-up or release cations).

This condition may be reflected by having elevated pH in subsequently taken groundwater samples.

Neat Cement Grout. Neat cement grout is one 42 kg (94 lb) bag of Portland cement mixed with 22–26 L (6–7 gal) of clean water (22). Grout mixed in this proportion shrinks approximately 17%. Additional water is sometimes added due to mixing difficulties, flash setting, and heat of hydration. Adding additional water may unacceptably increase shrinking and cracking of the grout. Shrinkage needs to be minimized to provide an effective annular seal.

As per ASTM C150 there are five types of cement which may be used to make grout. If neat cement is chosen for the grout, then in most monitoring situations Type I cement is used. Type I cement takes approximately 48–72 hours to cure, depending on the water content. The heat of hydration produced by neat cement grout while it is setting may compromise the structural integrity of PVC casing. This is especially true if a large void was created in the borehole wall when drilling. Adding bentonite powder (2–5%) to the grout may reduce shrinkage and provide plasticity. Adding sand or diatomaceous earth reduces the heat of hydration. Grouts (cement or bentonite slurry) need to be uniformly mixed. Lumpy grouts should not be used.

Bentonite. Some uncertainty exists as to whether bentonite should be used in the unsaturated zone. Bentonite used in the unsaturated zone may never fully hydrate and expand, or it may subsequently desiccate and shrink. Inadequately hydrated or desiccated bentonite may not provide an effective annular seal. Bentonite may, however, rehydrate when wetted, or stay sealed even in semiarid environments. Bentonite pellets may stick, hydrate, and bridge to any surface including drill flights and well casing protrusions. Therefore, if pellets are chosen, they should be emplaced with a clean tremmie pipe and funnel. Bentonite chips or pellets may have to be poured directly into the annulus from the surface if standing water is present.

When emplacing pellets, chips, or grout the tremmie should be rotated in a circular manner around the annulus as the tremmie pipe is raised upward to get an even coverage. Grouts must be emplaced with a grout pump with the discharge of the tremmie below the surface of the grout. The filter pack should be disturbed as little as possible while emplacing the annular seal and grout. To gauge the progress of grout emplacement a weighted tape should be used. The tremmie should be raised at the same rate that the grout is emplaced.

Surface Completion Details. Neat cement grout should extend from the top of the annular seal to the surface seal. Surface seals consist of either concrete or neat cement grout and should extend a minimum of one foot below the frost line (if applicable). Special surface seal designs are necessary if frost heave may become a problem. A protective outer casing equipped with a lockable cover is set into the surface seal. Sometimes vent holes (0.6 cm dia) are drilled into the base of the outer casing. Vent holes reduce the buildup of potentially explosive gases and allow water to drain between the casings. Pea gravel between the inner and outer casings may reduce the potential for frost heave and insect infestations. A hole drilled into the inner well cap is sometimes used to allow the well to vent with fluctuating water levels.

A case-hardened steel lock must be used to secure the well at all times to discourage vandalism. Locks are also analogous to a chain of custody seal. Lubrication of locks with oil or graphite is unacceptable. Steel or treated wood posts, concrete pyramids, berms, or underground vaults are used to protect monitoring wells from damage from vehicles and heavy equipment. Each monitoring well should have an aluminum identification tag affixed to it. The tag should contain the following information: well number, date of installation, well owner, driller, and elevation of the top of the innermost casing.

Surveying the Well

The elevation of the north side of the top of the innermost casing should be surveyed by a licensed surveyor to the nearest 1/100th of a foot above mean sea level. The survey should reference a National Geodetic Vertical Datum. The north side of the innermost well casing should be surveyed, as subsequent water level measurements made on the north side of the innermost well casing will reduce errors derived from uneven casing ends. Monitoring wells may need to be surveyed horizontally as well as vertically. For most monitoring situations third-order accuracy (1/5000) may be sufficient.

If the site benchmark is off then the groundwater flow direction and rate may still be *relatively* accurate, in terms of groundwater flow direction within the site. If individual wells are surveyed incorrectly groundwater flow directions and rates will be in error. In addition to errors introduced in the field, transcription errors are also possible. Spot-checking the accuracy of the survey using the original survey notes, the well logs, other site benchmarks, and a topographic map should be routinely undertaken. Resurveying potentially suspect wells with another survey crew may be warranted.

Developing the Well

Well development removes foreign materials introduced during drilling and installation and restores the natural hydraulic conductivity and water quality of the formation being sampled. At least during the initial development procedure a partial breakdown of the borehole wall should be undertaken.

A minimum of 72 hours should elapse before well development commences, to allow the grout to set up. To ascertain the efficacy of well development a nephelometer (measures turbidity) should be used. Wells constructed in fine-grained materials may have to be developed several times to reduce turbidity.

The most commonly used methods of monitoring well development include the following:

- overpumping
- surging
- bailing
- air lifting
- jetting
- combinations of the above

Overpumping. When developing a well by overpumping, the pumping rate is substantially greater than the ability of the formation to deliver water. If overpumping is used for well development the pump should be selected to hold up when pumping relatively large quantities of sand (23). Moreover, the pump must be capable of operating at various pumping rates. Overpumping may require adequate disposal of contaminated development water.

A variation of overpumping involves letting the pumped water column fall back into the casing (ie, rawhiding), causing a surging action. Overpumping a well (unless simultaneously rawhiding) is not usually effective in breaking down the borehole wall. Rawhiding is accomplished by starting and stopping a pump without a back-flow device. Pumps should not be allowed to operate in reverse as potentially contaminated air may be entrained into the filter pack. The pump should be raised and lowered in the screened section during development to ensure that the whole screened interval is subjected to the overpumping and/or surging.

Surging. Developing monitoring wells installed in consolidated materials or installed in sediments characterized by mixed sedimentary layers may be undertaken by *gently* surging the well. The use of excessive force when surging in unconsolidated formations will disrupt the natural orientations of the sediments, resulting in substantial reduction of groundwater flow in proximity to the well.

A surge block is similar to a piston attached to a rod. The rod is raised and lowered in the monitoring well which forces water back and forth through the screen. Some surge blocks have pressure relief holes, and flexible rings (like automobile engine piston rings) which are almost the same diameter as the well screen.

Numerous surge block designs are possible and they can be purchased or fabricated by sandwiching inert, flexible gasket material between wooden disks (slightly less than 5 cm in diameter). A length of PVC casing filled with sand (with end caps) tied to a clean length of rope may adequately function as a surge block. Tubing attached to a pump or a check valve may be incorporated into the design of a surge block, so water is brought to the surface simultaneously with surging (24).

The surge block is raised and lowered beginning at the top of the screen moving downward, with increasing vigor, periodically removing fine material from the well with either a bailer or a sand-resistant pump. The force of surging depends on the speed and length of stroke. Surging can collapse the screen and casing, and the surge block may become stuck (sand locked) if proper techniques are not employed.

Bailing. Bailing a well for well development may be effective in relatively clean, permeable formations where the water flows freely into the borehole (17). To develop a well by bailing, the bailer is agitated up and down at various depths of the well screen, and fine material should be collected from the bottom sump with the bailer. Care must be exercised when allowing the bailer to free fall, when agitating, or in rapid removal so as to not exceed the collapse strength of the screen or casing.

Air Lifting. Air lifting has generally been discouraged for developing monitoring wells due to concerns that this method will alter the chemistry of the groundwater (17). Air lifting involves introducing compressed air below the air–water interface within the monitoring well. The compressed air displaces water, causing it to flow to the surface. The air delivery tube must be at least 1.5–3 m above the top of the screen, and the pumping rate needs to less than or equal to the recharge rate into the well. Air bubbles or compressor oils forced into the filter pack or screen must be avoided when using this method.

Jetting. Jetting involves forcing water under pressure into the screened interval of the well. Jetting does not remove sediments from the well by itself. The water used to jet wells should not be sediment-ladened, or sediment may be forced back into the filter pack.

Combinations. Combining air lifting with jetting or surging may be one of the best ways to develop a well without disrupting the aquifer sediments, if the water is not excessively sediment-ladened (25). The relatively narrow size of the air lifting pipe allows simultaneous pumping and jetting. Jetting in conjunction with the use of a submersible pump may also be used to adequately develop wells (26). Lastly, carefully combining surging with overpumping may also effectively develop a well.

Hydraulic Conductivity Testing Methods

A pump or slug test are used to determine the hydraulic performance of the well, or the hydraulic conductivity of the formation. Laboratory-derived hydraulic conductivities have been shown to vary by orders of magnitude from values derived by field methods (27). As such, laboratory-derived values should only be regarded as gross estimations.

Pump tests are undertaken using a pumping well and observation well(s), or one well for both. Single-well pump tests are important to ascertain the maximum purge rate that can be used when sampling, without drawing water down into the screened interval. For monitoring wells a pump test may be difficult to perform. A pump test generally yields more accurate information about an aquifer's hydraulic characteristics, but slug tests are more widely used.

A slug test involves either quickly introducing or removing a solid object, or a known volume of water, into the well. A change of about a 0.3 m (one foot) in head is generally adequate. In highly permeable media a change in head can be accomplished by rapidly pressurizing the well (28). For longer tests with multiple wells, a manual water level indicator may be sufficient to obtain the change in water levels. A continuous data logger is recommended for observing drawdown or recovery when performing a slug test. There are several types of slug tests and analytical solutions that are used, depending on the hydrogeology of the site (29). Numerous assumptions, however, must be made that are specific to the slug test and analytical solution. These assumptions may or may not be valid.

Documentation

Adequate monitoring of well documentation should consist of the boring log, construction diagram, survey information, equipment decontamination report, a record of well development, laboratory-derived vertical permeabilities, and an accurate, scaled site map showing the location of the well. The "pedigree" of the well may be called into question if this information is not available.

Bullet Check List

The following is a "bullet" check list for monitoring well installation (30). Modifications to the following bullet check list may be appropriate. It is the responsibility of the consultant and permittee to ensure compliance with other laws, ordinances, rules, and requirements when installing monitoring wells, including those specified in the governing well code. The bullet check list should be provided to driller(s) when bids are requested.

1. The well driller must be a licensed driller, in conformance with the appropriate governing agency.
2. In accordance with OSHA regulation 1926.651 all utilities must be adequately notified where and when drilling is proposed.
3. Site-specific safety and health plan(s) in accordance with 29 CFR 1910.120 may be necessary, and this must be addressed in the work plan. A pre-drilling safety meeting should be held between the driller, consultant, and government regulators to define respective roles and responsibilities.
4. Wells proposed to be installed in flood-prone areas may require special design considerations, and this should be addressed in the work plan.
5. The exact method of drilling needs to be specified in the work plan. Mud rotary drilling should be discouraged unless an alternative method is not possible. If air rotary or percussion drilling is selected, then foams must not be used, and adequate filters must be used on the compressor so the potential for oil contamination is eliminated.
6. An individual with expertise in hydrogeology as defined by education and experience needs to be present on the drill rig to classify borings, and to possibly facilitate field changes to the approved work plan. Changes in the field include more frequent soil borings, changes in methodology, etc. Field changes may need approval from regulatory agencies, and this should be addressed in the work plan.
7. When installing monitoring wells soil samples must be taken a minimum of every five feet or with each change in lithology using ASTM D1586 (split barrel), D1587 (thin-walled tube), D3550 (ring-lined barrel), or an approved equivalent method. ASTM D2487 taxonomy must be used. Descriptions of the soil and lithologic units including textural classification, primary and secondary structures, voids, and soil color as per a Munsell chart must be included in the logs. In many drilling situations the most rapid change in lithology occurs in the first 4.5 m (15 ft), so continuously logging the first 4.5 m and the entire screened interval is recommended. In addition to anticipating when more frequent samples would be warranted in the work plan, the drill cuttings and the "pull" of the drill rig should be used to assess when more frequent soil samples are necessary.
8. Each soil unit identified should be laboratory tested for particle size, particle distribution, porosity, vertical permeability, and clay mineral content or cation-exchange capacity.
9. The filter pack for the proposed monitoring well(s) needs to be of uniform, well sorted grain size, and be based on the cumulative per cent retained determined from a grain size distribution curve. The well intake (screen) size should only be determined after the filter pack is specified, and should hold back between 95–100% of the filter pack material. In most situations this will require soil borings and laboratory analysis antecedent to monitoring well installation. The filter pack and screen size determinations needs to be submitted to the regulatory agency prior to well installation.
10. The monitoring well screens and riser pipe must be either factory cleaned (certified as cleaned) and delivered to the site in plastic bags removed form any potential source of contamination including gasoline, hydraulic fluids, etc, or adequately steam-cleaned in the field. The work plan must specify which procedure is proposed to be used.
11. The filter pack and grout must be emplaced from the bottom up using a tremmie, and not poured into the annulus from the surface. If bentonite chips or pellets are not used for the annular seal, 0.6 m of fine sand needs to be emplaced above the filter pack to prevent grout from invading the filter pack.
12. The type of annular sealant (ie, bentonite or cement) and grout must be specified, including the type of bentonite proposed (ie, chips, pellets, or a slurry) and the type of cement (ie, Type I–V). Additives to the cement or grout must be approved by the appropriate regulatory agency.
13. If bentonite pellets or chips are used for the annular seal then potable water may need to be added to hydrate the bentonite. Furthermore an adequate amount of time (1–4 h) must be allowed to elapse after emplacing the bentonite to allow proper hydration of the bentonite before allowing further construction of the well.
14. Down-hole drilling equipment including auger flights or drill bits must be steam cleaned or high pressure washed between drilling locations. The procedure must be specified in the work plan.
15. After the wells are installed, a minimum of 72 hours must elapse prior to well development to allow the grout to properly set up. A combination of surging with a surge block and/or overpumping is recommended for well development. To reduce the potential for altering ground water chemistry compressed air must not be used for well development. The method of well development must be submitted to the appropriate regulatory agency for review and approval. Turbidity of 10 nephelemetric turbidity units (NTUs) or less should be achieved in most monitoring situations.
16. Some of the newly installed monitoring wells may need to have the hydraulic conductivity ascertained (ie, slug tested). The number and locations of well(s) to be tested, the test method proposed, and the analytic solution (including assumptions) must be submitted to the appropriate regulatory agency for review and approval prior to initiating the slug test.
17. Newer, more accurate permanent site benchmarks need to be established if questions exist as to the accuracy and permanence of the existing site benchmark(s). This must be discussed in the work plan.
18. The north side of the innermost casing of the monitoring wells must be vertically surveyed to the nearest 1/100th of a foot, and the monitoring wells must be vertically surveyed (not using a measuring tape).

19. All monitoring wells in the monitoring system needs to be permanently labeled. An aluminum plate approximately 7.6 × 12.7 cm (3 × 5 in.) pop-riveted to the outermost casing with the field number, date of installation, and well owner inscribed with an awl or metal stamps is preferred. Adequate well labeling must be addressed in the work plan.
20. The well log, constructions diagram, formation log, laboratory analysis, surveying and slug testing results, and a (brief) written narrative discussing the drilling and installation of the monitoring well(s) including all deviations from the approved work plan must be submitted to the appropriate regulatory agency for review and approval subsequent to well installation.

GROUNDWATER SAMPLING EQUIPMENT

Groundwater sampling equipment falls into three broad categories:

1. Equipment used to measure the static water level or free product thickness
2. Sample acquisition devices (ie, pumps or bailers)
3. Field geochemical instrumentation

Groundwater sampling equipment needs to minimize sample alterations including degassing, volatilization, sorption, and leaching (31). Sampling equipment should also be reliable, durable, portable, and be easy to clean and operate. Data acquired in the field may be as important as the analytic data obtained in a laboratory. Therefore field sampling equipment should be of the best available quality and be maintained in optimal condition.

Measuring Static Water Levels

Static water levels are measured to ascertain the direction (and subsequently the rate) of groundwater flow. Static water levels are also used for well purging criteria. Static water levels need to be measured to the nearest one hundredth of a foot. Monitoring wells should be surveyed and subsequent static water elevations should be taken on the north side of the innermost casing due to uneven casing ends.

Steel Tapes. The two types of equipment most commonly used to measure static water levels in monitoring wells are steel tapes and electric water indicators. Steel tapes used for measuring static water levels are commonly 0.6 cm (0.25-in.) wide black steel surveying tapes on a hand-wound reel. Steel tapes may be purchased in lengths of up to 152 m (500 ft); for many applications a 60-m (200-ft) long tape should be adequate.

The black color of the tape facilitates a color contrast between the tape and the chalk or indicator paste that is applied to the tape. A slender heavy weight (usually a lead fishing sinker or surveyors plumb bob) is attached to the end of the tape to keep it under tension. Lead may be scraped off onto the casing, so a stainless steel weight may be preferred. It is recommended that the weight cause no more than a 0.015 m (0.05-ft) rise of water in the well (32).

Carpenters chalk or a special paste is applied to the end of the tape which is then lowered into the well. The water in the well washes off the chalk or changes the color of the paste, so a visible contrast appears on the tape. Most steel tapes are only calibrated in hundredths of a foot at the end of the tape, so it may be necessary to carry along a small piece of steel tape with the 0.003 m (0.01-ft) graduations. If light nonaqueous-phase liquids (LNAPLs) or dense nonaqueous-phase liquids (DNAPLs) are present special indicator pastes are available that can be applied to steel tapes to measure the free product thickness within the well bore. Extra health and safety precautions must be taken when LNAPLs and DNAPLs are anticipated.

It is important to only lower a chalked tape to the predetermined graduation (a stop mark) on the tape. If the tape is lowered and then raised to the stop mark on the tape a false reading will be imparted to the measurement. Several attempts may be necessary to obtain a static water elevation with a steel tape. Using a steel tape in subfreezing weather (the water freezes to the tape) or in the rain (the chalk washes off) may make using a steel tape a challenging task.

Electric Water Level Indicators. Electric water level indicators use ions in the water to complete a circuit, activating a tone and/or a light. Some of the newer electric water level indicators are made of a steel tape enclosed in Teflon making them as accurate as the steel tape and chalk method. Older electric water level indicators made of two strands of wire on a reel may not be as accurate as either a steel tape and chalk or an electric steel tape enclosed in Teflon because the wires may kink or stretch.

If nonpolar LNAPLs (or DNAPLs, when checking the bottom of the well) are present, electric water level indicators will not function. The thickness of LNAPLs or DNAPLs may be measured with an interface meter, a steel tape and indicator paste, or a special bailer. A correction factor should be applied to free product thickness measured in monitoring wells (33).

Tapes (both manual or electric) should be periodically checked for accuracy against a reliable standard. Tapes need to be adequately decontaminated between wells. It is especially important to decontaminate tapes used for domestic wells (usually with a bleach solution) to prevent the spread of iron bacteria. It is recommended that if an electric tape is selected that a steel tape (longer than the anticipated well depths) is also brought along, to provide a back-up in case of equipment failure.

Sample Acquisition Devices

The most commonly used devices to sample monitoring wells include grab mechanisms (bailers and syringe samplers), inertia pumps, peristaltic pumps, and positive displacement pumps (submersible and bladder pumps). Each device has its own advantages and limitations. Sampling devices may be permanently installed (ie, dedicated) or may be laboratory or field cleaned and used for all wells at a site. Dedicated equipment may have higher initial capitol costs with an attendant decrease in labor costs. Use of dedicated equipment has been shown to reduce sample alterations including turbidity. Dedicated equipment also greatly reduces sampling variations imparted between sampling events and between wells.

Grab Samplers

Bailers. The least expensive and perhaps the simplest method of acquiring groundwater samples is with a single-check valve bailer. Bailers are inexpensive, portable, and are relatively easy to clean and operate, depending on well depth. A bailer is a Teflon, polyethylene, stainless steel or PVC tube, with a ball check valve at the base. The ball check valve should have a specific gravity of between 1.4–2.0 (34). Bailers should be equipped with a bottom emptying device to reduce sample aeration.

Clean lengths of Teflon-coated wire cable or new lengths of braided polypropylene line are attached to the top of the bailer. The bailer is lowered to the air–water interface by hand or with a winch. As the bailer is allowed to slowly sink to the speficied depth in the well, the ball check valve is displaced upward by the water, allowing the bailer to fill. The bailer must be lowered and retrieved slowly to reduce turbidity. When the bailer is retrieved the ball check valve sinks to the bottom of the tube, closing off the bottom of the bailer.

Bailers may be laborious and time consuming to operate, especially in deeper wells. Samples obtained from a bailer are highly operator-dependent and are therefore quite variable. Bailers have been shown to be the sampling device that offers the greatest chance of altering the chemistry (by surging, mixing, and aeration) of any of the more commonly used sampling devices even when operated carefully (35). Several orders of magnitude increases in turbidity have been reported when using a bailer due to inconsistent operator usage and excessive purging (36). Large differences in colloid and polycyclic aromatic hydrocarbon (PAH) concentrations (100 and 750 times, respectively) have been observed between samples taken with a bailer and with a submersible pump (37).

Some field parameters (notably dissolved oxygen or ORP) need to be acquired in a flow-through cell. This is especially true when the dissolved oxygen concentrations are low, and/or the groundwater is reduced. This may exclude sampling with a bailer, if accurate dissolved oxygen and redox measurements are required.

Sampling LNAPLs may be facilitated by using a specially designed single-check valve bailer. The check valves of bailers used to sample LNAPLs have different specific gravities than bailers designed to sample water. DNAPLs may be best collected with special dual-check valve bailers (38).

Syringe Samplers. Syringe samplers operate much like a large medical syringe. The sampler is lowered to the desired sampling depth and the plunger is moved either mechanically or pneumatically to draw water into the device. Some syringe samplers can obtain a sample volume of up to 0.8 L. The sample may be shipped directly to the analytical laboratory inside the syringe. Field parameters may be difficult to obtain with this device. Suspended solids also may cause leakage around the seals of syringe samplers.

Inertia Pumps

An inertia pump consists of lengths of threaded pipe equipped with a check valve at the bottom. Sections of the pipe are screwed together and are lowered to the prescribed depth in the well. Some inertia pump designs incorporate one pipe inside another pipe. To operate a inertia pump a pipe is repeatedly raised and lowered either by hand or mechanically. When the pipe is lowered water moves upward past a check valve. As the pipe is raised the check valve rapidly sinks, which closes off the bottom of the device. The cycle is then repeated.

Manually operating these pumps may quickly become tiring, especially when sampling deeper wells. The pumping rate may be varied, but depending on the design a relatively large volume of water may be temporarily stored in the pipe. This water stored in the pipe could cause dewatering of the formation. This is especially true if the well is deep and is installed in less permeable formations. It may also be difficult for one person to operate and obtain a sample with a manually operated inertia pump.

Peristaltic Pumps

A peristaltic pump consists of a length of very flexible (usually silicone) tubing and a motor that turns two or three pinch rollers. The relatively small electric motor is usually powered by a 12 volt rechargeable battery or from a vehicle battery. The pinch rollers rotate in a circular direction around the length of tubing, squeezing and releasing it. This induces a vacuum into the sampling lines, drawing water upward. Peristaltic pump motors should have variable operating speeds; the pumping rate should not be controlled with pinch clamps on the sample delivery tubing.

Peristaltic pumps are easy to clean, transport, and operate. Peristaltic pumps have been shown to consistently provide the most reproducible results for trace metals, compared to other sampling devices (39). Peristaltic pumps are, however, limited to shallower applications (<7.6 m), because they operate by inducing a vacuum. Peristaltic pumps may also theoretically cause degassing by lowering the partial pressure of the sample. Lowering the partial pressure may cause the loss of volatile organic compounds (VOCs) and may also accelerate CO_2 evolution. Increased CO_2 concentrations may consume hydrogen ions, resulting in and increase of pH (40).

Positive Displacement Pumps

Bladder Pumps. A bladder pump consists of a pump and sample delivery lines, a flow-controller, and a source of compressed gas (compressor or cylinder). The pump design incorporates a flexible (usually Teflon) bladder within a stainless steel housing. Compressed gas, controlled by the flow-controller, alternately squeezes and releases the flexible bladder. Water flows into the pump past a screen and check valve, into the pump. The pumping rate at depth is controlled by adjusting the pressure exerted on the bladder, and the duration of the inflation/deflation cycle.

Some bladder pumps require the hydrostatic pressure of the overlying water to operate the pump. As such, some bladder pump designs may not operate if the pump is not deep enough into the water column. The sampling rate of bladder pumps may be somewhat misleading, as the flow rate is averaged over the two-stage cycle (41).

Electric Submersible Pumps. Electric (submersible) pumps use either meshing gears, staged impellers, or a helical rotor to positively displace water upward in the sample lines. A 12-volt battery or a generator (with filtered voltage) is required to operate submersible pumps. Submersible pumps should have variable flow rates and should not be controlled by flow restrictors. Overheating, stalling at low flow rates, sand locking, and excessive turbulence may be of concern with some submersible pumps. Submersible pumps are cooled by the surrounding water and may burn out if operated dry.

Multiport Flow-Through Cells

A multiport flow-through cell is necessary for obtaining reliable field parameters (42). Flow-through cells allow the probes to be in contact with the sample water, without the sample coming into contact with the atmosphere. Isolation of the sample from the atmosphere is especially important when the groundwater is near-anoxic or is reduced.

Clear acrylic or Plexiglas tubes are commonly used for flow-through cells. Silicone stoppers or threaded water-tight connectors are used to attach probes to the cell. A conical interior has been suggested to prevent inducing a streaming potential (static charge) on the probes. Flow-through cells should never be pressurized or the probes may be damaged. Multiport flow-through cells may be fabricated or purchased. Fabricating a flow-through cell gives the advantage of being able to use different sized probes from different manufacturers.

Geochemical Field Parameters

Geochemical parameters obtained in the field include temperature, pH, specific conductance, dissolved oxygen, ORP, turbidity, and alkalinity. These measurements need to be made in the field because of physical, chemical, and/or biological alterations that occur immediately after sampling. These parameters are useful for geochemical modeling. Most of these parameters are also used to ascertain when the well has stabilized, immediately prior to sample acquisition. The above field parameters may not accurately represent when a well has stabilized for VOC acquisition (43).

Temperature. Groundwater temperature provides information about

- stability and saturation of ions, and products of reactions
- rates of biotic and/or abiotic reactions
- residence times of groundwater (shallow groundwater may exhibit more pronounced seasonal swings)

Temperature is not a good primary indicator of well stability prior to sample acquisition (36). Temperatures of groundwater may be obtained either with an electronic (digital) thermometer or a mercury-filled glass thermometer. The thermometer should be accurate to the nearest 0.1 degree Celsius. At least once a day the thermometer should be checked against a thermometer certified by the National Institute of Standards and Technology.

pH. The pH in groundwater provides information about

- acidic or basic intensity of a solution, at a given temperature
- chemical speciation
- solubility of metals
- contaminant migration

pH is not a good primary indicator of well stability prior to sample acquisition (44).

Standard glass membrane pH electrodes rely on sodium ions on the outer surface of the glass bulb being (readily) replaced by hydrogen ions. This creates an electrical imbalance that is resolved by the reference electrode-generating electrons, which is measured as a voltage. pH is temperature-dependent. As such, the temperature of the solution needs to be reported with the pH measurement. pH meters need to be calibrated with two calibration solutions that bracket the anticipated sample pH.

Specific Conductance. Specific conductance (SC) in groundwater provides information about

- total dissolved solids: multiplying SC (in μmhos) by between 0.55–0.9 provides an estimation of total dissolved solids (45)
- major ion chemistry, as SC is less affected by small-scale heterogeneity (44)
- mobility of ionic species in the system

SC is a good primary indicator of well stability prior to sample acquisition, especially for inorganic species (44)

Conductivity is temperature-dependent so SC corrects the conductivity measurement to 25 degrees Celsius. A typical SC meter uses a platinum electrode that may need to be periodically replatinized. SC meters should be checked against known potassium chloride standards and should be accurate to between 0.1–1.0% of the meter's range.

Dissolved Oxygen. Dissolved oxygen (DO) in groundwater provides information about

- potential for attenuation of organic contaminants
- mobility of metals in the system
- residence times of groundwater

DO is a good primary indicator of well stability prior to sample acquisition (36). The mobility of trace metals and the potential for degradation of organic contaminants depends on DO concentrations. Therefore three-dimensional DO profiles should be considered for site investigations.

DO in (closed) groundwater systems is always consumed and is never produced. The solubility of oxygen increases with decreasing temperature and increasing hydrostatic pressure. DO concentrations are determined in the field either with a meter or with a titration method. DO meters rely on diffusion of oxygen across a semipermeable membrane in the probe, that in turn is influenced by pressure, temperature, and salinity.

Fouling of the membrane and inadequate replenishment of sample water to the probe may bias DO readings with a meter. DO meters need to be temperature compensated due to temperature-induced shrinking and swelling of the membrane. DO meters are also sensitive to changes in elevational or barometric pressure so may need to be recalibrated several times a day.

The Winkler titration method is an accurate method for determining DO in water. This method is, however, labor-intensive and does not provide continuous readings. Both methods for determining DO have precision and detection limits of 0.2 mg/L (46). DO meters should be checked against a Winkler titration at least once a day.

ORP. Oxidation reduction potential (ORP) in groundwater provides information about

- stability and occurrence of ions and compounds in the system

- potential of organic compounds to be co-metabolized by bacteria
- mobility of inorganic constituents in the system

ORP is a good primary indicator of well stability prior to sample acquisition (36).

Bacteria catalyze nearly all the important reduction–oxidation (redox) reactions in groundwater. Free oxygen, nitrate, sulfur, and ferric iron are all important electron acceptors. Organic matter is also necessary for redox reactions to occur at an appreciable rate. As oxygen, nitrate, sulfur, and ferric iron undergo reduction (ie, gain electrons), the groundwater becomes more reduced.

Redox buffering may enable the redox potential to remain relatively stable until, for example, all the free oxygen is consumed. Abrupt drops in ORP may occur as the oxygen, nitrate, sulfate, and ferric iron are consumed. Redox potentials are, however, usually mixed potentials, so it is impossible to relate the ORP to a single dominant redox couple in solution (47). Furthermore, redox couples are rarely in equilibrium with each other.

Using ORP measurements to identify zones along the groundwater path is essential for interpreting the geochemical dynamics of the site. ORP characterization is also necessary for ascertaining the applicability of certain remedial technologies, such as enhanced *in situ* bioremediation. The following sequence can be used to describe redox zones along groundwater flow paths (48,49):

1. Oxic (oxygen-rich): ORP potentials of 250–100 mV
2. Post-oxic (iron-rich): ORP potentials of 100–0 mV
3. Sulfidic (sulfate-rich): ORP potentials of 0– −200 mV
4. Methanic (methane-rich): ORP potentials of below −200 mV

There are three basic types of metallic electrodes used for ORP measurements:

1. A metal in contact with a solution of its own ions, for example, silver in a silver nitrate solution.
2. A metal coated with a barely soluble salt of this metal, in contact with a soluble salt with the same ion, for example, silver–silver chloride in a solution of potassium chloride.
3. An inert metal in contact with a solution containing both the oxidized and reduced states of an oxidation–reduction system, for example, platinum in contact with a solution of ferric–ferrous ions.

Each of these electrodes has its own specific application. For ORP measurements in groundwater a calomel electrode is commonly used. Calomel electrodes are made of mercury chloride and are filled with a potassium chloride solution. The indicator electrode may then be made of platinum.

Measuring ORP in the field may be undertaken using a pH meter with ORP probes. Light's or Zobell's solution are used as check solutions for ORP probes. Some manufacturers of pH calibration solutions use the same pH calibration solutions for ORP check solutions. ORP probes (following manufacturer instructions) may need to be routinely changed in the field and be carefully pretreated and stored in an oxygen scavenging solution of 0.2 *M* sodium sulfite (50).

Turbidity. Turbidity in groundwater provides information about

- groundwater flow conditions (fine-grained to conduit flow)
- groundwater chemistry, promoting colloidal stability
- monitoring well construction, development, and/or groundwater sampling techniques

Turbidity is a good primary indicator of well stability prior to sample acquisition (51,52).

Turbidity is an expression of the optical properties of the sample that causes light rays to be scattered and absorbed rather than being transmitted in straight lines thorough the sample (53). Turbidity is caused by dissolved or suspended particles of solids such as silts, clays, or organic materials. Turbidity is also caused by liquids and gases.

Turbidity is related to the debate on whether or not to field filter groundwater samples. Turbidity may be representative of natural groundwater conditions. Alternatively, turbidity may be attributable to an improperly sized filter pack or screen, to inadequate well development, or improper sampling techniques (eg, "slam" bailing). Unrepresentative turbidity should not be dealt with by ubiquitously field filtering groundwater samples.

Turbidity in groundwater is measured by a photoelectric nephelometer. Nephelometers may operate continuously using a probe (in a flow-through cell) or by taking and measuring discrete aliquots of the sample. It has been recommended that nephelometers use a tungsten-filament lamp operated at a temperature of between 2200–3000 K (45). Reagent-grade Formazin standard (a carcinogen) or styrene divinylbenzene beads are used to calibrate nephelometers. Sample tubes must not be handled where the light contacts them and must be discarded if they become scratched or etched.

Alkalinity. Alkalinity in groundwater provides information about

- capacity of the groundwater to neutralize acids
- contaminant migration (such as landfill leachate)
- solubilities and mobilities of some metals
- toxicities of some metals

Alkalinity is not used as a stabilization index for well purging criteria.

Alkalinity is measured by means of an electrometric titration (using a pH meter) or by colorometric titration. To perform an electrometric titration 100 mL of the sample is titrated to a pH of 4.5 using 0.02 *N* reagent-grade hydrochloric acid in 0.5 mL increments. With each 0.5 mL acid addition the pH is recorded. A titration curve is subsequently generated that defines inflection points and buffering capacity (53). The presence of suspended solids in the aliquot may interfere with electrometric alkalinity titrations.

Colorometric titrations involve the addition of an indicator (such as bromocresol green) that changes color at the end point of the titration. The amount of acid required to change the color of the indicator can be translated into the alkalinity of the sample. Colorometric titrations are not recommended if the sample is naturally colored, turbid, or contains waste material that could interfere with the indicator.

When determining alkalinity using either method the sample must not be filtered, diluted, concentrated, or altered in any way. Some parameters (eg, temperature) *must* be measured in the field. Some geochemical parameters may, however, be better acquired in a controlled laboratory setting. For example, which would be preferred: allowing a semiskilled sampler to perform an alkalinity titration in (sometimes) inclement conditions in the field, or performing the titration within 24 hours of sample acquisition in an analytical laboratory. These decisions are well-specific, based in part on the sampler's skills and specified holding times. Relatively large differences have been observed between alkalinity acquired in the field, and those taken in a laboratory 5 to 120 days subsequent to sampling due to chemical or biological alterations of the sample (54).

Field Filtering Groundwater Samples

The debate with respect to field filtering groundwater samples is ongoing. The U.S. EPA has advocated that groundwater samples remain unfiltered (55). Each state, and even individual programs within a state, have different approaches as to whether or not field filtering should be instituted. The debate may be summarized as follows.

For Filtering. Field filtering should be undertaken in some monitoring regimes (ie, silts and clays) because turbidity in the groundwater is not naturally occurring. Under these circumstances turbidity may be more attributable to the stratigraphic unit where the well is installed and is not representative of the groundwater flow regime.

Against Filtering. Oxygenation of groundwater that is anaerobic and reduced causes soluble ferrous iron to precipitate out of solution as ferric oxyhydroxide. Iron precipitation may change the pH, alkalinity, conductance, ionic strength, turbidity, and color of groundwater samples (56). Precipitated amorphous iron hydroxide absorbs trace metals. Furthermore a 0.45 μm filter size is somewhat arbitrary and as the filter becomes clogged it further reduces the filter's pore size.

Turbidity as it relates to field filtering may be evaluated by addressing the following questions:

1. Were the filter pack and screen properly sized in relation to the stratigraphic unit in which they were installed?
2. Has the monitoring well been properly developed in the recent past?
3. Are low flow purging and sampling techniques being used, possibly with dedicated equipment?

Trying to offset one mistake (ie, an improperly designed, installed, or sampled monitoring well) by instituting another (inappropriate field filtering) casts aspersions on the data. The differences between filtered and unfiltered groundwater samples, however, may not be statistically significant (even in fine-textured glacial till) if the answer is "yes" to all the above questions (36,57).

If the answer to any of the above questions is "no", then alternatives to field filtering should first be considered. This is because differences of greater than 20 times of metals (lead and chromium) between filtered and unfiltered samples have been observed (58). Alternatives to defaulting to inappropriate field filtering include well redevelopment, using dedicated sampling equipment operated at low flow rates, or in some situations replacing the well.

Sometimes even optimally designed, installed, developed, and sampled wells (installed in silts or clays) may produce samples that remain inherently turbid. Field filtering under these circumstances should be carefully evaluated in terms of the relative advantages or disadvantages of filtering. If filtering is decided on, the procedure should incorporate positive pressure (not a vacuum) and 5 μm filters vs 0.45 μm filters should be considered (36).

GROUNDWATER SAMPLING AND ANALYSIS PLANS

The equipment and techniques used to sample groundwater are selected based on costs, technical considerations, and program goals or requirements. Project costs include the costs incurred by preparing and implementing a specific protocol. If there are no regulatory requirements to use scientifically validated sampling techniques then a protocol (perhaps based on precedent) may be selected that is less than optimal. This is because the costs of preparing and executing a poor-quality protocol may be significantly less than the costs of implementing a more refined one. The costs incurred by using poor-quality data may, however, significantly outweigh the costs of obtaining good-quality data on which to base remedial decisions from the onset of the project.

Groundwater sampling protocols need to be formulated in terms of data quality objectives (DQOs). The DQO process helps define the goals of the project. The sampling equipment and techniques are subsequently selected to meet the goals specified by the DQO process. Technical considerations and limitations also need to be considered when designing a sampling protocol.

Data Quality Objectives

Some programs may be able to accept data of lesser accuracy than other programs. Quantifying contaminants in terms of their presence or absence may be acceptable to satisfy some program objectives. As such the sampling protocol may be less rigorous with potentially an attendant reduction in costs and data quality.

To ensure that the data quality reflects the needs of the program sampling plans need to be formulated in terms of DQOs. DQOs have at least three meanings: a process for defining the framework to be used for environmental investigations and cleanups, individual goals within the process, and levels of analytic uncertainty (59). The DQO process provides a structure for selecting the equipment and techniques used to collect data commensurate with the goals of the project.

The formal CERCLA (ie, EPA Superfund) DQO process may be tailored to other programs, even when the formal DQO process is not required. The output of the DQO process is a well-defined sampling and analysis plan with summary information, used to support decisions for remedial response activities.

The formal DQO process is comprised of three phases. Stage one involves identifying data users, evaluating existing data, developing a conceptual model, and specifying objectives for the project. Stage two identifies sample types, and data quality needs. Stage three involves preparation of a sampling and analysis plan (SAP) that specifies the numbers, locations, and frequency of each sample type required, and the sampling protocol used to obtain the samples.

Individual objectives within the DQO process include formulating statements which sets the tenor of the monitoring program. For example, a DQO statement for groundwater monitoring at a U.S. EPA RCRA Subtitle D landfill may be written as follows:

"Monitoring at this permitted solid waste landfill facility may be expected to continue for decades, and needs to accurately characterize contaminant concentrations to one quarter of the respective Maximum Contaminant Levels (MCLs) or lower. Monitoring data obtained at this facility influence decisions that may affect human and environmental health, and could necessitate expensive corrective actions. Therefore the ground water samples and subsequent analyses need to be of sufficient quality to characterize accurately the conditions at the site, without being prohibitively expensive. The following Sampling and Analysis Plan reflects the need for data of acceptable quality, while recognizing financial and technical constraints".

DQOs are also defined in terms of levels of accuracy in the analysis of the sample, as follows:

DQO Level 1: Field screening, such as using a pH meter, an OVM, ect

DQO Level 2: Field analysis (mobile laboratories)

DQO Level 3: Engineering level (level used for most programs)

DQO Level 4: Confirmational analysis (the Certified Laboratory Program)

DQO Level 5: Nonstandard analysis for less widely run analytes

The output of the DQO process may specify that long-term monitoring points need to be installed and sampled. Alternatively, temporary monitoring points (including the Hydropunch or Geoprobe) may provide data commensurate with the objectives of the project. The Geoprobe (usually considered DQO analytical Level 2) may be used to delineate subsurface contaminant plumes or to assist in locating permanent monitoring points. Laboratory confirmation (minimum of DQO analytical Level 3) of a fixed number of field-analyzed samples should be run to ascertain the precision and accuracy of the field analytical work.

Sampling and Analysis Plans

The formulated groundwater sampling and analysis plan (SAP) should state the objectives of the project and provide the detailed, sequential steps necessary to achieve the stated objectives. A site-specific sampling protocol is a part of a SAP. The following are components of a comprehensive SAP:

1. A title page and table of contents
2. An introduction that includes:
   - A brief overview of the contents of the sampling plan
   - The purpose and goals of groundwater monitoring at the site, including data quality objectives
   - The individuals who prepared the plan (with signatures) and date the plan was prepared
3. The names and phone numbers of those responsible for monitoring activities at the site, and those who can approve of field changes. This part should include a summary of the training and experience of the sampling leader and crew.
4. A site overview that includes a description of previous site activities and potential contaminants. The overview should summarize the geology and hydrogeology of the site, and provide groundwater flow direction and rates. This section should also include a USGS topographic map for site location, and a larger scale map showing the locations of any potentially concentrated contaminants, the monitoring points, property boundaries, etc.
5. A detailed sampling protocol that includes:
   - Well-specific sampling procedures that evaluates the drawdown and turbidity of each well
   - Sampling schedules
   - Parameters to be sampled
   - A table that includes well field numbers, relative location (upgradient or downgradient), casing diameters, top of riser elevations, and distances from measuring point to the top of the screen, and to the bottom of the well
   - The sequence of sampling, from the least to most contaminated well
6. Reference to the health and safety plan for sampling activities at the site
7. A quality assurance/quality control component including the proposed data validation and describing field audits
8. A description of how the samples are to be transported to the analytical laboratory including a chain-of-custody plan (if required)
9. A reference to and brief overview of the laboratory protocol and plan, including the title and date submitted and/or approved.

Sampling Protocols

A site-specific groundwater sampling protocol is a component of a groundwater SAP. Sampling protocols need to be specific about the equipment and techniques used for acquiring groundwater samples in individual wells at a site. The more specific the sampling plan, the less chance for introducing errors or erroneous assumptions (60). A site-specific sampling protocol needs to be a stand-alone document, taken out into the field with the sampling crew. The protocol should be written so it can be understood clearly by all sampling personnel.

Individual wells in a monitoring network need to be evaluated as to the optimal methodology (including placement) of the sampling device within the well to acquire samples. Placement of the sampling device within the well annulus needs to account for changes in stratigraphy along the length of the screened interval. This is especially true when the wells at a site are screened in vertically anisotropic or completely different stratigraphic units. Due to anisotropy vertical integration of solutes in a monitoring well using standard purging procedures may underestimate groundwater contamination by orders of magnitude (61).

A different sampling strategy may be appropriate for sampling a coarse outwash, than a fine-grained clayey till. Furthermore, the location of the screen in relation to the air–water interface, fluctuations of the potentiometric surface, and the chemical and physical characteristics of constituents of concern all need to be considered in terms of selecting an appropriate sampling methodology. The following are examples of how wells installed in different stratigraphic units (using positive displacement pumps) may be sampled for dissolved phase constituents:

Medium to High Yielding Wells may be stabilized at purge rates of 500 mL/min: for medium to high yield water table wells the U.S. EPA has recommended that the pump should be positioned as close to the air–water interface as possible. For deeper medium to high yield wells the U.S. EPA has recommended that the pump be positioned within the screened interval (62). The optimal purge rate should be evaluated in terms of reducing unrepresentative turbidity and minimizing draw-down in the well. The purge rate should not exceed aquifer yield so vertical adjustments of the pump intake should not be necessary.

Low Yield Wells With Recovery cannot be stabilized at purge rates of 500 mL/min: some monitoring wells are installed in stratigraphic units characterized by low recharge potential (low yield wells). If a low yield well is screened across the water table it should be purged to near dryness and allowed to at least partially recover before beginning the stabilization trials. If the low yield well is screened below the water table then evacuating the stagnant water above the screened interval and allowing partial recovery before beginning the stabilization trials could be considered. If water cascading down the screen in low yield wells is a concern, then after the first evacuation slow, partial evacuations should be undertaken.

Micro-Purging involves permanently installing a pump within the screened interval. The pump is then operated at a rate that is at or below the natural recharge rate to the well. Micropurging is based on the assumption that water flowing through the screen in proximity to the pump intake is representative of the formation and that any stagnant water above the screened interval will not be drawn into the pump. Furthermore, as the pump is permanently installed, turbidity induced by placement of the pump eliminated. As such traditional purging of between 3–5 casing volumes is obviated (63). Some purging after removing the dead volume in the pump and lines still appears to be warranted (64). Pump placement is based on a good-quality boring log. The pump is positioned at the mid-point of the screened interval in vertically isotropic aquifers or within targeted discrete zones in vertically anisotropic aquifers.

Stabilization Trials

For most monitoring programs, samples are obtained when the geochemical field parameters (specific conductance, dissolved oxygen, ORP, turbidity) have stabilized within predetermined ranges for three consecutive readings (a stabilization trial). The interval between taking a set of field parameters for a stabilization trial is usually based on evacuating a well-bore volume.

If micropurging is employed then stabilization trials may proceed immediately after purging the dead volume in the pump and lines. Furthermore, if micropurging is employed it is assumed that there is no stagnant water to be purged, subsequent to purging the pump and lines. As such, stabilization readings can be taken after evacuating partial (1/3–1/2 well volumes) instead of full well volumes.

If nondedicated equipment is selected, a minimum number of well-bore volumes should be evacuated prior to commencing stabilization trials. For aquifers with medium–high recharge rates usually three bore volumes are evacuated prior to initiating a stabilization trial. If micropurging is not employed each set of stabilization indices should be taken with each well volume that is evacuated. For less transmissive aquifers at least some purging is appropriate (65), but cascading of water down the screen (thus aerating it) should be avoided.

When field parameters have stabilized for three consecutive readings over the following values purging can stop, and groundwater samples may be acquired:

specific conductance: $\pm 10.0\ \mu S/cm$
dissolved oxygen: ± 0.2 mg O_2
turbidity: ≤ 10 NTUs or $\pm 5\%$ if >10 NTUs

These tolerances may need to be adjusted to individual programs.

Sampling Sequence

Several checklists should be prepared and be used for sampling wells. Sampling checklists should described the equipment maintenance required before going into the field, and the documentation, equipment, and supplies required for sampling. The following generic sequence is typically used for obtaining groundwater samples from monitoring wells:

1. Before Going Into the Field
   - Provide adequate notification to the laboratory, site owner, and regulatory agencies that a sampling event is going to occur
   - Check and calibrate the equipment
   - Assemble documents, equipment, and supplies
   - Review the most current site-specific sampling plan and the health and safety plan
   - Obtain the keys to the site and monitoring wells
2. Upon Arrival at the First Well in the Sampling Network
   - Inspect the well for damage, missing parts, labeling, and evidence of tampering (a secure lock in place?)
   - Document field conditions
   - Use an explosivity meter and/or organic vapor monitor to check the outer and inner casing, as necessary
   - Prepare the area around the well for sampling
   - Calibrate the equipment in the field, as required
   - Don Clean, disposable protective gloves
   - Take static water levels (and NAPL or DNAPL levels, if appropriate), calculate well volume
   - Measure field parameters while simultaneously purging the well
   - Adequately label the sample containers

- Document the field work including initial, stabilizing and final field parameters
- Acquire the sample, field filter (if appropriate), add preservatives
- Place the samples in a shipment cooler with ice
- Prepare QC samples
- Perform field analyses, such as alkalinity
- Complete the documentation for the well
- Clean the equipment and go to the next well
- Institute chain-of-custody controls as required
- Ship the samples to the laboratory for analysis, within prescribed holding times

Numerous permutations are possible with the above sequence. Many if not most sampling efforts, however, follow in the same general order the steps listed above. A detailed, step-by-step analysis of these tasks (including checklists) is available (66).

To ensure that samples were taken properly and are legally defensible if challenged in a court of law, each step of the process must be adequately documented by field personnel. Furthermore, the equipment and supplies used must be of the best available quality to ensure the quality of the data.

Summary for Obtaining Representative Groundwater Samples

The following may be summarized about sampling monitoring wells:

1. Monitoring wells need to be properly designed, installed, developed, and sampled to provide representative data.
2. The DQO process should be used for preparing groundwater sampling and analysis plans.
3. Each well in a network should be individually evaluated in terms of a proposed sampling methodology.
4. A muliport flow-though cell should be used for obtaining geochemical field parameters.
5. Bailers appear to be the device most prone to increasing unrepresentative turbidity by introducing a surging action within the screened interval (67). Multiport flow-through cells are also not compatible with bailer use. As such, routine bailer use should be discouraged.
6. Groundwater sampling is highly operator dependent. Sampling equipment that reduces sampling variability introduced by sampling crews in desirable. Equipment that reduces the number of sampling steps may therefore produce more representative samples.
7. Positive displacement pumps (bladder or submersible) should be encouraged for routine groundwater sampling.
8. Purge and sample rates should, whenever possible, not exceed the natural recharge rate to the well.
9. In isotropic or targeted anisotropic media micropurging should be encouraged. Micropurging has been demonstrated to reduce sample alterations (including turbidity), and may also be more cost-effective for long-term groundwater monitoring (68).
10. Samples should be obtained when the geochemical field parameters have stabilized within the predetermined ranges for three consecutive readings. If micropurging is employed then stabilization trials may proceed immediately subsequent to purging the dead volume in the pump and lines. If nondedicated equipment is selected, then a minimum number of well-bore volumes should be evacuated, prior to commencing stabilization trials. For aquifers with medium–high recharge rates typically three bore volumes are evacuated prior to initiating the stabilization trial. For less transmissive aquifers at least some purging is appropriate (65), but cascading of water down the screen should be avoided.

BIBLIOGRAPHY

1. S.J. Cullen, J.H. Kramer, L.G. Everett, and L.A. Eccles, "Is Our Ground Water Monitoring Strategy Illogical." *GWMR* (Summer 1992).
2. L.G., Everett, L.G. McMillion, and L.A. Eccles, *Suction Lysimeter Operations at Hazardous Waste Sites, Ground Water Contamination,* ASTM Field Methods, Philadelphia, Pa., 1988.
3. N. Wilson, *Soil Water and Ground Water Sampling,* Lewis Publishers, Inc., Boca Raton, Fla., 1995, pp. 15–30.
4. R.D. Morrison, and B. Lowery, *Soil Sci.* **149**(5), 308–316 (1990).
5. U.S. EPA, *Permit Guidance Document on Unsaturated Zone Monitoring for Hazardous Waste Land Treatment Units Section 4: Soil Pore Liquid Monitoring,* U.S.EPA, Washington, D.C., 1985.
6. J.M. Jemison, JR. and R.H. Fox, *Soil Sci.* **154**(2), 85–95 (1992).
7. B.R. Swistock, J.J. Yamona, D.R. Dewalle, and W.E. Sharp, *Water, Air, Soil Poll.* **50,** 387 (1990).
8. U.S. EPA, *Final Draft, Ground Water Monitoring,* Chap. 11, SW-846, Washington, D.C., Mar. 20, 1991.
9. American Society for Testing and Materials, *Standard Method for Penetration Test and Split-Barrel Sampling of Soils,* ASTM D1586-84, Philadelphia, Pa., 1984.
10. American Society for Testing and Materials,*Standard Guide for Soil Sampling in the Vadose Zone,* ASTM D4700-91, Philadelphia, Pa., Sept. 1991.
11. U.S. EPA, *Guide to Management of Investigation-Derived Wastes: Quick Reference Fact Sheet,* Publication No. 9345.3-03FS, U.S. EPA, Washington, D.C., 1992.
12. G.L. Hix, *Water Well J.* **47**(5) (May 1993).
13. K.D. Godsey, *Water Well J.* **47**(5) (May 1993).
14. American Society for Testing and Materials (ASTM) D5088-90, Standard Practice for Decontamination of Field Equipment Used at Nonradioactive Waste Sites, Sept. 1990.
15. D. Oakley, and N.E. Korte, *Ground Water Mon. Rev.* (Winter 1996).
16. F.G. Driscoll, *Ground Water and Wells,* Johnson Filtration Systems, Inc., 1986.
17. L. Aller, T.W. Bennett, G. Hackett, R.J. Petty, J.H. Lehr, H. Sedoris, D.M. Nielsen, and J.E. Denne, *Handbook of Suggested Practices for the Design and Installation of Ground-Water Monitoring Wells,* NWWA, Dublin, Ohio, 1989, 398 pp.
18. Technical data, ENV1091-10K, Johnson Environmental Well Products, 1991.
19. M.F. Glotfelty, *Water Well J.* **47**(5) (May 1993).
20. U.S.EPA, *RCRA Ground Water Monitoring Technical Enforcement Guidance Document,* Washington, D.C., Sept. 1986.
21. J.P. Gibb, *Ground Water Mon. Rev.* **7**(1) (Winter 1987).

22. American Society for Testing and Materials, *Standard Practice for Design and Installation of Ground Water Monitoring Wells in Aquifers,* ASTM D 5092-90, Philadelphia, Pa., Oct. 1990.
23. G.L. Hix, *Water Well J.* **7**(6) (June 1993).
24. T. Kwader, *National Drillers Buyers Guide,* Apr. 1993.
25. F.G. Driscoll, personel communication, Geraghty & Miller, Inc.
26. R.W. Reichart, *Water Well J.* (May 1996).
27. M.J. Barcelona, J.P. Gibb, J.A. Helfrich, and E.E. Garske, *Practical Guide for Ground Water Sampling: Illinois State Water Survey,* SWS Contract 374, Champaign, Ill., 1986.
28. D.W. Prosser, *Ground Water,* **19**(6) (1981).
29. American Society for Testing and Materials, *Standard Guide for Selection of Aquifer-Test Method in Determining of Hydraulic Properties by Well Techniques,* ASTM D4043-91, Dec. 1991.
30. N. Wilson, *Soil Water and Ground Water Sampling,* Lewis Publishers/CRC Press, Boca Raton, Fla., 1995.
31. M.J. Barcelona, J.P. Gibb, J.P. Helfrich, and E.E. Garske, *Practical Guide to Groundwater Sampling,* Contract Report 374, Illinois State Geological Survey, 1986.
32. *Standard Test Method for Determining Subsurface Liquid Levels in a Borehole or Monitoring Well (Observation Well),* ASTM D4750-87, Philadelphia, Pa., Jan. 1988.
33. P.F. Hudak, K.M. Clements, and H.A. Loaiciga, *J. Env. Eng.* **119** (1993).
34. *Standard Guide for Sampling Groundwater Monitoring Wells,* ASTM D4448-85a, Philadelphia, Pa., May 1986.
35. R.W. Puls, D.A. Clark, B. Bledsoe, R.M. Powel, and C.J. Paul, *Hazar. Waste Hazar. Mater.* **9**(2) (1992).
36. R.W. Puls, and R.M. Powell, *GWMR* (Summer 1992).
37. D.A. Backhus, J.N. Ryan, D.M. Groher, J.K. MacFarlane, and P.M. Gschwend, *Ground Water.* **31**(3) (May–June 1993).
38. U.S. EPA, *Final Draft Ground Water-Monitoring,* Chap. 11, SW-846, Washington, D.C., Mar. 1991.
39. R. Puls, *Improved Metals Sampling Techniques for Groundwater,* U.S. EPA Tech Trends, EPA/540/M-91/005 No. 7, Washington, D.C., Dec. 1991.
40. N.T. Loux, A.W. Garrison, and C.R. Chafin, *Intern. J. Environ. Anal Chem.* **38** 231–253 (1990).
41. C.J. Paul, and R.W. Puls, "Comparison of Ground-Water Sampling Devices Based on Equilibration of Water Quality Indicator Parameters," *National Ground Water Sampling Symposium Proceedings,* Washington, D.C., Nov. 30, 1992.
42. E.E. Garske, and M.R. Schock, *GWMR* (Summer 1986).
43. J. Gibs, and T.E. Imbrigiotta, *Ground Water,* **28**(1), 68–78 (1990).
44. M.J. Barcelona, H.A. Wehrmann, and M.D. Varlijen, *Ground Water,* **32**(1) (Jan.–Feb. 1994).
45. American Public Health Association, *Standard Methods for the Examination of Water and Waste Water,* 18th ed., APHA, Washington, D.C., 1992.
46. C.S. Rose, and A. Long, *GWMR* (Winter 1988).
47. R.D. Lindberg, and D.D. Runnells, *Science* **225,** 925–927 (Aug. 1984).
48. R.A. Berner, *J. Sedim. Petrol.* **51,** 359–365 (1981).
49. W.M. Edmunds, D.L. Miles, and J.M. Cook, in E. Eriksson, ed., *A Comparative Study of Sequential Redox Processes in Three British Aquifers: Hydrochemical Balances in Freshwater Systems,* International Association of Hydrological Sciences, 1984.
50. K. Walton-Day, D.L. Macalady, M.H. Brooks, and V.T. Tate, *GWMR* (Fall 1990).
51. R.W. Puls, R.M. Powel, D.A. Clark, and C.J. Paul, *Facilitated Transport of Inorganic Contaminants in Ground Water: Part II,* Colloidal Transport, EPA/600/M-91/040, 1991, 12 pp.
52. D.A. Backhus, J.N. Ryan, D.M. Groher, J.K. MacFarlane, and P.M. Gschwend, *Ground Water.* **31**(3) (May–June 1993).
53. *Standard Test Methods for Acidity or Alkalinity of Water,* ASTM D1067-88, Philadelphia, Pa., Feb. 1989.
54. M.R. Schock, and S. Schock, *Water Res.* **16,** 1455–1464 (1982).
55. *Fed. Reg.* **258.53,** 50978 (Oct. 9, 1991).
56. National Council of the Paper Industry for Air and Stream Improvement, Inc., *Tech. Bull.* **362,** 1–123 (1982).
57. S.B. Clark, N.M. Park, and R.C. Tuckfield, "Effects of Sample Collection Device and Filter Pore Size on Concentrations of Metals in Groundwater Sample," *National Ground Water Sampling Symposium Proceedings,* Nov. 30, 1992.
58. R.W. Puls, and M.J. Barcelona, *Hazar. Waste Hazar. Mate.* **6**(4) (1989).
59. U.S. EPA, *Data Quality Objectives for Remedial Response Activities,* Washington, D.C., 1987.
60. L. H. Keith, *Environmental Sampling and Analysis: A Practical Guide,* Lewis Publishers, Chelsea, Mich., 1991.
61. Robbins, G.A., and M.J. Martin-Hayden, *J. Contaminant Hydrol.* **8,** 203 (1991).
62. U.S.EPA, SW-846. *Ground Water Monitoring,* Chapt. 11, SW-846, Mar. 20, 1991.
63. Robin, M.J.L, and R.W. Gillham, *GWMR* (Fall 1987).
64. Barcelona, M.J., Wehrmann, H.A., and M.D. Varlijen, *Ground Water* **32**(1), 12–22 (1994).
65. Herzog, B., Chou, F.J., Valkenburg, J.R., and R.A. Grifin, *GWMR* **8**(4), 93 (1988).
66. N. Wilson, *Soil Water and Ground Water Sampling,* Lewis Publishers/CRC Press, Boca Raton, Fla., 1995.
67. Puls, R.W., Clark, D.A., Bledsoe, B., Powel, R.M., and C.J. Paul, *Hazar. Waste Hazar. Mater.* **9**(2), 149–161 (1992).
68. Parker, T.K., Eisen, M.F., Kopania, A.A., Robertson, A., Eades, R.L., and G. Thomas, "Selection, Design, Installation and Evaluation of Dedicated Groundwater Sampling Systems: A Case Study," *National Ground Water Sampling Symposium Proceedings,* Washington, D.C., Nov. 30, 1992.

SOLID-PHASE MICROEXTRACTION

TADEUSZ GÓRECKI
MARC ADAMS
MIN YANG
JANUSZ PAWLISZYN
University of Waterloo
Ontario, Canada

PRINCIPLES

Analysis of trace components in natural samples in most cases requires isolation and preconcentration of the analytes from the matrix. For organic compounds, this is usually done by liquid extraction, solid-phase extraction, or headspace extraction (1). All of these methods suffer from certain drawbacks. Liquid extraction (LE) requires large quantities of high purity solvents, is labor-intensive, time-consuming, and error-prone. Solid-phase extraction (SPE) requires only a small amount of solvent for analyte desorption from the sorbent, but the cartridges are easily plugged when suspended solids are in liquid samples, and analyte breakthrough might occur when the matrix composition is significantly different than that of the standards. Headspace extraction is limited to volatile analytes. Except for static headspace, the method is difficult to automate. Dynamic headspace techniques are multistage, thus potentially error-prone. Foaming can be a significant problem when analyzing natural samples.

Solid-phase microextraction (SPME), introduced as an alternative approach in 1990 (2), addresses most of the disadvantages of the methods mentioned above. SPME utilizes a small, fused silica fiber, usually coated with a polymeric stationary phase, for extracting an analyte from a matrix. The fiber is mounted in a syringe-like device for protection. The stationary phase can be a high viscosity liquid or a solid (porous polymer). The analytes are absorbed or adsorbed by the stationary phase (depending on the nature of the coating) until equilibrium is reached in the system. The amount of an analyte extracted by the coating at equilibrium is determined by the magnitude of the partition coefficient (distribution ratio) of the analyte between the sample and the coating. Because equilibrium rather than exhaustive extraction occurs, problems related to analyte breakthrough or limited sample capacity are virtually nonexistent in SPME.

Figure 1 shows a commercial SPME device, produced by Supelco, Inc. (Bellefonte, Pennsylvania). The fiber, glued into a piece of stainless steel tubing, is mounted in a special holder. The holder is equipped with an adjustable depth gauge, which controls repeatably how far the needle of the device penetrates the sample container (if any) or the injector. This is important, because the fibre is easily broken when it hits an obstacle. The movement of the plunger is limited by a small screw moving in the z-shaped slot of the device. For protection during storage or septum piercing, the fiber is withdrawn into the needle of the device, with the screw in the uppermost position. During extraction or desorption, the fibre is exposed by depressing the plunger, which can be locked in the lowered (middle) position by turning it clockwise (the position depicted in Fig. 1). The plunger is moved to its lowermost position only to replace the fiber assembly. Each type of fiber has a hub of a different color. The hub-viewing window enables a quick check of the type of fiber mounted in the device.

If the sample is placed in a vial, the septum of the vial is first pierced with the needle (with the fiber in the retracted position), and the plunger is lowered, which exposes the fiber to the sample. The analytes are allowed to partition into the coating for a predetermined time, and the fiber is then retracted back into the needle. The device is next transferred to the analytical instrument. When gas chromatography (GC) is used for analyte separation and quantitation, the fibre is inserted into a hot injector, where the trapped analytes are thermally desorbed. The process is easily automated by using a properly modified syringe autosampler commercially available from Varian Associates, Sunnyvale, California.

SPME combines sample isolation and preconcentration into a single step, eliminating many potential sources of error. The technique is completely solvent-free. Handling of the sample is minimal. Because of the simple, cylindrical geometry of the sorbent, no plugging occurs. The potential matrix effects are accounted for in most cases by using proper calibration methods. Examples of SPME applications include the determination of substituted benzene compounds in water (3,4), caffeine in beverages (5), volatile organic compounds in water (6), polycyclic aromatic hydrocarbons and polychlorinated biphenyls (7), chlorinated hydrocarbons (8), and phenols (9,10). A more detailed description of selected SPME applications is presented later.

The equilibrium nature of the technique facilitates speciation in natural systems because a minute fibre, which removes small amounts of target analytes, is not likely to disturb the system. Because of the small size, coated fibers are used to extract analytes from very small samples. For example, SPME is used to probe for substances emitted by a single flower bulb during its lifespan.

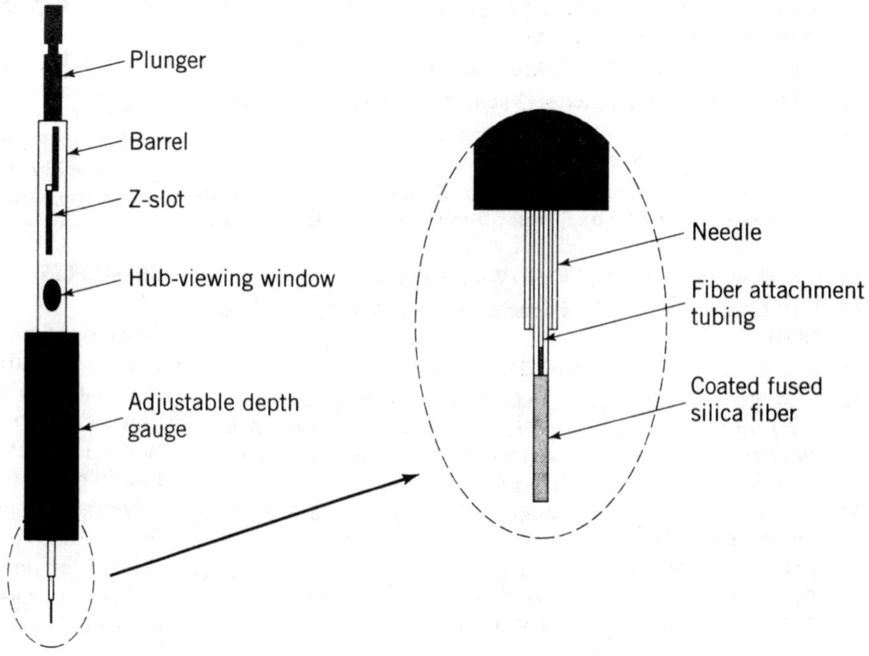

Figure 1. Design of the commercial SPME device.

When the sample volume is large enough (see below), the amount of an analyte extracted by the fiber does not depend on the volume of the sample. In other words, the fiber is exposed directly to ambient air, water, or a production stream, instead of collecting a fixed volume sample of any of these for analysis. By eliminating a separate sampling step, the entire analytical process is accelerated, and errors associated with analyte losses (eg, through decomposition or adsorption on the walls of the sampling container) are eliminated.

SPME sampling is performed in three basic modes: direct extraction, headspace extraction, and extraction with membrane protection. Figure 2 illustrates the differences between these modes. In the direct extraction mode (Fig. 2**a**), the coated fiber is inserted into the sample, and the analytes are transported directly from the sample matrix to the extracting phase. To facilitate rapid extraction, some level of agitation is required to transport the analytes from the bulk of the sample to the vicinity of the fiber. For gases the natural flow of air (eg, convection) is sufficient to facilitate rapid equilibration, but, for aqueous matrices, more efficient agitation techniques, such as fast sample flow, rapid fiber or vial movement, stirring, or sonication are required (11,12) to reduce the "depletion zone" effect produced close to the fiber from slow diffusional analyte transport through the stationary layer of liquid surrounding the fiber. A more detailed discussion of agitation effects is presented in the following sections.

In the headspace mode, the analytes are extracted from the gas phase equilibrated with the sample. The primary reason for this modification is to protect the fiber from adverse effects caused by nonvolatile, high molecular weight substances in the sample matrix (eg, humic acids or proteins). The headspace mode also allows matrix modifications, including pH adjustment, without affecting the fiber. In a system consisting of a liquid sample and its headspace, the amount of analyte extracted by the fiber coating does not depend on the location of the fiber in the liquid or gas phase. Therefore the sensitivity of headspace sampling is the same as the sensitivity of direct sampling as long as the volumes of the two phases are the same in both sampling modes. Even when no headspace is used in direct extraction, a significant sensitivity difference between direct and headspace sampling occurs only for very volatile analytes. However, the choice of sampling mode has a very significant impact on the extraction kinetics. When the fiber is in the headspace, the analytes are removed from the headspace first, followed by indirect extraction from the matrix. Therefore, volatile analytes are extracted faster than semivolatiles. Temperature significantly affects the process kinetics because it determines the vapor pressure of analytes. In general, the equilibration times for volatile compounds are shorter for headspace SPME extraction than for direct extraction under similar agitation, because a substantial portion of the analytes is in the headspace before the extraction process begins and because the diffusion coefficients in the gas phase are typically higher by four orders of magnitude than in liquids. However, since the concentration of semivolatile compounds in the gas phase at room temperature is small, headspace extraction rates for those compounds are substantially lower. They are improved by very efficient agitation or by increasing the extraction temperature (13). A more detailed discussion of the above phenomena is presented below.

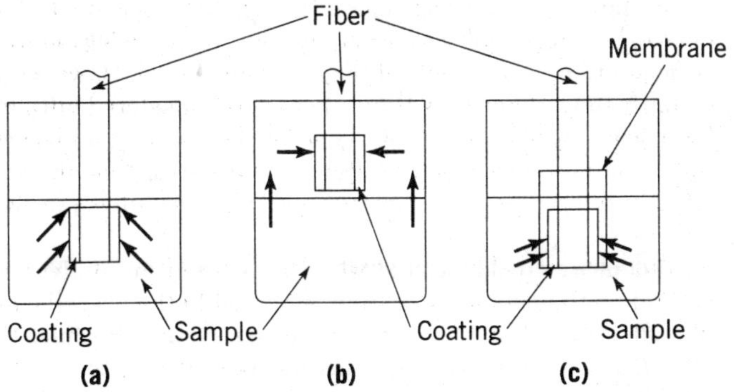

Figure 2. Modes of SPME operation: (**a**) direct extraction, (**b**) headspace extraction, and (**c**) membrane-protected SPME.

In the third mode (SPME extraction with membrane protection), the fiber is separated from the sample by a selective membrane, which lets the analytes through while blocking the intereferences. The main purpose for the membrane barrier is to protect the fiber against adverse effects caused by high molecular weight compounds when very dirty samples are analyzed. Although extraction from the headspace serves the same purpose, membrane protection enables the analysis of less volatile compounds. The extraction process is substantially slower than direct extraction because the analytes need to diffuse through the membrane before they reach the coating. Thin membranes and increased extraction temperature result in shorter extraction times (14).

Fiber coating should be selected accordingly to the application. Commercially available coatings are discussed below. The closer the properties of the coating to the properties of the analyte, the higher the affinity of the analyte for the coating. In general, nonpolar analytes should be extracted with nonpolar coatings, and vice versa. Coating selection is based on the chromatographic behavior of the target analytes. Compounds with long retention times on nonpolar GC columns are well extracted by nonpolar coatings (the opposite also applies).

SPME is easily interfaced to several types of instrumentation because it is solvent-free. The analytical instrument most frequently used with SPME is a gas chromatograph. Standard GC injectors, such as split/splitless, can be applied to SPME when a narrow insert with internal diameter close to the outside diameter of the needle is used. Narrow inserts are required to increase the linear carrier gas flow around the fiber, thus improving the desorption conditions by efficiently removing the analytes desorbed. The split should be turned off during SPME injection.

Coupling SPME with high performance liquid chromatography (HPLC) requires a special interface, but nevertheless it is analogous to the traditional loop injection. Figure 3 shows the design of a commercial SPME/HPLC interface (Supelco). It consists of a desorption chamber and a six-port injection valve. The desorption chamber is placed in the position where the injection loop normally resides on the injection valve. When the injection valve is in the "load" position, the fibre is introduced into the desorption chamber under ambient pressure. A different solvent can also be introduced at this stage to facilitate desorption. Once the valve is switched to the "inject" position, the analytes are desorbed from the fiber and introduced into the

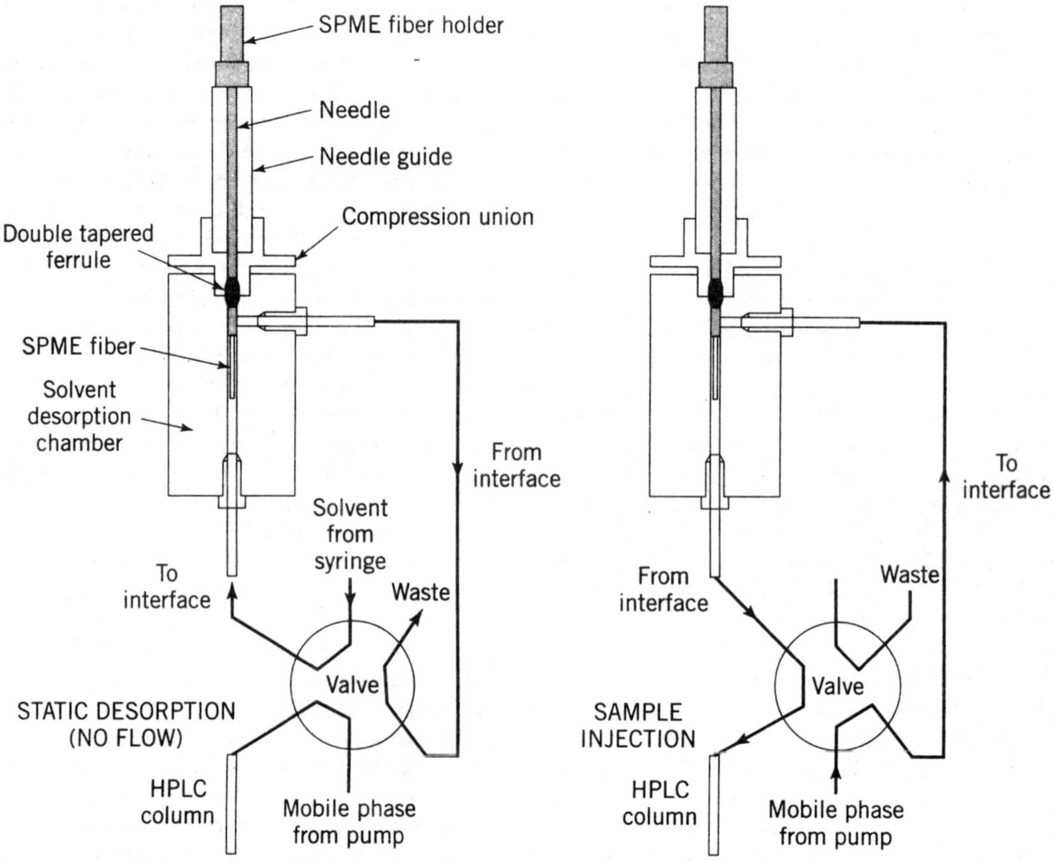

Figure 3. Schematic diagram of the commercial SPME/HPLC interface.

HPLC column. The same interface can also be used for coupling SPME with supercritical fluid chromatography (SFC). As of this time, coupling with capillary electrophoresis is still experimental.

The following sections present the theory, method development, and selected applications of SPME.

THEORY OF SOLID-PHASE MICROEXTRACTION

SPME theory provides insight and direction when developing methods and identifies parameters for rigorous control and optimization. Effective use of the theory minimizes the number of experiments that must be performed. Theory has been developed to understand the principal processes in SPME, in which fibers extract analytes by absorption (PDMS, PA), by applying the fundamentals of thermodynamics and mass transfer. To simplify mathematical relationships, the theory assumes ideal conditions. The theory for ideal extraction conditions is very accurate for trace concentrations in simple matrices like air or drinking water under ambient conditions, when secondary factors such as thermal expansion of polymers, changes in diffusion coefficients from solutes in polymers, heterogeneity of the matrix, etc., are neglected. When conditions are more complex, the theory for ideal cases still approximates well some of the parameters and general relationships between the parameters and extraction times or amounts extracted. The thermodynamics and the kinetics of the extraction process are described in this section. The amount of analyte extracted at equilibrium is calculated from thermodynamic principles, and the extraction time is estimated by solving differential equations describing mass transfer in the system.

Thermodynamics

Solid-phase microextraction is a multiphase equilibration process. Frequently, the extraction system is complex, as in a sample consisting of an aqueous phase with suspended solid particles, which have various adsorption interactions with analytes, plus a gaseous headspace. In some cases specific factors have to be considered, such as analyte losses by biodegradation or adsorption on the walls of the sampling vessel. However, to simplify the system, only three phases are considered initially, the fiber coating, the gas phase or headspace, and a homogeneous matrix, such as pure water. During extraction, analytes migrate between all three phases until equilibrium is reached.

Multiphase Equilibria in SPME. The mass of an analyte extracted by the polymeric coating is related to the overall equilibrium of the analyte in the three-phase system. Because the total mass of an analyte remains constant during the extraction,

$$C_0 V_s = C_f^\infty V_f + C_h^\infty V_h + C_s^\infty V_s \quad (1)$$

where C_0 is the initial concentration of the analyte in the matrix; C_f^∞, C_h^∞, and C_s^∞ are the equilibrium concentrations of the analyte in the coating, the headspace, and the matrix, respectively; and V_f, V_h, and V_s are the volumes of the coating, the headspace, and the matrix, respectively. If we define the coating/gas distribution constant as $K_{fh} = C_f^\infty/C_h^\infty$ and the gas/sample matrix distribution constant as $K_{hs} = C_h^\infty/C_s^\infty$, the mass of the analyte absorbed by the coating, $n = C_f^\infty V_f$, is expressed by

$$n = \frac{K_{fh}K_{hs}V_f C_0 V_s}{K_{fh}K_{hs}V_f + K_{hs}V_h + V_s} \tag{2}$$

Also,

$$K_{fs} = \frac{K_H}{K_F} = K_{fh}K_{hs} = K_{fg}K_{gs} \tag{3}$$

because the fiber/headspace distribution constant K_{fh} is approximated by the fiber/gas distribution constant K_{fg}, and the headspace/sample distribution constant K_{hs} by the gas/sample distribution constant K_{gs}, if the effect of moisture in the gaseous headspace is neglected. Thus, equation 2 is rewritten as

$$n = \frac{K_{fs}V_f C_0 V_s}{K_{fs}V_f + K_{hs}V_h + V_s} \tag{4}$$

As expected from the equilibrium conditions, the equation states that the amount of analyte extracted is independent of the location of the fiber in the system. The fiber may be placed in the headspace or directly in the sample as long as the volumes of the fiber coating, headspace, and sample remain constant. There are three terms in the denominator of equation 4 which measure the analyte capacity of each of the three phases, fiber ($K_{fs}V_f$), headspace ($K_{hs}V_h$), and the sample itself (V_s). If we assume that the vial containing the sample is completely filled (no headspace), the term $K_{hs}V_h$ in the denominator, which is related to the capacity (C_h^∞) of the headspace, is eliminated, resulting in

$$n = \frac{K_{fs}V_f C_0 V_s}{K_{fs}V_f + V_s} \tag{5}$$

Equations 4 and 5 describe the mass absorbed by the polymeric coating after equilibrium is reached in the system. For most analytes, K_{hs} is relatively small (eg, benzene has a K_{hs} value of 0.26), and sampling from the headspace does not affect the mass absorbed by the coating if the volume of the headspace is much smaller than that of the aqueous solution ($V_h \ll V_s$). Under these conditions, the detection limits of headspace SPME are therefore expected to be similar to those of direct SPME.

A proper expression for the above distribution constants should involve appropriate activities, but concentrations are a good approximation considering that the analytes are at trace levels in the sample and assuming that the matrix is relatively pure. The driving force for multiphase equilibrium is the initial difference between the chemical potentials of the analyte in the three phases. Although SPME is used to analyze compounds in various matrices and the fiber coating is an immobilized liquid polymer or porous solid material, in the following text, only multiphase systems including a liquid polymer, sample headspace, and an aqueous matrix are used to discuss equilibrium theory. General conclusions drawn from the discussions are valid for the majority of other types of coatings and matrices.

Prediction of Distribution Constants. In many cases, the distribution constants in equations 2–5, which determine the sensitivity of SPME extraction, are estimated from physicochemical data and chromatographic parameters. For example, distribution constants between a fiber coating and gaseous matrix (eg, air) are estimated from isothermal GC retention times on a column with a stationary phase identical to the fiber coating (15). This is possible because the partitioning process in gas chromatography is analogous to the partitioning process in solid phase microextraction, and there is a well-defined relationship between the distribution constant and the retention time. The nature of the gas phase does not affect the distribution constant, unless the components of the gas, such as moisture, swell the polymer, changing its properties. The formula which correlates the distribution constant and the retention time is as follows:

$$K_{fh} = K_{fg} = (t_R - t_A) \cdot F \cdot \frac{T}{T_m} \cdot \frac{p_m - p_w}{p_m} \cdot \frac{3}{2} \cdot \frac{(\frac{p_i}{p_0})^2 - 1}{(\frac{p_i}{p_0})^3 - 1} \cdot \frac{1}{V_L} \tag{6}$$

where t_R and t_A are the retention times of the solute and an unretained compound, respectively; F is the column flow measured by a soap-bubble flow meter; T and T_m are the temperatures of the column and the flow meter; p_m and p_w are the flow meter pressure and the saturated water vapor pressure; p_i and p_0 are the inlet and outlet pressures of the column; and V_L is the column stationary phase volume. Usually p_m and p_0 are equal to atmospheric pressure. The value of K_{fg} estimated by this method for partitioning of benzene between PDMS and the gas phase agrees within a few percent with the estimate obtained by SPME experimentation.

A very useful method for determining coating-to-gas distribution constants is the linear temperature programmed retention index (LTPRI) system, which indexes retention times of compounds relative to the retention times of n-alkanes. This system is applicable to retention times for temperature-programmed, gas–liquid chromatography. The logarithm of the coating-to-air distribution constants of n-alkanes is expressed as a linear function of their LTPRI values. For PDMS, this relationship is $\log K_{fg} = 0.00415{*}\text{LTPRI} - 0.188$ (16). Thus, the LTPRI system permits interpolating K_{fg} values from the plot of log K_{fg} versus retention index. The LTPRI values for many compounds are available in the literature. Hence this method allows estimating K_{fg} values without experimentation. If the LTPRI value for a compound is not available from published sources, it is determined from a GC run according to the definition

$$\text{LTPRI} = 100 \times N + 100 \times \left(\frac{t_{R(A)} - t_{R(N)}}{t_{R(N+1)} - t_{R(N)}}\right) \tag{7}$$

where N is the number of carbon atoms for the n-alkane eluting directly before the analyte at a retention time $t_{R(N)}$; $t_{R(A)}$ is the retention time of the analyte; and $t_{R(N+1)}$ is the retention time of the first n-alkane eluting after $t_{R(A)}$. Note that the GC column for determining LTPRI should be coated with the same material as the fiber coating.

The coating/water distribution constant is estimated from equation 5. The appropriate coating/gas distribution constant is found by applying techniques discussed above, and the gas/water distribution constant (Henry's constant) is obtained from physicochemical tables or is estimated by the structural unit contribution method (17).

Some correlations are used to anticipate trends in SPME coating/water distribution constants for analytes. For example, a number of investigators report the correlation between octanol/water distribution constant K_{ow} and K_{fw}. This is expected, because K_{ow} is a very general measure of the affinity of compounds to the organic phase. The trends are valid, however, only for compounds within homologous series, such as aliphatic hydrocarbons, aromatic hydrocarbons, or phenols. Trends should not be used for comparisons between different classes of compounds, because of different analyte activity coefficients in the polymer. Another, more universal approach, is to use the common fragment constants applied for a given coating phase based on theories developed for and tested by liquid chromatography (18–20). In this method, contributions of various functional groups in a given molecule are added to each other to estimate the appropriate constants.

Effect of Extraction Parameters. Thermodynamic theory predicts the effects of modifying certain extraction conditions on partitioning and indicates parameters to control reproducibility. The theory is used to optimize extraction conditions with a minimum number of experiments and to correct for variations in extraction conditions, without repeating calibration tests under the new conditions. For example, SPME analysis of outdoor air is done at ambient temperatures that vary significantly. The equation that predicts the effect of temperature on the amount of analyte extracted allows calibration without extensive experimentation. Extraction conditions that affect K_{fs} include temperature, salting, pH, and organic solvent content in water.

Temperature. If both the sample and the fiber change temperature from T_0 to T, the distribution constant changes according to

$$K_{fs} = K_0 e^{\frac{-\Delta H}{R}\left(\frac{1}{T}-\frac{1}{T_0}\right)} \tag{8}$$

where K_0 is the distribution constant when both the fibre and the sample are at temperature T_0 (K); ΔH is the molar change in enthalpy of analyte when it moves from the sample to the fiber coating; and R is the gas constant (21). The enthalpy change ΔH is considered constant over temperature ranges typical in SPME experiments. It is determined by measuring K_{fs} at two different temperatures. For coating/gas distribution constants, ΔH for a volatile compound is well approximated for PDMS by the heat of vaporization of the pure compound ΔH_v (19,22). Temperature effect must be considered when temperatures vary while sampling outdoors and when heating is used to increase extraction rates, stop metabolic activity, or enhance the release of analytes.

When the K_{fs} value for an analyte is greater than 1, the analyte has a lower potential energy in the fiber coating than in the sample, so the analyte partitioning into the fiber must be exothermic, which means that ΔH is greater than 0. Therefore equation 8 shows that raising the temperature decreases K_{fs}.

Equation 8 applies only to partitioning between two homogeneous phases. The equation does not apply to partitioning between a fiber and a multicomponent sample, but is still used to calculate the effect. For example, for poly(dimethylsiloxane) (PDMS) extraction of benzene from the headspace above an aqueous solution at 25°C $K_{fw} = 125, K_{fh} = 493$, ΔH_{fw} has been estimated as 13.9 kJ/Kmol (19), and ΔH_{fh} is estimated from the enthalpy change in vaporization (42.8 kJ/Kmol for benzene), where the subscript w represents water and the subscript h represents headspace. Therefore a 100-μm PDMS fiber extracting benzene from 3 mL of headspace above a 2-mL, 100-ppb aqueous solution extracts 5.5 ng at 25°C, according to equation 4. The amount extracted at 90°C is calculated from equations 4 and 8:

$$n = \frac{n_0}{1 + \frac{1}{V_f} V_w K_{wf} e^{\frac{\Delta H_{wf}}{R}\left(\frac{1}{T}-\frac{1}{T_0}\right)} + \frac{1}{V_f} V_h K_{hf} e^{\frac{\Delta H_{hf}}{R}\left(\frac{1}{T}-\frac{1}{T_0}\right)}} \tag{9}$$

which predicts 0.75 ng. Equation 8 does not apply directly to partitioning by adsorption.

Salting. Salting and pH adjustment are two common techniques for enhancing extraction of organics from aqueous solutions. Salting increases or decreases the amount extracted, depending on the compound and salt concentration. The effect of salting on SPME has been determined to date only by experiment. Figure **4a** illustrates the effect of salting on the extraction of benzene and toluene from water. A substantially increased amount of the analyte is extracted at salt concentrations above 1%, ultimately yielding about an order of magnitude increase in sensitivity at 30% salt. Saturation with salt is used to lower the detection limits and also to normalize variable salt concentrations in natural matrices. Note that salting lowers pH at high salt concentrations, because proton activity increases with increased solution ionic strength.

The effect of salting on SPME has not been examined theoretically, although a theory has been developed for the effect of salting on liquid–liquid extraction. It is described by Setchenow's equation:

$$K_{fs} = K_0 e^{k_s C_s} = K_0\left[1 + k_s C_s + \frac{(k_s C_s)^2}{2} + \frac{(k_s C_s)^3}{3} + \right] \tag{10}$$

where K_0 is the partition coefficient with no salt, k_s is a constant, and C_s is the salt concentration. This equation correctly predicts salt effects at least for low salt concentrations and sometimes also for high concentrations. An extensive review article summarizes the theory and tabulates most of the research published on salting-out and salting-in up to 1951 (23).

pH. Adjusting the pH of an aqueous solution changes K for dissociable species, assuming that only the neutral form of the acid or base is extracted by the fiber coating, according to

$$K = K_0 \frac{[H^+]}{K_a + [H^+]} \tag{11}$$

where K_0 is the distribution constant of the undissociated form between the sample and the fiber. This relationship is confirmed with results for extraction of acids as illustrated in Figure **4b** (24). As pH decreases, more acid is in the neutral form which partitions into the coating, resulting in higher sensitivity. To obtain the highest sensitivity in the analysis of acids, the pH must be two units lower than the pK_a of the acid.

Polarity of Sample Matrix and Coating Material. The presence of an organic solvent in water changes K according to

$$K_{fs} = K_{fw} \cdot 10^{(P_1 - P_2)/2} \tag{12}$$

where K_{fw} is the distribution constant for pure water, $P_1 = 10.2$ is the polarity parameter for water, and $P_2 = cP_s + (1 - c)P_1$ is the water/solvent mixture polarity parameter for a solvent of concentration c and polarity parameter P_s (20). This equation predicts the distribution constants for water heavily contaminated with miscible solvents, assuming that the solvent does not swell the coating. This relationship indicates that the concentration of the solvent must be above 1% to substantially change the properties of water and the distribution constant. Figure **4c** illustrates the decrease in the

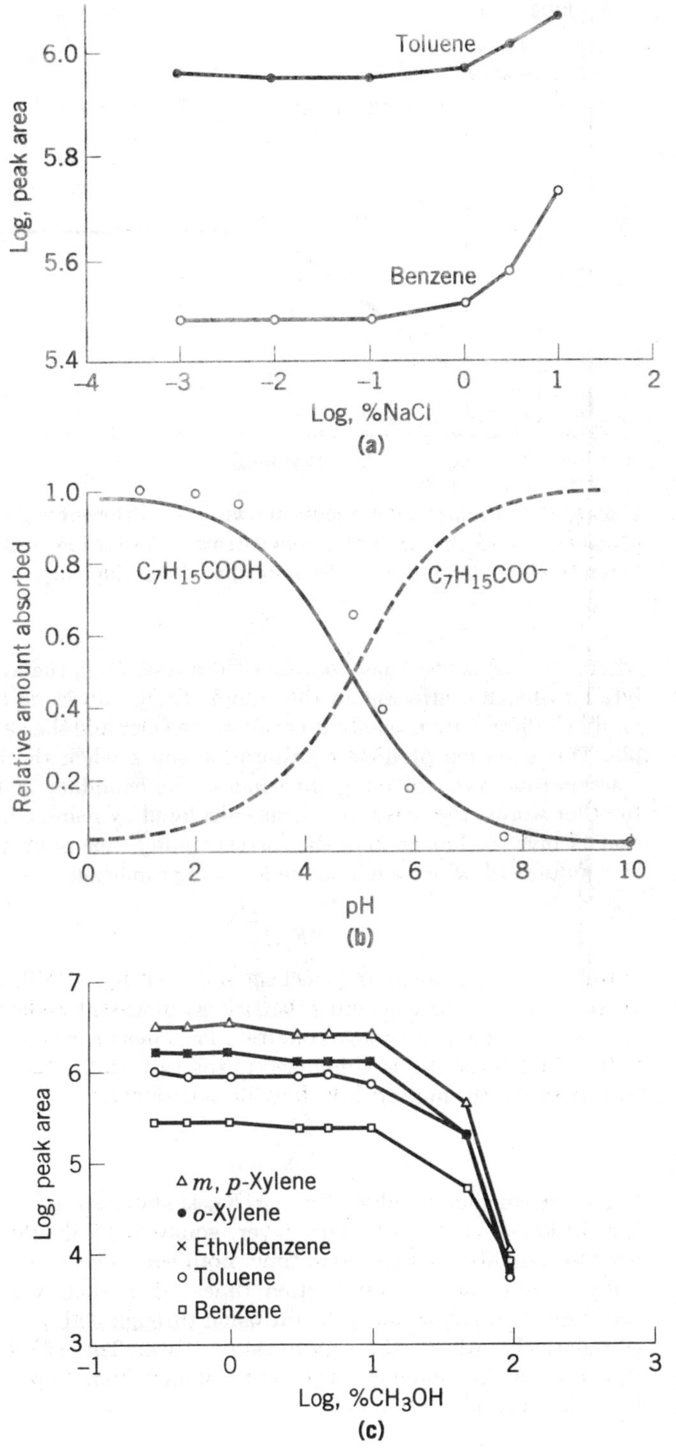

Figure 4. The effects of (**a**) salt, (**b**) pH, and (**c**) solvent on SPME.

amount of BTEX extracted into a PDMS coating with increasing methanol concentration in an aqueous matrix. This type of variation of the matrix composition can be compensated for by using internal standard calibrating techniques.

Compounds have affinity for phases of similar polarity. One study discussed the significance of polarity on fiber extraction of organics from water (25). The dielectric constant of PDMS varies from 2.6 to 2.8 depending on the molecular weight of the polymer, and the dielectric constant of PA varies from 2.6 to 3.6 (26). These polymer dielectric constants are similar to those for common organic compounds, for example, ϵ(toluene) = 2.4 and ϵ(acetic acid) = 4.1 (27). The polymer and organic ϵ values are very low compared to that of water at room temperature ($\epsilon(H_2O) = 78$) (23). The trends in dielectric constants indicate high K_{fs} values for typical organic compounds distributed between fiber coating and water.

Thermal Swelling. The theory developed for SPME assumes ideal conditions and neglects factors such as thermal expansion or interactions between the analytes or with interferences. This idealization is usually justified under ambient conditions and for solutes at trace concentration levels. The effect of thermal expansion on coated fiber extraction was considered (28). Thermal swelling changes the fiber coating radius according to

$$r = r_0(1 + \alpha T) \tag{13}$$

where r_0 is the radius at 0°C and α is the linear thermal expansion coefficient. For PDMS, α is listed as 2.7×10^{-4}°C^{-1}, which is negligible for typical SPME extraction conditions.

Kinetics

The extraction process kinetics determine the extraction speed. Most of the theory of mass transport is based on Fick's second law of diffusion describing mass balance in a dynamic system, which is expressed in one dimension by

$$\frac{\partial C}{\partial t} = D\frac{\partial^2 C}{\partial x^2} \tag{14}$$

where C is the concentration and D is the diffusion coefficient of the analyte. Considering the cylindrical geometry of the fiber and sampling system in three-dimensional space, equation 14 converts to

$$\frac{\partial C}{\partial t} = D\frac{1}{r}\left[\frac{\partial}{\partial r}\left(r\frac{\partial C}{\partial r}\right)\right] \tag{15}$$

Examining the kinetics based on this equation involves determining values for the diffusion coefficient, determining values for the boundary distribution constants, assuming boundary conditions, and solving the differential equations. Methods to solve these problems are well explained in one text (29). A discussion is presented below using graphs generated from analytical solutions.

The kinetic theory identifies "bottlenecks" of solid-phase microextraction and indicates strategies to increase extraction speed. It is assumed that diffusion proceeds according to Fick's law. The theory assumes zero interactions between analytes and vial surfaces or fiber core. Factors, such as thermal expansion, swelling, and analyte/analyte interactions are assumed negligible.

Direct Extraction. Let us first examine direct extraction of the analytes from a homogeneous water sample into a fiber's liquid polymer coating. No headspace is in the system. Figure 5 shows the geometry of the system investigated, where $b - a$ is the coating thickness.

Perfect Agitation. First consider the case where the water sample is perfectly agitated. In other words, the aqueous phase moves very rapidly with respect to the fiber, so that all analytes in the sample have access to the fiber coating. In this case, the equilibration time, defined as the time required to extract

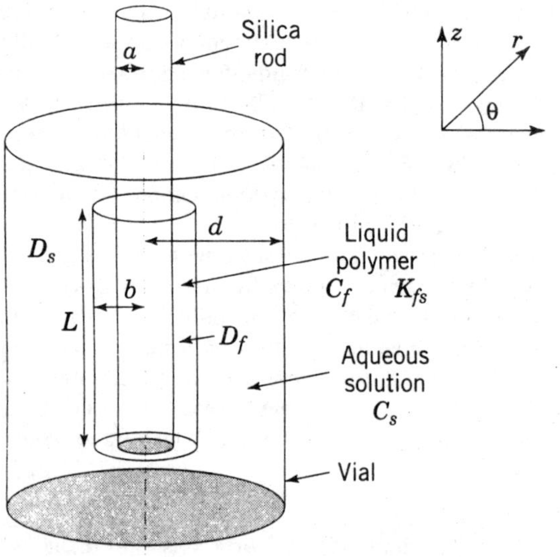

Figure 5. Graphic representation of the SPME/sample system with dimensions and parameters labeled as follows: *a*, fiber coating inner radius; *b*, fiber coating outer radius; *L*, fiber coating length; *d*, vial inner radius; C_f, analyte concentration in the fiber coating; D_f analyte diffusion coefficient in the fiber coating; C_s, analyte concentration in the sample; D_s, analyte diffusion coefficient in the sample; K_{fs}, analyte distribution coefficient between fiber coating and sample; $K_{fs} = C_f/C_s$.

95% of the equilibrium amount of an analyte from the sample, corresponds to

$$t_e = t_{95\%} = \frac{D_f}{2(b-a)^2} \quad (16)$$

Using this equation one estimates the shortest equilibration time possible for a practical system by substituting appropriate data for the diffusion coefficient of an analyte in the coating (D_f) and the fiber coating thickness ($b - a$). For example, the equilibration time for extracting benzene from a perfectly stirred solution with a 100-μm PDMS film is expected to be about 20 seconds. Equilibration times close to those predicted for perfectly agitated samples are obtained experimentally for extracting analytes from air samples (because of high diffusion coefficients in gas) or when very high sonication power facilitates mass transfer in aqueous samples. However, in practice, there is always a layer of unstirred water around the fiber. A higher stirring rate results in a thinner water layer around the fiber.

Practical Agitation. Independent of the agitation level, fluid contacting a fiber's surface is always stationary, and as the distance from the fiber surface increases, the fluid movement gradually increases until it corresponds to bulk flow in the sample. To model mass transport, the gradation in fluid motion and convection of molecules in the space surrounding the fiber surface are simplified by a zone of a defined thickness in which no convection occurs and perfect agitation in the bulk of the fluid everywhere else. This static layer zone is called the *Prandtl* boundary layer (Fig. 6) (30). Its thickness is determined by the agitation conditions and the viscosity of the fluid.

The equilibration time is estimated for practical cases from the following equation:

$$t_e = t_{95\%} = 3\,\frac{\delta K_{fs}(b-a)}{D_s} \quad (17)$$

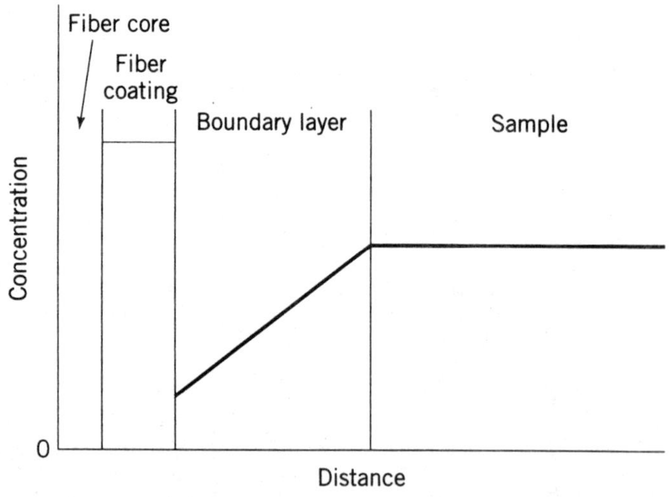

Figure 6. Boundary layer model showing the different regions considered and the assumed concentration vs radius profile when the boundary layer determines the extraction rate.

where $(b - a)$ is the fiber coating's thickness, D_s is the analyte's diffusion coefficient in the sample fluid, and K_{fs} is the analyte's distribution constant between the fiber and the sample. This equation predicts equilibration times when the extraction rate is controlled by diffusion in the boundary layer. In other words, the extraction time calculated by using equation 17 must be longer than the corresponding time predicted by equation 16, which leads to the following condition:

$$\delta > \frac{(b-a)D_s}{6K_{fs}D_f} \quad (18)$$

Diffusion coefficients for the most common coating, PDMS, are smaller than the corresponding coefficients in water by a factor of 5–6. Therefore, the above condition for aqueous extraction with a PDMS coating is simplified to the ratio of the coating thickness and the appropriate distribution constant:

$$\delta > \frac{b-a}{K_{fs}} \quad (19)$$

When the calculated value of $b - a/K_{fs}$ is small compared to the thickness of the boundary layer, equation 16 should be used to estimate the extraction time. Both equations substantially underestimate equilibration times when their values are close together, because the diffusion through both phases cumulatively affects the equilibration time. The effective thickness of the boundary layer is estimated from empirical formulas of fluid mechanics.

Headspace SPME. The geometry of SPME headspace extraction is illustrated in Figure 7**a** (13). An aqueous sample containing a dissolved organic compound is transferred into a closed container with a headspace. A fused silica fiber coated with a thin layer of a selected liquid organic polymer is inserted into the headspace portion of the container but is not in direct contact with the aqueous phase. The fiber's liquid coating starts to absorb the organic analyte from the headspace. The analyte undergoes a series of transport processes from water to gas phase (or headspace) and eventually to the coating, until the system finally reaches equilibrium. The diffusion process occurs in the axial direction and also in

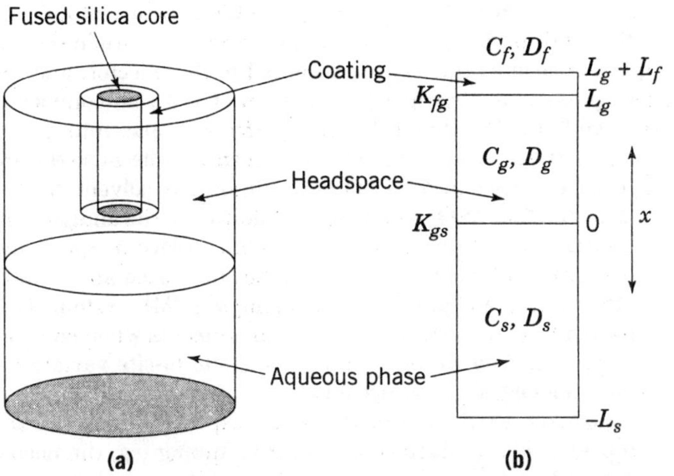

Figure 7. (**a**) Geometry for SPME headspace sampling; (**b**) one-dimensional model of the three-phase system: K_{fg} and K_{gs} are the coating/headspace and headspace/water partition coefficients; D_f, D_g, and D_s are the diffusion coefficients of the analyte in the coating, headspace, and water, respectively; C_f, C_h and C_s are the concentrations in the coating, headspace, and water, respectively; L_f, L_g, and L_s are the thicknesses of the coating, headspace, and aqueous phase, respectively.

the radial direction. Exact algebraic solutions of this diffusion problem can be derived. We are more interested in studying the main factors which control diffusion, however, rather than strict mathematical solutions. A simple one-dimensional diffusion model provides insight into this diffusion problem. In the model illustrated in Figure 7**b**, the diffusion occurs in only one direction (x-axis), L_f is the thickness of the polymeric coating, L_g is the length of the headspace, and L_s is the length of the aqueous solution. The solution to this problem is found by solving the appropriate mass balance differential equation.

For headspace SPME, the following equation estimates equilibration time:

$$t_e = t_{95\%} = 1.8\left(\frac{L_g}{K_{gw}D_g} + \frac{L_s}{1.6D_s}\right)K_{fw}L_f \qquad (20)$$

This equation is used when the amount of an analyte extracted by the fiber coating represents a small fraction of the analyte in the sample. The first term in the parentheses relates to mass transfer in the headspace and the second to mass transfer in the sample. As expected, the mass transfer in the sample is controlled by the diffusion coefficient D_s, in the sample, and the sample length L_s which analytes need to diffuse through. The rate of mass transfer in the headspace, on the other hand, is related to the gas/sample distribution constant K_{gw}, the diffusion coefficient of analytes in the headspace D_g, and the size of the headspace L_g. The equilibration time also increases with increasing fiber capacity, defined as a product of the fiber sample distribution constant K_{fs} and the fiber thickness L_f.

Equation 20 suggests that headspace SPME has a short equilibration time when the aqueous phase is static. The analyte should have a large Henry's constant, which occurs with high volatility and/or hydrophobicity of the analyte, and should not have a very large distribution constant between the fiber coating and the headspace. Because volatile organic compounds usually meet these requirements, this method is used to analyze them very effectively. The equilibration time when $V_g > 10K_{fg}V_f$ is estimated by

$$t_e = t_{95\%} = 100\,\frac{(K_{fg}V_f)^2}{D_g} \qquad (21)$$

This estimate of equilibration time is used for all static headspace extractions, including perfect agitation in the sample matrix, when the analyte extracted originates only from the headspace.

The discussion suggests that the extraction temperature is a very important parameter to optimize. When the temperature is high enough to move a significant fraction of the analyte to the headspace from the sample matrix, the extraction occurs only from the headspace, the equilibration time is very rapid, and it is independent of agitation.

Practical Agitation. In practice, perfect agitation of the matrix causes a similar condition in the headspace. When agitation in both phases is perfect, equation 16 is used to estimate equilibration times. For more realistic agitation, modified equations 20 and 21 estimate the equilibration time. A theoretical approach is to compute an eddy diffusion coefficient D_e, which describes the effect of agitation on mass transport. This eddy diffusion coefficient is added to the molecular diffusion coefficient to obtain an approximate "virtual" diffusion coefficient (31). For example, Fick's first law is adapted to describe mass flux in a stirred water sample as follows:

$$J = (D_e + D_s)\frac{\partial C}{\partial x} \qquad (22)$$

A relationship enabling the calculation of the eddy diffusion coefficient is expressed as,

$$D_e = 0.0075\ vR_d \qquad (23)$$

where v is the kinematic viscosity and R_d the Reynolds number of the fluid (32). For fluid stirred in a vial, the rotating Reynolds number is given by

$$R_d = 4\,R^2N\rho/v \qquad (24)$$

where R is the impeller radius, N is the rate of rotation, ρ is the fluid's density (33). The headspace of magnetically stirred sample can be considered a fluid in a cylindrical housing with a rotating disk at the bottom. The average rotational speed of the water surface is given by

$$N_w = \frac{1}{R}\int_0^R \frac{u(r)}{2\pi r}\,dr = 0.55\ N \qquad (25)$$

assuming that the stir bar radius is close to the radius of the vial. This relationship is used in equation 24 to calculate the headspace Reynolds number.

For example, for headspace SPME from magnetically stirred water in a vial, equation 20 becomes

$$t_e = t_{95\%} = 1.8\left(\frac{L_g}{K_{gw}(D_g + 2\cdot 10^{-5}NR^2)} + \frac{L_s}{1.6(D_s + .03\cdot NR^2)}\right)K_{fw}L_f \qquad (26)$$

where R is the radius of the vial and N is the revolution rate of the stir bar, whose radius is assumed close to that of the vial. Other equations with diffusion coefficients as parameters are modified for agitated systems similarly.

In-Tube SPME. The discussion above is limited to systems involving extraction phases dispersed on the surface of a rod which is exposed from the protective tubing (needle). In addition to this approach, the extraction phase in some solid phase microextraction systems, remains in the tubing during the extraction. These approaches involve coated rods retracted into the needles or an extracting phase dispersed onto the inner surface of the tubing. The theoretical aspects of the extraction processes which use these geometric arrangements are briefly described below.

Dynamic In-Tube SPME. Assume that the system consists of a piece of fused silica capillary, internally coated with a thin film of extracting phase (a piece of open tubular capillary GC column), or that the capillary is packed with an extracting phase dispersed on an inert supporting material (a piece of a micro-LC capillary column). In these geometric arrangements, the concentration profile along the x-axis of the tubing containing the extracting phase is described as a function of time t by adopting the expression for dispersion of the concentration front (34,35):

$$C(x,t) = \frac{1}{2}C_0\left[1 - erf\frac{x - \frac{ut}{1+k}}{\sigma\sqrt{2}}\right] \quad (27)$$

where u is the linear velocity of the fluid in the tube; k is the partition ratio defined as

$$k = K_{fs}\frac{V_f}{V_v} \quad (28)$$

where K_{sf} is a sample/coating distribution constant; V_f is the volume of the extracting phase and V_v is the void volume of the tubing containing the extracting phase. σ is the mean square root dispersion of the front defined as

$$\sigma = \sqrt{Ht\frac{u}{1+k}} \quad (29)$$

where H is equivalent to HETP (height equivalent to theoretical plate) in chromatographic systems. HETP is calculated as the sum of individual contributions to the front dispersion, which depend on the particular geometry of the extracting system (35).

Equation 27 indicates that the front of the analyte migrates through the capillary with a velocity proportional to the linear velocity of the sample and inversely proportional to the partition ratio. For short capillaries with low dispersion, the extraction time is assumed similar to the time required for the center of the band to reach the end of the capillary:

$$t_e = \frac{L(1 + K_{fs}\frac{V_f}{V_v})}{u} \quad (30)$$

where L is the length of the capillary holding the extracting phase. As expected, the extraction time is proportional to the length of the capillary and inversely proportional to the linear flow rate of the fluid. Extraction time also increases with an increase in the coating/sample distribution constant and with the volume of the extracting phase, but decreases with an increase in the void volume of the capillary. The equation above is used only for direct extraction when the sample matrix passes through the capillary. This approach is limited to particulate-free gas and clean water samples. The headspace SPME approach broadens the application of in-tube SPME. In that case, the mass transfer between the sample and its headspace is carefully considered to describe the process properly. Also, if the flow rate is very rapid and produces turbulent behavior and the coating/sample distribution constant is not very high, perfect agitation is assumed and equation 16 is used to estimate equilibration time.

Removal of analytes from a tube is an elution problem analogous to frontal chromatography and has been discussed in detail (35). In general, if the desorptive temperature of a GC is high and thin coatings are used, all the analytes are in the gas phase as soon as the coating is placed in the injector, and the desorption time corresponds to elution of two void volumes of the capillary. For liquid desorption (for example, into an LC system), the desorption volume is even smaller because the analytes are focused at the front of the desorptive solvent (36).

Static In-Tube SPME Sampling. In addition to the analyte concentration measurement at a well-defined place in space and time, obtained by using the approaches discussed above, integrated sampling is possible with a simple SPME system. This is particularly important in field measurements when changes in analyte concentration over time and site-to-site variations, must often be taken into account.

When the extracting phase is not exposed directly to the sample but is contained in a protective tubing (eg, the needle of the SPME device) without any flow of the sample through it (as discussed above), extraction occurs through the static gas phase in the needle. The integrating system consists of an extracting phase coating the interior of the tubing or an externally coated fiber withdrawn into the needle. These geometric arrangements represent a very powerful method for generating a response proportional to the integral of the analyte concentration over time and space (when the needle is moved through the space) (37). In these cases, the only mechanism of analyte transport to the extracting phase is diffusion through the gas phase contained in the tubing. During this process, a linear concentration profile is established in the tubing between the small needle opening, characterized by surface area A, and the distance Z between the needle opening and the position of the extracting phase. The amount of analyte dn extracted during time interval dt is calculated by considering Fick's first law of diffusion (37):

$$dn = AD_g\frac{dc}{dz}dt = AD_g\frac{\Delta C(t)}{Z}dt \quad (31)$$

where $\Delta C(t)/Z$ is the concentration gradient established in the needle between the needle opening and the position of the extracting phase, Z; $\Delta C(t) = C(t) - C_z$, where $C(t)$ is the time-dependent concentration of the analyte in the sample in the vicinity of the needle opening and C_z is the concentration of the analyte in the gas phase in the vicinity of the coating. C_z is close to zero for high coating/gas distribution constants; in this case, $\Delta C(t) = C(t)$. The concentration of the analyte C_z at the coating position in the needle increases with integration time, but it remains low compared to the sample concentration because of the sorbing coating. Therefore the amount accumulated over time is calculated as

$$n = D_g\frac{A}{Z}\int C(t)dt \quad (32)$$

As expected, the amount of analyte extracted is proportional to the integral of sample concentration over time, the diffusion coefficient of analytes in the gaseous phase D_g, and the surface area of the needle opening A, and inversely proportional to the distance between the coating and the needle opening Z. Equation 32 is valid only when the amount of analyte extracted onto the sorbent is a small fraction (below RSD of the measurement, typically 5%) of equilibrium amount with respect to the lowest concentration in the sample. To extend integration time, the coating is placed deeper inside the needle (larger Z), the opening of the needle is reduced by placing an additional orifice (smaller A), or a high capacity sorbent is used. The first two solutions result in low measurement sensitivity. An increase in sorbent capacity is a more attractive alternative. It is achieved by increasing the volume of the coating or its affinity toward the analyte. An increase in the coating volume requires an increase in the size of the device. The optimum approach to increased integration time is application of sorbents characterized by large coating/gas distribution constants.

METHOD DEVELOPMENT

Developing a new SPME method requires the following steps in most cases:

- selecting fiber coating
- selecting the derivatization reagent, if required
- selecting the extraction mode
- selecting the agitation method
- selecting the separation and/or detection technique
- optimizing desorption conditions
- optimizing sample volume
- determining the extraction time profile in a pure matrix
- determining extraction time
- calculating the distribution constant
- optimizing extraction conditions (pH, salt, temperature)
- determining the linear range for a pure matrix at optimum extraction conditions
- selecting the calibration method
- optimizing the extraction conditions for heterogeneous samples
- verifying the equilibration time, sensitivity, and linear dynamic range for complex samples
- determining method precision
- determining method detection limit
- validating the method
- automating the method

In most cases, not all the steps have to be performed. Knowledge gained from previous experiments and from literature, is often applied to the problem at hand. Most SPME methods developed to date are combined with gas chromatographic separation and an appropriate detection method. Hyphenation with HPLC and other techniques should be also considered. Following is a discussion of the particular steps in method development, focused mainly on aqueous samples.

Selecting the Fiber Coating

As of spring 1997, the following coatings are commercially available from Supelco, Inc. (Bellefonte, Pennsylvania): poly(dimethylsiloxane) (PDMS) 7, 30, and 100 μm thick; 85-μm poly(acrylate) (PA); 65-μm poly(dimenthylsiloxane)/divinylbenzene (PDMS/DVB); 60-μm PDMS/DVB optimized for HPLC; 65-μm poly(ethylene glycol)/divinylbenzene (CARBOWAX/DVB); 65-μm poly(ethylene glycol)/divinylbenzene template resin (CARBOWAX/TR); and 80-μm poly(dimethylsiloxane)/Carboxen. The PDMS and the PA coatings, still the most widely used, extract analytes by absorption. PDMS is a very viscous liquid, whereas PA is a low density solid polymer at room temperature. The low density allows analytes to diffuse into the coating, but the diffusion coefficients are lower than for PDMS. The remaining coatings are porous polymers, in which DVB or Carboxen porous microspheres are immobilized onto the fibre by using either Carbowax or PDMS as the glue to hold them together. These coatings extract analytes by adsorption rather than absorption. The pores in the template DVB polymer are uniform and result in less adsorptive discrimination as a function of analyte molecular weight (38). Figure 8 presents general guidelines for choosing a coating for a specific application. Typically, the chemical nature of the target analyte determines the type of coating used. A simple general rule "like dissolves like," applies very well for liquid coatings. Porous polymer coatings are mostly nonselective. Selection of the coating is based primarily on the polarity and volatility of the analyte. For example according to Figure 8, a 100-μm PDMS coating is used for compounds of high to medium volatility and low to medium polarity. In general, PDMS is the most useful coating and should be used whenever possible. It is very rugged and able to withstand high injector temperatures up to about 300°C. PDMS is nonpolar, thus it extracts nonpolar analytes very well. However, it is also applied successfully to more polar compounds, particularly after optimizing extraction conditions. An additional advantage of PDMS is the possibility of estimating distribution constants for organic compounds from retention parameters on PDMS-coated GC columns (15,16,39).

Both the coating thickness and the distribution constant determine the sensitivity of the method and the extraction time. Thick coatings offer increased sensitivity, but require much longer equilibration times. As a general rule, to speed up the sampling process, the thinnest coating offering the sensitivity required should be used. For 100-μm PDMS fibers, the equilibration time in direct aqueous extraction with magnetic stirring is less than 1 h for compounds characterized by

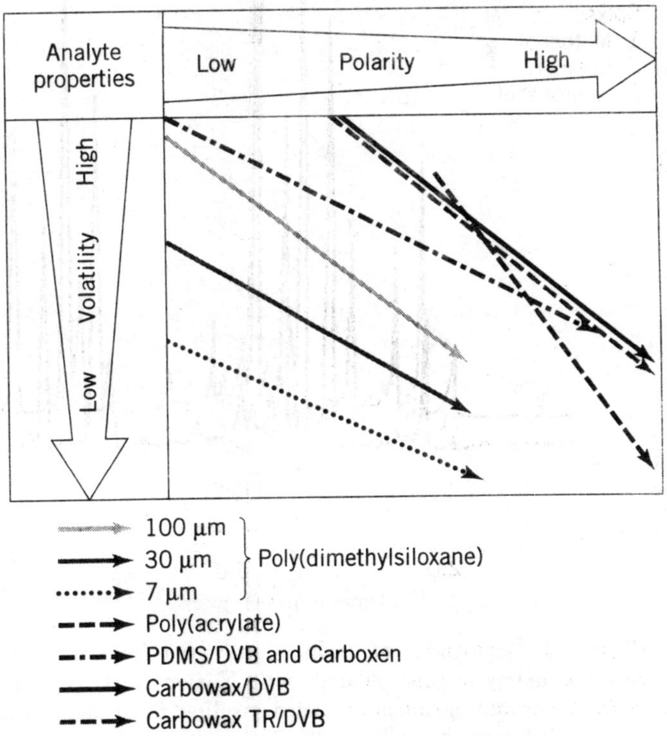

Figure 8. Coating selection guide.

distribution constants K lower than $\sim$10,000. Thinner PDMS coatings should be considered for compounds of larger K. For more polar compounds, such as phenols (9), the poly(acrylate) coating is more suitable.

Porous polymer coatings have complementary properties compared to PDMS and PA. They are more suitable for volatile species because the respective distribution ratios typically are larger than for PDMS. Changing the "glue" from PDMS to Carbowax results in different selectivity toward polar compounds, such as ketones and alcohols. The adsorption times are typically shorter for gaseous samples compared to 100-μm PDMS, since the analytes do not need to diffuse through the liquid polymeric phase.

The template resin coating is designed to reduce molecular weight discrimination for homologues differing in chain length. With regular DVB coatings characterized by nonuniform pore size distribution, the amount of analyte extracted from a solution of a given concentration varies as a function of molecular size. The template resin has uniform pore diameters and therefore uniform extraction efficiency.

When a group of analytes of different characteristics (eg, pesticides) is to be determined by SPME, primary consideration should be given to the analytes that are the most difficult to extract. A coating should be chosen that enables the determination of all the analytes with enough sensitivity. When none of the commercially available coatings meets the requirements, custom coatings are sometimes prepared. For example, Figure 9 shows the separation of alcohols extracted from unleaded gasoline using a fiber coated with commercially available Nafion perfluorinated resin (40).

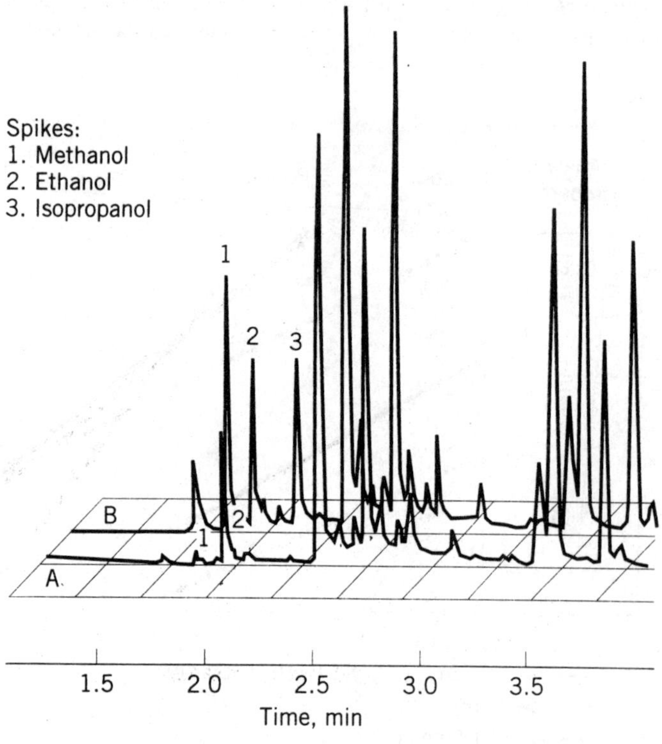

Figure 9. Separation of alcohols extracted from unleaded gasoline using a fiber coated with Nafion perfluorinated resin: A, chromatogram of unleaded gasoline, B chromatogram of unleaded gasoline spiked with 333 ppm each of methanol, ethanol, and isopropanol.

Selecting the Derivatizing Reagent

Derivatization performed before and/or during extraction enhances sensitivity and selectivity of both extraction and detection and enables SPME determination of analytes normally not amenable to analysis by this method, whereas postextraction methods only improve chromatographic behavior and detection. Incorporation of the derivatization step complicates the SPME procedure, and therefore should only be considered when necessary. Selective reactions producing specific analogues result in less interference during quantitation. This approach is used for analyte determination in complex matrices. Sensitivity is enhanced when the derivatizing reagent contains moieties that enhance detection. For example, the sensitivity of SPME determination of carboxylic acids in water is greatly improved by using 1-(pentafluorophenyl)diazoethane as the derivatizing reagent (41). This compound converts carboxylic acids directly in the aqueous matrix to pentafluorophenyl ester derivatives which are detected by electron capture detector (ECD) with very good sensitivity. Method selectivity is also greatly improved in this way.

In many cases, derivatization reagents designed specifically for liquid extraction procedures are used in SPME. For example, sodium tetraethylborate was used to enable the analysis of metal ions, including mercury and lead in water, and to perform partial speciation of organometallic compounds by SPME/GC (42,43).

The most interesting implementation of the derivatization procedure involves the use of doped fibers. Figure 10 presents a chromatogram obtained by exposing a poly(acrylate)-coated fiber doped with 1-pyrenyldiazomethane to the headspace of a sewage sample (44). All volatile carboxylic acids are detected. Without derivatization, SPME is not capable of extracting sufficient amounts of these acids because no appropriate selective coating yet exists. Derivatization results in high sensitivity and is compatible with field analysis requirements, because the derivatizing reagent and the derivatization products are nonvolatile. The doped fibers facilitate spot sampling and also can measure long-term exposure to a given contaminant, because analytes are continuously accumulated onto the fiber until all the reagent is consumed.

Selecting the Extraction Mode

Extraction mode selection is based on the sample matrix composition, analyte volatility, and its affinity to the matrix. For very dirty samples, the headspace or fiber protection mode should be selected. For clean matrices, both direct and headspace sampling are used. Headspace sampling is applicable to analytes of medium to high volatility. Headspace extraction is always preferential for volatile analytes because the equilibration times are shorter compared to direct extraction. Extraction conditions for many compounds, including polar and ionic materials, are improved by matrix modifications, as described later. Application of headspace SPME is extended to semivolatile compounds and analytes strongly bound to the matrix by increasing the extraction temperature, also discussed below. Fiber protection is used only for very dirty samples in cases where neither of the first two modes can

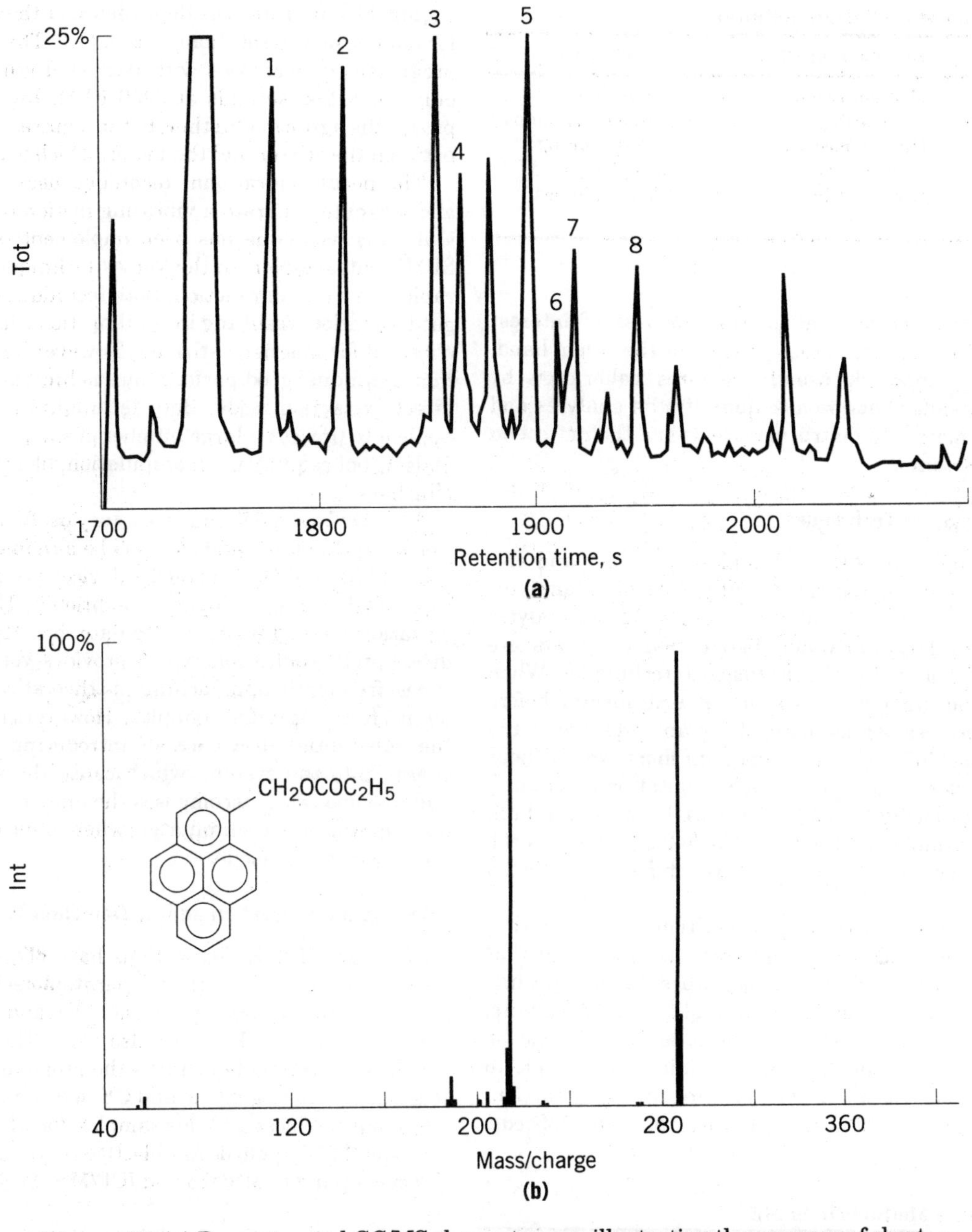

Figure 10. (**a**) Reconstructed GC/MS chromatogram illustrating the presence of short-chain fatty acids in a real sewage sample (SPME extraction with PA-coated fiber). Peak assignment: 1, acidic; 2, propionic; 3, isobutyric; 4, butyric; 5, pivalic; 6, isovaleric; 7, valeric; and 8, hexanoic acids. The peaks shown correspond to pyrenylmethyl esters of these acids. (**b**) An example of the mass fragmentogram and the structure of propionic acid/PDAM ester.

be applied. General sampling mode guidelines are presented in Table 1.

In SPME, analytes typically are not extracted quantitatively from the matrix. An equilibrium is established between the fiber coating and the sample instead. Equilibrium methods are more selective because they take full advantage of the differences in extracting-phase/matrix distribution constants to separate target analytes from interferences. Exhaustive extraction is achieved in SPME when the distribution constants are large enough. This is accomplished for most compounds by applying an internally cooled fiber (45). In exhaustive extraction, selectivity is sacrificed to obtain quantitative transfer of target analytes into the extracting phase. One advantage of this approach is that, in principle, it does not

Table 1. Sampling Mode Selection Criteria

Sampling Mode	Analyte Properties	Matrices
Direct	Medium to low volatility	Gaseous, liquid (preferably simple)
Headspace	High to medium volatility	Liquid (including complex), solid
Membrane protection	Low volatility	Complex samples

require calibration, because all of the analytes of interest are transferred to the extracting phase. On the other hand, the equilibrium approach usually requires calibration by surrogates or standard addition to quantify the analytes and compensate for matrix-to-matrix variations and their effect on distribution constants.

Selecting the Agitation Technique

Equilibration times in gaseous samples are short and frequently limited only by the rate of diffusion of the analytes in the coating. A similar situation occurs when analytes characterized by large air/water distribution constants are determined in water by the headspace technique. When the aqueous and gaseous phases are at equilibrium before the sampling process begins, most of the analytes are in the headspace. As a result, extraction times are short even without agitation. However, for aqueous samples, agitation is required in most cases to facilitate mass transport between the bulk of the aqueous sample and the fiber. Table 2 summarizes the properties of several agitation methods which have been tested with SPME.

Care must be taken with magnetic stirring to ensure that the rotational speed of the stirring bar is constant and that the base plate does not change temperature during stirring. This usually implies the use of high quality digital stirrers. Alternately, with cheaper stirrers, the base plate should be thermally insulated from the vial containing the sample to eliminate variations in sample temperature during extraction. Magnetic stirring is efficient with fast rotational speeds.

Table 2. Agitation Methods in SPME

Method	Advantages	Disadvantages
Static (no agitation)	Simple, performs well for gaseous samples	Limited to volatile analytes and headspace SPME
Magnetic stirring	Common equipment; good performance	Requires stirring bar in vial
Instrusive stirring	Very good performance	Sealing the sample vial difficult
Vortex/moving vial	Good performance; no need for stirring bar in vial	Stress on needle and fiber
Fiber movement	Good performance; no need for stirring bar in vial	Stress on needle and fiber; limited to small sample volumes
Flow through	Good agitation at rapid flows	Potential for cross-contamination; requires constant flow
Sonication	Very short extraction times	Noise; heating of sample

Figure 11 illustrates the dependence of the equilibration time for toluene on the rotational speed (12). The equilibration time progressively decreases with increased stirring speed and is only about 200 seconds at 3330 RPM. Intrusive stirring improves the agitation further, but it requires a direct connection between the stirrer and the motor, which is difficult to seal.

The needle vibrational technique uses an external motor and a cam to generate a vibrating motion of the fibre and the vial. This technique has been implemented by Varian in the SPME autosampler. In the Vortex technique the vial is moved rapidly in a circular motion. Both techniques generally provide good agitation, resulting in equilibration times similar to those obtained for magnetic stirring. However, for the needle vibration technique, good performance is limited to small vials and direct extraction mode. Both techniques can be conveniently applied to process a large number of samples since the sample vials do not require any manipulation, like introduction of stirring bars.

Flow through techniques are very useful in continuous monitoring applications and can also be automated (46). However, some additional flow metering devices are required to ensure reproducible mass-transfer conditions. The most efficient agitation method evaluated to date for SPME applications is direct probe sonication, which provides very short extraction times, frequently approaching the theoretical limits calculated for perfectly agitated samples. However, this technique has the substantial drawback of introducing large amounts of energy into the system, which cause the sample to heat up and in some cases decomposes the analyte (12). Most of these disadvantages are eliminated when sonication is applied in the flow-through mode.

Selecting the Separation and/or Detection Technique

So far, most SPME applications have been developed for gas chromatography, but other separation techniques, including HPLC, capillary electrophoresis (CE), and supercritical fluid chromatography (SFC), are also used. The complexity of the extraction mixture determines the proper quantitation device. Regular chromatographic and CE detectors are normally used for all but the most complex samples, for which mass spectrometry should be applied. As selective coatings become available, direct coupling to MS/MS and ICP/MS (47) becomes practical.

Optimizing Desorption Conditions

Standard gas chromatographic injectors, such as popular split-splitless types, are equipped with large volume inserts to accommodate the vapors of solvent introduced during liquid injections. As a result, the linear flow rates of the carrier gas in those injectors are very low in the splitless mode, and the transfer of the volatilized analytes onto the front of the GC column is very slow. No solvent is introduced during SPME injection, therefore a large insert volume is unnecessary. Opening the split line during SPME injection is not practical, because it reduces sensitivity. Efficient desorption and rapid transfer of the analytes from the injector to the column require high linear flow rates of the carrier gas around the coating. This is accomplished by reducing the internal diameter of the injector insert matching it as closely as possible to the outside diameter of the coated fiber. Narrow bore inserts for SPME are commercially available from Supelco for a range of GC instruments (48).

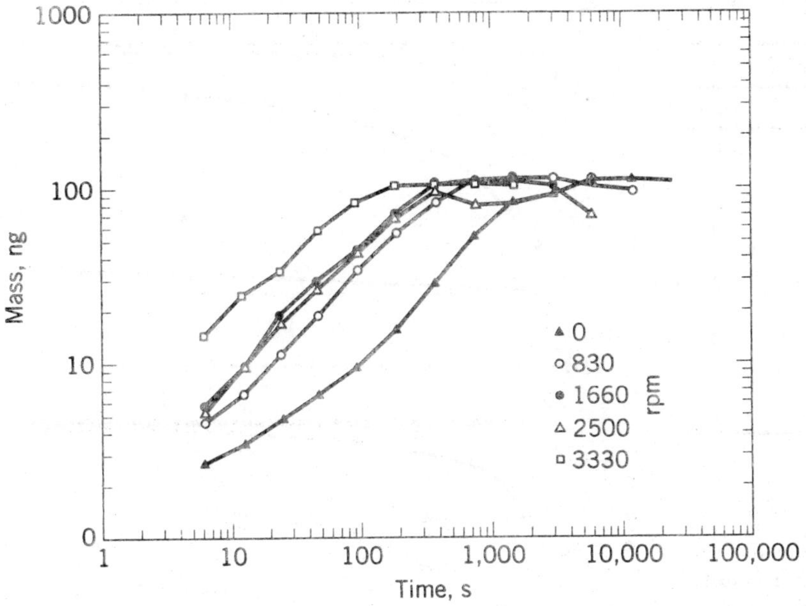

Figure 11. Extraction–time profiles obtained for direct extraction with magnetic stirring of 100 ppb toluene solution using 56-μm PDMS coating.

Other GC injectors, including programmed temperature vaporization (PTV) and septum-equipped programmable injector (SPI, Varian), are used with good results for SPME analysis. The use of on-column injectors is limited to those equipped with independent heating and capable of accommodating large diameter columns. Table 3 summarizes the characteristics of GC injectors suitable for SPME analysis.

Desorption time is determined by the temperature of the injector and the linear flow rate of the carrier gas around the fiber. Theoretical desorption times are very short because the diffusion coefficients of analytes in the coating increase and the gas/coating distribution constants rapidly decrease with increasing temperature. Table 4 illustrates the decrease in the gas/coating distribution constant for three analytes when the temperature is increased from ambient to 250°C, a typical injector temperature.

In practice, desorption temperature is determined by the thermal stability of the coating. It is advisable to use high desorption temperatures to achieve fast desorption. However, excessive temperatures adversely affect the coating's longevity and result in bleeding of the polymer, rendering separation and quantitation difficult.

To achieve good separation efficiency, it is crucial that the injection bandwidth is as small as possible. The few seconds needed to transfer the analytes onto the front of the column, when narrow bore inserts are used, might be still too long to produce narrow bands for very volatile analytes. It is necessary, therefore, to refocus the band after desorption is finished. A thick film column or cryofocusing is used for this purpose. It is always advisable to consider the application of a short length of deactivated narrow bore fused silica tubing as a retention gap. This eliminates the broadening effect created by the temperature gradient in the column close to the injector. Except for the most volatile analytes (gases), the combination of a thick film column and the retention gap is usually sufficient to refocus the injection band, and cryofocusing does not have to be applied.

Parameters which control the desorption process in the HPLC interface are analogous to those in GC applications. In addition to temperature and flow rate, the composition of the mobile phase also affects the process. In many cases, it is possible to use the mobile phase without any modifications, as in PAH analysis (49). In some instances, the addition of an appropriate solvent to the interface assists desorption. The linear flow rate of the mobile phase should be maximized by choosing a small ID tubing for the desorption chamber. Temperature is also important in accelerating the desorption process, as illustrated in Table 5. At ambient desorption temperatures, a significant carryover of Triton X-100 was observed, which was corrected when the temperature was raised to 135°C (50).

Table 3. GC Injectors and Their Compatibility with SPME

Split/Splitless	PTV	SPI
Can be used for SPME in splitless mode; low volume insert required	Can be used for SPME; low volume insert required	Low internal volume, best suitable for SPME

Table 4. Effect of Temperature on Gas/Coating Distribution Constant K_{fg}

	K_{fg}	
Compound	25°C	250°C
Benzene	300	0.4
Xylene	3,000	1.4
Undecane	26,000	13

Table 5. Effect of Desorption Chamber Temperature on Carryover of Triton X-100 Using Carbowax/DVB Fiber.

Number of Units in the Ethoxylate Chain (n =)	Carryover, %	
	21°C	135°C
4	15	0
5	20	2
6	30	2
7	25	3
8	26	1
9	26	2
10	27	2
11	27	3
12	26	2
13	28	3
14	26	0

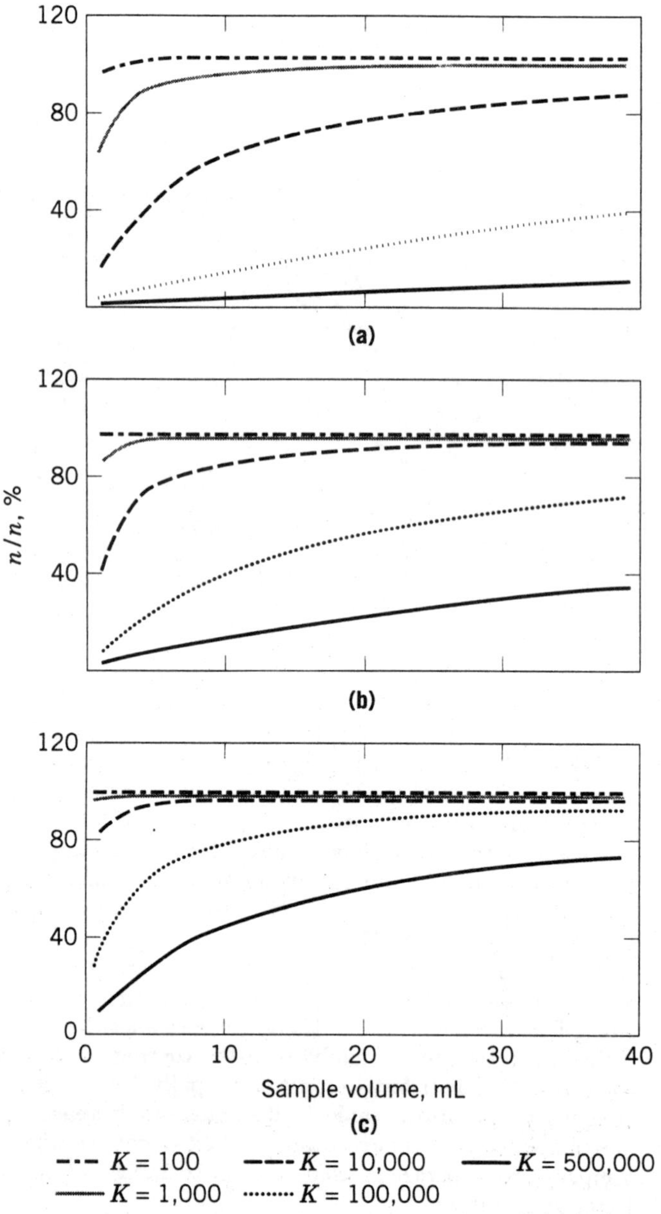

Figure 12. Dependence of n/n_I on V_s for (**a**) 100-μm fiber ($V_f = 0.65$ μL), (**b**) 30-μm fiber ($V_f = 0.14$ μL), and (**c**) 7-μm fiber ($V_f = 0.028$ μL) for direct SPME from small-volume samples (two-phase system).

Optimizing Sample Volume

The volume of the sample should be selected on the basis of the estimated distribution constant K_{fs}. The distribution constant is estimated by using literature values for the target analyte or a related compound with the coating selected. For the PDMS coating, it is also directly estimated from the chromatographic parameters (16). In all other cases, K_{fs} has to be determined experimentally by equilibrating the sample with the fiber and determining the amount of analyte extracted by the coating. Care must be taken to avoid analyte losses via adsorption, evaporation, microbial degradation, etc, when very long extraction times are required to reach equilibrium.

The sensitivity of the SPME method is proportional to n number of moles of analyte extracted from the sample which for the direct extraction method, is given by the following equation:

$$n = \frac{K_{fs}V_fV_sC_0}{V_s + K_{fs}V_f} \quad (33)$$

As the sample volume V_s increases, so does the amount of analyte extracted until the volume of the sample becomes significantly larger than the product of the distribution constant and volume of the coating (fiber capacity $K_{fs} \ll V_s$). At this point, the sensitivity of the method does not increase with further increases in volume. In practice, the limiting volume is calculated from the following relationship:

$$V_s = \frac{100K_{fs}V_f}{E} \quad (34)$$

where E is the magnitude of the relative measurement error (%). For example, for a 5% error, it is estimated that $V_s = 20K_{fs}V_f$. This means that for K_{fs} values up to about 200 and a 100-μm coating, a 2 mL vial is sufficient for maximum sensitivity, whereas a 40-mL vial should be used for K_{fs} values smaller than 4000, etc. Using sample volumes larger than the limiting value maximizes the sensitivity and also is more precise because variations in the sample volume do not affect the results. Figure 12 illustrates this relationship graphically. n_I corresponds to the amount of analyte extracted from an infinite sample volume. Even for a moderate distribution constant K of 1,000, characteristic of many volatiles, for the 100-μm PDMS coating (see Fig. 12**a**), there is a substantial difference in the amount extracted when the sample volume varies within a few mL range. Increasing the volume from 1 mL to 10 mL improves sensitivity by over 50%. The change is more dramatic for semivolatile analytes. The variation is less significant for thinner coatings (Fig. 12**b** and 12**c**), but the sample volume should still be considered when analytes with large distribution constants, such as PAHs, are analyzed. In practice, the sample size is determined by the vials available. The discussion above should assist in the choosing the most appropriate vial size.

In headspace SPME, the analytes partition to the gaseous phase and to the coating. Highly volatile compounds preferentially accumulate in the headspace resulting in a very substantial loss in sensitivity when the headspace volume is large. Figure 13 shows the dependence of the amount of analyte extracted by a 100-μm fiber on the headspace-to-

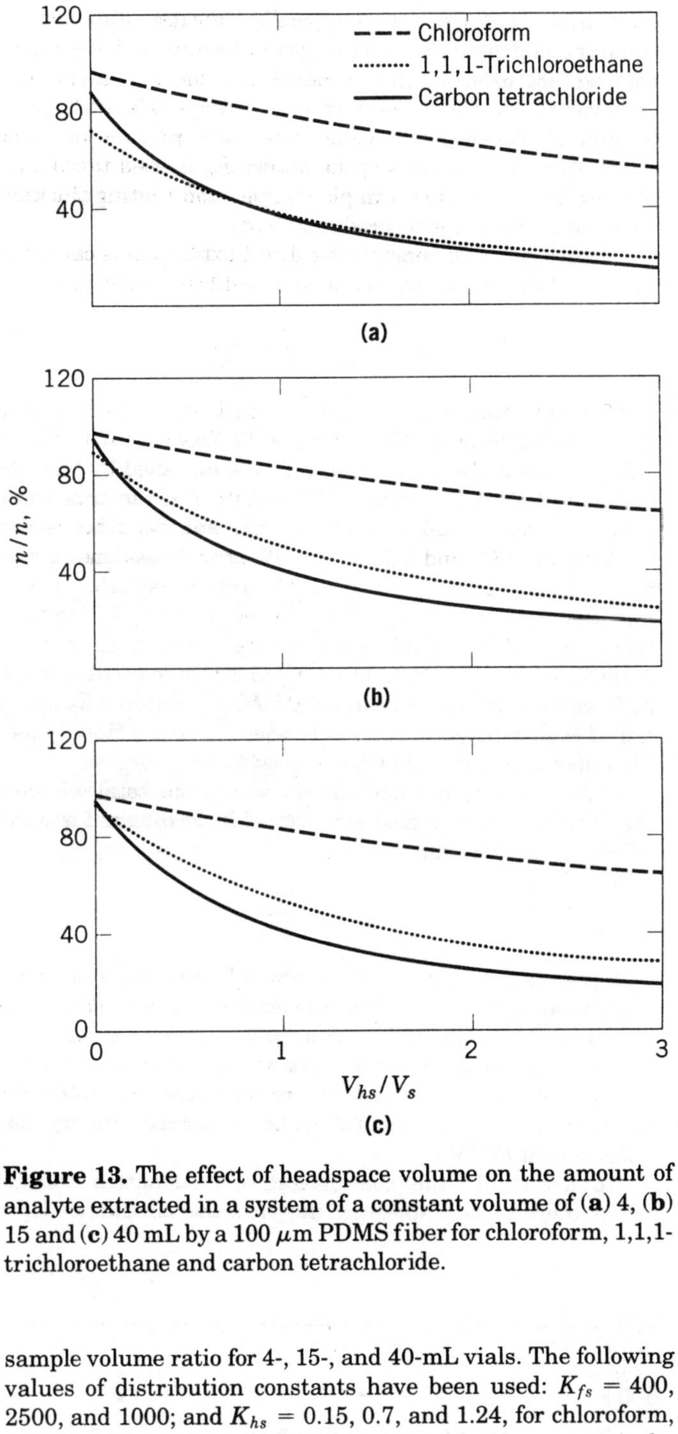

Figure 13. The effect of headspace volume on the amount of analyte extracted in a system of a constant volume of (**a**) 4, (**b**) 15 and (**c**) 40 mL by a 100 μm PDMS fiber for chloroform, 1,1,1-trichloroethane and carbon tetrachloride.

sample volume ratio for 4-, 15-, and 40-mL vials. The following values of distribution constants have been used: $K_{fs} = 400$, 2500, and 1000; and $K_{hs} = 0.15$, 0.7, and 1.24, for chloroform, 1,1,1,-trichloroethane and carbon tetrachloride, respectively. In this figure, n_I is the amount of analyte extracted from an infinite sample volume. The n/n_I ratio is always the largest for chloroform, which has the lowest K_{hs} value, even though its K_{fs} value is also the lowest because only a small fraction of the analyte is in the sample headspace, and therefore the equilibrium concentration of the analyte in the sample remains relatively high. For the two other curves and low headspace volumes, n/n_I is higher for carbon tetrachloride than for 1,1,1-trichloroethane, consistent with the values of their distribution constants K_{fs}. However, because carbon tetrachloride has the largest K_{hs}, n/n_I drops faster for this compound than for 1,1,1-trichloroethane, as a result of which the respective two lines cross each other. The biggest relative increase in n/n_I when moving from small to large vials (~44%) is observed for 1,1,1-trichloroethane, characterized by the highest K_{fs} value. For compounds with lower K_{fs} values, the relative increase in the amount extracted is much lower (~16 and ~12% for chloroform and carbon tetrachloride, respectively). It is therefore the combination of K_{fs} and K_{hs} for a given compound that determines the magnitude of the sample volume effect on the amount of analyte extracted by the fiber in a three-phase system involving headspace (51).

Headspace volume significantly affects equilibration time (extraction kinetics). If the amount of the analyte extracted by the fiber at equilibrium is negligibly small compared to the amount in the headspace equilibrated with the sample before extraction, the analyte is extracted almost exclusively from the gas phase, and the process is much faster compared to that when significant amounts of the analyte are transported from the liquid sample to the fiber via the headspace. Assuming that this occurs when 95% of the analyte is extracted exclusively from the headspace, the criterion that must be fulfilled is described by equation 35. The assumption is reasonable, because a 5% difference usually falls within the limits of experimental error for trace SPME/GC analysis:

$$\frac{K_{fs}V_f}{K_{hs}V_h} \leq 0.05 \qquad (35)$$

This criterion means that the headspace capacity ($K_{hs}V_{hs}$) must be at least 20 times larger than the capacity of the fiber (KV_f) for rapid extraction. For a given sample volume V_s, this is achieved by using a large enough headspace volume or by increasing K_{hs}. Increasing K_{hs} is accomplished by increasing the temperature or by salting the analyte out of the liquid phase. When criterion for equation 35 is fulfilled, equilibration takes as little as a few minutes and is almost independent of agitation (provided that the analyte is equilibrated between the liquid phase and its headspace before the extraction begins). The increase in headspace capacity, however, causes a significant loss in sensitivity. In any case, changing the vial size during the optimization process affects the sensitivity and also the equilibration time.

Determining the Extraction Time Profile in a Pure Matrix

Extraction time profiles for the analyte are determined first in a pure matrix, such as dry air, pure water, or sand, to basically understand the kinetics of analyte transfer from the matrix to the fiber under given mass-transfer conditions. The profile is determined by extracting samples of identical composition for progressively longer periods of time and plotting the resultant area counts vs time. The mass of the analyte extracted is quantified by proper calibration, eg, syringe injection of a standard solution of the analyte in an appropriate solvent. In gas chromatography, this type of calibration is reliable only when nonvaporizing injectors are used (PTV, SPI, heated on-column).

Determining the Extraction Time

An optimal approach to SPME analysis is to allow the analyte to reach equilibrium between the sample and the fiber coating. The equilibration time is defined as the time after which the amount of analyte extracted remains constant and, within the limits of experimental error, corresponds to the amount extracted after infinite time. Care should be taken when determining equilibration time, because in some cases a substantial reduction of the slope of the curve might be wrongly taken as the equilibrium point. Such a phenomenon often occurs in headspace SPME determinations of aqueous samples, where a rapid rise of the equilibration curve corresponding to extraction from the gas phase only is followed by a very slow increase related to analyte transfer from water through the headspace to the fiber. Figure 14 is an example of such a curve for phenanthrene. The amount extracted increases rapidly in the first five minutes, but the equilibrium is not reached until much later. Determining of the amount of analyte extracted at equilibrium allows calculating the distribution constants.

When an equilibration time is excessively long, the extraction time is shortened. However, in such cases the extraction time must be strictly controlled for good precision. Figure 15 presents equilibration time profiles for *p,p'*-DDT and Dichlorvos. The equilibration time for DDT is about two hours. At equilibrium, small variations in the extraction time do not affect the amount of the analyte extracted by the fiber. On the other hand, at the steep part of the curve, even small variations in the extraction time result in significant variations of the amount extracted. The shorter the extraction time the larger is the relative error. For a 10-minute extraction (Fig. 15**a**), the relative error caused by varying the extraction time by ±1 min is as high as 20%. It drops to 5% after 50 min of extraction because more analyte accumulates on the fiber. The timing is even more critical for short exposures when the equilibration curve rises rapidly. Figure 15**b** is an example where a 1-min deviation from a 10-min sampling time results in a 50% relative error of determination.

Autosamplers measure time very precisely, and the precision of analyte determination is very good, even when equilibrium is not reached in the system. However, this requires that the mass transfer conditions and the temperature remain constant during all experiments.

Calculating the Distribution Constant

The target analyte's distribution constant defines the sensitivity of the method. It is not necessary to calculate the fiber coating/sample matrix distribution constant K_{fs}, when the calibration is based on isotopically labelled standards or standard addition or when identical matrix and headspace volumes are used for the standard and for the sample with external calibration. However, it is always advisable to determine K_{fs} because this value gives more information about the experiment and aids optimization. K_{fs} is used to calculate the headspace volume, sample volume, and coating thickness required to reach the desired sensitivity.

The distribution constant for direct extraction is calculated from the following relationship obtained from equation 33:

$$K_{fs} = \frac{nV_s}{V_f(C_0V_s - n)} \tag{36}$$

It is important to consider the sample volume when calculating the value of K_{fs}. Table 6 presents the ranges of K values obtained when the amount of analyte n extracted by the fiber experimentally falls within ±5% (relative) of the true value, for two sample volumes (2 and 35 mL) and two fiber coating thicknesses (100 and 7 μm). A ±5% error is assumed typical for trace analysis by SPME/GC. To correctly estimate the distribution constants over 1,000 with 100-μm thick PDMS fiber, the volume of the sample must be larger than 2 mL. For very large K_{fs} values (>1, 000, 000), typical of high molecular weight hydrocarbons in water or air, no reliable estimates of K_{fs} are obtained even with a 35-mL sample when a 100-μm fiber is used. Therefore only the 7-μm fiber is used for this purpose.

If the distribution constant is low and the total volume of the sample is high, a good estimate of K_{fs} is obtained from the following relationship:

$$K_{fs} = \frac{n}{C_0V_f} \tag{37}$$

However, extreme caution is advised when using this equation, because it assumes that the analyte concentration in the sample does not change significantly after extraction. If this equation is used by mistake when the K_{fs} value is large or the sample volume is too small, the result of the calculation does not correspond to K_{fs}, but rather to the sample/coating phase volume ratio (V_s/V_f).

To determine the coating/matrix distribution constant when there is headspace in the vial, the matrix/headspace

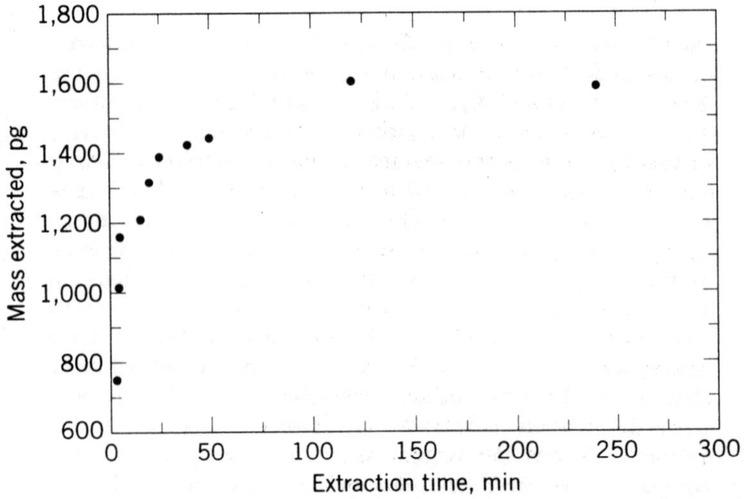

Figure 14. Extraction–time profile corresponding to extraction of phenanthrene-d10 from water at 50°C with magnetic stirring. Fiber coating: 7 μm PDMS.

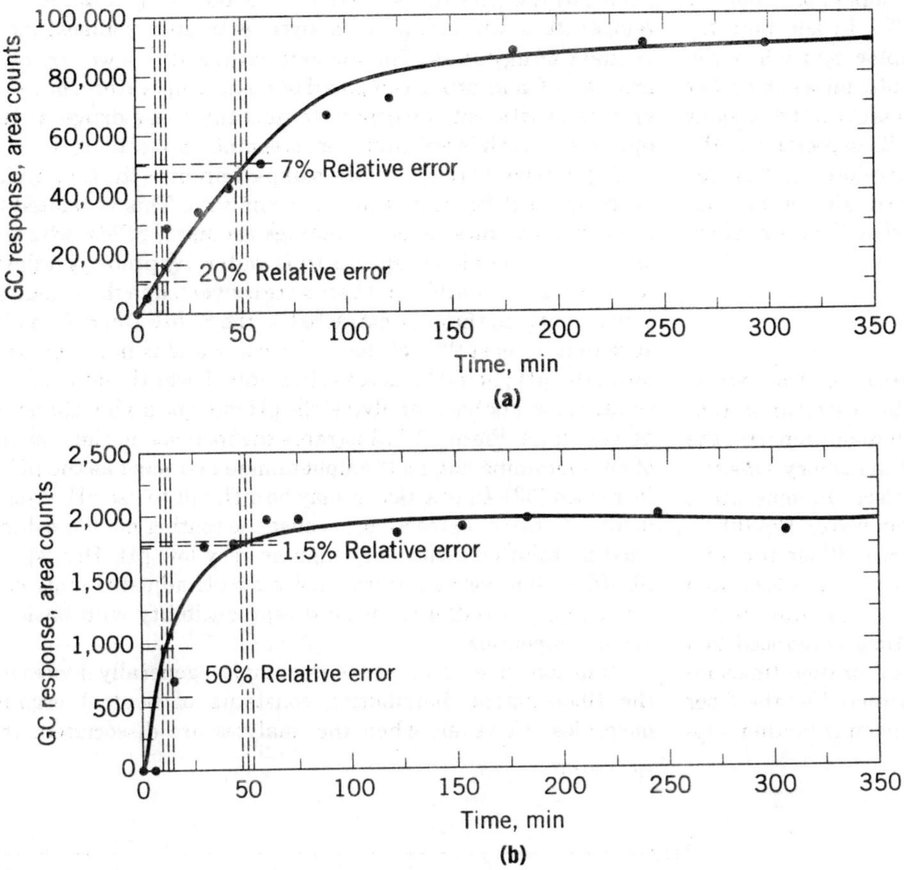

Figure 15. Selecting the extraction time based on extraction–time profiles.

Table 6. Ranges of K Values Obtained When the Amount of the Analyte n Extracted by the Fiber[a] Falls Within ±5% (relative) of the True Value, for Two Different Fiber Coating Thickness and Two Sample Volumes

2-mL sample		35-mL sample	
K	Range of K	K	Range of K
	100 μm		
100	105–95	100	105–95
1,000	1067–935	1,000	1051–949
10,000	12537–8172	10,000	10598–9413
100,000	∞–36190	100,000	115748–86928
1,000,000	∞–55072	1,000,000	4700000–492593
10,000,000	∞–58104	10,000,000	∞–923611
100,000,000	∞–58426	100,000,000	∞–1012177
	7 μm		
100	105–95	100	105–95
1,000	1051–949	1,000	1050–950
10,000	10574––9434	10,000	10504–9496
100,000	112903–88785	100,000	105422–94622
1,000,000	3500000–558824	1,000,000	1093750–913462
10,000,000	∞–1187500	10,000,000	17500000–6785714
100,000,000	∞–1338028	100,000,000	∞–19000000

[a] Determined experimentally.

distribution constant is required:

$$K_{fs} = \frac{n(K_{hs}V_h + V_s)}{V_f(C_0V_s - n)} \quad (38)$$

K_{fs} is obtained for aqueous samples from tables of Henry's constants (K_H):

$$K_{hs} = \frac{K_H}{RT} \quad (39)$$

For analytes with unknown Henry's constants, the following equation is used:

$$K_{fs} = \frac{nK_{fh}V_s}{K_{fh}V_fV_sC_0 - nK_{fh}V_f - nV_h} \quad (40)$$

Equation 40 is obtained by substituting $K_{hs} = K_{fs}/K_{fh}$ in equation 38 and rearranging it. The fiber/gas distribution constant is found by extracting target compounds from air mixtures by a simple bulb experiment (37). In addition, K_{fg} is obtained directly from a chromatographic run when the stationary phase is made of the same material as the fiber coating. For compounds with low vapor pressures, the amount of analyte in the headspace is very small, especially if the volume of the headspace is minimal. Therefore, in this situation, the headspace volume is neglected altogether, and equation 36, which corresponds to direct extraction, is applied.

Optimizing Extraction Conditions

An increase in extraction temperature increases the extraction rate, but simultaneously decreases the distribution constant. In general, if the extraction rate is of major concern, the highest temperature which still provides satisfactory sensitivity is used. Figure 16 demonstrates the effect of temperature on the extraction of methamphetamine from water. Equilibration takes several hours at room temperature. When the temperature is raised to 60°C, the equilibration time is less than 20 minutes for both compounds. When the temperature is further increased to 73°C, the equilibration time is reduced to a few minutes (Fig. 16**a**) (52). Very short equilibration times indicate that the majority of the analyte extracted by the fiber originates from the headspace. This conclusion is further supported by the stirring data obtained for 60°C (Fig. 16**b**). At this temperature, the amount of analytes extracted is almost independent of agitation. The amount extracted is lower by over an order of magnitude compared to room temperature, but still provides sufficient sensitivity to quantify these drugs at the ppb level, which is adequate for screening applications.

Adjustment of the pH of the sample improves the sensitivity of the method for basic and acidic analytes. This is related to the fact that unless special coatings are used, SPME extracts only neutral (nonionic) species from water. By properly adjusting the pH, weak acids and bases are converted to their neutral forms, in which they are extracted by the SPME fiber. To make sure that at least 99% of the acidic compound is in the neutral form, the pH should be at least two units lower than the pK_a of the analyte. For basic analytes, the pH must be higher than pK_b by two units. Figure 17 illustrates the increase in the amount of amphetamine and methamphetamine extracted as the pH is increased (53). In practice, it may be difficult to use pH adjustment with direct extraction, because the coating might be damaged by solutions with very high or very low pH. Headspace SPME is therefore a natural choice for pH-adjusted matrices. The sample is buffered for good reproducibility with basic or acidic compounds.

Addition of salt to aqueous samples generally increases the fiber/matrix distribution constants of neutral organic molecules. However, when the analytes are dissociated, the

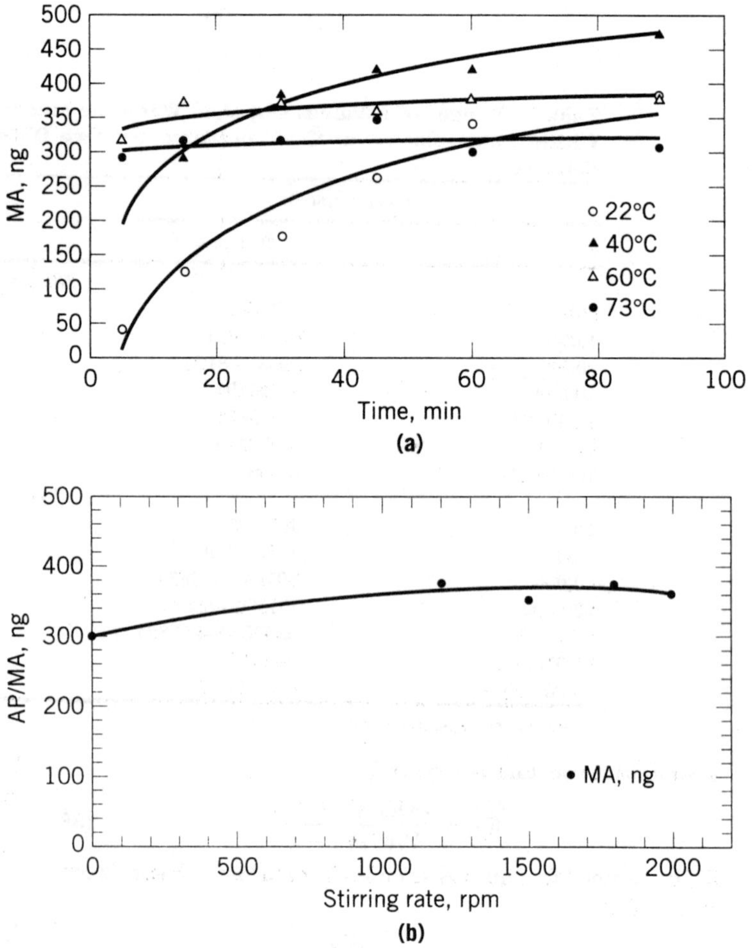

Figure 16. (**a**) Temperature dependence of the absorption–time profile obtained for methamphetamine: 22°C, 40°C, 60°C, and 73°C. (**b**) Amount of sample adsorbed as a function of the stirring rate at 60°C.

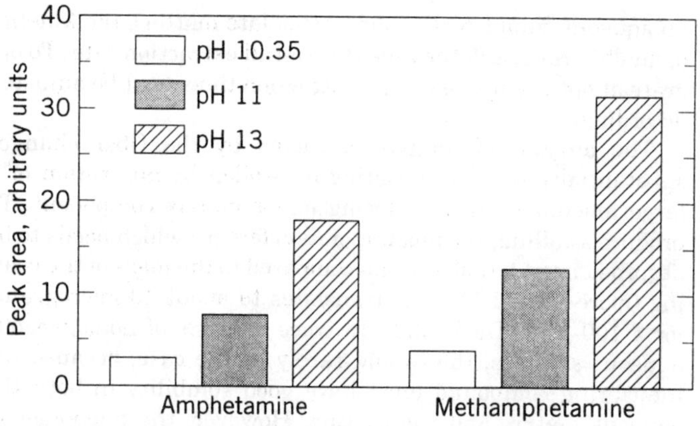

Figure 17. Effect of pH on amphetamine and methamphetamine extraction.

amount extracted decreases. This is related to the fact that the activity coefficients of ionic species in water decrease with an increase in the ionic strength of the solution. It is important, therefore, to convert the analytes to neutral forms first. Figure 18 shows the increase in the amount of amphetamines extracted from a solution when salt is added after the pH is lowered by KOH addition.

Determining the Linear Dynamic Range of the Method for a Pure Matrix at Optimum Extraction Conditions

Modifying of extraction conditions affects both the sensitivity and the equilibration time. It is advisable, therefore, to check the previously determined extraction time before determining the linear dynamic range. This step is required if the sensitivity changes substantially during optimization.

SPME coatings include polymeric liquids, such as PDMS, which, by definition, have a very broad linear range. For solid sorbents, such as Carbowax/DVB or PDMS/DVB, the linear range is narrower because of a limited number of sorption sites on the surface, but the range is still over several orders of magnitude for typical analytes in pure matrices. In some rare cases when the analyte has extremely high affinity to the surface (eg, basic protein adsorption on poly(acrylic acid) (54) or bioaffinity coatings), saturation occurs at low analyte concentrations. In such cases, the linear range is expanded by shortening the extraction time. In the majority of practical applications developed to date, the dynamic range of the detector limits the linear response of SPME methods.

Selecting the Calibration Method

Standard calibration procedures are used with SPME. A fiber blank is first checked to ensure that neither the fiber nor the instrument interfere with the determination. The fiber is conditioned before the first use by desorption in a GC injector or a specially designed conditioning device to ensure that the fiber coating itself does not introduce interferences. Fiber conditioning must be repeated after analysis of samples containing significant amounts of high molecular weight compounds, because such compounds require longer desorption times than the analytes of interest.

When the matrix is simple (eg, air or groundwater), the distribution constants are very similar to those for a pure matrix. Typical moisture levels in ambient air and the presence of salt and/or alcohol in water in concentrations lower than 1%, usually do not change the K values beyond the 5% RSD typical for SPME determinations. In many such instances, calibration is not necessary because the appropriate distribution constants, which define the external calibration curve, are available in the literature or are calculated from chromatographic retention parameters. External calibration is also successful when the matrix is more complex, but well defined (eg, process streams of relatively constant composition). Of course calibration standards must be prepared in such cases in a matrix of the same composition rather than in the pure medium (eg, water).

Special calibration procedures, such as isotopic dilution or standard addition, are used for more complex samples. In these methods, it is assumed that the target analytes behave similarly to spikes during extraction. This is usually valid when analyzing homogeneous samples. However, it might not be true when heterogeneous samples are analyzed, unless the native analytes are completely released from the matrix under the conditions applied. Moreover, whenever any of these methods is used, it is inherently assumed that the response is linear

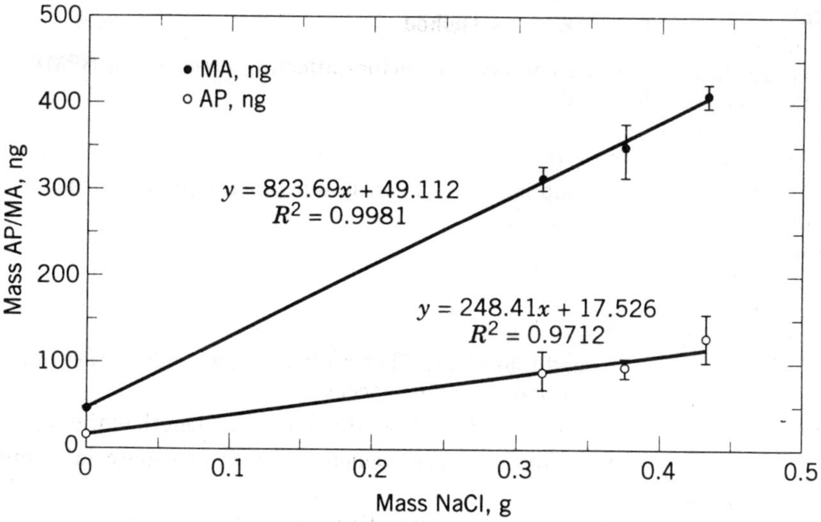

Figure 18. Effect of salt concentration on amphetamine and methamphetamine extraction (0.5 N KOH, 22°C).

in the concentration range between the original analyte concentration and the spiked concentration. Although this is usually true for fibers extracting analytes by absorption (PDMS, PA) and detectors with wide linear range (FID), problems arise when porous polymer fibers (PDMS/DVB, Carbowax/DVB) are used or when the detector has a narrow linear dynamic range. It is important, therefore, to check the linearity of the response with standard solutions before applying standard addition or isotopic dilution for calibration. To improve the accuracy and precision, multipoint standard addition is used whenever practical.

Optimizing Extraction Conditions for Heterogeneous Samples

Optimizing the process that releases native analytes from matrix components (eg, solid particles) is more difficult because the analyte/matrix association is usually poorly understood. Whenever a new type of complex matrix is considered, a study must be conducted to find the optimum extraction conditions for the fastest and most complete release and extraction of native analytes. Typically, an empirical approach is taken and several parameters are varied, such as temperature and type of additives. Change of extracting phase type, used in optimizing many solvent extraction methods, is of limited value in SPME because the fiber coating does not interact directly with solids in the matrix. Heating releases analytes from a solid matrix, but sometimes heating is inadequate because analytes are bound too strongly or the matrix or analytes are unstable at higher temperatures. Additives (matrix modifiers) which enhance the extraction of analytes are an alternative to heating. The additive selected should enhance the release of analytes from the matrix but not interfere with the partitioning of analytes into the fiber coating or with the analysis of the extracted compounds. Successful modifiers for SPME include nonvolatile acids, salts, and water (10,45). Better understanding of the analyte–matrix interaction allows a more rational choice of appropriate additives. In-depth investigation of sample matrices should be considered.

Frequently, conditions optimal for releasing the analytes from the matrix differ from conditions optimal for partitioning them into the coating. In most cases, high temperature removes native analytes from a matrix. Higher temperatures, however, reduce the distribution constant and thus the sensitivity of the method. To eliminate this disadvantage, a cooled fiber is used to condense analytes onto the fiber (45).

Optimizing extraction parameters is considered complete when the recoveries of analytes in native samples match those for spikes. This verification is based on extraction results obtained for certified standard reference materials or by comparison with standard extraction methods. Once the method is verified, quantitation compares extracted amounts of native and spiked analytes. Table 7 compares the results of PAH analysis by SPME with certified reference values obtained for Urban Air Particulate (NIST 1649) (55). The method consists of static hot water extraction at 250°C, followed by a cooldown to increase the amount of analytes absorbed by the PDMS coating.

Verifying the Equilibration Time, Sensitivity, and Linear Dynamic Range for Real Sample Matrices

The sample matrix affects the distribution constant and also the equilibration time. When the matrix is heterogeneous (eg, an aqueous sample with solid particulate matter), the kinetics of analyte release determine the overall extraction rate. To operate at optimum sensitivity, extraction time must be adjusted accordingly.

The amount of analyte extracted by the fiber changes substantially when the coating is swollen by absorption of a large amount of an interfering major matrix component. To produce swelling, the amount of interference which needs to be absorbed must be substantial compared to the mass of the coating (more than 1%). This translates to about 10 micrograms for a 100-μm PDMS fiber. In trace analysis of homogeneous aqueous samples, this would rarely be the case, because the interfering compound must have good solubility in both the aqueous matrix and the coating. However, the phenomenon occurs if the interference is a separate phase, such as an oily suspension or dispersed hydrophobic humic material (14). In such cases, SPME is of limited value as a quantitation tool, because the distribution constant and the volume of the coating change during extraction. The effect of high molecular weight compounds on the results of SPME analysis of aqueous samples is eliminated by using microporous hollow fibre membranes to isolate the coating from direct contact with the sample.

Besides extraction time, the linear dynamic range and method sensitivity should also be examined for real matrices. Typically, no differences between model solutions and real samples occur when fibers extracting the analytes by absorption (PDMS, PA) are used. When porous polymer fibers are used, however, competitive adsorption of interfering matrix components with large distribution ratios reduces the uptake of analytes of interest and causes nonlinearity of calibration curves. The adverse effect of competitive sorption is minimized by extracting the analytes under nonequilibrium conditions. For example, short extraction is performed with a well-agitated sample, or static extraction is used. In both cases, the uptake of volatile analytes with lower distribution ratios is only slightly affected, whereas the uptake of less volatile compounds with large distribution ratios is significantly reduced (40).

The sample matrix affects extraction and also separation. It should be verified that no matrix components coelute with the analytes of interest, thereby interfering with proper quantitation.

Precision of the Method

The most important factors affecting precision in SPME are listed below:

- agitation
- sampling time (if nonequilibrium conditions are used)
- temperature
- sample volume
- headspace volume
- vial shape
- condition of the fiber coating (cracks, adsorption of high molecular weights species)
- geometry of the fiber (thickness and length of the coating)
- sample matrix components (salt, organic material, humidity, etc)
- time between extraction and analysis

Table 7. Quantitation of PAHs from Urban Air Particulate (NIST 1649) by Static High Temperature Water Extraction

		Estimated Concentration as % of Certified Concentration (%RSD)		
		100 mg	50 mg	
Compound	Cert. Conc. µg/g, %RSD	270°C	270°C	300°C
Fluoranthene	7.1 (7)	137 (6)	149 (8)	128 (2)
Pyrene	7.2 (7)	113 (9)	121 (7)	105 (7)
Benzo(*a*)pyrene	2.9 (17)	49 (14)	72 (10)	70 (4)

- analyte losses (adsorption on the walls, permeation through Teflon, absorption by septa)
- geometry of the injector
- fiber positioning during injection
- condition of the injector (pieces of septa)
- stability of the detector response
- moisture in the needle

The majority of these factors and their effect on precision have already been discussed. Following is a brief explanation of the remaining parameters.

Reproducibility of the sample and headspace volumes affects the precision of SPME methods. The volume of the sample has an impact on the amount extracted when only a few-mL sample is used (see the discussion above). Therefore, small samples, such as 1 mL, must be measured very carefully. For large samples, small variations in the sample volume do not substantially affect the precision of the data. Obviously, the volume of all of the samples and standards must be the same for valid results.

Headspace volume is important in determining the precision of results in three-phase systems. It is relatively easy to measure sample volume accurately. However, vials are not manufactured with exactly the same total volume. Wall thicknesses and bottom shapes differ from vial to vial. Also, the shape of the septum in a closed vial varies from concave to convex. All of these factors affect the total volume of the system and therefore the headspace volume, which is the difference between the total volume and the sample volume. Such differences (especially related to septum shape) are usually more pronounced in small vials. Therefore worse precision can usually be expected when they are used. In addition to keeping the sample and standard volumes constant throughout all the experiments, vials of the same volume should be used for all samples and standards, so that the headspace volume is constant from vial to vial. This becomes a bigger challenge for small vials, when a small absolute variation in headspace volume translates to a large relative difference.

The condition of the fiber is very important. High molecular weight species, such as proteins and humic matter, adsorbed on the surface of the fiber, change its sorptive characteristics. In some cases, the substances adsorbed are removed by washing the fiber in appropriate solvents. When the coating is badly cracked, a small amount of the matrix itself held in the cracks by capillary forces is transferred into the analytical instrument and adversely affects its operation.

The volume of the coating determines the amount of analytes extracted. Fibers which begin to lose the coating phase should be replaced immediately. The length of the fiber should be monitored to ensure equivalency when changing the fiber. If the length of the fiber (thus the volume of the coating) differs, an appropriate correction factor is incorporated.

Time between preparation of standards or sample collection and extraction should be minimized. After placing aqueous samples or standards in a vial, analytes begin to interact with the walls of the container. Very hydrophobic substances are frequently strongly adsorbed on the surfaces and are not available for extraction by the coating. This results in lower than expected extracted amounts. One way to prevent this problem is to use vials with more inert walls, for example, silanized glass or Teflon. For volatile analytes, it is crucial that new Teflon-lined septa are always used. Once the Teflon lining is pierced, the analytes directly contact the septum material (typically polydimethylsiloxane), which usually results in losses through absorption. Finally, it should be remembered that many volatile compounds permeate through Teflon at relatively high rates. This phenomenon also contributes to analyte losses when the samples or standards are stored for prolonged periods.

Time between extraction and analysis should also be minimized. When this time is longer than a few seconds, the needle of the fiber assembly should be sealed with a septum and preferentially cooled to reduce losses and interferences. After the fiber is withdrawn from the sample, it begins to equilibrate with ambient air. This process is slow when the fiber is withdrawn into the needle. Therefore, the few seconds which are normally required to transfer the fiber to the injection port of a gas chromatograph do not result in any substantial losses of the analytes. However, when the fiber is to be stored for some time before the analysis, the needle must be sealed. In addition, this protection prevents contaminating the fiber with interferences or even target analytes in ambient air. Cooling the SPME device might be necessary to extend the stability of the volatile analytes in the coating (45).

Determining the Method Detection Limit

There are several methods described in the literature to determine the method detection limit. The most widely accepted definition is based on estimating the detection limit with low concentration spikes and calculating the standard deviation of the determination. The detection limit then is defined as three times the standard deviation for the measurement value not higher than 10 times the method detection limit.

Validating the Method

Validating the method includes comparing quantitative results with certified values obtained for standard reference materials which have similar matrix and target analytes. Another approach is to validate the method against officially accepted techniques for analyzing target samples and analytes. Table 8 summarizes the results of multilevel validation of SPME against the standard purge and trap technique for analyzing BTEX compounds in water. The regression line has a slope close to 1, with a very small intercept value and linear correlation coefficients better than 0.99 for all species. The very good agreement between both methods indicates that headspace SPME is suitable for analyzing volatile organic compounds in aqueous samples (56).

Finally, interlaboratory studies are frequently performed to validate a method. Table 9 summarizes the results from a round-robin test on pesticide analysis by SPME, which involved several laboratories in Europe and North America (57). In general, the results show good repeatability, which proves that SPME is valid for determining pesticides at trace levels. As expected, the interlaboratory and reproducibility standard deviations are higher because they include differences among laboratories. However, they are still satisfactory. The results also indicate that SPME is accurate. In all cases, the confidence intervals of the gross average and the "true" value overlap, which indicates that any differences between the two respective values are caused by random factors. Interestingly, for 10 out of 12 compounds, the values of the gross average are slightly lower than the "true" values. This might be caused in part by losses of analytes through adsorption (as described above) when the aqueous pesticide solutions were not prepared directly before analysis, as required by the test protocol.

Automating the Method

SPME is a very powerful investigative tool, but it can also be a technique chosen in many applications for processing large numbers of samples. To accomplish this task requires automation of the methods developed. As automated SPME devices with more advanced features and capabilities become available, automation of the methods developed becomes easier. A currently available SPME autosampler from Varian enables direct sampling with agitation of the sample by fiber vibration and static headspace sampling. In some cases, custom modifications to the commercially available systems facilitate operation of the method closer to optimum conditions.

APPLICATIONS OF SPME

A majority of the methods developed so far is applicable to clean matrix samples and only a few optimization parameters have been fully investigated. Many other potential applications are still waiting to be explored. The results obtained are very promising. A basic understanding is emerging regarding approaches which can be taken to the basic matrix types: gas, liquid and solid. The progress in this direction is summarized in the first three parts of this section. General approaches for successful extraction with these systems are emphasized. The objective of this section is not to demonstrate a comprehensive review, but to give examples of successful SPME applications in different areas.

Table 8. Regression Line Parameters for SPME vs Purge and Trap for Clean Water Analysis

	Regression Parameters: Purge and Trap on x-axis		
Compound	Slope	y-Intercept	Correlation
Benzene	1.07 ± 0.087	−5.9 ± 6.6	0.9951
Toluene	1.06 ± 0.060	−4.6 ± 4.6	0.9976
Ethylbenzene	1.06 ± 0.046	−3.8 ± 3.7	0.9986
m/p-Xylene	1.05 ± 0.073	−8.4 ± 11	0.9964
o-Xylene	1.07 ± 0.060	−4.0 ± 4.6	0.9977

Gaseous Matrices

All SPME methods developed to date for gas analysis are directed toward air analysis. However, the approaches described are also suitable for analyzing other gas mixtures.

Model studies on extraction from air have been performed by two methods: static and (the more realistic) dynamic. In the static method, the target compounds are introduced into a glass bulb equipped with a septum. After the analytes vaporize completely, an SPME fiber is introduced into the bulb through a septum (37). In practical ambient air measurements, the system is not static, because convection is always present. It is more appropriate, therefore, to use dynamic gas chambers for the modeling studies (22). Equilibration times for extracting trace contaminants from moving air are short, as predicted by the theory. Figure 19 shows the equilibration time profiles obtained for a 20-L chamber operated at 200 mL/min. The equilibration times are between 30 and 100 s, corresponding to pentane and undecane, respectively. They are close to theoretical values for perfect agitation. Increased air flow decreases the equilibration time for less volatile analytes.

The PDMS coating is generally unaffected by major air components. A small decrease in the response for organics was observed only for samples with relative humidity close to 100%. The primary experimental parameter which controls the response is temperature. However, the effect of temperature on the distribution constant is easily predicted, because $\log K_{fg}$ is linearly related to $1/T$ and the heat of vaporization of the pure solute ΔH^V (39):

$$\log_{10} K_{fg} = \frac{\Delta H^V}{2.303RT} + \left[\log_{10}\left(\frac{RT}{\gamma_i p^*}\right) - \frac{\Delta H^V}{2.303RT^*}\right] \quad (41)$$

where p^* is the analyte vapor pressure at a known temperature T^* for a pure solute and γ_i is the activity coefficient of the solute in the coating. In other words, K_{fg} can be calculated for given extraction conditions using the temperature measured and the heat of vaporization for the target compound. Figure 20 illustrates the above relationship graphically for a range of compounds varying in volatility. The linear relationship is clearly illustrated. In addition, as described in the section devoted to theory, the heat of vaporization and activity coefficient are related to the retention time of a compound on the GC column using the same coating material as the SPME fiber. Thus, the appropriate distribution constants are determined directly from the retention indices of the target analytes. For a PDMS coating, the following relationship between the gas/coating distribution constant and the linear temperature programmed retention index of a compound (LTPRI) has been found:

$$\log_{10} K_{fg} = 0.00415 \times \text{LPTRI} - 0.188 \quad (42)$$

The differences between the values of distribution constants calculated from the retention indices and determined directly were generally insignificant (16). This approach can be extended to other coatings when columns with appropriate

Table 9. Statistical Characteristics of the Results from the Round-Robin Test on Pesticide Analysis by SPME[a]

Compound	s_r[b]	s_L[c]	s_R[d]	r[e]	R[f]	G.A.[g]	C.I.[h]	T.V.[i]
Dichlorvos	2.06	5.04	5.44	5.83	15.4	27.3	27 ± 5.8	25 ± 1.35
EPTC	0.56	1.56	1.66	1.57	4.7	9.9	10 ± 1.6	10 ± 0.54
Ethoprofos	0.82	4.79	4.86	2.32	13.74	15.5	16 ± 2.3	17 ± 0.92
Trifluralin	0.27	0.57	0.63	0.76	1.79	1.6	1.6 ± 0.76	2 ± 0.11
Simazine	2.34	3.45	4.17	6.61	11.79	23.6	24 ± 6.6	25 ± 1.35
Propazine	1.21	2.04	2.37	3.42	6.71	9.5	10 ± 3.4	10 ± 0.54
Diazinon	0.63	2.13	2.22	1.79	6.29	8.2	8 ± 1.8	10 ± 0.54
M. chlorpyriphos	0.12	0.32	0.34	0.35	0.97	1.6	1.6 ± 0.35	2 ± 0.11
Heptachlor	2.03	2.89	3.53	5.75	10	8.9	9 ± 5.8	10 ± 0.54
Aldrin	0.54	0.73	0.91	1.53	2.58	2.0	2 ± 1.5	2 ± 0.11
Metolachlor	0.73	2.83	2.92	2.07	8.28	15.7	16 ± 2.1	17 ± 0.92
Endrin	0.87	3	3.13	2.47	8.85	8.8	9 ± 2.5	10 ± 0.54

[a] All values expressed in μg/L.
[b] s_r = repeatability standard deviation.
[c] s_r = interlaboratory standard deviation.
[d] s_L = reproducibility standard deviation.
[e] r = repeatability.
[f] R = reproducibility.
[g] G.A. = gross average.
[h] C.I. = confidence interval of the gross average.
[i] T.V. = confidence interval of the "true" value.

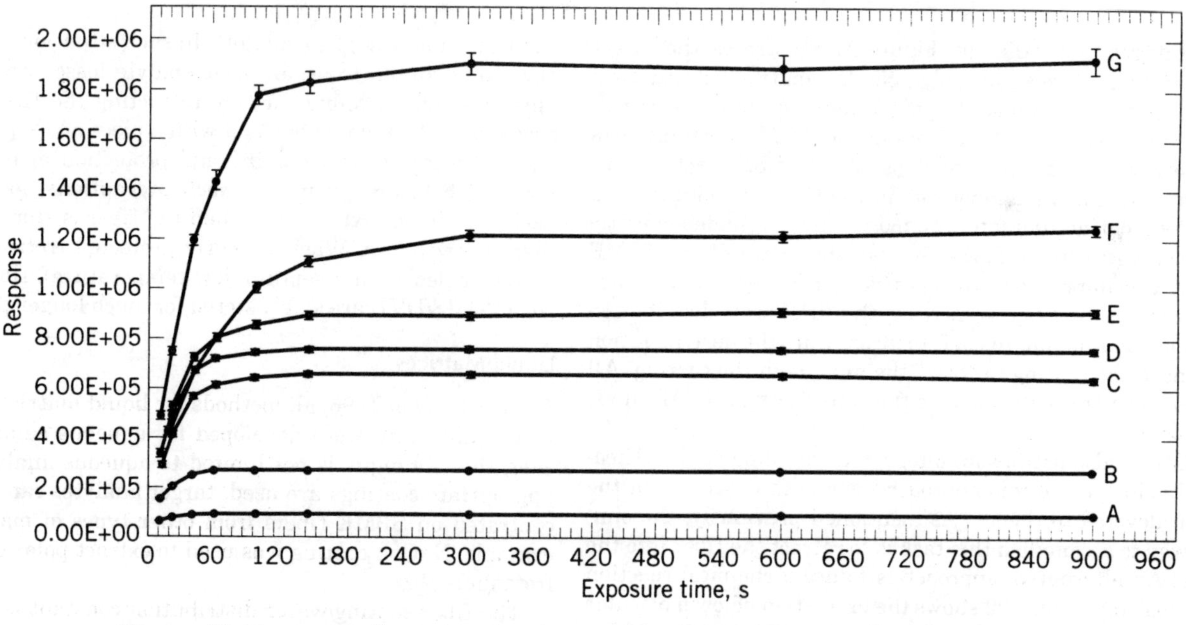

Figure 19. Representative extraction time profiles obtained for a standard gas mixture containing (A) benzene, (B) toluene, (C) ethylbenzene, (D) *p*-xylene, (E) *o*-xylene, (F) α-pinene, and (G) 1,3,5-trimethylbenzene. PDMS-coated fiber.

coatings are available. In addition, using this approach it is possible to determine the distribution constants of unidentified compounds. Therefore it is possible to use SPME to determine parameters like total petroleum hydrocarbons (TPH) in air. A comparison of SPME/PDMS with the standard charcoal tube technique obtained for an air sample drawn near a gas station resulted in 262 and 247 μg/L, respectively (16). The values agree well indicating that the approach described above quantifies analytes in gaseous samples without the need for identification. It should be pointed out, however, that this approach can be used only with detectors whose response is more or less independent of the nature of the analyte (eg, FID for analyzing hydrocarbons).

SPME has superior sensitivity for short term monitoring compared to traditional devices, which are limited by the gas throughput. SPME fibers can also be used at high

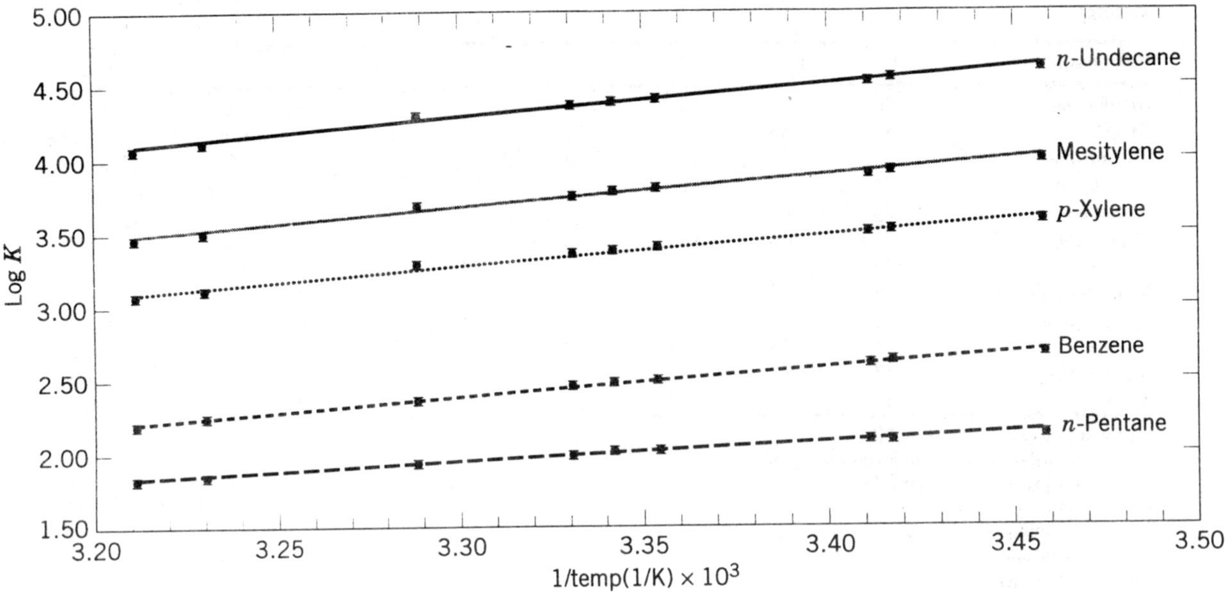

Figure 20. Representative plots for pentane, benzene, *p*-xylene, mesitylene, and *n*-undecane demonstrating the linear relationship between log K_{fg} and $1/T$.

temperatures. For example, Figure 21 illustrates the direct identification of several polycyclic aromatic hydrocarbons (PAHs) in diesel exhaust. A typical distribution of these pollutants is illustrated. The advantages of SPME for long-term sampling are not as clear now. Typically, the fiber reaches equilibrium with the air components in the first few minutes and then does not accumulate any more analytes, independently of the exposure time. However, the sensitivity of the SPME technique is improved by using thicker, more selective coatings or by cooling the fiber to increase the PDMS/gas distribution constant. Another approach is to incorporate the derivatization reagent in the coating to "trap" the analyte in the coating. All of these modifications enhance the sensitivity of SPME in air monitoring.

SPME is also used as an integrated sampling device. Theoretical principles of this approach have been described in the section devoted to theory. As mentioned previously, the simplest way to accomplish this task is to retract the fiber into the needle. An alternative approach is to use a chemical reaction in the coating. Figure 22 shows the extraction of low molecular weight carboxylic acids from a small volume of air using a fiber doped with 1-pyrenyldiazomethane (PDAM) (44). As expected, there is a linear relationship between the amount of acid extracted and time until about 60% of the acid is extracted from the sample. In practice, however, when the fiber is exposed to ambient air, analytes accumulate linearly with concentration as long as the fibre is exposed or the derivatization reagent is substantially consumed. This approach is particularly useful for integrated sampling of very low analyte concentrations in air.

Frequently, after the extraction, the fibers with analytes in the coating cannot be analyzed immediately. This is particularly true in field analysis, where an appropriate portable instrument might not be available. In such situations, the fibers have to be stored to assure that analyte losses are minimal. This is simply accomplished by retracting the fiber into the needle and sealing its open end with a piece of a septum. Cooling the needle provides additional protection against losses. For PDMS fibers, even with such precautions, losses of the most volatile analytes occur when the fiber is stored for more than a few hours. When properly protected (retracted fiber, needle sealed with a septum, low temperature), fibers coated with PDMS/DVB are safely stored for much longer periods.

Liquid Matrices

As of the end of 1996, all methods for liquid matrices reported in the literature were developed for aqueous samples. However, the technique is not limited to aqueous analysis. When appropriate coatings are used, target analytes can be preconcentrated on SPME fibers from other types of matrices. For example, Nafion coating was used to extract polar compounds from fuels (40).

The fiber coating/water distribution constant is calculated from the following equation:

$$K_{fw} = K_{fg}K_{aw} \tag{43}$$

where K_{fg} is determined from chromatographic data as discussed above and K_{aw} is the air/water distribution constant for a given analyte which can be found in tables of Henry's constants. For example, the equation for a PDMS coating and aqueous matrix is calculated from the equation below:

$$K_{fw} = 0.00415 \times \text{LTPRI} - 0.188 + \log_{10} K_{gw} \tag{44}$$

This equation is obtained after substituting the expression for K_{fg} given by equation 43 in equation 42. The K_{fw} values found from this equation agree very well with experimental data (39)

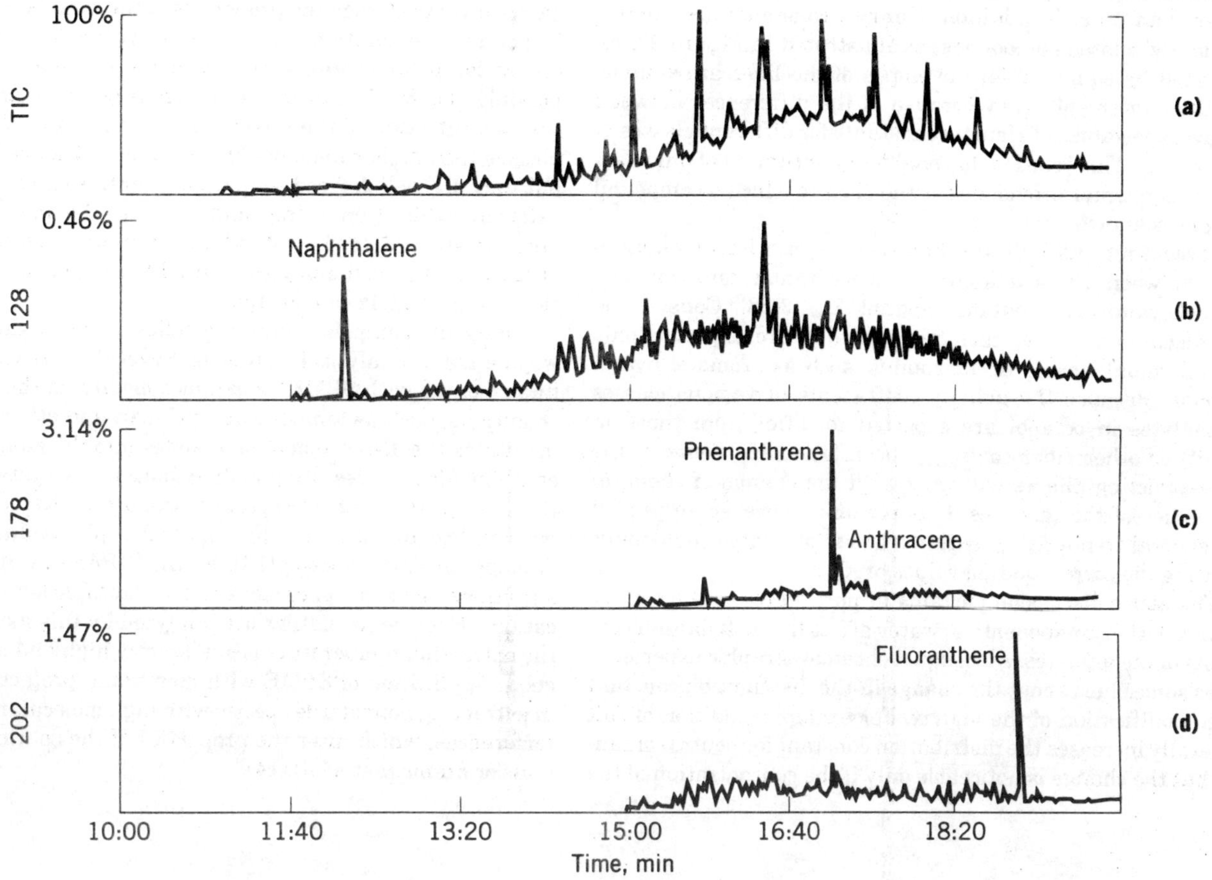

Figure 21. GC/MS traces corresponding to polycyclic aromatic hydrocarbons in diesel exhaust sampled by SPME/PDMS: (**a**) total ion chromatogram; selected ion traces at (**b**) m/z = 128, (**c**) 178, (**d**) 202.

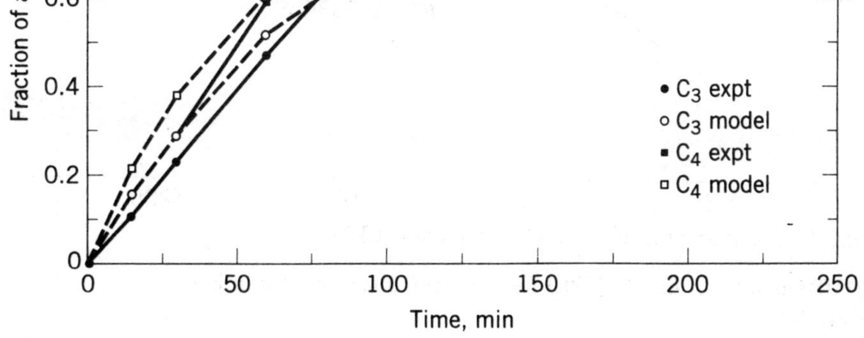

Figure 22. Comparison of experimental (expt) data vs theoretical model predictions for the reaction of gaseous propionic and butyric acids with PDAM in the PA-coated fiber.

considering that, typically, the errors in Henry's constants are larger than 10%. In addition, Henry's constants are similar for closely related compounds, as illustrated in Figure 23. As predicted by equation 44, the slopes of the lines are similar, but their intercepts vary because of the differences between the average values of Henry's constants for different classes of compounds. This relationship enables quantitation of unknown analytes in water provided that the class to which a compound belongs is known.

Several reports indicate the existence of a linear relationship between the coating/water distribution constant and octanol–water distribution constant K_{ow} (3,58). Considering the discussion above, such relationships are expected only for individual classes of compounds, such as aromatic hydrocarbons. Because the activity coefficients of various classes of analytes in octanol are expected to differ from those in PDMS or other fiber coatings, the relationship between the two extracting phases will vary with the change of chemical properties of the analytes. It is possible, however, to predict the general trends in K_{fw} within a group of related compounds by using the corresponding values of K_{ow}.

The above discussion pertains to pure water as the sample matrix. Other components in water affect the distribution constants of the analytes. Also, liquid chromatographic experience gives some clues about the change in the distribution constant with modification of the matrix. For example, addition of salt generally increases the distribution constant for neutral organics, but the change is noticeable only if the concentration of the salt exceeds 1%. Also, the presence of water miscible polar organic solvents change the properties of the matrix by reducing its polarity. In addition, swelling of the polymer by the solvent occurs for polar coatings, changing the coating volume and possibly also K_{fs}. However, the change is not substantial when the concentration of the solvent is below 1% (21). When samples contain higher amounts of salt and/or dissolved organics, but they are well defined so that pure matrix can be prepared, external calibration using standards in the matrix is still appropriate. Otherwise, standard addition (preferentially of isotopically labelled analytes) is used to compensate for variations in the matrix composition.

Analyzing samples containing solids, such as sludge, is always a major analytical challenge. Several approaches can be implemented with SPME. Sometimes modifying the extraction conditions, such as temperature, pH, salt and other additives, facilitates the displacement of analytes into the aqueous phase or the headspace, resulting in distribution constants similar to those for pure water. In certain cases, direct extraction is not possible because of a very dirty matrix or pH conditions which damage the fiber (above pH 12 for the PDMS coating). In such situations, the headspace mode is more suitable for many applications. Even semivolatiles are analyzed by this method when the extraction temperature is sufficiently high and agitation is good. Application of SPME with membrane protection assists in extracting nonvolatile species with high molecular weight interferences, which alter the properties of the coating (eg, proteins or humic material) (14).

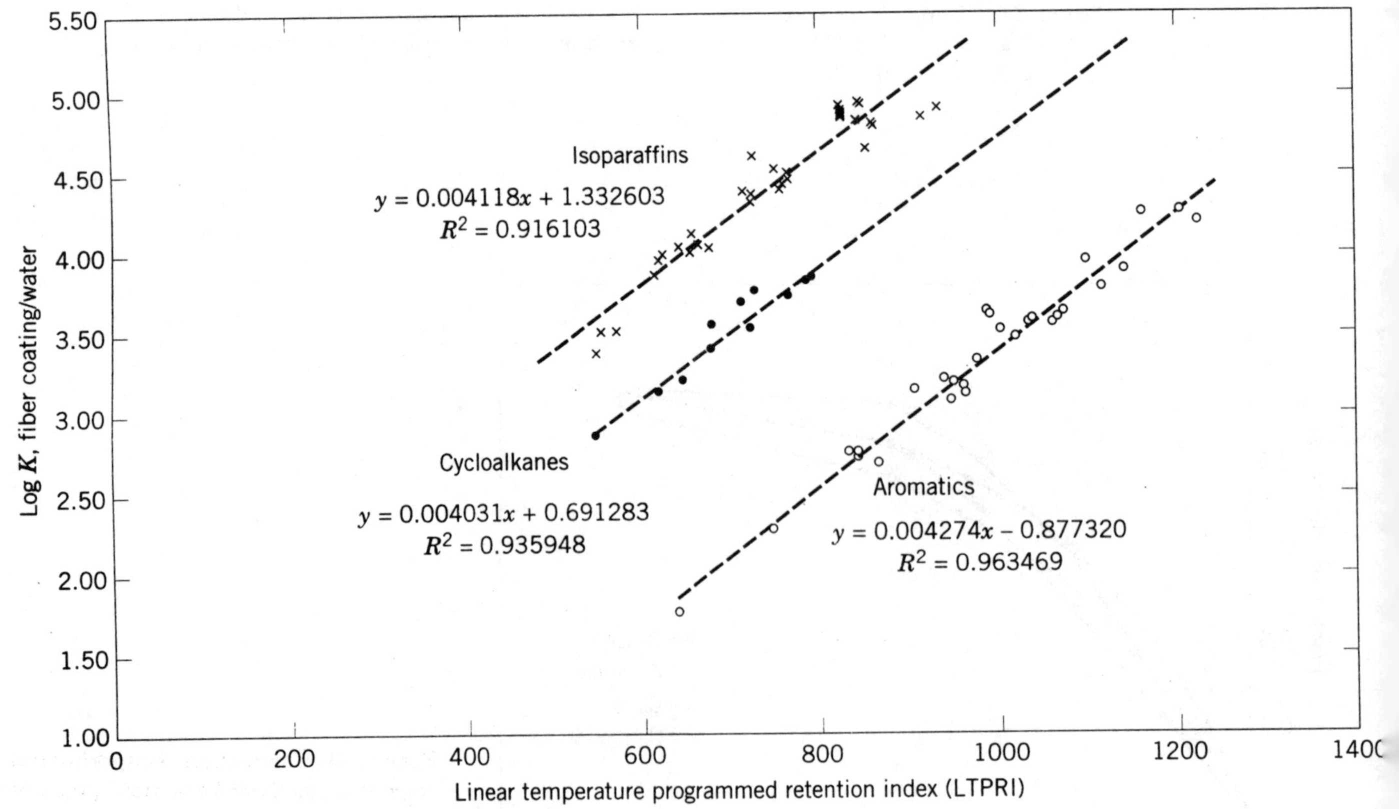

Figure 23. Log K_{fw} (PDMS) as a function of LTPRI for isoparaffins, substituted benzenes, and cycloalkanes at 25°C.

Solid Matrices

Accurate quantitation of target analytes in solids represents a very significant challenge to the analytical community. Although SPME cannot be used directly to extract analytes from solids, several approaches can be taken to facilitate simple sample preparation. For volatiles, the typical approach is to perform headspace analysis. To quantitatively release the analytes from the matrix, the temperature must be increased. This facilitates extraction and improves the kinetics of the process, but it also decreases the distribution constant. Typically, a maximum is observed for the relationship between the amount of analyte extracted by the fibre and the temperature (45). Initially, the loss of sensitivity from a decreased distribution ratio is more than compensated for by the increased concentration of the analyte in the headspace. At even higher temperatures, the loss of sensitivity predominates and the amount of analyte extracted by the fiber decreases. The inherent loss of sensitivity from the decreased distribution constant at high temperatures can be compensated for by cooling the fiber. However, this approach has not yet been commercialized. Alternatively, a matrix modifier is added to the solid sample. Water is very effective in displacing the analytes from solid surfaces for many types of samples, especially at elevated temperatures.

Another successful approach to SPME analysis of solids involves the use of water or a polar organic solvent, such as methanol, to extract analytes from a solid matrix before SPME analysis. When a polar solvent is used, the extract is spiked into pure water. Low temperature aqueous extraction followed by SPME is very useful for polar compounds, such as herbicides (59). Methanol with water spiking, on the other hand, is useful for analyzing volatile hydrocarbons (60). An interesting modification to the above procedure involves volatilizing the extract followed by fiber extraction from the gas phase (61). Quantitation of analytes in the gas phase is easier, as discussed above.

For less volatile analytes, the methanol approach described above gives good results. In addition, high pressure hot water extraction is a very suitable solvent-free alternative (55). Both static and dynamic hot water extraction have their advantages. The static method is very simple and inexpensive because it avoids high pressure pumps. The dynamic approach, on the other hand, provides an extract which is much cleaner than the original matrix. However, it might be possible to extract many semivolatile target analytes from very dirty matrices with the high temperature static system, when using headspace SPME.

Environmental Applications

A lot of research has been performed on samples of environmental origin, including air, water, sludge, and soil. The majority of applications has been developed for aqueous matrices. The results obtained for priority pollutants in water are very encouraging, indicating that in most cases the performance of SPME meets U.S. EPA method requirements. Table 10 summarizes the limits of detection for solid-phase microextraction of volatiles and semivolatiles, including phenols. In each case, the SPME method can be optimized to meet the requirements of the regulatory agency. Following is a brief discussion of the most important accomplishments for several groups of analytes.

Benzene, toluene, ethylbenzene and xylenes (BTEX) are the most investigated group of organic compounds of environmental interest. The sensitivity of the SPME/PDMS method for these compounds is very high, allowing mid to low ppt determinations in water. BTEX compounds are volatile, so the headspace SPME mode is successful in many applications. Validation research has been conducted to compare the performance of this method with more well-established alternatives. Figure 24 is an example of a multilevel validation for toluene. The correlation line indicates the equivalence between the purge and trap technique and SPME (56). The validation work has been extended to sludge and soils. Considering the complexity of the matrices, the correlation between the purge and trap and headspace SPME techniques was good. Validation work with other volatile organic compounds, such as chlorinated hydrocarbons, produced similar conclusions (61).

The measure of total petroleum hydrocarbons (TPH) is frequently used to estimate the contamination level of a sample. For air samples, this value is calculated directly from chromatographic data without identifying all individual components, as emphasized in the section devoted to gaseous matrices. TPH is then reported, together with more specific information about distribution and type of contaminants in the sample, without the need to calibrate. Determination of TPH of water samples is possible if the compounds in different classes are separated and identified by a chromatographic technique combined with a quantitative method, such as mass

Table 10. Summary of Detection limits for SPME Analysis of Various Classes of Organic Compounds

Analytes	LOD, pg/mL	EPA, pg/mL
BTEX	1–15	30–90
Polychlorinated HC solvents	1–100	10–100
PAHs	1.2–20	40
Chlorobiphenyls (PCBs)	3	60–100
Phenols	10–800	1500–42000

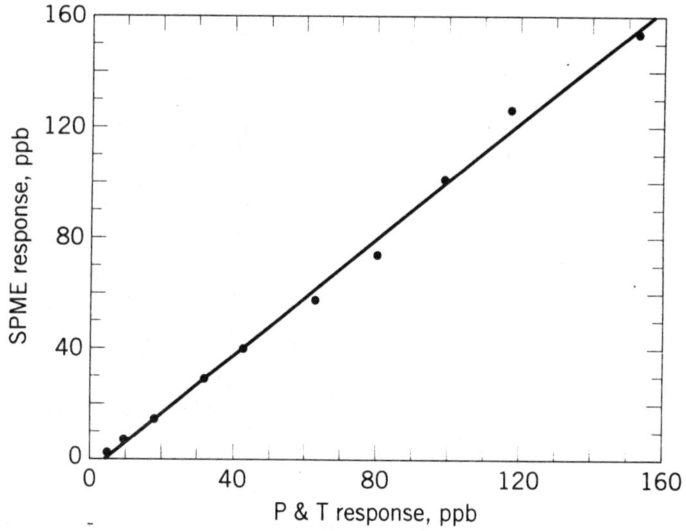

Figure 24. Regression plot for toluene. Each point on the graph represents a comparison between headspace SPME and purge and trap techniques for the same water sample.

spectrometry. For solids, TPH is obtained after the analytes are transferred into the gaseous or aqueous phase during extraction, as described in the section devoted to solid matrices.

Polar volatile solvents, like alcohols, ketones, and aldehydes, are difficult to quantify at trace levels in aqueous matrices. Extraction with nonpolar solvents leads to poor recoveries. More polar solvents are not used because they are miscible with water. SPME with a polar polymeric coating can provide the necessary alternative. Low ppb detection limits are possible even with FID when a PDMS/DVB coating and 35% salt concentration are used (40).

Nonpolar semivolatiles, such as PAHs and PCBs, have very large K_{fs} values, which results in very high sensitivities in SPME determination, frequently reaching low ppt levels (7,62). Figure 25 shows the major components in a pulp mill discharge detected and identified by SPME/GC/ITMS. The major advantage of SPME compared to alternative techniques is its portability. Frequently, it is very difficult to collect and preserve a representative sample because nonpolar semivolatiles adsorb on solid surfaces of the sampling vessel or on particulate matter. By sampling directly in the field, these phenomena are eliminated to a large extent.

Polar semivolatiles in natural matrices present a very significant analytical challenge. Adjustment of pH and salt addition help to reach the required detection limits. For example, the LOD obtained for phenols directly extracted from an acidified solution with a pH stable poly(acrylate) coating are lower than values specified by U.S. EPA. Derivatization with acetic anhydride further improves sensitivity and chromatographic performance. Derivatization also enhances extraction of carboxylic acids (41,44) and amines, particularly, of lower molecular weight species.

Direct extraction of very polar heteroaromatic compounds frequently present in creosote-contaminated groundwater with a PA coating combined with sensitive ion trap mass spectrometry detection provides low to sub ppb detection limits for these very water soluble compounds (63).

Pesticides are an analytically difficult group of compounds because of their wide range of chemical structures and properties. Some of them are classified as nonpolar (organochlorine pesticides) and others as very polar semivolatiles (herbicides). Despite this diversity, several research groups have very successfully applied SPME in their determination in aqueous matrices (64–68). Table 11 summarizes the SPME detection limits for several groups of pesticides. They are lower compared to U.S. EPA method LODs. It is even possible to develop single extraction conditions for rapidly screening 60 target pesticides combining three classes, organochlorine, organophosphorus and organonitrogen (68).

Application of sodium teraethylborate as an *in situ* derivatization reagent enables SPME analysis of some *metal ions and organometallic species* in aqueous samples. This approach was first used to determine methylmercury in water and biota (42), followed by the analysis of lead ion and organolead compounds (43). In this method, the analytes are first derivatized to their fully ethylated forms before they are extracted by the PDMS fiber. The derivatization products are volatile. Therefore headspace extraction by SPME is applied, facilitating rapid analysis of very complex matrices. The reaction time (about 15 minutes) typically determines the extraction time. Stirring

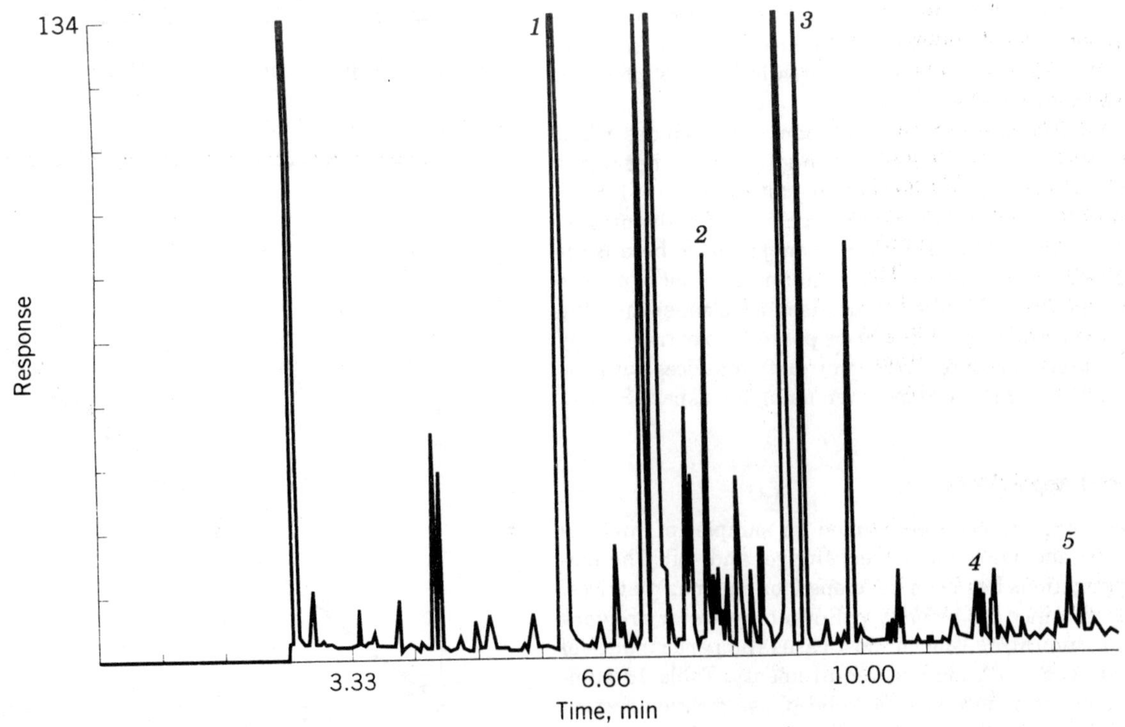

Figure 25. Total ion chromatogram showing organics detected and identified in a paper mill discharge wastewater sample extracted by headspace SPME at 25°C for 30 minutes: (*1*) naphthalene, (*2*) 1,1′-biphenyl, (*3*) dibenzofuran, (*4*) anthracene, (*5*) dibutylphthalate.

Table 11. Detection Limits for SPME Coupled with Various Detectors

Pesticide Classes	Target Analyte Subgroups	ECD, ng/L	NPD, ng/L	MS, ng/L	EPA methods, ng/L
Nitrogen-containing herbicides[a]	Thiocarbamates		20–60	0.05–8	100–200
	Triazines		40–6000	0.4–3	100–800
	Nitroanilines		10–30	0.02–0.4	200
	Substituted uracils		200–400	0.01–1	2500–4500
	Substituted amide		800	15	500
	Acetanilide		200	0.01	700
	Diphenyl ether		300	6	NA[b]
	Triazole		30	0.01	NA
Organochlorine pesticides[d]	HCHs	0.9–9[c]		0.01–0.04	10–25
	Hexachlorocyclodienes	0.06–1.6		0.02–0.6	2.5–50
	Diphenyl aliphatics	0.05–4.7		0.06–4.5	10–75
Phosphorus-containing pesticides[a]	Phosphate		500	6	100–2500
	Phosphorothiolate		130	2	N/A
	Phosphonothiate		15	8	9–200
	Pyrophospholate		16	1	6
	Phosphorodithioates		9–320	0.4–9	9–1500
	Phosphorothioates		11–280	0.7–100	4–2000

[a] Analyzed by SPME with 85 μm poly(acrylate) fiber.
[b] NA = not available
[c] Analyzed with 30% w/w salt added.
[d] Analyzed by SPME with 100 μm poly(dimethylsiloxane) fiber.

must be very efficient during derivatization, because the reagent decomposes rapidly after adjusting the pH. The affinity of neutral forms of organometallic compounds for the PDMS coating is very high, resulting in very low detection limits, approaching sub ppt levels for determining tetraethyllead in water by SPME/GC/ITMS. This method is applied to partial speciation of various forms of these metals in a sample, first by extracting native neutral organometallic species followed by the determination of the ionic forms (after derivatization).

Application of HPLC to separate species extracted by SPME opens new possibilities in determining nonvolatile and thermally unstable environmental pollutants. Several alkylphenol ethoxylate *nonionic surfactants* in water have been analyzed by SPME combined with HPLC (69). The detection limits obtained were at a low ppb level for individual ethoxamers with UV detection, which indicates that trace level determination in natural matrices is possible by this method.

Food and Pharmaceuticals

A unique feature of SPME in analyzing food and drugs is its ability, in some cases, to extract the analytes from the packaged product without opening the package. For example, flavors in wine can be checked before sale or purchase by introducing the fiber into the headspace of the wine through the bottle cork. Because only a small amount of flavor is extracted, the composition of the product does not change, and the extracting phase does not contaminate the product because the polymer is nonvolatile and insoluble. A similar process can be applied to on-line product monitoring to assure the best possible quality.

Food and pharmaceutical products are frequently contaminated by volatile *organic solvents,* used in manufacturing and processing. Considering the frequent complexity of their matrices, the analytical procedures are quite involved. Static headspace is frequently used for this purpose, but SPME is a good alternative for analyzing both contaminated pharmaceuticals (70) and foods (71) because of its superior sensitivity, especially for less volatile compounds. Samples are analyzed by dissolving the product in water before determination, or the analytes are extracted directly from the headspace above the product. Typically, standard addition is used to quantitate. The detection limits are similar to those obtained for environmental matrices.

Flavors constitute another very interesting subject for investigation. SPME has been applied to analyze a wide range of food products, spices, oils, beverages (72,73) and even black and white truffle aroma (74). Initial investigations indicate that good quantitation of volatile analytes in the matrix is obtained after the solid components of the sample matrix are separated from the aqueous phase by centrifugation. Combining SPME extraction with GC separation using a chiral stationary phase provides information about the composition of the products and also their natural purity (75). Figure 26 shows the separation of peppermint oil and peppermint-flavored candy using a β-cyclodextrin column.

Another interesting application of SPME for analyzing foods is quantifying *caffeine* in soft drinks (5). The most convenient quantitative method for this determination is based on isotopically labeled standards which are readily available. In addition to caffeine, flavor components are also detected in the same chromatographic run.

The number of research reports on semivolatiles in foods is more limited, but the scope of potential applications is as broad as in the environmental area. One of the interesting applications is determination of *herbicides* in wines (59).

Clinical and Forensic Analysis

The major advantage of SPME for clinical and forensic applications is its portability, which might eventually be translated into on-site sampling followed by immediate analysis with portable instruments. This approach would allow better monitoring of a patient's condition during treatment or therapy and

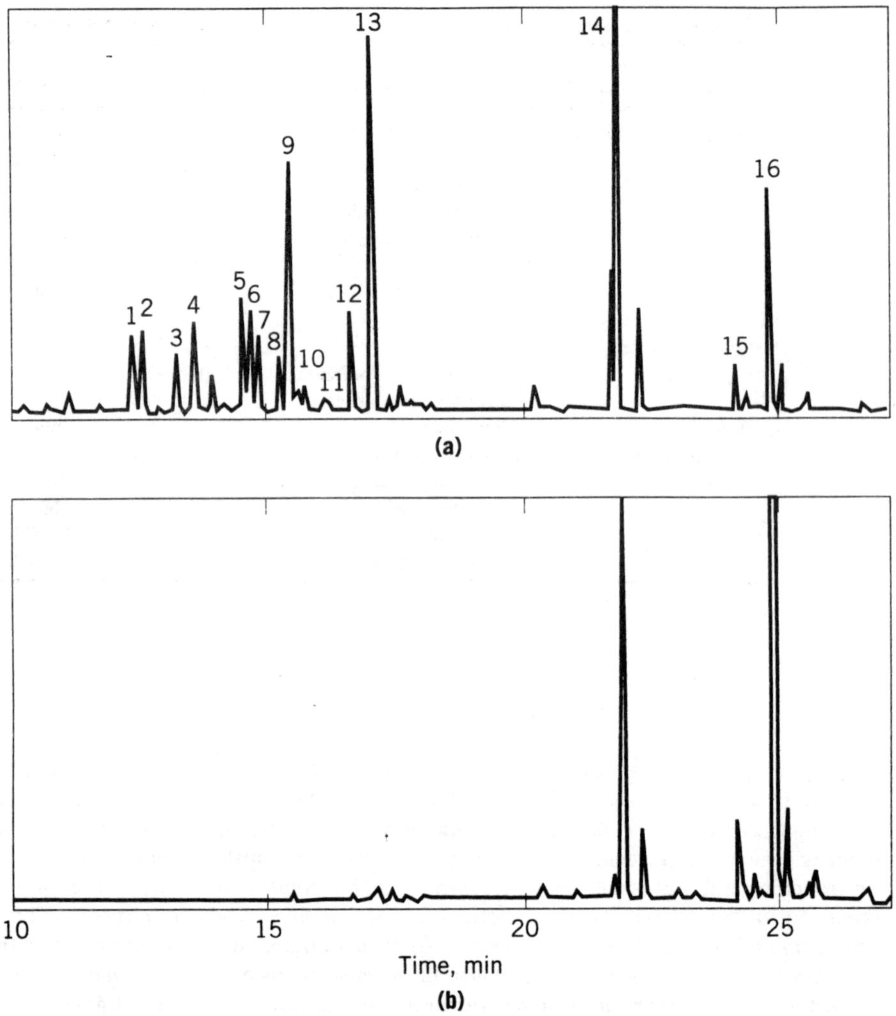

Figure 26. Headspace SPME extraction and chiral GC analysis of (**a**) peppermint oil and (**b**) peppermint-flavored candy using a β-cyclodextrin column. Oven temperature program: 40°C for 2 min, then 40–220°C at 4°C/min. Peaks: 1, (−)α-pinene; 2, (+)α-pinene; 3, β-myrcene; 4, (+)-sabinene; 5, (−)-camphene; 6, (−)β-pinene; 7, (+)-β-pinene; 8, α-terpinene; 9, 3-carene; 10, (−)-limonene; 11, (+)-limonene; 12, (±)β-phellandrene; 13, γ-terpinene; 14, cineole; 15, menthone; 16, (+)-menthol; 17, (−)-menthol; 18, isomenthol; 19, (±)-menthyl acetone.

better preservation of a crime scene because the objects would not be removed for analysis.

Monitoring of human health is achieved by noninvasive SPME methods. For example, breath contains the headspace of blood. Therefore, characterization of volatile components inside the human body is accomplished by adapting the SPME device to breath analysis. Figure 27 presents an example of breath analysis (76). The detection limits are at a low nmol/L level, which provides sufficient sensitivity for practical application to monitoring alcohol and acetone levels in blood. SPME is also used for rapid analysis of body fluids. Figure 28 illustrates the detection of several aromatic amines in the milk of a mother who smokes.

Analysis for *alcohol* and *drugs* in body fluids is frequently performed in clinical and forensic laboratories. SPME is suitable for monitoring alcohol levels in both urine and blood. Drugs in body fluids are more difficult to analyze, because typically they are semivolatile substances. However, a number of published preliminary results indicate that headspace SPME analysis at an elevated temperature or even direct extraction after dialysis of urine or blood serum is possible. Most of the work to date was done with *amphetamines* (53,77), but analysis of other drugs, including *valproic acid* (78), *cocaine, tricyclic antidepresants* (79), *nicotine,* and *local anaesthetics* is also reported.

In forensic analysis, headspace SPME has been applied to examine suspected *arson* cases (80,81). Figure 29 shows the total and selected ion chromatograms corresponding to a real arson sample and a gasoline standard, indicating the possible use of an accelerant.

Other Analytical Applications

SPME is used as a sampling and sample introduction technique for fast GC. Figure 30 demonstrates that, using this approach, volatile compounds in water are analyzed in less than 3 minutes (82), approximately ten times faster than the currently used purge and trap systems.

SPME is very well suited to field analysis. Field sampling and analysis reduce costs and eliminate many possible sources of error in handling and storing of samples. In addition, field analysis enables faster and better characterization of the problems because analytical information is available immediately for evaluation and decision. SPME is successfully used for fast GC analysis in the field with dedicated instrumentation (83).

Ion-exchange coating materials and crown ether ligands are used to selectively extract inorganic ions from aqueous matrices, allowing quantitation and also speciation (84,85). A polyacrylic acid coating has been developed for analyzing basic proteins (54).

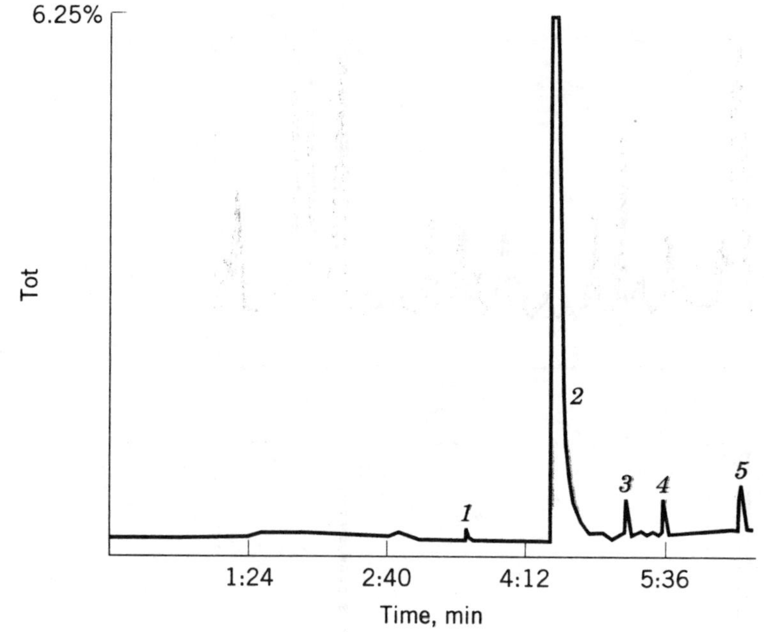

Figure 27. SPME/GC/ITMS analysis of a breath sample. The sample was taken after alcohol consumption. Fiber coating: Carbowax/DVB. Peak assignment: (*1*) acetaldehyde, (*2*) ethanol, (*3*) acetone, (*4*) isoprene, and (*5*) carbon disulfide.

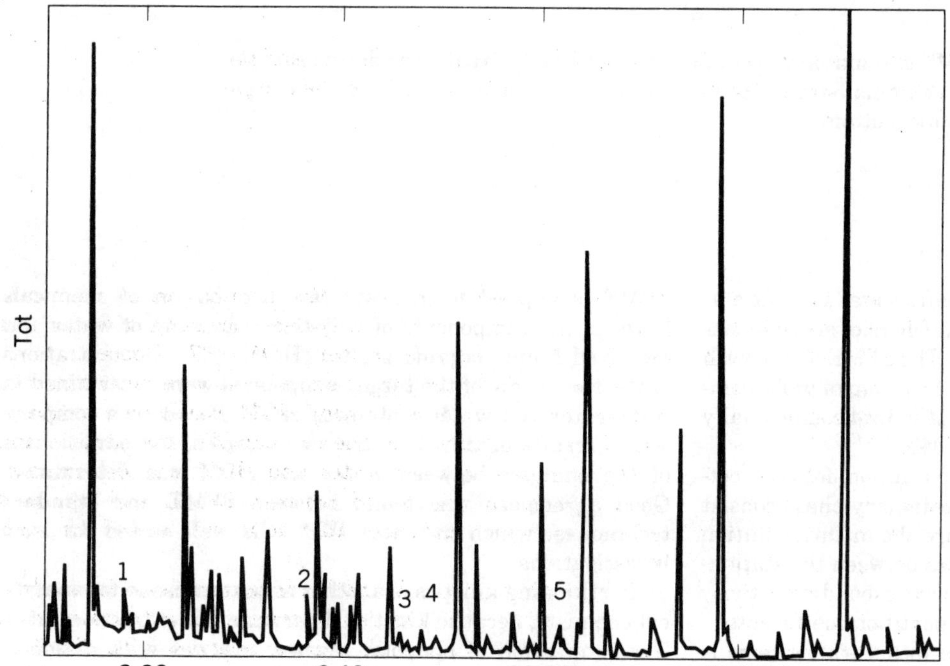

Figure 28. SPME/GC/ITMS analysis of human milk from a smoker. 1, aniline; 2, methylaniline; 3, dimethylaniline; 4, trimethylaniline; 5, nicotine.

Physicochemical Applications

Solid-phase microextraction is applied for extraction and also for measurements which better characterize the extraction system. The measurements include studying the properties of the fiber coating and investigating multiphase equilibria in the matrix. In addition, because detailed mathematical treatment of the solid-phase microextraction process is available, SPME is applied to study the parameters involved in the extraction, including diffusion coefficients in the coating and the extracted phase and various distribution constants.

The study of infinite-dilution activity coefficients of probe solutes in a polymer phase is fundamental to understanding solute–solvent interactions at the molecular level and the thermodynamic processes involved in forming the solution. SPME

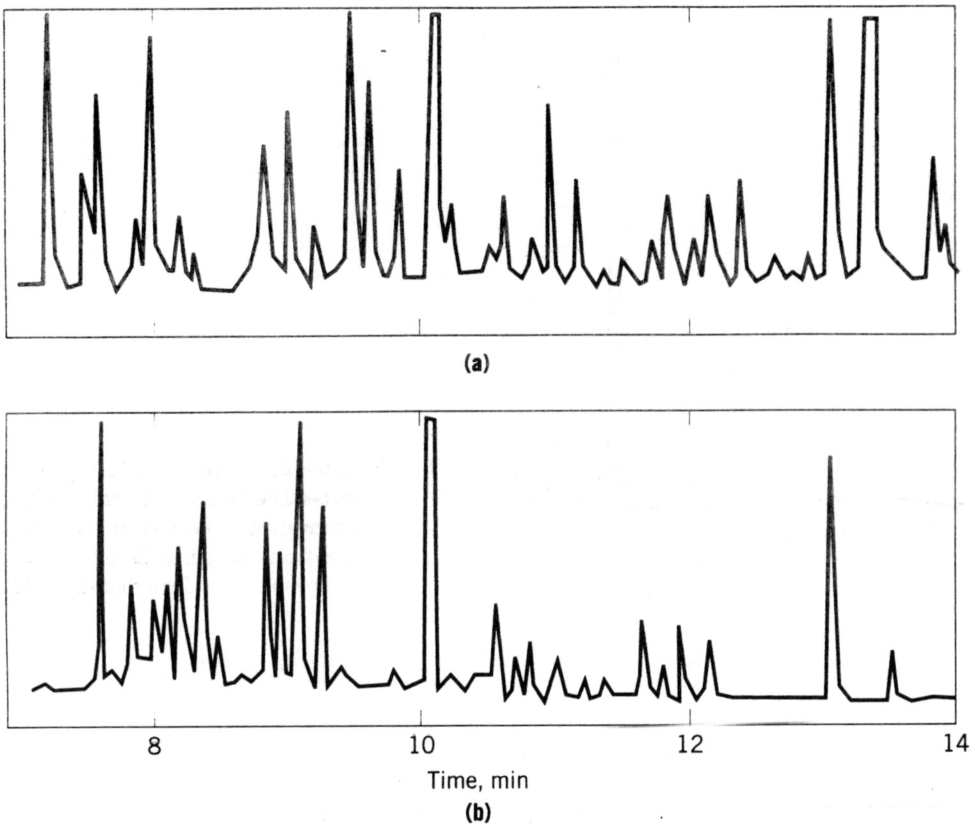

Figure 29. GC/ITMS chromatograms of headspace SPME of real fiber debris sample: (**a**) total ion current chromatogram; (**b**) selected ion trace for $m/z = 57$ for the sample (top) and pure gasoline (bottom).

is a useful tool for determining these coefficients (15). The stationary phases of interest are coated on fibers of suitable materials (fused silica, stainless steel, etc). The SPME device with a selected fiber coating is used to extract a group of probe compounds, which are then separated on a standard commercially available column and quantified by GC/MS.

The SPME measurements of infinite-dilution activity coefficients of probe solutes in a selected stationary phase consist of two steps. The first step is to measure the infinite-dilution partition coefficients of the probe solutes between the stationary phase and the gas phase. Then, the corresponding activity coefficients are determined from these partition coefficients.

Investigation of analyte distribution in natural systems is critical for many scientific disciplines, including environmental chemistry. Information gained during these studies is also very important to analytical chemists, because it facilitates the development of optimum sampling protocols and leads to more rational optimization of extraction parameters. One approach to studying multiphase equilibria involves separating the individual phases after equilibrium is reached and analyzing the individual phases to determine the concentrations of target compounds. For example, SPME is used to study the concentration of organic substances in octanol-saturated water (86). However, separating the phases is not always necessary. SPME is applied to measure the distribution of chemicals between the components of a system consisting of water and dissolved humic organic matter (HOM) (87). Concentrations of the free forms of the target compounds were determined in pure water and water containing HOM. Based on a comparison of results obtained for the two samples, the partitioning of the analytes between water and HOM was determined. Good agreement was found between SPME and standard techniques, which indicates that it is well suited for such investigations.

Partitioning kinetics is another important issue for analytical chemists, because kinetics determine the extraction rates. The interaction of nonpolar organic analytes with dissolved polymers is fast because physisorption does not require high activation energies. SPME is used to investigate the kinetics of the equilibration process of organic pollutants between water and dissolved HOM. In such a study, the sample containing native analytes is spiked with isotopically labeled analogues. Rapid extraction from a stirred HOM–water sample monitors concentration changes of target analytes over time. Figure 31 illustrates the change of the concentration of labeled compound in water vs native analyte as a function of time. The curve indicates rapid equilibration of the species between water and HOM.

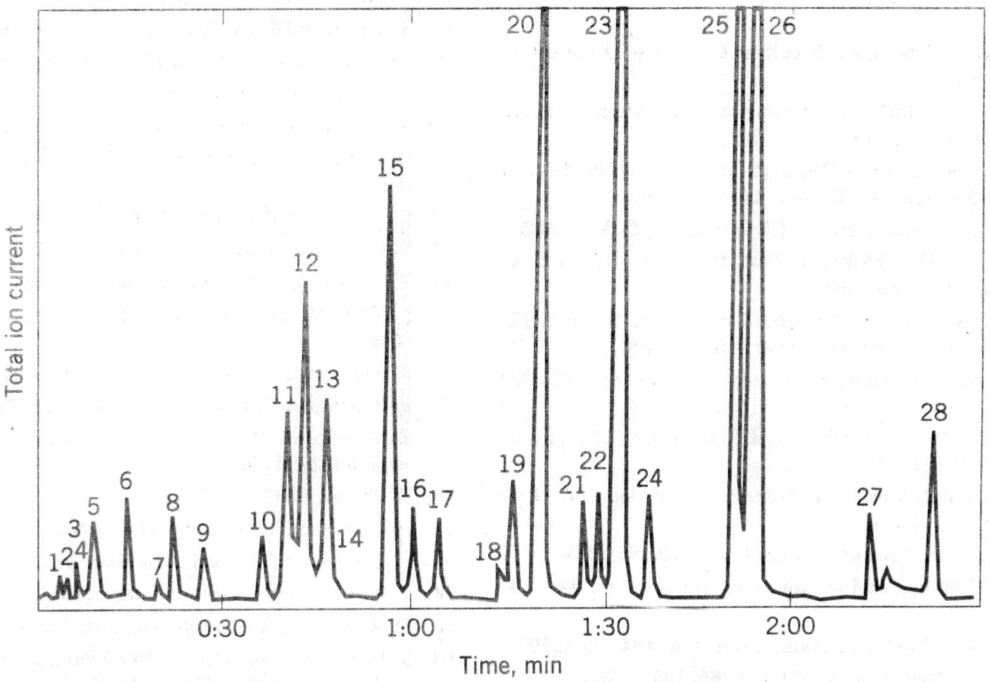

Figure 30. Separation of purgeables A, B and C on a VOCOL column. Conditions: 0°C–30°/min-70°; 2.1 atm, dedicated injector, capacitor voltage 24 V, MS detector, mass range 45–250. 1, chloromethane; 2, vinyl chloride; 3, bromomethane; 4, chloroethane; 5, trichlorofluoromethane; 6, 1,1-dichloroethene; 7, dichloromethane; 8, 1,2-dichloroethene; 9, 1,1-dichloroethane; 10, trichloromethane; 11, 1,1,1-trichloroethane; 12, tetrachloromethane; 13, benzene; 14, 1,2-dichloroethane; 15, trichloroethene; 16, 1,2-dichloropropane; 17, bromodichloromethane; 18, 2-chloroethyl vinyl ether; 19, *cis*-1,3-dichloropropene; 20, toluene; 21, *trans*-1,3-dichloropropene; 22, 1,1,2-trichloroethane; 23, tetrachloroethylene; 24, dibromochloromethane; 25, chlorobenzene; 26, ethylbenzene; 27, tribromomethane; 28, 1,1,2,2-tetrachloroethane.

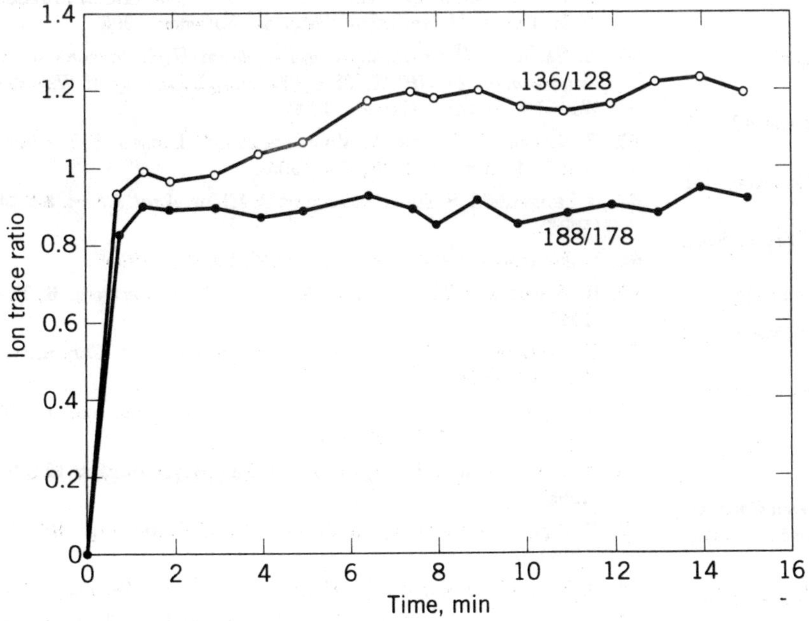

Figure 31. Kinetics of the sorption of naphthalene and anthracene onto DOM (100 ppm) from water measured by SPME. *y*-axis represents ratio of GC signals for deuterated spike and nondeuterated compound. Matrix: diluted wastewater with native PAHs.

BIBLIOGRAPHY

1. J. Namiesnik, T. Górecki, M. Biziuk, and L. Torres, *Anal. Chim. Acta.* **237,** 1 (1990).
2. J. Pawliszyn, *Solid-Phase Microextraction, Theory and Practice,* Wiley-VCH, New York, 1997.
3. C.L. Arthur, L.M. Killam, S. Motlagh, M. Lim, D.W. Potter, and J. Pawliszyn, *Environ. Sci. Technol.* **26,** 979 (1992)
4. D.W. Potter, and J. Pawliszyn, *J. Chromatogr.* **625,** 247 (1992).
5. S.B. Hawthorne, D.J. Miller, J. Pawliszyn, and C.L. Arthur, *J. Chromatogr.* **603,** 185 (1992).
6. C.L. Arthur, K. Pratt, S. Motlagh, J. Pawliszyn, and R.P. Belardi, *J. High Resolution Chromatogr.* **15,** 741 (1992).
7. D. Potter, and J. Pawliszyn, *Environ. Sci. Technol.* **28,** 298 (1994).
8. M. Chai, C.L. Arthur, J. Pawliszyn, R.P. Belardi, and K.F. Pratt, *Analyst* **118,** 1501 (1993).
9. K. Buchholz, and J. Pawliszyn, *Environ. Sci. Technol.* **27,** 2844 (1993).
10. K. Buchholz, and J. Pawliszyn, *Anal. Chem.* **66,** 160 (1994).
11. D. Louch, S. Motlagh, and J. Pawliszyn, *Anal. Chem.* **64,** 1187 (1992).
12. S. Motlagh and J. Pawliszyn, *Anal. Chim. Acta* **284,** 265 (1993).
13. Z. Zhang and J. Pawliszyn, *Anal. Chem.* **65,** 1843 (1993).
14. Z. Zhang, J. Poerschmann, and J. Pawliszyn, *Anal. Commun.* **33,** 219 (1996).
15. Z. Zhang and J. Pawliszyn, *J. Phys. Chem.* **100,** 17648 (1996).
16. P. Martos, A. Saraullo, and J. Pawliszyn, *Anal. Chem.* **69,** 402 (1997).
17. R. Schwarzenbach, P. Gschwend, and D. Imboden, *Environmental Organic Chemistry,* John Wiley & Sons, Inc., New York, 1993, pp. 109–123.
18. J. Li and P.W. Carr, *J. Chromatogr. A* **659,** 367 (1994).
19. J.H. Park, J.E. Lee, M.D. Jang, J. Li, and P.W. Carr. *J. Chromatogr.* **586,** 1 (1991).
20. A.J. Dallas and P.W. Carr, *J. Phys. Chem.* **98,** 4927 (1994).
21. C.L. Arthur, L.M. Killam, K.D. Buchholz, and J. Pawliszyn, *Anal. Chem.* **64,** 1960 (1992).
22. P. Martos, and J. Pawliszyn, *Anal. Chem.* (in press).
23. F.A. Long and W.F. McDevit, *Chem. Rev.* **51,** 119 (1952).
24. X. Yang and T. Peppard, *LC-GC* **13,** 882 (1995).
25. J.P. Conzen, J. Burck, and H.J. Ache, *Applied Spectrom.* **47,** 753 (1993).
26. J. Brandrup and E.H. Immergut, eds., *Polymer Handbook,* 3rd ed., John Wiley & Sons, Toronto, Canada, 1989.
27. *Handbook of Chemistry and Physics,* 70th ed. CRC Press, Boca Raton, Fla., 1989.
28. G.L. Klunder and R.E. Russo, *Appl. Spectrom.* **49,** 379 (1995).
29. H.S. Carslaw and J.C. Jaeger, *Conduction of Heat in Solids,* 2nd ed., Clarendon Press, Oxford, 1986.
30. A.D. Young, *Boundary Layers,* BSP Professional Books, Oxford, New York, 1989.
31. D.K. Edwards, V.E. Denny, and A.F. Mills, *Transfer Processes,* 2nd ed., Hemisphere Publishing, New York, 1979.
32. G.T. Orlob, *Contribution no. 19: Eddy Diffusion in Open Channel Flow.* Water Resources Center, University of California, Berkeley, Calif., 1958.
33. *Encyclopedia of Science and Technology,* 7th ed., McGraw–Hill, Toronto, Canada, 1992.
34. J. Crank, *Mathematics of Diffusion,* Clarendon Press, Oxford, 1986, p. 14.
35. J. Pawliszyn, *J. Chromatogr. Sci.* **31,** 31 (1993).
36. R. Eisert and J. Pawliszyn, *Anal. Chem.,* **69,** 3140 (1997).
37. M. Chai and J. Pawliszyn *Environ. Sci. Technol.* **29,** 693 (1995).
38. V. Mani, private communication, Supelco, Bellefonte, Pa., 1996.
39. A. Saraullo, P. Martos, and J. Pawliszyn, *Anal. Chem.* (in press).
40. T. Górecki, P. Martos, and J. Pawliszyn, *Anal. Chem.* (in press) (1997).
41. L. Pan and J. Pawliszyn, *Anal. Chem.* **69,** 196 (1997).
42. Y. Cai and J. Bayona, *J. Chromatogr.* **696,** 113 (1995).
43. T. Górecki and J. Pawliszyn, *Anal. Chem.* **68,** 3008 (1996).
44. L. Pan, M. Adams, and J. Pawliszyn, *Anal. Chem.* **67,** 4396 (1995).
45. Z. Zhang and J. Pawliszyn, *Anal. Chem.* **67,** 34 (1995).
46. R. Eisert and K. Levsen, *J. Chromatogr.* **737,** 59 (1996).
47. T. Górecki, M. Belkin, J. Caruso, and J. Pawliszyn, *Anal. Commun.* **34,** 275 (1997).
48. R. Shirey, *HRC* **18,** 495 (1995).
49. J. Chen and J. Pawliszyn, *Anal. Chem.* **67,** 2530 (1995).
50. H. Daimon and J. Pawliszyn, *Analyst* (in press) (1997).
51. T. Górecki and J. Pawliszyn, *Analyst,* **122,** 1079 (1997).
52. H. Lord and J. Pawliszyn, submitted to *Anal. Chem.*
53. L. Wang, *Determination of Amphetamine and Methamphetamine in Urine and Blood Plasma by Headspace Solid Phase Microextraction,* M.Sc. Thesis, University of Waterloo, Waterloo, Canada, 1996.
54. J–L. Liao, C–M. Zeng, S. Hjerten, and J. Pawliszyn *J. Microcol. Sep.* **8,** 1 (1996).
55. H. Daimon and J. Pawliszyn, *Anal. Commun.* **33,** 421 (1996).
56. B. MacGillivray, P. Fowlie, C. Sagara, and J. Pawliszyn *J. Chromatogr. Sci.* **32,** 317 (1994).
57. T. Górecki and J. Pawliszyn, *Analyst* **121,** 1381 (1996).
58. J. Dean, W. Tomlinson, V. Makovskaya, R. Cumming, M. Hetheridge, and M. Comber *Anal. Chem.* **68,** 130 (1996).
59. A. Boyd–Boland and J. Pawliszyn, *J. Chromatogr.* **704,** 163 (1995).
60. B. MacGillivray, *Analysis of Substituted Benzenes in Environmental Samples by Headspace Solid Phase Microextraction,* M.Sc. Thesis, University of Waterloo, Waterloo, 1996.
61. A. Saraullo, *Determination of Petroleum Hydrocarbons in the Environment by SPME,* M.Sc. Thesis, University of Waterloo, Waterloo, Ontario, Canada, 1996.
62. T. Nilsson, F. Pelusio, L. Montanarella, B. Larsen, S. Facchetti, and J. Madsen, *HRC* **18,** 617 (1995).
63. J. Langenfeld, S. Hawthorne, and D. Miller, *Anal. Chem.* **68,** 144 (1996).
64. S. Johanssen and J. Pawliszyn, *HRC,* **19,** 627 (1996).
65. R. Eisert and K. Levsen, *J. Am. Soc. Mass Spectrom.* **6,** 1119 (1995).
66. I. Barnabas, J. Dean, I. Fowlis, and S. Owen, *J. Chromatogr.* **705,** 305 (1995).
67. K. Graham, L. Sarna, G. Webster, J. Graynor, and H. Ng, *J. Chromatogr.* **725,** 129 (1996).
68. A. Boyd–Boland, S. Magdic, and J. Pawliszyn, *Analyst* **121,** 929 (1996).
69. P. Popp, K. Kalbitz, and G. Oppermann, *J. Chromatogr.* **687,** 133 (1994).
70. A. Boyd–Boland and J. Pawliszyn, *Anal. Chem.* **68,** 1521 (1996).
71. S. Scypinski, A.–M. Smith, L. Nelson, and S. Shaw, private communication.
72. D. Page and G. Lacroix, *J. Chromatogr.* **648,** 199 (1993).

73. X. Yang and T. Peppard, *J. Agric. Food Chem.* **42,** 1925 (1994).
74. A. Steffen and J. Pawliszyn, *J. Agric. Food Chem.* **44,** 2187 (1996).
75. F. Pelusio, T. Nilsson, L. Montanarella, R. Tilio, B. Larsen, S. Facchetti, and J. Madsen, *J. Agric. Food Chem.* **43,** 2138 (1995).
76. V. Mani and C. Woolley, *LC-GC* **13,** 734 (1995).
77. C. Grote and J. Pawliszyn, *Anal. Chem.,* **69,** 587 (1997).
78. N. Nagasawa, M. Yashiki, Y. Iwasaki, K. Hara, and T. Kojima, *Forensic Sci. Int.* **78,** 185 (1996).
79. M. Krogh, K. Johansen, F. Tonnesen, and K. Rasmussen, *J. Chromatogr.* **673,** 299 (1995).
80. T. Kumazawa, X–P. Lee, M–C. Tsai, H. Seno, A. Ishii, and K. Sato, *Jpn. J. Forensic Toxicol.* **13,** 25 (1996).
81. K. G Furton and J. Bruna, *HRC* **18,** 625 (1995).
82. A. Steffen and J. Pawliszyn, *Anal. Commun.* **33,** 129 (1996).
83. T. Górecki and J. Pawliszyn, *Anal. Chem.* **67,** 3265 (1995).
84. T. Górecki and J. Pawliszyn, *Field Analytical Chemistry and Technology,* **1**(5), 277 (1997).
85. E. Otu and J. Pawliszyn, *Microchim. Acta* **112,** 41 (1993).
86. J. Chongrong and J. Pawliszyn, *J. Microcol. Sep.* (in press) (1997).
87. P. Popp, A. Paschke, U. Schroter, and G. Oppermann, *Chemia Analityczna.* **40,** 897 (1995).
88. J. Poerschmann, Z. Zhang, F. Kopinke, and J. Pawliszyn, *Anal. Chem.* **69,** 597 (1997).

SOOT ANALYSIS. See BLACK CARBON IN SOIL, SEDIMENTS AND ICE CORES.

SORPTIVE BARRIERS FOR GROUNDWATER REMEDIATION

KIRK J. CANTRELL
DANIEL I. KAPLAN
Pacific Northwest National Laboratory
Richland, Washington

Alternatives for the remediation of contaminated groundwater can be grouped into four general categories (1):

1. isolation of the contaminated groundwater
2. removal and surface treatment of the contaminated groundwater and its subsequent disposal or reuse (pump-and-treat)
3. removal of contaminated soil or aquifer material, followed by treatment and/or disposal
4. *in situ* stabilization of aquifer contaminants

The first three categories can be classified as conventional alternatives to groundwater pollution treatment. The goals are either to cut the site off from groundwater flow and/or infiltration of meteoric water with impermeable barriers or to remove the contamination using pump-and-treat systems or excavating the soil (see PUMP-AND-TREAT TECHNOLOGY, AQUIFER REMEDIATION). The fourth alternative involves mostly innovative approaches which are attractive, but relatively new and untested.

The first remediation category, isolation and containment of groundwater pollutants, can be achieved through the use of physical barriers such as grout curtains and slurry walls (2) or hydraulic barriers resulting from a combination of pumping and injection wells (3). The use of physical barriers is most effective when groundwater is shallow, an impermeable soil or bedrock strata is available for contact with the barrier, and hydraulic gradients across the wall are small. The second remediation category usually involves pump-and-treat systems. This approach involves flushing the contaminated region with solvent, usually water, to mobilize and then remove the pollutants. More innovative, but also more risky approaches use surfactants, chelating agents, or other additives to enhance the mobilization of the contaminants. The flushing solvent is pumped to the surface, treated, and disposed. This method works best with highly soluble contaminants. Above-ground treatment alternatives for a wide variety of wastewaters have been developed (4). These treatments involve processes such as biological degradation, carbon adsorption, chemical oxidation and reduction, precipitation, flocculation, density separation, dialysis, distillation, evaporation, air stripping, ion exchange, and solvent extraction. Pollutants which are either sparingly soluble or insoluble in water, eg, PCBs or precipitated metal ions, are not easily brought to the surface by pump-and-treat methods. An alternative remediation option for these types of contaminants is the third remediation category, which is excavation and treatment of the contaminated soil.

Although *ex situ* treatment systems are generally available and straightforward for most contaminants, pump-and-treat as a remediation system has a poor success record and is generally viable only for a limited number of situations (5–7). Some of the practical problems associated with conventional *ex situ* groundwater treatment alternatives commonly include:

1. the need to manage large volumes of water containing very low concentrations of contaminants
2. the need to manage secondary waste generated from the treatment of the contaminated water
3. the significant disruptions of normal operations for long time periods
4. the unfavorable disruption of hydraulic characteristics in uncontaminated portions of the aquifer
5. the need to acquire discharge permits and purchase groundwater rights for the surface disposal of large volumes of treated groundwater
6. the high cost of operating a large-scale project for a long period of time

An innovative alternative to the conventional groundwater treatment processes is the fourth remediation category, *in situ* stabilization of aquifer contaminants (1,8–10). Here the goal is to halt the movement of contaminants without adversely affecting the movement of groundwater. This can be accomplished through the use of permeable reactive barriers (also referred to as *in situ* treatment wall, passive *in situ* chemical barrier, etc).

Techniques for emplacing permeable reactive barriers have been reviewed and described (11). Three major approaches are used to emplace permeable barriers. The first approach, referred to as trench-and-fill, involves digging a trench perpendicular to the flow path of the contaminant plume, then back-filling the trench with material that sequesters the contaminant (Fig. 1) (12,13). Alternatively, a permeable barrier may be formed by the placement of the media into the aquifer strata through injection wells (Fig. 2) (13,14). In this case, the media can either be introduced in the form of colloidal particles or soluble reactants which add material to or modify the chemical properties of the aquifer material. The third approach to sorptive barrier emplacement is broadly referred to as soil mixing (Fig. 3). In this case, the reactive material is pumped as a slurry into a hollow-stem auguring machine. As the machine augurs through the aquifer material, the reactive material is mixed with it, creating a reactive barrier.

These three basic methods can be adapted for emplacement of both horizontal and vertical reactive barriers. The major vertical barrier emplacement technologies include trenching back-filling, slurry trenching, and vertical drilling and injection, including boring (earth auguring), cable tool drilling, rotary drilling, sonic drilling, jetting methods, injection-mixing in drilled holes, and deep soil mixing. Major horizonal vertical barrier emplacement technologies include horizontal drilling, microtunneling, compaction boring, horizontal emplacement, longwall mining, hydraulic fracturing, and jetting methods.

Each of these methods has their own advantages and disadvantages (11). Selection of the appropriate method depends on the reactive material to be used, subsurface conditions, and the physical layout of the site and plume to be contained.

The processes that cause the retention of contaminants include physical and chemical adsorption, ion exchange, chemical oxidation or reduction, or precipitation. The scope of this article is to review barrier systems and materials which involve sorption, specifically physical and chemical adsorption and ion exchange. Additionally, some illustrustive examples and barrier efficiency calculations will be presented. Reactive barriers consisting of materials that sequester contaminants via precipitation will not be discussed herein (see CLAY BARRIERS, CHEMICAL AND MINERALOGICAL ANALYSIS; WASTE CONTAINMENT BARRIERS).

Laboratory studies (8,12,15–17) have demonstrated the effectiveness of various natural and synthetic sorbents in removing contaminants from aqueous solution. Selected materials have included metal oxides, peat, bottom ash, limestone, fly ash, vermiculite, illite, ottowa sand, activated carbon, kaolinite, natural zeolites, activated alumina, and cullite. These studies revealed that no single sorbent can reduce the concentration of all possible constituents in a contaminated groundwater to acceptable levels. However, combinations of different sorbents can often be used to reduce most contaminants to acceptable levels.

SELECTED SORBENTS AND TARGET CONTAMINANTS

Cost and effectiveness are the two most important factors to be considered when selecting a potential sorbent material for a reactive barrier. Cost of the material for *in situ* barrier technology is particularly important because much larger quantities are generally required than for *ex situ* applications. Another important factor to consider is the permeability of the material. The material, once installed as a reactive barrier, must have a permeability greater than or equal to the surrounding aquifer material. The barrier material must maintain its permeability throughout the life of the remediation. A final consideration

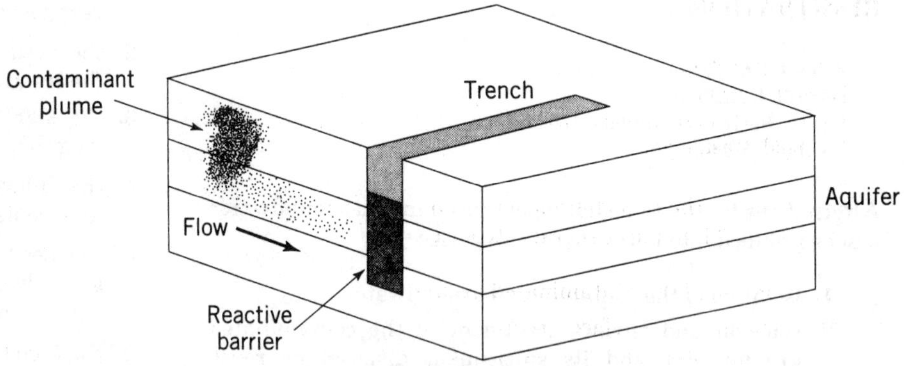

Figure 1. Trench-and-fill type permeable reactive barrier.

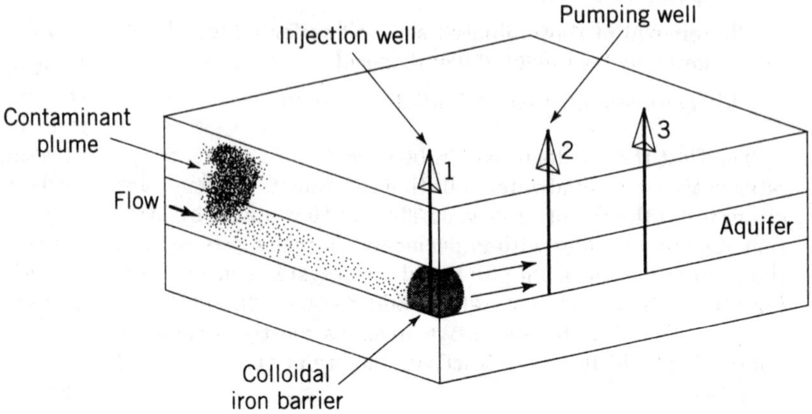

Figure 2. Permeable reactive barrier formed through injection of a solution or suspension. 1, 2, and 3 represent well numbers.

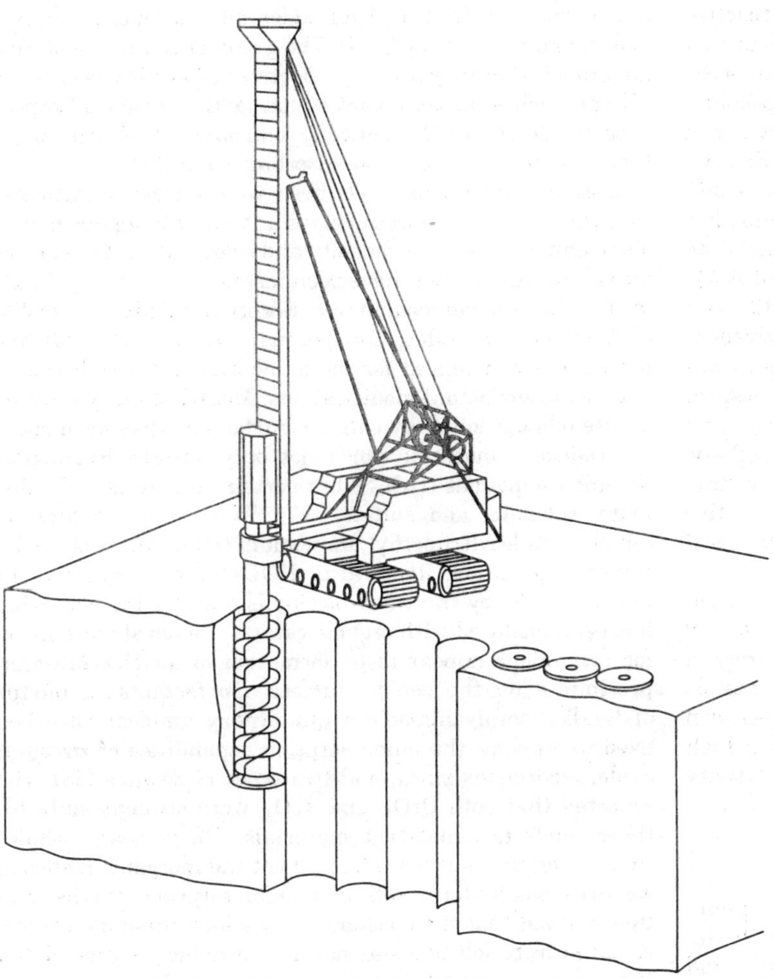

Figure 3. Permeable reactive barrier formed by soil mixing.

regarding the barrier material is that it must not produce any toxic or noxious by-products or leachates.

The selection of the appropriate material for a sorbent barrier is highly dependent on the contaminant(s) of concern, their concentrations, and the ambient geochemistry of the system. Not only will the unique chemistry of a contaminant be an important factor governing the effectiveness of a particular adsorbent to remove that contaminant, but the major ion chemistry of the water and other contaminants may also affect the sorption of the contaminant. Major ions or other contaminants may compete for adsorption sites on the barrier material or complex the targeted contaminant, resulting in decreased adsorption of the contaminant.

In the following discussion, various potential sorptive barrier materials are discussed with respect to their potential target contaminants and their effectiveness. High cost materials (eg, synthetic zeolites or ion-exchange resins) that are unlikely to be cost effective in *in situ* reactive barrier applications are omitted from this discussion.

Hydrous Ferric Oxide

Hydrous ferric oxides are well known for their ability to adsorb a variety of toxic substances including, metal ions, radionuclides, inorganic anions, and ionic organic compounds (18,19). The ability of hydrous ferric oxides to adsorb a particular compound is highly dependent on the pH. In general, cations are adsorbed more strongly at higher pH where the net charge on the surfaces of the hydrous ferric oxide is more negative and anions are adsorbed more strongly at lower pH where the net charge is more positive.

A number of factors should be considered before using hydrous ferric oxide. These include the potential influences of competing ions on adsorption of the contaminant of interest. Competition for adsorption sites with other ions can occur. The significance of the competitive effects depends on the major ion chemistry of the groundwater and the concentrations of co-contaminants. In addition to the previous factors, the length of time that the barrier is intended to last has to be considered. For example, hydrous ferric oxide is known for its high adsorption capacity resulting from its high surface area and high charge density; however, it is known that hydrous ferric oxide (ferrihydrite) transforms to goethite and/or hematite (20) which have lower surface areas and adsorption capacities.

Zeolites

Many zeolites, natural and synthetic, have been considered as adsorbents for a variety of metal ions and radionuclides. This discussion is limited to natural zeolites because synthetic zeolites are not likely to be cost effective in most *in situ* reactive barrier applications. Previous investigations have identified promising zeolites that can be used to restrict migration of Sr. It has been reported that the natural zeolites clinoptilolite, analcime, phillipsite, laumontite, erionite, mordenite, and chabazite all exhibit strong absorption properties for Sr from simulated groundwater (21). Data indicate that the zeolite chabazite is suited for the removal of Sr from waters that contain high concentrations of calcium and magnesium (22). A variety of synthetic zeolites (IE-95, IE-96, X-61, and A-51) and natural zeolites (clinoptilolite-3, clinoptilolite-12, and clinoptilolite-20) have been examined for their effectiveness as Sr sorbents (16). In addition to their effectiveness as adsorbents for Sr, these materials have been rated with respect to their relative cost, availability, long-term stability, and selectivity for Sr over competing cations such as Ca. Synthetic zeolite A-51 and clinoptilolite received the best overall rating. Further work on Sr adsorption on clinoptilolite as a reactive barrier material has been conducted (14,23) and is discussed in more detail below.

Natural zeolites have a greater affinity for Cs than Sr. However, because Cs adsorbs very strongly to several soil components, especially clay minerals, it is rarely a groundwater problem. The affinity of certain heavy metals (Ba^{2+}, Pb^{2+}, Cu^{2+}, Zn^{2+}, and Cd^{2+}) for clinoptilolite has been determined (24). It was found that clinoptilolite has a high selectivity for Ba^{2+} and Pb^{2+} and a somewhat lower selectivity for Cu^{2+}, Zn^{2+}, and Cd^{2+}.

Peat

Laboratory experiments have been conducted with sphagnum peat and mixtures of peat, $CaCO_3$, and hydrated lime to determine their effectiveness for immobilizing As, Mo, NO_3, and U from uranium-tailings leachate (25). In batch experiments, more than 97% of the U and As were removed from solutions which contained 5% peat (pH 3.6); 5% peat and 2.5% $CaCO_3$ (pH 6.3); 5% peat and 2.5% $Ca(OH)_2$ (pH 12); and 5% peat, 2.5% $CaCO_3$, and 2.5% $Ca(OH)_2$ (pH 10). Mo concentrations in solution significantly decreased when placed in contact with 5% peat and 5% peat and 2.5% $CaCO_3$, whereas Mo concentrations actually increased in solutions that were placed in contact with $Ca(OH)_2$. Nitrate was partially removed from the 5% peat solutions (80%) and a mixture containing 5% peat, 2.5% $CaCO_3$, and 2.5% $Ca(OH)_2$ (55%). Adsorption experiments were also conducted with U and Mo on peat (17). These experiments were conducted over a broad pH range. It was determined that U adsorption by peat was high at pH values between 4 and 8. Mo adsorption onto peat was only high at pH values below 6.

Anion Sorbing Agents

There are four broad classes of anionic sorbing materials that have been proposed: variable charge clays, surfactant-modified minerals, permanently charged positive minerals, and anion-precipitating minerals. Variable charge clays have a net positive charge at pH levels below their zero-point of charge, the pH value where the material has a net charge of zero coulomb/g. The anion-adsorbing capacity of the mineral increases as the pH decreases below the zero-point-of-charge. There are several pH-dependent charge minerals, the most common are Fe-oxyhydroxides. These minerals have a zero-point of charge between 7.5 to 8.3, depending on the specific iron oxide (26). Iron oxyhydroxides adsorb anionic forms of molybdenum and uranyl (17). The greatest limitation of these minerals is that they are only effective at low pH levels. As the pH approaches the zero-point-of-charge the anionic adsorption capacity decreases dramatically and above this critical pH, there is essentially no anion-adsorbing capacity.

Cationic surfactants have been widely used to form very thin films on the surfaces of clays and other inorganic materials to enhance sorption capacity and selectivity of the surfaces for anions. Anions that have been tested include CrO_4^- (27,28), 99-Tc (28), and radioiodine (29). Sorption affinities of zeolites (27), zirconium oxides, and porous carbon beads (30) have been greatly enhanced for inorganic anions through the use of cationic surfactant modifications. Modification of a natural zeolite (clinoptilolite) by cationic surfactants has been shown to produce a material which not only adsorbs hydrophobic organic compounds (HOCS) but certain anions as well (chromate, selenate, and sulfate) (27). In this case quaternary amine hexadecyltrimethyl ammonium (HDTMA) was used as the organic cation. Removal of HOCs by this organo-zeolite appears to be by the same mechanism as for the clay which has been modified with organic cations. Removal mechanisms for the anions appear to be formation of an HDTMA-anion precipitate on the zeolite surface. Surfactants (a mixture of trialkylmethlyammonium quaternary amines) have been used to enhance the anion sorption capabilities of zirconium oxide, amorphous silica, and two types of zeolites (28). They reported that both CrO_4^- and TcO_4^- were strongly sorbed by these surfactant-modified materials. They clearly showed that the organic surfactants, and not the inorganic materials, were responsible for the elevated anion sorption. Furthermore, they showed that the mechanism by which these anions were removed from solution was not due to reductive precipiation. It has been reported that for a number of clay minerals and natural soils treated with hexadecylpyridinium (HDPY) and benzethonium (BE), sorption rates and amounts for radioiodide (I^-) were increased several orders of magnitude above that of untreated material (29). Moderate increases in K_d values were observed for treatments using HDTMA, whereas treatments with trimethylpheny ammonium (TMPA) and tetramethyl ammonium (TMA) were ineffective. The fixation mechanism of the iodide was not clearly established. It was pointed out that the organic cations with relatively high molecular weights and long carbon chains caused swelling of the silicate layers in clay minerals, allowing the penetration of iodide. It was also indicated that the net negative charge on the clay minerals may have been shifted to positive values.

Most minerals in nature have a permanent negative charge, providing cation-exchange capacity to the sediment (31). These minerals have a negative charge, irrespective of the pH of the system. There are a few minerals, primarily man-made, that have a permanent positive charge. These minerals are presently being evaluated for use as anion exchangers, primarily for high pH systems where variable charge Fe-oxides have no anion adsorption capacity. Hydrotalcite is an example of such a permanent-positively charged mineral. It is rare in nature but relatively easily synthesized (32). It has been reported that although hydrotalcite minerals sorbed large quantities of I^- under some conditions, they have relatively low selectivity coefficients for I^- (32). Unfortunately, most contaminant anions are present in groundwater at concentrations that are several orders of magnitude less than those of background anionic constituents, such as carbonate, chloride, and sulfate. Another class of minerals are the cationic forms of clinoptilolite (a zeolite) with silver(I), thallium(I), mercury(I), and lead(II) on the exchange sites (33). These studies showed that the amount of I^- retained varies with the type of metal on the ion-exchange sites. Furthermore, the iodine was irreversibly sorbed (likely precipitated) onto these structures.

The final class of anionic sorbing materials is perhaps the most common and is referred to as anion-precipitating minerals. This class of material is outside the scope of this article; however, because it is quite commonly used, especially in the nuclear industry, a brief discussion is included here. The mechanism by which these minerals induce precipitation typically involve (*1*) reducing the contaminant from its mobile oxidized form to an immobile reduced form or (*2*) precipitating the anion as a sulfide or phosphate insoluble mineral. The ability of a number of sulfide minerals to retain 99-Tc and 129-I, both in anionic form, has been studied. The minerals evaluated included stibnite (Sb_2S_3), molybdenite (MoS_2), chalcopyrite ($FeCuS_2$), galena (PbS), and pyrite (FeS_2). It was reported that Tc is strongly sorbed by stibnite, the sorption ratio was about 2000 mL/g, and the sorption was irreversible. The sorption ratios for I on chalcopyrite and galena were about 100 mL/g and sorption was also irreversible. Anion retention by phosphate minerals has been described (35). Reductive precipitation of anions, including Cr, Mo, and Tc, have been demonstrated to be readily accomplished with zero-valent iron (36).

Surfactants Modified Minerals for Adsorption of Hydrophobic Organic Compounds

Because the sorption of hydrophobic organic compounds (HOCs) is largely controlled by the amount of organic matter in the aquifer material, an *in situ* reactive barrier for adsorption of HOCs could be constructed by increasing the organic matter content of the aquifer material or by constructing a permeable reactive barrier composed entirely of organic material. This could be accomplished in several ways using relatively inexpensive materials. For example, a trench could be dug and filled with a mixture of peat moss and aquifer materials. Soil mixing methods could also be used to incorporate peat moss into the aquifer materials. Other novel, low cost materials might also be used such as granulated rubber tire or sawdust. Another technique used to increase the organic content of aquifer material is to add clays which have been treated with organic cations or to treat the clays within the aquifer directly by injecting a solution of cationic organic compounds (37,38).

It has been observed that natural organic material, regardless of source, tends to be more or less comparable in their ability to accept nonionic organic compounds (39). The mineral components of soils and aquifer materials tend to be strongly hydrated while in contact with water. Most clays are negatively charged as a result of isomorphic substitution in the aluminosilicate structure. This negative charge is neutralized by exchangeable inorganic ions such as Na^+ or Ca^{2+}. These strongly hydrated ions impart a hydrophilic character to the mineral surface. HOCs cannot effectively compete with water for these adsorption sites and therefore interact primarily with the organic matter phase (40). By exchanging the inorganic ions with organic cations, the surface of the clay mineral can be changed from hydrophilic to organophilic. This hydrophobic adsorption tends to be driven by the incompatibility of HOCs with water, rather than the attraction of the HOCs to organic matter. As a result of this process, an aquifer could be altered such that its surfaces will have a strong sorptive capacity for HOCs. A number of organic cations have been studied for this purpose including quaternary ammonium cations and dodecylpyridinium (37,38,40–42). The adsorption of a variety of contaminant compounds were also studied, including benzene, toluene, ethylbenzene, and naphthalene (40,41), perchlorethylene (PCE) (38), diuron (42), and chlorobenzenes (37). In general, treatment of clays with organic cations greatly increases their ability to adsorb these organic compounds. It has also been found that the distribution coefficient (K_d = contaminant concentration on the solid phase/contaminant concentration in the liquid phase) is dependent on the particular organic cation used for the treatment. The results presented to date indicate that treatment of soils and subsurface materials with organic cations shows promise for promoting HOC sorption.

EFFECTS OF SORBENT CAPACITY, BARRIER THICKNESS, GROUNDWATER FLOW RATE, AND RESIDENCE TIME ON BARRIER EFFECTIVENESS

A number of parameters that influence the effectiveness of a sorptive barrier are evaluated in this section by using simplified calculations. Such "back-of-the-envelope" calculations, while being less accurate than those based on more sophisticated equations, provide important first approximations. For instance, by making a few simplifying assumptions, spreadsheet-type calculations can quickly provide a sensitivity analysis for a number of barrier parameters. Once the appropriate ranges of barrier parameters are determined using these simplified calculations, more sophisticated calculations, such as those described later in this article, must be conducted to accurately determine the design parameters of the sorptive barriers.

Kinetic Approach

Two general approaches are used here to describe the sorption of groundwater contaminants by a barrier: a kinetic and an equilibrium model. The kinetic model is used to describe sorptive barriers that sorb contaminants slowly with respect to the residence time of the contaminant in the barrier (ie, the reaction is rate limited). Kinetic models are commonly used to describe many redox or precipitation reactions, and to a lesser extent, adsorption reactions (43). The first-order kinetic model describes the rate of reaction as occurring at a fixed linear rate with respect to the concentration of the contaminant in solution. The rate constant, α (d^{-1}), is experimentally determined by monitoring the decrease in solution contaminant concentration as a function of time (44). It is described numerically in equation 1:

$$C_{\text{out}} = C_{\text{in}} e^{-\alpha t} \quad (1)$$

The symbols in equation 1 are described in Table 1. A useful parameter that can be calculated from the first-order rate constant is the half-life, $t_{0.5}$ (eq. 2).

$$t_{0.5} = \frac{-\ln(0.5)}{\alpha} \quad (2)$$

The half life, as it will be used in this section, can be thought of as the time required for the contaminant concentration to decrease by half. Calculation results using equations 1 and 2 are presented in Figure 4.

The residence time required to remove a given fraction of contaminant is very sensitive to the half-life of the reaction (see Fig. 4). For example, at a 10 day half-life, increasing the residence time from 0.7 to 1.1 days increases the ratio C_{in}/C_{out} from 10 to 1000. However, at a 100 day half-life, the residence time must be increased from 320 to 1000 days to obtain a similar increase in the C_{in}/C_{out} from 10 to 1000. Thus, very slight changes in the residence times for reactions with short half-lives can have a profound effect on the efficiency of the barrier.

The degradation of chlorinated ethenes by zero-valent iron reactive barriers provides an excellent example of the importance of a reaction half-life on barrier efficiency. Zero-valent iron barriers degrade this class of organic compounds by reductive dehalogenation and therefore are not sorptive barriers per se, yet the reactions are commonly described as occurring via first-order kinetics (45). The half-life for the degradation of tetrachloroethene (PCE) is 0.75 days and the degradation of vinyl chloride (VC) is 15.6 days (45).

Equilibrium Approach

The other approach to estimating sorptive barrier performance is through equilibrium models. Such models assume that (*1*) the reaction rates are essentially instantaneous and (*2*) all important reactions have gone to equilibrium while the contaminant is in contact with the barrier material. A mass balance equation describing groundwater contaminant adsorption to a barrier is presented in equation 3:

$$C_{in} \cdot q \cdot t = \phi \cdot M_{barrier} \quad (3)$$

Analysis of the units associated with equation 3 reveals that the number of equivalents of contaminant in solutions entering the sorptive barrier (left side of eq. 3) is equal to the number of equivalents of contaminant adsorbed to the barrier (right side of eq. 3). The mass of barrier material, $M_{barrier}$, can be expressed as:

$$M_{barrier} = \rho_b \cdot A \cdot L_{th} \quad (4)$$

Substituting equation 4 into equation 3 yields:

$$C_{inf} \cdot q \cdot t = \phi \cdot \rho_b \cdot A \cdot L_{th} \quad (5)$$

Finally, equation 5 can be rewritten to include the linear pore velocity, ν, a velocity term more commonly reported than the Darcy velocity (q),

$$\nu = \frac{q}{A \cdot n_e} \quad (6)$$

by substituting equation 6 into equation 5:

$$t = \frac{\phi \cdot \rho_b \cdot L_{th}}{C_{in} \cdot n_e \cdot \nu} \quad (7)$$

The capacity of a specific sorbent, ϕ is typically very sensitive to the contaminant concentration and the aqueous concentration of the background electrolytes. Consequently, it is important to measure the sorptive capacity of a barrier material under environmental conditions similar to those existing in the study site. For example, to determine the sorbent capacity of a proposed zeolite barrier for groundwater Sr-90 it is necessary to determine the sorpent capacity in the presence of naturally occurring strontium and competing cations such as calcium. To accomplish this goal, adsorption experiments were conducted with simulated groundwater using a range of initial strontium concentrations spiked with Sr-85 as a tracer (12). By saturating the zeolite adsorption sites with calcium prior to bringing the zeolite in contact with the Sr, the researchers were assured of providing a conservative estimate of the sorption capacity. Calcium, which exists in groundwater at several orders of magnitude greater concentrations than Sr^{2+}, is well known to compete with Sr^{2+} for cation-exchange sites. Additionally, by evaluating a range of Sr concentrations, researchers were able to determine how the sorbent capacity varied as a function of Sr concentrations. It was reported that the sorbent capacity varied as a function of strontium concentration in solution from 10^{-2} to 10^2 meq/kg. These experiments (12) will be discussed in more detail below.

Results of calculations using equation 7 are presented in Figure 5. For these calculations, barrier thickness, L_{th}, and linear pore velocity, ν, were varied and the contaminant concentration, C_{in}, bulk density, ρ_b, cross-sectional area, A, and sorbent capacity, ϕ, were held constant at the default

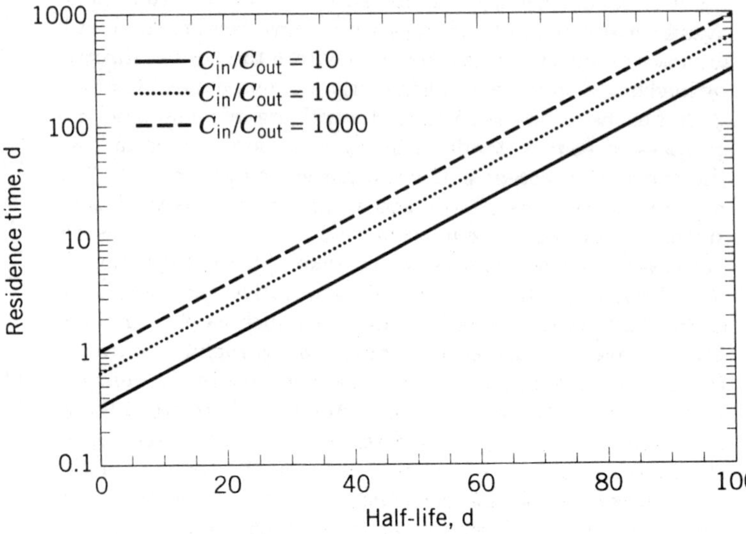

Figure 4. Required residence time of a contaminant plume in a sorptive barrier as a function of the reaction half-life and decrease in contaminant concentration (C_{in} = contaminant concentration going into barrier; C_{out} = contaminant concentration leaving barrier) (see eqs. 1 and 2).

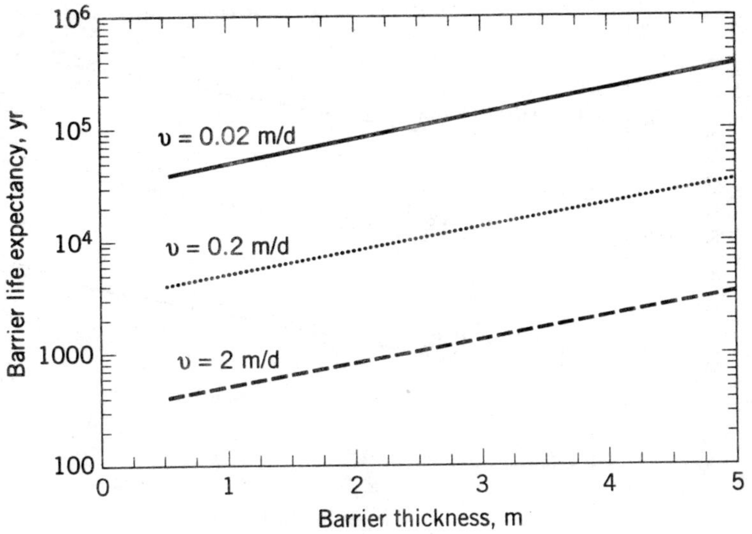

Figure 5. Barrier life expectancy as a function of barrier thickness and groundwater velocity (eq. 7).

Table 1. Symbol Definitions and Default Values Used in Calculations

Symbol	Definition	Default Value	Units
A	Cross-sectional area	1	m^2
C_{in}	Contaminant concentration going into barrier	10	meq/m^3
C_{out}	Contaminant concentration leaving barrier		meq/m^3
K_d	Distribution coefficient of barrier		m^3/kg
L_{th}	Barrier thickness		m
$M_{barrier}$	Mass of barrier material		kg
n_e	Effective porosity	0.2	m^3/m^3
q	Darcy flow velocity $= -K(dH/dx))$	0.04	m^3/d
t	Time		d
$t_{0.5}$	Half-life		d
α	First-order rate constant		1/d
ϕ	Sorbent capacity	700	meq/kg
ρ_b	Barrier bulk density	1600	kg/m^3
ν	Linear pore velocity $= q/(A^* n_e)$	0.2	m/d

values shown in Table 1. Perhaps the most striking conclusion that can be made from Figure 5 is that the groundwater velocity has a profound effect on the barrier life expectancy (the duration that the barrier is effective at sequestering the contaminant). An order of magnitude increase in the groundwater velocity results in an order of magnitude increase in the barrier life expectancy. These calculations emphasize the importance of having accurate groundwater velocity values for barrier efficiency calculations.

Figure 6 presents another set of calculations using equation 7. These calculations differ from those presented in Figure 5 in that sorbent capacity was varied and the linear pore velocity was held constant at 0.2 m/d. As expected, as sorbent capacity or barrier thickness increases, the barrier life expectancy increases.

The sorbent capacity, ϕ, can be expressed in terms of the commonly used distribution coefficient, K_d,

$$\phi = C_{in} \cdot K_d \tag{8}$$

Equation 8 is based on the following definition:

$$K_d = \frac{C_{ad}}{C_{eq}} \tag{9}$$

where C_{ad} is the concentration of the contaminant adsorbed (meq/kg) and C_{eq} is the concentration of the contaminant in solution in equilibrium with the adsorbent (meq/m^3). Assumptions underlying equation 8 include *(1)* C_{in} equals C_{eq} and *(2)* ϕ equals C_{ab}. These are reasonable assumptions for most sorptive barrier scenarios. Equation 8 can be substituted into equation 7 to yield an equation containing the commonly measured K_d parameter:

$$t = \frac{K_d \cdot \rho_b \cdot L_{th}}{N_e \cdot \nu} \tag{10}$$

The value of equation 10 is that it is in terms that are commonly used by environmental engineers and scientists (Fig. 7).

Using materials with large K_d values significantly extends the barrier life time. A K_d value of 3 m^3/kg (3000 mL/g) has

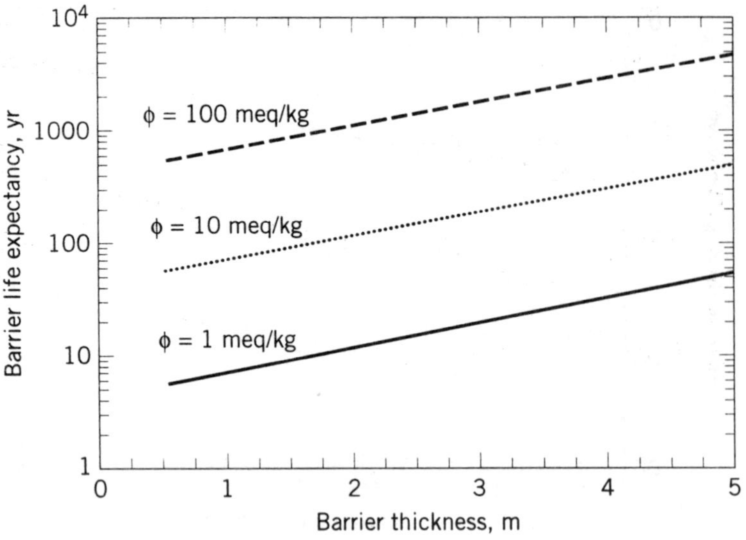

Figure 6. Barrier life expectancy as a function of barrier thickness and barrier material sorptive capacity (eq. 7).

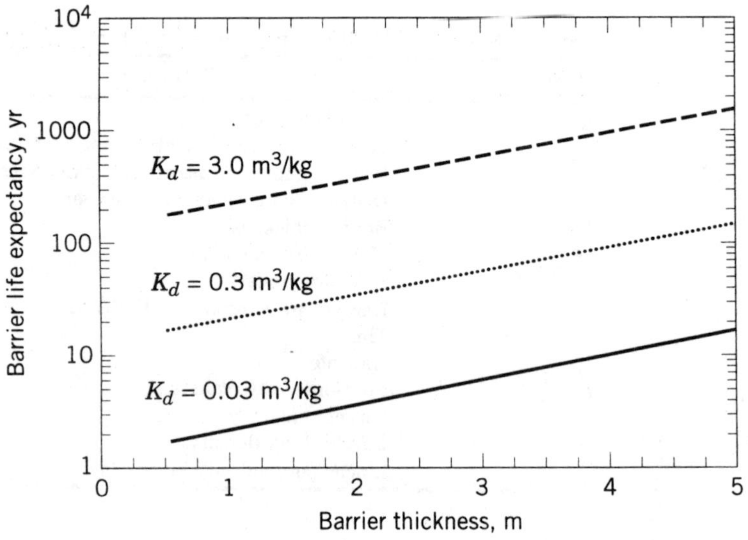

Figure 7. Barrier life expectancy as a function of barrier thickness and adsorption distribution coefficient (eq. 10).

been reported for several surface-modified clays such as zeolites and montmorillonites modified with quaternary amines (41,46). K_d values of greater than 1000 mg/mL have been reported for strontium onto zeolites (12).

CASE STUDIES OF LABORATORY TREATABILITY STUDIES FOR FIELD DEMONSTRATIONS

In this section two studies conducted to provide feasibility and design parameters for pilot-scale field demonstrations of *in situ* sorptive barriers are discussed. In order to fully evaluate a material's potential as an effective reactive barrier material, much more than just a simple determination of the ability of the material to adsorb the contaminant of interest is required. An equilibrium isotherm is required to determine the adsorption properties of the material as a function of contaminant concentration. From this data the adsorption capacity of the material as a function of the contaminant concentration can be determined. A simple exchange capacity determination or adsorption coefficient (K_d) is generally inadequate for this purpose. In addition, this adsorption isotherm should be conducted under geochemical conditions which are similar to those of the actual site where the barrier is to be installed. Another consideration, as discussed in the preceeding section, is the kinetics of the adsorption reaction. The reaction must proceed at a rate fast enough for the contaminant to be adsorbed before it passes through the barrier. As a result, the reaction kinetics and groundwater flow rate will determine how thick the barrier should be and its cost.

In addition to the important chemical properties mentioned above, the material must maintain a hydraulic conductivity that is greater than or equal to the aquifer material. If the reactive barrier is less permeable than the surrounding aquifer material, groundwater will be diverted around the barrier rendering it partially or completely ineffective. For some materials, reactivity or permeability of the barrier may change with time. Therefore, it is important to determine the permeability of the material and to evaluate any potential reactions that could alter the permeability of the barrier material with time.

Clinoptilolite Barrier for Adsorption of Sr-90 from Groundwater

Background. Past disposal practices in the 100-N area of the U.S. Department of Energy's (DOE) Hanford Site (Richland, Washington) have resulted in contamination of sediments and groundwater beneath the site. The source of the contamination was liquid wastes that were generated from the N-reactor coolant system, spent fuel storage basin, periphery coolant systems, and laboratory and radioactive drain systems in the reactor facility. Two trenches located parallel to the Columbia River at approximately 250 m and 600 m from the river bank were used for ground disposal. These trenches were operational from 1964 to 1991. The cumulative inventory of the major radionuclides disposed in the trenches is 5100 Ci Co-60, 2000 Ci Sr-90, and 2650 Ci Cs-137 (47). It has been determined that these two liquid waste disposal facilities are the source of Sr-90 contaminated groundwater which is entering the Columbia River adjacent to the site. Groundwater containing Sr-90 at levels above the drinking water standard (8 pCi/L) is discharging to the Columbia River in the 100-N area. Current geochemical and hydrogeologic data suggest that the remaining Sr-90 entering the river has desorbed from contaminated aquifer sediments in contact with groundwater. The maximum Sr-90 concentrations detected in the 100-N area in 1994 was 4030 pCi/L in a well between the trenches and the Columbia River.

A variety of remedial technologies were considered to prevent further migration of Sr-90 to the Columbia River. These include pump-and-treat, impermeable vertical barriers, and hydraulic control methods (48). The pump-and-treat and hydraulic control methods are expensive and require continuous maintenance. It is unlikely that the pump-and-treat option would be any more effective than the hydraulic control alternative because Sr-90 is adsorbed relatively strongly to the aquifer sediments, and the Sr-90 in the sediments will continue to desorb for a long time. With a 29-year half-life, it will require 300 years for the Sr-90 in the sediments to decay to less than 1% of its current inventory. In addition, these methods require treatment of the contaminated groundwater pumped to the surface and disposal of the radioactive waste created by the treatment method.

The impermeable vertical barrier methods have some significant disadvantages. The first is that an impermeable wall will raise the groundwater level. Because the concentrations of radionuclides in the sediments are greatest just above the groundwater (49), an increase in the groundwater level is likely to cause Sr-90 concentrations in the groundwater to increase. The other disadvantage is that it will not stop migration of Sr-90 to the river but only delay it.

Research was conducted to determine relevant physicochemical and hydrologic parameters for zeolites that can be applied to the design of a permeable barrier composed of clinoptilolite to effectively stop Sr-90 migration in groundwater (12,23). Both batch equilibrium isotherm data and kinetic data were collected. Particular attention was directed at understanding the competitive effects of Ca on Sr sorption to the zeolite. These data were used to develop a model for designing a barrier and determining its performance. Hydraulic conductivity data measurements were made on clinoptilolite samples of various particle size ranges. This information, along with the kinetic data determined using the various particle size ranges, was used to determine the best particle size range to use in the field demonstration. Experimental details have been presented (12 ,23).

Adsorption Isotherms. The adsorption data for Sr on clinoptilolite is shown in Figure 8. The data were fit to a Langmuir isotherm:

$$T = \frac{MKC}{(1 + KC)} \tag{11}$$

where T is the total average concentration of Sr adsorbed to the zeolite (mol/kg), C is the concentration of Sr in solution (mol/L), M is the adsorption maximum (mol/kg), and K is an affinity parameter (L/mol) (50). M was determined to be 0.136 mol/kg and K was 7030 L/mol using a nonlinear least-squares algorithm. The isotherm is linear at solution concentrations of less than 10^{-5} mol/L. In this region, an adsorption coefficient, K_d, of 956 L/kg (Langmuir K^*M) was determined. Because background Sr concentrations in Hanford groundwater are typically 3×10^{-6} mol/L, use of $K_d = 956$ L/kg for modeling purposes is appropriate.

Kinetic Experiments. In Figure 9, data determined from the following sources are compared: (*1*) 15-day data for 0.30 to 0.84 mm clinoptilolite determined in the previous equilibrium experiments, (*2*) 15-day data collected during the kinetic experiments for the 0.30 to 0.84 mm clinoptilolite, (*3*) 15-day kinetic data determined for 0.30 to 0.84 mm clinoptilolite that was soaked in solution containing 0.002 M $CaCl_2$ and 0.002 M NaCl before addition of Sr, and (*4*) 15-day kinetic data determined for 0.008 to 0.016 mm clinoptilolite. The 15-day data from the kinetic experiments are all in good agreement. However, it is apparent that the previous equilibrium data are slightly different from the data determined in the kinetic experiments. This disparity was determined to be the result of differences in the method of equilibration (23).

All the data shown in Figure 9, except the data from the first equilibrium isotherm data, were fit to a Langmuir isotherm. M was determined to be 0.0984 mol/kg and K was 26,500 L/mol using a nonlinear least-squares algorithm. The isotherm is linear at solution concentrations of less than 10^{-5} mol/L. In this region, an adsorption coefficient, K_d, of 2600 L/kg (Langmuir K^*M) was determined. Because background Sr concentrations in Hanford site groundwater are typically 3×10^{-6} mol/L, use of $K_d = 2600$ L/kg for modeling is appropriate (if equilibrium can be assumed).

An example of how the concentration of Sr in solution decreased with time for various particle size ranges of clinoptilolite particles is shown in Figure 9. It is clear that the rate of adsorption decreases as the particle size range increases. The most significant differences between particle sizes occurred during the first 24 hours. After this time, differences in the rate of adsorption between the particle sizes become relatively minor. The difference in the rate of adsorption between the 0.30 to 0.84 mm material and the 0.008 to 0.016 mm material is nearly insignificant after 0.5 h.

Figure 10 illustrates an example of the normalized concentration (C/C^0) of Sr in solution in contact with 0.30 to 0.84 mm clinoptilolite as a function of time for various initial strontium

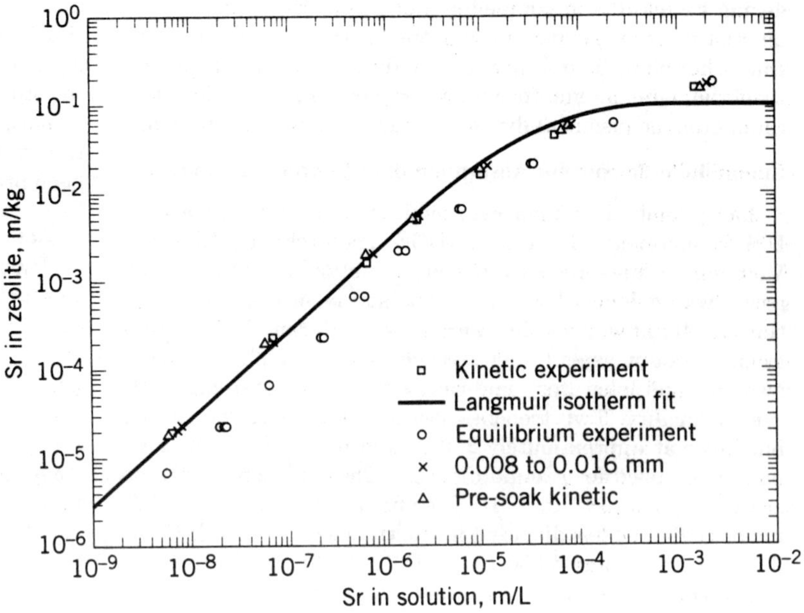

Figure 8. Comparison of adsorption data of Sr on 0.30 to 0.84 mm clinoptilolite (Ca-saturated).

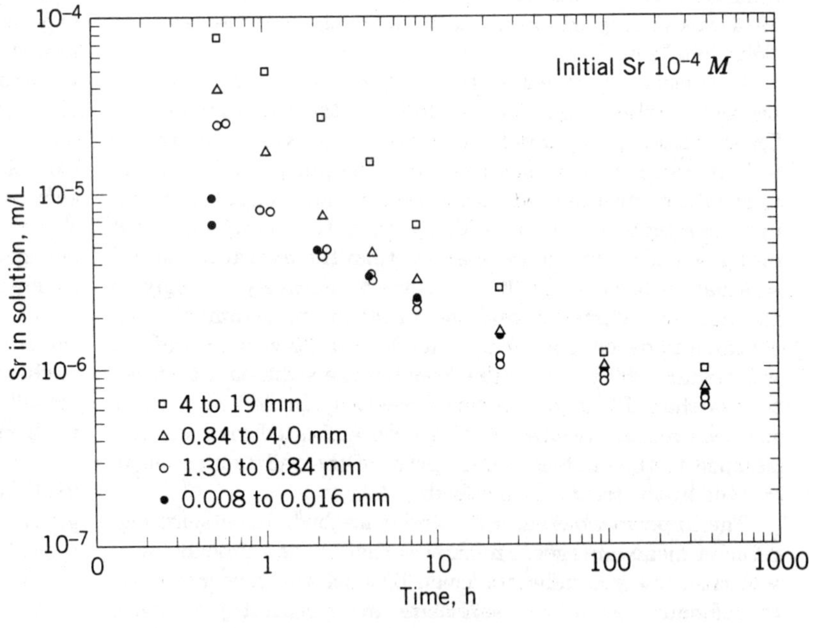

Figure 9. Sr concentration in solution in contact with various particle sizes of clinoptilolite (Ca-saturated) as a function of time.

values. These results indicate that the majority of adsorption occurs during the first 24 hours. At very low initial Sr concentrations, the change in the normalized concentration with time is essentially identical. As the concentration in solution increases to a point where the total strontium absorbed is a significant fraction of the total sites available, the decrease in the normalized solution concentration with time becomes significant.

A kinetic model for the process of adsorption of strontium on clinoptilolite was developed. The rate parameters were obtained by fitting the model to batch experimental data. The zeolite granules were taken to have a bipore distribution, micro (intracrystalline) and macro (intercrystalline) pores. Adsorption/desorption is assumed to occur as a two-stage process, first onto the surface offered by macropores followed by migration to micropores. Initially, the rate of adsorption is high but slows down quickly due to site saturation and desorption. The relatively slow migration to micropores prevents the rapid attainment of equilibrium, observed in the experiments. The details of the kinetic model are presented below.

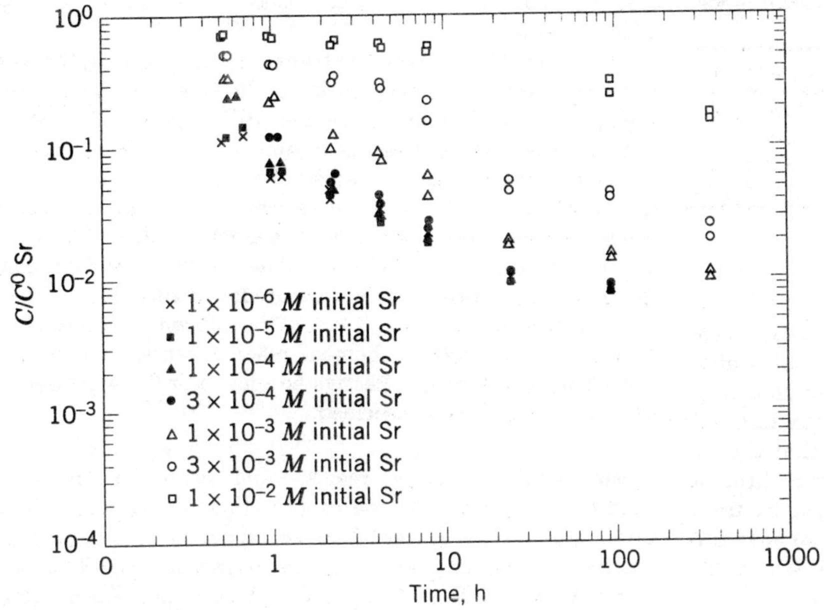

Figure 10. Normalized concentration of Sr in solution in contact with clinoptilolite (0.30 to 0.84 mm) as a function of time and initial Sr concentration.

Model fits for the kinetics of adsorption of Sr on clinoptilolite (0.30 to 0.84 mm, Ca-saturated) for various initial Sr concentrations are presented in Figure 11. In general, the model fits are excellent. For the 3 × $10^{-3}M$ initial Sr concentration data, the model fit deviates from the experimental data at 24 and 96 hours, but is otherwise in good agreement. Model fits for the other particle size ranges were as good or better than those illustrated.

Saturated hydraulic conductivities (K_{sat}) of various zeolite particle size ranges were measured in packed columns set up to operate as constant-head permeameters. Measurements were made for the following particle size ranges: 19–4.0 mm, 2.4–4.8 mm, 5 × 20 mesh, and 0.84–0.30 mm. During an experiment, water flows from the supply reservoir into the inlet port, upward into the zeolite, and out the outflow/overflow port at the upper end of the column. Upward flow is desirable because it helps remove trapped air from the pore space during initial wetting. The flow rate through the column was determined by collecting the column effluent in a plastic container; the hydraulic gradient across the sample was determined as the measured difference in water level in the two manometers. Saturated hydraulic conductivity, K (cm/s), was computed from these measurements using Darcy's law.

Results from the saturated hydraulic conductivity measurements are shown in Table 2. The mean and standard deviations were determined from the last three measurements taken from three separate tests (four in the case of the 2.4 to 4.8 mm material). A significant increase in conductivity is observed as the particle size increases. The design of a permeable reactive barrier must consider two objectives for

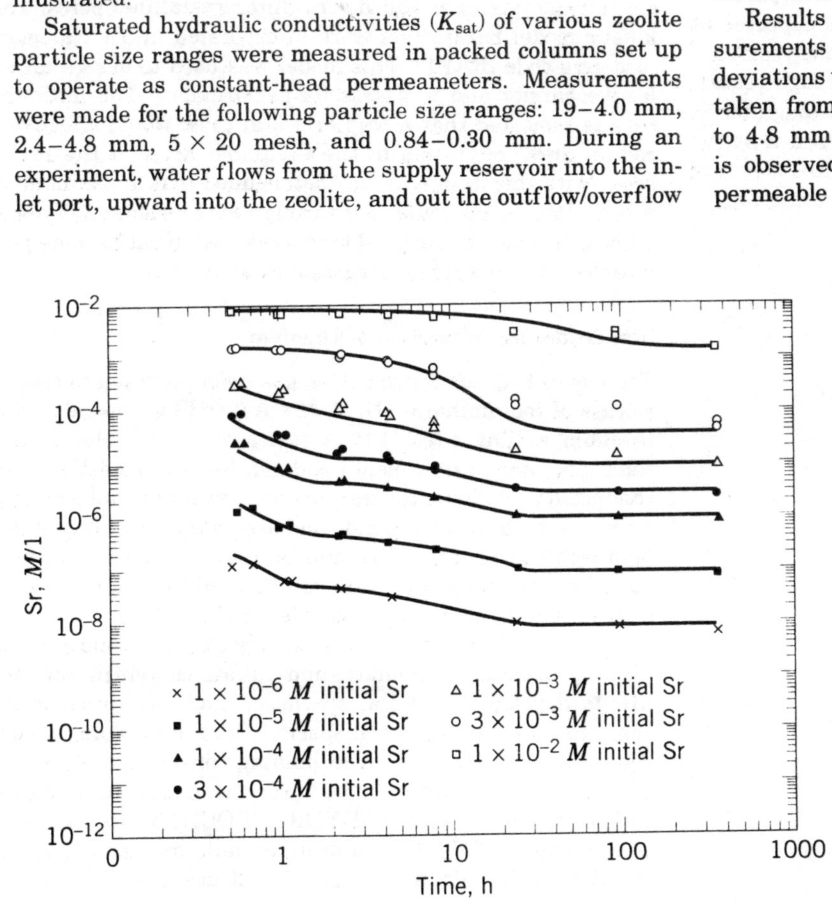

Figure 11. Sr concentrations in solution in contact with 0.30 to 0.84 mm clinoptilolite (Ca-saturated) and model estimates.

Table 2. Saturated Hydraulic Conductivity Measurements for Various Particle Size Ranges of Clinoptilolite

Particle Size Range, mm	K_{sat}, cm/min
0.30–0.84	3.9 ± 0.9
0.84–4.0	8.7 ± 2.5
2.4–4.8	18.4 ± 2.8
4.0–19	21.2 ± 5.8

selecting which particle size range of zeolite to use. The most important is that the barrier must remain more permeable then the surrounding aquifer material. If the permeable barrier does not remain more permeable then the surrounding aquifer material, groundwater may flow around the barrier, negating its effectiveness. The other objective is that the Sr must react with barrier material before it passes through the barrier material. The smallest particles remove Sr from solution most rapidly, but the smallest particles also have the lowest hydraulic conductivity. As a result, particle size range selection depends on site-specific hydraulic conductivity and groundwater velocity.

Performance Modeling of a Strontium Barrier Composed of Clinoptilolite. The breakthrough performance of a 1.0-m thick barrier of clinoptilolite was determined with a one-dimensional hydrologic model using the equilibrium adsorption isotherm data shown in Figure 12 (K_d = 2600 L/kg). The hydrologic parameters used in the model were hydraulic conductivity, 67 m/d; hydraulic gradient, 0.00037 m/m; porosity, 0.35; and bulk density, 1.03 g/cm^3. The parameters provide a pore velocity of 0.07 m/d. The values are typical of those expected at the 100-N area of the Hanford site. The hydraulic residence time within the barrier is 14 days. The breakthrough results are illustrated in Figure 12. Fifty percent breakthrough for Sr occurs at approximately 225 years. If it is assumed that the source term for Sr-90 entering the barrier remains constant and its radioactive half-life is 29 years, a maximum breakthrough occurs at 235 years at a concentration of 0.28% of its influent value.

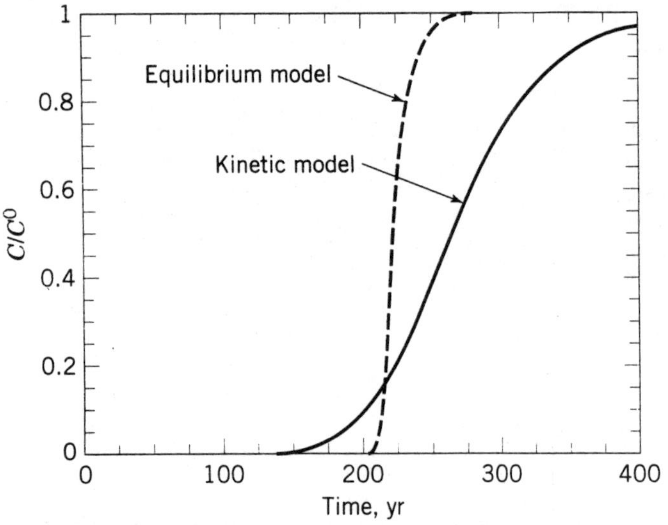

Figure 12. Transport of Sr through a permeable zeolite barrier.

The effect of kinetics on the transport of Sr through the barrier was assessed by incorporation of the kinetic model into the transport model. The kinetic results are compared with the results in which equilibrium is assumed in Figure 12. The effect of kinetics is relatively small. This is consistent with the results shown in Figure 9, when compared to the hydraulic residence time in the barrier. For the kinetic model, 50% breakthrough of Sr occurs at 265 years. The maximum Sr-90 breakthrough in the kinetic model also occurs at 235 years, but its maximum concentration is 0.10% of the influent concentration. Based on these results, a chemical reactive barrier composed of clinoptilolite will be an effective barrier to Sr-90 migration *in situ* for the conditions described.

Summary. Laboratory results and computer modeling (12,23) has shown that the zeolite clinoptilolite would be a very cost-effective material for construction of a permeable reactive barrier for preventing the migration of Sr-90 at the 100-N area of DOE's Hanford site in Richland, Washington. The Sr adsorption data for Ca-saturated clinoptilolite were found to fit to a Langmuir isotherm. The isotherm is linear at solution Sr concentrations of less than 10^{-5} mol/L. In this region, an adsorption coefficient (K_d) of 2600 L/kg was determined. Because Sr concentrations in Hanford groundwater are typically 3×10^{-6} mol/L, assuming linear adsorption (K_d = 2600 L/kg) for modeling purposes is appropriate.

Adsorption kinetics of Sr onto clinoptilolite were also determined. These data were used to develop a kinetic model. The zeolite granules were taken to have a bipore distribution, micro (intracrystalline) and macro (intercrystalline) pores. The kinetic model parameters were incorporated into a transport modeling code (RAFT). This model was used to design an effective barrier and to assess its performance. The modeling results indicated that a barrier 1.0-m thick would effectively reduce Sr-90 migration to the Columbia River at the 100-N area. With this design, it was determined that a maximum of 0.10% of the Sr-90 would pass through the barrier at 235 years. To be effective, the permeable reactive wall must be more permeable than the surrounding aquifer material.

Iron Oxides for Adsorption of Uranium

The use of hydrous ferric oxides has been proposed to control release of contaminants (U, V, Mo, As, and base metals) from uranium milling sites (13). A series of batch, column, and tank laboratory experiments and geochemical modeling were conducted to evaluate the interaction of uranium and molybdenum with hydrous ferric oxide and assess the feasibility of this approach (13,51–53). This information combined with site-specific characterization data can be used to design chemical barriers emplaced by injection of ferric chloride.

A series of batch adsorption experiments was conducted, at varying uranium, carbonate, and sulfate concentrations and pH, to develop a chemical speciation and adsorption model that accurately predicts transport of U(VI) for a wide range of chemical conditions (51). The binding constants and adsorption model developed from this work was used in a coupled hydrogeochemical model (HYDROGEOCHEM) (54) to simulate transport of U(VI) from uranium mill tailings (52). It was found that the adsorption behavior of uranium (VI) (uranyl ion and its complexes) is similar to other cations; a steep increase in adsorption occurs with increasing pH. A simple site complexation model, which included only one type of adsorption site and did not incorporate electrostatics, was found to provide a reasonable match to measured adsorption and proton-exchange data. This simple model has the advantage of efficient incorporation into the coupled reaction-transport model and lower run times for model simulations compared to that of a more complex model. Because uranium adsorption kinetics are known to be rapid (on the order of minutes) (53,55), determination of the rate of adsorption of U onto the hydrous ferric oxide was not necessary.

The coupled hydrogeochemical model was used to provide simulations for a variety of scenarios (52). The modeling results characterized the spatial and temporal evolution of the tailings-groundwater system and permitted delineation of the uranium(VI) plume in groundwater for the different scenarios. The model allowed investigators to optimize various design options with regard to regulatory requirements and economic constraints. It was found that adsorption of uranium(VI) onto hydrous ferric oxides can be an effective method for containing the release of uranium(VI) into groundwater, especially for neutral-pH mill tailings. It was also determined that essentially no retardation occurred in tailings with alkaline pore fluids.

Further work was conducted to evaluate the viability of injecting solutions of dissolved ferric chloride into the subsurface to form a hydrous ferric oxide reactive barrier (53). Dissolved ferric chloride hydrolyzes to hydrous ferric oxide ($Fe(OH)_3$) when it contacts calcite ($CaCO_3$) as follows:

$$FeCl_3 + 2\,CaCO_3 + 3\,H_2O \rightleftarrows Fe(OH)_3 + 2\,Ca^{2+} + 3\,Cl^- + H_2CO_3^0 + HCO_3^- \qquad (12)$$

As a result, injection of ferric chloride solutions through wells into aquifers which contain calcite may be a viable method to emplace a zone of hydrous ferric oxide. The work focused primarily on determining the effects of injecting ferric chloride into simulated aquifer material containing calcite. The factors which were evaluated during this study included (*1*) the effectiveness of the hydrous ferric oxide, formed by injection of $FeCl_3$ into media containing calcite, for removing U and Mo from contaminated groundwater, (*2*) the amount of decrease in hydraulic conductivity resulting from the introduction of the hydrous ferric oxide, and (*3*) the mobility of the hydrous ferric oxide subsequent to its formation in the porous media. Batch experiments were conducted to examine U adsorption onto hydrous ferric oxide prepared by mixing powdered calcite with $FeCl_3$ solutions.

Column and two-dimensional tank apparatus experiments were used to determine adsorption density (how effectively uranium utilized the available adsorption sites), changes in hydraulic conductivity resulting from the hydrous ferric oxide emplacement process, and the mobility of the hydrous ferric oxide within the porous media after it is formed.

Although the adsorption of U(VI) on hydrous ferric oxide is well understood (51), the phase of hydrous ferric oxide that forms can depend on the method of preparation (20). Therefore, experiments were conducted to determine if hydrous ferric oxide prepared with calcite has the same uranium adsorption properties as hydrous ferric oxide prepared with NaOH. The results of the batch experiments confirmed that uranium adsorption onto hydrous ferric oxide prepared with calcite increases sharply as pH increases from 6 to 8.5 in a manner which is nearly identical to that of hydrous ferric oxide prepared by hydrolysis of $FeCl_3$ with NaOH. These results indicate that both preparations result in hydrous ferric oxide.

The adsorption density determined in the tank experiment at breakthrough was found to approach the theoretical value to within 67 ± 22%. A significant observation regarding the adsorption of the uranium in the tank experiment was a gradual breakthrough of uranium. This gradual breakthrough is not the result of slow adsorption because it is known that the adsorption kinetics are fast relative to the flow rates used. One explanation was that the hydrous ferric oxide was deposited in a complex flow network consisting of wide (primary) channels through which most of the flow occurred, narrower (secondary, tertiary, etc) channels with progressively less flow, and pores containing stagnant water (53). As uranium-bearing fluid flows through the tank, flow occurs mainly through the primary channels, with a smaller portion accessing secondary channels and even less fluid reaching tertiary or higher order channels. Uranium could adsorb to hydrous ferric oxide in stagnant pores only after diffusing into the pore fluid. Initially, uranium adsorbs to hydrous ferric oxide in the primary flow channels. The proportion of U which adsorbs to sites in the secondary channels increases relative to the primary channels as more fluid passes through the tank, causing U retardation to decrease. If the flow rate is reduced, more of the dissolved U should access the secondary and higher order channels.

The hydrous ferric oxide was found to be immobile as demonstrated by low Fe concentrations measured in the effluent of two-column experiments and tank experiments. Hydrous ferric oxide remained immobile as water flowed through at average linear velocities of 25 cm/d in the columns and 256 cm/d in the tank. The velocity in the tank exceeded flow velocities of most groundwater aquifers, which ranged from about 0.3 to 100 cm/d.

SUMMARY AND CONCLUSIONS

Sorptive reactive barriers are a relatively recent innovation designed to mitigate or remediate groundwater contaminants. These barriers sorb contaminants, thereby retarding their migration, while permitting nontargeted groundwater constituents and water to pass through freely. They have been proposed for use or have been used for remediating metal or radionuclide contamination. They are less often used for organic contamination because it is more desirable to degrade the organic into benign by-products, an option not available for metals. For example, degradation of halogenated organic compounds can be accomplished through the use of reactive barriers made of zero-valent iron. Perhaps one of the best uses of sorptive barriers is for remediating short-lived radionuclides, such as 90-Sr or 137-Cs. The barriers sequester the radionuclides long enough for a sufficient amount of the radionuclide to decay into benign daughter products.

Sorptive barriers offer a number of advantages over conventional pump-and-treat technologies. Compared to pump-and-treat technologies, sorptive barriers are cheaper to operate, pose less health risks because contaminants are not brought above ground, do not require management of large volumes of treated water or secondary waste generated from the treated water, and do not require the acquisition of discharge permits. Perhaps the main disadvantages associated with sorptive barriers is that they have not been field tested and they have not been widely accepted by regulators and society. The reason for this lack of widespread acceptance is that sorptive barriers only delay the travel time of the contaminant, they do not remove the contaminant. The only technologies that permit removal of metals are pump-and-treat or excavation. However, pump-and-treat approaches are limited to easily extractable contaminants, whereas excavation is far too expensive for essentially all subsurface or large contaminated sites. Thus, sorptive barriers, along with precipitation-reactive barriers (such as zero-valent iron barriers), are among the few technologies presently available for mitigating large regions contaminated with sorbing contaminants (with K_d values greater than about 5 mL/g).

A number of sorptive barriers are under consideration for field demonstration. A great deal of performance and design information will likely be collected from these demonstrations. This information is expected to improve the current understanding of sorptive barriers and prove this to be a viable alternative to present-day technologies.

BIBLIOGRAPHY

1. B.M. Thomson and S.P. Shelton, "Permeable Barriers: A New Alternative for Treatment of Contaminated Ground Waters," *Proceedings of the FOCUS Conference on Southwestern Ground Water Issues,* Albuquerque, N.M., 1988, pp. 441–453.
2. P.A. Spooner, R.S. Wetzel, C.E. Spooner, C.A. Furman, and E.F. Tokarski, *Slurry Trench Construction for Pollution Migration Control,* Noyes Data Corp., Park Ridge, N.J., 1985.
3. J.F. Keely, *Ground Water Mon. Rev.* **4,** 63 (1984).
4. A.J. Schuckrow, A.P. Pajak, and C.J. Touhill, *Hazardous Waste Leachate Management Manual,* Noyes Publications, Park Ridge, N.J., 1982.
5. D.M. Mackay and J.A. Cherry, *Environ. Sci. Technol.* **23**(6), 631–636 (1989).
6. J. Bredehoeft, *Ground Water* **30**(6), 834 (1992).
7. J.A. MacDonald and M.C. Kavanaugh, *Environ. Sci. Technol.* **28**(8), 362A (1994).
8. J.E. Park, *Testing and Evaluation of Permeable Materials for Removing Pollutants from Leachates at Remedial Action Sites,* EPA/600/2-86/74, U.S. Environmental Protection Agency, Cincinnati, Ohio, 1986.
9. R. Sims, D. Sorensen, J. Sims, J. McLean, R. Mahmood, R. Dupont, and J. Jurinak, *Contaminated Surface Soils In-Place Treatment Techniques,* Noyes Publications, Park Ridge, N.J., 1986.
10. J.S. Devinny, L.G. Everett, J.C.S. Lu, and R.L. Stollar, *Subsurface Migration of Hazardous Wastes,* Van Nostrand Reinhold, New York, 1990.
11. W.H. Riggsbee, R.M. Schwarz, R.L. Treat, K.J. Cantrell, H.J. Stansfield, and S.J. Phillips, *Review of Potential Subsurface Permeable Barrier Emplacement and Monitoring Technologies,* PNL-9335, Pacific Northwest Laboratory, Richland, Wash., 1994.
12. K.J. Cantrell, P.F. Martin, and J.E. Szecsody, in G.W. Gee and N.R. Wing, eds., *In-Situ Remediation: Scientific Basis for Current and Future Technologies,* Battelle Press, Columbus, Ohio, 1994, pp. 839–850.
13. S.J. Morrison and R.R. Spangler, *Environ. Progress* **12**(3), 175 (1993).
14. K.J. Cantrell and D.I. Kaplan, *J. Environ. Eng.* **123**(5), 499 (1997).
15. J.W. Liskowitz, J.W., P.C. Chan, and R.B. Trattner, *Evaluation of Selected Sorbents for the Removal of Contaminants in Leachate from Industrial Sludges, Residual Management by Land Disposal,* EPA/600/9-76-015, U.S. Environmental Protection Agency, Cincinnati, Ohio, 1976.
16. H.D. Freeman, S.J. Mitchell, and J.L. Buelt, *Engineered Sorbent Barriers for Low-Level Waste Disposal,* PNL-6051, Pacific Northwest Laboratory, Richland, Wash., 1986.
17. S.J. Morrison and R.R. Spangler, *Environ. Sci. Technol.* **26**(10), 1922 (1992).
18. W. Stumm, *Chemistry of the Solid-Water Interface,* John Wiley & Sons, Inc., New York, 1992.
19. D.A. Dzombak and F.M.M. Morel, *Surface Complexation Modeling,* John Wiley & Sons New York, 1990, p. 393.
20. U. Schwertmann and R.M. Cornell, *Iron Oxides in the Laboratory,* VCH, New York, 1991, p. 89.
21. I. Neretnieks, *Retardation of Escaping Nuclides from a Final Repository,* KBS Technical Report, KBS-TR-30, Stockholm, Sweden, 1977.
22. S.M. Robinson, W.D. Arnold, and C.H. Byers, "Multicomponent Ion-Exchange Equilibria in Chabazite Zeolite," in D.W. Tedder and F.G. Pohland, eds., *ACS Symposium Series 468,* ACS, Washington, D.C., 1991, pp. 133–152.
23. K.J. Cantrell, "A Permeable Reactive Wall Composed of Clinoptilolite for Containment of Sr-90 in Hanford Groundwater," in *Nuclear and Hazardous Waste Management Spectrum '96,* American Nuclear Society, Inc., La Grange Park, Ill., 1996.
24. M.J. Semmens and M. Seyfarth, in L.B. Sand and F.A. Mumpton, eds., *Natural Zeolites: Occurrence, Properties and Use,* Pergamon, Oxford, U.K. 1978, pp. 517–526.
25. P.A. Longmire, D.G. Brookins, B.M. Thomson, and P.G. Eller, *Mat. Res. Soc. Symp. Proc.* **212,** 623–631 (1991).
26. W. Stumm and J.J. Morgan, *Aquatic Chemistry,* John Wiley & Sons, Inc., New York, 1981.
27. G.M. Haggerty and R.S. Bowman, *Environ. Sci. Technol.* **28**(3), 452 (1994).
28. X. Chen and J.E. Amonette, *Environ. Sci. Technol.* (in press) (1997).
29. J. Bors, *Radiochem. Acta* **51,** 139 (1990).
30. S.F. Marsh, Z.V. Svitra, and S.M. Bowen, *Distribution of 14 Elements on 63 Absorbers from Three Stimulant Solutions (Acid-Dissolved Sludge, Acidified Supernate, and Alkaline Supernate) for Hanford HLW Tank 102-SY,* LA-12654, Rev./UC-940, Los Alamos National Laboratory, Los Alamos, N.M., 1994, p. 86.
31. J.B. Dixon and S.B. Weed, eds., *Minerals in Soils Environments,* 2nd ed., Soil Science Society of America, Madison, Wis., 1989.
32. D.W. Oscarson, H.G. Miller, and R.L. Watson, *An Evaluation of Potential Additives to a Clay-Based Buffer Material for the Immobilization of 129-I,* AECL-9068, Whiteshell Nuclear Research Establishment, Pinawa, Manitoba, Canada, 1986.
33. G.D. Gradev, *J. Radioanal. Nuc. Chem.* **116,** 341 (1987).
34. Z. Huie, Z. Jishu, and Z. Lanying, *Radiochim. Acta* **44/45,** 143 (1988).
35. J.V. Wright, L.M. Peurrung, T.E. Moody, J.L. Conca, X. Chen, P.P. Didzerekis, and E. Wyse, *In Situ Immobilization of Heavy Metals in Apatite Mineral Formulations,* Pacific Northwest National Laboratory, Richland, Wash., 1995.
36. K.J. Cantrell, D.I. Kaplan, and T.W. Wietsma, *J. Haz. Mat.* **42,** 201 (1995).
37. J. Wagner, H. Chen, B.J. Brownawell, and J.C. Westall, *Environ. Sci. Technol.* **28**(2), 231 (1994).
38. S.R. Burris and C.P. Antworth, *J. Contam. Hydrol.* **10,** 325 (1992).

39. J.C. Westall, in W. Stumm, ed., *Aquatic Surface Chemistry*, John Wiley & Sons, Inc., New York, 1987, pp. 3–32.
40. S.A. Boyd, W.F. Jaynes, and B.S. Ross, in R.A. Baker, ed., *Organic Substances in Sediment and Water*, Lewis Publishers, Inc., Chelsea, Mich., 1991, pp. 181–200.
41. W.F. Jaynes and S.A. Boyd, *Soil Sci. Soc. Am. J.* **55**, 43 (1991).
42. N.A. Nzengung, E.A. Voudrias, P. Nkedi-Kizza, J.M. Wampler, and C.E. Weaver, *Environ. Sci. Technol.* **30**(1), 89 (1996).
43. D.L. Sparks, *Kinetics of Soil Chemical Processes*, Academic Press, Inc., New York, 1989.
44. M.C. Amnacher, "Methods of Obtaining and Analyzing Kinetic Data," in D.L. Sparks and D.L. Suarez, eds., *Rates of Soil Chemical Processes*, Special Publication Number 27, Soil Science Society of America, Inc., Madison, Wis., 1991.
45. R.W. Gillham and S.F. O'Hannesin, *Ground Water* **32**, 958 (1994).
46. W.F. Jaynes and S.A. Boyd, *J. Air Waste Mgmt. Assoc.* **40**, 1649 (1990).
47. DOE/RL, *RCRA Facility Investigation / Corrective Measures Study Work Plan for the 100-NR-2 Operable Unit, Hanford Site, Richland Washington*, DOE/RL-90-23, U.S. Dept. of Energy, Richland, Wash., 1992.
48. DOE/RL, *N Springs Expedited Response Action Proposal*, DOE/RL-93-23, Draft A, U.S. Dept. of Energy, Richland, Wash., 1992.
49. M.J. Hartman and K.A. Lindsey, *Hydrogeology of the 100-N Area, Hanford Site, Washington, Appendix G*, WHC-SD-EN-EV-027, Westinghouse Hanford Co., Richland, Wash., 1993.
50. D.G. Kinniburgh, *Environ. Sci. Technol.* **20**, 895 (1986).
51. S.J. Morrison, R.R. Spangler, and V.S. Tripathi, *J. Contam. Hydrol.* **17**, 333 (1995).
52. S.J. Morrison, V.S. Tripathi, and R.R. Spangler, *J. Contam. Hydrol.* **17**, 347 (1995).
53. S.J. Morrison, R.R. Spangler, and S.A. Morris, *Ground Water* **34**(1), 75 (1996).
54. G.T. Yeh and V.S. Tripathi, *HYDROGEOCHEM: A Coupled Model of Hydrological and Geochemical Equilibrium of Multi-Component Systems*, ORNL-6371, Oak Ridge National Laboratory, Oak Ridge, Tenn., 1990, p. 312.
55. C.D. Hsi and D. Langmuir, *Geochim. Cosmoshim. Acta*, **49**, 1931 (1985).

STABILITY CONSTANT PREDICTION

PETER WARWICK
PAUL DIMMOCK
REBECCA ROBBINS
Loughborough University
Leicestershire, United Kingdom

The transport of a pollutant or contaminant through the hydrosphere may be by advection (ie, the natural movement of an environmental water through a porous aquifer because of the presence of a hydraulic head) or by diffusion through, for example, a clay, if the water flow is very low or negligible. Irrespective of the mode of transport, the mobility of the pollutant is affected by its solubility and its interactions with the geological material through which it is transported. Both of these are fundamentally determined by the speciation of the pollutant after entering the surface or groundwater. Speciation may be defined as the chemical form or forms of a pollutant, and it is the most important determinant as to whether a pollutant is mobile through the environment (1–4). For example, uranium is relatively mobile when present in waters as uranyl carbonate [U(VI)] complexes, but is relatively immobile if it is converted to uranium (IV) in reducing conditions. Furthermore, speciation plays a vital role in determining the toxicological impact of, for example a metal pollutant, because its toxicity is determined by the activity of the metal species causing the toxic effect and not the total concentration of the metal contained in the water (5–8). A classical example of this is that chromium(III) is essential for life whereas chromium(VI) is toxic.

Speciation of a metal in relatively pure water may be easily predicted, using fundamental chemistry, to be an aquated cation or a simple hydrolysis product of the cation. However, environmental waters contain many agents capable of combining with a metal to form soluble, and therefore transportable, species. Inorganic ions such as sulfate or chloride may react with a metal to form complexes at low and neutral pHs. At higher pHs, carbonate and hydroxide ions are important complexants. The presence of anthropogenic complexing agents such as ethylenediamine tetraacetic acid (EDTA), which is extensively used in a range of consumer products, in the aquatic environment can also have significant environmental impact by mobilizing previously insoluble metals. Inorganic and organic colloids may also be present, and these may also be associated with pollutants. Thus, predicting the speciation of a pollutant in an environmental water is rather more difficult than predicting it in relatively pure water. So why should scientists want to predict speciation? Why not just experimentally determine the speciation of a pollutant? The answer to this question lies in the numerous variations in the chemistry of environmental waters which makes the determination of speciation in so many different waters impossible. For example, many countries are proposing to dispose of their nuclear waste in repositories deep underground. After closure, groundwater ingress into the repository may mobilize radionuclides through groundwater movement. In the near-field, that is the engineered part of the repository, groundwater is expected to be anoxic (reducing) and of high pH because of the presence of cement. As the water progresses to the far-field of the surrounding host geological formation, natural waters will dilute the near-field water producing more oxic water of lower pH. The experimental determination of radionuclide speciation in this range of water chemistries would require a prohibitive amount of time and effort.

Fortunately, there are computer programs available, such as PHREEQE (9), MINTEQA2 (10), and JESS (11–13) that allow us to predict speciation without going to the cost and effort of experimental work. These programs use a database of thermodynamic stability constants that are a measure of the strengths of reactions of the pollutant of interest with complexants present in the water. Unfortunately, although these databases are extensive, they are not fully comprehensive. As in speciation, missing stability constants could be measured by experiment, but this would be an onerous task even for the simplest natural systems. Even if the measurements were carried out, the accuracy of the measurements may be called into question when values are determined by a single investigator, or where difficult systems are complicated by hydrolysis, redox or solubility problems, or where values were determined in an unsuitable medium or under unsuitable conditions. Thus, the measurement of large numbers of stability constants of different chemical reactions under varying chemical conditions is not an attractive option. Fortunately, there are methods available that allow us to predict stability constants and therefore fill in the gaps in databases used to predict speciation. The general approaches are described below and concentrate on predicting metal-ligand stability constants in aqueous solutions because these are of major interest in environmental situations.

THE EQUILIBRIUM CONSTANT

The activity of a species a is related to its total concentration $[C]$ by its activity coefficient (γ).

$$\{a\} = \gamma[C]$$

For infinitely dilute solutions, γ approaches 1 so that

$$\{a\} = [C]$$

The *law of mass action* is only strictly valid when activities are used, and using these, the thermodynamic stability (formation) constant for a metal (M) reacting with a ligand (L) can be expressed in terms of the concentrations and activities of the species involved. For a polynuclear complex, the formation equilibrium is

$$xM + yL \rightleftharpoons MxLy$$

$$K^{\theta}_{xy} = \frac{\{M_xL_y\}}{\{M\}^x\{L\}^y}$$

$$K^{\theta}_{xy} = \frac{\gamma_{ML}[M_xL_y]}{\gamma^x_M[M]^x\gamma^y_L[L]^y}$$

The $\gamma_{ML}/\gamma^x{}_M\gamma^y{}_L$ term can be maintained effectively constant by having present a large excess of an inert background electrolyte. By using low concentrations of metal and ligand the stoichiometric stability constant, Kc, can be expressed as a function of the thermodynamic constant, K_θ as follows:

$$Kc_{xy} = \frac{[M_xL_y]}{[M]^x[L]^y} = K^{\theta}_{xy}\frac{\gamma^x_M\gamma^y_L}{\gamma_{ML}}$$

where Kc is a constant at given ionic strength.

Stepwise Formation Constants

Metal ions react with ligands in consecutive reactions that are characterized by equilibrium constants called stepwise formation constants.

$$M + L \rightleftharpoons ML$$

$$Kc_1 = \frac{[ML]}{[M][L]}$$

$$ML + L \rightleftharpoons ML_2$$

$$Kc_2 = \frac{[ML_2]}{[ML][L]}$$

$$ML_2 + L \rightleftharpoons ML_3$$

$$Kc_3 = \frac{[ML_3]}{[ML_2][L]}$$

The overall stability constant, β, is given by

$$\beta c_3 = \frac{[ML_3]}{[M][L]^3}$$

The relationship between the overall stability constant and the stepwise formation constants is

$$\beta c_3 = Kc_1 \times Kc_2 \times Kc_3$$

When a chemical reaction attains equilibrium, the rate of the forward reaction (rate constant equal to kf) is equal to the rate of the back reaction (rate constant equal to kb)

$$A + B \underset{kb}{\overset{kf}{\rightleftharpoons}} C + D$$

and the concentrations of the reactants and products are therefore constant.

$$kf[A][B] \rightleftharpoons kb[C][D]$$

rearranging this equation gives

$$\frac{kf}{kb} = Kc = \frac{[C][D]}{[A][B]}$$

in which Kc is the equilibrium constant.

pK_a

When an acid dissolves in water there is an equilibrium between the undissociated acid (HA), protons (H^+), and anions (A^-). If the solution is dilute, $\gamma_{H^+}\gamma_{A^-}/\gamma_{HA}$ can be considered as unity and the value of the acid dissociation constant (K_a) is given by

$$K_a = \frac{[H^+][A^-]}{[HA]}$$

The pK_a of an acid is defined as the negative log of K_a and its value is equal to the pH at which the activity of A^- is equal to that of HA:

$$pK_a = pH - \log_{10}\frac{[A^-]}{[HA]}$$

This expression is the Henderson–Hasselbach equation.

Effect of Ionic Strength on the Equilibrium Constant

As stated previously, the thermodynamic stability constant for a reaction may be calculated if the concentrations and activity coefficients of the various species involved are known. Unfortunately, activity coefficients depend upon the ionic strength of the solution and many stability constants are published at various ionic strengths. A noteable exception is Martell's critical compilation (14) of stability constants. Therefore, to convert published values of concentration-dependent stability constants into thermodynamic constants, activities must be calculated at the ionic strength of the published data. A number of methods exist (15–21) that relate activities to ionic strengths, but many of these are limited to a narrow range of ionic strengths and a unified method encompassing all ionic strengths is not yet available. A useful method for ionic strength corrections is a completely empirical approach (22). This approach is based on the trends in published data and represents a good, rapid, and surprisingly accurate method of consistent ionic strength correction.

Effect of Temperature on the Equilibrium Constant

At constant pressure, the thermodynamic equilibrium constant is dependent only upon temperature through

$$\Delta G = -RT \ln K$$

Using the Gibbs–Helmoltz equation

$$\frac{d}{dT}\left(\frac{\Delta G}{T}\right) = -\frac{\Delta H}{T^2}$$

the Van't Hoff equation, which relates the stability constant to the enthalpy of the reaction (ΔH), may be obtained

$$\frac{d(\ln K)}{dT} = \frac{\Delta H}{RT^2}$$

There are two ways of calculating the stability constant depending on whether or not ΔH is a function of temperature.

1. If ΔH is independent of temperature, the indefinite integral of the above equation gives
$$\ln K = \frac{-\Delta H}{RT} + C$$
and, if ΔH is known within the temperature range of interest, the definite integral of the equation becomes
$$\ln\left[\frac{K(T_2)}{K(T_1)}\right] = \frac{\Delta H}{R}\left(\frac{1}{T_1} - \frac{1}{T_2}\right)$$
2. If ΔH is temperature dependent, ΔH is related to the change in the heat capacity, ΔC_p, by:
$$\frac{d}{dT}(\Delta H) = \Delta C_p$$
This dependence is empirically modeled by a polynomial expression in T and, if ΔH is known at one temperature, an expression for K may be used as follows:
$$R \ln K = a + \frac{b}{T} + c \ln T + dT + eT^2$$
where a, b, c, d, and e are constants for a particular reaction. Again, there are few published studies on the temperature dependence of stability constants (23–25).

CORRELATIONS OF STABILITY CONSTANTS

Most approaches to predicting unknown metal-ligand stability constants are based on correlations of known metal-ligand stability constants to some chemical or physical parameter(s).

Linear-Free-Energy Relationships

Linear-free-energy relationships (LFERs) are among the most intrinsic concepts in physical organic chemistry (26–28) and have found many applications in the analysis of reactivity in coordination chemistry, especially in kinetics. The aim of LFER is to convert qualitative ideas into quantitative theories of reactivity. The validity of such correlations has been examined both experimentally and theoretically (29,30) and is well accepted. The basis for most LFERs is the existence of a linear relationship between the logarithm of the rate or equilibrium constant (log K) for one reaction and that for a second reaction (log K') subjected to some perturbation in reactant structure or conditions. This may be expressed as

$$\log K = a \log K' + b \tag{1}$$

in which a and b are constants.

The free energy of a reaction, ΔG, is related to its equilibrium constant by

$$\Delta G = -RT \ln K \tag{2}$$

and therefore a correlation between free energies of related reactions may be obtained through equations 1 and 2.

The interdependence need not be between two equilibrium constants. Log K has been correlated to other descriptive physical parameters of the reaction system such as ionization potential (31) or atomic number of the metal (32,33).

Hammett Correlations. Linear-free-energy relationships were extensively developed for physical organic chemistry by Hammett (34–37) for rationalizing aromatic substitution patterns and more generally by Taft (38–41) and others for aliphatic and alicyclic compounds.

Hammett derived a simple equation relating the rate constant, or equilibrium constant, for a series of meta- and para-substituted benzene ring side-chain reactions to that of the unsubstituted parent compound by two factors: the substituent constant σ, and the reaction constant ρ. The Hammett equation can be expressed as

$$\log k = \log k_0 + \rho\sigma \tag{3}$$

$$\log K = \log K_0 + \rho\sigma \tag{4}$$

where k is the predicted rate constant, K is the predicted equilibrium constant for the reaction of a m- or p-substituted benzene derivative, and k_0 and K_0 are the concomitant, known, reference values for the unsubstituted parent compound.

The substituent σ constant measures the effect of the substituent relative to hydrogen and is independent of the nature of the reaction; ρ, the reaction constant, which depends on the type of physical parameters of the reaction (solvent and temperature), measures the susceptibility of the reaction to polar effects. Extensive tabulations of ρ and σ values exist. These values have been used to predict stability constants of metal-ligand complexes by comparison with the respective σ values of the aromatic substituents.

Approaches to predicting stability constants have been developed along these lines by correlating stability constants to Hammett substituent constants (42–45). Specific examples for log K_1 of silver(I) complexes of various sulfur and selenium derivatized carboxylic acids and their dependence on σ values are shown in Figure 1 (42,43).

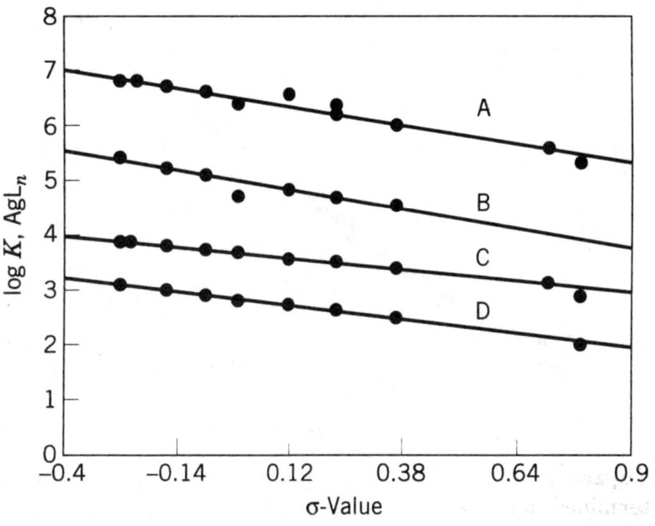

Figure 1. Correlation between the Hammett σ-value for a ligand and the stability constant for the complex formed between that ligand and silver(I), for a substituted chalcogenobenzoic acid (seleno analogue substituents H, m-/p-CH_3, m-/p-Cl, m-/p-$OCH_3$$m$-/$p$-$NO_2$, p-Br, p-OC_2H_5; thio analogue substituents H, m-/p-CH_3, m-/p-Cl, m-/p-OCH_3, p-NO_2, p-Br, p-CN, m-CF_3, where A is $Ag(L{-}Se)_2$; B, $Ag(L{-}S)_2$; C, $Ag(L{-}Se)$; and D, $Ag(L{-}S)$ (42,43).

pK_a Correlations. If pK_a values are unknown, accurate predictions of pK_a values have been available for many years using Hammett and Taft equations. Parameters for these equations are available for a wide range of functional groups (46). Substitution of these parameters into Hammett and Taft equations allows the dissociation constants, for example for organic acids, to be calculated. As Figure 2 demonstrates, pK_a prediction can be extremely accurate for a wide range of organic molecules.

More generally this approach could be used for predicting pK_a values for ligands and using these predicted values in equations used for estimating stability constants. K_a correlations can be viewed as a LFER, pK_a values are linked to thermodynamic quantities via equation 2, or as a subset of the acid–base discussion below.

Methods for predicting stability constants from pK_a values are of more limited success, many failing due to a lack of appreciation of the parameters of the metal within the model. Direct correlation to pK_a appears to be most successful where charged, monodentate ligands are involved, forming 1:1 complexes (47–51). Many examples of correlation to ligand pK_a exist, typical examples are shown in Figure 3. The correlation remains a useful technique for K_1 prediction in simple situations. However, the method is heavily reliant on an existing database of analogous data.

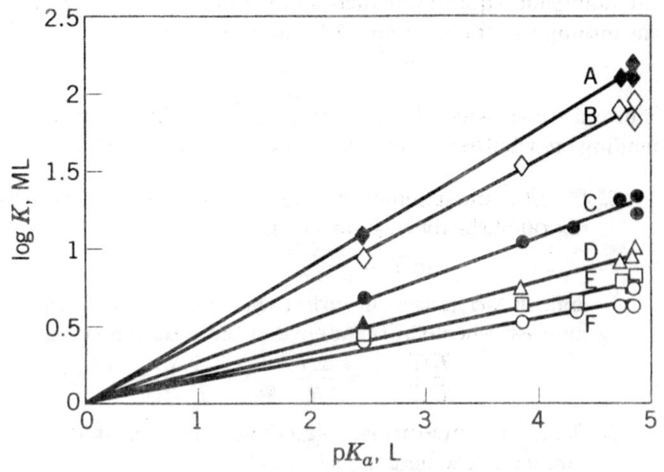

Figure 3. Correlation between pK_a and log K_1 for the complexes formed between the metals Co^{2+}, Ni^{2+}, Cu^{2+}, Zn^{2+}, Cd^{2+}, and Pb^{2+}, and the carboxylate ligands formate, acetate, butanoate, 2-cyanoacetate, 2-phenylacetate. A, Pb; B, Cu; C, Cd; D, Zn; E, Ni; and F, Co (49).

Log K(OH)$_M$ Correlations. Related to pK_a correlations for a ligand are correlations to the first hydrolysis product (or pK_a) of the metal to the stability constant for the metal-ligand reaction. Ligation by various oxygen-containing ligands show well-documented linear relationships (51), and examples of these are given in Figure 4.

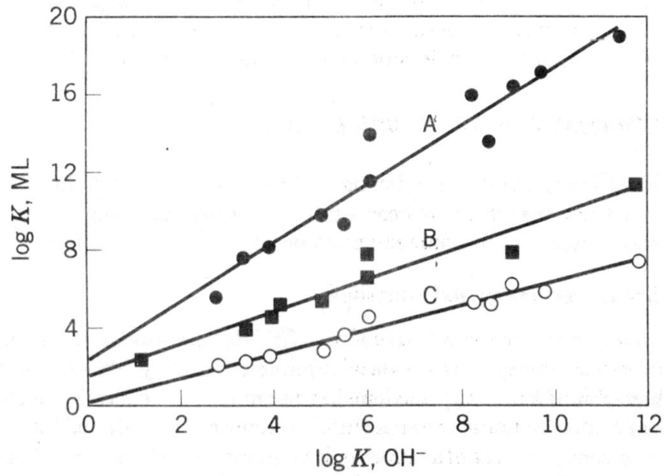

Figure 4. Correlation of log K_1 with log K_1, OH for metal–catechol (Mg^{2+}, Mn^{2+}, Cd^{2+}, Zn^{2+}, La^{3+}, Yb^{3+}, Cu^{2+}, Be^{2+}, UO_2^{2+}, Al^{3+}, Sc^{3+}, Ga^{3+}), metal–acetohyroxamate (Ca^{2+}, Mn^{2+}, Cd^{2+}, Ni^{2+}, Zn^{2+}, Pb^{2+}, Yb^{3+}, Cu^{2+}, Al^{3+}, Fe^{3+}), and metal–malonate (Mg^{2+}, Mn^{2+}, Cd^{2+}, Zn^{2+}, La^{3+}, Yb^{3+}, Be^{2+}, UO_2^{2+}, Al^{3+}, Sc^{3+}, Fe^{3+}) complexes. A is catechol; B, acetohydroxamate; and C, malonate (14).

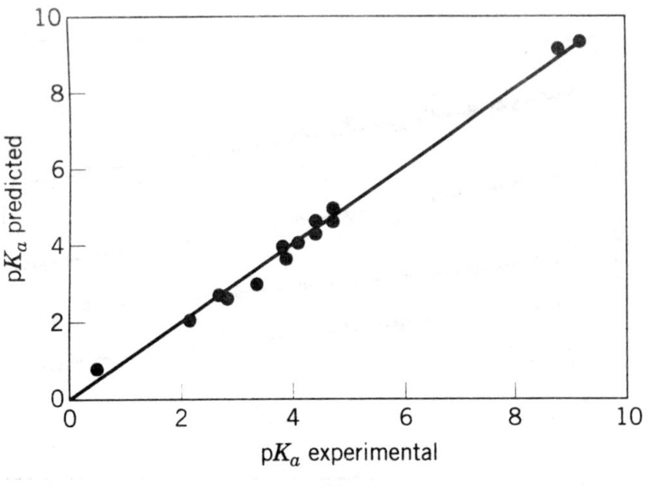

Figure 2. Correlation of predicted and experimentally determined pK_a values for a range of structurally diverse organic acids. The predicted values were calculated (31) and include 2-acetylbutandioic acid, methionine, 2,3-dihydroxy-2-hydroxymethylpropandioic acid, chlorodifluoroacetic acid, 3,4,5-trihydroxybenzoic acid, 4-bromo-1,2-benzenedicarboxylic acid, 4-hydroxy-3,5-dimethoxybenzoic acid, pyrimidine-2-carboxylic acid, glutamic acid, 4-nitrothiophen-2-carboxylic acid, 4-bromopyrrole-2-carboxylic acid, furan-2,4-dicarboxylic acid, 5-methylselenofuran-2-carboxylic acid, and pyrazole-3-carboxylic acid.

Also, related to the chelate effect, a general relationship between the stability constants of complexes formed by ligands with negatively charged oxygen atoms to the log $K(\mathrm{OH}^-)$ of the ligated metal has been derived (51,52). The intercept of the graph gives an indication of the denticity of the ligand. For example, a bidentate ligand would give an intercept on the y-axis of log 55.5, as is shown for catechol complexes in Figure 4.

$$\log K_1 = a \log K(\mathrm{OH}^-) + (n - 1) \log 55.5 \qquad (5)$$

where log $K(\mathrm{OH}^-)$ is defined as the stability constant for the formation of the complex M(OH). Plots of log K_1 vs log $K(\mathrm{OH}^-)$ give straight lines of intercept $(n - 1)$ log 55.5 relating the intercept to the denticity of the ligand.

Log K_{ML} and log $K_{ML'}$ Correlations. Many such relationships have been adequately demonstrated, and therefore these relationships provide useful methods for predicting unknown stability constants. The most obvious examples involve homologous series of the type $R(CH_2)_nX$, where stability constants for ligands with various values of n can be related to the stability constants of ligands for which most data are published.

A useful example of this relationship is the comparison between the metal complexes of EDTA and those of 1,2-diaminocyclohexanetetracarboxylic acid (CDTA) complexes (50) as is shown in Figure 5. This approach can be widely applied, especially to ligands that are closely structurally related. An example of this approach, as shown in Figure 6, is the linear relationship between metal-valine stability constants and metal–arginine constants, which may be used for predicting unknown metal–arginine constants. This relationship is easily made because of the large number of metal–valine stability constants published.

Humic and fulvic acids are present in all environmental waters and are known to form soluble complexes with contaminant or pollutant metals. Together with other organic species present in environmental waters, humics and fulvics are often described as *dissolved organic matter* or DOM (53,54). The stability constants of humic and fulvic reactions with a variety of metals have been determined (55–58). Linear-free-energy relationships have been used to predict unknown stability constants for metal–DOM complexes (59). Examples are the correlations between lactate and acetate metal complexes and DOM metal complexes and the correlation between pK_a of the metal and the DOM metal complex. The degree of correlation is shown in Figure 7.

Log K_{ML} and Log $K_{(ML+ML')}$ Correlations. The prediction of the stability constant of a multidentate ligand-metal interaction may be made easier by breaking down the ligand into monodentate ligands of known metal-ligand stability constants. Summation of these constants gives the stability constant of the metal–multidentate ligand reaction. An example of this approach is the prediction of picinolate stability constants using data from bipyridine and oxalate, as is shown in Figure 8 (51).

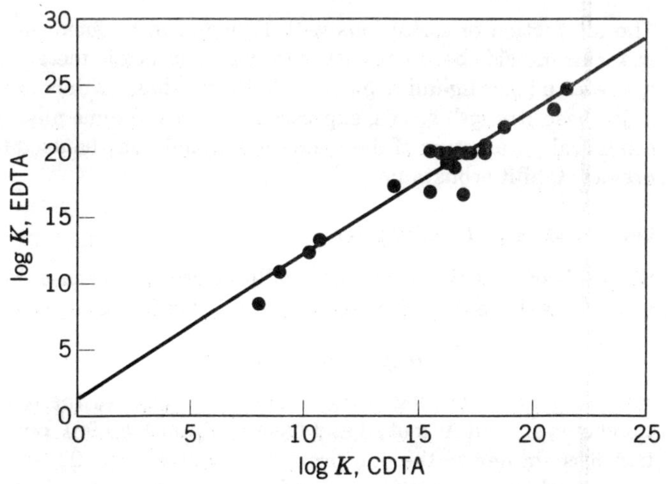

Figure 5. Correlation of log K_1 for metal–EDTA to log K_1 for metal–1,2-diaminocyclohexane tetraacetate complexes, where M = Ba^{2+}, Mg^{2+}, Ca^{2+}, Mn^{2+}, Al^{3+}, Cd^{2+}, Zn^{2+}, Co^{2+}, Gd^{3+}, Tb^{3+}, Pb^{2+}, Cu^{2+}, Yb^{3+}, Ga^{3+}, Hg^{2+}, La^{3+}, Ce^{3+}, Pr^{3+}, Nd^{3+}, Sm^{3+}, Eu^{3+}, Y^{3+}, VO^{2+}, Dy^{3+}, Tm^{3+}, and Lu^{3+} (14).

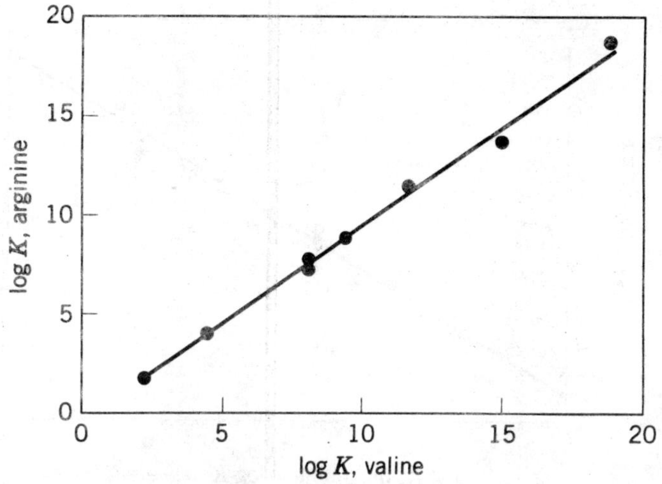

Figure 6. Correlation of log K_l for metal–valine to log K_1 for metal–arginine complexes, for $H^+(K_1, K_2)$, $Cu^{2+}(K_1, \beta_2)$, $Zn^{2+}(K_1, \beta_2)$, and $Hg^{2+}(K_1, \beta_2)$ (14).

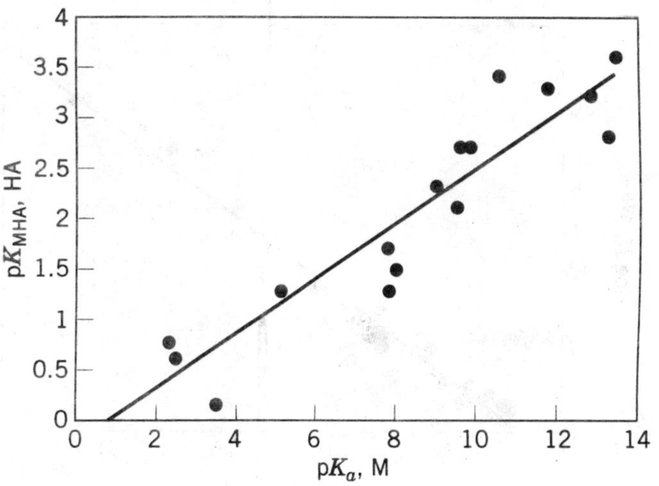

Figure 7. Correlation of pK_{MHA} for humic acid to pK_aM for Ba^{2+}, Mg^{2+}, Fe^{3+}, Ca^{2+}, Mn^{2+}, Cd^{2+}, Zn^{2+}, Co^{2+}, Pb^{2+}, Cu^{2+}, Hg^{2+}, Th^{4+}, Fe^{2+}, Eu^{3+}, and Sr^{2+} (59).

Log K_{ML} and Log $K_{M'L}$ Correlations. Intermetal correlations within groups and periods predate interligand correlations. A classic example of a metal–metal comparison is between Cu(II) and Ni(II) (50,60) as is shown in Figure 9. This correlation, between analogous complexes, is well documented, because a vast number of Cu(II) and Ni(II) stability constants have been determined (14,61–79).

Significant advances have been made in speciation studies of the heavy radioactive actinides using less hazardous elements (eg, Eu^{3+} as an analog for Am^{3+}), as is shown in Figure 10.

A summary of some useful LFERs between metal ions is given in Table 1. The general trends in stability constant values have been used to predict stability constants for carboxy-

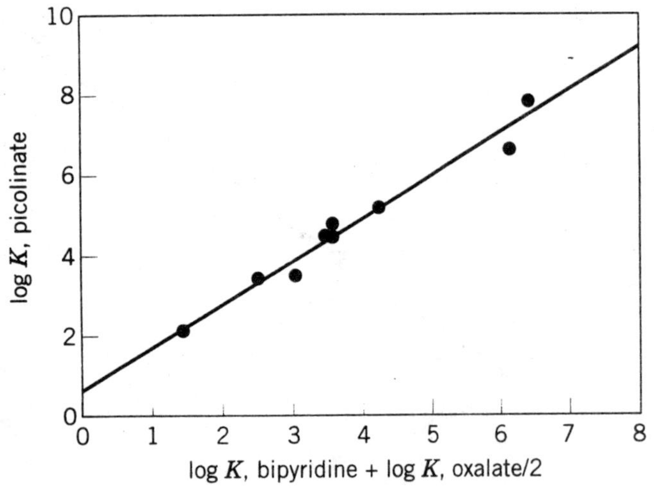

Figure 8. Prediction of metal–picinolate stability constants from bipyridine and oxalate metal stability constants for Mg^{2+}, La^{3+}, Mn^{2+}, Pb^{2+}, Cd^{2+}, Fe^{2+}, Zn^{2+}, Ni^{2+}, and Cu^{2+} (51).

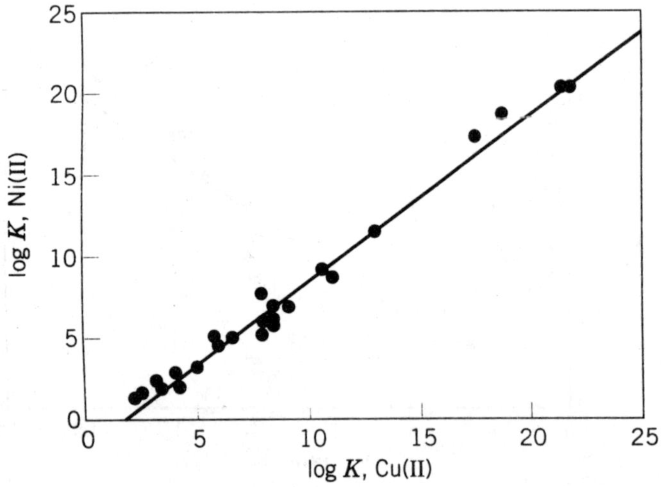

Figure 9. Relationship between log K_1 for a series of Ni(II) and Cu(II) complexes with ligands acetylacetonate, catechol, kojate, catechol-4-sulfonate, tiron, acetohydroxamate, EGTA, DTPA, DCTA, HEDTA, EDTA, picinolate, 2-aminobenzoate, 2-quinaldinate, 8-quinaldinate, NTA, MIDA, glycine–glycine, aspartate, IDA, alanine, glycine, phthalate, salicylic-5-sulfonate, salicylate, succinate, maleate, lactate, malonate, and acetate (14).

lates (80), alkylamino acid (81), and amino carboxylic acid (82) metal complexes.

Ternary Complexes. The emphasis of most experimental determinations of metal-ligand stability constants is on the calculation of simple binary complexes of the type $M(L)$, $M(L)_2$, and so on. However, in nature, the situation is more complicated, and mixed or ternary systems with different ligands attached to the same metal center, of the type $M(L)(L')$, may predominate. Little data have been published on such systems, with Cu^{2+} complexes being the most commonly studied (83–87). Mathematical considerations of the possible interactions in $M-L-L'$ mixtures allow estimates to be made of the ratios

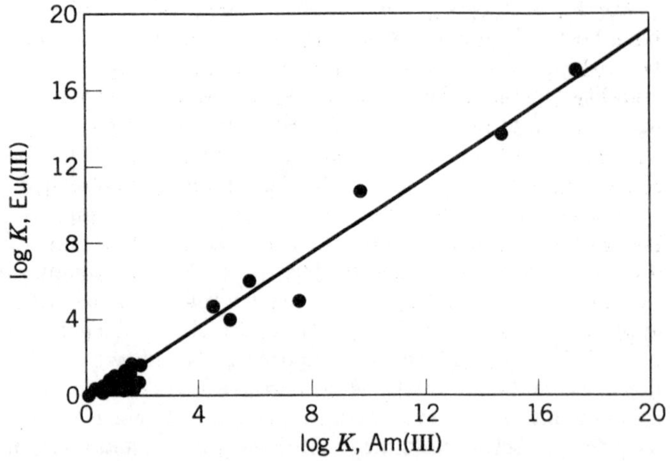

Figure 10. Relationship between log K_1 for a series of Eu(III) and Am(III) complexes with ligands carbonate, chloride, azide, hydroxide, thiocyanate, sulfate, mono- di-, and trichloroacetate, maleate, asparagine, and EDTA (14).

Table 1. Metal–Metal LFERs

Major Ion	Minor Ion
UO_2^{2+}	PuO_2^{2+}
Th^{4+}	Pu^{4+}
Cu^{2+}	Ni^{2+}
Eu^{3+}	Am^{3+}
CH_3Hg^+	Hg^{2+}
CH_3Hg^+	Ag_+

of ternary complexes to binary complexes present in solution. Furthermore, an estimation of the ternary complex stability constant can be made provided that the stability constants for the binary complexes are known.

ACID–BASE APPROACHES

The interaction of metal ions with ligands can be formalized in terms of acid–base chemistry in which an acidic metal reacts with a basic ligand to form a stable acid–base adduct (88). Acid–base strength scales, expressed in terms of some physicochemical parameters of the metal and ligand, may be used to predict stability constants.

Lewis Acid–Base Classification

Simple Brønsted classification is of little use in this case, but Lewis acid–base classification is useful. A useful example is

$$\log K_{AB} = E_A E_B + C_A C_B \tag{6}$$

where log K_{AB} is the log of the stability constant for the reaction between acid A (metal) and base B (ligand) (89,90). Equation 6 is similar to the one previously derived (91–93) (*vide infra*), and describes the complex formed in terms of ionic and covalent bonding factors E and C. Equation 6 may be extended to include desolvation factors, D_A and D_B, which accommodate inadequate correlation for ligands with second and third row donor atoms, which is caused by steric hindrance to solvation. The acid parameter E_A is equal to the log of the metal–fluoride stability constant, and C_A is equal to the log of the metal–hydroxide stability constant divided by 14:

For the metal fluoride $$E_A = \log K_1(F^-) \qquad (7)$$

For the metal hydroxide $$C_A = \log K_1(OH^-)/14 \qquad (8)$$

The base parameters E_B and C_B are derived from a graphical treatment of E_A, C_A, and K_{AB}. Manipulation of equation 6 gives

$$\log K_1/C_A = E_A E_B/C_A + C_B \qquad (9)$$

in which $\log K_1$ is the stability constant of the base in question with a particular metal. A plot of $\log K_1$ vs E_A/C_A gives a line of slope E_B and an intercept C_B. Some values of the parameters are shown in Tables 2 and 3.

Figure 11 shows the results for a range of metal ions with OH^-, CH_3COO^-, and Cl^- as ligands.

Earlier, a more refined approach to acid–base chemistry had been developed relating the polarizability and strength of an acid or base to its *hardness* or *softness*.

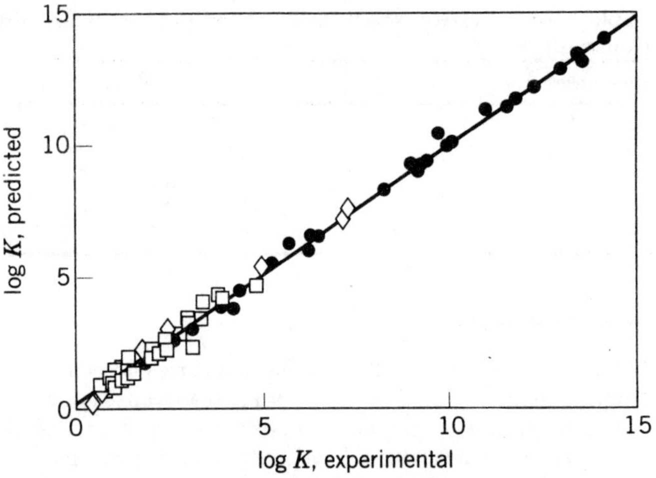

Figure 11. Correlation of predicted and experimental log K_1, OH^-, log K_1, Cl^-, and log K_1, CH_3COO^- for a series of metal ions using equation 7 for the metal ions Mg^{2+}, Mn^{2+}, Al^{3+}, Cd^{2+}, Zn^{2+}, Co^{2+}, Pb^{2+}, Cu^{2+}, Y^{3+}, Ga^{3+}, Hg^{2+}, La^{3+}, Sc^{3+}, Fe^{3+}, In^{3+}, Sn^{2+}, Co^{3+}, Bi^{3+}, VO^{2+}, Tl^{3+}, Cr^{3+}, Lu^{3+}, Ag^+, CH_3Hg^+, Ni^{2+}, U^{4+}, UO_2^{2+}, and Ca^{2+} (89,90).

Hard and Soft Acid–Base Classification

Pearson developed the concept of hard and soft acids and bases (HSAB) (94–98) from earlier empirical observations (99) of different classes of ligands reacting with different metal ions. The HSAB approach formalized many earlier observations (99–103), but quantification of the concept remains a problem. The dividing line between hard and soft behavior is not distinct (see Box 1), and subsequent approaches to quantification have resulted in confusing and contradictory results and predictions. However, dichotomies can often be rationalized in terms of reactivity with specific ligands. The prediction of stability constants for metal–ligand interactions, using the HSAB theory are documented in Table 4.

Table 2. Lewis Acid Parameters for a Range of Metal Ions[a]

Metal	E_a	C_a	Metal	E_a	C_a
Ag_+	−0.290	0.240	Pb^{2+}	2.763	0.413
Hg^{2+}	1.346	0.826	Mn^{2+}	1.581	0.223
Pd^{2+}	1.574	0.910	Cr^{3+}	5.148	0.721
CH_3Hg^+	1.359	0.708	Fe^{3+}	6.072	0.841
Tl^{3+}	1.938	0.914	UO_2^{2+}	4.755	0.646
Cu^{2+}	1.249	0.466	Ga^{3+}	6.060	0.788
H^+	3.067	1.009	U^{4+}	7.550	0.968
Cd^{2+}	1.139	0.344	Zr^{4+}	9.290	1.079
Ni^{2+}	1.011	0.300	Y^{3+}	4.451	0.475
Co^{3+}	3.299	0.875	La^{3+}	3.762	0.396
Zn^{2+}	1.329	0.312	Sc^{3+}	6.772	0.691
Co^{2+}	1.198	0.276	Mg^{2+}	1.862	0.178
Fe^{2+}	1.521	0.256	Al^{3+}	6.899	0.657
In^{3+}	5.153	0.818	Ca^{2+}	0.985	0.081
Bi^{3+}	5.777	0.904			

[a] Refs. 89 and 90.

Table 3. Lewis Base Parameters for Selected Bases[a]

Ligand	E_b	C_b
F^-	1.00	0
OH^-	0	14.00
NH_3	−0.088	12.34
Cl^-	−0.103	11.1
Br^-	−0.115	15.6
I^-	−0.112	23.1
$S{=}C(NH_2)_2$	−0.135	20.4
PPh_3	−0.118	25.0
CN−	−0.189	36

[a] Refs. 89 and 90.

BOX 1: Irving-Williams Series and Hard-Soft Classifications. Following is the Irving–Williams series describing the proclivities in stability constants for the first row transition metals, and the (**a**) and (**b**) classification for hard and soft metals. Metals can be classified by their reactivity with particular ligand moieties; the reactivity trends used to categorize hard and soft metals are shown.

$$Ba^{2+} < Sr^{2+} < Ca^{2+} < Mg^{2+} < Mn^{2+} < Fe^{2+} < Co^{2+} < Ni^{2+} < Cu^{2+} > Zn^{2+}$$

Hard—Class (**a**)	Borderline	Soft—class (**b**)
Acids		
H^+, Li^+, Na^+, K^+, Be^{2+}, Mg^{2+}, Ca^{2+}, BF_3, BCl_3, $B(OR)_3$, Al^{3+}, $AlCl_3$, $Al(CH_3)_3$, Sc^{3+}, Ti^{4+}, VO^{2+}, Cr^{3+}, Fe^{3+}, Co^{3+}	Fe^{2+}, Co^{2+}, Ni^{2+}, Cu^{2+}, Zn^{2+}, Rh^{3+}, $B(CH_3)_3$, R_3C^+, Pb^{2+}, Sn^{2+}	Cu^+, Ag^+, Au^+, Hg^{2+} Pt^{2+}, Pt^{4+}, MoO_2^{2+}, Pd^{2+}
Bases		
NH_3, RNH_2, N_2H_4, H_2O, OH^-, O^{2-}, ROH, RO^-, CO_3^{2-}, SO_4^{2-}, ClO_4^-, F^-	$C_6H_5NH_2$, N_3^-, N_2, Br^-, Cl^-	H^-, R^-, C_2H_4, C_6H_6, CN^-, CO, *SCN*$^-$, R_3P, R_2S, RSH

Table 4. Quantitative HSAB Models Used to Predict Stability Constants

References	Model
100, 101	$\log(K/K_0) = \alpha E_n + \beta H$
105, 106	$\log K = \alpha X + \beta Y + \gamma$
96	$\log K = S_A S_B + \sigma_A \sigma_B$
91–93	$\Delta H = E_A E_B + C_A C_B$
112, 113	$\log K = E_A E_B + C_A C_B D_A D_B$

Irving–Williams Series

Irving and Williams (32,33) developed a series describing the trend in relative stability of M^{2+} first-row transition metals with a particular ligand. Box 1 shows the order of the stability of complexes formed with a dipositive metal, for a given ligand, which is based on the radius of the metal ion and ligand field effects (104). This allows a reasonable estimation of unknown stability constants if a large number of stability constants exist for the reactions of a particular ligand with a series of metals. As this implies, it is heavily dependant on the existence of relevant data.

Edwards' Equation

It was early work (100–101) that led to development of the HSAB theory used to correlate kinetic data. The theory was later extended to include stability constant predictions

$$\log(K/K_0) = \alpha E_n + \beta H \qquad (10)$$

where

$$E_n = E_o + 2.60$$

$$H = pK_a + 1.74$$

and K/K_0 is the relative (to water) stability constant for the substitution reaction in question, E_o is the reduction potential of the donor ligand, and α and β are empirically derived scaling factors.

This progression of models was developed in an analogous manner to the Hammett equations. A notable feature of the Edwards' equation is its use of, in conjunction with empirical scaling factors α and β, physicochemical parameters to describe metal–ligand behavior. This approach of using independently determined physical properties rather than an existing database of stability constants was later exploited by Pearson.

Misono Equations

A dual parameter model is a common approach, linking basicity of the ligand to hardness or softness of the metal ion. An example of this approach is shown in equation 11 (105,106):

$$\log K = \alpha X + \beta Y + \gamma \qquad (11)$$

where X is a measure of the hardness of the metal, Y is a measure of the softness of the metal, α and β are the dual basicity factors (analogous to X and Y) for the ligand, and γ is a correlation factor. Figure 12 shows the degree of correlation for a series of metal–ligand complexes using this dual parameter model.

Pearson Equations

Pearson has also proposed a quantification of the HSAB principle (96):

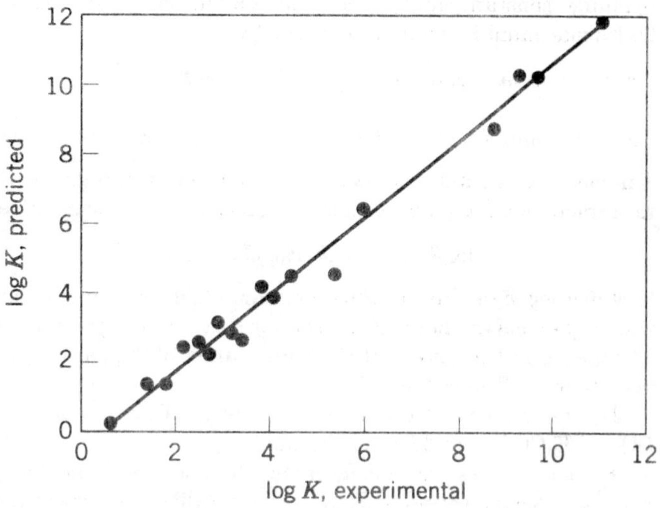

Figure 12. Correlation between predicted and experimental stability constants derived from equation 12 for ammine and hydroxo complexes with the metal ions Ag^+, Mg^{2+}, Fe^{2+}, Cu^{2+}, Zn^{2+}, Cd^{2+}, Hg^{2+}, Fe^{3+}, and In^{3+} (105,106).

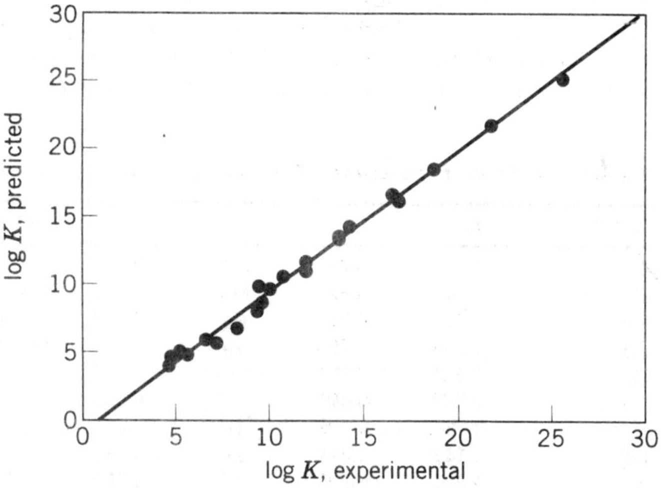

Figure 13. Correlation of predicted and experimental log K for metal–aminocarboxylate complexes using equation 16 for the aminocarboxylate ligands IDA, NTA, EDDA, EDDA′, and EDTA with the metal ions Ni^{2+}, Co^{2+}, Fe^{2+}, and Fe^{3+} (112).

$$\log K = S_a S_b + \sigma_a \sigma_b \qquad (12)$$

where S_a is a measure of the acid strength of the metal and S_b the base strength of the ligand. $\sigma_{a/b}$ is a measure of hardness/softness. Pearson's model has developed little, due to a reluctance to use hardness and softness as a quantitative scale, but it is used more as a guide to reactivity. Other scales of acid–base strength have been developed specifically to predict stability constant parameters based on the description of the strength of the interaction.

Drago Equation

Researchers (91–93) have published a large amount of predicted versus experimental data for the enthalpies of acid–base reactions, mainly involving uncharged organic, or

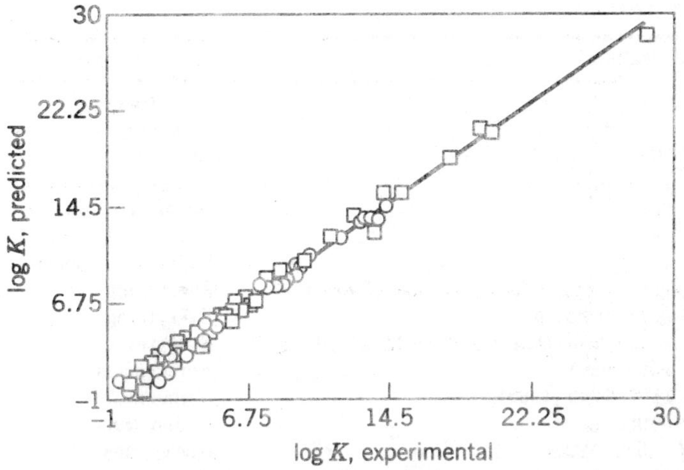

Figure 14. Correlation of predicted and experimental log K, OH and log K_1, ML using the equations in Box 2, for the metal ions $Li^+, Na^+, K^+, Be^{2+}, Mg^{2+}, Ca^{2+}, Sr^{2+}, Ba^{2+}, Sc^{3+}, Y^{3+}, La^{3+}, Ce^{3+}, Pr^{3+}, Nd^{3+}, Sm^{3+}, Eu^{3+}, Gd^{3+}, Er^{3+}, Yb^{3+}, Lu^{3+}, Zr^{4+}, Hf^{4+}, Th^{4+}, Al^{3+}, In^{3+}, Tl^{3+}, Tl^+, Pb^{2+}, Bi^{3+}, Zn^{2+}, Cd^{2+}, Hg^{2+}, Mn^{2+}, Fe^{2+}, Co^{2+}, Cu^{2+}, Ti^{3+}, V^{3+}, Fe^{3+}, CO^{3+}$, and Ag^+; and for the species $Na(SO_4)^-$, $TI(N_3)$, $TI(IO_3)$, BeF, $BaCO_3$, CoS_2O_3, CdI^+, CdI_2, $Pb(HPO_4)$, $Pb(HPO_4)_2$, $Am(SO_4)^+$, $Am(SO_4)_2^-$, $Am(SO_4)_3^{3-}$, $Bi(NO_3)^{2+}$, $Bi(NO_3)_2^+$, AlF^{2+}, AlF_2^+, AlF_3, $AlF_4^-, AlF_5^{2-}, AlF_6^{3-}$, $Mn(NH_3)^{2+}$, $Mn(NH_3)_2^{2+}$, $Mn(NH_3)_2^{2+}$, $Mn(NH_3)_3^{2+}, Zr(Cl)^{3+}$, $Zr(Cl)_2^{2+}$, $Zr(Cl)_3^+$, $UO_2(SO_4)$, $UO_2(SO_4)^{2-}, UO_2(SO_4)^{4-}$, $NpO_2(HPO_4)^-$, $Co(C_2O_4)_2^{2-}$, $Cd(en)^{2+}$, $Cd(en)_2^{2+}$, $Cd(en)_3^{2+}$, $Fe(gly)^+$, $Fe(gly)_2$, Pu(NTA), Ni(ida), $Ni(ida)_2^{2-}$, and $In(EDTA)^-$ (115–117).

organometallic, reactants largely in organic media, using a model derived from Pearson's HSAB concept,

$$-\Delta H = E_A E_B + C_A C_B \tag{13}$$

where E and C are electrostatic and covalent contributions, and ΔH is the enthalpy of the acid–base reaction. The relationship of enthalpy to log K is usually complicated by entropy and solvation effects in polar media such as water. However correlations of entropy of metal–fluoride interactions to ionic parameters have been reported (107) and experimentally confirmed (108). Although the Drago equation is very successful for predicting enthalpy data in organic media, it will not be further developed here because our major interest is in reactions in aqueous solutions.

Note that prior to Pearson's approach, stability constants were used to classify metals, thus requiring a significant amount of extant data before predictions were possible. The HSAB approach allows development of a model for the prediction of stability constant from physicochemical data, as opposed to large amounts of existing experimental data.

Partial Charge Model

A more recent extension of the concepts described in the HSAB theory is the partial charge model (PCM) (109,110). Briefly stated the PCM is as follows: *when two or more atoms initially different in electronegativity combine, they adjust to the same intermediate electronegativity in the compound.* This uses the concepts of hardness of metal ions together with electronegativity to describe the hydrolysis and ligation of metal ions with respect to pH. The approach is useful in describing the speciation of metal ions in aqueous solution with respect to hydrolysis, ionic dissociation, or ligation by a particular ligand. Although the model does not directly predict stability constants per se it has great utility in the qualitative description of the speciation of metal ions in solution.

Table 5. Electronicity Valves for Selected Ligands and Metals[a]

Ion	ϵ	Ion	ϵ	Ion	ϵ
F^-	−0.064	NH_3	1.569	Al^{3+}	0.015
HPO_4^{2-}	0.184	NO_3^-	1.639	Zr^{4+}	0.021
$H_2PO_4^-$	0.282	SCN^-	1.674	Sc^{3+}	0.085
SO_4^{2-}	0.699	$S_2O_3^{2-}$	1.865	Hf^{4+}	0.130
HCO_3^-	1.027	IO_3^-	1.927	Y^{3+}	0.321
NO_2^-	1.098	BrO_3^-	2.184	La^{3+}	0.488
CH_3COO^-	1.123	Cl^-	2.344	Mg^{2+}	1.049
ClO_4^-	1.260	Br^-	2.942	Co^{2+}	1.583
ClO_3^-	1.319	I^-	3.361	Ag^+	2.239

[a] Ref. 116.

EMPIRICAL MODELS

Correlated experimentally measured data, such as LFERs and pK_a correlations, may be condensed into mathematical equations, which can then be combined to obtain a general method of calculating unknown stability constants. Straightforward combinations of readily available physicochemical parameters, for example, charge and radius of metal ion, can be used to predict stability constants that are in excellent agreement with experimentally measured constants. Interpolating stability constants from a body of existing experimental results is a rapid and often useful way of predicting constants, but this approach is limited to reactions for which data exists.

Hancock Equations

Working from Adamson's model (111) of the chelate effect, Hancock and coworkers (112,113) derived an additive approach linking stability constants for reactions of metals with amines, aminocarboxylates, and carboxylates to the stability constants of the NH_3 and CH_3COO^- complexes:

$$\log K = 1.15 \log \beta_n(\text{NH}_3) - \sum_{i=1}^{n-1} i(0.50) + (n-1)\log 55.5 \tag{14}$$

$$\begin{aligned}\log K =& 1.15 \log \beta_n(\text{NH}_3) \\ &- \sum_{i=1}^{n-1} i(0.50) + m \log K(\text{CH}_3\text{COO}^-) \\ &- \sum_{i=1}^{m} i(0.19) + (n+m-1)\log 55.5\end{aligned} \tag{15}$$

where log K is the stability constant for the reaction between a metal and a polyamine or polyaminocarboxylate ligand, log $\beta_n(NH_3)$ is the log of the overall stability constant for the reaction of n NH_3 molecules with the metal in question, log $K(CH_3COO^-)$ is the log of the stability constant for the reaction of one acetate ligand with the metal in question, and

Table 6. Published Sources of Collated Stability Constant Values[a]

Author	Publication	Subject
Sillen and co-workers	*Stability Constants,* Special Publications 6, 7, 17, 25	Everything
Martell and co-workers	*Critical Stability Constants,* Vol. 1	Amino acids
Martell and co-workers	*Critical Stability Constants,* Vol. 2	Amines
Martell and co-workers	*Critical Stability Constants,* Vol. 3	Other organic ligands
Martell and co-workers	*Critical Stability Constants,* Vol. 4	Inorganic complexes
Martell and co-workers	*Critical Stability Constants,* Vol. 5	First supplement
Martell and co-workers	*Critical Stability Constants,* Vol. 6	Second supplement
Martell and co-workers	*NIST Critical Stability Constant Database* (computer based version of vols 1–6)	Everything
Pettit and Powell	*IUPAC Stability Constant Database* (computer-based)	Everything
IUPAC/various	*Stability Constants of Metal Ion Complexes, Chem. Data* Ser. Nos. 14, 17, 21, 22, 27	Everything
Yamauchi and co-workers	*Pure and Applied Chemistry,* **68,** 469–496 (1996)	Amino acids
Berthon	*Pure and And Applied Chemistry,* **67,** 1117–1240 (1995)	Amino acids
Sovago and co-workers	*Pure and Applied Chemistry,* **65,** 1029–1080 (1993)	Amino acids
Smith and Martell	*Pure and Applied Chemistry,* **63,** 1015–1080 (1991)	Nucleotides
Kiss and co-workers	*Pure and Applied Chemistry,* **63,** 597–638 (1991)	Glycine
Beck	*Pure and Applied Chemistry,* **59,** 1703–1720 (1987)	Cyanide
Pettit	*Pure and Applied Chemistry,* **56,** 247–292 (1984)	Amino Acids
Tuck	*Pure and Applied Chemistry* **55,** 1477–1528 (1983)	Indium
Anderegg	*Pure and Applied Chemistry* **54,** 2693–2758 (1982)	NTA
Sucha and Kotrly	*Handbook of Chemical Equilibria in Analytical Chemistry*	Everything

[a] The stability constants quoted vary from raw numbers lifted directly without comment from the literature to Martell's more discriminating collection of critical stability constants. Data are mostly collated by ligand, with the exception of the computer-based database which allows user-defined data extraction by metal, ligand, determination technique, and so on.

m is the number of carboxyl groups in the polyaminocarboxylate ligand. These equations allow the accurate prediction of stability constants for ubiquitous aminocarboxylate ligands, such as EDTA and NTA, as shown in Figure 13.

As an example of the use of equation 15, the stability constant for the formation of $[Ni(NTA)]^+$ is calculated as follows: Substituting $\log\beta_1(NH_3) = 2.72$, $\log\beta_1(CH_3COO^-) = 1.43$, $n = 1, m = 3$ into equation 15 gives

$$\log K = (1.15 \times 2.72) - 0 + (3 \times 1.43) - \{(1 + 2 + 3) \times 0.19\} + (3 \times \log 55.5)$$

the value of $\log K$ is calculated to be 11.51 as compared to the experimental value of 11.54.

Harris Equation

Another additive approach was developed (114) for a limited range of metals with nitrogen, oxygen, and sulfur ligands, allowing for the chelate effect of five- and six-membered rings. This approach is shown in equation 16:

$$\log K = \sum_{i=1}^{n} n_i X_i + \left[r_5 + \sum_{i=1}^{n_5} (r_5 f_5^{i-1}) \right]_{n_5 \neq 0} + \left[r_6 + \sum_{i=1}^{n_6} (r_6 f_6^{i-1}) \right]_{n_6 \neq 0} \quad (16)$$

where n is the number of ligating groups of a particular type, X is the stability increment of the ligating group of the ligand with a particular metal, r_5 and f_5 are increments for any five-membered rings formed by ligands at a metal, r_6 and f_6 are the corresponding increments for six-membered rings, and n_5 and n_6 are the numbers of five- and six-membered rings formed by the metal ligand complex. Equation 16 can be simplified:

$$\log K_{ML} = A + B + C \quad (17)$$

where A is a base value for the donor ligand involved. Typical values of the parameters defining A are available for the metals Ni^{2+}, Zn^{2+}, Cd^{2+}, and Fe^{2+} with various ligating moieties. Factors B and C allow for incremental adjustments to stability when five- and six-membered rings are formed during chelation, adding to the stability of the complex formed. The overall performance of the equation is good but is limited to the metals and ligands parameterized in its database.

As an example of the use of equation 16, the stability constant for the formation of $[Zn(EDTA)]^{2-}$ is calculated. The EDTA ligand contains two amine donors and four carboxylate donors and will form six five-membered rings with the central zinc atom. Therefore,

$$\log K([Zn(EDTA)]^{2-}) = (2 \times 2.5) + (4 \times 1.6) + 0.9 + \{(0.9 \times 1.1) + (0.9 \times 1.21) + (0.9 \times 1.33) + (0.9 \times 1.5) + (0.9 \times 1.65)\} = 18.4$$

The experimental value is 18.0.

Brown–Sylva–Ellis Equations

An empirical approach (115–117) enables predictions of metal species ligated by inorganic and simple organic ligands. Specific examples are given for hydroxo (115) species and dioxoactinides (117). The approach involves a highly parameterized empirical equation based on charge, radius and electronic structure of the metal ion, and the pK_a of the ligand (see Box 2).

Box 2: BSE Equations. Following are the equations derived to describe stability constants for M_pL_q complexes (115–117). Facile simplifications to allow predictions for the more common ML_q type complexes are easily accomplished; the inclusion of

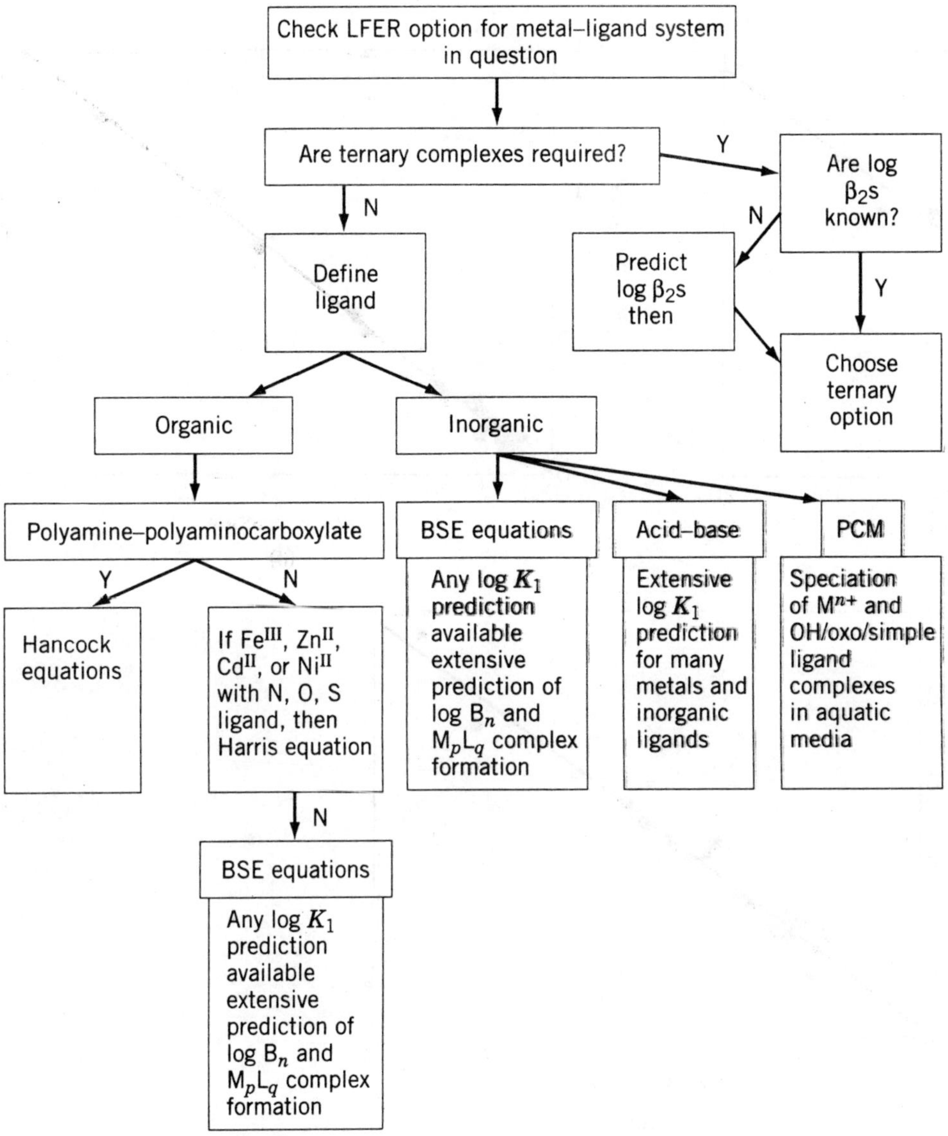

Figure 15. Flow chart describing the methodology of stability constant prediction.

Table 7. Analysis of the Organic Species Found in Landfill Leachate in Initial Acidification[a]

Organic Species	Concentration, mg/L
Acetic acid	11,000
Propanoic acid	3,760
2-Methylpropanoic acid	520
Butanoic acid	9,890
2-Methylbutanoic acid	350
3-Methylbutanoic acid	320
Pentanoic acid	2,510
4-Methylpentanoic acid	70
Hexanoic acid	5,770
Methylamine	6
Trimethylamine	83
t-Butylamine	41
sec-Butylamine	102
iso-Butylamine	32
Ethanol	277

[a] Ref. 132.

oligomeric terms relates to their origin in the description of hydroxide complexes.

$$\log \beta_{11} = \text{int}_1 + \text{slp}_1\left[g_1\left(\frac{z_M}{r_M^2} + g_2\right)\right]$$

or more generally

$$\log \beta_{pq} = [(p-1)(\text{int}_2) + q(\text{int}_1)] + [(p-1)(\text{slp}_2) + q(\text{slp}_1)] \times \left[g_1\left(\frac{z_M}{r_M^2} + g_2\right)\right] - \log U_{pq}$$

where

$$\text{slp}_1 = 0.031 + 0.0078\, \text{p}K_a$$

$$\text{int}_1 = 0.11 - 4.10[1 + (3 - z_L)(z_L + 1)]\text{slp}_1$$

$$\text{slp}_2 = 0.187\theta_L^2 - 0.814\theta_L + 0.78$$

$$\text{int}_2 = 3.58 - 62.67\text{slp}_2$$

$$g_1 = (1 + D + \varepsilon_L^2 S)(z_M + 2)$$

$$g_2 = g(n)(z_M - 1) - 0.05d[(\gamma - 1)(\varepsilon_L^4 - 3) - 1] \times [n - (3 + 2\gamma)]^2[1 - \gamma z_M](1 - S)$$

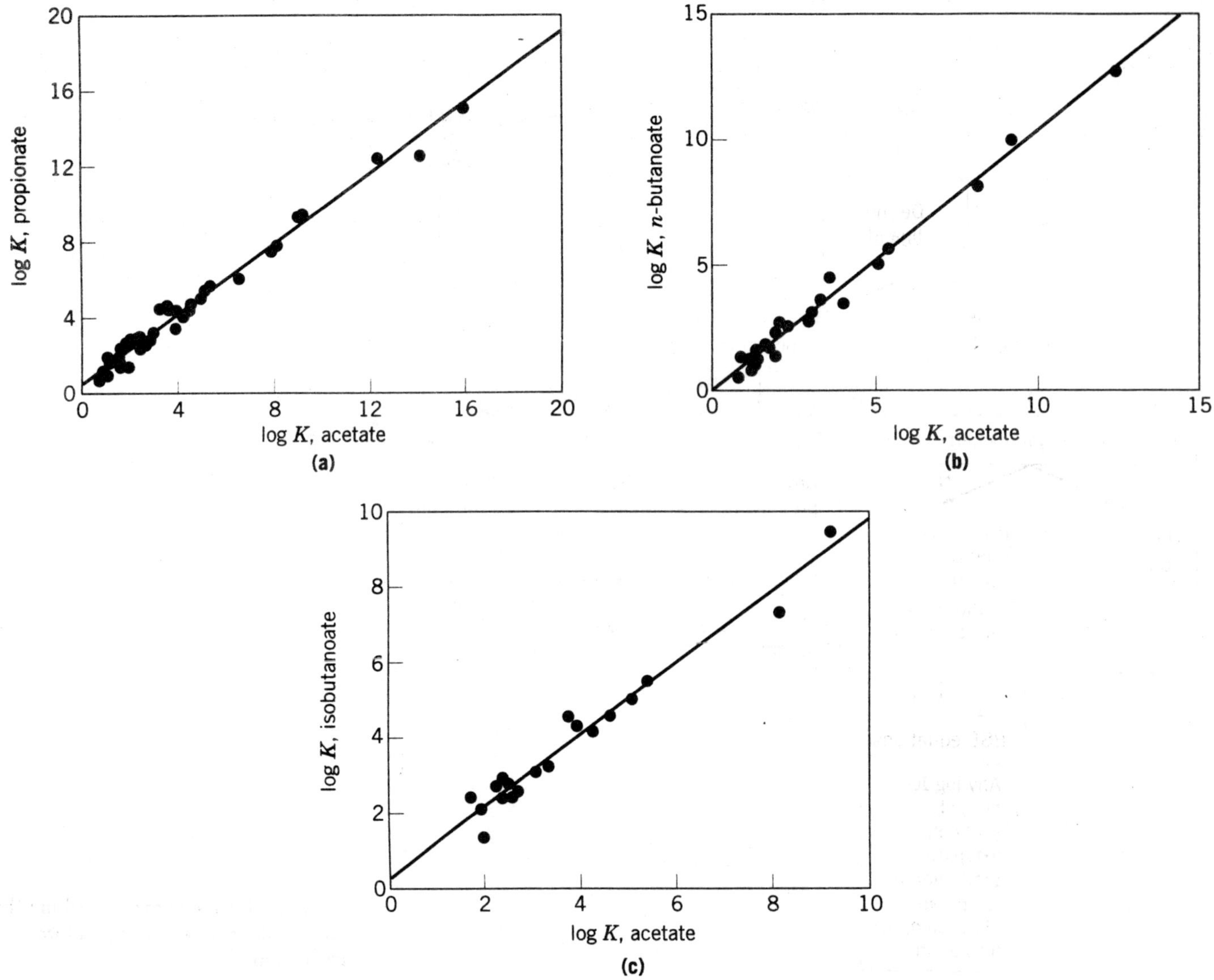

Figure 16. LFER for stability constants of metal–acetate complexes to (**a**) propionate, (**b**) 2-methyl butanoate, and (**c**) butanoate metal complexes.

and

$$\log k = -(2 - \theta_M)e^{-(\varepsilon_M - \varepsilon_L)^2}$$

$$\log U_{pq} = \log[(q + 1 - p)!] - 0.5(q - p)(q - p + 1)\log k$$

where $\log \beta_{11}$ is the overall stability constant for 1:1 metal ligand complex; z_M r_M are the charge and radius of the metal ion; $\log \beta_{pq}$ is the overall stability constant for a complex with p metals and q ligands; p and q are the number of metal ions and ligands respectively; pK_a is the negative log of the acid dissociation constant of the ligand; z_L is the charge on the ligand; θ_L is the number of ligands bonded to each metal; θ_M is the number of metal ions bonded to each ligand; ϵ_M is the electronicity of the metal; ϵ_L is the electronicity of the ligand; D is the presence or absence of d orbitals in outer shell of metal ion and is 0 or 1; d is the number of d electrons in the outermost shell of the metal; S is the presence or absence of s electrons in outer shell of metal ion and is 0 or 1; $g(n)$ is the Slater function and equal to 1 for all metals where $n > 1$; n is the principal quantum number of the metal; γ indicates if the ligand type oxo-anion type ligand and is 0 or 1; int_1, slp_1, int_2, slp_2, g_1, g_2, $\log U_{pq}$, and $\log k$ are all defined in the equations listed in terms of the above physical parameters.

The BSE model was developed from a series of studies on the hydrolysis of metal ions in aqueous solution. From these studies a general approach led to an equation describing the first hydrolysis product, which was capable of predicting a wide range of $\log \beta_{11}$ values (Fig. 14).

This approach was further developed to include higher oligomers, the production of which are closely associated with the hydrolysis of cations in aqueous solution. Extension of this approach to general unidentate ligands was accompanied by a conceptualization of metal and ligand bonding characteristics in the form of the *electronicity parameter,* which is a gauge of freeness of the valence electrons of both the metal and the ligand (Table 5).

Table 8. Predicted and Experimental Stability Constants for Plutonium, Amerieum, and Thorium with Major Organics in Landfill Leachate

Species	Ligand	Log K^1	log β_2	log β_3	log β_4	log β_5	log β_6
Th^{4+}	Acetate	2.68	5.03	6.60	7.60	8.40	10.10
	Propionate	4.74	8.85	11.70	14.10	17.00	19.00
	Butanoate	4.70	8.60	12.00	14.50	17.00	19.00
	Pentanoate	4.74	8.85	11.70	14.10	17.00	19.00
	Hexanoate	4.74	8.85	11.70	14.10	17.00	19.00
Pu^{3+}	Acetate	2.85	5.06	6.57	7.68	8.42	8.74
	Propionate	3.05	5.13	6.54	7.58	8.28	8.58
	Butanoate	2.98	5.30	6.88	8.04	8.82	9.16
	Pentanoate	3.27	5.47	6.73	7.10	6.60	
	Hexanoate	3.27	5.48	6.74	7.12	6.63	
Pu^{4+}	Acetate	6.10	10.6	16.2	21.2	26.0	
	Propionate	6.10	10.3	15.6	20.3	24.8	
	Butanoate	6.38	11.1	16.9	22.2	27.2	
	Pentanoate	5.89	11.1	15.6	19.6	23.1	26.1
	Hexanoate	5.90	11.1	15.6	19.7	23.2	26.2
PuO_2^+	Acetate	1.88	2.82	2.96			
	Propionate	2.14	3.03	3.16			
	Butanoate	1.96	2.95	3.10			
	Pentanoate	2.14	3.03	3.16			
	Hexanoate	2.14	3.03	3.16			
PuO_2^{2+}	Acetate	2.65	4.54	6.46			
	Propionate	3.02	5.47	7.28			
	Butanoate	2.61	5.04	7.27	9.31		
	Pentanoate	2.77	4.81	6.24			
	Hexanoate	2.77	4.82	6.26			
Am^{3+}	Acetate	2.89	4.78	5.90	6.10		
	Propionate	3.23	5.05	6.07	7.40		
	Butanoate	3.02	5.00	6.18	6.50		
	Pentanoate	3.23	5.05	6.07	7.40		
	Hexanoate	3.23	5.05	6.07			

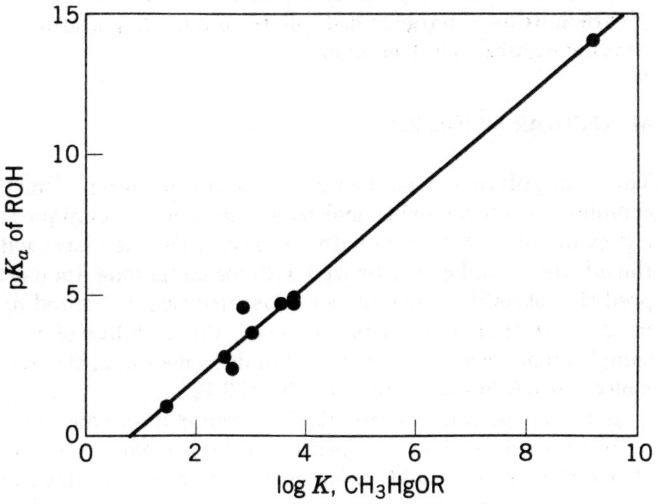

Figure 17. LFER for stability constants of log K_1, MMHgOR to pK_a of ROH ligands.

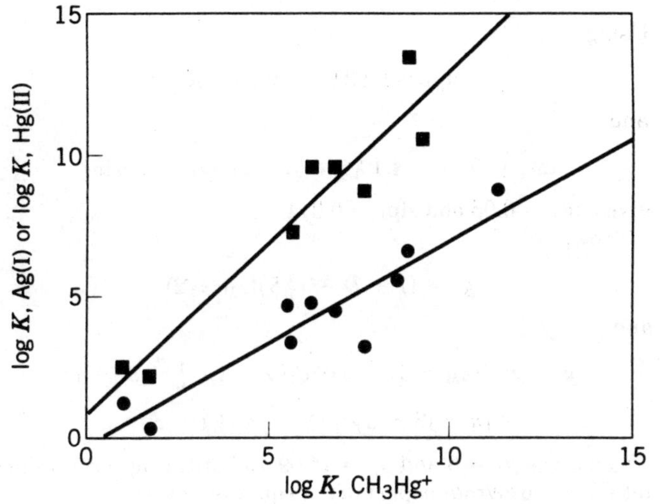

Figure 18. LFER for CH_3Hg^+, stability constants to Hg^{2+} and Ag^+ stability constants.

Electronicity can be thought of in terms of hardness (low electronicity/polarizability) and softness (high electronicity/polarizability). The equation describing metal–ligand interactions is an implicit function of z_M^2/r_M^2, where Z_M is the charge on the metal ion and r_M is the radius of the metal ion, because a factor of z_M is contained within the g terms. Plots of log β_{11} vs g terms gives int$_1$/slp$_1$ data for a particular ligand. However, int$_1$/slp$_1$ data may be calculated from z_L, the charge on the ligand and the pK_a of the ligand.

Table 9. Methyl Mercury Stability Constants for Potential Ligands in Landfill Leachate Analysis

Organic Species, L	Log $K_1(CH_3HgL)^a$
Acetic acid	**3.50**
Propanoic acid	**3.69**
2-Methylpropanoic acid	*3.66*
Butanoic acid	*3.66*
2-Methylbutanoic acid	*3.60*
3-Methylbutanoic acid	*3.57*
Pentanoic acid	*3.61*
4-Methylpentanoic acid	*3.61*
Hexanoic acid	*3.62*
Methylamine	**7.57**
Trimethylamine	**5.05**
t-Butylamine	**7.52**
sec-Butylamine	*6.56*
iso-Butylamine	*6.13*
Ethanol	*10.31*

[a] Experimental values in bold; predicted values in italic. Experimental values are from sources listed in Table 6.

The derived empirical equations shown in Box 2 have a wide applicability toward the prediction of stability constants as is shown in Figure 14.

As an example of the use of BSE equations in calculating stability constants, the constant for the formation of $CdCl_4^{2-}$ is calculated:
for the metal, Cd:

$$n = 4,\ g(n) = 1,\ D = 1,\ S = 0,\ d = 10,\ \epsilon_M = 2.731,$$

$$\theta_M = 1, z_M = 2, r_M = 0.97$$

and for the ligand, Cl:

$$\epsilon_L = 2.344, \theta_L = 0, z_L = -1, g = 0, pK_a = -2.19,$$

$$p = 1, q = 4$$

Using

$$slp_1 = 0.031 + 0.0078\ pK_a$$

and

$$int_1 = 0.11 - 4.10[1 + (3 - z_L)(z_L + 1)]slp_1$$

gives $int_1 = 0.05$ and $slp_1 = 0.014$.

Now,

$$g_1 = (1 + D + \varepsilon_L^2 S)(z_M + 2)$$

and

$$g_2 = g(n)(z_M - 1) - 0.05d[(\gamma - 1)(\varepsilon_L^4 - 3) - 1] \times [n - (3 + 2\gamma)]^2[1 - \gamma z_M](1 - S)$$

which gives $g_1 = 8$ and $g_2 = 15.09$. Substituting these values into the charge/radius part of the equation gives

$$\left[g_1\left(\frac{z_M}{r_M^2} + g_2\right)\right] = 137.73$$

$$\log k = -(2 - \theta_M)e^{-(\varepsilon_M - \varepsilon_L)^2}$$

gives $\log k = -0.861$ which can be substituted into

$$\log U_{pq} = \log[(q + 1 - p)!] - 0.5(q - p)(q - p + 1)\log k$$

to give $\log U_{pq} = 6.546$. Substituting all the above values into

$$\log \beta_{pq} = [(p - 1)(int_2) + q(int_1)] + [(p - 1)(slp_2) + q(slp_1)] \times \left[g_1\left(\frac{z_M}{r_M^2} + g_2\right)\right] - \log U_{pq}$$

gives a predicted $\log \beta_{14}$ of 1.37. The experimental value is 1.47.

OTHER EXAMPLES

Other more limited examples of equations derived for the behavior of specific metal-ligand systems are documented (118). For example, the complexation of actinide(III) metals by carbonate may be described by a simple quadratic dependence on the central metal ion radius (R). This dependence is sufficient to evaluate the stability constant for trivalent actinide carbonates. Equations 18 and 19 show this dependence

$$\log \beta_1 = -6.128 + 35.206R - 21.557R^2 \quad (18)$$

$$\log \beta_2 = 14.797 + 7.945R - 10.304R^2 \quad (19)$$

Actinide carbonates have also been described by a hard-sphere model of metal interaction with hydroxide, carbonate, and mixed hydroxycarbonates (119). The interaction of metal and ligand is described in purely electrostatic terms, treating the species as charged hard spheres with effective radii:

$$E = \sum_{i \neq j} \frac{Z_i Z_j}{\epsilon r_{ij}} \quad (20)$$

where E is the electrostatic potential energy, Z_i and Z_j are the effective electric charges of species i and j, ϵ is the dielectric constant, and r_{ij} is the distance between the hard spheres i and j. Good agreement with experimental results is obtained for hydroxy complexes, but the carbonate species are less successfully predicted. This is possibly due to the oversimplified model of carbonate as a charged hard sphere, and the hypothesis that bonding is purely electrostatic.

MOLECULAR MODELING

The manipulation, visualization, and optimization of metal complex structures using molecular mechanics techniques is a growing area of interest within coordination chemistry with the advent of well characterized data for metal ions. Its use to predict a stability constant per se is historically limited and no general approach exists. However, the stability of many complex macrocyclic systems is often described with the aid of molecular mechanics techniques (31,120,121).

It should be remembered that molecular mechanics calculations provide an idealized picture of the *gas-phase* structure of a molecule. Although this constraint frees the structure from any lattice interactions present in the solid state, solvent effects are completely neglected. An example of the difficulties that may be encountered in overlooking the importance of solvent effects is a consideration of the relative basicity of simple amines. The interplay of steric and electronic effects upon consecutive methyl substitution at N indicates that the basicity of the amine should increase: $NH_3 < CH_3NH_2 < (CH_3)_2NH < (CH_3)_3N$ but in fact, the experimental values in aqueous solution are as follows: $NH_3 < CH_3NH_2 < (CH_3)_2NH > (CH_3)_3N$. The values in the gas phase:

Table 10. Predicted Stability Constants for Complexes Formed Between Plutonium in the III, IV, V, and VI Oxidation States and Carboxylate and Alcoholic Moieties of Isosaccarinic Acid

Ligating Moiety	Reaction	Complex	Log β	Litigating Moiety	Reaction	Complex	Log β
COOH	$Pu^{3+} + HL^-$	$Pu(HL)^{2+}$	2.80	OH	$Pu^{3+} + L^{2-}$	$Pu(L)^+$	5.84
	$Pu^{3+} + 2HL^-$	$Pu(HL)_2^+$	4.61		$Pu^{3+} + 2L^{2-}$	$Pu(L)_2^-$	10.91
	$Pu^{3+} + 3HL^-$	$Pu(HL)_3$	5.57		$Pu^{3+} + 3L^{2-}$	$Pu(L)_3^{3-}$	15.35
	$Pu^{3+} + 4HL^-$	$Pu(HL)_4^-$	5.73		$Pu^{3+} + 4L^{2-}$	$Pu(L)_4^{5-}$	19.20
COOH	$Pu^{4+} + HL^-$	$Pu(HL)^{3+}$	5.03	OH	$Pu^{4+} + L^{2-}$	$Pu(L)^{2+}$	10.59
	$Pu^{4+} + 2HL^-$	$Pu(HL)_2^{2+}$	9.41		$Pu^{4+} + 2L^{2-}$	$Pu(L)_2$	20.42
	$Pu^{4+} + 3HL^-$	$Pu(HL)_3^+$	13.27		$Pu^{4+} + 3L^{2-}$	$Pu(L_3^{2-}$	29.61
	$Pu^{4+} + 4HL^-$	$Pu(HL)_4$	16.66		$Pu^{4+} + 4L^{2-}$	$Pu(L)_4^{4-}$	38.22
	$Pu^{4+} + 5HL^-$	$Pu(HL)_5^-$	19.60		$Pu^{4+} + 5L^{2-}$	$Pu(L)_5^{6-}$	46.26
COOH	$PuO_2^+ + HL^-$	$PuO_2(HL)$	0.65	OH	$PuO_2^+ + L^{2-}$	$PuO_2(L)^-$	2.46
	$PuO_2^+ + 2HL^-$	$PuO_2(HL)_2^-$	0.45		$PuO_2^+ + 2L^{2-}$	$PuO_2(L)_2^{3-}$	4.28
COOH	$PuO_2^{2+} + HL^-$	$PuO_2(HL)^+$	2.34	OH	$PuO_2^{2+} + L^{2-}$	$PuO_2(L)$	6.95
	$PuO_2^{2+} + 2HL^-$	$PuO_2(HL)_2$	4.03		$PuO_2^{2+} + 2L^{2-}$	$PuO_2(L)_2^{2-}$	13.41
	$PuO_2^{2+} + 3HL^-$	$PuO_2(HL)_3^-$	5.21		$PuO_2^{2+} + 3L^{4-}$	$PuO_2(L)_2^{3-}$	19.50

Table 11. Comparison of Experimental and Predicted Values of Stability Constants for Metal Reactions with 2-, 3-, 4-, 5-Tetrahydroxypentanoic Acid

Species	Experimental[a]	Predicted
CoL^+	1.76	1.92
EuL_2^+	1.94	2.41
SmL_2^+	1.96	2.36
CuL^+	5.37	2.60

[a] Experimental data from Ref. 138.

Table 12. Analysis of Cementitious Pore Water

Species	Concentration, *M*
Ca	4.7×10^{-2}
Mg	8.2×10^{-7}
Na	4.3×10^{-1}
K	4.4×10^{3}
Si	7.2×10^{-6}
Cl	4.0×10^{-1}
C	7.5×10^{-6}
S	7.0×10^{-3}
Pu	1.0×10^{-8}
pH	12.0
E_h	−400 mV

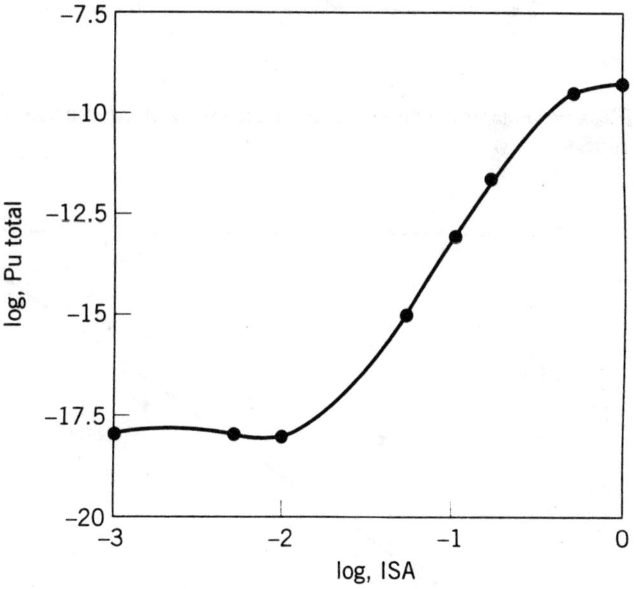

Figure 19. Speciation of plutonium and isosaccarinic acid in cementious water.

$NH_3 < CH_3NH_2 < (CH_3)_2NH < (CH_3)_3N$ confirm the electronic effect of repeated methyl substitution (inductive effect) and shows the anomaly to be the product of solvation effects in aqueous solution. Although a simple example, this demonstrates the importance of solvent effects, the incorporation of which in molecular mechanics and molecular orbital theory is still in its infancy (30).

An important area of macrocyclic chemistry of metal complexes considers the criteria for matching the size of the multidentate ligand cavity of molecules such as cryptands, with the ionic radius of the metal, which is closely linked with metal-ion recognition issues (51). Empirical rules derived from considerations of the steric requirements of chelation indicate that the size of the chelate ring, that is the ring formed by adjacent chelating atoms and the metal, is of greater importance than the macrocyclic ring size. This leads to an explanation for the poor correlation of ionic metal radii with macrocyclic cavity comparisons, which may be explained in terms of strain introduced by larger metal ions. This strain results in an increased chelate ring size (eg, five-membered to six-membered) giving a larger increase in stability constant for metals with ionic radii relative to those with larger radii. These rules are valid for both macrocyclic and nonmacrocyclic ligands.

MOLECULAR ORBITAL THEORY

Attempts to parameterize molecular orbital methods, such as CNDO (complete neglect of differential overlap) (122–125), to allow stability constant prediction have been attempted, but are limited in their application. The use of molecular electrostatic potentials (MEPs) as tools in interpreting and predicting reactivity trends in chemical systems is well established (126–

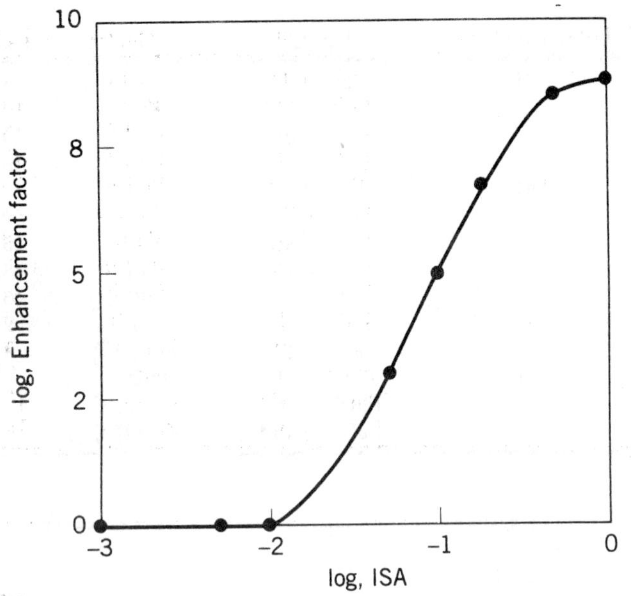

Figure 20. Enhancement factor of Pu solubility in the presence of ISA.

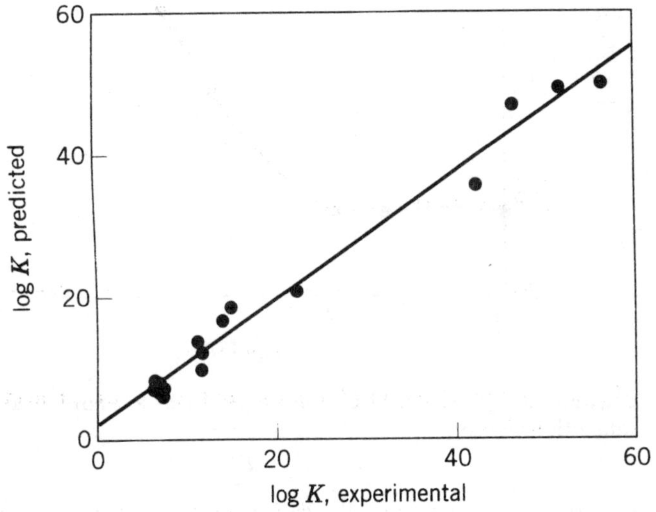

Figure 21. Correlation of predicted and experimental log K_n, $M(CO_3)_n$ for a range of 3+ and 4+ metal ions including Hf^{4+}, Np^{4+}, Zr^{4+}, Eu^{3+}, Y^{3+}, La^{3+}, Am^{3+}, Ce^{3+}, and Sc^{3+}; the data predicted cover 1,2, 1,3, 1,4, and 1,5 complexes where 1,n corresponds to $M_1(CO_3)_n$ (14).

128). A statistical ranking technique was used in a study of the correlation of Hammett σ-values to MEPs for a series of substituted benzoic acids and substituted anilines (128), in a move toward relating empirical measures of reactivity to basic molecular properties. This technique, which may prove applicable to metal-ligand interactions and stability constant predictions, considers the use of Hammett and other empirical descriptors as references to molecular properties.

STATISTICS

Correlation of a wide range of varied chemical parameters with experimentally determined stability constants would seem to be an area prime for analysis using multivariate statistical techniques, yet few applications exist. One such study is the lanthanide(III) promoted reaction of 4-oxo-4-1-benzopyran-3-carbaldehyde oxime (129). To critically evaluate the effect of Ln^{3+} on the reaction of oximes in acetone and methanol, the known stability constants for analogues of the reactants were collated, namely, a variety of oximes as analogues for 4-oxo-4-1-benzopyran-3-carbaldehyde oxime; acetylacetonate as an analogue for acetone; and 1,2-/1,3-diols and a triol as an analogue for methanol. Treating these data using factor analysis enabled conclusions to be rapidly drawn about the relative affinity of Ln^{3+} for the molecules in question. The statistical analysis gave valuable information about the bonding tendencies across the lanthanide series.

PREDICTION METHODOLOGY—A CONSISTENT APPROACH

Care must be taken when predicting stability constants, because, when using some of the methods which have been described, it is possible to predict constants for species that do not exist. For example, one can predict that the stability constant for a metal ion with twenty ligands attached to it may exist, but this of course would be physically impossible. Furthermore, some species may not exist due to kinetic factors. Some chemical intuition, or at the minimum a thorough perusal of the relevant literature of the metal ion chemistry, and the "typical" or usual binding mode(s) of the ligand is required.

Thus, all predictions must be correlated to existing data before use. This may be by comparison to similar systems for which published experimental data exist, or by using the same method of prediction on different systems where data exists.

Databases

A thorough search of the existing published data is essential. Table 6 details those works that present stability constant data in an easily accessible form. There are many compilations of data; one of the most comprehensive is available (14). In this text, data are presented in a clear form, indexed by ligand, and the only variables are ionic strength and occasionally temperature. A particular advantage of this series of texts is the large number of stability constants quoted at zero ionic strength, which allows immediate incorporation into speciation computer programs without ionic strength corrections. A computerized version of the database is available (79).

An alternative computerized database has been compiled as the SCD base (72). A series of pull-down menus allows searches by ligand class, metal class, H + /D+, experimental procedure, and so on. Data are easily extractable from the database and are downloadable to files for further processing. This is very convenient when constructing LFERs. However, data in the database are largely uncritical and all the published data, in whatever form, be it nonaqueous, high temperature, and so on, are included. However, the conditions under which the determinations were made, the experimental procedures, and literature references are included and therefore it is left to the user to assess the validity of the values. The criteria for

Table 13. Comparison of Experimental and Predictive Studies on Plutonium (IV) Carbonate Stability Constants[a]

Complex	Lierse	Kim	Nitsche	BSE	Hard Sphere
Ionic strength	0.3	1.0	0.5	0	1.0
$Pu(CO_3)^{2+}$	17.0	47.1	12.3	11.8	**15.6**
$Pu(CO_3)_2$	30.0	55.0	23.4	22.4	**28.0**
$Pu(CO_3)_3^{2-}$	39.1	57.9	30.0	32.2	**36.0**
$Pu(CO_3)_4^{4-}$	42.9	59.6	32.8	41.2	**41.2**
$Pu(CO_3)_5^{6-}$	44.5	62.4	33.9	49.5	**42.0**
$Pu(CO_3)_6^{8-}$			57.2	40.0	

[a] Refs. 139–142.

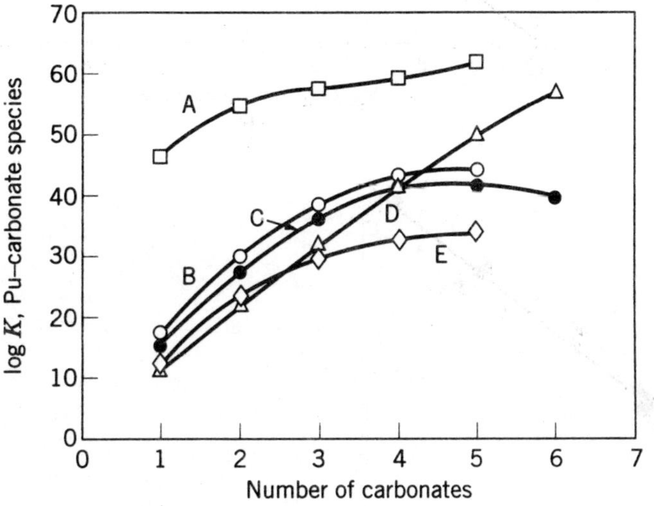

Figure 22. Comparison of plutonium carbonate predicted and experimental stability constants, where A is Kim; B, Lierse; C, hard sphere; D, BSE; and E, Nitsche.

critical stability constant selection here have been discussed (130).

Prediction Methodology

Figure 15 shows a useful approach for predicting an unknown stability constant and this approach has been used in the following examples.

Example 1: Landfill Leachate. A major environmental problem is that of waste disposal. A current option is to dispose of the waste in a landfill site by shallow burial of the waste (131). Water ingress into the site may create a zone of complicated chemistry which will mobilize pollutants and contaminants by producing soluble complexes. A typical example of the chemical constituents of landfill leachate, in the early stages of closure of the site, is shown in Table 7 (132).

Speciation of Actinide Ions in Landfill Leachate. In the United Kingdom, low level radioactive wastes contain a significant amount of cellulose which, in the chemical conditions of the repository, will largely degrade to produce simple carboxylic acids. Major organics present in the leachate are expected to be acetic, propanoic, butanoic, pentanoic, and hexanoic acids and their reactions with radionuclides of interest; Pu, U, Th and to a lesser extent Am, is of obvious interest in determining radionuclide mobility. Stability constants for the reactions of acetate with these radionuclides have been published (14,61–79), but unfortunately many of the other constants have not been measured. Two methods were used to predict these unknown constants. Firstly, for the lower carboxylic acids, propanoate, butanoate, and 2-methyl butanoate, correlations of stability constants of metal-propanoate, metal-butanoate, or metal-2-methyl butanoate with metal-acetate constants were made (Fig. 16).

From these correlations, stability constants of radionuclide-propanoate, radionuclide-butanoate, and radionuclide-2-methyl butanoate were determined. Secondly, because correlations of metal-pentanoate and metal-hexanoate could not be established because of the lack of data, the BSE equations were used to calculate the necessary stability constants. Examples of the results of the correlations and the calculations are shown in Table 8.

Methylmercury

Methylmercury, CH_3Hg^+, and other metal alkyl species are important in environmental toxicology because their lipophilic properties and volatility result in enhanced uptake across normally impermeable biological membranes (133). The CH_3Hg^+ subunit is kinetically stable to demethylation, and therefore exists as a separate entity as opposed to being in equilibrium with free mercury species. Often CH_3Hg^+ bioaccumulates in exposed organisms. The toxicity of CH_3Hg^+ makes it unpleasant and difficult to use, so much of the stability constant data, required to speciate mercury in such conditions, are missing. Table 9 shows the available stability constants (bold type) for the reactions of CH_3Hg^+ with possible complexing ligands present in a landfill leachate. To predict the other, non-existent, values LFERs were constructed as is shown in Figures 17 and 18. These correlations allow predictions of the unknown stability constants of the reactions between methyl mercury and the remaining carboxylic acids, ethanol and the amine ligands respectively. The predicted stability constants are shown in italics in Table 9.

Example 2: Cellulose Degradation Products. As stated in the introduction, groundwater in an intermediate level waste repository is expected to be of high pH and low E_h. As in low level wastes, intermediate level wastes will contain cellulose. Under the conditions of groundwater chemistry expected to be found in an intermediate level waste repository, cellulose is expected to degrade to species that may react with radionuclides forming water-soluble radionuclide complexes (134–137). One of the major degradation products of cellulose under these conditions is expected to be 2,4,5-trihydroxy 2-hydroxymethyl

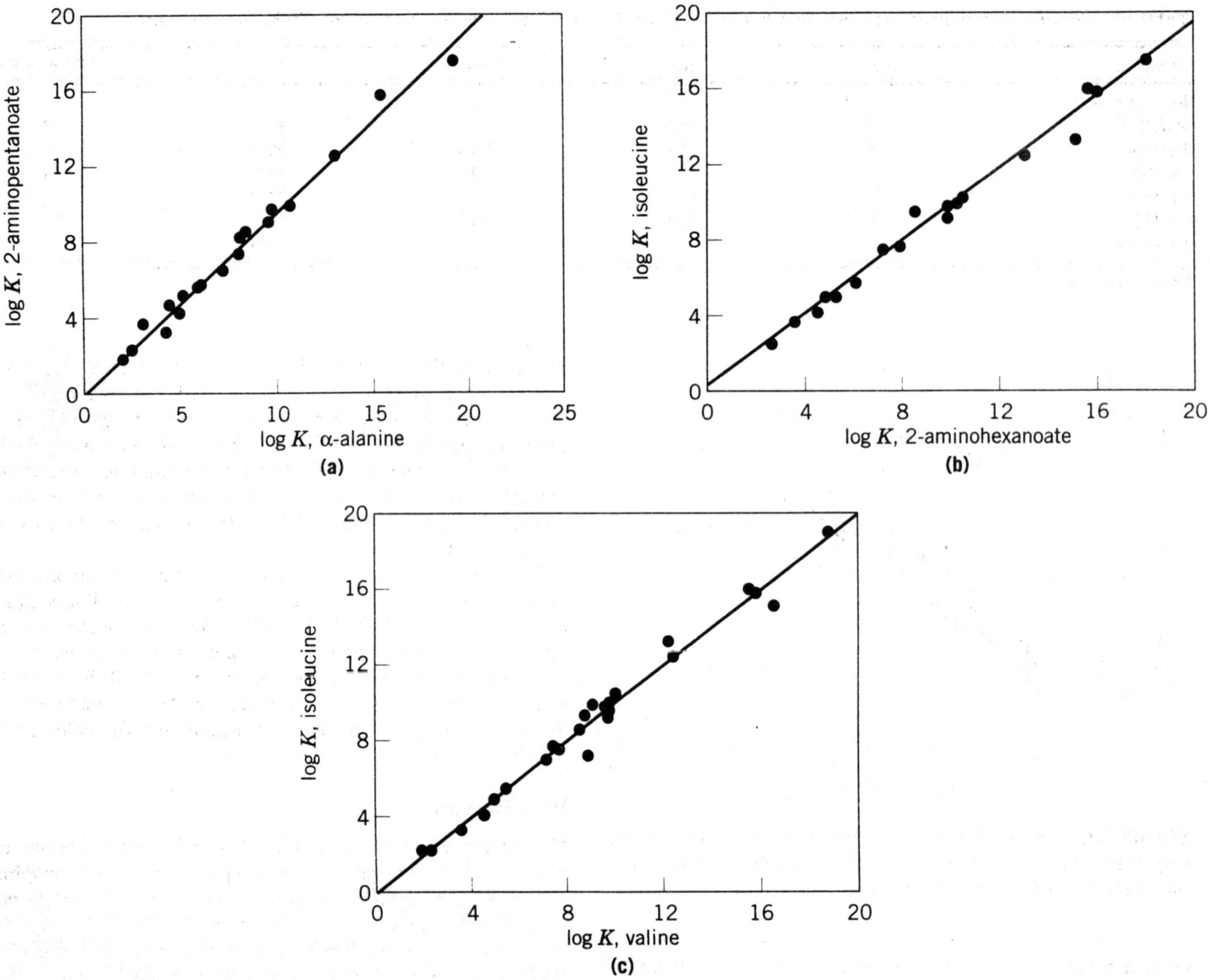

Figure 23. Typical amino acid LFERs (see Table 14): (**a**) 2-aminopentanoate and α-alanine, (**b**) isoileucine and 2-aminohexanoate and (**c**) isoleucine and valine.

pentanoic acid (iso-saccharinic acid (ISA)). At the expected pH of the near-field of the repository, the carboxylate group on the ISA will be deprotonated and therefore capable of ligating metal ions. Iso-saccharinic acid contains four alcohol groups and significant ionization of these groups may be expected at pH 12. Unfortunately, even though ISA is an important degradation product of cellulose, neither the pK_a values or the stability constants for radionuclide reactions with ISA exist. Therefore BSE equations were used to predict stability constants for reactions of plutonium and thorium with ISA.

The adopted approach to predicting stability constants was to firstly predict the pK_a values for the deprotonation of the acid using methods described previously. The predicted values were 3.53 for the COOH and 12.03, 13.79, 14.24, and 14.78 for the OH groups. Secondly, the stability constants for the reactions of the radionuclide of interest, eg, $Pu^{3+} + L^{2-} = (PuL)^{+}$, were calculated using BSE equations. The calculations were performed either for the reactions of the metal with the deprotonated carboxylate group or for the ionized OH group (pK_a = 12.03). Chelation was ignored in these calculations as there was no information available on the denticity of the reaction, or the possible chelation modes of ISA. The results of the calculations for the reaction of plutonium with ISA are shown in Table 10.

To obtain confidence in this approach, stability constants for reactions of cobalt, europium, samarium and copper with 2,3,4,5-tetrahydroxy pentanoic acid, which is an analogue of ISA, were calculated using BSE equations and compared to experimental values which were obtained by pH titrations and spectrophotometry (138). A comparison of the calculations to the experimental values is shown in Table 11.

Within experimental errors, good correlation is evident for cobalt, europium, and samarium but the results for copper show a large variation. Clearly, this variation requires further investigation and the lack of correlation encourages further analyses of the experimental or predictive methodologies used to obtain the values. To evaluate the importance of ISA in solubilizing and therefore possibly increasing the mobility of plutonium, calculations of plutonium solubility in the absence and presence of ISA were conducted using the speciation code PHREEQE. The predicted Pu-ISA values of the stability constants were converted into PHREEQE format and included in the database together with the cementitious porewater analysis shown in Table 12.

The results of the calculations are shown in Figures 19 and 20 Figure 19 shows the total concentration of plutonium in solution as a function of the concentration of ISA and Figure 20 shows the solubility enhancement factor calculated as the plutonium concentration in solution in the presence of ISA divided by the concentration of plutonium solution in the absence of ISA.

Example 3: Carbonate Speciation. The presence of CO_2 in the atmosphere in equilibrium with carbonic acid in aqueous solution results in metal–carbonate species being of special importance in environmental waters. Metal–carbonate species are thought to be especially important in the porewaters of a radioactive waste repository which contains cement because of the enhanced dissolution of CO_2 in alkaline solutions. A number of experimentally determined actinide-carbonate stability constants are available from the literature (139–142), but their values vary considerably. Comparing these values with predicted values may clarify the situation. To gain confidence in predicting metal–carbonate stability constants, BSE equations were used to predict stability constants for a range of M{3+} and M{4+} carbonate complexes and these values were compared with experimentally determined values reported in the literature.

The results shown in Figure 21 indicate a good correlation between the values, allowing a confident estimate of the values for the plutonium carbonate complexes shown in Table 13. Also shown in Table 13 is a comparison of all of the relevant published Pu(IV)–carbonate stability constants together with the predicted values using two of the methods described above, namely the BSE equations and the hard-sphere model. Although the use of BSE equations has been criticized for overestimating the stability constants for the higher Pu:CO_3complexes, if the values given in ref. 140 are omitted, Figure 22 demonstrates the degree of agreement for other carbonate species, up to $M(CO_3)_4^{n-}$.

Example 4: Amino Acid Stability Constants. Amino acids are the basic building blocks of life on earth and therefore their chemical reactions are of special importance. Because of their ubiquity in biological fluids and in the hydrosphere, a thorough knowledge of their reactions with metals is necessary to understand medical problems related to the presence of high concentrations of metals in biological fluids and environmental problems caused by the possibility of their enhancing the mobility of metals through the environment. As a demonstration of the use of LFERs to predict and correlate a wide number of metal–amino acid stability constants, Figure 23 shows examples of the very good correlations available if the correlations are restricted only to the aliphatic amino acids.

Table 14 shows the excellent agreement between a large number of correlations of experimentally determined stability constants. These correlations may be used to predict metal–amino acid stability constants for little-studied amino acids such as 2-amino hexanoate and 2-amino pentanoate.

Table 14. Tabulation of LFERS between Stability Constants for Amino Acid Metal Complexes with Estimate of Correlation

y-axis	*x* -Axis	Slope	Intercept	Correlation Coefficient, r^a
Leucine	Valine	1.016	0.067	0.963
Isoleucine	Leucine	0.970	−0.10225	0.966
β-Alanine	α-Alanine	0.928	0.408	0.932
β-Alanine	Valine	0.924	0.352	0.955
2-Aminohexanoate	Valine	1.029	0.016	0.983
Valine	2-Aminopentanoate	0.99741	0.168	0.958
Isoleucine	Valine	1.002	0.108	0.992
α-Alanine	2-Aminohexanoate	0.996	0.184	0.988
2-Aminopentanoate	α-Alanine	0.952	0.054	0.994
α-Alanine	Isoleucine	0.993	0.041	0.982
Leucine	α-Alanine	0.995	0.262	0.972
α-Alanine	Valine	0.968	0.378	0.966
2-Aminopentanoate	Leucine	0.885	0.169	0.946
Isoleucine	2-Aminohexanote	0.970	0.290	0.994
Isoleucine	2-Aminopentanoate	0.996	0.255	0.997
β-Alanine	Leucine	0.898	0.571	0.908
2-Aminohexanoate	Leucine	0.975	−0.167	0.964
β-Alanine	2-Aminohexanoate	0.984	−0.397	0.949
2-Aminopentanoate	β-Alanine	0.820	1.459	0.916
2-Aminopentanoate	2-Aminohexanoate	0.988	0.040	0.998
Isoleucine	β-Alanine	0.961	0.980	0.952

[a] $r = +1.00$, indicating a perfect positive correlation.

Acknowledgments

The Royal Society of Chemistry is thanked for giving permission to reproduce some of this article. The Company Research Laboratory, British Nuclear Fuels plc is thanked for financial support for much of the work reported.

BIBLIOGRAPHY

1. R. H. Neil and G. Sposito, *Soil Sci.* **142,** 164–172 (1986).
2. E. Matzner, D. Murach, and H. Fortmann, *Water, Air Soil Pollution* **31,** 273–282 (1986).
3. P. G. C. Campbell, P. M. Stokes, and J. N. Galloway, in *Heavy Metals in the Environment,* Proceedings of an International Conference, Heidelberg, **2,** 760–733, (1983).
4. P. Warwick and A. Hall, *Anal Proc.* **30,** 192 (1993).
5. D. C. Adrian, *Trace Metals in the Terrestrial Environment,* Springer Verlag, Berlin, 1986.
6. F. M. M. Morel, *Principles of Aquatic Chemistry,* John Wiley & Sons, Ltd., Chichester, U.K., 1983.
7. S. M. Ross, ed., *Toxic Metals in Soil-Plant Systems,* John Wiley & Sons, Ltd., Chichester, U.K., 1994.
8. A. Albert, *Selective Toxicity: The Physico-chemical Basis of Therapy,* 7th ed., Chapman and Hall, London, 1985.
9. D. L. Parkhurst, D. C. Thorstenson, and L. N. Plummer, *PHREEQE—A Computer Program for Geological Calculations,* U.S. Geological Survey Water Resources Investigations, No. 80-96, 1980.
10. J. D. Allison, D. S. Brown, and K. J. Novo-Gradac, Minteqa2/ *Prodefa2. A geochemical assessment model for environmental systems: version 3.0,* Environmental Research Laboratory, Office of Research and Development, U.S. Environmental Protection Agency, Athens, Ga., 1991.
11. P. M. May and K. Murray, *Talanta* **38,** 1409 (1991).
12. P. M. May and K. Murray, *Talanta* **38,** 11491 (1991).
13. P. M. May and K. Murray, *Talanta* **40,** 819 (1993).
14. R. M. Smith and A. E. Martell, *Critical Stability Constants,* Vols. 1–6, Plenum, New York, 1974–1976, 1982, 1989.
15. P. Debye and E. Huckel, *Phys. Z.* **24,** 185 (1923).
16. E. Huckel, *Phys. Z.* **26,** 93–147 (1925).
17. C. W. Davies, *Ionic Solution Theory,* Interscience Publishers, New York, 1962, p. 265.
18. P. Warwick, A. Hall, and D. Read, *Chemval II EC report* (in press) (1997).
19. G. Anderegg and S. Kholeif, *Talanta* **41,** 1507–1522 (1994).
20. I. Grenthe, J. Fuger, R. J. M. Konings, R. J. Lemire, A. B. Muller, C. Nguyen-Trung and H. Wanner, *Chemical Thermodynamics of Uranium,* OECD-NEA, North Holland, 1992.
21. K. S. Pitzer, *J. Phys. Chem.* **77,** 268 (1973).
22. A. E. Martell and R. J. Motekaitis, *Determination and Use of Stability Constants,* VCH Publishers, Weinheim, Germany, 1988.
23. E. L. Shock and C. M. Koretsky, *Geochim. Cosmochim. Acta,* **57,** 4899–4922 (1993).
24. E. L. Shock, E. H. Oelkers, D. A. Sverjensky, and H. C. Helgeson, *J. Chem. Soc. Faraday Trans.* **88,** 803–826 (1992).
25. E. L. Shock and H. C. Helgeson, *Geochim. Cosmochim. Acta* **52,** 2009–2036 (1988).
26. J. Shorter, *Correlation Analysis in Organic Chemistry: An Introduction to Linear Free Energy Relationships,* Clarendon Press, Oxford, 1973.
27. P. R. Wells, *Chem. Rev.* **63,** 171 (1963).
28. P. R. Wells, *Linear Free Energy Relationships,* Academic Press, London, 1968.
29. W. J. Albery, *Ann. Rev. Phys. Chem.* **31,** 227–263 (1980).
30. J. Aqvist, and A. Warshel, *Chem. Rev* **93,** 2523–2544 (1993).
31. M. Calvin, and N. C. Melchior, *J. Am. Chem. Soc.* **70,** 3270 (1948).
32. H. Irving and R. J. P. Williams, *Nature* **162,** 746 (1948).
33. H. Irving and R. J. P. Williams, *J. Chem. Soc.* 3192 (1953).
34. L. P. Hammett, *Chem. Rev.* **17,** 125 (1935).
35. L. P. Hammett, *Physical Organic Chemistry,* McGraw-Hill, New York, 1940, Chapt. 7.
36. L. P. Hammett, *J. Am. Chem. Soc.* **59,** 96 (1937).
37. S. Ehrenson, *Progr. Phys. Org. Chem.* **2,** 195 (1964).
38. R. W. Taft, in M. S. Newman, *Steric Effects in Organic Chemistry,* John Wiley & Sons, Inc., New York, 1956, Chapt. 13.
39. R. W. Taft, *J. Phys. Chem.* **64,** 1805 (1960).
40. R. W. Taft and I. C. Lewis, *J. Am. Chem. Soc.* **80,** 2436 (1958).
41. R. W. Taft, and I. C. Lewis, *J. Am. Chem. Soc.* **81,** 5343 (1959).
42. L. D. Pettit, C. Sherrington, and R. J. Whewell, *J. Chem. Soc. (A)* 2204 (1968).
43. L. D. Pettit, A. Royston, and R. J. Whewell, *J. Chem. Soc. (A),* 2009 (1968).
44. W. R. May and M. J. Jones, *J. Inorg. Nucl. Chem.* **24,** 511 (1962).
45. J. J. R. Frausto da Silva and J. Goncalves Calado, *J. Inorg. Nucl. Chem.* **28,** 125 (1966).
46. D. D. Perrin, B. Dempsey, and E. P. Serjeant, *pK_a Prediction for Organic Acids and Bases,* Chapman and Hall, London, 1981.
47. G. R. Choppin and L. H. J. Lajunan, *Inorg. Chem.* **25,** 3512 (1986).
48. G. R. Choppin, P. A. Bertrand, Y. Hasegawa, and E. N. Rizkella, *Inorg. Chem.* **21,** 3722 (1982).
49. I. Filipovic, M. Tkalcec, and B. S. Grabaric, *Inorg. Chem.* **29,** 1092 (1990).
50. H. Irving and H. Rossotti, *Acta Chem. Scand.* **10,** 72 (1956).
51. R. D. Hancock and A. E. Martell, *Chem. Rev.* **89,** 1875 (1989).
52. A. Evers, A. E. Martell, R. D. Hancock and R. M. Smith, *Inorg. Chem.* **28,** 2189 (1989).
53. *Sci. Total Env.* **117, 118;** special issue *Advances in Humic Substances Research* (1992).
54. M. H. B. Hayes, R. L. Malcom, P. MacCarthy, and R. S. Swift, eds., *Humic Substances II: In Search of Structure,* John Wiley & Sons, Ltd., Chichester, U.K., 1989.
55. J. Buffle and H. Sigel, *Metal Ions in Biological Systems,* Dekker, New York, 1984.
56. F. Frimmel and R. F. Christian, eds., *Humic Substances and Their Role in the Environment,* John Wiley & Sons, Ltd., Chichester, U.K., 1988.
57. J. Buffle, *Complexation Reactions in Aquatic Systems,* John Wiley & Sons, Ltd., Chichester, U.K., 1988.
58. J. J. R. Frausto da Silva and R. J. P. Williams, *The Biological Chemistry of the Elements: the Inorganic Chemistry of Life,* Clarendon Press, Oxford, 1991. p. 290.
59. M. A. Hurley and E. Tipping, *Geochim. Cosmochim. Acta* **56,** 3627 (1992).
60. A. E. Martell, R. J. Motekaitis, and R. M. Smith, *Env. Toxicol. Chem.* **7,** 417 (1988).
61. L. G. Sillen and A. E. Martell, *Stability Constants,* Special Publication No. 17, The Chemical Society, London, 1964.
62. L. G. Sillen and A. E. Martell, *Stability Constants,* Suppl. 1, Special Publication No. 25, The Chemical Society, London, 1971.
63. G. Anderegg, *Pure Appl. Chem.* **54,** 2693–2758 (1982).

64. L. D. Pettit, *Pure Appl. Chem.* **56**, 247–292 (1984).
65. M. T. Beck, *Pure Appl. Chem.* **59**, 1703–1720 (1987).
66. D. G. Tuck, *Pure. Appl. Chem.* **55**, 1477–1528 (1983).
67. T. Kiss, I. Sovago, and A. Gergely, *Pure. Appl. Chem.* **63**, 597–638 (1991).
68. R. M. Smith, A. E. Mantell, and Y. Chen, *Pure. Appl. Chem.* **63**, 1015–1080 (1991).
69. I. Sovago, T. Kiss, and A. Gergely, *Pure. Appl. Chem.* **65**, 1029–1080 (1993).
70. G. Berthon, *Pure. Appl. Chem.* **67**, 1117–1240 (1996).
71. O. Yamamuchi and A. Odani, *Pure. Appl. Chem.* **68**, 469–496 (1996).
72. L. D. Pettit and K. Powell, *IUPAC Stability Constants Database*, Academic Software, Otley, U.K., 1993.
73. S. Kotrly and L. Sucha, *Handbook of Chemical Equilibria in Analytical Chemistry*, John Wiley & Sons, Ltd., Chichester, U.K., 1985.
74. E. Hogfeldt, *Stability Constants of Metal Ion Complexes, Part A: Inorganic Ligands*, IUPAC, Chem. Data Ser. No. 21, Pergamon Press, Oxford, 1982.
75. D. D. Perrin, *Stability Constants of Metal Ion Complexes, Part A: Inorganic Ligands*, IUPAC, Chem. Data Ser. No. 22, Pergamon Press, Oxford, 1979.
76. W. A. E. McBryde, *A Critical Review of Equilibrium Data for Proton and Metal Complexes of 1,10-phenanthroline, 2,2'-Bipyridyl and Related Compounds*, IUPAC, Chem. Data Ser. No. 17, Pergamon Press, Oxford, 1975.
77. G. Anderegg, *A Critical Review of Stability Constants of EDTA Complexes*, IUPAC, Chem. Data Ser. No. 14, Pergamon Press, Oxford, 1977.
78. A. M. Bond and G. T. Hefter, *A Critical Review of Stability Constants and related Thermodynamic Data of Fluoride Complexes in Aqueous Solution*, IUPAC, Chem. Data Ser. No. 27, Pergamon Press, Oxford, 1980.
79. U. S. National Institute of Standards and Technology, *Critical Stability Constants of Metal Complexes, PC Based Database*, Gaithersburg, MD. Database 46, ver. 1.0 (1993), ver. 2.0 (1995), ver. 3.0 (1996).
80. R. M. Smith, A. E. Martell and R. J. Motekaitis, *Inorg. Chim. Acta* **99**, 207–216 (1985).
81. R. M. Smith, R. J. Motekaitis, and A. E. Martell, *Inorg. Chim. Acta*, **103**, 73–82 (1985).
82. R. M. Smith and A. E. Martell, *Total Env.* **64**, 125–147 (1987).
83. S. H. Laurie, *Inorg. Chim. Acta* **80**, L27–L28 (1983).
84. S. H. Laurie and C. James, *Inorg. Chim. Acta* **78**, 225–228 (1983).
85. V. S. Sharma and J. Schubert, *J. Chem. Ed.* **46**, 506–507 (1969).
86. H. Sigel, *Angew. Chem. Int. Ed. Engl.* **21**, 389–400 (1982).
87. H. Sigel, *Angew. Chem. Int. Ed. Engl.* **14**, 394–401 (1975).
88. J. E. Huheey, *Inorganic Chemistry*, 2nd ed., Harper and Row, London 1978, Chapt. 7.
89. R. D. Hancock and F. Marsicano, *Inorg. Chem.* **17**, 560 (1978).
90. R. D. Hancock and F. Marsicano, *Inorg. Chem.* **19**, 2709 (1980).
91. R. S. Drago, G. C. Vogel, and T. E. Needham, *J. Am. Chem. Soc.* **93**, 6014 (1971).
92. R. S. Drago and M. K. Kroeger, *J. Am. Chem. Soc.* **103**, 3250–3262 (1981).
93. R. S. Drago, *Struct. Bonding (Berlin)* **15**, 73 (1973).
94. R. G. Pearson, *J. Am. Chem. Soc.* **85**, 3533 (1963).
95. R. G. Pearson, *Science* **151**, 172 (1966).
96. R. G. Pearson, *Chem. Brit.* **3**, 103 (1967).
97. R. G. Pearson, *J. Chem. Ed.* **45**, 581–587 (1968).
98. R. G. Pearson, *J. Chem. Ed.* **45**, 643–648 (1968).
99. S. Ahrland, J. Chatt, and N. R. Davies, *Quart. Rev. Chem. Soc.* **12**, 265 (1958).
100. J. O. Edwards, *J. Am. Chem. Soc.* **76**, 1540 (1954).
101. J. O. Edwards, and R. G. Pearson, *J. Am. Chem. Soc.* **84**, 16 (1961).
102. G. Schwarzenbach, *Adv. Inorg. Radiochem.* **3**, 257 (1961).
103. R. G. Pearson, ed., *Hard and Soft Acids and Bases*, John Wiley & Sons, Ltd., Chichester, U.K., 1973.
104. R. J. P. Williams and J. D. Hale, *Struct. Bonding (Berlin)* **1**, 249 (1966).
105. M. Misono and Y. Saito, *Bull. Chem. Soc. Jpn.* **43**, 3680 (1970).
106. M. Misono, E. Ochiai, Y. Saito, and Y. Yoneda, *J. Inorg. Nucl. Chem.* **29**, 2685 (1967).
107. G. T. Hefter, *Coord. Chem. Rev.* **12**, 221 (1974).
108. P. M. Grant, P. A. Baisden, W. F. Kinand, and R. A. Torres, *Inorg. Chem.* **27**, 1156 (1988).
109. M. Henry and T. Merceron, *Radiochim. Acta* **66/67**, 57 (1994).
110. M. Henry, J. P. Jolivet, and J. Livage, *Struct. Bonding* **77**, 155 (1992).
111. A. W. Adamson, *J. Am. Chem. Soc.* **76**, 1578 (1954).
112. R. D. Hancock and F. Marsicano, *J. C. S. Dalton* **12**, 1096 (1976).
113. A. E. Martell, R. J. Motekaitis, and R. M. Smith, in *Environmental Inorganic Chemistry*, VCH Publishers, Deerfield Beach, Fla., 1985, pp. 89–115.
114. W. R. Harris, *J. Coord. Chem.* **13**, 16–28 (1983).
115. P. L. Brown, R. N. Sylva, and J. Ellis, *J. Chem. Soc. Dalton Trans.* 723–730 (1985).
116. P. L. Brown and R. N. Sylva, *J. Chem. Res. (S)* 4–5, 110–181, (1987).
117. P. L. Brown, *Talanta* **3**, 351 (1989).
118. K. J. Cantrell, *Polyhedron*, **7**, 573 (1988).
119. H. Moriyama, M. I. Pratopo, and K. Higashi, *Radiochim. Acta* **66/67**, 73 (1994).
120. G. R. Brubaker and D. W. Johnson, *Coord. Chem. Rev.* **53**, 1 (1984).
121. R. D. Hancock, *Prog. Inorg. Chem.* **37**, 187 (1989).
122. J. O. Morley, *Monatshefte Chem.* **119**, 1263 (1988).
123. J. O. Morley, *Rec. Trav. Chim. Pays-Bas* **107**, 89 (1988).
124. J. O. Morley *J. Chem. Soc., Perkin Trans. II*, 1243, (1987).
125. J. O. Morley and J. A. Hunter, *Polyhedron* **6**, 1215 (1987).
126. P. Politzer and D. G. Truhlar (eds.), *Chemical Applications of Atomic and Molecular Electrostatic Potentials*, Plenum, New York, 1981.
127. E. Scrocco and J. Tomasi, *Topics Curr. Chem.* **42**, 95 (1973).
128. H. B. Broughton, S. M. Green, and H. S. Rzepa. *J. Chem. Soc., Chem. Comm.*, 37, (1992).
129. O. Carugo and C. B. Castellani, *J. Chem. Res. (M)*, 1991, 1976–1994; (S) 1991.
130. R. M. Smith and A. E. Martell, in *Chemical Equilibrium and Reaction Models*, SSA Special Publication No. 42, 7–29, 1995.
131. E. Senior, *Microbiology of Landfill Sites*, CRC Press, Boca Raton, Fla., 1990.
132. J. Harmsen, *Water Res.* **17**, 699–705 (1983).
133. W. Stumm and J. J. Morgan, *Aquatic Chemistry*, 2nd ed., John Wiley & Sons, Ltd., Chichester, U.K., 1994, p. 386.
134. A. Atkinson and R. M. Guppy, UKAEA Report AERE R12961, 1988.

135. S. Bradshaw, S. C. Gaudie, B. F. Greenfield, C. E. Lyon, J. H. Rees, M. W. Spindler, and J. D. Wilkins, UKAEA Report AERE R12223, 1986.
136. M. J. Guest, UKAEA Report AERE R13083, 1987.
137. B. F. Greenfield, A. D. Moreton, M. W. Spindler, S. J. Williams, and D. R. Woodwark, *Mat. Res. Soc. Symp. Proc.* **257,** 299 (1992).
138. X. Bourbon and P. Toulhoat, *Migration 95,* St Malo, 1995.
139. T. W. Newton and J. C. Sullivan, in A. J. Freeman and C. Keller eds. *Handbook of the Physics and Chemistry of the Actinides,* North Holland Publishing Co., Amsterdam, 1985, pp. 387–406.
140. J. I. Kim, in A. J. Freeman and C. Keller eds, *Handbook of the Physics and Chemistry of the Actinides,* North Holland Publishing Co., Amsterdam, 1986, Chapt. 8.
141. C. Lierse, Ph. D. Thesis, Technische Universitat Munchen, Munchen, Germany, 1985.
142. H. Nitsche and R. J. Silva, *Radiochim. Acta* **72,** 65–72 (1996).

STABILIZATION OF HAZARDOUS WASTE. See HAZARDOUS WASTE, SOLIDIFICATION AND STABILIZATION.

STORMWATER MANAGEMENT, MUNICIPAL

THOMAS N. DEBO
Georgia Institute of Technology
Atlanta, Georgia

ANDREW J. REESE
OGDEN
Nashville, Tennessee

In recent years municipal stormwater management has taken on increased importance in many urban areas as drainage and flooding problems expand and increase in frequency and magnitude. In addition increased interest and regulations related to water quality issues have expanded urban stormwater management programs resulting in a greater need in the technical and engineering areas and increased pressures to fund expanding programs. However, to properly understand and manage stormwater requires more than a knowledge of design procedures and hydrologic or hydraulic practice. Actually making it happen in a real-world municipality, filled with the demands of regulations, financial pressures, stakeholder groups, and politics is also an important part of stormwater management. This article gives an overview of a larger publication that address many of the technical, institutional, and planning aspects of modern urban stormwater management issues and programs.

A hundred years ago the primary concern with surface water was draining it from wetlands and floodplains to make way for agricultural development. Later, as flooding of developed areas and demands for cheaper power increased, the concern broadened to include flood control and hydropower development. This was the heyday of the U.S. Corps of Engineers' and Bureau of Reclamation's river projects. Suburban growth in the 1950s through 1970s coupled with higher expectations for quality of development led to development of efficient curb and gutter drainage systems to quickly convey rainwater off streets and yards and into streams. Protection of roads and property was of paramount importance. Throughout the later part of this period it became apparent that localized flooding was often caused by this type of development and the importance of on-site and regional stormwater detention basins became recognized. Detention became a standard site development requirement in many of the more progressive urban areas. The development of personal computers enhanced the ability to actually plan drainage systems for existing and future development. Modern stormwater quantity management had begun.

Later, on the strength of an emerging body of information from Section 208 studies, the experience of various cities, and other data-gathering efforts such as the National Urban Runoff Program (NURP), the importance of urban areas as sources of nonpoint pollution grew. This culminated in the early 1990s with the publication of U.S. federal regulations mitigating and controlling pollution runoff in urban stormwater. Comprehensive stormwater quality management began.

Recently another paradigm shift from a purely chemical consideration of quality to a more holistic, ecologically based approach has emerged. In this approach both the riparian corridor and the land use throughout the watershed are considered as they impact the habitat and ecology of the stream and near-stream area. Further integration of all forms of water management, wastewater, drinking water, groundwater, and stormwater will occur.

In dealing with municipal stormwater management, it is important to understand the linkage between physical problems and deeper institutional root causes of those problems. Many municipalities have not understood this linkage and, as a result, wrestle continuously with the same problems, never arriving at permanent solutions. Figure 1 illustrates the dynamics of this technical–institutional relationship for stormwater quantity using a "five whys" methodology which comes out of a total quality management (TQM) consideration of stormwater management.

Typically, a stormwater administrator, public works engineer, or political leader gets a drainage complaint call: "I have a flooding problem and I want you to fix it." This is Level 1: the complaint. The complaint could just as easily have been an erosion or pollution complaint. Following the same methodology would eventually lead to the same conclusions.

When the question is asked: "Why is there a flooding problem?" (Level 2), there is usually one of (or a combination of) four reasons: (*1*) obstructed or damaged structures; (*2*) high risk residence location; (*3*) undersized structures; or (*4*) increased flow due to the impacts of urban development. The typical solution is to go out and fix the problem or, more commonly, to tell the resident that the problem is not the municipality's responsibility.

However, if the Level 3 question is asked: "Why did this flooding problem (and ones like it) occur in the first place?" a matching set of four more foundational reasons is uncovered.

1. Structures are obstructed or damaged because they are not inspected and maintained. Municipalities typically maintain very little of the drainage system.
2. Homes are located too close to streams because no one properly regulates the location. Because of the demands of the Federal Flood Insurance Program, most municipalities control the location and elevation of new construction within regulated floodplains.

3. Structures are undersized because of poor or inappropriate methodology and incomplete data. Most municipalities allow drainage structures to be sized using the Rational Method or other simplified technical methodologies.
4. Upstream development floods downstream development because it is not accounted for in the design of the downstream structures and/or it is not accounted for far enough downstream in the drainage system.

If again the "why" question (Level 4) is asked, three basic causative factors emerge.

1. Cities do not require appropriate levels of technical analysis because they are not sure what to require and how to implement these requirements.
2. Municipalities do not impose certain flood mitigation measures, development controls, or maintain off the public right-of-way because there is no legal authority to do so, and there is little impetus to establish such authority.
3. Even if these last two factors were solved, the bottom line is that there is no stable, adequate, and equitable funding source for stormwater management.

The more foundational factors emerge with the next "why" question (Level 5). Even when key stormwater staff understand the problems they must ask: who else is aware that flooding, erosion, or pollution problems exist? Who supports a growth in stormwater management? Who must support it for a successful program to be established?

The basic reason for lack of success in stormwater management is the same as for lack of success in anything else: lack of vision or direction (Level 6). In almost every case where a successful program has been developed, one or several individuals had or developed a vision for what their streams and creeks should and could be.

Notice that the first levels of assessment contain primarily physical and technical problems for which structural technical solutions are appropriate. Water is impacted by some physical means. However, when the later levels are considered, the solutions are institutional, programmatic, and nonstructural in nature. People are impacted by administrative means. These foundational problems allow or generate the more visible physical problems. If the root institutional problems are not eventually solved there will be a continual need to respond to an overwhelming number of flooding, erosion, and pollution complaints. The basic municipal stormwater management philosophy will continue to be that damage must be suffered before corrective action can be taken rather than taking preemptive action to avoid problems. Successful municipal stormwater management programs account for and deal with both the technical and institutional aspects.

This article introduces some of the more important aspects of municipal stormwater management. For a more complete description of these and other related areas, see Reference 1.

STORMWATER MANAGEMENT PROGRAMS

The term "stormwater management" can take on as many different meanings as there are stormwater management programs. Some are narrow, single-purpose programs whereas

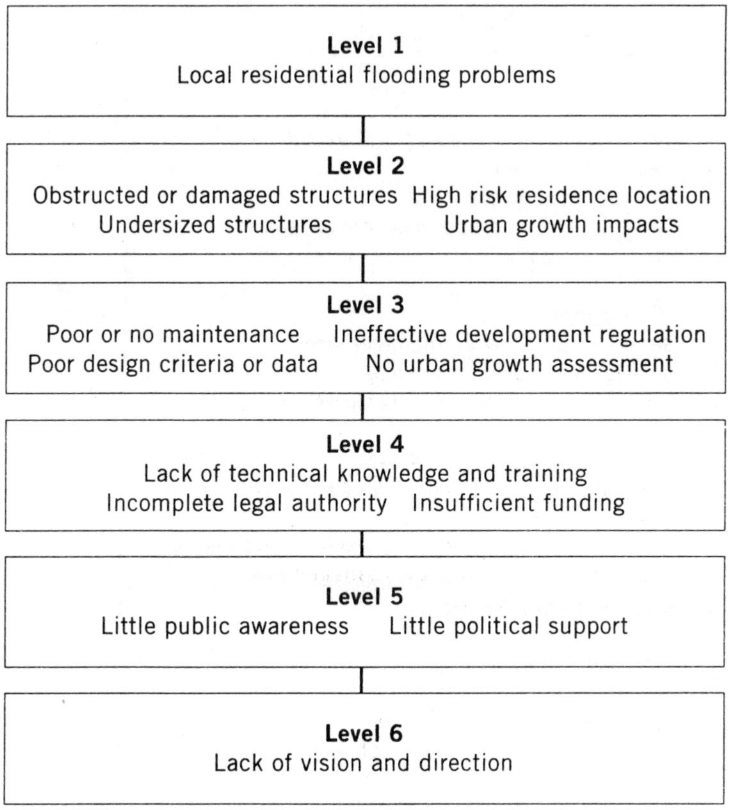

Figure 1. Multilevel stormwater problems.

others are broad, basin, or political entity-wide. There are also a number of general and specific ways to group the various functions and components of stormwater management for consideration. A stormwater management program may have many or few of these components and each of these components may be well or poorly developed. Several ways to understand and specify stormwater management are considered here.

The basic underlying purpose of stormwater management is to keep people from the water, to keep the water from people, and to protect or enhance the environment while doing so. Keeping people from the water involves a wide ranging number of nonstructural damage mitigation measures including zoning, floodplain management, home acquisition, flood-proofing, etc. Keeping the water from people involves direct interference with the water through the use of flow amount or flow elevation reduction measures such as retention or detention storage facilities or channel enlargement.

Maintenance or enhancement of water quality is taking on added importance through the mandates of the 1987 Water Quality Act and its implementing regulations. Aesthetics and multi-objective planning of water resources are also growing rapidly as expressed through stream restoration, greenway construction, and integration of flood control with parks or multi-use spaces.

FUNCTIONS IN STORMWATER MANAGEMENT

Figure 2 lists the basic functions of a typical stormwater management program from a management point of view. A typical municipality will have many of these functions dispersed throughout the municipality's governing structure.

Effective management of stormwater usually requires that a single staff position be identified as the central figure in stormwater control. This person should have primary control of all stormwater management functions (eg, drainage, floodplains, water quality, erosion, and sediment control).

Both large centralized organizations having almost all the functions in-house and small decentralized organizational structures hiring many of the functions both within municipal staff and outside have proven to be successful. Stormwater quality has been identified separately, although in practice it should be integrated with all the quantity functions in each functional area.

There are a large number of stormwater programs in the United States in various stages of development. One measure of the maturity of a stormwater program is the amount of money spent on it. Levels of funding for the basic programs (less any extraordinary items like large capital programs, large flood control maintenance programs, large flood control maintenance programs, costs shared with counties, etc) tend

1. Administration
   - General administration
   - General program planning and development

2. Special Programs
   - Public awareness and involvement
   - GIS and database management
   - Special program planning and development

3. Billing and Finance
   - Billing operations
   - Customer service
   - Financial management
   - Capital outlay

4. Indirect Cost Allocation
   - Overhead costs
   - Cost control
   - Support services

5. Stormwater Quality Management
   - Quality master planning
   - Retrofitting program
   - Comprehensive monitoring program
   - Structural and non-structural BMP programs
   - Pest, herb and fertilizer
   - Used oil and toxic materials
   - Street maintenance program
   - Spill response and cleanup

(continued)
   - Program for public education and reporting
   - Leakage and cross-connections
   - Industrial program
   - General commercial and residential program
   - Illicit con and illegal dumping
   - Landfills and other waste facilities

6. Engineering
   - Design criteria, standards and guidance
   - Field data collection
   - Quantity master planning
   - Design, field, and operations engineering
   - Hazard mitigation

7. Operations
   - General maintenance management
   - General routine maintenance
   - Emergency response maintenance

8. Regulation and Enforcement
   - General code development and enforcement
   - General permit administration
   - General drainage system inspection and regulation
   - multi-objective floodplain management
   - Erosion control program

9. Capital Improvements
   - Major capital improvements
   - Minor capital improvements
   - Land, easement, and right-of-way

Figure 2. Major stormwater management functional cost centers.

to be the same across the country. Table 1 indicates a typical range of program costs expressed in cost per acre (assuming a normal mix of urban development). Improvement or addition of the indicated capabilities are reflected in the cost ranges.

PUBLIC AWARENESS

Public awareness and education are the foundations of a successful urban stormwater management program. From before the program's inception and throughout its growth and service life, interaction with, and education of, the public is not an adjunct to the program's purpose; they are (or should be) a main purpose of the program. This is becoming especially important as the demands of the NPDES permit and its source control philosophy comes more into play in major and minor municipalities across the country.

Public awareness and education are carried out in stormwater management programs in two main ways: specific public awareness campaigns and ongoing "baseline" public information programs and activities. These two aspects differ in that a campaign has a beginning and an end whereas the ongoing program goes through transformations but does not envision an ending. For either case there are specific factors that should be kept in mind and methodologies that will make a stormwater management program successful.

Following are the major areas that must be included in a public awareness program.

- Defining the "public"
- Plans to include the public in the process
- Using public program techniques

Table 1. Typical Stormwater Program Costs

Program Type	Cost, Ac/Yr	Typical Program Features
Incidental	15–35	Reactive incidental maintenance and regulation primarily as part of other programs
Minimum	30–50	Add ROW maintenance, better regulation, and inspection, more staff, erosion control
Moderate	40–70	Add additional maintenance programs and levels of service, better regulation, and inspection, some planning, minor capital program, general upgrade of capabilities
Advanced	70–140	Add maintenance (of some sort) of the whole system, master planning, regional treatment, some water quality, data collection, multi-objective planning, strong control of development, and other programs, stormwater utility funding
Exceptional	>200	Add stormwater quality, ecological approaches, advanced flood control, and levels of service for maintenance, aesthetics become more important, public programs

- Development and use of volunteer programs
- Dealing with the media
- Risk communication in public awareness programs
- Communicating technical information

STORMWATER ORDINANCES AND REGULATIONS

Stormwater ordinances and regulations are the basic tool or method available to local governments for the control and/or management of urban drainage and flood control problems. Specifically, ordinances or regulations can be used for floodplain management, on-site detention and retention, erosion and sediment control, development regulation, and water quality control and enhancement. There is a difference between ordinances, resolutions, regulations, policies, etc. However, they are considered together in this article because the implementation and application of each is often similar. There are a number of regulatory drivers and resulting ordinance types used in municipalities throughout the country.

Passage of the National Flood Insurance Act of 1968 and subsequent amendments has provided an incentive for local governments to adopt floodplain regulations and ordinances. If a community wishes its citizens to be eligible for subsidized flood insurance and certain post-flood disaster assistance it must meet minimum requirements of the National Flood Insurance Program (NFIP) which includes the adoption of floodplain regulations for designated flood hazard areas. Most local governments with designated flood hazards are a part of the NFIP.

With regard to erosion and sediment control some states have passed legislation requiring local governments to enact erosion and sediment control regulations. Maryland was one of the first states to require such measures and continues to be a leader in this area. Another example is the state of Georgia which passed an Erosion and Sedimentation Act of 1975 that requires the governing authority of each county and municipality to adopt a comprehensive erosion and sedimentation ordinance establishing procedures governing land disturbing activities. The state of South Carolina has passed a comprehensive Stormwater Management and Sediment Regulation Act. This act mandates that stormwater management and water quality controls be enacted by all municipalities within the state. This is one of the most comprehensive stormwater management state laws.

Federally funded area-wide water quality planning activities in the late 1970s and early 1980s performed pursuant to Section 208, pl 92-500, The Federal Water Quality Control Act, resulted in recommending erosion and sediment control ordinances. Also, the U.S. Environmental Protection Agency (EPA) grants for sewage treatment and interceptor construction require municipalities to adopt erosion and sediment control ordinances as a grant condition. Recently, EPA has required municipalities to comply with water quality regulations related to the Clean Water Act.

The 1987 Water Quality Act and implementing regulations have led many municipalities into regulatory control of illicit discharges, illegal connections, industrial discharges, and a variety of other nonpoint discharge-related controls and ordinances.

These and other activities have resulted in the development and adoption of numerous municipal ordinances, regulations, and policies related to stormwater management quantity and quality.

Detention Ordinances

With regard to drainage and flood control, many local governments are requiring on-site detention facilities to reduce downstream runoff peaks because they feel it is good stormwater management practice to do so. Residents downstream of new developments are increasingly questioning the wisdom of allowing upstream developments to cast more water downstream for them to deal with. Such concern puts pressure on local governments to require on-site detention in new developments. In addition, some states have passed laws related to use of detention facilities. The state of Pennsylvania passed a Stormwater Management Act in 1978. This act requires the development of stormwater management plans for all watersheds within the State. Section 13 of this law, entitled Duty of Persons Engaged in the Development of Land, states:

Any landowner and any person engaged in the alteration or development of land which may affect stormwater runoff characteristics shall implement such measures consistent with the provisions of the applicable watershed stormwater plan as are reasonably necessary to prevent injury to health, safety, or other property. Such measures shall include such actions as are required:

1. *to assure that the maximum rate of stormwater runoff is no greater after development than prior to development activities; or*
2. *to manage the quantity, velocity, and direction of resulting stormwater runoff in a manner which otherwise adequately protects health and property from possible injury.*

Implementation of item (*1*) almost always involves the use of detention storage facilities. Such a requirement is the major focus of many municipal stormwater ordinances that have been adopted. This has resulted in the proliferation of many small detention storage facilities in many urban areas.

As a result of the increased use of ordinances and regulations to control quantity and quality problems related to storm runoff, municipalities have gained experience with implementing these regulations and some areas are experiencing the positive and negative effects resulting from their implementation. From this experience, professionals involved with urban stormwater management are posing several interesting questions. Is the adoption of ordinances which generally require the use of detention facilities on all developments the answer to flooding and drainage problems? Do sediment control regulations adequately protect water resources? What engineering and technical methods should be included in a community's stormwater management program? What controls should be included in stormwater quantity and quality control?

If a detention ordinance is implemented three additional factors should be present. First, the municipality should have a mechanism to encourage and implement regional storage facilities. Second, the detention pond should be designed to control downstream impacts, not just peak flows at the outlet. Third, a comprehensive inspection and maintenance program should be implemented.

Developing Policies

Ordinances state the policies of the local municipality related to stormwater management issues. Generally, the ordinance and supporting documents should state the "why" and "what" should be accomplished. Technical manuals and other guidance should provide the "how" information. However, the stated policies should be based on design parameters that are applicable and result in studies that are needed to solve or prevent local flooding and drainage problems. The following example shows how computer technology can be effectively utilized to develop policies that can be incorporated within a municipal ordinance.

After years of requiring engineers to analyze peak flows from proposed developments at the exit to these developments, municipalities are starting to realize that limiting the hydraulic analysis of stormwater management facilities to the exit of the proposed development or detention site is not sufficient to prevent increased downstream flooding and drainage problems. Thus there is a need for a policy to address this problem which could be incorporated into local ordinances and regulations. What is agreed is that hydraulic analysis should extend downstream from proposed developments but how far downstream the limits of this analysis should extend is difficult to determine. Requiring that all analysis extend a standard distance downstream (ie, one mile) may not be a sufficient distance for a very large development or may be much too far for very small developments. It must be remembered that any required analysis will cost money that must eventually be paid in higher development costs. Thus some procedure needs to be developed to account for developments of different size, location, and land use density. The policy should also not require expensive engineering and technical studies that will be a burden to the development process.

A study completed for the City and County of Greenville, South Carolina and Raleigh, North Carolina demonstrates how such a policy could be developed (2). This study used a hydrologic–hydraulic computer model to analyze the downstream effects of storm runoff from developments of different size, shape, physical characteristics, and location within larger drainage basins. This study also examined different sized flood events and different types of downstream drainage systems. The results of numerous computer runs of watersheds located throughout the United States were analyzed and Figure 3 shows a typical summary of these results. It can be seen from

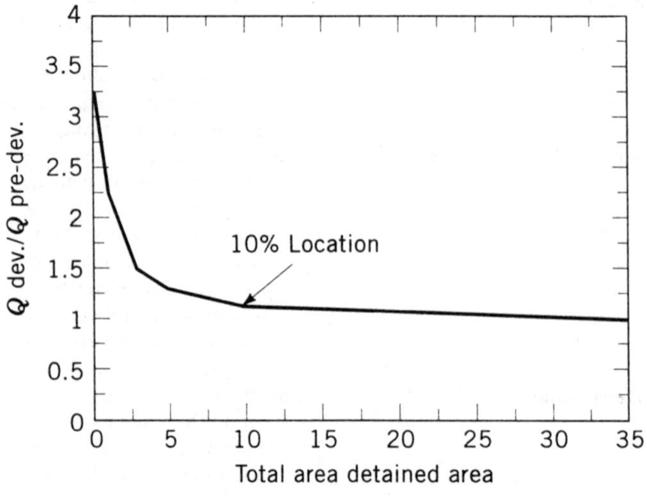

Figure 3. Determining downstream analysis limits.

Figure 3 that the effects of the development process stabilizes at the point where the proposed development represents from 5 to 10% of the drainage area, depending on the size of the development and the amount of increased impervious area. This analysis was then used as the basis for developing the following policy statement that has been incorporated in several municipal ordinances.

"In determining downstream effects from stormwater management structures and the development, hydrologic–hydraulic engineering studies shall extend downstream to a point where the proposed development represents less than ten (10%) percent of the total watershed draining to that point."

Similar computer model analyses can be effectively used to develop other reasonable and adequate policies which can provide stormwater management facilities designed to protect local residents without undue cost to the development process. Modeling the effects of different policies can provide valuable information in the policy development process.

Stormwater Management Design Manual

If the ordinance is designed to function as a policy statement then the municipality should publish a manual which would contain the engineering criteria and design standards needed for implementation of the stormwater management program. This manual would contain the technical information needed for both the flood and drainage aspects, and the water quality control aspects of the program.

The major thrust of a stormwater management design manual is to present criteria and examples such that a design engineer can determine generally what is expected by the municipality in different flood, drainage, and water quality designs. The manual should not be a limiting document forcing the engineer to use standard designs and procedures, but instead should present performance standards. Numerous design examples should be given. The manual should have the connotation of a document that suggests and informs rather than limits or prescribes. The manual should not replace good engineering judgment.

Stormwater management design manuals have recently been developed for several municipalities in North and South Carolina. An example of the general content of these manuals is given as the table of contents for the Greenville, South Carolina Stormwater Management Design Manual.

Chapter 1 - Introduction
Chapter 2 - Stormwater Management Ordinance
Chapter 3 - Data Availability
Chapter 4 - Documentation
Chapter 5 - Hydrology
Chapter 6 - Storm Drainage Systems
Chapter 7 - Design of Culverts
Chapter 8 - Open Channel Hydraulics
Chapter 9 - Storage Facilities
Chapter 10 - Energy Dissipation
Chapter 11 - Infiltration Facilities Design
Chapter 12 - Best Management Practices

FINANCING STORMWATER MANAGEMENT PROGRAMS

Municipal stormwater management programs have been funded using a number of mechanisms as the primary generator of funds including property taxes, sales taxes, state revolving funds, road funding, user fees, bonding, and surcharges on other utility fees. By far the most common current funding method is tax-based. However, the concept of a stormwater utility-based funding method is growing. In the early 1970s there were only one or two true stormwater utilities in existence. Today there are over 200. This number is expected to more than triple in the next decade as the financial impacts of stormwater quality legislation reach the many small municipalities.

A stormwater utility must be seen as an umbrella under which individual communities address their own specific needs in a manner consistent with local problems, priorities and practices. As such a stormwater utility provides a vehicle for:

- consolidating or coordinating responsibilities that were previously dispersed among several departments and divisions;
- generating funding that is adequate, stable, equitable, and dedicated solely to the stormwater function; and
- developing programs that are comprehensive, cohesive, and consistent year-to-year.

The more popular primary and secondary financing methods available to most municipalities to fund stormwater management programs include the following:

1. *Stormwater Drainage Services Charges:* This method of financing fits well with stormwater and flood control program needs in most municipalities. It is normally the primary financing mechanism for the utility and is often called "the utility". The functions and costs associated with stormwater management are similar to those for a water or sanitary sewer utility. Thus, this "stormwater utility" would be set up and function in a manner similar to other existing utilities. The principal difference between water and sanitary sewer services charges and stormwater service charges is the basis of measurement employed for setting rates. The typical basis for stormwater charges is the runoff production potential of a particular property. Usually, to avoid excessive utility administration and startup costs, only industrial/commercial/multifamily sites are considered on an individual basis whereas single family residences are covered under blanket charges.
2. *Revenue Bonding:* This method of municipal funding is most commonly used for facility construction, the purchase of equipment, or other major capital outlays. Unlike general obligation bonds, which are backed by the full faith and credit of the issuing governmental agency, revenue bonds are backed by revenues from specifically defined sources. Thus, it is intended to work in conjunction with some other ongoing funding system. "Anticipation notes" are somewhat similar with the anticipated revenues from a future utility covering the notes issued to generate funds to set up the utility.
3. *System Development Charges:* System development charges are also best used in conjunction with other

funding methods as a mechanism for balancing financial participation in the cost of services and facilities. They provide a funding mechanism through which owners of properties which develop in the future share in the cost of projects built in anticipation of their needs. They can properly be described as a deferral mechanism through which a property owner's financial participation in a project is delayed until development occurs which uses the additional capacity originally built into the system.

4. *Special Assessments And Improvement Districts:* A number of different methods of levying special assessments on benefitted properties have been used throughout the United States for stormwater and flood control improvements. Projects funded through special assessments must have a special benefit to the properties included in the assessment area and charges for each parcel must be consistent with the relative benefit to each property. Special assessment mechanisms are most often used for small local projects since they localize the cost of projects which serve a limited area and a limited constituency.
5. *Plan Review And Inspection Fees:* This method of funding is typically used by public works departments and utilities which operate stormwater drainage systems and must maintain and regulate their use. The fee is used to cover the cost of plan review and inspection of project sites during the construction of the proposed development or to fund ongoing inspection programs of private structural facilities such as detention ponds or water quality best management practices. Many municipalities charge only a token fee whereas others try to determine the actual costs involved and set a fee to cover these costs.
6. *In Lieu of Construction Fees:* Many municipalities require the construction of a stormwater detention facility for all nonresidential properties unless it can be demonstrated that construction of such a facility would adversely affect stormwater peak discharge rates downstream from the site. The *In Lieu of Construction Fee* provides a funding mechanism whereby construction of numerous individual on-site detention basins or other stormwater conveyance structures can be waived with developers contributing a fee to a fund to build a regional facility or make improvements in stormwater conveyance at locations remote from the development site. One drawback of this funding method is that the facility must be built ahead of the development and, therefore, an alternative funding source must be available. Rarely do the funds for construction become available prior to the construction start.
7. *General Facilities Charges:* Utility service charge rate structures sometimes incorporate surcharges for facilities which are necessary to provide adequate service to all rate payers. These charges are known by various names but the term *General Facilities Charges* describes their purpose. General facilities charges for stormwater and flood control improvements would be most appropriate for regional detention facilities or drainage systems serving only roads, parks, or other public properties used by the citizens in a municipality. Such charges could also be used for the development of a database or mapping information which serves all properties generally rather than just a certain sector or area.
8. *Impact Fees:* Impact fees are also designed to be used in conjunction with other funding mechanisms. They are most appropriate as a mechanism to balance financial participation with the cost of service, especially the short-term cost resulting from the development of private property. Impact fees must pass the "rational nexus" test and should only be used where impacts to be mitigated are specific rather than general; a consistent methodology is available for quantifying the impact; a separate accounting for each impact fee project can be maintained; and a mechanism exists for returning the unused portion of an impact fee after a mitigation period has passed. Impact fees have lately come under close legal scrutiny. In some cases, the cost of providing enough information to develop an appropriate impact fee method might make this funding method unattractive to the local municipality.
9. *Developer Extension/Late-Corner Fees:* Stormwater management facilities built as part of private developments often should be oversized to accommodate service beyond the immediate confines of a single project. This results in additional construction costs to serve the needs of these external areas. A financing method can be employed which allows an original developer to be compensated for these front-end expenses by developers subsequently building in the same area and served by the oversized facilities. These fees are particularly useful where an overall development master plan is available which defines future stormwater requirements.

Other secondary funding methods include sales taxes, surcharges, grants and loans, penalties, and many more.

MASTER PLANNING

Stormwater master planning is a term which has come to mean a wide variety of things. In some municipalities a stormwater master plan (SWMP) is nothing more than a collection of general specification drawings with some explanatory paragraphs as an introduction. In others it is a complex analysis of the total major and minor stormwater system including stormwater quality. In some locations master plans address existing problems while in some they seek to anticipate, and avoid, future problems.

Stormwater master planning plays a key role in identifying existing and future problems, opportunities for enhancement of stormwater use, cost estimates, impact assessment, and assigning priorities for change and improvement. Even master planning cannot address the core issues unless it explores, however summarily, factors other than strictly the technical ones.

Their are three major steps in the master planning process.

Step 1: Make A Needs Analysis

The first step in any master plan is to gain a clear understanding of the actual needs and goals of the municipality.

Step 2: Determine Constraints to Possible Solutions

There will always be constraints to solutions to problems. Constraints to any possible solutions fall naturally into the major categories of technical, political, social, physical, legal, and financial constraints.

Step 3: Formulate the Technical Approach

A master plan can concentrate on:

- the major drainage system and/or the minor system,
- quantity and/or quality and erosion control,
- existing and/or future conditions,
- urban and/or rural systems,
- correcting problems and/or planning to avoid problems,
- risk-based solutions or more deterministic approaches,
- "soft" solutions and/or "hard" solutions, or
- individual solutions and/or systems analysis.

The goals of the master plan study determine the level of effort and detail involved, as well as the degree of conservatism built into flow estimates. Flood hazard studies may require only an estimate of peak flows. Other studies may require the development of design stormwater runoff hydrographs and the timing of these hydrographs with respect to each other. There are varying complexity of studies and often studies are done in phases.

As important as master plans are to any comprehensive stormwater program, by themselves they will not solve problems or prevent flooding, drainage, or water quality problems. The master plans represent a blueprint for action that must be taken if these problems are to be solved or prevented. Too often people see the master plan as the end product and forget that if the plans are not implemented little good will result from the completed work. The real work begins when the master plan is completed.

Each master plan is undertaken in response to specific problems, goals, policies, etc, that relate to the watershed being planned and the implementation tools available. Detailed plans which analyze hydrologic–hydraulic and water quality parameters to a very small scale, say 10 to 20 acres, will be much more expensive to develop than more general plans which only analyze down to several hundred acres. However as a consequence, the general plans will not address flooding and drainage problems associated with small developments and single lots. Thus many of the limitations of any master plan will be related to the scale used for plan development and engineering analysis. Often large-scale planning is done over a whole area with key area small-scale plans done for certain areas of interest. These "nested" plans fit into the grid of the overall plan and feed data to it.

Other factors that may influence the limitations of specific master plans include data available for plan development, political acceptance of the final plan, acceptance and enforcement of the plan by local citizens and professional groups, and available funding. In addition, keeping the plan up-to-date is important if the plan is to remain a viable tool in dealing with stormwater management problems. The plan should not become a static document but rather it should be a part of a dynamic stormwater management program that changes as the problems and local circumstances change.

STORMWATER QUALITY MANAGEMENT PLANS

It has been known for years that urban runoff contains a wide variety of pollutants, often in concentrations and volumes to cause both acute and chronic environmental impairment of receiving streams and lakes. Numerous studies have been done over the last 20 years to attempt to quantify both the impacts of such runoff and to attempt to formulate tools to analyze the impacts and seek to mitigate them.

A profile of the impacts of stormwater discharges was developed by EPA in 1992 (3). It was found that between one-third and two-thirds of all designated use impairment for our nation's streams was a result of agricultural and urban runoff. Whereas agricultural nonpoint sources comprise the major source of pollution, urban runoff is also very significant. The pollutants found in urban runoff include sediment (total suspended solids (TSS) in the range of 35–400 mg/L), oxygen demanding substances (BOD in the range of 6–20 mg/L), heavy metals (typical range of 30–450 mg/L), nutrients (total Kjeldahl nitrogen (TKN) in the range of 1–4.5 mg/L), bacteria, oil, and grease and other more rare toxic chemicals, pesticides, etc (4).

Impacts on receiving waters are varied and often difficult to measure. Because of the transient and intermittent nature of the pollution loading, wastewater measurements and standards do not apply well. However, impairment is often measured in terms of:

- inability to attain a designated beneficial use (drinking water, fishable, body contact, etc);
- a violation of some chemical or biological numerical standard;
- an ecological impact such as a fish kill, lake eutrophication, reproductive disorders, etc; or
- a public perception impact such as offensive odor, litter, color, etc.

The water quality impacts of stormwater runoff to the receiving water depend on a number of factors including the magnitude and duration of rainfall events, soil types, time between storms, land use type and specific activities, illicit connections or illegal dumping, and the ratio of the runoff flow volume to the receiving water flow volumes. Although actual impacts of stormwater runoff are often difficult to ascertain, there is a general understanding that urban runoff contains significant levels of pollution comparable to primarily treated effluent from wastewater treatment plants and is a major source of receiving water pollution. Hard data were often scarce, not able to be compared, or contradictory in analysis and interpretation.

Water Quality Act

The Water Quality Act of 1987, Section 405 amends Section 402 of the Clean Water Act of 1972 to require the EPA to provide rules and regulations that establish a permit program for stormwater discharges associated with industrial activities and from municipal separate storm drain systems serving populations of 100,000 or more (5). This Act and the promulgated regulations have had a major impact on those municipalities affected.

The Act requires that dischargers to "Waters of the State" (WOS) be subject to the National Pollution Discharge Elimination System (NPDES) permit program. However, the requirements, and the philosophy behind these requirements, outlined in the draft regulation are far-reaching.

The Act, as it applies to municipalities, has three main thrusts.

1. It requires municipalities to effectively prohibit nonstormwater discharges into their publicly owned or operated separate storm drain system.
2. It requires municipalities to control discharge of pollution into their systems to the maximum extent practicable (MEP).
3. It defines one system-wide permit rather than permits for each individual discharge point.

These MEP plans consist of comprehensive stormwater quality management programs (SWQMP) which include management practices, control techniques and systems, design and engineering methods, and other such provisions as the Director of the program determines appropriate for control of such pollutants (6). The NPDES permit is in two parts. Part 1 concentrates on identification of and description of the system and dischargers. Part 2 concentrates on a program to reduce pollution discharges.

Defining Maximum Extent Practicable

Figure 4 depicts a simplified matrix schematic for MEP definition. For clarity of understanding and consideration, it is convenient to consider four major areas of polluters as identified in the regulations: (*1*) illicit connections or illegal dumping, (*2*) industrial, hazardous waste sites, and landfills, (*3*) erosion from construction sites, and (*4*) commercial and residential pollution. These areas are defined in this manner because the approach to deal with pollution from each of these areas is slightly different. Under each of the four major program areas it will be necessary to define a program which contains four major considerations: (*1*) technical approach adequacy and effectiveness, (*2*) legal authority, (*3*) financial sufficiency, and (*4*) administrative and organizational support. In actual analysis the matrix is greatly expanded to consider every facet of the proposed SWQMP to assure it is funded, staffed, legal, and will actually accomplish environmental or watershed objectives.

The MEP grid

	Tech	Legal	Money	Admin
Illicit and illegal				
Industry				
Construction sites				
Commercial and residence				

Figure 4. Defining MEP.

Evaluation of Best Management Practices

Best management practices (BMPs) are used to control and reduce the discharge of pollutants. There are many ways to categorize BMPs including structural or nonstructural, source control or treatment control, structural or programmatic; etc. However BMPs are categorized, there must be a consistent way to assess their effectiveness. The evaluation of best management practices is more an art than a science. EPA (7) and others describe several methods for the evaluation of BMPs. These methods are most often used together to assess various parts of a SWQMP.

The holistic approach is a combination of intuition, engineering judgment, basic analysis, and cost assessment which can lead to a set of BMPs and a strategy. This method can be very effective but requires the involvement of experienced personnel from a number of technical disciplines and from senior staff or political leadership. Often an ad hoc committee is selected which develops a set of objective criteria and then goes through a process to arrive at a final program.

Cost–benefit analysis is also an approach which can be used to develop SWQMPs. Such measures as cost per pound of pollutant removed or cost per day of activity can be used. This method is good for making decisions on which approach to use first and for how long. For example, it may be possible to achieve a reduction in a certain pollutant (such as petroleum hydrocarbons) using public education and regulatory processes. At some point though, structural controls may be necessary to complete the program. Cost–benefit analysis is typically used for more site or project specific activities rather than municipal-wide or more subjective nonstructural-type applications.

Optimization has been used in other sciences for decision making. Optimization consists of finding the maxima or minima of an objective function subject to a series of constraints each expressed in terms of decision variables. The objective function can be expressed in terms of cost or some complex combination of cost, benefit, and judgment. Nonlinear relationships among the variables and the fact that the numerical complexity can hide the subjectivity of many of the inputs often limits this method to narrow applications targeted toward a single site or pollution concern.

Matrix comparison is a more subjective method (unless numeric ratings are used) whereby a table is constructed with various BMP or BMP programs along the horizontal axis and evaluation criteria along the vertical axis. Each alternative program is then rated in terms of each of the criteria. A numeric rating (say 1 through 5) or a simple + or − rating can be used. Table 2 provides a list that has been used in various forms in several locations for BMP program development.

Some localities total an actual score and use the form for a numeric ranking of BMPs. Others simply use the pluses and minuses as flags to alert planners to particular BMP strengths or potential problems. Not all criteria apply to each BMP. Another list may be developed specifically for subcategories of BMPs.

Table 2. Best Management Practice Screening Criteria

BMP:	Criteria Description	+	0	−	CMTS
1.	Human risk, public safety, and potential liability				
2.	Environmental risk and implications				
3.	Ability to control key targeted pollutants				
4.	Costs to implement and continuing costs				
5.	Acceptability to public, stakeholders, staff, political leadership				
6.	Equitability to impacted persons				
7.	Reliability and consistency over time				
8.	Sustainability of maintenance or program management				
9.	Ability to be applied universally throughout the municipality, or, on a specific watershed basis				
10.	"Fit" with other operations and programs				
11.	Relationship to federal, state, or local requirements				
12.	Amenity or multi-use value				
Totals					

Nonstructural Best Management Practices

Best management practices (BMPs) are the basic mitigation measures used in the stormwater quality management plans to control pollutants within the municipality. In most cases it is much easier and less costly to prevent the pollutants from entering the drainage system than trying to control pollutants with structural BMPs. Thus within a "treatment train" concept, the nonstructural BMPs should be the first line of defense in protecting the receiving stream within the municipality. If used properly, the nonstructural BMPs can be very effective in controlling pollutants and greatly reduce the need for structural BMPs. In addition, nonstructural BMPs tend to be less costly, easier to design and implement, and easier to maintain than structural BMPs. BMPs included as nonstructural include the following:

- Public education/participation
- Land use planning/management
- Material use controls
- Material exposure controls
- Material disposal and recycling
- Spill prevention and cleanup
- Dumping controls
- Connection controls
- Street/storm drain maintenance
- Permanent erosion control

The field of municipal stormwater management is a very complex and rapidly expanding area combining many aspects of engineering design and analysis and policy formulation. Areas of interest span the range from complex engineering applications in culvert and channel design to citizen participation and ordinance development (1).

BIBLIOGRAPHY

1. T.N. Debo, and A.J. Reese, *Municipal Storm water Management,* Lewis Publishers, CRC Press, Inc., Boca Raton, Fla., 1995, and references therein.
2. T.N. Debo and A.J. Reese, "Downstream Impacts of Detention", *Proceedings of NOVATECH '92,* Nov. 3–5, 1992, Lyon, France.
3. U.S. EPA, *Environmental Impacts of Stormwater Discharges,* EPA 841-R-92-001, Washington, D.C., June 1992.
4. U.S. EPA, *Urban Targeting and BMP Selection,* Region V Water Division, Washington, D.C., Nov. 1990.
5. *Fed. Reg.,* 49416–49487 (Dec. 7, 1988).
6. U.S. EPA, *NPDES Permit Application Workshop For Storm Water Discharges From Municipal Separate Storm Sewer Systems and Industrial Sources—Speaker's Notebook,* U.S. EPA Region IV, Washington, D.C., 1992.
7. U.S. EPA, *Handbook—Urban Runoff Pollution Prevention and Control Planning,* EPA/625/R-93/004, Washington, D.C., Sept. 1993.

SUBSURFACE BARRIERS. See WASTE CONTAINMENT BARRIERS.

SUBSURFACE CONTAINMENT BARRIER INTEGRITY, VERIFICATION OF

JOHN H. HEISER
Brookhaven National Laboratory
Upton, New York

Extraordinary amounts of time, energy, and funds are being expended toward remediating contaminated sites or preventing future contamination of soils and groundwater. Contaminated soils, buried waste, and leaking underground storage tanks (UST) pose a threat to the environment through contaminant transport. One promising remediation option available to the waste management community is the construction of a subsurface barrier that consists of a wall of low permeability material. Subsurface barriers will improve remediation performance by removing pathways for contaminant transport due to groundwater movement, meteorological water infiltration, vapor- and gas-phase transport, transpiration, and so on. The use of subsurface barriers is a remediation option for many Department of Energy (DOE) defense sites including Hanford, Idaho National Engineering Laboratory, and Rocky Flats. Barriers are also considered an important remediation option by the U.S. Environmental Protection Agency (EPA) (1).

Subsurface barriers may be used to contain contaminant plumes, to direct contaminant movement to collection sumps or lysimeters in cases of unexpected remediation failures or transport mechanisms, to contain leakage from UST, and to restrict *in situ* soil cleanup operation and chemicals. They may be used alone or in combination with techniques, such as groundwater pumping, subsurface drains, and *in situ* biological or chemical treatment methods. Slurry walls are the most commonly used subsurface barriers because they are inexpensive, but grouted barriers such as grout curtains and sheet piling cut-off barriers are also used (2). Much of the recent interest in barriers is in the concept of a confinement barrier, which is analagous to putting a bowl underneath a waste site to catch any leakage (Fig. 1). For underground storage tanks, it was proposed that interim containment barriers be placed around the tanks. This would reduce or prevent future contamination of soil and groundwater in the event that further tank leakages occur before or during remediation. Use of interim subterranean barriers can provide sufficient time to evaluate and select appropriate remediation alternatives as well. To be most effective a barrier should be continuous, having no or as few holes or breaches as possible. A breach may be formed through numerous pathways including discontinuous grout application, from joints between panels, and from cracking due to grout curing or wet–dry cycling. The barrier material should be compatible with soil and waste conditions specific to the site. It should also have as low an effective diffusivity as is reasonably achievable to reduce or inhibit transport of moisture and contaminants.

The EPA has generally been flexible in its acceptance of any particular remediation technology for the cleanup of inactive hazardous waste disposal sites under the Comprehensive Environmental Response, Compensation, and Liability Act (CERCLA). The main consideration is whether the technology in question results in attainment of the remediation standards referred to in Section 121(d) of that Act. Section 121(b) of CERCLA provides the basis for this flexibility (3):

> *The President [presumably via the Administrator and staff of the EPA] shall conduct an assessment of permanent solutions and alternative treatment technologies or resource recovery technologies that, in whole or in part, will result in a permanent and significant decrease in the toxicity, mobility, or volume of the hazardous substance, pollutant, or contaminant. In making such an assessment, the President shall specifically address the long-term effectiveness of the various alternatives ... The President may select an alternative remedial action meeting the objectives of this subsection whether or not such action has been achieved in practice at any other facility or site that has similar characteristics. In making such a selection, the President may take into account the degree of support for such remedial action by parties interested in such site.*

As a practical matter, it is not the President but the EPA and the "parties interested in such site" (eg, governmental officials, private industry, and concerned citizens—namely, the stakeholders) who will be involved in assessing, selecting, and implementing the remediation technologies at any given facility or site. The EPA has considered subsurface barriers to be among the technologies potentially applicable to remediation of waste sites. For example, in an EPA handbook (2) intended for use as a basic reference tool on remedial action, the EPA devotes a whole section to subsurface barriers consisting of low permeability cut-off walls or diversions installed below ground to contain, capture, or redirect groundwater flow in the vicinity of a site.

To fully assess a technology the stakeholder must have tools available to measure the performance capabilities of the technology. The stakeholder cannot and should not take the technology at face value from a commercial contractor or the technology developer. The goal of any barrier installation should be a product that is continuous, having as few holes as possible. Unfortunately, no suitable method exists for the verification of an emplaced barrier's integrity. The large size and deep placement of subsurface barriers make detection of leaks challenging. This becomes magnified if the permissible leakage from the site is low. There is a unique need to be able to detect holes or breaches in a subsurface barrier without having to dig up the entire waste site. This report summarizes several commonly used or promising technologies that have been or may be applicable to *in situ* nondestructive barrier continuity/performance verification.

BACKGROUND

To understand why integrity and performance verification is required and how the verification might be accomplished it is helpful to understand some basics of the materials and installation processes that can be used in barrier applications. Performance, integrity, lifetime and failure mechanisms of the barrier will be a function of the emplacement method, barrier material, and the site characteristics (eg, geotechnical, geochemical, and contaminant characteristics) (see WASTE CONTAINMENT BARRIERS).

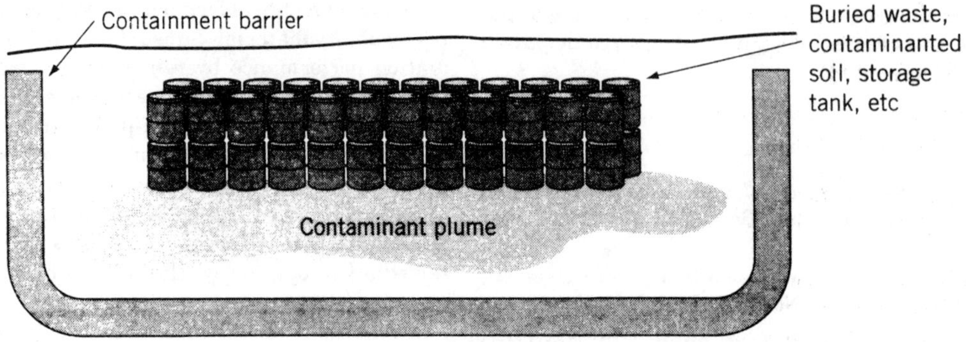

Figure 1. Subsurface barrier for containment of buried waste.

Barrier Materials

A variety of materials have been proposed as candidate barrier materials (4–6). Materials used in construction of the barrier will change the local properties of the formation including resistivity, compressibility, stiffness, porosity, shear and bulk modulus, and saturation. The combination of material chosen, its compatibility with the geotechnical, geochemical, and waste characteristics of the disposal site and the emplacement method places different requirements on verification and performance monitoring. The vast majority of subsurface barriers installed have been cement grout curtains or bentonite or cement/bentonite slurry walls. Recent innovations have brought a new selection of candidate barrier materials including thermosetting polymers (acrylics, siloxanes, etc) and montan was emulsions. Each material has specific installation requirements and failure mechanisms that may result in breaches or gaps in the final barrier.

Cement Grout. Most cement grouts (>99%) used for barriers are made with Portland cement; only these will be discussed in detail. Other hydraulic cement types such as phosphate, latex-modified, or magnesium cements behave similarly in terms of geophysical signal perturbations and breach formation pathways. Hydraulic cement is a generic term given to a class of inorganic materials that when mixed with water undergo hydration reactions and bind together to form a monolithic structure. Portland cement is prepared by heating clay minerals and lime together to form a clinker. Gypsum is then mixed with the clinker and ground into a fine powder to prepare the final cement. The major components of the cement are tricalcium silicate, dicalcium silicate, tricalcium aluminate, calcium aluminoferrite and gypsum. By adjusting the proportions of the raw materials final properties of the cement, such as set-time and strength, can be varied. Cured cement is an amorphous colloidal gel of calcium silicate hydrates. The reaction between the cement and water is exothermic and proceeds by first forming a gel. Curing continues from the gel and eventually the cement hardens. Hydration of cement is a complex set of chemical reactions and as such is subject to many interferences. The final properties of the cement barrier are dependent upon the composition of the cement, water to cement ratio, rate of curing, additives and any chemical or physical interactions of the cement with the waste or formation. It takes approximately 28 days for the cement to reach its final strength. The cured product is a hard brittle material that will change the compressibility, elasticity, density, consolidation, hydrogen index, and geochemistry of the local formation. Large castings of hydraulic cements invariably result in cracking due to shrinkage, thermal stresses induced by the hydration reactions, and wet–dry cycling prevalent at arid sites. Such cracking is an important breach formation pathway and may allow breach formations any time during the lifetime of the barrier. Cement grout barriers will require periodic reverification to detect if any failures have occurred due to aging and/or degradation of the cement.

Bentonite Slurries. Bentonite slurry walls are a commonly used method to produce cut-off walls. The barrier material is a bentonite clay mixed with soil. The bentonite concentration is typically 3–5% and the soil used can be site soil if the fines content (>25%) and blend are sufficient to produce a low permeability wall, otherwise soil must be trucked into the site. Bentonite slurries can also be jet grouted into the strata. Many organic contaminants adversely affect bentonite slurries by decreasing or collapsing the double layer of water surrounding the clay particles. Bentonite walls will increase the elasticity, change the resistivity/conductivity and compressibility, and increase the hydrogen index of the formation.

Polyacrylic Acid. Polyacrylic acid, under the trade name AC-400 (Geochemical Corporation) is among the most widely used hydraulic barrier materials for construction applications in the United States. This material is a magnesium acrylate and produces a soft gel rather than a hard solid with typical compressive strengths of 340 kPa (50 psi). AC-400 is commonly used to seal sewer pipe leaks and to stop water infiltration at construction sites. Acrylate grouts have been shown to have low water permeability and to be resistant to a large number of chemicals (7). Acrylate grout gels typically contain 80–90% water and are extremely sensitive to wet–dry and freeze–thaw cycling. Such high water content gels work best in 100% humidity or submerged conditions and will desiccate rapidly in arid conditions. The soft gels are also prone to creep fatigue failures and can fail with loading as low as 20% of the unconfined compressive strength (UCS) (8). The large water content of the gel makes properties such as electrical conductivity, acoustic properties, and hydrogen index of the barrier significantly different from the surrounding field.

Polymer Concrete. Organic polymers are innovative materials with desirable properties in both their fluid and solid states. This makes them suitable for applications where impermeability, high strength, and long-term durability are required. These materials have been used extensively in many commercial applications such as sewage and brine-handling systems and electrolytic baths. They have been used by the Army Corps of Engineers to repair dams and canal locks, and by the Federal Highway Administration for bridge deck and highway repairs. Polymer grouts are typically liquid organic chemicals that react with together in a polymerization to form a solid material made up of long-chain molecules of repeating units. Polymer grouts are candidates for high quality barrier materials because of their impermeability to gases and liquids, combined with their resistance to radiation, acidic, and alkaline environments. Plastic binders, which have insulating properties, generally increase the resistivity and lower the conductivity within the barrier wall (although some water-extended polymers decrease resistivity and increase conductivity). They also change the elastic properties, density, and consolidation of the soil.

Montan Wax Emulsion. Montan wax is the name given to a class of fossil plant waxes that are extracted from coal and lignite deposits. The wax is extracted from the coal or lignite using an organic solvent and recovered from the solvent by distillation. Montan wax is a complicated mixture of pure waxes (~70%), resins (~20%), and asphaltic materials (~10%) (9). The wax is a low melting point (85–90°C), low melt viscosity (60–350 cps at 90°C), hard, crystalline material. For subsurface barrier applications wax–bentonite emulsion is used. The emulsion is grouted into the strata to reduce the porosity. The grout was developed in Germany by Vereinigte Mitteldeutsche Braunkohlenwerke AG (MIBRAG) and DBI Gas-und, Umwelttechnik GmbH and has been tested in the United States by

Golder Associates (6). According to Golder the grout is a mixture of microscopic wax beads (formed by spraying molten wax into water) and water with a small amount of surfactant added to create the emulsion. Just before grouting the emulsion a small addition of bentonite clay is added. After grouting the bentonite slowly absorbs the surfactant breaking the emulsion. This allows the wax particles to begin to agglomerate. The aggregate mass grows and eventually blocks the porosity of the strata. There have been concerns that the water left behind is free to mobilize contaminants. MIBRAG claims the water is somehow held in the montan wax agglomerate as a gel. They have not offered an explanation of the chemistry that would allow this nor has any proof of this phenomena occurring been presented. In either case the montan wax grout greatly increases the hydrogen index and water content of the formation (although the more plausible emulsion-breaking mechanism would allow the water to eventually evaporate out of the formation). Geophysical perturbations in resistivity/conductivity should be readily apparent.

Silicate Grouts. Sodium silicate is one of the most widely used grouting materials worldwide. It is mainly used for waterproofing and consolidation of soils during construction operations. Proprietary silicate grouts have been suggested for use in containment barriers. Sodium silicate is available as a colloidal solution, typically 38% solids with a viscosity of 175–200 cps. The solution is generally diluted with additional water to reach the desired viscosity or silica content or reactivity parameters. Silicate grout is injected into the strata where reactive additives cause the formation and precipitation of polysilicic acids. Final characteristics of the hardened grout are dependent on the silica content of the grout, the rate of curing, the soil properties, and the catalyst/additives used. The degree of consolidation that the grout imparts to the soil is mainly dependent on the silicate content of the grout. Silicates are also subject to syneresis, that is, water will slowly migrate out of the gel causing shrinkage and degradation of the grout. The magnitude of the syneresis will decrease with increasing silicate content and shorter gel times. Shrinkage of the grout will cause an increase in the permeability of the formation. The degree of increase is related to the grain size (therefore pore size) of the soil. Coarse soils will show greater permeability losses than fine grained soils. Silicates are also subject to redissolution. If the polysilisic acid is contacted by a base, the gel formation process can be reversed. This happens to some extent to most silicate grouts due to residual hydroxides. In terms of geophysical parameters the silicate grouts change the density, the elastic modulus, conductivity/resistivity, and hydrogen index of the formation.

Installation of Subsurface Barriers

In 1992 a panel of experts for the DOE UST Integrated Demonstration was convened to identify technologies for placement of subterranean barriers around USTs. The group identified and ranked nine grouting technologies that have potential to place vertical barriers and five for horizontal barriers around the tank (10). These included jet-grouting, permeation grouting, sheet metal pilings, slurry walls, deep soil mixing, and cryogenics.

Jet Grouting. Conventional jet grout curtains are constructed by injecting the grout through the drilling tube of the grouting rig. The technique requires a pumpable grout that can be injected at pressures greater than 34 MPa (5000 psi) through a small orifice, typically 1–2 mm. This limits any aggregate additions to fine particle sizes. Most often, jet grouting uses a low viscosity grout (~50 cps), and incorporates only the existing soils for aggregate. The drill stem is drilled into the ground to the bottom where the barrier will start. The tube is then rotated and high pressure grout injection started. As the drill stem is turning, it is slowly withdrawn from the ground. The high velocity jet masticates and mixes the soil and grout, which results in a column approximately 1 m in diameter that resembles a pancake stack (Fig. 2). After the grout pumped into the primary holes has gelled, grout is injected into secondary holes to fill gaps in the primary grout injection. This results in a barrier 1.5 to 2 m thick. Jet grout curtains can be vertical using conventional drilling or may be angled or horizontal using directional drilling. There are often problems in forming a continuous grout barrier despite grouting via the secondary holes. Breaches may occur when cobble and other large buried objects interfere with the continuity of a jet-grouted column. These objects can block the pathway for the jet, thus preventing grout from reaching the downstream side of the obstruction. There may also be drill wander, which could cause misalignment of the columns and insufficient overlap of adjoining columns.

Jet grouting is the most widely adapted or modified placement method. Several promising jet-based grouting technologies that have been developed or are under development include panel jet grouting such as SoilSaw from Halliburton NUS, and ultrahigh pressure grouting from Quest Integrated Inc. or Superjet.

Panel jet grouting is a simple refinement to conventional jet grouting. The tool is turned back and forth only a few degrees, forming a bowtie-shaped panel, rather than rotating the jetting tool 360° and forming a cylindrical column. This forms a thin panel, typically 30–38 cm (12–15 in.) wide. Panels are laid side by side with a slight overlap to form a continuous barrier. A major advantage is the greatly reduced grout volumes resulting from thin panels, rather than thick columns.

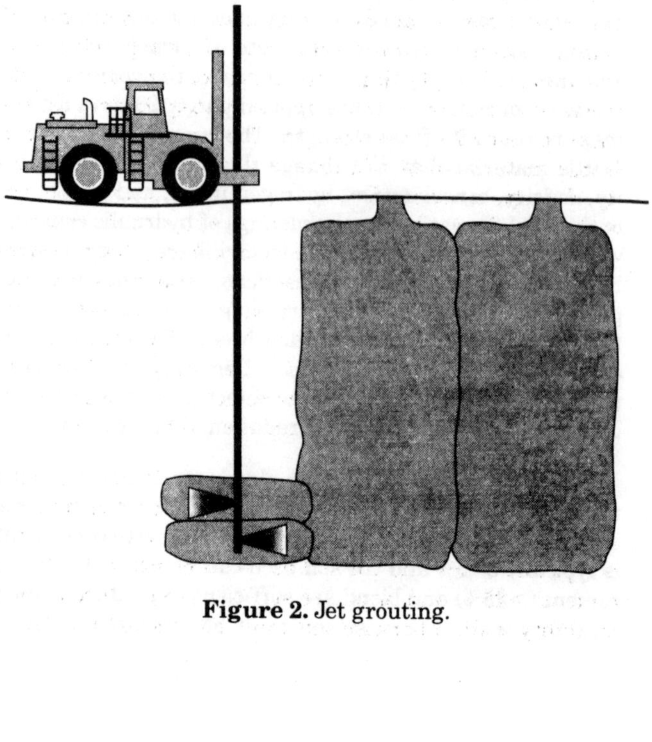

Figure 2. Jet grouting.

SoilSaw is a modification of panel jet grouting that consists of a long beam with multiple jets at regular intervals along a reciprocating beam. The beam is connected to a modified tracked excavator equipped with high-pressure pumps to deliver the grout to the jetting nozzles of the beam. Jetting action combined with the sawing of the beam liquifies the soil and cuts a pathway into the ground. Once the beam has reached the proper depth and cutting angle, the tracked excavator pulls the beam along the course intended for the barrier. The barrier is typically 30–38 cm (12–15 in.) wide, but the tool can be modified to form 10–15 cm (4–6 in.) wide walls. Grout requirements are similar to jet grouting, although the manufacturer suggests densifying the grout, to a density greater than the surrounding soils, by adding fine iron oxide. Pretesting of the grout would be required to assure compatibility with the iron oxide. This method uses more grout per cubic foot because the jetting bar must sink into the formation, which can only be done with a fully liquefied soil. In terms of verification this method is advantageous, because it has a positive placement; the grouting bar cannot proceed without delivering the grout to the formation. Breaches can still occur at the joint between panels (imperfect alignment of borehole) or through grout failures after solidification (eg, cracking). SoilSaw is also limited to vertical barriers. Halliburton NUS is developing a horizontal tool.

Ultrahigh pressure grouting uses pressures in excess of 170 MPa (25,000 psi) to greatly increase the affected field. This technique is used to make thin disks. The high pressure results in an affected area of up to 7.6 m (25 ft) in diameter. The method relies on overlapping disks to form continuous barriers. Advantages include fewer boreholes to be drilled, high pressure which can cut through cobble and buried objects, and thin barriers that conserve grout requirements. However, disadvantages exist as well. It may be difficult to align thin disks for proper overlap at depths of 30 m (100 ft) or more below grade. Grouts would also be limited to low viscosity, low solids content.

Permeation Grouting. A grouting material of low viscosity (typically <50 cps) is injected into the ground through a series of injection wells. Grout penetrates the soil where it then sets, forming the barrier in *in situ* (Fig. 3). The barrier is made continuous by overlapping the injection zones. Permeation is advantageous because it has little spoils, produces low stresses, and can produce low permeability barriers when used in conjunction with advanced polymer systems. One great disadvantage is that permeation grouting usually leads to hydrofracturing of the soil and/or short circuiting of the grout flow. Construction of a subterranean barrier by permeation injection requires the grout material to penetrate the soil completely, and remain in place until solidification is complete. Then, the solidified monolith must provide an impermeable and durable barrier. The ability of the grout to penetrate the soil is dependent on the grout viscosity and the permeability of the soil. A coarse sandy soil can be injected with moderate success; however, it is generally accepted that permeation grouting is limited to soils having hydraulic conductivities of more than 10^{-4} cm/s (8). It is easy to see that holes would occur in areas of hydrofractures or short circuiting where the grout takes the path of least resistance. The viscosity of the injection grout, the curing time, and the hydraulic conductivity of the soil will determine whether the binder remains in place or percolates away from the injection zone before hardening. A low viscosity material applied in coarse sand may run out, leaving a more permeable barrier after solidifying.

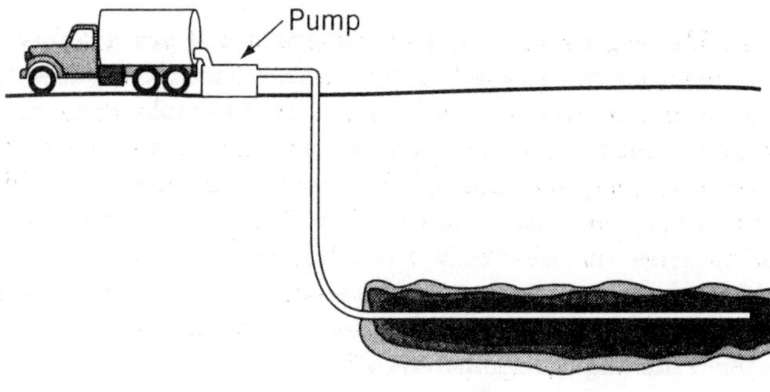

Figure 3. Permeation grouting.

Sheet Metal Piling. Sheet pile barriers, as the name implies, are made of interlocking sheets of the barrier material, typically a metal. Sheet piling barrier materials are generally installed by driving the sheets into the ground around the waste site. Corrosion of the steel by the soil limits the performance life of a sheet piling wall to between 7 and 40 years (2). Plastic (eg, high density polyethylene) sheet is a relatively new application and has had only limited actual field use. Sheets pose a problem for verification, because the seams are sealed using a grout, sealing of joints between sheets is often questionable, and seals cannot be verified. There may also be misalignment of adjoining sheets causing gaps.

Slurry Walls. To construct a slurry wall, a vertical trench is excavated, and the trench is filled with a suspension-grout slurry to form an engineered panel. The most common slurries are water–soil–bentonite and Portland cement–bentonite. Typically a bentonite–water slurry is used to hydraulically shore up the trench during construction and seal the pores in the trench walls. The construction of the wall advances by continuous excavation at one end of the trench while back-filling the other end. Back-filling is accomplished using a soil–bentonite, cement, cement–bentonite, or cement–soil–bentonite slurry. The back-fill slurry displaces the bentonite–water slurry. Overlapping is ensured by cutting into adjacent panels so the wall is *keyed.* Slurry walls are generally 6 to 24 m (20–80 ft) deep and 0.6 to 0.9 m (2–3 ft) wide. A major advantage of slurry walls is that they form a continuous unbroken barrier. The disadvantages include being limited to vertical barriers, the large amount of soil that must be excavated (and may be contaminated at a waste site), limited permeability reductions, and chemical resistance. One caution that the EPA literature on slurry walls mentions is the detrimental effect that certain chemicals can have on slurry walls containing bentonite mixtures (2,11,12). Certain chemicals can increase the permeability of bentonite mixtures or result in dissolution of the material. Some organic fluids result in desiccation and cracking of the bentonite material. The EPA literature emphasizes the necessity of compatibility testing between the slurry wall material and the chemicals it is likely to contact. Slurry walls are also susceptible to desiccation cracking.

Deep Soil Auger Mixing. Several large augers are used to bore into the soil followed by introduction of a grouting material. The augers mix the soil and grout and are gradually retrieved from the ground. The resultant panel, compared to jet-grouted curtains, generally has more favorable properties (eg, permeability) has fewer breaks, and has more consistent dimensions. Deep soil mixing is a proven technology but is limited to a depth of approximately 30 m (100 ft). It produces vertical barriers and results in a much larger quantity of spoils when compared to jet grouting. Breaches can occur if the barrier material cracks (eg, desiccation, chemical incompatibilities) or if the auger alignment is off.

Cryogenic or Frozen Soil Barriers Cryogenic barriers are formed by freezing the subsurface area where containment is needed. A series of steel casings is installed around the waste site approximately 3 to 6 m (10–20 ft) apart. These pipes are connected to a refrigeration unit. A nonhazardous refrigerant is circulated through the pipes, freezing the soil between pipes. Depending on-site conditions water may or may not have to be added to the subsurface and the barrier requires continuous, lifetime maintenance. Breaches could occur if the cooling pipes are not placed properly, if there is a loss of cooling to all or part of the barrier or if insufficient moisture exists to lower the permeability to the desired level. The matrix of steel piping running through the barrier will cause interference for many geophysical methods (eg, radar).

Regulatory Requirements on Barrier Integrity and Continuity

No suitable method exists for the verification of an emplaced barrier's integrity. The large size and deep placement of subsurface barriers make detection of leaks challenging. This becomes magnified if the permissible leakage from the site is low. Detection of small cracks (fractions of an inch) at depths of 30 m or more has not been possible using existing surface geophysical techniques. Such ability has been recognized as a need by the DOE in the FY1995 Needs Statement put out by the EM-50 In-Situ Remediation Integrated Program (13). Compounding the problem of locating flaws in a barrier is the fact that no placement technology can guarantee the completeness or integrity of the emplaced barrier.

The ability to verify barrier integrity is valuable to many government agencies (eg, DOE, EPA, DOD) and the commercial sector. Such ability will be required to gain full public acceptance of subsurface barriers as either primary or secondary confinement at waste sites. For subsurface barriers to be fully implemented and accepted, validation of the integrity and the ability to monitor performance over the lifetime of the barrier are required. At present there are no specific regulations under CERCLA or RCRA nor is there any regulatory guidance on subsurface barrier integrity or performance validation. The only regulatory standard for barriers is the RCRA (40 CFR 264, subpart N, Landfills) requirement of a 10^{-7} cm/s hydraulic conductivity at a thickness of 0.9 m (3 ft). This may change as the EPA Office of Emergency and Remedial Response has launched the Superfund Initiative on Long Term Reliability of Containment (14). The EPA will work with other agencies (eg, DOE, DOD) to develop technical guidance and methodologies to evaluate containment technologies.

The waste management community has a need to develop and refine barrier verification methods to determine the existence, size, and location of breaches in a subsurface barrier. After such determinations, the effect of the breaches may be factored into the performance assessment of the waste site. More appropriately the breaches could be repaired (and the repairs qualified with the same technology). Logically the methods must be nondestructive, because the barrier is required for continuing service after verification or performance monitoring. Depending on grout type and placement method, a subsurface barrier can affect the local density, porosity, electrical resistivity, and diffusivity. These local changes make the barrier detectable from the surrounding geology and waste structures (eg, drums or storage tanks). Present methods of subsurface characterization will need to be modified or new technologies developed to take advantage of this detectability.

TECHNOLOGIES FOR SUBSURFACE CHARACTERIZATION

Most methods presently being employed for *in situ* characterization of the subsurface can be classified as well logging technologies. They were developed for the oil, gas, and mining industries to characterize subsurface reservoirs and mineral deposits. Most are geophysical methods that react to changes in properties such as electrical conductivity or acoustic velocity caused by geological differences in the subsurface. When a barrier is emplaced in the subsurface it causes localized changes to the subsurface (15). Geophysical signals are perturbed by such changes and the logging device records an anomaly. If the resolution of the geophysical method is sufficient the anomaly can be identified and details (eg, a breach) may be distinguishable. Surface-deployed or large-field (ie, seismic) geophysical methods are best suited to directly image a barrier and develop a map of the subsurface. Geophysical methods deployed as borehole logging devices (ie, neutron) can be used to investigate discrete areas of the subsurface and to measure performance values. This article briefly discusses some commonly used logging techniques that may be applicable to integrity verification and performance monitoring of a subsurface containment barrier. The limitation(s) of the methods when used to verify grout continuity (barrier integrity) will also be discussed.

Other methods, such as gaseous tracers, measure flow pathways. Such measurements can be used to deduce information about a barrier. With a gaseous tracer if there is a hole in the barrier the tracer will be able to quickly move (eg, by diffusion or advection) across the containment zone. Knowing the tracer was transported across the barrier tells us there is a hole in the barrier. A new technology for characterization of the subsurface using gaseous tracers will be discussed. The technology is not only able to deduce a hole in a barrier but is also able to locate and size the hole.

Geophysical Technologies

Well logging began in the petroleum industries for locating oil and gas reserves. Logging quickly spread to other mining technologies including coal, uranium, and water. Early logging techniques focused on electrical properties and have since broadened to include gamma- and neutron-radiation, acoustic, thermal, and electromagnetic methods. Well logging can determine macrofeatures such as formation boundaries, local features such as water content, and microfeatures such as soil impurities or mineral content. For barrier continuity verification the determination of local features is most important. Well logging can determine water saturation, porosity, thermal conductivity, and many other parameters that may be related to barrier performance and continuity. Good discussions of well logging technologies can be found elsewhere (16,17). Traditional well logging consists of a sonde (probe) attached to a cable lowered into the borehole (Fig. 4). Some logging methods can be employed from the surface (Fig 5) or as a combination downhole–surface pairing of receiver and transmitter. The sonde receives and transmits signals in the formation. The type of logging is determined by the particular sonde used.

All logging methods measure distortions in a transmitted signal. Each method transmits some normal signal and receives back a distorted signal. How the signal is distorted and the magnitude of the distortion is related to the property of interest. If the formation is homogeneous then the received signal can be calibrated directly with no corrections for unwanted distortions. This is rarely the case. Subsurface formations are heterogeneous on the macro, local, and microscales. These heterogeneities can cause distortions in the logging signal. It is often impossible to differentiate between two different types of distortions or formation features. As an example a thin grout lens may give a similar resistivity as some ungrouted cobble. This is one reason logging technologies have limited use in barrier continuity verification. To be useful over a reasonable area requires correcting for unwanted distortions by solving partial differential equations relevant to each logging type and distortion type. Additionally, many logging techniques require numerous boreholes to provide full coverage of a barrier.

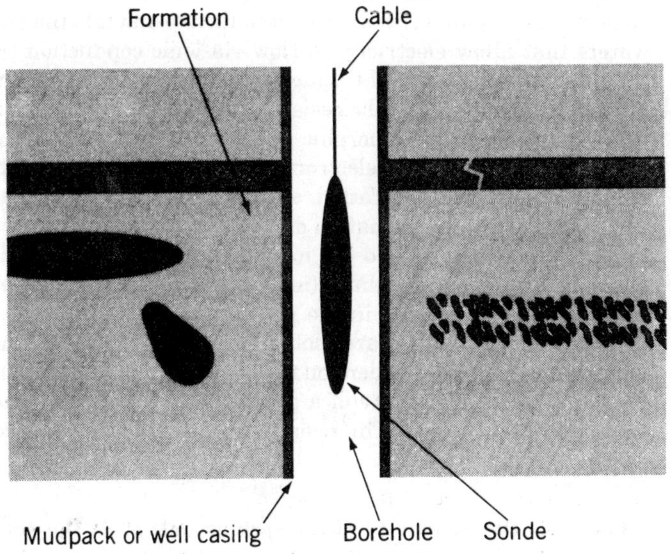

Figure 4. Typical borehole application of a geophysical survey.

Logging can however be very useful to determine barrier performance. Parameters such as porosity, water content, density, and brittleness can be measured and/or monitored. This is the traditional use of logging technologies and the methods are well developed and suited for such use.

Nuclear. Nuclear radiation can be used for subsurface characterization. There are many types of nuclear radiation including α, β, γ, and neutron. Alpha- and beta-particles do not penetrate deeply in soil and rock formations and therefore have not been utilized for well logging. Gamma- and neutron radiation have found wide-scale use for formation characterization. Neutron logging has been extensively used to determine formation saturation and densities. Gamma-rays are being used to determine subsurface compositions (minerals, contaminants, etc). Although these methods are valuable for mining and general characterization, they are of limited usefulness in barrier continuity determinations. The depth of penetration of radiation used in well logging is generally only 10 to 20 cm. With shielding and beam focusing, this may be extended but would still be less than 1 m. The number of boreholes required may be prohibitive (one every meter at best). The nuclear methods are not currently capable of resolutions of a fraction of an inch when the mean free path of the scanning radiation is a meter. Both neutron and gamma-radiation logging methods are not well suited to detect fractures in a formation. A fracture would simply raise or lower the average value of the parameter of interest (see NUCLEAR POWER).

Neutron Logging. Neutron well logging has become an important tool for estimating porosity and moisture content of a

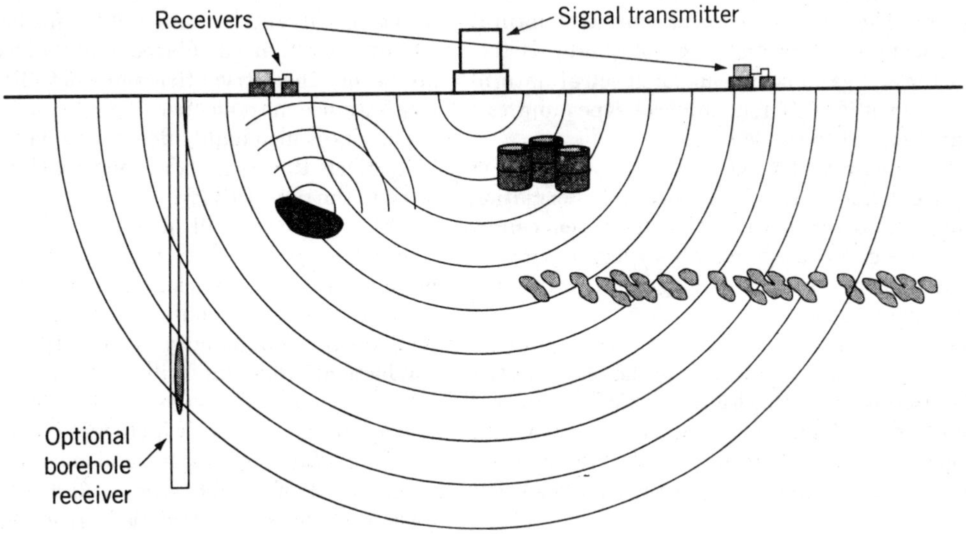

Figure 5. Surface deployment of a geophysical survey.

formation. Neutron logging is used to measure the hydrogen index I_h, which is defined as the equivalent volume fraction of fresh water containing the equivalent amount of hydrogen. Neutrons generally interact with atomic nuclei and therefore interactions are less frequent and neutron ranges are longer than other nuclear radiation. The most common neutron interaction is elastic scattering. Classical mechanics show that the neutron is moderated more efficiently by nuclei with a mass similar to the neutron such as hydrogen and other low mass elements. There are two types or neutron sources, chemical and accelerator. Both sources produce fast neutrons and the logging type is usually distinguished by the detector type, either thermal or epithermal. Chemical sources produce neutrons with an energy of approximately 4 MeV. Accelerators produce neutrons with an energy of about 14 MeV. Chemical sources have the advantage of being cheap and reliable, whereas accelerators have the advantage of being able to turn off the neutron source.

There are two common methods of measuring water content using neutron diffusion: the neutron log and the moisture gauge. An extended comparison of the two methods can be found elsewhere (18). Moisture gauges are sensitive to thermal neutrons and have an affected radius of 10 to 20 cm. Neutron logs measure thermal or epithermal neutrons depending on sonde configuration and have a measurement radius of 20 to 30 cm. Thermal neutron methods are used in fluid-filled boreholes and are of little value in air-filled borehole. Epithermal methods are preferred for air-filled boreholes and can also be used in fluid- or foam-filled boreholes. Neutron methods are dependent on many different parameters such as porosity, matrix, pore fluid, salinity, temperature, pressure, standoff, and borehole geometry. A complete mathematical solution incorporating all these parameters is not feasible, so numerical approximations are used and are field calibrated. Such approximations do not allow exact pictures of the subsurface but yield statistical equivalents that are averages over the affected range. Again the logging tool is very limited in the affected area and is not very useful as a continuity verification tool. The inadequacy of neutron logging for barrier-continuity verification has also been stated (19).

Gamma-Logging. Gamma-rays can be used in several ways to characterize a formation. Formation minerals contain naturally radioactive elements that can be detected by gamma-spectroscopy. There are three major sources of natural gamma-radiation: uranium, thorium, and potassium. Detection of the characteristic gamma-rays using spectral natural gamma-logging can identify different mineral types and can detect the concentrations of these elements.

Interactions of gamma-rays depend on the atomic number of the matrix and the gamma-energy. By exposing the formation to a source of gamma-rays the density of the formation can be estimated. This type of gamma-logging is $\gamma-\gamma$ logging or density logging. Measurements using $\gamma-\gamma$ logging utilize Compton scattering of gamma-rays.

In situ chemical analysis can be performed using gamma-logging. Gamma-ray exposure also results in the photoelectric effect and can be used to generate characteristic fluorescence X-rays. The energy of the X-rays is dependent on the atomic number of the incident matrix. Photoelectric effect interactions are the predominant interaction for gamma-ray energies below 100 keV.

Gamma-logging methods are of limited usefulness for barrier verification. The range of gamma-logging is only 5–12 cm. As mentioned for neutron logging, radioactive decay and subatomic interactions are statistical processes and are really only useful for bulk or average properties. It may be possible to detect components of a grout (20), but this would only indicate some grout in the local area not necessarily a homogeneous, breach-free mix of grout and soil.

Electrical. The earliest form of well logging was electrical resistance logging. This type of logging measures the resistivity (or the inverse, conductivity) of the formation. The variation in resistivity of rocks is larger than the variation of any other property commonly used for logging (eg, density or acoustic velocity). Electrical current may travel through the subsurface by electronic conduction as in metals where there are a large number of free electrons available; by electrolytic conduction as in an electrolyte solution by ionic transport; and by dielectric conduction as in an insulator with very few free electrons. Electronic conduction is by far the fastest method and is used in our homes using copper wires to carry electricity to appliances. Generally, formation rocks are nonconductive and it is the pore waters that allow electricity to flow via ionic conduction (although current can travel through metallic anomalies such as buried drums). Roughly the resistivity of rock formations is as follows: igneous > metamorphic > sedimentary. Ionic conduction is much slower than electronic transport and is a function of ionic mobility, concentration, and activity of the electrolyte. The resistivity of the formation depends on the quantity of water in the pore spaces and the ionic activity of the pore fluid, which is a function of salinity (ie, dissolved chlorides, carbonates, sulfates, etc). Salts in the pore water dissociate to form anions and cations that are mobile in an electric field. Since ionic conduction is dependent on the quantity of pore water and its salinity, the resistivity of a given class of rocks can vary with age and lithology. The resistivity of a formation is given by Archie's law:

$$\rho = \rho_w \phi^{-m} S^{-r}$$

where ρ is the resistivity of the rock or soil, ρ_w is the resistivity of the pore water, ϕ is the pore volume (porosity) of the rock or soil, m is a cementation factor, S is the fractional pore saturation, and r is a constant (≈ 2). Dielectric conduction occurs in insulators where the current changes with time as in alternating current. Electrons in the insulator, although not free to move about, are displaced slightly by a current causing polarization. Dielectric effects are not often used in subsurface logging since most electrical logging uses low frequencies. The resistivity is also highly dependent on temperature. Temperature affects the viscosity of the pore fluid, which is directly related to the conductivity.

Other artifacts will also affect the resistivity of a formation. Clays are strongly conductive due to the charged double layer of clay particles and will enhance the conductivity measured for the formulation. Minerals, such as sulfides, can increase the conductivity of the formation through electronic conduction. Fracturing will greatly affect resistivity logging by introducing nonconductive gaps in the current pathway. Electrical logging is severely influenced by surface anomalies such as weathering, topography, and moisture content. Even the geologic strike of the formation will affect the resistivity. Many formation rocks (eg, stratified layers of slate) are anisotropic and conduct better in the planar direction rather than perpendicular to it.

The equipment used for field measurement of resistivity includes a power source, electrodes, and meters for measuring resistivity or current and voltage. Constant reversal of current direction can be used to remove effects of polarization. Commonly DC power supplies are used and the current direction is reversed by switches, alternatively ac power sources at low frequencies can be used. Electrical resistivity logging requires a conductive coupling of the electrode to the formation. Electrodes can be metal rods (eg, stainless steel or brass) driven into the ground or contacts held against a borehole wall by packers or springs. The electrodes can be deployed at the surface with logging data limited to surface-related phenomena or one or more of the electrodes can be placed in a borehole to develop data at depth. The electrodes are usually placed in a collinear array to simplify the data interpretation. In boreholes where a nonconductive oil-based mud or an insulating casing is used induction sondes must be used for electrical measurements. Induction logging uses alternating current and coils to induce an eddy current in the formation. The strength of the eddy current is measured with another coil and is proportional to the formation resistivity. Induction probes measure the resistivity of the formation in a zone up to 1 m away from the borehole. This makes them poorly suited for large-scale barrier verification, but they might be useful for performance determinations (ie, quality of cement).

Electrical logging techniques measure bulk properties of the formation and are not well suited for barrier verification. An installed barrier will make local changes in the electrical properties of the formation. The porosity and pore fluid content will be altered. The premise behind a containment barrier is to locally reduce permeability (therefore porosity). Grouting may reduce the amount of conductive pore fluids (eg, polymer materials) or increase them (eg, cement grout). Depth of penetration of resistivity logging can be great (hundreds of feet) but spatial resolution is poor and variability in formation geology and geochemistry will contribute to background noise. Unless the changes that occur after barrier installation are great enough it may be extremely difficult to distinguish the barrier from the formation. One place that electrical logging may be useful in barrier verification and monitoring is for liquid leak detection. If a barrier contains water the conductivity inside the barrier will be quite different to that outside the barrier. If a leak should occur, the sudden influx of water and the resultant increase in conductivity to the outer soils should be readily apparent. It is doubtful though that the magnitude of the breach would be addressable. Most barriers are not designed to have standing water on the walls unless a failure of the waste package (eg, drums rust or tanks leak) occurs. It would be beneficial to know if the barrier is going to leak before it actually does. Electrical cross-borehole tomography (21) and borehole induction logs (22) have been used to map affected areas of remediation and to determine grout penetration in soils (19), but these studies have been limited to general locations and could not be used to infer homogeneity of a barrier.

Electrical Resistance Tomography. Included in electrical methods is electrical resistivity tomography (ERT) as developed at Lawrence Livermore National Laboratory. In Fernald, Ohio ERT will be employed to evaluate an *in situ* barrier placement (23). Electrical resistivity tomography is a cross-borehole method for measuring electrical resistivity. It uses a series of electrodes, placed within a borehole, that is in electrical contact with the formation. Direct current is applied to a pair of electrodes and the resultant voltages across other pairs of electrodes are measured. The current is then applied to a different pair of electrodes, and the voltages are measured again. This process is repeated until all the different combinations of electrode transmitter pairs are completed. The data is processed using sophisticated tomographic methods and provides a map of the area between boreholes. Researchers have reported (21) a spatial resolution of a few meters at best; "While we can identify the several factors controlling resolution, we cannot yet quantify the effect of each nor do we know the interplay between them" (21). Such poor resolution is typical of geophysical methods with borehole spacing on the order of 10 m (or signal travel distances of tens of meters for surface-deployed methods).

Electromagnetic. Electromagnetic (EM) logging responds to electrical conductors in the shallow subsurface using low-frequency (<1 MHz), time-varied electromagnetic fields. Generally EM is limited to near-surface detection and most often is used in mapping of buried metallic objects. A primary electromagnetic field is generated by passing an electrical current through a coil of wire. When the primary field encounters a conductor, a secondary field is emitted from the conductor that has the same frequency as the primary field but is phase shifted. Electromagnetism uses this secondary field to deduce the characteristics of the conductor. As mentioned earlier, EM responds to the secondary field induced in electrical conductors; therefore by measuring the in-phase and quadrature it is possible to differentiate between conductors and metallic objects. Quadrature (90 degrees out of phase component) measurements are sensitive to conductivity contrasts while in-phase (180° out of phase component) measurements are sensitive to metallic objects. The equipment consists of an ac power source, receiving and transmitting coils, and a recording system. An EM survey is done similarly to an electrical resistivity survey; the transmitting and receiving coils are moved to different grid points within the area to be mapped. Recordings are taken at each point and may be duplicated at varying frequencies to increase vertical resolution. Some EM systems run in a continuous mode, but data are not recorded continuously, rather it is sampled at regular intervals. Because EM methods are sensitive to electrical conductors they are not well suited for verification. The exception may be that aqueous leaks from a barrier might be detectable although the resolution will quickly diminish (to undetectable) with increasing depth. Even if the barrier material is conductive (ie, the pore waters of cement grouts) the secondary magnetic field set up by the barrier could not be resolved to see small cracks or holes. It has been shown (24) that even with conductors as large as 55 gallon (208 L) steel drums, EM can have extreme difficulties detecting the drum. A single drum buried 3 m (10 ft) below the surface was not detectable and three drums together (at 3 m) were barely detectable. If EM can be used in monitoring barrier placement, verification, and performance it is better suited as a general grout locating procedure to be used during grout injection for quality assurance/quality control purposes. It may also be useful to determine if water ingress has occurred after a rain storm. Percolation water will greatly increase the conductivity of the soil outside the containment area. Unless there is a hole in the barrier for the water to infiltrate, the soil inside the contained area should remain low in conductivity. If the affected area is large enough, EM may be able to see the water infiltration.

Ground-Penetrating Radar. Radar (radio detecting and ranging) was developed for detection of aircraft overhead. Ground-penetrating radar (GPR) is similar except that the antennas are pointed downward into the ground and a broader bandwidth (10 MHz to 1 GHz) is used. Ground-penetrating radar has been used to locate subsurface features such as old mine tunnels and buried waste (25) and is a surface-oriented characterization tool. A transmitting antenna is used to send an electromagnetic wave into the subsurface. The wave travels through the subsurface where parts of it are reflected by formation anomalies. The reflected waves are recorded at the surface and the results are analyzed to form a map of the anomalies. Thus GPR involves towing radar antennas across a site along a set of grid lines spaced about one half to one meter apart. At fixed distances along the grid a signal is transmitted from one antenna and the reflections recorded by a receiving antenna. The distance to an object can be determined by measuring the time the wave takes to travel to an object and back. A single recording is sufficient to determine the depth of an object but multiple recordings along the various grid lines allow definition of the object. Ground-penetrating radar depends on sharp contrasts in electrical properties of anomalies compared to the surrounding soil. Where less contrast is available (eg, a narrow crack in a barrier) GPR data are difficult if not impossible to interpret. Also GPR is subject to noise and clutter (eg, reflections from above-ground structures and signals from broadcast sources). Resolution of GPR is poor under ideal conditions and the penetration of the radar is severely limited through clay layers. Depth of penetration is related to formation electrical properties and frequencies of the electromagnetic wave and under ideal conditions is no more than 30 m. In clays, penetration can be as little as several inches. Penetration depth can be improved by lowering the frequency of the wave but resolution then becomes severely limited. In general GPR is useful in the top 3 m for locating buried objects. Formation features including water, rock, soil, plants, and buried human-made items result in strong radar absorption and reflection and a resultant high clutter in the signal. Depth and spacial resolution of GPR is greatly affected by the signal-to-clutter ratio.

Radio Imaging Method. Radio imaging method (RIM) consists of a low frequency electromagnetic transmitter and a remote field receiver. A transmitter is located below ground and emits a narrow-band radio signal through the ground. The receiver can be employed either a cross the borehole or on the surface. The speed of propagation through the formation and the rate of attenuation of the signal are functions of the frequency of the signal and the electrical conductivity of the formation. This method can be used similarly to GPR, for making ray traces, but also analyzes the phase change and attenuation of the wave, as in standard EM surveys. Boundaries of conductivity contrasts are determined using the phase shift and attenuation of the wave. Mapping is enhanced by making traces with multiple frequencies. Like ERT, sophisticated analysis tools are required to interpret the data. By measuring the electrical conductivity of the formation details on moisture variations, fracturing and lithology can be inferred. Radio imaging method technology was used at the Chemical Waste Landfill at Sandia National Laboratory and was stated to be able to detect soil units of 0.5 m (26). Such resolution is still not quite accurate enough for barrier continuity verification, although the method may be capable of finding leaks after water infiltration.

Acoustic Logging. Acoustic logging has been said to be "almost unique among logging methods because far more information is available than can be interpreted" (16). Acoustic waves are transmitted into the formation. The velocity and attenuation of these waves are affected by density, porosity, fracturing, pressure, pore fluid, temperature, and saturation. Two methods of acoustic logging can be used to map a formation: transmission or reflection. Transmission methods measure the velocity and/or attenuation of shear and compressional acoustic waves. Acoustic energy in the range of 20 kHz is transmitted from one borehole and the signal is received some distance away from the transmitter. Reflection methods generally operate around 500 kHz and measures the attenuation of the wave. The signal is reflected off formation features and the transmitter and the receiver can be in a single sonde. The higher frequency reflection methods have shown promise in the location of fractures, but have limited penetration depths.

Acoustic waves are well suited for measuring the elastic properties of materials and have been used in nondestructive testing of Portland cement concrete for paving (27). Researchers at Brookhaven National Laboratory believe that acoustic measurements could be used to assess cracks in barrier walls and to monitor the long-term integrity of the barrier. Inexpensive transducers could be emplaced in the barrier walls before solidification of the grout matrix. These transducers could then be used to look for fractures and gaps and to form a baseline acoustic signature of the barrier. The transducers could then be used in the future to recheck cracking and to compare the acoustic signature of the barrier. As mentioned earlier, there is often far too much information returned to the receiver in acoustic logging. Waves echo off the inhomogeneities of the formation and the echoes can reflect further. Acoustic techniques are much clearer in the more homogeneous grout media.

Seismic Tomography. Seismic tomography (a subset of acoustics) uses surface or borehole probes and low frequency acoustic energy pulses to transmit a signal. Early petroleum explorations used explosive charges to generate the seismic pulse, and this method is still used for petroleum and mineral exploration. Many of today's methods for shallower subsurface exploration (waste site characterization) use the simple ringing of a steel bar or the blast of a pneumatic hammer. Geophones then record the reflected waves as a function of time. The changes in the acoustic velocity of the seismic wave are dependent on the nature of the formation features. The velocity of the seismic wave is a function of shear modulus, bulk modulus, and formation density as given by

$$V_P = \sqrt{k/\rho}$$

for compressional waves, where V_P is the compressional wave velocity, k is the bulk modulus, and ρ is the density. For shear waves,

$$V_S = \sqrt{\mu/\rho}$$

where V_S is the shear wave velocity and μ is the shear modulus.

The low frequency of seismic logging allows great distances to be logged (ie, hundreds of meters as opposed to several feet for other logging techniques). The drawback to seismic evaluation is the resolution. Resolution is strongly dependent on frequency, the lower the frequency the less resolved, particularly in the vertical direction. To look hundreds of meters into a formation results in a resolution of tens of meters. It is generally accepted that minimum resolution is one-quarter to one-eighth the wavelength. From the above equations we see that the velocity of the wave depends on the medium it is traveling in but is a constant for a constant media. Typically, velocities in soils and rock range from 500 to 3000 m/s (1,600–10,000 ft/s). For a velocity of 1500 m/s (4900 ft/s) seismic reflection methods using frequencies on the order of 100 Hz results in wavelengths of 15 m (49 ft) and minimum (ideal) resolutions of approximately 1.9 to 3.8 m (6–12 ft). Evidence has been given that the use of shear waves (S-waves) in soil environments offers improved resolution over compressional (P-wave) waves (28). Depending on the soil, the S-wave velocity can be several times slower than P-wave while having a similar frequency. This yielded threefold resolution increases for the particular site used. However, this still only corresponds to detecting beds from 0.3 to 0.9 m (1–3 ft) thick. If methods can be developed to use higher frequencies (but lower than acoustics) the resolution of seismic tomography would increase. Of course the reflected waves must be distinguishable from background and signal noise and clutter. In the weathered soils where velocities are slower and for barrier detection where depth penetration of 30 m is adequate, it may be possible to use a compromise frequency.

Seismic tomography still represents one of the more promising of acoustic logging techniques and remains a popular geophysical tool for environmental characterization (29,30). As refinements to the resolution are made, the deep penetration of seismic waves will prove valuable to subsurface investigations. Cross-well seismics was evaluated at the Mixed Waste Landfill Integrated Demonstration (19), and although they were found useful in locating general grout penetrations, they were not considered adequate for verifying continuity of a barrier. In a study of high resolution seismic imaging of fractures in rock, it was concluded (31) that minor structures such as individual cracks were not detectable (at the frequencies they used).

Thermal Logging. Thermal logging looks at the temperature variation within a formation. Temperature measurements are most often used to correct data from other logging techniques but thermal logging can be used to gather useful information about the formation. Thermal conductivity is affected by the porosity and pore fluid of the formation rock. Water-filled porous rock exhibits lower thermal conductivities than solid rock, and air-filled pores further lower the conductivity. Injecting cooler or hotter water into the formation causes a thermal gradient to occur. Information about the porosity and water content of the formation can be obtained by observing the decay to thermal equilibrium.

Temperature logging can be used to track grouts in the subsurface. Measuring the heat of curing of cementitous or thermosetting grouts can be used to infer that grout has reached a certain area. This does not, however, yield data on the continuity of the barrier. Thermal logging is best suited to indicating positive grout injection into the desired zone.

Tracer Technologies

A tracer is any substance that can be easily or clearly monitored (traced) in the study media. Tracers for soil studies can be radioactive or nonradioactive liquids, gases, or solids. Tracer technologies can be used in transport and dispersion studies, leak detection studies, and material location.

In transport and dispersion studies, tracers are used to tag a flowing medium to determine where that flowing medium is going and how it is being dispersed in a surrounding matrix. Example applications are transport and dispersion of pollutants in air parcels on a continental scale; transport and dispersion of injected fluids in petroleum reservoirs for enhanced oil recovery, and transport and dispersion of air within homes and commercial buildings for energy-efficiency calculations. Typically, tracers for transport and dispersion are liquids and gases. They can be either radioactive or nonradioactive.

Leak detection studies use tracers to locate and estimate leak rates in various scenarios. These can be as simple as colored dyes used to visually locate cracks and holes in tanks or as complex as mass spectroscopy detection of helium to find leaks in vacuum systems.

Tracers can also be incorporated into a material (ie, grouts) to allow tracking of the material through a media or along a pathway. Radioactive tracers could be incorporated into the barrier grout, thus allowing nonintrusive detection of where the grout is placed. Chemical tracers could be used in the grouts but would have to be able to diffuse out of the grout to be detected by nonintrusive methods (eg, gas sampling tubes).

In this section several tracer technologies are investigated: nonradioactive, chemical tracers, including perfluorocarbons and sulfur hexafluoride, for leak/breach detection and radioactive tracers incorporated into the grouts for determining grout location.

Perfluorocarbon Tracers. Brookhaven National Laboratory (BNL) has developed a host of perfluorocarbon tracers (PFT) and has proposed using them for barrier continuity verification (32). These tracers were originally used in atmospheric and oceanographic studies and have since been applied to a great variety of problems including detecting leaks in buried natural gas pipelines and locating radon ingress pathways in residential basements (33). Prior accomplishments in determining the gas pathways in residential basements are readily applicable to barrier verification. The residential basements studied are essentially miniature barriers with vertical concrete walls and a horizontal concrete floor. Thus, PFT tracer technology should be directly applicable to barrier verification.

Perfluorocarbon tracers can be detected at extremely low levels. Parts per quadrillion are routinely measured. This allows detection of very small breaches in the barrier. A breach can be located by injecting a series of tracers on one side of a barrier wall and monitoring for those tracers on the other side. The injection and monitoring of the tracers can be accomplished using conventional low cost monitoring methods such as existing vadose zone monitoring wells or multilevel monitoring ports, placed using cone penetrometer techniques (eg, Hydropunch). The amount and type of tracer detected on the monitoring side of the barrier will determine the size and location of a breach. It is easy to see that the larger the opening in a barrier the greater the concentration of tracer is transported across the barrier. Locating the breach requires more sophistication in the tracer methodology. Multiple tracer types can be injected at different points along the barrier, in both vertical and horizontal directions (Fig. 6). Investigation of the spectra of tracers coming through a breach then gives a location relative to the various tracer injection points.

Perfluorocarbon tracer technology consists of the tracers themselves, injection techniques, samplers, and analyzers. This technology has several advantages over conventional tracers:

1. The background concentration of PFTs in the environment is negligible. Consequently, only small quantities are needed.
2. PFTs are nontoxic, nonreactive, nonflammable, environmentally safe (contain no chlorine), and are commercially available.
3. PFT technology is the most sensitive of all nonradioactive tracer technologies and concentrations in the range of 10 parts per quadrillion of air (ppq) can be routinely measured.
4. The PFTs technology is a multitracer technology allowing up to six PFTs to be simultaneously deployed, sampled, and analyzed with the same instrumentation. This results in a lower cost and flexibility in experimental design and data interpretation. All six PFTs can be analyzed in 15 min on a laboratory-based gas chromatograph.

Typically, the PFTs are measured by a capillary adsorbent tracer sampler (CATS), which is a small cigarette-sized glass tube containing a carbonaceous adsorbent specific for the PFTs. This sampler can be used dynamically (flowing a sample through the CATS) or passively (opening only one end to allow the CATS to sample by diffusion). The passive mode allows a

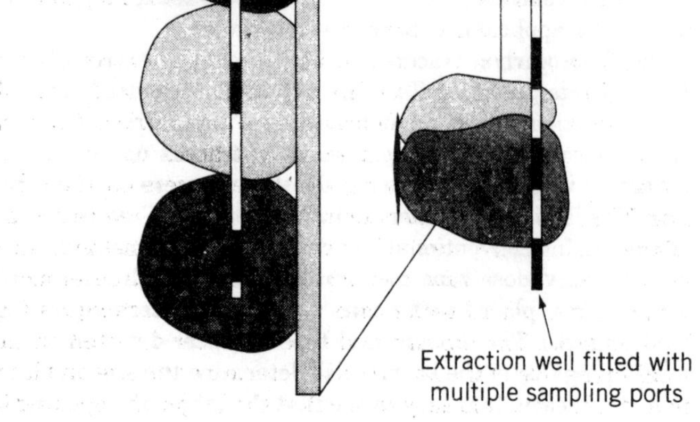

Figure 6. Perfluorocarbon tracer technology.

time-integrated PFT concentration to be measured in a simple manner. The CATS are shipped back to the laboratory for PFT analysis. Several real-time PFT analyzers are available, one that detects four different PFTs per 5-min sample down to the ambient background of the PFTs in air. Another real-time instrument can analyze PFT down to a part per trillion but cannot speciate the various PFTs.

One application relevant to barrier verification is the determination of radon flux into residential homes through cracks in the basement floor and surrounding subsurface walls (33). This can only be achieved with a multitracer technology. One type of PFT source (with a calibrated emission rate, most often prepared as a permeation device) is buried outside a house under study. Several other types of PFT sources are placed in the basement and the first floor of the house. From a time-averaged measurement of the various PFT concentrations in the basement and first floor and a measured radon concentration in these areas, it is possible to calculate a radon flux rate into the home. This technique of using a reference source of PFT to estimate the source rate of a compound of interest has also been used to estimate the following:

The source rate of dioxin into a commercial building from surrounding contaminated soil

The rate of leaking gasoline from underground storage tanks at gasoline stations

The rate of leaking dielectric fluid from subsurface electrical cables

Understanding mass transport through defects in barriers is central to the evaluation of barrier continuity. The migration of the tracers used in the experimental system will be analyzed using computer codes that predict the transport of the gas tracer through a porous soil and barrier with defects. Existing computer codes, such as DUST or TRACER3D, can be adapted as necessary for the problem.

Perfluorocarbon tracers may allow location and sizing of breaches at depth and should have a resolution of fractions of an inch. The technology has regulatory acceptance and is used commercially for nonwaste management practices (eg, detecting leaks in underground power cables). This technology has been used in a variety of soils and is applicable to the entire DOE complex as well as commercial waste sites. The major use of tracers is to verify placement continuity of a freshly emplaced barrier and to recheck corrective actions that may be used to seal or repair a breach. The PFTS also prove useful in measuring some performance parameters (eg, diffusion coefficient) of some higher permeability grouts (eg, Portland cement) and are useful to periodically check a barrier to determine the long-term integrity of the walls. This would be beneficial if a cementitous grout (Portland based) barrier were used. Cementitous grouts are prone to cracking from various degradation modes including wet–dry cycling that is prevalent at many arid DOE sites (eg, Sandia and Hanford). Tracers would allow determination of performance losses in containment over the life of the barrier.

Sulfur Hexafluoride. Sulfur hexafluoride (SF_6) is a chemical tracer used similarly to PFTs. The tracer is injected into the subsurface and then monitoring wells are used to detect the tracer as it moves through the subsurface. Typically, SF_6 is used in much higher concentrations that PFTs; pure gaseous

SF_6 versus 300 parts per million for PFT. In nature SF_6 can only be detected to about 10 parts per trillion (ppt) due to a natural background of approximately 3 ppt. In industrial areas were SF_6 is used in electrical equipment the background can be much higher. At Brookhaven National Laboratory where a particle accelerator is located the background is as high as 100 ppt. This would allow only eight to nine orders of magnitude dilution before the tracer blends into the background. For PFTs even starting at 300 ppm comfortably allows nine orders of magnitude dilution, whereas starting with a pure tracer would allow 13 to 14 orders of magnitude. It is however, advantageous to start with diluted tracer so that gas density differences are negligible. Another limiting feature of SF_6 is only having one form of tracer, it is not possible to use multiple tracers. This limits the ability of accurately and easily locating breaches and requires a long waiting period to recheck a barrier. Once a hole is found with any verification technology the logical next step is to fix the hole. Following the repair, it will be necessary to qualify the repair. With SF_6 you must wait until the tracer has dissipated to background levels, this could be a substantial time. For PFT-type tracers where multiple species exist and that can be detected independently, the only requirement to qualifying a repair is to change the tracer. SEAtrace, a monitoring system based on SF_6, is commercially available for impermeable barriers from Science & Engineering Associates, Inc. Probably SF_6 is limited to finding gross holes in a barrier and may be limited in reverification of repairs due to buildup of the tracer in the subsurface.

Radioisotopes. Radioactive tracers can be used to locate grout after application. A short-lived tracer is added to the grout prior to subsurface injection. Down-hole detectors are used to measure the radiation emitted by the tracer. The tracer is generally a gamma-emitter with a short half-life. The short half-life assures that the radioactivity will quickly decay away to innocuous levels. Multiple tracers and gamma-ray spectroscopy tools have been used to determine the diameter and vertical extent of grout columns for grouting well bore casings (34).

The use of radioactive tracers is limited to locating areas of grout penetration. Detection of gamma-energy in a local area suggests that the grout has reached that area but tells nothing of the continuity of the grout formation. A grout lens several feet from the detector may give a similar spectrograph as a homogeneous grout mix closer to the detector or as a thicker grout formation. Small (and perhaps large) holes and cracks would be invisible to present detection technologies.

RECENT CASE HISTORIES

Using Multiple Geophysical Methods to Map Disposal Pits at Brookhaven National Laboratory

The use of subsurface barriers is in its infancy; as such there is a lack of data on verification. One case at Brookhaven National Laboratory (BNL) uniquely demonstrates the intricacies of noninvasive geophysical surveys for subsurface characterization. Although this case involves locating and sizing buried waste pits and does not directly address a subsurface barrier, the requirements of delineating a subsurface structure are the same.

Brookhaven National Laboratory is located on Long Island in Upton, New York. It is a multidisciplinary scientific research center owned by the Department of Energy and operated by Associated Universities, Inc. In 1989 BNL was added to the U.S. EPA's National Priorities List. The site is divided into seven operable units (OU). OU1 includes the former landfill area. Associated with the former landfill area are the chemical/animal pits (CAP) and the glass holes. Beginning in the 1960s and continuing into the 1980s, BNL disposed of laboratory waste (glassware, chemicals, and animal carcasses) in numerous shallow pits. Individual pits were approximately 3 to 4.5 m (10–15 ft) in diameter and 4.5 to 6 m (15–20 ft) deep. Record keeping on the number of pits, location, and contents was poor by today's standards. This makes it difficult if not impossible to fully assess the problem and to develop a remediation plan. The drivers for remediating the pits are historical records that indicate hazardous materials may have been disposed of in the pits; groundwater contamination down-gradient of the pits; a test excavation of one of the glass holes that unearthed laboratory glass bottles with unidentified liquids still contained; and the fact that BNL rests atop an EPA designated sole-source aquifer system. In the summer of 1995 BNL commissioned a series of noninvasive geophysical surveys to characterize the site. This survey was performed to locate all the pits, determine the centers and boundaries, and define locations of future intrusive sampling by geoprobe techniques.

Geologically the area around the CAP and glass pits can best be described as a sand box. The surface layer (0–30 m below grade) is unconsolidated glacial sediments containing mostly fine to coarse quartz sand and some gravel; there are few cobbles or interfering geological anomalies. The closest geological structure to the pits is the discontinuous Gardiners Clay layer located between 24 and 30 m (80 and 110 ft) below sea level. Many geophysical measurements may be affected by the water table that is shallow and located approximately 7.6 to 9 m (25–30 ft) below grade. For this case study, the CAP surveys will be examined. Historical records suggested some 28 pits were dug and are contained within an area of approximately 9000 M^2. Laboratory waste was dumped into an unlined pit in layers separated by occasional layers of back-fill. Once filled a final layer of fill was used to cap the pit, but exact dimensions and contents were not recorded.

The logic used to survey this site was to use historical data (eg, aerial photographs showing excavations), topographical data, and a set of geophysical characterizations. Data from all these sets could be integrated together to produce an accurate map of the disposal areas. The geophysical methods included vertical ground-penetrating radar (GPR), an angled GPR denoted as side-looking underground radar (SLUR), low frequency electromagnetic induction (Geonics EM-31), time-domain electromagnetic induction (Geonics EM-61), and dense magnetic field gradient. The electromagnetic methods could be used to determine locations of pits, buried objects, and geologic anomalies and to distinguish ferrous versus nonferrous objects, but these methods are not as useful for making size and/or volume estimates. To better establish such estimates GPR and SLUR were used. Standard GPR would yield good data in the x–y plane (surface). To improve the resolution in the Z plane (vertical) SLUR was used to better define the bottoms of the pits.

Vertical Ground-Penetrating Radar. Two units were used for vertical GPR data collection. The first was a commercial GSSI radar operated by Soft Earth Geophysical Services. The second unit was an earth-penetrating radar imaging system (EPRIS) developed by the Coleman Research Corporation. A complete description of the system can be found in reports by Coleman Research (35,36). The EPRIS is a stepped frequency system operating between 100 and 1000 MHz in steps of 2 MHz. It uses a rapid imaging technique and coherent synthetic aperture imaging algorithms to enhance subsurface images. The antennas are 0.5 m in diameter and are positioned in a 2 m long, linear array. Imaging data was taken every 0.08 m along grid lines separated by 0.5 m. A detailed description of GPR theory was given previously.

Angled Ground-Penetrating Radar. Side-looking underground radar or SLUR provides an innovative twist on conventional GPR. The technique tows the GPR antenna array along the dig face of a trench. A trench is constructed such that the digface is at a 45-degree angle to the surface. This allows the GPR to be directed diagonally across the field, eliminating the vertical interferences and allowing better definition of the pit walls and bottoms. The unit used was a commercial GSSI system operated by Soft Earth Geophysical Services. A description of the SLUR technology is available (37).

Low Frequency Electromagnetic Induction. The Geonics EM-31 is a coplanar, dual coil, low frequency electromagnetic induction device used to detect near-surface contrasts in conductivity. A survey was conducted by Argonne National Laboratory (ANL) personnel along lines 5 m apart with data recorded every 0.5 m. The survey was conducted first with the boom parallel to the survey lines and then repeated with the boom perpendicular to the lines. This provided both validation of the data and evidence of any anisotropic conductivity effects and anomalies not directly under the survey lines. A description of the EM-31 equipment and results can be found in an ANL report (38).

Time-Domain Electromagnetic Induction. The Geonics EM-61 is a time-domain inductive electromagnetic device used to detect metallic objects near the surface. The EM-61 operates by producing a time varying magnetic field that induces a current flow within the subsurface. When the field is turned off the current flow in the subsurface begins to decay. This decay occurs slower in metallic objects allowing differentiation from nonmetallic structures. Data was recorded by ANL similarly to the EM-31 survey. A second survey was performed by Geonics with closer spacing to increase the spacial resolution. Details of the survey and equipment can be found in the ANL report (38).

Rapid Geophysical Surveyor. The Rapid Geophysical Surveyor (RGS) developed by Sage Earth Sciences uses an automated profile-based data acquisition system. It is said to provide higher resolution by allowing collection of closely spaced magnetic data. This was considered important at the BNL site where small laboratory scale items were expected. Magnetometer based systems are used to detect ferrous-based metals. When combined with the EM-31 and EM-61 data the RGS could be used to differentiate areas of ferrous and nonferrous metals. The data was collected every 0.15 m along lines spaced 0.5 m apart. Complete details of the equipment and survey are given in a report from Sage Earth Sciences (39).

Survey Results. As mentioned earlier, the objective of this characterization was to locate and size all of the pits in the CAP disposal area. Brookhaven National Laboratory recognized from the start that no one survey tool was likely to be sufficient and employed a suite of tools. Each survey was interpreted individually by the survey contractors, and these reports were combined and interpreted as a set (40). It is this combination of data that yields valuable information for BNL. At the same time this combination data set distinctly points out the shortcomings of geophysical techniques when applied to barrier verification, particularly when anomalies of fractions of an inch must be located. Figure 7 is shows the CAP area in 1964 and some excavation sites. This historical data was used to determine where to survey. Figure 8 is a topographical representation of the site (before clearing and leveling for the surveys). This map provided historical evidence of dig sites by identifying mounds that could be capped pits. Figures 9–13 show the results of the various surveys and Figure 14 shows the final determination of pit locations. None of the survey techniques could define every pit location. Ground-penetrating radar seems to be the best at finding anomalies, but still has so much associated noise that definition of every pit (or the boundaries of any one pit) is impossible. By combining these sophisticated technologies BNL could achieve their goal of locating and defining the boundaries of CAP disposal cells with a great deal of certainty. This was accomplished by examining historical data and physical measurements and by overlaying the suite of surveys. Anomalies that repeatedly occur on most of the surveys can be taken as pits. The size and volume of each pit were estimated by combining the survey results for each pit and making deductions based on knowledge of each geophysical method (eg, interferences to one technique might be absent in another).

The fact that BNL could not define all of the pits with any singular geophysical method is evidence that these technologies are not mature enough to locate breaches in a subsurface barrier, particularly if breach sizes of fractions of an inch are specified. These technologies are useful tools to define boundaries of a barrier or perhaps gaping holes (several square meters). They may be more useful as scoping quality assurance/quality control tools if used in symbiotic combinations. There was never a question that identifying individual bottles or cans in a specific pit was out of the resolution range of all these techniques. The future of these methods as barrier verification tools is uncertain. It appears to the author that magnetics and conductivity determinations are unlikely to mature to a point that they will be useful verification tools. The radar results are also severely limited in usefulness, but here the problem appears to be too much data (eg, reflections). Data handling must mature to the point that this data can be sifted or the superfluous data removed before such measurements are useful to verification. This is most likely a long-term problem, requiring experimental research and refinement of the mathematical and statistical evaluations, and will not be solved soon.

Proof-of-Concept Demonstration of Perfluorocarbon Tracers for Barrier Integrity Verification

A full-scale demonstration installation of a subsurface barrier was carried out at Hanford, Washington (41). Complementing the barrier demonstration is a continuing demonstration of the PFT verification technology discussed earlier. The objective of the PFT task was to demonstrate on a proof-of-concept

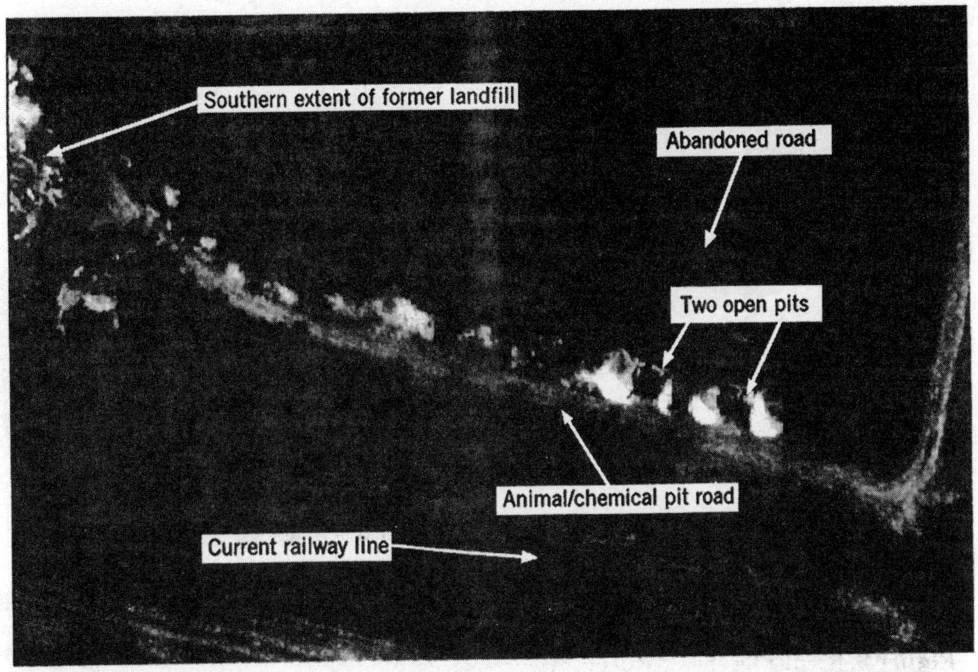

Figure 7. CAP site at Brookhaven National Laboratory. Courtesy of Brookhaven National Laboratory.

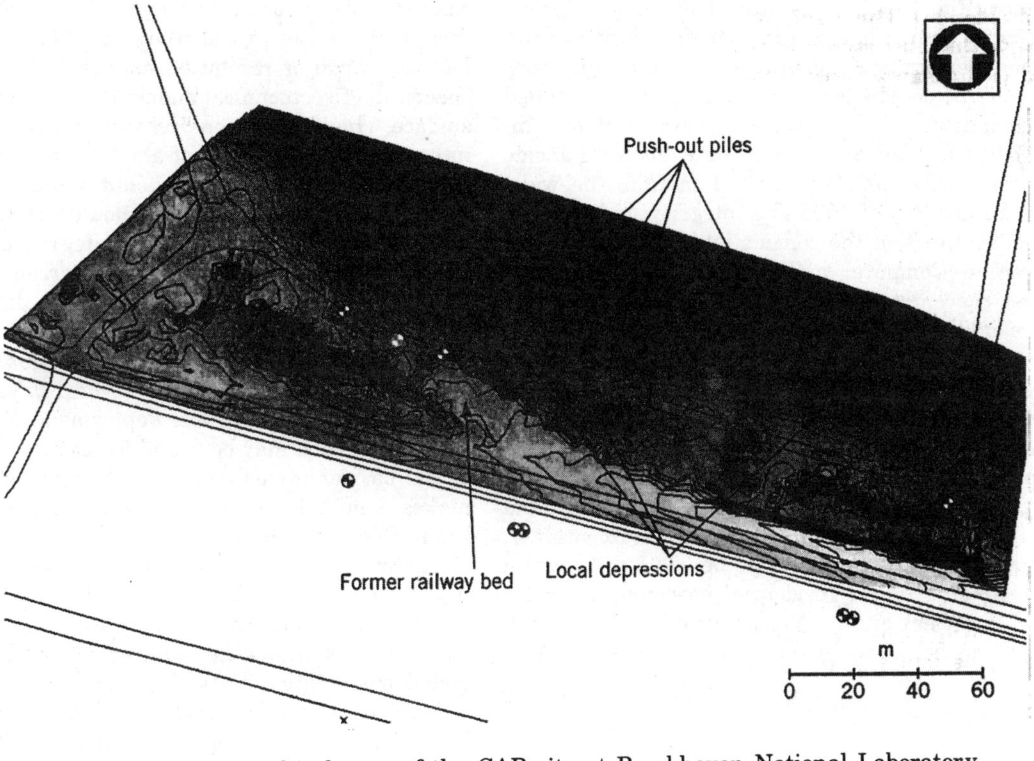

Figure 8. Topographical map of the CAP site at Brookhaven National Laboratory. Courtesy of Brookhaven National Laboratory.

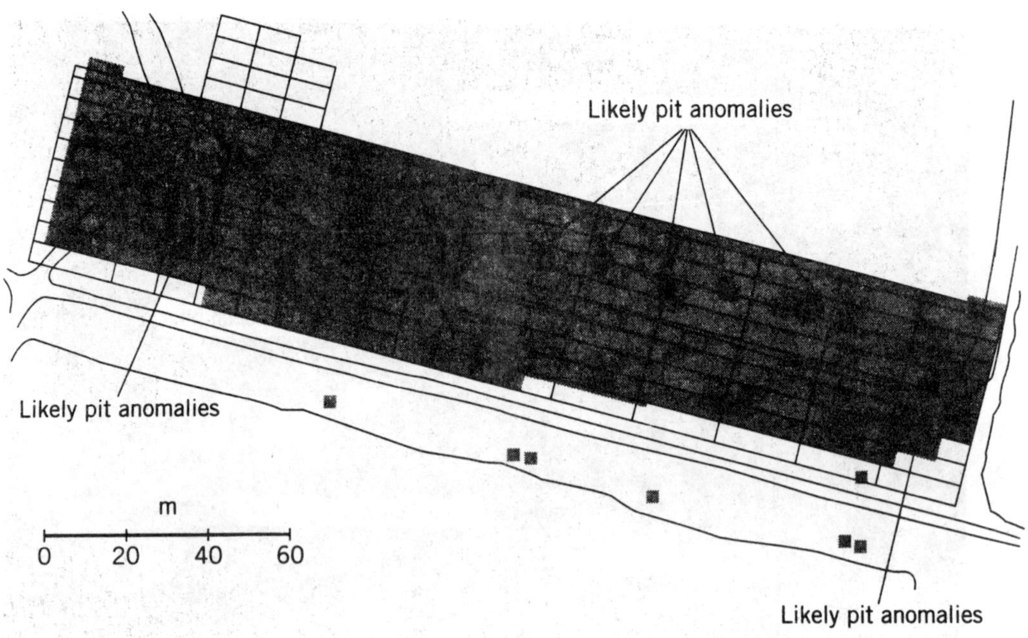

Figure 9. Results for the CAP using ground-penetrating radar. Courtesy of Brookhaven National Laboratory.

level that PFT technology can verify barrier integrity. A close-coupled barrier is produced by first installing a conventional cement grout curtain followed by a thin lining of a polymer grout (Fig. 15). The resultant barrier is a cement polymer composite that has economic benefits derived from the cement and performance benefits from the durable and resistant polymer layer. The Hanford demonstration installed a cone-shaped barrier using angled jet-grouting. Before installing the barrier, monitoring wells were installed inside and outside the barrier approximately 1 m from the walls (see Fig. 14). In October of 1995 the integrity and to some extent the performance of the cement layer were evaluated using the tracers. Computer modeling was used to decide starting tracer concentrations and to analyze the data (42). Because the cement layer was far more permeable than the polymer layer, we could estimate the gaseous permeability of this layer. An injection of PFTs was performed on the inside of the barrier. The injection point was at the bottom of the storage tank. The monitoring wells were then periodically sampled for tracers. Results were very promising; a uniform concentration of the tracer was detected around the barrier indicating an intact and breach-free container. Gaseous permeability was estimated to be 0.001 cm/s, which is within the range seen for residential basement concretes and high-quality cement grouts. A complete description of the demonstration can be found in the project Conceptual Test Plan (43).

CONCLUSIONS

Most subsurface characterization techniques have come from the petroleum and mining exploration industries. Scaled to subsurface barriers and faults of fractions of an inch, these geophysical techniques are short ranged or poorly resolved or both. The geophysical survey performed at Brookhaven National Laboratory is a very good example of proper use of these technologies. It was recognized early that no one method had the range or resolution nor capacity to cover the entire spectrum of geotechnical anomalies that can occur in the subsurface. This is a key consideration when attempting to verify integrity or performance of a subsurface barrier. The author believes that a suite of tools should be used to best address this issue, and the methods chosen should be site specific. Gaseous tracers can be used to verify the integrity of the barrier. This method will indicate whether, the barrier is intact and may yield performance data on the gaseous diffusion coefficient. This is fine for final verification of the barrier, but not as timely as needed for quality assurance/quality control during the installation. It may take several days to two weeks to get all the data from tracer deployment. Nondestructive geophysical method may be useful for monitoring the installation of the barrier in real time. Suppose the grouting contractor makes a mistake, such as skipping a column or having the grout flow interrupted for some portion of the installation. Here the large, gross errors in the installation are considered, not fraction-of-an-inch errors. Real-time geophysics allows the problem to be identified quickly before the contractor leaves the site. This allows the required repairs to be made without added mobilization costs. It is judicious to use well-resolved techniques such as tracers for regulatory proof of integrity and for long-term monitoring. During installation these techniques should be supplemented with real-time techniques for quality assurance/quality control. Geophysical methods are well suited for such broad-based characterizations. Further refinements in geophysical technologies may provide useful tools in the

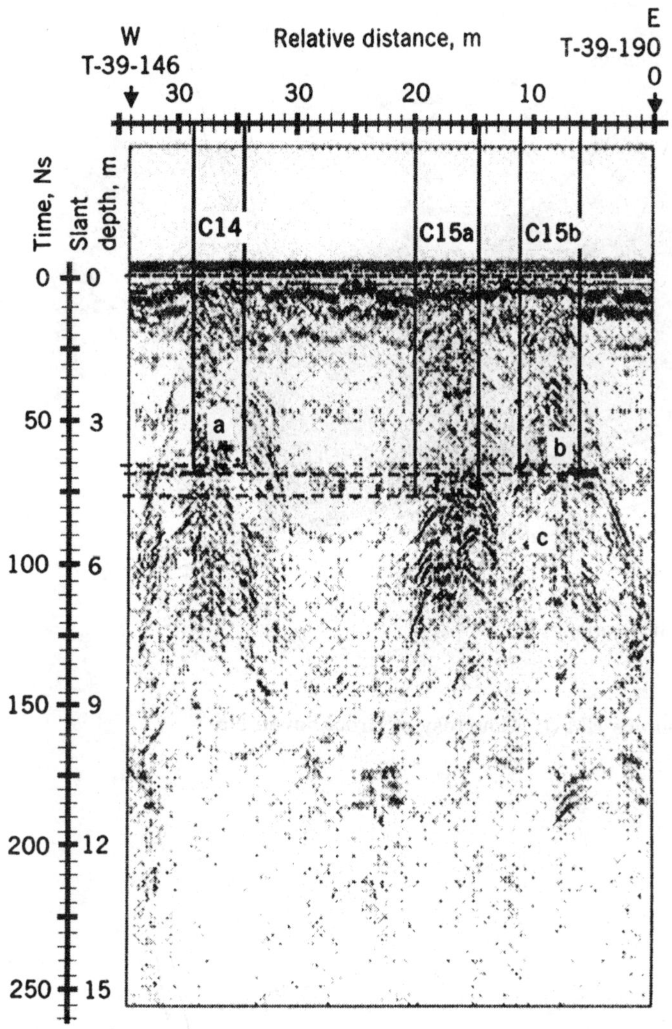

Figure 10. Results for the CAP using side-looking underground radar. Courtesy of Brookhaven National Laboratory.

future for verification and performance monitoring. Toward this, the following observations are made:

1. It does not appear that radiation-based (neutron and gamma logging) techniques will be viable methods for subsurface barrier continuity verification. Such methods, even if resolution were accurate enough would require a prohibitive number of boreholes to completely log the barrier. It would also be costly and time consuming to measure and analyze data from that many boreholes.
2. Acoustic and seismic logging methods have promise but must greatly increase resolution. As refinements to the resolution are made, the deep penetration of seismic waves will prove valuable to subsurface investigations. To date seismics have been best developed for deep penetration using low frequencies. Acoustics using high frequencies have been used to find cracks in brittle materials such as cement but have very limited penetration. For subsurface barriers it may be prudent to develop acoustic methods using frequencies better suited to the depths of typical waste sites. The use of embedded sonic transducers is also a worthwhile area for further research. Inexpensive transducers could be used to look for breaches and to form a baseline acoustic signature of the barrier. The transducers could then be used in the future to recheck cracking and to compare the acoustic signature of the barrier (a change in the signature signaling a change in the barrier).
3. Chemical tracers are an area that should prove valuable to subsurface barrier continuity verification. Perfluorocarbon tracers may allow locating and sizing of breaches at depth and should have a resolution of fractions of an inch. The technology has regulatory acceptance and is used commercially for nonwaste-management practices (eg, detecting leaks in underground power cables). This technology has been used in a variety of soils and will be applicable to the entire DOE complex and commercial waste sites. Tracers will allow final integrity verification of an emplaced barrier and rechecking of corrective actions used to repair a breach. They can also be used to periodically check a barrier to determine the long-term integrity of the walls.
4. Ground-penetrating radar technologies may be useful for real-time monitoring of the grouting procedure. Such techniques may be valuable in detecting operational mistakes and allowing timely corrective actions. The method requires further refinements in analysis and signal-to-noise ratio before becoming a useful monitoring tool. As a surface-oriented survey tool it is easy to move about the site and little site preparation is required.

It must also be clearly stated that the discussions of geophysical technologies in this report have only considered detection of a subsurface barrier through natural soil conditions. That is, no consideration has been given to the waste site contained within the barrier's boundaries and the introduced anomalies associated with typical waste disposal sites. If the site only contains contaminated soils from a spill then the waste will not affect geophysics appreciably. If the waste site contains buried waste such as drums, tanks, boxes, and debris, as would generally be expected, the geophysical methods will be greatly perturbed. The interference can be so great that many methods become useless for viewing the barrier. Ground-penetrating radar, electromagnetic, and resistivity/conductivity methods are overwhelmingly affected by buried metals. Seismic and acoustic methods will produce additional noise, reflections,and refractions of their signals due to the increased heterogeneity of the site.

For tracer technologies the waste site has less of an impact. The inner contained area may have greater heterogeneity in transport properties but the traces will still move about and be detectable outside the barrier (assuming a breach). Increased heterogeneity may require more closely spaced monitoring wells and ports or greater sampling frequencies. The important monitoring ports used to detect tracers that have escaped the barrier will be located outside the barrier. This will be an area where presumably no buried waste exists and the subsurface contains only natural anomalies.

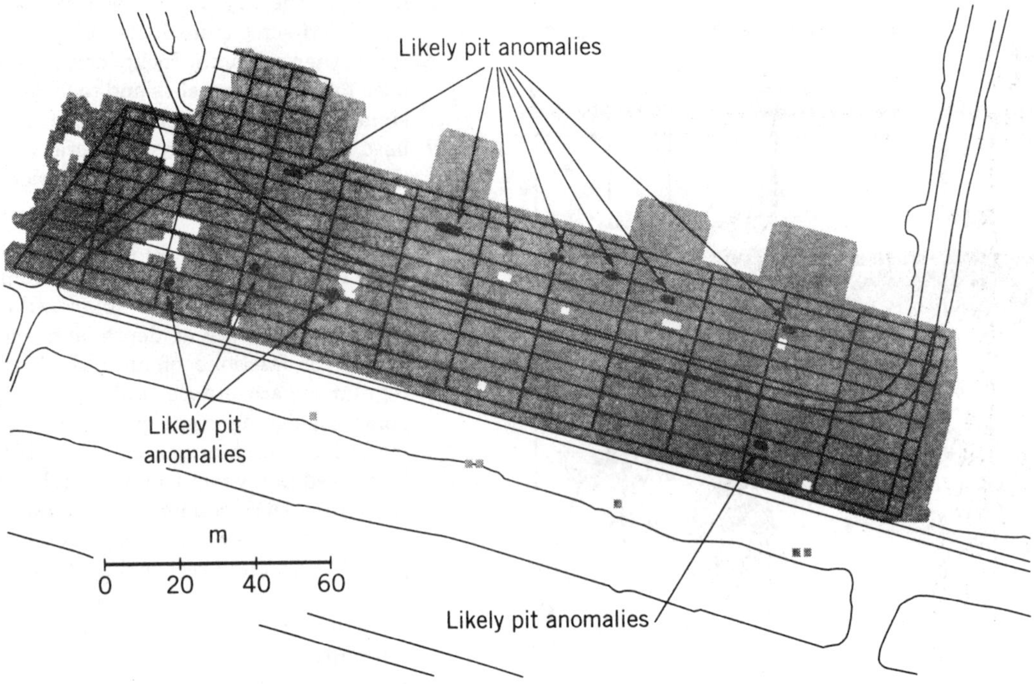

Figure 11. Results for the CAP using a Geonics EM 31. Courtesy of Brookhaven National Laboratory.

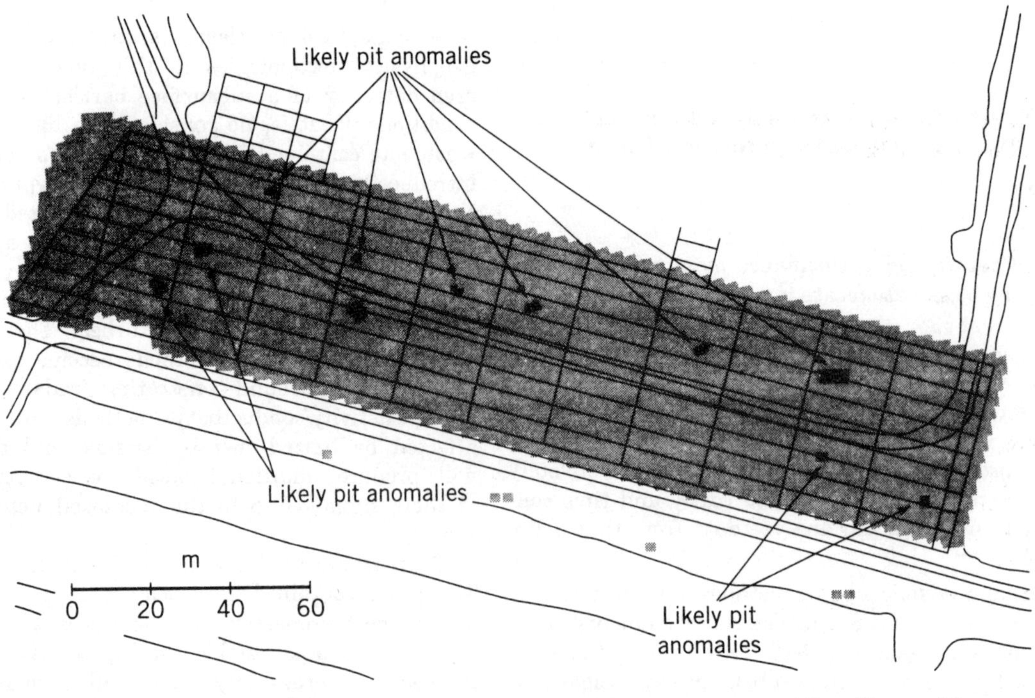

Figure 12. Results for the CAP using a Geonics EM-61. Courtesy of Brookhaven National Laboratory.

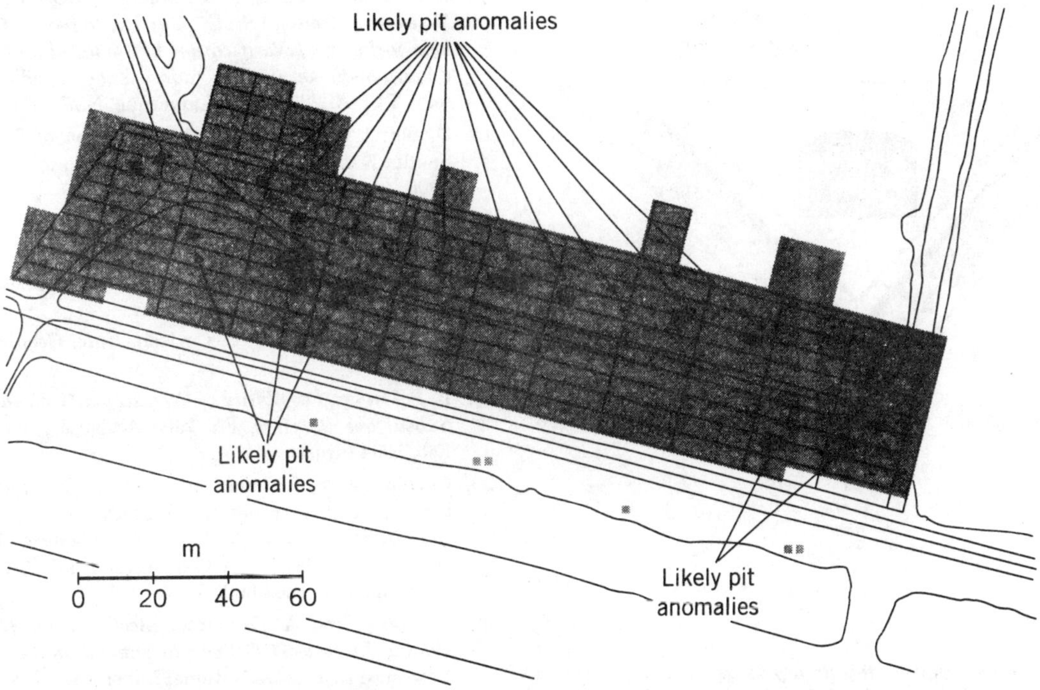

Figure 13. Results for the CAP using Rapid Geophysical Surveyor. Courtesy of Brookhaven National Laboratory.

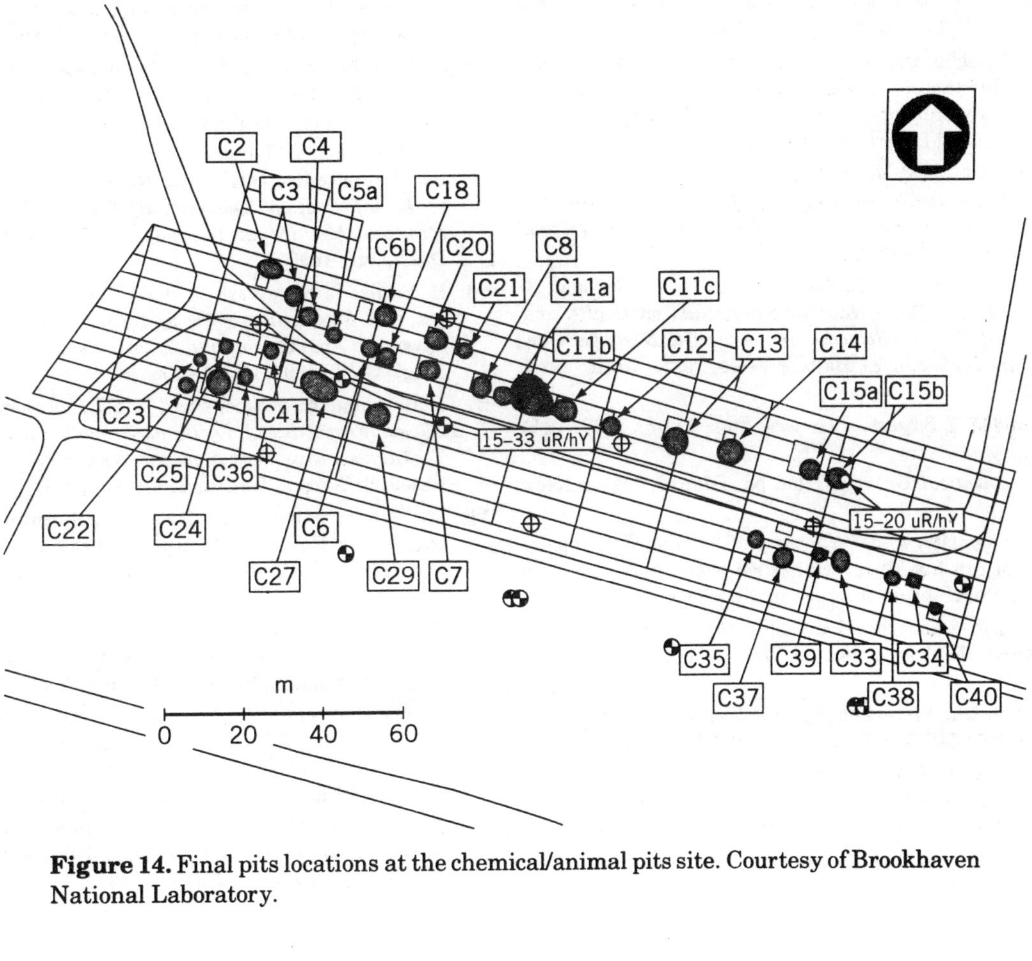

Figure 14. Final pits locations at the chemical/animal pits site. Courtesy of Brookhaven National Laboratory.

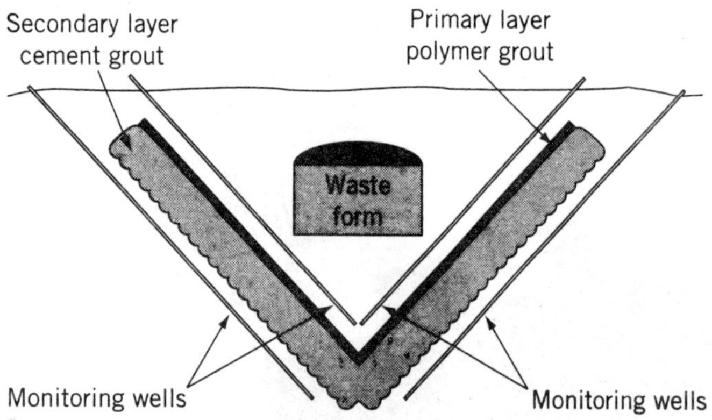

Figure 15. Schematic of close-coupled barrier installed at Hanford.

BIBLIOGRAPHY

1. B. Siskind, and J. Heiser, *Regulatory Issues and Assumptions Associated with Barriers in the Vadose Zone Surrounding Buried Waste,* Environmental and Waste Technology Center, Brookhaven National Laboratory, Feb. 1993, BNL-48749(I).
2. Science Applications International Corporation, *HANDBOOK: Remedial Action at Waste Disposal Sites (Revised),* EPA/625/6–85/006, Oct. 1985.
3. *Environmental Guidance Program Reference Book—Comprehensive Environmental Response, Compensation, and Liability Act,* ORNL/M-1572, Revision 11, Oct. 1, 1991.
4. J. Heiser, P. Colombo, and J. Clinton, *Polymers for Subterranean Containment Barriers for Underground Storage Tanks (USTs),* BNL-48750, Environmental and Waste Technology Center, Brookhaven National Laboratory, Upton, N.Y., Dec. 1992.
5. J. H. Heiser, and L. W Milian, *Laboratory Evaluation of Performance and Durability of Polymer Grouts for Subsurface Hydraulic/Diffusion Barriers,* BNL-61292, Brookhaven National Laboratory, Upton, N.Y., 1994.
6. C. Voss, M. Einberger, and R. Matalucci, In G. W. Gee and N. R. Wing, eds., *Thirty-Third Hanford Symposium on Health and the Environment, In-Situ Remediation: Scientific Basis for Current and Future Technologies,* Battelle Press, Pasco, Wash., 1994, pp. 113–134.
7. A. Bodocsi, and M. T. Bowers, *J. Geotech. Eng.,* **117**(8), 1227–1244 (Aug. 1991).
8. R. K. Karol, *Chemical Grouting,* 2nd ed., Marcel Dekker, New York, 1990.
9. D. Foedisch, *Chemistry and Technology of Montan Wax,* reprint from The American Ink Maker, translated from the German and distributed by, Strohmeyer & Arpe Co., New York.
10. *Interim Subsurface Barrier Technologies Workshop Report,* prepared by Bovay Northwest Inc., for Westinghouse Hanford Co., Sept. 10, 1992.
11. U.S. Environmental Protection Agency, *Engineering Bulletin—Slurry Walls,* EPA/540/S-92/008, Washington, D.C., Oct. 1992.
12. P. A. Spooner, R. S. Wetzel, C. E. Spooner, C. A. Furman, and E. F. Tokaraski, *Slurry Trench Construction for Pollution Migration Control,* EPA-540/2-84-001, Feb. 1984.
13. U.S. Dept. of Energy, EM-50 FY95 Needs Statement IS-9, *In-Situ Remediation Integrated Program,,* Washington, D.C., 1995.
14. J. D. Betsill, and R. D. Gruebel, *VAMOS The Verification and Monitoring Options Study, Current Research Options for In-Situ Monitoring and Verification of Contaminant Remediation and Containment within the Vadose Zone,* Sandia National Laboratory, TTP-AL221107, Albuquerque, N.M., May 1995.
15. D. Borns, *Geophysical Characterization of Subsurface Barriers,* Sandia National Laboratory, Albuquerque, N.M., SAND95-1461, 1995.
16. J. Hearst, and P. Nelson, *Well Logging for Physical Properties,* McGraw-Hill, Inc., New York, 1985.
17. D. Ellis, *Well Logging for Earth Scientists,* Elsevier Science Publishing, New York, 1987.
18. J. R., Hearst, and C. C. Carlson, *Nulc. Geophy.,* **8**,(2), 165–172, 1994.
19. B. P. Dwyer, *Feasibility of Permeation Grouting for Constructing Subsurface Barriers,* FY 1993 Activities, TTP No. AL-231004, Feb. 1994 (draft).
20. Golder Federal Services, Inc., *Multiple-Borehole Permeation Tests using Montan Wax and Sodium Silicate Barrier Materials in Unconsolidated Soils at the Mixed Waste Landfill Integrated Demonstration Site,* Draft Test Plan, Version 1.0, Redmond, Wash., Feb. 1994, pp. 913–1248.600c.
21. W. Daily, and A. Rameirez, *Electrical Resistivity Tomography During In Situ TCE Remediation at the Savannah River Site,* Lawrence Livermore National Laboratory, Livermore, Calif., Feb. 1993.
22. S. Boyd, R. Newmark, and M. Wilt, in R. Bell and C. Lepper, eds. *Proceedings of the Symposium on the Application of Geophysics to Engineering and Environmental Problems,* Vol. 1, Boston, Mass., Mar. 1994, pp. 295–308.
23. L. Murdoch, J. Dunlap, and J. Daniels, *Evaluation of an In Situ Horizontal Barrier,* Test Plan, University of Cincinnati, Engineering Research Division, Center for GeoEnvironmental Science and Technology, RF# 668148, project 728818, Apr. 1994.
24. O. Westphalen, and J. Rice, "Drum Detection: EM vs MAG Some Revealing Tests", *Proceedings of the Sixth National Outdoor Action Conference,* Las Vegas, Nev., May 11–13, 1992.
25. K. Davis, and G. Sandness, "Ground-Penetrating Radar for Remotely Controlled Site Characterization", Paper 31, *Proceedings of the Information Exchange Meeting on Characterization, Sensors, and Monitoring Technologies,* Dallas, Tex., CONF-920791 Draft, July 1992.
26. D. Borns, L. Stolarczyk, and W. Mondt, in R. Bell and C. Lepper, eds., *Proceedings of the Symposium on the Application of Geophysics to Engineering and Environmental Problems,* Vol. 2, San Diego, Calif., 1993, pp. 91–105.
27. J. Bay, and K. Stokoe, *Field and Laboratory Determination of Elastic Properties of Portland Cement Concrete Using Seismic Techniques,* Transportation and Research Record 1355, National Research Council, Washington, D.C., 1991, pp. 67–74.
28. W. Johnson, and J. Clark, "Improving Subsurface Resolution with the Seismic Reflection Method: Use S-Waves", *Proceeding of the Sixth National Outdoor Action Conference,* Las Vegas, Nev., May 1992.
29. G. Elbring, *Crosshole Shear-Wave Seismic Monitoring of an In Situ Air Stripping Waste Remediation Process,* SAND91-2742 Sandia National Laboratory, Albuquerque, N.M., 1992.
30. D. Steeples, and R. Miller, "Seismic Reflection Methods Applied to Engineering, Environmental, and Ground Water Problems", in S. Ward, ed., *Geotechnical and Environmental Geophysics,* Vol. 1, Society of Exploration Geophysicists, Tulsa, Okla., 1992.
31. E. Majer, L. Myer, J. Peterson, "High Resolution Imaging of Fractured Rock", *Third International Reservoir Characterization Technical Conference,* Nov. 3–5, 1991, Tulsa, Okla.

32. J. Heiser, *Verification of Subsurface Barrier Integrity using Perfluorocarbon Tracers,* TTP CH3-5-PR-19, proposal to U.S. Dept. of Energy, Office of Technology Development, In Situ Remediation Integrated Program, Washington, D.C., Mar. 1994.

33. T. W. D'Ottavio, and R. N. Dietz, "Radon Source Rate Measurements using Perfluorocarbon Tracers", *Indoor Air '87: the 4th International Conference on Indoor Air Quality and Climate,* Berlin, Germany, Aug. 17–21, 1987.

34. L. L. Gadeken, M. L. Gartner, and D. E. Sharbak, "Improved Evaluation Techniques for Multiple Radioactive Tracer Applications", *Twelfth International Logging Symposium of SAID,* Paris, France, Oct. 1989.

35. Coleman Research Corporation, *Brookhaven National Laboratory Ground Penetrating Radar Site Survey: Preliminary Survey Findings,* Orlando, Fla., 1995.

36. Coleman Research Corp., *Brookhaven National Laboratory Ground Penetrating Radar Site Survey: Final Hole Locations and Depth Estimates,* Orlando, Fla., 1995.

37. J. Daniels, *Preliminary Report, Ground Penetrating Radar Investigations at the Animal-Chemical Pit Area,* Soft Earth Associates, Westerville, Ohio, 1995.

38. R. Johnson, and J. Daniels, *Supplemental Characterization Report for the Animal/Chemical Pits and Glass Holes Area at Brookhaven National Laboratory, Upton, New York,* Argonne National Laboratory, Argonne, Ill., 1995.

39. Sage Earth Sciences, *Preliminary Report: RGS Magnetic Surveys of the Chemical/Animal Pits and Glass Holes Sites at Brookhaven National Laboratory, Upton, NY,* Idaho Falls, Idaho, 1995.

40. Argonne National Laboratory, Environmental Assessment Division, *Supplemental Characterization Report for the Animal/Chemical Pits and Glass Holes Area at Brookhaven National Laboratory, Upton, New York,* Argonne, Ill., Preliminary draft, Nov. 1995.

41. J. Heiser, and B. Dwyer, "Full-Scale Demonstration of Close-Coupled Subsurface Barrier Technology", *Waste Management '96,* Tucson, Ariz., Feb. 26–Mar. 1, 1996.

42. T. M. Sullivan, A. Gard, and J. Heiser, "Modeling of Subsurface Barrier Performance", *Waste Management '96,* Tucson, Ariz., Feb. 26–Mar. 1, 1996.

43. J. Heiser, and B. Dwyer, *Demonstration of Close-Coupled Barriers for Subsurface Containment of Buried Waste,* Conceptual Test Plan, Revision 2.0, Brookhaven National Laboratory, BNL-62573, Upton, N.Y., July 1995.